$192.95 per copy (in United States)
Price is subject to change without prior notice.

0164

The 2014 *Heavy Construction Cost Data,* 28th annual edition, and all of the RSMeans annual cost data books for 2014, are dedicated to the memory of our respected colleague and friend, Paula Reale-Camelio.

RSMeans Heavy Construction Cost Data

28TH ANNUAL EDITION

MW00674950

2014

RSMeans
A division of Reed Construction Data, LLC
Construction Publishers & Consultants
700 Longwater Drive
Norwell, MA 02061
USA
1-800-334-3509
www.rsmeans.com

Printed in the United States of America
ISSN 0893-5602
ISBN 978-1-940238-09-8

Senior Editor
Robert Fortier, PE *(2, 31, 32, 34, 35, 46)**

Contributing Editors
Christopher Babbitt
Adrian C. Charest, PE *(26, 27, 28, 48)*
Gary W. Christensen
David G. Drain, PE *(33)*
Cheryl Elsmore
David Fiske
Robert J. Kuchta *(8)*
Thomas Lane
Robert C. McNichols
Genevieve Medeiros
Melville J. Mossman, PE *(21, 22, 23)*
Jeannene D. Murphy
Marilyn Phelan, AIA *(9, 10, 11, 12)*
Stephen C. Plotner *(3, 5)*
Siobhan Soucie
Phillip R. Waier, PE *(1,13)*

Cover Design
Wendy McBay
Paul Robertson
Wayne Engebretson

Chief Operating Officer
Richard Remington

Vice President, Business Development
Steven Ritchie

Vice President, Finance & Operations
Renée Pinczes

Senior Director, Inside Sales
Derek F. Dean

Sales Manager
Reid Clanton

Director, Engineering
Bob Mewis, CCC *(4, 6, 7, 14, 41, 44)*

Senior Director, Product Management
Jennifer Johnson

Senior Product Marketing Manager
Andrea Sillah

Senior Director, Operations
Rodney L. Sharples

Production Manager
Michael Kokernak

Production
Jill Goodman
Jonathan Forgit
Debbie Panarelli
Sara Rutan
Mary Lou Geary

Technical Su
Christopher
Gary L. Hoi
Kathryn S.

** Numbers in italics are the divisional responsibilities for*
Please contact the designated editor directly with any

This book is printed on 90% recycled paper using soy-based ink and is entirely recyclable.

Related RSMeans Products and Services

The engineers at RSMeans suggest the following products and services as companion information resources to *RSMeans Heavy Construction Cost Data:*

Construction Cost Data Books

Building Construction Cost Data 2014
Site Work & Landscape Cost Data 2014
Concrete & Masonry Cost Data 2014

Reference Books

Landscape Estimating Methods
Unit Price Estimating Methods
Estimating Building Costs
RSMeans Estimating Handbook
Green Building: Project Planning & Estimating
How to Estimate with Means Data and CostWorks
Plan Reading & Material Takeoff
Project Scheduling and Management for Construction

Seminars and In-House Training

Site Work & Heavy Construction Estimating
RSMeans Online® Training
RSMeans Data for Job Order Contracting (JOC)
Plan Reading & Material Takeoff
Scheduling & Project Management
Unit Price Estimating

RSMeans Online Bookstore

Visit RSMeans at **www.rsmeans.com** for a one-stop portal for the most reliable and current resources available on the market. Here you can learn about our more than 20 annual cost data books, also available on CD and online versions, as well as our libary of reference books and RSMeans seminars.

RSMeans Electronic Data

Get the information found in RSMeans cost books electronically at **RSMeansOnline.com**.

RSMeans Business Solutions

Cost engineers and analysts conduct facility life cycle and benchmark research studies and predictive cost modeling, as well as offer consultation services for conceptual estimating, real property management, and job order contracting. Research studies are designed with the application of proprietary cost and project data from Reed Construction Data and RSMeans' extensive North American databases. Analysts offer building product manufacturers ualitative and quantitative market research, as well as time/motion udies for new products, and Web-based dashboards for market pportunity analysis. Clients are from both the public and private ctors, including federal, state, and municipal agencies; corporations; institutions; construction management firms; hospitals; and associations.

RSMeans for Job Order Contracting (JOC)

Best practice job order contracting (JOC) services support the owner in the development of their program. RSMeans develops all required contracting documents, works with stakeholders, and develops qualified and experienced contractor lists. RSMeans JOCWorks® software is the best of class in the functional tools that meet JOC specific requirements. RSMeans data is organized to meet JOC cost estimating requirements.

Construction Costs for Software Applications

More than 25 unit price and assemblies cost databases are available through a number of leading estimating and facilities management software providers (listed below). For more information see the "Other RSMeans Products" pages at the back of this publication.

RSMeansData™ is also available to federal, state, and local government agencies as multi-year, multi-seat licenses.

- 4Clicks-Solutions, LLC
- Assetworks
- Beck Technology
- BSD – Building Systems Design, Inc.
- CMS – Construction Management Software
- Corecon Technologies, Inc.
- CorVet Systems
- Hard Dollar Corporation
- Maxwell Systems
- MC²
- MOCA Systems
- Parsons Corporation
- ProEst
- Refined Data
- Sage Timberline Office
- UDA Technologies, Inc.
- US Cost, Inc.
- VFA – Vanderweil Facility Advisers
- WinEstimator, Inc.
- R & K Solutions
- Tririga

Table of Contents

Foreword

Our Mission

Since 1942, RSMeans has been actively engaged in construction cost publishing and consulting throughout North America.

Today, more than 70 years after RSMeans began, our primary objective remains the same: to provide you, the construction and facilities professional, with the most current and comprehensive construction cost data possible.

Whether you are a contractor, owner, architect, engineer, facilities manager, or anyone else who needs a reliable construction cost estimate, you'll find this publication to be a highly useful and necessary tool.

With the constant flow of new construction methods and materials today, it's difficult to find the time to look at and evaluate all the different construction cost possibilities. In addition, because labor and material costs keep changing, last year's cost information is not a reliable basis for today's estimate or budget.

That's why so many construction professionals turn to RSMeans. We keep track of the costs for you, along with a wide range of other key information, from city cost indexes . . . to productivity rates . . . to crew composition . . . to contractor's overhead and profit rates.

RSMeans performs these functions by collecting data from all facets of the industry and organizing it in a format that is instantly accessible to you. From the preliminary budget to the detailed unit price estimate, you'll find the data in this book useful for all phases of construction cost determination.

The Staff, the Organization, and Our Services

When you purchase one of RSMeans' publications, you are, in effect, hiring the services of a full-time staff of construction and engineering professionals.

Our thoroughly experienced and highly qualified staff works daily at collecting, analyzing, and disseminating comprehensive cost information for your needs. These staff members have years of practical construction experience and engineering training prior to joining the firm. As a result, you can count on them not only for accurate cost figures, but also for additional background reference information that will help you create a realistic estimate.

The RSMeans organization is always prepared to help you solve construction problems through its variety of data solutions, including online, CD, and print book formats, as well as cost estimating expertise available via our business solutions, training, and seminars.

Besides a full array of construction cost estimating books, RSMeans also publishes a number of other reference works for the construction industry. Subjects include construction estimating and project and business management, special topics such as green building and job order contracting, and a library of facility management references.

In addition, you can access all of our construction cost data electronically in convenient CD format or on the web. Visit **RSMeansOnline.com** for more information on our 24/7 online cost data.

What's more, you can increase your knowledge and improve your construction estimating and management performance with an RSMeans construction seminar or in-house training program. These two-day seminar programs offer unparalleled opportunities for everyone in your organization to become updated on a wide variety of construction-related issues.

RSMeans is also a worldwide provider of construction cost management and analysis services for commercial and government owners.

In short, RSMeans can provide you with the tools and expertise for constructing accurate and dependable construction estimates and budgets in a variety of ways.

Robert Snow Means Established a Tradition of Quality That Continues Today

Robert Snow Means spent years building RSMeans, making certain he always delivered a quality product.

Today, at RSMeans, we do more than talk about the quality of our data and the usefulness of our books. We stand behind all of our data, from historical cost indexes to construction materials and techniques to current costs.

If you have any questions about our products or services, please call us toll-free at 1-800-334-3509. Our customer service representatives will be happy to assist you. You can also visit our website at **www.rsmeans.com**

How the Book is Built: An Overview

The Construction Specifications Institute (CSI) and Construction Specifications Canada (CSC) have produced the 2012 edition of MasterFormat®, a system of titles and numbers used extensively to organize construction information.

All unit price data in the RSMeans cost data books is now arranged in the 50-division MasterFormat® 2012 system.

A Powerful Construction Tool

You have in your hands one of the most powerful construction tools available today. A successful project is built on the foundation of an accurate and dependable estimate. This book will enable you to construct just such an estimate.

For the casual user the book is designed to be:

- quickly and easily understood so you can get right to your estimate.
- filled with valuable information so you can understand the necessary factors that go into the cost estimate.

For the regular user, the book is designed to be:

- a handy desk reference that can be quickly referred to for key costs.
- a comprehensive, fully reliable source of current construction costs and productivity rates so you'll be prepared to estimate any project.
- a source book for preliminary project cost, product selections, and alternate materials and methods.

To meet all of these requirements, we have organized the book into the following clearly defined sections.

Quick Start

See our "Quick Start" instructions on the following page to get started right away.

Estimating with RSMeans Unit Price Cost Data

Please refer to these steps for guidance on completing an estimate using RSMeans unit price cost data.

How to Use the Book: The Details

This section contains an in-depth explanation of how the book is arranged . . . and how you can use it to determine a reliable construction cost estimate. It includes information about how we develop our cost figures and how to completely prepare your estimate.

Unit Price Section

All cost data has been divided into the 50 divisions according to the MasterFormat system of classification and numbering. For a listing of these divisions and an outline of their subdivisions, see the Unit Price Section Table of Contents.

Estimating tips are included at the beginning of each division.

Assemblies Section

The cost data in this section has been organized in an "Assemblies" format. These assemblies are the functional elements of a building and are arranged according to the 7 elements of the UNIFORMAT II classification system. For a complete explanation of a typical "Assemblies" page, see "How RSMeans Assemblies Data Works."

Reference Section

This section includes information on Equipment Rental Costs, Crew Listings, Historical Cost Indexes, City Cost Indexes, Location Factors, Reference Tables, Change Orders, Square Foot Costs, and a listing of Abbreviations.

Equipment Rental Costs: This section contains the average costs to rent and operate hundreds of pieces of construction equipment.

Crew Listings: This section lists all of the crews referenced in the book. For the purposes of this book, a crew is composed of more than one trade classification and/or the addition of power equipment to any trade classification. Power equipment is included in the cost of the crew. Costs are shown both with bare labor rates and with the installing contractor's overhead and profit added. For each, the total crew cost per eight-hour day and the composite cost per labor-hour are listed.

Historical Cost Indexes: These indexes provide you with data to adjust construction costs over time.

City Cost Indexes: All costs in this book are U.S. national averages. Costs vary because of the regional economy. You can adjust costs by CSI Division to over 700 locations throughout the U.S. and Canada by using the data in this section.

Location Factors: You can adjust total project costs to over 900 locations throughout the U.S. and Canada by using the data in this section.

Reference Tables: At the beginning of selected major classifications in the Unit Price section are reference numbers shown in a shaded box. These numbers refer you to related information in the Reference Section. In this section, you'll find reference tables, explanations, estimating information that support how we develop the unit price data, technical data, and estimating procedures.

Change Orders: This section includes information on the factors that influence the pricing of change orders.

Square Foot Costs: This section contains costs for 59 different building types that allow you to make a rough estimate for the overall cost of a project or its major components.

Abbreviations: A listing of abbreviations used throughout this book, along with the terms they represent, is included in this section.

Index

A comprehensive listing of all terms and subjects in this book will help you quickly find what you need when you are not sure where it occurs in MasterFormat.

The Scope of This Book

This book is designed to be as comprehensive and as easy to use as possible. To that end we have made certain assumptions and limited its scope in two key ways:

1. We have established material prices based on a national average.
2. We have computed labor costs based on a 30-city national average of union wage rates.

For a more detailed explanation of how the cost data is developed, see "How To Use the Book: The Details."

Project Size/Type

The material prices in RSMeans data cost books are "contractor's prices." They are the prices that contractors can expect to pay at the lumberyards, suppliers/distributers warehouses, etc. Small orders of speciality items would be higher than the costs shown, while very large orders, such as truckload lots, would be less. The variation would depend on the size, timing, and negotiating power of the contractor. The labor costs are primarily for new construction or major renovation rather than repairs or minor alterations.

With reasonable exercise of judgment, the figures can be used for any building work.

Absolute Essentials for a Quick Start

If you feel you are ready to use this book and don't think you will need the detailed instructions that begin on the following page, this Absolute Essentials for a Quick Start page is for you. These steps will allow you to get started estimating in a matter of minutes.

1 Scope

Think through the project that you will be estimating, and identify the many individual work tasks that will need to be covered in your estimate.

2 Quantify

Determine the number of units that will be required for each work task that you identified.

3 Pricing

Locate individual Unit Price line items that match the work tasks you identified. The Unit Price Section Table of Contents that begins on page 1 and the Index in the back of the book will help you find these line items.

4 Multiply

Multiply the Total Incl O&P cost for a Unit Price line item in the book by your quantity for that item. The price you calculate will be an estimate for a completed item of work performed by a subcontractor. Keep adding line items in this manner to build your estimate.

5 Project Overhead

Include project overhead items in your estimate. These items are needed to make the job run and are typically, but not always, provided by the General Contractor. They can be found in Division 1. An alternate method of estimating project overhead costs is to apply a percentage of the total project cost.

Include rented tools not included in crews, waste, rubbish handling, and cleanup.

6 Estimate Summary

Include General Contractor's markup on subcontractors, General Contractor's office overhead and profit, and sales tax on materials and equipment.

Adjust your estimate to the project's location by using the City Cost Indexes or Location Factors found in the Reference Section.

Editors' Note: We urge you to spend time reading and understanding the supporting material in the front of this book. An accurate estimate requires experience, knowledge, and careful calculation. The more you know about how we at RSMeans developed the data, the more accurate your estimate will be. In addition, it is important to take into consideration the reference material in the back of the book such as Equipment Listings, Crew Listings, City Cost Indexes, Location Factors, and Reference Tables.

Estimating with RSMeans Unit Price Cost Data

Following these steps will allow you to complete an accurate estimate using RSMeans Unit Price cost data.

1 Scope Out the Project

- Identify the individual work tasks that will need to be covered in your estimate.
- The Unit Price data in this book has been divided into 50 Divisions according to the CSI MasterFormat 2012—their titles are listed on the back cover of your book.
- Think through the project and identify those CSI Divisions needed in your estimate.
- The Unit Price Section Table of Contents on page 1 may also be helpful when scoping out your project.
- Experienced estimators find it helpful to begin with Division 2 and continue through completion. Division 1 can be estimated after the full project scope is known.

2 Quantify

- Determine the number of units required for each work task that you identified.
- Experienced estimators include an allowance for waste in their quantities. (Waste is not included in RSMeans Unit Price line items unless so stated.)

3 Price the Quantities

- Use the Unit Price Table of Contents, and the Index, to locate individual Unit Price line items for your estimate.
- Reference Numbers indicated within a Unit Price section refer to additional information that you may find useful.
- The crew indicates who is performing the work for that task. Crew codes are expanded in the Crew Listings in the Reference Section to include all trades and equipment that comprise the crew.
- The Daily Output is the amount of work the crew is expected to do in one day.
- The Labor-Hours value is the amount of time it will take for the crew to install one unit of work.
- The abbreviated Unit designation indicates the unit of measure upon which the crew, productivity, and prices are based.
- Bare Costs are shown for materials, labor, and equipment needed to complete the Unit Price line item. Bare costs do not include waste, project overhead, payroll insurance, payroll taxes, main office overhead, or profit.
- The Total Incl O&P cost is the billing rate or invoice amount of the installing contractor or subcontractor who performs the work for the Unit Price line item.

4 Multiply

- Multiply the total number of units needed for your project by the Total Incl O&P cost for each Unit Price line item.
- Be careful that your take off unit of measure matches the unit of measure in the Unit column.
- The price you calculate is an estimate for a completed item of work.
- Keep scoping individual tasks, determining the number of units required for those tasks, matching each task with individual Unit Price line items in the book, and multiply quantities by Total Incl O&P costs.
- An estimate completed in this manner is priced as if a subcontractor, or set of subcontractors, is performing the work. The estimate does not yet include Project Overhead or Estimate Summary components such as General Contractor markups on subcontracted work, General Contractor office overhead and profit, contingency, and location factor.

5 Project Overhead

- Include project overhead items from Division 1 – General Requirements.
- These items are needed to make the job run. They are typically, but not always, provided by the General Contractor. Items include, but are not limited to, field personnel,insurance, performance bond, permits, testing, temporary utilities, field office and storage facilities, temporary scaffolding and platforms, equipment mobilization and demobilization, temporary roads and sidewalks, winter protection, temporary barricades and fencing, temporary security, temporary signs, field engineering and layout, final cleaning and commissioning.
- Each item should be quantified, and matched to individual Unit Price line items in Division 1, and then priced and added to your estimate.
- An alternate method of estimating project overhead costs is to apply a percentage of the total project cost, usually 5% to 15% with an average of 10% (see General Conditions, page ix).
- Include other project related expenses in your estimate such as:
    - Rented equipment not itemized in the Crew Listings
    - Rubbish handling throughout the project (see 02 41 19.19)

6 Estimate Summary

- Include sales tax as required by laws of your state or county.
- Include the General Contractor's markup on self-performed work, usually 5% to 15% with an average of 10%.
- Include the General Contractor's markup on subcontracted work, usually 5% to 15% with an average of 10%.
- Include General Contractor's main office overhead and profit:
    - RSMeans gives general guidelines on the General Contractor's main office overhead (see section 01 31 13.40 and Reference Number R013113-50).
    - RSMeans gives no guidance on the General Contractor's profit.
    - Markups will depend on the size of the General Contractor's operations, his projected annual revenue, the level of risk he is taking on, and on the level of competition in the local area and for this project in particular.
- Include a contingency, usually 3% to 5%, if appropriate.
- Adjust your estimate to the project's location by using the City Cost Indexes or the Location Factors in the Reference Section:
    - Look at the rules on the pages for How to Use the City Cost Indexes to see how to apply the Indexes for your location.
    - When the proper Index or Factor has been identified for the project's location, convert it to a multiplier by dividing it by 100, and then multiply that multiplier by your estimated total cost. The original estimated total cost will now be adjusted up or down from the national average to a total that is appropriate for your location.

Editors' Notes:

1) *We urge you to spend time reading and understanding the supporting material in the front of this book. An accurate estimate requires experience, knowledge, and careful calculation. The more you know about how we at RSMeans developed the data, the more accurate your estimate will be. In addition, it is important to take into consideration the reference material in the back of the book such as Equipment Listings, Crew Listings, City Cost Indexes, Location Factors, and Reference Tables.*
2) *Contractors who are bidding or are involved in JOC, DOC, SABER, or IDIQ type contracts are cautioned that workers' compensation insurance, federal and state payroll taxes, waste, project supervision, project overhead, main office overhead, and profit are not included in bare costs. The coefficient or multiplier must cover these costs.*

How to Use the Book: The Details

What's Behind the Numbers? The Development of Cost Data

The staff at RSMeans continually monitors developments in the construction industry in order to ensure reliable, thorough, and up-to-date cost information. While overall construction costs may vary relative to general economic conditions, price fluctuations within the industry are dependent upon many factors. Individual price variations may, in fact, be opposite to overall economic trends. Therefore, costs are constantly tracked and complete updates are published yearly. Also, new items are frequently added in response to changes in materials and methods.

Costs—$ (U.S.)

All costs represent U.S. national averages and are given in U.S. dollars. The RSMeans City Cost Indexes can be used to adjust costs to a particular location. The City Cost Indexes for Canada can be used to adjust U.S. national averages to local costs in Canadian dollars. No exchange rate conversion is necessary.

G The processes or products identified by the green symbol in our publications have been determined to be environmentally responsible and/or resource-efficient solely by the RSMeans engineering staff. The inclusion of the green symbol does not represent compliance with any specific industry association or standard.

Material Costs

The RSMeans staff contacts manufacturers, dealers, distributors, and contractors all across the U.S. and Canada to determine national average material costs. If you have access to current material costs for your specific location, you may wish to make adjustments to reflect differences from the national average. Included within material costs are fasteners for a normal installation. RSMeans engineers use manufacturers' recommendations, written specifications, and/or standard construction practice for size and spacing of fasteners. Adjustments to material costs may be required for your specific application or location. The manufacturer's warranty is assumed. Extended warranties are not included in the material costs. Material costs do not include sales tax.

Labor Costs

Labor costs are based on the average of wage rates from 30 major U.S. cities. Rates are determined from labor union agreements or prevailing wages for construction trades for the current year. Rates, along with overhead and profit markups, are listed on the inside back cover of this book.

- If wage rates in your area vary from those used in this book, or if rate increases are expected within a given year, labor costs should be adjusted accordingly.

Labor costs reflect productivity based on actual working conditions. In addition to actual installation, these figures include time spent during a normal weekday on tasks such as, material receiving and handling, mobilization at site, site movement, breaks, and cleanup.

Productivity data is developed over an extended period so as not to be influenced by abnormal variations and reflects a typical average.

Equipment Costs

Equipment costs include not only rental, but also operating costs for equipment under normal use. The operating costs include parts and labor for routine servicing such as repair and replacement of pumps, filters, and worn lines. Normal operating expendables, such as fuel, lubricants, tires, and electricity (where applicable), are also included. Extraordinary operating expendables with highly variable wear patterns, such as diamond bits and blades, are excluded. These costs are included under materials. Equipment rental rates are obtained from industry sources throughout North America—contractors, suppliers, dealers, manufacturers, and distributors.

Equipment costs do not include operators' wages; nor do they include the cost to move equipment to a job site (mobilization) or from a job site (demobilization).

Equipment Cost/Day——The cost of power equipment required for each crew is included in the Crew Listings in the Reference Section (small tools that are considered as essential everyday tools are not listed out separately). The Crew Listings itemize specialized tools and heavy equipment along with labor trades. The daily cost of itemized equipment included in a crew is based on dividing the weekly bare rental rate by 5 (number of working days per week) and then adding the hourly operating cost times 8 (the number of hours per day). This Equipment Cost/Day is shown in the last column of the Equipment Rental Cost pages in the Reference Section.

Mobilization/Demobilization——The cost to move construction equipment from an equipment yard or rental company to the job site and back again is not included in equipment costs. Mobilization (to the site) and demobilization (from the site) costs can be found in the Unit Price Section. If a piece of equipment is already at the job site, it is not appropriate to utilize mob./demob. costs again in an estimate.

General Conditions

Cost data in this book is presented in two ways: Bare Costs and Total Cost including O&P (Overhead and Profit). General Conditions, when applicable, should also be added to the Total Cost including O&P. The costs for General Conditions are listed in Division 1 of the Unit Price Section and the Reference Section of this book. General Conditions for the *Installing Contractor* may range from 0% to 10% of the Total Cost including O&P. For the *General* or *Prime Contractor*, costs for General Conditions may range from 5% to 15% of the Total Cost including O&P, with a figure of 10% as the most typical allowance.

Overhead and Profit

Total Cost including O&P for the *Installing Contractor* is shown in the last column on both the Unit Price and the Assemblies pages of this book. This figure is the sum of the bare material cost plus 10% for profit, the bare labor cost plus total overhead and profit, and the bare equipment cost plus 10% for profit. Details for the calculation of Overhead and Profit on labor are shown on the inside back cover and in the Reference Section of this book. (See the "How RSMeans Data Works" for an example of this calculation.)

Factors Affecting Costs

Costs can vary depending upon a number of variables. Here's how we have handled the main factors affecting costs.

Quality——The prices for materials and the workmanship upon which productivity is based represent sound construction work. They are also in line with U.S. government specifications.

Overtime——We have made no allowance for overtime. If you anticipate premium time or work beyond normal working hours, be sure to make an appropriate adjustment to your labor costs.

Productivity——The productivity, daily output, and labor-hour figures for each line item are based on working an eight-hour day in daylight hours in moderate temperatures. For work that extends beyond normal work hours or is performed under adverse conditions, productivity may decrease. (See "How RSMeans Data Works" for more on productivity.)

Size of Project——The size, scope of work, and type of construction project will have a significant impact on cost. Economies of scale can reduce costs for large projects. Unit costs can often run higher for small projects.

Location——Material prices in this book are for metropolitan areas. However, in dense urban areas, traffic and site storage limitations may increase costs. Beyond a 20-mile radius of large cities, extra trucking or transportation charges may also increase the material costs slightly. On the other hand, lower wage rates may be in effect. Be sure to consider both of these factors when preparing an estimate, particularly if the job site is located in a central city or remote rural location. In addition, highly specialized subcontract items may require travel and per-diem expenses for mechanics.

Other Factors——

- season of year
- contractor management
- weather conditions
- local union restrictions
- building code requirements
- availability of:
  - adequate energy
  - skilled labor
  - building materials
- owner's special requirements/restrictions
- safety requirements
- environmental considerations
- traffic control

Unpredictable Factors——General business conditions influence "in-place" costs of all items. Substitute materials and construction methods may have to be employed. These may affect the installed cost and/or life cycle costs. Such factors may be difficult to evaluate and cannot necessarily be predicted on the basis of the job's location in a particular section of the country. Thus, where these factors apply, you may find significant but unavoidable cost variations for which you will have to apply a measure of judgment to your estimate.

Rounding of Costs

In general, all unit prices in excess of $5.00 have been rounded to make them easier to use and still maintain adequate precision of the results. The rounding rules we have chosen are in the following table.

Prices from...	Rounded to the nearest...
$.01 to $5.00	$.01
$5.01 to $20.00	$.05
$20.01 to $100.00	$.50
$100.01 to $300.00	$1.00
$300.01 to $1,000.00	$5.00
$1,000.01 to $10,000.00	$25.00
$10,000.01 to $50,000.00	$100.00
$50,000.01 and above	$500.00

How Subcontracted Items Affect Costs

A considerable portion of all large construction jobs is usually subcontracted. In fact, the percentage done by subcontractors is constantly increasing and may run over 90%. Since the workers employed by these companies do nothing else but install their particular product, they soon become expert in that line. The result is, installation by these firms is accomplished so efficiently that the total in-place cost, even adding the general contractor's overhead and profit, is no more, and often less, than if the principal contractor had handled the installation himself/herself. There is, moreover, the big advantage of having the work done correctly. Companies that deal with construction specialties are anxious to have their product perform well, and consequently, the installation will be the best possible.

Contingencies

The contractor should consider inflationary price trends and possible material shortages during the course of the job. These escalation factors are dependent upon both economic conditions and the anticipated time between the estimate and actual construction. If drawings are not complete or approved, or a budget cost is wanted, it is wise to add 5% to 10%. Contingencies, then, are a matter of judgment. Additional allowances for contingencies are shown in Division 1.

Estimating Precision

When making a construction cost estimate, ignore the cents column. Use only the total per-unit cost to the nearest dollar. The cents will average in a column of figures. A construction cost estimate of $257,323.37 is cumbersome. A figure of $257,325 is certainly more sensible, and $257,000 is better and just as likely to be correct.

If you follow this simple instruction, the time saved is tremendous, with an added important advantage. Using round figures leaves the professional estimator free to exercise judgment and common sense rather than being overwhelmed by a mass of computations.

Final Checklist

Estimating can be a straightforward process provided you remember the basics. Here's a checklist of some of the steps you should remember to complete before finalizing your estimate.

Did you remember to . . .

- factor in the City Cost Index for your locale?
- take into consideration which items have been marked up and by how much?
- mark up the entire estimate sufficiently for your purposes?
- read the background information on techniques and technical matters that could impact your project time span and cost?
- include all components of your project in the final estimate?
- double check your figures for accuracy?
- call RSMeans if you have any questions about your estimate or the data you've found in our publications?

For more information, please see "Tips for Accurate Estimating," R011105-05 in the Reference Section.

Remember, RSMeans stands behind its publications. If you have any questions about your estimate . . . about the costs you've used from our books . . . or even about the technical aspects of the job that may affect your estimate, feel free to call the RSMeans editors at 1-800-334-3509.

Free Quarterly Updates

Stay up-to-date throughout 2014 with RSMeans' free cost data updates four times a year. Sign up online to make sure you have access to the newest data. Every quarter we provide the city cost adjustment factors for hundreds of cities and key materials. Register at: **www.rsmeans2014updates.com**.

Unit Price Section

Table of Contents

Table of Contents (cont.)

How RSMeans Unit Price Data Works

All RSMeans unit price data is organized in the same way.

It is important to understand the structure, so that you can find information easily and use it correctly.

RSMeans **Line Numbers** consist of 12 characters, which identify a unique location in the database for each task. The first 6 or 8 digits conform to the Construction Specifications Institute MasterFormat® 2012. The remainder of the digits are a further breakdown by RSMeans in order to arrange items in understandable groups of similar tasks. Line numbers are consistent across all RSMeans publications, so a line number in any RSMeans product will always refer to the same unit of work.

RSMeans engineers have created **reference** information to assist you in your estimate. If there is information that applies to a section, it will be indicated at the start of the section. In this case, R033105-10 provides information on the proportionate quantities of formwork, reinforcing, and concrete used in cast-in-place concrete items such as footings, slabs, beams, and columns. The Reference Section is located in the back of the book on the pages with a gray edge.

RSMeans **Descriptions** are shown in a hierarchical structure to make them readable. In order to read a complete description, read up through the indents to the top of the section. Include everything that is above and to the left that is not contradicted by information below. For instance, the complete description for line 03 30 53.40 3550 is "Concrete in place, including forms (4 uses), Grade 60 rebar, concrete (Portland cement Type 1), placement and finishing unless otherwise indicated; Equipment pad (3000 psi), 4' x 4' x 6" thick."

When using **RSMeans data**, it is important to read through an entire section to ensure that you use the data that most closely matches your work. Note that sometimes there is additional information shown in the section that may improve your price. There are frequently lines that further describe, add to, or adjust data for specific situations.

03 30 Cast-In-Place Concrete

03 30 53 – Miscellaneous Cast-In-Place Concrete

03 30 53.40 Concrete In Place		
0010	**CONCRETE IN PLACE**	R033105-10
0020	Including forms (4 uses), Grade 60 rebar, concrete (Portland cement	R033105-20
0050	Type I), placement and finishing unless otherwise indicated	R033105-50
0300	Beams (3500 psi), 5 kip per L.F., 10' span	R033105-65
0350	25' span	R033105-70
0500	Chimney foundations (5000 psi), over 5 C.Y.	R033105-85
0510	(3500 psi), under 5 C.Y.	
0700	Columns, square (4000 psi), 12" x 12", less than 2% reinforcing	
3450	Over 10,000 S.F.	
3500	Add per floor for 3 to 6 stories high	
3520	For 7 to 20 stories high	
3540	Equipment pad (3000 psi), 3' x 3' x 6" thick	
3550	4' x 4' x 6" thick	
3560	5' x 5' x 8" thick	
3570	6' x 6' x 8" thick	

The data published in RSMeans print books represents a "national average" cost. This data should be modified to the project location using the **City Cost Indexes** or **Location Factors** tables found in the reference section (see pages 572–620). Use the location factors to adjust estimate totals if the project covers multiple trades. Use the city cost indexes (CCI) for single trade projects or projects where a more detailed analysis is required. All figures in the two tables are derived from the same research. The last row of data in the CCI, the weighted average, is the same as the numbers reported for each location in the location factor table.

Crews include labor or labor and equipment necessary to accomplish each task. In this case, Crew C-14H is used. RSMeans selects a crew to represent the workers and equipment that are typically used for that task. In this case, Crew C-14H consists of one carpenter foreman (outside), two carpenters, one rodman, one laborer, one cement finisher, and one gas engine vibrator. Details of all crews can be found in the reference section.

Crews

Crew No.	Bare Costs		Incl. Subs O&P		Cost Per Labor-Hour	
Crew C-14H	Hr.	Daily	Hr.	Daily	Bare Costs	Incl. O&P
1 Carpenter Foreman (outside)	$47.85	$382.80	$73.85	$590.80	$45.15	$69.42
2 Carpenters	45.85	733.60	70.75	1132.00		
1 Rodman (reinf.)	50.65	405.20	79.55	636.40		
1 Laborer	36.65	293.20	56.55	452.40		
1 Cement Finisher	44.05	352.40	65.10	520.80		
1 Gas Engine Vibrator		33.00		36.30	.69	.76
48 L.H., Daily Totals		$2200.20		$3368.70	$45.84	$70.18

The **Daily Output** is the amount of work that the crew can do in a normal 8-hour workday, including mobilization, layout, movement of materials, and cleanup. In this case, crew C-14H can install thirty 4' x 4' x 6" thick concrete pads in a day. Daily output is variable, based on many factors, including the size of the job, location, and environmental conditions. RSMeans data represents work done in daylight (or adequate lighting) and temperate conditions.

Bare Costs are the costs of materials, labor, and equipment that the installing contractor pays. They represent the cost, in U.S. dollars, for one unit of work. They do not include any markups for profit or labor burden.

Crew	Daily Output	Labor-Hours	Unit	2014 Bare Costs Material	Labor	Equipment	Total	Total Incl O&P
C-14A	15.62	12.804	C.Y.	3[illegible]5	590	47.50	952.50	1,325
"	18.55	10.782		335	495	40	870	1,175
C-14C	32.22	3.476		147	153	1.01	301.01	400
"	23.71	4.724		173	207	1.37	381.37	515
C-14A	11.96	16.722		355	770	62	1,187	1,675
	2200	.025		.80	1.04	.32	2.16	2.83
	31800	.002			.07	.02	.09	.13
	21200	.003			.11	.03	.14	.20
C-14H	45	1.067	Ea.	44.50	48	.74	93.24	124
	30	1.600		66.50	72	1.10	139.60	185
	18	2.667		119	120	1.84	240.84	320
	14	3.429		160	155	2.[illegible]7	317.37	415

The **Total Incl O&P column** is the total cost, including overhead and profit, that the installing contractor will charge the customer. This represents the cost of materials plus 10% profit, the cost of labor plus labor burden and 10% profit, and the cost of equipment plus 10% profit. It does not include the general contractor's overhead and profit. Note: See the inside back cover for details of how RSMeans calculates labor burden.

The **Total column** represents the total bare cost for the installing contractor, in U.S. dollars. In this case, the sum of $66.50 for material + $72.00 for labor + $1.10 for equipment is $139.60.

The figure in the **Labor Hours** column is the amount of labor required to perform one unit of work—in this case the amount of labor required to construct one 4' x 4' equipment pad. This figure is calculated by dividing the number of hours of labor in the crew by the daily output (48 labor hours divided by 30 pads = 1.6 hours of labor per pad). Multiply 1.600 times 60 to see the value in minutes: 60 x 1.6 = 96 minutes. Note: the labor hour figure is not dependent on the crew size. A change in crew size will result in a corresponding change in daily output, but the labor hours per unit of work will not change.

All RSMeans unit cost data includes the typical **Unit of Measure** used for estimating that item. For concrete-in-place the typical unit is cubic yards (C.Y.) or each (Ea.). For installing broadloom carpet it is square yard, and for gypsum board it is square foot. The estimator needs to take special care that the unit in the data matches the unit in the take-off. Unit conversions may be found in the Reference Section.

How RSMeans Unit Price Data Works (Continued)

Sample Estimate

This sample demonstrates the elements of an estimate, including a tally of the RSMeans data lines, and a summary of the markups on a contractor's work to arrive at a total cost to the owner. The RSMeans Location Factor is added at the bottom of the estimate to adjust the cost of the work to a specific location.

Work Performed: The body of the estimate shows the RSMeans data selected, including line number, a brief description of each item, its take-off unit and quantity, and the bare costs of materials, labor, and equipment. This estimate also includes a column titled "SubContract." This data is taken from the RSMeans column "Total Incl O&P," and represents the total that a subcontractor would charge a general contractor for the work, including the sub's markup for overhead and profit.

Division 1, General Requirements: This is the first division numerically, but the last division estimated. Division 1 includes project-wide needs provided by the general contractor. These requirements vary by project, but may include temporary facilities and utilities, security, testing, project cleanup, etc. For small projects a percentage can be used, typically between 5% and 15% of project cost. For large projects the costs may be itemized and priced individually.

Bonds: Bond costs should be added to the estimate. The figures here represent a typical performance bond, ensuring the owner that if the general contractor does not complete the obligations in the construction contract the bonding company will pay the cost for completion of the work.

Location Adjustment: RSMeans published data is based on national average costs. If necessary, adjust the total cost of the project using a location factor from the "Location Factor" table or the "City Cost Index" table. Use location factors if the work is general, covering multiple trades. If the work is by a single trade (e.g., masonry) use the more specific data found in the "City Cost Indexes."

Project Name: Pre-Engineered Steel Building		Architect: As Shown		
Location:	Anywhere, USA			
Line Number	Description	Qty	Unit	Material
03 30 53.40 3940	Strip footing, 12" x 24", reinforced	34	C.Y.	$4,624.00
03 30 53.40 3950	Strip footing, 12" x 36", reinforced	15	C.Y.	$1,950.00
03 11 13.65 3000	Concrete slab edge forms	500	L.F.	$155.00
03 22 11.10 0200	Welded wire fabric reinforcing	150	C.S.F.	$2,580.00
03 31 13.35 0300	Ready mix concrete, 4000 psi for slab on grade	278	C.Y.	$28,912.00
03 31 13.70 4300	Place, strike off & consolidate concrete slab	278	C.Y.	$0.00
03 35 13.30 0250	Machine float & trowel concrete slab	15,000	S.F.	$0.00
03 15 16.20 0140	Cut control joints in concrete slab	950	L.F.	$57.00
03 39 23.13 0300	Sprayed concrete curing membrane	150	C.S.F.	$1,642.50
Division 03	Subtotal			$39,920.50
08 36 13.10 2650	Manual 10' x 10' steel sectional overhead door	8	Ea.	$8,800.00
08 36 13.10 2860	Insulation and steel back panel for OH door	800	S.F.	$3,760.00
Division 08	Subtotal			$12,560.00
13 34 19.50 1100	Pre-Engineered Steel Building, 100' x 150' x 24'	15,000	SF Flr.	$0.00
13 34 19.50 6050	Framing for PESB door opening, 3' x 7'	4	Opng.	$0.00
13 34 19.50 6100	Framing for PESB door opening, 10' x 10'	8	Opng.	$0.00
13 34 19.50 6200	Framing for PESB window opening, 4' x 3'	6	Opng.	$0.00
13 34 19.50 5750	PESB door, 3' x 7', single leaf	4	Opng.	$2,340.00
13 34 19.50 7750	PESB sliding window, 4' x 3' with screen	6	Opng.	$2,250.00
13 34 19.50 6550	PESB gutter, eave type, 26 ga., painted	300	L.F.	$2,010.00
13 34 19.50 8650	PESB roof vent, 12" wide x 10' long	15	Ea.	$540.00
13 34 19.50 6900	PESB insulation, vinyl faced, 4" thick	27,400	S.F.	$10,960.00
Division 13	Subtotal			$18,100.00
		Subtotal		$70,580.50
Division 01	General Requirements @ 7%			4,940.64
		Estimate Subtotal		$75,521.14
		Sales Tax @ 5%		3,776.06
		Subtotal A		79,297.19
		GC O & P		7,929.72
		Subtotal B		87,226.91
		Contingency @ 5%		
		Subtotal C		
		Bond @ $12/1000 +10% O&P		
		Subtotal D		
		Location Adjustment Factor		
		Grand Total		

This estimate is based on an interactive spreadsheet.
A copy of this spreadsheet is located on the RSMeans website at **http://www.reedconstructiondata.com/rsmeans/extras/546011**.
You are free to download it and adjust it to your methodology.

Labor	Equipment	SubContract	Estimate Total	
		01/01/14	STD	
Labor	Equipment	SubContract	Estimate Total	
$3,468.00	$23.12	$0.00		
$1,230.00	$8.10	$0.00		
$1,160.00	$0.00	$0.00		
$3,900.00	$0.00	$0.00		
$0.00	$0.00	$0.00		
$4,642.60	$166.80	$0.00		
$8,700.00	$450.00	$0.00		
$370.50	$95.00	$0.00		
$922.50	$0.00	$0.00		
$24,393.60	**$743.02**	**$0.00**	**$65,057.12**	**Division 03**
$3,280.00	$0.00	$0.00		
$0.00	$0.00	$0.00		
$3,280.00	**$0.00**	**$0.00**	**$15,840.00**	**Division 08**
$0.00	$0.00	$345,000.00		
$0.00	$0.00	$2,260.00		
$0.00	$0.00	$9,200.00		
$0.00	$0.00	$3,300.00		
$656.00	$0.00	$0.00		
$558.00	$65.70	$0.00		
$768.00	$0.00	$0.00		
$3,060.00	$0.00	$0.00		
$8,768.00	$0.00	$0.00		
$13,810.00	**$65.70**	**$359,760.00**	**$391,735.70**	**Division 13**
$41,483.60	**$808.72**	**$359,760.00**	**$472,632.82**	**Subtotal**
2,903.85	**56.61**	**25,183.20**		**Gen. Requirements**
$44,387.45	**$865.33**	**$384,943.20**	**$472,632.82**	**Estimate Subtotal**
	43.27	**9,623.58**		**Sales tax**
44,387.45	**908.60**	**394,566.78**		**Subtotal**
24,368.71	**90.86**	**39,456.68**		**GC O & P**
68,756.16	**999.46**	**434,023.46**	**$591,005.99**	**Subtotal**
			29,550.30	**Contingency**
			$620,556.29	**Subtotal**
			8,191.34	**Bond**
			$628,747.63	**Subtotal**
102.30			**14,461.20**	**Location Adjustment**
			$643,208.83	**Grand Total**

Sales Tax: If the work is subject to state or local sales taxes, the amount must be added to the estimate. Sales tax may be added to material costs, equipment costs, and subcontracted work. In this case, sales tax was added in all three categories. It was assumed that approximately half the subcontracted work would be material cost, so the tax was applied to 50% of the subcontract total.

GC O&P: This entry represents the general contractor's markup on material, labor, equipment, and subcontractor costs. RSMeans' standard markup on materials, equipment, and subcontracted work is 10%. In this estimate, the markup on the labor performed by the GC's workers uses "Skilled Workers Average" shown in Column F on the table "Installing Contractor's Overhead & Profit," which can be found on the inside-back cover of the book.

Contingency: A factor for contingency may be added to any estimate to represent the cost of unknowns that may occur between the time that the estimate is performed and the time the project is constructed. The amount of the allowance will depend on the stage of design at which the estimate is done, and the contractor's assessment of the risk involved. Refer to section 01 21 16.50 for contigency allowances.

Estimating Tips

01 20 00 Price and Payment Procedures

- Allowances that should be added to estimates to cover contingencies and job conditions that are not included in the national average material and labor costs are shown in section 01 21.
- When estimating historic preservation projects (depending on the condition of the existing structure and the owner's requirements), a 15%-20% contingency or allowance is recommended, regardless of the stage of the drawings.

01 30 00 Administrative Requirements

- Before determining a final cost estimate, it is a good practice to review all the items listed in Subdivisions 01 31 and 01 32 to make final adjustments for items that may need customizing to specific job conditions.
- Requirements for initial and periodic submittals can represent a significant cost to the General Requirements of a job. Thoroughly check the submittal specifications when estimating a project to determine any costs that should be included.

01 40 00 Quality Requirements

- All projects will require some degree of Quality Control. This cost is not included in the unit cost of construction listed in each division. Depending upon the terms of the contract, the various costs of inspection and testing can be the responsibility of either the owner or the contractor. Be sure to include the required costs in your estimate.

01 50 00 Temporary Facilities and Controls

- Barricades, access roads, safety nets, scaffolding, security, and many more requirements for the execution of a safe project are elements of direct cost. These costs can easily be overlooked when preparing an estimate. When looking through the major classifications of this subdivision, determine which items apply to each division in your estimate.
- Construction Equipment Rental Costs can be found in the Reference Section in section 01 54 33. Operators' wages are not included in equipment rental costs.
- Equipment mobilization and demobilization costs are not included in equipment rental costs and must be considered separately in section 01 54 36.50.
- The cost of small tools provided by the installing contractor for his workers is covered in the "Overhead" column on the "Installing Contractor's Overhead and Profit" table that lists labor trades, base rates and markups and, therefore, is included in the "Total Incl. O&P" cost of any Unit Price line item. For those users who are constrained by contract terms to use only bare costs, there are two line items in section 01 54 39.70 for small tools as a percentage of bare labor cost. If some of those users are further constrained by contract terms to refrain from using line items from Division 1, you are advised to cover the cost of small tools within your coefficient or multiplier.

01 70 00 Execution and Closeout Requirements

- When preparing an estimate, thoroughly read the specifications to determine the requirements for Contract Closeout. Final cleaning, record documentation, operation and maintenance data, warranties and bonds, and spare parts and maintenance materials can all be elements of cost for the completion of a contract. Do not overlook these in your estimate.

Reference Numbers

Reference numbers are shown in shaded boxes at the beginning of some major classifications. These numbers refer to related items in the Reference Section. The reference information may be an estimating procedure, an alternate pricing method, or technical information.

Note: Not all subdivisions listed here necessarily appear in this publication.

01 11 Summary of Work

01 11 31 – Professional Consultants

01 11 31.30 Engineering Fees		Crew	Daily Output	Labor-Hours	Unit	2014 Bare Costs Material	Labor	Equipment	Total	Total Incl O&P
0010	**ENGINEERING FEES**									
0020	Educational planning consultant, minimum				Project				.50%	.50%
0100	Maximum				"				2.50%	2.50%
0200	Electrical, minimum				Contrct				4.10%	4.10%
0300	Maximum								10.10%	10.10%
0400	Elevator & conveying systems, minimum								2.50%	2.50%
0500	Maximum								5%	5%
0600	Food service & kitchen equipment, minimum								8%	8%
0700	Maximum								12%	12%
0800	Landscaping & site development, minimum								2.50%	2.50%
0900	Maximum								6%	6%
1000	Mechanical (plumbing & HVAC), minimum								4.10%	4.10%
1100	Maximum								10.10%	10.10%
1200	Structural, minimum				Project				1%	1%
1300	Maximum				"				2.50%	2.50%

01 21 Allowances

01 21 55 – Job Conditions Allowance

01 21 55.50 Job Conditions		Crew	Daily Output	Labor-Hours	Unit	Material	Labor	Equipment	Total	Total Incl O&P
0010	**JOB CONDITIONS** Modifications to applicable									
0020	cost summaries									
0100	Economic conditions, favorable, deduct				Project				2%	2%
0200	Unfavorable, add								5%	5%
0500	General Contractor management, experienced, deduct								2%	2%
0600	Inexperienced, add								10%	10%
0700	Labor availability, surplus, deduct								1%	1%
0800	Shortage, add								10%	10%
1100	Subcontractor availability, surplus, deduct								5%	5%
1200	Shortage, add								12%	12%

01 21 57 – Overtime Allowance

01 21 57.50 Overtime		Crew	Daily Output	Labor-Hours	Unit	Material	Labor	Equipment	Total	Total Incl O&P
0010	**OVERTIME** for early completion of projects or where R012909-90									
0020	labor shortages exist, add to usual labor, up to				Costs		100%			

01 21 61 – Cost Indexes

01 21 61.10 Construction Cost Index		Crew	Daily Output	Labor-Hours	Unit	Material	Labor	Equipment	Total	Total Incl O&P
0010	**CONSTRUCTION COST INDEX** (Reference) over 930 zip code locations in									
0020	the U.S. and Canada, total bldg. cost, min. (Longview, TX)				%					73.50%
0050	Average									100%
0100	Maximum (New York, NY)									131.10%

01 21 61.20 Historical Cost Indexes		Crew	Daily Output	Labor-Hours	Unit	Material	Labor	Equipment	Total	Total Incl O&P
0010	**HISTORICAL COST INDEXES** (See Reference Section)									

01 21 63 – Taxes

01 21 63.10 Taxes		Crew	Daily Output	Labor-Hours	Unit	Material	Labor	Equipment	Total	Total Incl O&P
0010	**TAXES** R012909-80									
0020	Sales tax, State, average				%	5.06%				
0050	Maximum R012909-85					7.25%				
0200	Social Security, on first $115,800 of wages						7.65%			
0300	Unemployment, combined Federal and State, minimum						.60%			
0350	Average						7.80%			

01 21 Allowances

01 21 63 – Taxes

01 21 63.10 Taxes		Crew	Daily Output	Labor-Hours	Unit	Material	2014 Bare Costs Labor	Equipment	Total	Total Incl O&P
0400	Maximum				%		13.50%			

01 31 Project Management and Coordination

01 31 13 – Project Coordination

01 31 13.20 Field Personnel

		Crew	Daily Output	Labor-Hours	Unit	Material	Labor	Equipment	Total	Total Incl O&P
0010	**FIELD PERSONNEL**									
0020	Clerk, average				Week		435		435	675
0100	Field engineer, minimum						1,025		1,025	1,600
0120	Average						1,350		1,350	2,100
0140	Maximum						1,525		1,525	2,375
0160	General purpose laborer, average						1,475		1,475	2,250
0180	Project manager, minimum						1,925		1,925	2,975
0200	Average						2,225		2,225	3,425
0220	Maximum						2,525		2,525	3,925
0240	Superintendent, minimum						1,875		1,875	2,900
0260	Average						2,050		2,050	3,200
0280	Maximum						2,350		2,350	3,625
0290	Timekeeper, average				↓		1,200		1,200	1,850

01 31 13.30 Insurance

		Crew	Daily Output	Labor-Hours	Unit	Material	Labor	Equipment	Total	Total Incl O&P
0010	**INSURANCE** R013113-40									
0020	Builders risk, standard, minimum				Job				.24%	.24%
0050	Maximum R013113-60								.64%	.64%
0200	All-risk type, minimum								.25%	.25%
0250	Maximum				↓				.62%	.62%
0400	Contractor's equipment floater, minimum				Value				.50%	.50%
0450	Maximum				"				1.50%	1.50%
0800	Workers' compensation & employer's liability, average									
0850	by trade, carpentry, general				Payroll		15.38%			
0900	Clerical						.51%			
0950	Concrete						13.15%			
1000	Electrical						5.83%			
1050	Excavation						9.69%			
1250	Masonry						13.49%			
1300	Painting & decorating						11.54%			
1350	Pile driving						14.65%			
1450	Plumbing						7.08%			
1500	Roofing						31.29%			
1600	Steel erection, structural						34.64%			
1700	Waterproofing, brush or hand caulking						6.86%			
1800	Wrecking						29.21%			
2000	Range of 35 trades in 50 states, excl. wrecking, min.						2.48%			
2100	Average						14%			
2200	Maximum				↓		150%			

01 31 13.40 Main Office Expense

		Crew	Daily Output	Labor-Hours	Unit	Material	Labor	Equipment	Total	Total Incl O&P
0010	**MAIN OFFICE EXPENSE** Average for General Contractors R013113-50									
0020	As a percentage of their annual volume									
0030	Annual volume to $300,000, minimum				% Vol.				20%	
0040	Maximum								30%	
0060	To $500,000, minimum								17%	
0070	Maximum				↓				22%	

01 31 Project Management and Coordination

01 31 13 – Project Coordination

01 31 13.40 Main Office Expense		Crew	Daily Output	Labor-Hours	Unit	2014 Bare Costs Material	Labor	Equipment	Total	Total Incl O&P
0080	To $1,000,000, minimum				% Vol.				16%	
0090	Maximum								19%	
0110	To $3,000,000, minimum								14%	
0120	Maximum								16%	
0125	Annual volume under 1 million dollars								17.50%	
0130	To $5,000,000, minimum								8%	
0140	Maximum								10%	
0150	Up to 4.0 million dollars								6.80%	
0200	Up to 7.0 million dollars								5.60%	
0250	Up to 10 million dollars								5.10%	
0300	Over 10 million dollars				↓				3.90%	

01 31 13.50 General Contractor's Mark-Up		Crew	Daily Output	Labor-Hours	Unit	Material	Labor	Equipment	Total	Total Incl O&P
0010	**GENERAL CONTRACTOR'S MARK-UP** on Change Orders									
0200	Extra work, by subcontractors, add R012909-80				%				10%	10%
0250	By General Contractor, add								15%	15%
0400	Omitted work, by subcontractors, deduct all but								5%	5%
0450	By General Contractor, deduct all but								7.50%	7.50%
0600	Overtime work, by subcontractors, add								15%	15%
0650	By General Contractor, add				↓				10%	10%
1150	Overhead markup, see Section 01 31 13.80									

01 31 13.80 Overhead and Profit		Crew	Daily Output	Labor-Hours	Unit	Material	Labor	Equipment	Total	Total Incl O&P
0010	**OVERHEAD & PROFIT** Allowance to add to items in this									
0020	book that do not include Subs O&P, average				%				25%	
0100	Allowance to add to items in this book that									
0110	do include Subs O&P, minimum				%				5%	5%
0150	Average								10%	10%
0200	Maximum								15%	15%
0300	Typical, by size of project, under $100,000								30%	
0350	$500,000 project								25%	
0400	$2,000,000 project								20%	
0450	Over $10,000,000 project				↓				15%	

01 31 13.90 Performance Bond		Crew	Daily Output	Labor-Hours	Unit	Material	Labor	Equipment	Total	Total Incl O&P
0010	**PERFORMANCE BOND**									
0020	For buildings, minimum				Job				.60%	.60%
0100	Maximum								2.50%	2.50%
0190	Highways & Bridges, new construction, minimum								1%	1%
0200	Maximum								1.50%	1.50%
0300	Highways & Bridges, resurfacing, minimum								.40%	.40%
0350	Maximum				↓				.94%	.94%

01 32 Construction Progress Documentation

01 32 13 – Scheduling of work

01 32 13.50 Scheduling		Crew	Daily Output	Labor-Hours	Unit	Material	Labor	Equipment	Total	Total Incl O&P
0010	**SCHEDULING**									
0020	Critical path, as % of architectural fee, minimum				%				.50%	.50%
0100	Maximum				"				1%	1%
0300	Computer-update, micro, no plots, minimum				Ea.				455	500
0400	Including plots, maximum				"				1,450	1,600
0600	Rule of thumb, CPM scheduling, small job ($10 Million)				Job				.05%	.05%
0650	Large job ($50 Million +)				↓				.03%	.03%

01 32 Construction Progress Documentation

01 32 13 – Scheduling of work

01 32 13.50 Scheduling		Crew	Daily Output	Labor-Hours	Unit	2014 Bare Costs Material	Labor	Equipment	Total	Total Incl O&P
0700	Including cost control, small job				Job				.08%	.08%
0750	Large job								.04%	.04%

01 32 33 – Photographic Documentation

01 32 33.50 Photographs		Crew	Daily Output	Labor-Hours	Unit	Material	Labor	Equipment	Total	Total Incl O&P
0010	**PHOTOGRAPHS**									
0020	8" x 10", 4 shots, 2 prints ea., std. mounting				Set	475			475	525
0100	Hinged linen mounts					530			530	580
0200	8" x 10", 4 shots, 2 prints each, in color					425			425	470
0300	For I.D. slugs, add to all above					5.30			5.30	5.85
0500	Aerial photos, initial fly-over, 6 shots, 1 print ea., 8" x 10"					865			865	950
0550	11" x 14" prints					980			980	1,075
0600	16" x 20" prints					1,175			1,175	1,300
0700	For full color prints, add					40%				
0750	Add for traffic control area					305			305	335
0900	For over 30 miles from airport, add per				Mile	5.55			5.55	6.15
1000	Vertical photography, 4 to 6 shots with									
1010	different scales, 1 print each				Set	1,150			1,150	1,275
1500	Time lapse equipment, camera and projector, buy				Ea.	2,650			2,650	2,925
1550	Rent per month				"	1,600			1,600	1,750
1700	Cameraman and film, including processing, B.&W.				Day	1,225			1,225	1,350
1720	Color				"	1,425			1,425	1,550

01 41 Regulatory Requirements

01 41 26 – Permit Requirements

01 41 26.50 Permits		Crew	Daily Output	Labor-Hours	Unit	Material	Labor	Equipment	Total	Total Incl O&P
0010	**PERMITS**									
0020	Rule of thumb, most cities, minimum				Job				.50%	.50%
0100	Maximum				"				2%	2%

01 45 Quality Control

01 45 23 – Testing and Inspecting Services

01 45 23.50 Testing		Crew	Daily Output	Labor-Hours	Unit	Material	Labor	Equipment	Total	Total Incl O&P
0010	**TESTING** and Inspecting Services									
0200	Asphalt testing, compressive strength Marshall stability, set of 3				Ea.				145	165
0220	Density, set of 3								86	95
0250	Extraction, individual tests on sample								136	150
0300	Penetration								41	45
0350	Mix design, 5 specimens								182	200
0360	Additional specimen								36	40
0400	Specific gravity								41	45
0420	Swell test								64	70
0450	Water effect and cohesion, set of 6								182	200
0470	Water effect and plastic flow								64	70
0600	Concrete testing, aggregates, abrasion, ASTM C 131								136	150
0650	Absorption, ASTM C 127								42	46
0800	Petrographic analysis, ASTM C 295								775	850
0900	Specific gravity, ASTM C 127								50	55
1000	Sieve analysis, washed, ASTM C 136								59	65
1050	Unwashed								59	65

01 45 Quality Control

01 45 23 – Testing and Inspecting Services

01 45 23.50 Testing		Crew	Daily Output	Labor-Hours	Unit	Material	Labor	Equipment	Total	Total Incl O&P
						2014 Bare Costs				
1200	Sulfate soundness				Ea.				114	125
1300	Weight per cubic foot								36	40
1500	Cement, physical tests, ASTM C 150								320	350
1600	Chemical tests, ASTM C 150								245	270
1800	Compressive test, cylinder, delivered to lab, ASTM C 39								12	13
1900	Picked up by lab, minimum								14	15
1950	Average								18	20
2000	Maximum								27	30
2200	Compressive strength, cores (not incl. drilling), ASTM C 42								36	40
2250	Core drilling, 4" diameter (plus technician)				Inch				23	25
2260	Technician for core drilling				Hr.				45	50
2300	Patching core holes				Ea.				22	24
2400	Drying shrinkage at 28 days								236	260
2500	Flexural test beams, ASTM C 78								59	65
2600	Mix design, one batch mix								259	285
2650	Added trial batches								120	132
2800	Modulus of elasticity, ASTM C 469								164	180
2900	Tensile test, cylinders, ASTM C 496								45	50
3000	Water-Cement ratio curve, 3 batches								141	155
3100	4 batches								186	205
3300	Masonry testing, absorption, per 5 brick, ASTM C 67								45	50
3350	Chemical resistance, per 2 brick								50	55
3400	Compressive strength, per 5 brick, ASTM C 67								68	75
3420	Efflorescence, per 5 brick, ASTM C 67								68	75
3440	Imperviousness, per 5 brick								87	96
3470	Modulus of rupture, per 5 brick								86	95
3500	Moisture, block only								32	35
3550	Mortar, compressive strength, set of 3								23	25
4100	Reinforcing steel, bend test								55	61
4200	Tensile test, up to #8 bar								36	40
4220	#9 to #11 bar								41	45
4240	#14 bar and larger								64	70
4400	Soil testing, Atterberg limits, liquid and plastic limits								59	65
4510	Hydrometer analysis								109	120
4530	Specific gravity, ASTM D 354								44	48
4600	Sieve analysis, washed, ASTM D 422								55	60
4700	Unwashed, ASTM D 422								59	65
4710	Consolidation test (ASTM D2435), minimum								250	275
4715	Maximum								430	475
4720	Density and classification of undisturbed sample								73	80
4735	Soil density, nuclear method, ASTM D2922								35	38.50
4740	Sand cone method ASTM D1556								27	30
4750	Moisture content, ASTM D 2216								9	10
4780	Permeability test, double ring infiltrometer								500	550
4800	Permeability, var. or constant head, undist., ASTM D 2434								227	250
4850	Recompacted								250	275
4900	Proctor compaction, 4" standard mold, ASTM D 698								123	135
4950	6" modified mold								68	75
5100	Shear tests, triaxial, minimum								410	450
5150	Maximum								545	600
5300	Direct shear, minimum, ASTM D 3080								320	350
5350	Maximum								410	450
5550	Technician for inspection, per day, earthwork								320	350

01 45 Quality Control

01 45 23 – Testing and Inspecting Services

01 45 23.50 Testing		Crew	Daily Output	Labor-Hours	Unit	2014 Bare Costs Material	Labor	Equipment	Total	Total Incl O&P
5570	Concrete				Ea.				280	310
5650	Bolting								400	440
5750	Roofing								480	530
5790	Welding								480	530
5820	Non-destructive metal testing, dye penetrant				Day				310	340
5840	Magnetic particle								310	340
5860	Radiography								450	495
5880	Ultrasonic								310	340
5900	Vibration monitoring, seismograph and technician								450	495
5910	Seismograph, rental only				Week				250	275
6000	Welding certification, minimum				Ea.				91	100
6100	Maximum				"				250	275
7000	Underground storage tank									
7500	Volumetric tightness test ,<=12,000 gal.				Ea.				435	480
7510	<=30,000 gal.				"				615	675
7600	Vadose zone (soil gas) sampling, 10-40 samples, min.				Day				1,375	1,500
7610	Maximum				"				2,275	2,500
7700	Ground water monitoring incl. drilling 3 wells, min.				Total				4,550	5,000
7710	Maximum				"				6,375	7,000
8000	X-ray concrete slabs				Ea.				182	200

01 51 Temporary Utilities

01 51 13 – Temporary Electricity

01 51 13.50 Temporary Power Equip (Pro-Rated Per Job)

		Crew	Daily Output	Labor-Hours	Unit	Material	Labor	Equipment	Total	Total Incl O&P
0010	**TEMPORARY POWER EQUIP (PRO-RATED PER JOB)**									
0020	Service, overhead feed, 3 use									
0030	100 Amp	1 Elec	1.25	6.400	Ea.	665	340		1,005	1,250
0040	200 Amp		1	8		860	425		1,285	1,600
0050	400 Amp		.75	10.667		1,575	570		2,145	2,600
0060	600 Amp		.50	16		2,300	855		3,155	3,800
0100	Underground feed, 3 use									
0110	100 Amp	1 Elec	2	4	Ea.	630	213		843	1,025
0120	200 Amp		1.15	6.957		855	370		1,225	1,500
0130	400 Amp		1	8		1,575	425		2,000	2,375
0140	600 Amp		.75	10.667		1,975	570		2,545	3,025
0150	800 Amp		.50	16		2,975	855		3,830	4,550
0160	1000 Amp		.35	22.857		3,325	1,225		4,550	5,475
0170	1200 Amp		.25	32		3,725	1,700		5,425	6,650
0180	2000 Amp		.20	40		4,525	2,125		6,650	8,175
0200	Transformers, 3 use									
0210	30 kVA	1 Elec	1	8	Ea.	1,650	425		2,075	2,475
0220	45 kVA		.75	10.667		1,975	570		2,545	3,025
0230	75 kVA		.50	16		3,325	855		4,180	4,925
0240	112.5 kVA		.40	20		3,600	1,075		4,675	5,575
0250	Feeder, PVC, CU wire in trench									
0260	60 Amp	1 Elec	96	.083	L.F.	4.09	4.45		8.54	11.15
0270	100 Amp		85	.094		7.80	5		12.80	16.10
0280	200 Amp		59	.136		18.75	7.25		26	31.50
0290	400 Amp		42	.190		43.50	10.15		53.65	62.50
0300	Feeder, PVC, aluminum wire in trench									
0310	60 Amp	1 Elec	96	.083	L.F.	4.49	4.45		8.94	11.60

01 51 13 – Temporary Electricity

01 51 13.50 Temporary Power Equip (Pro-Rated Per Job)		Crew	Daily Output	Labor-Hours	Unit	Material	2014 Bare Costs Labor	Equipment	Total	Total Incl O&P
0320	100 Amp	1 Elec	85	.094	L.F.	5	5		10	13
0330	200 Amp		59	.136		10.75	7.25		18	22.50
0340	400 Amp	↓	42	.190	↓	22.50	10.15		32.65	39.50
0350	Feeder, EMT, CU wire									
0360	60 Amp	1 Elec	90	.089	L.F.	3.43	4.74		8.17	10.85
0370	100 Amp		80	.100		8	5.35		13.35	16.80
0380	200 Amp		60	.133		17.15	7.10		24.25	29.50
0390	400 Amp	↓	35	.229	↓	50.50	12.20		62.70	74
0400	Feeder, EMT, alum. wire									
0410	60 Amp	1 Elec	90	.089	L.F.	4.71	4.74		9.45	12.30
0420	100 Amp		80	.100		5.90	5.35		11.25	14.50
0430	200 Amp		60	.133		16	7.10		23.10	28.50
0440	400 Amp	↓	35	.229	↓	28	12.20		40.20	49.50
0500	Equipment, 3 use									
0510	Spider box 50 Amp	1 Elec	8	1	Ea.	895	53.50		948.50	1,075
0520	Lighting cord 100′		8	1		116	53.50		169.50	208
0530	Light stanchion	↓	8	1	↓	70.50	53.50		124	158
0540	Temporary cords, 100′, 3 use									
0550	Feeder cord, 50 Amp	1 Elec	16	.500	Ea.	390	26.50		416.50	470
0560	Feeder cord, 100 Amp		12	.667		535	35.50		570.50	645
0570	Tap cord, 50 Amp		12	.667		221	35.50		256.50	296
0580	Tap cord, 100 Amp	↓	6	1.333	↓	535	71		606	695
0590	Temporary cords, 50′, 3 use									
0600	Feeder cord, 50 Amp	1 Elec	16	.500	Ea.	198	26.50		224.50	258
0610	Feeder cord, 100 Amp		12	.667		370	35.50		405.50	460
0620	Tap cord, 50 Amp		12	.667		109	35.50		144.50	173
0630	Tap cord, 100 Amp	↓	6	1.333	↓	370	71		441	510
0700	Connections									
0710	Compressor or pump									
0720	30 Amp	1 Elec	7	1.143	Ea.	23.50	61		84.50	117
0730	60 Amp		5.30	1.509		43.50	80.50		124	169
0740	100 Amp	↓	4	2	↓	97	107		204	267
0750	Tower crane									
0760	60 Amp	1 Elec	4.50	1.778	Ea.	43.50	95		138.50	190
0770	100 Amp	"	3	2.667	"	97	142		239	320
0780	Manlift									
0790	Single	1 Elec	3	2.667	Ea.	42	142		184	260
0800	Double	"	2	4	"	76.50	213		289.50	405
0810	Welder with disconnect									
0820	50 Amp	1 Elec	5	1.600	Ea.	271	85.50		356.50	425
0830	100 Amp		3.80	2.105		455	112		567	670
0840	200 Amp		2.50	3.200		820	171		991	1,150
0850	400 Amp	↓	1	8	↓	1,975	425		2,400	2,825

01 51 13.80 Temporary Utilities		Crew	Daily Output	Labor-Hours	Unit	Material	Labor	Equipment	Total	Total Incl O&P
0010	**TEMPORARY UTILITIES**									
0100	Heat, incl. fuel and operation, per week, 12 hrs. per day	1 Skwk	100	.080	CSF Flr	28	3.78		31.78	36.50
0200	24 hrs. per day	"	60	.133		53.50	6.30		59.80	68.50
0350	Lighting, incl. service lamps, wiring & outlets, minimum	1 Elec	34	.235		2.80	12.55		15.35	22
0360	Maximum	"	17	.471		5.85	25		30.85	44
0400	Power for temp lighting only, 6.6 KWH, per month								.92	1.01
0430	11.8 KWH, per month								1.65	1.82
0450	23.6 KWH, per month				↓				3.30	3.63

01 51 Temporary Utilities

01 51 13 – Temporary Electricity

01 51 13.80 Temporary Utilities		Crew	Daily Output	Labor-Hours	Unit	2014 Bare Costs Material	Labor	Equipment	Total	Total Incl O&P
0600	Power for job duration incl. elevator, etc., minimum				CSF Flr				47	51.50
0650	Maximum								110	121
0700	Temporary construction water bill per mo. average				Month	63.50			63.50	70
1000	Toilet, portable, see Equip. Rental 01 54 33 in Reference Section									

01 52 Construction Facilities

01 52 13 – Field Offices and Sheds

01 52 13.20 Office and Storage Space

		Crew	Daily Output	Labor-Hours	Unit	Material	Labor	Equipment	Total	Total Incl O&P
0010	**OFFICE AND STORAGE SPACE**									
0020	Office trailer, furnished, no hookups, 20′ x 8′, buy	2 Skwk	1	16	Ea.	9,150	755		9,905	11,300
0250	Rent per month					169			169	185
0300	32′ x 8′, buy	2 Skwk	.70	22.857		14,000	1,075		15,075	17,100
0350	Rent per month					203			203	223
0400	50′ x 10′, buy	2 Skwk	.60	26.667		22,400	1,250		23,650	26,600
0450	Rent per month					305			305	335
0500	50′ x 12′, buy	2 Skwk	.50	32		27,000	1,525		28,525	32,100
0550	Rent per month					350			350	385
0700	For air conditioning, rent per month, add					48.50			48.50	53.50
0800	For delivery, add per mile				Mile	11			11	12.10
0890	Delivery each way				Ea.	200			200	220
0900	Bunk house trailer, 8′ x 40′ duplex dorm with kitchen, no hookups, buy	2 Carp	1	16		37,300	735		38,035	42,100
0910	9 man with kitchen and bath, no hookups, buy		1	16		38,000	735		38,735	42,900
0920	18 man sleeper with bath, no hookups, buy		1	16		49,000	735		49,735	55,000
1000	Portable buildings, prefab, on skids, economy, 8′ x 8′		265	.060	S.F.	29.50	2.77		32.27	37
1100	Deluxe, 8′ x 12′		150	.107	"	24	4.89		28.89	34
1200	Storage boxes, 20′ x 8′, buy	2 Skwk	1.80	8.889	Ea.	2,775	420		3,195	3,700
1250	Rent per month					78.50			78.50	86
1300	40′ x 8′, buy	2 Skwk	1.40	11.429		3,700	540		4,240	4,900
1350	Rent per month					99			99	109

01 52 13.40 Field Office Expense

		Crew	Daily Output	Labor-Hours	Unit	Material	Labor	Equipment	Total	Total Incl O&P
0010	**FIELD OFFICE EXPENSE**									
0100	Office equipment rental average				Month	200			200	220
0120	Office supplies, average				"	75			75	82.50
0125	Office trailer rental, see Section 01 52 13.20									
0140	Telephone bill; avg. bill/month incl. long dist.				Month	81			81	89
0160	Lights & HVAC				"	152			152	167

01 54 Construction Aids

01 54 09 – Protection Equipment

01 54 09.60 Safety Nets

		Crew	Daily Output	Labor-Hours	Unit	Material	Labor	Equipment	Total	Total Incl O&P
0010	**SAFETY NETS**									
0020	No supports, stock sizes, nylon, 3 1/2″ mesh				S.F.	2.96			2.96	3.26
0100	Polypropylene, 6″ mesh					1.59			1.59	1.75
0200	Small mesh debris nets, 1/4″ mesh, stock sizes					.74			.74	.81
0220	Combined 3 1/2″ mesh and 1/4″ mesh, stock sizes					4.67			4.67	5.15
0300	Monthly rental, 4″ mesh, stock sizes, 1st month					.50			.50	.55
0320	2nd month rental					.25			.25	.28
0340	Maximum rental/year					1.15			1.15	1.27

01 54 Construction Aids

01 54 16 – Temporary Hoists

	01 54 16.50 Weekly Forklift Crew	Crew	Daily Output	Labor-Hours	Unit	Material	2014 Bare Costs Labor	Equipment	Total	Total Incl O&P
0010	**WEEKLY FORKLIFT CREW**									
0100	All-terrain forklift, 45′ lift, 35′ reach, 9000 lb. capacity	A-3P	.20	40	Week		1,875	2,650	4,525	5,775

01 54 19 – Temporary Cranes

	01 54 19.50 Daily Crane Crews	Crew	Daily Output	Labor-Hours	Unit	Material	Labor	Equipment	Total	Total Incl O&P
0010	**DAILY CRANE CREWS** for small jobs, portal to portal									
0100	12-ton truck-mounted hydraulic crane	A-3H	1	8	Day		400	875	1,275	1,575
0200	25-ton	A-3I	1	8			400	1,000	1,400	1,700
0300	40-ton	A-3J	1	8			400	1,250	1,650	1,975
0400	55-ton	A-3K	1	16			750	1,650	2,400	2,975
0500	80-ton	A-3L	1	16			750	2,400	3,150	3,775
0600	100-ton	A-3M	1	16			750	2,375	3,125	3,775
0900	If crane is needed on a Saturday, Sunday or Holiday									
0910	At time-and-a-half, add				Day		50%			
0920	At double time, add				"		100%			

	01 54 19.60 Monthly Tower Crane Crew	Crew	Daily Output	Labor-Hours	Unit	Material	Labor	Equipment	Total	Total Incl O&P
0010	**MONTHLY TOWER CRANE CREW**, excludes concrete footing									
0100	Static tower crane, 130′ high, 106′ jib, 6200 lb. capacity	A-3N	.05	176	Month		8,850	24,200	33,050	40,000

01 54 23 – Temporary Scaffolding and Platforms

	01 54 23.70 Scaffolding	Crew	Daily Output	Labor-Hours	Unit	Material	Labor	Equipment	Total	Total Incl O&P
0010	**SCAFFOLDING** R015423-10									
0906	Complete system for face of walls, no plank, material only rent/mo				C.S.F.	35			35	38.50
0910	Steel tubular, heavy duty shoring, buy									
0920	Frames 5′ high 2′ wide				Ea.	81.50			81.50	90
0925	5′ high 4′ wide					93			93	102
0930	6′ high 2′ wide					93.50			93.50	103
0935	6′ high 4′ wide					109			109	120
0940	Accessories									
0945	Cross braces				Ea.	15.50			15.50	17.05
0950	U-head, 8″ x 8″					19.10			19.10	21
0955	J-head, 4″ x 8″					13.90			13.90	15.30
0960	Base plate, 8″ x 8″					15.50			15.50	17.05
0965	Leveling jack					33.50			33.50	36.50
1000	Steel tubular, regular, buy									
1100	Frames 3′ high 5′ wide				Ea.	72.50			72.50	80
1150	5′ high 5′ wide					86.50			86.50	95
1200	6′-4″ high 5′ wide					118			118	129
1350	7′-6″ high 6′ wide					158			158	173
1500	Accessories cross braces					15.50			15.50	17.05
1550	Guardrail post					16.40			16.40	18.05
1600	Guardrail 7′ section					6.30			6.30	6.95
1650	Screw jacks & plates					21.50			21.50	24
1700	Sidearm brackets					25.50			25.50	28.50
1750	8″ casters					31			31	34
1800	Plank 2″ x 10″ x 16′-0″					52			52	57.50
1900	Stairway section					276			276	305
1910	Stairway starter bar					32			32	35
1920	Stairway inside handrail					53			53	58
1930	Stairway outside handrail					81.50			81.50	90
1940	Walk-thru frame guardrail					41.50			41.50	46
2000	Steel tubular, regular, rent/mo.									
2100	Frames 3′ high 5′ wide				Ea.	5			5	5.50

01 54 Construction Aids

01 54 23 – Temporary Scaffolding and Platforms

01 54 23.70 Scaffolding		Crew	Daily Output	Labor-Hours	Unit	2014 Bare Costs Material	Labor	Equipment	Total	Total Incl O&P
2150	5′ high 5′ wide				Ea.	5			5	5.50
2200	6′-4″ high 5′ wide					5.15			5.15	5.65
2250	7′-6″ high 6′ wide					7			7	7.70
2500	Accessories, cross braces					1			1	1.10
2550	Guardrail post					1			1	1.10
2600	Guardrail 7′ section					1			1	1.10
2650	Screw jacks & plates					2			2	2.20
2700	Sidearm brackets					2			2	2.20
2750	8″ casters					8			8	8.80
2800	Outrigger for rolling tower					3			3	3.30
2850	Plank 2″ x 10″ x 16′-0″					6			6	6.60
2900	Stairway section					35.50			35.50	39
2940	Walk-thru frame guardrail					2.50			2.50	2.75
3000	Steel tubular, heavy duty shoring, rent/mo.									
3250	5′ high 2′ & 4′ wide				Ea.	5			5	5.50
3300	6′ high 2′ & 4′ wide					5			5	5.50
3500	Accessories, cross braces					1			1	1.10
3600	U - head, 8″ x 8″					1			1	1.10
3650	J - head, 4″ x 8″					1			1	1.10
3700	Base plate, 8″ x 8″					1			1	1.10
3750	Leveling jack					2			2	2.20
4000	Scaffolding, stl. tubular, reg., no plank, labor only to erect & dismantle									
4100	Building exterior 2 stories	3 Carp	8	3	C.S.F.		138		138	212
4150	4 stories	″	8	3			138		138	212
4200	6 stories	4 Carp	8	4			183		183	283
4250	8 stories		8	4			183		183	283
4300	10 stories		7.50	4.267			196		196	300
4350	12 stories		7.50	4.267			196		196	300
5700	Planks, 2″ x 10″ x 16′-0″, labor only to erect & remove to 50′ H	3 Carp	72	.333	Ea.		15.30		15.30	23.50
5800	Over 50′ high	4 Carp	80	.400	″		18.35		18.35	28.50
6000	Heavy duty shoring for elevated slab forms to 8′-2″ high, floor area									
6100	Labor only to erect & dismantle	4 Carp	16	2	C.S.F.		91.50		91.50	142
6110	Materials only, rent.mo				″	29.50			29.50	32.50
6500	To 14′-8″ high									
6600	Labor only to erect & dismantle	4 Carp	10	3.200	C.S.F.		147		147	226
6610	Materials only, rent/mo				″	43			43	47.50

01 54 23.75 Scaffolding Specialties		Crew	Daily Output	Labor-Hours	Unit	Material	Labor	Equipment	Total	Total Incl O&P
0010	**SCAFFOLDING SPECIALTIES**									
1200	Sidewalk bridge, heavy duty steel posts & beams, including									
1210	parapet protection & waterproofing (material cost is rent/month)									
1230	3 posts	3 Carp	10	2.400	L.F.	56.50	110		166.50	232
1500	Sidewalk bridge using tubular steel scaffold frames including									
1510	planking (material cost is rent/month)	3 Carp	45	.533	L.F.	5.60	24.50		30.10	43.50
1600	For 2 uses per month, deduct from all above					50%				
1700	For 1 use every 2 months, add to all above					100%				
1900	Catwalks, 20″ wide, no guardrails, 7′ span, buy				Ea.	133			133	146
2000	10′ span, buy					186			186	205
3720	Putlog, standard, 8′ span, with hangers, buy					66.50			66.50	73
3730	Rent per month					10			10	11
3750	12′ span, buy					99			99	109
3755	Rent per month					15			15	16.50
3760	Trussed type, 16′ span, buy					229			229	252

01 54 Construction Aids

01 54 23 – Temporary Scaffolding and Platforms

01 54 23.75 Scaffolding Specialties		Crew	Daily Output	Labor-Hours	Unit	Material	2014 Bare Costs Labor	Equipment	Total	Total Incl O&P
3770	Rent per month				Ea.	20			20	22
3790	22′ span, buy					274			274	300
3795	Rent per month					30			30	33
3800	Rolling ladders with handrails, 30″ wide, buy, 2 step					231			231	254
4000	7 step					760			760	835
4050	10 step					1,050			1,050	1,150
4100	Rolling towers, buy, 5′ wide, 7′ long, 10′ high					1,150			1,150	1,275
4200	For 5′ high added sections, to buy, add					204			204	224
4300	Complete incl. wheels, railings, outriggers,									
4350	21′ high, to buy				Ea.	1,975			1,975	2,175
4400	Rent/month = 5% of purchase cost				″	170			170	187

01 54 26 – Temporary Swing Staging

01 54 26.50 Swing Staging

		Crew	Daily Output	Labor-Hours	Unit	Material	Labor	Equipment	Total	Total Incl O&P
0010	**SWING STAGING**, 500 lb. cap., 2′ wide to 24′ long, hand operated									
0020	steel cable type, with 60′ cables, buy				Ea.	4,900			4,900	5,375
0030	Rent per month				″	490			490	540
0600	Lightweight (not for masons) 24′ long for 150′ height,									
0610	manual type, buy				Ea.	10,200			10,200	11,200
0620	Rent per month					1,025			1,025	1,125
0700	Powered, electric or air, to 150′ high, buy					25,600			25,600	28,200
0710	Rent per month					1,800			1,800	1,975
0780	To 300′ high, buy					26,000			26,000	28,600
0800	Rent per month					1,825			1,825	2,000
1000	Bosun's chair or work basket 3′ x 3.5′, to 300′ high, electric, buy					10,400			10,400	11,400
1010	Rent per month					730			730	800
2200	Move swing staging (setup and remove)	E-4	2	16	Move		825	71	896	1,525

01 54 36 – Equipment Mobilization

01 54 36.50 Mobilization

		Crew	Daily Output	Labor-Hours	Unit	Material	Labor	Equipment	Total	Total Incl O&P
0010	**MOBILIZATION** (Use line item again for demobilization)									
0015	Up to 25 mi. haul dist. (50 mi. RT for mob/demob crew)									
0020	Dozer, loader, backhoe, excav., grader, paver, roller, 70 to 150 H.P.	B-34N	4	2	Ea.		75	142	217	272
0100	Above 150 HP	B-34K	3	2.667			100	320	420	510
0300	Scraper, towed type (incl. tractor), 6 C.Y. capacity		3	2.667			100	320	420	510
0400	10 C.Y.		2.50	3.200			120	385	505	610
0600	Self-propelled scraper, 15 C.Y.		2.50	3.200			120	385	505	610
0700	24 C.Y.		2	4			150	480	630	760
0900	Shovel or dragline, 3/4 C.Y.		3.60	2.222			83.50	267	350.50	420
1000	1-1/2 C.Y.		3	2.667			100	320	420	510
1100	Small equipment, placed in rear of, or towed by pickup truck	A-3A	8	1			36.50	21	57.50	78.50
1150	Equip up to 70 HP, on flatbed trailer behind pickup truck	A-3D	4	2			73	70.50	143.50	189
2000	Crane, truck-mounted, up to 75 ton, (driver only, one-way)	1 Eqhv	7.20	1.111			56		56	84.50
2100	Crane, truck-mounted, over 75 ton	A-3E	2.50	6.400			281	67.50	348.50	500
2200	Crawler-mounted, up to 75 ton	A-3F	2	8			350	500	850	1,075
2300	Over 75 ton	A-3G	1.50	10.667			470	755	1,225	1,550
2500	For each additional 5 miles haul distance, add						10%	10%		
3000	For large pieces of equipment, allow for assembly/knockdown									
3001	For mob/demob of vibrofloatation equip, see Section 31 45 13.10									
3100	For mob/demob of micro-tunneling equip, see Section 33 05 23.19									
3200	For mob/demob of pile driving equip, see Section 31 62 19.10									
3300	For mob/demob of caisson drilling equip, see Section 31 63 26.13									

01 54 Construction Aids

01 54 39 – Construction Equipment

01 54 39.70 Small Tools		Crew	Daily Output	Labor-Hours	Unit	Material	2014 Bare Costs Labor	Equipment	Total	Total Incl O&P
0010	**SMALL TOOLS**									
0020	As % of contractor's bare labor cost for project, minimum				Total		.50%			
0100	Maximum				″		2%			

01 55 Vehicular Access and Parking

01 55 23 – Temporary Roads

01 55 23.50 Roads and Sidewalks

		Crew	Daily Output	Labor-Hours	Unit	Material	Labor	Equipment	Total	Total Incl O&P
0010	**ROADS AND SIDEWALKS** Temporary									
0050	Roads, gravel fill, no surfacing, 4″ gravel depth	B-14	715	.067	S.Y.	3.98	2.60	.51	7.09	8.95
0100	8″ gravel depth	″	615	.078	″	7.95	3.02	.59	11.56	14.05
1000	Ramp, 3/4″ plywood on 2″ x 6″ joists, 16″ O.C.	2 Carp	300	.053	S.F.	1.44	2.45		3.89	5.35
1100	On 2″ x 10″ joists, 16″ O.C.	″	275	.058	″	1.99	2.67		4.66	6.30

01 56 Temporary Barriers and Enclosures

01 56 13 – Temporary Air Barriers

01 56 13.60 Tarpaulins

		Crew	Daily Output	Labor-Hours	Unit	Material	Labor	Equipment	Total	Total Incl O&P
0010	**TARPAULINS**									
0020	Cotton duck, 10 oz. to 13.13 oz. per S.Y., 6′x8′				S.F.	.75			.75	.83
0050	30′x30′					.43			.43	.47
0100	Polyvinyl coated nylon, 14 oz. to 18 oz., minimum					1.30			1.30	1.43
0150	Maximum					1.19			1.19	1.31
0200	Reinforced polyethylene 3 mils thick, white					.03			.03	.03
0300	4 mils thick, white, clear or black					.09			.09	.10
0400	5.5 mils thick, clear					.17			.17	.19
0500	White, fire retardant					.41			.41	.45
0600	12 mils, oil resistant, fire retardant					.27			.27	.30
0700	8.5 mils, black					.57			.57	.63
0710	Woven polyethylene, 6 mils thick					.17			.17	.19
0730	Polyester reinforced w/integral fastening system 11 mils thick					.24			.24	.26
0740	Mylar polyester, non-reinforced, 7 mils thick				↓	1.17			1.17	1.29

01 56 13.90 Winter Protection

		Crew	Daily Output	Labor-Hours	Unit	Material	Labor	Equipment	Total	Total Incl O&P
0010	**WINTER PROTECTION**									
0100	Framing to close openings	2 Clab	500	.032	S.F.	.41	1.17		1.58	2.26
0200	Tarpaulins hung over scaffolding, 8 uses, not incl. scaffolding		1500	.011		.25	.39		.64	.88
0300	Prefab fiberglass panels, steel frame, 8 uses	↓	1200	.013	↓	2.20	.49		2.69	3.17

01 56 16 – Temporary Dust Barriers

01 56 16.10 Dust Barriers, Temporary

		Crew	Daily Output	Labor-Hours	Unit	Material	Labor	Equipment	Total	Total Incl O&P
0010	**DUST BARRIERS, TEMPORARY**									
0020	Spring loaded telescoping pole & head, to 12′, erect and dismantle	1 Clab	240	.033	Ea.		1.22		1.22	1.88
0025	Cost per day (based upon 250 days)				Day	.24			.24	.26
0030	To 21′, erect and dismantle	1 Clab	240	.033	Ea.		1.22		1.22	1.88
0035	Cost per day (based upon 250 days)				Day	.38			.38	.42
0040	Accessories, caution tape reel, erect and dismantle	1 Clab	480	.017	Ea.		.61		.61	.94
0045	Cost per day (based upon 250 days)				Day	.36			.36	.40
0060	Foam rail and connector, erect and dismantle	1 Clab	240	.033	Ea.		1.22		1.22	1.88
0065	Cost per day (based upon 250 days)				Day	.10			.10	.11
0070	Caution tape	1 Clab	384	.021	C.L.F.	2.70	.76		3.46	4.15
0080	Zipper, standard duty	↓	60	.133	Ea.	15.95	4.89		20.84	25

01 56 Temporary Barriers and Enclosures

01 56 16 – Temporary Dust Barriers

01 56 16.10 Dust Barriers, Temporary		Crew	Daily Output	Labor-Hours	Unit	2014 Bare Costs Material	Labor	Equipment	Total	Total Incl O&P
0090	Heavy duty	1 Clab	48	.167	Ea.	18.90	6.10		25	30.50
0100	Polyethylene sheet, 4 mil		37	.216	Sq.	2.91	7.90		10.81	15.45
0110	6 mil		37	.216	″	3.98	7.90		11.88	16.65
1000	Dust partition, 6 mil polyethylene, 1″ x 3″ frame	2 Carp	2000	.008	S.F.	.29	.37		.66	.89
1080	2″ x 4″ frame	″	2000	.008	″	.32	.37		.69	.92

01 56 23 – Temporary Barricades

01 56 23.10 Barricades		Crew	Daily Output	Labor-Hours	Unit	Material	Labor	Equipment	Total	Total Incl O&P
0010	**BARRICADES**									
0020	5′ high, 3 rail @ 2″ x 8″, fixed	2 Carp	20	.800	L.F.	5.75	36.50		42.25	63
0150	Movable	″	30	.533	″	4.77	24.50		29.27	43
0300	Stock units, 6′ high, 8′ wide, plain, buy				Ea.	390			390	430
0350	With reflective tape, buy				″	405			405	445
0400	Break-a-way 3″ PVC pipe barricade									
0410	with 3 ea. 1′ x 4′ reflectorized panels, buy				Ea.	128			128	141
0500	Plywood with steel legs, 24″ wide					59			59	65
0600	Warning signal flag tree, 11′ high, 2 flags, buy					238			238	262
0800	Traffic cones, PVC, 18″ high					11			11	12.10
0850	28″ high					18.60			18.60	20.50
1000	Guardrail, wooden, 3′ high, 1″ x 6″, on 2″ x 4″ posts	2 Carp	200	.080	L.F.	1.29	3.67		4.96	7.05
1100	2″ x 6″, on 4″ x 4″ posts	″	165	.097		2.35	4.45		6.80	9.45
1200	Portable metal with base pads, buy					15.55			15.55	17.10
1250	Typical installation, assume 10 reuses	2 Carp	600	.027		2.55	1.22		3.77	4.70
1300	Barricade tape, polyethylene, 7 mil, 3″ wide x 500′ long roll				Ea.	25			25	27.50
3000	Detour signs, set up and remove									
3010	Reflective aluminum, MUTCD, 24″ x 24″, post mounted	1 Clab	20	.400	Ea.	2.35	14.65		17	25
5000	Barricades, see Section 01 54 33.40									

01 56 26 – Temporary Fencing

01 56 26.50 Temporary Fencing		Crew	Daily Output	Labor-Hours	Unit	Material	Labor	Equipment	Total	Total Incl O&P
0010	**TEMPORARY FENCING**									
0020	Chain link, 11 ga., 4′ high	2 Clab	400	.040	L.F.	2.95	1.47		4.42	5.50
0100	6′ high		300	.053		2.95	1.95		4.90	6.25
0200	Rented chain link, 6′ high, to 1000′ (up to 12 mo.)		400	.040		4.39	1.47		5.86	7.10
0250	Over 1000′ (up to 12 mo.)		300	.053		4.29	1.95		6.24	7.75
0350	Plywood, painted, 2″ x 4″ frame, 4′ high	A-4	135	.178		6.25	7.80		14.05	18.75
0400	4″ x 4″ frame, 8′ high	″	110	.218		11.45	9.55		21	27
0500	Wire mesh on 4″ x 4″ posts, 4′ high	2 Carp	100	.160		9.75	7.35		17.10	22
0550	8′ high	″	80	.200		14.70	9.15		23.85	30.50
0600	Plastic safety fence, light duty, 4′ high, posts at 10′	B-1	500	.048		1.18	1.79		2.97	4.06
0610	Medium duty, 4′ high, posts at 10′		500	.048		1.63	1.79		3.42	4.55
0620	Heavy duty, 4′ high, posts at 10′		500	.048		2.19	1.79		3.98	5.15
0630	Reflective heavy duty, 4′ high, posts at 10′		500	.048		4.64	1.79		6.43	7.85

01 56 29 – Temporary Protective Walkways

01 56 29.50 Protection		Crew	Daily Output	Labor-Hours	Unit	Material	Labor	Equipment	Total	Total Incl O&P
0010	**PROTECTION**									
0020	Stair tread, 2″ x 12″ planks, 1 use	1 Carp	75	.107	Tread	4.57	4.89		9.46	12.60
0100	Exterior plywood, 1/2″ thick, 1 use		65	.123		1.79	5.65		7.44	10.65
0200	3/4″ thick, 1 use		60	.133		2.64	6.10		8.74	12.35
2200	Sidewalks, 2″ x 12″ planks, 2 uses		350	.023	S.F.	.76	1.05		1.81	2.46
2300	Exterior plywood, 2 uses, 1/2″ thick		750	.011		.30	.49		.79	1.08
2400	5/8″ thick		650	.012		.37	.56		.93	1.28
2500	3/4″ thick		600	.013		.44	.61		1.05	1.42

01 56 Temporary Barriers and Enclosures

01 56 32 – Temporary Security

01 56 32.50 Watchman		Crew	Daily Output	Labor-Hours	Unit	2014 Bare Costs Material	Labor	Equipment	Total	Total Incl O&P
0010	**WATCHMAN**									
0020	Service, monthly basis, uniformed person, minimum				Hr.				25	27.50
0100	Maximum								45.50	50
0200	Person and command dog, minimum								31	34
0300	Maximum				↓				54.50	60
0500	Sentry dog, leased, with job patrol (yard dog), 1 dog				Week				290	320
0600	2 dogs				"				390	430
0800	Purchase, trained sentry dog, minimum				Ea.				1,375	1,500
0900	Maximum				"				2,725	3,000

01 58 Project Identification

01 58 13 – Temporary Project Signage

01 58 13.50 Signs		Crew	Daily Output	Labor-Hours	Unit	Material	Labor	Equipment	Total	Total Incl O&P
0010	**SIGNS**									
0020	High intensity reflectorized, no posts, buy				S.F.	31.50			31.50	34.50

01 71 Examination and Preparation

01 71 23 – Field Engineering

01 71 23.13 Construction Layout		Crew	Daily Output	Labor-Hours	Unit	Material	Labor	Equipment	Total	Total Incl O&P
0010	**CONSTRUCTION LAYOUT**									
1100	Crew for layout of building, trenching or pipe laying, 2 person crew	A-6	1	16	Day		730	71.50	801.50	1,200
1200	3 person crew	A-7	1	24			1,200	71.50	1,271.50	1,925
1400	Crew for roadway layout, 4 person crew	A-8	1	32	↓		1,575	71.50	1,646.50	2,475

01 71 23.19 Surveyor Stakes		Crew	Daily Output	Labor-Hours	Unit	Material	Labor	Equipment	Total	Total Incl O&P
0010	**SURVEYOR STAKES**									
0020	Hardwood, 1" x 1" x 48" long				C	66			66	72.50
0100	2" x 2" x 18" long					74			74	81.50
0150	2" x 2" x 24" long				↓	130			130	143

Division Notes

	CREW	DAILY OUTPUT	LABOR-HOURS	UNIT	BARE COSTS MAT.	LABOR	EQUIP.	TOTAL	TOTAL INCL O&P

Estimating Tips

02 30 00 Subsurface Investigation

In preparing estimates on structures involving earthwork or foundations, all information concerning soil characteristics should be obtained. Look particularly for hazardous waste, evidence of prior dumping of debris, and previous stream beds.

02 40 00 Demolition and Structure Moving

The costs shown for selective demolition do not include rubbish handling or disposal. These items should be estimated separately using RSMeans data or other sources.

- Historic preservation often requires that the contractor remove materials from the existing structure, rehab them, and replace them. The estimator must be aware of any related measures and precautions that must be taken when doing selective demolition and cutting and patching. Requirements may include special handling and storage, as well as security.
- In addition to Subdivision 02 41 00, you can find selective demolition items in each division. Example: Roofing demolition is in Division 7.

02 40 00 Building Deconstruction

This section provides costs for the careful dismantling and recycling of most of low-rise building materials.

02 50 00 Containment of Hazardous Waste

This section addresses on-site hazardous waste disposal costs.

02 80 00 Hazardous Material Disposal/ Remediation

This subdivision includes information on hazardous waste handling, asbestos remediation, lead remediation, and mold remediation. See reference R028213-20 and R028319-60 for further guidance in using these unit price lines.

02 90 00 Monitoring Chemical Sampling, Testing Analysis

This section provides costs for on-site sampling and testing hazardous waste.

Reference Numbers

Reference numbers are shown in shaded boxes at the beginning of some major classifications. These numbers refer to related items in the Reference Section. The reference information may be an estimating procedure, an alternate pricing method, or technical information.

Note: Not all subdivisions listed here necessarily appear in this publication.

02 21 Surveys

02 21 13 – Site Surveys

02 21 13.09 Topographical Surveys		Crew	Daily Output	Labor-Hours	Unit	Material	Labor	Equipment	Total	Total Incl O&P
						2014 Bare Costs				
0010	**TOPOGRAPHICAL SURVEYS**									
0020	Topographical surveying, conventional, minimum	A-7	3.30	7.273	Acre	19.20	365	21.50	405.70	610
0100	Maximum	A-8	.60	53.333	"	58.50	2,600	119	2,777.50	4,200

02 21 13.13 Boundary and Survey Markers

		Crew	Daily Output	Labor-Hours	Unit	Material	Labor	Equipment	Total	Total Incl O&P
0010	**BOUNDARY AND SURVEY MARKERS**									
0300	Lot location and lines, large quantities, minimum	A-7	2	12	Acre	34	605	35.50	674.50	1,000
0320	Average	"	1.25	19.200		54	970	57	1,081	1,600
0400	Small quantities, maximum	A-8	1	32		71.50	1,575	71.50	1,718	2,550
0600	Monuments, 3′ long	A-7	10	2.400	Ea.	34	121	7.15	162.15	231
0800	Property lines, perimeter, cleared land	"	1000	.024	L.F.	.04	1.21	.07	1.32	1.98
0900	Wooded land	A-8	875	.037	"	.06	1.79	.08	1.93	2.90

02 21 13.16 Aerial Surveys

		Crew	Daily Output	Labor-Hours	Unit	Material	Labor	Equipment	Total	Total Incl O&P
0010	**AERIAL SURVEYS**									
1500	Aerial surveying, including ground control, minimum fee, 10 acres				Total					4,700
1510	100 acres									9,400
1550	From existing photography, deduct									1,625
1600	2′ contours, 10 acres				Acre					470
1850	100 acres									94
2000	1000 acres									90
2050	10,000 acres									85

02 32 Geotechnical Investigations

02 32 13 – Subsurface Drilling and Sampling

02 32 13.10 Boring and Exploratory Drilling

		Crew	Daily Output	Labor-Hours	Unit	Material	Labor	Equipment	Total	Total Incl O&P
0010	**BORING AND EXPLORATORY DRILLING**									
0020	Borings, initial field stake out & determination of elevations	A-6	1	16	Day		730	71.50	801.50	1,200
0100	Drawings showing boring details				Total		335		335	425
0200	Report and recommendations from P.E.						775		775	970
0300	Mobilization and demobilization	B-55	4	6			220	280	500	650
0350	For over 100 miles, per added mile		450	.053	Mile		1.95	2.49	4.44	5.75
0600	Auger holes in earth, no samples, 2-1/2″ diameter		78.60	.305	L.F.		11.20	14.25	25.45	33
0650	4″ diameter		67.50	.356			13	16.60	29.60	38.50
0800	Cased borings in earth, with samples, 2-1/2″ diameter		55.50	.432		22	15.85	20	57.85	70.50
0850	4″ diameter		32.60	.736		33.50	27	34.50	95	116
1000	Drilling in rock, "BX" core, no sampling	B-56	34.90	.458			19.20	46	65.20	80
1050	With casing & sampling		31.70	.505		22	21	50.50	93.50	113
1200	"NX" core, no sampling		25.92	.617			26	62	88	108
1250	With casing and sampling		25	.640		26	27	64.50	117.50	141
1400	Borings, earth, drill rig and crew with truck mounted auger	B-55	1	24	Day		880	1,125	2,005	2,575
1450	Rock using crawler type drill	B-56	1	16	"		670	1,600	2,270	2,800
1500	For inner city borings add, minimum								10%	10%
1510	Maximum								20%	20%

02 32 19 – Exploratory Excavations

02 32 19.10 Test Pits

		Crew	Daily Output	Labor-Hours	Unit	Material	Labor	Equipment	Total	Total Incl O&P
0010	**TEST PITS**									
0020	Hand digging, light soil	1 Clab	4.50	1.778	C.Y.		65		65	101
0100	Heavy soil	"	2.50	3.200			117		117	181
0120	Loader-backhoe, light soil	B-11M	28	.571			24.50	14.20	38.70	53
0130	Heavy soil	"	20	.800			34	19.90	53.90	74.50
1000	Subsurface exploration, mobilization				Mile				6.75	8.40

02 32 Geotechnical Investigations

02 32 19 – Exploratory Excavations

02 32 19.10 Test Pits		Crew	Daily Output	Labor-Hours	Unit	Material	2014 Bare Costs Labor	Equipment	Total	Total Incl O&P
1010	Difficult access for rig, add				Hr.				260	320
1020	Auger borings, drill rig, incl. samples				L.F.				26.50	33
1030	Hand auger								31.50	40
1050	Drill and sample every 5′, split spoon								31.50	40
1060	Extra samples				Ea.				36	45.50

02 41 Demolition

02 41 13 – Selective Site Demolition

02 41 13.15 Hydrodemolition

		Crew	Daily Output	Labor-Hours	Unit	Material	Labor	Equipment	Total	Total Incl O&P
0010	**HYDRODEMOLITION** R024119-10									
0015	Hydrodemolition, concrete pavement, 4000 PSI, 2″ depth	B-5	500	.112	S.F.		4.53	2.85	7.38	10.10
0120	4″ depth		450	.124			5.05	3.17	8.22	11.20
0130	6″ depth		400	.140			5.65	3.56	9.21	12.60
0410	6000 PSI, 2″ depth		410	.137			5.50	3.47	8.97	12.25
0420	4″ depth		350	.160			6.45	4.07	10.52	14.40
0430	6″ depth		300	.187			7.55	4.75	12.30	16.80
0510	8000 PSI, 2″ depth		330	.170			6.85	4.32	11.17	15.25
0520	4″ depth		280	.200			8.10	5.10	13.20	18
0530	6″ depth		240	.233			9.45	5.95	15.40	21

02 41 13.17 Demolish, Remove Pavement and Curb

		Crew	Daily Output	Labor-Hours	Unit	Material	Labor	Equipment	Total	Total Incl O&P
0010	**DEMOLISH, REMOVE PAVEMENT AND CURB** R024119-10									
5010	Pavement removal, bituminous roads, up to 3″ thick	B-38	690	.058	S.Y.		2.41	1.86	4.27	5.75
5050	4″ to 6″ thick		420	.095			3.96	3.05	7.01	9.40
5100	Bituminous driveways		640	.063			2.60	2	4.60	6.20
5200	Concrete to 6″ thick, hydraulic hammer, mesh reinforced		255	.157			6.50	5	11.50	15.50
5300	Rod reinforced		200	.200			8.30	6.40	14.70	19.80
5400	Concrete, 7″ to 24″ thick, plain		33	1.212	C.Y.		50.50	39	89.50	120
5500	Reinforced		24	1.667	″		69.50	53.50	123	165
5600	With hand held air equipment, bituminous, to 6″ thick	B-39	1900	.025	S.F.		.98	.12	1.10	1.64
5700	Concrete to 6″ thick, no reinforcing		1600	.030			1.16	.15	1.31	1.95
5800	Mesh reinforced		1400	.034			1.33	.17	1.50	2.22
5900	Rod reinforced		765	.063			2.43	.31	2.74	4.08
6000	Curbs, concrete, plain	B-6	360	.067	L.F.		2.67	1.01	3.68	5.20
6100	Reinforced		275	.087			3.50	1.33	4.83	6.80
6200	Granite		360	.067			2.67	1.01	3.68	5.20
6300	Bituminous		528	.045			1.82	.69	2.51	3.55
6500	Site demo, berms under 4″ in height, bituminous		528	.045			1.82	.69	2.51	3.55
6600	4″ or over in height		300	.080			3.21	1.22	4.43	6.25

02 41 13.20 Selective Demo, Highway Guard Rails & Barriers

		Crew	Daily Output	Labor-Hours	Unit	Material	Labor	Equipment	Total	Total Incl O&P
0010	**SELECTIVE DEMOLITION, HIGHWAY GUARD RAILS & BARRIERS**									
0100	Guard rail, corrugated steel	B-6	600	.040	L.F.		1.60	.61	2.21	3.13
0200	End sections		40	.600	Ea.		24	9.15	33.15	47
0300	Wrap around		40	.600	″		24	9.15	33.15	47
0400	Timber 4″ x 8″		600	.040	L.F.		1.60	.61	2.21	3.13
0500	Three 3/4″ cables		600	.040	″		1.60	.61	2.21	3.13
0600	Wood posts		240	.100	Ea.		4.01	1.52	5.53	7.80
0700	Guide rail, 6″ x 6″ box beam	B-80B	120	.267	L.F.		10.45	2.07	12.52	18.35
0800	Median barrier, 6″ x 8″ box beam		240	.133			5.25	1.03	6.28	9.20
0850	Precast concrete 3′-6″ high x 2′ wide		300	.107			4.19	.83	5.02	7.35
0900	Impact barrier, UTMCD, barrel type	B-16	60	.533	Ea.		19.95	11.55	31.50	43

02 41 Demolition

02 41 13 – Selective Site Demolition

02 41 13.20 Selective Demo, Highway Guard Rails & Barriers		Crew	Daily Output	Labor-Hours	Unit	Material	2014 Bare Costs Labor	Equipment	Total	Total Incl O&P
1000	Resilient guide fence and light shield 6′ high	B-16	120	.267	L.F.		9.95	5.75	15.70	21.50
1100	Concrete posts, 6′-5″ triangular	B-6	200	.120	Ea.		4.81	1.83	6.64	9.40
1200	Speed bumps 10-1/2″ x 2-1/4″ x 48″		300	.080	L.F.		3.21	1.22	4.43	6.25
1300	Pavement marking channelizing		200	.120	Ea.		4.81	1.83	6.64	9.40
1400	Barrier and curb delineators		300	.080			3.21	1.22	4.43	6.25
1500	Rumble strips 24″ x 3-1/2″ x 1/2″		150	.160			6.40	2.44	8.84	12.55

02 41 13.33 Minor Site Demolition		Crew	Daily Output	Labor-Hours	Unit	Material	Labor	Equipment	Total	Total Incl O&P
0010	**MINOR SITE DEMOLITION** R024119-10									
0015	No hauling, abandon catch basin or manhole	B-6	7	3.429	Ea.		138	52	190	269
0020	Remove existing catch basin or manhole, masonry		4	6			241	91.50	332.50	470
0030	Catch basin or manhole frames and covers, stored		13	1.846			74	28	102	145
0040	Remove and reset		7	3.429			138	52	190	269
0100	Roadside delineators, remove only	B-80	175	.183			7.25	4.33	11.58	15.85
0110	Remove and reset	″	100	.320			12.70	7.60	20.30	28
0800	Guiderail, corrugated steel, remove only	B-80A	100	.240	L.F.		8.80	3.33	12.13	17.20
0850	Remove and reset	″	40	.600	″		22	8.35	30.35	43
0860	Guide posts, remove only	B-80B	120	.267	Ea.		10.45	2.07	12.52	18.35
0870	Remove and reset	B-55	50	.480			17.55	22.50	40.05	51.50
0900	Hydrants, fire, remove only	B-21A	5	8			365	96	461	665
0950	Remove and reset	″	2	20			915	241	1,156	1,675
1000	Masonry walls, block, solid	B-5	1800	.031	C.F.		1.26	.79	2.05	2.80
1200	Brick, solid		900	.062			2.52	1.58	4.10	5.60
1400	Stone, with mortar		900	.062			2.52	1.58	4.10	5.60
1500	Dry set		1500	.037			1.51	.95	2.46	3.36
1600	Median barrier, precast concrete, remove and store	B-3	430	.112	L.F.		4.39	6	10.39	13.35
1610	Remove and reset	″	390	.123			4.84	6.60	11.44	14.65
2900	Pipe removal, sewer/water, no excavation, 12″ diameter	B-6	175	.137			5.50	2.09	7.59	10.75
2930	15″-18″ diameter	B-12Z	150	.160			6.60	10.90	17.50	22
2960	21″-24″ diameter		120	.200			8.25	13.65	21.90	27.50
3000	27″-36″ diameter		90	.267			11	18.20	29.20	37
3200	Steel, welded connections, 4″ diameter	B-6	160	.150			6	2.28	8.28	11.70
3300	10″ diameter	″	80	.300			12.05	4.57	16.62	23.50
3500	Railroad track removal, ties and track	B-13	330	.170			6.75	2.26	9.01	12.90
3600	Ballast	B-14	500	.096	C.Y.		3.72	.73	4.45	6.50
3700	Remove and re-install, ties & track using new bolts & spikes		50	.960	L.F.		37	7.30	44.30	65
3800	Turnouts using new bolts and spikes		1	48	Ea.		1,850	365	2,215	3,250
4000	Sidewalk removal, bituminous, 2″ thick	B-6	350	.069	S.Y.		2.75	1.04	3.79	5.35
4010	2-1/2″ thick		325	.074			2.96	1.12	4.08	5.80
4050	Brick, set in mortar		185	.130			5.20	1.97	7.17	10.15
4100	Concrete, plain, 4″		160	.150			6	2.28	8.28	11.70
4110	Plain, 5″		140	.171			6.90	2.61	9.51	13.40
4120	Plain, 6″		120	.200			8	3.04	11.04	15.65
4200	Mesh reinforced, concrete, 4 ″		150	.160			6.40	2.44	8.84	12.55
4210	5 ″ thick		131	.183			7.35	2.79	10.14	14.30
4220	6 ″ thick		112	.214			8.60	3.26	11.86	16.75
4300	Slab on grade removal, plain	B-5	45	1.244	C.Y.		50.50	31.50	82	112
4310	Mesh reinforced		33	1.697			68.50	43	111.50	153
4320	Rod reinforced		25	2.240			90.50	57	147.50	202
4400	For congested sites or small quantities, add up to								200%	200%
4450	For disposal on site, add	B-11A	232	.069			2.95	5.70	8.65	10.80
4500	To 5 miles, add	B-34D	76	.105			3.95	9.90	13.85	16.95
6850	Runways, remove rubber skid marks, 4-6 passes	B-59A	35	.686	M.S.F.	36	25.50	14.50	76	95

02 41 13 – Selective Site Demolition

02 41 13.33 Minor Site Demolition		Crew	Daily Output	Labor-Hours	Unit	2014 Bare Costs Material	Labor	Equipment	Total	Total Incl O&P
6860	6-10 passes	B-59A	35	.686	M.S.F.	54.50	25.50	14.50	94.50	114

02 41 13.34 Selective Demolition, Utility Materials

		Crew	Daily Output	Labor-Hours	Unit	Material	Labor	Equipment	Total	Total Incl O&P
0010	**SELECTIVE DEMOLITION, UTILITY MATERIALS** R024119-10									
0015	Excludes excavation									
0020	See other utility items in Section 02 41 13.33									
0100	Fire Hydrant extensions	B-20	14	1.714	Ea.		70		70	108
0200	Precast Utility boxes up to 8′ x 14′ x 7′	B-13	2	28			1,125	375	1,500	2,100
0300	Handholes and meter pits	B-6	2	12			480	183	663	940
0400	Utility valves 4″-12″	B-20	4	6			245		245	380
0500	14″-24″	B-21	2	14			590	70	660	985

02 41 13.36 Selective Demolition, Utility Valves and Accessories

		Crew	Daily Output	Labor-Hours	Unit	Material	Labor	Equipment	Total	Total Incl O&P
0010	**SELECTIVE DEMOLITION, UTILITY VALVES & ACCESSORIES**									
0015	Excludes excavation									
0100	Utility valves 4″-12″ diam.	B-20	4	6	Ea.		245		245	380
0200	14″-24″ diam.	B-21	2	14			590	70	660	985
0300	Crosses 4″-12″	B-20	8	3			123		123	189
0400	14″-24″	B-21	4	7			295	35	330	495
0500	Utility cut-in valves 4″-12″ diam.	B-20	20	1.200			49		49	76
0600	Curb boxes	″	20	1.200			49		49	76

02 41 13.38 Selective Demo., Water & Sewer Piping & Fittings

		Crew	Daily Output	Labor-Hours	Unit	Material	Labor	Equipment	Total	Total Incl O&P
0010	**SELECTIVE DEMOLITION, WATER & SEWER PIPING AND FITTINGS**									
0015	Excludes excavation									
0020	See other utility items in Section 02 41 13.33									
0090	Concrete pipe 4″-10″ dia	B-6	250	.096	L.F.		3.85	1.46	5.31	7.50
0100	42″-48″ diameter	B-13B	96	.583			23.50	11.85	35.35	48.50
0200	60″-84″ diameter	″	80	.700			28	14.20	42.20	58.50
0300	96″ diameter	B-13C	80	.700			28	21.50	49.50	66.50
0400	108″-144″ diameter	″	64	.875			35	26.50	61.50	83
0450	Concrete fittings 12″ diameter	B-6	24	1	Ea.		40	15.20	55.20	78.50
0480	Concrete end pieces 12″ diameter		200	.120	L.F.		4.81	1.83	6.64	9.40
0485	15″ diameter		150	.160			6.40	2.44	8.84	12.55
0490	18″ diameter		150	.160			6.40	2.44	8.84	12.55
0500	24″ -36″ diameter		100	.240			9.65	3.65	13.30	18.75
0600	Concrete fittings 24″-36″ diameter		12	2	Ea.		80	30.50	110.50	157
0700	48″-84″ diameter	B-13B	12	4.667			186	94.50	280.50	390
0800	96″ diameter	″	8	7			279	142	421	585
0900	108″-144″ diameter	B-13C	4	14			560	425	985	1,325
1000	Ductile iron pipe 4″ diameter	B-21B	200	.200	L.F.		7.95	3.32	11.27	15.85
1100	6″-12″ diameter		175	.229			9.10	3.79	12.89	18.10
1200	14″-24″ diameter		120	.333			13.25	5.55	18.80	26.50
1300	Ductile iron fittings 4″-12″ diameter		24	1.667	Ea.		66.50	27.50	94	133
1400	14″-16″ diameter		18	2.222			88.50	37	125.50	177
1500	18″-24″ diameter		12	3.333			133	55.50	188.50	265
1600	Plastic pipe 3/4″-4″ diameter	B-20	700	.034	L.F.		1.40		1.40	2.17
1700	6″-8″ diameter		500	.048			1.96		1.96	3.03
1800	10″-18″ diameter		300	.080			3.27		3.27	5.05
1900	20″-36″ diameter		200	.120			4.90		4.90	7.60
1910	42″-48″ diameter		180	.133			5.45		5.45	8.40
1920	54″-60″ diameter		160	.150			6.15		6.15	9.45
2000	Plastic fittings 4″- 8″ diameter	B-6	75	.320	Ea.		12.85	4.87	17.72	25
2100	10″-14″ diameter		50	.480			19.25	7.30	26.55	37.50
2200	16″-24″ diameter		20	1.200			48	18.25	66.25	94

02 41 Demolition

02 41 13 – Selective Site Demolition

02 41 13.38 Selective Demo., Water & Sewer Piping & Fittings		Crew	Daily Output	Labor-Hours	Unit	Material	2014 Bare Costs Labor	Equipment	Total	Total Incl O&P
2210	30"-36" diameter	B-6	15	1.600	Ea.		64	24.50	88.50	126
2220	42"- 48" diameter		12	2			80	30.50	110.50	157
2300	Copper pipe 3/4"-2" diameter	Q-1	500	.032	L.F.		1.66		1.66	2.50
2400	2 1/2" - 3" diameter		300	.053			2.76		2.76	4.17
2500	4"- 6" diameter		200	.080			4.14		4.14	6.25
2600	Copper fittings 3/4"- 2" diameter		15	1.067	Ea.		55.50		55.50	83.50
2700	Cast iron pipe 4" diameter		200	.080	L.F.		4.14		4.14	6.25
2800	5"- 6" diameter	Q-2	200	.120			6.45		6.45	9.75
2900	8"- 12" diameter	Q-3	200	.160			8.75		8.75	13.25
3000	Cast iron fittings 4" diameter	Q-1	30	.533	Ea.		27.50		27.50	41.50
3100	5"- 6" diameter	Q-2	30	.800			43		43	65
3200	8"- 15" diameter	Q-3	30	1.067			58.50		58.50	88.50
3300	Vent cast iron pipe 4"- 8" diameter	Q-1	200	.080	L.F.		4.14		4.14	6.25
3400	10"- 15" diameter	Q-3	200	.160	"		8.75		8.75	13.25
3500	Vent cast iron fittings 4"- 8" diameter	Q-2	30	.800	Ea.		43		43	65
3600	10"- 15" diameter	"	20	1.200	"		64.50		64.50	97.50

02 41 13.40 Selective Demolition, Metal Drainage Piping

		Crew	Daily Output	Labor-Hours	Unit	Material	Labor	Equipment	Total	Total Incl O&P
0010	**SELECTIVE DEMOLITION, METAL DRAINAGE PIPING**									
0015	Excludes excavation									
0020	See other utility items in Sections 02 41 13.33, 02 41 13.36, 02 41 13.38									
0100	CMP pipe, aluminum, 6"-10" dia	B-21	800	.035	L.F.		1.48	.18	1.66	2.47
0110	12" dia		600	.047			1.97	.23	2.20	3.29
0120	18" dia		600	.047			1.97	.23	2.20	3.29
0140	Steel, 6"-10" dia		800	.035			1.48	.18	1.66	2.47
0150	12" dia		600	.047			1.97	.23	2.20	3.29
0160	18" dia		400	.070			2.95	.35	3.30	4.94
0170	24" dia	B-13	300	.187			7.45	2.49	9.94	14.15
0180	30" - 36" dia		250	.224			8.95	2.98	11.93	17
0190	48" - 60" dia		200	.280			11.15	3.73	14.88	21
0200	72" dia	B-13B	100	.560			22.50	11.35	33.85	46.50
0210	CMP end sections, steel, 10"-18" dia	B-21	40	.700	Ea.		29.50	3.50	33	49.50
0220	24"-36" dia	B-13	30	1.867			74.50	25	99.50	142
0230	48" dia		20	2.800			112	37.50	149.50	212
0240	60" dia		10	5.600			223	74.50	297.50	420
0250	72" dia	B-13B	10	5.600			223	114	337	465
0260	CMP fittings, 8"-12" dia	B-21	60	.467			19.70	2.33	22.03	33
0270	18" dia	"	40	.700			29.50	3.50	33	49.50
0280	24"-48" dia	B-13	30	1.867			74.50	25	99.50	142
0290	60" dia	"	20	2.800			112	37.50	149.50	212
0300	72" dia	B-13B	10	5.600			223	114	337	465
0310	Oval arch 17" x 13", 21" x 15", 15-18" equivalent	B-21	400	.070	L.F.		2.95	.35	3.30	4.94
0320	28" x 20", 24" equivalent	B-13	300	.187			7.45	2.49	9.94	14.15
0330	35" x 24", 42" x 29", 30-36" equivalent		250	.224			8.95	2.98	11.93	17
0340	49" x 33", 57" x 38", 42-48" equivalent		200	.280			11.15	3.73	14.88	21
0350	Oval arch 17" x 13" end piece, 15" equivalent	B-21	40	.700	Ea.		29.50	3.50	33	49.50
0360	42" x 29" end piece, 36" equivalent	B-13	30	1.867	"		74.50	25	99.50	142

02 41 13.42 Selective Demolition, Manholes and Catch Basins

		Crew	Daily Output	Labor-Hours	Unit	Material	Labor	Equipment	Total	Total Incl O&P
0010	**SELECTIVE DEMOLITION, MANHOLES & CATCH BASINS**									
0015	Excludes excavation									
0020	See other utility items in Section 02 41 13.33									
0100	Manholes, precast or brick over 8' deep	B-6	8	3	V.L.F.		120	45.50	165.50	234
0200	Cast in place 4'-8' deep	B-9	127	.315	SF Face		11.65	1.84	13.49	20

02 41 13.42 Selective Demolition, Manholes and Catch Basins		Crew	Daily Output	Labor-Hours	Unit	Material	2014 Bare Costs Labor	Equipment	Total	Total Incl O&P
0300	Over 8' deep	B-9	100	.400	SF Face		14.80	2.34	17.14	25.50
0400	Top, precast, 8" thick, 4'-6' dia	B-6	8	3	Ea.		120	45.50	165.50	234
0500	Steps	1 Clab	60	.133	"		4.89		4.89	7.55

02 41 13.43 Selective Demolition, Box Culvert

		Crew	Daily Output	Labor-Hours	Unit	Material	Labor	Equipment	Total	Total Incl O&P
0010	**SELECTIVE DEMOLITION, BOX CULVERT**									
0015	Excludes excavation									
0100	Box culvert 8' x 6' x 3' to 8' x 8' x 8'	B-69	300	.160	L.F.		6.45	5.50	11.95	15.95
0200	8' x 10' x 3' to 8' x 12' x 8'	"	200	.240	"		9.70	8.25	17.95	24

02 41 13.44 Selective Demolition, Septic Tanks and Related Components

		Crew	Daily Output	Labor-Hours	Unit	Material	Labor	Equipment	Total	Total Incl O&P
0010	**SELECTIVE DEMOLITION, SEPTIC TANKS & RELATED COMPONENTS**									
0020	Excludes excavation									
0100	Septic tanks, precast, 1000-1250 gal.	B-21	8	3.500	Ea.		148	17.50	165.50	247
0200	1500 gal.		7	4			169	20	189	282
0300	2000-2500 gal.		5	5.600			236	28	264	395
0400	4000 gal.		4	7			295	35	330	495
0500	Precast, 5000 gal., 4 piece	B-13	3	18.667			745	249	994	1,425
0600	15,000 gal.	B-13B	1.70	32.941			1,325	670	1,995	2,750
0700	25,000 gal.		1.10	50.909			2,025	1,025	3,050	4,250
0800	40,000 gal.		.80	70			2,800	1,425	4,225	5,850
0900	Precast, 50,000 gal., 5 piece	B-13C	.60	93.333			3,725	2,850	6,575	8,825
1000	Cast-in-place, 75,000 gal.	B-9	.06	666			24,700	3,900	28,600	42,400
1100	100,000 gal.	"	.05	800			29,600	4,675	34,275	51,000
1200	HDPE, 1000 gal.	B-21	9	3.111			131	15.55	146.55	219
1300	1500 gal.		8	3.500			148	17.50	165.50	247
1400	Galley, 4' x 4' x 4'		16	1.750			74	8.75	82.75	124
1500	Distribution boxes, concrete, 7 outlets	2 Clab	16	1			36.50		36.50	56.50
1600	9 outlets	"	8	2			73.50		73.50	113
1700	Leaching chambers 13' x 3'-7" x 1'-4", standard	B-13	16	3.500			140	46.50	186.50	266
1800	8' x 4' x 1'-6", heavy duty		14	4			159	53.50	212.50	305
1900	13' x 3'-9" x 1'-6"		12	4.667			186	62	248	355
2100	20' x 4' x 1'-6"		5	11.200			445	149	594	850
2200	Leaching pit 6'-6" x 6' deep	B-21	5	5.600			236	28	264	395
2300	6'-6" x 8' deep		4	7			295	35	330	495
2400	8' x 6' deep H20		4	7			295	35	330	495
2500	8' x 8' deep H20		3	9.333			395	46.50	441.50	655
2600	Velocity reducing pit, precast 6' x 3' deep		4.70	5.957			251	30	281	420

02 41 13.46 Selective Demolition, Steel Pipe With Insulation

		Crew	Daily Output	Labor-Hours	Unit	Material	Labor	Equipment	Total	Total Incl O&P
0010	**SELECTIVE DEMOLITION, STEEL PIPE WITH INSULATION**									
0020	Excludes excavation									
0100	Steel pipe, with insulation, 3/4"-4"	B-1A	400	.060	L.F.		2.24	.82	3.06	4.35
0200	5"-10"	B-1B	360	.089			3.60	2.75	6.35	8.55
0300	12"-16"		240	.133			5.40	4.12	9.52	12.85
0400	18"-24"		160	.200			8.10	6.20	14.30	19.25
0450	26"-36"		100	.320			13	9.90	22.90	31
0500	Steel gland seal, with insulation, 3/4"-4"	B-1A	100	.240	Ea.		8.95	3.27	12.22	17.40
0600	5"-10"	B-1B	75	.427			17.30	13.20	30.50	41
0700	12"-16"		60	.533			21.50	16.50	38	51
0800	18"-24"		50	.640			26	19.80	45.80	62
0850	26"-36"		40	.800			32.50	24.50	57	77
0900	Demo steel fittings with insulation 3/4"-4"	B-1A	60	.400			14.95	5.45	20.40	29
1000	5"-10"	B-1B	40	.800			32.50	24.50	57	77
1100	12"-16"		30	1.067			43.50	33	76.50	103

02 41 13 – Selective Site Demolition

02 41 13.46 Selective Demolition, Steel Pipe With Insulation		Crew	Daily Output	Labor-Hours	Unit	Material	2014 Bare Costs Labor	Equipment	Total	Total Incl O&P
1200	18"-24"	B-1B	20	1.600	Ea.		65	49.50	114.50	154
1300	26"-36"		15	2.133			86.50	66	152.50	206
1400	Steel pipe anchors, 5"-10"		40	.800			32.50	24.50	57	77
1500	12"-16"		30	1.067			43.50	33	76.50	103
1600	18"-24"		20	1.600			65	49.50	114.50	154
1700	26"-36"		15	2.133			86.50	66	152.50	206

02 41 13.48 Selective Demolition, Gasoline Containment Piping

		Crew	Daily Output	Labor-Hours	Unit	Material	Labor	Equipment	Total	Total Incl O&P
0010	**SELECTIVE DEMOLITION, GASOLINE CONTAINMENT PIPING**									
0020	Excludes excavation									
0030	Excludes environmental site remediation									
0100	Gasoline plastic primary containment piping 2" to 4"	Q-6	800	.030	L.F.		1.64		1.64	2.47
0200	Fittings 2" to 4"		40	.600	Ea.		33		33	49.50
0300	Gasoline plastic secondary containment piping 3" to 6"		800	.030	L.F.		1.64		1.64	2.47
0400	Fittings 3" to 6"		40	.600	Ea.		33		33	49.50

02 41 13.50 Selective Demolition, Natural Gas, PE Pipe

		Crew	Daily Output	Labor-Hours	Unit	Material	Labor	Equipment	Total	Total Incl O&P
0010	**SELECTIVE DEMOLITION, NATURAL GAS, PE PIPE**									
0020	Excludes excavation									
0100	Natural gas coils, PE, 1 1/4" to 3"	Q-6	800	.030	L.F.		1.64		1.64	2.47
0200	Joints, 40', PE, 3" - 4"		800	.030			1.64		1.64	2.47
0300	6" - 8"		600	.040			2.18		2.18	3.30

02 41 13.51 Selective Demolition, Natural Gas, Steel Pipe

		Crew	Daily Output	Labor-Hours	Unit	Material	Labor	Equipment	Total	Total Incl O&P
0010	**SELECTIVE DEMOLITION, NATURAL GAS, STEEL PIPE**									
0020	Excludes excavation									
0100	Natural gas steel pipe 1" - 4"	B-1A	800	.030	L.F.		1.12	.41	1.53	2.18
0200	5" - 10"	B-1B	360	.089			3.60	2.75	6.35	8.55
0300	12" - 16"		240	.133			5.40	4.12	9.52	12.85
0400	18" - 24"		160	.200			8.10	6.20	14.30	19.25
0500	Natural gas steel fittings 1" - 4"	B-1A	160	.150	Ea.		5.60	2.04	7.64	10.90
0600	5" - 10"	B-1B	160	.200			8.10	6.20	14.30	19.25
0700	12" - 16"		108	.296			12	9.15	21.15	28.50
0800	18" - 24"		70	.457			18.55	14.15	32.70	44

02 41 13.52 Selective Demo, Natural Gas, Valves, Fittings, Regulators

		Crew	Daily Output	Labor-Hours	Unit	Material	Labor	Equipment	Total	Total Incl O&P
0010	**SELECTIVE DEMO, NATURAL GAS, VALVES, FITTINGS, REGULATORS**									
0100	Gas stops 1 1/4" - 2"	1 Plum	22	.364	Ea.		21		21	31.50
0200	Gas regulator 1 1/2" - 2"	"	22	.364			21		21	31.50
0300	3" - 4"	Q-1	22	.727			37.50		37.50	57
0400	Gas plug valve, 3/4" - 2"	1 Plum	22	.364			21		21	31.50
0500	2 1/2" - 3"	Q-1	10	1.600			83		83	125

02 41 13.54 Selective Demolition, Electric Ducts and Fittings

		Crew	Daily Output	Labor-Hours	Unit	Material	Labor	Equipment	Total	Total Incl O&P
0010	**SELECTIVE DEMOLITION, ELECTRIC DUCTS & FITTINGS**									
0020	Excludes excavation									
0100	Plastic conduit, 1/2" - 2"	1 Elec	600	.013	L.F.		.71		.71	1.06
0200	3" - 6"	2 Elec	400	.040	"		2.13		2.13	3.19
0300	Fittings, 1/2" - 2"	1 Elec	50	.160	Ea.		8.55		8.55	12.80
0400	3" - 6"	"	40	.200	"		10.65		10.65	15.95

02 41 13.56 Selective Demolition, Electric Duct Banks

		Crew	Daily Output	Labor-Hours	Unit	Material	Labor	Equipment	Total	Total Incl O&P
0010	**SELECTIVE DEMOLITION, ELECTRIC DUCT BANKS**									
0020	Excludes excavation									
0100	Hand holes sized to 4' x 4' x 4'	R-3	7	2.857	Ea.		151	20	171	249
0200	Manholes sized to 6' x 10' x 7'	B-13	6	9.333	"		370	124	494	705
0300	Conduit 1 @ 2" diameter, EB plastic, no concrete	2 Elec	1000	.016	L.F.		.85		.85	1.28

02 41 Demolition

02 41 13 – Selective Site Demolition

02 41 13.56 Selective Demolition, Electric Duct Banks		Crew	Daily Output	Labor-Hours	Unit	Material	2014 Bare Costs Labor	Equipment	Total	Total Incl O&P
0400	2 @ 2″ diameter	2 Elec	500	.032	L.F.		1.71		1.71	2.56
0500	4 @ 2″ diameter		250	.064			3.41		3.41	5.10
0600	1 @ 3″ diameter		800	.020			1.07		1.07	1.60
0700	2 @ 3″ diameter		400	.040			2.13		2.13	3.19
0800	4 @ 3″ diameter		200	.080			4.27		4.27	6.40
0900	1 @ 4″ diameter		800	.020			1.07		1.07	1.60
1000	2 @ 4″ diameter		400	.040			2.13		2.13	3.19
1100	4 @ 4″ diameter		200	.080			4.27		4.27	6.40
1200	6 @ 4″ diameter		100	.160			8.55		8.55	12.80
1300	1 @ 5″ diameter		500	.032			1.71		1.71	2.56
1400	2 @ 5″ diameter		250	.064			3.41		3.41	5.10
1500	4 @ 5″ diameter		160	.100			5.35		5.35	8
1600	6 @ 5″ diameter		100	.160			8.55		8.55	12.80
1700	1 @ 6″ diameter		500	.032			1.71		1.71	2.56
1800	2 @ 6″ diameter		250	.064			3.41		3.41	5.10
1900	4 @ 6″ diameter		160	.100			5.35		5.35	8
2000	6 @ 6″ diameter		100	.160			8.55		8.55	12.80
2100	Conduit 1 EB plastic, with concrete, 0.92 C.F./L.F.	B-9	150	.267			9.90	1.56	11.46	16.95
2200	2 EB plastic 1.52 C.F./L.F.		100	.400			14.80	2.34	17.14	25.50
2300	2 x 2 EB plastic 2.51 C.F./L.F.		80	.500			18.55	2.92	21.47	31.50
2400	2 x 3 EB plastic 3.56 C.F./L.F.		60	.667			24.50	3.89	28.39	42.50
2500	Conduit 2 @ 2″ diameter, steel, no concrete	2 Elec	400	.040			2.13		2.13	3.19
2600	4 @ 2″ diameter		200	.080			4.27		4.27	6.40
2700	2 @ 3″ diameter		200	.080			4.27		4.27	6.40
2800	4 @ 3″ diameter		100	.160			8.55		8.55	12.80
2900	2 @ 4″ diameter		200	.080			4.27		4.27	6.40
3000	4 @ 4″ diameter		100	.160			8.55		8.55	12.80
3100	6 @ 4″ diameter		50	.320			17.05		17.05	25.50
3200	2 @ 5″ diameter		120	.133			7.10		7.10	10.65
3300	4 @ 5″ diameter		60	.267			14.25		14.25	21.50
3400	6 @ 5″ diameter		40	.400			21.50		21.50	32
3500	2 @ 6″ diameter		120	.133			7.10		7.10	10.65
3600	4 @ 6″ diameter		60	.267			14.25		14.25	21.50
3700	6 @ 6″ diameter		40	.400			21.50		21.50	32
3800	Conduit 2 steel, with concrete, 1.52 C.F./L.F.	B-9	80	.500			18.55	2.92	21.47	31.50
3900	2 x 2 EB steel 2.51 C.F./L.F.		60	.667			24.50	3.89	28.39	42.50
4000	2 x 3 steel 3.56 C.F./L.F.		50	.800			29.50	4.67	34.17	50.50
4100	Conduit fittings, PVC type EB, 2″ - 3″	1 Elec	30	.267	Ea.		14.25		14.25	21.50
4200	4″ - 6″	″	20	.400	″		21.50		21.50	32

02 41 13.60 Selective Demolition Fencing		Crew	Daily Output	Labor-Hours	Unit	Material	Labor	Equipment	Total	Total Incl O&P
0010	SELECTIVE DEMOLITION FENCING R024119-10									
0700	Snow fence, 4′ high	B-6	1000	.024	L.F.		.96	.37	1.33	1.88
1600	Fencing, barbed wire, 3 strand	2 Clab	430	.037			1.36		1.36	2.10
1650	5 strand	″	280	.057			2.09		2.09	3.23
1700	Chain link, posts & fabric, 8′ to 10′ high, remove only	B-6	445	.054			2.16	.82	2.98	4.22
1750	Remove and reset	″	70	.343			13.75	5.20	18.95	27
1775	Fencing, wood, average all types, 4′ to 6′ high	2 Clab	432	.037			1.36		1.36	2.09
1790	Remove and store	B-80	235	.136			5.40	3.23	8.63	11.85

02 41 13.62 Selective Demo., Chain Link Fences & Gates		Crew	Daily Output	Labor-Hours	Unit	Material	Labor	Equipment	Total	Total Incl O&P
0010	SELECTIVE DEMOLITION, CHAIN LINK FENCES & GATES									
0020	See other fence items in Section 02 41 13.60									
0100	Chain link, gates, 3′ - 4′ width	B-6	30	.800	Ea.		32	12.20	44.20	62.50

02 41 Demolition

02 41 13 – Selective Site Demolition

02 41 13.62 Selective Demo., Chain Link Fences & Gates		Crew	Daily Output	Labor-Hours	Unit	Material	2014 Bare Costs Labor	Equipment	Total	Total Incl O&P
0200	10-12′ width	B-6	16	1.500	Ea.		60	23	83	117
0300	14′ width		15	1.600			64	24.50	88.50	126
0400	20′ width		10	2.400			96.50	36.50	133	188
0500	18′ width with overhead & cantilever		80	.300	L.F.		12.05	4.57	16.62	23.50
0510	Sliding		80	.300			12.05	4.57	16.62	23.50
0520	Cantilever to 40′ wide		80	.300			12.05	4.57	16.62	23.50
0530	Motor operators	2 Skwk	1	16	Ea.		755		755	1,175
0540	Transmitter systems	"	15	1.067	"		50.50		50.50	78
0600	Chain link, fence, 5′ high	B-6	890	.027	L.F.		1.08	.41	1.49	2.11
0650	3′ - 4′ high		1000	.024			.96	.37	1.33	1.88
0675	12′ high		400	.060			2.41	.91	3.32	4.69
0800	Chain link, fence, braces	B-1	2000	.012	Ea.		.45		.45	.69
0900	Privacy slats	"	2000	.012			.45		.45	.69
1000	Fence posts, steel, in concrete	B-6	80	.300			12.05	4.57	16.62	23.50
1100	Fence fabric & accessories, fabric to 8′ high		800	.030	L.F.		1.20	.46	1.66	2.34
1200	Barbed wire		5000	.005	"		.19	.07	.26	.38
1300	Extension arms & eye tops		300	.080	Ea.		3.21	1.22	4.43	6.25
1400	Fence rails		2000	.012	L.F.		.48	.18	.66	.94
1500	Reinforcing wire		5000	.005	"		.19	.07	.26	.38

02 41 13.64 Selective Demolition, Vinyl Fences and Gates

		Crew	Daily Output	Labor-Hours	Unit	Material	Labor	Equipment	Total	Total Incl O&P
0010	**SELECTIVE DEMOLITION, VINYL FENCES & GATES**									
0100	Vinyl fence up to 6′ high	B-6	1000	.024	L.F.		.96	.37	1.33	1.88
0200	Gates, up to 6′ high	"	40	.600	Ea.		24	9.15	33.15	47

02 41 13.66 Selective Demolition, Misc Metal Fences and Gates

		Crew	Daily Output	Labor-Hours	Unit	Material	Labor	Equipment	Total	Total Incl O&P
0010	**SELECTIVE DEMOLITION, MISC METAL FENCES & GATES**									
0020	See other fence items in Section 02 41 13.60									
0100	Misc steel mesh fences, 4′ - 6′ high	B-6	600	.040	L.F.		1.60	.61	2.21	3.13
0200	Kennels, 6′ - 12′ long	2 Clab	8	2	Ea.		73.50		73.50	113
0300	Tops, 6′ - 12′ long	"	30	.533	"		19.55		19.55	30
0400	Security fences, 12′ - 16′ high	B-6	100	.240	L.F.		9.65	3.65	13.30	18.75
0500	Metal tubular picket fences 4′ - 6′ high		500	.048	"		1.93	.73	2.66	3.75
0600	Gates 3′ - 4′ wide		20	1.200	Ea.		48	18.25	66.25	94

02 41 13.68 Selective Demolition, Wood Fences and Gates

		Crew	Daily Output	Labor-Hours	Unit	Material	Labor	Equipment	Total	Total Incl O&P
0010	**SELECTIVE DEMOLITION, WOOD FENCES & GATES**									
0020	See other fence items in Section 02 41 13.60									
0100	Wood fence gates 3′ - 4′ wide	2 Clab	20	.800	Ea.		29.50		29.50	45
0200	Wood fence, open rail, to 4′ high		2560	.006	L.F.		.23		.23	.35
0300	to 8′ high		368	.043	"		1.59		1.59	2.46
0400	Post, in concrete		50	.320	Ea.		11.75		11.75	18.10

02 41 13.70 Selective Demolition, Rip-Rap and Rock Lining

		Crew	Daily Output	Labor-Hours	Unit	Material	Labor	Equipment	Total	Total Incl O&P
0010	**SELECTIVE DEMOLITION, RIP-RAP & ROCK LINING** R024119-10									
0100	Slope protection, broken stone	B-13	62	.903	C.Y.		36	12.05	48.05	68.50
0200	3/8 to 1/4 C.Y. pieces		60	.933	S.Y.		37	12.45	49.45	70.50
0300	18 inch depth		60	.933	"		37	12.45	49.45	70.50
0400	Dumped stone		93	.602	Ton		24	8	32	46
0500	Gabions, 6 – 12 inches deep		60	.933	S.Y.		37	12.45	49.45	70.50
0600	18 – 36 inches deep		30	1.867	"		74.50	25	99.50	142

02 41 13.72 Selective Demo., Shore Protect/Mooring Struct.

		Crew	Daily Output	Labor-Hours	Unit	Material	Labor	Equipment	Total	Total Incl O&P
0010	**SELECTIVE DEMOLITION, SHORE PROTECT/MOORING STRUCTURES**									
0100	Breakwaters, bulkheads, concrete, maximum	B-9	12	3.333	L.F.		124	19.45	143.45	213
0200	Breakwaters, bulkheads, concrete, 12′, minimum		10	4			148	23.50	171.50	255

02 41 Demolition

02 41 13 – Selective Site Demolition

02 41 13.72 Selective Demo., Shore Protect/Mooring Struct.		Crew	Daily Output	Labor-Hours	Unit	Material	2014 Bare Costs Labor	Equipment	Total	Total Incl O&P
0300	Maximum	B-9	9	4.444	L.F.		165	26	191	283
0400	Steel, from shore	B-40B	54	.889			36	22	58	79
0500	from barge	B-76A	30	2.133			84	70	154	206
0600	Jetties, docks, floating	B-21B	600	.067	S.F.		2.65	1.11	3.76	5.30
0700	Pier supported, 3" - 4" decking		300	.133			5.30	2.21	7.51	10.60
0800	Floating, prefab, small boat, minimum		600	.067			2.65	1.11	3.76	5.30
0900	Maximum		300	.133			5.30	2.21	7.51	10.60
1000	Floating, prefab, per slip, minimum		3.20	12.500	Ea.		495	207	702	995
1010	Maximum		2.80	14.286	"		570	237	807	1,125

02 41 13.74 Selective Demolition, Piles

		Crew	Daily Output	Labor-Hours	Unit	Material	Labor	Equipment	Total	Total Incl O&P
0010	**SELECTIVE DEMOLITION, PILES**									
0100	Cast in place piles, corrugated, 8" - 10"	B-19	600	.107	V.L.F.		4.91	3.07	7.98	11.05
0200	12"-14"		500	.128			5.90	3.68	9.58	13.25
0300	16"		400	.160			7.35	4.60	11.95	16.50
0400	fluted, 12"		600	.107			4.91	3.07	7.98	11.05
0500	14"-18"		500	.128			5.90	3.68	9.58	13.25
0600	end bearing, 12"		600	.107			4.91	3.07	7.98	11.05
0700	14"-18"		500	.128			5.90	3.68	9.58	13.25
0800	Precast prestressed piles, 12"-14" dia		700	.091			4.21	2.63	6.84	9.45
0900	16"-24" dia		600	.107			4.91	3.07	7.98	11.05
1000	36"-66" dia		300	.213			9.80	6.15	15.95	22
1100	10"-14" thick		600	.107			4.91	3.07	7.98	11.05
1200	16"-24" thick		500	.128			5.90	3.68	9.58	13.25
1300	Pressure grouted pile, 5"		150	.427			19.65	12.25	31.90	44
1400	Steel piles, 8"-12" tip		600	.107			4.91	3.07	7.98	11.05
1500	H sections HP8 to HP12		600	.107			4.91	3.07	7.98	11.05
1600	HP14	B-19A	600	.107			4.91	3.76	8.67	11.80
1700	Steel pipe piles, 8"-12"	B-19	600	.107			4.91	3.07	7.98	11.05
1800	14"-18" plain		500	.128			5.90	3.68	9.58	13.25
1900	concrete filled, 14"-18"		400	.160			7.35	4.60	11.95	16.50
2000	Timber piles to 14" dia		600	.107			4.91	3.07	7.98	11.05

02 41 13.76 Selective Demolition, Water Wells

		Crew	Daily Output	Labor-Hours	Unit	Material	Labor	Equipment	Total	Total Incl O&P
0010	**SELECTIVE DEMOLITION, WATER WELLS**									
0100	Well, 40' deep with casing & gravel pack, 24"-36" dia	B-23	.25	160	Ea.		5,925	11,500	17,425	21,900
0200	Riser pipe, 1-1/4", for observation well	"	300	.133	L.F.		4.94	9.60	14.54	18.15
0300	Pump, 1/2 to 5 HP up to 100' depth	Q-1	3	5.333	Ea.		276		276	415
0400	Up to 150' well 25 HP pump	Q-22	1.50	10.667			555	440	995	1,325
0500	Up to 500' well 30 HP pump	"	1	16			830	665	1,495	1,975
0600	Well screen 2" to 8"	B-23	300	.133	V.L.F.		4.94	9.60	14.54	18.15
0700	10" to 16"		200	.200			7.40	14.40	21.80	27.50
0800	18" to 26"		150	.267			9.90	19.25	29.15	36.50
0900	Slotted PVC for wells 1-1/4"-8"		600	.067			2.47	4.81	7.28	9.10
1000	Well screen and casing 6" to 16"		300	.133			4.94	9.60	14.54	18.15
1100	18" to 26"		150	.267			9.90	19.25	29.15	36.50
1200	30" to 36"		100	.400			14.80	29	43.80	54.50

02 41 13.78 Selective Demolition, Radio Towers

		Crew	Daily Output	Labor-Hours	Unit	Material	Labor	Equipment	Total	Total Incl O&P
0010	**SELECTIVE DEMOLITION, RADIO TOWERS**									
0100	Radio tower,guyed, 50'	2 Skwk	16	1	Ea.		47.50		47.50	73.50
0200	190', 40 lb. section	K-2	.70	34.286			1,600	475	2,075	3,275
0300	200', 70 lb. section		.70	34.286			1,600	475	2,075	3,275
0400	300', 70 lb. section		.40	60			2,825	835	3,660	5,725
0500	270', 90 lb. section		.40	60			2,825	835	3,660	5,725

02 41 Demolition

02 41 13 – Selective Site Demolition

02 41 13.78 Selective Demolition, Radio Towers		Crew	Daily Output	Labor-Hours	Unit	Material	2014 Bare Costs Labor	Equipment	Total	Total Incl O&P
0600	400′	K-2	.30	80	Ea.		3,750	1,100	4,850	7,625
0700	Self supported, 60′		.90	26.667			1,250	370	1,620	2,525
0800	120′		.80	30			1,400	415	1,815	2,850
0900	190′		.40	60			2,825	835	3,660	5,725

02 41 13.80 Selective Demolition, Utility Poles and Cross Arms

Line	Description	Crew	Daily Output	Labor-Hours	Unit	Material	Labor	Equipment	Total	Total Incl O&P
0010	**SELECTIVE DEMOLITION, UTILITY POLES & CROSS ARMS**									
0100	Utility poles, wood, 20′ - 30′ high	R-3	6	3.333	Ea.		176	23.50	199.50	291
0200	35′ - 45′ high	"	5	4			212	28	240	350
0300	Cross arms, wood, 4′ - 6′ long	1 Elec	5	1.600			85.50		85.50	128

02 41 13.82 Selective Removal, Pavement Lines and Markings

Line	Description	Crew	Daily Output	Labor-Hours	Unit	Material	Labor	Equipment	Total	Total Incl O&P
0010	**SELECTIVE REMOVAL, PAVEMENT LINES & MARKINGS**									
0015	Does not include traffic control costs									
0020	See other items in Section 32 17 23.13									
0100	Remove permanent painted traffic lines and markings	B-78A	500	.016	C.L.F.		.75	1.68	2.43	2.99
0200	Temporary traffic line tape	2 Clab	1500	.011	L.F.		.39		.39	.60
0300	Thermoplastic traffic lines and markings	B-79A	500	.024	C.L.F.		1.13	2.61	3.74	4.58
0400	Painted pavement markings	B-78B	500	.036	S.F.		1.36	.78	2.14	2.95

02 41 13.84 Selective Demolition, Walks, Steps and Pavers

Line	Description	Crew	Daily Output	Labor-Hours	Unit	Material	Labor	Equipment	Total	Total Incl O&P
0010	**SELECTIVE DEMOLITION, WALKS, STEPS AND PAVERS**									
0100	Splash blocks	1 Clab	300	.027	S.F.		.98		.98	1.51
0200	Tree grates	"	50	.160	Ea.		5.85		5.85	9.05
0300	Walks, limestone pavers	2 Clab	150	.107	S.F.		3.91		3.91	6.05
0400	Redwood sections		600	.027			.98		.98	1.51
0500	Redwood planks		480	.033			1.22		1.22	1.88
0600	Shale paver		300	.053			1.95		1.95	3.02
0700	Tile thinset paver		675	.024			.87		.87	1.34
0800	Wood round	B-1	350	.069	Ea.		2.56		2.56	3.95
0900	Asphalt block	2 Clab	450	.036	S.F.		1.30		1.30	2.01
1000	Bluestone		450	.036			1.30		1.30	2.01
1100	Slate, 1″ or thinner		675	.024			.87		.87	1.34
1200	Granite blocks		300	.053			1.95		1.95	3.02
1300	Precast patio blocks		450	.036			1.30		1.30	2.01
1400	Planter blocks		600	.027			.98		.98	1.51
1500	Brick paving, dry set		300	.053			1.95		1.95	3.02
1600	Mortar set		180	.089			3.26		3.26	5.05
1700	Dry set on edge		240	.067			2.44		2.44	3.77
1800	Steps, brick		200	.080	L.F.		2.93		2.93	4.52
1900	Railroad tie		150	.107			3.91		3.91	6.05
2000	Bluestone		180	.089			3.26		3.26	5.05
2100	Wood/steel edging for steps		1000	.016			.59		.59	.90
2200	Timber or railroad tie edging for steps		400	.040			1.47		1.47	2.26

02 41 13.86 Selective Demolition, Athletic Surfaces

Line	Description	Crew	Daily Output	Labor-Hours	Unit	Material	Labor	Equipment	Total	Total Incl O&P
0010	**SELECTIVE DEMOLITION, ATHLETIC SURFACES**									
0100	Synthetic grass	2 Clab	2000	.008	S.F.		.29		.29	.45
0200	Surface coat latex rubber	"	2000	.008	"		.29		.29	.45
0300	Tennis court posts	B-11C	16	1	Ea.		43	23	66	90.50

02 41 13.88 Selective Demolition, Lawn Sprinkler Systems

Line	Description	Crew	Daily Output	Labor-Hours	Unit	Material	Labor	Equipment	Total	Total Incl O&P
0010	**SELECTIVE DEMOLITION, LAWN SPRINKLER SYSTEMS**									
0100	Golf course sprinkler system, 9 hole	4 Skwk	.10	320	Ea.		15,100		15,100	23,400
0200	Sprinkler system, 24′ diam. @ 15′ O.C., per head	B-20	110	.218	Head		8.90		8.90	13.80
0300	60′ diam. @ 24′ O.C., per head	"	52	.462	"		18.85		18.85	29

02 41 Demolition

02 41 13 – Selective Site Demolition

02 41 13.88 Selective Demolition, Lawn Sprinkler Systems		Crew	Daily Output	Labor-Hours	Unit	Material	2014 Bare Costs Labor	Equipment	Total	Total Incl O&P
0400	Sprinkler heads, plastic	2 Skwk	150	.107	Ea.		5.05		5.05	7.80
0500	Impact circle pattern, 28′ - 76′ diam.		75	.213			10.10		10.10	15.65
0600	Pop-up, 42′ - 76′ diam.		50	.320			15.15		15.15	23.50
0700	39′ - 99′ diameter		50	.320			15.15		15.15	23.50
0800	Sprinkler valves		40	.400			18.90		18.90	29.50
0900	Valve boxes		40	.400			18.90		18.90	29.50
1000	Controls		2	8			380		380	585
1100	Backflow preventer		4	4			189		189	293
1200	Vacuum breaker		4	4			189		189	293

02 41 13.90 Selective Demolition, Retaining Walls

		Crew	Daily Output	Labor-Hours	Unit	Material	Labor	Equipment	Total	Total Incl O&P
0010	**SELECTIVE DEMOLITION, RETAINING WALLS**									
0020	See other retaining wall items in Section 02 41 13.33									
0100	Concrete retaining wall, 6′ high, no reinforcing	B-9	12.70	3.150	L.F.		117	18.40	135.40	200
0200	8′ high		10	4			148	23.50	171.50	255
0300	10′ high		7.80	5.128			190	30	220	325
0400	With reinforcing, 6′ high		11.50	3.478			129	20.50	149.50	222
0500	8′ high		9	4.444			165	26	191	283
0600	10′ high		7	5.714			212	33.50	245.50	360
0700	20′ high		4	10			370	58.50	428.50	635
0800	Concrete cribbing, 12′ high, open/closed face		126	.317	S.F.		11.75	1.85	13.60	20
0900	Interlocking segmental retaining wall	B-62	800	.030			1.20	.22	1.42	2.08
1000	Wall caps	"	600	.040			1.60	.29	1.89	2.78
1100	Metal bin retaining wall, 10′ wide, 4-12′ high	B-13	1200	.047			1.86	.62	2.48	3.53
1200	10′ wide, 16-28′ high		1000	.056			2.23	.75	2.98	4.24
1300	Stone filled gabions, 6′ x 3′ x 1′		170	.329	Ea.		13.15	4.39	17.54	25
1400	6′ x 3′ x 1′-6″		75	.747			30	9.95	39.95	56.50
1500	6′ x 3′ x 3′		25	2.240			89.50	30	119.50	170
1600	9′ x 3′ x 1′		75	.747			30	9.95	39.95	56.50
1700	9′ x 3′ x 1′-6″		33	1.697			67.50	22.50	90	129
1800	9′ x 3′ x 3′		12	4.667			186	62	248	355
1900	12′ x 3′ x 1′		42	1.333			53	17.75	70.75	101
2000	12′ x 3′ x 1′-6″		20	2.800			112	37.50	149.50	212
2100	12′ x 3′ x 3′		6	9.333			370	124	494	705

02 41 13.92 Selective Demolition, Parking Appurtenances

		Crew	Daily Output	Labor-Hours	Unit	Material	Labor	Equipment	Total	Total Incl O&P
0010	**SELECTIVE DEMOLITION, PARKING APPURTENANCES**									
0100	Bumper rails, garage, 6″ wide	B-6	300	.080	L.F.		3.21	1.22	4.43	6.25
0200	12″ channel rail		300	.080			3.21	1.22	4.43	6.25
0300	Parking bumper, timber		1000	.024			.96	.37	1.33	1.88
0400	Folding, with locks	B-1	100	.240	Ea.		8.95		8.95	13.80
0500	Flexible fixed garage stanchion	B-6	150	.160			6.40	2.44	8.84	12.55
0600	Wheel stops, precast concrete		120	.200			8	3.04	11.04	15.65
0700	Thermoplastic		120	.200			8	3.04	11.04	15.65
0800	Pipe bollards, 6″ - 12″ dia		80	.300			12.05	4.57	16.62	23.50

02 41 16 – Structure Demolition

02 41 16.13 Building Demolition

		Crew	Daily Output	Labor-Hours	Unit	Material	Labor	Equipment	Total	Total Incl O&P
0010	**BUILDING DEMOLITION** Large urban projects, incl. 20 mi. haul R024119-10									
0011	No foundation or dump fees, C.F. is vol. of building standing									
0020	Steel	B-8	21500	.003	C.F.		.12	.15	.27	.36
0050	Concrete		15300	.004			.17	.22	.39	.50
0080	Masonry		20100	.003			.13	.17	.30	.38
0100	Mixture of types		20100	.003			.13	.17	.30	.38
0500	Small bldgs, or single bldgs, no salvage included, steel	B-3	14800	.003			.13	.17	.30	.39

02 41 Demolition

02 41 16 – Structure Demolition

02 41 16.13 Building Demolition

	02 41 16.13 Building Demolition	Crew	Daily Output	Labor-Hours	Unit	Material	2014 Bare Costs Labor	Equipment	Total	Total Incl O&P
0600	Concrete	B-3	11300	.004	C.F.		.17	.23	.40	.51
0650	Masonry		14800	.003			.13	.17	.30	.39
0700	Wood		14800	.003			.13	.17	.30	.39
0750	For buildings with no interior walls, deduct								50%	50%
1000	Demoliton single family house, one story, wood 1600 S.F.	B-3	1	48	Ea.		1,900	2,575	4,475	5,725
1020	3200 S.F.		.50	96			3,775	5,150	8,925	11,500
1200	Demoliton two family house, two story, wood 2400 S.F.		.67	71.964			2,825	3,850	6,675	8,575
1220	4200 S.F.		.38	128			5,025	6,875	11,900	15,300
1300	Demoliton three family house, three story, wood 3200 S.F.		.50	96			3,775	5,150	8,925	11,500
1320	5400 S.F.		.30	160			6,300	8,575	14,875	19,100
5000	For buildings with no interior walls, deduct								50%	50%

02 41 16.15 Explosive/Implosive Demolition

	02 41 16.15 Explosive/Implosive Demolition	Crew	Daily Output	Labor-Hours	Unit	Material	Labor	Equipment	Total	Total Incl O&P
0010	**EXPLOSIVE/IMPLOSIVE DEMOLITION** R024119-10									
0011	Large projects,									
0020	No disposal fee based on building volume, steel building	B-5B	16900	.003	C.F.		.12	.17	.29	.38
0100	Concrete building		16900	.003			.12	.17	.29	.38
0200	Masonry building		16900	.003			.12	.17	.29	.38
0400	Disposal of material, minimum	B-3	445	.108	C.Y.		4.24	5.80	10.04	12.85
0500	Maximum	"	365	.132	"		5.15	7.05	12.20	15.65

02 41 16.17 Building Demolition Footings and Foundations

	02 41 16.17 Building Demolition Footings and Foundations	Crew	Daily Output	Labor-Hours	Unit	Material	Labor	Equipment	Total	Total Incl O&P
0010	**BUILDING DEMOLITION FOOTINGS AND FOUNDATIONS** R024119-10									
0200	Floors, concrete slab on grade,									
0240	4" thick, plain concrete	B-9	500	.080	S.F.		2.96	.47	3.43	5.10
0280	Reinforced, wire mesh		470	.085			3.15	.50	3.65	5.40
0300	Rods		400	.100			3.71	.58	4.29	6.35
0400	6" thick, plain concrete		375	.107			3.95	.62	4.57	6.80
0420	Reinforced, wire mesh		340	.118			4.36	.69	5.05	7.50
0440	Rods		300	.133			4.94	.78	5.72	8.45
1000	Footings, concrete, 1' thick, 2' wide	B-5	300	.187	L.F.		7.55	4.75	12.30	16.80
1080	1'-6" thick, 2' wide		250	.224			9.05	5.70	14.75	20
1120	3' wide		200	.280			11.30	7.10	18.40	25
1140	2' thick, 3' wide		175	.320			12.95	8.15	21.10	29
1200	Average reinforcing, add								10%	10%
1220	Heavy reinforcing, add								20%	20%
2000	Walls, block, 4" thick	1 Clab	180	.044	S.F.		1.63		1.63	2.51
2040	6" thick		170	.047			1.72		1.72	2.66
2080	8" thick		150	.053			1.95		1.95	3.02
2100	12" thick		150	.053			1.95		1.95	3.02
2200	For horizontal reinforcing, add								10%	10%
2220	For vertical reinforcing, add								20%	20%
2400	Concrete, plain concrete, 6" thick	B-9	160	.250			9.25	1.46	10.71	15.90
2420	8" thick		140	.286			10.60	1.67	12.27	18.20
2440	10" thick		120	.333			12.35	1.95	14.30	21
2500	12" thick		100	.400			14.80	2.34	17.14	25.50
2600	For average reinforcing, add								10%	10%
2620	For heavy reinforcing, add								20%	20%
4000	For congested sites or small quantities, add up to								200%	200%
4200	Add for disposal, on site	B-11A	232	.069	C.Y.		2.95	5.70	8.65	10.80
4250	To five miles	B-30	220	.109	"		4.51	11	15.51	18.95

02 41 Demolition

02 41 16 – Structure Demolition

02 41 16.33 Bridge Demolition		Crew	Daily Output	Labor-Hours	Unit	Material	2014 Bare Costs Labor	Equipment	Total	Total Incl O&P
0010	**BRIDGE DEMOLITION**									
0100	Bridges, pedestrian, precast, 60′ to 150′ long	B-21C	250	.224	S.F.		8.95	7.40	16.35	22
0200	Steel, 50′ to 160′ long, 8′ to 10′ wide	″	500	.112			4.46	3.71	8.17	10.95
0300	Laminated wood, 80′ to 130′ long	C-12	300	.160			7.25	2.21	9.46	13.60

02 41 19 – Selective Demolition

02 41 19.13 Selective Building Demolition

0010	**SELECTIVE BUILDING DEMOLITION**
0020	Costs related to selective demolition of specific building components
0025	are included under Common Work Results (XX 05)
0030	in the component's appropriate division.

02 41 19.16 Selective Demolition, Cutout

Line	Description	Crew	Daily Output	Labor-Hours	Unit	Material	Labor	Equipment	Total	Total Incl O&P
0010	**SELECTIVE DEMOLITION, CUTOUT** R024119-10									
0020	Concrete, elev. slab, light reinforcement, under 6 C.F.	B-9	65	.615	C.F.		23	3.59	26.59	39
0050	Light reinforcing, over 6 C.F.		75	.533	″		19.75	3.11	22.86	34
0200	Slab on grade to 6″ thick, not reinforced, under 8 S.F.		85	.471	S.F.		17.45	2.75	20.20	30
0250	8 – 16 S.F.		175	.229	″		8.45	1.33	9.78	14.50
0255	For over 16 S.F. see Line 02 41 16.17 0400									
0600	Walls, not reinforced, under 6 C.F.	B-9	60	.667	C.F.		24.50	3.89	28.39	42.50
0650	6 – 12 C.F.	″	80	.500	″		18.55	2.92	21.47	31.50
0655	For over 12 C.F. see Line 02 41 16.17 2500									
1000	Concrete, elevated slab, bar reinforced, under 6 C.F.	B-9	45	.889	C.F.		33	5.20	38.20	56.50
1050	Bar reinforced, over 6 C.F.		50	.800	″		29.50	4.67	34.17	50.50
1200	Slab on grade to 6″ thick, bar reinforced, under 8 S.F.		75	.533	S.F.		19.75	3.11	22.86	34
1250	8 – 16 S.F.		150	.267	″		9.90	1.56	11.46	16.95
1255	For over 16 S.F. see Line 02 41 16.17 0440									
1400	Walls, bar reinforced, under 6 C.F.	B-9	50	.800	C.F.		29.50	4.67	34.17	50.50
1450	6 – 12 C.F.	″	70	.571	″		21	3.34	24.34	36
1455	For over 12 C.F. see Lines 02 41 16.17 2500 and 2600									
2000	Brick, to 4 S.F. opening, not including toothing									
2040	4″ thick	B-9	30	1.333	Ea.		49.50	7.80	57.30	84.50
2060	8″ thick		18	2.222			82.50	13	95.50	141
2080	12″ thick		10	4			148	23.50	171.50	255
2400	Concrete block, to 4 S.F. opening, 2″ thick		35	1.143			42.50	6.65	49.15	73
2420	4″ thick		30	1.333			49.50	7.80	57.30	84.50
2440	8″ thick		27	1.481			55	8.65	63.65	94
2460	12″ thick		24	1.667			62	9.75	71.75	106
2600	Gypsum block, to 4 S.F. opening, 2″ thick		80	.500			18.55	2.92	21.47	31.50
2620	4″ thick		70	.571			21	3.34	24.34	36
2640	8″ thick		55	.727			27	4.25	31.25	46
2800	Terra cotta, to 4 S.F. opening, 4″ thick		70	.571			21	3.34	24.34	36
2840	8″ thick		65	.615			23	3.59	26.59	39
2880	12″ thick		50	.800			29.50	4.67	34.17	50.50
3000	Toothing masonry cutouts, brick, soft old mortar	1 Brhe	40	.200	V.L.F.		7.40		7.40	11.30
3100	Hard mortar		30	.267			9.85		9.85	15.05
3200	Block, soft old mortar		70	.114			4.23		4.23	6.45
3400	Hard mortar		50	.160			5.90		5.90	9
6000	Walls, interior, not including re-framing,									
6010	openings to 5 S.F.									
6100	Drywall to 5/8″ thick	1 Clab	24	.333	Ea.		12.20		12.20	18.85
6200	Paneling to 3/4″ thick		20	.400			14.65		14.65	22.50
6300	Plaster, on gypsum lath		20	.400			14.65		14.65	22.50

02 41 19.16 Selective Demolition, Cutout		Crew	Daily Output	Labor-Hours	Unit	Material	2014 Bare Costs Labor	Equipment	Total	Total Incl O&P
6340	On wire lath	1 Clab	14	.571	Ea.		21		21	32.50
7000	Wood frame, not including re-framing, openings to 5 S.F.									
7200	Floors, sheathing and flooring to 2″ thick	1 Clab	5	1.600	Ea.		58.50		58.50	90.50
7310	Roofs, sheathing to 1″ thick, not including roofing		6	1.333			49		49	75.50
7410	Walls, sheathing to 1″ thick, not including siding		7	1.143			42		42	64.50

02 41 19.18 Selective Demolition, Disposal Only

Code	Description	Crew	Daily Output	Labor-Hours	Unit	Material	Labor	Equipment	Total	Total Incl O&P
0010	**SELECTIVE DEMOLITION, DISPOSAL ONLY** R024119-10									
0015	Urban bldg w/salvage value allowed									
0020	Including loading and 5 mile haul to dump									
0200	Steel frame	B-3	430	.112	C.Y.		4.39	6	10.39	13.35
0300	Concrete frame		365	.132			5.15	7.05	12.20	15.65
0400	Masonry construction		445	.108			4.24	5.80	10.04	12.85
0500	Wood frame		247	.194			7.65	10.45	18.10	23

02 41 19.19 Selective Facility Services Demolition

Code	Description	Crew	Daily Output	Labor-Hours	Unit	Material	Labor	Equipment	Total	Total Incl O&P
0010	**SELECTIVE FACILITY SERVICES DEMOLITION**, Rubbish Handling R024119-10									
0020	The following are to be added to the demolition prices									
0050	The following are components for a complete chute system									
0100	Top chute circular steel, 4′ long, 18″ diameter R024119-30	B-1C	15	1.600	Ea.	256	59.50	30.50	346	405
0102	23″ diameter		15	1.600		277	59.50	30.50	367	430
0104	27″ diameter		15	1.600		299	59.50	30.50	389	455
0106	30″ diameter		15	1.600		320	59.50	30.50	410	475
0108	33″ diameter		15	1.600		340	59.50	30.50	430	500
0110	36″ diameter		15	1.600		365	59.50	30.50	455	525
0112	Regular chute, 18″ diameter		15	1.600		192	59.50	30.50	282	335
0114	23″ diameter		15	1.600		213	59.50	30.50	303	360
0116	27″ diameter		15	1.600		234	59.50	30.50	324	385
0118	30″ diameter		15	1.600		256	59.50	30.50	346	405
0120	33″ diameter		15	1.600		277	59.50	30.50	367	430
0122	36″ diameter		15	1.600		299	59.50	30.50	389	455
0124	Control door chute, 18″ diameter		15	1.600		365	59.50	30.50	455	525
0126	23″ diameter		15	1.600		385	59.50	30.50	475	545
0128	27″ diameter		15	1.600		405	59.50	30.50	495	570
0130	30″ diameter		15	1.600		425	59.50	30.50	515	595
0132	33″ diameter		15	1.600		445	59.50	30.50	535	615
0134	36″ diameter		15	1.600		470	59.50	30.50	560	640
0136	Chute liners, 14 ga., 18-30″ diameter		15	1.600		215	59.50	30.50	305	360
0138	33-36″ diameter		15	1.600		269	59.50	30.50	359	420
0140	17% thinner chute, 30″ diameter		15	1.600		213	59.50	30.50	303	360
0142	33% thinner chute, 30″ diameter		15	1.600		160	59.50	30.50	250	300
0144	Top chute cover	1 Clab	24	.333		147	12.20		159.20	181
0146	Door chute cover	″	24	.333		147	12.20		159.20	181
0148	Top chute trough	2 Clab	12	1.333		490	49		539	615
0150	Bolt down frame & counter weights, 250 lb.	B-1	4	6		4,075	224		4,299	4,825
0152	500 lb.		4	6		5,950	224		6,174	6,900
0154	750 lb.		4	6		8,750	224		8,974	9,975
0156	1000 lb.		2.67	8.989		9,775	335		10,110	11,300
0158	1500 lb.		2.67	8.989		12,500	335		12,835	14,300
0160	Chute warning light system, 5 stories	B-1C	4	6		8,625	224	114	8,963	9,950
0162	10 stories	″	2	12		13,700	450	228	14,378	16,000
0164	Dust control device for dumpsters	1 Clab	8	1		98	36.50		134.50	165
0166	Install or replace breakaway cord		8	1		26	36.50		62.50	85
0168	Install or replace warning sign		16	.500		11	18.35		29.35	40.50

02 41 19.19 Selective Facility Services Demolition		Crew	Daily Output	Labor-Hours	Unit	Material	2014 Bare Costs Labor	Equipment	Total	Total Incl O&P
0600	Dumpster, weekly rental, 1 dump/week, 6 C.Y. capacity (2 Tons)				Week	415			415	455
0700	10 C.Y. capacity (3 Tons)					480			480	530
0725	20 C.Y. capacity (5 Tons) R024119-20					565			565	625
0800	30 C.Y. capacity (7 Tons)					730			730	800
0840	40 C.Y. capacity (10 Tons)					775			775	850
0900	Alternate pricing for dumpsters									
0910	Delivery, average for all sizes				Ea.	75			75	82.50
0920	Haul, average for all sizes					235			235	259
0930	Rent per day, average for all sizes					20			20	22
0940	Rent per month, average for all sizes					80			80	88
0950	Disposal fee per ton, average for all sizes				Ton	88			88	97
2000	Load, haul, dump and return, 50′ haul, hand carried	2 Clab	24	.667	C.Y.		24.50		24.50	37.50
2005	Wheeled		37	.432			15.85		15.85	24.50
2040	51′ to 100′ haul, hand carried		16.50	.970			35.50		35.50	55
2045	Wheeled		25	.640			23.50		23.50	36
2080	Over 100′ haul, add per 100 L.F., hand carried		35.50	.451			16.50		16.50	25.50
2085	Wheeled		54	.296			10.85		10.85	16.75
2120	In elevators, per 10 floors, add		140	.114			4.19		4.19	6.45
2130	Load, haul, dump and return, up to 50′ haul, incl. up to 5 rsr stair, hand		23	.696			25.50		25.50	39.50
2135	Wheeled		35	.457			16.75		16.75	26
2140	6 – 10 riser stairs, hand carried		22	.727			26.50		26.50	41
2145	Wheeled		34	.471			17.25		17.25	26.50
2150	11 – 20 riser stairs, hand carried		20	.800			29.50		29.50	45
2155	Wheeled		31	.516			18.90		18.90	29
2160	21 – 40 riser stairs, hand carried		16	1			36.50		36.50	56.50
2165	Wheeled		24	.667			24.50		24.50	37.50
2170	100′ haul, incl. 5 riser stair, hand carried		15	1.067			39		39	60.50
2175	Wheeled		23	.696			25.50		25.50	39.50
2180	6 – 10 riser stair, hand carried		14	1.143			42		42	64.50
2185	Wheeled		21	.762			28		28	43
2190	11 – 20 riser stair, hand carried		12	1.333			49		49	75.50
2195	Wheeled		18	.889			32.50		32.50	50.50
2200	21 – 40 riser stair, hand carried		8	2			73.50		73.50	113
2205	Wheeled		12	1.333			49		49	75.50
2210	Over 100′ haul, add per 100 L.F., hand carried		35.50	.451			16.50		16.50	25.50
2215	Wheeled		54	.296			10.85		10.85	16.75
2220	For each additional flight of stairs, up to 5 risers, add		550	.029	Flight		1.07		1.07	1.65
2225	6 – 10 risers, add		275	.058			2.13		2.13	3.29
2230	11 – 20 risers, add		138	.116			4.25		4.25	6.55
2235	21 – 40 risers, add		69	.232			8.50		8.50	13.10
3000	Loading & trucking, including 2 mile haul, chute loaded	B-16	45	.711	C.Y.		26.50	15.40	41.90	58
3040	Hand loading truck, 50′ haul	"	48	.667			25	14.40	39.40	54.50
3080	Machine loading truck	B-17	120	.267			10.55	6.50	17.05	23.50
5000	Haul, per mile, up to 8 C.Y. truck	B-34B	1165	.007			.26	.59	.85	1.04
5100	Over 8 C.Y. truck	"	1550	.005			.19	.45	.64	.79

02 41 19.20 Selective Demolition, Dump Charges		Crew	Daily Output	Labor-Hours	Unit	Material	Labor	Equipment	Total	Total Incl O&P
0010	**SELECTIVE DEMOLITION, DUMP CHARGES** R024119-10									
0020	Dump charges, typical urban city, tipping fees only									
0100	Building construction materials				Ton	74			74	81
0200	Trees, brush, lumber					63			63	69.50
0300	Rubbish only					63			63	69.50
0500	Reclamation station, usual charge					74			74	81

02 41 Demolition

02 41 19 – Selective Demolition

02 41 19.21 Selective Demolition, Gutting		Crew	Daily Output	Labor-Hours	Unit	2014 Bare Costs Material	Labor	Equipment	Total	Total Incl O&P
0010	SELECTIVE DEMOLITION, GUTTING R024119-10									
0020	Building interior, including disposal, dumpster fees not included									
0500	Residential building									
0560	Minimum	B-16	400	.080	SF Flr.		2.99	1.73	4.72	6.50
0580	Maximum	"	360	.089	"		3.32	1.92	5.24	7.20
0900	Commercial building									
1000	Minimum	B-16	350	.091	SF Flr.		3.42	1.98	5.40	7.45
1020	Maximum	"	250	.128	"		4.78	2.77	7.55	10.40

02 41 19.25 Selective Demolition, Saw Cutting

Line	Description	Crew	Daily Output	Labor-Hours	Unit	Material	Labor	Equipment	Total	Total Incl O&P
0010	SELECTIVE DEMOLITION, SAW CUTTING R024119-10									
0015	Asphalt, up to 3" deep	B-89	1050	.015	L.F.	.14	.64	.50	1.28	1.67
0020	Each additional inch of depth	"	1800	.009		.05	.37	.29	.71	.93
1200	Masonry walls, hydraulic saw, brick, per inch of depth	B-89B	300	.053		.05	2.23	3.03	5.31	6.75
1220	Block walls, solid, per inch of depth	"	250	.064		.05	2.67	3.64	6.36	8.10
2000	Brick or masonry w/hand held saw, per inch of depth	A-1	125	.064		.06	2.35	.67	3.08	4.43
5000	Wood sheathing to 1" thick, on walls	1 Carp	200	.040			1.83		1.83	2.83
5020	On roof	"	250	.032			1.47		1.47	2.26

02 41 19.27 Selective Demolition, Torch Cutting

Line	Description	Crew	Daily Output	Labor-Hours	Unit	Material	Labor	Equipment	Total	Total Incl O&P
0010	SELECTIVE DEMOLITION, TORCH CUTTING R024119-10									
0020	Steel, 1" thick plate	E-25	333	.024	L.F.	.76	1.28	.03	2.07	3.12
0040	1" diameter bar	"	600	.013	Ea.	.13	.71	.02	.86	1.41
1000	Oxygen lance cutting, reinforced concrete walls									
1040	12" to 16" thick walls	1 Clab	10	.800	L.F.		29.50		29.50	45
1080	24" thick walls	"	6	1.333	"		49		49	75.50

02 43 Structure Moving

02 43 13 – Structure Relocation

02 43 13.13 Building Relocation

Line	Description	Crew	Daily Output	Labor-Hours	Unit	Material	Labor	Equipment	Total	Total Incl O&P
0010	BUILDING RELOCATION									
0011	One day move, up to 24' wide									
0020	Reset on new foundation, patch & hook-up, average move				Total					11,500
0040	Wood or steel frame bldg., based on ground floor area G	B-4	185	.259	S.F.		9.65	2.80	12.45	17.95
0060	Masonry bldg., based on ground floor area G	"	137	.350			13	3.78	16.78	24
0200	For 24' to 42' wide, add								15%	15%
0220	For each additional day on road, add G	B-4	1	48	Day		1,775	520	2,295	3,325
0240	Construct new basement, move building, 1 day									
0300	move, patch & hook-up, based on ground floor area G	B-3	155	.310	S.F.	10.25	12.20	16.60	39.05	48.50

02 58 Snow Control

02 58 13 – Snow Fencing

02 58 13.10 Snow Fencing System

Line	Description	Crew	Daily Output	Labor-Hours	Unit	Material	Labor	Equipment	Total	Total Incl O&P
0010	SNOW FENCING SYSTEM									
7001	Snow fence on steel posts 10' O.C., 4' high	B-1	500	.048	L.F.	.96	1.79		2.75	3.82

02 65 10 – Underground Tank and Contaminated Soil Removal

02 65 10.30 Removal of Underground Storage Tanks		Crew	Daily Output	Labor-Hours	Unit	Material	2014 Bare Costs Labor	Equipment	Total	Total Incl O&P
0010	**REMOVAL OF UNDERGROUND STORAGE TANKS**									
0011	Petroleum storage tanks, non-leaking									
0100	Excavate & load onto trailer									
0110	3000 gal. to 5000 gal. tank G	B-14	4	12	Ea.		465	91.50	556.50	815
0120	6000 gal. to 8000 gal. tank G	B-3A	3	13.333			520	345	865	1,175
0130	9000 gal. to 12000 gal. tank G	"	2	20			780	515	1,295	1,775
0190	Known leaking tank, add				%				100%	100%
0200	Remove sludge, water and remaining product from tank bottom									
0201	of tank with vacuum truck									
0300	3000 gal. to 5000 gal. tank G	A-13	5	1.600	Ea.		75.50	154	229.50	284
0310	6000 gal. to 8000 gal. tank G		4	2			94	193	287	355
0320	9000 gal. to 12000 gal. tank G		3	2.667			125	257	382	475
0390	Dispose of sludge off-site, average				Gal.				6.25	6.80
0400	Insert inert solid CO_2 "dry ice" into tank									
0401	For cleaning/transporting tanks (1.5 lb./100 gal. cap) G	1 Clab	500	.016	Lb.	1.08	.59		1.67	2.09
0403	Insert solid carbon dioxide, 1.5 lb./100 gal. G	"	400	.020	"	1.08	.73		1.81	2.32
0503	Disconnect and remove piping G	1 Plum	160	.050	L.F.		2.88		2.88	4.35
0603	Transfer liquids, 10% of volume G	"	1600	.005	Gal.		.29		.29	.43
0703	Cut accessway into underground storage tank G	1 Clab	5.33	1.501	Ea.		55		55	85
0813	Remove sludge, wash and wipe tank, 500 gal. G	1 Plum	8	1			57.50		57.50	87
0823	3,000 gal. G		6.67	1.199			69		69	104
0833	5,000 gal. G		6.15	1.301			75		75	113
0843	8,000 gal. G		5.33	1.501			86.50		86.50	130
0853	10,000 gal. G		4.57	1.751			101		101	152
0863	12,000 gal. G		4.21	1.900			109		109	165
1020	Haul tank to certified salvage dump, 100 miles round trip									
1023	3000 gal. to 5000 gal. tank				Ea.				760	830
1026	6000 gal. to 8000 gal. tank								880	960
1029	9,000 gal. to 12,000 gal. tank								1,050	1,150
1100	Disposal of contaminated soil to landfill									
1110	Minimum				C.Y.				145	160
1111	Maximum				"				400	440
1120	Disposal of contaminated soil to									
1121	bituminous concrete batch plant									
1130	Minimum				C.Y.				80	88
1131	Maximum				"				115	125
1203	Excavate, pull, & load tank, backfill hole, 8,000 gal. + G	B-12C	.50	32	Ea.		1,400	2,350	3,750	4,700
1213	Haul tank to certified dump, 100 miles rt, 8,000 gal. + G	B-34K	1	8			300	960	1,260	1,500
1223	Excavate, pull, & load tank, backfill hole, 500 gal. G	B-11C	1	16			685	365	1,050	1,450
1233	Excavate, pull, & load tank, backfill hole, 3,000 – 5,000 gal. G	B-11M	.50	32			1,375	795	2,170	2,975
1243	Haul tank to certified dump, 100 miles rt, 500 gal. G	B-34L	1	8			375	270	645	865
1253	Haul tank to certified dump, 100 miles rt, 3,000 – 5,000 gal. G	B-34M	1	8			375	335	710	935
2010	Decontamination of soil on site incl poly tarp on top/bottom									
2011	Soil containment berm, and chemical treatment									
2020	Minimum G	B-11C	100	.160	C.Y.	7.80	6.85	3.65	18.30	23
2021	Maximum G	"	100	.160		10.10	6.85	3.65	20.60	25.50
2050	Disposal of decontaminated soil, minimum								135	150
2055	Maximum								400	440

02 81 Transportation and Disposal of Hazardous Materials

02 81 20 – Hazardous Waste Handling

02 81 20.10 Hazardous Waste Cleanup/Pickup/Disposal		Crew	Daily Output	Labor-Hours	Unit	2014 Bare Costs Material	Labor	Equipment	Total	Total Incl O&P
0010	**HAZARDOUS WASTE CLEANUP/PICKUP/DISPOSAL**									
0100	For contractor rental equipment, i.e., Dozer,									
0110	Front end loader, Dump truck, etc., see 01 54 33 Reference Section									
1000	Solid pickup									
1100	55 gal. drums				Ea.				240	265
1120	Bulk material, minimum				Ton				190	210
1130	Maximum				"				595	655
1200	Transportation to disposal site									
1220	Truckload = 80 drums or 25 C.Y. or 18 tons									
1260	Minimum				Mile				3.95	4.45
1270	Maximum				"				7.25	7.35
3000	Liquid pickup, vacuum truck, stainless steel tank									
3100	Minimum charge, 4 hours									
3110	1 compartment, 2200 gallon				Hr.				140	155
3120	2 compartment, 5000 gallon				"				200	225
3400	Transportation in 6900 gallon bulk truck				Mile				7.95	8.75
3410	In teflon lined truck				"				10.20	11.25
5000	Heavy sludge or dry vacuumable material				Hr.				140	160
6000	Dumpsite disposal charge, minimum				Ton				140	155
6020	Maximum				"				415	455

Estimating Tips

General

- Carefully check all the plans and specifications. Concrete often appears on drawings other than structural drawings, including mechanical and electrical drawings for equipment pads. The cost of cutting and patching is often difficult to estimate. See Subdivision 03 81 for Concrete Cutting, Subdivision 02 41 19.16 for Cutout Demolition, Subdivision 03 05 05.10 for Concrete Demolition, and Subdivision 02 41 19.19 for Rubbish Handling (handling, loading and hauling of debris).
- Always obtain concrete prices from suppliers near the job site. A volume discount can often be negotiated, depending upon competition in the area. Remember to add for waste, particularly for slabs and footings on grade.

03 10 00 Concrete Forming and Accessories

- A primary cost for concrete construction is forming. Most jobs today are constructed with prefabricated forms. The selection of the forms best suited for the job and the total square feet of forms required for efficient concrete forming and placing are key elements in estimating concrete construction. Enough forms must be available for erection to make efficient use of the concrete placing equipment and crew.
- Concrete accessories for forming and placing depend upon the systems used. Study the plans and specifications to ensure that all special accessory requirements have been included in the cost estimate, such as anchor bolts, inserts, and hangers.
- Included within costs for forms-in-place are all necessary bracing and shoring.

03 20 00 Concrete Reinforcing

- Ascertain that the reinforcing steel supplier has included all accessories, cutting, bending, and an allowance for lapping, splicing, and waste. A good rule of thumb is 10% for lapping, splicing, and waste. Also, 10% waste should be allowed for welded wire fabric.
- The unit price items in the subdivisions for Reinforcing In Place, Glass Fiber Reinforcing, and Welded Wire Fabric include the labor to install accessories such as beam and slab bolsters, high chairs, and bar ties and tie wire. The material cost for these accessories is not included; they may be obtained from the Accessories Division.

03 30 00 Cast-In-Place Concrete

- When estimating structural concrete, pay particular attention to requirements for concrete additives, curing methods, and surface treatments. Special consideration for climate, hot or cold, must be included in your estimate. Be sure to include requirements for concrete placing equipment, and concrete finishing.
- For accurate concrete estimating, the estimator must consider each of the following major components individually: forms, reinforcing steel, ready-mix concrete, placement of the concrete, and finishing of the top surface. For faster estimating, Subdivision 03 30 53.40 for Concrete-In-Place can be used; here, various items of concrete work are presented that include the costs of all five major components (unless specifically stated otherwise).

03 40 00 Precast Concrete
03 50 00 Cast Decks and Underlayment

- The cost of hauling precast concrete structural members is often an important factor. For this reason, it is important to get a quote from the nearest supplier. It may become economically feasible to set up precasting beds on the site if the hauling costs are prohibitive.

Reference Numbers

Reference numbers are shown in shaded boxes at the beginning of some major classifications. These numbers refer to related items in the Reference Section. The reference information may be an estimating procedure, an alternate pricing method, or technical information.

Note: Not all subdivisions listed here necessarily appear in this publication.

03 01 Maintenance of Concrete

03 01 30 – Maintenance of Cast-In-Place Concrete

03 01 30.71 Concrete Crack Repair		Crew	Daily Output	Labor-Hours	Unit	Material	2014 Bare Costs Labor	Equipment	Total	Total Incl O&P
0010	**CONCRETE CRACK REPAIR**									
1000	Structural repair of concrete cracks by epoxy injection (ACI RAP-1)									
1001	suitable for horizontal, vertical and overhead repairs									
1010	Clean/grind concrete surface(s) free of contaminants	1 Cefi	400	.020	L.F.		.88		.88	1.30
1015	Rout crack with v-notch crack chaser, if needed	C-32	600	.027		.04	1.08	.15	1.27	1.83
1020	Blow out crack with oil-free dry compressed air (1 pass)	C-28	3000	.003			.12	.01	.13	.18
1030	Install surface-mounted entry ports (spacing = concrete depth)	1 Cefi	400	.020	Ea.	1.34	.88		2.22	2.77
1040	Cap crack at surface with epoxy gel (per side/face)		400	.020	L.F.	.33	.88		1.21	1.66
1050	Snap off ports, grind off epoxy cap residue after injection		200	.040	"		1.76		1.76	2.60
1100	Manual injection with 2-part epoxy cartridge, excludes prep									
1110	Up to 1/32" (0.03125") wide x 4" deep	1 Cefi	160	.050	L.F.	.17	2.20		2.37	3.44
1120	6" deep		120	.067		.25	2.94		3.19	4.61
1130	8" deep		107	.075		.33	3.29		3.62	5.25
1140	10" deep		100	.080		.41	3.52		3.93	5.65
1150	12" deep		96	.083		.50	3.67		4.17	5.95
1210	Up to 1/16" (0.0625") wide x 4" deep		160	.050		.33	2.20		2.53	3.62
1220	6" deep		120	.067		.50	2.94		3.44	4.89
1230	8" deep		107	.075		.66	3.29		3.95	5.60
1240	10" deep		100	.080		.83	3.52		4.35	6.10
1250	12" deep		96	.083		.99	3.67		4.66	6.50
1310	Up to 3/32" (0.09375") wide x 4" deep		160	.050		.50	2.20		2.70	3.81
1320	6" deep		120	.067		.74	2.94		3.68	5.15
1330	8" deep		107	.075		.99	3.29		4.28	5.95
1340	10" deep		100	.080		1.24	3.52		4.76	6.55
1350	12" deep		96	.083		1.49	3.67		5.16	7.05
1410	Up to 1/8" (0.125") wide x 4" deep		160	.050		.66	2.20		2.86	3.99
1420	6" deep		120	.067		.99	2.94		3.93	5.45
1430	8" deep		107	.075		1.32	3.29		4.61	6.30
1440	10" deep		100	.080		1.65	3.52		5.17	7
1450	12" deep		96	.083		1.98	3.67		5.65	7.60
1500	Pneumatic injection with 2-part bulk epoxy, excludes prep									
1510	Up to 5/32" (0.15625") wide x 4" deep	C-31	240	.033	L.F.	.23	1.47	1.54	3.24	4.13
1520	6" deep		180	.044		.35	1.96	2.06	4.37	5.55
1530	8" deep		160	.050		.47	2.20	2.32	4.99	6.30
1540	10" deep		150	.053		.58	2.35	2.47	5.40	6.85
1550	12" deep		144	.056		.70	2.45	2.57	5.72	7.20
1610	Up to 3/16" (0.1875") wide x 4" deep		240	.033		.28	1.47	1.54	3.29	4.18
1620	6" deep		180	.044		.42	1.96	2.06	4.44	5.60
1630	8" deep		160	.050		.56	2.20	2.32	5.08	6.45
1640	10" deep		150	.053		.70	2.35	2.47	5.52	6.95
1650	12" deep		144	.056		.84	2.45	2.57	5.86	7.40
1710	Up to 1/4" (0.1875") wide x 4" deep		240	.033		.37	1.47	1.54	3.38	4.28
1720	6" deep		180	.044		.56	1.96	2.06	4.58	5.75
1730	8" deep		160	.050		.75	2.20	2.32	5.27	6.65
1740	10" deep		150	.053		.94	2.35	2.47	5.76	7.20
1750	12" deep		144	.056		1.12	2.45	2.57	6.14	7.70
2000	Non-structural filling of concrete cracks by gravity-fed resin (ACI RAP-2)									
2001	suitable for individual cracks in stable horizontal surfaces only									
2010	Clean/grind concrete surface(s) free of contaminants	1 Cefi	400	.020	L.F.		.88		.88	1.30
2020	Rout crack with v-notch crack chaser, if needed	C-32	600	.027		.04	1.08	.15	1.27	1.83
2030	Blow out crack with oil-free dry compressed air (1 pass)	C-28	3000	.003			.12	.01	.13	.18
2040	Cap crack with epoxy gel at underside of elevated slabs, if needed	1 Cefi	400	.020		.33	.88		1.21	1.66
2050	Insert backer rod into crack, if needed		400	.020		.02	.88		.90	1.33

03 01 Maintenance of Concrete

03 01 30 – Maintenance of Cast-In-Place Concrete

03 01 30.71 Concrete Crack Repair		Crew	Daily Output	Labor-Hours	Unit	2014 Bare Costs Material	Labor	Equipment	Total	Total Incl O&P
2060	Partially fill crack with fine dry sand, if needed	1 Cefi	800	.010	L.F.	.02	.44		.46	.67
2070	Apply two beads of sealant alongside crack to form reservoir, if needed		400	.020		.25	.88		1.13	1.58
2100	Manual filling with squeeze bottle of 2-part epoxy resin									
2110	Full depth crack up to 1/16" (0.0625") wide x 4" deep	1 Cefi	300	.027	L.F.	.09	1.17		1.26	1.84
2120	6" deep		240	.033		.14	1.47		1.61	2.32
2130	8" deep		200	.040		.18	1.76		1.94	2.80
2140	10" deep		185	.043		.23	1.90		2.13	3.06
2150	12" deep		175	.046		.27	2.01		2.28	3.28
2210	Full depth crack up to 1/8" (0.125") wide x 4" deep		270	.030		.18	1.31		1.49	2.13
2220	6" deep		215	.037		.27	1.64		1.91	2.72
2230	8" deep		180	.044		.36	1.96		2.32	3.29
2240	10" deep		165	.048		.46	2.14		2.60	3.66
2250	12" deep		155	.052		.55	2.27		2.82	3.96
2310	Partial depth crack up to 3/16" (0.1875") wide x 1" deep		300	.027		.07	1.17		1.24	1.82
2410	Up to 1/4" (0.250") wide x 1" deep		270	.030		.09	1.31		1.40	2.03
2510	Up to 5/16" (0.3125") wide x 1" deep		250	.032		.11	1.41		1.52	2.21
2610	Up to 3/8" (0.375") wide x 1" deep		240	.033		.14	1.47		1.61	2.32

03 01 30.72 Concrete Surface Repairs		Crew	Daily Output	Labor-Hours	Unit	Material	Labor	Equipment	Total	Total Incl O&P
0010	**CONCRETE SURFACE REPAIRS**									
9000	Surface repair by Methacrylate flood coat (ACI RAP-13)									
9010	Suitable for healing and sealing horizontal surfaces only									
9020	Large cracks must previously have been repaired or filled									
9100	Shotblast entire surface to remove contaminants	A-1A	4000	.002	S.F.		.09	.05	.14	.21
9200	Blow off dust and debris with oil-free dry compressed air	C-28	16000	.001			.02		.02	.03
9300	Flood coat surface w/Methacrylate, distribute w/broom/squeegee, no prep.	3 Cefi	8000	.003		.96	.13		1.09	1.26
9400	Lightly broadcast even coat of dry silica sand while sealer coat is tacky	1 Cefi	8000	.001		.01	.04		.05	.08

03 05 Common Work Results for Concrete

03 05 05 – Selective Concrete Demolition

03 05 05.10 Selective Demolition, Concrete		Crew	Daily Output	Labor-Hours	Unit	Material	Labor	Equipment	Total	Total Incl O&P
0010	**SELECTIVE DEMOLITION, CONCRETE** R024119-10									
0012	Excludes saw cutting, torch cutting, loading or hauling									
0050	Break into small pieces, reinf. less than 1% of cross-sectional area	B-9	24	1.667	C.Y.		62	9.75	71.75	106
0060	Reinforcing 1% to 2% of cross-sectional area		16	2.500			92.50	14.60	107.10	159
0070	Reinforcing more than 2% of cross-sectional area		8	5			185	29	214	320
0150	Remove whole pieces, up to 2 tons per piece	E-18	36	1.111	Ea.		56.50	26.50	83	127
0160	2 – 5 tons per piece		30	1.333			68	32	100	152
0170	5 – 10 tons per piece		24	1.667			85	40	125	190
0180	10 – 15 tons per piece		18	2.222			113	53.50	166.50	254
0250	Precast unit embedded in masonry, up to 1 C.F.	D-1	16	1			41.50		41.50	63
0260	1 – 2 C.F.		12	1.333			55		55	84
0270	2 – 5 C.F.		10	1.600			66		66	101
0280	5 – 10 C.F.		8	2			82.50		82.50	126
0990	For hydrodemolition see Section 02 41 13.15									

03 05 13 – Basic Concrete Materials

03 05 13.20 Concrete Admixtures and Surface Treatments		Crew	Daily Output	Labor-Hours	Unit	Material	Labor	Equipment	Total	Total Incl O&P
0010	**CONCRETE ADMIXTURES AND SURFACE TREATMENTS**									
0040	Abrasives, aluminum oxide, over 20 tons				Lb.	1.82			1.82	2
0050	1 to 20 tons					1.95			1.95	2.14
0070	Under 1 ton					2.03			2.03	2.23

03 05 Common Work Results for Concrete

03 05 13 – Basic Concrete Materials

03 05 13.20 Concrete Admixtures and Surface Treatments		Crew	Daily Output	Labor-Hours	Unit	2014 Bare Costs Material	Labor	Equipment	Total	Total Incl O&P
0100	Silicon carbide, black, over 20 tons				Lb.	2.78			2.78	3.06
0110	1 to 20 tons					2.96			2.96	3.25
0120	Under 1 ton				↓	3.08			3.08	3.39
0200	Air entraining agent, .7 to 1.5 oz. per bag, 55 gallon drum				Gal.	13.35			13.35	14.65
0220	5 gallon pail					18.35			18.35	20
0300	Bonding agent, acrylic latex, 250 S.F. per gallon, 5 gallon pail					22			22	24.50
0320	Epoxy resin, 80 S.F. per gallon, 4 gallon case				↓	61.50			61.50	67.50
0400	Calcium chloride, 50 lb. bags, TL lots				Ton	770			770	845
0420	Less than truckload lots				Bag	23.50			23.50	25.50
0500	Carbon black, liquid, 2 to 8 lb. per bag of cement				Lb.	8.65			8.65	9.55
0600	Colored pigments, integral, 2 to 10 lb. per bag of cement, subtle colors					2.40			2.40	2.64
0610	Standard colors					3.25			3.25	3.58
0620	Premium colors				↓	5.40			5.40	5.95
0920	Dustproofing compound, 250 S.F./gal., 5 gallon pail				Gal.	6.20			6.20	6.80
1010	Epoxy based, 125 S.F./gal., 5 gallon pail				"	53			53	58
1100	Hardeners, metallic, 55 lb. bags, natural (grey)				Lb.	.66			.66	.73
1200	Colors					2.17			2.17	2.38
1300	Non-metallic, 55 lb. bags, natural grey					.41			.41	.45
1320	Colors				↓	.90			.90	.99
1550	Release agent, for tilt slabs, 5 gallon pail				Gal.	16.20			16.20	17.80
1570	For forms, 5 gallon pail					13.35			13.35	14.70
1590	Concrete release agent for forms, 100% biodegradeable, zero VOC, 5 gal pail G					19.20			19.20	21
1595	55 gallon drum G					15.75			15.75	17.35
1600	Sealer, hardener and dustproofer, epoxy-based, 125 S.F./gal., 5 gallon unit					53			53	58
1620	3 gallon unit					58			58	64
1630	Sealer, solvent-based, 250 S.F./gal., 55 gallon drum					23.50			23.50	25.50
1640	5 gallon pail					29.50			29.50	32.50
1650	Sealer, water based, 350 S.F., 55 gallon drum					22			22	24.50
1660	5 gallon pail					25			25	27.50
1900	Set retarder, 100 S.F./gal., 1 gallon pail				↓	22			22	24
2000	Waterproofing, integral 1 lb. per bag of cement				Lb.	2.02			2.02	2.22
2100	Powdered metallic, 40 lb. per 100 S.F., standard colors					2.40			2.40	2.64
2120	Premium colors				↓	3.36			3.36	3.70
3000	For integral colored pigments, 2500 psi (5 bag mix)									
3100	Standard colors, 1.8 lb. per bag, add				C.Y.	21.50			21.50	24
3200	9.4 lb. per bag, add					113			113	124
3400	Premium colors, 1.8 lb. per bag, add					29.50			29.50	32
3500	7.5 lb. per bag, add					122			122	134
3700	Ultra premium colors, 1.8 lb. per bag, add					48.50			48.50	53.50
3800	7.5 lb. per bag, add				↓	203			203	223
6000	Concrete ready mix additives, recycled coal fly ash, mixed at plant G				Ton	57			57	62.50
6010	Recycled blast furnace slag, mixed at plant G				"	87.50			87.50	96.50

03 05 13.25 Aggregate		Crew	Daily Output	Labor-Hours	Unit	Material	Labor	Equipment	Total	Total Incl O&P
0010	**AGGREGATE** R033105-20									
0100	Lightweight vermiculite or perlite, 4 C.F. bag, C.L. lots G				Bag	23.50			23.50	26
0150	L.C.L. lots G				"	26.50			26.50	29
0250	Sand & stone, loaded at pit, crushed bank gravel				Ton	21			21	23
0350	Sand, washed, for concrete					17.40			17.40	19.15
0400	For plaster or brick					17.40			17.40	19.15
0450	Stone, 3/4" to 1-1/2"					16.85			16.85	18.50
0470	Round, river stone					26.50			26.50	29
0500	3/8" roofing stone & 1/2" pea stone				↓	26			26	28.50

03 05 Common Work Results for Concrete

03 05 13 – Basic Concrete Materials

03 05 13.25 Aggregate		Crew	Daily Output	Labor-Hours	Unit	Material	2014 Bare Costs Labor	Equipment	Total	Total Incl O&P
0550	For trucking 10-mile round trip, add to the above	B-34B	117	.068	Ton		2.57	5.90	8.47	10.40
0600	For trucking 30-mile round trip, add to the above	"	72	.111	↓		4.17	9.60	13.77	16.95
0850	Sand & stone, loaded at pit, crushed bank gravel				C.Y.	29.50			29.50	32.50
0950	Sand, washed, for concrete					24			24	26.50
1000	For plaster or brick					24			24	26.50
1050	Stone, 3/4" to 1-1/2"					32			32	35.50
1055	Round, river stone					36			36	39.50
1100	3/8" roofing stone & 1/2" pea stone					24			24	26
1150	For trucking 10-mile round trip, add to the above	B-34B	78	.103			3.85	8.85	12.70	15.65
1200	For trucking 30-mile round trip, add to the above	"	48	.167	↓		6.25	14.40	20.65	25.50
1310	Onyx chips, 50 lb. bags				Cwt.	33			33	36.50
1330	Quartz chips, 50 lb. bags					12.30			12.30	13.55
1410	White marble, 3/8" to 1/2", 50 lb. bags				↓	8.25			8.25	9.10
1430	3/4", bulk				Ton	117			117	129

03 05 13.30 Cement		Crew	Daily Output	Labor-Hours	Unit	Material	Labor	Equipment	Total	Total Incl O&P
0010	**CEMENT** R033105-20									
0240	Portland, Type I/II, TL lots, 94 lb. bags				Bag	9.65			9.65	10.65
0250	LTL/LCL lots				"	10.70			10.70	11.80
0300	Trucked in bulk, per Cwt.				Cwt.	8.30			8.30	9.10
0400	Type III, high early strength, TL lots, 94 lb. bags				Bag	12.55			12.55	13.80
0420	L.T.L. or L.C.L. lots					13.95			13.95	15.35
0500	White, type III, high early strength, T.L. or C.L. lots, bags					25			25	27.50
0520	L.T.L. or L.C.L. lots					28			28	30.50
0600	White, type I, T.L. or C.L. lots, bags					23			23	25
0620	L.T.L. or L.C.L. lots				↓	24			24	26.50

03 05 13.80 Waterproofing and Dampproofing		Crew	Daily Output	Labor-Hours	Unit	Material	Labor	Equipment	Total	Total Incl O&P
0010	**WATERPROOFING AND DAMPPROOFING**									
0050	Integral waterproofing, add to cost of regular concrete				C.Y.	12.10			12.10	13.35

03 05 13.85 Winter Protection		Crew	Daily Output	Labor-Hours	Unit	Material	Labor	Equipment	Total	Total Incl O&P
0010	**WINTER PROTECTION**									
0012	For heated ready mix, add				C.Y.	4.50			4.50	4.95
0100	Temporary heat to protect concrete, 24 hours	2 Clab	50	.320	M.S.F.	535	11.75		546.75	605
0200	Temporary shelter for slab on grade, wood frame/polyethylene sheeting									
0201	Build or remove, light framing for short spans	2 Carp	10	1.600	M.S.F.	284	73.50		357.50	425
0210	Large framing for long spans	"	3	5.333	"	380	245		625	795
0500	Electrically heated pads, 110 volts, 15 watts per S.F., buy				S.F.	10.35			10.35	11.35
0600	20 watts per S.F., buy					13.75			13.75	15.10
0710	Electrically, heated pads, 15 watts/S.F., 20 uses				↓	.52			.52	.57

03 11 Concrete Forming

03 11 13 – Structural Cast-In-Place Concrete Forming

03 11 13.20 Forms In Place, Beams and Girders		Crew	Daily Output	Labor-Hours	Unit	Material	Labor	Equipment	Total	Total Incl O&P
0010	**FORMS IN PLACE, BEAMS AND GIRDERS**									
0500	Exterior spandrel, job-built plywood, 12" wide, 1 use	C-2	225	.213	SFCA	2.98	9.55		12.53	18
0550	2 use		275	.175		1.57	7.80		9.37	13.80
0600	3 use		295	.163		1.19	7.25		8.44	12.50
0650	4 use		310	.155		.97	6.90		7.87	11.70
1000	18" wide, 1 use		250	.192		2.69	8.55		11.24	16.20
1050	2 use		275	.175		1.48	7.80		9.28	13.65
1100	3 use	↓	305	.157	↓	1.07	7.05		8.12	12.05

03 11 Concrete Forming

03 11 13 – Structural Cast-In-Place Concrete Forming

03 11 13.20 Forms In Place, Beams and Girders		Crew	Daily Output	Labor-Hours	Unit	Material	Labor	Equipment	Total	Total Incl O&P
						2014 Bare Costs				
1150	4 use	C-2	315	.152	SFCA	.88	6.80		7.68	11.45
1500	24" wide, 1 use		265	.181		2.47	8.10		10.57	15.20
1550	2 use		290	.166		1.40	7.40		8.80	12.95
1600	3 use		315	.152		.99	6.80		7.79	11.60
1650	4 use		325	.148		.80	6.60		7.40	11.10
2000	Interior beam, job-built plywood, 12" wide, 1 use		300	.160		3.54	7.15		10.69	14.90
2050	2 use		340	.141		1.70	6.30		8	11.60
2100	3 use		364	.132		1.42	5.90		7.32	10.65
2150	4 use		377	.127		1.15	5.70		6.85	10
2500	24" wide, 1 use		320	.150		2.52	6.70		9.22	13.10
2550	2 use		365	.132		1.42	5.85		7.27	10.60
2600	3 use		385	.125		1	5.55		6.55	9.70
2650	4 use		395	.122		.81	5.45		6.26	9.25
3000	Encasing steel beam, hung, job-built plywood, 1 use		325	.148		3.09	6.60		9.69	13.60
3050	2 use		390	.123		1.70	5.50		7.20	10.35
3100	3 use		415	.116		1.24	5.15		6.39	9.30
3150	4 use		430	.112		1	4.98		5.98	8.80
3500	Bottoms only, to 30" wide, job-built plywood, 1 use		230	.209		3.98	9.30		13.28	18.80
3550	2 use		265	.181		2.24	8.10		10.34	14.95
3600	3 use		280	.171		1.59	7.65		9.24	13.55
3650	4 use		290	.166		1.29	7.40		8.69	12.80
4000	Sides only, vertical, 36" high, job-built plywood, 1 use		335	.143		5.10	6.40		11.50	15.45
4050	2 use		405	.119		2.80	5.30		8.10	11.25
4100	3 use		430	.112		2.04	4.98		7.02	9.95
4150	4 use		445	.108		1.65	4.82		6.47	9.25
4500	Sloped sides, 36" high, 1 use		305	.157		4.92	7.05		11.97	16.25
4550	2 use		370	.130		2.74	5.80		8.54	11.95
4600	3 use		405	.119		1.97	5.30		7.27	10.30
4650	4 use		425	.113		1.60	5.05		6.65	9.55
5000	Upstanding beams, 36" high, 1 use		225	.213		6.25	9.55		15.80	21.50
5050	2 use		255	.188		3.47	8.40		11.87	16.75
5100	3 use		275	.175		2.51	7.80		10.31	14.80
5150	4 use		280	.171		2.04	7.65		9.69	14.05

03 11 13.25 Forms In Place, Columns		Crew	Daily Output	Labor-Hours	Unit	Material	Labor	Equipment	Total	Total Incl O&P
0010	**FORMS IN PLACE, COLUMNS**									
0500	Round fiberglass, 4 use per mo., rent, 12" diameter	C-1	160	.200	L.F.	8.15	8.70		16.85	22.50
0550	16" diameter		150	.213		9.70	9.30		19	25
0600	18" diameter		140	.229		10.85	9.95		20.80	27.50
0650	24" diameter		135	.237		13.50	10.30		23.80	31
0700	28" diameter		130	.246		15.10	10.70		25.80	33
0800	30" diameter		125	.256		15.75	11.15		26.90	34.50
0850	36" diameter		120	.267		21	11.60		32.60	41
1500	Round fiber tube, recycled paper, 1 use, 8" diameter G		155	.206		1.66	9		10.66	15.70
1550	10" diameter G		155	.206		2.24	9		11.24	16.30
1600	12" diameter G		150	.213		2.60	9.30		11.90	17.20
1650	14" diameter G		145	.221		3.77	9.60		13.37	19
1700	16" diameter G		140	.229		4.52	9.95		14.47	20.50
1720	18" diameter G		140	.229		5.30	9.95		15.25	21
1750	20" diameter G		135	.237		7.30	10.30		17.60	24
1800	24" diameter G		130	.246		9.30	10.70		20	27
1850	30" diameter G		125	.256		13.60	11.15		24.75	32
1900	36" diameter G		115	.278		17.05	12.10		29.15	37.50

03 11 Concrete Forming

03 11 13 – Structural Cast-In-Place Concrete Forming

03 11 13.25 Forms In Place, Columns		Crew	Daily Output	Labor-Hours	Unit	2014 Bare Costs Material	Labor	Equipment	Total	Total Incl O&P
1950	42″ diameter G	C-1	100	.320	L.F.	42	13.95		55.95	68
2000	48″ diameter G		85	.376		48.50	16.40		64.90	79
2200	For seamless type, add					15%				
3000	Round, steel, 4 use per mo., rent, regular duty, 14″ diameter G	C-1	145	.221	L.F.	17.55	9.60		27.15	34
3050	16″ diameter G		125	.256		17.90	11.15		29.05	37
3100	Heavy duty, 20″ diameter G		105	.305		19.75	13.25		33	42
3150	24″ diameter G		85	.376		21.50	16.40		37.90	49
3200	30″ diameter G		70	.457		25	19.90		44.90	57.50
3250	36″ diameter G		60	.533		26.50	23		49.50	65
3300	48″ diameter G		50	.640		39.50	28		67.50	86.50
3350	60″ diameter G		45	.711		48.50	31		79.50	102
4500	For second and succeeding months, deduct					50%				
5000	Job-built plywood, 8″ x 8″ columns, 1 use	C-1	165	.194	SFCA	2.67	8.45		11.12	16
5050	2 use		195	.164		1.53	7.15		8.68	12.75
5100	3 use		210	.152		1.07	6.65		7.72	11.45
5150	4 use		215	.149		.88	6.50		7.38	10.95
5500	12″ x 12″ columns, 1 use		180	.178		2.56	7.75		10.31	14.75
5550	2 use		210	.152		1.41	6.65		8.06	11.80
5600	3 use		220	.145		1.02	6.35		7.37	10.90
5650	4 use		225	.142		.83	6.20		7.03	10.45
6000	16″ x 16″ columns, 1 use		185	.173		2.55	7.55		10.10	14.40
6050	2 use		215	.149		1.36	6.50		7.86	11.50
6100	3 use		230	.139		1.02	6.05		7.07	10.50
6150	4 use		235	.136		.83	5.95		6.78	10.05
6500	24″ x 24″ columns, 1 use		190	.168		2.87	7.35		10.22	14.45
6550	2 use		216	.148		1.58	6.45		8.03	11.70
6600	3 use		230	.139		1.15	6.05		7.20	10.60
6650	4 use		238	.134		.93	5.85		6.78	10.10
7000	36″ x 36″ columns, 1 use		200	.160		2.30	6.95		9.25	13.30
7050	2 use		230	.139		1.30	6.05		7.35	10.80
7100	3 use		245	.131		.92	5.70		6.62	9.80
7150	4 use		250	.128		.75	5.55		6.30	9.40
7400	Steel framed plywood, based on 50 uses of purchased									
7420	forms, and 4 uses of bracing lumber									
7500	8″ x 8″ column	C-1	340	.094	SFCA	2.11	4.10		6.21	8.60
7550	10″ x 10″		350	.091		1.84	3.98		5.82	8.15
7600	12″ x 12″		370	.086		1.56	3.77		5.33	7.50
7650	16″ x 16″		400	.080		1.22	3.48		4.70	6.75
7700	20″ x 20″		420	.076		1.08	3.32		4.40	6.30
7750	24″ x 24″		440	.073		.78	3.17		3.95	5.75
7755	30″ x 30″		440	.073		.99	3.17		4.16	6
7760	36″ x 36″		460	.070		.87	3.03		3.90	5.65

03 11 13.30 Forms In Place, Culvert		Crew	Daily Output	Labor-Hours	Unit	Material	Labor	Equipment	Total	Total Incl O&P
0010	**FORMS IN PLACE, CULVERT**									
0015	5′ to 8′ square or rectangular, 1 use	C-1	170	.188	SFCA	3.98	8.20		12.18	17
0050	2 use		180	.178		2.39	7.75		10.14	14.60
0100	3 use		190	.168		1.86	7.35		9.21	13.35
0150	4 use		200	.160		1.59	6.95		8.54	12.50

03 11 13.35 Forms In Place, Elevated Slabs		Crew	Daily Output	Labor-Hours	Unit	Material	Labor	Equipment	Total	Total Incl O&P
0010	**FORMS IN PLACE, ELEVATED SLABS**									
1000	Flat plate, job-built plywood, to 15′ high, 1 use	C-2	470	.102	S.F.	3.64	4.56		8.20	11.05
1050	2 use		520	.092		2	4.12		6.12	8.55

03 11 13 – Structural Cast-In-Place Concrete Forming

03 11 13.35 Forms In Place, Elevated Slabs		Crew	Daily Output	Labor-Hours	Unit	2014 Bare Costs Material	Labor	Equipment	Total	Total Incl O&P
1100	3 use	C-2	545	.088	S.F.	1.46	3.93		5.39	7.65
1150	4 use		560	.086		1.18	3.83		5.01	7.20
1500	15′ to 20′ high ceilings, 4 use		495	.097		1.25	4.33		5.58	8.05
1600	21′ to 35′ high ceilings, 4 use		450	.107		1.55	4.76		6.31	9.05
2000	Flat slab, drop panels, job-built plywood, to 15′ high, 1 use		449	.107		4.20	4.77		8.97	11.95
2050	2 use		509	.094		2.31	4.21		6.52	9.05
2100	3 use		532	.090		1.68	4.03		5.71	8.05
2150	4 use		544	.088		1.37	3.94		5.31	7.60
2250	15′ to 20′ high ceilings, 4 use		480	.100		2.43	4.47		6.90	9.55
2350	20′ to 35′ high ceilings, 4 use		435	.110		2.73	4.93		7.66	10.60
3000	Floor slab hung from steel beams, 1 use		485	.099		2.72	4.42		7.14	9.80
3050	2 use		535	.090		2.09	4.01		6.10	8.50
3100	3 use		550	.087		1.87	3.90		5.77	8.05
3150	4 use		565	.085		1.77	3.79		5.56	7.80
3500	Floor slab, with 1-way joist pans, 1 use		415	.116		5.95	5.15		11.10	14.50
3550	2 use		445	.108		4.07	4.82		8.89	11.95
3600	3 use		475	.101		3.44	4.51		7.95	10.75
3650	4 use		500	.096		3.12	4.29		7.41	10.05
4500	With 2-way waffle domes, 1 use		405	.119		6.10	5.30		11.40	14.85
4520	2 use		450	.107		4.20	4.76		8.96	11.95
4530	3 use		460	.104		3.57	4.66		8.23	11.10
4550	4 use		470	.102		3.25	4.56		7.81	10.65
5000	Box out for slab openings, over 16″ deep, 1 use		190	.253	SFCA	4.14	11.30		15.44	22
5050	2 use		240	.200	″	2.28	8.95		11.23	16.30
5500	Shallow slab box outs, to 10 S.F.		42	1.143	Ea.	10.50	51		61.50	90
5550	Over 10 S.F. (use perimeter)		600	.080	L.F.	1.41	3.57		4.98	7.05
6000	Bulkhead forms for slab, with keyway, 1 use, 2 piece		500	.096		1.95	4.29		6.24	8.75
6100	3 piece (see also edge forms)		460	.104		2.14	4.66		6.80	9.55
6200	Slab bulkhead form, 4-1/2″ high, exp metal, w/keyway & stakes	G	C-1	1200	.027		.88	1.16		2.04	2.76
6210	5-1/2″ high	G		1100	.029		1.01	1.27		2.28	3.06
6215	7-1/2″ high	G		960	.033		1.25	1.45		2.70	3.62
6220	9-1/2″ high	G		840	.038		1.36	1.66		3.02	4.06
6500	Curb forms, wood, 6″ to 12″ high, on elevated slabs, 1 use		180	.178	SFCA	1.51	7.75		9.26	13.60
6550	2 use		205	.156		.83	6.80		7.63	11.40
6600	3 use		220	.145		.60	6.35		6.95	10.40
6650	4 use		225	.142		.49	6.20		6.69	10.10
7000	Edge forms to 6″ high, on elevated slab, 4 use		500	.064	L.F.	.18	2.79		2.97	4.50
7070	7″ to 12″ high, 1 use		162	.198	SFCA	1.14	8.60		9.74	14.50
7080	2 use		198	.162		.62	7.05		7.67	11.55
7090	3 use		222	.144		.45	6.30		6.75	10.20
7101	4 use		350	.091		.18	3.98		4.16	6.35
7500	Depressed area forms to 12″ high, 4 use		300	.107	L.F.	.92	4.65		5.57	8.15
7550	12″ to 24″ high, 4 use		175	.183		1.26	7.95		9.21	13.70
8000	Perimeter deck and rail for elevated slabs, straight		90	.356		12	15.50		27.50	37
8050	Curved		65	.492		16.45	21.50		37.95	51
8500	Void forms, round plastic, 8″ high x 3″ diameter	G		450	.071	Ea.	2.08	3.10		5.18	7.05
8550	4″ diameter	G		425	.075		3.23	3.28		6.51	8.60
8600	6″ diameter	G		400	.080		5.05	3.48		8.53	10.95
8650	8″ diameter	G		375	.085		8.80	3.72		12.52	15.40

03 11 Concrete Forming

03 11 13 – Structural Cast-In-Place Concrete Forming

03 11 13.40 Forms In Place, Equipment Foundations		Crew	Daily Output	Labor-Hours	Unit	Material	2014 Bare Costs Labor	Equipment	Total	Total Incl O&P
0010	FORMS IN PLACE, EQUIPMENT FOUNDATIONS									
0020	1 use	C-2	160	.300	SFCA	3.09	13.40		16.49	24
0050	2 use		190	.253		1.70	11.30		13	19.25
0100	3 use		200	.240		1.24	10.70		11.94	17.90
0150	4 use		205	.234		1.01	10.45		11.46	17.25

03 11 13.45 Forms In Place, Footings

		Crew	Daily Output	Labor-Hours	Unit	Material	Labor	Equipment	Total	Total Incl O&P
0010	FORMS IN PLACE, FOOTINGS									
0020	Continuous wall, plywood, 1 use	C-1	375	.085	SFCA	6.25	3.72		9.97	12.60
0050	2 use		440	.073		3.42	3.17		6.59	8.65
0100	3 use		470	.068		2.49	2.97		5.46	7.30
0150	4 use		485	.066		2.03	2.87		4.90	6.65
0500	Dowel supports for footings or beams, 1 use		500	.064	L.F.	.84	2.79		3.63	5.20
1000	Integral starter wall, to 4″ high, 1 use		400	.080		.84	3.48		4.32	6.30
1500	Keyway, 4 use, tapered wood, 2″ x 4″	1 Carp	530	.015		.20	.69		.89	1.29
1550	2″ x 6″		500	.016		.28	.73		1.01	1.44
2000	Tapered plastic		530	.015		1.24	.69		1.93	2.43
2250	For keyway hung from supports, add		150	.053		.84	2.45		3.29	4.69
3000	Pile cap, square or rectangular, job-built plywood, 1 use	C-1	290	.110	SFCA	2.79	4.81		7.60	10.45
3050	2 use		346	.092		1.53	4.03		5.56	7.90
3100	3 use		371	.086		1.12	3.76		4.88	7.05
3150	4 use		383	.084		.91	3.64		4.55	6.60
4000	Triangular or hexagonal, 1 use		225	.142		3.26	6.20		9.46	13.15
4050	2 use		280	.114		1.79	4.98		6.77	9.65
4100	3 use		305	.105		1.30	4.57		5.87	8.50
4150	4 use		315	.102		1.06	4.42		5.48	8
5000	Spread footings, job-built lumber, 1 use		305	.105		2	4.57		6.57	9.25
5050	2 use		371	.086		1.11	3.76		4.87	7
5100	3 use		401	.080		.80	3.48		4.28	6.25
5150	4 use		414	.077		.65	3.37		4.02	5.90
6000	Supports for dowels, plinths or templates, 2′ x 2′ footing		25	1.280	Ea.	6.05	55.50		61.55	92.50
6050	4′ x 4′ footing		22	1.455		12.10	63.50		75.60	111
6100	8′ x 8′ footing		20	1.600		24	69.50		93.50	135
6150	12′ x 12′ footing		17	1.882		34	82		116	163
7000	Plinths, job-built plywood, 1 use		250	.128	SFCA	3.53	5.55		9.08	12.50
7100	4 use		270	.119	″	1.16	5.15		6.31	9.20

03 11 13.50 Forms In Place, Grade Beam

		Crew	Daily Output	Labor-Hours	Unit	Material	Labor	Equipment	Total	Total Incl O&P
0010	FORMS IN PLACE, GRADE BEAM									
0020	Job-built plywood, 1 use	C-2	530	.091	SFCA	2.82	4.04		6.86	9.35
0050	2 use		580	.083		1.55	3.70		5.25	7.40
0100	3 use		600	.080		1.13	3.57		4.70	6.75
0150	4 use		605	.079		.92	3.54		4.46	6.45

03 11 13.55 Forms In Place, Mat Foundation

		Crew	Daily Output	Labor-Hours	Unit	Material	Labor	Equipment	Total	Total Incl O&P
0010	FORMS IN PLACE, MAT FOUNDATION									
0020	Job-built plywood, 1 use	C-2	290	.166	SFCA	2.89	7.40		10.29	14.60
0050	2 use		310	.155		1.10	6.90		8	11.85
0100	3 use		330	.145		.71	6.50		7.21	10.80
0120	4 use		350	.137		.65	6.10		6.75	10.15

03 11 13.65 Forms In Place, Slab On Grade

		Crew	Daily Output	Labor-Hours	Unit	Material	Labor	Equipment	Total	Total Incl O&P
0010	FORMS IN PLACE, SLAB ON GRADE									
1000	Bulkhead forms w/keyway, wood, 6″ high, 1 use	C-1	510	.063	L.F.	.93	2.73		3.66	5.25
1050	2 uses		400	.080		.51	3.48		3.99	5.95

03 11 Concrete Forming

03 11 13 – Structural Cast-In-Place Concrete Forming

03 11 13.65 Forms In Place, Slab On Grade		Crew	Daily Output	Labor-Hours	Unit	2014 Bare Costs Material	Labor	Equipment	Total	Total Incl O&P
1100	4 uses	C-1	350	.091	L.F.	.30	3.98		4.28	6.50
1400	Bulkhead form for slab, 4-1/2" high, exp metal, incl keyway & stakes G		1200	.027		.88	1.16		2.04	2.76
1410	5-1/2" high G		1100	.029		1.01	1.27		2.28	3.06
1420	7-1/2" high G		960	.033		1.25	1.45		2.70	3.62
1430	9-1/2" high G		840	.038		1.36	1.66		3.02	4.06
2000	Curb forms, wood, 6" to 12" high, on grade, 1 use		215	.149	SFCA	2.43	6.50		8.93	12.65
2050	2 use		250	.128		1.34	5.55		6.89	10.10
2100	3 use		265	.121		.97	5.25		6.22	9.15
2150	4 use		275	.116		.79	5.05		5.84	8.65
3000	Edge forms, wood, 4 use, on grade, to 6" high		600	.053	L.F.	.31	2.32		2.63	3.92
3050	7" to 12" high		435	.074	SFCA	.77	3.20		3.97	5.80
3060	Over 12"		350	.091	"	.82	3.98		4.80	7.05
3500	For depressed slabs, 4 use, to 12" high		300	.107	L.F.	.67	4.65		5.32	7.90
3550	To 24" high		175	.183		.88	7.95		8.83	13.25
4000	For slab blockouts, to 12" high, 1 use		200	.160		.71	6.95		7.66	11.55
4050	To 24" high, 1 use		120	.267		.90	11.60		12.50	18.90
4100	Plastic (extruded), to 6" high, multiple use, on grade		800	.040		8.20	1.74		9.94	11.70
5020	Wood, incl. wood stakes, 1" x 3"		900	.036		.77	1.55		2.32	3.23
5050	2" x 4"		900	.036		.76	1.55		2.31	3.22
6000	Trench forms in floor, wood, 1 use		160	.200	SFCA	1.68	8.70		10.38	15.30
6050	2 use		175	.183		.92	7.95		8.87	13.30
6100	3 use		180	.178		.67	7.75		8.42	12.70
6150	4 use		185	.173		.55	7.55		8.10	12.20
8760	Void form, corrugated fiberboard, 4" x 12", 4' long G		3000	.011	S.F.	3.01	.46		3.47	4.03
8770	6" x 12", 4' long		3000	.011		3.59	.46		4.05	4.67
8780	1/4" thick hardboard protective cover for void form	2 Carp	1500	.011		.57	.49		1.06	1.38

03 11 13.85 Forms In Place, Walls		Crew	Daily Output	Labor-Hours	Unit	Material	Labor	Equipment	Total	Total Incl O&P
0010	**FORMS IN PLACE, WALLS**									
0100	Box out for wall openings, to 16" thick, to 10 S.F.	C-2	24	2	Ea.	25	89.50		114.50	166
0150	Over 10 S.F. (use perimeter)	"	280	.171	L.F.	2.13	7.65		9.78	14.15
0250	Brick shelf, 4" w, add to wall forms, use wall area above shelf									
0260	1 use	C-2	240	.200	SFCA	2.28	8.95		11.23	16.30
0300	2 use		275	.175		1.26	7.80		9.06	13.45
0350	4 use		300	.160		.91	7.15		8.06	12
0500	Bulkhead, wood with keyway, 1 use, 2 piece		265	.181	L.F.	1.90	8.10		10	14.60
0600	Bulkhead forms with keyway, 1 piece expanded metal, 8" wall G	C-1	1000	.032		1.25	1.39		2.64	3.53
0610	10" wall G		800	.040		1.36	1.74		3.10	4.19
0620	12" wall G		525	.061		1.63	2.65		4.28	5.90
0700	Buttress, to 8' high, 1 use	C-2	350	.137	SFCA	4.19	6.10		10.29	14.05
0750	2 use		430	.112		2.31	4.98		7.29	10.25
0800	3 use		460	.104		1.68	4.66		6.34	9.05
0850	4 use		480	.100		1.38	4.47		5.85	8.40
1000	Corbel or haunch, to 12" wide, add to wall forms, 1 use		150	.320	L.F.	2.28	14.30		16.58	24.50
1050	2 use		170	.282		1.26	12.60		13.86	21
1100	3 use		175	.274		.91	12.25		13.16	19.90
1150	4 use		180	.267		.74	11.90		12.64	19.15
2000	Wall, job-built plywood, to 8' high, 1 use		370	.130	SFCA	2.64	5.80		8.44	11.85
2050	2 use		435	.110		1.67	4.93		6.60	9.45
2150	4 use		505	.095		.99	4.24		5.23	7.65
2400	8' to 16' high, 1 use		280	.171		2.91	7.65		10.56	15
2450	2 use		345	.139		1.24	6.20		7.44	10.95
2500	3 use		375	.128		.89	5.70		6.59	9.75

03 11 13.85 Forms In Place, Walls		Crew	Daily Output	Labor-Hours	Unit	Material	2014 Bare Costs Labor	Equipment	Total	Total Incl O&P
2550	4 use	C-2	395	.122	SFCA	.73	5.45		6.18	9.15
2700	Over 16′ high, 1 use		235	.204		2.51	9.10		11.61	16.80
2750	2 use		290	.166		1.38	7.40		8.78	12.90
2800	3 use		315	.152		1	6.80		7.80	11.60
2850	4 use		330	.145		.82	6.50		7.32	10.90
4000	Radial, smooth curved, job-built plywood, 1 use		245	.196		2.41	8.75		11.16	16.15
4050	2 use		300	.160		1.33	7.15		8.48	12.45
4100	3 use		325	.148		.96	6.60		7.56	11.25
4150	4 use		335	.143		.78	6.40		7.18	10.70
4200	Below grade, job-built plywood, 1 use		225	.213		2.48	9.55		12.03	17.40
4210	2 use		225	.213		1.36	9.55		10.91	16.20
4220	3 use		225	.213		1.13	9.55		10.68	15.95
4230	4 use		225	.213		.81	9.55		10.36	15.60
4300	Curved, 2′ chords, job-built plywood, 1 use		290	.166		2	7.40		9.40	13.60
4350	2 use		355	.135		1.10	6.05		7.15	10.50
4400	3 use		385	.125		.80	5.55		6.35	9.50
4450	4 use		400	.120		.65	5.35		6	8.95
4500	Over 8′ high, 1 use		290	.166		.88	7.40		8.28	12.35
4525	2 use		355	.135		.48	6.05		6.53	9.85
4550	3 use		385	.125		.35	5.55		5.90	9
4575	4 use		400	.120		.29	5.35		5.64	8.55
4600	Retaining wall, battered, job-built plyw'd, to 8′ high, 1 use		300	.160		1.89	7.15		9.04	13.10
4650	2 use		355	.135		1.04	6.05		7.09	10.45
4700	3 use		375	.128		.76	5.70		6.46	9.65
4750	4 use		390	.123		.61	5.50		6.11	9.20
4900	Over 8′ to 16′ high, 1 use		240	.200		2.06	8.95		11.01	16.05
4950	2 use		295	.163		1.13	7.25		8.38	12.45
5000	3 use		305	.157		.82	7.05		7.87	11.75
5050	4 use		320	.150		.67	6.70		7.37	11.10
5100	Retaining wall form, plywood, smooth curve, 1 use		200	.240		3.03	10.70		13.73	19.90
5120	2 use		235	.204		1.67	9.10		10.77	15.90
5130	3 use		250	.192		1.21	8.55		9.76	14.60
5140	4 use	↓	260	.185		.99	8.25		9.24	13.80
5500	For gang wall forming, 192 S.F. sections, deduct					10%	10%			
5550	384 S.F. sections, deduct					20%	20%			
7500	Lintel or sill forms, 1 use	1 Carp	30	.267		3.06	12.25		15.31	22
7520	2 use		34	.235		1.68	10.80		12.48	18.50
7540	3 use		36	.222		1.23	10.20		11.43	17.05
7560	4 use	↓	37	.216	↓	1	9.90		10.90	16.40
7800	Modular prefabricated plywood, based on 20 uses of purchased									
7820	forms, and 4 uses of bracing lumber									
7860	To 8′ high	C-2	800	.060	SFCA	.85	2.68		3.53	5.05
8060	Over 8′ to 16′ high		600	.080		.90	3.57		4.47	6.50
8600	Pilasters, 1 use		270	.178		3.05	7.95		11	15.60
8620	2 use		330	.145		1.68	6.50		8.18	11.85
8640	3 use		370	.130		1.22	5.80		7.02	10.30
8660	4 use	↓	385	.125	↓	.99	5.55		6.54	9.70
9010	Steel framed plywood, based on 50 uses of purchased									
9020	forms, and 4 uses of bracing lumber									
9060	To 8′ high	C-2	600	.080	SFCA	.68	3.57		4.25	6.25
9260	Over 8′ to 16′ high		450	.107		.68	4.76		5.44	8.10
9460	Over 16′ to 20′ high	↓	400	.120	↓	.68	5.35		6.03	9
9475	For elevated walls, add						10%			

03 11 Concrete Forming

03 11 16 – Architectural Cast-in Place Concrete Forming

03 11 16.13 Concrete Form Liners		Crew	Daily Output	Labor-Hours	Unit	Material	2014 Bare Costs Labor	Equipment	Total	Total Incl O&P
0010	**CONCRETE FORM LINERS**									
5750	Liners for forms (add to wall forms), ABS plastic									
5800	Aged wood, 4" wide, 1 use	1 Carp	256	.031	SFCA	3.30	1.43		4.73	5.85
5820	2 use		256	.031		1.82	1.43		3.25	4.21
5830	3 use		256	.031		1.32	1.43		2.75	3.66
5840	4 use		256	.031		1.07	1.43		2.50	3.39
5900	Fractured rope rib, 1 use		192	.042		4.79	1.91		6.70	8.20
5925	2 use		192	.042		2.63	1.91		4.54	5.85
5950	3 use		192	.042		1.92	1.91		3.83	5.05
6000	4 use		192	.042		1.56	1.91		3.47	4.66
6100	Ribbed, 3/4" deep x 1-1/2" O.C., 1 use		224	.036		4.79	1.64		6.43	7.80
6125	2 use		224	.036		2.63	1.64		4.27	5.45
6150	3 use		224	.036		1.92	1.64		3.56	4.64
6200	4 use		224	.036		1.56	1.64		3.20	4.24
6300	Rustic brick pattern, 1 use		224	.036		3.30	1.64		4.94	6.15
6325	2 use		224	.036		1.82	1.64		3.46	4.53
6350	3 use		224	.036		1.32	1.64		2.96	3.98
6400	4 use		224	.036		1.07	1.64		2.71	3.71
6500	3/8" striated, random, 1 use		224	.036		3.30	1.64		4.94	6.15
6525	2 use		224	.036		1.82	1.64		3.46	4.53
6550	3 use		224	.036		1.32	1.64		2.96	3.98
6600	4 use		224	.036		1.07	1.64		2.71	3.71
6850	Random vertical rustication, 1 use		384	.021		6.25	.96		7.21	8.35
6900	2 use		384	.021		3.45	.96		4.41	5.25
6925	3 use		384	.021		2.51	.96		3.47	4.23
6950	4 use		384	.021		2.04	.96		3	3.71
7050	Wood, beveled edge, 3/4" deep, 1 use		384	.021	L.F.	.16	.96		1.12	1.65
7100	1" deep, 1 use		384	.021	"	.33	.96		1.29	1.83
7200	4" wide aged cedar, 1 use		256	.031	SFCA	3.30	1.43		4.73	5.85
7300	4" variable depth rough cedar		224	.036	"	4.79	1.64		6.43	7.80

03 11 19 – Insulating Concrete Forming

03 11 19.10 Insulating Forms, Left In Place		Crew	Daily Output	Labor-Hours	Unit	Material	Labor	Equipment	Total	Total Incl O&P
0010	**INSULATING FORMS, LEFT IN PLACE**									
0020	S.F. is for exterior face, but includes forms for both faces (total R22)									
2000	4" wall, straight block, 16" x 48" (5.33 S.F.) G	2 Carp	90	.178	Ea.	18.35	8.15		26.50	32.50
2010	90 corner block, exterior 16" x 38" x 22" (6.67 S.F.) G		75	.213		21.50	9.80		31.30	39
2020	45 corner block, exterior 16" x 34" x 18" (5.78 S.F.) G		75	.213		21.50	9.80		31.30	39
2100	6" wall, straight block, 16" x 48" (5.33 S.F.) G		90	.178		18.85	8.15		27	33
2110	90 corner block, exterior 16" x 32" x 24" (6.22 S.F.) G		75	.213		23.50	9.80		33.30	41
2120	45 corner block, exterior 16" x 26" x 18" (4.89 S.F.) G		75	.213		21	9.80		30.80	38
2130	Brick ledge block, 16" x 48" (5.33 S.F.) G		80	.200		24	9.15		33.15	40
2140	Taper top block, 16" x 48" (5.33 S.F.) G		80	.200		22	9.15		31.15	38.50
2200	8" wall, straight block, 16" x 48" (5.33 S.F.) G		90	.178		20	8.15		28.15	34.50
2210	90 corner block, exterior 16" x 34" x 26" (6.67 S.F.) G		75	.213		26	9.80		35.80	43.50
2220	45 corner block, exterior 16" x 28" x 20" (5.33 S.F.) G		75	.213		22	9.80		31.80	39
2230	Brick ledge block, 16" x 48" (5.33 S.F.) G		80	.200		25	9.15		34.15	41.50
2240	Taper top block, 16" x 48" (5.33 S.F.) G		80	.200		23	9.15		32.15	39.50

03 11 Concrete Forming

03 11 23 – Permanent Stair Forming

03 11 23.75 Forms In Place, Stairs		Crew	Daily Output	Labor-Hours	Unit	2014 Bare Costs Material	Labor	Equipment	Total	Total Incl O&P
0010	**FORMS IN PLACE, STAIRS**									
0015	(Slant length x width), 1 use	C-2	165	.291	S.F.	5.45	13		18.45	26
0050	2 use		170	.282		3.11	12.60		15.71	23
0100	3 use		180	.267		2.32	11.90		14.22	21
0150	4 use		190	.253		1.93	11.30		13.23	19.55
1000	Alternate pricing method (1.0 L.F./S.F.), 1 use		100	.480	LF Rsr	5.45	21.50		26.95	39
1050	2 use		105	.457		3.11	20.50		23.61	35
1100	3 use		110	.436		2.32	19.50		21.82	32.50
1150	4 use		115	.417		1.93	18.65		20.58	31
2000	Stairs, cast on sloping ground (length x width), 1 use		220	.218	S.F.	2.14	9.75		11.89	17.40
2025	2 use		232	.207		1.17	9.25		10.42	15.55
2050	3 use		244	.197		.85	8.80		9.65	14.50
2100	4 use		256	.188		.69	8.35		9.04	13.65

03 15 Concrete Accessories

03 15 05 – Concrete Forming Accessories

03 15 05.12 Chamfer Strips

		Crew	Daily Output	Labor-Hours	Unit	Material	Labor	Equipment	Total	Total Incl O&P
0010	**CHAMFER STRIPS**									
2000	Polyvinyl chloride, 1/2″ wide with leg	1 Carp	535	.015	L.F.	.38	.69		1.07	1.48
2200	3/4″ wide with leg		525	.015		.52	.70		1.22	1.65
2400	1″ radius with leg		515	.016		.38	.71		1.09	1.52
2800	2″ radius with leg		500	.016		.66	.73		1.39	1.86
5000	Wood, 1/2″ wide		535	.015		.13	.69		.82	1.20
5200	3/4″ wide		525	.015		.16	.70		.86	1.26
5400	1″ wide		515	.016		.33	.71		1.04	1.46

03 15 05.30 Hangers

		Crew	Daily Output	Labor-Hours	Unit	Material	Labor	Equipment	Total	Total Incl O&P
0010	**HANGERS**									
8500	Wire, black annealed, 15 gage G				Cwt.	147			147	162
8600	16 ga				″	199			199	219

03 15 05.70 Shores

		Crew	Daily Output	Labor-Hours	Unit	Material	Labor	Equipment	Total	Total Incl O&P
0010	**SHORES**									
0020	Erect and strip, by hand, horizontal members									
0500	Aluminum joists and stringers G	2 Carp	60	.267	Ea.		12.25		12.25	18.85
0600	Steel, adjustable beams G		45	.356			16.30		16.30	25
0700	Wood joists		50	.320			14.65		14.65	22.50
0800	Wood stringers		30	.533			24.50		24.50	37.50
1000	Vertical members to 10′ high G		55	.291			13.35		13.35	20.50
1050	To 13′ high G		50	.320			14.65		14.65	22.50
1100	To 16′ high G		45	.356			16.30		16.30	25
1500	Reshoring G		1400	.011	S.F.	.58	.52		1.10	1.45
1600	Flying truss system G	C-17D	9600	.009	SFCA		.42	.09	.51	.74
1760	Horizontal, aluminum joists, 6-1/4″ high x 5′ to 21′ span, buy G				L.F.	15.45			15.45	17
1770	Beams, 7-1/4″ high x 4′ to 30′ span G				″	18.10			18.10	19.90
1810	Horizontal, steel beam, W8x10, 7′ span, buy G				Ea.	54.50			54.50	60
1830	10′ span G					64.50			64.50	71
1920	15′ span G					111			111	122
1940	20′ span G					156			156	172
1970	Steel stringer, W8x10, 4′ to 16′ span, buy G				L.F.	6.45			6.45	7.10
3000	Rent for job duration, aluminum joist @ 2′ O.C., per mo. G				SF Flr.	.39			.39	.43
3050	Steel W8x10 G					.16			.16	.18

03 15 Concrete Accessories

03 15 05 – Concrete Forming Accessories

03 15 05.70 Shores

	03 15 05.70 Shores		Crew	Daily Output	Labor-Hours	Unit	Material (2014 Bare Costs)	Labor	Equipment	Total	Total Incl O&P
3060	Steel adjustable	G				SF Flr.	.16			.16	.18
3500	#1 post shore, steel, 5'-7" to 9'-6" high, 10000# cap., buy	G				Ea.	148			148	163
3550	#2 post shore, 7'-3" to 12'-10" high, 7800# capacity	G					171			171	188
3600	#3 post shore, 8'-10" to 16'-1" high, 3800# capacity	G					187			187	206
5010	Frame shoring systems, steel, 12000#/leg, buy										
5040	Frame, 2' wide x 6' high	G				Ea.	93.50			93.50	103
5250	X-brace	G					15.50			15.50	17.05
5550	Base plate	G					15.50			15.50	17.05
5600	Screw jack	G					33.50			33.50	36.50
5650	U-head, 8" x 8"	G					19.10			19.10	21

03 15 05.75 Sleeves and Chases

	03 15 05.75 Sleeves and Chases		Crew	Daily Output	Labor-Hours	Unit	Material	Labor	Equipment	Total	Total Incl O&P
0010	**SLEEVES AND CHASES**										
0100	Plastic, 1 use, 12" long, 2" diameter		1 Carp	100	.080	Ea.	2.44	3.67		6.11	8.35
0150	4" diameter			90	.089		6.65	4.08		10.73	13.60
0200	6" diameter			75	.107		14.25	4.89		19.14	23.50
0250	12" diameter			60	.133		46.50	6.10		52.60	60.50
5000	Sheet metal, 2" diameter	G		100	.080		1.36	3.67		5.03	7.15
5100	4" diameter	G		90	.089		1.70	4.08		5.78	8.15
5150	6" diameter	G		75	.107		1.90	4.89		6.79	9.65
5200	12" diameter	G		60	.133		3.63	6.10		9.73	13.45
6000	Steel pipe, 2" diameter	G		100	.080		4.65	3.67		8.32	10.75
6100	4" diameter	G		90	.089		16.70	4.08		20.78	24.50
6150	6" diameter	G		75	.107		36.50	4.89		41.39	47.50
6200	12" diameter	G		60	.133		86.50	6.10		92.60	104

03 15 05.80 Snap Ties

	03 15 05.80 Snap Ties		Crew	Daily Output	Labor-Hours	Unit	Material	Labor	Equipment	Total	Total Incl O&P
0010	**SNAP TIES**, 8-1/4" L&W (Lumber and wedge)										
0100	2250 lb., w/flat washer, 8" wall	G				C	87			87	95.50
0150	10" wall	G					127			127	140
0200	12" wall	G					132			132	145
0250	16" wall	G					146			146	161
0300	18" wall	G					151			151	166
0500	With plastic cone, 8" wall	G					78			78	86
0550	10" wall	G					81			81	89
0600	12" wall	G					86			86	94.50
0650	16" wall	G					95			95	105
0700	18" wall	G					98			98	108
1000	3350 lb., w/flat washer, 8" wall	G					158			158	174
1100	10" wall	G					174			174	191
1150	12" wall	G					178			178	196
1200	16" wall	G					204			204	224
1250	18" wall	G					212			212	233
1500	With plastic cone, 8" wall	G					128			128	141
1550	10" wall	G					141			141	155
1600	12" wall	G					145			145	160
1650	16" wall	G					165			165	182
1700	18" wall	G					171			171	188

03 15 05.85 Stair Tread Inserts

	03 15 05.85 Stair Tread Inserts		Crew	Daily Output	Labor-Hours	Unit	Material	Labor	Equipment	Total	Total Incl O&P
0010	**STAIR TREAD INSERTS**										
0105	Cast nosing insert, abrasive surface, pre-drilled, includes screws										
0110	Aluminum, 3" wide x 3' long		1 Cefi	32	.250	Ea.	52.50	11		63.50	74
0120	4' long			31	.258		67	11.35		78.35	90.50
0130	5' long			30	.267		82.50	11.75		94.25	108

03 15 Concrete Accessories

03 15 05 – Concrete Forming Accessories

03 15 05.85 Stair Tread Inserts

	03 15 05.85 Stair Tread Inserts	Crew	Daily Output	Labor-Hours	Unit	2014 Bare Costs Material	Labor	Equipment	Total	Total Incl O&P
0135	Extruded nosing insert, black abrasive strips, continuous anchor									
0140	Aluminum, 3″ wide x 3′ long	1 Cefi	64	.125	Ea.	34	5.50		39.50	45.50
0150	4′ long		60	.133		34	5.85		39.85	46
0160	5′ long		56	.143		52.50	6.30		58.80	67
0165	Extruded nosing insert, black abrasive strips, pre-drilled, incl. screws									
0170	Aluminum, 3″ wide x 3′ long	1 Cefi	32	.250	Ea.	46.50	11		57.50	67.50
0180	4′ long		31	.258		60.50	11.35		71.85	83.50
0190	5′ long		30	.267		71	11.75		82.75	95.50

03 15 05.95 Wall and Foundation Form Accessories

	03 15 05.95 Wall and Foundation Form Accessories	Crew	Daily Output	Labor-Hours	Unit	Material	Labor	Equipment	Total	Total Incl O&P
0010	**WALL AND FOUNDATION FORM ACCESSORIES**									
3000	Form oil, up to 1200 S.F. per gallon coverage				Gal.	13.15			13.15	14.45
3050	Up to 800 S.F. per gallon				"	19.85			19.85	22
3500	Form patches, 1-3/4″ diameter				C	26			26	28.50
3550	2-3/4″ diameter				"	45			45	49.50
4000	Nail stakes, 3/4″ diameter, 18″ long G				Ea.	3.08			3.08	3.39
4050	24″ long G					3.98			3.98	4.38
4200	30″ long G					5.10			5.10	5.60
4250	36″ long G					6.15			6.15	6.80

03 15 13 – Waterstops

03 15 13.50 Waterstops

	03 15 13.50 Waterstops	Crew	Daily Output	Labor-Hours	Unit	Material	Labor	Equipment	Total	Total Incl O&P
0010	**WATERSTOPS**, PVC and Rubber									
0020	PVC, ribbed 3/16″ thick, 4″ wide	1 Carp	155	.052	L.F.	1.22	2.37		3.59	4.99
0050	6″ wide		145	.055		2.16	2.53		4.69	6.30
0500	With center bulb, 6″ wide, 3/16″ thick		135	.059		1.88	2.72		4.60	6.25
0550	3/8″ thick		130	.062		3.35	2.82		6.17	8.05
0600	9″ wide x 3/8″ thick		125	.064		5.45	2.93		8.38	10.55
0800	Dumbbell type, 6″ wide, 3/16″ thick		150	.053		1.87	2.45		4.32	5.85
0850	3/8″ thick		145	.055		3.64	2.53		6.17	7.90
1000	9″ wide, 3/8″ thick, plain		130	.062		5.35	2.82		8.17	10.25
1050	Center bulb		130	.062		8.50	2.82		11.32	13.70
1250	Ribbed type, split, 3/16″ thick, 6″ wide		145	.055		1.84	2.53		4.37	5.90
1300	3/8″ thick		130	.062		4.11	2.82		6.93	8.85
2000	Rubber, flat dumbbell, 3/8″ thick, 6″ wide		145	.055		10.10	2.53		12.63	15
2050	9″ wide		135	.059		15.40	2.72		18.12	21
2500	Flat dumbbell split, 3/8″ thick, 6″ wide		145	.055		1.84	2.53		4.37	5.90
2550	9″ wide		135	.059		4.11	2.72		6.83	8.70
3000	Center bulb, 1/4″ thick, 6″ wide		145	.055		9.95	2.53		12.48	14.85
3050	9″ wide		135	.059		22	2.72		24.72	28
3500	Center bulb split, 3/8″ thick, 6″ wide		145	.055		14.10	2.53		16.63	19.40
3550	9″ wide		135	.059		24.50	2.72		27.22	30.50
5000	Waterstop fittings, rubber, flat									
5010	Dumbbell or center bulb, 3/8″ thick,									
5200	Field union, 6″ wide	1 Carp	50	.160	Ea.	34	7.35		41.35	49
5250	9″ wide		50	.160		45.50	7.35		52.85	61.50
5500	Flat cross, 6″ wide		30	.267		42.50	12.25		54.75	66
5550	9″ wide		30	.267		61.50	12.25		73.75	86.50
6000	Flat tee, 6″ wide		30	.267		41.50	12.25		53.75	64.50
6050	9″ wide		30	.267		56	12.25		68.25	81
6500	Flat ell, 6″ wide		40	.200		40.50	9.15		49.65	58.50
6550	9″ wide		40	.200		52	9.15		61.15	71
7000	Vertical tee, 6″ wide		25	.320		30	14.65		44.65	55.50
7050	9″ wide		25	.320		42.50	14.65		57.15	69.50

03 15 Concrete Accessories

03 15 13 – Waterstops

03 15 13.50 Waterstops		Crew	Daily Output	Labor-Hours	Unit	Material	Labor	Equipment	Total	Total Incl O&P
						2014 Bare Costs				
7500	Vertical ell, 6″ wide	1 Carp	35	.229	Ea.	28	10.50		38.50	47
7550	9″ wide		35	.229		36	10.50		46.50	55.50

03 15 16 – Concrete Construction Joints

03 15 16.20 Control Joints, Saw Cut		Crew	Daily Output	Labor-Hours	Unit	Material	Labor	Equipment	Total	Total Incl O&P
0010	**CONTROL JOINTS, SAW CUT**									
0100	Sawcut control joints in green concrete									
0120	1″ depth	C-27	2000	.008	L.F.	.04	.35	.09	.48	.66
0140	1-1/2″ depth		1800	.009		.06	.39	.10	.55	.75
0160	2″ depth		1600	.010		.07	.44	.11	.62	.85
0180	Sawcut joint reservoir in cured concrete									
0182	3/8″ wide x 3/4″ deep, with single saw blade	C-27	1000	.016	L.F.	.06	.70	.18	.94	1.29
0184	1/2″ wide x 1″ deep, with double saw blades		900	.018		.11	.78	.19	1.08	1.49
0186	3/4″ wide x 1-1/2″ deep, with double saw blades		800	.020		.22	.88	.22	1.32	1.79
0190	Water blast joint to wash away laitance, 2 passes	C-29	2500	.003			.12	.03	.15	.21
0200	Air blast joint to blow out debris and air dry, 2 passes	C-28	2000	.004			.18	.01	.19	.27
0344	For joint sealant, see Section 03 15 16.30									
0900	For replacement of joint sealant, see Section 07 01 90.81									

03 15 16.30 Expansion Joints		Crew	Daily Output	Labor-Hours	Unit	Material	Labor	Equipment	Total	Total Incl O&P
0010	**EXPANSION JOINTS**									
0020	Keyed, cold, 24 ga., incl. stakes, 3-1/2″ high G	1 Carp	200	.040	L.F.	.77	1.83		2.60	3.68
0050	4-1/2″ high G		200	.040		.88	1.83		2.71	3.80
0100	5-1/2″ high G		195	.041		1.01	1.88		2.89	4.01
0150	7-1/2″ high G		190	.042		1.25	1.93		3.18	4.36
0160	9-1/2″ high G		185	.043		1.36	1.98		3.34	4.56
0300	Poured asphalt, plain, 1/2″ x 1″	1 Clab	450	.018		.76	.65		1.41	1.85
0350	1″ x 2″		400	.020		3.02	.73		3.75	4.45
0500	Neoprene, liquid, cold applied, 1/2″ x 1″		450	.018		2.88	.65		3.53	4.18
0550	1″ x 2″		400	.020		11.50	.73		12.23	13.80
0700	Polyurethane, poured, 2 part, 1/2″ x 1″		400	.020		1.33	.73		2.06	2.59
0750	1″ x 2″		350	.023		5.30	.84		6.14	7.15
0900	Rubberized asphalt, hot or cold applied, 1/2″ x 1″		450	.018		.40	.65		1.05	1.45
0950	1″ x 2″		400	.020		1.60	.73		2.33	2.89
1100	Hot applied, fuel resistant, 1/2″ x 1″		450	.018		.60	.65		1.25	1.67
1150	1″ x 2″		400	.020		2.40	.73		3.13	3.77
2000	Premolded, bituminous fiber, 1/2″ x 6″	1 Carp	375	.021		.43	.98		1.41	1.98
2050	1″ x 12″		300	.027		1.95	1.22		3.17	4.04
2140	Concrete expansion joint, recycled paper and fiber, 1/2″ x 6″ G		390	.021		.42	.94		1.36	1.91
2150	1/2″ x 12″ G		360	.022		.83	1.02		1.85	2.49
2250	Cork with resin binder, 1/2″ x 6″		375	.021		.96	.98		1.94	2.57
2300	1″ x 12″		300	.027		2.70	1.22		3.92	4.86
2500	Neoprene sponge, closed cell, 1/2″ x 6″		375	.021		2.37	.98		3.35	4.12
2550	1″ x 12″		300	.027		8.25	1.22		9.47	11
2750	Polyethylene foam, 1/2″ x 6″		375	.021		.68	.98		1.66	2.26
2800	1″ x 12″		300	.027		2.19	1.22		3.41	4.30
3000	Polyethylene backer rod, 3/8″ diameter		460	.017		.03	.80		.83	1.26
3050	3/4″ diameter		460	.017		.06	.80		.86	1.29
3100	1″ diameter		460	.017		.09	.80		.89	1.33
3500	Polyurethane foam, with polybutylene, 1/2″ x 1/2″		475	.017		1.14	.77		1.91	2.44
3550	1″ x 1″		450	.018		2.88	.82		3.70	4.43
3750	Polyurethane foam, regular, closed cell, 1/2″ x 6″		375	.021		.84	.98		1.82	2.43
3800	1″ x 12″		300	.027		3	1.22		4.22	5.20
4000	Polyvinyl chloride foam, closed cell, 1/2″ x 6″		375	.021		2.28	.98		3.26	4.02

03 15 Concrete Accessories

03 15 16 – Concrete Construction Joints

03 15 16.30 Expansion Joints		Crew	Daily Output	Labor-Hours	Unit	Material	Labor (2014 Bare Costs)	Equipment	Total	Total Incl O&P
4050	1" x 12"	1 Carp	300	.027	L.F.	7.85	1.22		9.07	10.55
4250	Rubber, gray sponge, 1/2" x 6"		375	.021		1.92	.98		2.90	3.62
4300	1" x 12"		300	.027		6.90	1.22		8.12	9.50
4400	Redwood heartwood, 1" x 4"		400	.020		1.16	.92		2.08	2.70
4450	1" x 6"		375	.021		1.75	.98		2.73	3.44
5000	For installation in walls, add						75%			
5250	For installation in boxouts, add						25%			

03 15 19 – Cast-In Concrete Anchors

03 15 19.05 Anchor Bolt Accessories		Crew	Daily Output	Labor-Hours	Unit	Material	Labor (2014 Bare Costs)	Equipment	Total	Total Incl O&P
0010	**ANCHOR BOLT ACCESSORIES**									
0015	For anchor bolts set in fresh concrete, see Section 03 15 19.10									
8150	Anchor bolt sleeve, plastic, 1" diam. bolts	1 Carp	60	.133	Ea.	8.10	6.10		14.20	18.35
8500	1-1/2" diameter		28	.286		14.35	13.10		27.45	36
8600	2" diameter		24	.333		18.10	15.30		33.40	43.50
8650	3" diameter		20	.400		33	18.35		51.35	65
8800	Templates, steel, 8" bolt spacing G	2 Carp	16	1		10.80	46		56.80	83
8850	12" bolt spacing G		15	1.067		11.25	49		60.25	88
8900	16" bolt spacing G		14	1.143		13.50	52.50		66	96
8950	24" bolt spacing G		12	1.333		18	61		79	114
9100	Wood, 8" bolt spacing		16	1		.74	46		46.74	72
9150	12" bolt spacing		15	1.067		.98	49		49.98	76.50
9200	16" bolt spacing		14	1.143		1.23	52.50		53.73	82.50
9250	24" bolt spacing		16	1		1.72	46		47.72	73

03 15 19.10 Anchor Bolts		Crew	Daily Output	Labor-Hours	Unit	Material	Labor (2014 Bare Costs)	Equipment	Total	Total Incl O&P
0010	**ANCHOR BOLTS**									
0015	Made from recycled materials									
0025	Single bolts installed in fresh concrete, no templates									
0030	Hooked w/nut and washer, 1/2" diameter, 8" long G	1 Carp	132	.061	Ea.	1.49	2.78		4.27	5.95
0040	12" long G		131	.061		1.66	2.80		4.46	6.15
0050	5/8" diameter, 8" long G		129	.062		3.25	2.84		6.09	7.95
0060	12" long G		127	.063		4	2.89		6.89	8.85
0070	3/4" diameter, 8" long G		127	.063		4	2.89		6.89	8.85
0080	12" long G		125	.064		5	2.93		7.93	10.05
0090	2-bolt pattern, including job-built 2-hole template, per set									
0100	J-type, incl. hex nut & washer, 1/2" diameter x 6" long G	1 Carp	21	.381	Set	5.60	17.45		23.05	33
0110	12" long G		21	.381		6.30	17.45		23.75	34
0120	18" long G		21	.381		7.30	17.45		24.75	35
0130	3/4" diameter x 8" long G		20	.400		10.95	18.35		29.30	40.50
0140	12" long G		20	.400		12.95	18.35		31.30	43
0150	18" long G		20	.400		15.95	18.35		34.30	46
0160	1" diameter x 12" long G		19	.421		22	19.30		41.30	54.50
0170	18" long G		19	.421		26	19.30		45.30	58.50
0180	24" long G		19	.421		31	19.30		50.30	64
0190	36" long G		18	.444		41.50	20.50		62	77
0200	1-1/2" diameter x 18" long G		17	.471		65.50	21.50		87	106
0210	24" long G		16	.500		77.50	23		100.50	121
0300	L-type, incl. hex nut & washer, 3/4" diameter x 12" long G		20	.400		12.20	18.35		30.55	42
0310	18" long G		20	.400		14.80	18.35		33.15	45
0320	24" long G		20	.400		17.40	18.35		35.75	47.50
0330	30" long G		20	.400		21.50	18.35		39.85	52
0340	36" long G		20	.400		24	18.35		42.35	55
0350	1" diameter x 12" long G		19	.421		18.90	19.30		38.20	51

03 15 19.10 Anchor Bolts			Crew	Daily Output	Labor-Hours	Unit	Material	Labor	Equipment	Total	Total Incl O&P
							2014 Bare Costs				
0360	18″ long	G	1 Carp	19	.421	Set	23	19.30		42.30	55
0370	24″ long	G		19	.421		27.50	19.30		46.80	60
0380	30″ long	G		19	.421		32	19.30		51.30	65
0390	36″ long	G		18	.444		36	20.50		56.50	71
0400	42″ long	G		18	.444		43	20.50		63.50	79
0410	48″ long	G		18	.444		48	20.50		68.50	84.50
0420	1-1/4″ diameter x 18″ long	G		18	.444		33	20.50		53.50	68
0430	24″ long	G		18	.444		39	20.50		59.50	74
0440	30″ long	G		17	.471		44.50	21.50		66	82.50
0450	36″ long	G		17	.471		50	21.50		71.50	88.50
0460	42″ long	G	2 Carp	32	.500		56.50	23		79.50	97.50
0470	48″ long	G		32	.500		64	23		87	106
0480	54″ long	G		31	.516		74.50	23.50		98	119
0490	60″ long	G		31	.516		82	23.50		105.50	127
0500	1-1/2″ diameter x 18″ long	G		33	.485		48	22		70	87
0510	24″ long	G		32	.500		55.50	23		78.50	96.50
0520	30″ long	G		31	.516		62	23.50		85.50	105
0530	36″ long	G		30	.533		71	24.50		95.50	116
0540	42″ long	G		30	.533		80.50	24.50		105	126
0550	48″ long	G		29	.552		90	25.50		115.50	138
0560	54″ long	G		28	.571		109	26		135	161
0570	60″ long	G		28	.571		119	26		145	172
0580	1-3/4″ diameter x 18″ long	G		31	.516		67.50	23.50		91	111
0590	24″ long	G		30	.533		78.50	24.50		103	124
0600	30″ long	G		29	.552		91	25.50		116.50	139
0610	36″ long	G		28	.571		103	26		129	155
0620	42″ long	G		27	.593		116	27		143	169
0630	48″ long	G		26	.615		127	28		155	183
0640	54″ long	G		26	.615		156	28		184	216
0650	60″ long	G		25	.640		169	29.50		198.50	232
0660	2″ diameter x 24″ long	G		27	.593		105	27		132	157
0670	30″ long	G		27	.593		117	27		144	171
0680	36″ long	G		26	.615		129	28		157	186
0690	42″ long	G		25	.640		143	29.50		172.50	204
0700	48″ long	G		24	.667		164	30.50		194.50	228
0710	54″ long	G		23	.696		195	32		227	263
0720	60″ long	G		23	.696		209	32		241	279
0730	66″ long	G		22	.727		224	33.50		257.50	298
0740	72″ long	G		21	.762		245	35		280	325
1000	4-bolt pattern, including job-built 4-hole template, per set										
1100	J-type, incl. hex nut & washer, 1/2″ diameter x 6″ long	G	1 Carp	19	.421	Set	8.25	19.30		27.55	39
1110	12″ long	G		19	.421		9.60	19.30		28.90	40.50
1120	18″ long	G		18	.444		11.60	20.50		32.10	44.50
1130	3/4″ diameter x 8″ long	G		17	.471		18.95	21.50		40.45	54.50
1140	12″ long	G		17	.471		23	21.50		44.50	59
1150	18″ long	G		17	.471		29	21.50		50.50	65.50
1160	1″ diameter x 12″ long	G		16	.500		41	23		64	80.50
1170	18″ long	G		15	.533		49	24.50		73.50	91.50
1180	24″ long	G		15	.533		59	24.50		83.50	103
1190	36″ long	G		15	.533		80	24.50		104.50	126
1200	1-1/2″ diameter x 18″ long	G		13	.615		128	28		156	185
1210	24″ long	G		12	.667		152	30.50		182.50	215
1300	L-type, incl. hex nut & washer, 3/4″ diameter x 12″ long	G		17	.471		21.50	21.50		43	57

03 15 19 – Cast-In Concrete Anchors

03 15 19.10 Anchor Bolts		Crew	Daily Output	Labor-Hours	Unit	Material	Labor	Equipment	Total	Total Incl O&P
						2014 Bare Costs				
1310	18″ long G	1 Carp	17	.471	Set	26.50	21.50		48	63
1320	24″ long G		17	.471		32	21.50		53.50	68.50
1330	30″ long G		16	.500		39.50	23		62.50	79
1340	36″ long G		16	.500		45	23		68	85
1350	1″ diameter x 12″ long G		16	.500		35	23		58	74
1360	18″ long G		15	.533		42.50	24.50		67	84.50
1370	24″ long G		15	.533		52	24.50		76.50	94.50
1380	30″ long G		15	.533		61	24.50		85.50	105
1390	36″ long G		15	.533		69	24.50		93.50	114
1400	42″ long G		14	.571		83.50	26		109.50	132
1410	48″ long G		14	.571		93	26		119	143
1420	1-1/4″ diameter x 18″ long G		14	.571		63.50	26		89.50	110
1430	24″ long G		14	.571		74.50	26		100.50	123
1440	30″ long G		13	.615		86	28		114	138
1450	36″ long G	↓	13	.615		97.50	28		125.50	151
1460	42″ long G	2 Carp	25	.640		110	29.50		139.50	167
1470	48″ long G		24	.667		125	30.50		155.50	184
1480	54″ long G		23	.696		146	32		178	210
1490	60″ long G		23	.696		161	32		193	226
1500	1-1/2″ diameter x 18″ long G		25	.640		92.50	29.50		122	148
1510	24″ long G		24	.667		108	30.50		138.50	165
1520	30″ long G		23	.696		121	32		153	182
1530	36″ long G		22	.727		139	33.50		172.50	205
1540	42″ long G		22	.727		158	33.50		191.50	226
1550	48″ long G		21	.762		177	35		212	249
1560	54″ long G		20	.800		215	36.50		251.50	294
1570	60″ long G		20	.800		236	36.50		272.50	315
1580	1-3/4″ diameter x 18″ long G		22	.727		132	33.50		165.50	197
1590	24″ long G		21	.762		154	35		189	223
1600	30″ long G		21	.762		179	35		214	251
1610	36″ long G		20	.800		204	36.50		240.50	281
1620	42″ long G		19	.842		228	38.50		266.50	310
1630	48″ long G		18	.889		251	41		292	340
1640	54″ long G		18	.889		310	41		351	405
1650	60″ long G		17	.941		335	43		378	435
1660	2″ diameter x 24″ long G		19	.842		206	38.50		244.50	287
1670	30″ long G		18	.889		232	41		273	320
1680	36″ long G		18	.889		254	41		295	345
1690	42″ long G		17	.941		283	43		326	375
1700	48″ long G		16	1		325	46		371	430
1710	54″ long G		15	1.067		385	49		434	500
1720	60″ long G		15	1.067		415	49		464	530
1730	66″ long G		14	1.143		445	52.50		497.50	570
1740	72″ long G	↓	14	1.143	↓	485	52.50		537.50	615
1990	For galvanized, add				Ea.	75%				

03 15 19.20 Dovetail Anchor System		Crew	Daily Output	Labor-Hours	Unit	Material	Labor	Equipment	Total	Total Incl O&P
0010	**DOVETAIL ANCHOR SYSTEM**									
0500	Dovetail anchor slot, galvanized, foam-filled, 26 ga. G	1 Carp	425	.019	L.F.	.84	.86		1.70	2.25
0600	24 ga. G		400	.020		1.56	.92		2.48	3.14
0625	22 ga. G		400	.020		1.83	.92		2.75	3.43
0900	Stainless steel, foam-filled, 26 ga. G	↓	375	.021	↓	1.34	.98		2.32	2.99
1200	Dovetail brick anchor, corrugated, galvanized, 3-1/2″ long, 16 ga. G	1 Bric	10.50	.762	C	29	34.50		63.50	85

03 15 Concrete Accessories

03 15 19 – Cast-In Concrete Anchors

03 15 19.20 Dovetail Anchor System			Crew	Daily Output	Labor-Hours	Unit	2014 Bare Costs Material	Labor	Equipment	Total	Total Incl O&P
1300	12 ga.	G	1 Bric	10.50	.762	C	41	34.50		75.50	98
1500	Seismic, galvanized, 3-1/2" long, 16 ga.	G		10.50	.762		42	34.50		76.50	99.50
1600	12 ga.	G		10.50	.762		40	34.50		74.50	97
6000	Dovetail stone panel anchors, galvanized, 1/8" x 1" wide, 3-1/2" long	G		10.50	.762		92	34.50		126.50	154
6100	1/4" x 1" wide	G		10.50	.762		104	34.50		138.50	168

03 15 19.45 Machinery Anchors

			Crew	Daily Output	Labor-Hours	Unit	Material	Labor	Equipment	Total	Total Incl O&P
0010	**MACHINERY ANCHORS**, heavy duty, incl. sleeve, floating base nut,										
0020	lower stud & coupling nut, fiber plug, connecting stud, washer & nut.										
0030	For flush mounted embedment in poured concrete heavy equip. pads.										
0200	Stud & bolt, 1/2" diameter	G	E-16	40	.400	Ea.	72.50	21	3.55	97.05	121
0300	5/8" diameter	G		35	.457		80.50	24	4.06	108.56	135
0500	3/4" diameter	G		30	.533		93	28	4.74	125.74	156
0600	7/8" diameter	G		25	.640		101	33.50	5.70	140.20	176
0800	1" diameter	G		20	.800		117	41.50	7.10	165.60	210
0900	1-1/4" diameter	G		15	1.067		141	55.50	9.45	205.95	263

03 21 Reinforcement Bars

03 21 05 – Reinforcing Steel Accessories

03 21 05.10 Rebar Accessories

			Crew	Daily Output	Labor-Hours	Unit	Material	Labor	Equipment	Total	Total Incl O&P
0010	**REBAR ACCESSORIES**										
0030	Steel & plastic made from recycled materials										
0100	Beam bolsters (BB), lower, 1-1/2" high, plain steel	G				C.L.F.	28			28	31
0102	Galvanized	G					33.50			33.50	37
0104	Stainless tipped legs	G					455			455	500
0106	Plastic tipped legs	G					40			40	44
0108	Epoxy dipped	G					70			70	77
0110	2" high, plain	G					34			34	37.50
0120	Galvanized	G					41			41	45
0140	Stainless tipped legs	G					450			450	495
0160	Plastic tipped legs	G					47			47	51.50
0162	Epoxy dipped	G					79			79	87
0200	Upper (BBU), 1-1/2" high, plain steel	G					67			67	73.50
0210	3" high	G					76			76	83.50
0500	Slab bolsters, continuous (SB), 1" high, plain steel	G					28			28	31
0502	Galvanized	G					33.50			33.50	37
0504	Stainless tipped legs	G					455			455	500
0506	Plastic tipped legs	G					40			40	44
0510	2" high, plain steel	G					34			34	37.50
0515	Galvanized	G					41			41	45
0520	Stainless tipped legs	G					460			460	505
0525	Plastic tipped legs	G					47			47	51.50
0530	For bolsters with wire runners (SBR), add	G					37			37	40.50
0540	For bolsters with plates (SBP), add	G					88			88	97
0700	Bag ties, 16 ga., plain, 4" long	G				C	4			4	4.40
0710	5" long	G					5			5	5.50
0720	6" long	G					4			4	4.40
0730	7" long	G					5			5	5.50
1200	High chairs, individual (HC), 3" high, plain steel	G					56			56	61.50
1202	Galvanized	G					67			67	74
1204	Stainless tipped legs	G					480			480	530
1206	Plastic tipped legs	G					61			61	67

03 21 05.10 Rebar Accessories		Crew	Daily Output	Labor-Hours	Unit	2014 Bare Costs Material	Labor	Equipment	Total	Total Incl O&P
1210	5" high, plain	G				C	82			82	90
1212	Galvanized	G					98.50			98.50	108
1214	Stainless tipped legs	G					505			505	560
1216	Plastic tipped legs	G					90			90	99
1220	8" high, plain	G					121			121	133
1222	Galvanized	G					145			145	160
1224	Stainless tipped legs	G					545			545	600
1226	Plastic tipped legs	G					133			133	146
1230	12" high, plain	G					287			287	315
1232	Galvanized	G					345			345	380
1234	Stainless tipped legs	G					710			710	785
1236	Plastic tipped legs	G					315			315	345
1400	Individual high chairs, with plate (HCP), 5" high	G					171			171	188
1410	8" high	G					237			237	261
1500	Bar chair (BC), 1-1/2" high, plain steel	G					36			36	39.50
1520	Galvanized	G					41			41	45
1530	Stainless tipped legs	G					460			460	505
1540	Plastic tipped legs	G					39			39	43
1700	Continuous high chairs (CHC), legs 8" O.C., 4" high, plain steel	G				C.L.F.	49			49	54
1705	Galvanized	G					58.50			58.50	64.50
1710	Stainless tipped legs	G					475			475	520
1715	Plastic tipped legs	G					66			66	72.50
1718	Epoxy dipped	G					85			85	93.50
1720	6" high, plain	G					67			67	73.50
1725	Galvanized	G					80.50			80.50	88.50
1730	Stainless tipped legs	G					490			490	540
1735	Plastic tipped legs	G					90			90	99
1738	Epoxy dipped	G					116			116	128
1740	8" high, plain	G					75			75	82.50
1745	Galvanized	G					90			90	99
1750	Stainless tipped legs	G					500			500	550
1755	Plastic tipped legs	G					114			114	125
1758	Epoxy dipped	G					147			147	162
1900	For continuous bottom wire runners, add	G					50			50	55
1940	For continuous bottom plate, add	G					205			205	226
2200	Screed chair base, 1/2" coil thread diam., 2-1/2" high, plain steel	G				C	310			310	340
2210	Galvanized	G					370			370	405
2220	5-1/2" high, plain	G					370			370	410
2250	Galvanized	G					445			445	490
2300	3/4" coil thread diam., 2-1/2" high, plain steel	G					385			385	425
2310	Galvanized	G					465			465	510
2320	5-1/2" high, plain steel	G					475			475	525
2350	Galvanized	G					570			570	630
2400	Screed holder, 1/2" coil thread diam. for pipe screed, plain steel, 6" long	G					370			370	410
2420	12" long	G					570			570	630
2500	3/4" coil thread diam. for pipe screed, plain steel, 6" long	G					535			535	585
2520	12" long	G					855			855	940
2700	Screw anchor for bolts, plain steel, 3/4" diameter x 4" long	G					505			505	555
2720	1" diameter x 6" long	G					830			830	915
2740	1-1/2" diameter x 8" long	G					1,050			1,050	1,150
2800	Screw anchor eye bolts, 3/4" x 3" long	G					2,700			2,700	2,975
2820	1" x 3-1/2" long	G					3,625			3,625	3,975
2840	1-1/2" x 6" long	G					11,100			11,100	12,200

03 21 Reinforcement Bars

03 21 05 – Reinforcing Steel Accessories

03 21 05.10 Rebar Accessories			Crew	Daily Output	Labor-Hours	Unit	2014 Bare Costs Material	Labor	Equipment	Total	Total Incl O&P
2900	Screw anchor bolts, 3/4" x 9" long	G				C	1,525			1,525	1,675
2920	1" x 12" long	G					2,925			2,925	3,200
3001	Slab lifting inserts, single pickup, galv, 3/4" diam., 5" high	G					1,475			1,475	1,625
3010	6" high	G					1,500			1,500	1,650
3030	7" high	G					1,525			1,525	1,675
3100	1" diameter, 5-1/2" high	G					1,550			1,550	1,700
3120	7" high	G					1,600			1,600	1,750
3200	Double pickup lifting inserts, 1" diameter, 5-1/2" high	G					3,000			3,000	3,300
3220	7" high	G					3,325			3,325	3,650
3330	1-1/2" diameter, 8" high	G					4,175			4,175	4,600
3800	Subgrade chairs, #4 bar head, 3-1/2" high	G					35			35	38.50
3850	12" high	G					41			41	45
3900	#6 bar head, 3-1/2" high	G					35			35	38.50
3950	12" high	G					41			41	45
4200	Subgrade stakes, no nail holes, 3/4" diameter, 12" long	G					275			275	305
4250	24" long	G					335			335	370
4300	7/8" diameter, 12" long	G					350			350	385
4350	24" long	G				↓	585			585	645
4500	Tie wire, 16 ga. annealed steel	G				Cwt.	199			199	219

03 21 05.75 Splicing Reinforcing Bars			Crew	Daily Output	Labor-Hours	Unit	Material	Labor	Equipment	Total	Total Incl O&P
0010	**SPLICING REINFORCING BARS**										
0020	Including holding bars in place while splicing										
0100	Standard, self-aligning type, taper threaded, #4 bars	G	C-25	190	.168	Ea.	6.15	6.70		12.85	17.65
0105	#5 bars	G		170	.188		7.50	7.50		15	20.50
0110	#6 bars	G		150	.213		8.65	8.50		17.15	23.50
0120	#7 bars	G		130	.246		10.10	9.80		19.90	27
0300	#8 bars	G	↓	115	.278		17.10	11.10		28.20	37
0305	#9 bars	G	C-5	105	.533		18.70	26.50	7.10	52.30	70
0310	#10 bars	G		95	.589		21	29.50	7.85	58.35	77
0320	#11 bars	G		85	.659		22	33	8.80	63.80	85
0330	#14 bars	G		65	.862		33	43	11.50	87.50	116
0340	#18 bars	G		45	1.244		50.50	62	16.60	129.10	170
0500	Transition self-aligning, taper threaded, #18-14	G		45	1.244		52.50	62	16.60	131.10	172
0510	#18-11	G		45	1.244		53.50	62	16.60	132.10	173
0520	#14-11	G		65	.862		35	43	11.50	89.50	118
0540	#11-10	G		85	.659		24	33	8.80	65.80	87
0550	#10-9	G	↓	95	.589		23	29.50	7.85	60.35	79
0560	#9-8	G	C-25	105	.305		21	12.15		33.15	42.50
0580	#8-7	G		115	.278		19.20	11.10		30.30	39
0590	#7-6	G		130	.246		12.10	9.80		21.90	29
0600	Position coupler for curved bars, taper threaded, #4 bars	G		160	.200		28	8		36	44
0610	#5 bars	G		145	.221		29.50	8.80		38.30	46.50
0620	#6 bars	G		130	.246		35.50	9.80		45.30	55
0630	#7 bars	G		110	.291		37.50	11.60		49.10	60.50
0640	#8 bars	G	↓	100	.320		39	12.75		51.75	63.50
0650	#9 bars	G	C-5	90	.622		42.50	31	8.30	81.80	105
0660	#10 bars	G		80	.700		46	35	9.30	90.30	115
0670	#11 bars	G		70	.800		48	40	10.65	98.65	126
0680	#14 bars	G		55	1.018		60	51	13.55	124.55	160
0690	#18 bars	G		40	1.400		86	70	18.65	174.65	224
0700	Transition position coupler for curved bars, taper threaded, #18-14	G		40	1.400		88	70	18.65	176.65	227
0710	#18-11	G	↓	40	1.400	↓	89	70	18.65	177.65	228

03 21 05 – Reinforcing Steel Accessories

03 21 05.75 Splicing Reinforcing Bars			Crew	Daily Output	Labor-Hours	Unit	2014 Bare Costs Material	Labor	Equipment	Total	Total Incl O&P
0720	#14-11	G	C-5	55	1.018	Ea.	62	51	13.55	126.55	162
0730	#11-10	G		70	.800		50	40	10.65	100.65	129
0740	#10-9	G		80	.700		48	35	9.30	92.30	118
0750	#9-8	G	C-25	90	.356		45	14.20		59.20	72.50
0760	#8-7	G		100	.320		41	12.75		53.75	65.50
0770	#7-6	G		110	.291		40	11.60		51.60	63
0800	Sleeve type w/grout filler, for precast concrete, #6 bars	G		72	.444		22	17.75		39.75	53
0802	#7 bars	G		64	.500		26.50	19.95		46.45	61.50
0805	#8 bars	G		56	.571		31.50	23		54.50	72
0807	#9 bars	G		48	.667		37.50	26.50		64	84.50
0810	#10 bars	G	C-5	40	1.400		44.50	70	18.65	133.15	179
0900	#11 bars	G		32	1.750		49	87.50	23.50	160	216
0920	#14 bars	G		24	2.333		76.50	116	31	223.50	299
1000	Sleeve type w/ferrous filler, for critical structures, #6 bars	G	C-25	72	.444		60.50	17.75		78.25	95.50
1210	#7 bars	G		64	.500		61.50	19.95		81.45	100
1220	#8 bars	G		56	.571		64.50	23		87.50	108
1230	#9 bars	G	C-5	48	1.167		66.50	58	15.55	140.05	181
1240	#10 bars	G		40	1.400		71	70	18.65	159.65	208
1250	#11 bars	G		32	1.750		85.50	87.50	23.50	196.50	256
1260	#14 bars	G		24	2.333		107	116	31	254	335
1270	#18 bars	G		16	3.500		109	175	46.50	330.50	445
2000	Weldable half coupler, taper threaded, #4 bars	G	E-16	120	.133		9.15	6.95	1.18	17.28	23.50
2100	#5 bars	G		112	.143		10.80	7.45	1.27	19.52	26.50
2200	#6 bars	G		104	.154		17.10	8	1.37	26.47	34.50
2300	#7 bars	G		96	.167		19.90	8.70	1.48	30.08	39
2400	#8 bars	G		88	.182		21	9.45	1.61	32.06	41.50
2500	#9 bars	G		80	.200		23	10.40	1.78	35.18	45.50
2600	#10 bars	G		72	.222		23.50	11.60	1.97	37.07	48
2700	#11 bars	G		64	.250		25	13.05	2.22	40.27	53
2800	#14 bars	G		56	.286		29	14.90	2.54	46.44	61.50
2900	#18 bars	G		48	.333		47	17.35	2.96	67.31	85.50

03 21 11 – Plain Steel Reinforcement Bars

03 21 11.50 Reinforcing Steel, Mill Base Plus Extras			Crew	Daily Output	Labor-Hours	Unit	Material	Labor	Equipment	Total	Total Incl O&P
0010	**REINFORCING STEEL, MILL BASE PLUS EXTRAS**										
0150	Reinforcing, A615 grade 40, mill base	G				Ton	725			725	800
0200	Detailed, cut, bent, and delivered	G					1,000			1,000	1,100
0650	Reinforcing steel, A615 grade 60, mill base	G					725			725	800
0700	Detailed, cut, bent, and delivered	G					1,000			1,000	1,100
1000	Reinforcing steel, extras, included in delivered price										
1005	Mill extra, added for delivery to shop					Ton	34			34	37.50
1010	Shop extra, added for for handling & storage						43			43	47.50
1020	Shop extra, added for bending, limited percent of bars						32.50			32.50	36
1030	Average percent of bars						65			65	71.50
1050	Large percent of bars						130			130	143
1200	Shop extra, added for detailing, under 50 tons						50.50			50.50	55.50
1250	50 to 150 tons						38			38	42
1300	150 to 500 tons						36			36	39.50
1350	Over 500 tons						34			34	37.50
1700	Shop extra, added for listing						5			5	5.50
2000	Mill extra, added for quantity, under 20 tons						22			22	24
2100	Shop extra, added for quantity, under 20 tons						32			32	35
2200	20 to 49 tons						24			24	26.50

03 21 Reinforcement Bars

03 21 11 – Plain Steel Reinforcement Bars

03 21 11.50 Reinforcing Steel, Mill Base Plus Extras		Crew	Daily Output	Labor-Hours	Unit	2014 Bare Costs Material	Labor	Equipment	Total	Total Incl O&P
2250	50 to 99 tons				Ton	16			16	17.60
2300	100 to 300 tons					9.60			9.60	10.55
2500	Shop extra, added for size, #3					152			152	167
2550	#4					76			76	83.50
2600	#5					38			38	42
2650	#6					34			34	37.50
2700	#7 to #11					45.50			45.50	50
2750	#14					57			57	62.50
2800	#18					64.50			64.50	71
2900	Shop extra, added for delivery to job					17			17	18.70

03 21 11.60 Reinforcing In Place

Line	Description	Crew	Daily Output	Labor-Hours	Unit	Material	Labor	Equipment	Total	Total Incl O&P
0010	**REINFORCING IN PLACE**, 50-60 ton lots, A615 Grade 60									
0020	Includes labor, but not material cost, to install accessories									
0030	Made from recycled materials									
0100	Beams & Girders, #3 to #7 G	4 Rodm	1.60	20	Ton	1,000	1,025		2,025	2,700
0150	#8 to #18 G		2.70	11.852		1,000	600		1,600	2,050
0200	Columns, #3 to #7 G		1.50	21.333		1,000	1,075		2,075	2,800
0250	#8 to #18 G		2.30	13.913		1,000	705		1,705	2,200
0300	Spirals, hot rolled, 8″ to 15″ diameter G		2.20	14.545		1,575	735		2,310	2,875
0320	15″ to 24″ diameter G		2.20	14.545		1,500	735		2,235	2,800
0330	24″ to 36″ diameter G		2.30	13.913		1,425	705		2,130	2,675
0340	36″ to 48″ diameter G		2.40	13.333		1,350	675		2,025	2,550
0360	48″ to 64″ diameter G		2.50	12.800		1,500	650		2,150	2,675
0380	64″ to 84″ diameter G		2.60	12.308		1,575	625		2,200	2,700
0390	84″ to 96″ diameter G		2.70	11.852		1,650	600		2,250	2,775
0400	Elevated slabs, #4 to #7 G		2.90	11.034		1,000	560		1,560	1,975
0500	Footings, #4 to #7 G		2.10	15.238		1,000	770		1,770	2,300
0550	#8 to #18 G		3.60	8.889		1,000	450		1,450	1,800
0600	Slab on grade, #3 to #7 G		2.30	13.913		1,000	705		1,705	2,200
0700	Walls, #3 to #7 G		3	10.667		1,000	540		1,540	1,950
0750	#8 to #18 G		4	8		1,000	405		1,405	1,725
0900	For other than 50 – 60 ton lots									
1000	Under 10 ton job, #3 to #7, add					25%	10%			
1010	#8 to #18, add					20%	10%			
1050	10 – 50 ton job, #3 to #7, add					10%				
1060	#8 to #18, add					5%				
1100	60 – 100 ton job, #3 to #7, deduct					5%				
1110	#8 to #18, deduct					10%				
1150	Over 100 ton job, #3 to #7, deduct					10%				
1160	#8 to #18, deduct					15%				
1200	Reinforcing in place, A615 Grade 75, add G				Ton	92.50			92.50	102
1220	Grade 90, add					125			125	138
2000	Unloading & sorting, add to above	C-5	100	.560			28	7.45	35.45	51.50
2200	Crane cost for handling, 90 picks/day, up to 1.5 Tons/bundle, add to above		135	.415			20.50	5.55	26.05	38
2210	1.0 Ton/bundle		92	.609			30.50	8.10	38.60	56
2220	0.5 Ton/bundle		35	1.600			80	21.50	101.50	148
2400	Dowels, 2 feet long, deformed, #3 G	2 Rodm	520	.031	Ea.	.41	1.56		1.97	2.90
2410	#4 G		480	.033		.73	1.69		2.42	3.46
2420	#5 G		435	.037		1.15	1.86		3.01	4.19
2430	#6 G		360	.044		1.65	2.25		3.90	5.35
2450	Longer and heavier dowels, add G		725	.022	Lb.	.55	1.12		1.67	2.37
2500	Smooth dowels, 12″ long, 1/4″ or 3/8″ diameter G		140	.114	Ea.	.72	5.80		6.52	9.90

03 21 Reinforcement Bars

03 21 11 – Plain Steel Reinforcement Bars

03 21 11.60 Reinforcing In Place		Crew	Daily Output	Labor-Hours	Unit	Material	2014 Bare Costs Labor	Equipment	Total	Total Incl O&P
2520	5/8" diameter [G]	2 Rodm	125	.128	Ea.	1.26	6.50		7.76	11.60
2530	3/4" diameter [G]		110	.145		1.57	7.35		8.92	13.25
2600	Dowel sleeves for CIP concrete, 2-part system									
2610	Sleeve base, plastic, for 5/8" smooth dowel sleeve, fasten to edge form	1 Rodm	200	.040	Ea.	.52	2.03		2.55	3.75
2615	Sleeve, plastic, 12" long, for 5/8" smooth dowel, snap onto base		400	.020		1.16	1.01		2.17	2.87
2620	Sleeve base, for 3/4" smooth dowel sleeve		175	.046		.52	2.32		2.84	4.21
2625	Sleeve, 12" long, for 3/4" smooth dowel		350	.023		1.30	1.16		2.46	3.25
2630	Sleeve base, for 1" smooth dowel sleeve		150	.053		.66	2.70		3.36	4.97
2635	Sleeve, 12" long, for 1" smooth dowel		300	.027		1.37	1.35		2.72	3.63
2700	Dowel caps, visual warning only, plastic, #3 to #8	2 Rodm	800	.020		.27	1.01		1.28	1.89
2720	#8 to #18		750	.021		.68	1.08		1.76	2.45
2750	Impalement protective, plastic, #4 to #9		800	.020		1.50	1.01		2.51	3.24

03 21 13 – Galvanized Reinforcement Steel Bars

03 21 13.10 Galvanized Reinforcing

		Crew	Daily Output	Labor-Hours	Unit	Material	Labor	Equipment	Total	Total Incl O&P
0010	**GALVANIZED REINFORCING**									
0150	Add to plain steel rebar pricing for galvanized rebar				Ton	460			460	505

03 21 16 – Epoxy-Coated Reinforcement Steel Bars

03 21 16.10 Epoxy-Coated Reinforcing

		Crew	Daily Output	Labor-Hours	Unit	Material	Labor	Equipment	Total	Total Incl O&P
0010	**EPOXY-COATED REINFORCING**									
0100	Add to plain steel rebar pricing for epoxy-coated rebar				Ton	410			410	450

03 21 19 – Stainless Steel Reinforcement Bars

03 21 19.10 Stainless Steel Reinforcing

		Crew	Daily Output	Labor-Hours	Unit	Material	Labor	Equipment	Total	Total Incl O&P
0010	**STAINLESS STEEL REINFORCING**									
0100	Add to plain steel rebar pricing for stainless steel rebar					300%				

03 21 21 – Composite Reinforcement Bars

03 21 21.11 Glass Fiber-Reinforced Polymer Reinforcement Bars

		Crew	Daily Output	Labor-Hours	Unit	Material	Labor	Equipment	Total	Total Incl O&P
0010	**GLASS FIBER-REINFORCED POLYMER REINFORCEMENT BARS**									
0020	Includes labor, but not material cost, to install accessories									
0050	#2 bar, .043 lb./L.F.	4 Rodm	9500	.003	L.F.	.36	.17		.53	.67
0100	#3 bar, .092 lb./L.F.		9300	.003		.49	.17		.66	.81
0150	#4 bar, .160 lb./L.F.		9100	.004		.71	.18		.89	1.06
0200	#5 bar, .258 lb./L.F.		8700	.004		1.09	.19		1.28	1.49
0250	#6 bar, .372 lb./L.F.		8300	.004		1.41	.20		1.61	1.86
0300	#7 bar, .497 lb./L.F.		7900	.004		1.84	.21		2.05	2.34
0350	#8 bar, .620 lb./L.F.		7400	.004		2.50	.22		2.72	3.09
0400	#9 bar, .800 lb./L.F.		6800	.005		3.16	.24		3.40	3.85
0450	#10 bar, 1.08 lb./L.F.		5800	.006		3.93	.28		4.21	4.76
0500	For Bends, add per bend				Ea.	1.48			1.48	1.63

03 22 Fabric and Grid Reinforcing

03 22 11 – Plain Welded Wire Fabric Reinforcing

03 22 11.10 Plain Welded Wire Fabric			Crew	Daily Output	Labor-Hours	Unit	Material	2014 Bare Costs Labor	Equipment	Total	Total Incl O&P
0010	**PLAIN WELDED WIRE FABRIC** ASTM A185										
0020	Includes labor, but not material cost, to install accessories										
0030	Made from recycled materials										
0100	6 x 6 - W1.4 x W1.4 (10 x 10) 21 lb. per C.S.F.	G	2 Rodm	35	.457	C.S.F.	14.50	23		37.50	52.50
0200	6 x 6 - W2.1 x W2.1 (8 x 8) 30 lb. per C.S.F.	G		31	.516		17.20	26		43.20	60
0300	6 x 6 - W2.9 x W2.9 (6 x 6) 42 lb. per C.S.F.	G		29	.552		22.50	28		50.50	68.50
0400	6 x 6 - W4 x W4 (4 x 4) 58 lb. per C.S.F.	G		27	.593		31.50	30		61.50	81.50
0500	4 x 4 - W1.4 x W1.4 (10 x 10) 31 lb. per C.S.F.	G		31	.516		20	26		46	63
0600	4 x 4 - W2.1 x W2.1 (8 x 8) 44 lb. per C.S.F.	G		29	.552		25	28		53	71.50
0650	4 x 4 - W2.9 x W2.9 (6 x 6) 61 lb. per C.S.F.	G		27	.593		40.50	30		70.50	91.50
0700	4 x 4 - W4 x W4 (4 x 4) 85 lb. per C.S.F.	G		25	.640		50.50	32.50		83	107
0800	2 x 2 - #14 galv., 21 lb./C.S.F., beam & column wrap	G		6.50	2.462		41.50	125		166.50	242
0900	2 x 2 - #12 galv. for gunite reinforcing	G		6.50	2.462		62.50	125		187.50	265

03 22 13 – Galvanized Welded Wire Fabric Reinforcing

03 22 13.10 Galvanized Welded Wire Fabric		Crew	Daily Output	Labor-Hours	Unit	Material	Labor	Equipment	Total	Total Incl O&P
0010	**GALVANIZED WELDED WIRE FABRIC**									
0100	Add to plain welded wire pricing for galvanized welded wire				Lb.	.23			.23	.25

03 22 16 – Epoxy-Coated Welded Wire Fabric Reinforcing

03 22 16.10 Epoxy-Coated Welded Wire Fabric		Crew	Daily Output	Labor-Hours	Unit	Material	Labor	Equipment	Total	Total Incl O&P
0010	**EPOXY-COATED WELDED WIRE FABRIC**									
0100	Add to plain welded wire pricing for epoxy-coated welded wire				Lb.	.21			.21	.23

03 23 Stressed Tendon Reinforcing

03 23 05 – Prestressing Tendons

03 23 05.50 Prestressing Steel			Crew	Daily Output	Labor-Hours	Unit	Material	Labor	Equipment	Total	Total Incl O&P
0010	**PRESTRESSING STEEL**	R034136-90									
0100	Grouted strand, in beams, post-tensioned in field, 50′ span, 100 kip	G	C-3	1200	.053	Lb.	2.62	2.50	.09	5.21	6.90
0150	300 kip	G		2700	.024		1.12	1.11	.04	2.27	3.01
0300	100′ span, 100 kip	G		1700	.038		2.62	1.77	.07	4.46	5.70
0350	300 kip	G		3200	.020		2.25	.94	.04	3.23	3.98
0500	200′ span, 100 kip	G		2700	.024		2.62	1.11	.04	3.77	4.66
0550	300 kip	G		3500	.018		2.25	.86	.03	3.14	3.85
0800	Grouted bars, in beams, 50′ span, 42 kip	G		2600	.025		1.06	1.16	.04	2.26	3.02
0850	143 kip	G		3200	.020		1.02	.94	.04	2	2.62
1000	75′ span, 42 kip	G		3200	.020		1.08	.94	.04	2.06	2.68
1050	143 kip	G		4200	.015		.90	.72	.03	1.65	2.14
1200	Ungrouted strand, in beams, 50′ span, 100 kip	G	C-4	1275	.025		.61	1.28	.02	1.91	2.72
1250	300 kip	G		1475	.022		.61	1.11	.02	1.74	2.43
1400	100′ span, 100 kip	G		1500	.021		.61	1.09	.02	1.72	2.40
1450	300 kip	G		1650	.019		.61	.99	.02	1.62	2.25
1600	200′ span, 100 kip	G		1500	.021		.61	1.09	.02	1.72	2.40
1650	300 kip	G		1700	.019		.61	.96	.02	1.59	2.20
1800	Ungrouted bars, in beams, 50′ span, 42 kip	G		1400	.023		.50	1.17	.02	1.69	2.41
1850	143 kip	G		1700	.019		.50	.96	.02	1.48	2.08
2000	75′ span, 42 kip	G		1800	.018		.50	.91	.02	1.43	2
2050	143 kip	G		2200	.015		.50	.74	.01	1.25	1.74
2220	Ungrouted single strand, 100′ elevated slab, 25 kip	G		1200	.027		.61	1.36	.03	2	2.84
2250	35 kip	G		1475	.022		.61	1.11	.02	1.74	2.43
3000	Slabs on grade, 0.5-inch diam. non-bonded strands, HDPE sheathed,										
3050	attached dead-end anchors, loose stressing-end anchors										

03 23 Stressed Tendon Reinforcing

03 23 05 – Prestressing Tendons

03 23 05.50 Prestressing Steel		Crew	Daily Output	Labor-Hours	Unit	Material	Labor (2014 Bare Costs)	Equipment	Total	Total Incl O&P
3100	25′ x 30′ slab, strands @ 36″ O.C., placing	2 Rodm	2940	.005	S.F.	.60	.28		.88	1.09
3105	Stressing	C-4A	3750	.004			.22	.01	.23	.35
3110	42″ O.C., placing	2 Rodm	3200	.005		.53	.25		.78	.99
3115	Stressing	C-4A	4040	.004			.20	.01	.21	.33
3120	48″ O.C., placing	2 Rodm	3510	.005		.47	.23		.70	.87
3125	Stressing	C-4A	4390	.004			.18	.01	.19	.30
3150	25′ x 40′ slab, strands @ 36″ O.C., placing	2 Rodm	3370	.005		.58	.24		.82	1.02
3155	Stressing	C-4A	4360	.004			.19	.01	.20	.30
3160	42″ O.C., placing	2 Rodm	3760	.004		.50	.22		.72	.89
3165	Stressing	C-4A	4820	.003			.17	.01	.18	.27
3170	48″ O.C., placing	2 Rodm	4090	.004		.45	.20		.65	.80
3175	Stressing	C-4A	5190	.003			.16	.01	.17	.26
3200	30′ x 30′ slab, strands @ 36″ O.C., placing	2 Rodm	3260	.005		.58	.25		.83	1.03
3205	Stressing	C-4A	4190	.004			.19	.01	.20	.31
3210	42″ O.C., placing	2 Rodm	3530	.005		.52	.23		.75	.94
3215	Stressing	C-4A	4500	.004			.18	.01	.19	.29
3220	48″ O.C., placing	2 Rodm	3840	.004		.47	.21		.68	.84
3225	Stressing	C-4A	4850	.003			.17	.01	.18	.27
3230	30′ x 40′ slab, strands @ 36″ O.C., placing	2 Rodm	3780	.004		.56	.21		.77	.96
3235	Stressing	C-4A	4920	.003			.16	.01	.17	.27
3240	42″ O.C., placing	2 Rodm	4190	.004		.49	.19		.68	.84
3245	Stressing	C-4A	5410	.003			.15	.01	.16	.25
3250	48″ O.C., placing	2 Rodm	4520	.004		.45	.18		.63	.77
3255	Stressing	C-4A	5790	.003			.14	.01	.15	.23
3260	30′ x 50′ slab, strands @ 36″ O.C., placing	2 Rodm	4300	.004		.53	.19		.72	.89
3265	Stressing	C-4A	5650	.003			.14	.01	.15	.24
3270	42″ O.C., placing	2 Rodm	4720	.003		.47	.17		.64	.79
3275	Stressing	C-4A	6150	.003			.13	.01	.14	.22
3280	48″ O.C., placing	2 Rodm	5240	.003		.42	.15		.57	.70
3285	Stressing	C-4A	6760	.002	↓		.12	.01	.13	.20

03 24 Fibrous Reinforcing

03 24 05 – Reinforcing Fibers

03 24 05.30 Synthetic Fibers		Crew	Daily Output	Labor-Hours	Unit	Material	Labor	Equipment	Total	Total Incl O&P
0010	**SYNTHETIC FIBERS**									
0100	Synthetic fibers, add to concrete				Lb.	4.80			4.80	5.30
0110	1-1/2 lb. per C.Y.				C.Y.	7.40			7.40	8.15

03 24 05.70 Steel Fibers		Crew	Daily Output	Labor-Hours	Unit	Material	Labor	Equipment	Total	Total Incl O&P
0010	**STEEL FIBERS**									
0140	ASTM A850, Type V, continuously deformed, 1-1/2″ long x 0.045″ diam.									
0150	Add to price of ready mix concrete G				Lb.	1.25			1.25	1.38
0205	Alternate pricing, dosing at 5 lb. per C.Y., add to price of RMC G				C.Y.	6.25			6.25	6.90
0210	10 lb. per C.Y. G					12.50			12.50	13.75
0215	15 lb. per C.Y. G					18.75			18.75	20.50
0220	20 lb. per C.Y. G					25			25	27.50
0225	25 lb. per C.Y. G					31.50			31.50	34.50
0230	30 lb. per C.Y. G					37.50			37.50	41.50
0235	35 lb. per C.Y. G					44			44	48
0240	40 lb. per C.Y. G					50			50	55
0250	50 lb. per C.Y. G					62.50			62.50	69
0275	75 lb. per C.Y. G				↓	94			94	103

03 24 Fibrous Reinforcing

03 24 05 – Reinforcing Fibers

03 24 05.70 Steel Fibers		Crew	Daily Output	Labor-Hours	Unit	Material	2014 Bare Costs Labor	Equipment	Total	Total Incl O&P
0300	100 lb. per C.Y. G				C.Y.	125			125	138

03 30 Cast-In-Place Concrete

03 30 53 – Miscellaneous Cast-In-Place Concrete

03 30 53.40 Concrete In Place

03 30 53.40 Concrete In Place		Crew	Daily Output	Labor-Hours	Unit	Material	2014 Bare Costs Labor	Equipment	Total	Total Incl O&P
0010	**CONCRETE IN PLACE** R033053-10									
0020	Including forms (4 uses), Grade 60 rebar, concrete (Portland cement R033053-60									
0050	Type I), placement and finishing unless otherwise indicated R033105-10									
0300	Beams (3500 psi), 5 kip per L.F., 10′ span R033105-20	C-14A	15.62	12.804	C.Y.	315	590	47.50	952.50	1,325
0350	25′ span R033105-50		18.55	10.782		335	495	40	870	1,175
0700	Columns, square (4000 psi), 12″ x 12″, less than 2% reinforcing R033105-65		11.96	16.722		355	770	62	1,187	1,675
0720	2% to 3% reinforcing reinforcing R033105-80		10.13	19.743		565	910	73.50	1,548.50	2,100
0740	Over 3% reinforcing R033105-85		9.03	22.148		850	1,025	82.50	1,957.50	2,600
0800	16″ x 16″, less than 2% reinforcing		16.22	12.330		282	565	46	893	1,225
0820	2% to 3% reinforcing reinforcing		12.57	15.911		480	730	59	1,269	1,725
0840	Over 3% reinforcing		10.25	19.512		745	900	72.50	1,717.50	2,275
0900	24″ x 24″, less than 2% reinforcing		23.66	8.453		238	390	31.50	659.50	895
0920	2% to 3% reinforcing reinforcing		17.71	11.293		425	520	42	987	1,325
0940	Over 3% reinforcing		14.15	14.134		680	650	52.50	1,382.50	1,800
1000	36″ x 36″, less than 2% reinforcing		33.69	5.936		210	273	22	505	675
1020	2% to 3% reinforcing reinforcing		23.32	8.576		375	395	32	802	1,050
1040	Over 3% reinforcing		17.82	11.223		635	515	42	1,192	1,550
1100	Columns, round (4000 psi), tied, 12″ diameter, less than 2% reinforcing		20.97	9.537		305	440	35.50	780.50	1,050
1120	2% to 3% reinforcing reinforcing		15.27	13.098		515	605	48.50	1,168.50	1,550
1140	Over 3% reinforcing		12.11	16.515		785	760	61.50	1,606.50	2,100
1200	16″ diameter, less than 2% reinforcing		31.49	6.351		280	292	23.50	595.50	785
1220	2% to 3% reinforcing reinforcing		19.12	10.460		490	480	39	1,009	1,325
1240	Over 3% reinforcing		13.77	14.524		740	670	54	1,464	1,900
1300	20″ diameter, less than 2% reinforcing		41.04	4.873		283	224	18.15	525.15	675
1320	2% to 3% reinforcing reinforcing		24.05	8.316		475	385	31	891	1,150
1340	Over 3% reinforcing		17.01	11.758		740	540	43.50	1,323.50	1,700
1400	24″ diameter, less than 2% reinforcing		51.85	3.857		264	177	14.35	455.35	580
1420	2% to 3% reinforcing reinforcing		27.06	7.391		475	340	27.50	842.50	1,075
1440	Over 3% reinforcing		18.29	10.935		725	505	40.50	1,270.50	1,625
1500	36″ diameter, less than 2% reinforcing		75.04	2.665		265	123	9.90	397.90	490
1520	2% to 3% reinforcing reinforcing		37.49	5.335		450	245	19.85	714.85	895
1540	Over 3% reinforcing		22.84	8.757		705	405	32.50	1,142.50	1,425
1900	Elevated slab (4000 psi), flat slab with drops, 125 psf Sup. Load, 20′ span	C-14B	38.45	5.410		256	249	19.30	524.30	690
1950	30′ span		50.99	4.079		273	187	14.55	474.55	605
2100	Flat plate, 125 psf Sup. Load, 15′ span		30.24	6.878		234	315	24.50	573.50	770
2150	25′ span		49.60	4.194		245	193	14.95	452.95	585
2300	Waffle const., 30″ domes, 125 psf Sup. Load, 20′ span		37.07	5.611		254	258	20	532	695
2350	30′ span		44.07	4.720		237	217	16.85	470.85	615
2500	One way joists, 30″ pans, 125 psf Sup. Load, 15′ span		27.38	7.597		299	350	27	676	900
2550	25′ span		31.15	6.677		283	305	24	612	810
2700	One way beam & slab, 125 psf Sup. Load, 15′ span		20.59	10.102		253	465	36	754	1,025
2750	25′ span		28.36	7.334		241	335	26	602	815
2900	Two way beam & slab, 125 psf Sup. Load, 15′ span		24.04	8.652		245	395	31	671	920
2950	25′ span		35.87	5.799		212	266	20.50	498.50	665
3100	Elevated slabs, flat plate, including finish, not									
3110	including forms or reinforcing									

03 30 53 – Miscellaneous Cast-In-Place Concrete

03 30 53.40 Concrete In Place		Crew	Daily Output	Labor-Hours	Unit	2014 Bare Costs Material	Labor	Equipment	Total	Total Incl O&P
3150	Regular concrete (4000 psi), 4" slab	C-8	2613	.021	S.F.	1.39	.87	.27	2.53	3.16
3200	6" slab		2585	.022		2.03	.88	.27	3.18	3.87
3250	2-1/2" thick floor fill		2685	.021		.91	.85	.26	2.02	2.58
3300	Lightweight, 110# per C.F., 2-1/2" thick floor fill		2585	.022		1.24	.88	.27	2.39	3
3400	Cellular concrete, 1-5/8" fill, under 5000 S.F.		2000	.028		.84	1.14	.36	2.34	3.05
3450	Over 10,000 S.F.		2200	.025		.80	1.04	.32	2.16	2.83
3500	Add per floor for 3 to 6 stories high		31800	.002			.07	.02	.09	.13
3520	For 7 to 20 stories high		21200	.003			.11	.03	.14	.20
3540	Equipment pad (3000 psi), 3' x 3' x 6" thick	C-14H	45	1.067	Ea.	44.50	48	.74	93.24	124
3550	4' x 4' x 6" thick		30	1.600		66.50	72	1.10	139.60	185
3560	5' x 5' x 8" thick		18	2.667		119	120	1.84	240.84	320
3570	6' x 6' x 8" thick		14	3.429		160	155	2.37	317.37	415
3580	8' x 8' x 10" thick		8	6		335	271	4.14	610.14	790
3590	10' x 10' x 12" thick		5	9.600		575	435	6.60	1,016.60	1,300
3800	Footings (3000 psi), spread under 1 C.Y.	C-14C	28	4	C.Y.	164	176	1.16	341.16	455
3825	1 C.Y. to 5 C.Y.		43	2.605		198	114	.76	312.76	395
3850	Over 5 C.Y.		75	1.493		180	65.50	.43	245.93	299
3900	Footings, strip (3000 psi), 18" x 9", unreinforced	C-14L	40	2.400		120	103	.82	223.82	291
3920	18" x 9", reinforced	C-14C	35	3.200		144	141	.93	285.93	375
3925	20" x 10", unreinforced	C-14L	45	2.133		117	91.50	.73	209.23	269
3930	20" x 10", reinforced	C-14C	40	2.800		136	123	.81	259.81	340
3935	24" x 12", unreinforced	C-14L	55	1.745		116	74.50	.59	191.09	243
3940	24" x 12", reinforced	C-14C	48	2.333		136	102	.68	238.68	310
3945	36" x 12", unreinforced	C-14L	70	1.371		112	58.50	.47	170.97	214
3950	36" x 12", reinforced	C-14C	60	1.867		130	82	.54	212.54	271
4000	Foundation mat (3000 psi), under 10 C.Y.		38.67	2.896		204	127	.84	331.84	420
4050	Over 20 C.Y.		56.40	1.986		178	87	.58	265.58	330
4200	Wall, free-standing (3000 psi), 8" thick, 8' high	C-14D	45.83	4.364		152	199	16.25	367.25	490
4250	14' high		27.26	7.337		188	335	27.50	550.50	750
4260	12" thick, 8' high		64.32	3.109		142	142	11.55	295.55	390
4270	14' high		40.01	4.999		152	228	18.60	398.60	540
4300	15" thick, 8' high		80.02	2.499		137	114	9.30	260.30	335
4350	12' high		51.26	3.902		137	178	14.50	329.50	440
4500	18' high		48.85	4.094		155	187	15.25	357.25	475
4520	Handicap access ramp (4000 psi), railing both sides, 3' wide	C-14H	14.58	3.292	L.F.	315	149	2.27	466.27	580
4525	5' wide		12.22	3.928		330	177	2.71	509.71	635
4530	With 6" curb and rails both sides, 3' wide		8.55	5.614		325	253	3.87	581.87	755
4535	5' wide		7.31	6.566		330	296	4.53	630.53	825
4650	Slab on grade (3500 psi), not including finish, 4" thick	C-14E	60.75	1.449	C.Y.	120	65.50	.55	186.05	234
4700	6" thick	"	92	.957	"	115	43	.36	158.36	194
4701	Thickened slab edge (3500 psi), for slab on grade poured									
4702	monolithically with slab; depth is in addition to slab thickness;									
4703	formed vertical outside edge, earthen bottom and inside slope									
4705	8" deep x 8" wide bottom, unreinforced	C-14L	2190	.044	L.F.	3.33	1.88	.01	5.22	6.55
4710	8" x 8", reinforced	C-14C	1670	.067		5.70	2.95	.02	8.67	10.80
4715	12" deep x 12" wide bottom, unreinforced	C-14L	1800	.053		6.75	2.28	.02	9.05	11
4720	12" x 12", reinforced	C-14C	1310	.086		11.05	3.76	.02	14.83	18
4725	16" deep x 16" wide bottom, unreinforced	C-14L	1440	.067		11.40	2.85	.02	14.27	16.95
4730	16" x 16", reinforced	C-14C	1120	.100		16.50	4.39	.03	20.92	25
4735	20" deep x 20" wide bottom, unreinforced	C-14L	1150	.083		17.25	3.57	.03	20.85	24.50
4740	20" x 20", reinforced	C-14C	920	.122		24	5.35	.04	29.39	34.50
4745	24" deep x 24" wide bottom, unreinforced	C-14L	930	.103		24.50	4.42	.04	28.96	34
4750	24" x 24", reinforced	C-14C	740	.151		33	6.65	.04	39.69	46.50

03 30 Cast-In-Place Concrete

03 30 53 – Miscellaneous Cast-In-Place Concrete

03 30 53.40 Concrete In Place		Crew	Daily Output	Labor-Hours	Unit	2014 Bare Costs Material	Labor	Equipment	Total	Total Incl O&P
4751	Slab on grade (3500 psi), incl. troweled finish, not incl. forms									
4760	or reinforcing, over 10,000 S.F., 4" thick	C-14F	3425	.021	S.F.	1.31	.88	.01	2.20	2.77
4820	6" thick		3350	.021		1.91	.90	.01	2.82	3.46
4840	8" thick		3184	.023		2.62	.95	.01	3.58	4.31
4900	12" thick		2734	.026		3.92	1.10	.01	5.03	6
4950	15" thick		2505	.029		4.93	1.20	.01	6.14	7.20
5000	Slab on grade (3000 psi), incl. broom finish, not incl. forms									
5001	or reinforcing, 4" thick	C-14G	2873	.019	S.F.	1.29	.80	.01	2.10	2.63
5010	6" thick		2590	.022		2.01	.89	.01	2.91	3.56
5020	8" thick		2320	.024		2.62	.99	.01	3.62	4.40
5200	Lift slab in place above the foundation, incl. forms, reinforcing,									
5210	concrete (4000 psi) and columns, over 20,000 S.F. per floor	C-14B	2113	.098	S.F.	6.80	4.52	.35	11.67	14.85
5250	10,000 S.F. to 20,000 S.F. per floor		1650	.126		7.45	5.80	.45	13.70	17.65
5300	Under 10,000 S.F. per floor		1500	.139		8.10	6.35	.50	14.95	19.30
5500	Lightweight, ready mix, including screed finish only,									
5510	not including forms or reinforcing									
5550	1:4 (2500 psi) for structural roof decks	C-14B	260	.800	C.Y.	138	37	2.86	177.86	212
5600	1:6 (3000 psi) for ground slab with radiant heat	C-14F	92	.783		140	32.50	.36	172.86	203
5650	1:3:2 (2000 psi) with sand aggregate, roof deck	C-14B	260	.800		135	37	2.86	174.86	209
5700	Ground slab (2000 psi)	C-14F	107	.673		135	28	.31	163.31	191
5900	Pile caps (3000 psi), incl. forms and reinf., sq. or rect., under 10 C.Y.	C-14C	54.14	2.069		165	91	.60	256.60	325
5950	Over 10 C.Y.		75	1.493		155	65.50	.43	220.93	271
6000	Triangular or hexagonal, under 10 C.Y.		53	2.113		121	93	.61	214.61	277
6050	Over 10 C.Y.		85	1.318		137	58	.38	195.38	240
6200	Retaining walls (3000 psi), gravity, 4' high see Section 32 32	C-14D	66.20	3.021		137	138	11.25	286.25	375
6250	10' high		125	1.600		131	73	5.95	209.95	264
6300	Cantilever, level backfill loading, 8' high		70	2.857		147	130	10.65	287.65	375
6350	16' high		91	2.198		142	100	8.20	250.20	320
6800	Stairs (3500 psi), not including safety treads, free standing, 3'-6" wide	C-14H	83	.578	LF Nose	5.40	26	.40	31.80	46.50
6850	Cast on ground		125	.384	"	4.49	17.35	.27	22.11	31.50
7000	Stair landings, free standing		200	.240	S.F.	4.39	10.85	.17	15.41	21.50
7050	Cast on ground		475	.101	"	3.45	4.56	.07	8.08	10.85

03 31 Structural Concrete

03 31 13 – Heavyweight Structural Concrete

03 31 13.25 Concrete, Hand Mix

		Crew	Daily Output	Labor-Hours	Unit	Material	Labor	Equipment	Total	Total Incl O&P
0010	**CONCRETE, HAND MIX** for small quantities or remote areas									
0050	Includes bulk local aggregate, bulk sand, bagged Portland									
0060	cement (Type I) and water, using gas powered cement mixer									
0125	2500 psi	C-30	135	.059	C.F.	3.14	2.17	1.31	6.62	8.25
0130	3000 psi		135	.059		3.37	2.17	1.31	6.85	8.50
0135	3500 psi		135	.059		3.50	2.17	1.31	6.98	8.65
0140	4000 psi		135	.059		3.66	2.17	1.31	7.14	8.80
0145	4500 psi		135	.059		3.83	2.17	1.31	7.31	9
0150	5000 psi		135	.059		4.09	2.17	1.31	7.57	9.30
0300	Using pre-bagged dry mix and wheelbarrow (80-lb. bag = 0.6 C.F.)									
0340	4000 psi	1 Clab	48	.167	C.F.	6.25	6.10		12.35	16.35

03 31 Structural Concrete

03 31 13 – Heavyweight Structural Concrete

03 31 13.30 Concrete, Volumetric Site-Mixed		Crew	Daily Output	Labor-Hours	Unit	2014 Bare Costs Material	Labor	Equipment	Total	Total Incl O&P
0010	**CONCRETE, VOLUMETRIC SITE-MIXED**									
0015	Mixed on-site in volumetric truck									
0020	Includes local aggregate, sand, Portland cement (Type I) and water									
0025	Excludes all additives and treatments									
0100	3000 psi, 1 C.Y. mixed and discharged				C.Y.	159			159	175
0110	2 C.Y.					136			136	150
0120	3 C.Y.					122			122	134
0130	4 C.Y.					114			114	125
0140	5 C.Y.				↓	106			106	116
0200	For truck holding/waiting time past first 2 on-site hours, add				Hr.	72			72	79
0210	For trip charge beyond first 20 miles, each way, add				Mile	2.25			2.25	2.48
0220	For each additional increase of 500 psi, add				Ea.	4.17			4.17	4.59
03 31 13.35 Heavyweight Concrete, Ready Mix										
0010	**HEAVYWEIGHT CONCRETE, READY MIX,** delivered									
0012	Includes local aggregate, sand, Portland cement (Type I) and water									
0015	Excludes all additives and treatments R033105-20									
0020	2000 psi				C.Y.	93			93	102
0100	2500 psi					95.50			95.50	105
0150	3000 psi					99			99	109
0200	3500 psi					101			101	111
0300	4000 psi					104			104	114
0350	4500 psi					107			107	117
0400	5000 psi					110			110	121
0411	6000 psi					113			113	124
0412	8000 psi					119			119	130
0413	10,000 psi					125			125	137
0414	12,000 psi					131			131	144
1000	For high early strength (Portland cement Type III), add					10%				
1010	For structural lightweight with regular sand, add					25%				
1300	For winter concrete (hot water), add					4.50			4.50	4.95
1400	For hot weather concrete (ice), add					7.65			7.65	8.40
1410	For mid-range water reducer, add					3.69			3.69	4.06
1420	For high-range water reducer/superplasticizer, add					5.70			5.70	6.30
1430	For retarder, add					3.06			3.06	3.37
1440	For non-Chloride accelerator, add					5.65			5.65	6.20
1450	For Chloride accelerator, per 1%, add					3.32			3.32	3.65
1460	For fiber reinforcing, synthetic (1 lb./C.Y.), add					6.95			6.95	7.60
1500	For Saturday delivery, add				↓	4.93			4.93	5.40
1510	For truck holding/waiting time past 1st hour per load, add				Hr.	97			97	107
1520	For short load (less than 4 C.Y.), add per load				Ea.	77.50			77.50	85
2000	For all lightweight aggregate, add				C.Y.	45%				
4000	Flowable fill: ash, cement, aggregate, water									
4100	40 – 80 psi				C.Y.	78			78	85.50
4150	Structural: ash, cement, aggregate, water & sand									
4200	50 psi				C.Y.	78			78	85.50
4250	140 psi					78.50			78.50	86.50
4300	500 psi					81			81	89
4350	1000 psi				↓	84			84	92.50

03 31 13 – Heavyweight Structural Concrete

03 31 13.70 Placing Concrete		Crew	Daily Output	Labor-Hours	Unit	Material	2014 Bare Costs Labor	Equipment	Total	Total Incl O&P
0010	**PLACING CONCRETE**									
0020	Includes labor and equipment to place, level (strike off) and consolidate									
0050	Beams, elevated, small beams, pumped	C-20	60	1.067	C.Y.		42	12.95	54.95	78.50
0100	With crane and bucket	C-7	45	1.600			63.50	27.50	91	128
0200	Large beams, pumped	C-20	90	.711			28	8.65	36.65	52.50
0250	With crane and bucket	C-7	65	1.108			44	18.85	62.85	88.50
0400	Columns, square or round, 12" thick, pumped	C-20	60	1.067			42	12.95	54.95	78.50
0450	With crane and bucket	C-7	40	1.800			71.50	30.50	102	144
0600	18" thick, pumped	C-20	90	.711			28	8.65	36.65	52.50
0650	With crane and bucket	C-7	55	1.309			52	22.50	74.50	104
0800	24" thick, pumped	C-20	92	.696			27.50	8.45	35.95	51.50
0850	With crane and bucket	C-7	70	1.029			41	17.50	58.50	82
1000	36" thick, pumped	C-20	140	.457			18	5.55	23.55	33.50
1050	With crane and bucket	C-7	100	.720			28.50	12.25	40.75	57.50
1400	Elevated slabs, less than 6" thick, pumped	C-20	140	.457			18	5.55	23.55	33.50
1450	With crane and bucket	C-7	95	.758			30	12.90	42.90	60
1500	6" to 10" thick, pumped	C-20	160	.400			15.75	4.85	20.60	29.50
1550	With crane and bucket	C-7	110	.655			26	11.15	37.15	52.50
1600	Slabs over 10" thick, pumped	C-20	180	.356			14	4.31	18.31	26
1650	With crane and bucket	C-7	130	.554			22	9.45	31.45	44
1900	Footings, continuous, shallow, direct chute	C-6	120	.400			15.30	.55	15.85	24
1950	Pumped	C-20	150	.427			16.80	5.20	22	31
2000	With crane and bucket	C-7	90	.800			32	13.60	45.60	63.50
2100	Footings, continuous, deep, direct chute	C-6	140	.343			13.10	.47	13.57	20.50
2150	Pumped	C-20	160	.400			15.75	4.85	20.60	29.50
2200	With crane and bucket	C-7	110	.655			26	11.15	37.15	52.50
2400	Footings, spread, under 1 C.Y., direct chute	C-6	55	.873			33.50	1.20	34.70	52.50
2450	Pumped	C-20	65	.985			39	11.95	50.95	72.50
2500	With crane and bucket	C-7	45	1.600			63.50	27.50	91	128
2600	Over 5 C.Y., direct chute	C-6	120	.400			15.30	.55	15.85	24
2650	Pumped	C-20	150	.427			16.80	5.20	22	31
2700	With crane and bucket	C-7	100	.720			28.50	12.25	40.75	57.50
2900	Foundation mats, over 20 C.Y., direct chute	C-6	350	.137			5.25	.19	5.44	8.20
2950	Pumped	C-20	400	.160			6.30	1.94	8.24	11.80
3000	With crane and bucket	C-7	300	.240			9.55	4.09	13.64	19.10
3200	Grade beams, direct chute	C-6	150	.320			12.25	.44	12.69	19.20
3250	Pumped	C-20	180	.356			14	4.31	18.31	26
3300	With crane and bucket	C-7	120	.600			24	10.20	34.20	48
3500	High rise, for more than 5 stories, pumped, add per story	C-20	2100	.030			1.20	.37	1.57	2.25
3510	With crane and bucket, add per story	C-7	2100	.034			1.37	.58	1.95	2.73
3700	Pile caps, under 5 C.Y., direct chute	C-6	90	.533			20.50	.74	21.24	32
3750	Pumped	C-20	110	.582			23	7.05	30.05	43
3800	With crane and bucket	C-7	80	.900			36	15.35	51.35	72
3850	Pile cap, 5 C.Y. to 10 C.Y., direct chute	C-6	175	.274			10.50	.38	10.88	16.45
3900	Pumped	C-20	200	.320			12.60	3.88	16.48	23.50
3950	With crane and bucket	C-7	150	.480			19.10	8.15	27.25	38
4000	Over 10 C.Y., direct chute	C-6	215	.223			8.55	.31	8.86	13.40
4050	Pumped	C-20	240	.267			10.50	3.23	13.73	19.60
4100	With crane and bucket	C-7	185	.389			15.50	6.65	22.15	31
4300	Slab on grade, up to 6" thick, direct chute	C-6	110	.436			16.70	.60	17.30	26
4350	Pumped	C-20	130	.492			19.40	5.95	25.35	36
4400	With crane and bucket	C-7	110	.655			26	11.15	37.15	52.50

03 31 Structural Concrete

03 31 13 – Heavyweight Structural Concrete

03 31 13.70 Placing Concrete		Crew	Daily Output	Labor-Hours	Unit	2014 Bare Costs Material	Labor	Equipment	Total	Total Incl O&P
4600	Over 6" thick, direct chute	C-6	165	.291	C.Y.		11.10	.40	11.50	17.45
4650	Pumped	C-20	185	.346			13.60	4.20	17.80	25.50
4700	With crane and bucket	C-7	145	.497			19.75	8.45	28.20	39.50
4900	Walls, 8" thick, direct chute	C-6	90	.533			20.50	.74	21.24	32
4950	Pumped	C-20	100	.640			25	7.75	32.75	47
5000	With crane and bucket	C-7	80	.900			36	15.35	51.35	72
5050	12" thick, direct chute	C-6	100	.480			18.35	.66	19.01	28.50
5100	Pumped	C-20	110	.582			23	7.05	30.05	43
5200	With crane and bucket	C-7	90	.800			32	13.60	45.60	63.50
5300	15" thick, direct chute	C-6	105	.457			17.45	.63	18.08	27
5350	Pumped	C-20	120	.533			21	6.45	27.45	39
5400	With crane and bucket	C-7	95	.758			30	12.90	42.90	60
5600	Wheeled concrete dumping, add to placing costs above									
5610	Walking cart, 50' haul, add	C-18	32	.281	C.Y.		10.35	1.88	12.23	18.05
5620	150' haul, add		24	.375			13.85	2.51	16.36	24.50
5700	250' haul, add		18	.500			18.45	3.35	21.80	32
5800	Riding cart, 50' haul, add	C-19	80	.113			4.15	1.26	5.41	7.80
5810	150' haul, add		60	.150			5.55	1.69	7.24	10.40
5900	250' haul, add		45	.200			7.35	2.25	9.60	13.85
6000	Concrete in-fill for pan-type metal stairs and landings. Manual placement									
6010	includes up to 50' horizontal haul from point of concrete discharge.									
6100	Stair pan treads, 2" deep									
6110	Flights in 1st floor level up/down from discharge point	C-8A	3200	.015	S.F.		.59		.59	.90
6120	2nd floor level		2500	.019			.76		.76	1.15
6130	3rd floor level		2000	.024			.95		.95	1.44
6140	4th floor level		1800	.027			1.05		1.05	1.60
6200	Intermediate stair landings, pan-type 4" deep									
6210	Flights in 1st floor level up/down from discharge point	C-8A	2000	.024	S.F.		.95		.95	1.44
6220	2nd floor level		1500	.032			1.26		1.26	1.92
6230	3rd floor level		1200	.040			1.58		1.58	2.40
6240	4th floor level		1000	.048			1.89		1.89	2.88

03 35 Concrete Finishing

03 35 13 – High-Tolerance Concrete Floor Finishing

03 35 13.30 Finishing Floors, High Tolerance		Crew	Daily Output	Labor-Hours	Unit	Material	Labor	Equipment	Total	Total Incl O&P
0010	**FINISHING FLOORS, HIGH TOLERANCE**									
0012	Finishing requires that concrete first be placed, struck off & consolidated									
0015	Basic finishing for various unspecified flatwork									
0100	Bull float only	C-10	4000	.006	S.F.		.25		.25	.37
0125	Bull float & manual float		2000	.012			.50		.50	.75
0150	Bull float, manual float, & broom finish, w/edging & joints		1850	.013			.54		.54	.81
0200	Bull float, manual float & manual steel trowel		1265	.019			.79		.79	1.18
0210	For specified Random Access Floors in ACI Classes 1, 2, 3 and 4 to achieve									
0215	Composite Overall Floor Flatness and Levelness values up to FF35/FL25									
0250	Bull float, machine float & machine trowel (walk-behind)	C-10C	1715	.014	S.F.		.58	.03	.61	.90
0300	Power screed, bull float, machine float & trowel (walk-behind)	C-10D	2400	.010			.42	.05	.47	.67
0350	Power screed, bull float, machine float & trowel (ride-on)	C-10E	4000	.006			.25	.07	.32	.44
0352	For specified Random Access Floors in ACI Classes 5, 6, 7 and 8 to achieve									
0354	Composite Overall Floor Flatness and Levelness values up to FF50/FL50									
0356	Add for two-dimensional restraightening after power float	C-10	6000	.004	S.F.		.17		.17	.25
0358	For specified Random or Defined Access Floors in ACI Class 9 to achieve									

03 35 Concrete Finishing

03 35 13 – High-Tolerance Concrete Floor Finishing

03 35 13.30 Finishing Floors, High Tolerance		Crew	Daily Output	Labor-Hours	Unit	Material	2014 Bare Costs Labor	Equipment	Total	Total Incl O&P
0360	Composite Overall Floor Flatness and Levelness values up to FF100/FL100									
0362	Add for two-dimensional restraightening after bull float & power float	C-10	3000	.008	S.F.		.33		.33	.50
0364	For specified Superflat Defined Access Floors in ACI Class 9 to achieve									
0366	Minimum Floor Flatness and Levelness values of FF100/FL100									
0368	Add for 2-dim'l restraightening after bull float, power float, power trowel	C-10	2000	.012	S.F.		.50		.50	.75

03 35 16 – Heavy-Duty Concrete Floor Finishing

03 35 16.30 Finishing Floors, Heavy-Duty

		Crew	Daily Output	Labor-Hours	Unit	Material	Labor	Equipment	Total	Total Incl O&P
0010	**FINISHING FLOORS, HEAVY-DUTY**									
1800	Floor abrasives, .25 psf, aluminum oxide	1 Cefi	850	.009	S.F.	.51	.41		.92	1.17
1850	Silicon carbide		850	.009		.77	.41		1.18	1.46
2000	Floor hardeners, metallic, light service, .50 psf, add		850	.009		.51	.41		.92	1.17
2050	Medium service, .75 psf		750	.011		.77	.47		1.24	1.53
2100	Heavy service, 1.0 psf		650	.012		1.02	.54		1.56	1.92
2150	Extra heavy, 1.5 psf		575	.014		1.53	.61		2.14	2.59
2300	Non-metallic, light service, .50 psf		850	.009		.20	.41		.61	.83
2350	Medium service, .75 psf		750	.011		.31	.47		.78	1.03
2400	Heavy service, 1.00 psf		650	.012		.41	.54		.95	1.25
2450	Extra heavy, 1.50 psf		575	.014		.61	.61		1.22	1.58
2800	Trap rock wearing surface for monolithic floors									
2810	2.0 psf	C-10B	1250	.032	S.F.	.02	1.27	.23	1.52	2.19
3800	Dustproofing, solvent-based, 1 coat	1 Cefi	1900	.004		.18	.19		.37	.46
3850	2 coats		1300	.006		.63	.27		.90	1.10
4000	Epoxy-based, 1 coat		1500	.005		.14	.23		.37	.50
4050	2 coats		1500	.005		.28	.23		.51	.66

03 35 19 – Colored Concrete Finishing

03 35 19.30 Finishing Floors, Colored

		Crew	Daily Output	Labor-Hours	Unit	Material	Labor	Equipment	Total	Total Incl O&P
0010	**FINISHING FLOORS, COLORED**									
3000	Floor coloring, dusted on (0.6 psf), add to above	1 Cefi	1300	.006	S.F.	.44	.27		.71	.88
3050	(1.0 psf), add to above	"	625	.013	"	.73	.56		1.29	1.64
3100	Colored powder only				Lb.	.73			.73	.81
3600	1/2" topping using 0.6 psf powdered color	C-10B	590	.068	S.F.	4.81	2.69	.48	7.98	9.90
3650	1/2" topping using 1.0 psf powdered color	"	590	.068	"	5.10	2.69	.48	8.27	10.20

03 35 23 – Exposed Aggregate Concrete Finishing

03 35 23.30 Finishing Floors, Exposed Aggregate

		Crew	Daily Output	Labor-Hours	Unit	Material	Labor	Equipment	Total	Total Incl O&P
0010	**FINISHING FLOORS, EXPOSED AGGREGATE**									
1600	Exposed local aggregate finish, 3 lb. per S.F.	1 Cefi	625	.013	S.F.	.43	.56		.99	1.30
1650	4 lb. per S.F.	"	465	.017	"	.81	.76		1.57	2.01

03 35 29 – Tooled Concrete Finishing

03 35 29.30 Finishing Floors, Tooled

		Crew	Daily Output	Labor-Hours	Unit	Material	Labor	Equipment	Total	Total Incl O&P
0010	**FINISHING FLOORS, TOOLED**									
4400	Stair finish, float	1 Cefi	275	.029	S.F.		1.28		1.28	1.89
4500	Steel trowel finish		200	.040			1.76		1.76	2.60
4600	Silicon carbide finish, .25 psf		150	.053		.51	2.35		2.86	4.03

03 35 29.60 Finishing Walls

		Crew	Daily Output	Labor-Hours	Unit	Material	Labor	Equipment	Total	Total Incl O&P
0010	**FINISHING WALLS**									
0020	Break ties and patch voids	1 Cefi	540	.015	S.F.	.04	.65		.69	1
0050	Burlap rub with grout		450	.018		.04	.78		.82	1.20
0100	Carborundum rub, dry		270	.030			1.31		1.31	1.93
0150	Wet rub		175	.046			2.01		2.01	2.98
0300	Bush hammer, green concrete	B-39	1000	.048			1.86	.23	2.09	3.12

03 35 Concrete Finishing

03 35 29 – Tooled Concrete Finishing

03 35 29.60 Finishing Walls		Crew	Daily Output	Labor-Hours	Unit	Material	2014 Bare Costs Labor	Equipment	Total	Total Incl O&P
0350	Cured concrete	B-39	650	.074	S.F.		2.86	.36	3.22	4.80
0500	Acid etch	1 Cefi	575	.014		.14	.61		.75	1.06
0600	Float finish, 1/16" thick	"	300	.027		.36	1.17		1.53	2.13
0700	Sandblast, light penetration	E-11	1100	.029		.50	1.20	.22	1.92	2.78
0750	Heavy penetration	"	375	.085		1	3.51	.63	5.14	7.65
0850	Grind form fins flush	1 Clab	700	.011	L.F.		.42		.42	.65

03 35 33 – Stamped Concrete Finishing

03 35 33.50 Slab Texture Stamping		Crew	Daily Output	Labor-Hours	Unit	Material	Labor	Equipment	Total	Total Incl O&P
0010	**SLAB TEXTURE STAMPING**									
0050	Stamping requires that concrete first be placed, struck off, consolidated,									
0060	bull floated and free of bleed water. Decorative stamping tasks include:									
0100	Step 1 - first application of dry shake colored hardener	1 Cefi	6400	.001	S.F.	.43	.06		.49	.55
0110	Step 2 - bull float		6400	.001			.06		.06	.08
0130	Step 3 - second application of dry shake colored hardener		6400	.001		.21	.06		.27	.31
0140	Step 4 - bull float, manual float & steel trowel	3 Cefi	1280	.019			.83		.83	1.22
0150	Step 5 - application of dry shake colored release agent	1 Cefi	6400	.001		.10	.06		.16	.19
0160	Step 6 - place, tamp & remove mats	3 Cefi	2400	.010		1.43	.44		1.87	2.23
0170	Step 7 - touch up edges, mat joints & simulated grout lines	1 Cefi	1280	.006			.28		.28	.41
0300	Alternate stamping estimating method includes all tasks above	4 Cefi	800	.040		2.17	1.76		3.93	4.98
0400	Step 8 - pressure wash @ 3000 psi after 24 hours	1 Cefi	1600	.005			.22		.22	.33
0500	Step 9 - roll 2 coats cure/seal compound when dry	"	800	.010		.59	.44		1.03	1.30

03 37 Specialty Placed Concrete

03 37 13 – Shotcrete

03 37 13.30 Gunite (Dry-Mix)		Crew	Daily Output	Labor-Hours	Unit	Material	Labor	Equipment	Total	Total Incl O&P
0010	**GUNITE (DRY-MIX)**									
0020	Typical in place, 1" layers, no mesh included	C-16	2000	.028	S.F.	.32	1.14	.20	1.66	2.30
0100	Mesh for gunite 2 x 2, #12	2 Rodm	800	.020		.63	1.01		1.64	2.28
0150	#4 reinforcing bars @ 6" each way	"	500	.032		1.67	1.62		3.29	4.39
0300	Typical in place, including mesh, 2" thick, flat surfaces	C-16	1000	.056		1.26	2.28	.39	3.93	5.30
0350	Curved surfaces		500	.112		1.26	4.57	.79	6.62	9.20
0500	4" thick, flat surfaces		750	.075		1.90	3.05	.52	5.47	7.30
0550	Curved surfaces		350	.160		1.90	6.55	1.12	9.57	13.20
0900	Prepare old walls, no scaffolding, good condition	C-10	1000	.024			1		1	1.49
0950	Poor condition	"	275	.087			3.63		3.63	5.45
1100	For high finish requirement or close tolerance, add						50%			
1150	Very high						110%			

03 37 13.60 Shotcrete (Wet-Mix)		Crew	Daily Output	Labor-Hours	Unit	Material	Labor	Equipment	Total	Total Incl O&P
0010	**SHOTCRETE (WET-MIX)**									
0020	Wet mix, placed @ up to 12 C.Y. per hour, 3000 psi	C-8C	80	.600	C.Y.	103	24	5.70	132.70	156
0100	Up to 35 C.Y. per hour	C-8E	240	.200	"	92.50	8	2.23	102.73	117
1010	Fiber reinforced, 1" thick	C-8C	1740	.028	S.F.	.76	1.11	.26	2.13	2.82
1020	2" thick		900	.053		1.53	2.15	.51	4.19	5.50
1030	3" thick		825	.058		2.29	2.34	.55	5.18	6.70
1040	4" thick		750	.064		3.06	2.58	.61	6.25	7.95

03 37 23 – Roller-Compacted Concrete

03 37 23.50 Concrete, Roller-Compacted		Crew	Daily Output	Labor-Hours	Unit	Material	Labor	Equipment	Total	Total Incl O&P
0010	**CONCRETE, ROLLER-COMPACTED**									
0100	Mass placement, 1' lift, 1' layer	B-10C	1280	.009	C.Y.		.42	1.36	1.78	2.13
0200	2' lift, 6" layer	"	1600	.008			.34	1.09	1.43	1.70

03 37 Specialty Placed Concrete

03 37 23 – Roller-Compacted Concrete

03 37 23.50 Concrete, Roller-Compacted		Crew	Daily Output	Labor-Hours	Unit	Material	2014 Bare Costs Labor	Equipment	Total	Total Incl O&P
0210	Vertical face, formed, 1′ lift	B-11V	400	.060	C.Y.		2.20	.46	2.66	3.89
0220	6″ lift	″	200	.120			4.40	.91	5.31	7.80
0300	Sloped face, nonformed, 1′ lift	B-11L	384	.042			1.78	1.84	3.62	4.75
0360	6″ lift	″	192	.083			3.56	3.68	7.24	9.50
0400	Surface preparation, vacuum truck	B-6A	3280	.006	S.Y.		.26	.11	.37	.51
0450	Water clean	B-9A	3000	.008			.30	.17	.47	.64
0460	Water blast	B-9B	800	.030			1.11	.73	1.84	2.50
0500	Joint bedding placement, 1″ thick	B-11C	975	.016			.70	.37	1.07	1.48
0510	Conveyance of materials, 18 C.Y. truck, 5 min. cycle	B-34F	2048	.004	C.Y.		.15	.72	.87	1.01
0520	10 min. cycle		1024	.008			.29	1.44	1.73	2.03
0540	15 min. cycle		680	.012			.44	2.16	2.60	3.05
0550	With crane and bucket	C-23A	1600	.025			1.03	1.44	2.47	3.15
0560	With 4 C.Y. loader, 4 min. cycle	B-10U	480	.025			1.12	2.19	3.31	4.11
0570	8 min. cycle		240	.050			2.24	4.37	6.61	8.20
0580	12 min. cycle		160	.075			3.36	6.55	9.91	12.30
0590	With belt conveyor	C-7D	600	.093			3.61	.34	3.95	5.95
0600	With 17 C.Y. scraper, 5 min. cycle	B-33J	1440	.006			.27	1.27	1.54	1.81
0610	10 min. cycle		720	.011			.54	2.54	3.08	3.61
0620	15 min. cycle		480	.017			.82	3.81	4.63	5.45
0630	20 min. cycle		360	.022			1.09	5.10	6.19	7.25
0640	Water cure, small job, < 500 C.Y.	B-94C	8	1	Hr.		36.50	10.60	47.10	68
0650	Large job, over 500 C.Y.	B-59A	8	3	″		111	63.50	174.50	240
0660	RCC paving, with asphalt paver including material	B-25C	1000	.048	C.Y.	70	1.97	2.41	74.38	82.50
0670	8″ thick layers		4200	.011	S.Y.	18.35	.47	.57	19.39	21.50
0680	12″ thick layers		2800	.017	″	27.50	.70	.86	29.06	32
0700	Roller compacted concrete, 1.5″ - 2″ agg., 200 lb. cement/C.Y.				C.Y.	82.50			82.50	91

03 39 Concrete Curing

03 39 13 – Water Concrete Curing

03 39 13.50 Water Curing

		Crew	Daily Output	Labor-Hours	Unit	Material	Labor	Equipment	Total	Total Incl O&P
0010	**WATER CURING**									
0015	With burlap, 4 uses assumed, 7.5 oz.	2 Clab	55	.291	C.S.F.	14.70	10.65		25.35	32.50
0100	10 oz.	″	55	.291	″	26.50	10.65		37.15	45.50
0400	Curing blankets, 1″ to 2″ thick, buy				S.F.	.19			.19	.21

03 39 23 – Membrane Concrete Curing

03 39 23.13 Chemical Compound Membrane Concrete Curing

		Crew	Daily Output	Labor-Hours	Unit	Material	Labor	Equipment	Total	Total Incl O&P
0010	**CHEMICAL COMPOUND MEMBRANE CONCRETE CURING**									
0300	Sprayed membrane curing compound	2 Clab	95	.168	C.S.F.	10.95	6.15		17.10	21.50
0700	Curing compound, solvent based, 400 S.F./gal., 55 gallon lots				Gal.	23.50			23.50	25.50
0720	5 gallon lots					29.50			29.50	32.50
0800	Curing compound, water based, 250 S.F./gal., 55 gallon lots					22			22	24.50
0820	5 gallon lots					25			25	27.50

03 39 23.23 Sheet Membrane Concrete Curing

		Crew	Daily Output	Labor-Hours	Unit	Material	Labor	Equipment	Total	Total Incl O&P
0010	**SHEET MEMBRANE CONCRETE CURING**									
0200	Curing blanket, burlap/poly, 2-ply	2 Clab	70	.229	C.S.F.	19.05	8.40		27.45	34

03 41 Precast Structural Concrete

03 41 16 – Precast Concrete Slabs

03 41 16.20 Precast Concrete Channel Slabs		Crew	Daily Output	Labor-Hours	Unit	Material	2014 Bare Costs Labor	Equipment	Total	Total Incl O&P
0010	**PRECAST CONCRETE CHANNEL SLABS**									
0335	Lightweight concrete channel slab, long runs, 2-3/4" thick	C-12	1575	.030	S.F.	8	1.38	.42	9.80	11.40
0375	3-3/4" thick		1550	.031		8.30	1.41	.43	10.14	11.75
0475	4-3/4" thick		1525	.031		8.95	1.43	.43	10.81	12.55
1275	Short pieces, 2-3/4" thick		785	.061		12	2.78	.84	15.62	18.40
1375	3-3/4" thick		770	.062		12.45	2.83	.86	16.14	19
1475	4-3/4" thick		762	.063		13.45	2.86	.87	17.18	20

03 41 23 – Precast Concrete Stairs

03 41 23.50 Precast Stairs		Crew	Daily Output	Labor-Hours	Unit	Material	Labor	Equipment	Total	Total Incl O&P
0010	**PRECAST STAIRS**									
0020	Precast concrete treads on steel stringers, 3' wide	C-12	75	.640	Riser	134	29	8.85	171.85	201
0300	Front entrance, 5' wide with 48" platform, 2 risers		16	3	Flight	485	136	41.50	662.50	785
0350	5 risers		12	4		765	182	55	1,002	1,175
0500	6' wide, 2 risers		15	3.200		535	145	44	724	860
0550	5 risers		11	4.364		845	198	60.50	1,103.50	1,300
0700	7' wide, 2 risers		14	3.429		685	156	47.50	888.50	1,050
0750	5 risers		10	4.800		1,125	218	66.50	1,409.50	1,650
1200	Basement entrance stairwell, 6 steps, incl. steel bulkhead door	B-51	22	2.182		1,450	80.50	12.30	1,542.80	1,725
1250	14 steps	"	11	4.364		2,400	161	24.50	2,585.50	2,900

03 41 33 – Precast Structural Pretensioned Concrete

03 41 33.10 Precast Beams		Crew	Daily Output	Labor-Hours	Unit	Material	Labor	Equipment	Total	Total Incl O&P
0010	**PRECAST BEAMS**									
0011	L-shaped, 20' span, 12" x 20"	C-11	32	2.250	Ea.	2,900	113	57.50	3,070.50	3,450
0060	18" x 36"		24	3		3,975	151	76.50	4,202.50	4,725
0100	24" x 44"		22	3.273		4,775	165	83.50	5,023.50	5,625
0150	30' span, 12" x 36"		24	3		5,375	151	76.50	5,602.50	6,275
0200	18" x 44"		20	3.600		6,550	181	92	6,823	7,600
0250	24" x 52"		16	4.500		7,850	227	115	8,192	9,175
0400	40' span, 12" x 52"		20	3.600		8,350	181	92	8,623	9,575
0450	18" x 52"		16	4.500		9,325	227	115	9,667	10,800
0500	24" x 52"		12	6		10,500	300	153	10,953	12,200
1200	Rectangular, 20' span, 12" x 20"		32	2.250		2,825	113	57.50	2,995.50	3,350
1250	18" x 36"		24	3		3,500	151	76.50	3,727.50	4,200
1300	24" x 44"		22	3.273		4,175	165	83.50	4,423.50	4,975
1400	30' span, 12" x 36"		24	3		4,700	151	76.50	4,927.50	5,525
1450	18" x 44"		20	3.600		5,675	181	92	5,948	6,625
1500	24" x 52"		16	4.500		6,850	227	115	7,192	8,050
1600	40' span, 12" x 52"		20	3.600		6,975	181	92	7,248	8,075
1650	18" x 52"		16	4.500		7,950	227	115	8,292	9,275
1700	24" x 52"		12	6		9,125	300	153	9,578	10,700
2000	"T" shaped, 20' span, 12" x 20"		32	2.250		3,325	113	57.50	3,495.50	3,900
2050	18" x 36"		24	3		4,450	151	76.50	4,677.50	5,250
2100	24" x 44"		22	3.273		5,350	165	83.50	5,598.50	6,250
2200	30' span, 12" x 36"		24	3		6,125	151	76.50	6,352.50	7,075
2250	18" x 44"		20	3.600		7,425	181	92	7,698	8,550
2300	24" x 52"		16	4.500		8,875	227	115	9,217	10,300
2500	40' span, 12" x 52"		20	3.600		10,100	181	92	10,373	11,500
2550	18" x 52"		16	4.500		10,700	227	115	11,042	12,300
2600	24" x 52"		12	6		11,800	300	153	12,253	13,700

03 41 Precast Structural Concrete

03 41 33 – Precast Structural Pretensioned Concrete

03 41 33.15 Precast Columns		Crew	Daily Output	Labor-Hours	Unit	2014 Bare Costs Material	Labor	Equipment	Total	Total Incl O&P
0010	**PRECAST COLUMNS**									
0020	Rectangular to 12′ high, 16″ x 16″	C-11	120	.600	L.F.	166	30	15.30	211.30	252
0050	24″ x 24″		96	.750		228	38	19.15	285.15	335
0300	24′ high, 28″ x 28″		192	.375		265	18.90	9.55	293.45	335
0350	36″ x 36″		144	.500		355	25	12.75	392.75	445
0700	24′ high, 1 haunch, 12″ x 12″		32	2.250	Ea.	3,500	113	57.50	3,670.50	4,100
0800	20″ x 20″		28	2.571	″	4,825	130	65.50	5,020.50	5,600

03 41 33.25 Precast Joists

		Crew	Daily Output	Labor-Hours	Unit	Material	Labor	Equipment	Total	Total Incl O&P
0010	**PRECAST JOISTS**									
0015	40 psf L.L., 6″ deep for 12′ spans	C-12	600	.080	L.F.	23.50	3.63	1.10	28.23	32.50
0050	8″ deep for 16′ spans		575	.083		38.50	3.79	1.15	43.44	49.50
0100	10″ deep for 20′ spans		550	.087		67.50	3.96	1.21	72.67	82
0150	12″ deep for 24′ spans		525	.091		93	4.15	1.26	98.41	110

03 41 33.60 Precast Tees

		Crew	Daily Output	Labor-Hours	Unit	Material	Labor	Equipment	Total	Total Incl O&P
0010	**PRECAST TEES**									
0020	Quad tee, short spans, roof	C-11	7200	.010	S.F.	7.65	.50	.26	8.41	9.55
0050	Floor		7200	.010		7.65	.50	.26	8.41	9.55
0200	Double tee, floor members, 60′ span		8400	.009		9.10	.43	.22	9.75	11
0250	80′ span		8000	.009		11.80	.45	.23	12.48	14.05
0300	Roof members, 30′ span		4800	.015		7.65	.76	.38	8.79	10.10
0350	50′ span		6400	.011		8.40	.57	.29	9.26	10.55
0400	Wall members, up to 55′ high		3600	.020		11.35	1.01	.51	12.87	14.80
0500	Single tee roof members, 40′ span		3200	.023		11.30	1.13	.57	13	14.95
0550	80′ span		5120	.014		11.70	.71	.36	12.77	14.45
0600	100′ span		6000	.012		17.15	.60	.31	18.06	20
0650	120′ span		6000	.012		18.80	.60	.31	19.71	22
1000	Double tees, floor members									
1100	Lightweight, 20″ x 8′ wide, 45′ span	C-11	20	3.600	Ea.	3,025	181	92	3,298	3,725
1150	24″ x 8′ wide, 50′ span		18	4		3,350	202	102	3,654	4,150
1200	32″ x 10′ wide, 60′ span		16	4.500		5,050	227	115	5,392	6,075
1250	Standard weight, 12″ x 8′ wide, 20′ span		22	3.273		1,225	165	83.50	1,473.50	1,725
1300	16″ x 8′ wide, 25′ span		20	3.600		1,525	181	92	1,798	2,075
1350	18″ x 8′ wide, 30′ span		20	3.600		1,825	181	92	2,098	2,425
1400	20″ x 8′ wide, 45′ span		18	4		2,750	202	102	3,054	3,475
1450	24″ x 8′ wide, 50′ span		16	4.500		3,050	227	115	3,392	3,875
1500	32″ x 10′ wide, 60′ span		14	5.143		4,575	259	131	4,965	5,650
2000	Roof members									
2050	Lightweight, 20″ x 8′ wide, 40′ span	C-11	20	3.600	Ea.	2,700	181	92	2,973	3,350
2100	24″ x 8′ wide, 50′ span		18	4		3,350	202	102	3,654	4,150
2150	32″ x 10′ wide, 60′ span		16	4.500		5,050	227	115	5,392	6,075
2200	Standard weight, 12″ x 8′ wide, 30′ span		22	3.273		1,825	165	83.50	2,073.50	2,400
2250	16″ x 8′ wide, 30′ span		20	3.600		1,925	181	92	2,198	2,525
2300	18″ x 8′ wide, 30′ span		20	3.600		2,025	181	92	2,298	2,625
2350	20″ x 8′ wide, 40′ span		18	4		2,450	202	102	2,754	3,150
2400	24″ x 8′ wide, 50′ span		16	4.500		3,050	227	115	3,392	3,875
2450	32″ x 10′ wide, 60′ span		14	5.143		4,575	259	131	4,965	5,650

03 47 Site-Cast Concrete

03 47 13 – Tilt-Up Concrete

03 47 13.50 Tilt-Up Wall Panels		Crew	Daily Output	Labor-Hours	Unit	2014 Bare Costs Material	Labor	Equipment	Total	Total Incl O&P
0010	**TILT-UP WALL PANELS**									
0015	Wall panel construction, walls only, 5-1/2" thick	C-14	1600	.090	S.F.	5.40	4.04	1.03	10.47	13.30
0100	7-1/2" thick		1550	.093		6.75	4.17	1.07	11.99	14.95
0500	Walls and columns, 5-1/2" thick walls, 12" x 12" columns		1565	.092		8.15	4.13	1.06	13.34	16.45
0550	7-1/2" thick wall, 12" x 12" columns		1370	.105	↓	9.95	4.72	1.21	15.88	19.55
0800	Columns only, site precast, 12" x 12"		200	.720	L.F.	19.65	32.50	8.25	60.40	80.50
0850	16" x 16"	↓	105	1.371	"	28.50	61.50	15.75	105.75	144

03 48 Precast Concrete Specialties

03 48 43 – Precast Concrete Trim

03 48 43.90 Precast Window Sills

		Crew	Daily Output	Labor-Hours	Unit	Material	Labor	Equipment	Total	Total Incl O&P
0010	**PRECAST WINDOW SILLS**									
0600	Precast concrete, 4" tapers to 3", 9" wide	D-1	70	.229	L.F.	11.90	9.45		21.35	27.50
0650	11" wide		60	.267		15.55	11		26.55	34
0700	13" wide, 3 1/2" tapers to 2 1/2", 12" wall	↓	50	.320	↓	15.50	13.20		28.70	37

03 52 Lightweight Concrete Roof Insulation

03 52 16 – Lightweight Insulating Concrete

03 52 16.13 Lightweight Cellular Insulating Concrete

		Crew	Daily Output	Labor-Hours	Unit	Material	Labor	Equipment	Total	Total Incl O&P
0010	**LIGHTWEIGHT CELLULAR INSULATING CONCRETE**									
0020	Portland cement and foaming agent G	C-8	50	1.120	C.Y.	115	45.50	14.20	174.70	211

03 52 16.16 Lightweight Aggregate Insulating Concrete

		Crew	Daily Output	Labor-Hours	Unit	Material	Labor	Equipment	Total	Total Incl O&P
0010	**LIGHTWEIGHT AGGREGATE INSULATING CONCRETE**									
0100	Poured vermiculite or perlite, field mix,									
0110	1:6 field mix G	C-8	50	1.120	C.Y.	251	45.50	14.20	310.70	360
0200	Ready mix, 1:6 mix, roof fill, 2" thick G		10000	.006	S.F.	1.39	.23	.07	1.69	1.96
0250	3" thick G	↓	7700	.007		2.09	.30	.09	2.48	2.85
0400	Expanded volcanic glass rock, 1" thick G	2 Carp	1500	.011		.55	.49		1.04	1.35
0450	3" thick G	"	1200	.013	↓	1.64	.61		2.25	2.75

03 62 Non-Shrink Grouting

03 62 13 – Non-Metallic Non-Shrink Grouting

03 62 13.50 Grout, Non-Metallic Non-shrink

		Crew	Daily Output	Labor-Hours	Unit	Material	Labor	Equipment	Total	Total Incl O&P
0010	**GROUT, NON-METALLIC NON-SHRINK**									
0300	Non-shrink, non-metallic, 1" deep	1 Cefi	35	.229	S.F.	6.15	10.05		16.20	21.50
0350	2" deep	"	25	.320	"	12.30	14.10		26.40	34.50

03 62 16 – Metallic Non-Shrink Grouting

03 62 16.50 Grout, Metallic Non-Shrink

		Crew	Daily Output	Labor-Hours	Unit	Material	Labor	Equipment	Total	Total Incl O&P
0010	**GROUT, METALLIC NON-SHRINK**									
0020	Column & machine bases, non-shrink, metallic, 1" deep	1 Cefi	35	.229	S.F.	8.50	10.05		18.55	24.50
0050	2" deep	"	25	.320	"	16.95	14.10		31.05	39.50

03 63 Epoxy Grouting

03 63 05 – Grouting of Dowels and Fasteners

03 63 05.10 Epoxy Only		Crew	Daily Output	Labor-Hours	Unit	2014 Bare Costs Material	Labor	Equipment	Total	Total Incl O&P
0010	**EPOXY ONLY**									
1500	Chemical anchoring, epoxy cartridge, excludes layout, drilling, fastener									
1530	For fastener 3/4" diam. x 6" embedment	2 Skwk	72	.222	Ea.	5.55	10.50		16.05	22.50
1535	1" diam. x 8" embedment		66	.242		8.30	11.45		19.75	27
1540	1-1/4" diam. x 10" embedment		60	.267		16.60	12.60		29.20	38
1545	1-3/4" diam. x 12" embedment		54	.296		27.50	14		41.50	52
1550	14" embedment		48	.333		33	15.75		48.75	61
1555	2" diam. x 12" embedment		42	.381		44.50	18		62.50	76.50
1560	18" embedment		32	.500		55.50	23.50		79	97.50

03 81 Concrete Cutting

03 81 13 – Flat Concrete Sawing

03 81 13.50 Concrete Floor/Slab Cutting

		Crew	Daily Output	Labor-Hours	Unit	Material	Labor	Equipment	Total	Total Incl O&P
0010	**CONCRETE FLOOR/SLAB CUTTING**									
0050	Includes blade cost, layout and set-up time									
0300	Saw cut concrete slabs, plain, up to 3" deep	B-89	1060	.015	L.F.	.13	.63	.50	1.26	1.66
0320	Each additional inch of depth		3180	.005		.04	.21	.17	.42	.55
0400	Mesh reinforced, up to 3" deep		980	.016		.15	.68	.54	1.37	1.79
0420	Each additional inch of depth		2940	.005		.05	.23	.18	.46	.60
0500	Rod reinforced, up to 3" deep		800	.020		.19	.84	.66	1.69	2.19
0520	Each additional inch of depth		2400	.007		.06	.28	.22	.56	.73

03 81 13.75 Concrete Saw Blades

		Crew	Daily Output	Labor-Hours	Unit	Material	Labor	Equipment	Total	Total Incl O&P
0010	**CONCRETE SAW BLADES**									
3000	Blades for saw cutting, included in cutting line items									
3020	Diamond, 12" diameter				Ea.	244			244	269
3040	18" diameter					445			445	490
3080	24" diameter					755			755	830
3120	30" diameter					1,100			1,100	1,200
3160	36" diameter					1,425			1,425	1,575
3200	42" diameter					2,700			2,700	2,975

03 81 16 – Track Mounted Concrete Wall Sawing

03 81 16.50 Concrete Wall Cutting

		Crew	Daily Output	Labor-Hours	Unit	Material	Labor	Equipment	Total	Total Incl O&P
0010	**CONCRETE WALL CUTTING**									
0750	Includes blade cost, layout and set-up time									
0800	Concrete walls, hydraulic saw, plain, per inch of depth	B-89B	250	.064	L.F.	.04	2.67	3.64	6.35	8.10
0820	Rod reinforcing, per inch of depth	"	150	.107	"	.06	4.46	6.05	10.57	13.50

03 82 Concrete Boring

03 82 13 – Concrete Core Drilling

03 82 13.10 Core Drilling

		Crew	Daily Output	Labor-Hours	Unit	Material	Labor	Equipment	Total	Total Incl O&P
0010	**CORE DRILLING**									
0015	Includes bit cost, layout and set-up time									
0020	Reinforced concrete slab, up to 6" thick									
0100	1" diameter core	B-89A	17	.941	Ea.	.18	39.50	6.85	46.53	68.50
0150	For each additional inch of slab thickness in same hole, add		1440	.011		.03	.47	.08	.58	.84
0200	2" diameter core		16.50	.970		.29	40.50	7.05	47.84	71
0250	For each additional inch of slab thickness in same hole, add		1080	.015		.05	.62	.11	.78	1.13
0300	3" diameter core		16	1		.39	42	7.25	49.64	73.50

03 82 13 – Concrete Core Drilling

03 82 13.10 Core Drilling		Crew	Daily Output	Labor-Hours	Unit	Material	2014 Bare Costs Labor	Equipment	Total	Total Incl O&P
0350	For each additional inch of slab thickness in same hole, add	B-89A	720	.022	Ea.	.07	.93	.16	1.16	1.69
0500	4″ diameter core		15	1.067		.50	45	7.75	53.25	78
0550	For each additional inch of slab thickness in same hole, add		480	.033		.08	1.40	.24	1.72	2.52
0700	6″ diameter core		14	1.143		.78	48	8.30	57.08	84
0750	For each additional inch of slab thickness in same hole, add		360	.044		.13	1.87	.32	2.32	3.38
0900	8″ diameter core		13	1.231		1.08	51.50	8.95	61.53	91
0950	For each additional inch of slab thickness in same hole, add		288	.056		.18	2.33	.40	2.91	4.25
1100	10″ diameter core		12	1.333		1.51	56	9.70	67.21	99
1150	For each additional inch of slab thickness in same hole, add		240	.067		.25	2.80	.48	3.53	5.15
1300	12″ diameter core		11	1.455		1.79	61	10.55	73.34	108
1350	For each additional inch of slab thickness in same hole, add		206	.078		.30	3.26	.56	4.12	6
1500	14″ diameter core		10	1.600		2.11	67	11.60	80.71	119
1550	For each additional inch of slab thickness in same hole, add		180	.089		.35	3.73	.65	4.73	6.85
1700	18″ diameter core		9	1.778		2.84	74.50	12.90	90.24	132
1750	For each additional inch of slab thickness in same hole, add		144	.111		.47	4.66	.81	5.94	8.60
1754	24″ diameter core		8	2		3.99	84	14.50	102.49	150
1756	For each additional inch of slab thickness in same hole, add		120	.133		.67	5.60	.97	7.24	10.45
1760	For horizontal holes, add to above						20%	20%		
1770	Prestressed hollow core plank, 8″ thick									
1780	1″ diameter core	B-89A	17.50	.914	Ea.	.24	38.50	6.65	45.39	67
1790	For each additional inch of plank thickness in same hole, add		3840	.004		.03	.18	.03	.24	.33
1794	2″ diameter core		17.25	.928		.38	39	6.75	46.13	68
1796	For each additional inch of plank thickness in same hole, add		2880	.006		.05	.23	.04	.32	.45
1800	3″ diameter core		17	.941		.52	39.50	6.85	46.87	69
1810	For each additional inch of plank thickness in same hole, add		1920	.008		.07	.35	.06	.48	.68
1820	4″ diameter core		16.50	.970		.66	40.50	7.05	48.21	71.50
1830	For each additional inch of plank thickness in same hole, add		1280	.013		.08	.52	.09	.69	1
1840	6″ diameter core		15.50	1.032		1.04	43.50	7.50	52.04	76.50
1850	For each additional inch of plank thickness in same hole, add		960	.017		.13	.70	.12	.95	1.35
1860	8″ diameter core		15	1.067		1.44	45	7.75	54.19	79
1870	For each additional inch of plank thickness in same hole, add		768	.021		.18	.87	.15	1.20	1.72
1880	10″ diameter core		14	1.143		2.01	48	8.30	58.31	85.50
1890	For each additional inch of plank thickness in same hole, add		640	.025		.25	1.05	.18	1.48	2.10
1900	12″ diameter core		13.50	1.185		2.39	50	8.60	60.99	89
1910	For each additional inch of plank thickness in same hole, add		548	.029		.30	1.23	.21	1.74	2.46
3000	Bits for core drilling, included in drilling line items									
3010	Diamond, premium, 1″ diameter				Ea.	72			72	79.50
3020	2″ diameter					115			115	127
3030	3″ diameter					157			157	173
3040	4″ diameter					199			199	219
3060	6″ diameter					315			315	345
3080	8″ diameter					435			435	475
3110	10″ diameter					605			605	665
3120	12″ diameter					715			715	790
3140	14″ diameter					845			845	930
3180	18″ diameter					1,125			1,125	1,250
3240	24″ diameter					1,600			1,600	1,750

03 82 Concrete Boring

03 82 16 – Concrete Drilling

03 82 16.10 Concrete Impact Drilling		Crew	Daily Output	Labor-Hours	Unit	Material	2014 Bare Costs Labor	Equipment	Total	Total Incl O&P
0010	**CONCRETE IMPACT DRILLING**									
0020	Includes bit cost, layout and set-up time, no anchors									
0050	Up to 4" deep in concrete/brick floors/walls									
0100	Holes, 1/4" diameter	1 Carp	75	.107	Ea.	.06	4.89		4.95	7.60
0150	For each additional inch of depth in same hole, add		430	.019		.02	.85		.87	1.34
0200	3/8" diameter		63	.127		.05	5.80		5.85	9.05
0250	For each additional inch of depth in same hole, add		340	.024		.01	1.08		1.09	1.67
0300	1/2" diameter		50	.160		.05	7.35		7.40	11.35
0350	For each additional inch of depth in same hole, add		250	.032		.01	1.47		1.48	2.27
0400	5/8" diameter		48	.167		.08	7.65		7.73	11.90
0450	For each additional inch of depth in same hole, add		240	.033		.02	1.53		1.55	2.38
0500	3/4" diameter		45	.178		.11	8.15		8.26	12.70
0550	For each additional inch of depth in same hole, add		220	.036		.03	1.67		1.70	2.60
0600	7/8" diameter		43	.186		.15	8.55		8.70	13.30
0650	For each additional inch of depth in same hole, add		210	.038		.04	1.75		1.79	2.74
0700	1" diameter		40	.200		.16	9.15		9.31	14.35
0750	For each additional inch of depth in same hole, add		190	.042		.04	1.93		1.97	3.02
0800	1-1/4" diameter		38	.211		.23	9.65		9.88	15.15
0850	For each additional inch of depth in same hole, add		180	.044		.06	2.04		2.10	3.20
0900	1-1/2" diameter		35	.229		.32	10.50		10.82	16.50
0950	For each additional inch of depth in same hole, add	▼	165	.048	▼	.08	2.22		2.30	3.52
1000	For ceiling installations, add						40%			

Estimating Tips

04 05 00 Common Work Results for Masonry

- The terms *mortar* and *grout* are often used interchangeably, and incorrectly. Mortar is used to bed masonry units, seal the entry of air and moisture, provide architectural appearance, and allow for size variations in the units. Grout is used primarily in reinforced masonry construction and is used to bond the masonry to the reinforcing steel. Common mortar types are M(2500 psi), S(1800 psi), N(750 psi), and O(350 psi), and conform to ASTM C270. Grout is either fine or coarse and conforms to ASTM C476, and in-place strengths generally exceed 2500 psi. Mortar and grout are different components of masonry construction and are placed by entirely different methods. An estimator should be aware of their unique uses and costs.
- Mortar is included in all assembled masonry line items. The mortar cost, part of the assembled masonry material cost, includes all ingredients, all labor, and all equipment required. Please see reference number R040513-10.
- Waste, specifically the loss/droppings of mortar and the breakage of brick and block, is included in all masonry assemblies in this division. A factor of 25% is added for mortar and 3% for brick and concrete masonry units.
- Scaffolding or staging is not included in any of the Division 4 costs. Refer to Subdivision 01 54 23 for scaffolding and staging costs.

04 20 00 Unit Masonry

- The most common types of unit masonry are brick and concrete masonry. The major classifications of brick are building brick (ASTM C62), facing brick (ASTM C216), glazed brick, fire brick, and pavers. Many varieties of texture and appearance can exist within these classifications, and the estimator would be wise to check local custom and availability within the project area. For repair and remodeling jobs, matching the existing brick may be the most important criteria.
- Brick and concrete block are priced by the piece and then converted into a price per square foot of wall. Openings less than two square feet are generally ignored by the estimator because any savings in units used is offset by the cutting and trimming required.
- It is often difficult and expensive to find and purchase small lots of historic brick. Costs can vary widely. Many design issues affect costs, selection of mortar mix, and repairs or replacement of masonry materials. Cleaning techniques must be reflected in the estimate.
- All masonry walls, whether interior or exterior, require bracing. The cost of bracing walls during construction should be included by the estimator, and this bracing must remain in place until permanent bracing is complete. Permanent bracing of masonry walls is accomplished by masonry itself, in the form of pilasters or abutting wall corners, or by anchoring the walls to the structural frame. Accessories in the form of anchors, anchor slots, and ties are used, but their supply and installation can be by different trades. For instance, anchor slots on spandrel beams and columns are supplied and welded in place by the steel fabricator, but the ties from the slots into the masonry are installed by the bricklayer. Regardless of the installation method, the estimator must be certain that these accessories are accounted for in pricing.

Reference Numbers

Reference numbers are shown in shaded boxes at the beginning of some major classifications. These numbers refer to related items in the Reference Section. The reference information may be an estimating procedure, an alternate pricing method, or technical information.

Note: Not all subdivisions listed here necessarily appear in this publication.

04 01 Maintenance of Masonry

04 01 20 – Maintenance of Unit Masonry

04 01 20.20 Pointing Masonry

	04 01 20.20 Pointing Masonry	Crew	Daily Output	Labor-Hours	Unit	2014 Bare Costs Material	Labor	Equipment	Total	Total Incl O&P
0010	**POINTING MASONRY**									
0300	Cut and repoint brick, hard mortar, running bond	1 Bric	80	.100	S.F.	.51	4.56		5.07	7.50
0320	Common bond		77	.104		.51	4.74		5.25	7.75
0360	Flemish bond		70	.114		.54	5.20		5.74	8.55
0400	English bond		65	.123		.54	5.60		6.14	9.15
0600	Soft old mortar, running bond		100	.080		.51	3.65		4.16	6.10
0620	Common bond		96	.083		.51	3.80		4.31	6.35
0640	Flemish bond		90	.089		.54	4.05		4.59	6.80
0680	English bond		82	.098		.54	4.45		4.99	7.40
0700	Stonework, hard mortar		140	.057	L.F.	.68	2.61		3.29	4.72
0720	Soft old mortar		160	.050	"	.68	2.28		2.96	4.23
1000	Repoint, mask and grout method, running bond		95	.084	S.F.	.68	3.84		4.52	6.60
1020	Common bond		90	.089		.68	4.05		4.73	6.95
1040	Flemish bond		86	.093		.72	4.24		4.96	7.25
1060	English bond		77	.104		.72	4.74		5.46	8
2000	Scrub coat, sand grout on walls, thin mix, brushed		120	.067		2.72	3.04		5.76	7.60
2020	Troweled		98	.082		3.79	3.72		7.51	9.80

04 01 20.40 Sawing Masonry

	04 01 20.40 Sawing Masonry	Crew	Daily Output	Labor-Hours	Unit	Material	Labor	Equipment	Total	Total Incl O&P
0010	**SAWING MASONRY**									
0050	Brick or block by hand, per inch depth	A-1	125	.064	L.F.	.06	2.35	.67	3.08	4.43

04 01 20.41 Unit Masonry Stabilization

	04 01 20.41 Unit Masonry Stabilization	Crew	Daily Output	Labor-Hours	Unit	Material	Labor	Equipment	Total	Total Incl O&P
0010	**UNIT MASONRY STABILIZATION**									
0100	Structural repointing method									
0110	Cut / grind mortar joint	1 Bric	240	.033	L.F.		1.52		1.52	2.32
0120	Clean and mask joint		2500	.003		.13	.15		.28	.36
0130	Epoxy paste and 1/4" FRP rod		240	.033		1.07	1.52		2.59	3.50
0132	3/8" FRP rod		160	.050		1.63	2.28		3.91	5.25
0134	1/4" CFRP rod		240	.033		13.90	1.52		15.42	17.55
0136	3/8" CFRP rod		160	.050		17.50	2.28		19.78	22.50
0140	Remove masking		14400	.001			.03		.03	.04
0300	Structural fabric method									
0310	Primer	1 Bric	600	.013	S.F.	.46	.61		1.07	1.43
0320	Apply filling/leveling paste		720	.011		.43	.51		.94	1.24
0330	Epoxy, glass fiber fabric		720	.011		5.65	.51		6.16	6.95
0340	Carbon fiber fabric		720	.011		15.45	.51		15.96	17.75

04 01 20.50 Toothing Masonry

	04 01 20.50 Toothing Masonry	Crew	Daily Output	Labor-Hours	Unit	Material	Labor	Equipment	Total	Total Incl O&P
0010	**TOOTHING MASONRY**									
0500	Brickwork, soft old mortar	1 Clab	40	.200	V.L.F.		7.35		7.35	11.30
0520	Hard mortar		30	.267			9.75		9.75	15.10
0700	Blockwork, soft old mortar		70	.114			4.19		4.19	6.45
0720	Hard mortar		50	.160			5.85		5.85	9.05

04 01 30 – Unit Masonry Cleaning

04 01 30.20 Cleaning Masonry

	04 01 30.20 Cleaning Masonry	Crew	Daily Output	Labor-Hours	Unit	Material	Labor	Equipment	Total	Total Incl O&P
0010	**CLEANING MASONRY**									
0200	By chemical, brush and rinse, new work, light construction dust	D-1	1000	.016	S.F.	.06	.66		.72	1.07
0220	Medium construction dust		800	.020		.08	.83		.91	1.35
0240	Heavy construction dust, drips or stains		600	.027		.11	1.10		1.21	1.80
0260	Low pressure wash and rinse, light restoration, light soil		800	.020		.12	.83		.95	1.40
0270	Average soil, biological staining		400	.040		.18	1.65		1.83	2.72
0280	Heavy soil, biological and mineral staining, paint		330	.048		.25	2		2.25	3.32
0300	High pressure wash and rinse, heavy restoration, light soil		600	.027		.11	1.10		1.21	1.80

04 01 Maintenance of Masonry

04 01 30 – Unit Masonry Cleaning

04 01 30.20 Cleaning Masonry		Crew	Daily Output	Labor-Hours	Unit	Material	2014 Bare Costs Labor	Equipment	Total	Total Incl O&P
0310	Average soil, biological staining	D-1	400	.040	S.F.	.16	1.65		1.81	2.70
0320	Heavy soil, biological and mineral staining, paint		250	.064		.22	2.64		2.86	4.27
0400	High pressure wash, water only, light soil	C-29	500	.016			.59	.14	.73	1.05
0420	Average soil, biological staining		375	.021			.78	.19	.97	1.41
0440	Heavy soil, biological and mineral staining, paint		250	.032			1.17	.28	1.45	2.12
0800	High pressure water and chemical, light soil		450	.018		.15	.65	.15	.95	1.34
0820	Average soil, biological staining		300	.027		.22	.98	.23	1.43	2
0840	Heavy soil, biological and mineral staining, paint		200	.040		.30	1.47	.35	2.12	2.97
1200	Sandblast, wet system, light soil	J-6	1750	.018		.33	.74	.14	1.21	1.65
1220	Average soil, biological staining		1100	.029		.50	1.18	.22	1.90	2.59
1240	Heavy soil, biological and mineral staining, paint		700	.046		.67	1.86	.34	2.87	3.93
1400	Dry system, light soil		2500	.013		.33	.52	.09	.94	1.26
1420	Average soil, biological staining		1750	.018		.50	.74	.14	1.38	1.83
1440	Heavy soil, biological and mineral staining, paint		1000	.032		.67	1.30	.24	2.21	2.98
1800	For walnut shells, add					.82			.82	.90
1820	For corn chips, add					.82			.82	.90
2000	Steam cleaning, light soil	A-1H	750	.011			.39	.10	.49	.71
2020	Average soil, biological staining		625	.013			.47	.12	.59	.85
2040	Heavy soil, biological and mineral staining		375	.021			.78	.20	.98	1.43
4000	Add for masking doors and windows	1 Clab	800	.010		.04	.37		.41	.61
4200	Add for pedestrian protection				Job				10%	10%

04 01 30.60 Brick Washing		Crew	Daily Output	Labor-Hours	Unit	Material	Labor	Equipment	Total	Total Incl O&P
0010	**BRICK WASHING**									
0012	Acid cleanser, smooth brick surface	1 Bric	560	.014	S.F.	.06	.65		.71	1.06
0050	Rough brick		400	.020		.09	.91		1	1.49
0060	Stone, acid wash		600	.013		.10	.61		.71	1.04
1000	Muriatic acid, price per gallon in 5 gallon lots				Gal.	13			13	14.30

04 05 Common Work Results for Masonry

04 05 05 – Selective Masonry Demolition

04 05 05.10 Selective Demolition		Crew	Daily Output	Labor-Hours	Unit	Material	Labor	Equipment	Total	Total Incl O&P
0010	**SELECTIVE DEMOLITION** R024119-10									
1000	Chimney, 16″ x 16″, soft old mortar	1 Clab	55	.145	C.F.		5.35		5.35	8.25
1020	Hard mortar		40	.200			7.35		7.35	11.30
1030	16″ x 20″, soft old mortar		55	.145			5.35		5.35	8.25
1040	Hard mortar		40	.200			7.35		7.35	11.30
1050	16″ x 24″, soft old mortar		55	.145			5.35		5.35	8.25
1060	Hard mortar		40	.200			7.35		7.35	11.30
1080	20″ x 20″, soft old mortar		55	.145			5.35		5.35	8.25
1100	Hard mortar		40	.200			7.35		7.35	11.30
1110	20″ x 24″, soft old mortar		55	.145			5.35		5.35	8.25
1120	Hard mortar		40	.200			7.35		7.35	11.30
1140	20″ x 32″, soft old mortar		55	.145			5.35		5.35	8.25
1160	Hard mortar		40	.200			7.35		7.35	11.30
1200	48″ x 48″, soft old mortar		55	.145			5.35		5.35	8.25
1220	Hard mortar		40	.200			7.35		7.35	11.30
1250	Metal, high temp steel jacket, 24″ diameter	E-2	130	.431	V.L.F.		21.50	11.75	33.25	49.50
1260	60″ diameter	″	60	.933			47	25.50	72.50	108
1280	Flue lining, up to 12″ x 12″	1 Clab	200	.040			1.47		1.47	2.26
1282	Up to 24″ x 24″		150	.053			1.95		1.95	3.02
2000	Columns, 8″ x 8″, soft old mortar		48	.167			6.10		6.10	9.45

04 05 Common Work Results for Masonry

04 05 05 – Selective Masonry Demolition

04 05 05.10 Selective Demolition		Crew	Daily Output	Labor-Hours	Unit	Material	2014 Bare Costs Labor	Equipment	Total	Total Incl O&P
2020	Hard mortar	1 Clab	40	.200	V.L.F.		7.35		7.35	11.30
2060	16" x 16", soft old mortar		16	.500			18.35		18.35	28.50
2100	Hard mortar		14	.571			21		21	32.50
2140	24" x 24", soft old mortar		8	1			36.50		36.50	56.50
2160	Hard mortar		6	1.333			49		49	75.50
2200	36" x 36", soft old mortar		4	2			73.50		73.50	113
2220	Hard mortar		3	2.667			97.50		97.50	151
2230	Alternate pricing method, soft old mortar		30	.267	C.F.		9.75		9.75	15.10
2240	Hard mortar		23	.348	"		12.75		12.75	19.65
3000	Copings, precast or masonry, to 8" wide									
3020	Soft old mortar	1 Clab	180	.044	L.F.		1.63		1.63	2.51
3040	Hard mortar	"	160	.050	"		1.83		1.83	2.83
3100	To 12" wide									
3120	Soft old mortar	1 Clab	160	.050	L.F.		1.83		1.83	2.83
3140	Hard mortar	"	140	.057	"		2.09		2.09	3.23
4000	Fireplace, brick, 30" x 24" opening									
4020	Soft old mortar	1 Clab	2	4	Ea.		147		147	226
4040	Hard mortar		1.25	6.400			235		235	360
4100	Stone, soft old mortar		1.50	5.333			195		195	300
4120	Hard mortar		1	8			293		293	450
5000	Veneers, brick, soft old mortar		140	.057	S.F.		2.09		2.09	3.23
5020	Hard mortar		125	.064			2.35		2.35	3.62
5050	Glass block, up to 4" thick		500	.016			.59		.59	.90
5100	Granite and marble, 2" thick		180	.044			1.63		1.63	2.51
5120	4" thick		170	.047			1.72		1.72	2.66
5140	Stone, 4" thick		180	.044			1.63		1.63	2.51
5160	8" thick		175	.046			1.68		1.68	2.58
5400	Alternate pricing method, stone, 4" thick		60	.133	C.F.		4.89		4.89	7.55
5420	8" thick		85	.094	"		3.45		3.45	5.30

04 05 13 – Masonry Mortaring

04 05 13.10 Cement

		Crew	Daily Output	Labor-Hours	Unit	Material	Labor	Equipment	Total	Total Incl O&P
0010	**CEMENT**									
0100	Masonry, 70 lb. bag, T.L. lots				Bag	11.05			11.05	12.15
0150	L.T.L. lots					11.70			11.70	12.90
0200	White, 70 lb. bag, T.L. lots					15.65			15.65	17.25
0250	L.T.L. lots					18.30			18.30	20

04 05 13.20 Lime

		Crew	Daily Output	Labor-Hours	Unit	Material	Labor	Equipment	Total	Total Incl O&P
0010	**LIME**									
0020	Masons, hydrated, 50 lb. bag, T.L. lots				Bag	10.10			10.10	11.10
0050	L.T.L. lots					11.10			11.10	12.20
0200	Finish, double hydrated, 50 lb. bag, T.L. lots					8.65			8.65	9.55
0250	L.T.L. lots					9.55			9.55	10.50

04 05 13.23 Surface Bonding Masonry Mortaring

		Crew	Daily Output	Labor-Hours	Unit	Material	Labor	Equipment	Total	Total Incl O&P
0010	**SURFACE BONDING MASONRY MORTARING**									
0020	Gray or white colors, not incl. block work	1 Bric	540	.015	S.F.	.13	.68		.81	1.18

04 05 13.30 Mortar

		Crew	Daily Output	Labor-Hours	Unit	Material	Labor	Equipment	Total	Total Incl O&P
0010	**MORTAR**									
0020	With masonry cement									
0100	Type M, 1:1:6 mix	1 Brhe	143	.056	C.F.	4.91	2.07		6.98	8.55
0200	Type N, 1:3 mix		143	.056		4.94	2.07		7.01	8.60
0300	Type O, 1:3 mix		143	.056		4.78	2.07		6.85	8.40

04 05 Common Work Results for Masonry

04 05 13 – Masonry Mortaring

04 05 13.30 Mortar		Crew	Daily Output	Labor-Hours	Unit	Material	Labor	Equipment	Total	Total Incl O&P
						2014 Bare Costs				
0400	Type PM, 1:1:6 mix, 2500 psi	1 Brhe	143	.056	C.F.	3.99	2.07		6.06	7.55
0500	Type S, 1/2:1:4 mix	↓	143	.056	↓	5.35	2.07		7.42	9.05
2000	With portland cement and lime									
2100	Type M, 1:1/4:3 mix	1 Brhe	143	.056	C.F.	8.45	2.07		10.52	12.45
2200	Type N, 1:1:6 mix, 750 psi		143	.056		6.60	2.07		8.67	10.40
2300	Type O, 1:2:9 mix (Pointing Mortar)		143	.056		7	2.07		9.07	10.85
2400	Type PL, 1:1/2:4 mix, 2500 psi		143	.056		3.99	2.07		6.06	7.55
2600	Type S, 1:1/2:4 mix, 1800 psi	↓	143	.056		7.75	2.07		9.82	11.65
2650	Pre-mixed, type S or N					5.05			5.05	5.55
2700	Mortar for glass block	1 Brhe	143	.056	↓	10.40	2.07		12.47	14.55
2900	Mortar for fire brick, dry mix, 10 lb. pail				Ea.	26			26	28.50

04 05 13.91 Masonry Restoration Mortaring

		Crew	Daily Output	Labor-Hours	Unit	Material	Labor	Equipment	Total	Total Incl O&P
0010	**MASONRY RESTORATION MORTARING**									
0020	Masonry restoration mix				Lb.	.23			.23	.25
0050	White				"	.23			.23	.25

04 05 13.93 Mortar Pigments

		Crew	Daily Output	Labor-Hours	Unit	Material	Labor	Equipment	Total	Total Incl O&P
0010	**MORTAR PIGMENTS**, 50 lb. bags (2 bags per M bricks)									
0020	Color admixture, range 2 to 10 lb. per bag of cement, light colors				Lb.	4.25			4.25	4.68
0050	Medium colors					6.20			6.20	6.80
0100	Dark colors				↓	12.90			12.90	14.20

04 05 13.95 Sand

		Crew	Daily Output	Labor-Hours	Unit	Material	Labor	Equipment	Total	Total Incl O&P
0010	**SAND**, screened and washed at pit									
0020	For mortar, per ton				Ton	17.40			17.40	19.15
0050	With 10 mile haul					32			32	35
0100	With 30 mile haul				↓	52			52	57
0200	Screened and washed, at the pit				C.Y.	24			24	26.50
0250	With 10 mile haul					44			44	48.50
0300	With 30 mile haul				↓	72			72	79.50

04 05 13.98 Mortar Admixtures

		Crew	Daily Output	Labor-Hours	Unit	Material	Labor	Equipment	Total	Total Incl O&P
0010	**MORTAR ADMIXTURES**									
0020	Waterproofing admixture, per quart (1 qt. to 2 bags of masonry cement)				Qt.	3.50			3.50	3.85

04 05 16 – Masonry Grouting

04 05 16.30 Grouting

		Crew	Daily Output	Labor-Hours	Unit	Material	Labor	Equipment	Total	Total Incl O&P
0010	**GROUTING**									
0011	Bond beams & lintels, 8" deep, 6" thick, 0.15 C.F. per L.F.	D-4	1480	.022	L.F.	.66	.90	.09	1.65	2.19
0020	8" thick, 0.2 C.F. per L.F.		1400	.023		1.04	.95	.10	2.09	2.70
0050	10" thick, 0.25 C.F. per L.F.		1200	.027		1.09	1.11	.11	2.31	3.01
0060	12" thick, 0.3 C.F. per L.F.	↓	1040	.031	↓	1.31	1.28	.13	2.72	3.53
0200	Concrete block cores, solid, 4" thk., by hand, 0.067 C.F./S.F. of wall	D-8	1100	.036	S.F.	.29	1.53		1.82	2.66
0210	6" thick, pumped, 0.175 C.F. per S.F.	D-4	720	.044		.76	1.85	.19	2.80	3.87
0250	8" thick, pumped, 0.258 C.F. per S.F.		680	.047		1.13	1.96	.20	3.29	4.44
0300	10" thick, pumped, 0.340 C.F. per S.F.		660	.048		1.49	2.02	.20	3.71	4.92
0350	12" thick, pumped, 0.422 C.F. per S.F.		640	.050		1.84	2.08	.21	4.13	5.45
0500	Cavity walls, 2" space, pumped, 0.167 C.F./S.F. of wall		1700	.019		.73	.78	.08	1.59	2.08
0550	3" space, 0.250 C.F./S.F.		1200	.027		1.09	1.11	.11	2.31	3.01
0600	4" space, 0.333 C.F. per S.F.		1150	.028		1.46	1.16	.12	2.74	3.49
0700	6" space, 0.500 C.F. per S.F.		800	.040	↓	2.19	1.67	.17	4.03	5.15
0800	Door frames, 3' x 7' opening, 2.5 C.F. per opening		60	.533	Opng.	10.95	22	2.25	35.20	48.50
0850	6' x 7' opening, 3.5 C.F. per opening		45	.711	"	15.30	29.50	2.99	47.79	65
2000	Grout, C476, for bond beams, lintels and CMU cores	↓	350	.091	C.F.	4.37	3.81	.38	8.56	11.05

04 05 19.05 Anchor Bolts		Crew	Daily Output	Labor-Hours	Unit	Material	2014 Bare Costs Labor	Equipment	Total	Total Incl O&P
0010	**ANCHOR BOLTS**									
0015	Installed in fresh grout in CMU bond beams or filled cores, no templates									
0020	Hooked, with nut and washer, 1/2" diam., 8" long	1 Bric	132	.061	Ea.	1.49	2.76		4.25	5.85
0030	12" long		131	.061		1.66	2.78		4.44	6.05
0040	5/8" diameter, 8" long		129	.062		3.25	2.83		6.08	7.90
0050	12" long		127	.063		4	2.87		6.87	8.80
0060	3/4" diameter, 8" long		127	.063		4	2.87		6.87	8.80
0070	12" long		125	.064		5	2.92		7.92	9.95

04 05 19.16 Masonry Anchors		Crew	Daily Output	Labor-Hours	Unit	Material	Labor	Equipment	Total	Total Incl O&P
0010	**MASONRY ANCHORS**									
0020	For brick veneer, galv., corrugated, 7/8" x 7", 22 Ga.	1 Bric	10.50	.762	C	13.45	34.50		47.95	68
0100	24 Ga.		10.50	.762		9.50	34.50		44	63.50
0150	16 Ga.		10.50	.762		26.50	34.50		61	82
0200	Buck anchors, galv., corrugated, 16 ga., 2" bend, 8" x 2"		10.50	.762		49	34.50		83.50	107
0250	8" x 3"		10.50	.762		51	34.50		85.50	109
0300	Adjustable, rectangular, 4-1/8" wide									
0350	Anchor and tie, 3/16" wire, mill galv.									
0400	2-3/4" eye, 3-1/4" tie	1 Bric	1.05	7.619	M	430	345		775	1,000
0500	4-3/4" tie		1.05	7.619		465	345		810	1,050
0520	5-1/2" tie		1.05	7.619		505	345		850	1,075
0550	4-3/4" eye, 3-1/4" tie		1.05	7.619		475	345		820	1,050
0570	4-3/4" tie		1.05	7.619		515	345		860	1,100
0580	5-1/2" tie		1.05	7.619		560	345		905	1,150
0660	Cavity wall, Z-type, galvanized, 6" long, 1/8" diam.		10.50	.762	C	23.50	34.50		58	79
0670	3/16" diameter		10.50	.762		27	34.50		61.50	82.50
0680	1/4" diameter		10.50	.762		40.50	34.50		75	97.50
0850	8" long, 3/16" diameter		10.50	.762		25	34.50		59.50	80.50
0855	1/4" diameter		10.50	.762		48.50	34.50		83	107
1000	Rectangular type, galvanized, 1/4" diameter, 2" x 6"		10.50	.762		70	34.50		104.50	130
1050	4" x 6"		10.50	.762		85	34.50		119.50	147
1100	3/16" diameter, 2" x 6"		10.50	.762		42	34.50		76.50	99.50
1150	4" x 6"		10.50	.762		48.50	34.50		83	106
1200	Mesh wall tie, 1/2" mesh, hot dip galvanized									
1400	16 ga., 12" long, 3" wide	1 Bric	9	.889	C	84.50	40.50		125	155
1420	6" wide		9	.889		122	40.50		162.50	197
1440	12" wide		8.50	.941		198	43		241	283
1500	Rigid partition anchors, plain, 8" long, 1" x 1/8"		10.50	.762		235	34.50		269.50	310
1550	1" x 1/4"		10.50	.762		277	34.50		311.50	360
1580	1-1/2" x 1/8"		10.50	.762		259	34.50		293.50	340
1600	1-1/2" x 1/4"		10.50	.762		325	34.50		359.50	410
1650	2" x 1/8"		10.50	.762		305	34.50		339.50	390
1700	2" x 1/4"		10.50	.762		405	34.50		439.50	500
2000	Column flange ties, wire, galvanized									
2300	3/16" diameter, up to 3" wide	1 Bric	10.50	.762	C	87	34.50		121.50	149
2350	To 5" wide		10.50	.762		94.50	34.50		129	157
2400	To 7" wide		10.50	.762		102	34.50		136.50	165
2600	To 9" wide		10.50	.762		109	34.50		143.50	172
2650	1/4" diameter, up to 3" wide		10.50	.762		110	34.50		144.50	174
2700	To 5" wide		10.50	.762		118	34.50		152.50	183
2800	To 7" wide		10.50	.762		133	34.50		167.50	199
2850	To 9" wide		10.50	.762		143	34.50		177.50	210
2900	For hot dip galvanized, add					35%				

04 05 19 – Masonry Anchorage and Reinforcing

04 05 19.16 Masonry Anchors

	04 05 19.16 Masonry Anchors	Crew	Daily Output	Labor-Hours	Unit	2014 Bare Costs Material	Labor	Equipment	Total	Total Incl O&P
4000	Channel slots, 1-3/8" x 1/2" x 8"									
4100	12 ga., plain	1 Bric	10.50	.762	C	212	34.50		246.50	286
4150	16 ga., galvanized	"	10.50	.762	"	145	34.50		179.50	212
4200	Channel slot anchors									
4300	16 ga., galvanized, 1-1/4" x 3-1/2"				C	56.50			56.50	62
4350	1-1/4" x 5-1/2"					66			66	72.50
4400	1-1/4" x 7-1/2"					74.50			74.50	82
4500	1/8" plain, 1-1/4" x 3-1/2"					143			143	157
4550	1-1/4" x 5-1/2"					152			152	167
4600	1-1/4" x 7-1/2"					166			166	183
4700	For corrugation, add					76.50			76.50	84.50
4750	For hot dip galvanized, add				↓	35%				
5000	Dowels									
5100	Plain, 1/4" diameter, 3" long				C	55.50			55.50	61
5150	4" long					61.50			61.50	67.50
5200	6" long					75			75	82.50
5300	3/8" diameter, 3" long					73			73	80.50
5350	4" long					91			91	100
5400	6" long					105			105	115
5500	1/2" diameter, 3" long					107			107	117
5550	4" long					127			127	139
5600	6" long					164			164	180
5700	5/8" diameter, 3" long					146			146	161
5750	4" long					180			180	198
5800	6" long					247			247	272
6000	3/4" diameter, 3" long					182			182	200
6100	4" long					231			231	254
6150	6" long					330			330	365
6300	For hot dip galvanized, add				↓	35%				

04 05 19.26 Masonry Reinforcing Bars

	04 05 19.26 Masonry Reinforcing Bars	Crew	Daily Output	Labor-Hours	Unit	Material	Labor	Equipment	Total	Total Incl O&P
0010	**MASONRY REINFORCING BARS**									
0015	Steel bars A615, placed horiz., #3 & #4 bars	1 Bric	450	.018	Lb.	.50	.81		1.31	1.79
0020	#5 & #6 bars		800	.010		.50	.46		.96	1.25
0050	Placed vertical, #3 & #4 bars		350	.023		.50	1.04		1.54	2.14
0060	#5 & #6 bars		650	.012	↓	.50	.56		1.06	1.41
0200	Joint reinforcing, regular truss, to 6" wide, mill std galvanized		30	.267	C.L.F.	21	12.15		33.15	41.50
0250	12" wide		20	.400		24.50	18.25		42.75	55
0400	Cavity truss with drip section, to 6" wide		30	.267		22.50	12.15		34.65	43
0450	12" wide	↓	20	.400	↓	26	18.25		44.25	56.50
0500	Joint reinforcing, ladder type, mill std galvanized									
0600	9 ga. sides, 9 ga. ties, 4" wall	1 Bric	30	.267	C.L.F.	15.30	12.15		27.45	35.50
0650	6" wall		30	.267		17.70	12.15		29.85	38
0700	8" wall		25	.320		18.95	14.60		33.55	43
0750	10" wall		20	.400		16.25	18.25		34.50	46
0800	12" wall	↓	20	.400	↓	16.45	18.25		34.70	46
1000	Truss type									
1100	9 ga. sides, 9 ga. ties, 4" wall	1 Bric	30	.267	C.L.F.	20.50	12.15		32.65	41
1150	6" wall		30	.267		21	12.15		33.15	41.50
1200	8" wall		25	.320		22	14.60		36.60	46
1250	10" wall		20	.400		20.50	18.25		38.75	50.50
1300	12" wall		20	.400		21	18.25		39.25	51.50
1500	3/16" sides, 9 ga. ties, 4" wall	↓	30	.267	↓	23.50	12.15		35.65	44.50

04 05 Common Work Results for Masonry

04 05 19 – Masonry Anchorage and Reinforcing

04 05 19.26 Masonry Reinforcing Bars		Crew	Daily Output	Labor-Hours	Unit	Material	2014 Bare Costs Labor	Equipment	Total	Total Incl O&P
1550	6″ wall	1 Bric	30	.267	C.L.F.	29.50	12.15		41.65	51
1600	8″ wall		25	.320		30.50	14.60		45.10	56
1650	10″ wall		20	.400		31.50	18.25		49.75	62.50
1700	12″ wall		20	.400		32	18.25		50.25	63.50
2000	3/16″ sides, 3/16″ ties, 4″ wall		30	.267		33	12.15		45.15	55
2050	6″ wall		30	.267		34	12.15		46.15	56
2100	8″ wall		25	.320		35	14.60		49.60	60.50
2150	10″ wall		20	.400		37	18.25		55.25	68.50
2200	12″ wall		20	.400		39	18.25		57.25	71
2500	Cavity truss type, galvanized									
2600	9 ga. sides, 9 ga. ties, 4″ wall	1 Bric	25	.320	C.L.F.	35.50	14.60		50.10	61
2650	6″ wall		25	.320		31	14.60		45.60	56.50
2700	8″ wall		20	.400		37.50	18.25		55.75	69
2750	10″ wall		15	.533		39.50	24.50		64	80
2800	12″ wall		15	.533		36	24.50		60.50	76.50
3000	3/16″ sides, 9 ga. ties, 4″ wall		25	.320		33.50	14.60		48.10	58.50
3050	6″ wall		25	.320		30.50	14.60		45.10	55.50
3100	8″ wall		20	.400		44.50	18.25		62.75	77
3150	10″ wall		15	.533		29.50	24.50		54	69
3200	12″ wall		15	.533		35	24.50		59.50	75.50
3500	For hot dip galvanizing, add				Ton	460			460	505

04 05 23 – Masonry Accessories

04 05 23.13 Masonry Control and Expansion Joints		Crew	Daily Output	Labor-Hours	Unit	Material	Labor	Equipment	Total	Total Incl O&P
0010	**MASONRY CONTROL AND EXPANSION JOINTS**									
0020	Rubber, for double wythe 8″ minimum wall (Brick/CMU)	1 Bric	400	.020	L.F.	1.77	.91		2.68	3.34
0025	"T" shaped		320	.025		1.16	1.14		2.30	3.02
0030	Cross-shaped for CMU units		280	.029		1.42	1.30		2.72	3.55
0050	PVC, for double wythe 8″ minimum wall (Brick/CMU)		400	.020		1.27	.91		2.18	2.79
0120	"T" shaped		320	.025		.73	1.14		1.87	2.54
0160	Cross-shaped for CMU units		280	.029		.92	1.30		2.22	3

04 05 23.95 Wall Plugs		Crew	Daily Output	Labor-Hours	Unit	Material	Labor	Equipment	Total	Total Incl O&P
0010	**WALL PLUGS** (for nailing to brickwork)									
0020	25 ga., galvanized, plain	1 Bric	10.50	.762	C	26.50	34.50		61	82
0050	Wood filled	"	10.50	.762	"	63.50	34.50		98	123

04 21 Clay Unit Masonry

04 21 13 – Brick Masonry

04 21 13.13 Brick Veneer Masonry		Crew	Daily Output	Labor-Hours	Unit	Material	Labor	Equipment	Total	Total Incl O&P
0010	**BRICK VENEER MASONRY**, T.L. lots, excl. scaff., grout & reinforcing									
0015	Material costs incl. 3% brick and 25% mortar waste									
0020	Standard, select common, 4″ x 2-2/3″ x 8″ (6.75/S.F.)	D-8	1.50	26.667	M	655	1,125		1,780	2,450
0601	Buff or gray face, running bond, (6.75/S.F.)		1.50	26.667		595	1,125		1,720	2,375
0700	Glazed face, 4″ x 2-2/3″ x 8″, running bond		1.40	28.571		1,775	1,200		2,975	3,775
0750	Full header every 6th course (7.88/S.F.)		1.35	29.630		1,700	1,250		2,950	3,750
1999	Alternate method of figuring by square foot									
2000	Standard, sel. common, 4″ x 2-2/3″ x 8″, (6.75/S.F.)	D-8	230	.174	S.F.	4.43	7.35		11.78	16.10
2020	Red, 4″ x 2-2/3″ x 8″, running bond		220	.182		4	7.65		11.65	16.10
2600	Buff or gray face, running bond, (6.75/S.F.)		220	.182		4.23	7.65		11.88	16.35
2700	Glazed face brick, running bond		210	.190		11.40	8.05		19.45	25
2750	Full header every 6th course (7.88/S.F.)		170	.235		13.30	9.90		23.20	30

04 21 13 – Brick Masonry

04 21 13.13 Brick Veneer Masonry		Crew	Daily Output	Labor-Hours	Unit	2014 Bare Costs Material	Labor	Equipment	Total	Total Incl O&P
3550	For curved walls, add						30%			

04 21 13.14 Thin Brick Veneer

		Crew	Daily Output	Labor-Hours	Unit	Material	Labor	Equipment	Total	Total Incl O&P
0010	**THIN BRICK VENEER**									
0015	Material costs incl. 3% brick and 25% mortar waste									
0020	On & incl. metal panel support sys, modular, 2-2/3" x 5/8" x 8", red	D-7	92	.174	S.F.	8.75	6.55		15.30	19.30
0100	Closure, 4" x 5/8" x 8"		110	.145		8.75	5.45		14.20	17.70
0110	Norman, 2-2/3" x 5/8" x 12"		110	.145		8.95	5.45		14.40	17.90
0120	Utility, 4" x 5/8" x 12"		125	.128		8.95	4.81		13.76	16.95
0130	Emperor, 4" x 3/4" x 16"		175	.091		10.15	3.44		13.59	16.25
0140	Super emperor, 8" x 3/4" x 16"		195	.082		10.15	3.09		13.24	15.75
0150	For L shaped corners with 4" return, add				L.F.	9.25			9.25	10.20
0200	On masonry/plaster back-up, modular, 2-2/3" x 5/8" x 8", red	D-7	137	.117	S.F.	3.95	4.39		8.34	10.80
0210	Closure, 4" x 5/8" x 8"		165	.097		3.95	3.65		7.60	9.75
0220	Norman, 2-2/3" x 5/8" x 12"		165	.097		4.15	3.65		7.80	9.95
0230	Utility, 4" x 5/8" x 12"		185	.086		4.15	3.25		7.40	9.35
0240	Emperor, 4" x 3/4" x 16"		260	.062		5.35	2.31		7.66	9.30
0250	Super emperor, 8" x 3/4" x 16"		285	.056		5.35	2.11		7.46	9
0260	For L shaped corners with 4" return, add				L.F.	9.25			9.25	10.20
0270	For embedment into pre-cast concrete panels, add				S.F.	14.40			14.40	15.85

04 21 13.15 Chimney

		Crew	Daily Output	Labor-Hours	Unit	Material	Labor	Equipment	Total	Total Incl O&P
0010	**CHIMNEY**, excludes foundation, scaffolding, grout and reinforcing									
0100	Brick, 16" x 16", 8" flue	D-1	18.20	.879	V.L.F.	25	36.50		61.50	83
0150	16" x 20" with one 8" x 12" flue		16	1		38.50	41.50		80	106
0200	16" x 24" with two 8" x 8" flues		14	1.143		55.50	47		102.50	133
0250	20" x 20" with one 12" x 12" flue		13.70	1.168		50	48		98	128
0300	20" x 24" with two 8" x 12" flues		12	1.333		63	55		118	153
0350	20" x 32" with two 12" x 12" flues		10	1.600		87.50	66		153.50	198

04 21 13.45 Face Brick

		Crew	Daily Output	Labor-Hours	Unit	Material	Labor	Equipment	Total	Total Incl O&P
0010	**FACE BRICK** Material Only, C216, TL lots									
0300	Standard modular, 4" x 2-2/3" x 8", minimum				M	490			490	540
0350	Maximum					695			695	760
0450	Economy, 4" x 4" x 8", minimum					765			765	840
0500	Maximum					990			990	1,100
0510	Economy, 4" x 4" x 12", minimum					1,200			1,200	1,325
0520	Maximum					1,575			1,575	1,725
0550	Jumbo, 6" x 4" x 12", minimum					1,375			1,375	1,500
0600	Maximum					1,825			1,825	2,000
0610	Jumbo, 8" x 4" x 12", minimum					1,375			1,375	1,500
0620	Maximum					1,825			1,825	2,000
0650	Norwegian, 4" x 3-1/5" x 12", minimum					1,125			1,125	1,225
0700	Maximum					1,450			1,450	1,600
0710	Norwegian, 6" x 3-1/5" x 12", minimum					1,475			1,475	1,600
0720	Maximum					1,900			1,900	2,100
0850	Standard glazed, plain colors, 4" x 2-2/3" x 8", minimum					1,550			1,550	1,725
0900	Maximum					2,025			2,025	2,225
1000	Deep trim shades, 4" x 2-2/3" x 8", minimum					1,975			1,975	2,150
1050	Maximum					2,200			2,200	2,425
1080	Jumbo utility, 4" x 4" x 12"					1,375			1,375	1,525
1120	4" x 8" x 8"					1,725			1,725	1,875
1140	4" x 8" x 16"					5,050			5,050	5,550
1260	Engineer, 4" x 3-1/5" x 8", minimum					515			515	565
1270	Maximum					715			715	785

04 21 Clay Unit Masonry

04 21 13 – Brick Masonry

04 21 13.45 Face Brick		Crew	Daily Output	Labor-Hours	Unit	2014 Bare Costs Material	Labor	Equipment	Total	Total Incl O&P
1350	King, 4" x 2-3/4" x 10", minimum				M	465			465	515
1360	Maximum					575			575	630
1770	Standard modular, double glazed, 4" x 2-2/3" x 8"					2,325			2,325	2,550
1850	Jumbo, colored glazed ceramic, 6" x 4" x 12"					2,450			2,450	2,700
2050	Jumbo utility, glazed, 4" x 4" x 12"					4,600			4,600	5,050
2100	4" x 8" x 8"					5,400			5,400	5,950
2150	4" x 16" x 8"					6,325			6,325	6,950
2170	For less than truck load lots, add					15			15	16.50
2180	For buff or gray brick, add					16			16	17.60
3050	Used brick, minimum					405			405	445
3150	Add for brick to match existing work, minimum					5%				
3200	Maximum				▼	50%				

04 22 Concrete Unit Masonry

04 22 10 – Concrete Masonry Units

04 22 10.10 Concrete Block

		Crew	Daily Output	Labor-Hours	Unit	Material	Labor	Equipment	Total	Total Incl O&P
0010	**CONCRETE BLOCK** Material Only									
0020	2" x 8" x 16" solid, normal-weight, 2,000 psi				Ea.	.98			.98	1.08
0050	3,500 psi					1.06			1.06	1.17
0100	5,000 psi					1.22			1.22	1.34
0150	Lightweight, std.					.99			.99	1.09
0300	3" x 8" x 16" solid, normal-weight, 2000 psi					1.02			1.02	1.12
0350	3,500 psi					1.27			1.27	1.40
0400	5,000 psi					1.44			1.44	1.58
0450	Lightweight, std.					1.13			1.13	1.24
0600	4" x 8" x 16" hollow, normal-weight, 2000 psi					1.11			1.11	1.22
0650	3,500 psi					1.24			1.24	1.36
0700	5000 psi					1.48			1.48	1.63
0750	Lightweight, std.					1.23			1.23	1.35
1300	Solid, normal-weight, 2,000 psi					1.40			1.40	1.54
1350	3,500 psi					1.31			1.31	1.44
1400	5,000 psi					1.34			1.34	1.47
1450	Lightweight, std.					1.13			1.13	1.24
1600	6" x 8" x 16" hollow, normal-weight, 2,000 psi					1.46			1.46	1.61
1650	3,500 psi					1.65			1.65	1.82
1700	5,000 psi					1.80			1.80	1.98
1750	Lightweight, std.					1.77			1.77	1.95
2300	Solid, normal-weight, 2,000 psi					1.61			1.61	1.77
2350	3,500 psi					1.50			1.50	1.65
2400	5,000 psi					2			2	2.20
2450	Lightweight, std.					2.14			2.14	2.35
2600	8" x 8" x 16" hollow, normal-weight, 2000 psi					1.59			1.59	1.75
2650	3500 psi					1.66			1.66	1.83
2700	5,000 psi					2.19			2.19	2.41
2750	Lightweight, std.					2.13			2.13	2.34
3200	Solid, normal-weight, 2,000 psi					2.64			2.64	2.90
3250	3,500 psi					2.84			2.84	3.12
3300	5,000 psi					2.94			2.94	3.23
3350	Lightweight, std.					2.75			2.75	3.03
3400	10" x 8" x 16" hollow, normal-weight, 2000 psi					1.59			1.59	1.75
3410	3500 psi				▼	1.66			1.66	1.83

04 22 Concrete Unit Masonry

04 22 10 – Concrete Masonry Units

04 22 10.10 Concrete Block		Crew	Daily Output	Labor-Hours	Unit	Material	Labor	Equipment	Total	Total Incl O&P
						2014 Bare Costs				
3420	5,000 psi				Ea.	2.19			2.19	2.41
3430	Lightweight, std.					2.13			2.13	2.34
3480	Solid, normal-weight, 2,000 psi					2.64			2.64	2.90
3490	3,500 psi					2.84			2.84	3.12
3500	5,000 psi					2.94			2.94	3.23
3510	Lightweight, std.					2.75			2.75	3.03
3600	12" x 8" x 16" hollow, normal-weight, 2000 psi					2.53			2.53	2.78
3650	3,500 psi					2.59			2.59	2.85
3700	5,000 psi					2.94			2.94	3.23
3750	Lightweight, std.					2.65			2.65	2.92
4300	Solid, normal-weight, 2,000 psi					4.04			4.04	4.44
4350	3,500 psi					3.31			3.31	3.64
4400	5,000 psi					3.23			3.23	3.55
4500	Lightweight, std.				↓	2.71			2.71	2.98

04 22 10.11 Autoclave Aerated Concrete Block		Crew	Daily Output	Labor-Hours	Unit	Material	Labor	Equipment	Total	Total Incl O&P
0010	**AUTOCLAVE AERATED CONCRETE BLOCK**, excl. scaffolding, grout & reinforcing									
0050	Solid, 4" x 8" x 24", incl. mortar G	D-8	600	.067	S.F.	1.48	2.81		4.29	5.90
0060	6" x 8" x 24" G		600	.067		2.23	2.81		5.04	6.75
0070	8" x 8" x 24" G		575	.070		2.97	2.93		5.90	7.75
0080	10" x 8" x 24" G		575	.070		3.63	2.93		6.56	8.45
0090	12" x 8" x 24" G	↓	550	.073	↓	4.45	3.07		7.52	9.55

04 22 10.14 Concrete Block, Back-Up		Crew	Daily Output	Labor-Hours	Unit	Material	Labor	Equipment	Total	Total Incl O&P
0010	**CONCRETE BLOCK, BACK-UP**, C90, 2000 psi									
0020	Normal weight, 8" x 16" units, tooled joint 1 side									
0050	Not-reinforced, 2000 psi, 2" thick	D-8	475	.084	S.F.	1.40	3.55		4.95	6.95
0200	4" thick		460	.087		1.66	3.67		5.33	7.45
0300	6" thick		440	.091		2.18	3.83		6.01	8.25
0350	8" thick		400	.100		2.45	4.22		6.67	9.15
0400	10" thick	↓	330	.121		2.95	5.10		8.05	11.05
0450	12" thick	D-9	310	.155		3.76	6.40		10.16	13.90
1000	Reinforced, alternate courses, 4" thick	D-8	450	.089		1.82	3.75		5.57	7.70
1100	6" thick		430	.093		2.34	3.92		6.26	8.55
1150	8" thick		395	.101		2.62	4.27		6.89	9.40
1200	10" thick	↓	320	.125		3.10	5.25		8.35	11.45
1250	12" thick	D-9	300	.160		3.92	6.60		10.52	14.35
2000	Lightweight, not reinforced, 4" thick	D-8	460	.087		1.80	3.67		5.47	7.60
2100	6" thick		445	.090		2.53	3.79		6.32	8.60
2150	8" thick		435	.092		3.06	3.88		6.94	9.25
2200	10" thick	↓	410	.098		3.69	4.11		7.80	10.30
2250	12" thick	D-9	390	.123		3.89	5.10		8.99	12.05
3000	Reinforced, alternate courses, 4" thick	D-8	450	.089		1.95	3.75		5.70	7.85
3100	6" thick		430	.093		2.69	3.92		6.61	8.95
3150	8" thick		420	.095		3.23	4.02		7.25	9.65
3200	10" thick	↓	400	.100		3.85	4.22		8.07	10.70
3250	12" thick	D-9	380	.126	↓	4.05	5.20		9.25	12.40

04 22 10.16 Concrete Block, Bond Beam		Crew	Daily Output	Labor-Hours	Unit	Material	Labor	Equipment	Total	Total Incl O&P
0010	**CONCRETE BLOCK, BOND BEAM**, C90, 2000 psi									
0020	Not including grout or reinforcing									
0125	Regular block, 6" thick	D-8	584	.068	L.F.	2.25	2.89		5.14	6.90
0130	8" high, 8" thick	"	565	.071		2.63	2.98		5.61	7.45
0150	12" thick	D-9	510	.094		3.91	3.89		7.80	10.20
0525	Lightweight, 6" thick	D-8	592	.068	↓	2.67	2.85		5.52	7.30

04 22 Concrete Unit Masonry

04 22 10 – Concrete Masonry Units

04 22 10.16 Concrete Block, Bond Beam

	04 22 10.16 Concrete Block, Bond Beam	Crew	Daily Output	Labor-Hours	Unit	2014 Bare Costs Material	Labor	Equipment	Total	Total Incl O&P
0530	8″ high, 8″ thick	D-8	575	.070	L.F.	3.22	2.93		6.15	8
0550	12″ thick	D-9	520	.092		4.33	3.81		8.14	10.55
2000	Including grout and 2 #5 bars									
2100	Regular block, 8″ high, 8″ thick	D-8	300	.133	L.F.	4.72	5.60		10.32	13.75
2150	12″ thick	D-9	250	.192		6.55	7.95		14.50	19.30
2500	Lightweight, 8″ high, 8″ thick	D-8	305	.131		5.30	5.55		10.85	14.30
2550	12″ thick	D-9	255	.188		7	7.75		14.75	19.55

04 22 10.18 Concrete Block, Column

	04 22 10.18 Concrete Block, Column	Crew	Daily Output	Labor-Hours	Unit	Material	Labor	Equipment	Total	Total Incl O&P
0010	**CONCRETE BLOCK, COLUMN** or pilaster									
0050	Including vertical reinforcing (4-#4 bars) and grout									
0160	1 piece unit, 16″ x 16″	D-1	26	.615	V.L.F.	15.85	25.50		41.35	56
0170	2 piece units, 16″ x 20″		24	.667		20.50	27.50		48	64.50
0180	20″ x 20″		22	.727		26	30		56	74.50
0190	22″ x 24″		18	.889		38.50	36.50		75	98.50
0200	20″ x 32″		14	1.143		45	47		92	122

04 22 10.19 Concrete Block, Insulation Inserts

	04 22 10.19 Concrete Block, Insulation Inserts	Crew	Daily Output	Labor-Hours	Unit	Material	Labor	Equipment	Total	Total Incl O&P
0010	**CONCRETE BLOCK, INSULATION INSERTS**									
0100	Styrofoam, plant installed, add to block prices									
0200	8″ x 16″ units, 6″ thick				S.F.	1.20			1.20	1.32
0250	8″ thick					1.35			1.35	1.49
0300	10″ thick					1.50			1.50	1.65
0350	12″ thick					1.65			1.65	1.82
0500	8″ x 8″ units, 8″ thick					1.20			1.20	1.32
0550	12″ thick					1.20			1.20	1.32

04 22 10.24 Concrete Block, Exterior

	04 22 10.24 Concrete Block, Exterior	Crew	Daily Output	Labor-Hours	Unit	Material	Labor	Equipment	Total	Total Incl O&P
0010	**CONCRETE BLOCK, EXTERIOR**, C90, 2000 psi									
0020	Reinforced alt courses, tooled joints 2 sides									
0100	Normal weight, 8″ x 16″ x 6″ thick	D-8	395	.101	S.F.	2.27	4.27		6.54	9
0200	8″ thick		360	.111		3.92	4.68		8.60	11.45
0250	10″ thick		290	.138		4.08	5.80		9.88	13.35
0300	12″ thick	D-9	250	.192		4.68	7.95		12.63	17.25
0500	Lightweight, 8″ x 16″ x 6″ thick	D-8	450	.089		3	3.75		6.75	9
0600	8″ thick		430	.093		3.82	3.92		7.74	10.20
0650	10″ thick		395	.101		3.93	4.27		8.20	10.85
0700	12″ thick	D-9	350	.137		4	5.65		9.65	13.05

04 22 10.26 Concrete Block Foundation Wall

	04 22 10.26 Concrete Block Foundation Wall	Crew	Daily Output	Labor-Hours	Unit	Material	Labor	Equipment	Total	Total Incl O&P
0010	**CONCRETE BLOCK FOUNDATION WALL**, C90/C145									
0050	Normal-weight, cut joints, horiz joint reinf, no vert reinf.									
0200	Hollow, 8″ x 16″ x 6″ thick	D-8	455	.088	S.F.	2.64	3.71		6.35	8.55
0250	8″ thick		425	.094		2.93	3.97		6.90	9.30
0300	10″ thick		350	.114		3.43	4.82		8.25	11.10
0350	12″ thick	D-9	300	.160		4.25	6.60		10.85	14.75
0500	Solid, 8″ x 16″ block, 6″ thick	D-8	440	.091		2.81	3.83		6.64	8.95
0550	8″ thick	"	415	.096		4.12	4.06		8.18	10.75
0600	12″ thick	D-9	350	.137		5.95	5.65		11.60	15.20
1000	Reinforced, #4 vert @ 48″									
1125	6″ thick	D-8	445	.090	S.F.	3.51	3.79		7.30	9.65
1150	8″ thick		415	.096		4.16	4.06		8.22	10.80
1200	10″ thick		340	.118		5	4.96		9.96	13.05
1250	12″ thick	D-9	290	.166		6.20	6.85		13.05	17.20
1500	Solid, 8″ x 16″ block, 6″ thick	D-8	430	.093		2.82	3.92		6.74	9.10
1600	8″ thick	"	405	.099		4.12	4.16		8.28	10.90

04 22 10.26 Concrete Block Foundation Wall

	04 22 10.26 Concrete Block Foundation Wall	Crew	Daily Output	Labor-Hours	Unit	Material	Labor	Equipment	Total	Total Incl O&P
						2014 Bare Costs				
1650	12" thick	D-9	340	.141	S.F.	5.95	5.85		11.80	15.45

04 22 10.28 Concrete Block, High Strength

	04 22 10.28 Concrete Block, High Strength	Crew	Daily Output	Labor-Hours	Unit	Material	Labor	Equipment	Total	Total Incl O&P
0010	**CONCRETE BLOCK, HIGH STRENGTH**									
0050	Hollow, reinforced alternate courses, 8" x 16" units									
0200	3500 psi, 4" thick	D-8	440	.091	S.F.	1.88	3.83		5.71	7.90
0250	6" thick		395	.101		2.20	4.27		6.47	8.90
0300	8" thick		360	.111		3.85	4.68		8.53	11.40
0350	12" thick	D-9	250	.192		4.57	7.95		12.52	17.15
0500	5000 psi, 4" thick	D-8	440	.091		1.95	3.83		5.78	8
0550	6" thick		395	.101		2.84	4.27		7.11	9.60
0600	8" thick		360	.111		4.04	4.68		8.72	11.60
0650	12" thick	D-9	300	.160		4.59	6.60		11.19	15.10
1000	For 75% solid block, add					30%				
1050	For 100% solid block, add					50%				

04 22 10.38 Concrete Brick

	04 22 10.38 Concrete Brick	Crew	Daily Output	Labor-Hours	Unit	Material	Labor	Equipment	Total	Total Incl O&P
0010	**CONCRETE BRICK**, C55, grade N, type 1									
0100	Regular, 4 x 2-1/4 x 8	D-8	660	.061	Ea.	.52	2.56		3.08	4.46
0125	Rusticated, 4 x 2-1/4 x 8		660	.061		.58	2.56		3.14	4.53
0150	Frog, 4 x 2-1/4 x 8		660	.061		.56	2.56		3.12	4.51
0200	Double, 4 x 4-7/8 x 8		535	.075		.91	3.15		4.06	5.80

04 22 10.44 Glazed Concrete Block

	04 22 10.44 Glazed Concrete Block	Crew	Daily Output	Labor-Hours	Unit	Material	Labor	Equipment	Total	Total Incl O&P
0010	**GLAZED CONCRETE BLOCK** C744									
0100	Single face, 8" x 16" units, 2" thick	D-8	360	.111	S.F.	9.55	4.68		14.23	17.65
0200	4" thick		345	.116		9.80	4.89		14.69	18.25
0250	6" thick		330	.121		10.60	5.10		15.70	19.45
0300	8" thick		310	.129		11.30	5.45		16.75	21
0350	10" thick		295	.136		12.95	5.70		18.65	23
0400	12" thick	D-9	280	.171		13.80	7.10		20.90	26
0700	Double face, 8" x 16" units, 4" thick	D-8	340	.118		13.85	4.96		18.81	23
0750	6" thick		320	.125		16.85	5.25		22.10	26.50
0800	8" thick		300	.133		17.65	5.60		23.25	28
1000	Jambs, bullnose or square, single face, 8" x 16", 2" thick		315	.127	Ea.	18.25	5.35		23.60	28
1050	4" thick		285	.140	"	21.50	5.90		27.40	32.50
1200	Caps, bullnose or square, 8" x 16", 2" thick		420	.095	L.F.	16.95	4.02		20.97	25
1250	4" thick		380	.105	"	18.45	4.44		22.89	27.50
1256	Corner, bullnose or square, 2" thick		280	.143	Ea.	19.80	6		25.80	31
1258	4" thick		270	.148		21.50	6.25		27.75	33
1260	6" thick		260	.154		26.50	6.50		33	39.50
1270	8" thick		250	.160		32.50	6.75		39.25	46.50
1280	10" thick		240	.167		33	7.05		40.05	47
1290	12" thick		230	.174		34.50	7.35		41.85	49
1500	Cove base, 8" x 16", 2" thick		315	.127	L.F.	8.55	5.35		13.90	17.55
1550	4" thick		285	.140		8.65	5.90		14.55	18.50
1600	6" thick		265	.151		9.25	6.35		15.60	19.90
1650	8" thick		245	.163		9.70	6.90		16.60	21

04 24 Adobe Unit Masonry

04 24 16 – Manufactured Adobe Unit Masonry

04 24 16.06 Adobe Brick		Crew	Daily Output	Labor-Hours	Unit	Material	Labor	Equipment	Total	Total Incl O&P
						2014 Bare Costs				
0010	**ADOBE BRICK**, Semi-stabilized, with cement mortar									
0060	Brick, 10" x 4" x 14", 2.6/S.F. G	D-8	560	.071	S.F.	4.17	3.01		7.18	9.20
0080	12" x 4" x 16", 2.3/S.F. G		580	.069		6.85	2.91		9.76	12
0100	10" x 4" x 16", 2.3/S.F. G		590	.068		6.30	2.86		9.16	11.30
0120	8" x 4" x 16", 2.3/S.F. G		560	.071		4.82	3.01		7.83	9.90
0140	4" x 4" x 16", 2.3/S.F. G		540	.074		4.75	3.12		7.87	10
0160	6" x 4" x 16", 2.3/S.F. G		540	.074		4.39	3.12		7.51	9.60
0180	4" x 4" x 12", 3.0/S.F. G		520	.077		5	3.24		8.24	10.45
0200	8" x 4" x 12", 3.0/S.F. G		520	.077		3.98	3.24		7.22	9.30

04 25 Unit Masonry Panels

04 25 20 – Pre-Fabricated Masonry Panels

04 25 20.10 Brick and Epoxy Mortar Panels

		Crew	Daily Output	Labor-Hours	Unit	Material	Labor	Equipment	Total	Total Incl O&P
0010	**BRICK AND EPOXY MORTAR PANELS**									
0020	Prefabricated brick & epoxy mortar, 4" thick, minimum	C-11	775	.093	S.F.	5.45	4.68	2.37	12.50	16.60
0100	Maximum	"	500	.144		6.65	7.25	3.68	17.58	24
0200	For 2" concrete back-up, add					50%				
0300	For 1" urethane & 3" concrete back-up, add					70%				

04 27 Multiple-Wythe Unit Masonry

04 27 10 – Multiple-Wythe Masonry

04 27 10.30 Brick Walls

		Crew	Daily Output	Labor-Hours	Unit	Material	Labor	Equipment	Total	Total Incl O&P
0010	**BRICK WALLS**, including mortar, excludes scaffolding									
0020	Estimating by number of brick									
0140	Face brick, 4" thick wall, 6.75 brick/S.F.	D-8	1.45	27.586	M	585	1,175		1,760	2,425
0150	Common brick, 4" thick wall, 6.75 brick/S.F.		1.60	25		585	1,050		1,635	2,250
0204	8" thick, 13.50 bricks per S.F.		1.80	22.222		600	935		1,535	2,075
0250	12" thick, 20.25 bricks per S.F.		1.90	21.053		605	890		1,495	2,025
0304	16" thick, 27.00 bricks per S.F.		2	20		610	845		1,455	1,950
0500	Reinforced, face brick, 4" thick wall, 6.75 brick/S.F.		1.40	28.571		610	1,200		1,810	2,500
0520	Common brick, 4" thick wall, 6.75 brick/S.F.		1.55	25.806		610	1,100		1,710	2,325
0550	8" thick, 13.50 bricks per S.F.		1.75	22.857		625	965		1,590	2,175
0600	12" thick, 20.25 bricks per S.F.		1.85	21.622		630	910		1,540	2,100
0650	16" thick, 27.00 bricks per S.F.		1.95	20.513		635	865		1,500	2,025
0790	Alternate method of figuring by square foot									
0800	Face brick, 4" thick wall, 6.75 brick/S.F.	D-8	215	.186	S.F.	3.93	7.85		11.78	16.30
0850	Common brick, 4" thick wall, 6.75 brick/S.F.		240	.167		3.94	7.05		10.99	15.05
0900	8" thick, 13.50 bricks per S.F.		135	.296		8.10	12.50		20.60	28
1000	12" thick, 20.25 bricks per S.F.		95	.421		12.20	17.75		29.95	40.50
1050	16" thick, 27.00 bricks per S.F.		75	.533		16.50	22.50		39	52.50
1200	Reinforced, face brick, 4" thick wall, 6.75 brick/S.F.		210	.190		4.10	8.05		12.15	16.75
1220	Common brick, 4" thick wall, 6.75 brick/S.F.		235	.170		4.11	7.20		11.31	15.45
1250	8" thick, 13.50 bricks per S.F.		130	.308		8.45	12.95		21.40	29
1300	12" thick, 20.25 bricks per S.F.		90	.444		12.75	18.75		31.50	42.50
1350	16" thick, 27.00 bricks per S.F.		70	.571		17.20	24		41.20	55.50

04 27 Multiple-Wythe Unit Masonry

04 27 10 – Multiple-Wythe Masonry

04 27 10.40 Steps		Crew	Daily Output	Labor-Hours	Unit	2014 Bare Costs Material	Labor	Equipment	Total	Total Incl O&P
0010	**STEPS**									
0012	Entry steps, select common brick	D-1	.30	53.333	M	550	2,200		2,750	3,950

04 41 Dry-Placed Stone

04 41 10 – Dry Placed Stone

04 41 10.10 Rough Stone Wall

			Crew	Daily Output	Labor-Hours	Unit	Material	Labor	Equipment	Total	Total Incl O&P
0011	**ROUGH STONE WALL**, Dry										
0012	Dry laid (no mortar), under 18″ thick	G	D-1	60	.267	C.F.	12.40	11		23.40	30.50
0150	Over 18″ thick	G	D-12	63	.508	″	14.85	21		35.85	49
0500	Field stone veneer	G	D-8	120	.333	S.F.	11.45	14.05		25.50	34

04 43 Stone Masonry

04 43 10 – Masonry with Natural and Processed Stone

04 43 10.05 Ashlar Veneer

		Crew	Daily Output	Labor-Hours	Unit	Material	Labor	Equipment	Total	Total Incl O&P
0011	**ASHLAR VENEER** 4″ + or - thk, random or random rectangular									
0150	Sawn face, split joints, low priced stone	D-8	140	.286	S.F.	11.55	12.05		23.60	31
0200	Medium priced stone		130	.308		13.30	12.95		26.25	34.50
0300	High priced stone		120	.333		17.85	14.05		31.90	41
0600	Seam face, split joints, medium price stone		125	.320		20	13.50		33.50	42.50
0700	High price stone		120	.333		20.50	14.05		34.55	44
1000	Split or rock face, split joints, medium price stone		125	.320		14.70	13.50		28.20	36.50
1100	High price stone		120	.333		19.05	14.05		33.10	42.50

04 43 10.10 Bluestone

		Crew	Daily Output	Labor-Hours	Unit	Material	Labor	Equipment	Total	Total Incl O&P
0010	**BLUESTONE**, cut to size									
0100	Paving, natural cleft, to 4′, 1″ thick	D-8	150	.267	S.F.	6.50	11.25		17.75	24.50
0150	1-1/2″ thick		145	.276		6.50	11.65		18.15	25
0200	Smooth finish, 1″ thick		150	.267		7	11.25		18.25	25
0250	1-1/2″ thick		145	.276		7.50	11.65		19.15	26
0300	Thermal finish, 1″ thick		150	.267		7	11.25		18.25	25
0350	1-1/2″ thick		145	.276		7.50	11.65		19.15	26
1000	Stair treads, natural cleft, 12″ wide, 6′ long, 1-1/2″ thick	D-10	115	.278	L.F.	12	12.55	4.19	28.74	37
1050	2″ thick		105	.305		13	13.75	4.58	31.33	40.50
1100	Smooth finish, 1-1/2″ thick		115	.278		12	12.55	4.19	28.74	37
1150	2″ thick		105	.305		13	13.75	4.58	31.33	40.50
1300	Thermal finish, 1-1/2″ thick		115	.278		12	12.55	4.19	28.74	37
1350	2″ thick		105	.305		13	13.75	4.58	31.33	40.50

04 43 10.45 Granite

		Crew	Daily Output	Labor-Hours	Unit	Material	Labor	Equipment	Total	Total Incl O&P
0010	**GRANITE**, cut to size									
0050	Veneer, polished face, 3/4″ to 1-1/2″ thick									
0150	Low price, gray, light gray, etc.	D-10	130	.246	S.F.	26.50	11.10	3.70	41.30	50.50
0180	Medium price, pink, brown, etc.		130	.246		29.50	11.10	3.70	44.30	53.50
0220	High price, red, black, etc.		130	.246		42	11.10	3.70	56.80	67.50
0300	1-1/2″ to 2-1/2″ thick, veneer									
0350	Low price, gray, light gray, etc.	D-10	130	.246	S.F.	28.50	11.10	3.70	43.30	52.50
0500	Medium price, pink, brown, etc.		130	.246		33.50	11.10	3.70	48.30	58
0550	High price, red, black, etc.		130	.246		52.50	11.10	3.70	67.30	78.50
0700	2-1/2″ to 4″ thick, veneer									
0750	Low price, gray, light gray, etc.	D-10	110	.291	S.F.	38.50	13.10	4.38	55.98	67.50

04 43 Stone Masonry

04 43 10 – Masonry with Natural and Processed Stone

04 43 10.45 Granite		Crew	Daily Output	Labor-Hours	Unit	Material	2014 Bare Costs Labor	Equipment	Total	Total Incl O&P
0850	Medium price, pink, brown, etc.	D-10	110	.291	S.F.	44	13.10	4.38	61.48	73.50
0950	High price, red, black, etc.		110	.291		63	13.10	4.38	80.48	94.50
1000	For bush hammered finish, deduct					5%				
1050	Coarse rubbed finish, deduct					10%				
1100	Honed finish, deduct					5%				
1150	Thermal finish, deduct					18%				
1800	Carving or bas-relief, from templates or plaster molds									
1850	Minimum	D-10	80	.400	C.F.	178	18.05	6	202.05	230
1900	Maximum	"	80	.400	"	520	18.05	6	544.05	605
2000	Intricate or hand finished pieces									
2010	Mouldings, radius cuts, bullnose edges, etc.									
2050	Add, minimum					30%				
2100	Add, maximum					300%				
2450	For radius under 5′, add				L.F.	100%				
2500	Steps, copings, etc., finished on more than one surface									
2550	Minimum	D-10	50	.640	C.F.	91	29	9.65	129.65	155
2600	Maximum	"	50	.640	"	146	29	9.65	184.65	215
2700	Pavers, Belgian block, 8″-13″ long, 4″-6″ wide, 4″-6″ deep	D-11	120	.200	S.F.	27	8.70		35.70	43
2800	Pavers, 4″ x 4″ x 4″ blocks, split face and joints									
2850	Minimum	D-11	80	.300	S.F.	13.10	13		26.10	34.50
2900	Maximum	"	80	.300	"	29	13		42	52
3000	Pavers, 4″ x 4″ x 4″, thermal face, sawn joints									
3050	Minimum	D-11	65	.369	S.F.	24.50	16		40.50	51
3100	Maximum	"	65	.369		32	16		48	60
4000	Soffits, 2″ thick, minimum	D-13	35	1.371		37.50	60	13.75	111.25	149
4100	Maximum		35	1.371		91.50	60	13.75	165.25	207
4200	4″ thick, minimum		35	1.371		63	60	13.75	136.75	177
4300	Maximum		35	1.371		119	60	13.75	192.75	238
04 43 10.50 Lightweight Natural Stone										
0011	**LIGHTWEIGHT NATURAL STONE** Lava type									
0100	Veneer, rubble face, sawed back, irregular shapes G	D-10	130	.246	S.F.	6.25	11.10	3.70	21.05	28
0200	Sawed face and back, irregular shapes G	"	130	.246	"	6.25	11.10	3.70	21.05	28
04 43 10.80 Slate										
0010	**SLATE**									
0040	Pennsylvania - blue gray to black									
0050	Vermont - unfading green, mottled green & purple, gray & purple									
0100	Virginia - blue black									
0200	Exterior paving, natural cleft, 1″ thick									
0250	6″ x 6″ Pennsylvania	D-12	100	.320	S.F.	6.75	13.40		20.15	28
0300	Vermont		100	.320		11.25	13.40		24.65	33
0350	Virginia		100	.320		14.10	13.40		27.50	36
0500	24″ x 24″, Pennsylvania		120	.267		13	11.15		24.15	31.50
0550	Vermont		120	.267		28	11.15		39.15	47.50
0600	Virginia		120	.267		20.50	11.15		31.65	39.50
0700	18″ x 30″ Pennsylvania		120	.267		14.75	11.15		25.90	33.50
0750	Vermont		120	.267		28	11.15		39.15	47.50
0800	Virginia		120	.267		18.20	11.15		29.35	37
3100	Stair landings, 1″ thick, black, clear	D-1	65	.246		20	10.15		30.15	37.50
3200	Ribbon	"	65	.246		22	10.15		32.15	39.50
3500	Stair treads, sand finish, 1″ thick x 12″ wide									
3550	Under 3 L.F.	D-10	85	.376	L.F.	22	17	5.65	44.65	56.50
3600	3 L.F. to 6 L.F.	"	120	.267	"	24	12.05	4.01	40.06	48.50

04 43 Stone Masonry

04 43 10 – Masonry with Natural and Processed Stone

04 43 10.80 Slate		Crew	Daily Output	Labor-Hours	Unit	2014 Bare Costs Material	Labor	Equipment	Total	Total Incl O&P
3700	Ribbon, sand finish, 1" thick x 12" wide									
3750	To 6 L.F.	D-10	120	.267	L.F.	20	12.05	4.01	36.06	44.50

04 43 10.85 Window Sill

		Crew	Daily Output	Labor-Hours	Unit	Material	Labor	Equipment	Total	Total Incl O&P
0010	**WINDOW SILL**									
0020	Bluestone, thermal top, 10" wide, 1-1/2" thick	D-1	85	.188	S.F.	9.25	7.75		17	22
0050	2" thick		75	.213	"	8.85	8.80		17.65	23
0100	Cut stone, 5" x 8" plain		48	.333	L.F.	11.90	13.75		25.65	34
0200	Face brick on edge, brick, 8" wide		80	.200		5.30	8.25		13.55	18.40
0400	Marble, 9" wide, 1" thick		85	.188		8.50	7.75		16.25	21
0900	Slate, colored, unfading, honed, 12" wide, 1" thick		85	.188		8.50	7.75		16.25	21
0950	2" thick	▼	70	.229	▼	8.50	9.45		17.95	24

04 51 Flue Liner Masonry

04 51 10 – Clay Flue Lining

04 51 10.10 Flue Lining

		Crew	Daily Output	Labor-Hours	Unit	Material	Labor	Equipment	Total	Total Incl O&P
0010	**FLUE LINING**, including mortar									
0020	Clay, 8" x 8"	D-1	125	.128	V.L.F.	5.30	5.30		10.60	13.85
0100	8" x 12"		103	.155		7.85	6.40		14.25	18.45
0200	12" x 12"		93	.172		10.20	7.10		17.30	22
0300	12" x 18"		84	.190		20	7.85		27.85	34
0400	18" x 18"		75	.213		26	8.80		34.80	42.50
0500	20" x 20"		66	.242		40	10		50	59.50
0600	24" x 24"		56	.286		51.50	11.80		63.30	74.50
1000	Round, 18" diameter		66	.242		37	10		47	56
1100	24" diameter	▼	47	.340	▼	72	14.05		86.05	101

04 54 Refractory Brick Masonry

04 54 10 – Refractory Brick Work

04 54 10.10 Fire Brick

		Crew	Daily Output	Labor-Hours	Unit	Material	Labor	Equipment	Total	Total Incl O&P
0010	**FIRE BRICK**									
0012	Low duty, 2000°F, 9" x 2-1/2" x 4-1/2"	D-1	.60	26.667	M	1,375	1,100		2,475	3,200
0050	High duty, 3000°F	"	.60	26.667	"	1,800	1,100		2,900	3,650

04 54 10.20 Fire Clay

		Crew	Daily Output	Labor-Hours	Unit	Material	Labor	Equipment	Total	Total Incl O&P
0010	**FIRE CLAY**									
0020	Gray, high duty, 100 lb. bag				Bag	30			30	33
0050	100 lb. drum, premixed (400 brick per drum)				Drum	41			41	45

Division Notes

		DAILY OUTPUT	LABOR-HOURS		BARE COSTS				TOTAL
	CREW			UNIT	MAT.	LABOR	EQUIP.	TOTAL	INCL O&P

Estimating Tips

05 05 00 Common Work Results for Metals

- Nuts, bolts, washers, connection angles, and plates can add a significant amount to both the tonnage of a structural steel job and the estimated cost. As a rule of thumb, add 10% to the total weight to account for these accessories.
- Type 2 steel construction, commonly referred to as "simple construction," consists generally of field-bolted connections with lateral bracing supplied by other elements of the building, such as masonry walls or x-bracing. The estimator should be aware, however, that shop connections may be accomplished by welding or bolting. The method may be particular to the fabrication shop and may have an impact on the estimated cost.

05 10 00 Structural Steel

- Steel items can be obtained from two sources: a fabrication shop or a metals service center. Fabrication shops can fabricate items under more controlled conditions than can crews in the field. They are also more efficient and can produce items more economically. Metal service centers serve as a source of long mill shapes to both fabrication shops and contractors.
- Most line items in this structural steel subdivision, and most items in 05 50 00 Metal Fabrications, are indicated as being shop fabricated. The bare material cost for these shop fabricated items is the "Invoice Cost" from the shop and includes the mill base price of steel plus mill extras, transportation to the shop, shop drawings and detailing where warranted, shop fabrication and handling, sandblasting and a shop coat of primer paint, all necessary structural bolts, and delivery to the job site. The bare labor cost and bare equipment cost for these shop fabricated items is for field installation or erection.
- Line items in Subdivision 05 12 23.40 Lightweight Framing, and other items scattered in Division 5, are indicated as being field fabricated. The bare material cost for these field fabricated items is the "Invoice Cost" from the metals service center and includes the mill base price of steel plus mill extras, transportation to the metals service center, material handling, and delivery of long lengths of mill shapes to the job site. Material costs for structural bolts and welding rods should be added to the estimate. The bare labor cost and bare equipment cost for these items is for both field fabrication and field installation or erection, and include time for cutting, welding and drilling in the fabricated metal items. Drilling into concrete and fasteners to fasten field fabricated items to other work are not included and should be added to the estimate.

05 20 00 Steel Joist Framing

- In any given project the total weight of open web steel joists is determined by the loads to be supported and the design. However, economies can be realized in minimizing the amount of labor used to place the joists. This is done by maximizing the joist spacing, and therefore minimizing the number of joists required to be installed on the job. Certain spacings and locations may be required by the design, but in other cases maximizing the spacing and keeping it as uniform as possible will keep the costs down.

05 30 00 Steel Decking

- The takeoff and estimating of metal deck involves more than simply the area of the floor or roof and the type of deck specified or shown on the drawings. Many different sizes and types of openings may exist. Small openings for individual pipes or conduits may be drilled after the floor/roof is installed, but larger openings may require special deck lengths as well as reinforcing or structural support. The estimator should determine who will be supplying this reinforcing. Additionally, some deck terminations are part of the deck package, such as screed angles and pour stops, and others will be part of the steel contract, such as angles attached to structural members and cast-in-place angles and plates. The estimator must ensure that all pieces are accounted for in the complete estimate.

05 50 00 Metal Fabrications

- The most economical steel stairs are those that use common materials, standard details, and most importantly, a uniform and relatively simple method of field assembly. Commonly available A36 channels and plates are very good choices for the main stringers of the stairs, as are angles and tees for the carrier members. Risers and treads are usually made by specialty shops, and it is most economical to use a typical detail in as many places as possible. The stairs should be pre-assembled and shipped directly to the site. The field connections should be simple and straightforward to be accomplished efficiently, and with minimum equipment and labor.

Reference Numbers

Reference numbers are shown in shaded boxes at the beginning of some major classifications. These numbers refer to related items in the Reference Section. The reference information may be an estimating procedure, an alternate pricing method, or technical information. *Note:* Not all subdivisions listed here necessarily appear in this publication.

05 01 Maintenance of Metals

05 01 10 – Maintenance of Structural Metal Framing

05 01 10.51 Cleaning of Structural Metal Framing		Crew	Daily Output	Labor-Hours	Unit	Material	2014 Bare Costs Labor	Equipment	Total	Total Incl O&P
0010	CLEANING OF STRUCTURAL METAL FRAMING									
6125	Steel surface treatments, PDCA guidelines									
6170	Wire brush, hand (SSPC-SP2)	1 Psst	400	.020	S.F.	.02	.81		.83	1.49
6180	Power tool (SSPC-SP3)	"	700	.011		.09	.46		.55	.93
6215	Pressure washing, up to 5000 psi, 5000-15,000 S.F./day	1 Pord	10000	.001			.03		.03	.05
6220	Steam cleaning, 600 psi @ 300 F, 1250 – 2500 S.F./day		2000	.004			.16		.16	.24
6225	Water blasting, up to 25,000 psi, 1750 – 3500 S.F./day		2500	.003			.13		.13	.19
6230	Brush-off blast (SSPC-SP7)	E-11	1750	.018		.17	.75	.14	1.06	1.58
6235	Com'l blast (SSPC-SP6), loose scale, fine pwder rust, 2.0#/S.F. sand		1200	.027		.33	1.10	.20	1.63	2.42
6240	Tight mill scale, little/no rust, 3.0#/S.F. sand		1000	.032		.50	1.32	.24	2.06	3
6245	Exist coat blistered/pitted, 4.0#/S.F. sand		875	.037		.67	1.51	.27	2.45	3.55
6250	Exist coat badly pitted/nodules, 6.7#/S.F. sand		825	.039		1.12	1.60	.29	3.01	4.21
6255	Near white blast (SSPC-SP10), loose scale, fine rust, 5.6#/S.F. sand		450	.071		.94	2.93	.53	4.40	6.50
6260	Tight mill scale, little/no rust, 6.9#/S.F. sand		325	.098		1.16	4.05	.73	5.94	8.80
6265	Exist coat blistered/pitted, 9.0#/S.F. sand		225	.142		1.51	5.85	1.05	8.41	12.55
6270	Exist coat badly pitted/nodules, 11.3#/S.F. sand		150	.213		1.89	8.80	1.58	12.27	18.40

05 05 Common Work Results for Metals

05 05 05 – Selective Metals Demolition

05 05 05.10 Selective Demolition, Metals

		Crew	Daily Output	Labor-Hours	Unit	Material	Labor	Equipment	Total	Total Incl O&P
0010	SELECTIVE DEMOLITION, METALS R024119-10									
0015	Excludes shores, bracing, cutting, loading, hauling, dumping									
0020	Remove nuts only up to 3/4" diameter	1 Sswk	480	.017	Ea.		.85		.85	1.50
0030	7/8" to 1-1/4" diameter		240	.033			1.70		1.70	3.01
0040	1-3/8" to 2" diameter		160	.050			2.56		2.56	4.51
0060	Unbolt and remove structural bolts up to 3/4" diameter		240	.033			1.70		1.70	3.01
0070	7/8" to 2" diameter		160	.050			2.56		2.56	4.51
0140	Light weight framing members, remove whole or cut up, up to 20 lb.		240	.033			1.70		1.70	3.01
0150	21 – 40 lb.	2 Sswk	210	.076			3.89		3.89	6.85
0160	41 – 80 lb.	3 Sswk	180	.133			6.80		6.80	12.05
0170	81 – 120 lb.	4 Sswk	150	.213			10.90		10.90	19.25
0230	Structural members, remove whole or cut up, up to 500 lb.	E-19	48	.500			25	20	45	64.50
0240	1/4 – 2 tons	E-18	36	1.111			56.50	26.50	83	127
0250	2 – 5 tons	E-24	30	1.067			54	25	79	120
0260	5 – 10 tons	E-20	24	2.667			134	49	183	283
0270	10 – 15 tons	E-2	18	3.111			156	85	241	360
0340	Fabricated item, remove whole or cut up, up to 20 lb.	1 Sswk	96	.083			4.26		4.26	7.50
0350	21 – 40 lb.	2 Sswk	84	.190			9.75		9.75	17.20
0360	41 – 80 lb.	3 Sswk	72	.333			17.05		17.05	30
0370	81 – 120 lb.	4 Sswk	60	.533			27.50		27.50	48
0380	121 – 500 lb.	E-19	48	.500			25	20	45	64.50
0390	501 – 1000 lb.	"	36	.667			33.50	26.50	60	86
0500	Steel roof decking, uncovered, bare	B-2	5000	.008	S.F.		.30		.30	.46

05 05 13 – Shop-Applied Coatings for Metal

05 05 13.50 Paints and Protective Coatings

		Crew	Daily Output	Labor-Hours	Unit	Material	Labor	Equipment	Total	Total Incl O&P
0010	PAINTS AND PROTECTIVE COATINGS									
5900	Galvanizing structural steel in shop, under 1 ton				Ton	550			550	605
5950	1 ton to 20 tons					505			505	555
6000	Over 20 tons					460			460	505

05 05 19 – Post-Installed Concrete Anchors

05 05 19.10 Chemical Anchors

	05 05 19.10 Chemical Anchors	Crew	Daily Output	Labor-Hours	Unit	2014 Bare Costs Material	Labor	Equipment	Total	Total Incl O&P
0010	**CHEMICAL ANCHORS**									
0020	Includes layout & drilling									
1430	Chemical anchor, w/rod & epoxy cartridge, 3/4" diam. x 9-1/2" long	B-89A	27	.593	Ea.	9.75	25	4.30	39.05	54
1435	1" diameter x 11-3/4" long		24	.667		16.70	28	4.84	49.54	67
1440	1-1/4" diameter x 14" long		21	.762		36	32	5.55	73.55	95.50
1445	1-3/4" diameter x 15" long		20	.800		64.50	33.50	5.80	103.80	129
1450	18" long		17	.941		77.50	39.50	6.85	123.85	154
1455	2" diameter x 18" long		16	1		102	42	7.25	151.25	185
1460	24" long		15	1.067		132	45	7.75	184.75	224

05 05 19.20 Expansion Anchors

	05 05 19.20 Expansion Anchors		Crew	Daily Output	Labor-Hours	Unit	Material	Labor	Equipment	Total	Total Incl O&P
0010	**EXPANSION ANCHORS**										
0100	Anchors for concrete, brick or stone, no layout and drilling										
0200	Expansion shields, zinc, 1/4" diameter, 1-5/16" long, single	G	1 Carp	90	.089	Ea.	.48	4.08		4.56	6.85
0300	1-3/8" long, double	G		85	.094		.59	4.32		4.91	7.30
0400	3/8" diameter, 1-1/2" long, single	G		85	.094		.81	4.32		5.13	7.55
0500	2" long, double	G		80	.100		1.32	4.59		5.91	8.55
0600	1/2" diameter, 2-1/16" long, single	G		80	.100		1.62	4.59		6.21	8.90
0700	2-1/2" long, double	G		75	.107		2.10	4.89		6.99	9.85
0800	5/8" diameter, 2-5/8" long, single	G		75	.107		2.85	4.89		7.74	10.70
0900	2-3/4" long, double	G		70	.114		2.99	5.25		8.24	11.40
1000	3/4" diameter, 2-3/4" long, single	G		70	.114		3.25	5.25		8.50	11.70
1100	3-15/16" long, double	G		65	.123		5.80	5.65		11.45	15.10
5700	Lag screw shields, 1/4" diameter, short	G		90	.089		.37	4.08		4.45	6.70
5800	Long	G		85	.094		.41	4.32		4.73	7.10
5900	3/8" diameter, short	G		85	.094		.65	4.32		4.97	7.35
6000	Long	G		80	.100		.78	4.59		5.37	7.95
6100	1/2" diameter, short	G		80	.100		1.05	4.59		5.64	8.25
6200	Long	G		75	.107		1.18	4.89		6.07	8.85
6300	5/8" diameter, short	G		70	.114		1.81	5.25		7.06	10.10
6400	Long	G		65	.123		2	5.65		7.65	10.90
6600	Lead, #6 & #8, 3/4" long	G		260	.031		.18	1.41		1.59	2.38
6700	#10 - #14, 1-1/2" long	G		200	.040		.29	1.83		2.12	3.15
6800	#16 & #18, 1-1/2" long	G		160	.050		.39	2.29		2.68	3.97
6900	Plastic, #6 & #8, 3/4" long			260	.031		.04	1.41		1.45	2.22
7000	#8 & #10, 7/8" long			240	.033		.04	1.53		1.57	2.40
7100	#10 & #12, 1" long			220	.036		.05	1.67		1.72	2.63
7200	#14 & #16, 1-1/2" long			160	.050		.07	2.29		2.36	3.62
8000	Wedge anchors, not including layout or drilling										
8050	Carbon steel, 1/4" diameter, 1-3/4" long	G	1 Carp	150	.053	Ea.	.40	2.45		2.85	4.21
8100	3-1/4" long	G		140	.057		.53	2.62		3.15	4.62
8150	3/8" diameter, 2-1/4" long	G		145	.055		.49	2.53		3.02	4.44
8200	5" long	G		140	.057		.86	2.62		3.48	4.99
8250	1/2" diameter, 2-3/4" long	G		140	.057		.97	2.62		3.59	5.10
8300	7" long	G		125	.064		1.66	2.93		4.59	6.35
8350	5/8" diameter, 3-1/2" long	G		130	.062		1.69	2.82		4.51	6.20
8400	8-1/2" long	G		115	.070		3.60	3.19		6.79	8.90
8450	3/4" diameter, 4-1/4" long	G		115	.070		2.81	3.19		6	8
8500	10" long	G		95	.084		6.40	3.86		10.26	13
8550	1" diameter, 6" long	G		100	.080		9.05	3.67		12.72	15.60
8575	9" long	G		85	.094		11.75	4.32		16.07	19.60
8600	12" long	G		75	.107		12.70	4.89		17.59	21.50
8650	1-1/4" diameter, 9" long	G		70	.114		24	5.25		29.25	34

05 05 Common Work Results for Metals

05 05 19 – Post-Installed Concrete Anchors

05 05 19.20 Expansion Anchors			Crew	Daily Output	Labor-Hours	Unit	2014 Bare Costs Material	Labor	Equipment	Total	Total Incl O&P
8700	12″ long	G	1 Carp	60	.133	Ea.	30.50	6.10		36.60	43
8750	For type 303 stainless steel, add						350%				
8800	For type 316 stainless steel, add						450%				
8950	Self-drilling concrete screw, hex washer head, 3/16″ diam. x 1-3/4″ long	G	1 Carp	300	.027	Ea.	.20	1.22		1.42	2.11
8960	2-1/4″ long	G		250	.032		.23	1.47		1.70	2.51
8970	Phillips flat head, 3/16″ diam. x 1-3/4″ long	G		300	.027		.20	1.22		1.42	2.11
8980	2-1/4″ long	G		250	.032		.23	1.47		1.70	2.51

05 05 21 – Fastening Methods for Metal

05 05 21.10 Cutting Steel

		Crew	Daily Output	Labor-Hours	Unit	Material	Labor	Equipment	Total	Total Incl O&P
0010	**CUTTING STEEL**									
0020	Hand burning, incl. preparation, torch cutting & grinding, no staging									
0050	Steel to 1/4″ thick	E-25	400	.020	L.F.	.17	1.06	.03	1.26	2.09
0100	1/2″ thick		320	.025		.32	1.33	.04	1.69	2.73
0150	3/4″ thick		260	.031		.53	1.63	.04	2.20	3.51
0200	1″ thick		200	.040		.76	2.12	.06	2.94	4.64

05 05 21.15 Drilling Steel

		Crew	Daily Output	Labor-Hours	Unit	Material	Labor	Equipment	Total	Total Incl O&P
0010	**DRILLING STEEL**									
1910	Drilling & layout for steel, up to 1/4″ deep, no anchor									
1920	Holes, 1/4″ diameter	1 Sswk	112	.071	Ea.	.08	3.65		3.73	6.55
1925	For each additional 1/4″ depth, add		336	.024		.08	1.22		1.30	2.24
1930	3/8″ diameter		104	.077		.09	3.93		4.02	7.05
1935	For each additional 1/4″ depth, add		312	.026		.09	1.31		1.40	2.41
1940	1/2″ diameter		96	.083		.09	4.26		4.35	7.60
1945	For each additional 1/4″ depth, add		288	.028		.09	1.42		1.51	2.61
1950	5/8″ diameter		88	.091		.13	4.65		4.78	8.35
1955	For each additional 1/4″ depth, add		264	.030		.13	1.55		1.68	2.88
1960	3/4″ diameter		80	.100		.17	5.10		5.27	9.20
1965	For each additional 1/4″ depth, add		240	.033		.17	1.70		1.87	3.19
1970	7/8″ diameter		72	.111		.22	5.70		5.92	10.25
1975	For each additional 1/4″ depth, add		216	.037		.22	1.89		2.11	3.58
1980	1″ diameter		64	.125		.23	6.40		6.63	11.55
1985	For each additional 1/4″ depth, add		192	.042		.23	2.13		2.36	4.01
1990	For drilling up, add						40%			

05 05 21.90 Welding Steel

		Crew	Daily Output	Labor-Hours	Unit	Material	Labor	Equipment	Total	Total Incl O&P
0010	**WELDING STEEL**, Structural R050521-20									
0020	Field welding, 1/8″ E6011, cost per welder, no operating engineer	E-14	8	1	Hr.	4.33	53	17.75	75.08	118
0200	With 1/2 operating engineer	E-13	8	1.500		4.33	76.50	17.75	98.58	153
0300	With 1 operating engineer	E-12	8	2		4.33	100	17.75	122.08	189
0500	With no operating engineer, 2# weld rod per ton	E-14	8	1	Ton	4.33	53	17.75	75.08	118
0600	8# E6011 per ton	"	2	4		17.30	212	71	300.30	470
0800	With one operating engineer per welder, 2# E6011 per ton	E-12	8	2		4.33	100	17.75	122.08	189
0900	8# E6011 per ton	"	2	8		17.30	400	71	488.30	755
1200	Continuous fillet, down welding									
1300	Single pass, 1/8″ thick, 0.1#/L.F.	E-14	150	.053	L.F.	.22	2.83	.95	4	6.30
1400	3/16″ thick, 0.2#/L.F.		75	.107		.43	5.65	1.89	7.97	12.55
1500	1/4″ thick, 0.3#/L.F.		50	.160		.65	8.50	2.84	11.99	18.85
1610	5/16″ thick, 0.4#/L.F.		38	.211		.87	11.20	3.74	15.81	25
1800	3 passes, 3/8″ thick, 0.5#/L.F.		30	.267		1.08	14.15	4.73	19.96	31.50
2010	4 passes, 1/2″ thick, 0.7#/L.F.		22	.364		1.52	19.30	6.45	27.27	43
2200	5 to 6 passes, 3/4″ thick, 1.3#/L.F.		12	.667		2.81	35.50	11.85	50.16	78.50
2400	8 to 11 passes, 1″ thick, 2.4#/L.F.		6	1.333		5.20	71	23.50	99.70	157
2600	For vertical joint welding, add						20%			

05 05 21 – Fastening Methods for Metal

05 05 21.90 Welding Steel			Crew	Daily Output	Labor-Hours	Unit	Material	2014 Bare Costs Labor	Equipment	Total	Total Incl O&P
2700	Overhead joint welding, add					L.F.		300%			
2900	For semi-automatic welding, obstructed joints, deduct							5%			
3000	Exposed joints, deduct							15%			
4000	Cleaning and welding plates, bars, or rods										
4010	to existing beams, columns, or trusses		E-14	12	.667	L.F.	1.08	35.50	11.85	48.43	76.50

05 05 23 – Metal Fastenings

05 05 23.10 Bolts and Hex Nuts			Crew	Daily Output	Labor-Hours	Unit	Material	Labor	Equipment	Total	Total Incl O&P
0010	**BOLTS & HEX NUTS**, Steel, A307										
0100	1/4" diameter, 1/2" long	G	1 Sswk	140	.057	Ea.	.06	2.92		2.98	5.20
0200	1" long	G		140	.057		.07	2.92		2.99	5.25
0300	2" long	G		130	.062		.10	3.14		3.24	5.65
0400	3" long	G		130	.062		.15	3.14		3.29	5.70
0500	4" long	G		120	.067		.17	3.41		3.58	6.20
0600	3/8" diameter, 1" long	G		130	.062		.14	3.14		3.28	5.70
0700	2" long	G		130	.062		.18	3.14		3.32	5.75
0800	3" long	G		120	.067		.24	3.41		3.65	6.25
0900	4" long	G		120	.067		.30	3.41		3.71	6.35
1000	5" long	G		115	.070		.38	3.56		3.94	6.70
1100	1/2" diameter, 1-1/2" long	G		120	.067		.40	3.41		3.81	6.45
1200	2" long	G		120	.067		.46	3.41		3.87	6.50
1300	4" long	G		115	.070		.75	3.56		4.31	7.15
1400	6" long	G		110	.073		1.05	3.72		4.77	7.70
1500	8" long	G		105	.076		1.38	3.89		5.27	8.35
1600	5/8" diameter, 1-1/2" long	G		120	.067		.98	3.41		4.39	7.10
1700	2" long	G		120	.067		1.09	3.41		4.50	7.20
1800	4" long	G		115	.070		1.59	3.56		5.15	8.05
1900	6" long	G		110	.073		2.05	3.72		5.77	8.80
2000	8" long	G		105	.076		3.06	3.89		6.95	10.20
2100	10" long	G		100	.080		3.87	4.09		7.96	11.45
2200	3/4" diameter, 2" long	G		120	.067		1.15	3.41		4.56	7.25
2300	4" long	G		110	.073		1.65	3.72		5.37	8.35
2400	6" long	G		105	.076		2.12	3.89		6.01	9.20
2500	8" long	G		95	.084		3.20	4.30		7.50	11.10
2600	10" long	G		85	.094		4.20	4.81		9.01	13.10
2700	12" long	G		80	.100		4.92	5.10		10.02	14.40
2800	1" diameter, 3" long	G		105	.076		2.69	3.89		6.58	9.80
2900	6" long	G		90	.089		3.94	4.54		8.48	12.35
3000	12" long	G		75	.107		7.10	5.45		12.55	17.40
3100	For galvanized, add						75%				
3200	For stainless, add						350%				

05 05 23.30 Lag Screws			Crew	Daily Output	Labor-Hours	Unit	Material	Labor	Equipment	Total	Total Incl O&P
0010	**LAG SCREWS**										
0020	Steel, 1/4" diameter, 2" long	G	1 Carp	200	.040	Ea.	.09	1.83		1.92	2.93
0100	3/8" diameter, 3" long	G		150	.053		.29	2.45		2.74	4.09
0200	1/2" diameter, 3" long	G		130	.062		.63	2.82		3.45	5.05
0300	5/8" diameter, 3" long	G		120	.067		1.13	3.06		4.19	5.95

05 05 23.50 Powder Actuated Tools and Fasteners			Crew	Daily Output	Labor-Hours	Unit	Material	Labor	Equipment	Total	Total Incl O&P
0010	**POWDER ACTUATED TOOLS & FASTENERS**										
0020	Stud driver, .22 caliber, single shot					Ea.	147			147	161
0100	.27 caliber, semi automatic, strip					"	435			435	475
0300	Powder load, single shot, .22 cal, power level 2, brown					C	5.30			5.30	5.80
0400	Strip, .27 cal, power level 4, red						7.60			7.60	8.35

05 05 Common Work Results for Metals

05 05 23 – Metal Fastenings

05 05 23.50 Powder Actuated Tools and Fasteners			Crew	Daily Output	Labor-Hours	Unit	Material	Labor	Equipment	Total	Total Incl O&P
							2014 Bare Costs				
0600	Drive pin, .300 x 3/4" long	G	1 Carp	4.80	1.667	C	4.86	76.50		81.36	123
0700	.300 x 3" long with washer	G	"	4	2		12.50	91.50		104	156

05 05 23.55 Rivets			Crew	Daily Output	Labor-Hours	Unit	Material	Labor	Equipment	Total	Total Incl O&P
0010	**RIVETS**										
0100	Aluminum rivet & mandrel, 1/2" grip length x 1/8" diameter	G	1 Carp	4.80	1.667	C	7.65	76.50		84.15	126
0200	3/16" diameter	G		4	2		11.40	91.50		102.90	155
0300	Aluminum rivet, steel mandrel, 1/8" diameter	G		4.80	1.667		10.25	76.50		86.75	129
0400	3/16" diameter	G		4	2		16.45	91.50		107.95	160
0500	Copper rivet, steel mandrel, 1/8" diameter	G		4.80	1.667		9.10	76.50		85.60	128
0800	Stainless rivet & mandrel, 1/8" diameter	G		4.80	1.667		26	76.50		102.50	147
0900	3/16" diameter	G		4	2		36	91.50		127.50	182
1000	Stainless rivet, steel mandrel, 1/8" diameter	G		4.80	1.667		15.65	76.50		92.15	135
1100	3/16" diameter	G		4	2		25.50	91.50		117	170
1200	Steel rivet and mandrel, 1/8" diameter	G		4.80	1.667		7.40	76.50		83.90	126
1300	3/16" diameter	G		4	2		12	91.50		103.50	155
1400	Hand riveting tool, standard					Ea.	71.50			71.50	78.50
1500	Deluxe						380			380	420
1600	Power riveting tool, standard						570			570	625
1700	Deluxe						3,600			3,600	3,950

05 05 23.70 Structural Blind Bolts			Crew	Daily Output	Labor-Hours	Unit	Material	Labor	Equipment	Total	Total Incl O&P
0010	**STRUCTURAL BLIND BOLTS**										
0100	1/4" diameter x 1/4" grip	G	1 Sswk	240	.033	Ea.	1.24	1.70		2.94	4.37
0150	1/2" grip	G		216	.037		1.33	1.89		3.22	4.80
0200	3/8" diameter x 1/2" grip	G		232	.034		1.75	1.76		3.51	5.05
0250	3/4" grip	G		208	.038		1.84	1.97		3.81	5.50
0300	1/2" diameter x 1/2" grip	G		224	.036		3.99	1.82		5.81	7.60
0350	3/4" grip	G		200	.040		5.60	2.04		7.64	9.75
0400	5/8" diameter x 3/4" grip	G		216	.037		8.25	1.89		10.14	12.45
0450	1" grip	G		192	.042		9.50	2.13		11.63	14.20

05 05 23.80 Vibration and Bearing Pads			Crew	Daily Output	Labor-Hours	Unit	Material	Labor	Equipment	Total	Total Incl O&P
0010	**VIBRATION & BEARING PADS**										
0300	Laminated synthetic rubber impregnated cotton duck, 1/2" thick		2 Sswk	24	.667	S.F.	69	34		103	136
0400	1" thick			20	.800		135	41		176	221
0600	Neoprene bearing pads, 1/2" thick			24	.667		26.50	34		60.50	89
0700	1" thick			20	.800		52.50	41		93.50	130
0900	Fabric reinforced neoprene, 5000 psi, 1/2" thick			24	.667		11.50	34		45.50	72.50
1000	1" thick			20	.800		23	41		64	97.50
1200	Felt surfaced vinyl pads, cork and sisal, 5/8" thick			24	.667		29	34		63	92
1300	1" thick			20	.800		52.50	41		93.50	130
1500	Teflon bonded to 10 ga. carbon steel, 1/32" layer			24	.667		51.50	34		85.50	117
1600	3/32" layer			24	.667		77	34		111	145
1800	Bonded to 10 ga. stainless steel, 1/32" layer			24	.667		91	34		125	160
1900	3/32" layer			24	.667		120	34		154	192
2100	Circular machine leveling pad & stud					Kip	6.45			6.45	7.10

05 05 23.85 Weld Shear Connectors			Crew	Daily Output	Labor-Hours	Unit	Material	Labor	Equipment	Total	Total Incl O&P
0010	**WELD SHEAR CONNECTORS**										
0020	3/4" diameter, 3-3/16" long	G	E-10	960	.017	Ea.	.53	.87	.49	1.89	2.66
0030	3-3/8" long	G		950	.017		.56	.88	.50	1.94	2.71
0200	3-7/8" long	G		945	.017		.60	.88	.50	1.98	2.77
0300	4-3/16" long	G		935	.017		.63	.89	.51	2.03	2.82
0500	4-7/8" long	G		930	.017		.70	.90	.51	2.11	2.91
0600	5-3/16" long	G		920	.017		.73	.91	.52	2.16	2.97

05 05 23 – Metal Fastenings

05 05 23.85 Weld Shear Connectors			Crew	Daily Output	Labor-Hours	Unit	Material	Labor	Equipment	Total	Total Incl O&P
							2014 Bare Costs				
0800	5-3/8" long	G	E-10	910	.018	Ea.	.74	.92	.52	2.18	3
0900	6-3/16" long	G		905	.018		.81	.92	.52	2.25	3.10
1000	7-3/16" long	G		895	.018		1	.93	.53	2.46	3.32
1100	8-3/16" long	G		890	.018		1.10	.94	.53	2.57	3.45
1500	7/8" diameter, 3-11/16" long	G		920	.017		.86	.91	.52	2.29	3.12
1600	4-3/16" long	G		910	.018		.93	.92	.52	2.37	3.21
1700	5-3/16" long	G		905	.018		1.05	.92	.52	2.49	3.37
1800	6-3/16" long	G		895	.018		1.17	.93	.53	2.63	3.51
1900	7-3/16" long	G		890	.018		1.30	.94	.53	2.77	3.67
2000	8-3/16" long	G		880	.018		1.42	.95	.54	2.91	3.83

05 05 23.87 Weld Studs			Crew	Daily Output	Labor-Hours	Unit	Material	Labor	Equipment	Total	Total Incl O&P
0010	**WELD STUDS**										
0020	1/4" diameter, 2-11/16" long	G	E-10	1120	.014	Ea.	.35	.74	.42	1.51	2.17
0100	4-1/8" long	G		1080	.015		.33	.77	.44	1.54	2.20
0200	3/8" diameter, 4-1/8" long	G		1080	.015		.38	.77	.44	1.59	2.26
0300	6-1/8" long	G		1040	.015		.49	.80	.46	1.75	2.45
0400	1/2" diameter, 2-1/8" long	G		1040	.015		.35	.80	.46	1.61	2.30
0500	3-1/8" long	G		1025	.016		.43	.81	.46	1.70	2.42
0600	4-1/8" long	G		1010	.016		.50	.83	.47	1.80	2.53
0700	5-5/16" long	G		990	.016		.62	.84	.48	1.94	2.70
0800	6-1/8" long	G		975	.016		.67	.86	.49	2.02	2.79
0900	8-1/8" long	G		960	.017		.95	.87	.49	2.31	3.11
1000	5/8" diameter, 2-11/16" long	G		1000	.016		.61	.83	.48	1.92	2.67
1010	4-3/16" long	G		990	.016		.76	.84	.48	2.08	2.86
1100	6-9/16" long	G		975	.016		.99	.86	.49	2.34	3.14
1200	8-3/16" long	G		960	.017		1.33	.87	.49	2.69	3.53

05 05 23.90 Welding Rod		Crew	Daily Output	Labor-Hours	Unit	Material	Labor	Equipment	Total	Total Incl O&P
0010	**WELDING ROD**									
0020	Steel, type 6011, 1/8" diam., less than 500#				Lb.	2.16			2.16	2.38
0100	500# to 2,000#					1.95			1.95	2.15
0200	2,000# to 5,000#					1.83			1.83	2.02
0300	5/32" diameter, less than 500#					2.20			2.20	2.42
0310	500# to 2,000#					1.98			1.98	2.18
0320	2,000# to 5,000#					1.86			1.86	2.05
0400	3/16" diam., less than 500#					2.46			2.46	2.71
0500	500# to 2,000#					2.22			2.22	2.44
0600	2,000# to 5,000#					2.09			2.09	2.30
0620	Steel, type 6010, 1/8" diam., less than 500#					2.22			2.22	2.44
0630	500# to 2,000#					2			2	2.20
0640	2,000# to 5,000#					1.88			1.88	2.07
0650	Steel, type 7018 Low Hydrogen, 1/8" diam., less than 500#					2.15			2.15	2.37
0660	500# to 2,000#					1.94			1.94	2.13
0670	2,000# to 5,000#					1.82			1.82	2.01
0700	Steel, type 7024 Jet Weld, 1/8" diam., less than 500#					2.43			2.43	2.67
0710	500# to 2,000#					2.19			2.19	2.41
0720	2,000# to 5,000#					2.06			2.06	2.26
1550	Aluminum, type 4043 TIG, 1/8" diam., less than 10#					5.10			5.10	5.65
1560	10# to 60#					4.61			4.61	5.05
1570	Over 60#					4.33			4.33	4.77
1600	Aluminum, type 5356 TIG, 1/8" diam., less than 10#					5.45			5.45	6
1610	10# to 60#					4.90			4.90	5.40
1620	Over 60#					4.61			4.61	5.05

05 05 Common Work Results for Metals

05 05 23 – Metal Fastenings

05 05 23.90 Welding Rod			Crew	Daily Output	Labor-Hours	Unit	2014 Bare Costs Material	Labor	Equipment	Total	Total Incl O&P
1900	Cast iron, type 8 Nickel, 1/8″ diam., less than 500#					Lb.	22			22	24
1910	500# to 1,000#						19.75			19.75	21.50
1920	Over 1,000#						18.55			18.55	20.50
2000	Stainless steel, type 316/316L, 1/8″ diam., less than 500#						7.10			7.10	7.85
2100	500# to 1000#						6.40			6.40	7.05
2220	Over 1000#						6.05			6.05	6.65

05 12 Structural Steel Framing

05 12 23 – Structural Steel for Buildings

05 12 23.15 Columns, Lightweight

Code	Description		Crew	Daily Output	Labor-Hours	Unit	Material	Labor	Equipment	Total	Total Incl O&P
0010	**COLUMNS, LIGHTWEIGHT**										
1000	Lightweight units (lally), 3-1/2″ diameter		E-2	780	.072	L.F.	8.25	3.60	1.96	13.81	17.35
1050	4″ diameter		″	900	.062	″	10.15	3.12	1.70	14.97	18.30
5800	Adjustable jack post, 8′ maximum height, 2-3/4″ diameter	G				Ea.	52			52	57
5850	4″ diameter	G				″	83			83	91.50

05 12 23.17 Columns, Structural

Code	Description		Crew	Daily Output	Labor-Hours	Unit	Material	Labor	Equipment	Total	Total Incl O&P
0010	**COLUMNS, STRUCTURAL**										
0015	Made from recycled materials										
0020	Shop fab'd for 100-ton, 1-2 story project, bolted connections										
0800	Steel, concrete filled, extra strong pipe, 3-1/2″ diameter		E-2	660	.085	L.F.	43.50	4.26	2.32	50.08	58
0830	4″ diameter			780	.072		48.50	3.60	1.96	54.06	62
0890	5″ diameter			1020	.055		58	2.76	1.50	62.26	70.50
0930	6″ diameter			1200	.047		77	2.34	1.27	80.61	90
0940	8″ diameter			1100	.051		77	2.56	1.39	80.95	90.50
1100	For galvanizing, add					Lb.	.25			.25	.28
1300	For web ties, angles, etc., add per added lb.		1 Sswk	945	.008		1.33	.43		1.76	2.22
1500	Steel pipe, extra strong, no concrete, 3″ to 5″ diameter	G	E-2	16000	.004		1.33	.18	.10	1.61	1.87
1600	6″ to 12″ diameter	G		14000	.004		1.33	.20	.11	1.64	1.92
1700	Steel pipe, extra strong, no concrete, 3″ diameter x 12′-0″	G		60	.933	Ea.	163	47	25.50	235.50	287
1750	4″ diameter x 12′-0″	G		58	.966		238	48.50	26.50	313	375
1800	6″ diameter x 12′-0″	G		54	1.037		455	52	28.50	535.50	620
1850	8″ diameter x 14′-0″	G		50	1.120		805	56	30.50	891.50	1,025
1900	10″ diameter x 16′-0″	G		48	1.167		1,150	58.50	32	1,240.50	1,400
1950	12″ diameter x 18′-0″	G		45	1.244		1,550	62.50	34	1,646.50	1,875
3300	Structural tubing, square, A500GrB, 4″ to 6″ square, light section	G		11270	.005	Lb.	1.33	.25	.14	1.72	2.03
3600	Heavy section	G		32000	.002	″	1.33	.09	.05	1.47	1.66
4000	Concrete filled, add					L.F.	4.16			4.16	4.57
4500	Structural tubing, square, 4″ x 4″ x 1/4″ x 12′-0″	G	E-2	58	.966	Ea.	219	48.50	26.50	294	350
4550	6″ x 6″ x 1/4″ x 12′-0″	G		54	1.037		360	52	28.50	440.50	515
4600	8″ x 8″ x 3/8″ x 14′-0″	G		50	1.120		775	56	30.50	861.50	985
4650	10″ x 10″ x 1/2″ x 16′-0″	G		48	1.167		1,450	58.50	32	1,540.50	1,700
5100	Structural tubing, rect., 5″ to 6″ wide, light section	G		8000	.007	Lb.	1.33	.35	.19	1.87	2.27
5200	Heavy section	G		12000	.005		1.33	.23	.13	1.69	2
5300	7″ to 10″ wide, light section	G		15000	.004		1.33	.19	.10	1.62	1.89
5400	Heavy section	G		18000	.003		1.33	.16	.08	1.57	1.82
5500	Structural tubing, rect., 5″ x 3″ x 1/4″ x 12′-0″	G		58	.966	Ea.	212	48.50	26.50	287	345
5550	6″ x 4″ x 5/16″ x 12′-0″	G		54	1.037		330	52	28.50	410.50	485
5600	8″ x 4″ x 3/8″ x 12′-0″	G		54	1.037		485	52	28.50	565.50	650
5650	10″ x 6″ x 3/8″ x 14′-0″	G		50	1.120		775	56	30.50	861.50	985
5700	12″ x 8″ x 1/2″ x 16′-0″	G		48	1.167		1,425	58.50	32	1,515.50	1,700
6800	W Shape, A992 steel, 2 tier, W8 x 24	G		1080	.052	L.F.	35	2.60	1.42	39.02	44.50

05 12 Structural Steel Framing

05 12 23 – Structural Steel for Buildings

05 12 23.17 Columns, Structural			Crew	Daily Output	Labor-Hours	Unit	2014 Bare Costs Material	Labor	Equipment	Total	Total Incl O&P
6850	W8 x 31	G	E-2	1080	.052	L.F.	45	2.60	1.42	49.02	55.50
6900	W8 x 48	G		1032	.054		70	2.72	1.48	74.20	83.50
6950	W8 x 67	G		984	.057		97.50	2.86	1.55	101.91	114
7000	W10 x 45	G		1032	.054		65.50	2.72	1.48	69.70	78.50
7050	W10 x 68	G		984	.057		99	2.86	1.55	103.41	116
7100	W10 x 112	G		960	.058		163	2.93	1.59	167.52	187
7150	W12 x 50	G		1032	.054		73	2.72	1.48	77.20	86.50
7200	W12 x 87	G		984	.057		127	2.86	1.55	131.41	146
7250	W12 x 120	G		960	.058		175	2.93	1.59	179.52	199
7300	W12 x 190	G		912	.061		277	3.08	1.68	281.76	310
7350	W14 x 74	G		984	.057		108	2.86	1.55	112.41	126
7400	W14 x 120	G		960	.058		175	2.93	1.59	179.52	199
7450	W14 x 176	G		912	.061		257	3.08	1.68	261.76	289
8090	For projects 75 to 99 tons, add					All	10%				
8092	50 to 74 tons, add						20%				
8094	25 to 49 tons, add						30%	10%			
8096	10 to 24 tons, add						50%	25%			
8098	2 to 9 tons, add						75%	50%			
8099	Less than 2 tons, add						100%	100%			

05 12 23.20 Curb Edging

Line	Description		Crew	Daily Output	Labor-Hours	Unit	Material	Labor	Equipment	Total	Total Incl O&P
0010	**CURB EDGING**										
0020	Steel angle w/anchors, shop fabricated, on forms, 1" x 1", 0.8#/L.F.	G	E-4	350	.091	L.F.	1.67	4.72	.41	6.80	10.65
0100	2" x 2" angles, 3.92#/L.F.	G		330	.097		6.65	5	.43	12.08	16.60
0200	3" x 3" angles, 6.1#/L.F.	G		300	.107		10.50	5.50	.47	16.47	22
0300	4" x 4" angles, 8.2#/L.F.	G		275	.116		13.85	6	.52	20.37	26.50
1000	6" x 4" angles, 12.3#/L.F.	G		250	.128		20.50	6.60	.57	27.67	35
1050	Steel channels with anchors, on forms, 3" channel, 5#/L.F.	G		290	.110		8.35	5.70	.49	14.54	19.80
1100	4" channel, 5.4#/L.F.	G		270	.119		9	6.10	.53	15.63	21.50
1200	6" channel, 8.2#/L.F.	G		255	.125		13.85	6.50	.56	20.91	27.50
1300	8" channel, 11.5#/L.F.	G		225	.142		19.10	7.35	.63	27.08	34.50
1400	10" channel, 15.3#/L.F.	G		180	.178		25	9.15	.79	34.94	44.50
1500	12" channel, 20.7#/L.F.	G		140	.229		33.50	11.80	1.01	46.31	59
2000	For curved edging, add						35%	10%			

05 12 23.40 Lightweight Framing

Line	Description		Crew	Daily Output	Labor-Hours	Unit	Material	Labor	Equipment	Total	Total Incl O&P
0010	**LIGHTWEIGHT FRAMING**										
0015	Made from recycled materials										
0400	Angle framing, field fabricated, 4" and larger	G	E-3	440	.055	Lb.	.77	2.82	.32	3.91	6.20
0450	Less than 4" angles	G		265	.091		.80	4.69	.54	6.03	9.75
0600	Channel framing, field fabricated, 8" and larger	G		500	.048		.80	2.49	.28	3.57	5.55
0650	Less than 8" channels	G		335	.072		.80	3.71	.42	4.93	7.90
1000	Continuous slotted channel framing system, shop fab, simple framing	G	2 Sswk	2400	.007		4.11	.34		4.45	5.10
1200	Complex framing	G	"	1600	.010		4.64	.51		5.15	6
1250	Plate & bar stock for reinforcing beams and trusses	G					1.46			1.46	1.60
1300	Cross bracing, rods, shop fabricated, 3/4" diameter	G	E-3	700	.034		1.59	1.78	.20	3.57	5.10
1310	7/8" diameter	G		850	.028		1.59	1.46	.17	3.22	4.51
1320	1" diameter	G		1000	.024		1.59	1.24	.14	2.97	4.10
1330	Angle, 5" x 5" x 3/8"	G		2800	.009		1.59	.44	.05	2.08	2.59
1350	Hanging lintels, shop fabricated	G		850	.028		1.59	1.46	.17	3.22	4.51
1380	Roof frames, shop fabricated, 3'-0" square, 5' span	G	E-2	4200	.013		1.59	.67	.36	2.62	3.29
1400	Tie rod, not upset, 1-1/2" to 4" diameter, with turnbuckle	G	2 Sswk	800	.020		1.72	1.02		2.74	3.69
1420	No turnbuckle	G		700	.023		1.66	1.17		2.83	3.88
1500	Upset, 1-3/4" to 4" diameter, with turnbuckle	G		800	.020		1.72	1.02		2.74	3.69

05 12 Structural Steel Framing

05 12 23 – Structural Steel for Buildings

05 12 23.40 Lightweight Framing			Crew	Daily Output	Labor-Hours	Unit	2014 Bare Costs Material	Labor	Equipment	Total	Total Incl O&P
1520	No turnbuckle	G	2 Sswk	700	.023	Lb.	1.66	1.17		2.83	3.88

05 12 23.45 Lintels

			Crew	Daily Output	Labor-Hours	Unit	Material	Labor	Equipment	Total	Total Incl O&P
0010	**LINTELS**										
0015	Made from recycled materials										
0020	Plain steel angles, shop fabricated, under 500 lb.	G	1 Bric	550	.015	Lb.	1.02	.66		1.68	2.13
0100	500 to 1000 lb.	G		640	.013		.99	.57		1.56	1.96
0200	1,000 to 2,000 lb.	G		640	.013		.97	.57		1.54	1.93
0300	2,000 to 4,000 lb.	G		640	.013		.94	.57		1.51	1.90
0500	For built-up angles and plates, add to above	G					.33			.33	.36
0700	For engineering, add to above						.13			.13	.15
0900	For galvanizing, add to above, under 500 lb.						.30			.30	.33
0950	500 to 2,000 lb.						.28			.28	.30
1000	Over 2,000 lb.						.25			.25	.28

05 12 23.60 Pipe Support Framing

			Crew	Daily Output	Labor-Hours	Unit	Material	Labor	Equipment	Total	Total Incl O&P
0010	**PIPE SUPPORT FRAMING**										
0020	Under 10#/L.F., shop fabricated	G	E-4	3900	.008	Lb.	1.78	.42	.04	2.24	2.74
0200	10.1 to 15#/L.F.	G		4300	.007		1.75	.38	.03	2.16	2.64
0400	15.1 to 20#/L.F.	G		4800	.007		1.72	.34	.03	2.09	2.53
0600	Over 20#/L.F.	G		5400	.006		1.70	.31	.03	2.04	2.44

05 12 23.65 Plates

			Crew	Daily Output	Labor-Hours	Unit	Material	Labor	Equipment	Total	Total Incl O&P
0010	**PLATES**										
0015	Made from recycled materials										
0020	For connections & stiffener plates, shop fabricated										
0050	1/8" thick (5.1 lb./S.F.)	G				S.F.	6.75			6.75	7.45
0100	1/4" thick (10.2 lb./S.F.)	G					13.50			13.50	14.85
0300	3/8" thick (15.3 lb./S.F.)	G					20.50			20.50	22.50
0400	1/2" thick (20.4 lb./S.F.)	G					27			27	29.50
0450	3/4" thick (30.6 lb./S.F.)	G					40.50			40.50	44.50
0500	1" thick (40.8 lb./S.F.)	G					54			54	59.50
2000	Steel plate, warehouse prices, no shop fabrication										
2100	1/4" thick (10.2 lb./S.F.)	G				S.F.	8.05			8.05	8.85

05 12 23.70 Stressed Skin Steel Roof and Ceiling System

			Crew	Daily Output	Labor-Hours	Unit	Material	Labor	Equipment	Total	Total Incl O&P
0010	**STRESSED SKIN STEEL ROOF & CEILING SYSTEM**										
0020	Double panel flat roof, spans to 100'	G	E-2	1150	.049	S.F.	10.60	2.44	1.33	14.37	17.25
0100	Double panel convex roof, spans to 200'	G		960	.058		17.25	2.93	1.59	21.77	25.50
0200	Double panel arched roof, spans to 300'	G		760	.074		26.50	3.70	2.01	32.21	37.50

05 12 23.75 Structural Steel Members

			Crew	Daily Output	Labor-Hours	Unit	Material	Labor	Equipment	Total	Total Incl O&P
0010	**STRUCTURAL STEEL MEMBERS**										
0015	Made from recycled materials										
0020	Shop fab'd for 100-ton, 1-2 story project, bolted connections										
0100	Beam or girder, W 6 x 9	G	E-2	600	.093	L.F.	13.10	4.68	2.55	20.33	25
0120	x 15	G		600	.093		22	4.68	2.55	29.23	35
0140	x 20	G		600	.093		29	4.68	2.55	36.23	43
0300	W 8 x 10	G		600	.093		14.60	4.68	2.55	21.83	27
0320	x 15	G		600	.093		22	4.68	2.55	29.23	35
0350	x 21	G		600	.093		30.50	4.68	2.55	37.73	44.50
0360	x 24	G		550	.102		35	5.10	2.78	42.88	50.50
0370	x 28	G		550	.102		41	5.10	2.78	48.88	57
0500	x 31	G		550	.102		45	5.10	2.78	52.88	61.50
0520	x 35	G		550	.102		51	5.10	2.78	58.88	68
0540	x 48	G		550	.102		70	5.10	2.78	77.88	89

05 12 Structural Steel Framing

05 12 23 – Structural Steel for Buildings

05 12 23.75 Structural Steel Members			Crew	Daily Output	Labor-Hours	Unit	Material	2014 Bare Costs Labor	Equipment	Total	Total Incl O&P
0600	W 10 x 12	G	E-2	600	.093	L.F.	17.50	4.68	2.55	24.73	30
0620	x 15	G		600	.093		22	4.68	2.55	29.23	35
0700	x 22	G		600	.093		32	4.68	2.55	39.23	46.50
0720	x 26	G		600	.093		38	4.68	2.55	45.23	52.50
0740	x 33	G		550	.102		48	5.10	2.78	55.88	65
0900	x 49	G		550	.102		71.50	5.10	2.78	79.38	90.50
1100	W 12 x 16	G		880	.064		23.50	3.19	1.74	28.43	33
1300	x 22	G		880	.064		32	3.19	1.74	36.93	43
1500	x 26	G		880	.064		38	3.19	1.74	42.93	49
1520	x 35	G		810	.069		51	3.47	1.89	56.36	64
1560	x 50	G		750	.075		73	3.75	2.04	78.79	88.50
1580	x 58	G		750	.075		84.50	3.75	2.04	90.29	102
1700	x 72	G		640	.088		105	4.39	2.39	111.78	125
1740	x 87	G		640	.088		127	4.39	2.39	133.78	149
1900	W 14 x 26	G		990	.057		38	2.84	1.54	42.38	48
2100	x 30	G		900	.062		43.50	3.12	1.70	48.32	55
2300	x 34	G		810	.069		49.50	3.47	1.89	54.86	62.50
2320	x 43	G		810	.069		62.50	3.47	1.89	67.86	77
2340	x 53	G		800	.070		77.50	3.51	1.91	82.92	93
2360	x 74	G		760	.074		108	3.70	2.01	113.71	128
2380	x 90	G		740	.076		131	3.80	2.07	136.87	153
2500	x 120	G		720	.078		175	3.90	2.12	181.02	201
2700	W 16 x 26	G		1000	.056		38	2.81	1.53	42.34	48
2900	x 31	G		900	.062		45	3.12	1.70	49.82	56.50
3100	x 40	G		800	.070		58.50	3.51	1.91	63.92	72
3120	x 50	G		800	.070		73	3.51	1.91	78.42	88
3140	x 67	G		760	.074		97.50	3.70	2.01	103.21	116
3300	W 18 x 35	G	E-5	960	.083		51	4.22	1.74	56.96	65
3500	x 40	G		960	.083		58.50	4.22	1.74	64.46	73
3520	x 46	G		960	.083		67	4.22	1.74	72.96	83
3700	x 50	G		912	.088		73	4.44	1.83	79.27	89.50
3900	x 55	G		912	.088		80	4.44	1.83	86.27	97.50
3920	x 65	G		900	.089		94.50	4.50	1.86	100.86	114
3940	x 76	G		900	.089		111	4.50	1.86	117.36	132
3960	x 86	G		900	.089		125	4.50	1.86	131.36	148
3980	x 106	G		900	.089		155	4.50	1.86	161.36	180
4100	W 21 x 44	G		1064	.075		64	3.81	1.57	69.38	79
4300	x 50	G		1064	.075		73	3.81	1.57	78.38	88.50
4500	x 62	G		1036	.077		90.50	3.91	1.61	96.02	108
4700	x 68	G		1036	.077		99	3.91	1.61	104.52	117
4720	x 83	G		1000	.080		121	4.05	1.67	126.72	142
4740	x 93	G		1000	.080		136	4.05	1.67	141.72	158
4760	x 101	G		1000	.080		147	4.05	1.67	152.72	171
4780	x 122	G		1000	.080		178	4.05	1.67	183.72	205
4900	W 24 x 55	G		1110	.072		80	3.65	1.51	85.16	96
5100	x 62	G		1110	.072		90.50	3.65	1.51	95.66	107
5300	x 68	G		1110	.072		99	3.65	1.51	104.16	117
5500	x 76	G		1110	.072		111	3.65	1.51	116.16	130
5700	x 84	G		1080	.074		122	3.75	1.55	127.30	143
5720	x 94	G		1080	.074		137	3.75	1.55	142.30	159
5740	x 104	G		1050	.076		152	3.86	1.59	157.45	175
5760	x 117	G		1050	.076		171	3.86	1.59	176.45	196
5780	x 146	G		1050	.076		213	3.86	1.59	218.45	242

05 12 Structural Steel Framing

05 12 23 – Structural Steel for Buildings

05 12 23.75 Structural Steel Members			Crew	Daily Output	Labor-Hours	Unit	2014 Bare Costs Material	Labor	Equipment	Total	Total Incl O&P
5800	W 27 x 84	G	E-5	1190	.067	L.F.	122	3.41	1.40	126.81	142
5900	x 94	G		1190	.067		137	3.41	1.40	141.81	158
5920	x 114	G		1150	.070		166	3.52	1.45	170.97	191
5940	x 146	G		1150	.070		213	3.52	1.45	217.97	242
5960	x 161	G		1150	.070		235	3.52	1.45	239.97	266
6100	W 30 x 99	G		1200	.067		144	3.38	1.39	148.77	166
6300	x 108	G		1200	.067		157	3.38	1.39	161.77	180
6500	x 116	G		1160	.069		169	3.49	1.44	173.93	194
6520	x 132	G		1160	.069		192	3.49	1.44	196.93	220
6540	x 148	G		1160	.069		216	3.49	1.44	220.93	245
6560	x 173	G		1120	.071		252	3.62	1.49	257.11	285
6580	x 191	G		1120	.071		278	3.62	1.49	283.11	315
6700	W 33 x 118	G		1176	.068		172	3.45	1.42	176.87	196
6900	x 130	G		1134	.071		189	3.57	1.47	194.04	216
7100	x 141	G		1134	.071		206	3.57	1.47	211.04	234
7120	x 169	G		1100	.073		246	3.68	1.52	251.20	279
7140	x 201	G		1100	.073		293	3.68	1.52	298.20	330
7300	W 36 x 135	G		1170	.068		197	3.46	1.43	201.89	224
7500	x 150	G		1170	.068		219	3.46	1.43	223.89	248
7600	x 170	G		1150	.070		248	3.52	1.45	252.97	281
7700	x 194	G		1125	.071		283	3.60	1.49	288.09	320
7900	x 231	G		1125	.071		335	3.60	1.49	340.09	380
7920	x 262	G		1035	.077		380	3.92	1.61	385.53	430
8100	x 302	G		1035	.077		440	3.92	1.61	445.53	495
8490	For projects 75 to 99 tons, add						10%				
8492	50 to 74 tons, add						20%				
8494	25 to 49 tons, add						30%	10%			
8496	10 to 24 tons, add						50%	25%			
8498	2 to 9 tons, add						75%	50%			
8499	Less than 2 tons, add						100%	100%			

05 12 23.77 Structural Steel Projects			Crew	Daily Output	Labor-Hours	Unit	Material	Labor	Equipment	Total	Total Incl O&P
0010	**STRUCTURAL STEEL PROJECTS**										
0015	Made from recycled materials										
0020	Shop fab'd for 100-ton, 1-2 story project, bolted connections										
1300	Industrial bldgs., 1 story, beams & girders, steel bearing R050521-20	G	E-5	12.90	6.202	Ton	2,650	315	130	3,095	3,600
1400	Masonry bearing	G	"	10	8	"	2,650	405	167	3,222	3,800
1500	Industrial bldgs., 1 story, under 10 tons,	G									
1510	steel from warehouse, trucked	G	E-2	7.50	7.467	Ton	3,175	375	204	3,754	4,350
1600	1 story with roof trusses, steel bearing	G	E-5	10.60	7.547		3,125	380	158	3,663	4,275
1700	Masonry bearing	G	"	8.30	9.639		3,125	490	201	3,816	4,500
1900	Monumental structures, banks, stores, etc., simple connections	G	E-6	13	9.846		2,650	500	143	3,293	3,950
2000	Moment/composite connections	G		9	14.222		4,400	720	207	5,327	6,325
2800	Power stations, fossil fuels, simple connections	G		11	11.636		2,650	590	170	3,410	4,125
2900	Moment/composite connections	G		5.70	22.456		3,975	1,150	325	5,450	6,675
2950	Nuclear fuels, non-safety steel, simple connections	G		7	18.286		2,650	925	266	3,841	4,825
3000	Moment/composite connections	G		5.50	23.273		3,975	1,175	340	5,490	6,775
3040	Safety steel, simple connections	G		2.50	51.200		3,875	2,600	745	7,220	9,550
3070	Moment/composite connections	G		1.50	85.333		5,100	4,325	1,250	10,675	14,400
3100	Roof trusses, simple connections	G	E-5	13	6.154		3,700	310	129	4,139	4,750
3200	Moment/composite connections	G	"	8.30	9.639		4,500	490	201	5,191	6,000
3810	Fabrication shop costs (incl in project bare material cost, above)										
3820	Mini mill base price, Grade A992	G				Ton	770			770	845

05 12 23.77 Structural Steel Projects			Crew	Daily Output	Labor-Hours	Unit	2014 Bare Costs Material	Labor	Equipment	Total	Total Incl O&P
3830	Mill extras plus delivery to warehouse					Ton	275			275	305
3835	Delivery from warehouse to fabrication shop						85			85	93.50
3840	Shop extra for shop drawings and detailing						295			295	325
3850	Shop fabricating and handling						950			950	1,050
3860	Shop sandblasting and primer coat of paint						155			155	171
3870	Shop delivery to the job site						120			120	132
3880	Total material cost, shop fabricated, primed, delivered					↓	2,650			2,650	2,925
3900	High strength steel mill spec extras:										
3950	A529, A572 (50 ksi) and A36: same as A992 steel (no extra)										
4000	Add to A992 price for A572 (60, 65 ksi)	G				Ton	80			80	88
4100	A242 and A588 Weathering	G				"	80			80	88
4200	Mill size extras for W-Shapes: 0 to 30 plf: no extra charge										
4210	Member sizes 31 to 65 plf, deduct	G				Ton	.01			.01	.01
4220	Member sizes 66 to 100 plf, deduct	G					5.65			5.65	6.25
4230	Member sizes 101 to 387 plf, add	G				↓	55.50			55.50	61
4300	Column base plates, light, up to 150 lb.	G	2 Sswk	2000	.008	Lb.	1.46	.41		1.87	2.32
4400	Heavy, over 150 lb.	G	E-2	7500	.007	"	1.52	.37	.20	2.09	2.54
5390	For projects 75 to 99 tons, add					Ton	10%				
5392	50 to 74 tons, add						20%				
5394	25 to 49 tons, add						30%	10%			
5396	10 to 24 tons, add						50%	25%			
5398	2 to 9 tons, add						75%	50%			
5399	Less than 2 tons, add					↓	100%	100%			

05 12 23.78 Structural Steel Secondary Members		Crew	Daily Output	Labor-Hours	Unit	Material	Labor	Equipment	Total	Total Incl O&P
0010	**STRUCTURAL STEEL SECONDARY MEMBERS**									
0015	Made from recycled materials									
0020	Shop fabricated for 20-ton girt/purlin framing package, materials only									
0100	Girts/purlins, C/Z-shapes, includes clips and bolts									
0110	6" x 2-1/2" x 2-1/2", 16 ga., 3.0 lb./L.F.				L.F.	3.58			3.58	3.94
0115	14 ga., 3.5 lb./L.F.					4.17			4.17	4.59
0120	8" x 2-3/4" x 2-3/4", 16 ga., 3.4 lb./L.F.					4.05			4.05	4.46
0125	14 ga., 4.1 lb./L.F.					4.89			4.89	5.40
0130	12 ga., 5.6 lb./L.F.					6.70			6.70	7.35
0135	10" x 3-1/2" x 3-1/2", 14 ga., 4.7 lb./L.F.					5.60			5.60	6.15
0140	12 ga., 6.7 lb./L.F.					8			8	8.80
0145	12" x 3-1/2" x 3-1/2", 14 ga., 5.3 lb./L.F.					6.30			6.30	6.95
0150	12 ga., 7.4 lb./L.F.				↓	8.80			8.80	9.70
0200	Eave struts, C-shape, includes clips and bolts									
0210	6" x 4" x 3", 16 ga., 3.1 lb./L.F.				L.F.	3.70			3.70	4.07
0215	14 ga., 3.9 lb./L.F.					4.65			4.65	5.10
0220	8" x 4" x 3", 16 ga., 3.5 lb./L.F.					4.17			4.17	4.59
0225	14 ga., 4.4 lb./L.F.					5.25			5.25	5.75
0230	12 ga., 6.2 lb./L.F.					7.40			7.40	8.15
0235	10" x 5" x 3", 14 ga., 5.2 lb./L.F.					6.20			6.20	6.80
0240	12 ga., 7.3 lb./L.F.					8.70			8.70	9.60
0245	12" x 5" x 4", 14 ga., 6.0 lb./L.F.					7.15			7.15	7.85
0250	12 ga., 8.4 lb./L.F.				↓	10			10	11
0300	Rake/base angle, excludes concrete drilling and expansion anchors									
0310	2" x 2", 14 ga., 1.0 lb./L.F.	2 Sswk	640	.025	L.F.	1.19	1.28		2.47	3.57
0315	3" x 2", 14 ga., 1.3 lb./L.F.		535	.030		1.55	1.53		3.08	4.41
0320	3" x 3", 14 ga., 1.6 lb./L.F.		500	.032		1.91	1.64		3.55	4.99
0325	4" x 3", 14 ga., 1.8 lb./L.F.	↓	480	.033	↓	2.15	1.70		3.85	5.35

05 12 Structural Steel Framing

05 12 23 – Structural Steel for Buildings

05 12 23.78 Structural Steel Secondary Members		Crew	Daily Output	Labor-Hours	Unit	Material	2014 Bare Costs Labor	Equipment	Total	Total Incl O&P
0600	Installation of secondary members, erection only									
0610	Girts, purlins, eave struts, 16 ga., 6" deep	E-18	100	.400	Ea.		20.50	9.60	30.10	45.50
0615	8" deep		80	.500			25.50	12	37.50	57
0620	14 ga., 6" deep		80	.500			25.50	12	37.50	57
0625	8" deep		65	.615			31.50	14.80	46.30	70.50
0630	10" deep		55	.727			37	17.45	54.45	83
0635	12" deep		50	.800			41	19.20	60.20	91
0640	12 ga., 8" deep		50	.800			41	19.20	60.20	91
0645	10" deep		45	.889			45.50	21.50	67	102
0650	12" deep		40	1			51	24	75	114
0900	For less than 20-ton job lots									
0905	For 15 to 19 tons, add					10%				
0910	For 10 to 14 tons, add					25%				
0915	For 5 to 9 tons, add					50%	50%	50%		
0920	For 1 to 4 tons, add					75%	75%	75%		
0925	For less than 1 ton, add					100%	100%	100%		

05 14 Structural Aluminum Framing

05 14 23 – Non-Exposed Structural Aluminum Framing

05 14 23.05 Aluminum Shapes

			Crew	Daily Output	Labor-Hours	Unit	Material	Labor	Equipment	Total	Total Incl O&P
0010	**ALUMINUM SHAPES**										
0015	Made from recycled materials										
0020	Structural shapes, 1" to 10" members, under 1 ton	G	E-2	4000	.014	Lb.	4.18	.70	.38	5.26	6.20
0050	1 to 5 tons	G		4300	.013		3.83	.65	.36	4.84	5.70
0100	Over 5 tons	G		4600	.012		3.65	.61	.33	4.59	5.45
0300	Extrusions, over 5 tons, stock shapes	G		1330	.042		3.48	2.11	1.15	6.74	8.70
0400	Custom shapes	G		1330	.042		3.61	2.11	1.15	6.87	8.80

05 15 Wire Rope Assemblies

05 15 16 – Steel Wire Rope Assemblies

05 15 16.05 Accessories for Steel Wire Rope

			Crew	Daily Output	Labor-Hours	Unit	Material	Labor	Equipment	Total	Total Incl O&P
0010	**ACCESSORIES FOR STEEL WIRE ROPE**										
0015	Made from recycled materials										
1500	Thimbles, heavy duty, 1/4"	G	E-17	160	.100	Ea.	.51	5.20		5.71	9.75
1510	1/2"	G		160	.100		2.23	5.20		7.43	11.65
1520	3/4"	G		105	.152		5.05	7.95		13	19.55
1530	1"	G		52	.308		10.15	16.05		26.20	39.50
1540	1-1/4"	G		38	.421		15.60	22		37.60	55.50
1550	1-1/2"	G		13	1.231		44	64		108	161
1560	1-3/4"	G		8	2		90.50	104		194.50	284
1570	2"	G		6	2.667		132	139		271	390
1580	2-1/4"	G		4	4		178	208		386	565
1600	Clips, 1/4" diameter	G		160	.100		2.64	5.20		7.84	12.10
1610	3/8" diameter	G		160	.100		2.90	5.20		8.10	12.40
1620	1/2" diameter	G		160	.100		4.66	5.20		9.86	14.30
1630	3/4" diameter	G		102	.157		7.55	8.15		15.70	22.50
1640	1" diameter	G		64	.250		12.60	13.05		25.65	37
1650	1-1/4" diameter	G		35	.457		20.50	24		44.50	64.50
1670	1-1/2" diameter	G		26	.615		28	32		60	87

05 15 16 – Steel Wire Rope Assemblies

05 15 16.05 Accessories for Steel Wire Rope			Crew	Daily Output	Labor-Hours	Unit	2014 Bare Costs Material	Labor	Equipment	Total	Total Incl O&P
1680	1-3/4" diameter	G	E-17	16	1	Ea.	65	52		117	164
1690	2" diameter	G		12	1.333		72	69.50		141.50	203
1700	2-1/4" diameter	G		10	1.600		106	83.50		189.50	264
1800	Sockets, open swage, 1/4" diameter	G		160	.100		45.50	5.20		50.70	59
1810	1/2" diameter	G		77	.208		65.50	10.85		76.35	91.50
1820	3/4" diameter	G		19	.842		102	44		146	190
1830	1" diameter	G		9	1.778		183	92.50		275.50	365
1840	1-1/4" diameter	G		5	3.200		254	167		421	575
1850	1-1/2" diameter	G		3	5.333		555	278		833	1,100
1860	1-3/4" diameter	G		3	5.333		985	278		1,263	1,575
1870	2" diameter	G		1.50	10.667		1,500	555		2,055	2,625
1900	Closed swage, 1/4" diameter	G		160	.100		27	5.20		32.20	38.50
1910	1/2" diameter	G		104	.154		46.50	8		54.50	65
1920	3/4" diameter	G		32	.500		69.50	26		95.50	123
1930	1" diameter	G		15	1.067		122	55.50		177.50	232
1940	1-1/4" diameter	G		7	2.286		183	119		302	410
1950	1-1/2" diameter	G		4	4		330	208		538	735
1960	1-3/4" diameter	G		3	5.333		490	278		768	1,025
1970	2" diameter	G		2	8		955	415		1,370	1,775
2000	Open spelter, galv., 1/4" diameter	G		160	.100		58	5.20		63.20	73
2010	1/2" diameter	G		70	.229		60.50	11.90		72.40	87.50
2020	3/4" diameter	G		26	.615		90.50	32		122.50	157
2030	1" diameter	G		10	1.600		252	83.50		335.50	425
2040	1-1/4" diameter	G		5	3.200		360	167		527	690
2050	1-1/2" diameter	G		4	4		765	208		973	1,200
2060	1-3/4" diameter	G		2	8		1,325	415		1,740	2,200
2070	2" diameter	G		1.20	13.333		1,525	695		2,220	2,925
2080	2-1/2" diameter	G		1	16		2,825	835		3,660	4,575
2100	Closed spelter, galv., 1/4" diameter	G		160	.100		48.50	5.20		53.70	62
2110	1/2" diameter	G		88	.182		51.50	9.45		60.95	73.50
2120	3/4" diameter	G		30	.533		78.50	28		106.50	135
2130	1" diameter	G		13	1.231		167	64		231	296
2140	1-1/4" diameter	G		7	2.286		267	119		386	505
2150	1-1/2" diameter	G		6	2.667		575	139		714	875
2160	1-3/4" diameter	G		2.80	5.714		765	298		1,063	1,375
2170	2" diameter	G		2	8		945	415		1,360	1,775
2200	Jaw & jaw turnbuckles, 1/4" x 4"	G		160	.100		16.35	5.20		21.55	27
2250	1/2" x 6"	G		96	.167		20.50	8.70		29.20	38
2260	1/2" x 9"	G		77	.208		27.50	10.85		38.35	49.50
2270	1/2" x 12"	G		66	.242		31	12.65		43.65	56.50
2300	3/4" x 6"	G		38	.421		40.50	22		62.50	83
2310	3/4" x 9"	G		30	.533		45	28		73	98.50
2320	3/4" x 12"	G		28	.571		57.50	30		87.50	116
2330	3/4" x 18"	G		23	.696		69	36		105	140
2350	1" x 6"	G		17	.941		78.50	49		127.50	173
2360	1" x 12"	G		13	1.231		86	64		150	208
2370	1" x 18"	G		10	1.600		129	83.50		212.50	289
2380	1" x 24"	G		9	1.778		142	92.50		234.50	320
2400	1-1/4" x 12"	G		7	2.286		145	119		264	370
2410	1-1/4" x 18"	G		6.50	2.462		179	128		307	425
2420	1-1/4" x 24"	G		5.60	2.857		241	149		390	530
2450	1-1/2" x 12"	G		5.20	3.077		425	160		585	750
2460	1-1/2" x 18"	G		4	4		450	208		658	870

05 15 Wire Rope Assemblies

05 15 16 – Steel Wire Rope Assemblies

05 15 16.05 Accessories for Steel Wire Rope			Crew	Daily Output	Labor-Hours	Unit	2014 Bare Costs Material	Labor	Equipment	Total	Total Incl O&P
2470	1-1/2" x 24"	G	E-17	3.20	5	Ea.	610	261		871	1,125
2500	1-3/4" x 18"	G		3.20	5		920	261		1,181	1,450
2510	1-3/4" x 24"	G		2.80	5.714		1,050	298		1,348	1,675
2550	2" x 24"	G		1.60	10		1,425	520		1,945	2,475

05 15 16.50 Steel Wire Rope

			Crew	Daily Output	Labor-Hours	Unit	Material	Labor	Equipment	Total	Total Incl O&P
0010	**STEEL WIRE ROPE**										
0015	Made from recycled materials										
0020	6 x 19, bright, fiber core, 5000' rolls, 1/2" diameter	G				L.F.	.88			.88	.96
0050	Steel core	G					1.15			1.15	1.27
0100	Fiber core, 1" diameter	G					2.95			2.95	3.25
0150	Steel core	G					3.37			3.37	3.70
0300	6 x 19, galvanized, fiber core, 1/2" diameter	G					1.29			1.29	1.42
0350	Steel core	G					1.48			1.48	1.62
0400	Fiber core, 1" diameter	G					3.78			3.78	4.16
0450	Steel core	G					3.96			3.96	4.36
0500	6 x 7, bright, IPS, fiber core, <500 L.F. w/acc., 1/4" diameter	G	E-17	6400	.003		1.47	.13		1.60	1.85
0510	1/2" diameter	G		2100	.008		3.59	.40		3.99	4.65
0520	3/4" diameter	G		960	.017		6.50	.87		7.37	8.70
0550	6 x 19, bright, IPS, IWRC, <500 L.F. w/acc., 1/4" diameter	G		5760	.003		.97	.14		1.11	1.32
0560	1/2" diameter	G		1730	.009		1.57	.48		2.05	2.57
0570	3/4" diameter	G		770	.021		2.72	1.08		3.80	4.90
0580	1" diameter	G		420	.038		4.61	1.99		6.60	8.55
0590	1-1/4" diameter	G		290	.055		7.65	2.87		10.52	13.45
0600	1-1/2" diameter	G		192	.083		9.40	4.34		13.74	18
0610	1-3/4" diameter	G	E-18	240	.167		15	8.50	4	27.50	35.50
0620	2" diameter	G		160	.250		19.20	12.75	6	37.95	49.50
0630	2-1/4" diameter	G		160	.250		25.50	12.75	6	44.25	57
0650	6 x 37, bright, IPS, IWRC, <500 L.F. w/acc., 1/4" diameter	G	E-17	6400	.003		1.15	.13		1.28	1.50
0660	1/2" diameter	G		1730	.009		1.96	.48		2.44	3
0670	3/4" diameter	G		770	.021		3.16	1.08		4.24	5.40
0680	1" diameter	G		430	.037		5	1.94		6.94	8.90
0690	1-1/4" diameter	G		290	.055		7.60	2.87		10.47	13.40
0700	1-1/2" diameter	G		190	.084		10.85	4.39		15.24	19.70
0710	1-3/4" diameter	G	E-18	260	.154		17.20	7.85	3.69	28.74	36.50
0720	2" diameter	G		200	.200		22.50	10.20	4.80	37.50	47.50
0730	2-1/4" diameter	G		160	.250		29.50	12.75	6	48.25	61
0800	6 x 19 & 6 x 37, swaged, 1/2" diameter	G	E-17	1220	.013		2.43	.68		3.11	3.88
0810	9/16" diameter	G		1120	.014		2.83	.74		3.57	4.42
0820	5/8" diameter	G		930	.017		3.35	.90		4.25	5.25
0830	3/4" diameter	G		640	.025		4.27	1.30		5.57	7
0840	7/8" diameter	G		480	.033		5.40	1.74		7.14	9
0850	1" diameter	G		350	.046		6.55	2.38		8.93	11.45
0860	1-1/8" diameter	G		288	.056		8.10	2.89		10.99	14
0870	1-1/4" diameter	G		230	.070		9.80	3.62		13.42	17.15
0880	1-3/8" diameter	G		192	.083		11.30	4.34		15.64	20
0890	1-1/2" diameter	G	E-18	300	.133		13.75	6.80	3.20	23.75	30.50

05 15 16.60 Galvanized Steel Wire Rope and Accessories

			Crew	Daily Output	Labor-Hours	Unit	Material	Labor	Equipment	Total	Total Incl O&P
0010	**GALVANIZED STEEL WIRE ROPE & ACCESSORIES**										
0015	Made from recycled materials										
3000	Aircraft cable, galvanized, 7 x 7 x 1/8"	G	E-17	5000	.003	L.F.	.17	.17		.34	.48
3100	Clamps, 1/8"	G	"	125	.128	Ea.	1.89	6.65		8.54	13.85

05 15 Wire Rope Assemblies

05 15 16 – Steel Wire Rope Assemblies

05 15 16.70 Temporary Cable Safety Railing		Crew	Daily Output	Labor-Hours	Unit	2014 Bare Costs Material	Labor	Equipment	Total	Total Incl O&P
0010	**TEMPORARY CABLE SAFETY RAILING**, Each 100′ strand incl.									
0020	2 eyebolts, 1 turnbuckle, 100′ cable, 2 thimbles, 6 clips									
0025	Made from recycled materials									
0100	One strand using 1/4″ cable & accessories G	2 Sswk	4	4	C.L.F.	205	204		409	585
0200	1/2″ cable & accessories G	″	2	8	″	430	410		840	1,200

05 21 Steel Joist Framing

05 21 13 – Deep Longspan Steel Joist Framing

05 21 13.50 Deep Longspan Joists

		Crew	Daily Output	Labor-Hours	Unit	Material	Labor	Equipment	Total	Total Incl O&P
0010	**DEEP LONGSPAN JOISTS**									
3010	DLH series, 40-ton job lots, bolted cross bridging, shop primer									
3015	Made from recycled materials									
3040	Spans to 144′ (shipped in 2 pieces) G	E-7	13	6.154	Ton	1,925	310	139	2,374	2,825
3500	For less than 40-ton job lots									
3502	For 30 to 39 tons, add					10%				
3504	20 to 29 tons, add					20%				
3506	10 to 19 tons, add					30%				
3507	5 to 9 tons, add					50%	25%			
3508	1 to 4 tons, add					75%	50%			
3509	Less than 1 ton, add					100%	100%			
4010	SLH series, 40-ton job lots, bolted cross bridging, shop primer									
4040	Spans to 200′ (shipped in 3 pieces) G	E-7	13	6.154	Ton	1,975	310	139	2,424	2,875
6100	For less than 40-ton job lots									
6102	For 30 to 39 tons, add					10%				
6104	20 to 29 tons, add					20%				
6106	10 to 19 tons, add					30%				
6107	5 to 9 tons, add					50%	25%			
6108	1 to 4 tons, add					75%	50%			
6109	Less than 1 ton, add					100%	100%			

05 21 16 – Longspan Steel Joist Framing

05 21 16.50 Longspan Joists

		Crew	Daily Output	Labor-Hours	Unit	Material	Labor	Equipment	Total	Total Incl O&P
0010	**LONGSPAN JOISTS**									
2000	LH series, 40-ton job lots, bolted cross bridging, shop primer									
2015	Made from recycled materials									
2040	Longspan joists, LH series, up to 96′ G	E-7	13	6.154	Ton	1,825	310	139	2,274	2,725
2600	For less than 40-ton job lots									
2602	For 30 to 39 tons, add					10%				
2604	20 to 29 tons, add					20%				
2606	10 to 19 tons, add					30%				
2607	5 to 9 tons, add					50%	25%			
2608	1 to 4 tons, add					75%	50%			
2609	Less than 1 ton, add					100%	100%			
6000	For welded cross bridging, add						30%			

05 21 19 – Open Web Steel Joist Framing

05 21 19.10 Open Web Joists

		Crew	Daily Output	Labor-Hours	Unit	Material	Labor	Equipment	Total	Total Incl O&P
0010	**OPEN WEB JOISTS**									
0015	Made from recycled materials									
0050	K series, 40-ton lots, horiz. bridging, spans to 30′, shop primer G	E-7	12	6.667	Ton	1,650	340	151	2,141	2,575
0440	K series, 30′ to 50′ spans G	″	17	4.706	″	1,625	238	107	1,970	2,325

05 21 Steel Joist Framing

05 21 19 – Open Web Steel Joist Framing

05 21 19.10 Open Web Joists			Crew	Daily Output	Labor-Hours	Unit	2014 Bare Costs Material	Labor	Equipment	Total	Total Incl O&P
0800	For less than 40-ton job lots										
0802	For 30 to 39 tons, add						10%				
0804	20 to 29 tons, add						20%				
0806	10 to 19 tons, add						30%				
0807	5 to 9 tons, add						50%	25%			
0808	1 to 4 tons, add						75%	50%			
0809	Less than 1 ton, add						100%	100%			
1010	CS series, 40-ton job lots, horizontal bridging, shop primer										
1040	Spans to 30′	G	E-7	12	6.667	Ton	1,700	340	151	2,191	2,625
1500	For less than 40-ton job lots										
1502	For 30 to 39 tons, add						10%				
1504	20 to 29 tons, add						20%				
1506	10 to 19 tons, add						30%				
1507	5 to 9 tons, add						50%	25%			
1508	1 to 4 tons, add						75%	50%			
1509	Less than 1 ton, add						100%	100%			
6200	For shop prime paint other than mfrs. standard, add						20%				
6300	For bottom chord extensions, add per chord	G				Ea.	36			36	39.50
6400	Individual steel bearing plate, 6″ x 6″ x 1/4″ with J-hook	G	1 Bric	160	.050	″	7.95	2.28		10.23	12.25

05 21 23 – Steel Joist Girder Framing

05 21 23.50 Joist Girders			Crew	Daily Output	Labor-Hours	Unit	Material	Labor	Equipment	Total	Total Incl O&P
0010	**JOIST GIRDERS**										
0015	Made from recycled materials										
7020	Joist girders, 40-ton job lots, shop primer	G	E-5	13	6.154	Ton	1,650	310	129	2,089	2,500
7100	For less than 40-ton job lots										
7102	For 30 to 39 tons, add						10%				
7104	20 to 29 tons, add						20%				
7106	10 to 19 tons, add						30%				
7107	5 to 9 tons, add						50%	25%			
7108	1 to 4 tons, add						75%	50%			
7109	Less than 1 ton, add						100%	100%			
8000	Trusses, 40-ton job lots, shop fabricated WT chords, shop primer	G	E-5	11	7.273	Ton	5,425	370	152	5,947	6,750
8100	For less than 40-ton job lots										
8102	For 30 to 39 tons, add						10%				
8104	20 to 29 tons, add						20%				
8106	10 to 19 tons, add						30%				
8107	5 to 9 tons, add						50%	25%			
8108	1 to 4 tons, add						75%	50%			
8109	Less than 1 ton, add						100%	100%			

05 31 Steel Decking

05 31 13 – Steel Floor Decking

05 31 13.50 Floor Decking			Crew	Daily Output	Labor-Hours	Unit	Material	Labor	Equipment	Total	Total Incl O&P
0010	**FLOOR DECKING**	R053100-10									
0015	Made from recycled materials										
5100	Non-cellular composite decking, galvanized, 1-1/2″ deep, 16 ga.	G	E-4	3500	.009	S.F.	3.31	.47	.04	3.82	4.51
5120	18 ga.	G		3650	.009		2.68	.45	.04	3.17	3.79
5140	20 ga.	G		3800	.008		2.13	.43	.04	2.60	3.16
5200	2″ deep, 22 ga.	G		3860	.008		1.86	.43	.04	2.33	2.85
5300	20 ga.	G	↓	3600	.009	↓	2.05	.46	.04	2.55	3.11

05 31 Steel Decking

05 31 13 – Steel Floor Decking

05 31 13.50 Floor Decking		Crew	Daily Output	Labor-Hours	Unit	Material	2014 Bare Costs Labor	Equipment	Total	Total Incl O&P
5400	18 ga. G	E-4	3380	.009	S.F.	2.62	.49	.04	3.15	3.79
5500	16 ga. G		3200	.010		3.27	.52	.04	3.83	4.56
5700	3" deep, 22 ga. G		3200	.010		2.03	.52	.04	2.59	3.19
5800	20 ga. G		3000	.011		2.26	.55	.05	2.86	3.50
5900	18 ga. G		2850	.011		2.79	.58	.05	3.42	4.14
6000	16 ga. G	↓	2700	.012	↓	3.72	.61	.05	4.38	5.25

05 31 23 – Steel Roof Decking

05 31 23.50 Roof Decking

		Crew	Daily Output	Labor-Hours	Unit	Material	Labor	Equipment	Total	Total Incl O&P
0010	**ROOF DECKING**									
0015	Made from recycled materials									
2100	Open type, 1-1/2" deep, Type B, wide rib, galv., 22 ga., under 50 sq. G	E-4	4500	.007	S.F.	1.97	.37	.03	2.37	2.85
2400	Over 500 squares G		5100	.006		1.42	.32	.03	1.77	2.16
2600	20 ga., under 50 squares G		3865	.008		2.30	.43	.04	2.77	3.33
2700	Over 500 squares G		4300	.007		1.66	.38	.03	2.07	2.55
2900	18 ga., under 50 squares G		3800	.008		2.97	.43	.04	3.44	4.08
3000	Over 500 squares G		4300	.007		2.14	.38	.03	2.55	3.08
3050	16 ga., under 50 squares G		3700	.009		4	.45	.04	4.49	5.25
3100	Over 500 squares G		4200	.008		2.88	.39	.03	3.30	3.90
3200	3" deep, Type N, 22 ga., under 50 squares G		3600	.009		2.90	.46	.04	3.40	4.04
3300	20 ga., under 50 squares G		3400	.009		3.13	.49	.04	3.66	4.35
3400	18 ga., under 50 squares G		3200	.010		4.04	.52	.04	4.60	5.40
3500	16 ga., under 50 squares G		3000	.011		5.35	.55	.05	5.95	6.90
3700	4-1/2" deep, Type J, 20 ga., over 50 squares G		2700	.012		3.50	.61	.05	4.16	4.99
3800	18 ga. G		2460	.013		4.62	.67	.06	5.35	6.35
3900	16 ga. G		2350	.014		6	.70	.06	6.76	7.90
4100	6" deep, Type H, 18 ga., over 50 squares G		2000	.016		5.55	.83	.07	6.45	7.65
4200	16 ga. G		1930	.017		6.90	.86	.07	7.83	9.20
4300	14 ga. G		1860	.017		8.90	.89	.08	9.87	11.45
4500	7-1/2" deep, Type H, 18 ga., over 50 squares G		1690	.019		6.55	.98	.08	7.61	9.05
4600	16 ga. G		1590	.020		8.20	1.04	.09	9.33	10.95
4700	14 ga. G	↓	1490	.021	↓	10.20	1.11	.10	11.41	13.30
4800	For painted instead of galvanized, deduct					5%				
5000	For acoustical perforated with fiberglass insulation, add				S.F.	25%				
5100	For type F intermediate rib instead of type B wide rib, add G					25%				
5150	For type A narrow rib instead of type B wide rib, add G				↓	25%				

05 31 33 – Steel Form Decking

05 31 33.50 Form Decking

		Crew	Daily Output	Labor-Hours	Unit	Material	Labor	Equipment	Total	Total Incl O&P
0010	**FORM DECKING**									
0015	Made from recycled materials									
6100	Slab form, steel, 28 ga., 9/16" deep, Type UFS, uncoated G	E-4	4000	.008	S.F.	1.49	.41	.04	1.94	2.41
6200	Galvanized G		4000	.008		1.32	.41	.04	1.77	2.22
6220	24 ga., 1" deep, Type UF1X, uncoated G		3900	.008		1.43	.42	.04	1.89	2.36
6240	Galvanized G		3900	.008		1.68	.42	.04	2.14	2.64
6300	24 ga., 1-5/16" deep, Type UFX, uncoated G		3800	.008		1.52	.43	.04	1.99	2.48
6400	Galvanized G		3800	.008		1.79	.43	.04	2.26	2.78
6500	22 ga., 1-5/16" deep, uncoated G		3700	.009		1.92	.45	.04	2.41	2.94
6600	Galvanized G		3700	.009		1.96	.45	.04	2.45	2.99
6700	22 ga., 2" deep, uncoated G		3600	.009		2.50	.46	.04	3	3.60
6800	Galvanized G	↓	3600	.009	↓	2.45	.46	.04	2.95	3.55
7000	Sheet metal edge closure form, 12" wide with 2 bends, galvanized									
7100	18 ga. G	E-14	360	.022	L.F.	4.04	1.18	.39	5.61	6.95
7200	16 ga. G	"	360	.022	"	5.45	1.18	.39	7.02	8.50

05 35 Raceway Decking Assemblies

05 35 13 – Steel Cellular Decking

05 35 13.50 Cellular Decking

			Crew	Daily Output	Labor-Hours	Unit	2014 Bare Costs Material	Labor	Equipment	Total	Total Incl O&P
0010	**CELLULAR DECKING**										
0015	Made from recycled materials										
0200	Cellular units, galv, 1-1/2" deep, Type BC, 20-20 ga., over 15 squares	G	E-4	1460	.022	S.F.	7.75	1.13	.10	8.98	10.65
0250	18-20 ga.	G		1420	.023		8.85	1.16	.10	10.11	11.85
0300	18-18 ga.	G		1390	.023		9.05	1.19	.10	10.34	12.15
0320	16-18 ga.	G		1360	.024		10.80	1.21	.10	12.11	14.10
0340	16-16 ga.	G		1330	.024		12	1.24	.11	13.35	15.50
0400	3" deep, Type NC, galvanized, 20-20 ga.	G		1375	.023		8.55	1.20	.10	9.85	11.65
0500	18-20 ga.	G		1350	.024		10.35	1.22	.11	11.68	13.65
0600	18-18 ga.	G		1290	.025		10.30	1.28	.11	11.69	13.70
0700	16-18 ga.	G		1230	.026		11.60	1.34	.12	13.06	15.25
0800	16-16 ga.	G		1150	.028		12.65	1.44	.12	14.21	16.55
1000	4-1/2" deep, Type JC, galvanized, 18-20 ga.	G		1100	.029		11.90	1.50	.13	13.53	15.90
1100	18-18 ga.	G		1040	.031		11.85	1.59	.14	13.58	15.95
1200	16-18 ga.	G		980	.033		13.35	1.68	.15	15.18	17.80
1300	16-16 ga.	G		935	.034		14.55	1.77	.15	16.47	19.30
1500	For acoustical deck, add						15%				
1700	For cells used for ventilation, add						15%				
1900	For multi-story or congested site, add							50%			
8000	Metal deck and trench, 2" thick, 20 ga., combination										
8010	60% cellular, 40% non-cellular, inserts and trench	G	R-4	1100	.036	S.F.	15.55	1.89	.13	17.57	20.50

05 51 Metal Stairs

05 51 13 – Metal Pan Stairs

05 51 13.50 Pan Stairs

			Crew	Daily Output	Labor-Hours	Unit	Material	Labor	Equipment	Total	Total Incl O&P
0010	**PAN STAIRS**, shop fabricated, steel stringers										
0015	Made from recycled materials										
0200	Cement fill metal pan, picket rail, 3'-6" wide	G	E-4	35	.914	Riser	500	47	4.06	551.06	640
0300	4'-0" wide	G		30	1.067		560	55	4.74	619.74	715
0350	Wall rail, both sides, 3'-6" wide	G		53	.604		380	31	2.68	413.68	480
1500	Landing, steel pan, conventional	G		160	.200	S.F.	66	10.30	.89	77.19	92
1600	Pre-erected	G		255	.125	"	118	6.50	.56	125.06	142
1700	Pre-erected, steel pan tread, 3'-6" wide, 2 line pipe rail	G	E-2	87	.644	Riser	550	32.50	17.55	600.05	680

05 51 16 – Metal Floor Plate Stairs

05 51 16.50 Floor Plate Stairs

			Crew	Daily Output	Labor-Hours	Unit	Material	Labor	Equipment	Total	Total Incl O&P
0010	**FLOOR PLATE STAIRS**, shop fabricated, steel stringers										
0015	Made from recycled materials										
0400	Cast iron tread and pipe rail, 3'-6" wide	G	E-4	35	.914	Riser	535	47	4.06	586.06	675
0500	Checkered plate tread, industrial, 3'-6" wide	G		28	1.143		330	59	5.05	394.05	470
0550	Circular, for tanks, 3'-0" wide	G		33	.970		370	50	4.31	424.31	500
0600	For isolated stairs, add							100%			
0800	Custom steel stairs, 3'-6" wide, economy	G	E-4	35	.914		500	47	4.06	551.06	640
0810	Medium priced	G		30	1.067		660	55	4.74	719.74	830
0900	Deluxe	G		20	1.600		825	82.50	7.10	914.60	1,075
1100	For 4' wide stairs, add						5%	5%			
1300	For 5' wide stairs, add						10%	10%			

05 51 Metal Stairs

05 51 19 – Metal Grating Stairs

05 51 19.50 Grating Stairs			Crew	Daily Output	Labor-Hours	Unit	2014 Bare Costs Material	Labor	Equipment	Total	Total Incl O&P
0010	**GRATING STAIRS**, shop fabricated, steel stringers, safety nosing on treads										
0015	Made from recycled materials										
0020	Grating tread and pipe railing, 3'-6" wide	G	E-4	35	.914	Riser	330	47	4.06	381.06	450
0100	4'-0" wide	G	"	30	1.067	"	430	55	4.74	489.74	575

05 51 33 – Metal Ladders

05 51 33.13 Vertical Metal Ladders			Crew	Daily Output	Labor-Hours	Unit	Material	Labor	Equipment	Total	Total Incl O&P
0010	**VERTICAL METAL LADDERS**, shop fabricated										
0015	Made from recycled materials										
0020	Steel, 20" wide, bolted to concrete, with cage	G	E-4	50	.640	V.L.F.	64	33	2.84	99.84	132
0100	Without cage	G		85	.376		37	19.45	1.67	58.12	77
0300	Aluminum, bolted to concrete, with cage	G		50	.640		115	33	2.84	150.84	188
0400	Without cage	G		85	.376		49.50	19.45	1.67	70.62	91

05 52 Metal Railings

05 52 13 – Pipe and Tube Railings

05 52 13.50 Railings, Pipe			Crew	Daily Output	Labor-Hours	Unit	Material	Labor	Equipment	Total	Total Incl O&P
0010	**RAILINGS, PIPE**, shop fab'd, 3'-6" high, posts @ 5' O.C.										
0015	Made from recycled materials										
0020	Aluminum, 2 rail, satin finish, 1-1/4" diameter	G	E-4	160	.200	L.F.	36	10.30	.89	47.19	58.50
0030	Clear anodized	G		160	.200		44	10.30	.89	55.19	67.50
0040	Dark anodized	G		160	.200		49	10.30	.89	60.19	73
0080	1-1/2" diameter, satin finish	G		160	.200		42.50	10.30	.89	53.69	65.50
0090	Clear anodized	G		160	.200		47.50	10.30	.89	58.69	71
0100	Dark anodized	G		160	.200		52.50	10.30	.89	63.69	76.50
0140	Aluminum, 3 rail, 1-1/4" diam., satin finish	G		137	.234		54.50	12.05	1.04	67.59	82
0150	Clear anodized	G		137	.234		67.50	12.05	1.04	80.59	97
0160	Dark anodized	G		137	.234		75	12.05	1.04	88.09	105
0200	1-1/2" diameter, satin finish	G		137	.234		64.50	12.05	1.04	77.59	93.50
0210	Clear anodized	G		137	.234		74	12.05	1.04	87.09	104
0220	Dark anodized	G		137	.234		80.50	12.05	1.04	93.59	111
0500	Steel, 2 rail, on stairs, primed, 1-1/4" diameter	G		160	.200		25	10.30	.89	36.19	46.50
0520	1-1/2" diameter	G		160	.200		27	10.30	.89	38.19	49
0540	Galvanized, 1-1/4" diameter	G		160	.200		34	10.30	.89	45.19	56.50
0560	1-1/2" diameter	G		160	.200		38.50	10.30	.89	49.69	61
0580	Steel, 3 rail, primed, 1-1/4" diameter	G		137	.234		37.50	12.05	1.04	50.59	63.50
0600	1-1/2" diameter	G		137	.234		39	12.05	1.04	52.09	65.50
0620	Galvanized, 1-1/4" diameter	G		137	.234		52.50	12.05	1.04	65.59	80
0640	1-1/2" diameter	G		137	.234		60.50	12.05	1.04	73.59	89
0700	Stainless steel, 2 rail, 1-1/4" diam. #4 finish	G		137	.234		110	12.05	1.04	123.09	144
0720	High polish	G		137	.234		178	12.05	1.04	191.09	219
0740	Mirror polish	G		137	.234		223	12.05	1.04	236.09	268
0760	Stainless steel, 3 rail, 1-1/2" diam., #4 finish	G		120	.267		166	13.75	1.18	180.93	209
0770	High polish	G		120	.267		275	13.75	1.18	289.93	330
0780	Mirror finish	G		120	.267		335	13.75	1.18	349.93	395
0900	Wall rail, alum. pipe, 1-1/4" diam., satin finish	G		213	.150		20	7.75	.67	28.42	36.50
0905	Clear anodized	G		213	.150		25	7.75	.67	33.42	42
0910	Dark anodized	G		213	.150		29.50	7.75	.67	37.92	47
0915	1-1/2" diameter, satin finish	G		213	.150		22.50	7.75	.67	30.92	39.50
0920	Clear anodized	G		213	.150		28	7.75	.67	36.42	45.50
0925	Dark anodized	G		213	.150		35	7.75	.67	43.42	53

05 52 Metal Railings

05 52 13 – Pipe and Tube Railings

05 52 13.50 Railings, Pipe			Crew	Daily Output	Labor-Hours	Unit	Material	2014 Bare Costs Labor	Equipment	Total	Total Incl O&P
0930	Steel pipe, 1-1/4" diameter, primed	G	E-4	213	.150	L.F.	15	7.75	.67	23.42	31
0935	Galvanized	G		213	.150		22	7.75	.67	30.42	38.50
0940	1-1/2" diameter	G		176	.182		15.50	9.40	.81	25.71	34.50
0945	Galvanized	G		213	.150		22	7.75	.67	30.42	38.50
0955	Stainless steel pipe, 1-1/2" diam., #4 finish	G		107	.299		88	15.45	1.33	104.78	125
0960	High polish	G		107	.299		179	15.45	1.33	195.78	225
0965	Mirror polish	G		107	.299		212	15.45	1.33	228.78	261
2000	2-line pipe rail (1-1/2" T&B) with 1/2" pickets @ 4-1/2" O.C.,										
2005	attached handrail on brackets										
2010	42" high aluminum, satin finish, straight & level	G	E-4	120	.267	L.F.	198	13.75	1.18	212.93	244
2050	42" high steel, primed, straight & level	G	"	120	.267		127	13.75	1.18	141.93	166
4000	For curved and level rails, add						10%	10%			
4100	For sloped rails for stairs, add						30%	30%			

05 52 16 – Industrial Railings

05 52 16.50 Railings, Industrial			Crew	Daily Output	Labor-Hours	Unit	Material	Labor	Equipment	Total	Total Incl O&P
0010	**RAILINGS, INDUSTRIAL**, shop fab'd, 3'-6" high, posts @ 5' O.C.										
0020	2 rail, 3'-6" high, 1-1/2" pipe	G	E-4	255	.125	L.F.	27	6.50	.56	34.06	42
0100	2" angle rail	G	"	255	.125		25	6.50	.56	32.06	39.50
0200	For 4" high kick plate, 10 ga., add	G					5.65			5.65	6.20
0300	1/4" thick, add	G					7.40			7.40	8.15
0500	For curved level rails, add						10%	10%			
0550	For sloped rails for stairs, add						30%	30%			

05 53 Metal Gratings

05 53 13 – Bar Gratings

05 53 13.10 Floor Grating, Aluminum			Crew	Daily Output	Labor-Hours	Unit	Material	Labor	Equipment	Total	Total Incl O&P
0010	**FLOOR GRATING, ALUMINUM**, field fabricated from panels										
0015	Made from recycled materials										
0110	Bearing bars @ 1-3/16" O.C., cross bars @ 4" O.C.,										
0111	Up to 300 S.F., 1" x 1/8" bar	G	E-4	900	.036	S.F.	16.90	1.83	.16	18.89	22
0112	Over 300 S.F.	G		850	.038		15.35	1.94	.17	17.46	20.50
0113	1-1/4" x 1/8" bar, up to 300 S.F.	G		800	.040		17.55	2.06	.18	19.79	23
0114	Over 300 S.F.	G		1000	.032		15.95	1.65	.14	17.74	20.50
0122	1-1/4" x 3/16" bar, up to 300 S.F.	G		750	.043		27	2.20	.19	29.39	33.50
0124	Over 300 S.F.	G		1000	.032		24.50	1.65	.14	26.29	30
0132	1-1/2" x 3/16" bar, up to 300 S.F.	G		700	.046		31	2.36	.20	33.56	38.50
0134	Over 300 S.F.	G		1000	.032		28.50	1.65	.14	30.29	34
0136	1-3/4" x 3/16" bar, up to 300 S.F.	G		500	.064		34.50	3.30	.28	38.08	44
0138	Over 300 S.F.	G		1000	.032		31.50	1.65	.14	33.29	37.50
0146	2-1/4" x 3/16" bar, up to 300 S.F.	G		600	.053		44	2.75	.24	46.99	53
0148	Over 300 S.F.	G		1000	.032		40	1.65	.14	41.79	47
0162	Cross bars @ 2" O.C., 1" x 1/8", up to 300 S.F.	G		600	.053		29.50	2.75	.24	32.49	37.50
0164	Over 300 S.F.	G		1000	.032		27	1.65	.14	28.79	32.50
0172	1-1/4" x 3/16" bar, up to 300 S.F.	G		600	.053		48.50	2.75	.24	51.49	58
0174	Over 300 S.F.	G		1000	.032		44	1.65	.14	45.79	51.50
0182	1-1/2" x 3/16" bar, up to 300 S.F.	G		600	.053		55.50	2.75	.24	58.49	66
0184	Over 300 S.F.	G		1000	.032		50.50	1.65	.14	52.29	58.50
0186	1-3/4" x 3/16" bar, up to 300 S.F.	G		600	.053		60	2.75	.24	62.99	71
0188	Over 300 S.F.	G		1000	.032		54.50	1.65	.14	56.29	63
0200	For straight cuts, add					L.F.	4.30			4.30	4.73

05 53 Metal Gratings

05 53 13 – Bar Gratings

05 53 13.10 Floor Grating, Aluminum

			Crew	Daily Output	Labor-Hours	Unit	2014 Bare Costs Material	Labor	Equipment	Total	Total Incl O&P
0212	Bearing bars @ 15/16" O.C., 1" x 1/8", up to 300 S.F.	G	E-4	520	.062	S.F.	30	3.18	.27	33.45	39
0214	Over 300 S.F.	G		920	.035		27	1.79	.15	28.94	33.50
0222	1-1/4" x 3/16", up to 300 S.F.	G		520	.062		49	3.18	.27	52.45	60
0224	Over 300 S.F.	G		920	.035		44.50	1.79	.15	46.44	52.50
0232	1-1/2" x 3/16", up to 300 S.F.	G		520	.062		56.50	3.18	.27	59.95	68
0234	Over 300 S.F.	G		920	.035		51.50	1.79	.15	53.44	60
0300	For curved cuts, add					L.F.	5.30			5.30	5.85
0400	For straight banding, add	G					5.50			5.50	6.05
0500	For curved banding, add	G					6.60			6.60	7.25
0600	For aluminum checkered plate nosings, add	G					7.15			7.15	7.85
0700	For straight toe plate, add	G					10.80			10.80	11.90
0800	For curved toe plate, add	G					12.60			12.60	13.85
1000	For cast aluminum abrasive nosings, add	G					10.50			10.50	11.55
1200	Expanded aluminum, .65# per S.F.	G	E-4	1050	.030	S.F.	11.90	1.57	.14	13.61	16.05
1400	Extruded I bars are 10% less than 3/16" bars										
1600	Heavy duty, all extruded plank, 3/4" deep, 1.8# per S.F.	G	E-4	1100	.029	S.F.	26	1.50	.13	27.63	31.50
1700	1-1/4" deep, 2.9# per S.F.	G		1000	.032		26	1.65	.14	29.79	34
1800	1-3/4" deep, 4.2# per S.F.	G		925	.035		39.50	1.78	.15	41.43	47
1900	2-1/4" deep, 5.0# per S.F.	G		875	.037		60.50	1.89	.16	62.55	70
2100	For safety serrated surface, add						15%				

05 53 13.70 Floor Grating, Steel

			Crew	Daily Output	Labor-Hours	Unit	Material	Labor	Equipment	Total	Total Incl O&P
0010	**FLOOR GRATING, STEEL**, field fabricated from panels										
0015	Made from recycled materials										
0300	Platforms, to 12' high, rectangular	G	E-4	3150	.010	Lb.	3.21	.52	.05	3.78	4.51
0400	Circular	G	"	2300	.014	"	4.01	.72	.06	4.79	5.75
0410	Painted bearing bars @ 1-3/16"										
0412	Cross bars @ 4" O.C., 3/4" x 1/8" bar, up to 300 S.F.	G	E-2	500	.112	S.F.	7.70	5.60	3.06	16.36	21.50
0414	Over 300 S.F.	G		750	.075		7	3.75	2.04	12.79	16.30
0422	1-1/4" x 3/16", up to 300 S.F.	G		400	.140		11.85	7.05	3.82	22.72	29
0424	Over 300 S.F.	G		600	.093		10.80	4.68	2.55	18.03	22.50
0432	1-1/2" x 3/16", up to 300 S.F.	G		400	.140		13.60	7.05	3.82	24.47	31
0434	Over 300 S.F.	G		600	.093		12.35	4.68	2.55	19.58	24.50
0436	1-3/4" x 3/16", up to 300 S.F.	G		400	.140		17.80	7.05	3.82	28.67	35.50
0438	Over 300 S.F.	G		600	.093		16.15	4.68	2.55	23.38	28.50
0452	2-1/4" x 3/16", up to 300 S.F.	G		300	.187		22	9.35	5.10	36.45	46
0454	Over 300 S.F.	G		450	.124		20	6.25	3.40	29.65	36.50
0462	Cross bars @ 2" O.C., 3/4" x 1/8", up to 300 S.F.	G		500	.112		15.15	5.60	3.06	23.81	29.50
0464	Over 300 S.F.	G		750	.075		12.60	3.75	2.04	18.39	22.50
0472	1-1/4" x 3/16", up to 300 S.F.	G		400	.140		19.80	7.05	3.82	30.67	38
0474	Over 300 S.F.	G		600	.093		16.50	4.68	2.55	23.73	29
0482	1-1/2" x 3/16", up to 300 S.F.	G		400	.140		22.50	7.05	3.82	33.37	40.50
0484	Over 300 S.F.	G		600	.093		18.60	4.68	2.55	25.83	31.50
0486	1-3/4" x 3/16", up to 300 S.F.	G		400	.140		33.50	7.05	3.82	44.37	53
0488	Over 300 S.F.	G		600	.093		28	4.68	2.55	35.23	42
0502	2-1/4" x 3/16", up to 300 S.F.	G		300	.187		33	9.35	5.10	47.45	57.50
0504	Over 300 S.F.	G		450	.124		27.50	6.25	3.40	37.15	44.50
0601	Painted bearing bars @ 15/16" O.C., cross bars @ 4" O.C.,										
0612	Up to 300 S.F., 3/4" x 3/16" bars	G	E-4	850	.038	S.F.	11.35	1.94	.17	13.46	16.10
0622	1-1/4" x 3/16" bars	G		600	.053		15.15	2.75	.24	18.14	22
0632	1-1/2" x 3/16" bars	G		550	.058		17.40	3	.26	20.66	24.50
0636	1-3/4" x 3/16" bars	G		450	.071		24.50	3.67	.32	28.49	34
0652	2-1/4" x 3/16" bars	G	E-2	300	.187		25.50	9.35	5.10	39.95	49.50

05 53 Metal Gratings

05 53 13 – Bar Gratings

05 53 13.70 Floor Grating, Steel			Crew	Daily Output	Labor-Hours	Unit	Material	2014 Bare Costs Labor	Equipment	Total	Total Incl O&P
0662	Cross bars @ 2" O.C., up to 300 S.F., 3/4" x 3/16"	G	E-2	500	.112	S.F.	15.80	5.60	3.06	24.46	30.50
0672	1-1/4" x 3/16" bars	G		400	.140		20	7.05	3.82	30.87	38
0682	1-1/2" x 3/16" bars	G		400	.140		23	7.05	3.82	33.87	41
0686	1-3/4" x 3/16" bars	G		300	.187		29	9.35	5.10	43.45	53
0690	For galvanized grating, add						25%				
0800	For straight cuts, add					L.F.	6.15			6.15	6.75
0900	For curved cuts, add						7.80			7.80	8.60
1000	For straight banding, add	G					6.60			6.60	7.25
1100	For curved banding, add	G					8.70			8.70	9.55
1200	For checkered plate nosings, add	G					7.65			7.65	8.40
1300	For straight toe or kick plate, add	G					13.70			13.70	15.05
1400	For curved toe or kick plate, add	G					15.50			15.50	17.05
1500	For abrasive nosings, add	G					10.15			10.15	11.15
1600	For safety serrated surface, bearing bars @ 1-3/16" O.C., add						15%				
1700	Bearing bars @ 15/16" O.C., add						25%				
2000	Stainless steel gratings, close spaced, 1" x 1/8" bars, up to 300 S.F.	G	E-4	450	.071	S.F.	73	3.67	.32	76.99	87.50
2100	Standard spacing, 3/4" x 1/8" bars	G		500	.064		84	3.30	.28	87.58	98
2200	1-1/4" x 3/16" bars	G		400	.080		80	4.13	.36	84.49	95.50
2400	Expanded steel grating, at ground, 3.0# per S.F.	G		900	.036		8.05	1.83	.16	10.04	12.30
2500	3.14# per S.F.	G		900	.036		7.15	1.83	.16	9.14	11.30
2600	4.0# per S.F.	G		850	.038		8.60	1.94	.17	10.71	13.10
2650	4.27# per S.F.	G		850	.038		9.30	1.94	.17	11.41	13.85
2700	5.0# per S.F.	G		800	.040		13.80	2.06	.18	16.04	19
2800	6.25# per S.F.	G		750	.043		17.80	2.20	.19	20.19	23.50
2900	7.0# per S.F.	G		700	.046		19.65	2.36	.20	22.21	26
3100	For flattened expanded steel grating, add						8%				
3300	For elevated installation above 15', add							15%			

05 53 16 – Plank Gratings

05 53 16.50 Grating Planks			Crew	Daily Output	Labor-Hours	Unit	Material	Labor	Equipment	Total	Total Incl O&P
0010	**GRATING PLANKS**, field fabricated from planks										
0020	Aluminum, 9-1/2" wide, 14 ga., 2" rib	G	E-4	950	.034	L.F.	25.50	1.74	.15	27.39	31.50
0200	Galvanized steel, 9-1/2" wide, 14 ga., 2-1/2" rib	G		950	.034		16.80	1.74	.15	18.69	21.50
0300	4" rib	G		950	.034		18.50	1.74	.15	20.39	23.50
0500	12 ga., 2-1/2" rib	G		950	.034		16.55	1.74	.15	18.44	21.50
0600	3" rib	G		950	.034		22.50	1.74	.15	24.39	27.50
0800	Stainless steel, type 304, 16 ga., 2" rib	G		950	.034		39	1.74	.15	40.89	46
0900	Type 316	G		950	.034		57	1.74	.15	58.89	66

05 53 19 – Floor Grating Frame

05 53 19.30 Grating Frame			Crew	Daily Output	Labor-Hours	Unit	Material	Labor	Equipment	Total	Total Incl O&P
0010	**GRATING FRAME**, field fabricated										
0020	Aluminum, for gratings 1" to 1-1/2" deep	G	1 Sswk	70	.114	L.F.	3.72	5.85		9.57	14.40
0100	For each corner, add	G				Ea.	5.60			5.60	6.15

05 54 Metal Floor Plates

05 54 13 – Floor Plates

05 54 13.20 Checkered Plates		Crew	Daily Output	Labor-Hours	Unit	Material	2014 Bare Costs Labor	Equipment	Total	Total Incl O&P
0010	**CHECKERED PLATES**, steel, field fabricated									
0015	Made from recycled materials									
0020	1/4" & 3/8", 2000 to 5000 S.F., bolted G	E-4	2900	.011	Lb.	.87	.57	.05	1.49	2.01
0100	Welded G		4400	.007	"	.83	.38	.03	1.24	1.61
0300	Pit or trench cover and frame, 1/4" plate, 2' to 3' wide G		100	.320	S.F.	11.05	16.50	1.42	28.97	42.50
0400	For galvanizing, add G				Lb.	.29			.29	.32
0500	Platforms, 1/4" plate, no handrails included, rectangular G	E-4	4200	.008		3.21	.39	.03	3.63	4.26
0600	Circular G	"	2500	.013		4.01	.66	.06	4.73	5.65

05 54 13.70 Trench Covers		Crew	Daily Output	Labor-Hours	Unit	Material	Labor	Equipment	Total	Total Incl O&P
0010	**TRENCH COVERS**, field fabricated									
0020	Cast iron grating with bar stops and angle frame, to 18" wide G	1 Sswk	20	.400	L.F.	235	20.50		255.50	294
0100	Frame only (both sides of trench), 1" grating G		45	.178		1.38	9.10		10.48	17.55
0150	2" grating G		35	.229		3.27	11.70		14.97	24
0200	Aluminum, stock units, including frames and									
0210	3/8" plain cover plate, 4" opening G	E-4	205	.156	L.F.	22.50	8.05	.69	31.24	39.50
0300	6" opening G		185	.173		27	8.95	.77	36.72	46
0400	10" opening G		170	.188		36	9.70	.84	46.54	57.50
0500	16" opening G		155	.206		50	10.65	.92	61.57	75
0700	Add per inch for additional widths to 24" G					2.06			2.06	2.27
0900	For custom fabrication, add					50%				
1100	For 1/4" plain cover plate, deduct					12%				
1500	For cover recessed for tile, 1/4" thick, deduct					12%				
1600	3/8" thick, add					5%				
1800	For checkered plate cover, 1/4" thick, deduct					12%				
1900	3/8" thick, add					2%				
2100	For slotted or round holes in cover, 1/4" thick, add					3%				
2200	3/8" thick, add					4%				
2300	For abrasive cover, add					12%				

05 55 Metal Stair Treads and Nosings

05 55 19 – Metal Stair Tread Covers

05 55 19.50 Stair Tread Covers for Renovation		Crew	Daily Output	Labor-Hours	Unit	Material	Labor	Equipment	Total	Total Incl O&P
0010	**STAIR TREAD COVERS FOR RENOVATION**									
0205	Extruded tread cover with nosing, pre-drilled, includes screws									
0210	Aluminum with black abrasive strips, 9" wide x 3' long	1 Carp	24	.333	Ea.	106	15.30		121.30	140
0220	4' long		22	.364		140	16.65		156.65	179
0230	5' long		20	.400		174	18.35		192.35	220
0240	11" wide x 3' long		24	.333		136	15.30		151.30	174
0250	4' long		22	.364		175	16.65		191.65	219
0260	5' long		20	.400		225	18.35		243.35	277
0305	Black abrasive strips with yellow front strips									
0310	Aluminum, 9" wide x 3' long	1 Carp	24	.333	Ea.	114	15.30		129.30	149
0320	4' long		22	.364		152	16.65		168.65	193
0330	5' long		20	.400		193	18.35		211.35	241
0340	11" wide x 3' long		24	.333		146	15.30		161.30	184
0350	4' long		22	.364		188	16.65		204.65	233
0360	5' long		20	.400		242	18.35		260.35	296
0405	Black abrasive strips with photoluminescent front strips									
0410	Aluminum, 9" wide x 3' long	1 Carp	24	.333	Ea.	151	15.30		166.30	190
0420	4' long		22	.364		172	16.65		188.65	215
0430	5' long		20	.400		215	18.35		233.35	265

05 55 Metal Stair Treads and Nosings

05 55 19 – Metal Stair Tread Covers

05 55 19.50 Stair Tread Covers for Renovation			Crew	Daily Output	Labor-Hours	Unit	Material	Labor	Equipment	Total	Total Incl O&P
							2014 Bare Costs				
0440	11″ wide x 3′ long		1 Carp	24	.333	Ea.	153	15.30		168.30	192
0450	4′ long			22	.364		204	16.65		220.65	250
0460	5′ long		↓	20	.400	↓	255	18.35		273.35	310

05 56 Metal Castings

05 56 13 – Metal Construction Castings

05 56 13.50 Construction Castings

			Crew	Daily Output	Labor-Hours	Unit	Material	Labor	Equipment	Total	Total Incl O&P
0010	**CONSTRUCTION CASTINGS**										
0020	Manhole covers and frames, see Section 33 44 13.13										
0100	Column bases, cast iron, 16″ x 16″, approx. 65 lb.	G	E-4	46	.696	Ea.	153	36	3.09	192.09	235
0200	32″ x 32″, approx. 256 lb.	G	″	23	1.391		570	72	6.20	648.20	765
0400	Cast aluminum for wood columns, 8″ x 8″	G	1 Carp	32	.250		45.50	11.45		56.95	67.50
0500	12″ x 12″	G	″	32	.250	↓	99	11.45		110.45	127
0600	Miscellaneous C.I. castings, light sections, less than 150 lb.	G	E-4	3200	.010	Lb.	2.38	.52	.04	2.94	3.58
1100	Heavy sections, more than 150 lb.	G		4200	.008		1.15	.39	.03	1.57	2
1300	Special low volume items	G	↓	3200	.010		4.13	.52	.04	4.69	5.50
1500	For ductile iron, add					↓	100%				

05 58 Formed Metal Fabrications

05 58 21 – Formed Chain

05 58 21.05 Alloy Steel Chain

			Crew	Daily Output	Labor-Hours	Unit	Material	Labor	Equipment	Total	Total Incl O&P
0010	**ALLOY STEEL CHAIN**, Grade 80, for lifting										
0015	Self-colored, cut lengths, 1/4″	G	E-17	4	4	C.L.F.	725	208		933	1,175
0020	3/8″	G		2	8		920	415		1,335	1,750
0030	1/2″	G		1.20	13.333		1,475	695		2,170	2,850
0040	5/8″	G	↓	.72	22.222		2,375	1,150		3,525	4,675
0050	3/4″	G	E-18	.48	83.333		3,125	4,250	2,000	9,375	13,000
0060	7/8″	G		.40	100		5,525	5,100	2,400	13,025	17,500
0070	1″	G		.35	114		7,100	5,825	2,750	15,675	20,900
0080	1-1/4″	G	↓	.24	166	↓	13,400	8,500	4,000	25,900	33,700
0110	Hook, Grade 80, Clevis slip, 1/4″	G				Ea.	26			26	28.50
0120	3/8″	G					30.50			30.50	33.50
0130	1/2″	G					49			49	54
0140	5/8″	G					72			72	79.50
0150	3/4″	G					105			105	116
0160	Hook, Grade 80, eye/sling w/hammerlock coupling, 15 Ton	G					335			335	365
0170	22 Ton	G					905			905	995
0180	37 Ton	G				↓	2,650			2,650	2,925

05 75 Decorative Formed Metal

05 75 13 – Columns

05 75 13.10 Aluminum Columns			Crew	Daily Output	Labor-Hours	Unit	Material	2014 Bare Costs Labor	Equipment	Total	Total Incl O&P
0010	**ALUMINUM COLUMNS**										
0015	Made from recycled materials										
0020	Aluminum, extruded, stock units, no cap or base, 6″ diameter	G	E-4	240	.133	L.F.	10.85	6.90	.59	18.34	25
0100	8″ diameter	G		170	.188		14.25	9.70	.84	24.79	33.50
0200	10″ diameter	G		150	.213		19.20	11	.95	31.15	41.50
0300	12″ diameter	G		140	.229		36	11.80	1.01	48.81	62
0400	15″ diameter	G	↓	120	.267	↓	48	13.75	1.18	62.93	78.50
0410	Caps and bases, plain, 6″ diameter	G				Set	22.50			22.50	25
0420	8″ diameter	G					29.50			29.50	32.50
0430	10″ diameter	G					41			41	45.50
0440	12″ diameter	G					59.50			59.50	65
0450	15″ diameter	G					110			110	121
0460	Caps, ornamental, plain	G					340			340	370
0470	Fancy	G				↓	1,675			1,675	1,850
0500	For square columns, add to column prices above					L.F.	50%				

05 75 13.20 Columns, Ornamental			Crew	Daily Output	Labor-Hours	Unit	Material	Labor	Equipment	Total	Total Incl O&P
0010	**COLUMNS, ORNAMENTAL**, shop fabricated										
6400	Mild steel, flat, 9″ wide, stock units, painted, plain	G	E-4	160	.200	V.L.F.	9.15	10.30	.89	20.34	29
6450	Fancy	G		160	.200		17.75	10.30	.89	28.94	38.50
6500	Corner columns, painted, plain	G		160	.200		15.75	10.30	.89	26.94	36.50
6550	Fancy	G	↓	160	.200	↓	31	10.30	.89	42.19	53.50

Division Notes

	CREW	DAILY OUTPUT	LABOR-HOURS	UNIT	BARE COSTS MAT.	LABOR	EQUIP.	TOTAL	TOTAL INCL O&P

Estimating Tips

06 05 00 Common Work Results for Wood, Plastics, and Composites

- Common to any wood-framed structure are the accessory connector items such as screws, nails, adhesives, hangers, connector plates, straps, angles, and hold-downs. For typical wood-framed buildings, such as residential projects, the aggregate total for these items can be significant, especially in areas where seismic loading is a concern. For floor and wall framing, the material cost is based on 10 to 25 lbs. per MBF. Hold-downs, hangers, and other connectors should be taken off by the piece.

  Included with material costs are fasteners for a normal installation. RSMeans engineers use manufacturer's recommendations, written specifications, and/or standard construction practice for size and spacing of fasteners. Prices for various fasteners are shown for informational purposes only. Adjustments should be made if unusual fastening conditions exist.

06 10 00 Carpentry

- Lumber is a traded commodity and therefore sensitive to supply and demand in the marketplace. Even in "budgetary" estimating of wood-framed projects, it is advisable to call local suppliers for the latest market pricing.
- Common quantity units for wood-framed projects are "thousand board feet" (MBF). A board foot is a volume of wood, 1" x 1' x 1', or 144 cubic inches. Board-foot quantities are generally calculated using nominal material dimensions—dressed sizes are ignored. Board foot per lineal foot of any stick of lumber can be calculated by dividing the nominal cross-sectional area by 12. As an example, 2,000 lineal feet of 2 x 12 equates to 4 MBF by dividing the nominal area, 2 x 12, by 12, which equals 2, and multiplying by 2,000 to give 4,000 board feet. This simple rule applies to all nominal dimensioned lumber.
- Waste is an issue of concern at the quantity takeoff for any area of construction. Framing lumber is sold in even foot lengths, i.e., 10', 12', 14', 16', and depending on spans, wall heights, and the grade of lumber, waste is inevitable. A rule of thumb for lumber waste is 5%–10% depending on material quality and the complexity of the framing.
- Wood in various forms and shapes is used in many projects, even where the main structural framing is steel, concrete, or masonry. Plywood as a back-up partition material and 2x boards used as blocking and cant strips around roof edges are two common examples. The estimator should ensure that the costs of all wood materials are included in the final estimate.

06 20 00 Finish Carpentry

- It is necessary to consider the grade of workmanship when estimating labor costs for erecting millwork and interior finish. In practice, there are three grades: premium, custom, and economy. The RSMeans daily output for base and case moldings is in the range of 200 to 250 L.F. per carpenter per day. This is appropriate for most average custom-grade projects. For premium projects, an adjustment to productivity of 25%–50% should be made, depending on the complexity of the job.

Reference Numbers

Reference numbers are shown in shaded boxes at the beginning of some major classifications. These numbers refer to related items in the Reference Section. The reference information may be an estimating procedure, an alternate pricing method, or technical information.

Note: Not all subdivisions listed here necessarily appear in this publication.

06 05 Common Work Results for Wood, Plastics, and Composites

06 05 05 – Selective Wood and Plastics Demolition

06 05 05.10 Selective Demolition Wood Framing		Crew	Daily Output	Labor-Hours	Unit	Material	2014 Bare Costs Labor	Equipment	Total	Total Incl O&P
0010	**SELECTIVE DEMOLITION WOOD FRAMING** R024119-10									
0100	Timber connector, nailed, small	1 Clab	96	.083	Ea.		3.05		3.05	4.71
0110	Medium		60	.133			4.89		4.89	7.55
0120	Large		48	.167			6.10		6.10	9.45
0130	Bolted, small		48	.167			6.10		6.10	9.45
0140	Medium		32	.250			9.15		9.15	14.15
0150	Large	↓	24	.333	↓		12.20		12.20	18.85
3162	Alternate pricing method	B-1	1.10	21.818	M.B.F.		815		815	1,250

06 05 23 – Wood, Plastic, and Composite Fastenings

06 05 23.60 Timber Connectors		Crew	Daily Output	Labor-Hours	Unit	Material	Labor	Equipment	Total	Total Incl O&P
0010	**TIMBER CONNECTORS**									
0020	Add up cost of each part for total cost of connection									
0100	Connector plates, steel, with bolts, straight	2 Carp	75	.213	Ea.	30.50	9.80		40.30	48.50
0110	Tee, 7 ga.		50	.320		35	14.65		49.65	61
0120	T-Strap, 14 ga., 12" x 8" x 2"		50	.320		35	14.65		49.65	61
0150	Anchor plates, 7 ga., 9" x 7"	↓	75	.213		30.50	9.80		40.30	48.50
0200	Bolts, machine, sq. hd. with nut & washer, 1/2" diameter, 4" long	1 Carp	140	.057		.75	2.62		3.37	4.87
0300	7-1/2" long		130	.062		1.37	2.82		4.19	5.85
0500	3/4" diameter, 7-1/2" long		130	.062		3.20	2.82		6.02	7.85
0610	Machine bolts, w/nut, washer, 3/4" diam., 15" L, HD's & beam hangers		95	.084	↓	5.95	3.86		9.81	12.50
0800	Drilling bolt holes in timber, 1/2" diameter		450	.018	Inch		.82		.82	1.26
0900	1" diameter	↓	350	.023	"		1.05		1.05	1.62

06 05 23.70 Rough Hardware		Crew	Daily Output	Labor-Hours	Unit	Material	Labor	Equipment	Total	Total Incl O&P
0010	**ROUGH HARDWARE**, average percent of carpentry material									
0020	Minimum				Job	.50%				
0200	Maximum					1.50%				
0210	In seismic or hurricane areas, up to				↓	10%				

06 11 Wood Framing

06 11 10 – Framing with Dimensional, Engineered or Composite Lumber

06 11 10.14 Posts and Columns		Crew	Daily Output	Labor-Hours	Unit	Material	Labor	Equipment	Total	Total Incl O&P
0010	**POSTS AND COLUMNS**									
0100	4" x 4"	2 Carp	390	.041	L.F.	1.71	1.88		3.59	4.78
0150	4" x 6"		275	.058		2.59	2.67		5.26	6.95
0200	4" x 8"		220	.073		3.49	3.33		6.82	9
0250	6" x 6"		215	.074		4.76	3.41		8.17	10.50
0300	6" x 8"		175	.091		6.35	4.19		10.54	13.45
0350	6" x 10"	↓	150	.107	↓	7.95	4.89		12.84	16.30

06 12 Structural Panels

06 12 10 – Structural Insulated Panels

06 12 10.10 OSB Faced Panels			Crew	Daily Output	Labor-Hours	Unit	Material	Labor	Equipment	Total	Total Incl O&P
							2014 Bare Costs				
0010	**OSB FACED PANELS**										
0500	Structural insulated panel, 7/16" OSB both sides, straw core										
0510	4-3/8" T, walls (w/sill, splines, plates)	G	F-6	2400	.017	S.F.	7.40	.72	.28	8.40	9.50
0520	Floors (w/splines)	G		2400	.017		7.40	.72	.28	8.40	9.50
0530	Roof (w/splines)	G		2400	.017		7.40	.72	.28	8.40	9.50
0550	7-7/8" T, walls (w/sill, splines, plates)	G		2400	.017		11.20	.72	.28	12.20	13.70
0560	Floors (w/splines)	G		2400	.017		11.20	.72	.28	12.20	13.70
0570	Roof (w/splines)	G	↓	2400	.017	↓	11.20	.72	.28	12.20	13.70

06 13 Heavy Timber Construction

06 13 23 – Heavy Timber Framing

06 13 23.10 Heavy Framing

			Crew	Daily Output	Labor-Hours	Unit	Material	Labor	Equipment	Total	Total Incl O&P
0010	**HEAVY FRAMING**										
0020	Beams, single 6" x 10"		2 Carp	1.10	14.545	M.B.F.	1,600	665		2,265	2,775
0100	Single 8" x 16"			1.20	13.333		2,000	610		2,610	3,150
0200	Built from 2" lumber, multiple 2" x 14"	R061110-30		.90	17.778		760	815		1,575	2,100
0210	Built from 3" lumber, multiple 3" x 6"			.70	22.857		930	1,050		1,980	2,650
0220	Multiple 3" x 8"			.80	20		1,275	915		2,190	2,825
0230	Multiple 3" x 10"			.90	17.778		1,275	815		2,090	2,650
0240	Multiple 3" x 12"			1	16		1,275	735		2,010	2,525
0250	Built from 4" lumber, multiple 4" x 6"			.80	20		1,300	915		2,215	2,850
0260	Multiple 4" x 8"			.90	17.778		1,300	815		2,115	2,675
0270	Multiple 4" x 10"			1	16		1,275	735		2,010	2,525
0280	Multiple 4" x 12"			1.10	14.545		1,275	665		1,940	2,425
0290	Columns, structural grade, 1500f, 4" x 4"			.60	26.667		1,275	1,225		2,500	3,275
0300	6" x 6"			.65	24.615		1,400	1,125		2,525	3,300
0400	8" x 8"			.70	22.857		1,350	1,050		2,400	3,100
0500	10" x 10"			.75	21.333		1,450	980		2,430	3,100
0600	12" x 12"			.80	20		1,525	915		2,440	3,100
0800	Floor planks, 2" thick, T & G, 2" x 6"			1.05	15.238		1,600	700		2,300	2,825
0900	2" x 10"			1.10	14.545		1,600	665		2,265	2,775
1100	3" thick, 3" x 6"			1.05	15.238		1,600	700		2,300	2,825
1200	3" x 10"			1.10	14.545		1,600	665		2,265	2,775
1400	Girders, structural grade, 12" x 12"			.80	20		1,525	915		2,440	3,100
1500	10" x 16"			1	16		2,400	735		3,135	3,775
2300	Roof purlins, 4" thick, structural grade		↓	1.05	15.238	↓	1,300	700		2,000	2,500

06 13 33 – Heavy Timber Pier Construction

06 13 33.50 Jetties, Docks, Fixed

		Crew	Daily Output	Labor-Hours	Unit	Material	Labor	Equipment	Total	Total Incl O&P
0010	**JETTIES, DOCKS, FIXED**									
0020	Pile supported, treated wood,									
0030	5' x 20' platform, freshwater	F-3	115	.348	S.F.	18.35	16.25	5.75	40.35	51.50
0060	Saltwater		115	.348		38.50	16.25	5.75	60.50	73.50
0100	6' x 20' platform, freshwater		110	.364		16	17	6.05	39.05	50.50
0160	Saltwater		110	.364		34	17	6.05	57.05	70
0200	8' x 20' platform, freshwater		105	.381		13.25	17.80	6.30	37.35	49
0260	Saltwater		105	.381		29	17.80	6.30	53.10	66.50
0420	5' x 30' platform, freshwater		100	.400		17.70	18.70	6.65	43.05	55.50
0460	Saltwater	↓	100	.400	↓	53.50	18.70	6.65	78.85	94.50
0500	For Greenhart lumber, add					40%				
0550	Diagonal planking, add								25%	25%

06 13 Heavy Timber Construction

06 13 33 – Heavy Timber Pier Construction

06 13 33.52 Jetties, Piers		Crew	Daily Output	Labor-Hours	Unit	Material	2014 Bare Costs Labor	Equipment	Total	Total Incl O&P
0010	**JETTIES, PIERS**, Municipal with 3" x 12" framing and 3" decking,									
0020	wood piles and cross bracing, alternate bents battered	B-76	60	1.200	S.F.	74.50	55	56	185.50	230
0200	Treated piles, not including mobilization									
0210	50' long, 20 lb. creosote, shore driven	B-19	540	.119	V.L.F.	20.50	5.45	3.41	29.36	35
0220	Barge driven	B-76	320	.225		20.50	10.30	10.45	41.25	50
0230	2.5 lb. treatment, shore driven	B-19	540	.119		17.90	5.45	3.41	26.76	32
0240	Barge driven	B-76	320	.225		17.90	10.30	10.45	38.65	47.50
0250	30' long, 20 lb. creosote, shore driven	B-19	540	.119		18.20	5.45	3.41	27.06	32.50
0260	Barge driven	B-76	320	.225		18.20	10.30	10.45	38.95	47.50
0270	2.5 lb. treatment, shore driven	B-19	540	.119		17.90	5.45	3.41	26.76	32
0280	Barge driven	B-76	320	.225	↓	17.90	10.30	10.45	38.65	47.50
0300	Mobilization, barge, by tug boat	B-83	25	.640	Mile		27.50	35.50	63	81
0350	Standby time for shore pile driving crew				Hr.				600	750
0360	Standby time for barge driving rig				"				840	1,050

06 17 Shop-Fabricated Structural Wood

06 17 33 – Wood I-Joists

06 17 33.10 Wood and Composite I-Joists		Crew	Daily Output	Labor-Hours	Unit	Material	Labor	Equipment	Total	Total Incl O&P
0010	**WOOD AND COMPOSITE I-JOISTS**									
0100	Plywood webs, incl. bridging & blocking, panels 24" O.C.									
1200	15' to 24' span, 50 psf live load	F-5	2400	.013	SF Flr.	1.95	.62		2.57	3.09
1300	55 psf live load		2250	.014		2.21	.66		2.87	3.45
1400	24' to 30' span, 45 psf live load		2600	.012		2.61	.57		3.18	3.76
1500	55 psf live load	↓	2400	.013	↓	3.89	.62		4.51	5.25

06 18 Glued-Laminated Construction

06 18 13 – Glued-Laminated Beams

06 18 13.20 Laminated Framing		Crew	Daily Output	Labor-Hours	Unit	Material	Labor	Equipment	Total	Total Incl O&P
0010	**LAMINATED FRAMING**									
0020	30 lb., short term live load, 15 lb. dead load									
0200	Straight roof beams, 20' clear span, beams 8' O.C.	F-3	2560	.016	SF Flr.	1.91	.73	.26	2.90	3.50
0300	Beams 16' O.C.		3200	.013		1.38	.58	.21	2.17	2.65
0500	40' clear span, beams 8' O.C.		3200	.013		3.65	.58	.21	4.44	5.15
0600	Beams 16' O.C.	↓	3840	.010		2.99	.49	.17	3.65	4.23
0800	60' clear span, beams 8' O.C.	F-4	2880	.017		6.30	.77	.39	7.46	8.50
0900	Beams 16' O.C.	"	3840	.013		4.68	.58	.30	5.56	6.35
1100	Tudor arches, 30' to 40' clear span, frames 8' O.C.	F-3	1680	.024		8.20	1.11	.39	9.70	11.15
1200	Frames 16' O.C.	"	2240	.018		6.40	.83	.30	7.53	8.65
1400	50' to 60' clear span, frames 8' O.C.	F-4	2200	.022		8.85	1.01	.52	10.38	11.80
1500	Frames 16' O.C.		2640	.018		7.50	.84	.43	8.77	10
1700	Radial arches, 60' clear span, frames 8' O.C.		1920	.025		8.25	1.16	.59	10	11.50
1800	Frames 16' O.C.		2880	.017		6.35	.77	.39	7.51	8.60
2000	100' clear span, frames 8' O.C.		1600	.030		8.55	1.39	.71	10.65	12.30
2100	Frames 16' O.C.		2400	.020		7.50	.92	.47	8.89	10.20
2300	120' clear span, frames 8' O.C.		1440	.033		11.35	1.54	.79	13.68	15.75
2400	Frames 16' O.C.	↓	1920	.025		10.35	1.16	.59	12.10	13.80
2600	Bowstring trusses, 20' O.C., 40' clear span	F-3	2400	.017		5.10	.78	.28	6.16	7.15
2700	60' clear span	F-4	3600	.013		4.60	.62	.32	5.54	6.35
2800	100' clear span	↓	4000	.012	↓	6.50	.55	.28	7.33	8.30

06 18 Glued-Laminated Construction

06 18 13 – Glued-Laminated Beams

06 18 13.20 Laminated Framing		Crew	Daily Output	Labor-Hours	Unit	2014 Bare Costs Material	Labor	Equipment	Total	Total Incl O&P
2900	120′ clear span	F-4	3600	.013	SF Flr.	7	.62	.32	7.94	9
3100	For premium appearance, add to S.F. prices					5%				
3300	For industrial type, deduct					15%				
3500	For stain and varnish, add					5%				
3900	For 3/4″ laminations, add to straight					25%				
4100	Add to curved					15%				
4300	Alternate pricing method: (use nominal footage of									
4310	components). Straight beams, camber less than 6″	F-3	3.50	11.429	M.B.F.	2,825	535	189	3,549	4,150
4400	Columns, including hardware		2	20		3,050	935	330	4,315	5,150
4600	Curved members, radius over 32′		2.50	16		3,125	750	265	4,140	4,875
4700	Radius 10′ to 32′		3	13.333		3,075	625	221	3,921	4,600
4900	For complicated shapes, add maximum					100%				
5100	For pressure treating, add to straight					35%				
5200	Add to curved					45%				
6000	Laminated veneer members, southern pine or western species									
6050	1-3/4″ wide x 5-1/2″ deep	2 Carp	480	.033	L.F.	2.94	1.53		4.47	5.60
6100	9-1/2″ deep		480	.033		4.35	1.53		5.88	7.15
6150	14″ deep		450	.036		6.75	1.63		8.38	9.90
6200	18″ deep		450	.036		9.50	1.63		11.13	12.95
6300	Parallel strand members, southern pine or western species									
6350	1-3/4″ wide x 9-1/4″ deep	2 Carp	480	.033	L.F.	4.35	1.53		5.88	7.15
6400	11-1/4″ deep		450	.036		4.93	1.63		6.56	7.90
6450	14″ deep		400	.040		6.75	1.83		8.58	10.25
6500	3-1/2″ wide x 9-1/4″ deep		480	.033		15.15	1.53		16.68	19
6550	11-1/4″ deep		450	.036		18.10	1.63		19.73	22.50
6600	14″ deep		400	.040		22.50	1.83		24.33	27.50
6650	7″ wide x 9-1/4″ deep		450	.036		30.50	1.63		32.13	36
6700	11-1/4″ deep		420	.038		38	1.75		39.75	44
6750	14″ deep		400	.040		44.50	1.83		46.33	52

Division Notes

	CREW	DAILY OUTPUT	LABOR-HOURS	UNIT	BARE COSTS MAT.	LABOR	EQUIP.	TOTAL	TOTAL INCL O&P

Estimating Tips

07 10 00 Dampproofing and Waterproofing

- Be sure of the job specifications before pricing this subdivision. The difference in cost between waterproofing and dampproofing can be great. Waterproofing will hold back standing water. Dampproofing prevents the transmission of water vapor. Also included in this section are vapor retarding membranes.

07 20 00 Thermal Protection

- Insulation and fireproofing products are measured by area, thickness, volume or R-value. Specifications may give only what the specific R-value should be in a certain situation. The estimator may need to choose the type of insulation to meet that R-value.

07 30 00 Steep Slope Roofing
07 40 00 Roofing and Siding Panels

- Many roofing and siding products are bought and sold by the square. One square is equal to an area that measures 100 square feet.

This simple change in unit of measure could create a large error if the estimator is not observant. Accessories necessary for a complete installation must be figured into any calculations for both material and labor.

07 50 00 Membrane Roofing
07 60 00 Flashing and Sheet Metal
07 70 00 Roofing and Wall Specialties and Accessories

- The items in these subdivisions compose a roofing system. No one component completes the installation, and all must be estimated. Built-up or single-ply membrane roofing systems are made up of many products and installation trades. Wood blocking at roof perimeters or penetrations, parapet coverings, reglets, roof drains, gutters, downspouts, sheet metal flashing, skylights, smoke vents, and roof hatches all need to be considered along with the roofing material. Several different installation trades will need to work together on the roofing system. Inherent difficulties in the scheduling and coordination of various trades must be accounted for when estimating labor costs.

07 90 00 Joint Protection

- To complete the weather-tight shell, the sealants and caulkings must be estimated. Where different materials meet—at expansion joints, at flashing penetrations, and at hundreds of other locations throughout a construction project—they provide another line of defense against water penetration. Often, an entire system is based on the proper location and placement of caulking or sealants. The detailed drawings that are included as part of a set of architectural plans show typical locations for these materials. When caulking or sealants are shown at typical locations, this means the estimator must include them for all the locations where this detail is applicable. Be careful to keep different types of sealants separate, and remember to consider backer rods and primers if necessary.

Reference Numbers

Reference numbers are shown in shaded boxes at the beginning of some major classifications. These numbers refer to related items in the Reference Section. The reference information may be an estimating procedure, an alternate pricing method, or technical information.

Note: Not all subdivisions listed here necessarily appear in this publication.

07 01 Operation and Maint. of Thermal and Moisture Protection

07 01 90 – Maintenance of Joint Protection

07 01 90.81 Joint Sealant Replacement		Crew	Daily Output	Labor-Hours	Unit	Material	Labor	Equipment	Total	Total Incl O&P
						2014 Bare Costs				
0010	**JOINT SEALANT REPLACEMENT**									
0050	Control joints in concrete floors/slabs									
0100	Option 1 for joints with hard dry sealant									
0110	Step 1: Sawcut to remove 95% of old sealant									
0112	1/4" wide x 1/2" deep, with single saw blade	C-27	4800	.003	L.F.	.02	.15	.04	.21	.28
0114	3/8" wide x 3/4" deep, with single saw blade		4000	.004		.03	.18	.04	.25	.34
0116	1/2" wide x 1" deep, with double saw blades		3600	.004		.06	.20	.05	.31	.41
0118	3/4" wide x 1-1/2" deep, with double saw blades		3200	.005		.13	.22	.05	.40	.53
0120	Step 2: Water blast joint faces and edges	C-29	2500	.003			.12	.03	.15	.21
0130	Step 3: Air blast joint faces and edges	C-28	2000	.004			.18	.01	.19	.27
0140	Step 4: Sand blast joint faces and edges	E-11	2000	.016			.66	.12	.78	1.23
0150	Step 5: Air blast joint faces and edges	C-28	2000	.004			.18	.01	.19	.27
0200	Option 2 for joints with soft pliable sealant									
0210	Step 1: Plow joint with rectangular blade	B-62	2600	.009	L.F.		.37	.07	.44	.64
0220	Step 2: Sawcut to re-face joint faces									
0222	1/4" wide x 1/2" deep, with single saw blade	C-27	2400	.007	L.F.	.02	.29	.07	.38	.53
0224	3/8" wide x 3/4" deep, with single saw blade		2000	.008		.04	.35	.09	.48	.67
0226	1/2" wide x 1" deep, with double saw blades		1800	.009		.08	.39	.10	.57	.78
0228	3/4" wide x 1-1/2" deep, with double saw blades		1600	.010		.17	.44	.11	.72	.95
0230	Step 3: Water blast joint faces and edges	C-29	2500	.003			.12	.03	.15	.21
0240	Step 4: Air blast joint faces and edges	C-28	2000	.004			.18	.01	.19	.27
0250	Step 5: Sand blast joint faces and edges	E-11	2000	.016			.66	.12	.78	1.23
0260	Step 6: Air blast joint faces and edges	C-28	2000	.004			.18	.01	.19	.27
0290	For saw cutting new control joints, see Section 03 15 16.20									
8920	For joint sealant, see Sections 03 15 16.30									

07 25 Weather Barriers

07 25 10 – Weather Barriers or Wraps

07 25 10.10 Weather Barriers		Crew	Daily Output	Labor-Hours	Unit	Material	Labor	Equipment	Total	Total Incl O&P
0010	**WEATHER BARRIERS**									
0400	Asphalt felt paper, 15#	1 Carp	37	.216	Sq.	5.40	9.90		15.30	21.50
0401	Per square foot	"	3700	.002	S.F.	.05	.10		.15	.21
0450	Housewrap, exterior, spun bonded polypropylene									
0470	Small roll	1 Carp	3800	.002	S.F.	.14	.10		.24	.31
0480	Large roll	"	4000	.002	"	.14	.09		.23	.29
2100	Asphalt felt roof deck vapor barrier, class 1 metal decks	1 Rofc	37	.216	Sq.	22	8.45		30.45	38.50
2200	For all other decks	"	37	.216		16.40	8.45		24.85	32.50
2800	Asphalt felt, 50% recycled content, 15 lb., 4 sq. per roll	1 Carp	36	.222		5.40	10.20		15.60	21.50
2810	30 lb., 2 sq. per roll	"	36	.222		10.80	10.20		21	27.50
3000	Building wrap, spunbonded polyethylene	2 Carp	8000	.002	S.F.	.14	.09		.23	.29

Estimating Tips

General

- Room Finish Schedule: A complete set of plans should contain a room finish schedule. If one is not available, it would be well worth the time and effort to obtain one.

09 20 00 Plaster and Gypsum Board

- Lath is estimated by the square yard plus a 5% allowance for waste. Furring, channels, and accessories are measured by the linear foot. An extra foot should be allowed for each accessory miter or stop.
- Plaster is also estimated by the square yard. Deductions for openings vary by preference, from zero deduction to 50% of all openings over 2 feet in width. The estimator should allow one extra square foot for each linear foot of horizontal interior or exterior angle located below the ceiling level. Also, double the areas of small radius work.
- Drywall accessories, studs, track, and acoustical caulking are all measured by the linear foot. Drywall taping is figured by the square foot. Gypsum wallboard is estimated by the square foot. No material deductions should be made for door or window openings under 32 S.F.

09 60 00 Flooring

- Tile and terrazzo areas are taken off on a square foot basis. Trim and base materials are measured by the linear foot. Accent tiles are listed per each. Two basic methods of installation are used. Mud set is approximately 30% more expensive than thin set. In terrazzo work, be sure to include the linear footage of embedded decorative strips, grounds, machine rubbing, and power cleanup.
- Wood flooring is available in strip, parquet, or block configuration. The latter two types are set in adhesives with quantities estimated by the square foot. The laying pattern will influence labor costs and material waste. In addition to the material and labor for laying wood floors, the estimator must make allowances for sanding and finishing these areas unless the flooring is prefinished.
- Sheet flooring is measured by the square yard. Roll widths vary, so consideration should be given to use the most economical width, as waste must be figured into the total quantity. Consider also the installation methods available, direct glue down or stretched.

09 70 00 Wall Finishes

- Wall coverings are estimated by the square foot. The area to be covered is measured, length by height of wall above baseboards, to calculate the square footage of each wall. This figure is divided by the number of square feet in the single roll which is being used. Deduct, in full, the areas of openings such as doors and windows. Where a pattern match is required allow 25%–30% waste.

09 80 00 Acoustic Treatment

- Acoustical systems fall into several categories. The takeoff of these materials should be by the square foot of area with a 5% allowance for waste. Do not forget about scaffolding, if applicable, when estimating these systems.

09 90 00 Painting and Coating

- A major portion of the work in painting involves surface preparation. Be sure to include cleaning, sanding, filling, and masking costs in the estimate.
- Protection of adjacent surfaces is not included in painting costs. When considering the method of paint application, an important factor is the amount of protection and masking required. These must be estimated separately and may be the determining factor in choosing the method of application.

Reference Numbers

Reference numbers are shown in shaded boxes at the beginning of some major classifications. These numbers refer to related items in the Reference Section. The reference information may be an estimating procedure, an alternate pricing method, or technical information.

Note: Not all subdivisions listed here necessarily appear in this publication.

09 97 Special Coatings

09 97 13 – Steel Coatings

09 97 13.23 Exterior Steel Coatings		Crew	Daily Output	Labor-Hours	Unit	2014 Bare Costs Material	Labor	Equipment	Total	Total Incl O&P
0010	**EXTERIOR STEEL COATINGS**									
6100	Cold galvanizing, brush in field	1 Psst	1100	.007	S.F.	.12	.29		.41	.67
6510	Paints & protective coatings, sprayed in field									
6520	Alkyds, primer	2 Psst	3600	.004	S.F.	.08	.18		.26	.41
6540	Gloss topcoats		3200	.005		.08	.20		.28	.46
6560	Silicone alkyd		3200	.005		.15	.20		.35	.53
6610	Epoxy, primer		3000	.005		.29	.22		.51	.71
6630	Intermediate or topcoat		2800	.006		.26	.23		.49	.71
6650	Enamel coat		2800	.006		.33	.23		.56	.78
6810	Latex primer		3600	.004		.06	.18		.24	.39
6830	Topcoats		3200	.005		.06	.20		.26	.44
6910	Universal primers, one part, phenolic, modified alkyd		2000	.008		.40	.32		.72	1.02
6940	Two part, epoxy spray		2000	.008		.30	.32		.62	.91
7000	Zinc rich primers, self cure, spray, inorganic		1800	.009		.86	.36		1.22	1.60
7010	Epoxy, spray, organic	↓	1800	.009	↓	.26	.36		.62	.94
7020	Above one story, spray painting simple structures, add						25%			
7030	Intricate structures, add						50%			

Estimating Tips

General

- The items in this division are usually priced per square foot or each.
- Many items in Division 10 require some type of support system or special anchors that are not usually furnished with the item. The required anchors must be added to the estimate in the appropriate division.
- Some items in Division 10, such as lockers, may require assembly before installation. Verify the amount of assembly required. Assembly can often exceed installation time.

10 20 00 Interior Specialties

- Support angles and blocking are not included in the installation of toilet compartments, shower/dressing compartments, or cubicles. Appropriate line items from Divisions 5 or 6 may need to be added to support the installations.
- Toilet partitions are priced by the stall. A stall consists of a side wall, pilaster, and door with hardware. Toilet tissue holders and grab bars are extra.
- The required acoustical rating of a folding partition can have a significant impact on costs. Verify the sound transmission coefficient rating of the panel priced to the specification requirements.
- Grab bar installation does not include supplemental blocking or backing to support the required load. When grab bars are installed at an existing facility, provisions must be made to attach the grab bars to solid structure.

Reference Numbers

Reference numbers are shown in shaded boxes at the beginning of some major classifications. These numbers refer to related items in the Reference Section. The reference information may be an estimating procedure, an alternate pricing method, or technical information.

Note: Not all subdivisions listed here necessarily appear in this publication.

10 05 Common Work Results for Specialties

10 05 05 – Selective Specialties Demolition

10 05 05.10 Selective Demolition, Specialties		Crew	Daily Output	Labor-Hours	Unit	Material	2014 Bare Costs Labor	Equipment	Total	Total Incl O&P
0010	**SELECTIVE DEMOLITION, SPECIALTIES**									
4000	Removal of traffic signs, including supports									
4020	To 10 S.F.	B-80B	16	2	Ea.		78.50	15.50	94	138
4030	11 S.F. to 20 S.F.	"	5	6.400			251	49.50	300.50	440
4040	21 S.F. to 40 S.F.	B-14	1.80	26.667			1,025	203	1,228	1,800
4050	41 S.F. to 100 S.F.	B-13	1.30	43.077			1,725	575	2,300	3,250
4070	Remove traffic posts to 12′-0″ high	B-6	100	.240	↓		9.65	3.65	13.30	18.75

10 14 Signage

10 14 53 – Traffic Signage

10 14 53.20 Traffic Signs		Crew	Daily Output	Labor-Hours	Unit	Material	2014 Bare Costs Labor	Equipment	Total	Total Incl O&P
0010	**TRAFFIC SIGNS**									
0012	Stock, 24″ x 24″, no posts, .080″ alum. reflectorized	B-80	70	.457	Ea.	83	18.15	10.85	112	131
0100	High intensity		70	.457		95.50	18.15	10.85	124.50	145
0300	30″ x 30″, reflectorized		70	.457		121	18.15	10.85	150	173
0400	High intensity		70	.457		133	18.15	10.85	162	186
0600	Guide and directional signs, 12″ x 18″, reflectorized		70	.457		34.50	18.15	10.85	63.50	78
0700	High intensity		70	.457		51	18.15	10.85	80	96.50
0900	18″ x 24″, stock signs, reflectorized		70	.457		46	18.15	10.85	75	91
1000	High intensity		70	.457		51	18.15	10.85	80	96.50
1200	24″ x 24″, stock signs, reflectorized		70	.457		56	18.15	10.85	85	101
1300	High intensity		70	.457		61	18.15	10.85	90	107
1500	Add to above for steel posts, galvanized, 10′-0″ upright, bolted		200	.160		32.50	6.35	3.79	42.64	50
1600	12′-0″ upright, bolted		140	.229	↓	39	9.10	5.40	53.50	63
1800	Highway road signs, aluminum, over 20 S.F., reflectorized		350	.091	S.F.	33.50	3.63	2.17	39.30	45
2000	High intensity		350	.091		33.50	3.63	2.17	39.30	45
2200	Highway, suspended over road, 80 S.F. min., reflectorized		165	.194		31	7.70	4.59	43.29	51
2300	High intensity		165	.194	↓	29.50	7.70	4.59	41.79	49.50
2350	Roadway delineators and reference markers		500	.064	Ea.	15.95	2.54	1.52	20.01	23
2360	Delineator post only, 6′	↓	500	.064		18	2.54	1.52	22.06	25.50
2400	Highway sign bridge structure, 45′ to 80′								30,800	33,900
2410	Cantilever structure, add				↓				20%	20%
5200	Remove and relocate signs, including supports									
5210	To 10 S.F.	B-80B	5	6.400	Ea.	350	251	49.50	650.50	825
5220	11 S.F. to 20 S.F.	"	1.70	18.824		785	740	146	1,671	2,150
5230	21 S.F. to 40 S.F.	B-14	.56	85.714		830	3,325	650	4,805	6,725
5240	41 S.F. to 100 S.F.	B-13	.32	175	↓	1,375	6,975	2,325	10,675	14,800
8000	For temporary barricades and lights, see Section 01 56 23.10									

Estimating Tips

General

- The items in this division are usually priced per square foot or each. Most of these items are purchased by the owner and installed by the contractor. Do not assume the items in Division 12 will be purchased and installed by the contractor. Check the specifications for responsibilities and include receiving, storage, installation, and mechanical and electrical hookups in the appropriate divisions.
- Some items in this division require some type of support system that is not usually furnished with the item. Examples of these systems include blocking for the attachment of casework and heavy drapery rods. The required blocking must be added to the estimate in the appropriate division.

Reference Numbers

Reference numbers are shown in shaded boxes at the beginning of some major classifications. These numbers refer to related items in the Reference Section. The reference information may be an estimating procedure, an alternate pricing method, or technical information.

Note: Not all subdivisions listed here necessarily appear in this publication.

12 93 Site Furnishings

12 93 43 – Site Seating and Tables

12 93 43.13 Site Seating		Crew	Daily Output	Labor-Hours	Unit	2014 Bare Costs Material	Labor	Equipment	Total	Total Incl O&P
0010	**SITE SEATING**									
0012	Seating, benches, park, precast conc., w/backs, wood rails, 4′ long	2 Clab	5	3.200	Ea.	580	117		697	815
0100	8′ long		4	4		935	147		1,082	1,250
0300	Fiberglass, without back, one piece, 4′ long		10	1.600		605	58.50		663.50	760
0400	8′ long		7	2.286		780	84		864	990
0500	Steel barstock pedestals w/backs, 2″ x 3″ wood rails, 4′ long		10	1.600		1,100	58.50		1,158.50	1,300
0510	8′ long		7	2.286		1,400	84		1,484	1,675
0520	3″ x 8″ wood plank, 4′ long		10	1.600		1,175	58.50		1,233.50	1,400
0530	8′ long		7	2.286		1,450	84		1,534	1,725
0540	Backless, 4″ x 4″ wood plank, 4′ square		10	1.600		920	58.50		978.50	1,125
0550	8′ long		7	2.286		990	84		1,074	1,225
0600	Aluminum pedestals, with backs, aluminum slats, 8′ long		8	2		470	73.50		543.50	630
0610	15′ long		5	3.200		955	117		1,072	1,225
0620	Portable, aluminum slats, 8′ long		8	2		455	73.50		528.50	615
0630	15′ long		5	3.200		550	117		667	785
0800	Cast iron pedestals, back & arms, wood slats, 4′ long		8	2		415	73.50		488.50	575
0820	8′ long		5	3.200		920	117		1,037	1,200
0840	Backless, wood slats, 4′ long		8	2		580	73.50		653.50	755
0860	8′ long		5	3.200		1,150	117		1,267	1,425
1700	Steel frame, fir seat, 10′ long	▼	10	1.600	▼	345	58.50		403.50	465

Estimating Tips

General

- The items and systems in this division are usually estimated, purchased, supplied, and installed as a unit by one or more subcontractors. The estimator must ensure that all parties are operating from the same set of specifications and assumptions, and that all necessary items are estimated and will be provided. Many times the complex items and systems are covered, but the more common ones, such as excavation or a crane, are overlooked for the very reason that everyone assumes nobody could miss them. The estimator should be the central focus and be able to ensure that all systems are complete.
- Another area where problems can develop in this division is at the interface between systems. The estimator must ensure, for instance, that anchor bolts, nuts, and washers are estimated and included for the air-supported structures and pre-engineered buildings to be bolted to their foundations. Utility supply is a common area where essential items or pieces of equipment can be missed or overlooked due to the fact that each subcontractor may feel it is another's responsibility. The estimator should also be aware of certain items which may be supplied as part of a package but installed by others, and ensure that the installing contractor's estimate includes the cost of installation. Conversely, the estimator must also ensure that items are not costed by two different subcontractors, resulting in an inflated overall estimate.

13 30 00 Special Structures

- The foundations and floor slab, as well as rough mechanical and electrical, should be estimated, as this work is required for the assembly and erection of the structure. Generally, as noted in the book, the pre-engineered building comes as a shell. Pricing is based on the size and structural design parameters stated in the reference section. Additional features, such as windows and doors with their related structural framing, must also be included by the estimator. Here again, the estimator must have a clear understanding of the scope of each portion of the work and all the necessary interfaces.

Reference Numbers

Reference numbers are shown in shaded boxes at the beginning of some major classifications. These numbers refer to related items in the Reference Section. The reference information may be an estimating procedure, an alternate pricing method, or technical information.

Note: Not all subdivisions listed here necessarily appear in this publication.

13 05 05 – Selective Special Construction Demolition

13 05 05.45 Selective Demolition, Lightning Protection		Crew	Daily Output	Labor-Hours	Unit	2014 Bare Costs Material	Labor	Equipment	Total	Total Incl O&P
0010	**SELECTIVE DEMOLITION, LIGHTNING PROTECTION**									
0020	Air terminal & base, copper, 3/8" diam. x 10", to 75' h	1 Clab	16	.500	Ea.		18.35		18.35	28.50
0030	1/2" diam. x 12", over 75' h		16	.500			18.35		18.35	28.50
0050	Aluminum, 1/2" diam. x 12", to 75' h		16	.500			18.35		18.35	28.50
0060	5/8" diam. x 12", over 75' h		16	.500			18.35		18.35	28.50
0070	Cable, copper, 220 lb. per thousand feet, to 75' high		640	.013	L.F.		.46		.46	.71
0080	375 lb. per thousand feet, over 75' high		460	.017			.64		.64	.98
0090	Aluminum, 101 lb. per thousand feet, to 75' high		560	.014			.52		.52	.81
0100	199 lb. per thousand feet, over 75' high		480	.017			.61		.61	.94
0110	Arrester, 175 V AC, to ground		16	.500	Ea.		18.35		18.35	28.50
0120	650 V AC, to ground		13	.615	"		22.50		22.50	35

13 05 05.50 Selective Demolition, Pre-Engineered Steel Buildings

		Crew	Daily Output	Labor-Hours	Unit	Material	Labor	Equipment	Total	Total Incl O&P
0010	**SELECTIVE DEMOLITION, PRE-ENGINEERED STEEL BUILDINGS**									
0500	Pre-engd. steel bldgs., rigid frame, clear span & multi post, excl. salvage									
0550	3,500 to 7,500 S.F.	L-10	1000	.024	SF Flr.		1.24	.66	1.90	2.81
0600	7,501 to 12,500 S.F.		1500	.016			.82	.44	1.26	1.88
0650	12,500 S.F. or greater		1650	.015			.75	.40	1.15	1.70
0700	Pre-engd. steel building components									
0710	Entrance canopy, including frame 4' x 4'	E-24	8	4	Ea.		202	93	295	450
0720	4' x 8'	"	7	4.571			231	107	338	510
0730	HM doors, self framing, single leaf	2 Skwk	8	2			94.50		94.50	147
0740	Double leaf		5	3.200			151		151	234
0760	Gutter, eave type		600	.027	L.F.		1.26		1.26	1.95
0770	Sash, single slide, double slide or fixed		24	.667	Ea.		31.50		31.50	49
0780	Skylight, fiberglass, to 30 S.F.		16	1			47.50		47.50	73.50
0785	Roof vents, circular, 12" to 24" diameter		12	1.333			63		63	97.50
0790	Continuous, 10' long		8	2			94.50		94.50	147
0900	Shelters, aluminum frame									
0910	Acrylic glazing, 3' x 9' x 8' high	2 Skwk	2	8	Ea.		380		380	585
0920	9' x 12' x 8' high	"	1.50	10.667	"		505		505	780

13 05 05.60 Selective Demolition, Silos

		Crew	Daily Output	Labor-Hours	Unit	Material	Labor	Equipment	Total	Total Incl O&P
0010	**SELECTIVE DEMOLITION, SILOS**									
0020	Conc stave, indstrl, conical/sloping bott, excl fndtn, 12' diam., 35' h	E-24	.18	177	Ea.		8,975	4,150	13,125	19,900
0030	16' diam., 45' h		.12	266			13,500	6,225	19,725	29,800
0040	25' diam., 75' h		.08	400			20,200	9,325	29,525	44,800
0050	Steel, factory fabricated, 30,000 gal. cap, painted or epoxy lined	L-5	2	28			1,425	375	1,800	2,875

13 05 05.75 Selective Demolition, Storage Tanks

		Crew	Daily Output	Labor-Hours	Unit	Material	Labor	Equipment	Total	Total Incl O&P
0010	**SELECTIVE DEMOLITION, STORAGE TANKS**									
0500	Steel tank, single wall, above ground, not incl. fdn., pumps or piping									
0510	Single wall, 275 gallon R024119-10	Q-1	3	5.333	Ea.		276		276	415
0520	550 thru 2,000 gallon	B-34P	2	12			575	350	925	1,250
0530	5,000 thru 10,000 gallon	B-34Q	2	12			580	640	1,220	1,575
0540	15,000 thru 30,000 gallon	B-34S	2	16			820	1,800	2,620	3,225
0600	Steel tank, double wall, above ground not incl. fdn., pumps & piping									
0620	500 thru 2,000 gallon	B-34P	2	12	Ea.		575	350	925	1,250

13 31 Fabric Structures

13 31 13 – Air-Supported Fabric Structures

13 31 13.09 Air Supported Tank Covers		Crew	Daily Output	Labor-Hours	Unit	2014 Bare Costs Material	Labor	Equipment	Total	Total Incl O&P
0010	**AIR SUPPORTED TANK COVERS**, vinyl polyester									
0100	Scrim, double layer, with hardware, blower, standby & controls									
0200	Round, 75′ diameter	B-2	4500	.009	S.F.	10.20	.33		10.53	11.75
0300	100′ diameter		5000	.008		9.30	.30		9.60	10.65
0400	150′ diameter		5000	.008		7.35	.30		7.65	8.55
0500	Rectangular, 20′ x 20′		4500	.009		23	.33		23.33	26
0600	30′ x 40′		4500	.009		23	.33		23.33	26
0700	50′ x 60′		4500	.009		23	.33		23.33	26
0800	For single wall construction, deduct, minimum					.79			.79	.87
0900	Maximum					2.33			2.33	2.56
1000	For maximum resistance to atmosphere or cold, add					1.14			1.14	1.25
1100	For average shipping charges, add				Total	1,975			1,975	2,175

13 31 13.13 Single-Walled Air-Supported Structures		Crew	Daily Output	Labor-Hours	Unit	Material	Labor	Equipment	Total	Total Incl O&P
0010	**SINGLE-WALLED AIR-SUPPORTED STRUCTURES**									
0020	Site preparation, incl. anchor placement and utilities	B-11B	1000	.016	SF Flr.	1.16	.67	.30	2.13	2.63
0030	For concrete, see Section 03 30 53.40									
0050	Warehouse, polyester/vinyl fabric, 28 oz., over 10 yr. life, welded									
0060	Seams, tension cables, primary & auxiliary inflation system,									
0070	airlock, personnel doors and liner									
0100	5,000 S.F.	4 Clab	5000	.006	SF Flr.	25	.23		25.23	28
0250	12,000 S.F.	"	6000	.005		18	.20		18.20	20
0400	24,000 S.F.	8 Clab	12000	.005		12.65	.20		12.85	14.25
0500	50,000 S.F.	"	12500	.005		11.75	.19		11.94	13.20
0700	12 oz. reinforced vinyl fabric, 5 yr. life, sewn seams,									
0710	accordion door, including liner									
0750	3000 S.F.	4 Clab	3000	.011	SF Flr.	12	.39		12.39	13.80
0800	12,000 S.F.	"	6000	.005		10.20	.20		10.40	11.55
0850	24,000 S.F.	8 Clab	12000	.005		8.65	.20		8.85	9.80
0950	Deduct for single layer					1.03			1.03	1.13
1000	Add for welded seams					1.01			1.01	1.11
1050	Add for double layer, welded seams included					2.08			2.08	2.29
1250	Tedlar/vinyl fabric, 28 oz., with liner, over 10 yr. life,									
1260	incl. overhead and personnel doors									
1300	3000 S.F.	4 Clab	3000	.011	SF Flr.	24.50	.39		24.89	27.50
1450	12,000 S.F.	"	6000	.005		17.20	.20		17.40	19.25
1550	24,000 S.F.	8 Clab	12000	.005		13.30	.20		13.50	14.95
1700	Deduct for single layer					1.44			1.44	1.58
2860	For low temperature conditions, add					1.14			1.14	1.25
2870	For average shipping charges, add				Total	5,600			5,600	6,150
2900	Thermal liner, translucent reinforced vinyl				SF Flr.	1.14			1.14	1.25
2950	Metalized mylar fabric and mesh, double liner				"	2.33			2.33	2.56
3050	Stadium/convention center, teflon coated fiberglass, heavy weight,									
3060	over 20 yr. life, incl. thermal liner and heating system									
3100	Minimum	9 Clab	26000	.003	SF Flr.	57.50	.10		57.60	63
3110	Maximum	"	19000	.004	"	68	.14		68.14	75
3400	Doors, air lock, 15′ long, 10′ x 10′	2 Carp	.80	20	Ea.	20,400	915		21,315	23,900
3600	15′ x 15′	"	.50	32		30,300	1,475		31,775	35,700
3700	For each added 5′ length, add					5,425			5,425	5,950
3900	Revolving personnel door, 6′ diameter, 6′-6″ high	2 Carp	.80	20		15,200	915		16,115	18,100
4200	Double wall, self supporting, shell only, minimum				SF Flr.				19.10	21
4300	Maximum				"				35.50	39

13 31 Fabric Structures

13 31 23 – Tensioned Fabric Structures

13 31 23.50 Tension Structures		Crew	Daily Output	Labor-Hours	Unit	Material	Labor (2014 Bare Costs)	Equipment (2014 Bare Costs)	Total (2014 Bare Costs)	Total Incl O&P
0010	**TENSION STRUCTURES** Rigid steel/alum. frame, vinyl coated poly									
0100	Fabric shell, 60′ clear span, not incl. foundations or floors									
0200	6,000 S.F.	B-41	1000	.044	SF Flr.	13.80	1.67	.29	15.76	18.10
0300	12,000 S.F.		1100	.040		13.20	1.52	.27	14.99	17.20
0400	80′ to 99′ clear span, 20,800 S.F.		1220	.036		13	1.37	.24	14.61	16.70
0410	100′ to 119′ clear span, 10,000 S.F.	L-5	2175	.026		13.70	1.32	.34	15.36	17.70
0430	26,000 S.F.		2300	.024		12.75	1.25	.32	14.32	16.55
0450	36,000 S.F.		2500	.022		12.60	1.15	.30	14.05	16.15
0460	120′ to 149′ clear span, 24,000 S.F.		3000	.019		13.90	.96	.25	15.11	17.20
0470	150′ to 199′ clear span, 30,000 S.F.		6000	.009		14.45	.48	.12	15.05	16.80
0480	200′ clear span, 40,000 S.F.	E-6	8000	.016		17.75	.81	.23	18.79	21
0500	For roll-up door, 12′ x 14′, add	L-2	1	16	Ea.	5,500	645		6,145	7,050
0600	For personnel doors, add, minimum				SF Flr.	5%				
0700	Add, maximum					15%				
0800	For site work, simple foundation, etc., add, minimum								1.25	1.95
0900	Add, maximum								2.75	3.05

13 34 Fabricated Engineered Structures

13 34 16 – Grandstands and Bleachers

13 34 16.53 Bleachers		Crew	Daily Output	Labor-Hours	Unit	Material	Labor	Equipment	Total	Total Incl O&P
0010	**BLEACHERS**									
0020	Bleachers, outdoor, portable, 5 tiers, 42 seats	2 Sswk	120	.133	Seat	90	6.80		96.80	111
0100	5 tiers, 54 seats		80	.200		80.50	10.20		90.70	107
0200	10 tiers, 104 seats		120	.133		91	6.80		97.80	112
0300	10 tiers, 144 seats		80	.200		82	10.20		92.20	108
0500	Permanent bleachers, aluminum seat, steel frame, 24″ row									
0600	8 tiers, 80 seats	2 Sswk	60	.267	Seat	66	13.65		79.65	96.50
0700	8 tiers, 160 seats		48	.333		57	17.05		74.05	93
0925	15 tiers, 154 to 165 seats		60	.267		99	13.65		112.65	133
0975	15 tiers, 214 to 225 seats		60	.267		90	13.65		103.65	123
1050	15 tiers, 274 to 285 seats		60	.267		81	13.65		94.65	114
1200	Seat backs only, 30″ row, fiberglass		160	.100		40	5.10		45.10	53
1300	Steel and wood		160	.100		18.15	5.10		23.25	29
1400	NOTE: average seating is 1.5′ in width									

13 34 19 – Metal Building Systems

13 34 19.50 Pre-Engineered Steel Buildings		Crew	Daily Output	Labor-Hours	Unit	Material	Labor	Equipment	Total	Total Incl O&P
0010	**PRE-ENGINEERED STEEL BUILDINGS**									
0100	Clear span rigid frame, 26 ga. colored roofing and siding									
0150	20′ to 29′ wide, 10′ eave height	E-2	425	.132	SF Flr.	8.50	6.60	3.60	18.70	24.50
0160	14′ eave height		350	.160		9.20	8.05	4.37	21.62	28.50
0170	16′ eave height		320	.175		9.85	8.80	4.78	23.43	31
0180	20′ eave height		275	.204		10.75	10.20	5.55	26.50	35.50
0190	24′ eave height		240	.233		11.85	11.70	6.35	29.90	40
0200	30′ to 49′ wide, 10′ eave height		535	.105		6.45	5.25	2.86	14.56	19.15
0300	14′ eave height		450	.124		7	6.25	3.40	16.65	22
0400	16′ eave height		415	.135		7.45	6.75	3.68	17.88	24
0500	20′ eave height		360	.156		8.10	7.80	4.25	20.15	27
0600	24′ eave height		320	.175		8.90	8.80	4.78	22.48	30
0700	50′ to 100′ wide, 10′ eave height		770	.073		5.50	3.65	1.99	11.14	14.45
0900	16′ eave height		600	.093		6.30	4.68	2.55	13.53	17.70

13 34 19 – Metal Building Systems

13 34 19.50 Pre-Engineered Steel Buildings		Crew	Daily Output	Labor-Hours	Unit	Material	Labor	Equipment	Total	Total Incl O&P
						2014 Bare Costs				
1000	20′ eave height	E-2	490	.114	SF Flr.	6.85	5.75	3.12	15.72	20.50
1100	24′ eave height	↓	435	.129	↓	7.55	6.45	3.51	17.51	23
1200	Clear span tapered beam frame, 26 ga. colored roofing/siding									
1300	30′ to 39′ wide, 10′ eave height	E-2	535	.105	SF Flr.	7.35	5.25	2.86	15.46	20
1400	14′ eave height		450	.124		8.10	6.25	3.40	17.75	23
1500	16′ eave height		415	.135		8.50	6.75	3.68	18.93	25
1600	20′ eave height		360	.156		9.35	7.80	4.25	21.40	28
1700	40′ wide, 10′ eave height		600	.093		6.50	4.68	2.55	13.73	17.90
1800	14′ eave height		510	.110		7.20	5.50	3	15.70	20.50
1900	16′ eave height		475	.118		7.55	5.90	3.22	16.67	22
2000	20′ eave height		415	.135		8.30	6.75	3.68	18.73	24.50
2100	50′ to 79′ wide, 10′ eave height		770	.073		6.05	3.65	1.99	11.69	15.05
2200	14′ eave height		675	.083		6.60	4.16	2.26	13.02	16.80
2300	16′ eave height		635	.088		6.85	4.43	2.41	13.69	17.70
2400	20′ eave height		490	.114		7.45	5.75	3.12	16.32	21.50
2410	80′ to 100′ wide, 10′ eave height		935	.060		5.40	3.01	1.64	10.05	12.80
2420	14′ eave height		750	.075		5.90	3.75	2.04	11.69	15.10
2430	16′ eave height		685	.082		6.20	4.10	2.23	12.53	16.20
2440	20′ eave height		560	.100		6.60	5	2.73	14.33	18.80
2460	101′ to 120′ wide, 10′ eave height		950	.059		4.94	2.96	1.61	9.51	12.20
2470	14′ eave height		770	.073		5.50	3.65	1.99	11.14	14.45
2480	16′ eave height		675	.083		5.85	4.16	2.26	12.27	15.95
2490	20′ eave height	↓	560	.100	↓	6.25	5	2.73	13.98	18.35
2500	Single post 2-span frame, 26 ga. colored roofing and siding									
2600	80′ wide, 14′ eave height	E-2	740	.076	SF Flr.	5.50	3.80	2.07	11.37	14.75
2700	16′ eave height		695	.081		5.85	4.04	2.20	12.09	15.65
2800	20′ eave height		625	.090		6.30	4.50	2.45	13.25	17.30
2900	24′ eave height		570	.098		6.90	4.93	2.68	14.51	18.90
3000	100′ wide, 14′ eave height		835	.067		5.35	3.37	1.83	10.55	13.55
3100	16′ eave height		795	.070		4.95	3.54	1.92	10.41	13.55
3200	20′ eave height		730	.077		6	3.85	2.09	11.94	15.45
3300	24′ eave height		670	.084		6.65	4.19	2.28	13.12	16.95
3400	120′ wide, 14′ eave height		870	.064		6.25	3.23	1.76	11.24	14.30
3500	16′ eave height		830	.067		5.55	3.39	1.84	10.78	13.90
3600	20′ eave height		765	.073		6	3.67	2	11.67	15.05
3700	24′ eave height	↓	705	.079	↓	6.60	3.99	2.17	12.76	16.40
3800	Double post 3-span frame, 26 ga. colored roofing and siding									
3900	150′ wide, 14′ eave height	E-2	925	.061	SF Flr.	4.35	3.04	1.65	9.04	11.75
4000	16′ eave height		890	.063		4.53	3.16	1.72	9.41	12.20
4100	20′ eave height		820	.068		4.93	3.43	1.86	10.22	13.25
4200	24′ eave height	↓	765	.073	↓	5.45	3.67	2	11.12	14.45
4300	Triple post 4-span frame, 26 ga. colored roofing and siding									
4400	160′ wide, 14′ eave height	E-2	970	.058	SF Flr.	4.28	2.90	1.58	8.76	11.35
4500	16′ eave height		930	.060		4.46	3.02	1.64	9.12	11.85
4600	20′ eave height		870	.064		4.37	3.23	1.76	9.36	12.25
4700	24′ eave height		815	.069		4.97	3.45	1.88	10.30	13.35
4800	200′ wide, 14′ eave height		1030	.054		3.90	2.73	1.48	8.11	10.55
4900	16′ eave height		995	.056		4.05	2.82	1.54	8.41	10.95
5000	20′ eave height		935	.060		4.47	3.01	1.64	9.12	11.80
5100	24′ eave height	↓	885	.063	↓	5	3.18	1.73	9.91	12.80
5200	Accessory items: add to the basic building cost above									
5250	Eave overhang, 2′ wide, 26 ga., with soffit	E-2	360	.156	L.F.	29.50	7.80	4.25	41.55	50.50
5300	4′ wide, without soffit	↓	300	.187	↓	26	9.35	5.10	40.45	50

13 34 Fabricated Engineered Structures

13 34 19 – Metal Building Systems

13 34 19.50 Pre-Engineered Steel Buildings		Crew	Daily Output	Labor-Hours	Unit	2014 Bare Costs Material	Labor	Equipment	Total	Total Incl O&P
5350	With soffit	E-2	250	.224	L.F.	38	11.25	6.10	55.35	67.50
5400	6′ wide, without soffit		250	.224		34	11.25	6.10	51.35	63.50
5450	With soffit		200	.280		45.50	14.05	7.65	67.20	82.50
5500	Entrance canopy, incl. frame, 4′ x 4′		25	2.240	Ea.	475	112	61	648	780
5550	4′ x 8′		19	2.947	″	550	148	80.50	778.50	945
5600	End wall roof overhang, 4′ wide, without soffit		850	.066	L.F.	16.55	3.31	1.80	21.66	26
5650	With soffit		500	.112	″	27	5.60	3.06	35.66	42.50
5700	Doors, HM self-framing, incl. butts, lockset and trim									
5750	Single leaf, 3070 (3′ x 7′), economy	2 Sswk	5	3.200	Opng.	585	164		749	935
5800	Deluxe		4	4		640	204		844	1,075
5825	Glazed		4	4		745	204		949	1,175
5850	3670 (3′-6″ x 7′)		4	4		815	204		1,019	1,250
5900	4070 (4′ x 7′)		3	5.333		885	273		1,158	1,450
5950	Double leaf, 6070 (6′ x 7′)		2	8		1,100	410		1,510	1,925
6000	Glazed		2	8		1,400	410		1,810	2,275
6050	Framing only, for openings, 3′ x 7′		4	4		185	204		389	565
6100	10′ x 10′		3	5.333		610	273		883	1,150
6150	For windows below, 2020 (2′ x 2′)		6	2.667		196	136		332	455
6200	4030 (4′ x 3′)		5	3.200		239	164		403	550
6250	Flashings, 26 ga., corner or eave, painted		240	.067	L.F.	4.30	3.41		7.71	10.75
6300	Galvanized		240	.067		4.10	3.41		7.51	10.50
6350	Rake flashing, painted		240	.067		4.64	3.41		8.05	11.10
6400	Galvanized		240	.067		4.40	3.41		7.81	10.85
6450	Ridge flashing, 18″ wide, painted		240	.067		6.20	3.41		9.61	12.80
6500	Galvanized		240	.067		7.10	3.41		10.51	13.80
6550	Gutter, eave type, 26 ga., painted		320	.050		6.70	2.56		9.26	11.90
6650	Valley type, between buildings, painted		120	.133		12.85	6.80		19.65	26
6710	Insulation, rated .6 lb. density, unfaced 4″ thick, R13	2 Carp	2300	.007	S.F.	.39	.32		.71	.92
6730	10″ thick, R30	″	2300	.007	″	1	.32		1.32	1.59
6750	Insulation, rated .6 lb. density, poly/scrim/foil (PSF) faced									
6760	4″ thick R13	2 Carp	2300	.007	S.F.	.58	.32		.90	1.13
6770	6″ thick, R19		2300	.007		.82	.32		1.14	1.39
6780	9 1/2″ thick, R30		2300	.007		.90	.32		1.22	1.48
6800	Insulation, rated .6 lb. density, vinyl faced 1-1/2″ thick, R5		2300	.007		.28	.32		.60	.80
6850	3″ thick, R10		2300	.007		.30	.32		.62	.82
6900	4″ thick, R13		2300	.007		.40	.32		.72	.93
6920	6″ thick, R19		2300	.007		.53	.32		.85	1.07
6930	10″ thick, R30		2300	.007		1.32	.32		1.64	1.94
6950	Foil/scrim/kraft (FSK) faced, 1-1/2″ thick, R5		2300	.007		.30	.32		.62	.82
7000	2″ thick, R6		2300	.007		.39	.32		.71	.92
7050	3″ thick, R10		2300	.007		.41	.32		.73	.94
7100	4″ thick, R13		2300	.007		.43	.32		.75	.96
7110	6″ thick, R19		2300	.007		.60	.32		.92	1.15
7120	10″ thick, R30		2300	.007		.84	.32		1.16	1.41
7150	Metalized polyester/scrim/kraft (PSK) facing,1-1/2″ thk, R5		2300	.007		.49	.32		.81	1.03
7200	2″ thick, R6		2300	.007		.58	.32		.90	1.13
7250	3″ thick, R11		2300	.007		.66	.32		.98	1.22
7300	4″ thick, R13		2300	.007		.75	.32		1.07	1.32
7310	6″ thick, R19		2300	.007		.97	.32		1.29	1.56
7320	10″ thick, R30		2300	.007		1.08	.32		1.40	1.68
7350	Vinyl/scrim/foil (VSF), 1-1/2″ thick, R5		2300	.007		.43	.32		.75	.96
7400	2″ thick, R6		2300	.007		.55	.32		.87	1.10
7450	3″ thick, R10		2300	.007		.59	.32		.91	1.14

13 34 Fabricated Engineered Structures

13 34 19 – Metal Building Systems

13 34 19.50 Pre-Engineered Steel Buildings		Crew	Daily Output	Labor-Hours	Unit	2014 Bare Costs Material	Labor	Equipment	Total	Total Incl O&P
7500	4" thick, R13	2 Carp	2300	.007	S.F.	.73	.32		1.05	1.29
7510	Vinyl/scrim/vinyl (VSV) 4" thick, R13		2300	.007		.48	.32		.80	1.02
7585	Vinyl/scrim/polyester (VSP), 4" thick, R13		2300	.007		.52	.32		.84	1.06
7650	Sash, single slide, glazed, with screens, 2020 (2' x 2')	E-1	22	1.091	Opng.	125	55	6.45	186.45	238
7700	3030 (3' x 3')		14	1.714		282	86.50	10.15	378.65	465
7750	4030 (4' x 3')		13	1.846		375	93	10.95	478.95	585
7800	6040 (6' x 4')		12	2		750	101	11.85	862.85	1,000
7850	Double slide sash, 3030 (3' x 3')		14	1.714		220	86.50	10.15	316.65	400
7900	6040 (6' x 4')		12	2		585	101	11.85	697.85	830
7950	Fixed glass, no screens, 3030 (3' x 3')		14	1.714		215	86.50	10.15	311.65	395
8000	6040 (6' x 4')		12	2		575	101	11.85	687.85	815
8050	Prefinished storm sash, 3030 (3' x 3')		70	.343		75	17.30	2.03	94.33	114
8200	Skylight, fiberglass panels, to 30 S.F.		10	2.400	Ea.	107	121	14.20	242.20	335
8250	Larger sizes, add for excess over 30 S.F.		300	.080	S.F.	3.55	4.03	.47	8.05	11.25
8300	Roof vents, turbine ventilator, wind driven									
8350	No damper, includes base, galvanized									
8400	12" diameter	Q-9	10	1.600	Ea.	91	79		170	220
8450	20" diameter		8	2		257	98.50		355.50	430
8500	24" diameter		8	2		405	98.50		503.50	595
8600	Continuous, 26 ga., 10' long, 9" wide	2 Sswk	4	4		36	204		240	400
8650	12" wide	"	4	4		36	204		240	400

13 34 23 – Fabricated Structures

13 34 23.15 Domes		Crew	Daily Output	Labor-Hours	Unit	Material	Labor	Equipment	Total	Total Incl O&P
0010	**DOMES**									
1500	Domes, bulk storage, shell only, dual radius hemisphere, arch, steel									
1600	framing, corrugated steel covering, 150' diameter	E-2	550	.102	SF Flr.	28	5.10	2.78	35.88	42.50
1700	400' diameter	"	720	.078		22.50	3.90	2.12	28.52	34
1800	Wood framing, wood decking, to 400' diameter	F-4	400	.120		37.50	5.55	2.84	45.89	53
1900	Radial framed wood (2" x 6"), 1/2" thick									
2000	plywood, asphalt shingles, 50' diameter	F-3	2000	.020	SF Flr.	72.50	.93	.33	73.76	81.50
2100	60' diameter		1900	.021		62	.98	.35	63.33	70
2200	72' diameter		1800	.022		51.50	1.04	.37	52.91	58.50
2300	116' diameter		1730	.023		35	1.08	.38	36.46	40.50
2400	150' diameter		1500	.027		37.50	1.25	.44	39.19	44

13 34 43 – Aircraft Hangars

13 34 43.50 Hangars		Crew	Daily Output	Labor-Hours	Unit	Material	Labor	Equipment	Total	Total Incl O&P
0010	**HANGARS** Prefabricated steel T hangars, Galv. steel roof &									
0100	walls, incl. electric bi-folding doors									
0110	not including floors or foundations, 4 unit	E-2	1275	.044	SF Flr.	12.45	2.20	1.20	15.85	18.75
0130	8 unit		1063	.053		11.35	2.64	1.44	15.43	18.50
0900	With bottom rolling doors, 4 unit		1386	.040		11.50	2.03	1.10	14.63	17.30
1000	8 unit		966	.058		10.30	2.91	1.58	14.79	18.05
1200	Alternate pricing method:									
1300	Galv. roof and walls, electric bi-folding doors, 4 plane	E-2	1.06	52.830	Plane	16,500	2,650	1,450	20,600	24,300
1500	8 plane		.91	61.538		13,500	3,100	1,675	18,275	22,000
1600	With bottom rolling doors, 4 plane		1.25	44.800		15,300	2,250	1,225	18,775	22,000
1800	8 plane		.97	57.732		12,300	2,900	1,575	16,775	20,300
2000	Circular type, prefab., steel frame, plastic skin, electric									
2010	door, including foundations, 80' diameter,									

13 34 Fabricated Engineered Structures

13 34 53 – Agricultural Structures

13 34 53.50 Silos		Crew	Daily Output	Labor-Hours	Unit	2014 Bare Costs Material	Labor	Equipment	Total	Total Incl O&P
0010	**SILOS**									
0500	Steel, factory fab., 30,000 gallon cap., painted, minimum	L-5	1	56	Ea.	22,100	2,875	745	25,720	30,100
0700	Maximum		.50	112		35,100	5,750	1,500	42,350	50,000
0800	Epoxy lined, minimum		1	56		36,100	2,875	745	39,720	45,500
1000	Maximum	↓	.50	112	↓	45,700	5,750	1,500	52,950	62,000

13 36 Towers

13 36 13 – Metal Towers

13 36 13.50 Control Towers

		Crew	Daily Output	Labor-Hours	Unit	Material	Labor	Equipment	Total	Total Incl O&P
0010	**CONTROL TOWERS**									
0020	Modular 12′ x 10′, incl. instruments				Ea.	810,000			810,000	891,000
0100	Maximum								970,000	1,067,000
0500	With standard 40′ tower				↓	1,400,000			1,400,000	1,540,000
1000	Temporary portable control towers, 8′ x 12′,									
1010	complete with one position communications, minimum				Ea.				266,000	293,000
2000	For fixed facilities, depending on height, minimum								57,000	63,000
2010	Maximum				↓				115,000	127,000

13 53 Meteorological Instrumentation

13 53 09 – Weather Instrumentation

13 53 09.50 Weather Station

		Crew	Daily Output	Labor-Hours	Unit	Material	Labor	Equipment	Total	Total Incl O&P
0010	**WEATHER STATION**									
0020	Remote recording, solar powered, with rain gauge & display, 400 ft range				Ea.	1,625			1,625	1,775
0100	1 mile range				"	1,825			1,825	2,025

Estimating Tips

22 10 00 Plumbing Piping and Pumps

This subdivision is primarily basic pipe and related materials. The pipe may be used by any of the mechanical disciplines, i.e., plumbing, fire protection, heating, and air conditioning.

Note: CPVC plastic piping approved for fire protection is located in 21 11 13.

- The labor adjustment factors listed in Subdivision 22 01 02.20 apply throughout Divisions 21, 22, and 23. CAUTION: the correct percentage may vary for the same items. For example, the percentage add for the basic pipe installation should be based on the maximum height that the craftsman must install for that particular section. If the pipe is to be located 14′ above the floor but it is suspended on threaded rod from beams, the bottom flange of which is 18′ high (4′ rods), then the height is actually 18′ and the add is 20%. The pipe coverer, however, does not have to go above the 14′, and so his or her add should be 10%.
- Most pipe is priced first as straight pipe with a joint (coupling, weld, etc.) every 10′ and a hanger usually every 10′. There are exceptions with hanger spacing such as for cast iron pipe (5′) and plastic pipe (3 per 10′). Following each type of pipe there are several lines listing sizes and the amount to be subtracted to delete couplings and hangers. This is for pipe that is to be buried or supported together on trapeze hangers. The reason that the couplings are deleted is that these runs are usually long, and frequently longer lengths of pipe are used. By deleting the couplings, the estimator is expected to look up and add back the correct reduced number of couplings.
- When preparing an estimate, it may be necessary to approximate the fittings. Fittings usually run between 25% and 50% of the cost of the pipe. The lower percentage is for simpler runs, and the higher number is for complex areas, such as mechanical rooms.
- For historic restoration projects, the systems must be as invisible as possible, and pathways must be sought for pipes, conduit, and ductwork. While installations in accessible spaces (such as basements and attics) are relatively straightforward to estimate, labor costs may be more difficult to determine when delivery systems must be concealed.

22 40 00 Plumbing Fixtures

- Plumbing fixture costs usually require two lines: the fixture itself and its "rough-in, supply, and waste."
- In the Assemblies Section (Plumbing D2010) for the desired fixture, the System Components Group at the center of the page shows the fixture on the first line. The rest of the list (fittings, pipe, tubing, etc.) will total up to what we refer to in the Unit Price section as "Rough-in, supply, waste, and vent." Note that for most fixtures we allow a nominal 5′ of tubing to reach from the fixture to a main or riser.
- Remember that gas- and oil-fired units need venting.

Reference Numbers

Reference numbers are shown in shaded boxes at the beginning of some major classifications. These numbers refer to related items in the Reference Section. The reference information may be an estimating procedure, an alternate pricing method, or technical information.

Note: Not all subdivisions listed here necessarily appear in this publication.

Note: **Trade Service,** *in part, has been used as a reference source for some of the material prices used in Division 22.*

22 01 Operation and Maintenance of Plumbing

22 01 02 – Labor Adjustments

22 01 02.20 Labor Adjustment Factors		Crew	Daily Output	Labor-Hours	Unit	2014 Bare Costs Material	Labor	Equipment	Total	Total Incl O&P
0010	**LABOR ADJUSTMENT FACTORS**, (For Div. 21, 22 and 23) R220102-20									
0100	Labor factors, The below are reasonable suggestions, however									
0110	each project must be evaluated for its own peculiarities, and									
0120	the adjustments be increased or decreased depending on the									
0130	severity of the special conditions.									
1000	Add to labor for elevated installation (Above floor level)									
1080	10′ to 14.5′ high						10%			
1100	15′ to 19.5′ high						20%			
1120	20′ to 24.5′ high						25%			
1140	25′ to 29.5′ high						35%			
1160	30′ to 34.5′ high						40%			
1180	35′ to 39.5′ high						50%			
1200	40′ and higher						55%			
2000	Add to labor for crawl space									
2100	3′ high						40%			
2140	4′ high						30%			
3000	Add to labor for multi-story building									
3100	Add per floor for floors 3 thru 19						2%			
3140	Add per floor for floors 20 and up						4%			
4000	Add to labor for working in existing occupied buildings									
4100	Hospital						35%			
4140	Office building						25%			
4180	School						20%			
4220	Factory or warehouse						15%			
4260	Multi dwelling						15%			
5000	Add to labor, miscellaneous									
5100	Cramped shaft						35%			
5140	Congested area						15%			
5180	Excessive heat or cold						30%			
9000	Labor factors, The above are reasonable suggestions, however									
9010	each project should be evaluated for its own peculiarities.									
9100	Other factors to be considered are:									
9140	Movement of material and equipment through finished areas									
9180	Equipment room									
9220	Attic space									
9260	No service road									
9300	Poor unloading/storage area									
9340	Congested site area/heavy traffic									

22 05 Common Work Results for Plumbing

22 05 05 – Selective Plumbing Demolition

22 05 05.10 Plumbing Demolition		Crew	Daily Output	Labor-Hours	Unit	Material	Labor	Equipment	Total	Total Incl O&P
0010	**PLUMBING DEMOLITION**									
2000	Piping, metal, up thru 1-1/2″ diameter	1 Plum	200	.040	L.F.		2.30		2.30	3.48
2050	2″ thru 3-1/2″ diameter	″	150	.053			3.07		3.07	4.63
2100	4″ thru 6″ diameter	2 Plum	100	.160			9.20		9.20	13.90
2150	8″ thru 14″ diameter	″	60	.267			15.35		15.35	23
2153	16″ thru 20″ diameter	Q-18	70	.343			18.70	.79	19.49	29.50
2155	24″ thru 26″ diameter		55	.436			24	1	25	37
2156	30″ thru 36″ diameter		40	.600			33	1.38	34.38	51
2160	Plastic pipe with fittings, up thru 1-1/2″ diameter	1 Plum	250	.032			1.84		1.84	2.78

22 05 Common Work Results for Plumbing

22 05 05 – Selective Plumbing Demolition

22 05 05.10 Plumbing Demolition		Crew	Daily Output	Labor-Hours	Unit	Material	2014 Bare Costs Labor	Equipment	Total	Total Incl O&P
2162	2″ thru 3″ diameter	1 Plum	200	.040	L.F.		2.30		2.30	3.48
2164	4″ thru 6″ diameter	Q-1	200	.080			4.14		4.14	6.25
2166	8″ thru 14″ diameter		150	.107			5.55		5.55	8.35
2168	16″ diameter		100	.160			8.30		8.30	12.50
2212	Deduct for salvage, aluminum scrap				Ton				700	770
2214	Brass scrap								2,450	2,675
2216	Copper scrap								3,200	3,525
2218	Lead scrap								520	570
2220	Steel scrap								180	200
2250	Water heater, 40 gal.	1 Plum	6	1.333	Ea.		76.50		76.50	116

22 05 23 – General-Duty Valves for Plumbing Piping

22 05 23.20 Valves, Bronze		Crew	Daily Output	Labor-Hours	Unit	Material	2014 Bare Costs Labor	Equipment	Total	Total Incl O&P
0010	**VALVES, BRONZE**									
1020	Angle, 150 lb., rising stem, threaded									
1030	1/8″	1 Plum	24	.333	Ea.	135	19.20		154.20	177
1040	1/4″		24	.333		135	19.20		154.20	177
1050	3/8″		24	.333		135	19.20		154.20	177
1060	1/2″		22	.364		135	21		156	180
1070	3/4″		20	.400		184	23		207	237
1080	1″		19	.421		265	24		289	330
1100	1-1/2″		13	.615		445	35.50		480.50	545
1110	2″		11	.727		715	42		757	850
1510	2-1/2″		9	.889		295	51		346	400
1520	3″		8	1		450	57.50		507.50	575
1750	Check, swing, class 150, regrinding disc, threaded									
1800	1/8″	1 Plum	24	.333	Ea.	63	19.20		82.20	98.50
1830	1/4″		24	.333		63	19.20		82.20	98.50
1840	3/8″		24	.333		67	19.20		86.20	103
1850	1/2″		24	.333		72	19.20		91.20	108
1860	3/4″		20	.400		95	23		118	139
1870	1″		19	.421		137	24		161	187
1880	1-1/4″		15	.533		198	30.50		228.50	265
1890	1-1/2″		13	.615		230	35.50		265.50	305
1900	2″		11	.727		340	42		382	435
1910	2-1/2″	Q-1	15	1.067		760	55.50		815.50	920
2000	For 200 lb., add					5%	10%			
2040	For 300 lb., add					15%	15%			
2850	Gate, N.R.S., soldered, 125 psi									
2900	3/8″	1 Plum	24	.333	Ea.	54.50	19.20		73.70	89
2920	1/2″		24	.333		46.50	19.20		65.70	80.50
2940	3/4″		20	.400		53	23		76	93
2950	1″		19	.421		75	24		99	119
2960	1-1/4″		15	.533		115	30.50		145.50	173
2970	1-1/2″		13	.615		129	35.50		164.50	196
2980	2″		11	.727		182	42		224	264
2990	2-1/2″	Q-1	15	1.067		455	55.50		510.50	590
3000	3″	″	13	1.231		595	64		659	750
3850	Rising stem, soldered, 300 psi									
3950	1″	1 Plum	19	.421	Ea.	189	24		213	245
3980	2″	″	11	.727		505	42		547	625
4000	3″	Q-1	13	1.231		1,675	64		1,739	1,950
4250	Threaded, class 150									

22 05 23.20 Valves, Bronze		Crew	Daily Output	Labor-Hours	Unit	2014 Bare Costs Material	Labor	Equipment	Total	Total Incl O&P
4310	1/4"	1 Plum	24	.333	Ea.	71	19.20		90.20	108
4320	3/8"		24	.333		71	19.20		90.20	108
4330	1/2"		24	.333		65	19.20		84.20	101
4340	3/4"		20	.400		76	23		99	119
4350	1"		19	.421		102	24		126	149
4360	1-1/4"		15	.533		139	30.50		169.50	200
4370	1-1/2"		13	.615		175	35.50		210.50	246
4380	2"		11	.727		235	42		277	320
4390	2-1/2"	Q-1	15	1.067		550	55.50		605.50	690
4400	3"	"	13	1.231		765	64		829	935
4500	For 300 psi, threaded, add					100%	15%			
4540	For chain operated type, add					15%				
4850	Globe, class 150, rising stem, threaded									
4920	1/4"	1 Plum	24	.333	Ea.	102	19.20		121.20	141
4940	3/8" size		24	.333		101	19.20		120.20	140
4950	1/2"		24	.333		101	19.20		120.20	140
4960	3/4"		20	.400		135	23		158	183
4970	1"		19	.421		212	24		236	270
4980	1-1/4"		15	.533		335	30.50		365.50	415
4990	1-1/2"		13	.615		410	35.50		445.50	505
5000	2"		11	.727		615	42		657	740
5010	2-1/2"	Q-1	15	1.067		1,225	55.50		1,280.50	1,425
5020	3"	"	13	1.231		1,750	64		1,814	2,025
5120	For 300 lb. threaded, add					50%	15%			
5600	Relief, pressure & temperature, self-closing, ASME, threaded									
5640	3/4"	1 Plum	28	.286	Ea.	253	16.45		269.45	305
5650	1"		24	.333		405	19.20		424.20	475
5660	1-1/4"		20	.400		895	23		918	1,025
5670	1-1/2"		18	.444		1,250	25.50		1,275.50	1,425
5680	2"		16	.500		1,350	29		1,379	1,550
5950	Pressure, poppet type, threaded									
6000	1/2"	1 Plum	30	.267	Ea.	68.50	15.35		83.85	98.50
6040	3/4"	"	28	.286	"	73.50	16.45		89.95	106
6400	Pressure, water, ASME, threaded									
6440	3/4"	1 Plum	28	.286	Ea.	116	16.45		132.45	152
6450	1"		24	.333		260	19.20		279.20	315
6460	1-1/4"		20	.400		390	23		413	465
6470	1-1/2"		18	.444		570	25.50		595.50	665
6480	2"		16	.500		820	29		849	950
6490	2-1/2"		15	.533		3,150	30.50		3,180.50	3,525
6900	Reducing, water pressure									
6920	300 psi to 25-75 psi, threaded or sweat									
6940	1/2"	1 Plum	24	.333	Ea.	395	19.20		414.20	465
6950	3/4"		20	.400		395	23		418	470
6960	1"		19	.421		615	24		639	710
6970	1-1/4"		15	.533		1,100	30.50		1,130.50	1,250
6980	1-1/2"		13	.615		1,650	35.50		1,685.50	1,875
8350	Tempering, water, sweat connections									
8400	1/2"	1 Plum	24	.333	Ea.	109	19.20		128.20	149
8440	3/4"	"	20	.400	"	126	23		149	174
8650	Threaded connections									
8700	1/2"	1 Plum	24	.333	Ea.	126	19.20		145.20	168
8740	3/4"		20	.400		770	23		793	880

22 05 Common Work Results for Plumbing

22 05 23 – General-Duty Valves for Plumbing Piping

22 05 23.20 Valves, Bronze		Crew	Daily Output	Labor-Hours	Unit	2014 Bare Costs Material	Labor	Equipment	Total	Total Incl O&P
8750	1"	1 Plum	19	.421	Ea.	865	24		889	990
8760	1-1/4"		15	.533		1,350	30.50		1,380.50	1,525
8770	1-1/2"		13	.615		1,475	35.50		1,510.50	1,650
8780	2"		11	.727		2,200	42		2,242	2,500

22 05 23.60 Valves, Plastic		Crew	Daily Output	Labor-Hours	Unit	Material	Labor	Equipment	Total	Total Incl O&P
0010	**VALVES, PLASTIC**									
1150	Ball, PVC, socket or threaded, single union									
1230	1/2"	1 Plum	26	.308	Ea.	28	17.70		45.70	57.50
1240	3/4"		25	.320		32	18.40		50.40	63.50
1250	1"		23	.348		40	20		60	74.50
1260	1-1/4"		21	.381		54	22		76	92.50
1270	1-1/2"		20	.400		64.50	23		87.50	106
1280	2"		17	.471		93.50	27		120.50	144
1290	2-1/2"	Q-1	26	.615		395	32		427	485
1300	3"		24	.667		480	34.50		514.50	575
1310	4"		20	.800		810	41.50		851.50	955
1360	For PVC, flanged, add					100%	15%			
3150	Ball check, PVC, socket or threaded									
3200	1/4"	1 Plum	26	.308	Ea.	44	17.70		61.70	74.50
3220	3/8"		26	.308		44	17.70		61.70	74.50
3240	1/2"		26	.308		44	17.70		61.70	74.50
3250	3/4"		25	.320		49	18.40		67.40	82
3260	1"		23	.348		61.50	20		81.50	97.50
3270	1-1/4"		21	.381		103	22		125	147
3280	1-1/2"		20	.400		103	23		126	149
3290	2"		17	.471		140	27		167	195
3310	3"	Q-1	24	.667		390	34.50		424.50	480
3320	4"	"	20	.800		550	41.50		591.50	670
3360	For PVC, flanged, add					50%	15%			

22 05 76 – Facility Drainage Piping Cleanouts

22 05 76.10 Cleanouts		Crew	Daily Output	Labor-Hours	Unit	Material	Labor	Equipment	Total	Total Incl O&P
0010	**CLEANOUTS**									
0060	Floor type									
0080	Round or square, scoriated nickel bronze top									
0100	2" pipe size	1 Plum	10	.800	Ea.	183	46		229	271
0120	3" pipe size		8	1		274	57.50		331.50	385
0140	4" pipe size		6	1.333		274	76.50		350.50	415
0160	5" pipe size		4	2		345	115		460	555
0180	6" pipe size	Q-1	6	2.667		345	138		483	590
0200	8" pipe size	"	4	4		615	207		822	990
0340	Recessed for tile, same price									
0980	Round top, recessed for terrazzo									
1000	2" pipe size	1 Plum	9	.889	Ea.	183	51		234	278
1080	3" pipe size		6	1.333		274	76.50		350.50	415
1100	4" pipe size		4	2		274	115		389	475
1120	5" pipe size	Q-1	6	2.667		345	138		483	590
1140	6" pipe size		5	3.200		345	166		511	630
1160	8" pipe size		4	4		615	207		822	990
2000	Round scoriated nickel bronze top, extra heavy duty									
2060	2" pipe size	1 Plum	9	.889	Ea.	249	51		300	350
2080	3" pipe size		6	1.333		340	76.50		416.50	490
2100	4" pipe size		4	2		340	115		455	550

22 05 76 – Facility Drainage Piping Cleanouts

22 05 76.10 Cleanouts		Crew	Daily Output	Labor-Hours	Unit	Material	2014 Bare Costs Labor	Equipment	Total	Total Incl O&P
2120	5″ pipe size	Q-1	6	2.667	Ea.	415	138		553	665
2140	6″ pipe size		5	3.200		415	166		581	705
2160	8″ pipe size		4	4		680	207		887	1,075
4000	Wall type, square smooth cover, over wall frame									
4060	2″ pipe size	1 Plum	14	.571	Ea.	298	33		331	380
4080	3″ pipe size		12	.667		320	38.50		358.50	415
4100	4″ pipe size		10	.800		345	46		391	450
4120	5″ pipe size		9	.889		500	51		551	625
4140	6″ pipe size		8	1		555	57.50		612.50	695
4160	8″ pipe size	Q-1	11	1.455		740	75.50		815.50	925
5000	Extension, C.I.; bronze countersunk plug, 8″ long									
5040	2″ pipe size	1 Plum	16	.500	Ea.	157	29		186	216
5060	3″ pipe size		14	.571		182	33		215	250
5080	4″ pipe size		13	.615		185	35.50		220.50	257
5100	5″ pipe size		12	.667		289	38.50		327.50	380
5120	6″ pipe size		11	.727		350	42		392	450

22 05 76.20 Cleanout Tees		Crew	Daily Output	Labor-Hours	Unit	Material	Labor	Equipment	Total	Total Incl O&P
0010	**CLEANOUT TEES**									
0100	Cast iron, B&S, with countersunk plug									
0200	2″ pipe size	1 Plum	4	2	Ea.	248	115		363	445
0220	3″ pipe size		3.60	2.222		270	128		398	490
0240	4″ pipe size		3.30	2.424		335	140		475	580
0260	5″ pipe size	Q-1	5.50	2.909		730	151		881	1,025
0280	6″ pipe size	″	5	3.200		910	166		1,076	1,250
0300	8″ pipe size	Q-3	5	6.400		1,050	350		1,400	1,675
0500	For round smooth access cover, same price									
0600	For round scoriated access cover, same price									
0700	For square smooth access cover, add				Ea.	60%				
4000	Plastic, tees and adapters. Add plugs									
4010	ABS, DWV									
4020	Cleanout tee, 1-1/2″ pipe size	1 Plum	15	.533	Ea.	20	30.50		50.50	68.50
4030	2″ pipe size	Q-1	27	.593		22.50	30.50		53	71.50
4040	3″ pipe size		21	.762		43.50	39.50		83	108
4050	4″ pipe size		16	1		87	52		139	174
4100	Cleanout plug, 1-1/2″ pipe size	1 Plum	32	.250		2.97	14.40		17.37	25
4110	2″ pipe size	Q-1	56	.286		3.90	14.80		18.70	27
4120	3″ pipe size		36	.444		6.30	23		29.30	42
4130	4″ pipe size		30	.533		11	27.50		38.50	53.50
4180	Cleanout adapter fitting, 1-1/2″ pipe size	1 Plum	32	.250		4.70	14.40		19.10	26.50
4190	2″ pipe size	Q-1	56	.286		7.20	14.80		22	30.50
4200	3″ pipe size		36	.444		18.45	23		41.45	55.50
4210	4″ pipe size		30	.533		34	27.50		61.50	79
5000	PVC, DWV									
5010	Cleanout tee, 1-1/2″ pipe size	1 Plum	15	.533	Ea.	12.75	30.50		43.25	60.50
5020	2″ pipe size	Q-1	27	.593		14.85	30.50		45.35	63
5030	3″ pipe size		21	.762		29	39.50		68.50	91
5040	4″ pipe size		16	1		51.50	52		103.50	135
5090	Cleanout plug, 1-1/2″ pipe size	1 Plum	32	.250		3.04	14.40		17.44	25
5100	2″ pipe size	Q-1	56	.286		3.30	14.80		18.10	26
5110	3″ pipe size		36	.444		6.05	23		29.05	41.50
5120	4″ pipe size		30	.533		8.90	27.50		36.40	51.50
5130	6″ pipe size		24	.667		29.50	34.50		64	84.50

22 05 Common Work Results for Plumbing

22 05 76 – Facility Drainage Piping Cleanouts

22 05 76.20 Cleanout Tees		Crew	Daily Output	Labor-Hours	Unit	Material	Labor	Equipment	Total	Total Incl O&P
						2014 Bare Costs				
5170	Cleanout adapter fitting, 1-1/2″ pipe size	1 Plum	32	.250	Ea.	4.06	14.40		18.46	26
5180	2″ pipe size	Q-1	56	.286		5.20	14.80		20	28.50
5190	3″ pipe size		36	.444		14.70	23		37.70	51
5200	4″ pipe size		30	.533		23.50	27.50		51	67.50
5210	6″ pipe size		24	.667		70.50	34.50		105	130

22 11 Facility Water Distribution

22 11 13 – Facility Water Distribution Piping

22 11 13.23 Pipe/Tube, Copper

Line	Description	Crew	Daily Output	Labor-Hours	Unit	Material	Labor	Equipment	Total	Total Incl O&P
0010	**PIPE/TUBE, COPPER**, Solder joints									
1000	Type K tubing, couplings & clevis hanger assemblies 10′ O.C.									
1100	1/4″ diameter	1 Plum	84	.095	L.F.	4.62	5.50		10.12	13.40
1120	3/8″ diameter		82	.098		3.99	5.60		9.59	12.90
1140	1/2″ diameter		78	.103		4.47	5.90		10.37	13.80
1160	5/8″ diameter		77	.104		5.95	6		11.95	15.55
1180	3/4″ diameter		74	.108		7.80	6.20		14	18
1200	1″ diameter		66	.121		10.60	7		17.60	22
1220	1-1/4″ diameter		56	.143		13.25	8.20		21.45	27
1240	1-1/2″ diameter		50	.160		17.15	9.20		26.35	33
1260	2″ diameter		40	.200		26.50	11.50		38	46.50
1280	2-1/2″ diameter	Q-1	60	.267		41	13.80		54.80	66.50
1300	3″ diameter		54	.296		58	15.35		73.35	86.50
1320	3-1/2″ diameter		42	.381		82	19.75		101.75	120
1330	4″ diameter		38	.421		101	22		123	144
1340	5″ diameter		32	.500		203	26		229	262
1360	6″ diameter	Q-2	38	.632		305	34		339	385
1380	8″ diameter	″	34	.706		480	38		518	590
1390	For other than full hard temper, add					13%				
1440	For silver solder, add						15%			
1800	For medical clean, (oxygen class), add					12%				
1950	To delete cplgs. & hngrs., 1/4″-1″ pipe, subtract					27%	60%			
1960	1-1/4″ -3″ pipe, subtract					14%	52%			
1970	3-1/2″-5″ pipe, subtract					10%	60%			
1980	6″-8″ pipe, subtract					19%	53%			

22 11 13.25 Pipe/Tube Fittings, Copper

Line	Description	Crew	Daily Output	Labor-Hours	Unit	Material	Labor	Equipment	Total	Total Incl O&P
0010	**PIPE/TUBE FITTINGS, COPPER**, Wrought unless otherwise noted									
0040	Solder joints, copper x copper									
0070	90° elbow, 1/4″	1 Plum	22	.364	Ea.	9.70	21		30.70	42
0090	3/8″		22	.364		9.25	21		30.25	41.50
0100	1/2″		20	.400		3.08	23		26.08	38.50
0110	5/8″		19	.421		10.65	24		34.65	48
0120	3/4″		19	.421		6.95	24		30.95	44
0130	1″		16	.500		17	29		46	62
0140	1-1/4″		15	.533		25.50	30.50		56	74.50
0150	1-1/2″		13	.615		40.50	35.50		76	98
0160	2″		11	.727		72.50	42		114.50	143
0170	2-1/2″	Q-1	13	1.231		146	64		210	258
0180	3″		11	1.455		195	75.50		270.50	330
0190	3-1/2″		10	1.600		685	83		768	875
0200	4″		9	1.778		500	92		592	690
0210	5″		6	2.667		2,100	138		2,238	2,500

22 11 13.25 Pipe/Tube Fittings, Copper		Crew	Daily Output	Labor-Hours	Unit	2014 Bare Costs Material	Labor	Equipment	Total	Total Incl O&P
0220	6"	Q-2	9	2.667	Ea.	2,800	143		2,943	3,300
0230	8"	"	8	3		10,300	161		10,461	11,600
0250	45° elbow, 1/4"	1 Plum	22	.364		17.30	21		38.30	50.50
0270	3/8"		22	.364		14.80	21		35.80	48
0280	1/2"		20	.400		5.65	23		28.65	41.50
0290	5/8"		19	.421		27	24		51	66
0300	3/4"		19	.421		9.90	24		33.90	47.50
0310	1"		16	.500		25	29		54	71
0320	1-1/4"		15	.533		33.50	30.50		64	83.50
0330	1-1/2"		13	.615		40.50	35.50		76	98
0340	2"		11	.727		67.50	42		109.50	138
0350	2-1/2"	Q-1	13	1.231		144	64		208	256
0360	3"		13	1.231		214	64		278	335
0370	3-1/2"		10	1.600		600	83		683	785
0380	4"		9	1.778		455	92		547	640
0390	5"		6	2.667		1,775	138		1,913	2,150
0400	6"	Q-2	9	2.667		2,775	143		2,918	3,300
0410	8"	"	8	3		9,450	161		9,611	10,600
0450	Tee, 1/4"	1 Plum	14	.571		19.40	33		52.40	71
0470	3/8"		14	.571		15.60	33		48.60	66.50
0480	1/2"		13	.615		5.25	35.50		40.75	59.50
0490	5/8"		12	.667		34	38.50		72.50	95.50
0500	3/4"		12	.667		12.70	38.50		51.20	72
0510	1"		10	.800		39.50	46		85.50	113
0520	1-1/4"		9	.889		53.50	51		104.50	136
0530	1-1/2"		8	1		82.50	57.50		140	178
0540	2"		7	1.143		129	66		195	242
0550	2-1/2"	Q-1	8	2		260	104		364	440
0560	3"		7	2.286		395	118		513	615
0570	3-1/2"		6	2.667		1,150	138		1,288	1,475
0580	4"		5	3.200		960	166		1,126	1,300
0590	5"		4	4		3,150	207		3,357	3,775
0600	6"	Q-2	6	4		4,300	215		4,515	5,050
0610	8"	"	5	4.800		16,600	258		16,858	18,700
0612	Tee, reducing on the outlet, 1/4"	1 Plum	15	.533		33	30.50		63.50	82.50
0613	3/8"		15	.533		31	30.50		61.50	80.50
0614	1/2"		14	.571		27	33		60	79.50
0615	5/8"		13	.615		54.50	35.50		90	114
0616	3/4"		12	.667		15	38.50		53.50	74.50
0617	1"		11	.727		39.50	42		81.50	107
0618	1-1/4"		10	.800		58	46		104	133
0619	1-1/2"		9	.889		61.50	51		112.50	145
0620	2"		8	1		97.50	57.50		155	194
0621	2-1/2"	Q-1	9	1.778		305	92		397	475
0622	3"		8	2		365	104		469	560
0623	4"		6	2.667		685	138		823	960
0624	5"		5	3.200		3,150	166		3,316	3,700
0625	6"	Q-2	7	3.429		4,075	184		4,259	4,775
0626	8"	"	6	4		16,600	215		16,815	18,500
0630	Tee, reducing on the run, 1/4"	1 Plum	15	.533		41	30.50		71.50	92
0631	3/8"		15	.533		47	30.50		77.50	98
0632	1/2"		14	.571		37.50	33		70.50	90.50
0633	5/8"		13	.615		55	35.50		90.50	114

22 11 13 – Facility Water Distribution Piping

22 11 13.25 Pipe/Tube Fittings, Copper		Crew	Daily Output	Labor-Hours	Unit	Material	Labor (2014 Bare Costs)	Equipment	Total	Total Incl O&P
0634	3/4"	1 Plum	12	.667	Ea.	15	38.50		53.50	74.50
0635	1"		11	.727		47	42		89	115
0636	1-1/4"		10	.800		74.50	46		120.50	152
0637	1-1/2"		9	.889		123	51		174	212
0638	2"		8	1		168	57.50		225.50	271
0639	2-1/2"	Q-1	9	1.778		415	92		507	600
0640	3"		8	2		560	104		664	770
0641	4"		6	2.667		1,275	138		1,413	1,600
0642	5"		5	3.200		2,975	166		3,141	3,525
0643	6"	Q-2	7	3.429		4,550	184		4,734	5,275
0644	8"	"	6	4		16,100	215		16,315	18,000
0650	Coupling, 1/4"	1 Plum	24	.333		2.25	19.20		21.45	31.50
0670	3/8"		24	.333		2.98	19.20		22.18	32.50
0680	1/2"		22	.364		2.34	21		23.34	34
0690	5/8"		21	.381		7.05	22		29.05	41
0700	3/4"		21	.381		4.70	22		26.70	38
0710	1"		18	.444		9.60	25.50		35.10	49
0715	1-1/4"		17	.471		16.60	27		43.60	59.50
0716	1-1/2"		15	.533		22	30.50		52.50	70.50
0718	2"		13	.615		36.50	35.50		72	94
0721	2-1/2"	Q-1	15	1.067		78	55.50		133.50	169
0722	3"		13	1.231		117	64		181	226
0724	3-1/2"		8	2		224	104		328	405
0726	4"		7	2.286		246	118		364	450
0728	5"		6	2.667		605	138		743	875
0731	6"	Q-2	8	3		1,000	161		1,161	1,350
0732	8"	"	7	3.429		3,250	184		3,434	3,850
0850	Unions, 1/4"	1 Plum	21	.381		70.50	22		92.50	111
0870	3/8"		21	.381		71	22		93	111
0880	1/2"		19	.421		38	24		62	78
0890	5/8"		18	.444		161	25.50		186.50	216
0900	3/4"		18	.444		47.50	25.50		73	91
0910	1"		15	.533		81.50	30.50		112	137
0920	1-1/4"		14	.571		142	33		175	206
0930	1-1/2"		12	.667		154	38.50		192.50	228
0940	2"		10	.800		320	46		366	420
0950	2-1/2"	Q-1	12	1.333		695	69		764	870
0960	3"	"	10	1.600		1,800	83		1,883	2,100
0980	Adapter, copper x male IPS, 1/4"	1 Plum	20	.400		32	23		55	70.50
0990	3/8"		20	.400		15.95	23		38.95	52.50
1000	1/2"		18	.444		6.45	25.50		31.95	45.50
1010	3/4"		17	.471		10.85	27		37.85	53
1020	1"		15	.533		28	30.50		58.50	77.50
1030	1-1/4"		13	.615		41.50	35.50		77	99
1040	1-1/2"		12	.667		47.50	38.50		86	111
1050	2"		11	.727		80.50	42		122.50	152
1060	2-1/2"	Q-1	10.50	1.524		310	79		389	460
1070	3"		10	1.600		410	83		493	575
1080	3-1/2"		9	1.778		485	92		577	675
1090	4"		8	2		555	104		659	765
1200	5"		6	2.667		3,275	138		3,413	3,800
1210	6"	Q-2	8.50	2.824		3,650	152		3,802	4,225
1250	Cross, 1/2"	1 Plum	10	.800		51.50	46		97.50	126

22 11 13.25 Pipe/Tube Fittings, Copper		Crew	Daily Output	Labor-Hours	Unit	2014 Bare Costs Material	Labor	Equipment	Total	Total Incl O&P
1260	3/4″	1 Plum	9.50	.842	Ea.	99.50	48.50		148	183
1270	1″		8	1		169	57.50		226.50	273
1280	1-1/4″		7.50	1.067		243	61.50		304.50	360
1290	1-1/2″		6.50	1.231		345	71		416	485
1300	2″		5.50	1.455		655	83.50		738.50	845
1310	2-1/2″	Q-1	6.50	2.462		1,500	128		1,628	1,850
1320	3″	″	5.50	2.909		1,775	151		1,926	2,175
1500	Tee fitting, mechanically formed,(Type 1, 'branch sizes up to 2 in.')									
1520	1/2″ run size, 3/8″ to 1/2″ branch size	1 Plum	80	.100	Ea.		5.75		5.75	8.70
1530	3/4″ run size, 3/8″ to 3/4″ branch size		60	.133			7.65		7.65	11.60
1540	1″ run size, 3/8″ to 1″ branch size		54	.148			8.55		8.55	12.85
1550	1-1/4″ run size, 3/8″ to 1-1/4″ branch size		48	.167			9.60		9.60	14.50
1560	1-1/2″ run size, 3/8″ to 1-1/2″ branch size		40	.200			11.50		11.50	17.40
1570	2″ run size, 3/8″ to 2″ branch size		35	.229			13.15		13.15	19.85
1580	2-1/2″ run size, 1/2″ to 2″ branch size		32	.250			14.40		14.40	21.50
1590	3″ run size, 1″ to 2″ branch size		26	.308			17.70		17.70	26.50
1600	4″ run size, 1″ to 2″ branch size		24	.333			19.20		19.20	29
1640	Tee fitting, mechanically formed, (Type 2, branches 2-1/2″ thru 4″)									
1650	2-1/2″ run size, 2-1/2″ branch size	1 Plum	12.50	.640	Ea.		37		37	55.50
1660	3″ run size, 2-1/2″ to 3″ branch size		12	.667			38.50		38.50	58
1670	3-1/2″ run size, 2-1/2″ to 3-1/2″ branch size		11	.727			42		42	63
1680	4″ run size, 2-1/2″ to 4″ branch size		10.50	.762			44		44	66
1698	5″ run size, 2″ to 4″ branch size		9.50	.842			48.50		48.50	73
1700	6″ run size, 2″ to 4″ branch size		8.50	.941			54		54	82
1710	8″ run size, 2″ to 4″ branch size		7	1.143			66		66	99.50
2000	DWV, solder joints, copper x copper									
2030	90° Elbow, 1-1/4″	1 Plum	13	.615	Ea.	29.50	35.50		65	86
2050	1-1/2″		12	.667		44.50	38.50		83	107
2070	2″		10	.800		58.50	46		104.50	134
2090	3″	Q-1	10	1.600		144	83		227	284
2100	4″	″	9	1.778		1,000	92		1,092	1,250
2150	45° Elbow, 1-1/4″	1 Plum	13	.615		27.50	35.50		63	83.50
2170	1-1/2″		12	.667		22.50	38.50		61	83
2180	2″		10	.800		52	46		98	127
2190	3″	Q-1	10	1.600		111	83		194	247
2200	4″	″	9	1.778		515	92		607	705
2250	Tee, Sanitary, 1-1/4″	1 Plum	9	.889		59	51		110	142
2270	1-1/2″		8	1		73.50	57.50		131	168
2290	2″		7	1.143		85.50	66		151.50	194
2310	3″	Q-1	7	2.286		370	118		488	590
2330	4″	″	6	2.667		945	138		1,083	1,250
2400	Coupling, 1-1/4″	1 Plum	14	.571		13.95	33		46.95	65
2420	1-1/2″		13	.615		17.35	35.50		52.85	72.50
2440	2″		11	.727		24	42		66	89.50
2460	3″	Q-1	11	1.455		49	75.50		124.50	168
2480	4″	″	10	1.600		149	83		232	288

22 11 13.44 Pipe, Steel		Crew	Daily Output	Labor-Hours	Unit	Material	Labor	Equipment	Total	Total Incl O&P
0010	**PIPE, STEEL**									
0020	All pipe sizes are to Spec. A-53 unless noted otherwise									
0050	Schedule 40, threaded, with couplings, and clevis hanger									
0060	assemblies sized for covering, 10′ O.C.									
0540	Black, 1/4″ diameter	1 Plum	66	.121	L.F.	4.51	7		11.51	15.50

22 11 Facility Water Distribution

22 11 13 – Facility Water Distribution Piping

22 11 13.44 Pipe, Steel		Crew	Daily Output	Labor-Hours	Unit	Material	2014 Bare Costs Labor	Equipment	Total	Total Incl O&P
0550	3/8" diameter	1 Plum	65	.123	L.F.	5.10	7.10		12.20	16.30
0560	1/2" diameter		63	.127		2.73	7.30		10.03	14.05
0570	3/4" diameter		61	.131		3.17	7.55		10.72	14.90
0580	1" diameter		53	.151		4.64	8.70		13.34	18.20
0590	1-1/4" diameter	Q-1	89	.180		5.75	9.30		15.05	20.50
0600	1-1/2" diameter		80	.200		6.70	10.35		17.05	23
0610	2" diameter		64	.250		8.95	12.95		21.90	29.50
0620	2-1/2" diameter		50	.320		14.05	16.60		30.65	40.50
0630	3" diameter		43	.372		18	19.25		37.25	49
0640	3-1/2" diameter		40	.400		24.50	20.50		45	58.50
0650	4" diameter		36	.444		27	23		50	65
0809	A-106, gr. A/B, seamless w/cplgs. & clevis hanger assemblies									
0811	1/4" diameter	1 Plum	66	.121	L.F.	7.75	7		14.75	19.05
0812	3/8" diameter		65	.123		7.70	7.10		14.80	19.15
0813	1/2" diameter		63	.127		8.90	7.30		16.20	21
0814	3/4" diameter		61	.131		10.80	7.55		18.35	23.50
0815	1" diameter		53	.151		11.90	8.70		20.60	26
0816	1-1/4" diameter	Q-1	89	.180		14.45	9.30		23.75	30
0817	1-1/2" diameter		80	.200		22	10.35		32.35	40
0819	2" diameter		64	.250		24	12.95		36.95	46
0821	2-1/2" diameter		50	.320		29	16.60		45.60	57
0822	3" diameter		43	.372		32	19.25		51.25	64
0823	4" diameter		36	.444		49	23		72	89
1220	To delete coupling & hanger, subtract									
1230	1/4" diam. to 3/4" diam.					31%	56%			
1240	1" diam. to 1-1/2" diam.					23%	51%			
1250	2" diam. to 4" diam.					23%	41%			
1280	All pipe sizes are to Spec. A-53 unless noted otherwise									
1281	Schedule 40, threaded, with couplings and clevis hanger									
1282	assemblies sized for covering, 10' O. C.									
1290	Galvanized, 1/4" diameter	1 Plum	66	.121	L.F.	6.40	7		13.40	17.60
1300	3/8" diameter		65	.123		7.05	7.10		14.15	18.45
1310	1/2" diameter		63	.127		3.76	7.30		11.06	15.20
1320	3/4" diameter		61	.131		4.45	7.55		12	16.30
1330	1" diameter		53	.151		6.10	8.70		14.80	19.80
1340	1-1/4" diameter	Q-1	89	.180		7.70	9.30		17	22.50
1350	1-1/2" diameter		80	.200		9.05	10.35		19.40	25.50
1360	2" diameter		64	.250		12.15	12.95		25.10	33
1370	2-1/2" diameter		50	.320		20	16.60		36.60	47
1380	3" diameter		43	.372		25.50	19.25		44.75	57.50
1390	3-1/2" diameter		40	.400		33	20.50		53.50	68
1400	4" diameter		36	.444		38	23		61	76.50
1750	To delete coupling & hanger, subtract									
1760	1/4" diam. to 3/4" diam.					31%	56%			
1770	1" diam. to 1-1/2" diam.					23%	51%			
1780	2" diam. to 4" diam.					23%	41%			

22 11 13.45 Pipe Fittings, Steel, Threaded		Crew	Daily Output	Labor-Hours	Unit	Material	Labor	Equipment	Total	Total Incl O&P
0010	**PIPE FITTINGS, STEEL, THREADED**									
0020	Cast Iron									
1300	Extra heavy weight, black									
1310	Couplings, steel straight									
1320	1/4"	1 Plum	19	.421	Ea.	3.77	24		27.77	40.50

22 11 13.45	Pipe Fittings, Steel, Threaded	Crew	Daily Output	Labor-Hours	Unit	2014 Bare Costs Material	Labor	Equipment	Total	Total Incl O&P
1330	3/8"	1 Plum	19	.421	Ea.	4.10	24		28.10	41
1340	1/2"		19	.421		5.55	24		29.55	42.50
1350	3/4"		18	.444		5.90	25.50		31.40	45
1360	1"		15	.533		7.55	30.50		38.05	55
1370	1-1/4"	Q-1	26	.615		12.10	32		44.10	61.50
1380	1-1/2"		24	.667		12.10	34.50		46.60	65.50
1390	2"		21	.762		18.45	39.50		57.95	80
1400	2-1/2"		18	.889		27.50	46		73.50	99.50
1410	3"		14	1.143		32.50	59		91.50	126
1420	3-1/2"		12	1.333		43.50	69		112.50	152
1430	4"		10	1.600		51.50	83		134.50	182
1510	90° Elbow, straight									
1520	1/2"	1 Plum	15	.533	Ea.	29.50	30.50		60	79
1530	3/4"		14	.571		30	33		63	82.50
1540	1"		13	.615		36	35.50		71.50	93
1550	1-1/4"	Q-1	22	.727		54	37.50		91.50	116
1560	1-1/2"		20	.800		66.50	41.50		108	136
1580	2"		18	.889		82.50	46		128.50	160
1590	2-1/2"		14	1.143		200	59		259	310
1600	3"		10	1.600		264	83		347	415
1610	4"		6	2.667		540	138		678	805
1650	45° Elbow, straight									
1660	1/2"	1 Plum	15	.533	Ea.	42	30.50		72.50	92.50
1670	3/4"		14	.571		40.50	33		73.50	94
1680	1"		13	.615		48.50	35.50		84	107
1690	1-1/4"	Q-1	22	.727		80	37.50		117.50	145
1700	1-1/2"		20	.800		88	41.50		129.50	159
1710	2"		18	.889		125	46		171	207
1720	2-1/2"		14	1.143		216	59		275	325
1800	Tee, straight									
1810	1/2"	1 Plum	9	.889	Ea.	46	51		97	128
1820	3/4"		9	.889		46	51		97	128
1830	1"		8	1		55.50	57.50		113	148
1840	1-1/4"	Q-1	14	1.143		82.50	59		141.50	180
1850	1-1/2"		13	1.231		106	64		170	214
1860	2"		11	1.455		132	75.50		207.50	259
1870	2-1/2"		9	1.778		281	92		373	450
1880	3"		6	2.667		385	138		523	635
1890	4"		4	4		750	207		957	1,150
4000	Standard weight, black									
4010	Couplings, steel straight, merchants									
4030	1/4"	1 Plum	19	.421	Ea.	1.01	24		25.01	37.50
4040	3/8"		19	.421		1.22	24		25.22	38
4050	1/2"		19	.421		1.30	24		25.30	38
4060	3/4"		18	.444		1.64	25.50		27.14	40.50
4070	1"		15	.533		2.31	30.50		32.81	49
4080	1-1/4"	Q-1	26	.615		2.95	32		34.95	51.50
4090	1-1/2"		24	.667		3.73	34.50		38.23	56
4100	2"		21	.762		5.35	39.50		44.85	65.50
4110	2-1/2"		18	.889		16.75	46		62.75	88
4120	3"		14	1.143		23.50	59		82.50	116
4130	3-1/2"		12	1.333		41.50	69		110.50	150
4140	4"		10	1.600		41.50	83		124.50	171

22 11 13 – Facility Water Distribution Piping

22 11 13.45 Pipe Fittings, Steel, Threaded		Crew	Daily Output	Labor-Hours	Unit	Material	Labor	Equipment	Total	Total Incl O&P
						2014 Bare Costs				
4200	Standard weight, galvanized									
4210	Couplings, steel straight, merchants									
4230	1/4"	1 Plum	19	.421	Ea.	1.16	24		25.16	38
4240	3/8"		19	.421		1.49	24		25.49	38
4250	1/2"		19	.421		1.58	24		25.58	38
4260	3/4"		18	.444		1.99	25.50		27.49	40.50
4270	1"	↓	15	.533		2.78	30.50		33.28	49.50
4280	1-1/4"	Q-1	26	.615		3.55	32		35.55	52
4290	1-1/2"		24	.667		4.41	34.50		38.91	57
4300	2"		21	.762		6.60	39.50		46.10	67
4310	2-1/2"		18	.889		20.50	46		66.50	92
4320	3"		14	1.143		27.50	59		86.50	120
4330	3-1/2"		12	1.333		48	69		117	157
4340	4"	↓	10	1.600	↓	48	83		131	178

22 11 13.47 Pipe Fittings, Steel

Line	Description	Crew	Daily Output	Labor-Hours	Unit	Material	Labor	Equipment	Total	Total Incl O&P
0010	**PIPE FITTINGS, STEEL**, Flanged, Welded & Special									
0620	Gasket and bolt set, 150#, 1/2" pipe size	1 Plum	20	.400	Ea.	2.50	23		25.50	38
0622	3/4" pipe size		19	.421		2.64	24		26.64	39.50
0624	1" pipe size		18	.444		2.67	25.50		28.17	41.50
0626	1-1/4" pipe size		17	.471		2.80	27		29.80	44
0628	1-1/2" pipe size		15	.533		2.98	30.50		33.48	50
0630	2" pipe size		13	.615		4.85	35.50		40.35	59
0640	2-1/2" pipe size		12	.667		5.60	38.50		44.10	64
0650	3" pipe size		11	.727		5.80	42		47.80	69.50
0660	3-1/2" pipe size		9	.889		9.40	51		60.40	87.50
0670	4" pipe size		8	1		10.40	57.50		67.90	98.50
0680	5" pipe size		7	1.143		17.10	66		83.10	118
0690	6" pipe size		6	1.333		18.10	76.50		94.60	136
0700	8" pipe size		5	1.600		19.55	92		111.55	161
0710	10" pipe size		4.50	1.778		35.50	102		137.50	193
0720	12" pipe size		4.20	1.905		37.50	110		147.50	208
0730	14" pipe size		4	2		39.50	115		154.50	218
0740	16" pipe size		3	2.667		41.50	153		194.50	278
0750	18" pipe size		2.70	2.963		80.50	171		251.50	345
0760	20" pipe size		2.30	3.478		131	200		331	445
0780	24" pipe size	↓	1.90	4.211	↓	164	242		406	545

22 11 13.48 Pipe, Fittings and Valves, Steel, Grooved-Joint

Line	Description	Crew	Daily Output	Labor-Hours	Unit	Material	Labor	Equipment	Total	Total Incl O&P
0010	**PIPE, FITTINGS AND VALVES, STEEL, GROOVED-JOINT**									
0012	Fittings are ductile iron. Steel fittings noted.									
0020	Pipe includes coupling & clevis type hanger assemblies, 10' O.C.									
1000	Schedule 40, black									
1040	3/4" diameter	1 Plum	71	.113	L.F.	4.62	6.50		11.12	14.90
1050	1" diameter		63	.127		4.41	7.30		11.71	15.90
1060	1-1/4" diameter		58	.138		5.70	7.95		13.65	18.25
1070	1-1/2" diameter		51	.157		6.50	9.05		15.55	21
1080	2" diameter	↓	40	.200		7.85	11.50		19.35	26
1090	2-1/2" diameter	Q-1	57	.281		13.45	14.55		28	37
1100	3" diameter		50	.320		16.55	16.60		33.15	43
1110	4" diameter		45	.356		22.50	18.40		40.90	53
1120	5" diameter	↓	37	.432		36.50	22.50		59	74.50
1130	6" diameter	Q-2	42	.571		47	30.50		77.50	98.50
1140	8" diameter	↓	37	.649	↓	77	35		112	137

22 11 Facility Water Distribution

22 11 13 – Facility Water Distribution Piping

22 11 13.48	Pipe, Fittings and Valves, Steel, Grooved-Joint	Crew	Daily Output	Labor-Hours	Unit	2014 Bare Costs Material	Labor	Equipment	Total	Total Incl O&P
1150	10″ diameter	Q-2	31	.774	L.F.	104	41.50		145.50	177
1160	12″ diameter		27	.889		116	48		164	200
1170	14″ diameter		20	1.200		119	64.50		183.50	229
1180	16″ diameter		17	1.412		179	76		255	310
1190	18″ diameter		14	1.714		178	92		270	335
1200	20″ diameter		12	2		208	107		315	390
1210	24″ diameter		10	2.400		234	129		363	450
1740	To delete coupling & hanger, subtract									
1750	3/4″ diam. to 2″ diam.					65%	27%			
1760	2-1/2″ diam. to 5″ diam.					41%	18%			
1770	6″ diam. to 12″ diam.					31%	13%			
1780	14″ diam. to 24″ diam.					35%	10%			
1800	Galvanized									
1840	3/4″ diameter	1 Plum	71	.113	L.F.	5.85	6.50		12.35	16.25
1850	1″ diameter		63	.127		5.55	7.30		12.85	17.15
1860	1-1/4″ diameter		58	.138		7.20	7.95		15.15	19.95
1870	1-1/2″ diameter		51	.157		8.40	9.05		17.45	23
1880	2″ diameter		40	.200		12	11.50		23.50	30.50
1890	2-1/2″ diameter	Q-1	57	.281		18.20	14.55		32.75	42
1900	3″ diameter		50	.320		23	16.60		39.60	50.50
1910	4″ diameter		45	.356		32	18.40		50.40	63.50
1920	5″ diameter		37	.432		58.50	22.50		81	98.50
1930	6″ diameter	Q-2	42	.571		63.50	30.50		94	117
1940	8″ diameter		37	.649		74.50	35		109.50	135
1950	10″ diameter		31	.774		120	41.50		161.50	194
1960	12″ diameter		27	.889		146	48		194	233
2540	To delete coupling & hanger, subtract									
2550	3/4″ diam. to 2″ diam.					36%	27%			
2560	2-1/2″ diam. to 5″ diam.					19%	18%			
2570	6″ diam. to 12″ diam.					14%	13%			
4690	Tee, painted									
4700	3/4″ diameter	1 Plum	38	.211	Ea.	60.50	12.10		72.60	85
4740	1″ diameter		33	.242		46.50	13.95		60.45	72.50
4750	1-1/4″ diameter		27	.296		46.50	17.05		63.55	77.50
4760	1-1/2″ diameter		22	.364		46.50	21		67.50	83
4770	2″ diameter		17	.471		46.50	27		73.50	92.50
4780	2-1/2″ diameter	Q-1	27	.593		46.50	30.50		77	98
4790	3″ diameter		22	.727		64	37.50		101.50	127
4800	4″ diameter		17	.941		97	49		146	181
4810	5″ diameter		13	1.231		226	64		290	345
4820	6″ diameter	Q-2	17	1.412		261	76		337	400
4830	8″ diameter		14	1.714		570	92		662	770
4840	10″ diameter		12	2		1,150	107		1,257	1,400
4850	12″ diameter		10	2.400		1,600	129		1,729	1,950
4851	14″ diameter		9	2.667		1,500	143		1,643	1,900
4852	16″ diameter		8	3		1,550	161		1,711	1,950
4853	18″ diameter	Q-3	11	2.909		1,625	159		1,784	2,050
4854	20″ diameter		10	3.200		2,325	175		2,500	2,850
4855	24″ diameter		8	4		3,550	219		3,769	4,225
4900	For galvanized tees, add					24%				
4939	Couplings									
4940	Flexible, standard, painted									
4950	3/4″ diameter	1 Plum	100	.080	Ea.	16.60	4.60		21.20	25

22 11 13 – Facility Water Distribution Piping

22 11 13.48 Pipe, Fittings and Valves, Steel, Grooved-Joint		Crew	Daily Output	Labor-Hours	Unit	2014 Bare Costs Material	Labor	Equipment	Total	Total Incl O&P
4960	1" diameter	1 Plum	100	.080	Ea.	16.60	4.60		21.20	25
4970	1-1/4" diameter		80	.100		22	5.75		27.75	32.50
4980	1-1/2" diameter		67	.119		24	6.85		30.85	36.50
4990	2" diameter		50	.160		25.50	9.20		34.70	42
5000	2-1/2" diameter	Q-1	80	.200		29.50	10.35		39.85	48
5010	3" diameter		67	.239		33	12.35		45.35	55
5020	3-1/2" diameter		57	.281		47	14.55		61.55	73.50
5030	4" diameter		50	.320		47.50	16.60		64.10	77
5040	5" diameter		40	.400		72	20.50		92.50	111
5050	6" diameter	Q-2	50	.480		85	26		111	133
5070	8" diameter		42	.571		138	30.50		168.50	199
5090	10" diameter		35	.686		225	37		262	305
5110	12" diameter		32	.750		256	40.50		296.50	340
5120	14" diameter		24	1		365	53.50		418.50	480
5130	16" diameter		20	1.200		480	64.50		544.50	625
5140	18" diameter		18	1.333		560	71.50		631.50	725
5150	20" diameter		16	1.500		880	80.50		960.50	1,100
5160	24" diameter		13	1.846		965	99		1,064	1,225
5200	For galvanized couplings, add					33%				
5750	Flange, w/groove gasket, black steel									
5754	See Line 22 11 13.47 0620 for gasket & bolt set									
5760	ANSI class 125 and 150, painted									
5780	2" pipe size	1 Plum	23	.348	Ea.	99.50	20		119.50	139
5790	2-1/2" pipe size	Q-1	37	.432		123	22.50		145.50	170
5800	3" pipe size		31	.516		133	26.50		159.50	188
5820	4" pipe size		23	.696		178	36		214	251
5830	5" pipe size		19	.842		206	43.50		249.50	293
5840	6" pipe size	Q-2	23	1.043		225	56		281	335
5850	8" pipe size		17	1.412		254	76		330	395
5860	10" pipe size		14	1.714		400	92		492	580
5870	12" pipe size		12	2		525	107		632	740
5880	14" pipe size		10	2.400		1,025	129		1,154	1,325
5890	16" pipe size		9	2.667		1,200	143		1,343	1,525
5900	18" pipe size		6	4		1,475	215		1,690	1,950
5910	20" pipe size		5	4.800		1,775	258		2,033	2,350
5920	24" pipe size		4.50	5.333		2,250	287		2,537	2,900
8000	Butterfly valve, 2 position handle, with standard trim									
8010	1-1/2" pipe size	1 Plum	50	.160	Ea.	242	9.20		251.20	280
8020	2" pipe size	"	38	.211		242	12.10		254.10	284
8030	3" pipe size	Q-1	50	.320		345	16.60		361.60	405
8050	4" pipe size	"	38	.421		380	22		402	455
8070	6" pipe size	Q-2	38	.632		770	34		804	895
8080	8" pipe size		27	.889		1,025	48		1,073	1,225
8090	10" pipe size		20	1.200		1,700	64.50		1,764.50	1,975
8200	With stainless steel trim									
8240	1-1/2" pipe size	1 Plum	50	.160	Ea.	305	9.20		314.20	350
8250	2" pipe size	"	38	.211		305	12.10		317.10	355
8270	3" pipe size	Q-1	50	.320		410	16.60		426.60	480
8280	4" pipe size	"	38	.421		445	22		467	525
8300	6" pipe size	Q-2	38	.632		835	34		869	970
8310	8" pipe size		27	.889		2,075	48		2,123	2,350
8320	10" pipe size		20	1.200		3,450	64.50		3,514.50	3,900
9000	Cut one groove, labor									

22 11 Facility Water Distribution

22 11 13 – Facility Water Distribution Piping

22 11 13.48 Pipe, Fittings and Valves, Steel, Grooved-Joint		Crew	Daily Output	Labor-Hours	Unit	Material	2014 Bare Costs Labor	Equipment	Total	Total Incl O&P
9010	3/4" pipe size	Q-1	152	.105	Ea.		5.45		5.45	8.25
9020	1" pipe size		140	.114			5.90		5.90	8.95
9030	1-1/4" pipe size		124	.129			6.70		6.70	10.10
9040	1-1/2" pipe size		114	.140			7.25		7.25	11
9050	2" pipe size		104	.154			7.95		7.95	12.05
9060	2-1/2" pipe size		96	.167			8.65		8.65	13.05
9070	3" pipe size		88	.182			9.40		9.40	14.20
9080	3-1/2" pipe size		83	.193			10		10	15.10
9090	4" pipe size		78	.205			10.65		10.65	16.05
9100	5" pipe size		72	.222			11.50		11.50	17.40
9110	6" pipe size		70	.229			11.85		11.85	17.90
9120	8" pipe size		54	.296			15.35		15.35	23
9130	10" pipe size		38	.421			22		22	33
9140	12" pipe size		30	.533			27.50		27.50	41.50
9150	14" pipe size		20	.800			41.50		41.50	62.50
9160	16" pipe size		19	.842			43.50		43.50	66
9170	18" pipe size		18	.889			46		46	69.50
9180	20" pipe size		17	.941			49		49	73.50
9190	24" pipe size	↓	15	1.067	↓		55.50		55.50	83.50
9210	Roll one groove									
9220	3/4" pipe size	Q-1	266	.060	Ea.		3.12		3.12	4.71
9230	1" pipe size		228	.070			3.64		3.64	5.50
9240	1-1/4" pipe size		200	.080			4.14		4.14	6.25
9250	1-1/2" pipe size		178	.090			4.66		4.66	7.05
9260	2" pipe size		116	.138			7.15		7.15	10.80
9270	2-1/2" pipe size		110	.145			7.55		7.55	11.40
9280	3" pipe size		100	.160			8.30		8.30	12.50
9290	3-1/2" pipe size		94	.170			8.80		8.80	13.30
9300	4" pipe size		86	.186			9.65		9.65	14.55
9310	5" pipe size		84	.190			9.85		9.85	14.90
9320	6" pipe size		80	.200			10.35		10.35	15.65
9330	8" pipe size		66	.242			12.55		12.55	18.95
9340	10" pipe size		58	.276			14.30		14.30	21.50
9350	12" pipe size		46	.348			18		18	27
9360	14" pipe size		30	.533			27.50		27.50	41.50
9370	16" pipe size		28	.571			29.50		29.50	44.50
9380	18" pipe size		27	.593			30.50		30.50	46.50
9390	20" pipe size		25	.640			33		33	50
9400	24" pipe size	↓	23	.696	↓		36		36	54.50

22 11 13.74 Pipe, Plastic		Crew	Daily Output	Labor-Hours	Unit	Material	Labor	Equipment	Total	Total Incl O&P
0010	**PIPE, PLASTIC**									
4100	DWV type, schedule 40, couplings 10' O.C., clevis hanger assy's, 3 per 10'									
4210	ABS, schedule 40, foam core type									
4212	Plain end black									
4214	1-1/2" diameter	1 Plum	39	.205	L.F.	2.96	11.80		14.76	21
4216	2" diameter	Q-1	62	.258		3.40	13.35		16.75	23.50
4218	3" diameter		56	.286		6.10	14.80		20.90	29
4220	4" diameter		51	.314		8.70	16.25		24.95	34
4222	6" diameter	↓	42	.381	↓	20.50	19.75		40.25	52.50
4240	To delete coupling & hangers, subtract									
4244	1-1/2" diam. to 6" diam.					43%	48%			
4400	PVC									

22 11 13.74 Pipe, Plastic		Crew	Daily Output	Labor-Hours	Unit	Material (2014 Bare Costs)	Labor (2014 Bare Costs)	Equipment (2014 Bare Costs)	Total (2014 Bare Costs)	Total Incl O&P
4410	1-1/4" diameter	1 Plum	42	.190	L.F.	3.41	10.95		14.36	20.50
4420	1-1/2" diameter	"	36	.222		2.98	12.80		15.78	22.50
4460	2" diameter	Q-1	59	.271		3.39	14.05		17.44	24.50
4470	3" diameter		53	.302		6.05	15.65		21.70	30
4480	4" diameter		48	.333		8.35	17.25		25.60	35
4490	6" diameter		39	.410		17.25	21.50		38.75	51
4500	8" diameter	Q-2	48	.500		27.50	27		54.50	71
4510	To delete coupling & hangers, subtract									
4520	1-1/4" diam. to 1-1/2" diam.					48%	60%			
4530	2" diam. to 8" diam.					42%	54%			
4550	PVC, schedule 40, foam core type									
4552	Plain end, white									
4554	1-1/2" diameter	1 Plum	39	.205	L.F.	2.72	11.80		14.52	21
4556	2" diameter	Q-1	62	.258		3.07	13.35		16.42	23.50
4558	3" diameter		56	.286		5.45	14.80		20.25	28.50
4560	4" diameter		51	.314		7.40	16.25		23.65	32.50
4562	6" diameter		42	.381		15.40	19.75		35.15	47
4564	8" diameter	Q-2	51	.471		24.50	25.50		50	65
4568	10" diameter		48	.500		29.50	27		56.50	73
4570	12" diameter		46	.522		34	28		62	80
4580	To delete coupling & hangers, subtract									
4582	1-1/2" dia to 2" dia					58%	54%			
4584	3" dia to 12" dia					46%	42%			
5300	CPVC, socket joint, couplings 10' O.C., clevis hanger assemblies, 3 per 10'									
5302	Schedule 40									
5304	1/2" diameter	1 Plum	54	.148	L.F.	3.65	8.55		12.20	16.85
5305	3/4" diameter		51	.157		4.30	9.05		13.35	18.40
5306	1" diameter		46	.174		5.50	10		15.50	21
5307	1-1/4" diameter		42	.190		6.90	10.95		17.85	24
5308	1-1/2" diameter		36	.222		8.05	12.80		20.85	28
5309	2" diameter	Q-1	59	.271		9.90	14.05		23.95	32
5310	2-1/2" diameter		56	.286		16.60	14.80		31.40	41
5311	3" diameter		53	.302		20	15.65		35.65	45.50
5312	4" diameter		48	.333		27.50	17.25		44.75	56.50
5314	6" diameter		43	.372		53	19.25		72.25	87
5318	To delete coupling & hangers, subtract									
5319	1/2" diam. to 3/4" diam.					37%	77%			
5320	1" diam. to 1-1/4" diam.					27%	70%			
5321	1-1/2" diam. to 3" diam.					21%	57%			
5322	4" diam. to 6" diam.					16%	57%			
5360	CPVC, threaded, couplings 10' O.C., clevis hanger assemblies, 3 per 10'									
5380	Schedule 40									
5460	1/2" diameter	1 Plum	54	.148	L.F.	4.45	8.55		13	17.75
5470	3/4" diameter		51	.157		5.70	9.05		14.75	19.90
5480	1" diameter		46	.174		6.95	10		16.95	22.50
5490	1-1/4" diameter		42	.190		8	10.95		18.95	25.50
5500	1-1/2" diameter		36	.222		9	12.80		21.80	29
5510	2" diameter	Q-1	59	.271		11.10	14.05		25.15	33
5520	2-1/2" diameter		56	.286		17.85	14.80		32.65	42
5530	3" diameter		53	.302		22	15.65		37.65	47.50
5540	4" diameter		48	.333		34.50	17.25		51.75	64
5550	6" diameter		43	.372		56.50	19.25		75.75	91
5730	To delete coupling & hangers, subtract									

22 11 13 – Facility Water Distribution Piping

22 11 13.74 Pipe, Plastic		Crew	Daily Output	Labor-Hours	Unit	2014 Bare Costs Material	Labor	Equipment	Total	Total Incl O&P
5740	1/2" diam. to 3/4" diam.					37%	77%			
5750	1" diam. to 1-1/4" diam.					27%	70%			
5760	1-1/2" diam. to 3" diam.					21%	57%			
5770	4" diam. to 6" diam.					16%	57%			
7280	PEX, flexible, no couplings or hangers									
7282	Note: For labor costs add 25% to the couplings and fittings labor total.									
7300	Non-barrier type, hot/cold tubing rolls									
7310	1/4" diameter x 100'				L.F.	.47			.47	.52
7350	3/8" diameter x 100'					.52			.52	.57
7360	1/2" diameter x 100'					.58			.58	.64
7370	1/2" diameter x 500'					.58			.58	.64
7380	1/2" diameter x 1000'					.58			.58	.64
7400	3/4" diameter x 100'					1.05			1.05	1.16
7410	3/4" diameter x 500'					1.05			1.05	1.16
7420	3/4" diameter x 1000'					1.05			1.05	1.16
7460	1" diameter x 100'					1.82			1.82	2
7470	1" diameter x 300'					1.82			1.82	2
7480	1" diameter x 500'					1.82			1.82	2
7500	1-1/4" diameter x 100'					3.08			3.08	3.39
7510	1-1/4" diameter x 300'					3.08			3.08	3.39
7540	1-1/2" diameter x 100'					4.19			4.19	4.61
7550	1-1/2" diameter x 300'				↓	4.19			4.19	4.61
7596	Most sizes available in red or blue									

22 11 13.76 Pipe Fittings, Plastic		Crew	Daily Output	Labor-Hours	Unit	Material	Labor	Equipment	Total	Total Incl O&P
0010	**PIPE FITTINGS, PLASTIC**									
0030	Epoxy resin, fiberglass reinforced, general service									
0090	Elbow, 90°, 2"	Q-1	33.10	.483	Ea.	102	25		127	150
0100	3"		20.80	.769		113	40		153	184
0110	4"		16.50	.970		115	50		165	202
0120	6"	↓	10.10	1.584		225	82		307	370
0130	8"	Q-2	9.30	2.581		415	139		554	665
0140	10"		8.50	2.824		520	152		672	805
0150	12"	↓	7.60	3.158	↓	750	170		920	1,075
0160	45° Elbow, same as 90°									
0290	Tee, 2"	Q-1	20	.800	Ea.	54	41.50		95.50	122
0300	3"		13.90	1.151		64.50	59.50		124	161
0310	4"		11	1.455		89.50	75.50		165	212
0320	6"	↓	6.70	2.388		225	124		349	435
0330	8"	Q-2	6.20	3.871		1,600	208		1,808	2,075
0340	10"		5.70	4.211		1,800	226		2,026	2,350
0350	12"	↓	5.10	4.706	↓	2,200	253		2,453	2,800
0380	Couplings									
0410	2"	Q-1	33.10	.483	Ea.	22.50	25		47.50	63
0420	3"		20.80	.769		24.50	40		64.50	87
0430	4"		16.50	.970		33	50		83	113
0440	6"	↓	10.10	1.584		78.50	82		160.50	210
0450	8"	Q-2	9.30	2.581		133	139		272	355
0460	10"		8.50	2.824		195	152		347	445
0470	12"	↓	7.60	3.158	↓	262	170		432	545
0473	High corrosion resistant couplings, add					30%				
2100	PVC schedule 80, socket joint									
2110	90° elbow, 1/2"	1 Plum	30.30	.264	Ea.	2.48	15.20		17.68	25.50

22 11 13.76 Pipe Fittings, Plastic		Crew	Daily Output	Labor-Hours	Unit	Material	Labor	Equipment	Total	Total Incl O&P
						2014 Bare Costs				
2130	3/4"	1 Plum	26	.308	Ea.	3.17	17.70		20.87	30
2140	1"		22.70	.352		4.95	20.50		25.45	36
2150	1-1/4"		20.20	.396		6.85	23		29.85	42
2160	1-1/2"		18.20	.440		7.30	25.50		32.80	46
2170	2"	Q-1	33.10	.483		8.85	25		33.85	47.50
2180	3"		20.80	.769		23.50	40		63.50	85.50
2190	4"		16.50	.970		35.50	50		85.50	115
2200	6"		10.10	1.584		101	82		183	235
2210	8"	Q-2	9.30	2.581		277	139		416	515
2250	45° elbow, 1/2"	1 Plum	30.30	.264		4.67	15.20		19.87	28
2270	3/4"		26	.308		7.15	17.70		24.85	34.50
2280	1"		22.70	.352		10.70	20.50		31.20	42.50
2290	1-1/4"		20.20	.396		13.60	23		36.60	49.50
2300	1-1/2"		18.20	.440		16.10	25.50		41.60	55.50
2310	2"	Q-1	33.10	.483		21	25		46	61
2320	3"		20.80	.769		53.50	40		93.50	119
2330	4"		16.50	.970		96	50		146	182
2340	6"		10.10	1.584		121	82		203	257
2350	8"	Q-2	9.30	2.581		262	139		401	495
2400	Tee, 1/2"	1 Plum	20.20	.396		7	23		30	42
2420	3/4"		17.30	.462		7.35	26.50		33.85	48
2430	1"		15.20	.526		9.15	30.50		39.65	55.50
2440	1-1/4"		13.50	.593		25.50	34		59.50	79.50
2450	1-1/2"		12.10	.661		25.50	38		63.50	85.50
2460	2"	Q-1	20	.800		31.50	41.50		73	97
2470	3"		13.90	1.151		43	59.50		102.50	137
2480	4"		11	1.455		49.50	75.50		125	169
2490	6"		6.70	2.388		169	124		293	375
2500	8"	Q-2	6.20	3.871		395	208		603	745
2510	Flange, socket, 150 lb., 1/2"	1 Plum	55.60	.144		13.60	8.30		21.90	27.50
2514	3/4"		47.60	.168		14.50	9.65		24.15	30.50
2518	1"		41.70	.192		16.15	11.05		27.20	34.50
2522	1-1/2"		33.30	.240		16.65	13.85		30.50	39.50
2526	2"	Q-1	60.60	.264		22.50	13.70		36.20	45.50
2530	4"		30.30	.528		49	27.50		76.50	95.50
2534	6"		18.50	.865		77	45		122	152
2538	8"	Q-2	17.10	1.404		138	75.50		213.50	265
2550	Coupling, 1/2"	1 Plum	30.30	.264		4.49	15.20		19.69	28
2570	3/4"		26	.308		6.05	17.70		23.75	33
2580	1"		22.70	.352		6.25	20.50		26.75	37.50
2590	1-1/4"		20.20	.396		9.50	23		32.50	45
2600	1-1/2"		18.20	.440		10.25	25.50		35.75	49.50
2610	2"	Q-1	33.10	.483		11	25		36	50
2620	3"		20.80	.769		31	40		71	94
2630	4"		16.50	.970		39	50		89	119
2640	6"		10.10	1.584		84	82		166	216
2650	8"	Q-2	9.30	2.581		114	139		253	335
2660	10"		8.50	2.824		400	152		552	670
2670	12"		7.60	3.158		450	170		620	750
4500	DWV, ABS, non pressure, socket joints									
4540	1/4 Bend, 1-1/4"	1 Plum	20.20	.396	Ea.	6.15	23		29.15	41.50
4560	1-1/2"	"	18.20	.440		4.69	25.50		30.19	43
4570	2"	Q-1	33.10	.483		7.45	25		32.45	46

22 11 13 – Facility Water Distribution Piping

22 11 13.76 Pipe Fittings, Plastic		Crew	Daily Output	Labor-Hours	Unit	2014 Bare Costs Material	Labor	Equipment	Total	Total Incl O&P
4580	3″	Q-1	20.80	.769	Ea.	18.30	40		58.30	80
4590	4″		16.50	.970		37.50	50		87.50	118
4600	6″		10.10	1.584		164	82		246	305
4650	1/8 Bend, same as 1/4 Bend									
4800	Tee, sanitary									
4820	1-1/4″	1 Plum	13.50	.593	Ea.	8.15	34		42.15	60.50
4830	1-1/2″	″	12.10	.661		7.15	38		45.15	65.50
4840	2″	Q-1	20	.800		11	41.50		52.50	74.50
4850	3″		13.90	1.151		30.50	59.50		90	124
4860	4″		11	1.455		53.50	75.50		129	173
4862	Tee, sanitary, reducing, 2″ x 1-1/2″		22	.727		11.15	37.50		48.65	69.50
4864	3″ x 2″		15.30	1.046		22	54		76	107
4868	4″ x 3″		12.10	1.322		52.50	68.50		121	161
4870	Combination Y and 1/8 bend									
4872	1-1/2″	1 Plum	12.10	.661	Ea.	17.80	38		55.80	77
4874	2″	Q-1	20	.800		22	41.50		63.50	87
4876	3″		13.90	1.151		41	59.50		100.50	136
4878	4″		11	1.455		87	75.50		162.50	210
4880	3″ x 1-1/2″		15.50	1.032		53.50	53.50		107	140
4882	4″ x 3″		12.10	1.322		82.50	68.50		151	194
4900	Wye, 1-1/4″	1 Plum	13.50	.593		9.40	34		43.40	62
4902	1-1/2″	″	12.10	.661		11.60	38		49.60	70.50
4904	2″	Q-1	20	.800		14.30	41.50		55.80	78
4906	3″		13.90	1.151		33.50	59.50		93	127
4908	4″		11	1.455		68.50	75.50		144	190
4910	6″		6.70	2.388		204	124		328	410
4918	3″ x 1-1/2″		15.50	1.032		27	53.50		80.50	111
4920	4″ x 3″		12.10	1.322		54	68.50		122.50	163
4922	6″ x 4″		6.90	2.319		168	120		288	365
4930	Double Wye, 1-1/2″	1 Plum	9.10	.879		33	50.50		83.50	113
4932	2″	Q-1	16.60	.964		39.50	50		89.50	119
4934	3″		10.40	1.538		89.50	79.50		169	219
4936	4″		8.25	1.939		175	100		275	345
4940	2″ x 1-1/2″		16.80	.952		34.50	49.50		84	112
4942	3″ x 2″		10.60	1.509		66	78		144	191
4944	4″ x 3″		8.45	1.893		139	98		237	300
4946	6″ x 4″		7.25	2.207		214	114		328	410
4950	Reducer bushing, 2″ x 1-1/2″		36.40	.440		3.84	23		26.84	38.50
4952	3″ x 1-1/2″		27.30	.586		16.20	30.50		46.70	64
4954	4″ x 2″		18.20	.879		31	45.50		76.50	103
4956	6″ x 4″		11.10	1.441		87	74.50		161.50	209
4960	Couplings, 1-1/2″	1 Plum	18.20	.440		2.33	25.50		27.83	40.50
4962	2″	Q-1	33.10	.483		3.14	25		28.14	41.50
4963	3″		20.80	.769		9.05	40		49.05	70
4964	4″		16.50	.970		16.35	50		66.35	94
4966	6″		10.10	1.584		68	82		150	199
4970	2″ x 1-1/2″		33.30	.480		6.70	25		31.70	45
4972	3″ x 1-1/2″		21	.762		18.55	39.50		58.05	80
4974	4″ x 3″		16.70	.958		29.50	49.50		79	108
4978	Closet flange, 4″	1 Plum	32	.250		16.20	14.40		30.60	39.50
4980	4″ x 3″	″	34	.235		18.50	13.55		32.05	41
5000	DWV, PVC, schedule 40, socket joints									
5040	1/4 bend, 1-1/4″	1 Plum	20.20	.396	Ea.	12.55	23		35.55	48.50

22 11 13.76 Pipe Fittings, Plastic		Crew	Daily Output	Labor-Hours	Unit	2014 Bare Costs Material	Labor	Equipment	Total	Total Incl O&P
5060	1-1/2"	1 Plum	18.20	.440	Ea.	3.45	25.50		28.95	42
5070	2"	Q-1	33.10	.483		5.45	25		30.45	44
5080	3"		20.80	.769		16	40		56	77.50
5090	4"		16.50	.970		31.50	50		81.50	111
5100	6"		10.10	1.584		111	82		193	246
5105	8"	Q-2	9.30	2.581		224	139		363	455
5110	1/4 bend, long sweep, 1-1/2"	1 Plum	18.20	.440		8.30	25.50		33.80	47
5112	2"	Q-1	33.10	.483		8.95	25		33.95	48
5114	3"		20.80	.769		21	40		61	83.50
5116	4"		16.50	.970		40	50		90	121
5150	1/8 bend, 1-1/4"	1 Plum	20.20	.396		8.20	23		31.20	43.50
5170	1-1/2"	"	18.20	.440		3.38	25.50		28.88	41.50
5180	2"	Q-1	33.10	.483		4.87	25		29.87	43.50
5190	3"		20.80	.769		14.35	40		54.35	76
5200	4"		16.50	.970		25	50		75	104
5210	6"		10.10	1.584		102	82		184	236
5215	8"	Q-2	9.30	2.581		158	139		297	380
5250	Tee, sanitary 1-1/4"	1 Plum	13.50	.593		12.30	34		46.30	65
5254	1-1/2"	"	12.10	.661		6.25	38		44.25	64.50
5255	2"	Q-1	20	.800		9.20	41.50		50.70	72.50
5256	3"		13.90	1.151		23	59.50		82.50	115
5257	4"		11	1.455		43	75.50		118.50	161
5259	6"		6.70	2.388		179	124		303	385
5261	8"	Q-2	6.20	3.871		535	208		743	905
5264	2" x 1-1/2"	Q-1	22	.727		8.10	37.50		45.60	66
5266	3" x 1-1/2"		15.50	1.032		17.05	53.50		70.55	100
5268	4" x 3"		12.10	1.322		52	68.50		120.50	161
5271	6" x 4"		6.90	2.319		173	120		293	370
5314	Combination Y & 1/8 bend, 1-1/2"	1 Plum	12.10	.661		15.10	38		53.10	74
5315	2"	Q-1	20	.800		16.15	41.50		57.65	80.50
5317	3"		13.90	1.151		40	59.50		99.50	134
5318	4"		11	1.455		79.50	75.50		155	202
5324	Combination Y & 1/8 bend, reducing									
5325	2" x 2" x 1-1/2"	Q-1	22	.727	Ea.	21	37.50		58.50	80.50
5327	3" x 3" x 1-1/2"		15.50	1.032		38	53.50		91.50	123
5328	3" x 3" x 2"		15.30	1.046		26.50	54		80.50	111
5329	4" x 4" x 2"		12.20	1.311		43	68		111	151
5331	Wye, 1-1/4"	1 Plum	13.50	.593		16.10	34		50.10	69
5332	1-1/2"	"	12.10	.661		11.80	38		49.80	70.50
5333	2"	Q-1	20	.800		11.20	41.50		52.70	75
5334	3"		13.90	1.151		30	59.50		89.50	123
5335	4"		11	1.455		55	75.50		130.50	175
5336	6"		6.70	2.388		165	124		289	370
5337	8"	Q-2	6.20	3.871		385	208		593	740
5341	2" x 1-1/2"	Q-1	22	.727		13.85	37.50		51.35	72.50
5342	3" x 1-1/2"		15.50	1.032		19.85	53.50		73.35	103
5343	4" x 3"		12.10	1.322		44.50	68.50		113	152
5344	6" x 4"		6.90	2.319		122	120		242	315
5345	8" x 6"	Q-2	6.40	3.750		291	201		492	625
5347	Double wye, 1-1/2"	1 Plum	9.10	.879		25.50	50.50		76	105
5348	2"	Q-1	16.60	.964		29.50	50		79.50	108
5349	3"		10.40	1.538		60	79.50		139.50	186
5350	4"		8.25	1.939		122	100		222	286

22 11 13 – Facility Water Distribution Piping

22 11 13.76 Pipe Fittings, Plastic		Crew	Daily Output	Labor-Hours	Unit	2014 Bare Costs Material	Labor	Equipment	Total	Total Incl O&P
5353	Double wye, reducing									
5354	2" x 2" x 1-1/2" x 1-1/2"	Q-1	16.80	.952	Ea.	26	49.50		75.50	103
5355	3" x 3" x 2" x 2"		10.60	1.509		44.50	78		122.50	167
5356	4" x 4" x 3" x 3"		8.45	1.893		96.50	98		194.50	254
5357	6" x 6" x 4" x 4"		7.25	2.207		325	114		439	530
5374	Coupling, 1-1/4"	1 Plum	20.20	.396		7.60	23		30.60	43
5376	1-1/2"	"	18.20	.440		1.72	25.50		27.22	40
5378	2"	Q-1	33.10	.483		2.24	25		27.24	40.50
5380	3"		20.80	.769		7.70	40		47.70	68.50
5390	4"		16.50	.970		13.65	50		63.65	91
5400	6"		10.10	1.584		44	82		126	173
5402	8"	Q-2	9.30	2.581		109	139		248	330
5404	2" x 1-1/2"	Q-1	33.30	.480		4.99	25		29.99	43
5406	3" x 1-1/2"		21	.762		15.15	39.50		54.65	76
5408	4" x 3"		16.70	.958		24.50	49.50		74	102
5410	Reducer bushing, 2" x 1-1/4"		36.50	.438		7.05	22.50		29.55	42.50
5412	3" x 1-1/2"		27.30	.586		14.15	30.50		44.65	61.50
5414	4" x 2"		18.20	.879		24.50	45.50		70	96
5416	6" x 4"		11.10	1.441		65.50	74.50		140	185
5418	8" x 6"	Q-2	10.20	2.353		142	126		268	345
5425	Closet flange 4"	Q-1	32	.500		17.35	26		43.35	58
5426	4" x 3"	"	34	.471		17.50	24.50		42	56.50
5450	Solvent cement for PVC, industrial grade, per quart				Qt.	35.50			35.50	39
7340	PVC flange, slip-on, Sch 80 std., 1/2"	1 Plum	22	.364	Ea.	13.60	21		34.60	46.50
7350	3/4"		21	.381		14.50	22		36.50	49
7360	1"		18	.444		16.15	25.50		41.65	56.50
7370	1-1/4"		17	.471		16.65	27		43.65	59.50
7380	1-1/2"		16	.500		17	29		46	62
7390	2"	Q-1	26	.615		22.50	32		54.50	73
7400	2-1/2"		24	.667		35	34.50		69.50	90.50
7410	3"		18	.889		38.50	46		84.50	112
7420	4"		15	1.067		49	55.50		104.50	138
7430	6"		10	1.600		77	83		160	210
7440	8"	Q-2	11	2.182		138	117		255	330
7550	Union, schedule 40, socket joints, 1/2"	1 Plum	19	.421		5.20	24		29.20	42.50
7560	3/4"		18	.444		5.35	25.50		30.85	44.50
7570	1"		15	.533		5.55	30.50		36.05	52.50
7580	1-1/4"		14	.571		16.65	33		49.65	68
7590	1-1/2"		13	.615		18.60	35.50		54.10	74
7600	2"	Q-1	20	.800		25	41.50		66.50	90

22 11 19 – Domestic Water Piping Specialties

22 11 19.10 Flexible Connectors

		Crew	Daily Output	Labor-Hours	Unit	Material	Labor	Equipment	Total	Total Incl O&P
0010	**FLEXIBLE CONNECTORS**, Corrugated, 7/8" O.D., 1/2" I.D.									
0050	Gas, seamless brass, steel fittings									
0200	12" long	1 Plum	36	.222	Ea.	17.40	12.80		30.20	38.50
0220	18" long		36	.222		21.50	12.80		34.30	43.50
0240	24" long		34	.235		25.50	13.55		39.05	48.50
0260	30" long		34	.235		27.50	13.55		41.05	51
0280	36" long		32	.250		30.50	14.40		44.90	55
0320	48" long		30	.267		38.50	15.35		53.85	65.50
0340	60" long		30	.267		46	15.35		61.35	73.50
0360	72" long		30	.267		53	15.35		68.35	81.50

22 11 Facility Water Distribution

22 11 19 – Domestic Water Piping Specialties

22 11 19.10 Flexible Connectors		Crew	Daily Output	Labor-Hours	Unit	2014 Bare Costs Material	Labor	Equipment	Total	Total Incl O&P
2000	Water, copper tubing, dielectric separators									
2100	12″ long	1 Plum	36	.222	Ea.	17.35	12.80		30.15	38.50
2220	15″ long		36	.222		19.30	12.80		32.10	40.50
2240	18″ long		36	.222		21	12.80		33.80	42.50
2260	24″ long		34	.235		26	13.55		39.55	49

22 11 19.38 Water Supply Meters		Crew	Daily Output	Labor-Hours	Unit	Material	Labor	Equipment	Total	Total Incl O&P
0010	**WATER SUPPLY METERS**									
1000	Detector, serves dual systems such as fire and domestic or									
1020	process water, wide range cap., UL and FM approved									
1100	3″ mainline x 2″ by-pass, 400 GPM	Q-1	3.60	4.444	Ea.	7,225	230		7,455	8,300
1140	4″ mainline x 2″ by-pass, 700 GPM	″	2.50	6.400		7,225	330		7,555	8,450
1180	6″ mainline x 3″ by-pass, 1600 GPM	Q-2	2.60	9.231		11,100	495		11,595	13,000
1220	8″ mainline x 4″ by-pass, 2800 GPM		2.10	11.429		16,400	615		17,015	18,900
1260	10″ mainline x 6″ by-pass, 4400 GPM		2	12		23,400	645		24,045	26,800
1300	10″ x 12″ mainlines x 6″ by-pass, 5400 GPM		1.70	14.118		31,800	760		32,560	36,200
2000	Domestic/commercial, bronze									
2020	Threaded									
2060	5/8″ diameter, to 20 GPM	1 Plum	16	.500	Ea.	48.50	29		77.50	97
2080	3/4″ diameter, to 30 GPM		14	.571		88.50	33		121.50	147
2100	1″ diameter, to 50 GPM		12	.667		134	38.50		172.50	206
2300	Threaded/flanged									
2340	1-1/2″ diameter, to 100 GPM	1 Plum	8	1	Ea.	330	57.50		387.50	445
2360	2″ diameter, to 160 GPM	″	6	1.333	″	445	76.50		521.50	605
2600	Flanged, compound									
2640	3″ diameter, 320 GPM	Q-1	3	5.333	Ea.	3,025	276		3,301	3,750
2660	4″ diameter, to 500 GPM		1.50	10.667		4,850	555		5,405	6,150
2680	6″ diameter, to 1,000 GPM		1	16		7,725	830		8,555	9,750
2700	8″ diameter, to 1,800 GPM		.80	20		12,100	1,025		13,125	14,900
7000	Turbine									
7260	Flanged									
7300	2″ diameter, to 160 GPM	1 Plum	7	1.143	Ea.	590	66		656	750
7320	3″ diameter, to 450 GPM	Q-1	3.60	4.444		1,225	230		1,455	1,700
7340	4″ diameter, to 650 GPM	″	2.50	6.400		2,050	330		2,380	2,750
7360	6″ diameter, to 1800 GPM	Q-2	2.60	9.231		3,650	495		4,145	4,775
7380	8″ diameter, to 2500 GPM		2.10	11.429		5,800	615		6,415	7,300
7400	10″ diameter, to 5500 GPM		1.70	14.118		7,825	760		8,585	9,750

22 11 19.42 Backflow Preventers		Crew	Daily Output	Labor-Hours	Unit	Material	Labor	Equipment	Total	Total Incl O&P
0010	**BACKFLOW PREVENTERS**, Includes valves									
0020	and four test cocks, corrosion resistant, automatic operation									
1000	Double check principle									
1010	Threaded, with ball valves									
1020	3/4″ pipe size	1 Plum	16	.500	Ea.	231	29		260	298
1030	1″ pipe size		14	.571		264	33		297	340
1040	1-1/2″ pipe size		10	.800		560	46		606	690
1050	2″ pipe size		7	1.143		655	66		721	820
1080	Threaded, with gate valves									
1100	3/4″ pipe size	1 Plum	16	.500	Ea.	1,075	29		1,104	1,250
1120	1″ pipe size		14	.571		1,100	33		1,133	1,250
1140	1-1/2″ pipe size		10	.800		1,400	46		1,446	1,625
1160	2″ pipe size		7	1.143		1,725	66		1,791	2,000
1200	Flanged, valves are gate									
1210	3″ pipe size	Q-1	4.50	3.556	Ea.	2,800	184		2,984	3,350

22 11 Facility Water Distribution

22 11 19 – Domestic Water Piping Specialties

22 11 19.42 Backflow Preventers		Crew	Daily Output	Labor-Hours	Unit	2014 Bare Costs Material	Labor	Equipment	Total	Total Incl O&P
1220	4″ pipe size	Q-1	3	5.333	Ea.	2,900	276		3,176	3,625
1230	6″ pipe size	Q-2	3	8		4,275	430		4,705	5,350
1240	8″ pipe size		2	12		7,775	645		8,420	9,525
1250	10″ pipe size		1	24		11,500	1,300		12,800	14,700
1300	Flanged, valves are OS&Y									
1380	3″ pipe size	Q-1	4.50	3.556	Ea.	2,875	184		3,059	3,425
1400	4″ pipe size	″	3	5.333		3,475	276		3,751	4,250
1420	6″ pipe size	Q-2	3	8		5,450	430		5,880	6,650
4000	Reduced pressure principle									
4100	Threaded, bronze, valves are ball									
4120	3/4″ pipe size	1 Plum	16	.500	Ea.	445	29		474	535
4140	1″ pipe size		14	.571		480	33		513	575
4150	1-1/4″ pipe size		12	.667		845	38.50		883.50	990
4160	1-1/2″ pipe size		10	.800		960	46		1,006	1,125
4180	2″ pipe size		7	1.143		1,075	66		1,141	1,300
5000	Flanged, bronze, valves are OS&Y									
5060	2-1/2″ pipe size	Q-1	5	3.200	Ea.	4,025	166		4,191	4,675
5080	3″ pipe size		4.50	3.556		4,650	184		4,834	5,375
5100	4″ pipe size		3	5.333		5,500	276		5,776	6,475
5120	6″ pipe size	Q-2	3	8		8,750	430		9,180	10,300
5200	Flanged, iron, valves are gate									
5210	2-1/2″ pipe size	Q-1	5	3.200	Ea.	2,650	166		2,816	3,175
5220	3″ pipe size		4.50	3.556		2,775	184		2,959	3,325
5230	4″ pipe size		3	5.333		3,750	276		4,026	4,550
5240	6″ pipe size	Q-2	3	8		5,275	430		5,705	6,475
5250	8″ pipe size		2	12		9,475	645		10,120	11,400
5260	10″ pipe size		1	24		13,400	1,300		14,700	16,700
5600	Flanged, iron, valves are OS&Y									
5660	2-1/2″ pipe size	Q-1	5	3.200	Ea.	3,025	166		3,191	3,575
5680	3″ pipe size		4.50	3.556		3,175	184		3,359	3,775
5700	4″ pipe size		3	5.333		3,975	276		4,251	4,800
5720	6″ pipe size	Q-2	3	8		5,775	430		6,205	7,000
5740	8″ pipe size		2	12		10,100	645		10,745	12,100
5760	10″ pipe size		1	24		13,600	1,300		14,900	16,900

22 11 19.64 Hydrants

Line	Description	Crew	Daily Output	Labor-Hours	Unit	Material	Labor	Equipment	Total	Total Incl O&P
0010	**HYDRANTS**									
0050	Wall type, moderate climate, bronze, encased									
0200	3/4″ IPS connection	1 Plum	16	.500	Ea.	705	29		734	825
0300	1″ IPS connection		14	.571		810	33		843	940
0500	Anti-siphon type, 3/4″ connection		16	.500		610	29		639	715
1000	Non-freeze, bronze, exposed									
1100	3/4″ IPS connection, 4″ to 9″ thick wall	1 Plum	14	.571	Ea.	480	33		513	575
1120	10″ to 14″ thick wall	″	12	.667		515	38.50		553.50	630
1280	For anti-siphon type, add					126			126	139
2000	Non-freeze bronze, encased, anti-siphon type									
2100	3/4″ IPS connection, 5″ to 9″ thick wall	1 Plum	14	.571	Ea.	1,225	33		1,258	1,375
2120	10″ to 14″ thick wall		12	.667		1,250	38.50		1,288.50	1,425
2140	15″ to 19″ thick wall		12	.667		1,300	38.50		1,338.50	1,500
3000	Ground box type, bronze frame, 3/4″ IPS connection									
3080	Non-freeze, all bronze, polished face, set flush									
3100	2 feet depth of bury	1 Plum	8	1	Ea.	905	57.50		962.50	1,075
3140	4 feet depth of bury		8	1		1,050	57.50		1,107.50	1,225

22 11 Facility Water Distribution

22 11 19 – Domestic Water Piping Specialties

22 11 19.64 Hydrants		Crew	Daily Output	Labor-Hours	Unit	Material	2014 Bare Costs Labor	Equipment	Total	Total Incl O&P
3180	6 feet depth of bury	1 Plum	7	1.143	Ea.	1,175	66		1,241	1,400
3220	8 feet depth of bury		5	1.600		1,325	92		1,417	1,600
3400	For 1" IPS connection, add					15%	10%			
3550	For 2" connection, add					445%	24%			
3600	For tapped drain port in box, add					82			82	90
5000	Moderate climate, all bronze, polished face									
5020	and scoriated cover, set flush									
5100	3/4" IPS connection	1 Plum	16	.500	Ea.	625	29		654	735
5120	1" IPS connection	"	14	.571		770	33		803	895
5200	For tapped drain port in box, add					82			82	90

22 11 23 – Domestic Water Pumps

22 11 23.10 General Utility Pumps		Crew	Daily Output	Labor-Hours	Unit	Material	Labor	Equipment	Total	Total Incl O&P
0010	**GENERAL UTILITY PUMPS**									
2000	Single stage									
3000	Double suction,									
3190	75 HP, to 2500 GPM	Q-3	.28	114	Ea.	20,100	6,275		26,375	31,600
3220	100 HP, to 3000 GPM		.26	123		25,500	6,750		32,250	38,300
3240	150 HP, to 4000 GPM		.24	133		36,000	7,300		43,300	50,500

22 13 Facility Sanitary Sewerage

22 13 16 – Sanitary Waste and Vent Piping

22 13 16.20 Pipe, Cast Iron		Crew	Daily Output	Labor-Hours	Unit	Material	Labor	Equipment	Total	Total Incl O&P
0010	**PIPE, CAST IRON**, Soil, on clevis hanger assemblies, 5' O.C.									
0020	Single hub, service wt., lead & oakum joints 10' O.C.									
2120	2" diameter	Q-1	63	.254	L.F.	9.30	13.15		22.45	30
2140	3" diameter		60	.267		12.95	13.80		26.75	35.50
2160	4" diameter		55	.291		16.85	15.05		31.90	41.50
2180	5" diameter	Q-2	76	.316		23	16.95		39.95	51
2200	6" diameter	"	73	.329		29	17.65		46.65	58.50
2220	8" diameter	Q-3	59	.542		44	29.50		73.50	93.50
2240	10" diameter		54	.593		71	32.50		103.50	127
2260	12" diameter		48	.667		101	36.50		137.50	166
2320	For service weight, double hub, add					10%				
2340	For extra heavy, single hub, add					48%	4%			
2360	For extra heavy, double hub, add					71%	4%			
2400	Lead for caulking, (1#/diam. in.)	Q-1	160	.100	Lb.	1.04	5.20		6.24	8.95
2420	Oakum for caulking, (1/8#/diam. in.)	"	40	.400	"	4.80	20.50		25.30	37
2960	To delete hangers, subtract									
2970	2" diam. to 4" diam.					16%	19%			
2980	5" diam. to 8" diam.					14%	14%			
2990	10" diam. to 15" diam.					13%	19%			
3000	Single hub, service wt., push-on gasket joints 10' O.C.									
3010	2" diameter	Q-1	66	.242	L.F.	10.25	12.55		22.80	30
3020	3" diameter		63	.254		14.20	13.15		27.35	35.50
3030	4" diameter		57	.281		18.40	14.55		32.95	42
3040	5" diameter	Q-2	79	.304		25.50	16.30		41.80	52.50
3050	6" diameter	"	75	.320		31.50	17.20		48.70	61
3060	8" diameter	Q-3	62	.516		49.50	28.50		78	97
3070	10" diameter		56	.571		80.50	31.50		112	136
3080	12" diameter		49	.653		113	36		149	178

22 13 16 – Sanitary Waste and Vent Piping

22 13 16.20 Pipe, Cast Iron		Crew	Daily Output	Labor-Hours	Unit	2014 Bare Costs Material	Labor	Equipment	Total	Total Incl O&P
3082	15" diameter	Q-3	40	.800	L.F.	164	44		208	247
3100	For service weight, double hub, add					65%				
3110	For extra heavy, single hub, add					48%	4%			
3120	For extra heavy, double hub, add					29%	4%			
3130	To delete hangers, subtract									
3140	2" diam. to 4" diam.					12%	21%			
3150	5" diam. to 8" diam.					10%	16%			
3160	10" diam. to 15" diam.					9%	21%			
4000	No hub, couplings 10' O.C.									
4100	1-1/2" diameter	Q-1	71	.225	L.F.	9.15	11.65		20.80	27.50
4120	2" diameter		67	.239		9.40	12.35		21.75	29
4140	3" diameter		64	.250		13.35	12.95		26.30	34
4160	4" diameter	↓	58	.276		16.95	14.30		31.25	40
4180	5" diameter	Q-2	83	.289		23	15.55		38.55	49
4200	6" diameter	"	79	.304		30	16.30		46.30	57.50
4220	8" diameter	Q-3	69	.464		52	25.50		77.50	96
4240	10" diameter		61	.525		85	29		114	137
4244	12" diameter		58	.552		105	30		135	162
4248	15" diameter	↓	52	.615	↓	156	33.50		189.50	222
4280	To delete hangers, subtract									
4290	1-1/2" diam. to 6" diam.					22%	47%			
4300	8" diam. to 10" diam.					21%	44%			
4310	12" diam. to 15" diam.					19%	40%			

22 13 16.30 Pipe Fittings, Cast Iron		Crew	Daily Output	Labor-Hours	Unit	Material	Labor	Equipment	Total	Total Incl O&P
0010	**PIPE FITTINGS, CAST IRON**, Soil									
0040	Hub and spigot, service weight, lead & oakum joints									
0080	1/4 bend, 2"	Q-1	16	1	Ea.	19.25	52		71.25	99
0120	3"		14	1.143		25.50	59		84.50	118
0140	4"	↓	13	1.231		40	64		104	141
0160	5"	Q-2	18	1.333		56	71.50		127.50	170
0180	6"	"	17	1.412		69.50	76		145.50	192
0200	8"	Q-3	11	2.909		208	159		367	470
0220	10"		10	3.200		305	175		480	600
0224	12"	↓	9	3.556		415	195		610	750
0266	Closet bend, 3" diameter with flange 10" x 16"	Q-1	14	1.143		125	59		184	228
0268	16" x 16"		12	1.333		138	69		207	255
0270	Closet bend, 4" diameter, 2-1/2" x 4" ring, 6" x 16"		13	1.231		112	64		176	220
0280	8" x 16"		13	1.231		100	64		164	207
0290	10" x 12"		12	1.333		94.50	69		163.50	208
0300	10" x 18"		11	1.455		130	75.50		205.50	257
0310	12" x 16"		11	1.455		112	75.50		187.50	238
0330	16" x 16"		10	1.600		141	83		224	280
0340	1/8 bend, 2"		16	1		13.65	52		65.65	93
0350	3"		14	1.143		21.50	59		80.50	113
0360	4"	↓	13	1.231		31.50	64		95.50	131
0380	5"	Q-2	18	1.333		43.50	71.50		115	156
0400	6"	"	17	1.412		53	76		129	173
0420	8"	Q-3	11	2.909		158	159		317	415
0440	10"		10	3.200		227	175		402	515
0460	12"	↓	9	3.556		430	195		625	770
0500	Sanitary tee, 2"	Q-1	10	1.600		26.50	83		109.50	155
0540	3"	↓	9	1.778	↓	43.50	92		135.50	187

22 13 16.30 Pipe Fittings, Cast Iron		Crew	Daily Output	Labor-Hours	Unit	Material	2014 Bare Costs Labor	Equipment	Total	Total Incl O&P
0620	4"	Q-1	8	2	Ea.	53	104		157	214
0700	5"	Q-2	12	2		105	107		212	278
0800	6"	"	11	2.182		119	117		236	310
0880	8"	Q-3	7	4.571		315	251		566	730
1000	Tee, 2"	Q-1	10	1.600		38.50	83		121.50	168
1060	3"		9	1.778		57.50	92		149.50	203
1120	4"		8	2		74	104		178	237
1200	5"	Q-2	12	2		157	107		264	335
1300	6"	"	11	2.182		154	117		271	345
1380	8"	Q-3	7	4.571		272	251		523	680
1400	Combination Y and 1/8 bend									
1420	2"	Q-1	10	1.600	Ea.	33.50	83		116.50	162
1460	3"		9	1.778		51	92		143	195
1520	4"		8	2		70	104		174	233
1540	5"	Q-2	12	2		133	107		240	310
1560	6"		11	2.182		168	117		285	360
1580	8"		7	3.429		415	184		599	735
1582	12"	Q-3	6	5.333		850	292		1,142	1,375
1600	Double Y, 2"	Q-1	8	2		58.50	104		162.50	221
1610	3"		7	2.286		74	118		192	261
1620	4"		6.50	2.462		97	128		225	300
1630	5"	Q-2	9	2.667		199	143		342	435
1640	6"	"	8	3		249	161		410	515
1650	8"	Q-3	5.50	5.818		610	320		930	1,150
1660	10"		5	6.400		1,350	350		1,700	2,000
1670	12"		4.50	7.111		1,550	390		1,940	2,300
1740	Reducer, 3" x 2"	Q-1	15	1.067		18.75	55.50		74.25	104
1750	4" x 2"		14.50	1.103		21.50	57		78.50	110
1760	4" x 3"		14	1.143		24.50	59		83.50	117
1770	5" x 2"		14	1.143		51	59		110	146
1780	5" x 3"		13.50	1.185		54	61.50		115.50	152
1790	5" x 4"		13	1.231		31.50	64		95.50	131
1800	6" x 2"		13.50	1.185		48.50	61.50		110	146
1810	6" x 3"		13	1.231		50	64		114	152
1830	6" x 4"		12.50	1.280		49	66.50		115.50	154
1840	6" x 5"		11	1.455		53	75.50		128.50	172
1880	8" x 3"	Q-2	13.50	1.778		94	95.50		189.50	247
1900	8" x 4"		13	1.846		82	99		181	240
1920	8" x 5"		12	2		85	107		192	256
1940	8" x 6"		12	2		84.50	107		191.50	255
1960	Increaser, 2" x 3"	Q-1	15	1.067		45	55.50		100.50	133
1980	2" x 4"		14	1.143		45	59		104	139
2000	2" x 5"		13	1.231		55	64		119	157
2020	3" x 4"		13	1.231		49.50	64		113.50	151
2040	3" x 5"		13	1.231		55	64		119	157
2060	3" x 6"		12	1.333		66	69		135	177
2070	4" x 5"		13	1.231		58.50	64		122.50	161
2080	4" x 6"		12	1.333		67.50	69		136.50	178
2090	4" x 8"	Q-2	13	1.846		136	99		235	300
2100	5" x 6"	Q-1	11	1.455		99	75.50		174.50	223
2110	5" x 8"	Q-2	12	2		160	107		267	340
2120	6" x 8"		12	2		160	107		267	340
2130	6" x 10"		8	3		288	161		449	560

22 13 16 – Sanitary Waste and Vent Piping

22 13 16.30	Pipe Fittings, Cast Iron	Crew	Daily Output	Labor-Hours	Unit	2014 Bare Costs Material	Labor	Equipment	Total	Total Incl O&P
2140	8" x 10"	Q-2	6.50	3.692	Ea.	290	198		488	620
2150	10" x 12"		5.50	4.364		515	234		749	920
2500	Y, 2"	Q-1	10	1.600		24.50	83		107.50	152
2510	3"		9	1.778		45	92		137	189
2520	4"		8	2		60.50	104		164.50	223
2530	5"	Q-2	12	2		107	107		214	280
2540	6"	"	11	2.182		139	117		256	330
2550	8"	Q-3	7	4.571		340	251		591	755
2560	10"		6	5.333		550	292		842	1,050
2570	12"		5	6.400		1,250	350		1,600	1,900
2580	15"		4	8		2,725	440		3,165	3,650
3000	For extra heavy, add					44%	4%			
3600	Hub and spigot, service weight gasket joint									
3605	Note: gaskets and joint labor have									
3606	been included with all listed fittings.									
3610	1/4 bend, 2"	Q-1	20	.800	Ea.	28.50	41.50		70	94
3620	3"		17	.941		38	49		87	115
3630	4"		15	1.067		55.50	55.50		111	145
3640	5"	Q-2	21	1.143		80	61.50		141.50	181
3650	6"	"	19	1.263		94.50	68		162.50	206
3660	8"	Q-3	12	2.667		263	146		409	510
3670	10"		11	2.909		400	159		559	680
3680	12"		10	3.200		535	175		710	855
3700	Closet bend, 3" diameter with ring 10" x 16"	Q-1	17	.941		138	49		187	225
3710	16" x 16"		15	1.067		150	55.50		205.50	249
3730	Closet bend, 4" diameter, 1" x 4" ring, 6" x 16"		15	1.067		127	55.50		182.50	224
3740	8" x 16"		15	1.067		115	55.50		170.50	211
3750	10" x 12"		14	1.143		110	59		169	211
3760	10" x 18"		13	1.231		145	64		209	257
3770	12" x 16"		13	1.231		128	64		192	238
3780	16" x 16"		12	1.333		156	69		225	276
3800	1/8 bend, 2"		20	.800		23	41.50		64.50	88
3810	3"		17	.941		33.50	49		82.50	111
3820	4"		15	1.067		46.50	55.50		102	135
3830	5"	Q-2	21	1.143		67.50	61.50		129	167
3840	6"	"	19	1.263		77.50	68		145.50	188
3850	8"	Q-3	12	2.667		213	146		359	455
3860	10"		11	2.909		325	159		484	595
3870	12"		10	3.200		555	175		730	875
3900	Sanitary Tee, 2"	Q-1	12	1.333		45.50	69		114.50	154
3910	3"		10	1.600		67.50	83		150.50	200
3920	4"		9	1.778		83.50	92		175.50	231
3930	5"	Q-2	13	1.846		153	99		252	320
3940	6"	"	11	2.182		169	117		286	365
3950	8"	Q-3	8.50	3.765		425	206		631	780
3980	Tee, 2"	Q-1	12	1.333		57.50	69		126.50	167
3990	3"		10	1.600		82	83		165	215
4000	4"		9	1.778		104	92		196	254
4010	5"	Q-2	13	1.846		205	99		304	375
4020	6"	"	11	2.182		204	117		321	400
4030	8"	Q-3	8	4		380	219		599	750
4060	Combination Y and 1/8 bend									
4070	2"	Q-1	12	1.333	Ea.	52.50	69		121.50	162

22 13 Facility Sanitary Sewerage

22 13 16 – Sanitary Waste and Vent Piping

	22 13 16.30 Pipe Fittings, Cast Iron	Crew	Daily Output	Labor-Hours	Unit	Material (2014 Bare Costs)	Labor (2014 Bare Costs)	Equipment (2014 Bare Costs)	Total (2014 Bare Costs)	Total Incl O&P
4080	3″	Q-1	10	1.600	Ea.	75.50	83		158.50	208
4090	4″		9	1.778		101	92		193	250
4100	5″	Q-2	13	1.846		181	99		280	350
4110	6″	″	11	2.182		218	117		335	415
4120	8″	Q-3	8	4		525	219		744	910
4121	12″	″	7	4.571		1,100	251		1,351	1,575
4160	Double Y, 2″	Q-1	10	1.600		86.50	83		169.50	221
4170	3″		8	2		111	104		215	278
4180	4″		7	2.286		143	118		261	335
4190	5″	Q-2	10	2.400		271	129		400	495
4200	6″	″	9	2.667		325	143		468	570
4210	8″	Q-3	6	5.333		775	292		1,067	1,300
4220	10″		5	6.400		1,625	350		1,975	2,325
4230	12″		4.50	7.111		1,900	390		2,290	2,700
4260	Reducer, 3″ x 2″	Q-1	17	.941		40.50	49		89.50	118
4270	4″ x 2″		16.50	.970		46	50		96	127
4280	4″ x 3″		16	1		52	52		104	135
4290	5″ x 2″		16	1		84	52		136	171
4300	5″ x 3″		15.50	1.032		90	53.50		143.50	180
4310	5″ x 4″		15	1.067		70.50	55.50		126	161
4320	6″ x 2″		15.50	1.032		82.50	53.50		136	172
4330	6″ x 3″		15	1.067		87	55.50		142.50	179
4336	6″ x 4″		14	1.143		89.50	59		148.50	188
4340	6″ x 5″		13	1.231		102	64		166	209
4360	8″ x 3″	Q-2	15	1.600		161	86		247	305
4370	8″ x 4″		15	1.600		152	86		238	298
4380	8″ x 5″		14	1.714		164	92		256	320
4390	8″ x 6″		14	1.714		164	92		256	320
4430	Increaser, 2″ x 3″	Q-1	17	.941		57	49		106	137
4440	2″ x 4″		16	1		60	52		112	144
4450	2″ x 5″		15	1.067		79	55.50		134.50	171
4460	3″ x 4″		15	1.067		64.50	55.50		120	155
4470	3″ x 5″		15	1.067		79	55.50		134.50	171
4480	3″ x 6″		14	1.143		91	59		150	190
4490	4″ x 5″		15	1.067		82.50	55.50		138	175
4500	4″ x 6″		14	1.143		92.50	59		151.50	192
4510	4″ x 8″	Q-2	15	1.600		191	86		277	340
4520	5″ x 6″	Q-1	13	1.231		124	64		188	233
4530	5″ x 8″	Q-2	14	1.714		215	92		307	375
4540	6″ x 8″		14	1.714		215	92		307	375
4550	6″ x 10″		10	2.400		385	129		514	615
4560	8″ x 10″		8.50	2.824		385	152		537	655
4570	10″ x 12″		7.50	3.200		635	172		807	960
4600	Y, 2″	Q-1	12	1.333		43	69		112	152
4610	3″		10	1.600		69.50	83		152.50	202
4620	4″		9	1.778		91	92		183	239
4630	5″	Q-2	13	1.846		155	99		254	320
4640	6″	″	11	2.182		189	117		306	385
4650	8″	Q-3	8	4		450	219		669	825
4660	10″		7	4.571		740	251		991	1,200
4670	12″		6	5.333		1,500	292		1,792	2,100
4672	15″		5	6.400		3,025	350		3,375	3,850
4900	For extra heavy, add					44%	4%			

22 13 16.30 Pipe Fittings, Cast Iron		Crew	Daily Output	Labor-Hours	Unit	2014 Bare Costs Material	Labor	Equipment	Total	Total Incl O&P
4940	Gasket and making push-on joint									
4950	2"	Q-1	40	.400	Ea.	9.40	20.50		29.90	42
4960	3"		35	.457		12.20	23.50		35.70	49.50
4970	4"		32	.500		15.30	26		41.30	56
4980	5"	Q-2	43	.558		24	30		54	72
4990	6"	"	40	.600		25	32		57	76
5000	8"	Q-3	32	1		55	55		110	144
5010	10"		29	1.103		95.50	60.50		156	197
5020	12"		25	1.280		122	70		192	240
5022	15"		21	1.524		146	83.50		229.50	287
5030	Note: gaskets and joint labor have									
5040	Been included with all listed fittings.									
5990	No hub									
6000	Cplg. & labor required at joints not incl. in fitting									
6010	price. Add 1 coupling per joint for installed price									
6020	1/4 Bend, 1-1/2"				Ea.	9.85			9.85	10.85
6060	2"					10.70			10.70	11.80
6080	3"					15			15	16.50
6120	4"					22			22	24.50
6140	5"					53.50			53.50	58.50
6160	6"					53.50			53.50	59
6180	8"					151			151	166
6184	1/4 Bend, long sweep, 1-1/2"					24.50			24.50	27
6186	2"					23.50			23.50	26
6188	3"					28.50			28.50	31.50
6189	4"					45			45	49.50
6190	5"					87.50			87.50	96.50
6191	6"					99.50			99.50	110
6192	8"					269			269	296
6193	10"					550			550	605
6200	1/8 Bend, 1-1/2"					8.35			8.35	9.15
6210	2"					9.25			9.25	10.15
6212	3"					12.40			12.40	13.65
6214	4"					16.25			16.25	17.85
6216	5"					34			34	37
6218	6"					36			36	39.50
6220	8"					104			104	114
6222	10"					197			197	217
6380	Sanitary Tee, tapped, 1-1/2"					19.60			19.60	21.50
6382	2" x 1-1/2"					17.35			17.35	19.05
6384	2"					18.55			18.55	20.50
6386	3" x 2"					27.50			27.50	30.50
6388	3"					47.50			47.50	52.50
6390	4" x 1-1/2"					24.50			24.50	27
6392	4" x 2"					28			28	30.50
6393	4"					28			28	30.50
6394	6" x 1-1/2"					64			64	70.50
6396	6" x 2"					65			65	71.50
6459	Sanitary Tee, 1-1/2"					13.85			13.85	15.25
6460	2"					14.75			14.75	16.25
6470	3"					18.25			18.25	20
6472	4"					34.50			34.50	38
6474	5"					81			81	89

22 13 16.30 Pipe Fittings, Cast Iron		Crew	Daily Output	Labor-Hours	Unit	2014 Bare Costs Material	Labor	Equipment	Total	Total Incl O&P
6476	6"				Ea.	82.50			82.50	91
6478	8"					335			335	365
6730	Y, 1-1/2"					14			14	15.40
6740	2"					13.70			13.70	15.10
6750	3"					19.95			19.95	22
6760	4"					32			32	35
6762	5"					75.50			75.50	83
6764	6"					85			85	93.50
6768	8"					200			200	220
6769	10"					445			445	490
6770	12"					875			875	960
6771	15"					1,950			1,950	2,150
6791	Y, reducing, 3" x 2"					14.85			14.85	16.30
6792	4" x 2"					21.50			21.50	23.50
6793	5" x 2"					47			47	51.50
6794	6" x 2"					52			52	57.50
6795	6" x 4"					68			68	74.50
6796	8" x 4"					116			116	128
6797	8" x 6"					143			143	157
6798	10" x 6"					320			320	355
6799	10" x 8"					385			385	425
6800	Double Y, 2"					21.50			21.50	24
6920	3"					40			40	44
7000	4"					81.50			81.50	89.50
7100	6"					144			144	158
7120	8"					410			410	450
7200	Combination Y and 1/8 Bend									
7220	1-1/2"				Ea.	14.90			14.90	16.40
7260	2"					15.60			15.60	17.15
7320	3"					24.50			24.50	27
7400	4"					47.50			47.50	52
7480	5"					97			97	107
7500	6"					130			130	143
7520	8"					305			305	335
7800	Reducer, 3" x 2"					7.55			7.55	8.30
7820	4" x 2"					11.70			11.70	12.85
7840	4" x 3"					11.70			11.70	12.85
7842	6" x 3"					31.50			31.50	34.50
7844	6" x 4"					31.50			31.50	34.50
7846	6" x 5"					32			32	35.50
7848	8" x 2"					50			50	55
7850	8" x 3"					46			46	50.50
7852	8" x 4"					48.50			48.50	53.50
7854	8" x 5"					55			55	60.50
7856	8" x 6"					54			54	59.50
7858	10" x 4"					95.50			95.50	105
7860	10" x 6"					101			101	111
7862	10" x 8"					118			118	130
7864	12" x 4"					197			197	217
7866	12" x 6"					211			211	232
7868	12" x 8"					217			217	239
7870	12" x 10"					221			221	243
7872	15" x 4"					410			410	455

22 13 Facility Sanitary Sewerage

22 13 16 – Sanitary Waste and Vent Piping

22 13 16.30 Pipe Fittings, Cast Iron		Crew	Daily Output	Labor-Hours	Unit	Material	2014 Bare Costs Labor	Equipment	Total	Total Incl O&P
7874	15" x 6"				Ea.	390			390	430
7876	15" x 8"					450			450	495
7878	15" x 10"					465			465	510
7880	15" x 12"					470			470	515
8000	Coupling, standard (by CISPI Mfrs.)									
8020	1-1/2"	Q-1	48	.333	Ea.	13.95	17.25		31.20	41.50
8040	2"		44	.364		15.25	18.85		34.10	45.50
8080	3"		38	.421		17.10	22		39.10	52
8120	4"		33	.485		20	25		45	60
8160	5"	Q-2	44	.545		41	29.50		70.50	89.50
8180	6"	"	40	.600		46.50	32		78.50	100
8200	8"	Q-3	33	.970		97	53		150	187
8220	10"	"	26	1.231		124	67.50		191.50	238
8300	Coupling, cast iron clamp & neoprene gasket (by MG)									
8310	1-1/2"	Q-1	48	.333	Ea.	6.40	17.25		23.65	33
8320	2"		44	.364		8.35	18.85		27.20	37.50
8330	3"		38	.421		12.65	22		34.65	47
8340	4"		33	.485		14.60	25		39.60	54
8350	5"	Q-2	44	.545		20.50	29.50		50	67
8360	6"	"	40	.600		32.50	32		64.50	84.50
8380	8"	Q-3	33	.970		116	53		169	208
8400	10"	"	26	1.231		189	67.50		256.50	310
8600	Coupling, Stainless steel, heavy duty									
8620	1-1/2"	Q-1	48	.333	Ea.	9.55	17.25		26.80	36.50
8630	2"		44	.364		9.90	18.85		28.75	39.50
8640	2" x 1-1/2"		44	.364		12.05	18.85		30.90	42
8650	3"		38	.421		10.75	22		32.75	45
8660	4"		33	.485		12.15	25		37.15	51.50
8670	4" x 3"		33	.485		17.50	25		42.50	57.50
8680	5"	Q-2	44	.545		26	29.50		55.50	73
8690	6"	"	40	.600		29.50	32		61.50	80.50
8700	8"	Q-3	33	.970		49.50	53		102.50	135
8710	10"		26	1.231		62.50	67.50		130	171
8712	12"		22	1.455		97	79.50		176.50	226
8715	15"		18	1.778		114	97.50		211.50	273

22 13 19 – Sanitary Waste Piping Specialties

22 13 19.13 Sanitary Drains		Crew	Daily Output	Labor-Hours	Unit	Material	Labor	Equipment	Total	Total Incl O&P
0010	**SANITARY DRAINS**									
0400	Deck, auto park, C.I., 13" top									
0440	3", 4", 5", and 6" pipe size	Q-1	8	2	Ea.	1,325	104		1,429	1,625
0480	For galvanized body, add				"	720			720	790
2000	Floor, medium duty, C.I., deep flange, 7" diam. top									
2040	2" and 3" pipe size	Q-1	12	1.333	Ea.	186	69		255	310
2080	For galvanized body, add					89.50			89.50	98.50
2120	With polished bronze top					291			291	320
2400	Heavy duty, with sediment bucket, C.I., 12" diam. loose grate									
2420	2", 3", 4", 5", and 6" pipe size	Q-1	9	1.778	Ea.	640	92		732	845
2460	With polished bronze top				"	905			905	995
2500	Heavy duty, cleanout & trap w/bucket, C.I., 15" top									
2540	2", 3", and 4" pipe size	Q-1	6	2.667	Ea.	6,075	138		6,213	6,900
2560	For galvanized body, add					1,550			1,550	1,725
2580	With polished bronze top					6,575			6,575	7,250

22 13 29 – Sanitary Sewerage Pumps

22 13 29.13 Wet-Pit-Mounted, Vertical Sewerage Pumps		Crew	Daily Output	Labor-Hours	Unit	Material	2014 Bare Costs Labor	Equipment	Total	Total Incl O&P
0010	**WET-PIT-MOUNTED, VERTICAL SEWERAGE PUMPS**									
0020	Controls incl. alarm/disconnect panel w/wire. Excavation not included									
0260	Simplex, 9 GPM at 60 PSIG, 91 gal. tank				Ea.	3,325			3,325	3,650
0300	Unit with manway, 26" I.D., 18" high					3,700			3,700	4,050
0340	26" I.D., 36" high					3,750			3,750	4,125
0380	43" I.D., 4' high					3,975			3,975	4,375
0600	Simplex, 9 GPM at 60 PSIG, 150 gal. tank, indoor					3,525			3,525	3,875
0700	Unit with manway, 26" I.D., 36" high					4,225			4,225	4,650
0740	26" I.D., 4' high					4,450			4,450	4,900
2000	Duplex, 18 GPM at 60 PSIG, 150 gal. tank, indoor					7,025			7,025	7,750
2060	Unit with manway, 43" I.D., 4' high					8,100			8,100	8,900
2400	For core only					1,875			1,875	2,050
3000	Indoor residential type installation									
3020	Simplex, 9 GPM at 60 PSIG, 91 gal. HDPE tank				Ea.	3,350			3,350	3,675

22 13 29.14 Sewage Ejector Pumps

		Crew	Daily Output	Labor-Hours	Unit	Material	Labor	Equipment	Total	Total Incl O&P
0010	**SEWAGE EJECTOR PUMPS**, With operating and level controls									
0100	Simplex system incl. tank, cover, pump 15' head									
0500	37 gal. PE tank, 12 GPM, 1/2 HP, 2" discharge	Q-1	3.20	5	Ea.	480	259		739	915
0510	3" discharge		3.10	5.161		520	267		787	975
0530	87 GPM, .7 HP, 2" discharge		3.20	5		735	259		994	1,200
0540	3" discharge		3.10	5.161		795	267		1,062	1,275
0600	45 gal. coated stl. tank, 12 GPM, 1/2 HP, 2" discharge		3	5.333		855	276		1,131	1,350
0610	3" discharge		2.90	5.517		890	286		1,176	1,400
0630	87 GPM, .7 HP, 2" discharge		3	5.333		1,100	276		1,376	1,625
0640	3" discharge		2.90	5.517		1,150	286		1,436	1,700
0660	134 GPM, 1 HP, 2" discharge		2.80	5.714		1,175	296		1,471	1,750
0680	3" discharge		2.70	5.926		1,250	305		1,555	1,850
0700	70 gal. PE tank, 12 GPM, 1/2 HP, 2" discharge		2.60	6.154		920	320		1,240	1,500
0710	3" discharge		2.40	6.667		980	345		1,325	1,600
0730	87 GPM, 0.7 HP, 2" discharge		2.50	6.400		1,200	330		1,530	1,800
0740	3" discharge		2.30	6.957		1,275	360		1,635	1,950
0760	134 GPM, 1 HP, 2" discharge		2.20	7.273		1,300	375		1,675	2,000
0770	3" discharge		2	8		1,375	415		1,790	2,150
0800	75 gal. coated stl. tank, 12 GPM, 1/2 HP, 2" discharge		2.40	6.667		1,025	345		1,370	1,650
0810	3" discharge		2.20	7.273		1,075	375		1,450	1,750
0830	87 GPM, .7 HP, 2" discharge		2.30	6.957		1,300	360		1,660	1,975
0840	3" discharge		2.10	7.619		1,350	395		1,745	2,100
0860	134 GPM, 1 HP, 2" discharge		2	8		1,375	415		1,790	2,150
0880	3" discharge		1.80	8.889		1,450	460		1,910	2,300
1040	Duplex system incl. tank, covers, pumps									
1060	110 gal. fiberglass tank, 24 GPM, 1/2 HP, 2" discharge	Q-1	1.60	10	Ea.	1,850	520		2,370	2,800
1080	3" discharge		1.40	11.429		1,950	590		2,540	3,025
1100	174 GPM, .7 HP, 2" discharge		1.50	10.667		2,400	555		2,955	3,450
1120	3" discharge		1.30	12.308		2,475	640		3,115	3,700
1140	268 GPM, 1 HP, 2" discharge		1.20	13.333		2,600	690		3,290	3,900
1160	3" discharge		1	16		2,700	830		3,530	4,200
1260	135 gal. coated stl. tank, 24 GPM, 1/2 HP, 2" discharge	Q-2	1.70	14.118		1,900	760		2,660	3,250
2000	3" discharge		1.60	15		2,025	805		2,830	3,450
2640	174 GPM, .7 HP, 2" discharge		1.60	15		2,500	805		3,305	3,975
2660	3" discharge		1.50	16		2,625	860		3,485	4,200
2700	268 GPM, 1 HP, 2" discharge		1.30	18.462		2,700	990		3,690	4,475
3040	3" discharge		1.10	21.818		2,875	1,175		4,050	4,950

22 13 Facility Sanitary Sewerage

22 13 29 – Sanitary Sewerage Pumps

22 13 29.14 Sewage Ejector Pumps		Crew	Daily Output	Labor-Hours	Unit	2014 Bare Costs Material	Labor	Equipment	Total	Total Incl O&P
3060	275 gal. coated stl. tank, 24 GPM, 1/2 HP, 2" discharge	Q-2	1.50	16	Ea.	2,375	860		3,235	3,925
3080	3" discharge		1.40	17.143		2,425	920		3,345	4,050
3100	174 GPM, .7 HP, 2" discharge		1.40	17.143		3,075	920		3,995	4,775
3120	3" discharge		1.30	18.462		3,250	990		4,240	5,075
3140	268 GPM, 1 HP, 2" discharge		1.10	21.818		3,375	1,175		4,550	5,475
3160	3" discharge	↓	.90	26.667	↓	3,550	1,425		4,975	6,075
3260	Pump system accessories, add									
3300	Alarm horn and lights, 115 V mercury switch	Q-1	8	2	Ea.	97	104		201	262
3340	Switch, mag. contactor, alarm bell, light, 3 level control		5	3.200		475	166		641	775
3380	Alternator, mercury switch activated	↓	4	4	↓	855	207		1,062	1,250

22 14 Facility Storm Drainage

22 14 23 – Storm Drainage Piping Specialties

22 14 23.33 Backwater Valves		Crew	Daily Output	Labor-Hours	Unit	Material	Labor	Equipment	Total	Total Incl O&P
0010	**BACKWATER VALVES**, C.I. Body									
6980	Bronze gate and automatic flapper valves									
7000	3" and 4" pipe size	Q-1	13	1.231	Ea.	1,900	64		1,964	2,175
7100	5" and 6" pipe size	"	13	1.231	"	2,900	64		2,964	3,300
7240	Bronze flapper valve, bolted cover									
7260	2" pipe size	Q-1	16	1	Ea.	555	52		607	690
7300	4" pipe size	"	13	1.231		1,075	64		1,139	1,275
7340	6" pipe size	Q-2	17	1.412	↓	1,525	76		1,601	1,800

22 14 26 – Facility Storm Drains

22 14 26.19 Facility Trench Drains		Crew	Daily Output	Labor-Hours	Unit	Material	Labor	Equipment	Total	Total Incl O&P
0010	**FACILITY TRENCH DRAINS**									
5980	Trench, floor, heavy duty, modular, C.I., 12" x 12" top									
6000	2", 3", 4", 5", & 6" pipe size	Q-1	8	2	Ea.	830	104		934	1,075
6100	For unit with polished bronze top	"	8	2	"	1,250	104		1,354	1,500
6600	Trench, floor, for cement concrete encasement									
6610	Not including trenching or concrete									
6640	Polyester polymer concrete									
6650	4" internal width, with grate									
6660	Light duty steel grate	Q-1	120	.133	L.F.	34	6.90		40.90	48
6670	Medium duty steel grate		115	.139		39	7.20		46.20	54
6680	Heavy duty iron grate	↓	110	.145	↓	60	7.55		67.55	77.50
6700	12" internal width, with grate									
6770	Heavy duty galvanized grate	Q-1	80	.200	L.F.	164	10.35		174.35	196
6800	Fiberglass									
6810	8" internal width, with grate									
6820	Medium duty galvanized grate	Q-1	115	.139	L.F.	96.50	7.20		103.70	117
6830	Heavy duty iron grate	"	110	.145	"	102	7.55		109.55	124

22 14 29 – Sump Pumps

22 14 29.13 Wet-Pit-Mounted, Vertical Sump Pumps		Crew	Daily Output	Labor-Hours	Unit	Material	Labor	Equipment	Total	Total Incl O&P
0010	**WET-PIT-MOUNTED, VERTICAL SUMP PUMPS**									
0400	Molded PVC base, 21 GPM at 15' head, 1/3 HP	1 Plum	5	1.600	Ea.	135	92		227	288
0800	Iron base, 21 GPM at 15' head, 1/3 HP		5	1.600		164	92		256	320
1200	Solid brass, 21 GPM at 15' head, 1/3 HP	↓	5	1.600	↓	289	92		381	460
2000	Sump pump, single stage									
2010	25 GPM, 1 HP, 1-1/2" discharge	Q-1	1.80	8.889	Ea.	3,625	460		4,085	4,675
2020	75 GPM, 1-1/2 HP, 2" discharge	↓	1.50	10.667	↓	3,825	555		4,380	5,025

22 14 Facility Storm Drainage

22 14 29 – Sump Pumps

22 14 29.13 Wet-Pit-Mounted, Vertical Sump Pumps		Crew	Daily Output	Labor-Hours	Unit	Material	2014 Bare Costs Labor	Equipment	Total	Total Incl O&P
2030	100 GPM, 2 HP, 2-1/2" discharge	Q-1	1.30	12.308	Ea.	3,900	640		4,540	5,250
2040	150 GPM, 3 HP, 3" discharge		1.10	14.545		3,900	755		4,655	5,425
2050	200 GPM, 3 HP, 3" discharge		1	16		4,125	830		4,955	5,800
2060	300 GPM, 10 HP, 4" discharge	Q-2	1.20	20		4,450	1,075		5,525	6,525
2070	500 GPM, 15 HP, 5" discharge		1.10	21.818		5,075	1,175		6,250	7,350
2080	800 GPM, 20 HP, 6" discharge		1	24		6,000	1,300		7,300	8,550
2090	1000 GPM, 30 HP, 6" discharge		.85	28.235		6,575	1,525		8,100	9,550
2100	1600 GPM, 50 HP, 8" discharge		.72	33.333		10,300	1,800		12,100	14,000
2110	2000 GPM, 60 HP, 8" discharge	Q-3	.85	37.647		10,500	2,075		12,575	14,700
2202	For general purpose float switch, copper coated float, add	Q-1	5	3.200		108	166		274	370

22 14 29.16 Submersible Sump Pumps		Crew	Daily Output	Labor-Hours	Unit	Material	Labor	Equipment	Total	Total Incl O&P
0010	**SUBMERSIBLE SUMP PUMPS**									
7000	Sump pump, automatic									
7100	Plastic, 1-1/4" discharge, 1/4 HP	1 Plum	6	1.333	Ea.	134	76.50		210.50	263
7140	1/3 HP		5	1.600		194	92		286	350
7160	1/2 HP		5	1.600		238	92		330	400
7180	1-1/2" discharge, 1/2 HP		4	2		272	115		387	475
7500	Cast iron, 1-1/4" discharge, 1/4 HP		6	1.333		189	76.50		265.50	325
7540	1/3 HP		6	1.333		222	76.50		298.50	360
7560	1/2 HP		5	1.600		269	92		361	435

22 31 Domestic Water Softeners

22 31 16 – Commercial Domestic Water Softeners

22 31 16.10 Water Softeners		Crew	Daily Output	Labor-Hours	Unit	Material	Labor	Equipment	Total	Total Incl O&P
0010	**WATER SOFTENERS**									
5800	Softener systems, automatic, intermediate sizes									
5820	available, may be used in multiples.									
6000	Hardness capacity between regenerations and flow									
6100	150,000 grains, 37 GPM cont., 51 GPM peak	Q-1	1.20	13.333	Ea.	5,300	690		5,990	6,900
6200	300,000 grains, 81 GPM cont., 113 GPM peak		1	16		9,375	830		10,205	11,600
6300	750,000 grains, 160 GPM cont., 230 GPM peak		.80	20		12,200	1,025		13,225	15,000
6400	900,000 grains, 185 GPM cont., 270 GPM peak		.70	22.857		19,700	1,175		20,875	23,400

22 52 Fountain Plumbing Systems

22 52 16 – Fountain Pumps

22 52 16.10 Fountain Water Pumps		Crew	Daily Output	Labor-Hours	Unit	Material	Labor	Equipment	Total	Total Incl O&P
0010	**FOUNTAIN WATER PUMPS**									
0100	Pump w/controls									
0200	Single phase, 100' cord, 1/2 H.P. pump	2 Skwk	4.40	3.636	Ea.	1,275	172		1,447	1,675
0300	3/4 H.P. pump		4.30	3.721		2,175	176		2,351	2,675
0400	1 H.P. pump		4.20	3.810		2,350	180		2,530	2,875
0500	1-1/2 H.P. pump		4.10	3.902		2,800	185		2,985	3,350
0600	2 H.P. pump		4	4		3,775	189		3,964	4,475
0700	Three phase, 200' cord, 5 H.P. pump		3.90	4.103		5,175	194		5,369	6,000
0800	7-1/2 H.P. pump		3.80	4.211		9,000	199		9,199	10,200
0900	10 H.P. pump		3.70	4.324		13,300	205		13,505	15,000
1000	15 H.P. pump		3.60	4.444		16,800	210		17,010	18,800
2000	DESIGN NOTE: Use two horsepower per surface acre.									

22 52 Fountain Plumbing Systems

22 52 33 – Fountain Ancillary

22 52 33.10 Fountain Miscellaneous		Crew	Daily Output	Labor-Hours	Unit	Material	Labor	Equipment	Total	Total Incl O&P
						2014 Bare Costs				
0010	**FOUNTAIN MISCELLANEOUS**									
1100	Nozzles, minimum	2 Skwk	8	2	Ea.	149	94.50		243.50	310
1200	Maximum		8	2		320	94.50		414.50	495
1300	Lights w/mounting kits, 200 watt		18	.889		1,050	42		1,092	1,225
1400	300 watt		18	.889		1,275	42		1,317	1,475
1500	500 watt		18	.889		1,425	42		1,467	1,650
1600	Color blender		12	1.333		555	63		618	710

22 66 Chemical-Waste Systems for Lab. and Healthcare Facilities

22 66 53 – Laboratory Chemical-Waste and Vent Piping

22 66 53.30 Glass Pipe

		Crew	Daily Output	Labor-Hours	Unit	Material	Labor	Equipment	Total	Total Incl O&P
0010	**GLASS PIPE**, Borosilicate, couplings & clevis hanger assemblies, 10′ O.C.									
0020	Drainage									
1100	1-1/2″ diameter	Q-1	52	.308	L.F.	10.60	15.95		26.55	35.50
1120	2″ diameter		44	.364		13.85	18.85		32.70	44
1140	3″ diameter		39	.410		18.70	21.50		40.20	52.50
1160	4″ diameter		30	.533		33.50	27.50		61	78
1180	6″ diameter		26	.615		58	32		90	112
1870	To delete coupling & hanger, subtract									
1880	1-1/2″ diam. to 2″ diam.					19%	22%			
1890	3″ diam. to 6″ diam.					20%	17%			
2000	Process supply (pressure), beaded joints									
2040	1/2″ diameter	1 Plum	36	.222	L.F.	5.20	12.80		18	25
2060	3/4″ diameter		31	.258		5.85	14.85		20.70	29
2080	1″ diameter		27	.296		15.75	17.05		32.80	43.50
2100	1-1/2″ diameter	Q-1	47	.340		13.90	17.65		31.55	42
2120	2″ diameter		39	.410		18.90	21.50		40.40	53
2140	3″ diameter		34	.471		25	24.50		49.50	64.50
2160	4″ diameter		25	.640		38	33		71	92
2180	6″ diameter		21	.762		107	39.50		146.50	178
2860	To delete coupling & hanger, subtract									
2870	1/2″ diam. to 1″ diam.					25%	33%			
2880	1-1/2″ diam. to 3″ diam.					22%	21%			
2890	4″ diam. to 6″ diam.					23%	15%			
3800	Conical joint, transparent									
3980	6″ diameter	Q-1	21	.762	L.F.	143	39.50		182.50	217
4500	To delete couplings & hangers, subtract									
4530	6″ diam.					22%	26%			

22 66 53.40 Pipe Fittings, Glass

		Crew	Daily Output	Labor-Hours	Unit	Material	Labor	Equipment	Total	Total Incl O&P
0010	**PIPE FITTINGS, GLASS**									
0020	Drainage, beaded ends									
0040	Coupling & labor required at joints not incl. in fitting									
0050	price. Add 1 per joint for installed price									
0070	90° Bend or sweep, 1-1/2″				Ea.	31.50			31.50	35
0090	2″					40			40	44
0100	3″					66			66	72.50
0110	4″					105			105	116
0120	6″ (sweep only)					330			330	360
0200	45° Bend or sweep same as 90°									
0350	Tee, single sanitary, 1-1/2″				Ea.	51			51	56.50

22 66 53 – Laboratory Chemical-Waste and Vent Piping

22 66 53.40 Pipe Fittings, Glass		Crew	Daily Output	Labor-Hours	Unit	2014 Bare Costs Material	Labor	Equipment	Total	Total Incl O&P
0370	2"				Ea.	51			51	56.50
0380	3"					76.50			76.50	84
0390	4"					137			137	150
0400	6"					365			365	405
0410	Tee, straight, 1-1/2"					63.50			63.50	69.50
0430	2"					63.50			63.50	69.50
0440	3"					91.50			91.50	101
0450	4"					107			107	117
0460	6"					395			395	435
0500	Coupling, stainless steel, TFE seal ring									
0520	1-1/2"	Q-1	32	.500	Ea.	23.50	26		49.50	64.50
0530	2"		30	.533		29.50	27.50		57	73.50
0540	3"		25	.640		39.50	33		72.50	93.50
0550	4"		23	.696		67.50	36		103.50	129
0560	6"		20	.800		152	41.50		193.50	230
0600	Coupling, stainless steel, bead to plain end									
0610	1-1/2"	Q-1	36	.444	Ea.	30	23		53	68
0620	2"		34	.471		37	24.50		61.50	78
0630	3"		29	.552		63	28.50		91.50	113
0640	4"		27	.593		94	30.50		124.50	151
0650	6"		24	.667		320	34.50		354.50	400
2350	Coupling, Viton liner, for temperatures to 400°F									
2370	1/2"	Q-1	40	.400	Ea.	46.50	20.50		67	83
2380	3/4"		37	.432		53	22.50		75.50	92.50
2390	1"		35	.457		63	23.50		86.50	106
2400	1-1/2"		32	.500		23	26		49	64.50
2410	2"		30	.533		29.50	27.50		57	74
2420	3"		25	.640		39.50	33		72.50	93.50
2430	4"		23	.696		67.50	36		103.50	129
2440	6"		20	.800		152	41.50		193.50	230
2550	For beaded joint armored fittings, add					200%				
2600	Conical ends. Flange set, gasket & labor not incl. in fitting									
2620	price. Add 1 per joint for installed price.									
2650	90° Sweep elbow, 1"				Ea.	105			105	116
2670	1-1/2"					211			211	232
2680	2"					219			219	241
2690	3"					340			340	375
2700	4"					605			605	665
2710	6"					1,050			1,050	1,175
2750	Cross (straight), add					55%				
2850	Tee, add					20%				

22 66 53.60 Corrosion Resistant Pipe		Crew	Daily Output	Labor-Hours	Unit	Material	Labor	Equipment	Total	Total Incl O&P
0010	**CORROSION RESISTANT PIPE**, No couplings or hangers									
0020	Iron alloy, drain, mechanical joint									
1000	1-1/2" diameter	Q-1	70	.229	L.F.	46	11.85		57.85	68.50
1100	2" diameter		66	.242		47	12.55		59.55	70.50
1120	3" diameter		60	.267		60.50	13.80		74.30	87.50
1140	4" diameter		52	.308		77.50	15.95		93.45	110
1980	Iron alloy, drain, B&S joint									
2000	2" diameter	Q-1	54	.296	L.F.	55.50	15.35		70.85	84.50
2100	3" diameter		52	.308		71	15.95		86.95	102
2120	4" diameter		48	.333		98	17.25		115.25	134

22 66 Chemical-Waste Systems for Lab. and Healthcare Facilities

22 66 53 – Laboratory Chemical-Waste and Vent Piping

22 66 53.60 Corrosion Resistant Pipe		Crew	Daily Output	Labor-Hours	Unit	2014 Bare Costs Material	Labor	Equipment	Total	Total Incl O&P
2140	6″ diameter	Q-2	59	.407	L.F.	158	22		180	206
2160	8″ diameter	″	54	.444		297	24		321	360
2980	Plastic, epoxy, fiberglass filament wound, B&S joint									
3000	2″ diameter	Q-1	62	.258	L.F.	12	13.35		25.35	33
3100	3″ diameter		51	.314		14	16.25		30.25	40
3120	4″ diameter		45	.356		20	18.40		38.40	50
3140	6″ diameter		32	.500		28	26		54	70
3160	8″ diameter	Q-2	38	.632		44	34		78	99.50
3180	10″ diameter		32	.750		60	40.50		100.50	127
3200	12″ diameter		28	.857		72	46		118	149
3980	Polyester, fiberglass filament wound, B&S joint									
4000	2″ diameter	Q-1	62	.258	L.F.	13.05	13.35		26.40	34.50
4100	3″ diameter		51	.314		17	16.25		33.25	43
4120	4″ diameter		45	.356		25	18.40		43.40	55.50
4140	6″ diameter		32	.500		36	26		62	78.50
4160	8″ diameter	Q-2	38	.632		85.50	34		119.50	145
4180	10″ diameter		32	.750		105	40.50		145.50	176
4200	12″ diameter		28	.857		125	46		171	208
4980	Polypropylene, acid resistant, fire retardant, schedule 40									
5000	1-1/2″ diameter	Q-1	68	.235	L.F.	7.90	12.20		20.10	27
5100	2″ diameter		62	.258		12.45	13.35		25.80	33.50
5120	3″ diameter		51	.314		22	16.25		38.25	49
5140	4″ diameter		45	.356		28	18.40		46.40	59
5160	6″ diameter		32	.500		56.50	26		82.50	101
5980	Proxylene, fire retardant, Schedule 40									
6000	1-1/2″ diameter	Q-1	68	.235	L.F.	12.40	12.20		24.60	32
6100	2″ diameter		62	.258		17	13.35		30.35	38.50
6120	3″ diameter		51	.314		30.50	16.25		46.75	58.50
6140	4″ diameter		45	.356		43.50	18.40		61.90	75.50
6160	6″ diameter		32	.500		73.50	26		99.50	120
6820	For Schedule 80, add					35%	2%			

22 66 53.70 Pipe Fittings, Corrosion Resistant		Crew	Daily Output	Labor-Hours	Unit	Material	Labor	Equipment	Total	Total Incl O&P
0010	**PIPE FITTINGS, CORROSION RESISTANT**									
0030	Iron alloy									
0050	Mechanical joint									
0060	1/4 Bend, 1-1/2″	Q-1	12	1.333	Ea.	77	69		146	189
0080	2″		10	1.600		126	83		209	264
0090	3″		9	1.778		151	92		243	305
0100	4″		8	2		174	104		278	345
0110	1/8 Bend, 1-1/2″		12	1.333		49	69		118	158
0130	2″		10	1.600		84	83		167	218
0140	3″		9	1.778		112	92		204	262
0150	4″		8	2		150	104		254	320
0160	Tee and Y, sanitary, straight									
0170	1-1/2″	Q-1	8	2	Ea.	84	104		188	249
0180	2″		7	2.286		112	118		230	300
0190	3″		6	2.667		174	138		312	400
0200	4″		5	3.200		320	166		486	600
0360	Coupling, 1-1/2″		14	1.143		47.50	59		106.50	142
0380	2″		12	1.333		54	69		123	164
0390	3″		11	1.455		56.50	75.50		132	177
0400	4″		10	1.600		64.50	83		147.50	196

22 66 Chemical-Waste Systems for Lab. and Healthcare Facilities

22 66 53 – Laboratory Chemical-Waste and Vent Piping

22 66 53.70 Pipe Fittings, Corrosion Resistant		Crew	Daily Output	Labor-Hours	Unit	2014 Bare Costs Material	Labor	Equipment	Total	Total Incl O&P
0500	Bell & Spigot									
0510	1/4 and 1/16 bend, 2"	Q-1	16	1	Ea.	99.50	52		151.50	187
0520	3"		14	1.143		232	59		291	345
0530	4"		13	1.231		238	64		302	360
0540	6"	Q-2	17	1.412		470	76		546	635
0550	8"	"	12	2		1,950	107		2,057	2,275
0620	1/8 bend, 2"	Q-1	16	1		112	52		164	201
0640	3"		14	1.143		207	59		266	320
0650	4"		13	1.231		238	64		302	360
0660	6"	Q-2	17	1.412		395	76		471	550
0680	8"	"	12	2		1,550	107		1,657	1,850
0700	Tee, sanitary, 2"	Q-1	10	1.600		213	83		296	360
0710	3"		9	1.778		690	92		782	900
0720	4"		8	2		575	104		679	785
0730	6"	Q-2	11	2.182		715	117		832	965
0740	8"	"	8	3		1,950	161		2,111	2,400
1800	Y, sanitary, 2"	Q-1	10	1.600		224	83		307	370
1820	3"		9	1.778		400	92		492	580
1830	4"		8	2		360	104		464	550
1840	6"	Q-2	11	2.182		1,200	117		1,317	1,500
1850	8"	"	8	3		3,300	161		3,461	3,875
3000	Epoxy, filament wound									
3030	Quick-lock joint									
3040	90° Elbow, 2"	Q-1	28	.571	Ea.	97.50	29.50		127	152
3060	3"		16	1		112	52		164	201
3070	4"		13	1.231		153	64		217	265
3080	6"		8	2		223	104		327	400
3090	8"	Q-2	9	2.667		410	143		553	665
3100	10"		7	3.429		520	184		704	850
3110	12"		6	4		740	215		955	1,150
3120	45° Elbow, 2"	Q-1	28	.571		75	29.50		104.50	127
3130	3"		16	1		106	52		158	195
3140	4"		13	1.231		109	64		173	217
3150	6"		8	2		223	104		327	400
3160	8"	Q-2	9	2.667		410	143		553	665
3170	10"		7	3.429		515	184		699	850
3180	12"		6	4		740	215		955	1,150
3190	Tee, 2"	Q-1	19	.842		233	43.50		276.50	320
3200	3"		11	1.455		280	75.50		355.50	425
3210	4"		9	1.778		335	92		427	510
3220	6"		5	3.200		575	166		741	880
3230	8"	Q-2	6	4		645	215		860	1,025
3240	10"		5	4.800		900	258		1,158	1,375
3250	12"		4	6		1,275	320		1,595	1,875
4000	Polypropylene, acid resistant									
4020	Non-pressure, electrofusion joints									
4050	1/4 bend, 1-1/2"	1 Plum	16	.500	Ea.	13.55	29		42.55	58.50
4060	2"	Q-1	28	.571		26	29.50		55.50	73
4080	3"		17	.941		28.50	49		77.50	105
4090	4"		14	1.143		46.50	59		105.50	141
4110	6"		8	2		110	104		214	277
4150	1/4 Bend, long sweep									
4170	1-1/2"	1 Plum	16	.500	Ea.	14.95	29		43.95	60

22 66 53 – Laboratory Chemical-Waste and Vent Piping

22 66 53.70 Pipe Fittings, Corrosion Resistant		Crew	Daily Output	Labor-Hours	Unit	2014 Bare Costs Material	Labor	Equipment	Total	Total Incl O&P
4180	2"	Q-1	28	.571	Ea.	26.50	29.50		56	73.50
4200	3"		17	.941		33	49		82	110
4210	4"		14	1.143		47.50	59		106.50	142
4250	1/8 bend, 1-1/2"	1 Plum	16	.500		14	29		43	59
4260	2"	Q-1	28	.571		16.40	29.50		45.90	62.50
4280	3"		17	.941		30	49		79	107
4290	4"		14	1.143		33.50	59		92.50	127
4310	6"		8	2		92.50	104		196.50	258
4400	Tee, sanitary									
4420	1-1/2"	1 Plum	10	.800	Ea.	17.10	46		63.10	88.50
4430	2"	Q-1	17	.941		19.95	49		68.95	95.50
4450	3"		11	1.455		40	75.50		115.50	158
4460	4"		9	1.778		60	92		152	205
4480	6"		5	3.200		400	166		566	690
4490	Tee, sanitary reducing, 2" x 2" x 1-1/2"		17	.941		19.95	49		68.95	95.50
4492	3" x 3" x 2"		11	1.455		40	75.50		115.50	158
4494	4" x 4" x 3"		9	1.778		58.50	92		150.50	203
4496	6" x 6" x 4"		5	3.200		196	166		362	465
4650	Wye 45°, 1-1/2"	1 Plum	10	.800		18.50	46		64.50	90
4652	2"	Q-1	17	.941		25.50	49		74.50	102
4653	3"		11	1.455		42.50	75.50		118	161
4654	4"		9	1.778		62	92		154	207
4656	6"		5	3.200		160	166		326	425
4678	Combination Y & 1/8 bend									
4681	1-1/2"	1 Plum	10	.800	Ea.	27	46		73	99.50
4683	2"	Q-1	17	.941		28.50	49		77.50	105
4684	3"		11	1.455		48.50	75.50		124	168
4685	4"		9	1.778		66	92		158	212

Estimating Tips

The labor adjustment factors listed in Subdivision 22 01 02.20 also apply to Division 23.

23 10 00 Facility Fuel Systems

- The prices in this subdivision for above- and below-ground storage tanks do not include foundations or hold-down slabs, unless noted. The estimator should refer to Divisions 3 and 31 for foundation system pricing. In addition to the foundations, required tank accessories, such as tank gauges, leak detection devices, and additional manholes and piping, must be added to the tank prices.

23 50 00 Central Heating Equipment

- When estimating the cost of an HVAC system, check to see who is responsible for providing and installing the temperature control system. It is possible to overlook controls, assuming that they would be included in the electrical estimate.
- When looking up a boiler, be careful on specified capacity. Some manufacturers rate their products on output while others use input.
- Include HVAC insulation for pipe, boiler, and duct (wrap and liner).
- Be careful when looking up mechanical items to get the correct pressure rating and connection type (thread, weld, flange).

23 70 00 Central HVAC Equipment

- Combination heating and cooling units are sized by the air conditioning requirements. (See Reference No. R236000-20 for preliminary sizing guide.)
- A ton of air conditioning is nominally 400 CFM.
- Rectangular duct is taken off by the linear foot for each size, but its cost is usually estimated by the pound. Remember that SMACNA standards now base duct on internal pressure.
- Prefabricated duct is estimated and purchased like pipe: straight sections and fittings.
- Note that cranes or other lifting equipment are not included on any lines in Division 23. For example, if a crane is required to lift a heavy piece of pipe into place high above a gym floor, or to put a rooftop unit on the roof of a four-story building, etc., it must be added. Due to the potential for extreme variation—from nothing additional required to a major crane or helicopter—we feel that including a nominal amount for "lifting contingency" would be useless and detract from the accuracy of the estimate. When using equipment rental cost data from RSMeans, do not forget to include the cost of the operator(s).

Reference Numbers

Reference numbers are shown in shaded boxes at the beginning of some major classifications. These numbers refer to related items in the Reference Section. The reference information may be an estimating procedure, an alternate pricing method, or technical information.

Note: Not all subdivisions listed here necessarily appear in this publication.

Note: **Trade Service,** *in part, has been used as a reference source for some of the material prices used in Division 23.*

23 05 05 – Selective HVAC Demolition

23 05 05.10 HVAC Demolition		Crew	Daily Output	Labor-Hours	Unit	Material	2014 Bare Costs Labor	Equipment	Total	Total Incl O&P
0010	**HVAC DEMOLITION**									
0100	Air conditioner, split unit, 3 ton	Q-5	2	8	Ea.		420		420	635
0150	Package unit, 3 ton	Q-6	3	8	"		435		435	660
0298	Boilers									
0300	Electric, up thru 148 kW	Q-19	2	12	Ea.		635		635	955
0310	150 thru 518 kW	"	1	24			1,275		1,275	1,900
0320	550 thru 2000 kW	Q-21	.40	80			4,350		4,350	6,550
0330	2070 kW and up	"	.30	106			5,800		5,800	8,725
0340	Gas and/or oil, up thru 150 MBH	Q-7	2.20	14.545			810		810	1,225
0350	160 thru 2000 MBH		.80	40			2,225		2,225	3,375
0360	2100 thru 4500 MBH		.50	64			3,575		3,575	5,375
0370	4600 thru 7000 MBH		.30	106			5,950		5,950	8,975
0380	7100 thru 12,000 MBH		.16	200			11,100		11,100	16,800
0390	12,200 thru 25,000 MBH		.12	266			14,900		14,900	22,400
1000	Ductwork, 4" high, 8" wide	1 Clab	200	.040	L.F.		1.47		1.47	2.26
1100	6" high, 8" wide		165	.048			1.78		1.78	2.74
1200	10" high, 12" wide		125	.064			2.35		2.35	3.62
1300	12"-14" high, 16"-18" wide		85	.094			3.45		3.45	5.30
1400	18" high, 24" wide		67	.119			4.38		4.38	6.75
1500	30" high, 36" wide		56	.143			5.25		5.25	8.10
1540	72" wide		50	.160			5.85		5.85	9.05
3000	Mechanical equipment, light items. Unit is weight, not cooling.	Q-5	.90	17.778	Ton		935		935	1,425
3600	Heavy items	"	1.10	14.545	"		765		765	1,150
5090	Remove refrigerant from system	1 Stpi	40	.200	Lb.		11.70		11.70	17.65

23 05 23 – General-Duty Valves for HVAC Piping

23 05 23.30 Valves, Iron Body		Crew	Daily Output	Labor-Hours	Unit	Material	Labor	Equipment	Total	Total Incl O&P
0010	**VALVES, IRON BODY**									
1020	Butterfly, wafer type, gear actuator, 200 lb.									
1030	2"	1 Plum	14	.571	Ea.	96	33		129	155
1040	2-1/2"	Q-1	9	1.778		97.50	92		189.50	246
1050	3"		8	2		101	104		205	267
1060	4"		5	3.200		113	166		279	375
1070	5"	Q-2	5	4.800		126	258		384	530
1080	6"	"	5	4.800		143	258		401	550
1650	Gate, 125 lb., N.R.S.									
2150	Flanged									
2200	2"	1 Plum	5	1.600	Ea.	680	92		772	890
2240	2-1/2"	Q-1	5	3.200		700	166		866	1,025
2260	3"		4.50	3.556		780	184		964	1,150
2280	4"		3	5.333		1,125	276		1,401	1,650
2300	6"	Q-2	3	8		1,900	430		2,330	2,750
3550	OS&Y, 125 lb., flanged									
3600	2"	1 Plum	5	1.600	Ea.	455	92		547	640
3660	3"	Q-1	4.50	3.556		495	184		679	825
3680	4"	"	3	5.333		730	276		1,006	1,225
3700	6"	Q-2	3	8		1,175	430		1,605	1,950
3900	For 175 lb., flanged, add					200%	10%			
5450	Swing check, 125 lb., threaded									
5470	1"	1 Plum	13	.615	Ea.	635	35.50		670.50	755
5500	2"	"	11	.727		410	42		452	515
5540	2-1/2"	Q-1	15	1.067		525	55.50		580.50	665
5550	3"		13	1.231		570	64		634	720

23 05 Common Work Results for HVAC

23 05 23 – General-Duty Valves for HVAC Piping

23 05 23.30 Valves, Iron Body		Crew	Daily Output	Labor-Hours	Unit	2014 Bare Costs Material	Labor	Equipment	Total	Total Incl O&P
5560	4"	Q-1	10	1.600	Ea.	905	83		988	1,125
5950	Flanged									
6000	2"	1 Plum	5	1.600	Ea.	395	92		487	575
6040	2-1/2"	Q-1	5	3.200		365	166		531	650
6050	3"		4.50	3.556		390	184		574	705
6060	4"		3	5.333		610	276		886	1,075
6070	6"	Q-2	3	8		1,050	430		1,480	1,800

23 05 23.80 Valves, Steel

Line	Description	Crew	Daily Output	Labor-Hours	Unit	Material	Labor	Equipment	Total	Total Incl O&P
0010	**VALVES, STEEL**									
0800	Cast									
1350	Check valve, swing type, 150 lb., flanged									
1370	1"	1 Plum	10	.800	Ea.	390	46		436	500
1400	2"	"	8	1		705	57.50		762.50	860
1440	2-1/2"	Q-1	5	3.200		820	166		986	1,150
1450	3"		4.50	3.556		835	184		1,019	1,200
1460	4"		3	5.333		1,200	276		1,476	1,750
1540	For 300 lb., flanged, add					50%	15%			
1548	For 600 lb., flanged, add					110%	20%			
1950	Gate valve, 150 lb., flanged									
2000	2"	1 Plum	8	1	Ea.	760	57.50		817.50	920
2040	2-1/2"	Q-1	5	3.200		1,075	166		1,241	1,425
2050	3"		4.50	3.556		1,075	184		1,259	1,450
2060	4"		3	5.333		1,325	276		1,601	1,900
2070	6"	Q-2	3	8		2,075	430		2,505	2,925
3650	Globe valve, 150 lb., flanged									
3700	2"	1 Plum	8	1	Ea.	955	57.50		1,012.50	1,125
3740	2-1/2"	Q-1	5	3.200		1,200	166		1,366	1,575
3750	3"		4.50	3.556		1,200	184		1,384	1,600
3760	4"		3	5.333		1,775	276		2,051	2,375
3770	6"	Q-2	3	8		2,775	430		3,205	3,700
5150	Forged									
5650	Check valve, class 800, horizontal, socket									
5698	Threaded									
5700	1/4"	1 Plum	24	.333	Ea.	94.50	19.20		113.70	133
5720	3/8"		24	.333		94.50	19.20		113.70	133
5730	1/2"		24	.333		94.50	19.20		113.70	133
5740	3/4"		20	.400		101	23		124	146
5750	1"		19	.421		119	24		143	168
5760	1-1/4"		15	.533		233	30.50		263.50	305

23 05 93 – Testing, Adjusting, and Balancing for HVAC

23 05 93.50 Piping, Testing

Line	Description	Crew	Daily Output	Labor-Hours	Unit	Material	Labor	Equipment	Total	Total Incl O&P
0010	**PIPING, TESTING**									
0100	Nondestructive testing									
0110	Nondestructive hydraulic pressure test, isolate & 1 hr. hold									
0120	1" - 4" pipe									
0140	0 – 250 L.F.	1 Stpi	1.33	6.015	Ea.		350		350	530
0160	250 – 500 L.F.	"	.80	10			585		585	885
0180	500 – 1000 L.F.	Q-5	1.14	14.035			740		740	1,125
0200	1000 – 2000 L.F.	"	.80	20			1,050		1,050	1,600
0300	6" - 10" pipe									
0320	0 – 250 L.F.	Q-5	1	16	Ea.		840		840	1,275
0340	250 – 500 L.F.		.73	21.918			1,150		1,150	1,750

23 05 Common Work Results for HVAC

23 05 93 – Testing, Adjusting, and Balancing for HVAC

23 05 93.50 Piping, Testing		Crew	Daily Output	Labor-Hours	Unit	Material	2014 Bare Costs Labor	Equipment	Total	Total Incl O&P
0360	500 – 1000 L.F.	Q-5	.53	30.189	Ea.		1,600		1,600	2,400
0380	1000 – 2000 L.F.	↓	.38	42.105	↓		2,225		2,225	3,350
1000	Pneumatic pressure test, includes soaping joints									
1120	1" - 4" pipe									
1140	0 – 250 L.F.	Q-5	2.67	5.993	Ea.	12.20	315		327.20	490
1160	250 – 500 L.F.		1.33	12.030		24.50	635		659.50	980
1180	500 – 1000 L.F.		.80	20		36.50	1,050		1,086.50	1,650
1200	1000 – 2000 L.F.	↓	.50	32	↓	49	1,675		1,724	2,600
1300	6" - 10" pipe									
1320	0 – 250 L.F.	Q-5	1.33	12.030	Ea.	12.20	635		647.20	970
1340	250 – 500 L.F.		.67	23.881		24.50	1,250		1,274.50	1,925
1360	500 – 1000 L.F.		.40	40		49	2,100		2,149	3,225
1380	1000 – 2000 L.F.	↓	.25	64	↓	61	3,375		3,436	5,175
2000	X-Ray of welds									
2110	2" diam.	1 Stpi	8	1	Ea.	12.60	58.50		71.10	102
2120	3" diam.		8	1		12.60	58.50		71.10	102
2130	4" diam.		8	1		18.90	58.50		77.40	110
2140	6" diam.		8	1		18.90	58.50		77.40	110
2150	8" diam.		6.60	1.212		18.90	71		89.90	128
2160	10" diam.	↓	6	1.333	↓	25	78		103	146
3000	Liquid penetration of welds									
3110	2" diam.	1 Stpi	14	.571	Ea.	3.95	33.50		37.45	55
3120	3" diam.		13.60	.588		3.95	34.50		38.45	56.50
3130	4" diam.		13.40	.597		3.95	35		38.95	57.50
3140	6" diam.		13.20	.606		3.95	35.50		39.45	58
3150	8" diam.		13	.615		5.95	36		41.95	61
3160	10" diam.	↓	12.80	.625	↓	5.95	36.50		42.45	61.50

23 13 Facility Fuel-Storage Tanks

23 13 23 – Facility Aboveground Fuel-Oil, Storage Tanks

23 13 23.13 Vertical, Steel, Abvground Fuel-Oil, Stor. Tanks		Crew	Daily Output	Labor-Hours	Unit	Material	Labor	Equipment	Total	Total Incl O&P
0010	**VERTICAL, STEEL, ABOVEGROUND FUEL-OIL, STORAGE TANKS**									
4000	Fixed roof oil storage tanks, steel, (1 BBL=42 gal. w/foundation 3'D x 1'W)									
4200	5,000 barrels				Ea.				194,000	213,500
4300	24,000 barrels								333,500	367,000
4500	56,000 barrels								729,000	802,000
4600	110,000 barrels								1,060,000	1,166,000
4800	143,000 barrels								1,250,000	1,375,000
4900	225,000 barrels								1,360,000	1,496,000
5100	Floating roof gasoline tanks, steel, 5,000 barrels (w/foundation 3'D x 1'W)								204,000	225,000
5200	25,000 barrels								381,000	419,000
5400	55,000 barrels								839,000	923,000
5500	100,000 barrels								1,253,000	1,379,000
5700	150,000 barrels								1,532,000	1,685,000
5800	225,000 barrels				↓				2,300,000	2,783,000

Estimating Tips

26 05 00 Common Work Results for Electrical

- Conduit should be taken off in three main categories—power distribution, branch power, and branch lighting—so the estimator can concentrate on systems and components, therefore making it easier to ensure all items have been accounted for.
- For cost modifications for elevated conduit installation, add the percentages to labor according to the height of installation, and only to the quantities exceeding the different height levels, not to the total conduit quantities.
- Remember that aluminum wiring of equal ampacity is larger in diameter than copper and may require larger conduit.
- If more than three wires at a time are being pulled, deduct percentages from the labor hours of that grouping of wires.
- When taking off grounding systems, identify separately the type and size of wire, and list each unique type of ground connection.
- The estimator should take the weights of materials into consideration when completing a takeoff. Topics to consider include: How will the materials be supported? What methods of support are available? How high will the support structure have to reach? Will the final support structure be able to withstand the total burden? Is the support material included or separate from the fixture, equipment, and material specified?
- Do not overlook the costs for equipment used in the installation. If scaffolding or highlifts are available in the field, contractors may use them in lieu of the proposed ladders and rolling staging.

26 20 00 Low-Voltage Electrical Transmission

- Supports and concrete pads may be shown on drawings for the larger equipment, or the support system may be only a piece of plywood for the back of a panelboard. In either case, it must be included in the costs.

26 40 00 Electrical and Cathodic Protection

- When taking off cathodic protections systems, identify the type and size of cable, and list each unique type of anode connection.

26 50 00 Lighting

- Fixtures should be taken off room by room, using the fixture schedule, specifications, and the ceiling plan. For large concentrations of lighting fixtures in the same area, deduct the percentages from labor hours.

Reference Numbers

Reference numbers are shown in shaded boxes at the beginning of some major classifications. These numbers refer to related items in the Reference Section. The reference information may be an estimating procedure, an alternate pricing method, or technical information.

Note: Not all subdivisions listed here necessarily appear in this publication.

Note: **Trade Service,** *in part, has been used as a reference source for some of the material prices used in Division 26.*

26 05 05.10 Electrical Demolition		Crew	Daily Output	Labor-Hours	Unit	Material	2014 Bare Costs Labor	Equipment	Total	Total Incl O&P
0010	**ELECTRICAL DEMOLITION**									
0020	Conduit to 15′ high, including fittings & hangers									
0100	Rigid galvanized steel, 1/2″ to 1″ diameter	1 Elec	242	.033	L.F.		1.76		1.76	2.64
0120	1-1/4″ to 2″	″	200	.040			2.13		2.13	3.19
0140	2-1/2″ to 3-1/2″	2 Elec	302	.053			2.83		2.83	4.23
0160	4″ to 6″	″	160	.100			5.35		5.35	8
0200	Electric metallic tubing (EMT), 1/2″ to 1″	1 Elec	394	.020			1.08		1.08	1.62
0220	1-1/4″ to 1-1/2″		326	.025			1.31		1.31	1.96
0240	2″ to 3″		236	.034			1.81		1.81	2.71
0260	3-1/2″ to 4″	2 Elec	310	.052			2.75		2.75	4.12
0270	Armored cable, (BX) avg. 50′ runs									
0280	#14, 2 wire	1 Elec	690	.012	L.F.		.62		.62	.93
0290	#14, 3 wire		571	.014			.75		.75	1.12
0300	#12, 2 wire		605	.013			.71		.71	1.06
0310	#12, 3 wire		514	.016			.83		.83	1.24
0320	#10, 2 wire		514	.016			.83		.83	1.24
0330	#10, 3 wire		425	.019			1		1	1.50
0340	#8, 3 wire		342	.023			1.25		1.25	1.87
0350	Non metallic sheathed cable (Romex)									
0360	#14, 2 wire	1 Elec	720	.011	L.F.		.59		.59	.89
0370	#14, 3 wire		657	.012			.65		.65	.97
0380	#12, 2 wire		629	.013			.68		.68	1.02
0390	#10, 3 wire		450	.018			.95		.95	1.42
0400	Wiremold raceway, including fittings & hangers									
0420	No. 3000	1 Elec	250	.032	L.F.		1.71		1.71	2.56
0440	No. 4000		217	.037			1.97		1.97	2.94
0460	No. 6000		166	.048			2.57		2.57	3.85
0500	Channels, steel, including fittings & hangers									
0520	3/4″ x 1-1/2″	1 Elec	308	.026	L.F.		1.39		1.39	2.07
0540	1-1/2″ x 1-1/2″		269	.030			1.59		1.59	2.37
0560	1-1/2″ x 1-7/8″		229	.035			1.86		1.86	2.79
0600	Copper bus duct, indoor, 3 phase									
0610	Including hangers & supports									
0620	225 amp	2 Elec	135	.119	L.F.		6.30		6.30	9.45
0640	400 amp		106	.151			8.05		8.05	12.05
0660	600 amp		86	.186			9.95		9.95	14.85
0680	1000 amp		60	.267			14.25		14.25	21.50
0700	1600 amp		40	.400			21.50		21.50	32
0720	3000 amp		10	1.600			85.50		85.50	128
1300	Transformer, dry type, 1 phase, incl. removal of									
1320	supports, wire & conduit terminations									
1340	1 kVA	1 Elec	7.70	1.039	Ea.		55.50		55.50	83
1420	75 kVA	2 Elec	2.50	6.400	″		340		340	510
1440	3 phase to 600V, primary									
1460	3 kVA	1 Elec	3.85	2.078	Ea.		111		111	166
1520	75 kVA	2 Elec	2.70	5.926			315		315	475
1550	300 kVA	R-3	1.80	11.111			590	78	668	965
1570	750 kVA	″	1.10	18.182			960	127	1,087	1,600
1800	Wire, THW-THWN-THHN, removed from									
1810	in place conduit, to 15′ high									
1830	#14	1 Elec	65	.123	C.L.F.		6.55		6.55	9.85
1840	#12		55	.145			7.75		7.75	11.60
1850	#10		45.50	.176			9.40		9.40	14.05

26 05 Common Work Results for Electrical

26 05 05 – Selective Electrical Demolition

26 05 05.10 Electrical Demolition		Crew	Daily Output	Labor-Hours	Unit	Material	2014 Bare Costs Labor	Equipment	Total	Total Incl O&P
1860	#8	1 Elec	40.40	.198	C.L.F.		10.55		10.55	15.80
1870	#6		32.60	.245			13.10		13.10	19.60
1880	#4	2 Elec	53	.302			16.10		16.10	24
1890	#3		50	.320			17.05		17.05	25.50
1900	#2		44.60	.359			19.15		19.15	28.50
1910	1/0		33.20	.482			25.50		25.50	38.50
1920	2/0		29.20	.548			29		29	44
1930	3/0		25	.640			34		34	51
1940	4/0		22	.727			39		39	58
1950	250 kcmil		20	.800			42.50		42.50	64
1960	300 kcmil		19	.842			45		45	67
1970	350 kcmil		18	.889			47.50		47.50	71
1980	400 kcmil		17	.941			50		50	75
1990	500 kcmil		16.20	.988			52.50		52.50	79
2000	Interior fluorescent fixtures, incl. supports									
2010	& whips, to 15′ high									
2100	Recessed drop-in 2′ x 2′, 2 lamp	2 Elec	35	.457	Ea.		24.50		24.50	36.50
2120	2′ x 4′, 2 lamp		33	.485			26		26	38.50
2140	2′ x 4′, 4 lamp		30	.533			28.50		28.50	42.50
2160	4′ x 4′, 4 lamp		20	.800			42.50		42.50	64
2180	Surface mount, acrylic lens & hinged frame									
2200	1′ x 4′, 2 lamp	2 Elec	44	.364	Ea.		19.40		19.40	29
2220	2′ x 2′, 2 lamp		44	.364			19.40		19.40	29
2260	2′ x 4′, 4 lamp		33	.485			26		26	38.50
2280	4′ x 4′, 4 lamp		23	.696			37		37	55.50
2300	Strip fixtures, surface mount									
2320	4′ long, 1 lamp	2 Elec	53	.302	Ea.		16.10		16.10	24
2340	4′ long, 2 lamp		50	.320			17.05		17.05	25.50
2360	8′ long, 1 lamp		42	.381			20.50		20.50	30.50
2380	8′ long, 2 lamp		40	.400			21.50		21.50	32
2400	Pendant mount, industrial, incl. removal									
2410	of chain or rod hangers, to 15′ high									
2420	4′ long, 2 lamp	2 Elec	35	.457	Ea.		24.50		24.50	36.50
2440	8′ long, 2 lamp	"	27	.593	"		31.50		31.50	47.50

26 05 13 – Medium-Voltage Cables

26 05 13.16 Medium-Voltage, Single Cable

Line	Description	Crew	Daily Output	Labor-Hours	Unit	Material	Labor	Equipment	Total	Total Incl O&P
0010	**MEDIUM-VOLTAGE, SINGLE CABLE** Splicing & terminations not included									
0040	Copper, XLP shielding, 5 kV, #6	2 Elec	4.40	3.636	C.L.F.	158	194		352	465
0050	#4		4.40	3.636		205	194		399	515
0100	#2		4	4		242	213		455	585
0200	#1		4	4		278	213		491	625
0400	1/0		3.80	4.211		305	225		530	670
0600	2/0		3.60	4.444		385	237		622	780
0800	4/0		3.20	5		500	267		767	950
1000	250 kcmil	3 Elec	4.50	5.333		620	285		905	1,100
1200	350 kcmil		3.90	6.154		800	330		1,130	1,375
1400	500 kcmil		3.60	6.667		975	355		1,330	1,600
1600	15 kV, ungrounded neutral, #1	2 Elec	4	4		340	213		553	690
1800	1/0		3.80	4.211		405	225		630	780
2000	2/0		3.60	4.444		460	237		697	860
2200	4/0		3.20	5		610	267		877	1,075
2400	250 kcmil	3 Elec	4.50	5.333		675	285		960	1,175

26 05 Common Work Results for Electrical

26 05 13 – Medium-Voltage Cables

26 05 13.16 Medium-Voltage, Single Cable		Crew	Daily Output	Labor-Hours	Unit	Material	2014 Bare Costs Labor	Equipment	Total	Total Incl O&P
2600	350 kcmil	3 Elec	3.90	6.154	C.L.F.	855	330		1,185	1,425
2800	500 kcmil		3.60	6.667		1,050	355		1,405	1,675
3000	25 kV, grounded neutral, #1/0	2 Elec	3.60	4.444		560	237		797	970
3200	2/0		3.40	4.706		615	251		866	1,050
3400	4/0		3	5.333		770	285		1,055	1,275
3600	250 kcmil	3 Elec	4.20	5.714		960	305		1,265	1,500
3800	350 kcmil		3.60	6.667		1,125	355		1,480	1,750
3900	500 kcmil		3.30	7.273		1,325	390		1,715	2,025
4000	35 kV, grounded neutral, #1/0	2 Elec	3.40	4.706		590	251		841	1,025
4200	2/0		3.20	5		695	267		962	1,175
4400	4/0		2.80	5.714		875	305		1,180	1,425
4600	250 kcmil	3 Elec	3.90	6.154		1,025	330		1,355	1,625
4800	350 kcmil		3.30	7.273		1,225	390		1,615	1,925
5000	500 kcmil		3	8		1,450	425		1,875	2,250
5050	Aluminum, XLP shielding, 5 kV, #2	2 Elec	5	3.200		168	171		339	440
5070	#1		4.40	3.636		174	194		368	480
5090	1/0		4	4		202	213		415	540
5100	2/0		3.80	4.211		226	225		451	585
5150	4/0		3.60	4.444		268	237		505	650
5200	250 kcmil	3 Elec	4.80	5		325	267		592	755
5220	350 kcmil		4.50	5.333		380	285		665	845
5240	500 kcmil		3.90	6.154		490	330		820	1,025
5260	750 kcmil		3.60	6.667		650	355		1,005	1,250
5300	15 kV aluminum, XLP, #1	2 Elec	4.40	3.636		216	194		410	530
5320	1/0		4	4		223	213		436	565
5340	2/0		3.80	4.211		266	225		491	630
5360	4/0		3.60	4.444		295	237		532	680
5380	250 kcmil	3 Elec	4.80	5		355	267		622	790
5400	350 kcmil		4.50	5.333		395	285		680	860
5420	500 kcmil		3.90	6.154		545	330		875	1,100
5440	750 kcmil		3.60	6.667		725	355		1,080	1,325

26 05 19 – Low-Voltage Electrical Power Conductors and Cables

26 05 19.55 Non-Metallic Sheathed Cable		Crew	Daily Output	Labor-Hours	Unit	Material	Labor	Equipment	Total	Total Incl O&P
0010	**NON-METALLIC SHEATHED CABLE** 600 volt									
1000	URD - triplex underground distribution cable, alum. 2 #4 + #4 neutral	2 Elec	2.80	5.714	C.L.F.	101	305		406	565
1010	2 #2 + #4 neutral		2.65	6.038		123	320		443	615
1020	2 #2 + #2 neutral		2.55	6.275		125	335		460	640
1030	2 1/0 + #2 neutral		2.40	6.667		155	355		510	700
1040	2 1/0 + 1/0 neutral		2.30	6.957		170	370		540	740
1050	2 2/0 + #1 neutral		2.20	7.273		183	390		573	780
1060	2 2/0 + 2/0 neutral		2.10	7.619		198	405		603	825
1070	2 3/0 + 1/0 neutral		1.95	8.205		207	440		647	885
1080	2 3/0 + 3/0 neutral		1.95	8.205		237	440		677	915
1090	2 4/0 + 2/0 neutral		1.85	8.649		252	460		712	970
1100	2 4/0 + 4/0 neutral		1.85	8.649		280	460		740	1,000
1450	UF underground feeder cable, copper with ground, #14, 2 conductor	1 Elec	4	2		29.50	107		136.50	193
1500	#12, 2 conductor		3.50	2.286		44.50	122		166.50	232
1550	#10, 2 conductor		3	2.667		69.50	142		211.50	290
1600	#14, 3 conductor		3.50	2.286		41	122		163	228
1650	#12, 3 conductor		3	2.667		62	142		204	281
1700	#10, 3 conductor		2.50	3.200		97	171		268	365

26 05 33 – Raceway and Boxes for Electrical Systems

26 05 33.18 Pull Boxes		Crew	Daily Output	Labor-Hours	Unit	Material	Labor	Equipment	Total	Total Incl O&P
						2014 Bare Costs				
0010	**PULL BOXES**									
2100	Pull box, NEMA 3R, type SC, raintight & weatherproof									
2150	6" L x 6" W x 6" D	1 Elec	10	.800	Ea.	18.85	42.50		61.35	84.50
2200	8" L x 6" W x 6" D		8	1		28	53.50		81.50	111
2250	10" L x 6" W x 6" D		7	1.143		37	61		98	132
2300	12" L x 12" W x 6" D		5	1.600		44	85.50		129.50	177
2350	16" L x 16" W x 6" D		4.50	1.778		115	95		210	268
2400	20" L x 20" W x 6" D		4	2		142	107		249	315
2450	24" L x 18" W x 8" D		3	2.667		125	142		267	350
2500	24" L x 24" W x 10" D		2.50	3.200		161	171		332	435
2550	30" L x 24" W x 12" D		2	4		380	213		593	735
2600	36" L x 36" W x 12" D		1.50	5.333		430	285		715	895
2800	Cast iron, pull boxes for surface mounting									
3000	NEMA 4, watertight & dust tight									
3050	6" L x 6" W x 6" D	1 Elec	4	2	Ea.	248	107		355	430
3100	8" L x 6" W x 6" D		3.20	2.500		345	133		478	580
3150	10" L x 6" W x 6" D		2.50	3.200		385	171		556	680
3200	12" L x 12" W x 6" D		2.30	3.478		705	186		891	1,050
3250	16" L x 16" W x 6" D		1.30	6.154		1,125	330		1,455	1,750
3300	20" L x 20" W x 6" D		.80	10		2,100	535		2,635	3,125
3350	24" L x 18" W x 8" D		.70	11.429		2,600	610		3,210	3,800
3400	24" L x 24" W x 10" D		.50	16		4,925	855		5,780	6,700
3450	30" L x 24" W x 12" D		.40	20		9,025	1,075		10,100	11,500
3500	36" L x 36" W x 12" D		.20	40		7,025	2,125		9,150	10,900

26 05 39 – Underfloor Raceways for Electrical Systems

26 05 39.30 Conduit In Concrete Slab

		Crew	Daily Output	Labor-Hours	Unit	Material	Labor	Equipment	Total	Total Incl O&P
0010	**CONDUIT IN CONCRETE SLAB** Including terminations,									
0020	fittings and supports									
3230	PVC, schedule 40, 1/2" diameter	1 Elec	270	.030	L.F.	.64	1.58		2.22	3.07
3250	3/4" diameter		230	.035		.75	1.86		2.61	3.60
3270	1" diameter		200	.040		.94	2.13		3.07	4.23
3300	1-1/4" diameter		170	.047		1.36	2.51		3.87	5.25
3330	1-1/2" diameter		140	.057		1.63	3.05		4.68	6.35
3350	2" diameter		120	.067		2.07	3.56		5.63	7.55
3370	2-1/2" diameter		90	.089		3.53	4.74		8.27	11
3400	3" diameter	2 Elec	160	.100		4.47	5.35		9.82	12.90
3430	3-1/2" diameter		120	.133		5.70	7.10		12.80	16.95
3440	4" diameter		100	.160		6.25	8.55		14.80	19.65
3450	5" diameter		80	.200		9.55	10.65		20.20	26.50
3460	6" diameter		60	.267		13.65	14.25		27.90	36.50
3530	Sweeps, 1" diameter, 30" radius	1 Elec	32	.250	Ea.	37	13.35		50.35	60.50
3550	1-1/4" diameter		24	.333		44	17.80		61.80	74.50
3570	1-1/2" diameter		21	.381		46	20.50		66.50	81
3600	2" diameter		18	.444		50	23.50		73.50	90.50
3630	2-1/2" diameter		14	.571		66.50	30.50		97	119
3650	3" diameter		10	.800		84.50	42.50		127	157
3670	3-1/2" diameter		8	1		103	53.50		156.50	194
3700	4" diameter		7	1.143		141	61		202	248
3710	5" diameter		6	1.333		165	71		236	287
3730	Couplings, 1/2" diameter					.25			.25	.28
3750	3/4" diameter					.28			.28	.31
3770	1" diameter					.42			.42	.46

26 05 Common Work Results for Electrical

26 05 39 – Underfloor Raceways for Electrical Systems

26 05 39.30 Conduit In Concrete Slab		Crew	Daily Output	Labor-Hours	Unit	Material	Labor	Equipment	Total	Total Incl O&P
						2014 Bare Costs				
3800	1-1/4" diameter				Ea.	.65			.65	.72
3830	1-1/2" diameter					.80			.80	.88
3850	2" diameter					1.04			1.04	1.14
3870	2-1/2" diameter					1.85			1.85	2.04
3900	3" diameter					2.99			2.99	3.29
3930	3-1/2" diameter					3.77			3.77	4.15
3950	4" diameter					4.21			4.21	4.63
3960	5" diameter					10.70			10.70	11.75
3970	6" diameter					14.55			14.55	16
4030	End bells 1" diameter, PVC	1 Elec	60	.133		4.59	7.10		11.69	15.70
4050	1-1/4" diameter		53	.151		5.65	8.05		13.70	18.30
4100	1-1/2" diameter		48	.167		5.65	8.90		14.55	19.50
4150	2" diameter		34	.235		8.40	12.55		20.95	28
4170	2-1/2" diameter		27	.296		9.30	15.80		25.10	34
4200	3" diameter		20	.400		9.80	21.50		31.30	43
4250	3-1/2" diameter		16	.500		11.65	26.50		38.15	53
4300	4" diameter		14	.571		12.70	30.50		43.20	59.50
4310	5" diameter		12	.667		13.40	35.50		48.90	68
4320	6" diameter		9	.889		15.90	47.50		63.40	88.50
4350	Rigid galvanized steel, 1/2" diameter		200	.040	L.F.	2.31	2.13		4.44	5.75
4400	3/4" diameter		170	.047		2.44	2.51		4.95	6.45
4450	1" diameter		130	.062		3.36	3.28		6.64	8.60
4500	1-1/4" diameter		110	.073		4.59	3.88		8.47	10.85
4600	1-1/2" diameter		100	.080		5.15	4.27		9.42	12.05
4800	2" diameter		90	.089		6.40	4.74		11.14	14.15

26 05 39.40 Conduit In Trench		Crew	Daily Output	Labor-Hours	Unit	Material	Labor	Equipment	Total	Total Incl O&P
0010	**CONDUIT IN TRENCH** Includes terminations and fittings									
0020	Does not include excavation or backfill, see Section 31 23 16.00									
0200	Rigid galvanized steel, 2" diameter	1 Elec	150	.053	L.F.	6	2.85		8.85	10.85
0400	2-1/2" diameter	"	100	.080		11.75	4.27		16.02	19.35
0600	3" diameter	2 Elec	160	.100		13.90	5.35		19.25	23.50
0800	3-1/2" diameter		140	.114		18.35	6.10		24.45	29
1000	4" diameter		100	.160		20.50	8.55		29.05	35.50
1200	5" diameter		80	.200		42.50	10.65		53.15	63
1400	6" diameter		60	.267		62.50	14.25		76.75	90.50

26 12 Medium-Voltage Transformers

26 12 19 – Pad-Mounted, Liquid-Filled, Medium-Voltage Transformers

26 12 19.10 Transformer, Oil-Filled		Crew	Daily Output	Labor-Hours	Unit	Material	Labor	Equipment	Total	Total Incl O&P
0010	**TRANSFORMER, OIL-FILLED** primary delta or Y,									
0050	Pad mounted 5 kV or 15 kV, with taps, 277/480 V secondary, 3 phase									
0100	150 kVA	R-3	.65	30.769	Ea.	8,875	1,625	215	10,715	12,500
0110	225 kVA		.55	36.364		10,100	1,925	255	12,280	14,300
0200	300 kVA		.45	44.444		12,600	2,350	310	15,260	17,800
0300	500 kVA		.40	50		18,000	2,650	350	21,000	24,200
0400	750 kVA		.38	52.632		22,700	2,775	370	25,845	29,600
0500	1000 kVA		.26	76.923		27,000	4,075	540	31,615	36,400

26 12 Medium-Voltage Transformers

26 12 19 – Pad-Mounted, Liquid-Filled, Medium-Voltage Transformers

26 12 19.20 Transformer, Liquid-Filled		Crew	Daily Output	Labor-Hours	Unit	2014 Bare Costs Material	Labor	Equipment	Total	Total Incl O&P
0010	**TRANSFORMER, LIQUID-FILLED** Pad mounted									
0020	5 kV or 15 kV primary, 277/480 volt secondary, 3 phase									
0050	225 kVA	R-3	.55	36.364	Ea.	13,500	1,925	255	15,680	18,000
0100	300 kVA		.45	44.444		16,000	2,350	310	18,660	21,500
0200	500 kVA		.40	50		20,200	2,650	350	23,200	26,600
0250	750 kVA		.38	52.632		26,100	2,775	370	29,245	33,300
0300	1000 kVA		.26	76.923		30,300	4,075	540	34,915	40,100

26 13 Medium-Voltage Switchgear

26 13 16 – Medium-Voltage Fusible Interrupter Switchgear

26 13 16.10 Switchgear

		Crew	Daily Output	Labor-Hours	Unit	Material	Labor	Equipment	Total	Total Incl O&P
0010	**SWITCHGEAR**, Incorporate switch with cable connections, transformer,									
0100	& Low Voltage section									
0200	Load interrupter switch, 600 amp, 2 position									
0300	NEMA 1, 4.8 kV, 300 kVA & below w/CLF fuses	R-3	.40	50	Ea.	21,200	2,650	350	24,200	27,800
0400	400 kVA & above w/CLF fuses		.38	52.632		23,700	2,775	370	26,845	30,700
0500	Non fusible		.41	48.780		17,200	2,575	340	20,115	23,200
0600	13.8 kV, 300 kVA & below w/CLF fuses		.38	52.632		26,700	2,775	370	29,845	33,900
0700	400 kVA & above w/CLF fuses		.36	55.556		26,700	2,950	390	30,040	34,100
0800	Non fusible		.40	50		20,200	2,650	350	23,200	26,600
0900	Cable lugs for 2 feeders 4.8 kV or 13.8 kV	1 Elec	8	1		645	53.50		698.50	790
1000	Pothead, one 3 conductor or three 1 conductor		4	2		3,100	107		3,207	3,575
1100	Two 3 conductor or six 1 conductor		2	4		6,125	213		6,338	7,075
1200	Key interlocks		8	1		715	53.50		768.50	870
1300	Lightning arresters, Distribution class (no charge)									
1400	Intermediate class or line type 4.8 kV	1 Elec	2.70	2.963	Ea.	3,475	158		3,633	4,050
1500	13.8 kV		2	4		4,625	213		4,838	5,400
1600	Station class, 4.8 kV		2.70	2.963		5,975	158		6,133	6,800
1700	13.8 kV		2	4		10,300	213		10,513	11,600
1800	Transformers, 4800 volts to 480/277 volts, 75 kVA	R-3	.68	29.412		18,200	1,550	206	19,956	22,600
1900	112.5 kVA		.65	30.769		22,200	1,625	215	24,040	27,100
2000	150 kVA		.57	35.088		25,200	1,850	246	27,296	30,800
2100	225 kVA		.48	41.667		29,000	2,200	292	31,492	35,500
2200	300 kVA		.41	48.780		32,400	2,575	340	35,315	39,900
2300	500 kVA		.36	55.556		42,600	2,950	390	45,940	51,500
2400	750 kVA		.29	68.966		48,400	3,650	485	52,535	59,000
2500	13,800 volts to 480/277 volts, 75 kVA		.61	32.787		25,600	1,725	230	27,555	31,000
2600	112.5 kVA		.55	36.364		33,900	1,925	255	36,080	40,500
2700	150 kVA		.49	40.816		34,300	2,150	286	36,736	41,300
2800	225 kVA		.41	48.780		39,600	2,575	340	42,515	47,800
2900	300 kVA		.37	54.054		40,400	2,850	380	43,630	49,100
3000	500 kVA		.31	64.516		44,700	3,425	450	48,575	55,000
3100	750 kVA		.26	76.923		49,200	4,075	540	53,815	60,500
3200	Forced air cooling & temperature alarm	1 Elec	1	8		3,950	425		4,375	5,000
3300	Low voltage components									
3400	Maximum panel height 49-1/2", single or twin row									
3500	Breaker heights, type FA or FH, 6"									
3600	type KA or KH, 8"									
3700	type LA, 11"									
3800	type MA, 14"									

26 13 Medium-Voltage Switchgear

26 13 16 – Medium-Voltage Fusible Interrupter Switchgear

26 13 16.10 Switchgear		Crew	Daily Output	Labor-Hours	Unit	Material	2014 Bare Costs Labor	Equipment	Total	Total Incl O&P
3900	Breakers, 2 pole, 15 to 60 amp, type FA	1 Elec	5.60	1.429	Ea.	340	76		416	490
4000	70 to 100 amp, type FA		4.20	1.905		430	102		532	620
4100	15 to 60 amp, type FH		5.60	1.429		560	76		636	730
4200	70 to 100 amp, type FH		4.20	1.905		650	102		752	870
4300	125 to 225 amp, type KA		3.40	2.353		1,000	126		1,126	1,300
4400	125 to 225 amp, type KH		3.40	2.353		2,525	126		2,651	3,000
4500	125 to 400 amp, type LA		2.50	3.200		2,275	171		2,446	2,775
4600	125 to 600 amp, type MA		1.80	4.444		2,975	237		3,212	3,625
4700	700 & 800 amp, type MA		1.50	5.333		3,850	285		4,135	4,675
4800	3 pole, 15 to 60 amp, type FA		5.30	1.509		435	80.50		515.50	600
4900	70 to 100 amp, type FA		4	2		540	107		647	750
5000	15 to 60 amp, type FH		5.30	1.509		655	80.50		735.50	840
5100	70 to 100 amp, type FH		4	2		745	107		852	975
5200	125 to 225 amp, type KA		3.20	2.500		1,250	133		1,383	1,575
5300	125 to 225 amp, type KH		3.20	2.500		2,825	133		2,958	3,300
5400	125 to 400 amp, type LA		2.30	3.478		2,825	186		3,011	3,375
5500	125 to 600 amp, type MA		1.60	5		4,075	267		4,342	4,900
5600	700 & 800 amp, type MA	↓	1.30	6.154	↓	5,300	330		5,630	6,325

26 24 Switchboards and Panelboards

26 24 16 – Panelboards

26 24 16.30 Panelboards Commercial Applications

		Crew	Daily Output	Labor-Hours	Unit	Material	Labor	Equipment	Total	Total Incl O&P
0010	**PANELBOARDS COMMERCIAL APPLICATIONS**									
0050	NQOD, w/20 amp 1 pole bolt-on circuit breakers									
0100	3 wire, 120/240 volts, 100 amp main lugs									
0150	10 circuits	1 Elec	1	8	Ea.	535	425		960	1,225
0200	14 circuits		.88	9.091		655	485		1,140	1,450
0250	18 circuits		.75	10.667		710	570		1,280	1,625
0300	20 circuits	↓	.65	12.308		795	655		1,450	1,850
0350	225 amp main lugs, 24 circuits	2 Elec	1.20	13.333		900	710		1,610	2,075
0400	30 circuits		.90	17.778		1,050	950		2,000	2,575
0450	36 circuits		.80	20		1,175	1,075		2,250	2,900
0500	38 circuits		.72	22.222		1,275	1,175		2,450	3,175
0550	42 circuits	↓	.66	24.242		1,325	1,300		2,625	3,375
0600	4 wire, 120/208 volts, 100 amp main lugs, 12 circuits	1 Elec	1	8		630	425		1,055	1,325
0650	16 circuits		.75	10.667		725	570		1,295	1,650
0700	20 circuits		.65	12.308		835	655		1,490	1,900
0750	24 circuits		.60	13.333		905	710		1,615	2,075
0800	30 circuits	↓	.53	15.094		1,025	805		1,830	2,350
0850	225 amp main lugs, 32 circuits	2 Elec	.90	17.778		1,150	950		2,100	2,700
0900	34 circuits		.84	19.048		1,200	1,025		2,225	2,825
0950	36 circuits		.80	20		1,225	1,075		2,300	2,950
1000	42 circuits	↓	.68	23.529	↓	1,350	1,250		2,600	3,375

26 27 13 – Electricity Metering

26 27 13.10 Meter Centers and Sockets		Crew	Daily Output	Labor-Hours	Unit	2014 Bare Costs Material	Labor	Equipment	Total	Total Incl O&P
0010	**METER CENTERS AND SOCKETS**									
0100	Sockets, single position, 4 terminal, 100 amp	1 Elec	3.20	2.500	Ea.	42	133		175	246
0200	150 amp		2.30	3.478		54.50	186		240.50	340
0300	200 amp		1.90	4.211		89.50	225		314.50	435
2000	Meter center, main fusible switch, 1P 3W 120/240 volt									
2030	400 amp	2 Elec	1.60	10	Ea.	1,600	535		2,135	2,550
2040	600 amp		1.10	14.545		2,775	775		3,550	4,225
2050	800 amp		.90	17.778		4,350	950		5,300	6,225
2060	Rainproof 1P 3W 120/240 volt, 400 amp		1.60	10		1,600	535		2,135	2,550
2070	600 amp		1.10	14.545		2,775	775		3,550	4,225
2080	800 amp		.90	17.778		4,350	950		5,300	6,225
2100	3P 4W 120/208 V, 400 amp		1.60	10		1,825	535		2,360	2,825
2110	600 amp		1.10	14.545		3,425	775		4,200	4,900
2120	800 amp		.90	17.778		6,350	950		7,300	8,400
2130	Rainproof 3P 4W 120/208 V, 400 amp		1.60	10		1,825	535		2,360	2,825
2140	600 amp		1.10	14.545		3,425	775		4,200	4,900
2150	800 amp		.90	17.778		6,350	950		7,300	8,400
2170	Main circuit breaker, 1P 3W 120/240 V									
2180	400 amp	2 Elec	1.60	10	Ea.	2,875	535		3,410	3,950
2190	600 amp		1.10	14.545		3,850	775		4,625	5,400
2200	800 amp		.90	17.778		4,500	950		5,450	6,375
2210	1000 amp		.80	20		6,200	1,075		7,275	8,425
2220	1200 amp		.76	21.053		8,350	1,125		9,475	10,900
2230	1600 amp		.68	23.529		18,700	1,250		19,950	22,500
2240	Rainproof 1P 3W 120/240 V, 400 amp		1.60	10		2,875	535		3,410	3,950
2250	600 amp		1.10	14.545		3,850	775		4,625	5,400
2260	800 amp		.90	17.778		4,500	950		5,450	6,375
2270	1000 amp		.80	20		6,200	1,075		7,275	8,425
2280	1200 amp		.76	21.053		8,350	1,125		9,475	10,900
2300	3P 4W 120/208 V, 400 amp		1.60	10		3,250	535		3,785	4,375
2310	600 amp		1.10	14.545		4,600	775		5,375	6,200
2320	800 amp		.90	17.778		5,450	950		6,400	7,425
2330	1000 amp		.80	20		7,175	1,075		8,250	9,475
2340	1200 amp		.76	21.053		9,175	1,125		10,300	11,800
2350	1600 amp		.68	23.529		18,700	1,250		19,950	22,500
2360	Rainproof 3P 4W 120/208 V, 400 amp		1.60	10		3,250	535		3,785	4,375
2370	600 amp		1.10	14.545		4,600	775		5,375	6,200
2380	800 amp		.90	17.778		5,450	950		6,400	7,425
2390	1000 amp		.76	21.053		7,175	1,125		8,300	9,550
2400	1200 amp		.68	23.529		9,175	1,250		10,425	12,000

26 28 Low-Voltage Circuit Protective Devices

26 28 16 – Enclosed Switches and Circuit Breakers

26 28 16.20 Safety Switches		Crew	Daily Output	Labor-Hours	Unit	2014 Bare Costs Material	Labor	Equipment	Total	Total Incl O&P
0010	**SAFETY SWITCHES**									
0100	General duty 240 volt, 3 pole NEMA 1, fusible, 30 amp	1 Elec	3.20	2.500	Ea.	86	133		219	295
0200	60 amp		2.30	3.478		146	186		332	440
0300	100 amp		1.90	4.211		250	225		475	610
0400	200 amp		1.30	6.154		535	330		865	1,075
0500	400 amp	2 Elec	1.80	8.889		1,350	475		1,825	2,200
0600	600 amp	"	1.20	13.333		2,550	710		3,260	3,875
5510	Heavy duty, 600 volt, 3 pole 3ph. NEMA 3R fusible, 30 amp	1 Elec	3.10	2.581		395	138		533	640
5520	60 amp		2.20	3.636		465	194		659	800
5530	100 amp		1.80	4.444		725	237		962	1,150
5540	200 amp		1.20	6.667		1,000	355		1,355	1,625
5550	400 amp	2 Elec	1.60	10		2,325	535		2,860	3,375

26 32 Packaged Generator Assemblies

26 32 13 – Engine Generators

26 32 13.13 Diesel-Engine-Driven Generator Sets

		Crew	Daily Output	Labor-Hours	Unit	Material	Labor	Equipment	Total	Total Incl O&P
0010	**DIESEL-ENGINE-DRIVEN GENERATOR SETS**									
2000	Diesel engine, including battery, charger,									
2010	muffler, & day tank, 30 kW	R-3	.55	36.364	Ea.	16,400	1,925	255	18,580	21,200
2100	50 kW		.42	47.619		19,400	2,525	335	22,260	25,500
2200	75 kW		.35	57.143		25,100	3,025	400	28,525	32,700
2300	100 kW		.31	64.516		28,000	3,425	450	31,875	36,400
2400	125 kW		.29	68.966		29,500	3,650	485	33,635	38,400
2500	150 kW		.26	76.923		33,900	4,075	540	38,515	44,000
2600	175 kW		.25	80		36,900	4,225	560	41,685	47,600
2700	200 kW		.24	83.333		38,200	4,400	585	43,185	49,300
2800	250 kW		.23	86.957		44,800	4,600	610	50,010	57,000
2900	300 kW		.22	90.909		48,600	4,800	635	54,035	61,500
3000	350 kW		.20	100		55,000	5,300	700	61,000	69,000
3100	400 kW		.19	105		68,000	5,575	735	74,310	83,500
3200	500 kW		.18	111		85,500	5,875	780	92,155	103,500
3220	600 kW		.17	117		111,500	6,225	825	118,550	133,500
3240	750 kW	R-13	.38	110		138,500	5,675	500	144,675	161,500

26 32 13.16 Gas-Engine-Driven Generator Sets

		Crew	Daily Output	Labor-Hours	Unit	Material	Labor	Equipment	Total	Total Incl O&P
0010	**GAS-ENGINE-DRIVEN GENERATOR SETS**									
0020	Gas or gasoline operated, includes battery,									
0050	charger, & muffler									
0200	3 phase 4 wire, 277/480 volt, 7.5 kW	R-3	.83	24.096	Ea.	7,275	1,275	169	8,719	10,100
0300	11.5 kW		.71	28.169		10,300	1,500	197	11,997	13,700
0400	20 kW		.63	31.746		12,200	1,675	222	14,097	16,200
0500	35 kW		.55	36.364		14,500	1,925	255	16,680	19,100
0520	60 kW		.50	40		19,100	2,125	280	21,505	24,500
0600	80 kW		.40	50		23,700	2,650	350	26,700	30,500
0700	100 kW		.33	60.606		26,000	3,200	425	29,625	33,900
0800	125 kW		.28	71.429		53,000	3,775	500	57,275	64,500
0900	185 kW		.25	80		70,500	4,225	560	75,285	84,500

26 51 Interior Lighting

26 51 13 – Interior Lighting Fixtures, Lamps, and Ballasts

	26 51 13.90 Ballast, Replacement HID	Crew	Daily Output	Labor-Hours	Unit	2014 Bare Costs Material	Labor	Equipment	Total	Total Incl O&P
0010	**BALLAST, REPLACEMENT HID**									
7510	Multi-tap 120/208/240/277 volt									
7550	High pressure sodium, 70 watt	1 Elec	10	.800	Ea.	160	42.50		202.50	240
7560	100 watt		9.40	.851		167	45.50		212.50	251
7570	150 watt		9	.889		180	47.50		227.50	269
7580	250 watt		8.50	.941		268	50		318	370
7590	400 watt		7	1.143		305	61		366	425
7600	1000 watt		6	1.333		420	71		491	565
7610	Metal halide, 175 watt		8	1		98.50	53.50		152	188
7620	250 watt		8	1		127	53.50		180.50	220
7630	400 watt		7	1.143		159	61		220	267
7640	1000 watt		6	1.333		272	71		343	405
7650	1500 watt	↓	5	1.600	↓	335	85.50		420.50	500

26 56 Exterior Lighting

26 56 13 – Lighting Poles and Standards

26 56 13.10 Lighting Poles

	26 56 13.10 Lighting Poles	Crew	Daily Output	Labor-Hours	Unit	Material	Labor	Equipment	Total	Total Incl O&P
0010	**LIGHTING POLES**									
2800	Light poles, anchor base									
2820	not including concrete bases									
2840	Aluminum pole, 8' high	1 Elec	4	2	Ea.	695	107		802	925
2850	10' high		4	2		735	107		842	970
2860	12' high		3.80	2.105		765	112		877	1,025
2870	14' high		3.40	2.353		795	126		921	1,075
2880	16' high	↓	3	2.667		875	142		1,017	1,175
3000	20' high	R-3	2.90	6.897		925	365	48.50	1,338.50	1,625
3200	30' high		2.60	7.692		1,750	405	54	2,209	2,600
3400	35' high		2.30	8.696		1,900	460	61	2,421	2,850
3600	40' high	↓	2	10		2,175	530	70	2,775	3,275
3800	Bracket arms, 1 arm	1 Elec	8	1		119	53.50		172.50	211
4000	2 arms		8	1		240	53.50		293.50	345
4200	3 arms		5.30	1.509		360	80.50		440.50	515
4400	4 arms		5.30	1.509		480	80.50		560.50	650
4500	Steel pole, galvanized, 8' high		3.80	2.105		600	112		712	835
4510	10' high		3.70	2.162		630	115		745	870
4520	12' high		3.40	2.353		680	126		806	940
4530	14' high		3.10	2.581		725	138		863	1,000
4540	16' high		2.90	2.759		765	147		912	1,075
4550	18' high	↓	2.70	2.963		810	158		968	1,125
4600	20' high	R-3	2.60	7.692		1,100	405	54	1,559	1,875
4800	30' high		2.30	8.696		1,300	460	61	1,821	2,175
5000	35' high		2.20	9.091		1,425	480	63.50	1,968.50	2,350
5200	40' high	↓	1.70	11.765		1,750	625	82.50	2,457.50	2,950
5400	Bracket arms, 1 arm	1 Elec	8	1		179	53.50		232.50	276
5600	2 arms		8	1		277	53.50		330.50	385
5800	3 arms		5.30	1.509		300	80.50		380.50	450
6000	4 arms	↓	5.30	1.509		415	80.50		495.50	580
6100	Fiberglass pole, 1 or 2 fixtures, 20' high	R-3	4	5		665	265	35	965	1,175
6200	30' high		3.60	5.556		825	294	39	1,158	1,400
6300	35' high		3.20	6.250		1,300	330	44	1,674	1,975
6400	40' high	↓	2.80	7.143	↓	1,500	380	50	1,930	2,275

26 56 Exterior Lighting

26 56 13 – Lighting Poles and Standards

26 56 13.10 Lighting Poles			Crew	Daily Output	Labor-Hours	Unit	Material	Labor	Equipment	Total	Total Incl O&P
							2014 Bare Costs				
6420	Wood pole, 4-1/2" x 5-1/8", 8' high		1 Elec	6	1.333	Ea.	325	71		396	465
6430	10' high			6	1.333		375	71		446	515
6440	12' high			5.70	1.404		470	75		545	630
6450	15' high			5	1.600		550	85.50		635.50	735
6460	20' high			4	2		665	107		772	890
7300	Transformer bases, not including concrete bases										
7320	Maximum pole size, steel, 40' high		1 Elec	2	4	Ea.	1,400	213		1,613	1,875
7340	Cast aluminum, 30' high			3	2.667		740	142		882	1,025
7350	40' high			2.50	3.200		1,125	171		1,296	1,500

26 56 16 – Parking Lighting

26 56 16.55 Parking LED Lighting			Crew	Daily Output	Labor-Hours	Unit	Material	Labor	Equipment	Total	Total Incl O&P
0010	**PARKING LED LIGHTING**										
0100	Round pole mounting, 88 lamp watts	G	1 Elec	2	4	Ea.	1,975	213		2,188	2,500
0110	Square pole mounting, 223 lamp watts	G	"	2	4	"	3,475	213		3,688	4,150

26 56 19 – Roadway Lighting

26 56 19.20 Roadway Luminaire			Crew	Daily Output	Labor-Hours	Unit	Material	Labor	Equipment	Total	Total Incl O&P
0010	**ROADWAY LUMINAIRE**										
2650	Roadway area luminaire, low pressure sodium, 135 watt		1 Elec	2	4	Ea.	645	213		858	1,025
2700	180 watt		"	2	4		690	213		903	1,075
2750	Metal halide, 400 watt		2 Elec	4.40	3.636		550	194		744	895
2760	1000 watt			4	4		620	213		833	1,000
2780	High pressure sodium, 400 watt			4.40	3.636		575	194		769	920
2790	1000 watt			4	4		655	213		868	1,050

26 56 19.55 Roadway LED Luminaire			Crew	Daily Output	Labor-Hours	Unit	Material	Labor	Equipment	Total	Total Incl O&P
0010	**ROADWAY LED LUMINAIRE**										
0100	LED fixture, 72 LEDs, 120 V AC or 12 V DC, equal to 60 watt	G	1 Elec	2.70	2.963	Ea.	585	158		743	880
0110	108 LEDs, 120 V AC or 12 V DC, equal to 90 watt	G		2.70	2.963		690	158		848	995
0120	144 LEDs, 120 V AC or 12 V DC, equal to 120 watt	G		2.70	2.963		845	158		1,003	1,175
0130	252 LEDs, 120 V AC or 12 V DC, equal to 210 watt	G	2 Elec	4.40	3.636		1,175	194		1,369	1,600
0140	Replaces high pressure sodium fixture, 75 watt	G	1 Elec	2.70	2.963		715	158		873	1,025
0150	125 watt	G		2.70	2.963		815	158		973	1,125
0160	150 watt	G		2.70	2.963		1,025	158		1,183	1,350
0170	175 watt	G		2.70	2.963		1,225	158		1,383	1,575
0180	200 watt	G		2.70	2.963		1,425	158		1,583	1,800
0190	250 watt	G	2 Elec	4.40	3.636		1,625	194		1,819	2,100
0200	320 watt	G	"	4.40	3.636		1,775	194		1,969	2,275

26 56 23 – Area Lighting

26 56 23.10 Exterior Fixtures			Crew	Daily Output	Labor-Hours	Unit	Material	Labor	Equipment	Total	Total Incl O&P
0010	**EXTERIOR FIXTURES** With lamps										
0200	Wall mounted, incandescent, 100 watt		1 Elec	8	1	Ea.	34.50	53.50		88	118
0400	Quartz, 500 watt			5.30	1.509		49.50	80.50		130	176
0420	1500 watt			4.20	1.905		102	102		204	264
1100	Wall pack, low pressure sodium, 35 watt			4	2		225	107		332	410
1150	55 watt			4	2		268	107		375	455
1160	High pressure sodium, 70 watt			4	2		221	107		328	405
1170	150 watt			4	2		242	107		349	425
1180	Metal Halide, 175 watt			4	2		245	107		352	430
1190	250 watt			4	2		279	107		386	465
1195	400 watt			4	2		330	107		437	525
1250	Induction lamp, 40 watt			4	2		340	107		447	535
1260	80 watt			4	2		575	107		682	795

26 56 Exterior Lighting

26 56 23 – Area Lighting

26 56 23.10 Exterior Fixtures		Crew	Daily Output	Labor-Hours	Unit	2014 Bare Costs Material	Labor	Equipment	Total	Total Incl O&P
1278	LED, poly lens, 26 watt	1 Elec	4	2	Ea.	325	107		432	520
1280	110 watt		4	2		1,300	107		1,407	1,575
1500	LED, glass lens, 13 watt	↓	4	2	↓	325	107		432	520

26 56 26 – Landscape Lighting

26 56 26.20 Landscape Fixtures		Crew	Daily Output	Labor-Hours	Unit	Material	Labor	Equipment	Total	Total Incl O&P
0010	**LANDSCAPE FIXTURES**									
0012	Incl. conduit, wire, trench									
0030	Bollards									
0040	Incandescent, 24″	1 Elec	2.50	3.200	Ea.	370	171		541	660
0050	36″		2	4		480	213		693	845
0060	42″		2	4		510	213		723	880
0070	H.I.D., 24″		2.50	3.200		430	171		601	725
0080	36″		2	4		480	213		693	845
0090	42″		2	4		620	213		833	1,000
0100	Concrete, 18″ diam.		1.20	6.667		1,425	355		1,780	2,075
0110	24″ diam.	↓	.75	10.667	↓	1,775	570		2,345	2,800
0120	Dry niche									
0130	300 W 120 Volt	1 Elec	4	2	Ea.	935	107		1,042	1,175
0140	1000 W 120 Volt		2	4		1,250	213		1,463	1,700
0150	300 W 12 Volt	↓	4	2	↓	935	107		1,042	1,175
0160	Low voltage									
0170	Recessed uplight	1 Elec	2.20	3.636	Ea.	355	194		549	680
0180	Walkway		4	2		325	107		432	515
0190	Malibu - 5 light set		3	2.667		235	142		377	470
0200	Mushroom 24″ pier	↓	4	2	↓	241	107		348	425
0210	Recessed, adjustable									
0220	Incandescent, 150 W	1 Elec	2.50	3.200	Ea.	550	171		721	860
0230	300 W	″	2.50	3.200	″	635	171		806	950
0250	Recessed uplight									
0260	Incandescent, 50 W	1 Elec	2.50	3.200	Ea.	450	171		621	750
0270	150 W		2.50	3.200		465	171		636	770
0280	300 W		2.50	3.200		520	171		691	830
0310	Quartz 500 W	↓	2.50	3.200	↓	515	171		686	820
0400	Recessed wall light									
0410	Incandescent 100 W	1 Elec	4	2	Ea.	225	107		332	410
0420	Fluorescent		4	2		204	107		311	385
0430	H.I.D. 100 W	↓	3	2.667	↓	420	142		562	675
0500	Step lights									
0510	Incandescent	1 Elec	5	1.600	Ea.	201	85.50		286.50	350
0520	Fluorescent	″	5	1.600	″	213	85.50		298.50	360
0600	Tree lights, surface adjustable									
0610	Incandescent 50 W	1 Elec	3	2.667	Ea.	475	142		617	740
0620	Incandescent 100 W		3	2.667		495	142		637	760
0630	Incandescent 150 W	↓	2	4	↓	525	213		738	900
0700	Underwater lights									
0710	150 W 120 Volt	1 Elec	6	1.333	Ea.	845	71		916	1,025
0720	300 W 120 Volt		6	1.333		1,000	71		1,071	1,225
0730	1000 W 120 Volt		4	2		1,275	107		1,382	1,550
0740	50 W 12 Volt		6	1.333		960	71		1,031	1,150
0750	300 W 12 Volt	↓	6	1.333	↓	865	71		936	1,050
0800	Walkway, adjustable									
0810	Fluorescent, 2′	1 Elec	3	2.667	Ea.	375	142		517	625

26 56 Exterior Lighting

26 56 26 – Landscape Lighting

26 56 26.20 Landscape Fixtures		Crew	Daily Output	Labor-Hours	Unit	2014 Bare Costs Material	Labor	Equipment	Total	Total Incl O&P
0820	Fluorescent, 4′	1 Elec	3	2.667	Ea.	400	142		542	655
0830	Fluorescent, 8′		2	4		775	213		988	1,175
0840	Incandescent, 50 W		4	2		355	107		462	550
0850	150 W		4	2		380	107		487	575
0900	Wet niche									
0910	300 W 120 Volt	1 Elec	2.50	3.200	Ea.	770	171		941	1,100
0920	1000 W 120 Volt		1.50	5.333		930	285		1,215	1,450
0930	300 W 12 Volt		2.50	3.200		885	171		1,056	1,225
7380	Landscape recessed uplight, incl. housing, ballast, transformer									
7390	& reflector, not incl. conduit, wire, trench									
7420	Incandescent, 250 watt	1 Elec	5	1.600	Ea.	585	85.50		670.50	775
7440	Quartz, 250 watt		5	1.600		555	85.50		640.50	740
7460	500 watt		4	2		570	107		677	790

26 56 33 – Walkway Lighting

26 56 33.10 Walkway Luminaire

		Crew	Daily Output	Labor-Hours	Unit	Material	Labor	Equipment	Total	Total Incl O&P
0010	**WALKWAY LUMINAIRE**									
6500	Bollard light, lamp & ballast, 42″ high with polycarbonate lens									
6800	Metal halide, 175 watt	1 Elec	3	2.667	Ea.	805	142		947	1,100
6900	High pressure sodium, 70 watt		3	2.667		825	142		967	1,125
7000	100 watt		3	2.667		825	142		967	1,125
7100	150 watt		3	2.667		805	142		947	1,100
7200	Incandescent, 150 watt		3	2.667		590	142		732	860
7810	Walkway luminaire, square 16″, metal halide 250 watt		2.70	2.963		630	158		788	930
7820	High pressure sodium, 70 watt		3	2.667		720	142		862	1,000
7830	100 watt		3	2.667		735	142		877	1,025
7840	150 watt		3	2.667		735	142		877	1,025
7850	200 watt		3	2.667		740	142		882	1,025
7910	Round 19″, metal halide, 250 watt		2.70	2.963		920	158		1,078	1,225
7920	High pressure sodium, 70 watt		3	2.667		1,000	142		1,142	1,325
7930	100 watt		3	2.667		1,000	142		1,142	1,325
7940	150 watt		3	2.667		1,025	142		1,167	1,350
7950	250 watt		2.70	2.963		1,075	158		1,233	1,400
8000	Sphere 14″ opal, incandescent, 200 watt		4	2		292	107		399	480
8020	Sphere 18″ opal, incandescent, 300 watt		3.50	2.286		355	122		477	575
8040	Sphere 16″ clear, high pressure sodium, 70 watt		3	2.667		610	142		752	885
8050	100 watt		3	2.667		650	142		792	930
8100	Cube 16″ opal, incandescent, 300 watt		3.50	2.286		390	122		512	610
8120	High pressure sodium, 70 watt		3	2.667		565	142		707	840
8130	100 watt		3	2.667		580	142		722	855
8230	Lantern, high pressure sodium, 70 watt		3	2.667		505	142		647	775
8240	100 watt		3	2.667		545	142		687	815
8250	150 watt		3	2.667		510	142		652	775
8260	250 watt		2.70	2.963		715	158		873	1,025
8270	Incandescent, 300 watt		3.50	2.286		375	122		497	600
8330	Reflector 22″ w/globe, high pressure sodium, 70 watt		3	2.667		485	142		627	750
8340	100 watt		3	2.667		495	142		637	760
8350	150 watt		3	2.667		500	142		642	765
8360	250 watt		2.70	2.963		640	158		798	940

26 56 33.55 Walkway LED Luminaire

		Crew	Daily Output	Labor-Hours	Unit	Material	Labor	Equipment	Total	Total Incl O&P
0010	**WALKWAY LED LUMINAIRE**									
0100	Pole mounted, 86 watts, 4350 lumens G	1 Elec	3	2.667	Ea.	1,350	142		1,492	1,725
0110	4630 lumens G		3	2.667		2,275	142		2,417	2,725

26 56 Exterior Lighting

26 56 33 – Walkway Lighting

26 56 33.55 Walkway LED Luminaire		Crew	Daily Output	Labor-Hours	Unit	2014 Bare Costs Material	Labor	Equipment	Total	Total Incl O&P
0120	80 watts, 4000 lumens G	1 Elec	3	2.667	Ea.	2,150	142		2,292	2,575

26 56 36 – Flood Lighting

26 56 36.20 Floodlights

		Crew	Daily Output	Labor-Hours	Unit	Material	Labor	Equipment	Total	Total Incl O&P
0010	**FLOODLIGHTS** with ballast and lamp,									
1290	floor mtd, mount with swivel bracket									
1300	Induction lamp, 40 watt	1 Elec	3	2.667	Ea.	445	142		587	705
1310	80 watt		3	2.667		680	142		822	960
1320	150 watt		3	2.667		1,225	142		1,367	1,575
1400	Pole mounted, pole not included									
1950	Metal halide, 175 watt	1 Elec	2.70	2.963	Ea.	335	158		493	605
2000	400 watt	2 Elec	4.40	3.636		415	194		609	750
2200	1000 watt		4	4		575	213		788	950
2210	1500 watt		3.70	4.324		595	231		826	1,000
2250	Low pressure sodium, 55 watt	1 Elec	2.70	2.963		580	158		738	870
2270	90 watt		2	4		640	213		853	1,025
2290	180 watt		2	4		810	213		1,023	1,225
2340	High pressure sodium, 70 watt		2.70	2.963		243	158		401	505
2360	100 watt		2.70	2.963		250	158		408	510
2380	150 watt		2.70	2.963		287	158		445	550
2400	400 watt	2 Elec	4.40	3.636		375	194		569	705
2600	1000 watt	"	4	4		645	213		858	1,025

26 56 36.55 LED Floodlights

		Crew	Daily Output	Labor-Hours	Unit	Material	Labor	Equipment	Total	Total Incl O&P
0010	**LED FLOODLIGHTS** with ballast and lamp,									
0020	Pole mounted, pole not included									
0100	11 watt G	1 Elec	4	2	Ea.	345	107		452	540
0110	46 watt G		4	2		1,025	107		1,132	1,300
0120	90 watt G		4	2		1,500	107		1,607	1,800
0130	288 watt G		4	2		2,325	107		2,432	2,700

26 61 Lighting Systems and Accessories

26 61 23 – Lamps Applications

26 61 23.10 Lamps

		Crew	Daily Output	Labor-Hours	Unit	Material	Labor	Equipment	Total	Total Incl O&P
0010	**LAMPS**									
0600	Mercury vapor, mogul base, deluxe white, 100 watt	1 Elec	.30	26.667	C	3,650	1,425		5,075	6,150
0650	175 watt		.30	26.667		1,725	1,425		3,150	4,000
0700	250 watt		.30	26.667		3,550	1,425		4,975	6,025
0800	400 watt		.30	26.667		5,100	1,425		6,525	7,725
0900	1000 watt		.20	40		7,350	2,125		9,475	11,300
1000	Metal halide, mogul base, 175 watt		.30	26.667		3,000	1,425		4,425	5,425
1100	250 watt		.30	26.667		3,400	1,425		4,825	5,875
1200	400 watt		.30	26.667		5,600	1,425		7,025	8,275
1300	1000 watt		.20	40		7,800	2,125		9,925	11,800
1320	1000 watt, 125,000 initial lumens		.20	40		22,100	2,125		24,225	27,500
1330	1500 watt		.20	40		20,400	2,125		22,525	25,700
1350	High pressure sodium, 70 watt		.30	26.667		3,650	1,425		5,075	6,150
1360	100 watt		.30	26.667		3,825	1,425		5,250	6,325
1370	150 watt		.30	26.667		3,925	1,425		5,350	6,450
1380	250 watt		.30	26.667		5,800	1,425		7,225	8,525
1400	400 watt		.30	26.667		4,275	1,425		5,700	6,825
1450	1000 watt		.20	40		11,800	2,125		13,925	16,100

26 61 Lighting Systems and Accessories

26 61 23 – Lamps Applications

26 61 23.10 Lamps		Crew	Daily Output	Labor-Hours	Unit	2014 Bare Costs Material	Labor	Equipment	Total	Total Incl O&P
1500	Low pressure sodium, 35 watt	1 Elec	.30	26.667	C	12,000	1,425		13,425	15,300
1550	55 watt		.30	26.667		11,900	1,425		13,325	15,200
1600	90 watt		.30	26.667		13,300	1,425		14,725	16,700
1650	135 watt		.20	40		20,000	2,125		22,125	25,200
1700	180 watt		.20	40		23,100	2,125		25,225	28,600
1750	Quartz line, clear, 500 watt		1.10	7.273		1,375	390		1,765	2,100
1760	1500 watt		.20	40		3,575	2,125		5,700	7,150
1762	Spot, MR 16, 50 watt	▼	1.30	6.154	▼	1,050	330		1,380	1,675

Estimating Tips

27 20 00 Data Communications
27 30 00 Voice Communications
27 40 00 Audio-Video Communications

When estimating material costs for special systems, it is always prudent to obtain manufacturers' quotations for equipment prices and special installation requirements which will affect the total costs.

Reference Numbers

Reference numbers are shown in shaded boxes at the beginning of some major classifications. These numbers refer to related items in the Reference Section. The reference information may be an estimating procedure, an alternate pricing method, or technical information.

Note: Not all subdivisions listed here necessarily appear in this publication.

Note: **Trade Service,** *in part, has been used as a reference source for some of the material prices used in Division 27.*

27 13 Communications Backbone Cabling

27 13 23 – Communications Optical Fiber Backbone Cabling

27 13 23.13 Communications Optical Fiber		Crew	Daily Output	Labor-Hours	Unit	2014 Bare Costs Material	Labor	Equipment	Total	Total Incl O&P
0010	**COMMUNICATIONS OPTICAL FIBER**									
0040	Specialized tools & techniques cause installation costs to vary.									
0070	Fiber optic, cable, bulk simplex, single mode	1 Elec	8	1	C.L.F.	23	53.50		76.50	106
0080	Multi mode		8	1		49.50	53.50		103	135
0090	4 strand, single mode		7.34	1.090		44	58		102	136
0095	Multi mode		7.34	1.090		50.50	58		108.50	143
0100	12 strand, single mode		6.67	1.199		105	64		169	212
0105	Multi mode		6.67	1.199		106	64		170	213
0150	Jumper				Ea.	33.50			33.50	37
0200	Pigtail					33.50			33.50	37
0300	Connector	1 Elec	24	.333		20.50	17.80		38.30	49
0350	Finger splice		32	.250		32.50	13.35		45.85	55.50
0400	Transceiver (low cost bi-directional)		8	1		415	53.50		468.50	540
0450	Rack housing, 4 rack spaces, 12 panels (144 fibers)		2	4		495	213		708	865
0500	Patch panel, 12 ports		6	1.333		245	71		316	375
1000	Cable, 62.5 microns, direct burial, 4 fiber	R-15	1200	.040	L.F.	.63	2.10	.27	3	4.13
1020	Indoor, 2 fiber	R-19	1000	.020		.36	1.07		1.43	2
1040	Outdoor, aerial/duct	"	1670	.012		.72	.64		1.36	1.75
1060	50 microns, direct burial, 8 fiber	R-22	4000	.009		1.36	.42		1.78	2.14
1080	12 fiber		4000	.009		1.52	.42		1.94	2.31
1100	Indoor, 12 fiber		759	.049		1.46	2.23		3.69	4.99
1120	Connectors, 62.5 micron cable, transmission	R-19	40	.500	Ea.	15.40	26.50		41.90	57
1140	Cable splice		40	.500		15.45	26.50		41.95	57
1160	125 micron cable, transmission		16	1.250		15.40	67		82.40	117
1180	Receiver, 1.2 mile range		20	1		355	53.50		408.50	475
1200	1.9 mile range		20	1		425	53.50		478.50	545
1220	6.2 mile range		5	4		685	214		899	1,075
1240	Transmitter, 1.2 mile range		20	1		355	53.50		408.50	475
1260	1.9 mile range		20	1		425	53.50		478.50	545
1280	6.2 mile range		5	4		685	214		899	1,075
1300	Modem, 1.2 mile range		5	4		400	214		614	760
1320	6.2 mile range		5	4		795	214		1,009	1,200
1340	1.9 mile range, 12 channel		5	4		1,825	214		2,039	2,325
1360	Repeater, 1.2 mile range		10	2		530	107		637	745
1380	1.9 mile range		10	2		530	107		637	745
1400	6.2 mile range		5	4		910	214		1,124	1,325
1420	1.2 mile range, digital		5	4		530	214		744	905
1440	Transceiver, 1.9 mile range		5	4		435	214		649	795
1460	1.2 mile range, digital		5	4		355	214		569	715
1480	Cable enclosure, interior NEMA 13		7	2.857		150	153		303	395
1500	Splice w/enclosure encapsulant		16	1.250		230	67		297	355

Estimating Tips

31 05 00 Common Work Results for Earthwork

- Estimating the actual cost of performing earthwork requires careful consideration of the variables involved. This includes items such as type of soil, whether water will be encountered, dewatering, whether banks need bracing, disposal of excavated earth, and length of haul to fill or spoil sites, etc. If the project has large quantities of cut or fill, consider raising or lowering the site to reduce costs, while paying close attention to the effect on site drainage and utilities.
- If the project has large quantities of fill, creating a borrow pit on the site can significantly lower the costs.
- It is very important to consider what time of year the project is scheduled for completion. Bad weather can create large cost overruns from dewatering, site repair, and lost productivity from cold weather.

Reference Numbers

Reference numbers are shown in shaded boxes at the beginning of some major classifications. These numbers refer to related items in the Reference Section. The reference information may be an estimating procedure, an alternate pricing method, or technical information.

Note: Not all subdivisions listed here necessarily appear in this publication.

31 05 Common Work Results for Earthwork

31 05 23 – Cement and Concrete for Earthwork

31 05 23.30 Plant Mixed Bituminous Concrete		Crew	Daily Output	Labor-Hours	Unit	2014 Bare Costs Material	Labor	Equipment	Total	Total Incl O&P
0010	**PLANT MIXED BITUMINOUS CONCRETE**									
0020	Asphaltic concrete plant mix (145 lb. per C.F.)				Ton	68			68	75
0040	Asphaltic concrete less than 300 tons add trucking costs									
0050	See Section 31 23 23.20 for hauling costs									
0200	All weather patching mix, hot				Ton	68			68	75
0250	Cold patch					78			78	86
0300	Berm mix					68			68	75
0400	Base mix					68			68	75
0500	Binder mix					68			68	75
0600	Sand or sheet mix				↓	67			67	73.50

31 05 23.40 Recycled Plant Mixed Bituminous Concrete		Crew	Daily Output	Labor-Hours	Unit	Material	Labor	Equipment	Total	Total Incl O&P
0010	**RECYCLED PLANT MIXED BITUMINOUS CONCRETE**									
0200	Reclaimed pavement in stockpile G				Ton	52			52	57
0400	Recycled pavement, at plant, ratio old: new, 70:30 G					57			57	62.50
0600	Ratio old: new, 30:70 G				↓	63			63	69.50

31 06 Schedules for Earthwork

31 06 60 – Schedules for Special Foundations and Load Bearing Elements

31 06 60.14 Piling Special Costs		Crew	Daily Output	Labor-Hours	Unit	Material	Labor	Equipment	Total	Total Incl O&P
0010	**PILING SPECIAL COSTS**									
0011	Piling special costs, pile caps, see Section 03 30 53.40									
0500	Cutoffs, concrete piles, plain	1 Pile	5.50	1.455	Ea.		64.50		64.50	102
0600	With steel thin shell, add		38	.211			9.35		9.35	14.80
0700	Steel pile or "H" piles		19	.421			18.70		18.70	29.50
0800	Wood piles	↓	38	.211	↓		9.35		9.35	14.80
0900	Pre-augering up to 30′ deep, average soil, 24″ diameter	B-43	180	.267	L.F.		10.75	14.15	24.90	32
0920	36″ diameter		115	.417			16.85	22	38.85	50.50
0960	48″ diameter		70	.686			27.50	36.50	64	82.50
0980	60″ diameter	↓	50	.960	↓		39	51	90	116
1000	Testing, any type piles, test load is twice the design load									
1050	50 ton design load, 100 ton test				Ea.				14,000	15,500
1100	100 ton design load, 200 ton test								20,000	22,000
1150	150 ton design load, 300 ton test								26,000	28,500
1200	200 ton design load, 400 ton test								28,000	31,000
1250	400 ton design load, 800 ton test				↓				32,000	35,000
1500	Wet conditions, soft damp ground									
1600	Requiring mats for crane, add								40%	40%
1700	Barge mounted driving rig, add								30%	30%

31 06 60.15 Mobilization		Crew	Daily Output	Labor-Hours	Unit	Material	Labor	Equipment	Total	Total Incl O&P
0010	**MOBILIZATION**									
0020	Set up & remove, air compressor, 600 CFM	A-5	3.30	5.455	Ea.		200	20.50	220.50	335
0100	1200 CFM	"	2.20	8.182			300	30.50	330.50	495
0200	Crane, with pile leads and pile hammer, 75 ton	B-19	.60	106			4,900	3,075	7,975	11,000
0300	150 ton	"	.36	177			8,175	5,125	13,300	18,300
0500	Drill rig, for caissons, to 36″, minimum	B-43	2	24			970	1,275	2,245	2,875
0520	Maximum		.50	96			3,875	5,100	8,975	11,600
0600	Up to 84″	↓	1	48			1,950	2,550	4,500	5,775
0800	Auxiliary boiler, for steam small	A-5	1.66	10.843			395	40.50	435.50	655
0900	Large	"	.83	21.687			795	81.50	876.50	1,325
1100	Rule of thumb: complete pile driving set up, small	B-19	.45	142	↓		6,550	4,100	10,650	14,700

31 06 Schedules for Earthwork

31 06 60 – Schedules for Special Foundations and Load Bearing Elements

31 06 60.15 Mobilization		Crew	Daily Output	Labor-Hours	Unit	Material	2014 Bare Costs Labor	Equipment	Total	Total Incl O&P
1200	Large	B-19	.27	237	Ea.		10,900	6,825	17,725	24,500
1300	Mobilization by water for barge driving rig									
1310	Minimum				Ea.				7,000	7,700
1320	Maximum				″				45,000	49,500
1500	Mobilization, barge, by tug boat	B-83	25	.640	Mile		27.50	35.50	63	81
1600	Standby time for shore pile driving crew				Hr.				715	890
1700	Standby time for barge driving rig				″				1,000	1,250

31 11 Clearing and Grubbing

31 11 10 – Clearing and Grubbing Land

31 11 10.10 Clear and Grub Site		Crew	Daily Output	Labor-Hours	Unit	Material	2014 Bare Costs Labor	Equipment	Total	Total Incl O&P
0010	**CLEAR AND GRUB SITE**									
0020	Cut & chip light trees to 6″ diam.	B-7	1	48	Acre		1,875	1,675	3,550	4,725
0150	Grub stumps and remove	B-30	2	12			495	1,200	1,695	2,075
0200	Cut & chip medium, trees to 12″ diam.	B-7	.70	68.571			2,675	2,400	5,075	6,750
0250	Grub stumps and remove	B-30	1	24			990	2,425	3,415	4,150
0300	Cut & chip heavy, trees to 24″ diam.	B-7	.30	160			6,250	5,600	11,850	15,800
0350	Grub stumps and remove	B-30	.50	48			1,975	4,825	6,800	8,350
0400	If burning is allowed, deduct cut & chip								40%	40%
3000	Chipping stumps, to 18″ deep, 12″ diam.	B-86	20	.400	Ea.		19.55	8.10	27.65	38.50
3040	18″ diameter		16	.500			24.50	10.15	34.65	48
3080	24″ diameter		14	.571			28	11.60	39.60	55.50
3100	30″ diameter		12	.667			32.50	13.50	46	64.50
3120	36″ diameter		10	.800			39	16.20	55.20	77.50
3160	48″ diameter		8	1			49	20.50	69.50	96.50
5000	Tree thinning, feller buncher, conifer									
5080	Up to 8″ diameter	B-93	240	.033	Ea.		1.63	3.04	4.67	5.80
5120	12″ diameter		160	.050			2.45	4.56	7.01	8.70
5240	Hardwood, up to 4″ diameter		240	.033			1.63	3.04	4.67	5.80
5280	8″ diameter		180	.044			2.17	4.05	6.22	7.75
5320	12″ diameter		120	.067			3.26	6.10	9.36	11.65
7000	Tree removal, congested area, aerial lift truck									
7040	8″ diameter	B-85	7	5.714	Ea.		224	149	373	510
7080	12″ diameter		6	6.667			262	174	436	590
7120	18″ diameter		5	8			315	209	524	710
7160	24″ diameter		4	10			395	262	657	890
7240	36″ diameter		3	13.333			525	350	875	1,200
7280	48″ diameter		2	20			785	525	1,310	1,775
9000	Site clearing with 335 H.P. dozer, trees to 6″ diameter	B-10M	280	.043			1.92	6.50	8.42	10.10
9010	To 12″ diameter		150	.080			3.59	12.10	15.69	18.75
9020	To 24″ diameter		100	.120			5.40	18.15	23.55	28
9030	To 36″ diameter		50	.240			10.75	36.50	47.25	56.50
9100	Grub stumps, trees to 6″ diameter		400	.030			1.34	4.54	5.88	7.05
9103	To 36″ diameter		195	.062			2.76	9.30	12.06	14.45

31 13 Selective Tree and Shrub Removal and Trimming

31 13 13 – Selective Tree and Shrub Removal

31 13 13.10 Selective Clearing		Crew	Daily Output	Labor-Hours	Unit	Material	2014 Bare Costs Labor	Equipment	Total	Total Incl O&P
0010	**SELECTIVE CLEARING**									
0020	Clearing brush with brush saw	A-1C	.25	32	Acre		1,175	125	1,300	1,925
0100	By hand	1 Clab	.12	66.667			2,450		2,450	3,775
0300	With dozer, ball and chain, light clearing	B-11A	2	8			340	660	1,000	1,250
0400	Medium clearing		1.50	10.667			455	885	1,340	1,675
0500	With dozer and brush rake, light		10	1.600			68.50	133	201.50	251
0550	Medium brush to 4" diameter		8	2			85.50	166	251.50	315
0600	Heavy brush to 4" diameter		6.40	2.500			107	207	314	390
1000	Brush mowing, tractor w/rotary mower, no removal									
1020	Light density	B-84	2	4	Acre		196	182	378	495
1040	Medium density		1.50	5.333			261	243	504	660
1080	Heavy density		1	8			390	365	755	995

31 13 13.20 Selective Tree Removal		Crew	Daily Output	Labor-Hours	Unit	Material	2014 Bare Costs Labor	Equipment	Total	Total Incl O&P
0010	**SELECTIVE TREE REMOVAL**									
0011	With tractor, large tract, firm									
0020	level terrain, no boulders, less than 12" diam. trees									
0300	300 HP dozer, up to 400 trees/acre, 0 to 25% hardwoods	B-10M	.75	16	Acre		715	2,425	3,140	3,775
0340	25% to 50% hardwoods		.60	20			895	3,025	3,920	4,700
0370	75% to 100% hardwoods		.45	26.667			1,200	4,050	5,250	6,275
0400	500 trees/acre, 0% to 25% hardwoods		.60	20			895	3,025	3,920	4,700
0440	25% to 50% hardwoods		.48	25			1,125	3,775	4,900	5,875
0470	75% to 100% hardwoods		.36	33.333			1,500	5,050	6,550	7,825
0500	More than 600 trees/acre, 0 to 25% hardwoods		.52	23.077			1,025	3,500	4,525	5,425
0540	25% to 50% hardwoods		.42	28.571			1,275	4,325	5,600	6,700
0570	75% to 100% hardwoods		.31	38.710			1,725	5,850	7,575	9,100
0900	Large tract clearing per tree									
1500	300 HP dozer, to 12" diameter, softwood	B-10M	320	.038	Ea.		1.68	5.70	7.38	8.80
1550	Hardwood		100	.120			5.40	18.15	23.55	28
1600	12" to 24" diameter, softwood		200	.060			2.69	9.10	11.79	14.10
1650	Hardwood		80	.150			6.70	22.50	29.20	35.50
1700	24" to 36" diameter, softwood		100	.120			5.40	18.15	23.55	28
1750	Hardwood		50	.240			10.75	36.50	47.25	56.50
1800	36" to 48" diameter, softwood		70	.171			7.70	26	33.70	40
1850	Hardwood		35	.343			15.35	52	67.35	80.50
2000	Stump removal on site by hydraulic backhoe, 1-1/2 C.Y.									
2040	4" to 6" diameter	B-17	60	.533	Ea.		21	13.05	34.05	46.50
2050	8" to 12" diameter	B-30	33	.727			30	73	103	127
2100	14" to 24" diameter		25	.960			39.50	96.50	136	167
2150	26" to 36" diameter		16	1.500			62	151	213	261
3000	Remove selective trees, on site using chain saws and chipper,									
3050	not incl. stumps, up to 6" diameter	B-7	18	2.667	Ea.		104	93	197	263
3100	8" to 12" diameter		12	4			156	140	296	395
3150	14" to 24" diameter		10	4.800			187	168	355	475
3200	26" to 36" diameter		8	6			234	210	444	590
3300	Machine load, 2 mile haul to dump, 12" diam. tree	A-3B	8	2			86.50	152	238.50	298

31 14 Earth Stripping and Stockpiling

31 14 13 – Soil Stripping and Stockpiling

31 14 13.23 Topsoil Stripping and Stockpiling		Crew	Daily Output	Labor-Hours	Unit	Material	2014 Bare Costs Labor	Equipment	Total	Total Incl O&P
0010	TOPSOIL STRIPPING AND STOCKPILING									
0020	200 H.P. dozer, ideal conditions	B-10B	2300	.005	C.Y.		.23	.58	.81	.99
0100	Adverse conditions	″	1150	.010			.47	1.15	1.62	1.98
0200	300 H.P. dozer, ideal conditions	B-10M	3000	.004			.18	.61	.79	.94
0300	Adverse conditions	″	1650	.007			.33	1.10	1.43	1.71
1400	Loam or topsoil, remove and stockpile on site									
1420	6″ deep, 200′ haul	B-10B	865	.014	C.Y.		.62	1.53	2.15	2.63
1430	300′ haul		520	.023			1.03	2.55	3.58	4.38
1440	500′ haul		225	.053			2.39	5.90	8.29	10.15
1450	Alternate method: 6″ deep, 200′ haul		5090	.002	S.Y.		.11	.26	.37	.45
1460	500′ haul		1325	.009	″		.41	1	1.41	1.72
1500	Loam or topsoil, remove/stockpile on site									
1510	By hand, 6″ deep, 50′ haul, less than 100 S.Y.	B-1	100	.240	S.Y.		8.95		8.95	13.80
1520	By skid steer, 6″ deep, 100′ haul, 101-500 S.Y.	B-62	500	.048			1.93	.35	2.28	3.33
1530	100′ haul, 501-900 S.Y.	″	900	.027			1.07	.19	1.26	1.85
1540	200′ haul, 901-1100 S.Y.	B-63	1000	.040			1.55	.17	1.72	2.57
1550	By dozer, 200′ haul, 1101-4000 S.Y.	B-10B	4000	.003			.13	.33	.46	.56

31 22 Grading

31 22 13 – Rough Grading

31 22 13.20 Rough Grading Sites		Crew	Daily Output	Labor-Hours	Unit	Material	Labor	Equipment	Total	Total Incl O&P
0010	ROUGH GRADING SITES									
0100	Rough grade sites 400 S.F. or less	B-1	2	12	Ea.		450		450	690
0120	410-1000 S.F.	″	1	24			895		895	1,375
0130	1100-3000 S.F.	B-62	1.50	16			640	116	756	1,125
0140	3100-5000 S.F.	″	1	24			965	174	1,139	1,675
0150	5100-8000 S.F.	B-63	1	40			1,550	174	1,724	2,575
0160	8100-10000 S.F.	″	.75	53.333			2,075	233	2,308	3,425
0170	8100-10000 S.F.	B-10L	1	12			540	480	1,020	1,350
0200	Rough grade open sites 10000-20000 S.F.	B-11L	1.80	8.889			380	395	775	1,000
0210	20100-25000 S.F.		1.40	11.429			490	505	995	1,300
0220	25100-30000 S.F.		1.20	13.333			570	590	1,160	1,525
0230	30100-35000 S.F.		1	16			685	705	1,390	1,825
0240	35100-40000 S.F.		.90	17.778			760	785	1,545	2,025
0250	40100-45000 S.F.		.80	20			855	885	1,740	2,275
0260	45100-50000 S.F.		.72	22.222			950	980	1,930	2,525
0270	50100-75000 S.F.		.50	32			1,375	1,425	2,800	3,650
0280	75100-100000 S.F.		.36	44.444			1,900	1,975	3,875	5,050

31 22 16 – Fine Grading

31 22 16.10 Finish Grading		Crew	Daily Output	Labor-Hours	Unit	Material	Labor	Equipment	Total	Total Incl O&P
0010	FINISH GRADING									
0012	Finish grading area to be paved with grader, small area	B-11L	400	.040	S.Y.		1.71	1.77	3.48	4.55
0100	Large area		2000	.008			.34	.35	.69	.91
0200	Grade subgrade for base course, roadways		3500	.005			.20	.20	.40	.52
1020	For large parking lots	B-32C	5000	.010			.41	.45	.86	1.13
1050	For small irregular areas	″	2000	.024			1.03	1.14	2.17	2.83
1100	Fine grade for slab on grade, machine	B-11L	1040	.015			.66	.68	1.34	1.76
1150	Hand grading	B-18	700	.034			1.28	.07	1.35	2.04
1200	Fine grade granular base for sidewalks and bikeways	B-62	1200	.020			.80	.15	.95	1.39
2550	Hand grade select gravel	2 Clab	60	.267	C.S.F.		9.75		9.75	15.10

31 22 Grading

31 22 16 – Fine Grading

	31 22 16.10 Finish Grading	Crew	Daily Output	Labor-Hours	Unit	Material	2014 Bare Costs Labor	Equipment	Total	Total Incl O&P
3000	Hand grade select gravel, including compaction, 4″ deep	B-18	555	.043	S.Y.		1.61	.08	1.69	2.58
3100	6″ deep		400	.060			2.24	.12	2.36	3.58
3120	8″ deep		300	.080			2.99	.15	3.14	4.78
3300	Finishing grading slopes, gentle	B-11L	8900	.002			.08	.08	.16	.21
3310	Steep slopes		7100	.002			.10	.10	.20	.26
3312	Steep slopes, large quantities		64	.250	M.S.F.		10.70	11.05	21.75	28.50
3500	Finish grading lagoon bottoms		4	4			171	177	348	455
3600	Fine grade, top of lagoon banks for compaction		30	.533			23	23.50	46.50	61

31 23 Excavation and Fill

31 23 16 – Excavation

31 23 16.13 Excavating, Trench

	Description	Crew	Daily Output	Labor-Hours	Unit	Material	Labor	Equipment	Total	Total Incl O&P
0010	**EXCAVATING, TRENCH**									
0011	Or continuous footing									
0020	Common earth with no sheeting or dewatering included									
0050	1′ to 4′ deep, 3/8 C.Y. excavator	B-11C	150	.107	B.C.Y.		4.56	2.44	7	9.65
0060	1/2 C.Y. excavator	B-11M	200	.080			3.42	1.99	5.41	7.45
0062	3/4 C.Y. excavator	B-12F	270	.059			2.57	2.45	5.02	6.65
0090	4′ to 6′ deep, 1/2 C.Y. excavator	B-11M	200	.080			3.42	1.99	5.41	7.45
0100	5/8 C.Y. excavator	B-12Q	250	.064			2.78	2.38	5.16	6.85
0110	3/4 C.Y. excavator	B-12F	300	.053			2.32	2.21	4.53	5.95
0120	1 C.Y. hydraulic excavator	B-12A	400	.040			1.74	2.04	3.78	4.90
0130	1-1/2 C.Y. excavator	B-12B	540	.030			1.29	1.91	3.20	4.07
0300	1/2 C.Y. excavator, truck mounted	B-12J	200	.080			3.48	4.41	7.89	10.15
0500	6′ to 10′ deep, 3/4 C.Y. excavator	B-12F	225	.071			3.09	2.94	6.03	7.95
0510	1 C.Y. excavator	B-12A	400	.040			1.74	2.04	3.78	4.90
0600	1 C.Y. excavator, truck mounted	B-12K	400	.040			1.74	2.51	4.25	5.40
0610	1-1/2 C.Y. excavator	B-12B	600	.027			1.16	1.72	2.88	3.66
0620	2-1/2 C.Y. excavator	B-12S	1000	.016			.70	1.64	2.34	2.86
0900	10′ to 14′ deep, 3/4 C.Y. excavator	B-12F	200	.080			3.48	3.31	6.79	8.95
0910	1 C.Y. excavator	B-12A	360	.044			1.93	2.26	4.19	5.45
1000	1-1/2 C.Y. excavator	B-12B	540	.030			1.29	1.91	3.20	4.07
1020	2-1/2 C.Y. excavator	B-12S	900	.018			.77	1.82	2.59	3.18
1030	3 C.Y. excavator	B-12D	1400	.011			.50	1.74	2.24	2.68
1040	3-1/2 C.Y. excavator	″	1800	.009			.39	1.36	1.75	2.08
1300	14′ to 20′ deep, 1 C.Y. excavator	B-12A	320	.050			2.17	2.55	4.72	6.10
1310	1-1/2 C.Y. excavator	B-12B	480	.033			1.45	2.15	3.60	4.57
1320	2-1/2 C.Y. excavator	B-12S	765	.021			.91	2.14	3.05	3.74
1330	3 C.Y. excavator	B-12D	1000	.016			.70	2.44	3.14	3.75
1335	3-1/2 C.Y. excavator	″	1150	.014			.60	2.12	2.72	3.26
1340	20′ to 24′ deep, 1 C.Y. excavator	B-12A	288	.056			2.41	2.83	5.24	6.80
1342	1-1/2 C.Y. excavator	B-12B	432	.037			1.61	2.39	4	5.10
1344	2-1/2 C.Y. excavator	B-12S	685	.023			1.02	2.39	3.41	4.18
1346	3 C.Y. excavator	B-12D	900	.018			.77	2.71	3.48	4.17
1348	3-1/2 C.Y. excavator	″	1050	.015			.66	2.33	2.99	3.57
1352	4′ to 6′ deep, 1/2 C.Y. excavator w/trench box	B-13H	188	.085			3.70	5.15	8.85	11.30
1354	5/8 C.Y. excavator	″	235	.068			2.96	4.11	7.07	9.05
1356	3/4 C.Y. excavator	B-13G	282	.057			2.47	2.65	5.12	6.70
1358	1 C.Y. excavator	B-13D	376	.043			1.85	2.39	4.24	5.45
1360	1-1/2 C.Y. excavator	B-13E	508	.032			1.37	2.20	3.57	4.50
1362	6′ to 10′ deep, 3/4 C.Y. excavator w/trench box	B-13G	212	.075			3.28	3.52	6.80	8.85

31 23 16.13 Excavating, Trench		Crew	Daily Output	Labor-Hours	Unit	Material	2014 Bare Costs Labor	Equipment	Total	Total Incl O&P
1370	1 C.Y. excavator	B-13D	376	.043	B.C.Y.		1.85	2.39	4.24	5.45
1371	1-1/2 C.Y. excavator	B-13E	564	.028			1.23	1.98	3.21	4.05
1372	2-1/2 C.Y. excavator	B-13J	940	.017			.74	1.83	2.57	3.14
1374	10′ to 14′ deep, 3/4 C.Y. excavator w/trench box	B-13G	188	.085			3.70	3.97	7.67	10
1375	1 C.Y. excavator	B-13D	338	.047			2.06	2.66	4.72	6.05
1376	1-1/2 C.Y. excavator	B-13E	508	.032			1.37	2.20	3.57	4.50
1377	2-1/2 C.Y. excavator	B-13J	845	.019			.82	2.04	2.86	3.50
1378	3 C.Y. excavator	B-13F	1316	.012			.53	1.92	2.45	2.92
1380	3-1/2 C.Y. excavator	"	1692	.009			.41	1.49	1.90	2.27
1381	14′ to 20′ deep, 1 C.Y. excavator w/trench box	B-13D	301	.053			2.31	2.99	5.30	6.80
1382	1-1/2 C.Y. excavator	B-13E	451	.035			1.54	2.47	4.01	5.10
1383	2-1/2 C.Y. excavator	B-13J	720	.022			.97	2.39	3.36	4.11
1384	3 C.Y. excavator	B-13F	940	.017			.74	2.69	3.43	4.09
1385	3-1/2 C.Y. excavator	"	1081	.015			.64	2.34	2.98	3.55
1386	20′ to 24′ deep, 1 C.Y. excavator w/trench box	B-13D	271	.059			2.57	3.32	5.89	7.55
1387	1-1/2 C.Y. excavator	B-13E	406	.039			1.71	2.75	4.46	5.65
1388	2-1/2 C.Y. excavator	B-13J	645	.025			1.08	2.67	3.75	4.59
1389	3 C.Y. excavator	B-13F	846	.019			.82	2.99	3.81	4.54
1390	3-1/2 C.Y. excavator	"	987	.016			.70	2.56	3.26	3.90
1391	Shoring by S.F./day trench wall protected loose mat., 4′ W	B-6	3200	.008	SF Wall	.55	.30	.11	.96	1.19
1392	Rent shoring per week per S.F. wall protected, loose mat., 4′ W					1.52			1.52	1.68
1395	Hydraulic shoring, S.F. trench wall protected stable mat., 4′ W	2 Clab	2700	.006		.17	.22		.39	.53
1397	semi-stable material, 4′ W	"	2400	.007		.24	.24		.48	.65
1398	Rent hydraulic shoring per day/S.F. wall, stable mat., 4′ W					.29			.29	.32
1399	semi-stable material					.36			.36	.39
1400	By hand with pick and shovel 2′ to 6′ deep, light soil	1 Clab	8	1	B.C.Y.		36.50		36.50	56.50
1500	Heavy soil	"	4	2	"		73.50		73.50	113
1700	For tamping backfilled trenches, air tamp, add	A-1G	100	.080	E.C.Y.		2.93	.56	3.49	5.15
1900	Vibrating plate, add	B-18	180	.133	"		4.98	.26	5.24	8
2100	Trim sides and bottom for concrete pours, common earth		1500	.016	S.F.		.60	.03	.63	.95
2300	Hardpan		600	.040	"		1.49	.08	1.57	2.38
2400	Pier and spread footing excavation, add to above				B.C.Y.				30%	30%
3000	Backfill trench, F.E. loader, wheel mtd., 1 C.Y. bucket									
3020	Minimal haul	B-10R	400	.030	L.C.Y.		1.34	.74	2.08	2.86
3040	100′ haul		200	.060			2.69	1.48	4.17	5.70
3060	200′ haul		100	.120			5.40	2.95	8.35	11.45
3080	2-1/4 C.Y. bucket, minimum haul	B-10T	600	.020			.90	.87	1.77	2.33
3090	100′ haul		300	.040			1.79	1.74	3.53	4.65
3100	200′ haul		150	.080			3.59	3.49	7.08	9.30
4000	For backfill with dozer, see Section 31 23 23.14									
4010	For compaction of backfill, see Section 31 23 23.23									
5020	Loam & Sandy clay with no sheeting or dewatering included									
5050	1′ to 4′ deep, 3/8 C.Y. tractor loader/backhoe	B-11C	162	.099	B.C.Y.		4.23	2.25	6.48	8.95
5060	1/2 C.Y. excavator	B-11M	216	.074			3.17	1.84	5.01	6.85
5070	3/4 C.Y. excavator	B-12F	292	.055			2.38	2.27	4.65	6.15
5080	4′ to 6′ deep, 1/2 C.Y. excavator	B-11M	216	.074			3.17	1.84	5.01	6.85
5090	5/8 C.Y. excavator	B-12Q	276	.058			2.52	2.15	4.67	6.20
5100	3/4 C.Y. excavator	B-12F	324	.049			2.15	2.04	4.19	5.55
5110	1 C.Y. excavator	B-12A	432	.037			1.61	1.89	3.50	4.54
5120	1-1/2 C.Y. excavator	B-12B	583	.027			1.19	1.77	2.96	3.76
5130	1/2 C.Y. excavator, truck mounted	B-12J	216	.074			3.22	4.08	7.30	9.40
5140	6′ to 10′ deep, 3/4 C.Y. excavator	B-12F	243	.066			2.86	2.72	5.58	7.35
5150	1 C.Y. excavator	B-12A	432	.037			1.61	1.89	3.50	4.54

31 23 16.13 Excavating, Trench		Crew	Daily Output	Labor-Hours	Unit	Material	2014 Bare Costs Labor	Equipment	Total	Total Incl O&P
5160	1 C.Y. excavator, truck mounted	B-12K	432	.037	B.C.Y.		1.61	2.33	3.94	5
5170	1-1/2 C.Y. excavator	B-12B	648	.025			1.07	1.59	2.66	3.39
5180	2-1/2 C.Y. excavator	B-12S	1080	.015			.64	1.52	2.16	2.65
5190	10′ to 14′ deep, 3/4 C.Y. excavator	B-12F	216	.074			3.22	3.07	6.29	8.30
5200	1 C.Y. excavator	B-12A	389	.041			1.79	2.09	3.88	5.05
5210	1-1/2 C.Y. excavator	B-12B	583	.027			1.19	1.77	2.96	3.76
5220	2-1/2 C.Y. hydraulic backhoe	B-12S	970	.016			.72	1.69	2.41	2.95
5230	3 C.Y. excavator	B-12D	1512	.011			.46	1.61	2.07	2.48
5240	3-1/2 C.Y. excavator	″	1944	.008			.36	1.26	1.62	1.93
5250	14′ to 20′ deep, 1 C.Y. excavator	B-12A	346	.046			2.01	2.36	4.37	5.65
5260	1-1/2 C.Y. excavator	B-12B	518	.031			1.34	1.99	3.33	4.24
5270	2-1/2 C.Y. excavator	B-12S	826	.019			.84	1.98	2.82	3.47
5280	3 C.Y. excavator	B-12D	1080	.015			.64	2.26	2.90	3.47
5290	3-1/2 C.Y. excavator	″	1242	.013			.56	1.97	2.53	3.02
5300	20′ to 24′ deep, 1 C.Y. excavator	B-12A	311	.051			2.24	2.62	4.86	6.30
5310	1-1/2 C.Y. excavator	B-12B	467	.034			1.49	2.21	3.70	4.70
5320	2-1/2 C.Y. excavator	B-12S	740	.022			.94	2.21	3.15	3.87
5330	3 C.Y. excavator	B-12D	972	.016			.72	2.51	3.23	3.85
5340	3-1/2 C.Y. excavator	″	1134	.014			.61	2.15	2.76	3.31
5352	4′ to 6′ deep, 1/2 C.Y. excavator w/trench box	B-13H	205	.078			3.39	4.71	8.10	10.40
5354	5/8 C.Y. excavator	″	257	.062			2.71	3.76	6.47	8.25
5356	3/4 C.Y. excavator	B-13G	308	.052			2.26	2.42	4.68	6.10
5358	1 C.Y. excavator	B-13D	410	.039			1.70	2.19	3.89	5
5360	1-1/2 C.Y. excavator	B-13E	554	.029			1.25	2.01	3.26	4.13
5362	6′ to 10′ deep, 3/4 C.Y. excavator w/trench box	B-13G	231	.069			3.01	3.23	6.24	8.15
5364	1 C.Y. excavator	B-13D	410	.039			1.70	2.19	3.89	5
5366	1-1/2 C.Y. excavator	B-13E	616	.026			1.13	1.81	2.94	3.71
5368	2-1/2 C.Y. excavator	B-13J	1015	.016			.68	1.70	2.38	2.91
5370	10′ to 14′ deep, 3/4 C.Y. excavator w/trench box	B-13G	205	.078			3.39	3.64	7.03	9.20
5372	1 C.Y. excavator	B-13D	370	.043			1.88	2.43	4.31	5.55
5374	1-1/2 C.Y. excavator	B-13E	554	.029			1.25	2.01	3.26	4.13
5376	2-1/2 C.Y. excavator	B-13J	914	.018			.76	1.88	2.64	3.23
5378	3 C.Y. excavator	B-13F	1436	.011			.48	1.76	2.24	2.67
5380	3-1/2 C.Y. excavator	″	1847	.009			.38	1.37	1.75	2.07
5382	14′ to 20′ deep, 1 C.Y. excavator w/trench box	B-13D	329	.049			2.11	2.73	4.84	6.25
5384	1-1/2 C.Y. excavator	B-13E	492	.033			1.41	2.27	3.68	4.65
5386	2-1/2 C.Y. excavator	B-13J	780	.021			.89	2.21	3.10	3.79
5388	3 C.Y. excavator	B-13F	1026	.016			.68	2.46	3.14	3.74
5390	3-1/2 C.Y. excavator	″	1180	.014			.59	2.14	2.73	3.26
5392	20′ to 24′ deep, 1 C.Y. excavator w/trench box	B-13D	295	.054			2.36	3.05	5.41	6.95
5394	1-1/2 C.Y. excavator	B-13E	444	.036			1.57	2.51	4.08	5.15
5396	2-1/2 C.Y. excavator	B-13J	695	.023			1	2.48	3.48	4.25
5398	3 C.Y. excavator	B-13F	923	.017			.75	2.74	3.49	4.16
5400	3-1/2 C.Y. excavator	″	1077	.015			.65	2.35	3	3.57
6020	Sand & gravel with no sheeting or dewatering included									
6050	1′ to 4′ deep, 3/8 C.Y. excavator	B-11C	165	.097	B.C.Y.		4.15	2.21	6.36	8.80
6060	1/2 C.Y. excavator	B-11M	220	.073			3.11	1.81	4.92	6.75
6070	3/4 C.Y. excavator	B-12F	297	.054			2.34	2.23	4.57	6.05
6080	4′ to 6′ deep, 1/2 C.Y. excavator	B-11M	220	.073			3.11	1.81	4.92	6.75
6090	5/8 C.Y. excavator	B-12Q	275	.058			2.53	2.16	4.69	6.25
6100	3/4 C.Y. excavator	B-12F	330	.048			2.11	2.01	4.12	5.45
6110	1 C.Y. excavator	B-12A	440	.036			1.58	1.85	3.43	4.45
6120	1-1/2 C.Y. excavator	B-12B	594	.027			1.17	1.74	2.91	3.70

31 23 Excavation and Fill

31 23 16 – Excavation

31 23 16.13 Excavating, Trench		Crew	Daily Output	Labor-Hours	Unit	Material	2014 Bare Costs Labor	Equipment	Total	Total Incl O&P
6130	1/2 C.Y. excavator, truck mounted	B-12J	220	.073	B.C.Y.		3.16	4.01	7.17	9.25
6140	6′ to 10′ deep, 3/4 C.Y. excavator	B-12F	248	.065			2.80	2.67	5.47	7.20
6150	1 C.Y. excavator	B-12A	440	.036			1.58	1.85	3.43	4.45
6160	1 C.Y. excavator, truck mounted	B-12K	440	.036			1.58	2.28	3.86	4.92
6170	1-1/2 C.Y. excavator	B-12B	660	.024			1.05	1.56	2.61	3.33
6180	2-1/2 C.Y. excavator	B-12S	1100	.015			.63	1.49	2.12	2.61
6190	10′ to 14′ deep, 3/4 C.Y. excavator	B-12F	220	.073			3.16	3.01	6.17	8.15
6200	1 C.Y. excavator	B-12A	396	.040			1.76	2.06	3.82	4.94
6210	1-1/2 C.Y. excavator	B-12B	594	.027			1.17	1.74	2.91	3.70
6220	2-1/2 C.Y. excavator	B-12S	990	.016			.70	1.65	2.35	2.89
6230	3 C.Y. excavator	B-12D	1540	.010			.45	1.59	2.04	2.43
6240	3-1/2 C.Y. excavator	"	1980	.008			.35	1.23	1.58	1.90
6250	14′ to 20′ deep, 1 C.Y. excavator	B-12A	352	.045			1.97	2.31	4.28	5.55
6260	1-1/2 C.Y. excavator	B-12B	528	.030			1.32	1.95	3.27	4.16
6270	2-1/2 C.Y. excavator	B-12S	840	.019			.83	1.95	2.78	3.40
6280	3 C.Y. excavator	B-12D	1100	.015			.63	2.22	2.85	3.41
6290	3-1/2 C.Y. excavator	"	1265	.013			.55	1.93	2.48	2.96
6300	20′ to 24′ deep, 1 C.Y. excavator	B-12A	317	.050			2.19	2.57	4.76	6.20
6310	1-1/2 C.Y. excavator	B-12B	475	.034			1.46	2.17	3.63	4.63
6320	2-1/2 C.Y. excavator	B-12S	755	.021			.92	2.17	3.09	3.79
6330	3 C.Y. excavator	B-12D	990	.016			.70	2.47	3.17	3.78
6340	3-1/2 C.Y. excavator	"	1155	.014			.60	2.11	2.71	3.25
6352	4′ to 6′ deep, 1/2 C.Y. excavator w/trench box	B-13H	209	.077			3.33	4.62	7.95	10.20
6354	5/8 C.Y. excavator	"	261	.061			2.66	3.70	6.36	8.15
6356	3/4 C.Y. excavator	B-13G	314	.051			2.21	2.38	4.59	6
6358	1 C.Y. excavator	B-13D	418	.038			1.66	2.15	3.81	4.91
6360	1-1/2 C.Y. excavator	B-13E	564	.028			1.23	1.98	3.21	4.05
6362	6′ to 10′ deep, 3/4 C.Y. excavator w/trench box	B-13G	236	.068			2.95	3.16	6.11	8
6364	1 C.Y. excavator	B-13D	418	.038			1.66	2.15	3.81	4.91
6366	1-1/2 C.Y. excavator	B-13E	627	.026			1.11	1.78	2.89	3.65
6368	2-1/2 C.Y. excavator	B-13J	1035	.015			.67	1.66	2.33	2.86
6370	10′ to 14′ deep, 3/4 C.Y. excavator w/trench box	B-13G	209	.077			3.33	3.57	6.90	9.05
6372	1 C.Y. excavator	B-13D	376	.043			1.85	2.39	4.24	5.45
6374	1-1/2 C.Y. excavator	B-13E	564	.028			1.23	1.98	3.21	4.05
6376	2-1/2 C.Y. excavator	B-13J	930	.017			.75	1.85	2.60	3.18
6378	3 C.Y. excavator	B-13F	1463	.011			.48	1.73	2.21	2.63
6380	3-1/2 C.Y. excavator	"	1881	.009			.37	1.34	1.71	2.04
6382	14′ to 20′ deep, 1 C.Y. excavator w/trench box	B-13D	334	.048			2.08	2.69	4.77	6.15
6384	1-1/2 C.Y. excavator	B-13E	502	.032			1.38	2.22	3.60	4.56
6386	2-1/2 C.Y. excavator	B-13J	790	.020			.88	2.18	3.06	3.74
6388	3 C.Y. excavator	B-13F	1045	.015			.67	2.42	3.09	3.68
6390	3-1/2 C.Y. excavator	"	1202	.013			.58	2.10	2.68	3.19
6392	20′ to 24′ deep, 1 C.Y. excavator w/trench box	B-13D	301	.053			2.31	2.99	5.30	6.80
6394	1-1/2 C.Y. excavator	B-13E	452	.035			1.54	2.47	4.01	5.05
6396	2-1/2 C.Y. excavator	B-13J	710	.023			.98	2.42	3.40	4.17
6398	3 C.Y. excavator	B-13F	941	.017			.74	2.68	3.42	4.08
6400	3-1/2 C.Y. excavator	"	1097	.015			.63	2.30	2.93	3.50
7020	Dense hard clay with no sheeting or dewatering included									
7050	1′ to 4′ deep, 3/8 C.Y. excavator	B-11C	132	.121	B.C.Y.		5.20	2.77	7.97	10.95
7060	1/2 C.Y. excavator	B-11M	176	.091			3.89	2.26	6.15	8.45
7070	3/4 C.Y. excavator	B-12F	238	.067			2.92	2.78	5.70	7.50
7080	4′ to 6′ deep, 1/2 C.Y. excavator	B-11M	176	.091			3.89	2.26	6.15	8.45
7090	5/8 C.Y. excavator	B-12Q	220	.073			3.16	2.70	5.86	7.80

31 23 16.13 Excavating, Trench		Crew	Daily Output	Labor-Hours	Unit	Material	2014 Bare Costs Labor	Equipment	Total	Total Incl O&P
7100	3/4 C.Y. excavator	B-12F	264	.061	B.C.Y.		2.63	2.51	5.14	6.80
7110	1 C.Y. excavator	B-12A	352	.045			1.97	2.31	4.28	5.55
7120	1-1/2 C.Y. excavator	B-12B	475	.034			1.46	2.17	3.63	4.63
7130	1/2 C.Y. excavator, truck mounted	B-12J	176	.091			3.95	5	8.95	11.55
7140	6′ to 10′ deep, 3/4 C.Y. excavator	B-12F	198	.081			3.51	3.34	6.85	9.05
7150	1 C.Y. excavator	B-12A	352	.045			1.97	2.31	4.28	5.55
7160	1 C.Y. excavator, truck mounted	B-12K	352	.045			1.97	2.85	4.82	6.15
7170	1-1/2 C.Y. excavator	B-12B	528	.030			1.32	1.95	3.27	4.16
7180	2-1/2 C.Y. excavator	B-12S	880	.018			.79	1.86	2.65	3.26
7190	10′ to 14′ deep, 3/4 C.Y. excavator	B-12F	176	.091			3.95	3.76	7.71	10.20
7200	1 C.Y. excavator	B-12A	317	.050			2.19	2.57	4.76	6.20
7210	1-1/2 C.Y. excavator	B-12B	475	.034			1.46	2.17	3.63	4.63
7220	2-1/2 C.Y. excavator	B-12S	790	.020			.88	2.07	2.95	3.62
7230	3 C.Y. excavator	B-12D	1232	.013			.56	1.98	2.54	3.04
7240	3-1/2 C.Y. excavator	"	1584	.010			.44	1.54	1.98	2.37
7250	14′ to 20′ deep, 1 C.Y. excavator	B-12A	282	.057			2.47	2.89	5.36	6.95
7260	1-1/2 C.Y. excavator	B-12B	422	.038			1.65	2.44	4.09	5.20
7270	2-1/2 C.Y. excavator	B-12S	675	.024			1.03	2.42	3.45	4.24
7280	3 C.Y. excavator	B-12D	880	.018			.79	2.77	3.56	4.26
7290	3-1/2 C.Y. excavator	"	1012	.016			.69	2.41	3.10	3.70
7300	20′ to 24′ deep, 1 C.Y. excavator	B-12A	254	.063			2.74	3.21	5.95	7.70
7310	1-1/2 C.Y. excavator	B-12B	380	.042			1.83	2.71	4.54	5.80
7320	2-1/2 C.Y. excavator	B-12S	605	.026			1.15	2.71	3.86	4.74
7330	3 C.Y. excavator	B-12D	792	.020			.88	3.08	3.96	4.73
7340	3-1/2 C.Y. excavator	"	924	.017			.75	2.64	3.39	4.06

31 23 16.14 Excavating, Utility Trench		Crew	Daily Output	Labor-Hours	Unit	Material	Labor	Equipment	Total	Total Incl O&P
0010	**EXCAVATING, UTILITY TRENCH**									
0011	Common earth									
0050	Trenching with chain trencher, 12 H.P., operator walking									
0100	4″ wide trench, 12″ deep	B-53	800	.010	L.F.		.47	.09	.56	.81
0150	18″ deep		750	.011			.50	.10	.60	.86
0200	24″ deep		700	.011			.54	.10	.64	.93
0300	6″ wide trench, 12″ deep		650	.012			.58	.11	.69	1
0350	18″ deep		600	.013			.63	.12	.75	1.08
0400	24″ deep		550	.015			.68	.13	.81	1.18
0450	36″ deep		450	.018			.84	.16	1	1.44
0600	8″ wide trench, 12″ deep		475	.017			.79	.15	.94	1.36
0650	18″ deep		400	.020			.94	.18	1.12	1.63
0700	24″ deep		350	.023			1.08	.20	1.28	1.85
0750	36″ deep		300	.027			1.25	.24	1.49	2.16
0830	Fly wheel trencher, 18″ wide trench, 6′ deep, light soil	B-54A	1992	.005	B.C.Y.		.22	.61	.83	1.01
0840	Medium soil		1594	.006			.28	.76	1.04	1.25
0850	Heavy soil		1295	.007			.34	.93	1.27	1.54
0860	24″ wide trench, 9′ deep, light soil	B-54B	4981	.002			.09	.38	.47	.56
0870	Medium soil		4000	.003			.12	.47	.59	.70
0880	Heavy soil		3237	.003			.14	.59	.73	.87
1000	Backfill by hand including compaction, add									
1050	4″ wide trench, 12″ deep	A-1G	800	.010	L.F.		.37	.07	.44	.65
1100	18″ deep		530	.015			.55	.11	.66	.97
1150	24″ deep		400	.020			.73	.14	.87	1.28
1300	6″ wide trench, 12″ deep		540	.015			.54	.10	.64	.95
1350	18″ deep		405	.020			.72	.14	.86	1.27

31 23 16.14 Excavating, Utility Trench

	31 23 16.14 Excavating, Utility Trench	Crew	Daily Output	Labor-Hours	Unit	Material	2014 Bare Costs Labor	Equipment	Total	Total Incl O&P
1400	24″ deep	A-1G	270	.030	L.F.		1.09	.21	1.30	1.91
1450	36″ deep		180	.044			1.63	.31	1.94	2.85
1600	8″ wide trench, 12″ deep		400	.020			.73	.14	.87	1.28
1650	18″ deep		265	.030			1.11	.21	1.32	1.94
1700	24″ deep		200	.040			1.47	.28	1.75	2.57
1750	36″ deep		135	.059			2.17	.41	2.58	3.81
2000	Chain trencher, 40 H.P. operator riding									
2050	6″ wide trench and backfill, 12″ deep	B-54	1200	.007	L.F.		.31	.28	.59	.79
2100	18″ deep		1000	.008			.38	.34	.72	.94
2150	24″ deep		975	.008			.39	.35	.74	.97
2200	36″ deep		900	.009			.42	.38	.80	1.04
2250	48″ deep		750	.011			.50	.45	.95	1.26
2300	60″ deep		650	.012			.58	.52	1.10	1.45
2400	8″ wide trench and backfill, 12″ deep		1000	.008			.38	.34	.72	.94
2450	18″ deep		950	.008			.40	.36	.76	.99
2500	24″ deep		900	.009			.42	.38	.80	1.04
2550	36″ deep		800	.010			.47	.42	.89	1.18
2600	48″ deep		650	.012			.58	.52	1.10	1.45
2700	12″ wide trench and backfill, 12″ deep		975	.008			.39	.35	.74	.97
2750	18″ deep		860	.009			.44	.39	.83	1.09
2800	24″ deep		800	.010			.47	.42	.89	1.18
2850	36″ deep		725	.011			.52	.47	.99	1.30
3000	16″ wide trench and backfill, 12″ deep		835	.010			.45	.41	.86	1.13
3050	18″ deep		750	.011			.50	.45	.95	1.26
3100	24″ deep		700	.011			.54	.49	1.03	1.35
3200	Compaction with vibratory plate, add								35%	35%
5100	Hand excavate and trim for pipe bells after trench excavation									
5200	8″ pipe	1 Clab	155	.052	L.F.		1.89		1.89	2.92
5300	18″ pipe	″	130	.062	″		2.26		2.26	3.48
5400	For clay or till, add up to								150%	150%
6000	Excavation utility trench, rock material									
6050	Note: assumption teeth change every 100 feet of trench									
6100	Chain trencher, 6″ wide, 30″ depth, rock saw	B-54D	400	.040	L.F.	21	1.71	.92	23.63	27
6200	Chain trencher, 18″ wide, 8′ depth, rock trencher	B-54E	280	.057	″	36	2.44	9.25	47.69	53.50

31 23 16.15 Excavating, Utility Trench, Plow

	31 23 16.15 Excavating, Utility Trench, Plow	Crew	Daily Output	Labor-Hours	Unit	Material	Labor	Equipment	Total	Total Incl O&P
0010	**EXCAVATING, UTILITY TRENCH, PLOW**									
0100	Single cable, plowed into fine material	B-54C	3800	.004	L.F.		.18	.32	.50	.63
0200	Two cable		3200	.005			.21	.38	.59	.74
0300	Single cable, plowed into coarse material		2000	.008			.34	.60	.94	1.18

31 23 16.16 Structural Excavation for Minor Structures

	31 23 16.16 Structural Excavation for Minor Structures	Crew	Daily Output	Labor-Hours	Unit	Material	Labor	Equipment	Total	Total Incl O&P
0010	**STRUCTURAL EXCAVATION FOR MINOR STRUCTURES**									
0015	Hand, pits to 6′ deep, sandy soil	1 Clab	8	1	B.C.Y.		36.50		36.50	56.50
0100	Heavy soil or clay	″	4	2			73.50		73.50	113
0200	Pits to 2′ deep, normal soil	B-2	24	1.667			62		62	95.50
0300	Pits 6′ to 12′ deep, sandy soil	1 Clab	5	1.600			58.50		58.50	90.50
0500	Heavy soil or clay		3	2.667			97.50		97.50	151
0700	Pits 12′ to 18′ deep, sandy soil		4	2			73.50		73.50	113
0900	Heavy soil or clay		2	4			147		147	226
1000	Hand trimming, bottom of excavation	B-2	2400	.017	S.F.		.62		.62	.95
1010	Slopes and sides		2400	.017	″		.62		.62	.95
1030	Around obstructions		8	5	B.C.Y.		185		185	286
1100	Hand loading trucks from stock pile, sandy soil	1 Clab	12	.667			24.50		24.50	37.50

31 23 16 – Excavation

31 23 16.16 Structural Excavation for Minor Structures		Crew	Daily Output	Labor-Hours	Unit	Material	2014 Bare Costs Labor	Equipment	Total	Total Incl O&P
1300	Heavy soil or clay	1 Clab	8	1	B.C.Y.		36.50		36.50	56.50
1500	For wet or muck hand excavation, add to above								50%	50%
1550	Excavation rock by hand/air tool	B-9	3.40	11.765	B.C.Y.		435	68.50	503.50	750
6000	Machine excavation, for spread and mat footings, elevator pits,									
6001	and small building foundations									
6030	Common earth, hydraulic backhoe, 1/2 C.Y. bucket	B-12E	55	.291	B.C.Y.		12.65	8.15	20.80	28.50
6035	3/4 C.Y. bucket	B-12F	90	.178			7.70	7.35	15.05	19.90
6040	1 C.Y. bucket	B-12A	108	.148			6.45	7.55	14	18.15
6050	1-1/2 C.Y. bucket	B-12B	144	.111			4.83	7.15	11.98	15.30
6060	2 C.Y. bucket	B-12C	200	.080			3.48	5.90	9.38	11.75
6070	Sand and gravel, 3/4 C.Y. bucket	B-12F	100	.160			6.95	6.60	13.55	17.90
6080	1 C.Y. bucket	B-12A	120	.133			5.80	6.80	12.60	16.30
6090	1-1/2 C.Y. bucket	B-12B	160	.100			4.35	6.45	10.80	13.75
6100	2 C.Y. bucket	B-12C	220	.073			3.16	5.35	8.51	10.75
6110	Clay, till, or blasted rock, 3/4 C.Y. bucket	B-12F	80	.200			8.70	8.30	17	22.50
6120	1 C.Y. bucket	B-12A	95	.168			7.30	8.60	15.90	20.50
6130	1-1/2 C.Y. bucket	B-12B	130	.123			5.35	7.95	13.30	16.85
6140	2 C.Y. bucket	B-12C	175	.091			3.97	6.70	10.67	13.45
6230	Sandy clay & loam, hydraulic backhoe, 1/2 C.Y. bucket	B-12E	60	.267			11.60	7.45	19.05	26
6235	3/4 C.Y. bucket	B-12F	98	.163			7.10	6.75	13.85	18.30
6240	1 C.Y. bucket	B-12A	116	.138			6	7	13	16.90
6250	1-1/2 C.Y. bucket	B-12B	156	.103			4.46	6.60	11.06	14.05
9010	For mobilization or demobilization, see Section 01 54 36.50									
9020	For dewatering, see Section 31 23 19.20									
9022	For larger structures, see Bulk Excavation, Section 31 23 16.42									
9024	For loading onto trucks, add								15%	15%
9026	For hauling, see Section 31 23 23.20									
9030	For sheeting or soldier bms/lagging, see Section 31 52 16.10									
9040	For trench excavation of strip ftgs, see Section 31 23 16.13									

31 23 16.26 Rock Removal		Crew	Daily Output	Labor-Hours	Unit	Material	Labor	Equipment	Total	Total Incl O&P
0010	**ROCK REMOVAL**									
0015	Drilling only rock, 2" hole for rock bolts	B-47	316	.076	L.F.		3.10	5.15	8.25	10.40
0800	2-1/2" hole for pre-splitting		600	.040			1.63	2.70	4.33	5.45
4600	Quarry operations, 2-1/2" to 3-1/2" diameter		715	.034			1.37	2.27	3.64	4.60
4610	6" diameter drill holes	B-47A	1350	.018			.79	.93	1.72	2.22

31 23 16.30 Drilling and Blasting Rock		Crew	Daily Output	Labor-Hours	Unit	Material	Labor	Equipment	Total	Total Incl O&P
0010	**DRILLING AND BLASTING ROCK**									
0020	Rock, open face, under 1500 C.Y.	B-47	225	.107	B.C.Y.	3.15	4.35	7.20	14.70	18.05
0100	Over 1500 C.Y.		300	.080		3.15	3.26	5.40	11.81	14.40
0200	Areas where blasting mats are required, under 1500 C.Y.		175	.137		3.15	5.60	9.25	18	22
0250	Over 1500 C.Y.		250	.096		3.15	3.91	6.50	13.56	16.60
0300	Bulk drilling and blasting, can vary greatly, average								9.65	12.20
0500	Pits, average								25.50	31.50
1300	Deep hole method, up to 1500 C.Y.	B-47	50	.480		3.15	19.55	32.50	55.20	69
1400	Over 1500 C.Y.		66	.364		3.15	14.85	24.50	42.50	53
1900	Restricted areas, up to 1500 C.Y.		13	1.846		3.15	75.50	125	203.65	255
2000	Over 1500 C.Y.		20	1.200		3.15	49	81	133.15	167
2200	Trenches, up to 1500 C.Y.		22	1.091		9.15	44.50	74	127.65	159
2300	Over 1500 C.Y.		26	.923		9.15	37.50	62.50	109.15	136
2500	Pier holes, up to 1500 C.Y.		22	1.091		3.15	44.50	74	121.65	152
2600	Over 1500 C.Y.		31	.774		3.15	31.50	52.50	87.15	109
2800	Boulders under 1/2 C.Y., loaded on truck, no hauling	B-100	80	.150			6.70	12	18.70	23.50

31 23 16.30 Drilling and Blasting Rock

	31 23 16.30 Drilling and Blasting Rock	Crew	Daily Output	Labor-Hours	Unit	Material	2014 Bare Costs Labor	Equipment	Total	Total Incl O&P
2900	Boulders, drilled, blasted	B-47	100	.240	B.C.Y.	3.15	9.80	16.20	29.15	36.50
3100	Jackhammer operators with foreman compressor, air tools	B-9	1	40	Day		1,475	234	1,709	2,525
3300	Track drill, compressor, operator and foreman	B-47	1	24	"		980	1,625	2,605	3,275
3500	Blasting caps				Ea.	6.25			6.25	6.85
3700	Explosives					.46			.46	.51
3900	Blasting mats, rent, for first day					144			144	158
4000	Per added day					48			48	53
4200	Preblast survey for 6 room house, individual lot, minimum	A-6	2.40	6.667			305	29.50	334.50	505
4300	Maximum	"	1.35	11.852			540	53	593	890
4500	City block within zone of influence, minimum	A-8	25200	.001	S.F.		.06		.06	.10
4600	Maximum	"	15100	.002	"		.10		.10	.17
5000	Excavate and load boulders, less than 0.5 C.Y.	B-10T	80	.150	B.C.Y.		6.70	6.55	13.25	17.45
5020	0.5 C.Y. to 1 C.Y.	B-10U	100	.120			5.40	10.50	15.90	19.75
5200	Excavate and load blasted rock, 3 C.Y. power shovel	B-12T	1530	.010			.45	1.04	1.49	1.84
5400	Haul boulders, 25 Ton off-highway dump, 1 mile round trip	B-34E	330	.024			.91	3.98	4.89	5.75
5420	2 mile round trip		275	.029			1.09	4.78	5.87	6.90
5440	3 mile round trip		225	.036			1.34	5.85	7.19	8.50
5460	4 mile round trip		200	.040			1.50	6.55	8.05	9.55
5600	Bury boulders on site, less than 0.5 C.Y., 300 H.P. dozer									
5620	150′ haul	B-10M	310	.039	B.C.Y.		1.74	5.85	7.59	9.10
5640	300′ haul		210	.057			2.56	8.65	11.21	13.40
5800	0.5 to 1 C.Y., 300 H.P. dozer, 150′ haul		300	.040			1.79	6.05	7.84	9.40
5820	300′ haul		200	.060			2.69	9.10	11.79	14.10

31 23 16.32 Ripping

	31 23 16.32 Ripping	Crew	Daily Output	Labor-Hours	Unit	Material	Labor	Equipment	Total	Total Incl O&P
0010	**RIPPING**									
0020	Ripping, trap rock, soft, 300 HP dozer, ideal conditions	B-11S	700	.017	B.C.Y.		.77	2.71	3.48	4.15
1500	Adverse conditions		660	.018			.81	2.88	3.69	4.40
1600	Medium hard, 300 HP dozer, ideal conditons		600	.020			.90	3.16	4.06	4.85
1700	Adverse conditions		540	.022			1	3.52	4.52	5.40
2000	Very hard, 410 HP dozer, ideal conditions	B-11T	350	.034			1.54	7.15	8.69	10.20
2100	Adverse conditions	"	310	.039			1.74	8.05	9.79	11.50
2200	Shale, soft, 300 HP dozer, ideal conditons	B-11S	1500	.008			.36	1.27	1.63	1.94
2300	Adverse conditions	"	1350	.009			.40	1.41	1.81	2.16
2310	Grader rear ripper, 180 H.P. ideal conditions	B-11J	740	.022			.92	1.07	1.99	2.58
2320	Adverse conditions	"	630	.025			1.09	1.25	2.34	3.04
2400	Medium hard, 300 HP dozer, ideal conditons	B-11S	1200	.010			.45	1.58	2.03	2.42
2500	Adverse conditions	"	1080	.011			.50	1.76	2.26	2.69
2510	Grader rear ripper, 180 H.P. ideal conditions	B-11J	625	.026			1.10	1.26	2.36	3.06
2520	Adverse conditions	"	530	.030			1.29	1.49	2.78	3.61
2600	Very hard, 410 HP dozer, ideal conditons	B-11T	800	.015			.67	3.13	3.80	4.46
2700	Adverse conditions	"	720	.017			.75	3.47	4.22	4.96
2800	Till, boulder clay/hardpan, soft, 300 H.P. dozer, ideal conditions	B-11S	7000	.002			.08	.27	.35	.42
2810	Adverse conditions	"	6300	.002			.09	.30	.39	.46
2815	Grader rear ripper, 180 H.P. ideal conditions	B-11J	1500	.011			.46	.53	.99	1.28
2816	Adverse conditions	"	1275	.013			.54	.62	1.16	1.50
2820	Medium hard, 300 H.P. dozer, ideal conditions	B-11S	6000	.002			.09	.32	.41	.49
2830	Adverse conditions	"	5400	.002			.10	.35	.45	.54
2835	Grader rear ripper, 180 H.P. ideal conditions	B-11J	740	.022			.92	1.07	1.99	2.58
2836	Adverse conditions	"	630	.025			1.09	1.25	2.34	3.04
2840	Very hard, 410 H.P. dozer, ideal conditions	B-11T	5000	.002			.11	.50	.61	.71
2850	Adverse conditions	"	4500	.003			.12	.56	.68	.79
3000	Dozing ripped material, 200 HP, 100′ haul	B-10B	700	.017			.77	1.89	2.66	3.25

31 23 16.32 Ripping		Crew	Daily Output	Labor-Hours	Unit	Material	2014 Bare Costs Labor	Equipment	Total	Total Incl O&P
3050	300′ haul	B-10B	250	.048	B.C.Y.		2.15	5.30	7.45	9.15
3200	300 HP, 100′ haul	B-10M	1150	.010			.47	1.58	2.05	2.45
3250	300′ haul	"	400	.030			1.34	4.54	5.88	7.05
3400	410 HP, 100′ haul	B-10X	1680	.007			.32	1.43	1.75	2.07
3450	300′ haul	"	600	.020			.90	4.02	4.92	5.80
31 23 16.42 Excavating, Bulk Bank Measure										
0010	**EXCAVATING, BULK BANK MEASURE** R312316-40									
0011	Common earth piled									
0020	For loading onto trucks, add								15%	15%
0050	For mobilization and demobilization, see Section 01 54 36.50									
0100	For hauling, see Section 31 23 23.20									
0200	Excavator, hydraulic, crawler mtd., 1 C.Y. cap. = 100 C.Y./hr.	B-12A	800	.020	B.C.Y.		.87	1.02	1.89	2.45
0250	1-1/2 C.Y. cap. = 125 C.Y./hr.	B-12B	1000	.016			.70	1.03	1.73	2.19
0260	2 C.Y. cap. = 165 C.Y./hr.	B-12C	1320	.012			.53	.89	1.42	1.78
0300	3 C.Y. cap. = 260 C.Y./hr.	B-12D	2080	.008			.33	1.17	1.50	1.80
0305	3-1/2 C.Y. cap. = 300 C.Y./hr.	"	2400	.007			.29	1.02	1.31	1.56
0310	Wheel mounted, 1/2 C.Y. cap. = 40 C.Y./hr.	B-12E	320	.050			2.17	1.40	3.57	4.86
0360	3/4 C.Y. cap. = 60 C.Y./hr.	B-12F	480	.033			1.45	1.38	2.83	3.73
0500	Clamshell, 1/2 C.Y. cap. = 20 C.Y./hr.	B-12G	160	.100			4.35	4.51	8.86	11.60
0550	1 C.Y. cap. = 35 C.Y./hr.	B-12H	280	.057			2.48	4.36	6.84	8.60
0950	Dragline, 1/2 C.Y. cap. = 30 C.Y./hr.	B-12I	240	.067			2.90	3.75	6.65	8.55
1000	3/4 C.Y. cap. = 35 C.Y./hr.	"	280	.057			2.48	3.22	5.70	7.35
1050	1-1/2 C.Y. cap. = 65 C.Y./hr.	B-12P	520	.031			1.34	2.33	3.67	4.61
1100	3 C.Y. cap. = 112 C.Y./hr.	B-12V	900	.018			.77	1.72	2.49	3.07
1200	Front end loader, track mtd., 1-1/2 C.Y. cap. = 70 C.Y./hr.	B-10N	560	.021			.96	.93	1.89	2.49
1250	2-1/2 C.Y. cap. = 95 C.Y./hr.	B-10O	760	.016			.71	1.26	1.97	2.47
1300	3 C.Y. cap. = 130 C.Y./hr.	B-10P	1040	.012			.52	1.15	1.67	2.05
1350	5 C.Y. cap. = 160 C.Y./hr.	B-10Q	1280	.009			.42	1.20	1.62	1.96
1500	Wheel mounted, 3/4 C.Y. cap. = 45 C.Y./hr.	B-10R	360	.033			1.49	.82	2.31	3.18
1550	1-1/2 C.Y. cap. = 80 C.Y./hr.	B-10S	640	.019			.84	.58	1.42	1.92
1600	2-1/4 C.Y. cap. = 100 C.Y./hr.	B-10T	800	.015			.67	.65	1.32	1.74
1601	3 C.Y. cap. = 140 C.Y./hr.	"	1120	.011			.48	.47	.95	1.24
1650	5 C.Y. cap. = 185 C.Y./hr.	B-10U	1480	.008			.36	.71	1.07	1.33
1800	Hydraulic excavator, truck mtd. 1/2 C.Y. = 30 C.Y./hr.	B-12J	240	.067			2.90	3.67	6.57	8.45
1850	48 inch bucket, 1 C.Y. = 45 C.Y./hr.	B-12K	360	.044			1.93	2.79	4.72	6
3700	Shovel, 1/2 C.Y. capacity = 55 C.Y./hr.	B-12L	440	.036			1.58	1.69	3.27	4.27
3750	3/4 C.Y. capacity = 85 C.Y./hr.	B-12M	680	.024			1.02	1.39	2.41	3.09
3800	1 C.Y. capacity = 120 C.Y./hr.	B-12N	960	.017			.72	1.30	2.02	2.54
3850	1-1/2 C.Y. capacity = 160 C.Y./hr.	B-12O	1280	.013			.54	.98	1.52	1.91
3900	3 C.Y. cap. = 250 C.Y./hr.	B-12T	2000	.008			.35	.80	1.15	1.41
4000	For soft soil or sand, deduct								15%	15%
4100	For heavy soil or stiff clay, add								60%	60%
4200	For wet excavation with clamshell or dragline, add								100%	100%
4250	All other equipment, add								50%	50%
4400	Clamshell in sheeting or cofferdam, minimum	B-12H	160	.100			4.35	7.65	12	15.05
4450	Maximum	"	60	.267			11.60	20.50	32.10	40
5000	Excavating, bulk bank measure, sandy clay & loam piled									
5020	For loading onto trucks, add								15%	15%
5100	Excavator, hydraulic, crawler mtd., 1 C.Y. cap. = 120 C.Y./hr.	B-12A	960	.017	B.C.Y.		.72	.85	1.57	2.04
5150	1-1/2 C.Y. cap. = 150 C.Y./hr.	B-12B	1200	.013			.58	.86	1.44	1.82
5300	2 C.Y. cap. = 195 C.Y./hr.	B-12C	1560	.010			.45	.75	1.20	1.51
5400	3 C.Y. cap. = 300 C.Y./hr.	B-12D	2400	.007			.29	1.02	1.31	1.56

31 23 16 – Excavation

31 23 16.42 Excavating, Bulk Bank Measure		Crew	Daily Output	Labor-Hours	Unit	Material	2014 Bare Costs Labor	Equipment	Total	Total Incl O&P
5500	3.5 C.Y. cap. = 350 C.Y./hr.	B-12D	2800	.006	B.C.Y.		.25	.87	1.12	1.34
5610	Wheel mounted, 1/2 C.Y. cap. = 44 C.Y./hr.	B-12E	352	.045			1.97	1.27	3.24	4.42
5660	3/4 C.Y. cap. = 66 C.Y./hr.	B-12F	528	.030	↓		1.32	1.25	2.57	3.39
8000	For hauling excavated material, see Section 31 23 23.20									

31 23 16.43 Excavating, Large Volume Projects		Crew	Daily Output	Labor-Hours	Unit	Material	Labor	Equipment	Total	Total Incl O&P
0010	**EXCAVATING, LARGE VOLUME PROJECTS**, various materials									
0050	Minimum project size 200,000 B.C.Y.									
0080	For hauling, see Section 31 23 23.20									
0100	Unrestricted means continous loading into hopper or railroad cars									
0200	Loader, 8 C.Y. bucket, 110% fill factor, unrestricted operation	B-14J	5280	.002	L.C.Y.		.10	.38	.48	.57
0250	10 C.Y. bucket	B-14K	7040	.002			.08	.38	.46	.54
0300	8 C.Y. bucket, 105% fill factor	B-14J	5040	.002			.11	.39	.50	.59
0350	10 C.Y. bucket	B-14K	6720	.002			.08	.40	.48	.56
0400	8 C.Y. bucket, 100% fill factor	B-14J	4800	.003			.11	.41	.52	.63
0450	10 C.Y. bucket	B-14K	6400	.002			.08	.42	.50	.60
0500	8 C.Y. bucket, 95% fill factor	B-14J	4560	.003			.12	.44	.56	.66
0550	10 C.Y. bucket	B-14K	6080	.002			.09	.44	.53	.62
0600	8 C.Y. bucket, 90% fill factor	B-14J	4320	.003			.12	.46	.58	.70
0650	10 C.Y. bucket	B-14K	5760	.002			.09	.47	.56	.66
0700	8 C.Y. bucket, 85% fill factor	B-14J	4080	.003			.13	.49	.62	.74
0750	10 C.Y. bucket	B-14K	5440	.002			.10	.50	.60	.70
0800	8 C.Y. bucket, 80% fill factor	B-14J	3840	.003			.14	.52	.66	.78
0850	10 C.Y. bucket	B-14K	5120	.002			.10	.53	.63	.74
0900	8 C.Y. bucket, 75% fill factor	B-14J	3600	.003			.15	.55	.70	.84
0950	10 C.Y. bucket	B-14K	4800	.003			.11	.56	.67	.79
1000	8 C.Y. bucket, 70% fill factor	B-14J	3360	.004			.16	.59	.75	.89
1050	10 C.Y. bucket	B-14K	4480	.003			.12	.60	.72	.84
1100	8 C.Y. bucket, 65% fill factor	B-14J	3120	.004			.17	.64	.81	.96
1150	10 C.Y. bucket	B-14K	4160	.003			.13	.65	.78	.91
1200	8 C.Y. bucket, 60% fill factor	B-14J	2880	.004			.19	.69	.88	1.04
1250	10 C.Y. bucket	B-14K	3840	.003	↓		.14	.71	.85	.99
2100	Restricted loading trucks									
2200	Loader, 8 C.Y. bucket, 110% fill factor, loading trucks	B-14J	4400	.003	L.C.Y.		.12	.45	.57	.69
2250	10 C.Y. bucket	B-14K	5940	.002			.09	.46	.55	.64
2300	8 C.Y. bucket, 105% fill factor	B-14J	4200	.003			.13	.47	.60	.72
2350	10 C.Y. bucket	B-14K	5670	.002			.10	.48	.58	.67
2400	8 C.Y. bucket, 100% fill factor	B-14J	4000	.003			.13	.50	.63	.75
2450	10 C.Y. bucket	B-14K	5400	.002			.10	.50	.60	.70
2500	8 C.Y. bucket, 95% fill factor	B-14J	3800	.003			.14	.52	.66	.80
2550	10 C.Y. bucket	B-14K	5130	.002			.10	.53	.63	.74
2600	8 C.Y. bucket, 90% fill factor	B-14J	3600	.003			.15	.55	.70	.84
2650	10 C.Y. bucket	B-14K	4860	.002			.11	.56	.67	.78
2700	8 C.Y. bucket, 85% fill factor	B-14J	3400	.004			.16	.59	.75	.88
2750	10 C.Y. bucket	B-14K	4590	.003			.12	.59	.71	.83
2800	8 C.Y. bucket, 80% fill factor	B-14J	3200	.004			.17	.62	.79	.94
2850	10 C.Y. bucket	B-14K	4320	.003			.12	.63	.75	.88
2900	8 C.Y. bucket, 75% fill factor	B-14J	3000	.004			.18	.66	.84	1
2950	10 C.Y. bucket	B-14K	4050	.003			.13	.67	.80	.93
4000	8 C.Y. bucket, 70% fill factor	B-14J	2800	.004			.19	.71	.90	1.07
4050	10 C.Y. bucket	B-14K	3780	.003			.14	.71	.85	1.01
4100	8 C.Y. bucket, 65% fill factor	B-14J	2600	.005			.21	.77	.98	1.16
4150	10 C.Y. bucket	B-14K	3510	.003	↓		.15	.77	.92	1.08

31 23 Excavation and Fill

31 23 16 – Excavation

31 23 16.43 Excavating, Large Volume Projects		Crew	Daily Output	Labor-Hours	Unit	Material	2014 Bare Costs Labor	Equipment	Total	Total Incl O&P
4200	8 C.Y. bucket, 60% fill factor	B-14J	2400	.005	L.C.Y.		.22	.83	1.05	1.25
4250	10 C.Y. bucket	B-14K	3240	.004			.17	.83	1	1.17
4290	Continuous excavation with no truck loading									
4300	Excavator, 4.5 C.Y. bucket, 100% fill factor, no truck loading	B-14A	4000	.003	B.C.Y.		.14	.76	.90	1.05
4320	6 C.Y. bucket	B-14B	5900	.002			.09	.60	.69	.80
4330	7 C.Y. bucket	B-14C	6900	.002			.08	.52	.60	.69
4340	Shovel, 7 C.Y. bucket	B-14F	8900	.001			.06	.42	.48	.55
4360	12 C.Y. bucket	B-14G	14800	.001			.04	.35	.39	.45
4400	Excavator, 4.5 C.Y. bucket, 95% fill factor	B-14A	3800	.003			.14	.80	.94	1.10
4420	6 C.Y. bucket	B-14B	5605	.002			.10	.63	.73	.84
4430	7 C.Y. bucket	B-14C	6555	.002			.08	.55	.63	.73
4440	Shovel, 7 C.Y. bucket	B-14F	8455	.001			.06	.44	.50	.58
4460	12 C.Y. bucket	B-14G	14060	.001			.04	.37	.41	.47
4500	Excavator, 4.5 C.Y. bucket, 90% fill factor	B-14A	3600	.003			.15	.84	.99	1.16
4520	6 C.Y. bucket	B-14B	5310	.002			.10	.66	.76	.89
4530	7 C.Y. bucket	B-14C	6210	.002			.09	.58	.67	.77
4540	Shovel, 7 C.Y. bucket	B-14F	8010	.002			.07	.46	.53	.61
4560	12 C.Y. bucket	B-14G	13320	.001			.04	.39	.43	.49
4600	Excavator, 4.5 C.Y. bucket, 85% fill factor	B-14A	3400	.004			.16	.90	1.06	1.23
4620	6 C.Y. bucket	B-14B	5015	.002			.11	.70	.81	.94
4630	7 C.Y. bucket	B-14C	5865	.002			.09	.61	.70	.82
4640	Shovel, 7 C.Y. bucket	B-14F	7565	.002			.07	.49	.56	.65
4660	12 C.Y. bucket	B-14G	12580	.001			.04	.41	.45	.52
4700	Excavator, 4.5 C.Y. bucket, 80% fill factor	B-14A	3200	.004			.17	.95	1.12	1.31
4720	6 C.Y. bucket	B-14B	4720	.003			.12	.75	.87	1
4730	7 C.Y. bucket	B-14C	5520	.002			.10	.65	.75	.87
4740	Shovel, 7 C.Y. bucket	B-14F	7120	.002			.08	.52	.60	.70
4760	12 C.Y. bucket	B-14G	11840	.001			.05	.44	.49	.55
5290	Excavation with truck loading									
5300	Excavator, 4.5 C.Y. bucket, 100% fill factor, with truck loading	B-14A	3400	.004	B.C.Y.		.16	.90	1.06	1.23
5320	6 C.Y. bucket	B-14B	5000	.002			.11	.70	.81	.95
5330	7 C.Y. bucket	B-14C	5800	.002			.09	.62	.71	.82
5340	Shovel, 7 C.Y. bucket	B-14F	7500	.002			.07	.50	.57	.65
5360	12 C.Y. bucket	B-14G	12500	.001			.04	.42	.46	.53
5400	Excavator, 4.5 C.Y. bucket, 95% fill factor	B-14A	3230	.004			.17	.94	1.11	1.30
5420	6 C.Y. bucket	B-14B	4750	.003			.12	.74	.86	1
5430	7 C.Y. bucket	B-14C	5510	.002			.10	.65	.75	.87
5440	Shovel, 7 C.Y. bucket	B-14F	7125	.002			.08	.52	.60	.69
5460	12 C.Y. bucket	B-14G	11875	.001			.05	.44	.49	.55
5500	Excavator, 4.5 C.Y. bucket, 90% fill factor	B-14A	3060	.004			.18	.99	1.17	1.36
5520	6 C.Y. bucket	B-14B	4500	.003			.12	.78	.90	1.05
5530	7 C.Y. bucket	B-14C	5220	.002			.11	.69	.80	.92
5540	Shovel, 7 C.Y. bucket	B-14F	6750	.002			.08	.55	.63	.73
5560	12 C.Y. bucket	B-14G	11250	.001			.05	.46	.51	.58
5600	Excavator, 4.5 C.Y. bucket, 85% fill factor	B-14A	2890	.004			.19	1.05	1.24	1.45
5620	6 C.Y. bucket	B-14B	4250	.003			.13	.83	.96	1.11
5630	7 C.Y. bucket	B-14C	4930	.002			.11	.73	.84	.97
5640	Shovel, 7 C.Y. bucket	B-14F	6375	.002			.09	.58	.67	.77
5660	12 C.Y. bucket	B-14G	10625	.001			.05	.49	.54	.62
5700	Excavator, 4.5 C.Y. bucket, 80% fill factor	B-14A	2720	.004			.20	1.12	1.32	1.54
5720	6 C.Y. bucket	B-14B	4000	.003			.14	.88	1.02	1.18
5740	Shovel, 7 C.Y. bucket	B-14F	6000	.002			.09	.62	.71	.82
5760	12 C.Y. bucket	B-14G	10000	.001			.05	.52	.57	.65

31 23 16 – Excavation

31 23 16.43 Excavating, Large Volume Projects		Crew	Daily Output	Labor-Hours	Unit	Material	2014 Bare Costs Labor	Equipment	Total	Total Incl O&P
6000	Self propelled scraper, 1/4 push dozer, average productivity									
6100	31 C.Y., 1500′ haul	B-33K	2046	.007	B.C.Y.		.31	2.04	2.35	2.71
6200	44 C.Y.	B-33H	2640	.005	"		.24	1.93	2.17	2.49

31 23 16.46 Excavating, Bulk, Dozer		Crew	Daily Output	Labor-Hours	Unit	Material	Labor	Equipment	Total	Total Incl O&P
0010	**EXCAVATING, BULK, DOZER**									
0011	Open site									
2000	80 H.P., 50′ haul, sand & gravel	B-10L	460	.026	B.C.Y.		1.17	1.05	2.22	2.93
2010	Sandy clay & loam		440	.027			1.22	1.09	2.31	3.06
2020	Common earth		400	.030			1.34	1.20	2.54	3.37
2040	Clay		250	.048			2.15	1.92	4.07	5.40
2200	150′ haul, sand & gravel		230	.052			2.34	2.09	4.43	5.85
2210	Sandy clay & loam		220	.055			2.44	2.19	4.63	6.15
2220	Common earth		200	.060			2.69	2.40	5.09	6.75
2240	Clay		125	.096			4.30	3.85	8.15	10.80
2400	300′ haul, sand & gravel		120	.100			4.48	4.01	8.49	11.25
2410	Sandy clay & loam		115	.104			4.68	4.18	8.86	11.75
2420	Common earth		100	.120			5.40	4.81	10.21	13.50
2440	Clay	↓	65	.185			8.25	7.40	15.65	21
3000	105 H.P., 50′ haul, sand & gravel	B-10W	700	.017			.77	.85	1.62	2.10
3010	Sandy clay & loam		680	.018			.79	.87	1.66	2.17
3020	Common earth		610	.020			.88	.97	1.85	2.41
3040	Clay		385	.031			1.40	1.54	2.94	3.82
3200	150′ haul, sand & gravel		310	.039			1.74	1.91	3.65	4.74
3210	Sandy clay & loam		300	.040			1.79	1.97	3.76	4.90
3220	Common earth		270	.044			1.99	2.19	4.18	5.45
3240	Clay		170	.071			3.16	3.48	6.64	8.65
3300	300′ haul, sand & gravel		140	.086			3.84	4.23	8.07	10.50
3310	Sandy clay & loam		135	.089			3.98	4.38	8.36	10.85
3320	Common earth		120	.100			4.48	4.93	9.41	12.30
3340	Clay	↓	100	.120			5.40	5.90	11.30	14.70
4000	200 H.P., 50′ haul, sand & gravel	B-10B	1400	.009			.38	.95	1.33	1.63
4010	Sandy clay & loam		1360	.009			.40	.97	1.37	1.67
4020	Common earth		1230	.010			.44	1.08	1.52	1.86
4040	Clay		770	.016			.70	1.72	2.42	2.95
4200	150′ haul, sand & gravel		595	.020			.90	2.23	3.13	3.83
4210	Sandy clay & loam		580	.021			.93	2.28	3.21	3.92
4220	Common earth		516	.023			1.04	2.57	3.61	4.42
4240	Clay		325	.037			1.65	4.08	5.73	7
4400	300′ haul, sand & gravel		310	.039			1.74	4.27	6.01	7.35
4410	Sandy clay & loam		300	.040			1.79	4.42	6.21	7.60
4420	Common earth		270	.044			1.99	4.91	6.90	8.45
4440	Clay	↓	170	.071			3.16	7.80	10.96	13.35
5000	300 H.P., 50′ haul, sand & gravel	B-10M	1900	.006			.28	.96	1.24	1.48
5010	Sandy clay & loam		1850	.006			.29	.98	1.27	1.52
5020	Common earth		1650	.007			.33	1.10	1.43	1.71
5040	Clay		1025	.012			.52	1.77	2.29	2.75
5200	150′ haul, sand & gravel		920	.013			.58	1.97	2.55	3.06
5210	Sandy clay & loam		895	.013			.60	2.03	2.63	3.15
5220	Common earth		800	.015			.67	2.27	2.94	3.52
5240	Clay		500	.024			1.08	3.63	4.71	5.65
5400	300′ haul, sand & gravel		470	.026			1.14	3.87	5.01	6
5410	Sandy clay & loam	↓	455	.026	↓		1.18	3.99	5.17	6.20

31 23 16.46 Excavating, Bulk, Dozer

31 23 16.46 Excavating, Bulk, Dozer		Crew	Daily Output	Labor-Hours	Unit	Material	2014 Bare Costs Labor	Equipment	Total	Total Incl O&P
5420	Common earth	B-10M	410	.029	B.C.Y.		1.31	4.43	5.74	6.90
5440	Clay	↓	250	.048			2.15	7.25	9.40	11.30
5500	460 H.P., 50′ haul, sand & gravel	B-10X	1930	.006			.28	1.25	1.53	1.79
5506	Sandy clay & loam		1880	.006			.29	1.28	1.57	1.85
5510	Common earth		1680	.007			.32	1.43	1.75	2.07
5520	Clay		1050	.011			.51	2.29	2.80	3.30
5530	150′ haul, sand & gravel		1290	.009			.42	1.87	2.29	2.69
5535	Sandy clay & loam		1250	.010			.43	1.93	2.36	2.78
5540	Common earth		1120	.011			.48	2.15	2.63	3.10
5550	Clay		700	.017			.77	3.44	4.21	4.96
5560	300′ haul, sand & gravel		660	.018			.81	3.65	4.46	5.25
5565	Sandy clay & loam		640	.019			.84	3.76	4.60	5.40
5570	Common earth		575	.021			.94	4.19	5.13	6.05
5580	Clay	↓	350	.034			1.54	6.90	8.44	9.90
6000	700 H.P., 50′ haul, sand & gravel	B-10V	3500	.003			.15	1.28	1.43	1.64
6006	Sandy clay & loam		3400	.004			.16	1.32	1.48	1.69
6010	Common earth		3035	.004			.18	1.48	1.66	1.89
6020	Clay		1925	.006			.28	2.33	2.61	2.99
6030	150′ haul, sand & gravel		2025	.006			.27	2.22	2.49	2.84
6035	Sandy clay & loam		1960	.006			.27	2.29	2.56	2.94
6040	Common earth		1750	.007			.31	2.56	2.87	3.29
6050	Clay		1100	.011			.49	4.08	4.57	5.25
6060	300′ haul, sand & gravel		1030	.012			.52	4.35	4.87	5.60
6065	Sandy clay & loam		1005	.012			.54	4.46	5	5.75
6070	Common earth		900	.013			.60	4.98	5.58	6.40
6080	Clay	↓	550	.022	↓		.98	8.15	9.13	10.45
6090	For dozer with ripper, see Section 31 23 16.32									

31 23 16.48 Excavation, Bulk, Dragline

31 23 16.48 Excavation, Bulk, Dragline		Crew	Daily Output	Labor-Hours	Unit	Material	Labor	Equipment	Total	Total Incl O&P
0010	**EXCAVATION, BULK, DRAGLINE**									
0011	Excavate and load on truck, bank measure									
0012	Bucket drag line, 3/4 C.Y., sand/gravel	B-12I	440	.036	B.C.Y.		1.58	2.05	3.63	4.66
0100	Light clay		310	.052			2.24	2.91	5.15	6.65
0110	Heavy clay		280	.057			2.48	3.22	5.70	7.35
0120	Unclassified soil	↓	250	.064			2.78	3.60	6.38	8.20
0200	1-1/2 C.Y. bucket, sand/gravel	B-12P	575	.028			1.21	2.11	3.32	4.17
0210	Light clay		440	.036			1.58	2.76	4.34	5.45
0220	Heavy clay		352	.045			1.97	3.44	5.41	6.80
0230	Unclassified soil	↓	300	.053			2.32	4.04	6.36	8
0300	3 C.Y., sand/gravel	B-12V	720	.022			.97	2.15	3.12	3.84
0310	Light clay		700	.023			.99	2.21	3.20	3.95
0320	Heavy clay		600	.027			1.16	2.57	3.73	4.60
0330	Unclassified soil	↓	550	.029	↓		1.26	2.81	4.07	5

31 23 16.50 Excavation, Bulk, Scrapers

31 23 16.50 Excavation, Bulk, Scrapers		Crew	Daily Output	Labor-Hours	Unit	Material	Labor	Equipment	Total	Total Incl O&P
0010	**EXCAVATION, BULK, SCRAPERS**									
0100	Elev. scraper 11 C.Y., sand & gravel 1500′ haul, 1/4 dozer	B-33F	690	.020	B.C.Y.		.92	2.37	3.29	4.01
0150	3000′ haul		610	.023			1.04	2.68	3.72	4.54
0200	5000′ haul		505	.028			1.26	3.24	4.50	5.50
0300	Common earth, 1500′ haul		600	.023			1.06	2.72	3.78	4.61
0350	3000′ haul		530	.026			1.20	3.08	4.28	5.20
0400	5000′ haul		440	.032			1.44	3.71	5.15	6.30
0410	Sandy clay & loam, 1500′ haul		648	.022			.98	2.52	3.50	4.26
0420	3000′ haul	↓	572	.024	↓		1.11	2.86	3.97	4.83

31 23 16 – Excavation

31 23 16.50 Excavation, Bulk, Scrapers		Crew	Daily Output	Labor-Hours	Unit	Material	2014 Bare Costs Labor	Equipment	Total	Total Incl O&P
0430	5000' haul	B-33F	475	.029	B.C.Y.		1.34	3.44	4.78	5.80
0500	Clay, 1500' haul		375	.037			1.69	4.36	6.05	7.35
0550	3000' haul		330	.042			1.93	4.95	6.88	8.40
0600	5000' haul		275	.051			2.31	5.95	8.26	10.05
1000	Self propelled scraper, 14 C.Y. 1/4 push dozer, sand									
1050	Sand and gravel, 1500' haul	B-33D	920	.015	B.C.Y.		.69	2.48	3.17	3.78
1100	3000' haul		805	.017			.79	2.84	3.63	4.32
1200	5000' haul		645	.022			.99	3.54	4.53	5.40
1300	Common earth, 1500' haul		800	.018			.79	2.85	3.64	4.35
1350	3000' haul		700	.020			.91	3.26	4.17	4.97
1400	5000' haul		560	.025			1.14	4.08	5.22	6.20
1420	Sandy clay & loam, 1500' haul		864	.016			.74	2.64	3.38	4.03
1430	3000' haul		786	.018			.81	2.90	3.71	4.43
1440	5000' haul		605	.023			1.05	3.77	4.82	5.75
1500	Clay, 1500' haul		500	.028			1.27	4.57	5.84	6.95
1550	3000' haul		440	.032			1.44	5.20	6.64	7.90
1600	5000' haul		350	.040			1.82	6.50	8.32	9.95
2000	21 C.Y., 1/4 push dozer, sand & gravel, 1500' haul	B-33E	1180	.012			.54	2.59	3.13	3.67
2100	3000' haul		910	.015			.70	3.35	4.05	4.75
2200	5000' haul		750	.019			.85	4.07	4.92	5.75
2300	Common earth, 1500' haul		1030	.014			.62	2.96	3.58	4.20
2350	3000' haul		790	.018			.80	3.86	4.66	5.45
2400	5000' haul		650	.022			.98	4.70	5.68	6.65
2420	Sandy clay & loam, 1500' haul		1112	.013			.57	2.75	3.32	3.89
2430	3000' haul		854	.016			.74	3.57	4.31	5.05
2440	5000' haul		702	.020			.91	4.35	5.26	6.15
2500	Clay, 1500' haul		645	.022			.99	4.73	5.72	6.70
2550	3000' haul		495	.028			1.28	6.15	7.43	8.75
2600	5000' haul		405	.035			1.57	7.55	9.12	10.70
2700	Towed, 10 C.Y., 1/4 push dozer, sand & gravel, 1500' haul	B-33B	560	.025			1.14	4.31	5.45	6.45
2720	3000' haul		450	.031			1.41	5.35	6.76	8.05
2730	5000' haul		365	.038			1.74	6.60	8.34	9.90
2750	Common earth, 1500' haul		420	.033			1.51	5.75	7.26	8.60
2770	3000' haul		400	.035			1.59	6.05	7.64	9.05
2780	5000' haul		310	.045			2.05	7.80	9.85	11.65
2785	Sandy clay & Loam, 1500' haul		454	.031			1.40	5.30	6.70	8
2790	3000' haul		432	.032			1.47	5.60	7.07	8.40
2795	5000' haul		340	.041			1.87	7.10	8.97	10.65
2800	Clay, 1500' haul		315	.044			2.02	7.65	9.67	11.45
2820	3000' haul		300	.047			2.12	8.05	10.17	12.10
2840	5000' haul		225	.062			2.82	10.70	13.52	16.10
2900	15 C.Y., 1/4 push dozer, sand & gravel, 1500' haul	B-33C	800	.018			.79	3.04	3.83	4.55
2920	3000' haul		640	.022			.99	3.80	4.79	5.70
2940	5000' haul		520	.027			1.22	4.67	5.89	7
2960	Common earth, 1500' haul		600	.023			1.06	4.05	5.11	6.05
2980	3000' haul		560	.025			1.14	4.34	5.48	6.50
3000	5000' haul		440	.032			1.44	5.50	6.94	8.30
3005	Sandy clay & Loam, 1500' haul		648	.022			.98	3.75	4.73	5.60
3010	3000' haul		605	.023			1.05	4.02	5.07	6
3015	5000' haul		475	.029			1.34	5.10	6.44	7.70
3020	Clay, 1500' haul		450	.031			1.41	5.40	6.81	8.10
3040	3000' haul		420	.033			1.51	5.80	7.31	8.65
3060	5000' haul		320	.044			1.99	7.60	9.59	11.35

31 23 19 – Dewatering

31 23 19.10 Cut Drainage Ditch		Crew	Daily Output	Labor-Hours	Unit	2014 Bare Costs Material	Labor	Equipment	Total	Total Incl O&P
0010	**CUT DRAINAGE DITCH**									
0020	Cut drainage ditch common earth, 30′ wide x 1′ deep	B-11L	6000	.003	L.F.		.11	.12	.23	.30
0200	Clay and till		4200	.004			.16	.17	.33	.44
0250	Clean wet drainage ditch, 30′ wide		10000	.002			.07	.07	.14	.18

31 23 19.20 Dewatering Systems

		Crew	Daily Output	Labor-Hours	Unit	Material	Labor	Equipment	Total	Total Incl O&P
0010	**DEWATERING SYSTEMS**									
0020	Excavate drainage trench, 2′ wide, 2′ deep	B-11C	90	.178	C.Y.		7.60	4.06	11.66	16.05
0100	2′ wide, 3′ deep, with backhoe loader	″	135	.119			5.05	2.71	7.76	10.75
0200	Excavate sump pits by hand, light soil	1 Clab	7.10	1.127			41.50		41.50	63.50
0300	Heavy soil	″	3.50	2.286			84		84	129
0500	Pumping 8 hr., attended 2 hrs. per day, including 20 L.F.									
0550	of suction hose & 100 L.F. discharge hose									
0600	2″ diaphragm pump used for 8 hours	B-10H	4	3	Day		134	18.95	152.95	226
0620	Add per additional pump							72	72	79
0650	4″ diaphragm pump used for 8 hours	B-10I	4	3			134	31	165	239
0670	Add per additional pump							116	116	127
0800	8 hrs. attended, 2″ diaphragm pump	B-10H	1	12			540	75.50	615.50	905
0820	Add per additional pump							72	72	79
0900	3″ centrifugal pump	B-10J	1	12			540	84.50	624.50	915
0920	Add per additional pump							79	79	86.50
1000	4″ diaphragm pump	B-10I	1	12			540	124	664	955
1020	Add per additional pump							116	116	127
1100	6″ centrifugal pump	B-10K	1	12			540	365	905	1,225
1120	Add per additional pump							340	340	375
1300	CMP, incl. excavation 3′ deep, 12″ diameter	B-6	115	.209	L.F.	8.35	8.35	3.18	19.88	25.50
1400	18″ diameter		100	.240	″	12.55	9.65	3.65	25.85	32.50
1600	Sump hole construction, incl. excavation and gravel, pit		1250	.019	C.F.	1.12	.77	.29	2.18	2.73
1700	With 12″ gravel collar, 12″ pipe, corrugated, 16 ga.		70	.343	L.F.	21.50	13.75	5.20	40.45	51
1800	15″ pipe, corrugated, 16 ga.		55	.436		28	17.50	6.65	52.15	65
1900	18″ pipe, corrugated, 16 ga.		50	.480		32.50	19.25	7.30	59.05	73.50
2000	24″ pipe, corrugated, 14 ga.		40	.600		39	24	9.15	72.15	90
2200	Wood lining, up to 4′ x 4′, add		300	.080	SFCA	16.25	3.21	1.22	20.68	24
9950	See Section 31 23 19.40 for wellpoints									
9960	See Section 31 23 19.30 for deep well systems									
9970	See Section 22 11 23 for pumps									

31 23 19.30 Wells

		Crew	Daily Output	Labor-Hours	Unit	Material	Labor	Equipment	Total	Total Incl O&P
0010	**WELLS**									
0011	For dewatering 10′ to 20′ deep, 2′ diameter									
0020	with steel casing, minimum	B-6	165	.145	V.L.F.	38	5.85	2.21	46.06	53
0050	Average		98	.245		43	9.85	3.73	56.58	66.50
0100	Maximum		49	.490		48.50	19.65	7.45	75.60	91.50
0300	For dewatering pumps see 01 54 33 in Reference Section									
0500	For domestic water wells, see Section 33 21 13.10									

31 23 19.40 Wellpoints

		Crew	Daily Output	Labor-Hours	Unit	Material	Labor	Equipment	Total	Total Incl O&P
0010	**WELLPOINTS**									
0011	For equipment rental, see 01 54 33 in Reference Section									
0100	Installation and removal of single stage system									
0110	Labor only, .75 labor-hours per L.F.	1 Clab	10.70	.748	LF Hdr		27.50		27.50	42.50
0200	2.0 labor-hours per L.F.	″	4	2	″		73.50		73.50	113
0400	Pump operation, 4 @ 6 hr. shifts									
0410	Per 24 hour day	4 Eqlt	1.27	25.197	Day		1,175		1,175	1,800

31 23 Excavation and Fill

31 23 19 – Dewatering

31 23 19.40 Wellpoints		Crew	Daily Output	Labor-Hours	Unit	Material	2014 Bare Costs Labor	Equipment	Total	Total Incl O&P
0500	Per 168 hour week, 160 hr. straight, 8 hr. double time	4 Eqlt	.18	177	Week		8,375		8,375	12,700
0550	Per 4.3 week month		.04	800	Month		37,600		37,600	57,000
0600	Complete installation, operation, equipment rental, fuel &									
0610	removal of system with 2" wellpoints 5' O.C.									
0700	100' long header, 6" diameter, first month	4 Eqlt	3.23	9.907	LF Hdr	159	465		624	880
0800	Thereafter, per month		4.13	7.748		127	365		492	695
1000	200' long header, 8" diameter, first month		6	5.333		144	251		395	540
1100	Thereafter, per month		8.39	3.814		71.50	179		250.50	350
1300	500' long header, 8" diameter, first month		10.63	3.010		55.50	142		197.50	276
1400	Thereafter, per month		20.91	1.530		40	72		112	153
1600	1,000' long header, 10" diameter, first month		11.62	2.754		47.50	130		177.50	249
1700	Thereafter, per month		41.81	.765		24	36		60	80.50
1900	Note: above figures include pumping 168 hrs. per week									
1910	and include the pump operator and one stand-by pump.									

31 23 23 – Fill

31 23 23.13 Backfill

Line	Description	Crew	Daily Output	Labor-Hours	Unit	Material	Labor	Equipment	Total	Total Incl O&P
0010	**BACKFILL**									
0015	By hand, no compaction, light soil	1 Clab	14	.571	L.C.Y.		21		21	32.50
0100	Heavy soil		11	.727	"		26.50		26.50	41
0300	Compaction in 6" layers, hand tamp, add to above		20.60	.388	E.C.Y.		14.25		14.25	22
0400	Roller compaction operator walking, add	B-10A	100	.120			5.40	1.83	7.23	10.20
0500	Air tamp, add	B-9D	190	.211			7.80	1.40	9.20	13.60
0600	Vibrating plate, add	A-1D	60	.133			4.89	.60	5.49	8.20
0800	Compaction in 12" layers, hand tamp, add to above	1 Clab	34	.235			8.60		8.60	13.30
0900	Roller compaction operator walking, add	B-10A	150	.080			3.59	1.22	4.81	6.80
1000	Air tamp, add	B-9	285	.140			5.20	.82	6.02	8.90
1100	Vibrating plate, add	A-1E	90	.089			3.26	.51	3.77	5.60
3000	For flowable fill, see Section 03 31 13.35									

31 23 23.14 Backfill, Structural

Line	Description	Crew	Daily Output	Labor-Hours	Unit	Material	Labor	Equipment	Total	Total Incl O&P
0010	**BACKFILL, STRUCTURAL**									
0011	Dozer or F.E. loader									
0020	From existing stockpile, no compaction									
2000	80 H.P., 50' haul, sand & gravel	B-10L	1100	.011	L.C.Y.		.49	.44	.93	1.22
2010	Sandy clay & loam		1070	.011			.50	.45	.95	1.26
2020	Common earth		975	.012			.55	.49	1.04	1.38
2040	Clay		850	.014			.63	.57	1.20	1.58
2200	150' haul, sand & gravel		550	.022			.98	.87	1.85	2.45
2210	Sandy clay & loam		535	.022			1.01	.90	1.91	2.52
2220	Common earth		490	.024			1.10	.98	2.08	2.75
2240	Clay		425	.028			1.27	1.13	2.40	3.18
2400	300' haul, sand & gravel		370	.032			1.45	1.30	2.75	3.64
2410	Sandy clay & loam		360	.033			1.49	1.34	2.83	3.75
2420	Common earth		330	.036			1.63	1.46	3.09	4.08
2440	Clay		290	.041			1.85	1.66	3.51	4.65
3000	105 H.P., 50' haul, sand & gravel	B-10W	1350	.009			.40	.44	.84	1.09
3010	Sandy clay & loam		1325	.009			.41	.45	.86	1.11
3020	Common earth		1225	.010			.44	.48	.92	1.20
3040	Clay		1100	.011			.49	.54	1.03	1.33
3200	150' haul, sand & gravel		670	.018			.80	.88	1.68	2.19
3210	Sandy clay & loam		655	.018			.82	.90	1.72	2.24
3220	Common earth		610	.020			.88	.97	1.85	2.41
3240	Clay		550	.022			.98	1.08	2.06	2.67

31 23 23 – Fill

31 23 23.14 Backfill, Structural		Crew	Daily Output	Labor-Hours	Unit	Material (2014 Bare Costs)	Labor (2014 Bare Costs)	Equipment (2014 Bare Costs)	Total (2014 Bare Costs)	Total Incl O&P
3300	300' haul, sand & gravel	B-10W	465	.026	L.C.Y.		1.16	1.27	2.43	3.16
3310	Sandy clay & loam		455	.026			1.18	1.30	2.48	3.23
3320	Common earth		415	.029			1.30	1.43	2.73	3.54
3340	Clay		370	.032			1.45	1.60	3.05	3.97
4000	200 H.P., 50' haul, sand & gravel	B-10B	2500	.005			.22	.53	.75	.91
4010	Sandy clay & loam		2435	.005			.22	.54	.76	.94
4020	Common earth		2200	.005			.24	.60	.84	1.03
4040	Clay		1950	.006			.28	.68	.96	1.17
4200	150' haul, sand & gravel		1225	.010			.44	1.08	1.52	1.86
4210	Sandy clay & loam		1200	.010			.45	1.10	1.55	1.89
4220	Common earth		1100	.011			.49	1.20	1.69	2.07
4240	Clay		975	.012			.55	1.36	1.91	2.34
4400	300' haul, sand & gravel		805	.015			.67	1.65	2.32	2.83
4410	Sandy clay & loam		790	.015			.68	1.68	2.36	2.89
4420	Common earth		735	.016			.73	1.80	2.53	3.10
4440	Clay		660	.018			.81	2.01	2.82	3.45
5000	300 H.P., 50' haul, sand & gravel	B-10M	3170	.004			.17	.57	.74	.89
5010	Sandy clay & loam		3110	.004			.17	.58	.75	.90
5020	Common earth		2900	.004			.19	.63	.82	.97
5040	Clay		2700	.004			.20	.67	.87	1.04
5200	150' haul, sand & gravel		2200	.005			.24	.83	1.07	1.28
5210	Sandy clay & loam		2150	.006			.25	.84	1.09	1.31
5220	Common earth		1950	.006			.28	.93	1.21	1.44
5240	Clay		1700	.007			.32	1.07	1.39	1.66
5400	300' haul, sand & gravel		1500	.008			.36	1.21	1.57	1.88
5410	Sandy clay & loam		1470	.008			.37	1.24	1.61	1.92
5420	Common earth		1350	.009			.40	1.35	1.75	2.09
5440	Clay		1225	.010			.44	1.48	1.92	2.30
6000	For compaction, see Section 31 23 23.23									
6010	For trench backfill, see Section 31 23 16.13 and 31 23 16.14									

31 23 23.15 Borrow, Loading And/Or Spreading		Crew	Daily Output	Labor-Hours	Unit	Material	Labor	Equipment	Total	Total Incl O&P
0010	**BORROW, LOADING AND/OR SPREADING**									
4000	Common earth, shovel, 1 C.Y. bucket	B-12N	840	.019	B.C.Y.	16.65	.83	1.48	18.96	21
4010	1-1/2 C.Y. bucket	B-12O	1135	.014		16.65	.61	1.11	18.37	20.50
4020	3 C.Y. bucket	B-12T	1800	.009		16.65	.39	.89	17.93	19.85
4030	Front end loader, wheel mounted									
4050	3/4 C.Y. bucket	B-10R	550	.022	B.C.Y.	16.65	.98	.54	18.17	20.50
4060	1-1/2 C.Y. bucket	B-10S	970	.012		16.65	.55	.39	17.59	19.55
4070	3 C.Y. bucket	B-10T	1575	.008		16.65	.34	.33	17.32	19.20
4080	5 C.Y. bucket	B-10U	2600	.005		16.65	.21	.40	17.26	19.05
5000	Select granular fill, shovel, 1 C.Y. bucket	B-12N	925	.017		21	.75	1.34	23.09	25.50
5010	1-1/2 C.Y. bucket	B-12O	1250	.013		21	.56	1.01	22.57	25
5020	3 C.Y. bucket	B-12T	1980	.008		21	.35	.81	22.16	24.50
5030	Front end loader, wheel mounted									
5050	3/4 C.Y. bucket	B-10R	800	.015	B.C.Y.	21	.67	.37	22.04	24.50
5060	1-1/2 C.Y. bucket	B-10S	1065	.011		21	.51	.35	21.86	24
5070	3 C.Y. bucket	B-10T	1735	.007		21	.31	.30	21.61	24
5080	5 C.Y. bucket	B-10U	2850	.004		21	.19	.37	21.56	23.50
6000	Clay, till, or blasted rock, shovel, 1 C.Y. bucket	B-12N	715	.022		12.35	.97	1.74	15.06	16.95
6010	1-1/2 C.Y. bucket	B-12O	965	.017		12.35	.72	1.30	14.37	16.10
6020	3 C.Y. bucket	B-12T	1530	.010		12.35	.45	1.04	13.84	15.40
6030	Front end loader, wheel mounted									

31 23 23.15 Borrow, Loading And/Or Spreading

	31 23 23.15 Borrow, Loading And/Or Spreading	Crew	Daily Output	Labor-Hours	Unit	2014 Bare Costs Material	Labor	Equipment	Total	Total Incl O&P
6035	3/4 C.Y. bucket	B-10R	465	.026	B.C.Y.	12.35	1.16	.63	14.14	16
6040	1-1/2 C.Y. bucket	B-10S	825	.015		12.35	.65	.45	13.45	15.05
6045	3 C.Y. bucket	B-10T	1340	.009		12.35	.40	.39	13.14	14.60
6050	5 C.Y. bucket	B-10U	2200	.005		12.35	.24	.48	13.07	14.45
6060	Front end loader, track mounted									
6065	1-1/2 C.Y. bucket	B-10N	715	.017	B.C.Y.	12.35	.75	.73	13.83	15.50
6070	3 C.Y. bucket	B-10P	1190	.010		12.35	.45	1	13.80	15.35
6075	5 C.Y. bucket	B-10Q	1835	.007		12.35	.29	.84	13.48	14.90
7000	Topsoil or loam from stockpile, shovel, 1 C.Y. bucket	B-12N	840	.019		24.50	.83	1.48	26.81	30
7010	1-1/2 C.Y. bucket	B-12O	1135	.014		24.50	.61	1.11	26.22	29
7020	3 C.Y. bucket	B-12T	1800	.009		24.50	.39	.89	25.78	28.50
7030	Front end loader, wheel mounted									
7050	3/4 C.Y. bucket	B-10R	550	.022	B.C.Y.	24.50	.98	.54	26.02	29
7060	1-1/2 C.Y. bucket	B-10S	970	.012		24.50	.55	.39	25.44	28.50
7070	3 C.Y. bucket	B-10T	1575	.008		24.50	.34	.33	25.17	28
7080	5 C.Y. bucket	B-10U	2600	.005		24.50	.21	.40	25.11	28
8900	For larger hauling units, deduct from above								30%	30%
9000	Hauling only, excavated or borrow material, see Section 31 23 23.20									

31 23 23.16 Fill By Borrow and Utility Bedding

	31 23 23.16 Fill By Borrow and Utility Bedding	Crew	Daily Output	Labor-Hours	Unit	Material	Labor	Equipment	Total	Total Incl O&P
0010	**FILL BY BORROW AND UTILITY BEDDING**									
0049	Utility bedding, for pipe & conduit, not incl. compaction									
0050	Crushed or screened bank run gravel	B-6	150	.160	L.C.Y.	25	6.40	2.44	33.84	40
0100	Crushed stone 3/4" to 1/2"		150	.160		28.50	6.40	2.44	37.34	43.50
0200	Sand, dead or bank		150	.160		17.55	6.40	2.44	26.39	32
0500	Compacting bedding in trench	A-1D	90	.089	E.C.Y.		3.26	.40	3.66	5.50
0600	If material source exceeds 2 miles, add for extra mileage.									
0610	See Section 31 23 23.20 for hauling mileage add.									

31 23 23.17 General Fill

	31 23 23.17 General Fill	Crew	Daily Output	Labor-Hours	Unit	Material	Labor	Equipment	Total	Total Incl O&P
0010	**GENERAL FILL**									
0011	Spread dumped material, no compaction									
0020	By dozer, no compaction	B-10B	1000	.012	L.C.Y.		.54	1.33	1.87	2.28
0100	By hand	1 Clab	12	.667	"		24.50		24.50	37.50
0150	Spread fill, from stockpile with 2-1/2 C.Y. F.E. loader									
0170	130 H.P., 300' haul	B-10P	600	.020	L.C.Y.		.90	1.99	2.89	3.55
0190	With dozer 300 H.P., 300' haul	B-10M	600	.020	"		.90	3.03	3.93	4.70
0400	For compaction of embankment, see Section 31 23 23.23									
0500	Gravel fill, compacted, under floor slabs, 4" deep	B-37	10000	.005	S.F.	.42	.19	.02	.63	.77
0600	6" deep		8600	.006		.63	.22	.02	.87	1.05
0700	9" deep		7200	.007		1.05	.26	.02	1.33	1.58
0800	12" deep		6000	.008		1.48	.31	.03	1.82	2.13
1000	Alternate pricing method, 4" deep		120	.400	E.C.Y.	31.50	15.50	1.32	48.32	60.50
1100	6" deep		160	.300		31.50	11.60	.99	44.09	54
1200	9" deep		200	.240		31.50	9.30	.79	41.59	50
1300	12" deep		220	.218		31.50	8.45	.72	40.67	49
1500	For fill under exterior paving, see Section 32 11 23.23									
1600	For flowable fill, see Section 03 31 13.35									

31 23 23.19 Backfill, Airport Subgrade

	31 23 23.19 Backfill, Airport Subgrade	Crew	Daily Output	Labor-Hours	Unit	Material	Labor	Equipment	Total	Total Incl O&P
0010	**BACKFILL, AIRPORT SUBGRADE**									
0100	Backfill, dozer, sand & gravel, 12" depth ,300' haul, no compaction	B-10M	4500	.003	S.Y.		.12	.40	.52	.62
0200	13" depth		4155	.003			.13	.44	.57	.68
0300	14" depth		3860	.003			.14	.47	.61	.73
0400	15" depth		3600	.003			.15	.50	.65	.78

31 23 23 – Fill

31 23 23.19 Backfill, Airport Subgrade		Crew	Daily Output	Labor-Hours	Unit	Material	2014 Bare Costs Labor	Equipment	Total	Total Incl O&P
0500	16″ depth	B-10M	3375	.004	S.Y.		.16	.54	.70	.83
0600	17″ depth		3175	.004			.17	.57	.74	.89
0700	18″ depth		3000	.004			.18	.61	.79	.94
0800	19″ depth		2845	.004			.19	.64	.83	.99
0900	20″ depth		2700	.004			.20	.67	.87	1.04
1100	22″ depth		2455	.005			.22	.74	.96	1.14
1200	23″ depth		2350	.005			.23	.77	1	1.20
1300	24″ depth		2250	.005			.24	.81	1.05	1.25
1400	26″ depth		2075	.006			.26	.88	1.14	1.35
1500	28″ depth		1930	.006			.28	.94	1.22	1.46
1600	30″ depth		1800	.007			.30	1.01	1.31	1.57
1700	32″ depth		1690	.007			.32	1.08	1.40	1.66
1800	34″ depth		1590	.008			.34	1.14	1.48	1.78
1900	36″ depth		1500	.008			.36	1.21	1.57	1.88
2000	38″ depth		1420	.008			.38	1.28	1.66	1.99
2100	40″ depth		1350	.009			.40	1.35	1.75	2.09
2200	42″ depth		1285	.009			.42	1.41	1.83	2.20
2300	44″ depth		1230	.010			.44	1.48	1.92	2.30
2400	46″ depth		1175	.010			.46	1.55	2.01	2.40
2500	48″ depth		1125	.011			.48	1.62	2.10	2.51
2600	54″ depth		1000	.012			.54	1.82	2.36	2.82
2700	60″ depth		900	.013			.60	2.02	2.62	3.13
3100	Backfill, dozer, common earth, 12″ depth ,300′ haul, no compaction		4050	.003			.13	.45	.58	.69
3200	13″ depth		3740	.003			.14	.49	.63	.75
3300	14″ depth		3470	.003			.16	.52	.68	.82
3400	15″ depth		3240	.004			.17	.56	.73	.87
3500	16″ depth		3040	.004			.18	.60	.78	.93
3600	17″ depth		2860	.004			.19	.64	.83	.99
3700	18″ depth		2700	.004			.20	.67	.87	1.04
3800	19″ depth		2560	.005			.21	.71	.92	1.10
3900	20″ depth		2430	.005			.22	.75	.97	1.16
4000	21″ depth		2315	.005			.23	.78	1.01	1.21
4100	22″ depth		2210	.005			.24	.82	1.06	1.27
4200	23″ depth		2115	.006			.25	.86	1.11	1.33
4300	24″ depth		2025	.006			.27	.90	1.17	1.39
4400	30″ depth		1620	.007			.33	1.12	1.45	1.74
4500	36″ depth		900	.013			.60	2.02	2.62	3.13
9900	For compaction of subgrade see Section 31 23 23.25									

31 23 23.20 Hauling		Crew	Daily Output	Labor-Hours	Unit	Material	Labor	Equipment	Total	Total Incl O&P
0010	**HAULING**									
0011	Excavated or borrow, loose cubic yards									
0012	no loading equipment, including hauling,waiting, loading/dumping									
0013	time per cycle (wait, load, travel, unload or dump & return)									
0014	8 C.Y. truck, 15 MPH ave, cycle 0.5 miles, 10 min. wait/Ld./Uld.	B-34A	320	.025	L.C.Y.		.94	1.30	2.24	2.86
0016	cycle 1 mile		272	.029			1.10	1.53	2.63	3.38
0018	cycle 2 miles		208	.038			1.44	2.01	3.45	4.41
0020	cycle 4 miles		144	.056			2.09	2.90	4.99	6.35
0022	cycle 6 miles		112	.071			2.68	3.73	6.41	8.20
0024	cycle 8 miles		88	.091			3.41	4.74	8.15	10.40
0026	20 MPH ave,cycle 0.5 mile		336	.024			.89	1.24	2.13	2.73
0028	cycle 1 mile		296	.027			1.02	1.41	2.43	3.10
0030	cycle 2 miles		240	.033			1.25	1.74	2.99	3.82

31 23 23.20 Hauling		Crew	Daily Output	Labor-Hours	Unit	Material	2014 Bare Costs Labor	Equipment	Total	Total Incl O&P
0032	cycle 4 miles	B-34A	176	.045	L.C.Y.		1.71	2.37	4.08	5.20
0034	cycle 6 miles		136	.059			2.21	3.07	5.28	6.75
0036	cycle 8 miles		112	.071			2.68	3.73	6.41	8.20
0044	25 MPH ave, cycle 4 miles		192	.042			1.56	2.17	3.73	4.78
0046	cycle 6 miles		160	.050			1.88	2.61	4.49	5.75
0048	cycle 8 miles		128	.063			2.35	3.26	5.61	7.15
0050	30 MPH ave, cycle 4 miles		216	.037			1.39	1.93	3.32	4.25
0052	cycle 6 miles		176	.045			1.71	2.37	4.08	5.20
0054	cycle 8 miles		144	.056			2.09	2.90	4.99	6.35
0114	15 MPH ave, cycle 0.5 mile, 15 min. wait/Ld./Uld.		224	.036			1.34	1.86	3.20	4.10
0116	cycle 1 mile		200	.040			1.50	2.09	3.59	4.58
0118	cycle 2 miles		168	.048			1.79	2.48	4.27	5.45
0120	cycle 4 miles		120	.067			2.50	3.48	5.98	7.65
0122	cycle 6 miles		96	.083			3.13	4.35	7.48	9.55
0124	cycle 8 miles		80	.100			3.76	5.20	8.96	11.50
0126	20 MPH ave, cycle 0.5 mile		232	.034			1.29	1.80	3.09	3.96
0128	cycle 1 mile		208	.038			1.44	2.01	3.45	4.41
0130	cycle 2 miles		184	.043			1.63	2.27	3.90	4.98
0132	cycle 4 miles		144	.056			2.09	2.90	4.99	6.35
0134	cycle 6 miles		112	.071			2.68	3.73	6.41	8.20
0136	cycle 8 miles		96	.083			3.13	4.35	7.48	9.55
0144	25 MPH ave, cycle 4 miles		152	.053			1.98	2.74	4.72	6.05
0146	cycle 6 miles		128	.063			2.35	3.26	5.61	7.15
0148	cycle 8 miles		112	.071			2.68	3.73	6.41	8.20
0150	30 MPH ave, cycle 4 miles		168	.048			1.79	2.48	4.27	5.45
0152	cycle 6 miles		144	.056			2.09	2.90	4.99	6.35
0154	cycle 8 miles		120	.067			2.50	3.48	5.98	7.65
0214	15 MPH ave, cycle 0.5 mile, 20 min wait/Ld./Uld.		176	.045			1.71	2.37	4.08	5.20
0216	cycle 1 mile		160	.050			1.88	2.61	4.49	5.75
0218	cycle 2 miles		136	.059			2.21	3.07	5.28	6.75
0220	cycle 4 miles		104	.077			2.89	4.01	6.90	8.80
0222	cycle 6 miles		88	.091			3.41	4.74	8.15	10.40
0224	cycle 8 miles		72	.111			4.17	5.80	9.97	12.70
0226	20 MPH ave, cycle 0.5 mile		176	.045			1.71	2.37	4.08	5.20
0228	cycle 1 mile		168	.048			1.79	2.48	4.27	5.45
0230	cycle 2 miles		144	.056			2.09	2.90	4.99	6.35
0232	cycle 4 miles		120	.067			2.50	3.48	5.98	7.65
0234	cycle 6 miles		96	.083			3.13	4.35	7.48	9.55
0236	cycle 8 miles		88	.091			3.41	4.74	8.15	10.40
0244	25 MPH ave, cycle 4 miles		128	.063			2.35	3.26	5.61	7.15
0246	cycle 6 miles		112	.071			2.68	3.73	6.41	8.20
0248	cycle 8 miles		96	.083			3.13	4.35	7.48	9.55
0250	30 MPH ave, cycle 4 miles		136	.059			2.21	3.07	5.28	6.75
0252	cycle 6 miles		120	.067			2.50	3.48	5.98	7.65
0254	cycle 8 miles		104	.077			2.89	4.01	6.90	8.80
0314	15 MPH ave, cycle 0.5 mile, 25 min wait/Ld./Uld.		144	.056			2.09	2.90	4.99	6.35
0316	cycle 1 mile		128	.063			2.35	3.26	5.61	7.15
0318	cycle 2 miles		112	.071			2.68	3.73	6.41	8.20
0320	cycle 4 miles		96	.083			3.13	4.35	7.48	9.55
0322	cycle 6 miles		80	.100			3.76	5.20	8.96	11.50
0324	cycle 8 miles		64	.125			4.69	6.50	11.19	14.30
0326	20 MPH ave, cycle 0.5 mile		144	.056			2.09	2.90	4.99	6.35
0328	cycle 1 mile		136	.059			2.21	3.07	5.28	6.75

31 23 23.20 Hauling		Crew	Daily Output	Labor-Hours	Unit	Material	2014 Bare Costs Labor	Equipment	Total	Total Incl O&P
0330	cycle 2 miles	B-34A	120	.067	L.C.Y.		2.50	3.48	5.98	7.65
0332	cycle 4 miles		104	.077			2.89	4.01	6.90	8.80
0334	cycle 6 miles		88	.091			3.41	4.74	8.15	10.40
0336	cycle 8 miles		80	.100			3.76	5.20	8.96	11.50
0344	25 MPH ave, cycle 4 miles		112	.071			2.68	3.73	6.41	8.20
0346	cycle 6 miles		96	.083			3.13	4.35	7.48	9.55
0348	cycle 8 miles		88	.091			3.41	4.74	8.15	10.40
0350	30 MPH ave, cycle 4 miles		112	.071			2.68	3.73	6.41	8.20
0352	cycle 6 miles		104	.077			2.89	4.01	6.90	8.80
0354	cycle 8 miles		96	.083			3.13	4.35	7.48	9.55
0414	15 MPH ave, cycle 0.5 mile, 30 min wait/Ld./Uld.		120	.067			2.50	3.48	5.98	7.65
0416	cycle 1 mile		112	.071			2.68	3.73	6.41	8.20
0418	cycle 2 miles		96	.083			3.13	4.35	7.48	9.55
0420	cycle 4 miles		80	.100			3.76	5.20	8.96	11.50
0422	cycle 6 miles		72	.111			4.17	5.80	9.97	12.70
0424	cycle 8 miles		64	.125			4.69	6.50	11.19	14.30
0426	20 MPH ave, cycle 0.5 mile		120	.067			2.50	3.48	5.98	7.65
0428	cycle 1 mile		112	.071			2.68	3.73	6.41	8.20
0430	cycle 2 miles		104	.077			2.89	4.01	6.90	8.80
0432	cycle 4 miles		88	.091			3.41	4.74	8.15	10.40
0434	cycle 6 miles		80	.100			3.76	5.20	8.96	11.50
0436	cycle 8 miles		72	.111			4.17	5.80	9.97	12.70
0444	25 MPH ave, cycle 4 miles		96	.083			3.13	4.35	7.48	9.55
0446	cycle 6 miles		88	.091			3.41	4.74	8.15	10.40
0448	cycle 8 miles		80	.100			3.76	5.20	8.96	11.50
0450	30 MPH ave, cycle 4 miles		96	.083			3.13	4.35	7.48	9.55
0452	cycle 6 miles		88	.091			3.41	4.74	8.15	10.40
0454	cycle 8 miles		80	.100			3.76	5.20	8.96	11.50
0514	15 MPH ave, cycle 0.5 mile, 35 min wait/Ld./Uld.		104	.077			2.89	4.01	6.90	8.80
0516	cycle 1 mile		96	.083			3.13	4.35	7.48	9.55
0518	cycle 2 miles		88	.091			3.41	4.74	8.15	10.40
0520	cycle 4 miles		72	.111			4.17	5.80	9.97	12.70
0522	cycle 6 miles		64	.125			4.69	6.50	11.19	14.30
0524	cycle 8 miles		56	.143			5.35	7.45	12.80	16.40
0526	20 MPH ave, cycle 0.5 mile		104	.077			2.89	4.01	6.90	8.80
0528	cycle 1 mile		96	.083			3.13	4.35	7.48	9.55
0530	cycle 2 miles		96	.083			3.13	4.35	7.48	9.55
0532	cycle 4 miles		80	.100			3.76	5.20	8.96	11.50
0534	cycle 6 miles		72	.111			4.17	5.80	9.97	12.70
0536	cycle 8 miles		64	.125			4.69	6.50	11.19	14.30
0544	25 MPH ave, cycle 4 miles		88	.091			3.41	4.74	8.15	10.40
0546	cycle 6 miles		80	.100			3.76	5.20	8.96	11.50
0548	cycle 8 miles		72	.111			4.17	5.80	9.97	12.70
0550	30 MPH ave, cycle 4 miles		88	.091			3.41	4.74	8.15	10.40
0552	cycle 6 miles		80	.100			3.76	5.20	8.96	11.50
0554	cycle 8 miles		72	.111			4.17	5.80	9.97	12.70
1014	12 C.Y. truck, cycle 0.5 mile, 15 MPH ave, 15 min. wait/Ld./Uld.	B-34B	336	.024			.89	2.06	2.95	3.63
1016	cycle 1 mile		300	.027			1	2.31	3.31	4.07
1018	cycle 2 miles		252	.032			1.19	2.75	3.94	4.84
1020	cycle 4 miles		180	.044			1.67	3.85	5.52	6.80
1022	cycle 6 miles		144	.056			2.09	4.81	6.90	8.50
1024	cycle 8 miles		120	.067			2.50	5.75	8.25	10.15
1025	cycle 10 miles		96	.083			3.13	7.20	10.33	12.70

31 23 23.20 Hauling		Crew	Daily Output	Labor-Hours	Unit	Material	2014 Bare Costs Labor	Equipment	Total	Total Incl O&P
1026	20 MPH ave, cycle 0.5 mile	B-34B	348	.023	L.C.Y.		.86	1.99	2.85	3.51
1028	cycle 1 mile		312	.026			.96	2.22	3.18	3.91
1030	cycle 2 miles		276	.029			1.09	2.51	3.60	4.42
1032	cycle 4 miles		216	.037			1.39	3.21	4.60	5.65
1034	cycle 6 miles		168	.048			1.79	4.12	5.91	7.25
1036	cycle 8 miles		144	.056			2.09	4.81	6.90	8.50
1038	cycle 10 miles		120	.067			2.50	5.75	8.25	10.15
1040	25 MPH ave, cycle 4 miles		228	.035			1.32	3.04	4.36	5.35
1042	cycle 6 miles		192	.042			1.56	3.61	5.17	6.35
1044	cycle 8 miles		168	.048			1.79	4.12	5.91	7.25
1046	cycle 10 miles		144	.056			2.09	4.81	6.90	8.50
1050	30 MPH ave, cycle 4 miles		252	.032			1.19	2.75	3.94	4.84
1052	cycle 6 miles		216	.037			1.39	3.21	4.60	5.65
1054	cycle 8 miles		180	.044			1.67	3.85	5.52	6.80
1056	cycle 10 miles		156	.051			1.93	4.44	6.37	7.80
1060	35 MPH ave, cycle 4 miles		264	.030			1.14	2.62	3.76	4.62
1062	cycle 6 miles		228	.035			1.32	3.04	4.36	5.35
1064	cycle 8 miles		204	.039			1.47	3.39	4.86	6
1066	cycle 10 miles		180	.044			1.67	3.85	5.52	6.80
1068	cycle 20 miles		120	.067			2.50	5.75	8.25	10.15
1069	cycle 30 miles		84	.095			3.58	8.25	11.83	14.50
1070	cycle 40 miles		72	.111			4.17	9.60	13.77	16.95
1072	40 MPH ave, cycle 6 miles		240	.033			1.25	2.88	4.13	5.10
1074	cycle 8 miles		216	.037			1.39	3.21	4.60	5.65
1076	cycle 10 miles		192	.042			1.56	3.61	5.17	6.35
1078	cycle 20 miles		120	.067			2.50	5.75	8.25	10.15
1080	cycle 30 miles		96	.083			3.13	7.20	10.33	12.70
1082	cycle 40 miles		72	.111			4.17	9.60	13.77	16.95
1084	cycle 50 miles		60	.133			5	11.55	16.55	20.50
1094	45 MPH ave, cycle 8 miles		216	.037			1.39	3.21	4.60	5.65
1096	cycle 10 miles		204	.039			1.47	3.39	4.86	6
1098	cycle 20 miles		132	.061			2.28	5.25	7.53	9.20
1100	cycle 30 miles		108	.074			2.78	6.40	9.18	11.30
1102	cycle 40 miles		84	.095			3.58	8.25	11.83	14.50
1104	cycle 50 miles		72	.111			4.17	9.60	13.77	16.95
1106	50 MPH ave, cycle 10 miles		216	.037			1.39	3.21	4.60	5.65
1108	cycle 20 miles		144	.056			2.09	4.81	6.90	8.50
1110	cycle 30 miles		108	.074			2.78	6.40	9.18	11.30
1112	cycle 40 miles		84	.095			3.58	8.25	11.83	14.50
1114	cycle 50 miles		72	.111			4.17	9.60	13.77	16.95
1214	15 MPH ave, cycle 0.5 mile, 20 min. wait/Ld./Uld.		264	.030			1.14	2.62	3.76	4.62
1216	cycle 1 mile		240	.033			1.25	2.88	4.13	5.10
1218	cycle 2 miles		204	.039			1.47	3.39	4.86	6
1220	cycle 4 miles		156	.051			1.93	4.44	6.37	7.80
1222	cycle 6 miles		132	.061			2.28	5.25	7.53	9.20
1224	cycle 8 miles		108	.074			2.78	6.40	9.18	11.30
1225	cycle 10 miles		96	.083			3.13	7.20	10.33	12.70
1226	20 MPH ave, cycle 0.5 mile		264	.030			1.14	2.62	3.76	4.62
1228	cycle 1 mile		252	.032			1.19	2.75	3.94	4.84
1230	cycle 2 miles		216	.037			1.39	3.21	4.60	5.65
1232	cycle 4 miles		180	.044			1.67	3.85	5.52	6.80
1234	cycle 6 miles		144	.056			2.09	4.81	6.90	8.50
1236	cycle 8 miles		132	.061			2.28	5.25	7.53	9.20

31 23 23.20 Hauling		Crew	Daily Output	Labor-Hours	Unit	Material	2014 Bare Costs Labor	Equipment	Total	Total Incl O&P
1238	cycle 10 miles	B-34B	108	.074	L.C.Y.		2.78	6.40	9.18	11.30
1240	25 MPH ave, cycle 4 miles		192	.042			1.56	3.61	5.17	6.35
1242	cycle 6 miles		168	.048			1.79	4.12	5.91	7.25
1244	cycle 8 miles		144	.056			2.09	4.81	6.90	8.50
1246	cycle 10 miles		132	.061			2.28	5.25	7.53	9.20
1250	30 MPH ave, cycle 4 miles		204	.039			1.47	3.39	4.86	6
1252	cycle 6 miles		180	.044			1.67	3.85	5.52	6.80
1254	cycle 8 miles		156	.051			1.93	4.44	6.37	7.80
1256	cycle 10 miles		144	.056			2.09	4.81	6.90	8.50
1260	35 MPH ave, cycle 4 miles		216	.037			1.39	3.21	4.60	5.65
1262	cycle 6 miles		192	.042			1.56	3.61	5.17	6.35
1264	cycle 8 miles		168	.048			1.79	4.12	5.91	7.25
1266	cycle 10 miles		156	.051			1.93	4.44	6.37	7.80
1268	cycle 20 miles		108	.074			2.78	6.40	9.18	11.30
1269	cycle 30 miles		72	.111			4.17	9.60	13.77	16.95
1270	cycle 40 miles		60	.133			5	11.55	16.55	20.50
1272	40 MPH ave, cycle 6 miles		192	.042			1.56	3.61	5.17	6.35
1274	cycle 8 miles		180	.044			1.67	3.85	5.52	6.80
1276	cycle 10 miles		156	.051			1.93	4.44	6.37	7.80
1278	cycle 20 miles		108	.074			2.78	6.40	9.18	11.30
1280	cycle 30 miles		84	.095			3.58	8.25	11.83	14.50
1282	cycle 40 miles		72	.111			4.17	9.60	13.77	16.95
1284	cycle 50 miles		60	.133			5	11.55	16.55	20.50
1294	45 MPH ave, cycle 8 miles		180	.044			1.67	3.85	5.52	6.80
1296	cycle 10 miles		168	.048			1.79	4.12	5.91	7.25
1298	cycle 20 miles		120	.067			2.50	5.75	8.25	10.15
1300	cycle 30 miles		96	.083			3.13	7.20	10.33	12.70
1302	cycle 40 miles		72	.111			4.17	9.60	13.77	16.95
1304	cycle 50 miles		60	.133			5	11.55	16.55	20.50
1306	50 MPH ave, cycle 10 miles		180	.044			1.67	3.85	5.52	6.80
1308	cycle 20 miles		132	.061			2.28	5.25	7.53	9.20
1310	cycle 30 miles		96	.083			3.13	7.20	10.33	12.70
1312	cycle 40 miles		84	.095			3.58	8.25	11.83	14.50
1314	cycle 50 miles		72	.111			4.17	9.60	13.77	16.95
1414	15 MPH ave, cycle 0.5 mile, 25 min. wait/Ld./Uld.		204	.039			1.47	3.39	4.86	6
1416	cycle 1 mile		192	.042			1.56	3.61	5.17	6.35
1418	cycle 2 miles		168	.048			1.79	4.12	5.91	7.25
1420	cycle 4 miles		132	.061			2.28	5.25	7.53	9.20
1422	cycle 6 miles		120	.067			2.50	5.75	8.25	10.15
1424	cycle 8 miles		96	.083			3.13	7.20	10.33	12.70
1425	cycle 10 miles		84	.095			3.58	8.25	11.83	14.50
1426	20 MPH ave, cycle 0.5 mile		216	.037			1.39	3.21	4.60	5.65
1428	cycle 1 mile		204	.039			1.47	3.39	4.86	6
1430	cycle 2 miles		180	.044			1.67	3.85	5.52	6.80
1432	cycle 4 miles		156	.051			1.93	4.44	6.37	7.80
1434	cycle 6 miles		132	.061			2.28	5.25	7.53	9.20
1436	cycle 8 miles		120	.067			2.50	5.75	8.25	10.15
1438	cycle 10 miles		96	.083			3.13	7.20	10.33	12.70
1440	25 MPH ave, cycle 4 miles		168	.048			1.79	4.12	5.91	7.25
1442	cycle 6 miles		144	.056			2.09	4.81	6.90	8.50
1444	cycle 8 miles		132	.061			2.28	5.25	7.53	9.20
1446	cycle 10 miles		108	.074			2.78	6.40	9.18	11.30
1450	30 MPH ave, cycle 4 miles	↓	168	.048	↓		1.79	4.12	5.91	7.25

31 23 23.20 Hauling		Crew	Daily Output	Labor-Hours	Unit	Material	2014 Bare Costs Labor	Equipment	Total	Total Incl O&P
1452	cycle 6 miles	B-34B	156	.051	L.C.Y.		1.93	4.44	6.37	7.80
1454	cycle 8 miles		132	.061			2.28	5.25	7.53	9.20
1456	cycle 10 miles		120	.067			2.50	5.75	8.25	10.15
1460	35 MPH ave, cycle 4 miles		180	.044			1.67	3.85	5.52	6.80
1462	cycle 6 miles		156	.051			1.93	4.44	6.37	7.80
1464	cycle 8 miles		144	.056			2.09	4.81	6.90	8.50
1466	cycle 10 miles		132	.061			2.28	5.25	7.53	9.20
1468	cycle 20 miles		96	.083			3.13	7.20	10.33	12.70
1469	cycle 30 miles		72	.111			4.17	9.60	13.77	16.95
1470	cycle 40 miles		60	.133			5	11.55	16.55	20.50
1472	40 MPH ave, cycle 6 miles		168	.048			1.79	4.12	5.91	7.25
1474	cycle 8 miles		156	.051			1.93	4.44	6.37	7.80
1476	cycle 10 miles		144	.056			2.09	4.81	6.90	8.50
1478	cycle 20 miles		96	.083			3.13	7.20	10.33	12.70
1480	cycle 30 miles		84	.095			3.58	8.25	11.83	14.50
1482	cycle 40 miles		60	.133			5	11.55	16.55	20.50
1484	cycle 50 miles		60	.133			5	11.55	16.55	20.50
1494	45 MPH ave, cycle 8 miles		156	.051			1.93	4.44	6.37	7.80
1496	cycle 10 miles		144	.056			2.09	4.81	6.90	8.50
1498	cycle 20 miles		108	.074			2.78	6.40	9.18	11.30
1500	cycle 30 miles		84	.095			3.58	8.25	11.83	14.50
1502	cycle 40 miles		72	.111			4.17	9.60	13.77	16.95
1504	cycle 50 miles		60	.133			5	11.55	16.55	20.50
1506	50 MPH ave, cycle 10 miles		156	.051			1.93	4.44	6.37	7.80
1508	cycle 20 miles		120	.067			2.50	5.75	8.25	10.15
1510	cycle 30 miles		96	.083			3.13	7.20	10.33	12.70
1512	cycle 40 miles		72	.111			4.17	9.60	13.77	16.95
1514	cycle 50 miles		60	.133			5	11.55	16.55	20.50
1614	15 MPH, cycle 0.5 mile, 30 min. wait/Ld./Uld.		180	.044			1.67	3.85	5.52	6.80
1616	cycle 1 mile		168	.048			1.79	4.12	5.91	7.25
1618	cycle 2 miles		144	.056			2.09	4.81	6.90	8.50
1620	cycle 4 miles		120	.067			2.50	5.75	8.25	10.15
1622	cycle 6 miles		108	.074			2.78	6.40	9.18	11.30
1624	cycle 8 miles		84	.095			3.58	8.25	11.83	14.50
1625	cycle 10 miles		84	.095			3.58	8.25	11.83	14.50
1626	20 MPH ave, cycle 0.5 mile		180	.044			1.67	3.85	5.52	6.80
1628	cycle 1 mile		168	.048			1.79	4.12	5.91	7.25
1630	cycle 2 miles		156	.051			1.93	4.44	6.37	7.80
1632	cycle 4 miles		132	.061			2.28	5.25	7.53	9.20
1634	cycle 6 miles		120	.067			2.50	5.75	8.25	10.15
1636	cycle 8 miles		108	.074			2.78	6.40	9.18	11.30
1638	cycle 10 miles		96	.083			3.13	7.20	10.33	12.70
1640	25 MPH ave, cycle 4 miles		144	.056			2.09	4.81	6.90	8.50
1642	cycle 6 miles		132	.061			2.28	5.25	7.53	9.20
1644	cycle 8 miles		108	.074			2.78	6.40	9.18	11.30
1646	cycle 10 miles		108	.074			2.78	6.40	9.18	11.30
1650	30 MPH ave, cycle 4 miles		144	.056			2.09	4.81	6.90	8.50
1652	cycle 6 miles		132	.061			2.28	5.25	7.53	9.20
1654	cycle 8 miles		120	.067			2.50	5.75	8.25	10.15
1656	cycle 10 miles		108	.074			2.78	6.40	9.18	11.30
1660	35 MPH ave, cycle 4 miles		156	.051			1.93	4.44	6.37	7.80
1662	cycle 6 miles		144	.056			2.09	4.81	6.90	8.50
1664	cycle 8 miles		132	.061			2.28	5.25	7.53	9.20

31 23 23.20 Hauling		Crew	Daily Output	Labor-Hours	Unit	Material	2014 Bare Costs Labor	Equipment	Total	Total Incl O&P
1666	cycle 10 miles	B-34B	120	.067	L.C.Y.		2.50	5.75	8.25	10.15
1668	cycle 20 miles		84	.095			3.58	8.25	11.83	14.50
1669	cycle 30 miles		72	.111			4.17	9.60	13.77	16.95
1670	cycle 40 miles		60	.133			5	11.55	16.55	20.50
1672	40 MPH, cycle 6 miles		144	.056			2.09	4.81	6.90	8.50
1674	cycle 8 miles		132	.061			2.28	5.25	7.53	9.20
1676	cycle 10 miles		120	.067			2.50	5.75	8.25	10.15
1678	cycle 20 miles		96	.083			3.13	7.20	10.33	12.70
1680	cycle 30 miles		72	.111			4.17	9.60	13.77	16.95
1682	cycle 40 miles		60	.133			5	11.55	16.55	20.50
1684	cycle 50 miles		48	.167			6.25	14.40	20.65	25.50
1694	45 MPH ave, cycle 8 miles		144	.056			2.09	4.81	6.90	8.50
1696	cycle 10 miles		132	.061			2.28	5.25	7.53	9.20
1698	cycle 20 miles		96	.083			3.13	7.20	10.33	12.70
1700	cycle 30 miles		84	.095			3.58	8.25	11.83	14.50
1702	cycle 40 miles		60	.133			5	11.55	16.55	20.50
1704	cycle 50 miles		60	.133			5	11.55	16.55	20.50
1706	50 MPH ave, cycle 10 miles		132	.061			2.28	5.25	7.53	9.20
1708	cycle 20 miles		108	.074			2.78	6.40	9.18	11.30
1710	cycle 30 miles		84	.095			3.58	8.25	11.83	14.50
1712	cycle 40 miles		72	.111			4.17	9.60	13.77	16.95
1714	cycle 50 miles		60	.133			5	11.55	16.55	20.50
2000	Hauling, 8 C.Y. truck, small project cost per hour	B-34A	8	1	Hr.		37.50	52	89.50	115
2100	12 C.Y. Truck	B-34B	8	1			37.50	86.50	124	153
2150	16.5 C.Y. Truck	B-34C	8	1			37.50	92.50	130	160
2175	18 C.Y. 8 wheel Truck	B-34I	8	1			37.50	108	145.50	177
2200	20 C.Y. Truck	B-34D	8	1			37.50	94	131.50	162
2300	Grading at dump, or embankment if required, by dozer	B-10B	1000	.012	L.C.Y.		.54	1.33	1.87	2.28
2310	Spotter at fill or cut, if required	1 Clab	8	1	Hr.		36.50		36.50	56.50
2500	Dust control, light	B-59	1	8	Day		300	510	810	1,025
2510	Heavy	"	.50	16			600	1,025	1,625	2,050
2600	Haul road maintenance	B-86A	1	8			390	705	1,095	1,375
3014	16.5 C.Y. truck, 15 min. wait/Ld./Uld., 15 MPH, cycle 0.5 mile	B-34C	462	.017	L.C.Y.		.65	1.60	2.25	2.75
3016	cycle 1 mile		413	.019			.73	1.79	2.52	3.08
3018	cycle 2 miles		347	.023			.87	2.13	3	3.66
3020	cycle 4 miles		248	.032			1.21	2.98	4.19	5.15
3022	cycle 6 miles		198	.040			1.52	3.73	5.25	6.40
3024	cycle 8 miles		165	.048			1.82	4.48	6.30	7.70
3025	cycle 10 miles		132	.061			2.28	5.60	7.88	9.60
3026	20 MPH ave, cycle 0.5 mile		479	.017			.63	1.54	2.17	2.66
3028	cycle 1 mile		429	.019			.70	1.72	2.42	2.97
3030	cycle 2 miles		380	.021			.79	1.94	2.73	3.35
3032	cycle 4 miles		281	.028			1.07	2.63	3.70	4.52
3034	cycle 6 miles		231	.035			1.30	3.20	4.50	5.50
3036	cycle 8 miles		198	.040			1.52	3.73	5.25	6.40
3038	cycle 10 miles		165	.048			1.82	4.48	6.30	7.70
3040	25 MPH ave, cycle 4 miles		314	.025			.96	2.35	3.31	4.05
3042	cycle 6 miles		264	.030			1.14	2.80	3.94	4.82
3044	cycle 8 miles		231	.035			1.30	3.20	4.50	5.50
3046	cycle 10 miles		198	.040			1.52	3.73	5.25	6.40
3050	30 MPH ave, cycle 4 miles		347	.023			.87	2.13	3	3.66
3052	cycle 6 miles		281	.028			1.07	2.63	3.70	4.52
3054	cycle 8 miles		248	.032			1.21	2.98	4.19	5.15

31 23 23 – Fill

31 23 23.20 Hauling		Crew	Daily Output	Labor-Hours	Unit	Material	2014 Bare Costs Labor	Equipment	Total	Total Incl O&P
3056	cycle 10 miles	B-34C	215	.037	L.C.Y.		1.40	3.44	4.84	5.90
3060	35 MPH ave, cycle 4 miles		363	.022			.83	2.04	2.87	3.50
3062	cycle 6 miles		314	.025			.96	2.35	3.31	4.05
3064	cycle 8 miles		264	.030			1.14	2.80	3.94	4.82
3066	cycle 10 miles		248	.032			1.21	2.98	4.19	5.15
3068	cycle 20 miles		149	.054			2.02	4.96	6.98	8.55
3070	cycle 30 miles		116	.069			2.59	6.35	8.94	10.95
3072	cycle 40 miles		83	.096			3.62	8.90	12.52	15.30
3074	40 MPH ave, cycle 6 miles		330	.024			.91	2.24	3.15	3.85
3076	cycle 8 miles		281	.028			1.07	2.63	3.70	4.52
3078	cycle 10 miles		264	.030			1.14	2.80	3.94	4.82
3080	cycle 20 miles		165	.048			1.82	4.48	6.30	7.70
3082	cycle 30 miles		132	.061			2.28	5.60	7.88	9.60
3084	cycle 40 miles		99	.081			3.03	7.45	10.48	12.85
3086	cycle 50 miles		83	.096			3.62	8.90	12.52	15.30
3094	45 MPH ave, cycle 8 miles		297	.027			1.01	2.49	3.50	4.28
3096	cycle 10 miles		281	.028			1.07	2.63	3.70	4.52
3098	cycle 20 miles		182	.044			1.65	4.06	5.71	7
3100	cycle 30 miles		132	.061			2.28	5.60	7.88	9.60
3102	cycle 40 miles		116	.069			2.59	6.35	8.94	10.95
3104	cycle 50 miles		99	.081			3.03	7.45	10.48	12.85
3106	50 MPH ave, cycle 10 miles		281	.028			1.07	2.63	3.70	4.52
3108	cycle 20 miles		198	.040			1.52	3.73	5.25	6.40
3110	cycle 30 miles		149	.054			2.02	4.96	6.98	8.55
3112	cycle 40 miles		116	.069			2.59	6.35	8.94	10.95
3114	cycle 50 miles		99	.081			3.03	7.45	10.48	12.85
3214	20 min. wait/Ld./Uld., 15 MPH, cycle 0.5 mile		363	.022			.83	2.04	2.87	3.50
3216	cycle 1 mile		330	.024			.91	2.24	3.15	3.85
3218	cycle 2 miles		281	.028			1.07	2.63	3.70	4.52
3220	cycle 4 miles		215	.037			1.40	3.44	4.84	5.90
3222	cycle 6 miles		182	.044			1.65	4.06	5.71	7
3224	cycle 8 miles		149	.054			2.02	4.96	6.98	8.55
3225	cycle 10 miles		132	.061			2.28	5.60	7.88	9.60
3226	20 MPH ave, cycle 0.5 mile		363	.022			.83	2.04	2.87	3.50
3228	cycle 1 mile		347	.023			.87	2.13	3	3.66
3230	cycle 2 miles		297	.027			1.01	2.49	3.50	4.28
3232	cycle 4 miles		248	.032			1.21	2.98	4.19	5.15
3234	cycle 6 miles		198	.040			1.52	3.73	5.25	6.40
3236	cycle 8 miles		182	.044			1.65	4.06	5.71	7
3238	cycle 10 miles		149	.054			2.02	4.96	6.98	8.55
3240	25 MPH ave, cycle 4 miles		264	.030			1.14	2.80	3.94	4.82
3242	cycle 6 miles		231	.035			1.30	3.20	4.50	5.50
3244	cycle 8 miles		198	.040			1.52	3.73	5.25	6.40
3246	cycle 10 miles		182	.044			1.65	4.06	5.71	7
3250	30 MPH ave, cycle 4 miles		281	.028			1.07	2.63	3.70	4.52
3252	cycle 6 miles		248	.032			1.21	2.98	4.19	5.15
3254	cycle 8 miles		215	.037			1.40	3.44	4.84	5.90
3256	cycle 10 miles		198	.040			1.52	3.73	5.25	6.40
3260	35 MPH ave, cycle 4 miles		297	.027			1.01	2.49	3.50	4.28
3262	cycle 6 miles		264	.030			1.14	2.80	3.94	4.82
3264	cycle 8 miles		231	.035			1.30	3.20	4.50	5.50
3266	cycle 10 miles		215	.037			1.40	3.44	4.84	5.90
3268	cycle 20 miles	↓	149	.054	↓		2.02	4.96	6.98	8.55

31 23 23.20 Hauling		Crew	Daily Output	Labor-Hours	Unit	Material	2014 Bare Costs Labor	Equipment	Total	Total Incl O&P
3270	cycle 30 miles	B-34C	99	.081	L.C.Y.		3.03	7.45	10.48	12.85
3272	cycle 40 miles		83	.096			3.62	8.90	12.52	15.30
3274	40 MPH ave, cycle 6 miles		264	.030			1.14	2.80	3.94	4.82
3276	cycle 8 miles		248	.032			1.21	2.98	4.19	5.15
3278	cycle 10 miles		215	.037			1.40	3.44	4.84	5.90
3280	cycle 20 miles		149	.054			2.02	4.96	6.98	8.55
3282	cycle 30 miles		116	.069			2.59	6.35	8.94	10.95
3284	cycle 40 miles		99	.081			3.03	7.45	10.48	12.85
3286	cycle 50 miles		83	.096			3.62	8.90	12.52	15.30
3294	45 MPH ave, cycle 8 miles		248	.032			1.21	2.98	4.19	5.15
3296	cycle 10 miles		231	.035			1.30	3.20	4.50	5.50
3298	cycle 20 miles		165	.048			1.82	4.48	6.30	7.70
3300	cycle 30 miles		132	.061			2.28	5.60	7.88	9.60
3302	cycle 40 miles		99	.081			3.03	7.45	10.48	12.85
3304	cycle 50 miles		83	.096			3.62	8.90	12.52	15.30
3306	50 MPH ave, cycle 10 miles		248	.032			1.21	2.98	4.19	5.15
3308	cycle 20 miles		182	.044			1.65	4.06	5.71	7
3310	cycle 30 miles		132	.061			2.28	5.60	7.88	9.60
3312	cycle 40 miles		116	.069			2.59	6.35	8.94	10.95
3314	cycle 50 miles		99	.081			3.03	7.45	10.48	12.85
3414	25 min. wait/Ld./Uld., 15 MPH, cycle 0.5 mile		281	.028			1.07	2.63	3.70	4.52
3416	cycle 1 mile		264	.030			1.14	2.80	3.94	4.82
3418	cycle 2 miles		231	.035			1.30	3.20	4.50	5.50
3420	cycle 4 miles		182	.044			1.65	4.06	5.71	7
3422	cycle 6 miles		165	.048			1.82	4.48	6.30	7.70
3424	cycle 8 miles		132	.061			2.28	5.60	7.88	9.60
3425	cycle 10 miles		116	.069			2.59	6.35	8.94	10.95
3426	20 MPH ave, cycle 0.5 mile		297	.027			1.01	2.49	3.50	4.28
3428	cycle 1 mile		281	.028			1.07	2.63	3.70	4.52
3430	cycle 2 miles		248	.032			1.21	2.98	4.19	5.15
3432	cycle 4 miles		215	.037			1.40	3.44	4.84	5.90
3434	cycle 6 miles		182	.044			1.65	4.06	5.71	7
3436	cycle 8 miles		165	.048			1.82	4.48	6.30	7.70
3438	cycle 10 miles		132	.061			2.28	5.60	7.88	9.60
3440	25 MPH ave, cycle 4 miles		231	.035			1.30	3.20	4.50	5.50
3442	cycle 6 miles		198	.040			1.52	3.73	5.25	6.40
3444	cycle 8 miles		182	.044			1.65	4.06	5.71	7
3446	cycle 10 miles		165	.048			1.82	4.48	6.30	7.70
3450	30 MPH ave, cycle 4 miles		231	.035			1.30	3.20	4.50	5.50
3452	cycle 6 miles		215	.037			1.40	3.44	4.84	5.90
3454	cycle 8 miles		182	.044			1.65	4.06	5.71	7
3456	cycle 10 miles		165	.048			1.82	4.48	6.30	7.70
3460	35 MPH ave, cycle 4 miles		248	.032			1.21	2.98	4.19	5.15
3462	cycle 6 miles		215	.037			1.40	3.44	4.84	5.90
3464	cycle 8 miles		198	.040			1.52	3.73	5.25	6.40
3466	cycle 10 miles		182	.044			1.65	4.06	5.71	7
3468	cycle 20 miles		132	.061			2.28	5.60	7.88	9.60
3470	cycle 30 miles		99	.081			3.03	7.45	10.48	12.85
3472	cycle 40 miles		83	.096			3.62	8.90	12.52	15.30
3474	40 MPH, cycle 6 miles		231	.035			1.30	3.20	4.50	5.50
3476	cycle 8 miles		215	.037			1.40	3.44	4.84	5.90
3478	cycle 10 miles		198	.040			1.52	3.73	5.25	6.40
3480	cycle 20 miles		132	.061			2.28	5.60	7.88	9.60

31 23 23.20 Hauling		Crew	Daily Output	Labor-Hours	Unit	Material	2014 Bare Costs Labor	Equipment	Total	Total Incl O&P
3482	cycle 30 miles	B-34C	116	.069	L.C.Y.		2.59	6.35	8.94	10.95
3484	cycle 40 miles		83	.096			3.62	8.90	12.52	15.30
3486	cycle 50 miles		83	.096			3.62	8.90	12.52	15.30
3494	45 MPH ave, cycle 8 miles		215	.037			1.40	3.44	4.84	5.90
3496	cycle 10 miles		198	.040			1.52	3.73	5.25	6.40
3498	cycle 20 miles		149	.054			2.02	4.96	6.98	8.55
3500	cycle 30 miles		116	.069			2.59	6.35	8.94	10.95
3502	cycle 40 miles		99	.081			3.03	7.45	10.48	12.85
3504	cycle 50 miles		83	.096			3.62	8.90	12.52	15.30
3506	50 MPH ave, cycle 10 miles		215	.037			1.40	3.44	4.84	5.90
3508	cycle 20 miles		165	.048			1.82	4.48	6.30	7.70
3510	cycle 30 miles		132	.061			2.28	5.60	7.88	9.60
3512	cycle 40 miles		99	.081			3.03	7.45	10.48	12.85
3514	cycle 50 miles		83	.096			3.62	8.90	12.52	15.30
3614	30 min. wait/Ld./Uld., 15 MPH, cycle 0.5 mile		248	.032			1.21	2.98	4.19	5.15
3616	cycle 1 mile		231	.035			1.30	3.20	4.50	5.50
3618	cycle 2 miles		198	.040			1.52	3.73	5.25	6.40
3620	cycle 4 miles		165	.048			1.82	4.48	6.30	7.70
3622	cycle 6 miles		149	.054			2.02	4.96	6.98	8.55
3624	cycle 8 miles		116	.069			2.59	6.35	8.94	10.95
3625	cycle 10 miles		116	.069			2.59	6.35	8.94	10.95
3626	20 MPH ave, cycle 0.5 mile		248	.032			1.21	2.98	4.19	5.15
3628	cycle 1 mile		231	.035			1.30	3.20	4.50	5.50
3630	cycle 2 miles		215	.037			1.40	3.44	4.84	5.90
3632	cycle 4 miles		182	.044			1.65	4.06	5.71	7
3634	cycle 6 miles		165	.048			1.82	4.48	6.30	7.70
3636	cycle 8 miles		149	.054			2.02	4.96	6.98	8.55
3638	cycle 10 miles		132	.061			2.28	5.60	7.88	9.60
3640	25 MPH ave, cycle 4 miles		198	.040			1.52	3.73	5.25	6.40
3642	cycle 6 miles		182	.044			1.65	4.06	5.71	7
3644	cycle 8 miles		149	.054			2.02	4.96	6.98	8.55
3646	cycle 10 miles		149	.054			2.02	4.96	6.98	8.55
3650	30 MPH ave, cycle 4 miles		198	.040			1.52	3.73	5.25	6.40
3652	cycle 6 miles		182	.044			1.65	4.06	5.71	7
3654	cycle 8 miles		165	.048			1.82	4.48	6.30	7.70
3656	cycle 10 miles		149	.054			2.02	4.96	6.98	8.55
3660	35 MPH ave, cycle 4 miles		215	.037			1.40	3.44	4.84	5.90
3662	cycle 6 miles		198	.040			1.52	3.73	5.25	6.40
3664	cycle 8 miles		182	.044			1.65	4.06	5.71	7
3666	cycle 10 miles		165	.048			1.82	4.48	6.30	7.70
3668	cycle 20 miles		116	.069			2.59	6.35	8.94	10.95
3670	cycle 30 miles		99	.081			3.03	7.45	10.48	12.85
3672	cycle 40 miles		83	.096			3.62	8.90	12.52	15.30
3674	40 MPH, cycle 6 miles		198	.040			1.52	3.73	5.25	6.40
3676	cycle 8 miles		182	.044			1.65	4.06	5.71	7
3678	cycle 10 miles		165	.048			1.82	4.48	6.30	7.70
3680	cycle 20 miles		132	.061			2.28	5.60	7.88	9.60
3682	cycle 30 miles		99	.081			3.03	7.45	10.48	12.85
3684	cycle 40 miles		83	.096			3.62	8.90	12.52	15.30
3686	cycle 50 miles		66	.121			4.55	11.20	15.75	19.25
3694	45 MPH ave, cycle 8 miles		198	.040			1.52	3.73	5.25	6.40
3696	cycle 10 miles		182	.044			1.65	4.06	5.71	7
3698	cycle 20 miles		132	.061			2.28	5.60	7.88	9.60

31 23 23.20 Hauling		Crew	Daily Output	Labor-Hours	Unit	Material	2014 Bare Costs Labor	Equipment	Total	Total Incl O&P
3700	cycle 30 miles	B-34C	116	.069	L.C.Y.		2.59	6.35	8.94	10.95
3702	cycle 40 miles		83	.096			3.62	8.90	12.52	15.30
3704	cycle 50 miles		83	.096			3.62	8.90	12.52	15.30
3706	50 MPH ave, cycle 10 miles		182	.044			1.65	4.06	5.71	7
3708	cycle 20 miles		149	.054			2.02	4.96	6.98	8.55
3710	cycle 30 miles		116	.069			2.59	6.35	8.94	10.95
3712	cycle 40 miles		99	.081			3.03	7.45	10.48	12.85
3714	cycle 50 miles		83	.096			3.62	8.90	12.52	15.30
4014	20 C.Y. truck, 15 min. wait/Ld./Uld., 15 MPH, cycle 0.5 mile	B-34D	560	.014			.54	1.35	1.89	2.30
4016	cycle 1 mile		500	.016			.60	1.51	2.11	2.58
4018	cycle 2 miles		420	.019			.72	1.79	2.51	3.06
4020	cycle 4 miles		300	.027			1	2.51	3.51	4.29
4022	cycle 6 miles		240	.033			1.25	3.14	4.39	5.35
4024	cycle 8 miles		200	.040			1.50	3.77	5.27	6.45
4025	cycle 10 miles		160	.050			1.88	4.71	6.59	8.05
4026	20 MPH ave, cycle 0.5 mile		580	.014			.52	1.30	1.82	2.22
4028	cycle 1 mile		520	.015			.58	1.45	2.03	2.47
4030	cycle 2 miles		460	.017			.65	1.64	2.29	2.80
4032	cycle 4 miles		340	.024			.88	2.22	3.10	3.79
4034	cycle 6 miles		280	.029			1.07	2.69	3.76	4.60
4036	cycle 8 miles		240	.033			1.25	3.14	4.39	5.35
4038	cycle 10 miles		200	.040			1.50	3.77	5.27	6.45
4040	25 MPH ave, cycle 4 miles		380	.021			.79	1.98	2.77	3.39
4042	cycle 6 miles		320	.025			.94	2.36	3.30	4.02
4044	cycle 8 miles		280	.029			1.07	2.69	3.76	4.60
4046	cycle 10 miles		240	.033			1.25	3.14	4.39	5.35
4050	30 MPH ave, cycle 4 miles		420	.019			.72	1.79	2.51	3.06
4052	cycle 6 miles		340	.024			.88	2.22	3.10	3.79
4054	cycle 8 miles		300	.027			1	2.51	3.51	4.29
4056	cycle 10 miles		260	.031			1.16	2.90	4.06	4.95
4060	35 MPH ave, cycle 4 miles		440	.018			.68	1.71	2.39	2.92
4062	cycle 6 miles		380	.021			.79	1.98	2.77	3.39
4064	cycle 8 miles		320	.025			.94	2.36	3.30	4.02
4066	cycle 10 miles		300	.027			1	2.51	3.51	4.29
4068	cycle 20 miles		180	.044			1.67	4.19	5.86	7.15
4070	cycle 30 miles		140	.057			2.15	5.40	7.55	9.15
4072	cycle 40 miles		100	.080			3	7.55	10.55	12.90
4074	40 MPH ave, cycle 6 miles		400	.020			.75	1.88	2.63	3.22
4076	cycle 8 miles		340	.024			.88	2.22	3.10	3.79
4078	cycle 10 miles		320	.025			.94	2.36	3.30	4.02
4080	cycle 20 miles		200	.040			1.50	3.77	5.27	6.45
4082	cycle 30 miles		160	.050			1.88	4.71	6.59	8.05
4084	cycle 40 miles		120	.067			2.50	6.30	8.80	10.70
4086	cycle 50 miles		100	.080			3	7.55	10.55	12.90
4094	45 MPH ave, cycle 8 miles		360	.022			.83	2.09	2.92	3.57
4096	cycle 10 miles		340	.024			.88	2.22	3.10	3.79
4098	cycle 20 miles		220	.036			1.37	3.43	4.80	5.85
4100	cycle 30 miles		160	.050			1.88	4.71	6.59	8.05
4102	cycle 40 miles		140	.057			2.15	5.40	7.55	9.15
4104	cycle 50 miles		120	.067			2.50	6.30	8.80	10.70
4106	50 MPH ave, cycle 10 miles		340	.024			.88	2.22	3.10	3.79
4108	cycle 20 miles		240	.033			1.25	3.14	4.39	5.35
4110	cycle 30 miles		180	.044			1.67	4.19	5.86	7.15

31 23 23 – Fill

31 23 23.20 Hauling		Crew	Daily Output	Labor-Hours	Unit	Material	2014 Bare Costs Labor	Equipment	Total	Total Incl O&P
4112	cycle 40 miles	B-34D	140	.057	L.C.Y.		2.15	5.40	7.55	9.15
4114	cycle 50 miles		120	.067			2.50	6.30	8.80	10.70
4214	20 min. wait/Ld./Uld., 15 MPH, cycle 0.5 mile		440	.018			.68	1.71	2.39	2.92
4216	cycle 1 mile		400	.020			.75	1.88	2.63	3.22
4218	cycle 2 miles		340	.024			.88	2.22	3.10	3.79
4220	cycle 4 miles		260	.031			1.16	2.90	4.06	4.95
4222	cycle 6 miles		220	.036			1.37	3.43	4.80	5.85
4224	cycle 8 miles		180	.044			1.67	4.19	5.86	7.15
4225	cycle 10 miles		160	.050			1.88	4.71	6.59	8.05
4226	20 MPH ave, cycle 0.5 mile		440	.018			.68	1.71	2.39	2.92
4228	cycle 1 mile		420	.019			.72	1.79	2.51	3.06
4230	cycle 2 miles		360	.022			.83	2.09	2.92	3.57
4232	cycle 4 miles		300	.027			1	2.51	3.51	4.29
4234	cycle 6 miles		240	.033			1.25	3.14	4.39	5.35
4236	cycle 8 miles		220	.036			1.37	3.43	4.80	5.85
4238	cycle 10 miles		180	.044			1.67	4.19	5.86	7.15
4240	25 MPH ave, cycle 4 miles		320	.025			.94	2.36	3.30	4.02
4242	cycle 6 miles		280	.029			1.07	2.69	3.76	4.60
4244	cycle 8 miles		240	.033			1.25	3.14	4.39	5.35
4246	cycle 10 miles		220	.036			1.37	3.43	4.80	5.85
4250	30 MPH ave, cycle 4 miles		340	.024			.88	2.22	3.10	3.79
4252	cycle 6 miles		300	.027			1	2.51	3.51	4.29
4254	cycle 8 miles		260	.031			1.16	2.90	4.06	4.95
4256	cycle 10 miles		240	.033			1.25	3.14	4.39	5.35
4260	35 MPH ave, cycle 4 miles		360	.022			.83	2.09	2.92	3.57
4262	cycle 6 miles		320	.025			.94	2.36	3.30	4.02
4264	cycle 8 miles		280	.029			1.07	2.69	3.76	4.60
4266	cycle 10 miles		260	.031			1.16	2.90	4.06	4.95
4268	cycle 20 miles		180	.044			1.67	4.19	5.86	7.15
4270	cycle 30 miles		120	.067			2.50	6.30	8.80	10.70
4272	cycle 40 miles		100	.080			3	7.55	10.55	12.90
4274	40 MPH ave, cycle 6 miles		320	.025			.94	2.36	3.30	4.02
4276	cycle 8 miles		300	.027			1	2.51	3.51	4.29
4278	cycle 10 miles		260	.031			1.16	2.90	4.06	4.95
4280	cycle 20 miles		180	.044			1.67	4.19	5.86	7.15
4282	cycle 30 miles		140	.057			2.15	5.40	7.55	9.15
4284	cycle 40 miles		120	.067			2.50	6.30	8.80	10.70
4286	cycle 50 miles		100	.080			3	7.55	10.55	12.90
4294	45 MPH ave, cycle 8 miles		300	.027			1	2.51	3.51	4.29
4296	cycle 10 miles		280	.029			1.07	2.69	3.76	4.60
4298	cycle 20 miles		200	.040			1.50	3.77	5.27	6.45
4300	cycle 30 miles		160	.050			1.88	4.71	6.59	8.05
4302	cycle 40 miles		120	.067			2.50	6.30	8.80	10.70
4304	cycle 50 miles		100	.080			3	7.55	10.55	12.90
4306	50 MPH ave, cycle 10 miles		300	.027			1	2.51	3.51	4.29
4308	cycle 20 miles		220	.036			1.37	3.43	4.80	5.85
4310	cycle 30 miles		180	.044			1.67	4.19	5.86	7.15
4312	cycle 40 miles		140	.057			2.15	5.40	7.55	9.15
4314	cycle 50 miles		120	.067			2.50	6.30	8.80	10.70
4414	25 min. wait/Ld./Uld., 15 MPH, cycle 0.5 mile		340	.024			.88	2.22	3.10	3.79
4416	cycle 1 mile		320	.025			.94	2.36	3.30	4.02
4418	cycle 2 miles		280	.029			1.07	2.69	3.76	4.60
4420	cycle 4 miles		220	.036			1.37	3.43	4.80	5.85

31 23 23.20 Hauling		Crew	Daily Output	Labor-Hours	Unit	Material	2014 Bare Costs Labor	Equipment	Total	Total Incl O&P
4422	cycle 6 miles	B-34D	200	.040	L.C.Y.		1.50	3.77	5.27	6.45
4424	cycle 8 miles		160	.050			1.88	4.71	6.59	8.05
4425	cycle 10 miles		140	.057			2.15	5.40	7.55	9.15
4426	20 MPH ave, cycle 0.5 mile		360	.022			.83	2.09	2.92	3.57
4428	cycle 1 mile		340	.024			.88	2.22	3.10	3.79
4430	cycle 2 miles		300	.027			1	2.51	3.51	4.29
4432	cycle 4 miles		260	.031			1.16	2.90	4.06	4.95
4434	cycle 6 miles		220	.036			1.37	3.43	4.80	5.85
4436	cycle 8 miles		200	.040			1.50	3.77	5.27	6.45
4438	cycle 10 miles		160	.050			1.88	4.71	6.59	8.05
4440	25 MPH ave, cycle 4 miles		280	.029			1.07	2.69	3.76	4.60
4442	cycle 6 miles		240	.033			1.25	3.14	4.39	5.35
4444	cycle 8 miles		220	.036			1.37	3.43	4.80	5.85
4446	cycle 10 miles		200	.040			1.50	3.77	5.27	6.45
4450	30 MPH ave, cycle 4 miles		280	.029			1.07	2.69	3.76	4.60
4452	cycle 6 miles		260	.031			1.16	2.90	4.06	4.95
4454	cycle 8 miles		220	.036			1.37	3.43	4.80	5.85
4456	cycle 10 miles		200	.040			1.50	3.77	5.27	6.45
4460	35 MPH ave, cycle 4 miles		300	.027			1	2.51	3.51	4.29
4462	cycle 6 miles		260	.031			1.16	2.90	4.06	4.95
4464	cycle 8 miles		240	.033			1.25	3.14	4.39	5.35
4466	cycle 10 miles		220	.036			1.37	3.43	4.80	5.85
4468	cycle 20 miles		160	.050			1.88	4.71	6.59	8.05
4470	cycle 30 miles		120	.067			2.50	6.30	8.80	10.70
4472	cycle 40 miles		100	.080			3	7.55	10.55	12.90
4474	40 MPH, cycle 6 miles		280	.029			1.07	2.69	3.76	4.60
4476	cycle 8 miles		260	.031			1.16	2.90	4.06	4.95
4478	cycle 10 miles		240	.033			1.25	3.14	4.39	5.35
4480	cycle 20 miles		160	.050			1.88	4.71	6.59	8.05
4482	cycle 30 miles		140	.057			2.15	5.40	7.55	9.15
4484	cycle 40 miles		100	.080			3	7.55	10.55	12.90
4486	cycle 50 miles		100	.080			3	7.55	10.55	12.90
4494	45 MPH ave, cycle 8 miles		260	.031			1.16	2.90	4.06	4.95
4496	cycle 10 miles		240	.033			1.25	3.14	4.39	5.35
4498	cycle 20 miles		180	.044			1.67	4.19	5.86	7.15
4500	cycle 30 miles		140	.057			2.15	5.40	7.55	9.15
4502	cycle 40 miles		120	.067			2.50	6.30	8.80	10.70
4504	cycle 50 miles		100	.080			3	7.55	10.55	12.90
4506	50 MPH ave, cycle 10 miles		260	.031			1.16	2.90	4.06	4.95
4508	cycle 20 miles		200	.040			1.50	3.77	5.27	6.45
4510	cycle 30 miles		160	.050			1.88	4.71	6.59	8.05
4512	cycle 40 miles		120	.067			2.50	6.30	8.80	10.70
4514	cycle 50 miles		100	.080			3	7.55	10.55	12.90
4614	30 min. wait/Ld./Uld., 15 MPH, cycle 0.5 mile		300	.027			1	2.51	3.51	4.29
4616	cycle 1 mile		280	.029			1.07	2.69	3.76	4.60
4618	cycle 2 miles		240	.033			1.25	3.14	4.39	5.35
4620	cycle 4 miles		200	.040			1.50	3.77	5.27	6.45
4622	cycle 6 miles		180	.044			1.67	4.19	5.86	7.15
4624	cycle 8 miles		140	.057			2.15	5.40	7.55	9.15
4625	cycle 10 miles		140	.057			2.15	5.40	7.55	9.15
4626	20 MPH ave, cycle 0.5 mile		300	.027			1	2.51	3.51	4.29
4628	cycle 1 mile		280	.029			1.07	2.69	3.76	4.60
4630	cycle 2 miles		260	.031			1.16	2.90	4.06	4.95

31 23 23.20 Hauling		Crew	Daily Output	Labor-Hours	Unit	Material	2014 Bare Costs Labor	Equipment	Total	Total Incl O&P
4632	cycle 4 miles	B-34D	220	.036	L.C.Y.		1.37	3.43	4.80	5.85
4634	cycle 6 miles		200	.040			1.50	3.77	5.27	6.45
4636	cycle 8 miles		180	.044			1.67	4.19	5.86	7.15
4638	cycle 10 miles		160	.050			1.88	4.71	6.59	8.05
4640	25 MPH ave, cycle 4 miles		240	.033			1.25	3.14	4.39	5.35
4642	cycle 6 miles		220	.036			1.37	3.43	4.80	5.85
4644	cycle 8 miles		180	.044			1.67	4.19	5.86	7.15
4646	cycle 10 miles		180	.044			1.67	4.19	5.86	7.15
4650	30 MPH ave, cycle 4 miles		240	.033			1.25	3.14	4.39	5.35
4652	cycle 6 miles		220	.036			1.37	3.43	4.80	5.85
4654	cycle 8 miles		200	.040			1.50	3.77	5.27	6.45
4656	cycle 10 miles		180	.044			1.67	4.19	5.86	7.15
4660	35 MPH ave, cycle 4 miles		260	.031			1.16	2.90	4.06	4.95
4662	cycle 6 miles		240	.033			1.25	3.14	4.39	5.35
4664	cycle 8 miles		220	.036			1.37	3.43	4.80	5.85
4666	cycle 10 miles		200	.040			1.50	3.77	5.27	6.45
4668	cycle 20 miles		140	.057			2.15	5.40	7.55	9.15
4670	cycle 30 miles		120	.067			2.50	6.30	8.80	10.70
4672	cycle 40 miles		100	.080			3	7.55	10.55	12.90
4674	40 MPH, cycle 6 miles		240	.033			1.25	3.14	4.39	5.35
4676	cycle 8 miles		220	.036			1.37	3.43	4.80	5.85
4678	cycle 10 miles		200	.040			1.50	3.77	5.27	6.45
4680	cycle 20 miles		160	.050			1.88	4.71	6.59	8.05
4682	cycle 30 miles		120	.067			2.50	6.30	8.80	10.70
4684	cycle 40 miles		100	.080			3	7.55	10.55	12.90
4686	cycle 50 miles		80	.100			3.76	9.40	13.16	16.10
4694	45 MPH ave, cycle 8 miles		220	.036			1.37	3.43	4.80	5.85
4696	cycle 10 miles		220	.036			1.37	3.43	4.80	5.85
4698	cycle 20 miles		160	.050			1.88	4.71	6.59	8.05
4700	cycle 30 miles		140	.057			2.15	5.40	7.55	9.15
4702	cycle 40 miles		100	.080			3	7.55	10.55	12.90
4704	cycle 50 miles		100	.080			3	7.55	10.55	12.90
4706	50 MPH ave, cycle 10 miles		220	.036			1.37	3.43	4.80	5.85
4708	cycle 20 miles		180	.044			1.67	4.19	5.86	7.15
4710	cycle 30 miles		140	.057			2.15	5.40	7.55	9.15
4712	cycle 40 miles		120	.067			2.50	6.30	8.80	10.70
4714	cycle 50 miles		100	.080			3	7.55	10.55	12.90
5000	22 C.Y. off-road, 15 min. wait/Ld./Uld., 5 MPH, cycle 2000 ft	B-34F	528	.015			.57	2.78	3.35	3.93
5010	cycle 3000 ft		484	.017			.62	3.04	3.66	4.29
5020	cycle 4000 ft		440	.018			.68	3.34	4.02	4.71
5030	cycle 0.5 mile		506	.016			.59	2.91	3.50	4.11
5040	cycle 1 mile		374	.021			.80	3.93	4.73	5.55
5050	cycle 2 miles		264	.030			1.14	5.55	6.69	7.85
5060	10 MPH, cycle 2000 ft		594	.013			.51	2.48	2.99	3.49
5070	cycle 3000 ft		572	.014			.53	2.57	3.10	3.63
5080	cycle 4000 ft		528	.015			.57	2.78	3.35	3.93
5090	cycle 0.5 mile		572	.014			.53	2.57	3.10	3.63
5100	cycle 1 mile		506	.016			.59	2.91	3.50	4.11
5110	cycle 2 miles		374	.021			.80	3.93	4.73	5.55
5120	cycle 4 miles		264	.030			1.14	5.55	6.69	7.85
5130	15 MPH, cycle 2000 ft		638	.013			.47	2.30	2.77	3.25
5140	cycle 3000 ft		594	.013			.51	2.48	2.99	3.49
5150	cycle 4000 ft		572	.014			.53	2.57	3.10	3.63

31 23 23.20 Hauling		Crew	Daily Output	Labor-Hours	Unit	Material	2014 Bare Costs Labor	Equipment	Total	Total Incl O&P
5160	cycle 0.5 mile	B-34F	616	.013	L.C.Y.		.49	2.39	2.88	3.37
5170	cycle 1 mile		550	.015			.55	2.67	3.22	3.77
5180	cycle 2 miles		462	.017			.65	3.18	3.83	4.49
5190	cycle 4 miles		330	.024			.91	4.45	5.36	6.30
5200	20 MPH, cycle 2 miles		506	.016			.59	2.91	3.50	4.11
5210	cycle 4 miles		374	.021			.80	3.93	4.73	5.55
5220	25 MPH, cycle 2 miles		528	.015			.57	2.78	3.35	3.93
5230	cycle 4 miles		418	.019			.72	3.52	4.24	4.97
5300	20 min. wait/Ld./Uld., 5 MPH, cycle 2000 ft		418	.019			.72	3.52	4.24	4.97
5310	cycle 3000 ft		396	.020			.76	3.71	4.47	5.25
5320	cycle 4000 ft		352	.023			.85	4.18	5.03	5.90
5330	cycle 0.5 mile		396	.020			.76	3.71	4.47	5.25
5340	cycle 1 mile		330	.024			.91	4.45	5.36	6.30
5350	cycle 2 miles		242	.033			1.24	6.05	7.29	8.60
5360	10 MPH, cycle 2000 ft		462	.017			.65	3.18	3.83	4.49
5370	cycle 3000 ft		440	.018			.68	3.34	4.02	4.71
5380	cycle 4000 ft		418	.019			.72	3.52	4.24	4.97
5390	cycle 0.5 mile		462	.017			.65	3.18	3.83	4.49
5400	cycle 1 mile		396	.020			.76	3.71	4.47	5.25
5410	cycle 2 miles		330	.024			.91	4.45	5.36	6.30
5420	cycle 4 miles		242	.033			1.24	6.05	7.29	8.60
5430	15 MPH, cycle 2000 ft		484	.017			.62	3.04	3.66	4.29
5440	cycle 3000 ft		462	.017			.65	3.18	3.83	4.49
5450	cycle 4000 ft		462	.017			.65	3.18	3.83	4.49
5460	cycle 0.5 mile		484	.017			.62	3.04	3.66	4.29
5470	cycle 1 mile		440	.018			.68	3.34	4.02	4.71
5480	cycle 2 miles		374	.021			.80	3.93	4.73	5.55
5490	cycle 4 miles		286	.028			1.05	5.15	6.20	7.25
5500	20 MPH, cycle 2 miles		396	.020			.76	3.71	4.47	5.25
5510	cycle 4 miles		330	.024			.91	4.45	5.36	6.30
5520	25 MPH, cycle 2 miles		418	.019			.72	3.52	4.24	4.97
5530	cycle 4 miles		352	.023			.85	4.18	5.03	5.90
5600	25 min. wait/Ld./Uld., 5 MPH, cycle 2000 ft		352	.023			.85	4.18	5.03	5.90
5610	cycle 3000 ft		330	.024			.91	4.45	5.36	6.30
5620	cycle 4000 ft		308	.026			.98	4.77	5.75	6.75
5630	cycle 0.5 mile		330	.024			.91	4.45	5.36	6.30
5640	cycle 1 mile		286	.028			1.05	5.15	6.20	7.25
5650	cycle 2 miles		220	.036			1.37	6.70	8.07	9.45
5660	10 MPH, cycle 2000 ft		374	.021			.80	3.93	4.73	5.55
5670	cycle 3000 ft		374	.021			.80	3.93	4.73	5.55
5680	cycle 4000 ft		352	.023			.85	4.18	5.03	5.90
5690	cycle 0.5 mile		374	.021			.80	3.93	4.73	5.55
5700	cycle 1 mile		330	.024			.91	4.45	5.36	6.30
5710	cycle 2 miles		286	.028			1.05	5.15	6.20	7.25
5720	cycle 4 miles		220	.036			1.37	6.70	8.07	9.45
5730	15 MPH, cycle 2000 ft		396	.020			.76	3.71	4.47	5.25
5740	cycle 3000 ft		374	.021			.80	3.93	4.73	5.55
5750	cycle 4000 ft		374	.021			.80	3.93	4.73	5.55
5760	cycle 0.5 mile		374	.021			.80	3.93	4.73	5.55
5770	cycle 1 mile		352	.023			.85	4.18	5.03	5.90
5780	cycle 2 miles		308	.026			.98	4.77	5.75	6.75
5790	cycle 4 miles		242	.033			1.24	6.05	7.29	8.60
5800	20 MPH, cycle 2 miles		330	.024			.91	4.45	5.36	6.30

31 23 23.20 Hauling		Crew	Daily Output	Labor-Hours	Unit	Material	2014 Bare Costs Labor	Equipment	Total	Total Incl O&P
5810	cycle 4 miles	B-34F	286	.028	L.C.Y.		1.05	5.15	6.20	7.25
5820	25 MPH, cycle 2 miles		352	.023			.85	4.18	5.03	5.90
5830	cycle 4 miles		308	.026			.98	4.77	5.75	6.75
6000	34 C.Y. off-road, 15 min. wait/Ld./Uld., 5 MPH, cycle 2000 ft	B-34G	816	.010			.37	2.19	2.56	2.97
6010	cycle 3000 ft		748	.011			.40	2.39	2.79	3.24
6020	cycle 4000 ft		680	.012			.44	2.63	3.07	3.56
6030	cycle 0.5 mile		782	.010			.38	2.29	2.67	3.10
6040	cycle 1 mile		578	.014			.52	3.09	3.61	4.19
6050	cycle 2 miles		408	.020			.74	4.38	5.12	5.95
6060	10 MPH, cycle 2000 ft		918	.009			.33	1.95	2.28	2.64
6070	cycle 3000 ft		884	.009			.34	2.02	2.36	2.74
6080	cycle 4000 ft		816	.010			.37	2.19	2.56	2.97
6090	cycle 0.5 mile		884	.009			.34	2.02	2.36	2.74
6100	cycle 1 mile		782	.010			.38	2.29	2.67	3.10
6110	cycle 2 miles		578	.014			.52	3.09	3.61	4.19
6120	cycle 4 miles		408	.020			.74	4.38	5.12	5.95
6130	15 MPH, cycle 2000 ft		986	.008			.30	1.81	2.11	2.45
6140	cycle 3000 ft		918	.009			.33	1.95	2.28	2.64
6150	cycle 4000 ft		884	.009			.34	2.02	2.36	2.74
6160	cycle 0.5 mile		952	.008			.32	1.88	2.20	2.54
6170	cycle 1 mile		850	.009			.35	2.10	2.45	2.85
6180	cycle 2 miles		714	.011			.42	2.50	2.92	3.39
6190	cycle 4 miles		510	.016			.59	3.50	4.09	4.76
6200	20 MPH, cycle 2 miles		782	.010			.38	2.29	2.67	3.10
6210	cycle 4 miles		578	.014			.52	3.09	3.61	4.19
6220	25 MPH, cycle 2 miles		816	.010			.37	2.19	2.56	2.97
6230	cycle 4 miles		646	.012			.46	2.77	3.23	3.75
6300	20 min. wait/Ld./Uld., 5 MPH, cycle 2000 ft		646	.012			.46	2.77	3.23	3.75
6310	cycle 3000 ft		612	.013			.49	2.92	3.41	3.96
6320	cycle 4000 ft		544	.015			.55	3.29	3.84	4.45
6330	cycle 0.5 mile		612	.013			.49	2.92	3.41	3.96
6340	cycle 1 mile		510	.016			.59	3.50	4.09	4.76
6350	cycle 2 miles		374	.021			.80	4.78	5.58	6.50
6360	10 MPH, cycle 2000 ft		714	.011			.42	2.50	2.92	3.39
6370	cycle 3000 ft		680	.012			.44	2.63	3.07	3.56
6380	cycle 4000 ft		646	.012			.46	2.77	3.23	3.75
6390	cycle 0.5 mile		714	.011			.42	2.50	2.92	3.39
6400	cycle 1 mile		612	.013			.49	2.92	3.41	3.96
6410	cycle 2 miles		510	.016			.59	3.50	4.09	4.76
6420	cycle 4 miles		374	.021			.80	4.78	5.58	6.50
6430	15 MPH, cycle 2000 ft		748	.011			.40	2.39	2.79	3.24
6440	cycle 3000 ft		714	.011			.42	2.50	2.92	3.39
6450	cycle 4000 ft		714	.011			.42	2.50	2.92	3.39
6460	cycle 0.5 mile		748	.011			.40	2.39	2.79	3.24
6470	cycle 1 mile		680	.012			.44	2.63	3.07	3.56
6480	cycle 2 miles		578	.014			.52	3.09	3.61	4.19
6490	cycle 4 miles		442	.018			.68	4.04	4.72	5.50
6500	20 MPH, cycle 2 miles		612	.013			.49	2.92	3.41	3.96
6510	cycle 4 miles		510	.016			.59	3.50	4.09	4.76
6520	25 MPH, cycle 2 miles		646	.012			.46	2.77	3.23	3.75
6530	cycle 4 miles		544	.015			.55	3.29	3.84	4.45
6600	25 min. wait/Ld./Uld., 5 MPH, cycle 2000 ft		544	.015			.55	3.29	3.84	4.45
6610	cycle 3000 ft		510	.016			.59	3.50	4.09	4.76

31 23 23.20 Hauling		Crew	Daily Output	Labor-Hours	Unit	Material	2014 Bare Costs Labor	Equipment	Total	Total Incl O&P
6620	cycle 4000 ft	B-34G	476	.017	L.C.Y.		.63	3.76	4.39	5.10
6630	cycle 0.5 mile		510	.016			.59	3.50	4.09	4.76
6640	cycle 1 mile		442	.018			.68	4.04	4.72	5.50
6650	cycle 2 miles		340	.024			.88	5.25	6.13	7.15
6660	10 MPH, cycle 2000 ft		578	.014			.52	3.09	3.61	4.19
6670	cycle 3000 ft		578	.014			.52	3.09	3.61	4.19
6680	cycle 4000 ft		544	.015			.55	3.29	3.84	4.45
6690	cycle 0.5 mile		578	.014			.52	3.09	3.61	4.19
6700	cycle 1 mile		510	.016			.59	3.50	4.09	4.76
6710	cycle 2 miles		442	.018			.68	4.04	4.72	5.50
6720	cycle 4 miles		340	.024			.88	5.25	6.13	7.15
6730	15 MPH, cycle 2000 ft		612	.013			.49	2.92	3.41	3.96
6740	cycle 3000 ft		578	.014			.52	3.09	3.61	4.19
6750	cycle 4000 ft		578	.014			.52	3.09	3.61	4.19
6760	cycle 0.5 mile		612	.013			.49	2.92	3.41	3.96
6770	cycle 1 mile		544	.015			.55	3.29	3.84	4.45
6780	cycle 2 miles		476	.017			.63	3.76	4.39	5.10
6790	cycle 4 miles		374	.021			.80	4.78	5.58	6.50
6800	20 MPH, cycle 2 miles		510	.016			.59	3.50	4.09	4.76
6810	cycle 4 miles		442	.018			.68	4.04	4.72	5.50
6820	25 MPH, cycle 2 miles		544	.015			.55	3.29	3.84	4.45
6830	cycle 4 miles		476	.017			.63	3.76	4.39	5.10
7000	42 C.Y. off-road, 20 min. wait/Ld./Uld., 5 MPH, cycle 2000 ft	B-34H	798	.010			.38	2.25	2.63	3.05
7010	cycle 3000 ft		756	.011			.40	2.38	2.78	3.22
7020	cycle 4000 ft		672	.012			.45	2.67	3.12	3.62
7030	cycle 0.5 mile		756	.011			.40	2.38	2.78	3.22
7040	cycle 1 mile		630	.013			.48	2.85	3.33	3.87
7050	cycle 2 miles		462	.017			.65	3.89	4.54	5.25
7060	10 MPH, cycle 2000 ft		882	.009			.34	2.04	2.38	2.76
7070	cycle 3000 ft		840	.010			.36	2.14	2.50	2.90
7080	cycle 4000 ft		798	.010			.38	2.25	2.63	3.05
7090	cycle 0.5 mile		882	.009			.34	2.04	2.38	2.76
7100	cycle 1 mile		798	.010			.38	2.25	2.63	3.05
7110	cycle 2 miles		630	.013			.48	2.85	3.33	3.87
7120	cycle 4 miles		462	.017			.65	3.89	4.54	5.25
7130	15 MPH, cycle 2000 ft		924	.009			.33	1.95	2.28	2.64
7140	cycle 3000 ft		882	.009			.34	2.04	2.38	2.76
7150	cycle 4000 ft		882	.009			.34	2.04	2.38	2.76
7160	cycle 0.5 mile		882	.009			.34	2.04	2.38	2.76
7170	cycle 1 mile		840	.010			.36	2.14	2.50	2.90
7180	cycle 2 miles		714	.011			.42	2.52	2.94	3.41
7190	cycle 4 miles		546	.015			.55	3.29	3.84	4.46
7200	20 MPH, cycle 2 miles		756	.011			.40	2.38	2.78	3.22
7210	cycle 4 miles		630	.013			.48	2.85	3.33	3.87
7220	25 MPH, cycle 2 miles		798	.010			.38	2.25	2.63	3.05
7230	cycle 4 miles		672	.012			.45	2.67	3.12	3.62
7300	25 min. wait/Ld./Uld., 5 MPH, cycle 2000 ft		672	.012			.45	2.67	3.12	3.62
7310	cycle 3000 ft		630	.013			.48	2.85	3.33	3.87
7320	cycle 4000 ft		588	.014			.51	3.06	3.57	4.14
7330	cycle 0.5 mile		630	.013			.48	2.85	3.33	3.87
7340	cycle 1 mile		546	.015			.55	3.29	3.84	4.46
7350	cycle 2 miles		378	.021			.79	4.75	5.54	6.45
7360	10 MPH, cycle 2000 ft		714	.011			.42	2.52	2.94	3.41

31 23 23 – Fill

31 23 23.20 Hauling		Crew	Daily Output	Labor-Hours	Unit	Material	2014 Bare Costs Labor	Equipment	Total	Total Incl O&P
7370	cycle 3000 ft	B-34H	714	.011	L.C.Y.		.42	2.52	2.94	3.41
7380	cycle 4000 ft		672	.012			.45	2.67	3.12	3.62
7390	cycle 0.5 mile		714	.011			.42	2.52	2.94	3.41
7400	cycle 1 mile		630	.013			.48	2.85	3.33	3.87
7410	cycle 2 miles		546	.015			.55	3.29	3.84	4.46
7420	cycle 4 miles		378	.021			.79	4.75	5.54	6.45
7430	15 MPH, cycle 2000 ft		756	.011			.40	2.38	2.78	3.22
7440	cycle 3000 ft		714	.011			.42	2.52	2.94	3.41
7450	cycle 4000 ft		714	.011			.42	2.52	2.94	3.41
7460	cycle 0.5 mile		714	.011			.42	2.52	2.94	3.41
7470	cycle 1 mile		672	.012			.45	2.67	3.12	3.62
7480	cycle 2 miles		588	.014			.51	3.06	3.57	4.14
7490	cycle 4 miles		462	.017			.65	3.89	4.54	5.25
7500	20 MPH, cycle 2 miles		630	.013			.48	2.85	3.33	3.87
7510	cycle 4 miles		546	.015			.55	3.29	3.84	4.46
7520	25 MPH, cycle 2 miles		672	.012			.45	2.67	3.12	3.62
7530	cycle 4 miles		588	.014			.51	3.06	3.57	4.14
8000	60 C.Y. off-road, 20 min. wait/Ld./Uld., 5 MPH, cycle 2000 ft	B-34J	1140	.007			.26	2.54	2.80	3.20
8010	cycle 3000 ft		1080	.007			.28	2.68	2.96	3.37
8020	cycle 4000 ft		960	.008			.31	3.02	3.33	3.80
8030	cycle 0.5 mile		1080	.007			.28	2.68	2.96	3.37
8040	cycle 1 mile		900	.009			.33	3.22	3.55	4.05
8050	cycle 2 miles		660	.012			.46	4.39	4.85	5.50
8060	10 MPH, cycle 2000 ft		1260	.006			.24	2.30	2.54	2.89
8070	cycle 3000 ft		1200	.007			.25	2.42	2.67	3.04
8080	cycle 4000 ft		1140	.007			.26	2.54	2.80	3.20
8090	cycle 0.5 mile		1260	.006			.24	2.30	2.54	2.89
8100	cycle 1 mile		1080	.007			.28	2.68	2.96	3.37
8110	cycle 2 miles		900	.009			.33	3.22	3.55	4.05
8120	cycle 4 miles		660	.012			.46	4.39	4.85	5.50
8130	15 MPH, cycle 2000 ft		1320	.006			.23	2.20	2.43	2.76
8140	cycle 3000 ft		1260	.006			.24	2.30	2.54	2.89
8150	cycle 4000 ft		1260	.006			.24	2.30	2.54	2.89
8160	cycle 0.5 mile		1320	.006			.23	2.20	2.43	2.76
8170	cycle 1 mile		1200	.007			.25	2.42	2.67	3.04
8180	cycle 2 miles		1020	.008			.29	2.84	3.13	3.57
8190	cycle 4 miles		780	.010			.39	3.72	4.11	4.68
8200	20 MPH, cycle 2 miles		1080	.007			.28	2.68	2.96	3.37
8210	cycle 4 miles		900	.009			.33	3.22	3.55	4.05
8220	25 MPH, cycle 2 miles		1140	.007			.26	2.54	2.80	3.20
8230	cycle 4 miles		960	.008			.31	3.02	3.33	3.80
8300	25 min. wait/Ld./Uld., 5 MPH, cycle 2000 ft		960	.008			.31	3.02	3.33	3.80
8310	cycle 3000 ft		900	.009			.33	3.22	3.55	4.05
8320	cycle 4000 ft		840	.010			.36	3.45	3.81	4.34
8330	cycle 0.5 mile		900	.009			.33	3.22	3.55	4.05
8340	cycle 1 mile		780	.010			.39	3.72	4.11	4.68
8350	cycle 2 miles		600	.013			.50	4.83	5.33	6.05
8360	10 MPH, cycle 2000 ft		1020	.008			.29	2.84	3.13	3.57
8370	cycle 3000 ft		1020	.008			.29	2.84	3.13	3.57
8380	cycle 4000 ft		960	.008			.31	3.02	3.33	3.80
8390	cycle 0.5 mile		1020	.008			.29	2.84	3.13	3.57
8400	cycle 1 mile		900	.009			.33	3.22	3.55	4.05
8410	cycle 2 miles		780	.010			.39	3.72	4.11	4.68

31 23 23.20 Hauling		Crew	Daily Output	Labor-Hours	Unit	Material	2014 Bare Costs Labor	Equipment	Total	Total Incl O&P
8420	cycle 4 miles	B-34J	600	.013	L.C.Y.		.50	4.83	5.33	6.05
8430	15 MPH, cycle 2000 ft		1080	.007			.28	2.68	2.96	3.37
8440	cycle 3000 ft		1020	.008			.29	2.84	3.13	3.57
8450	cycle 4000 ft		1020	.008			.29	2.84	3.13	3.57
8460	cycle 0.5 mile		1080	.007			.28	2.68	2.96	3.37
8470	cycle 1 mile		960	.008			.31	3.02	3.33	3.80
8480	cycle 2 miles		840	.010			.36	3.45	3.81	4.34
8490	cycle 4 miles		660	.012			.46	4.39	4.85	5.50
8500	20 MPH, cycle 2 miles		900	.009			.33	3.22	3.55	4.05
8510	cycle 4 miles		780	.010			.39	3.72	4.11	4.68
8520	25 MPH, cycle 2 miles		960	.008			.31	3.02	3.33	3.80
8530	cycle 4 miles		840	.010			.36	3.45	3.81	4.34
9014	18 C.Y. truck, 8 wheels,15 min. wait/Ld./Uld.,15 MPH, cycle 0.5 mi.	B-34I	504	.016			.60	1.72	2.32	2.80
9016	cycle 1 mile		450	.018			.67	1.93	2.60	3.14
9018	cycle 2 miles		378	.021			.79	2.29	3.08	3.73
9020	cycle 4 miles		270	.030			1.11	3.21	4.32	5.25
9022	cycle 6 miles		216	.037			1.39	4.02	5.41	6.55
9024	cycle 8 miles		180	.044			1.67	4.82	6.49	7.85
9025	cycle 10 miles		144	.056			2.09	6.05	8.14	9.85
9026	20 MPH ave, cycle 0.5 mile		522	.015			.58	1.66	2.24	2.71
9028	cycle 1 mile		468	.017			.64	1.85	2.49	3.02
9030	cycle 2 miles		414	.019			.73	2.10	2.83	3.41
9032	cycle 4 miles		324	.025			.93	2.68	3.61	4.36
9034	cycle 6 miles		252	.032			1.19	3.44	4.63	5.60
9036	cycle 8 miles		216	.037			1.39	4.02	5.41	6.55
9038	cycle 10 miles		180	.044			1.67	4.82	6.49	7.85
9040	25 MPH ave, cycle 4 miles		342	.023			.88	2.54	3.42	4.13
9042	cycle 6 miles		288	.028			1.04	3.01	4.05	4.90
9044	cycle 8 miles		252	.032			1.19	3.44	4.63	5.60
9046	cycle 10 miles		216	.037			1.39	4.02	5.41	6.55
9050	30 MPH ave, cycle 4 miles		378	.021			.79	2.29	3.08	3.73
9052	cycle 6 miles		324	.025			.93	2.68	3.61	4.36
9054	cycle 8 miles		270	.030			1.11	3.21	4.32	5.25
9056	cycle 10 miles		234	.034			1.28	3.71	4.99	6.05
9060	35 MPH ave, cycle 4 miles		396	.020			.76	2.19	2.95	3.57
9062	cycle 6 miles		342	.023			.88	2.54	3.42	4.13
9064	cycle 8 miles		288	.028			1.04	3.01	4.05	4.90
9066	cycle 10 miles		270	.030			1.11	3.21	4.32	5.25
9068	cycle 20 miles		162	.049			1.85	5.35	7.20	8.75
9070	cycle 30 miles		126	.063			2.38	6.90	9.28	11.20
9072	cycle 40 miles		90	.089			3.34	9.65	12.99	15.70
9074	40 MPH ave, cycle 6 miles		360	.022			.83	2.41	3.24	3.92
9076	cycle 8 miles		324	.025			.93	2.68	3.61	4.36
9078	cycle 10 miles		288	.028			1.04	3.01	4.05	4.90
9080	cycle 20 miles		180	.044			1.67	4.82	6.49	7.85
9082	cycle 30 miles		144	.056			2.09	6.05	8.14	9.85
9084	cycle 40 miles		108	.074			2.78	8.05	10.83	13.10
9086	cycle 50 miles		90	.089			3.34	9.65	12.99	15.70
9094	45 MPH ave, cycle 8 miles		324	.025			.93	2.68	3.61	4.36
9096	cycle 10 miles		306	.026			.98	2.83	3.81	4.62
9098	cycle 20 miles		198	.040			1.52	4.38	5.90	7.15
9100	cycle 30 miles		144	.056			2.09	6.05	8.14	9.85
9102	cycle 40 miles		126	.063			2.38	6.90	9.28	11.20

31 23 23.20 Hauling		Crew	Daily Output	Labor-Hours	Unit	Material	2014 Bare Costs Labor	Equipment	Total	Total Incl O&P
9104	cycle 50 miles	B-34I	108	.074	L.C.Y.		2.78	8.05	10.83	13.10
9106	50 MPH ave, cycle 10 miles		324	.025			.93	2.68	3.61	4.36
9108	cycle 20 miles		216	.037			1.39	4.02	5.41	6.55
9110	cycle 30 miles		162	.049			1.85	5.35	7.20	8.75
9112	cycle 40 miles		126	.063			2.38	6.90	9.28	11.20
9114	cycle 50 miles		108	.074			2.78	8.05	10.83	13.10
9214	20 min. wait/Ld./Uld.,15 MPH, cycle 0.5 mi.		396	.020			.76	2.19	2.95	3.57
9216	cycle 1 mile		360	.022			.83	2.41	3.24	3.92
9218	cycle 2 miles		306	.026			.98	2.83	3.81	4.62
9220	cycle 4 miles		234	.034			1.28	3.71	4.99	6.05
9222	cycle 6 miles		198	.040			1.52	4.38	5.90	7.15
9224	cycle 8 miles		162	.049			1.85	5.35	7.20	8.75
9225	cycle 10 miles		144	.056			2.09	6.05	8.14	9.85
9226	20 MPH ave, cycle 0.5 mile		396	.020			.76	2.19	2.95	3.57
9228	cycle 1 mile		378	.021			.79	2.29	3.08	3.73
9230	cycle 2 miles		324	.025			.93	2.68	3.61	4.36
9232	cycle 4 miles		270	.030			1.11	3.21	4.32	5.25
9234	cycle 6 miles		216	.037			1.39	4.02	5.41	6.55
9236	cycle 8 miles		198	.040			1.52	4.38	5.90	7.15
9238	cycle 10 miles		162	.049			1.85	5.35	7.20	8.75
9240	25 MPH ave, cycle 4 miles		288	.028			1.04	3.01	4.05	4.90
9242	cycle 6 miles		252	.032			1.19	3.44	4.63	5.60
9244	cycle 8 miles		216	.037			1.39	4.02	5.41	6.55
9246	cycle 10 miles		198	.040			1.52	4.38	5.90	7.15
9250	30 MPH ave, cycle 4 miles		306	.026			.98	2.83	3.81	4.62
9252	cycle 6 miles		270	.030			1.11	3.21	4.32	5.25
9254	cycle 8 miles		234	.034			1.28	3.71	4.99	6.05
9256	cycle 10 miles		216	.037			1.39	4.02	5.41	6.55
9260	35 MPH ave, cycle 4 miles		324	.025			.93	2.68	3.61	4.36
9262	cycle 6 miles		288	.028			1.04	3.01	4.05	4.90
9264	cycle 8 miles		252	.032			1.19	3.44	4.63	5.60
9266	cycle 10 miles		234	.034			1.28	3.71	4.99	6.05
9268	cycle 20 miles		162	.049			1.85	5.35	7.20	8.75
9270	cycle 30 miles		108	.074			2.78	8.05	10.83	13.10
9272	cycle 40 miles		90	.089			3.34	9.65	12.99	15.70
9274	40 MPH ave, cycle 6 miles		288	.028			1.04	3.01	4.05	4.90
9276	cycle 8 miles		270	.030			1.11	3.21	4.32	5.25
9278	cycle 10 miles		234	.034			1.28	3.71	4.99	6.05
9280	cycle 20 miles		162	.049			1.85	5.35	7.20	8.75
9282	cycle 30 miles		126	.063			2.38	6.90	9.28	11.20
9284	cycle 40 miles		108	.074			2.78	8.05	10.83	13.10
9286	cycle 50 miles		90	.089			3.34	9.65	12.99	15.70
9294	45 MPH ave, cycle 8 miles		270	.030			1.11	3.21	4.32	5.25
9296	cycle 10 miles		252	.032			1.19	3.44	4.63	5.60
9298	cycle 20 miles		180	.044			1.67	4.82	6.49	7.85
9300	cycle 30 miles		144	.056			2.09	6.05	8.14	9.85
9302	cycle 40 miles		108	.074			2.78	8.05	10.83	13.10
9304	cycle 50 miles		90	.089			3.34	9.65	12.99	15.70
9306	50 MPH ave, cycle 10 miles		270	.030			1.11	3.21	4.32	5.25
9308	cycle 20 miles		198	.040			1.52	4.38	5.90	7.15
9310	cycle 30 miles		144	.056			2.09	6.05	8.14	9.85
9312	cycle 40 miles		126	.063			2.38	6.90	9.28	11.20
9314	cycle 50 miles		108	.074			2.78	8.05	10.83	13.10

	31 23 23.20 Hauling	Crew	Daily Output	Labor-Hours	Unit	Material	2014 Bare Costs Labor	Equipment	Total	Total Incl O&P
9414	25 min. wait/Ld./Uld.,15 MPH, cycle 0.5 mi.	B-34I	306	.026	L.C.Y.		.98	2.83	3.81	4.62
9416	cycle 1 mile		288	.028			1.04	3.01	4.05	4.90
9418	cycle 2 miles		252	.032			1.19	3.44	4.63	5.60
9420	cycle 4 miles		198	.040			1.52	4.38	5.90	7.15
9422	cycle 6 miles		180	.044			1.67	4.82	6.49	7.85
9424	cycle 8 miles		144	.056			2.09	6.05	8.14	9.85
9425	cycle 10 miles		126	.063			2.38	6.90	9.28	11.20
9426	20 MPH ave, cycle 0.5 mile		324	.025			.93	2.68	3.61	4.36
9428	cycle 1 mile		306	.026			.98	2.83	3.81	4.62
9430	cycle 2 miles		270	.030			1.11	3.21	4.32	5.25
9432	cycle 4 miles		234	.034			1.28	3.71	4.99	6.05
9434	cycle 6 miles		198	.040			1.52	4.38	5.90	7.15
9436	cycle 8 miles		180	.044			1.67	4.82	6.49	7.85
9438	cycle 10 miles		144	.056			2.09	6.05	8.14	9.85
9440	25 MPH ave, cycle 4 miles		252	.032			1.19	3.44	4.63	5.60
9442	cycle 6 miles		216	.037			1.39	4.02	5.41	6.55
9444	cycle 8 miles		198	.040			1.52	4.38	5.90	7.15
9446	cycle 10 miles		180	.044			1.67	4.82	6.49	7.85
9450	30 MPH ave, cycle 4 miles		252	.032			1.19	3.44	4.63	5.60
9452	cycle 6 miles		234	.034			1.28	3.71	4.99	6.05
9454	cycle 8 miles		198	.040			1.52	4.38	5.90	7.15
9456	cycle 10 miles		180	.044			1.67	4.82	6.49	7.85
9460	35 MPH ave, cycle 4 miles		270	.030			1.11	3.21	4.32	5.25
9462	cycle 6 miles		234	.034			1.28	3.71	4.99	6.05
9464	cycle 8 miles		216	.037			1.39	4.02	5.41	6.55
9466	cycle 10 miles		198	.040			1.52	4.38	5.90	7.15
9468	cycle 20 miles		144	.056			2.09	6.05	8.14	9.85
9470	cycle 30 miles		108	.074			2.78	8.05	10.83	13.10
9472	cycle 40 miles		90	.089			3.34	9.65	12.99	15.70
9474	40 MPH ave, cycle 6 miles		252	.032			1.19	3.44	4.63	5.60
9476	cycle 8 miles		234	.034			1.28	3.71	4.99	6.05
9478	cycle 10 miles		216	.037			1.39	4.02	5.41	6.55
9480	cycle 20 miles		144	.056			2.09	6.05	8.14	9.85
9482	cycle 30 miles		126	.063			2.38	6.90	9.28	11.20
9484	cycle 40 miles		90	.089			3.34	9.65	12.99	15.70
9486	cycle 50 miles		90	.089			3.34	9.65	12.99	15.70
9494	45 MPH ave, cycle 8 miles		234	.034			1.28	3.71	4.99	6.05
9496	cycle 10 miles		216	.037			1.39	4.02	5.41	6.55
9498	cycle 20 miles		162	.049			1.85	5.35	7.20	8.75
9500	cycle 30 miles		126	.063			2.38	6.90	9.28	11.20
9502	cycle 40 miles		108	.074			2.78	8.05	10.83	13.10
9504	cycle 50 miles		90	.089			3.34	9.65	12.99	15.70
9506	50 MPH ave, cycle 10 miles		234	.034			1.28	3.71	4.99	6.05
9508	cycle 20 miles		180	.044			1.67	4.82	6.49	7.85
9510	cycle 30 miles		144	.056			2.09	6.05	8.14	9.85
9512	cycle 40 miles		108	.074			2.78	8.05	10.83	13.10
9514	cycle 50 miles		90	.089			3.34	9.65	12.99	15.70
9614	30 min. wait/Ld./Uld.,15 MPH, cycle 0.5 mi.		270	.030			1.11	3.21	4.32	5.25
9616	cycle 1 mile		252	.032			1.19	3.44	4.63	5.60
9618	cycle 2 miles		216	.037			1.39	4.02	5.41	6.55
9620	cycle 4 miles		180	.044			1.67	4.82	6.49	7.85
9622	cycle 6 miles		162	.049			1.85	5.35	7.20	8.75
9624	cycle 8 miles		126	.063			2.38	6.90	9.28	11.20

31 23 23.20 Hauling		Crew	Daily Output	Labor-Hours	Unit	Material	2014 Bare Costs Labor	Equipment	Total	Total Incl O&P
9625	cycle 10 miles	B-34I	126	.063	L.C.Y.		2.38	6.90	9.28	11.20
9626	20 MPH ave, cycle 0.5 mile		270	.030			1.11	3.21	4.32	5.25
9628	cycle 1 mile		252	.032			1.19	3.44	4.63	5.60
9630	cycle 2 miles		234	.034			1.28	3.71	4.99	6.05
9632	cycle 4 miles		198	.040			1.52	4.38	5.90	7.15
9634	cycle 6 miles		180	.044			1.67	4.82	6.49	7.85
9636	cycle 8 miles		162	.049			1.85	5.35	7.20	8.75
9638	cycle 10 miles		144	.056			2.09	6.05	8.14	9.85
9640	25 MPH ave, cycle 4 miles		216	.037			1.39	4.02	5.41	6.55
9642	cycle 6 miles		198	.040			1.52	4.38	5.90	7.15
9644	cycle 8 miles		180	.044			1.67	4.82	6.49	7.85
9646	cycle 10 miles		162	.049			1.85	5.35	7.20	8.75
9650	30 MPH ave, cycle 4 miles		216	.037			1.39	4.02	5.41	6.55
9652	cycle 6 miles		198	.040			1.52	4.38	5.90	7.15
9654	cycle 8 miles		180	.044			1.67	4.82	6.49	7.85
9656	cycle 10 miles		162	.049			1.85	5.35	7.20	8.75
9660	35 MPH ave, cycle 4 miles		234	.034			1.28	3.71	4.99	6.05
9662	cycle 6 miles		216	.037			1.39	4.02	5.41	6.55
9664	cycle 8 miles		198	.040			1.52	4.38	5.90	7.15
9666	cycle 10 miles		180	.044			1.67	4.82	6.49	7.85
9668	cycle 20 miles		126	.063			2.38	6.90	9.28	11.20
9670	cycle 30 miles		108	.074			2.78	8.05	10.83	13.10
9672	cycle 40 miles		90	.089			3.34	9.65	12.99	15.70
9674	40 MPH ave, cycle 6 miles		216	.037			1.39	4.02	5.41	6.55
9676	cycle 8 miles		198	.040			1.52	4.38	5.90	7.15
9678	cycle 10 miles		180	.044			1.67	4.82	6.49	7.85
9680	cycle 20 miles		144	.056			2.09	6.05	8.14	9.85
9682	cycle 30 miles		108	.074			2.78	8.05	10.83	13.10
9684	cycle 40 miles		90	.089			3.34	9.65	12.99	15.70
9686	cycle 50 miles		72	.111			4.17	12.05	16.22	19.60
9694	45 MPH ave, cycle 8 miles		216	.037			1.39	4.02	5.41	6.55
9696	cycle 10 miles		198	.040			1.52	4.38	5.90	7.15
9698	cycle 20 miles		144	.056			2.09	6.05	8.14	9.85
9700	cycle 30 miles		126	.063			2.38	6.90	9.28	11.20
9702	cycle 40 miles		108	.074			2.78	8.05	10.83	13.10
9704	cycle 50 miles		90	.089			3.34	9.65	12.99	15.70
9706	50 MPH ave, cycle 10 miles		198	.040			1.52	4.38	5.90	7.15
9708	cycle 20 miles		162	.049			1.85	5.35	7.20	8.75
9710	cycle 30 miles		126	.063			2.38	6.90	9.28	11.20
9712	cycle 40 miles		108	.074			2.78	8.05	10.83	13.10
9714	cycle 50 miles		90	.089			3.34	9.65	12.99	15.70

31 23 23.23 Compaction

		Crew	Daily Output	Labor-Hours	Unit	Material	Labor	Equipment	Total	Total Incl O&P
0010	**COMPACTION**									
5000	Riding, vibrating roller, 6" lifts, 2 passes	B-10Y	3000	.004	E.C.Y.		.18	.19	.37	.48
5020	3 passes		2300	.005			.23	.24	.47	.63
5040	4 passes		1900	.006			.28	.29	.57	.75
5050	8" lifts, 2 passes		4100	.003			.13	.14	.27	.35
5060	12" lifts, 2 passes		5200	.002			.10	.11	.21	.28
5080	3 passes		3500	.003			.15	.16	.31	.41
5100	4 passes		2600	.005			.21	.22	.43	.56
5600	Sheepsfoot or wobbly wheel roller, 6" lifts, 2 passes	B-10G	2400	.005			.22	.51	.73	.90
5620	3 passes		1735	.007			.31	.70	1.01	1.24

31 23 23 – Fill

31 23 23.23 Compaction		Crew	Daily Output	Labor-Hours	Unit	Material	2014 Bare Costs Labor	Equipment	Total	Total Incl O&P
5640	4 passes	B-10G	1300	.009	E.C.Y.		.41	.94	1.35	1.66
5680	12" lifts, 2 passes		5200	.002			.10	.23	.33	.42
5700	3 passes		3500	.003			.15	.35	.50	.61
5720	4 passes		2600	.005			.21	.47	.68	.84
6000	Towed sheepsfoot or wobbly wheel roller, 6" lifts, 2 passes	B-10D	10000	.001			.05	.18	.23	.27
6020	3 passes		2000	.006			.27	.88	1.15	1.38
6030	4 passes		1500	.008			.36	1.18	1.54	1.85
6050	12" lifts, 2 passes		6000	.002			.09	.29	.38	.46
6060	3 passes		4000	.003			.13	.44	.57	.69
6070	4 passes		3000	.004			.18	.59	.77	.92
6200	Vibrating roller, 6" lifts, 2 passes	B-10C	2600	.005			.21	.67	.88	1.06
6210	3 passes		1735	.007			.31	1	1.31	1.57
6220	4 passes		1300	.009			.41	1.34	1.75	2.10
6250	12" lifts, 2 passes		5200	.002			.10	.33	.43	.53
6260	3 passes		3465	.003			.16	.50	.66	.79
6270	4 passes		2600	.005			.21	.67	.88	1.06
7000	Walk behind, vibrating plate 18" wide, 6" lifts, 2 passes	A-1D	200	.040			1.47	.18	1.65	2.46
7020	3 passes		185	.043			1.58	.19	1.77	2.66
7040	4 passes		140	.057			2.09	.26	2.35	3.51
7200	12" lifts, 2 passes, 21" wide	A-1E	560	.014			.52	.08	.60	.90
7220	3 passes		375	.021			.78	.12	.90	1.35
7240	4 passes		280	.029			1.05	.16	1.21	1.80
7500	Vibrating roller 24" wide, 6" lifts, 2 passes	B-10A	420	.029			1.28	.44	1.72	2.43
7520	3 passes		280	.043			1.92	.65	2.57	3.65
7540	4 passes		210	.057			2.56	.87	3.43	4.86
7600	12" lifts, 2 passes		840	.014			.64	.22	.86	1.22
7620	3 passes		560	.021			.96	.33	1.29	1.82
7640	4 passes		420	.029			1.28	.44	1.72	2.43
8000	Rammer tamper, 6" to 11", 4" lifts, 2 passes	A-1F	130	.062			2.26	.38	2.64	3.90
8050	3 passes		97	.082			3.02	.51	3.53	5.20
8100	4 passes		65	.123			4.51	.76	5.27	7.80
8200	8" lifts, 2 passes		260	.031			1.13	.19	1.32	1.95
8250	3 passes		195	.041			1.50	.25	1.75	2.60
8300	4 passes		130	.062			2.26	.38	2.64	3.90
8400	13" to 18", 4" lifts, 2 passes	A-1G	390	.021			.75	.14	.89	1.32
8450	3 passes		290	.028			1.01	.19	1.20	1.77
8500	4 passes		195	.041			1.50	.29	1.79	2.64
8600	8" lifts, 2 passes		780	.010			.38	.07	.45	.66
8650	3 passes		585	.014			.50	.10	.60	.88
8700	4 passes		390	.021			.75	.14	.89	1.32
9000	Water, 3000 gal. truck, 3 mile haul	B-45	1888	.008		1.20	.37	.49	2.06	2.42
9010	6 mile haul		1444	.011		1.20	.48	.64	2.32	2.75
9020	12 mile haul		1000	.016		1.20	.69	.92	2.81	3.38
9030	6000 gal. wagon, 3 mile haul	B-59	2000	.004		1.20	.15	.25	1.60	1.83
9040	6 mile haul	"	1600	.005		1.20	.19	.32	1.71	1.96

31 23 23.25 Compaction, Airports		Crew	Daily Output	Labor-Hours	Unit	Material	Labor	Equipment	Total	Total Incl O&P
0010	**COMPACTION, AIRPORTS**									
0100	Airport subgrade, compaction									
0110	non cohesive soils, 85% proctor, 12" depth	B-10G	15600	.001	S.Y.		.03	.08	.11	.14
0200	24" depth		7800	.002			.07	.16	.23	.28
0300	36" depth		5200	.002			.10	.23	.33	.42
0400	48" depth		3900	.003			.14	.31	.45	.55

31 23 Excavation and Fill

31 23 23 – Fill

31 23 23.25 Compaction, Airports		Crew	Daily Output	Labor-Hours	Unit	Material	2014 Bare Costs Labor	Equipment	Total	Total Incl O&P
0500	60″ depth	B-10G	3120	.004	S.Y.		.17	.39	.56	.69
0600	90% proctor, 12″ depth		7200	.002			.07	.17	.24	.30
0700	24″ depth		3600	.003			.15	.34	.49	.60
0800	36″ depth		2400	.005			.22	.51	.73	.90
0900	48″ depth		1800	.007			.30	.68	.98	1.20
1000	60″ depth		1440	.008			.37	.85	1.22	1.50
1100	95% proctor, 12″ depth	B-10D	6000	.002			.09	.29	.38	.46
1200	24″ depth		3000	.004			.18	.59	.77	.92
1300	36″ depth		2000	.006			.27	.88	1.15	1.38
1400	42″ depth		1715	.007			.31	1.03	1.34	1.62
1500	100% proctor, 12″ depth		4500	.003			.12	.39	.51	.61
1550	18″ depth		3000	.004			.18	.59	.77	.92
1600	24″ depth		2250	.005			.24	.79	1.03	1.22
1700	cohesive soils, 80% proctor, 12″ depth	B-10G	7200	.002			.07	.17	.24	.30
1800	24″ depth	B-10D	15000	.001			.04	.12	.16	.18
1900	36″ depth		10000	.001			.05	.18	.23	.27
2000	85% proctor, 12″ depth		6000	.002			.09	.29	.38	.46
2100	18″ depth		4000	.003			.13	.44	.57	.69
2200	24″ depth		3000	.004			.18	.59	.77	.92
2300	27″ depth		2670	.004			.20	.66	.86	1.04
2400	90% proctor, 6″ depth		10400	.001			.05	.17	.22	.27
2500	9″ depth		6935	.002			.08	.26	.34	.40
2600	12″ depth		5200	.002			.10	.34	.44	.53
2700	15″ depth		4160	.003			.13	.42	.55	.67
2800	18″ depth		3470	.003			.16	.51	.67	.80
2900	95% proctor, 6″ depth		1500	.008			.36	1.18	1.54	1.85
3000	7″ depth		1285	.009			.42	1.38	1.80	2.15
3100	8″ depth		1125	.011			.48	1.57	2.05	2.46
3200	9″ depth		1000	.012			.54	1.77	2.31	2.77

31 25 Erosion and Sedimentation Controls

31 25 14 – Stabilization Measures for Erosion and Sedimentation Control

31 25 14.16 Rolled Erosion Control Mats and Blankets

		Crew	Daily Output	Labor-Hours	Unit	Material	Labor	Equipment	Total	Total Incl O&P
0010	**ROLLED EROSION CONTROL MATS AND BLANKETS**									
0020	Jute mesh, 100 S.Y. per roll, 4′ wide, stapled G	B-80A	2400	.010	S.Y.	1.11	.37	.14	1.62	1.94
0060	Polyethylene 3 dimensional geomatrix, 50 mil thick G		700	.034		2.40	1.26	.48	4.14	5.10
0062	120 mil thick G		515	.047		2.63	1.71	.65	4.99	6.25
0070	Paper biodegradable mesh G	B-1	2500	.010		.10	.36		.46	.66
0080	Paper mulch G	B-64	20000	.001		.10	.03	.02	.15	.17
0100	Plastic netting, stapled, 2″ x 1″ mesh, 20 mil G	B-1	2500	.010		.23	.36		.59	.80
0120	Revegetation mat, webbed G	2 Clab	1000	.016		4.55	.59		5.14	5.90
0200	Polypropylene mesh, stapled, 6.5 oz./S.Y. G	B-1	2500	.010		2.33	.36		2.69	3.11
0300	Tobacco netting, or jute mesh #2, stapled G	″	2500	.010		.18	.36		.54	.75
0400	Soil sealant, liquid sprayed from truck G	B-81	5000	.005		.42	.20	.14	.76	.92
1000	Silt fence, polypropylene, 3′ high, ideal conditions G	2 Clab	1600	.010	L.F.	.24	.37		.61	.83
1100	Adverse conditions G	″	950	.017	″	.24	.62		.86	1.21
1130	Cellular confinement, poly, 3-dimen, 8′ x 20′ panels, 4″ deep cell G	B-6	1600	.015	S.F.	2.05	.60	.23	2.88	3.43
1140	8″ deep cells G	″	1200	.020	″	2.82	.80	.30	3.92	4.66
1200	Place and remove hay bales G	A-2	3	8	Ton	241	293	90	624	815
1250	Hay bales, staked G	″	2500	.010	L.F.	9.65	.35	.11	10.11	11.25

31 31 Soil Treatment

31 31 16 – Termite Control

31 31 16.13 Chemical Termite Control		Crew	Daily Output	Labor-Hours	Unit	2014 Bare Costs Material	Labor	Equipment	Total	Total Incl O&P
0010	**CHEMICAL TERMITE CONTROL**									
0020	Slab and walls, residential	1 Skwk	1200	.007	SF Flr.	.37	.32		.69	.90
0100	Commercial, minimum		2496	.003		.32	.15		.47	.59
0200	Maximum		1645	.005		.48	.23		.71	.89
0400	Insecticides for termite control, minimum		14.20	.563	Gal.	63	26.50		89.50	111
0500	Maximum		11	.727	"	108	34.50		142.50	173
3000	Soil poisoning (sterilization)	1 Clab	4496	.002	S.F.	.45	.07		.52	.60
3100	Herbicide application from truck	B-59	19000	.001	S.Y.		.02	.03	.05	.05

31 32 Soil Stabilization

31 32 13 – Soil Mixing Stabilization

31 32 13.13 Asphalt Soil Stabilization		Crew	Daily Output	Labor-Hours	Unit	Material	Labor	Equipment	Total	Total Incl O&P
0010	**ASPHALT SOIL STABILIZATION**									
0011	Including scarifying and compaction									
0020	Asphalt, 1-1/2" deep, 1/2 gal/S.Y.	B-75	4000	.014	S.Y.	.94	.62	1.51	3.07	3.64
0040	3/4 gal/S.Y.		4000	.014		1.41	.62	1.51	3.54	4.16
0100	3" deep, 1 gal/S.Y.		3500	.016		1.88	.71	1.73	4.32	5.05
0140	1-1/2 gal/S.Y.		3500	.016		2.82	.71	1.73	5.26	6.05
0200	6" deep, 2 gal/S.Y.		3000	.019		3.76	.82	2.02	6.60	7.60
0240	3 gal/S.Y.		3000	.019		5.65	.82	2.02	8.49	9.65
0300	8" deep, 2-2/3 gal/S.Y.		2800	.020		5	.88	2.16	8.04	9.20
0340	4 gal/S.Y.		2800	.020		7.50	.88	2.16	10.54	11.95
0500	12" deep, 4 gal/S.Y.		5000	.011		7.50	.49	1.21	9.20	10.35
0540	6 gal/S.Y.		2600	.022		11.30	.95	2.33	14.58	16.40

31 32 13.16 Cement Soil Stabilization		Crew	Daily Output	Labor-Hours	Unit	Material	Labor	Equipment	Total	Total Incl O&P
0010	**CEMENT SOIL STABILIZATION**									
0011	Including scarifying and compaction									
1020	Cement, 4% mix, by volume, 6" deep	B-74	1100	.058	S.Y.	1.90	2.52	5.50	9.92	12
1030	8" deep		1050	.061		2.48	2.64	5.75	10.87	13.05
1060	12" deep		960	.067		3.73	2.88	6.30	12.91	15.40
1100	6% mix, 6" deep		1100	.058		2.73	2.52	5.50	10.75	12.90
1120	8" deep		1050	.061		3.56	2.64	5.75	11.95	14.25
1160	12" deep		960	.067		5.40	2.88	6.30	14.58	17.20
1200	9% mix, 6" deep		1100	.058		4.14	2.52	5.50	12.16	14.45
1220	8" deep		1050	.061		5.40	2.64	5.75	13.79	16.20
1260	12" deep		960	.067		8.10	2.88	6.30	17.28	20.50
1300	12% mix, 6" deep		1100	.058		5.40	2.52	5.50	13.42	15.80
1320	8" deep		1050	.061		7.20	2.64	5.75	15.59	18.20
1360	12" deep		960	.067		10.75	2.88	6.30	19.93	23

31 32 13.19 Lime Soil Stabilization		Crew	Daily Output	Labor-Hours	Unit	Material	Labor	Equipment	Total	Total Incl O&P
0010	**LIME SOIL STABILIZATION**									
0011	Including scarifying and compaction									
2020	Hydrated lime, for base, 2% mix by weight, 6" deep	B-74	1800	.036	S.Y.	2.39	1.54	3.35	7.28	8.65
2030	8" deep		1700	.038		3.42	1.63	3.55	8.60	10.15
2060	12" deep		1550	.041		4.84	1.79	3.89	10.52	12.30
2100	4% mix, 6" deep		1800	.036		3.57	1.54	3.35	8.46	9.95
2120	8" deep		1700	.038		4.82	1.63	3.55	10	11.70
2160	12" deep		1550	.041		7.20	1.79	3.89	12.88	14.95
2200	6% mix, 6" deep		1800	.036		5.40	1.54	3.35	10.29	12
2220	8" deep		1700	.038		7.20	1.63	3.55	12.38	14.35

31 32 Soil Stabilization

31 32 13 – Soil Mixing Stabilization

31 32 13.19 Lime Soil Stabilization		Crew	Daily Output	Labor-Hours	Unit	Material	Labor	Equipment	Total	Total Incl O&P
						2014 Bare Costs				
2260	12" deep	B-74	1550	.041	S.Y.	10.75	1.79	3.89	16.43	18.80

31 32 13.30 Calcium Chloride

		Crew	Daily Output	Labor-Hours	Unit	Material	Labor	Equipment	Total	Total Incl O&P
0010	**CALCIUM CHLORIDE**									
0020	Calcium chloride delivered, 100 lb. bags, truckload lots				Ton	385			385	425
0030	Solution, 4 lb. flake per gallon, tank truck delivery				Gal.	1.44			1.44	1.58

31 32 19 – Geosynthetic Soil Stabilization and Layer Separation

31 32 19.16 Geotextile Soil Stabilization

		Crew	Daily Output	Labor-Hours	Unit	Material	Labor	Equipment	Total	Total Incl O&P
0010	**GEOTEXTILE SOIL STABILIZATION**									
1500	Geotextile fabric, woven, 200 lb. tensile strength	2 Clab	2500	.006	S.Y.	1.88	.23		2.11	2.43
1510	Heavy Duty, 600 lb. tensile strength		2400	.007		1.80	.24		2.04	2.36
1550	Non-woven, 120 lb. tensile strength		2500	.006		1.12	.23		1.35	1.59

31 32 36 – Soil Nailing

31 32 36.16 Grouted Soil Nailing

		Crew	Daily Output	Labor-Hours	Unit	Material	Labor	Equipment	Total	Total Incl O&P
0010	**GROUTED SOIL NAILING**									
0020	Soil nailing does not include guniting of surfaces									
0030	See Section 03 37 13.30 for guniting of surfaces									
0035	Layout and vertical and horizontal control per day	A-6	1	16	Day		730	71.50	801.50	1,200
0038	Layout and vertical and horizontal control average holes per day	"	60	.267	Ea.		12.20	1.19	13.39	20
0050	For Grade 150 soil nail add $ 1.37/L.F. to base material cost									
0060	Material delivery add $ 1.83 to $ 2.00 per truck mile for shipping									
0090	Average Soil nailing, grade 75, 15 min setup per hole & 80'/hr. drilling									
0100	Soil nailing , drill hole, install # 8 nail, grout 20' depth average	B-47G	16	2	Ea.	133	79.50	125	337.50	405
0110	25' depth average		14.20	2.254		166	89.50	140	395.50	475
0120	30' depth average		12.80	2.500		275	99.50	156	530.50	625
0130	35' depth average		11.60	2.759		320	110	172	602	705
0140	40' depth average		10.70	2.991		365	119	186	670	790
0150	45' depth average		9.90	3.232		410	128	201	739	870
0160	50' depth average		9.10	3.516		475	140	219	834	975
0170	55' depth average		8.50	3.765		520	150	234	904	1,050
0180	60' depth average		8	4		565	159	249	973	1,150
0190	65' depth average		7.50	4.267		610	170	266	1,046	1,225
0200	70' depth average		7.10	4.507		655	179	281	1,115	1,300
0210	75' depth average		6.70	4.776		700	190	297	1,187	1,375
0290	Average Soil nailing, grade 75, 15 min setup per hole & 90'/hr. drilling									
0300	Soil nailing , drill hole, install # 8 nail, grout 20' depth average	B-47G	16.60	1.928	Ea.	133	76.50	120	329.50	395
0310	25' depth average		15	2.133		166	85	133	384	460
0320	30' depth average		13.70	2.336		275	93	145	513	605
0330	35' depth average		12.30	2.602		320	103	162	585	685
0340	40' depth average		11.40	2.807		365	112	175	652	765
0350	45' depth average		10.70	2.991		410	119	186	715	840
0360	50' depth average		9.80	3.265		475	130	203	808	945
0370	55' depth average		9.20	3.478		520	138	217	875	1,025
0380	60' depth average		8.70	3.678		565	146	229	940	1,100
0390	65' depth average		8.10	3.951		610	157	246	1,013	1,175
0400	70' depth average		7.70	4.156		655	165	259	1,079	1,250
0410	75' depth average		7.40	4.324		700	172	269	1,141	1,325
0490	Average Soil nailing, grade 75, 15 min setup per hole & 100'/hr. drilling									
0500	Soil nailing , drill hole, install # 8 nail, grout 20' depth average	B-47G	17.80	1.798	Ea.	133	71.50	112	316.50	380
0510	25' depth average		16	2		166	79.50	125	370.50	440
0520	30' depth average		14.60	2.192		275	87	137	499	585
0530	35' depth average		13.30	2.406		320	95.50	150	565.50	660

31 32 36.16 Grouted Soil Nailing		Crew	Daily Output	Labor-Hours	Unit	2014 Bare Costs Material	Labor	Equipment	Total	Total Incl O&P
0540	40′ depth average	B-47G	12.30	2.602	Ea.	365	103	162	630	735
0550	45′ depth average		11.40	2.807		410	112	175	697	815
0560	50′ depth average		10.70	2.991		475	119	186	780	910
0570	55′ depth average		10	3.200		520	127	199	846	985
0580	60′ depth average		9.40	3.404		565	135	212	912	1,050
0590	65′ depth average		8.90	3.596		610	143	224	977	1,125
0600	70′ depth average		8.40	3.810		655	151	237	1,043	1,225
0610	75′ depth average		8	4		700	159	249	1,108	1,300
0690	Average Soil nailing, grade 75, 15 min setup per hole & 110′/hr. drilling									
0700	Soil nailing , drill hole, install # 8 nail, grout 20′ depth average	B-47G	18.50	1.730	Ea.	133	69	108	310	370
0710	25′ depth average		16.60	1.928		166	76.50	120	362.50	435
0720	30′ depth average		15.50	2.065		275	82	129	486	565
0730	35′ depth average		14.10	2.270		320	90	141	551	645
0740	40′ depth average		13	2.462		365	98	153	616	720
0750	45′ depth average		12	2.667		410	106	166	682	795
0760	50′ depth average		11.40	2.807		475	112	175	762	885
0770	55′ depth average		10.70	2.991		520	119	186	825	960
0780	60′ depth average		10	3.200		565	127	199	891	1,025
0790	65′ depth average		9.40	3.404		610	135	212	957	1,100
0800	70′ depth average		9.10	3.516		655	140	219	1,014	1,175
0810	75′ depth average		8.60	3.721		700	148	232	1,080	1,250
0890	Average Soil nailing, grade 75, 15 min setup per hole & 120′/hr. drilling									
0900	Soil nailing , drill hole, install # 8 nail, grout 20′ depth average	B-47G	19.20	1.667	Ea.	133	66.50	104	303.50	365
0910	25′ depth average		17.10	1.871		166	74.50	117	357.50	425
0920	30′ depth average		16	2		275	79.50	125	479.50	560
0930	35′ depth average		14.60	2.192		320	87	137	544	635
0940	40′ depth average		13.70	2.336		365	93	145	603	705
0950	45′ depth average		12.60	2.540		410	101	158	669	780
0960	50′ depth average		12	2.667		475	106	166	747	865
0970	55′ depth average		11.20	2.857		520	114	178	812	940
0980	60′ depth average		10.70	2.991		565	119	186	870	1,000
0990	65′ depth average		10	3.200		610	127	199	936	1,075
1000	70′ depth average		9.60	3.333		655	133	208	996	1,150
1010	75′ depth average		9.10	3.516		700	140	219	1,059	1,225
1190	Difficult Soil nailing, grade 75, 20 min setup per hole & 80′/hr. drilling									
1200	Soil nailing , drill hole, install # 8 nail, grout 20′ depth difficult	B-47G	13.70	2.336	Ea.	133	93	145	371	450
1210	25′ depth difficult		12.30	2.602		166	103	162	431	520
1220	30′ depth difficult		11.20	2.857		275	114	178	567	670
1230	35′ depth difficult		10.40	3.077		320	122	192	634	750
1240	40′ depth difficult		9.60	3.333		365	133	208	706	830
1250	45′ depth difficult		8.90	3.596		410	143	224	777	915
1260	50′ depth difficult		8.30	3.855		475	153	240	868	1,025
1270	55′ depth difficult		7.90	4.051		520	161	252	933	1,100
1280	60′ depth difficult		7.40	4.324		565	172	269	1,006	1,175
1290	65′ depth difficult		7	4.571		610	182	285	1,077	1,275
1300	70′ depth difficult		6.60	4.848		655	193	300	1,148	1,350
1310	75′ depth difficult		6.30	5.079		700	202	315	1,217	1,425
1390	Difficult soil nailing, grade 75, 20 min setup per hole & 90′/hr. drilling									
1400	Soil nailing , drill hole, install # 8 nail, grout 20′ depth difficult	B-47G	14.10	2.270	Ea.	133	90	141	364	440
1410	25′ depth difficult		13	2.462		166	98	153	417	500
1420	30′ depth difficult		12	2.667		275	106	166	547	645
1430	35′ depth difficult		10.90	2.936		320	117	183	620	730
1440	40′ depth difficult		10.20	3.137		365	125	195	685	805

31 32 Soil Stabilization

31 32 36 – Soil Nailing

31 32 36.16 Grouted Soil Nailing		Crew	Daily Output	Labor-Hours	Unit	2014 Bare Costs Material	Labor	Equipment	Total	Total Incl O&P
1450	45′ depth difficult	B-47G	9.60	3.333	Ea.	410	133	208	751	880
1460	50′ depth difficult		8.90	3.596		475	143	224	842	985
1470	55′ depth difficult		8.40	3.810		520	151	237	908	1,075
1480	60′ depth difficult		8	4		565	159	249	973	1,150
1490	65′ depth difficult		7.50	4.267		610	170	266	1,046	1,225
1500	70′ depth difficult		7.20	4.444		655	177	277	1,109	1,300
1510	75′ depth difficult		6.90	4.638		700	184	289	1,173	1,375
1590	Difficult soil nailing, grade 75, 20 min setup per hole & 100′/hr. drilling									
1600	Soil nailing , drill hole, install # 8 nail, grout 20′ depth difficult	B-47G	15	2.133	Ea.	133	85	133	351	425
1610	25′ depth difficult		13.70	2.336		166	93	145	404	485
1620	30′ depth difficult		12.60	2.540		275	101	158	534	630
1630	35′ depth difficult		11.70	2.735		320	109	170	599	705
1640	40′ depth difficult		10.90	2.936		365	117	183	665	780
1650	45′ depth difficult		10.20	3.137		410	125	195	730	855
1660	50′ depth difficult		9.60	3.333		475	133	208	816	950
1670	55′ depth difficult		9.10	3.516		520	140	219	879	1,025
1680	60′ depth difficult		8.60	3.721		565	148	232	945	1,100
1690	65′ depth difficult		8.10	3.951		610	157	246	1,013	1,175
1700	70′ depth difficult		7.70	4.156		655	165	259	1,079	1,250
1710	75′ depth difficult		7.40	4.324		700	172	269	1,141	1,325
1790	Difficult soil nailing, grade 75, 20 min setup per hole & 110′/hr. drilling									
1800	Soil nailing , drill hole, install # 8 nail, grout 20′ depth difficult	B-47G	15.50	2.065	Ea.	133	82	129	344	415
1810	25′ depth difficult		14.10	2.270		166	90	141	397	475
1820	30′ depth difficult		13.30	2.406		275	95.50	150	520.50	610
1830	35′ depth difficult		12.30	2.602		320	103	162	585	685
1840	40′ depth difficult		11.40	2.807		365	112	175	652	765
1850	45′ depth difficult		10.70	2.991		410	119	186	715	840
1860	50′ depth difficult		10.20	3.137		475	125	195	795	925
1870	55′ depth difficult		9.60	3.333		520	133	208	861	1,000
1880	60′ depth difficult		9.10	3.516		565	140	219	924	1,075
1890	65′ depth difficult		8.60	3.721		610	148	232	990	1,150
1900	70′ depth difficult		8.30	3.855		655	153	240	1,048	1,225
1910	75′ depth difficult		7.90	4.051		700	161	252	1,113	1,300
1990	Difficult soil nailing, grade 75, 20 min setup per hole & 120′/hr. drilling									
2000	Soil nailing , drill hole, install # 8 nail, grout 20′ depth difficult	B-47G	16	2	Ea.	133	79.50	125	337.50	405
2010	25′ depth difficult		14.60	2.192		166	87	137	390	465
2020	30′ depth difficult		13.70	2.336		275	93	145	513	605
2030	35′ depth difficult		12.60	2.540		320	101	158	579	680
2040	40′ depth difficult		12	2.667		365	106	166	637	745
2050	45′ depth difficult		11.20	2.857		410	114	178	702	820
2060	50′ depth difficult		10.70	2.991		475	119	186	780	910
2070	55′ depth difficult		10	3.200		520	127	199	846	985
2080	60′ depth difficult		9.60	3.333		565	133	208	906	1,050
2090	65′ depth difficult		9.10	3.516		610	140	219	969	1,125
2100	70′ depth difficult		8.70	3.678		655	146	229	1,030	1,200
2110	75′ depth difficult		8.30	3.855		700	153	240	1,093	1,275
2990	Very difficult soil nailing, grade 75, 25 min setup & 80′/hr. drilling									
3000	Soil nailing, drill, install # 8 nail, grout 20′ depth very difficult	B-47G	12	2.667	Ea.	133	106	166	405	495
3010	25′ depth very difficult		10.90	2.936		166	117	183	466	565
3020	30′ depth very difficult		10	3.200		275	127	199	601	715
3030	35′ depth very difficult		9.40	3.404		320	135	212	667	790
3040	40′ depth very difficult		8.70	3.678		365	146	229	740	875
3050	45′ depth very difficult		8.10	3.951		410	157	246	813	960

31 32 36.16 Grouted Soil Nailing		Crew	Daily Output	Labor-Hours	Unit	2014 Bare Costs Material	Labor	Equipment	Total	Total Incl O&P
3060	50′ depth very difficult	B-47G	7.60	4.211	Ea.	475	167	262	904	1,075
3070	55′ depth very difficult		7.30	4.384		520	174	273	967	1,150
3080	60′ depth very difficult		6.90	4.638		565	184	289	1,038	1,225
3090	65′ depth very difficult		6.50	4.923		610	196	305	1,111	1,300
3100	70′ depth very difficult		6.20	5.161		655	205	320	1,180	1,400
3110	75′ depth very difficult		5.90	5.424		700	216	340	1,256	1,475
3190	Very difficult soil nailing, grade 75, 25 min setup & 90′/hr. drilling									
3200	Soil nailing, drill, install # 8 nail, grout 20′ depth very difficult	B-47G	12.30	2.602	Ea.	133	103	162	398	485
3210	25′ depth very difficult		11.40	2.807		166	112	175	453	545
3220	30′ depth very difficult		10.70	2.991		275	119	186	580	690
3230	35′ depth very difficult		9.80	3.265		320	130	203	653	775
3240	40′ depth very difficult		9.20	3.478		365	138	217	720	850
3250	45′ depth very difficult		8.70	3.678		410	146	229	785	925
3260	50′ depth very difficult		8.10	3.951		475	157	246	878	1,025
3270	55′ depth very difficult		7.70	4.156		520	165	259	944	1,100
3280	60′ depth very difficult		7.40	4.324		565	172	269	1,006	1,175
3290	65′ depth very difficult		7	4.571		610	182	285	1,077	1,275
3300	70′ depth very difficult		6.70	4.776		655	190	297	1,142	1,325
3310	75′ depth very difficult		6.40	5		700	199	310	1,209	1,425
3390	Very difficult soil nailing, grade 75, 25 min setup & 100′/hr. drilling									
3400	Soil nailing, drill, install # 8 nail, grout 20′ depth very difficult	B-47G	13	2.462	Ea.	133	98	153	384	465
3410	25′ depth very difficult		12	2.667		166	106	166	438	530
3420	30′ depth very difficult		11.20	2.857		275	114	178	567	670
3430	35′ depth very difficult		10.40	3.077		320	122	192	634	750
3440	40′ depth very difficult		9.80	3.265		365	130	203	698	825
3450	45′ depth very difficult		9.20	3.478		410	138	217	765	900
3460	50′ depth very difficult		8.70	3.678		475	146	229	850	995
3470	55′ depth very difficult		8.30	3.855		520	153	240	913	1,075
3480	60′ depth very difficult		7.90	4.051		565	161	252	978	1,150
3490	65′ depth very difficult		7.50	4.267		610	170	266	1,046	1,225
3500	70′ depth very difficult		7.20	4.444		655	177	277	1,109	1,300
3510	75′ depth very difficult		6.90	4.638		700	184	289	1,173	1,375
3590	Very difficult soil nailing, grade 75, 25 min setup & 110′/hr. drilling									
3600	Soil nailing, drill, install # 8 nail, grout 20′ depth very difficult	B-47G	13.30	2.406	Ea.	133	95.50	150	378.50	460
3610	25′ depth very difficult		13.70	2.336		166	93	145	404	485
3620	30′ depth very difficult		11.70	2.735		275	109	170	554	655
3630	35′ depth very difficult		10.90	2.936		320	117	183	620	730
3640	40′ depth very difficult		10.20	3.137		365	125	195	685	805
3650	45′ depth very difficult		9.60	3.333		410	133	208	751	880
3660	50′ depth very difficult		9.20	3.478		475	138	217	830	970
3670	55′ depth very difficult		8.70	3.678		520	146	229	895	1,050
3680	60′ depth very difficult		8.30	3.855		565	153	240	958	1,125
3690	65′ depth very difficult		7.90	4.051		610	161	252	1,023	1,200
3700	70′ depth very difficult		7.60	4.211		655	167	262	1,084	1,275
3710	75′ depth very difficult		7.30	4.384		700	174	273	1,147	1,350
3790	Very difficult soil nailing, grade 75, 25 min setup & 120′/hr. drilling									
3800	Soil nailing, drill, install # 8 nail, grout 20′ depth very difficult	B-47G	13.70	2.336	Ea.	133	93	145	371	450
3810	25′ depth very difficult		12.60	2.540		166	101	158	425	510
3820	30′ depth very difficult		12	2.667		275	106	166	547	645
3830	35′ depth very difficult		11.20	2.857		320	114	178	612	720
3840	40′ depth very difficult		10.70	2.991		365	119	186	670	790
3850	45′ depth very difficult		10	3.200		410	127	199	736	865
3860	50′ depth very difficult		9.60	3.333		475	133	208	816	950

31 32 36.16 Grouted Soil Nailing		Crew	Daily Output	Labor-Hours	Unit	Material	Labor	Equipment	Total	Total Incl O&P
						2014 Bare Costs				
3870	55′ depth very difficult	B-47G	9.10	3.516	Ea.	520	140	219	879	1,025
3880	60′ depth very difficult		8.70	3.678		565	146	229	940	1,100
3890	65′ depth very difficult		8.30	3.855		610	153	240	1,003	1,175
3900	70′ depth very difficult		8	4		655	159	249	1,063	1,250
3910	75′ depth very difficult		7.60	4.211		700	167	262	1,129	1,325
3990	Severe soil nailing, grade 75, 30 min setup per hole & 80′/hr. drilling									
4000	Soil nailing , drill hole, install # 8 nail, grout 20′ depth severe	B-47G	10.70	2.991	Ea.	133	119	186	438	535
4010	25′ depth severe		9.80	3.265		166	130	203	499	605
4020	30′ depth severe		9.10	3.516		275	140	219	634	755
4030	35′ depth severe		8.60	3.721		320	148	232	700	830
4040	40′ depth severe		8	4		365	159	249	773	920
4050	45′ depth severe		7.50	4.267		410	170	266	846	1,000
4060	50′ depth severe		7.10	4.507		475	179	281	935	1,100
4070	55′ depth severe		6.80	4.706		520	187	293	1,000	1,175
4080	60′ depth severe		6.40	5		565	199	310	1,074	1,275
4090	65′ depth severe		6.10	5.246		610	209	325	1,144	1,350
4100	70′ depth severe		5.80	5.517		655	219	345	1,219	1,425
4110	75′ depth severe		5.60	5.714		700	227	355	1,282	1,500
4190	Severe soil nailing, grade 75, 30 min setup per hole & 90′/hr. drilling									
4200	Soil nailing , drill hole, install # 8 nail, grout 20′ depth severe	B-47G	10.90	2.936	Ea.	133	117	183	433	525
4210	25′ depth severe		10.20	3.137		166	125	195	486	590
4220	30′ depth severe		9.60	3.333		275	133	208	616	730
4230	35′ depth severe		8.90	3.596		320	143	224	687	815
4240	40′ depth severe		8.40	3.810		365	151	237	753	895
4250	45′ depth severe		8	4		410	159	249	818	970
4260	50′ depth severe		7.50	4.267		475	170	266	911	1,075
4270	55′ depth severe		7.20	4.444		520	177	277	974	1,150
4280	60′ depth severe		6.90	4.638		565	184	289	1,038	1,225
4290	65′ depth severe		6.50	4.923		610	196	305	1,111	1,300
4300	70′ depth severe		6.20	5.161		655	205	320	1,180	1,400
4310	75′ depth severe		6	5.333		700	212	330	1,242	1,450
4390	Severe soil nailing, grade 75, 30 min setup per hole & 100′/hr. drilling									
4400	Soil nailing , drill hole, install # 8 nail, grout 20′ depth severe	B-47G	11.40	2.807	Ea.	133	112	175	420	510
4410	25′ depth severe		10.70	2.991		166	119	186	471	570
4420	30′ depth severe		10	3.200		275	127	199	601	715
4430	35′ depth severe		9.40	3.404		320	135	212	667	790
4440	40′ depth severe		8.90	3.596		365	143	224	732	865
4450	45′ depth severe		8.40	3.810		410	151	237	798	945
4460	50′ depth severe		8	4		475	159	249	883	1,050
4470	55′ depth severe		7.60	4.211		520	167	262	949	1,125
4480	60′ depth severe		7.30	4.384		565	174	273	1,012	1,200
4490	65′ depth severe		7	4.571		610	182	285	1,077	1,275
4500	70′ depth severe		6.70	4.776		655	190	297	1,142	1,325
4510	75′ depth severe		6.40	5		700	199	310	1,209	1,425
4590	Severe soil nailing, grade 75, 30 min setup per hole & 110′/hr. drilling									
4600	Soil nailing , drill hole, install # 8 nail, grout 20′ depth severe	B-47G	11.70	2.735	Ea.	133	109	170	412	500
4610	25′ depth severe		10.90	2.936		166	117	183	466	565
4620	30′ depth severe		10.40	3.077		275	122	192	589	700
4630	35′ depth severe		9.80	3.265		320	130	203	653	775
4640	40′ depth severe		9.20	3.478		365	138	217	720	850
4650	45′ depth severe		8.70	3.678		410	146	229	785	925
4660	50′ depth severe		8.40	3.810		475	151	237	863	1,025
4670	55′ depth severe		8	4		520	159	249	928	1,100

31 32 Soil Stabilization

31 32 36 – Soil Nailing

31 32 36.16 Grouted Soil Nailing		Crew	Daily Output	Labor-Hours	Unit	2014 Bare Costs Material	Labor	Equipment	Total	Total Incl O&P
4680	60′ depth severe	B-47G	7.60	4.211	Ea.	565	167	262	994	1,175
4690	65′ depth severe		7.30	4.384		610	174	273	1,057	1,250
4700	70′ depth severe		7.10	4.507		655	179	281	1,115	1,300
4710	75′ depth severe		6.80	4.706		700	187	293	1,180	1,375
4790	Severe soil nailing, grade 75, 30 min setup per hole & 120′/hr. drilling									
4800	Soil nailing , drill hole, install # 8 nail, grout 20′ depth severe	B-47G	12	2.667	Ea.	133	106	166	405	495
4810	25′ depth severe		11.20	2.857		166	114	178	458	555
4820	30′ depth severe		10.70	2.991		275	119	186	580	690
4830	35′ depth severe		10	3.200		320	127	199	646	765
4840	40′ depth severe		9.60	3.333		365	133	208	706	830
4850	45′ depth severe		9.10	3.516		410	140	219	769	905
4860	50′ depth severe		8.70	3.678		475	146	229	850	995
4870	55′ depth severe		8.30	3.855		520	153	240	913	1,075
4880	60′ depth severe		8	4		565	159	249	973	1,150
4890	65′ depth severe		7.60	4.211		610	167	262	1,039	1,225
4900	70′ depth severe		7.40	4.324		655	172	269	1,096	1,275
4910	75′ depth severe		7.10	4.507		700	179	281	1,160	1,350

31 33 Rock Stabilization

31 33 13 – Rock Bolting and Grouting

31 33 13.10 Rock Bolting

		Crew	Daily Output	Labor-Hours	Unit	Material	Labor	Equipment	Total	Total Incl O&P
0010	**ROCK BOLTING**									
2020	Hollow core, prestressable anchor, 1″ diameter, 5′ long	2 Skwk	32	.500	Ea.	104	23.50		127.50	151
2025	10′ long		24	.667		208	31.50		239.50	277
2060	2″ diameter, 5′ long		32	.500		395	23.50		418.50	470
2065	10′ long		24	.667		775	31.50		806.50	905
2100	Super high-tensile, 3/4″ diameter, 5′ long		32	.500		23.50	23.50		47	62.50
2105	10′ long		24	.667		47	31.50		78.50	101
2160	2″ diameter, 5′ long		32	.500		210	23.50		233.50	268
2165	10′ long		24	.667		420	31.50		451.50	510
4400	Drill hole for rock bolt, 1-3/4″ diam., 5′ long (for 3/4″ bolt)	B-56	17	.941			39.50	94.50	134	164
4405	10′ long		9	1.778			74.50	179	253.50	310
4420	2″ diameter, 5′ long (for 1″ bolt)		13	1.231			51.50	124	175.50	215
4425	10′ long		7	2.286			95.50	230	325.50	400
4460	3-1/2″ diameter, 5′ long (for 2″ bolt)		10	1.600			67	161	228	279
4465	10′ long		5	3.200			134	320	454	560

31 36 Gabions

31 36 13 – Gabion Boxes

31 36 13.10 Gabion Box Systems

		Crew	Daily Output	Labor-Hours	Unit	Material	Labor	Equipment	Total	Total Incl O&P
0010	**GABION BOX SYSTEMS**									
0400	Gabions, galvanized steel mesh mats or boxes, stone filled, 6″ deep	B-13	200	.280	S.Y.	26.50	11.15	3.73	41.38	50
0500	9″ deep		163	.344		38.50	13.70	4.58	56.78	68
0600	12″ deep		153	.366		52	14.60	4.88	71.48	85
0700	18″ deep		102	.549		73	22	7.30	102.30	122
0800	36″ deep		60	.933		97.50	37	12.45	146.95	178

31 37 Riprap

31 37 13 – Machined Riprap

31 37 13.10 Riprap and Rock Lining		Crew	Daily Output	Labor-Hours	Unit	2014 Bare Costs Material	Labor	Equipment	Total	Total Incl O&P
0010	**RIPRAP AND ROCK LINING**									
0011	Random, broken stone									
0100	Machine placed for slope protection	B-12G	62	.258	L.C.Y.	29.50	11.20	11.65	52.35	62.50
0110	3/8 to 1/4 C.Y. pieces, grouted	B-13	80	.700	S.Y.	56	28	9.30	93.30	115
0200	18" minimum thickness, not grouted	"	53	1.057	"	18.45	42	14.05	74.50	101
0300	Dumped, 50 lb. average	B-11A	800	.020	Ton	25	.86	1.66	27.52	30.50
0350	100 lb. average		700	.023		25	.98	1.89	27.87	31
0370	300 lb. average		600	.027		25	1.14	2.21	28.35	31.50

31 41 Shoring

31 41 13 – Timber Shoring

31 41 13.10 Building Shoring		Crew	Daily Output	Labor-Hours	Unit	Material	Labor	Equipment	Total	Total Incl O&P
0010	**BUILDING SHORING**									
0020	Shoring, existing building, with timber, no salvage allowance	B-51	2.20	21.818	M.B.F.	760	805	123	1,688	2,225
1000	On cribbing with 35 ton screw jacks, per box and jack	"	3.60	13.333	Jack	71	495	75	641	920
1100	Masonry openings in walls, see Section 02 41 19.16									

31 41 16 – Sheet Piling

31 41 16.10 Sheet Piling Systems		Crew	Daily Output	Labor-Hours	Unit	Material	Labor	Equipment	Total	Total Incl O&P
0010	**SHEET PILING SYSTEMS**									
0020	Sheet piling steel, not incl. wales, 22 psf, 15' excav., left in place	B-40	10.81	5.920	Ton	1,450	272	355	2,077	2,425
0100	Drive, extract & salvage		6	10.667		505	490	635	1,630	2,025
0300	20' deep excavation, 27 psf, left in place		12.95	4.942		1,450	227	295	1,972	2,275
0400	Drive, extract & salvage		6.55	9.771		505	450	585	1,540	1,900
0600	25' deep excavation, 38 psf, left in place		19	3.368		1,450	155	201	1,806	2,075
0700	Drive, extract & salvage		10.50	6.095		505	280	365	1,150	1,400
0900	40' deep excavation, 38 psf, left in place		21.20	3.019		1,450	139	180	1,769	2,025
1000	Drive, extract & salvage		12.25	5.224		505	240	310	1,055	1,275
1200	15' deep excavation, 22 psf, left in place		983	.065	S.F.	16.95	3	3.89	23.84	27.50
1300	Drive, extract & salvage		545	.117		5.65	5.40	7	18.05	22.50
1500	20' deep excavation, 27 psf, left in place		960	.067		21.50	3.07	3.98	28.55	32.50
1600	Drive, extract & salvage		485	.132		7.35	6.05	7.85	21.25	26
1800	25' deep excavation, 38 psf, left in place		1000	.064		31.50	2.94	3.82	38.26	43.50
1900	Drive, extract & salvage		553	.116		10.05	5.30	6.90	22.25	27
2100	Rent steel sheet piling and wales, first month				Ton	305			305	335
2200	Per added month					30.50			30.50	33.50
2300	Rental piling left in place, add to rental					1,150			1,150	1,275
2500	Wales, connections & struts, 2/3 salvage					475			475	525
2700	High strength piling, 50,000 psi, add					63			63	69.50
2800	55,000 psi, add					81			81	89
3000	Tie rod, not upset, 1-1/2" to 4" diameter with turnbuckle					2,100			2,100	2,300
3100	No turnbuckle					1,625			1,625	1,800
3300	Upset, 1-3/4" to 4" diameter with turnbuckle					2,375			2,375	2,600
3400	No turnbuckle					2,050			2,050	2,250
3600	Lightweight, 18" to 28" wide, 7 ga., 9.22 psf, and									
3610	9 ga., 8.6 psf, minimum				Lb.	.79			.79	.87
3700	Average					.90			.90	.99
3750	Maximum					1.05			1.05	1.16
3900	Wood, solid sheeting, incl. wales, braces and spacers,									
3910	drive, extract & salvage, 8' deep excavation	B-31	330	.121	S.F.	1.76	4.71	.66	7.13	9.90
4000	10' deep, 50 S.F./hr. in & 150 S.F./hr. out		300	.133		1.81	5.20	.73	7.74	10.80

31 41 Shoring

31 41 16 – Sheet Piling

31 41 16.10 Sheet Piling Systems		Crew	Daily Output	Labor-Hours	Unit	Material	Labor	Equipment	Total	Total Incl O&P
						2014 Bare Costs				
4100	12′ deep, 45 S.F./hr. in & 135 S.F./hr. out	B-31	270	.148	S.F.	1.87	5.75	.81	8.43	11.85
4200	14′ deep, 42 S.F./hr. in & 126 S.F./hr. out		250	.160		1.92	6.20	.88	9	12.70
4300	16′ deep, 40 S.F./hr. in & 120 S.F./hr. out		240	.167		1.98	6.50	.91	9.39	13.20
4400	18′ deep, 38 S.F./hr. in & 114 S.F./hr. out		230	.174		2.05	6.75	.95	9.75	13.75
4500	20′ deep, 35 S.F./hr. in & 105 S.F./hr. out		210	.190		2.12	7.40	1.04	10.56	14.95
4520	Left in place, 8′ deep, 55 S.F./hr.		440	.091		3.18	3.54	.50	7.22	9.50
4540	10′ deep, 50 S.F./hr.		400	.100		3.34	3.89	.55	7.78	10.30
4560	12′ deep, 45 S.F./hr.		360	.111		3.53	4.32	.61	8.46	11.20
4565	14′ deep, 42 S.F./hr.		335	.119		3.74	4.64	.65	9.03	12
4570	16′ deep, 40 S.F./hr.		320	.125		3.97	4.86	.69	9.52	12.60
4580	18′ deep, 38 S.F./hr.		305	.131		4.23	5.10	.72	10.05	13.30
4590	20′ deep, 35 S.F./hr.		280	.143	▼	4.54	5.55	.78	10.87	14.40
4700	Alternate pricing, left in place, 8′ deep		1.76	22.727	M.B.F.	715	885	125	1,725	2,300
4800	Drive, extract and salvage, 8′ deep	▼	1.32	30.303	"	635	1,175	166	1,976	2,700
5000	For treated lumber add cost of treatment to lumber									

31 43 Concrete Raising

31 43 13 – Pressure Grouting

31 43 13.13 Concrete Pressure Grouting

		Crew	Daily Output	Labor-Hours	Unit	Material	Labor	Equipment	Total	Total Incl O&P
0010	**CONCRETE PRESSURE GROUTING**									
0020	Grouting, pressure, cement & sand, 1:1 mix, minimum	B-61	124	.323	Bag	13.95	12.60	2.81	29.36	38
0100	Maximum		51	.784	"	13.95	30.50	6.85	51.30	70
0200	Cement and sand, 1:1 mix, minimum		250	.160	C.F.	28	6.25	1.40	35.65	41.50
0300	Maximum		100	.400		42	15.65	3.49	61.14	74
0310	Grouting, cement and sand, 1:2 mix, geothermal wells		250	.160		11.15	6.25	1.40	18.80	23.50
0320	Bentonite mix 1:4		250	.160		8.10	6.25	1.40	15.75	20
0400	Epoxy cement grout, minimum		137	.292		720	11.40	2.55	733.95	815
0500	Maximum	▼	57	.702	▼	720	27.50	6.10	753.60	845
0700	Alternate pricing method: (Add for materials)									
0710	5 person crew and equipment	B-61	1	40	Day		1,575	350	1,925	2,775

31 43 19 – Mechanical Jacking

31 43 19.10 Slabjacking

		Crew	Daily Output	Labor-Hours	Unit	Material	Labor	Equipment	Total	Total Incl O&P
0010	**SLABJACKING**									
0100	4″ thick slab	D-4	1500	.021	S.F.	.31	.89	.09	1.29	1.79
0150	6″ thick slab		1200	.027		.43	1.11	.11	1.65	2.28
0200	8″ thick slab		1000	.032		.52	1.33	.13	1.98	2.75
0250	10″ thick slab		900	.036		.55	1.48	.15	2.18	3.02
0300	12″ thick slab	▼	850	.038	▼	.60	1.57	.16	2.33	3.22

31 45 Vibroflotation and Densification

31 45 13 – Vibroflotation

31 45 13.10 Vibroflotation Densification		Crew	Daily Output	Labor-Hours	Unit	Material	2014 Bare Costs Labor	Equipment	Total	Total Incl O&P
0010	**VIBROFLOTATION DENSIFICATION**									
0900	Vibroflotation compacted sand cylinder, minimum	B-60	750	.075	V.L.F.		3.20	2.94	6.14	8.10
0950	Maximum		325	.172			7.40	6.80	14.20	18.70
1100	Vibro replacement compacted stone cylinder, minimum		500	.112			4.80	4.41	9.21	12.15
1150	Maximum		250	.224			9.60	8.85	18.45	24.50
1300	Mobilization and demobilization, minimum		.47	119	Total		5,100	4,700	9,800	13,000
1400	Maximum		.14	400	"		17,100	15,800	32,900	43,500

31 46 Needle Beams

31 46 13 – Cantilever Needle Beams

31 46 13.10 Needle Beams		Crew	Daily Output	Labor-Hours	Unit	Material	Labor	Equipment	Total	Total Incl O&P
0010	**NEEDLE BEAMS**									
0011	Incl. wood shoring 10′ x 10′ opening									
0400	Block, concrete, 8″ thick	B-9	7.10	5.634	Ea.	50	209	33	292	410
0420	12″ thick		6.70	5.970		58	221	35	314	445
0800	Brick, 4″ thick with 8″ backup block		5.70	7.018		58	260	41	359	510
1000	Brick, solid, 8″ thick		6.20	6.452		50	239	37.50	326.50	465
1040	12″ thick		4.90	8.163		58	300	47.50	405.50	580
1080	16″ thick		4.50	8.889		74	330	52	456	650
2000	Add for additional floors of shoring	B-1	6	4		50	149		199	285

31 48 Underpinning

31 48 13 – Underpinning Piers

31 48 13.10 Underpinning Foundations		Crew	Daily Output	Labor-Hours	Unit	Material	Labor	Equipment	Total	Total Incl O&P
0010	**UNDERPINNING FOUNDATIONS**									
0011	Including excavation,									
0020	forming, reinforcing, concrete and equipment									
0100	5′ to 16′ below grade, 100 to 500 C.Y.	B-52	2.30	24.348	C.Y.	274	1,025	259	1,558	2,150
0200	Over 500 C.Y.		2.50	22.400		247	950	238	1,435	1,975
0400	16′ to 25′ below grade, 100 to 500 C.Y.		2	28		300	1,200	298	1,798	2,475
0500	Over 500 C.Y.		2.10	26.667		285	1,125	283	1,693	2,375
0700	26′ to 40′ below grade, 100 to 500 C.Y.		1.60	35		330	1,500	370	2,200	3,050
0800	Over 500 C.Y.		1.80	31.111		300	1,325	330	1,955	2,725
0900	For under 50 C.Y., add					10%	40%			
1000	For 50 C.Y. to 100 C.Y., add					5%	20%			

31 52 Cofferdams

31 52 16 – Timber Cofferdams

31 52 16.10 Cofferdams		Crew	Daily Output	Labor-Hours	Unit	Material	Labor	Equipment	Total	Total Incl O&P
0010	**COFFERDAMS**									
0011	Incl. mobilization and temporary sheeting									
0020	Shore driven	B-40	960	.067	S.F.	21	3.07	3.98	28.05	32
0060	Barge driven	"	550	.116	"	21	5.35	6.95	33.30	39
0080	Soldier beams & lagging H piles with 3″ wood sheeting									
0090	horizontal between piles, including removal of wales & braces									
0100	No hydrostatic head, 15′ deep, 1 line of braces, minimum	B-50	545	.206	S.F.	7.60	9	4.53	21.13	27.50
0200	Maximum		495	.226		8.45	9.90	4.99	23.34	30.50

31 52 Cofferdams

31 52 16 – Timber Cofferdams

31 52 16.10 Cofferdams		Crew	Daily Output	Labor-Hours	Unit	Material (2014 Bare Costs)	Labor (2014 Bare Costs)	Equipment (2014 Bare Costs)	Total (2014 Bare Costs)	Total Incl O&P
0400	15′ to 22′ deep with 2 lines of braces, 10″ H, minimum	B-50	360	.311	S.F.	8.95	13.65	6.85	29.45	39
0500	Maximum		330	.339		10.15	14.85	7.50	32.50	42.50
0700	23′ to 35′ deep with 3 lines of braces, 12″ H, minimum		325	.345		11.70	15.10	7.60	34.40	45
0800	Maximum		295	.380		12.70	16.65	8.35	37.70	49
1000	36′ to 45′ deep with 4 lines of braces, 14″ H, minimum		290	.386		13.15	16.90	8.50	38.55	50.50
1100	Maximum		265	.423		13.85	18.50	9.30	41.65	54.50
1300	No hydrostatic head, left in place, 15′ dp., 1 line of braces, min.		635	.176		10.15	7.75	3.89	21.79	27.50
1400	Maximum		575	.195		10.90	8.55	4.29	23.74	30
1600	15′ to 22′ deep with 2 lines of braces, minimum		455	.246		15.25	10.80	5.45	31.50	39.50
1700	Maximum		415	.270		16.95	11.80	5.95	34.70	43.50
1900	23′ to 35′ deep with 3 lines of braces, minimum		420	.267		18.15	11.70	5.90	35.75	44.50
2000	Maximum		380	.295		20	12.90	6.50	39.40	49
2200	36′ to 45′ deep with 4 lines of braces, minimum		385	.291		22	12.75	6.40	41.15	51
2300	Maximum		350	.320		25.50	14	7.05	46.55	58
2350	Lagging only, 3″ thick wood between piles 8′ O.C., minimum	B-46	400	.120		1.69	4.90	.11	6.70	9.70
2370	Maximum		250	.192		2.54	7.85	.18	10.57	15.30
2400	Open sheeting no bracing, for trenches to 10′ deep, min.		1736	.028		.76	1.13	.03	1.92	2.64
2450	Maximum		1510	.032		.85	1.30	.03	2.18	2.99
2500	Tie-back method, add to open sheeting, add, minimum								20%	20%
2550	Maximum								60%	60%
2700	Tie-backs only, based on tie-backs total length, minimum	B-46	86.80	.553	L.F.	14.50	22.50	.52	37.52	52
2750	Maximum		38.50	1.247	″	25.50	51	1.17	77.67	109
3500	Tie-backs only, typical average, 25′ long		2	24	Ea.	640	980	22.50	1,642.50	2,250
3600	35′ long		1.58	30.380	″	850	1,250	28.50	2,128.50	2,925
4500	Trench box, 7′ deep, 16′ x 8′, see 01 54 33 in Reference Section				Day				191	210
4600	20′ x 10′, see 01 54 33 in Reference Section				″				240	265
5200	Wood sheeting, in trench, jacks at 4′ O.C., 8′ deep	B-1	800	.030	S.F.	.59	1.12		1.71	2.37
5250	12′ deep		700	.034		.69	1.28		1.97	2.73
5300	15′ deep		600	.040		.95	1.49		2.44	3.35
6000	See also Section 31 41 16.10									

31 56 Slurry Walls

31 56 23 – Lean Concrete Slurry Walls

31 56 23.20 Slurry Trench

31 56 23.20 Slurry Trench		Crew	Daily Output	Labor-Hours	Unit	Material	Labor	Equipment	Total	Total Incl O&P
0010	**SLURRY TRENCH**									
0011	Excavated slurry trench in wet soils									
0020	backfilled with 3000 PSI concrete, no reinforcing steel									
0050	Minimum	C-7	333	.216	C.F.	7.35	8.60	3.68	19.63	25.50
0100	Maximum		200	.360	″	12.30	14.35	6.15	32.80	42.50
0200	Alternate pricing method, minimum		150	.480	S.F.	14.65	19.10	8.15	41.90	54
0300	Maximum		120	.600		22	24	10.20	56.20	72
0500	Reinforced slurry trench, minimum	B-48	177	.316		11	13.10	16.45	40.55	50
0600	Maximum	″	69	.812		36.50	33.50	42	112	139
0800	Haul for disposal, 2 mile haul, excavated material, add	B-34B	99	.081	C.Y.		3.03	7	10.03	12.35
0900	Haul bentonite castings for disposal, add	″	40	.200	″		7.50	17.30	24.80	30.50

31 62 Driven Piles

31 62 13 – Concrete Piles

31 62 13.23 Prestressed Concrete Piles		Crew	Daily Output	Labor-Hours	Unit	Material	Labor	Equipment	Total	Total Incl O&P
						2014 Bare Costs				
0010	**PRESTRESSED CONCRETE PILES**, 200 piles									
0020	Unless specified otherwise, not incl. pile caps or mobilization									
2200	Precast, prestressed, 50′ long, 12″ diam., 2-3/8″ wall	B-19	720	.089	V.L.F.	23.50	4.09	2.56	30.15	35
2300	14″ diameter, 2-1/2″ wall		680	.094		31.50	4.33	2.71	38.54	44
2500	16″ diameter, 3″ wall		640	.100		44.50	4.60	2.88	51.98	59.50
2600	18″ diameter, 3″ wall	B-19A	600	.107		56	4.91	3.76	64.67	74
2800	20″ diameter, 3-1/2″ wall		560	.114		67	5.25	4.03	76.28	86.50
2900	24″ diameter, 3-1/2″ wall		520	.123		75.50	5.65	4.34	85.49	97
2920	36″ diameter, 4-1/2″ wall	B-19	400	.160		101	7.35	4.60	112.95	128
2940	54″ diameter, 5″ wall		340	.188		132	8.65	5.40	146.05	164
2960	66″ diameter, 6″ wall		220	.291		206	13.40	8.35	227.75	257
3100	Precast, prestressed, 40′ long, 10″ thick, square		700	.091		18.05	4.21	2.63	24.89	29.50
3200	12″ thick, square		680	.094		21.50	4.33	2.71	28.54	33.50
3275	Shipping for 60 foot long concrete piles, add to material cost					.92			.92	1.01
3400	14″ thick, square	B-19	600	.107		25	4.91	3.07	32.98	38.50
3500	Octagonal		640	.100		34.50	4.60	2.88	41.98	48
3700	16″ thick, square		560	.114		40.50	5.25	3.29	49.04	56.50
3800	Octagonal		600	.107		41	4.91	3.07	48.98	56.50
4000	18″ thick, square	B-19A	520	.123		47	5.65	4.34	56.99	65
4100	Octagonal	B-19	560	.114		48.50	5.25	3.29	57.04	65.50
4300	20″ thick, square	B-19A	480	.133		59.50	6.15	4.71	70.36	80.50
4400	Octagonal	B-19	520	.123		54	5.65	3.54	63.19	71.50
4600	24″ thick, square	B-19A	440	.145		77	6.70	5.15	88.85	101
4700	Octagonal	B-19	480	.133		77.50	6.15	3.83	87.48	99.50
4750	Mobilization for 10,000 L.F. pile job, add		3300	.019			.89	.56	1.45	2
4800	25,000 L.F. pile job, add		8500	.008			.35	.22	.57	.78
4850	Mobilization by water for barge driving rig, add								100%	100%

31 62 16 – Steel Piles

31 62 16.13 Steel Piles		Crew	Daily Output	Labor-Hours	Unit	Material	Labor	Equipment	Total	Total Incl O&P
0010	**STEEL PILES**									
0100	Step tapered, round, concrete filled									
0110	8″ tip, 60 ton capacity, 30′ depth	B-19	760	.084	V.L.F.	11.35	3.87	2.42	17.64	21
0120	60′ depth		740	.086		12.85	3.98	2.49	19.32	23
0130	80′ depth		700	.091		13.30	4.21	2.63	20.14	24
0150	10″ tip, 90 ton capacity, 30′ depth		700	.091		13.65	4.21	2.63	20.49	24.50
0160	60′ depth		690	.093		14.05	4.27	2.67	20.99	25
0170	80′ depth		670	.096		15.15	4.39	2.75	22.29	26.50
0190	12″ tip, 120 ton capacity, 30′ depth		660	.097		18.10	4.46	2.79	25.35	30
0200	60′ depth, 12″ diameter		630	.102		17.45	4.67	2.92	25.04	29.50
0210	80′ depth		590	.108		16.95	4.99	3.12	25.06	30
0250	"H" Sections, 50′ long, HP8 x 36		640	.100		16.15	4.60	2.88	23.63	28
0400	HP10 X 42		610	.105		18.75	4.83	3.02	26.60	31.50
0500	HP10 X 57		610	.105		26	4.83	3.02	33.85	39.50
0700	HP12 X 53		590	.108		24	4.99	3.12	32.11	37.50
0800	HP12 X 74	B-19A	590	.108		33	4.99	3.83	41.82	48.50
1000	HP14 X 73		540	.119		32.50	5.45	4.18	42.13	49
1100	HP14 X 89		540	.119		39.50	5.45	4.18	49.13	56.50
1300	HP14 X 102		510	.125		45	5.75	4.43	55.18	63.50
1400	HP14 X 117		510	.125		51.50	5.75	4.43	61.68	71
1600	Splice on standard points, not in leads, 8″ or 10″	1 Sswl	5	1.600	Ea.	118	82		200	274

31 62 Driven Piles

31 62 19 – Timber Piles

31 62 19.10 Wood Piles

	31 62 19.10 Wood Piles	Crew	Daily Output	Labor-Hours	Unit	2014 Bare Costs Material	Labor	Equipment	Total	Total Incl O&P
0010	**WOOD PILES**									
0011	Friction or end bearing, not including									
0050	mobilization or demobilization									
0100	Untreated piles, up to 30′ long, 12″ butts, 8″ points	B-19	625	.102	V.L.F.	17.40	4.71	2.95	25.06	29.50
0200	30′ to 39′ long, 12″ butts, 8″ points		700	.091		17.40	4.21	2.63	24.24	28.50
0300	40′ to 49′ long, 12″ butts, 7″ points		720	.089		17.40	4.09	2.56	24.05	28.50
0400	50′ to 59′ long, 13″butts, 7″ points		800	.080		17.40	3.68	2.30	23.38	27.50
0500	60′ to 69′ long, 13″ butts, 7″ points		840	.076		19.60	3.51	2.19	25.30	29.50
0600	70′ to 80′ long, 13″ butts, 6″ points		840	.076		22	3.51	2.19	27.70	32
0700	70′ to 80′ long, 16″ butts, 6″ points		800	.080	Ea.	27.50	3.68	2.30	33.48	39
0800	Treated piles, 12 lb. per C.F.,									
0810	friction or end bearing, ASTM class B									
1000	Up to 30′ long, 12″ butts, 8″ points	B-19	625	.102	V.L.F.	16.70	4.71	2.95	24.36	29
1100	30′ to 39′ long, 12″ butts, 8″ points		700	.091		18.20	4.21	2.63	25.04	29.50
1200	40′ to 49′ long, 12″ butts, 7″ points		720	.089		20.50	4.09	2.56	27.15	31.50
1300	50′ to 59′ long, 13″ butts, 7″ points		800	.080		21.50	3.68	2.30	27.48	32
1400	60′ to 69′ long, 13″ butts, 6″ points	B-19A	840	.076		26.50	3.51	2.69	32.70	38
1500	70′ to 80′ long, 13″ butts, 6″ points	″	840	.076		28	3.51	2.69	34.20	39.50
1600	Treated piles, C.C.A., 2.5# per C.F.									
1610	8″ butts, 10′ long	B-19	400	.160	V.L.F.	17.90	7.35	4.60	29.85	36
1620	11′ to 16′ long		500	.128		17.90	5.90	3.68	27.48	33
1630	17′ to 20′ long		575	.111		17.90	5.10	3.20	26.20	31
1640	10″ butts, 10′ to 16′ long		500	.128		17.90	5.90	3.68	27.48	33
1650	17′ to 20′ long		575	.111		17.90	5.10	3.20	26.20	31
1660	21′ to 40′ long		700	.091		17.90	4.21	2.63	24.74	29
1670	12″ butts, 10′ to 20′ long		575	.111		17.90	5.10	3.20	26.20	31
1680	21′ to 35′ long		650	.098		17.90	4.53	2.83	25.26	30
1690	36′ to 40′ long		700	.091		17.90	4.21	2.63	24.74	29
1695	14″ butts, to 40′ long		700	.091		17.90	4.21	2.63	24.74	29
1700	Boot for pile tip, minimum	1 Pile	27	.296	Ea.	23	13.15		36.15	46.50
1800	Maximum		21	.381		69	16.90		85.90	103
2000	Point for pile tip, minimum		20	.400		23	17.75		40.75	53.50
2100	Maximum		15	.533		83	23.50		106.50	129
2300	Splice for piles over 50′ long, minimum	B-46	35	1.371		56.50	56	1.29	113.79	151
2400	Maximum		20	2.400		68	98	2.26	168.26	231
2600	Concrete encasement with wire mesh and tube		331	.145	V.L.F.	10.50	5.95	.14	16.59	21
2700	Mobilization for 10,000 L.F. pile job, add	B-19	3300	.019			.89	.56	1.45	2
2800	25,000 L.F. pile job, add	″	8500	.008			.35	.22	.57	.78
2900	Mobilization by water for barge driving rig, add								100%	100%

31 62 19.30 Marine Wood Piles

	31 62 19.30 Marine Wood Piles	Crew	Daily Output	Labor-Hours	Unit	Material	Labor	Equipment	Total	Total Incl O&P
0010	**MARINE WOOD PILES**									
0040	Friction or end bearing, not including									
0050	mobilization or demobilization									
0055	Marine piles includes production adjusted									
0060	due to thickness of silt layers									
0090	Shore driven piles with silt layer									
1000	Shore driven 50′ untreated pile, 13″ butts, 7″ points, 10′ silt	B-19	952	.067	V.L.F.	17.40	3.09	1.93	22.42	26
1050	55′ untreated pile		936	.068		17.40	3.15	1.97	22.52	26
1075	60′ untreated pile		969	.066		19.60	3.04	1.90	24.54	28.50
1100	65′ untreated pile		958	.067		19.60	3.07	1.92	24.59	28.50
1125	70′ untreated pile, 6″ points		948	.068		22	3.11	1.94	27.05	31
1150	75′ untreated pile, 6″ points		940	.068		22	3.13	1.96	27.09	31

31 62 19 – Timber Piles

31 62 19.30 Marine Wood Piles		Crew	Daily Output	Labor-Hours	Unit	Material	Labor	Equipment	Total	Total Incl O&P
						2014 Bare Costs				
1175	80′ untreated pile, 6″ points	B-19	933	.069	V.L.F.	22	3.16	1.97	27.13	31
1200	50′ treated pile, 7″ points		952	.067		21.50	3.09	1.93	26.52	30.50
1225	55′ treated pile, 7″ points		936	.068		21.50	3.15	1.97	26.62	30.50
1250	60′ treated pile, 6″ points		969	.066		26.50	3.04	1.90	31.44	36.50
1275	65′ treated pile, 6″ points		958	.067		26.50	3.07	1.92	31.49	36.50
1300	70′ treated pile, 6″ points		948	.068		28	3.11	1.94	33.05	38
1325	75′ treated pile, 6″ points		940	.068		28	3.13	1.96	33.09	38
1350	80′ treated pile, 6″ points		933	.069		28	3.16	1.97	33.13	38
1400	Shore driven 50′ untreated pile, 13″ butts, 7″ points, 15′ silt		1053	.061		17.40	2.80	1.75	21.95	25.50
1450	55′ untreated pile		1023	.063		17.40	2.88	1.80	22.08	25.50
1475	60′ untreated pile		1050	.061		19.60	2.80	1.75	24.15	28
1500	65′ untreated pile		1030	.062		19.60	2.86	1.79	24.25	28
1525	70′ untreated pile, 6″ points		1013	.063		22	2.91	1.82	26.73	30.50
1550	75′ untreated pile, 6″ points		1000	.064		22	2.94	1.84	26.78	30.50
1575	80′ untreated pile, 6″ points		988	.065		22	2.98	1.86	26.84	30.50
1600	50′ treated pile, 7″ points		1053	.061		21.50	2.80	1.75	26.05	30
1625	55′ treated pile, 7″ points		1023	.063		21.50	2.88	1.80	26.18	30
1650	60′ treated pile, 6″ points		1050	.061		26.50	2.80	1.75	31.05	36
1675	65′ treated pile, 6″ points		1030	.062		26.50	2.86	1.79	31.15	36
1700	70′ treated pile, 6″ points		1014	.063		28	2.90	1.82	32.72	37.50
1725	75′ treated pile, 6″ points		1000	.064		28	2.94	1.84	32.78	37.50
1750	80′ treated pile, 6″ points		988	.065		28	2.98	1.86	32.84	37.50
1800	Shore driven 50′ untreated pile, 13″ butts, 7″ points, 20′ silt		1177	.054		17.40	2.50	1.56	21.46	25
1850	55′ untreated pile		1128	.057		17.40	2.61	1.63	21.64	25
1875	60′ untreated pile		1145	.056		19.60	2.57	1.61	23.78	27.50
1900	65′ untreated pile		1114	.057		19.60	2.64	1.65	23.89	27.50
1925	70′ untreated pile, 6″ points		1089	.059		22	2.70	1.69	26.39	30
1950	75′ untreated pile, 6″ points		1068	.060		22	2.76	1.72	26.48	30
1975	80′ untreated pile, 6″ points		1050	.061		22	2.80	1.75	26.55	30.50
2000	50′ treated pile, 7″ points		1177	.054		21.50	2.50	1.56	25.56	29
2025	55′ treated pile, 7″ points		1128	.057		21.50	2.61	1.63	25.74	29.50
2050	60′ treated pile, 6″ points		1145	.056		26.50	2.57	1.61	30.68	35.50
2075	65′ treated pile, 6″ points		1114	.057		26.50	2.64	1.65	30.79	35.50
2100	70′ treated pile, 6″ points		1089	.059		28	2.70	1.69	32.39	37
2125	75′ treated pile, 6″ points		1068	.060		28	2.76	1.72	32.48	37
2150	80′ treated pile, 6″ points		1050	.061		28	2.80	1.75	32.55	37.50
2200	Shore driven 60′ untreated pile, 13″ butts, 7″ points, 25′ silt		1260	.051		19.60	2.34	1.46	23.40	27
2225	65′ untreated pile		1213	.053		19.60	2.43	1.52	23.55	27
2250	70′ untreated pile, 6″ points		1176	.054		22	2.50	1.57	26.07	29.50
2275	75′ untreated pile, 6″ points		1146	.056		22	2.57	1.61	26.18	30
2300	80′ untreated pile, 6″ points		1120	.057		22	2.63	1.64	26.27	30
2350	60′ treated pile, 6″ points		1260	.051		26.50	2.34	1.46	30.30	35
2375	65′ treated pile, 6″ points		1213	.053		26.50	2.43	1.52	30.45	35
2400	70′ treated pile, 6″ points		1176	.054		28	2.50	1.57	32.07	36.50
2425	75′ treated pile, 6″ points		1146	.056		28	2.57	1.61	32.18	37
2450	80′ treated pile, 6″ points		1120	.057		28	2.63	1.64	32.27	37
2500	Shore driven 70′ untreated pile, 13″ butts, 6″ points, 30′ silt		1278	.050		22	2.30	1.44	25.74	29
2525	75′ untreated pile, 6″ points		1235	.052		22	2.38	1.49	25.87	29.50
2550	80′ untreated pile, 6″ points		1200	.053		22	2.45	1.53	25.98	29.50
2600	70′ treated pile, 6″ points		1278	.050		28	2.30	1.44	31.74	36
2625	75′ treated pile, 6″ points		1235	.052		28	2.38	1.49	31.87	36.50
2650	80′ treated pile, 6″ points		1200	.053		28	2.45	1.53	31.98	36.50
2700	Shore driven 70′ untreated pile, 13″ butts, 6″ points, 35′ silt	↓	1400	.046	↓	22	2.10	1.31	25.41	28.50

31 62 19.30 Marine Wood Piles		Crew	Daily Output	Labor-Hours	Unit	Material	2014 Bare Costs Labor	Equipment	Total	Total Incl O&P
2725	75′ untreated pile, 6″ points	B-19	1340	.048	V.L.F.	22	2.20	1.37	25.57	29
2750	80′ untreated pile, 6″ points		1292	.050		22	2.28	1.42	25.70	29
2800	70′ treated pile, 6″ points		1400	.046		28	2.10	1.31	31.41	35.50
2825	75′ treated pile, 6″ points		1340	.048		28	2.20	1.37	31.57	36
2850	80′ treated pile, 6″ points		1292	.050		28	2.28	1.42	31.70	36
2875	Shore driven 75′ untreated pile, 13″ butts, 6″ points, 40′ silt		1465	.044		22	2.01	1.26	25.27	28.50
2900	80′ untreated pile, 6″ points		1400	.046		22	2.10	1.31	25.41	28.50
2925	75′ treated pile, 6″ points		1465	.044		28	2.01	1.26	31.27	35.50
2950	80′ treated pile, 6″ points		1400	.046		28	2.10	1.31	31.41	35.50
2990	Barge driven piles with silt layer									
3000	Barge driven 50′ untreated pile, 13″ butts, 7″ points, 10′ silt	B-19B	762	.084	V.L.F.	17.40	3.86	3.44	24.70	29
3050	55′ untreated pile		749	.085		17.40	3.93	3.50	24.83	29
3075	60′ untreated pile		775	.083		19.60	3.80	3.38	26.78	31
3100	65′ untreated pile		766	.084		19.60	3.84	3.42	26.86	31.50
3125	70′ untreated pile, 6″ points		759	.084		22	3.88	3.45	29.33	34
3150	75′ untreated pile, 6″ points		752	.085		22	3.92	3.49	29.41	34
3175	80′ untreated pile, 6″ points		747	.086		22	3.94	3.51	29.45	34
3200	50′ treated pile, 7″ points		762	.084		21.50	3.86	3.44	28.80	33.50
3225	55′ treated pile, 7″ points		749	.085		21.50	3.93	3.50	28.93	33.50
3250	60′ treated pile, 6″ points		775	.083		26.50	3.80	3.38	33.68	39
3275	65′ treated pile, 6″ points		766	.084		26.50	3.84	3.42	33.76	39.50
3300	70′ treated pile, 6″ points		759	.084		28	3.88	3.45	35.33	41
3325	75′ treated pile, 6″ points		752	.085		28	3.92	3.49	35.41	41
3350	80′ treated pile, 6″ points		747	.086		28	3.94	3.51	35.45	41
3400	Barge driven 50′ untreated pile, 13″ butts, 7″ points, 15′ silt		842	.076		17.40	3.50	3.11	24.01	28
3450	55′ untreated pile		819	.078		17.40	3.60	3.20	24.20	28.50
3475	60′ untreated pile		840	.076		19.60	3.51	3.12	26.23	30.50
3500	65′ untreated pile		824	.078		19.60	3.57	3.18	26.35	30.50
3525	70′ untreated pile, 6″ points		811	.079		22	3.63	3.23	28.86	33
3550	75′ untreated pile, 6″ points		800	.080		22	3.68	3.28	28.96	33.50
3575	80′ untreated pile, 6″ points		791	.081		22	3.72	3.31	29.03	33.50
3600	50′ treated pile, 7″ points		842	.076		21.50	3.50	3.11	28.11	32.50
3625	55′ treated pile, 7″ points		819	.078		21.50	3.60	3.20	28.30	32.50
3650	60′ treated pile, 6″ points		840	.076		26.50	3.51	3.12	33.13	38.50
3675	65′ treated pile, 6″ points		824	.078		26.50	3.57	3.18	33.25	38.50
3700	70′ treated pile, 6″ points		811	.079		28	3.63	3.23	34.86	40
3725	75′ treated pile, 6″ points		800	.080		28	3.68	3.28	34.96	40.50
3750	80′ treated pile, 6″ points		791	.081		28	3.72	3.31	35.03	40.50
3800	Barge driven 50′ untreated pile, 13″ butts, 7″ points, 20′ silt		941	.068		17.40	3.13	2.79	23.32	27
3850	55′ untreated pile		903	.071		17.40	3.26	2.90	23.56	27.50
3875	60′ untreated pile		916	.070		19.60	3.21	2.86	25.67	29.50
3900	65′ untreated pile		891	.072		19.60	3.30	2.94	25.84	30
3925	70′ untreated pile, 6″ points		871	.073		22	3.38	3.01	28.39	32.50
3950	75′ untreated pile, 6″ points		854	.075		22	3.45	3.07	28.52	32.50
3975	80′ untreated pile, 6″ points		840	.076		22	3.51	3.12	28.63	33
4000	50′ treated pile, 7″ points		941	.068		21.50	3.13	2.79	27.42	31.50
4025	55′ treated pile, 7″ points		903	.071		21.50	3.26	2.90	27.66	32
4050	60′ treated pile, 6″ points		916	.070		26.50	3.21	2.86	32.57	37.50
4075	65′ treated pile, 6″ points		891	.072		26.50	3.30	2.94	32.74	38
4100	70′ treated pile, 6″ points		871	.073		28	3.38	3.01	34.39	39.50
4125	75′ treated pile, 6″ points		854	.075		28	3.45	3.07	34.52	39.50
4150	80′ treated pile, 6″ points		840	.076		28	3.51	3.12	34.63	40
4200	Barge driven 60′ untreated pile, 13″ butts, 7″ points, 25′ silt		1008	.063		19.60	2.92	2.60	25.12	29

31 62 Driven Piles

31 62 19 – Timber Piles

31 62 19.30 Marine Wood Piles		Crew	Daily Output	Labor-Hours	Unit	2014 Bare Costs Material	Labor	Equipment	Total	Total Incl O&P
4225	65′ untreated pile	B-19B	971	.066	V.L.F.	19.60	3.03	2.70	25.33	29
4250	70′ untreated pile, 6″ points		941	.068		22	3.13	2.79	27.92	32
4275	75′ untreated pile, 6″ points		916	.070		22	3.21	2.86	28.07	32
4300	80′ untreated pile, 6″ points		896	.071		22	3.29	2.93	28.22	32.50
4350	60′ treated pile, 6″ points		1008	.063		26.50	2.92	2.60	32.02	37
4375	65′ treated pile, 6″ points		971	.066		26.50	3.03	2.70	32.23	37
4400	70′ treated pile, 6″ points		941	.068		28	3.13	2.79	33.92	39
4425	75′ treated pile, 6″ points		916	.070		28	3.21	2.86	34.07	39
4450	80′ treated pile, 6″ points		896	.071		28	3.29	2.93	34.22	39.50
4500	Barge driven 70′ untreated pile, 13″ butts, 6″ points, 30′ silt		1023	.063		22	2.88	2.56	27.44	31.50
4525	75′ untreated pile, 6″ points		988	.065		22	2.98	2.65	27.63	31.50
4550	80′ untreated pile, 6″ points		960	.067		22	3.07	2.73	27.80	32
4600	70′ treated pile, 6″ points		1023	.063		28	2.88	2.56	33.44	38.50
4625	75′ treated pile, 6″ points		988	.065		28	2.98	2.65	33.63	38.50
4650	80′ treated pile, 6″ points		960	.067		28	3.07	2.73	33.80	39
4700	Barge driven 70′ untreated pile, 13″ butts, 6″ points, 35′ silt		1120	.057		22	2.63	2.34	26.97	30.50
4725	75′ untreated pile, 6″ points		1072	.060		22	2.75	2.45	27.20	31
4750	80′ untreated pile, 6″ points		1034	.062		22	2.85	2.54	27.39	31
4800	70′ treated pile, 6″ points		1120	.057		28	2.63	2.34	32.97	37.50
4825	75′ treated pile, 6″ points		1072	.060		28	2.75	2.45	33.20	38
4850	80′ treated pile, 6″ points		1034	.062		28	2.85	2.54	33.39	38
4875	Barge driven 75′ untreated pile, 13″ butts, 6″ points, 40′ silt		1172	.055		22	2.51	2.24	26.75	30.50
4900	80′ untreated pile, 6″ points		1120	.057		22	2.63	2.34	26.97	30.50
4925	75′ treated pile, 6″ points		1172	.055		28	2.51	2.24	32.75	37.50
4950	80′ treated pile, 6″ points		1120	.057		28	2.63	2.34	32.97	37.50
5000	Piles, wood, treated, added undriven length, 13″ butt					28			28	31
5020	Untreated pile, 13″ butts					22			22	24

31 62 23 – Composite Piles

31 62 23.10 Recycled Plastic and Steel Piles

			Crew	Daily Output	Labor-Hours	Unit	Material	Labor	Equipment	Total	Total Incl O&P
0010	**RECYCLED PLASTIC AND STEEL PILES**, 200 PILES										
5000	Marine pilings, recycled plastic w/steel reinf, up to 90′ long, 8″ dia	G	B-19	500	.128	V.L.F.	30.50	5.90	3.68	40.08	47
5010	10″ dia	G		500	.128		38.50	5.90	3.68	48.08	55.50
5020	13″ dia	G		400	.160		54.50	7.35	4.60	66.45	76.50
5030	16″ dia	G		400	.160		68	7.35	4.60	79.95	91.50
5040	20″ dia	G		350	.183		96.50	8.40	5.25	110.15	125
5050	24″ dia	G		350	.183		121	8.40	5.25	134.65	152

31 62 23.12 Marine Recycled Plastic and Steel Piles, 200 Piles

			Crew	Daily Output	Labor-Hours	Unit	Material	Labor	Equipment	Total	Total Incl O&P
0010	**MARINE RECYCLED PLASTIC AND STEEL PILES, 200 PILES**										
0040	Friction or end bearing, not including										
0050	mobilization or demobilization										
0055	Marine piles includes production adjusted										
0060	due to thickness of silt layers										
0090	Shore driven piles with silt layer										
0100	Marine pilings, recycled plastic w/steel, 50′ long, 8″ dia,10′ silt	G	B-19	595	.108	V.L.F.	30.50	4.95	3.09	38.54	44.50
0110	50′ long, 10″ dia	G		595	.108		38.50	4.95	3.09	46.54	53
0120	50′ long, 13″ dia	G		476	.134		54.50	6.20	3.87	64.57	74
0130	50′ long, 16″ dia	G		476	.134		68	6.20	3.87	78.07	89
0140	50′ long, 20″ dia	G		417	.153		96.50	7.05	4.41	107.96	122
0150	50′ long, 24″ dia	G		417	.153		121	7.05	4.41	132.46	149
0210	55′ long, 8″ dia	G		585	.109		30.50	5.05	3.15	38.70	45
0220	55′ long, 10″ dia	G		585	.109		38.50	5.05	3.15	46.70	53.50
0230	55′ long, 13″ dia	G		468	.137		54.50	6.30	3.93	64.73	74

31 62 23 – Composite Piles

31 62 23.12 Marine Recycled Plastic and Steel Piles, 200 Piles			Crew	Daily Output	Labor-Hours	Unit	Material	Labor	Equipment	Total	Total Incl O&P
							2014 Bare Costs				
0240	55′ long, 16″ dia	G	B-19	468	.137	V.L.F.	68	6.30	3.93	78.23	89
0250	55′ long, 20″ dia	G		410	.156		96.50	7.20	4.49	108.19	122
0260	55′ long, 24″ dia	G		410	.156		121	7.20	4.49	132.69	149
0310	60′ long, 8″ dia	G		576	.111		30.50	5.10	3.20	38.80	45
0320	60′ long, 10″ dia	G		577	.111		38.50	5.10	3.19	46.79	53.50
0330	60′ long, 13″ dia	G		462	.139		54.50	6.35	3.98	64.83	74.50
0340	60′ long, 16″ dia	G		462	.139		68	6.35	3.98	78.33	89.50
0350	60′ long, 20″ dia	G		404	.158		96.50	7.30	4.56	108.36	122
0360	60′ long, 24″ dia	G		404	.158		121	7.30	4.56	132.86	149
0410	65′ long, 8″ dia	G		570	.112		30.50	5.15	3.23	38.88	45
0420	65′ long, 10″ dia	G		570	.112		38.50	5.15	3.23	46.88	53.50
0430	65′ long, 13″ dia	G		456	.140		54.50	6.45	4.04	64.99	74.50
0440	65′ long, 16″ dia	G		456	.140		68	6.45	4.04	78.49	89.50
0450	65′ long, 20″ dia	G		399	.160		96.50	7.40	4.61	108.51	123
0460	65′ long, 24″ dia	G		399	.160		121	7.40	4.61	133.01	150
0510	70′ long, 8″ dia	G		565	.113		30.50	5.20	3.26	38.96	45
0520	70′ long, 10″ dia	G		565	.113		38.50	5.20	3.26	46.96	53.50
0530	70′ long, 13″ dia	G		452	.142		54.50	6.50	4.07	65.07	74.50
0540	70′ long, 16″ dia	G		452	.142		68	6.50	4.07	78.57	89.50
0550	70′ long, 20″ dia	G		395	.162		96.50	7.45	4.66	108.61	123
0560	70′ long, 24″ dia	G		399	.160		121	7.40	4.61	133.01	150
0610	75′ long, 8″ dia	G		560	.114		30.50	5.25	3.29	39.04	45.50
0620	75′ long, 10″ dia	G		560	.114		38.50	5.25	3.29	47.04	54
0630	75′ long, 13″ dia	G		448	.143		54.50	6.55	4.11	65.16	75
0640	75′ long, 16″ dia	G		448	.143		68	6.55	4.11	78.66	90
0650	75′ long, 20″ dia	G		392	.163		96.50	7.50	4.70	108.70	123
0660	75′ long, 24″ dia	G		392	.163		121	7.50	4.70	133.20	150
0710	80′ long, 8″ dia	G		556	.115		30.50	5.30	3.31	39.11	45.50
0720	80′ long, 10″ dia	G		556	.115		38.50	5.30	3.31	47.11	54
0730	80′ long, 13″ dia	G		444	.144		54.50	6.65	4.15	65.30	75
0740	80′ long, 16″ dia	G		444	.144		68	6.65	4.15	78.80	90
0750	80′ long, 20″ dia	G		389	.165		96.50	7.55	4.73	108.78	123
0760	80′ long, 24″ dia	G		389	.165		121	7.55	4.73	133.28	150
1000	Marine pilings, recycled plastic w/steel, 50′ long, 8″ dia, 15′ silt	G		658	.097		30.50	4.47	2.80	37.77	43.50
1010	50′ long, 10″ dia	G		658	.097		38.50	4.47	2.80	45.77	52
1020	50′ long, 13″ dia	G		526	.122		54.50	5.60	3.50	63.60	72.50
1030	50′ long, 16″ dia	G		526	.122		68	5.60	3.50	77.10	87.50
1040	50′ long, 20″ dia	G		461	.139		96.50	6.40	3.99	106.89	120
1050	50′ long, 24″ dia	G		461	.139		121	6.40	3.99	131.39	147
1210	55′ long, 8″ dia	G		640	.100		30.50	4.60	2.88	37.98	44
1220	55′ long, 10″ dia	G		640	.100		38.50	4.60	2.88	45.98	52.50
1230	55′ long, 13″ dia	G		512	.125		54.50	5.75	3.60	63.85	73
1240	55′ long, 16″ dia	G		512	.125		68	5.75	3.60	77.35	88
1250	55′ long, 20″ dia	G		448	.143		96.50	6.55	4.11	107.16	121
1260	55′ long, 24″ dia	G		448	.143		121	6.55	4.11	131.66	148
1310	60′ long, 8″ dia	G		625	.102		30.50	4.71	2.95	38.16	44
1320	60′ long, 10″ dia	G		625	.102		38.50	4.71	2.95	46.16	52.50
1330	60′ long, 13″ dia	G		500	.128		54.50	5.90	3.68	64.08	73.50
1340	60′ long, 16″ dia	G		500	.128		68	5.90	3.68	77.58	88.50
1350	60′ long, 20″ dia	G		438	.146		96.50	6.70	4.20	107.40	121
1360	60′ long, 24″ dia	G		438	.146		121	6.70	4.20	131.90	148
1410	65′ long, 8″ dia	G		613	.104		30.50	4.80	3	38.30	44.50
1420	65′ long, 10″ dia	G		613	.104		38.50	4.80	3	46.30	53

31 62 23 – Composite Piles

31 62 23.12 Marine Recycled Plastic and Steel Piles, 200 Piles		Crew	Daily Output	Labor-Hours	Unit	Material	Labor	Equipment	Total	Total Incl O&P
						2014 Bare Costs				
1430	65′ long, 13″ dia G	B-19	491	.130	V.L.F.	54.50	6	3.75	64.25	73.50
1440	65′ long, 16″ dia G		491	.130		68	6	3.75	77.75	88.50
1450	65′ long, 20″ dia G		429	.149		96.50	6.85	4.29	107.64	121
1460	65′ long, 24″ dia G		429	.149		121	6.85	4.29	132.14	148
1510	70′ long, 8″ dia G		603	.106		30.50	4.88	3.05	38.43	44.50
1520	70′ long, 10″ dia G		603	.106		38.50	4.88	3.05	46.43	53
1530	70′ long, 13″ dia G		483	.133		54.50	6.10	3.81	64.41	73.50
1540	70′ long, 16″ dia G		483	.133		68	6.10	3.81	77.91	88.50
1550	70′ long, 20″ dia G		422	.152		96.50	7	4.36	107.86	122
1560	70′ long, 24″ dia G		422	.152		121	7	4.36	132.36	149
1610	75′ long, 8″ dia G		595	.108		30.50	4.95	3.09	38.54	44.50
1620	75′ long, 10″ dia G		595	.108		38.50	4.95	3.09	46.54	53
1630	75′ long, 13″ dia G		476	.134		54.50	6.20	3.87	64.57	74
1640	75′ long, 16″ dia G		476	.134		68	6.20	3.87	78.07	89
1650	75′ long, 20″ dia G		417	.153		96.50	7.05	4.41	107.96	122
1660	75′ long, 24″ dia G		417	.153		121	7.05	4.41	132.46	149
1710	80′ long, 8″ dia G		588	.109		30.50	5	3.13	38.63	44.50
1720	80′ long, 10″ dia G		588	.109		38.50	5	3.13	46.63	53
1730	80′ long, 13″ dia G		471	.136		54.50	6.25	3.91	64.66	74
1740	80′ long, 16″ dia G		471	.136		68	6.25	3.91	78.16	89
1750	80′ long, 20″ dia G		412	.155		96.50	7.15	4.47	108.12	122
1760	80′ long, 24″ dia G		412	.155		121	7.15	4.47	132.62	149
2000	Marine pilings, recycled plastic w/steel, 50′ long, 8″ dia, 20′ silt G		735	.087		30.50	4.01	2.50	37.01	42.50
2010	50′ long, 10″ dia G		735	.087		38.50	4.01	2.50	45.01	51
2020	50′ long, 13″ dia G		588	.109		54.50	5	3.13	62.63	71
2030	50′ long, 16″ dia G		588	.109		68	5	3.13	76.13	86
2040	50′ long, 20″ dia G		515	.124		96.50	5.70	3.57	105.77	119
2050	50′ long, 24″ dia G		515	.124		121	5.70	3.57	130.27	146
2210	55′ long, 8″ dia G		705	.091		30.50	4.18	2.61	37.29	43
2220	55′ long, 10″ dia G		705	.091		38.50	4.18	2.61	45.29	51.50
2230	55′ long, 13″ dia G		564	.113		54.50	5.20	3.26	62.96	71.50
2240	55′ long, 16″ dia G		564	.113		68	5.20	3.26	76.46	86.50
2250	55′ long, 20″ dia G		494	.130		96.50	5.95	3.73	106.18	119
2260	55′ long, 24″ dia G		494	.130		121	5.95	3.73	130.68	146
2310	60′ long, 8″ dia G		682	.094		30.50	4.32	2.70	37.52	43
2320	60′ long, 10″ dia G		682	.094		38.50	4.32	2.70	45.52	51.50
2330	60′ long, 13″ dia G		546	.117		54.50	5.40	3.37	63.27	72
2340	60′ long, 16″ dia G		546	.117		68	5.40	3.37	76.77	87
2350	60′ long, 20″ dia G		477	.134		96.50	6.15	3.86	106.51	120
2360	60′ long, 24″ dia G		477	.134		121	6.15	3.86	131.01	147
2410	65′ long, 8″ dia G		663	.097		30.50	4.44	2.78	37.72	43.50
2420	65′ long, 10″ dia G		663	.097		38.50	4.44	2.78	45.72	52
2430	65′ long, 13″ dia G		531	.121		54.50	5.55	3.47	63.52	72.50
2440	65′ long, 16″ dia G		531	.121		68	5.55	3.47	77.02	87.50
2450	65′ long, 20″ dia G		464	.138		96.50	6.35	3.97	106.82	120
2460	65′ long, 24″ dia G		464	.138		121	6.35	3.97	131.32	147
2510	70′ long, 8″ dia G		648	.099		30.50	4.54	2.84	37.88	43.50
2520	70′ long, 10″ dia G		648	.099		38.50	4.54	2.84	45.88	52
2530	70′ long, 13″ dia G		518	.124		54.50	5.70	3.55	63.75	73
2540	70′ long, 16″ dia G		518	.124		68	5.70	3.55	77.25	88
2550	70′ long, 20″ dia G		454	.141		96.50	6.50	4.05	107.05	121
2560	70′ long, 24″ dia G		454	.141		121	6.50	4.05	131.55	148
2610	75′ long, 8″ dia G		636	.101		30.50	4.63	2.89	38.02	44

31 62 23 – Composite Piles

31 62 23.12 Marine Recycled Plastic and Steel Piles, 200 Piles		Crew	Daily Output	Labor-Hours	Unit	Material	Labor	Equipment	Total	Total Incl O&P
						2014 Bare Costs				
2620	75′ long, 10″ dia G	B-19	636	.101	V.L.F.	38.50	4.63	2.89	46.02	52.50
2630	75′ long, 13″ dia G		508	.126		54.50	5.80	3.62	63.92	73
2640	75′ long, 16″ dia G		508	.126		68	5.80	3.62	77.42	88
2650	75′ long, 20″ dia G		445	.144		96.50	6.60	4.14	107.24	121
2660	75′ long, 24″ dia G		445	.144		121	6.60	4.14	131.74	148
2710	80′ long, 8″ dia G		625	.102		30.50	4.71	2.95	38.16	44
2720	80′ long, 10″ dia G		625	.102		38.50	4.71	2.95	46.16	52.50
2730	80′ long, 13″ dia G		500	.128		54.50	5.90	3.68	64.08	73.50
2740	80′ long, 16″ dia G		500	.128		68	5.90	3.68	77.58	88.50
2750	80′ long, 20″ dia G		437	.146		96.50	6.75	4.21	107.46	121
2760	80′ long, 24″ dia G		437	.146		121	6.75	4.21	131.96	148
4990	Barge driven piles with silt layer									
5100	Marine pilings, recycled plastic w/steel, 50′ long, 8″ dia,10′ silt G	B-19B	476	.134	V.L.F.	30.50	6.20	5.50	42.20	49
5110	50′ long, 10″ dia G		476	.134		38.50	6.20	5.50	50.20	57.50
5120	50′ long, 13″ dia G		381	.168		54.50	7.75	6.90	69.15	79.50
5130	50′ long, 16″ dia G		381	.168		68	7.75	6.90	82.65	94.50
5140	50′ long, 20″ dia G		333	.192		96.50	8.85	7.85	113.20	128
5150	50′ long, 24″ dia G		333	.192		121	8.85	7.85	137.70	155
5210	55′ long, 8″ dia G		468	.137		30.50	6.30	5.60	42.40	49.50
5220	55′ long, 10″ dia G		468	.137		38.50	6.30	5.60	50.40	58
5230	55′ long, 13″ dia G		374	.171		54.50	7.85	7	69.35	80
5240	55′ long, 16″ dia G		374	.171		68	7.85	7	82.85	95
5250	55′ long, 20″ dia G		328	.195		96.50	9	8	113.50	129
5260	55′ long, 24″ dia G		328	.195		121	9	8	138	156
5310	60′ long, 8″ dia G		461	.139		30.50	6.40	5.70	42.60	49.50
5320	60′ long, 10″ dia G		461	.139		38.50	6.40	5.70	50.60	58
5330	60′ long, 13″ dia G		369	.173		54.50	8	7.10	69.60	80.50
5340	60′ long, 16″ dia G		369	.173		68	8	7.10	83.10	95.50
5350	60′ long, 20″ dia G		323	.198		96.50	9.10	8.10	113.70	129
5360	60′ long, 24″ dia G		323	.198		121	9.10	8.10	138.20	156
5410	65′ long, 8″ dia G		456	.140		30.50	6.45	5.75	42.70	50
5420	65′ long, 10″ dia G		456	.140		38.50	6.45	5.75	50.70	58.50
5430	65′ long, 13″ dia G		365	.175		54.50	8.05	7.20	69.75	80.50
5440	65′ long, 16″ dia G		365	.175		68	8.05	7.20	83.25	95.50
5450	65′ long, 20″ dia G		319	.201		96.50	9.25	8.20	113.95	129
5460	65′ long, 24″ dia G		319	.201		121	9.25	8.20	138.45	156
5510	70′ long, 8″ dia G		452	.142		30.50	6.50	5.80	42.80	50
5520	70′ long, 10″ dia G		452	.142		38.50	6.50	5.80	50.80	58.50
5530	70′ long, 13″ dia G		361	.177		54.50	8.15	7.25	69.90	80.50
5540	70′ long, 16″ dia G		361	.177		68	8.15	7.25	83.40	95.50
5550	70′ long, 20″ dia G		316	.203		96.50	9.30	8.30	114.10	130
5560	70′ long, 24″ dia G		316	.203		121	9.30	8.30	138.60	157
5610	75′ long, 8″ dia G		448	.143		30.50	6.55	5.85	42.90	50
5620	75′ long, 10″ dia G		448	.143		38.50	6.55	5.85	50.90	58.50
5630	75′ long, 13″ dia G		358	.179		54.50	8.25	7.30	70.05	81
5640	75′ long, 16″ dia G		358	.179		68	8.25	7.30	83.55	96
5650	75′ long, 20″ dia G		313	.204		96.50	9.40	8.40	114.30	130
5660	75′ long, 24″ dia G		313	.204		121	9.40	8.40	138.80	157
5710	80′ long, 8″ dia G		444	.144		30.50	6.65	5.90	43.05	50.50
5720	80′ long, 10″ dia G		444	.144		38.50	6.65	5.90	51.05	59
5730	80′ long, 13″ dia G		356	.180		54.50	8.25	7.35	70.10	81
5740	80′ long, 16″ dia G		356	.180		68	8.25	7.35	83.60	96
5750	80′ long, 20″ dia G		311	.206		96.50	9.45	8.45	114.40	130

31 62 23 – Composite Piles

31 62 23.12 Marine Recycled Plastic and Steel Piles, 200 Piles			Crew	Daily Output	Labor-Hours	Unit	Material	Labor	Equipment	Total	Total Incl O&P
							2014 Bare Costs				
5760	80′ long, 24″ dia	G	B-19B	311	.206	V.L.F.	121	9.45	8.45	138.90	157
6100	Marine pilings, recycled plastic w/steel, 50′ long, 8″ dia,15′ silt	G		526	.122		30.50	5.60	4.98	41.08	47.50
6110	50′ long, 10″ dia	G		526	.122		38.50	5.60	4.98	49.08	56
6120	50′ long, 13″ dia	G		421	.152		54.50	7	6.25	67.75	78
6130	50′ long, 16″ dia	G		421	.152		68	7	6.25	81.25	93
6140	50′ long, 20″ dia	G		368	.174		96.50	8	7.15	111.65	126
6150	50′ long, 24″ dia	G		368	.174		121	8	7.15	136.15	153
6210	55′ long, 8″ dia	G		512	.125		30.50	5.75	5.10	41.35	48
6220	55′ long, 10″ dia	G		512	.125		38.50	5.75	5.10	49.35	56.50
6230	55′ long, 13″ dia	G		409	.156		54.50	7.20	6.40	68.10	78.50
6240	55′ long, 16″ dia	G		409	.156		68	7.20	6.40	81.60	93.50
6250	55′ long, 20″ dia	G		358	.179		96.50	8.25	7.30	112.05	127
6260	55′ long, 24″ dia	G		358	.179		121	8.25	7.30	136.55	154
6310	60′ long, 8″ dia	G		500	.128		30.50	5.90	5.25	41.65	48.50
6320	60′ long, 10″ dia	G		500	.128		38.50	5.90	5.25	49.65	57
6330	60′ long, 13″ dia	G		400	.160		54.50	7.35	6.55	68.40	78.50
6340	60′ long, 16″ dia	G		400	.160		68	7.35	6.55	81.90	93.50
6350	60′ long, 20″ dia	G		350	.183		96.50	8.40	7.50	112.40	127
6360	60′ long, 24″ dia	G		350	.183		121	8.40	7.50	136.90	154
6410	65′ long, 8″ dia	G		491	.130		30.50	6	5.35	41.85	48.50
6420	65′ long, 10″ dia	G		491	.130		38.50	6	5.35	49.85	57
6430	65′ long, 13″ dia	G		392	.163		54.50	7.50	6.70	68.70	79
6440	65′ long, 16″ dia	G		392	.163		68	7.50	6.70	82.20	94
6450	65′ long, 20″ dia	G		343	.187		96.50	8.60	7.65	112.75	128
6460	65′ long, 24″ dia	G		343	.187		121	8.60	7.65	137.25	155
6510	70′ long, 8″ dia	G		483	.133		30.50	6.10	5.45	42.05	49
6520	70′ long, 10″ dia	G		483	.133		38.50	6.10	5.45	50.05	57.50
6530	70′ long, 13″ dia	G		386	.166		54.50	7.65	6.80	68.95	79.50
6540	70′ long, 16″ dia	G		386	.166		68	7.65	6.80	82.45	94.50
6550	70′ long, 20″ dia	G		338	.189		96.50	8.70	7.75	112.95	128
6560	70′ long, 24″ dia	G		338	.189		121	8.70	7.75	137.45	155
6610	75′ long, 8″ dia	G		476	.134		30.50	6.20	5.50	42.20	49
6620	75′ long, 10″ dia	G		476	.134		38.50	6.20	5.50	50.20	57.50
6630	75′ long, 13″ dia	G		381	.168		54.50	7.75	6.90	69.15	79.50
6640	75′ long, 16″ dia	G		381	.168		68	7.75	6.90	82.65	94.50
6650	75′ long, 20″ dia	G		333	.192		96.50	8.85	7.85	113.20	128
6660	75′ long, 24″ dia	G		333	.192		121	8.85	7.85	137.70	155
6710	80′ long, 8″ dia	G		471	.136		30.50	6.25	5.55	42.30	49.50
6720	80′ long, 10″ dia	G		471	.136		38.50	6.25	5.55	50.30	58
6730	80′ long, 13″ dia	G		376	.170		54.50	7.85	6.95	69.30	80
6740	80′ long, 16″ dia	G		376	.170		68	7.85	6.95	82.80	95
6750	80′ long, 20″ dia	G		329	.195		96.50	8.95	7.95	113.40	129
6760	80′ long, 24″ dia	G		329	.195		121	8.95	7.95	137.90	156
7100	Marine pilings, recycled plastic w/steel, 50′ long, 8″ dia,20′ silt	G		588	.109		30.50	5	4.46	39.96	46
7110	50′ long, 10″ dia	G		588	.109		38.50	5	4.46	47.96	54.50
7120	50′ long, 13″ dia	G		471	.136		54.50	6.25	5.55	66.30	76
7130	50′ long, 16″ dia	G		471	.136		68	6.25	5.55	79.80	91
7140	50′ long, 20″ dia	G		412	.155		96.50	7.15	6.35	110	124
7150	50′ long, 24″ dia	G		412	.155		121	7.15	6.35	134.50	151
7210	55′ long, 8″ dia	G		564	.113		30.50	5.20	4.65	40.35	47
7220	55′ long, 10″ dia	G		564	.113		38.50	5.20	4.65	48.35	55.50
7230	55′ long, 13″ dia	G		451	.142		54.50	6.55	5.80	66.85	76.50
7240	55′ long, 16″ dia	G		451	.142		68	6.55	5.80	80.35	91.50

31 62 23 – Composite Piles

31 62 23.12 Marine Recycled Plastic and Steel Piles, 200 Piles		Crew	Daily Output	Labor-Hours	Unit	2014 Bare Costs Material	Labor	Equipment	Total	Total Incl O&P
7250	55′ long, 20″ dia G	B-19B	395	.162	V.L.F.	96.50	7.45	6.65	110.60	125
7260	55′ long, 24″ dia G		395	.162		121	7.45	6.65	135.10	152
7310	60′ long, 8″ dia G		545	.117		30.50	5.40	4.81	40.71	47
7320	60′ long, 10″ dia G		545	.117		38.50	5.40	4.81	48.71	55.50
7330	60′ long, 13″ dia G		436	.147		54.50	6.75	6	67.25	77
7340	60′ long, 16″ dia G		436	.147		68	6.75	6	80.75	92
7350	60′ long, 20″ dia G		382	.168		96.50	7.70	6.85	111.05	126
7360	60′ long, 24″ dia G		382	.168		121	7.70	6.85	135.55	153
7410	65′ long, 8″ dia G		531	.121		30.50	5.55	4.94	40.99	47.50
7420	65′ long, 10″ dia G		531	.121		38.50	5.55	4.94	48.99	56
7430	65′ long, 13″ dia G		424	.151		54.50	6.95	6.20	67.65	77.50
7440	65′ long, 16″ dia G		424	.151		68	6.95	6.20	81.15	92.50
7450	65′ long, 20″ dia G		371	.173		96.50	7.95	7.05	111.50	126
7460	65′ long, 24″ dia G		371	.173		121	7.95	7.05	136	153
7510	70′ long, 8″ dia G		518	.124		30.50	5.70	5.05	41.25	48
7520	70′ long, 10″ dia G		518	.124		38.50	5.70	5.05	49.25	56.50
7530	70′ long, 13″ dia G		415	.154		54.50	7.10	6.30	67.90	78
7540	70′ long, 16″ dia G		415	.154		68	7.10	6.30	81.40	93
7550	70′ long, 20″ dia G		363	.176		96.50	8.10	7.20	111.80	127
7560	70′ long, 24″ dia G		363	.176		121	8.10	7.20	136.30	154
7610	75′ long, 8″ dia G		508	.126		30.50	5.80	5.15	41.45	48.50
7620	75′ long, 10″ dia G		508	.126		38.50	5.80	5.15	49.45	57
7630	75′ long, 13″ dia G		407	.157		54.50	7.25	6.45	68.20	78.50
7640	75′ long, 16″ dia G		407	.157		68	7.25	6.45	81.70	93.50
7650	75′ long, 20″ dia G		356	.180		96.50	8.25	7.35	112.10	127
7660	75′ long, 24″ dia G		356	.180		121	8.25	7.35	136.60	154
7710	80′ long, 8″ dia G		500	.128		30.50	5.90	5.25	41.65	48.50
7720	80′ long, 10″ dia G		500	.128		38.50	5.90	5.25	49.65	57
7730	80′ long, 13″ dia G		400	.160		54.50	7.35	6.55	68.40	78.50
7740	80′ long, 16″ dia G		400	.160		68	7.35	6.55	81.90	93.50
7750	80′ long, 20″ dia G		350	.183		96.50	8.40	7.50	112.40	127
7760	80′ long, 24″ dia G	▼	350	.183	▼	121	8.40	7.50	136.90	154

31 62 23.13 Concrete-Filled Steel Piles

		Crew	Daily Output	Labor-Hours	Unit	Material	Labor	Equipment	Total	Total Incl O&P
0010	**CONCRETE-FILLED STEEL PILES** no mobilization or demobilization									
2600	Pipe piles, 50′ lg. 8″ diam., 29 lb. per L.F., no concrete	B-19	500	.128	V.L.F.	20.50	5.90	3.68	30.08	36.50
2700	Concrete filled		460	.139		21.50	6.40	4	31.90	38.50
2900	10″ diameter, 34 lb. per L.F., no concrete		500	.128		27	5.90	3.68	36.58	43
3000	Concrete filled		450	.142		28.50	6.55	4.09	39.14	46
3200	12″ diameter, 44 lb. per L.F., no concrete		475	.135		33.50	6.20	3.88	43.58	50.50
3300	Concrete filled		415	.154		36.50	7.10	4.44	48.04	56.50
3500	14″ diameter, 46 lb. per L.F., no concrete		430	.149		35.50	6.85	4.28	46.63	54.50
3600	Concrete filled		355	.180		40.50	8.30	5.20	54	63
3800	16″ diameter, 52 lb. per L.F., no concrete		385	.166		39.50	7.65	4.78	51.93	60.50
3900	Concrete filled		335	.191		44.50	8.80	5.50	58.80	69
4100	18″ diameter, 59 lb. per L.F., no concrete		355	.180		50.50	8.30	5.20	64	74
4200	Concrete filled	▼	310	.206	▼	56.50	9.50	5.95	71.95	83.50
4400	Splices for pipe piles, not in leads, 8″ diameter	1 Sswl	4.67	1.713	Ea.	90.50	87.50		178	255
4500	14″ diameter		3.79	2.111		118	108		226	320
4600	16″ diameter		3.03	2.640		146	135		281	400
4650	18″ diameter		4.50	1.778		193	91		284	375
4800	Points, standard, 8″ diameter		4.61	1.735		150	88.50		238.50	320
4900	14″ diameter	▼	4.05	1.975	▼	209	101		310	405

31 62 Driven Piles

31 62 23 – Composite Piles

31 62 23.13 Concrete-Filled Steel Piles		Crew	Daily Output	Labor-Hours	Unit	2014 Bare Costs Material	Labor	Equipment	Total	Total Incl O&P
5000	16″ diameter	1 Sswl	3.37	2.374	Ea.	254	121		375	495
5050	18″ diameter		5	1.600		400	82		482	580
5200	Points, heavy duty, 10″ diameter		2.89	2.768		297	141		438	575
5300	14″ or 16″ diameter		2.02	3.960		475	202		677	875
5500	For reinforcing steel, add		1150	.007	Lb.	.90	.36		1.26	1.62
5700	For thick wall sections, add				"	.94			.94	1.03
6020	Steel pipe pile end plates, 8″ diameter	1 Sswl	14	.571	Ea.	75.50	29		104.50	135
6050	10″ diameter		14	.571		79.50	29		108.50	139
6100	12″ diameter		12	.667		84	34		118	153
6150	14″ diameter		10	.800		87	41		128	168
6200	16″ diameter		9	.889		93	45.50		138.50	182
6250	18″ diameter		8	1		105	51		156	205
6300	Steel pipe pile shoes, 8″ diameter		12	.667		162	34		196	238
6350	10″ diameter		12	.667		168	34		202	245
6400	12″ diameter		10	.800		179	41		220	269
6450	14″ diameter		9	.889		207	45.50		252.50	305
6500	16″ diameter		8	1		230	51		281	340
6550	18″ diameter		6	1.333		257	68		325	405

31 62 33 – Drilled Micropiles

31 62 33.10 Drilled Micropiles Metal Pipe		Crew	Daily Output	Labor-Hours	Unit	Material	Labor	Equipment	Total	Total Incl O&P
0010	**DRILLED MICROPILES METAL PIPE**									
0011	No mobilization or demobilization									
5000	Pressure grouted pin pile, 5″ diam., cased, up to 50 ton,									
5040	End bearing, less than 20′	B-48	90	.622	V.L.F.	49.50	25.50	32.50	107.50	129
5080	More than 40′		135	.415		29.50	17.15	21.50	68.15	82
5120	Friction, loose sand and gravel		107	.523		49.50	21.50	27	98	117
5160	Dense sand and gravel		135	.415		29.50	17.15	21.50	68.15	82
5200	Uncased, up to 10 ton capacity, 20′		135	.415		29.50	17.15	21.50	68.15	82

31 63 Bored Piles

31 63 26 – Drilled Caissons

31 63 26.13 Fixed End Caisson Piles		Crew	Daily Output	Labor-Hours	Unit	Material	Labor	Equipment	Total	Total Incl O&P
0010	**FIXED END CAISSON PILES**									
0015	Including excavation, concrete, 50 lb. reinforcing									
0020	per C.Y., not incl. mobilization, boulder removal, disposal									
0100	Open style, machine drilled, to 50′ deep, in stable ground, no									
0110	casings or ground water, 18″ diam., 0.065 C.Y./L.F.	B-43	200	.240	V.L.F.	8.05	9.70	12.75	30.50	38
0200	24″ diameter, 0.116 C.Y./L.F.		190	.253		14.40	10.20	13.45	38.05	46
0300	30″ diameter, 0.182 C.Y./L.F.		150	.320		22.50	12.95	17	52.45	63.50
0400	36″ diameter, 0.262 C.Y./L.F.		125	.384		32.50	15.50	20.50	68.50	82
0500	48″ diameter, 0.465 C.Y./L.F.		100	.480		57.50	19.40	25.50	102.40	121
0600	60″ diameter, 0.727 C.Y./L.F.		90	.533		90	21.50	28.50	140	163
0700	72″ diameter, 1.05 C.Y./L.F.		80	.600		130	24	32	186	215
0800	84″ diameter, 1.43 C.Y./L.F.		75	.640		177	26	34	237	272
1000	For bell excavation and concrete, add									
1020	4′ bell diameter, 24″ shaft, 0.444 C.Y.	B-43	20	2.400	Ea.	44	97	128	269	340
1040	6′ bell diameter, 30″ shaft, 1.57 C.Y.		5.70	8.421		155	340	450	945	1,175
1060	8′ bell diameter, 36″ shaft, 3.72 C.Y.		2.40	20		370	810	1,075	2,255	2,825
1080	9′ bell diameter, 48″ shaft, 4.48 C.Y.		2	24		445	970	1,275	2,690	3,375
1100	10′ bell diameter, 60″ shaft, 5.24 C.Y.		1.70	28.235		520	1,150	1,500	3,170	3,975

31 63 26.13 Fixed End Caisson Piles		Crew	Daily Output	Labor-Hours	Unit	2014 Bare Costs Material	Labor	Equipment	Total	Total Incl O&P
1120	12′ bell diameter, 72″ shaft, 8.74 C.Y.	B-43	1	48	Ea.	865	1,950	2,550	5,365	6,725
1140	14′ bell diameter, 84″ shaft, 13.6 C.Y.		.70	68.571		1,350	2,775	3,650	7,775	9,725
1200	Open style, machine drilled, to 50′ deep, in wet ground, pulled									
1300	casing and pumping, 18″ diameter, 0.065 C.Y./L.F.	B-48	160	.350	V.L.F.	8.05	14.45	18.20	40.70	51
1400	24″ diameter, 0.116 C.Y./L.F.		125	.448		14.40	18.50	23.50	56.40	70
1500	30″ diameter, 0.182 C.Y./L.F.		85	.659		22.50	27	34	83.50	104
1600	36″ diameter, 0.262 C.Y./L.F.		60	.933		32.50	38.50	48.50	119.50	148
1700	48″ diameter, 0.465 C.Y./L.F.	B-49	55	1.600		57.50	68.50	66.50	192.50	243
1800	60″ diameter, 0.727 C.Y./L.F.		35	2.514		90	108	104	302	380
1900	72″ diameter, 1.05 C.Y./L.F.		30	2.933		130	126	122	378	470
2000	84″ diameter, 1.43 C.Y./L.F.		25	3.520		177	151	146	474	590
2100	For bell excavation and concrete, add									
2120	4′ bell diameter, 24″ shaft, 0.444 C.Y.	B-48	19.80	2.828	Ea.	44	117	147	308	390
2140	6′ bell diameter, 30″ shaft, 1.57 C.Y.		5.70	9.825		155	405	510	1,070	1,350
2160	8′ bell diameter, 36″ shaft, 3.72 C.Y.		2.40	23.333		370	965	1,200	2,535	3,200
2180	9′ bell diameter, 48″ shaft, 4.48 C.Y.	B-49	3.30	26.667		445	1,150	1,100	2,695	3,475
2200	10′ bell diameter, 60″ shaft, 5.24 C.Y.		2.80	31.429		520	1,350	1,300	3,170	4,075
2220	12′ bell diameter, 72″ shaft, 8.74 C.Y.		1.60	55		865	2,350	2,275	5,490	7,100
2240	14′ bell diameter, 84″ shaft, 13.6 C.Y.		1	88		1,350	3,775	3,650	8,775	11,300
2300	Open style, machine drilled, to 50′ deep, in soft rocks and									
2400	medium hard shales, 18″ diameter, 0.065 C.Y./L.F.	B-49	50	1.760	V.L.F.	8.05	75.50	73	156.55	205
2500	24″ diameter, 0.116 C.Y./L.F.		30	2.933		14.40	126	122	262.40	345
2600	30″ diameter, 0.182 C.Y./L.F.		20	4.400		22.50	189	183	394.50	515
2700	36″ diameter, 0.262 C.Y./L.F.		15	5.867		32.50	252	244	528.50	690
2800	48″ diameter, 0.465 C.Y./L.F.		10	8.800		57.50	380	365	802.50	1,050
2900	60″ diameter, 0.727 C.Y./L.F.		7	12.571		90	540	520	1,150	1,500
3000	72″ diameter, 1.05 C.Y./L.F.		6	14.667		130	630	610	1,370	1,775
3100	84″ diameter, 1.43 C.Y./L.F.		5	17.600		177	755	730	1,662	2,150
3200	For bell excavation and concrete, add									
3220	4′ bell diameter, 24″ shaft, 0.444 C.Y.	B-49	10.90	8.073	Ea.	44	345	335	724	955
3240	6′ bell diameter, 30″ shaft, 1.57 C.Y.		3.10	28.387		155	1,225	1,175	2,555	3,350
3260	8′ bell diameter, 36″ shaft, 3.72 C.Y.		1.30	67.692		370	2,900	2,800	6,070	7,975
3280	9′ bell diameter, 48″ shaft, 4.48 C.Y.		1.10	80		445	3,425	3,325	7,195	9,425
3300	10′ bell diameter, 60″ shaft, 5.24 C.Y.		.90	97.778		520	4,200	4,050	8,770	11,500
3320	12′ bell diameter, 72″ shaft, 8.74 C.Y.		.60	146		865	6,300	6,100	13,265	17,300
3340	14′ bell diameter, 84″ shaft, 13.6 C.Y.		.40	220		1,350	9,450	9,150	19,950	26,100
3600	For rock excavation, sockets, add, minimum		120	.733	C.F.		31.50	30.50	62	82
3650	Average		95	.926			40	38.50	78.50	104
3700	Maximum		48	1.833			78.50	76	154.50	205
3900	For 50′ to 100′ deep, add				V.L.F.				7%	7%
4000	For 100′ to 150′ deep, add								25%	25%
4100	For 150′ to 200′ deep, add								30%	30%
4200	For casings left in place, add				Lb.	1.13			1.13	1.24
4300	For other than 50 lb. reinf. per C.Y., add or deduct				"	1.10			1.10	1.21
4400	For steel "I" beam cores, add	B-49	8.30	10.602	Ton	2,125	455	440	3,020	3,500
4500	Load and haul excess excavation, 2 miles	B-34B	178	.045	L.C.Y.		1.69	3.89	5.58	6.85
4600	For mobilization, 50 mile radius, rig to 36″	B-43	2	24	Ea.		970	1,275	2,245	2,875
4650	Rig to 84″	B-48	1.75	32			1,325	1,650	2,975	3,850
4700	For low headroom, add								50%	50%
4750	For difficult access, add								25%	25%
5000	Bottom inspection	1 Skwk	1.20	6.667			315		315	490

31 63 26 – Drilled Caissons

	31 63 26.16 Concrete Caissons for Marine Construction	Crew	Daily Output	Labor-Hours	Unit	Material	2014 Bare Costs Labor	Equipment	Total	Total Incl O&P
0010	**CONCRETE CAISSONS FOR MARINE CONSTRUCTION**									
0100	Caissons, incl. mobilization and demobilization, up to 50 miles									
0200	Uncased shafts, 30 to 80 tons cap., 17" diam., 10' depth	B-44	88	.727	V.L.F.	19.80	33	21	73.80	96.50
0300	25' depth		165	.388		14.15	17.50	11.15	42.80	55.50
0400	80-150 ton capacity, 22" diameter, 10' depth		80	.800		25	36	23	84	109
0500	20' depth		130	.492		19.80	22	14.15	55.95	72
0700	Cased shafts, 10 to 30 ton capacity, 10-5/8" diam., 20' depth		175	.366		14.15	16.50	10.50	41.15	53
0800	30' depth		240	.267		13.20	12.05	7.65	32.90	42
0850	30 to 60 ton capacity, 12" diameter, 20' depth		160	.400		19.80	18.05	11.50	49.35	62.50
0900	40' depth		230	.278		15.25	12.55	8	35.80	45
1000	80 to 100 ton capacity, 16" diameter, 20' depth		160	.400		28.50	18.05	11.50	58.05	71.50
1100	40' depth		230	.278		26.50	12.55	8	47.05	57.50
1200	110 to 140 ton capacity, 17-5/8" diameter, 20' depth		160	.400		30.50	18.05	11.50	60.05	74
1300	40' depth		230	.278		28.50	12.55	8	49.05	59.50
1400	140 to 175 ton capacity, 19" diameter, 20' depth		130	.492		33	22	14.15	69.15	86.50
1500	40' depth		210	.305		30.50	13.75	8.75	53	64.50
1700	Over 30' long, L.F. cost tends to be lower									
1900	Maximum depth is about 90'									

31 63 29 – Drilled Concrete Piers and Shafts

31 63 29.13 Uncased Drilled Concrete Piers

		Crew	Daily Output	Labor-Hours	Unit	Material	Labor	Equipment	Total	Total Incl O&P
0010	**UNCASED DRILLED CONCRETE PIERS**									
0020	Unless specified otherwise, not incl. pile caps or mobilization									
0050	Cast in place augered piles, no casing or reinforcing									
0060	8" diameter	B-43	540	.089	V.L.F.	4.04	3.59	4.72	12.35	15.15
0065	10" diameter		480	.100		6.40	4.04	5.30	15.74	19.10
0070	12" diameter		420	.114		9.05	4.62	6.05	19.72	24
0075	14" diameter		360	.133		12.20	5.40	7.10	24.70	29.50
0080	16" diameter		300	.160		16.45	6.45	8.50	31.40	37.50
0085	18" diameter		240	.200		20.50	8.10	10.65	39.25	46.50
0100	Cast in place, thin wall shell pile, straight sided,									
0110	not incl. reinforcing, 8" diam., 16 ga., 5.8 lb./L.F.	B-19	700	.091	V.L.F.	9.10	4.21	2.63	15.94	19.50
0200	10" diameter, 16 ga. corrugated, 7.3 lb./L.F.		650	.098		11.95	4.53	2.83	19.31	23.50
0300	12" diameter, 16 ga. corrugated, 8.7 lb./L.F.		600	.107		15.50	4.91	3.07	23.48	28
0400	14" diameter, 16 ga. corrugated, 10.0 lb./L.F.		550	.116		18.25	5.35	3.35	26.95	32
0500	16" diameter, 16 ga. corrugated, 11.6 lb./L.F.		500	.128		22.50	5.90	3.68	32.08	38
0800	Cast in place friction pile, 50' long, fluted,									
0810	tapered steel, 4000 psi concrete, no reinforcing									
0900	12" diameter, 7 ga.	B-19	600	.107	V.L.F.	29.50	4.91	3.07	37.48	43.50
1000	14" diameter, 7 ga.		560	.114		32	5.25	3.29	40.54	47
1100	16" diameter, 7 ga.		520	.123		37.50	5.65	3.54	46.69	54
1200	18" diameter, 7 ga.		480	.133		44	6.15	3.83	53.98	62.50
1300	End bearing, fluted, constant diameter,									
1320	4000 psi concrete, no reinforcing									
1340	12" diameter, 7 ga.	B-19	600	.107	V.L.F.	30.50	4.91	3.07	38.48	44.50
1360	14" diameter, 7 ga.		560	.114		38.50	5.25	3.29	47.04	54
1380	16" diameter, 7 ga.		520	.123		44.50	5.65	3.54	53.69	61.50
1400	18" diameter, 7 ga.		480	.133		49	6.15	3.83	58.98	68

31 63 29.20 Cast In Place Piles, Adds

		Crew	Daily Output	Labor-Hours	Unit	Material	Labor	Equipment	Total	Total Incl O&P
0010	**CAST IN PLACE PILES, ADDS**									
1500	For reinforcing steel, add				Lb.	1			1	1.10
1700	For ball or pedestal end, add	B-19	11	5.818	C.Y.	148	268	167	583	760

31 63 Bored Piles

31 63 29 – Drilled Concrete Piers and Shafts

31 63 29.20 Cast In Place Piles, Adds		Crew	Daily Output	Labor-Hours	Unit	Material	2014 Bare Costs Labor	Equipment	Total	Total Incl O&P
1900	For lengths above 60′, concrete, add	B-19	11	5.818	C.Y.	154	268	167	589	770
2000	For steel thin shell, pipe only				Lb.	1.25			1.25	1.38

31 71 Tunnel Excavation

31 71 13 – Shield Driving Tunnel Excavation

31 71 13.10 Soft Ground Shield Driven Boring

		Crew	Daily Output	Labor-Hours	Unit	Material	Labor	Equipment	Total	Total Incl O&P
0010	**SOFT GROUND SHIELD DRIVEN BORING**									
0300	Earth excavation, minimum				L.F.				350	385
0310	Average								855	940
0320	Maximum								2,150	2,375

31 71 16 – Tunnel Excavation by Drilling and Blasting

31 71 16.20 Shaft Construction

		Crew	Daily Output	Labor-Hours	Unit	Material	Labor	Equipment	Total	Total Incl O&P
0010	**SHAFT CONSTRUCTION**									
0400	Shaft excavation, rock, minimum				C.Y.				72	79
0410	Average								100	110
0420	Maximum								200	220
0430	Earth, minimum								41	45
0440	Average								63.50	69.50
0450	Maximum								179	196
3000	Ventilation for tunnel construction									
3100	Fiberglass Ductwork 46 inch diameter for mine	Q-10	32	.750	L.F.	167	38.50		205.50	243
3200	28,000 CFM, 20 HP	Q-20	.40	50	Ea.	16,300	2,500		18,800	21,700

31 71 19 – Tunnel Excavation by Tunnel Boring Machine

31 71 19.10 Rock Excavation, Tunnel Boring General

		Crew	Daily Output	Labor-Hours	Unit	Material	Labor	Equipment	Total	Total Incl O&P
0010	**ROCK EXCAVATION, TUNNEL BORING GENERAL**									
0020	Bored Tunnels, incl. mucking 20′ diameter				L.F.				475	525
0100	Rock excavation, minimum								475	525
0110	Average								1,025	1,125
0120	Maximum								2,775	3,050
0200	Mixed rock and earth, minimum								790	870
0210	Average								1,675	1,850
0220	Maximum								2,550	2,800

31 71 21 – Tunnel Excavation by Cut and Cover

31 71 21.10 Cut and Cover Tunnels

		Crew	Daily Output	Labor-Hours	Unit	Material	Labor	Equipment	Total	Total Incl O&P
0010	**CUT AND COVER TUNNELS**									
2000	Cut and cover, excavation, not incl. hauling or backfill	B-57	80	.600	C.Y.		25.50	21	46.50	62
2100	Decking on steel frames, concrete				S.F.				36	39.50
2110	Steel plates								46.50	51
2120	Wood								26	28.50
2400	Gunite, dry mix, tunnel walls, including mesh, 4″ thick	C-16	500	.112		1.90	4.57	.79	7.26	9.90

31 73 Tunnel Grouting

31 73 13 – Cement Tunnel Grouting

31 73 13.10 Tunnel Liner Grouting		Crew	Daily Output	Labor-Hours	Unit	2014 Bare Costs Material	Labor	Equipment	Total	Total Incl O&P
0010	**TUNNEL LINER GROUTING**									
0800	Contact grouting incl. drilling and connecting, minimum				C.F.				31	34
0820	Maximum				"				65	71.50

31 74 Tunnel Construction

31 74 13 – Cast-in-place concrete tunnel lining

31 74 13.10 Cast-in-Place Concrete Tunnel Linings		Crew	Daily Output	Labor-Hours	Unit	Material	Labor	Equipment	Total	Total Incl O&P
0010	**CAST-IN-PLACE CONCRETE TUNNEL LININGS**									
0500	Tunnel liner invert, incl. reinf., for bored tunnels, 20′ diam., rock				L.F.				535	590
0510	Earth								360	395
0520	Arch, incl. reinf., for bored tunnels, rock								535	590
0530	Earth				↓				510	560

Division Notes

	CREW	DAILY OUTPUT	LABOR-HOURS	UNIT	BARE COSTS MAT.	LABOR	EQUIP.	TOTAL	TOTAL INCL O&P

Estimating Tips

32 01 00 Operations and Maintenance of Exterior Improvements

- Recycling of asphalt pavement is becoming very popular and is an alternative to removal and replacement. It can be a good value engineering proposal if removed pavement can be recycled, either at the project site or at another site that is reasonably close to the project site. Sections on repair of flexible and rigid pavement are included.

32 10 00 Bases, Ballasts, and Paving

- When estimating paving, keep in mind the project schedule. Also note that prices for asphalt and concrete are generally higher in the cold seasons. Lines for pavement markings, including tactile warning systems and fence lines, are included.

32 90 00 Planting

- The timing of planting and guarantee specifications often dictate the costs for establishing tree and shrub growth and a stand of grass or ground cover. Establish the work performance schedule to coincide with the local planting season. Maintenance and growth guarantees can add from 20%–100% to the total landscaping cost and can be contractually cumbersome. The cost to replace trees and shrubs can be as high as 5% of the total cost, depending on the planting zone, soil conditions, and time of year.

Reference Numbers

Reference numbers are shown in shaded boxes at the beginning of some major classifications. These numbers refer to related items in the Reference Section. The reference information may be an estimating procedure, an alternate pricing method, or technical information.

Note: Not all subdivisions listed here necessarily appear in this publication.

32 01 13 – Flexible Paving Surface Treatment

32 01 13.61 Slurry Seal (Latex Modified)		Crew	Daily Output	Labor-Hours	Unit	2014 Bare Costs Material	Labor	Equipment	Total	Total Incl O&P
0010	**SLURRY SEAL (LATEX MODIFIED)**									
3600	Waterproofing, membrane, tar and fabric, small area	B-63	233	.172	S.Y.	13.90	6.65	.75	21.30	26.50
3640	Large area		1435	.028		12.80	1.08	.12	14	15.90
3680	Preformed rubberized asphalt, small area		100	.400		18.95	15.50	1.74	36.19	47
3720	Large area	↓	367	.109		17.25	4.22	.48	21.95	26
3780	Rubberized asphalt (latex) seal	B-45	5000	.003	↓	2.78	.14	.18	3.10	3.47

32 01 13.62 Asphalt Surface Treatment		Crew	Daily Output	Labor-Hours	Unit	Material	Labor	Equipment	Total	Total Incl O&P
0010	**ASPHALT SURFACE TREATMENT**									
3000	Pavement overlay, polypropylene									
3040	6 oz. per S.Y., ideal conditions	B-63	10000	.004	S.Y.	2.16	.15	.02	2.33	2.64
3080	Adverse conditions		1000	.040		2.89	1.55	.17	4.61	5.75
3120	4 oz. per S.Y., ideal conditions		10000	.004		1.68	.15	.02	1.85	2.11
3160	Adverse conditions	↓	1000	.040		2.17	1.55	.17	3.89	4.95
3200	Tack coat, emulsion, .05 gal. per S.Y., 1000 S.Y.	B-45	2500	.006		.28	.28	.37	.93	1.14
3240	10,000 S.Y.		10000	.002		.22	.07	.09	.38	.46
3270	.10 gal. per S.Y., 1000 S.Y.		2500	.006		.53	.28	.37	1.18	1.41
3275	10,000 S.Y.		10000	.002		.42	.07	.09	.58	.67
3280	.15 gal. per S.Y., 1000 S.Y.		2500	.006		.77	.28	.37	1.42	1.68
3320	10,000 S.Y.	↓	10000	.002	↓	.62	.07	.09	.78	.89

32 01 13.64 Sand Seal		Crew	Daily Output	Labor-Hours	Unit	Material	Labor	Equipment	Total	Total Incl O&P
0010	**SAND SEAL**									
2080	Sand sealing, sharp sand, asphalt emulsion, small area	B-91	10000	.006	S.Y.	1.47	.28	.24	1.99	2.30
2120	Roadway or large area	"	18000	.004	"	1.26	.15	.13	1.54	1.76
3000	Sealing random cracks, min 1/2" wide, to 1-1/2", 1,000 L.F.	B-77	2800	.014	L.F.	1.50	.53	.21	2.24	2.70
3040	10,000 L.F.		4000	.010	"	1.04	.37	.15	1.56	1.88
3080	Alternate method, 1,000 L.F.		200	.200	Gal.	39.50	7.40	3.01	49.91	57.50
3120	10,000 L.F.	↓	325	.123	"	31.50	4.56	1.85	37.91	43.50
3200	Multi-cracks (flooding), 1 coat, small area	B-92	460	.070	S.Y.	2.32	2.58	1.39	6.29	8.05
3240	Large area		2850	.011		2.06	.42	.22	2.70	3.16
3280	2 coat, small area		230	.139		14.40	5.15	2.78	22.33	27
3320	Large area		1425	.022	↓	13.05	.83	.45	14.33	16.15
3360	Alternate method, small area		115	.278	Gal.	14.40	10.35	5.55	30.30	38
3400	Large area	↓	715	.045	"	13.50	1.66	.89	16.05	18.40

32 01 13.66 Fog Seal		Crew	Daily Output	Labor-Hours	Unit	Material	Labor	Equipment	Total	Total Incl O&P
0010	**FOG SEAL**									
0012	Sealcoating, 2 coat coal tar pitch emulsion over 10,000 S.Y.	B-45	5000	.003	S.Y.	1.29	.14	.18	1.61	1.83
0030	1000 to 10,000 S.Y.	"	3000	.005		1.29	.23	.31	1.83	2.11
0100	Under 1000 S.Y.	B-1	1050	.023		1.29	.85		2.14	2.74
0300	Petroleum resistant, over 10,000 S.Y.	B-45	5000	.003		1.29	.14	.18	1.61	1.83
0320	1000 to 10,000 S.Y.	"	3000	.005		1.29	.23	.31	1.83	2.11
0400	Under 1000 S.Y.	B-1	1050	.023		1.29	.85		2.14	2.74
0600	Non-skid pavement renewal, over 10,000 S.Y.	B-45	5000	.003		1.36	.14	.18	1.68	1.91
0620	1000 to 10,000 S.Y.	"	3000	.005		1.36	.23	.31	1.90	2.19
0700	Under 1000 S.Y.	B-1	1050	.023		1.36	.85		2.21	2.82
0800	Prepare and clean surface for above	A-2	8545	.003	↓		.10	.03	.13	.19
1000	Hand seal asphalt curbing	B-1	4420	.005	L.F.	.61	.20		.81	.98
1900	Asphalt surface treatment, single course, small area									
1901	0.30 gal/S.Y. asphalt material, 20#/S.Y. aggregate	B-91	5000	.013	S.Y.	1.29	.55	.47	2.31	2.78
1910	Roadway or large area		10000	.006		1.19	.28	.24	1.71	1.99
1950	Asphalt surface treatment, dbl. course for small area		3000	.021		2.91	.92	.78	4.61	5.45
1960	Roadway or large area		6000	.011		2.62	.46	.39	3.47	4.01
1980	Asphalt surface treatment, single course, for shoulders	↓	7500	.009	↓	1.45	.37	.31	2.13	2.51

32 01 13 – Flexible Paving Surface Treatment

32 01 13.68 Slurry Seal

	32 01 13.68 Slurry Seal	Crew	Daily Output	Labor-Hours	Unit	Material	2014 Bare Costs Labor	Equipment	Total	Total Incl O&P
0010	**SLURRY SEAL**									
0100	Slurry seal, type I, 8 lb. agg./S.Y., 1 coat, small or irregular area	B-90	2800	.023	S.Y.	1.84	.91	.79	3.54	4.28
0150	Roadway or large area		10000	.006		1.84	.25	.22	2.31	2.65
0200	Type II, 12 lb. aggregate/S.Y., 2 coats,small or irregular area		2000	.032		3.68	1.27	1.11	6.06	7.20
0250	Roadway or large area		8000	.008		3.68	.32	.28	4.28	4.85
0300	Type III, 20 lb. aggregate/S.Y., 2 coats, small or irregular area		1800	.036		4.34	1.41	1.23	6.98	8.30
0350	Roadway or large area		6000	.011		4.34	.42	.37	5.13	5.85
0400	Slurry seal, thermoplastic coal-tar, type I, small or irregular area		2400	.027		3.73	1.06	.93	5.72	6.75
0450	Roadway or large area		8000	.008		3.73	.32	.28	4.33	4.90
0500	Type II, small or irregular area		2400	.027		4.77	1.06	.93	6.76	7.90
0550	Roadway or large area		7800	.008		4.77	.33	.28	5.38	6.05

32 01 16 – Flexible Paving Rehabilitation

32 01 16.71 Cold Milling Asphalt Paving

	32 01 16.71 Cold Milling Asphalt Paving	Crew	Daily Output	Labor-Hours	Unit	Material	Labor	Equipment	Total	Total Incl O&P
0010	**COLD MILLING ASPHALT PAVING**									
5200	Cold planing & cleaning, 1″ to 3″ asphalt pavmt., over 25,000 S.Y.	B-71	6000	.009	S.Y.		.39	1.15	1.54	1.87
5280	5,000 S.Y. to 10,000 S.Y.	″	4000	.014	″		.59	1.73	2.32	2.80
5300	Asphalt pavement removal from conc. base, no haul									
5320	Rip, load & sweep 1″ to 3″	B-70	8000	.007	S.Y.		.30	.23	.53	.70
5330	3″ to 6″ deep	″	5000	.011			.47	.37	.84	1.12
5340	Profile grooving, asphalt pavement load & sweep, 1″ deep	B-71	12500	.004			.19	.55	.74	.90
5350	3″ deep		9000	.006			.26	.77	1.03	1.25
5360	6″ deep		5000	.011			.47	1.38	1.85	2.24
5400	Mixing material in windrow, 180 H.P. grader	B-11L	9400	.002	C.Y.		.07	.08	.15	.19
5450	For cold laid asphalt pavement, see Section 32 12 16.19									

32 01 16.73 In Place Cold Reused Asphalt Paving

	32 01 16.73 In Place Cold Reused Asphalt Paving	Crew	Daily Output	Labor-Hours	Unit	Material	Labor	Equipment	Total	Total Incl O&P
0010	**IN PLACE COLD REUSED ASPHALT PAVING**									
5000	Reclamation, pulverizing and blending with existing base									
5040	Aggregate base, 4″ thick pavement, over 15,000 S.Y.	B-73	2400	.027	S.Y.		1.19	2.16	3.35	4.18
5080	5,000 S.Y. to 15,000 S.Y.		2200	.029			1.30	2.35	3.65	4.57
5120	8″ thick pavement, over 15,000 S.Y.		2200	.029			1.30	2.35	3.65	4.57
5160	5,000 S.Y. to 15,000 S.Y.		2000	.032			1.43	2.59	4.02	5

32 01 16.74 In Place Hot Reused Asphalt Paving

	32 01 16.74 In Place Hot Reused Asphalt Paving	Crew	Daily Output	Labor-Hours	Unit	Material	Labor	Equipment	Total	Total Incl O&P
0010	**IN PLACE HOT REUSED ASPHALT PAVING**									
5500	Recycle asphalt pavement at site									
5520	Remove, rejuvenate and spread 4″ deep [G]	B-72	2500	.026	S.Y.	4.59	1.10	4.60	10.29	11.80
5521	6″ deep [G]	″	2000	.032	″	6.75	1.38	5.75	13.88	15.80

32 01 17 – Flexible Paving Repair

32 01 17.10 Repair of Asphalt Pavement Holes

	32 01 17.10 Repair of Asphalt Pavement Holes	Crew	Daily Output	Labor-Hours	Unit	Material	Labor	Equipment	Total	Total Incl O&P
0010	**REPAIR OF ASPHALT PAVEMENT HOLES** (cold patch)									
0100	Flexible pavement repair holes, roadway,light traffic, 1 C.F. size	B-37A	24	1	Ea.	9.05	36.50	15.60	61.15	83.50
0150	Group of two, 1 C.F. size each		16	1.500	Set	18.05	55	23.50	96.55	130
0200	Group of three, 1 C.F. size each		12	2	″	27	73	31	131	178
0300	Medium traffic, 1 C.F. size each	B-37B	24	1.333	Ea.	9.05	49	15.60	73.65	102
0350	Group of two, 1 C.F. size each		16	2	Set	18.05	73	23.50	114.55	159
0400	Group of three, 1 C.F. size each		12	2.667	″	27	97.50	31	155.50	215
0500	Highway/heavy traffic, 1 C.F. size each	B-37C	24	1.333	Ea.	9.05	49	27	85.05	114
0550	Group of two, 1 C.F. size each		16	2	Set	18.05	73	40.50	131.55	176
0600	Group of three, 1 C.F. size each		12	2.667	″	27	97.50	54	178.50	239
0700	Add police officer and car for traffic control				Hr.	45.50			45.50	50
1000	Flexible pavement repair holes,parking lot, bag material, 1 C.F. size each	B-37D	18	.889	Ea.	64.50	32.50	8.70	105.70	131

32 01 17.10 Repair of Asphalt Pavement Holes		Crew	Daily Output	Labor-Hours	Unit	Material	Labor	Equipment	Total	Total Incl O&P
						2014 Bare Costs				
1010	Economy bag material, 1 C.F. each	B-37D	18	.889	Ea.	43	32.50	8.70	84.20	107
1100	Group of two, 1 C.F. size each		12	1.333	Set	129	49	13.10	191.10	231
1110	Group of two, economy bag, 1 C.F. size each		12	1.333		86	49	13.10	148.10	184
1200	Group of three 1 C.F. size each		10	1.600		194	58.50	15.70	268.20	320
1210	Economy bag, group of three, 1 C.F. size each		10	1.600		129	58.50	15.70	203.20	249
1300	Flexible pavement repair holes, parking lot, 1 C.F. single hole	A-3A	4	2	Ea.	64.50	73	42	179.50	229
1310	Economy material, 1 C.F. single hole		4	2	"	43	73	42	158	205
1400	Flexible pavement repair holes, parking lot, 1 C.F. four holes		2	4	Set	259	146	84	489	600
1410	Economy material, 1 C.F. four holes		2	4	"	172	146	84	402	505
1500	Flexible pavement repair holes,large parking lot, bulk matl , 1 C.F. size	B-37A	32	.750	Ea.	9.05	27.50	11.70	48.25	65

32 01 17.20 Repair of Asphalt Pavement Patches		Crew	Daily Output	Labor-Hours	Unit	Material	Labor	Equipment	Total	Total Incl O&P
0010	**REPAIR OF ASPHALT PAVEMENT PATCHES**									
0100	Flexible pavement patches, roadway,light traffic,sawcut 10-25 S.F.	B-89	12	1.333	Ea.		55.50	44	99.50	133
0150	Sawcut 26-60 S.F.		10	1.600			67	52.50	119.50	160
0200	Sawcut 61-100 S.F.		8	2			83.50	65.50	149	200
0210	Sawcut groups of small size patches		24	.667			28	22	50	66.50
0220	Large size patches		16	1			42	33	75	99.50
0300	Flexible pavement patches, roadway,light traffic,digout 10-25 S.F.	B-6	16	1.500			60	23	83	117
0350	Digout 26-60 S.F.		12	2			80	30.50	110.50	157
0400	Digout 61-100 S.F.		8	3			120	45.50	165.50	234
0450	Add 8 C.Y. truck, small project debris haulaway	B-34A	8	1	Hr.		37.50	52	89.50	115
0460	Add 12 C.Y. truck, small project debris haulaway	B-34B	8	1			37.50	86.50	124	153
0480	Add flagger for non-intersection medium traffic	1 Clab	8	1			36.50		36.50	56.50
0490	Add flasher truck for intersection medium traffic or heavy traffic	A-2B	8	1			36.50	34	70.50	92.50
0500	Flexible pavement patches, roadway, repave,cold, 15 S.F., 4" D	B-37	8	6	Ea.	43	232	19.85	294.85	425
0510	6" depth		8	6		65	232	19.85	316.85	450
0520	Repave,cold, 20 S.F., 4" depth		8	6		57.50	232	19.85	309.35	440
0530	6" depth		8	6		87	232	19.85	338.85	475
0540	Repave,cold, 25 S.F., 4" depth		8	6		71.50	232	19.85	323.35	455
0550	6" depth		8	6		108	232	19.85	359.85	495
0600	Repave,cold, 30 S.F., 4" depth		8	6		86	232	19.85	337.85	470
0610	6" depth		8	6		130	232	19.85	381.85	520
0640	Repave,cold, 40 S.F., 4" depth		8	6		114	232	19.85	365.85	505
0650	6" depth		8	6		173	232	19.85	424.85	570
0680	Repave,cold, 50 S.F., 4" depth		8	6		143	232	19.85	394.85	535
0690	6" depth		8	6		217	232	19.85	468.85	615
0720	Repave,cold, 60 S.F., 4" depth		8	6		172	232	19.85	423.85	565
0730	6" depth		8	6		260	232	19.85	511.85	665
0800	Repave,cold, 70 S.F., 4" depth		7	6.857		200	266	22.50	488.50	655
0810	6" depth		7	6.857		305	266	22.50	593.50	770
0820	Repave,cold, 75 S.F., 4" depth		7	6.857		213	266	22.50	501.50	670
0830	6" depth		7	6.857		325	266	22.50	613.50	795
0900	Repave,cold, 80 S.F., 4" depth		6	8		229	310	26.50	565.50	755
0910	6" depth		6	8		345	310	26.50	681.50	885
0940	Repave,cold, 90 S.F., 4" depth		6	8		257	310	26.50	593.50	785
0950	6" depth		6	8		390	310	26.50	726.50	935
0980	Repave,cold, 100 S.F., 4" depth		6	8		286	310	26.50	622.50	820
0990	6" depth		6	8		435	310	26.50	771.50	980
1000	Add flasher truck for paving operations in medium or heavy traffic	A-2B	8	1	Hr.		36.50	34	70.50	92.50
1100	Prime coat for repair 15-40 S.F., 25% overspray	B-37A	48	.500	Ea.	7	18.30	7.80	33.10	44.50
1150	41-60 S.F.		40	.600		14	22	9.35	45.35	59.50
1175	61-80 S.F.		32	.750		18.95	27.50	11.70	58.15	76

32 01 Operation and Maintenance of Exterior Improvements

32 01 17 – Flexible Paving Repair

32 01 17.20 Repair of Asphalt Pavement Patches		Crew	Daily Output	Labor-Hours	Unit	2014 Bare Costs Material	Labor	Equipment	Total	Total Incl O&P
1200	81-100 S.F.	B-37A	32	.750	Ea.	23.50	27.50	11.70	62.70	80.50
1210	Groups of patches w/25% overspray		3600	.007	S.F.	.23	.24	.10	.57	.75
1300	Flexible pavement repair patches,street repave,hot, 60 S.F., 4″ D	B-37	8	6	Ea.	100	232	19.85	351.85	485
1310	6″ depth		8	6		151	232	19.85	402.85	545
1320	Repave,hot, 70 S.F., 4″ depth		7	6.857		116	266	22.50	404.50	565
1330	6″ depth		7	6.857		176	266	22.50	464.50	630
1340	Repave,hot, 75 S.F., 4″ depth		7	6.857		124	266	22.50	412.50	570
1350	6″ depth		7	6.857		189	266	22.50	477.50	645
1360	Repave,hot, 80 S.F., 4″ depth		6	8		133	310	26.50	469.50	650
1370	6″ depth		6	8		202	310	26.50	538.50	725
1380	Repave,hot, 90 S.F., 4″ depth		6	8		150	310	26.50	486.50	670
1390	6″ depth		6	8		227	310	26.50	563.50	755
1400	Repave,hot, 100 S.F., 4″ depth		6	8		166	310	26.50	502.50	685
1410	6″ depth		6	8		252	310	26.50	588.50	780
1420	Pave hot groups of patches, 4″ depth		900	.053	S.F.	1.66	2.06	.18	3.90	5.20
1430	6″ depth		900	.053	″	2.52	2.06	.18	4.76	6.15
1500	Add 8 C.Y. truck for hot asphalt paving operations	B-34A	1	8	Day		300	415	715	920
1550	Add flasher truck for hot paving operations in med/heavy traffic	A-2B	8	1	Hr.		36.50	34	70.50	92.50
2000	Add police officer and car for traffic control				″	45.50			45.50	50

32 01 17.61 Sealing Cracks In Asphalt Paving		Crew	Daily Output	Labor-Hours	Unit	Material	Labor	Equipment	Total	Total Incl O&P
0010	**SEALING CRACKS IN ASPHALT PAVING**									
0100	Sealing cracks in asphalt paving, 1/8″ wide x 1/2″ depth, slow set	B-37A	2000	.012	L.F.	.17	.44	.19	.80	1.08
0110	1/4″ wide x 1/2″ depth		2000	.012		.20	.44	.19	.83	1.11
0130	3/8″ wide x 1/2″ depth		2000	.012		.22	.44	.19	.85	1.14
0140	1/2″ wide x 1/2″ depth		1800	.013		.25	.49	.21	.95	1.26
0150	3/4″ wide x 1/2″ depth		1600	.015		.31	.55	.23	1.09	1.44
0160	1″ wide x 1/2″ depth		1600	.015		.36	.55	.23	1.14	1.50
0165	1/8″ wide x 1″ depth, rapid set	B-37F	2000	.016		.25	.59	.30	1.14	1.51
0170	1/4″ wide x 1″ depth		2000	.016		.48	.59	.30	1.37	1.77
0175	3/8″ wide x 1″ depth		1800	.018		.73	.65	.34	1.72	2.17
0180	1/2″ wide x 1″ depth		1800	.018		.97	.65	.34	1.96	2.44
0185	5/8″ wide x 1″ depth		1800	.018		1.21	.65	.34	2.20	2.70
0190	3/4″ wide x 1″ depth		1600	.020		1.46	.73	.38	2.57	3.15
0195	1″ wide x 1″ depth		1600	.020		1.94	.73	.38	3.05	3.68
0200	1/8″ wide x 2″ depth, rapid set		2000	.016		.48	.59	.30	1.37	1.77
0210	1/4″ wide x 2″ depth		2000	.016		.97	.59	.30	1.86	2.31
0220	3/8″ wide x 2″ depth		1800	.018		1.46	.65	.34	2.45	2.97
0230	1/2″ wide x 2″ depth		1800	.018		1.94	.65	.34	2.93	3.50
0240	5/8″ wide x 2″ depth		1800	.018		2.43	.65	.34	3.42	4.04
0250	3/4″ wide x 2″ depth		1600	.020		2.91	.73	.38	4.02	4.75
0260	1″ wide x 2″ depth		1600	.020		3.88	.73	.38	4.99	5.80
0300	Add flagger for non-intersection medium traffic	1 Clab	8	1	Hr.		36.50		36.50	56.50
0400	Add flasher truck for intersection medium traffic or heavy traffic	A-2B	8	1	″		36.50	34	70.50	92.50

32 01 29 – Rigid Paving Repair

32 01 29.61 Partial Depth Patching of Rigid Pavement		Crew	Daily Output	Labor-Hours	Unit	Material	Labor	Equipment	Total	Total Incl O&P
0010	**PARTIAL DEPTH PATCHING OF RIGID PAVEMENT**									
0100	Rigid pavement repair, roadway,light traffic, 25% pitting 15 S.F.	B-37F	16	2	Ea.	27.50	73	38	138.50	185
0110	Pitting 20 S.F.		16	2		36.50	73	38	147.50	195
0120	Pitting 25 S.F.		16	2		45.50	73	38	156.50	205
0130	Pitting 30 S.F.		16	2		54.50	73	38	165.50	215
0140	Pitting 40 S.F.		16	2		73	73	38	184	235
0150	Pitting 50 S.F.		12	2.667		91	97.50	51	239.50	305

32 01 29.61 Partial Depth Patching of Rigid Pavement		Crew	Daily Output	Labor-Hours	Unit	2014 Bare Costs Material	Labor	Equipment	Total	Total Incl O&P
0160	Pitting 60 S.F.	B-37F	12	2.667	Ea.	109	97.50	51	257.50	325
0170	Pitting 70 S.F.		12	2.667		128	97.50	51	276.50	345
0180	Pitting 75 S.F.		12	2.667		137	97.50	51	285.50	355
0190	Pitting 80 S.F.		8	4		146	146	76	368	470
0200	Pitting 90 S.F.		8	4		164	146	76	386	490
0210	Pitting 100 S.F.		8	4		182	146	76	404	510
0300	Roadway,light traffic, 50% pitting 15 S.F.		12	2.667		54.50	97.50	51	203	266
0310	Pitting 20 S.F.		12	2.667		98	97.50	51	246.50	315
0320	Pitting 25 S.F.		12	2.667		91	97.50	51	239.50	305
0330	Pitting 30 S.F.		12	2.667		109	97.50	51	257.50	325
0340	Pitting 40 S.F.		12	2.667		146	97.50	51	294.50	365
0350	Pitting 50 S.F.		8	4		182	146	76	404	510
0360	Pitting 60 S.F.		8	4		219	146	76	441	550
0370	Pitting 70 S.F.		8	4		255	146	76	477	590
0380	Pitting 75 S.F.		8	4		273	146	76	495	610
0390	Pitting 80 S.F.		6	5.333		292	195	101	588	730
0400	Pitting 90 S.F.		6	5.333		330	195	101	626	770
0410	Pitting 100 S.F.		6	5.333		365	195	101	661	810
1000	Rigid pavement repair,light traffic,surface patch, 2" deep, 15 S.F.		8	4		109	146	76	331	430
1010	Surface patch, 2" deep, 20 S.F.		8	4		146	146	76	368	470
1020	Surface patch, 2" deep, 25 S.F.		8	4		182	146	76	404	510
1030	Surface patch, 2" deep, 30 S.F.		8	4		219	146	76	441	550
1040	Surface patch, 2" deep, 40 S.F.		8	4		292	146	76	514	630
1050	Surface patch, 2" deep, 50 S.F.		6	5.333		365	195	101	661	810
1060	Surface patch, 2" deep, 60 S.F.		6	5.333		440	195	101	736	890
1070	Surface patch, 2" deep, 70 S.F.		6	5.333		510	195	101	806	970
1080	Surface patch, 2" deep, 75 S.F.		6	5.333		545	195	101	841	1,000
1090	Surface patch, 2" deep, 80 S.F.		5	6.400		585	234	122	941	1,125
1100	Surface patch, 2" deep, 90 S.F.		5	6.400		655	234	122	1,011	1,225
1200	Surface patch, 2" deep, 100 S.F.		5	6.400		730	234	122	1,086	1,300
2000	Add flagger for non-intersection medium traffic	1 Clab	8	1	Hr.		36.50		36.50	56.50
2100	Add flasher truck for intersection medium traffic or heavy traffic	A-2B	8	1			36.50	34	70.50	92.50
2200	Add police officer and car for traffic control					45.50			45.50	50

32 01 29.70 Full Depth Patching of Rigid Pavement		Crew	Daily Output	Labor-Hours	Unit	Material	Labor	Equipment	Total	Total Incl O&P
0010	**FULL DEPTH PATCHING OF RIGID PAVEMENT**									
0015	Pavement preparation includes sawcut, remove pavement and replace									
0020	6 inches of subbase and one layer of reinforcement									
0030	Trucking and haulaway of debris excluded									
0100	Rigid pavement replace, light traffic,prep, 15 S.F.,6" depth	B-37E	16	3.500	Ea.	21.50	139	67.50	228	310
0110	Replacement preparation 20 S.F., 6" depth		16	3.500		28.50	139	67.50	235	320
0115	25 S.F., 6" depth		16	3.500		35.50	139	67.50	242	325
0120	30 S.F., 6" depth		14	4		42.50	159	77	278.50	375
0125	35 S.F., 6" depth		14	4		50	159	77	286	385
0130	40 S.F., 6" depth		14	4		57	159	77	293	390
0135	45 S.F., 6" depth		14	4		64	159	77	300	400
0140	50 S.F., 6" depth		12	4.667		71	186	90	347	460
0145	55 S.F., 6" depth		12	4.667		78.50	186	90	354.50	470
0150	60 S.F., 6" depth		10	5.600		85.50	223	108	416.50	555
0155	65 S.F., 6" depth		10	5.600		92.50	223	108	423.50	560
0160	70 S.F., 6" depth		10	5.600		99.50	223	108	430.50	570
0165	75 S.F., 6" depth		10	5.600		107	223	108	438	575
0170	80 S.F., 6" depth		8	7		114	279	135	528	700

32 01 29 – Rigid Paving Repair

32 01 29.70 Full Depth Patching of Rigid Pavement		Crew	Daily Output	Labor-Hours	Unit	2014 Bare Costs Material	Labor	Equipment	Total	Total Incl O&P
0175	85 S.F., 6" depth	B-37E	8	7	Ea.	121	279	135	535	705
0180	90 S.F., 6" depth		8	7		128	279	135	542	715
0185	95 S.F., 6" depth		8	7		135	279	135	549	725
0190	100 S.F., 6" depth		8	7		142	279	135	556	730
0200	Pavement preparation for 8 inch rigid paving same as 6 inch									
0290	Pavement preparation for 9 inch rigid paving same as 10 inch									
0300	Rigid pavement replace, light traffic,prep, 15 S.F.,10" depth	B-37E	16	3.500	Ea.	35	139	67.50	241.50	325
0302	Pavement preparation 10 inch includes sawcut, remove pavement and									
0304	replace 6 inches of subbase and two layers of reinforcement									
0306	Trucking and haulaway of debris excluded									
0310	Replacement preparation 20 S.F., 10" depth	B-37E	16	3.500	Ea.	46.50	139	67.50	253	340
0315	25 S.F., 10" depth		16	3.500		58	139	67.50	264.50	350
0320	30 S.F., 10" depth		14	4		69.50	159	77	305.50	405
0325	35 S.F., 10" depth		14	4		54.50	159	77	290.50	390
0330	40 S.F., 10" depth		14	4		93	159	77	329	430
0335	45 S.F., 10" depth		12	4.667		105	186	90	381	500
0340	50 S.F., 10" depth		12	4.667		116	186	90	392	510
0345	55 S.F., 10" depth		12	4.667		128	186	90	404	525
0350	60 S.F., 10" depth		10	5.600		139	223	108	470	610
0355	65 S.F., 10" depth		10	5.600		151	223	108	482	625
0360	70 S.F., 10" depth		10	5.600		163	223	108	494	640
0365	75 S.F., 10" depth		8	7		174	279	135	588	765
0370	80 S.F., 10" depth		8	7		186	279	135	600	780
0375	85 S.F., 10" depth		8	7		198	279	135	612	790
0380	90 S.F., 10" depth		8	7		209	279	135	623	805
0385	95 S.F., 10" depth		8	7		221	279	135	635	815
0390	100 S.F., 10" depth		8	7		232	279	135	646	830
0500	Add 8 C.Y. truck, small project debris haulaway	B-34A	8	1	Hr.		37.50	52	89.50	115
0550	Add 12 C.Y. truck, small project debris haulaway	B-34B	8	1			37.50	86.50	124	153
0600	Add flagger for non-intersection medium traffic	1 Clab	8	1			36.50		36.50	56.50
0650	Add flasher truck for intersection medium traffic or heavy traffic	A-2B	8	1			36.50	34	70.50	92.50
0700	Add police officer and car for traffic control					45.50			45.50	50
0990	Concrete will replaced using quick set mix with aggregate									
1000	Rigid pavement replace, light traffic,repour, 15 S.F.,6" depth	B-37F	18	1.778	Ea.	163	65	34	262	315
1010	20 S.F., 6" depth		18	1.778		217	65	34	316	375
1015	25 S.F., 6" depth		18	1.778		272	65	34	371	435
1020	30 S.F., 6" depth		16	2		325	73	38	436	515
1025	35 S.F., 6" depth		16	2		380	73	38	491	575
1030	40 S.F., 6" depth		16	2		435	73	38	546	635
1035	45 S.F., 6" depth		16	2		490	73	38	601	695
1040	50 S.F., 6" depth		16	2		545	73	38	656	750
1045	55 S.F., 6" depth		12	2.667		595	97.50	51	743.50	860
1050	60 S.F., 6" depth		12	2.667		650	97.50	51	798.50	920
1055	65 S.F., 6" depth		12	2.667		705	97.50	51	853.50	980
1060	70 S.F., 6" depth		12	2.667		760	97.50	51	908.50	1,050
1065	75 S.F., 6" depth		10	3.200		815	117	61	993	1,150
1070	80 S.F., 6" depth		10	3.200		870	117	61	1,048	1,200
1075	85 S.F., 6" depth		8	4		925	146	76	1,147	1,325
1080	90 S.F., 6" depth		8	4		980	146	76	1,202	1,375
1085	95 S.F., 6" depth		8	4		1,025	146	76	1,247	1,425
1090	100 S.F., 6" depth		8	4		1,075	146	76	1,297	1,500
1099	Concrete will be replaced using 4500 PSI Concrete ready mix									
1100	Rigid pavement replace, light traffic,repour, 15 S.F.,6" depth	A-2	12	2	Ea.	289	73	22.50	384.50	460

32 01 29 – Rigid Paving Repair

	32 01 29.70 Full Depth Patching of Rigid Pavement	Crew	Daily Output	Labor-Hours	Unit	2014 Bare Costs Material	Labor	Equipment	Total	Total Incl O&P
1110	20-50 S.F., 6″ depth	A-2	12	2	Ea.	289	73	22.50	384.50	460
1120	55-65 S.F., 6″ depth		10	2.400		320	88	27	435	515
1130	70-80 S.F., 6″ depth		10	2.400		345	88	27	460	545
1140	85-100 S.F., 6″ depth		8	3		405	110	34	549	650
1200	Repour 15-40 S.F.,8″ depth		12	2		289	73	22.50	384.50	460
1210	45-50 S.F.,8″ depth		12	2		320	73	22.50	415.50	490
1220	55-60 S.F.,8″ depth		10	2.400		345	88	27	460	545
1230	65-80 S.F.,8″ depth		10	2.400		405	88	27	520	610
1240	85-90 S.F.,8″ depth		8	3		435	110	34	579	685
1250	95-100 S.F.,8″ depth		8	3		460	110	34	604	715
1300	Repour 15-30 S.F.,10″ depth		12	2		289	73	22.50	384.50	460
1310	35-40 S.F.,10″ depth		12	2		320	73	22.50	415.50	490
1320	45 S.F.,10″ depth		12	2		345	73	22.50	440.50	520
1330	50-65 S.F.,10″ depth		10	2.400		405	88	27	520	610
1340	70 S.F.,10″ depth		10	2.400		435	88	27	550	645
1350	75-80 S.F.,10″ depth		10	2.400		460	88	27	575	675
1360	85-95 S.F.,10″ depth		8	3		520	110	34	664	775
1370	100 S.F.,10″ depth		8	3		550	110	34	694	810

32 01 30 – Operation and Maintenance of Site Improvements

	32 01 30.10 Site Maintenance	Crew	Daily Output	Labor-Hours	Unit	Material	Labor	Equipment	Total	Total Incl O&P
0010	**SITE MAINTENANCE**									
1550	General site work maintenance									
1560	Clearing brush with brush saw & rake	1 Clab	565	.014	S.Y.		.52		.52	.80
1570	By hand	"	280	.029			1.05		1.05	1.62
1580	With dozer, ball and chain, light clearing	B-11A	3675	.004			.19	.36	.55	.68
1590	Medium clearing	"	3110	.005			.22	.43	.65	.81
3000	Lawn maintenance									
3010	Aerate lawn, 18″ cultivating width, walk behind	A-1K	95	.084	M.S.F.		3.09	.86	3.95	5.70
3040	48″ cultivating width	B-66	750	.011			.50	.35	.85	1.14
3060	72″ cultivating width	"	1100	.007			.34	.24	.58	.78
3100	Edge lawn, by hand at walks	1 Clab	16	.500	C.L.F.		18.35		18.35	28.50
3150	At planting beds		7	1.143			42		42	64.50
3200	Using gas powered edger at walks		88	.091			3.33		3.33	5.15
3250	At planting beds		24	.333			12.20		12.20	18.85
3260	Vacuum, 30″ gas, outdoors with hose		96	.083	M.L.F.		3.05		3.05	4.71
3400	Weed lawn, by hand		3	2.667	M.S.F.		97.50		97.50	151
4500	Rake leaves or lawn, by hand		7.50	1.067			39		39	60.50
4510	Power rake		45	.178			6.50		6.50	10.05
4700	Seeding lawn, see Section 32 92 19.14									
4750	Sodding, see Section 32 92 23.10									
5900	Road & walk maintenance									
5915	De-icing roads and walks									
5920	Calcium Chloride in truckload lots see Section 31 32 13.30									
6000	Ice melting comp., 90% Calc. Chlor., effec. to -30°F									
6010	50-80 lb. poly bags, med. applic. 19 lb./M.S.F., by hand	1 Clab	60	.133	M.S.F.	18.20	4.89		23.09	27.50
6050	With hand operated rotary spreader		110	.073		18.20	2.67		20.87	24
6100	Rock salt, med. applic. on road & walkway, by hand		60	.133		4.50	4.89		9.39	12.50
6110	With hand operated rotary spreader		110	.073		4.50	2.67		7.17	9.05
6600	Shrub maintenance									
6640	Shrub bed fertilize dry granular 3 lb./M.S.F.	1 Clab	85	.094	M.S.F.	6.55	3.45		10	12.50
6800	Weed, by handhoe		8	1			36.50		36.50	56.50
6810	Spray out		32	.250			9.15		9.15	14.15

32 01 Operation and Maintenance of Exterior Improvements

32 01 30 – Operation and Maintenance of Site Improvements

32 01 30.10 Site Maintenance		Crew	Daily Output	Labor-Hours	Unit	Material	2014 Bare Costs Labor	Equipment	Total	Total Incl O&P
6820	Spray after mulch	1 Clab	48	.167	M.S.F.		6.10		6.10	9.45
7100	Tree maintenance									
7140	Clear and grub trees, see Section 31 11 10.10									
7160	Cutting and piling trees, see Section 31 13 13.20									
7200	Fertilize, tablets, slow release, 30 gram/tree	1 Clab	100	.080	Ea.	.57	2.93		3.50	5.15
7280	Guying, including stakes, guy wire & wrap, see Section 32 94 50.10									
7300	Planting, trees, Deciduous, in prep. beds, see Section 32 93 43.20 or .30									
7400	Removal, trees see Section 32 96 43.20									
7420	Pest control, spray	1 Clab	24	.333	Ea.	26	12.20		38.20	47.50
7430	Systemic	"	48	.167	"	26.50	6.10		32.60	38.50

32 01 90 – Operation and Maintenance of Planting

32 01 90.13 Fertilizing

Code	Description	Crew	Daily Output	Labor-Hours	Unit	Material	Labor	Equipment	Total	Total Incl O&P
0010	**FERTILIZING**									
0100	Dry granular, 4#/M.S.F., hand spread	1 Clab	24	.333	M.S.F.	2.75	12.20		14.95	22
0110	Push rotary		140	.057	"	2.75	2.09		4.84	6.25
0112	Push rotary, per 1076 feet squared		130	.062	Ea.	2.75	2.26		5.01	6.50
0120	Tractor towed spreader, 8′	B-66	500	.016	M.S.F.	2.75	.75	.52	4.02	4.74
0130	12′ spread		800	.010		2.75	.47	.32	3.54	4.10
0140	Truck whirlwind spreader		1200	.007		2.75	.31	.22	3.28	3.75
0180	Water soluable, hydro spread, 1.5#/M.S.F.	B-64	600	.027		2.82	.98	.72	4.52	5.40
0190	Add for weed control					.47			.47	.52

32 01 90.19 Mowing

Code	Description	Crew	Daily Output	Labor-Hours	Unit	Material	Labor	Equipment	Total	Total Incl O&P
0010	**MOWING**									
1650	Mowing brush, tractor with rotary mower									
1660	Light density	B-84	22	.364	M.S.F.		17.80	16.55	34.35	45
1670	Medium density		13	.615			30	28	58	76.50
1680	Heavy density		9	.889			43.50	40.50	84	111
2000	Mowing, brush/grass, tractor, rotary mower, highway/airport median		13	.615			30	28	58	76.50
2010	Traffic safety flashing truck for highway/airport median mowing	A-2B	1	8	Day		292	270	562	740
4050	Lawn mowing, power mower, 18″ - 22″	1 Clab	65	.123	M.S.F.		4.51		4.51	6.95
4100	22″ - 30″		110	.073			2.67		2.67	4.11
4150	30″ - 32″		140	.057			2.09		2.09	3.23
4160	Riding mower, 36″ - 44″	B-66	300	.027			1.25	.87	2.12	2.85
4170	48″ - 58″	"	480	.017			.78	.54	1.32	1.79
4175	Mowing with tractor & attachments									
4180	3 gang reel, 7′	B-66	930	.009	M.S.F.		.40	.28	.68	.92
4190	5 gang reel, 12′		1200	.007			.31	.22	.53	.72
4200	Cutter or sickle-bar, 5′, rough terrain		210	.038			1.79	1.24	3.03	4.08
4210	Cutter or sickle-bar, 5′, smooth terrain		340	.024			1.11	.76	1.87	2.52
4220	Drainage channel, 5′ sickle bar		5	1.600	Mile		75.50	52	127.50	171
4250	Lawnmower, rotary type, sharpen (all sizes)	1 Clab	10	.800	Ea.		29.50		29.50	45
4260	Repair or replace part		7	1.143	"		42		42	64.50
5000	Edge trimming with weed whacker		5760	.001	L.F.		.05		.05	.08

32 01 90.23 Pruning

Code	Description	Crew	Daily Output	Labor-Hours	Unit	Material	Labor	Equipment	Total	Total Incl O&P
0010	**PRUNING**									
0020	1-1/2″ caliper	1 Clab	84	.095	Ea.		3.49		3.49	5.40
0030	2″ caliper		70	.114			4.19		4.19	6.45
0040	2-1/2″ caliper		50	.160			5.85		5.85	9.05
0050	3″ caliper		30	.267			9.75		9.75	15.10
0060	4″ caliper, by hand	2 Clab	21	.762			28		28	43
0070	Aerial lift equipment	B-85	38	1.053			41.50	27.50	69	94
0100	6″ caliper, by hand	2 Clab	12	1.333			49		49	75.50

32 01 Operation and Maintenance of Exterior Improvements

32 01 90 – Operation and Maintenance of Planting

32 01 90.23 Pruning		Crew	Daily Output	Labor-Hours	Unit	Material	2014 Bare Costs Labor	Equipment	Total	Total Incl O&P
0110	Aerial lift equipment	B-85	20	2	Ea.		78.50	52.50	131	178
0200	9″ caliper, by hand	2 Clab	7.50	2.133			78		78	121
0210	Aerial lift equipment	B-85	12.50	3.200			126	83.50	209.50	285
0300	12″ caliper, by hand	2 Clab	6.50	2.462			90		90	139
0310	Aerial lift equipment	B-85	10.80	3.704			145	97	242	330
0400	18″ caliper by hand	2 Clab	5.60	2.857			105		105	162
0410	Aerial lift equipment	B-85	9.30	4.301			169	112	281	385
0500	24″ caliper, by hand	2 Clab	4.60	3.478			127		127	197
0510	Aerial lift equipment	B-85	7.70	5.195			204	136	340	465
0600	30″ caliper, by hand	2 Clab	3.70	4.324			158		158	245
0610	Aerial lift equipment	B-85	6.20	6.452			253	169	422	575
0700	36″ caliper, by hand	2 Clab	2.70	5.926			217		217	335
0710	Aerial lift equipment	B-85	4.50	8.889			350	232	582	790
0800	48″ caliper, by hand	2 Clab	1.70	9.412			345		345	530
0810	Aerial lift equipment	B-85	2.80	14.286			560	375	935	1,275
32 01 90.24 Shrub Pruning										
0010	**SHRUB PRUNING**									
6700	Prune, shrub bed	1 Clab	7	1.143	M.S.F.		42		42	64.50
6710	Shrub under 3′ height		190	.042	Ea.		1.54		1.54	2.38
6720	4′ height		90	.089			3.26		3.26	5.05
6730	Over 6′		50	.160			5.85		5.85	9.05
7350	Prune trees from ground		20	.400			14.65		14.65	22.50
7360	High work		8	1			36.50		36.50	56.50
32 01 90.26 Watering										
0010	**WATERING**									
4900	Water lawn or planting bed with hose, 1″ of water	1 Clab	16	.500	M.S.F.		18.35		18.35	28.50
4910	50′ soaker hoses, in place		82	.098			3.58		3.58	5.50
4920	60′ soaker hoses, in place		89	.090			3.29		3.29	5.10
7500	Water trees or shrubs, under 1″ caliper		32	.250	Ea.		9.15		9.15	14.15
7550	1″ - 3″ caliper		17	.471			17.25		17.25	26.50
7600	3″ - 4″ caliper		12	.667			24.50		24.50	37.50
7650	Over 4″ caliper		10	.800			29.50		29.50	45
9000	For sprinkler irrigation systems, see Section 32 84 23.10									
32 01 90.29 Topsoil Preservation										
0010	**TOPSOIL PRESERVATION**									
0100	Weed planting bed	1 Clab	800	.010	S.Y.		.37		.37	.57

32 06 Schedules for Exterior Improvements

32 06 10 – Schedules for Bases, Ballasts, and Paving

32 06 10.10 Sidewalks, Driveways and Patios		Crew	Daily Output	Labor-Hours	Unit	Material	Labor	Equipment	Total	Total Incl O&P
0010	**SIDEWALKS, DRIVEWAYS AND PATIOS** No base									
0020	Asphaltic concrete, 2″ thick	B-37	720	.067	S.Y.	7.20	2.58	.22	10	12.10
0100	2-1/2″ thick	″	660	.073	″	9.15	2.82	.24	12.21	14.65
0110	Bedding for brick or stone, mortar, 1″ thick	D-1	300	.053	S.F.	.76	2.20		2.96	4.20
0120	2″ thick	″	200	.080		1.90	3.30		5.20	7.15
0130	Sand, 2″ thick	B-18	8000	.003		.26	.11	.01	.38	.47
0140	4″ thick	″	4000	.006		.52	.22	.01	.75	.94
0300	Concrete, 3000 psi, CIP, 6 x 6 - W1.4 x W1.4 mesh,									
0310	broomed finish, no base, 4″ thick	B-24	600	.040	S.F.	1.65	1.69		3.34	4.39
0350	5″ thick		545	.044		2.20	1.86		4.06	5.25

32 06 Schedules for Exterior Improvements

32 06 10 – Schedules for Bases, Ballasts, and Paving

32 06 10.10 Sidewalks, Driveways and Patios		Crew	Daily Output	Labor-Hours	Unit	Material	Labor	Equipment	Total	Total Incl O&P
						2014 Bare Costs				
0400	6″ thick	B-24	510	.047	S.F.	2.57	1.99		4.56	5.85
0450	For bank run gravel base, 4″ thick, add	B-18	2500	.010		.51	.36	.02	.89	1.13
0520	8″ thick, add	″	1600	.015		1.03	.56	.03	1.62	2.02
0550	Exposed aggregate finish, add to above, minimum	B-24	1875	.013		.10	.54		.64	.93
0600	Maximum		455	.053		.34	2.23		2.57	3.75
0700	Patterned surface, add to above min.		1200	.020			.84		.84	1.28
0950	Concrete tree grate, 5′ square	B-6	25	.960	Ea.	420	38.50	14.60	473.10	535
0955	Tree well & cover, concrete, 3′ square		25	.960		315	38.50	14.60	368.10	420
0960	Cast iron tree grate with frame, 2 piece, round, 5′ diameter		25	.960		1,100	38.50	14.60	1,153.10	1,275
0980	Square, 5′ side		25	.960		1,100	38.50	14.60	1,153.10	1,275
1000	Crushed stone, 1″ thick, white marble	2 Clab	1700	.009	S.F.	.40	.34		.74	.97
1050	Bluestone		1700	.009		.22	.34		.56	.77
1070	Granite chips		1700	.009		.22	.34		.56	.77
1200	For 2″ asphaltic conc. base and tack coat, add to above	B-37	7200	.007		1.05	.26	.02	1.33	1.57
1660	Limestone pavers, 3″ thick	D-1	72	.222		9.95	9.20		19.15	25
1670	4″ thick		70	.229		13.20	9.45		22.65	29
1680	5″ thick		68	.235		16.50	9.70		26.20	33
1700	Redwood, prefabricated, 4′ x 4′ sections	2 Carp	316	.051		4.77	2.32		7.09	8.85
1750	Redwood planks, 1″ thick, on sleepers	″	240	.067		4.77	3.06		7.83	9.95
1830	1-1/2″ thick	B-28	167	.144		7.55	6.15		13.70	17.80
1840	2″ thick		167	.144		9.75	6.15		15.90	20
1850	3″ thick		150	.160		14.40	6.85		21.25	26.50
1860	4″ thick		150	.160		18.85	6.85		25.70	31
1870	5″ thick		150	.160		24	6.85		30.85	37
2100	River or beach stone, stock	B-1	18	1.333	Ton	29.50	50		79.50	110
2150	Quarried	″	18	1.333	″	51.50	50		101.50	134
2160	Load, dump, and spread stone with skid steer, 100′ haul	B-62	24	1	C.Y.		40	7.25	47.25	69.50
2165	200′ haul		18	1.333			53.50	9.70	63.20	92.50
2168	300′ haul		12	2			80	14.50	94.50	139
2170	Shale paver, 2-1/4″ thick	D-1	200	.080	S.F.	3.10	3.30		6.40	8.45
2200	Coarse washed sand bed, 1″	B-62	1350	.018	S.Y.	1.75	.71	.13	2.59	3.16
2250	Stone dust, 4″ thick	″	900	.027	″	3.89	1.07	.19	5.15	6.10
2300	Tile thinset pavers, 3/8″ thick	D-1	300	.053	S.F.	3.50	2.20		5.70	7.20
2350	3/4″ thick	″	280	.057	″	5.60	2.36		7.96	9.75
2400	Wood rounds, cypress	B-1	175	.137	Ea.	9.85	5.10		14.95	18.75
8000	For temporary barricades, see Section 01 56 23.10									

32 06 10.20 Steps		Crew	Daily Output	Labor-Hours	Unit	Material	Labor	Equipment	Total	Total Incl O&P
0010	**STEPS**									
0011	Incl. excav., borrow & concrete base as required									
0100	Brick steps	B-24	35	.686	LF Riser	15	29		44	60.50
0200	Railroad ties	2 Clab	25	.640		3.55	23.50		27.05	40
0300	Bluestone treads, 12″ x 2″ or 12″ x 1-1/2″	B-24	30	.800		33	33.50		66.50	87.50
0500	Concrete, cast in place, see Section 03 30 53.40									
0600	Precast concrete, see Section 03 41 23.50									

32 11 Base Courses

32 11 23 – Aggregate Base Courses

32 11 23.23 Base Course Drainage Layers		Crew	Daily Output	Labor-Hours	Unit	2014 Bare Costs Material	Labor	Equipment	Total	Total Incl O&P
0010	**BASE COURSE DRAINAGE LAYERS**									
0011	For roadways and large areas									
0050	Crushed 3/4" stone base, compacted, 3" deep	B-36C	5200	.008	S.Y.	2.76	.34	.79	3.89	4.42
0100	6" deep		5000	.008		5.50	.36	.82	6.68	7.50
0200	9" deep		4600	.009		8.30	.39	.89	9.58	10.65
0300	12" deep		4200	.010		11.05	.42	.98	12.45	13.85
0301	Crushed 1-1/2" stone base, compacted to 4" deep	B-36B	6000	.011		3.80	.46	.79	5.05	5.75
0302	6" deep		5400	.012		5.70	.51	.88	7.09	8
0303	8" deep		4500	.014		7.60	.61	1.05	9.26	10.45
0304	12" deep		3800	.017		11.40	.73	1.24	13.37	15.05
0350	Bank run gravel, spread and compacted									
0370	6" deep	B-32	6000	.005	S.Y.	4.26	.24	.38	4.88	5.50
0390	9" deep		4900	.007		6.40	.30	.46	7.16	8
0400	12" deep		4200	.008		8.55	.35	.54	9.44	10.55
0600	Cold laid asphalt pavement, see Section 32 12 16.19									
1500	Alternate method to figure base course									
1510	Crushed stone, 3/4", compacted, 3" deep	B-36C	435	.092	E.C.Y.	28.50	4.10	9.45	42.05	47.50
1511	6" deep	B-36B	835	.077		28.50	3.31	5.65	37.46	42.50
1512	9" deep		1150	.056		28.50	2.40	4.11	35.01	39
1513	12" deep		1400	.046		28.50	1.97	3.38	33.85	37.50
1520	Crushed stone, 1-1/2", compacted 4" deep		665	.096		28.50	4.15	7.10	39.75	45
1521	6" deep		900	.071		28.50	3.07	5.25	36.82	41.50
1522	8" deep		1000	.064		28.50	2.76	4.73	35.99	40.50
1523	12" deep		1265	.051		28.50	2.18	3.74	34.42	38.50
1530	Gravel, bank run, compacted, 6" deep	B-36C	835	.048		22	2.14	4.91	29.05	32.50
1531	9" deep		1150	.035		22	1.55	3.57	27.12	30.50
1532	12" deep		1400	.029		22	1.27	2.93	26.20	29
2010	Crushed stone, 3/4" maximum size, 3" deep	B-36	540	.074	Ton	17.10	3.11	2.98	23.19	27
2011	6" deep		1625	.025		17.10	1.03	.99	19.12	21.50
2012	9" deep		1785	.022		17.10	.94	.90	18.94	21
2013	12" deep		1950	.021		17.10	.86	.82	18.78	21
2020	Crushed stone, 1-1/2" maximum size, 4" deep		720	.056		17.10	2.33	2.23	21.66	25
2021	6" deep		815	.049		17.10	2.06	1.97	21.13	24
2022	8" deep		835	.048		17.10	2.01	1.93	21.04	24
2023	12" deep		975	.041		17.10	1.72	1.65	20.47	23
2030	Bank run gravel, 6" deep	B-32A	875	.027		14.75	1.23	1.60	17.58	19.85
2031	9" deep		970	.025		14.75	1.11	1.44	17.30	19.50
2032	12" deep		1060	.023		14.75	1.01	1.32	17.08	19.20
6000	Stabilization fabric, polypropylene, 6 oz./S.Y.	B-6	10000	.002	S.Y.	1.27	.10	.04	1.41	1.59
6900	For small and irregular areas, add						50%	50%		
7000	Prepare and roll sub-base, small areas to 2500 S.Y.	B-32A	1500	.016	S.Y.		.72	.93	1.65	2.12
8000	Large areas over 2500 S.Y.	"	3500	.007			.31	.40	.71	.91
8050	For roadways	B-32	4000	.008			.37	.57	.94	1.19

32 11 26 – Asphaltic Base Courses

32 11 26.13 Plant Mix Asphaltic Base Courses		Crew	Daily Output	Labor-Hours	Unit	Material	Labor	Equipment	Total	Total Incl O&P
0010	**PLANT MIX ASPHALTIC BASE COURSES**									
0011	Roadways and large paved areas									
0500	Bituminous concrete, 4" thick	B-25	4545	.019	S.Y.	15.30	.78	.61	16.69	18.70
0550	6" thick		3700	.024		22.50	.96	.75	24.21	27
0560	8" thick		3000	.029		30	1.18	.92	32.10	36
0570	10" thick		2545	.035		37	1.39	1.08	39.47	44.50
1600	Macadam base, crushed stone or slag, dry-bound	B-36D	1400	.023	E.C.Y.	67	1.06	2.30	70.36	77.50

32 11 Base Courses

32 11 26 – Asphaltic Base Courses

	32 11 26.13 Plant Mix Asphaltic Base Courses	Crew	Daily Output	Labor-Hours	Unit	Material	2014 Bare Costs Labor	Equipment	Total	Total Incl O&P
1610	Water-bound	B-36C	1400	.029	E.C.Y.	67	1.27	2.93	71.20	78.50
2000	Alternate method to figure base course									
2005	Bituminous concrete, 4" thick	B-25	1000	.088	Ton	68	3.54	2.76	74.30	83.50
2006	6" thick		1220	.072		68	2.90	2.26	73.16	82
2007	8" thick		1320	.067		68	2.68	2.09	72.77	81.50
2008	10" thick		1400	.063		68	2.53	1.97	72.50	81
8900	For small and irregular areas, add						50%	50%		

	32 11 26.19 Bituminous-Stabilized Base Courses	Crew	Daily Output	Labor-Hours	Unit	Material	Labor	Equipment	Total	Total Incl O&P
0010	**BITUMINOUS-STABILIZED BASE COURSES**									
0020	And large paved areas									
0700	Liquid application to gravel base, asphalt emulsion	B-45	6000	.003	Gal.	4.25	.12	.15	4.52	5.05
0800	Prime and seal, cut back asphalt		6000	.003	"	5	.12	.15	5.27	5.85
1000	Macadam penetration crushed stone, 2 gal. per S.Y., 4" thick		6000	.003	S.Y.	8.50	.12	.15	8.77	9.70
1100	6" thick, 3 gal. per S.Y.		4000	.004		12.75	.17	.23	13.15	14.55
1200	8" thick, 4 gal. per S.Y.		3000	.005		17	.23	.31	17.54	19.40
8900	For small and irregular areas, add						50%	50%		

32 12 Flexible Paving

32 12 16 – Asphalt Paving

	32 12 16.13 Plant-Mix Asphalt Paving	Crew	Daily Output	Labor-Hours	Unit	Material	Labor	Equipment	Total	Total Incl O&P
0010	**PLANT-MIX ASPHALT PAVING**									
0020	And large paved areas with no hauling included									
0025	See Section 31 23 23.20 for hauling costs									
0080	Binder course, 1-1/2" thick	B-25	7725	.011	S.Y.	5.55	.46	.36	6.37	7.20
0120	2" thick		6345	.014		7.40	.56	.44	8.40	9.50
0130	2-1/2" thick		5620	.016		9.25	.63	.49	10.37	11.65
0160	3" thick		4905	.018		11.10	.72	.56	12.38	13.95
0170	3-1/2" thick		4520	.019		12.95	.78	.61	14.34	16.10
0200	4" thick		4140	.021		14.80	.85	.67	16.32	18.30
0300	Wearing course, 1" thick	B-25B	10575	.009		3.67	.37	.28	4.32	4.92
0340	1-1/2" thick		7725	.012		6.15	.51	.39	7.05	8
0380	2" thick		6345	.015		8.30	.62	.47	9.39	10.55
0420	2-1/2" thick		5480	.018		10.20	.72	.55	11.47	12.95
0460	3" thick		4900	.020		12.20	.80	.61	13.61	15.30
0470	3-1/2" thick		4520	.021		14.30	.87	.66	15.83	17.75
0480	4" thick		4140	.023		16.35	.95	.73	18.03	20
0500	Open graded friction course	B-25C	5000	.010		2.04	.39	.48	2.91	3.38
0800	Alternate method of figuring paving costs									
0810	Binder course, 1-1/2" thick	B-25	630	.140	Ton	68	5.60	4.38	77.98	88.50
0811	2" thick		690	.128		68	5.10	4	77.10	87.50
0812	3" thick		800	.110		68	4.42	3.45	75.87	85.50
0813	4" thick		900	.098		68	3.93	3.07	75	84.50
0850	Wearing course, 1" thick	B-25B	575	.167		68	6.85	5.20	80.05	91
0851	1-1/2" thick		630	.152		68	6.25	4.76	79.01	90
0852	2" thick		690	.139		68	5.70	4.35	78.05	88.50
0853	2-1/2" thick		765	.125		68	5.15	3.92	77.07	87
0854	3" thick		800	.120		68	4.91	3.75	76.66	86.50
1000	Pavement replacement over trench, 2" thick	B-37	90	.533	S.Y.	7.65	20.50	1.77	29.92	42.50
1050	4" thick		70	.686		15.10	26.50	2.27	43.87	60
1080	6" thick		55	.873		24	34	2.89	60.89	81.50
1100	Turnouts and driveway entrances to highways									

32 12 Flexible Paving

32 12 16 – Asphalt Paving

32 12 16.13	Plant-Mix Asphalt Paving	Crew	Daily Output	Labor-Hours	Unit	Material	2014 Bare Costs Labor	Equipment	Total	Total Incl O&P
1110	Binder course, 1-1/2" thick	B-25	315	.279	Ton	68	11.20	8.75	87.95	102
1120	2" thick		345	.255		68	10.25	8	86.25	99.50
1130	3" thick		400	.220		68	8.85	6.90	83.75	96
1140	4" thick		450	.196		68	7.85	6.15	82	94
1150	Binder course, 1-1/2" thick		3860	.023	S.Y.	5.55	.92	.72	7.19	8.30
1160	2" thick		3175	.028		7.40	1.11	.87	9.38	10.80
1170	2-1/2" thick		2810	.031		9.25	1.26	.98	11.49	13.15
1180	3" thick		2455	.036		11.10	1.44	1.12	13.66	15.65
1190	3-1/2" thick		2260	.039		12.95	1.56	1.22	15.73	18
1195	4" thick		2070	.043		14.80	1.71	1.33	17.84	20.50
1200	Wearing course, 1" thick	B-25B	290	.331	Ton	68	13.55	10.35	91.90	107
1210	1-1/2" thick		315	.305		68	12.45	9.55	90	105
1220	2" thick		345	.278		68	11.40	8.70	88.10	102
1230	2-1/2" thick		385	.249		68	10.20	7.80	86	99.50
1240	3" thick		400	.240		68	9.80	7.50	85.30	98.50
1250	Wearing course, 1" thick		5290	.018	S.Y.	3.67	.74	.57	4.98	5.80
1260	1-1/2" thick		3860	.025		6.15	1.02	.78	7.95	9.20
1270	2" thick		3175	.030		8.30	1.24	.95	10.49	12.05
1280	2-1/2" thick		2740	.035		10.20	1.43	1.10	12.73	14.65
1285	3" thick		2450	.039		12.20	1.60	1.23	15.03	17.20
1290	3-1/2" thick		2260	.042		14.30	1.74	1.33	17.37	19.80
1295	4" thick		2070	.046		16.35	1.90	1.45	19.70	22.50
1400	Bridge approach less than 300 tons									
1410	Binder course, 1-1/2" thick	B-25	330	.267	Ton	68	10.70	8.35	87.05	101
1420	2" thick		360	.244		68	9.80	7.65	85.45	98.50
1430	3" thick		420	.210		68	8.40	6.55	82.95	95
1440	4" thick		470	.187		68	7.50	5.85	81.35	93
1450	Binder course, 1-1/2" thick		4055	.022	S.Y.	5.55	.87	.68	7.10	8.20
1460	2" thick		3330	.026		7.40	1.06	.83	9.29	10.70
1470	2-1/2" thick		2950	.030		9.25	1.20	.94	11.39	13
1480	3" thick		2575	.034		11.10	1.37	1.07	13.54	15.50
1490	3-1/2" thick		2370	.037		12.95	1.49	1.16	15.60	17.80
1495	4" thick		2175	.040		14.80	1.63	1.27	17.70	20
1500	Wearing course, 1" thick	B-25B	300	.320	Ton	68	13.10	10	91.10	106
1510	1-1/2" thick		330	.291		68	11.90	9.10	89	103
1520	2" thick		360	.267		68	10.90	8.35	87.25	101
1530	2-1/2" thick		400	.240		68	9.80	7.50	85.30	98.50
1540	3" thick		420	.229		68	9.35	7.15	84.50	97
1550	Wearing course, 1" thick		5555	.017	S.Y.	3.67	.71	.54	4.92	5.70
1560	1-1/2" thick		4055	.024		6.15	.97	.74	7.86	9.10
1570	2" thick		3330	.029		8.30	1.18	.90	10.38	11.90
1580	2-1/2" thick		2875	.033		10.20	1.37	1.04	12.61	14.50
1585	3" thick		2570	.037		12.20	1.53	1.17	14.90	17
1590	3-1/2" thick		2375	.040		14.30	1.65	1.26	17.21	19.60
1595	4" thick		2175	.044		16.35	1.81	1.38	19.54	22
1600	Bridge approach 300-800 tons									
1610	Binder course, 1-1/2" thick	B-25	390	.226	Ton	68	9.05	7.10	84.15	96.50
1620	2" thick		430	.205		68	8.20	6.40	82.60	94.50
1630	3" thick		500	.176		68	7.05	5.50	80.55	92
1640	4" thick		560	.157		68	6.30	4.93	79.23	90
1650	Binder course, 1-1/2" thick		4815	.018	S.Y.	5.55	.73	.57	6.85	7.85
1660	2" thick		3955	.022		7.40	.89	.70	8.99	10.30
1670	2-1/2" thick		3500	.025		9.25	1.01	.79	11.05	12.55

32 12 16 – Asphalt Paving

32 12 16.13 Plant-Mix Asphalt Paving		Crew	Daily Output	Labor-Hours	Unit	Material	Labor (2014 Bare Costs)	Equipment (2014 Bare Costs)	Total (2014 Bare Costs)	Total Incl O&P
1680	3" thick	B-25	3060	.029	S.Y.	11.10	1.16	.90	13.16	14.95
1690	3-1/2" thick		2820	.031		12.95	1.25	.98	15.18	17.25
1695	4" thick		2580	.034		14.80	1.37	1.07	17.24	19.55
1700	Wearing course, 1" thick	B-25B	360	.267	Ton	68	10.90	8.35	87.25	101
1710	1-1/2" thick		390	.246		68	10.05	7.70	85.75	99
1720	2" thick		430	.223		68	9.15	7	84.15	96.50
1730	2-1/2" thick		475	.202		68	8.25	6.30	82.55	94.50
1740	3" thick		500	.192		68	7.85	6	81.85	93.50
1750	Wearing course, 1" thick		6590	.015	S.Y.	3.67	.60	.46	4.73	5.45
1760	1-1/2" thick		4815	.020		6.15	.82	.62	7.59	8.75
1770	2" thick		3955	.024		8.30	.99	.76	10.05	11.45
1780	2-1/2" thick		3415	.028		10.20	1.15	.88	12.23	14
1785	3" thick		3055	.031		12.20	1.29	.98	14.47	16.45
1790	3-1/2" thick		2820	.034		14.30	1.39	1.06	16.75	19
1795	4" thick		2580	.037		16.35	1.52	1.16	19.03	21.50
1800	Bridge approach over 800 tons									
1810	Binder course, 1-1/2" thick	B-25	525	.168	Ton	68	6.75	5.25	80	91
1820	2" thick		575	.153		68	6.15	4.80	78.95	90
1830	3" thick		670	.131		68	5.30	4.12	77.42	87.50
1840	4" thick		750	.117		68	4.71	3.68	76.39	86.50
1850	Binder course, 1-1/2" thick		6450	.014	S.Y.	5.55	.55	.43	6.53	7.40
1860	2" thick		5300	.017		7.40	.67	.52	8.59	9.75
1870	2-1/2" thick		4690	.019		9.25	.75	.59	10.59	11.95
1880	3" thick		4095	.021		11.10	.86	.67	12.63	14.25
1890	3-1/2" thick		3775	.023		12.95	.94	.73	14.62	16.50
1895	4" thick		3455	.025		14.80	1.02	.80	16.62	18.70
1900	Wearing course, 1" thick	B-25B	480	.200	Ton	68	8.20	6.25	82.45	94.50
1910	1-1/2" thick		525	.183		68	7.50	5.70	81.20	93
1920	2" thick		575	.167		68	6.85	5.20	80.05	91
1930	2-1/2" thick		640	.150		68	6.15	4.69	78.84	89.50
1940	3" thick		670	.143		68	5.85	4.48	78.33	89
1950	Wearing course, 1" thick		8830	.011	S.Y.	3.67	.44	.34	4.45	5.10
1960	1-1/2" thick		6450	.015		6.15	.61	.47	7.23	8.25
1970	2" thick		5300	.018		8.30	.74	.57	9.61	10.85
1980	2-1/2" thick		4575	.021		10.20	.86	.66	11.72	13.30
1985	3" thick		4090	.023		12.20	.96	.73	13.89	15.70
1990	3-1/2" thick		3775	.025		14.30	1.04	.80	16.14	18.15
1995	4" thick		3455	.028		16.35	1.14	.87	18.36	20.50
2000	Intersections									
2010	Binder course, 1-1/2" thick	B-25	440	.200	Ton	68	8.05	6.25	82.30	94.50
2020	2" thick		485	.181		68	7.30	5.70	81	92.50
2030	3" thick		560	.157		68	6.30	4.93	79.23	90
2040	4" thick		630	.140		68	5.60	4.38	77.98	88.50
2050	Binder course, 1-1/2" thick		5410	.016	S.Y.	5.55	.65	.51	6.71	7.65
2060	2" thick		4440	.020		7.40	.80	.62	8.82	10.05
2070	2-1/2" thick		3935	.022		9.25	.90	.70	10.85	12.30
2080	3" thick		3435	.026		11.10	1.03	.80	12.93	14.65
2090	3-1/2" thick		3165	.028		12.95	1.12	.87	14.94	16.90
2095	4" thick		2900	.030		14.80	1.22	.95	16.97	19.15
2100	Wearing course, 1" thick	B-25B	400	.240	Ton	68	9.80	7.50	85.30	98.50
2110	1-1/2" thick		440	.218		68	8.90	6.80	83.70	96
2120	2" thick		485	.198		68	8.10	6.20	82.30	94
2130	2-1/2" thick		535	.179		68	7.35	5.60	80.95	92.50

32 12 Flexible Paving

32 12 16 – Asphalt Paving

32 12 16.13 Plant-Mix Asphalt Paving		Crew	Daily Output	Labor-Hours	Unit	2014 Bare Costs Material	Labor	Equipment	Total	Total Incl O&P
2140	3″ thick	B-25B	560	.171	Ton	68	7	5.35	80.35	91.50
2150	Wearing course, 1″ thick		7400	.013	S.Y.	3.67	.53	.41	4.61	5.30
2160	1-1/2″ thick		5410	.018		6.15	.73	.55	7.43	8.50
2170	2″ thick		4440	.022		8.30	.88	.68	9.86	11.20
2180	2-1/2″ thick		3835	.025		10.20	1.02	.78	12	13.70
2185	3″ thick		3430	.028		12.20	1.14	.88	14.22	16.10
2190	3-1/2″ thick		3165	.030		14.30	1.24	.95	16.49	18.65
2195	4″ thick		2900	.033		16.35	1.35	1.04	18.74	21
3000	Prime coat, emulsion, .30 gal. per S.Y., 1000 S.Y.	B-45	2500	.006		1.68	.28	.37	2.33	2.68
3100	Tack coat, emulsion, .10 gal. per S.Y., 1000 S.Y.	″	2500	.006		.56	.28	.37	1.21	1.45

32 12 16.14 Asphaltic Concrete Paving		Crew	Daily Output	Labor-Hours	Unit	Material	Labor	Equipment	Total	Total Incl O&P
0011	**ASPHALTIC CONCRETE PAVING**, parking lots & driveways									
0015	No asphalt hauling included									
0018	Use 6.05 C.Y. per inch per M.S.F. for hauling									
0020	6″ stone base, 2″ binder course, 1″ topping	B-25C	9000	.005	S.F.	1.74	.22	.27	2.23	2.55
0025	2″ binder course, 2″ topping		9000	.005		2.16	.22	.27	2.65	3
0030	3″ binder course, 2″ topping		9000	.005		2.57	.22	.27	3.06	3.46
0035	4″ binder course, 2″ topping		9000	.005		2.98	.22	.27	3.47	3.91
0040	1.5″ binder course, 1″ topping		9000	.005		1.54	.22	.27	2.03	2.32
0042	3″ binder course, 1″ topping		9000	.005		2.16	.22	.27	2.65	3
0045	3″ binder course, 3″ topping		9000	.005		2.98	.22	.27	3.47	3.91
0050	4″ binder course, 3″ topping		9000	.005		3.39	.22	.27	3.88	4.36
0055	4″ binder course, 4″ topping		9000	.005		3.80	.22	.27	4.29	4.81
0300	Binder course, 1-1/2″ thick		35000	.001		.62	.06	.07	.75	.85
0400	2″ thick		25000	.002		.80	.08	.10	.98	1.11
0500	3″ thick		15000	.003		1.24	.13	.16	1.53	1.74
0600	4″ thick		10800	.004		1.62	.18	.22	2.02	2.31
0800	Sand finish course, 3/4″ thick		41000	.001		.32	.05	.06	.43	.48
0900	1″ thick		34000	.001		.39	.06	.07	.52	.60
1000	Fill pot holes, hot mix, 2″ thick	B-16	4200	.008		.84	.28	.16	1.28	1.54
1100	4″ thick		3500	.009		1.23	.34	.20	1.77	2.10
1120	6″ thick		3100	.010		1.65	.39	.22	2.26	2.66
1140	Cold patch, 2″ thick	B-51	3000	.016		.98	.59	.09	1.66	2.08
1160	4″ thick		2700	.018		1.86	.66	.10	2.62	3.16
1180	6″ thick		1900	.025		2.89	.93	.14	3.96	4.78
2000	From 60% recycled content, base course 3″ thick G	B-25	800	.110	Ton	45.50	4.42	3.45	53.37	60.50
3000	Prime coat, emulsion, .30 gal. per S.Y., 1000 S.Y.	B-45	2500	.006	S.Y.	1.68	.28	.37	2.33	2.68
3100	Tack coat, emulsion, .10 gal. per S.Y., 1000 S.Y.	″	2500	.006	″	.56	.28	.37	1.21	1.45

32 12 16.19 Cold-Mix Asphalt Paving		Crew	Daily Output	Labor-Hours	Unit	Material	Labor	Equipment	Total	Total Incl O&P
0010	**COLD-MIX ASPHALT PAVING** 0.5 gal. asphalt/S.Y.per in depth									
0020	Well graded granular aggregate									
0100	Blade mixed in windrows, spread & compacted 4″ course	B-90A	1600	.035	S.Y.	15.90	1.54	1.25	18.69	21
0200	Traveling plant mixed in windrows, compacted 4″ course	B-90B	3000	.016		15.90	.69	.76	17.35	19.40
0300	Rotary plant mixed in place, compacted 4″ course	″	3500	.014		15.90	.59	.65	17.14	19.10
0400	Central stationary plant, mixed, compacted 4″ course	B-36	7200	.006		32	.23	.22	32.45	35.50

32 13 Rigid Paving

32 13 13 – Concrete Paving

32 13 13.23 Concrete Paving Surface Treatment		Crew	Daily Output	Labor-Hours	Unit	Material	2014 Bare Costs Labor	Equipment	Total	Total Incl O&P
0010	**CONCRETE PAVING SURFACE TREATMENT**									
0015	Including joints, finishing and curing									
0020	Fixed form, 12′ pass, unreinforced, 6″ thick	B-26	3000	.029	S.Y.	21	1.20	1.12	23.32	26.50
0030	7″ thick		2850	.031		26	1.27	1.18	28.45	31.50
0100	8″ thick		2750	.032		29	1.31	1.22	31.53	35.50
0110	8″ thick, small area		1375	.064		29	2.62	2.45	34.07	38.50
0200	9″ thick		2500	.035		33	1.44	1.35	35.79	40
0300	10″ thick		2100	.042		36	1.72	1.60	39.32	44.50
0310	10″ thick, small area		1050	.084		36	3.44	3.20	42.64	49
0400	12″ thick		1800	.049		41.50	2	1.87	45.37	50.50
0410	Conc. pavement, w/jt.,fnsh.&curing,fix form,24′ pass,unreinforced,6″T		6000	.015		20	.60	.56	21.16	23.50
0420	7″ thick		5700	.015		24.50	.63	.59	25.72	28.50
0430	8″ thick		5500	.016		27.50	.66	.61	28.77	31.50
0440	9″ thick		5000	.018		31.50	.72	.67	32.89	36.50
0450	10″ thick		4200	.021		34.50	.86	.80	36.16	40
0460	12″ thick		3600	.024		40	1	.93	41.93	46.50
0470	15″ thick		3000	.029		52	1.20	1.12	54.32	60.50
0500	Fixed form 12′ pass 15″ thick		1500	.059		52.50	2.41	2.24	57.15	64
0510	For small irregular areas, add				%	10%	100%	100%		
0520	Welded wire fabric, sheets for rigid paving 2.33 lb./S.Y.	2 Rodm	389	.041	S.Y.	1.31	2.08		3.39	4.71
0530	Reinforcing steel for rigid paving 12 lb./S.Y.		666	.024		6.30	1.22		7.52	8.85
0540	Reinforcing steel for rigid paving 18 lb./S.Y.		444	.036		9.45	1.83		11.28	13.25
0610	For under 10′ pass, add				%	10%	100%	100%		
0620	Slip form, 12′ pass, unreinforced, 6″ thick	B-26A	5600	.016	S.Y.	20.50	.64	.63	21.77	24
0622	7″ thick		5600	.016		25	.64	.63	26.27	28.50
0624	8″ thick		5300	.017		28	.68	.67	29.35	32.50
0626	9″ thick		4820	.018		32	.75	.73	33.48	37
0628	10″ thick		4050	.022		35	.89	.87	36.76	41
0630	12″ thick		3470	.025		40.50	1.04	1.02	42.56	47
0632	15″ thick		2890	.030		51	1.25	1.22	53.47	59.50
0640	Slip form, 24′ pass, unreinforced, 6″ thick		11200	.008		20	.32	.32	20.64	23
0642	7″ thick		11200	.008		24	.32	.32	24.64	27
0644	8″ thick		10600	.008		27	.34	.33	27.67	30.50
0646	9″ thick		9640	.009		31	.37	.37	31.74	35
0648	10″ thick		8100	.011		34	.45	.44	34.89	38.50
0650	12″ thick		6940	.013		39.50	.52	.51	40.53	45
0652	15″ thick		5780	.015		49.50	.62	.61	50.73	56
0700	Finishing, broom finish small areas	2 Cefi	120	.133			5.85		5.85	8.70
0710	Transverse joint support dowels	C-1	350	.091	Ea.	5.60	3.98		9.58	12.30
0720	Transverse contraction joints, saw cut & grind	A-1B	120	.067	L.F.		2.44	1.46	3.90	5.40
0730	Transverse expansion joints, incl. premolded bit. jt. filler	C-1	150	.213		2.37	9.30		11.67	16.95
0740	Transverse construction joint using bulkhead	"	73	.438		3.44	19.10		22.54	33.50
0750	Longitudinal joint tie bars, grouted	B-23	70	.571	Ea.	5.35	21	41	67.35	84
1000	Curing, with sprayed membrane by hand	2 Clab	1500	.011	S.Y.	.99	.39		1.38	1.69
1650	For integral coloring, see Section 03 05 13.20									
3200	Concrete grooving, continuous for roadways	B-71	700	.080	S.Y.		3.38	9.90	13.28	16

32 13 13.24 Plain Cement Concrete Pavement, Airports

		Crew	Daily Output	Labor-Hours	Unit	Material	Labor	Equipment	Total	Total Incl O&P
0010	**PLAIN CEMENT CONCRETE PAVEMENT, AIRPORTS**									
0050	includes joints, finishing, and curing									
0100	fixed form, 12′ pass, unreinforced, 8″ thick	B-26	2750	.032	S.Y.	29.50	1.31	1.22	32.03	36
0200	9″ thick		2500	.035		32.50	1.44	1.35	35.29	39.50
0300	10″ thick		2100	.042		36	1.72	1.60	39.32	44

32 13 Rigid Paving

32 13 13 – Concrete Paving

32 13 13.24	Plain Cement Concrete Pavement, Airports	Crew	Daily Output	Labor-Hours	Unit	Material	Labor	Equipment	Total	Total Incl O&P
						2014 Bare Costs				
0400	11″ thick	B-26	1910	.046	S.Y.	39	1.89	1.76	42.65	48
0500	12″ thick		1800	.049		41	2	1.87	44.87	50.50
0600	13″ thick		1670	.053		46	2.16	2.01	50.17	56
0700	14″ thick		1550	.057		49	2.33	2.17	53.50	60
0800	15″ thick		1500	.059		52	2.41	2.24	56.65	63.50
0900	16″ thick		1410	.062		54.50	2.56	2.39	59.45	66.50
1000	17″ thick		1325	.066		57.50	2.72	2.54	62.76	70.50
1100	18″ thick		1250	.070		61.50	2.89	2.69	67.08	75
1200	19″ thick		1190	.074		65.50	3.03	2.83	71.36	80
1300	20″ thick		1125	.078		68.50	3.21	2.99	74.70	83.50
1400	21″ thick		1075	.082		70.50	3.36	3.13	76.99	86
1500	22″ thick		1025	.086		74	3.52	3.28	80.80	90
1600	23″ thick		1000	.088		77	3.61	3.36	83.97	94.50
1700	24″ thick		1000	.088		80.50	3.61	3.36	87.47	98
1800	25″ thick		1000	.088		84	3.61	3.36	90.97	101
1900	26″ thick		1000	.088		87	3.61	3.36	93.97	105
2000	27″ thick		1000	.088		90	3.61	3.36	96.97	109
2100	fixed form, 24′ pass, unreinforced, 8″ thick		5500	.016		28.50	.66	.61	29.77	33
2110	9″ thick		5000	.018		31.50	.72	.67	32.89	36.50
2120	10″ thick		4200	.021		34.50	.86	.80	36.16	40
2130	11″ thick		3820	.023		37.50	.94	.88	39.32	44
2140	12″ thick		3600	.024		39.50	1	.93	41.43	46
2150	13″ thick		3340	.026		43.50	1.08	1.01	45.59	51
2160	14″ thick		3100	.028		46.50	1.16	1.09	48.75	54.50
2170	15″ thick		3000	.029		49.50	1.20	1.12	51.82	57.50
2180	16″ thick		2820	.031		52	1.28	1.19	54.47	60.50
2190	17″ thick		2650	.033		55	1.36	1.27	57.63	64
2200	18″ thick		2500	.035		59	1.44	1.35	61.79	68.50
2210	19″ thick		2380	.037		62	1.52	1.41	64.93	72.50
2220	20″ thick		2250	.039		65.50	1.60	1.50	68.60	76
2230	21″ thick		2150	.041		67.50	1.68	1.56	70.74	78.50
2240	22″ thick		2050	.043		70.50	1.76	1.64	73.90	82
2250	23″ thick		2000	.044		74	1.80	1.68	77.48	85.50
2260	24″ thick		2000	.044		77	1.80	1.68	80.48	89
2270	25″ thick		2000	.044		80	1.80	1.68	83.48	92
2280	26″ thick		2000	.044		83	1.80	1.68	86.48	95.50
2290	27″ thick		2000	.044		86	1.80	1.68	89.48	99.50
2300	Welded wire fabric, sheets for rigid paving 2.33 lb./S.Y.	2 Rodm	389	.041		1.31	2.08		3.39	4.71
2310	Reinforcing steel for rigid paving 12 lb./S.Y.		666	.024		6.30	1.22		7.52	8.85
2320	Reinforcing steel for rigid paving 18 lb./S.Y.		444	.036		9.45	1.83		11.28	13.25
3000	5/8 inch reinforced deformed tie dowel, 30″ long		300	.053	Ea.	1.95	2.70		4.65	6.40
3100	3/4″ diameter smooth dowel 18 inch long		225	.071		4.17	3.60		7.77	10.25
3200	1″ diameter, 19″ long		200	.080		4.64	4.05		8.69	11.45
3300	1-1/4″ diameter, 20″ long		180	.089		6.65	4.50		11.15	14.35
3400	1-1/2″ diameter dowel 20″ long		160	.100		8.35	5.05		13.40	17.15
3500	2″ diameter, 24″ long		120	.133		16.30	6.75		23.05	28.50
4000	6″ Keyed expansion joint, 24 ga	1 Carp	195	.041	L.F.	1.25	1.88		3.13	4.28
4100	9″		190	.042		2.16	1.93		4.09	5.35
4200	12″		185	.043		3.73	1.98		5.71	7.15
4300	15″		180	.044		4.65	2.04		6.69	8.25
4400	18″		175	.046		5.60	2.10		7.70	9.40
5000	Expansion joint, premolded, bituminous fiber, type B, per S.F.		300	.027	S.F.	1.95	1.22		3.17	4.04

	32 13 13.26 Slip Form Concrete Pavement, Airports	Crew	Daily Output	Labor-Hours	Unit	Material	Labor (2014 Bare Costs)	Equipment	Total	Total Incl O&P
0010	**SLIP FORM CONCRETE PAVEMENT, AIRPORTS**									
0015	Note: Slip forming method is not desirable if pavement may freeze									
0020	Unit price includes finish, curing, and green sawed joints									
0100	Slip form concrete pavement, 12′ pass, unreinforced, 8″ thick	B-26A	5300	.017	S.Y.	28.50	.68	.67	29.85	33.50
0200	9″ thick		4820	.018		32	.75	.73	33.48	37
0300	10″ thick		4050	.022		35	.89	.87	36.76	41
0400	11″ thick		3680	.024		38	.98	.96	39.94	44.50
0500	12″ thick		3470	.025		40.50	1.04	1.02	42.56	47
0600	13″ thick		3220	.027		44.50	1.12	1.10	46.72	52
0700	14″ thick		2990	.029		48	1.21	1.18	50.39	55.50
0800	15″ thick		2890	.030		51	1.25	1.22	53.47	59.50
0900	16″ thick		2720	.032		53	1.33	1.30	55.63	62
1000	17″ thick		2555	.034		56.50	1.41	1.39	59.30	65.50
1100	18″ thick		2410	.037		60.50	1.50	1.47	63.47	70.50
1200	19″ thick		2310	.038		63.50	1.56	1.53	66.59	74
1300	20″ thick		2170	.041		67	1.66	1.63	70.29	78
1400	21″ thick		2070	.043		69	1.74	1.71	72.45	80.50
1500	22″ thick		1975	.045		72	1.83	1.79	75.62	84.50
1600	23″ thick		1900	.046		75.50	1.90	1.86	79.26	88
1700	24″ thick		1900	.046		78.50	1.90	1.86	82.26	91.50
1800	25″ thick		1900	.046		82	1.90	1.86	85.76	95
1900	26″ thick		1900	.046		85	1.90	1.86	88.76	98.50
2000	27″ thick		1900	.046		88	1.90	1.86	91.76	102
2100	Slip form Conc pavement, 24′ pass, unreinforced, 8″ thick		10600	.008		28	.34	.33	28.67	32
2110	9″ thick		9640	.009		31	.37	.37	31.74	35
2120	10″ thick		8100	.011		34	.45	.44	34.89	38.50
2130	11″ thick		7360	.012		37.50	.49	.48	38.47	42.50
2140	12″ thick		6940	.013		39.50	.52	.51	40.53	45
2150	13″ thick		6440	.014		43	.56	.55	44.11	48.50
2160	14″ thick		5980	.015		46	.60	.59	47.19	52
2170	15″ thick		5780	.015		49	.62	.61	50.23	55.50
2180	16″ thick		5440	.016		51.50	.66	.65	52.81	58
2190	17″ thick		5110	.017		54.50	.71	.69	55.90	62
2200	18″ thick		4820	.018		58.50	.75	.73	59.98	66.50
2210	19″ thick		4620	.019		61.50	.78	.77	63.05	69.50
2220	20″ thick		4340	.020		64.50	.83	.82	66.15	73
2230	21″ thick		4140	.021		66.50	.87	.86	68.23	76
2240	22″ thick		3950	.022		70	.91	.90	71.81	79.50
2250	23″ thick		3800	.023		73	.95	.93	74.88	83
2260	24″ thick		3800	.023		76	.95	.93	77.88	86.50
2270	25″ thick		3800	.023		78.50	.95	.93	80.38	89
2280	26″ thick		3800	.023		82	.95	.93	83.88	92.50
2290	27″ thick		3800	.023		85	.95	.93	86.88	96
2300	Welded wire fabric, sheets, 6 x 6 - W1.4 x W1.4 , 21 lb./C.S.F.	2 Rodm	389	.041		1.31	2.08		3.39	4.71
2310	Reinforcing steel for rigid paving 12 lb./S.Y.		666	.024		6.30	1.22		7.52	8.85
2320	Reinforcing steel for rigid paving 18 lb./S.Y.		444	.036		9.45	1.83		11.28	13.25
3000	5/8″ reinforced deformed tie dowel, 30″ long		300	.053	Ea.	1.95	2.70		4.65	6.40
3100	3/4″ diameter smooth dowel 18 inch long		225	.071		4.17	3.60		7.77	10.25
3200	1″ diameter, 19 inch long		200	.080		4.64	4.05		8.69	11.45
3300	1-1/4″ diameter, 20 inch long		180	.089		6.65	4.50		11.15	14.35
3400	1-1/2″ diameter dowel 20 inch long		160	.100		8.35	5.05		13.40	17.15
3500	2″ diameter, 24 inch long		120	.133		16.30	6.75		23.05	28.50

32 13 Rigid Paving

32 13 13 – Concrete Paving

32 13 13.26 Slip Form Concrete Pavement, Airports		Crew	Daily Output	Labor-Hours	Unit	2014 Bare Costs Material	Labor	Equipment	Total	Total Incl O&P
4000	6″ Keyed expansion joint	1 Carp	195	.041	L.F.	1.25	1.88		3.13	4.28
4100	9″		190	.042		2.16	1.93		4.09	5.35
4200	12″		185	.043		3.73	1.98		5.71	7.15
4300	15″		180	.044		4.65	2.04		6.69	8.25
4400	18″		175	.046		5.60	2.10		7.70	9.40
5000	Expansion joint, premolded, bituminous fiber, type B, per S.F.		300	.027	S.F.	1.95	1.22		3.17	4.04

32 13 13.28 Slip Form Cement Concrete Pavement, Canals

		Crew	Daily Output	Labor-Hours	Unit	Material	Labor	Equipment	Total	Total Incl O&P
0010	**SLIP FORM CEMENT CONCRETE PAVEMENT, CANALS**									
0015	Note: Slip forming method is not desirable if pavement may freeze									
0020	Unit price includes finish, curing, and green sawed joints									
0110	Slip form Conc canal lining, unreinforced, 6″ thick	B-26B	6000	.016	S.Y.	21.50	.67	.72	22.89	26
0120	8″ thick	″	5500	.017		28	.73	.78	29.51	32.50
2300	Welded wire fabric, sheets for rigid paving 2.33 lb./S.Y.	2 Rodm	389	.041		1.31	2.08		3.39	4.71
2310	Reinforcing steel for rigid paving 12 lb./S.Y.		666	.024		6.30	1.22		7.52	8.85
2320	Reinforcing steel for rigid paving 18 lb./S.Y.		444	.036		9.45	1.83		11.28	13.25

32 14 Unit Paving

32 14 13 – Precast Concrete Unit Paving

32 14 13.16 Precast Concrete Unit Paving Slabs

		Crew	Daily Output	Labor-Hours	Unit	Material	Labor	Equipment	Total	Total Incl O&P
0010	**PRECAST CONCRETE UNIT PAVING SLABS**									
0710	Precast concrete patio blocks, 2-3/8″ thick, colors, 8″ x 16″	D-1	265	.060	S.F.	2.14	2.49		4.63	6.15
0715	12″ x 12″		300	.053		2.78	2.20		4.98	6.40
0720	16″ x 16″		335	.048		3.06	1.97		5.03	6.40
0730	24″ x 24″		510	.031		3.81	1.30		5.11	6.15
0740	Green, 8″ x 16″		265	.060		2.76	2.49		5.25	6.85
0750	Exposed local aggregate, natural	2 Bric	250	.064		7.60	2.92		10.52	12.80
0800	Colors		250	.064		7.60	2.92		10.52	12.80
0850	Exposed granite or limestone aggregate		250	.064		7.35	2.92		10.27	12.55
0900	Exposed white tumblestone aggregate		250	.064		6.55	2.92		9.47	11.70

32 14 13.18 Precast Concrete Plantable Pavers

		Crew	Daily Output	Labor-Hours	Unit	Material	Labor	Equipment	Total	Total Incl O&P
0010	**PRECAST CONCRETE PLANTABLE PAVERS** (50% grass)									
0015	Subgrade preparation and grass planting not included									
0100	Precast concrete plantable pavers with topsoil, 24″ x 16″	B-63	800	.050	S.F.	3.71	1.94	.22	5.87	7.30
0200	Less than 600 Square Feet or irregular area	″	500	.080	″	3.71	3.10	.35	7.16	9.20
0300	3/4″ crushed stone base for plantable pavers, 6 inch depth	B-62	1000	.024	S.Y.	5.50	.96	.17	6.63	7.70
0400	8 inch depth		900	.027		7.40	1.07	.19	8.66	10
0500	10 inch depth		800	.030		9.20	1.20	.22	10.62	12.20
0600	12 inch depth		700	.034		11.05	1.38	.25	12.68	14.55
0700	Hydro seeding plantable pavers	B-81A	20	.800	M.S.F.	19.35	29.50	23.50	72.35	92
0800	Apply fertilizer and seed to plantable pavers	1 Clab	8	1	″	51.50	36.50		88	113

32 14 16 – Brick Unit Paving

32 14 16.10 Brick Paving

		Crew	Daily Output	Labor-Hours	Unit	Material	Labor	Equipment	Total	Total Incl O&P
0010	**BRICK PAVING**									
0012	4″ x 8″ x 1-1/2″, without joints (4.5 brick/S.F.)	D-1	110	.145	S.F.	2.21	6		8.21	11.60
0100	Grouted, 3/8″ joint (3.9 brick/S.F.)		90	.178		2.15	7.35		9.50	13.55
0200	4″ x 8″ x 2-1/4″, without joints (4.5 bricks/S.F.)		110	.145		2.48	6		8.48	11.90
0300	Grouted, 3/8″ joint (3.9 brick/S.F.)		90	.178		2.15	7.35		9.50	13.55
0500	Bedding, asphalt, 3/4″ thick	B-25	5130	.017		.68	.69	.54	1.91	2.40
0540	Course washed sand bed, 1″ thick	B-18	5000	.005		.29	.18	.01	.48	.61
0580	Mortar, 1″ thick	D-1	300	.053		.63	2.20		2.83	4.06

32 14 Unit Paving

32 14 16 – Brick Unit Paving

32 14 16.10 Brick Paving		Crew	Daily Output	Labor-Hours	Unit	2014 Bare Costs Material	Labor	Equipment	Total	Total Incl O&P
0620	2" thick	D-1	200	.080	S.F.	1.27	3.30		4.57	6.45
1500	Brick on 1" thick sand bed laid flat, 4.5 per S.F.		100	.160		3.12	6.60		9.72	13.50
2000	Brick pavers, laid on edge, 7.2 per S.F.		70	.229		3.83	9.45		13.28	18.60
2500	For 4" thick concrete bed and joints, add		595	.027		1.15	1.11		2.26	2.95
2800	For steam cleaning, add	A-1H	950	.008		.09	.31	.08	.48	.67

32 14 23 – Asphalt Unit Paving

32 14 23.10 Asphalt Blocks		Crew	Daily Output	Labor-Hours	Unit	Material	Labor	Equipment	Total	Total Incl O&P
0010	**ASPHALT BLOCKS**									
0020	Rectangular, 6" x 12" x 1-1/4", w/bed & neopr. adhesive	D-1	135	.119	S.F.	9.20	4.89		14.09	17.55
0100	3" thick		130	.123		12.90	5.10		18	22
0300	Hexagonal tile, 8" wide, 1-1/4" thick		135	.119		9.20	4.89		14.09	17.55
0400	2" thick		130	.123		12.90	5.10		18	22
0500	Square, 8" x 8", 1-1/4" thick		135	.119		9.20	4.89		14.09	17.55
0600	2" thick		130	.123		12.90	5.10		18	22
0900	For exposed aggregate (ground finish) add					.61			.61	.67
0910	For colors, add					.46			.46	.51

32 14 40 – Stone Paving

32 14 40.10 Stone Pavers		Crew	Daily Output	Labor-Hours	Unit	Material	Labor	Equipment	Total	Total Incl O&P
0010	**STONE PAVERS**									
1100	Flagging, bluestone, irregular, 1" thick,	D-1	81	.198	S.F.	6.35	8.15		14.50	19.45
1110	1-1/2" thick		90	.178		7.50	7.35		14.85	19.45
1120	Pavers, 1/2" thick		110	.145		10.55	6		16.55	21
1130	3/4" thick		95	.168		13.45	6.95		20.40	25.50
1140	1" thick		81	.198		14.40	8.15		22.55	28.50
1150	Snapped random rectangular, 1" thick		92	.174		9.60	7.20		16.80	21.50
1200	1-1/2" thick		85	.188		11.55	7.75		19.30	24.50
1250	2" thick		83	.193		13.45	7.95		21.40	27
1300	Slate, natural cleft, irregular, 3/4" thick		92	.174		9.15	7.20		16.35	21
1310	1" thick		85	.188		10.65	7.75		18.40	23.50
1350	Random rectangular, gauged, 1/2" thick		105	.152		19.80	6.30		26.10	31.50
1400	Random rectangular, butt joint, gauged, 1/4" thick		150	.107		21.50	4.41		25.91	30
1450	For sand rubbed finish, add					9.20			9.20	10.10
1550	Granite blocks, 3-1/2" x 3-1/2" x 3-1/2"	D-1	92	.174		11.60	7.20		18.80	23.50
1560	4" x 4" x 4"		95	.168		12.25	6.95		19.20	24
1600	4" to 12" long, 3" to 5" wide, 3" to 5" thick		98	.163		9.70	6.75		16.45	21
1650	6" to 15" long, 3" to 6" wide, 3" to 5" thick		105	.152		5.15	6.30		11.45	15.30

32 16 Curbs, Gutters, Sidewalks, and Driveways

32 16 13 – Curbs and Gutters

32 16 13.13 Cast-in-Place Concrete Curbs and Gutters		Crew	Daily Output	Labor-Hours	Unit	Material	Labor	Equipment	Total	Total Incl O&P
0010	**CAST-IN-PLACE CONCRETE CURBS AND GUTTERS**									
0290	Forms only, no concrete									
0300	Concrete, wood forms, 6" x 18", straight	C-2	500	.096	L.F.	2.95	4.29		7.24	9.85
0400	6" x 18", radius	"	200	.240	"	3.06	10.70		13.76	19.90
0402	Forms and concrete complete									
0404	Concrete, wood forms, 6" x 18", straight & concrete	C-2A	500	.096	L.F.	5.70	4.26		9.96	12.80
0406	6" x 18", radius		200	.240		5.85	10.65		16.50	22.50
0410	Steel forms, 6" x 18", straight		700	.069		4.37	3.04		7.41	9.45
0411	6" x 18", radius		400	.120		3.64	5.30		8.94	12.15
0415	Machine formed, 6" x 18", straight	B-69A	2000	.024		3.54	.97	.45	4.96	5.85

32 16 Curbs, Gutters, Sidewalks, and Driveways

32 16 13 – Curbs and Gutters

32 16 13.13 Cast-in-Place Concrete Curbs and Gutters		Crew	Daily Output	Labor-Hours	Unit	2014 Bare Costs Material	Labor	Equipment	Total	Total Incl O&P
0416	6" x 18", radius	B-69A	900	.053	L.F.	3.57	2.15	1	6.72	8.30
0421	Curb and gutter, straight									
0422	with 6" high curb and 6" thick gutter, wood forms									
0430	24" wide, .055 C.Y. per L.F.	C-2A	375	.128	L.F.	15.20	5.70		20.90	25.50
0435	30" wide, .066 C.Y. per L.F.		340	.141		16.70	6.25		22.95	28
0440	Steel forms, 24" wide, straight		700	.069		6.80	3.04		9.84	12.10
0441	Radius		500	.096		6.60	4.26		10.86	13.80
0442	30" wide, straight		700	.069		7.95	3.04		10.99	13.35
0443	Radius		500	.096		7.50	4.26		11.76	14.75
0445	Machine formed, 24" wide, straight	B-69A	2000	.024		5.55	.97	.45	6.97	8.05
0446	Radius		900	.053		5.55	2.15	1	8.70	10.50
0447	30" wide, straight		2000	.024		6.45	.97	.45	7.87	9.05
0448	Radius		900	.053		6.45	2.15	1	9.60	11.50
0451	Median mall, machine formed, 2' x 9" high, straight		2200	.022		5.55	.88	.41	6.84	7.90
0452	Radius	B-69B	900	.053		5.55	2.15	.86	8.56	10.30
0453	4' x 9" high, straight		2000	.024		11.10	.97	.39	12.46	14.10
0454	Radius		800	.060		11.10	2.42	.96	14.48	16.95

32 16 13.23 Precast Concrete Curbs and Gutters		Crew	Daily Output	Labor-Hours	Unit	Material	Labor	Equipment	Total	Total Incl O&P
0010	**PRECAST CONCRETE CURBS AND GUTTERS**									
0550	Precast, 6" x 18", straight	B-29	700	.080	L.F.	9.90	3.19	1.26	14.35	17.20
0600	6" x 18", radius	"	325	.172	"	10.40	6.85	2.71	19.96	25

32 16 13.33 Asphalt Curbs		Crew	Daily Output	Labor-Hours	Unit	Material	Labor	Equipment	Total	Total Incl O&P
0010	**ASPHALT CURBS**									
0012	Curbs, asphaltic, machine formed, 8" wide, 6" high, 40 L.F./ton	B-27	1000	.032	L.F.	1.74	1.19	.29	3.22	4.07
0100	8" wide, 8" high, 30 L.F. per ton		900	.036		2.33	1.32	.33	3.98	4.96
0150	Asphaltic berm, 12" W, 3"-6" H, 35 L.F./ton, before pavement		700	.046		.04	1.70	.42	2.16	3.12
0200	12" W, 1-1/2" to 4" H, 60 L.F. per ton, laid with pavement	B-2	1050	.038		.02	1.41		1.43	2.21

32 16 13.43 Stone Curbs		Crew	Daily Output	Labor-Hours	Unit	Material	Labor	Equipment	Total	Total Incl O&P
0010	**STONE CURBS**									
1000	Granite, split face, straight, 5" x 16"	D-13	275	.175	L.F.	12.30	7.65	1.75	21.70	27
1100	6" x 18"	"	250	.192		16.15	8.45	1.93	26.53	33
1300	Radius curbing, 6" x 18", over 10' radius	B-29	260	.215		19.80	8.60	3.39	31.79	39
1400	Corners, 2' radius	"	80	.700	Ea.	66.50	28	11	105.50	128
1600	Edging, 4-1/2" x 12", straight	d-13	300	.160	L.F.	6.15	7	1.60	14.75	19.20
1800	Curb inlets, (guttermouth) straight	B-29	41	1.366	Ea.	148	54.50	21.50	224	269
2000	Indian granite (belgian block)									
2100	Jumbo, 10-1/2" x 7-1/2" x 4", grey	D-1	150	.107	L.F.	6.05	4.41		10.46	13.35
2150	Pink		150	.107		8.20	4.41		12.61	15.70
2200	Regular, 9" x 4-1/2" x 4-1/2", grey		160	.100		4.53	4.13		8.66	11.30
2250	Pink		160	.100		5.80	4.13		9.93	12.65
2300	Cubes, 4" x 4" x 4", grey		175	.091		3.30	3.78		7.08	9.40
2350	Pink		175	.091		3.58	3.78		7.36	9.70
2400	6" x 6" x 6", pink		155	.103		11.85	4.26		16.11	19.50
2500	Alternate pricing method for indian granite									
2550	Jumbo, 10-1/2" x 7-1/2" x 4" (30 lb.), grey				Ton	345			345	380
2600	Pink					475			475	525
2650	Regular, 9" x 4-1/2" x 4-1/2" (20 lb.), grey					320			320	350
2700	Pink					405			405	445
2750	Cubes, 4" x 4" x 4" (5 lb.), grey					400			400	440
2800	Pink					460			460	505
2850	6" x 6" x 6" (25 lb.), pink					460			460	505
2900	For pallets, add					22			22	24

32 17 Paving Specialties

32 17 13 – Parking Bumpers

32 17 13.13 Metal Parking Bumpers

	32 17 13.13 Metal Parking Bumpers	Crew	Daily Output	Labor-Hours	Unit	Material	2014 Bare Costs Labor	Equipment	Total	Total Incl O&P
0010	**METAL PARKING BUMPERS**									
0015	Bumper rails for garages, 12 Ga. rail, 6" wide, with steel									
0020	posts 12'-6" O.C., minimum	E-4	190	.168	L.F.	17.30	8.70	.75	26.75	35
0030	Average		165	.194		21.50	10	.86	32.36	42.50
0100	Maximum		140	.229		26	11.80	1.01	38.81	50.50
0300	12" channel rail, minimum		160	.200		21.50	10.30	.89	32.69	43
0400	Maximum		120	.267		32.50	13.75	1.18	47.43	61.50
1300	Pipe bollards, conc. filled/paint, 8' L x 4' D hole, 6" diam.	B-6	20	1.200	Ea.	660	48	18.25	726.25	820
1400	8" diam.		15	1.600		890	64	24.50	978.50	1,100
1500	12" diam.		12	2		1,125	80	30.50	1,235.50	1,375
2030	Folding with individual padlocks	B-2	50	.800		600	29.50		629.50	705

32 17 13.16 Plastic Parking Bumpers

	32 17 13.16 Plastic Parking Bumpers	Crew	Daily Output	Labor-Hours	Unit	Material	Labor	Equipment	Total	Total Incl O&P
0010	**PLASTIC PARKING BUMPERS**									
1200	Thermoplastic, 6" x 10" x 6'-0"	B-2	120	.333	Ea.	52	12.35		64.35	76

32 17 13.19 Precast Concrete Parking Bumpers

	32 17 13.19 Precast Concrete Parking Bumpers	Crew	Daily Output	Labor-Hours	Unit	Material	Labor	Equipment	Total	Total Incl O&P
0010	**PRECAST CONCRETE PARKING BUMPERS**									
1000	Wheel stops, precast concrete incl. dowels, 6" x 10" x 6'-0"	B-2	120	.333	Ea.	39.50	12.35		51.85	62
1100	8" x 13" x 6'-0"	"	120	.333	"	45.50	12.35		57.85	69

32 17 13.26 Wood Parking Bumpers

	32 17 13.26 Wood Parking Bumpers	Crew	Daily Output	Labor-Hours	Unit	Material	Labor	Equipment	Total	Total Incl O&P
0010	**WOOD PARKING BUMPERS**									
0020	Parking barriers, timber w/saddles, treated type									
0100	4" x 4" for cars	B-2	520	.077	L.F.	2.79	2.85		5.64	7.45
0200	6" x 6" for trucks		520	.077	"	5.80	2.85		8.65	10.75
0600	Flexible fixed stanchion, 2' high, 3" diameter		100	.400	Ea.	40	14.80		54.80	67

32 17 23 – Pavement Markings

32 17 23.13 Painted Pavement Markings

	32 17 23.13 Painted Pavement Markings	Crew	Daily Output	Labor-Hours	Unit	Material	Labor	Equipment	Total	Total Incl O&P
0010	**PAINTED PAVEMENT MARKINGS**									
0020	Acrylic waterborne, white or yellow, 4" wide, less than 3000 L.F.	B-78	20000	.002	L.F.	.15	.09	.03	.27	.34
0030	3000-16000 L.F.		20000	.002		.09	.09	.03	.21	.27
0040	over 16000 L.F.		20000	.002		.09	.09	.03	.21	.27
0200	6" wide, less than 3000 L.F.		11000	.004		.23	.16	.06	.45	.56
0220	3000-16000 L.F.		11000	.004		.13	.16	.06	.35	.45
0230	over 16000 L.F.		11000	.004		.13	.16	.06	.35	.45
0500	8" wide, less than 3000 L.F.		10000	.005		.30	.18	.06	.54	.67
0520	3000-16000 L.F.		10000	.005		.17	.18	.06	.41	.53
0530	over 16000 L.F.		10000	.005		.17	.18	.06	.41	.53
0600	12" wide, less than 3000 L.F.		4000	.012		.45	.44	.16	1.05	1.36
0604	3000-16000 L.F.		4000	.012		.26	.44	.16	.86	1.15
0608	over 16000 L.F.		4000	.012		.26	.44	.16	.86	1.15
0620	Arrows or gore lines		2300	.021	S.F.	.21	.77	.28	1.26	1.73
0640	Temporary paint, white or yellow, less than 3000 L.F.		15000	.003	L.F.	.07	.12	.04	.23	.31
0642	3000-16000 L.F.		15000	.003		.06	.12	.04	.22	.30
0644	Over 16000 L.F.		15000	.003		.05	.12	.04	.21	.29
0660	Removal	1 Clab	300	.027			.98		.98	1.51
0710	Thermoplastic, white or yellow, 4" wide, less than 6000 L.F.	B-79	15000	.003		.30	.10	.10	.50	.60
0715	6000 L.F. or more		15000	.003		.28	.10	.10	.48	.58
0730	6" wide, less than 6000 L.F.		14000	.003		.45	.11	.11	.67	.78
0735	6000 L.F. or more		14000	.003		.42	.11	.11	.64	.74
0740	8" wide, less than 6000 L.F.		12000	.003		.60	.12	.13	.85	.99
0745	6000 L.F. or more		12000	.003		.56	.12	.13	.81	.95
0750	12" wide, less than 6000 L.F.		6000	.007		.89	.25	.26	1.40	1.64

32 17 Paving Specialties

32 17 23 – Pavement Markings

32 17 23.13 Painted Pavement Markings

	32 17 23.13 Painted Pavement Markings	Crew	Daily Output	Labor-Hours	Unit	2014 Bare Costs Material	Labor	Equipment	Total	Total Incl O&P
0755	6000 L.F. or more	B-79	6000	.007	L.F.	.83	.25	.26	1.34	1.58
0760	Arrows		660	.061	S.F.	.60	2.24	2.38	5.22	6.75
0770	Gore lines		2500	.016		.60	.59	.63	1.82	2.26
0780	Letters		660	.061		.60	2.24	2.38	5.22	6.75
0782	Thermoplastic material, small users				Ton	1,775			1,775	1,950
0784	Glass beads, highway use, add				Lb.	.61			.61	.67
0786	Thermoplastic material, highway departments				Ton	1,550			1,550	1,700
1000	Airport painted markings									
1050	Traffic safety flashing truck for airport painting	A-2B	1	8	Day		292	270	562	740
1100	Painting, white or yellow, taxiway markings	B-78	4000	.012	S.F.	.26	.44	.16	.86	1.15
1110	with 12 lb. beads per 100 S.F.		4000	.012		.53	.44	.16	1.13	1.44
1200	Runway markings		3500	.014		.26	.51	.18	.95	1.27
1210	with 12 lb. beads per 100 S.F.		3500	.014		.53	.51	.18	1.22	1.56
1300	Pavement location or direction signs		2500	.019		.26	.71	.26	1.23	1.66
1310	with 12 lb. beads per 100 S.F.		2500	.019		.53	.71	.26	1.50	1.95
1350	Mobilization airport pavement painting		4	12	Ea.		445	161	606	860
1400	Paint markings or pavement signs removal daytime	B-78B	400	.045	S.F.		1.70	.98	2.68	3.70
1500	Removal nighttime		335	.054	"		2.03	1.17	3.20	4.42
1600	Mobilization pavement paint removal		4	4.500	Ea.		170	98	268	370

32 17 23.14 Pavement Parking Markings

	32 17 23.14 Pavement Parking Markings	Crew	Daily Output	Labor-Hours	Unit	Material	Labor	Equipment	Total	Total Incl O&P
0010	**PAVEMENT PARKING MARKINGS**									
0790	Layout of pavement marking	A-2	25000	.001	L.F.		.04	.01	.05	.06
0800	Lines on pvmt., parking stall, paint, white, 4" wide	B-78B	400	.045	Stall	4.62	1.70	.98	7.30	8.80
0825	Parking stall, small quantities	2 Pord	80	.200		9.25	7.90		17.15	22
0830	Lines on pvmt., parking stall, thermoplastic, white, 4" wide	B-79	300	.133		13	4.94	5.25	23.19	27.50
1000	Street letters and numbers	B-78B	1600	.011	S.F.	.70	.43	.24	1.37	1.69
1100	Pavement marking letter, 6"	2 Pord	400	.040	Ea.	11	1.58		12.58	14.50
1110	12" letter		272	.059		13.90	2.33		16.23	18.80
1120	24" letter		160	.100		30.50	3.96		34.46	40
1130	36" letter		84	.190		40	7.55		47.55	55.50
1140	42" letter		84	.190		53	7.55		60.55	70
1150	72" letter		40	.400		43.50	15.80		59.30	72
1200	Handicap symbol		40	.400		32	15.80		47.80	59
1210	Handicap Parking sign 12" x 18 " and post	A-2	12	2		128	73	22.50	223.50	279
1300	Pavement marking, thermoplastic tape including layout, 4 inch width	B-79B	320	.025	L.F.	2.50	.92	.48	3.90	4.68
1310	12 inch width		192	.042	"	6.75	1.53	.79	9.07	10.70
1320	Letters including layout, 4 inch		240	.033	Ea.	12.25	1.22	.63	14.10	16.10
1330	6 inch		160	.050		15	1.83	.95	17.78	20.50
1340	12 inch		120	.067		19	2.44	1.27	22.71	26
1350	48 inch		64	.125		71	4.58	2.38	77.96	87.50
1360	96 inch		32	.250		90	9.15	4.75	103.90	118
1380	4 letter words, 8 feet tall		8	1		335	36.50	19	390.50	450
1385	Sample words: "BUMP, FIRE, LANE"									

32 17 23.33 Plastic Pavement Markings

	32 17 23.33 Plastic Pavement Markings	Crew	Daily Output	Labor-Hours	Unit	Material	Labor	Equipment	Total	Total Incl O&P
0010	**PLASTIC PAVEMENT MARKINGS**									
0020	Thermoplastic markings with glass beads									
0100	Thermoplastic striping, 6# beads per 100 S.F, white or yellow, 4" wide	B-79	15000	.003	L.F.	.31	.10	.10	.51	.61
0110	8# beads per 100 S.F		15000	.003		.32	.10	.10	.52	.62
0120	10# beads per 100 S.F		15000	.003		.32	.10	.10	.52	.62
0130	12# beads per 100 S.F		15000	.003		.32	.10	.10	.52	.63
0200	6# beads per 100 S.F, white or yellow, 6" wide		14000	.003		.47	.11	.11	.69	.80
0210	8# beads per 100 S.F		14000	.003		.47	.11	.11	.69	.80

32 17 Paving Specialties

32 17 23 – Pavement Markings

32 17 23.33 Plastic Pavement Markings		Crew	Daily Output	Labor-Hours	Unit	2014 Bare Costs Material	Labor	Equipment	Total	Total Incl O&P
0220	10# beads per 100 S.F	B-79	14000	.003	L.F.	.48	.11	.11	.70	.81
0230	12# beads per 100 S.F		14000	.003		.49	.11	.11	.71	.82
0300	6# beads per 100 S.F, white or yellow, 8" wide		12000	.003		.62	.12	.13	.87	1.02
0310	8# beads per 100 S.F		12000	.003		.63	.12	.13	.88	1.03
0320	10# beads per 100 S.F		12000	.003		.64	.12	.13	.89	1.03
0330	12# beads per 100 S.F		12000	.003		.65	.12	.13	.90	1.04
0400	6# beads per 100 S.F, white or yellow, 12" wide		6000	.007		.94	.25	.26	1.45	1.70
0410	8# beads per 100 S.F		6000	.007		.95	.25	.26	1.46	1.71
0420	10# beads per 100 S.F		6000	.007		.96	.25	.26	1.47	1.73
0430	12# beads per 100 S.F		6000	.007		.97	.25	.26	1.48	1.74
0500	Gore lines with 6# beads per 100 S.F.		2500	.016	S.F.	.64	.59	.63	1.86	2.30
0510	8# beads per 100 S.F.		2500	.016		.65	.59	.63	1.87	2.31
0520	10# beads per 100 S.F.		2500	.016		.66	.59	.63	1.88	2.33
0530	12# beads per 100 S.F.		2500	.016		.67	.59	.63	1.89	2.34
0600	Arrows with 6# beads per100 S.F.		660	.061		.64	2.24	2.38	5.26	6.75
0610	8# beads per100 S.F.		660	.061		.65	2.24	2.38	5.27	6.80
0620	10# beads per100 S.F.		660	.061		.66	2.24	2.38	5.28	6.80
0630	12# beads per100 S.F.		660	.061		.67	2.24	2.38	5.29	6.80
0700	Letters with 6# beads per 100 S.F.		660	.061		.64	2.24	2.38	5.26	6.75
0710	8# beads per 100 S.F.		660	.061		.65	2.24	2.38	5.27	6.80
0720	10# beads per 100 S.F.		660	.061		.66	2.24	2.38	5.28	6.80
0730	12# beads per 100 S.F.		660	.061		.67	2.24	2.38	5.29	6.80
1000	Airport thermoplastic markings									
1050	Traffic safety flashing truck for airport markings	A-2B	1	8	Day		292	270	562	740
1100	Thermoplastic taxiway markings with 24# beads per 100 S.F.	B-79	2500	.016	S.F.	1.16	.59	.63	2.38	2.88
1200	Runway markings		2100	.019		1.16	.71	.75	2.62	3.19
1250	Location or direction signs		1500	.027		1.16	.99	1.05	3.20	3.95
1350	Mobilization airport pavement marking		4	10	Ea.		370	395	765	1,000
1400	Thermostatic markings or pavement signs removal daytime	B-78B	400	.045	S.F.		1.70	.98	2.68	3.70
1500	Removal nighttime		335	.054	"		2.03	1.17	3.20	4.42
1600	Mobilization thermostatic marking removal		4	4.500	Ea.		170	98	268	370
1700	Thermoplastic road marking material, small users				Ton	1,775			1,775	1,950
1710	Highway departments					1,550			1,550	1,700
1720	Thermoplastic airport marking material, small users					1,875			1,875	2,075
1730	Large users					1,650			1,650	1,800
1800	Glass beads material for highway use, type II				Lb.	.61			.61	.67
1900	Glass beads material for airport use, type III				"	2.25			2.25	2.48

32 18 Athletic and Recreational Surfacing

32 18 23 – Athletic Surfacing

32 18 23.33 Running Track Surfacing

Line	Description	Crew	Daily Output	Labor-Hours	Unit	Material	Labor	Equipment	Total	Total Incl O&P
0010	**RUNNING TRACK SURFACING**									
0020	Running track, asphalt, incl base, 3" thick	B-37	300	.160	S.Y.	24	6.20	.53	30.73	36
0102	Surface, latex rubber system, 1/2" thick, black	B-20	115	.209		39	8.55		47.55	56
0152	Colors		115	.209		48	8.55		56.55	66
0302	Urethane rubber system, 1/2" thick, black		110	.218		29	8.90		37.90	45.50
0402	Color coating		110	.218		35.50	8.90		44.40	53

32 31 Fences and Gates

32 31 13 – Chain Link Fences and Gates

32 31 13.20 Fence, Chain Link Industrial		Crew	Daily Output	Labor-Hours	Unit	Material	Labor (2014 Bare Costs)	Equipment	Total	Total Incl O&P
0010	**FENCE, CHAIN LINK INDUSTRIAL**									
0011	Schedule 40, including concrete									
0020	3 strands barb wire, 2" post @ 10' O.C., set in concrete, 6' H									
0200	9 ga. wire, galv. steel, in concrete	B-80C	240	.100	L.F.	19.10	3.66	1.16	23.92	28
0248	Fence, add for vinyl coated fabric				S.F.	.68			.68	.75
0300	Aluminized steel	B-80C	240	.100	L.F.	19.70	3.66	1.16	24.52	28.50
0500	6 ga. wire, galv. steel		240	.100		20.50	3.66	1.16	25.32	30
0600	Aluminized steel		240	.100		29.50	3.66	1.16	34.32	39.50
0800	6 ga. wire, 6' high but omit barbed wire, galv. steel		250	.096		19.40	3.51	1.11	24.02	28
0900	Aluminized steel, in concrete		250	.096		23.50	3.51	1.11	28.12	32.50
0920	8' H, 6 ga. wire, 2-1/2" line post, galv. steel, in concrete		180	.133		31	4.88	1.55	37.43	43
0940	Aluminized steel, in concrete		180	.133		37.50	4.88	1.55	43.93	50.50
1400	Gate for 6' high fence, 1-5/8" frame, 3' wide, galv. steel		10	2.400	Ea.	191	88	28	307	375
1500	Aluminized steel, in concrete		10	2.400	"	191	88	28	307	375
2000	5'-0" high fence, 9 ga., no barbed wire, 2" line post, in concrete									
2010	10' O.C., 1-5/8" top rail, in concrete									
2100	Galvanized steel, in concrete	B-80C	300	.080	L.F.	18.10	2.93	.93	21.96	25.50
2200	Aluminized steel, in concrete		300	.080	"	18.75	2.93	.93	22.61	26
2400	Gate, 4' wide, 5' high, 2" frame, galv. steel, in concrete		10	2.400	Ea.	176	88	28	292	360
2500	Aluminized steel, in concrete		10	2.400	"	192	88	28	308	380
3100	Overhead slide gate, chain link, 6' high, to 18' wide, in concrete		38	.632	L.F.	92.50	23	7.30	122.80	146
3105	8' high, in concrete	B-80	30	1.067		92.50	42.50	25.50	160.50	195
3108	10' high, in concrete		24	1.333		154	53	31.50	238.50	285
3110	Cantilever type, in concrete		48	.667		118	26.50	15.80	160.30	188
3120	8' high, in concrete		24	1.333		157	53	31.50	241.50	288
3130	10' high, in concrete		18	1.778		198	70.50	42	310.50	370
5000	Double swing gates, incl. posts & hardware, in concrete									
5010	5' high, 12' opening, in concrete	B-80C	3.40	7.059	Opng.	365	258	82	705	885
5020	20' opening, in concrete		2.80	8.571		485	315	99.50	899.50	1,125
5060	6' high, 12' opening, in concrete		3.20	7.500		435	275	87	797	990
5070	20' opening, in concrete		2.60	9.231		605	340	107	1,052	1,300
5080	8' high, 12' opening, in concrete	B-80	2.13	15.002		435	595	355	1,385	1,775
5090	20' opening, in concrete		1.45	22.069		655	875	525	2,055	2,650
5100	10' high, 12' opening, in concrete		1.31	24.427		770	970	580	2,320	2,950
5110	20' opening, in concrete		1.03	31.068		835	1,225	735	2,795	3,625
5120	12' high, 12' opening, in concrete		1.05	30.476		1,250	1,200	720	3,170	4,025
5130	20' opening, in concrete		.85	37.647		1,325	1,500	890	3,715	4,725
5190	For aluminized steel add					20%				
7055	Braces, galv. steel	B-80A	960	.025	L.F.	2.79	.92	.35	4.06	4.85
7056	Aluminized steel	"	960	.025	"	3.34	.92	.35	4.61	5.45
7075	Fence, for small jobs 100 L.F. or less fence w/or wo gate, add				S.F.	20%				

32 31 13.25 Fence, Chain Link Residential		Crew	Daily Output	Labor-Hours	Unit	Material	Labor (2014 Bare Costs)	Equipment	Total	Total Incl O&P
0010	**FENCE, CHAIN LINK RESIDENTIAL**									
0011	Schedule 20, 11 ga. wire, 1-5/8" post									
0020	10' O.C., 1-3/8" top rail, 2" corner post, galv. stl. 3' high	B-80C	500	.048	L.F.	2	1.76	.56	4.32	5.50
0050	4' high		400	.060		7.05	2.20	.70	9.95	11.90
0100	6' high		200	.120		9.70	4.39	1.39	15.48	18.95
0150	Add for gate 3' wide, 1-3/8" frame, 3' high		12	2	Ea.	81	73	23	177	228
0170	4' high		10	2.400		87	88	28	203	261
0190	6' high		10	2.400		110	88	28	226	287
0200	Add for gate 4' wide, 1-3/8" frame, 3' high		9	2.667		88	97.50	31	216.50	281
0220	4' high		9	2.667		96	97.50	31	224.50	290

32 31 13 – Chain Link Fences and Gates

32 31 13.25 Fence, Chain Link Residential		Crew	Daily Output	Labor-Hours	Unit	2014 Bare Costs Material	Labor	Equipment	Total	Total Incl O&P
0240	6′ high	B-80C	8	3	Ea.	120	110	35	265	340
0350	Aluminized steel, 11 ga. wire, 3′ high		500	.048	L.F.	8.05	1.76	.56	10.37	12.15
0380	4′ high		400	.060		9	2.20	.70	11.90	14.05
0400	6′ high		200	.120		10.75	4.39	1.39	16.53	20
0450	Add for gate 3′ wide, 1-3/8″ frame, 3′ high		12	2	Ea.	89	73	23	185	237
0470	4′ high		10	2.400		97.50	88	28	213.50	273
0490	6′ high		10	2.400		122	88	28	238	300
0500	Add for gate 4′ wide, 1-3/8″ frame, 3′ high		10	2.400		98.50	88	28	214.50	274
0520	4′ high		9	2.667		118	97.50	31	246.50	315
0540	6′ high		8	3		131	110	35	276	350
0620	Vinyl covered, 9 ga. wire, 3′ high		500	.048	L.F.	7.15	1.76	.56	9.47	11.15
0640	4′ high		400	.060		8.15	2.20	.70	11.05	13.10
0660	6′ high		200	.120		10.05	4.39	1.39	15.83	19.35
0720	Add for gate 3′ wide, 1-3/8″ frame, 3′ high		12	2	Ea.	89	73	23	185	237
0740	4′ high		10	2.400		96	88	28	212	272
0760	6′ high		10	2.400		120	88	28	236	298
0780	Add for gate 4′ wide, 1-3/8″ frame, 3′ high		10	2.400		96.50	88	28	212.50	272
0800	4′ high		9	2.667		101	97.50	31	229.50	295
0820	6′ high		8	3		128	110	35	273	350
7076	Fence, for small jobs 100 L.F. fence or less w/or wo gate, add				S.F.	20%				

32 31 13.26 Tennis Court Fences and Gates		Crew	Daily Output	Labor-Hours	Unit	Material	Labor	Equipment	Total	Total Incl O&P
0010	**TENNIS COURT FENCES AND GATES**									
0860	Tennis courts, 11 ga. wire, 2-1/2″ post set									
0870	in concrete, 10′ O.C., 1-5/8″ top rail									
0900	10′ high	B-80	190	.168	L.F.	21.50	6.70	3.99	32.19	38
0920	12′ high		170	.188	″	22.50	7.50	4.46	34.46	41.50
1000	Add for gate 4′ wide, 1-5/8″ frame 7′ high		10	3.200	Ea.	228	127	76	431	530
1040	Aluminized steel, 11 ga. wire 10′ high		190	.168	L.F.	19.35	6.70	3.99	30.04	36
1100	12′ high		170	.188	″	21.50	7.50	4.46	33.46	40
1140	Add for gate 4′ wide, 1-5/8″ frame, 7′ high		10	3.200	Ea.	258	127	76	461	565
1250	Vinyl covered, 9 ga. wire, 10′ high		190	.168	L.F.	19.80	6.70	3.99	30.49	36.50
1300	12′ high		170	.188	″	23	7.50	4.46	34.96	42
1310	Fence, CL, tennis court, transom gate, single, galv., 4′ x 7′	B-80A	8.72	2.752	Ea.	295	101	38	434	525
1400	Add for gate 4′ wide, 1-5/8″ frame, 7′ high	B-80	10	3.200	″	310	127	76	513	620

32 31 13.30 Fence, Chain Link, Gates and Posts		Crew	Daily Output	Labor-Hours	Unit	Material	Labor	Equipment	Total	Total Incl O&P
0010	**FENCE, CHAIN LINK, GATES & POSTS**									
0011	(1/3 post length in ground)									
6580	Line posts, galvanized, 2-1/2″ OD, set in conc., 4′	B-80	80	.400	Ea.	29	15.90	9.50	54.40	66.50
6585	5′		76	.421		33	16.70	9.95	59.65	73
6590	6′		74	.432		35.50	17.15	10.25	62.90	77
6595	7′		72	.444		43	17.65	10.55	71.20	86
6600	8′		69	.464		46.50	18.40	11	75.90	91
6610	H-beam, 1-7/8″, 4′		83	.386		32	15.30	9.15	56.45	68.50
6615	5′		81	.395		37	15.70	9.35	62.05	75
6620	6′		78	.410		43	16.30	9.70	69	82.50
6625	7′		75	.427		47	16.95	10.10	74.05	88.50
6630	8′		73	.438		53	17.40	10.40	80.80	96
6635	Vinyl coated, 2-1/2″ OD, set in conc., 4′		79	.405		37.50	16.10	9.60	63.20	76.50
6640	5′		77	.416		37	16.50	9.85	63.35	77
6645	6′		74	.432		50	17.15	10.25	77.40	93
6650	7′		72	.444		62.50	17.65	10.55	90.70	108
6655	8′		69	.464		69	18.40	11	98.40	116

32 31 13.30 Fence, Chain Link, Gates and Posts		Crew	Daily Output	Labor-Hours	Unit	2014 Bare Costs Material	Labor	Equipment	Total	Total Incl O&P
6660	End gate post, steel, 3″ OD, set in conc.,4′	B-80	68	.471	Ea.	42.50	18.70	11.15	72.35	88
6665	5′		65	.492		46	19.55	11.65	77.20	93.50
6670	6′		63	.508		50	20	12.05	82.05	99.50
6675	7′		61	.525		57	21	12.45	90.45	108
6680	8′		59	.542		65.50	21.50	12.85	99.85	119
6685	Vinyl, 4′		68	.471		47	18.70	11.15	76.85	92.50
6690	5′		65	.492		54.50	19.55	11.65	85.70	103
6695	6′		63	.508		87	20	12.05	119.05	140
6700	7′		61	.525		103	21	12.45	136.45	160
6705	8′		59	.542		115	21.50	12.85	149.35	174
6710	Corner post, galv. steel, 4″ OD, set in conc., 4′		65	.492		90	19.55	11.65	121.20	141
6715	6′		63	.508		99.50	20	12.05	131.55	153
6720	7′		61	.525		125	21	12.45	158.45	183
6725	8′		65	.492		129	19.55	11.65	160.20	185
6730	Vinyl, 5′		65	.492		85.50	19.55	11.65	116.70	137
6735	6′		63	.508		128	20	12.05	160.05	185
6740	7′		61	.525		152	21	12.45	185.45	213
6745	8′		59	.542		168	21.50	12.85	202.35	232
7031	For corner, end, & pull post bracing, add					20%	15%			
7795	Cantilever, manual, exp. roller, (pr), 40′ wide x 8′ high	B-22	1	30	Ea.	5,600	1,275	210	7,085	8,350
7800	30′ wide x 8′ high		1	30		4,275	1,275	210	5,760	6,900
7805	24′ wide x 8′ high		1	30		3,475	1,275	210	4,960	6,025
7810	Motor operators for gates, (no elec wiring), 3′ wide swing	2 Skwk	.50	32		1,200	1,525		2,725	3,650
7815	Up to 20′ wide swing		.50	32		1,500	1,525		3,025	4,000
7820	Up to 45′ sliding		.50	32		2,775	1,525		4,300	5,400
7825	Overhead gate, 6′ to 18′ wide, sliding/cantilever		45	.356	L.F.	310	16.80		326.80	365
7830	Gate operators, digital receiver		7	2.286	Ea.	73	108		181	248
7835	Two button transmitter		24	.667		22.50	31.50		54	74
7840	3 button station		14	1.143		20	54		74	106
7845	Master slave system		4	4		158	189		347	465
7900	Auger fence post hole, 3′ deep, medium soil, by hand	1 Clab	30	.267			9.75		9.75	15.10
7925	By machine	B-80	175	.183			7.25	4.33	11.58	15.85
7950	Rock, with jackhammer	B-9	32	1.250			46.50	7.30	53.80	79.50
7975	With rock drill	B-47C	65	.246			10.30	25	35.30	43.50

32 31 13.33 Chain Link Backstops

		Crew	Daily Output	Labor-Hours	Unit	Material	Labor	Equipment	Total	Total Incl O&P
0010	**CHAIN LINK BACKSTOPS**									
0015	Backstops, baseball, prefabricated, 30′ wide, 12′ high & 1 overhang	B-1	1	24	Ea.	2,550	895		3,445	4,175
0100	40′ wide, 12′ high & 2 overhangs	"	.75	32		6,825	1,200		8,025	9,350
0300	Basketball, steel, single goal	B-13	3.04	18.421		1,425	735	245	2,405	2,975
0400	Double goal	"	1.92	29.167		1,950	1,175	390	3,515	4,350
0600	Tennis, wire mesh with pair of ends	B-1	2.48	9.677	Set	2,675	360		3,035	3,475
0700	Enclosed court	"	1.30	18.462	Ea.	9,225	690		9,915	11,200

32 31 13.40 Fence, Fabric and Accessories

		Crew	Daily Output	Labor-Hours	Unit	Material	Labor	Equipment	Total	Total Incl O&P
0010	**FENCE, FABRIC & ACCESSORIES**									
1000	Fabric, 9 ga., galv., 1.2 oz. coat, 2″ chain link, 4′	B-80A	304	.079	L.F.	3.36	2.89	1.10	7.35	9.35
1150	5′		285	.084		4.08	3.09	1.17	8.34	10.55
1200	6′		266	.090		7.75	3.31	1.25	12.31	15.05
1250	7′		247	.097		9.40	3.56	1.35	14.31	17.35
1300	8′		228	.105		12.10	3.86	1.46	17.42	21
1400	9 ga., fused, 4′		304	.079		3.79	2.89	1.10	7.78	9.85
1450	5′		285	.084		4.40	3.09	1.17	8.66	10.90
1500	6′		266	.090		4.61	3.31	1.25	9.17	11.55

32 31 Fences and Gates

32 31 13 – Chain Link Fences and Gates

32 31 13.40 Fence, Fabric and Accessories		Crew	Daily Output	Labor-Hours	Unit	2014 Bare Costs Material	Labor	Equipment	Total	Total Incl O&P
1550	7′	B-80A	247	.097	L.F.	5.95	3.56	1.35	10.86	13.55
1600	8′		228	.105		9.75	3.86	1.46	15.07	18.25
1650	Barbed wire, galv., cost per strand		2280	.011		.12	.39	.15	.66	.89
1700	Vinyl coated		2280	.011		.16	.39	.15	.70	.94
1750	Extension arms, 3 strands		143	.168	Ea.	4.36	6.15	2.33	12.84	16.85
1800	6 strands, 2-3/8″		119	.202		10.65	7.40	2.80	20.85	26
1850	Eye tops, 2-3/8″		143	.168		1.77	6.15	2.33	10.25	14
1900	Top rail, incl. tie wires, 1-5/8″, galv.		912	.026	L.F.	4.61	.96	.37	5.94	6.95
1950	Vinyl coated		912	.026		5.25	.96	.37	6.58	7.65
2100	Rail, middle/bottom, w/tie wire, 1-5/8″, galv.		912	.026		4.61	.96	.37	5.94	6.95
2150	Vinyl coated		912	.026		5.25	.96	.37	6.58	7.65
2200	Reinforcing wire, coiled spring, 7 ga. galv.		2279	.011		.16	.39	.15	.70	.94
2250	9 ga., vinyl coated		2282	.011		.49	.39	.15	1.03	1.29
2300	Steel T-post, galvanized with clips, 5′, common earth, flat		200	.120	Ea.	8.80	4.40	1.67	14.87	18.35
2310	Clay		176	.136		8.80	5	1.89	15.69	19.50
2320	Soil & rock		144	.167		8.80	6.10	2.31	17.21	21.50
2330	5.5′, common earth, flat		200	.120		9.95	4.40	1.67	16.02	19.60
2340	Clay		176	.136		9.95	5	1.89	16.84	20.50
2350	Soil & rock		144	.167		9.95	6.10	2.31	18.36	23
2360	6′, common earth, flat		200	.120		10.75	4.40	1.67	16.82	20.50
2370	Clay		176	.136		10.75	5	1.89	17.64	21.50
2375	Soil & rock		144	.167		10.75	6.10	2.31	19.16	24
2600	Steel T-post, galvanized with clips, 5′, common earth, hills		180	.133		8.80	4.89	1.85	15.54	19.30
2610	Clay		160	.150		8.80	5.50	2.08	16.38	20.50
2620	Soil & rock		130	.185		8.80	6.75	2.56	18.11	23
2630	5.5′, common earth, hills		180	.133		9.95	4.89	1.85	16.69	20.50
2640	Clay		160	.150		9.95	5.50	2.08	17.53	21.50
2650	Soil & rock		130	.185		9.95	6.75	2.56	19.26	24
2660	6′, common earth, hills		180	.133		10.75	4.89	1.85	17.49	21.50
2670	Clay		160	.150		10.75	5.50	2.08	18.33	22.50
2675	Soil & rock		130	.185		10.75	6.75	2.56	20.06	25

32 31 23 – Plastic Fences and Gates

32 31 23.10 Fence, Vinyl		Crew	Daily Output	Labor-Hours	Unit	Material	Labor	Equipment	Total	Total Incl O&P
0010	**FENCE, VINYL**									
0011	White, steel reinforced, stainless steel fasteners									
0020	Picket, 4″ x 4″ posts @ 6′ - 0″ OC, 3′ high	B-1	140	.171	L.F.	21	6.40		27.40	33
0030	4′ high		130	.185		23.50	6.90		30.40	36
0040	5′ high		120	.200		26	7.45		33.45	40
0100	Board (semi-privacy), 5″ x 5″ posts @ 7′ - 6″ OC, 5′ high		130	.185		23	6.90		29.90	36
0120	6′ high		125	.192		27	7.15		34.15	40.50
0200	Basketweave, 5″ x 5″ posts @ 7′ - 6″ OC, 5′ high		160	.150		23.50	5.60		29.10	34.50
0220	6′ high		150	.160		26.50	5.95		32.45	38.50
0300	Privacy, 5″ x 5″ posts @ 7′ - 6″ OC, 5′ high		130	.185		23	6.90		29.90	36
0320	6′ high		150	.160		27	5.95		32.95	38.50
0350	Gate, 5′ high		9	2.667	Ea.	320	99.50		419.50	510
0360	6′ high		9	2.667		370	99.50		469.50	560
0400	For posts set in concrete, add		25	.960		8.10	36		44.10	64.50
0500	Post and rail fence, 2 rail		150	.160	L.F.	5.40	5.95		11.35	15.15
0510	3 rail		150	.160		6.95	5.95		12.90	16.85
0515	4 rail		150	.160		9.05	5.95		15	19.15

32 31 26 – Wire Fences and Gates

32 31 26.10 Fences, Misc. Metal		Crew	Daily Output	Labor-Hours	Unit	Material	Labor (2014 Bare Costs)	Equipment	Total	Total Incl O&P
0010	**FENCES, MISC. METAL**									
0012	Chicken wire, posts @ 4′, 1″ mesh, 4′ high	B-80C	410	.059	L.F.	3.61	2.14	.68	6.43	8
0100	2″ mesh, 6′ high		350	.069		3.80	2.51	.79	7.10	8.90
0200	Galv. steel, 12 ga., 2″ x 4″ mesh, posts 5′ O.C., 3′ high		300	.080		2.75	2.93	.93	6.61	8.55
0300	5′ high		300	.080		3.35	2.93	.93	7.21	9.20
0400	14 ga., 1″ x 2″ mesh, 3′ high		300	.080		3.40	2.93	.93	7.26	9.25
0500	5′ high		300	.080		4.55	2.93	.93	8.41	10.50
1000	Kennel fencing, 1-1/2″ mesh, 6′ long, 3′-6″ wide, 6′-2″ high	2 Clab	4	4	Ea.	500	147		647	775
1050	12′ long		4	4		715	147		862	1,000
1200	Top covers, 1-1/2″ mesh, 6′ long		15	1.067		135	39		174	210
1250	12′ long		12	1.333		190	49		239	285
4500	Security fence, prison grade, set in concrete, 12′ high	B-80	25	1.280	L.F.	61	51	30.50	142.50	179
4600	16′ high	"	20	1.600	"	79	63.50	38	180.50	226

32 31 26.20 Wire Fencing, General		Crew	Daily Output	Labor-Hours	Unit	Material	Labor	Equipment	Total	Total Incl O&P
0010	**WIRE FENCING, GENERAL**									
0015	Barbed wire, galvanized, domestic steel, hi-tensile 15-1/2 ga.				M.L.F.	91			91	100
0020	Standard, 12-3/4 ga.					103			103	113
0210	Barbless wire, 2-strand galvanized, 12-1/2 ga.					103			103	113
0500	Helical razor ribbon, stainless steel, 18″ dia x 18″ spacing				C.L.F.	168			168	185
0600	Hardware cloth galv., 1/4″ mesh, 23 ga., 2′ wide				C.S.F.	65			65	71.50
0700	3′ wide					48			48	53
0900	1/2″ mesh, 19 ga., 2′ wide					37.50			37.50	41
1000	4′ wide					23.50			23.50	26
1200	Chain link fabric, steel, 2″ mesh, 6 ga., galvanized					247			247	271
1300	9 ga., galvanized					86.50			86.50	95.50
1350	Vinyl coated					88.50			88.50	97
1360	Aluminized					80			80	88
1400	2-1/4″ mesh, 11.5 ga., galvanized					54			54	59.50
1600	1-3/4″ mesh (tennis courts), 11.5 ga. (core), vinyl coated					54			54	59.50
1700	9 ga., galvanized					84			84	92.50
2100	Welded wire fabric, galvanized, 1″ x 2″, 14 ga.					60			60	66
2200	2″ x 4″, 12-1/2 ga.					60			60	66

32 31 29 – Wood Fences and Gates

32 31 29.10 Fence, Wood		Crew	Daily Output	Labor-Hours	Unit	Material	Labor	Equipment	Total	Total Incl O&P
0010	**FENCE, WOOD**									
0011	Basket weave, 3/8″ x 4″ boards, 2″ x 4″									
0020	stringers on spreaders, 4″ x 4″ posts									
0050	No. 1 cedar, 6′ high	B-80C	160	.150	L.F.	25.50	5.50	1.74	32.74	38.50
0070	Treated pine, 6′ high	"	150	.160	"	34	5.85	1.85	41.70	48.50
0200	Board fence, 1″ x 4″ boards, 2″ x 4″ rails, 4″ x 4″ post									
0220	Preservative treated, 2 rail, 3′ high	B-80C	145	.166	L.F.	9.05	6.05	1.92	17.02	21.50
0240	4′ high		135	.178		10.45	6.50	2.06	19.01	24
0260	3 rail, 5′ high		130	.185		11.55	6.75	2.14	20.44	25.50
0300	6′ high		125	.192		13.25	7.05	2.23	22.53	28
0320	No. 2 grade western cedar, 2 rail, 3′ high		145	.166		9.80	6.05	1.92	17.77	22
0340	4′ high		135	.178		10.80	6.50	2.06	19.36	24
0360	3 rail, 5′ high		130	.185		12.25	6.75	2.14	21.14	26.50
0400	6′ high		125	.192		13.15	7.05	2.23	22.43	27.50
0420	No. 1 grade cedar, 2 rail, 3′ high		145	.166		11.85	6.05	1.92	19.82	24.50
0440	4′ high		135	.178		13.80	6.50	2.06	22.36	27.50
0460	3 rail, 5′ high		130	.185		16.95	6.75	2.14	25.84	31.50

32 31 Fences and Gates

32 31 29 – Wood Fences and Gates

32 31 29.10 Fence, Wood		Crew	Daily Output	Labor-Hours	Unit	2014 Bare Costs Material	Labor	Equipment	Total	Total Incl O&P
0500	6′ high	B-80C	125	.192	L.F.	18.45	7.05	2.23	27.73	34
0860	Open rail fence, split rails, 2 rail 3′ high, no. 1 cedar		160	.150		11.70	5.50	1.74	18.94	23
0870	No. 2 cedar		160	.150		10.90	5.50	1.74	18.14	22.50
0880	3 rail, 4′ high, no. 1 cedar		150	.160		12.10	5.85	1.85	19.80	24.50
0890	No. 2 cedar		150	.160		8.70	5.85	1.85	16.40	20.50
0920	Rustic rails, 2 rail 3′ high, no. 1 cedar		160	.150		9.35	5.50	1.74	16.59	20.50
0930	No. 2 cedar		160	.150		8.70	5.50	1.74	15.94	19.95
0940	3 rail, 4′ high		150	.160		9	5.85	1.85	16.70	21
0950	No. 2 cedar		150	.160		7.30	5.85	1.85	15	19.05
1240	Stockade fence, no. 1 cedar, 3-1/4″ rails, 6′ high		160	.150		11.95	5.50	1.74	19.19	23.50
1260	8′ high		155	.155		16.50	5.65	1.79	23.94	29
1270	Gate, 3′-6″ wide		9	2.667	Ea.	235	97.50	31	363.50	445
1300	No. 2 cedar, treated wood rails, 6′ high		160	.150	L.F.	12.30	5.50	1.74	19.54	24
1320	Gate, 3′-6″ wide		8	3	Ea.	80	110	35	225	296
1360	Treated pine, treated rails, 6′ high		160	.150	L.F.	12.80	5.50	1.74	20.04	24.50
1400	8′ high		150	.160	″	18.65	5.85	1.85	26.35	31.50
1420	Gate, 3′-6″ wide		9	2.667	Ea.	90	97.50	31	218.50	283

32 31 29.20 Fence, Wood Rail		Crew	Daily Output	Labor-Hours	Unit	Material	Labor	Equipment	Total	Total Incl O&P
0010	**FENCE, WOOD RAIL**									
0012	Picket, No. 2 cedar, Gothic, 2 rail, 3′ high	B-1	160	.150	L.F.	7.55	5.60		13.15	16.95
0050	Gate, 3′-6″ wide	B-80C	9	2.667	Ea.	76.50	97.50	31	205	268
0400	3 rail, 4′ high		150	.160	L.F.	8.45	5.85	1.85	16.15	20.50
0500	Gate, 3′-6″ wide		9	2.667	Ea.	94.50	97.50	31	223	288
5000	Fence rail, redwood, 2″ x 4″, merch. grade 8′	B-1	2400	.010	L.F.	2.39	.37		2.76	3.21
6000	Fence post, select redwood, earthpacked & treated, 4″ x 4″ x 6′		96	.250	Ea.	13.50	9.35		22.85	29.50
6010	4″ x 4″ x 8′		96	.250		18.50	9.35		27.85	35
6020	Set in concrete, 4″ x 4″ x 6′		50	.480		21	17.90		38.90	50.50
6030	4″ x 4″ x 8′		50	.480		22	17.90		39.90	51.50
6040	Wood post, 4′ high, set in concrete, incl. concrete		50	.480		13.50	17.90		31.40	42.50
6050	Earth packed		96	.250		16.50	9.35		25.85	32.50
6060	6′ high, set in concrete, incl. concrete		50	.480		17	17.90		34.90	46
6070	Earth packed		96	.250		12.90	9.35		22.25	28.50

32 32 Retaining Walls

32 32 13 – Cast-in-Place Concrete Retaining Walls

32 32 13.10 Retaining Walls, Cast Concrete		Crew	Daily Output	Labor-Hours	Unit	Material	Labor	Equipment	Total	Total Incl O&P
0010	**RETAINING WALLS, CAST CONCRETE**									
1800	Concrete gravity wall with vertical face including excavation & backfill									
1850	No reinforcing									
1900	6′ high, level embankment	C-17C	36	2.306	L.F.	71.50	110	17.20	198.70	269
2000	33° slope embankment		32	2.594		84	124	19.35	227.35	305
2200	8′ high, no surcharge		27	3.074		90	147	23	260	350
2300	33° slope embankment		24	3.458		109	165	26	300	405
2500	10′ high, level embankment		19	4.368		129	209	32.50	370.50	505
2600	33° slope embankment		18	4.611		178	220	34.50	432.50	575
2800	Reinforced concrete cantilever, incl. excavation, backfill & reinf.									
2900	6′ high, 33° slope embankment	C-17C	35	2.371	L.F.	66	113	17.70	196.70	267
3000	8′ high, 33° slope embankment		29	2.862		76	137	21.50	234.50	320
3100	10′ high, 33° slope embankment		20	4.150		99	198	31	328	450
3200	20′ high, 500 lb. per L.F. surcharge		7.50	11.067		297	530	82.50	909.50	1,225
3500	Concrete cribbing, incl. excavation and backfill									

32 32 Retaining Walls

32 32 13 – Cast-in-Place Concrete Retaining Walls

32 32 13.10 Retaining Walls, Cast Concrete		Crew	Daily Output	Labor-Hours	Unit	Material	2014 Bare Costs Labor	Equipment	Total	Total Incl O&P
3700	12′ high, open face	B-13	210	.267	S.F.	33	10.65	3.55	47.20	56.50
3900	Closed face	″	210	.267	″	31	10.65	3.55	45.20	54
4100	Concrete filled slurry trench, see Section 31 56 23.20									

32 32 23 – Segmental Retaining Walls

32 32 23.13 Segmental Conc. Unit Masonry Retaining Walls

		Crew	Daily Output	Labor-Hours	Unit	Material	Labor	Equipment	Total	Total Incl O&P
0010	**SEGMENTAL CONC. UNIT MASONRY RETAINING WALLS**									
7100	Segmental Retaining Wall system, incl. pins, and void fill									
7120	base not included									
7140	Large unit, 8″ high x 18″ wide x 20″ deep, 3 plane split	B-62	300	.080	S.F.	13.60	3.21	.58	17.39	20.50
7150	Straight split		300	.080		13.50	3.21	.58	17.29	20.50
7160	Medium, lt. wt., 8″ high x 18″ wide x 12″ deep, 3 plane split		400	.060		10.45	2.41	.44	13.30	15.65
7170	Straight split		400	.060		9.85	2.41	.44	12.70	15
7180	Small unit, 4″ x 18″ x 10″ deep, 3 plane split		400	.060		13.35	2.41	.44	16.20	18.80
7190	Straight split		400	.060		13	2.41	.44	15.85	18.45
7200	Cap unit, 3 plane split		300	.080		13.60	3.21	.58	17.39	20.50
7210	Cap unit, straight split		300	.080		13.60	3.21	.58	17.39	20.50
7250	Geo-grid soil reinforcement 4′ x 50′	2 Clab	22500	.001		.70	.03		.73	.81
7255	Geo-grid soil reinforcement 6′ x 150′	″	22500	.001		.56	.03		.59	.66
8000	For higher walls, add components as necessary									

32 32 26 – Metal Crib Retaining Walls

32 32 26.10 Metal Bin Retaining Walls

		Crew	Daily Output	Labor-Hours	Unit	Material	Labor	Equipment	Total	Total Incl O&P
0010	**METAL BIN RETAINING WALLS**									
0011	Aluminized steel bin, excavation									
0020	and backfill not included, 10′ wide									
0100	4′ high, 5.5′ deep	B-13	650	.086	S.F.	27.50	3.43	1.15	32.08	36.50
0200	8′ high, 5.5′ deep		615	.091		31.50	3.63	1.21	36.34	41.50
0300	10′ high, 7.7′ deep		580	.097		34.50	3.85	1.29	39.64	45.50
0400	12′ high, 7.7′ deep		530	.106		37.50	4.21	1.41	43.12	49
0500	16′ high, 7.7′ deep		515	.109		39.50	4.33	1.45	45.28	51.50
0600	16′ high, 9.9′ deep		500	.112		42.50	4.46	1.49	48.45	55.50
0700	20′ high, 9.9′ deep		470	.119		48	4.75	1.59	54.34	62
0800	20′ high, 12.1′ deep		460	.122		43.50	4.85	1.62	49.97	56.50
0900	24′ high, 12.1′ deep		455	.123		46	4.91	1.64	52.55	60.50
1000	24′ high, 14.3′ deep		450	.124		54	4.96	1.66	60.62	68.50
1100	28′ high, 14.3′ deep		440	.127		56	5.05	1.70	62.75	71
1300	For plain galvanized bin type walls, deduct					10%				

32 32 29 – Timber Retaining Walls

32 32 29.10 Landscape Timber Retaining Walls

		Crew	Daily Output	Labor-Hours	Unit	Material	Labor	Equipment	Total	Total Incl O&P
0010	**LANDSCAPE TIMBER RETAINING WALLS**									
0100	Treated timbers, 6″ x 6″	1 Clab	265	.030	L.F.	2	1.11		3.11	3.91
0110	6″ x 8″	″	200	.040	″	2.62	1.47		4.09	5.15
0120	Drilling holes in timbers for fastening, 1/2″	1 Carp	450	.018	Inch		.82		.82	1.26
0130	5/8″	″	450	.018	″		.82		.82	1.26
0140	Reinforcing rods for fastening, 1/2″	1 Clab	312	.026	L.F.	.37	.94		1.31	1.85
0150	5/8″	″	312	.026	″	.57	.94		1.51	2.08
0160	Reinforcing fabric	2 Clab	2500	.006	S.Y.	1.88	.23		2.11	2.43
0170	Gravel backfill		28	.571	C.Y.	19.90	21		40.90	54.50
0180	Perforated pipe, 4″ diameter with silt sock		1200	.013	L.F.	1.29	.49		1.78	2.17
0190	Galvanized 60d common nails	1 Clab	625	.013	Ea.	.16	.47		.63	.90
0200	20d common nails	″	3800	.002	″	.04	.08		.12	.16

32 32 Retaining Walls

32 32 36 – Gabion Retaining Walls

32 32 36.10 Stone Gabion Retaining Walls		Crew	Daily Output	Labor-Hours	Unit	2014 Bare Costs Material	Labor	Equipment	Total	Total Incl O&P
0010	**STONE GABION RETAINING WALLS**									
4300	Stone filled gabions, not incl. excavation,									
4310	Stone, delivered, 3′ wide									
4340	Galvanized, 6′ long, 1′ high	B-13	113	.496	Ea.	106	19.75	6.60	132.35	155
4400	1′-6″ high		50	1.120		165	44.50	14.90	224.40	267
4490	3′-0″ high		13	4.308		245	172	57.50	474.50	595
4590	9′ long, 1′ high		50	1.120		200	44.50	14.90	259.40	305
4650	1′-6″ high		22	2.545		213	101	34	348	430
4690	3′-0″ high		6	9.333		385	370	124	879	1,125
4890	12′ long, 1′ high		28	2		245	79.50	26.50	351	420
4950	1′-6″ high		13	4.308		325	172	57.50	554.50	680
4990	3′-0″ high		3	18.667		460	745	249	1,454	1,925
5200	PVC coated, 6′ long, 1′ high		113	.496		115	19.75	6.60	141.35	165
5250	1′-6″ high		50	1.120		175	44.50	14.90	234.40	278
5300	3′ high		13	4.308		265	172	57.50	494.50	620
5500	9′ long, 1′ high		50	1.120		220	44.50	14.90	279.40	325
5550	1′-6″ high		22	2.545		234	101	34	369	450
5600	3′ high		6	9.333		415	370	124	909	1,150
5800	12′ long, 1′ high		28	2		287	79.50	26.50	393	465
5850	1′-6″ high		13	4.308		350	172	57.50	579.50	710
5900	3′ high		3	18.667		490	745	249	1,484	1,975
6000	Galvanized, 6′ long, 1′ high		75	.747	C.Y.	96	30	9.95	135.95	162
6010	1′-6″ high		50	1.120		165	44.50	14.90	224.40	267
6020	3′-0″ high		25	2.240		123	89.50	30	242.50	305
6030	9′ long, 1′ high		50	1.120		200	44.50	14.90	259.40	305
6040	1′-6″ high		33.30	1.682		142	67	22.50	231.50	284
6050	3′-0″ high		16.70	3.353		128	134	44.50	306.50	395
6060	12′ long, 1′ high		37.50	1.493		184	59.50	19.90	263.40	315
6070	1′-6″ high		25	2.240		162	89.50	30	281.50	350
6080	3′-0″ high		12.50	4.480		115	179	59.50	353.50	465
6100	PVC coated, 6′ long, 1′ high		75	.747		173	30	9.95	212.95	246
6110	1′-6″ high		50	1.120		175	44.50	14.90	234.40	278
6120	3′ high		25	2.240		133	89.50	30	252.50	315
6130	9′ long, 1′ high		50	1.120		220	44.50	14.90	279.40	325
6140	1′-6″ high		33.30	1.682		156	67	22.50	245.50	300
6150	3′ high		16.67	3.359		138	134	45	317	405
6160	12′ long, 1′ high		37.50	1.493		215	59.50	19.90	294.40	350
6170	1′-6″ high		25	2.240		154	89.50	30	273.50	340
6180	3′ high		12.50	4.480		123	179	59.50	361.50	475

32 32 60 – Stone Retaining Walls

32 32 60.10 Retaining Walls, Stone		Crew	Daily Output	Labor-Hours	Unit	Material	Labor	Equipment	Total	Total Incl O&P
0010	**RETAINING WALLS, STONE**									
0015	Including excavation, concrete footing and									
0020	stone 3′ below grade. Price is exposed face area.									
0200	Decorative random stone, to 6′ high, 1′-6″ thick, dry set	D-1	35	.457	S.F.	30.50	18.90		49.40	62.50
0300	Mortar set		40	.400		33.50	16.50		50	62
0500	Cut stone, to 6′ high, 1′-6″ thick, dry set		35	.457		31.50	18.90		50.40	64
0600	Mortar set		40	.400		32.50	16.50		49	61
0800	Random stone, 6′ to 10′ high, 2′ thick, dry set		45	.356		41	14.70		55.70	67.50
0900	Mortar set		50	.320		42	13.20		55.20	66
1100	Cut stone, 6′ to 10′ high, 2′ thick, dry set		45	.356		40	14.70		54.70	66.50
1200	Mortar set		50	.320		42	13.20		55.20	66

32 32 Retaining Walls

32 32 60 – Stone Retaining Walls

32 32 60.10 Retaining Walls, Stone		Crew	Daily Output	Labor-Hours	Unit	Material	Labor	Equipment	Total	Total Incl O&P
						2014 Bare Costs				
5100	Setting stone, dry	D-1	100	.160	C.F.		6.60		6.60	10.05
5600	With mortar	↓	120	.133	"		5.50		5.50	8.40

32 34 Fabricated Bridges

32 34 10 – Fabricated Highway Bridges

32 34 10.10 Bridges, Highway		Crew	Daily Output	Labor-Hours	Unit	Material	Labor	Equipment	Total	Total Incl O&P
0010	**BRIDGES, HIGHWAY**									
0020	Structural steel, rolled beams	E-5	8.50	9.412	Ton	2,650	475	197	3,322	3,950
0500	Built up, plate girders	E-6	10.50	12.190	"	3,125	620	178	3,923	4,700
1000	Concrete in place, no reinforcing, abutment footings	C-17B	30	2.733	C.Y.	221	131	14.20	366.20	460
1050	Abutment		23	3.565		221	170	18.55	409.55	530
1100	Walls, stems and wing walls		20	4.100		298	196	21.50	515.50	655
1150	Parapets		12	6.833		298	325	35.50	658.50	870
1170	Pier footings	↓	40	2.050	↓	221	98	10.65	329.65	405
1180	Piers and columns, see Section 03 30 53.40									
1190	Pier caps including shoring	C-17B	10	8.200	C.Y.	350	390	42.50	782.50	1,025
1210	Beams, including shoring	C-14	15	9.600		430	430	110	970	1,250
1220	Haunches		10	14.400	↓	500	645	165	1,310	1,725
1230	Bridge sidewalks	↓	750	.192	S.F.	9.10	8.60	2.20	19.90	25.50
1250	Decks, including finish and cure, 8″ thick									
1260	Shored forms	C-14	3000	.048	S.F.	7.40	2.16	.55	10.11	12.10
1270	Beam supported forms		3500	.041		5.45	1.85	.47	7.77	9.35
1280	Stay in place metal slab forms, 20 ga.	↓	5000	.029	↓	7.05	1.29	.33	8.67	10.10
1500	Precast, prestressed concrete									
1510	Box girders, 35′ to 50′ span				Ea.				13,900	16,000
1520	50′ to 65′ span								18,100	20,800
1530	65′ to 85′ span				↓				23,700	27,300
1540	Deck beams, 12″ deep				S.F.				67	77
1541	15″ deep								76	87.50
1542	18″ deep								84.50	97
1543	20″ deep				↓				93.50	108
1550	Double T, 40′ to 45′ span				Ea.				11,600	13,400
1555	45′ to 55′ span								14,200	16,300
1560	55′ to 65′ span				↓				16,800	18,200
1580	Price per S.F.				S.F.				31	35.50
1600	I beams, 60′ to 80′ span				Ea.				12,400	14,200
1610	80′ to 100′ span								15,500	17,800
1620	100′ to 120′ span								18,500	21,300
1630	120′ to 140′ span				↓				21,600	24,900
2000	Reinforcing, in place	4 Rodm	3	10.667	Ton	2,100	540		2,640	3,175
2050	Galvanized coated		3	10.667		2,775	540		3,315	3,925
2100	Epoxy coated	↓	3	10.667	↓	2,650	540		3,190	3,750
2110	See also Section 03 21 11.50 & Section 03 21 11.60									
3000	Expansion dams, steel, double upset 4″ x 8″ angles welded to									
3010	double 8″ x 8″ angles, 1-3/4″ compression seal	C-22	30	1.400	L.F.	570	71	3.11	644.11	740
3040	Double 8″ x 8″ angles only, 1-3/4″ compression seal		35	1.200		430	61	2.66	493.66	570
3050	Galvanized		35	1.200		525	61	2.66	588.66	680
3060	Double 8″ x 6″ angles only, 1-3/4″ compression seal		35	1.200		330	61	2.66	393.66	460
3100	Double 10″ channels, 1-3/4″ compression seal	↓	35	1.200		299	61	2.66	362.66	430
3420	For 3″ compression seal, add					61.50			61.50	67.50
3440	For double slotted extrusions with seal strip, add				↓	275			275	305

32 34 Fabricated Bridges

32 34 10 – Fabricated Highway Bridges

32 34 10.10 Bridges, Highway		Crew	Daily Output	Labor-Hours	Unit	Material	Labor	Equipment	Total	Total Incl O&P
						2014 Bare Costs				
3490	For galvanizing, add				Lb.	.42			.42	.46
4000	Approach railings, steel, galv. pipe, 2 line	C-22	140	.300	L.F.	210	15.25	.67	225.92	256
4200	Bridge railings, steel, galv. pipe, 3 line w/screen		85	.494		300	25	1.10	326.10	370
4220	4 line w/screen		75	.560		350	28.50	1.24	379.74	430
4300	Aluminum, pipe, 3 line w/screen		95	.442		200	22.50	.98	223.48	257
8000	For structural excavation, see Section 31 23 16.16									
8010	For dewatering, see Section 31 23 19.20									

32 34 20 – Fabricated Pedestrian Bridges

32 34 20.10 Bridges, Pedestrian		Crew	Daily Output	Labor-Hours	Unit	Material	Labor	Equipment	Total	Total Incl O&P
0010	**BRIDGES, PEDESTRIAN**									
0011	Spans over streams, roadways, etc.									
0020	including erection, not including foundations									
0050	Precast concrete, complete in place, 8′ wide, 60′ span	E-2	215	.260	S.F.	114	13.05	7.10	134.15	155
0100	100′ span		185	.303		125	15.20	8.25	148.45	172
0150	120′ span		160	.350		136	17.55	9.55	163.10	190
0200	150′ span		145	.386		141	19.40	10.55	170.95	200
0300	Steel, trussed or arch spans, compl. in place, 8′ wide, 40′ span		320	.175		115	8.80	4.78	128.58	147
0400	50′ span		395	.142		103	7.10	3.87	113.97	129
0500	60′ span		465	.120		103	6.05	3.29	112.34	127
0600	80′ span		570	.098		123	4.93	2.68	130.61	146
0700	100′ span		465	.120		173	6.05	3.29	182.34	204
0800	120′ span		365	.153		218	7.70	4.19	229.89	258
0900	150′ span		310	.181		232	9.05	4.93	245.98	276
1000	160′ span		255	.220		232	11	6	249	280
1100	10′ wide, 80′ span		640	.088		123	4.39	2.39	129.78	145
1200	120′ span		415	.135		159	6.75	3.68	169.43	191
1300	150′ span		445	.126		178	6.30	3.44	187.74	211
1400	200′ span		205	.273		190	13.70	7.45	211.15	241
1600	Wood, laminated type, complete in place, 80′ span	C-12	203	.236		85	10.75	3.27	99.02	114
1700	130′ span	"	153	.314		88.50	14.25	4.33	107.08	124

32 35 Screening Devices

32 35 16 – Sound Barriers

32 35 16.10 Traffic Barriers, Highway Sound Barriers		Crew	Daily Output	Labor-Hours	Unit	Material	Labor	Equipment	Total	Total Incl O&P
0010	**TRAFFIC BARRIERS, HIGHWAY SOUND BARRIERS**									
0020	Highway sound barriers, not including footing									
0100	Precast concrete, concrete columns @ 30′ OC, 8″ T, 8′ H	C-12	400	.120	L.F.	130	5.45	1.66	137.11	153
0110	12′ H		265	.181		195	8.20	2.50	205.70	230
0120	16′ H		200	.240		260	10.90	3.31	274.21	305
0130	20′ H		160	.300		325	13.60	4.14	342.74	385
0400	Lt. Wt. composite panel, cementitious face, St. posts @ 12′ OC, 8′ H	B-80B	190	.168		154	6.60	1.31	161.91	181
0410	12′ H		125	.256		231	10.05	1.99	243.04	272
0420	16′ H		95	.337		310	13.20	2.61	325.81	365
0430	20′ H		75	.427		385	16.75	3.31	405.06	455

32 84 Planting Irrigation

32 84 23 – Underground Sprinklers

32 84 23.10 Sprinkler Irrigation System		Crew	Daily Output	Labor-Hours	Unit	2014 Bare Costs Material	Labor	Equipment	Total	Total Incl O&P
0010	**SPRINKLER IRRIGATION SYSTEM**									
0011	For lawns									
0100	Golf course with fully automatic system	C-17	.05	1600	9 holes	100,000	76,500		176,500	228,000
0200	24′ diam. head at 15′ O.C. incl. piping, auto oper., minimum	B-20	70	.343	Head	23.50	14		37.50	47
0300	Maximum		40	.600		56	24.50		80.50	99.50
0500	60′ diam. head at 40′ O.C. incl. piping, auto oper., minimum		28	.857		70.50	35		105.50	132
0600	Maximum		23	1.043		190	42.50		232.50	275
0800	Residential system, custom, 1″ supply		2000	.012	S.F.	.26	.49		.75	1.05
0900	1-1/2″ supply		1800	.013	″	.49	.54		1.03	1.38
1020	Pop up spray head w/risers, hi-pop, full circle pattern, 4″	2 Skwk	76	.211	Ea.	4.48	9.95		14.43	20.50
1030	1/2 circle pattern, 4″		76	.211		4.48	9.95		14.43	20.50
1040	6″, full circle pattern		76	.211		9.85	9.95		19.80	26.50
1050	1/2 circle pattern, 6″		76	.211		8.50	9.95		18.45	25
1060	12″, full circle pattern		76	.211		11.60	9.95		21.55	28
1070	1/2 circle pattern, 12″		76	.211		11.50	9.95		21.45	28
1080	Pop up bubbler head w/risers, hi-pop bubbler head, 4″		76	.211		4.35	9.95		14.30	20
1090	6″		76	.211		9.10	9.95		19.05	25.50
1100	12″		76	.211		10.90	9.95		20.85	27.50
1110	Impact full/part circle sprinklers, 28′-54′ 25-60 PSI		37	.432		19.60	20.50		40.10	53
1120	Spaced 37′-49′ @ 25-50 PSI		37	.432		27.50	20.50		48	62
1130	Spaced 43′-61′ @ 30-60 PSI		37	.432		77	20.50		97.50	116
1140	Spaced 54′-78′ @ 40-80 PSI		37	.432		133	20.50		153.50	178
1145	Impact rotor pop-up full/part commercial circle sprinklers									
1150	Spaced 42′-65′ 35-80 PSI	2 Skwk	25	.640	Ea.	17	30.50		47.50	65.50
1160	Spaced 48′-76′ 45-85 PSI	″	25	.640	″	18.70	30.50		49.20	67.50
1165	Impact rotor pop-up part. circle comm., 53′-75′, 55-100 PSI, w/accessories									
1170	Plastic case, metal cover	2 Skwk	25	.640	Ea.	82	30.50		112.50	137
1180	Rubber cover		25	.640		63	30.50		93.50	117
1190	Iron case, metal cover		22	.727		135	34.50		169.50	203
1200	Rubber cover		22	.727		143	34.50		177.50	211
1250	Plastic case, 2 nozzle, metal cover		25	.640		128	30.50		158.50	187
1260	Rubber cover		25	.640		131	30.50		161.50	191
1270	Iron case, 2 nozzle, metal cover		22	.727		185	34.50		219.50	258
1280	Rubber cover		22	.727		185	34.50		219.50	258
1282	Impact rotor pop-up full circle commercial, 39′-99′, 30-100 PSI									
1284	Plastic case, metal cover	2 Skwk	25	.640	Ea.	138	30.50		168.50	199
1286	Rubber cover		25	.640		158	30.50		188.50	220
1288	Iron case, metal cover		22	.727		203	34.50		237.50	277
1290	Rubber cover		22	.727		210	34.50		244.50	285
1292	Plastic case, 2 nozzle, metal cover		22	.727		150	34.50		184.50	219
1294	Rubber cover		22	.727		150	34.50		184.50	219
1296	Iron case, 2 nozzle, metal cover		20	.800		198	38		236	276
1298	Rubber cover		20	.800		205	38		243	285
1305	Electric remote control valve, plastic, 3/4″		18	.889		17.30	42		59.30	84
1310	1″		18	.889		33	42		75	102
1320	1-1/2″		18	.889		60.50	42		102.50	132
1330	2″		18	.889		86	42		128	160
1335	Quick coupling valves, brass, locking cover									
1340	Inlet coupling valve, 3/4″	2 Skwk	18.75	.853	Ea.	22	40.50		62.50	86.50
1350	1″		18.75	.853		30.50	40.50		71	96.50
1360	Controller valve boxes, 6″ round boxes		18.75	.853		7	40.50		47.50	70
1370	10″ round boxes		14.25	1.123		11.30	53		64.30	95
1380	12″ square box		9.75	1.641		15.90	77.50		93.40	137

32 84 Planting Irrigation

32 84 23 – Underground Sprinklers

32 84 23.10 Sprinkler Irrigation System		Crew	Daily Output	Labor-Hours	Unit	Material	2014 Bare Costs Labor	Equipment	Total	Total Incl O&P
1388	Electromech. control, 14 day 3-60 min., auto start to 23/day									
1390	4 station	2 Skwk	1.04	15.385	Ea.	185	730		915	1,325
1400	7 station		.64	25		201	1,175		1,376	2,050
1410	12 station		.40	40		220	1,900		2,120	3,175
1420	Dual programs, 18 station		.24	66.667		610	3,150		3,760	5,550
1430	23 station	↓	.16	100	↓	690	4,725		5,415	8,075
1435	Backflow preventer, bronze, 0-175 PSI, w/valves, test cocks									
1440	3/4"	2 Skwk	2	8	Ea.	103	380		483	700
1450	1"		2	8		118	380		498	715
1460	1-1/2"		2	8		280	380		660	895
1470	2"	↓	2	8	↓	345	380		725	965
1475	Pressure vacuum breaker, brass, 15-150 PSI									
1480	3/4"	2 Skwk	2	8	Ea.	60.50	380		440.50	650
1490	1"		2	8		65.50	380		445.50	655
1500	1-1/2"		2	8		270	380		650	880
1510	2"	↓	2	8	↓	325	380		705	945

32 91 Planting Preparation

32 91 13 – Soil Preparation

32 91 13.16 Mulching		Crew	Daily Output	Labor-Hours	Unit	Material	Labor	Equipment	Total	Total Incl O&P
0010	**MULCHING**									
0200	Hay, 1" deep, hand spread	1 Clab	475	.017	S.Y.	.48	.62		1.10	1.48
0250	Power mulcher, small	B-64	180	.089	M.S.F.	53.50	3.25	2.39	59.14	66
0350	Large	B-65	530	.030	"	53.50	1.10	1.12	55.72	61.50
0400	Humus peat, 1" deep, hand spread	1 Clab	700	.011	S.Y.	2.56	.42		2.98	3.47
0450	Push spreader	"	2500	.003	"	2.56	.12		2.68	3
0550	Tractor spreader	B-66	700	.011	M.S.F.	284	.54	.37	284.91	315
0600	Oat straw, 1" deep, hand spread	1 Clab	475	.017	S.Y.	.55	.62		1.17	1.56
0650	Power mulcher, small	B-64	180	.089	M.S.F.	61	3.25	2.39	66.64	74.50
0700	Large	B-65	530	.030	"	61	1.10	1.12	63.22	70
0750	Add for asphaltic emulsion	B-45	1770	.009	Gal.	5.75	.39	.52	6.66	7.50
0800	Peat moss, 1" deep, hand spread	1 Clab	900	.009	S.Y.	2.63	.33		2.96	3.39
0850	Push spreader	"	2500	.003	"	2.63	.12		2.75	3.07
0950	Tractor spreader	B-66	700	.011	M.S.F.	292	.54	.37	292.91	320
1000	Polyethylene film, 6 mil	2 Clab	2000	.008	S.Y.	.44	.29		.73	.93
1010	4 mil		2300	.007		.35	.26		.61	.78
1020	1-1/2 mil		2500	.006		.19	.23		.42	.57
1050	Filter fabric weed barrier	↓	2000	.008		.70	.29		.99	1.22
1100	Redwood nuggets, 3" deep, hand spread	1 Clab	150	.053	↓	2.77	1.95		4.72	6.05
1150	Skid steer loader	B-63	13.50	2.963	M.S.F.	310	115	12.90	437.90	530
1200	Stone mulch, hand spread, ceramic chips, economy	1 Clab	125	.064	S.Y.	6.85	2.35		9.20	11.15
1250	Deluxe	"	95	.084	"	10.50	3.09		13.59	16.30
1300	Granite chips	B-1	10	2.400	C.Y.	69.50	89.50		159	215
1400	Marble chips		10	2.400		152	89.50		241.50	305
1600	Pea gravel		28	.857		75.50	32		107.50	133
1700	Quartz	↓	10	2.400	↓	188	89.50		277.50	345
1800	Tar paper, 15 lb. felt	1 Clab	800	.010	S.Y.	.49	.37		.86	1.10
1900	Wood chips, 2" deep, hand spread	"	220	.036	"	1.45	1.33		2.78	3.66
1950	Skid steer loader	B-63	20.30	1.970	M.S.F.	161	76.50	8.60	246.10	305

32 91 Planting Preparation

32 91 13 – Soil Preparation

	32 91 13.23 Structural Soil Mixing	Crew	Daily Output	Labor-Hours	Unit	2014 Bare Costs Material	Labor	Equipment	Total	Total Incl O&P
0010	**STRUCTURAL SOIL MIXING**									
0100	Rake topsoil, site material, harley rock rake, ideal	B-6	33	.727	M.S.F.		29	11.05	40.05	56.50
0200	Adverse	"	7	3.429			138	52	190	269
0300	Screened loam, york rake and finish, ideal	B-62	24	1			40	7.25	47.25	69.50
0400	Adverse	"	20	1.200			48	8.70	56.70	83.50
1000	Remove topsoil & stock pile on site, 75 HP dozer, 6" deep, 50' haul	B-10L	30	.400			17.95	16.05	34	45
1050	300' haul		6.10	1.967			88	79	167	221
1100	12" deep, 50' haul		15.50	.774			34.50	31	65.50	87
1150	300' haul		3.10	3.871			174	155	329	435
1200	200 HP dozer, 6" deep, 50' haul	B-10B	125	.096			4.30	10.60	14.90	18.20
1250	300' haul		30.70	.391			17.50	43	60.50	74
1300	12" deep, 50' haul		62	.194			8.65	21.50	30.15	36.50
1350	300' haul		15.40	.779			35	86	121	148
1400	Alternate method, 75 HP dozer, 50' haul	B-10L	860	.014	C.Y.		.63	.56	1.19	1.57
1450	300' haul	"	114	.105			4.72	4.22	8.94	11.85
1500	200 HP dozer, 50' haul	B-10B	2660	.005			.20	.50	.70	.86
1600	300' haul	"	570	.021			.94	2.32	3.26	4
1800	Rolling topsoil, hand push roller	1 Clab	3200	.003	S.F.		.09		.09	.14
1850	Tractor drawn roller	B-66	10666	.001	"		.04	.02	.06	.08
1900	Remove rocks & debris from grade, by hand	B-62	80	.300	M.S.F.		12.05	2.18	14.23	21
1920	With rock picker	B-10S	140	.086			3.84	2.67	6.51	8.80
2000	Root raking and loading, residential, no boulders	B-6	53.30	.450			18.05	6.85	24.90	35
2100	With boulders		32	.750			30	11.40	41.40	58.50
2200	Municipal, no boulders		200	.120			4.81	1.83	6.64	9.40
2300	With boulders		120	.200			8	3.04	11.04	15.65
2400	Large commercial, no boulders	B-10B	400	.030			1.34	3.31	4.65	5.70
2500	With boulders	"	240	.050			2.24	5.50	7.74	9.45
2600	Rough grade & scarify subsoil to receive topsoil, common earth									
2610	200 H.P. dozer with scarifier	B-11A	80	.200	M.S.F.		8.55	16.55	25.10	31.50
2620	180 H.P. grader with scarifier	B-11L	110	.145			6.20	6.45	12.65	16.55
2700	Clay and till, 200 H.P. dozer with scarifier	B-11A	50	.320			13.70	26.50	40.20	50
2710	180 H.P. grader with scarifier	B-11L	40	.400			17.10	17.70	34.80	45.50
3000	Scarify subsoil, residential, skid steer loader w/scarifiers, 50 HP	B-66	32	.250			11.75	8.10	19.85	27
3050	Municipal, skid steer loader w/scarifiers, 50 HP	"	120	.067			3.14	2.16	5.30	7.15
3100	Large commercial, 75 HP, dozer w/scarifier	B-10L	240	.050			2.24	2	4.24	5.60
3200	Grader with scarifier, 135 H.P.	B-11L	280	.057			2.44	2.53	4.97	6.50
3500	Screen topsoil from stockpile, vibrating screen, wet material (organic)	B-10P	200	.060	C.Y.		2.69	5.95	8.64	10.65
3550	Dry material	"	300	.040			1.79	3.97	5.76	7.10
3600	Mixing with conditioners, manure and peat	B-10R	550	.022			.98	.54	1.52	2.08
3650	Mobilization add for 2 days or less operation	B-34K	3	2.667	Job		100	320	420	510
3800	Spread conditioned topsoil, 6" deep, by hand	B-1	360	.067	S.Y.	5.45	2.49		7.94	9.85
3850	300 HP dozer	B-10M	27	.444	M.S.F.	590	19.90	67.50	677.40	755
3900	4" deep, by hand	B-1	470	.051	S.Y.	4.92	1.91		6.83	8.35
3920	300 H.P. dozer	B-10M	34	.353	M.S.F.	445	15.80	53.50	514.30	570
3940	180 H.P. grader	B-11L	37	.432	"	445	18.50	19.10	482.60	535
4000	Spread soil conditioners, alum. sulfate, 1#/S.Y., hand push spreader	1 Clab	17500	.001	S.Y.	19.50	.02		19.52	21.50
4050	Tractor spreader	B-66	700	.011	M.S.F.	2,175	.54	.37	2,175.91	2,375
4100	Fertilizer, 0.2#/S.Y., push spreader	1 Clab	17500	.001	S.Y.	.49	.02		.51	.57
4150	Tractor spreader	B-66	700	.011	M.S.F.	54.50	.54	.37	55.41	61
4200	Ground limestone, 1#/S.Y., push spreader	1 Clab	17500	.001	S.Y.	.18	.02		.20	.23
4250	Tractor spreader	B-66	700	.011	M.S.F.	20	.54	.37	20.91	23
4400	Manure, 18#/S.Y., push spreader	1 Clab	2500	.003	S.Y.	7.85	.12		7.97	8.85

32 91 Planting Preparation

32 91 13 – Soil Preparation

32 91 13.23 Structural Soil Mixing		Crew	Daily Output	Labor-Hours	Unit	Material	2014 Bare Costs Labor	Equipment	Total	Total Incl O&P
4450	Tractor spreader	B-66	280	.029	M.S.F.	870	1.34	.93	872.27	965
4500	Perlite, 1" deep, push spreader	1 Clab	17500	.001	S.Y.	9.25	.02		9.27	10.25
4550	Tractor spreader	B-66	700	.011	M.S.F.	1,025	.54	.37	1,025.91	1,125
4600	Vermiculite, push spreader	1 Clab	17500	.001	S.Y.	8.75	.02		8.77	9.70
4650	Tractor spreader	B-66	700	.011	M.S.F.	970	.54	.37	970.91	1,075
5000	Spread topsoil, skid steer loader and hand dress	B-62	270	.089	C.Y.	24.50	3.57	.65	28.72	33
5100	Articulated loader and hand dress	B-100	320	.038		24.50	1.68	3	29.18	33
5200	Articulated loader and 75HP dozer	B-10M	500	.024		24.50	1.08	3.63	29.21	32.50
5300	Road grader and hand dress	B-11L	1000	.016		24.50	.68	.71	25.89	29
6000	Tilling topsoil, 20 HP tractor, disk harrow, 2" deep	B-66	450	.018	M.S.F.		.84	.58	1.42	1.90
6050	4" deep		360	.022			1.05	.72	1.77	2.38
6100	6" deep		270	.030			1.39	.96	2.35	3.17
6150	26" rototiller, 2" deep	A-1J	1250	.006	S.Y.		.23	.05	.28	.41
6200	4" deep		1000	.008			.29	.06	.35	.51
6250	6" deep		750	.011			.39	.08	.47	.68

32 91 13.26 Planting Beds		Crew	Daily Output	Labor-Hours	Unit	Material	Labor	Equipment	Total	Total Incl O&P
0010	**PLANTING BEDS**									
0100	Backfill planting pit, by hand, on site topsoil	2 Clab	18	.889	C.Y.		32.50		32.50	50.50
0200	Prepared planting mix, by hand	"	24	.667			24.50		24.50	37.50
0300	Skid steer loader, on site topsoil	B-62	340	.071			2.83	.51	3.34	4.90
0400	Prepared planting mix	"	410	.059			2.35	.43	2.78	4.07
1000	Excavate planting pit, by hand, sandy soil	2 Clab	16	1			36.50		36.50	56.50
1100	Heavy soil or clay	"	8	2			73.50		73.50	113
1200	1/2 C.Y. backhoe, sandy soil	B-11C	150	.107			4.56	2.44	7	9.65
1300	Heavy soil or clay	"	115	.139			5.95	3.18	9.13	12.60
2000	Mix planting soil, incl. loam, manure, peat, by hand	2 Clab	60	.267		43	9.75		52.75	62.50
2100	Skid steer loader	B-62	150	.160		43	6.40	1.16	50.56	58.50
3000	Pile sod, skid steer loader	"	2800	.009	S.Y.		.34	.06	.40	.60
3100	By hand	2 Clab	400	.040			1.47		1.47	2.26
4000	Remove sod, F.E. loader	B-10S	2000	.006			.27	.19	.46	.62
4100	Sod cutter	B-12K	3200	.005			.22	.31	.53	.68
4200	By hand	2 Clab	240	.067			2.44		2.44	3.77

32 91 19 – Landscape Grading

32 91 19.13 Topsoil Placement and Grading		Crew	Daily Output	Labor-Hours	Unit	Material	Labor	Equipment	Total	Total Incl O&P
0010	**TOPSOIL PLACEMENT AND GRADING**									
0300	Fine grade, base course for paving, see Section 32 11 23.23									
0400	Spread from pile to rough finish grade, F.E. loader, 1.5 C.Y.	B-10S	200	.060	C.Y.		2.69	1.87	4.56	6.15
0500	Up to 200' radius, by hand	1 Clab	14	.571			21		21	32.50
0600	Top dress by hand, 1 C.Y. for 600 S.F.	"	11.50	.696		28	25.50		53.50	70
0700	Furnish and place, truck dumped, screened, 4" deep	B-10S	1300	.009	S.Y.	3.49	.41	.29	4.19	4.79
0800	6" deep	"	820	.015	"	4.46	.66	.46	5.58	6.40

32 92 Turf and Grasses

32 92 19 – Seeding

32 92 19.14 Seeding, Athletic Fields		Crew	Daily Output	Labor-Hours	Unit	2014 Bare Costs Material	Labor	Equipment	Total	Total Incl O&P
0010	**SEEDING, ATHLETIC FIELDS**									
0020	Seeding, athletic fields, athletic field mix, 8#/M.S.F. push spreader	1 Clab	8	1	M.S.F.	17.60	36.50		54.10	76
0100	Tractor spreader	B-66	52	.154		17.60	7.25	4.99	29.84	36
0200	Hydro or air seeding, with mulch and fertilizer	B-81	80	.300		19.35	12.30	8.85	40.50	50
0400	Birdsfoot trefoil, .45#/M.S.F., push spreader	1 Clab	8	1		12	36.50		48.50	69.50
0500	Tractor spreader	B-66	52	.154		12	7.25	4.99	24.24	29.50
0600	Hydro or air seeding, with mulch and fertilizer	B-81	80	.300		23	12.30	8.85	44.15	54
0800	Bluegrass, 4#/M.S.F., common, push spreader	1 Clab	8	1		14.30	36.50		50.80	72.50
0900	Tractor spreader	B-66	52	.154		14.30	7.25	4.99	26.54	32.50
1000	Hydro or air seeding, with mulch and fertilizer	B-81	80	.300		23.50	12.30	8.85	44.65	54.50
1100	Baron, push spreader	1 Clab	8	1		16.80	36.50		53.30	75
1200	Tractor spreader	B-66	52	.154		16.80	7.25	4.99	29.04	35
1300	Hydro or air seeding, with mulch and fertilizer	B-81	80	.300		23	12.30	8.85	44.15	54
1500	Clover, 0.67#/M.S.F., white, push spreader	1 Clab	8	1		2.13	36.50		38.63	59
1600	Tractor spreader	B-66	52	.154		2.13	7.25	4.99	14.37	18.85
1700	Hydro or air seeding, with mulch and fertilizer	B-81	80	.300		11.70	12.30	8.85	32.85	41.50
1800	Ladino, push spreader	1 Clab	8	1		2.21	36.50		38.71	59
1900	Tractor spreader	B-66	52	.154		2.21	7.25	4.99	14.45	18.95
2000	Hydro or air seeding, with mulch and fertilizer	B-81	80	.300		9.70	12.30	8.85	30.85	39
2200	Fescue 5.5#/M.S.F., tall, push spreader	1 Clab	8	1		10.80	36.50		47.30	68.50
2300	Tractor spreader	B-66	52	.154		10.80	7.25	4.99	23.04	28.50
2400	Hydro or air seeding, with mulch and fertilizer	B-81	80	.300		35.50	12.30	8.85	56.65	67.50
2500	Chewing, push spreader	1 Clab	8	1		6	36.50		42.50	63
2600	Tractor spreader	B-66	52	.154		6	7.25	4.99	18.24	23
2700	Hydro or air seeding, with mulch and fertilizer	B-81	80	.300		19.80	12.30	8.85	40.95	50.50
2900	Crown vetch, 4#/M.S.F., push spreader	1 Clab	8	1		80	36.50		116.50	145
3000	Tractor spreader	B-66	52	.154		80	7.25	4.99	92.24	105
3100	Hydro or air seeding, with mulch and fertilizer	B-81	80	.300		110	12.30	8.85	131.15	150
3300	Rye, 10#/M.S.F., annual, push spreader	1 Clab	8	1		8.80	36.50		45.30	66
3400	Tractor spreader	B-66	52	.154		8.80	7.25	4.99	21.04	26
3500	Hydro or air seeding, with mulch and fertilizer	B-81	80	.300		19.35	12.30	8.85	40.50	50
3600	Fine textured, push spreader	1 Clab	8	1		12	36.50		48.50	69.50
3700	Tractor spreader	B-66	52	.154		12	7.25	4.99	24.24	29.50
3800	Hydro or air seeding, with mulch and fertilizer	B-81	80	.300		26.50	12.30	8.85	47.65	57.50
4000	Shade mix, 6#/M.S.F., push spreader	1 Clab	8	1		9.30	36.50		45.80	67
4100	Tractor spreader	B-66	52	.154		9.30	7.25	4.99	21.54	27
4200	Hydro or air seeding, with mulch and fertilizer	B-81	80	.300		20.50	12.30	8.85	41.65	51
4400	Slope mix, 6#/M.S.F., push spreader	1 Clab	8	1		10.40	36.50		46.90	68
4500	Tractor spreader	B-66	52	.154		10.40	7.25	4.99	22.64	28
4600	Hydro or air seeding, with mulch and fertilizer	B-81	80	.300		26	12.30	8.85	47.15	57
4800	Turf mix, 4#/M.S.F., push spreader	1 Clab	8	1		10.90	36.50		47.40	68.50
4900	Tractor spreader	B-66	52	.154		10.90	7.25	4.99	23.14	28.50
5000	Hydro or air seeding, with mulch and fertilizer	B-81	80	.300		27.50	12.30	8.85	48.65	58.50
5200	Utility mix, 7#/M.S.F., push spreader	1 Clab	8	1		9.30	36.50		45.80	67
5300	Tractor spreader	B-66	52	.154		9.30	7.25	4.99	21.54	27
5400	Hydro or air seeding, with mulch and fertilizer	B-81	80	.300		35	12.30	8.85	56.15	67
5600	Wildflower, .10#/M.S.F., push spreader	1 Clab	8	1		1.80	36.50		38.30	58.50
5700	Tractor spreader	B-66	52	.154		1.80	7.25	4.99	14.04	18.50
5800	Hydro or air seeding, with mulch and fertilizer	B-81	80	.300		9.90	12.30	8.85	31.05	39.50
7000	Apply fertilizer, 800 lb./acre	B-66	4	2	Ton	1,000	94	65	1,159	1,325
7025	Fertilizer, mechanical spread	1 Clab	1.75	4.571	Acre	5.70	168		173.70	265
7100	Apply mulch, see Section 32 91 13.16									

32 92 Turf and Grasses

32 92 23 – Sodding

32 92 23.10 Sodding Systems		Crew	Daily Output	Labor-Hours	Unit	Material	2014 Bare Costs Labor	Equipment	Total	Total Incl O&P
0010	**SODDING SYSTEMS**									
0020	Sodding, 1" deep, bluegrass sod, on level ground, over 8 M.S.F.	B-63	22	1.818	M.S.F.	250	70.50	7.95	328.45	390
0200	4 M.S.F.		17	2.353		265	91	10.25	366.25	445
0300	1000 S.F.		13.50	2.963		290	115	12.90	417.90	510
0500	Sloped ground, over 8 M.S.F.		6	6.667		250	258	29	537	700
0600	4 M.S.F.		5	8		265	310	35	610	805
0700	1000 S.F.		4	10		290	385	43.50	718.50	965
1000	Bent grass sod, on level ground, over 6 M.S.F.		20	2		253	77.50	8.70	339.20	410
1100	3 M.S.F.		18	2.222		265	86	9.70	360.70	435
1200	Sodding 1000 S.F. or less		14	2.857		293	111	12.45	416.45	510
1500	Sloped ground, over 6 M.S.F.		15	2.667		253	103	11.65	367.65	450
1600	3 M.S.F.		13.50	2.963		265	115	12.90	392.90	480
1700	1000 S.F.		12	3.333		293	129	14.55	436.55	540

32 92 26 – Sprigging

32 92 26.13 Stolonizing		Crew	Daily Output	Labor-Hours	Unit	Material	Labor	Equipment	Total	Total Incl O&P
0010	**STOLONIZING**									
0100	6" O.C., by hand	1 Clab	4	2	M.S.F.	26	73.50		99.50	142
0110	Walk behind sprig planter	"	80	.100		26	3.67		29.67	34
0120	Towed sprig planter	B-66	350	.023		26	1.08	.74	27.82	31
0130	9" O.C., by hand	1 Clab	5.20	1.538		17.25	56.50		73.75	106
0140	Walk behind sprig planter	"	92	.087		17.25	3.19		20.44	24
0150	Towed sprig planter	B-66	420	.019		17.25	.90	.62	18.77	21
0160	12" O.C., by hand	1 Clab	6	1.333		12.85	49		61.85	89.50
0170	Walk behind sprig planter	"	110	.073		12.85	2.67		15.52	18.25
0180	Towed sprig planter	B-66	500	.016		12.85	.75	.52	14.12	15.85
0200	Broadcast, by hand, 2 Bu. per M.S.F.	1 Clab	15	.533		5.15	19.55		24.70	35.50
0210	4 Bu. per M.S.F.		10	.800		10.30	29.50		39.80	56.50
0220	6 Bu. per M.S.F.		6.50	1.231		15.50	45		60.50	86.50
0300	Hydro planter, 6 Bu. per M.S.F.	B-64	100	.160		15.50	5.85	4.30	25.65	31
0320	Manure spreader planting 6 Bu. per M.S.F.	B-66	200	.040		15.50	1.88	1.30	18.68	21.50

32 93 Plants

32 93 43 – Trees

32 93 43.10 Planting		Crew	Daily Output	Labor-Hours	Unit	Material	Labor	Equipment	Total	Total Incl O&P
0010	**PLANTING**									
0011	Trees, shrubs and ground cover									
0100	Light soil									
0110	Bare root seedlings, 3" to 5" height	1 Clab	960	.008	Ea.		.31		.31	.47
0120	6" to 10"		520	.015			.56		.56	.87
0130	11" to 16"		370	.022			.79		.79	1.22
0140	17" to 24"		210	.038			1.40		1.40	2.15
0200	Potted, 2-1/4" diameter		840	.010			.35		.35	.54
0210	3" diameter		700	.011			.42		.42	.65
0220	4" diameter		620	.013			.47		.47	.73
0300	Container, 1 gallon	2 Clab	84	.190			7		7	10.75
0310	2 gallon		52	.308			11.30		11.30	17.40
0320	3 gallon		40	.400			14.65		14.65	22.50
0330	5 gallon		29	.552			20		20	31
0400	Bagged and burlapped, 12" diameter ball, by hand		19	.842			31		31	47.50
0410	Backhoe/loader, 48 H.P.	B-6	40	.600			24	9.15	33.15	47

32 93 Plants

32 93 43 – Trees

32 93 43.10 Planting		Crew	Daily Output	Labor-Hours	Unit	Material	2014 Bare Costs Labor	Equipment	Total	Total Incl O&P
0415	15″ diameter, by hand	2 Clab	16	1	Ea.		36.50		36.50	56.50
0416	Backhoe/loader, 48 H.P.	B-6	30	.800			32	12.20	44.20	62.50
0420	18″ diameter by hand	2 Clab	12	1.333			49		49	75.50
0430	Backhoe/loader, 48 H.P.	B-6	27	.889			35.50	13.55	49.05	69.50
0440	24″ diameter by hand	2 Clab	9	1.778			65		65	101
0450	Backhoe/loader 48 H.P.	B-6	21	1.143			46	17.40	63.40	89.50
0470	36″ diameter, backhoe/loader, 48 H.P.	″	17	1.412			56.50	21.50	78	111
0550	Medium soil									
0560	Bare root seedlings, 3″ to 5″	1 Clab	672	.012	Ea.		.44		.44	.67
0561	6″ to 10″		364	.022			.81		.81	1.24
0562	11″ to 16″		260	.031			1.13		1.13	1.74
0563	17″ to 24″		145	.055			2.02		2.02	3.12
0570	Potted, 2-1/4″ diameter		590	.014			.50		.50	.77
0572	3″ diameter		490	.016			.60		.60	.92
0574	4″ diameter		435	.018			.67		.67	1.04
0590	Container, 1 gallon	2 Clab	59	.271			9.95		9.95	15.35
0592	2 gallon		36	.444			16.30		16.30	25
0594	3 gallon		28	.571			21		21	32.50
0595	5 gallon		20	.800			29.50		29.50	45
0600	Bagged and burlapped, 12″ diameter ball, by hand		13	1.231			45		45	69.50
0605	Backhoe/loader, 48 H.P.	B-6	28	.857			34.50	13.05	47.55	67
0607	15″ diameter, by hand	2 Clab	11.20	1.429			52.50		52.50	81
0608	Backhoe/loader, 48 H.P.	B-6	21	1.143			46	17.40	63.40	89.50
0610	18″ diameter, by hand	2 Clab	8.50	1.882			69		69	106
0615	Backhoe/loader, 48 H.P.	B-6	19	1.263			50.50	19.25	69.75	98.50
0620	24″ diameter, by hand	2 Clab	6.30	2.540			93		93	144
0625	Backhoe/loader, 48 H.P.	B-6	14.70	1.633			65.50	25	90.50	128
0630	36″ diameter, backhoe/loader, 48 H.P.	″	12	2			80	30.50	110.50	157
0700	Heavy or stoney soil									
0710	Bare root seedlings, 3″ to 5″	1 Clab	470	.017	Ea.		.62		.62	.96
0711	6″ to 10″		255	.031			1.15		1.15	1.77
0712	11″ to 16″		182	.044			1.61		1.61	2.49
0713	17″ to 24″		101	.079			2.90		2.90	4.48
0720	Potted, 2-1/4″ diameter		360	.022			.81		.81	1.26
0722	3″ diameter		343	.023			.85		.85	1.32
0724	4″ diameter		305	.026			.96		.96	1.48
0730	Container, 1 gallon	2 Clab	41.30	.387			14.20		14.20	22
0732	2 gallon		25.20	.635			23.50		23.50	36
0734	3 gallon		19.60	.816			30		30	46
0735	5 gallon		14	1.143			42		42	64.50
0750	Bagged and burlapped, 12″ diameter ball, by hand		9.10	1.758			64.50		64.50	99.50
0751	Backhoe/loader	B-6	19.60	1.224			49	18.65	67.65	96
0752	15″ diameter, by hand	2 Clab	7.80	2.051			75		75	116
0753	Backhoe/loader, 48 H.P.	B-6	14.70	1.633			65.50	25	90.50	128
0754	18″ diameter, by hand	2 Clab	5.60	2.857			105		105	162
0755	Backhoe/loader, 48 H.P.	B-6	13.30	1.805			72.50	27.50	100	141
0756	24″ diameter, by hand	2 Clab	4.40	3.636			133		133	206
0757	Backhoe/loader, 48 H.P.	B-6	10.30	2.330			93.50	35.50	129	182
0758	36″ diameter backhoe/loader, 48 H.P.	″	8.40	2.857			115	43.50	158.50	224
2000	Stake out tree and shrub locations	2 Clab	220	.073			2.67		2.67	4.11

32 93 Plants

32 93 43 – Trees

32 93 43.30 Trees, Deciduous		Crew	Daily Output	Labor-Hours	Unit	Material	Labor	Equipment	Total	Total Incl O&P
0010	**TREES, DECIDUOUS**									
0011	Zones 2 – 6									
0100	Acer campestre, (Hedge Maple), Z4, B&B									
0110	4′ to 5′				Ea.	80			80	88
0120	5′ to 6′					91			91	100
0130	1-1/2″ to 2″ Cal.					225			225	248
0140	2″ to 2-1/2″ Cal.					300			300	330
0150	2-1/2″ to 3″ Cal.				↓	350			350	385
0200	Acer ginnala, (Amur Maple), Z2, cont/BB									
0210	8′ to 10′				Ea.	160			160	176
0220	10′ to 12′					239			239	262
0230	12′ to 14′				↓	234			234	257
0600	Acer platanoides, (Norway Maple), Z4, B&B									
0610	8′ to 10′				Ea.	215			215	237
0620	1-1/2″ to 2″ Cal.					200			200	220
0630	2″ to 2-1/2″ Cal.					315			315	345
0640	2-1/2″ to 3″ Cal.					375			375	415
0650	3″ to 3-1/2″ Cal.					450			450	495
0660	Bare root, 8′ to 10′					345			345	375
0670	10′ to 12′					375			375	415
0680	12′ to 14′				↓	415			415	455
0700	Acer platanoides columnare, (Column maple), Z4, B&B									
0710	2″ to 2-1/2″ Cal.				Ea.	315			315	345
0720	2-1/2″ to 3″ Cal.					365			365	400
0730	3″ to 3-1/2″ Cal.					440			440	485
0740	3-1/2″ to 4″ Cal.					600			600	660
0750	4″ to 4-1/2″ Cal.					1,050			1,050	1,150
0760	4-1/2″ to 5″ Cal.					1,475			1,475	1,625
0770	5″ to 5-1/2″ Cal.				↓	1,925			1,925	2,125
0800	Acer rubrum, (Red Maple), Z4, B&B									
0810	1-1/2″ to 2″ Cal.				Ea.	200			200	220
0820	2″ to 2-1/2″ Cal.					250			250	275
0830	2-1/2″ to 3″ Cal.					300			300	330
0840	Bare Root, 8′ to 10′					250			250	275
0850	10′ to 12′				↓	320			320	350
0900	Acer saccharum, (Sugar Maple), Z3, B&B									
0910	1-1/2″ to 2″ Cal.				Ea.	255			255	281
0920	2″ to 2-1/2″ Cal.					305			305	335
0930	2-1/2″ to 3″ Cal.					355			355	395
0940	3″ to 3-1/2″ Cal.					460			460	505
0950	3-1/2″ to 4″ Cal.					560			560	615
0960	Bare Root, 8′ to 10′					143			143	157
0970	10′ to 12′					219			219	241
0980	12′ to 14′				↓	280			280	310

32 94 Planting Accessories

32 94 50 – Tree Guying

32 94 50.10 Tree Guying Systems		Crew	Daily Output	Labor-Hours	Unit	Material	Labor	Equipment	Total	Total Incl O&P
						2014 Bare Costs				
0010	**TREE GUYING SYSTEMS**									
0015	Tree guying Including stakes, guy wire and wrap									
0100	Less than 3" caliper, 2 stakes	2 Clab	35	.457	Ea.	14.20	16.75		30.95	41.50
0200	3" to 4" caliper, 3 stakes	"	21	.762		19.20	28		47.20	64
0300	6" to 8" caliper, 3 stakes	B-1	8	3	↓	36.50	112		148.50	213
1000	Including arrowhead anchor, cable, turnbuckles and wrap									
1100	Less than 3" caliper, 3" anchors	2 Clab	20	.800	Ea.	44.50	29.50		74	94
1200	3" to 6" caliper, 4" anchors		15	1.067		38	39		77	103
1300	6" caliper, 6" anchors		12	1.333		44.50	49		93.50	125
1400	8" caliper, 8" anchors	↓	9	1.778	↓	128	65		193	242

32 96 Transplanting

32 96 23 – Plant and Bulb Transplanting

32 96 23.23 Planting		Crew	Daily Output	Labor-Hours	Unit	Material	Labor	Equipment	Total	Total Incl O&P
0010	**PLANTING**									
0012	Moving shrubs on site, 12" ball	B-62	28	.857	Ea.		34.50	6.20	40.70	59.50
0100	24" ball	"	22	1.091	"		44	7.90	51.90	75.50

32 96 23.43 Moving Trees		Crew	Daily Output	Labor-Hours	Unit	Material	Labor	Equipment	Total	Total Incl O&P
0010	**MOVING TREES**, On site									
0300	Moving trees on site, 36" ball	B-6	3.75	6.400	Ea.		257	97.50	354.50	500
0400	60" ball	"	1	24	"		965	365	1,330	1,875

32 96 43 – Tree Transplanting

32 96 43.20 Tree Removal		Crew	Daily Output	Labor-Hours	Unit	Material	Labor	Equipment	Total	Total Incl O&P
0010	**TREE REMOVAL**									
0100	Dig & lace, shrubs, broadleaf evergreen, 18"-24" high	B-1	55	.436	Ea.		16.30		16.30	25
0200	2'-3'	"	35	.686			25.50		25.50	39.50
0300	3'-4'	B-6	30	.800			32	12.20	44.20	62.50
0400	4'-5'	"	20	1.200			48	18.25	66.25	94
1000	Deciduous, 12"-15"	B-1	110	.218			8.15		8.15	12.55
1100	18"-24"		65	.369			13.80		13.80	21.50
1200	2'-3'	↓	55	.436			16.30		16.30	25
1300	3'-4'	B-6	50	.480			19.25	7.30	26.55	37.50
2000	Evergreen, 18"-24"	B-1	55	.436			16.30		16.30	25
2100	2'-0" to 2'-6"		50	.480			17.90		17.90	27.50
2200	2'-6" to 3'-0"		35	.686			25.50		25.50	39.50
2300	3'-0" to 3'-6"		20	1.200			45		45	69
3000	Trees, deciduous, small, 2'-3'	↓	55	.436			16.30		16.30	25
3100	3'-4'	B-6	50	.480			19.25	7.30	26.55	37.50
3200	4'-5'		35	.686			27.50	10.45	37.95	53.50
3300	5'-6'		30	.800			32	12.20	44.20	62.50
4000	Shade, 5'-6'		50	.480			19.25	7.30	26.55	37.50
4100	6'-8'		35	.686			27.50	10.45	37.95	53.50
4200	8'-10'		25	.960			38.50	14.60	53.10	75
4300	2" caliper		12	2			80	30.50	110.50	157
5000	Evergreen, 4'-5'		35	.686			27.50	10.45	37.95	53.50
5100	5'-6'		25	.960			38.50	14.60	53.10	75
5200	6'-7'		19	1.263			50.50	19.25	69.75	98.50
5300	7'-8'		15	1.600			64	24.50	88.50	126
5400	8'-10'	↓	11	2.182	↓		87.50	33	120.50	171

Estimating Tips

33 10 00 Water Utilities
33 30 00 Sanitary Sewerage Utilities
33 40 00 Storm Drainage Utilities

- Never assume that the water, sewer, and drainage lines will go in at the early stages of the project. Consider the site access needs before dividing the site in half with open trenches, loose pipe, and machinery obstructions. Always inspect the site to establish that the site drawings are complete. Check off all existing utilities on your drawings as you locate them. Be especially careful with underground utilities because appurtenances are sometimes buried during regrading or repaving operations. If you find any discrepancies, mark up the site plan for further research. Differing site conditions can be very costly if discovered later in the project.
- See also Section 33 01 00 for restoration of pipe where removal/replacement may be undesirable. Use of new types of piping materials can reduce the overall project cost. Owners/design engineers should consider the installing contractor as a valuable source of current information on utility products and local conditions that could lead to significant cost savings.

Reference Numbers

Reference numbers are shown in shaded boxes at the beginning of some major classifications. These numbers refer to related items in the Reference Section. The reference information may be an estimating procedure, an alternate pricing method, or technical information.

Note: Not all subdivisions listed here necessarily appear in this publication.

Note: **Trade Service,** *in part, has been used as a reference source for some of the material prices used in Division 33.*

33 01 10.10 Corrosion Resistance

	33 01 10.10 Corrosion Resistance	Crew	Daily Output	Labor-Hours	Unit	2014 Bare Costs Material	Labor	Equipment	Total	Total Incl O&P
0010	**CORROSION RESISTANCE**									
0012	Wrap & coat, add to pipe, 4″ diameter				L.F.	2.53			2.53	2.78
0020	5″ diameter					3.18			3.18	3.50
0040	6″ diameter					3.81			3.81	4.19
0060	8″ diameter					4.53			4.53	4.98
0080	10″ diameter					5.70			5.70	6.25
0100	12″ diameter					7.45			7.45	8.20
0120	14″ diameter					8.60			8.60	9.45
0140	16″ diameter					9.75			9.75	10.70
0160	18″ diameter					10.70			10.70	11.80
0180	20″ diameter					11.85			11.85	13
0200	24″ diameter					14.40			14.40	15.85
0220	Small diameter pipe, 1″ diameter, add					1.65			1.65	1.82
0240	2″ diameter					1.81			1.81	1.99
0260	2-1/2″ diameter					1.99			1.99	2.19
0280	3″ diameter					2.39			2.39	2.63
0300	Fittings, field covered, add				S.F.	10.45			10.45	11.45
0500	Coating, bituminous, per diameter inch, 1 coat, add				L.F.	.64			.64	.70
0540	3 coat					1.93			1.93	2.12
0560	Coal tar epoxy, per diameter inch, 1 coat, add					.23			.23	.25
0600	3 coat					.69			.69	.76
1000	Polyethylene H.D. extruded, .025″ thk., 1/2″ diameter add					.26			.26	.29
1020	3/4″ diameter					.29			.29	.32
1040	1″ diameter					.33			.33	.36
1060	1-1/4″ diameter					.41			.41	.45
1080	1-1/2″ diameter					.42			.42	.46
1100	.030″ thk., 2″ diameter					.46			.46	.51
1120	2-1/2″ diameter					.52			.52	.57
1140	.035″ thk., 3″ diameter					.75			.75	.83
1160	3-1/2″ diameter					.80			.80	.88
1180	4″ diameter					.85			.85	.94
1200	.040″ thk, 5″ diameter					1.04			1.04	1.14
1220	6″ diameter					1.10			1.10	1.21
1240	8″ diameter					1.44			1.44	1.58
1260	10″ diameter					1.80			1.80	1.98
1280	12″ diameter					2.16			2.16	2.38
1300	.060″ thk., 14″ diameter					2.99			2.99	3.29
1320	16″ diameter					3.56			3.56	3.92
1340	18″ diameter					4.12			4.12	4.53
1360	20″ diameter					4.68			4.68	5.15
1380	Fittings, field wrapped, add				S.F.	4.80			4.80	5.30

33 01 10.20 Pipe Repair

	33 01 10.20 Pipe Repair	Crew	Daily Output	Labor-Hours	Unit	Material	Labor	Equipment	Total	Total Incl O&P
0010	**PIPE REPAIR**									
0020	Not including excavation or backfill									
0100	Clamp, stainless steel, lightweight, for steel pipe									
0110	3″ long, 1/2″ diameter pipe	1 Plum	34	.235	Ea.	11.90	13.55		25.45	33.50
0120	3/4″ diameter pipe		32	.250		12.55	14.40		26.95	35.50
0130	1″ diameter pipe		30	.267		13.20	15.35		28.55	37.50
0140	1-1/4″ diameter pipe		28	.286		13.95	16.45		30.40	40.50
0150	1-1/2″ diameter pipe		26	.308		14.65	17.70		32.35	42.50
0160	2″ diameter pipe		24	.333		15.75	19.20		34.95	46.50
0170	2-1/2″ diameter pipe		23	.348		21.50	20		41.50	53.50

33 01 10 – Operation and Maintenance of Water Utilities

33 01 10.20 Pipe Repair		Crew	Daily Output	Labor-Hours	Unit	Material	Labor	Equipment	Total	Total Incl O&P
						2014 Bare Costs				
0180	3" diameter pipe	1 Plum	22	.364	Ea.	19.30	21		40.30	52.50
0190	3-1/2" diameter pipe	"	21	.381	"	25	22		47	60.50
0200	4" diameter pipe	B-20	44	.545		25	22.50		47.50	62
0210	5" diameter pipe		42	.571		29.50	23.50		53	68.50
0220	6" diameter pipe		38	.632		34	26		60	77.50
0230	8" diameter pipe		30	.800		40.50	32.50		73	95
0240	10" diameter pipe		28	.857		130	35		165	197
0250	12" diameter pipe		24	1		180	41		221	261
0260	14" diameter pipe		22	1.091		260	44.50		304.50	355
0270	16" diameter pipe		20	1.200		280	49		329	385
0280	18" diameter pipe		18	1.333		310	54.50		364.50	425
0290	20" diameter pipe		16	1.500		330	61.50		391.50	460
0300	24" diameter pipe	↓	14	1.714		350	70		420	495
0360	For 6" long, add					100%	40%			
0370	For 9" long, add					200%	100%			
0380	For 12" long, add					300%	150%			
0390	For 18" long, add				↓	500%	200%			
0400	Pipe Freezing for live repairs of systems 3/8 inch to 6 inch									
0410	Note: Pipe Freezing can also be used to install a valve into a live system									
0420	Pipe Freezing each side 3/8 inch	2 Skwk	8	2	Ea.	530	94.50		624.50	730
0425	Pipe Freezing each side 3/8 inch, second location same kit		8	2		22.50	94.50		117	172
0430	Pipe Freezing each side 3/4 inch		8	2		510	94.50		604.50	705
0435	Pipe Freezing each side 3/4 inch, second location same kit		8	2		22.50	94.50		117	172
0440	Pipe Freezing each side 1-1/2 inch		6	2.667		510	126		636	755
0445	Pipe Freezing each side 1-1/2 inch , second location same kit		6	2.667		22.50	126		148.50	220
0450	Pipe Freezing each side 2 inch		6	2.667		890	126		1,016	1,175
0455	Pipe Freezing each side 2 inch, second location same kit		6	2.667		22.50	126		148.50	220
0460	Pipe Freezing each side 2-1/2 inch-3 inch		6	2.667		925	126		1,051	1,225
0465	Pipe Freeze each side 2-1/2 -3 inch, second location same kit		6	2.667		45.50	126		171.50	245
0470	Pipe Freezing each side 4 inch		4	4		1,475	189		1,664	1,925
0475	Pipe Freezing each side 4 inch, second location same kit		4	4		72	189		261	375
0480	Pipe Freezing each side 5-6 inch		4	4		4,000	189		4,189	4,700
0485	Pipe Freezing each side 5-6 inch, second location same kit	↓	4	4		216	189		405	530
0490	Pipe Freezing extra 20 lb. CO_2 cylinders (3/8" to 2" - 1 ea, 3" -2 ea)					223			223	245
0500	Pipe Freezing extra 50 lb. CO_2 cylinders (4" - 2 ea, 5"-6" -6 ea)				↓	440			440	485
1000	Clamp, stainless steel, with threaded service tap									
1040	Full seal for iron, steel, PVC pipe									
1100	6" long, 2" diameter pipe	1 Plum	17	.471	Ea.	87	27		114	137
1110	2-1/2" diameter pipe		16	.500		90	29		119	143
1120	3" diameter pipe		15.60	.513		152	29.50		181.50	212
1130	3-1/2" diameter pipe	↓	15	.533		157	30.50		187.50	219
1140	4" diameter pipe	B-20	32	.750		164	30.50		194.50	228
1150	6" diameter pipe		28	.857		172	35		207	243
1160	8" diameter pipe		21	1.143		179	46.50		225.50	269
1170	10" diameter pipe		20	1.200		224	49		273	320
1180	12" diameter pipe	↓	17	1.412		261	57.50		318.50	375
1205	For 9" long, add					20%	45%			
1210	For 12" long, add					40%	80%			
1220	For 18" long, add				↓	70%	110%			
1600	Clamp, stainless steel, single section									
1640	Full seal for iron, steel, PVC pipe									
1700	6" long, 2" diameter pipe	1 Plum	17	.471	Ea.	87	27		114	137
1710	2-1/2" diameter pipe	↓	16	.500	↓	90.50	29		119.50	143

33 01 Operation and Maintenance of Utilities

33 01 10 – Operation and Maintenance of Water Utilities

33 01 10.20 Pipe Repair		Crew	Daily Output	Labor-Hours	Unit	Material	Labor	Equipment	Total	Total Incl O&P
						2014 Bare Costs				
1720	3″ diameter pipe	1 Plum	15.60	.513	Ea.	105	29.50		134.50	160
1730	3-1/2″ diameter pipe		15	.533		110	30.50		140.50	168
1740	4″ diameter pipe	B-20	32	.750		117	30.50		147.50	176
1750	6″ diameter pipe		27	.889		145	36.50		181.50	216
1760	8″ diameter pipe		21	1.143		171	46.50		217.50	260
1770	10″ diameter pipe		20	1.200		224	49		273	320
1780	12″ diameter pipe		17	1.412		261	57.50		318.50	375
1800	For 9″ long, add					40%	45%			
1810	For 12″ long, add					60%	80%			
1820	For 18″ long, add					120%	110%			
2000	Clamp, stainless steel, two section									
2040	Full seal, for iron, steel, PVC pipe									
2100	6″ long, 4″ diameter pipe	B-20	24	1	Ea.	230	41		271	315
2110	6″ diameter pipe		20	1.200		264	49		313	365
2120	8″ diameter pipe		13	1.846		300	75.50		375.50	450
2130	10″ diameter pipe		12	2		305	81.50		386.50	460
2140	12″ diameter pipe		10	2.400		390	98		488	580
2200	9″ long, 4″ diameter pipe		16	1.500		300	61.50		361.50	425
2210	6″ diameter pipe		13	1.846		340	75.50		415.50	490
2220	8″ diameter pipe		9	2.667		375	109		484	585
2230	10″ diameter pipe		8	3		495	123		618	735
2240	12″ diameter pipe		7	3.429		565	140		705	840
2250	14″ diameter pipe		6.40	3.750		760	153		913	1,075
2260	16″ diameter pipe		6	4		710	163		873	1,025
2270	18″ diameter pipe		5	4.800		800	196		996	1,175
2280	20″ diameter pipe		4.60	5.217		885	213		1,098	1,300
2290	24″ diameter pipe		4	6		1,175	245		1,420	1,675
2320	For 12″ long, add to 9″					15%	25%			
2330	For 18″ long, add to 9″					70%	55%			
8000	For internal cleaning and inspection, see Section 33 01 30.16									
8100	For pipe testing, see Section 23 05 93.50									

33 01 30 – Operation and Maintenance of Sewer Utilities

33 01 30.16 TV Inspection of Sewer Pipelines		Crew	Daily Output	Labor-Hours	Unit	Material	Labor	Equipment	Total	Total Incl O&P
0010	**TV INSPECTION OF SEWER PIPELINES**									
0100	Pipe internal cleaning & inspection, cleaning, pressure pipe systems									
0120	Pig method, lengths 1000′ to 10,000′									
0140	4″ diameter thru 24″ diameter, minimum				L.F.				3.60	4.14
0160	Maximum				″				18	21
6000	Sewage/sanitary systems									
6100	Power rodder with header & cutters									
6110	Mobilization charge, minimum				Total				695	800
6120	Mobilization charge, maximum				″				9,125	10,600
6140	Cleaning 4″-12″ diameter				L.F.				3.39	3.90
6190	14″-24″ diameter								3.98	4.59
6240	30″ diameter								5.80	6.65
6250	36″ diameter								6.75	7.75
6260	48″ diameter								7.70	8.85
6270	60″ diameter								8.70	9.95
6280	72″ diameter								9.65	11.10
9000	Inspection, television camera with video									
9060	up to 500 linear feet				Total				715	820

33 01 30 – Operation and Maintenance of Sewer Utilities

33 01 30.71 Pipebursting

	33 01 30.71 Pipebursting	Crew	Daily Output	Labor-Hours	Unit	2014 Bare Costs Material	Labor	Equipment	Total	Total Incl O&P
0010	**PIPEBURSTING**									
0011	300′ runs, replace with HDPE pipe									
0020	Not including excavation, backfill, shoring, or dewatering									
0100	6″ to 15″ diameter, minimum				L.F.				104	114
0200	Maximum								214	236
0300	18″ to 36″ diameter, minimum								208	229
0400	Maximum								435	510
0500	Mobilize and demobilize, minimum				Job				3,000	3,300
0600	Maximum				″				32,100	35,400

33 01 30.72 Relining Sewers

	33 01 30.72 Relining Sewers	Crew	Daily Output	Labor-Hours	Unit	Material	Labor	Equipment	Total	Total Incl O&P
0010	**RELINING SEWERS**									
0011	With cement incl. bypass & cleaning									
0020	Less than 10,000 L.F., urban, 6″ to 10″	C-17E	130	.615	L.F.	9.55	29.50	.73	39.78	57
0050	10″ to 12″		125	.640		11.75	30.50	.76	43.01	61
0070	12″ to 16″		115	.696		12.05	33	.82	45.87	65.50
0100	16″ to 20″		95	.842		14.15	40	.99	55.14	78.50
0200	24″ to 36″		90	.889		15.25	42.50	1.05	58.80	83.50
0300	48″ to 72″		80	1		24.50	47.50	1.18	73.18	102
0500	Rural, 6″ to 10″		180	.444		9.55	21	.52	31.07	44
0550	10″ to 12″		175	.457		11.75	22	.54	34.29	47.50
0570	12″ to 16″		160	.500		12.05	24	.59	36.64	51
0600	16″ to 20″		135	.593		14.15	28.50	.70	43.35	60.50
0700	24″ to 36″		125	.640		15.25	30.50	.76	46.51	65
0800	48″ to 72″		100	.800		24.50	38	.94	63.44	86.50
1000	Greater than 10,000 L.F., urban, 6″ to 10″		160	.500		9.55	24	.59	34.14	48
1050	10″ to 12″		155	.516		11.75	24.50	.61	36.86	51.50
1070	12″ to 16″		140	.571		12.05	27.50	.67	40.22	56
1100	16″ to 20″		120	.667		14.15	32	.79	46.94	66
1200	24″ to 36″		115	.696		15.25	33	.82	49.07	69
1300	48″ to 72″		95	.842		24.50	40	.99	65.49	89.50
1500	Rural, 6″ to 10″		215	.372		9.55	17.75	.44	27.74	38.50
1550	10″ to 12″		210	.381		11.75	18.15	.45	30.35	41.50
1570	12″ to 16″		185	.432		12.05	20.50	.51	33.06	46
1600	16″ to 20″		150	.533		14.15	25.50	.63	40.28	55.50
1700	24″ to 36″		140	.571		15.25	27.50	.67	43.42	59.50
1800	48″ to 72″		120	.667		24.50	32	.79	57.29	77

33 01 30.74 HDPE Pipe Lining

	33 01 30.74 HDPE Pipe Lining	Crew	Daily Output	Labor-Hours	Unit	Material	Labor	Equipment	Total	Total Incl O&P
0010	**HDPE PIPE LINING**, excludes cleaning and video inspection									
0020	Pipe relined with one pipe size smaller than original (4″ for 6″)									
0100	6″ diameter, original size	B-6B	600	.080	L.F.	4.13	2.99	1.55	8.67	10.85
0150	8″ diameter, original size		600	.080		8.20	2.99	1.55	12.74	15.30
0200	10″ diameter, original size		600	.080		10.65	2.99	1.55	15.19	18
0250	12″ diameter, original size		400	.120		20	4.48	2.33	26.81	31.50
0300	14″ diameter, original size		400	.120		22.50	4.48	2.33	29.31	34

33 05 16 – Utility Structures

33 05 16.13 Precast Concrete Utility Boxes		Crew	Daily Output	Labor-Hours	Unit	Material	Labor	Equipment	Total	Total Incl O&P
						2014 Bare Costs				
0010	**PRECAST CONCRETE UTILITY BOXES**, 6" thick									
0040	4' x 6' x 6' high, I.D.	B-13	2	28	Ea.	2,950	1,125	375	4,450	5,350
0050	5' x 10' x 6' high, I.D.		2	28		3,675	1,125	375	5,175	6,125
0100	6' x 10' x 6' high, I.D.		2	28		3,800	1,125	375	5,300	6,300
0150	5' x 12' x 6' high, I.D.		2	28		4,025	1,125	375	5,525	6,525
0200	6' x 12' x 6' high, I.D.		1.80	31.111		4,525	1,250	415	6,190	7,325
0250	6' x 13' x 6' high, I.D.		1.50	37.333		5,925	1,500	495	7,920	9,350
0300	8' x 14' x 7' high, I.D.	↓	1	56	↓	6,400	2,225	745	9,370	11,300
0350	Hand hole, precast concrete, 1-1/2" thick									
0400	1'-0" x 2'-0" x 1'-9", I.D., light duty	B-1	4	6	Ea.	410	224		634	795
0450	4'-6" x 3'-2" x 2'-0", O.D., heavy duty	B-6	3	8		1,475	320	122	1,917	2,225
0460	Meter pit, 4' x 4', 4' deep		2	12		1,200	480	183	1,863	2,250
0470	6' deep		1.60	15		1,700	600	228	2,528	3,025
0480	8' deep		1.40	17.143		2,225	690	261	3,176	3,775
0490	10' deep		1.20	20		2,850	800	305	3,955	4,675
0500	15' deep		1	24		4,150	965	365	5,480	6,425
0510	6' x 6', 4' deep		1.40	17.143		2,200	690	261	3,151	3,750
0520	6' deep		1.20	20		3,300	800	305	4,405	5,200
0530	8' deep		1	24		4,425	965	365	5,755	6,725
0540	10' deep		.80	30		5,525	1,200	455	7,180	8,425
0550	15' deep	↓	.60	40	↓	8,400	1,600	610	10,610	12,300

33 05 23 – Trenchless Utility Installation

33 05 23.19 Microtunneling		Crew	Daily Output	Labor-Hours	Unit	Material	Labor	Equipment	Total	Total Incl O&P
0010	**MICROTUNNELING**									
0011	Not including excavation, backfill, shoring,									
0020	or dewatering, average 50'/day, slurry method									
0100	24" to 48" outside diameter, minimum				L.F.				875	965
0110	Adverse conditions, add				%				50%	50%
1000	Rent microtunneling machine, average monthly lease				Month				97,500	107,000
1010	Operating technician				Day				630	705
1100	Mobilization and demobilization, minimum				Job				41,200	45,900
1110	Maximum				"				445,500	490,500

33 05 23.20 Horizontal Boring		Crew	Daily Output	Labor-Hours	Unit	Material	Labor	Equipment	Total	Total Incl O&P
0010	**HORIZONTAL BORING**									
0011	Casing only, 100' minimum,									
0020	not incl. jacking pits or dewatering									
0100	Roadwork, 1/2" thick wall, 24" diameter casing	B-42	20	3.200	L.F.	121	132	68.50	321.50	415
0200	36" diameter		16	4		223	165	85.50	473.50	600
0300	48" diameter		15	4.267		310	176	91	577	715
0500	Railroad work, 24" diameter		15	4.267		121	176	91	388	510
0600	36" diameter		14	4.571		223	189	97.50	509.50	650
0700	48" diameter	↓	12	5.333		310	220	114	644	810
0900	For ledge, add								20%	20%
1000	Small diameter boring, 3", sandy soil	B-82	900	.018		22	.74	.09	22.83	25
1040	Rocky soil	"	500	.032	↓	22	1.34	.17	23.51	26
1100	Prepare jacking pits, incl. mobilization & demobilization, minimum				Ea.				3,225	3,700
1101	Maximum				"				22,000	25,500

33 05 23.22 Directional Drilling		Crew	Daily Output	Labor-Hours	Unit	Material	Labor	Equipment	Total	Total Incl O&P
0010	**DIRECTIONAL DRILLING**									
0011	Excluding cost of conduit, 100' minimum									
0100	Small equipment to 300', not to exceed 12" diam.									
0102	small unit mobilization to site	B-82A	2	8	Ea.		335	550	885	1,125

33 05 Common Work Results for Utilities

33 05 23 – Trenchless Utility Installation

33 05 23.22 Directional Drilling		Crew	Daily Output	Labor-Hours	Unit	Material	2014 Bare Costs Labor	Equipment	Total	Total Incl O&P
0105	small unit setup per drill	B-82A	4	4	Ea.		167	276	443	560
0109	minimum charge gravel, sand & silt, up to 12" diam.		3.20	5		10.15	209	345	564.15	710
0110	gravel, sand & silt, up to 12" diam.		300	.053	L.F.	.07	2.23	3.68	5.98	7.55
0118	min charge , clay & soft sandstone, up to 10" diam.		3.20	5	Ea.	20.50	209	345	574.50	725
0120	Clay & soft sandstone, up to 10" diam.		300	.053	L.F.	.14	2.23	3.68	6.05	7.60
0128	min charge , hard clay & cobble, up to 8" diam.		1	16	Ea.	54	670	1,100	1,824	2,300
0130	Hard clay & cobble, up to 8" diam.		100	.160	L.F.	.54	6.70	11.05	18.29	23
0200	Medium equipment to 600', not to exceed 12" diam.									
0202	medium unit mobilization to site	B-82B	2	12	Ea.		480	690	1,170	1,500
0205	medium unit setup per drill		4	6			241	345	586	750
0209	minimum charge gravel, sand & silt, up to 12" diam.		3.20	7.500		10.15	300	430	740.15	945
0210	gravel, sand & silt, up to 12" diam.		350	.069	L.F.	.07	2.75	3.95	6.77	8.65
0218	min charge , clay & soft sandstone, up to 10" diam.		3.20	7.500	Ea.	20.50	300	430	750.50	960
0220	Clay & soft sandstone, up to 10" diam.		300	.080	L.F.	.14	3.21	4.61	7.96	10.10
0228	min charge , hard clay & cobble, up to 8" diam.		1.78	13.483	Ea.	54	540	775	1,369	1,750
0230	Hard clay & cobble, up to 8" diam.		150	.160	L.F.	.54	6.40	9.20	16.14	20.50
0300	Large equipment to 1000', not to exceed 12" diam.									
0302	large unit mobilization to site	B-82C	2	12	Ea.		480	845	1,325	1,675
0305	large unit setup per drill		4	6			241	425	666	835
0309	minimum charge gravel, sand & silt, up to 12" diam.		4	6		10.15	241	425	676.15	845
0310	gravel, sand & silt, up to 12" diam.		400	.060	L.F.	.07	2.41	4.23	6.71	8.40
0318	min charge , clay & soft sandstone, up to 12" diam.		3.20	7.500	Ea.	20.50	300	530	850.50	1,075
0320	Clay & soft sandstone, up to 12" diam.		350	.069	L.F.	.14	2.75	4.84	7.73	9.65
0328	min charge , hard clay & cobble, up to 10" diam.		1.78	13.483	Ea.	54	540	950	1,544	1,950
0330	Rock & cobble, up to 10" diam.		150	.160	L.F.	.54	6.40	11.30	18.24	23
0400	Directional drilling, mud trailer per day	B-82D	1	8	Day		375	375	750	980
1000	Additional charges for mobilization may apply at some locations									

33 05 26 – Utility Identification

33 05 26.10 Utility Accessories		Crew	Daily Output	Labor-Hours	Unit	Material	Labor	Equipment	Total	Total Incl O&P
0010	**UTILITY ACCESSORIES**									
0400	Underground tape, detectable, reinforced, alum. foil core, 2"	1 Clab	150	.053	C.L.F.	5.65	1.95		7.60	9.25
0500	6"	"	140	.057	"	27.50	2.09		29.59	33.50

33 11 Water Utility Distribution Piping

33 11 13 – Public Water Utility Distribution Piping

33 11 13.10 Water Supply, Concrete Pipe		Crew	Daily Output	Labor-Hours	Unit	Material	Labor	Equipment	Total	Total Incl O&P
0010	**WATER SUPPLY, CONCRETE PIPE**									
0020	Not including excavation or backfill, without gaskets									
3000	Conc. cylinder pipe (CCP), 150 PSI, 12" diam.	B-13	192	.292	L.F.	95.50	11.65	3.89	111.04	127
3010	24" diameter	"	128	.438		118	17.45	5.85	141.30	163
3040	36" diameter	B-13B	96	.583		147	23.50	11.85	182.35	210
3050	48" diameter		64	.875		201	35	17.75	253.75	294
3060	Prestressed (PCCP), 150 PSI, 60" diameter		60	.933		330	37	18.95	385.95	440
3070	72" diameter		60	.933		415	37	18.95	470.95	535
3080	84" diameter		40	1.400		540	56	28.50	624.50	705
3090	96" diameter	B-13C	40	1.400		795	56	42.50	893.50	1,000
3100	108" diameter		32	1.750		1,100	70	53	1,223	1,375
3102	120" diameter		16	3.500		1,600	140	106	1,846	2,100
3104	144" diameter		16	3.500		1,975	140	106	2,221	2,500
3110	Conc. cylinder pipe (CCP), 150 PSI, elbow, 90°, 12" diameter	B-13	24	2.333	Ea.	1,275	93	31	1,399	1,575

33 11 Water Utility Distribution Piping

33 11 13 – Public Water Utility Distribution Piping

33 11 13.10 Water Supply, Concrete Pipe		Crew	Daily Output	Labor-Hours	Unit	Material	Labor	Equipment	Total	Total Incl O&P
						2014 Bare Costs				
3140	24″ diameter	B-13	6	9.333	Ea.	2,500	370	124	2,994	3,450
3150	36″ diameter	B-13B	4	14		4,875	560	284	5,719	6,525
3160	48″ diameter		3	18.667		9,625	745	380	10,750	12,200
3170	Prestressed (PCCP), 150 PSI, elbow, 90°, 60″ diameter		2	28		17,500	1,125	570	19,195	21,600
3180	72″ diameter		1.60	35		27,400	1,400	710	29,510	33,000
3190	84″ diameter		1.30	43.077		33,300	1,725	875	35,900	40,300
3200	96″ diameter		1	56		44,700	2,225	1,125	48,050	54,000
3210	108″ diameter	B-13C	.66	84.848		66,500	3,375	2,575	72,450	81,500
3220	120″ diameter		.40	140		67,500	5,575	4,250	77,325	87,000
3225	144″ diameter		.30	184		106,500	7,350	5,600	119,450	134,500
3230	Concrete cylinder pipe (CCP), 150 PSI, elbow, 45°, 12″ diameter	B-13	24	2.333		795	93	31	919	1,050
3250	24″ diameter	″	6	9.333		1,600	370	124	2,094	2,475
3260	36″ diameter	B-13B	4	14		2,725	560	284	3,569	4,175
3270	48″ diameter		3	18.667		4,950	745	380	6,075	7,025
3280	Prestressed, (PCCP), 150 PSI, elbow, 45°, 60″ diameter		2	28		10,900	1,125	570	12,595	14,300
3290	72″ diameter		1.60	35		17,000	1,400	710	19,110	21,600
3300	84″ diameter		1.30	42.945		26,000	1,700	870	28,570	32,200
3310	96″ diameter		1	56		32,200	2,225	1,125	35,550	40,100
3320	108″ diameter	B-13C	.66	84.337		40,600	3,350	2,575	46,525	52,500
3330	120″ diameter		.40	140		41,000	5,575	4,250	50,825	58,500
3340	144″ diameter		.30	184		65,500	7,350	5,600	78,450	89,500

33 11 13.15 Water Supply, Ductile Iron Pipe		Crew	Daily Output	Labor-Hours	Unit	Material	Labor	Equipment	Total	Total Incl O&P
0010	**WATER SUPPLY, DUCTILE IRON PIPE**									
0011	Cement lined									
0020	Not including excavation or backfill									
2000	Pipe, class 50 water piping, 18′ lengths									
2020	Mechanical joint, 4″ diameter	B-21A	200	.200	L.F.	15.60	9.15	2.41	27.16	34
2040	6″ diameter		160	.250		17.75	11.45	3.01	32.21	40.50
2060	8″ diameter		133.33	.300		24.50	13.75	3.61	41.86	52
2080	10″ diameter		114.29	.350		35	16.05	4.21	55.26	67.50
2100	12″ diameter		105.26	.380		42.50	17.40	4.57	64.47	78.50
2120	14″ diameter		100	.400		60	18.35	4.81	83.16	99.50
2140	16″ diameter		72.73	.550		63.50	25	6.60	95.10	116
2160	18″ diameter		68.97	.580		74.50	26.50	7	108	130
2170	20″ diameter		57.14	.700		85.50	32	8.40	125.90	153
2180	24″ diameter		47.06	.850		108	39	10.25	157.25	190
3000	Tyton, push-on joint, 4″ diameter		400	.100		17	4.58	1.20	22.78	27
3020	6″ diameter		333.33	.120		17.85	5.50	1.44	24.79	29.50
3040	8″ diameter		200	.200		23.50	9.15	2.41	35.06	42
3060	10″ diameter		181.82	.220		35	10.10	2.65	47.75	57
3080	12″ diameter		160	.250		37	11.45	3.01	51.46	62
3100	14″ diameter		133.33	.300		41	13.75	3.61	58.36	70
3120	16″ diameter		114.29	.350		49.50	16.05	4.21	69.76	83.50
3140	18″ diameter		100	.400		55	18.35	4.81	78.16	94
3160	20″ diameter		88.89	.450		57	20.50	5.40	82.90	100
3180	24″ diameter		76.92	.520		64	24	6.25	94.25	114
6170	Cap, 4″ diameter	B-20A	32	1	Ea.	109	44.50		153.50	188
6180	6″ diameter		25.60	1.250		166	56		222	268
6190	8″ diameter		21.33	1.500		227	67		294	350
6200	12″ diameter		16.84	1.900		380	85		465	550
6210	18″ diameter		11	2.909		1,200	130		1,330	1,500
6220	24″ diameter	B-21A	9.41	4.251		2,425	195	51	2,671	3,025

33 11 13.15 Water Supply, Ductile Iron Pipe		Crew	Daily Output	Labor-Hours	Unit	Material	Labor	Equipment	Total	Total Incl O&P
						2014 Bare Costs				
6230	30" diameter	B-21A	8	5	Ea.	5,000	229	60	5,289	5,925
6240	36" diameter	↓	6.90	5.797	↓	6,000	266	69.50	6,335.50	7,050
8000	Fittings, mechanical joint									
8006	90° bend, 4" diameter	B-20A	16	2	Ea.	273	89.50		362.50	435
8020	6" diameter		12.80	2.500		405	112		517	615
8040	8" diameter	↓	10.67	2.999		725	134		859	1,000
8060	10" diameter	B-21A	11.43	3.500		980	160	42	1,182	1,375
8080	12" diameter		10.53	3.799		1,375	174	45.50	1,594.50	1,825
8100	14" diameter		10	4		1,850	183	48	2,081	2,350
8120	16" diameter		7.27	5.502		2,400	252	66	2,718	3,075
8140	18" diameter		6.90	5.797		3,225	266	69.50	3,560.50	4,025
8160	20" diameter		5.71	7.005		4,050	320	84.50	4,454.50	5,025
8180	24" diameter	↓	4.70	8.511		6,275	390	102	6,767	7,600
8200	Wye or tee, 4" diameter	B-20A	10.67	2.999		440	134		574	690
8220	6" diameter		8.53	3.751		670	168		838	990
8240	8" diameter	↓	7.11	4.501		1,075	201		1,276	1,475
8260	10" diameter	B-21A	7.62	5.249		1,550	241	63	1,854	2,125
8280	12" diameter		7.02	5.698		2,025	261	68.50	2,354.50	2,700
8300	14" diameter		6.67	5.997		3,250	275	72	3,597	4,075
8320	16" diameter		4.85	8.247		3,600	380	99	4,079	4,650
8340	18" diameter		4.60	8.696		4,875	400	105	5,380	6,100
8360	20" diameter		3.81	10.499		6,825	480	126	7,431	8,400
8380	24" diameter	↓	3.14	12.739		11,600	585	153	12,338	13,800
8398	45° bends, 4" diameter	B-20A	16	2		254	89.50		343.50	415
8400	6" diameter		12.80	2.500		365	112		477	575
8405	8" diameter	↓	10.67	2.999		520	134		654	780
8410	12" diameter	B-21A	10.53	3.799		1,175	174	45.50	1,394.50	1,625
8420	16" diameter		7.27	5.502		2,100	252	66	2,418	2,750
8430	20" diameter		5.71	7.005		3,275	320	84.50	3,679.50	4,200
8440	24" diameter	↓	4.70	8.511		4,575	390	102	5,067	5,750
8450	Decreaser, 6" x 4" diameter	B-20A	14.22	2.250		360	101		461	550
8460	8" x 6" diameter	"	11.64	2.749		525	123		648	765
8470	10" x 6" diameter	B-21A	13.33	3.001		645	138	36	819	960
8480	12" x 6" diameter		12.70	3.150		795	144	38	977	1,125
8490	16" x 6" diameter		10	4		1,425	183	48	1,656	1,900
8500	20" x 6" diameter	↓	8.42	4.751	↓	3,125	218	57	3,400	3,850
8552	For water utility valves see Section 33 12 16									
8700	Joint restraint, ductile iron mechanical joints									
8710	4" diameter	B-20A	32	1	Ea.	36.50	44.50		81	109
8720	6" diameter		25.60	1.250		43.50	56		99.50	133
8730	8" diameter		21.33	1.500		64	67		131	172
8740	10" diameter		18.28	1.751		92	78.50		170.50	220
8750	12" diameter		16.84	1.900		131	85		216	275
8760	14" diameter		16	2		170	89.50		259.50	325
8770	16" diameter		11.64	2.749		227	123		350	435
8780	18" diameter		11.03	2.901		315	130		445	545
8785	20" diameter		9.14	3.501		385	157		542	665
8790	24" diameter	↓	7.53	4.250		505	190		695	845
9600	Steel sleeve with tap, 4" diameter	B-20	3	8		430	325		755	980
9620	6" diameter		2	12		470	490		960	1,275
9630	8" diameter	↓	2	12	↓	515	490		1,005	1,325

33 11 Water Utility Distribution Piping

33 11 13 – Public Water Utility Distribution Piping

33 11 13.20 Water Supply, Polyethylene Pipe, C901		Crew	Daily Output	Labor-Hours	Unit	2014 Bare Costs Material	Labor	Equipment	Total	Total Incl O&P
0010	**WATER SUPPLY, POLYETHYLENE PIPE, C901**									
0020	Not including excavation or backfill									
1000	Piping, 160 PSI, 3/4" diameter	Q-1A	525	.019	L.F.	.38	1.10		1.48	2.09
1120	1" diameter		485	.021		.52	1.19		1.71	2.37
1140	1-1/2" diameter		450	.022		1.27	1.29		2.56	3.34
1160	2" diameter		365	.027		2.10	1.59		3.69	4.71
2000	Fittings, insert type, nylon, 160 & 250 psi, cold water									
2220	Clamp ring, stainless steel, 3/4" diameter	Q-1A	345	.029	Ea.	1.15	1.68		2.83	3.81
2240	1" diameter		321	.031		1.34	1.81		3.15	4.20
2260	1-1/2" diameter		285	.035		2.40	2.03		4.43	5.70
2280	2" diameter		255	.039		3.15	2.27		5.42	6.90
2300	Coupling, 3/4" diameter		66	.152		1.27	8.80		10.07	14.65
2320	1" diameter		57	.175		1.48	10.15		11.63	17
2340	1-1/2" diameter		51	.196		3	11.35		14.35	20.50
2360	2" diameter		48	.208		4.70	12.05		16.75	23.50
2400	Elbow, 90°, 3/4" diameter		66	.152		2.06	8.80		10.86	15.50
2420	1" diameter		57	.175		2.49	10.15		12.64	18.10
2440	1-1/2" diameter		51	.196		4.10	11.35		15.45	21.50
2460	2" diameter		48	.208		5.75	12.05		17.80	24.50
2500	Tee, 3/4" diameter		42	.238		2.49	13.80		16.29	23.50
2520	1" diameter		39	.256		3.45	14.85		18.30	26.50
2540	1-1/2" diameter		33	.303		6.75	17.55		24.30	34
2560	2" diameter		30	.333		12.80	19.30		32.10	43

33 11 13.25 Water Supply, Polyvinyl Chloride Pipe

33 11 13.25 Water Supply, Polyvinyl Chloride Pipe		Crew	Daily Output	Labor-Hours	Unit	Material	Labor	Equipment	Total	Total Incl O&P
0010	**WATER SUPPLY, POLYVINYL CHLORIDE PIPE**									
0020	Not including excavation or backfill, unless specified									
2100	PVC pipe, Class 150, 1-1/2" diameter	Q-1A	750	.013	L.F.	.43	.77		1.20	1.64
2120	2" diameter		686	.015		.67	.84		1.51	2.02
2140	2-1/2" diameter		500	.020		1.28	1.16		2.44	3.16
2160	3" diameter	B-20	430	.056		1.38	2.28		3.66	5.05
3010	AWWA C905, PR 100, DR 25									
3030	14" diameter	B-20A	213	.150	L.F.	13.65	6.70		20.35	25.50
3040	16" diameter		200	.160		19.20	7.15		26.35	32
3050	18" diameter		160	.200		24	8.95		32.95	40
3060	20" diameter		133	.241		30	10.75		40.75	49.50
3070	24" diameter		107	.299		43	13.40		56.40	67.50
3080	30" diameter		80	.400		81	17.90		98.90	117
3090	36" diameter		80	.400		126	17.90		143.90	166
3100	42" diameter		60	.533		168	24		192	222
3200	48" diameter		60	.533		220	24		244	279
4520	Pressure pipe Class 150, SDR 18, AWWA C900, 4" diameter		380	.084		2.61	3.77		6.38	8.60
4530	6" diameter		316	.101		5.25	4.53		9.78	12.70
4540	8" diameter		264	.121		8.90	5.40		14.30	18
4550	10" diameter		220	.145		15.45	6.50		21.95	27
4560	12" diameter		186	.172		20.50	7.70		28.20	34.50
8000	Fittings with rubber gasket									
8003	Class 150, DR 18									
8006	90° Bend , 4" diameter	B-20	100	.240	Ea.	42.50	9.80		52.30	61.50
8020	6" diameter		90	.267		74.50	10.90		85.40	99
8040	8" diameter		80	.300		144	12.25		156.25	178
8060	10" diameter		50	.480		330	19.60		349.60	390
8080	12" diameter		30	.800		425	32.50		457.50	515

33 11 13.25 Water Supply, Polyvinyl Chloride Pipe		Crew	Daily Output	Labor-Hours	Unit	Material	2014 Bare Costs Labor	Equipment	Total	Total Incl O&P
8100	Tee, 4" diameter	B-20	90	.267	Ea.	58	10.90		68.90	80.50
8120	6" diameter		80	.300		144	12.25		156.25	177
8140	8" diameter		70	.343		184	14		198	224
8160	10" diameter		40	.600		590	24.50		614.50	690
8180	12" diameter		20	1.200		765	49		814	915
8200	45° Bend, 4" diameter		100	.240		42	9.80		51.80	61
8220	6" diameter		90	.267		73	10.90		83.90	97
8240	8" diameter		50	.480		144	19.60		163.60	190
8260	10" diameter		50	.480		330	19.60		349.60	390
8280	12" diameter		30	.800		425	32.50		457.50	515
8300	Reducing tee 6" x 4"		100	.240		126	9.80		135.80	153
8320	8" x 6"		90	.267		224	10.90		234.90	263
8330	10" x 6"		90	.267		380	10.90		390.90	430
8340	10" x 8"		90	.267		395	10.90		405.90	445
8350	12" x 6"		90	.267		450	10.90		460.90	505
8360	12" x 8"		90	.267		470	10.90		480.90	530
8400	Tapped service tee (threaded type) 6" x 6" x 3/4"		100	.240		95	9.80		104.80	120
8430	6" x 6" x 1"		90	.267		95	10.90		105.90	122
8440	6" x 6" x 1-1/2"		90	.267		95	10.90		105.90	122
8450	6" x 6" x 2"		90	.267		95	10.90		105.90	122
8460	8" x 8" x 3/4"		90	.267		140	10.90		150.90	171
8470	8" x 8" x 1"		90	.267		140	10.90		150.90	171
8480	8" x 8" x 1-1/2"		90	.267		140	10.90		150.90	171
8490	8" x 8" x 2"		90	.267		140	10.90		150.90	171
8500	Repair coupling 4"		100	.240		22.50	9.80		32.30	40
8520	6" diameter		90	.267		34.50	10.90		45.40	55
8540	8" diameter		50	.480		83	19.60		102.60	122
8560	10" diameter		50	.480		208	19.60		227.60	260
8580	12" diameter		50	.480		233	19.60		252.60	288
8600	Plug end 4"		100	.240		22.50	9.80		32.30	40
8620	6" diameter		90	.267		40.50	10.90		51.40	61.50
8640	8" diameter		50	.480		69	19.60		88.60	107
8660	10" diameter		50	.480		96.50	19.60		116.10	137
8680	12" diameter	↓	50	.480	↓	119	19.60		138.60	162
8700	PVC pipe, joint restraint									
8710	4" diameter	B-20A	32	1	Ea.	46	44.50		90.50	119
8720	6" diameter		25.60	1.250		57	56		113	148
8730	8" diameter		21.33	1.500		79.50	67		146.50	190
8740	10" diameter		18.28	1.751		141	78.50		219.50	275
8750	12" diameter		16.84	1.900		148	85		233	293
8760	14" diameter		16	2		221	89.50		310.50	380
8770	16" diameter		11.64	2.749		274	123		397	485
8780	18" diameter		11.03	2.901		330	130		460	565
8785	20" diameter		9.14	3.501		400	157		557	680
8790	24" diameter	↓	7.53	4.250	↓	465	190		655	800

33 11 13.35 Water Supply, HDPE		Crew	Daily Output	Labor-Hours	Unit	Material	Labor	Equipment	Total	Total Incl O&P
0010	**WATER SUPPLY, HDPE**									
0011	Butt fusion joints, SDR 21 40' lengths not including excavation or backfill									
0100	4" diameter	B-22A	400	.100	L.F.	4.13	4.19	1.66	9.98	12.80
0200	6" diameter		380	.105		8.20	4.41	1.74	14.35	17.70
0300	8" diameter		320	.125		10.65	5.25	2.07	17.97	22
0400	10" diameter	↓	300	.133	↓	20	5.60	2.21	27.81	33

33 11 13 – Public Water Utility Distribution Piping

33 11 13.35 Water Supply, HDPE

		Crew	Daily Output	Labor-Hours	Unit	Material	Labor	Equipment	Total	Total Incl O&P
						2014 Bare Costs				
0500	12" diameter	B-22A	260	.154	L.F.	22.50	6.45	2.55	31.50	37
0600	14" diameter	B-22B	220	.182		38.50	7.60	5.05	51.15	60
0700	16" diameter		180	.222		44	9.30	6.15	59.45	69
0800	18" diameter		140	.286		49.50	11.95	7.90	69.35	81
0900	24" diameter		100	.400		70.50	16.75	11.10	98.35	116
1000	Fittings									
1100	Elbows, 90 degrees									
1200	4" diameter	B-22A	32	1.250	Ea.	23.50	52.50	20.50	96.50	129
1300	6" diameter		28	1.429		55	60	23.50	138.50	179
1400	8" diameter		24	1.667		148	70	27.50	245.50	300
1500	10" diameter		18	2.222		550	93	37	680	790
1600	12" diameter		12	3.333		550	140	55	745	880
1700	14" diameter	B-22B	9	4.444		760	186	123	1,069	1,250
1800	16" diameter		6	6.667		870	279	185	1,334	1,600
1900	18" diameter		4	10		985	420	277	1,682	2,025
2000	24" diameter		3	13.333		1,325	560	370	2,255	2,725
2100	Tees									
2200	4" diameter	B-22A	30	1.333	Ea.	28	56	22	106	142
2300	6" diameter		26	1.538		68.50	64.50	25.50	158.50	202
2400	8" diameter		22	1.818		175	76	30	281	345
2500	10" diameter		15	2.667		231	112	44	387	475
2600	12" diameter		10	4		865	168	66	1,099	1,275
2700	14" diameter	B-22B	8	5		1,025	210	139	1,374	1,600
2800	16" diameter		6	6.667		1,200	279	185	1,664	1,950
2900	18" diameter		4	10		1,400	420	277	2,097	2,500
3000	24" diameter		2	20		1,875	840	555	3,270	3,975
4100	Caps									
4110	4" diameter	B-22A	34	1.176	Ea.	17.65	49.50	19.45	86.60	117
4120	6" diameter		30	1.333		41	56	22	119	156
4130	8" diameter		26	1.538		71	64.50	25.50	161	205
4150	10" diameter		20	2		310	84	33	427	510
4160	12" diameter		14	2.857		320	120	47.50	487.50	585

33 11 13.40 Water Supply, Black Steel Pipe

		Crew	Daily Output	Labor-Hours	Unit	Material	Labor	Equipment	Total	Total Incl O&P
0010	**WATER SUPPLY, BLACK STEEL PIPE**									
0011	Not including excavation or backfill									
1000	Pipe, black steel, plain end, welded, 1/4" wall thk, 8" diam.	B-35A	208	.269	L.F.	32	11.95	7.85	51.80	62
1010	10" diameter		204	.275		40	12.20	8	60.20	71.50
1020	12" diameter		195	.287		49.50	12.75	8.40	70.65	83
1030	18" diameter		175	.320		72	14.20	9.35	95.55	111
1040	5/16" wall thickness, 12" diameter		195	.287		60	12.75	8.40	81.15	94.50
1050	18" diameter		175	.320		89.50	14.20	9.35	113.05	130
1060	36" diameter		28.96	1.934		179	86	56.50	321.50	390
1070	3/8" wall thickness, 18" diameter		43.20	1.296		108	57.50	38	203.50	248
1080	24" diameter		36	1.556		148	69	45.50	262.50	320
1090	30" diameter		30.40	1.842		180	81.50	54	315.50	380
1100	1/2" wall thickness, 36" diameter		26.08	2.147		283	95.50	62.50	441	525
1110	48" diameter		21.68	2.583		420	115	75.50	610.50	725
1135	7/16" wall thickness, 48" diameter		20.80	2.692		385	119	78.50	582.50	690
1140	5/8" wall thickness, 48" diameter		21.68	2.583		545	115	75.50	735.50	860

33 11 Water Utility Distribution Piping

33 11 13 – Public Water Utility Distribution Piping

33 11 13.45 Water Supply, Copper Pipe		Crew	Daily Output	Labor-Hours	Unit	Material	Labor	Equipment	Total	Total Incl O&P
						2014 Bare Costs				
0010	**WATER SUPPLY, COPPER PIPE**									
0020	Not including excavation or backfill									
2000	Tubing, type K, 20′ joints, 3/4″ diameter	Q-1	400	.040	L.F.	6.75	2.07		8.82	10.60
2200	1″ diameter		320	.050		9	2.59		11.59	13.80
3000	1-1/2″ diameter		265	.060		14.35	3.13		17.48	20.50
3020	2″ diameter		230	.070		22	3.60		25.60	30
3040	2-1/2″ diameter		146	.110		32.50	5.70		38.20	44.50
3060	3″ diameter		134	.119		45	6.20		51.20	59.50
4012	4″ diameter		95	.168		75	8.70		83.70	95
4016	6″ diameter	Q-2	80	.300		161	16.10		177.10	202
5000	Tubing, type L									
5108	2″ diameter	Q-1	230	.070	L.F.	17.65	3.60		21.25	25
6010	3″ diameter		134	.119		36	6.20		42.20	49
6012	4″ diameter		95	.168		59.50	8.70		68.20	78.50
6016	6″ diameter	Q-2	80	.300		142	16.10		158.10	181
7020	Fittings, brass, corporation stops, 3/4″ diameter	1 Plum	19	.421	Ea.	45	24		69	86
7040	1″ diameter		16	.500		45.50	29		74.50	93.50
7060	1-1/2″ diameter		13	.615		143	35.50		178.50	211
7080	2″ diameter		11	.727		211	42		253	295
7100	Curb stops, 3/4″ diameter		19	.421		48	24		72	89.50
7120	1″ diameter		16	.500		77	29		106	128
7140	1-1/2″ diameter		13	.615		220	35.50		255.50	296
7160	2″ diameter		11	.727		330	42		372	430
7165	Fittings, brass, corporation stops, no lead, 3/4″ diameter		19	.421		76	24		100	120
7166	1″ diameter		16	.500		100	29		129	154
7167	1-1/2″ diameter		13	.615		220	35.50		255.50	296
7168	2″ diameter		11	.727		330	42		372	430
7170	Curb stops, no lead, 3/4″ diameter		19	.421		103	24		127	151
7171	1″ diameter		16	.500		147	29		176	206
7172	1-1/2″ diameter		13	.615		275	35.50		310.50	360
7173	2″ diameter		11	.727		290	42		332	385
7180	Curb box, cast iron, 1/2″ to 1″ curb stops		12	.667		52	38.50		90.50	115
7200	1-1/4″ to 2″ curb stops		8	1		84.50	57.50		142	180
7220	Saddles, 3/4″ & 1″ diameter, add					74			74	81.50
7240	1-1/2″ to 2″ diameter, add					107			107	117
7250	For copper fittings, see Section 22 11 13.25									

33 11 13.90 Water Supply, Thrust Blocks		Crew	Daily Output	Labor-Hours	Unit	Material	Labor	Equipment	Total	Total Incl O&P
0010	**WATER SUPPLY, THRUST BLOCKS**									
0015	Piping, not including excavation or backfill									
0110	Thrust block for 90 elbow, 4″ Diameter	C-30	41	.195	Ea.	16.05	7.15	4.31	27.51	33.50
0115	6″ Diameter		23	.348		27.50	12.75	7.70	47.95	58
0120	8″ Diameter		14	.571		42.50	21	12.65	76.15	93
0125	10″ Diameter		9	.889		61	32.50	19.65	113.15	139
0130	12″ Diameter		7	1.143		83	42	25.50	150.50	184
0135	14″ Diameter		5	1.600		111	58.50	35.50	205	252
0140	16″ Diameter		4	2		140	73.50	44	257.50	315
0145	18″ Diameter		3	2.667		173	97.50	59	329.50	405
0150	20″ Diameter		2.50	3.200		209	117	70.50	396.50	490
0155	24″ Diameter		2	4		300	147	88.50	535.50	655
0210	Thrust block for tee or deadend, 4″ Diameter		65	.123		11	4.51	2.72	18.23	22
0215	6″ Diameter		35	.229		19.45	8.40	5.05	32.90	40
0220	8″ Diameter		21	.381		30.50	13.95	8.40	52.85	64.50

33 11 Water Utility Distribution Piping

33 11 13 – Public Water Utility Distribution Piping

33 11 13.90 Water Supply, Thrust Blocks		Crew	Daily Output	Labor-Hours	Unit	2014 Bare Costs Material	Labor	Equipment	Total	Total Incl O&P
0225	10″ Diameter	C-30	14	.571	Ea.	44	21	12.65	77.65	94.50
0230	12″ Diameter		10	.800		59.50	29.50	17.70	106.70	130
0235	14″ Diameter		7	1.143		80	42	25.50	147.50	181
0240	16″ Diameter		5	1.600		101	58.50	35.50	195	241
0245	18″ Diameter		4	2		124	73.50	44	241.50	299
0250	20″ Diameter		3.50	2.286		150	84	50.50	284.50	350
0255	24″ Diameter		2.50	3.200		212	117	70.50	399.50	495

33 12 Water Utility Distribution Equipment

33 12 13 – Water Service Connections

33 12 13.15 Tapping, Crosses and Sleeves		Crew	Daily Output	Labor-Hours	Unit	Material	Labor	Equipment	Total	Total Incl O&P
0010	**TAPPING, CROSSES AND SLEEVES**									
4000	Drill and tap pressurized main (labor only)									
4100	6″ main, 1″ to 2″ service	Q-1	3	5.333	Ea.		276		276	415
4150	8″ main, 1″ to 2″ service	″	2.75	5.818	″		300		300	455
4500	Tap and insert gate valve									
4600	8″ main, 4″ branch	B-21	3.20	8.750	Ea.		370	44	414	620
4650	6″ branch		2.70	10.370			440	52	492	730
4700	10″ Main, 4″ branch		2.70	10.370			440	52	492	730
4750	6″ branch		2.35	11.915			505	59.50	564.50	840
4800	12″ main, 6″ branch		2.35	11.915			505	59.50	564.50	840
4850	8″ branch		2.35	11.915			505	59.50	564.50	840
7000	Tapping crosses, sleeves, valves; with rubber gaskets									
7020	Crosses, 4″ x 4″	B-21	37	.757	Ea.	1,550	32	3.78	1,585.78	1,750
7060	8″ x 6″		21	1.333		1,600	56.50	6.65	1,663.15	1,875
7080	8″ x 8″		21	1.333		1,675	56.50	6.65	1,738.15	1,950
7100	10″ x 6″		21	1.333		2,625	56.50	6.65	2,688.15	3,000
7160	12″ x 12″		18	1.556		3,675	65.50	7.80	3,748.30	4,125
7180	14″ x 6″		16	1.750		7,150	74	8.75	7,232.75	7,975
7240	16″ x 10″		14	2		7,750	84.50	10	7,844.50	8,700
7280	18″ x 6″		10	2.800		11,500	118	14	11,632	12,800
7320	18″ x 18″		10	2.800		10,500	118	14	10,632	11,800
7340	20″ x 6″		8	3.500		9,250	148	17.50	9,415.50	10,400
7360	20″ x 12″		8	3.500		9,925	148	17.50	10,090.50	11,100
7420	24″ x 12″		6	4.667		12,100	197	23.50	12,320.50	13,700
7440	24″ x 18″		6	4.667		18,100	197	23.50	18,320.50	20,200
7600	Cut-in sleeves with rubber gaskets, 4″		18	1.556		435	65.50	7.80	508.30	585
7640	8″		10	2.800		640	118	14	772	900
7680	12″		9	3.111		1,200	131	15.55	1,346.55	1,550
7800	Cut-in valves with rubber gaskets, 4″		18	1.556		475	65.50	7.80	548.30	630
7840	8″		10	2.800		1,000	118	14	1,132	1,300
7880	12″		9	3.111		1,250	131	15.55	1,396.55	1,600
7900	Tapping Valve 4 inch, MJ, ductile iron		18	1.556		570	65.50	7.80	643.30	740
7920	6 inch, MJ, ductile iron		12	2.333		790	98.50	11.65	900.15	1,025
8000	Sleeves with rubber gaskets, 4″ x 4″		37	.757		870	32	3.78	905.78	1,025
8030	8″ x 4″		21	1.333		880	56.50	6.65	943.15	1,075
8040	8″ x 6″		21	1.333		1,025	56.50	6.65	1,088.15	1,225
8060	8″ x 8″		21	1.333		1,350	56.50	6.65	1,413.15	1,575
8070	10″ x 4″		21	1.333		970	56.50	6.65	1,033.15	1,175
8080	10″ x 6″		21	1.333		1,050	56.50	6.65	1,113.15	1,250
8090	10″ x 8″		21	1.333		1,525	56.50	6.65	1,588.15	1,775

33 12 Water Utility Distribution Equipment

33 12 13 – Water Service Connections

33 12 13.15 Tapping, Crosses and Sleeves		Crew	Daily Output	Labor-Hours	Unit	Material	2014 Bare Costs Labor	Equipment	Total	Total Incl O&P
8140	12" x 12"	B-21	18	1.556	Ea.	2,900	65.50	7.80	2,973.30	3,275
8160	14" x 6"		16	1.750		1,775	74	8.75	1,857.75	2,075
8220	16" x 10"		14	2		3,325	84.50	10	3,419.50	3,825
8260	18" x 6"		10	2.800		2,125	118	14	2,257	2,525
8300	18" x 18"		10	2.800		12,000	118	14	12,132	13,400
8320	20" x 6"		8	3.500		2,475	148	17.50	2,640.50	2,950
8340	20" x 12"		8	3.500		4,175	148	17.50	4,340.50	4,850
8400	24" x 12"		6	4.667		4,325	197	23.50	4,545.50	5,100
8420	24" x 18"		6	4.667		10,700	197	23.50	10,920.50	12,100
8800	Hydrant valve box, 6' long	B-20	20	1.200		259	49		308	360
8820	8' long		18	1.333		335	54.50		389.50	450
8830	Valve box w/lid 4' deep		14	1.714		116	70		186	236
8840	Valve box and large base w/lid		14	1.714		325	70		395	470

33 12 16 – Water Utility Distribution Valves

33 12 16.10 Valves		Crew	Daily Output	Labor-Hours	Unit	Material	Labor	Equipment	Total	Total Incl O&P
0010	**VALVES**, water distribution									
0011	See Sections 22 05 23.20 and 22 05 23.60									
3000	Butterfly valves with boxes, cast iron, mech. jt.									
3100	4" diameter	B-6	6	4	Ea.	830	160	61	1,051	1,225
3180	8" diameter		6	4		1,150	160	61	1,371	1,575
3340	12" diameter		6	4		1,675	160	61	1,896	2,175
3400	14" diameter		4	6		4,025	241	91.50	4,357.50	4,900
3460	18" diameter		4	6		6,150	241	91.50	6,482.50	7,225
3480	20" diameter		4	6		5,375	241	91.50	5,707.50	6,375
3500	24" diameter		4	6		7,500	241	91.50	7,832.50	8,725
3600	With lever operator									
3610	4" diameter	B-6	6	4	Ea.	715	160	61	936	1,100
3616	8" diameter		6	4		1,025	160	61	1,246	1,450
3620	12" diameter		6	4		1,550	160	61	1,771	2,050
3624	16" diameter		4	6		3,225	241	91.50	3,557.50	4,025
3630	24" diameter		4	6		7,175	241	91.50	7,507.50	8,350
3700	Check valves, flanged									
3710	4" diameter	B-6	6	4	Ea.	940	160	61	1,161	1,350
3714	6" diameter		6	4		1,375	160	61	1,596	1,825
3716	8" diameter		6	4		2,300	160	61	2,521	2,850
3720	12" diameter		6	4		3,825	160	61	4,046	4,550
3724	16" diameter		4	6		6,600	241	91.50	6,932.50	7,725
3726	18" diameter		4	6		9,775	241	91.50	10,107.50	11,300
3730	24" diameter		4	6		16,300	241	91.50	16,632.50	18,400
3800	Gate valves, C.I., 250 PSI, mechanical joint, w/boxes									
3810	4" diameter	B-6	6	4	Ea.	395	160	61	616	750
3814	6" diameter		6	4		520	160	61	741	890
3816	8" diameter		6	4		920	160	61	1,141	1,325
3820	12" diameter		6	4		1,475	160	61	1,696	1,950
3824	16" diameter		4	6		5,075	241	91.50	5,407.50	6,075
3828	20" diameter		4	6		11,700	241	91.50	12,032.50	13,300
3830	24" diameter		4	6		14,600	241	91.50	14,932.50	16,600
3831	30" diameter		4	6		38,400	241	91.50	38,732.50	42,700
3832	36" diameter		4	6		56,000	241	91.50	56,332.50	62,000
3880	Sleeve, for tapping mains, 8" x 4", add					880			880	970
3884	10" x 6", add					1,050			1,050	1,150
3888	12" x 6", add					1,675			1,675	1,825

33 12 Water Utility Distribution Equipment

33 12 16 – Water Utility Distribution Valves

33 12 16.10 Valves		Crew	Daily Output	Labor-Hours	Unit	Material	Labor	Equipment	Total	Total Incl O&P
						2014 Bare Costs				
3892	12" x 8", add				Ea.	1,350			1,350	1,500

33 12 16.20 Valves		Crew	Daily Output	Labor-Hours	Unit	Material	Labor	Equipment	Total	Total Incl O&P
0010	**VALVES**									
0011	Special trim or use									
9000	Valves, gate valve, N.R.S. PIV with post, 4" diameter	B-6	6	4	Ea.	1,400	160	61	1,621	1,850
9040	8" diameter		6	4		1,925	160	61	2,146	2,425
9080	12" diameter		6	4		2,475	160	61	2,696	3,050
9120	OS&Y, 4" diameter		6	4		730	160	61	951	1,125
9160	8" diameter		6	4		1,550	160	61	1,771	2,025
9200	12" diameter		6	4		3,725	160	61	3,946	4,425
9220	14" diameter		4	6		5,075	241	91.50	5,407.50	6,075
9400	Check valves, rubber disc, 2-1/2" diameter		6	4		710	160	61	931	1,100
9440	4" diameter		6	4		940	160	61	1,161	1,350
9500	8" diameter		6	4		2,300	160	61	2,521	2,850
9540	12" diameter		6	4		3,825	160	61	4,046	4,550
9700	Detector check valves, reducing, 4" diameter		6	4		1,325	160	61	1,546	1,775
9740	8" diameter		6	4		2,975	160	61	3,196	3,575
9800	Galvanized, 4" diameter		6	4		1,550	160	61	1,771	2,025
9840	8" diameter		6	4		3,250	160	61	3,471	3,900

33 12 19 – Water Utility Distribution Fire Hydrants

33 12 19.10 Fire Hydrants		Crew	Daily Output	Labor-Hours	Unit	Material	Labor	Equipment	Total	Total Incl O&P
0010	**FIRE HYDRANTS**									
0020	Mechanical joints unless otherwise noted									
1000	Fire hydrants, two way; excavation and backfill not incl.									
1100	4-1/2" valve size, depth 2'-0"	B-21	10	2.800	Ea.	1,650	118	14	1,782	2,025
1120	2'-6"		10	2.800		1,750	118	14	1,882	2,125
1140	3'-0"		10	2.800		1,875	118	14	2,007	2,250
1160	3'-6"		9	3.111		1,900	131	15.55	2,046.55	2,325
1170	4'-0"		9	3.111		1,950	131	15.55	2,096.55	2,375
1200	4'-6"		9	3.111		1,975	131	15.55	2,121.55	2,400
1220	5'-0"		8	3.500		2,025	148	17.50	2,190.50	2,475
1240	5'-6"		8	3.500		2,050	148	17.50	2,215.50	2,500
1260	6'-0"		7	4		2,100	169	20	2,289	2,575
1280	6'-6"		7	4		2,100	169	20	2,289	2,575
1300	7'-0"		6	4.667		2,150	197	23.50	2,370.50	2,700
1340	8'-0"		6	4.667		2,225	197	23.50	2,445.50	2,775
1420	10'-0"		5	5.600		2,375	236	28	2,639	3,025
2000	5-1/4" valve size, depth 2'-0"		10	2.800		1,775	118	14	1,907	2,175
2080	4'-0"		9	3.111		1,900	131	15.55	2,046.55	2,300
2160	6'-0"		7	4		2,050	169	20	2,239	2,550
2240	8'-0"		6	4.667		2,175	197	23.50	2,395.50	2,700
2320	10'-0"		5	5.600		2,500	236	28	2,764	3,150
2350	For threeway valves, add					7%				
2400	Lower barrel extensions with stems, 1'-0"	B-20	14	1.714		355	70		425	500
2440	2'-0"		13	1.846		430	75.50		505.50	590
2480	3'-0"		12	2		770	81.50		851.50	970
2520	4'-0"		10	2.400		815	98		913	1,050
5000	Indicator post									
5020	Adjustable, valve size 4" to 14", 4' bury	B-21	10	2.800	Ea.	800	118	14	932	1,075
5060	8' bury		7	4		1,100	169	20	1,289	1,500
5080	10' bury		6	4.667		1,150	197	23.50	1,370.50	1,600
5100	12' bury		5	5.600		1,400	236	28	1,664	1,950

33 12 Water Utility Distribution Equipment

33 12 19 – Water Utility Distribution Fire Hydrants

33 12 19.10 Fire Hydrants		Crew	Daily Output	Labor-Hours	Unit	Material	2014 Bare Costs Labor	Equipment	Total	Total Incl O&P
5120	14′ bury	B-21	4	7	Ea.	1,525	295	35	1,855	2,175
5500	Non-adjustable, valve size 4″ to 14″, 3′ bury		10	2.800		800	118	14	932	1,075
5520	3′-6″ bury		10	2.800		800	118	14	932	1,075
5540	4′ bury	↓	9	3.111	↓	800	131	15.55	946.55	1,100

33 16 Water Utility Storage Tanks

33 16 13 – Aboveground Water Utility Storage Tanks

33 16 13.13 Steel Water Storage Tanks

		Crew	Daily Output	Labor-Hours	Unit	Material	Labor	Equipment	Total	Total Incl O&P
0010	**STEEL WATER STORAGE TANKS**									
0910	Steel, ground level, ht./diam. less than 1, not incl. fdn., 100,000 gallons				Ea.				202,000	244,500
1000	250,000 gallons								295,500	324,000
1200	500,000 gallons								417,000	458,500
1250	750,000 gallons								538,000	591,500
1300	1,000,000 gallons								558,000	725,500
1500	2,000,000 gallons								1,043,000	1,148,000
1600	4,000,000 gallons								2,121,000	2,333,000
1800	6,000,000 gallons								3,095,000	3,405,000
1850	8,000,000 gallons								4,068,000	4,475,000
1910	10,000,000 gallons				↓				5,050,000	5,554,500
2100	Steel standpipes, ht./diam. more than 1, 100′ to overflow, no fdn.									
2200	500,000 gallons				Ea.				546,500	600,500
2400	750,000 gallons								722,500	794,500
2500	1,000,000 gallons								1,060,500	1,167,000
2700	1,500,000 gallons								1,749,000	1,923,000
2800	2,000,000 gallons				↓				2,327,000	2,559,000

33 16 13.16 Prestressed Conc. Water Storage Tanks

		Crew	Daily Output	Labor-Hours	Unit	Material	Labor	Equipment	Total	Total Incl O&P
0010	**PRESTRESSED CONC. WATER STORAGE TANKS**									
0020	Not including fdn., pipe or pumps, 250,000 gallons				Ea.				299,000	329,500
0100	500,000 gallons								487,000	536,000
0300	1,000,000 gallons								707,000	807,500
0400	2,000,000 gallons								1,072,000	1,179,000
0600	4,000,000 gallons								1,706,000	1,877,000
0700	6,000,000 gallons								2,266,000	2,493,000
0750	8,000,000 gallons								2,924,000	3,216,000
0800	10,000,000 gallons				↓				3,533,000	3,886,000

33 16 13.23 Plastic-Coated Fabric Pillow Water Tanks

		Crew	Daily Output	Labor-Hours	Unit	Material	Labor	Equipment	Total	Total Incl O&P
0010	**PLASTIC-COATED FABRIC PILLOW WATER TANKS**									
7000	Water tanks, vinyl coated fabric pillow tanks, freestanding, 5,000 gallons	4 Clab	4	8	Ea.	3,600	293		3,893	4,400
7100	Supporting embankment not included, 25,000 gallons	6 Clab	2	24		13,000	880		13,880	15,600
7200	50,000 gallons	8 Clab	1.50	42.667		18,100	1,575		19,675	22,300
7300	100,000 gallons	9 Clab	.90	80		41,500	2,925		44,425	50,000
7400	150,000 gallons		.50	144		59,500	5,275		64,775	73,500
7500	200,000 gallons		.40	180		73,500	6,600		80,100	91,000
7600	250,000 gallons	↓	.30	240	↓	103,500	8,800		112,300	127,500

33 16 19 – Elevated Water Utility Storage Tanks

33 16 19.50 Elevated Water Storage Tanks

		Crew	Daily Output	Labor-Hours	Unit	Material	Labor	Equipment	Total	Total Incl O&P
0010	**ELEVATED WATER STORAGE TANKS**									
0011	Not incl. pipe, pumps or foundation									
3000	Elevated water tanks, 100′ to bottom capacity line, incl. painting									
3010	50,000 gallons				Ea.				185,000	204,000

33 16 Water Utility Storage Tanks

33 16 19 – Elevated Water Utility Storage Tanks

33 16 19.50 Elevated Water Storage Tanks		Crew	Daily Output	Labor-Hours	Unit	2014 Bare Costs Material	Labor	Equipment	Total	Total Incl O&P
3300	100,000 gallons				Ea.				280,000	307,500
3400	250,000 gallons								751,500	826,500
3600	500,000 gallons								1,336,000	1,470,000
3700	750,000 gallons								1,622,000	1,783,500
3900	1,000,000 gallons				↓				2,322,000	2,556,000

33 21 Water Supply Wells

33 21 13 – Public Water Supply Wells

33 21 13.10 Wells and Accessories		Crew	Daily Output	Labor-Hours	Unit	Material	Labor	Equipment	Total	Total Incl O&P
0010	**WELLS & ACCESSORIES**									
0011	Domestic									
0100	Drilled, 4" to 6" diameter	B-23	120	.333	L.F.		12.35	24	36.35	45.50
0200	8" diameter	"	95.20	.420	"		15.55	30.50	46.05	57.50
0400	Gravel pack well, 40' deep, incl. gravel & casing, complete									
0500	24" diameter casing x 18" diameter screen	B-23	.13	307	Total	38,800	11,400	22,200	72,400	84,500
0600	36" diameter casing x 18" diameter screen		.12	333	"	40,000	12,400	24,000	76,400	89,500
0800	Observation wells, 1-1/4" riser pipe	↓	163	.245	V.L.F.	20.50	9.10	17.70	47.30	56
0900	For flush Buffalo roadway box, add	1 Skwk	16.60	.482	Ea.	51.50	23		74.50	92
1200	Test well, 2-1/2" diameter, up to 50' deep (15 to 50 GPM)	B-23	1.51	26.490	"	810	980	1,900	3,690	4,525
1300	Over 50' deep, add	"	121.80	.328	L.F.	21.50	12.15	23.50	57.15	68.50
1500	Pumps, installed in wells to 100' deep, 4" submersible									
1510	1/2 H.P.	Q-1	3.22	4.969	Ea.	450	257		707	885
1520	3/4 H.P.		2.66	6.015		590	310		900	1,125
1600	1 H.P.	↓	2.29	6.987		680	360		1,040	1,300
1700	1-1/2 H.P.	Q-22	1.60	10		735	520	415	1,670	2,050
1800	2 H.P.		1.33	12.030		890	625	500	2,015	2,475
1900	3 H.P.		1.14	14.035		1,375	725	580	2,680	3,250
2000	5 H.P.		1.14	14.035		2,300	725	580	3,605	4,275
2050	Remove and install motor only, 4 H.P.		1.14	14.035		1,100	725	580	2,405	2,950
3000	Pump, 6" submersible, 25' to 150' deep, 25 H.P., 249 to 297 GPM		.89	17.978		7,175	930	745	8,850	10,100
3100	25' to 500' deep, 30 H.P., 100 to 300 GPM	↓	.73	21.918	↓	8,600	1,125	910	10,635	12,200
8000	Steel well casing	B-23A	3020	.008	Lb.	1.22	.33	.90	2.45	2.83
8110	Well screen assembly, stainless steel, 2" diameter		273	.088	L.F.	81	3.64	9.90	94.54	106
8120	3" diameter		253	.095		141	3.93	10.70	155.63	174
8130	4" diameter		200	.120		177	4.97	13.55	195.52	217
8140	5" diameter		168	.143		187	5.90	16.10	209	233
8150	6" diameter		126	.190		212	7.90	21.50	241.40	269
8160	8" diameter		98.50	.244		288	10.10	27.50	325.60	360
8170	10" diameter		73	.329		360	13.60	37	410.60	455
8180	12" diameter		62.50	.384		410	15.90	43.50	469.40	520
8190	14" diameter		54.30	.442		460	18.30	50	528.30	590
8200	16" diameter		48.30	.497		510	20.50	56	586.50	655
8210	18" diameter		39.20	.612		635	25.50	69	729.50	815
8220	20" diameter		31.20	.769		710	32	87	829	930
8230	24" diameter		23.80	1.008		855	42	114	1,011	1,125
8240	26" diameter		21	1.143		930	47.50	129	1,106.50	1,250
8250	PVC well casing, 2" diameter		280	.086		1.66	3.55	9.65	14.86	17.95
8252	3" diameter		260	.092		3.40	3.82	10.40	17.62	21
8254	4" diameter		205	.117		4.89	4.85	13.20	22.94	27.50
8256	6" diameter		130	.185		8.50	7.65	21	37.15	44
8258	8" diameter	↓	100	.240	↓	12.95	9.95	27	49.90	59.50

33 21 Water Supply Wells

33 21 13 – Public Water Supply Wells

33 21 13.10 Wells and Accessories

	33 21 13.10 Wells and Accessories	Crew	Daily Output	Labor-Hours	Unit	2014 Bare Costs Material	Labor	Equipment	Total	Total Incl O&P
8300	Slotted PVC, 1-1/4″ diameter	B-23A	521	.046	L.F.	2.41	1.91	5.20	9.52	11.25
8310	1-1/2″ diameter		488	.049		2.95	2.04	5.55	10.54	12.45
8320	2″ diameter		273	.088		4.03	3.64	9.90	17.57	21
8330	3″ diameter		253	.095		5.85	3.93	10.70	20.48	24
8340	4″ diameter		200	.120		6.75	4.97	13.55	25.27	30
8350	5″ diameter		168	.143		14.05	5.90	16.10	36.05	42.50
8360	6″ diameter		126	.190		16.70	7.90	21.50	46.10	54
8370	8″ diameter		98.50	.244		26	10.10	27.50	63.60	74
8400	Artificial gravel pack, 2″ screen, 6″ casing	B-23B	174	.138		3.61	5.70	17.50	26.81	32
8405	8″ casing		111	.216		4.63	8.95	27.50	41.08	49
8410	10″ casing		74.50	.322		6.60	13.35	41	60.95	73
8415	12″ casing		60	.400		9.25	16.55	51	76.80	91.50
8420	14″ casing		50.20	.478		10.80	19.80	60.50	91.10	109
8425	16″ casing		40.70	.590		13.60	24.50	75	113.10	135
8430	18″ casing		36	.667		16.35	27.50	84.50	128.35	154
8435	20″ casing		29.50	.814		19.25	33.50	103	155.75	187
8440	24″ casing		25.70	.934		22	38.50	119	179.50	215
8445	26″ casing		24.60	.976		24.50	40.50	124	189	225
8450	30″ casing		20	1.200		28	49.50	152	229.50	275
8455	36″ casing		16.40	1.463		31.50	60.50	186	278	335
8500	Develop well		8	3	Hr.	243	124	380	747	875
8550	Pump test well		8	3		81	124	380	585	700
8560	Standby well	B-23A	8	3		76.50	124	340	540.50	645
8570	Standby, drill rig		8	3			124	340	464	560
8580	Surface seal well, concrete filled		1	24	Ea.	835	995	2,700	4,530	5,425
8590	Well test pump, install & remove	B-23	1	40			1,475	2,875	4,350	5,450
8600	Well sterilization, chlorine	2 Clab	1	16		120	585		705	1,025
8610	Well water pressure switch	1 Clab	12	.667		49.50	24.50		74	92
8630	Well, water pressure switch with manual reset	″	12	.667		35	24.50		59.50	76
9950	See Section 31 23 19.40 for wellpoints									
9960	See Section 31 23 19.30 for drainage wells									

33 21 13.20 Water Supply Wells, Pumps

	33 21 13.20 Water Supply Wells, Pumps	Crew	Daily Output	Labor-Hours	Unit	Material	Labor	Equipment	Total	Total Incl O&P
0010	**WATER SUPPLY WELLS, PUMPS**									
0011	With pressure control									
1000	Deep well, jet, 42 gal. galvanized tank									
1040	3/4 HP	1 Plum	.80	10	Ea.	1,100	575		1,675	2,100
3000	Shallow well, jet, 30 gal. galvanized tank									
3040	1/2 HP	1 Plum	2	4	Ea.	895	230		1,125	1,325

33 31 Sanitary Utility Sewerage Piping

33 31 13 – Public Sanitary Utility Sewerage Piping

33 31 13.13 Sewage Collection, Vent Cast Iron Pipe

	33 31 13.13 Sewage Collection, Vent Cast Iron Pipe	Crew	Daily Output	Labor-Hours	Unit	Material	Labor	Equipment	Total	Total Incl O&P
0010	**SEWAGE COLLECTION, VENT CAST IRON PIPE**									
0020	Not including excavation or backfill									
2022	Sewage vent cast iron, B&S, 4″ diameter	Q-1	66	.242	L.F.	15.95	12.55		28.50	36.50
2024	5″ diameter	Q-2	88	.273		22.50	14.65		37.15	47
2026	6″ diameter	″	84	.286		27	15.35		42.35	53
2028	8″ diameter	Q-3	70	.457		44	25		69	86.50
2030	10″ diameter		66	.485		73.50	26.50		100	121
2032	12″ diameter		57	.561		105	31		136	163
2034	15″ diameter		49	.653		151	36		187	220

33 31 13.13 Sewage Collection, Vent Cast Iron Pipe		Crew	Daily Output	Labor-Hours	Unit	Material	Labor	Equipment	Total	Total Incl O&P
						2014 Bare Costs				
8001	Fittings, bends and elbows									
8110	4" diameter	Q-1	13	1.231	Ea.	55.50	64		119.50	158
8112	5" diameter	Q-2	18	1.333		80	71.50		151.50	196
8114	6" diameter	"	17	1.412		94.50	76		170.50	219
8116	8" diameter	Q-3	11	2.909		263	159		422	530
8118	10" diameter		10	3.200		400	175		575	705
8120	12" diameter		9	3.556		535	195		730	885
8122	15" diameter		7	4.571		1,575	251		1,826	2,100
8500	Wyes and tees									
8510	4" diameter	Q-1	8	2	Ea.	91	104		195	256
8512	5" diameter	Q-2	12	2		155	107		262	335
8514	6" diameter	"	11	2.182		189	117		306	385
8516	8" diameter	Q-3	7	4.571		450	251		701	875
8518	10" diameter		6	5.333		740	292		1,032	1,250
8520	12" diameter		4	8		1,500	440		1,940	2,300
8522	15" diameter		3	10.667		3,025	585		3,610	4,200

33 31 13.15 Sewage Collection, Concrete Pipe										
0010	**SEWAGE COLLECTION, CONCRETE PIPE**									
0020	See Section 33 41 13.60 for sewage/drainage collection, concrete pipe									

33 31 13.20 Sewage Collection, Plastic Pipe		Crew	Daily Output	Labor-Hours	Unit	Material	Labor	Equipment	Total	Total Incl O&P
0010	**SEWAGE COLLECTION, PLASTIC PIPE**									
0020	Not including excavation & backfill									
3000	Piping, HDPE Corrugated Type S with watertight gaskets, 4" diameter	B-20	425	.056	L.F.	.94	2.31		3.25	4.61
3020	6" diameter		400	.060		2.24	2.45		4.69	6.25
3040	8" diameter		380	.063		4.59	2.58		7.17	9.05
3060	10" diameter		370	.065		6.60	2.65		9.25	11.35
3080	12" diameter		340	.071		7.35	2.89		10.24	12.55
3100	15" diameter		300	.080		8.75	3.27		12.02	14.70
3120	18" diameter	B-21	275	.102		13.95	4.30	.51	18.76	22.50
3140	24" diameter		250	.112		18.50	4.73	.56	23.79	28.50
3160	30" diameter		200	.140		24	5.90	.70	30.60	36.50
3180	36" diameter		180	.156		32.50	6.55	.78	39.83	46.50
3200	42" diameter		175	.160		43.50	6.75	.80	51.05	59.50
3220	48" diameter		170	.165		53.50	6.95	.82	61.27	70.50
3240	54" diameter		160	.175		97	7.40	.88	105.28	119
3260	60" diameter		150	.187		142	7.90	.93	150.83	170
3300	Watertight elbows 12" diameter	B-20	11	2.182	Ea.	72.50	89		161.50	218
3320	15" diameter	"	9	2.667		110	109		219	289
3340	18" diameter	B-21	9	3.111		181	131	15.55	327.55	420
3360	24" diameter		9	3.111		380	131	15.55	526.55	640
3380	30" diameter		8	3.500		610	148	17.50	775.50	915
3400	36" diameter		8	3.500		785	148	17.50	950.50	1,100
3420	42" diameter		6	4.667		990	197	23.50	1,210.50	1,425
3440	48" diameter		6	4.667		1,775	197	23.50	1,995.50	2,300
3460	Watertight tee 12" diameter	B-20	7	3.429		123	140		263	350
3480	15" diameter	"	6	4		184	163		347	455
3500	18" diameter	B-21	6	4.667		258	197	23.50	478.50	615
3520	24" diameter		5	5.600		355	236	28	619	785
3540	30" diameter		5	5.600		705	236	28	969	1,175
3560	36" diameter		4	7		795	295	35	1,125	1,375
3580	42" diameter		4	7		870	295	35	1,200	1,450
3600	48" diameter		4	7		1,500	295	35	1,830	2,150

	33 31 13.25 Sewage Collection, Polyvinyl Chloride Pipe	Crew	Daily Output	Labor-Hours	Unit	Material	2014 Bare Costs Labor	Equipment	Total	Total Incl O&P
0010	**SEWAGE COLLECTION, POLYVINYL CHLORIDE PIPE**									
0020	Not including excavation or backfill									
2000	20′ lengths, SDR 35, B&S, 4″ diameter	B-20	375	.064	L.F.	1.46	2.62		4.08	5.65
2040	6″ diameter		350	.069		3.29	2.80		6.09	7.95
2080	13′ lengths , SDR 35, B&S, 8″ diameter		335	.072		6.95	2.93		9.88	12.10
2120	10″ diameter	B-21	330	.085		11.40	3.58	.42	15.40	18.50
2160	12″ diameter		320	.088		12.75	3.69	.44	16.88	20
2200	15″ diameter		240	.117		13	4.92	.58	18.50	22.50
2300	18″ diameter		200	.140		15.05	5.90	.70	21.65	26.50
2400	21″ diameter		190	.147		21	6.20	.74	27.94	33.50
2500	24″ diameter		180	.156		27	6.55	.78	34.33	40.50
3040	Fittings, bends or elbows, 4″ diameter	2 Skwk	24	.667	Ea.	9.85	31.50		41.35	60
3080	6″ diameter		24	.667		32	31.50		63.50	84
3120	Tees, 4″ diameter		16	1		13.50	47.50		61	88.50
3160	6″ diameter		16	1		43	47.50		90.50	121
3200	Wyes, 4″ diameter		16	1		16.70	47.50		64.20	92
3240	6″ diameter		16	1		46	47.50		93.50	124
3661	Cap, 4″		48	.333		7.70	15.75		23.45	33
3670	6″		48	.333		12.25	15.75		28	38
3671	8″		40	.400		28.50	18.90		47.40	61
3672	10″		24	.667		88	31.50		119.50	146
3673	12″		20	.800		125	38		163	197
3674	15″		18	.889		215	42		257	300
3675	18″		18	.889		315	42		357	410
3676	21″		18	.889		680	42		722	815
3677	24″		18	.889		785	42		827	930
3745	Elbow, 90 degree, 8″ diameter		20	.800		55	38		93	119
3750	10″ diameter		12	1.333		184	63		247	300
3755	12″ diameter		10	1.600		238	75.50		313.50	380
3760	15″ diameter		9	1.778		500	84		584	680
3765	18″ diameter		9	1.778		850	84		934	1,075
3770	21″ diameter		9	1.778		1,450	84		1,534	1,725
3775	24″ diameter		9	1.778		1,950	84		2,034	2,275
3795	Elbow, 45 degree, 8″ diameter		20	.800		49	38		87	113
3805	12″ diameter		10	1.600		183	75.50		258.50	320
3810	15″ diameter		9	1.778		410	84		494	580
3815	18″ diameter		9	1.778		655	84		739	850
3820	21″ diameter		9	1.778		1,100	84		1,184	1,350
3825	24″ diameter		9	1.778		1,525	84		1,609	1,800
3875	Elbow, 22 1/2 degree, 8″ diameter		20	.800		49	38		87	113
3880	10″ diameter		12	1.333		134	63		197	245
3885	12″ diameter		12	1.333		175	63		238	291
3890	15″ diameter		9	1.778		460	84		544	640
3895	18″ diameter		9	1.778		675	84		759	875
3900	21″ diameter		9	1.778		960	84		1,044	1,175
3905	24″ diameter		9	1.778		1,375	84		1,459	1,650
3910	Tee, 8″ diameter		13	1.231		80	58		138	178
3915	10″ diameter		8	2		290	94.50		384.50	465
3920	12″ diameter		7	2.286		415	108		523	620
3925	15″ diameter		6	2.667		685	126		811	950
3930	18″ diameter		6	2.667		1,125	126		1,251	1,450
3935	21″ diameter		6	2.667		2,750	126		2,876	3,225

33 31 Sanitary Utility Sewerage Piping

33 31 13 – Public Sanitary Utility Sewerage Piping

33 31 13.25 Sewage Collection, Polyvinyl Chloride Pipe		Crew	Daily Output	Labor-Hours	Unit	2014 Bare Costs Material	Labor	Equipment	Total	Total Incl O&P
3940	24″ diameter	2 Skwk	6	2.667	Ea.	4,450	126		4,576	5,100
4000	Piping, DWV PVC, no exc./bkfill., 10′ L, Sch 40, 4″ diameter	B-20	375	.064	L.F.	4.64	2.62		7.26	9.15
4010	6″ diameter		350	.069		10.20	2.80		13	15.55
4020	8″ diameter		335	.072		19.70	2.93		22.63	26

33 32 Wastewater Utility Pumping Stations

33 32 13 – Packaged Utility Lift Stations

33 32 13.13 Packaged Sewage Lift Stations

		Crew	Daily Output	Labor-Hours	Unit	Material	Labor	Equipment	Total	Total Incl O&P
0010	**PACKAGED SEWAGE LIFT STATIONS**									
2500	Sewage lift station, 200,000 GPD	E-8	.20	520	Ea.	225,500	26,200	10,500	262,200	304,500
2510	500,000 GPD		.15	684		251,000	34,500	13,800	299,300	350,000
2520	800,000 GPD		.13	812		302,000	41,000	16,400	359,400	420,000

33 36 Utility Septic Tanks

33 36 13 – Utility Septic Tank and Effluent Wet Wells

33 36 13.13 Concrete Utility Septic Tank

		Crew	Daily Output	Labor-Hours	Unit	Material	Labor	Equipment	Total	Total Incl O&P
0010	**CONCRETE UTILITY SEPTIC TANK**									
0011	Not including excavation or piping									
0015	Septic tanks, precast, 1,000 gallon	B-21	8	3.500	Ea.	1,075	148	17.50	1,240.50	1,450
0020	1,250 gallon		8	3.500		1,100	148	17.50	1,265.50	1,475
0060	1,500 gallon		7	4		1,500	169	20	1,689	1,925
0100	2,000 gallon		5	5.600		1,975	236	28	2,239	2,550
0140	2,500 gallon		5	5.600		2,500	236	28	2,764	3,125
0180	4,000 gallon		4	7		5,575	295	35	5,905	6,625
0200	5,000 gallon	B-13	3.50	16		9,475	640	213	10,328	11,600
0220	5,000 gal., 4 piece	"	3	18.667		10,400	745	249	11,394	12,800
0300	15,000 gallon, 4 piece	B-13B	1.70	32.941		21,700	1,325	670	23,695	26,700
0400	25,000 gallon, 4 piece		1.10	50.909		42,200	2,025	1,025	45,250	50,500
0500	40,000 gallon, 4 piece		.80	70		54,000	2,800	1,425	58,225	65,500
0520	50,000 gallon, 5 piece	B-13C	.60	93.333		62,000	3,725	2,850	68,575	77,500
0640	75,000 gallon, cast in place	C-14C	.25	448		75,500	19,700	130	95,330	113,500
0660	100,000 gallon	"	.15	746		93,500	32,800	217	126,517	153,500
0800	Septic tanks, precast, 2 compartment, 1,000 gallon	B-21	8	3.500		1,225	148	17.50	1,390.50	1,575
0805	1,250 gallon		8	3.500		1,550	148	17.50	1,715.50	1,950
0810	1,500 gallon		7	4		1,900	169	20	2,089	2,350
0815	2,000 gallon		5	5.600		2,400	236	28	2,664	3,025
0820	2,500 gallon		5	5.600		3,000	236	28	3,264	3,700
1150	Leaching field chambers, 13′ x 3′-7″ x 1′-4″, standard	B-13	16	3.500		480	140	46.50	666.50	790
1200	Heavy duty, 8′ x 4′ x 1′-6″		14	4		284	159	53.50	496.50	615
1300	13′ x 3′-9″ x 1′-6″		12	4.667		1,125	186	62	1,373	1,600
1350	20′ x 4′ x 1′-6″		5	11.200		1,175	445	149	1,769	2,125
1400	Leaching pit, precast concrete, 3′ diameter, 3′ deep	B-21	8	3.500		710	148	17.50	875.50	1,025
1500	6′ diameter, 3′ section		4.70	5.957		885	251	30	1,166	1,400
1600	Leaching pit, 6′-6″ diameter, 6′ deep		5	5.600		1,025	236	28	1,289	1,525
1620	8′ deep		4	7		1,200	295	35	1,530	1,825
1700	8′ diameter, H-20 load, 6′ deep		4	7		1,500	295	35	1,830	2,150
1720	8′ deep		3	9.333		2,500	395	46.50	2,941.50	3,400
2000	Velocity reducing pit, precast conc., 6′ diameter, 3′ deep		4.70	5.957		1,600	251	30	1,881	2,175

33 36 Utility Septic Tanks

33 36 13 – Utility Septic Tank and Effluent Wet Wells

	33 36 13.19 Polyethylene Utility Septic Tank	Crew	Daily Output	Labor-Hours	Unit	Material	2014 Bare Costs Labor	Equipment	Total	Total Incl O&P
0010	**POLYETHYLENE UTILITY SEPTIC TANK**									
0015	High density polyethylene, 1,000 gallon	B-21	8	3.500	Ea.	1,300	148	17.50	1,465.50	1,675
0020	1,250 gallon		8	3.500		1,400	148	17.50	1,565.50	1,800
0025	1,500 gallon		7	4		1,650	169	20	1,839	2,100
1015	Septic tanks, HDPE, 2 compartment, 1,000 gallon		8	3.500		1,275	148	17.50	1,440.50	1,650
1020	1,500 gallon		7	4		1,825	169	20	2,014	2,275

33 36 19 – Utility Septic Tank Effluent Filter

	33 36 19.13 Utility Septic Tank Effluent Tube Filter	Crew	Daily Output	Labor-Hours	Unit	Material	Labor	Equipment	Total	Total Incl O&P
0010	**UTILITY SEPTIC TANK EFFLUENT TUBE FILTER**									
3000	Effluent filter, 4" diameter	1 Skwk	8	1	Ea.	55.50	47.50		103	135
3020	6" diameter	"	7	1.143	"	230	54		284	335

33 36 33 – Utility Septic Tank Drainage Field

	33 36 33.13 Utility Septic Tank Tile Drainage Field	Crew	Daily Output	Labor-Hours	Unit	Material	Labor	Equipment	Total	Total Incl O&P
0010	**UTILITY SEPTIC TANK TILE DRAINAGE FIELD**									
0015	Distribution box, concrete, 5 outlets	2 Clab	20	.800	Ea.	78.50	29.50		108	131
0020	7 outlets		16	1		78.50	36.50		115	143
0025	9 outlets		8	2		475	73.50		548.50	640
0115	Distribution boxes, HDPE, 5 outlets		20	.800		64.50	29.50		94	116
0120	8 outlets		10	1.600		68.50	58.50		127	166
0240	Distribution boxes, Outlet Flow Leveler	1 Clab	50	.160		2.22	5.85		8.07	11.50
0300	Precast concrete, galley, 4' x 4' x 4'	B-21	16	1.750		241	74	8.75	323.75	390

33 36 50 – Drainage Field Systems

	33 36 50.10 Drainage Field Excavation and Fill	Crew	Daily Output	Labor-Hours	Unit	Material	Labor	Equipment	Total	Total Incl O&P
0010	**DRAINAGE FIELD EXCAVATION AND FILL**									
2200	Septic tank & drainage field excavation with 3/4 c.y backhoe	B-12F	145	.110	C.Y.		4.79	4.57	9.36	12.30
2400	4' trench for disposal field, 3/4 C.Y. backhoe	"	335	.048	L.F.		2.08	1.98	4.06	5.35
2600	Gravel fill, run of bank	B-6	150	.160	C.Y.	19.90	6.40	2.44	28.74	34.50
2800	Crushed stone, 3/4"	"	150	.160	"	34.50	6.40	2.44	43.34	50

33 41 Storm Utility Drainage Piping

33 41 13 – Public Storm Utility Drainage Piping

	33 41 13.40 Piping, Storm Drainage, Corrugated Metal	Crew	Daily Output	Labor-Hours	Unit	Material	Labor	Equipment	Total	Total Incl O&P
0010	**PIPING, STORM DRAINAGE, CORRUGATED METAL**									
0020	Not including excavation or backfill									
2000	Corrugated metal pipe, galvanized									
2020	Bituminous coated with paved invert, 20' lengths									
2040	8" diameter, 16 ga.	B-14	330	.145	L.F.	8.70	5.65	1.11	15.46	19.40
2060	10" diameter, 16 ga.		260	.185		9.05	7.15	1.41	17.61	22.50
2080	12" diameter, 16 ga.		210	.229		11.10	8.85	1.74	21.69	27.50
2100	15" diameter, 16 ga.		200	.240		15.25	9.30	1.83	26.38	33
2120	18" diameter, 16 ga.		190	.253		16.75	9.80	1.92	28.47	35.50
2140	24" diameter, 14 ga.		160	.300		21.50	11.60	2.28	35.38	44
2160	30" diameter, 14 ga.	B-13	120	.467		28	18.60	6.20	52.80	66
2180	36" diameter, 12 ga.		120	.467		35.50	18.60	6.20	60.30	74.50
2200	48" diameter, 12 ga.		100	.560		53	22.50	7.45	82.95	100
2220	60" diameter, 10 ga.	B-13B	75	.747		80	30	15.15	125.15	150
2240	72" diameter, 8 ga.	"	45	1.244		95.50	49.50	25.50	170.50	209
2250	End sections, 8" diameter, 16 ga.	B-14	20	2.400	Ea.	43.50	93	18.25	154.75	211
2255	10" diameter, 16 ga.		20	2.400		60	93	18.25	171.25	229

33 41 13.40 Piping, Storm Drainage, Corrugated Metal		Crew	Daily Output	Labor-Hours	Unit	2014 Bare Costs Material	Labor	Equipment	Total	Total Incl O&P
2260	12" diameter, 16 ga.	B-14	18	2.667	Ea.	120	103	20.50	243.50	315
2265	15" diameter, 16 ga.		18	2.667		205	103	20.50	328.50	410
2270	18" diameter, 16 ga.		16	3		240	116	23	379	470
2275	24" diameter, 16 ga.	B-13	16	3.500		280	140	46.50	466.50	575
2280	30" diameter, 16 ga.		14	4		490	159	53.50	702.50	840
2285	36" diameter, 14 ga.		14	4		730	159	53.50	942.50	1,100
2290	48" diameter, 14 ga.		10	5.600		1,850	223	74.50	2,147.50	2,450
2292	60" diameter, 14 ga.		6	9.333		1,950	370	124	2,444	2,850
2294	72" diameter, 14 ga.	B-13B	5	11.200		3,125	445	227	3,797	4,375
2300	Bends or elbows, 8" diameter	B-14	28	1.714		117	66.50	13.05	196.55	245
2320	10" diameter		25	1.920		146	74.50	14.60	235.10	290
2340	12" diameter, 16 ga.		23	2.087		169	81	15.90	265.90	325
2342	18" diameter, 16 ga.		20	2.400		244	93	18.25	355.25	430
2344	24" diameter, 14 ga.		16	3		335	116	23	474	575
2346	30" diameter, 14 ga.		15	3.200		430	124	24.50	578.50	690
2348	36" diameter, 14 ga.	B-13	15	3.733		580	149	49.50	778.50	920
2350	48" diameter, 12 ga.	"	12	4.667		805	186	62	1,053	1,250
2352	60" diameter, 10 ga.	B-13B	10	5.600		1,075	223	114	1,412	1,650
2354	72" diameter, 10 ga.	"	6	9.333		1,325	370	189	1,884	2,250
2360	Wyes or tees, 8" diameter	B-14	25	1.920		164	74.50	14.60	253.10	310
2380	10" diameter		21	2.286		205	88.50	17.40	310.90	380
2400	12" diameter, 16 ga.		19	2.526		244	98	19.25	361.25	440
2410	18" diameter, 16 ga.		16	3		325	116	23	464	560
2412	24" diameter, 14 ga.		16	3		485	116	23	624	740
2414	30" diameter, 14 ga.	B-13	12	4.667		645	186	62	893	1,075
2416	36" diameter, 14 ga.		11	5.091		795	203	68	1,066	1,250
2418	48" diameter, 12 ga.		10	5.600		1,175	223	74.50	1,472.50	1,700
2420	60" diameter, 10 ga.	B-13B	8	7		1,700	279	142	2,121	2,450
2422	72" diameter, 10 ga.	"	5	11.200		2,050	445	227	2,722	3,175
2500	Galvanized, uncoated, 20' lengths									
2520	8" diameter, 16 ga.	B-14	355	.135	L.F.	7.85	5.25	1.03	14.13	17.85
2540	10" diameter, 16 ga.		280	.171		9	6.65	1.30	16.95	21.50
2560	12" diameter, 16 ga.		220	.218		10	8.45	1.66	20.11	26
2580	15" diameter, 16 ga.		220	.218		12.50	8.45	1.66	22.61	28.50
2600	18" diameter, 16 ga.		205	.234		15.10	9.05	1.78	25.93	32.50
2620	24" diameter, 14 ga.		175	.274		19	10.60	2.09	31.69	39.50
2640	30" diameter, 14 ga.	B-13	130	.431		25	17.15	5.75	47.90	60.50
2660	36" diameter, 12 ga.		130	.431		32	17.15	5.75	54.90	68
2680	48" diameter, 12 ga.		110	.509		47.50	20.50	6.80	74.80	91
2690	60" diameter, 10 ga.	B-13B	78	.718		72	28.50	14.55	115.05	139
2695	72" diameter, 10 ga.	"	60	.933		86	37	18.95	141.95	173
2711	Bends or elbows, 12" diameter, 16 ga.	B-14	30	1.600	Ea.	146	62	12.20	220.20	269
2712	15" diameter, 16 ga.		25.04	1.917		180	74	14.60	268.60	330
2714	18" diameter, 16 ga.		20	2.400		202	93	18.25	313.25	385
2716	24" diameter, 14 ga.		16	3		293	116	23	432	525
2718	30" diameter, 14 ga.		15	3.200		340	124	24.50	488.50	595
2720	36" diameter, 14 ga.	B-13	15	3.733		515	149	49.50	713.50	850
2722	48" diameter, 12 ga.		12	4.667		680	186	62	928	1,100
2724	60" diameter, 10 ga.		10	5.600		1,050	223	74.50	1,347.50	1,575
2726	72" diameter, 10 ga.		6	9.333		1,350	370	124	1,844	2,175
2728	Wyes or tees, 12" diameter, 16 ga.	B-14	22.48	2.135		193	82.50	16.25	291.75	360
2730	18" diameter, 16 ga.		15	3.200		282	124	24.50	430.50	530
2732	24" diameter, 14 ga.		15	3.200		450	124	24.50	598.50	715

33 41 Storm Utility Drainage Piping

33 41 13 – Public Storm Utility Drainage Piping

33 41 13.40 Piping, Storm Drainage, Corrugated Metal		Crew	Daily Output	Labor-Hours	Unit	Material	2014 Bare Costs Labor	Equipment	Total	Total Incl O&P
2734	30″ diameter, 14 ga.	B-14	14	3.429	Ea.	580	133	26	739	870
2736	36″ diameter, 14 ga.	B-13	14	4		755	159	53.50	967.50	1,125
2738	48″ diameter, 12 ga.		12	4.667		1,075	186	62	1,323	1,550
2740	60″ diameter, 10 ga.		10	5.600		1,575	223	74.50	1,872.50	2,150
2742	72″ diameter, 10 ga.		6	9.333		1,850	370	124	2,344	2,750
2780	End sections, 8″ diameter	B-14	35	1.371		73	53	10.45	136.45	173
2785	10″ diameter		35	1.371		77	53	10.45	140.45	178
2790	12″ diameter		35	1.371		114	53	10.45	177.45	218
2800	18″ diameter		30	1.600		115	62	12.20	189.20	236
2810	24″ diameter	B-13	25	2.240		215	89.50	30	334.50	405
2820	30″ diameter		25	2.240		330	89.50	30	449.50	530
2825	36″ diameter		20	2.800		480	112	37.50	629.50	740
2830	48″ diameter		10	5.600		950	223	74.50	1,247.50	1,475
2835	60″ diameter	B-13B	5	11.200		1,650	445	227	2,322	2,725
2840	72″ diameter	″	4	14		1,975	560	284	2,819	3,325
2850	Couplings, 12″ diameter					10.60			10.60	11.65
2855	18″ diameter					15.15			15.15	16.65
2860	24″ diameter					20			20	22
2865	30″ diameter					25			25	27.50
2870	36″ diameter					32			32	35
2875	48″ diameter					47.50			47.50	52.50
2880	60″ diameter					60			60	66
2885	72″ diameter					72.50			72.50	80

33 41 13.50 Piping, Drainage & Sewage, Corrug. HDPE Type S		Crew	Daily Output	Labor-Hours	Unit	Material	Labor	Equipment	Total	Total Incl O&P
0010	**PIPING, DRAINAGE & SEWAGE, CORRUGATED HDPE TYPE S**									
0020	Not including excavation & backfill, bell & spigot									
1000	With gaskets, 4″ diameter	B-20	425	.056	L.F.	.82	2.31		3.13	4.47
1010	6″ diameter		400	.060		1.95	2.45		4.40	5.95
1020	8″ diameter		380	.063		3.99	2.58		6.57	8.40
1030	10″ diameter		370	.065		5.75	2.65		8.40	10.45
1040	12″ diameter		340	.071		6.40	2.89		9.29	11.50
1050	15″ diameter		300	.080		7.60	3.27		10.87	13.45
1060	18″ diameter	B-21	275	.102		12.10	4.30	.51	16.91	20.50
1070	24″ diameter		250	.112		16.10	4.73	.56	21.39	25.50
1080	30″ diameter		200	.140		21	5.90	.70	27.60	33
1090	36″ diameter		180	.156		28	6.55	.78	35.33	42
1100	42″ diameter		175	.160		38	6.75	.80	45.55	53.50
1110	48″ diameter		170	.165		46.50	6.95	.82	54.27	62.50
1120	54″ diameter		160	.175		84.50	7.40	.88	92.78	105
1130	60″ diameter		150	.187		124	7.90	.93	132.83	149
1135	Add 15% to material pipe cost for water tight connection bell & spigot									
1140	HDPE type S, elbows 12″ diameter	B-20	11	2.182	Ea.	63	89		152	207
1150	15″ diameter	″	9	2.667		96	109		205	273
1160	18″ diameter	B-21	9	3.111		158	131	15.55	304.55	390
1170	24″ diameter		9	3.111		335	131	15.55	481.55	585
1180	30″ diameter		8	3.500		530	148	17.50	695.50	825
1190	36″ diameter		8	3.500		680	148	17.50	845.50	995
1200	42″ diameter		6	4.667		860	197	23.50	1,080.50	1,275
1220	48″ diameter		6	4.667		1,550	197	23.50	1,770.50	2,025
1240	HDPE type S, Tee 12″ diameter	B-20	7	3.429		107	140		247	335
1260	15″ diameter	″	6	4		160	163		323	430
1280	18″ diameter	B-21	6	4.667		224	197	23.50	444.50	580

33 41 13.50 Piping, Drainage & Sewage, Corrug. HDPE Type S		Crew	Daily Output	Labor-Hours	Unit	2014 Bare Costs Material	Labor	Equipment	Total	Total Incl O&P
1300	24″ diameter	B-21	5	5.600	Ea.	310	236	28	574	735
1320	30″ diameter		5	5.600		615	236	28	879	1,075
1340	36″ diameter		4	7		690	295	35	1,020	1,250
1360	42″ diameter		4	7		755	295	35	1,085	1,325
1380	48″ diameter		4	7		1,300	295	35	1,630	1,925
1400	Add to basic installation cost for each split coupling joint									
1402	HDPE type S, split coupling, 12″ diameter	B-20	17	1.412	Ea.	7.50	57.50		65	97.50
1420	15″ diameter		15	1.600		12.50	65.50		78	115
1440	18″ diameter		13	1.846		21.50	75.50		97	141
1460	24″ diameter		12	2		31.50	81.50		113	161
1480	30″ diameter		10	2.400		70	98		168	229
1500	36″ diameter		9	2.667		133	109		242	315
1520	42″ diameter		8	3		157	123		280	360
1540	48″ diameter		8	3		168	123		291	375

33 41 13.60 Sewage/Drainage Collection, Concrete Pipe

		Crew	Daily Output	Labor-Hours	Unit	Material	Labor	Equipment	Total	Total Incl O&P
0010	**SEWAGE/DRAINAGE COLLECTION, CONCRETE PIPE**									
0020	Not including excavation or backfill									
0050	Box culvert, cast in place, 6′ x 6′	C-15	16	4.500	L.F.	232	194		426	555
0060	8′ x 8′		14	5.143		340	222		562	715
0070	12′ x 12′		10	7.200		670	310		980	1,225
0100	Box culvert, precast, base price, 8′ long, 6′ x 3′	B-69	140	.343		254	13.85	11.80	279.65	315
0150	6′ x 7′		125	.384		320	15.50	13.20	348.70	390
0200	8′ x 3′		113	.425		315	17.15	14.60	346.75	390
0250	8′ x 8′		100	.480		385	19.40	16.50	420.90	470
0300	10′ x 3′		110	.436		440	17.65	15	472.65	530
0350	10′ x 8′		80	.600		510	24	20.50	554.50	620
0400	12′ x 3′		100	.480		660	19.40	16.50	695.90	775
0450	12′ x 8′		67	.716		840	29	24.50	893.50	995
0500	Set up charge at plant, add to base price				Job	5,875			5,875	6,475
0510	Inserts and keyway, add				Ea.	545			545	595
0520	Sloped or skewed end, add				″	830			830	915
1000	Non-reinforced pipe, extra strength, B&S or T&G joints									
1010	6″ diameter	B-14	265.04	.181	L.F.	5.60	7	1.38	13.98	18.45
1020	8″ diameter		224	.214		6.15	8.30	1.63	16.08	21.50
1040	12″ diameter		200	.240		8.15	9.30	1.83	19.28	25.50
1050	15″ diameter		180	.267		11.25	10.35	2.03	23.63	30.50
1060	18″ diameter		144	.333		13.20	12.90	2.54	28.64	37
1070	21″ diameter		112	.429		16.30	16.60	3.26	36.16	47
1080	24″ diameter		100	.480		22	18.60	3.65	44.25	56.50
1560	Reinforced culvert, class 2, no gaskets									
1590	27″ diameter	B-21	88	.318	L.F.	36	13.45	1.59	51.04	62
1592	30″ diameter	B-13	80	.700		39.50	28	9.30	76.80	97
1594	36″ diameter	″	72	.778		55.50	31	10.35	96.85	120
2000	Reinforced culvert, class 3, no gaskets									
2010	12″ diameter	B-14	150	.320	L.F.	10	12.40	2.44	24.84	32.50
2020	15″ diameter		150	.320		13	12.40	2.44	27.84	36
2030	18″ diameter		132	.364		16.75	14.10	2.77	33.62	43
2035	21″ diameter		120	.400		21	15.50	3.04	39.54	50.50
2040	24″ diameter		100	.480		25	18.60	3.65	47.25	60
2045	27″ diameter	B-13	92	.609		36	24.50	8.10	68.60	85.50
2050	30″ diameter		88	.636		40.50	25.50	8.50	74.50	93
2060	36″ diameter		72	.778		55	31	10.35	96.35	119

33 41 13.60 Sewage/Drainage Collection, Concrete Pipe		Crew	Daily Output	Labor-Hours	Unit	Material	2014 Bare Costs Labor	Equipment	Total	Total Incl O&P
2070	42″ diameter	B-13B	72	.778	L.F.	74	31	15.80	120.80	146
2080	48″ diameter		64	.875		88	35	17.75	140.75	170
2090	60″ diameter		48	1.167		135	46.50	23.50	205	247
2100	72″ diameter		40	1.400		205	56	28.50	289.50	345
2120	84″ diameter		32	1.750		250	70	35.50	355.50	420
2140	96″ diameter		24	2.333		300	93	47.50	440.50	525
2200	With gaskets, class 3, 12″ diameter	B-21	168	.167		11	7.05	.83	18.88	24
2220	15″ diameter		160	.175		14.30	7.40	.88	22.58	28
2230	18″ diameter		152	.184		18.45	7.80	.92	27.17	33.50
2240	24″ diameter		136	.206		31.50	8.70	1.03	41.23	49
2260	30″ diameter	B-13	88	.636		48	25.50	8.50	82	101
2270	36″ diameter	″	72	.778		64.50	31	10.35	105.85	129
2290	48″ diameter	B-13B	64	.875		100	35	17.75	152.75	183
2310	72″ diameter	″	40	1.400		224	56	28.50	308.50	365
2330	Flared ends, 12″ diameter	B-21	31	.903	Ea.	225	38	4.52	267.52	310
2340	15″ diameter		25	1.120		266	47.50	5.60	319.10	370
2400	18″ diameter		20	1.400		305	59	7	371	435
2420	24″ diameter		14	2		370	84.50	10	464.50	545
2440	36″ diameter	B-13	10	5.600		790	223	74.50	1,087.50	1,300
2500	Class 4									
2510	12″ diameter	B-21	168	.167	L.F.	12	7.05	.83	19.88	25
2512	15″ diameter		160	.175		16	7.40	.88	24.28	30
2514	18″ diameter		152	.184		20	7.80	.92	28.72	35
2516	21″ diameter		144	.194		25.50	8.20	.97	34.67	41.50
2518	24″ diameter		136	.206		31.50	8.70	1.03	41.23	49
2520	27″ diameter		120	.233		42	9.85	1.17	53.02	62.50
2522	30″ diameter	B-13	88	.636		48	25.50	8.50	82	101
2524	36″ diameter	″	72	.778		65	31	10.35	106.35	130
2600	Class 5									
2610	12″ diameter	B-21	168	.167	L.F.	14	7.05	.83	21.88	27
2612	15″ diameter		160	.175		18	7.40	.88	26.28	32
2614	18″ diameter		152	.184		22.50	7.80	.92	31.22	38
2616	21″ diameter		144	.194		28.50	8.20	.97	37.67	45
2618	24″ diameter		136	.206		35.50	8.70	1.03	45.23	53.50
2620	27″ diameter		120	.233		46.50	9.85	1.17	57.52	67.50
2622	30″ diameter	B-13	88	.636		53.50	25.50	8.50	87.50	107
2624	36″ diameter	″	72	.778		73	31	10.35	114.35	139
2800	Add for rubber joints 12″- 36″ diameter					12%				
3080	Radius pipe, add to pipe prices, 12″ to 60″ diameter					50%				
3090	Over 60″ diameter, add					20%				
3500	Reinforced elliptical, 8′ lengths, C507 class 3									
3520	14″ x 23″ inside, round equivalent 18″ diameter	B-21	82	.341	L.F.	40.50	14.40	1.71	56.61	68.50
3530	24″ x 38″ inside, round equivalent 30″ diameter	B-13	58	.966		61.50	38.50	12.85	112.85	141
3540	29″ x 45″ inside, round equivalent 36″ diameter		52	1.077		78	43	14.35	135.35	167
3550	38″ x 60″ inside, round equivalent 48″ diameter		38	1.474		133	58.50	19.65	211.15	258
3560	48″ x 76″ inside, round equivalent 60″ diameter		26	2.154		185	86	28.50	299.50	370
3570	58″ x 91″ inside, round equivalent 72″ diameter		22	2.545		270	101	34	405	490
3780	Concrete slotted pipe, class 4 mortar joint									
3800	12″ diameter	B-21	168	.167	L.F.	27	7.05	.83	34.88	41.50
3840	18″ diameter	″	152	.184	″	31.50	7.80	.92	40.22	47.50
3900	Concrete slotted pipe, Class 4 O-ring joint									
3940	12″ diameter	B-21	168	.167	L.F.	27	7.05	.83	34.88	41.50
3960	18″ diameter	″	152	.184	″	31.50	7.80	.92	40.22	47.50

33 41 Storm Utility Drainage Piping

33 41 13 – Public Storm Utility Drainage Piping

33 41 13.60 Sewage/Drainage Collection, Concrete Pipe		Crew	Daily Output	Labor-Hours	Unit	2014 Bare Costs Material	Labor	Equipment	Total	Total Incl O&P
6200	Gasket, conc. pipe joint, 12"				Ea.	3.94			3.94	4.33
6220	24"					6.50			6.50	7.15
6240	36"					9.30			9.30	10.25
6260	48"					12.40			12.40	13.60
6270	60"					15.40			15.40	16.95
6280	72"					18.70			18.70	20.50

33 42 Culverts

33 42 16 – Concrete Culverts

33 42 16.15 Oval Arch Culverts

		Crew	Daily Output	Labor-Hours	Unit	Material	Labor	Equipment	Total	Total Incl O&P
0010	**OVAL ARCH CULVERTS**									
3000	Corrugated galvanized or aluminum, coated & paved									
3020	17" x 13", 16 ga., 15" equivalent	B-14	200	.240	L.F.	13.25	9.30	1.83	24.38	31
3040	21" x 15", 16 ga., 18" equivalent		150	.320		16.05	12.40	2.44	30.89	39.50
3060	28" x 20", 14 ga., 24" equivalent		125	.384		24	14.85	2.92	41.77	52
3080	35" x 24", 14 ga., 30" equivalent		100	.480		29.50	18.60	3.65	51.75	65
3100	42" x 29", 12 ga., 36" equivalent	B-13	100	.560		35.50	22.50	7.45	65.45	81
3120	49" x 33", 12 ga., 42" equivalent		90	.622		41	25	8.30	74.30	92.50
3140	57" x 38", 12 ga., 48" equivalent		75	.747		57	30	9.95	96.95	119
3160	Steel, plain oval arch culverts, plain									
3180	17" x 13", 16 ga., 15" equivalent	B-14	225	.213	L.F.	11.95	8.25	1.62	21.82	27.50
3200	21" x 15", 16 ga., 18" equivalent		175	.274		14.45	10.60	2.09	27.14	34.50
3220	28" x 20", 14 ga., 24" equivalent		150	.320		21.50	12.40	2.44	36.34	45
3240	35" x 24", 14 ga., 30" equivalent	B-13	108	.519		26.50	20.50	6.90	53.90	68
3260	42" x 29", 12 ga., 36" equivalent		108	.519		32	20.50	6.90	59.40	74.50
3280	49" x 33", 12 ga., 42" equivalent		92	.609		37	24.50	8.10	69.60	87
3300	57" x 38", 12 ga., 48" equivalent		75	.747		51.50	30	9.95	91.45	113
3320	End sections, 17" x 13"		22	2.545	Ea.	144	101	34	279	355
3340	42" x 29"		17	3.294	"	395	131	44	570	685
3360	Multi-plate arch, steel	B-20	1690	.014	Lb.	1.30	.58		1.88	2.33

33 44 Storm Utility Water Drains

33 44 13 – Utility Area Drains

33 44 13.13 Catchbasins

		Crew	Daily Output	Labor-Hours	Unit	Material	Labor	Equipment	Total	Total Incl O&P
0010	**CATCHBASINS**									
0011	Not including footing & excavation									
1580	Curb inlet frame, grate, and curb box									
1582	Large 24" x 36" heavy duty	B-24	2	12	Ea.	520	505		1,025	1,350
1590	Small 10" x 21" medium duty	"	2	12		365	505		870	1,175
1600	Frames & grates, C.I., 24" square, 500 lb.	B-6	7.80	3.077		340	123	47	510	615
1700	26" D shape, 600 lb.		7	3.429		505	138	52	695	825
1800	Light traffic, 18" diameter, 100 lb.		10	2.400		122	96.50	36.50	255	320
1900	24" diameter, 300 lb.		8.70	2.759		196	111	42	349	430
2000	36" diameter, 900 lb.		5.80	4.138		570	166	63	799	950
2100	Heavy traffic, 24" diameter, 400 lb.		7.80	3.077		244	123	47	414	510
2200	36" diameter, 1150 lb.		3	8		795	320	122	1,237	1,500
2300	Mass. State standard, 26" diameter, 475 lb.		7	3.429		265	138	52	455	560
2400	30" diameter, 620 lb.		7	3.429		345	138	52	535	650
2500	Watertight, 24" diameter, 350 lb.		7.80	3.077		320	123	47	490	590

33 44 Storm Utility Water Drains

33 44 13 – Utility Area Drains

33 44 13.13 Catchbasins		Crew	Daily Output	Labor-Hours	Unit	Material	Labor	Equipment	Total	Total Incl O&P
						2014 Bare Costs				
2600	26″ diameter, 500 lb.	B-6	7	3.429	Ea.	425	138	52	615	735
2700	32″ diameter, 575 lb.	↓	6	4	↓	850	160	61	1,071	1,250
2800	3 piece cover & frame, 10″ deep,									
2900	1200 lb., for heavy equipment	B-6	3	8	Ea.	1,050	320	122	1,492	1,775
3000	Raised for paving 1-1/4″ to 2″ high									
3100	4 piece expansion ring									
3200	20″ to 26″ diameter	1 Clab	3	2.667	Ea.	157	97.50		254.50	325
3300	30″ to 36″ diameter	″	3	2.667	″	216	97.50		313.50	390
3320	Frames and covers, existing, raised for paving, 2″, including									
3340	row of brick, concrete collar, up to 12″ wide frame	B-6	18	1.333	Ea.	44.50	53.50	20.50	118.50	154
3360	20″ to 26″ wide frame		11	2.182		66.50	87.50	33	187	244
3380	30″ to 36″ wide frame	↓	9	2.667		82.50	107	40.50	230	300
3400	Inverts, single channel brick	D-1	3	5.333		96	220		316	440
3500	Concrete		5	3.200		103	132		235	315
3600	Triple channel, brick		2	8		147	330		477	665
3700	Concrete	↓	3	5.333	↓	139	220		359	485

33 46 Subdrainage

33 46 16 – Subdrainage Piping

33 46 16.25 Piping, Subdrainage, Corrugated Metal

		Crew	Daily Output	Labor-Hours	Unit	Material	Labor	Equipment	Total	Total Incl O&P
0010	**PIPING, SUBDRAINAGE, CORRUGATED METAL**									
0021	Not including excavation and backfill									
2010	Aluminum, perforated									
2020	6″ diameter, 18 ga.	B-20	380	.063	L.F.	6.55	2.58		9.13	11.20
2200	8″ diameter, 16 ga.	″	370	.065		8.60	2.65		11.25	13.55
2220	10″ diameter, 16 ga.	B-21	360	.078		10.75	3.28	.39	14.42	17.30
2240	12″ diameter, 16 ga.		285	.098		12	4.15	.49	16.64	20
2260	18″ diameter, 16 ga.	↓	205	.137	↓	18	5.75	.68	24.43	29.50
3000	Uncoated galvanized, perforated									
3020	6″ diameter, 18 ga.	B-20	380	.063	L.F.	6.05	2.58		8.63	10.65
3200	8″ diameter, 16 ga.	″	370	.065		8.35	2.65		11	13.25
3220	10″ diameter, 16 ga.	B-21	360	.078		8.85	3.28	.39	12.52	15.20
3240	12″ diameter, 16 ga.		285	.098		9.85	4.15	.49	14.49	17.75
3260	18″ diameter, 16 ga.	↓	205	.137	↓	15.05	5.75	.68	21.48	26
4000	Steel, perforated, asphalt coated									
4020	6″ diameter 18 ga.	B-20	380	.063	L.F.	6.55	2.58		9.13	11.20
4030	8″ diameter 18 ga.	″	370	.065		8.60	2.65		11.25	13.55
4040	10″ diameter 16 ga.	B-21	360	.078		10.10	3.28	.39	13.77	16.60
4050	12″ diameter 16 ga.		285	.098		11.10	4.15	.49	15.74	19.15
4060	18″ diameter 16 ga.	↓	205	.137	↓	17.15	5.75	.68	23.58	28.50

33 46 16.30 Piping, Subdrainage, Plastic

		Crew	Daily Output	Labor-Hours	Unit	Material	Labor	Equipment	Total	Total Incl O&P
0010	**PIPING, SUBDRAINAGE, PLASTIC**									
0020	Not including excavation and backfill									
2100	Perforated PVC, 4″ diameter	B-14	314	.153	L.F.	1.46	5.90	1.16	8.52	12
2110	6″ diameter		300	.160		3.29	6.20	1.22	10.71	14.45
2120	8″ diameter		290	.166		6.30	6.40	1.26	13.96	18.20
2130	10″ diameter		280	.171		9.15	6.65	1.30	17.10	21.50
2140	12″ diameter	↓	270	.178	↓	12.75	6.90	1.35	21	26

33 46 Subdrainage

33 46 16 – Subdrainage Piping

33 46 16.35 Piping, Subdrain., Corr. Plas. Tubing, Perf. or Plain		Crew	Daily Output	Labor-Hours	Unit	Material	2014 Bare Costs Labor	Equipment	Total	Total Incl O&P
0010	PIPING, SUBDRAINAGE, CORR. PLASTIC TUBING, PERF. OR PLAIN									
0020	In rolls, not including excavation and backfill									
0030	3″ diameter	2 Clab	1200	.013	L.F.	.46	.49		.95	1.26
0040	4″ diameter		1200	.013		.51	.49		1	1.31
0041	With silt sock		1200	.013		1.29	.49		1.78	2.17
0060	6″ diameter		900	.018		1.47	.65		2.12	2.63
0080	8″ diameter		700	.023		2.75	.84		3.59	4.32
0200	Fittings									
0230	Elbows, 3″ diameter	1 Clab	32	.250	Ea.	4.70	9.15		13.85	19.30
0240	4″ diameter		32	.250		5.30	9.15		14.45	20
0250	5″ diameter		32	.250		6.60	9.15		15.75	21.50
0260	6″ diameter		32	.250		9.30	9.15		18.45	24.50
0280	8″ diameter		32	.250		12.30	9.15		21.45	27.50
0330	Tees, 3″ diameter		27	.296		3.96	10.85		14.81	21
0340	4″ diameter		27	.296		5.45	10.85		16.30	23
0350	5″ diameter		27	.296		6.50	10.85		17.35	24
0360	6″ diameter		27	.296		10.05	10.85		20.90	28
0370	6″ x 6″ x 4″		27	.296		10	10.85		20.85	28
0380	8″ diameter		27	.296		16.05	10.85		26.90	34.50
0390	8″ x 8″ x 6″		27	.296		16.05	10.85		26.90	34.50
0430	End cap, 3″ diameter		32	.250		2.01	9.15		11.16	16.35
0440	4″ diameter		32	.250		2.51	9.15		11.66	16.90
0460	6″ diameter		32	.250		5.70	9.15		14.85	20.50
0480	8″ diameter		32	.250		6.45	9.15		15.60	21.50
0530	Coupler, 3″ diameter		32	.250		2.26	9.15		11.41	16.65
0540	4″ diameter		32	.250		2.03	9.15		11.18	16.40
0550	5″ diameter		32	.250		2.53	9.15		11.68	16.95
0560	6″ diameter		32	.250		4.27	9.15		13.42	18.85
0580	8″ diameter		32	.250		5.95	9.15		15.10	20.50
0590	Heavy duty highway type, add					10%				
0660	Reducer, 6″ to 4″	1 Clab	32	.250		6.05	9.15		15.20	21
0680	8″ to 6″		32	.250		10.30	9.15		19.45	25.50
0730	"Y" fitting, 3″ diameter		27	.296		5	10.85		15.85	22.50
0740	4″ diameter		27	.296		7.75	10.85		18.60	25.50
0750	5″ diameter		27	.296		9.90	10.85		20.75	27.50
0760	6″ diameter		27	.296		12.30	10.85		23.15	30.50
0780	8″ diameter		27	.296		25	10.85		35.85	44.50
0860	Silt sock only for above tubing, 6″ diameter				L.F.	3.70			3.70	4.07
0880	8″ diameter				″	6.10			6.10	6.70

33 46 26 – Geotextile Subsurface Drainage Filtration

33 46 26.10 Geotextiles for Subsurface Drainage		Crew	Daily Output	Labor-Hours	Unit	Material	Labor	Equipment	Total	Total Incl O&P
0010	GEOTEXTILES FOR SUBSURFACE DRAINAGE									
0100	Fabric, laid in trench, polypropylene, ideal conditions	2 Clab	2400	.007	S.Y.	1.85	.24		2.09	2.42
0110	Adverse conditions		1600	.010	″	1.85	.37		2.22	2.61
0170	Fabric ply bonded to 3 dimen. nylon mat, .4″ thick, ideal conditions		2000	.008	S.F.	.23	.29		.52	.70
0180	Adverse conditions		1200	.013	″	.28	.49		.77	1.06
0185	Soil drainage mat on vertical wall, 0.44″ thick		265	.060	S.Y.	1.80	2.21		4.01	5.40
0188	0.25″ thick		300	.053	″	.77	1.95		2.72	3.87
0190	0.8″ thick, ideal conditions		2400	.007	S.F.	.23	.24		.47	.63
0200	Adverse conditions		1600	.010	″	.37	.37		.74	.98
0300	Drainage material, 3/4″ gravel fill in trench	B-6	260	.092	C.Y.	24	3.70	1.41	29.11	34
0400	Pea stone	″	260	.092	″	22	3.70	1.41	27.11	31.50

33 47 Ponds and Reservoirs

33 47 13 – Pond and Reservoir Liners

33 47 13.53 Reservoir Liners HDPE		Crew	Daily Output	Labor-Hours	Unit	Material	Labor	Equipment	Total	Total Incl O&P
						2014 Bare Costs				
0010	**RESERVOIR LINERS HDPE**									
0011	Membrane lining									
1100	30 mil thick	3 Skwk	1850	.013	S.F.	.52	.61		1.13	1.52
1200	60 mil thick		1600	.015	"	.58	.71		1.29	1.74
1220	60 mil thick		1.60	15	M.S.F.	580	710		1,290	1,750
1300	120 mil thick	↓	1440	.017	S.F.	.71	.79		1.50	2

33 49 Storm Drainage Structures

33 49 13 – Storm Drainage Manholes, Frames, and Covers

33 49 13.10 Storm Drainage Manholes, Frames and Covers		Crew	Daily Output	Labor-Hours	Unit	Material	Labor	Equipment	Total	Total Incl O&P
0010	**STORM DRAINAGE MANHOLES, FRAMES & COVERS**									
0020	Excludes footing, excavation, backfill (See line items for frame & cover)									
0050	Brick, 4′ inside diameter, 4′ deep	D-1	1	16	Ea.	510	660		1,170	1,550
0100	6′ deep		.70	22.857		725	945		1,670	2,250
0150	8′ deep		.50	32	↓	935	1,325		2,260	3,050
0200	For depths over 8′, add		4	4	V.L.F.	83.50	165		248.50	345
0400	Concrete blocks (radial), 4′ I.D., 4′ deep		1.50	10.667	Ea.	370	440		810	1,075
0500	6′ deep		1	16		495	660		1,155	1,550
0600	8′ deep		.70	22.857	↓	620	945		1,565	2,125
0700	For depths over 8′, add	↓	5.50	2.909	V.L.F.	65	120		185	255
0800	Concrete, cast in place, 4′ x 4′, 8″ thick, 4′ deep	C-14H	2	24	Ea.	505	1,075	16.55	1,596.55	2,250
0900	6′ deep		1.50	32		725	1,450	22	2,197	3,050
1000	8′ deep		1	48	↓	1,050	2,175	33	3,258	4,500
1100	For depths over 8′, add	↓	8	6	V.L.F.	118	271	4.14	393.14	550
1110	Precast, 4′ I.D., 4′ deep	B-22	4.10	7.317	Ea.	775	315	51	1,141	1,375
1120	6′ deep		3	10		970	425	70	1,465	1,800
1130	8′ deep		2	15	↓	1,100	640	105	1,845	2,325
1140	For depths over 8′, add	↓	16	1.875	V.L.F.	125	80	13.15	218.15	275
1150	5′ I.D., 4′ deep	B-6	3	8	Ea.	1,625	320	122	2,067	2,400
1160	6′ deep		2	12		1,675	480	183	2,338	2,775
1170	8′ deep		1.50	16	↓	2,350	640	244	3,234	3,850
1180	For depths over 8′, add		12	2	V.L.F.	277	80	30.50	387.50	460
1190	6′ I.D., 4′ deep		2	12	Ea.	2,125	480	183	2,788	3,275
1200	6′ deep		1.50	16		2,550	640	244	3,434	4,050
1210	8′ deep		1	24	↓	3,150	965	365	4,480	5,325
1220	For depths over 8′, add	↓	8	3	V.L.F.	375	120	45.50	540.50	645
1250	Slab tops, precast, 8″ thick									
1300	4′ diameter manhole	B-6	8	3	Ea.	245	120	45.50	410.50	505
1400	5′ diameter manhole		7.50	3.200		405	128	48.50	581.50	695
1500	6′ diameter manhole	↓	7	3.429		625	138	52	815	955
3800	Steps, heavyweight cast iron, 7″ x 9″	1 Bric	40	.200		19.05	9.10		28.15	35
3900	8″ x 9″		40	.200		23	9.10		32.10	39
3928	12″ x 10-1/2″		40	.200		26.50	9.10		35.60	43.50
4000	Standard sizes, galvanized steel		40	.200		21.50	9.10		30.60	37.50
4100	Aluminum		40	.200		24	9.10		33.10	40
4150	Polyethylene	↓	40	.200		24.50	9.10		33.60	41
4210	Rubber boot 6″ diam. or smaller	1 Clab	32	.250		80	9.15		89.15	102
4215	8″ diam.		24	.333		91	12.20		103.20	119
4220	10″ diam.		19	.421		104	15.45		119.45	139
4225	12″ diam.		16	.500		133	18.35		151.35	175
4230	16″ diam.	↓	15	.533	↓	169	19.55		188.55	216

33 49 Storm Drainage Structures

33 49 13 – Storm Drainage Manholes, Frames, and Covers

33 49 13.10 Storm Drainage Manholes, Frames and Covers		Crew	Daily Output	Labor-Hours	Unit	2014 Bare Costs Material	Labor	Equipment	Total	Total Incl O&P
4235	18″ diam.	1 Clab	15	.533	Ea.	196	19.55		215.55	246
4240	24″ diam.		14	.571		228	21		249	284
4245	30″ diam.		12	.667		298	24.50		322.50	370

33 51 Natural-Gas Distribution

33 51 13 – Natural-Gas Piping

33 51 13.10 Piping, Gas Service and Distribution, P.E.

		Crew	Daily Output	Labor-Hours	Unit	Material	Labor	Equipment	Total	Total Incl O&P
0010	**PIPING, GAS SERVICE AND DISTRIBUTION, POLYETHYLENE**									
0020	Not including excavation or backfill									
1000	60 psi coils, compression coupling @ 100′, 1/2″ diameter, SDR 11	B-20A	608	.053	L.F.	.61	2.35		2.96	4.26
1010	1″ diameter, SDR 11		544	.059		1.41	2.63		4.04	5.55
1040	1-1/4″ diameter, SDR 11		544	.059		1.76	2.63		4.39	5.95
1100	2″ diameter, SDR 11		488	.066		2.85	2.93		5.78	7.60
1160	3″ diameter, SDR 11		408	.078		5.60	3.51		9.11	11.50
1500	60 PSI 40′ joints with coupling, 3″ diameter, SDR 11	B-21A	408	.098		5.55	4.49	1.18	11.22	14.25
1540	4″ diameter, SDR 11		352	.114		12.15	5.20	1.37	18.72	23
1600	6″ diameter, SDR 11		328	.122		33.50	5.60	1.47	40.57	47
1640	8″ diameter, SDR 11		272	.147		51.50	6.75	1.77	60.02	68.50

33 51 13.20 Piping, Gas Service and Distribution, Steel

		Crew	Daily Output	Labor-Hours	Unit	Material	Labor	Equipment	Total	Total Incl O&P
0010	**PIPING, GAS SERVICE & DISTRIBUTION, STEEL**									
0020	Not including excavation or backfill, tar coated and wrapped									
4000	Pipe schedule 40, plain end									
4040	1″ diameter	Q-4	300	.107	L.F.	5.25	5.85	.18	11.28	14.85
4080	2″ diameter		280	.114		8.25	6.25	.20	14.70	18.70
4120	3″ diameter		260	.123		13.65	6.75	.21	20.61	25.50
4160	4″ diameter	B-35	255	.188		17.75	8.60	2.81	29.16	35.50
4200	5″ diameter		220	.218		25.50	9.95	3.26	38.71	47.50
4240	6″ diameter		180	.267		31.50	12.20	3.98	47.68	57.50
4280	8″ diameter		140	.343		50	15.65	5.10	70.75	84.50
4320	10″ diameter		100	.480		117	22	7.15	146.15	170
4360	12″ diameter		80	.600		130	27.50	8.95	166.45	195
4400	14″ diameter		75	.640		139	29	9.55	177.55	207
4440	16″ diameter		70	.686		152	31.50	10.25	193.75	226
4480	18″ diameter		65	.738		195	33.50	11.05	239.55	278
4520	20″ diameter		60	.800		305	36.50	11.95	353.45	405
4560	24″ diameter		50	.960		345	44	14.35	403.35	465
6000	Schedule 80, plain end									
6002	4″ diameter	B-35	144	.333	L.F.	39.50	15.20	4.98	59.68	72.50
6006	6″ diameter		126	.381		87.50	17.40	5.70	110.60	129
6008	8″ diameter		108	.444		117	20.50	6.65	144.15	167
6012	12″ diameter		72	.667		234	30.50	9.95	274.45	315
8008	Elbow, weld joint, standard weight									
8020	4″ diameter	Q-16	6.80	3.529	Ea.	96	190	8.10	294.10	400
8026	8″ diameter		3.40	7.059		400	380	16.25	796.25	1,025
8030	12″ diameter		2.30	10.435		885	560	24	1,469	1,850
8034	16″ diameter		1.50	16		2,000	860	37	2,897	3,550
8038	20″ diameter		1.20	20		3,675	1,075	46	4,796	5,725
8040	24″ diameter		1.02	23.529		5,200	1,275	54	6,529	7,675
8100	Extra heavy									
8102	4″ diameter	Q-16	5.30	4.528	Ea.	192	243	10.40	445.40	585
8108	8″ diameter		2.60	9.231		600	495	21	1,116	1,425

33 51 Natural-Gas Distribution

33 51 13 – Natural-Gas Piping

33 51 13.20	Piping, Gas Service and Distribution, Steel	Crew	Daily Output	Labor-Hours	Unit	Material	2014 Bare Costs Labor	Equipment	Total	Total Incl O&P
8112	12" diameter	Q-16	1.80	13.333	Ea.	1,175	715	30.50	1,920.50	2,400
8116	16" diameter		1.20	20		2,675	1,075	46	3,796	4,625
8120	20" diameter		.94	25.532		4,900	1,375	58.50	6,333.50	7,525
8122	24" diameter		.80	30		6,925	1,600	69	8,594	10,100
8200	Malleable, standard weight									
8202	4" diameter	B-20	12	2	Ea.	510	81.50		591.50	685
8208	8" diameter		6	4		1,400	163		1,563	1,775
8212	12" diameter		4	6		1,475	245		1,720	2,000
8300	Extra heavy									
8302	4" diameter	B-20	12	2	Ea.	1,025	81.50		1,106.50	1,250
8308	8" diameter	B-21	6	4.667		1,750	197	23.50	1,970.50	2,250
8312	12" diameter	"	4	7		2,350	295	35	2,680	3,100
8500	Tee weld, standard weight									
8510	4" diameter	Q-16	4.50	5.333	Ea.	177	287	12.25	476.25	645
8514	6" diameter		3	8		305	430	18.40	753.40	1,000
8516	8" diameter		2.30	10.435		530	560	24	1,114	1,450
8520	12" diameter		1.50	16		1,450	860	37	2,347	2,950
8524	16" diameter		1	24		2,875	1,300	55	4,230	5,150
8528	20" diameter		.80	30		7,150	1,600	69	8,819	10,400
8530	24" diameter		.70	34.286		9,225	1,850	79	11,154	13,000
8810	Malleable, standard weight									
8812	4" diameter	B-20	8	3	Ea.	870	123		993	1,150
8818	8" diameter	B-21	4	7	"	1,500	295	35	1,830	2,150
8900	Extra heavy									
8902	4" diameter	B-20	8	3	Ea.	1,300	123		1,423	1,625
8908	8" diameter	B-21	4	7		1,875	295	35	2,205	2,550
8912	12" diameter	"	2.70	10.370		2,725	440	52	3,217	3,725

33 51 33 – Natural-Gas Metering

33 51 33.10 Piping, Valves and Meters, Gas Distribution

Line	Description	Crew	Daily Output	Labor-Hours	Unit	Material	Labor	Equipment	Total	Total Incl O&P
0010	**PIPING, VALVES & METERS, GAS DISTRIBUTION**									
0020	Not including excavation or backfill									
0100	Gas stops, with or without checks									
0140	1-1/4" size	1 Plum	12	.667	Ea.	61	38.50		99.50	126
0180	1-1/2" size		10	.800		86.50	46		132.50	165
0200	2" size		8	1		115	57.50		172.50	214
0600	Pressure regulator valves, iron and bronze									
0680	2" diameter	1 Plum	11	.727	Ea.	275	42		317	370
0700	3" diameter	Q-1	13	1.231		555	64		619	705
0740	4" diameter	"	8	2		1,875	104		1,979	2,225
2000	Lubricated semi-steel plug valve									
2040	3/4" diameter	1 Plum	16	.500	Ea.	90	29		119	143
2080	1" diameter		14	.571		115	33		148	176
2100	1-1/4" diameter		12	.667		137	38.50		175.50	208
2140	1-1/2" diameter		11	.727		147	42		189	225
2180	2" diameter		8	1		175	57.50		232.50	279
2300	2-1/2" diameter	Q-1	5	3.200		269	166		435	545
2340	3" diameter	"	4.50	3.556		330	184		514	645

33 52 Liquid Fuel Distribution

33 52 13 – Fuel-Oil Distribution

33 52 13.14 Petroleum Products		Crew	Daily Output	Labor-Hours	Unit	2014 Bare Costs Material	Labor	Equipment	Total	Total Incl O&P
0010	**PETROLEUM PRODUCTS** steel distribution piping									
0020	Major steel pipeline transmission lines									
1000	Piping, steel pipeline distribution, 10 foot lenghs, 24" dia, 3/8 thick	B-35A	80	.700	L.F.	149	31	20.50	200.50	234
1100	8 foot lenghs, 36" dia, 1/2" thick		24	2.333		285	104	68	457	550
1200	6 foot lenghs, 48" dia, 1/2" thick		18	3.111		500	138	91	729	860

33 52 16 – Gasoline Distribution

33 52 16.13 Gasoline Piping

		Crew	Daily Output	Labor-Hours	Unit	Material	Labor	Equipment	Total	Total Incl O&P
0010	**GASOLINE PIPING**									
0020	Primary containment pipe, fiberglass-reinforced									
0030	Plastic pipe 15' & 30' lengths									
0040	2" diameter	Q-6	425	.056	L.F.	5.95	3.08		9.03	11.20
0050	3" diameter		400	.060		10.40	3.28		13.68	16.35
0060	4" diameter		375	.064		13.70	3.49		17.19	20.50
0100	Fittings									
0110	Elbows, 90° & 45°, bell-ends, 2"	Q-6	24	1	Ea.	43.50	54.50		98	130
0120	3" diameter		22	1.091		55	59.50		114.50	151
0130	4" diameter		20	1.200		70	65.50		135.50	176
0200	Tees, bell ends, 2"		21	1.143		60.50	62.50		123	161
0210	3" diameter		18	1.333		64	73		137	181
0220	4" diameter		15	1.600		84	87.50		171.50	225
0230	Flanges bell ends, 2"		24	1		33.50	54.50		88	120
0240	3" diameter		22	1.091		39	59.50		98.50	133
0250	4" diameter		20	1.200		45	65.50		110.50	149
0260	Sleeve couplings, 2"		21	1.143		12.40	62.50		74.90	108
0270	3" diameter		18	1.333		17.80	73		90.80	130
0280	4" diameter		15	1.600		23	87.50		110.50	158
0290	Threaded adapters 2"		21	1.143		18	62.50		80.50	114
0300	3" diameter		18	1.333		34	73		107	148
0310	4" diameter		15	1.600		38	87.50		125.50	174
0320	Reducers, 2"		27	.889		27.50	48.50		76	104
0330	3" diameter		22	1.091		27.50	59.50		87	120
0340	4" diameter		20	1.200		37	65.50		102.50	140
1010	Gas station product line for secondary containment (double wall)									
1100	Fiberglass reinforced plastic pipe 25' lengths									
1120	Pipe, plain end, 3" diameter	Q-6	375	.064	L.F.	26.50	3.49		29.99	35
1130	4" diameter		350	.069		32	3.74		35.74	41
1140	5" diameter		325	.074		35.50	4.03		39.53	45
1150	6" diameter		300	.080		39	4.37		43.37	49.50
1200	Fittings									
1230	Elbows, 90° & 45°, 3" diameter	Q-6	18	1.333	Ea.	136	73		209	260
1240	4" diameter		16	1.500		167	82		249	310
1250	5" diameter		14	1.714		184	93.50		277.50	345
1260	6" diameter		12	2		204	109		313	390
1270	Tees, 3" diameter		15	1.600		166	87.50		253.50	315
1280	4" diameter		12	2		202	109		311	385
1290	5" diameter		9	2.667		315	146		461	570
1300	6" diameter		6	4		375	218		593	745
1310	Couplings, 3" diameter		18	1.333		55	73		128	171
1320	4" diameter		16	1.500		119	82		201	255
1330	5" diameter		14	1.714		214	93.50		307.50	375
1340	6" diameter		12	2		315	109		424	515
1350	Cross-over nipples, 3" diameter		18	1.333		10.60	73		83.60	122

33 52 Liquid Fuel Distribution

33 52 16 – Gasoline Distribution

33 52 16.13 Gasoline Piping		Crew	Daily Output	Labor-Hours	Unit	Material	Labor (2014 Bare Costs)	Equipment	Total	Total Incl O&P
1360	4″ diameter	Q-6	16	1.500	Ea.	12.75	82		94.75	138
1370	5″ diameter		14	1.714		15.90	93.50		109.40	159
1380	6″ diameter		12	2		19.10	109		128.10	186
1400	Telescoping, reducers, concentric 4″ x 3″		18	1.333		47.50	73		120.50	163
1410	5″ x 4″		17	1.412		95.50	77		172.50	221
1420	6″ x 5″		16	1.500		234	82		316	380

33 61 Hydronic Energy Distribution

33 61 13 – Underground Hydronic Energy Distribution

33 61 13.10 Chilled/HVAC Hot Water Distribution

		Crew	Daily Output	Labor-Hours	Unit	Material	Labor	Equipment	Total	Total Incl O&P
0010	**CHILLED/HVAC HOT WATER DISTRIBUTION**									
1005	Pipe, black steel w/2″ polyurethane insul, 20′ lengths									
1010	Align & tackweld on sleepers (NIC), 1-1/4″ diameter	B-35	864	.056	L.F.	30	2.54	.83	33.37	38
1020	1-1/2″		824	.058		33.50	2.66	.87	37.03	42
1030	2″		680	.071		35	3.22	1.05	39.27	44.50
1040	2-1/2″		560	.086		38	3.91	1.28	43.19	49
1050	3″		528	.091		39	4.15	1.36	44.51	50.50
1060	4″		384	.125		42.50	5.70	1.87	50.07	58
1070	5″		360	.133		63	6.10	1.99	71.09	80.50
1080	6″		296	.162		68.50	7.40	2.42	78.32	89.50
1090	8″		264	.182		91.50	8.30	2.72	102.52	117
1100	12″		216	.222		153	10.15	3.32	166.47	187
1110	On trench bottom, 18″ diameter		176.13	.273		236	12.45	4.07	252.52	283
1120	24″	B-35A	145	.386		355	17.15	11.25	383.40	430
1130	30″		121	.463		510	20.50	13.50	544	605
1140	36″		100	.560		680	25	16.35	721.35	805
1150	Elbows, on sleepers, 1-1/2″ diameter	Q-17	21	.762	Ea.	780	40	2.64	822.64	925
1160	3″		9.36	1.709		1,025	90	5.90	1,120.90	1,275
1170	4″		8	2		1,250	105	6.90	1,361.90	1,550
1180	6″		6	2.667		1,625	140	9.25	1,774.25	2,000
1190	8″		4.64	3.448		1,975	182	11.95	2,168.95	2,425
1200	Tees, 1-1/2″ diameter		17	.941		1,550	49.50	3.26	1,602.76	1,775
1210	3″		8.50	1.882		1,925	99	6.50	2,030.50	2,275
1220	4″		6	2.667		2,075	140	9.25	2,224.25	2,500
1230	6″	Q-17A	6.72	3.571		2,600	185	107	2,892	3,250
1240	8″	″	6.40	3.750		3,050	194	112	3,356	3,775
1250	Reducer, 3″ diameter	Q-17	16	1		410	52.50	3.46	465.96	535
1260	4″		12	1.333		490	70	4.61	564.61	645
1270	6″		12	1.333		695	70	4.61	769.61	875
1280	8″		10	1.600		900	84	5.55	989.55	1,125
1290	Anchor, 4″ diameter	Q-17A	12	2		1,200	104	60	1,364	1,525
1300	6″		10.50	2.286		1,375	119	68.50	1,562.50	1,750
1310	8″		10	2.400		1,525	124	72	1,721	1,975
1320	Cap, 1-1/2″ diameter	Q-17	42	.381		125	20	1.32	146.32	169
1330	3″		14.64	1.093		169	57.50	3.78	230.28	277
1340	4″		16	1		194	52.50	3.46	249.96	296
1350	6″		16	1		335	52.50	3.46	390.96	450
1360	8″		13.50	1.185		425	62.50	4.10	491.60	565
1365	12″		11	1.455		625	76.50	5.05	706.55	805
1370	Elbow, fittings on trench bottom, 12″ diameter	Q-17A	12	2		3,000	104	60	3,164	3,525
1380	18″		8	3		5,375	156	90	5,621	6,225

33 61 Hydronic Energy Distribution

33 61 13 – Underground Hydronic Energy Distribution

33 61 13.10 Chilled/HVAC Hot Water Distribution

	33 61 13.10 Chilled/HVAC Hot Water Distribution	Crew	Daily Output	Labor-Hours	Unit	Material	2014 Bare Costs Labor	Equipment	Total	Total Incl O&P
1390	24"	Q-17A	6	4	Ea.	7,650	207	120	7,977	8,850
1400	30"		5.36	4.478		10,500	232	134	10,866	12,000
1410	36"		4	6		14,200	310	180	14,690	16,300
1420	Tee, 12" diameter		6.72	3.571		3,950	185	107	4,242	4,750
1430	18"		6	4		6,600	207	120	6,927	7,700
1440	24"		4.64	5.172		10,300	268	155	10,723	11,900
1450	30"		4.16	5.769		15,800	299	173	16,272	17,900
1460	36"		3.36	7.143		23,200	370	214	23,784	26,300
1470	Reducer, 12" diameter		10.64	2.256		1,250	117	67.50	1,434.50	1,625
1480	18"		9.68	2.479		2,000	129	74	2,203	2,475
1490	24"		8	3		3,075	156	90	3,321	3,700
1500	30"		7.04	3.409		4,775	177	102	5,054	5,625
1510	36"		5.36	4.478		6,800	232	134	7,166	7,975
1520	Anchor, 12" diameter		11	2.182		1,600	113	65.50	1,778.50	2,000
1530	18"		6.72	3.571		1,700	185	107	1,992	2,275
1540	24"		6.32	3.797		2,375	197	114	2,686	3,025
1550	30"		4.64	5.172		2,875	268	155	3,298	3,750
1560	36"		4	6		3,525	310	180	4,015	4,550
1565	Weld in place and install shrink collar									
1570	On sleepers, 1-1/2" diameter	Q-17A	18.50	1.297	Ea.	29	67.50	39	135.50	177
1580	3"		6.72	3.571		49.50	185	107	341.50	455
1590	4"		5.36	4.478		49.50	232	134	415.50	550
1600	6"		4	6		58	310	180	548	730
1610	8"		3.36	7.143		89.50	370	214	673.50	895
1620	12"		2.64	9.091		116	470	272	858	1,150
1630	On trench bottom, 18" diameter		2	12		174	620	360	1,154	1,525
1640	24"		1.36	17.647		232	915	530	1,677	2,200
1650	30"		1.04	23.077		290	1,200	690	2,180	2,875
1660	36"		1	24		350	1,250	720	2,320	3,050

33 61 13.20 Pipe Conduit, Prefabricated/Preinsulated

	33 61 13.20 Pipe Conduit, Prefabricated/Preinsulated	Crew	Daily Output	Labor-Hours	Unit	Material	Labor	Equipment	Total	Total Incl O&P
0010	**PIPE CONDUIT, PREFABRICATED/PREINSULATED**									
0020	Does not include trenching, fittings or crane.									
0300	For cathodic protection, add 12 to 14%									
0310	of total built-up price (casing plus service pipe)									
0580	Polyurethane insulated system, 250°F. max. temp.									
0620	Black steel service pipe, standard wt., 1/2" insulation									
0660	3/4" diam. pipe size	Q-17	54	.296	L.F.	57	15.60	1.03	73.63	87.50
0670	1" diam. pipe size		50	.320		63	16.85	1.11	80.96	95.50
0680	1-1/4" diam. pipe size		47	.340		70	17.90	1.18	89.08	105
0690	1-1/2" diam. pipe size		45	.356		76	18.70	1.23	95.93	113
0700	2" diam. pipe size		42	.381		79	20	1.32	100.32	118
0710	2-1/2" diam. pipe size		34	.471		80	25	1.63	106.63	127
0720	3" diam. pipe size		28	.571		93	30	1.98	124.98	150
0730	4" diam. pipe size		22	.727		117	38.50	2.52	158.02	190
0740	5" diam. pipe size		18	.889		149	47	3.08	199.08	238
0750	6" diam. pipe size	Q-18	23	1.043		173	57	2.40	232.40	280
0760	8" diam. pipe size		19	1.263		254	69	2.91	325.91	385
0770	10" diam. pipe size		16	1.500		320	82	3.45	405.45	485
0780	12" diam. pipe size		13	1.846		400	101	4.25	505.25	595
0790	14" diam. pipe size		11	2.182		445	119	5	569	675
0800	16" diam. pipe size		10	2.400		510	131	5.50	646.50	770
0810	18" diam. pipe size		8	3		590	164	6.90	760.90	905

33 61 13 – Underground Hydronic Energy Distribution

33 61 13.20 Pipe Conduit, Prefabricated/Preinsulated		Crew	Daily Output	Labor-Hours	Unit	Material	2014 Bare Costs Labor	Equipment	Total	Total Incl O&P
0820	20" diam. pipe size	Q-18	7	3.429	L.F.	655	187	7.90	849.90	1,000
0830	24" diam. pipe size		6	4		805	218	9.20	1,032.20	1,225
0900	For 1" thick insulation, add					10%				
0940	For 1-1/2" thick insulation, add					13%				
0980	For 2" thick insulation, add					20%				
1500	Gland seal for system, 3/4" diam. pipe size	Q-17	32	.500	Ea.	775	26.50	1.73	803.23	890
1510	1" diam. pipe size		32	.500		775	26.50	1.73	803.23	890
1540	1-1/4" diam. pipe size		30	.533		825	28	1.85	854.85	955
1550	1-1/2" diam. pipe size		30	.533		825	28	1.85	854.85	955
1560	2" diam. pipe size		28	.571		980	30	1.98	1,011.98	1,125
1570	2-1/2" diam. pipe size		26	.615		1,050	32.50	2.13	1,084.63	1,225
1580	3" diam. pipe size		24	.667		1,125	35	2.31	1,162.31	1,300
1590	4" diam. pipe size		22	.727		1,350	38.50	2.52	1,391.02	1,525
1600	5" diam. pipe size		19	.842		1,650	44.50	2.91	1,697.41	1,900
1610	6" diam. pipe size	Q-18	26	.923		1,750	50.50	2.12	1,802.62	2,000
1620	8" diam. pipe size		25	.960		2,050	52.50	2.21	2,104.71	2,325
1630	10" diam. pipe size		23	1.043		2,375	57	2.40	2,434.40	2,725
1640	12" diam. pipe size		21	1.143		2,625	62.50	2.63	2,690.13	3,000
1650	14" diam. pipe size		19	1.263		2,950	69	2.91	3,021.91	3,350
1660	16" diam. pipe size		18	1.333		3,500	73	3.07	3,576.07	3,975
1670	18" diam. pipe size		16	1.500		3,700	82	3.45	3,785.45	4,200
1680	20" diam. pipe size		14	1.714		4,200	93.50	3.94	4,297.44	4,775
1690	24" diam. pipe size		12	2		4,675	109	4.60	4,788.60	5,325
2000	Elbow, 45° for system									
2020	3/4" diam. pipe size	Q-17	14	1.143	Ea.	490	60	3.95	553.95	630
2040	1" diam. pipe size		13	1.231		500	65	4.26	569.26	655
2050	1-1/4" diam. pipe size		11	1.455		570	76.50	5.05	651.55	745
2060	1-1/2" diam. pipe size		9	1.778		595	93.50	6.15	694.65	805
2070	2" diam. pipe size		6	2.667		620	140	9.25	769.25	905
2080	2-1/2" diam. pipe size		4	4		670	211	13.85	894.85	1,075
2090	3" diam. pipe size		3.50	4.571		775	241	15.80	1,031.80	1,225
2100	4" diam. pipe size		3	5.333		905	281	18.45	1,204.45	1,450
2110	5" diam. pipe size		2.80	5.714		1,175	300	19.75	1,494.75	1,750
2120	6" diam. pipe size	Q-18	4	6		1,325	330	13.80	1,668.80	1,950
2130	8" diam. pipe size		3	8		1,925	435	18.40	2,378.40	2,775
2140	10" diam. pipe size		2.40	10		2,350	545	23	2,918	3,425
2150	12" diam. pipe size		2	12		3,075	655	27.50	3,757.50	4,425
2160	14" diam. pipe size		1.80	13.333		3,825	730	30.50	4,585.50	5,350
2170	16" diam. pipe size		1.60	15		4,550	820	34.50	5,404.50	6,300
2180	18" diam. pipe size		1.30	18.462		5,725	1,000	42.50	6,767.50	7,875
2190	20" diam. pipe size		1	24		7,200	1,300	55	8,555	9,950
2200	24" diam. pipe size		.70	34.286		9,050	1,875	79	11,004	12,900
2260	For elbow, 90°, add					25%				
2300	For tee, straight, add					85%	30%			
2340	For tee, reducing, add					170%	30%			
2380	For weldolet, straight, add					50%				

33 63 13 – Underground Steam and Condensate Distribution Piping

33 63 13.10 Calcium Silicate Insulated System		Crew	Daily Output	Labor-Hours	Unit	2014 Bare Costs Material	Labor	Equipment	Total	Total Incl O&P
0010	**CALCIUM SILICATE INSULATED SYSTEM**									
0011	High temp. (1200 degrees F)									
2840	Steel casing with protective exterior coating									
2850	6-5/8" diameter	Q-18	52	.462	L.F.	106	25	1.06	132.06	155
2860	8-5/8" diameter		50	.480		115	26	1.10	142.10	168
2870	10-3/4" diameter		47	.511		135	28	1.17	164.17	191
2880	12-3/4" diameter		44	.545		147	30	1.25	178.25	208
2890	14" diameter		41	.585		166	32	1.35	199.35	232
2900	16" diameter		39	.615		178	33.50	1.42	212.92	248
2910	18" diameter		36	.667		196	36.50	1.53	234.03	273
2920	20" diameter		34	.706		221	38.50	1.62	261.12	305
2930	22" diameter		32	.750		305	41	1.73	347.73	400
2940	24" diameter		29	.828		350	45	1.90	396.90	455
2950	26" diameter		26	.923		400	50.50	2.12	452.62	520
2960	28" diameter		23	1.043		495	57	2.40	554.40	635
2970	30" diameter		21	1.143		530	62.50	2.63	595.13	675
2980	32" diameter		19	1.263		595	69	2.91	666.91	760
2990	34" diameter		18	1.333		600	73	3.07	676.07	775
3000	36" diameter		16	1.500		645	82	3.45	730.45	840
3040	For multi-pipe casings, add					10%				
3060	For oversize casings, add					2%				
3400	Steel casing gland seal, single pipe									
3420	6-5/8" diameter	Q-18	25	.960	Ea.	1,475	52.50	2.21	1,529.71	1,700
3440	8-5/8" diameter		23	1.043		1,725	57	2.40	1,784.40	2,000
3450	10-3/4" diameter		21	1.143		1,925	62.50	2.63	1,990.13	2,225
3460	12-3/4" diameter		19	1.263		2,275	69	2.91	2,346.91	2,600
3470	14" diameter		17	1.412		2,525	77	3.25	2,605.25	2,900
3480	16" diameter		16	1.500		2,950	82	3.45	3,035.45	3,375
3490	18" diameter		15	1.600		3,275	87.50	3.68	3,366.18	3,725
3500	20" diameter		13	1.846		3,625	101	4.25	3,730.25	4,125
3510	22" diameter		12	2		4,075	109	4.60	4,188.60	4,650
3520	24" diameter		11	2.182		4,575	119	5	4,699	5,200
3530	26" diameter		10	2.400		5,225	131	5.50	5,361.50	5,925
3540	28" diameter		9.50	2.526		5,900	138	5.80	6,043.80	6,700
3550	30" diameter		9	2.667		6,000	146	6.15	6,152.15	6,825
3560	32" diameter		8.50	2.824		6,750	154	6.50	6,910.50	7,675
3570	34" diameter		8	3		7,350	164	6.90	7,520.90	8,350
3580	36" diameter		7	3.429		7,825	187	7.90	8,019.90	8,900
3620	For multi-pipe casings, add					5%				
4000	Steel casing anchors, single pipe									
4020	6-5/8" diameter	Q-18	8	3	Ea.	1,325	164	6.90	1,495.90	1,700
4040	8-5/8" diameter		7.50	3.200		1,375	175	7.35	1,557.35	1,800
4050	10-3/4" diameter		7	3.429		1,800	187	7.90	1,994.90	2,300
4060	12-3/4" diameter		6.50	3.692		1,925	202	8.50	2,135.50	2,450
4070	14" diameter		6	4		2,275	218	9.20	2,502.20	2,850
4080	16" diameter		5.50	4.364		2,625	238	10.05	2,873.05	3,275
4090	18" diameter		5	4.800		2,975	262	11.05	3,248.05	3,675
4100	20" diameter		4.50	5.333		3,275	291	12.25	3,578.25	4,050
4110	22" diameter		4	6		3,650	330	13.80	3,993.80	4,525
4120	24" diameter		3.50	6.857		3,975	375	15.75	4,365.75	4,950
4130	26" diameter		3	8		4,500	435	18.40	4,953.40	5,625
4140	28" diameter		2.50	9.600		4,950	525	22	5,497	6,250
4150	30" diameter		2	12		5,250	655	27.50	5,932.50	6,800

	33 63 13.10 Calcium Silicate Insulated System	Crew	Daily Output	Labor-Hours	Unit	2014 Bare Costs Material	Labor	Equipment	Total	Total Incl O&P
4160	32″ diameter	Q-18	1.50	16	Ea.	6,275	875	37	7,187	8,300
4170	34″ diameter		1	24		7,050	1,300	55	8,405	9,775
4180	36″ diameter		1	24		7,675	1,300	55	9,030	10,500
4220	For multi-pipe, add					5%	20%			
4800	Steel casing elbow									
4820	6-5/8″ diameter	Q-18	15	1.600	Ea.	1,800	87.50	3.68	1,891.18	2,125
4830	8-5/8″ diameter		15	1.600		1,925	87.50	3.68	2,016.18	2,250
4850	10-3/4″ diameter		14	1.714		2,325	93.50	3.94	2,422.44	2,700
4860	12-3/4″ diameter		13	1.846		2,750	101	4.25	2,855.25	3,175
4870	14″ diameter		12	2		2,950	109	4.60	3,063.60	3,425
4880	16″ diameter		11	2.182		3,225	119	5	3,349	3,725
4890	18″ diameter		10	2.400		3,675	131	5.50	3,811.50	4,250
4900	20″ diameter		9	2.667		3,925	146	6.15	4,077.15	4,550
4910	22″ diameter		8	3		4,225	164	6.90	4,395.90	4,900
4920	24″ diameter		7	3.429		4,650	187	7.90	4,844.90	5,425
4930	26″ diameter		6	4		5,100	218	9.20	5,327.20	5,950
4940	28″ diameter		5	4.800		5,500	262	11.05	5,773.05	6,450
4950	30″ diameter		4	6		5,525	330	13.80	5,868.80	6,575
4960	32″ diameter		3	8		6,275	435	18.40	6,728.40	7,600
4970	34″ diameter		2	12		6,900	655	27.50	7,582.50	8,625
4980	36″ diameter		2	12		7,350	655	27.50	8,032.50	9,125
5500	Black steel service pipe, std. wt., 1″ thick insulation									
5510	3/4″ diameter pipe size	Q-17	54	.296	L.F.	48	15.60	1.03	64.63	77
5540	1″ diameter pipe size		50	.320		49.50	16.85	1.11	67.46	81
5550	1-1/4″ diameter pipe size		47	.340		56	17.90	1.18	75.08	90
5560	1-1/2″ diameter pipe size		45	.356		61	18.70	1.23	80.93	97
5570	2″ diameter pipe size		42	.381		67.50	20	1.32	88.82	106
5580	2-1/2″ diameter pipe size		34	.471		71	25	1.63	97.63	118
5590	3″ diameter pipe size		28	.571		81	30	1.98	112.98	137
5600	4″ diameter pipe size		22	.727		104	38.50	2.52	145.02	175
5610	5″ diameter pipe size		18	.889		140	47	3.08	190.08	228
5620	6″ diameter pipe size	Q-18	23	1.043		153	57	2.40	212.40	257
6000	Black steel service pipe, std. wt., 1-1/2″ thick insul.									
6010	3/4″ diameter pipe size	Q-17	54	.296	L.F.	48.50	15.60	1.03	65.13	78
6040	1″ diameter pipe size		50	.320		54	16.85	1.11	71.96	86
6050	1-1/4″ diameter pipe size		47	.340		61	17.90	1.18	80.08	95.50
6060	1-1/2″ diameter pipe size		45	.356		66.50	18.70	1.23	86.43	103
6070	2″ diameter pipe size		42	.381		72	20	1.32	93.32	111
6080	2-1/2″ diameter pipe size		34	.471		77	25	1.63	103.63	124
6090	3″ diameter pipe size		28	.571		86	30	1.98	117.98	143
6100	4″ diameter pipe size		22	.727		110	38.50	2.52	151.02	182
6110	5″ diameter pipe size		18	.889		140	47	3.08	190.08	228
6120	6″ diameter pipe size	Q-18	23	1.043		160	57	2.40	219.40	265
6130	8″ diameter pipe size		19	1.263		230	69	2.91	301.91	360
6140	10″ diameter pipe size		16	1.500		288	82	3.45	373.45	445
6150	12″ diameter pipe size		13	1.846		345	101	4.25	450.25	535
6190	For 2″ thick insulation, add					15%				
6220	For 2-1/2″ thick insulation, add					25%				
6260	For 3″ thick insulation, add					30%				
6800	Black steel service pipe, ex. hvy. wt., 1″ thick insul.									
6820	3/4″ diameter pipe size	Q-17	50	.320	L.F.	50.50	16.85	1.11	68.46	82
6840	1″ diameter pipe size		47	.340		53.50	17.90	1.18	72.58	87.50
6850	1-1/4″ diameter pipe size		44	.364		62	19.15	1.26	82.41	98.50

33 63 Steam Energy Distribution

33 63 13 – Underground Steam and Condensate Distribution Piping

33 63 13.10 Calcium Silicate Insulated System		Crew	Daily Output	Labor-Hours	Unit	2014 Bare Costs Material	Labor	Equipment	Total	Total Incl O&P
6860	1-1/2" diameter pipe size	Q-17	42	.381	L.F.	65.50	20	1.32	86.82	104
6870	2" diameter pipe size		40	.400		69.50	21	1.38	91.88	110
6880	2-1/2" diameter pipe size		31	.516		86	27	1.79	114.79	138
6890	3" diameter pipe size		27	.593		97	31	2.05	130.05	156
6900	4" diameter pipe size		21	.762		128	40	2.64	170.64	203
6910	5" diameter pipe size		17	.941		179	49.50	3.26	231.76	275
6920	6" diameter pipe size	Q-18	22	1.091		181	59.50	2.51	243.01	292
7400	Black steel service pipe, ex. hvy. wt., 1-1/2" thick insul.									
7420	3/4" diameter pipe size	Q-17	50	.320	L.F.	42	16.85	1.11	59.96	72.50
7440	1" diameter pipe size		47	.340		48	17.90	1.18	67.08	81
7450	1-1/4" diameter pipe size		44	.364		55.50	19.15	1.26	75.91	91.50
7460	1-1/2" diameter pipe size		42	.381		61	20	1.32	82.32	99
7470	2" diameter pipe size		40	.400		67.50	21	1.38	89.88	108
7480	2-1/2" diameter pipe size		31	.516		73	27	1.79	101.79	123
7490	3" diameter pipe size		27	.593		85.50	31	2.05	118.55	143
7500	4" diameter pipe size		21	.762		112	40	2.64	154.64	186
7510	5" diameter pipe size		17	.941		154	49.50	3.26	206.76	249
7520	6" diameter pipe size	Q-18	22	1.091		170	59.50	2.51	232.01	280
7530	8" diameter pipe size		18	1.333		255	73	3.07	331.07	395
7540	10" diameter pipe size		15	1.600		305	87.50	3.68	396.18	470
7550	12" diameter pipe size		13	1.846		370	101	4.25	475.25	565
7590	For 2" thick insulation, add					13%				
7640	For 2-1/2" thick insulation, add					18%				
7680	For 3" thick insulation, add					24%				

33 71 Electrical Utility Transmission and Distribution

33 71 13 – Electrical Utility Towers

33 71 13.23 Steel Electrical Utility Towers		Crew	Daily Output	Labor-Hours	Unit	Material	Labor	Equipment	Total	Total Incl O&P
0010	**STEEL ELECTRICAL UTILITY TOWERS**									
0100	Excavation and backfill, earth	R-5	135.38	.650	C.Y.		30.50	11.20	41.70	58.50
0105	Rock		21.46	4.101	"		191	70.50	261.50	365
0200	Steel footings (grillage) in earth		3.91	22.506	Ton	2,175	1,050	385	3,610	4,400
0205	In rock		3.20	27.500	"	2,175	1,275	475	3,925	4,850
0290	See also Section 33 05 23.19									
0300	Rock anchors	R-5	5.87	14.991	Ea.	575	700	258	1,533	1,975
0400	Concrete foundations	"	12.85	6.848	C.Y.	131	320	118	569	755
0490	See also Section 03 30 53.40									
0500	Towers-material handling and spotting	R-7	22.56	2.128	Ton		80.50	9.95	90.45	135
0540	Steel tower erection	R-5	7.65	11.503		2,125	535	198	2,858	3,350
0550	Lace and box		7.10	12.394		2,125	580	213	2,918	3,425
0560	Painting total structure		1.47	59.864	Ea.	460	2,800	1,025	4,285	5,850
0570	Disposal of surplus material	R-7	20.87	2.300	Mile		87	10.75	97.75	146
0600	Special towers-material handling and spotting	"	12.31	3.899	Ton		148	18.25	166.25	247
0640	Special steel structure erection	R-6	6.52	13.497		2,650	630	525	3,805	4,450
0650	Special steel lace and box	"	6.29	13.990		2,650	650	545	3,845	4,500
0670	Disposal of surplus material	R-7	7.87	6.099	Mile		231	28.50	259.50	385

33 71 13.80 Transmission Line Right of Way		Crew	Daily Output	Labor-Hours	Unit	Material	Labor	Equipment	Total	Total Incl O&P
0010	**TRANSMISSION LINE RIGHT OF WAY**									
0100	Clearing right of way	B-87	6.67	5.997	Acre		279	450	729	920
0200	Restoration & seeding	B-10D	4	3	"	1,075	134	440	1,649	1,875

	33 71 16.23 Steel Electrical Utility Poles	Crew	Daily Output	Labor-Hours	Unit	2014 Bare Costs Material	Labor	Equipment	Total	Total Incl O&P
0010	**STEEL ELECTRICAL UTILITY POLES**									
0100	Steel, square, tapered shaft, 20′	R-15A	6	8	Ea.	370	375	54	799	1,025
0120	25′		5.35	8.972		390	420	60.50	870.50	1,125
0140	30′		4.80	10		455	470	67.50	992.50	1,275
0160	35′		4.35	11.034		470	515	74.50	1,059.50	1,375
0180	40′		4	12		665	560	81	1,306	1,675
0300	Galvanized, round, tapered shaft, 40′		4	12		1,975	560	81	2,616	3,125
0320	45′		3.70	12.973		2,200	605	88	2,893	3,450
0340	50′		3.45	13.913		2,425	650	94	3,169	3,750
0360	55′		3	16		2,550	750	108	3,408	4,050
0380	60′		2.65	18.113		2,850	850	123	3,823	4,525
0400	65′	R-13	2.20	19.091		3,525	980	86.50	4,591.50	5,450
0420	70′		1.90	22.105		3,850	1,125	100	5,075	6,025
0440	75′		1.70	24.706		5,050	1,275	112	6,437	7,575
0460	80′		1.50	28		4,550	1,450	127	6,127	7,325
0480	85′		1.40	30		4,950	1,550	136	6,636	7,925
0500	90′		1.25	33.600		5,300	1,725	152	7,177	8,625
0520	95′		1.15	36.522		5,725	1,875	165	7,765	9,300
0540	100′		1	42		6,175	2,150	190	8,515	10,300
0700	12 sided, tapered shaft, 40′	R-15A	4	12		1,100	560	81	1,741	2,150
0720	45′		3.70	12.973		1,225	605	88	1,918	2,375
0740	50′		3.45	13.913		1,400	650	94	2,144	2,650
0760	55′		3	16		1,625	750	108	2,483	3,025
0780	60′		2.65	18.113		1,675	850	123	2,648	3,250
0800	65′	R-13	2.20	19.091		1,975	980	86.50	3,041.50	3,750
0820	70′		1.90	22.105		2,150	1,125	100	3,375	4,150
0840	75′		1.70	24.706		2,325	1,275	112	3,712	4,575
0860	80′		1.50	28		3,950	1,450	127	5,527	6,650
0880	85′		1.40	30		6,400	1,550	136	8,086	9,525
0900	90′		1.25	33.600		8,375	1,725	152	10,252	12,000
0920	95′		1.15	36.522		10,500	1,875	165	12,540	14,600
0940	100′		1	42		12,800	2,150	190	15,140	17,600
0960	105′		.90	46.667		14,200	2,400	211	16,811	19,400
0980	Ladder clips					226			226	249
1000	Galvanized steel, round, tapered, w/one 6′ arm, 20′	R-15A	9.20	5.217		890	244	35.50	1,169.50	1,400
1020	25′		7.65	6.275		970	294	42.50	1,306.50	1,575
1040	30′		6.55	7.328		1,075	345	49.50	1,469.50	1,750
1060	35′		5.75	8.348		1,150	390	56.50	1,596.50	1,925
1080	40′		5.15	9.320		1,225	435	63	1,723	2,075
1200	Two 6′ arms, 20′		6.55	7.328		1,150	345	49.50	1,544.50	1,850
1220	25′		5.75	8.348		1,175	390	56.50	1,621.50	1,950
1240	30′		5.15	9.320		1,300	435	63	1,798	2,150
1260	35′		4.60	10.435		1,400	490	70.50	1,960.50	2,350
1280	40′		4.20	11.429		1,600	535	77.50	2,212.50	2,675
1400	One 12′ truss arm, 25′		7.65	6.275		1,100	294	42.50	1,436.50	1,725
1420	30′		6.55	7.328		1,150	345	49.50	1,544.50	1,825
1440	35′		5.75	8.348		1,250	390	56.50	1,696.50	2,025
1460	40′		5.15	9.320		1,400	435	63	1,898	2,250
1480	45′		4.60	10.435		1,600	490	70.50	2,160.50	2,600
1600	Two 12′ truss arms, 25′		5.75	8.348		1,325	390	56.50	1,771.50	2,100
1620	30′		5.15	9.320		1,350	435	63	1,848	2,225
1640	35′		4.60	10.435		1,450	490	70.50	2,010.50	2,425

33 71 16 – Electrical Utility Poles

33 71 16.23 Steel Electrical Utility Poles		Crew	Daily Output	Labor-Hours	Unit	Material	2014 Bare Costs Labor	Equipment	Total	Total Incl O&P
1660	40′	R-15A	4.20	11.429	Ea.	1,600	535	77.50	2,212.50	2,675
1680	45′		3.85	12.468		1,825	585	84.50	2,494.50	2,975
3400	Galvanized steel, tapered, 10′		15.50	3.097		277	145	21	443	545
3420	12′		11.50	4.174		405	195	28.50	628.50	770
3440	14′		9.20	5.217		299	244	35.50	578.50	740
3460	16′		8.45	5.680		310	266	38.50	614.50	785
3480	18′		7.65	6.275		320	294	42.50	656.50	840
3500	20′		7.10	6.761		300	315	46	661	860
6000	Digging holes in earth, average	R-5	25.14	3.500			163	60	223	315
6010	In rock, average	"	4.51	19.512			910	335	1,245	1,750
6020	Formed plate pole structure									
6030	Material handling and spotting	R-7	2.40	20	Ea.		755	93.50	848.50	1,275
6040	Erect steel plate pole	R-5	1.95	45.128		9,050	2,100	775	11,925	14,000
6050	Guys, anchors and hardware for pole, in earth		7.04	12.500		550	585	215	1,350	1,725
6060	In rock		17.96	4.900		655	228	84.50	967.50	1,150
6070	Foundations for line poles									
6080	Excavation, in earth	R-5	135.38	.650	C.Y.		30.50	11.20	41.70	58.50
6090	In rock		20	4.400			205	75.50	280.50	395
6110	Concrete foundations		11	8		131	375	138	644	860

33 71 16.33 Wood Electrical Utility Poles		Crew	Daily Output	Labor-Hours	Unit	Material	Labor	Equipment	Total	Total Incl O&P
0010	**WOOD ELECTRICAL UTILITY POLES**									
0011	Excludes excavation, backfill and cast-in-place concrete									
1020	12″ Ponderosa Pine Poles treated 0.40 ACQ, 16′	R-3	3.20	6.250	Ea.	880	330	44	1,254	1,500
5000	Wood, class 3 yellow pine, penta-treated, 25′	R-15A	8.60	5.581		199	261	38	498	655
5020	30′		7.70	6.234		221	292	42	555	730
5040	35′		5.80	8.276		292	385	56	733	965
5060	40′		5.30	9.057		365	425	61.50	851.50	1,100
5080	45′		4.70	10.213		405	480	69	954	1,250
5100	50′		4.20	11.429		505	535	77.50	1,117.50	1,450
5120	55′		3.80	12.632		605	590	85.50	1,280.50	1,650
5140	60′		3.50	13.714		840	640	93	1,573	2,000
5160	65′		3.20	15		1,125	700	102	1,927	2,400
5180	70′		3	16		1,600	750	108	2,458	3,000
5200	75′		2.80	17.143		2,075	805	116	2,996	3,625
6000	Wood, class 1 type C, CCA/ACA-treated, 25′		8.60	5.581		231	261	38	530	690
6020	30′		7.70	6.234		267	292	42	601	780
6040	35′		5.80	8.276		355	385	56	796	1,050
6060	40′		5.30	9.057		450	425	61.50	936.50	1,200
6080	45′		4.70	10.213		545	480	69	1,094	1,400
6100	50′		4.20	11.429		600	535	77.50	1,212.50	1,550
6120	55′		3.80	12.632		690	590	85.50	1,365.50	1,750
6200	Electric & tel sitework, 20′ high, treated wd., see Section 26 56 13.10	R-3	3.10	6.452		223	340	45	608	805
6400	25′ high		2.90	6.897		263	365	48.50	676.50	895
6600	30′ high		2.60	7.692		430	405	54	889	1,150
6800	35′ high		2.40	8.333		495	440	58.50	993.50	1,275
7000	40′ high		2.30	8.696		710	460	61	1,231	1,525
7200	45′ high		1.70	11.765		865	625	82.50	1,572.50	1,975
7400	Cross arms with hardware & insulators									
7600	4′ long	1 Elec	2.50	3.200	Ea.	148	171		319	420
7800	5′ long		2.40	3.333		170	178		348	455
8000	6′ long		2.20	3.636		165	194		359	470
9000	Disposal of pole & hardware surplus material	R-7	20.87	2.300	Mile		87	10.75	97.75	146

33 71 16 – Electrical Utility Poles

33 71 16.33 Wood Electrical Utility Poles		Crew	Daily Output	Labor-Hours	Unit	Material	2014 Bare Costs Labor	Equipment	Total	Total Incl O&P
9100	Disposal of crossarms & hardware surplus material	R-7	40	1.200	Mile		45.50	5.60	51.10	76

33 71 19 – Electrical Underground Ducts and Manholes

33 71 19.15 Underground Ducts and Manholes		Crew	Daily Output	Labor-Hours	Unit	Material	2014 Bare Costs Labor	Equipment	Total	Total Incl O&P
0010	UNDERGROUND DUCTS AND MANHOLES									
0011	Not incl. excavation, backfill and concrete, in slab or duct bank									
1000	Direct burial									
1010	PVC, schedule 40, w/coupling, 1/2" diameter	1 Elec	340	.024	L.F.	.41	1.26		1.67	2.33
1020	3/4" diameter		290	.028		.55	1.47		2.02	2.81
1030	1" diameter		260	.031		.81	1.64		2.45	3.35
1040	1-1/2" diameter		210	.038		1.33	2.03		3.36	4.50
1050	2" diameter		180	.044		1.74	2.37		4.11	5.45
1060	3" diameter	2 Elec	240	.067		3.33	3.56		6.89	8.95
1070	4" diameter		160	.100		4.66	5.35		10.01	13.15
1080	5" diameter		120	.133		6.75	7.10		13.85	18.10
1090	6" diameter		90	.178		8.95	9.50		18.45	24
1110	Elbows, 1/2" diameter	1 Elec	48	.167	Ea.	1.06	8.90		9.96	14.45
1120	3/4" diameter		38	.211		1.04	11.25		12.29	17.95
1130	1" diameter		32	.250		1.57	13.35		14.92	21.50
1140	1-1/2" diameter		21	.381		3.11	20.50		23.61	34
1150	2" diameter		16	.500		4.22	26.50		30.72	44.50
1160	3" diameter		12	.667		12.70	35.50		48.20	67
1170	4" diameter		9	.889		20.50	47.50		68	93.50
1180	5" diameter		8	1		33	53.50		86.50	117
1190	6" diameter		5	1.600		76	85.50		161.50	212
1210	Adapters, 1/2" diameter		52	.154		.30	8.20		8.50	12.65
1220	3/4" diameter		43	.186		.49	9.95		10.44	15.40
1230	1" diameter		39	.205		.66	10.95		11.61	17.15
1240	1-1/2" diameter		35	.229		1.14	12.20		13.34	19.50
1250	2" diameter		26	.308		1.44	16.40		17.84	26
1260	3" diameter		20	.400		3.50	21.50		25	36
1270	4" diameter		14	.571		6.20	30.50		36.70	52.50
1280	5" diameter		12	.667		13.20	35.50		48.70	67.50
1290	6" diameter		9	.889		22	47.50		69.50	95
1340	Bell end & cap, 1-1/2" diameter		35	.229		8.20	12.20		20.40	27.50
1350	Bell end & plug, 2" diameter		26	.308		10.85	16.40		27.25	36.50
1360	3" diameter		20	.400		13.20	21.50		34.70	46.50
1370	4" diameter		14	.571		16.65	30.50		47.15	64
1380	5" diameter		12	.667		18.80	35.50		54.30	73.50
1390	6" diameter		9	.889		22.50	47.50		70	96
1450	Base spacer, 2" diameter		56	.143		1.54	7.60		9.14	13.10
1460	3" diameter		46	.174		1.58	9.30		10.88	15.65
1470	4" diameter		41	.195		1.63	10.40		12.03	17.40
1480	5" diameter		37	.216		2.04	11.55		13.59	19.50
1490	6" diameter		34	.235		2.77	12.55		15.32	22
1550	Intermediate spacer, 2" diameter		60	.133		1.51	7.10		8.61	12.30
1560	3" diameter		46	.174		1.70	9.30		11	15.75
1570	4" diameter		41	.195		1.56	10.40		11.96	17.30
1580	5" diameter		37	.216		1.84	11.55		13.39	19.25
1590	6" diameter		34	.235		2.69	12.55		15.24	22

33 71 19.17 Electric and Telephone Underground		Crew	Daily Output	Labor-Hours	Unit	Material	2014 Bare Costs Labor	Equipment	Total	Total Incl O&P
0010	**ELECTRIC AND TELEPHONE UNDERGROUND**									
0011	Not including excavation									
0200	backfill and cast in place concrete									
0400	Hand holes, precast concrete, with concrete cover									
0600	2′ x 2′ x 3′ deep	R-3	2.40	8.333	Ea.	415	440	58.50	913.50	1,175
0800	3′ x 3′ x 3′ deep		1.90	10.526		535	555	73.50	1,163.50	1,500
1000	4′ x 4′ x 4′ deep		1.40	14.286		1,200	755	100	2,055	2,525
1200	Manholes, precast with iron racks & pulling irons, C.I. frame									
1400	and cover, 4′ x 6′ x 7′ deep	B-13	2	28	Ea.	5,925	1,125	375	7,425	8,625
1600	6′ x 8′ x 7′ deep		1.90	29.474		6,650	1,175	395	8,220	9,550
1800	6′ x 10′ x 7′ deep		1.80	31.111		7,450	1,250	415	9,115	10,600
4200	Underground duct, banks ready for concrete fill, min. of 7.5″									
4400	between conduits, center to center									
4580	PVC, type EB, 1 @ 2″ diameter	2 Elec	480	.033	L.F.	.72	1.78		2.50	3.46
4600	2 @ 2″ diameter		240	.067		1.45	3.56		5.01	6.90
4800	4 @ 2″ diameter		120	.133		2.90	7.10		10	13.85
4900	1 @ 3″ diameter		400	.040		.99	2.13		3.12	4.28
5000	2 @ 3″ diameter		200	.080		1.98	4.27		6.25	8.55
5200	4 @ 3″ diameter		100	.160		3.95	8.55		12.50	17.15
5300	1 @ 4″ diameter		320	.050		1.51	2.67		4.18	5.65
5400	2 @ 4″ diameter		160	.100		3.03	5.35		8.38	11.35
5600	4 @ 4″ diameter		80	.200		6.05	10.65		16.70	22.50
5800	6 @ 4″ diameter		54	.296		9.10	15.80		24.90	33.50
5810	1 @ 5″ diameter		260	.062		2.25	3.28		5.53	7.40
5820	2 @ 5″ diameter		130	.123		4.51	6.55		11.06	14.80
5840	4 @ 5″ diameter		70	.229		9	12.20		21.20	28
5860	6 @ 5″ diameter		50	.320		13.50	17.05		30.55	40.50
5870	1 @ 6″ diameter		200	.080		3.22	4.27		7.49	9.95
5880	2 @ 6″ diameter		100	.160		6.45	8.55		15	19.90
5900	4 @ 6″ diameter		50	.320		12.85	17.05		29.90	39.50
5920	6 @ 6″ diameter		30	.533		19.30	28.50		47.80	63.50
6200	Rigid galvanized steel, 2 @ 2″ diameter		180	.089		12.10	4.74		16.84	20.50
6400	4 @ 2″ diameter		90	.178		24	9.50		33.50	40.50
6800	2 @ 3″ diameter		100	.160		27.50	8.55		36.05	43
7000	4 @ 3″ diameter		50	.320		55	17.05		72.05	86
7200	2 @ 4″ diameter		70	.229		40	12.20		52.20	62.50
7400	4 @ 4″ diameter		34	.471		80	25		105	126
7600	6 @ 4″ diameter		22	.727		120	39		159	190
7620	2 @ 5″ diameter		60	.267		83	14.25		97.25	113
7640	4 @ 5″ diameter		30	.533		166	28.50		194.50	225
7660	6 @ 5″ diameter		18	.889		248	47.50		295.50	345
7680	2 @ 6″ diameter		40	.400		119	21.50		140.50	163
7700	4 @ 6″ diameter		20	.800		238	42.50		280.50	325
7720	6 @ 6″ diameter		14	1.143		360	61		421	485
8000	Fittings, PVC type EB, elbow, 2″ diameter	1 Elec	16	.500	Ea.	15.90	26.50		42.40	57.50
8200	3″ diameter		14	.571		20.50	30.50		51	68
8400	4″ diameter		12	.667		28.50	35.50		64	84.50
8420	5″ diameter		10	.800		152	42.50		194.50	231
8440	6″ diameter		9	.889		173	47.50		220.50	261
8500	Coupling, 2″ diameter					.70			.70	.77
8600	3″ diameter					3.22			3.22	3.54
8700	4″ diameter					3.52			3.52	3.87

33 71 Electrical Utility Transmission and Distribution

33 71 19 – Electrical Underground Ducts and Manholes

33 71 19.17 Electric and Telephone Underground		Crew	Daily Output	Labor-Hours	Unit	2014 Bare Costs Material	Labor	Equipment	Total	Total Incl O&P
8720	5″ diameter				Ea.	6.65			6.65	7.35
8740	6″ diameter					22.50			22.50	24.50
8800	Adapter, 2″ diameter	1 Elec	26	.308		1.03	16.40		17.43	25.50
9000	3″ diameter		20	.400		2.88	21.50		24.38	35
9200	4″ diameter		16	.500		4.25	26.50		30.75	44.50
9220	5″ diameter		13	.615		10.25	33		43.25	60.50
9240	6″ diameter		10	.800		13.30	42.50		55.80	78.50
9400	End bell, 2″ diameter		16	.500		1.31	26.50		27.81	41.50
9600	3″ diameter		14	.571		6.65	30.50		37.15	53
9800	4″ diameter		12	.667		4.27	35.50		39.77	57.50
9810	5″ diameter		10	.800		13	42.50		55.50	78.50
9820	6″ diameter		8	1		66	53.50		119.50	153
9830	5° angle coupling, 2″ diameter		26	.308		19.55	16.40		35.95	46
9840	3″ diameter		20	.400		29.50	21.50		51	64.50
9850	4″ diameter		16	.500		15.25	26.50		41.75	57
9860	5″ diameter		13	.615		14.75	33		47.75	65
9870	6″ diameter		10	.800		23	42.50		65.50	89
9880	Expansion joint, 2″ diameter		16	.500		33.50	26.50		60	77
9890	3″ diameter		18	.444		64	23.50		87.50	106
9900	4″ diameter		12	.667		85.50	35.50		121	147
9910	5″ diameter		10	.800		114	42.50		156.50	190
9920	6″ diameter		8	1		143	53.50		196.50	237
9930	Heat bender, 2″ diameter					545			545	600
9940	6″ diameter					1,850			1,850	2,025
9950	Cement, quart					19.55			19.55	21.50

33 71 23 – Insulators and Fittings

33 71 23.16 Post Insulators		Crew	Daily Output	Labor-Hours	Unit	Material	Labor	Equipment	Total	Total Incl O&P
0010	**POST INSULATORS**									
7400	Insulators, pedestal type	R-11	112	.500	Ea.		25.50	7.95	33.45	46.50
7490	See also line 33 71 39.13 1000									

33 71 26 – Transmission and Distribution Equipment

33 71 26.13 Capacitor Banks		Crew	Daily Output	Labor-Hours	Unit	Material	Labor	Equipment	Total	Total Incl O&P
0010	**CAPACITOR BANKS**									
1300	Station capacitors									
1350	Synchronous, 13 to 26 kV	R-11	3.11	18.006	MVAR	6,650	910	285	7,845	9,025
1360	46 kV		3.33	16.817		8,500	850	267	9,617	10,900
1370	69 kV		3.81	14.698		8,350	745	233	9,328	10,600
1380	161 kV		6.51	8.602		7,800	435	136	8,371	9,400
1390	500 kV		10.37	5.400		6,800	273	85.50	7,158.50	7,975
1450	Static, 13 to 26 kV		3.11	18.006		5,650	910	285	6,845	7,900
1460	46 kV		3.01	18.605		7,125	940	295	8,360	9,600
1470	69 kV		3.81	14.698		6,925	745	233	7,903	9,000
1480	161 kV		6.51	8.602		6,425	435	136	6,996	7,875
1490	500 kV		10.37	5.400		5,875	273	85.50	6,233.50	6,950
1600	Voltage regulators, 13 to 26 kV		.75	74.667	Ea.	252,500	3,775	1,175	257,450	285,000

33 71 26.23 Current Transformers		Crew	Daily Output	Labor-Hours	Unit	Material	Labor	Equipment	Total	Total Incl O&P
0010	**CURRENT TRANSFORMERS**									
4050	Current transformers, 13 to 26 kV	R-11	14	4	Ea.	3,100	202	63.50	3,365.50	3,775
4060	46 kV		9.33	6.002		9,075	305	95	9,475	10,500
4070	69 kV		7	8		9,400	405	127	9,932	11,100
4080	161 kV		1.87	29.947		30,500	1,525	475	32,500	36,300

33 71 26 – Transmission and Distribution Equipment

33 71 26.26 Potential Transformers		Crew	Daily Output	Labor-Hours	Unit	2014 Bare Costs Material	Labor	Equipment	Total	Total Incl O&P
0010	**POTENTIAL TRANSFORMERS**									
4100	Potential transformers, 13 to 26 kV	R-11	11.20	5	Ea.	4,450	253	79.50	4,782.50	5,350
4110	46 kV		8	7		9,125	355	111	9,591	10,700
4120	69 kV		6.22	9.003		9,675	455	143	10,273	11,400
4130	161 kV		2.24	25		20,900	1,275	395	22,570	25,300
4140	500 kV		1.40	40		62,500	2,025	635	65,160	72,500

33 71 39 – High-Voltage Wiring

33 71 39.13 Overhead High-Voltage Wiring		Crew	Daily Output	Labor-Hours	Unit	Material	Labor	Equipment	Total	Total Incl O&P
0010	**OVERHEAD HIGH-VOLTAGE WIRING**									
0100	Conductors, primary circuits									
0110	Material handling and spotting	R-5	9.78	8.998	W.Mile		420	155	575	805
0120	For river crossing, add		11	8			375	138	513	715
0150	Conductors, per wire, 210 to 636 kcmil		1.96	44.898		10,300	2,100	775	13,175	15,300
0160	795 to 954 kcmil		1.87	47.059		20,500	2,200	810	23,510	26,700
0170	1000 to 1600 kcmil		1.47	59.864		35,100	2,800	1,025	38,925	44,000
0180	Over 1600 kcmil		1.35	65.185		49,100	3,050	1,125	53,275	60,000
0200	For river crossing, add, 210 to 636 kcmil		1.24	70.968			3,300	1,225	4,525	6,350
0220	795 to 954 kcmil		1.09	80.734			3,750	1,400	5,150	7,200
0230	1000 to 1600 kcmil		.97	90.722			4,225	1,550	5,775	8,125
0240	Over 1600 kcmil		.87	101			4,725	1,750	6,475	9,050
0300	Joints and dead ends	R-8	6	8	Ea.	1,400	380	63.50	1,843.50	2,200
0400	Sagging	R-5	7.33	12.001	W.Mile		560	207	767	1,075
0500	Clipping, per structure, 69 kV	R-10	9.60	5	Ea.		252	65	317	450
0510	161 kV		5.33	9.006			455	117	572	810
0520	345 to 500 kV		2.53	18.972			955	247	1,202	1,700
0600	Make and install jumpers, per structure, 69 kV	R-8	3.20	15		380	710	119	1,209	1,625
0620	161 kV		1.20	40		765	1,900	320	2,985	4,050
0640	345 to 500 kV		.32	150		1,275	7,075	1,200	9,550	13,500
0700	Spacers	R-10	68.57	.700		76	35	9.10	120.10	147
0720	For river crossings, add	"	60	.800			40.50	10.40	50.90	72
0800	Installing pulling line (500 kV only)	R-9	1.45	44.138	W.Mile	755	1,950	264	2,969	4,075
0810	Disposal of surplus material, high voltage conductors	R-7	6.96	6.897	Mile		261	32.50	293.50	435
0820	With trailer mounted reel stands	"	13.71	3.501	"		133	16.40	149.40	222
0900	Insulators and hardware, primary circuits									
0920	Material handling and spotting, 69 kV	R-7	480	.100	Ea.		3.79	.47	4.26	6.30
0930	161 kV		685.71	.070			2.65	.33	2.98	4.43
0950	345 to 500 kV		960	.050			1.89	.23	2.12	3.17
1000	Disk insulators, 69 kV	R-5	880	.100		77.50	4.66	1.72	83.88	94.50
1020	161 kV		977.78	.090		89	4.19	1.55	94.74	106
1040	345 to 500 kV		1100	.080		89	3.73	1.38	94.11	105
1060	See Section 33 71 23.16 for pin or pedestal insulator									
1100	Install disk insulator at river crossing, add									
1110	69 kV	R-5	586.67	.150	Ea.		7	2.58	9.58	13.40
1120	161 kV		880	.100			4.66	1.72	6.38	8.95
1140	345 to 500 kV		880	.100			4.66	1.72	6.38	8.95
1150	Disposal of surplus material, high voltage insulators	R-7	41.74	1.150	Mile		43.50	5.40	48.90	73
1300	Overhead ground wire installation									
1320	Material handling and spotting	R-7	5.65	8.496	W.Mile		320	40	360	540
1340	Overhead ground wire	R-5	1.76	50		2,975	2,325	860	6,160	7,750
1350	At river crossing, add		1.17	75.214			3,500	1,300	4,800	6,725
1360	Disposal of surplus material, grounding wire		41.74	2.108	Mile		98.50	36.50	135	188
1400	Installing conductors, underbuilt circuits									

33 71 Electrical Utility Transmission and Distribution

33 71 39 – High-Voltage Wiring

33 71 39.13 Overhead High-Voltage Wiring		Crew	Daily Output	Labor-Hours	Unit	Material	2014 Bare Costs Labor	Equipment	Total	Total Incl O&P
1420	Material handling and spotting	R-7	5.65	8.496	W.Mile		320	40	360	540
1440	Conductors, per wire, 210 to 636 kcmil	R-5	1.96	44.898		10,300	2,100	775	13,175	15,300
1450	795 to 954 kcmil		1.87	47.059		20,500	2,200	810	23,510	26,700
1460	1000 to 1600 kcmil		1.47	59.864		35,100	2,800	1,025	38,925	44,000
1470	Over 1600 kcmil		1.35	65.185		49,100	3,050	1,125	53,275	60,000
1500	Joints and dead ends	R-8	6	8	Ea.	1,400	380	63.50	1,843.50	2,200
1550	Sagging	R-5	8.80	10	W.Mile		465	172	637	895
1600	Clipping, per structure, 69 kV	R-10	9.60	5	Ea.		252	65	317	450
1620	161 kV		5.33	9.006			455	117	572	810
1640	345 to 500 kV		2.53	18.972			955	247	1,202	1,700
1700	Making and installing jumpers, per structure, 69 kV	R-8	5.87	8.177		380	385	65	830	1,075
1720	161 kV		.96	50		765	2,350	400	3,515	4,850
1740	345 to 500 kV		.32	150		1,275	7,075	1,200	9,550	13,500
1800	Spacers	R-10	96	.500		76	25	6.50	107.50	129
1810	Disposal of surplus material, conductors & hardware	R-7	6.96	6.897	Mile		261	32.50	293.50	435
2000	Insulators and hardware for underbuilt circuits									
2100	Material handling and spotting	R-7	1200	.040	Ea.		1.51	.19	1.70	2.54
2150	Disk insulators, 69 kV	R-8	600	.080		77.50	3.78	.64	81.92	92
2160	161 kV		686	.070		89	3.30	.56	92.86	103
2170	345 to 500 kV		800	.060		89	2.83	.48	92.31	102
2180	Disposal of surplus material, insulators & hardware	R-7	41.74	1.150	Mile		43.50	5.40	48.90	73
2300	Sectionalizing switches, 69 kV	R-5	1.26	69.841	Ea.	20,000	3,250	1,200	24,450	28,300
2310	161 kV		.80	110		22,600	5,125	1,900	29,625	34,700
2500	Protective devices		5.50	16		6,550	745	275	7,570	8,625
2600	Clearance poles, 8 poles per mile									
2650	In earth, 69 kV	R-5	1.16	75.862	Mile	5,400	3,525	1,300	10,225	12,700
2660	161 kV	"	.64	137		8,850	6,400	2,375	17,625	22,000
2670	345 to 500 kV	R-6	.48	183		10,600	8,550	7,150	26,300	32,500
2800	In rock, 69 kV	R-5	.69	127		5,400	5,950	2,200	13,550	17,300
2820	161 kV	"	.35	251		8,850	11,700	4,325	24,875	32,200
2840	345 to 500 kV	R-6	.24	366		10,600	17,100	14,300	42,000	53,000

33 72 Utility Substations

33 72 26 – Substation Bus Assemblies

33 72 26.13 Aluminum Substation Bus Assemblies

		Crew	Daily Output	Labor-Hours	Unit	Material	Labor	Equipment	Total	Total Incl O&P
0010	**ALUMINUM SUBSTATION BUS ASSEMBLIES**									
7300	Bus	R-11	590	.095	Lb.	2.73	4.80	1.50	9.03	11.85

33 72 33 – Control House Equipment

33 72 33.33 Raceway/Boxes for Utility Substations

		Crew	Daily Output	Labor-Hours	Unit	Material	Labor	Equipment	Total	Total Incl O&P
0010	**RACEWAY/BOXES FOR UTILITY SUBSTATIONS**									
7000	Conduit, conductors, and insulators									
7100	Conduit, metallic	R-11	560	.100	Lb.	2.50	5.05	1.59	9.14	12.10
7110	Non-metallic		800	.070		1.75	3.54	1.11	6.40	8.50
7200	Wire and cable		700	.080		8.20	4.05	1.27	13.52	16.50

33 72 33.36 Cable Trays for Utility Substations

		Crew	Daily Output	Labor-Hours	Unit	Material	Labor	Equipment	Total	Total Incl O&P
0010	**CABLE TRAYS FOR UTILITY SUBSTATIONS**									
7700	Cable tray	R-11	40	1.400	L.F.	23.50	71	22	116.50	157

33 72 Utility Substations

33 72 33 – Control House Equipment

33 72 33.43 Substation Backup Batteries		Crew	Daily Output	Labor-Hours	Unit	Material	Labor	Equipment	Total	Total Incl O&P
0010	SUBSTATION BACKUP BATTERIES									
9120	Battery chargers	R-11	11.20	5	Ea.	3,775	253	79.50	4,107.50	4,650
9200	Control batteries	"	14	4	K.A.H.	87	202	63.50	352.50	470

(2014 Bare Costs: Material, Labor, Equipment, Total)

33 72 33.46 Substation Converter Stations		Crew	Daily Output	Labor-Hours	Unit	Material	Labor	Equipment	Total	Total Incl O&P
0010	SUBSTATION CONVERTER STATIONS									
9000	Station service equipment									
9100	Conversion equipment									
9110	Station service transformers	R-11	5.60	10	Ea.	91,500	505	159	92,164	101,500

33 73 Utility Transformers

33 73 23 – Dry-Type Utility Transformers

33 73 23.20 Primary Transformers		Crew	Daily Output	Labor-Hours	Unit	Material	Labor	Equipment	Total	Total Incl O&P
0010	PRIMARY TRANSFORMERS									
1000	Main conversion equipment									
1050	Power transformers, 13 to 26 kV	R-11	1.72	32.558	MVA	23,900	1,650	515	26,065	29,300
1060	46 kV		3.50	16		22,400	810	254	23,464	26,100
1070	69 kV		3.11	18.006		19,100	910	285	20,295	22,700
1080	110 kV		3.29	17.021		18,100	860	270	19,230	21,600
1090	161 kV		4.31	12.993		16,800	655	206	17,661	19,700
1100	500 kV		7	8		16,800	405	127	17,332	19,200
1200	Grounding transformers		3.11	18.006	Ea.	105,500	910	285	106,695	117,500

33 75 High-Voltage Switchgear and Protection Devices

33 75 13 – Air High-Voltage Circuit Breaker

33 75 13.13 High-Voltage Circuit Breaker, Air		Crew	Daily Output	Labor-Hours	Unit	Material	Labor	Equipment	Total	Total Incl O&P
0010	HIGH-VOLTAGE CIRCUIT BREAKER, AIR									
2100	Air circuit breakers, 13 to 26 kV	R-11	.56	100	Ea.	59,000	5,050	1,575	65,625	74,500
2110	161 kV	"	.24	233	"	253,500	11,800	3,700	269,000	301,000

33 75 16 – Oil High-Voltage Circuit Breaker

33 75 16.13 Oil Circuit Breaker		Crew	Daily Output	Labor-Hours	Unit	Material	Labor	Equipment	Total	Total Incl O&P
0010	OIL CIRCUIT BREAKER									
2000	Power circuit breakers									
2050	Oil circuit breakers, 13 to 26 kV	R-11	1.12	50	Ea.	56,500	2,525	795	59,820	66,500
2060	46 kV		.75	74.667		82,500	3,775	1,175	87,450	97,500
2070	69 kV		.45	124		191,500	6,300	1,975	199,775	222,000
2080	161 kV		.16	350		292,500	17,700	5,550	315,750	354,000
2090	500 kV		.06	933		1,094,500	47,200	14,800	1,156,500	1,291,500

33 75 19 – Gas High-Voltage Circuit Breaker

33 75 19.13 Gas Circuit Breaker		Crew	Daily Output	Labor-Hours	Unit	Material	Labor	Equipment	Total	Total Incl O&P
0010	GAS CIRCUIT BREAKER									
2150	Gas circuit breakers, 13 to 26 kV	R-11	.56	100	Ea.	221,500	5,050	1,575	228,125	253,000
2160	161 kV		.08	700		291,000	35,400	11,100	337,500	385,500
2170	500 kV		.04	1400		1,024,500	71,000	22,200	1,117,700	1,258,000

33 75 23 – Vacuum High-Voltage Circuit Breaker

33 75 23.13 Vacuum Circuit Breaker		Crew	Daily Output	Labor-Hours	Unit	Material	Labor	Equipment	Total	Total Incl O&P
0010	VACUUM CIRCUIT BREAKER									
2200	Vacuum circuit breakers, 13 to 26 kV	R-11	.56	100	Ea.	48,300	5,050	1,575	54,925	62,500

33 75 36 – High-Voltage Utility Fuses

33 75 36.13 Fuses		Crew	Daily Output	Labor-Hours	Unit	Material	Labor	Equipment	Total	Total Incl O&P
						2014 Bare Costs				
0010	**FUSES**									
8250	Fuses, 13 to 26 kV	R-11	18.67	2.999	Ea.	2,050	152	47.50	2,249.50	2,525
8260	46 kV		11.20	5		2,325	253	79.50	2,657.50	3,050
8270	69 kV		8	7		2,425	355	111	2,891	3,325
8280	161 kV		4.67	11.991		3,075	605	190	3,870	4,500

33 75 39 – High-Voltage Surge Arresters

33 75 39.13 Surge Arresters		Crew	Daily Output	Labor-Hours	Unit	Material	Labor	Equipment	Total	Total Incl O&P
0010	**SURGE ARRESTERS**									
8000	Protective equipment									
8050	Lightning arresters, 13 to 26 kV	R-11	18.67	2.999	Ea.	1,525	152	47.50	1,724.50	1,950
8060	46 kV		14	4		4,125	202	63.50	4,390.50	4,925
8070	69 kV		11.20	5		5,300	253	79.50	5,632.50	6,300
8080	161 kV		5.60	10		7,300	505	159	7,964	8,975
8090	500 kV		1.40	40		24,800	2,025	635	27,460	31,000

33 75 43 – Shunt Reactors

33 75 43.13 Reactors		Crew	Daily Output	Labor-Hours	Unit	Material	Labor	Equipment	Total	Total Incl O&P
0010	**REACTORS**									
8150	Reactors and resistors, 13 to 26 kV	R-11	28	2	Ea.	3,450	101	31.50	3,582.50	3,975
8160	46 kV		4.31	12.993		10,400	655	206	11,261	12,700
8170	69 kV		2.80	20		17,000	1,000	315	18,315	20,600
8180	161 kV		2.24	25		19,400	1,275	395	21,070	23,700
8190	500 kV		.08	700		78,000	35,400	11,100	124,500	151,000

33 75 53 – High-Voltage Switches

33 75 53.13 Switches		Crew	Daily Output	Labor-Hours	Unit	Material	Labor	Equipment	Total	Total Incl O&P
0010	**SWITCHES**									
3000	Disconnecting switches									
3050	Gang operated switches									
3060	Manual operation, 13 to 26 kV	R-11	1.65	33.939	Ea.	13,600	1,725	540	15,865	18,100
3070	46 kV		1.12	50		22,600	2,525	795	25,920	29,500
3080	69 kV		.80	70		25,300	3,550	1,100	29,950	34,500
3090	161 kV		.56	100		30,500	5,050	1,575	37,125	42,900
3100	500 kV		.14	400		83,000	20,200	6,350	109,550	129,000
3110	Motor operation, 161 kV		.51	109		45,900	5,550	1,750	53,200	61,000
3120	500 kV		.28	200		127,000	10,100	3,175	140,275	158,000
3250	Circuit switches, 161 kV		.41	136		91,500	6,900	2,175	100,575	113,500
3300	Single pole switches									
3350	Disconnecting switches, 13 to 26 kV	R-11	28	2	Ea.	11,900	101	31.50	12,032.50	13,300
3360	46 kV		8	7		21,000	355	111	21,466	23,800
3370	69 kV		5.60	10		23,300	505	159	23,964	26,600
3380	161 kV		2.80	20		89,000	1,000	315	90,315	99,500
3390	500 kV		.22	254		253,000	12,900	4,025	269,925	302,500
3450	Grounding switches, 46 kV		5.60	10		32,100	505	159	32,764	36,200
3460	69 kV		3.73	15.013		33,300	760	238	34,298	38,100
3470	161 kV		2.24	25		35,300	1,275	395	36,970	41,100
3480	500 kV		.62	90.323		43,800	4,575	1,425	49,800	56,500

33 79 Site Grounding

33 79 83 – Site Grounding Conductors

33 79 83.13 Grounding Wire, Bar, and Rod		Crew	Daily Output	Labor-Hours	Unit	2014 Bare Costs Material	Labor	Equipment	Total	Total Incl O&P
0010	**GROUNDING WIRE, BAR, AND ROD**									
7500	Grounding systems	R-11	280	.200	Lb.	10.80	10.10	3.17	24.07	30.50

33 81 Communications Structures

33 81 13 – Communications Transmission Towers

33 81 13.10 Radio Towers

		Crew	Daily Output	Labor-Hours	Unit	Material	Labor	Equipment	Total	Total Incl O&P
0010	**RADIO TOWERS**									
0020	Guyed, 50′ H, 40 lb. sect., 70 MPH basic wind spd.	2 Sswk	1	16	Ea.	2,750	820		3,570	4,475
0100	Wind load 90 MPH basic wind speed	"	1	16		3,700	820		4,520	5,525
0300	190′ high, 40 lb. section, wind load 70 MPH basic wind speed	K-2	.33	72.727		9,550	3,400	1,000	13,950	17,400
0400	200′ high, 70 lb. section, wind load 90 MPH basic wind speed		.33	72.727		15,900	3,400	1,000	20,300	24,400
0600	300′ high, 70 lb. section, wind load 70 MPH basic wind speed		.20	120		25,500	5,625	1,675	32,800	39,400
0700	270′ high, 90 lb. section, wind load 90 MPH basic wind speed		.20	120		27,600	5,625	1,675	34,900	41,700
0800	400′ high, 100 lb. section, wind load 70 MPH basic wind speed		.14	171		38,100	8,050	2,375	48,525	58,000
0900	Self-supporting, 60′ high, wind load 70 MPH basic wind speed		.80	30		4,500	1,400	415	6,315	7,800
0910	60′ high, wind load 90 MPH basic wind speed		.45	53.333		5,275	2,500	740	8,515	10,900
1000	120′ high, wind load 70 MPH basic wind speed		.40	60		10,000	2,825	835	13,660	16,700
1200	190′ high, wind load 90 MPH basic wind speed	↓	.20	120		28,300	5,625	1,675	35,600	42,500
2000	For states west of Rocky Mountains, add for shipping				↓	10%				

Estimating Tips

34 11 00 Rail Tracks

This subdivision includes items that may involve either repair of existing, or construction of new, railroad tracks. Additional preparation work, such as the roadbed earthwork, would be found in Division 31. Additional new construction siding and turnouts are found in Subdivision 34 72. Maintenance of railroads is found under 34 01 23 Operation and Maintenance of Railways.

34 40 00 Traffic Signals

This subdivision includes traffic signal systems. Other traffic control devices such as traffic signs are found in Subdivision 10 14 53 Traffic Signage.

34 70 00 Vehicle Barriers

This subdivision includes security vehicle barriers, guide and guard rails, crash barriers, and delineators. The actual maintenance and construction of concrete and asphalt pavement is found in Division 32.

Reference Numbers

Reference numbers are shown in shaded boxes at the beginning of some major classifications. These numbers refer to related items in the Reference Section. The reference information may be an estimating procedure, an alternate pricing method, or technical information.

Note: Not all subdivisions listed here necessarily appear in this publication.

34 01 13.10 Maintenance Grading of Roadways		Crew	Daily Output	Labor-Hours	Unit	Material	2014 Bare Costs Labor	Equipment	Total	Total Incl O&P
0010	**MAINTENANCE GRADING OF ROADWAYS**									
0100	Maintenance grading mob/demob of equipment per hour	B-11L	8	2	Hr.		85.50	88.50	174	228
0200	Maintenance grading of ditches one ditch slope only 1.0 MPH average		5	3.200	Mile		137	141	278	365
0210	1.5 MPH		7.50	2.133			91.50	94.50	186	243
0220	2.0 MPH		10	1.600			68.50	70.50	139	183
0230	2.5 MPH		12.50	1.280			55	56.50	111.50	146
0240	3.0 MPH		15	1.067			45.50	47	92.50	122
0300	Maintenance grading of roadway, 3 passes, 3.0 MPH average		5	3.200			137	141	278	365
0310	4 passes		3.80	4.211			180	186	366	480
0320	3 passes, 3.5 MPH average		5.80	2.759			118	122	240	315
0330	4 passes		4.40	3.636			156	161	317	415
0340	3 passes, 4 MPH average		6.70	2.388			102	106	208	272
0350	4 passes		5	3.200			137	141	278	365
0360	3 passes, 4.5 MPH average		7.50	2.133			91.50	94.50	186	243
0370	4 passes		5.60	2.857			122	126	248	325
0380	3 passes, 5 MPH average		8.30	1.928			82.50	85	167.50	220
0390	4 passes		6.30	2.540			109	112	221	289
0400	3 passes, 5.5 MPH average		9.20	1.739			74.50	77	151.50	199
0410	4 passes		6.90	2.319			99	102	201	265
0420	3 passes, 6 MPH average		10	1.600			68.50	70.50	139	183
0430	4 passes		7.50	2.133			91.50	94.50	186	243
0440	3 passes, 6.5 MPH average		10.80	1.481			63.50	65.50	129	169
0450	4 passes		8.10	1.975			84.50	87.50	172	225
0460	3 passes, 7 MPH average		11.70	1.368			58.50	60.50	119	156
0470	4 passes		8.80	1.818			78	80.50	158.50	208
0480	3 passes, 7.5 MPH average		12.50	1.280			55	56.50	111.50	146
0490	4 passes		9.40	1.702			73	75	148	194
0500	3 passes, 8 MPH average		13.30	1.203			51.50	53	104.50	137
0510	4 passes		10	1.600			68.50	70.50	139	183
0520	3 passes, 8.5 MPH average		14.20	1.127			48	50	98	129
0530	4 passes		10.60	1.509			64.50	66.50	131	172
0540	3 passes, 9 MPH average		15	1.067			45.50	47	92.50	122
0550	4 passes		11.30	1.416			60.50	62.50	123	162
0560	3 passes, 9.5 MPH average		15.80	1.013			43.50	45	88.50	115
0570	4 passes		11.90	1.345			57.50	59.50	117	154
0600	Maintenance grading of roadway, 8′ wide pass, 3.0 MPH average		15	1.067			45.50	47	92.50	122
0610	3.5 MPH average		17.50	.914			39	40.50	79.50	105
0620	4 MPH average		20	.800			34	35.50	69.50	91.50
0630	4.5 MPH average		22.50	.711			30.50	31.50	62	81
0640	5 MPH average		25	.640			27.50	28.50	56	73
0650	5.5 MPH average		27.50	.582			25	25.50	50.50	66.50
0660	6 MPH average		30	.533			23	23.50	46.50	61
0670	6.5 MPH average		32.50	.492			21	22	43	56
0680	7 MPH average		35	.457			19.55	20	39.55	52
0690	7.5 MPH average		37.50	.427			18.25	18.85	37.10	48.50
0700	8 MPH average		40	.400			17.10	17.70	34.80	45.50
0710	8.5 MPH average		42.50	.376			16.10	16.65	32.75	43
0720	9 MPH average		45	.356			15.20	15.70	30.90	40.50
0730	9.5 MPH average		47.50	.337			14.40	14.90	29.30	38.50

34 01 Operation and Maintenance of Transportation

34 01 23 – Operation and Maintenance of Railways

34 01 23.51 Maintenance of Railroads		Crew	Daily Output	Labor-Hours	Unit	Material	2014 Bare Costs Labor	Equipment	Total	Total Incl O&P
0010	MAINTENANCE OF RAILROADS									
0400	Resurface and realign existing track	B-14	200	.240	L.F.		9.30	1.83	11.13	16.30
0600	For crushed stone ballast, add	"	500	.096	"	11.40	3.72	.73	15.85	19.05

34 11 Rail Tracks

34 11 13 – Track Rails

34 11 13.23 Heavy Rail Track

		Crew	Daily Output	Labor-Hours	Unit	Material	Labor	Equipment	Total	Total Incl O&P
0010	HEAVY RAIL TRACK									
1000	Rail, 100 lb. prime grade				L.F.	35			35	38
1500	Relay rail				"	17.40			17.40	19.10

34 11 33 – Track Cross Ties

34 11 33.13 Concrete Track Cross Ties

		Crew	Daily Output	Labor-Hours	Unit	Material	Labor	Equipment	Total	Total Incl O&P
0010	CONCRETE TRACK CROSS TIES									
1400	Ties, concrete, 8'-6" long, 30" O.C.	B-14	80	.600	Ea.	172	23	4.57	199.57	230

34 11 33.16 Timber Track Cross Ties

		Crew	Daily Output	Labor-Hours	Unit	Material	Labor	Equipment	Total	Total Incl O&P
0010	TIMBER TRACK CROSS TIES									
1600	Wood, pressure treated, 6" x 8" x 8'-6", C.L. lots	B-14	90	.533	Ea.	51.50	20.50	4.06	76.06	93.50
1700	L.C.L. lots		90	.533		54	20.50	4.06	78.56	96
1900	Heavy duty, 7" x 9" x 8'-6", C.L. lots		70	.686		57	26.50	5.20	88.70	109
2000	L.C.L. lots		70	.686		57	26.50	5.20	88.70	109

34 11 33.17 Timber Switch Ties

		Crew	Daily Output	Labor-Hours	Unit	Material	Labor	Equipment	Total	Total Incl O&P
0010	TIMBER SWITCH TIES									
1200	Switch timber, for a #8 switch, pressure treated	B-14	3.70	12.973	M.B.F.	3,100	500	98.50	3,698.50	4,275
1300	Complete set of timbers, 3.7 MBF for #8 switch	"	1	48	Total	11,900	1,850	365	14,115	16,400

34 11 93 – Track Appurtenances and Accessories

34 11 93.50 Track Accessories

		Crew	Daily Output	Labor-Hours	Unit	Material	Labor	Equipment	Total	Total Incl O&P
0010	TRACK ACCESSORIES									
0020	Car bumpers, test	B-14	2	24	Ea.	3,875	930	183	4,988	5,900
0100	Heavy duty		2	24		7,350	930	183	8,463	9,700
0200	Derails hand throw (sliding)		10	4.800		1,225	186	36.50	1,447.50	1,675
0300	Hand throw with standard timbers, open stand & target		8	6		1,325	232	45.50	1,602.50	1,850
2400	Wheel stops, fixed		18	2.667	Pr.	955	103	20.50	1,078.50	1,225
2450	Hinged		14	3.429	"	1,300	133	26	1,459	1,650

34 11 93.60 Track Material

		Crew	Daily Output	Labor-Hours	Unit	Material	Labor	Equipment	Total	Total Incl O&P
0010	TRACK MATERIAL									
0020	Track bolts				Ea.	3.98			3.98	4.38
0100	Joint bars				Pr.	92.50			92.50	101
0200	Spikes				Ea.	1.31			1.31	1.44
0300	Tie plates				"	15.45			15.45	16.95

34 41 Roadway Signaling and Control Equipment

34 41 13 – Traffic Signals

34 41 13.10 Traffic Signals Systems		Crew	Daily Output	Labor-Hours	Unit	Material	2014 Bare Costs Labor	Equipment	Total	Total Incl O&P
0010	**TRAFFIC SIGNALS SYSTEMS**									
0020	Mid block pedestrian crosswalk with pushbutton and mast arms	R-11	.30	186	Total	68,000	9,450	2,950	80,400	92,500
0600	Traffic signals, school flashing system, solar powered, remote controlled	"	1	56	Signal	12,000	2,825	890	15,715	18,400
1000	Intersection traffic signals, LED, mast, programmable, no lane control									
1010	Includes all labor, material and equip. for complete installation	R-11	1	56	Signal	158,000	2,825	890	161,715	179,000
1100	Intersection traffic signals, LED, mast, programmable, R/L lane control									
1110	Includes all labor, material and equip. for complete installation	R-11	1	56	Signal	216,000	2,825	890	219,715	242,500
1200	Add protective/permissive left turns to existing traffic light									
1210	Includes all labor, material and equip. for complete installation	R-11	1	56	Signal	66,000	2,825	890	69,715	77,500
1300	Replace existing light heads with LED heads wire hung									
1310	Includes all labor, material and equip. for complete installation	R-11	1	56	Signal	41,000	2,825	890	44,715	50,500
1400	Replace existing light heads with LED heads mast arm hung									
1410	Includes all labor, material and equip. for complete installation	R-11	1	56	Signal	77,000	2,825	890	80,715	89,500

34 43 Airfield Signaling and Control Equipment

34 43 13 – Airfield Signals

34 43 13.10 Airport Lighting		Crew	Daily Output	Labor-Hours	Unit	Material	Labor	Equipment	Total	Total Incl O&P
0010	**AIRPORT LIGHTING**									
0100	Runway centerline, bidir., semi-flush, 200 W, w/shallow insert base	R-22	12.40	3.006	Ea.	1,625	136		1,761	1,975
0110	for mounting in base housing		18.60	2.004		815	91		906	1,050
0120	Flush, 200 W, w/shallow insert base		12.40	3.006		1,600	136		1,736	1,975
0130	for mounting in base housing		18.64	2		855	91		946	1,075
0140	45 W, for mounting in base housing		18.64	2		850	91		941	1,075
0150	Touchdown zone light, unidirectional, 200 W, w/shallow insert base		12.40	3.006		1,350	136		1,486	1,675
0160	115 W		12.40	3.006		1,300	136		1,436	1,625
0170	Bi-directional, 62 W		12.40	3.006		1,350	136		1,486	1,700
0180	Unidirectional, 200 W, for mounting in base housing		18.60	2.004		600	91		691	800
0190	115 W		18.60	2.004		645	91		736	850
0200	Bi-directional, 62 W		18.60	2.004		720	91		811	930
0210	Runway edge & threshold light, bidir., 200 W, for base housing		9.36	3.983		920	181		1,101	1,300
0220	300 W		9.36	3.983		1,300	181		1,481	1,700
0230	499 W		7.44	5.011		2,150	227		2,377	2,725
0240	Threshold & approach light, unidir., 200 W, for base housing		9.36	3.983		535	181		716	865
0250	499 W		7.44	5.011		650	227		877	1,050
0260	Runway edge, bi-directional, 2-115 W, for base housing		12.40	3.006		1,425	136		1,561	1,750
0270	2-185 W		12.40	3.006		1,575	136		1,711	1,925
0280	Runway threshold & end, bidir., 2-115 W, for base housing		12.40	3.006		1,550	136		1,686	1,925
0370	45 W, flush, for mounting in base housing		18.64	2		765	91		856	975
0380	115 W		18.64	2		780	91		871	995
0400	Transformer, 45 W		37.20	1.002		167	45.50		212.50	253
0410	65 W		37.20	1.002		172	45.50		217.50	258
0420	100 W		37.20	1.002		213	45.50		258.50	305
0430	200 W, 6.6/6.6 A		37.20	1.002		209	45.50		254.50	298
0440	20/6.6 A		24.80	1.503		223	68.50		291.50	350
0450	500 W		24.80	1.503		292	68.50		360.50	425
0460	Base housing, 24" deep, 12" diameter		18.64	2		470	91		561	655
0470	15" diameter		18.64	2		1,025	91		1,116	1,250
0480	Connection kit for 1 conductor cable		24.80	1.503		33.50	68.50		102	140
0500	Elevated runway marker, high intensity quartz, edge, 115 W		14.88	2.505		330	114		444	530
0510	175 W		14.88	2.505		340	114		454	540
0520	Threshold, 115 W		14.88	2.505		365	114		479	575

34 43 Airfield Signaling and Control Equipment

34 43 13 – Airfield Signals

34 43 13.10 Airport Lighting		Crew	Daily Output	Labor-Hours	Unit	Material	2014 Bare Costs Labor	Equipment	Total	Total Incl O&P
0530	Edge, 200 W, w/base and transformer	R-22	11.85	3.146	Ea.	1,050	143		1,193	1,375
0540	Elevated runway light, quartz, runway edge, 45 W		14.88	2.505		222	114		336	415
0550	Threshold, 115 W		14.88	2.505		365	114		479	575
0560	Taxiway, 30 W		14.88	2.505		203	114		317	395
0570	Fixture, 200 W, w/base and transformer		11.85	3.146		1,050	143		1,193	1,375
0580	Elevated threshold marker, quartz, 115 W		14.88	2.505		465	114		579	680
0590	175 W		14.88	2.505		465	114		579	680

34 43 23 – Weather Observation Equipment

34 43 23.16 Airfield Wind Cones		Crew	Daily Output	Labor-Hours	Unit	Material	Labor	Equipment	Total	Total Incl O&P
0010	**AIRFIELD WIND CONES**									
1200	Wind cone, 12′ lighted assembly, rigid, w/obstruction light	R-21	1.36	24.118	Ea.	9,025	1,275	44.50	10,344.50	11,900
1210	Without obstruction light		1.52	21.579		7,575	1,150	40	8,765	10,100
1220	Unlighted assembly, w/obstruction light		1.68	19.524		8,250	1,050	36	9,336	10,700
1230	Without obstruction light		1.84	17.826		7,575	950	33	8,558	9,775
1240	Wind cone slip fitter, 2-1/2″ pipe		21.84	1.502		100	80	2.78	182.78	233
1250	Wind cone sock, 12′ x 3′, cotton		6.56	5		545	267	9.25	821.25	1,000
1260	Nylon		6.56	5		555	267	9.25	831.25	1,025

34 71 Roadway Construction

34 71 13 – Vehicle Barriers

34 71 13.26 Vehicle Guide Rails		Crew	Daily Output	Labor-Hours	Unit	Material	Labor	Equipment	Total	Total Incl O&P
0010	**VEHICLE GUIDE RAILS**									
0012	Corrugated stl., galv. stl. posts, 6′-3″ O.C.	B-80	850	.038	L.F.	21.50	1.50	.89	23.89	27.50
0100	Double face		570	.056	″	32.50	2.23	1.33	36.06	41
0200	End sections, galvanized, flared		50	.640	Ea.	94	25.50	15.15	134.65	160
0300	Wrap around end		50	.640		137	25.50	15.15	177.65	207
0350	Anchorage units		15	2.133		1,175	84.50	50.50	1,310	1,475
0365	End section, flared	B-80A	78	.308		44	11.30	4.27	59.57	70.50
0370	End section, wrap-around, corrugated steel	″	78	.308		54	11.30	4.27	69.57	81
0400	Timber guide rail, 4″ x 8″ with 6″ x 8″ wood posts, treated	B-80	960	.033	L.F.	11.80	1.32	.79	13.91	15.90
0600	Cable guide rail, 3 at 3/4″ cables, steel posts, single face		900	.036		10.60	1.41	.84	12.85	14.75
0650	Double face		635	.050		21.50	2	1.19	24.69	28
0700	Wood posts		950	.034		12.70	1.34	.80	14.84	16.90
0750	Double face		650	.049		22	1.95	1.17	25.12	29
0760	Breakaway wood posts		195	.164	Ea.	345	6.50	3.89	355.39	395
0800	Anchorage units, breakaway		15	2.133	″	1,525	84.50	50.50	1,660	1,850
0900	Guide rail, steel box beam, 6″ x 6″		120	.267	L.F.	33	10.60	6.30	49.90	59.50
0950	End assembly	B-80A	48	.500	Ea.	335	18.35	6.95	360.30	405
1100	Median barrier, steel box beam, 6″ x 8″	B-80	215	.149	L.F.	44	5.90	3.53	53.43	61.50
1120	Shop curved	B-80A	92	.261	″	56	9.55	3.62	69.17	80.50
1140	End assembly		48	.500	Ea.	510	18.35	6.95	535.30	595
1150	Corrugated beam		400	.060	L.F.	42	2.20	.83	45.03	50.50
1400	Resilient guide fence and light shield, 6′ high	B-2	130	.308	″	37.50	11.40		48.90	58.50
1500	Concrete posts, individual, 6′-5″, triangular	B-80	110	.291	Ea.	71	11.55	6.90	89.45	104
1550	Square		110	.291		76.50	11.55	6.90	94.95	110
1600	Wood guide posts		150	.213		50	8.45	5.05	63.50	73.50
2000	Median, precast concrete, 3′-6″ high, 2′ wide, single face	B-29	380	.147	L.F.	53	5.85	2.32	61.17	70
2200	Double face	″	340	.165		61	6.55	2.59	70.14	80
2300	Cast in place, steel forms	C-2	170	.282		109	12.60		121.60	139
2320	Slipformed	C-7	352	.205		45	8.15	3.48	56.63	66

34 71 Roadway Construction

34 71 13 – Vehicle Barriers

34 71 13.26 Vehicle Guide Rails		Crew	Daily Output	Labor-Hours	Unit	2014 Bare Costs Material	Labor	Equipment	Total	Total Incl O&P
2400	Speed bumps, thermoplastic, 10-1/2" x 2-1/4" x 48" long	B-2	120	.333	Ea.	135	12.35		147.35	168
3000	Energy absorbing terminal, 10 bay, 3' wide	B-80B	.10	320		42,300	12,600	2,475	57,375	68,500
3010	7 bay, 2' - 6" wide		.20	160		31,600	6,275	1,250	39,125	45,800
3020	5 bay, 2' wide		.20	160		24,200	6,275	1,250	31,725	37,600
3030	Impact barrier, UTMCD, barrel type	B-16	30	1.067		500	40	23	563	635
3100	Wide hazard protection, foam sandwich, 7 bay, 7' - 6" wide	B-80B	.14	228		31,600	8,975	1,775	42,350	50,500
3110	5' wide		.15	213		26,300	8,375	1,650	36,325	43,600
3120	3' wide		.18	177		24,200	6,975	1,375	32,550	38,800
5000	For bridge railing, see Section 32 34 10.10									

34 71 19 – Vehicle Delineators

34 71 19.13 Fixed Vehicle Delineators

		Crew	Daily Output	Labor-Hours	Unit	Material	Labor	Equipment	Total	Total Incl O&P
0010	**FIXED VEHICLE DELINEATORS**									
0020	Crash barriers									
0100	Traffic channelizing pavement markers, layout only	A-7	2000	.012	Ea.		.61	.04	.65	.97
0110	13" x 7-1/2" x 2-1/2" high, non-plowable install	2 Clab	96	.167		24	6.10		30.10	35.50
0200	8" x 8" x 3-1/4" high, non-plowable, install		96	.167		22.50	6.10		28.60	34
0230	4" x 4" x 3/4" high, non-plowable, install		120	.133		1.75	4.89		6.64	9.50
0240	9-1/4" x 5-7/8" x 1/4" high, plowable, concrete pavmt.	A-2A	70	.343		17.85	12.55	6.35	36.75	46
0250	9-1/4" x 5-7/8" x 1/4" high, plowable, asphalt pav't	"	120	.200		3.75	7.30	3.71	14.76	19.45
0300	Barrier and curb delineators, reflectorized, 2" x 4"	2 Clab	150	.107		2	3.91		5.91	8.25
0310	3" x 5"	"	150	.107		3.90	3.91		7.81	10.35
0500	Rumble strip, polycarbonate									
0510	24" x 3-1/2" x 1/2" high	2 Clab	50	.320	Ea.	8.25	11.75		20	27

34 72 Railway Construction

34 72 16 – Railway Siding

34 72 16.50 Railroad Sidings

		Crew	Daily Output	Labor-Hours	Unit	Material	Labor	Equipment	Total	Total Incl O&P
0010	**RAILROAD SIDINGS**									
0800	Siding, yard spur, level grade									
0808	Wood ties and ballast, 80 lb. new rail	B-14	57	.842	L.F.	112	32.50	6.40	150.90	181
0809	80 lb. relay rail		57	.842		79	32.50	6.40	117.90	144
0812	90 lb. new rail		57	.842		120	32.50	6.40	158.90	189
0813	90 lb. relay rail		57	.842		81	32.50	6.40	119.90	146
0820	100 lb. new rail		57	.842		129	32.50	6.40	167.90	198
0822	100 lb. relay rail		57	.842		85	32.50	6.40	123.90	151
0830	110 lb. new rail		57	.842		138	32.50	6.40	176.90	209
0832	110 lb. relay rail		57	.842		90.50	32.50	6.40	129.40	157
1002	Steel ties in concrete, incl. fasteners & plates									
1003	80 lb. new rail	B-14	22	2.182	L.F.	165	84.50	16.60	266.10	330
1005	80 lb. relay rail		22	2.182		136	84.50	16.60	237.10	298
1012	90 lb. new rail		22	2.182		172	84.50	16.60	273.10	335
1015	90 lb. relay rail		22	2.182		139	84.50	16.60	240.10	300
1020	100 lb. new rail		22	2.182		179	84.50	16.60	280.10	345
1025	100 lb. relay rail		22	2.182		143	84.50	16.60	244.10	305
1030	110 lb. new rail		22	2.182		188	84.50	16.60	289.10	355
1035	110 lb. relay rail		22	2.182		147	84.50	16.60	248.10	310

34 72 Railway Construction

34 72 16 – Railway Siding

34 72 16.60 Railroad Turnouts		Crew	Daily Output	Labor-Hours	Unit	2014 Bare Costs Material	Labor	Equipment	Total	Total Incl O&P
0010	**RAILROAD TURNOUTS**									
2200	Turnout, #8 complete, w/rails, plates, bars, frog, switch point,									
2250	timbers, and ballast to 6" below bottom of ties									
2280	90 lb. rails	B-13	.25	224	Ea.	39,900	8,925	2,975	51,800	61,000
2290	90 lb. relay rails		.25	224		26,800	8,925	2,975	38,700	46,500
2300	100 lb. rails		.25	224		44,500	8,925	2,975	56,400	66,000
2310	100 lb. relay rails		.25	224		29,900	8,925	2,975	41,800	49,900
2320	110 lb. rails		.25	224		49,000	8,925	2,975	60,900	71,000
2330	110 lb. relay rails		.25	224		33,000	8,925	2,975	44,900	53,500
2340	115 lb. rails		.25	224		53,500	8,925	2,975	65,400	76,000
2350	115 lb. relay rails		.25	224		35,800	8,925	2,975	47,700	56,500
2360	132 lb. rails		.25	224		61,500	8,925	2,975	73,400	84,500
2370	132 lb. relay rails	▼	.25	224	▼	40,700	8,925	2,975	52,600	61,500

34 77 Transportation Equipment

34 77 13 – Passenger Boarding Bridges

34 77 13.16 Movable Aircraft Passenger Boarding Bridges

		Crew	Daily Output	Labor-Hours	Unit	Material	Labor	Equipment	Total	Total Incl O&P
0010	**MOVABLE AIRCRAFT PASSENGER BOARDING BRIDGES**									
0100	Aircraft movable passenger bridge	L-5B	.05	1440	Ea.	1,086,000	75,500	33,000	1,194,500	1,351,500

Division Notes

	CREW	DAILY OUTPUT	LABOR-HOURS	UNIT	BARE COSTS MAT.	BARE COSTS LABOR	BARE COSTS EQUIP.	BARE COSTS TOTAL	TOTAL INCL O&P

Estimating Tips

35 01 50 Operation and Maintenance of Marine Construction

Includes unit price lines for pile cleaning and pile wrapping for protection.

35 20 16 Hydraulic Gates

This subdivision includes various types of gates that are commonly used in waterway and canal construction. Various earthwork items and structural support is found in Division 31, and concrete work in Division 3.

35 20 23 Dredging

This subdivision includes barge and shore dredging systems for rivers, canals, and channels.

35 31 00 Shoreline Protection

This subdivision includes breakwaters, bulkheads, and revetments for ocean and river inlets. Additional earthwork may be required from Division 31, and concrete work from Division 3.

35 41 00 Levees

Information on levee construction, including estimated cost of clay cone material.

35 49 00 Waterway Structures

This subdivision includes breakwaters and bulkheads for canals.

35 51 00 Floating Construction

This section includes floating piers, docks, and dock accessories. Fixed Pier Timber Construction is found in 06 13 33. Driven piles are found in Division 31, as well as sheet piling, cofferdams, and riprap.

Reference Numbers

Reference numbers are shown in shaded boxes at the beginning of some major classifications. These numbers refer to related items in the Reference Section. The reference information may be an estimating procedure, an alternate pricing method, or technical information.

Note: Not all subdivisions listed here necessarily appear in this publication.

35 01 50 – Operation and Maintenance of Marine Construction

35 01 50.10 Cleaning of Marine Pier Piles		Crew	Daily Output	Labor-Hours	Unit	Material	2014 Bare Costs Labor	Equipment	Total	Total Incl O&P
0010	**CLEANING OF MARINE PIER PILES**									
0015	Exposed piles pressure washing from boat									
0100	Clean wood piles from boat, 8″ diam., 5′ length G	B-1D	17	.941	Ea.		34.50	16.20	50.70	71
0110	6-10′ long G		9	1.778			65	30.50	95.50	135
0120	11-15′ long G		6.50	2.462			90	42.50	132.50	186
0130	10″ diam., 5′ length G		13.75	1.164			42.50	20	62.50	88
0140	6-10′ long G		7.25	2.207			81	38	119	167
0150	11-15′ long G		5	3.200			117	55	172	242
0160	12″ diam., 5′ length G		11.50	1.391			51	24	75	105
0170	6-10′ long G		6	2.667			97.50	46	143.50	202
0180	11-15′ long G		4.25	3.765			138	65	203	284
0190	13″ diam., 5′ length G		10.50	1.524			56	26	82	115
0200	6-10′ long G		5.50	2.909			107	50	157	220
0210	11-15′ long G		4	4			147	69	216	300
0220	14″ diam., 5′ length G		9.75	1.641			60	28	88	124
0230	6-10′ long G		5.25	3.048			112	52.50	164.50	230
0240	11-15′ long G		3.75	4.267			156	73.50	229.50	320
0300	Clean concrete piles from boat, 12″ diam., 5′ length G		14	1.143			42	19.65	61.65	86
0310	6-10′ long G		7.25	2.207			81	38	119	167
0320	11-15′ long G		5	3.200			117	55	172	242
0330	14″ diam., 5′ length G		12	1.333			49	23	72	101
0340	6-10′ long G		6.25	2.560			94	44	138	194
0350	11-15′ long G		4.50	3.556			130	61	191	269
0360	16″ diam., 5′ length G		10.50	1.524			56	26	82	115
0370	6-10′ long G		5.50	2.909			107	50	157	220
0380	11-15′ long G		4	4			147	69	216	300
0390	18″ diam., 5′ length G		9.25	1.730			63.50	30	93.50	131
0400	6-10′ long G		5	3.200			117	55	172	242
0410	11-15′ long G		3.50	4.571			168	78.50	246.50	345
0420	20″ diam., 5′ length G		8.50	1.882			69	32.50	101.50	142
0430	6-10′ long G		4.50	3.556			130	61	191	269
0440	11-15′ long G		3	5.333			195	91.50	286.50	400
0450	24″ diam., 5′ length G		7	2.286			84	39.50	123.50	173
0460	6-10′ long G		3.75	4.267			156	73.50	229.50	320
0470	11-15′ long G		2.50	6.400			235	110	345	480
0480	36″ diam., 5′ length G		4.75	3.368			123	58	181	254
0490	6-10′ long G		2.50	6.400			235	110	345	480
0500	11-15′ long G		1.75	9.143			335	157	492	690
0510	54″ diam., 5′ length G		3	5.333			195	91.50	286.50	400
0520	6-10′ long G		1.75	9.143			335	157	492	690
0530	11-15′ long G		1.25	12.800			470	220	690	965
0540	66″ diam., 5′ length G		2.50	6.400			235	110	345	480
0550	6-10′ long G		1.50	10.667			390	183	573	805
0560	11-15′ long G		1	16			585	275	860	1,200
0600	10″ square pile, 5′ length G		13.25	1.208			44.50	21	65.50	91.50
0610	6-10′ long G		7	2.286			84	39.50	123.50	173
0620	11-15′ long G		4.75	3.368			123	58	181	254
0630	12″ square pile, 5′ length G		11	1.455			53.50	25	78.50	110
0640	6-10′ long G		5.75	2.783			102	48	150	210
0650	11-15′ long G		4	4			147	69	216	300
0660	14″ square pile, 5′ length G		9.50	1.684			61.50	29	90.50	127
0670	6-10′ long G		5	3.200			117	55	172	242
0680	11-15′ long G	▼	3.50	4.571	▼		168	78.50	246.50	345

35 01 50 – Operation and Maintenance of Marine Construction

35 01 50.10 Cleaning of Marine Pier Piles			Crew	Daily Output	Labor-Hours	Unit	Material	2014 Bare Costs Labor	Equipment	Total	Total Incl O&P
0690	16″ square pile, 5′ length	G	B-1D	8.25	1.939	Ea.		71	33.50	104.50	147
0700	6-10′ long	G		4.25	3.765			138	65	203	284
0710	11-15′ long	G		3	5.333			195	91.50	286.50	400
0720	18″ square pile, 5′ length	G		7.25	2.207			81	38	119	167
0730	6-10′ long	G		3.75	4.267			156	73.50	229.50	320
0740	11-15′ long	G		2.75	5.818			213	100	313	440
0750	20″ square pile, 5′ length	G		6.50	2.462			90	42.50	132.50	186
0760	6-10′ long	G		3.50	4.571			168	78.50	246.50	345
0770	11-15′ long	G		2.50	6.400			235	110	345	480
0780	24″ square pile, 5′ length	G		5.50	2.909			107	50	157	220
0790	6-10′ long	G		3	5.333			195	91.50	286.50	400
0800	11-15′ long	G		2	8			293	138	431	600
0810	14″ diameter Octagonal piles, 5′ length	G		12.50	1.280			47	22	69	96.50
0820	6-10′ long	G		6.50	2.462			90	42.50	132.50	186
0830	11-15′ long	G		4.50	3.556			130	61	191	269
0840	16″ diameter Octagonal piles, 5′ length	G		11	1.455			53.50	25	78.50	110
0850	6-10′ long	G		5.75	2.783			102	48	150	210
0860	11-15′ long	G		4	4			147	69	216	300
0870	18″ diameter Octagonal piles, 5′ length	G		9.75	1.641			60	28	88	124
0880	6-10′ long	G		5	3.200			117	55	172	242
0890	11-15′ long	G		3.50	4.571			168	78.50	246.50	345
0900	20″ diameter Octagonal piles, 5′ length	G		8.75	1.829			67	31.50	98.50	138
0910	6-10′ long	G		4.50	3.556			130	61	191	269
0920	11-15′ long	G		3.25	4.923			180	84.50	264.50	370
0930	24″ diameter Octagonal piles, 5′ length	G		7.25	2.207			81	38	119	167
0940	6-10′ long	G		3.75	4.267			156	73.50	229.50	320
0950	11-15′ long	G		2.75	5.818			213	100	313	440
1000	Clean steel piles from boat, 8″ diam., 5′ length	G		14.50	1.103			40.50	19	59.50	83.50
1010	6-10′ long	G		7.75	2.065			75.50	35.50	111	156
1020	11-15′ long	G		5.50	2.909			107	50	157	220
1030	10″ diam., 5′ length	G		11.50	1.391			51	24	75	105
1040	6-10′ long	G		6	2.667			97.50	46	143.50	202
1050	11-15′ long	G		4.50	3.556			130	61	191	269
1060	12″ diam., 5′ length	G		9.50	1.684			61.50	29	90.50	127
1070	6-10′ long	G		5	3.200			117	55	172	242
1080	11-15′ long	G		3.75	4.267			156	73.50	229.50	320
1100	HP 8 x 36, 5′ length	G		7.50	2.133			78	36.50	114.50	162
1110	6-10′ long	G		4	4			147	69	216	300
1120	11-15′ long	G		3	5.333			195	91.50	286.50	400
1130	HP 10 x 42, 5′ length	G		6.25	2.560			94	44	138	194
1140	6-10′ long	G		3.25	4.923			180	84.50	264.50	370
1150	11-15′ long	G		2.30	6.957			255	120	375	525
1160	HP 10 x 57, 5′ length	G		6	2.667			97.50	46	143.50	202
1170	6-10′ long	G		3.25	4.923			180	84.50	264.50	370
1180	11-15′ long	G		2.30	6.957			255	120	375	525
1200	HP 12 x 53, 5′ length	G		5	3.200			117	55	172	242
1210	6-10′ long	G		2.75	5.818			213	100	313	440
1220	11-15′ long	G		2	8			293	138	431	600
1230	HP 12 x 74, 5′ length	G		5	3.200			117	55	172	242
1240	6-10′ long	G		2.75	5.818			213	100	313	440
1250	11-15′ long	G		1.90	8.421			310	145	455	635
1300	HP 14 x 73, 5′ length	G		4.25	3.765			138	65	203	284
1310	6-10′ long	G		2.33	6.867			252	118	370	520

35 01 Operation and Maint. of Waterway & Marine Construction

35 01 50 – Operation and Maintenance of Marine Construction

35 01 50.10 Cleaning of Marine Pier Piles			Crew	Daily Output	Labor-Hours	Unit	Material	Labor	Equipment	Total	Total Incl O&P
							2014 Bare Costs				
1320	11-15′ long	G	B-1D	1.60	10	Ea.		365	172	537	755
1330	HP 14 x 89, 5′ length	G		4.25	3.765			138	65	203	284
1340	6-10′ long	G		2.33	6.867			252	118	370	520
1350	11-15′ long	G		1.60	10			365	172	537	755
1360	HP 14 x 102, 5′ length	G		4.25	3.765			138	65	203	284
1370	6-10′ long	G		2.33	6.867			252	118	370	520
1380	11-15′ long	G		1.60	10			365	172	537	755
1385	HP 14 x 117, 5′ length	G		4.25	3.765			138	65	203	284
1390	6-10′ long	G		2.33	6.867			252	118	370	520
1395	11-15′ long	G		1.60	10			365	172	537	755

35 01 50.20 Protective Wrapping of Marine Pier Piles

Line	Description		Crew	Daily Output	Labor-Hours	Unit	Material	Labor	Equipment	Total	Total Incl O&P
0010	**PROTECTIVE WRAPPING OF MARINE PIER PILES**										
0015	Exposed piles wrapped using boat										
0020	Note: piles must be cleaned before wrapping										
0100	Wrap protective material on wood piles with nails 8″ diameter	G	B-1G	162	.099	V.L.F.	18.25	3.62	1.17	23.04	27
0110	10″ diameter	G		130	.123		23	4.51	1.46	28.97	33.50
0120	12″ diameter	G		108	.148		27.50	5.45	1.75	34.70	40.50
0130	13″ diameter	G		99	.162		29.50	5.90	1.91	37.31	44
0140	14″ diameter	G		93	.172		32	6.30	2.04	40.34	47
0150	Wrap protective material on wood piles, straps, 8″ diam.	G		180	.089		28.50	3.26	1.05	32.81	37.50
0160	10″ diameter	G		144	.111		36	4.07	1.31	41.38	47.50
0170	12″ diameter	G		120	.133		43	4.89	1.58	49.47	56.50
0180	13″ diameter	G		110	.145		46.50	5.35	1.72	53.57	61
0190	14″ diameter	G		103	.155		50	5.70	1.84	57.54	66
0200	Wrap protective material on concrete piles, straps, 12″ diam.	G		120	.133		43	4.89	1.58	49.47	56.50
0210	14″ diam.	G		103	.155		50	5.70	1.84	57.54	66
0220	16″ diam.	G		90	.178		57	6.50	2.10	65.60	75.50
0230	18″ diam.	G		80	.200		64.50	7.35	2.37	74.22	84.50
0240	20″ diam.	G		72	.222		71.50	8.15	2.63	82.28	94
0250	24″ diam.	G		60	.267		85.50	9.75	3.15	98.40	113
0260	36″ diam.	G		40	.400		129	14.65	4.73	148.38	169
0270	54″ diam.	G		27	.593		193	21.50	7	221.50	253
0280	66″ diam.	G		22	.727		236	26.50	8.60	271.10	310
0300	Wrap protective material on concrete piles,straps, 10″ square	G		113	.142		45.50	5.20	1.68	52.38	60
0310	12″ square	G		94	.170		54.50	6.25	2.01	62.76	72
0320	14″ square	G		81	.198		63.50	7.25	2.34	73.09	83.50
0330	16″ square	G		71	.225		73	8.25	2.67	83.92	95.50
0340	18″ square	G		63	.254		82	9.30	3	94.30	108
0350	20″ square	G		63	.254		91	9.30	3	103.30	118
0360	24″ square	G		47	.340		109	12.50	4.03	125.53	144
0400	Wrap protective material, concrete piles,straps, 14″ octagon	G		107	.150		48	5.50	1.77	55.27	63.50
0410	16″ octagon	G		94	.170		55	6.25	2.01	63.26	72.50
0420	18″ octagon	G		83	.193		62	7.05	2.28	71.33	81.50
0430	20″ octagon	G		75	.213		68.50	7.80	2.52	78.82	90.50
0440	24″ octagon	G		62	.258		82.50	9.45	3.05	95	108
0500	Wrap protective material, steel piles,straps, HP 8 x 36	G		90	.178		72	6.50	2.10	80.60	91.50
0510	HP 10 x 42	G		72	.222		88.50	8.15	2.63	99.28	113
0520	HP 10 x 57	G		72	.222		90	8.15	2.63	100.78	114
0530	HP 12 x 53	G		60	.267		106	9.75	3.15	118.90	136
0540	HP 12 x 74	G		60	.267		108	9.75	3.15	120.90	138
0550	HP 14 x 73	G		52	.308		126	11.30	3.64	140.94	160
0560	HP 14 x 89	G		52	.308		128	11.30	3.64	142.94	162

35 01 Operation and Maint. of Waterway & Marine Construction

35 01 50 – Operation and Maintenance of Marine Construction

35 01 50.20 Protective Wrapping of Marine Pier Piles			Crew	Daily Output	Labor-Hours	Unit	Material	2014 Bare Costs Labor	Equipment	Total	Total Incl O&P
0570	HP 14 x 102	G	B-1G	52	.308	V.L.F.	128	11.30	3.64	142.94	162
0580	HP 14 x 117	G	↓	52	.308	↓	130	11.30	3.64	144.94	164

35 20 Waterway and Marine Construction and Equipment

35 20 16 – Hydraulic Gates

35 20 16.26 Hydraulic Sluice Gates		Crew	Daily Output	Labor-Hours	Unit	Material	Labor	Equipment	Total	Total Incl O&P
0010	**HYDRAULIC SLUICE GATES**									
0100	Heavy duty, self contained w/crank oper. gate, 18" x 18"	L-5A	1.70	18.824	Ea.	9,200	965	355	10,520	12,100
0110	24" x 24"		1.20	26.667		13,500	1,375	505	15,380	17,700
0120	30" x 30"		1	32		14,100	1,650	605	16,355	19,000
0130	36" x 36"		.90	35.556		17,300	1,825	675	19,800	22,900
0140	42" x 42"		.80	40		19,800	2,050	760	22,610	26,100
0150	48" x 48"		.50	64		21,900	3,300	1,225	26,425	31,000
0160	54" x 54"		.40	80		29,700	4,100	1,525	35,325	41,400
0170	60" x 60"		.30	106		32,500	5,475	2,025	40,000	47,300
0180	66" x 66"		.30	106		38,100	5,475	2,025	45,600	53,500
0190	72" x 72"		.20	160		42,300	8,225	3,025	53,550	64,000
0200	78" x 78"		.20	160		54,000	8,225	3,025	65,250	77,000
0210	84" x 84"	↓	.10	320		59,000	16,400	6,075	81,475	99,500
0220	90" x 90"	E-20	.30	213		70,500	10,700	3,925	85,125	100,000
0230	96" x 96"		.30	213		75,000	10,700	3,925	89,625	105,000
0240	108" x 108"		.20	320		85,500	16,100	5,875	107,475	128,000
0250	120" x 120"		.10	640		94,500	32,200	11,800	138,500	171,500
0260	132" x 132"	↓	.10	640	↓	116,000	32,200	11,800	160,000	195,500

35 20 16.63 Canal Gates		Crew	Daily Output	Labor-Hours	Unit	Material	Labor	Equipment	Total	Total Incl O&P
0010	**CANAL GATES**									
0011	Cast iron body, fabricated frame									
0100	12" diameter	L-5A	4.60	6.957	Ea.	810	360	132	1,302	1,650
0110	18" diameter		4	8		1,525	410	152	2,087	2,575
0120	24" diameter		3.50	9.143		2,825	470	174	3,469	4,100
0130	30" diameter		2.80	11.429		3,225	585	217	4,027	4,800
0140	36" diameter		2.30	13.913		3,825	715	264	4,804	5,750
0150	42" diameter		1.70	18.824		6,575	965	355	7,895	9,275
0160	48" diameter		1.20	26.667		8,325	1,375	505	10,205	12,100
0170	54" diameter		.90	35.556		14,400	1,825	675	16,900	19,700
0180	60" diameter		.50	64		17,100	3,300	1,225	21,625	25,700
0190	66" diameter		.50	64		12,800	3,300	1,225	17,325	21,000
0200	72" diameter	↓	.40	80	↓	23,700	4,100	1,525	29,325	34,700

35 20 16.66 Flap Gates		Crew	Daily Output	Labor-Hours	Unit	Material	Labor	Equipment	Total	Total Incl O&P
0010	**FLAP GATES**									
0100	Aluminum, 18" diameter	L-5A	5	6.400	Ea.	2,100	330	121	2,551	3,000
0110	24" diameter		4	8		2,800	410	152	3,362	3,975
0120	30" diameter		3.50	9.143		3,400	470	174	4,044	4,750
0130	36" diameter		2.80	11.429		4,475	585	217	5,277	6,175
0140	42" diameter		2.30	13.913		6,525	715	264	7,504	8,700
0150	48" diameter		1.70	18.824		7,750	965	355	9,070	10,600
0160	54" diameter		1.20	26.667		9,650	1,375	505	11,530	13,500
0170	60" diameter		.80	40		12,500	2,050	760	15,310	18,000
0180	66" diameter		.50	64		15,700	3,300	1,225	20,225	24,200
0190	72" diameter	↓	.40	80	↓	18,600	4,100	1,525	24,225	29,200

35 20 16 – Hydraulic Gates

35 20 16.69 Knife Gates		Crew	Daily Output	Labor-Hours	Unit	2014 Bare Costs Material	Labor	Equipment	Total	Total Incl O&P
0010	**KNIFE GATES**									
0100	Incl. handwheel operator for hub, 6″ diameter	Q-23	7.70	3.117	Ea.	1,450	172	125	1,747	2,000
0110	8″ diameter		7.20	3.333		1,825	184	133	2,142	2,425
0120	10″ diameter		4.80	5		2,650	277	200	3,127	3,575
0130	12″ diameter		3.60	6.667		3,700	370	267	4,337	4,925
0140	14″ diameter		3.40	7.059		5,125	390	282	5,797	6,525
0150	16″ diameter		3.20	7.500		6,850	415	300	7,565	8,475
0160	18″ diameter		3	8		9,300	445	320	10,065	11,200
0170	20″ diameter		2.70	8.889		12,900	490	355	13,745	15,300
0180	24″ diameter		2.40	10		16,700	555	400	17,655	19,600
0190	30″ diameter		1.80	13.333		34,600	740	535	35,875	39,700
0200	36″ diameter		1.20	20		45,500	1,100	800	47,400	52,500

35 20 16.73 Slide Gates		Crew	Daily Output	Labor-Hours	Unit	Material	Labor	Equipment	Total	Total Incl O&P
0010	**SLIDE GATES**									
0100	Steel, self contained incl. anchor bolts and grout, 12″ x 12″	L-5A	4.60	6.957	Ea.	4,850	360	132	5,342	6,075
0110	18″ x 18″		4	8		5,025	410	152	5,587	6,400
0120	24″ x 24″		3.50	9.143		5,725	470	174	6,369	7,275
0130	30″ x 30″		2.80	11.429		6,125	585	217	6,927	7,975
0140	36″ x 36″		2.30	13.913		6,450	715	264	7,429	8,625
0150	42″ x 42″		1.70	18.824		7,075	965	355	8,395	9,825
0160	48″ x 48″		1.20	26.667		7,575	1,375	505	9,455	11,200
0170	54″ x 54″		.90	35.556		8,225	1,825	675	10,725	12,900
0180	60″ x 60″		.55	58.182		8,450	3,000	1,100	12,550	15,600
0190	72″ x 72″		.36	88.889		10,700	4,575	1,675	16,950	21,400

35 20 23 – Dredging

35 20 23.13 Mechanical Dredging		Crew	Daily Output	Labor-Hours	Unit	Material	Labor	Equipment	Total	Total Incl O&P
0010	**MECHANICAL DREDGING**									
0020	Dredging mobilization and demobilization, add to below, minimum	B-8	.53	120	Total		4,950	6,275	11,225	14,500
0100	Maximum	″	.10	640	″		26,300	33,200	59,500	76,500
0300	Barge mounted clamshell excavation into scows									
0310	Dumped 20 miles at sea, minimum	B-57	310	.155	B.C.Y.		6.50	5.45	11.95	15.95
0400	Maximum	″	213	.225	″		9.50	7.95	17.45	23.50
0500	Barge mounted dragline or clamshell, hopper dumped,									
0510	pumped 1000′ to shore dump, minimum	B-57	340	.141	B.C.Y.		5.95	4.99	10.94	14.60
0525	All pumping uses 2000 gallons of water per cubic yard									
0600	Maximum	B-57	243	.198	B.C.Y.		8.30	6.95	15.25	20.50

35 20 23.23 Hydraulic Dredging		Crew	Daily Output	Labor-Hours	Unit	Material	Labor	Equipment	Total	Total Incl O&P
0010	**HYDRAULIC DREDGING**									
1000	Hydraulic method, pumped 1000′ to shore dump, minimum	B-57	460	.104	B.C.Y.		4.40	3.68	8.08	10.75
1100	Maximum		310	.155			6.50	5.45	11.95	15.95
1400	Into scows dumped 20 miles, minimum		425	.113			4.76	3.99	8.75	11.65
1500	Maximum		243	.198			8.30	6.95	15.25	20.50
1600	For inland rivers and canals in South, deduct								30%	30%

35 31 Shoreline Protection

35 31 16 – Seawalls

35 31 16.13 Concrete Seawalls		Crew	Daily Output	Labor-Hours	Unit	Material	Labor	Equipment	Total	Total Incl O&P
						2014 Bare Costs				
0010	**CONCRETE SEAWALLS**									
0011	Reinforced concrete									
0015	include footing and tie-backs									
0020	Up to 6' high, minimum	C-17C	28	2.964	L.F.	49.50	142	22	213.50	298
0060	Maximum		24.25	3.423		79	164	25.50	268.50	370
0100	12' high, minimum		20	4.150		129	198	31	358	480
0160	Maximum	↓	18.50	4.486	↓	149	214	33.50	396.50	530
0180	Precast bulkhead, complete, including									
0190	vertical and battered piles, face panels, and cap									
0195	Using 16' vertical piles				L.F.				445	515
0196	Using 20' vertical piles				"				475	550

35 31 16.19 Steel Sheet Piling Seawalls		Crew	Daily Output	Labor-Hours	Unit	Material	Labor	Equipment	Total	Total Incl O&P
0010	**STEEL SHEET PILING SEAWALLS**									
0200	Steel sheeting, with 4' x 4' x 8" concrete deadmen, @ 10' O.C.									
0210	12' high, shore driven	B-40	27	2.370	L.F.	97	109	141	347	435
0260	Barge driven	B-76	15	4.800	"	146	220	223	589	750
6000	Crushed stone placed behind bulkhead by clam bucket	B-12H	120	.133	L.C.Y.	17.10	5.80	10.15	33.05	39

35 31 19 – Revetments

35 31 19.18 Revetments, Concrete		Crew	Daily Output	Labor-Hours	Unit	Material	Labor	Equipment	Total	Total Incl O&P
0010	**REVETMENTS, CONCRETE**									
0100	Concrete revetment matt 8' x 20' x 4 1/2" excluding site preparation									
0110	Includes all labor, material and equip. for complete installation				Ea.	3,300			3,300	3,625

35 41 Levees

35 41 13 – Landside Levee Berms

35 41 13.10 Landside Levee Total Clay Cost		Crew	Daily Output	Labor-Hours	Unit	Material	Labor	Equipment	Total	Total Incl O&P
0010	**LANDSIDE LEVEE TOTAL CLAY COST**									
0015	Assumption swell factor of clay 40%									
0400	Low cost clay for levee, 3 :1 slope, 16 feet wide, 8 feet high				M.L.F.	115,500			115,500	127,000
0410	9 feet high					139,500			139,500	153,500
0420	10 feet high					166,000			166,000	182,500
0430	12 feet high					225,000			225,000	247,500
0440	14 feet high					293,000			293,000	322,000
0450	16 feet high					369,500			369,500	406,500
0460	18 feet high					454,500			454,500	500,000
0470	20 feet high					548,500			548,500	603,000
0480	22 feet high					652,000			652,000	717,000
0490	24 feet high				↓	762,000			762,000	838,000
0500	Low cost clay per mile, 8 feet high				Mile	609,500			609,500	670,500
0510	9 feet high					737,000			737,000	811,000
0520	10 feet high					876,000			876,000	964,000
0530	12 feet high					1,188,500			1,188,500	1,307,500
0540	14 feet high					1,546,500			1,546,500	1,701,500
0550	16 feet high					1,950,500			1,950,500	2,145,500
0560	18 feet high					2,400,000			2,400,000	2,640,000
0570	20 feet high					2,895,500			2,895,500	3,185,000
0580	22 feet high					3,436,500			3,436,500	3,780,000
0590	24 feet high				↓	4,023,000			4,023,000	4,425,000
0600	Medium cost clay per M.L.F., 8 feet high				M.L.F.	211,500			211,500	232,500
0610	9 feet high				↓	255,500			255,500	281,000

35 41 Levees

35 41 13 – Landside Levee Berms

	35 41 13.10 Landside Levee Total Clay Cost	Crew	Daily Output	Labor-Hours	Unit	2014 Bare Costs Material	Labor	Equipment	Total	Total Incl O&P
0620	10 feet high				M.L.F.	303,500			303,500	334,000
0630	12 feet high					412,000			412,000	453,000
0640	14 feet high					536,000			536,000	589,500
0650	16 feet high					676,000			676,000	743,500
0660	18 feet high					832,000			832,000	915,000
0670	20 feet high					1,003,500			1,003,500	1,104,000
0680	22 feet high					1,191,000			1,191,000	1,310,000
0690	24 feet high					1,394,500			1,394,500	1,534,000
0700	Medium cost clay per mile, 8 feet high				Mile	1,115,500			1,115,500	1,227,000
0710	9 feet high					1,349,000			1,349,000	1,484,000
0720	10 feet high					1,603,500			1,603,500	1,764,000
0730	12 feet high					2,175,000			2,175,000	2,392,500
0740	14 feet high					2,830,500			2,830,500	3,113,500
0750	16 feet high					3,569,500			3,569,500	3,926,500
0760	18 feet high					4,392,000			4,392,000	4,831,500
0770	20 feet high					5,298,500			5,298,500	5,828,500
0780	22 feet high					6,288,500			6,288,500	6,917,500
0790	24 feet high					7,362,000			7,362,000	8,098,500
0800	High cost clay per M.L.F., 8 feet high				M.L.F.	255,000			255,000	280,500
0810	9 feet high					308,500			308,500	339,500
0820	10 feet high					366,500			366,500	403,500
0830	12 feet high					497,500			497,500	547,000
0840	14 feet high					647,000			647,000	712,000
0850	16 feet high					816,000			816,000	898,000
0860	18 feet high					1,004,500			1,004,500	1,105,000
0870	20 feet high					1,211,500			1,211,500	1,332,500
0880	22 feet high					1,438,000			1,438,000	1,581,500
0890	24 feet high					1,683,500			1,683,500	1,852,000
0900	High cost clay per mile, 8 feet high				Mile	1,347,000			1,347,000	1,481,500
0910	9 feet high					1,629,000			1,629,000	1,791,500
0920	10 feet high					1,936,000			1,936,000	2,129,500
0930	12 feet high					2,626,000			2,626,000	2,889,000
0940	14 feet high					3,417,500			3,417,500	3,759,000
0950	16 feet high					4,309,500			4,309,500	4,740,500
0960	18 feet high					5,303,000			5,303,000	5,833,000
1000	Levee low clay core, 12′ width, 4′ below grade to 2′ below top, 8′ high				M.L.F.	78,500			78,500	86,500
1010	9 feet high levee					86,500			86,500	95,000
1020	10 feet high levee					94,500			94,500	104,000
1030	12 feet high levee					110,000			110,000	121,000
1040	14 feet high levee					126,000			126,000	138,500
1050	16 feet high levee					141,500			141,500	156,000
1060	18 feet high levee					157,500			157,500	173,000
1070	20 feet high levee					173,000			173,000	190,500
1080	22 feet high levee					189,000			189,000	208,000
1090	24 feet high levee					204,500			204,500	225,000
1100	Levee med clay core, 12′ width, 4′ below grade to 2′ below top, 8′ high					144,000			144,000	158,500
1110	9 feet high levee					158,500			158,500	174,500
1120	10 feet high levee					173,000			173,000	190,000
1130	12 feet high levee					201,500			201,500	222,000
1140	14 feet high levee					230,500			230,500	253,500
1150	16 feet high levee					259,500			259,500	285,000
1160	18 feet high levee					288,000			288,000	317,000
1170	20 feet high levee					317,000			317,000	348,500

35 41 13 – Landside Levee Berms

35 41 13.10 Landside Levee Total Clay Cost		Crew	Daily Output	Labor-Hours	Unit	Material	2014 Bare Costs Labor	Equipment	Total	Total Incl O&P
1180	22 feet high levee				M.L.F.	345,500			345,500	380,500
1190	24 feet high levee					374,500			374,500	412,000
1200	High cost clay core, 8′ high levee					174,000			174,000	191,500
1210	9 feet high levee					191,500			191,500	210,500
1220	10 feet high levee					208,500			208,500	229,500
1230	12 feet high levee					243,500			243,500	268,000
1240	14 feet high levee					278,500			278,500	306,000
1250	16 feet high levee					313,000			313,000	344,500
1260	18 feet high levee					348,000			348,000	382,500
1270	20 feet high levee					382,500			382,500	421,000
1280	22 feet high levee					417,500			417,500	459,000
1290	24 feet high levee					452,000			452,000	497,500
2000	Levee low clay core, 12′ width, 5′ below grade to 2′ below top, 8′ high					86,500			86,500	95,000
2010	9 feet high levee					94,500			94,500	104,000
2020	10 feet high levee					102,500			102,500	112,500
2030	12 feet high levee					118,000			118,000	130,000
2040	14 feet high levee					134,000			134,000	147,000
2050	16 feet high levee					149,500			149,500	164,500
2060	18 feet high levee					165,500			165,500	182,000
2070	20 feet high levee					181,000			181,000	199,000
2080	22 feet high levee					197,000			197,000	216,500
2090	24 feet high levee				↓	212,500			212,500	234,000

35 41 13.20 Landside Levees Unit Price Items		Crew	Daily Output	Labor-Hours	Unit	Material	Labor	Equipment	Total	Total Incl O&P
0010	**LANDSIDE LEVEES UNIT PRICE ITEMS**									
0020	Clay material cost delivered 20 miles from site									
0030	Clay backfill material delivered up to 20 miles from site low cost				L.C.Y.	12.65			12.65	13.90
0040	Medium cost					23			23	25.50
0050	High cost				↓	28			28	31
0100	Compaction of levee Clay material									
0200	Open area, 3 passes	B-10G	4380	.003	E.C.Y.		.12	.28	.40	.50
0220	Restricted area		3290	.004			.16	.37	.53	.66
0240	Slope area		2190	.005			.25	.56	.81	.98
0300	Open area, 4 passes		2800	.004			.19	.44	.63	.77
0320	Restricted area		2100	.006			.26	.58	.84	1.03
0340	Slope area		1400	.009			.38	.87	1.25	1.55
0400	Open area, 5 passes		2240	.005			.24	.54	.78	.97
0420	Restricted area		1700	.007			.32	.72	1.04	1.27
0440	Slope area		1120	.011			.48	1.09	1.57	1.93
0500	Open area, 6 passes		1870	.006			.29	.65	.94	1.16
0520	Restricted area		1400	.009			.38	.87	1.25	1.55
0540	Slope area	↓	935	.013	↓		.58	1.30	1.88	2.31
0600	Backfill of levee with loader, Clay material									
0620	Open area	B-10U	3280	.004	L.C.Y.		.16	.32	.48	.60
0640	Limited target area		2460	.005			.22	.43	.65	.80
0660	Restricted and limited target area	↓	1640	.007	↓		.33	.64	.97	1.20

35 49 Waterway Structures

35 49 13 – Floodwalls

35 49 13.30 Breakwaters, Bulkheads, Residential Canal		Crew	Daily Output	Labor-Hours	Unit	Material	Labor	Equipment	Total	Total Incl O&P
						2014 Bare Costs				
0010	**BREAKWATERS, BULKHEADS, RESIDENTIAL CANAL**									
0020	Aluminum panel sheeting, incl. concrete cap and anchor									
0030	Coarse compact sand, 4′-0″ high, 2′-0″ embedment	B-40	200	.320	L.F.	61.50	14.70	19.10	95.30	112
0040	3′-6″ embedment		140	.457		72.50	21	27.50	121	143
0060	6′-0″ embedment		90	.711		92.50	32.50	42.50	167.50	200
0120	6′-0″ high, 2′-6″ embedment		170	.376		78.50	17.30	22.50	118.30	138
0140	4′-0″ embedment		125	.512		92.50	23.50	30.50	146.50	172
0160	5′-6″ embedment		95	.674		122	31	40	193	227
0220	8′-0″ high, 3′-6″ embedment		140	.457		105	21	27.50	153.50	179
0240	5′-0″ embedment		100	.640		105	29.50	38	172.50	204
0420	Medium compact sand, 3′-0″ high, 2′-0″ embedment		235	.272		136	12.55	16.25	164.80	186
0440	4′-0″ embedment		150	.427		167	19.65	25.50	212.15	243
0460	5′-6″ embedment		115	.557		202	25.50	33	260.50	299
0520	5′-0″ high, 3′-6″ embedment		165	.388		188	17.85	23	228.85	261
0540	5′-0″ embedment		120	.533		217	24.50	32	273.50	310
0560	6′-6″ embedment		105	.610		282	28	36.50	346.50	395
0620	7′-0″ high, 4′-6″ embedment		135	.474		245	22	28.50	295.50	335
0640	6′-0″ embedment		110	.582		271	27	34.50	332.50	380
0720	Loose silty sand, 3′-0″ high, 3′-0″ embedment		205	.312		136	14.35	18.65	169	192
0740	4′-6″ embedment		155	.413		164	19	24.50	207.50	237
0760	6′-0″ embedment		125	.512		191	23.50	30.50	245	280
0820	4′-6″ high, 4′-6″ embedment		155	.413		188	19	24.50	231.50	264
0840	6′-0″ embedment		125	.512		218	23.50	30.50	272	310
0860	7′-0″ embedment		115	.557		264	25.50	33	322.50	365
0920	6′-0″ high, 5′-6″ embedment		130	.492		245	22.50	29.50	297	340
0940	7′-0″ embedment	↓	115	.557	↓	271	25.50	33	329.50	375

35 51 Floating Construction

35 51 13 – Floating Piers

35 51 13.23 Floating Wood Piers

		Crew	Daily Output	Labor-Hours	Unit	Material	Labor	Equipment	Total	Total Incl O&P
0010	**FLOATING WOOD PIERS**									
0020	Polyethylene encased polystyrene, no pilings included	F-3	330	.121	S.F.	28.50	5.65	2.01	36.16	42
0030	See Section 06 13 33.50 or Section 06 13 33.52 for fixed docks									
0200	Pile supported, shore constructed, bare, 3″ decking	F-3	130	.308	S.F.	27.50	14.40	5.10	47	57.50
0250	4″ decking		120	.333		28.50	15.60	5.55	49.65	61
0400	Floating, small boat, prefab, no shore facilities, minimum		250	.160		24	7.50	2.65	34.15	41
0500	Maximum		150	.267	↓	51	12.45	4.42	67.87	80
0700	Per slip, minimum (180 S.F. each)		1.59	25.157	Ea.	4,900	1,175	415	6,490	7,625
0800	Maximum	↓	1.40	28.571	"	8,250	1,325	475	10,050	11,600

35 51 13.24 Jetties, Docks

		Crew	Daily Output	Labor-Hours	Unit	Material	Labor	Equipment	Total	Total Incl O&P
0010	**JETTIES, DOCKS**									
0011	Floating including anchors									
0030	See Section 06 13 33.50 or Section 06 13 33.52 for fixed docks									
1000	Polystyrene flotation, minimum	F-3	200	.200	S.F.	27	9.35	3.32	39.67	47.50
1040	Maximum		135	.296	"	35	13.85	4.91	53.76	65.50
1100	Alternate method of figuring, minimum		1.13	35.398	Slip	4,925	1,650	585	7,160	8,600
1140	Maximum		.70	57.143	"	7,050	2,675	945	10,670	12,900
1200	Galv. steel frame and wood deck, 3′ wide, minimum		320	.125	S.F.	16.75	5.85	2.07	24.67	29.50
1240	Maximum		200	.200		27	9.35	3.32	39.67	47.50
1300	4′ wide, minimum		320	.125		17	5.85	2.07	24.92	30
1340	Maximum	↓	200	.200	↓	27	9.35	3.32	39.67	48

35 51 Floating Construction

35 51 13 – Floating Piers

35 51 13.24 Jetties, Docks		Crew	Daily Output	Labor-Hours	Unit	Material	2014 Bare Costs Labor	Equipment	Total	Total Incl O&P
1500	8′ wide, minimum	F-3	250	.160	S.F.	18.55	7.50	2.65	28.70	35
1540	Maximum		160	.250		28	11.70	4.15	43.85	53.50
1700	Treated wood frames and deck, 3′ wide, minimum		250	.160		18.80	7.50	2.65	28.95	35
1740	Maximum		125	.320		64.50	14.95	5.30	84.75	100
2000	Polyethylene drums, treated wood frame and deck									
2100	6′ wide, minimum	F-3	250	.160	S.F.	32	7.50	2.65	42.15	49.50
2140	Maximum		125	.320		42.50	14.95	5.30	62.75	75.50
2200	8′ wide, minimum		233	.172		26	8	2.85	36.85	44
2240	Maximum		120	.333		35.50	15.60	5.55	56.65	69
2300	10′ wide, minimum		200	.200		28.50	9.35	3.32	41.17	49
2340	Maximum		110	.364		39	17	6.05	62.05	75.50
2400	Concrete pontoons, treated wood frame and deck									
2500	Breakwater, concrete pontoon, 50′ L x 8′ W x 6.5′ D	F-4	4	12	Ea.	45,300	555	284	46,139	51,000
2600	Inland breakwater, concrete pontoon, 50′ L x 8′ W x 4′ D		4	12	"	43,300	555	284	44,139	48,800
2700	Docks, concrete pontoon w/treated wood deck		2400	.020	S.F.	108	.92	.47	109.39	121
2800	Docks,concrete pontoons docks 400 S.F. minimum order									

35 51 13.28 Jetties, Floating Dock Accessories		Crew	Daily Output	Labor-Hours	Unit	Material	Labor	Equipment	Total	Total Incl O&P
0010	**JETTIES, FLOATING DOCK ACCESSORIES**									
0200	Dock connectors, stressed cables with rubber spacers									
0220	25″ long, 3′ wide dock	1 Clab	2	4	Joint	205	147		352	450
0240	5′ wide dock		2	4		239	147		386	490
0400	38″ long, 4′ wide dock		1.75	4.571		224	168		392	505
0440	6′ wide dock		1.45	5.517		255	202		457	590
1000	Gangway, aluminum, one end rolling, no hand rails									
1020	3′ wide, minimum	1 Clab	67	.119	L.F.	195	4.38		199.38	222
1040	Maximum		32	.250		224	9.15		233.15	261
1100	4′ wide, minimum		40	.200		207	7.35		214.35	238
1140	Maximum		24	.333		238	12.20		250.20	281
1180	For handrails, add					70			70	76.50
2000	Pile guides, beads on stainless cable	1 Clab	4	2	Ea.	178	73.50		251.50	310
2020	Rod type, 8″ diameter piles, minimum		4	2		48.50	73.50		122	167
2040	Maximum		2	4		71	147		218	305
2100	10″ to 14″ diameter piles minimum		3.20	2.500		56	91.50		147.50	203
2140	Maximum		1.75	4.571		98.50	168		266.50	365
2200	Roller type, 4 rollers, minimum		4	2		161	73.50		234.50	290
2240	Maximum		1.75	4.571		229	168		397	510

35 59 Marine Specialties

35 59 33 – Marine Bollards and Cleats

35 59 33.50 Jetties, Dock Accessories		Crew	Daily Output	Labor-Hours	Unit	Material	Labor	Equipment	Total	Total Incl O&P
0010	**JETTIES, DOCK ACCESSORIES**									
0100	Cleats, aluminum, "S" type, 12″ long	1 Clab	8	1	Ea.	25	36.50		61.50	84
0140	10″ long		6.70	1.194		60.50	44		104.50	134
0180	15″ long		6	1.333		79.50	49		128.50	163
0400	Dock wheel for corners and piles, vinyl, 12″ diameter		4	2		73	73.50		146.50	193
1000	Electrical receptacle with circuit breaker,									
1020	Pile mounted, double 30 amp, 125 volt	1 Elec	2	4	Unit	790	213		1,003	1,175
1060	Double 50 amp, 125/240 volt		1.60	5		1,075	267		1,342	1,575
1120	Free standing, add		4	2		194	107		301	375
1140	Double free standing, add		2.70	2.963		180	158		338	435
1160	Light, 2 louvered, with photo electric switch, add		8	1		197	53.50		250.50	296

35 59 Marine Specialties

35 59 33 – Marine Bollards and Cleats

35 59 33.50 Jetties, Dock Accessories		Crew	Daily Output	Labor-Hours	Unit	2014 Bare Costs Material	Labor	Equipment	Total	Total Incl O&P
1180	Telephone jack on stanchion	1 Elec	8	1	Unit	196	53.50		249.50	295
1300	Fender, Vinyl, 4" high	1 Clab	160	.050	L.F.	13.10	1.83		14.93	17.30
1380	Corner piece		80	.100	Ea.	16.50	3.67		20.17	24
1400	Hose holder, cast aluminum		16	.500	"	50	18.35		68.35	83.50
1500	Ladder, aluminum, heavy duty									
1520	Crown top, 5 to 7 step, minimum	1 Clab	5.30	1.509	Ea.	160	55.50		215.50	262
1560	Maximum		2	4		258	147		405	510
1580	Bracket for portable clamp mounting		8	1		9.15	36.50		45.65	66.50
1800	Line holder, treated wood, small		16	.500		8.60	18.35		26.95	38
1840	Large		13.30	.602		14.60	22		36.60	50
2000	Mooring whip, fiberglass bolted to dock,									
2020	1200 lb. boat	1 Clab	8.80	.909	Pr.	340	33.50		373.50	420
2040	10,000 lb. boat		6.70	1.194		685	44		729	825
2080	60,000 lb. boat		4	2		800	73.50		873.50	995
2400	Shock absorbing tubing, vertical bumpers									
2420	3" diam., vinyl, white	1 Clab	80	.100	L.F.	4.67	3.67		8.34	10.80
2440	Polybutyl, clear		80	.100	"	5.20	3.67		8.87	11.35
2480	Mounts, polybutyl		20	.400	Ea.	5.45	14.65		20.10	28.50
2490	Deluxe		20	.400	"	12.20	14.65		26.85	36

Estimating Tips

Products such as conveyors, material handling cranes and hoists, as well as other items specified in this division, require trained installers. The general contractor may not have any choice as to who will perform the installation or when it will be performed. Long lead times are often required for these products, making early decisions in purchasing and scheduling necessary.

The installation of this type of equipment may require the embedment of mounting hardware during construction of floors, structural walls, or interior walls/partitions. Electrical connections will require coordination with the electrical contractor.

Reference Numbers

Reference numbers are shown in shaded boxes at the beginning of some major classifications. These numbers refer to related items in the Reference Section. The reference information may be an estimating procedure, an alternate pricing method, or technical information.

Note: Not all subdivisions listed here necessarily appear in this publication.

41 21 Conveyors

41 21 23 – Piece Material Conveyors

41 21 23.16 Container Piece Material Conveyors		Crew	Daily Output	Labor-Hours	Unit	2014 Bare Costs Material	Labor	Equipment	Total	Total Incl O&P
0010	**CONTAINER PIECE MATERIAL CONVEYORS**									
0020	Gravity fed, 2″ rollers, 3″ O.C.									
0050	10′ sections with 2 supports, 600 lb. capacity, 18″ wide				Ea.	450			450	495
0100	24″ wide					515			515	570
0150	1400 lb. capacity, 18″ wide					415			415	455
0200	24″ wide					460			460	505
0350	Horizontal belt, center drive and takeup, 60 fpm									
0400	16″ belt, 26.5′ length	2 Mill	.50	32	Ea.	3,150	1,550		4,700	5,750
0450	24″ belt, 41.5′ length		.40	40		4,900	1,925		6,825	8,200
0500	61.5′ length		.30	53.333		6,525	2,575		9,100	11,000
0600	Inclined belt, 10′ rise with horizontal loader and									
0620	End idler assembly, 27.5′ length, 18″ belt	2 Mill	.30	53.333	Ea.	7,700	2,575		10,275	12,300
0700	24″ belt	″	.15	106	″	8,600	5,125		13,725	17,000
3600	Monorail, overhead, manual, channel type									
3700	125 lb. per L.F.	1 Mill	26	.308	L.F.	14.45	14.80		29.25	38
3900	500 lb. per L.F.	″	21	.381	″	20.50	18.30		38.80	49.50
4000	Trolleys for above, 2 wheel, 125 lb. capacity				Ea.	74			74	81.50
4200	4 wheel, 250 lb. capacity					290			290	320
4300	8 wheel, 500 lb. capacity					675			675	745

41 22 Cranes and Hoists

41 22 13 – Cranes

41 22 13.10 Crane Rail

		Crew	Daily Output	Labor-Hours	Unit	Material	Labor	Equipment	Total	Total Incl O&P
0010	**CRANE RAIL**									
0020	Box beam bridge, no equipment included	E-4	3400	.009	Lb.	1.33	.49	.04	1.86	2.37
0210	Running track only, 104 lb. per yard, 20′ piece	″	160	.200	L.F.	23	10.30	.89	34.19	44.50

41 22 23 – Hoists

41 22 23.10 Material Handling

		Crew	Daily Output	Labor-Hours	Unit	Material	Labor	Equipment	Total	Total Incl O&P
0010	**MATERIAL HANDLING**, cranes, hoists and lifts									
1500	Cranes, portable hydraulic, floor type, 2,000 lb. capacity				Ea.	3,500			3,500	3,875
1600	4,000 lb. capacity					4,025			4,025	4,425
1800	Movable gantry type, 12′ to 15′ range, 2,000 lb. capacity					2,050			2,050	2,250
1900	6,000 lb. capacity					4,825			4,825	5,300
2100	Hoists, electric overhead, chain, hook hung, 15′ lift, 1 ton cap.					2,875			2,875	3,175
2200	3 ton capacity					3,375			3,375	3,725
2500	5 ton capacity					7,800			7,800	8,575
2600	For hand-pushed trolley, add					15%				
2700	For geared trolley, add					30%				
2800	For motor trolley, add					75%				
3000	For lifts over 15′, 1 ton, add				L.F.	25			25	27.50
3100	5 ton, add				″	52			52	57
3300	Lifts, scissor type, portable, electric, 36″ high, 2,000 lb.				Ea.	3,475			3,475	3,825
3400	48″ high, 4,000 lb.				″	4,175			4,175	4,600

Division 46 - Water & Wastewater Equipment

Estimating Tips

This division contains information about water and wastewater equipment and systems, which was formerly located in Division 44.

The main areas of focus are total wastewater treatment plants and components of wastewater treatment plants. In addition, there are assemblies such as sewage treatment lagoons that can be found in some publications under G30 Site Mechanical Utilities.

Also included in this section are oil/water separators for wastewater treatment.

Reference Numbers

Reference numbers are shown in shaded boxes at the beginning of some major classifications. These numbers refer to related items in the Reference Section. The reference information may be an estimating procedure, an alternate pricing method, or technical information.

Note: Not all subdivisions listed here necessarily appear in this publication.

46 07 Packaged Water and Wastewater Treatment Equipment

46 07 53 – Packaged Wastewater Treatment Equipment

46 07 53.10 Biological Packaged Water Treatment Plants		Crew	Daily Output	Labor-Hours	Unit	2014 Bare Costs Material	Labor	Equipment	Total	Total Incl O&P
0010	BIOLOGICAL PACKAGED WATER TREATMENT PLANTS									
0011	Not including fencing or external piping									
0020	Steel packaged, blown air aeration plants									
0100	1,000 GPD				Gal.				55	60.50
0200	5,000 GPD								22	24
0300	15,000 GPD								22	24
0400	30,000 GPD								15.40	16.95
0500	50,000 GPD								11	12.10
0600	100,000 GPD								9.90	10.90
0700	200,000 GPD								8.80	9.70
0800	500,000 GPD				↓				7.70	8.45
1000	Concrete, extended aeration, primary and secondary treatment									
1010	10,000 GPD				Gal.				22	24
1100	30,000 GPD								15.40	16.95
1200	50,000 GPD								11	12.10
1400	100,000 GPD								9.90	10.90
1500	500,000 GPD				↓				7.70	8.45
1700	Municipal wastewater treatment facility									
1720	1.0 MGD				Gal.				11	12.10
1740	1.5 MGD								10.60	11.65
1760	2.0 MGD								10	11
1780	3.0 MGD								7.80	8.60
1800	5.0 MGD				↓				5.80	6.70
2000	Holding tank system, not incl. excavation or backfill									
2010	Recirculating chemical water closet	2 Plum	4	4	Ea.	520	230		750	920
2100	For voltage converter, add	"	16	1		295	57.50		352.50	410
2200	For high level alarm, add	1 Plum	7.80	1.026	↓	114	59		173	215

46 07 53.20 Wastewater Treatment System		Crew	Daily Output	Labor-Hours	Unit	Material	Labor	Equipment	Total	Total Incl O&P
0010	WASTEWATER TREATMENT SYSTEM									
0020	Fiberglass, 1,000 gallon	B-21	1.29	21.705	Ea.	4,075	915	109	5,099	6,025
0100	1,500 gallon	"	1.03	27.184	"	8,150	1,150	136	9,436	10,900

46 25 Oil and Grease Separation and Removal Equipment

46 25 16 – API Oil-Water Separators

46 25 16.10 Oil/Water Separators		Crew	Daily Output	Labor-Hours	Unit	Material	Labor	Equipment	Total	Total Incl O&P
0010	OIL/WATER SEPARATORS									
0020	Complete system, not including excavation and backfill									
0100	Treated capacity 0.2 cubic feet per second	B-22	1	30	Ea.	4,950	1,275	210	6,435	7,650
0200	0.5 cubic foot per second	B-13	.75	74.667		7,875	2,975	995	11,845	14,400
0300	1.0 cubic foot per second		.60	93.333		11,300	3,725	1,250	16,275	19,500
0400	2.4 – 3 cubic feet per second		.30	186		21,900	7,450	2,475	31,825	38,200
0500	11 cubic feet per second		.17	329		24,800	13,100	4,400	42,300	52,000
0600	22 cubic feet per second	↓	.10	560	↓	37,100	22,300	7,450	66,850	83,000

46 51 Air and Gas Diffusion Equipment

46 51 13 – Floating Mechanical Aerators

	46 51 13.10 Aeration Equipment	Crew	Daily Output	Labor-Hours	Unit	2014 Bare Costs Material	Labor	Equipment	Total	Total Incl O&P
0010	**AERATION EQUIPMENT**									
0020	Aeration equipment includes floats and excludes power supply and anchorage									
4320	Surface aerator, 50 lb. oxygen/hr, 30 HP, 900 RPM, inc floats, no anchoring	B-21B	1	40	Ea.	33,300	1,600	665	35,565	39,800
4322	Anchoring aerators per cell, 6 anchors per cell	B-6	.50	48		3,675	1,925	730	6,330	7,775
4324	Anchoring aerators per cell, 8 anchors per cell	"	.33	72.727		5,050	2,925	1,100	9,075	11,300

46 51 20 – Air and Gas Handling Equipment

46 51 20.10 Blowers and System Components

	46 51 20.10 Blowers and System Components	Crew	Daily Output	Labor-Hours	Unit	Material	Labor	Equipment	Total	Total Incl O&P
0010	**BLOWERS AND SYSTEM COMPONENTS**									
0020	Rotary Lobe Blowers									
0030	Medium Pressure (7 to 14 psig)									
0120	38CFM, 2.1 BHP	Q-2	2.50	9.600	Ea.	1,200	515		1,715	2,100
0130	125 CFM, 5.5 BHP		2.40	10		1,350	535		1,885	2,275
0140	245 CFM, 10.2 BHP		2	12		1,675	645		2,320	2,825
0150	363 CFM, 14.5 BHP		1.80	13.333		2,375	715		3,090	3,700
0160	622 CFM, 24.5 BHP		1.40	17.143		3,950	920		4,870	5,750
0170	1125 CFM, 42.8 BHP		1.10	21.818		6,275	1,175		7,450	8,675
0180	1224 CFM, 47.4 BHP		1	24		8,900	1,300		10,200	11,800
0400	Prepackaged Medium Pressure (7 to 14 PSIG) Blower									
0405	incl. 3 ph. motor, filter, silencer, valves, check valve,press. gage									
0420	38CFM, 2.1 BHP	Q-2	2.50	9.600	Ea.	2,850	515		3,365	3,925
0430	125 CFM, 5.5 BHP		2.40	10		3,650	535		4,185	4,800
0440	245 CFM, 10.2 BHP		2	12		4,875	645		5,520	6,350
0450	363 CFM, 14.5 BHP		1.80	13.333		7,550	715		8,265	9,375
0460	521 CFM, 20 BHP		1.50	16		7,600	860		8,460	9,650
1010	Filters and Silencers									
1100	Silencer with paper filter									
1105	1" Connection	1 Plum	14	.571	Ea.	32	33		65	84.50
1110	1.5" Connection		11	.727		38	42		80	105
1115	2" Connection		9	.889		130	51		181	220
1120	2.5" Connection		8	1		140	57.50		197.50	241
1125	3" Connection	Q-1	8	2		137	104		241	305
1130	4" Connection	"	5	3.200		285	166		451	565
1135	5" Connection	Q-2	5	4.800		330	258		588	755
1140	6" Connection	"	5	4.800		345	258		603	770
1300	Silencer with polyester filter									
1305	1" Connection	1 Plum	14	.571	Ea.	43	33		76	96.50
1310	1.5" Connection		11	.727		47	42		89	115
1315	2" Connection		9	.889		163	51		214	257
1320	2.5" Connection		8	1		163	57.50		220.50	267
1325	3" Connection	Q-1	8	2		168	104		272	340
1330	4" Connection	"	5	3.200		315	166		481	600
1335	5" Connection	Q-2	5	4.800		385	258		643	810
1340	6" Connection	"	5	4.800		435	258		693	865
1500	Chamber Silencers									
1505	1" Connection	1 Plum	14	.571	Ea.	80	33		113	138
1510	1.5" Connection		11	.727		106	42		148	180
1515	2" Connection		9	.889		116	51		167	205
1520	2.5" Connection		8	1		217	57.50		274.50	325
1525	3" Connection	Q-1	8	2		286	104		390	470
1530	4" Connection	"	5	3.200		400	166		566	690
1700	Blower Couplings									
1701	Blower Flexible Coupling									

46 51 Air and Gas Diffusion Equipment

46 51 20 – Air and Gas Handling Equipment

46 51 20.10 Blowers and System Components		Crew	Daily Output	Labor-Hours	Unit	Material	2014 Bare Costs Labor	Equipment	Total	Total Incl O&P
1710	1.5" Connection	Q-1	71	.225	Ea.	30	11.65		41.65	50.50
1715	2" Connection		67	.239		33	12.35		45.35	55
1720	2.5" Connection		65	.246		38	12.75		50.75	61.50
1725	3" Connection		64	.250		51	12.95		63.95	75.50
1730	4" Connection		58	.276		61	14.30		75.30	88.50
1735	5" Connection	Q-2	83	.289		66	15.55		81.55	96
1740	6" Connection		79	.304		91	16.30		107.30	125
1745	8" Connection		69	.348		150	18.70		168.70	193

46 51 20.20 Aeration System Air Process Piping		Crew	Daily Output	Labor-Hours	Unit	Material	Labor	Equipment	Total	Total Incl O&P
0010	**AERATION SYSTEM AIR PROCESS PIPING**									
0100	Blower Pressure Relief Valves-adjustable									
0105	1" Diameter	1 Plum	14	.571	Ea.	110	33		143	171
0110	2" Diameter		9	.889		120	51		171	209
0115	2.5" Diameter		8	1		170	57.50		227.50	274
0120	3" Diameter	Q-1	8	2		170	104		274	345
0125	4" Diameter	"	5	3.200		190	166		356	460
0200	Blower Pressure Relief Valves-weight loaded									
0205	1" Diameter	1 Plum	14	.571	Ea.	191	33		224	260
0210	2" Diameter	"	9	.889		340	51		391	450
0220	3" Diameter	Q-1	8	2		975	104		1,079	1,225
0225	4" Diameter	"	5	3.200		1,375	166		1,541	1,750
0300	Pressure Relief Valves-preset high flow									
0310	2" Diameter	1 Plum	9	.889	Ea.	260	51		311	360
0320	3" Diameter	Q-1	8	2	"	560	104		664	770
1000	Check Valves, Wafer Style									
1110	2" Diameter	1 Plum	9	.889	Ea.	119	51		170	208
1120	3" Diameter	Q-1	8	2		140	104		244	310
1125	4" Diameter	"	5	3.200		175	166		341	440
1130	5" Diameter	Q-2	6	4		265	215		480	615
1135	6" Diameter		5	4.800		320	258		578	740
1140	8" Diameter		4.50	5.333		440	287		727	920
1200	Check Valves, Flanged Steel									
1210	2" Diameter	1 Plum	8	1	Ea.	274	57.50		331.50	385
1220	3" Diameter	Q-1	4.50	3.556		330	184		514	645
1225	4" Diameter	"	3	5.333		400	276		676	855
1230	5" Diameter	Q-2	3	8		590	430		1,020	1,300
1235	6" Diameter		3	8		590	430		1,020	1,300
1240	8" Diameter		2.50	9.600		920	515		1,435	1,775
2100	Butterfly Valves, Lever Operated-Wafer Style									
2110	2" Diameter	1 Plum	14	.571	Ea.	46.50	33		79.50	101
2120	3" Diameter	Q-1	8	2		58.50	104		162.50	220
2125	4" Diameter	"	5	3.200		81	166		247	340
2130	5" Diameter	Q-2	6	4		102	215		317	435
2135	6" Diameter		5	4.800		119	258		377	520
2140	8" Diameter		4.50	5.333		179	287		466	630
2145	10" Diameter		4	6		235	320		555	745
2200	Butterfly Valves, Gear Operated-Wafer Style									
2210	2" Diameter	1 Plum	14	.571	Ea.	83	33		116	141
2220	3" Diameter	Q-1	8	2		93	104		197	258
2225	4" Diameter	"	5	3.200		112	166		278	375
2230	5" Diameter	Q-2	6	4		135	215		350	475
2235	6" Diameter		5	4.800		153	258		411	560

46 51 Air and Gas Diffusion Equipment

46 51 20 – Air and Gas Handling Equipment

46 51 20.20 Aeration System Air Process Piping		Crew	Daily Output	Labor-Hours	Unit	2014 Bare Costs Material	Labor	Equipment	Total	Total Incl O&P
2240	8″ Diameter	Q-2	4.50	5.333	Ea.	234	287		521	695
2245	10″ Diameter		4	6		273	320		593	785
2250	12″ Diameter		3	8		355	430		785	1,050

46 51 20.30 Aeration System Blower Control Panels

Line	Description	Crew	Daily Output	Labor-Hours	Unit	Material	Labor	Equipment	Total	Total Incl O&P
0010	**AERATION SYSTEM BLOWER CONTROL PANELS**									
0020	Single Phase simplex									
0030	7 to 10 overload amp range	1 Elec	2.70	2.963	Ea.	770	158		928	1,075
0040	9 to 13 overload amp range		2.70	2.963		770	158		928	1,075
0050	12 to 18 overload amp range		2.70	2.963		770	158		928	1,075
0060	16 to 24 overload amp range		2.70	2.963		770	158		928	1,075
0070	23 to 32 overload amp range		2.70	2.963		830	158		988	1,150
0080	30 to 40 overload amp range		2.70	2.963		830	158		988	1,150
0090	Single Phase duplex									
0100	7 to 10 overload amp range	1 Elec	2	4	Ea.	1,250	213		1,463	1,700
0110	9 to 13 overload amp range		2	4		1,250	213		1,463	1,700
0120	12 to 18 overload amp range		2	4		1,250	213		1,463	1,700
0130	16 to 24 overload amp range		2	4		1,375	213		1,588	1,850
0140	23 to 32 overload amp range		2	4		1,400	213		1,613	1,875
0150	30 to 40 overload amp range		2	4		1,400	213		1,613	1,875
0160	Three Phase simplex									
0170	1.6 to 2.5 overload amp range	1 Elec	2.50	3.200	Ea.	925	171		1,096	1,275
0180	2.5 to 4 overload amp range		2.50	3.200		925	171		1,096	1,275
0190	4 to 6.3 overload amp range		2.50	3.200		925	171		1,096	1,275
0200	6 to 10 overload amp range		2.50	3.200		925	171		1,096	1,275
0210	9 to 14 overload amp range		2.50	3.200		960	171		1,131	1,300
0220	13 to 18 overload amp range		2.50	3.200		960	171		1,131	1,300
0230	17 to 23 overload amp range		2.50	3.200		970	171		1,141	1,325
0240	20 to 25 overload amp range		2.50	3.200		970	171		1,141	1,325
0250	23 to 32 overload amp range		2.50	3.200		1,050	171		1,221	1,425
0260	37 to 50 overload amp range		2.50	3.200		1,175	171		1,346	1,550
0270	Three Phase duplex									
0280	1.6 to 2.5 overload amp range	1 Elec	1.90	4.211	Ea.	1,250	225		1,475	1,700
0290	2.5 to 4 overload amp range		1.90	4.211		1,325	225		1,550	1,800
0300	4 to 6.3 overload amp range		1.90	4.211		1,325	225		1,550	1,800
0310	6 to 10 overload amp range		1.90	4.211		1,325	225		1,550	1,800
0320	9 to 14 overload amp range		1.90	4.211		1,325	225		1,550	1,800
0330	13 to 18 overload amp range		1.90	4.211		1,375	225		1,600	1,825
0340	17 to 23 overload amp range		1.90	4.211		1,375	225		1,600	1,850
0350	20 to 25 overload amp range		1.90	4.211		1,375	225		1,600	1,850
0360	23 to 32 overload amp range		1.90	4.211		1,400	225		1,625	1,875
0370	37 to 50 overload amp range		1.90	4.211		1,825	225		2,050	2,325

46 51 36 – Ceramic Disc Fine Bubble Diffusers

46 51 36.10 Ceramic Disc Air Diffuser Systems

Line	Description	Crew	Daily Output	Labor-Hours	Unit	Material	Labor	Equipment	Total	Total Incl O&P
0010	**CERAMIC DISC AIR DIFFUSER SYSTEMS**									
0020	Price for air diffuser pipe system by cell size excluding concrete work									
0030	Price for air diffuser pipe system by cell size excluding air supply									
0040	Depth of 12 feet is the waste depth not the cell dimensions									
0100	Ceramic disc air diffuser system, cell size 20′ x 20′ x 12′	2 Plum	.38	42.667	Ea.	11,900	2,450		14,350	16,800
0120	20′ x 30′ x 12′		.26	61.326		17,700	3,525		21,225	24,700
0140	20′ x 40′ x 12′		.20	80		23,400	4,600		28,000	32,800
0160	20′ x 50′ x 12′		.16	98.644		29,200	5,675		34,875	40,700
0180	20′ x 60′ x 12′		.14	117		34,900	6,750		41,650	48,600

46 51 Air and Gas Diffusion Equipment

46 51 36 – Ceramic Disc Fine Bubble Diffusers

46 51 36.10 Ceramic Disc Air Diffuser Systems		Crew	Daily Output	Labor-Hours	Unit	2014 Bare Costs Material	Labor	Equipment	Total	Total Incl O&P
0200	20′ x 70′ x 12′	2 Plum	.12	136	Ea.	40,700	7,825		48,525	56,500
0220	20′ x 80′ x 12′		.10	154		46,400	8,900		55,300	64,500
0240	20′ x 90′ x 12′		.09	173		52,000	9,975		61,975	72,500
0260	20′ x 100′ x 12′		.08	192		58,000	11,100		69,100	80,500
0280	20′ x 120′ x 12′		.07	229		69,500	13,200		82,700	96,500
0300	20′ x 140′ x 12′		.06	266		81,000	15,300		96,300	112,000
0320	20′ x 160′ x 12′		.05	304		92,500	17,500		110,000	128,500
0340	20′ x 180′ x 12′		.05	341		104,000	19,600		123,600	144,000
0360	20′ x 200′ x 12′		.04	378		115,500	21,800		137,300	160,000
0380	20′ x 250′ x 12′		.03	471		144,500	27,200		171,700	200,000
0400	20′ x 300′ x 12′		.03	565		173,000	32,500		205,500	239,500
0420	20′ x 350′ x 12′		.02	658		202,000	37,900		239,900	279,000
0440	20′ x 400′ x 12′		.02	751		230,500	43,200		273,700	319,000
0460	20′ x 450′ x 12′		.02	846		259,500	48,700		308,200	359,000
0480	20′ x 500′ x 12′		.02	941		288,500	54,000		342,500	399,000

46 53 Biological Treatment Systems

46 53 17 – Activated Sludge Treatment

46 53 17.10 Activated Sludge Treatment Cells		Crew	Daily Output	Labor-Hours	Unit	Material	Labor	Equipment	Total	Total Incl O&P
0010	**ACTIVATED SLUDGE TREATMENT CELLS**									
0015	Price for cell construction excluding aerator piping & drain									
0020	Cell construction by dimensions									
0100	Treatment cell, end or single, 20′ x 20′ x 14′ high (2′ freeboard)	C-14D	.46	434	Ea.	16,700	19,800	1,625	38,125	51,000
0120	20′ x 30′ x 14′ high		.35	568		21,700	25,900	2,125	49,725	66,000
0140	20′ x 40′ x 14′ high		.29	701		26,700	32,000	2,600	61,300	81,500
0160	20′ x 50′ x 14′ high		.24	835		31,700	38,100	3,100	72,900	97,500
0180	20′ x 60′ x 14′ high		.21	969		36,700	44,200	3,600	84,500	112,500
0200	20′ x 70′ x 14′ high		.18	1103		41,700	50,500	4,100	96,300	128,000
0220	20′ x 80′ x 14′ high		.16	1236		46,700	56,500	4,600	107,800	143,500
0240	20′ x 90′ x 14′ high		.15	1370		51,500	62,500	5,100	119,100	159,000
0260	20′ x 100′ x 14′ high		.13	1503		56,500	68,500	5,600	130,600	174,500
0280	20′ x 120′ x 14′ high		.11	1771		66,500	81,000	6,600	154,100	205,500
0300	20′ x 140′ x 14′ high		.10	2038		76,500	93,000	7,575	177,075	236,000
0320	20′ x 160′ x 14′ high		.09	2306		86,500	105,500	8,575	200,575	267,000
0340	20′ x 180′ x 14′ high		.08	2574		96,500	117,500	9,575	223,575	297,500
0360	20′ x 200′ x 14′ high		.07	2840		106,500	129,500	10,600	246,600	328,500
0380	20′ x 250′ x 14′ high		.06	3508		131,500	160,000	13,100	304,600	406,000
0400	20′ x 300′ x 14′ high		.05	4175		156,500	190,500	15,500	362,500	482,500
0420	20′ x 350′ x 14′ high		.04	4842		181,500	221,000	18,000	420,500	560,000
0440	20′ x 400′ x 14′ high		.04	5509		206,500	251,500	20,500	478,500	637,000
0460	20′ x 450′ x 14′ high		.03	6191		231,000	282,500	23,000	536,500	715,500
0480	20′ x 500′ x 14′ high		.03	6849		256,000	312,500	25,500	594,000	792,000
0500	Treatment cell, end or single, 20′ x 20′ x 12′ high (2′ freeboard)		.48	414		15,800	18,900	1,550	36,250	48,200
0520	20′ x 30′ x 12′ high		.37	543		20,600	24,800	2,025	47,425	63,000
0540	20′ x 40′ x 12′ high		.30	671		25,300	30,600	2,500	58,400	78,000
0560	20′ x 50′ x 12′ high		.25	800		30,100	36,500	2,975	69,575	93,000
0580	20′ x 60′ x 12′ high		.22	928		34,800	42,400	3,450	80,650	107,500
0600	20′ x 70′ x 12′ high		.19	1057		39,600	48,300	3,925	91,825	122,500
0620	20′ x 80′ x 12′ high		.17	1186		44,400	54,000	4,425	102,825	137,000
0640	20′ x 90′ x 12′ high		.15	1314		49,100	60,000	4,900	114,000	152,000
0660	20′ x 100′ x 12′ high		.14	1444		54,000	66,000	5,375	125,375	167,000

46 53 Biological Treatment Systems

46 53 17 – Activated Sludge Treatment

46 53 17.10 Activated Sludge Treatment Cells		Crew	Daily Output	Labor-Hours	Unit	2014 Bare Costs Material	Labor	Equipment	Total	Total Incl O&P
0680	20′ x 120′ x 12′ high	C-14D	.12	1700	Ea.	63,500	77,500	6,325	147,325	196,000
0700	20′ x 140′ x 12′ high		.10	1958		73,000	89,500	7,275	169,775	226,000
0720	20′ x 160′ x 12′ high		.09	2214		82,500	101,000	8,250	191,750	255,500
0740	20′ x 180′ x 12′ high		.08	2472		92,000	113,000	9,200	214,200	285,000
0760	20′ x 200′ x 12′ high		.07	2732		101,500	124,500	10,200	236,200	314,500
0780	20′ x 250′ x 12′ high		.06	3372		125,500	154,000	12,500	292,000	389,500
0800	20′ x 300′ x 12′ high		.05	4016		149,000	183,500	14,900	347,400	463,000
0820	20′ x 350′ x 12′ high		.04	4662		173,000	212,500	17,300	402,800	537,000
0840	20′ x 400′ x 12′ high		.04	5305		196,500	242,000	19,700	458,200	611,000
0860	20′ x 450′ x 12′ high		.03	5952		220,500	271,500	22,100	514,100	686,000
0880	20′ x 500′ x 12′ high		.03	6600		244,500	301,000	24,600	570,100	760,000
1100	Treatment cell, connecting, 20′ x 20′ x 14′ high (2′ freeboard)		.57	350		13,300	16,000	1,300	30,600	40,800
1120	20′ x 30′ x 14′ high		.45	442		16,700	20,200	1,650	38,550	51,500
1140	20′ x 40′ x 14′ high		.37	534		20,000	24,400	2,000	46,400	62,000
1160	20′ x 50′ x 14′ high		.32	626		23,300	28,600	2,325	54,225	72,500
1180	20′ x 60′ x 14′ high		.28	718		26,600	32,800	2,675	62,075	82,500
1200	20′ x 70′ x 14′ high		.25	810		29,900	37,000	3,025	69,925	93,000
1220	20′ x 80′ x 14′ high		.22	902		33,200	41,200	3,350	77,750	103,500
1240	20′ x 90′ x 14′ high		.20	994		36,500	45,400	3,700	85,600	114,000
1260	20′ x 100′ x 14′ high		.18	1086		39,800	49,600	4,050	93,450	125,000
1280	20′ x 120′ x 14′ high		.16	1270		46,400	58,000	4,725	109,125	145,500
1300	20′ x 140′ x 14′ high		.14	1454		53,000	66,500	5,400	124,900	167,000
1320	20′ x 160′ x 14′ high		.12	1638		59,500	74,500	6,100	140,100	187,000
1340	20′ x 180′ x 14′ high		.11	1823		66,000	83,000	6,775	155,775	209,000
1360	20′ x 200′ x 14′ high		.10	2006		73,000	91,500	7,450	171,950	229,000
1380	20′ x 250′ x 14′ high		.08	2466		89,500	112,500	9,175	211,175	282,000
1400	20′ x 300′ x 14′ high		.07	2923		106,000	133,500	10,900	250,400	334,000
1420	20′ x 350′ x 14′ high		.06	3384		122,500	154,500	12,600	289,600	386,500
1440	20′ x 400′ x 14′ high		.05	3846		139,000	175,500	14,300	328,800	439,000
1460	20′ x 450′ x 14′ high		.05	4301		155,500	196,500	16,000	368,000	491,000
1480	20′ x 500′ x 14′ high		.04	4761		172,000	217,500	17,700	407,200	543,500
1500	Treatment cell, connecting, 20′ x 20′ x 12′ high (2′ freeboard)		.60	335		12,700	15,300	1,250	29,250	38,900
1520	20′ x 30′ x 12′ high		.47	425		15,800	19,400	1,575	36,775	49,100
1540	20′ x 40′ x 12′ high		.39	514		19,000	23,500	1,925	44,425	59,000
1560	20′ x 50′ x 12′ high		.33	604		22,200	27,600	2,250	52,050	69,500
1580	20′ x 60′ x 12′ high		.29	693		25,400	31,700	2,575	59,675	79,500
1600	20′ x 70′ x 12′ high		.26	783		28,600	35,700	2,925	67,225	89,500
1620	20′ x 80′ x 12′ high		.23	872		31,800	39,800	3,250	74,850	100,000
1640	20′ x 90′ x 12′ high		.21	962		35,000	43,900	3,575	82,475	110,000
1660	20′ x 100′ x 12′ high		.19	1051		38,200	48,000	3,900	90,100	120,500
1680	20′ x 120′ x 12′ high		.16	1230		44,500	56,000	4,575	105,075	140,500
1700	20′ x 140′ x 12′ high		.14	1409		51,000	64,500	5,250	120,750	161,000
1720	20′ x 160′ x 12′ high		.13	1588		57,500	72,500	5,900	135,900	181,500
1740	20′ x 180′ x 12′ high		.11	1766		63,500	80,500	6,575	150,575	201,500
1760	20′ x 200′ x 12′ high		.10	1945		70,000	89,000	7,225	166,225	222,000
1780	20′ x 250′ x 12′ high		.08	2392		86,000	109,000	8,900	203,900	273,000
1800	20′ x 300′ x 12′ high		.07	2840		102,000	129,500	10,600	242,100	323,500
1820	20′ x 350′ x 12′ high		.06	3289		118,000	150,000	12,200	280,200	374,500
1840	20′ x 400′ x 12′ high		.05	3738		134,000	170,500	13,900	318,400	425,500
1860	20′ x 450′ x 12′ high		.05	4184		150,000	191,000	15,600	356,600	476,000
1880	20′ x 500′ x 12′ high		.04	5571		165,500	254,000	20,700	440,200	597,500
1996	Price for aerator curb per cell excluding aerator piping & drain									
2000	Treatment cell, 20′ x 20′ aerator curb	F-7	1.88	17.067	Ea.	565	705		1,270	1,700

46 53 Biological Treatment Systems

46 53 17 – Activated Sludge Treatment

46 53 17.10 Activated Sludge Treatment Cells		Crew	Daily Output	Labor-Hours	Unit	2014 Bare Costs Material	Labor	Equipment	Total	Total Incl O&P
2010	20′ x 30′	F-7	1.25	25.600	Ea.	845	1,050		1,895	2,550
2020	20′ x 40′		.94	34.133		1,125	1,400		2,525	3,425
2030	20′ x 50′		.75	42.667		1,400	1,750		3,150	4,275
2040	20′ x 60′		.63	51.200		1,700	2,100		3,800	5,100
2050	20′ x 70′		.54	59.735		1,975	2,475		4,450	5,975
2060	20′ x 80′		.47	68.259		2,250	2,825		5,075	6,825
2070	20′ x 90′		.42	76.794		2,525	3,175		5,700	7,675
2080	20′ x 100′		.38	85.333		2,825	3,525		6,350	8,525
2090	20′ x 120′		.31	102		3,375	4,225		7,600	10,300
2100	20′ x 140′		.27	119		3,950	4,925		8,875	11,900
2110	20′ x 160′		.23	136		4,500	5,625		10,125	13,700
2120	20′ x 180′		.21	153		5,075	6,325		11,400	15,400
2130	20′ x 200′		.19	170		5,625	7,050		12,675	17,100
2160	20′ x 250′		.15	213		7,050	8,800		15,850	21,400
2190	20′ x 300′		.13	256		8,450	10,600		19,050	25,600
2220	20′ x 350′		.11	298		9,850	12,300		22,150	29,800
2250	20′ x 400′		.09	341		11,300	14,100		25,400	34,100
2280	20′ x 450′		.08	384		12,700	15,800		28,500	38,400
2310	20′ x 500′	▼	.08	426	▼	14,100	17,600		31,700	42,700

Estimating Tips

- When estimating costs for the installation of electrical power generation equipment, factors to review include access to the job site, access and setting at the installation site, required connections, uncrating pads, anchors, leveling, final assembly of the components, and temporary protection from physical damage, including from exposure to the environment.
- Be aware of the cost of equipment supports, concrete pads, and vibration isolators, and cross-reference to other trades' specifications. Also, review site and structural drawings for items that must be included in the estimates.
- It is important to include items that are not documented in the plans and specifications but must be priced. These items include, but are not limited to, testing, dust protection, roof penetration, core drilling concrete floors and walls, patching, cleanup, and final adjustments. Add a contingency or allowance for utility company fees for power hookups, if needed.
- The project size and scope of electrical power generation equipment will have a significant impact on cost. The intent of RSMeans cost data is to provide a benchmark cost so that owners, engineers, and electrical contractors will have a comfortable number with which to start a project. Additionally, there are many websites available to use for research and to obtain a vendor's quote to finalize costs.

Reference Numbers

Reference numbers are shown in shaded boxes at the beginning of some major classifications. These numbers refer to related items in the Reference Section. The reference information may be an estimating procedure, an alternate pricing method, or technical information.

Note: Not all subdivisions listed here necessarily appear in this publication.

48 15 Wind Energy Electrical Power Generation Equipment

48 15 13 – Wind Turbines

48 15 13.50 Wind Turbines and Components			Crew	Daily Output	Labor-Hours	Unit	2014 Bare Costs Material	Labor	Equipment	Total	Total Incl O&P
0010	**WIND TURBINES & COMPONENTS**										
0500	Complete system, grid connected										
1000	20 kW, 31′ dia, incl. labor & material, min	G				System					49,900
1010	Maximum	G									92,000
1500	10 kW, 23′ dia, incl. labor & material	G									74,000
2000	2.4 kW, 12′ dia, incl. labor & material	G				↓					18,000

48 18 Fuel Cell Electrical Power Generation Equipment

48 18 13 – Electrical Power Generation Fuel Cells

48 18 13.10 Fuel Cells			Crew	Daily Output	Labor-Hours	Unit	Material	Labor	Equipment	Total	Total Incl O&P
0010	**FUEL CELLS**										
2001	Complete system, natural gas, installed, 200 kW	G				System					1,150,000
2002	400 kW	G									2,168,000
2003	600 kW	G									3,060,000
2004	800 kW	G									4,000,000
2005	1000 kW	G				↓					5,005,000
2010	Comm type, for battery charging, uses hydrogen, 100 watt, 12 V	G				Ea.	2,075			2,075	2,300
2020	500 watt, 12 V	G					4,275			4,275	4,700
2030	1000 watt, 48 V	G					7,000			7,000	7,700
2040	Small, for demonstration or battery charging units, 12 V	G					188			188	207
2050	Spare anode set for	G					42.50			42.50	46.50
2060	3 watts, uses hydrogen gas	G					435			435	475
2070	6 watts, uses hydrogen gas	G					780			780	860
2080	10 watts, uses hydrogen gas	G					1,125			1,125	1,250
2090	One month rental of 261 C.F. hydrogen gas cylinder for, min	G					21.50			21.50	23.50
2100	Maximum	G				↓	585			585	645

Assemblies Section

Table of Contents

How RSMeans Assemblies Data Works

Assemblies estimating provides a fast and reasonably accurate way to develop construction costs. An assembly is the grouping of individual work items, with appropriate quantities, to provide a cost for a major construction component in a convenient unit of measure.

An assemblies estimate is often used during early stages of design development to compare the cost impact of various design alternatives on total building cost.

Assemblies estimates are also used as an efficient tool to verify construction estimates.

Assemblies estimates do not require a completed design or detailed drawings. Instead, they are based on the general size of the structure and other known parameters of the project. The degree of accuracy of an assemblies estimate is generally within +/- 15%.

NARRATIVE FORMAT

G20 Site Improvements

G2040 Site Development

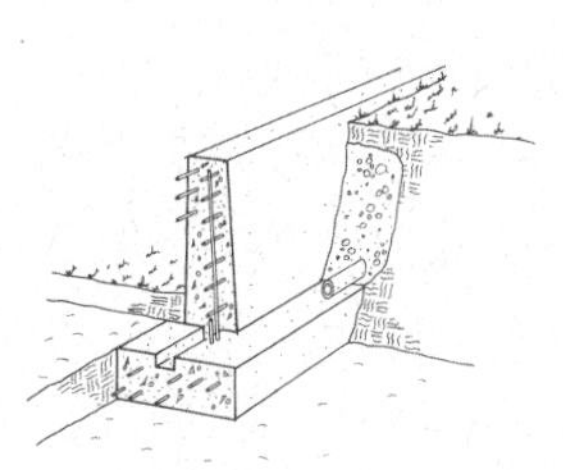

There are four basic types of Concrete Retaining Wall Systems: reinforced concrete with level backfill; reinforced concrete with sloped backfill or surcharge; unreinforced with level backfill; and unreinforced with sloped backfill or surcharge. System elements include: all necessary forms (4 uses); 3,000 p.s.i. concrete with an 8″ chute; all necessary reinforcing steel; and underdrain. Exposed concrete is patched and rubbed.

The Expanded System Listing shows walls that range in thickness from 10″ to 18″ for reinforced concrete walls with level backfill and 12″ to 24″ for reinforced walls with sloped backfill. Walls range from a height of 4′ to 20′. Unreinforced level and sloped backfill walls range from a height of 3′ to 10′.

RSMeans assemblies are identified by a **unique 12-character identifier**. The assemblies are numbered using UNIFORMAT II, ASTM Standard E1557. The first 6 characters represent this system to Level 4. The last 6 characters represent further breakdown by RSMeans in order to arrange items in understandable groups of similar tasks. Line numbers are consistent across all RSMeans publications, so a line number in any RSMeans assemblies data set will always refer to the same work.

System Components	QUANTITY	UNIT	COST PER L.F. MAT.	INST.	TOTAL
SYSTEM G2040 210 1000					
CONC. RETAIN. WALL REINFORCED, LEVEL BACKFILL, 4′ HIGH					
Forms in place, cont. wall footing & keyway, 4 uses	2.000	S.F.	4.46	8.86	13.32
Forms in place, retaining wall forms, battered to 8′ high, 4 uses	8.000	SFCA	5.44	68	73.44
Reinforcing in place, walls, #3 to #7	.004	Ton	4.40	3.40	7.80
Concrete ready mix, regular weight, 3000 psi	.204	C.Y.	22.24		22.24
Placing concrete and vibrating footing con., shallow direct chute	.074	C.Y.		1.78	1.78
Placing concrete and vibrating walls, 8″ thick, direct chute	.130	C.Y.		4.14	4.14
Pipe bedding, crushed or screened bank run gravel	1.000	L.F.	3.03	1.37	4.40
Pipe, subdrainage, corrugated plastic, 4″ diameter	1.000	L.F.	.56	.75	1.31
Finish walls and break ties, patch walls	4.000	S.F.	.16	3.84	4
TOTAL			40.29	92.14	132.43

G2040 210	Concrete Retaining Walls	COST PER L.F. MAT.	INST.	TOTAL
1000	Conc. retain. wall, reinforced, level backfill, 4′ high x 2′-2″ base, 10″ th	40.50	92	132.50
1200	6′ high x 3′-3″ base, 10″ thick	58	134	192
1400	8′ high x 4′-3″ base, 10″ thick	75	175	250
1600	10′ high x 5′-4″ base, 13″ thick	98	256	354
1800	12′ high x 6′-6″ base, 14″ thick	121	305	426
2200	16′ high x 8′-6″ base, 16″ thick	184	415	599
2600	20′ high x 10′-5″ base, 18″ thick	270	535	805
3000	Sloped backfill, 4′ high x 3′-2″ base, 12″ thick	49	95.50	144.50
3200	6′ high x 4′-6″ base, 12″ thick	69	138	207
3400	8′ high x 5′-11″ base, 12″ thick	91.50	181	272.50
3600	10′ high x 7′-5″ base, 16″ thick	130	268	398
3800	12′ high x 8′-10″ base, 18″ thick	170	325	495
4200	16′ high x 11′-10″ base, 21″ thick	285	460	745
4600	20′ high x 15′-0″ base, 24″ thick	445	625	1,070
5000	Unreinforced, level backfill, 3′-0″ high x 1′-6″ base	21	62	83
5200	4′-0″ high x 2′-0″ base	32.50	82.50	115
5400	6′-0″ high x 3′-0″ base	60.50	128	188.50
5600	8′-0″ high x 4′-0″ base	96.50	170	266.50

RSMeans assemblies descriptions appear in two formats: narrative and table. **Narrative descriptions** are shown in a hierarchical structure to make them readable. In order to read a complete description, read up through the indents to the top of the section. Include everything that is above and to the left that is not contradicted by information below.

For supplemental customizable square foot estimating forms, visit: **http:/www.reedconstructiondata.com/rsmeans/assemblies**

Most assemblies consist of three major elements: a graphic, the system components, and the cost data itself. The **Graphic** is a visual representation showing the typical appearance of the assembly in question, frequently accompanied by additional explanatory technical information describing the class of items. The **System Components** is a listing of the individual tasks that make up the assembly, including the quantity and unit of measure for each item, along with the cost of material and installation. The **Assemblies Data** below lists prices for other similar systems with dimensional and/or size variations.

All RSMeans assemblies costs represent the cost for the installing contractor. An allowance for profit has been added to all material, labor, and equipment rental costs. A markup for labor burdens, including workers' compensation, fixed overhead, and business overhead, is included with installation costs.

The data published in RSMeans print books represents a "national average" cost. This data should be modified to the project location using the **City Cost Indexes** or **Location Factors** tables found in the Reference Section in the back of the book.

TABLE FORMAT

A20 Basement Construction

A2020 Basement Walls

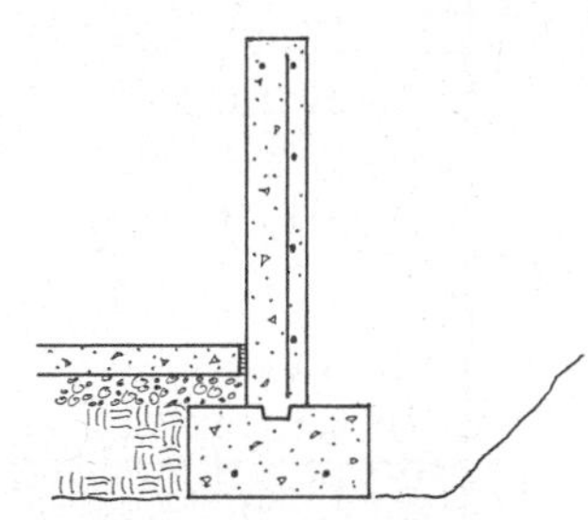

The Foundation Bearing Wall System includes: forms up to 16′ high (four uses); 3,000 p.s.i. concrete placed and vibrated; and form removal with breaking form ties and patching walls. The wall systems list walls from 6″ to 16″ thick and are designed with minimum reinforcement.

Excavation and backfill are not included.

Please see the reference section for further design and cost information.

All RSMeans assemblies data includes a typical **Unit of Measure** used for estimating that item. For instance, for continuous footings or foundation walls the unit is linear feet (L.F.). For spread footings the unit is each (Ea.). The estimator needs to take special care that the unit in the data matches the unit in the takeoff. Abbreviations and unit conversions can be found in the Reference Section.

System Components	QUANTITY	UNIT	COST PER L.F. MAT.	COST PER L.F. INST.	COST PER L.F. TOTAL
SYSTEM A2020 110 1500					
FOUNDATION WALL, CAST IN PLACE, DIRECT CHUTE, 4′ HIGH, 6″ THICK					
Formwork	8.000	SFCA	6	44	50
Reinforcing	3.300	Lb.	1.82	1.40	3.22
Unloading & sorting reinforcing	3.300	Lb.		.08	.08
Concrete, 3,000 psi	.074	C.Y.	8.07		8.07
Place concrete, direct chute	.074	C.Y.		2.35	2.35
Finish walls, break ties and patch voids, one side	4.000	S.F.	.16	3.84	4
TOTAL			16.05	51.67	67.72

System components are listed separately to detail what is included in the development of the total system price.

A2020 110	Walls, Cast in Place							
	WALL HEIGHT (FT.)	PLACING METHOD	CONCRETE (C.Y. per L.F.)	REINFORCING (LBS. per L.F.)	WALL THICKNESS (IN.)	COST PER L.F. MAT.	COST PER L.F. INST.	COST PER L.F. TOTAL
1500	4′	direct chute	.074	3.3	6	16.05	51.50	67.55
1520			.099	4.8	8	19.60	53	72.60
1540			.123	6.0	10	23	54	77
1560			.148	7.2	12	26.50	55	81.50
1580			.173	8.1	14	29.50	56	85.50
1600			.197	9.44	16	33	57.50	90.50
1700	4′	pumped	.074	3.3	6	16.05	52.50	68.55
1720			.099	4.8	8	19.60	55	74.60
1740			.123	6.0	10	23	56	79
1760			.148	7.2	12	26.50	57	83.50
1780			.173	8.1	14	29.50	58.50	88
1800			.197	9.44	16	33	60	93
3000	6′	direct chute	.111	4.95	6	24	77.50	101.50
3020			.149	7.20	8	29.50	79.50	109
3040			.184	9.00	10	34.50	81	115.50
3060			.222	10.8	12	39.50	83	122.50
3080			.260	12.15	14	44.50	84	128.50
3100			.300	14.39	16	50	86.50	136.50

Table descriptions work in a similar fashion, except that if there is a blank in the column at a particular line number, read up to the description above in the same column.

How RSMeans Assemblies Data Works (Continued)

Sample Estimate

This sample demonstrates the elements of an estimate, including a tally of the RSMeans data lines. Published assemblies costs include all markups for labor burden and profit for the installing contractor. This estimate adds a summary of the markups applied by a general contractor on the installing contractors' work. These figures represent the total cost to the owner. The RSMeans location factor is added at the bottom of the estimate to adjust the cost of the work to a specific location. **NOTE: The following assemblies are not found in this book and are used here for example only.**

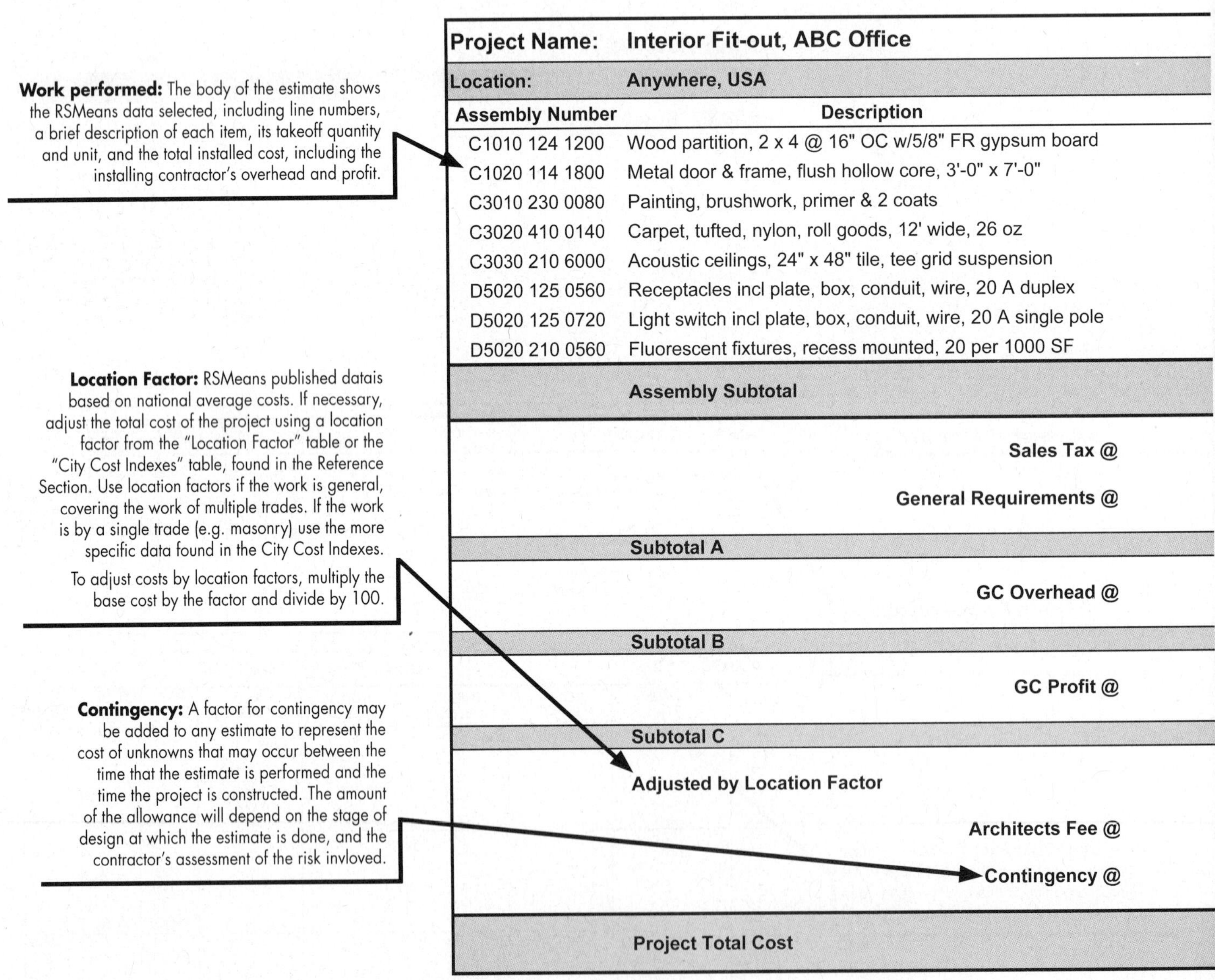

Project Name:	Interior Fit-out, ABC Office
Location:	Anywhere, USA
Assembly Number	**Description**
C1010 124 1200	Wood partition, 2 x 4 @ 16" OC w/5/8" FR gypsum board
C1020 114 1800	Metal door & frame, flush hollow core, 3'-0" x 7'-0"
C3010 230 0080	Painting, brushwork, primer & 2 coats
C3020 410 0140	Carpet, tufted, nylon, roll goods, 12' wide, 26 oz
C3030 210 6000	Acoustic ceilings, 24" x 48" tile, tee grid suspension
D5020 125 0560	Receptacles incl plate, box, conduit, wire, 20 A duplex
D5020 125 0720	Light switch incl plate, box, conduit, wire, 20 A single pole
D5020 210 0560	Fluorescent fixtures, recess mounted, 20 per 1000 SF
	Assembly Subtotal
	Sales Tax @
	General Requirements @
	Subtotal A
	GC Overhead @
	Subtotal B
	GC Profit @
	Subtotal C
	Adjusted by Location Factor
	Architects Fee @
	Contingency @
	Project Total Cost

Work performed: The body of the estimate shows the RSMeans data selected, including line numbers, a brief description of each item, its takeoff quantity and unit, and the total installed cost, including the installing contractor's overhead and profit.
Location Factor: RSMeans published datais based on national average costs. If necessary, adjust the total cost of the project using a location factor from the "Location Factor" table or the "City Cost Indexes" table, found in the Reference Section. Use location factors if the work is general, covering the work of multiple trades. If the work is by a single trade (e.g. masonry) use the more specific data found in the City Cost Indexes.
To adjust costs by location factors, multiply the base cost by the factor and divide by 100.
Contingency: A factor for contingency may be added to any estimate to represent the cost of unknowns that may occur between the time that the estimate is performed and the time the project is constructed. The amount of the allowance will depend on the stage of design at which the estimate is done, and the contractor's assessment of the risk invloved.

This estimate is based on an interactive spreadsheet. A copy of this spreadsheet is located on the RSMeans website at **http://www.reedconstructiondata.com/rsmeans/extras/546011.** You are free to download it and adjust it to your methodology.

Date: 1/1/2014		STD
Qty.	Unit	Subtotal
560.000	S.F.	$2,592.80
2.000	Ea.	$2,610.00
1,120.000	S.F.	$1,321.60
240.000	S.F.	$1,147.20
200.000	S.F.	$914.00
8.000	Ea.	$2,104.00
2.000	Ea.	$527.00
200.000	S.F.	$2,092.00
		$13,308.60
5 %		$ 332.72
7 %		$ 931.60
		$14,572.92
5 %		$ 728.65
		$15,301.56
5 %		$ 765.08
		$16,066.64
118.1		$ 18,974.70
8 %		$ 1,517.98
15 %		$ 2,846.21
		$ 23,338.88

Sales Tax: If the work is subject to state or local sales taxes, the amount must be added to the estimate. In a conceptual estimate it can be assumed that one half of the total represents material costs. Therefore, apply the sales tax rate to 50% of the assembly subtotal.

General Requirements: This item covers project-wide needs provided by the General Contractor. These items vary by project, but may include temporary facilities and utilities, security, testing, project cleanup, etc. In assemblies estimates a percentage is used, typically between 5% and 15% of project cost.

General Contractor Overhead: This entry represents the General Contractor's markup on all work to cover project administration costs.

General Contractor Profit: This entry represents the GC's profit on all work performed. The value included here can vary widely by project, and is influenced by the GC's perception of the project's financial risk and market conditions.

Architects Fee: If appropriate, add the design cost to the project estimate. These fees vary based on project complexity and size. Typical design and engineering fees can be found in the reference section.

A Substructure

A1010 Standard Foundations

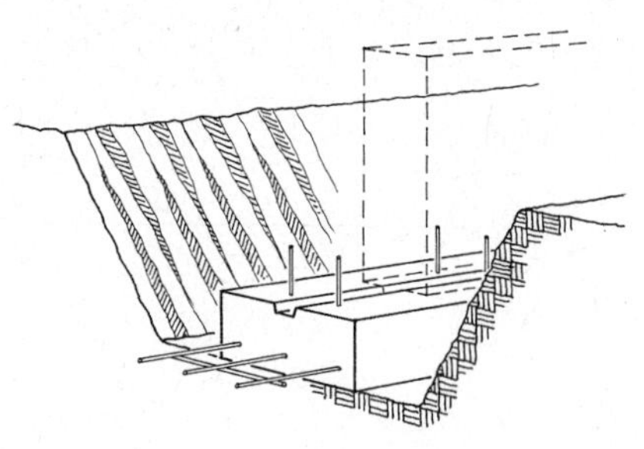

The Strip Footing System includes: excavation; hand trim; all forms needed for footing placement; forms for 2″ x 6″ keyway (four uses); dowels; and 3,000 p.s.i. concrete.

The footing size required varies for different soils. Soil bearing capacities are listed for 3 KSF and 6 KSF. Depths of the system range from 8″ and deeper. Widths range from 16″ and wider. Smaller strip footings may not require reinforcement.

Please see the reference section for further design and cost information.

System Components	QUANTITY	UNIT	COST PER L.F.		
			MAT.	INST.	TOTAL
SYSTEM A1010 110 2500					
STRIP FOOTING, LOAD 5.1 KLF, SOIL CAP. 3 KSF, 24″ WIDE X 12″ DEEP, REINF.					
Trench excavation	.148	C.Y.		1.43	1.43
Hand trim	2.000	S.F.		1.90	1.90
Compacted backfill	.074	C.Y.		.29	.29
Formwork, 4 uses	2.000	S.F.	4.46	8.86	13.32
Keyway form, 4 uses	1.000	L.F.	.31	1.13	1.44
Reinforcing, fy = 60000 psi	3.000	Lb.	1.71	1.83	3.54
Dowels	2.000	Ea.	1.62	5.30	6.92
Concrete, f'c = 3000 psi	.074	C.Y.	8.07		8.07
Place concrete, direct chute	.074	C.Y.		1.78	1.78
Screed finish	2.000	S.F.		.74	.74
TOTAL			16.17	23.26	39.43

A1010 110	Strip Footings	MAT.	INST.	TOTAL
		COST PER L.F.		
2100	Strip footing, load 2.6 KLF, soil capacity 3 KSF, 16″ wide x 8″ deep, plain	8.30	13.95	22.25
2300	Load 3.9 KLF, soil capacity, 3 KSF, 24″ wide x 8″ deep, plain	10.15	15.35	25.50
2500	Load 5.1 KLF, soil capacity 3 KSF, 24″ wide x 12″ deep, reinf.	16.15	23	39.15
2700	Load 11.1 KLF, soil capacity 6 KSF, 24″ wide x 12″ deep, reinf.	16.15	23	39.15
2900	Load 6.8 KLF, soil capacity 3 KSF, 32″ wide x 12″ deep, reinf.	19.55	25	44.55
3100	Load 14.8 KLF, soil capacity 6 KSF, 32″ wide x 12″ deep, reinf.	19.55	25	44.55
3300	Load 9.3 KLF, soil capacity 3 KSF, 40″ wide x 12″ deep, reinf.	23	27	50
3500	Load 18.4 KLF, soil capacity 6 KSF, 40″ wide x 12″ deep, reinf.	23	27.50	50.50
3700	Load 10.1 KLF, soil capacity 3 KSF, 48″ wide x 12″ deep, reinf.	25.50	29.50	55
3900	Load 22.1 KLF, soil capacity 6 KSF, 48″ wide x 12″ deep, reinf.	27	31.50	58.50
4100	Load 11.8 KLF, soil capacity 3 KSF, 56″ wide x 12″ deep, reinf.	29.50	32.50	62
4300	Load 25.8 KLF, soil capacity 6 KSF, 56″ wide x 12″ deep, reinf.	32	35.50	67.50
4500	Load 10 KLF, soil capacity 3 KSF, 48″ wide x 16″ deep, reinf.	32	34.50	66.50
4700	Load 22 KLF, soil capacity 6 KSF, 48″ wide, 16″ deep, reinf.	33	35.50	68.50
4900	Load 11.6 KLF, soil capacity 3 KSF, 56″ wide x 16″ deep, reinf.	36.50	50	86.50
5100	Load 25.6 KLF, soil capacity 6 KSF, 56″ wide x 16″ deep, reinf.	38.50	52	90.50
5300	Load 13.3 KLF, soil capacity 3 KSF, 64″ wide x 16″ deep, reinf.	41.50	41	82.50
5500	Load 29.3 KLF, soil capacity 6 KSF, 64″ wide x 16″ deep, reinf.	44.50	44	88.50
5700	Load 15 KLF, soil capacity 3 KSF, 72″ wide x 20″ deep, reinf.	55.50	49	104.50
5900	Load 33 KLF, soil capacity 6 KSF, 72″ wide x 20″ deep, reinf.	58.50	52.50	111
6100	Load 18.3 KLF, soil capacity 3 KSF, 88″ wide x 24″ deep, reinf.	78	63	141
6300	Load 40.3 KLF, soil capacity 6 KSF, 88″ wide x 24″ deep, reinf.	84.50	69.50	154
6500	Load 20 KLF, soil capacity 3 KSF, 96″ wide x 24″ deep, reinf.	84.50	66.50	151
6700	Load 44 KLF, soil capacity 6 KSF, 96″ wide x 24″ deep, reinf.	89.50	72	161.50

A1010 Standard Foundations

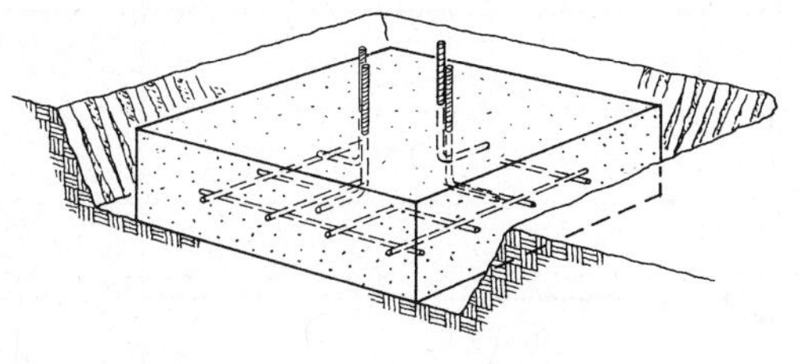

The Spread Footing System includes: excavation; backfill; forms (four uses); all reinforcement; 3,000 p.s.i. concrete (chute placed); and float finish.

Footing systems are priced per individual unit. The Expanded System Listing at the bottom shows various footing sizes. It is assumed that excavation is done by a truck mounted hydraulic excavator with an operator and oiler.

Backfill is with a dozer, and compaction by air tamp. The excavation and backfill equipment is assumed to operate at 30 C.Y. per hour.

Please see the reference section for further design and cost information.

System Components	QUANTITY	UNIT	COST EACH MAT.	COST EACH INST.	COST EACH TOTAL
SYSTEM A1010 210 7100					
SPREAD FOOTINGS, LOAD 25K, SOIL CAPACITY 3 KSF, 3′ SQ X 12″ DEEP					
Bulk excavation	.590	C.Y.		4.99	4.99
Hand trim	9.000	S.F.		8.55	8.55
Compacted backfill	.260	C.Y.		1.01	1.01
Formwork, 4 uses	12.000	S.F.	8.52	62.40	70.92
Reinforcing, fy = 60,000 psi	.006	Ton	6.60	7.20	13.80
Dowel or anchor bolt templates	6.000	L.F.	5.52	25.80	31.32
Concrete, f′c = 3,000 psi	.330	C.Y.	35.97		35.97
Place concrete, direct chute	.330	C.Y.		7.96	7.96
Float finish	9.000	S.F.		3.33	3.33
TOTAL			56.61	121.24	177.85

A1010 210	Spread Footings	COST EACH MAT.	COST EACH INST.	COST EACH TOTAL
7090	Spread footings, 3000 psi concrete, chute delivered			
7100	Load 25K, soil capacity 3 KSF, 3′-0″ sq. x 12″ deep	56.50	122	178.50
7150	Load 50K, soil capacity 3 KSF, 4′-6″ sq. x 12″ deep	122	209	331
7200	Load 50K, soil capacity 6 KSF, 3′-0″ sq. x 12″ deep	56.50	122	178.50
7250	Load 75K, soil capacity 3 KSF, 5′-6″ sq. x 13″ deep	194	294	488
7300	Load 75K, soil capacity 6 KSF, 4′-0″ sq. x 12″ deep	98.50	179	277.50
7350	Load 100K, soil capacity 3 KSF, 6′-0″ sq. x 14″ deep	245	355	600
7410	Load 100K, soil capacity 6 KSF, 4′-6″ sq. x 15″ deep	150	246	396
7450	Load 125K, soil capacity 3 KSF, 7′-0″ sq. x 17″ deep	390	505	895
7500	Load 125K, soil capacity 6 KSF, 5′-0″ sq. x 16″ deep	193	295	488
7550	Load 150K, soil capacity 3 KSF 7′-6″ sq. x 18″ deep	470	590	1,060
7610	Load 150K, soil capacity 6 KSF, 5′-6″ sq. x 18″ deep	257	370	627
7650	Load 200K, soil capacity 3 KSF, 8′-6″ sq. x 20″ deep	670	775	1,445
7700	Load 200K, soil capacity 6 KSF, 6′-0″ sq. x 20″ deep	340	455	795
7750	Load 300K, soil capacity 3 KSF, 10′-6″ sq. x 25″ deep	1,225	1,275	2,500
7810	Load 300K, soil capacity 6 KSF, 7′-6″ sq. x 25″ deep	640	765	1,405
7850	Load 400K, soil capacity 3 KSF, 12′-6″ sq. x 28″ deep	1,975	1,900	3,875
7900	Load 400K, soil capacity 6 KSF, 8′-6″ sq. x 27″ deep	895	995	1,890
7950	Load 500K, soil capacity 3 KSF, 14′-0″ sq. x 31″ deep	2,725	2,450	5,175
8010	Load 500K, soil capacity 6 KSF, 9′-6″ sq. x 30″ deep	1,225	1,300	2,525
8050	Load 600K, soil capacity 3 KSF, 16′-0″ sq. x 35″ deep	4,000	3,375	7,375
8100	Load 600K, soil capacity 6 KSF, 10′-6″ sq. x 33″ deep	1,675	1,675	3,350

A1010 Standard Foundations

A1010 210	Spread Footings	COST EACH		
		MAT.	INST.	TOTAL
8150	Load 700K, soil capacity 3 KSF, 17'-0" sq. x 37" deep	4,700	3,875	8,575
8200	Load 700K, soil capacity 6 KSF, 11'-6" sq. x 36" deep	2,125	2,025	4,150
8250	Load 800K, soil capacity 3 KSF, 18'-0" sq. x 39" deep	5,575	4,475	10,050
8300	Load 800K, soil capacity 6 KSF, 12'-0" sq. x 37" deep	2,400	2,250	4,650
8350	Load 900K, soil capacity 3 KSF, 19'-0" sq. x 40" deep	6,475	5,125	11,600
8400	Load 900K, soil capacity 6 KSF, 13'-0" sq. x 39" deep	2,950	2,675	5,625
8450	Load 1000K, soil capacity 3 KSF, 20'-0" sq. x 42" deep	7,450	5,775	13,225
8500	Load 1000K, soil capacity 6 KSF, 13'-6" sq. x 41" deep	3,350	2,975	6,325
8550	Load 1200K, soil capacity 6 KSF, 15'-0" sq. x 48" deep	4,450	3,775	8,225
8600	Load 1400K, soil capacity 6 KSF, 16'-0" sq. x 47" deep	5,450	4,475	9,925
8650	Load 1600K, soil capacity 6 KSF, 18'-0" sq. x 52" deep	7,550	5,925	13,475

A1010 Standard Foundations

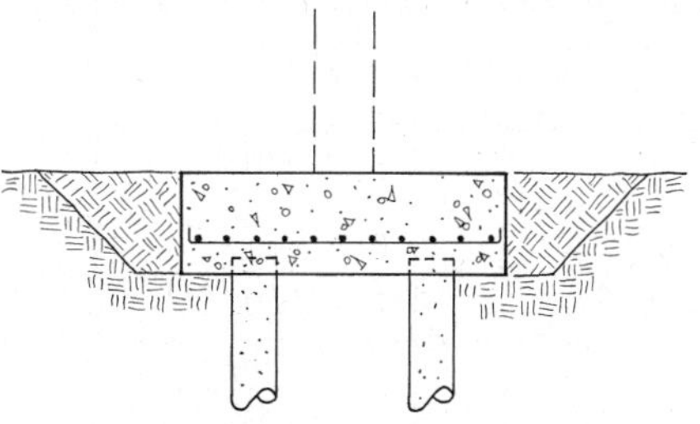

These pile cap systems include excavation with a truck mounted hydraulic excavator, hand trimming, compacted backfill, forms for concrete, templates for dowels or anchor bolts, reinforcing steel and concrete placed and floated.

Pile embedment is assumed as 6″. Design is consistent with the Concrete Reinforcing Steel Institute Handbook f'c = 3000 psi, fy = 60,000.

Please see the reference section for further design and cost information.

System Components	QUANTITY	UNIT	COST EACH MAT.	COST EACH INST.	COST EACH TOTAL
SYSTEM A1010 250 5100					
CAP FOR 2 PILES, 6'-6″X3'-6″X20″, 15 TON PILE, 8″ MIN. COL., 45K COL. LOAD					
Excavation, bulk, hyd excavator, truck mtd. 30″ bucket 1/2 CY	2.890	C.Y.		24.48	24.48
Trim sides and bottom of trench, regular soil	23.000	S.F.		21.85	21.85
Dozer backfill & roller compaction	1.500	C.Y.		5.86	5.86
Forms in place pile cap, square or rectangular, 4 uses	33.000	SFCA	33	184.80	217.80
Templates for dowels or anchor bolts	8.000	Ea.	7.36	34.40	41.76
Reinforcing in place footings, #8 to #14	.025	Ton	27.50	17.63	45.13
Concrete ready mix, regular weight, 3000 psi	1.400	C.Y.	152.60		152.60
Place and vibrate concrete for pile caps, under 5 CY, direct chute	1.400	C.Y.		44.53	44.53
Float finish	23.000	S.F.		8.51	8.51
TOTAL			220.46	342.06	562.52

A1010 250	Pile Caps						
	NO. PILES	SIZE FT-IN X FT-IN X IN	PILE CAPACITY (TON)	COLUMN SIZE (IN)	COLUMN LOAD (K)	COST EACH MAT.	COST EACH INST.	COST EACH TOTAL

	NO. PILES	SIZE FT-IN X FT-IN X IN	PILE CAPACITY (TON)	COLUMN SIZE (IN)	COLUMN LOAD (K)	COST EACH MAT.	COST EACH INST.	COST EACH TOTAL
5100	2	6-6x3-6x20	15	8	45	220	345	565
5150		26	40	8	155	270	420	690
5200		34	80	11	314	365	540	905
5250		37	120	14	473	390	575	965
5300	3	5-6x5-1x23	15	8	75	265	395	660
5350		28	40	10	232	295	445	740
5400		32	80	14	471	340	505	845
5450		38	120	17	709	395	580	975
5500	4	5-6x5-6x18	15	10	103	300	395	695
5550		30	40	11	308	435	565	1,000
5600		36	80	16	626	510	660	1,170
5650		38	120	19	945	540	695	1,235
5700	6	8-6x5-6x18	15	12	156	505	570	1,075
5750		37	40	14	458	805	845	1,650
5800		40	80	19	936	925	940	1,865
5850		45	120	24	1413	1,025	1,050	2,075
5900	8	8-6x7-9x19	15	12	205	770	775	1,545
5950		36	40	16	610	1,075	1,025	2,100
6000		44	80	22	1243	1,350	1,225	2,575
6050		47	120	27	1881	1,450	1,325	2,775

A1010 Standard Foundations

A1010 250 Pile Caps

	NO. PILES	SIZE FT-IN X FT-IN X IN	PILE CAPACITY (TON)	COLUMN SIZE (IN)	COLUMN LOAD (K)	COST EACH		
						MAT.	INST.	TOTAL
6100	10	11-6x7-9x21	15	14	250	1,075	945	2,020
6150		39	40	17	756	1,625	1,350	2,975
6200		47	80	25	1547	1,975	1,625	3,600
6250		49	120	31	2345	2,100	1,700	3,800
6300	12	11-6x8-6x22	15	15	316	1,375	1,150	2,525
6350		49	40	19	900	2,125	1,725	3,850
6400		52	80	27	1856	2,375	1,875	4,250
6450		55	120	34	2812	2,575	2,025	4,600
6500	14	11-6x10-9x24	15	16	345	1,775	1,425	3,200
6550		41	40	21	1056	2,300	1,775	4,075
6600		55	80	29	2155	3,025	2,225	5,250
6700	16	11-6x11-6x26	15	18	400	2,175	1,625	3,800
6750		48	40	22	1200	2,825	2,100	4,925
6800		60	80	31	2460	3,525	2,550	6,075
6900	18	13-0x11-6x28	15	20	450	2,450	1,800	4,250
6950		49	40	23	1349	3,300	2,325	5,625
7000		56	80	33	2776	3,900	2,750	6,650
7100	20	14-6x11-6x30	15	20	510	3,000	2,125	5,125
7150		52	40	24	1491	3,975	2,725	6,700

A1010 Standard Foundations

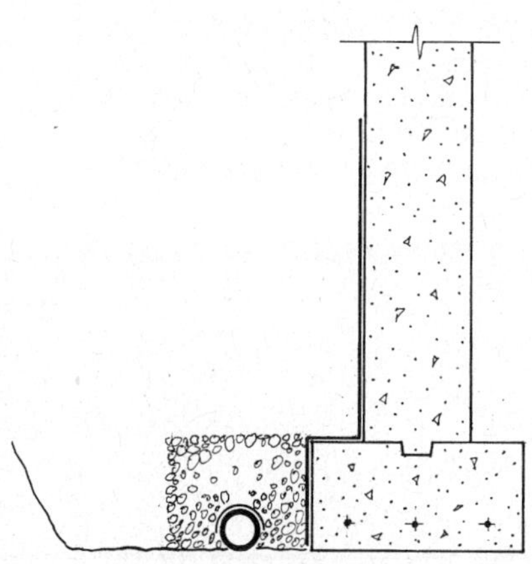

General: Footing drains can be placed either inside or outside of foundation walls depending upon the source of water to be intercepted. If the source of subsurface water is principally from grade or a subsurface stream above the bottom of the footing, outside drains should be used. For high water tables, use inside drains or both inside and outside.

The effectiveness of underdrains depends on good waterproofing. This must be carefully installed and protected during construction.

Costs below include the labor and materials for the pipe and 6″ of crushed stone around pipe. Excavation and backfill are not included.

System Components	QUANTITY	UNIT	COST PER L.F. MAT.	INST.	TOTAL
SYSTEM A1010 310 1000					
FOUNDATION UNDERDRAIN, OUTSIDE ONLY, PVC, 4″ DIAM.					
PVC pipe 4″ diam. S.D.R. 35	1.000	L.F.	1.61	4.04	5.65
Crushed stone 3/4″ to 1/2″	.070	C.Y.	2.17	.88	3.05
TOTAL			3.78	4.92	8.70

A1010 310	Foundation Underdrain	COST PER L.F. MAT.	INST.	TOTAL
1000	Foundation underdrain, outside only, PVC, 4″ diameter	3.78	4.92	8.70
1100	6″ diameter	6.40	5.45	11.85
1400	Perforated HDPE, 6″ diameter	4.41	2.14	6.55
1450	8″ diameter	6.45	2.66	9.11
1500	12″ diameter	12	6.50	18.50
1600	Corrugated metal, 16 ga. asphalt coated, 6″ diameter	10	5.10	15.10
1650	8″ diameter	12.85	5.50	18.35
1700	10″ diameter	15.85	7.15	23
3000	Outside and inside, PVC, 4″ diameter	7.55	9.85	17.40
3100	6″ diameter	12.80	10.95	23.75
3400	Perforated HDPE, 6″ diameter	8.80	4.27	13.07
3450	8″ diameter	12.90	5.35	18.25
3500	12″ diameter	24	12.90	36.90
3600	Corrugated metal, 16 ga., asphalt coated, 6″ diameter	20	10.25	30.25
3650	8″ diameter	25.50	10.95	36.45
3700	10″ diameter	31.50	14.20	45.70

A1010 Standard Foundations

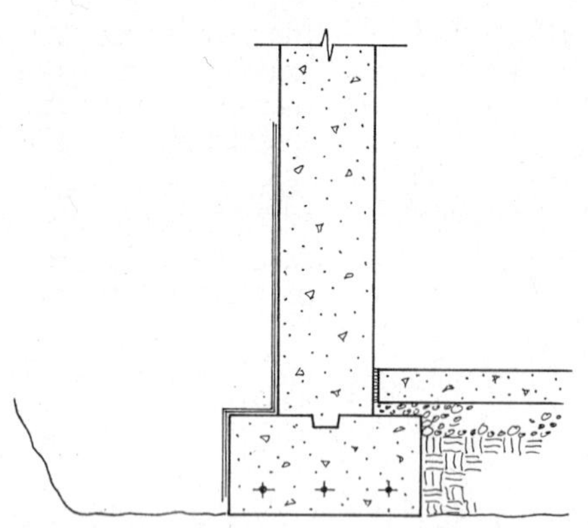

General: Apply foundation wall dampproofing over clean concrete giving particular attention to the joint between the wall and the footing. Use care in backfilling to prevent damage to the dampproofing.

Costs for four types of dampproofing are listed below.

System Components	QUANTITY	UNIT	COST PER L.F. MAT.	INST.	TOTAL
SYSTEM A1010 320 1000 FOUNDATION DAMPPROOFING, BITUMINOUS, 1 COAT, 4′ HIGH					
Bituminous asphalt dampproofing brushed on below grade, 1 coat	4.000	S.F.	1	3.20	4.20
Labor for protection of dampproofing during backfilling	4.000	S.F.		1.35	1.35
TOTAL			1	4.55	5.55

A1010 320	Foundation Dampproofing	COST PER L.F. MAT.	INST.	TOTAL
1000	Foundation dampproofing, bituminous, 1 coat, 4′ high	1	4.55	5.55
1400	8′ high	2	9.10	11.10
1800	12′ high	3	14.10	17.10
2000	2 coats, 4′ high	2	5.65	7.65
2400	8′ high	4	11.25	15.25
2800	12′ high	6	17.35	23.35
3000	Asphalt with fibers, 1/16″ thick, 4′ high	1.48	5.65	7.13
3400	8′ high	2.96	11.25	14.21
3800	12′ high	4.44	17.35	21.79
4000	1/8″ thick, 4′ high	2.60	6.65	9.25
4400	8′ high	5.20	13.35	18.55
4800	12′ high	7.80	20.50	28.30
5000	Asphalt coated board and mastic, 1/4″ thick, 4′ high	4.40	6.10	10.50
5400	8′ high	8.80	12.20	21
5800	12′ high	13.20	18.80	32
6000	1/2″ thick, 4′ high	6.60	8.50	15.10
6400	8′ high	13.20	17	30.20
6800	12′ high	19.80	26	45.80
7000	Cementitious coating, on walls, 1/8″ thick coating, 4′ high	2.64	8.80	11.44
7400	8′ high	5.30	17.60	22.90
7800	12′ high	7.90	26.50	34.40
8000	Cementitious/metallic slurry, 4 coat, 1/2″thick, 2′ high	.67	8.70	9.37
8400	4′ high	1.34	17.40	18.74
8800	6′ high	2.01	26	28.01

A1020 Special Foundations

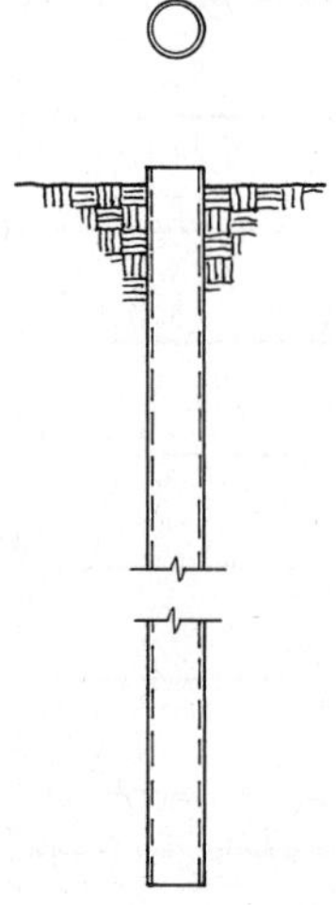

The Cast-in-Place Concrete Pile System includes: a defined number of 4,000 p.s.i. concrete piles with thin-wall, straight-sided, steel shells that have a standard steel plate driving point. An allowance for cutoffs is included.

The Expanded System Listing shows costs per cluster of piles. Clusters range from one pile to twenty piles. Loads vary from 50 Kips to 1,600 Kips. Both end-bearing and friction-type piles are shown.

Please see the reference section for cost of mobilization of the pile driving equipment and other design and cost information.

System Components	QUANTITY	UNIT	COST EACH MAT.	COST EACH INST.	COST EACH TOTAL
SYSTEM A1020 110 2220					
CIP SHELL CONCRETE PILE, 25′ LONG, 50K LOAD, END BEARING, 1 PILE					
7 Ga. shell, 12″ diam.	27.000	V.L.F.	904.50	297.81	1,202.31
Steel pipe pile standard point, 12″ or 14″ diameter pile	1.000	Ea.	114.50	89	203.50
Pile cutoff, conc. pile with thin steel shell	1.000	Ea.		14.80	14.80
TOTAL			1,019	401.61	1,420.61

A1020 110	C.I.P. Concrete Piles	COST EACH MAT.	COST EACH INST.	COST EACH TOTAL
2220	CIP shell concrete pile, 25′ long, 50K load, end bearing, 1 pile	1,025	400	1,425
2240	100K load, end bearing, 2 pile cluster	2,050	805	2,855
2260	200K load, end bearing, 4 pile cluster	4,075	1,625	5,700
2280	400K load, end bearing, 7 pile cluster	7,125	2,825	9,950
2300	10 pile cluster	10,200	4,025	14,225
2320	800K load, end bearing, 13 pile cluster	22,300	7,325	29,625
2340	17 pile cluster	29,200	9,575	38,775
2360	1200K load, end bearing, 14 pile cluster	24,100	7,900	32,000
2380	19 pile cluster	32,600	10,700	43,300
2400	1600K load, end bearing, 19 pile cluster	32,600	10,700	43,300
2420	50′ long, 50K load, end bearing, 1 pile	1,900	690	2,590
2440	Friction type, 2 pile cluster	3,675	1,375	5,050
2460	3 pile cluster	5,500	2,050	7,550
2480	100K load, end bearing, 2 pile cluster	3,775	1,375	5,150
2500	Friction type, 4 pile cluster	7,350	2,750	10,100
2520	6 pile cluster	11,000	4,125	15,125
2540	200K load, end bearing, 4 pile cluster	7,550	2,750	10,300
2560	Friction type, 8 pile cluster	14,700	5,500	20,200
2580	10 pile cluster	18,400	6,900	25,300
2600	400K load, end bearing, 7 pile cluster	13,200	4,825	18,025
2620	Friction type, 16 pile cluster	29,400	11,000	40,400
2640	19 pile cluster	34,900	13,100	48,000
2660	800K load, end bearing, 14 pile cluster	43,700	12,900	56,600
2680	20 pile cluster	62,500	18,500	81,000
2700	1200K load, end bearing, 15 pile cluster	46,800	13,900	60,700
2720	1600K load, end bearing, 20 pile cluster	62,500	18,500	81,000
3740	75′ long, 50K load, end bearing, 1 pile	2,900	1,175	4,075
3760	Friction type, 2 pile cluster	5,625	2,325	7,950

A1020 Special Foundations

A1020 110	C.I.P. Concrete Piles	COST EACH		
		MAT.	INST.	TOTAL
3780	3 pile cluster	8,425	3,500	11,925
3800	100K load, end bearing, 2 pile cluster	5,775	2,325	8,100
3820	Friction type, 3 pile cluster	8,425	3,500	11,925
3840	5 pile cluster	14,100	5,825	19,925
3860	200K load, end bearing, 4 pile cluster	11,600	4,675	16,275
3880	6 pile cluster	17,300	7,000	24,300
3900	Friction type, 6 pile cluster	16,900	7,000	23,900
3910	7 pile cluster	19,700	8,150	27,850
3920	400K load, end bearing, 7 pile cluster	20,200	8,150	28,350
3930	11 pile cluster	31,800	12,800	44,600
3940	Friction type, 12 pile cluster	33,700	14,000	47,700
3950	14 pile cluster	39,400	16,400	55,800
3960	800K load, end bearing, 15 pile cluster	70,500	22,800	93,300
3970	20 pile cluster	93,500	30,400	123,900
3980	1200K load, end bearing, 17 pile cluster	79,500	25,800	105,300

A1020 Special Foundations

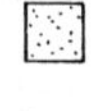

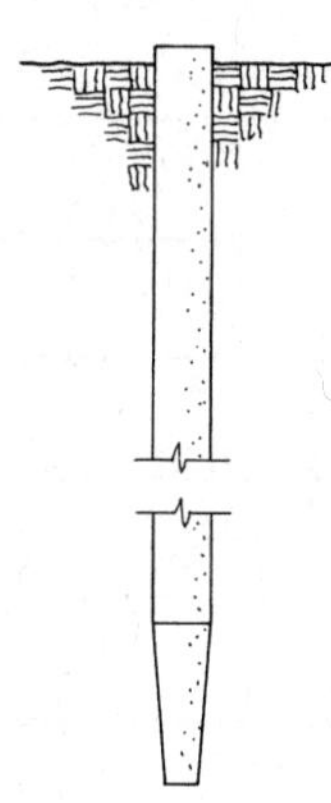

The Precast Concrete Pile System includes: pre-stressed concrete piles; standard steel driving point; and an allowance for cutoffs.

The Expanded System Listing shows costs per cluster of piles. Clusters range from one pile to twenty piles. Loads vary from 50 Kips to 1,600 Kips. Both end-bearing and friction type piles are listed.

Please see the reference section for cost of mobilization of the pile driving equipment and other design and cost information.

System Components	QUANTITY	UNIT	COST EACH MAT.	COST EACH INST.	COST EACH TOTAL
SYSTEM A1020 120 2220					
PRECAST CONCRETE PILE, 50′ LONG, 50K LOAD, END BEARING, 1 PILE					
Precast, prestressed conc. piles, 10″ square, no mobil.	53.000	V.L.F.	1,054.70	500.32	1,555.02
Steel pipe pile standard point, 8″ to 10″ diameter	1.000	Ea.	82.50	78.50	161
Piling special costs cutoffs concrete piles plain	1.000	Ea.		102	102
TOTAL			1,137.20	680.82	1,818.02

A1020 120	Precast Concrete Piles	COST EACH MAT.	COST EACH INST.	COST EACH TOTAL
2220	Precast conc pile, 50′ long, 50K load, end bearing, 1 pile	1,125	685	1,810
2240	Friction type, 2 pile cluster	2,775	1,425	4,200
2260	4 pile cluster	5,550	2,825	8,375
2280	100K load, end bearing, 2 pile cluster	2,275	1,350	3,625
2300	Friction type, 2 pile cluster	2,775	1,425	4,200
2320	4 pile cluster	5,550	2,825	8,375
2340	7 pile cluster	9,700	4,950	14,650
2360	200K load, end bearing, 3 pile cluster	3,400	2,025	5,425
2380	4 pile cluster	4,550	2,725	7,275
2400	Friction type, 8 pile cluster	11,100	5,675	16,775
2420	9 pile cluster	12,500	6,375	18,875
2440	14 pile cluster	19,400	9,875	29,275
2460	400K load, end bearing, 6 pile cluster	6,825	4,100	10,925
2480	8 pile cluster	9,100	5,450	14,550
2500	Friction type, 14 pile cluster	19,400	9,875	29,275
2520	16 pile cluster	22,200	11,300	33,500
2540	18 pile cluster	25,000	12,700	37,700
2560	800K load, end bearing, 12 pile cluster	14,800	8,850	23,650
2580	16 pile cluster	18,200	10,900	29,100
2600	1200K load, end bearing, 19 pile cluster	56,500	15,900	72,400
2620	20 pile cluster	59,500	16,700	76,200
2640	1600K load, end bearing, 19 pile cluster	56,500	15,900	72,400
4660	100′ long, 50K load, end bearing, 1 pile	2,175	1,175	3,350
4680	Friction type, 1 pile	2,625	1,225	3,850

A1020 Special Foundations

A1020 120	Precast Concrete Piles	COST EACH		
		MAT.	INST.	TOTAL
4700	2 pile cluster	5,275	2,425	7,700
4720	100K load, end bearing, 2 pile cluster	4,350	2,325	6,675
4740	Friction type, 2 pile cluster	5,275	2,425	7,700
4760	3 pile cluster	7,900	3,650	11,550
4780	4 pile cluster	10,500	4,850	15,350
4800	200K load, end bearing, 3 pile cluster	6,525	3,500	10,025
4820	4 pile cluster	8,700	4,700	13,400
4840	Friction type, 3 pile cluster	7,900	3,650	11,550
4860	5 pile cluster	13,200	6,075	19,275
4880	400K load, end bearing, 6 pile cluster	13,000	7,025	20,025
4900	8 pile cluster	17,400	9,375	26,775
4910	Friction type, 8 pile cluster	21,100	9,700	30,800
4920	10 pile cluster	26,300	12,100	38,400
4930	800K load, end bearing, 13 pile cluster	28,200	15,300	43,500
4940	16 pile cluster	34,800	18,800	53,600
4950	1200K load, end bearing, 19 pile cluster	109,500	27,500	137,000
4960	20 pile cluster	115,000	29,000	144,000
4970	1600K load, end bearing, 19 pile cluster	109,500	27,500	137,000

A1020 Special Foundations

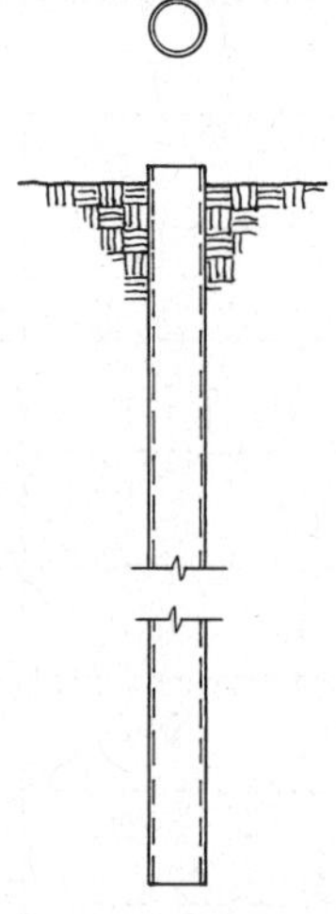

The Steel Pipe Pile System includes: steel pipe sections filled with 4,000 p.s.i. concrete; a standard steel driving point; splices when required and an allowance for cutoffs.

The Expanded System Listing shows costs per cluster of piles. Clusters range from one pile to twenty piles. Loads vary from 50 Kips to 1,600 Kips. Both end-bearing and friction-type piles are shown.

Please see the reference section for cost of mobilization of the pile driving equipment and other design and cost information.

System Components	QUANTITY	UNIT	COST EACH MAT.	INST.	TOTAL
SYSTEM A1020 130 2220					
CONC. FILL STEEL PIPE PILE, 50′ LONG, 50K LOAD, END BEARING, 1 PILE					
Piles, steel, pipe, conc. filled, 12″ diameter	53.000	V.L.F.	2,146.50	844.29	2,990.79
Steel pipe pile, standard point, for 12″ or 14″ diameter pipe	1.000	Ea.	229	178	407
Pile cut off, concrete pile, thin steel shell	1.000	Ea.		14.80	14.80
TOTAL			2,375.50	1,037.09	3,412.59

A1020 130	Steel Pipe Piles	COST EACH MAT.	INST.	TOTAL
2220	Conc. fill steel pipe pile, 50′ long, 50K load, end bearing, 1 pile	2,375	1,050	3,425
2240	Friction type, 2 pile cluster	4,750	2,075	6,825
2250	100K load, end bearing, 2 pile cluster	4,750	2,075	6,825
2260	3 pile cluster	7,125	3,100	10,225
2300	Friction type, 4 pile cluster	9,500	4,150	13,650
2320	5 pile cluster	11,900	5,200	17,100
2340	10 pile cluster	23,800	10,400	34,200
2360	200K load, end bearing, 3 pile cluster	7,125	3,100	10,225
2380	4 pile cluster	9,500	4,150	13,650
2400	Friction type, 4 pile cluster	9,500	4,150	13,650
2420	8 pile cluster	19,000	8,300	27,300
2440	9 pile cluster	21,400	9,325	30,725
2460	400K load, end bearing, 6 pile cluster	14,300	6,225	20,525
2480	7 pile cluster	16,600	7,250	23,850
2500	Friction type, 9 pile cluster	21,400	9,325	30,725
2520	16 pile cluster	38,000	16,700	54,700
2540	19 pile cluster	45,100	19,700	64,800
2560	800K load, end bearing, 11 pile cluster	26,100	11,400	37,500
2580	14 pile cluster	33,300	14,500	47,800
2600	15 pile cluster	35,600	15,600	51,200
2620	Friction type, 17 pile cluster	40,400	17,600	58,000
2640	1200K load, end bearing, 16 pile cluster	38,000	16,700	54,700
2660	20 pile cluster	47,500	20,800	68,300
2680	1600K load, end bearing, 17 pile cluster	40,400	17,600	58,000
3700	100′ long, 50K load, end bearing, 1 pile	4,600	2,050	6,650
3720	Friction type, 1 pile	4,600	2,050	6,650

A1020 Special Foundations

A1020 130	Steel Pipe Piles	COST EACH		
		MAT.	INST.	TOTAL
3740	2 pile cluster	9,225	4,100	13,325
3760	100K load, end bearing, 2 pile cluster	9,225	4,100	13,325
3780	Friction type, 2 pile cluster	9,225	4,100	13,325
3800	3 pile cluster	13,800	6,150	19,950
3820	200K load, end bearing, 3 pile cluster	13,800	6,150	19,950
3840	4 pile cluster	18,400	8,225	26,625
3860	Friction type, 3 pile cluster	13,800	6,150	19,950
3880	4 pile cluster	18,400	8,225	26,625
3900	400K load, end bearing, 6 pile cluster	27,700	12,300	40,000
3910	7 pile cluster	32,300	14,400	46,700
3920	Friction type, 5 pile cluster	23,100	10,300	33,400
3930	8 pile cluster	36,900	16,400	53,300
3940	800K load, end bearing, 11 pile cluster	50,500	22,600	73,100
3950	14 pile cluster	64,500	28,800	93,300
3960	15 pile cluster	69,000	30,800	99,800
3970	1200K load, end bearing, 16 pile cluster	74,000	32,900	106,900
3980	20 pile cluster	92,000	41,100	133,100
3990	1600K load, end bearing, 17 pile cluster	78,500	34,900	113,400

A1020 Special Foundations

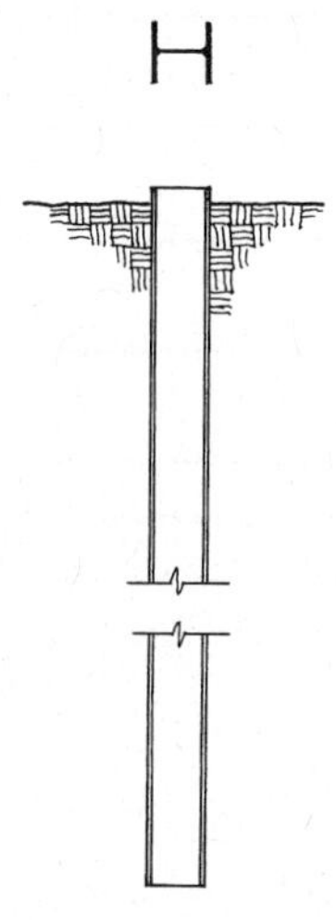

A Steel "H" Pile System includes: steel H sections; heavy duty driving point; splices where applicable and allowance for cutoffs.

The Expanded System Listing shows costs per cluster of piles. Clusters range from one pile to eighteen piles. Loads vary from 50 Kips to 2,000 Kips. All loads for Steel H Pile systems are given in terms of end bearing capacity.

Steel sections range from 10″ x 10″ to 14″ x 14″ in the Expanded System Listing. The 14″ x 14″ steel section is used for all H piles used in applications requiring a working load over 800 Kips.

Please see the reference section for cost of mobilization of the pile driving equipment and other design and cost information.

System Components	QUANTITY	UNIT	COST EACH MAT.	COST EACH INST.	COST EACH TOTAL
SYSTEM A1020 140 2220					
STEEL H PILES, 50′ LONG, 100K LOAD, END BEARING, 1 PILE					
Steel H piles 10″ x 10″, 42 #/L.F.	53.000	V.L.F.	1,086.50	573.46	1,659.96
Heavy duty point, 10″	1.000	Ea.	203	180	383
Pile cut off, steel pipe or H piles	1.000	Ea.		29.50	29.50
TOTAL			1,289.50	782.96	2,072.46

A1020 140	Steel H Piles	COST EACH MAT.	COST EACH INST.	COST EACH TOTAL
2220	Steel H piles, 50′ long, 100K load, end bearing, 1 pile	1,300	780	2,080
2260	2 pile cluster	2,575	1,575	4,150
2280	200K load, end bearing, 2 pile cluster	2,575	1,575	4,150
2300	3 pile cluster	3,875	2,350	6,225
2320	400K load, end bearing, 3 pile cluster	3,875	2,350	6,225
2340	4 pile cluster	5,150	3,125	8,275
2360	6 pile cluster	7,725	4,700	12,425
2380	800K load, end bearing, 5 pile cluster	6,450	3,900	10,350
2400	7 pile cluster	9,025	5,475	14,500
2420	12 pile cluster	15,500	9,375	24,875
2440	1200K load, end bearing, 8 pile cluster	10,300	6,250	16,550
2460	11 pile cluster	14,200	8,600	22,800
2480	17 pile cluster	21,900	13,300	35,200
2500	1600K load, end bearing, 10 pile cluster	16,300	8,225	24,525
2520	14 pile cluster	22,800	11,500	34,300
2540	2000K load, end bearing, 12 pile cluster	19,500	9,875	29,375
2560	18 pile cluster	29,300	14,800	44,100
3580	100′ long, 50K load, end bearing, 1 pile	4,225	1,775	6,000
3600	100K load, end bearing, 1 pile	4,225	1,775	6,000
3620	2 pile cluster	8,475	3,600	12,075
3640	200K load, end bearing, 2 pile cluster	8,475	3,600	12,075
3660	3 pile cluster	12,700	5,375	18,075
3680	400K load, end bearing, 3 pile cluster	12,700	5,375	18,075
3700	4 pile cluster	16,900	7,150	24,050
3720	6 pile cluster	25,400	10,800	36,200
3740	800K load, end bearing, 5 pile cluster	21,200	8,975	30,175

A1020 Special Foundations

A1020 140	Steel H Piles	COST EACH		
		MAT.	INST.	TOTAL
3760	7 pile cluster	29,600	12,500	42,100
3780	12 pile cluster	51,000	21,500	72,500
3800	1200K load, end bearing, 8 pile cluster	33,900	14,400	48,300
3820	11 pile cluster	46,600	19,700	66,300
3840	17 pile cluster	72,000	30,400	102,400
3860	1600K load, end bearing, 10 pile cluster	42,300	17,900	60,200
3880	14 pile cluster	59,000	25,100	84,100
3900	2000K load, end bearing, 12 pile cluster	51,000	21,500	72,500
3920	18 pile cluster	76,000	32,200	108,200

A1020 Special Foundations

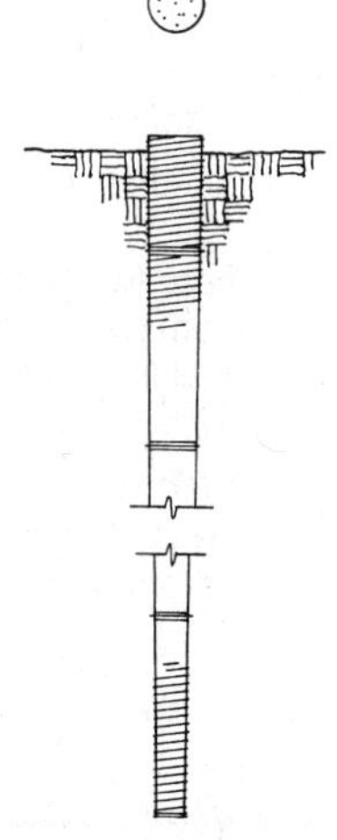

The Step Tapered Steel Pile System includes: step tapered piles filled with 4,000 p.s.i. concrete. The cost for splices and pile cutoffs is included.

The Expanded System Listing shows costs per cluster of piles. Clusters range from one pile to twenty-four piles. Both end bearing piles and friction piles are listed. Loads vary from 50 Kips to 1,600 Kips.

Please see the reference section for cost of mobilization of the pile driving equipment and other design and cost information.

System Components	QUANTITY	UNIT	COST EACH MAT.	COST EACH INST.	COST EACH TOTAL
SYSTEM A1020 150 1000					
STEEL PILE, STEP TAPERED, 50′ LONG, 50K LOAD, END BEARING, 1 PILE					
Steel shell step tapered conc. filled piles, 8″ tip 60 ton capacity to 60′	53.000	V.L.F.	747.30	473.82	1,221.12
Pile cutoff, steel pipe or H piles	1.000	Ea.		29.50	29.50
TOTAL			747.30	503.32	1,250.62

A1020 150	Step-Tapered Steel Piles	COST EACH MAT.	COST EACH INST.	COST EACH TOTAL
1000	Steel pile, step tapered, 50′ long, 50K load, end bearing, 1 pile	745	505	1,250
1200	Friction type, 3 pile cluster	2,250	1,500	3,750
1400	100K load, end bearing, 2 pile cluster	1,500	1,000	2,500
1600	Friction type, 4 pile cluster	3,000	2,000	5,000
1800	200K load, end bearing, 4 pile cluster	3,000	2,000	5,000
2000	Friction type, 6 pile cluster	4,475	3,025	7,500
2200	400K load, end bearing, 7 pile cluster	5,225	3,525	8,750
2400	Friction type, 10 pile cluster	7,475	5,025	12,500
2600	800K load, end bearing, 14 pile cluster	10,500	7,050	17,550
2800	Friction type, 18 pile cluster	13,500	9,075	22,575
3000	1200K load, end bearing, 16 pile cluster	13,100	8,575	21,675
3200	Friction type, 21 pile cluster	17,200	11,300	28,500
3400	1600K load, end bearing, 18 pile cluster	14,700	9,675	24,375
3600	Friction type, 24 pile cluster	19,700	12,900	32,600
5000	100′ long, 50K load, end bearing, 1 pile	1,500	1,000	2,500
5200	Friction type, 2 pile cluster	3,000	2,000	5,000
5400	100K load, end bearing, 2 pile cluster	3,000	2,000	5,000
5600	Friction type, 3 pile cluster	4,500	3,000	7,500
5800	200K load, end bearing, 4 pile cluster	6,025	4,025	10,050
6000	Friction type, 5 pile cluster	7,525	5,025	12,550
6200	400K load, end bearing, 7 pile cluster	10,500	7,000	17,500
6400	Friction type, 8 pile cluster	12,000	8,000	20,000
6600	800K load, end bearing, 15 pile cluster	22,600	15,100	37,700
6800	Friction type, 16 pile cluster	24,100	16,100	40,200
7000	1200K load, end bearing, 17 pile cluster	43,300	26,200	69,500
7200	Friction type, 19 pile cluster	32,600	19,900	52,500
7400	1600K load, end bearing, 20 pile cluster	34,300	20,900	55,200
7600	Friction type, 22 pile cluster	37,700	23,100	60,800

A1020 Special Foundations

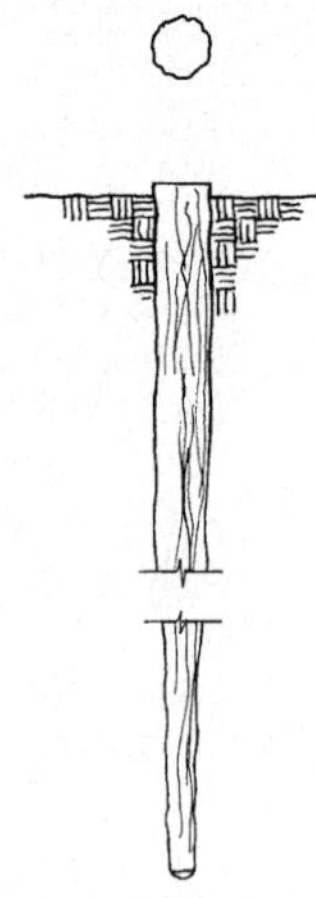

The Treated Wood Pile System includes: creosoted wood piles; a standard steel driving point; and an allowance for cutoffs.

The Expanded System Listing shows costs per cluster of piles. Clusters range from two piles to twenty piles. Loads vary from 50 Kips to 400 Kips. Both end-bearing and friction type piles are listed.

Please see the reference section for cost of mobilization of the pile driving equipment and other design and cost information.

System Components	QUANTITY	UNIT	COST EACH MAT.	COST EACH INST.	COST EACH TOTAL
SYSTEM A1020 160 2220					
WOOD PILES, 25′ LONG, 50K LOAD, END BEARING, 3 PILE CLUSTER					
Wood piles, treated, 12″ butt, 8″ tip, up to 30′ long	81.000	V.L.F.	1,486.35	857.79	2,344.14
Point for driving wood piles	3.000	Ea.	76.50	84	160.50
Pile cutoff, wood piles	3.000	Ea.		44.40	44.40
TOTAL			1,562.85	986.19	2,549.04

A1020 160	Treated Wood Piles	COST EACH MAT.	COST EACH INST.	COST EACH TOTAL
2220	Wood piles, 25′ long, 50K load, end bearing, 3 pile cluster	1,575	985	2,560
2240	Friction type, 3 pile cluster	1,475	900	2,375
2260	5 pile cluster	2,475	1,500	3,975
2280	100K load, end bearing, 4 pile cluster	2,075	1,325	3,400
2300	5 pile cluster	2,600	1,625	4,225
2320	6 pile cluster	3,125	1,975	5,100
2340	Friction type, 5 pile cluster	2,475	1,500	3,975
2360	6 pile cluster	2,975	1,800	4,775
2380	10 pile cluster	4,950	3,000	7,950
2400	200K load, end bearing, 8 pile cluster	4,175	2,625	6,800
2420	10 pile cluster	5,200	3,300	8,500
2440	12 pile cluster	6,250	3,950	10,200
2460	Friction type, 10 pile cluster	4,950	3,000	7,950
2480	400K load, end bearing, 16 pile cluster	8,325	5,250	13,575
2500	20 pile cluster	10,400	6,575	16,975
4520	50′ long, 50K load, end bearing, 3 pile cluster	3,775	1,475	5,250
4540	4 pile cluster	5,050	1,950	7,000
4560	Friction type, 2 pile cluster	2,450	895	3,345
4580	3 pile cluster	3,675	1,325	5,000
4600	100K load, end bearing, 5 pile cluster	6,300	2,425	8,725
4620	8 pile cluster	10,100	3,900	14,000
4640	Friction type, 3 pile cluster	3,675	1,325	5,000
4660	5 pile cluster	6,100	2,225	8,325
4680	200K load, end bearing, 9 pile cluster	11,300	4,375	15,675
4700	10 pile cluster	12,600	4,875	17,475
4720	15 pile cluster	18,900	7,300	26,200
4740	Friction type, 5 pile cluster	6,100	2,225	8,325
4760	6 pile cluster	7,325	2,675	10,000

A1020 Special Foundations

A1020 160	Treated Wood Piles	COST EACH		
		MAT.	INST.	TOTAL
4780	10 pile cluster	12,200	4,475	16,675
4800	400K load, end bearing, 18 pile cluster	22,700	8,775	31,475
4820	20 pile cluster	25,200	9,750	34,950
4840	Friction type, 9 pile cluster	11,000	4,000	15,000
4860	10 pile cluster	12,200	4,475	16,675
9000	Add for boot for driving tip, each pile	25.50	21	46.50

A1020 Special Foundations

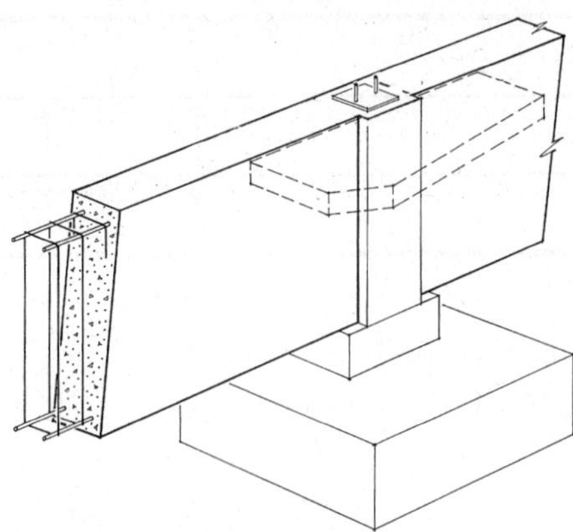

The Grade Beam System includes: excavation with a truck mounted backhoe; hand trim; backfill; forms (four uses); reinforcing steel; and 3,000 p.s.i. concrete placed from chute.

Superimposed loads vary in the listing from 1 Kip per linear foot (KLF) and above. In the Expanded System Listing, the span of the beams varies from 15′ to 40′. Depth varies from 28″ to 52″. Width varies from 12″ and wider.

Please see the reference section for further design and cost information.

System Components	QUANTITY	UNIT	COST PER L.F. MAT.	INST.	TOTAL
SYSTEM A1020 210 2220					
GRADE BEAM, 15′ SPAN, 28″ DEEP, 12″ WIDE, 8 KLF LOAD					
Excavation, trench, hydraulic backhoe, 3/8 CY bucket	.260	C.Y.		2.51	2.51
Trim sides and bottom of trench, regular soil	2.000	S.F.		1.90	1.90
Backfill, by hand, compaction in 6″ layers, using vibrating plate	.170	C.Y.		1.39	1.39
Forms in place, grade beam, 4 uses	4.700	SFCA	4.75	25.62	30.37
Reinforcing in place, beams & girders, #8 to #14	.019	Ton	20.90	17.96	38.86
Concrete ready mix, regular weight, 3000 psi	.090	C.Y.	9.81		9.81
Place and vibrate conc. for grade beam, direct chute	.090	C.Y.		1.72	1.72
TOTAL			35.46	51.10	86.56

A1020 210	Grade Beams	COST PER L.F. MAT.	INST.	TOTAL
2220	Grade beam, 15′ span, 28″ deep, 12″ wide, 8 KLF load	35.50	51	86.50
2240	14″ wide, 12 KLF load	36.50	51.50	88
2260	40″ deep, 12″ wide, 16 KLF load	36.50	64	100.50
2280	20 KLF load	42	68.50	110.50
2300	52″ deep, 12″ wide, 30 KLF load	48	86	134
2320	40 KLF load	59	95.50	154.50
2340	50 KLF load	70	104	174
3360	20′ span, 28″ deep, 12″ wide, 2 KLF load	22.50	40	62.50
3380	16″ wide, 4 KLF load	29.50	44.50	74
3400	40″ deep, 12″ wide, 8 KLF load	36.50	64	100.50
3420	12 KLF load	46.50	71.50	118
3440	14″ wide, 16 KLF load	56	79.50	135.50
3460	52″ deep, 12″ wide, 20 KLF load	60.50	96.50	157
3480	14″ wide, 30 KLF load	83.50	114	197.50
3500	20″ wide, 40 KLF load	97.50	120	217.50
3520	24″ wide, 50 KLF load	121	137	258
4540	30′ span, 28″ deep, 12″ wide, 1 KLF load	23.50	41	64.50
4560	14″ wide, 2 KLF load	37.50	56	93.50
4580	40″ deep, 12″ wide, 4 KLF load	44	67.50	111.50
4600	18″ wide, 8 KLF load	62.50	81.50	144
4620	52″ deep, 14″ wide, 12 KLF load	80	111	191
4640	20″ wide, 16 KLF load	97.50	121	218.50
4660	24″ wide, 20 KLF load	121	138	259
4680	36″ wide, 30 KLF load	173	173	346
4700	48″ wide, 40 KLF load	228	209	437
5720	40′ span, 40″ deep, 12″ wide, 1 KLF load	32	60	92

A1020 Special Foundations

A1020 210	Grade Beams	COST PER L.F.		
		MAT.	INST.	TOTAL
5740	2 KLF load	39.50	66.50	106
5760	52" deep, 12" wide, 4 KLF load	59	95.50	154.50
5780	20" wide, 8 KLF load	94.50	118	212.50
5800	28" wide, 12 KLF load	129	143	272
5820	38" wide, 16 KLF load	178	174	352
5840	46" wide, 20 KLF load	233	214	447

A1020 Special Foundations

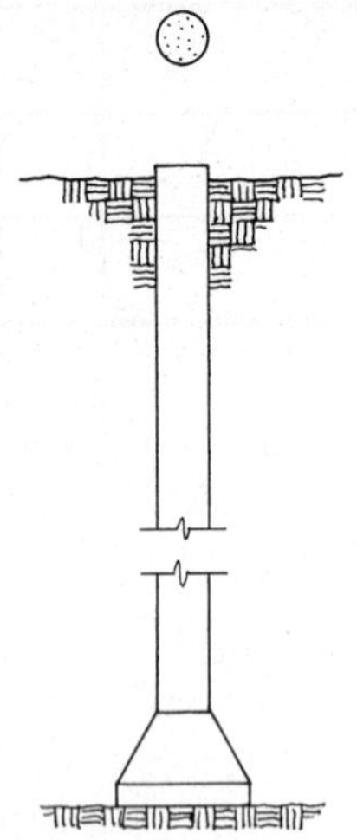

Caisson Systems are listed for three applications: stable ground, wet ground and soft rock. Concrete used is 3,000 p.s.i. placed from chute. Included are a bell at the bottom of the caisson shaft (if applicable) along with required excavation and disposal of excess excavated material up to two miles from job site.

The Expanded System lists cost per caisson. End-bearing loads vary from 200 Kips to 3,200 Kips. The dimensions of the caissons range from 2′ x 50′ to 7′ x 200′.

Please see the reference section for further design and cost information.

System Components	QUANTITY	UNIT	COST EACH MAT.	COST EACH INST.	COST EACH TOTAL
SYSTEM A1020 310 2200					
CAISSON, STABLE GROUND, 3000 PSI CONC., 10 KSF BRNG, 200K LOAD, 2′X50′					
Caissons, drilled, to 50′, 24″ shaft diameter, .116 C.Y./L.F.	50.000	V.L.F.	790	1,520	2,310
Reinforcing in place, columns, #3 to #7	.060	Ton	66	102	168
4′ bell diameter, 24″ shaft, 0.444 C.Y.	1.000	Ea.	48.50	289	337.50
Load & haul excess excavation, 2 miles	6.240	C.Y.		42.81	42.81
TOTAL			904.50	1,953.81	2,858.31

A1020 310	Caissons	COST EACH MAT.	COST EACH INST.	COST EACH TOTAL
2200	Caisson, stable ground, 3000 PSI conc, 10 KSF brng, 200K load, 2′-0″x50′-0	905	1,950	2,855
2400	400K load, 2′-6″x50′-0″	1,500	3,125	4,625
2600	800K load, 3′-0″x100′-0″	4,075	7,475	11,550
2800	1200K load, 4′-0″x100′-0″	7,075	9,350	16,425
3000	1600K load, 5′-0″x150′-0″	15,900	14,500	30,400
3200	2400K load, 6′-0″x150′-0″	23,200	18,900	42,100
3400	3200K load, 7′-0″x200′-0″	41,700	27,600	69,300
5000	Wet ground, 3000 PSI conc., 10 KSF brng, 200K load, 2′-0″x50′-0″	785	2,775	3,560
5200	400K load, 2′-6″x50′-0″	1,325	4,750	6,075
5400	800K load, 3′-0″x100′-0″	3,550	12,800	16,350
5600	1200K load, 4′-0″x100′-0″	6,125	19,000	25,125
5800	1600K load, 5′-0″x150′-0″	13,600	40,800	54,400
6000	2400K load, 6′-0″x150′-0″	20,000	50,500	70,500
6200	3200K load, 7′-0″x200′-0″	35,800	80,500	116,300
7800	Soft rock, 3000 PSI conc., 10 KSF brng, 200K load, 2′-0″x50′-0″	785	15,000	15,785
8000	400K load, 2′-6″x50′-0″	1,325	24,300	25,625
8200	800K load, 3′-0″x100′-0″	3,550	63,500	67,050
8400	1200K load, 4′-0″x100′-0″	6,125	93,000	99,125
8600	1600K load, 5′-0″x150′-0″	13,600	192,000	205,600
8800	2400K load, 6′-0″x150′-0″	20,000	228,000	248,000
9000	3200K load, 7′-0″x200′-0″	35,800	360,500	396,300

A1020 Special Foundations

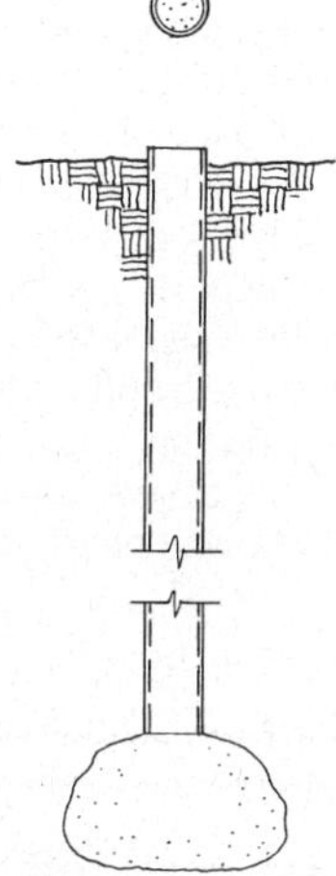

Pressure Injected Piles are usually uncased up to 25′ and cased over 25′ depending on soil conditions.

These costs include excavation and hauling of excess materials; steel casing over 25′; reinforcement; 3,000 p.s.i. concrete; plus mobilization and demobilization of equipment for a distance of up to fifty miles to and from the job site.

The Expanded System lists cost per cluster of piles. Clusters range from one pile to eight piles. End-bearing loads range from 50 Kips to 1,600 Kips.

Please see the reference section for further design and cost information.

System Components	QUANTITY	UNIT	COST EACH MAT.	COST EACH INST.	COST EACH TOTAL
SYSTEM A1020 710 4200					
PRESSURE INJECTED FOOTING, END BEARING, 50′ LONG, 50K LOAD, 1 PILE					
Pressure injected footings, cased, 30-60 ton cap., 12″ diameter	50.000	V.L.F.	837.50	1,420	2,257.50
Pile cutoff, concrete pile with thin steel shell	1.000	Ea.		14.80	14.80
TOTAL			837.50	1,434.80	2,272.30

A1020 710	Pressure Injected Footings	COST EACH MAT.	COST EACH INST.	COST EACH TOTAL
2200	Pressure injected footing, end bearing, 25′ long, 50K load, 1 pile	390	1,000	1,390
2400	100K load, 1 pile	390	1,000	1,390
2600	2 pile cluster	780	2,025	2,805
2800	200K load, 2 pile cluster	780	2,025	2,805
3200	400K load, 4 pile cluster	1,550	4,025	5,575
3400	7 pile cluster	2,725	7,075	9,800
3800	1200K load, 6 pile cluster	3,300	7,625	10,925
4000	1600K load, 7 pile cluster	3,850	8,875	12,725
4200	50′ long, 50K load, 1 pile	840	1,425	2,265
4400	100K load, 1 pile	1,450	1,425	2,875
4600	2 pile cluster	2,900	2,875	5,775
4800	200K load, 2 pile cluster	2,900	2,875	5,775
5000	4 pile cluster	5,800	5,725	11,525
5200	400K load, 4 pile cluster	5,800	5,725	11,525
5400	8 pile cluster	11,600	11,500	23,100
5600	800K load, 7 pile cluster	10,200	10,100	20,300
5800	1200K load, 6 pile cluster	9,300	8,625	17,925
6000	1600K load, 7 pile cluster	10,900	10,100	21,000

A1030 Slab on Grade

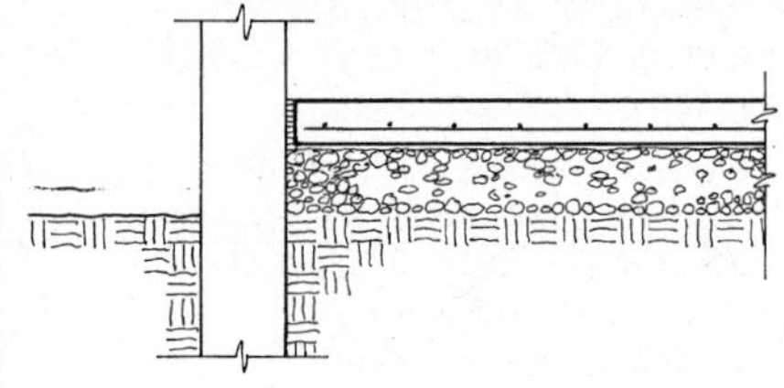

There are four types of Slab on Grade Systems listed: Non-industrial, Light industrial, Industrial and Heavy industrial. Each type is listed two ways: reinforced and non-reinforced. A Slab on Grade system includes three passes with a grader; 6″ of compacted gravel fill; polyethylene vapor barrier; 3500 p.s.i. concrete placed by chute; bituminous fibre expansion joint; all necessary edge forms (4 uses); steel trowel finish; and sprayed-on membrane curing compound.

The Expanded System Listing shows costs on a per square foot basis. Thicknesses of the slabs range from 4″ and above. Non-industrial applications are for foot traffic only with negligible abrasion. Light industrial applications are for pneumatic wheels and light abrasion. Industrial applications are for solid rubber wheels and moderate abrasion. Heavy industrial applications are for steel wheels and severe abrasion. All slabs are either shown unreinforced or reinforced with welded wire fabric.

System Components	QUANTITY	UNIT	COST PER S.F.		
			MAT.	INST.	TOTAL
SYSTEM A1030 120 2220					
SLAB ON GRADE, 4″ THICK, NON INDUSTRIAL, NON REINFORCED					
Fine grade, 3 passes with grader and roller	.110	S.Y.		.50	.50
Gravel under floor slab, 6″ deep, compacted	1.000	S.F.	.46	.31	.77
Polyethylene vapor barrier, standard, .006″ thick	1.000	S.F.	.04	.15	.19
Concrete ready mix, regular weight, 3500 psi	.012	C.Y.	1.33		1.33
Place and vibrate concrete for slab on grade, 4″ thick, direct chute	.012	C.Y.		.32	.32
Expansion joint, premolded bituminous fiber, 1/2″ x 6″	.100	L.F.	.10	.33	.43
Edge forms in place for slab on grade to 6″ high, 4 uses	.030	L.F.	.01	.11	.12
Cure with sprayed membrane curing compound	1.000	S.F.	.12	.10	.22
Finishing floor, monolithic steel trowel	1.000	S.F.		.90	.90
TOTAL			2.06	2.72	4.78

A1030 120	Plain & Reinforced	COST PER S.F.		
		MAT.	INST.	TOTAL
2220	Slab on grade, 4″ thick, non industrial, non reinforced	2.06	2.72	4.78
2240	Reinforced	2.22	3.09	5.31
2260	Light industrial, non reinforced	2.66	3.33	5.99
2280	Reinforced	2.82	3.70	6.52
2300	Industrial, non reinforced	3.27	6.95	10.22
2320	Reinforced	3.43	7.35	10.78
3340	5″ thick, non industrial, non reinforced	2.40	2.79	5.19
3360	Reinforced	2.56	3.16	5.72
3380	Light industrial, non reinforced	3	3.40	6.40
3400	Reinforced	3.16	3.77	6.93
3420	Heavy industrial, non reinforced	4.25	8.30	12.55
3440	Reinforced	4.35	8.70	13.05
4460	6″ thick, non industrial, non reinforced	2.84	2.73	5.57
4480	Reinforced	3.07	3.22	6.29
4500	Light industrial, non reinforced	3.46	3.34	6.80
4520	Reinforced	3.82	4	7.82
4540	Heavy industrial, non reinforced	4.71	8.40	13.11
4560	Reinforced	4.94	8.90	13.84

A2010 Basement Excavation

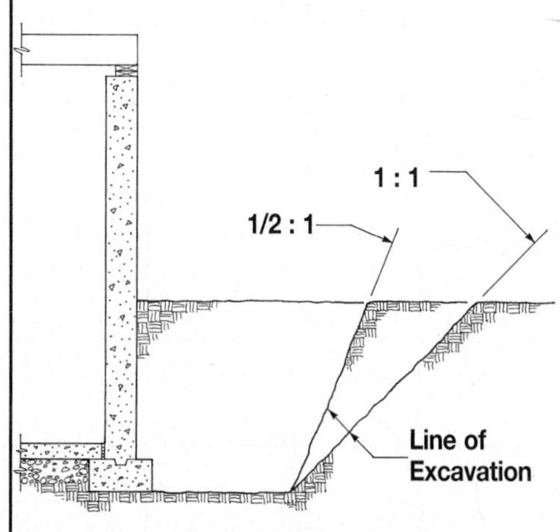

Pricing Assumptions: Two-thirds of excavation is by 2-1/2 C.Y. wheel mounted front end loader and one-third by 1-1/2 C.Y. hydraulic excavator.

Two-mile round trip haul by 12 C.Y. tandem trucks is included for excavation wasted and storage of suitable fill from excavated soil. For excavation in clay, all is wasted and the cost of suitable backfill with two-mile haul is included.

Sand and gravel assumes 15% swell and compaction; common earth assumes 25% swell and 15% compaction; clay assumes 40% swell and 15% compaction (non-clay).

In general, the following items are accounted for in the costs in the table below.

1. Excavation for building or other structure to depth and extent indicated.
2. Backfill compacted in place.
3. Haul of excavated waste.
4. Replacement of unsuitable material with bank run gravel.

Note: Additional excavation and fill beyond this line of general excavation for the building (as required for isolated spread footings, strip footings, etc.) are included in the cost of the appropriate component systems.

System Components	QUANTITY	UNIT	COST PER S.F. MAT.	INST.	TOTAL
SYSTEM A2010 110 2280					
EXCAVATE & FILL, 1000 S.F., 8′ DEEP, SAND, ON SITE STORAGE					
Excavating bulk shovel, 1.5 C.Y. bucket, 150 cy/hr	.262	C.Y.		.50	.50
Excavation, front end loader, 2-1/2 C.Y.	.523	C.Y.		1.29	1.29
Haul earth, 12 C.Y. dump truck	.341	L.C.Y.		2.31	2.31
Backfill, dozer bulk push 300′, including compaction	.562	C.Y.		2.19	2.19
TOTAL				6.29	6.29

A2010 110	Building Excavation & Backfill	COST PER S.F. MAT.	INST.	TOTAL
2220	Excav & fill, 1000 S.F., 4′ sand, gravel, or common earth, on site storage		1.04	1.04
2240	Off site storage		1.57	1.57
2260	Clay excavation, bank run gravel borrow for backfill	2.40	2.42	4.82
2280	8′ deep, sand, gravel, or common earth, on site storage		6.30	6.30
2300	Off site storage		13.90	13.90
2320	Clay excavation, bank run gravel borrow for backfill	9.75	11.30	21.05
2340	16′ deep, sand, gravel, or common earth, on site storage		17.20	17.20
2350	Off site storage		34	34
2360	Clay excavation, bank run gravel borrow for backfill	27	29	56
3380	4000 S.F., 4′ deep, sand, gravel, or common earth, on site storage		.57	.57
3400	Off site storage		1.11	1.11
3420	Clay excavation, bank run gravel borrow for backfill	1.22	1.22	2.44
3440	8′ deep, sand, gravel, or common earth, on site storage		4.44	4.44
3460	Off site storage		7.80	7.80
3480	Clay excavation, bank run gravel borrow for backfill	4.44	6.80	11.24
3500	16′ deep, sand, gravel, or common earth, on site storage		10.65	10.65
3520	Off site storage		21.50	21.50
3540	Clay, excavation, bank run gravel borrow for backfill	11.90	16.15	28.05
4560	10,000 S.F., 4′ deep, sand gravel, or common earth, on site storage		.32	.32
4580	Off site storage		.65	.65
4600	Clay excavation, bank run gravel borrow for backfill	.74	.74	1.48
4620	8′ deep, sand, gravel, or common earth, on site storage		3.85	3.85
4640	Off site storage		5.85	5.85
4660	Clay excavation, bank run gravel borrow for backfill	2.69	5.30	7.99
4680	16′ deep, sand, gravel, or common earth, on site storage		8.75	8.75
4700	Off site storage		14.90	14.90
4720	Clay excavation, bank run gravel borrow for backfill	7.15	12.10	19.25

A2010 Basement Excavation

A2010 110	Building Excavation & Backfill	COST PER S.F.		
		MAT.	INST.	TOTAL
5740	30,000 S.F., 4′ deep, sand, gravel, or common earth, on site storage		.18	.18
5760	Off site storage		.38	.38
5780	Clay excavation, bank run gravel borrow for backfill	.43	.43	.86
5800	8′ deep, sand & gravel, or common earth, on site storage		3.46	3.46
5820	Off site storage		4.58	4.58
5840	Clay excavation, bank run gravel borrow for backfill	1.51	4.28	5.79
5860	16′ deep, sand, gravel, or common earth, on site storage		7.50	7.50
5880	Off site storage		10.85	10.85
5900	Clay excavation, bank run gravel borrow for backfill	3.96	9.35	13.31
6910	100,000 S.F., 4′ deep, sand, gravel, or common earth, on site storage		.11	.11
6920	Off site storage		.22	.22
6930	Clay excavation, bank run gravel borrow for backfill	.22	.22	.44
6940	8′ deep, sand, gravel, or common earth, on site storage		3.24	3.24
6950	Off site storage		3.83	3.83
6960	Clay excavation, bank run gravel borrow for backfill	.82	3.69	4.51
6970	16′ deep, sand, gravel, or common earth, on site storage		6.80	6.80
6980	Off site storage		8.55	8.55
6990	Clay excavation, bank run gravel borrow for backfill	2.14	7.80	9.94

A20 Basement Construction

A2020 Basement Walls

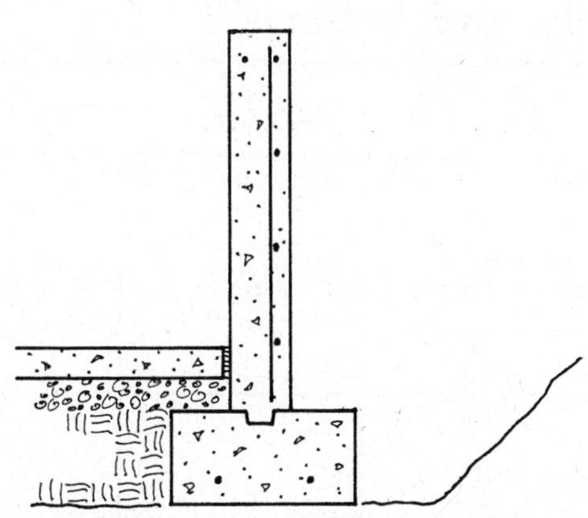

The Foundation Bearing Wall System includes: forms up to 16′ high (four uses); 3,000 p.s.i. concrete placed and vibrated; and form removal with breaking form ties and patching walls. The wall systems list walls from 6″ to 16″ thick and are designed with minimum reinforcement.

Excavation and backfill are not included.

Please see the reference section for further design and cost information.

System Components	QUANTITY	UNIT	COST PER L.F. MAT.	INST.	TOTAL
SYSTEM A2020 110 1500					
FOUNDATION WALL, CAST IN PLACE, DIRECT CHUTE, 4′ HIGH, 6″ THICK					
Formwork	8.000	SFCA	6	44	50
Reinforcing	3.300	Lb.	1.82	1.40	3.22
Unloading & sorting reinforcing	3.300	Lb.		.08	.08
Concrete, 3,000 psi	.074	C.Y.	8.07		8.07
Place concrete, direct chute	.074	C.Y.		2.35	2.35
Finish walls, break ties and patch voids, one side	4.000	S.F.	.16	3.84	4
TOTAL			16.05	51.67	67.72

A2020 110 Walls, Cast in Place

	WALL HEIGHT (FT.)	PLACING METHOD	CONCRETE (C.Y. per L.F.)	REINFORCING (LBS. per L.F.)	WALL THICKNESS (IN.)	COST PER L.F. MAT.	INST.	TOTAL
1500	4′	direct chute	.074	3.3	6	16.05	51.50	67.55
1520			.099	4.8	8	19.60	53	72.60
1540			.123	6.0	10	23	54	77
1560			.148	7.2	12	26.50	55	81.50
1580			.173	8.1	14	29.50	56	85.50
1600			.197	9.44	16	33	57.50	90.50
1700	4′	pumped	.074	3.3	6	16.05	52.50	68.55
1720			.099	4.8	8	19.60	55	74.60
1740			.123	6.0	10	23	56	79
1760			.148	7.2	12	26.50	57	83.50
1780			.173	8.1	14	29.50	58.50	88
1800			.197	9.44	16	33	60	93
3000	6′	direct chute	.111	4.95	6	24	77.50	101.50
3020			.149	7.20	8	29.50	79.50	109
3040			.184	9.00	10	34.50	81	115.50
3060			.222	10.8	12	39.50	83	122.50
3080			.260	12.15	14	44.50	84	128.50
3100			.300	14.39	16	50	86.50	136.50
3200	6′	pumped	.111	4.95	6	24	79	103
3220			.149	7.20	8	29.50	82	111.50
3240			.184	9.00	10	34.50	83.50	118
3260			.222	10.8	12	39.50	86.50	126
3280			.260	12.15	14	44.50	87.50	132
3300			.300	14.39	16	50	90	140

A2020 Basement Walls

A2020 110 Walls, Cast in Place

	WALL HEIGHT (FT.)	PLACING METHOD	CONCRETE (C.Y. per L.F.)	REINFORCING (LBS. per L.F.)	WALL THICKNESS (IN.)	COST PER L.F.		
						MAT.	INST.	TOTAL
5000	8′	direct chute	.148	6.6	6	32	103	135
5020			.199	9.6	8	39.50	106	145.50
5040			.250	12	10	46	108	154
5060			.296	14.39	12	52.50	110	162.50
5080			.347	16.19	14	54.50	111	165.50
5100			.394	19.19	16	66	115	181
5200	8′	pumped	.148	6.6	6	32	105	137
5220			.199	9.6	8	39.50	110	149.50
5240			.250	12	10	46	112	158
5260			.296	14.39	12	52.50	114	166.50
5280			.347	16.19	14	54.50	115	169.50
5300			.394	19.19	16	66	120	186
6020	10′	direct chute	.248	12	8	49	133	182
6040			.307	14.99	10	57	135	192
6060			.370	17.99	12	65.50	138	203.50
6080			.433	20.24	14	73.50	140	213.50
6100			.493	23.99	16	82.50	143	225.50
6220	10′	pumped	.248	12	8	49	137	186
6240			.307	14.99	10	57	139	196
6260			.370	17.99	12	65.50	144	209.50
6280			.433	20.24	14	73.50	146	219.50
6300			.493	23.99	16	82.50	150	232.50
7220	12′	pumped	.298	14.39	8	59	164	223
7240			.369	17.99	10	68.50	167	235.50
7260			.444	21.59	12	79	173	252
7280			.52	24.29	14	88.50	175	263.50
7300			.591	28.79	16	98.50	179	277.50
7420	12′	crane & bucket	.298	14.39	8	59	171	230
7440			.369	17.99	10	68.50	175	243.50
7460			.444	21.59	12	79	182	261
7480			.52	24.29	14	88.50	186	274.50
7500			.591	28.79	16	98.50	194	292.50
8220	14′	pumped	.347	16.79	8	68.50	191	259.50
8240			.43	20.99	10	80	195	275
8260			.519	25.19	12	92	201	293
8280			.607	28.33	14	103	204	307
8300			.69	33.59	16	115	210	325
8420	14′	crane & bucket	.347	16.79	8	68.50	200	268.50
8440			.43	20.99	10	80	205	285
8460			.519	25.19	12	92	212	304
8480			.607	28.33	14	103	217	320
8500			.69	33.59	16	115	224	339
9220	16′	pumped	.397	19.19	8	78.50	218	296.50
9240			.492	23.99	10	91.50	223	314.50
9260			.593	28.79	12	105	230	335
9280			.693	32.39	14	118	233	351
9300			.788	38.38	16	132	240	372
9420	16′	crane & bucket	.397	19.19	8	78.50	229	307.50
9440			.492	23.99	10	91.50	234	325.50
9460			.593	28.79	12	105	242	347
9480			.693	32.39	14	118	248	366
9500			.788	38.38	16	132	256	388

B1010 Floor Construction

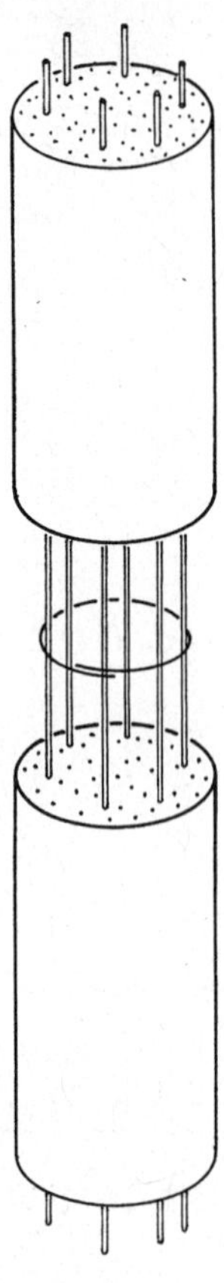

CONCRETE COLUMNS

General: It is desirable for purposes of consistency and simplicity to maintain constant column sizes throughout the building height. To do this, concrete strength may be varied (higher strength concrete at lower stories and lower strength concrete at upper stories), as well as varying the amount of reinforcing.

The first portion of the table provides probable minimum column sizes with related costs and weights per lineal foot of story height for bottom level columns.

The second portion of the table provides costs by column size for top level columns with minimum code reinforcement. Probable maximum loads for these columns are also given.

How to Use Table:

1. Enter the second portion (minimum reinforcing) of the table with the minimum allowable column size from the selected cast in place floor system.

   If the total load on the column does not exceed the allowable working load shown, use the cost per L.F. multiplied by the length of columns required to obtain the column cost.

2. If the total load on the column exceeds the allowable working load shown in the second portion of the table, enter the first portion of the table with the total load on the column and the minimum allowable column size from the selected cast in place floor system.

   Select a cost per L.F. for bottom level columns by total load or minimum allowable column size.

   Select a cost per L.F. for top level columns using the column size required for bottom level columns from the second portion of the table.

$$\frac{\text{Btm. + Top Col. Costs/L.F.}}{2} = \text{Avg. Col. Cost/L.F.}$$

Column Cost = Average Col. Cost/L.F. x Length of Cols. Required.

See reference section to determine total loads.

Design and Pricing Assumptions:
Normal wt. concrete, f'c = 4 or 6 KSI, placed by pump.
Steel, fy = 60 KSI, spliced every other level.
Minimum design eccentricity of 0.1t.
Assumed load level depth is 8″ (weights prorated to full story basis).
Gravity loads only (no frame or lateral loads included).

Please see the reference section for further design and cost information.

System Components	QUANTITY	UNIT	COST PER V.L.F. MAT.	INST.	TOTAL
SYSTEM B1010 201 1050					
ROUND TIED COLUMNS, 4 KSI CONCRETE, 100K MAX. LOAD, 10′ STORY, 12″ SIZE					
Forms in place, columns, round fiber tube, 12″ diam, 1 use	1.000	L.F.	2.86	14.35	17.21
Reinforcing in place, column ties	1.393	Lb.	4.38	6.54	10.92
Concrete ready mix, regular weight, 4000 psi	.029	C.Y.	3.31		3.31
Placing concrete, incl. vibrating, 12″ sq./round columns, pumped	.029	C.Y.		2.27	2.27
Finish, burlap rub w/grout	3.140	S.F.	.13	3.64	3.77
TOTAL			10.68	26.80	37.48

B1010 201 — C.I.P. Column - Round Tied

	LOAD (KIPS)	STORY HEIGHT (FT.)	COLUMN SIZE (IN.)	COLUMN WEIGHT (P.L.F.)	CONCRETE STRENGTH (PSI)	COST PER V.L.F. MAT.	INST.	TOTAL
1050	100	10	12	110	4000	10.70	27	37.70
1060		12	12	111	4000	10.90	27	37.90
1070		14	12	112	4000	11.10	27.50	38.60
1080	150	10	12	110	4000	11.80	28.50	40.30
1090		12	12	111	4000	12	29	41
1100		14	14	153	4000	14.85	31	45.85
1120	200	10	14	150	4000	15.10	31.50	46.60
1140		12	14	152	4000	15.40	32	47.40
1160		14	14	153	4000	15.70	32.50	48.20

B1010 Floor Construction

B1010 201 — C.I.P. Column - Round Tied

	LOAD (KIPS)	STORY HEIGHT (FT.)	COLUMN SIZE (IN.)	COLUMN WEIGHT (P.L.F.)	CONCRETE STRENGTH (PSI)	COST PER V.L.F.		
						MAT.	INST.	TOTAL
1180	300	10	16	194	4000	18.75	34.50	53.25
1190		12	18	250	4000	25.50	39.50	65
1200		14	18	252	4000	26	40	66
1220	400	10	20	306	4000	29.50	44.50	74
1230		12	20	310	4000	30	45.50	75.50
1260		14	20	313	4000	31	46.50	77.50
1280	500	10	22	368	4000	36.50	50.50	87
1300		12	22	372	4000	37	51.50	88.50
1325		14	22	375	4000	38	52.50	90.50
1350	600	10	24	439	4000	41.50	56	97.50
1375		12	24	445	4000	42.50	57	99.50
1400		14	24	448	4000	43	58.50	101.50
1420	700	10	26	517	4000	51.50	63	114.50
1430		12	26	524	4000	52.50	64.50	117
1450		14	26	528	4000	53.50	66	119.50
1460	800	10	28	596	4000	57	69.50	126.50
1480		12	28	604	4000	58.50	71	129.50
1490		14	28	609	4000	59.50	73	132.50
1500	900	10	28	596	4000	61	74.50	135.50
1510		12	28	604	4000	62	76.50	138.50
1520		14	28	609	4000	63	78.50	141.50
1530	1000	10	30	687	4000	66	77	143
1540		12	30	695	4000	67.50	79.50	147
1620		14	30	701	4000	69	81.50	150.50
1640	100	10	12	110	6000	10.95	27	37.95
1660		12	12	111	6000	11.20	27.50	38.70
1680		14	12	112	6000	11.40	27.50	38.90
1700	150	10	12	110	6000	11.85	28	39.85
1710		12	12	111	6000	12.05	28.50	40.55
1720		14	12	112	6000	12.30	29	41.30
1730	200	10	12	110	6000	12.50	29	41.50
1740		12	12	111	6000	12.75	29.50	42.25
1760		14	12	112	6000	12.95	30	42.95
1780	300	10	14	150	6000	15.80	32	47.80
1790		12	14	152	6000	16.10	32.50	48.60
1800		14	16	153	6000	18.55	34	52.55
1810	400	10	16	194	6000	20	35.50	55.50
1820		12	16	196	6000	20.50	36.50	57
1830		14	18	252	6000	26	39.50	65.50
1850	500	10	18	247	6000	27	41	68
1870		12	18	250	6000	27.50	41.50	69
1880		14	20	252	6000	30.50	44.50	75
1890	600	10	20	306	6000	31.50	46.50	78
1900		12	20	310	6000	32.50	47.50	80
1905		14	20	313	6000	33	48.50	81.50
1910	700	10	22	368	6000	37	50.50	87.50
1915		12	22	372	6000	38	51.50	89.50
1920		14	22	375	6000	39	52.50	91.50
1925	800	10	24	439	6000	42.50	56	98.50
1930		12	24	445	6000	43.50	57	100.50
1935		14	24	448	6000	44.50	58.50	103
1940	900	10	24	439	6000	45	60	105
1945		12	24	445	6000	46	61	107

B1010 Floor Construction

B1010 201 — C.I.P. Column - Round Tied

	LOAD (KIPS)	STORY HEIGHT (FT.)	COLUMN SIZE (IN.)	COLUMN WEIGHT (P.L.F.)	CONCRETE STRENGTH (PSI)	COST PER V.L.F.		
						MAT.	INST.	TOTAL
1970	1000	10	26	517	6000	54	64.50	118.50
1980		12	26	524	6000	55	66	121
1995		14	26	528	6000	56	67.50	123.50

B1010 202 — C.I.P. Columns, Round Tied - Minimum Reinforcing

	LOAD (KIPS)	STORY HEIGHT (FT.)	COLUMN SIZE (IN.)	COLUMN WEIGHT (P.L.F.)	CONCRETE STRENGTH (PSI)	COST PER V.L.F.		
						MAT.	INST.	TOTAL
2500	100	10-14	12	107	4000	9.60	25.50	35.10
2510	200	10-14	16	190	4000	16.80	31.50	48.30
2520	400	10-14	20	295	4000	26.50	40	66.50
2530	600	10-14	24	425	4000	37	49.50	86.50
2540	800	10-14	28	580	4000	51	60.50	111.50
2550	1100	10-14	32	755	4000	66	70.50	136.50
2560	1400	10-14	36	960	4000	78.50	82.50	161
2570								

B1010 Floor Construction

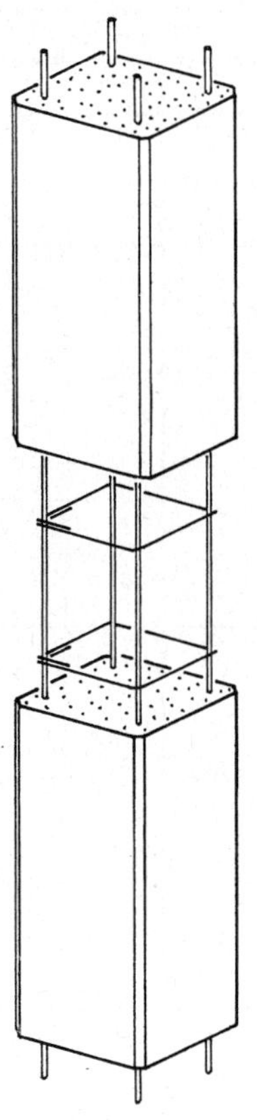

CONCRETE COLUMNS

General: It is desirable for purposes of consistency and simplicity to maintain constant column sizes throughout the building height. To do this, concrete strength may be varied (higher strength concrete at lower stories and lower strength concrete at upper stories), as well as varying the amount of reinforcing.

The first portion of the table provides probable minimum column sizes with related costs and weights per lineal foot of story height for bottom level columns.

The second portion of the table provides costs by column size for top level columns with minimum code reinforcement. Probable maximum loads for these columns are also given.

How to Use Table:

1. Enter the second portion (minimum reinforcing) of the table with the minimum allowable column size from the selected cast in place floor system.

   If the total load on the column does not exceed the allowable working load shown, use the cost per L.F. multiplied by the length of columns required to obtain the column cost.

2. If the total load on the column exceeds the allowable working load shown in the second portion of the table, enter the first portion of the table with the total load on the column and the minimum allowable column size from the selected cast in place floor system.

   Select a cost per L.F. for bottom level columns by total load or minimum allowable column size.

   Select a cost per L.F. for top level columns using the column size required for bottom level columns from the second portion of the table.

$$\frac{\text{Btm. + Top Col. Costs/L.F.}}{2} = \text{Avg. Col. Cost/L.F.}$$

Column Cost = Average Col. Cost/L.F. x Length of Cols. Required.

See reference section in back of book to determine total loads.

Design and Pricing Assumptions:
Normal wt. concrete, f'c = 4 or 6 KSI, placed by pump.
Steel, fy = 60 KSI, spliced every other level.
Minimum design eccentricity of 0.1t.
Assumed load level depth is 8″ (weights prorated to full story basis).
Gravity loads only (no frame or lateral loads included).

Please see the reference section for further design and cost information.

System Components	QUANTITY	UNIT	COST PER V.L.F. MAT.	INST.	TOTAL
SYSTEM B1010 203 0640					
SQUARE COLUMNS, 100K LOAD, 10′ STORY, 10″ SQUARE					
Forms in place, columns, plywood, 10″ x 10″, 4 uses	3.323	SFCA	3.03	31.83	34.86
Chamfer strip, wood, 3/4″ wide	4.000	L.F.	.72	4.32	5.04
Reinforcing in place, column ties	1.405	Lb.	3.88	5.78	9.66
Concrete ready mix, regular weight, 4000 psi	.026	C.Y.	2.96		2.96
Placing concrete, incl. vibrating, 12″ sq./round columns, pumped	.026	C.Y.		2.03	2.03
Finish, break ties, patch voids, burlap rub w/grout	3.323	S.F.	.13	3.87	4
TOTAL			10.72	47.83	58.55

B1010 203 C.I.P. Column, Square Tied

	LOAD (KIPS)	STORY HEIGHT (FT.)	COLUMN SIZE (IN.)	COLUMN WEIGHT (P.L.F.)	CONCRETE STRENGTH (PSI)	COST PER V.L.F. MAT.	INST.	TOTAL
0640	100	10	10	96	4000	10.70	48	58.70
0680		12	10	97	4000	10.90	48.50	59.40
0700		14	12	142	4000	14.30	58.50	72.80
0710								

B1010 Floor Construction

B1010 203 C.I.P. Column, Square Tied

	LOAD (KIPS)	STORY HEIGHT (FT.)	COLUMN SIZE (IN.)	COLUMN WEIGHT (P.L.F.)	CONCRETE STRENGTH (PSI)	COST PER V.L.F.		
						MAT.	INST.	TOTAL
0740	150	10	10	96	4000	12.65	51	63.65
0780		12	12	142	4000	14.90	59	73.90
0800		14	12	143	4000	15.15	59.50	74.65
0840	200	10	12	140	4000	15.70	60.50	76.20
0860		12	12	142	4000	16	61	77
0900		14	14	196	4000	18.65	67.50	86.15
0920	300	10	14	192	4000	19.75	69.50	89.25
0960		12	14	194	4000	20	70	90
0980		14	16	253	4000	23.50	77.50	101
1020	400	10	16	248	4000	25	79.50	104.50
1060		12	16	251	4000	25.50	80.50	106
1080		14	16	253	4000	25.50	81	106.50
1200	500	10	18	315	4000	29	89	118
1250		12	20	394	4000	35	101	136
1300		14	20	397	4000	36	102	138
1350	600	10	20	388	4000	36.50	103	139.50
1400		12	20	394	4000	37.50	105	142.50
1600		14	20	397	4000	38	106	144
1900	700	10	20	388	4000	42	114	156
2100		12	22	474	4000	42.50	115	157.50
2300		14	22	478	4000	43.50	116	159.50
2600	800	10	22	388	4000	44.50	117	161.50
2900		12	22	474	4000	45	119	164
3200		14	22	478	4000	46	120	166
3400	900	10	24	560	4000	49.50	128	177.50
3800		12	24	567	4000	50.50	130	180.50
4000		14	24	571	4000	52	132	184
4250	1000	10	24	560	4000	54	135	189
4500		12	26	667	4000	57	141	198
4750		14	26	673	4000	58	143	201
5600	100	10	10	96	6000	10.95	48	58.95
5800		12	10	97	6000	11.15	48	59.15
6000		14	12	142	6000	14.70	58.50	73.20
6200	150	10	10	96	6000	12.90	50.50	63.40
6400		12	12	98	6000	15.25	59	74.25
6600		14	12	143	6000	15.55	59.50	75.05
6800	200	10	12	140	6000	16.10	60.50	76.60
7000		12	12	142	6000	16.35	61	77.35
7100		14	14	196	6000	19.15	67.50	86.65
7300	300	10	14	192	6000	19.90	69	88.90
7500		12	14	194	6000	20.50	69.50	90
7600		14	14	196	6000	20.50	70	90.50
7700	400	10	14	192	6000	21.50	71	92.50
7800		12	14	194	6000	22	71.50	93.50
7900		14	16	253	6000	24.50	77.50	102
8000	500	10	16	248	6000	25.50	79.50	105
8050		12	16	251	6000	26	80.50	106.50
8100		14	16	253	6000	26.50	81	107.50
8200	600	10	18	315	6000	29.50	89.50	119
8300		12	18	319	6000	30	90.50	120.50
8400		14	18	321	6000	30.50	91.50	122

B1010 Floor Construction

B1010 203 — C.I.P. Column, Square Tied

	LOAD (KIPS)	STORY HEIGHT (FT.)	COLUMN SIZE (IN.)	COLUMN WEIGHT (P.L.F.)	CONCRETE STRENGTH (PSI)	COST PER V.L.F.		
						MAT.	INST.	TOTAL
8500	700	10	18	315	6000	31	92.50	123.50
8600		12	18	319	6000	32	93	125
8700		14	18	321	6000	32.50	94	126.50
8800	800	10	20	388	6000	36	101	137
8900		12	20	394	6000	37	102	139
9000		14	20	397	6000	37.50	103	140.50
9100	900	10	20	388	6000	39	106	145
9300		12	20	394	6000	40	107	147
9600		14	20	397	6000	41	108	149
9800	1000	10	22	469	6000	44.50	116	160.50
9840		12	22	474	6000	45.50	117	162.50
9900		14	22	478	6000	46.50	119	165.50

B1010 204 — C.I.P. Column, Square Tied-Minimum Reinforcing

	LOAD (KIPS)	STORY HEIGHT (FT.)	COLUMN SIZE (IN.)	COLUMN WEIGHT (P.L.F.)	CONCRETE STRENGTH (PSI)	COST PER V.L.F.		
						MAT.	INST.	TOTAL
9913	150	10-14	12	135	4000	12.90	56.50	69.40
9918	300	10-14	16	240	4000	21	73.50	94.50
9924	500	10-14	20	375	4000	31	95.50	126.50
9930	700	10-14	24	540	4000	43	118	161
9936	1000	10-14	28	740	4000	56.50	144	200.50
9942	1400	10-14	32	965	4000	69.50	162	231.50
9948	1800	10-14	36	1220	4000	86.50	189	275.50
9954	2300	10-14	40	1505	4000	106	218	324

B1010 Floor Construction

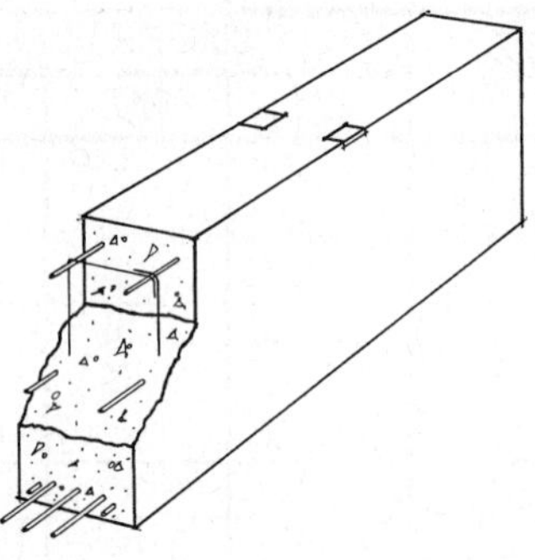

General: Beams priced in the following table are plant produced prestressed members transported to the site and erected.

Pricing Assumptions: Prices are based upon 10,000 S.F. to 20,000 S.F. projects and 50 mile to 100 mile transport.

Normal steel for connections is included in price. Deduct 20% from prices for Southern states. Add 10% to prices for Western states.

Design Assumptions: Normal weight concrete to 150 lbs./C.F. f'c = 5 KSI. Prestressing Steel: 250 KSI straight strand. Non-Prestressing Steel: fy = 60 KSI.

B1010 213 Rectangular Precast Beams

	SPAN (FT.)	SUPERIMPOSED LOAD (K.L.F.)	SIZE W X D (IN.)	BEAM WEIGHT (P.L.F.)	TOTAL LOAD (K.L.F.)	COST PER L.F.		
						MAT.	INST.	TOTAL
2200	15	2.32	12x16	200	2.52	150	19.60	169.60
2250		3.80	12x20	250	4.05	155	19.60	174.60
2300		5.60	12x24	300	5.90	160	20.50	180.50
2350		7.65	12x28	350	8.00	165	20.50	185.50
2400		5.85	18x20	375	6.73	165	20.50	185.50
2450		8.26	18x24	450	8.71	171	22	193
2500		11.39	18x28	525	11.91	182	22	204
2550		18.90	18x36	675	19.58	192	22	214
2600		10.78	24x24	600	11.38	187	22	209
2700		15.12	24x28	700	15.82	197	22	219
2750		25.23	24x36	900	26.13	213	23	236
2800	20	1.22	12x16	200	1.44	150	14.70	164.70
2850		2.03	12x20	250	2.28	155	15.25	170.25
2900		3.02	12x24	300	3.32	160	15.25	175.25
2950		4.15	12x28	350	4.50	165	15.25	180.25
3000		6.85	12x36	450	7.30	173	16.50	189.50
3050		3.13	18x20	375	3.50	165	15.25	180.25
3100		4.45	18x24	450	4.90	171	16.50	187.50
3150		6.18	18x28	525	6.70	182	16.50	198.50
3200		10.33	18x36	675	11.00	192	17.15	209.15
3400		15.48	18x44	825	16.30	208	18.75	226.75
3450		21.63	18x52	975	22.60	219	18.75	237.75
3500		5.80	24x24	600	6.40	187	16.50	203.50
3600		8.20	24x28	700	8.90	197	17.15	214.15
3700		13.80	24x36	900	14.70	213	18.75	231.75
3800		20.70	24x44	1100	21.80	230	20.50	250.50
3850		29.20	24x52	1300	30.50	251	26	277
3900	25	1.82	12x24	300	2.12	160	12.20	172.20
4000		2.53	12x28	350	2.88	165	13.20	178.20
4050		5.18	12x36	450	5.63	173	13.20	186.20
4100		6.55	12x44	550	7.10	182	13.20	195.20
4150		9.08	12x52	650	9.73	192	13.70	205.70
4200		1.86	18x20	375	2.24	165	13.20	178.20
4250		2.69	18x24	450	3.14	171	13.20	184.20
4300		3.76	18x28	525	4.29	182	13.20	195.20
4350		6.37	18x36	675	7.05	192	15	207
4400		9.60	18x44	872	10.43	208	16.50	224.50
4450		13.49	18x52	975	14.47	219	16.50	235.50

B1010 Floor Construction

B1010 213 Rectangular Precast Beams

	SPAN (FT.)	SUPERIMPOSED LOAD (K.L.F.)	SIZE W X D (IN.)	BEAM WEIGHT (P.L.F.)	TOTAL LOAD (K.L.F.)	COST PER L.F.		
						MAT.	INST.	TOTAL
4500	25	18.40	18x60	1125	19.53	235	18.30	253.30
4600		3.50	24x24	600	4.10	187	15	202
4700		5.00	24x28	700	5.70	197	15	212
4800		8.50	24x36	900	9.40	213	16.50	229.50
4900		12.85	24x44	1100	13.95	230	18.30	248.30
5000		18.22	24x52	1300	19.52	251	20.50	271.50
5050		24.48	24x60	1500	26.00	267	23.50	290.50
5100	30	1.65	12x28	350	2.00	165	11.45	176.45
5150		2.79	12x36	450	3.24	173	11.45	184.45
5200		4.38	12x44	550	4.93	182	12.45	194.45
5250		6.10	12x52	650	6.75	192	12.45	204.45
5300		8.36	12x60	750	9.11	202	13.70	215.70
5400		1.72	18x24	450	2.17	171	11.45	182.45
5450		2.45	18x28	525	2.98	182	12.45	194.45
5750		4.21	18x36	675	4.89	192	12.45	204.45
6000		6.42	18x44	825	7.25	208	13.70	221.70
6250		9.07	18x52	975	10.05	219	15.25	234.25
6500		12.43	18x60	1125	13.56	235	17.15	252.15
6750		2.24	24x24	600	2.84	187	12.45	199.45
7000		3.26	24x28	700	3.96	197	13.70	210.70
7100		5.63	24x36	900	6.53	213	15.25	228.25
7200	35	8.59	24x44	1100	9.69	230	16.85	246.85
7300		12.25	24x52	1300	13.55	251	16.85	267.85
7400		16.54	24x60	1500	18.04	267	16.85	283.85
7500	40	2.23	12x44	550	2.78	182	10.30	192.30
7600		3.15	12x52	650	3.80	192	10.30	202.30
7700		4.38	12x60	750	5.13	202	11.45	213.45
7800		2.08	18x36	675	2.76	192	11.45	203.45
7900		3.25	18x44	825	4.08	208	11.45	219.45
8000		4.68	18x52	975	5.66	219	12.85	231.85
8100		6.50	18x60	1175	7.63	235	12.85	247.85
8200		2.78	24x36	900	3.60	213	15.85	228.85
8300		4.35	24x44	1100	5.45	230	15.85	245.85
8400		6.35	24x52	1300	7.65	251	17.15	268.15
8500		8.65	24x60	1500	10.15	267	17.15	284.15
8600	45	2.35	12x52	650	3.00	192	10.15	202.15
8800		3.29	12x60	750	4.04	202	10.15	212.15
9000		2.39	18x44	825	3.22	208	13.05	221.05
9020		3.49	18x52	975	4.47	219	13.05	232.05
9200		4.90	18x60	1125	6.03	235	15.25	250.25
9250		3.21	24x44	1100	4.31	230	16.65	246.65
9300		4.72	24x52	1300	6.02	251	16.65	267.65
9500		6.52	24x60	1500	8.02	267	16.65	283.65
9600	50	2.64	18x52	975	3.62	219	13.70	232.70
9700		3.75	18x60	1125	4.88	235	16.50	251.50
9750		3.58	24x52	1300	4.88	251	16.50	267.50
9800		5.00	24x60	1500	6.50	267	18.30	285.30
9950	55	3.86	24x60	1500	5.36	267	16.65	283.65

B1010 Floor Construction

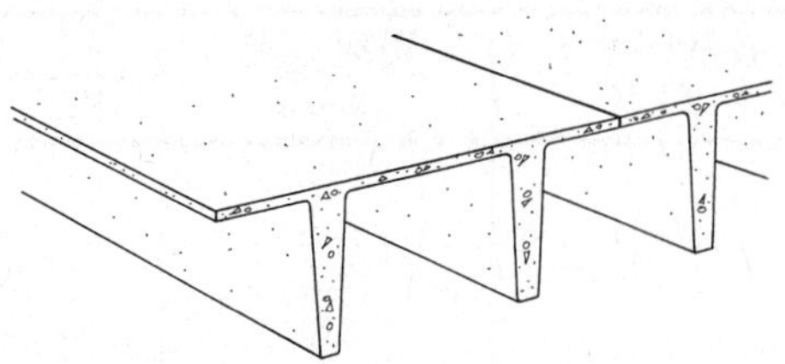

Most widely used for moderate span floors and roofs. At shorter spans, they tend to be competitive with hollow core slabs. They are also used as wall panels.

System Components	QUANTITY	UNIT	COST PER S.F. MAT.	INST.	TOTAL
SYSTEM B1010 235 6700					
PRECAST, DOUBLE "T", 2" TOPPING, 30' SPAN, 30 PSF SUP. LOAD, 18" X 8'					
Double "T" beams, reg. wt, 18" x 8' w, 30' span	1.000	S.F.	9.27	1.71	10.98
Edge forms to 6" high on elevated slab, 4 uses	.050	L.F.	.01	.22	.23
Concrete, ready mix, regular weight, 3000 psi	.250	C.F.	1.01		1.01
Place and vibrate concrete, elevated slab less than 6", pumped	.250	C.F.		.40	.40
Finishing floor, monolithic steel trowel finish for finish floor	1.000	S.F.		.90	.90
Curing with sprayed membrane curing compound	.010	C.S.F.	.12	.10	.22
TOTAL			10.41	3.33	13.74

B1010 234 Precast Double "T" Beams with No Topping

	SPAN (FT.)	SUPERIMPOSED LOAD (P.S.F.)	DBL. "T" SIZE D (IN.) W (FT.)	CONCRETE "T" TYPE	TOTAL LOAD (P.S.F.)	COST PER S.F. MAT.	INST.	TOTAL
1500	30	30	18x8	Reg. Wt.	92	9.25	1.71	10.96
1600		40	18x8	Reg. Wt.	102	9.45	2.21	11.66
1700		50	18x8	Reg. Wt	112	9.45	2.21	11.66
1800		75	18x8	Reg. Wt.	137	9.45	2.29	11.74
1900		100	18x8	Reg. Wt.	162	9.45	2.29	11.74
2000	40	30	20x8	Reg. Wt.	87	8.45	1.43	9.88
2100		40	20x8	Reg. Wt.	97	8.55	1.71	10.26
2200		50	20x8	Reg. Wt.	107	8.55	1.71	10.26
2300		75	20x8	Reg. Wt.	132	8.55	1.84	10.39
2400		100	20x8	Reg. Wt.	157	8.70	2.25	10.95
2500	50	30	24x8	Reg. Wt.	103	8.40	1.30	9.70
2600		40	24x8	Reg. Wt.	113	8.45	1.58	10.03
2700		50	24x8	Reg. Wt.	123	8.50	1.69	10.19
2800		75	24x8	Reg. Wt.	148	8.50	1.71	10.21
2900		100	24x8	Reg. Wt.	173	8.65	2.11	10.76
3000	60	30	24x8	Reg. Wt.	82	8.50	1.71	10.21
3100		40	32x10	Reg. Wt.	104	8.55	1.39	9.94
3150		50	32x10	Reg. Wt.	114	8.50	1.20	9.70
3200		75	32x10	Reg. Wt.	139	8.55	1.30	9.85
3250		100	32x10	Reg. Wt.	164	8.65	1.63	10.28
3300	70	30	32x10	Reg. Wt.	94	8.50	1.28	9.78
3350		40	32x10	Reg. Wt.	104	8.55	1.30	9.85
3400		50	32x10	Reg. Wt.	114	8.65	1.62	10.27
3450		75	32x10	Reg. Wt.	139	8.75	1.94	10.69
3500		100	32x10	Reg. Wt.	164	8.95	2.59	11.54
3550	80	30	32x10	Reg. Wt.	94	8.65	1.62	10.27
3600		40	32x10	Reg. Wt.	104	8.85	2.26	11.11
3900		50	32x10	Reg. Wt.	114	8.95	2.58	11.53

B1010 Floor Construction

B1010 234 Precast Double "T" Beams with No Topping

	SPAN (FT.)	SUPERIMPOSED LOAD (P.S.F.)	DBL. "T" SIZE D (IN.) W (FT.)	CONCRETE "T" TYPE	TOTAL LOAD (P.S.F.)	COST PER S.F. MAT.	INST.	TOTAL
4300	50	30	20x8	Lt. Wt.	66	9.30	1.57	10.87
4400		40	20x8	Lt. Wt.	76	9.35	1.42	10.77
4500		50	20x8	Lt. Wt.	86	9.45	1.71	11.16
4600		75	20x8	Lt. Wt.	111	9.50	1.95	11.45
4700		100	20x8	Lt. Wt.	136	9.65	2.36	12.01
4800	60	30	24x8	Lt. Wt.	70	9.40	1.55	10.95
4900		40	32x10	Lt. Wt.	88	9.30	1.08	10.38
5000		50	32x10	Lt. Wt.	98	9.40	1.31	10.71
5200		75	32x10	Lt. Wt.	123	9.45	1.51	10.96
5400		100	32x10	Lt. Wt.	148	9.55	1.83	11.38
5600	70	30	32x10	Lt. Wt.	78	9.30	1.08	10.38
5750		40	32x10	Lt. Wt.	88	9.40	1.31	10.71
5900		50	32x10	Lt. Wt.	98	9.45	1.50	10.95
6000		75	32x10	Lt. Wt.	123	9.55	1.82	11.37
6100		100	32x10	Lt. Wt.	148	9.80	2.47	12.27
6200	80	30	32x10	Lt. Wt.	78	9.45	1.50	10.95
6300		40	32x10	Lt. Wt.	88	9.55	1.82	11.37
6400		50	32x10	Lt. Wt.	98	9.65	2.14	11.79

B1010 235 Precast Double "T" Beams With 2″ Topping

	SPAN (FT.)	SUPERIMPOSED LOAD (P.S.F.)	DBL. "T" SIZE D (IN.) W (FT.)	CONCRETE "T" TYPE	TOTAL LOAD (P.S.F.)	COST PER S.F. MAT.	INST.	TOTAL
6700	30	30	18x8	Reg. Wt.	117	10.40	3.33	13.73
6750		40	18x8	Reg. Wt.	127	10.40	3.33	13.73
6800		50	18x8	Reg. Wt.	137	10.50	3.61	14.11
6900		75	18x8	Reg. Wt.	162	10.50	3.61	14.11
7000		100	18x8	Reg. Wt.	187	10.55	3.75	14.30
7100	40	30	18x8	Reg. Wt.	120	8.20	3.19	11.39
7200		40	20x8	Reg. Wt.	130	9.60	3.05	12.65
7300		50	20x8	Reg. Wt.	140	9.65	3.33	12.98
7400		75	20x8	Reg. Wt.	165	9.70	3.46	13.16
7500		100	20x8	Reg. Wt.	190	9.85	3.87	13.72
7550	50	30	24x8	Reg. Wt.	120	9.60	3.20	12.80
7600		40	24x8	Reg. Wt.	130	9.65	3.31	12.96
7750		50	24x8	Reg. Wt.	140	9.65	3.33	12.98
7800		75	24x8	Reg. Wt.	165	9.80	3.73	13.53
7900		100	32x10	Reg. Wt.	189	9.70	3.05	12.75
8000	60	30	32x10	Reg. Wt.	118	9.55	2.60	12.15
8100		40	32x10	Reg. Wt.	129	9.65	2.82	12.47
8200		50	32x10	Reg. Wt.	139	9.70	3.05	12.75
8300		75	32x10	Reg. Wt.	164	9.75	3.25	13
8350		100	32x10	Reg. Wt.	189	9.90	3.57	13.47
8400	70	30	32x10	Reg. Wt.	119	9.65	2.92	12.57
8450		40	32x10	Reg. Wt.	129	9.75	3.24	12.99
8500		50	32x10	Reg. Wt.	139	9.90	3.56	13.46
8550		75	32x10	Reg. Wt.	164	10.10	4.21	14.31
8600	80	30	32x10	Reg. Wt.	119	10.10	4.20	14.30
8800	50	30	20x8	Lt. Wt.	105	10.50	3.32	13.82
8850		40	24x8	Lt. Wt.	121	10.55	3.48	14.03
8900		50	24x8	Lt. Wt.	131	10.65	3.72	14.37
8950		75	24x8	Lt. Wt.	156	10.65	3.72	14.37
9000		100	24x8	Lt. Wt.	181	10.75	4.13	14.88

B1010 Floor Construction

B1010 235 Precast Double "T" Beams With 2″ Topping

	SPAN (FT.)	SUPERIMPOSED LOAD (P.S.F.)	DBL. "T" SIZE D (IN.) W (FT.)	CONCRETE "T" TYPE	TOTAL LOAD (P.S.F.)	COST PER S.F.		
						MAT.	INST.	TOTAL
9200	60	30	32x10	Lt. Wt.	103	10.45	2.70	13.15
9300		40	32x10	Lt. Wt.	113	10.55	2.93	13.48
9350		50	32x10	Lt. Wt.	123	10.55	2.93	13.48
9400		75	32x10	Lt. Wt.	148	10.60	3.13	13.73
9450		100	32x10	Lt. Wt.	173	10.70	3.45	14.15
9500	70	30	32x10	Lt. Wt.	103	10.60	3.12	13.72
9550		40	32x10	Lt. Wt.	113	10.60	3.12	13.72
9600		50	32x10	Lt. Wt.	123	10.70	3.44	14.14
9650		75	32x10	Lt. Wt.	148	10.80	3.77	14.57
9700	80	30	32x10	Lt. Wt.	103	10.70	3.44	14.14
9800		40	32x10	Lt. Wt.	113	10.80	3.76	14.56
9900		50	32x10	Lt. Wt.	123	10.90	4.08	14.98

B2010 Exterior Walls

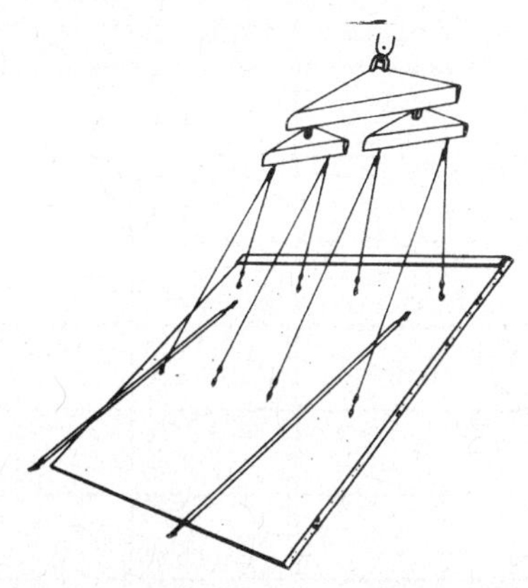

The advantage of tilt up construction is in the low cost of forms and placing of concrete and reinforcing. Tilt up has been used for several types of buildings, including warehouses, stores, offices, and schools. The panels are cast in forms on the ground, or floor slab. Most jobs use 5-1/2″ thick solid reinforced concrete panels.

Design Assumptions:
Conc. f'c = 3000 psi
Reinf. fy = 60,000

System Components	QUANTITY	UNIT	COST EACH		
			MAT.	INST.	TOTAL
SYSTEM B2010 106 3200					
TILT-UP PANELS, 20′ X 25′, BROOM FINISH, 5-1/2″ THICK, 3000 PSI					
Apply liquid bond release agent	500.000	S.F.	10	70	80
Edge forms in place for slab on grade	120.000	L.F.	40.80	429.60	470.40
Reinforcing in place	.350	Ton	385	385	770
Footings, form braces, steel	1.000	Set	499.80		499.80
Slab lifting inserts	1.000	Set	69		69
Framing, less than 4″ angles	1.000	Set	82.65	844.55	927.20
Concrete, ready mix, regular weight, 3000 psi	8.550	C.Y.	931.95		931.95
Place and vibrate concrete for slab on grade, 4″ thick, direct chute	8.550	C.Y.		223.67	223.67
Finish floor, monolithic broom finish	500.000	S.F.		405	405
Cure with curing compound, sprayed	500.000	S.F.	60.25	47.50	107.75
Erection crew	.058	Day		682.95	682.95
TOTAL	500.000		2,079.45	3,088.27	5,167.72
COST PER S.F.			4.16	6.18	10.34

B2010 106	Tilt-Up Concrete Panel	COST PER S.F.		
		MAT.	INST.	TOTAL
3200	Tilt-up conc panels, broom finish, 5-1/2″ thick, 3000 PSI	4.15	6.20	10.35
3250	5000 PSI	4.26	6.10	10.36
3300	6″ thick, 3000 PSI	4.58	6.35	10.93
3350	5000 PSI	4.69	6.25	10.94
3400	7-1/2″ thick, 3000 PSI	5.85	6.55	12.40
3450	5000 PSI	6	6.45	12.45
3500	8″ thick, 3000 PSI	6.25	6.70	12.95
3550	5000 PSI	6.45	6.60	13.05
3700	Steel trowel finish, 5-1/2″ thick, 3000 PSI	4.16	6.25	10.41
3750	5000 PSI	4.26	6.15	10.41
3800	6″ thick, 3000 PSI	4.58	6.40	10.98
3850	5000 PSI	4.69	6.30	10.99
3900	7-1/2″ thick, 3000 PSI	5.85	6.65	12.50
3950	5000 PSI	6	6.50	12.50
4000	8″ thick, 3000 PSI	6.25	6.80	13.05
4050	5000 PSI	6.45	6.65	13.10
4200	Exp. aggregate finish, 5-1/2″ thick, 3000 PSI	4.74	6.35	11.09
4250	5000 PSI	4.84	6.20	11.04
4300	6″ thick, 3000 PSI	5.15	6.45	11.60
4350	5000 PSI	5.25	6.35	11.60
4400	7-1/2″ thick, 3000 PSI	6.40	6.70	13.10
4450	5000 PSI	6.60	6.60	13.20

B2010 Exterior Walls

B2010 106	Tilt-Up Concrete Panel	COST PER S.F.		
		MAT.	INST.	TOTAL
4500	8" thick, 3000 PSI	6.85	6.85	13.70
4550	5000 PSI	7.05	6.75	13.80
4600	Exposed aggregate & vert. rustication 5-1/2" thick, 3000 PSI	7	7.80	14.80
4650	5000 PSI	7.10	7.70	14.80
4700	6" thick, 3000 PSI	7.40	7.95	15.35
4750	5000 PSI	7.50	7.85	15.35
4800	7-1/2" thick, 3000 PSI	8.65	8.15	16.80
4850	5000 PSI	8.80	8.05	16.85
4900	8" thick, 3000 PSI	9.10	8.30	17.40
4950	5000 PSI	9.25	8.20	17.45
5000	Vertical rib & light sandblast, 5-1/2" thick, 3000 PSI	6.55	10.20	16.75
5050	5000 PSI	6.65	10.10	16.75
5100	6" thick, 3000 PSI	6.95	10.35	17.30
5150	5000 PSI	7.05	10.25	17.30
5200	7-1/2" thick, 3000 PSI	8.20	10.65	18.85
5250	5000 PSI	8.35	10.55	18.90
5300	8" thick, 3000 PSI	8.65	10.80	19.45
5350	5000 PSI	8.80	10.70	19.50
6000	Broom finish w/2" polystyrene insulation, 6" thick, 3000 PSI	3.95	7.65	11.60
6050	5000 PSI	4.10	7.65	11.75
6100	Broom finish 2" fiberplank insulation, 6" thick, 3000 PSI	4.54	7.55	12.09
6150	5000 PSI	4.69	7.55	12.24
6200	Exposed aggregate w/2" polystyrene insulation, 6" thick, 3000 PSI	4.42	7.60	12.02
6250	5000 PSI	4.57	7.60	12.17
6300	Exposed aggregate 2" fiberplank insulation, 6" thick, 3000 PSI	5	7.55	12.55
6350	5000 PSI	5.15	7.55	12.70

G Building Sitework

G1010 Site Clearing

System Components	QUANTITY	UNIT	COST PER ACRE MAT.	COST PER ACRE INST.	COST PER ACRE TOTAL
SYSTEM G1010 120 1000					
REMOVE TREES & STUMPS UP TO 6" DIA. BY CUT & CHIP & STUMP HAUL AWAY					
Cut & chip light trees to 6" diameter	1.000	Acre		4,725	4,725
Grub stumps & remove, ideal conditions	1.000	Acre		2,080	2,080
TOTAL				6,805	6,805

G1010 120	Clear & Grub Site	COST PER ACRE MAT.	COST PER ACRE INST.	COST PER ACRE TOTAL
0980	Clear & grub includes removal and haulaway of tree stumps			
1000	Remove trees & stumps up to 6 " in dia. by cut & chip & remove stumps		6,800	6,800
1100	Up to 12 " in diameter		10,900	10,900
1990	Removal of brush includes haulaway of cut material			
1992	Percentage of vegetation is actual plant coverage			
2000	Remove brush by saw, 4' tall, 5 % vegetation coverage, 10 mile haul cycle		4,125	4,125
2010	10 % vegetation coverage		8,250	8,250
2020	15 % vegetation coverage		12,400	12,400
2030	20 % vegetation coverage		16,500	16,500
2040	25 % vegetation coverage		20,600	20,600

G10 Site Preparation

G1010 Site Clearing

Stripping and stock piling top soil can be done with a variety of equipment depending on the size of the job. Typically this would be a skid steer, tracked, or wheeled loader. Larger jobs can be done with a loader or skid steer. Even larger jobs can be done by a dozer.

System Components	QUANTITY	UNIT	COST PER S.Y. MAT.	COST PER S.Y. INST.	COST PER S.Y. TOTAL
SYSTEM G1010 122 1000					
STRIP TOPSOIL AND STOCKPILE, 6 INCH DEEP BY HAND, 50′ HAUL					
By hand, 6″ deep, 50′ haul, less than 100 S.Y.	1.000	S.Y.		13.80	13.80
TOTAL				13.80	13.80

G1010 122	Strip Topsoil	COST PER S.Y. MAT.	COST PER S.Y. INST.	COST PER S.Y. TOTAL
1000	Strip topsoil & stockpile, 6 inch deep by hand, 50′ haul		13.80	13.80
1100	100′ haul, 101-500 S.Y., by skid steer		3.33	3.33
1200	100′ haul, 501-900 S.Y.		1.85	1.85
1300	200′ haul , 901-1100 S.Y.		2.57	2.57
1400	200′ haul, 1101-4000 S.Y., by dozer		.56	.56

G1030 Site Earthwork

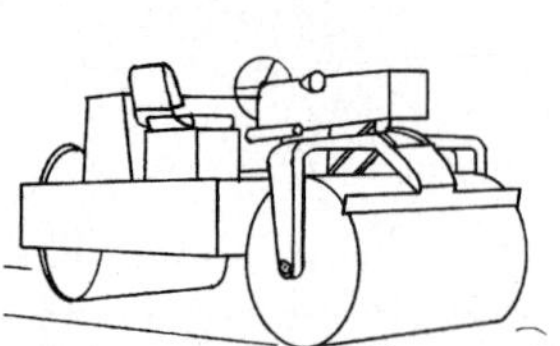

The Cut and Fill Gravel System includes: moving gravel cut from an area above the specified grade to an area below the specified grade utilizing a bulldozer and/or scraper with the addition of compaction equipment, plus a water wagon to adjust the moisture content of the soil.

The Expanded System Listing shows Cut and Fill operations with hauling distances that vary from 50′ to 5000′. Lifts for compaction in the filled area vary from 6″ to 12″. There is no waste included in the assumptions.

System Components	QUANTITY	UNIT	COST PER C.Y. EQUIP.	COST PER C.Y. LABOR	COST PER C.Y. TOTAL
SYSTEM G1030 105 1000					
GRAVEL CUT & FILL, 80 HP DOZER & COMPACT, 50′ HAUL, 6″ LIFT, 2 PASSES					
Excavating, bulk, dozer, 80 H.P. 50′ haul sand and gravel	1.000	C.Y.	1.15	1.78	2.93
Water wagon rent per day	.003	Hr.	.21	.17	.38
Backfill, dozer, 50′ haul sand and gravel	1.000	C.Y.	.55	.85	1.40
Compaction, vibrating roller, 6″ lifts, 2 passes	1.000	C.Y.	.21	.27	.48
TOTAL			2.12	3.07	5.19

G1030 105	Cut & Fill Gravel	COST PER C.Y. EQUIP.	COST PER C.Y. LABOR	COST PER C.Y. TOTAL
1000	Gravel cut & fill, 80 HP dozer & roller compact, 50′ haul, 6″ lift, 2 passes	2.12	3.07	5.19
1050	4 passes	2.23	3.23	5.46
1100	12″ lift, 2 passes	2.03	2.96	4.99
1150	4 passes	2.15	3.12	5.27
1200	150′ haul, 6″ lift, 2 passes	3.82	5.70	9.52
1250	4 passes	3.93	5.85	9.78
1300	12″ lift, 2 passes	3.73	5.60	9.33
1350	4 passes	3.85	5.75	9.60
1400	300′ haul, 6″ lift, 2 passes	6.45	9.85	16.30
1450	4 passes	6.60	10	16.60
1500	12″ lift, 2 passes	6.40	9.70	16.10
1550	4 passes	6.50	9.90	16.40
1600	105 HP dozer & roller compactor, 50′ haul, 6″ lift, 2 passes	1.90	2.31	4.21
1650	4 passes	2.01	2.47	4.48
1700	12″ lift, 2 passes	1.81	2.20	4.01
1750	4 passes	1.93	2.36	4.29
1800	150′ haul, 6″ lift, 2 passes	3.64	4.48	8.12
1850	4 passes	3.75	4.64	8.39
1900	12″ lift, 2 passes	3.55	4.37	7.92
1950	4 passes	3.67	4.53	8.20
2000	300′ haul, 6″ lift, 2 passes	6.70	8.30	15
2050	4 passes	6.80	8.45	15.25
2100	12′ lift, 2 passes	6.60	8.20	14.80
2150	4 passes	6.70	8.35	15.05
2200	200 HP dozer & roller compactor, 150′ haul, 6″ lift, 2 passes	4.24	2.59	6.83
2250	4 passes	4.35	2.75	7.10
2300	12″ lift, 2 passes	4.15	2.48	6.63
2350	4 passes	4.27	2.64	6.91

G1030 Site Earthwork

G1030 105	Cut & Fill Gravel	COST PER C.Y.		
		EQUIP.	LABOR	TOTAL
2600	300' haul, 6" lift, 2 passes	7.20	4.25	11.45
2650	4 passes	7.30	4.41	11.71
2700	12" lift, 2 passes	7.10	4.14	11.24
2750	4 passes	7.25	4.30	11.55
3000	300 HP dozer & roller compactors, 150' haul, 6" lift, 2 passes	3.64	1.76	5.40
3050	4 passes	3.75	1.92	5.67
3100	12" lift, 2 passes	3.55	1.65	5.20
3150	4 passes	3.67	1.81	5.48
3200	300' haul, 6" lift, 2 passes	6.20	2.81	9.01
3250	4 passes	6.30	2.97	9.27
3300	12" lift, 2 passes	6.10	2.70	8.80
3350	4 passes	6.25	2.86	9.11
4200	10 C.Y. elevating scraper & roller compacter, 1500' haul, 6" lift, 2 passes	4.62	3.84	8.46
4250	4 passes	5	4.25	9.25
4300	12" lift, 2 passes	4.47	3.76	8.23
4350	4 passes	4.61	3.87	8.48
4800	5000' haul, 6" lift, 2 passes	6.20	5.15	11.35
4850	4 passes	6.45	5.40	11.85
4900	12" lift, 2 passes	5.90	4.89	10.79
4950	4 passes	6.20	5.20	11.40
5000	15 C.Y. S.P. scraper & roller compactor, 1500' haul, 6" lift, 2 passes	4.22	2.83	7.05
5050	4 passes	4.48	3.06	7.54
5100	12" lift, 2 passes	4.07	2.74	6.81
5150	4 passes	4.21	2.85	7.06
5400	5000' haul, 6" lift, 2 passes	5.90	3.93	9.83
5450	4 passes	6.15	4.17	10.32
5500	12" lift, 2 passes	5.75	3.85	9.60
5550	4 passes	5.90	3.96	9.86
5600	21 C.Y. S.P. scraper & roller compactor, 1500' haul, 6" lift, 2 passes	4.07	2.26	6.33
5650	4 passes	4.34	2.49	6.83
5700	12" lift, 2 passes	3.93	2.16	6.09
5750	4 passes	4.07	2.27	6.34
6000	5000' haul, 6" lift, 2 passes	6.25	3.41	9.66
6050	4 passes	6.50	3.65	10.15
6100	12" lift, 2 passes	6.10	3.32	9.42
6150	4 passes	6.25	3.43	9.68

G1030 Site Earthwork

The Excavation of Sand and Gravel System balances the productivity of the Excavating equipment to the hauling equipment. It is assumed that the hauling equipment will encounter light traffic and will move up no considerable grades on the haul route. No mobilization cost is included. All costs given in these systems include a swell factor of 15%.

The Expanded System Listing shows excavation systems using backhoes ranging from 1/2 Cubic Yard capacity to 3-1/2 Cubic Yards. Power shovels indicated range from 1/2 Cubic Yard to 3 Cubic Yards. Dragline bucket rigs used range from 1/2 Cubic Yard to 3 Cubic Yards. Truck capacities indicated range from 8 Cubic Yards to 20 Cubic Yards. Each system lists the number of trucks involved and the distance (round trip) that each must travel.

System Components	QUANTITY	UNIT	COST PER C.Y. EQUIP.	LABOR	TOTAL
SYSTEM G1030 110 1000					
EXCAVATE SAND & GRAVEL, 1/2 CY BACKHOE, TWO 8 CY DUMP TRUCKS 2 MRT					
Excavating, bulk, hyd. backhoe, wheel mtd., 1-1/2 CY	1.000	B.C.Y.	1.13	2.45	3.58
Haul earth, 8 CY dump truck, 2 mile round trip, 2.6 loads/hr	1.150	L.C.Y.	4.39	4.39	8.78
Spotter at earth fill dump or in cut	.020	Hr.		1.02	1.02
TOTAL			5.52	7.86	13.38

G1030 110	Excavate Sand & Gravel	COST PER C.Y. EQUIP.	LABOR	TOTAL
1000	Excavate sand & gravel, 1/2 C.Y. backhoe, two 8 C.Y. dump trucks, 2 MRT	5.50	7.85	13.35
1200	Three 8 C.Y. dumptruck, 4 mile round trip	7.70	10	17.70
1400	Two 12 C.Y. dumptrucks, 4 mile round trip	7	6.50	13.50
1600	3/4 C.Y.backhoe, four 8 C.Y. dump trucks, 2 mile round trip	5.25	6.65	11.90
1700	Six 8 C.Y. dump trucks, 4 mile round trip	7.40	8.80	16.20
1800	Three 12 C.Y. dump trucks, 4 mile round trip	7	5.55	12.55
1900	Two 16 C.Y. dump trailers, 3 mile round trip	5.60	4.51	10.11
2000	Two 20 C.Y. dump trailers, 4 mile round trip	5.20	4.23	9.43
2200	1-1/2 C.Y. backhoe, four 12 C.Y. dump trucks, 2 mile round trip	5.60	3.90	9.50
2300	Six 12 C.Y. dump trucks, 4 mile round trip	6.80	4.36	11.16
2400	Three 16 C.Y. dump trailers, 3 mile round trip	5.60	3.82	9.42
2600	Three 20 C.Y. dump trailers, 4 mile round trip	5.20	3.54	8.74
2800	2-1/2 C.Y. backhoe, six 12 C.Y. dump trucks, 2 mile round trip	5.75	4.07	9.82
2900	Eight 12 C.Y. dump trucks, 4 mile round trip	7.05	4.82	11.87
3000	Four 16 C.Y. dump trailers, 2 mile round trip	4.80	3.21	8.01
3100	Six 16 C.Y. dump trailers, 4 mile round trip	6.05	4.04	10.09
3200	Four 20 C.Y. dump trailers, 3 mile round trip	5	3.29	8.29
3400	3-1/2 C.Y. backhoe, eight 16 C.Y. dump trailers, 3 mile round trip	5.65	3.31	8.96
3500	Ten 16 C.Y. dump trailers, 4 mile round trip	6	3.46	9.46
3700	Six 20 C.Y. dump trailers, 2 mile round trip	4.10	2.41	6.51
3800	Nine 20 C.Y. dump trailers, 4 mile round trip	5.15	2.95	8.10
4000	1/2 C.Y. pwr. shovel, four 8 C.Y. dump trucks, 2 mile round trip	5.40	6.25	11.65
4100	Six 8 C.Y. dump trucks, 3 mile round trip	6.35	7.45	13.80
4200	Two 12 C.Y. dump trucks, 1 mile round trip	4.46	3.88	8.34
4300	Four 12 C.Y. dump trucks, 4 mile round trip	6.95	5.15	12.10
4400	Two 16 C.Y. dump trailers, 2 mile round trip	4.70	3.78	8.48
4500	Two 20 C.Y. dump trailers, 3 mile round trip	4.91	3.87	8.78
4600	Three 20 C.Y. dump trailers, 4 mile round trip	5.05	3.65	8.70
4800	3/4 C.Y. pwr. shovel, six 8 C.Y. dump trucks, 2 mile round trip	5.35	6.05	11.40
4900	Four 12 C.Y. dump trucks, 2 mile round trip	5.70	4.06	9.76
5000	Six 12 C.Y. dump trucks, 4 mile round trip	6.85	5	11.85
5100	Four 16 C.Y. dump trailers, 4 mile round trip	5.95	4.03	9.98
5200	Two 20 C.Y. dump trailers, 1 mile round trip	3.69	2.79	6.48
5400	1-1/2 C.Y. pwr. shovel, six 12 C.Y. dump trucks, 2 mile round trip	5.55	3.87	9.42

G1030 Site Earthwork

G1030 110	Excavate Sand & Gravel	COST PER C.Y.		
		EQUIP.	LABOR	TOTAL
5600	Four 16 C.Y. dump trailers, 1 mile round trip	3.97	2.54	6.51
5700	Six 16 C.Y. dump trailers, 3 mile round trip	5.35	3.48	8.83
5800	Six 20 C.Y. dump trailers, 4 mile round trip	4.96	3.21	8.17
6000	3 C.Y. pwr. shovel, eight 16 C.Y. dump trailers, 1 mile round trip	3.76	2.31	6.07
6200	Eight 20 C.Y. dump trailers, 2 mile round trip	3.64	2.28	5.92
6400	Twelve 20 C.Y. dump trailers, 4 mile round trip	4.77	2.82	7.59
6600	1/2 C.Y. dragline bucket, three 8 C.Y. dump trucks, 1 mile round trip	4.68	5.40	10.08
6700	Five 8 C.Y.dump trucks, 4 mile round trip	9.25	10.65	19.90
6800	Two 12 C.Y. dump trucks, 2 mile round trip	7.55	6.45	14
6900	Three 12 C.Y. dump trucks, 4 mile round trip	8.60	6.90	15.50
7000	Two 16 C.Y. dump trailers, 4 mile round trip	7.85	6.40	14.25
7200	3/4 C.Y. dragline bucket, four 8 C.Y. dump trucks, 1 mile round trip	4.98	5.80	10.78
7300	Six 8 C.Y. dump trucks, 3 mile round trip	7.40	8.35	15.75
7400	Four 12 C.Y. dump trucks, 4 mile round trip	7.95	6.50	14.45
7500	Two 16 C.Y. dump trailers, 2 mile round trip	5.85	4.78	10.63
7600	Three 16 C.Y. dump trailers, 4 mile round trip	6.95	5.25	12.20
7800	1-1/2 C.Y. dragline bucket, six 8 C.Y. dump trucks, 1 mile round trip	3.90	4.14	8.04
7900	Ten 8 C.Y. dump trucks, 3 mile round trip	7.05	7.20	14.25
8000	Six 12 C.Y. dump trucks, 2 mile round trip	6.10	4.29	10.39
8100	Eight 12 C.Y. dump trucks, 4 mile round trip	7.35	5.05	12.40
8200	Four 16 C.Y. dump trailers, 2 mile round trip	5.15	3.46	8.61
8300	Three 20 C.Y. dump trailers, 2 mile round trip	4.77	3.60	8.37
8400	3 C.Y. dragline bucket, eight 12 C.Y. dump trucks, 2 mile round trip	6	3.93	9.93
8500	Six 16 C.Y. dump trailers, 2 mile round trip	4.95	3.21	8.16
8600	Nine 16 C.Y. dump trailers, 4 mile round trip	6.15	3.82	9.97
8700	Eight 20 C.Y. dump trailers, 4 mile round trip	5.35	3.31	8.66

G1030 Site Earthwork

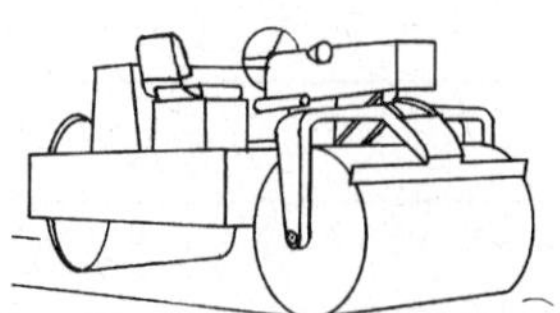

The Cut and Fill Common Earth System includes: moving common earth cut from an area above the specified grade to an area below the specified grade utilizing a bulldozer and/or scraper, with the addition of compaction equipment, plus a water wagon to adjust the moisture content of the soil.

The Expanded System Listing shows Cut and Fill operations with hauling distances that vary from 50′ to 5000′. Lifts for compaction in the filled area vary from 4″ to 8″. There is no waste included in the assumptions.

System Components	QUANTITY	UNIT	COST PER C.Y. EQUIP.	LABOR	TOTAL
SYSTEM G1030 115 1000					
EARTH CUT & FILL, 80 HP DOZER & COMPACTOR, 50′ HAUL, 4″ LIFT, 2 PASSES					
Excavating, bulk, dozer, 50′ haul, common earth	1.000	C.Y.	1.88	2.91	4.79
Water wagon, rent per day	.008	Hr.	.56	.46	1.02
Backfill dozer, from existing stockpile, 75 H.P., 50′ haul	1.000	C.Y.	.71	1.10	1.81
Compaction, roller, 4″ lifts, 2 passes	1.000	C.Y.	.09	.41	.50
TOTAL			3.24	4.88	8.12

G1030 115	Cut & Fill Common Earth	COST PER C.Y. EQUIP.	LABOR	TOTAL
1000	Earth cut & fill, 80 HP dozer & roller compact, 50′ haul, 4″ lift, 2 passes	3.24	4.88	8.10
1050	4 passes	2.77	4.43	7.20
1100	8″ lift, 2 passes	2.68	4.02	6.70
1150	4 passes	2.77	4.43	7.20
1200	150′ haul, 4″ lift, 2 passes	5.20	7.80	13
1250	4 passes	5.35	8.55	13.90
1300	8″ lift, 2 passes	5.20	7.80	13
1350	4 passes	5.35	8.55	13.90
1400	300′ haul, 4″ lift, 2 passes	9.35	14.05	23.50
1450	4 passes	9.60	15.30	25
1500	8″ lift, 2 passes	9.35	14.05	23.50
1550	4 passes	9.60	15.30	25
1600	105 H.P. dozer and roller compactor, 50′ haul, 4″ lift, 2 passes	2.08	2.65	4.73
1650	4 passes	2.13	2.86	4.99
1700	8″ lift, 2 passes	2.08	2.65	4.73
1750	4 passes	2.13	2.86	4.99
1800	150′ haul, 4″ lift, 2 passes	4.55	5.80	10.35
1850	4 passes	4.64	6.25	10.90
1900	8″ lift, 2 passes	4.53	5.70	10.25
1950	4 passes	4.62	6.15	10.75
2000	300′ haul, 4″ lift, 2 passes	8.45	11.05	19.50
2050	4 passes	8.65	12.05	20.50
2100	8″ lift, 2 passes	8.10	10.75	18.85
2150	4 passes	8.45	11.05	19.50
2200	200 H.P. dozer & roller compactor, 150′ haul, 4″ lift, 2 passes	4.99	3.05	8.05
2250	4 passes	5.05	3.26	8.30
2300	8″ lift, 2 passes	4.95	2.95	7.90
2350	4 passes	4.99	3.05	8.05
2600	300′ haul, 4″ lift, 2 passes	8.80	5.55	14.35
2650	4 passes	8.90	5.95	14.85

G1030 115	Cut & Fill Common Earth	COST PER C.Y.		
		EQUIP.	LABOR	TOTAL
2700	8″ lift, 2 passes	8.70	5.35	14.05
2750	4 passes	8.65	5.15	13.80
3000	300 H.P. dozer & roller compactor, 150′ haul, 4″ lifts, 2 passes	4.36	2.23	6.60
3050	4 passes	4.46	2.53	7
3100	8″ lift, 2 passes	4.31	2.07	6.40
3150	4 passes	4.36	2.23	6.60
3200	300′ haul, 4″ lifts, 2 passes	7.50	3.57	11.05
3250	4 passes	7.60	3.87	11.45
3300	8″ lifts, 2 passes	7.45	3.41	10.85
3350	4 passes	7.50	3.57	11.05
4200	10 C.Y. elevating scraper & roller compact, 1500′ haul, 4″ lifts, 2 passes	4.29	3.31	7.60
4250	4 passes	4.38	3.72	8.10
4300	8″ lifts, 2 passes	4.24	3.10	7.35
4350	4 passes	4.29	3.31	7.60
4800	5000′ haul, 4″ lifts, 2 passes	5.65	4.23	9.90
4850	4 passes	5.75	4.64	10.40
4900	8″ lifts, 2 passes	5.60	4.02	9.60
4950	4 passes	5.65	4.23	9.90
5000	15 C.Y. S.P. scraper & roller compact, 1500′ haul, 4″ lifts, 2 passes	4.45	2.72	7.15
5050	4 passes	4.54	2.97	7.50
5100	8″ lifts, 2 passes	4.54	2.97	7.50
5150	4 passes	4.41	2.59	7
5400	5000′ haul, 4″ lifts, 2 passes	5.50	2.91	8.40
5450	4 passes	5.55	3.12	8.65
5500	8″ lifts, 2 passes	5.95	3.44	9.40
5550	4 passes	5.95	3.39	9.35
5600	21 C.Y. S.P. scraper & roller compact, 1500′ haul, 4″ lifts, 2 passes	4.46	2.31	6.75
5650	4 passes	4.52	2.52	7.05
5700	8″ lifts, 2 passes	4.60	2.45	7.05
5750	4 passes	4.64	2.55	7.20
6000	5000′ haul, 4″ lifts, 2 passes	6.35	2.98	9.35
6050	4 passes	6.45	3.28	9.75
6100	8″ lifts, 2 passes	6.30	2.82	9.10
6150	4 passes	6.35	2.98	9.35

G1030 Site Earthwork

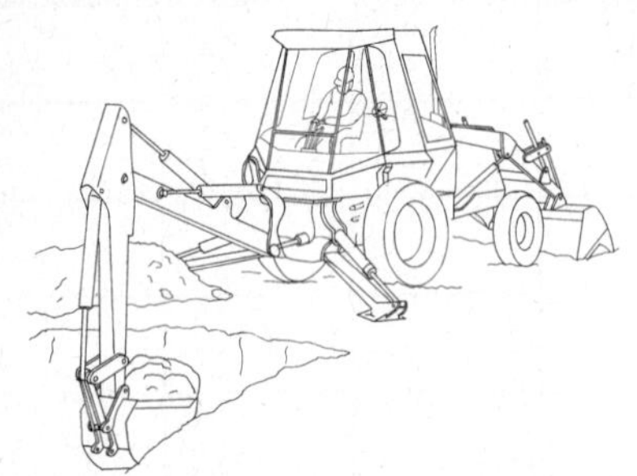

The Excavation of Common Earth System balances the productivity of the excavating equipment to the hauling equipment. It is assumed that the hauling equipment will encounter light traffic and will move up no considerable grades on the haul route. No mobilization cost is included. All costs given in these systems include a swell factor of 25% for hauling.

The Expanded System Listing shows Excavation systems using backhoes ranging from 1/2 Cubic Yard capacity to 3-1/2 Cubic Yards. Power shovels indicated range from 1/2 Cubic Yard to 3 Cubic Yards. Dragline bucket rigs range from 1/2 Cubic Yard to 3 Cubic Yards. Truck capacities range from 8 Cubic Yards to 20 Cubic Yards. Each system lists the number of trucks involved and the distance (round trip) that each must travel.

System Components	QUANTITY	UNIT	COST PER C.Y.		
			EQUIP.	LABOR	TOTAL
SYSTEM G1030 120 1000					
EXCAVATE COMMON EARTH, 1/2 CY BACKHOE, TWO 8 CY DUMP TRUCKS, 1 MRT					
Excavating, bulk hyd. backhoe wheel mtd., 1/2 C.Y.	1.000	B.C.Y.	.97	2.10	3.07
Hauling, 8 CY truck, cycle 0.5 mile, 20 MPH, 15 min. wait/Ld./Uld.	1.280	L.C.Y.	2.53	2.53	5.06
Spotter at earth fill dump or in cut	.020	Hr.		.90	.90
TOTAL			3.50	5.53	9.03

G1030 120	Excavate Common Earth	COST PER C.Y.		
		EQUIP.	LABOR	TOTAL
1000	Excavate common earth, 1/2 C.Y. backhoe, two 8 C.Y. dump trucks, 1 mile RT	3.50	5.55	9.05
1200	Three 8 C.Y. dump trucks, 3 mile round trip	7.15	9.30	16.45
1400	Two 12 C.Y. dump trucks, 4 mile round trip	7.80	7.15	14.95
1600	3/4 C.Y. backhoe, three 8 C.Y. dump trucks, 1 mile round trip	3.49	4.55	8.04
1700	Five 8 C.Y. dump trucks, 3 mile round trip	7.05	8.55	15.60
1800	Two 12 C.Y. dump trucks, 2 mile round trip	6.50	5.50	12
1900	Two 16 C.Y. dump trailers, 3 mile round trip	6.10	4.67	10.77
2000	Two 20 C.Y. dump trailers, 4 mile round trip	5.80	4.61	10.41
2200	1-1/2 C.Y. backhoe, eight 8 C.Y. dump trucks, 3 mile round trip	6.95	7.55	14.50
2300	Four 12 C.Y. dump trucks, 2 mile round trip	6.20	4.69	10.89
2400	Six 12 C.Y. dump trucks, 4 mile round trip	7.50	5.40	12.90
2500	Three 16 C.Y. dump trailers, 2 mile round trip	5	3.48	8.48
2600	Two 20 C.Y. dump trailers, 1 mile round trip	3.97	2.87	6.84
2700	Three 20 C.Y. dump trailer, 3 mile round trip	5.25	3.56	8.81
2800	2-1/2 C.Y. excavator, six 12 C.Y. dump trucks, 1 mile round trip	4.47	3.16	7.63
2900	Eight 12 C.Y. dump trucks, 3 mile round trip	6.45	4.41	10.86
3000	Four 16 C.Y. dump trailers, 1 mile round trip	4.38	2.88	7.26
3100	Six 16 C.Y. dump trailers, 3 mile round trip	5.90	3.90	9.80
3200	Six 20 C.Y. dump trailers, 4 mile round trip	5.50	3.59	9.09
3400	3-1/2 C.Y. backhoe, six 16 C.Y. dump trailers, 1 mile round trip	4.64	2.74	7.38
3600	Ten 16 C.Y. dump trailers, 4 mile round trip	6.65	3.87	10.52
3800	Eight 20 C.Y. dump trailers, 3 mile round trip	5.30	3.07	8.37
4000	1/2 C.Y. pwr. shovel, four 8 C.Y. dump trucks, 2 mile round trip	6	6.90	12.90
4100	Two 12 C.Y. dump trucks, 1 mile round trip	4.95	4.29	9.24
4200	Four 12 C.Y. dump trucks, 4 mile round trip	7.70	5.70	13.40
4300	Two 16 C.Y. dump trailers, 2 mile round trip	5.20	4.19	9.39
4400	Two 20 C.Y. dump trailers, 4 mile round trip	6	4.80	10.80
4800	3/4 C.Y. pwr. shovel, six 8 C.Y. dump trucks, 2 mile round trip	5.90	6.65	12.55
4900	Three 12 C.Y. dump trucks, 1 mile round trip	4.86	3.68	8.54
5000	Five 12 C.Y. dump trucks, 4 mile round trip	7.75	5.50	13.25
5100	Three 16 C.Y. dump trailers, 3 mile round trip	6.35	4.46	10.81
5200	Three 20 C.Y. dump trailers, 4 mile round trip	5.90	4.15	10.05
5400	1-1/2 C.Y. pwr. shovel, six 12 C.Y. dump trucks, 1 mile round trip	4.49	3.15	7.64
5500	Ten 12 C.Y. dump trucks, 4 mile round trip	7.40	4.96	12.36

G1030 Site Earthwork

G1030 120	Excavate Common Earth	COST PER C.Y.		
		EQUIP.	LABOR	TOTAL
5600	Six 16 C.Y. dump trailers, 3 mile round trip	5.95	3.88	9.83
5700	Four 20 C.Y. dump trailers, 2 mile round trip	4.29	2.99	7.28
5800	Six 20 C.Y. dump trailers, 4 mile round trip	5.50	3.58	9.08
6000	3 C.Y. pwr. shovel, ten 12 C.Y. dump trucks, 1 mile round trip	4.38	2.87	7.25
6200	Twelve 16 C.Y. dump trailers, 4 mile round trip	6.25	3.77	10.02
6400	Eight 20 C.Y. dump trailers, 2 mile round trip	4.04	2.50	6.54
6600	1/2 C.Y. dragline bucket, three 8 C.Y. dump trucks, 2 mile round trip	8.20	9.15	17.35
6700	Two 12 C.Y. dump trucks, 3 mile round trip	9.20	7.95	17.15
6800	Three 12 C.Y. dump trucks, 4 mile round trip	9.50	7.60	17.10
7000	Two 16 C.Y. dump trailers, 4 mile round trip	8.65	7.05	15.70
7200	3/4 C.Y. dragline bucket, four 8 C.Y. dump trucks, 1 mile round trip	4.49	5.10	9.59
7300	Six 8 C.Y. dump trucks, 3 mile round trip	8.15	9.20	17.35
7400	Two 12 C.Y. dump trucks, 1 mile round trip	6.25	5.45	11.70
7500	Four 12 C.Y. dump trucks, 4 mile round trip	8.80	6.70	15.50
7600	Two 16 C.Y. dump trailers, 2 mile round trip	6.45	5.30	11.75
7800	1-1/2 C.Y. dragline bucket, four 12 C.Y. dump trucks, 1 mile round trip	5.50	3.89	9.39
7900	Six 12 C.Y. dump trucks, 3 mile round trip	7.25	5.15	12.40
8000	Four 16 C.Y. dump trucks, 2 mile round trip	5.70	3.85	9.55
8200	Five 16 C.Y. dump trucks, 4 mile round trip	7.30	4.77	12.07
8400	3 C.Y. dragline bucket, eight 12 C.Y. dump trucks, 2 mile round trip	6.65	4.39	11.04
8500	Twelve 12 C.Y. dump trucks, 4 mile round trip	7.95	5.05	13
8600	Eight 16 C.Y. dump trailers, 3 mile round trip	6.40	3.94	10.34
8700	Six 20 C.Y. dump trailers, 2 mile round trip	4.64	2.93	7.57

G1030 Site Earthwork

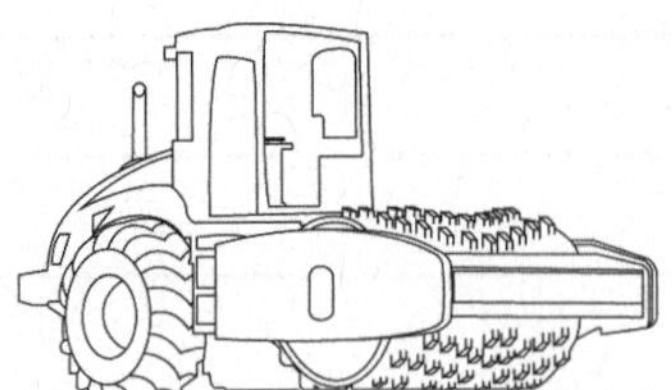

The Cut and Fill Clay System includes: moving clay from an area above the specified grade to an area below the specified grade utilizing a bulldozer and compaction equipment; plus a water wagon to adjust the moisture content of the soil.

The Expanded System Listing shows Cut and Fill operations with hauling distances that vary from 50′ to 5000′. Lifts for compaction in the filled areas vary from 4″ to 8″. There is no waste included in the assumptions.

System Components	QUANTITY	UNIT	COST PER C.Y. EQUIP.	COST PER C.Y. LABOR	COST PER C.Y. TOTAL
SYSTEM G1030 125 1000					
CLAY CUT & FILL, 80 HP DOZER & COMPACTOR, 50′ HAUL, 4″ LIFTS, 2 PASSES					
Excavating , bulk, dozer, 50′ haul, clay	1.000	C.Y.	2.12	3.28	5.40
Water wagon, rent per day	.004	Hr.	.28	.23	.51
Backfill dozer, from existing, stock pile, 75 H.P., 50′ haul, clay	1.330	C.Y.	.82	1.28	2.10
Compaction, tamper, 4″ lifts, 2 passes	245.000	S.F.	.56	.34	.90
TOTAL			3.78	5.13	8.91

G1030 125	Cut & Fill Clay	COST PER C.Y. EQUIP.	COST PER C.Y. LABOR	COST PER C.Y. TOTAL
1000	Clay cut & fill, 80 HP dozer & compactor, 50′ haul, 4″ lifts, 2 passes	3.78	5.15	8.93
1050	4 passes	4.53	5.65	10.18
1100	8″ lifts, 2 passes	3.34	4.84	8.18
1150	4 passes	3.74	5.10	8.84
1200	150′ haul, 4″ lifts, 2 passes	6.75	9.70	16.45
1250	4 passes	7.50	10.20	17.70
1300	8″ lifts, 2 passes	6.30	9.40	15.70
1350	4 passes	6.70	9.65	16.35
1400	300′ haul, 4″ lifts, 2 passes	11.40	16.95	28.35
1450	4 passes	12.15	17.45	29.60
1500	8″ lifts, 2 passes	10.95	16.65	27.60
1550	4 passes	11.35	16.90	28.25
1600	105 HP dozer & sheeps foot compactors, 50′ haul, 4″ lifts, 2 passes	3.31	3.68	6.99
1650	4 passes	4.06	4.20	8.26
1700	8″ lifts, 2 passes	2.87	3.39	6.26
1750	4 passes	3.27	3.66	6.93
1800	150′ haul, 4″ lifts, 2 passes	6.25	7.35	13.60
1850	4 passes	7	7.90	14.90
1900	8″ lifts, 2 passes	5.80	7.10	12.90
1950	4 passes	6.20	7.35	13.55
2000	300′ haul, 4″ lifts, 2 passes	9.70	11.70	21.40
2050	4 passes	10.45	12.25	22.70
2100	8″ lifts, 2 passes	9.25	11.40	20.65
2150	4 passes	9.65	11.70	21.35
2200	200 HP dozer & sheepsfoot compactors, 150′ haul, 4″ lifts, 2 passes	7.30	4.21	11.51
2250	4 passes	8.05	4.73	12.78
2300	8″ lifts, 2 passes	6.90	3.92	10.82
2350	4 passes	7.30	4.19	11.49
2600	300′ haul, 4″ lifts, 2 passes	12.35	7.05	19.40
2650	4 passes	13.10	7.55	20.65

G1030 Site Earthwork

G1030 125	Cut & Fill Clay	COST PER C.Y.		
		EQUIP.	LABOR	TOTAL
2700	8″ lifts, 2 passes	11.90	6.75	18.65
2750	4 passes	12.30	7	19.30
3000	300 HP dozer & sheepsfoot compactors, 150′ haul, 4″ lifts, 2 passes	6.40	2.85	9.25
3050	4 passes	7.15	3.37	10.52
3100	8″ lifts, 2 passes	5.95	2.56	8.51
3150	4 passes	6.35	2.83	9.18
3200	300′ haul, 4″ lifts, 2 passes	11	4.74	15.74
3250	4 passes	11.75	5.25	17
3300	8″ lifts, 2 passes	10.55	4.45	15
3350	4 passes	10.95	4.72	15.67
4200	10 C.Y. elev. scraper & sheepsfoot rollers, 1500′ haul, 4″ lifts, 2 passes	10.30	5.90	16.20
4250	4 passes	10.55	6.15	16.70
4300	8″ lifts, 2 passes	10.20	5.85	16.05
4350	4 passes	10.35	5.95	16.30
4800	5000′ haul, 4″ lifts, 2 passes	14	8	22
4850	4 passes	14.25	8.20	22.45
4900	8″ lifts, 2 passes	13.85	7.95	21.80
4950	4 passes	14	8.05	22.05
5000	15 C.Y. SP scraper & sheepsfoot rollers, 1500′ haul, 4″ lifts, 2 passes	8.55	4.09	12.64
5050	4 passes	8.80	4.32	13.12
5100	8″ lifts, 2 passes	8.40	4	12.40
5150	4 passes	8.55	4.11	12.66
5400	5000′ haul, 4″ lifts, 2 passes	12.20	5.75	17.95
5450	4 passes	12.50	6	18.50
5500	8″ lifts, 2 passes	12.10	5.70	17.80
5550	4 passes	12.25	5.80	18.05
5600	21 C.Y. SP scraper & sheepsfoot rollers, 1500′ haul, 4″ lifts, 2 passes	8	3.22	11.22
5650	4 passes	8.30	3.46	11.76
5700	8″ lifts, 2 passes	7.90	3.13	11.03
5750	4 passes	8	3.24	11.24
6000	5000′ haul, 4″ lift, 2 passes	12.60	4.99	17.59
6050	4 passes	12.85	5.25	18.10
6100	8″ lift, 2 passes	12.50	4.91	17.41
6150	4 passes	12.60	5	17.60

G1030 Site Earthwork

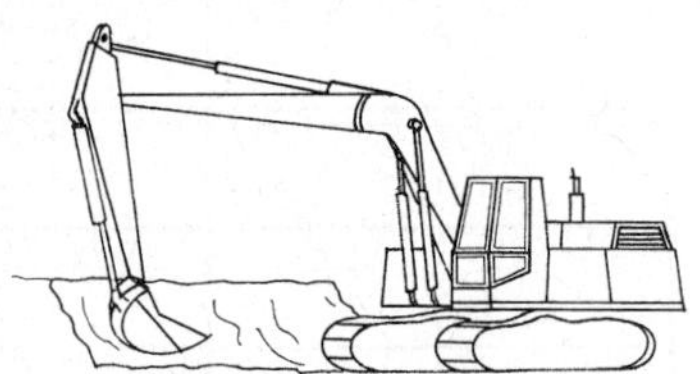

The Excavation of Clay System balances the productivity of excavating equipment to hauling equipment. It is assumed that the hauling equipment will encounter light traffic and will move up no considerable grades on the haul route. No mobilization cost is included. All costs given in these systems include a swell factor of 40%.

The Expanded System Listing shows Excavation systems using backhoes ranging from 1/2 Cubic Yard capacity to 3-1/2 Cubic Yards. Power shovels indicated range from 1/2 Cubic Yard to 3 Cubic Yards. Dragline bucket rigs range from 1/2 Cubic Yard to 3 Cubic Yards. Truck capacities range from 8 Cubic Yards to 20 Cubic Yards. Each system lists the number of trucks involved and the distance (round trip) that each must travel.

System Components	QUANTITY	UNIT	COST PER C.Y. EQUIP.	COST PER C.Y. LABOR	COST PER C.Y. TOTAL
SYSTEM G1030 130 1000					
EXCAVATE CLAY, 1/2 CY BACKHOE, TWO 8 CY DUMP TRUCKS, 2 MI ROUND TRIP					
Excavating bulk hyd. backhoe, wheel mtd. 1/2 C.Y.	1.000	B.C.Y.	1.38	2.98	4.36
Haul earth, 8 C.Y. dump truck, 2 mile round trip, 2.6 loads/hr	1.350	L.C.Y.	5.16	5.16	10.32
Spotter at earth fill dump or in cut	.020	Hr.		1.30	1.30
TOTAL			6.54	9.44	15.98

G1030 130	Excavate Clay	COST PER C.Y. EQUIP.	COST PER C.Y. LABOR	COST PER C.Y. TOTAL
1000	Excavate clay, 1/2 C.Y. backhoe, two 8 C.Y. dump trucks, 2 mile round trip	6.55	9.45	16
1200	Three 8 C.Y. dump trucks, 4 mile round trip	9.10	11.85	20.95
1400	Two 12 C.Y. dump trucks, 4 mile round trip	8.30	7.75	16.05
1600	3/4 C.Y. backhoe, three 8 C.Y. dump trucks, 2 mile round trip	6.55	8	14.55
1700	Five 8 C.Y. dump trucks, 4 mile round trip	8.95	10.20	19.15
1800	Two 12 C.Y. dump trucks, 2 mile round trip	6.90	5.95	12.85
1900	Three 12 C.Y. dump trucks, 4 mile round trip	8.25	6.60	14.85
2000	1-1/2 C.Y. backhoe, eight 8 C.Y. dump trucks, 3 mile round trip	7.40	7.70	15.10
2100	Three 12 C.Y. dump trucks, 1 mile round trip	5.05	3.78	8.83
2200	Six 12 C.Y. dump trucks, 4 mile round trip	8	5.75	13.75
2300	Three 16 C.Y. dump trailers, 2 mile round trip	5.35	3.76	9.11
2400	Four 16 C.Y. dump trailers, 4 mile round trip	6.95	4.64	11.59
2600	Two 20 C.Y. dump trailers, 1 mile round trip	4.24	3.11	7.35
2800	2-1/2 C.Y. backhoe, eight 12 C.Y. dump trucks, 3 mile round trip	6.85	4.73	11.58
2900	Four 16 C.Y. dump trailers, 1 mile round trip	4.66	3.05	7.71
3000	Six 16 C.Y. dump trailers, 3 mile round trip	6.30	4.17	10.47
3100	Four 20 C.Y. dump trailers, 2 mile round trip	4.45	3.12	7.57
3200	Six 20 C.Y. dump trailers, 4 mile round trip	5.80	3.86	9.66
3400	3-1/2 C.Y. backhoe, ten 12 C.Y. dump trucks, 2 mile round trip	6.75	4.13	10.88
3500	Six 16 C.Y. dump trailers, 1 mile round trip	4.95	2.90	7.85
3600	Ten 16 C.Y. dump trailers, 4 mile round trip	7.05	4.09	11.14
3800	Eight 20 C.Y. dump trailers, 3 mile round trip	5.65	3.25	8.90
4000	1/2 C.Y. pwr. shovel, three 8 C.Y. dump trucks, 1 mile round trip	5.80	6.80	12.60
4100	Six 8 C.Y. dump trucks, 4 mile round trip	8.95	10.55	19.50
4200	Two 12 C.Y. dump trucks, 1 mile round trip	5.30	4.64	9.94
4400	Three 12 C.Y. dump trucks, 3 mile round trip	7.25	5.75	13
4600	3/4 C.Y. pwr. shovel, six 8 C.Y. dump trucks, 2 mile round trip	6.30	7.15	13.45
4700	Nine 8 C.Y. dump trucks, 4 mile round trip	8.85	9.65	18.50
4800	Three 12 C.Y. dump trucks, 1 mile round trip	5.20	3.98	9.18
4900	Four 12 C.Y. dump trucks, 3 mile round trip	7.30	5.35	12.65
5000	Two 16 C.Y. dump trailers, 1 mile round trip	5.15	3.92	9.07
5200	Three 16 C.Y. dump trailers, 3 mile round trip	6.75	4.80	11.55

G1030 Site Earthwork

G1030 130	Excavate Clay	COST PER C.Y.		
		EQUIP.	LABOR	TOTAL
5400	1-1/2 C.Y. pwr. shovel, six 12 C.Y. dump trucks, 2 mile round trip	6.55	4.61	11.16
5500	Four 16 C.Y. dump trailers, 1 mile round trip	4.69	3.04	7.73
5600	Seven 16 C.Y. dump trailers, 4 mile round trip	6.75	4.42	11.17
5800	Five 20 C.Y. dump trailers, 3 mile round trip	5.45	3.43	8.88
6000	3 C.Y. pwr. shovel, nine 16 C.Y. dump trailers, 2 mile round trip	5	3.09	8.09
6100	Twelve 16 C.Y. dump trailers, 4 mile round trip	6.60	3.98	10.58
6200	Six 20 C.Y. dump trailers, 1 mile round trip	3.91	2.47	6.38
6400	Nine 20 C.Y. dump trailers, 3 mile round trip	5.25	3.22	8.47
6600	1/2 C.Y. dragline bucket, two 8 C.Y. dump trucks, 1 mile round trip	8.95	10.25	19.20
6800	Four 8 C.Y. dump trucks, 4 mile round trip	11.75	12.95	24.70
7000	Two 12 C.Y. dump trucks, 3 mile round trip	9.95	8.65	18.60
7200	3/4 C.Y. dragline bucket, four 8 C.Y. dump trucks, 2 mile round trip	7.95	8.75	16.70
7300	Six 8 C.Y. dump trucks, 4 mile round trip	10.45	11.85	22.30
7400	Two 12 C.Y. dump trucks, 2 mile round trip	9	7.70	16.70
7500	Three 12 C.Y. dump trucks, 4 mile round trip	10.15	8.20	18.35
7600	Two 16 C.Y. dump trailers, 3 mile round trip	8.45	6.90	15.35
7800	1-1/2 C.Y. dragline bucket, five 12 C.Y. dump trucks, 3 mile round trip	8.20	5.65	13.85
7900	Three 16 C.Y. dump trailers, 1 mile round trip	5.70	3.92	9.62
8000	Five 16 C.Y. dump trailers, 4 mile round trip	7.80	5.10	12.90
8100	Three 20 C.Y. dump trailers, 2 mile round trip	5.70	3.96	9.66
8400	3 C.Y. dragline bucket, eight 16 C.Y. dump trailers, 3 mile round trip	6.60	4.07	10.67
8500	Six 20 C.Y. dump trailers, 2 mile round trip	4.97	3.19	8.16
8600	Eight 20 C.Y. dump trailers, 4 mile round trip	6.35	3.93	10.28

G1030 Site Earthwork

The Loading and Hauling of Sand and Gravel System balances the productivity of loading equipment to hauling equipment. It is assumed that the hauling equipment will encounter light traffic and will move up no considerable grades on the haul route.

The Expanded System Listing shows Loading and Hauling systems that use either a track or wheel front-end loader. Track loaders indicated range from 1-1/2 Cubic Yards capacity to 4-1/2 Cubic Yards capacity. Wheel loaders range from 1-1/2 Cubic Yards to 5 Cubic Yards. Trucks for hauling range from 12 Cubic Yards capacity to 20 Cubic Yards capacity. Each system lists the number of trucks involved and the distance (round trip) that each must travel.

System Components	QUANTITY	UNIT	COST PER C.Y. EQUIP.	LABOR	TOTAL
SYSTEM G1030 135 1000					
LOAD & HAUL SAND & GRAVEL, 1-1/2 CY LOADER, FOUR 12 CY TRUCKS, 1 MRT					
Excavating bulk, F.E. loader, track mtd., 1/2 C.Y.	1.000	B.C.Y.	.64	.91	1.55
Haul earth, 12 C.Y. dump truck, 1 mile round trip, 2.7 loads/hr	1.150	L.C.Y.	3.47	2.09	5.56
Spotter at earth fill dump or in cut	.040	Hr.		.23	.23
TOTAL			4.11	3.23	7.34

G1030 135	Load & Haul Sand & Gravel	COST PER C.Y. EQUIP.	LABOR	TOTAL
1000	Load & haul sand & gravel, 1-1/2 CY tr.loader, four 12 CY dump trucks,1 MRT	4.11	3.23	7.34
1200	Six 12 C.Y. dump trucks, 3 mile round trip	5.80	4.45	10.25
1400	Four 16 C.Y. dump trailers, 3 mile round trip	5.45	3.93	9.38
1600	Three 20 C.Y. dump trailers, 2 mile round trip	3.86	3.02	6.88
1800	Four 20 C.Y.dump trailers, 4 mile round trip	5.05	3.65	8.70
2000	2-1/2 C.Y. track loader, six 12 C.Y. dump trucks, 2 mile round trip	5.80	4.08	9.88
2200	Eight 12 C.Y. dump trucks, 4 mile round trip	7.10	4.82	11.92
2400	Five 16 C.Y. dump trailers, 3 mile round trip	5.75	3.69	9.44
2600	Three 20 C.Y. dump trailers, 1 mile round trip	3.82	2.57	6.39
3000	3-1/2 C.Y. track loader, six 12 C.Y. dump trucks, 1 mile round trip	4.44	3.04	7.48
3200	Six 16 C.Y. dump trailers, 2 mile round trip	4.68	3.04	7.72
3600	Eight 16 C.Y. dump trailers, 4 mile round trip	6.05	3.74	9.79
4000	4-1/2 C.Y. track loader, six 16 C.Y. dump trailers, 1 mile round trip	4.26	2.56	6.82
4200	Eight 16 C.Y. dump trailers, 2 mile round trip	4.65	2.78	7.43
4400	Eight 20 C.Y. dump trailers, 3 mile round trip	4.86	2.86	7.72
4600	Nine 20 C.Y. dump trailers, 4 mile round trip	5.25	3.06	8.31
5000	1-1/2 C.Y. wheel loader, four 12 C.Y. dump trucks, 1 mile round trip	3.93	3.23	7.16
5200	Six 12 C.Y. dump trucks, 3 mile round trip	5.65	4.45	10.10
5400	Four 16 C.Y. dump trailers, 2 mile round trip	4.19	3.21	7.40
5600	Five 16 C.Y. dump trailers, 4 mile round trip	5.60	4.02	9.62
6000	3 C.Y. wheel loader, ten 12 C.Y.dump trucks, 3 mile round trip	5.45	3.79	9.24
6200	Five 16 C.Y. dump trailers, 1 mile round trip	3.59	2.43	6.02
6400	Six 16 C.Y. dump trailers, 2 mile round trip	4.05	2.87	6.92
6600	Seven 20 C.Y. dump trailers, 4 mile round trip	4.62	3.07	7.69
7000	5 C.Y. wheel loader, eight 16 C.Y. dump trailers, 1 mile round trip	3.80	2.38	6.18
7200	Twelve 16 C.Y. dump trailers, 3 mile round trip	5.20	3.16	8.36
7400	Nine 20 C.Y. dump trailers, 2 mile round trip	3.63	2.26	5.89
7600	Twelve 20 C.Y. dump trailers, 4 mile round	4.80	2.89	7.69

G1030 Site Earthwork

The Loading and Hauling of Common Earth System balances the productivity of loading equipment to hauling equipment. It is assumed that the hauling equipment will encounter light traffic and will move up no considerable grades on the haul route.

The Expanded System Listing shows Loading and Hauling systems that use either a track or wheel front–end loader. Track loaders indicated range from 1-1/2 Cubic Yards capacity to 4-1/2 Cubic Yards capacity. Wheel loaders range from 1-1/2 Cubic Yards to 5 Cubic Yards. Trucks for hauling range from 8 Cubic Yards capacity to 20 Cubic Yards capacity. Each system lists the number of trucks involved and the distance (round trip) that each must travel.

System Components	QUANTITY	UNIT	COST PER C.Y. EQUIP.	LABOR	TOTAL
SYSTEM G1030 140 1000					
LOAD & HAUL COMMON EARTH, 1-1/2 CY LOADER, SIX 8 CY TRUCKS, 1 MRT					
Excavating bulk, F.E. loader track mtd., 1.5 C.Y.	1.000	B.C.Y.	.76	1.08	1.84
8 C.Y. truck, cycle 2 miles	1.280	L.C.Y.	4.31	4.31	8.62
Spotter at earth fill dump or in cut	.010	Hr.		.57	.57
TOTAL			5.07	5.96	11.03

G1030 140	Load & Haul Common Earth	COST PER C.Y. EQUIP.	LABOR	TOTAL
1000	Load & haul common earth, 1-1/2 C.Y. tr. loader, six 8 C.Y. trucks, 1 MRT	5.05	5.95	11
1200	Four 12 C.Y. dump trucks, 2 mile round trip	6.25	4.81	11.06
1400	Three 16 C.Y. dump trailers, 2 mile round trip	5.05	3.94	8.99
1600	Four 16 C.Y. dump trailers, 4 mile round trip	6.60	4.80	11.40
2000	2-1/2 C.Y. track loader, six 12 C.Y. dump trucks, 3 mile round trip	7	4.94	11.94
2200	Four 16 C.Y. dump trailers, 2 mile round trip	5.40	3.58	8.98
2400	Five 16 C.Y. dump trailers, 4 mile round trip	6.95	4.48	11.43
2600	Three 20 C.Y. dump trailers, 1 mile round trip	4.22	2.82	7.04
3000	3-1/2 C.Y. track loader, six 12 C.Y. dump trucks, 1 mile round trip	4.92	3.33	8.25
3200	Seven 16 C.Y. dump trailers, 4 mile round trip	6.85	4.32	11.17
3400	Four 20 C.Y. dump trailers, 1 mile round trip	4.14	2.54	6.68
3600	Six 20 C.Y. dump trailers, 4 mile round trip	5.95	3.77	9.72
4000	4-1/2 C.Y. track loader, eight 12 C.Y. dump trucks, 1 mile round trip	4.89	3.11	8
4200	Six 16 C.Y. dump trailers, 1 mile round trip	4.73	2.87	7.60
4400	Six 20 C.Y. dump trailers, 2 mile round trip	4.63	2.77	7.40
4600	Eight 20 C.Y. dump trailers, 4 mile round trip	5.95	3.47	9.42
5000	1-1/2 C.Y. wheel loader, eight 8 C.Y. dump trucks, 2 mile round trip	5.40	6.50	11.90
5200	Four 12 C.Y. dump trucks, 1 mile round trip	4.36	3.60	7.96
5400	Six 12 C.Y. dump trucks, 3 mile round trip	6.25	4.94	11.19
5600	Five 16 C.Y. dump trailers, 4 mile round trip	6.20	4.48	10.68
6000	3 C.Y. wheel loader, eight 12 C.Y. dump trucks, 2 mile round trip	5.70	4.03	9.73
6200	Five 16 C.Y. dump trailers, 1 mile round trip	4	2.68	6.68
6400	Eight 16 C.Y. dump trailers, 3 mile round trip	5.55	3.62	9.17
6600	Six 20 C.Y. dump trailers, 2 mile round trip	3.80	2.62	6.42
7000	5 C.Y. wheel loader, eight 16 C.Y. dump trailers, 1 mile round trip	4.23	2.69	6.92
7200	Twelve 16 C.Y. dump trailers, 3 mile round trip	5.75	3.55	9.30
7400	Nine 20 C.Y. dump trailers, 2 mile round trip	4.03	2.48	6.51
7600	Twelve 20 C.Y. dump trailers, 4 mile round trip	5.35	3.18	8.53

G1030 Site Earthwork

The Loading and Hauling of Clay System balances the productivity of loading equipment to hauling equipment. It is assumed that the hauling equipment will encounter light traffic and will move up no considerable grades on the haul route.

The Expanded System Listing shows Loading and Hauling systems that use either a track or wheel front-end loader. Track loaders indicated range from 1-1/2 Cubic Yards capacity to 4-1/2 Cubic Yards capacity. Wheel loaders range from 1-1/2 Cubic Yards to 5 Cubic Yards. Trucks for hauling range from 8 Cubic Yards capacity to 20 Cubic Yards capacity. Each system lists the number of trucks involved and the distance (round trip) that each must travel.

System Components	QUANTITY	UNIT	COST PER C.Y. EQUIP.	LABOR	TOTAL
SYSTEM G1030 145 1000					
LOAD & HAUL CLAY, 1-1/2 CY LOADER, EIGHT 8 CY DUMP TRUCKS, 3 MRT					
Excavating bulk, F.E. loader, track mtd., 1.5 C.Y.	1.000	B.C.Y.	1	1.42	2.42
Haul earth, 8 C.Y. dump truck, 3 mile round trip, 2.1 loads/hr	1.350	L.C.Y.	6.45	6.44	12.89
Spotter at earth fill dump or in cut	.010	Hr.		.79	.79
TOTAL			7.45	8.65	16.10

G1030 145	Load & Haul Clay	COST PER C.Y. EQUIP.	LABOR	TOTAL
1000	Load & haul clay, 1-1/2 C.Y. loader, eight 8 C.Y. dump trucks, 3 MRT	7.45	8.65	16.10
1200	Four 12 C.Y. dump trucks, 2 mile round trip	6.65	5.20	11.85
1400	Six 12 C.Y. dump trucks, 4 mile round trip	8.05	6.20	14.25
1600	Three 16 C.Y. dump trailers, 2 mile round trip	5.40	4.28	9.68
2000	2-1/2 C.Y. track loader, four 12 C.Y. dump trucks, 1 mile round trip	5.50	3.85	9.35
2200	Six 12 C.Y. dump trucks, 3 mile round trip	7.45	5.30	12.75
2400	Four 16 C.Y. dump trailers, 2 mile round trip	5.75	3.82	9.57
2600	Five 16 C.Y. dump trailers, 4 mile round trip	7.45	4.77	12.22
3000	3-1/2 C.Y. track loader, six 12 C.Y. dump trucks, 1 mile round trip	5.25	3.60	8.85
3200	Six 16 C.Y. dump trailers, 2 mile round trip	5.55	3.59	9.14
3400	Eight 16 C.Y. dump trailers, 4 mile round trip	7.10	4.48	11.58
3600	Four 20 C.Y. dump trailers, 1 mile round trip	4.44	2.94	7.38
4000	4-1/2 C.Y. track loader, six 16 C.Y. dump trailers, 1 mile round trip	5.05	3.04	8.09
4200	Ten 16 C.Y. dump trailers, 4 mile round trip	7.15	4.22	11.37
4400	Five 20 C.Y. dump trailers, 1 mile round trip	4.47	2.56	7.03
4600	Eight 20 C.Y. dump trailers, 3 mile round trip	5.75	3.39	9.14
5000	1-1/2 C.Y. wheel loader, eight 8 C.Y. dump trucks, 2 mile round trip	5.75	6.95	12.70
5200	Four 12 C.Y. dump trucks, 1 mile round trip	4.63	3.85	8.48
5400	Six 12 C.Y. dump trucks, 3 mile round trip	6.65	5.30	11.95
5600	Five 16 C.Y. dump trailers, 4 mile round trip	6.55	4.77	11.32
6000	3 C.Y. wheel loader, eight 12 C.Y. dump trucks, 2 mile round trip	6.05	4.32	10.37
6200	Six 16 C.Y. dump trailers, 2 mile round trip	4.77	3.39	8.16
6400	Eight 16 C.Y. dump trailers, 4 mile round trip	6.35	4.28	10.63
6600	Four 20 C.Y. dump trailers, 1 mile round trip	3.65	2.50	6.15
7000	5 C.Y. wheel loader, eight 16 C.Y. dump trailers, 1 mile round trip	4.49	2.85	7.34
7200	Twelve 16 C.Y. dump trailers, 3 mile round trip	6.10	3.75	9.85
7400	Nine 20 C.Y. dump trailers, 2 mile round trip	4.29	2.70	6.99
7600	Twelve 20 C.Y. dump trailers, 4 mile round trip	5.65	3.44	9.09

G1030 Site Earthwork

The Loading and Hauling of Rock System balances the productivity of loading equipment to hauling equipment. It is assumed that the hauling equipment will encounter light traffic and will move up no considerable grades on the haul route.

The Expanded System Listing shows Loading and Hauling systems that use either a track or wheel front-end loader. Track loaders indicated range from 1-1/2 Cubic Yards capacity to 4-1/2 Cubic Yards capacity. Wheel loaders range from 1-1/2 Cubic Yards to 5 Cubic Yards. Trucks for hauling range from 8 Cubic Yards capacity to 20 Cubic Yards capacity. Each system lists the number of trucks involved and the distance (round trip) that each must travel.

System Components	QUANTITY	UNIT	COST PER C.Y. EQUIP.	LABOR	TOTAL
SYSTEM G1030 150 1000					
LOAD & HAUL ROCK, 1-1/2 C.Y. TRACK LOADER, SIX 8 C.Y. TRUCKS, 1 MRT					
Excavating bulk, F.E. loader, track mtd., 1.5 C.Y.	1.000	B.C.Y.	1	1.42	2.42
8 C.Y. truck, cycle 2 miles	1.650	L.C.Y.	5.56	5.56	11.12
Spotter at earth fill dump or in cut	.010	Hr.		.79	.79
TOTAL			6.56	7.77	14.33

G1030 150	Load & Haul Rock	COST PER C.Y. EQUIP.	LABOR	TOTAL
1000	Load & haul rock, 1-1/2 C.Y. track loader, six 8 C.Y. trucks, 1 MRT	6.55	7.75	14.30
1200	Nine 8 C.Y. dump trucks, 3 mile round trip	8.95	10.20	19.15
1400	Six 12 C.Y. dump trucks, 4 mile round trip	9.80	7.50	17.30
1600	Three 16 C.Y. dump trucks, 2 mile round trip	6.55	5.10	11.65
2000	2-1/2 C.Y. track loader, twelve 8 C.Y. dump trucks, 3 mile round trip	9.35	9.60	18.95
2200	Five 12 C.Y. dump trucks, 1 mile round trip	6.35	4.33	10.68
2400	Eight 12 C.Y. dump trucks, 4 mile round trip	10.15	7	17.15
2600	Four 16 C.Y. dump trailers, 2 mile round trip	7	4.61	11.61
3000	3-1/2 C.Y. track loader, eight 12 C.Y. dump trucks, 2 mile round trip	8.25	5.45	13.70
3200	Five 16 C.Y. dump trucks, 1 mile round trip	6.05	3.71	9.76
3400	Seven 16 C.Y. dump trailers, 3 mile round trip	8.10	5.10	13.20
3600	Seven 20 C.Y. dump trailers, 4 mile round trip	7.50	4.69	12.19
4000	4-1/2 C.Y. track loader, nine 12 C.Y. dump trucks, 1 mile round trip	6.20	3.92	10.12
4200	Eight 16 C.Y. dump trailers, 2 mile round trip	6.65	4	10.65
4400	Eleven 16 C.Y. dump trailers, 4 mile round trip	8.60	5.10	13.70
4600	Seven 20 C.Y. dump trailers, 2 mile round trip	5.80	3.46	9.26
5000	1-1/2 C.Y. wheel loader, nine 8 C.Y. dump trucks, 2 mile round trip	6.90	8.20	15.10
5200	Four 12 C.Y. dump trucks, 1 mile round trip	5.65	4.64	10.29
5400	Seven 12 C.Y. dump trucks, 4 mile round trip	9.40	7.20	16.60
5600	Five 16 C.Y. dump trailers, 4 mile round trip	8	5.80	13.80
6000	3 C.Y. wheel loader, eight 12 C.Y. dump trucks, 2 mile round trip	7.40	5.25	12.65
6200	Five 16 C.Y. dump trailers, 1 mile round trip	5.15	3.48	8.63
6400	Seven 16 C.Y. dump trailers, 3 mile round trip	6.20	4.30	10.50
6600	Seven 20 C.Y. dump trailers, 4 mile round trip	6.60	4.45	11.05
7000	5 C.Y. wheel loader, twelve 12 C.Y. dump trucks, 1 mile round trip	5.60	3.66	9.26
7200	Nine 16 C.Y. dump trailers, 1 mile round trip	5.40	3.36	8.76
7400	Eight 20 C.Y. dump trailers, 1 mile round trip	4.63	2.88	7.51
7600	Twelve 20 C.Y. dump trailers, 3 mile round trip	6.25	3.79	10.04

G1030 Site Earthwork

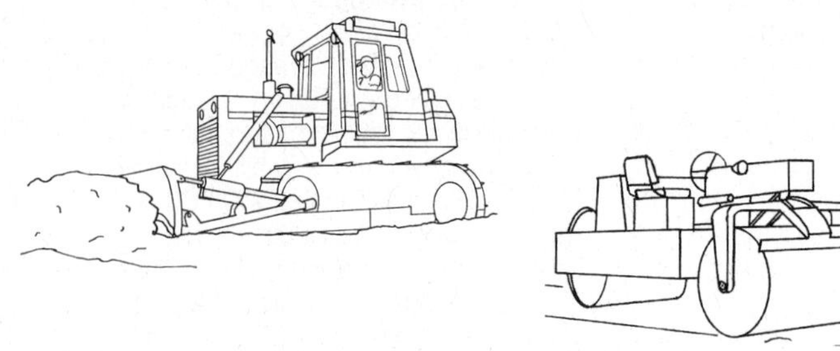

The Cut and Fill Common Earth System includes: moving common earth cut from an area above the specified grade to an area below the specified grade utilizing a bulldozer and/or scraper, with the addition of compaction equipment, plus a water wagon to adjust the moisture content of the soil.

The Expanded System Listing shows Cut and Fill operations with hauling distances that vary from 50′ to 5000′. Lifts for compaction in the filled area vary from 4″ to 8″. There is no waste included in the assumptions.

System Components	QUANTITY	UNIT	COST PER C.Y. EQUIP.	LABOR	TOTAL
SYSTEM G1030 155 1000					
SANDY CLAY CUT & FILL, 80 HP DOZER & COMPACTOR, 50′ HAUL, 4″ LIFT, 2 PASSES					
Excavating, bulk dozer, 50′ haul, sandy clay & loam	1.000	C.Y.	1.20	1.86	3.06
Water wagon, rent per day	.008	Hr.	.56	.46	1.02
Backfill dozer, from existing stockpile, 75 H.P., 50′ haul	1.000	C.Y.	.69	1.08	1.77
Compaction, roller, 4″ lifts, 2 passes	1.000	C.Y.	.09	.41	.50
TOTAL			2.54	3.81	6.35

G1030 155	Cut & Fill Sandy Clay/Loam	COST PER C.Y. EQUIP.	LABOR	TOTAL
1000	Sandy Clay cut and fill, 80 HP dozer & roller, 50′ haul, 4″ lift, 2 passes	2.54	3.81	6.35
1050	4 passes	2.88	4.52	7.40
1100	8″ lift, 2 passes	2.49	3.60	6.10
1150	4 passes	2.53	3.79	6.30
1200	150′ haul, 4″ lift, 2 passes	4.93	7.40	12.35
1250	4 passes	5.05	7.90	12.95
1300	8″ lift, 2 passes	4.86	7.05	11.90
1350	4 passes	4.91	7.30	12.20
1400	300′ haul, 4″ lift, 2 passes	8.60	12.95	21.50
1450	4 passes	8.80	13.75	22.50
1500	8″ lift, 2 passes	8.45	12.35	21
1550	4 passes	8.55	12.75	21.50
1600	105 H.P. dozer and roller compactor, 50′ haul, 4″ lift, 2 passes	1.97	2.51	4.48
1650	4 passes	2	2.66	4.66
1700	8″ lift, 2 passes	1.95	2.41	4.36
1750	4 passes	1.97	2.48	4.45
1800	150′ haul, 4″ lift, 2 passes	4.30	5.50	9.80
1850	4 passes	4.39	5.90	10.30
1900	8″ lift, 2 passes	4.28	5.40	9.70
1950	4 passes	4.37	5.80	10.15
2000	300′ haul, 4″ lift, 2 passes	7.75	10.15	17.90
2050	4 passes	7.95	11.20	19.15
2100	8″ lift, 2 passes	7.40	9.90	17.30
2150	4 passes	7.75	10.15	17.90
2200	200 H.P. dozer & roller compactor, 150′ haul, 4″ lift, 2 passes	4.62	2.85	7.45
2250	4 passes	4.68	3.06	7.75
2300	8″ lift, 2 passes	4.58	2.75	7.35
2350	4 passes	4.62	2.85	7.45
2600	300′ haul, 4″ lift, 2 passes	8.25	5.20	13.45
2650	4 passes	8.40	5.65	14.05

G1030 Site Earthwork

G1030 155	Cut & Fill Sandy Clay/Loam	COST PER C.Y.		
		EQUIP.	LABOR	TOTAL
2700	8″ lift, 2 passes	8.20	5	13.20
2750	4 passes	8.30	5.30	13.60
3000	300 H.P. dozer & roller compactor, 150′ haul, 4″ lifts, 2 passes	4.05	2.11	6.15
3050	4 passes	4.15	2.41	6.55
3100	8″ lift, 2 passes	4	1.95	5.95
3150	4 passes	4.05	2.11	6.15
3200	300′ haul, 4″ lifts, 2 passes	6.95	3.35	10.30
3250	4 passes	7.05	3.65	10.70
3300	8″ lifts, 2 passes	6.90	3.19	10.10
3350	4 passes	6.95	3.35	10.30
4200	10 C.Y. elevating scraper & roller compact, 1500′ haul, 4″ lifts, 2 passes	3.97	3.07	7.05
4250	4 passes	4.03	3.37	7.40
4300	8″ lifts, 2 passes	3.92	2.86	6.80
4350	4 passes	3.96	3.02	7
4800	5000′ haul, 4″ lifts, 2 passes	4.94	3.52	8.45
4850	4 passes	4.99	3.73	8.70
4900	8″ lifts, 2 passes	4.92	3.42	8.35
4950	4 passes	4.94	3.52	8.45
5000	15 C.Y. S.P. scraper & roller compact, 1500′ haul, 4″ lifts, 2 passes	4.38	2.83	7.20
5050	4 passes	4.47	3.08	7.55
5100	8″ lifts, 2 passes	4.70	3.37	8.05
5150	4 passes	4.57	2.99	7.55
5400	5000′ haul, 4″ lifts, 2 passes	5.30	2.98	8.30
5450	4 passes	5.35	3.19	8.55
5500	8″ lifts, 2 passes	5.25	2.88	8.15
5550	4 passes	5.25	2.83	8.10
5600	21 C.Y. S.P. scraper & roller compact, 1500′ haul, 4″ lifts, 2 passes	4.19	2.20	6.40
5650	4 passes	4.25	2.41	6.65
5700	8″ lifts, 2 passes	4.33	2.33	6.65
5750	4 passes	4.37	2.43	6.80
6000	5000′ haul, 4″ lifts, 2 passes	5.85	2.75	8.60
6050	4 passes	5.95	3.05	9
6100	8″ lifts, 2 passes	5.80	2.59	8.40
6150	4 passes	5.85	2.75	8.60

G1030 Site Earthwork

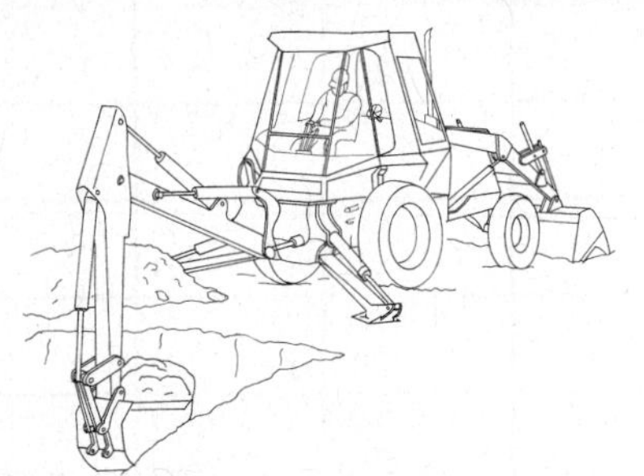

The Excavation of Sandy Clay/Loam System balances the productivity of the excavating equipment to the hauling equipment. It is assumed that the hauling equipment will encounter light traffic and will move up no considerable grades on the haul route. No mobilization cost is included. All costs given in these systems include a swell factor of 25% for hauling.

The Expanded System Listing shows Excavation systems using backhoes ranging from 1/2 Cubic Yard capacity to 3-1/2 Cubic Yards. Truck capacities range from 8 Cubic Yards to 20 Cubic Yards. Each system lists the number of trucks involved and the distance (round trip) that each must travel.

System Components	QUANTITY	UNIT	COST PER C.Y. EQUIP.	LABOR	TOTAL
SYSTEM G1030 160 1000					
EXCAVATE SANDY CLAY/LOAM, 1/2 CY BACKHOE, TWO 8 CY DUMP TRUCKS, 1 MRT					
Excavating, bulk hyd. backhoe wheel mtd., 1/2 C.Y.	1.000	B.C.Y.	.97	2.10	3.07
Hauling, 8 CY truck, cycle 0.5 mile, 20 MPH, 15 min. wait/Ld./Uld.	1.280	L.C.Y.	2.53	2.53	5.06
Spotter at earth fill dump or in cut	.020	Hr.		.90	.90
TOTAL			5.26	7.19	12.45

G1030 160	Excavate Sandy Clay/Loam	COST PER C.Y. EQUIP.	LABOR	TOTAL
1000	Excavate sandy clay/loam, 1/2 C.Y. backhoe, two 8 C.Y. dump trucks, 1 MRT	5.25	7.20	12.45
1200	Three 8 C.Y. dump trucks, 3 mile round trip	7.15	9.15	16.30
1400	Two 12 C.Y. dump trucks, 4 mile round trip	7.80	7	14.80
1600	3/4 C.Y. backhoe, three 8 C.Y. dump trucks, 1 mile round trip	5.25	6.25	11.50
1700	Five 8 C.Y. dump trucks, 3 mile round trip	7.05	8.55	15.60
1800	Two 12 C.Y. dump trucks, 2 mile round trip	6.50	5.45	11.95
1900	Two 16 C.Y. dump trailers, 3 mile round trip	6.10	4.60	10.70
2000	Two 20 C.Y. dump trailers, 4 mile round trip	5.75	4.53	10.30
2200	1-1/2 C.Y. excavator, eight 8 C.Y. dump trucks, 3 mile round trip	6.90	7.55	14.45
2300	Four 12 C.Y. dump trucks, 2 mile round trip	6.15	4.62	10.75
2400	Six 12 C.Y. dump trucks, 4 mile round trip	7.50	5.30	12.80
2500	Three 16 C.Y. dump trailers, 2 mile round trip	4.94	3.39	8.35
2600	Two 20 C.Y. dump trailers, 1 mile round trip	3.88	2.76	6.65
2700	Three 20 C.Y. dump trailer, 3 mile round trip	5.20	3.47	8.65
2800	2-1/2 C.Y. excavator, six 12 C.Y. dump trucks, 1 mile round trip	4.33	3.04	7.35
2900	Eight 12 C.Y. dump trucks, 3 mile round trip	6.30	4.30	10.60
3000	Four 16 C.Y. dump trailers, 1 mile round trip	4.21	2.72	6.95
3100	Six 16 C.Y. dump trailers, 3 mile round trip	5.80	3.76	9.55
3200	Six 20 C.Y. dump trailers, 4 mile round trip	5.35	3.45	8.80
3400	3-1/2 C.Y. excavator, six 16 C.Y. dump trailers, 1 mile round trip	4.46	2.68	7.15
3600	Ten 16 C.Y. dump trailers, 4 mile round trip	6.50	3.83	10.35
3800	Eight 20 C.Y. dump trailers, 3 mile round trip	5.15	3.02	8.15

G1030 Site Earthwork

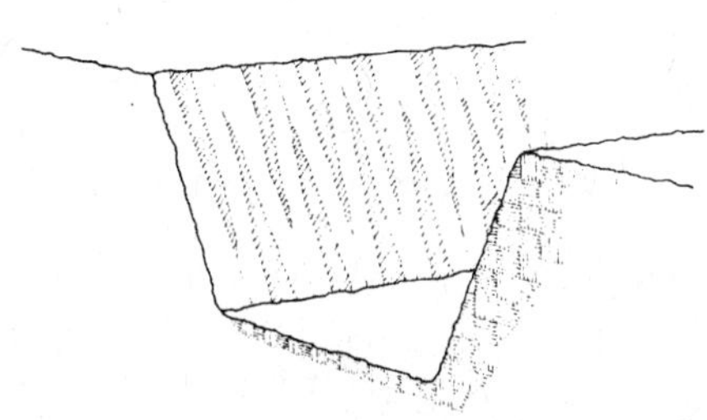

Trenching Systems are shown on a cost per linear foot basis. The systems include: excavation; backfill and removal of spoil; and compaction for various depths and trench bottom widths. The backfill has been reduced to accommodate a pipe of suitable diameter and bedding.

The slope for trench sides varies from none to 1:1 and shoring equipment is not included.

The Expanded System Listing shows Trenching Systems that range from 2′ to 12′ in width. Depths range from 2′ to 25′.

System Components	QUANTITY	UNIT	COST PER L.F. EQUIP.	COST PER L.F. LABOR	COST PER L.F. TOTAL
SYSTEM G1030 805 1310					
TRENCHING COMMON EARTH, NO SLOPE, 2′ WIDE, 2′ DP, 3/8 C.Y. BUCKET					
Excavation, trench, hyd. backhoe, track mtd., 3/8 C.Y. bucket	.148	B.C.Y.	.40	1.03	1.43
Backfill and load spoil, from stockpile	.153	L.C.Y.	.12	.31	.43
Compaction by vibrating plate, 6″ lifts, 4 passes	.118	E.C.Y.	.03	.38	.41
Remove excess spoil, 8 C.Y. dump truck, 2 mile roundtrip	.040	L.C.Y.	.15	.15	.30
TOTAL			.70	1.87	2.57

G1030 805	Trenching Common Earth	COST PER L.F. EQUIP.	COST PER L.F. LABOR	COST PER L.F. TOTAL
1310	Trenching, common earth, no slope, 2′ wide, 2′ deep, 3/8 C.Y. bucket	.70	1.87	2.57
1320	3′ deep, 3/8 C.Y. bucket	.99	2.82	3.81
1330	4′ deep, 3/8 C.Y. bucket	1.29	3.78	5.07
1340	6′ deep, 3/8 C.Y. bucket	1.68	4.92	6.60
1350	8′ deep, 1/2 C.Y. bucket	2.22	6.55	8.77
1360	10′ deep, 1 C.Y. bucket	3.48	7.75	11.23
1400	4′ wide, 2′ deep, 3/8 C.Y. bucket	1.61	3.70	5.31
1410	3′ deep, 3/8 C.Y. bucket	2.21	5.60	7.81
1420	4′ deep, 1/2 C.Y. bucket	2.57	6.30	8.87
1430	6′ deep, 1/2 C.Y. bucket	4.15	10.15	14.30
1440	8′ deep, 1/2 C.Y. bucket	6.65	13.05	19.70
1450	10′ deep, 1 C.Y. bucket	8.05	16.20	24.25
1460	12′ deep, 1 C.Y. bucket	10.35	20.50	30.85
1470	15′ deep, 1-1/2 C.Y. bucket	9.40	18.50	27.90
1480	18′ deep, 2-1/2 C.Y. bucket	13.05	26	39.05
1520	6′ wide, 6′ deep, 5/8 C.Y. bucket w/trench box	8.60	14.95	23.55
1530	8′ deep, 3/4 C.Y. bucket	11.40	19.75	31.15
1540	10′ deep, 1 C.Y. bucket	11.55	20.50	32.05
1550	12′ deep, 1-1/2 C.Y. bucket	12.40	22	34.40
1560	16′ deep, 2-1/2 C.Y. bucket	17.10	27.50	44.60
1570	20′ deep, 3-1/2 C.Y. bucket	22	33	55
1580	24′ deep, 3-1/2 C.Y. bucket	26	39.50	65.50
1640	8′ wide, 12′ deep, 1-1/2 C.Y. bucket w/trench box	17.40	28	45.40
1650	15′ deep, 1-1/2 C.Y. bucket	22.50	36.50	59
1660	18′ deep, 2-1/2 C.Y. bucket	25	36.50	61.50
1680	24′ deep, 3-1/2 C.Y. bucket	35.50	51	86.50
1730	10′ wide, 20′ deep, 3-1/2 C.Y. bucket w/trench box	28.50	48.50	77
1740	24′ deep, 3-1/2 C.Y. bucket	42.50	58.50	101
1800	1/2 to 1 slope, 2′ wide, 2′ deep, 3/8 C.Y. bucket	.99	2.82	3.81
1810	3′ deep, 3/8 C.Y. bucket	1.67	4.97	6.64
1820	4′ deep, 3/8 C.Y. bucket	2.49	7.60	10.09
1840	6′ deep, 3/8 C.Y. bucket	4.02	12.35	16.37
1860	8′ deep, 1/2 C.Y. bucket	6.40	19.75	26.15
1880	10′ deep, 1 C.Y. bucket	11.95	27.50	39.45

G1030 805	Trenching Common Earth	COST PER L.F.		
		EQUIP.	LABOR	TOTAL
2300	4' wide, 2' deep, 3/8 C.Y. bucket	1.91	4.65	6.56
2310	3' deep, 3/8 C.Y. bucket	2.88	7.75	10.63
2320	4' deep, 1/2 C.Y. bucket	3.63	9.65	13.28
2340	6' deep, 1/2 C.Y. bucket	6.95	18	24.95
2360	8' deep, 1/2 C.Y. bucket	12.85	26.50	39.35
2380	10' deep, 1 C.Y. bucket	17.80	37	54.80
2400	12' deep, 1 C.Y. bucket	24.50	50.50	75
2430	15' deep, 1-1/2 C.Y. bucket	26.50	54	80.50
2460	18' deep, 2-1/2 C.Y. bucket	46.50	84.50	131
2840	6' wide, 6' deep, 5/8 C.Y. bucket w/trench box	12.70	22.50	35.20
2860	8' deep, 3/4 C.Y. bucket	18.35	33.50	51.85
2880	10' deep, 1 C.Y. bucket	18.60	34	52.60
2900	12' deep, 1-1/2 C.Y. bucket	23.50	44.50	68
2940	16' deep, 2-1/2 C.Y. bucket	39	65.50	104.50
2980	20' deep, 3-1/2 C.Y. bucket	55	88.50	143.50
3020	24' deep, 3-1/2 C.Y. bucket	77.50	121	198.50
3100	8' wide, 12' deep, 1-1/2 C.Y. bucket w/trench box	29	51	80
3120	15' deep, 1-1/2 C.Y. bucket	42.50	74	116.50
3140	18' deep, 2-1/2 C.Y. bucket	54	88	142
3180	24' deep, 3-1/2 C.Y. bucket	87	133	220
3270	10' wide, 20' deep, 3-1/2 C.Y. bucket w/trench box	56	102	158
3280	24' deep, 3-1/2 C.Y. bucket	96	145	241
3370	12' wide, 20' deep, 3-1/2 C.Y. bucket w/trench box	81	118	199
3380	25' deep, 3-1/2 C.Y. bucket	112	169	281
3500	1 to 1 slope, 2' wide, 2' deep, 3/8 C.Y. bucket	1.29	3.78	5.07
3520	3' deep, 3/8 C.Y. bucket	3.72	8.50	12.22
3540	4' deep, 3/8 C.Y. bucket	3.68	11.40	15.08
3560	6' deep, 1/2 C.Y. bucket	4.02	12.35	16.37
3580	8' deep, 1/2 C.Y. bucket	7.95	24.50	32.45
3600	10' deep, 1 C.Y. bucket	20.50	47	67.50
3800	4' wide, 2' deep, 3/8 C.Y. bucket	2.21	5.60	7.81
3820	3' deep, 3/8 C.Y. bucket	3.55	9.90	13.45
3840	4' deep, 1/2 C.Y. bucket	4.67	12.90	17.57
3860	6' deep, 1/2 C.Y. bucket	9.75	26	35.75
3880	8' deep, 1/2 C.Y. bucket	19.10	40	59.10
3900	10' deep, 1 C.Y. bucket	27.50	58	85.50
3920	12' deep, 1 C.Y. bucket	40.50	84	124.50
3940	15' deep, 1-1/2 C.Y. bucket	43.50	89.50	133
3960	18' deep, 2-1/2 C.Y. bucket	64	116	180
4030	6' wide, 6' deep, 5/8 C.Y. bucket w/trench box	16.60	30	46.60
4040	8' deep, 3/4 C.Y. bucket	24	41.50	65.50
4050	10' deep, 1 C.Y. bucket	27	50	77
4060	12' deep, 1-1/2 C.Y. bucket	36	68	104
4070	16' deep, 2-1/2 C.Y. bucket	61	104	165
4080	20' deep, 3-1/2 C.Y. bucket	88.50	144	232.50
4090	24' deep, 3-1/2 C.Y. bucket	129	203	332
4500	8' wide, 12' deep, 1-1/2 C.Y. bucket w/trench box	41	74	115
4550	15' deep, 1-1/2 C.Y. bucket	62	111	173
4600	18' deep, 2-1/2 C.Y. bucket	81.50	136	217.50
4650	24' deep, 3-1/2 C.Y. bucket	139	214	353
4800	10' wide, 20' deep, 3-1/2 C.Y. bucket w/trench box	83	155	238
4850	24' deep, 3-1/2 C.Y. bucket	148	226	374
4950	12' wide, 20' deep, 3-1/2 C.Y. bucket w/trench box	117	175	292
4980	25' deep, 3-1/2 C.Y. bucket	168	256	424

G1030 Site Earthwork

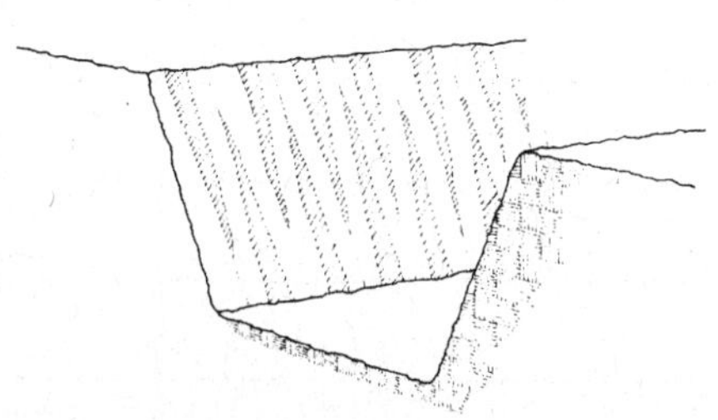

Trenching Systems are shown on a cost per linear foot basis. The systems include: excavation; backfill and removal of spoil; and compaction for various depths and trench bottom widths. The backfill has been reduced to accommodate a pipe of suitable diameter and bedding.

The slope for trench sides varies from none to 1:1.

The Expanded System Listing shows Trenching Systems that range from 2′ to 12′ in width. Depths range from 2′ to 25′.

System Components	QUANTITY	UNIT	COST PER L.F. EQUIP.	LABOR	TOTAL
SYSTEM G1030 806 1310					
TRENCHING LOAM & SANDY CLAY, NO SLOPE, 2′ WIDE, 2′ DP, 3/8 C.Y. BUCKET					
Excavation, trench, hyd. backhoe, track mtd., 3/8 C.Y. bucket	.148	B.C.Y.	.37	.95	1.32
Backfill and load spoil, from stockpile	.165	L.C.Y.	.13	.34	.47
Compaction by vibrating plate 18″ wide, 6″ lifts, 4 passes	.118	E.C.Y.	.03	.38	.41
Remove excess spoil, 8 C.Y. dump truck, 2 mile roundtrip	.042	L.C.Y.	.16	.16	.32
TOTAL			.70	1.87	2.57

G1030 806	Trenching Loam & Sandy Clay	COST PER L.F. EQUIP.	LABOR	TOTAL
1310	Trenching, loam & sandy clay, no slope, 2′ wide, 2′ deep, 3/8 C.Y. bucket	.69	1.83	2.52
1320	3′ deep, 3/8 C.Y. bucket	1.08	3.03	4.11
1330	4′ deep, 3/8 C.Y. bucket	1.26	3.69	4.95
1340	6′ deep, 3/8 C.Y. bucket	1.80	4.40	6.20
1350	8′ deep, 1/2 C.Y. bucket	2.39	5.80	8.20
1360	10′ deep, 1 C.Y. bucket	2.69	6.25	8.95
1400	4′ wide, 2′ deep, 3/8 C.Y. bucket	1.60	3.63	5.25
1410	3′ deep, 3/8 C.Y. bucket	2.18	5.50	7.70
1420	4′ deep, 1/2 C.Y. bucket	2.56	6.20	8.75
1430	6′ deep, 1/2 C.Y. bucket	4.42	9.05	13.45
1440	8′ deep, 1/2 C.Y. bucket	6.50	12.85	19.35
1450	10′ deep, 1 C.Y. bucket	6.45	13.20	19.65
1460	12′ deep, 1 C.Y. bucket	8.10	16.40	24.50
1470	15′ deep, 1-1/2 C.Y. bucket	9.80	19.10	29
1480	18′ deep, 2-1/2 C.Y. bucket	11.60	21	32.50
1520	6′ wide, 6′ deep, 5/8 C.Y. bucket w/trench box	8.35	14.75	23
1530	8′ deep, 3/4 C.Y. bucket	11	19.35	30.50
1540	10′ deep, 1 C.Y. bucket	10.60	19.70	30.50
1550	12′ deep, 1-1/2 C.Y. bucket	12.10	22	34
1560	16′ deep, 2-1/2 C.Y. bucket	16.70	28	44.50
1570	20′ deep, 3-1/2 C.Y. bucket	20	33	53
1580	24′ deep, 3-1/2 C.Y. bucket	25.50	40	65.50
1640	8′ wide, 12′ deep, 1-1/4 C.Y. bucket w/trench box	17.15	28	45
1650	15′ deep, 1-1/2 C.Y. bucket	21	35.50	56.50
1660	18′ deep, 2-1/2 C.Y. bucket	25.50	40	65.50
1680	24′ deep, 3-1/2 C.Y. bucket	34.50	52	86.50
1730	10′ wide, 20′ deep, 3-1/2 C.Y. bucket w/trench box	35	52	87
1740	24′ deep, 3-1/2 C.Y. bucket	43.50	64	108
1780	12′ wide, 20′ deep, 3-1/2 C.Y. bucket w/trench box	42	61.50	104
1790	25′ deep, 3-1/2 C.Y. bucket	54.50	79.50	134
1800	1/2:1 slope, 2′ wide, 2′ deep, 3/8 C.Y. bucket	.98	2.76	3.74
1810	3′ deep, 3/8 C.Y. bucket	1.63	4.86	6.50
1820	4′ deep, 3/8 C.Y. bucket	2.43	7.40	9.85
1840	6′ deep, 3/8 C.Y. bucket	4.32	11	15.30

G1030 Site Earthwork

G1030 806	Trenching Loam & Sandy Clay	COST PER L.F.		
		EQUIP.	LABOR	TOTAL
1860	8' deep, 1/2 C.Y. bucket	6.85	17.65	24.50
1880	10' deep, 1 C.Y. bucket	9.15	22	31
2300	4' wide, 2' deep, 3/8 C.Y. bucket	1.90	4.56	6.45
2310	3' deep, 3/8 C.Y. bucket	2.84	7.60	10.45
2320	4' deep, 1/2 C.Y. bucket	3.59	9.45	13.05
2340	6' deep, 1/2 C.Y. bucket	7.40	16.15	23.50
2360	8' deep, 1/2 C.Y. bucket	12.50	26	38.50
2380	10' deep, 1 C.Y. bucket	14.20	30.50	44.50
2400	12' deep, 1 C.Y. bucket	23.50	50	73.50
2430	15' deep, 1-1/2 C.Y. bucket	27.50	56	83.50
2460	18' deep, 2-1/2 C.Y. bucket	46	85.50	132
2840	6' wide, 6' deep, 5/8 C.Y. bucket w/trench box	12.10	22.50	34.50
2860	8' deep, 3/4 C.Y. bucket	17.70	33	50.50
2880	10' deep, 1 C.Y. bucket	18.95	37	56
2900	12' deep, 1-1/2 C.Y. bucket	23.50	45	68.50
2940	16' deep, 2-1/2 C.Y. bucket	38	66.50	105
2980	20' deep, 3-1/2 C.Y. bucket	53	89.50	143
3020	24' deep, 3-1/2 C.Y. bucket	75	123	198
3100	8' wide, 12' deep, 1-1/2 C.Y. bucket w/trench box	28.50	51	79.50
3120	15' deep, 1-1/2 C.Y. bucket	39	72	111
3140	18' deep, 2-1/2 C.Y. bucket	52.50	89	142
3180	24' deep, 3-1/2 C.Y. bucket	84.50	135	220
3270	10' wide, 20' deep, 3-1/2 C.Y. bucket w/trench box	68	109	177
3280	24' deep, 3-1/2 C.Y. bucket	93.50	147	241
3320	12' wide, 20' deep, 3-1/2 C.Y. bucket w/trench box	75	118	193
3380	25' deep, 3-1/2 C.Y. bucket w/trench box	102	159	261
3500	1:1 slope, 2' wide, 2' deep, 3/8 C.Y. bucket	1.26	3.69	4.95
3520	3' deep, 3/8 C.Y. bucket	2.29	6.95	9.25
3540	4' deep, 3/8 C.Y. bucket	3.57	11.10	14.65
3560	6' deep, 1/2 C.Y. bucket	4.32	11	15.30
3580	8' deep, 1/2 C.Y. bucket	11.35	29.50	41
3600	10' deep, 1 C.Y. bucket	15.65	38	53.50
3800	4' wide, 2' deep, 3/8 C.Y. bucket	2.18	5.50	7.70
3820	3' deep, 1/2 C.Y. bucket	3.49	9.70	13.20
3840	4' deep, 1/2 C.Y. bucket	4.61	12.65	17.25
3860	6' deep, 1/2 C.Y. bucket	10.35	23	33.50
3880	8' deep, 1/2 C.Y. bucket	18.55	39.50	58
3900	10' deep, 1 C.Y. bucket	22	47.50	69.50
3920	12' deep, 1 C.Y. bucket	31.50	67	98.50
3940	15' deep, 1-1/2 C.Y. bucket	45.50	92.50	138
3960	18' deep, 2-1/2 C.Y. bucket	63	118	181
4030	6' wide, 6' deep, 5/8 C.Y. bucket w/trench box	15.85	30	46
4040	8' deep, 3/4 C.Y. bucket	24.50	46.50	71
4050	10' deep, 1 C.Y. bucket	27.50	54.50	82
4060	12' deep, 1-1/2 C.Y. bucket	35	68.50	104
4070	16' deep, 2-1/2 C.Y. bucket	59.50	105	165
4080	20' deep, 3-1/2 C.Y. bucket	86	146	232
4090	24' deep, 3-1/2 C.Y. bucket	125	206	330
4500	8' wide, 12' deep, 1-1/4 C.Y. bucket w/trench box	40	74	114
4550	15' deep, 1-1/2 C.Y. bucket	57	108	165
4600	18' deep, 2-1/2 C.Y. bucket	79.50	138	218
4650	24' deep, 3-1/2 C.Y. bucket	134	217	350
4800	10' wide, 20' deep, 3-1/2 C.Y. bucket w/trench box	101	166	267
4850	24' deep, 3-1/2 C.Y. bucket	143	230	375
4950	12' wide, 20' deep, 3-1/2 C.Y. bucket w/trench box	108	175	283
4980	25' deep, 3-1/2 C.Y. bucket	163	259	420

G1030 Site Earthwork

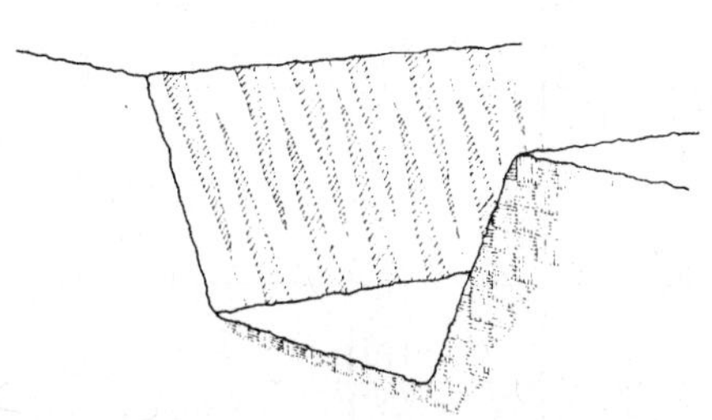

Trenching Systems are shown on a cost per linear foot basis. The systems include: excavation; backfill and removal of spoil; and compaction for various depths and trench bottom widths. The backfill has been reduced to accommodate a pipe of suitable diameter and bedding.

The slope for trench sides varies from none to 1:1.

The Expanded System Listing shows Trenching Systems that range from 2′ to 12′ in width. Depths range from 2′ to 25′.

System Components	QUANTITY	UNIT	COST PER L.F. EQUIP.	LABOR	TOTAL
SYSTEM G1030 807 1310					
TRENCHING SAND & GRAVEL, NO SLOPE, 2′ WIDE, 2′ DEEP, 3/8 C.Y. BUCKET					
Excavation, trench, hyd. backhoe, track mtd., 3/8 C.Y. bucket	.148	B.C.Y.	.36	.94	1.30
Backfill and load spoil, from stockpile	.140	L.C.Y.	.11	.29	.40
Compaction by vibrating plate 18″ wide, 6″ lifts, 4 passes	.118	E.C.Y.	.03	.38	.41
Remove excess spoil, 8 C.Y. dump truck, 2 mile roundtrip	.035	L.C.Y.	.14	.14	.28
TOTAL			.64	1.75	2.39

G1030 807	Trenching Sand & Gravel	COST PER L.F. EQUIP.	LABOR	TOTAL
1310	Trenching, sand & gravel, no slope, 2′ wide, 2′ deep, 3/8 C.Y. bucket	.64	1.75	2.39
1320	3′ deep, 3/8 C.Y. bucket	1.02	2.93	3.95
1330	4′ deep, 3/8 C.Y. bucket	1.18	3.52	4.70
1340	6′ deep, 3/8 C.Y. bucket	1.72	4.19	5.90
1350	8′ deep, 1/2 C.Y. bucket	2.26	5.55	7.80
1360	10′ deep, 1 C.Y. bucket	2.51	5.90	8.40
1400	4′ wide, 2′ deep, 3/8 C.Y. bucket	1.46	3.42	4.88
1410	3′ deep, 3/8 C.Y. bucket	2	5.20	7.20
1420	4′ deep, 1/2 C.Y. bucket	2.35	5.85	8.20
1430	6′ deep, 1/2 C.Y. bucket	4.18	8.65	12.85
1440	8′ deep, 1/2 C.Y. bucket	6.15	12.20	18.35
1450	10′ deep, 1 C.Y. bucket	6.10	12.45	18.55
1460	12′ deep, 1 C.Y. bucket	7.65	15.55	23
1470	15′ deep, 1-1/2 C.Y. bucket	9.25	18	27.50
1480	18′ deep, 2-1/2 C.Y. bucket	11	19.70	30.50
1520	6′ wide, 6′ deep, 5/8 C.Y. bucket w/trench box	7.85	13.95	22
1530	8′ deep, 3/4 C.Y. bucket	10.40	18.40	29
1540	10′ deep, 1 C.Y. bucket	10	18.60	28.50
1550	12′ deep, 1-1/2 C.Y. bucket	11.40	21	32.50
1560	16′ deep, 2 C.Y. bucket	16	26	42
1570	20′ deep, 3-1/2 C.Y. bucket	19.15	31	50
1580	24′ deep, 3-1/2 C.Y. bucket	24	37.50	61.50
1640	8′ wide, 12′ deep, 1-1/2 C.Y. bucket w/trench box	16.10	26.50	42.50
1650	15′ deep, 1-1/2 C.Y. bucket	19.55	33.50	53
1660	18′ deep, 2-1/2 C.Y. bucket	24	37.50	61.50
1680	24′ deep, 3-1/2 C.Y. bucket	32.50	48.50	81
1730	10′ wide, 20′ deep, 3-1/2 C.Y. bucket w/trench box	32.50	48.50	81
1740	24′ deep, 3-1/2 C.Y. bucket	41	60	101
1780	12′ wide, 20′ deep, 3-1/2 C.Y. bucket w/trench box	39.50	57.50	97
1790	25′ deep, 3-1/2 C.Y. bucket	51.50	74.50	126
1800	1/2:1 slope, 2′ wide, 2′ deep, 3/8 C.Y. bucket	.91	2.63	3.54
1810	3′ deep, 3/8 C.Y. bucket	1.53	4.64	6.15
1820	4′ deep, 3/8 C.Y. bucket	2.28	7.10	9.40
1840	6′ deep, 3/8 C.Y. bucket	4.11	10.55	14.65

G1030 Site Earthwork

G1030 807	Trenching Sand & Gravel	COST PER L.F.		
		EQUIP.	LABOR	TOTAL
1860	8' deep, 1/2 C.Y. bucket	6.55	16.85	23.50
1880	10' deep, 1 C.Y. bucket	8.60	21	29.50
2300	4' wide, 2' deep, 3/8 C.Y. bucket	1.73	4.31	6.05
2310	3' deep, 3/8 C.Y. bucket	2.61	7.20	9.80
2320	4' deep, 1/2 C.Y. bucket	3.31	8.95	12.25
2340	6' deep, 1/2 C.Y. bucket	7.05	15.45	22.50
2360	8' deep, 1/2 C.Y. bucket	11.85	24.50	36.50
2380	10' deep, 1 C.Y. bucket	13.40	28.50	42
2400	12' deep, 1 C.Y. bucket	22.50	47.50	70
2430	15' deep, 1-1/2 C.Y. bucket	26	52.50	78.50
2460	18' deep, 2-1/2 C.Y. bucket	43.50	80.50	124
2840	6' wide, 6' deep, 5/8 C.Y. bucket w/trench box	11.45	21	32.50
2860	8' deep, 3/4 C.Y. bucket	16.80	31	48
2880	10' deep, 1 C.Y. bucket	17.95	35	53
2900	12' deep, 1-1/2 C.Y. bucket	22.50	42.50	65
2940	16' deep, 2 C.Y. bucket	36.50	62.50	99
2980	20' deep, 3-1/2 C.Y. bucket	50.50	84	135
3020	24' deep, 3-1/2 C.Y. bucket	71.50	115	187
3100	8' wide, 12' deep, 1-1/4 C.Y. bucket w/trench box	27	48.50	75.50
3120	15' deep, 1-1/2 C.Y. bucket	36.50	67.50	104
3140	18' deep, 2-1/2 C.Y. bucket	50	83.50	134
3180	24' deep, 3-1/2 C.Y. bucket	80	126	206
3270	10' wide, 20' deep, 3-1/2 C.Y. bucket w/trench box	64	102	166
3280	24' deep, 3-1/2 C.Y. bucket	88.50	138	227
3370	12' wide, 20' deep, 3-1/2 C.Y. bucket w/trench box	71.50	113	185
3380	25' deep, 3-1/2 C.Y. bucket	103	159	262
3500	1:1 slope, 2' wide, 2' deep, 3/8 C.Y. bucket	2.05	4.29	6.35
3520	3' deep, 3/8 C.Y. bucket	2.14	6.65	8.80
3540	4' deep, 3/8 C.Y. bucket	3.36	10.65	14
3560	6' deep, 3/8 C.Y. bucket	4.11	10.55	14.65
3580	8' deep, 1/2 C.Y. bucket	10.80	28	39
3600	10' deep, 1 C.Y. bucket	14.65	35.50	50
3800	4' wide, 2' deep, 3/8 C.Y. bucket	2	5.20	7.20
3820	3' deep, 3/8 C.Y. bucket	3.23	9.20	12.45
3840	4' deep, 1/2 C.Y. bucket	4.26	12	16.25
3860	6' deep, 1/2 C.Y. bucket	9.90	22	32
3880	8' deep, 1/2 C.Y. bucket	17.60	37.50	55
3900	10' deep, 1 C.Y. bucket	21	45	66
3920	12' deep, 1 C.Y. bucket	30	63.50	93.50
3940	15' deep, 1-1/2 C.Y. bucket	43	87.50	131
3960	18' deep, 2-1/2 C.Y. bucket	59.50	111	171
4030	6' wide, 6' deep, 5/8 C.Y. bucket w/trench box	15.05	28.50	43.50
4040	8' deep, 3/4 C.Y. bucket	23	44	67
4050	10' deep, 1 C.Y. bucket	26	51.50	77.50
4060	12' deep, 1-1/2 C.Y. bucket	33	64.50	97.50
4070	16' deep, 2 C.Y. bucket	57	99	156
4080	20' deep, 3-1/2 C.Y. bucket	81.50	137	219
4090	24' deep, 3-1/2 C.Y. bucket	119	193	310
4500	8' wide, 12' deep, 1-1/2 C.Y. bucket w/trench box	38	70	108
4550	15' deep, 1-1/2 C.Y. bucket	53.50	102	156
4600	18' deep, 2-1/2 C.Y. bucket	76	129	205
4650	24' deep, 3-1/2 C.Y. bucket	127	204	330
4800	10' wide, 20' deep, 3-1/2 C.Y. bucket w/trench box	95.50	155	251
4850	24' deep, 3-1/2 C.Y. bucket	136	216	350
4950	12' wide, 20' deep, 3-1/2 C.Y. bucket w/trench box	102	164	266
4980	25' deep, 3-1/2 C.Y. bucket	154	243	395

G1030 Site Earthwork

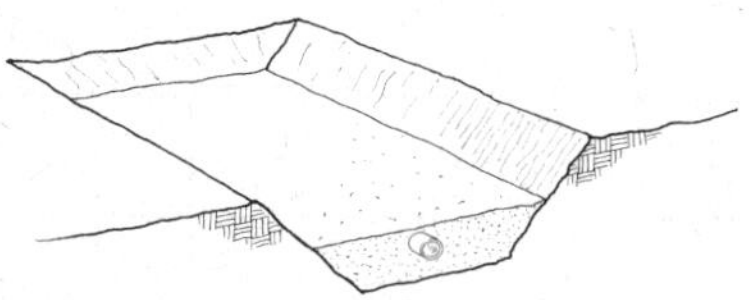

The Pipe Bedding System is shown for various pipe diameters. Compacted bank sand is used for pipe bedding and to fill 12″ over the pipe. No backfill is included. Various side slopes are shown to accommodate different soil conditions. Pipe sizes vary from 6″ to 84″ diameter.

System Components	QUANTITY	UNIT	COST PER L.F. MAT.	INST.	TOTAL
SYSTEM G1030 815 1440					
PIPE BEDDING, SIDE SLOPE 0 TO 1, 1′ WIDE, PIPE SIZE 6″ DIAMETER					
Borrow, bank sand, 2 mile haul, machine spread	.086	C.Y.	1.65	.67	2.32
Compaction, vibrating plate	.086	C.Y.		.21	.21
TOTAL			1.65	.88	2.53

G1030 815	Pipe Bedding	COST PER L.F. MAT.	INST.	TOTAL
1440	Pipe bedding, side slope 0 to 1, 1′ wide, pipe size 6″ diameter	1.65	.88	2.53
1460	2′ wide, pipe size 8″ diameter	3.57	1.91	5.48
1480	Pipe size 10″ diameter	3.65	1.96	5.61
1500	Pipe size 12″ diameter	3.72	2	5.72
1520	3′ wide, pipe size 14″ diameter	6.05	3.23	9.28
1540	Pipe size 15″ diameter	6.10	3.25	9.35
1560	Pipe size 16″ diameter	6.15	3.29	9.44
1580	Pipe size 18″ diameter	6.25	3.36	9.61
1600	4′ wide, pipe size 20″ diameter	8.90	4.77	13.67
1620	Pipe size 21″ diameter	9	4.81	13.81
1640	Pipe size 24″ diameter	9.20	4.93	14.13
1660	Pipe size 30″ diameter	9.35	5	14.35
1680	6′ wide, pipe size 32″ diameter	16	8.60	24.60
1700	Pipe size 36″ diameter	16.40	8.75	25.15
1720	7′ wide, pipe size 48″ diameter	26	13.95	39.95
1740	8′ wide, pipe size 60″ diameter	31.50	16.95	48.45
1760	10′ wide, pipe size 72″ diameter	44	23.50	67.50
1780	12′ wide, pipe size 84″ diameter	58.50	31	89.50
2140	Side slope 1/2 to 1, 1′ wide, pipe size 6″ diameter	3.08	1.64	4.72
2160	2′ wide, pipe size 8″ diameter	5.25	2.79	8.04
2180	Pipe size 10″ diameter	5.60	3.01	8.61
2200	Pipe size 12″ diameter	5.95	3.18	9.13
2220	3′ wide, pipe size 14″ diameter	8.60	4.60	13.20
2240	Pipe size 15″ diameter	8.80	4.69	13.49
2260	Pipe size 16″ diameter	9	4.82	13.82
2280	Pipe size 18″ diameter	9.50	5.05	14.55
2300	4′ wide, pipe size 20″ diameter	12.50	6.70	19.20
2320	Pipe size 21″ diameter	12.80	6.85	19.65
2340	Pipe size 24″ diameter	13.60	7.25	20.85
2360	Pipe size 30″ diameter	15.10	8.05	23.15
2380	6′ wide, pipe size 32″ diameter	22	11.90	33.90
2400	Pipe size 36″ diameter	23.50	12.65	36.15
2420	7′ wide, pipe size 48″ diameter	37	19.75	56.75
2440	8′ wide, pipe size 60″ diameter	47	25	72
2460	10′ wide, pipe size 72″ diameter	64	34.50	98.50
2480	12′ wide, pipe size 84″ diameter	84	45	129
2620	Side slope 1 to 1, 1′ wide, pipe size 6″ diameter	4.51	2.41	6.92
2640	2′ wide, pipe size 8″ diameter	6.95	3.71	10.66

G1030 Site Earthwork

G1030 815	Pipe Bedding	COST PER L.F. MAT.	INST.	TOTAL
2660	Pipe size 10″ diameter	7.55	4.03	11.58
2680	Pipe size 12″ diameter	8.20	4.40	12.60
2700	3′ wide, pipe size 14″ diameter	11.15	5.95	17.10
2720	Pipe size 15″ diameter	11.50	6.15	17.65
2740	Pipe size 16″ diameter	11.85	6.35	18.20
2760	Pipe size 18″ diameter	12.70	6.80	19.50
2780	4′ wide, pipe size 20″ diameter	16.10	8.60	24.70
2800	Pipe size 21″ diameter	16.55	8.85	25.40
2820	Pipe size 24″ diameter	17.95	9.60	27.55
2840	Pipe size 30″ diameter	20.50	11.10	31.60
2860	6′ wide, pipe size 32″ diameter	28.50	15.25	43.75
2880	Pipe size 36″ diameter	31	16.50	47.50
2900	7′ wide, pipe size 48″ diameter	47.50	25.50	73
2920	8′ wide, pipe size 60″ diameter	62	33	95
2940	10′ wide, pipe size 72″ diameter	84.50	45	129.50
2960	12′ wide, pipe size 84″ diameter	110	59	169

G2030 Pedestrian Paving

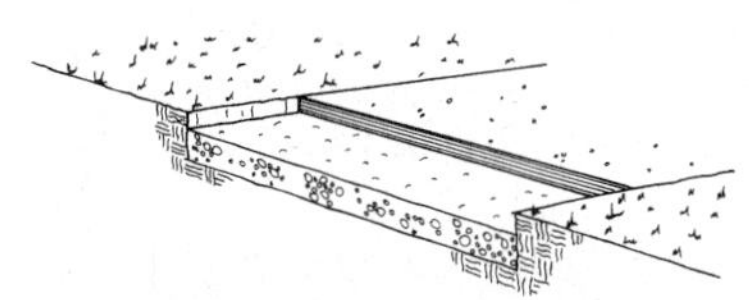

The Bituminous Sidewalk System includes: excavation; compacted gravel base (hand graded), bituminous surface; and hand grading along the edge of the completed walk.

The Expanded System Listing shows Bituminous Sidewalk systems with wearing course depths ranging from 1″ to 2-1/2″ of bituminous material. The gravel base ranges from 4″ to 8″. Sidewalk widths are shown ranging from 3′ to 5′. Costs are on a linear foot basis.

System Components	QUANTITY	UNIT	COST PER L.F. MAT.	INST.	TOTAL
SYSTEM G2030 110 1580					
BITUMINOUS SIDEWALK, 1″ THICK PAVING, 4″ GRAVEL BASE, 3′ WIDTH					
Excavation, bulk, dozer, push 50′	.046	B.C.Y.		.09	.09
Borrow, bank run gravel, haul 2 mi., spread w/dozer, no compaction	.037	L.C.Y.	.89	.29	1.18
Compact w/vib. plate, 8″ lifts	.037	E.C.Y.		.09	.09
Fine grade, area to be paved, small area	.333	S.Y.		3.14	3.14
Asphaltic concrete, 2″ thick	.333	S.Y.	1.32	.70	2.02
Backfill by hand, no compaction, light soil	.006	L.C.Y.		.20	.20
TOTAL			2.21	4.51	6.72

G2030 110	Bituminous Sidewalks	COST PER L.F. MAT.	INST.	TOTAL
1580	Bituminous sidewalk, 1″ thick paving, 4″ gravel base, 3′ width	2.21	4.51	6.72
1600	4′ width	2.93	4.90	7.83
1620	5′ width	3.73	5.15	8.88
1640	6″ gravel base, 3′ width	2.66	4.76	7.42
1660	4′ width	3.53	5.20	8.73
1680	5′ width	4.43	5.70	10.13
1700	8″ gravel base, 3′ width	3.10	4.98	8.08
1720	4′ width	4.13	5.50	9.63
1740	5′ width	5.20	6.10	11.30
1800	1-1/2″ thick paving, 4″ gravel base, 3′ width	2.94	4.72	7.66
1820	4′ width	3.88	5.20	9.08
1840	5′ width	4.78	6	10.78
1860	6″ gravel base, 3′ width	3.39	4.95	8.34
1880	4′ width	4.41	5.80	10.21
1900	5′ width	5.60	6.15	11.75
1920	8″ gravel base, 3′ width	3.76	5.40	9.16
1940	4′ width	5.10	5.85	10.95
1960	5′ width	6.25	6.80	13.05
2120	2″ thick paving, 4″ gravel base, 3′ width	3.52	5.45	8.97
2140	4′ width	4.69	6.05	10.74
2160	5′ width	5.90	6.70	12.60
2180	6″ gravel base, 3′ width	3.97	5.65	9.62
2200	4′ width	5.30	6.35	11.65
2240	8″ gravel base, 3′ width	4.41	5.90	10.31
2280	5′ width	7.35	7.45	14.80
2400	2-1/2″ thick paving, 4″ gravel base, 3′ width	5.10	6.05	11.15
2420	4′ width	6.75	6.85	13.60
2460	6″ gravel base, 3′ width	5.55	6.30	11.85
2500	5′ width	9.25	8.05	17.30
2520	8″ gravel base, 3′ width	5.95	6.50	12.45
2540	4′ width	7.95	7.45	15.40
2560	5′ width	10	8.45	18.45

G2030 Pedestrian Paving

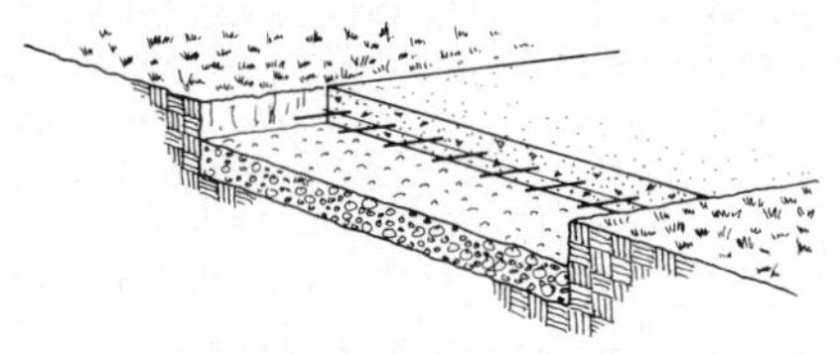

The Concrete Sidewalk System includes: excavation; compacted gravel base (hand graded); forms; welded wire fabric; and 3,000 p.s.i. air-entrained concrete (broom finish).

The Expanded System Listing shows Concrete Sidewalk systems with wearing course depths ranging from 4″ to 6″. The gravel base ranges from 4″ to 8″. Sidewalk widths are shown ranging from 3′ to 5′. Costs are on a linear foot basis.

System Components	QUANTITY	UNIT	COST PER L.F. MAT.	INST.	TOTAL
SYSTEM G2030 120 1580					
CONCRETE SIDEWALK 4″ THICK, 4″ GRAVEL BASE, 3′ WIDE					
Excavation, box out with dozer	.100	B.C.Y.		.21	.21
Gravel base, haul 2 miles, spread with dozer	.037	L.C.Y.	.89	.29	1.18
Compaction with vibrating plate	.037	E.C.Y.		.09	.09
Fine grade by hand	.333	S.Y.		3.19	3.19
Concrete in place including forms and reinforcing	.037	C.Y.	5.46	7.71	13.17
Backfill edges by hand	.010	L.C.Y.		.33	.33
TOTAL			6.35	11.82	18.17

G2030 120	Concrete Sidewalks	COST PER L.F. MAT.	INST.	TOTAL
1580	Concrete sidewalk, 4″ thick, 4″ gravel base, 3′ wide	6.35	11.85	18.20
1600	4′ wide	8.45	14.60	23.05
1620	5′ wide	10.60	17.30	27.90
1640	6″ gravel base, 3′ wide	6.80	12.05	18.85
1660	4′ wide	9.05	14.90	23.95
1680	5′ wide	11.35	17.70	29.05
1700	8″ gravel base, 3′ wide	7.25	12.25	19.50
1720	4′ wide	9.65	15.20	24.85
1740	5′ wide	12.05	18.10	30.15
1800	5″ thick concrete, 4″ gravel base, 3′ wide	8.15	12.75	20.90
1820	4′ wide	10.85	15.75	26.60
1840	5′ wide	13.60	18.80	32.40
1860	6″ gravel base, 3′ wide	8.60	13	21.60
1880	4′ wide	11.45	16.10	27.55
1900	5′ wide	14.35	18.50	32.85
1920	8″ gravel base, 3′ wide	9.05	13.25	22.30
1940	4′ wide	12.05	17.20	29.25
1960	5′ wide	15.05	18.90	33.95
2120	6″ thick concrete, 4″ gravel base, 3′ wide	9.35	13.45	22.80
2140	4′ wide	12.45	16.70	29.15
2160	5′ wide	15.60	19.90	35.50
2180	6″ gravel base, 3′ wide	9.80	13.75	23.55
2200	4′ wide	13.05	17	30.05
2220	5′ wide	16.35	20.50	36.85
2240	8″ gravel base, 3′ wide	10.25	13.95	24.20
2260	4′ wide	13.65	17.30	30.95
2280	5′ wide	17.05	20.50	37.55

G2030 Pedestrian Paving

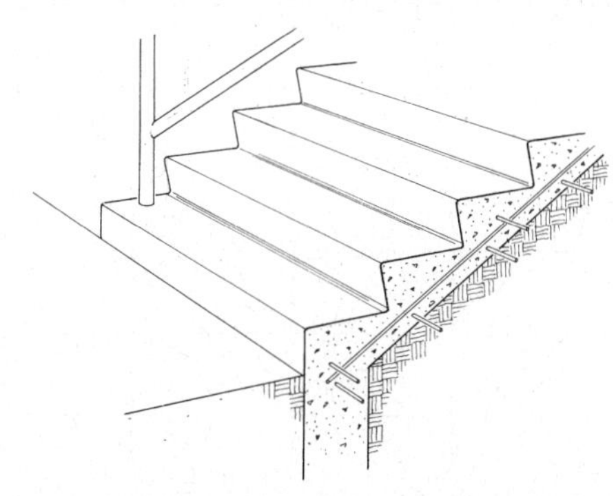

The Step System has three basic types: railroad tie, cast-in-place concrete or brick with a concrete base. System elements include: gravel base compaction; and backfill.

Wood Step Systems use 6″ x 6″ railroad ties that produce 3′ to 6′ wide steps that range from 2-riser to 5-riser configurations. Cast in Place Concrete Step Systems are either monolithic or aggregate finish. They range from 3′ to 6′ in width. Concrete Steps Systems are either in a 2-riser or 5-riser configuration. Precast Concrete Step Systems are listed for 4′ to 7′ widths with both 2-riser and 5-riser models. Brick Step Systems are placed on a 12″ concrete base. The size range is the same as cost in place concrete. Costs are on a per unit basis. All systems are assumed to include a full landing at the top 4′ long.

System Components	QUANTITY	UNIT	COST PER EACH		
			MAT.	INST.	TOTAL
SYSTEM G2030 310 0960					
STAIRS, RAILROAD TIES, 6″ X 8″, 3′ WIDE, 2 RISERS					
Excavation, by hand, sandy soil	1.327	C.Y.		74.98	74.98
Borrow fill, bank run gravel	.664	C.Y.	14.61		14.61
Delivery charge	.664	C.Y.		19.29	19.29
Backfill compaction, 6″ layers, hand tamp	.664	C.Y.		14.61	14.61
Railroad ties, wood, creosoted, 6″ x 8″ x 8′-6″, C.L. lots	5.000	Ea.	285	182.30	467.30
Backfill by hand, no compaction, light soil	.800	C.Y.		26	26
Remove excess spoil, 12 C.Y. dump truck	1.200	L.C.Y.		8.14	8.14
TOTAL			299.61	325.32	624.93

G2030 310	Stairs	COST PER EACH		
		MAT.	INST.	TOTAL
0960	Stairs, railroad ties, 6″ x 8″, 3′ wide, 2 risers	300	325	625
0980	5 risers	705	660	1,365
1000	4′ wide, 2 risers	305	360	665
1020	5 risers	710	690	1,400
1040	5′ wide, 2 risers	360	420	780
1060	5 risers	885	850	1,735
1080	6′ wide, 2 risers	365	455	820
1100	5 risers	890	875	1,765
2520	Concrete, cast in place, 3′ wide, 2 risers	192	450	642
2540	5 risers	255	635	890
2560	4′ wide, 2 risers	237	585	822
2580	5 risers	310	820	1,130
2600	5′ wide, 2 risers	283	730	1,013
2620	5 risers	360	980	1,340
2640	6′ wide, 2 risers	325	850	1,175
2660	5 risers	415	1,150	1,565
2800	Exposed aggregate finish, 3′ wide, 2 risers	199	465	664
2820	5 risers	267	635	902
2840	4′ wide, 2 risers	246	585	831
2860	5 risers	325	820	1,145
2880	5′ wide, 2 risers	295	730	1,025
2900	5 risers	380	980	1,360
2920	6′ wide, 2 risers	335	850	1,185
2940	5 risers	440	1,150	1,590
3200	Precast, 4′ wide, 2 risers	480	340	820
3220	5 risers	740	505	1,245

G2030 Pedestrian Paving

G2030 310	Stairs	COST PER EACH		
		MAT.	INST.	TOTAL
3240	5′ wide, 2 risers	545	405	950
3260	5 risers	860	540	1,400
3280	6′ wide, 2 risers	605	430	1,035
3300	5 risers	950	580	1,530
3320	7′ wide, 2 risers	770	455	1,225
3340	5 risers	1,275	625	1,900
4100	Brick, incl 12″ conc base, 3′ wide, 2 risers	325	1,125	1,450
4120	5 risers	475	1,700	2,175
4140	4′ wide, 2 risers	410	1,400	1,810
4160	5 risers	585	2,100	2,685
4180	5′ wide, 2 risers	490	1,650	2,140
4200	5 risers	695	2,450	3,145
4220	6′ wide, 2 risers	565	1,950	2,515
4240	5 risers	810	2,825	3,635

G2040 Site Development

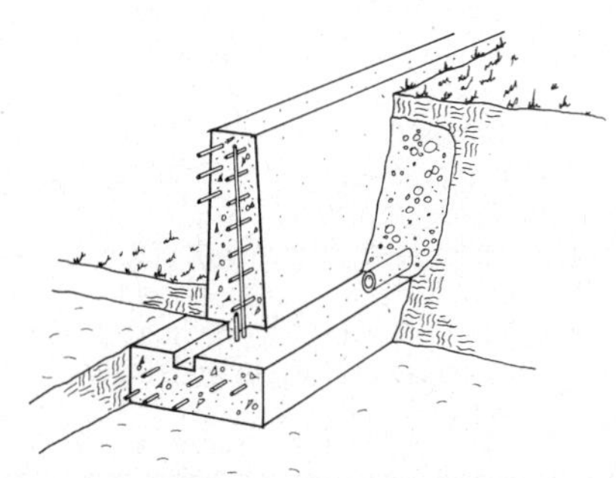

There are four basic types of Concrete Retaining Wall Systems: reinforced concrete with level backfill; reinforced concrete with sloped backfill or surcharge; unreinforced with level backfill; and unreinforced with sloped backfill or surcharge. System elements include: all necessary forms (4 uses); 3,000 p.s.i. concrete with an 8″ chute; all necessary reinforcing steel; and underdrain. Exposed concrete is patched and rubbed.

The Expanded System Listing shows walls that range in thickness from 10″ to 18″ for reinforced concrete walls with level backfill and 12″ to 24″ for reinforced walls with sloped backfill. Walls range from a height of 4′ to 20′. Unreinforced level and sloped backfill walls range from a height of 3′ to 10′.

System Components	QUANTITY	UNIT	COST PER L.F. MAT.	INST.	TOTAL
SYSTEM G2040 210 1000					
CONC. RETAIN. WALL REINFORCED, LEVEL BACKFILL, 4′ HIGH					
Forms in place, cont. wall footing & keyway, 4 uses	2.000	S.F.	4.46	8.86	13.32
Forms in place, retaining wall forms, battered to 8′ high, 4 uses	8.000	SFCA	5.44	68	73.44
Reinforcing in place, walls, #3 to #7	.004	Ton	4.40	3.40	7.80
Concrete ready mix, regular weight, 3000 psi	.204	C.Y.	22.24		22.24
Placing concrete and vibrating footing con., shallow direct chute	.074	C.Y.		1.78	1.78
Placing concrete and vibrating walls, 8″ thick, direct chute	.130	C.Y.		4.14	4.14
Pipe bedding, crushed or screened bank run gravel	1.000	L.F.	3.03	1.37	4.40
Pipe, subdrainage, corrugated plastic, 4″ diameter	1.000	L.F.	.56	.75	1.31
Finish walls and break ties, patch walls	4.000	S.F.	.16	3.84	4
TOTAL			40.29	92.14	132.43

G2040 210	Concrete Retaining Walls	COST PER L.F. MAT.	INST.	TOTAL
1000	Conc. retain. wall, reinforced, level backfill, 4′ high x 2′-2″ base,10″ thick	40.50	92	132.50
1200	6′ high x 3′-3″ base, 10″ thick	58	134	192
1400	8′ high x 4′-3″ base, 10″ thick	75	175	250
1600	10′ high x 5′-4″ base, 13″ thick	98	256	354
1800	12′ high x 6′-6″ base, 14″ thick	121	305	426
2200	16′ high x 8′-6″ base, 16″ thick	184	415	599
2600	20′ high x 10′-5″ base, 18″ thick	270	535	805
3000	Sloped backfill, 4′ high x 3′-2″ base, 12″ thick	49	95.50	144.50
3200	6′ high x 4′-6″ base, 12″ thick	69	138	207
3400	8′ high x 5′-11″ base, 12″ thick	91.50	181	272.50
3600	10′ high x 7′-5″ base, 16″ thick	130	268	398
3800	12′ high x 8′-10″ base, 18″ thick	170	325	495
4200	16′ high x 11′-10″ base, 21″ thick	285	460	745
4600	20′ high x 15′-0″ base, 24″ thick	445	625	1,070
5000	Unreinforced, level backfill, 3′-0″ high x 1′-6″ base	21	62	83
5200	4′-0″ high x 2′-0″ base	32.50	82.50	115
5400	6′-0″ high x 3′-0″ base	60.50	128	188.50
5600	8′-0″ high x 4′-0″ base	96.50	170	266.50
5800	10′-0″ high x 5′-0″ base	143	263	406
7000	Sloped backfill, 3′-0″ high x 2′-0″ base	25	64	89
7200	4′-0″ high x 3′-0″ base	42	86	128
7400	6′-0″ high x 5′-0″ base	87.50	136	223.50
7600	8′-0″ high x 7′-0″ base	148	184	332
7800	10′-0″ high x 9′-0″ base	227	285	512

G2040 Site Development

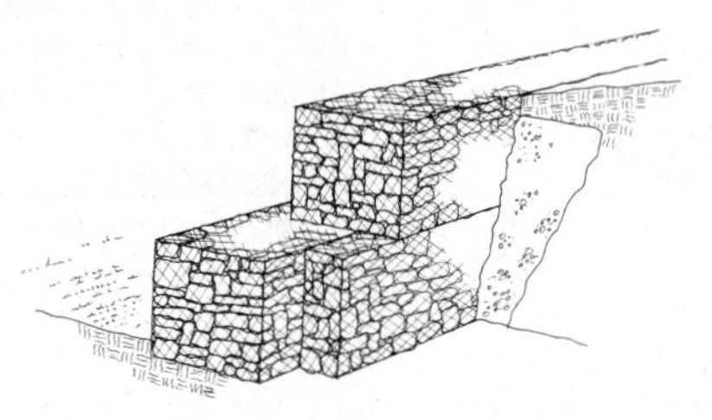

The Gabion Retaining Wall Systems list three types of surcharge conditions: level, sloped and highway, for two different facing configurations: stepped and straight with batter. Costs are expressed per L.F. for heights ranging from 6′ to 18′ for retaining sandy or clay soil. For protection against sloughing in wet clay materials counterforts have been added for these systems. Drainage stone has been added behind the walls to avoid any additional lateral pressures.

System Components	QUANTITY	UNIT	COST PER L.F. MAT.	INST.	TOTAL
SYSTEM G2040 270 1000					
GABION RET. WALL, LEVEL BACKFILL, STEPPED FACE, 4′ BASE, 6′ HIGH					
3′ x 3′ cross section gabion	2.000	Ea.	94.44	157.11	251.55
3′ x 1′ cross section gabion	1.000	Ea.	24.44	9.43	33.87
Crushed stone drainage	.220	C.Y.	7.75	1.96	9.71
TOTAL			126.63	168.50	295.13

G2040 270	Gabion Retaining Walls	COST PER L.F. MAT.	INST.	TOTAL
1000	Gabion ret. wall, level backfill, stepped face, 4′ base, 6′ high, sandy soil	127	169	296
1040	Clay soil with counterforts @ 16′ O.C.	190	345	535
1100	5′ base, 9′ high, sandy soil	226	267	493
1140	Clay soil with counterforts @ 16′ O.C.	385	535	920
1200	6′ base, 12′ high, sandy soil	325	425	750
1240	Clay soil with counterforts @ 16′ O.C.	575	1,125	1,700
1300	7′-6″ base, 15′ high, sandy soil	460	600	1,060
1340	Clay soil with counterforts @ 16′ O.C.	1,100	1,575	2,675
1400	9′ base, 18′ high, sandy soil	605	835	1,440
1440	Clay soil with counterforts @ 16′ O.C.	1,250	1,900	3,150
2000	Straight face w/1:6 batter, 3′ base, 6′ high, sandy soil	110	161	271
2040	Clay soil with counterforts @ 16′ O.C.	173	340	513
2100	4′-6″ base, 9′ high, sandy soil	189	263	452
2140	Clay soil with counterforts @ 16′ O.C.	350	530	880
2200	6′ base, 12′ high, sandy soil	290	420	710
2240	Clay soil with counterforts @ 16′ O.C.	545	1,125	1,670
2300	7′-6″ base, 15′ high, sandy soil	415	600	1,015
2340	Clay soil with counterforts @ 16′ O.C.	850	1,250	2,100
2400	18′ high, sandy soil	540	780	1,320
2440	Clay soil with counterforts @ 16′ O.C.	870	910	1,780
3000	Backfill sloped 1-1/2:1, stepped face, 4′-6″ base, 6′ high, sandy soil	128	181	309
3040	Clay soil with counterforts @ 16′ O.C.	191	360	551
3100	6′ base, 9′ high, sandy soil	226	340	566
3140	Clay soil with counterforts @ 16′ O.C.	385	605	990
3200	7′-6″ base, 12′ high, sandy soil	350	515	865
3240	Clay soil with counterforts @ 16′ O.C.	605	1,225	1,830
3300	9′ base, 15′ high, sandy soil	495	755	1,250
3340	Clay soil with counterforts @ 16′ O.C.	1,125	1,750	2,875
3400	10′-6″ base, 18′ high, sandy soil	670	1,025	1,695
3440	Clay soil with counterforts @ 16′ O.C.	1,300	2,075	3,375
4000	Straight face with 1:6 batter, 4′-6″ base, 6′ high, sandy soil	136	182	318
4040	Clay soil with counterforts @ 16′ O.C.	199	360	559
4100	6′ base, 9′ high, sandy soil	237	340	577
4140	Clay soil with counterforts @ 16′ O.C.	395	605	1,000

G2040 Site Development

G2040 270	Gabion Retaining Walls	COST PER L.F.		
		MAT.	INST.	TOTAL
4200	7'-6" base, 12' high, sandy soil	365	520	885
4240	Clay soil with counterforts @ 16' O.C.	615	1,225	1,840
4300	9' base, 15' high, sandy soil	510	760	1,270
4340	Clay soil with counterforts @ 16' O.C.	1,150	1,750	2,900
4400	18' high, sandy soil	655	995	1,650
4440	Clay soil with counterforts @ 16' O.C.	1,300	2,050	3,350
5000	Highway surcharge, straight face, 6' base, 6' high, sandy soil	197	315	512
5040	Clay soil with counterforts @ 16' O.C.	260	495	755
5100	9' base, 9' high, sandy soil	340	555	895
5140	Clay soil with counterforts @ 16' O.C.	500	820	1,320
5200	12' high, sandy soil	625	825	1,450
5240	Clay soil with counterforts @ 16' O.C.	935	2,050	2,985
5300	12' base, 15' high, sandy soil	680	1,100	1,780
5340	Clay soil with counterforts @ 16' O.C.	1,125	1,750	2,875
5400	18' high, sandy soil	875	1,425	2,300
5440	Clay soil with counterforts @ 16' O.C.	1,500	2,475	3,975

G3020 Sanitary Sewer

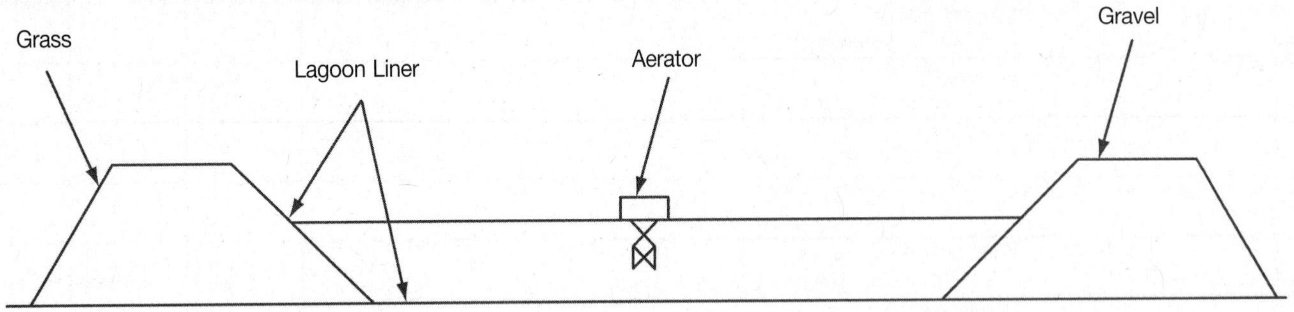

Grass
Lagoon Liner
Aerator
Gravel

System Components	QUANTITY	UNIT	COST EACH MAT.	COST EACH INST.	COST EACH TOTAL
SYSTEM G3020 710 1000					
AERATED DOMESTIC SEWAGE LAGOON, COMMON EARTH, 500,000 GPD					
Supervision of lagoon construction	45.000	Day		26,325	26,325
Quality Control of Earthwork per day	20.000	Day		9,000	9,000
Anchoring aerators per cell, 6 anchors per cell	2.000	Ea.	8,050	7,510	15,560
Excavate lagoon	7778.000	B.C.Y.		18,744.98	18,744.98
Finish grade slopes for seeding	25.000	M.S.F.		712.50	712.50
Spread & grade top of lagoon banks for compaction	454.000	M.S.F.		27,694	27,694
Compact lagoon banks	7127.000	E.C.Y.		5,345.25	5,345.25
Install pond liner	71.000	M.S.F.	45,440	78,100	123,540
Provide water for compaction and dust control	1556.000	E.C.Y.	2,053.92	3,205.36	5,259.28
Dispose of excess material on-site	846.000	L.C.Y.		2,994.84	2,994.84
Seed non-lagoon side of banks	25.000	M.S.F.	712.50	712.50	1,425
Place gravel on top of banks	1269.000	S.Y.	7,677.45	1,827.36	9,504.81
Provide 6 ' high fence around lagoon	1076.000	L.F.	22,596	7,456.68	30,052.68
Provide 12' wide fence gate	1.000	Opng.	475	515.50	990.50
Surface aerator, 30 HP, 50 lbs/HR, 900 RPM	4.000	Ea.	146,400	12,720	159,120
Temporary Fencing	1269.000	L.F.	4,124.25	2,867.94	6,992.19
TOTAL			237,529.12	205,731.91	443,261.03

G3020 710	Aerated Sewage Lagoon	COST EACH MAT.	COST EACH INST.	COST EACH TOTAL
1000	500,000 GPD Domestic Sewage Aerated Lagoon, common earth, excluding power	237,500	205,500	443,000
1100	600,000 GPD	319,000	238,500	557,500
1200	700,000 GPD	325,000	257,000	582,000
1300	800,000 GPD	405,000	286,000	691,000
1400	900,000 GPD	410,000	306,000	716,000
1500	1 MGD	459,500	386,000	845,500
2000	500,000 GPD sandy clay & loam	237,000	203,500	440,500
2100	600,000 GPD	319,000	236,000	555,000
2200	700,000 GPD	325,000	252,000	577,000
2300	800,000 GPD	405,000	282,500	687,500
2400	900,000 GPD	410,000	302,500	712,500
2500	1 MGD	564,000	478,000	1,042,000
3000	500,000 GPD sand & gravel	237,000	201,500	438,500
3100	600,000 GPD	319,000	234,500	553,500
3200	700,000 GPD	325,000	251,500	576,500
3300	800,000 GPD	411,000	280,500	691,500
3400	900,000 GPD	410,000	297,500	707,500
3500	1 MGD	460,000	381,000	841,000

G3020 Sanitary Sewer

G3020 710	Aerated Sewage Lagoon	COST EACH		
		MAT.	INST.	TOTAL
5000	500,000 GPD Lagoon aerators power supply	8,500	11,400	19,900
5100	600,000 GPD	10,200	14,000	24,200
5200	700,000 GPD	10,200	14,000	24,200
5300	800,000 GPD	17,000	22,800	39,800
5400	900,000 GPD	17,000	22,800	39,800
5500	1 MGD	17,000	22,800	39,800
5600	Lagoons, aerated without liner			
6000	500,000 GPD Domestic Sewage Aerated Lagoon, common earth, no power/liner	191,500	121,500	313,000
6100	600,000 GPD	269,000	146,500	415,500
6200	700,000 GPD	270,500	155,500	426,000
6300	800,000 GPD	345,500	177,500	523,000
6400	900,000 GPD	345,500	188,000	533,500
6500	1 MGD	368,500	218,000	586,500
7000	500,000 GPD sandy clay & loam	191,500	119,000	310,500
7100	600,000 GPD	269,000	144,500	413,500
7200	700,000 GPD	270,500	152,500	423,000
7300	800,000 GPD	345,500	174,500	520,000
7400	900,000 GPD	355,000	187,500	542,500
7500	1 MGD	368,500	214,500	583,000

G3020 Sanitary Sewer

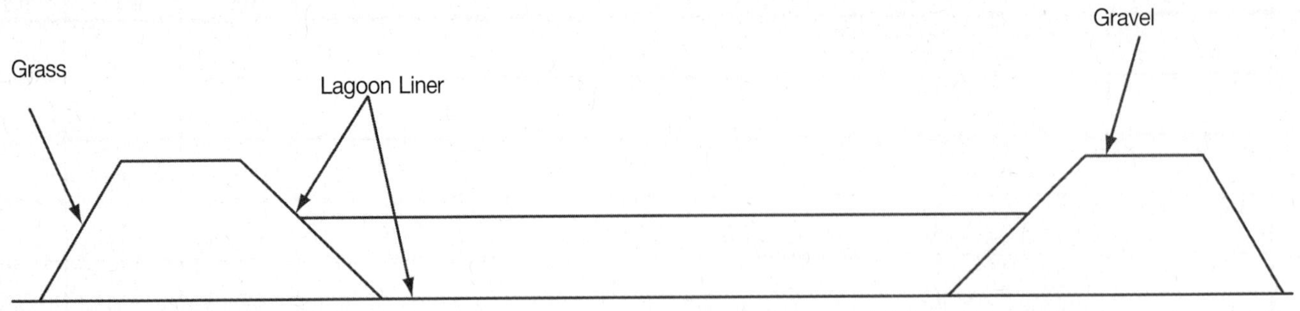

Gravel

Grass

Lagoon Liner

System Components	QUANTITY	UNIT	COST EACH MAT.	COST EACH INST.	COST EACH TOTAL
SYSTEM G3020 720 1000					
500,000 GPD, PRIMARY DOMESTIC SEWAGE LAGOON, COMMON EARTH					
Supervision of lagoon construction	125.000	Day		73,125	73,125
Quality Control of Earthwork per day	50.000	Day		22,500	22,500
Excavate lagoon	22000.000	B.C.Y.		53,020	53,020
Finish grade slopes for seeding	54.000	M.S.F.		1,539	1,539
Spread & grade top of lagoon banks for compaction	1013.000	M.S.F.		61,793	61,793
Compact lagoon banks	14437.000	E.C.Y.		10,827.75	10,827.75
Install pond liner	369.000	M.S.F.	236,160	405,900	642,060
Provide water for compaction and dust control	4400.000	E.C.Y.	5,808	9,064	14,872
Dispose of excess material on-site	9832.000	L.C.Y.		34,805.28	34,805.28
Seed non-lagoon side of banks	54.000	M.S.F.	1,539	1,539	3,078
Place gravel on top of banks	3111.000	S.Y.	18,821.55	4,479.84	23,301.39
Provide 6 ' high fence around lagoon	2012.000	L.F.	42,252	13,943.16	56,195.16
Provide 12′ wide fence gate	3.000	Opng.	1,425	1,546.50	2,971.50
Temporary Fencing	2012.000	L.F.	6,539	4,547.12	11,086.12
TOTAL			312,544.55	698,629.65	1,011,174.20

G3020 720	Primary Sewage Lagoon	COST EACH MAT.	COST EACH INST.	COST EACH TOTAL
0980	Lagoons, Primary, 20 day retention, 5.1 foot depth, no mechanical aeration			
1000	500,000 GPD, Primary Domestic Sewage Lagoon, common earth, 20 day retention	312,500	698,500	1,011,000
1100	600,000 GPD	350,500	798,000	1,148,500
1200	700,000 GPD	395,500	1,067,000	1,462,500
1300	800,000 GPD	462,000	1,003,500	1,465,500
1400	900,000 GPD	520,500	1,148,000	1,668,500
1500	1 MGD	569,000	1,220,000	1,789,000
2000	500,000 GPD sandy clay & loam	312,500	697,500	1,010,000
2100	600,000 GPD	350,500	796,000	1,146,500
2200	700,000 GPD	395,500	862,500	1,258,000
2300	800,000 GPD	462,000	1,004,000	1,466,000
2400	900,000 GPD	521,000	1,149,500	1,670,500
2500	1 MGD	570,000	1,269,000	1,839,000
3000	500,000 GPD sand & gravel	312,000	678,500	990,500
3100	600,000 GPD	350,500	774,500	1,125,000
3200	700,000 GPD	395,500	839,500	1,235,000
3300	800,000 GPD	462,000	978,000	1,440,000
3400	900,000 GPD	521,000	1,119,000	1,640,000
3500	1 MGD	570,000	1,235,000	1,805,000
5600	Lagoons, Primary without liner			

G3020 Sanitary Sewer

G3020 720	Primary Sewage Lagoon	COST EACH		
		MAT.	INST.	TOTAL
6000	500,000 GPD, Domestic Sewage Primary Lagoon, common earth, 20 day, no liner	76,500	292,500	369,000
6100	600,000 GPD	81,500	336,000	417,500
6200	700,000 GPD	87,500	538,000	625,500
6300	800,000 GPD	103,500	387,500	491,000
6400	900,000 GPD	111,000	444,000	555,000
6500	1 MGD	116,500	490,500	607,000
7000	500,000 GPD sandy clay & loam	76,500	291,500	368,000
7100	600,000 GPD	81,500	334,000	415,500
7200	700,000 GPD	87,500	333,500	421,000
7300	800,000 GPD	103,500	388,000	491,500
7400	900,000 GPD	111,000	444,500	555,500
7500	1 MGD	116,500	490,000	606,500

G3020 Sanitary Sewer

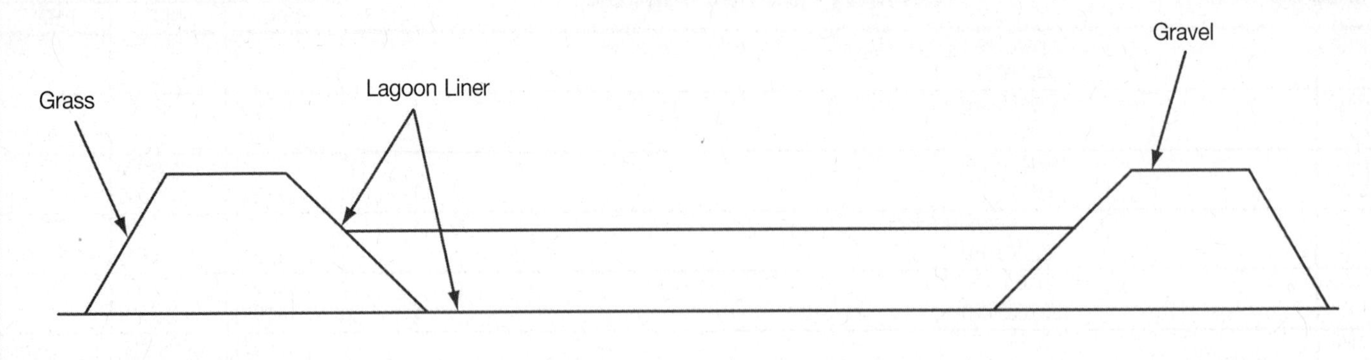

System Components	QUANTITY	UNIT	COST EACH MAT.	COST EACH INST.	COST EACH TOTAL
SYSTEM G3020 730 1000					
500,000 GPD, SECONDARY DOMESTIC SEWAGE LAGOON, COMMON EARTH					
Supervision of lagoon construction	60.000	Day		35,100	35,100
Quality Control of Earthwork per day	30.000	Day		13,500	13,500
Excavate lagoon	10780.000	B.C.Y.		25,979.80	25,979.80
Finish grade slopes for seeding	40.000	M.S.F.		1,140	1,140
Spread & grade top of lagoon banks for compaction	753.000	M.S.F.		45,933	45,933
Compact lagoon banks	10725.000	E.C.Y.		8,043.75	8,043.75
Install pond liner	202.000	M.S.F.	129,280	222,200	351,480
Provide water for compaction and dust control	2156.000	E.C.Y.	2,845.92	4,441.36	7,287.28
Dispose of excess material on-site	72.000	L.C.Y.		254.88	254.88
Seed non-lagoon side of banks	40.000	M.S.F.	1,140	1,140	2,280
Place gravel on top of banks	2311.000	S.Y.	13,981.55	3,327.84	17,309.39
Provide 6 ' high fence around lagoon	1562.000	L.F.	32,802	10,824.66	43,626.66
Provide 12' wide fence gate	3.000	Opng.	1,425	1,546.50	2,971.50
Temporary Fencing	1562.000	L.F.	5,076.50	3,530.12	8,606.62
TOTAL			186,550.97	376,961.91	563,512.88

G3020 730	Secondary Sewage Lagoon	COST EACH MAT.	COST EACH INST.	COST EACH TOTAL
0980	Lagoons, Secondary, 10 day retention, 5.1 foot depth, no mechanical aeration			
1000	500,000 GPD Secondary Domestic Sewage Lagoon, common earth,10 day retention	186,500	377,000	563,500
1100	600,000 GPD	212,000	442,000	654,000
1200	700,000 GPD	235,500	585,000	820,500
1300	800,000 GPD	260,500	534,000	794,500
1400	900,000 GPD	284,500	592,500	877,000
1500	1 MGD	331,500	685,000	1,016,500
1600	2 MGD	549,500	1,200,000	1,749,500
1700	3 MGD	887,500	1,855,500	2,743,000
2000	500,000 GPD sandy clay & loam	186,500	378,000	564,500
2100	600,000 GPD	212,000	443,000	655,000
2200	700,000 GPD	235,500	479,000	714,500
2300	800,000 GPD	260,500	535,500	796,000
2400	900,000 GPD	284,500	594,000	878,500
2500	1 MGD	331,500	686,500	1,018,000
2600	2 MGD	549,500	1,200,000	1,749,500
2700	3 MGD	887,500	1,857,500	2,745,000
3000	500,000 GPD sand & gravel	186,500	369,000	555,500
3100	600,000 GPD	212,000	432,500	644,500
3200	700,000 GPD	235,500	468,000	703,500
3300	800,000 GPD	260,500	522,000	782,500
3400	900,000 GPD	284,500	579,000	863,500

G3020 Sanitary Sewer

G3020 730	Secondary Sewage Lagoon	COST EACH		
		MAT.	INST.	TOTAL
3500	1 MGD	331,500	669,500	1,001,000
3600	2 MGD	549,500	1,168,000	1,717,500
3700	3 MGD	887,500	1,808,000	2,695,500
6000	500,000 GPD, Domestic Sewage Secondary Lagoon, common earth, 10 day, no liner	48,100	152,500	200,600
6100	600,000 GPD	61,500	183,500	245,000
6200	700,000 GPD	65,500	191,500	257,000
6300	800,000 GPD	69,000	212,000	281,000
6400	900,000 GPD	72,500	228,500	301,000
6500	1 MGD	85,500	262,500	348,000
6600	2 MGD	114,500	452,000	566,500
6700	3 MGD	201,000	675,500	876,500
7000	500,000 GPD sandy clay & loam	57,500	156,000	213,500
7100	600,000 GPD	61,500	184,500	246,000
7200	700,000 GPD	65,500	186,500	252,000
7300	800,000 GPD	69,000	206,500	275,500
7400	900,000 GPD	72,500	229,500	302,000
7500	1 MGD	85,500	264,000	349,500
7600	2 MGD	114,500	452,000	566,500
7700	3 MGD	201,000	677,500	878,500

G3020 Sanitary Sewer

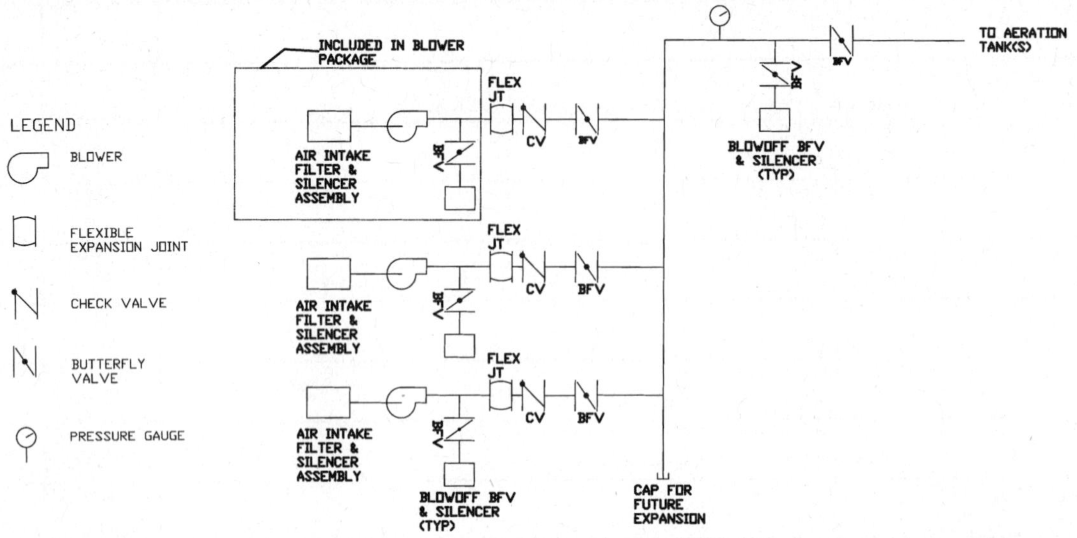

System Components	QUANTITY	UNIT	COST EACH		
			MAT.	INST.	TOTAL
SYSTEM G3020 740 1000					
WASTEWATER TREATMENT, AERATION, 75 CFM PREPACKAGED BLOWER SYSTEM					
Prepacked blower, 38 CFM	3.000	Ea.	9,450	2,340	11,790
Flexible coupling-2" diameter	3.000	Ea.	109.50	56.10	165.60
Wafer style check valve-2" diameter	3.000	Ea.	393	231	624
Wafer style butterfly valve-2" diameter	3.000	Ea.	273	148.50	421.50
Wafer style butterfly valve-3" diameter	1.000	Ea.	102	156	258
Blower Pressure Relief Valves-adjustable-3" Diameter	1.000	Ea.	187	156	343
Silencer with polyester filter-3" Connection	1.000	Ea.	185	156	341
Pipe-2" diameter	30.000	L.F.	384	690	1,074
Pipe-3" diameter	50.000	L.F.	1,225	1,250	2,475
3 x 3 x 2 Tee	3.000	Ea.	300	270	570
Pipe cap-3" diameter	1.000	Ea.	87	41.50	128.50
Pressure gage	1.000	Ea.	39.50	22	61.50
Master control panel	1.000	Ea.	1,025	256	1,281
TOTAL			13,760	5,773.10	19,533.10

G3020 740	Blower System for Wastewater Aeration	COST EACH		
		MAT.	INST.	TOTAL
1000	75 CFM Prepackaged Blower System	13,800	5,775	19,575
1100	250 CFM	13,100	8,250	21,350
1200	450 CFM	24,800	8,750	33,550
1300	700 CFM	40,300	11,000	51,300
1400	1000 CFM	40,400	11,700	52,100

G3030 Storm Sewer

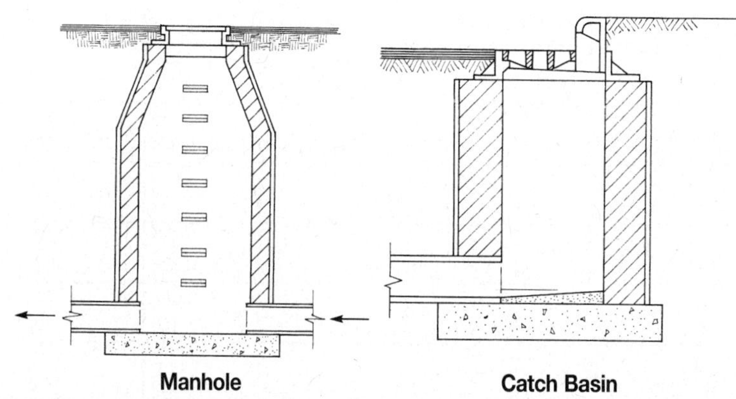

The Manhole and Catch Basin System includes: excavation with a backhoe; a formed concrete footing; frame and cover; cast iron steps and compacted backfill.

The Expanded System Listing shows manholes that have a 4′, 5′ and 6′ inside diameter riser. Depths range from 4′ to 14′. Construction material shown is either concrete, concrete block, precast concrete, or brick.

System Components	QUANTITY	UNIT	COST PER EACH MAT.	COST PER EACH INST.	COST PER EACH TOTAL
SYSTEM G3030 210 1920					
MANHOLE/CATCH BASIN, BRICK, 4′ I.D. RISER, 4′ DEEP					
Excavation, hydraulic backhoe, 3/8 C.Y. bucket	14.815	B.C.Y.		117.93	117.93
Trim sides and bottom of excavation	64.000	S.F.		60.80	60.80
Forms in place, manhole base, 4 uses	20.000	SFCA	14.20	104	118.20
Reinforcing in place footings, #4 to #7	.019	Ton	20.90	22.80	43.70
Concrete, 3000 psi	.925	C.Y.	100.83		100.83
Place and vibrate concrete, footing, direct chute	.925	C.Y.		48.40	48.40
Catch basin or MH, brick, 4′ ID, 4′ deep	1.000	Ea.	560	1,000	1,560
Catch basin or MH steps; heavy galvanized cast iron	1.000	Ea.	21	13.90	34.90
Catch basin or MH frame and cover	1.000	Ea.	268	240.50	508.50
Fill, granular	12.954	L.C.Y.	297.94		297.94
Backfill, spread with wheeled front end loader	12.954	L.C.Y.		32.78	32.78
Backfill compaction, 12″ lifts, air tamp	12.954	E.C.Y.		115.29	115.29
TOTAL			1,282.87	1,756.40	3,039.27

G3030 210	Manholes & Catch Basins	COST PER EACH MAT.	COST PER EACH INST.	COST PER EACH TOTAL
1920	Manhole/catch basin, brick, 4′ I.D. riser, 4′ deep	1,275	1,750	3,025
1940	6′ deep	1,825	2,450	4,275
1960	8′ deep	2,425	3,375	5,800
1980	10′ deep	2,875	4,150	7,025
3000	12′ deep	3,575	4,500	8,075
3020	14′ deep	4,425	6,300	10,725
3200	Block, 4′ I.D. riser, 4′ deep	1,125	1,425	2,550
3220	6′ deep	1,575	2,000	3,575
3240	8′ deep	2,075	2,800	4,875
3260	10′ deep	2,475	3,425	5,900
3280	12′ deep	3,150	4,350	7,500
3300	14′ deep	3,950	5,325	9,275
4620	Concrete, cast-in-place, 4′ I.D. riser, 4′ deep	1,275	2,450	3,725
4640	6′ deep	1,825	3,250	5,075
4660	8′ deep	2,550	4,700	7,250
4680	10′ deep	3,075	5,825	8,900
4700	12′ deep	3,875	7,225	11,100
4720	14′ deep	4,775	8,650	13,425
5820	Concrete, precast, 4′ I.D. riser, 4′ deep	1,575	1,300	2,875
5840	6′ deep	2,100	1,750	3,850

G3030 Storm Sewer

G3030 210	Manholes & Catch Basins	COST PER EACH		
		MAT.	INST.	TOTAL
5860	8' deep	2,625	2,425	5,050
5880	10' deep	3,150	3,000	6,150
5900	12' deep	3,975	3,700	7,675
5920	14' deep	4,900	4,700	9,600
6000	5' I.D. riser, 4' deep	2,575	1,475	4,050
6020	6' deep	2,925	2,075	5,000
6040	8' deep	4,100	2,750	6,850
6060	10' deep	5,225	3,475	8,700
6080	12' deep	6,450	4,450	10,900
6100	14' deep	7,800	5,400	13,200
6200	6' I.D. riser, 4' deep	3,250	1,950	5,200
6220	6' deep	4,100	2,575	6,675
6240	8' deep	5,225	3,600	8,825
6260	10' deep	6,625	4,550	11,175
6280	12' deep	8,125	5,725	13,850
6300	14' deep	9,775	6,950	16,725

G3030 Storm Sewer

The Headwall Systems are listed in concrete and different stone wall materials for two different backfill slope conditions. The backfill slope directly affects the length of the wing walls. Walls are listed for different culvert sizes starting at 30″ diameter. Excavation and backfill are included in the system components, and are figured from an elevation 2′ below the bottom of the pipe.

System Components	QUANTITY	UNIT	COST PER EACH		
			MAT.	INST.	TOTAL
SYSTEM G3030 310 2000					
HEADWALL, C.I.P. CONCRETE FOR 30″ PIPE, 3′ LONG WING WALLS					
Excavation, hydraulic backhoe, 3/8 C.Y. bucket	2.500	B.C.Y.		222.88	222.88
Formwork, 2 uses	157.000	SFCA	412.25	1,873.16	2,285.41
Reinforcing in place, A615 Gr 60, longer and heavier dowels, add	45.000	Lb.	27.45	79.20	106.65
Concrete, 3000 psi	2.600	C.Y.	283.40		283.40
Place concrete, spread footings, direct chute	2.600	C.Y.		136.03	136.03
Backfill, dozer	2.500	L.C.Y.		52.92	52.92
TOTAL			723.10	2,364.19	3,087.29

G3030 310	Headwalls	COST PER EACH		
		MAT.	INST.	TOTAL
2000	Headwall, 1-1/2 to 1 slope soil, C.I.P. conc, 30″ pipe, 3′ long wing walls	725	2,350	3,075
2020	Pipe size 36″, 3′-6″ long wing walls	905	2,825	3,730
2040	Pipe size 42″, 4′ long wing walls	1,100	3,325	4,425
2060	Pipe size 48″, 4′-6″ long wing walls	1,300	3,875	5,175
2080	Pipe size 54″, 5′-0″ long wing walls	1,550	4,500	6,050
2100	Pipe size 60″, 5′-6″ long wing walls	1,800	5,100	6,900
2120	Pipe size 72″, 6′-6″ long wing walls	2,400	6,600	9,000
2140	Pipe size 84″, 7′-6″ long wing walls	3,075	8,125	11,200
3000	$350/ton stone, pipe size 30″, 3′ long wing walls	1,600	865	2,465
3020	Pipe size 36″, 3′-6″ long wing walls	2,050	1,025	3,075
3040	Pipe size 42″, 4′ long wing walls	2,575	1,225	3,800
3060	Pipe size 48″, 4′-6″ long wing walls	3,200	1,450	4,650
3080	Pipe size 54″, 5′ long wing walls	3,850	1,700	5,550
3100	Pipe size 60″, 5′-6″ long wing walls	4,575	1,950	6,525
3120	Pipe size 72″, 6′-6″ long wing walls	6,300	2,600	8,900
3140	Pipe size 84″, 7′-6″ long wing walls	8,300	3,325	11,625
4500	2 to 1 slope soil, C.I.P. concrete, pipe size 30″, 4′-3″ long wing walls	860	2,725	3,585
4520	Pipe size 36″, 5′ long wing walls	1,100	3,325	4,425
4540	Pipe size 42″, 5′-9″ long wing walls	1,350	4,000	5,350
4560	Pipe size 48″, 6′-6″ long wing walls	1,625	4,625	6,250
4580	Pipe size 54″, 7′-3″ long wing walls	1,925	5,425	7,350
4600	Pipe size 60″, 8′-0″ long wing walls	2,250	6,200	8,450
4620	Pipe size 72″, 9′-6″ long wing walls	3,000	8,075	11,075
4640	Pipe size 84″, 11′-0″ long wing walls	3,875	10,000	13,875
5500	$350/ton stone, pipe size 30″, 4′-3″ long wing walls	1,950	995	2,945
5520	Pipe size 36″, 5′ long wing walls	2,550	1,225	3,775
5540	Pipe size 42″, 5′-9″ long wing walls	3,225	1,475	4,700
5560	Pipe size 48″, 6′-6″ long wing walls	4,000	1,750	5,750
5580	Pipe size 54″, 7′-3″ long wing walls	4,850	2,075	6,925
5600	Pipe size 60″, 8′-0″ long wing walls	5,775	2,400	8,175
5620	Pipe size 72″, 9′-6″ long wing walls	8,000	3,225	11,225
5640	Pipe size 84″, 11′-0″ long wing walls	10,700	4,200	14,900

G3060 Fuel Distribution

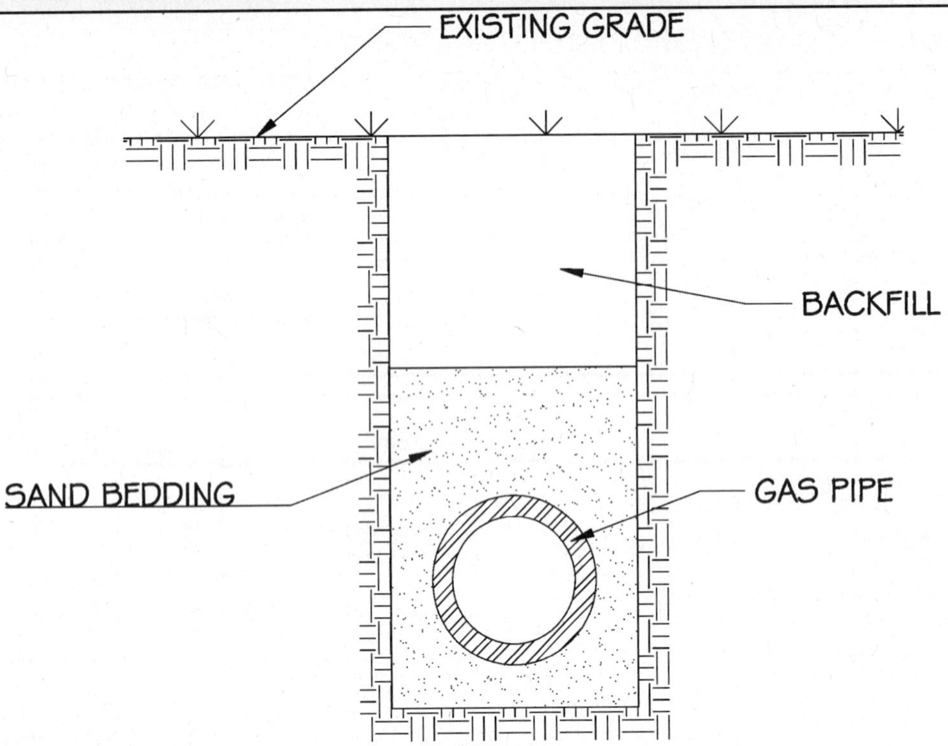

System Components	QUANTITY	UNIT	COST L.F. MAT.	COST L.F. INST.	COST L.F. TOTAL
SYSTEM G3060 112 1000					
GASLINE 1/2" PE, 60 PSI COILS, COMPRESSION COUPLING, INCLUDING COMMON EARTH EXCAVATION, BEDDING, BACKFILL AND COMPACTION					
Excavate Trench	.170	B.C.Y.		1.13	1.13
Sand bedding material	.090	L.C.Y.	1.74	1.13	2.87
Compact bedding	.070	E.C.Y.		.38	.38
Install 1/2" gas pipe	1.000	L.F.	.67	3.59	4.26
Backfill trench	.120	L.C.Y.		.35	.35
Compact fill material in trench	.100	E.C.Y.		.49	.49
Dispose of excess fill material on-site	.090	L.C.Y.		1.03	1.03
TOTAL			2.41	8.10	10.51

G3060 112	Gasline (Common Earth Excavation)	COST L.F. MAT.	COST L.F. INST.	COST L.F. TOTAL
0950	Gasline, including common earth excavation, bedding, backfill and compaction			
1000	1/2" diameter PE, 60 psi coils, compression coupling, SDR 11, 2' deep	2.41	8.10	10.51
1010	4' deep	2.41	10.35	12.76
1020	6' deep	2.41	12.60	15.01
1050	1" diameter PE, 60 psi coils, compression coupling, SDR 11, 2' deep	3.29	8.50	11.79
1060	4' deep	3.29	10.75	14.04
1070	6' deep	3.29	13	16.29
1100	1 1/4" diameter PE, 60 psi coils, compression coupling, SDR 11, 2' deep	3.67	8.50	12.17
1110	4' deep	3.67	10.75	14.42
1120	6' deep	3.67	13	16.67
1150	2" diameter PE, 60 psi coils, compression coupling, SDR 11, 2' deep	4.88	8.95	13.83
1160	4' deep	4.88	11.25	16.13
1180	6' deep	4.88	13.45	18.33
1200	3" diameter PE, 60 psi coils, compression coupling, SDR 11, 2' deep	8.85	10.95	19.80
1210	4' deep	8.85	13.20	22.05
1220	6' deep	8.85	15.45	24.30

G3060 Fuel Distribution

G3060 112	Gasline (Common Earth Excavation)	COST L.F. MAT.	COST L.F. INST.	COST L.F. TOTAL
1250	4″ diameter PE, 60 psi 40′ length, compression coupling, SDR 11, 2′ deep	16.30	15.25	31.55
1260	4′ deep	16.30	17.50	33.80
1270	6′ deep	16.30	19.75	36.05
1300	6″ diameter PE, 60 psi 40′ length w/coupling, SDR 11, 2′ deep	40	16.15	56.15
1310	4′ deep	40	18.35	58.35
1320	6′ deep	40	20.50	60.50
1350	8″ diameter PE, 60 psi 40′ length w/coupling, SDR 11, 4′ deep	60	21	81
1360	6′ deep	60	23	83
2000	1″ diameter steel plain end, Schedule 40, 2′ deep	7.55	13.55	21.10
2010	4′ deep	7.55	15.80	23.35
2020	6′ deep	7.55	18.05	25.60
2050	2″ diameter, Steel plain end, Schedule 40, 2′ deep	10.80	14.20	25
2060	4′ deep	10.80	16.45	27.25
2070	6′ deep	10.80	18.70	29.50
2100	3″ diameter, steel plain end, Schedule 40, 2′ deep	17.70	16.05	33.75
2110	4′ deep	17.70	18.30	36
2120	6′ deep	17.70	20.50	38.20
2150	4″ diameter, steel plain end, Schedule 40, 2′ deep	22.50	22	44.50
2160	4′ deep	22.50	24.50	47
2170	6′ deep	22.50	26.50	49
2200	5″ diameter, steel plain end, Schedule 40, 2′ deep	31.50	24.50	56
2210	4′ deep	31.50	27	58.50
2220	6′ deep	31.50	29	60.50
2250	6″ diameter, steel plain end, Schedule 40, 2′ deep	37.50	29	66.50
2260	4′ deep	37.50	31	68.50
2270	6′ deep	37.50	33.50	71
2300	8″ diameter, steel plain end, Schedule 40, 2′ deep	58.50	35.50	94
2310	4′ deep	58.50	38	96.50
2320	6′ deep	58.50	40	98.50
2350	10″ diameter, steel plain end, Schedule 40, 2′ deep	132	47.50	179.50
2360	4′ deep	132	49.50	181.50
2370	6′ deep	132	52	184
2400	12″ diameter, steel plain end, Schedule 40, 2′ deep	146	58	204
2410	4′ deep	146	60	206
2420	6′ deep	146	62.50	208.50
2460	14″ diameter, steel plain end, Schedule 40, 4′ deep	156	64.50	220.50
2470	6′ deep	156	67	223
2510	16″ diameter, Schedule 40, 4′ deep	171	69.50	240.50
2520	6′ deep	171	72	243
2560	18″ diameter, Schedule 40, 4′ deep	219	74.50	293.50
2570	6′ deep	219	77.50	296.50
2600	20″ diameter, steel plain end, Schedule 40, 4′ deep	340	81	421
2610	6′ deep	340	83.50	423.50
2650	24″ diameter, steel plain end, Schedule 40, 4′ deep	385	96	481
2660	6′ deep	385	99.50	484.50
3150	4″ diameter, steel plain end, Schedule 80, 2′ deep	46.50	35	81.50
3160	4′ deep	46.50	37	83.50
3170	6′ deep	46.50	39.50	86
3200	5″ diameter, steel plain end, Schedule 80, 2′ deep	73	35.50	108.50
3210	4′ deep	73	38	111
3220	6′ deep	73	40	113
3250	6″ diameter, steel plain end, Schedule 80, 2′ deep	99.50	39	138.50
3260	4′ deep	99.50	41	140.50
3270	6′ deep	99.50	43	142.50
3300	8″ diameter, steel plain end, Schedule 80, 2′ deep	132	44.50	176.50
3310	4′ deep	132	46.50	178.50
3320	6′ deep	132	49	181
3350	10″ diameter, steel plain end, Schedule 80, 2′ deep	196	52	248

G3060 Fuel Distribution

G3060 112	Gasline (Common Earth Excavation)	COST L.F.		
		MAT.	INST.	TOTAL
3360	4′ deep	196	54	250
3370	6′ deep	196	56	252
3400	12″ diameter, steel plain end, Schedule 80, 2′ deep	260	63.50	323.50
3410	4′ deep	260	65.50	325.50
3420	6′ deep	260	68	328

G4010 Electrical Distribution

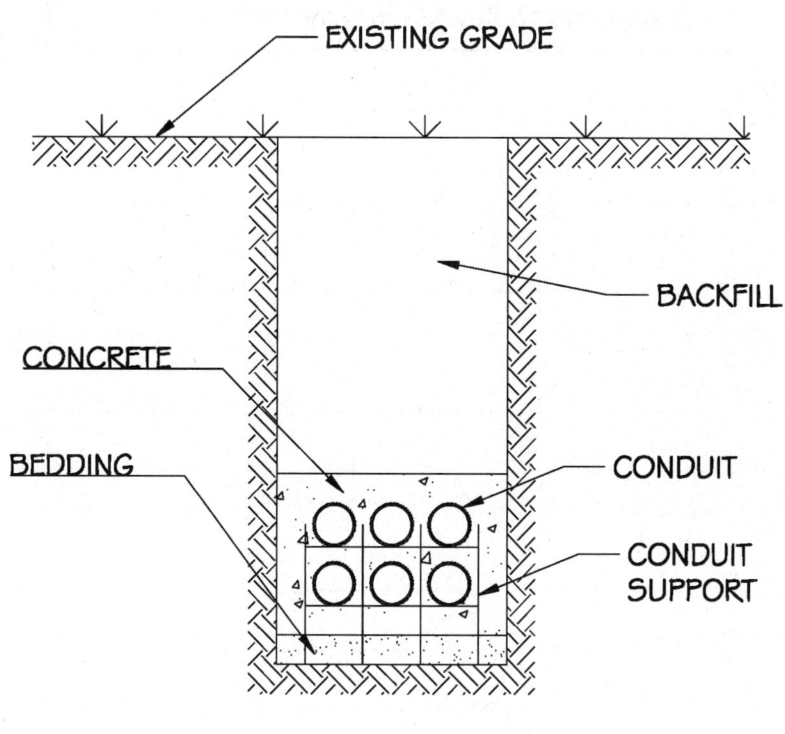

System Components	QUANTITY	UNIT	COST L.F. MAT.	COST L.F. INST.	COST L.F. TOTAL
SYSTEM G4010 320 1012					
UNDERGROUND ELECTRICAL CONDUIT, 2″ DIAMETER, INCLUDING EXCAVATION, CONCRETE, BEDDING, BACKFILL, AND COMPACTION					
Excavate Trench	.150	B.C.Y.		1	1
Base spacer	.200	Ea.	.34	2.28	2.62
Conduit	1.000	L.F.	1.91	3.55	5.46
Concrete material	.050	C.Y.	5.45		5.45
Concrete placement	.050	C.Y.		1.21	1.21
Backfill trench	.120	L.C.Y.		.35	.35
Compact fill material in trench	.100	E.C.Y.		.49	.49
Dispose of excess fill material on-site	.120	L.C.Y.		1.37	1.37
Marking tape	.010	C.L.F.	.31	.03	.34
TOTAL			8.01	10.28	18.29

G4010 320	Underground Electrical Conduit	COST L.F. MAT.	COST L.F. INST.	COST L.F. TOTAL
0950	Underground electrical conduit, including common earth excavation,			
0952	Concrete, bedding, backfill and compaction			
1012	2″ dia. Schedule 40 PVC, 2′ deep	8	10.30	18.30
1014	4′ deep	8	14.70	22.70
1016	6′ deep	8	18.90	26.90
1022	2 @ 2″ dia. Schedule 40 PVC, 2′ deep	10.25	16.15	26.40
1024	4′ deep	10.25	20.50	30.75
1026	6′ deep	10.25	24.50	34.75
1032	3 @ 2″ dia. Schedule 40 PVC, 2′ deep	12.50	22	34.50
1034	4′ deep	12.50	26.50	39
1036	6′ deep	12.50	30.50	43
1042	4 @ 2″ dia. Schedule 40 PVC, 2′ deep	16.90	27.50	44.40
1044	4′ deep	16.90	31.50	48.40
1046	6′ deep	16.90	36	52.90

G4010 Electrical Distribution

G4010 320	Underground Electrical Conduit	COST L.F.		
		MAT.	INST.	TOTAL
1062	6 @ 2" dia. Schedule 40 PVC, 2' deep	21.50	39	60.50
1064	4' deep	21.50	43	64.50
1066	6' deep	21.50	47.50	69
1082	8 @ 2" dia. Schedule 40 PVC, 2' deep	26	50.50	76.50
1084	4' deep	26	54.50	80.50
1086	6' deep	26	59	85
1092	9 @ 2" dia. Schedule 40 PVC, 2' deep	28	56	84
1094	4' deep	30.50	60	90.50
1096	6' deep	30.50	64.50	95
1212	3" dia. Schedule 40 PVC, 2' deep	10.85	12.60	23.45
1214	4' deep	10.85	17	27.85
1216	6' deep	10.85	21.50	32.35
1222	2 @ 3" dia. Schedule 40 PVC, 2' deep	13.80	20.50	34.30
1224	4' deep	13.80	25	38.80
1226	6' deep	13.80	29	42.80
1232	3 @ 3" dia. Schedule 40 PVC, 2' deep	17.80	28.50	46.30
1234	4' deep	17.80	33	50.80
1236	6' deep	17.80	37	54.80
1242	4 @ 3" dia. Schedule 40 PVC, 2' deep	25	37	62
1244	4' deep	25	41	66
1246	6' deep	25	45.50	70.50
1262	6 @ 3" dia. Schedule 40 PVC, 2' deep	32	52.50	84.50
1264	4' deep	32	57	89
1266	6' deep	32	61.50	93.50
1282	8 @ 3" dia. Schedule 40 PVC, 2' deep	40	69	109
1284	4' deep	40	73	113
1286	6' deep	40	78	118
1292	9 @ 3" dia. Schedule 40 PVC, 2' deep	46.50	78	124.50
1294	4' deep	47.50	81	128.50
1296	6' deep	47.50	85	132.50
1312	4" dia. Schedule 40 PVC, 2' deep	12.35	15.45	27.80
1314	4' deep	12.35	19.85	32.20
1316	6' deep	12.35	24	36.35
1322	2 @ 4" dia. Schedule 40 PVC, 2' deep	17.85	26.50	44.35
1324	4' deep	17.85	30.50	48.35
1326	6' deep	17.85	35	52.85
1332	3 @ 4" dia. Schedule 40 PVC, 2' deep	22.50	37.50	60
1334	4' deep	22.50	41.50	64
1336	6' deep	22.50	46	68.50
1342	4 @ 4" dia. Schedule 40 PVC, 2' deep	32	49	81
1344	4' deep	32	53	85
1346	6' deep	32	57.50	89.50
1362	6 @ 4" dia. Schedule 40 PVC, 2' deep	42	71	113
1364	4' deep	42	75	117
1366	6' deep	42	79.50	121.50
1382	8 @ 4" dia. Schedule 40 PVC, 2' deep	54	93.50	147.50
1384	4' deep	54	99	153
1386	6' deep	54	104	158
1392	9 @ 4" dia. Schedule 40 PVC, 2' deep	60.50	105	165.50
1394	4' deep	60.50	108	168.50
1396	6' deep	60.50	112	172.50
1412	5" dia. Schedule 40 PVC, 2' deep	14.75	18.40	33.15
1414	4' deep	14.75	22.50	37.25
1416	6' deep	14.75	27	41.75
1422	2 @ 5" dia. Schedule 40 PVC, 2' deep	22.50	32.50	55
1424	4' deep	22.50	36.50	59
1426	6' deep	22.50	41	63.50
1432	3 @ 5" dia. Schedule 40 PVC, 2' deep	30.50	46.50	77

G4010 Electrical Distribution

G4010 320	Underground Electrical Conduit	COST L.F.		
		MAT.	INST.	TOTAL
1434	4′ deep	30.50	51.50	82
1436	6′ deep	30.50	56	86.50
1442	4 @ 5″ dia. Schedule 40 PVC, 2′ deep	41.50	60	101.50
1444	4′ deep	41.50	65	106.50
1446	6′ deep	41.50	69	110.50
1462	6 @ 5″ dia. Schedule 40 PVC, 2′ deep	56.50	88.50	145
1464	4′ deep	56.50	93	149.50
1466	6′ deep	56.50	97.50	154
1482	8 @ 5″ dia. Schedule 40 PVC, 2′ deep	74	118	192
1484	4′ deep	74	124	198
1486	6′ deep	74	129	203
1492	9 @ 5″ dia. Schedule 40 PVC, 2′ deep	83	132	215
1494	4′ deep	83	135	218
1496	6′ deep	83	139	222
1512	6″ dia. Schedule 40 PVC, 2′ deep	18.40	22.50	40.90
1514	4′ deep	18.40	26.50	44.90
1516	6′ deep	18.40	31	49.40
1522	2 @ 6″ dia. Schedule 40 PVC, 2′ deep	28	40	68
1524	4′ deep	28	44	72
1526	6′ deep	28	48.50	76.50
1532	3 @ 6″ dia. Schedule 40 PVC, 2′ deep	39.50	59	98.50
1534	4′ deep	39.50	64	103.50
1536	6′ deep	39.50	69	108.50
1542	4 @ 6″ dia. Schedule 40 PVC, 2′ deep	52	75.50	127.50
1544	4′ deep	52	80	132
1546	6′ deep	52	84	136
1562	6 @ 6″ dia. Schedule 40 PVC, 2′ deep	74	112	186
1564	4′ deep	74	117	191
1566	6′ deep	74	123	197
1582	8 @ 6″ dia. Schedule 40 PVC, 2′ deep	98	149	247
1584	4′ deep	98	156	254
1586	6′ deep	98	163	261
1592	9 @ 6″ dia. Schedule 40 PVC, 2′ deep	110	168	278
1594	4′ deep	110	171	281
1596	6′ deep	110	175	285

G9020 Site Repair & Maintenance

G9020 100	Clean and Wrap Marine Piles	COST EACH		
		MAT.	INST.	TOTAL
1000	Clean & wrap 5 foot long, 8 inch diameter wood pile using nails from boat	100	106	206
1010	6 foot long	120	176	296
1020	7 foot long	140	183	323
1030	8 foot long	160	190	350
1040	9 foot long	180	196	376
1050	10 foot long	200	204	404
1060	11 foot long	220	262	482
1070	12 foot long	240	268	508
1075	13 foot long	260	276	536
1080	14 foot long	280	282	562
1090	15 foot long	300	289	589
1100	Clean & wrap 5 foot long, 10 inch diameter	125	131	256
1110	6 foot long	150	219	369
1120	7 foot long	175	227	402
1130	8 foot long	200	236	436
1140	9 foot long	225	245	470
1150	10 foot long	250	253	503
1160	11 foot long	275	335	610
1170	12 foot long	300	345	645
1175	13 foot long	325	355	680
1180	14 foot long	350	360	710
1190	15 foot long	375	370	745
1200	Clean & wrap 5 foot long, 12 inch diameter	150	157	307
1210	6 foot long	180	263	443
1220	7 foot long	210	274	484
1230	8 foot long	240	284	524
1240	9 foot long	270	295	565
1250	10 foot long	300	305	605
1260	11 foot long	330	395	725
1270	12 foot long	360	410	770
1275	13 foot long	390	415	805
1280	14 foot long	420	430	850
1290	15 foot long	450	440	890
1300	Clean & wrap 5 foot long, 13 inch diameter	163	172	335
1310	6 foot long	195	288	483
1320	7 foot long	228	299	527
1330	8 foot long	260	310	570
1340	9 foot long	293	320	613
1350	10 foot long	325	335	660
1360	11 foot long	360	425	785
1370	12 foot long	390	435	825
1375	13 foot long	425	450	875
1380	14 foot long	455	460	915
1390	15 foot long	490	470	960
1400	Clean & wrap 5 foot long, 14 inch diameter	175	184	359
1410	6 foot long	210	300	510
1420	7 foot long	245	315	560
1430	8 foot long	280	325	605
1440	9 foot long	315	340	655
1450	10 foot long	350	350	700
1460	11 foot long	385	455	840
1470	12 foot long	420	465	885
1475	13 foot long	455	480	935
1480	14 foot long	490	490	980
1490	15 foot long	525	500	1,025
1500	Clean & wrap 5 foot long, 8 inch diameter wood pile using straps from boat	158	102	260
1510	6 foot long	189	172	361
1520	7 foot long	221	178	399

G9020 Site Repair & Maintenance

G9020 100	Clean and Wrap Marine Piles	COST EACH		
		MAT.	INST.	TOTAL
1530	8 foot long	252	184	436
1540	9 foot long	284	190	474
1550	10 foot long	315	197	512
1560	11 foot long	345	255	600
1570	12 foot long	380	261	641
1575	13 foot long	410	267	677
1580	14 foot long	440	273	713
1590	15 foot long	475	279	754
1600	Clean & wrap 5 foot long, 10 inch diameter	198	127	325
1610	6 foot long	237	214	451
1620	7 foot long	277	221	498
1630	8 foot long	315	229	544
1640	9 foot long	355	237	592
1650	10 foot long	395	245	640
1660	11 foot long	435	325	760
1670	12 foot long	475	335	810
1675	13 foot long	515	345	860
1680	14 foot long	555	350	905
1690	15 foot long	595	360	955
1700	Clean & wrap 5 foot long, 12 inch diameter	235	151	386
1710	6 foot long	282	257	539
1720	7 foot long	330	267	597
1730	8 foot long	375	276	651
1740	9 foot long	425	285	710
1750	10 foot long	470	295	765
1760	11 foot long	515	385	900
1770	12 foot long	565	395	960
1775	13 foot long	610	405	1,015
1780	14 foot long	660	415	1,075
1790	15 foot long	705	420	1,125
1800	Clean & wrap 5 foot long, 13 inch diameter	255	166	421
1810	6 foot long	305	282	587
1820	7 foot long	355	291	646
1830	8 foot long	410	300	710
1840	9 foot long	460	310	770
1850	10 foot long	510	320	830
1860	11 foot long	560	410	970
1870	12 foot long	610	425	1,035
1875	13 foot long	665	435	1,100
1880	14 foot long	715	440	1,155
1890	15 foot long	765	455	1,220
1900	Clean & wrap 5 foot long, 14 inch diameter	275	178	453
1910	6 foot long	330	295	625
1920	7 foot long	385	305	690
1930	8 foot long	440	315	755
1940	9 foot long	495	325	820
1950	10 foot long	550	340	890
1960	11 foot long	605	445	1,050
1970	12 foot long	660	450	1,110
1975	13 foot long	715	460	1,175
1980	14 foot long	770	475	1,245
1990	15 foot long	825	485	1,310
2000	Clean & wrap 5 foot long, 12 inch diameter concrete pile from boat w/straps	235	132	367
2010	6 foot long	282	223	505
2020	7 foot long	330	232	562
2030	8 foot long	375	241	616
2040	9 foot long	425	251	676
2050	10 foot long	470	261	731

G9020 Site Repair & Maintenance

G9020 100	Clean and Wrap Marine Piles	COST EACH		
		MAT.	INST.	TOTAL
2060	11 foot long	515	345	860
2070	12 foot long	565	355	920
2080	13 foot long	610	360	970
2085	14 foot long	660	370	1,030
2090	15 foot long	705	380	1,085
2100	Clean & wrap 5 foot long, 14 inch diameter	275	155	430
2110	6 foot long	330	259	589
2120	7 foot long	385	270	655
2130	8 foot long	440	280	720
2140	9 foot long	495	291	786
2150	10 foot long	550	300	850
2160	11 foot long	605	390	995
2170	12 foot long	660	395	1,055
2180	13 foot long	715	410	1,125
2185	14 foot long	770	420	1,190
2190	15 foot long	825	435	1,260

Reference Section

All the reference information is in one section, making it easy to find what you need to know . . . and easy to use the book on a daily basis. This section is visually identified by a vertical gray bar on the page edges.

In this Reference Section, we've included Equipment Rental Costs, a listing of rental and operating costs; Crew Listings, a full listing of all crews and equipment, and their costs; Historical Cost Indexes for cost comparisons over time; City Cost Indexes and Location Factors for adjusting costs to the region you are in; Reference Tables, where you will find explanations, estimating information and procedures, or technical data; Change Orders, information on pricing changes to contract documents; Square Foot Costs that allow you to make a rough estimate for the overall cost of a project; and an explanation of all the Abbreviations in the book.

Table of Contents

Equipment Rental Costs

Estimating Tips

- This section contains the average costs to rent and operate hundreds of pieces of construction equipment. This is useful information when estimating the time and material requirements of any particular operation in order to establish a unit or total cost. Equipment costs include not only rental, but also operating costs for equipment under normal use.

Rental Costs

- Equipment rental rates are obtained from industry sources throughout North America–contractors, suppliers, dealers, manufacturers, and distributors.
- Rental rates vary throughout the country, with larger cities generally having lower rates. Lease plans for new equipment are available for periods in excess of six months, with a percentage of payments applying toward purchase.
- Monthly rental rates vary from 2% to 5% of the purchase price of the equipment depending on the anticipated life of the equipment and its wearing parts.
- Weekly rental rates are about 1/3 the monthly rates, and daily rental rates are about 1/3 the weekly rate.
- Rental rates can also be treated as reimbursement costs for contractor-owned equipment. Owned equipment costs include depreciation, loan payments, interest, taxes, insurance, storage, and major repairs.

Operating Costs

- The operating costs include parts and labor for routine servicing, such as repair and replacement of pumps, filters and worn lines. Normal operating expendables, such as fuel, lubricants, tires and electricity (where applicable), are also included.
- Extraordinary operating expendables with highly variable wear patterns, such as diamond bits and blades, are excluded. These costs can be found as material costs in the Unit Price section.
- The hourly operating costs listed do not include the operator's wages.

Equipment Cost/Day

- Any power equipment required by a crew is shown in the Crew Listings with a daily cost.
- The daily cost of equipment needed by a crew is based on dividing the weekly rental rate by 5 (number of working days in the week), and then adding the hourly operating cost times 8 (the number of hours in a day). This "Equipment Cost/ Day" is shown in the far right column of the Equipment Rental pages.
- If equipment is needed for only one or two days, it is best to develop your own cost by including components for daily rent and hourly operating cost. This is important when the listed Crew for a task does not contain the equipment needed, such as a crane for lifting mechanical heating/cooling equipment up onto a roof.
- If the quantity of work is less than the crew's Daily Output shown for a Unit Price line item that includes a bare unit equipment cost, it is recommended to estimate one day's rental cost and operating cost for equipment shown in the Crew Listing for that line item.

Mobilization/ Demobilization

- The cost to move construction equipment from an equipment yard or rental company to the jobsite and back again is not included in equipment rental costs listed in the Reference section, nor in the bare equipment cost of any Unit Price line item, nor in any equipment costs shown in the Crew listings.
- Mobilization (to the site) and demobilization (from the site) costs can be found in the Unit Price section.
- If a piece of equipment is already at the jobsite, it is not appropriate to utilize mobil./ demob. costs again in an estimate.

01 54 33 | Equipment Rental

			UNIT	HOURLY OPER. COST	RENT PER DAY	RENT PER WEEK	RENT PER MONTH	EQUIPMENT COST/DAY	
10	0010	**CONCRETE EQUIPMENT RENTAL** without operators R015433-10							10
	0200	Bucket, concrete lightweight, 1/2 C.Y.	Ea.	.80	23	69	207	20.20	
	0300	1 C.Y.		.85	27	81	243	23	
	0400	1-1/2 C.Y.		1.10	36.50	110	330	30.80	
	0500	2 C.Y.		1.20	43.50	130	390	35.60	
	0580	8 C.Y.		5.95	252	755	2,275	198.60	
	0600	Cart, concrete, self-propelled, operator walking, 10 C.F.		3.40	55	165	495	60.20	
	0700	Operator riding, 18 C.F.		5.65	93.50	280	840	101.20	
	0800	Conveyer for concrete, portable, gas, 16″ wide, 26′ long		13.50	122	365	1,100	181	
	0900	46′ long		13.90	147	440	1,325	199.20	
	1000	56′ long		14.05	155	465	1,400	205.40	
	1100	Core drill, electric, 2-1/2 H.P., 1″ to 8″ bit diameter		1.50	57	171	515	46.20	
	1150	11 H.P., 8″ to 18″ cores		5.90	115	345	1,025	116.20	
	1200	Finisher, concrete floor, gas, riding trowel, 96″ wide		14.05	142	425	1,275	197.40	
	1300	Gas, walk-behind, 3 blade, 36″ trowel		2.35	20	60	180	30.80	
	1400	4 blade, 48″ trowel		4.70	26.50	80	240	53.60	
	1500	Float, hand-operated (Bull float) 48″ wide		.08	13.35	40	120	8.65	
	1570	Curb builder, 14 H.P., gas, single screw		14.65	250	750	2,250	267.20	
	1590	Double screw		15.30	287	860	2,575	294.40	
	1600	Floor grinder, concrete and terrazzo, electric, 22″ path		2.19	134	402	1,200	97.90	
	1700	Edger, concrete, electric, 7″ path		1.02	51.50	155	465	39.15	
	1750	Vacuum pick-up system for floor grinders, wet/dry		1.47	81.50	245	735	60.75	
	1800	Mixer, powered, mortar and concrete, gas, 6 C.F., 18 H.P.		9.20	117	350	1,050	143.60	
	1900	10 C.F., 25 H.P.		11.60	140	420	1,250	176.80	
	2000	16 C.F.		11.95	162	485	1,450	192.60	
	2100	Concrete, stationary, tilt drum, 2 C.Y.		6.80	228	685	2,050	191.40	
	2120	Pump, concrete, truck mounted 4″ line 80′ boom		24.55	855	2,570	7,700	710.40	
	2140	5″ line, 110′ boom		31.75	1,125	3,400	10,200	934	
	2160	Mud jack, 50 C.F. per hr.		7.35	127	380	1,150	134.80	
	2180	225 C.F. per hr.		9.65	145	435	1,300	164.20	
	2190	Shotcrete pump rig, 12 C.Y./hr.		15.10	220	660	1,975	252.80	
	2200	35 C.Y./hr.		17.80	237	710	2,125	284.40	
	2600	Saw, concrete, manual, gas, 18 H.P.		7.25	43.50	130	390	84	
	2650	Self-propelled, gas, 30 H.P.		14.40	100	300	900	175.20	
	2675	V-groove crack chaser, manual, gas, 6 H.P.		2.55	17.35	52	156	30.80	
	2700	Vibrators, concrete, electric, 60 cycle, 2 H.P.		.43	8.65	26	78	8.65	
	2800	3 H.P.		.61	11.65	35	105	11.90	
	2900	Gas engine, 5 H.P.		2.20	16	48	144	27.20	
	3000	8 H.P.		3.00	15	45	135	33	
	3050	Vibrating screed, gas engine, 8 H.P.		2.78	68.50	205	615	63.25	
	3120	Concrete transit mixer, 6 x 4, 250 H.P., 8 C.Y., rear discharge		62.50	565	1,690	5,075	838	
	3200	Front discharge		73.60	690	2,065	6,200	1,002	
	3300	6 x 6, 285 H.P., 12 C.Y., rear discharge		72.65	650	1,950	5,850	971.20	
	3400	Front discharge		75.95	695	2,085	6,250	1,025	
20	0010	**EARTHWORK EQUIPMENT RENTAL** without operators R015433-10							20
	0040	Aggregate spreader, push type 8′ to 12′ wide	Ea.	3.25	25.50	76	228	41.20	
	0045	Tailgate type, 8′ wide		3.10	32.50	98	294	44.40	
	0055	Earth auger, truck-mounted, for fence & sign posts, utility poles		19.75	445	1,335	4,000	425	
	0060	For borings and monitoring wells		47.30	680	2,040	6,125	786.40	
	0070	Portable, trailer mounted		3.15	32.50	98	294	44.80	
	0075	Truck-mounted, for caissons, water wells		100.90	2,900	8,720	26,200	2,551	
	0080	Horizontal boring machine, 12″ to 36″ diameter, 45 H.P.		24.60	193	580	1,750	312.80	
	0090	12″ to 48″ diameter, 65 H.P.		34.85	335	1,005	3,025	479.80	
	0095	Auger, for fence posts, gas engine, hand held		.55	6	18	54	8	
	0100	Excavator, diesel hydraulic, crawler mounted, 1/2 C.Y. cap.		25.00	415	1,240	3,725	448	
	0120	5/8 C.Y. capacity		33.05	550	1,650	4,950	594.40	
	0140	3/4 C.Y. capacity		36.00	625	1,870	5,600	662	
	0150	1 C.Y. capacity		50.60	685	2,050	6,150	814.80	

01 54 33 | Equipment Rental

			UNIT	HOURLY OPER. COST	RENT PER DAY	RENT PER WEEK	RENT PER MONTH	EQUIPMENT COST/DAY	
20	0200	1-1/2 C.Y. capacity	Ea.	60.80	910	2,725	8,175	1,031	20
	0300	2 C.Y. capacity		68.50	1,050	3,135	9,400	1,175	
	0320	2-1/2 C.Y. capacity		104.75	1,325	3,995	12,000	1,637	
	0325	3-1/2 C.Y. capacity		147.35	2,100	6,315	18,900	2,442	
	0330	4-1/2 C.Y. capacity		176.65	2,725	8,150	24,500	3,043	
	0335	6 C.Y. capacity		224.40	2,875	8,640	25,900	3,523	
	0340	7 C.Y. capacity		226.60	2,975	8,915	26,700	3,596	
	0342	Excavator attachments, bucket thumbs		3.20	245	735	2,200	172.60	
	0345	Grapples		2.75	193	580	1,750	138	
	0347	Hydraulic hammer for boom mounting, 5000 ft lb.		14.30	425	1,280	3,850	370.40	
	0349	11,000 ft lb.		23.20	750	2,250	6,750	635.60	
	0350	Gradall type, truck mounted, 3 ton @ 15′ radius, 5/8 C.Y.		43.45	890	2,670	8,000	881.60	
	0370	1 C.Y. capacity		47.65	1,050	3,120	9,350	1,005	
	0400	Backhoe-loader, 40 to 45 H.P., 5/8 C.Y. capacity		14.95	233	700	2,100	259.60	
	0450	45 H.P. to 60 H.P., 3/4 C.Y. capacity		23.90	290	870	2,600	365.20	
	0460	80 H.P., 1-1/4 C.Y. capacity		26.20	315	940	2,825	397.60	
	0470	112 H.P., 1-1/2 C.Y. capacity		41.65	645	1,935	5,800	720.20	
	0482	Backhoe-loader attachment, compactor, 20,000 lb.		5.75	142	425	1,275	131	
	0485	Hydraulic hammer, 750 ft lb.		3.25	96.50	290	870	84	
	0486	Hydraulic hammer, 1200 ft lb.		6.30	217	650	1,950	180.40	
	0500	Brush chipper, gas engine, 6″ cutter head, 35 H.P.		11.95	107	320	960	159.60	
	0550	Diesel engine, 12″ cutter head, 130 H.P.		28.20	285	855	2,575	396.60	
	0600	15″ cutter head, 165 H.P.		33.90	330	985	2,950	468.20	
	0750	Bucket, clamshell, general purpose, 3/8 C.Y.		1.25	38.50	115	345	33	
	0800	1/2 C.Y.		1.35	45	135	405	37.80	
	0850	3/4 C.Y.		1.50	53.50	160	480	44	
	0900	1 C.Y.		1.55	58.50	175	525	47.40	
	0950	1-1/2 C.Y.		2.55	80	240	720	68.40	
	1000	2 C.Y.		2.65	88.50	265	795	74.20	
	1010	Bucket, dragline, medium duty, 1/2 C.Y.		.75	23.50	70	210	20	
	1020	3/4 C.Y.		.75	24.50	74	222	20.80	
	1030	1 C.Y.		.80	26	78	234	22	
	1040	1-1/2 C.Y.		1.20	40	120	360	33.60	
	1050	2 C.Y.		1.25	45	135	405	37	
	1070	3 C.Y.		1.95	61.50	185	555	52.60	
	1200	Compactor, manually guided 2-drum vibratory smooth roller, 7.5 H.P.		7.35	207	620	1,850	182.80	
	1250	Rammer/tamper, gas, 8″		2.70	46.50	140	420	49.60	
	1260	15″		3.00	53.50	160	480	56	
	1300	Vibratory plate, gas, 18″ plate, 3000 lb. blow		2.65	24.50	73	219	35.80	
	1350	21″ plate, 5000 lb. blow		3.30	32.50	98	294	46	
	1370	Curb builder/extruder, 14 H.P., gas, single screw		14.65	250	750	2,250	267.20	
	1390	Double screw		15.30	287	860	2,575	294.40	
	1500	Disc harrow attachment, for tractor		.43	72.50	217	650	46.85	
	1810	Feller buncher, shearing & accumulating trees, 100 H.P.		42.50	650	1,945	5,825	729	
	1860	Grader, self-propelled, 25,000 lb.		39.55	595	1,780	5,350	672.40	
	1910	30,000 lb.		43.25	600	1,805	5,425	707	
	1920	40,000 lb.		64.65	1,000	3,015	9,050	1,120	
	1930	55,000 lb.		84.65	1,600	4,800	14,400	1,637	
	1950	Hammer, pavement breaker, self-propelled, diesel, 1000 to 1250 lb.		30.25	355	1,060	3,175	454	
	2000	1300 to 1500 lb.		45.38	705	2,120	6,350	787.05	
	2050	Pile driving hammer, steam or air, 4150 ft lb. @ 225 bpm		10.35	475	1,425	4,275	367.80	
	2100	8750 ft lb. @ 145 bpm		12.35	660	1,980	5,950	494.80	
	2150	15,000 ft lb. @ 60 bpm		13.95	795	2,380	7,150	587.60	
	2200	24,450 ft lb. @ 111 bpm		14.95	880	2,640	7,925	647.60	
	2250	Leads, 60′ high for pile driving hammers up to 20,000 ft lb.		3.30	80.50	242	725	74.80	
	2300	90′ high for hammers over 20,000 ft lb.		4.95	141	424	1,275	124.40	
	2350	Diesel type hammer, 22,400 ft lb.		26.93	535	1,610	4,825	537.45	
	2400	41,300 ft lb.		36.30	585	1,760	5,275	642.40	

01 54 | Construction Aids

01 54 33 | Equipment Rental

			UNIT	HOURLY OPER. COST	RENT PER DAY	RENT PER WEEK	RENT PER MONTH	EQUIPMENT COST/DAY	
20	2450	141,000 ft lb.	Ea.	55.55	1,050	3,160	9,475	1,076	20
	2500	Vib. elec. hammer/extractor, 200 kW diesel generator, 34 H.P.		56.25	635	1,910	5,725	832	
	2550	80 H.P.		103.45	930	2,785	8,350	1,385	
	2600	150 H.P.		196.65	1,775	5,335	16,000	2,640	
	2800	Log chipper, up to 22" diameter, 600 H.P.		64.95	630	1,895	5,675	898.60	
	2850	Logger, for skidding & stacking logs, 150 H.P.		53.45	820	2,460	7,375	919.60	
	2860	Mulcher, diesel powered, trailer mounted		24.70	213	640	1,925	325.60	
	2900	Rake, spring tooth, with tractor		18.31	335	1,007	3,025	347.90	
	3000	Roller, vibratory, tandem, smooth drum, 20 H.P.		8.75	148	445	1,325	159	
	3050	35 H.P.		11.25	253	760	2,275	242	
	3100	Towed type vibratory compactor, smooth drum, 50 H.P.		26.25	335	1,010	3,025	412	
	3150	Sheepsfoot, 50 H.P.		27.60	370	1,115	3,350	443.80	
	3170	Landfill compactor, 220 H.P.		91.85	1,500	4,485	13,500	1,632	
	3200	Pneumatic tire roller, 80 H.P.		17.00	355	1,065	3,200	349	
	3250	120 H.P.		25.45	605	1,815	5,450	566.60	
	3300	Sheepsfoot vibratory roller, 240 H.P.		69.65	1,100	3,305	9,925	1,218	
	3320	340 H.P.		101.85	1,600	4,830	14,500	1,781	
	3350	Smooth drum vibratory roller, 75 H.P.		25.15	595	1,790	5,375	559.20	
	3400	125 H.P.		32.85	715	2,150	6,450	692.80	
	3410	Rotary mower, brush, 60", with tractor		22.65	305	915	2,750	364.20	
	3420	Rototiller, walk-behind, gas, 5 H.P.		2.23	66	198	595	57.45	
	3422	8 H.P.		3.30	91.50	275	825	81.40	
	3440	Scrapers, towed type, 7 C.Y. capacity		5.70	110	330	990	111.60	
	3450	10 C.Y. capacity		6.45	150	450	1,350	141.60	
	3500	15 C.Y. capacity		6.85	173	520	1,550	158.80	
	3525	Self-propelled, single engine, 14 C.Y. capacity		113.60	1,525	4,600	13,800	1,829	
	3550	Dual engine, 21 C.Y. capacity		174.50	2,000	6,015	18,000	2,599	
	3600	31 C.Y. capacity		232.10	2,850	8,545	25,600	3,566	
	3640	44 C.Y. capacity		287.55	3,650	10,960	32,900	4,492	
	3650	Elevating type, single engine, 11 C.Y. capacity		71.20	1,025	3,050	9,150	1,180	
	3700	22 C.Y. capacity		138.15	2,225	6,690	20,100	2,443	
	3710	Screening plant 110 H.P. w/5' x 10' screen		31.63	435	1,300	3,900	513.05	
	3720	5' x 16' screen		34.85	540	1,615	4,850	601.80	
	3850	Shovel, crawler-mounted, front-loading, 7 C.Y. capacity		246.75	2,900	8,700	26,100	3,714	
	3855	12 C.Y. capacity		354.50	3,950	11,840	35,500	5,204	
	3860	Shovel/backhoe bucket, 1/2 C.Y.		2.45	66.50	200	600	59.60	
	3870	3/4 C.Y.		2.50	73.50	220	660	64	
	3880	1 C.Y.		2.60	83.50	250	750	70.80	
	3890	1-1/2 C.Y.		2.75	96.50	290	870	80	
	3910	3 C.Y.		3.10	130	390	1,175	102.80	
	3950	Stump chipper, 18" deep, 30 H.P.		7.06	176	528	1,575	162.10	
	4110	Dozer, crawler, torque converter, diesel 80 H.P.		29.75	405	1,215	3,650	481	
	4150	105 H.P.		36.10	505	1,515	4,550	591.80	
	4200	140 H.P.		52.10	780	2,335	7,000	883.80	
	4260	200 H.P.		77.35	1,175	3,530	10,600	1,325	
	4310	300 H.P.		100.55	1,700	5,065	15,200	1,817	
	4360	410 H.P.		135.65	2,200	6,620	19,900	2,409	
	4370	500 H.P.		174.30	2,875	8,610	25,800	3,116	
	4380	700 H.P.		261.35	4,000	11,970	35,900	4,485	
	4400	Loader, crawler, torque conv., diesel, 1-1/2 C.Y., 80 H.P.		31.35	455	1,360	4,075	522.80	
	4450	1-1/2 to 1-3/4 C.Y., 95 H.P.		34.15	585	1,760	5,275	625.20	
	4510	1-3/4 to 2-1/4 C.Y., 130 H.P.		54.60	870	2,615	7,850	959.80	
	4530	2-1/2 to 3-1/4 C.Y., 190 H.P.		67.00	1,100	3,275	9,825	1,191	
	4560	3-1/2 to 5 C.Y., 275 H.P.		86.10	1,400	4,215	12,600	1,532	
	4610	Front end loader, 4WD, articulated frame, diesel, 1 to 1-1/4 C.Y., 70 H.P.		19.90	227	680	2,050	295.20	
	4620	1-1/2 to 1-3/4 C.Y., 95 H.P.		25.30	285	855	2,575	373.40	
	4650	1-3/4 to 2 C.Y., 130 H.P.		29.45	365	1,100	3,300	455.60	
	4710	2-1/2 to 3-1/2 C.Y., 145 H.P.	↓	33.35	425	1,280	3,850	522.80	

01 54 33 | Equipment Rental

		01 54 33 Equipment Rental	UNIT	HOURLY OPER. COST	RENT PER DAY	RENT PER WEEK	RENT PER MONTH	EQUIPMENT COST/DAY	
20	4730	3 to 4-1/2 C.Y., 185 H.P.	Ea.	42.85	550	1,655	4,975	673.80	20
	4760	5-1/4 to 5-3/4 C.Y., 270 H.P.		67.80	845	2,535	7,600	1,049	
	4810	7 to 9 C.Y., 475 H.P.		116.25	1,775	5,295	15,900	1,989	
	4870	9 - 11 C.Y., 620 H.P.		155.75	2,425	7,285	21,900	2,703	
	4880	Skid steer loader, wheeled, 10 C.F., 30 H.P. gas		10.90	145	435	1,300	174.20	
	4890	1 C.Y., 78 H.P., diesel		20.35	243	730	2,200	308.80	
	4892	Skid-steer attachment, auger		.50	83.50	251	755	54.20	
	4893	Backhoe		.66	111	332	995	71.70	
	4894	Broom		.74	124	372	1,125	80.30	
	4895	Forks		.23	38.50	115	345	24.85	
	4896	Grapple		.56	92.50	278	835	60.10	
	4897	Concrete hammer		1.06	177	531	1,600	114.70	
	4898	Tree spade		1.03	172	515	1,550	111.25	
	4899	Trencher		.67	112	335	1,000	72.35	
	4900	Trencher, chain, boom type, gas, operator walking, 12 H.P.		5.40	46.50	140	420	71.20	
	4910	Operator riding, 40 H.P.		20.55	292	875	2,625	339.40	
	5000	Wheel type, diesel, 4′ deep, 12″ wide		87.50	845	2,530	7,600	1,206	
	5100	6′ deep, 20″ wide		93.25	1,925	5,760	17,300	1,898	
	5150	Chain type, diesel, 5′ deep, 8″ wide		38.85	560	1,685	5,050	647.80	
	5200	Diesel, 8′ deep, 16″ wide		156.45	3,550	10,620	31,900	3,376	
	5202	Rock trencher, wheel type, 6″ wide x 18″ deep		21.50	330	985	2,950	369	
	5206	Chain type, 18″ wide x 7′ deep		112.60	2,825	8,445	25,300	2,590	
	5210	Tree spade, self-propelled		14.00	267	800	2,400	272	
	5250	Truck, dump, 2-axle, 12 ton, 8 C.Y. payload, 220 H.P.		35.15	227	680	2,050	417.20	
	5300	Three axle dump, 16 ton, 12 C.Y. payload, 400 H.P.		62.40	320	965	2,900	692.20	
	5310	Four axle dump, 25 ton, 18 C.Y. payload, 450 H.P.		73.20	470	1,410	4,225	867.60	
	5350	Dump trailer only, rear dump, 16-1/2 C.Y.		5.40	138	415	1,250	126.20	
	5400	20 C.Y.		5.85	157	470	1,400	140.80	
	5450	Flatbed, single axle, 1-1/2 ton rating		28.65	68.50	205	615	270.20	
	5500	3 ton rating		34.25	98.50	295	885	333	
	5550	Off highway rear dump, 25 ton capacity		74.55	1,200	3,590	10,800	1,314	
	5600	35 ton capacity		83.10	1,350	4,025	12,100	1,470	
	5610	50 ton capacity		103.40	1,600	4,800	14,400	1,787	
	5620	65 ton capacity		106.30	1,575	4,735	14,200	1,797	
	5630	100 ton capacity		154.35	2,775	8,315	24,900	2,898	
	6000	Vibratory plow, 25 H.P., walking		9.25	60	180	540	110	
40	0010	**GENERAL EQUIPMENT RENTAL** without operators R015433-10							40
	0150	Aerial lift, scissor type, to 15′ high, 1000 lb. cap., electric	Ea.	3.10	51.50	155	465	55.80	
	0160	To 25′ high, 2000 lb. capacity		3.55	66.50	200	600	68.40	
	0170	Telescoping boom to 40′ high, 500 lb. capacity, diesel		17.00	315	945	2,825	325	
	0180	To 45′ high, 500 lb. capacity		18.85	355	1,070	3,200	364.80	
	0190	To 60′ high, 600 lb. capacity		21.25	475	1,430	4,300	456	
	0195	Air compressor, portable, 6.5 CFM, electric		.56	12.65	38	114	12.10	
	0196	Gasoline		.79	19	57	171	17.70	
	0200	Towed type, gas engine, 60 CFM		13.55	48.50	145	435	137.40	
	0300	160 CFM		15.80	51.50	155	465	157.40	
	0400	Diesel engine, rotary screw, 250 CFM		17.05	110	330	990	202.40	
	0500	365 CFM		23.10	133	400	1,200	264.80	
	0550	450 CFM		29.35	167	500	1,500	334.80	
	0600	600 CFM		51.70	232	695	2,075	552.60	
	0700	750 CFM		51.90	240	720	2,150	559.20	
	0800	For silenced models, small sizes, add to rent		3%	5%	5%	5%		
	0900	Large sizes, add to rent		5%	7%	7%	7%		
	0930	Air tools, breaker, pavement, 60 lb.	Ea.	.50	9.65	29	87	9.80	
	0940	80 lb.		.50	10.35	31	93	10.20	
	0950	Drills, hand (jackhammer) 65 lb.		.60	17.35	52	156	15.20	
	0960	Track or wagon, swing boom, 4″ drifter		63.45	885	2,660	7,975	1,040	
	0970	5″ drifter		75.50	1,075	3,255	9,775	1,255	

01 54 33 | Equipment Rental

			UNIT	HOURLY OPER. COST	RENT PER DAY	RENT PER WEEK	RENT PER MONTH	EQUIPMENT COST/DAY	
40	0975	Track mounted quarry drill, 6" diameter drill	Ea.	129.75	1,600	4,830	14,500	2,004	40
	0980	Dust control per drill		.97	24	72	216	22.15	
	0990	Hammer, chipping, 12 lb.		.55	26	78	234	20	
	1000	Hose, air with couplings, 50' long, 3/4" diameter		.03	5	15	45	3.25	
	1100	1" diameter		.04	6.35	19	57	4.10	
	1200	1-1/2" diameter		.05	9	27	81	5.80	
	1300	2" diameter		.07	12	36	108	7.75	
	1400	2-1/2" diameter		.11	19	57	171	12.30	
	1410	3" diameter		.14	23	69	207	14.90	
	1450	Drill, steel, 7/8" x 2'		.05	8.65	26	78	5.60	
	1460	7/8" x 6'		.05	9	27	81	5.80	
	1520	Moil points		.02	3.33	10	30	2.15	
	1525	Pneumatic nailer w/accessories		.47	31	93	279	22.35	
	1530	Sheeting driver for 60 lb. breaker		.03	5.65	17	51	3.65	
	1540	For 90 lb. breaker		.11	7.35	22	66	5.30	
	1550	Spade, 25 lb.		.45	7	21	63	7.80	
	1560	Tamper, single, 35 lb.		.55	36.50	109	325	26.20	
	1570	Triple, 140 lb.		.82	54.50	164	490	39.35	
	1580	Wrenches, impact, air powered, up to 3/4" bolt		.40	12.65	38	114	10.80	
	1590	Up to 1-1/4" bolt		.50	23.50	70	210	18	
	1600	Barricades, barrels, reflectorized, 1 to 99 barrels		.03	4.60	13.80	41.50	3	
	1610	100 to 200 barrels		.02	3.53	10.60	32	2.30	
	1620	Barrels with flashers, 1 to 99 barrels		.03	5.25	15.80	47.50	3.40	
	1630	100 to 200 barrels		.03	4.20	12.60	38	2.75	
	1640	Barrels with steady burn type C lights		.04	7	21	63	4.50	
	1650	Illuminated board, trailer mounted, with generator		3.55	130	390	1,175	106.40	
	1670	Portable barricade, stock, with flashers, 1 to 6 units		.03	5.25	15.80	47.50	3.40	
	1680	25 to 50 units		.03	4.90	14.70	44	3.20	
	1685	Butt fusion machine, wheeled, 1.5 HP electric, 2" - 8" diameter pipe		2.62	167	500	1,500	120.95	
	1690	Tracked, 20 HP diesel, 4"-12" diameter pipe		11.10	490	1,465	4,400	381.80	
	1695	83 HP diesel, 8" - 24" diameter pipe		30.30	975	2,930	8,800	828.40	
	1700	Carts, brick, hand powered, 1000 lb. capacity		.44	72.50	218	655	47.10	
	1800	Gas engine, 1500 lb., 7-1/2' lift		4.10	115	345	1,025	101.80	
	1822	Dehumidifier, medium, 6 lb./hr., 150 CFM		.99	62	186	560	45.10	
	1824	Large, 18 lb./hr., 600 CFM		2.01	126	378	1,125	91.70	
	1830	Distributor, asphalt, trailer mounted, 2000 gal., 38 H.P. diesel		10.15	330	985	2,950	278.20	
	1840	3000 gal., 38 H.P. diesel		11.65	360	1,075	3,225	308.20	
	1850	Drill, rotary hammer, electric		1.00	26.50	80	240	24	
	1860	Carbide bit, 1-1/2" diameter, add to electric rotary hammer		.02	3.61	10.84	32.50	2.35	
	1865	Rotary, crawler, 250 H.P.		150.45	2,100	6,285	18,900	2,461	
	1870	Emulsion sprayer, 65 gal., 5 H.P. gas engine		2.90	98	294	880	82	
	1880	200 gal., 5 H.P. engine		7.85	163	490	1,475	160.80	
	1900	Floor auto-scrubbing machine, walk-behind, 28" path		4.52	293	880	2,650	212.15	
	1930	Floodlight, mercury vapor, or quartz, on tripod, 1000 watt		.41	19.35	58	174	14.90	
	1940	2000 watt		.78	39	117	350	29.65	
	1950	Floodlights, trailer mounted with generator, 1 - 300 watt light		3.65	71.50	215	645	72.20	
	1960	2 - 1000 watt lights		4.90	98.50	295	885	98.20	
	2000	4 - 300 watt lights		4.60	93.50	280	840	92.80	
	2005	Foam spray rig, incl. box trailer, compressor, generator, proportioner		36.53	490	1,465	4,400	585.25	
	2020	Forklift, straight mast, 12' lift, 5000 lb., 2 wheel drive, gas		25.35	203	610	1,825	324.80	
	2040	21' lift, 5000 lb., 4 wheel drive, diesel		21.15	242	725	2,175	314.20	
	2050	For rough terrain, 42' lift, 35' reach, 9000 lb., 110 H.P.		29.95	485	1,460	4,375	531.60	
	2060	For plant, 4 ton capacity, 80 H.P., 2 wheel drive, gas		15.50	93.50	280	840	180	
	2080	10 ton capacity, 120 H.P., 2 wheel drive, diesel		24.30	162	485	1,450	291.40	
	2100	Generator, electric, gas engine, 1.5 kW to 3 kW		3.95	11.35	34	102	38.40	
	2200	5 kW		5.10	15.35	46	138	50	
	2300	10 kW		9.65	36.50	110	330	99.20	
	2400	25 kW	↓	11.00	86.50	260	780	140	

01 54 33 | Equipment Rental

			UNIT	HOURLY OPER. COST	RENT PER DAY	RENT PER WEEK	RENT PER MONTH	EQUIPMENT COST/DAY	
40	2500	Diesel engine, 20 kW	Ea.	12.55	68.50	205	615	141.40	40
	2600	50 kW		24.35	103	310	930	256.80	
	2700	100 kW		45.15	130	390	1,175	439.20	
	2800	250 kW		88.85	237	710	2,125	852.80	
	2850	Hammer, hydraulic, for mounting on boom, to 500 ft lb.		2.55	75	225	675	65.40	
	2860	1000 ft lb.		4.35	127	380	1,150	110.80	
	2900	Heaters, space, oil or electric, 50 MBH		1.93	7.65	23	69	20.05	
	3000	100 MBH		3.61	10.65	32	96	35.30	
	3100	300 MBH		10.58	38.50	115	345	107.65	
	3150	500 MBH		17.40	45	135	405	166.20	
	3200	Hose, water, suction with coupling, 20′ long, 2″ diameter		.02	3	9	27	1.95	
	3210	3″ diameter		.03	4.33	13	39	2.85	
	3220	4″ diameter		.03	5	15	45	3.25	
	3230	6″ diameter		.11	17.65	53	159	11.50	
	3240	8″ diameter		.20	33.50	100	300	21.60	
	3250	Discharge hose with coupling, 50′ long, 2″ diameter		.01	1.33	4	12	.90	
	3260	3″ diameter		.01	2.33	7	21	1.50	
	3270	4″ diameter		.02	3.67	11	33	2.35	
	3280	6″ diameter		.06	9.35	28	84	6.10	
	3290	8″ diameter		.20	33.50	100	300	21.60	
	3295	Insulation blower		.77	6	18	54	9.75	
	3300	Ladders, extension type, 16′ to 36′ long		.14	22.50	68	204	14.70	
	3400	40′ to 60′ long		.19	31	93	279	20.10	
	3405	Lance for cutting concrete		2.52	89	267	800	73.55	
	3407	Lawn mower, rotary, 22″, 5 H.P.		2.10	59	177	530	52.20	
	3408	48″ self propelled		2.92	75	225	675	68.35	
	3410	Level, electronic, automatic, with tripod and leveling rod		1.49	99	297	890	71.30	
	3430	Laser type, for pipe and sewer line and grade		.73	48.50	145	435	34.85	
	3440	Rotating beam for interior control		.97	64	192	575	46.15	
	3460	Builder's optical transit, with tripod and rod		.10	16.65	50	150	10.80	
	3500	Light towers, towable, with diesel generator, 2000 watt		4.60	93.50	280	840	92.80	
	3600	4000 watt		4.90	98.50	295	885	98.20	
	3700	Mixer, powered, plaster and mortar, 6 C.F., 7 H.P.		2.90	19	57	171	34.60	
	3800	10 C.F., 9 H.P.		3.05	30.50	91	273	42.60	
	3850	Nailer, pneumatic		.47	31	93	279	22.35	
	3900	Paint sprayers complete, 8 CFM		.94	62.50	187	560	44.90	
	4000	17 CFM		1.57	104	313	940	75.15	
	4020	Pavers, bituminous, rubber tires, 8′ wide, 50 H.P., diesel		31.85	500	1,500	4,500	554.80	
	4030	10′ wide, 150 H.P.		107.95	1,825	5,460	16,400	1,956	
	4050	Crawler, 8′ wide, 100 H.P., diesel		90.45	1,825	5,500	16,500	1,824	
	4060	10′ wide, 150 H.P.		115.10	2,250	6,715	20,100	2,264	
	4070	Concrete paver, 12′ to 24′ wide, 250 H.P.		102.05	1,525	4,600	13,800	1,736	
	4080	Placer-spreader-trimmer, 24′ wide, 300 H.P.		147.05	2,475	7,405	22,200	2,657	
	4100	Pump, centrifugal gas pump, 1-1/2″ diam., 65 GPM		3.95	50	150	450	61.60	
	4200	2″ diameter, 130 GPM		5.30	55	165	495	75.40	
	4300	3″ diameter, 250 GPM		5.60	56.50	170	510	78.80	
	4400	6″ diameter, 1500 GPM		29.35	177	530	1,600	340.80	
	4500	Submersible electric pump, 1-1/4″ diameter, 55 GPM		.38	16.65	50	150	13.05	
	4600	1-1/2″ diameter, 83 GPM		.42	19	57	171	14.75	
	4700	2″ diameter, 120 GPM		1.43	23.50	71	213	25.65	
	4800	3″ diameter, 300 GPM		2.55	41.50	125	375	45.40	
	4900	4″ diameter, 560 GPM		10.65	160	480	1,450	181.20	
	5000	6″ diameter, 1590 GPM		15.75	215	645	1,925	255	
	5100	Diaphragm pump, gas, single, 1-1/2″ diameter		1.18	50.50	152	455	39.85	
	5200	2″ diameter		4.25	63.50	190	570	72	
	5300	3″ diameter		4.25	63.50	190	570	72	
	5400	Double, 4″ diameter		6.35	108	325	975	115.80	
	5450	Pressure washer 5 GPM, 3000 psi	↓	4.80	51.50	155	465	69.40	

01 54 33 | Equipment Rental

		01 54 33 | Equipment Rental	UNIT	HOURLY OPER. COST	RENT PER DAY	RENT PER WEEK	RENT PER MONTH	EQUIPMENT COST/DAY	
40	5460	7 GPM, 3000 psi	Ea.	6.25	60	180	540	86	40
	5500	Trash pump, self-priming, gas, 2" diameter		4.50	21.50	64	192	48.80	
	5600	Diesel, 4" diameter		9.30	90	270	810	128.40	
	5650	Diesel, 6" diameter		25.55	150	450	1,350	294.40	
	5655	Grout Pump		26.80	260	780	2,350	370.40	
	5700	Salamanders, L.P. gas fired, 100,000 Btu		3.81	13.65	41	123	38.70	
	5705	50,000 Btu		2.39	10.35	31	93	25.30	
	5720	Sandblaster, portable, open top, 3 C.F. capacity		.55	26.50	80	240	20.40	
	5730	6 C.F. capacity		.95	40	120	360	31.60	
	5740	Accessories for above		.13	21.50	64	192	13.85	
	5750	Sander, floor		.71	17	51	153	15.90	
	5760	Edger		.57	18.35	55	165	15.55	
	5800	Saw, chain, gas engine, 18" long		2.25	22	66	198	31.20	
	5900	Hydraulic powered, 36" long		.75	65	195	585	45	
	5950	60" long		.75	66.50	200	600	46	
	6000	Masonry, table mounted, 14" diameter, 5 H.P.		1.31	56.50	170	510	44.50	
	6050	Portable cut-off, 8 H.P.		2.40	33.50	100	300	39.20	
	6100	Circular, hand held, electric, 7-1/4" diameter		.22	4.67	14	42	4.55	
	6200	12" diameter		.22	8	24	72	6.55	
	6250	Wall saw, w/hydraulic power, 10 H.P.		10.45	60	180	540	119.60	
	6275	Shot blaster, walk-behind, 20" wide		4.80	293	880	2,650	214.40	
	6280	Sidewalk broom, walk-behind		2.32	70	210	630	60.55	
	6300	Steam cleaner, 100 gallons per hour		3.70	76.50	230	690	75.60	
	6310	200 gallons per hour		5.25	95	285	855	99	
	6340	Tar Kettle/Pot, 400 gallons		10.50	76.50	230	690	130	
	6350	Torch, cutting, acetylene-oxygen, 150' hose, excludes gases		.30	15	45	135	11.40	
	6360	Hourly operating cost includes tips and gas		19.00				152	
	6410	Toilet, portable chemical		.12	20.50	62	186	13.35	
	6420	Recycle flush type		.15	24.50	74	222	16	
	6430	Toilet, fresh water flush, garden hose,		.18	29.50	89	267	19.25	
	6440	Hoisted, non-flush, for high rise		.15	24.50	73	219	15.80	
	6465	Tractor, farm with attachment		21.90	277	830	2,500	341.20	
	6500	Trailers, platform, flush deck, 2 axle, 25 ton capacity		5.45	117	350	1,050	113.60	
	6600	40 ton capacity		7.00	160	480	1,450	152	
	6700	3 axle, 50 ton capacity		7.55	177	530	1,600	166.40	
	6800	75 ton capacity		9.35	232	695	2,075	213.80	
	6810	Trailer mounted cable reel for high voltage line work		5.35	255	764	2,300	195.60	
	6820	Trailer mounted cable tensioning rig		10.64	505	1,520	4,550	389.10	
	6830	Cable pulling rig		71.33	2,825	8,510	25,500	2,273	
	6900	Water tank trailer, engine driven discharge, 5000 gallons		6.95	143	430	1,300	141.60	
	6925	10,000 gallons		9.50	198	595	1,775	195	
	6950	Water truck, off highway, 6000 gallons		89.80	775	2,330	7,000	1,184	
	7010	Tram car for high voltage line work, powered, 2 conductor		6.27	138	415	1,250	133.15	
	7020	Transit (builder's level) with tripod		.10	16.65	50	150	10.80	
	7030	Trench box, 3000 lb., 6' x 8'		.56	93.50	280	840	60.50	
	7040	7200 lb., 6' x 20'		.78	130	389	1,175	84.05	
	7050	8000 lb., 8' x 16'		1.08	180	540	1,625	116.65	
	7060	9500 lb., 8' x 20'		1.21	201	603	1,800	130.30	
	7065	11,000 lb., 8' x 24'		1.27	211	633	1,900	136.75	
	7070	12,000 lb., 10' x 20'		1.36	227	680	2,050	146.90	
	7100	Truck, pickup, 3/4 ton, 2 wheel drive		15.25	58.50	175	525	157	
	7200	4 wheel drive		15.55	73.50	220	660	168.40	
	7250	Crew carrier, 9 passenger		21.60	86.50	260	780	224.80	
	7290	Flat bed truck, 20,000 lb. GVW		22.30	125	375	1,125	253.40	
	7300	Tractor, 4 x 2, 220 H.P.		31.05	197	590	1,775	366.40	
	7410	330 H.P.		46.05	272	815	2,450	531.40	
	7500	6 x 4, 380 H.P.		52.85	315	950	2,850	612.80	
	7600	450 H.P.	↓	64.05	385	1,150	3,450	742.40	

01 54 33 | Equipment Rental

		01 54 33 Equipment Rental	UNIT	HOURLY OPER. COST	RENT PER DAY	RENT PER WEEK	RENT PER MONTH	EQUIPMENT COST/DAY
40	7610	Tractor, with A frame, boom and winch, 225 H.P.	Ea.	33.95	272	815	2,450	434.60
	7620	Vacuum truck, hazardous material, 2500 gallons		12.00	292	875	2,625	271
	7625	5,000 gallons		14.27	410	1,230	3,700	360.15
	7650	Vacuum, HEPA, 16 gallon, wet/dry		.83	18	54	162	17.45
	7655	55 gallon, wet/dry		.77	27	81	243	22.35
	7660	Water tank, portable		.16	26.50	80	240	17.30
	7690	Sewer/catch basin vacuum, 14 C.Y., 1500 gallons		18.61	615	1,850	5,550	518.90
	7700	Welder, electric, 200 amp		3.64	16.35	49	147	38.90
	7800	300 amp		5.41	20	60	180	55.30
	7900	Gas engine, 200 amp		13.90	23.50	70	210	125.20
	8000	300 amp		15.90	24.50	74	222	142
	8100	Wheelbarrow, any size		.08	13	39	117	8.45
	8200	Wrecking ball, 4000 lb.	▼	2.30	68.50	205	615	59.40
50	0010	**HIGHWAY EQUIPMENT RENTAL** without operators R015433-10						
	0050	Asphalt batch plant, portable drum mixer, 100 ton/hr.	Ea.	79.15	1,450	4,320	13,000	1,497
	0060	200 ton/hr.		90.32	1,525	4,590	13,800	1,641
	0070	300 ton/hr.		107.39	1,800	5,390	16,200	1,937
	0100	Backhoe attachment, long stick, up to 185 H.P., 10.5′ long		.35	23	69	207	16.60
	0140	Up to 250 H.P., 12′ long		.37	24.50	74	222	17.75
	0180	Over 250 H.P., 15′ long		.52	34.50	103	310	24.75
	0200	Special dipper arm, up to 100 H.P., 32′ long		1.06	70.50	211	635	50.70
	0240	Over 100 H.P., 33′ long		1.32	87.50	263	790	63.15
	0280	Catch basin/sewer cleaning truck, 3 ton, 9 C.Y., 1000 gal.		46.25	390	1,170	3,500	604
	0300	Concrete batch plant, portable, electric, 200 C.Y./hr.		22.08	510	1,525	4,575	481.65
	0520	Grader/dozer attachment, ripper/scarifier, rear mounted, up to 135 H.P.		3.45	65	195	585	66.60
	0540	Up to 180 H.P.		4.05	81.50	245	735	81.40
	0580	Up to 250 H.P.		4.45	93.50	280	840	91.60
	0700	Pvmt. removal bucket, for hyd. excavator, up to 90 H.P.		1.90	51.50	155	465	46.20
	0740	Up to 200 H.P.		2.10	73.50	220	660	60.80
	0780	Over 200 H.P.		2.25	85	255	765	69
	0900	Aggregate spreader, self-propelled, 187 H.P.		53.75	685	2,055	6,175	841
	1000	Chemical spreader, 3 C.Y.		3.40	43.50	130	390	53.20
	1900	Hammermill, traveling, 250 H.P.		77.73	2,050	6,120	18,400	1,846
	2000	Horizontal borer, 3″ diameter, 13 H.P. gas driven		6.25	55	165	495	83
	2150	Horizontal directional drill, 20,000 lb. thrust, 78 H.P. diesel		31.55	675	2,030	6,100	658.40
	2160	30,000 lb. thrust, 115 H.P.		39.40	1,025	3,105	9,325	936.20
	2170	50,000 lb. thrust, 170 H.P.		56.45	1,325	3,970	11,900	1,246
	2190	Mud trailer for HDD, 1500 gallons, 175 H.P., gas		35.35	153	460	1,375	374.80
	2200	Hydromulcher, diesel, 3000 gallon, for truck mounting		24.05	247	740	2,225	340.40
	2300	Gas, 600 gallon		9.20	98.50	295	885	132.60
	2400	Joint & crack cleaner, walk behind, 25 H.P.		4.35	50	150	450	64.80
	2500	Filler, trailer mounted, 400 gallons, 20 H.P.		9.60	212	635	1,900	203.80
	3000	Paint striper, self-propelled, 40 gallon, 22 H.P.		7.60	157	470	1,400	154.80
	3100	120 gallon, 120 H.P.		24.80	400	1,195	3,575	437.40
	3200	Post drivers, 6″ I-Beam frame, for truck mounting		19.65	380	1,145	3,425	386.20
	3400	Road sweeper, self-propelled, 8′ wide, 90 H.P.		40.30	570	1,715	5,150	665.40
	3450	Road sweeper, vacuum assisted, 4 C.Y., 220 gallons		77.55	625	1,870	5,600	994.40
	4000	Road mixer, self-propelled, 130 H.P.		47.80	775	2,320	6,950	846.40
	4100	310 H.P.		84.40	2,100	6,335	19,000	1,942
	4220	Cold mix paver, incl. pug mill and bitumen tank, 165 H.P.		98.80	2,325	6,965	20,900	2,183
	4240	Pavement brush, towed		3.20	93.50	280	840	81.60
	4250	Paver, asphalt, wheel or crawler, 130 H.P., diesel		98.45	2,300	6,905	20,700	2,169
	4300	Paver, road widener, gas 1′ to 6′, 67 H.P.		49.40	905	2,715	8,150	938.20
	4400	Diesel, 2′ to 14′, 88 H.P.		63.70	1,050	3,170	9,500	1,144
	4600	Slipform pavers, curb and gutter, 2 track, 75 H.P.		52.00	805	2,410	7,225	898
	4700	4 track, 165 H.P.		41.65	730	2,185	6,550	770.20
	4800	Median barrier, 215 H.P.	▼	54.00	890	2,670	8,000	966

01 54 33 | Equipment Rental

			UNIT	HOURLY OPER. COST	RENT PER DAY	RENT PER WEEK	RENT PER MONTH	EQUIPMENT COST/DAY	
50	4901	Trailer, low bed, 75 ton capacity	Ea.	10.10	232	695	2,075	219.80	50
	5000	Road planer, walk behind, 10" cutting width, 10 H.P.		4.00	33.50	100	300	52	
	5100	Self-propelled, 12" cutting width, 64 H.P.		11.45	113	340	1,025	159.60	
	5120	Traffic line remover, metal ball blaster, truck mounted, 115 H.P.		48.75	750	2,245	6,725	839	
	5140	Grinder, truck mounted, 115 H.P.		55.45	810	2,435	7,300	930.60	
	5160	Walk-behind, 11 H.P.		4.55	53.50	160	480	68.40	
	5200	Pavement profiler, 4′ to 6′ wide, 450 H.P.		258.10	3,350	10,045	30,100	4,074	
	5300	8′ to 10′ wide, 750 H.P.		405.10	4,400	13,180	39,500	5,877	
	5400	Roadway plate, steel, 1" x 8′ x 20′		.08	13	39	117	8.45	
	5600	Stabilizer, self-propelled, 150 H.P.		50.10	610	1,835	5,500	767.80	
	5700	310 H.P.		100.25	1,700	5,125	15,400	1,827	
	5800	Striper, truck mounted, 120 gallon paint, 460 H.P.		71.05	485	1,460	4,375	860.40	
	5900	Thermal paint heating kettle, 115 gallons		5.95	25.50	77	231	63	
	6000	Tar kettle, 330 gallon, trailer mounted		8.70	58.50	175	525	104.60	
	7000	Tunnel locomotive, diesel, 8 to 12 ton		32.40	585	1,760	5,275	611.20	
	7005	Electric, 10 ton		26.55	670	2,010	6,025	614.40	
	7010	Muck cars, 1/2 C.Y. capacity		2.05	24.50	74	222	31.20	
	7020	1 C.Y. capacity		2.30	32.50	98	294	38	
	7030	2 C.Y. capacity		2.45	36.50	110	330	41.60	
	7040	Side dump, 2 C.Y. capacity		2.65	45	135	405	48.20	
	7050	3 C.Y. capacity		3.55	51.50	155	465	59.40	
	7060	5 C.Y. capacity		5.10	65	195	585	79.80	
	7100	Ventilating blower for tunnel, 7-1/2 H.P.		2.05	51.50	155	465	47.40	
	7110	10 H.P.		2.22	53.50	160	480	49.75	
	7120	20 H.P.		3.43	69.50	208	625	69.05	
	7140	40 H.P.		5.59	98.50	295	885	103.70	
	7160	60 H.P.		8.54	152	455	1,375	159.30	
	7175	75 H.P.		11.25	207	620	1,850	214	
	7180	200 H.P.		22.56	305	910	2,725	362.50	
	7800	Windrow loader, elevating	↓	55.95	1,325	3,960	11,900	1,240	
60	0010	**LIFTING AND HOISTING EQUIPMENT RENTAL** without operators (R015433-10)							60
	0120	Aerial lift truck, 2 person, to 80′	Ea.	26.75	715	2,145	6,425	643	
	0140	Boom work platform, 40′ snorkel (R015433-15)		16.10	277	830	2,500	294.80	
	0150	Crane, flatbed mounted, 3 ton capacity		16.55	193	580	1,750	248.40	
	0200	Crane, climbing, 106′ jib, 6000 lb. capacity, 410 fpm (R312316-45)		37.76	1,625	4,860	14,600	1,274	
	0300	101′ jib, 10,250 lb. capacity, 270 fpm		44.26	2,050	6,160	18,500	1,586	
	0500	Tower, static, 130′ high, 106′ jib, 6200 lb. capacity at 400 fpm		41.56	1,875	5,620	16,900	1,456	
	0600	Crawler mounted, lattice boom, 1/2 C.Y., 15 tons at 12′ radius		37.52	640	1,920	5,750	684.15	
	0700	3/4 C.Y., 20 tons at 12′ radius		50.03	800	2,400	7,200	880.25	
	0800	1 C.Y., 25 tons at 12′ radius		66.70	1,075	3,195	9,575	1,173	
	0900	1-1/2 C.Y., 40 tons at 12′ radius		66.70	1,075	3,225	9,675	1,179	
	1000	2 C.Y., 50 tons at 12′ radius		70.70	1,250	3,765	11,300	1,319	
	1100	3 C.Y., 75 tons at 12′ radius		75.65	1,475	4,435	13,300	1,492	
	1200	100 ton capacity, 60′ boom		85.65	1,700	5,090	15,300	1,703	
	1300	165 ton capacity, 60′ boom		109.15	2,000	5,980	17,900	2,069	
	1400	200 ton capacity, 70′ boom		132.10	2,500	7,480	22,400	2,553	
	1500	350 ton capacity, 80′ boom		184.60	3,725	11,210	33,600	3,719	
	1600	Truck mounted, lattice boom, 6 x 4, 20 tons at 10′ radius		37.31	1,100	3,310	9,925	960.50	
	1700	25 tons at 10′ radius		40.35	1,200	3,610	10,800	1,045	
	1800	8 x 4, 30 tons at 10′ radius		43.85	1,275	3,840	11,500	1,119	
	1900	40 tons at 12′ radius		46.92	1,325	4,010	12,000	1,177	
	2000	60 tons at 15′ radius		53.24	1,425	4,240	12,700	1,274	
	2050	82 tons at 15′ radius		60.06	1,525	4,540	13,600	1,388	
	2100	90 tons at 15′ radius		67.57	1,650	4,940	14,800	1,529	
	2200	115 tons at 15′ radius		76.34	1,850	5,520	16,600	1,715	
	2300	150 tons at 18′ radius		84.40	1,950	5,815	17,400	1,838	
	2350	165 tons at 18′ radius		90.25	2,050	6,160	18,500	1,954	
	2400	Truck mounted, hydraulic, 12 ton capacity	↓	43.00	530	1,595	4,775	663	

01 54 33 | Equipment Rental

			UNIT	HOURLY OPER. COST	RENT PER DAY	RENT PER WEEK	RENT PER MONTH	EQUIPMENT COST/DAY	
60	2500	25 ton capacity	Ea.	45.35	640	1,915	5,750	745.80	60
	2550	33 ton capacity		45.90	655	1,960	5,875	759.20	
	2560	40 ton capacity		59.35	765	2,290	6,875	932.80	
	2600	55 ton capacity		77.50	860	2,585	7,750	1,137	
	2700	80 ton capacity		101.35	1,400	4,205	12,600	1,652	
	2720	100 ton capacity		94.95	1,450	4,325	13,000	1,625	
	2740	120 ton capacity		109.95	1,550	4,665	14,000	1,813	
	2760	150 ton capacity		128.55	2,050	6,135	18,400	2,255	
	2800	Self-propelled, 4 x 4, with telescoping boom, 5 ton		17.75	230	690	2,075	280	
	2900	12-1/2 ton capacity		32.65	365	1,100	3,300	481.20	
	3000	15 ton capacity		33.35	385	1,160	3,475	498.80	
	3050	20 ton capacity		36.30	450	1,350	4,050	560.40	
	3100	25 ton capacity		37.90	505	1,520	4,550	607.20	
	3150	40 ton capacity		46.50	570	1,705	5,125	713	
	3200	Derricks, guy, 20 ton capacity, 60′ boom, 75′ mast		27.18	395	1,190	3,575	455.45	
	3300	100′ boom, 115′ mast		42.63	685	2,050	6,150	751.05	
	3400	Stiffleg, 20 ton capacity, 70′ boom, 37′ mast		29.70	515	1,550	4,650	547.60	
	3500	100′ boom, 47′ mast		45.64	825	2,480	7,450	861.10	
	3550	Helicopter, small, lift to 1250 lb. maximum, w/pilot		100.87	3,200	9,610	28,800	2,729	
	3600	Hoists, chain type, overhead, manual, 3/4 ton		.10	.33	1	3	1	
	3900	10 ton		.75	6	18	54	9.60	
	4000	Hoist and tower, 5000 lb. cap., portable electric, 40′ high		4.77	228	685	2,050	175.15	
	4100	For each added 10′ section, add		.11	18	54	162	11.70	
	4200	Hoist and single tubular tower, 5000 lb. electric, 100′ high		6.47	320	957	2,875	243.15	
	4300	For each added 6′-6″ section, add		.19	31	93	279	20.10	
	4400	Hoist and double tubular tower, 5000 lb., 100′ high		6.95	350	1,054	3,150	266.40	
	4500	For each added 6′-6″ section, add		.21	34.50	103	310	22.30	
	4550	Hoist and tower, mast type, 6000 lb., 100′ high		7.48	365	1,093	3,275	278.45	
	4570	For each added 10′ section, add		.13	21.50	64	192	13.85	
	4600	Hoist and tower, personnel, electric, 2000 lb., 100′ @ 125 fpm		15.90	970	2,910	8,725	709.20	
	4700	3000 lb., 100′ @ 200 fpm		18.13	1,100	3,290	9,875	803.05	
	4800	3000 lb., 150′ @ 300 fpm		20.13	1,225	3,690	11,100	899.05	
	4900	4000 lb., 100′ @ 300 fpm		20.82	1,250	3,760	11,300	918.55	
	5000	6000 lb., 100′ @ 275 fpm		22.44	1,325	3,950	11,900	969.50	
	5100	For added heights up to 500′, add	L.F.	.01	1.67	5	15	1.10	
	5200	Jacks, hydraulic, 20 ton	Ea.	.05	2	6	18	1.60	
	5500	100 ton		.40	11.65	35	105	10.20	
	6100	Jacks, hydraulic, climbing w/50′ jackrods, control console, 30 ton cap.		1.97	131	394	1,175	94.55	
	6150	For each added 10′ jackrod section, add		.05	3.33	10	30	2.40	
	6300	50 ton capacity		3.17	211	633	1,900	151.95	
	6350	For each added 10′ jackrod section, add		.06	4	12	36	2.90	
	6500	125 ton capacity		8.30	555	1,660	4,975	398.40	
	6550	For each added 10′ jackrod section, add		.57	37.50	113	340	27.15	
	6600	Cable jack, 10 ton capacity with 200′ cable		1.65	110	329	985	79	
	6650	For each added 50′ of cable, add		.20	13	39	117	9.40	
70	0010	**WELLPOINT EQUIPMENT RENTAL** without operators R015433-10							70
	0020	Based on 2 months rental							
	0100	Combination jetting & wellpoint pump, 60 H.P. diesel	Ea.	18.14	325	976	2,925	340.30	
	0200	High pressure gas jet pump, 200 H.P., 300 psi	″	43.54	278	834	2,500	515.10	
	0300	Discharge pipe, 8″ diameter	L.F.	.01	.53	1.59	4.77	.40	
	0350	12″ diameter		.01	.78	2.33	7	.55	
	0400	Header pipe, flows up to 150 GPM, 4″ diameter		.01	.48	1.44	4.32	.35	
	0500	400 GPM, 6″ diameter		.01	.56	1.69	5.05	.40	
	0600	800 GPM, 8″ diameter		.01	.78	2.33	7	.55	
	0700	1500 GPM, 10″ diameter		.01	.82	2.46	7.40	.55	
	0800	2500 GPM, 12″ diameter		.02	1.55	4.65	13.95	1.10	
	0900	4500 GPM, 16″ diameter		.03	1.98	5.95	17.85	1.45	

01 54 33 | Equipment Rental

		01 54 33 | Equipment Rental	UNIT	HOURLY OPER. COST	RENT PER DAY	RENT PER WEEK	RENT PER MONTH	EQUIPMENT COST/DAY	
70	0950	For quick coupling aluminum and plastic pipe, add	L.F.	.03	2.05	6.16	18.50	1.45	70
	1100	Wellpoint, 25′ long, with fittings & riser pipe, 1-1/2″ or 2″ diameter		.06	4.10	12.29	37	2.95	
	1200	Wellpoint pump, diesel powered, 4″ suction, 20 H.P.		7.71	188	563	1,700	174.30	
	1300	6″ suction, 30 H.P.		10.54	233	698	2,100	223.90	
	1400	8″ suction, 40 H.P.		14.24	320	957	2,875	305.30	
	1500	10″ suction, 75 H.P.		21.97	375	1,119	3,350	399.55	
	1600	12″ suction, 100 H.P.		31.24	590	1,770	5,300	603.90	
	1700	12″ suction, 175 H.P.	↓	46.78	655	1,970	5,900	768.25	
80	0010	**MARINE EQUIPMENT RENTAL** without operators R015433-10							80
	0200	Barge, 400 Ton, 30′ wide x 90′ long	Ea.	17.50	1,075	3,205	9,625	781	
	0240	800 Ton, 45′ wide x 90′ long		21.20	1,300	3,900	11,700	949.60	
	2000	Tugboat, diesel, 100 H.P.		38.95	217	650	1,950	441.60	
	2040	250 H.P.		81.55	395	1,185	3,550	889.40	
	2080	380 H.P.		160.00	1,175	3,545	10,600	1,989	
	3000	Small work boat, gas, 16-foot, 50 H.P.		19.15	60	180	540	189.20	
	4000	Large, diesel, 48-foot, 200 H.P.	↓	90.80	1,250	3,745	11,200	1,475	

Crew No.	Bare Costs		Incl. Subs O&P		Cost Per Labor-Hour	
Crew A-1	**Hr.**	**Daily**	**Hr.**	**Daily**	**Bare Costs**	**Incl. O&P**
1 Building Laborer	$36.65	$293.20	$56.55	$452.40	$36.65	$56.55
1 Concrete Saw, Gas Manual		84.00		92.40	10.50	11.55
8 L.H., Daily Totals		$377.20		$544.80	$47.15	$68.10
Crew A-1A	**Hr.**	**Daily**	**Hr.**	**Daily**	**Bare Costs**	**Incl. O&P**
1 Skilled Worker	$47.30	$378.40	$73.25	$586.00	$47.30	$73.25
1 Shot Blaster, 20"		214.40		235.84	26.80	29.48
8 L.H., Daily Totals		$592.80		$821.84	$74.10	$102.73
Crew A-1B	**Hr.**	**Daily**	**Hr.**	**Daily**	**Bare Costs**	**Incl. O&P**
1 Building Laborer	$36.65	$293.20	$56.55	$452.40	$36.65	$56.55
1 Concrete Saw		175.20		192.72	21.90	24.09
8 L.H., Daily Totals		$468.40		$645.12	$58.55	$80.64
Crew A-1C	**Hr.**	**Daily**	**Hr.**	**Daily**	**Bare Costs**	**Incl. O&P**
1 Building Laborer	$36.65	$293.20	$56.55	$452.40	$36.65	$56.55
1 Chain Saw, Gas, 18"		31.20		34.32	3.90	4.29
8 L.H., Daily Totals		$324.40		$486.72	$40.55	$60.84
Crew A-1D	**Hr.**	**Daily**	**Hr.**	**Daily**	**Bare Costs**	**Incl. O&P**
1 Building Laborer	$36.65	$293.20	$56.55	$452.40	$36.65	$56.55
1 Vibrating Plate, Gas, 18"		35.80		39.38	4.47	4.92
8 L.H., Daily Totals		$329.00		$491.78	$41.13	$61.47
Crew A-1E	**Hr.**	**Daily**	**Hr.**	**Daily**	**Bare Costs**	**Incl. O&P**
1 Building Laborer	$36.65	$293.20	$56.55	$452.40	$36.65	$56.55
1 Vibrating Plate, Gas, 21"		46.00		50.60	5.75	6.33
8 L.H., Daily Totals		$339.20		$503.00	$42.40	$62.88
Crew A-1F	**Hr.**	**Daily**	**Hr.**	**Daily**	**Bare Costs**	**Incl. O&P**
1 Building Laborer	$36.65	$293.20	$56.55	$452.40	$36.65	$56.55
1 Rammer/Tamper, Gas, 8"		49.60		54.56	6.20	6.82
8 L.H., Daily Totals		$342.80		$506.96	$42.85	$63.37
Crew A-1G	**Hr.**	**Daily**	**Hr.**	**Daily**	**Bare Costs**	**Incl. O&P**
1 Building Laborer	$36.65	$293.20	$56.55	$452.40	$36.65	$56.55
1 Rammer/Tamper, Gas, 15"		56.00		61.60	7.00	7.70
8 L.H., Daily Totals		$349.20		$514.00	$43.65	$64.25
Crew A-1H	**Hr.**	**Daily**	**Hr.**	**Daily**	**Bare Costs**	**Incl. O&P**
1 Building Laborer	$36.65	$293.20	$56.55	$452.40	$36.65	$56.55
1 Exterior Steam Cleaner		75.60		83.16	9.45	10.40
8 L.H., Daily Totals		$368.80		$535.56	$46.10	$66.94
Crew A-1J	**Hr.**	**Daily**	**Hr.**	**Daily**	**Bare Costs**	**Incl. O&P**
1 Building Laborer	$36.65	$293.20	$56.55	$452.40	$36.65	$56.55
1 Cultivator, Walk-Behind, 5 H.P.		57.45		63.20	7.18	7.90
8 L.H., Daily Totals		$350.65		$515.60	$43.83	$64.45
Crew A-1K	**Hr.**	**Daily**	**Hr.**	**Daily**	**Bare Costs**	**Incl. O&P**
1 Building Laborer	$36.65	$293.20	$56.55	$452.40	$36.65	$56.55
1 Cultivator, Walk-Behind, 8 H.P.		81.40		89.54	10.18	11.19
8 L.H., Daily Totals		$374.60		$541.94	$46.83	$67.74
Crew A-1M	**Hr.**	**Daily**	**Hr.**	**Daily**	**Bare Costs**	**Incl. O&P**
1 Building Laborer	$36.65	$293.20	$56.55	$452.40	$36.65	$56.55
1 Snow Blower, Walk-Behind		60.55		66.61	7.57	8.33
8 L.H., Daily Totals		$353.75		$519.01	$44.22	$64.88

Crew No.	Bare Costs		Incl. Subs O&P		Cost Per Labor-Hour	
Crew A-2	**Hr.**	**Daily**	**Hr.**	**Daily**	**Bare Costs**	**Incl. O&P**
2 Laborers	$36.65	$586.40	$56.55	$904.80	$36.60	$56.27
1 Truck Driver (light)	36.50	292.00	55.70	445.60		
1 Flatbed Truck, Gas, 1.5 Ton		270.20		297.22	11.26	12.38
24 L.H., Daily Totals		$1148.60		$1647.62	$47.86	$68.65
Crew A-2A	**Hr.**	**Daily**	**Hr.**	**Daily**	**Bare Costs**	**Incl. O&P**
2 Laborers	$36.65	$586.40	$56.55	$904.80	$36.60	$56.27
1 Truck Driver (light)	36.50	292.00	55.70	445.60		
1 Flatbed Truck, Gas, 1.5 Ton		270.20		297.22		
1 Concrete Saw		175.20		192.72	18.56	20.41
24 L.H., Daily Totals		$1323.80		$1840.34	$55.16	$76.68
Crew A-2B	**Hr.**	**Daily**	**Hr.**	**Daily**	**Bare Costs**	**Incl. O&P**
1 Truck Driver (light)	$36.50	$292.00	$55.70	$445.60	$36.50	$55.70
1 Flatbed Truck, Gas, 1.5 Ton		270.20		297.22	33.77	37.15
8 L.H., Daily Totals		$562.20		$742.82	$70.28	$92.85
Crew A-3A	**Hr.**	**Daily**	**Hr.**	**Daily**	**Bare Costs**	**Incl. O&P**
1 Truck Driver (light)	$36.50	$292.00	$55.70	$445.60	$36.50	$55.70
1 Pickup Truck, 4 x 4, 3/4 Ton		168.40		185.24	21.05	23.16
8 L.H., Daily Totals		$460.40		$630.84	$57.55	$78.86
Crew A-3B	**Hr.**	**Daily**	**Hr.**	**Daily**	**Bare Costs**	**Incl. O&P**
1 Equip. Oper. (medium)	$48.90	$391.20	$74.15	$593.20	$43.23	$65.72
1 Truck Driver (heavy)	37.55	300.40	57.30	458.40		
1 Dump Truck, 12 C.Y., 400 H.P.		692.20		761.42		
1 F.E. Loader, W.M.,2.5 C.Y.		522.80		575.08	75.94	83.53
16 L.H., Daily Totals		$1906.60		$2388.10	$119.16	$149.26
Crew A-3C	**Hr.**	**Daily**	**Hr.**	**Daily**	**Bare Costs**	**Incl. O&P**
1 Equip. Oper. (light)	$47.05	$376.40	$71.35	$570.80	$47.05	$71.35
1 Loader, Skid Steer, 78 H.P.		308.80		339.68	38.60	42.46
8 L.H., Daily Totals		$685.20		$910.48	$85.65	$113.81
Crew A-3D	**Hr.**	**Daily**	**Hr.**	**Daily**	**Bare Costs**	**Incl. O&P**
1 Truck Driver (light)	$36.50	$292.00	$55.70	$445.60	$36.50	$55.70
1 Pickup Truck, 4 x 4, 3/4 Ton		168.40		185.24		
1 Flatbed Trailer, 25 Ton		113.60		124.96	35.25	38.77
8 L.H., Daily Totals		$574.00		$755.80	$71.75	$94.47
Crew A-3E	**Hr.**	**Daily**	**Hr.**	**Daily**	**Bare Costs**	**Incl. O&P**
1 Equip. Oper. (crane)	$50.25	$402.00	$76.20	$609.60	$43.90	$66.75
1 Truck Driver (heavy)	37.55	300.40	57.30	458.40		
1 Pickup Truck, 4 x 4, 3/4 Ton		168.40		185.24	10.53	11.58
16 L.H., Daily Totals		$870.80		$1253.24	$54.42	$78.33
Crew A-3F	**Hr.**	**Daily**	**Hr.**	**Daily**	**Bare Costs**	**Incl. O&P**
1 Equip. Oper. (crane)	$50.25	$402.00	$76.20	$609.60	$43.90	$66.75
1 Truck Driver (heavy)	37.55	300.40	57.30	458.40		
1 Pickup Truck, 4 x 4, 3/4 Ton		168.40		185.24		
1 Truck Tractor, 6x4, 380 H.P.		612.80		674.08		
1 Lowbed Trailer, 75 Ton		219.80		241.78	62.56	68.82
16 L.H., Daily Totals		$1703.40		$2169.10	$106.46	$135.57

Crew No.	Bare Costs		Incl. Subs O&P		Cost Per Labor-Hour	
Crew A-3G	**Hr.**	**Daily**	**Hr.**	**Daily**	**Bare Costs**	**Incl. O&P**
1 Equip. Oper. (crane)	$50.25	$402.00	$76.20	$609.60	$43.90	$66.75
1 Truck Driver (heavy)	37.55	300.40	57.30	458.40		
1 Pickup Truck, 4 x 4, 3/4 Ton		168.40		185.24		
1 Truck Tractor, 6x4, 450 H.P.		742.40		816.64		
1 Lowbed Trailer, 75 Ton		219.80		241.78	70.66	77.73
16 L.H., Daily Totals		$1833.00		$2311.66	$114.56	$144.48
Crew A-3H	**Hr.**	**Daily**	**Hr.**	**Daily**	**Bare Costs**	**Incl. O&P**
1 Equip. Oper. (crane)	$50.25	$402.00	$76.20	$609.60	$50.25	$76.20
1 Hyd. Crane, 12 Ton (Daily)		874.00		961.40	109.25	120.18
8 L.H., Daily Totals		$1276.00		$1571.00	$159.50	$196.38
Crew A-3I	**Hr.**	**Daily**	**Hr.**	**Daily**	**Bare Costs**	**Incl. O&P**
1 Equip. Oper. (crane)	$50.25	$402.00	$76.20	$609.60	$50.25	$76.20
1 Hyd. Crane, 25 Ton (Daily)		1003.00		1103.30	125.38	137.91
8 L.H., Daily Totals		$1405.00		$1712.90	$175.63	$214.11
Crew A-3J	**Hr.**	**Daily**	**Hr.**	**Daily**	**Bare Costs**	**Incl. O&P**
1 Equip. Oper. (crane)	$50.25	$402.00	$76.20	$609.60	$50.25	$76.20
1 Hyd. Crane, 40 Ton (Daily)		1240.00		1364.00	155.00	170.50
8 L.H., Daily Totals		$1642.00		$1973.60	$205.25	$246.70
Crew A-3K	**Hr.**	**Daily**	**Hr.**	**Daily**	**Bare Costs**	**Incl. O&P**
1 Equip. Oper. (crane)	$50.25	$402.00	$76.20	$609.60	$46.90	$71.10
1 Equip. Oper. (oiler)	43.55	348.40	66.00	528.00		
1 Hyd. Crane, 55 Ton (Daily)		1480.00		1628.00		
1 P/U Truck, 3/4 Ton (Daily)		182.00		200.20	103.88	114.26
16 L.H., Daily Totals		$2412.40		$2965.80	$150.78	$185.36
Crew A-3L	**Hr.**	**Daily**	**Hr.**	**Daily**	**Bare Costs**	**Incl. O&P**
1 Equip. Oper. (crane)	$50.25	$402.00	$76.20	$609.60	$46.90	$71.10
1 Equip. Oper. (oiler)	43.55	348.40	66.00	528.00		
1 Hyd. Crane, 80 Ton (Daily)		2211.00		2432.10		
1 P/U Truck, 3/4 Ton (Daily)		182.00		200.20	149.56	164.52
16 L.H., Daily Totals		$3143.40		$3769.90	$196.46	$235.62
Crew A-3M	**Hr.**	**Daily**	**Hr.**	**Daily**	**Bare Costs**	**Incl. O&P**
1 Equip. Oper. (crane)	$50.25	$402.00	$76.20	$609.60	$46.90	$71.10
1 Equip. Oper. (oiler)	43.55	348.40	66.00	528.00		
1 Hyd. Crane, 100 Ton (Daily)		2200.00		2420.00		
1 P/U Truck, 3/4 Ton (Daily)		182.00		200.20	148.88	163.76
16 L.H., Daily Totals		$3132.40		$3757.80	$195.78	$234.86
Crew A-3N	**Hr.**	**Daily**	**Hr.**	**Daily**	**Bare Costs**	**Incl. O&P**
1 Equip. Oper. (crane)	$50.25	$402.00	$76.20	$609.60	$50.25	$76.20
1 Tower Cane (monthly)		1100.00		1210.00	137.50	151.25
8 L.H., Daily Totals		$1502.00		$1819.60	$187.75	$227.45
Crew A-3P	**Hr.**	**Daily**	**Hr.**	**Daily**	**Bare Costs**	**Incl. O&P**
1 Equip. Oper. (light)	$47.05	$376.40	$71.35	$570.80	$47.05	$71.35
1 A.T. Forklift, 42' lift		531.60		584.76	66.45	73.09
8 L.H., Daily Totals		$908.00		$1155.56	$113.50	$144.44
Crew A-4	**Hr.**	**Daily**	**Hr.**	**Daily**	**Bare Costs**	**Incl. O&P**
2 Carpenters	$45.85	$733.60	$70.75	$1132.00	$43.75	$67.00
1 Painter, Ordinary	39.55	316.40	59.50	476.00		
24 L.H., Daily Totals		$1050.00		$1608.00	$43.75	$67.00

Crew No.	Bare Costs		Incl. Subs O&P		Cost Per Labor-Hour	
Crew A-5	**Hr.**	**Daily**	**Hr.**	**Daily**	**Bare Costs**	**Incl. O&P**
2 Laborers	$36.65	$586.40	$56.55	$904.80	$36.63	$56.46
.25 Truck Driver (light)	36.50	73.00	55.70	111.40		
.25 Flatbed Truck, Gas, 1.5 Ton		67.55		74.31	3.75	4.13
18 L.H., Daily Totals		$726.95		$1090.51	$40.39	$60.58
Crew A-6	**Hr.**	**Daily**	**Hr.**	**Daily**	**Bare Costs**	**Incl. O&P**
1 Instrument Man	$47.30	$378.40	$73.25	$586.00	$45.67	$70.13
1 Rodman/Chainman	44.05	352.40	67.00	536.00		
1 Level, Electronic		71.30		78.43	4.46	4.90
16 L.H., Daily Totals		$802.10		$1200.43	$50.13	$75.03
Crew A-7	**Hr.**	**Daily**	**Hr.**	**Daily**	**Bare Costs**	**Incl. O&P**
1 Chief of Party	$59.90	$479.20	$92.15	$737.20	$50.42	$77.47
1 Instrument Man	47.30	378.40	73.25	586.00		
1 Rodman/Chainman	44.05	352.40	67.00	536.00		
1 Level, Electronic		71.30		78.43	2.97	3.27
24 L.H., Daily Totals		$1281.30		$1937.63	$53.39	$80.73
Crew A-8	**Hr.**	**Daily**	**Hr.**	**Daily**	**Bare Costs**	**Incl. O&P**
1 Chief of Party	$59.90	$479.20	$92.15	$737.20	$48.83	$74.85
1 Instrument Man	47.30	378.40	73.25	586.00		
2 Rodmen/Chainmen	44.05	704.80	67.00	1072.00		
1 Level, Electronic		71.30		78.43	2.23	2.45
32 L.H., Daily Totals		$1633.70		$2473.63	$51.05	$77.30
Crew A-9	**Hr.**	**Daily**	**Hr.**	**Daily**	**Bare Costs**	**Incl. O&P**
1 Asbestos Foreman	$51.65	$413.20	$80.35	$642.80	$51.21	$79.69
7 Asbestos Workers	51.15	2864.40	79.60	4457.60		
64 L.H., Daily Totals		$3277.60		$5100.40	$51.21	$79.69
Crew A-10A	**Hr.**	**Daily**	**Hr.**	**Daily**	**Bare Costs**	**Incl. O&P**
1 Asbestos Foreman	$51.65	$413.20	$80.35	$642.80	$51.32	$79.85
2 Asbestos Workers	51.15	818.40	79.60	1273.60		
24 L.H., Daily Totals		$1231.60		$1916.40	$51.32	$79.85
Crew A-10B	**Hr.**	**Daily**	**Hr.**	**Daily**	**Bare Costs**	**Incl. O&P**
1 Asbestos Foreman	$51.65	$413.20	$80.35	$642.80	$51.27	$79.79
3 Asbestos Workers	51.15	1227.60	79.60	1910.40		
32 L.H., Daily Totals		$1640.80		$2553.20	$51.27	$79.79
Crew A-10C	**Hr.**	**Daily**	**Hr.**	**Daily**	**Bare Costs**	**Incl. O&P**
3 Asbestos Workers	$51.15	$1227.60	$79.60	$1910.40	$51.15	$79.60
1 Flatbed Truck, Gas, 1.5 Ton		270.20		297.22	11.26	12.38
24 L.H., Daily Totals		$1497.80		$2207.62	$62.41	$91.98
Crew A-10D	**Hr.**	**Daily**	**Hr.**	**Daily**	**Bare Costs**	**Incl. O&P**
2 Asbestos Workers	$51.15	$818.40	$79.60	$1273.60	$49.02	$75.35
1 Equip. Oper. (crane)	50.25	402.00	76.20	609.60		
1 Equip. Oper. (oiler)	43.55	348.40	66.00	528.00		
1 Hydraulic Crane, 33 Ton		759.20		835.12	23.73	26.10
32 L.H., Daily Totals		$2328.00		$3246.32	$72.75	$101.45
Crew A-11	**Hr.**	**Daily**	**Hr.**	**Daily**	**Bare Costs**	**Incl. O&P**
1 Asbestos Foreman	$51.65	$413.20	$80.35	$642.80	$51.21	$79.69
7 Asbestos Workers	51.15	2864.40	79.60	4457.60		
2 Chip. Hammers, 12 Lb., Elec.		40.00		44.00	.63	.69
64 L.H., Daily Totals		$3317.60		$5144.40	$51.84	$80.38

Crews

Crew No.	Bare Costs		Incl. Subs O&P		Cost Per Labor-Hour	

Crew A-12	Hr.	Daily	Hr.	Daily	Bare Costs	Incl. O&P
1 Asbestos Foreman	$51.65	$413.20	$80.35	$642.80	$51.21	$79.69
7 Asbestos Workers	51.15	2864.40	79.60	4457.60		
1 Trk-Mtd Vac, 14 CY, 1500 Gal.		518.90		570.79		
1 Flatbed Truck, 20,000 GVW		253.40		278.74	12.07	13.27
64 L.H., Daily Totals		$4049.90		$5949.93	$63.28	$92.97

Crew A-13	Hr.	Daily	Hr.	Daily	Bare Costs	Incl. O&P
1 Equip. Oper. (light)	$47.05	$376.40	$71.35	$570.80	$47.05	$71.35
1 Trk-Mtd Vac, 14 CY, 1500 Gal.		518.90		570.79		
1 Flatbed Truck, 20,000 GVW		253.40		278.74	96.54	106.19
8 L.H., Daily Totals		$1148.70		$1420.33	$143.59	$177.54

Crew B-1	Hr.	Daily	Hr.	Daily	Bare Costs	Incl. O&P
1 Labor Foreman (outside)	$38.65	$309.20	$59.65	$477.20	$37.32	$57.58
2 Laborers	36.65	586.40	56.55	904.80		
24 L.H., Daily Totals		$895.60		$1382.00	$37.32	$57.58

Crew B-1A	Hr.	Daily	Hr.	Daily	Bare Costs	Incl. O&P
1 Labor Foreman (outside)	$38.65	$309.20	$59.65	$477.20	$37.32	$57.58
2 Laborers	36.65	586.40	56.55	904.80		
2 Cutting Torches		22.80		25.08		
2 Sets of Gases		304.00		334.40	13.62	14.98
24 L.H., Daily Totals		$1222.40		$1741.48	$50.93	$72.56

Crew B-1B	Hr.	Daily	Hr.	Daily	Bare Costs	Incl. O&P
1 Labor Foreman (outside)	$38.65	$309.20	$59.65	$477.20	$40.55	$62.24
2 Laborers	36.65	586.40	56.55	904.80		
1 Equip. Oper. (crane)	50.25	402.00	76.20	609.60		
2 Cutting Torches		22.80		25.08		
2 Sets of Gases		304.00		334.40		
1 Hyd. Crane, 12 Ton		663.00		729.30	30.93	34.02
32 L.H., Daily Totals		$2287.40		$3080.38	$71.48	$96.26

Crew B-1C	Hr.	Daily	Hr.	Daily	Bare Costs	Incl. O&P
1 Labor Foreman (outside)	$38.65	$309.20	$59.65	$477.20	$37.32	$57.58
2 Laborers	36.65	586.40	56.55	904.80		
1 Aerial Lift Truck, 60' Boom		456.00		501.60	19.00	20.90
24 L.H., Daily Totals		$1351.60		$1883.60	$56.32	$78.48

Crew B-1D	Hr.	Daily	Hr.	Daily	Bare Costs	Incl. O&P
2 Laborers	$36.65	$586.40	$56.55	$904.80	$36.65	$56.55
1 Small Work Boat, Gas, 50 H.P.		189.20		208.12		
1 Pressure Washer, 7 GPM		86.00		94.60	17.20	18.92
16 L.H., Daily Totals		$861.60		$1207.52	$53.85	$75.47

Crew B-1E	Hr.	Daily	Hr.	Daily	Bare Costs	Incl. O&P
1 Labor Foreman (outside)	$38.65	$309.20	$59.65	$477.20	$37.15	$57.33
3 Laborers	36.65	879.60	56.55	1357.20		
1 Work Boat, Diesel, 200 H.P.		1475.00		1622.50		
2 Pressure Washer, 7 GPM		172.00		189.20	51.47	56.62
32 L.H., Daily Totals		$2835.80		$3646.10	$88.62	$113.94

Crew B-1F	Hr.	Daily	Hr.	Daily	Bare Costs	Incl. O&P
2 Skilled Workers	$47.30	$756.80	$73.25	$1172.00	$43.75	$67.68
1 Laborer	36.65	293.20	56.55	452.40		
1 Small Work Boat, Gas, 50 H.P.		189.20		208.12		
1 Pressure Washer, 7 GPM		86.00		94.60	11.47	12.61
24 L.H., Daily Totals		$1325.20		$1927.12	$55.22	$80.30

Crew B-1G	Hr.	Daily	Hr.	Daily	Bare Costs	Incl. O&P
2 Laborers	$36.65	$586.40	$56.55	$904.80	$36.65	$56.55
1 Small Work Boat, Gas, 50 H.P.		189.20		208.12	11.82	13.01
16 L.H., Daily Totals		$775.60		$1112.92	$48.48	$69.56

Crew B-1H	Hr.	Daily	Hr.	Daily	Bare Costs	Incl. O&P
2 Skilled Workers	$47.30	$756.80	$73.25	$1172.00	$43.75	$67.68
1 Laborer	36.65	293.20	56.55	452.40		
1 Small Work Boat, Gas, 50 H.P.		189.20		208.12	7.88	8.67
24 L.H., Daily Totals		$1239.20		$1832.52	$51.63	$76.36

Crew B-1J	Hr.	Daily	Hr.	Daily	Bare Costs	Incl. O&P
1 Labor Foreman (inside)	$37.15	$297.20	$57.30	$458.40	$36.90	$56.92
1 Laborer	36.65	293.20	56.55	452.40		
16 L.H., Daily Totals		$590.40		$910.80	$36.90	$56.92

Crew B-1K	Hr.	Daily	Hr.	Daily	Bare Costs	Incl. O&P
1 Carpenter Foreman (inside)	$46.35	$370.80	$71.50	$572.00	$46.10	$71.13
1 Carpenter	45.85	366.80	70.75	566.00		
16 L.H., Daily Totals		$737.60		$1138.00	$46.10	$71.13

Crew B-2	Hr.	Daily	Hr.	Daily	Bare Costs	Incl. O&P
1 Labor Foreman (outside)	$38.65	$309.20	$59.65	$477.20	$37.05	$57.17
4 Laborers	36.65	1172.80	56.55	1809.60		
40 L.H., Daily Totals		$1482.00		$2286.80	$37.05	$57.17

Crew B-2A	Hr.	Daily	Hr.	Daily	Bare Costs	Incl. O&P
1 Labor Foreman (outside)	$38.65	$309.20	$59.65	$477.20	$37.32	$57.58
2 Laborers	36.65	586.40	56.55	904.80		
1 Aerial Lift Truck, 60' Boom		456.00		501.60	19.00	20.90
24 L.H., Daily Totals		$1351.60		$1883.60	$56.32	$78.48

Crew B-3	Hr.	Daily	Hr.	Daily	Bare Costs	Incl. O&P
1 Labor Foreman (outside)	$38.65	$309.20	$59.65	$477.20	$39.33	$60.25
2 Laborers	36.65	586.40	56.55	904.80		
1 Equip. Oper. (medium)	48.90	391.20	74.15	593.20		
2 Truck Drivers (heavy)	37.55	600.80	57.30	916.80		
1 Crawler Loader, 3 C.Y.		1191.00		1310.10		
2 Dump Trucks, 12 C.Y., 400 H.P.		1384.40		1522.84	53.65	59.02
48 L.H., Daily Totals		$4463.00		$5724.94	$92.98	$119.27

Crew B-3A	Hr.	Daily	Hr.	Daily	Bare Costs	Incl. O&P
4 Laborers	$36.65	$1172.80	$56.55	$1809.60	$39.10	$60.07
1 Equip. Oper. (medium)	48.90	391.20	74.15	593.20		
1 Hyd. Excavator, 1.5 C.Y.		1031.00		1134.10	25.77	28.35
40 L.H., Daily Totals		$2595.00		$3536.90	$64.88	$88.42

Crew B-3B	Hr.	Daily	Hr.	Daily	Bare Costs	Incl. O&P
2 Laborers	$36.65	$586.40	$56.55	$904.80	$39.94	$61.14
1 Equip. Oper. (medium)	48.90	391.20	74.15	593.20		
1 Truck Driver (heavy)	37.55	300.40	57.30	458.40		
1 Backhoe Loader, 80 H.P.		397.60		437.36		
1 Dump Truck, 12 C.Y., 400 H.P.		692.20		761.42	34.06	37.46
32 L.H., Daily Totals		$2367.80		$3155.18	$73.99	$98.60

Crew B-3C	Hr.	Daily	Hr.	Daily	Bare Costs	Incl. O&P
3 Laborers	$36.65	$879.60	$56.55	$1357.20	$39.71	$60.95
1 Equip. Oper. (medium)	48.90	391.20	74.15	593.20		
1 Crawler Loader, 4 C.Y.		1532.00		1685.20	47.88	52.66
32 L.H., Daily Totals		$2802.80		$3635.60	$87.59	$113.61

Crew B-4	Bare Costs Hr.	Bare Costs Daily	Incl. Subs O&P Hr.	Incl. Subs O&P Daily	Cost Per Labor-Hour Bare Costs	Cost Per Labor-Hour Incl. O&P
1 Labor Foreman (outside)	$38.65	$309.20	$59.65	$477.20	$37.13	$57.19
4 Laborers	36.65	1172.80	56.55	1809.60		
1 Truck Driver (heavy)	37.55	300.40	57.30	458.40		
1 Truck Tractor, 220 H.P.		366.40		403.04		
1 Flatbed Trailer, 40 Ton		152.00		167.20	10.80	11.88
48 L.H., Daily Totals		$2300.80		$3315.44	$47.93	$69.07

Crew B-5	Hr.	Daily	Hr.	Daily	Bare Costs	Incl. O&P
1 Labor Foreman (outside)	$38.65	$309.20	$59.65	$477.20	$40.44	$62.02
4 Laborers	36.65	1172.80	56.55	1809.60		
2 Equip. Oper. (medium)	48.90	782.40	74.15	1186.40		
1 Air Compressor, 250 cfm		202.40		222.64		
2 Breakers, Pavement, 60 lb.		19.60		21.56		
2 -50' Air Hoses, 1.5"		11.60		12.76		
1 Crawler Loader, 3 C.Y.		1191.00		1310.10	25.44	27.98
56 L.H., Daily Totals		$3689.00		$5040.26	$65.88	$90.00

Crew B-5A	Hr.	Daily	Hr.	Daily	Bare Costs	Incl. O&P
1 Labor Foreman (outside)	$38.65	$309.20	$59.65	$477.20	$39.88	$61.10
6 Laborers	36.65	1759.20	56.55	2714.40		
2 Equip. Oper. (medium)	48.90	782.40	74.15	1186.40		
1 Equip. Oper. (light)	47.05	376.40	71.35	570.80		
2 Truck Drivers (heavy)	37.55	600.80	57.30	916.80		
1 Air Compressor, 365 cfm		264.80		291.28		
2 Breakers, Pavement, 60 lb.		19.60		21.56		
8 -50' Air Hoses, 1"		32.80		36.08		
2 Dump Trucks, 8 C.Y., 220 H.P.		834.40		917.84	12.00	13.20
96 L.H., Daily Totals		$4979.60		$7132.36	$51.87	$74.30

Crew B-5B	Hr.	Daily	Hr.	Daily	Bare Costs	Incl. O&P
1 Powderman	$47.30	$378.40	$73.25	$586.00	$42.96	$65.58
2 Equip. Oper. (medium)	48.90	782.40	74.15	1186.40		
3 Truck Drivers (heavy)	37.55	901.20	57.30	1375.20		
1 F.E. Loader, W.M.,2.5 C.Y.		522.80		575.08		
3 Dump Trucks, 12 C.Y., 400 H.P.		2076.60		2284.26		
1 Air Compressor, 365 CFM		264.80		291.28	59.67	65.64
48 L.H., Daily Totals		$4926.20		$6298.22	$102.63	$131.21

Crew B-5C	Hr.	Daily	Hr.	Daily	Bare Costs	Incl. O&P
3 Laborers	$36.65	$879.60	$56.55	$1357.20	$40.97	$62.58
1 Equip. Oper. (medium)	48.90	391.20	74.15	593.20		
2 Truck Drivers (heavy)	37.55	600.80	57.30	916.80		
1 Equip. Oper. (crane)	50.25	402.00	76.20	609.60		
1 Equip. Oper. (oiler)	43.55	348.40	66.00	528.00		
2 Dump Trucks, 12 C.Y., 400 H.P.		1384.40		1522.84		
1 Crawler Loader, 4 C.Y.		1532.00		1685.20		
1 S.P. Crane, 4x4, 25 Ton		607.20		667.92	55.06	60.56
64 L.H., Daily Totals		$6145.60		$7880.76	$96.03	$123.14

Crew B-5D	Hr.	Daily	Hr.	Daily	Bare Costs	Incl. O&P
1 Labor Foreman (outside)	$38.65	$309.20	$59.65	$477.20	$40.08	$61.43
4 Laborers	36.65	1172.80	56.55	1809.60		
2 Equip. Oper. (medium)	48.90	782.40	74.15	1186.40		
1 Truck Driver (heavy)	37.55	300.40	57.30	458.40		
1 Air Compressor, 250 cfm		202.40		222.64		
2 Breakers, Pavement, 60 lb.		19.60		21.56		
2 -50' Air Hoses, 1.5"		11.60		12.76		
1 Crawler Loader, 3 C.Y.		1191.00		1310.10		
1 Dump Truck, 12 C.Y., 400 H.P.		692.20		761.42	33.08	36.38
64 L.H., Daily Totals		$4681.60		$6260.08	$73.15	$97.81

Crew B-6	Hr.	Daily	Hr.	Daily	Bare Costs	Incl. O&P
2 Laborers	$36.65	$586.40	$56.55	$904.80	$40.12	$61.48
1 Equip. Oper. (light)	47.05	376.40	71.35	570.80		
1 Backhoe Loader, 48 H.P.		365.20		401.72	15.22	16.74
24 L.H., Daily Totals		$1328.00		$1877.32	$55.33	$78.22

Crew B-6A	Hr.	Daily	Hr.	Daily	Bare Costs	Incl. O&P
.5 Labor Foreman (outside)	$38.65	$154.60	$59.65	$238.60	$41.95	$64.21
1 Laborer	36.65	293.20	56.55	452.40		
1 Equip. Oper. (medium)	48.90	391.20	74.15	593.20		
1 Vacuum Truck, 5000 Gal.		360.15		396.17	18.01	19.81
20 L.H., Daily Totals		$1199.15		$1680.37	$59.96	$84.02

Crew B-6B	Hr.	Daily	Hr.	Daily	Bare Costs	Incl. O&P
2 Labor Foremen (outside)	$38.65	$618.40	$59.65	$954.40	$37.32	$57.58
4 Laborers	36.65	1172.80	56.55	1809.60		
1 S.P. Crane, 4x4, 5 Ton		280.00		308.00		
1 Flatbed Truck, Gas, 1.5 Ton		270.20		297.22		
1 Butt Fusion Mach., 4"-12" diam.		381.80		419.98	19.42	21.36
48 L.H., Daily Totals		$2723.20		$3789.20	$56.73	$78.94

Crew B-6C	Hr.	Daily	Hr.	Daily	Bare Costs	Incl. O&P
2 Labor Foremen (outside)	$38.65	$618.40	$59.65	$954.40	$37.32	$57.58
4 Laborers	36.65	1172.80	56.55	1809.60		
1 S.P. Crane, 4x4, 12 Ton		481.20		529.32		
1 Flatbed Truck, Gas, 3 Ton		333.00		366.30		
1 Butt Fusion Mach., 8"-24" diam.		828.40		911.24	34.22	37.64
48 L.H., Daily Totals		$3433.80		$4570.86	$71.54	$95.23

Crew B-7	Hr.	Daily	Hr.	Daily	Bare Costs	Incl. O&P
1 Labor Foreman (outside)	$38.65	$309.20	$59.65	$477.20	$39.02	$60.00
4 Laborers	36.65	1172.80	56.55	1809.60		
1 Equip. Oper. (medium)	48.90	391.20	74.15	593.20		
1 Brush Chipper, 12", 130 H.P.		396.60		436.26		
1 Crawler Loader, 3 C.Y.		1191.00		1310.10		
2 Chain Saws, Gas, 36" Long		90.00		99.00	34.95	38.45
48 L.H., Daily Totals		$3550.80		$4725.36	$73.97	$98.44

Crew B-7A	Hr.	Daily	Hr.	Daily	Bare Costs	Incl. O&P
2 Laborers	$36.65	$586.40	$56.55	$904.80	$40.12	$61.48
1 Equip. Oper. (light)	47.05	376.40	71.35	570.80		
1 Rake w/Tractor		347.90		382.69		
2 Chain Saw, Gas, 18"		62.40		68.64	17.10	18.81
24 L.H., Daily Totals		$1373.10		$1926.93	$57.21	$80.29

Crew B-7B	Hr.	Daily	Hr.	Daily	Bare Costs	Incl. O&P
1 Labor Foreman (outside)	$38.65	$309.20	$59.65	$477.20	$38.81	$59.61
4 Laborers	36.65	1172.80	56.55	1809.60		
1 Equip. Oper. (medium)	48.90	391.20	74.15	593.20		
1 Truck Driver (heavy)	37.55	300.40	57.30	458.40		
1 Brush Chipper, 12", 130 H.P.		396.60		436.26		
1 Crawler Loader, 3 C.Y.		1191.00		1310.10		
2 Chain Saws, Gas, 36" Long		90.00		99.00		
1 Dump Truck, 8 C.Y., 220 H.P.		417.20		458.92	37.41	41.15
56 L.H., Daily Totals		$4268.40		$5642.68	$76.22	$100.76

Crew No.	Bare Costs		Incl. Subs O&P		Cost Per Labor-Hour	

Crew B-7C	Hr.	Daily	Hr.	Daily	Bare Costs	Incl. O&P
1 Labor Foreman (outside)	$38.65	$309.20	$59.65	$477.20	$38.81	$59.61
4 Laborers	36.65	1172.80	56.55	1809.60		
1 Equip. Oper. (medium)	48.90	391.20	74.15	593.20		
1 Truck Driver (heavy)	37.55	300.40	57.30	458.40		
1 Brush Chipper, 12", 130 H.P.		396.60		436.26		
1 Crawler Loader, 3 C.Y.		1191.00		1310.10		
2 Chain Saws, Gas, 36" Long		90.00		99.00		
1 Dump Truck, 12 C.Y., 400 H.P.		692.20		761.42	42.32	46.55
56 L.H., Daily Totals		$4543.40		$5945.18	$81.13	$106.16

Crew B-8	Hr.	Daily	Hr.	Daily	Bare Costs	Incl. O&P
1 Labor Foreman (outside)	$38.65	$309.20	$59.65	$477.20	$41.05	$62.71
2 Laborers	36.65	586.40	56.55	904.80		
2 Equip. Oper. (medium)	48.90	782.40	74.15	1186.40		
1 Equip. Oper. (oiler)	43.55	348.40	66.00	528.00		
2 Truck Drivers (heavy)	37.55	600.80	57.30	916.80		
1 Hyd. Crane, 25 Ton		745.80		820.38		
1 Crawler Loader, 3 C.Y.		1191.00		1310.10		
2 Dump Trucks, 12 C.Y., 400 H.P.		1384.40		1522.84	51.89	57.08
64 L.H., Daily Totals		$5948.40		$7666.52	$92.94	$119.79

Crew B-9	Hr.	Daily	Hr.	Daily	Bare Costs	Incl. O&P
1 Labor Foreman (outside)	$38.65	$309.20	$59.65	$477.20	$37.05	$57.17
4 Laborers	36.65	1172.80	56.55	1809.60		
1 Air Compressor, 250 cfm		202.40		222.64		
2 Breakers, Pavement, 60 lb.		19.60		21.56		
2 -50' Air Hoses, 1.5"		11.60		12.76	5.84	6.42
40 L.H., Daily Totals		$1715.60		$2543.76	$42.89	$63.59

Crew B-9A	Hr.	Daily	Hr.	Daily	Bare Costs	Incl. O&P
2 Laborers	$36.65	$586.40	$56.55	$904.80	$36.95	$56.80
1 Truck Driver (heavy)	37.55	300.40	57.30	458.40		
1 Water Tank Trailer, 5000 Gal.		141.60		155.76		
1 Truck Tractor, 220 H.P.		366.40		403.04		
2 -50' Discharge Hoses, 3"		3.00		3.30	21.29	23.42
24 L.H., Daily Totals		$1397.80		$1925.30	$58.24	$80.22

Crew B-9B	Hr.	Daily	Hr.	Daily	Bare Costs	Incl. O&P
2 Laborers	$36.65	$586.40	$56.55	$904.80	$36.95	$56.80
1 Truck Driver (heavy)	37.55	300.40	57.30	458.40		
2 -50' Discharge Hoses, 3"		3.00		3.30		
1 Water Tank Trailer, 5000 Gal.		141.60		155.76		
1 Truck Tractor, 220 H.P.		366.40		403.04		
1 Pressure Washer		69.40		76.34	24.18	26.60
24 L.H., Daily Totals		$1467.20		$2001.64	$61.13	$83.40

Crew B-9D	Hr.	Daily	Hr.	Daily	Bare Costs	Incl. O&P
1 Labor Foreman (outside)	$38.65	$309.20	$59.65	$477.20	$37.05	$57.17
4 Common Laborers	36.65	1172.80	56.55	1809.60		
1 Air Compressor, 250 cfm		202.40		222.64		
2 -50' Air Hoses, 1.5"		11.60		12.76		
2 Air Powered Tampers		52.40		57.64	6.66	7.33
40 L.H., Daily Totals		$1748.40		$2579.84	$43.71	$64.50

Crew B-10	Hr.	Daily	Hr.	Daily	Bare Costs	Incl. O&P
1 Equip. Oper. (medium)	$48.90	$391.20	$74.15	$593.20	$44.82	$68.28
.5 Laborer	36.65	146.60	56.55	226.20		
12 L.H., Daily Totals		$537.80		$819.40	$44.82	$68.28

Crew B-10A	Hr.	Daily	Hr.	Daily	Bare Costs	Incl. O&P
1 Equip. Oper. (medium)	$48.90	$391.20	$74.15	$593.20	$44.82	$68.28
.5 Laborer	36.65	146.60	56.55	226.20		
1 Roller, 2-Drum, W.B., 7.5 H.P.		182.80		201.08	15.23	16.76
12 L.H., Daily Totals		$720.60		$1020.48	$60.05	$85.04

Crew B-10B	Hr.	Daily	Hr.	Daily	Bare Costs	Incl. O&P
1 Equip. Oper. (medium)	$48.90	$391.20	$74.15	$593.20	$44.82	$68.28
.5 Laborer	36.65	146.60	56.55	226.20		
1 Dozer, 200 H.P.		1325.00		1457.50	110.42	121.46
12 L.H., Daily Totals		$1862.80		$2276.90	$155.23	$189.74

Crew B-10C	Hr.	Daily	Hr.	Daily	Bare Costs	Incl. O&P
1 Equip. Oper. (medium)	$48.90	$391.20	$74.15	$593.20	$44.82	$68.28
.5 Laborer	36.65	146.60	56.55	226.20		
1 Dozer, 200 H.P.		1325.00		1457.50		
1 Vibratory Roller, Towed, 23 Ton		412.00		453.20	144.75	159.22
12 L.H., Daily Totals		$2274.80		$2730.10	$189.57	$227.51

Crew B-10D	Hr.	Daily	Hr.	Daily	Bare Costs	Incl. O&P
1 Equip. Oper. (medium)	$48.90	$391.20	$74.15	$593.20	$44.82	$68.28
.5 Laborer	36.65	146.60	56.55	226.20		
1 Dozer, 200 H.P.		1325.00		1457.50		
1 Sheepsft. Roller, Towed		443.80		488.18	147.40	162.14
12 L.H., Daily Totals		$2306.60		$2765.08	$192.22	$230.42

Crew B-10E	Hr.	Daily	Hr.	Daily	Bare Costs	Incl. O&P
1 Equip. Oper. (medium)	$48.90	$391.20	$74.15	$593.20	$44.82	$68.28
.5 Laborer	36.65	146.60	56.55	226.20		
1 Tandem Roller, 5 Ton		159.00		174.90	13.25	14.57
12 L.H., Daily Totals		$696.80		$994.30	$58.07	$82.86

Crew B-10F	Hr.	Daily	Hr.	Daily	Bare Costs	Incl. O&P
1 Equip. Oper. (medium)	$48.90	$391.20	$74.15	$593.20	$44.82	$68.28
.5 Laborer	36.65	146.60	56.55	226.20		
1 Tandem Roller, 10 Ton		242.00		266.20	20.17	22.18
12 L.H., Daily Totals		$779.80		$1085.60	$64.98	$90.47

Crew B-10G	Hr.	Daily	Hr.	Daily	Bare Costs	Incl. O&P
1 Equip. Oper. (medium)	$48.90	$391.20	$74.15	$593.20	$44.82	$68.28
.5 Laborer	36.65	146.60	56.55	226.20		
1 Sheepsfoot Roller, 240 H.P.		1218.00		1339.80	101.50	111.65
12 L.H., Daily Totals		$1755.80		$2159.20	$146.32	$179.93

Crew B-10H	Hr.	Daily	Hr.	Daily	Bare Costs	Incl. O&P
1 Equip. Oper. (medium)	$48.90	$391.20	$74.15	$593.20	$44.82	$68.28
.5 Laborer	36.65	146.60	56.55	226.20		
1 Diaphragm Water Pump, 2"		72.00		79.20		
1 -20' Suction Hose, 2"		1.95		2.15		
2 -50' Discharge Hoses, 2"		1.80		1.98	6.31	6.94
12 L.H., Daily Totals		$613.55		$902.73	$51.13	$75.23

Crew B-10I	Hr.	Daily	Hr.	Daily	Bare Costs	Incl. O&P
1 Equip. Oper. (medium)	$48.90	$391.20	$74.15	$593.20	$44.82	$68.28
.5 Laborer	36.65	146.60	56.55	226.20		
1 Diaphragm Water Pump, 4"		115.80		127.38		
1 -20' Suction Hose, 4"		3.25		3.58		
2 -50' Discharge Hoses, 4"		4.70		5.17	10.31	11.34
12 L.H., Daily Totals		$661.55		$955.52	$55.13	$79.63

Crew No.	Bare Costs		Incl. Subs O&P		Cost Per Labor-Hour	

Crew B-10J	Hr.	Daily	Hr.	Daily	Bare Costs	Incl. O&P
1 Equip. Oper. (medium)	$48.90	$391.20	$74.15	$593.20	$44.82	$68.28
.5 Laborer	36.65	146.60	56.55	226.20		
1 Centrifugal Water Pump, 3"		78.80		86.68		
1 -20' Suction Hose, 3"		2.85		3.13		
2 -50' Discharge Hoses, 3"		3.00		3.30	7.05	7.76
12 L.H., Daily Totals		$622.45		$912.51	$51.87	$76.04

Crew B-10K	Hr.	Daily	Hr.	Daily	Bare Costs	Incl. O&P
1 Equip. Oper. (medium)	$48.90	$391.20	$74.15	$593.20	$44.82	$68.28
.5 Laborer	36.65	146.60	56.55	226.20		
1 Centr. Water Pump, 6"		340.80		374.88		
1 -20' Suction Hose, 6"		11.50		12.65		
2 -50' Discharge Hoses, 6"		12.20		13.42	30.38	33.41
12 L.H., Daily Totals		$902.30		$1220.35	$75.19	$101.70

Crew B-10L	Hr.	Daily	Hr.	Daily	Bare Costs	Incl. O&P
1 Equip. Oper. (medium)	$48.90	$391.20	$74.15	$593.20	$44.82	$68.28
.5 Laborer	36.65	146.60	56.55	226.20		
1 Dozer, 80 H.P.		481.00		529.10	40.08	44.09
12 L.H., Daily Totals		$1018.80		$1348.50	$84.90	$112.38

Crew B-10M	Hr.	Daily	Hr.	Daily	Bare Costs	Incl. O&P
1 Equip. Oper. (medium)	$48.90	$391.20	$74.15	$593.20	$44.82	$68.28
.5 Laborer	36.65	146.60	56.55	226.20		
1 Dozer, 300 H.P.		1817.00		1998.70	151.42	166.56
12 L.H., Daily Totals		$2354.80		$2818.10	$196.23	$234.84

Crew B-10N	Hr.	Daily	Hr.	Daily	Bare Costs	Incl. O&P
1 Equip. Oper. (medium)	$48.90	$391.20	$74.15	$593.20	$44.82	$68.28
.5 Laborer	36.65	146.60	56.55	226.20		
1 F.E. Loader, T.M., 1.5 C.Y		522.80		575.08	43.57	47.92
12 L.H., Daily Totals		$1060.60		$1394.48	$88.38	$116.21

Crew B-10O	Hr.	Daily	Hr.	Daily	Bare Costs	Incl. O&P
1 Equip. Oper. (medium)	$48.90	$391.20	$74.15	$593.20	$44.82	$68.28
.5 Laborer	36.65	146.60	56.55	226.20		
1 F.E. Loader, T.M., 2.25 C.Y.		959.80		1055.78	79.98	87.98
12 L.H., Daily Totals		$1497.60		$1875.18	$124.80	$156.26

Crew B-10P	Hr.	Daily	Hr.	Daily	Bare Costs	Incl. O&P
1 Equip. Oper. (medium)	$48.90	$391.20	$74.15	$593.20	$44.82	$68.28
.5 Laborer	36.65	146.60	56.55	226.20		
1 Crawler Loader, 3 C.Y.		1191.00		1310.10	99.25	109.18
12 L.H., Daily Totals		$1728.80		$2129.50	$144.07	$177.46

Crew B-10Q	Hr.	Daily	Hr.	Daily	Bare Costs	Incl. O&P
1 Equip. Oper. (medium)	$48.90	$391.20	$74.15	$593.20	$44.82	$68.28
.5 Laborer	36.65	146.60	56.55	226.20		
1 Crawler Loader, 4 C.Y.		1532.00		1685.20	127.67	140.43
12 L.H., Daily Totals		$2069.80		$2504.60	$172.48	$208.72

Crew B-10R	Hr.	Daily	Hr.	Daily	Bare Costs	Incl. O&P
1 Equip. Oper. (medium)	$48.90	$391.20	$74.15	$593.20	$44.82	$68.28
.5 Laborer	36.65	146.60	56.55	226.20		
1 F.E. Loader, W.M., 1 C.Y.		295.20		324.72	24.60	27.06
12 L.H., Daily Totals		$833.00		$1144.12	$69.42	$95.34

Crew No.	Bare Costs		Incl. Subs O&P		Cost Per Labor-Hour	

Crew B-10S	Hr.	Daily	Hr.	Daily	Bare Costs	Incl. O&P
1 Equip. Oper. (medium)	$48.90	$391.20	$74.15	$593.20	$44.82	$68.28
.5 Laborer	36.65	146.60	56.55	226.20		
1 F.E. Loader, W.M., 1.5 C.Y.		373.40		410.74	31.12	34.23
12 L.H., Daily Totals		$911.20		$1230.14	$75.93	$102.51

Crew B-10T	Hr.	Daily	Hr.	Daily	Bare Costs	Incl. O&P
1 Equip. Oper. (medium)	$48.90	$391.20	$74.15	$593.20	$44.82	$68.28
.5 Laborer	36.65	146.60	56.55	226.20		
1 F.E. Loader, W.M.,2.5 C.Y.		522.80		575.08	43.57	47.92
12 L.H., Daily Totals		$1060.60		$1394.48	$88.38	$116.21

Crew B-10U	Hr.	Daily	Hr.	Daily	Bare Costs	Incl. O&P
1 Equip. Oper. (medium)	$48.90	$391.20	$74.15	$593.20	$44.82	$68.28
.5 Laborer	36.65	146.60	56.55	226.20		
1 F.E. Loader, W.M., 5.5 C.Y.		1049.00		1153.90	87.42	96.16
12 L.H., Daily Totals		$1586.80		$1973.30	$132.23	$164.44

Crew B-10V	Hr.	Daily	Hr.	Daily	Bare Costs	Incl. O&P
1 Equip. Oper. (medium)	$48.90	$391.20	$74.15	$593.20	$44.82	$68.28
.5 Laborer	36.65	146.60	56.55	226.20		
1 Dozer, 700 H.P.		4485.00		4933.50	373.75	411.13
12 L.H., Daily Totals		$5022.80		$5752.90	$418.57	$479.41

Crew B-10W	Hr.	Daily	Hr.	Daily	Bare Costs	Incl. O&P
1 Equip. Oper. (medium)	$48.90	$391.20	$74.15	$593.20	$44.82	$68.28
.5 Laborer	36.65	146.60	56.55	226.20		
1 Dozer, 105 H.P.		591.80		650.98	49.32	54.25
12 L.H., Daily Totals		$1129.60		$1470.38	$94.13	$122.53

Crew B-10X	Hr.	Daily	Hr.	Daily	Bare Costs	Incl. O&P
1 Equip. Oper. (medium)	$48.90	$391.20	$74.15	$593.20	$44.82	$68.28
.5 Laborer	36.65	146.60	56.55	226.20		
1 Dozer, 410 H.P.		2409.00		2649.90	200.75	220.82
12 L.H., Daily Totals		$2946.80		$3469.30	$245.57	$289.11

Crew B-10Y	Hr.	Daily	Hr.	Daily	Bare Costs	Incl. O&P
1 Equip. Oper. (medium)	$48.90	$391.20	$74.15	$593.20	$44.82	$68.28
.5 Laborer	36.65	146.60	56.55	226.20		
1 Vibr. Roller, Towed, 12 Ton		559.20		615.12	46.60	51.26
12 L.H., Daily Totals		$1097.00		$1434.52	$91.42	$119.54

Crew B-11A	Hr.	Daily	Hr.	Daily	Bare Costs	Incl. O&P
1 Equipment Oper. (med.)	$48.90	$391.20	$74.15	$593.20	$42.77	$65.35
1 Laborer	36.65	293.20	56.55	452.40		
1 Dozer, 200 H.P.		1325.00		1457.50	82.81	91.09
16 L.H., Daily Totals		$2009.40		$2503.10	$125.59	$156.44

Crew B-11B	Hr.	Daily	Hr.	Daily	Bare Costs	Incl. O&P
1 Equipment Oper. (light)	$47.05	$376.40	$71.35	$570.80	$41.85	$63.95
1 Laborer	36.65	293.20	56.55	452.40		
1 Air Powered Tamper		26.20		28.82		
1 Air Compressor, 365 cfm		264.80		291.28		
2 -50' Air Hoses, 1.5"		11.60		12.76	18.91	20.80
16 L.H., Daily Totals		$972.20		$1356.06	$60.76	$84.75

Crew B-11C	Hr.	Daily	Hr.	Daily	Bare Costs	Incl. O&P
1 Equipment Oper. (med.)	$48.90	$391.20	$74.15	$593.20	$42.77	$65.35
1 Laborer	36.65	293.20	56.55	452.40		
1 Backhoe Loader, 48 H.P.		365.20		401.72	22.82	25.11
16 L.H., Daily Totals		$1049.60		$1447.32	$65.60	$90.46

Crew No.	Bare Costs		Incl. Subs O&P		Cost Per Labor-Hour	

Crew B-11J	Hr.	Daily	Hr.	Daily	Bare Costs	Incl. O&P
1 Equipment Oper. (med.)	$48.90	$391.20	$74.15	$593.20	$42.77	$65.35
1 Laborer	36.65	293.20	56.55	452.40		
1 Grader, 30,000 Lbs.		707.00		777.70		
1 Ripper, Beam & 1 Shank		81.40		89.54	49.27	54.20
16 L.H., Daily Totals		$1472.80		$1912.84	$92.05	$119.55

Crew B-11K	Hr.	Daily	Hr.	Daily	Bare Costs	Incl. O&P
1 Equipment Oper. (med.)	$48.90	$391.20	$74.15	$593.20	$42.77	$65.35
1 Laborer	36.65	293.20	56.55	452.40		
1 Trencher, Chain Type, 8' D		3376.00		3713.60	211.00	232.10
16 L.H., Daily Totals		$4060.40		$4759.20	$253.78	$297.45

Crew B-11L	Hr.	Daily	Hr.	Daily	Bare Costs	Incl. O&P
1 Equipment Oper. (med.)	$48.90	$391.20	$74.15	$593.20	$42.77	$65.35
1 Laborer	36.65	293.20	56.55	452.40		
1 Grader, 30,000 Lbs.		707.00		777.70	44.19	48.61
16 L.H., Daily Totals		$1391.40		$1823.30	$86.96	$113.96

Crew B-11M	Hr.	Daily	Hr.	Daily	Bare Costs	Incl. O&P
1 Equipment Oper. (med.)	$48.90	$391.20	$74.15	$593.20	$42.77	$65.35
1 Laborer	36.65	293.20	56.55	452.40		
1 Backhoe Loader, 80 H.P.		397.60		437.36	24.85	27.34
16 L.H., Daily Totals		$1082.00		$1482.96	$67.63	$92.69

Crew B-11N	Hr.	Daily	Hr.	Daily	Bare Costs	Incl. O&P
1 Labor Foreman (outside)	$38.65	$309.20	$59.65	$477.20	$40.19	$61.31
2 Equipment Operators (med.)	48.90	782.40	74.15	1186.40		
6 Truck Drivers (heavy)	37.55	1802.40	57.30	2750.40		
1 F.E. Loader, W.M., 5.5 C.Y.		1049.00		1153.90		
1 Dozer, 410 H.P.		2409.00		2649.90		
6 Dump Trucks, Off Hwy., 50 Ton		10722.00		11794.20	196.94	216.64
72 L.H., Daily Totals		$17074.00		$20012.00	$237.14	$277.94

Crew B-11Q	Hr.	Daily	Hr.	Daily	Bare Costs	Incl. O&P
1 Equipment Operator (med.)	$48.90	$391.20	$74.15	$593.20	$44.82	$68.28
.5 Laborer	36.65	146.60	56.55	226.20		
1 Dozer, 140 H.P.		883.80		972.18	73.65	81.02
12 L.H., Daily Totals		$1421.60		$1791.58	$118.47	$149.30

Crew B-11R	Hr.	Daily	Hr.	Daily	Bare Costs	Incl. O&P
1 Equipment Operator (med.)	$48.90	$391.20	$74.15	$593.20	$44.82	$68.28
.5 Laborer	36.65	146.60	56.55	226.20		
1 Dozer, 200 H.P.		1325.00		1457.50	110.42	121.46
12 L.H., Daily Totals		$1862.80		$2276.90	$155.23	$189.74

Crew B-11S	Hr.	Daily	Hr.	Daily	Bare Costs	Incl. O&P
1 Equipment Operator (med.)	$48.90	$391.20	$74.15	$593.20	$44.82	$68.28
.5 Laborer	36.65	146.60	56.55	226.20		
1 Dozer, 300 H.P.		1817.00		1998.70		
1 Ripper, Beam & 1 Shank		81.40		89.54	158.20	174.02
12 L.H., Daily Totals		$2436.20		$2907.64	$203.02	$242.30

Crew B-11T	Hr.	Daily	Hr.	Daily	Bare Costs	Incl. O&P
1 Equipment Operator (med.)	$48.90	$391.20	$74.15	$593.20	$44.82	$68.28
.5 Laborer	36.65	146.60	56.55	226.20		
1 Dozer, 410 H.P.		2409.00		2649.90		
1 Ripper, Beam & 2 Shanks		91.60		100.76	208.38	229.22
12 L.H., Daily Totals		$3038.40		$3570.06	$253.20	$297.51

Crew B-11U	Hr.	Daily	Hr.	Daily	Bare Costs	Incl. O&P
1 Equipment Operator (med.)	$48.90	$391.20	$74.15	$593.20	$44.82	$68.28
.5 Laborer	36.65	146.60	56.55	226.20		
1 Dozer, 520 H.P.		3116.00		3427.60	259.67	285.63
12 L.H., Daily Totals		$3653.80		$4247.00	$304.48	$353.92

Crew B-11V	Hr.	Daily	Hr.	Daily	Bare Costs	Incl. O&P
3 Laborers	$36.65	$879.60	$56.55	$1357.20	$36.65	$56.55
1 Roller, 2-Drum, W.B., 7.5 H.P.		182.80		201.08	7.62	8.38
24 L.H., Daily Totals		$1062.40		$1558.28	$44.27	$64.93

Crew B-11W	Hr.	Daily	Hr.	Daily	Bare Costs	Incl. O&P
1 Equipment Operator (med.)	$48.90	$391.20	$74.15	$593.20	$38.42	$58.64
1 Common Laborer	36.65	293.20	56.55	452.40		
10 Truck Drivers (heavy)	37.55	3004.00	57.30	4584.00		
1 Dozer, 200 H.P.		1325.00		1457.50		
1 Vibratory Roller, Towed, 23 Ton		412.00		453.20		
10 Dump Trucks, 8 C.Y., 220 H.P.		4172.00		4589.20	61.55	67.71
96 L.H., Daily Totals		$9597.40		$12129.50	$99.97	$126.35

Crew B-11Y	Hr.	Daily	Hr.	Daily	Bare Costs	Incl. O&P
1 Labor Foreman (outside)	$38.65	$309.20	$59.65	$477.20	$40.96	$62.76
5 Common Laborers	36.65	1466.00	56.55	2262.00		
3 Equipment Operators (med.)	48.90	1173.60	74.15	1779.60		
1 Dozer, 80 H.P.		481.00		529.10		
2 Roller, 2-Drum, W.B., 7.5 H.P.		365.60		402.16		
4 Vibrating Plate, Gas, 21"		184.00		202.40	14.31	15.75
72 L.H., Daily Totals		$3979.40		$5652.46	$55.27	$78.51

Crew B-12A	Hr.	Daily	Hr.	Daily	Bare Costs	Incl. O&P
1 Equip. Oper. (crane)	$50.25	$402.00	$76.20	$609.60	$43.45	$66.38
1 Laborer	36.65	293.20	56.55	452.40		
1 Hyd. Excavator, 1 C.Y.		814.80		896.28	50.92	56.02
16 L.H., Daily Totals		$1510.00		$1958.28	$94.38	$122.39

Crew B-12B	Hr.	Daily	Hr.	Daily	Bare Costs	Incl. O&P
1 Equip. Oper. (crane)	$50.25	$402.00	$76.20	$609.60	$43.45	$66.38
1 Laborer	36.65	293.20	56.55	452.40		
1 Hyd. Excavator, 1.5 C.Y.		1031.00		1134.10	64.44	70.88
16 L.H., Daily Totals		$1726.20		$2196.10	$107.89	$137.26

Crew B-12C	Hr.	Daily	Hr.	Daily	Bare Costs	Incl. O&P
1 Equip. Oper. (crane)	$50.25	$402.00	$76.20	$609.60	$43.45	$66.38
1 Laborer	36.65	293.20	56.55	452.40		
1 Hyd. Excavator, 2 C.Y.		1175.00		1292.50	73.44	80.78
16 L.H., Daily Totals		$1870.20		$2354.50	$116.89	$147.16

Crew B-12D	Hr.	Daily	Hr.	Daily	Bare Costs	Incl. O&P
1 Equip. Oper. (crane)	$50.25	$402.00	$76.20	$609.60	$43.45	$66.38
1 Laborer	36.65	293.20	56.55	452.40		
1 Hyd. Excavator, 3.5 C.Y.		2442.00		2686.20	152.63	167.89
16 L.H., Daily Totals		$3137.20		$3748.20	$196.07	$234.26

Crew B-12E	Hr.	Daily	Hr.	Daily	Bare Costs	Incl. O&P
1 Equip. Oper. (crane)	$50.25	$402.00	$76.20	$609.60	$43.45	$66.38
1 Laborer	36.65	293.20	56.55	452.40		
1 Hyd. Excavator, .5 C.Y.		448.00		492.80	28.00	30.80
16 L.H., Daily Totals		$1143.20		$1554.80	$71.45	$97.17

Crew No.	Bare Costs		Incl. Subs O&P		Cost Per Labor-Hour	

Crew B-12F	Hr.	Daily	Hr.	Daily	Bare Costs	Incl. O&P
1 Equip. Oper. (crane)	$50.25	$402.00	$76.20	$609.60	$43.45	$66.38
1 Laborer	36.65	293.20	56.55	452.40		
1 Hyd. Excavator, .75 C.Y.		662.00		728.20	41.38	45.51
16 L.H., Daily Totals		$1357.20		$1790.20	$84.83	$111.89

Crew B-12G	Hr.	Daily	Hr.	Daily	Bare Costs	Incl. O&P
1 Equip. Oper. (crane)	$50.25	$402.00	$76.20	$609.60	$43.45	$66.38
1 Laborer	36.65	293.20	56.55	452.40		
1 Crawler Crane, 15 Ton		684.15		752.57		
1 Clamshell Bucket, .5 C.Y.		37.80		41.58	45.12	49.63
16 L.H., Daily Totals		$1417.15		$1856.15	$88.57	$116.01

Crew B-12H	Hr.	Daily	Hr.	Daily	Bare Costs	Incl. O&P
1 Equip. Oper. (crane)	$50.25	$402.00	$76.20	$609.60	$43.45	$66.38
1 Laborer	36.65	293.20	56.55	452.40		
1 Crawler Crane, 25 Ton		1173.00		1290.30		
1 Clamshell Bucket, 1 C.Y.		47.40		52.14	76.28	83.90
16 L.H., Daily Totals		$1915.60		$2404.44	$119.72	$150.28

Crew B-12I	Hr.	Daily	Hr.	Daily	Bare Costs	Incl. O&P
1 Equip. Oper. (crane)	$50.25	$402.00	$76.20	$609.60	$43.45	$66.38
1 Laborer	36.65	293.20	56.55	452.40		
1 Crawler Crane, 20 Ton		880.25		968.27		
1 Dragline Bucket, .75 C.Y.		20.80		22.88	56.32	61.95
16 L.H., Daily Totals		$1596.25		$2053.16	$99.77	$128.32

Crew B-12J	Hr.	Daily	Hr.	Daily	Bare Costs	Incl. O&P
1 Equip. Oper. (crane)	$50.25	$402.00	$76.20	$609.60	$43.45	$66.38
1 Laborer	36.65	293.20	56.55	452.40		
1 Gradall, 5/8 C.Y.		881.60		969.76	55.10	60.61
16 L.H., Daily Totals		$1576.80		$2031.76	$98.55	$126.99

Crew B-12K	Hr.	Daily	Hr.	Daily	Bare Costs	Incl. O&P
1 Equip. Oper. (crane)	$50.25	$402.00	$76.20	$609.60	$43.45	$66.38
1 Laborer	36.65	293.20	56.55	452.40		
1 Gradall, 3 Ton, 1 C.Y.		1005.00		1105.50	62.81	69.09
16 L.H., Daily Totals		$1700.20		$2167.50	$106.26	$135.47

Crew B-12L	Hr.	Daily	Hr.	Daily	Bare Costs	Incl. O&P
1 Equip. Oper. (crane)	$50.25	$402.00	$76.20	$609.60	$43.45	$66.38
1 Laborer	36.65	293.20	56.55	452.40		
1 Crawler Crane, 15 Ton		684.15		752.57		
1 F.E. Attachment, .5 C.Y.		59.60		65.56	46.48	51.13
16 L.H., Daily Totals		$1438.95		$1880.13	$89.93	$117.51

Crew B-12M	Hr.	Daily	Hr.	Daily	Bare Costs	Incl. O&P
1 Equip. Oper. (crane)	$50.25	$402.00	$76.20	$609.60	$43.45	$66.38
1 Laborer	36.65	293.20	56.55	452.40		
1 Crawler Crane, 20 Ton		880.25		968.27		
1 F.E. Attachment, .75 C.Y.		64.00		70.40	59.02	64.92
16 L.H., Daily Totals		$1639.45		$2100.68	$102.47	$131.29

Crew B-12N	Hr.	Daily	Hr.	Daily	Bare Costs	Incl. O&P
1 Equip. Oper. (crane)	$50.25	$402.00	$76.20	$609.60	$43.45	$66.38
1 Laborer	36.65	293.20	56.55	452.40		
1 Crawler Crane, 25 Ton		1173.00		1290.30		
1 F.E. Attachment, 1 C.Y.		70.80		77.88	77.74	85.51
16 L.H., Daily Totals		$1939.00		$2430.18	$121.19	$151.89

Crew B-12O	Hr.	Daily	Hr.	Daily	Bare Costs	Incl. O&P
1 Equip. Oper. (crane)	$50.25	$402.00	$76.20	$609.60	$43.45	$66.38
1 Laborer	36.65	293.20	56.55	452.40		
1 Crawler Crane, 40 Ton		1179.00		1296.90		
1 F.E. Attachment, 1.5 C.Y.		80.00		88.00	78.69	86.56
16 L.H., Daily Totals		$1954.20		$2446.90	$122.14	$152.93

Crew B-12P	Hr.	Daily	Hr.	Daily	Bare Costs	Incl. O&P
1 Equip. Oper. (crane)	$50.25	$402.00	$76.20	$609.60	$43.45	$66.38
1 Laborer	36.65	293.20	56.55	452.40		
1 Crawler Crane, 40 Ton		1179.00		1296.90		
1 Dragline Bucket, 1.5 C.Y.		33.60		36.96	75.79	83.37
16 L.H., Daily Totals		$1907.80		$2395.86	$119.24	$149.74

Crew B-12Q	Hr.	Daily	Hr.	Daily	Bare Costs	Incl. O&P
1 Equip. Oper. (crane)	$50.25	$402.00	$76.20	$609.60	$43.45	$66.38
1 Laborer	36.65	293.20	56.55	452.40		
1 Hyd. Excavator, 5/8 C.Y.		594.40		653.84	37.15	40.87
16 L.H., Daily Totals		$1289.60		$1715.84	$80.60	$107.24

Crew B-12S	Hr.	Daily	Hr.	Daily	Bare Costs	Incl. O&P
1 Equip. Oper. (crane)	$50.25	$402.00	$76.20	$609.60	$43.45	$66.38
1 Laborer	36.65	293.20	56.55	452.40		
1 Hyd. Excavator, 2.5 C.Y.		1637.00		1800.70	102.31	112.54
16 L.H., Daily Totals		$2332.20		$2862.70	$145.76	$178.92

Crew B-12T	Hr.	Daily	Hr.	Daily	Bare Costs	Incl. O&P
1 Equip. Oper. (crane)	$50.25	$402.00	$76.20	$609.60	$43.45	$66.38
1 Laborer	36.65	293.20	56.55	452.40		
1 Crawler Crane, 75 Ton		1492.00		1641.20		
1 F.E. Attachment, 3 C.Y.		102.80		113.08	99.67	109.64
16 L.H., Daily Totals		$2290.00		$2816.28	$143.13	$176.02

Crew B-12V	Hr.	Daily	Hr.	Daily	Bare Costs	Incl. O&P
1 Equip. Oper. (crane)	$50.25	$402.00	$76.20	$609.60	$43.45	$66.38
1 Laborer	36.65	293.20	56.55	452.40		
1 Crawler Crane, 75 Ton		1492.00		1641.20		
1 Dragline Bucket, 3 C.Y.		52.60		57.86	96.54	106.19
16 L.H., Daily Totals		$2239.80		$2761.06	$139.99	$172.57

Crew B-12Y	Hr.	Daily	Hr.	Daily	Bare Costs	Incl. O&P
1 Equip. Oper. (crane)	$50.25	$402.00	$76.20	$609.60	$41.18	$63.10
2 Laborers	36.65	586.40	56.55	904.80		
1 Hyd. Excavator, 3.5 C.Y.		2442.00		2686.20	101.75	111.93
24 L.H., Daily Totals		$3430.40		$4200.60	$142.93	$175.03

Crew B-12Z	Hr.	Daily	Hr.	Daily	Bare Costs	Incl. O&P
1 Equip. Oper. (crane)	$50.25	$402.00	$76.20	$609.60	$41.18	$63.10
2 Laborers	36.65	586.40	56.55	904.80		
1 Hyd. Excavator, 2.5 C.Y.		1637.00		1800.70	68.21	75.03
24 L.H., Daily Totals		$2625.40		$3315.10	$109.39	$138.13

Crew B-13	Hr.	Daily	Hr.	Daily	Bare Costs	Incl. O&P
1 Labor Foreman (outside)	$38.65	$309.20	$59.65	$477.20	$39.86	$61.15
4 Laborers	36.65	1172.80	56.55	1809.60		
1 Equip. Oper. (crane)	50.25	402.00	76.20	609.60		
1 Equip. Oper. (oiler)	43.55	348.40	66.00	528.00		
1 Hyd. Crane, 25 Ton		745.80		820.38	13.32	14.65
56 L.H., Daily Totals		$2978.20		$4244.78	$53.18	$75.80

Crews

Crew B-13A	Bare Costs Hr.	Bare Costs Daily	Incl. Subs O&P Hr.	Incl. Subs O&P Daily	Bare Costs (Per Labor-Hour)	Incl. O&P (Per Labor-Hour)
1 Labor Foreman (outside)	$38.65	$309.20	$59.65	$477.20	$40.69	$62.24
2 Laborers	36.65	586.40	56.55	904.80		
2 Equipment Operators (med.)	48.90	782.40	74.15	1186.40		
2 Truck Drivers (heavy)	37.55	600.80	57.30	916.80		
1 Crawler Crane, 75 Ton		1492.00		1641.20		
1 Crawler Loader, 4 C.Y.		1532.00		1685.20		
2 Dump Trucks, 8 C.Y., 220 H.P.		834.40		917.84	68.90	75.79
56 L.H., Daily Totals		$6137.20		$7729.44	$109.59	$138.03

Crew B-13B	Hr.	Daily	Hr.	Daily	Bare Costs	Incl. O&P
1 Labor Foreman (outside)	$38.65	$309.20	$59.65	$477.20	$39.86	$61.15
4 Laborers	36.65	1172.80	56.55	1809.60		
1 Equip. Oper. (crane)	50.25	402.00	76.20	609.60		
1 Equip. Oper. (oiler)	43.55	348.40	66.00	528.00		
1 Hyd. Crane, 55 Ton		1137.00		1250.70	20.30	22.33
56 L.H., Daily Totals		$3369.40		$4675.10	$60.17	$83.48

Crew B-13C	Hr.	Daily	Hr.	Daily	Bare Costs	Incl. O&P
1 Labor Foreman (outside)	$38.65	$309.20	$59.65	$477.20	$39.86	$61.15
4 Laborers	36.65	1172.80	56.55	1809.60		
1 Equip. Oper. (crane)	50.25	402.00	76.20	609.60		
1 Equip. Oper. (oiler)	43.55	348.40	66.00	528.00		
1 Crawler Crane, 100 Ton		1703.00		1873.30	30.41	33.45
56 L.H., Daily Totals		$3935.40		$5297.70	$70.28	$94.60

Crew B-13D	Hr.	Daily	Hr.	Daily	Bare Costs	Incl. O&P
1 Laborer	$36.65	$293.20	$56.55	$452.40	$43.45	$66.38
1 Equip. Oper. (crane)	50.25	402.00	76.20	609.60		
1 Hyd. Excavator, 1 C.Y.		814.80		896.28		
1 Trench Box		84.05		92.45	56.18	61.80
16 L.H., Daily Totals		$1594.05		$2050.74	$99.63	$128.17

Crew B-13E	Hr.	Daily	Hr.	Daily	Bare Costs	Incl. O&P
1 Laborer	$36.65	$293.20	$56.55	$452.40	$43.45	$66.38
1 Equip. Oper. (crane)	50.25	402.00	76.20	609.60		
1 Hyd. Excavator, 1.5 C.Y.		1031.00		1134.10		
1 Trench Box		84.05		92.45	69.69	76.66
16 L.H., Daily Totals		$1810.25		$2288.55	$113.14	$143.03

Crew B-13F	Hr.	Daily	Hr.	Daily	Bare Costs	Incl. O&P
1 Laborer	$36.65	$293.20	$56.55	$452.40	$43.45	$66.38
1 Equip. Oper. (crane)	50.25	402.00	76.20	609.60		
1 Hyd. Excavator, 3.5 C.Y.		2442.00		2686.20		
1 Trench Box		84.05		92.45	157.88	173.67
16 L.H., Daily Totals		$3221.25		$3840.66	$201.33	$240.04

Crew B-13G	Hr.	Daily	Hr.	Daily	Bare Costs	Incl. O&P
1 Laborer	$36.65	$293.20	$56.55	$452.40	$43.45	$66.38
1 Equip. Oper. (crane)	50.25	402.00	76.20	609.60		
1 Hyd. Excavator, .75 C.Y.		662.00		728.20		
1 Trench Box		84.05		92.45	46.63	51.29
16 L.H., Daily Totals		$1441.25		$1882.66	$90.08	$117.67

Crew B-13H	Hr.	Daily	Hr.	Daily	Bare Costs	Incl. O&P
1 Laborer	$36.65	$293.20	$56.55	$452.40	$43.45	$66.38
1 Equip. Oper. (crane)	50.25	402.00	76.20	609.60		
1 Gradall, 5/8 C.Y.		881.60		969.76		
1 Trench Box		84.05		92.45	60.35	66.39
16 L.H., Daily Totals		$1660.85		$2124.22	$103.80	$132.76

Crew B-13I	Hr.	Daily	Hr.	Daily	Bare Costs	Incl. O&P
1 Laborer	$36.65	$293.20	$56.55	$452.40	$43.45	$66.38
1 Equip. Oper. (crane)	50.25	402.00	76.20	609.60		
1 Gradall, 3 Ton, 1 C.Y.		1005.00		1105.50		
1 Trench Box		84.05		92.45	68.07	74.87
16 L.H., Daily Totals		$1784.25		$2259.95	$111.52	$141.25

Crew B-13J	Hr.	Daily	Hr.	Daily	Bare Costs	Incl. O&P
1 Laborer	$36.65	$293.20	$56.55	$452.40	$43.45	$66.38
1 Equip. Oper. (crane)	50.25	402.00	76.20	609.60		
1 Hyd. Excavator, 2.5 C.Y.		1637.00		1800.70		
1 Trench Box		84.05		92.45	107.57	118.32
16 L.H., Daily Totals		$2416.25		$2955.16	$151.02	$184.70

Crew B-14	Hr.	Daily	Hr.	Daily	Bare Costs	Incl. O&P
1 Labor Foreman (outside)	$38.65	$309.20	$59.65	$477.20	$38.72	$59.53
4 Laborers	36.65	1172.80	56.55	1809.60		
1 Equip. Oper. (light)	47.05	376.40	71.35	570.80		
1 Backhoe Loader, 48 H.P.		365.20		401.72	7.61	8.37
48 L.H., Daily Totals		$2223.60		$3259.32	$46.33	$67.90

Crew B-14A	Hr.	Daily	Hr.	Daily	Bare Costs	Incl. O&P
1 Equip. Oper. (crane)	$50.25	$402.00	$76.20	$609.60	$45.72	$69.65
.5 Laborer	36.65	146.60	56.55	226.20		
1 Hyd. Excavator, 4.5 C.Y.		3043.00		3347.30	253.58	278.94
12 L.H., Daily Totals		$3591.60		$4183.10	$299.30	$348.59

Crew B-14B	Hr.	Daily	Hr.	Daily	Bare Costs	Incl. O&P
1 Equip. Oper. (crane)	$50.25	$402.00	$76.20	$609.60	$45.72	$69.65
.5 Laborer	36.65	146.60	56.55	226.20		
1 Hyd. Excavator, 6 C.Y.		3523.00		3875.30	293.58	322.94
12 L.H., Daily Totals		$4071.60		$4711.10	$339.30	$392.59

Crew B-14C	Hr.	Daily	Hr.	Daily	Bare Costs	Incl. O&P
1 Equip. Oper. (crane)	$50.25	$402.00	$76.20	$609.60	$45.72	$69.65
.5 Laborer	36.65	146.60	56.55	226.20		
1 Hyd. Excavator, 7 C.Y.		3596.00		3955.60	299.67	329.63
12 L.H., Daily Totals		$4144.60		$4791.40	$345.38	$399.28

Crew B-14F	Hr.	Daily	Hr.	Daily	Bare Costs	Incl. O&P
1 Equip. Oper. (crane)	$50.25	$402.00	$76.20	$609.60	$45.72	$69.65
.5 Laborer	36.65	146.60	56.55	226.20		
1 Hyd. Shovel, 7 C.Y.		3714.00		4085.40	309.50	340.45
12 L.H., Daily Totals		$4262.60		$4921.20	$355.22	$410.10

Crew B-14G	Hr.	Daily	Hr.	Daily	Bare Costs	Incl. O&P
1 Equip. Oper. (crane)	$50.25	$402.00	$76.20	$609.60	$45.72	$69.65
.5 Laborer	36.65	146.60	56.55	226.20		
1 Hyd. Shovel, 12 C.Y.		5204.00		5724.40	433.67	477.03
12 L.H., Daily Totals		$5752.60		$6560.20	$479.38	$546.68

Crew B-14J	Hr.	Daily	Hr.	Daily	Bare Costs	Incl. O&P
1 Equip. Oper. (medium)	$48.90	$391.20	$74.15	$593.20	$44.82	$68.28
.5 Laborer	36.65	146.60	56.55	226.20		
1 F.E. Loader, 8 C.Y.		1989.00		2187.90	165.75	182.32
12 L.H., Daily Totals		$2526.80		$3007.30	$210.57	$250.61

Crew No.	Bare Costs		Incl. Subs O&P		Cost Per Labor-Hour	
Crew B-14K	**Hr.**	**Daily**	**Hr.**	**Daily**	**Bare Costs**	**Incl. O&P**
1 Equip. Oper. (medium)	$48.90	$391.20	$74.15	$593.20	$44.82	$68.28
.5 Laborer	36.65	146.60	56.55	226.20		
1 F.E. Loader, 10 C.Y.		2703.00		2973.30	225.25	247.78
12 L.H., Daily Totals		$3240.80		$3792.70	$270.07	$316.06
Crew B-15	**Hr.**	**Daily**	**Hr.**	**Daily**	**Bare Costs**	**Incl. O&P**
1 Equipment Oper. (med.)	$48.90	$391.20	$74.15	$593.20	$40.66	$62.01
.5 Laborer	36.65	146.60	56.55	226.20		
2 Truck Drivers (heavy)	37.55	600.80	57.30	916.80		
2 Dump Trucks, 12 C.Y., 400 H.P.		1384.40		1522.84		
1 Dozer, 200 H.P.		1325.00		1457.50	96.76	106.44
28 L.H., Daily Totals		$3848.00		$4716.54	$137.43	$168.45
Crew B-16	**Hr.**	**Daily**	**Hr.**	**Daily**	**Bare Costs**	**Incl. O&P**
1 Labor Foreman (outside)	$38.65	$309.20	$59.65	$477.20	$37.38	$57.51
2 Laborers	36.65	586.40	56.55	904.80		
1 Truck Driver (heavy)	37.55	300.40	57.30	458.40		
1 Dump Truck, 12 C.Y., 400 H.P.		692.20		761.42	21.63	23.79
32 L.H., Daily Totals		$1888.20		$2601.82	$59.01	$81.31
Crew B-17	**Hr.**	**Daily**	**Hr.**	**Daily**	**Bare Costs**	**Incl. O&P**
2 Laborers	$36.65	$586.40	$56.55	$904.80	$39.48	$60.44
1 Equip. Oper. (light)	47.05	376.40	71.35	570.80		
1 Truck Driver (heavy)	37.55	300.40	57.30	458.40		
1 Backhoe Loader, 48 H.P.		365.20		401.72		
1 Dump Truck, 8 C.Y., 220 H.P.		417.20		458.92	24.45	26.90
32 L.H., Daily Totals		$2045.60		$2794.64	$63.92	$87.33
Crew B-17A	**Hr.**	**Daily**	**Hr.**	**Daily**	**Bare Costs**	**Incl. O&P**
2 Labor Foremen (outside)	$38.65	$618.40	$59.65	$954.40	$39.38	$60.82
6 Laborers	36.65	1759.20	56.55	2714.40		
1 Skilled Worker Foreman (out)	49.30	394.40	76.35	610.80		
1 Skilled Worker	47.30	378.40	73.25	586.00		
80 L.H., Daily Totals		$3150.40		$4865.60	$39.38	$60.82
Crew B-17B	**Hr.**	**Daily**	**Hr.**	**Daily**	**Bare Costs**	**Incl. O&P**
2 Laborers	$36.65	$586.40	$56.55	$904.80	$39.48	$60.44
1 Equip. Oper. (light)	47.05	376.40	71.35	570.80		
1 Truck Driver (heavy)	37.55	300.40	57.30	458.40		
1 Backhoe Loader, 48 H.P.		365.20		401.72		
1 Dump Truck, 12 C.Y., 400 H.P.		692.20		761.42	33.04	36.35
32 L.H., Daily Totals		$2320.60		$3097.14	$72.52	$96.79
Crew B-18	**Hr.**	**Daily**	**Hr.**	**Daily**	**Bare Costs**	**Incl. O&P**
1 Labor Foreman (outside)	$38.65	$309.20	$59.65	$477.20	$37.32	$57.58
2 Laborers	36.65	586.40	56.55	904.80		
1 Vibrating Plate, Gas, 21"		46.00		50.60	1.92	2.11
24 L.H., Daily Totals		$941.60		$1432.60	$39.23	$59.69
Crew B-19	**Hr.**	**Daily**	**Hr.**	**Daily**	**Bare Costs**	**Incl. O&P**
1 Pile Driver Foreman (outside)	$46.40	$371.20	$73.60	$588.80	$46.01	$71.70
4 Pile Drivers	44.40	1420.80	70.40	2252.80		
2 Equip. Oper. (crane)	50.25	804.00	76.20	1219.20		
1 Equip. Oper. (oiler)	43.55	348.40	66.00	528.00		
1 Crawler Crane, 40 Ton		1179.00		1296.90		
1 Lead, 90' High		124.40		136.84		
1 Hammer, Diesel, 22k ft-lb		537.45		591.20	28.76	31.64
64 L.H., Daily Totals		$4785.25		$6613.73	$74.77	$103.34

Crew No.	Bare Costs		Incl. Subs O&P		Cost Per Labor-Hour	
Crew B-19A	**Hr.**	**Daily**	**Hr.**	**Daily**	**Bare Costs**	**Incl. O&P**
1 Pile Driver Foreman (outside)	$46.40	$371.20	$73.60	$588.80	$46.01	$71.70
4 Pile Drivers	44.40	1420.80	70.40	2252.80		
2 Equip. Oper. (crane)	50.25	804.00	76.20	1219.20		
1 Equip. Oper. (oiler)	43.55	348.40	66.00	528.00		
1 Crawler Crane, 75 Ton		1492.00		1641.20		
1 Lead, 90' high		124.40		136.84		
1 Hammer, Diesel, 41k ft-lb		642.40		706.64	35.29	38.82
64 L.H., Daily Totals		$5203.20		$7073.48	$81.30	$110.52
Crew B-19B	**Hr.**	**Daily**	**Hr.**	**Daily**	**Bare Costs**	**Incl. O&P**
1 Pile Driver Foreman (outside)	$46.40	$371.20	$73.60	$588.80	$46.01	$71.70
4 Pile Drivers	44.40	1420.80	70.40	2252.80		
2 Equip. Oper. (crane)	50.25	804.00	76.20	1219.20		
1 Equip. Oper. (oiler)	43.55	348.40	66.00	528.00		
1 Crawler Crane, 40 Ton		1179.00		1296.90		
1 Lead, 90' High		124.40		136.84		
1 Hammer, Diesel, 22k ft-lb		537.45		591.20		
1 Barge, 400 Ton		781.00		859.10	40.97	45.06
64 L.H., Daily Totals		$5566.25		$7472.84	$86.97	$116.76
Crew B-19C	**Hr.**	**Daily**	**Hr.**	**Daily**	**Bare Costs**	**Incl. O&P**
1 Pile Driver Foreman (outside)	$46.40	$371.20	$73.60	$588.80	$46.01	$71.70
4 Pile Drivers	44.40	1420.80	70.40	2252.80		
2 Equip. Oper. (crane)	50.25	804.00	76.20	1219.20		
1 Equip. Oper. (oiler)	43.55	348.40	66.00	528.00		
1 Crawler Crane, 75 Ton		1492.00		1641.20		
1 Lead, 90' High		124.40		136.84		
1 Hammer, Diesel, 41k ft-lb		642.40		706.64		
1 Barge, 400 Ton		781.00		859.10	47.50	52.25
64 L.H., Daily Totals		$5984.20		$7932.58	$93.50	$123.95
Crew B-20	**Hr.**	**Daily**	**Hr.**	**Daily**	**Bare Costs**	**Incl. O&P**
1 Labor Foreman (outside)	$38.65	$309.20	$59.65	$477.20	$40.87	$63.15
1 Skilled Worker	47.30	378.40	73.25	586.00		
1 Laborer	36.65	293.20	56.55	452.40		
24 L.H., Daily Totals		$980.80		$1515.60	$40.87	$63.15
Crew B-20A	**Hr.**	**Daily**	**Hr.**	**Daily**	**Bare Costs**	**Incl. O&P**
1 Labor Foreman (outside)	$38.65	$309.20	$59.65	$477.20	$44.73	$68.16
1 Laborer	36.65	293.20	56.55	452.40		
1 Plumber	57.55	460.40	86.90	695.20		
1 Plumber Apprentice	46.05	368.40	69.55	556.40		
32 L.H., Daily Totals		$1431.20		$2181.20	$44.73	$68.16
Crew B-21	**Hr.**	**Daily**	**Hr.**	**Daily**	**Bare Costs**	**Incl. O&P**
1 Labor Foreman (outside)	$38.65	$309.20	$59.65	$477.20	$42.21	$65.01
1 Skilled Worker	47.30	378.40	73.25	586.00		
1 Laborer	36.65	293.20	56.55	452.40		
.5 Equip. Oper. (crane)	50.25	201.00	76.20	304.80		
.5 S.P. Crane, 4x4, 5 Ton		140.00		154.00	5.00	5.50
28 L.H., Daily Totals		$1321.80		$1974.40	$47.21	$70.51
Crew B-21A	**Hr.**	**Daily**	**Hr.**	**Daily**	**Bare Costs**	**Incl. O&P**
1 Labor Foreman (outside)	$38.65	$309.20	$59.65	$477.20	$45.83	$69.77
1 Laborer	36.65	293.20	56.55	452.40		
1 Plumber	57.55	460.40	86.90	695.20		
1 Plumber Apprentice	46.05	368.40	69.55	556.40		
1 Equip. Oper. (crane)	50.25	402.00	76.20	609.60		
1 S.P. Crane, 4x4, 12 Ton		481.20		529.32	12.03	13.23
40 L.H., Daily Totals		$2314.40		$3320.12	$57.86	$83.00

Crew No.	Bare Costs		Incl. Subs O&P		Cost Per Labor-Hour	
Crew B-21B	**Hr.**	**Daily**	**Hr.**	**Daily**	**Bare Costs**	**Incl. O&P**
1 Labor Foreman (outside)	$38.65	$309.20	$59.65	$477.20	$39.77	$61.10
3 Laborers	36.65	879.60	56.55	1357.20		
1 Equip. Oper. (crane)	50.25	402.00	76.20	609.60		
1 Hyd. Crane, 12 Ton		663.00		729.30	16.57	18.23
40 L.H., Daily Totals		$2253.80		$3173.30	$56.34	$79.33
Crew B-21C	**Hr.**	**Daily**	**Hr.**	**Daily**	**Bare Costs**	**Incl. O&P**
1 Labor Foreman (outside)	$38.65	$309.20	$59.65	$477.20	$39.86	$61.15
4 Laborers	36.65	1172.80	56.55	1809.60		
1 Equip. Oper. (crane)	50.25	402.00	76.20	609.60		
1 Equip. Oper. (oiler)	43.55	348.40	66.00	528.00		
2 Cutting Torches		22.80		25.08		
2 Sets of Gases		304.00		334.40		
1 Lattice Boom Crane, 90 Ton		1529.00		1681.90	33.14	36.45
56 L.H., Daily Totals		$4088.20		$5465.78	$73.00	$97.60
Crew B-22	**Hr.**	**Daily**	**Hr.**	**Daily**	**Bare Costs**	**Incl. O&P**
1 Labor Foreman (outside)	$38.65	$309.20	$59.65	$477.20	$42.74	$65.76
1 Skilled Worker	47.30	378.40	73.25	586.00		
1 Laborer	36.65	293.20	56.55	452.40		
.75 Equip. Oper. (crane)	50.25	301.50	76.20	457.20		
.75 S.P. Crane, 4x4, 5 Ton		210.00		231.00	7.00	7.70
30 L.H., Daily Totals		$1492.30		$2203.80	$49.74	$73.46
Crew B-22A	**Hr.**	**Daily**	**Hr.**	**Daily**	**Bare Costs**	**Incl. O&P**
1 Labor Foreman (outside)	$38.65	$309.20	$59.65	$477.20	$41.90	$64.44
1 Skilled Worker	47.30	378.40	73.25	586.00		
2 Laborers	36.65	586.40	56.55	904.80		
1 Equipment Operator, Crane	50.25	402.00	76.20	609.60		
1 S.P. Crane, 4x4, 5 Ton		280.00		308.00		
1 Butt Fusion Mach., 4"-12" diam.		381.80		419.98	16.55	18.20
40 L.H., Daily Totals		$2337.80		$3305.58	$58.45	$82.64
Crew B-22B	**Hr.**	**Daily**	**Hr.**	**Daily**	**Bare Costs**	**Incl. O&P**
1 Labor Foreman (outside)	$38.65	$309.20	$59.65	$477.20	$41.90	$64.44
1 Skilled Worker	47.30	378.40	73.25	586.00		
2 Laborers	36.65	586.40	56.55	904.80		
1 Equip. Oper. (crane)	50.25	402.00	76.20	609.60		
1 S.P. Crane, 4x4, 5 Ton		280.00		308.00		
1 Butt Fusion Mach., 8"-24" diam.		828.40		911.24	27.71	30.48
40 L.H., Daily Totals		$2784.40		$3796.84	$69.61	$94.92
Crew B-22C	**Hr.**	**Daily**	**Hr.**	**Daily**	**Bare Costs**	**Incl. O&P**
1 Skilled Worker	$47.30	$378.40	$73.25	$586.00	$41.98	$64.90
1 Laborer	36.65	293.20	56.55	452.40		
1 Butt Fusion Mach., 2"-8" diam.		120.95		133.04	7.56	8.32
16 L.H., Daily Totals		$792.55		$1171.44	$49.53	$73.22
Crew B-23	**Hr.**	**Daily**	**Hr.**	**Daily**	**Bare Costs**	**Incl. O&P**
1 Labor Foreman (outside)	$38.65	$309.20	$59.65	$477.20	$37.05	$57.17
4 Laborers	36.65	1172.80	56.55	1809.60		
1 Drill Rig, Truck-Mounted		2551.00		2806.10		
1 Flatbed Truck, Gas, 3 Ton		333.00		366.30	72.10	79.31
40 L.H., Daily Totals		$4366.00		$5459.20	$109.15	$136.48

Crew No.	Bare Costs		Incl. Subs O&P		Cost Per Labor-Hour	
Crew B-23A	**Hr.**	**Daily**	**Hr.**	**Daily**	**Bare Costs**	**Incl. O&P**
1 Labor Foreman (outside)	$38.65	$309.20	$59.65	$477.20	$41.40	$63.45
1 Laborer	36.65	293.20	56.55	452.40		
1 Equip. Oper. (medium)	48.90	391.20	74.15	593.20		
1 Drill Rig, Truck-Mounted		2551.00		2806.10		
1 Pickup Truck, 3/4 Ton		157.00		172.70	112.83	124.12
24 L.H., Daily Totals		$3701.60		$4501.60	$154.23	$187.57
Crew B-23B	**Hr.**	**Daily**	**Hr.**	**Daily**	**Bare Costs**	**Incl. O&P**
1 Labor Foreman (outside)	$38.65	$309.20	$59.65	$477.20	$41.40	$63.45
1 Laborer	36.65	293.20	56.55	452.40		
1 Equip. Oper. (medium)	48.90	391.20	74.15	593.20		
1 Drill Rig, Truck-Mounted		2551.00		2806.10		
1 Pickup Truck, 3/4 Ton		157.00		172.70		
1 Centr. Water Pump, 6"		340.80		374.88	127.03	139.74
24 L.H., Daily Totals		$4042.40		$4876.48	$168.43	$203.19
Crew B-24	**Hr.**	**Daily**	**Hr.**	**Daily**	**Bare Costs**	**Incl. O&P**
1 Cement Finisher	$44.05	$352.40	$65.10	$520.80	$42.18	$64.13
1 Laborer	36.65	293.20	56.55	452.40		
1 Carpenter	45.85	366.80	70.75	566.00		
24 L.H., Daily Totals		$1012.40		$1539.20	$42.18	$64.13
Crew B-25	**Hr.**	**Daily**	**Hr.**	**Daily**	**Bare Costs**	**Incl. O&P**
1 Labor Foreman (outside)	$38.65	$309.20	$59.65	$477.20	$40.17	$61.63
7 Laborers	36.65	2052.40	56.55	3166.80		
3 Equip. Oper. (medium)	48.90	1173.60	74.15	1779.60		
1 Asphalt Paver, 130 H.P.		2169.00		2385.90		
1 Tandem Roller, 10 Ton		242.00		266.20		
1 Roller, Pneum. Whl., 12 Ton		349.00		383.90	31.36	34.50
88 L.H., Daily Totals		$6295.20		$8459.60	$71.54	$96.13
Crew B-25B	**Hr.**	**Daily**	**Hr.**	**Daily**	**Bare Costs**	**Incl. O&P**
1 Labor Foreman (outside)	$38.65	$309.20	$59.65	$477.20	$40.90	$62.67
7 Laborers	36.65	2052.40	56.55	3166.80		
4 Equip. Oper. (medium)	48.90	1564.80	74.15	2372.80		
1 Asphalt Paver, 130 H.P.		2169.00		2385.90		
2 Tandem Rollers, 10 Ton		484.00		532.40		
1 Roller, Pneum. Whl., 12 Ton		349.00		383.90	31.27	34.40
96 L.H., Daily Totals		$6928.40		$9319.00	$72.17	$97.07
Crew B-25C	**Hr.**	**Daily**	**Hr.**	**Daily**	**Bare Costs**	**Incl. O&P**
1 Labor Foreman (outside)	$38.65	$309.20	$59.65	$477.20	$41.07	$62.93
3 Laborers	36.65	879.60	56.55	1357.20		
2 Equip. Oper. (medium)	48.90	782.40	74.15	1186.40		
1 Asphalt Paver, 130 H.P.		2169.00		2385.90		
1 Tandem Roller, 10 Ton		242.00		266.20	50.23	55.25
48 L.H., Daily Totals		$4382.20		$5672.90	$91.30	$118.19
Crew B-25D	**Hr.**	**Daily**	**Hr.**	**Daily**	**Bare Costs**	**Incl. O&P**
1 Labor Foreman (outside)	$38.65	$309.20	$59.65	$477.20	$41.15	$63.05
3 Laborers	36.65	879.60	56.55	1357.20		
2.125 Equip. Oper. (medium)	48.90	831.30	74.15	1260.55		
.125 Truck Driver (heavy)	37.55	37.55	57.30	57.30		
.125 Truck Tractor, 6x4, 380 H.P.		76.60		84.26		
.125 Dist. Tanker, 3000 Gallon		38.52		42.38		
1 Asphalt Paver, 130 H.P.		2169.00		2385.90		
1 Tandem Roller, 10 Ton		242.00		266.20	50.52	55.57
50 L.H., Daily Totals		$4583.77		$5930.99	$91.68	$118.62

Crew B-25E	Hr.	Daily	Hr.	Daily	Bare Costs	Incl. O&P
1 Labor Foreman (outside)	$38.65	$309.20	$59.65	$477.20	$41.23	$63.15
3 Laborers	36.65	879.60	56.55	1357.20		
2.250 Equip. Oper. (medium)	48.90	880.20	74.15	1334.70		
.25 Truck Driver (heavy)	37.55	75.10	57.30	114.60		
.25 Truck Tractor, 6x4, 380 H.P.		153.20		168.52		
.25 Dist. Tanker, 3000 Gallon		77.05		84.75		
1 Asphalt Paver, 130 H.P.		2169.00		2385.90		
1 Tandem Roller, 10 Ton		242.00		266.20	50.79	55.87
52 L.H., Daily Totals		$4785.35		$6189.07	$92.03	$119.02

Crew B-26	Hr.	Daily	Hr.	Daily	Bare Costs	Incl. O&P
1 Labor Foreman (outside)	$38.65	$309.20	$59.65	$477.20	$41.00	$62.90
6 Laborers	36.65	1759.20	56.55	2714.40		
2 Equip. Oper. (medium)	48.90	782.40	74.15	1186.40		
1 Rodman (reinf.)	50.65	405.20	79.55	636.40		
1 Cement Finisher	44.05	352.40	65.10	520.80		
1 Grader, 30,000 Lbs.		707.00		777.70		
1 Paving Mach. & Equip.		2657.00		2922.70	38.23	42.05
88 L.H., Daily Totals		$6972.40		$9235.60	$79.23	$104.95

Crew B-26A	Hr.	Daily	Hr.	Daily	Bare Costs	Incl. O&P
1 Labor Foreman (outside)	$38.65	$309.20	$59.65	$477.20	$41.00	$62.90
6 Laborers	36.65	1759.20	56.55	2714.40		
2 Equip. Oper. (medium)	48.90	782.40	74.15	1186.40		
1 Rodman (reinf.)	50.65	405.20	79.55	636.40		
1 Cement Finisher	44.05	352.40	65.10	520.80		
1 Grader, 30,000 Lbs.		707.00		777.70		
1 Paving Mach. & Equip.		2657.00		2922.70		
1 Concrete Saw		175.20		192.72	40.22	44.24
88 L.H., Daily Totals		$7147.60		$9428.32	$81.22	$107.14

Crew B-26B	Hr.	Daily	Hr.	Daily	Bare Costs	Incl. O&P
1 Labor Foreman (outside)	$38.65	$309.20	$59.65	$477.20	$41.66	$63.84
6 Laborers	36.65	1759.20	56.55	2714.40		
3 Equip. Oper. (medium)	48.90	1173.60	74.15	1779.60		
1 Rodman (reinf.)	50.65	405.20	79.55	636.40		
1 Cement Finisher	44.05	352.40	65.10	520.80		
1 Grader, 30,000 Lbs.		707.00		777.70		
1 Paving Mach. & Equip.		2657.00		2922.70		
1 Concrete Pump, 110' Boom		934.00		1027.40	44.77	49.25
96 L.H., Daily Totals		$8297.60		$10856.20	$86.43	$113.09

Crew B-26C	Hr.	Daily	Hr.	Daily	Bare Costs	Incl. O&P
1 Labor Foreman (outside)	$38.65	$309.20	$59.65	$477.20	$40.22	$61.77
6 Laborers	36.65	1759.20	56.55	2714.40		
1 Equip. Oper. (medium)	48.90	391.20	74.15	593.20		
1 Rodman (reinf.)	50.65	405.20	79.55	636.40		
1 Cement Finisher	44.05	352.40	65.10	520.80		
1 Paving Mach. & Equip.		2657.00		2922.70		
1 Concrete Saw		175.20		192.72	35.40	38.94
80 L.H., Daily Totals		$6049.40		$8057.42	$75.62	$100.72

Crew B-27	Hr.	Daily	Hr.	Daily	Bare Costs	Incl. O&P
1 Labor Foreman (outside)	$38.65	$309.20	$59.65	$477.20	$37.15	$57.33
3 Laborers	36.65	879.60	56.55	1357.20		
1 Berm Machine		294.40		323.84	9.20	10.12
32 L.H., Daily Totals		$1483.20		$2158.24	$46.35	$67.44

Crew B-28	Hr.	Daily	Hr.	Daily	Bare Costs	Incl. O&P
2 Carpenters	$45.85	$733.60	$70.75	$1132.00	$42.78	$66.02
1 Laborer	36.65	293.20	56.55	452.40		
24 L.H., Daily Totals		$1026.80		$1584.40	$42.78	$66.02

Crew B-29	Hr.	Daily	Hr.	Daily	Bare Costs	Incl. O&P
1 Labor Foreman (outside)	$38.65	$309.20	$59.65	$477.20	$39.86	$61.15
4 Laborers	36.65	1172.80	56.55	1809.60		
1 Equip. Oper. (crane)	50.25	402.00	76.20	609.60		
1 Equip. Oper. (oiler)	43.55	348.40	66.00	528.00		
1 Gradall, 5/8 C.Y.		881.60		969.76	15.74	17.32
56 L.H., Daily Totals		$3114.00		$4394.16	$55.61	$78.47

Crew B-30	Hr.	Daily	Hr.	Daily	Bare Costs	Incl. O&P
1 Equip. Oper. (medium)	$48.90	$391.20	$74.15	$593.20	$41.33	$62.92
2 Truck Drivers (heavy)	37.55	600.80	57.30	916.80		
1 Hyd. Excavator, 1.5 C.Y.		1031.00		1134.10		
2 Dump Trucks, 12 C.Y., 400 H.P.		1384.40		1522.84	100.64	110.71
24 L.H., Daily Totals		$3407.40		$4166.94	$141.97	$173.62

Crew B-31	Hr.	Daily	Hr.	Daily	Bare Costs	Incl. O&P
1 Labor Foreman (outside)	$38.65	$309.20	$59.65	$477.20	$38.89	$60.01
3 Laborers	36.65	879.60	56.55	1357.20		
1 Carpenter	45.85	366.80	70.75	566.00		
1 Air Compressor, 250 cfm		202.40		222.64		
1 Sheeting Driver		5.30		5.83		
2 -50' Air Hoses, 1.5"		11.60		12.76	5.48	6.03
40 L.H., Daily Totals		$1774.90		$2641.63	$44.37	$66.04

Crew B-32	Hr.	Daily	Hr.	Daily	Bare Costs	Incl. O&P
1 Laborer	$36.65	$293.20	$56.55	$452.40	$45.84	$69.75
3 Equip. Oper. (medium)	48.90	1173.60	74.15	1779.60		
1 Grader, 30,000 Lbs.		707.00		777.70		
1 Tandem Roller, 10 Ton		242.00		266.20		
1 Dozer, 200 H.P.		1325.00		1457.50	71.06	78.17
32 L.H., Daily Totals		$3740.80		$4733.40	$116.90	$147.92

Crew B-32A	Hr.	Daily	Hr.	Daily	Bare Costs	Incl. O&P
1 Laborer	$36.65	$293.20	$56.55	$452.40	$44.82	$68.28
2 Equip. Oper. (medium)	48.90	782.40	74.15	1186.40		
1 Grader, 30,000 Lbs.		707.00		777.70		
1 Roller, Vibratory, 25 Ton		692.80		762.08	58.33	64.16
24 L.H., Daily Totals		$2475.40		$3178.58	$103.14	$132.44

Crew B-32B	Hr.	Daily	Hr.	Daily	Bare Costs	Incl. O&P
1 Laborer	$36.65	$293.20	$56.55	$452.40	$44.82	$68.28
2 Equip. Oper. (medium)	48.90	782.40	74.15	1186.40		
1 Dozer, 200 H.P.		1325.00		1457.50		
1 Roller, Vibratory, 25 Ton		692.80		762.08	84.08	92.48
24 L.H., Daily Totals		$3093.40		$3858.38	$128.89	$160.77

Crew B-32C	Hr.	Daily	Hr.	Daily	Bare Costs	Incl. O&P
1 Labor Foreman (outside)	$38.65	$309.20	$59.65	$477.20	$43.11	$65.87
2 Laborers	36.65	586.40	56.55	904.80		
3 Equip. Oper. (medium)	48.90	1173.60	74.15	1779.60		
1 Grader, 30,000 Lbs.		707.00		777.70		
1 Tandem Roller, 10 Ton		242.00		266.20		
1 Dozer, 200 H.P.		1325.00		1457.50	47.38	52.11
48 L.H., Daily Totals		$4343.20		$5663.00	$90.48	$117.98

Crew No.	Bare Costs		Incl. Subs O&P		Cost Per Labor-Hour	
Crew B-33A	**Hr.**	**Daily**	**Hr.**	**Daily**	**Bare Costs**	**Incl. O&P**
1 Equip. Oper. (medium)	$48.90	$391.20	$74.15	$593.20	$45.40	$69.12
.5 Laborer	36.65	146.60	56.55	226.20		
.25 Equip. Oper. (medium)	48.90	97.80	74.15	148.30		
1 Scraper, Towed, 7 C.Y.		111.60		122.76		
1.25 Dozers, 300 H.P.		2271.25		2498.38	170.20	187.22
14 L.H., Daily Totals		$3018.45		$3588.84	$215.60	$256.35

Crew B-33B	Hr.	Daily	Hr.	Daily	Bare Costs	Incl. O&P
1 Equip. Oper. (medium)	$48.90	$391.20	$74.15	$593.20	$45.40	$69.12
.5 Laborer	36.65	146.60	56.55	226.20		
.25 Equip. Oper. (medium)	48.90	97.80	74.15	148.30		
1 Scraper, Towed, 10 C.Y.		141.60		155.76		
1.25 Dozers, 300 H.P.		2271.25		2498.38	172.35	189.58
14 L.H., Daily Totals		$3048.45		$3621.84	$217.75	$258.70

Crew B-33C	Hr.	Daily	Hr.	Daily	Bare Costs	Incl. O&P
1 Equip. Oper. (medium)	$48.90	$391.20	$74.15	$593.20	$45.40	$69.12
.5 Laborer	36.65	146.60	56.55	226.20		
.25 Equip. Oper. (medium)	48.90	97.80	74.15	148.30		
1 Scraper, Towed, 15 C.Y.		158.80		174.68		
1.25 Dozers, 300 H.P.		2271.25		2498.38	173.57	190.93
14 L.H., Daily Totals		$3065.65		$3640.76	$218.97	$260.05

Crew B-33D	Hr.	Daily	Hr.	Daily	Bare Costs	Incl. O&P
1 Equip. Oper. (medium)	$48.90	$391.20	$74.15	$593.20	$45.40	$69.12
.5 Laborer	36.65	146.60	56.55	226.20		
.25 Equip. Oper. (medium)	48.90	97.80	74.15	148.30		
1 S.P. Scraper, 14 C.Y.		1829.00		2011.90		
.25 Dozer, 300 H.P.		454.25		499.68	163.09	179.40
14 L.H., Daily Totals		$2918.85		$3479.28	$208.49	$248.52

Crew B-33E	Hr.	Daily	Hr.	Daily	Bare Costs	Incl. O&P
1 Equip. Oper. (medium)	$48.90	$391.20	$74.15	$593.20	$45.40	$69.12
.5 Laborer	36.65	146.60	56.55	226.20		
.25 Equip. Oper. (medium)	48.90	97.80	74.15	148.30		
1 S.P. Scraper, 21 C.Y.		2599.00		2858.90		
.25 Dozer, 300 H.P.		454.25		499.68	218.09	239.90
14 L.H., Daily Totals		$3688.85		$4326.27	$263.49	$309.02

Crew B-33F	Hr.	Daily	Hr.	Daily	Bare Costs	Incl. O&P
1 Equip. Oper. (medium)	$48.90	$391.20	$74.15	$593.20	$45.40	$69.12
.5 Laborer	36.65	146.60	56.55	226.20		
.25 Equip. Oper. (medium)	48.90	97.80	74.15	148.30		
1 Elev. Scraper, 11 C.Y.		1180.00		1298.00		
.25 Dozer, 300 H.P.		454.25		499.68	116.73	128.41
14 L.H., Daily Totals		$2269.85		$2765.38	$162.13	$197.53

Crew B-33G	Hr.	Daily	Hr.	Daily	Bare Costs	Incl. O&P
1 Equip. Oper. (medium)	$48.90	$391.20	$74.15	$593.20	$45.40	$69.12
.5 Laborer	36.65	146.60	56.55	226.20		
.25 Equip. Oper. (medium)	48.90	97.80	74.15	148.30		
1 Elev. Scraper, 22 C.Y.		2443.00		2687.30		
.25 Dozer, 300 H.P.		454.25		499.68	206.95	227.64
14 L.H., Daily Totals		$3532.85		$4154.68	$252.35	$296.76

Crew No.	Bare Costs		Incl. Subs O&P		Cost Per Labor-Hour	
Crew B-33H	**Hr.**	**Daily**	**Hr.**	**Daily**	**Bare Costs**	**Incl. O&P**
.5 Laborer	$36.65	$146.60	$56.55	$226.20	$45.40	$69.12
1 Equipment Operator (med.)	48.90	391.20	74.15	593.20		
.25 Equipment Operator (med.)	48.90	97.80	74.15	148.30		
1 S.P. Scraper, 44 C.Y.		4492.00		4941.20		
.25 Dozer, 410 H.P.		602.25		662.48	363.88	400.26
14 L.H., Daily Totals		$5729.85		$6571.38	$409.27	$469.38

Crew B-33J	Hr.	Daily	Hr.	Daily	Bare Costs	Incl. O&P
1 Equipment Operator (med.)	$48.90	$391.20	$74.15	$593.20	$48.90	$74.15
1 S.P. Scraper, 14 C.Y.		1829.00		2011.90	228.63	251.49
8 L.H., Daily Totals		$2220.20		$2605.10	$277.52	$325.64

Crew B-33K	Hr.	Daily	Hr.	Daily	Bare Costs	Incl. O&P
1 Equipment Operator (med.)	$48.90	$391.20	$74.15	$593.20	$45.40	$69.12
.25 Equipment Operator (med.)	48.90	97.80	74.15	148.30		
.5 Laborer	36.65	146.60	56.55	226.20		
1 S.P. Scraper, 31 C.Y.		3566.00		3922.60		
.25 Dozer, 410 H.P.		602.25		662.48	297.73	327.51
14 L.H., Daily Totals		$4803.85		$5552.77	$343.13	$396.63

Crew B-34A	Hr.	Daily	Hr.	Daily	Bare Costs	Incl. O&P
1 Truck Driver (heavy)	$37.55	$300.40	$57.30	$458.40	$37.55	$57.30
1 Dump Truck, 8 C.Y., 220 H.P.		417.20		458.92	52.15	57.37
8 L.H., Daily Totals		$717.60		$917.32	$89.70	$114.67

Crew B-34B	Hr.	Daily	Hr.	Daily	Bare Costs	Incl. O&P
1 Truck Driver (heavy)	$37.55	$300.40	$57.30	$458.40	$37.55	$57.30
1 Dump Truck, 12 C.Y., 400 H.P.		692.20		761.42	86.53	95.18
8 L.H., Daily Totals		$992.60		$1219.82	$124.08	$152.48

Crew B-34C	Hr.	Daily	Hr.	Daily	Bare Costs	Incl. O&P
1 Truck Driver (heavy)	$37.55	$300.40	$57.30	$458.40	$37.55	$57.30
1 Truck Tractor, 6x4, 380 H.P.		612.80		674.08		
1 Dump Trailer, 16.5 C.Y.		126.20		138.82	92.38	101.61
8 L.H., Daily Totals		$1039.40		$1271.30	$129.93	$158.91

Crew B-34D	Hr.	Daily	Hr.	Daily	Bare Costs	Incl. O&P
1 Truck Driver (heavy)	$37.55	$300.40	$57.30	$458.40	$37.55	$57.30
1 Truck Tractor, 6x4, 380 H.P.		612.80		674.08		
1 Dump Trailer, 20 C.Y.		140.80		154.88	94.20	103.62
8 L.H., Daily Totals		$1054.00		$1287.36	$131.75	$160.92

Crew B-34E	Hr.	Daily	Hr.	Daily	Bare Costs	Incl. O&P
1 Truck Driver (heavy)	$37.55	$300.40	$57.30	$458.40	$37.55	$57.30
1 Dump Truck, Off Hwy., 25 Ton		1314.00		1445.40	164.25	180.68
8 L.H., Daily Totals		$1614.40		$1903.80	$201.80	$237.97

Crew B-34F	Hr.	Daily	Hr.	Daily	Bare Costs	Incl. O&P
1 Truck Driver (heavy)	$37.55	$300.40	$57.30	$458.40	$37.55	$57.30
1 Dump Truck, Off Hwy., 35 Ton		1470.00		1617.00	183.75	202.13
8 L.H., Daily Totals		$1770.40		$2075.40	$221.30	$259.43

Crew B-34G	Hr.	Daily	Hr.	Daily	Bare Costs	Incl. O&P
1 Truck Driver (heavy)	$37.55	$300.40	$57.30	$458.40	$37.55	$57.30
1 Dump Truck, Off Hwy., 50 Ton		1787.00		1965.70	223.38	245.71
8 L.H., Daily Totals		$2087.40		$2424.10	$260.93	$303.01

Crew B-34H	Hr.	Daily	Hr.	Daily	Bare Costs	Incl. O&P
1 Truck Driver (heavy)	$37.55	$300.40	$57.30	$458.40	$37.55	$57.30
1 Dump Truck, Off Hwy., 65 Ton		1797.00		1976.70	224.63	247.09
8 L.H., Daily Totals		$2097.40		$2435.10	$262.18	$304.39

Crew B-34I	Hr.	Daily	Hr.	Daily	Bare Costs	Incl. O&P
1 Truck Driver (heavy)	$37.55	$300.40	$57.30	$458.40	$37.55	$57.30
1 Dump Truck, 18 C.Y., 450 H.P.		867.60		954.36	108.45	119.30
8 L.H., Daily Totals		$1168.00		$1412.76	$146.00	$176.60

Crew B-34J	Hr.	Daily	Hr.	Daily	Bare Costs	Incl. O&P
1 Truck Driver (heavy)	$37.55	$300.40	$57.30	$458.40	$37.55	$57.30
1 Dump Truck, Off Hwy., 100 Ton		2898.00		3187.80	362.25	398.48
8 L.H., Daily Totals		$3198.40		$3646.20	$399.80	$455.77

Crew B-34K	Hr.	Daily	Hr.	Daily	Bare Costs	Incl. O&P
1 Truck Driver (heavy)	$37.55	$300.40	$57.30	$458.40	$37.55	$57.30
1 Truck Tractor, 6x4, 450 H.P.		742.40		816.64		
1 Lowbed Trailer, 75 Ton		219.80		241.78	120.28	132.30
8 L.H., Daily Totals		$1262.60		$1516.82	$157.82	$189.60

Crew B-34L	Hr.	Daily	Hr.	Daily	Bare Costs	Incl. O&P
1 Equip. Oper. (light)	$47.05	$376.40	$71.35	$570.80	$47.05	$71.35
1 Flatbed Truck, Gas, 1.5 Ton		270.20		297.22	33.77	37.15
8 L.H., Daily Totals		$646.60		$868.02	$80.83	$108.50

Crew B-34M	Hr.	Daily	Hr.	Daily	Bare Costs	Incl. O&P
1 Equip. Oper. (light)	$47.05	$376.40	$71.35	$570.80	$47.05	$71.35
1 Flatbed Truck, Gas, 3 Ton		333.00		366.30	41.63	45.79
8 L.H., Daily Totals		$709.40		$937.10	$88.67	$117.14

Crew B-34N	Hr.	Daily	Hr.	Daily	Bare Costs	Incl. O&P
1 Truck Driver (heavy)	$37.55	$300.40	$57.30	$458.40	$37.55	$57.30
1 Dump Truck, 8 C.Y., 220 H.P.		417.20		458.92		
1 Flatbed Trailer, 40 Ton		152.00		167.20	71.15	78.27
8 L.H., Daily Totals		$869.60		$1084.52	$108.70	$135.57

Crew B-34P	Hr.	Daily	Hr.	Daily	Bare Costs	Incl. O&P
1 Pipe Fitter	$58.50	$468.00	$88.35	$706.80	$47.97	$72.73
1 Truck Driver (light)	36.50	292.00	55.70	445.60		
1 Equip. Oper. (medium)	48.90	391.20	74.15	593.20		
1 Flatbed Truck, Gas, 3 Ton		333.00		366.30		
1 Backhoe Loader, 48 H.P.		365.20		401.72	29.09	32.00
24 L.H., Daily Totals		$1849.40		$2513.62	$77.06	$104.73

Crew B-34Q	Hr.	Daily	Hr.	Daily	Bare Costs	Incl. O&P
1 Pipe Fitter	$58.50	$468.00	$88.35	$706.80	$48.42	$73.42
1 Truck Driver (light)	36.50	292.00	55.70	445.60		
1 Equip. Oper. (crane)	50.25	402.00	76.20	609.60		
1 Flatbed Trailer, 25 Ton		113.60		124.96		
1 Dump Truck, 8 C.Y., 220 H.P.		417.20		458.92		
1 Hyd. Crane, 25 Ton		745.80		820.38	53.19	58.51
24 L.H., Daily Totals		$2438.60		$3166.26	$101.61	$131.93

Crew B-34R	Hr.	Daily	Hr.	Daily	Bare Costs	Incl. O&P
1 Pipe Fitter	$58.50	$468.00	$88.35	$706.80	$48.42	$73.42
1 Truck Driver (light)	36.50	292.00	55.70	445.60		
1 Equip. Oper. (crane)	50.25	402.00	76.20	609.60		
1 Flatbed Trailer, 25 Ton		113.60		124.96		
1 Dump Truck, 8 C.Y., 220 H.P.		417.20		458.92		
1 Hyd. Crane, 25 Ton		745.80		820.38		
1 Hyd. Excavator, 1 C.Y.		814.80		896.28	87.14	95.86
24 L.H., Daily Totals		$3253.40		$4062.54	$135.56	$169.27

Crew B-34S	Hr.	Daily	Hr.	Daily	Bare Costs	Incl. O&P
2 Pipe Fitters	$58.50	$936.00	$88.35	$1413.60	$51.20	$77.55
1 Truck Driver (heavy)	37.55	300.40	57.30	458.40		
1 Equip. Oper. (crane)	50.25	402.00	76.20	609.60		
1 Flatbed Trailer, 40 Ton		152.00		167.20		
1 Truck Tractor, 6x4, 380 H.P.		612.80		674.08		
1 Hyd. Crane, 80 Ton		1652.00		1817.20		
1 Hyd. Excavator, 2 C.Y.		1175.00		1292.50	112.24	123.47
32 L.H., Daily Totals		$5230.20		$6432.58	$163.44	$201.02

Crew B-34T	Hr.	Daily	Hr.	Daily	Bare Costs	Incl. O&P
2 Pipe Fitters	$58.50	$936.00	$88.35	$1413.60	$51.20	$77.55
1 Truck Driver (heavy)	37.55	300.40	57.30	458.40		
1 Equip. Oper. (crane)	50.25	402.00	76.20	609.60		
1 Flatbed Trailer, 40 Ton		152.00		167.20		
1 Truck Tractor, 6x4, 380 H.P.		612.80		674.08		
1 Hyd. Crane, 80 Ton		1652.00		1817.20	75.53	83.08
32 L.H., Daily Totals		$4055.20		$5140.08	$126.72	$160.63

Crew B-35	Hr.	Daily	Hr.	Daily	Bare Costs	Incl. O&P
1 Labor Foreman (outside)	$38.65	$309.20	$59.65	$477.20	$45.66	$69.76
1 Skilled Worker	47.30	378.40	73.25	586.00		
1 Welder (plumber)	57.55	460.40	86.90	695.20		
1 Laborer	36.65	293.20	56.55	452.40		
1 Equip. Oper. (crane)	50.25	402.00	76.20	609.60		
1 Equip. Oper. (oiler)	43.55	348.40	66.00	528.00		
1 Welder, Electric, 300 amp		55.30		60.83		
1 Hyd. Excavator, .75 C.Y.		662.00		728.20	14.94	16.44
48 L.H., Daily Totals		$2908.90		$4137.43	$60.60	$86.20

Crew B-35A	Hr.	Daily	Hr.	Daily	Bare Costs	Incl. O&P
1 Labor Foreman (outside)	$38.65	$309.20	$59.65	$477.20	$44.37	$67.87
2 Laborers	36.65	586.40	56.55	904.80		
1 Skilled Worker	47.30	378.40	73.25	586.00		
1 Welder (plumber)	57.55	460.40	86.90	695.20		
1 Equip. Oper. (crane)	50.25	402.00	76.20	609.60		
1 Equip. Oper. (oiler)	43.55	348.40	66.00	528.00		
1 Welder, Gas Engine, 300 amp		142.00		156.20		
1 Crawler Crane, 75 Ton		1492.00		1641.20	29.18	32.10
56 L.H., Daily Totals		$4118.80		$5598.20	$73.55	$99.97

Crew B-36	Hr.	Daily	Hr.	Daily	Bare Costs	Incl. O&P
1 Labor Foreman (outside)	$38.65	$309.20	$59.65	$477.20	$41.95	$64.21
2 Laborers	36.65	586.40	56.55	904.80		
2 Equip. Oper. (medium)	48.90	782.40	74.15	1186.40		
1 Dozer, 200 H.P.		1325.00		1457.50		
1 Aggregate Spreader		41.20		45.32		
1 Tandem Roller, 10 Ton		242.00		266.20	40.20	44.23
40 L.H., Daily Totals		$3286.20		$4337.42	$82.16	$108.44

Crew B-36A	Hr.	Daily	Hr.	Daily	Bare Costs	Incl. O&P
1 Labor Foreman (outside)	$38.65	$309.20	$59.65	$477.20	$43.94	$67.05
2 Laborers	36.65	586.40	56.55	904.80		
4 Equip. Oper. (medium)	48.90	1564.80	74.15	2372.80		
1 Dozer, 200 H.P.		1325.00		1457.50		
1 Aggregate Spreader		41.20		45.32		
1 Tandem Roller, 10 Ton		242.00		266.20		
1 Roller, Pneum. Whl., 12 Ton		349.00		383.90	34.95	38.45
56 L.H., Daily Totals		$4417.60		$5907.72	$78.89	$105.50

Crew B-36B	Hr.	Daily	Hr.	Daily	Bare Costs	Incl. O&P
1 Labor Foreman (outside)	$38.65	$309.20	$59.65	$477.20	$43.14	$65.83
2 Laborers	36.65	586.40	56.55	904.80		
4 Equip. Oper. (medium)	48.90	1564.80	74.15	2372.80		
1 Truck Driver (heavy)	37.55	300.40	57.30	458.40		
1 Grader, 30,000 Lbs.		707.00		777.70		
1 F.E. Loader, Crl, 1.5 C.Y.		625.20		687.72		
1 Dozer, 300 H.P.		1817.00		1998.70		
1 Roller, Vibratory, 25 Ton		692.80		762.08		
1 Truck Tractor, 6x4, 450 H.P.		742.40		816.64		
1 Water Tank Trailer, 5000 Gal.		141.60		155.76	73.84	81.23
64 L.H., Daily Totals		$7486.80		$9411.80	$116.98	$147.06

Crew B-36C	Hr.	Daily	Hr.	Daily	Bare Costs	Incl. O&P
1 Labor Foreman (outside)	$38.65	$309.20	$59.65	$477.20	$44.58	$67.88
3 Equip. Oper. (medium)	48.90	1173.60	74.15	1779.60		
1 Truck Driver (heavy)	37.55	300.40	57.30	458.40		
1 Grader, 30,000 Lbs.		707.00		777.70		
1 Dozer, 300 H.P.		1817.00		1998.70		
1 Roller, Vibratory, 25 Ton		692.80		762.08		
1 Truck Tractor, 6x4, 450 H.P.		742.40		816.64		
1 Water Tank Trailer, 5000 Gal.		141.60		155.76	102.52	112.77
40 L.H., Daily Totals		$5884.00		$7226.08	$147.10	$180.65

Crew B-36D	Hr.	Daily	Hr.	Daily	Bare Costs	Incl. O&P
1 Labor Foreman (outside)	$38.65	$309.20	$59.65	$477.20	$46.34	$70.53
3 Equip. Oper. (medium)	48.90	1173.60	74.15	1779.60		
1 Grader, 30,000 Lbs.		707.00		777.70		
1 Dozer, 300 H.P.		1817.00		1998.70		
1 Roller, Vibratory, 25 Ton		692.80		762.08	100.53	110.58
32 L.H., Daily Totals		$4699.60		$5795.28	$146.86	$181.10

Crew B-37	Hr.	Daily	Hr.	Daily	Bare Costs	Incl. O&P
1 Labor Foreman (outside)	$38.65	$309.20	$59.65	$477.20	$38.72	$59.53
4 Laborers	36.65	1172.80	56.55	1809.60		
1 Equip. Oper. (light)	47.05	376.40	71.35	570.80		
1 Tandem Roller, 5 Ton		159.00		174.90	3.31	3.64
48 L.H., Daily Totals		$2017.40		$3032.50	$42.03	$63.18

Crew B-37A	Hr.	Daily	Hr.	Daily	Bare Costs	Incl. O&P
2 Laborers	$36.65	$586.40	$56.55	$904.80	$36.60	$56.27
1 Truck Driver (light)	36.50	292.00	55.70	445.60		
1 Flatbed Truck, Gas, 1.5 Ton		270.20		297.22		
1 Tar Kettle, T.M.		104.60		115.06	15.62	17.18
24 L.H., Daily Totals		$1253.20		$1762.68	$52.22	$73.44

Crew B-37B	Hr.	Daily	Hr.	Daily	Bare Costs	Incl. O&P
3 Laborers	$36.65	$879.60	$56.55	$1357.20	$36.61	$56.34
1 Truck Driver (light)	36.50	292.00	55.70	445.60		
1 Flatbed Truck, Gas, 1.5 Ton		270.20		297.22		
1 Tar Kettle, T.M.		104.60		115.06	11.71	12.88
32 L.H., Daily Totals		$1546.40		$2215.08	$48.33	$69.22

Crew B-37C	Hr.	Daily	Hr.	Daily	Bare Costs	Incl. O&P
2 Laborers	$36.65	$586.40	$56.55	$904.80	$36.58	$56.13
2 Truck Drivers (light)	36.50	584.00	55.70	891.20		
2 Flatbed Trucks, Gas, 1.5 Ton		540.40		594.44		
1 Tar Kettle, T.M.		104.60		115.06	20.16	22.17
32 L.H., Daily Totals		$1815.40		$2505.50	$56.73	$78.30

Crew B-37D	Hr.	Daily	Hr.	Daily	Bare Costs	Incl. O&P
1 Laborer	$36.65	$293.20	$56.55	$452.40	$36.58	$56.13
1 Truck Driver (light)	36.50	292.00	55.70	445.60		
1 Pickup Truck, 3/4 Ton		157.00		172.70	9.81	10.79
16 L.H., Daily Totals		$742.20		$1070.70	$46.39	$66.92

Crew B-37E	Hr.	Daily	Hr.	Daily	Bare Costs	Incl. O&P
3 Laborers	$36.65	$879.60	$56.55	$1357.20	$39.84	$60.94
1 Equip. Oper. (light)	47.05	376.40	71.35	570.80		
1 Equip. Oper. (medium)	48.90	391.20	74.15	593.20		
2 Truck Drivers (light)	36.50	584.00	55.70	891.20		
4 Barrels w/ Flasher		13.60		14.96		
1 Concrete Saw		175.20		192.72		
1 Rotary Hammer Drill		24.00		26.40		
1 Hammer Drill Bit		2.35		2.59		
1 Loader, Skid Steer, 30 H.P.		174.20		191.62		
1 Conc. Hammer Attach.		114.70		126.17		
1 Vibrating Plate, Gas, 18"		35.80		39.38		
2 Flatbed Trucks, Gas, 1.5 Ton		540.40		594.44	19.29	21.22
56 L.H., Daily Totals		$3311.45		$4600.68	$59.13	$82.15

Crew B-37F	Hr.	Daily	Hr.	Daily	Bare Costs	Incl. O&P
3 Laborers	$36.65	$879.60	$56.55	$1357.20	$36.61	$56.34
1 Truck Driver (light)	36.50	292.00	55.70	445.60		
4 Barrels w/ Flasher		13.60		14.96		
1 Concrete Mixer, 10 C.F.		176.80		194.48		
1 Air Compressor, 60 cfm		137.40		151.14		
1 -50' Air Hose, 3/4"		3.25		3.58		
1 Spade (Chipper)		7.80		8.58		
1 Flatbed Truck, Gas, 1.5 Ton		270.20		297.22	19.03	20.94
32 L.H., Daily Totals		$1780.65		$2472.76	$55.65	$77.27

Crew B-37G	Hr.	Daily	Hr.	Daily	Bare Costs	Incl. O&P
1 Labor Foreman (outside)	$38.65	$309.20	$59.65	$477.20	$38.72	$59.53
4 Laborers	36.65	1172.80	56.55	1809.60		
1 Equip. Oper. (light)	47.05	376.40	71.35	570.80		
1 Berm Machine		294.40		323.84		
1 Tandem Roller, 5 Ton		159.00		174.90	9.45	10.39
48 L.H., Daily Totals		$2311.80		$3356.34	$48.16	$69.92

Crew B-37H	Hr.	Daily	Hr.	Daily	Bare Costs	Incl. O&P
1 Labor Foreman (outside)	$38.65	$309.20	$59.65	$477.20	$38.72	$59.53
4 Laborers	36.65	1172.80	56.55	1809.60		
1 Equip. Oper. (light)	47.05	376.40	71.35	570.80		
1 Tandem Roller, 5 Ton		159.00		174.90		
1 Flatbed Trucks, Gas, 1.5 Ton		270.20		297.22		
1 Tar Kettle, T.M.		104.60		115.06	11.12	12.23
48 L.H., Daily Totals		$2392.20		$3444.78	$49.84	$71.77

Crew B-37I	Bare Costs Hr.	Bare Costs Daily	Incl. Subs O&P Hr.	Incl. Subs O&P Daily	Cost Per Labor-Hour Bare Costs	Cost Per Labor-Hour Incl. O&P
3 Laborers	$36.65	$879.60	$56.55	$1357.20	$39.84	$60.94
1 Equip. Oper. (light)	47.05	376.40	71.35	570.80		
1 Equip. Oper. (medium)	48.90	391.20	74.15	593.20		
2 Truck Drivers (light)	36.50	584.00	55.70	891.20		
4 Barrels w/ Flasher		13.60		14.96		
1 Concrete Saw		175.20		192.72		
1 Rotary Hammer Drill		24.00		26.40		
1 Hammer Drill Bit		2.35		2.59		
1 Air Compressor, 60 cfm		137.40		151.14		
1 -50' Air Hose, 3/4"		3.25		3.58		
1 Spade (Chipper)		7.80		8.58		
1 Loader, Skid Steer, 30 H.P.		174.20		191.62		
1 Conc. Hammer Attach.		114.70		126.17		
1 Concrete Mixer, 10 C.F.		176.80		194.48		
1 Vibrating Plate, Gas, 18"		35.80		39.38		
2 Flatbed Trucks, Gas, 1.5 Ton		540.40		594.44	25.10	27.61
56 L.H., Daily Totals		$3636.70		$4958.45	$64.94	$88.54

Crew B-37J	Hr.	Daily	Hr.	Daily	Bare Costs	Incl. O&P
1 Labor Foreman (outside)	$38.65	$309.20	$59.65	$477.20	$38.72	$59.53
4 Laborers	36.65	1172.80	56.55	1809.60		
1 Equip. Oper. (light)	47.05	376.40	71.35	570.80		
1 Air Compressor, 60 cfm		137.40		151.14		
1 -50' Air Hose, 3/4"		3.25		3.58		
2 Concrete Mixer, 10 C.F.		353.60		388.96		
2 Flatbed Trucks, Gas, 1.5 Ton		540.40		594.44		
1 Shot Blaster, 20"		214.40		235.84	26.02	28.62
48 L.H., Daily Totals		$3107.45		$4231.56	$64.74	$88.16

Crew B-37K	Hr.	Daily	Hr.	Daily	Bare Costs	Incl. O&P
1 Labor Foreman (outside)	$38.65	$309.20	$59.65	$477.20	$38.72	$59.53
4 Laborers	36.65	1172.80	56.55	1809.60		
1 Equip. Oper. (light)	47.05	376.40	71.35	570.80		
1 Air Compressor, 60 cfm		137.40		151.14		
1 -50' Air Hose, 3/4"		3.25		3.58		
2 Flatbed Trucks, Gas, 1.5 Ton		540.40		594.44		
1 Shot Blaster, 20"		214.40		235.84	18.66	20.52
48 L.H., Daily Totals		$2753.85		$3842.59	$57.37	$80.05

Crew B-38	Hr.	Daily	Hr.	Daily	Bare Costs	Incl. O&P
1 Labor Foreman (outside)	$38.65	$309.20	$59.65	$477.20	$41.58	$63.65
2 Laborers	36.65	586.40	56.55	904.80		
1 Equip. Oper. (light)	47.05	376.40	71.35	570.80		
1 Equip. Oper. (medium)	48.90	391.20	74.15	593.20		
1 Backhoe Loader, 48 H.P.		365.20		401.72		
1 Hyd. Hammer, (1200 lb.)		180.40		198.44		
1 F.E. Loader, W.M., 4 C.Y.		673.80		741.18		
1 Pvmt. Rem. Bucket		60.80		66.88	32.01	35.21
40 L.H., Daily Totals		$2943.40		$3954.22	$73.58	$98.86

Crew B-39	Hr.	Daily	Hr.	Daily	Bare Costs	Incl. O&P
1 Labor Foreman (outside)	$38.65	$309.20	$59.65	$477.20	$38.72	$59.53
4 Laborers	36.65	1172.80	56.55	1809.60		
1 Equip. Oper. (light)	47.05	376.40	71.35	570.80		
1 Air Compressor, 250 cfm		202.40		222.64		
2 Breakers, Pavement, 60 lb.		19.60		21.56		
2 -50' Air Hoses, 1.5"		11.60		12.76	4.87	5.35
48 L.H., Daily Totals		$2092.00		$3114.56	$43.58	$64.89

Crew B-40	Hr.	Daily	Hr.	Daily	Bare Costs	Incl. O&P
1 Pile Driver Foreman (outside)	$46.40	$371.20	$73.60	$588.80	$46.01	$71.70
4 Pile Drivers	44.40	1420.80	70.40	2252.80		
2 Equip. Oper. (crane)	50.25	804.00	76.20	1219.20		
1 Equip. Oper. (oiler)	43.55	348.40	66.00	528.00		
1 Crawler Crane, 40 Ton		1179.00		1296.90		
1 Vibratory Hammer & Gen.		2640.00		2904.00	59.67	65.64
64 L.H., Daily Totals		$6763.40		$8789.70	$105.68	$137.34

Crew B-40B	Hr.	Daily	Hr.	Daily	Bare Costs	Incl. O&P
1 Labor Foreman (outside)	$38.65	$309.20	$59.65	$477.20	$40.40	$61.92
3 Laborers	36.65	879.60	56.55	1357.20		
1 Equip. Oper. (crane)	50.25	402.00	76.20	609.60		
1 Equip. Oper. (oiler)	43.55	348.40	66.00	528.00		
1 Lattice Boom Crane, 40 Ton		1177.00		1294.70	24.52	26.97
48 L.H., Daily Totals		$3116.20		$4266.70	$64.92	$88.89

Crew B-41	Hr.	Daily	Hr.	Daily	Bare Costs	Incl. O&P
1 Labor Foreman (outside)	$38.65	$309.20	$59.65	$477.20	$37.95	$58.44
4 Laborers	36.65	1172.80	56.55	1809.60		
.25 Equip. Oper. (crane)	50.25	100.50	76.20	152.40		
.25 Equip. Oper. (oiler)	43.55	87.10	66.00	132.00		
.25 Crawler Crane, 40 Ton		294.75		324.23	6.70	7.37
44 L.H., Daily Totals		$1964.35		$2895.43	$44.64	$65.81

Crew B-42	Hr.	Daily	Hr.	Daily	Bare Costs	Incl. O&P
1 Labor Foreman (outside)	$38.65	$309.20	$59.65	$477.20	$41.27	$64.78
4 Laborers	36.65	1172.80	56.55	1809.60		
1 Equip. Oper. (crane)	50.25	402.00	76.20	609.60		
1 Equip. Oper. (oiler)	43.55	348.40	66.00	528.00		
1 Welder	51.10	408.80	90.20	721.60		
1 Hyd. Crane, 25 Ton		745.80		820.38		
1 Welder, Gas Engine, 300 amp		142.00		156.20		
1 Horz. Boring Csg. Mch.		479.80		527.78	21.37	23.51
64 L.H., Daily Totals		$4008.80		$5650.36	$62.64	$88.29

Crew B-43	Hr.	Daily	Hr.	Daily	Bare Costs	Incl. O&P
1 Labor Foreman (outside)	$38.65	$309.20	$59.65	$477.20	$40.40	$61.92
3 Laborers	36.65	879.60	56.55	1357.20		
1 Equip. Oper. (crane)	50.25	402.00	76.20	609.60		
1 Equip. Oper. (oiler)	43.55	348.40	66.00	528.00		
1 Drill Rig, Truck-Mounted		2551.00		2806.10	53.15	58.46
48 L.H., Daily Totals		$4490.20		$5778.10	$93.55	$120.38

Crew B-44	Hr.	Daily	Hr.	Daily	Bare Costs	Incl. O&P
1 Pile Driver Foreman (outside)	$46.40	$371.20	$73.60	$588.80	$45.14	$70.52
4 Pile Drivers	44.40	1420.80	70.40	2252.80		
2 Equip. Oper. (crane)	50.25	804.00	76.20	1219.20		
1 Laborer	36.65	293.20	56.55	452.40		
1 Crawler Crane, 40 Ton		1179.00		1296.90		
1 Lead, 60' High		74.80		82.28		
1 Hammer, Diesel, 15K ft.-lbs.		587.60		646.36	28.77	31.65
64 L.H., Daily Totals		$4730.60		$6538.74	$73.92	$102.17

Crew B-45	Hr.	Daily	Hr.	Daily	Bare Costs	Incl. O&P
1 Equip. Oper. (medium)	$48.90	$391.20	$74.15	$593.20	$43.23	$65.72
1 Truck Driver (heavy)	37.55	300.40	57.30	458.40		
1 Dist. Tanker, 3000 Gallon		308.20		339.02		
1 Truck Tractor, 6x4, 380 H.P.		612.80		674.08	57.56	63.32
16 L.H., Daily Totals		$1612.60		$2064.70	$100.79	$129.04

Crew No.	Bare Costs		Incl. Subs O&P		Cost Per Labor-Hour	
Crew B-46	**Hr.**	**Daily**	**Hr.**	**Daily**	**Bare Costs**	**Incl. O&P**
1 Pile Driver Foreman (outside)	$46.40	$371.20	$73.60	$588.80	$40.86	$64.01
2 Pile Drivers	44.40	710.40	70.40	1126.40		
3 Laborers	36.65	879.60	56.55	1357.20		
1 Chain Saw, Gas, 36" Long		45.00		49.50	.94	1.03
48 L.H., Daily Totals		$2006.20		$3121.90	$41.80	$65.04
Crew B-47	**Hr.**	**Daily**	**Hr.**	**Daily**	**Bare Costs**	**Incl. O&P**
1 Blast Foreman (outside)	$38.65	$309.20	$59.65	$477.20	$40.78	$62.52
1 Driller	36.65	293.20	56.55	452.40		
1 Equip. Oper. (light)	47.05	376.40	71.35	570.80		
1 Air Track Drill, 4"		1040.00		1144.00		
1 Air Compressor, 600 cfm		552.60		607.86		
2 -50' Air Hoses, 3"		29.80		32.78	67.60	74.36
24 L.H., Daily Totals		$2601.20		$3285.04	$108.38	$136.88
Crew B-47A	**Hr.**	**Daily**	**Hr.**	**Daily**	**Bare Costs**	**Incl. O&P**
1 Drilling Foreman (outside)	$38.65	$309.20	$59.65	$477.20	$44.15	$67.28
1 Equip. Oper. (heavy)	50.25	402.00	76.20	609.60		
1 Equip. Oper. (oiler)	43.55	348.40	66.00	528.00		
1 Air Track Drill, 5"		1255.00		1380.50	52.29	57.52
24 L.H., Daily Totals		$2314.60		$2995.30	$96.44	$124.80
Crew B-47C	**Hr.**	**Daily**	**Hr.**	**Daily**	**Bare Costs**	**Incl. O&P**
1 Laborer	$36.65	$293.20	$56.55	$452.40	$41.85	$63.95
1 Equip. Oper. (light)	47.05	376.40	71.35	570.80		
1 Air Compressor, 750 cfm		559.20		615.12		
2 -50' Air Hoses, 3"		29.80		32.78		
1 Air Track Drill, 4"		1040.00		1144.00	101.81	111.99
16 L.H., Daily Totals		$2298.60		$2815.10	$143.66	$175.94
Crew B-47E	**Hr.**	**Daily**	**Hr.**	**Daily**	**Bare Costs**	**Incl. O&P**
1 Labor Foreman (outside)	$38.65	$309.20	$59.65	$477.20	$37.15	$57.33
3 Laborers	36.65	879.60	56.55	1357.20		
1 Flatbed Truck, Gas, 3 Ton		333.00		366.30	10.41	11.45
32 L.H., Daily Totals		$1521.80		$2200.70	$47.56	$68.77
Crew B-47G	**Hr.**	**Daily**	**Hr.**	**Daily**	**Bare Costs**	**Incl. O&P**
1 Labor Foreman (outside)	$38.65	$309.20	$59.65	$477.20	$39.75	$61.02
2 Laborers	36.65	586.40	56.55	904.80		
1 Equip. Oper. (light)	47.05	376.40	71.35	570.80		
1 Air Track Drill, 4"		1040.00		1144.00		
1 Air Compressor, 600 cfm		552.60		607.86		
2 -50' Air Hoses, 3"		29.80		32.78		
1 Gunite Pump Rig		370.40		407.44	62.27	68.50
32 L.H., Daily Totals		$3264.80		$4144.88	$102.03	$129.53
Crew B-47H	**Hr.**	**Daily**	**Hr.**	**Daily**	**Bare Costs**	**Incl. O&P**
1 Skilled Worker Foreman (out)	$49.30	$394.40	$76.35	$610.80	$47.80	$74.03
3 Skilled Workers	47.30	1135.20	73.25	1758.00		
1 Flatbed Truck, Gas, 3 Ton		333.00		366.30	10.41	11.45
32 L.H., Daily Totals		$1862.60		$2735.10	$58.21	$85.47

Crew No.	Bare Costs		Incl. Subs O&P		Cost Per Labor-Hour	
Crew B-48	**Hr.**	**Daily**	**Hr.**	**Daily**	**Bare Costs**	**Incl. O&P**
1 Labor Foreman (outside)	$38.65	$309.20	$59.65	$477.20	$41.35	$63.26
3 Laborers	36.65	879.60	56.55	1357.20		
1 Equip. Oper. (crane)	50.25	402.00	76.20	609.60		
1 Equip. Oper. (oiler)	43.55	348.40	66.00	528.00		
1 Equip. Oper. (light)	47.05	376.40	71.35	570.80		
1 Centr. Water Pump, 6"		340.80		374.88		
1 -20' Suction Hose, 6"		11.50		12.65		
1 -50' Discharge Hose, 6"		6.10		6.71		
1 Drill Rig, Truck-Mounted		2551.00		2806.10	51.95	57.15
56 L.H., Daily Totals		$5225.00		$6743.14	$93.30	$120.41
Crew B-49	**Hr.**	**Daily**	**Hr.**	**Daily**	**Bare Costs**	**Incl. O&P**
1 Labor Foreman (outside)	$38.65	$309.20	$59.65	$477.20	$42.91	$65.99
3 Laborers	36.65	879.60	56.55	1357.20		
2 Equip. Oper. (crane)	50.25	804.00	76.20	1219.20		
2 Equip. Oper. (oilers)	43.55	696.80	66.00	1056.00		
1 Equip. Oper. (light)	47.05	376.40	71.35	570.80		
2 Pile Drivers	44.40	710.40	70.40	1126.40		
1 Hyd. Crane, 25 Ton		745.80		820.38		
1 Centr. Water Pump, 6"		340.80		374.88		
1 -20' Suction Hose, 6"		11.50		12.65		
1 -50' Discharge Hose, 6"		6.10		6.71		
1 Drill Rig, Truck-Mounted		2551.00		2806.10	41.54	45.69
88 L.H., Daily Totals		$7431.60		$9827.52	$84.45	$111.68
Crew B-50	**Hr.**	**Daily**	**Hr.**	**Daily**	**Bare Costs**	**Incl. O&P**
2 Pile Driver Foremen (outside)	$46.40	$742.40	$73.60	$1177.60	$43.80	$68.40
6 Pile Drivers	44.40	2131.20	70.40	3379.20		
2 Equip. Oper. (crane)	50.25	804.00	76.20	1219.20		
1 Equip. Oper. (oiler)	43.55	348.40	66.00	528.00		
3 Laborers	36.65	879.60	56.55	1357.20		
1 Crawler Crane, 40 Ton		1179.00		1296.90		
1 Lead, 60' High		74.80		82.28		
1 Hammer, Diesel, 15K ft.-lbs.		587.60		646.36		
1 Air Compressor, 600 cfm		552.60		607.86		
2 -50' Air Hoses, 3"		29.80		32.78		
1 Chain Saw, Gas, 36" Long		45.00		49.50	22.04	24.25
112 L.H., Daily Totals		$7374.40		$10376.88	$65.84	$92.65
Crew B-51	**Hr.**	**Daily**	**Hr.**	**Daily**	**Bare Costs**	**Incl. O&P**
1 Labor Foreman (outside)	$38.65	$309.20	$59.65	$477.20	$36.96	$56.92
4 Laborers	36.65	1172.80	56.55	1809.60		
1 Truck Driver (light)	36.50	292.00	55.70	445.60		
1 Flatbed Truck, Gas, 1.5 Ton		270.20		297.22	5.63	6.19
48 L.H., Daily Totals		$2044.20		$3029.62	$42.59	$63.12
Crew B-52	**Hr.**	**Daily**	**Hr.**	**Daily**	**Bare Costs**	**Incl. O&P**
1 Carpenter Foreman (outside)	$47.85	$382.80	$73.85	$590.80	$42.50	$65.17
1 Carpenter	45.85	366.80	70.75	566.00		
3 Laborers	36.65	879.60	56.55	1357.20		
1 Cement Finisher	44.05	352.40	65.10	520.80		
.5 Rodman (reinf.)	50.65	202.60	79.55	318.20		
.5 Equip. Oper. (medium)	48.90	195.60	74.15	296.60		
.5 Crawler Loader, 3 C.Y.		595.50		655.05	10.63	11.70
56 L.H., Daily Totals		$2975.30		$4304.65	$53.13	$76.87
Crew B-53	**Hr.**	**Daily**	**Hr.**	**Daily**	**Bare Costs**	**Incl. O&P**
1 Equip. Oper. (light)	$47.05	$376.40	$71.35	$570.80	$47.05	$71.35
1 Trencher, Chain, 12 H.P.		71.20		78.32	8.90	9.79
8 L.H., Daily Totals		$447.60		$649.12	$55.95	$81.14

Crew B-54	Bare Costs Hr.	Bare Costs Daily	Incl. Subs O&P Hr.	Incl. Subs O&P Daily	Cost Per Labor-Hour Bare Costs	Cost Per Labor-Hour Incl. O&P
1 Equip. Oper. (light)	$47.05	$376.40	$71.35	$570.80	$47.05	$71.35
1 Trencher, Chain, 40 H.P.		339.40		373.34	42.42	46.67
8 L.H., Daily Totals		$715.80		$944.14	$89.47	$118.02

Crew B-54A	Hr.	Daily	Hr.	Daily	Bare Costs	Incl. O&P
.17 Labor Foreman (outside)	$38.65	$52.56	$59.65	$81.12	$47.41	$72.04
1 Equipment Operator (med.)	48.90	391.20	74.15	593.20		
1 Wheel Trencher, 67 H.P.		1206.00		1326.60	128.85	141.73
9.36 L.H., Daily Totals		$1649.76		$2000.92	$176.26	$213.77

Crew B-54B	Hr.	Daily	Hr.	Daily	Bare Costs	Incl. O&P
.25 Labor Foreman (outside)	$38.65	$77.30	$59.65	$119.30	$46.85	$71.25
1 Equipment Operator (med.)	48.90	391.20	74.15	593.20		
1 Wheel Trencher, 150 H.P.		1898.00		2087.80	189.80	208.78
10 L.H., Daily Totals		$2366.50		$2800.30	$236.65	$280.03

Crew B-54C	Hr.	Daily	Hr.	Daily	Bare Costs	Incl. O&P
1 Laborer	$36.65	$293.20	$56.55	$452.40	$42.77	$65.35
1 Equipment Operator (med.)	48.90	391.20	74.15	593.20		
1 Wheel Trencher, 67 H.P.		1206.00		1326.60	75.38	82.91
16 L.H., Daily Totals		$1890.40		$2372.20	$118.15	$148.26

Crew B-54D	Hr.	Daily	Hr.	Daily	Bare Costs	Incl. O&P
1 Laborer	$36.65	$293.20	$56.55	$452.40	$42.77	$65.35
1 Equipment Operator (med.)	48.90	391.20	74.15	593.20		
1 Rock Trencher, 6" Width		369.00		405.90	23.06	25.37
16 L.H., Daily Totals		$1053.40		$1451.50	$65.84	$90.72

Crew B-54E	Hr.	Daily	Hr.	Daily	Bare Costs	Incl. O&P
1 Laborer	$36.65	$293.20	$56.55	$452.40	$42.77	$65.35
1 Equipment Operator (med.)	48.90	391.20	74.15	593.20		
1 Rock Trencher, 18" Width		2590.00		2849.00	161.88	178.06
16 L.H., Daily Totals		$3274.40		$3894.60	$204.65	$243.41

Crew B-55	Hr.	Daily	Hr.	Daily	Bare Costs	Incl. O&P
2 Laborers	$36.65	$586.40	$56.55	$904.80	$36.60	$56.27
1 Truck Driver (light)	36.50	292.00	55.70	445.60		
1 Truck-Mounted Earth Auger		786.40		865.04		
1 Flatbed Truck, Gas, 3 Ton		333.00		366.30	46.64	51.31
24 L.H., Daily Totals		$1997.80		$2581.74	$83.24	$107.57

Crew B-56	Hr.	Daily	Hr.	Daily	Bare Costs	Incl. O&P
1 Laborer	$36.65	$293.20	$56.55	$452.40	$41.85	$63.95
1 Equip. Oper. (light)	47.05	376.40	71.35	570.80		
1 Air Track Drill, 4"		1040.00		1144.00		
1 Air Compressor, 600 cfm		552.60		607.86		
1 -50' Air Hose, 3"		14.90		16.39	100.47	110.52
16 L.H., Daily Totals		$2277.10		$2791.45	$142.32	$174.47

Crew B-57	Hr.	Daily	Hr.	Daily	Bare Costs	Incl. O&P
1 Labor Foreman (outside)	$38.65	$309.20	$59.65	$477.20	$42.13	$64.38
2 Laborers	36.65	586.40	56.55	904.80		
1 Equip. Oper. (crane)	50.25	402.00	76.20	609.60		
1 Equip. Oper. (light)	47.05	376.40	71.35	570.80		
1 Equip. Oper. (oiler)	43.55	348.40	66.00	528.00		
1 Crawler Crane, 25 Ton		1173.00		1290.30		
1 Clamshell Bucket, 1 C.Y.		47.40		52.14		
1 Centr. Water Pump, 6"		340.80		374.88		
1 -20' Suction Hose, 6"		11.50		12.65		
20 -50' Discharge Hoses, 6"		122.00		134.20	35.31	38.84
48 L.H., Daily Totals		$3717.10		$4954.57	$77.44	$103.22

Crew B-58	Hr.	Daily	Hr.	Daily	Bare Costs	Incl. O&P
2 Laborers	$36.65	$586.40	$56.55	$904.80	$40.12	$61.48
1 Equip. Oper. (light)	47.05	376.40	71.35	570.80		
1 Backhoe Loader, 48 H.P.		365.20		401.72		
1 Small Helicopter, w/ Pilot		2729.00		3001.90	128.93	141.82
24 L.H., Daily Totals		$4057.00		$4879.22	$169.04	$203.30

Crew B-59	Hr.	Daily	Hr.	Daily	Bare Costs	Incl. O&P
1 Truck Driver (heavy)	$37.55	$300.40	$57.30	$458.40	$37.55	$57.30
1 Truck Tractor, 220 H.P.		366.40		403.04		
1 Water Tank Trailer, 5000 Gal.		141.60		155.76	63.50	69.85
8 L.H., Daily Totals		$808.40		$1017.20	$101.05	$127.15

Crew B-59A	Hr.	Daily	Hr.	Daily	Bare Costs	Incl. O&P
2 Laborers	$36.65	$586.40	$56.55	$904.80	$36.95	$56.80
1 Truck Driver (heavy)	37.55	300.40	57.30	458.40		
1 Water Tank Trailer, 5000 Gal.		141.60		155.76		
1 Truck Tractor, 220 H.P.		366.40		403.04	21.17	23.28
24 L.H., Daily Totals		$1394.80		$1922.00	$58.12	$80.08

Crew B-60	Hr.	Daily	Hr.	Daily	Bare Costs	Incl. O&P
1 Labor Foreman (outside)	$38.65	$309.20	$59.65	$477.20	$42.84	$65.38
2 Laborers	36.65	586.40	56.55	904.80		
1 Equip. Oper. (crane)	50.25	402.00	76.20	609.60		
2 Equip. Oper. (light)	47.05	752.80	71.35	1141.60		
1 Equip. Oper. (oiler)	43.55	348.40	66.00	528.00		
1 Crawler Crane, 40 Ton		1179.00		1296.90		
1 Lead, 60' High		74.80		82.28		
1 Hammer, Diesel, 15K ft.-lbs.		587.60		646.36		
1 Backhoe Loader, 48 H.P.		365.20		401.72	39.40	43.34
56 L.H., Daily Totals		$4605.40		$6088.46	$82.24	$108.72

Crew B-61	Hr.	Daily	Hr.	Daily	Bare Costs	Incl. O&P
1 Labor Foreman (outside)	$38.65	$309.20	$59.65	$477.20	$39.13	$60.13
3 Laborers	36.65	879.60	56.55	1357.20		
1 Equip. Oper. (light)	47.05	376.40	71.35	570.80		
1 Cement Mixer, 2 C.Y.		191.40		210.54		
1 Air Compressor, 160 cfm		157.40		173.14	8.72	9.59
40 L.H., Daily Totals		$1914.00		$2788.88	$47.85	$69.72

Crew B-62	Hr.	Daily	Hr.	Daily	Bare Costs	Incl. O&P
2 Laborers	$36.65	$586.40	$56.55	$904.80	$40.12	$61.48
1 Equip. Oper. (light)	47.05	376.40	71.35	570.80		
1 Loader, Skid Steer, 30 H.P.		174.20		191.62	7.26	7.98
24 L.H., Daily Totals		$1137.00		$1667.22	$47.38	$69.47

Crew No.	Bare Costs		Incl. Subs O&P		Cost Per Labor-Hour	
Crew B-63	**Hr.**	**Daily**	**Hr.**	**Daily**	**Bare Costs**	**Incl. O&P**
4 Laborers	$36.65	$1172.80	$56.55	$1809.60	$38.73	$59.51
1 Equip. Oper. (light)	47.05	376.40	71.35	570.80		
1 Loader, Skid Steer, 30 H.P.		174.20		191.62	4.36	4.79
40 L.H., Daily Totals		$1723.40		$2572.02	$43.09	$64.30
Crew B-63B	**Hr.**	**Daily**	**Hr.**	**Daily**	**Bare Costs**	**Incl. O&P**
1 Labor Foreman (inside)	$37.15	$297.20	$57.30	$458.40	$39.38	$60.44
2 Laborers	36.65	586.40	56.55	904.80		
1 Equip. Oper. (light)	47.05	376.40	71.35	570.80		
1 Loader, Skid Steer, 78 H.P.		308.80		339.68	9.65	10.62
32 L.H., Daily Totals		$1568.80		$2273.68	$49.02	$71.05
Crew B-64	**Hr.**	**Daily**	**Hr.**	**Daily**	**Bare Costs**	**Incl. O&P**
1 Laborer	$36.65	$293.20	$56.55	$452.40	$36.58	$56.13
1 Truck Driver (light)	36.50	292.00	55.70	445.60		
1 Power Mulcher (small)		159.60		175.56		
1 Flatbed Truck, Gas, 1.5 Ton		270.20		297.22	26.86	29.55
16 L.H., Daily Totals		$1015.00		$1370.78	$63.44	$85.67
Crew B-65	**Hr.**	**Daily**	**Hr.**	**Daily**	**Bare Costs**	**Incl. O&P**
1 Laborer	$36.65	$293.20	$56.55	$452.40	$36.58	$56.13
1 Truck Driver (light)	36.50	292.00	55.70	445.60		
1 Power Mulcher (Large)		325.60		358.16		
1 Flatbed Truck, Gas, 1.5 Ton		270.20		297.22	37.24	40.96
16 L.H., Daily Totals		$1181.00		$1553.38	$73.81	$97.09
Crew B-66	**Hr.**	**Daily**	**Hr.**	**Daily**	**Bare Costs**	**Incl. O&P**
1 Equip. Oper. (light)	$47.05	$376.40	$71.35	$570.80	$47.05	$71.35
1 Loader-Backhoe, 40 H.P.		259.60		285.56	32.45	35.70
8 L.H., Daily Totals		$636.00		$856.36	$79.50	$107.05
Crew B-67	**Hr.**	**Daily**	**Hr.**	**Daily**	**Bare Costs**	**Incl. O&P**
1 Millwright	$48.10	$384.80	$70.85	$566.80	$47.58	$71.10
1 Equip. Oper. (light)	47.05	376.40	71.35	570.80		
1 Forklift, R/T, 4,000 Lb.		314.20		345.62	19.64	21.60
16 L.H., Daily Totals		$1075.40		$1483.22	$67.21	$92.70
Crew B-67B	**Hr.**	**Daily**	**Hr.**	**Daily**	**Bare Costs**	**Incl. O&P**
1 Millwright Foreman (inside)	$48.60	$388.80	$71.60	$572.80	$48.35	$71.22
1 Millwright	48.10	384.80	70.85	566.80		
16 L.H., Daily Totals		$773.60		$1139.60	$48.35	$71.22
Crew B-68	**Hr.**	**Daily**	**Hr.**	**Daily**	**Bare Costs**	**Incl. O&P**
2 Millwrights	$48.10	$769.60	$70.85	$1133.60	$47.75	$71.02
1 Equip. Oper. (light)	47.05	376.40	71.35	570.80		
1 Forklift, R/T, 4,000 Lb.		314.20		345.62	13.09	14.40
24 L.H., Daily Totals		$1460.20		$2050.02	$60.84	$85.42
Crew B-68A	**Hr.**	**Daily**	**Hr.**	**Daily**	**Bare Costs**	**Incl. O&P**
1 Millwright Foreman (inside)	$48.60	$388.80	$71.60	$572.80	$48.27	$71.10
2 Millwrights	48.10	769.60	70.85	1133.60		
1 Forklift, 8,000 Lb.		180.00		198.00	7.50	8.25
24 L.H., Daily Totals		$1338.40		$1904.40	$55.77	$79.35

Crew No.	Bare Costs		Incl. Subs O&P		Cost Per Labor-Hour	
Crew B-68B	**Hr.**	**Daily**	**Hr.**	**Daily**	**Bare Costs**	**Incl. O&P**
1 Millwright Foreman (inside)	$48.60	$388.80	$71.60	$572.80	$52.37	$78.11
2 Millwrights	48.10	769.60	70.85	1133.60		
2 Electricians	53.35	853.60	79.85	1277.60		
2 Plumbers	57.55	920.80	86.90	1390.40		
1 Forklift, 5,000 Lb.		324.80		357.28	5.80	6.38
56 L.H., Daily Totals		$3257.60		$4731.68	$58.17	$84.49
Crew B-68C	**Hr.**	**Daily**	**Hr.**	**Daily**	**Bare Costs**	**Incl. O&P**
1 Millwright Foreman (inside)	$48.60	$388.80	$71.60	$572.80	$51.90	$77.30
1 Millwright	48.10	384.80	70.85	566.80		
1 Electrician	53.35	426.80	79.85	638.80		
1 Plumber	57.55	460.40	86.90	695.20		
1 Forklift, 5,000 Lb.		324.80		357.28	10.15	11.16
32 L.H., Daily Totals		$1985.60		$2830.88	$62.05	$88.47
Crew B-68D	**Hr.**	**Daily**	**Hr.**	**Daily**	**Bare Costs**	**Incl. O&P**
1 Labor Foreman (inside)	$37.15	$297.20	$57.30	$458.40	$40.28	$61.73
1 Laborer	36.65	293.20	56.55	452.40		
1 Equip. Oper. (light)	47.05	376.40	71.35	570.80		
1 Forklift, 5,000 Lb.		324.80		357.28	13.53	14.89
24 L.H., Daily Totals		$1291.60		$1838.88	$53.82	$76.62
Crew B-68E	**Hr.**	**Daily**	**Hr.**	**Daily**	**Bare Costs**	**Incl. O&P**
1 Struc. Steel Foreman (inside)	$51.60	$412.80	$91.05	$728.40	$51.20	$90.37
3 Struc. Steel Workers	51.10	1226.40	90.20	2164.80		
1 Welder	51.10	408.80	90.20	721.60		
1 Forklift, 8,000 Lb.		180.00		198.00	4.50	4.95
40 L.H., Daily Totals		$2228.00		$3812.80	$55.70	$95.32
Crew B-68F	**Hr.**	**Daily**	**Hr.**	**Daily**	**Bare Costs**	**Incl. O&P**
1 Skilled Worker Foreman (out)	$49.30	$394.40	$76.35	$610.80	$47.97	$74.28
2 Skilled Workers	47.30	756.80	73.25	1172.00		
1 Forklift, 5,000 Lb.		324.80		357.28	13.53	14.89
24 L.H., Daily Totals		$1476.00		$2140.08	$61.50	$89.17
Crew B-69	**Hr.**	**Daily**	**Hr.**	**Daily**	**Bare Costs**	**Incl. O&P**
1 Labor Foreman (outside)	$38.65	$309.20	$59.65	$477.20	$40.40	$61.92
3 Laborers	36.65	879.60	56.55	1357.20		
1 Equip. Oper. (crane)	50.25	402.00	76.20	609.60		
1 Equip. Oper. (oiler)	43.55	348.40	66.00	528.00		
1 Hyd. Crane, 80 Ton		1652.00		1817.20	34.42	37.86
48 L.H., Daily Totals		$3591.20		$4789.20	$74.82	$99.78
Crew B-69A	**Hr.**	**Daily**	**Hr.**	**Daily**	**Bare Costs**	**Incl. O&P**
1 Labor Foreman (outside)	$38.65	$309.20	$59.65	$477.20	$40.26	$61.42
3 Laborers	36.65	879.60	56.55	1357.20		
1 Equip. Oper. (medium)	48.90	391.20	74.15	593.20		
1 Concrete Finisher	44.05	352.40	65.10	520.80		
1 Curb/Gutter Paver, 2-Track		898.00		987.80	18.71	20.58
48 L.H., Daily Totals		$2830.40		$3936.20	$58.97	$82.00
Crew B-69B	**Hr.**	**Daily**	**Hr.**	**Daily**	**Bare Costs**	**Incl. O&P**
1 Labor Foreman (outside)	$38.65	$309.20	$59.65	$477.20	$40.26	$61.42
3 Laborers	36.65	879.60	56.55	1357.20		
1 Equip. Oper. (medium)	48.90	391.20	74.15	593.20		
1 Cement Finisher	44.05	352.40	65.10	520.80		
1 Curb/Gutter Paver, 4-Track		770.20		847.22	16.05	17.65
48 L.H., Daily Totals		$2702.60		$3795.62	$56.30	$79.08

Crew No.	Bare Costs		Incl. Subs O&P		Cost Per Labor-Hour	
Crew B-70	**Hr.**	**Daily**	**Hr.**	**Daily**	**Bare Costs**	**Incl. O&P**
1 Labor Foreman (outside)	$38.65	$309.20	$59.65	$477.20	$42.19	$64.54
3 Laborers	36.65	879.60	56.55	1357.20		
3 Equip. Oper. (medium)	48.90	1173.60	74.15	1779.60		
1 Grader, 30,000 Lbs.		707.00		777.70		
1 Ripper, Beam & 1 Shank		81.40		89.54		
1 Road Sweeper, S.P., 8' wide		665.40		731.94		
1 F.E. Loader, W.M., 1.5 C.Y.		373.40		410.74	32.63	35.89
56 L.H., Daily Totals		$4189.60		$5623.92	$74.81	$100.43
Crew B-70A	**Hr.**	**Daily**	**Hr.**	**Daily**	**Bare Costs**	**Incl. O&P**
1 Laborer	$36.65	$293.20	$56.55	$452.40	$46.45	$70.63
4 Equip. Oper. (medium)	48.90	1564.80	74.15	2372.80		
1 Grader, 40,000 Lbs.		1120.00		1232.00		
1 F.E. Loader, W.M.,2.5 C.Y.		522.80		575.08		
1 Dozer, 80 H.P.		481.00		529.10		
1 Roller, Pneum. Whl., 12 Ton		349.00		383.90	61.82	68.00
40 L.H., Daily Totals		$4330.80		$5545.28	$108.27	$138.63
Crew B-71	**Hr.**	**Daily**	**Hr.**	**Daily**	**Bare Costs**	**Incl. O&P**
1 Labor Foreman (outside)	$38.65	$309.20	$59.65	$477.20	$42.19	$64.54
3 Laborers	36.65	879.60	56.55	1357.20		
3 Equip. Oper. (medium)	48.90	1173.60	74.15	1779.60		
1 Pvmt. Profiler, 750 H.P.		5877.00		6464.70		
1 Road Sweeper, S.P., 8' wide		665.40		731.94		
1 F.E. Loader, W.M., 1.5 C.Y.		373.40		410.74	123.50	135.85
56 L.H., Daily Totals		$9278.20		$11221.38	$165.68	$200.38
Crew B-72	**Hr.**	**Daily**	**Hr.**	**Daily**	**Bare Costs**	**Incl. O&P**
1 Labor Foreman (outside)	$38.65	$309.20	$59.65	$477.20	$43.02	$65.74
3 Laborers	36.65	879.60	56.55	1357.20		
4 Equip. Oper. (medium)	48.90	1564.80	74.15	2372.80		
1 Pvmt. Profiler, 750 H.P.		5877.00		6464.70		
1 Hammermill, 250 H.P.		1846.00		2030.60		
1 Windrow Loader		1240.00		1364.00		
1 Mix Paver 165 H.P.		2183.00		2401.30		
1 Roller, Pneum. Whl., 12 Ton		349.00		383.90	179.61	197.57
64 L.H., Daily Totals		$14248.60		$16851.70	$222.63	$263.31
Crew B-73	**Hr.**	**Daily**	**Hr.**	**Daily**	**Bare Costs**	**Incl. O&P**
1 Labor Foreman (outside)	$38.65	$309.20	$59.65	$477.20	$44.56	$67.94
2 Laborers	36.65	586.40	56.55	904.80		
5 Equip. Oper. (medium)	48.90	1956.00	74.15	2966.00		
1 Road Mixer, 310 H.P.		1942.00		2136.20		
1 Tandem Roller, 10 Ton		242.00		266.20		
1 Hammermill, 250 H.P.		1846.00		2030.60		
1 Grader, 30,000 Lbs.		707.00		777.70		
.5 F.E. Loader, W.M., 1.5 C.Y.		186.70		205.37		
.5 Truck Tractor, 220 H.P.		183.20		201.52		
.5 Water Tank Trailer, 5000 Gal.		70.80		77.88	80.90	88.99
64 L.H., Daily Totals		$8029.30		$10043.47	$125.46	$156.93

Crew No.	Bare Costs		Incl. Subs O&P		Cost Per Labor-Hour	
Crew B-74	**Hr.**	**Daily**	**Hr.**	**Daily**	**Bare Costs**	**Incl. O&P**
1 Labor Foreman (outside)	$38.65	$309.20	$59.65	$477.20	$43.25	$65.92
1 Laborer	36.65	293.20	56.55	452.40		
4 Equip. Oper. (medium)	48.90	1564.80	74.15	2372.80		
2 Truck Drivers (heavy)	37.55	600.80	57.30	916.80		
1 Grader, 30,000 Lbs.		707.00		777.70		
1 Ripper, Beam & 1 Shank		81.40		89.54		
2 Stabilizers, 310 H.P.		3654.00		4019.40		
1 Flatbed Truck, Gas, 3 Ton		333.00		366.30		
1 Chem. Spreader, Towed		53.20		58.52		
1 Roller, Vibratory, 25 Ton		692.80		762.08		
1 Water Tank Trailer, 5000 Gal.		141.60		155.76		
1 Truck Tractor, 220 H.P.		366.40		403.04	94.21	103.63
64 L.H., Daily Totals		$8797.40		$10851.54	$137.46	$169.56
Crew B-75	**Hr.**	**Daily**	**Hr.**	**Daily**	**Bare Costs**	**Incl. O&P**
1 Labor Foreman (outside)	$38.65	$309.20	$59.65	$477.20	$44.06	$67.16
1 Laborer	36.65	293.20	56.55	452.40		
4 Equip. Oper. (medium)	48.90	1564.80	74.15	2372.80		
1 Truck Driver (heavy)	37.55	300.40	57.30	458.40		
1 Grader, 30,000 Lbs.		707.00		777.70		
1 Ripper, Beam & 1 Shank		81.40		89.54		
2 Stabilizers, 310 H.P.		3654.00		4019.40		
1 Dist. Tanker, 3000 Gallon		308.20		339.02		
1 Truck Tractor, 6x4, 380 H.P.		612.80		674.08		
1 Roller, Vibratory, 25 Ton		692.80		762.08	108.15	118.96
56 L.H., Daily Totals		$8523.80		$10422.62	$152.21	$186.12
Crew B-76	**Hr.**	**Daily**	**Hr.**	**Daily**	**Bare Costs**	**Incl. O&P**
1 Dock Builder Foreman (outside)	$46.40	$371.20	$73.60	$588.80	$45.83	$71.56
5 Dock Builders	44.40	1776.00	70.40	2816.00		
2 Equip. Oper. (crane)	50.25	804.00	76.20	1219.20		
1 Equip. Oper. (oiler)	43.55	348.40	66.00	528.00		
1 Crawler Crane, 50 Ton		1319.00		1450.90		
1 Barge, 400 Ton		781.00		859.10		
1 Hammer, Diesel, 15K ft.-lbs.		587.60		646.36		
1 Lead, 60' High		74.80		82.28		
1 Air Compressor, 600 cfm		552.60		607.86		
2 -50' Air Hoses, 3"		29.80		32.78	46.46	51.10
72 L.H., Daily Totals		$6644.40		$8831.28	$92.28	$122.66
Crew B-76A	**Hr.**	**Daily**	**Hr.**	**Daily**	**Bare Costs**	**Incl. O&P**
1 Labor Foreman (outside)	$38.65	$309.20	$59.65	$477.20	$39.46	$60.58
5 Laborers	36.65	1466.00	56.55	2262.00		
1 Equip. Oper. (crane)	50.25	402.00	76.20	609.60		
1 Equip. Oper. (oiler)	43.55	348.40	66.00	528.00		
1 Crawler Crane, 50 Ton		1319.00		1450.90		
1 Barge, 400 Ton		781.00		859.10	32.81	36.09
64 L.H., Daily Totals		$4625.60		$6186.80	$72.28	$96.67
Crew B-77	**Hr.**	**Daily**	**Hr.**	**Daily**	**Bare Costs**	**Incl. O&P**
1 Labor Foreman (outside)	$38.65	$309.20	$59.65	$477.20	$37.02	$57.00
3 Laborers	36.65	879.60	56.55	1357.20		
1 Truck Driver (light)	36.50	292.00	55.70	445.60		
1 Crack Cleaner, 25 H.P.		64.80		71.28		
1 Crack Filler, Trailer Mtd.		203.80		224.18		
1 Flatbed Truck, Gas, 3 Ton		333.00		366.30	15.04	16.54
40 L.H., Daily Totals		$2082.40		$2941.76	$52.06	$73.54

Crew No.	Bare Costs		Incl. Subs O&P		Cost Per Labor-Hour	
Crew B-78	**Hr.**	**Daily**	**Hr.**	**Daily**	**Bare Costs**	**Incl. O&P**
1 Labor Foreman (outside)	$38.65	$309.20	$59.65	$477.20	$36.96	$56.92
4 Laborers	36.65	1172.80	56.55	1809.60		
1 Truck Driver (light)	36.50	292.00	55.70	445.60		
1 Paint Striper, S.P., 40 Gallon		154.80		170.28		
1 Flatbed Truck, Gas, 3 Ton		333.00		366.30		
1 Pickup Truck, 3/4 Ton		157.00		172.70	13.43	14.78
48 L.H., Daily Totals		$2418.80		$3441.68	$50.39	$71.70
Crew B-78A	**Hr.**	**Daily**	**Hr.**	**Daily**	**Bare Costs**	**Incl. O&P**
1 Equip. Oper. (light)	$47.05	$376.40	$71.35	$570.80	$47.05	$71.35
1 Line Rem. (Metal Balls) 115 H.P.		839.00		922.90	104.88	115.36
8 L.H., Daily Totals		$1215.40		$1493.70	$151.93	$186.71
Crew B-78B	**Hr.**	**Daily**	**Hr.**	**Daily**	**Bare Costs**	**Incl. O&P**
2 Laborers	$36.65	$586.40	$56.55	$904.80	$37.81	$58.19
.25 Equip. Oper. (light)	47.05	94.10	71.35	142.70		
1 Pickup Truck, 3/4 Ton		157.00		172.70		
1 Line Rem.,11 H.P.,Walk Behind		68.40		75.24		
.25 Road Sweeper, S.P., 8' wide		166.35		182.99	21.76	23.94
18 L.H., Daily Totals		$1072.25		$1478.43	$59.57	$82.13
Crew B-78C	**Hr.**	**Daily**	**Hr.**	**Daily**	**Bare Costs**	**Incl. O&P**
1 Labor Foreman (outside)	$38.65	$309.20	$59.65	$477.20	$36.96	$56.92
4 Laborers	36.65	1172.80	56.55	1809.60		
1 Truck Driver (light)	36.50	292.00	55.70	445.60		
1 Paint Striper, T.M., 120 Gal.		860.40		946.44		
1 Flatbed Truck, Gas, 3 Ton		333.00		366.30		
1 Pickup Truck, 3/4 Ton		157.00		172.70	28.13	30.95
48 L.H., Daily Totals		$3124.40		$4217.84	$65.09	$87.87
Crew B-78D	**Hr.**	**Daily**	**Hr.**	**Daily**	**Bare Costs**	**Incl. O&P**
2 Labor Foremen (outside)	$38.65	$618.40	$59.65	$954.40	$37.03	$57.09
7 Laborers	36.65	2052.40	56.55	3166.80		
1 Truck Driver (light)	36.50	292.00	55.70	445.60		
1 Paint Striper, T.M., 120 Gal.		860.40		946.44		
1 Flatbed Truck, Gas, 3 Ton		333.00		366.30		
3 Pickup Trucks, 3/4 Ton		471.00		518.10		
1 Air Compressor, 60 cfm		137.40		151.14		
1 -50' Air Hose, 3/4"		3.25		3.58		
1 Breakers, Pavement, 60 lb.		9.80		10.78	22.69	24.95
80 L.H., Daily Totals		$4777.65		$6563.14	$59.72	$82.04
Crew B-78E	**Hr.**	**Daily**	**Hr.**	**Daily**	**Bare Costs**	**Incl. O&P**
2 Labor Foremen (outside)	$38.65	$618.40	$59.65	$954.40	$36.97	$57.00
9 Laborers	36.65	2638.80	56.55	4071.60		
1 Truck Driver (light)	36.50	292.00	55.70	445.60		
1 Paint Striper, T.M., 120 Gal.		860.40		946.44		
1 Flatbed Truck, Gas, 3 Ton		333.00		366.30		
4 Pickup Trucks, 3/4 Ton		628.00		690.80		
2 Air Compressor, 60 cfm		274.80		302.28		
2 -50' Air Hose, 3/4"		6.50		7.15		
2 Breakers, Pavement, 60 lb.		19.60		21.56	22.11	24.32
96 L.H., Daily Totals		$5671.50		$7806.13	$59.08	$81.31

Crew No.	Bare Costs		Incl. Subs O&P		Cost Per Labor-Hour	
Crew B-78F	**Hr.**	**Daily**	**Hr.**	**Daily**	**Bare Costs**	**Incl. O&P**
2 Labor Foremen (outside)	$38.65	$618.40	$59.65	$954.40	$36.92	$56.93
11 Laborers	36.65	3225.20	56.55	4976.40		
1 Truck Driver (light)	36.50	292.00	55.70	445.60		
1 Paint Striper, T.M., 120 Gal.		860.40		946.44		
1 Flatbed Truck, Gas, 3 Ton		333.00		366.30		
7 Pickup Trucks, 3/4 Ton		1099.00		1208.90		
3 Air Compressor, 60 cfm		412.20		453.42		
3 -50' Air Hose, 3/4"		9.75		10.73		
3 Breakers, Pavement, 60 lb.		29.40		32.34	24.50	26.95
112 L.H., Daily Totals		$6879.35		$9394.52	$61.42	$83.88
Crew B-79	**Hr.**	**Daily**	**Hr.**	**Daily**	**Bare Costs**	**Incl. O&P**
1 Labor Foreman (outside)	$38.65	$309.20	$59.65	$477.20	$37.02	$57.00
3 Laborers	36.65	879.60	56.55	1357.20		
1 Truck Driver (light)	36.50	292.00	55.70	445.60		
1 Paint Striper, T.M., 120 Gal.		860.40		946.44		
1 Heating Kettle, 115 Gallon		63.00		69.30		
1 Flatbed Truck, Gas, 3 Ton		333.00		366.30		
2 Pickup Trucks, 3/4 Ton		314.00		345.40	39.26	43.19
40 L.H., Daily Totals		$3051.20		$4007.44	$76.28	$100.19
Crew B-79A	**Hr.**	**Daily**	**Hr.**	**Daily**	**Bare Costs**	**Incl. O&P**
1.5 Equip. Oper. (light)	$47.05	$564.60	$71.35	$856.20	$47.05	$71.35
.5 Line Remov. (Grinder) 115 H.P.		465.30		511.83		
1 Line Rem. (Metal Balls) 115 H.P.		839.00		922.90	108.69	119.56
12 L.H., Daily Totals		$1868.90		$2290.93	$155.74	$190.91
Crew B-79B	**Hr.**	**Daily**	**Hr.**	**Daily**	**Bare Costs**	**Incl. O&P**
1 Laborer	$36.65	$293.20	$56.55	$452.40	$36.65	$56.55
1 Set of Gases		152.00		167.20	19.00	20.90
8 L.H., Daily Totals		$445.20		$619.60	$55.65	$77.45
Crew B-79C	**Hr.**	**Daily**	**Hr.**	**Daily**	**Bare Costs**	**Incl. O&P**
1 Labor Foreman (outside)	$38.65	$309.20	$59.65	$477.20	$36.91	$56.87
5 Laborers	36.65	1466.00	56.55	2262.00		
1 Truck Driver (light)	36.50	292.00	55.70	445.60		
1 Paint Striper, T.M., 120 Gal.		860.40		946.44		
1 Heating Kettle, 115 Gallon		63.00		69.30		
1 Flatbed Truck, Gas, 3 Ton		333.00		366.30		
3 Pickup Trucks, 3/4 Ton		471.00		518.10		
1 Air Compressor, 60 cfm		137.40		151.14		
1 -50' Air Hose, 3/4"		3.25		3.58		
1 Breakers, Pavement, 60 lb.		9.80		10.78	33.53	36.89
56 L.H., Daily Totals		$3945.05		$5250.44	$70.45	$93.76
Crew B-79D	**Hr.**	**Daily**	**Hr.**	**Daily**	**Bare Costs**	**Incl. O&P**
2 Labor Foremen (outside)	$38.65	$618.40	$59.65	$954.40	$37.13	$57.22
5 Laborers	36.65	1466.00	56.55	2262.00		
1 Truck Driver (light)	36.50	292.00	55.70	445.60		
1 Paint Striper, T.M., 120 Gal.		860.40		946.44		
1 Heating Kettle, 115 Gallon		63.00		69.30		
1 Flatbed Truck, Gas, 3 Ton		333.00		366.30		
4 Pickup Trucks, 3/4 Ton		628.00		690.80		
1 Air Compressor, 60 cfm		137.40		151.14		
1 -50' Air Hose, 3/4"		3.25		3.58		
1 Breakers, Pavement, 60 lb.		9.80		10.78	31.79	34.97
64 L.H., Daily Totals		$4411.25		$5900.34	$68.93	$92.19

Crew No.	Bare Costs		Incl. Subs O&P		Cost Per Labor-Hour	
Crew B-79E	**Hr.**	**Daily**	**Hr.**	**Daily**	**Bare Costs**	**Incl. O&P**
2 Labor Foremen (outside)	$38.65	$618.40	$59.65	$954.40	$37.03	$57.09
7 Laborers	36.65	2052.40	56.55	3166.80		
1 Truck Driver (light)	36.50	292.00	55.70	445.60		
1 Paint Striper, T.M., 120 Gal.		860.40		946.44		
1 Heating Kettle, 115 Gallon		63.00		69.30		
1 Flatbed Truck, Gas, 3 Ton		333.00		366.30		
5 Pickup Trucks, 3/4 Ton		785.00		863.50		
2 Air Compressors, 60 cfm		274.80		302.28		
2 -50' Air Hoses, 3/4"		6.50		7.15		
2 Breakers, Pavement, 60 lb.		19.60		21.56	29.28	32.21
80 L.H., Daily Totals		$5305.10		$7143.33	$66.31	$89.29
Crew B-80	**Hr.**	**Daily**	**Hr.**	**Daily**	**Bare Costs**	**Incl. O&P**
1 Labor Foreman (outside)	$38.65	$309.20	$59.65	$477.20	$39.71	$60.81
1 Laborer	36.65	293.20	56.55	452.40		
1 Truck Driver (light)	36.50	292.00	55.70	445.60		
1 Equip. Oper. (light)	47.05	376.40	71.35	570.80		
1 Flatbed Truck, Gas, 3 Ton		333.00		366.30		
1 Earth Auger, Truck-Mtd.		425.00		467.50	23.69	26.06
32 L.H., Daily Totals		$2028.80		$2779.80	$63.40	$86.87
Crew B-80A	**Hr.**	**Daily**	**Hr.**	**Daily**	**Bare Costs**	**Incl. O&P**
3 Laborers	$36.65	$879.60	$56.55	$1357.20	$36.65	$56.55
1 Flatbed Truck, Gas, 3 Ton		333.00		366.30	13.88	15.26
24 L.H., Daily Totals		$1212.60		$1723.50	$50.52	$71.81
Crew B-80B	**Hr.**	**Daily**	**Hr.**	**Daily**	**Bare Costs**	**Incl. O&P**
3 Laborers	$36.65	$879.60	$56.55	$1357.20	$39.25	$60.25
1 Equip. Oper. (light)	47.05	376.40	71.35	570.80		
1 Crane, Flatbed Mounted, 3 Ton		248.40		273.24	7.76	8.54
32 L.H., Daily Totals		$1504.40		$2201.24	$47.01	$68.79
Crew B-80C	**Hr.**	**Daily**	**Hr.**	**Daily**	**Bare Costs**	**Incl. O&P**
2 Laborers	$36.65	$586.40	$56.55	$904.80	$36.60	$56.27
1 Truck Driver (light)	36.50	292.00	55.70	445.60		
1 Flatbed Truck, Gas, 1.5 Ton		270.20		297.22		
1 Manual Fence Post Auger, Gas		8.00		8.80	11.59	12.75
24 L.H., Daily Totals		$1156.60		$1656.42	$48.19	$69.02
Crew B-81	**Hr.**	**Daily**	**Hr.**	**Daily**	**Bare Costs**	**Incl. O&P**
1 Laborer	$36.65	$293.20	$56.55	$452.40	$41.03	$62.67
1 Equip. Oper. (medium)	48.90	391.20	74.15	593.20		
1 Truck Driver (heavy)	37.55	300.40	57.30	458.40		
1 Hydromulcher, T.M., 3000 Gal.		340.40		374.44		
1 Truck Tractor, 220 H.P.		366.40		403.04	29.45	32.40
24 L.H., Daily Totals		$1691.60		$2281.48	$70.48	$95.06
Crew B-81A	**Hr.**	**Daily**	**Hr.**	**Daily**	**Bare Costs**	**Incl. O&P**
1 Laborer	$36.65	$293.20	$56.55	$452.40	$36.58	$56.13
1 Truck Driver (light)	36.50	292.00	55.70	445.60		
1 Hydromulcher, T.M., 600 Gal.		132.60		145.86		
1 Flatbed Truck, Gas, 3 Ton		333.00		366.30	29.10	32.01
16 L.H., Daily Totals		$1050.80		$1410.16	$65.67	$88.14
Crew B-82	**Hr.**	**Daily**	**Hr.**	**Daily**	**Bare Costs**	**Incl. O&P**
1 Laborer	$36.65	$293.20	$56.55	$452.40	$41.85	$63.95
1 Equip. Oper. (light)	47.05	376.40	71.35	570.80		
1 Horiz. Borer, 6 H.P.		83.00		91.30	5.19	5.71
16 L.H., Daily Totals		$752.60		$1114.50	$47.04	$69.66

Crew No.	Bare Costs		Incl. Subs O&P		Cost Per Labor-Hour	
Crew B-82A	**Hr.**	**Daily**	**Hr.**	**Daily**	**Bare Costs**	**Incl. O&P**
1 Laborer	$36.65	$293.20	$56.55	$452.40	$41.85	$63.95
1 Equip. Oper. (light)	47.05	376.40	71.35	570.80		
1 Flatbed Truck, Gas, 3 Ton		333.00		366.30		
1 Flatbed Trailer, 25 Ton		113.60		124.96		
1 Horiz. Dir. Drill, 20k lb. Thrust		658.40		724.24	69.06	75.97
16 L.H., Daily Totals		$1774.60		$2238.70	$110.91	$139.92
Crew B-82B	**Hr.**	**Daily**	**Hr.**	**Daily**	**Bare Costs**	**Incl. O&P**
2 Laborers	$36.65	$586.40	$56.55	$904.80	$40.12	$61.48
1 Equip. Oper. (light)	47.05	376.40	71.35	570.80		
1 Flatbed Truck, Gas, 3 Ton		333.00		366.30		
1 Flatbed Trailer, 25 Ton		113.60		124.96		
1 Horiz. Dir. Drill, 30k lb. Thrust		936.20		1029.82	57.62	63.38
24 L.H., Daily Totals		$2345.60		$2996.68	$97.73	$124.86
Crew B-82C	**Hr.**	**Daily**	**Hr.**	**Daily**	**Bare Costs**	**Incl. O&P**
2 Laborers	$36.65	$586.40	$56.55	$904.80	$40.12	$61.48
1 Equip. Oper. (light)	47.05	376.40	71.35	570.80		
1 Flatbed Truck, Gas, 3 Ton		333.00		366.30		
1 Flatbed Trailer, 25 Ton		113.60		124.96		
1 Horiz. Dir. Drill, 50k lb. Thrust		1246.00		1370.60	70.53	77.58
24 L.H., Daily Totals		$2655.40		$3337.46	$110.64	$139.06
Crew B-82D	**Hr.**	**Daily**	**Hr.**	**Daily**	**Bare Costs**	**Incl. O&P**
1 Equip. Oper. (light)	$47.05	$376.40	$71.35	$570.80	$47.05	$71.35
1 Mud Trailer for HDD, 1500 Gal.		374.80		412.28	46.85	51.53
8 L.H., Daily Totals		$751.20		$983.08	$93.90	$122.89
Crew B-83	**Hr.**	**Daily**	**Hr.**	**Daily**	**Bare Costs**	**Incl. O&P**
1 Tugboat Captain	$48.90	$391.20	$74.15	$593.20	$42.77	$65.35
1 Tugboat Hand	36.65	293.20	56.55	452.40		
1 Tugboat, 250 H.P.		889.40		978.34	55.59	61.15
16 L.H., Daily Totals		$1573.80		$2023.94	$98.36	$126.50
Crew B-84	**Hr.**	**Daily**	**Hr.**	**Daily**	**Bare Costs**	**Incl. O&P**
1 Equip. Oper. (medium)	$48.90	$391.20	$74.15	$593.20	$48.90	$74.15
1 Rotary Mower/Tractor		364.20		400.62	45.52	50.08
8 L.H., Daily Totals		$755.40		$993.82	$94.42	$124.23
Crew B-85	**Hr.**	**Daily**	**Hr.**	**Daily**	**Bare Costs**	**Incl. O&P**
3 Laborers	$36.65	$879.60	$56.55	$1357.20	$39.28	$60.22
1 Equip. Oper. (medium)	48.90	391.20	74.15	593.20		
1 Truck Driver (heavy)	37.55	300.40	57.30	458.40		
1 Aerial Lift Truck, 80'		643.00		707.30		
1 Brush Chipper, 12", 130 H.P.		396.60		436.26		
1 Pruning Saw, Rotary		6.55		7.21	26.15	28.77
40 L.H., Daily Totals		$2617.35		$3559.57	$65.43	$88.99
Crew B-86	**Hr.**	**Daily**	**Hr.**	**Daily**	**Bare Costs**	**Incl. O&P**
1 Equip. Oper. (medium)	$48.90	$391.20	$74.15	$593.20	$48.90	$74.15
1 Stump Chipper, S.P.		162.10		178.31	20.26	22.29
8 L.H., Daily Totals		$553.30		$771.51	$69.16	$96.44
Crew B-86A	**Hr.**	**Daily**	**Hr.**	**Daily**	**Bare Costs**	**Incl. O&P**
1 Equip. Oper. (medium)	$48.90	$391.20	$74.15	$593.20	$48.90	$74.15
1 Grader, 30,000 Lbs.		707.00		777.70	88.38	97.21
8 L.H., Daily Totals		$1098.20		$1370.90	$137.28	$171.36

Crew B-86B	Bare Costs Hr.	Bare Costs Daily	Incl. Subs O&P Hr.	Incl. Subs O&P Daily	Cost Per Labor-Hour Bare Costs	Cost Per Labor-Hour Incl. O&P
1 Equip. Oper. (medium)	$48.90	$391.20	$74.15	$593.20	$48.90	$74.15
1 Dozer, 200 H.P.		1325.00		1457.50	165.63	182.19
8 L.H., Daily Totals		$1716.20		$2050.70	$214.53	$256.34

Crew B-87	Hr.	Daily	Hr.	Daily	Bare Costs	Incl. O&P
1 Laborer	$36.65	$293.20	$56.55	$452.40	$46.45	$70.63
4 Equip. Oper. (medium)	48.90	1564.80	74.15	2372.80		
2 Feller Bunchers, 100 H.P.		1458.00		1603.80		
1 Log Chipper, 22" Tree		898.60		988.46		
1 Dozer, 105 H.P.		591.80		650.98		
1 Chain Saw, Gas, 36" Long		45.00		49.50	74.83	82.32
40 L.H., Daily Totals		$4851.40		$6117.94	$121.29	$152.95

Crew B-88	Hr.	Daily	Hr.	Daily	Bare Costs	Incl. O&P
1 Laborer	$36.65	$293.20	$56.55	$452.40	$47.15	$71.64
6 Equip. Oper. (medium)	48.90	2347.20	74.15	3559.20		
2 Feller Bunchers, 100 H.P.		1458.00		1603.80		
1 Log Chipper, 22" Tree		898.60		988.46		
2 Log Skidders, 50 H.P.		1839.20		2023.12		
1 Dozer, 105 H.P.		591.80		650.98		
1 Chain Saw, Gas, 36" Long		45.00		49.50	86.30	94.93
56 L.H., Daily Totals		$7473.00		$9327.46	$133.45	$166.56

Crew B-89	Hr.	Daily	Hr.	Daily	Bare Costs	Incl. O&P
1 Equip. Oper. (light)	$47.05	$376.40	$71.35	$570.80	$41.77	$63.52
1 Truck Driver (light)	36.50	292.00	55.70	445.60		
1 Flatbed Truck, Gas, 3 Ton		333.00		366.30		
1 Concrete Saw		175.20		192.72		
1 Water Tank, 65 Gal.		17.30		19.03	32.84	36.13
16 L.H., Daily Totals		$1193.90		$1594.45	$74.62	$99.65

Crew B-89A	Hr.	Daily	Hr.	Daily	Bare Costs	Incl. O&P
1 Skilled Worker	$47.30	$378.40	$73.25	$586.00	$41.98	$64.90
1 Laborer	36.65	293.20	56.55	452.40		
1 Core Drill (Large)		116.20		127.82	7.26	7.99
16 L.H., Daily Totals		$787.80		$1166.22	$49.24	$72.89

Crew B-89B	Hr.	Daily	Hr.	Daily	Bare Costs	Incl. O&P
1 Equip. Oper. (light)	$47.05	$376.40	$71.35	$570.80	$41.77	$63.52
1 Truck Driver (light)	36.50	292.00	55.70	445.60		
1 Wall Saw, Hydraulic, 10 H.P.		119.60		131.56		
1 Generator, Diesel, 100 kW		439.20		483.12		
1 Water Tank, 65 Gal.		17.30		19.03		
1 Flatbed Truck, Gas, 3 Ton		333.00		366.30	56.82	62.50
16 L.H., Daily Totals		$1577.50		$2016.41	$98.59	$126.03

Crew B-90	Hr.	Daily	Hr.	Daily	Bare Costs	Incl. O&P
1 Labor Foreman (outside)	$38.65	$309.20	$59.65	$477.20	$39.73	$60.83
3 Laborers	36.65	879.60	56.55	1357.20		
2 Equip. Oper. (light)	47.05	752.80	71.35	1141.60		
2 Truck Drivers (heavy)	37.55	600.80	57.30	916.80		
1 Road Mixer, 310 H.P.		1942.00		2136.20		
1 Dist. Truck, 2000 Gal.		278.20		306.02	34.69	38.16
64 L.H., Daily Totals		$4762.60		$6335.02	$74.42	$98.98

Crew B-90A	Hr.	Daily	Hr.	Daily	Bare Costs	Incl. O&P
1 Labor Foreman (outside)	$38.65	$309.20	$59.65	$477.20	$43.94	$67.05
2 Laborers	36.65	586.40	56.55	904.80		
4 Equip. Oper. (medium)	48.90	1564.80	74.15	2372.80		
2 Graders, 30,000 Lbs.		1414.00		1555.40		
1 Tandem Roller, 10 Ton		242.00		266.20		
1 Roller, Pneum. Whl., 12 Ton		349.00		383.90	35.80	39.38
56 L.H., Daily Totals		$4465.40		$5960.30	$79.74	$106.43

Crew B-90B	Hr.	Daily	Hr.	Daily	Bare Costs	Incl. O&P
1 Labor Foreman (outside)	$38.65	$309.20	$59.65	$477.20	$43.11	$65.87
2 Laborers	36.65	586.40	56.55	904.80		
3 Equip. Oper. (medium)	48.90	1173.60	74.15	1779.60		
1 Roller, Pneum. Whl., 12 Ton		349.00		383.90		
1 Road Mixer, 310 H.P.		1942.00		2136.20	47.73	52.50
48 L.H., Daily Totals		$4360.20		$5681.70	$90.84	$118.37

Crew B-90C	Hr.	Daily	Hr.	Daily	Bare Costs	Incl. O&P
1 Labor Foreman (outside)	$38.65	$309.20	$59.65	$477.20	$40.42	$61.84
4 Laborers	36.65	1172.80	56.55	1809.60		
3 Equip. Oper. (medium)	48.90	1173.60	74.15	1779.60		
3 Truck Drivers (heavy)	37.55	901.20	57.30	1375.20		
3 Road Mixers, 310 H.P.		5826.00		6408.60	66.20	72.83
88 L.H., Daily Totals		$9382.80		$11850.20	$106.62	$134.66

Crew B-90D	Hr.	Daily	Hr.	Daily	Bare Costs	Incl. O&P
1 Labor Foreman (outside)	$38.65	$309.20	$59.65	$477.20	$39.84	$61.02
6 Laborers	36.65	1759.20	56.55	2714.40		
3 Equip. Oper. (medium)	48.90	1173.60	74.15	1779.60		
3 Truck Drivers (heavy)	37.55	901.20	57.30	1375.20		
3 Road Mixers, 310 H.P.		5826.00		6408.60	56.02	61.62
104 L.H., Daily Totals		$9969.20		$12755.00	$95.86	$122.64

Crew B-90E	Hr.	Daily	Hr.	Daily	Bare Costs	Incl. O&P
1 Labor Foreman (outside)	$38.65	$309.20	$59.65	$477.20	$41.06	$62.84
4 Laborers	36.65	1172.80	56.55	1809.60		
3 Equip. Oper. (medium)	48.90	1173.60	74.15	1779.60		
1 Truck Driver (heavy)	37.55	300.40	57.30	458.40		
1 Road Mixers, 310 H.P.		1942.00		2136.20	26.97	29.67
72 L.H., Daily Totals		$4898.00		$6661.00	$68.03	$92.51

Crew B-91	Hr.	Daily	Hr.	Daily	Bare Costs	Incl. O&P
1 Labor Foreman (outside)	$38.65	$309.20	$59.65	$477.20	$43.14	$65.83
2 Laborers	36.65	586.40	56.55	904.80		
4 Equip. Oper. (medium)	48.90	1564.80	74.15	2372.80		
1 Truck Driver (heavy)	37.55	300.40	57.30	458.40		
1 Dist. Tanker, 3000 Gallon		308.20		339.02		
1 Truck Tractor, 6x4, 380 H.P.		612.80		674.08		
1 Aggreg. Spreader, S.P.		841.00		925.10		
1 Roller, Pneum. Whl., 12 Ton		349.00		383.90		
1 Tandem Roller, 10 Ton		242.00		266.20	36.77	40.44
64 L.H., Daily Totals		$5113.80		$6801.50	$79.90	$106.27

Crew B-91B	Hr.	Daily	Hr.	Daily	Bare Costs	Incl. O&P
1 Laborer	$36.65	$293.20	$56.55	$452.40	$42.77	$65.35
1 Equipment Oper. (med.)	48.90	391.20	74.15	593.20		
1 Road Sweeper, Vac. Assist.		994.40		1093.84	62.15	68.36
16 L.H., Daily Totals		$1678.80		$2139.44	$104.93	$133.72

Crew No.	Bare Costs		Incl. Subs O&P		Cost Per Labor-Hour	
Crew B-91C	**Hr.**	**Daily**	**Hr.**	**Daily**	**Bare Costs**	**Incl. O&P**
1 Laborer	$36.65	$293.20	$56.55	$452.40	$36.58	$56.13
1 Truck Driver (light)	36.50	292.00	55.70	445.60		
1 Catch Basin Cleaning Truck		604.00		664.40	37.75	41.52
16 L.H., Daily Totals		$1189.20		$1562.40	$74.33	$97.65

Crew B-91D	Hr.	Daily	Hr.	Daily	Bare Costs	Incl. O&P
1 Labor Foreman (outside)	$38.65	$309.20	$59.65	$477.20	$41.65	$63.67
5 Laborers	36.65	1466.00	56.55	2262.00		
5 Equip. Oper. (medium)	48.90	1956.00	74.15	2966.00		
2 Truck Drivers (heavy)	37.55	600.80	57.30	916.80		
1 Aggreg. Spreader, S.P.		841.00		925.10		
2 Truck Tractor, 6x4, 380 H.P.		1225.60		1348.16		
2 Dist. Tanker, 3000 Gallon		616.40		678.04		
2 Pavement Brush, Towed		163.20		179.52		
2 Roller, Pneum. Whl., 12 Ton		698.00		767.80	34.08	37.49
104 L.H., Daily Totals		$7876.20		$10520.62	$75.73	$101.16

Crew B-92	Hr.	Daily	Hr.	Daily	Bare Costs	Incl. O&P
1 Labor Foreman (outside)	$38.65	$309.20	$59.65	$477.20	$37.15	$57.33
3 Laborers	36.65	879.60	56.55	1357.20		
1 Crack Cleaner, 25 H.P.		64.80		71.28		
1 Air Compressor, 60 cfm		137.40		151.14		
1 Tar Kettle, T.M.		104.60		115.06		
1 Flatbed Truck, Gas, 3 Ton		333.00		366.30	19.99	21.99
32 L.H., Daily Totals		$1828.60		$2538.18	$57.14	$79.32

Crew B-93	Hr.	Daily	Hr.	Daily	Bare Costs	Incl. O&P
1 Equip. Oper. (medium)	$48.90	$391.20	$74.15	$593.20	$48.90	$74.15
1 Feller Buncher, 100 H.P.		729.00		801.90	91.13	100.24
8 L.H., Daily Totals		$1120.20		$1395.10	$140.03	$174.39

Crew B-94A	Hr.	Daily	Hr.	Daily	Bare Costs	Incl. O&P
1 Laborer	$36.65	$293.20	$56.55	$452.40	$36.65	$56.55
1 Diaphragm Water Pump, 2"		72.00		79.20		
1 -20' Suction Hose, 2"		1.95		2.15		
2 -50' Discharge Hoses, 2"		1.80		1.98	9.47	10.42
8 L.H., Daily Totals		$368.95		$535.73	$46.12	$66.97

Crew B-94B	Hr.	Daily	Hr.	Daily	Bare Costs	Incl. O&P
1 Laborer	$36.65	$293.20	$56.55	$452.40	$36.65	$56.55
1 Diaphragm Water Pump, 4"		115.80		127.38		
1 -20' Suction Hose, 4"		3.25		3.58		
2 -50' Discharge Hoses, 4"		4.70		5.17	15.47	17.02
8 L.H., Daily Totals		$416.95		$588.52	$52.12	$73.57

Crew B-94C	Hr.	Daily	Hr.	Daily	Bare Costs	Incl. O&P
1 Laborer	$36.65	$293.20	$56.55	$452.40	$36.65	$56.55
1 Centrifugal Water Pump, 3"		78.80		86.68		
1 -20' Suction Hose, 3"		2.85		3.13		
2 -50' Discharge Hoses, 3"		3.00		3.30	10.58	11.64
8 L.H., Daily Totals		$377.85		$545.51	$47.23	$68.19

Crew B-94D	Hr.	Daily	Hr.	Daily	Bare Costs	Incl. O&P
1 Laborer	$36.65	$293.20	$56.55	$452.40	$36.65	$56.55
1 Centr. Water Pump, 6"		340.80		374.88		
1 -20' Suction Hose, 6"		11.50		12.65		
2 -50' Discharge Hoses, 6"		12.20		13.42	45.56	50.12
8 L.H., Daily Totals		$657.70		$853.35	$82.21	$106.67

Crew No.	Bare Costs		Incl. Subs O&P		Cost Per Labor-Hour	
Crew C-1	**Hr.**	**Daily**	**Hr.**	**Daily**	**Bare Costs**	**Incl. O&P**
3 Carpenters	$45.85	$1100.40	$70.75	$1698.00	$43.55	$67.20
1 Laborer	36.65	293.20	56.55	452.40		
32 L.H., Daily Totals		$1393.60		$2150.40	$43.55	$67.20

Crew C-2	Hr.	Daily	Hr.	Daily	Bare Costs	Incl. O&P
1 Carpenter Foreman (outside)	$47.85	$382.80	$73.85	$590.80	$44.65	$68.90
4 Carpenters	45.85	1467.20	70.75	2264.00		
1 Laborer	36.65	293.20	56.55	452.40		
48 L.H., Daily Totals		$2143.20		$3307.20	$44.65	$68.90

Crew C-2A	Hr.	Daily	Hr.	Daily	Bare Costs	Incl. O&P
1 Carpenter Foreman (outside)	$47.85	$382.80	$73.85	$590.80	$44.35	$67.96
3 Carpenters	45.85	1100.40	70.75	1698.00		
1 Cement Finisher	44.05	352.40	65.10	520.80		
1 Laborer	36.65	293.20	56.55	452.40		
48 L.H., Daily Totals		$2128.80		$3262.00	$44.35	$67.96

Crew C-3	Hr.	Daily	Hr.	Daily	Bare Costs	Incl. O&P
1 Rodman Foreman (outside)	$52.65	$421.20	$82.70	$661.60	$46.95	$73.17
4 Rodmen (reinf.)	50.65	1620.80	79.55	2545.60		
1 Equip. Oper. (light)	47.05	376.40	71.35	570.80		
2 Laborers	36.65	586.40	56.55	904.80		
3 Stressing Equipment		30.60		33.66		
.5 Grouting Equipment		82.10		90.31	1.76	1.94
64 L.H., Daily Totals		$3117.50		$4806.77	$48.71	$75.11

Crew C-4	Hr.	Daily	Hr.	Daily	Bare Costs	Incl. O&P
1 Rodman Foreman (outside)	$52.65	$421.20	$82.70	$661.60	$51.15	$80.34
3 Rodmen (reinf.)	50.65	1215.60	79.55	1909.20		
3 Stressing Equipment		30.60		33.66	.96	1.05
32 L.H., Daily Totals		$1667.40		$2604.46	$52.11	$81.39

Crew C-4A	Hr.	Daily	Hr.	Daily	Bare Costs	Incl. O&P
2 Rodmen (reinf.)	$50.65	$810.40	$79.55	$1272.80	$50.65	$79.55
4 Stressing Equipment		40.80		44.88	2.55	2.81
16 L.H., Daily Totals		$851.20		$1317.68	$53.20	$82.36

Crew C-5	Hr.	Daily	Hr.	Daily	Bare Costs	Incl. O&P
1 Rodman Foreman (outside)	$52.65	$421.20	$82.70	$661.60	$49.86	$77.59
4 Rodmen (reinf.)	50.65	1620.80	79.55	2545.60		
1 Equip. Oper. (crane)	50.25	402.00	76.20	609.60		
1 Equip. Oper. (oiler)	43.55	348.40	66.00	528.00		
1 Hyd. Crane, 25 Ton		745.80		820.38	13.32	14.65
56 L.H., Daily Totals		$3538.20		$5165.18	$63.18	$92.24

Crew C-6	Hr.	Daily	Hr.	Daily	Bare Costs	Incl. O&P
1 Labor Foreman (outside)	$38.65	$309.20	$59.65	$477.20	$38.22	$58.49
4 Laborers	36.65	1172.80	56.55	1809.60		
1 Cement Finisher	44.05	352.40	65.10	520.80		
2 Gas Engine Vibrators		66.00		72.60	1.38	1.51
48 L.H., Daily Totals		$1900.40		$2880.20	$39.59	$60.00

Crew No.	Bare Costs		Incl. Subs O&P		Cost Per Labor-Hour	
Crew C-7	**Hr.**	**Daily**	**Hr.**	**Daily**	**Bare Costs**	**Incl. O&P**
1 Labor Foreman (outside)	$38.65	$309.20	$59.65	$477.20	$39.82	$60.85
5 Laborers	36.65	1466.00	56.55	2262.00		
1 Cement Finisher	44.05	352.40	65.10	520.80		
1 Equip. Oper. (medium)	48.90	391.20	74.15	593.20		
1 Equip. Oper. (oiler)	43.55	348.40	66.00	528.00		
2 Gas Engine Vibrators		66.00		72.60		
1 Concrete Bucket, 1 C.Y.		23.00		25.30		
1 Hyd. Crane, 55 Ton		1137.00		1250.70	17.03	18.73
72 L.H., Daily Totals		$4093.20		$5729.80	$56.85	$79.58
Crew C-7A	**Hr.**	**Daily**	**Hr.**	**Daily**	**Bare Costs**	**Incl. O&P**
1 Labor Foreman (outside)	$38.65	$309.20	$59.65	$477.20	$37.13	$57.13
5 Laborers	36.65	1466.00	56.55	2262.00		
2 Truck Drivers (heavy)	37.55	600.80	57.30	916.80		
2 Conc. Transit Mixers		2050.00		2255.00	32.03	35.23
64 L.H., Daily Totals		$4426.00		$5911.00	$69.16	$92.36
Crew C-7B	**Hr.**	**Daily**	**Hr.**	**Daily**	**Bare Costs**	**Incl. O&P**
1 Labor Foreman (outside)	$38.65	$309.20	$59.65	$477.20	$39.46	$60.58
5 Laborers	36.65	1466.00	56.55	2262.00		
1 Equipment Operator, Crane	50.25	402.00	76.20	609.60		
1 Equipment Oiler	43.55	348.40	66.00	528.00		
1 Conc. Bucket, 2 C.Y.		35.60		39.16		
1 Lattice Boom Crane, 165 Ton		1954.00		2149.40	31.09	34.20
64 L.H., Daily Totals		$4515.20		$6065.36	$70.55	$94.77
Crew C-7C	**Hr.**	**Daily**	**Hr.**	**Daily**	**Bare Costs**	**Incl. O&P**
1 Labor Foreman (outside)	$38.65	$309.20	$59.65	$477.20	$39.96	$61.34
5 Laborers	36.65	1466.00	56.55	2262.00		
2 Equipment Operators (med.)	48.90	782.40	74.15	1186.40		
2 F.E. Loaders, W.M., 4 C.Y.		1347.60		1482.36	21.06	23.16
64 L.H., Daily Totals		$3905.20		$5407.96	$61.02	$84.50
Crew C-7D	**Hr.**	**Daily**	**Hr.**	**Daily**	**Bare Costs**	**Incl. O&P**
1 Labor Foreman (outside)	$38.65	$309.20	$59.65	$477.20	$38.69	$59.51
5 Laborers	36.65	1466.00	56.55	2262.00		
1 Equip. Oper. (medium)	48.90	391.20	74.15	593.20		
1 Concrete Conveyer		205.40		225.94	3.67	4.03
56 L.H., Daily Totals		$2371.80		$3558.34	$42.35	$63.54
Crew C-8	**Hr.**	**Daily**	**Hr.**	**Daily**	**Bare Costs**	**Incl. O&P**
1 Labor Foreman (outside)	$38.65	$309.20	$59.65	$477.20	$40.80	$61.95
3 Laborers	36.65	879.60	56.55	1357.20		
2 Cement Finishers	44.05	704.80	65.10	1041.60		
1 Equip. Oper. (medium)	48.90	391.20	74.15	593.20		
1 Concrete Pump (Small)		710.40		781.44	12.69	13.95
56 L.H., Daily Totals		$2995.20		$4250.64	$53.49	$75.90
Crew C-8A	**Hr.**	**Daily**	**Hr.**	**Daily**	**Bare Costs**	**Incl. O&P**
1 Labor Foreman (outside)	$38.65	$309.20	$59.65	$477.20	$39.45	$59.92
3 Laborers	36.65	879.60	56.55	1357.20		
2 Cement Finishers	44.05	704.80	65.10	1041.60		
48 L.H., Daily Totals		$1893.60		$2876.00	$39.45	$59.92

Crew No.	Bare Costs		Incl. Subs O&P		Cost Per Labor-Hour	
Crew C-8B	**Hr.**	**Daily**	**Hr.**	**Daily**	**Bare Costs**	**Incl. O&P**
1 Labor Foreman (outside)	$38.65	$309.20	$59.65	$477.20	$39.50	$60.69
3 Laborers	36.65	879.60	56.55	1357.20		
1 Equip. Oper. (medium)	48.90	391.20	74.15	593.20		
1 Vibrating Power Screed		63.25		69.58		
1 Roller, Vibratory, 25 Ton		692.80		762.08		
1 Dozer, 200 H.P.		1325.00		1457.50	52.03	57.23
40 L.H., Daily Totals		$3661.05		$4716.76	$91.53	$117.92
Crew C-8C	**Hr.**	**Daily**	**Hr.**	**Daily**	**Bare Costs**	**Incl. O&P**
1 Labor Foreman (outside)	$38.65	$309.20	$59.65	$477.20	$40.26	$61.42
3 Laborers	36.65	879.60	56.55	1357.20		
1 Cement Finisher	44.05	352.40	65.10	520.80		
1 Equip. Oper. (medium)	48.90	391.20	74.15	593.20		
1 Shotcrete Rig, 12 C.Y./hr		252.80		278.08		
1 Air Compressor, 160 cfm		157.40		173.14		
4 -50' Air Hoses, 1"		16.40		18.04		
4 -50' Air Hoses, 2"		31.00		34.10	9.53	10.49
48 L.H., Daily Totals		$2390.00		$3451.76	$49.79	$71.91
Crew C-8D	**Hr.**	**Daily**	**Hr.**	**Daily**	**Bare Costs**	**Incl. O&P**
1 Labor Foreman (outside)	$38.65	$309.20	$59.65	$477.20	$41.60	$63.16
1 Laborer	36.65	293.20	56.55	452.40		
1 Cement Finisher	44.05	352.40	65.10	520.80		
1 Equipment Oper. (light)	47.05	376.40	71.35	570.80		
1 Air Compressor, 250 cfm		202.40		222.64		
2 -50' Air Hoses, 1"		8.20		9.02	6.58	7.24
32 L.H., Daily Totals		$1541.80		$2252.86	$48.18	$70.40
Crew C-8E	**Hr.**	**Daily**	**Hr.**	**Daily**	**Bare Costs**	**Incl. O&P**
1 Labor Foreman (outside)	$38.65	$309.20	$59.65	$477.20	$39.95	$60.96
3 Laborers	36.65	879.60	56.55	1357.20		
1 Cement Finisher	44.05	352.40	65.10	520.80		
1 Equipment Oper. (light)	47.05	376.40	71.35	570.80		
1 Shotcrete Rig, 35 C.Y./hr		284.40		312.84		
1 Air Compressor, 250 cfm		202.40		222.64		
4 -50' Air Hoses, 1"		16.40		18.04		
4 -50' Air Hoses, 2"		31.00		34.10	11.13	12.24
48 L.H., Daily Totals		$2451.80		$3513.62	$51.08	$73.20
Crew C-10	**Hr.**	**Daily**	**Hr.**	**Daily**	**Bare Costs**	**Incl. O&P**
1 Laborer	$36.65	$293.20	$56.55	$452.40	$41.58	$62.25
2 Cement Finishers	44.05	704.80	65.10	1041.60		
24 L.H., Daily Totals		$998.00		$1494.00	$41.58	$62.25
Crew C-10B	**Hr.**	**Daily**	**Hr.**	**Daily**	**Bare Costs**	**Incl. O&P**
3 Laborers	$36.65	$879.60	$56.55	$1357.20	$39.61	$59.97
2 Cement Finishers	44.05	704.80	65.10	1041.60		
1 Concrete Mixer, 10 C.F.		176.80		194.48		
2 Trowels, 48" Walk-Behind		107.20		117.92	7.10	7.81
40 L.H., Daily Totals		$1868.40		$2711.20	$46.71	$67.78
Crew C-10C	**Hr.**	**Daily**	**Hr.**	**Daily**	**Bare Costs**	**Incl. O&P**
1 Laborer	$36.65	$293.20	$56.55	$452.40	$41.58	$62.25
2 Cement Finishers	44.05	704.80	65.10	1041.60		
1 Trowel, 48" Walk-Behind		53.60		58.96	2.23	2.46
24 L.H., Daily Totals		$1051.60		$1552.96	$43.82	$64.71

Crews

Crew C-10D	Bare Costs Hr.	Bare Costs Daily	Incl. Subs O&P Hr.	Incl. Subs O&P Daily	Cost Per Labor-Hour Bare Costs	Cost Per Labor-Hour Incl. O&P
1 Laborer	$36.65	$293.20	$56.55	$452.40	$41.58	$62.25
2 Cement Finishers	44.05	704.80	65.10	1041.60		
1 Vibrating Power Screed		63.25		69.58		
1 Trowel, 48" Walk-Behind		53.60		58.96	4.87	5.36
24 L.H., Daily Totals		$1114.85		$1622.54	$46.45	$67.61

Crew C-10E	Hr.	Daily	Hr.	Daily	Bare Costs	Incl. O&P
1 Laborer	$36.65	$293.20	$56.55	$452.40	$41.58	$62.25
2 Cement Finishers	44.05	704.80	65.10	1041.60		
1 Vibrating Power Screed		63.25		69.58		
1 Cement Trowel, 96" Ride-On		197.40		217.14	10.86	11.95
24 L.H., Daily Totals		$1258.65		$1780.71	$52.44	$74.20

Crew C-10F	Hr.	Daily	Hr.	Daily	Bare Costs	Incl. O&P
1 Laborer	$36.65	$293.20	$56.55	$452.40	$41.58	$62.25
2 Cement Finishers	44.05	704.80	65.10	1041.60		
1 Aerial Lift Truck, 60' Boom		456.00		501.60	19.00	20.90
24 L.H., Daily Totals		$1454.00		$1995.60	$60.58	$83.15

Crew C-11	Hr.	Daily	Hr.	Daily	Bare Costs	Incl. O&P
1 Struc. Steel Foreman (outside)	$53.10	$424.80	$93.70	$749.60	$50.39	$86.34
6 Struc. Steel Workers	51.10	2452.80	90.20	4329.60		
1 Equip. Oper. (crane)	50.25	402.00	76.20	609.60		
1 Equip. Oper. (oiler)	43.55	348.40	66.00	528.00		
1 Lattice Boom Crane, 150 Ton		1838.00		2021.80	25.53	28.08
72 L.H., Daily Totals		$5466.00		$8238.60	$75.92	$114.43

Crew C-12	Hr.	Daily	Hr.	Daily	Bare Costs	Incl. O&P
1 Carpenter Foreman (outside)	$47.85	$382.80	$73.85	$590.80	$45.38	$69.81
3 Carpenters	45.85	1100.40	70.75	1698.00		
1 Laborer	36.65	293.20	56.55	452.40		
1 Equip. Oper. (crane)	50.25	402.00	76.20	609.60		
1 Hyd. Crane, 12 Ton		663.00		729.30	13.81	15.19
48 L.H., Daily Totals		$2841.40		$4080.10	$59.20	$85.00

Crew C-13	Hr.	Daily	Hr.	Daily	Bare Costs	Incl. O&P
1 Struc. Steel Worker	$51.10	$408.80	$90.20	$721.60	$49.35	$83.72
1 Welder	51.10	408.80	90.20	721.60		
1 Carpenter	45.85	366.80	70.75	566.00		
1 Welder, Gas Engine, 300 amp		142.00		156.20	5.92	6.51
24 L.H., Daily Totals		$1326.40		$2165.40	$55.27	$90.22

Crew C-14	Hr.	Daily	Hr.	Daily	Bare Costs	Incl. O&P
1 Carpenter Foreman (outside)	$47.85	$382.80	$73.85	$590.80	$44.90	$69.13
5 Carpenters	45.85	1834.00	70.75	2830.00		
4 Laborers	36.65	1172.80	56.55	1809.60		
4 Rodmen (reinf.)	50.65	1620.80	79.55	2545.60		
2 Cement Finishers	44.05	704.80	65.10	1041.60		
1 Equip. Oper. (crane)	50.25	402.00	76.20	609.60		
1 Equip. Oper. (oiler)	43.55	348.40	66.00	528.00		
1 Hyd. Crane, 80 Ton		1652.00		1817.20	11.47	12.62
144 L.H., Daily Totals		$8117.60		$11772.40	$56.37	$81.75

Crew C-14A	Hr.	Daily	Hr.	Daily	Bare Costs	Incl. O&P
1 Carpenter Foreman (outside)	$47.85	$382.80	$73.85	$590.80	$46.01	$71.06
16 Carpenters	45.85	5868.80	70.75	9056.00		
4 Rodmen (reinf.)	50.65	1620.80	79.55	2545.60		
2 Laborers	36.65	586.40	56.55	904.80		
1 Cement Finisher	44.05	352.40	65.10	520.80		
1 Equip. Oper. (medium)	48.90	391.20	74.15	593.20		
1 Gas Engine Vibrator		33.00		36.30		
1 Concrete Pump (Small)		710.40		781.44	3.72	4.09
200 L.H., Daily Totals		$9945.80		$15028.94	$49.73	$75.14

Crew C-14B	Hr.	Daily	Hr.	Daily	Bare Costs	Incl. O&P
1 Carpenter Foreman (outside)	$47.85	$382.80	$73.85	$590.80	$45.94	$70.83
16 Carpenters	45.85	5868.80	70.75	9056.00		
4 Rodmen (reinf.)	50.65	1620.80	79.55	2545.60		
2 Laborers	36.65	586.40	56.55	904.80		
2 Cement Finishers	44.05	704.80	65.10	1041.60		
1 Equip. Oper. (medium)	48.90	391.20	74.15	593.20		
1 Gas Engine Vibrator		33.00		36.30		
1 Concrete Pump (Small)		710.40		781.44	3.57	3.93
208 L.H., Daily Totals		$10298.20		$15549.74	$49.51	$74.76

Crew C-14C	Hr.	Daily	Hr.	Daily	Bare Costs	Incl. O&P
1 Carpenter Foreman (outside)	$47.85	$382.80	$73.85	$590.80	$43.92	$67.77
6 Carpenters	45.85	2200.80	70.75	3396.00		
2 Rodmen (reinf.)	50.65	810.40	79.55	1272.80		
4 Laborers	36.65	1172.80	56.55	1809.60		
1 Cement Finisher	44.05	352.40	65.10	520.80		
1 Gas Engine Vibrator		33.00		36.30	.29	.32
112 L.H., Daily Totals		$4952.20		$7626.30	$44.22	$68.09

Crew C-14D	Hr.	Daily	Hr.	Daily	Bare Costs	Incl. O&P
1 Carpenter Foreman (outside)	$47.85	$382.80	$73.85	$590.80	$45.63	$70.35
18 Carpenters	45.85	6602.40	70.75	10188.00		
2 Rodmen (reinf.)	50.65	810.40	79.55	1272.80		
2 Laborers	36.65	586.40	56.55	904.80		
1 Cement Finisher	44.05	352.40	65.10	520.80		
1 Equip. Oper. (medium)	48.90	391.20	74.15	593.20		
1 Gas Engine Vibrator		33.00		36.30		
1 Concrete Pump (Small)		710.40		781.44	3.72	4.09
200 L.H., Daily Totals		$9869.00		$14888.14	$49.34	$74.44

Crew C-14E	Hr.	Daily	Hr.	Daily	Bare Costs	Incl. O&P
1 Carpenter Foreman (outside)	$47.85	$382.80	$73.85	$590.80	$45.10	$69.85
2 Carpenters	45.85	733.60	70.75	1132.00		
4 Rodmen (reinf.)	50.65	1620.80	79.55	2545.60		
3 Laborers	36.65	879.60	56.55	1357.20		
1 Cement Finisher	44.05	352.40	65.10	520.80		
1 Gas Engine Vibrator		33.00		36.30	.38	.41
88 L.H., Daily Totals		$4002.20		$6182.70	$45.48	$70.26

Crew C-14F	Hr.	Daily	Hr.	Daily	Bare Costs	Incl. O&P
1 Labor Foreman (outside)	$38.65	$309.20	$59.65	$477.20	$41.81	$62.59
2 Laborers	36.65	586.40	56.55	904.80		
6 Cement Finishers	44.05	2114.40	65.10	3124.80		
1 Gas Engine Vibrator		33.00		36.30	.46	.50
72 L.H., Daily Totals		$3043.00		$4543.10	$42.26	$63.10

Crew No.	Bare Costs		Incl. Subs O&P		Cost Per Labor-Hour	
Crew C-14G	**Hr.**	**Daily**	**Hr.**	**Daily**	**Bare Costs**	**Incl. O&P**
1 Labor Foreman (outside)	$38.65	$309.20	$59.65	$477.20	$41.16	$61.88
2 Laborers	36.65	586.40	56.55	904.80		
4 Cement Finishers	44.05	1409.60	65.10	2083.20		
1 Gas Engine Vibrator		33.00		36.30	.59	.65
56 L.H., Daily Totals		$2338.20		$3501.50	$41.75	$62.53

Crew No.	Bare Costs		Incl. Subs O&P		Cost Per Labor-Hour	
Crew C-14H	**Hr.**	**Daily**	**Hr.**	**Daily**	**Bare Costs**	**Incl. O&P**
1 Carpenter Foreman (outside)	$47.85	$382.80	$73.85	$590.80	$45.15	$69.42
2 Carpenters	45.85	733.60	70.75	1132.00		
1 Rodman (reinf.)	50.65	405.20	79.55	636.40		
1 Laborer	36.65	293.20	56.55	452.40		
1 Cement Finisher	44.05	352.40	65.10	520.80		
1 Gas Engine Vibrator		33.00		36.30	.69	.76
48 L.H., Daily Totals		$2200.20		$3368.70	$45.84	$70.18

Crew No.	Bare Costs		Incl. Subs O&P		Cost Per Labor-Hour	
Crew C-14L	**Hr.**	**Daily**	**Hr.**	**Daily**	**Bare Costs**	**Incl. O&P**
1 Carpenter Foreman (outside)	$47.85	$382.80	$73.85	$590.80	$42.80	$65.80
6 Carpenters	45.85	2200.80	70.75	3396.00		
4 Laborers	36.65	1172.80	56.55	1809.60		
1 Cement Finisher	44.05	352.40	65.10	520.80		
1 Gas Engine Vibrator		33.00		36.30	.34	.38
96 L.H., Daily Totals		$4141.80		$6353.50	$43.14	$66.18

Crew No.	Bare Costs		Incl. Subs O&P		Cost Per Labor-Hour	
Crew C-14M	**Hr.**	**Daily**	**Hr.**	**Daily**	**Bare Costs**	**Incl. O&P**
1 Carpenter Foreman (outside)	$47.85	$382.80	$73.85	$590.80	$44.56	$68.41
2 Carpenters	45.85	733.60	70.75	1132.00		
1 Rodman (reinf.)	50.65	405.20	79.55	636.40		
2 Laborers	36.65	586.40	56.55	904.80		
1 Cement Finisher	44.05	352.40	65.10	520.80		
1 Equip. Oper. (medium)	48.90	391.20	74.15	593.20		
1 Gas Engine Vibrator		33.00		36.30		
1 Concrete Pump (Small)		710.40		781.44	11.62	12.78
64 L.H., Daily Totals		$3595.00		$5195.74	$56.17	$81.18

Crew No.	Bare Costs		Incl. Subs O&P		Cost Per Labor-Hour	
Crew C-15	**Hr.**	**Daily**	**Hr.**	**Daily**	**Bare Costs**	**Incl. O&P**
1 Carpenter Foreman (outside)	$47.85	$382.80	$73.85	$590.80	$43.14	$66.08
2 Carpenters	45.85	733.60	70.75	1132.00		
3 Laborers	36.65	879.60	56.55	1357.20		
2 Cement Finishers	44.05	704.80	65.10	1041.60		
1 Rodman (reinf.)	50.65	405.20	79.55	636.40		
72 L.H., Daily Totals		$3106.00		$4758.00	$43.14	$66.08

Crew No.	Bare Costs		Incl. Subs O&P		Cost Per Labor-Hour	
Crew C-16	**Hr.**	**Daily**	**Hr.**	**Daily**	**Bare Costs**	**Incl. O&P**
1 Labor Foreman (outside)	$38.65	$309.20	$59.65	$477.20	$40.80	$61.95
3 Laborers	36.65	879.60	56.55	1357.20		
2 Cement Finishers	44.05	704.80	65.10	1041.60		
1 Equip. Oper. (medium)	48.90	391.20	74.15	593.20		
1 Gunite Pump Rig		370.40		407.44		
2 -50' Air Hoses, 3/4"		6.50		7.15		
2 -50' Air Hoses, 2"		15.50		17.05	7.01	7.71
56 L.H., Daily Totals		$2677.20		$3900.84	$47.81	$69.66

Crew No.	Bare Costs		Incl. Subs O&P		Cost Per Labor-Hour	
Crew C-16A	**Hr.**	**Daily**	**Hr.**	**Daily**	**Bare Costs**	**Incl. O&P**
1 Laborer	$36.65	$293.20	$56.55	$452.40	$43.41	$65.22
2 Cement Finishers	44.05	704.80	65.10	1041.60		
1 Equip. Oper. (medium)	48.90	391.20	74.15	593.20		
1 Gunite Pump Rig		370.40		407.44		
2 -50' Air Hoses, 3/4"		6.50		7.15		
2 -50' Air Hoses, 2"		15.50		17.05		
1 Aerial Lift Truck, 60' Boom		456.00		501.60	26.51	29.16
32 L.H., Daily Totals		$2237.60		$3020.44	$69.92	$94.39

Crew No.	Bare Costs		Incl. Subs O&P		Cost Per Labor-Hour	
Crew C-17	**Hr.**	**Daily**	**Hr.**	**Daily**	**Bare Costs**	**Incl. O&P**
2 Skilled Worker Foremen (out)	$49.30	$788.80	$76.35	$1221.60	$47.70	$73.87
8 Skilled Workers	47.30	3027.20	73.25	4688.00		
80 L.H., Daily Totals		$3816.00		$5909.60	$47.70	$73.87

Crew No.	Bare Costs		Incl. Subs O&P		Cost Per Labor-Hour	
Crew C-17A	**Hr.**	**Daily**	**Hr.**	**Daily**	**Bare Costs**	**Incl. O&P**
2 Skilled Worker Foremen (out)	$49.30	$788.80	$76.35	$1221.60	$47.73	$73.90
8 Skilled Workers	47.30	3027.20	73.25	4688.00		
.125 Equip. Oper. (crane)	50.25	50.25	76.20	76.20		
.125 Hyd. Crane, 80 Ton		206.50		227.15	2.55	2.80
81 L.H., Daily Totals		$4072.75		$6212.95	$50.28	$76.70

Crew No.	Bare Costs		Incl. Subs O&P		Cost Per Labor-Hour	
Crew C-17B	**Hr.**	**Daily**	**Hr.**	**Daily**	**Bare Costs**	**Incl. O&P**
2 Skilled Worker Foremen (out)	$49.30	$788.80	$76.35	$1221.60	$47.76	$73.93
8 Skilled Workers	47.30	3027.20	73.25	4688.00		
.25 Equip. Oper. (crane)	50.25	100.50	76.20	152.40		
.25 Hyd. Crane, 80 Ton		413.00		454.30		
.25 Trowel, 48" Walk-Behind		13.40		14.74	5.20	5.72
82 L.H., Daily Totals		$4342.90		$6531.04	$52.96	$79.65

Crew No.	Bare Costs		Incl. Subs O&P		Cost Per Labor-Hour	
Crew C-17C	**Hr.**	**Daily**	**Hr.**	**Daily**	**Bare Costs**	**Incl. O&P**
2 Skilled Worker Foremen (out)	$49.30	$788.80	$76.35	$1221.60	$47.79	$73.95
8 Skilled Workers	47.30	3027.20	73.25	4688.00		
.375 Equip. Oper. (crane)	50.25	150.75	76.20	228.60		
.375 Hyd. Crane, 80 Ton		619.50		681.45	7.46	8.21
83 L.H., Daily Totals		$4586.25		$6819.65	$55.26	$82.16

Crew No.	Bare Costs		Incl. Subs O&P		Cost Per Labor-Hour	
Crew C-17D	**Hr.**	**Daily**	**Hr.**	**Daily**	**Bare Costs**	**Incl. O&P**
2 Skilled Worker Foremen (out)	$49.30	$788.80	$76.35	$1221.60	$47.82	$73.98
8 Skilled Workers	47.30	3027.20	73.25	4688.00		
.5 Equip. Oper. (crane)	50.25	201.00	76.20	304.80		
.5 Hyd. Crane, 80 Ton		826.00		908.60	9.83	10.82
84 L.H., Daily Totals		$4843.00		$7123.00	$57.65	$84.80

Crew No.	Bare Costs		Incl. Subs O&P		Cost Per Labor-Hour	
Crew C-17E	**Hr.**	**Daily**	**Hr.**	**Daily**	**Bare Costs**	**Incl. O&P**
2 Skilled Worker Foremen (out)	$49.30	$788.80	$76.35	$1221.60	$47.70	$73.87
8 Skilled Workers	47.30	3027.20	73.25	4688.00		
1 Hyd. Jack with Rods		94.55		104.01	1.18	1.30
80 L.H., Daily Totals		$3910.55		$6013.60	$48.88	$75.17

Crew No.	Bare Costs		Incl. Subs O&P		Cost Per Labor-Hour	
Crew C-18	**Hr.**	**Daily**	**Hr.**	**Daily**	**Bare Costs**	**Incl. O&P**
.125 Labor Foreman (outside)	$38.65	$38.65	$59.65	$59.65	$36.87	$56.89
1 Laborer	36.65	293.20	56.55	452.40		
1 Concrete Cart, 10 C.F.		60.20		66.22	6.69	7.36
9 L.H., Daily Totals		$392.05		$578.27	$43.56	$64.25

Crew No.	Bare Costs		Incl. Subs O&P		Cost Per Labor-Hour	
Crew C-19	**Hr.**	**Daily**	**Hr.**	**Daily**	**Bare Costs**	**Incl. O&P**
.125 Labor Foreman (outside)	$38.65	$38.65	$59.65	$59.65	$36.87	$56.89
1 Laborer	36.65	293.20	56.55	452.40		
1 Concrete Cart, 18 C.F.		101.20		111.32	11.24	12.37
9 L.H., Daily Totals		$433.05		$623.37	$48.12	$69.26

Crew No.	Bare Costs		Incl. Subs O&P		Cost Per Labor-Hour	
Crew C-20	**Hr.**	**Daily**	**Hr.**	**Daily**	**Bare Costs**	**Incl. O&P**
1 Labor Foreman (outside)	$38.65	$309.20	$59.65	$477.20	$39.36	$60.21
5 Laborers	36.65	1466.00	56.55	2262.00		
1 Cement Finisher	44.05	352.40	65.10	520.80		
1 Equip. Oper. (medium)	48.90	391.20	74.15	593.20		
2 Gas Engine Vibrators		66.00		72.60		
1 Concrete Pump (Small)		710.40		781.44	12.13	13.34
64 L.H., Daily Totals		$3295.20		$4707.24	$51.49	$73.55

Crew No.	Bare Costs		Incl. Subs O&P		Cost Per Labor-Hour	
Crew C-21	**Hr.**	**Daily**	**Hr.**	**Daily**	**Bare Costs**	**Incl. O&P**
1 Labor Foreman (outside)	$38.65	$309.20	$59.65	$477.20	$39.36	$60.21
5 Laborers	36.65	1466.00	56.55	2262.00		
1 Cement Finisher	44.05	352.40	65.10	520.80		
1 Equip. Oper. (medium)	48.90	391.20	74.15	593.20		
2 Gas Engine Vibrators		66.00		72.60		
1 Concrete Conveyer		205.40		225.94	4.24	4.66
64 L.H., Daily Totals		$2790.20		$4151.74	$43.60	$64.87

Crew C-22	Hr.	Daily	Hr.	Daily	Bare Costs	Incl. O&P
1 Rodman Foreman (outside)	$52.65	$421.20	$82.70	$661.60	$50.85	$79.75
4 Rodmen (reinf.)	50.65	1620.80	79.55	2545.60		
.125 Equip. Oper. (crane)	50.25	50.25	76.20	76.20		
.125 Equip. Oper. (oiler)	43.55	43.55	66.00	66.00		
.125 Hyd. Crane, 25 Ton		93.22		102.55	2.22	2.44
42 L.H., Daily Totals		$2229.03		$3451.95	$53.07	$82.19

Crew C-23	Hr.	Daily	Hr.	Daily	Bare Costs	Incl. O&P
2 Skilled Worker Foremen (out)	$49.30	$788.80	$76.35	$1221.60	$47.62	$73.44
6 Skilled Workers	47.30	2270.40	73.25	3516.00		
1 Equip. Oper. (crane)	50.25	402.00	76.20	609.60		
1 Equip. Oper. (oiler)	43.55	348.40	66.00	528.00		
1 Lattice Boom Crane, 90 Ton		1529.00		1681.90	19.11	21.02
80 L.H., Daily Totals		$5338.60		$7557.10	$66.73	$94.46

Crew C-23A	Hr.	Daily	Hr.	Daily	Bare Costs	Incl. O&P
1 Labor Foreman (outside)	$38.65	$309.20	$59.65	$477.20	$41.15	$62.99
2 Laborers	36.65	586.40	56.55	904.80		
1 Equip. Oper. (crane)	50.25	402.00	76.20	609.60		
1 Equip. Oper. (oiler)	43.55	348.40	66.00	528.00		
1 Crawler Crane, 100 Ton		1703.00		1873.30		
3 Conc. Buckets, 8 C.Y.		595.80		655.38	57.47	63.22
40 L.H., Daily Totals		$3944.80		$5048.28	$98.62	$126.21

Crew C-24	Hr.	Daily	Hr.	Daily	Bare Costs	Incl. O&P
2 Skilled Worker Foremen (out)	$49.30	$788.80	$76.35	$1221.60	$47.62	$73.44
6 Skilled Workers	47.30	2270.40	73.25	3516.00		
1 Equip. Oper. (crane)	50.25	402.00	76.20	609.60		
1 Equip. Oper. (oiler)	43.55	348.40	66.00	528.00		
1 Lattice Boom Crane, 150 Ton		1838.00		2021.80	22.98	25.27
80 L.H., Daily Totals		$5647.60		$7897.00	$70.59	$98.71

Crew C-25	Hr.	Daily	Hr.	Daily	Bare Costs	Incl. O&P
2 Rodmen (reinf.)	$50.65	$810.40	$79.55	$1272.80	$39.90	$64.58
2 Rodmen Helpers	29.15	466.40	49.60	793.60		
32 L.H., Daily Totals		$1276.80		$2066.40	$39.90	$64.58

Crew C-27	Hr.	Daily	Hr.	Daily	Bare Costs	Incl. O&P
2 Cement Finishers	$44.05	$704.80	$65.10	$1041.60	$44.05	$65.10
1 Concrete Saw		175.20		192.72	10.95	12.05
16 L.H., Daily Totals		$880.00		$1234.32	$55.00	$77.14

Crew C-28	Hr.	Daily	Hr.	Daily	Bare Costs	Incl. O&P
1 Cement Finisher	$44.05	$352.40	$65.10	$520.80	$44.05	$65.10
1 Portable Air Compressor, Gas		17.70		19.47	2.21	2.43
8 L.H., Daily Totals		$370.10		$540.27	$46.26	$67.53

Crew C-29	Hr.	Daily	Hr.	Daily	Bare Costs	Incl. O&P
1 Laborer	$36.65	$293.20	$56.55	$452.40	$36.65	$56.55
1 Pressure Washer		69.40		76.34	8.68	9.54
8 L.H., Daily Totals		$362.60		$528.74	$45.33	$66.09

Crew C-30	Hr.	Daily	Hr.	Daily	Bare Costs	Incl. O&P
1 Laborer	$36.65	$293.20	$56.55	$452.40	$36.65	$56.55
1 Concrete Mixer, 10 C.F.		176.80		194.48	22.10	24.31
8 L.H., Daily Totals		$470.00		$646.88	$58.75	$80.86

Crew C-31	Hr.	Daily	Hr.	Daily	Bare Costs	Incl. O&P
1 Cement Finisher	$44.05	$352.40	$65.10	$520.80	$44.05	$65.10
1 Grout Pump		370.40		407.44	46.30	50.93
8 L.H., Daily Totals		$722.80		$928.24	$90.35	$116.03

Crew C-32	Hr.	Daily	Hr.	Daily	Bare Costs	Incl. O&P
1 Cement Finisher	$44.05	$352.40	$65.10	$520.80	$40.35	$60.83
1 Laborer	36.65	293.20	56.55	452.40		
1 Crack Chaser Saw, Gas, 6 H.P.		30.80		33.88		
1 Vacuum Pick-Up System		60.75		66.83	5.72	6.29
16 L.H., Daily Totals		$737.15		$1073.91	$46.07	$67.12

Crew D-1	Hr.	Daily	Hr.	Daily	Bare Costs	Incl. O&P
1 Bricklayer	$45.60	$364.80	$69.50	$556.00	$41.30	$62.95
1 Bricklayer Helper	37.00	296.00	56.40	451.20		
16 L.H., Daily Totals		$660.80		$1007.20	$41.30	$62.95

Crew D-2	Hr.	Daily	Hr.	Daily	Bare Costs	Incl. O&P
3 Bricklayers	$45.60	$1094.40	$69.50	$1668.00	$42.50	$64.85
2 Bricklayer Helpers	37.00	592.00	56.40	902.40		
.5 Carpenter	45.85	183.40	70.75	283.00		
44 L.H., Daily Totals		$1869.80		$2853.40	$42.50	$64.85

Crew D-3	Hr.	Daily	Hr.	Daily	Bare Costs	Incl. O&P
3 Bricklayers	$45.60	$1094.40	$69.50	$1668.00	$42.34	$64.57
2 Bricklayer Helpers	37.00	592.00	56.40	902.40		
.25 Carpenter	45.85	91.70	70.75	141.50		
42 L.H., Daily Totals		$1778.10		$2711.90	$42.34	$64.57

Crew D-4	Hr.	Daily	Hr.	Daily	Bare Costs	Incl. O&P
1 Bricklayer	$45.60	$364.80	$69.50	$556.00	$41.66	$63.41
2 Bricklayer Helpers	37.00	592.00	56.40	902.40		
1 Equip. Oper. (light)	47.05	376.40	71.35	570.80		
1 Grout Pump, 50 C.F./hr.		134.80		148.28	4.21	4.63
32 L.H., Daily Totals		$1468.00		$2177.48	$45.88	$68.05

Crew D-5	Hr.	Daily	Hr.	Daily	Bare Costs	Incl. O&P
1 Bricklayer	$45.60	$364.80	$69.50	$556.00	$45.60	$69.50
8 L.H., Daily Totals		$364.80		$556.00	$45.60	$69.50

Crew D-6	Hr.	Daily	Hr.	Daily	Bare Costs	Incl. O&P
3 Bricklayers	$45.60	$1094.40	$69.50	$1668.00	$41.48	$63.26
3 Bricklayer Helpers	37.00	888.00	56.40	1353.60		
.25 Carpenter	45.85	91.70	70.75	141.50		
50 L.H., Daily Totals		$2074.10		$3163.10	$41.48	$63.26

Crew D-7	Hr.	Daily	Hr.	Daily	Bare Costs	Incl. O&P
1 Tile Layer	$41.95	$335.60	$61.85	$494.80	$37.60	$55.42
1 Tile Layer Helper	33.25	266.00	49.00	392.00		
16 L.H., Daily Totals		$601.60		$886.80	$37.60	$55.42

Crew D-8	Hr.	Daily	Hr.	Daily	Bare Costs	Incl. O&P
3 Bricklayers	$45.60	$1094.40	$69.50	$1668.00	$42.16	$64.26
2 Bricklayer Helpers	37.00	592.00	56.40	902.40		
40 L.H., Daily Totals		$1686.40		$2570.40	$42.16	$64.26

Crew No.	Bare Costs		Incl. Subs O&P		Cost Per Labor-Hour	
Crew D-9	**Hr.**	**Daily**	**Hr.**	**Daily**	**Bare Costs**	**Incl. O&P**
3 Bricklayers	$45.60	$1094.40	$69.50	$1668.00	$41.30	$62.95
3 Bricklayer Helpers	37.00	888.00	56.40	1353.60		
48 L.H., Daily Totals		$1982.40		$3021.60	$41.30	$62.95

Crew D-10	Hr.	Daily	Hr.	Daily	Bare Costs	Incl. O&P
1 Bricklayer Foreman (outside)	$47.60	$380.80	$72.55	$580.40	$45.11	$68.66
1 Bricklayer	45.60	364.80	69.50	556.00		
1 Bricklayer Helper	37.00	296.00	56.40	451.20		
1 Equip. Oper. (crane)	50.25	402.00	76.20	609.60		
1 S.P. Crane, 4x4, 12 Ton		481.20		529.32	15.04	16.54
32 L.H., Daily Totals		$1924.80		$2726.52	$60.15	$85.20

Crew D-11	Hr.	Daily	Hr.	Daily	Bare Costs	Incl. O&P
1 Bricklayer Foreman (outside)	$47.60	$380.80	$72.55	$580.40	$43.40	$66.15
1 Bricklayer	45.60	364.80	69.50	556.00		
1 Bricklayer Helper	37.00	296.00	56.40	451.20		
24 L.H., Daily Totals		$1041.60		$1587.60	$43.40	$66.15

Crew D-12	Hr.	Daily	Hr.	Daily	Bare Costs	Incl. O&P
1 Bricklayer Foreman (outside)	$47.60	$380.80	$72.55	$580.40	$41.80	$63.71
1 Bricklayer	45.60	364.80	69.50	556.00		
2 Bricklayer Helpers	37.00	592.00	56.40	902.40		
32 L.H., Daily Totals		$1337.60		$2038.80	$41.80	$63.71

Crew D-13	Hr.	Daily	Hr.	Daily	Bare Costs	Incl. O&P
1 Bricklayer Foreman (outside)	$47.60	$380.80	$72.55	$580.40	$43.88	$66.97
1 Bricklayer	45.60	364.80	69.50	556.00		
2 Bricklayer Helpers	37.00	592.00	56.40	902.40		
1 Carpenter	45.85	366.80	70.75	566.00		
1 Equip. Oper. (crane)	50.25	402.00	76.20	609.60		
1 S.P. Crane, 4x4, 12 Ton		481.20		529.32	10.03	11.03
48 L.H., Daily Totals		$2587.60		$3743.72	$53.91	$77.99

Crew E-1	Hr.	Daily	Hr.	Daily	Bare Costs	Incl. O&P
1 Welder Foreman (outside)	$53.10	$424.80	$93.70	$749.60	$50.42	$85.08
1 Welder	51.10	408.80	90.20	721.60		
1 Equip. Oper. (light)	47.05	376.40	71.35	570.80		
1 Welder, Gas Engine, 300 amp		142.00		156.20	5.92	6.51
24 L.H., Daily Totals		$1352.00		$2198.20	$56.33	$91.59

Crew E-2	Hr.	Daily	Hr.	Daily	Bare Costs	Incl. O&P
1 Struc. Steel Foreman (outside)	$53.10	$424.80	$93.70	$749.60	$50.19	$85.24
4 Struc. Steel Workers	51.10	1635.20	90.20	2886.40		
1 Equip. Oper. (crane)	50.25	402.00	76.20	609.60		
1 Equip. Oper. (oiler)	43.55	348.40	66.00	528.00		
1 Lattice Boom Crane, 90 Ton		1529.00		1681.90	27.30	30.03
56 L.H., Daily Totals		$4339.40		$6455.50	$77.49	$115.28

Crew E-3	Hr.	Daily	Hr.	Daily	Bare Costs	Incl. O&P
1 Struc. Steel Foreman (outside)	$53.10	$424.80	$93.70	$749.60	$51.77	$91.37
1 Struc. Steel Worker	51.10	408.80	90.20	721.60		
1 Welder	51.10	408.80	90.20	721.60		
1 Welder, Gas Engine, 300 amp		142.00		156.20	5.92	6.51
24 L.H., Daily Totals		$1384.40		$2349.00	$57.68	$97.88

Crew E-3A	Hr.	Daily	Hr.	Daily	Bare Costs	Incl. O&P
1 Struc. Steel Foreman (outside)	$53.10	$424.80	$93.70	$749.60	$51.77	$91.37
1 Struc. Steel Worker	51.10	408.80	90.20	721.60		
1 Welder	51.10	408.80	90.20	721.60		
1 Welder, Gas Engine, 300 amp		142.00		156.20		
1 Aerial Lift Truck, 40' Boom		325.00		357.50	19.46	21.40
24 L.H., Daily Totals		$1709.40		$2706.50	$71.22	$112.77

Crew E-4	Hr.	Daily	Hr.	Daily	Bare Costs	Incl. O&P
1 Struc. Steel Foreman (outside)	$53.10	$424.80	$93.70	$749.60	$51.60	$91.08
3 Struc. Steel Workers	51.10	1226.40	90.20	2164.80		
1 Welder, Gas Engine, 300 amp		142.00		156.20	4.44	4.88
32 L.H., Daily Totals		$1793.20		$3070.60	$56.04	$95.96

Crew E-5	Hr.	Daily	Hr.	Daily	Bare Costs	Incl. O&P
2 Struc. Steel Foremen (outside)	$53.10	$849.60	$93.70	$1499.20	$50.66	$87.08
5 Struc. Steel Workers	51.10	2044.00	90.20	3608.00		
1 Equip. Oper. (crane)	50.25	402.00	76.20	609.60		
1 Welder	51.10	408.80	90.20	721.60		
1 Equip. Oper. (oiler)	43.55	348.40	66.00	528.00		
1 Lattice Boom Crane, 90 Ton		1529.00		1681.90		
1 Welder, Gas Engine, 300 amp		142.00		156.20	20.89	22.98
80 L.H., Daily Totals		$5723.80		$8804.50	$71.55	$110.06

Crew E-6	Hr.	Daily	Hr.	Daily	Bare Costs	Incl. O&P
3 Struc. Steel Foremen (outside)	$53.10	$1274.40	$93.70	$2248.80	$50.70	$87.29
9 Struc. Steel Workers	51.10	3679.20	90.20	6494.40		
1 Equip. Oper. (crane)	50.25	402.00	76.20	609.60		
1 Welder	51.10	408.80	90.20	721.60		
1 Equip. Oper. (oiler)	43.55	348.40	66.00	528.00		
1 Equip. Oper. (light)	47.05	376.40	71.35	570.80		
1 Lattice Boom Crane, 90 Ton		1529.00		1681.90		
1 Welder, Gas Engine, 300 amp		142.00		156.20		
1 Air Compressor, 160 cfm		157.40		173.14		
2 Impact Wrenches		36.00		39.60	14.57	16.02
128 L.H., Daily Totals		$8353.60		$13224.04	$65.26	$103.31

Crew E-7	Hr.	Daily	Hr.	Daily	Bare Costs	Incl. O&P
1 Struc. Steel Foreman (outside)	$53.10	$424.80	$93.70	$749.60	$50.66	$87.08
4 Struc. Steel Workers	51.10	1635.20	90.20	2886.40		
1 Equip. Oper. (crane)	50.25	402.00	76.20	609.60		
1 Equip. Oper. (oiler)	43.55	348.40	66.00	528.00		
1 Welder Foreman (outside)	53.10	424.80	93.70	749.60		
2 Welders	51.10	817.60	90.20	1443.20		
1 Lattice Boom Crane, 90 Ton		1529.00		1681.90		
2 Welder, Gas Engine, 300 amp		284.00		312.40	22.66	24.93
80 L.H., Daily Totals		$5865.80		$8960.70	$73.32	$112.01

Crew E-8	Hr.	Daily	Hr.	Daily	Bare Costs	Incl. O&P
1 Struc. Steel Foreman (outside)	$53.10	$424.80	$93.70	$749.60	$50.45	$86.35
4 Struc. Steel Workers	51.10	1635.20	90.20	2886.40		
1 Welder Foreman (outside)	53.10	424.80	93.70	749.60		
4 Welders	51.10	1635.20	90.20	2886.40		
1 Equip. Oper. (crane)	50.25	402.00	76.20	609.60		
1 Equip. Oper. (oiler)	43.55	348.40	66.00	528.00		
1 Equip. Oper. (light)	47.05	376.40	71.35	570.80		
1 Lattice Boom Crane, 90 Ton		1529.00		1681.90		
4 Welder, Gas Engine, 300 amp		568.00		624.80	20.16	22.18
104 L.H., Daily Totals		$7343.80		$11287.10	$70.61	$108.53

Crew E-9	Bare Costs Hr.	Bare Costs Daily	Incl. Subs O&P Hr.	Incl. Subs O&P Daily	Cost Per Labor-Hour Bare Costs	Cost Per Labor-Hour Incl. O&P
2 Struc. Steel Foremen (outside)	$53.10	$849.60	$93.70	$1499.20	$50.70	$87.29
5 Struc. Steel Workers	51.10	2044.00	90.20	3608.00		
1 Welder Foreman (outside)	53.10	424.80	93.70	749.60		
5 Welders	51.10	2044.00	90.20	3608.00		
1 Equip. Oper. (crane)	50.25	402.00	76.20	609.60		
1 Equip. Oper. (oiler)	43.55	348.40	66.00	528.00		
1 Equip. Oper. (light)	47.05	376.40	71.35	570.80		
1 Lattice Boom Crane, 90 Ton		1529.00		1681.90		
5 Welder, Gas Engine, 300 amp		710.00		781.00	17.49	19.24
128 L.H., Daily Totals		$8728.20		$13636.10	$68.19	$106.53

Crew E-10	Hr.	Daily	Hr.	Daily	Bare Costs	Incl. O&P
1 Welder Foreman (outside)	$53.10	$424.80	$93.70	$749.60	$52.10	$91.95
1 Welder	51.10	408.80	90.20	721.60		
1 Welder, Gas Engine, 300 amp		142.00		156.20		
1 Flatbed Truck, Gas, 3 Ton		333.00		366.30	29.69	32.66
16 L.H., Daily Totals		$1308.60		$1993.70	$81.79	$124.61

Crew E-11	Hr.	Daily	Hr.	Daily	Bare Costs	Incl. O&P
2 Painters, Struc. Steel	$40.50	$648.00	$73.05	$1168.80	$41.17	$68.50
1 Building Laborer	36.65	293.20	56.55	452.40		
1 Equip. Oper. (light)	47.05	376.40	71.35	570.80		
1 Air Compressor, 250 cfm		202.40		222.64		
1 Sandblaster, Portable, 3 C.F.		20.40		22.44		
1 Set Sand Blasting Accessories		13.85		15.23	7.40	8.13
32 L.H., Daily Totals		$1554.25		$2452.32	$48.57	$76.63

Crew E-11A	Hr.	Daily	Hr.	Daily	Bare Costs	Incl. O&P
2 Painters, Struc. Steel	$40.50	$648.00	$73.05	$1168.80	$41.17	$68.50
1 Building Laborer	36.65	293.20	56.55	452.40		
1 Equip. Oper. (light)	47.05	376.40	71.35	570.80		
1 Air Compressor, 250 cfm		202.40		222.64		
1 Sandblaster, Portable, 3 C.F.		20.40		22.44		
1 Set Sand Blasting Accessories		13.85		15.23		
1 Aerial Lift Truck, 60' Boom		456.00		501.60	21.65	23.81
32 L.H., Daily Totals		$2010.25		$2953.92	$62.82	$92.31

Crew E-11B	Hr.	Daily	Hr.	Daily	Bare Costs	Incl. O&P
2 Painters, Struc. Steel	$40.50	$648.00	$73.05	$1168.80	$39.22	$67.55
1 Building Laborer	36.65	293.20	56.55	452.40		
2 Paint Sprayer, 8 C.F.M.		89.80		98.78		
1 Aerial Lift Truck, 60' Boom		456.00		501.60	22.74	25.02
24 L.H., Daily Totals		$1487.00		$2221.58	$61.96	$92.57

Crew E-12	Hr.	Daily	Hr.	Daily	Bare Costs	Incl. O&P
1 Welder Foreman (outside)	$53.10	$424.80	$93.70	$749.60	$50.08	$82.53
1 Equip. Oper. (light)	47.05	376.40	71.35	570.80		
1 Welder, Gas Engine, 300 amp		142.00		156.20	8.88	9.76
16 L.H., Daily Totals		$943.20		$1476.60	$58.95	$92.29

Crew E-13	Hr.	Daily	Hr.	Daily	Bare Costs	Incl. O&P
1 Welder Foreman (outside)	$53.10	$424.80	$93.70	$749.60	$51.08	$86.25
.5 Equip. Oper. (light)	47.05	188.20	71.35	285.40		
1 Welder, Gas Engine, 300 amp		142.00		156.20	11.83	13.02
12 L.H., Daily Totals		$755.00		$1191.20	$62.92	$99.27

Crew E-14	Hr.	Daily	Hr.	Daily	Bare Costs	Incl. O&P
1 Welder Foreman (outside)	$53.10	$424.80	$93.70	$749.60	$53.10	$93.70
1 Welder, Gas Engine, 300 amp		142.00		156.20	17.75	19.52
8 L.H., Daily Totals		$566.80		$905.80	$70.85	$113.22

Crew E-16	Hr.	Daily	Hr.	Daily	Bare Costs	Incl. O&P
1 Welder Foreman (outside)	$53.10	$424.80	$93.70	$749.60	$52.10	$91.95
1 Welder	51.10	408.80	90.20	721.60		
1 Welder, Gas Engine, 300 amp		142.00		156.20	8.88	9.76
16 L.H., Daily Totals		$975.60		$1627.40	$60.98	$101.71

Crew E-17	Hr.	Daily	Hr.	Daily	Bare Costs	Incl. O&P
1 Struc. Steel Foreman (outside)	$53.10	$424.80	$93.70	$749.60	$52.10	$91.95
1 Structural Steel Worker	51.10	408.80	90.20	721.60		
16 L.H., Daily Totals		$833.60		$1471.20	$52.10	$91.95

Crew E-18	Hr.	Daily	Hr.	Daily	Bare Costs	Incl. O&P
1 Struc. Steel Foreman (outside)	$53.10	$424.80	$93.70	$749.60	$51.06	$87.69
3 Structural Steel Workers	51.10	1226.40	90.20	2164.80		
1 Equipment Operator (med.)	48.90	391.20	74.15	593.20		
1 Lattice Boom Crane, 20 Ton		960.50		1056.55	24.01	26.41
40 L.H., Daily Totals		$3002.90		$4564.15	$75.07	$114.10

Crew E-19	Hr.	Daily	Hr.	Daily	Bare Costs	Incl. O&P
1 Struc. Steel Foreman (outside)	$53.10	$424.80	$93.70	$749.60	$50.42	$85.08
1 Structural Steel Worker	51.10	408.80	90.20	721.60		
1 Equip. Oper. (light)	47.05	376.40	71.35	570.80		
1 Lattice Boom Crane, 20 Ton		960.50		1056.55	40.02	44.02
24 L.H., Daily Totals		$2170.50		$3098.55	$90.44	$129.11

Crew E-20	Hr.	Daily	Hr.	Daily	Bare Costs	Incl. O&P
1 Struc. Steel Foreman (outside)	$53.10	$424.80	$93.70	$749.60	$50.30	$85.86
5 Structural Steel Workers	51.10	2044.00	90.20	3608.00		
1 Equip. Oper. (crane)	50.25	402.00	76.20	609.60		
1 Equip. Oper. (oiler)	43.55	348.40	66.00	528.00		
1 Lattice Boom Crane, 40 Ton		1177.00		1294.70	18.39	20.23
64 L.H., Daily Totals		$4396.20		$6789.90	$68.69	$106.09

Crew E-22	Hr.	Daily	Hr.	Daily	Bare Costs	Incl. O&P
1 Skilled Worker Foreman (out)	$49.30	$394.40	$76.35	$610.80	$47.97	$74.28
2 Skilled Workers	47.30	756.80	73.25	1172.00		
24 L.H., Daily Totals		$1151.20		$1782.80	$47.97	$74.28

Crew E-24	Hr.	Daily	Hr.	Daily	Bare Costs	Incl. O&P
3 Structural Steel Workers	$51.10	$1226.40	$90.20	$2164.80	$50.55	$86.19
1 Equipment Operator (med.)	48.90	391.20	74.15	593.20		
1 Hyd. Crane, 25 Ton		745.80		820.38	23.31	25.64
32 L.H., Daily Totals		$2363.40		$3578.38	$73.86	$111.82

Crew E-25	Hr.	Daily	Hr.	Daily	Bare Costs	Incl. O&P
1 Welder Foreman (outside)	$53.10	$424.80	$93.70	$749.60	$53.10	$93.70
1 Cutting Torch		11.40		12.54	1.43	1.57
8 L.H., Daily Totals		$436.20		$762.14	$54.52	$95.27

Crew F-3	Hr.	Daily	Hr.	Daily	Bare Costs	Incl. O&P
4 Carpenters	$45.85	$1467.20	$70.75	$2264.00	$46.73	$71.84
1 Equip. Oper. (crane)	50.25	402.00	76.20	609.60		
1 Hyd. Crane, 12 Ton		663.00		729.30	16.57	18.23
40 L.H., Daily Totals		$2532.20		$3602.90	$63.31	$90.07

Crew F-4	Hr.	Daily	Hr.	Daily	Bare Costs	Incl. O&P
4 Carpenters	$45.85	$1467.20	$70.75	$2264.00	$46.20	$70.87
1 Equip. Oper. (crane)	50.25	402.00	76.20	609.60		
1 Equip. Oper. (oiler)	43.55	348.40	66.00	528.00		
1 Hyd. Crane, 55 Ton		1137.00		1250.70	23.69	26.06
48 L.H., Daily Totals		$3354.60		$4652.30	$69.89	$96.92

Crew F-5	Bare Costs Hr.	Bare Costs Daily	Incl. Subs O&P Hr.	Incl. Subs O&P Daily	Cost Per Labor-Hour Bare Costs	Cost Per Labor-Hour Incl. O&P
1 Carpenter Foreman (outside)	$47.85	$382.80	$73.85	$590.80	$46.35	$71.53
3 Carpenters	45.85	1100.40	70.75	1698.00		
32 L.H., Daily Totals		$1483.20		$2288.80	$46.35	$71.53

Crew F-6	Hr.	Daily	Hr.	Daily	Bare Costs	Incl. O&P
2 Carpenters	$45.85	$733.60	$70.75	$1132.00	$43.05	$66.16
2 Building Laborers	36.65	586.40	56.55	904.80		
1 Equip. Oper. (crane)	50.25	402.00	76.20	609.60		
1 Hyd. Crane, 12 Ton		663.00		729.30	16.57	18.23
40 L.H., Daily Totals		$2385.00		$3375.70	$59.63	$84.39

Crew F-7	Hr.	Daily	Hr.	Daily	Bare Costs	Incl. O&P
2 Carpenters	$45.85	$733.60	$70.75	$1132.00	$41.25	$63.65
2 Building Laborers	36.65	586.40	56.55	904.80		
32 L.H., Daily Totals		$1320.00		$2036.80	$41.25	$63.65

Crew G-1	Hr.	Daily	Hr.	Daily	Bare Costs	Incl. O&P
1 Roofer Foreman (outside)	$41.15	$329.20	$70.05	$560.40	$36.58	$62.26
4 Roofers Composition	39.15	1252.80	66.65	2132.80		
2 Roofer Helpers	29.15	466.40	49.60	793.60		
1 Application Equipment		199.20		219.12		
1 Tar Kettle/Pot		130.00		143.00		
1 Crew Truck		224.80		247.28	9.89	10.88
56 L.H., Daily Totals		$2602.40		$4096.20	$46.47	$73.15

Crew G-2	Hr.	Daily	Hr.	Daily	Bare Costs	Incl. O&P
1 Plasterer	$41.95	$335.60	$63.15	$505.20	$38.60	$58.57
1 Plasterer Helper	37.20	297.60	56.00	448.00		
1 Building Laborer	36.65	293.20	56.55	452.40		
1 Grout Pump, 50 C.F./hr.		134.80		148.28	5.62	6.18
24 L.H., Daily Totals		$1061.20		$1553.88	$44.22	$64.75

Crew G-2A	Hr.	Daily	Hr.	Daily	Bare Costs	Incl. O&P
1 Roofer Composition	$39.15	$313.20	$66.65	$533.20	$34.98	$57.60
1 Roofer Helper	29.15	233.20	49.60	396.80		
1 Building Laborer	36.65	293.20	56.55	452.40		
1 Foam Spray Rig, Trailer-Mtd.		585.25		643.77		
1 Pickup Truck, 3/4 Ton		157.00		172.70	30.93	34.02
24 L.H., Daily Totals		$1581.85		$2198.88	$65.91	$91.62

Crew G-3	Hr.	Daily	Hr.	Daily	Bare Costs	Incl. O&P
2 Sheet Metal Workers	$54.70	$875.20	$83.55	$1336.80	$45.67	$70.05
2 Building Laborers	36.65	586.40	56.55	904.80		
32 L.H., Daily Totals		$1461.60		$2241.60	$45.67	$70.05

Crew G-4	Hr.	Daily	Hr.	Daily	Bare Costs	Incl. O&P
1 Labor Foreman (outside)	$38.65	$309.20	$59.65	$477.20	$37.32	$57.58
2 Building Laborers	36.65	586.40	56.55	904.80		
1 Flatbed Truck, Gas, 1.5 Ton		270.20		297.22		
1 Air Compressor, 160 cfm		157.40		173.14	17.82	19.60
24 L.H., Daily Totals		$1323.20		$1852.36	$55.13	$77.18

Crew G-5	Hr.	Daily	Hr.	Daily	Bare Costs	Incl. O&P
1 Roofer Foreman (outside)	$41.15	$329.20	$70.05	$560.40	$35.55	$60.51
2 Roofers Composition	39.15	626.40	66.65	1066.40		
2 Roofer Helpers	29.15	466.40	49.60	793.60		
1 Application Equipment		199.20		219.12	4.98	5.48
40 L.H., Daily Totals		$1621.20		$2639.52	$40.53	$65.99

Crew G-6A	Hr.	Daily	Hr.	Daily	Bare Costs	Incl. O&P
2 Roofers Composition	$39.15	$626.40	$66.65	$1066.40	$39.15	$66.65
1 Small Compressor, Electric		12.10		13.31		
2 Pneumatic Nailers		44.70		49.17	3.55	3.90
16 L.H., Daily Totals		$683.20		$1128.88	$42.70	$70.56

Crew G-7	Hr.	Daily	Hr.	Daily	Bare Costs	Incl. O&P
1 Carpenter	$45.85	$366.80	$70.75	$566.00	$45.85	$70.75
1 Small Compressor, Electric		12.10		13.31		
1 Pneumatic Nailer		22.35		24.59	4.31	4.74
8 L.H., Daily Totals		$401.25		$603.89	$50.16	$75.49

Crew H-1	Hr.	Daily	Hr.	Daily	Bare Costs	Incl. O&P
2 Glaziers	$44.05	$704.80	$67.00	$1072.00	$47.58	$78.60
2 Struc. Steel Workers	51.10	817.60	90.20	1443.20		
32 L.H., Daily Totals		$1522.40		$2515.20	$47.58	$78.60

Crew H-2	Hr.	Daily	Hr.	Daily	Bare Costs	Incl. O&P
2 Glaziers	$44.05	$704.80	$67.00	$1072.00	$41.58	$63.52
1 Building Laborer	36.65	293.20	56.55	452.40		
24 L.H., Daily Totals		$998.00		$1524.40	$41.58	$63.52

Crew H-3	Hr.	Daily	Hr.	Daily	Bare Costs	Incl. O&P
1 Glazier	$44.05	$352.40	$67.00	$536.00	$39.35	$60.35
1 Helper	34.65	277.20	53.70	429.60		
16 L.H., Daily Totals		$629.60		$965.60	$39.35	$60.35

Crew H-4	Hr.	Daily	Hr.	Daily	Bare Costs	Incl. O&P
1 Carpenter	$45.85	$366.80	$70.75	$566.00	$42.87	$65.75
1 Carpenter Helper	34.65	277.20	53.70	429.60		
.5 Electrician	53.35	213.40	79.85	319.40		
20 L.H., Daily Totals		$857.40		$1315.00	$42.87	$65.75

Crew J-1	Hr.	Daily	Hr.	Daily	Bare Costs	Incl. O&P
3 Plasterers	$41.95	$1006.80	$63.15	$1515.60	$40.05	$60.29
2 Plasterer Helpers	37.20	595.20	56.00	896.00		
1 Mixing Machine, 6 C.F.		143.60		157.96	3.59	3.95
40 L.H., Daily Totals		$1745.60		$2569.56	$43.64	$64.24

Crew J-2	Hr.	Daily	Hr.	Daily	Bare Costs	Incl. O&P
3 Plasterers	$41.95	$1006.80	$63.15	$1515.60	$40.14	$60.19
2 Plasterer Helpers	37.20	595.20	56.00	896.00		
1 Lather	40.60	324.80	59.70	477.60		
1 Mixing Machine, 6 C.F.		143.60		157.96	2.99	3.29
48 L.H., Daily Totals		$2070.40		$3047.16	$43.13	$63.48

Crew J-3	Hr.	Daily	Hr.	Daily	Bare Costs	Incl. O&P
1 Terrazzo Worker	$41.65	$333.20	$61.40	$491.20	$38.23	$56.35
1 Terrazzo Helper	34.80	278.40	51.30	410.40		
1 Floor Grinder, 22" Path		97.90		107.69		
1 Terrazzo Mixer		192.60		211.86	18.16	19.97
16 L.H., Daily Totals		$902.10		$1221.15	$56.38	$76.32

Crew J-4	Hr.	Daily	Hr.	Daily	Bare Costs	Incl. O&P
2 Cement Finishers	$44.05	$704.80	$65.10	$1041.60	$41.58	$62.25
1 Laborer	36.65	293.20	56.55	452.40		
1 Floor Grinder, 22" Path		97.90		107.69		
1 Floor Edger, 7" Path		39.15		43.06		
1 Vacuum Pick-Up System		60.75		66.83	8.24	9.07
24 L.H., Daily Totals		$1195.80		$1711.58	$49.83	$71.32

Crew No.	Bare Costs		Incl. Subs O&P		Cost Per Labor-Hour	
Crew J-4A	**Hr.**	**Daily**	**Hr.**	**Daily**	**Bare Costs**	**Incl. O&P**
2 Cement Finishers	$44.05	$704.80	$65.10	$1041.60	$40.35	$60.83
2 Laborers	36.65	586.40	56.55	904.80		
1 Floor Grinder, 22" Path		97.90		107.69		
1 Floor Edger, 7" Path		39.15		43.06		
1 Vacuum Pick-Up System		60.75		66.83		
1 Floor Auto Scrubber		212.15		233.37	12.81	14.09
32 L.H., Daily Totals		$1701.15		$2397.34	$53.16	$74.92
Crew J-4B	**Hr.**	**Daily**	**Hr.**	**Daily**	**Bare Costs**	**Incl. O&P**
1 Laborer	$36.65	$293.20	$56.55	$452.40	$36.65	$56.55
1 Floor Auto Scrubber		212.15		233.37	26.52	29.17
8 L.H., Daily Totals		$505.35		$685.76	$63.17	$85.72
Crew J-6	**Hr.**	**Daily**	**Hr.**	**Daily**	**Bare Costs**	**Incl. O&P**
2 Painters	$39.55	$632.80	$59.50	$952.00	$40.70	$61.73
1 Building Laborer	36.65	293.20	56.55	452.40		
1 Equip. Oper. (light)	47.05	376.40	71.35	570.80		
1 Air Compressor, 250 cfm		202.40		222.64		
1 Sandblaster, Portable, 3 C.F.		20.40		22.44		
1 Set Sand Blasting Accessories		13.85		15.23	7.40	8.13
32 L.H., Daily Totals		$1539.05		$2235.51	$48.10	$69.86
Crew J-7	**Hr.**	**Daily**	**Hr.**	**Daily**	**Bare Costs**	**Incl. O&P**
2 Painters	$39.55	$632.80	$59.50	$952.00	$39.55	$59.50
1 Floor Belt Sander		15.90		17.49		
1 Floor Sanding Edger		15.55		17.11	1.97	2.16
16 L.H., Daily Totals		$664.25		$986.60	$41.52	$61.66
Crew K-1	**Hr.**	**Daily**	**Hr.**	**Daily**	**Bare Costs**	**Incl. O&P**
1 Carpenter	$45.85	$366.80	$70.75	$566.00	$41.17	$63.23
1 Truck Driver (light)	36.50	292.00	55.70	445.60		
1 Flatbed Truck, Gas, 3 Ton		333.00		366.30	20.81	22.89
16 L.H., Daily Totals		$991.80		$1377.90	$61.99	$86.12
Crew K-2	**Hr.**	**Daily**	**Hr.**	**Daily**	**Bare Costs**	**Incl. O&P**
1 Struc. Steel Foreman (outside)	$53.10	$424.80	$93.70	$749.60	$46.90	$79.87
1 Struc. Steel Worker	51.10	408.80	90.20	721.60		
1 Truck Driver (light)	36.50	292.00	55.70	445.60		
1 Flatbed Truck, Gas, 3 Ton		333.00		366.30	13.88	15.26
24 L.H., Daily Totals		$1458.60		$2283.10	$60.77	$95.13
Crew L-1	**Hr.**	**Daily**	**Hr.**	**Daily**	**Bare Costs**	**Incl. O&P**
1 Electrician	$53.35	$426.80	$79.85	$638.80	$55.45	$83.38
1 Plumber	57.55	460.40	86.90	695.20		
16 L.H., Daily Totals		$887.20		$1334.00	$55.45	$83.38
Crew L-2	**Hr.**	**Daily**	**Hr.**	**Daily**	**Bare Costs**	**Incl. O&P**
1 Carpenter	$45.85	$366.80	$70.75	$566.00	$40.25	$62.23
1 Carpenter Helper	34.65	277.20	53.70	429.60		
16 L.H., Daily Totals		$644.00		$995.60	$40.25	$62.23
Crew L-3	**Hr.**	**Daily**	**Hr.**	**Daily**	**Bare Costs**	**Incl. O&P**
1 Carpenter	$45.85	$366.80	$70.75	$566.00	$49.94	$76.22
.5 Electrician	53.35	213.40	79.85	319.40		
.5 Sheet Metal Worker	54.70	218.80	83.55	334.20		
16 L.H., Daily Totals		$799.00		$1219.60	$49.94	$76.22

Crew No.	Bare Costs		Incl. Subs O&P		Cost Per Labor-Hour	
Crew L-3A	**Hr.**	**Daily**	**Hr.**	**Daily**	**Bare Costs**	**Incl. O&P**
1 Carpenter Foreman (outside)	$47.85	$382.80	$73.85	$590.80	$50.13	$77.08
.5 Sheet Metal Worker	54.70	218.80	83.55	334.20		
12 L.H., Daily Totals		$601.60		$925.00	$50.13	$77.08
Crew L-4	**Hr.**	**Daily**	**Hr.**	**Daily**	**Bare Costs**	**Incl. O&P**
2 Skilled Workers	$47.30	$756.80	$73.25	$1172.00	$43.08	$66.73
1 Helper	34.65	277.20	53.70	429.60		
24 L.H., Daily Totals		$1034.00		$1601.60	$43.08	$66.73
Crew L-5	**Hr.**	**Daily**	**Hr.**	**Daily**	**Bare Costs**	**Incl. O&P**
1 Struc. Steel Foreman (outside)	$53.10	$424.80	$93.70	$749.60	$51.26	$88.70
5 Struc. Steel Workers	51.10	2044.00	90.20	3608.00		
1 Equip. Oper. (crane)	50.25	402.00	76.20	609.60		
1 Hyd. Crane, 25 Ton		745.80		820.38	13.32	14.65
56 L.H., Daily Totals		$3616.60		$5787.58	$64.58	$103.35
Crew L-5A	**Hr.**	**Daily**	**Hr.**	**Daily**	**Bare Costs**	**Incl. O&P**
1 Struc. Steel Foreman (outside)	$53.10	$424.80	$93.70	$749.60	$51.39	$87.58
2 Structural Steel Workers	51.10	817.60	90.20	1443.20		
1 Equip. Oper. (crane)	50.25	402.00	76.20	609.60		
1 S.P. Crane, 4x4, 25 Ton		607.20		667.92	18.98	20.87
32 L.H., Daily Totals		$2251.60		$3470.32	$70.36	$108.45
Crew L-5B	**Hr.**	**Daily**	**Hr.**	**Daily**	**Bare Costs**	**Incl. O&P**
1 Struc. Steel Foreman (outside)	$53.10	$424.80	$93.70	$749.60	$52.53	$83.63
2 Structural Steel Workers	51.10	817.60	90.20	1443.20		
2 Electricians	53.35	853.60	79.85	1277.60		
2 Steamfitters/Pipefitters	58.50	936.00	88.35	1413.60		
1 Equip. Oper. (crane)	50.25	402.00	76.20	609.60		
1 Equip. Oper. (oiler)	43.55	348.40	66.00	528.00		
1 Hyd. Crane, 80 Ton		1652.00		1817.20	22.94	25.24
72 L.H., Daily Totals		$5434.40		$7838.80	$75.48	$108.87
Crew L-6	**Hr.**	**Daily**	**Hr.**	**Daily**	**Bare Costs**	**Incl. O&P**
1 Plumber	$57.55	$460.40	$86.90	$695.20	$56.15	$84.55
.5 Electrician	53.35	213.40	79.85	319.40		
12 L.H., Daily Totals		$673.80		$1014.60	$56.15	$84.55
Crew L-7	**Hr.**	**Daily**	**Hr.**	**Daily**	**Bare Costs**	**Incl. O&P**
2 Carpenters	$45.85	$733.60	$70.75	$1132.00	$44.29	$67.99
1 Building Laborer	36.65	293.20	56.55	452.40		
.5 Electrician	53.35	213.40	79.85	319.40		
28 L.H., Daily Totals		$1240.20		$1903.80	$44.29	$67.99
Crew L-8	**Hr.**	**Daily**	**Hr.**	**Daily**	**Bare Costs**	**Incl. O&P**
2 Carpenters	$45.85	$733.60	$70.75	$1132.00	$48.19	$73.98
.5 Plumber	57.55	230.20	86.90	347.60		
20 L.H., Daily Totals		$963.80		$1479.60	$48.19	$73.98
Crew L-9	**Hr.**	**Daily**	**Hr.**	**Daily**	**Bare Costs**	**Incl. O&P**
1 Labor Foreman (inside)	$37.15	$297.20	$57.30	$458.40	$41.83	$66.78
2 Building Laborers	36.65	586.40	56.55	904.80		
1 Struc. Steel Worker	51.10	408.80	90.20	721.60		
.5 Electrician	53.35	213.40	79.85	319.40		
36 L.H., Daily Totals		$1505.80		$2404.20	$41.83	$66.78

Crew No.	Bare Costs		Incl. Subs O&P		Cost Per Labor-Hour	

Crew L-10	Hr.	Daily	Hr.	Daily	Bare Costs	Incl. O&P
1 Struc. Steel Foreman (outside)	$53.10	$424.80	$93.70	$749.60	$51.48	$86.70
1 Structural Steel Worker	51.10	408.80	90.20	721.60		
1 Equip. Oper. (crane)	50.25	402.00	76.20	609.60		
1 Hyd. Crane, 12 Ton		663.00		729.30	27.63	30.39
24 L.H., Daily Totals		$1898.60		$2810.10	$79.11	$117.09

Crew L-11	Hr.	Daily	Hr.	Daily	Bare Costs	Incl. O&P
2 Wreckers	$36.65	$586.40	$61.60	$985.60	$42.65	$67.69
1 Equip. Oper. (crane)	50.25	402.00	76.20	609.60		
1 Equip. Oper. (light)	47.05	376.40	71.35	570.80		
1 Hyd. Excavator, 2.5 C.Y.		1637.00		1800.70		
1 Loader, Skid Steer, 78 H.P.		308.80		339.68	60.81	66.89
32 L.H., Daily Totals		$3310.60		$4306.38	$103.46	$134.57

Crew M-1	Hr.	Daily	Hr.	Daily	Bare Costs	Incl. O&P
3 Elevator Constructors	$74.15	$1779.60	$110.95	$2662.80	$70.44	$105.39
1 Elevator Apprentice	59.30	474.40	88.70	709.60		
5 Hand Tools		48.00		52.80	1.50	1.65
32 L.H., Daily Totals		$2302.00		$3425.20	$71.94	$107.04

Crew M-3	Hr.	Daily	Hr.	Daily	Bare Costs	Incl. O&P
1 Electrician Foreman (outside)	$55.35	$442.80	$82.85	$662.80	$55.92	$84.14
1 Common Laborer	36.65	293.20	56.55	452.40		
.25 Equipment Operator (med.)	48.90	97.80	74.15	148.30		
1 Elevator Constructor	74.15	593.20	110.95	887.60		
1 Elevator Apprentice	59.30	474.40	88.70	709.60		
.25 S.P. Crane, 4x4, 20 Ton		140.10		154.11	4.12	4.53
34 L.H., Daily Totals		$2041.50		$3014.81	$60.04	$88.67

Crew M-4	Hr.	Daily	Hr.	Daily	Bare Costs	Incl. O&P
1 Electrician Foreman (outside)	$55.35	$442.80	$82.85	$662.80	$55.31	$83.24
1 Common Laborer	36.65	293.20	56.55	452.40		
.25 Equipment Operator, Crane	50.25	100.50	76.20	152.40		
.25 Equip. Oper. (oiler)	43.55	87.10	66.00	132.00		
1 Elevator Constructor	74.15	593.20	110.95	887.60		
1 Elevator Apprentice	59.30	474.40	88.70	709.60		
.25 S.P. Crane, 4x4, 40 Ton		178.25		196.07	4.95	5.45
36 L.H., Daily Totals		$2169.45		$3192.88	$60.26	$88.69

Crew Q-1	Hr.	Daily	Hr.	Daily	Bare Costs	Incl. O&P
1 Plumber	$57.55	$460.40	$86.90	$695.20	$51.80	$78.22
1 Plumber Apprentice	46.05	368.40	69.55	556.40		
16 L.H., Daily Totals		$828.80		$1251.60	$51.80	$78.22

Crew Q-1A	Hr.	Daily	Hr.	Daily	Bare Costs	Incl. O&P
.25 Plumber Foreman (outside)	$59.55	$119.10	$89.90	$179.80	$57.95	$87.50
1 Plumber	57.55	460.40	86.90	695.20		
10 L.H., Daily Totals		$579.50		$875.00	$57.95	$87.50

Crew Q-1C	Hr.	Daily	Hr.	Daily	Bare Costs	Incl. O&P
1 Plumber	$57.55	$460.40	$86.90	$695.20	$50.83	$76.87
1 Plumber Apprentice	46.05	368.40	69.55	556.40		
1 Equip. Oper. (medium)	48.90	391.20	74.15	593.20		
1 Trencher, Chain Type, 8' D		3376.00		3713.60	140.67	154.73
24 L.H., Daily Totals		$4596.00		$5558.40	$191.50	$231.60

Crew Q-2	Hr.	Daily	Hr.	Daily	Bare Costs	Incl. O&P
2 Plumbers	$57.55	$920.80	$86.90	$1390.40	$53.72	$81.12
1 Plumber Apprentice	46.05	368.40	69.55	556.40		
24 L.H., Daily Totals		$1289.20		$1946.80	$53.72	$81.12

Crew No.	Bare Costs		Incl. Subs O&P		Cost Per Labor-Hour	

Crew Q-3	Hr.	Daily	Hr.	Daily	Bare Costs	Incl. O&P
1 Plumber Foreman (inside)	$58.05	$464.40	$87.65	$701.20	$54.80	$82.75
2 Plumbers	57.55	920.80	86.90	1390.40		
1 Plumber Apprentice	46.05	368.40	69.55	556.40		
32 L.H., Daily Totals		$1753.60		$2648.00	$54.80	$82.75

Crew Q-4	Hr.	Daily	Hr.	Daily	Bare Costs	Incl. O&P
1 Plumber Foreman (inside)	$58.05	$464.40	$87.65	$701.20	$54.80	$82.75
1 Plumber	57.55	460.40	86.90	695.20		
1 Welder (plumber)	57.55	460.40	86.90	695.20		
1 Plumber Apprentice	46.05	368.40	69.55	556.40		
1 Welder, Electric, 300 amp		55.30		60.83	1.73	1.90
32 L.H., Daily Totals		$1808.90		$2708.83	$56.53	$84.65

Crew Q-5	Hr.	Daily	Hr.	Daily	Bare Costs	Incl. O&P
1 Steamfitter	$58.50	$468.00	$88.35	$706.80	$52.65	$79.50
1 Steamfitter Apprentice	46.80	374.40	70.65	565.20		
16 L.H., Daily Totals		$842.40		$1272.00	$52.65	$79.50

Crew Q-6	Hr.	Daily	Hr.	Daily	Bare Costs	Incl. O&P
2 Steamfitters	$58.50	$936.00	$88.35	$1413.60	$54.60	$82.45
1 Steamfitter Apprentice	46.80	374.40	70.65	565.20		
24 L.H., Daily Totals		$1310.40		$1978.80	$54.60	$82.45

Crew Q-7	Hr.	Daily	Hr.	Daily	Bare Costs	Incl. O&P
1 Steamfitter Foreman (inside)	$59.00	$472.00	$89.10	$712.80	$55.70	$84.11
2 Steamfitters	58.50	936.00	88.35	1413.60		
1 Steamfitter Apprentice	46.80	374.40	70.65	565.20		
32 L.H., Daily Totals		$1782.40		$2691.60	$55.70	$84.11

Crew Q-8	Hr.	Daily	Hr.	Daily	Bare Costs	Incl. O&P
1 Steamfitter Foreman (inside)	$59.00	$472.00	$89.10	$712.80	$55.70	$84.11
1 Steamfitter	58.50	468.00	88.35	706.80		
1 Welder (steamfitter)	58.50	468.00	88.35	706.80		
1 Steamfitter Apprentice	46.80	374.40	70.65	565.20		
1 Welder, Electric, 300 amp		55.30		60.83	1.73	1.90
32 L.H., Daily Totals		$1837.70		$2752.43	$57.43	$86.01

Crew Q-9	Hr.	Daily	Hr.	Daily	Bare Costs	Incl. O&P
1 Sheet Metal Worker	$54.70	$437.60	$83.55	$668.40	$49.23	$75.17
1 Sheet Metal Apprentice	43.75	350.00	66.80	534.40		
16 L.H., Daily Totals		$787.60		$1202.80	$49.23	$75.17

Crew Q-10	Hr.	Daily	Hr.	Daily	Bare Costs	Incl. O&P
2 Sheet Metal Workers	$54.70	$875.20	$83.55	$1336.80	$51.05	$77.97
1 Sheet Metal Apprentice	43.75	350.00	66.80	534.40		
24 L.H., Daily Totals		$1225.20		$1871.20	$51.05	$77.97

Crew Q-11	Hr.	Daily	Hr.	Daily	Bare Costs	Incl. O&P
1 Sheet Metal Foreman (inside)	$55.20	$441.60	$84.30	$674.40	$52.09	$79.55
2 Sheet Metal Workers	54.70	875.20	83.55	1336.80		
1 Sheet Metal Apprentice	43.75	350.00	66.80	534.40		
32 L.H., Daily Totals		$1666.80		$2545.60	$52.09	$79.55

Crew Q-12	Hr.	Daily	Hr.	Daily	Bare Costs	Incl. O&P
1 Sprinkler Installer	$55.40	$443.20	$83.70	$669.60	$49.85	$75.33
1 Sprinkler Apprentice	44.30	354.40	66.95	535.60		
16 L.H., Daily Totals		$797.60		$1205.20	$49.85	$75.33

Crew Q-13	Bare Costs Hr.	Bare Costs Daily	Incl. Subs O&P Hr.	Incl. Subs O&P Daily	Cost Per Labor-Hour Bare Costs	Cost Per Labor-Hour Incl. O&P
1 Sprinkler Foreman (inside)	$55.90	$447.20	$84.45	$675.60	$52.75	$79.70
2 Sprinkler Installers	55.40	886.40	83.70	1339.20		
1 Sprinkler Apprentice	44.30	354.40	66.95	535.60		
32 L.H., Daily Totals		$1688.00		$2550.40	$52.75	$79.70

Crew Q-14	Hr.	Daily	Hr.	Daily	Bare Costs	Incl. O&P
1 Asbestos Worker	$51.15	$409.20	$79.60	$636.80	$46.02	$71.63
1 Asbestos Apprentice	40.90	327.20	63.65	509.20		
16 L.H., Daily Totals		$736.40		$1146.00	$46.02	$71.63

Crew Q-15	Hr.	Daily	Hr.	Daily	Bare Costs	Incl. O&P
1 Plumber	$57.55	$460.40	$86.90	$695.20	$51.80	$78.22
1 Plumber Apprentice	46.05	368.40	69.55	556.40		
1 Welder, Electric, 300 amp		55.30		60.83	3.46	3.80
16 L.H., Daily Totals		$884.10		$1312.43	$55.26	$82.03

Crew Q-16	Hr.	Daily	Hr.	Daily	Bare Costs	Incl. O&P
2 Plumbers	$57.55	$920.80	$86.90	$1390.40	$53.72	$81.12
1 Plumber Apprentice	46.05	368.40	69.55	556.40		
1 Welder, Electric, 300 amp		55.30		60.83	2.30	2.53
24 L.H., Daily Totals		$1344.50		$2007.63	$56.02	$83.65

Crew Q-17	Hr.	Daily	Hr.	Daily	Bare Costs	Incl. O&P
1 Steamfitter	$58.50	$468.00	$88.35	$706.80	$52.65	$79.50
1 Steamfitter Apprentice	46.80	374.40	70.65	565.20		
1 Welder, Electric, 300 amp		55.30		60.83	3.46	3.80
16 L.H., Daily Totals		$897.70		$1332.83	$56.11	$83.30

Crew Q-17A	Hr.	Daily	Hr.	Daily	Bare Costs	Incl. O&P
1 Steamfitter	$58.50	$468.00	$88.35	$706.80	$51.85	$78.40
1 Steamfitter Apprentice	46.80	374.40	70.65	565.20		
1 Equip. Oper. (crane)	50.25	402.00	76.20	609.60		
1 Hyd. Crane, 12 Ton		663.00		729.30		
1 Welder, Electric, 300 amp		55.30		60.83	29.93	32.92
24 L.H., Daily Totals		$1962.70		$2671.73	$81.78	$111.32

Crew Q-18	Hr.	Daily	Hr.	Daily	Bare Costs	Incl. O&P
2 Steamfitters	$58.50	$936.00	$88.35	$1413.60	$54.60	$82.45
1 Steamfitter Apprentice	46.80	374.40	70.65	565.20		
1 Welder, Electric, 300 amp		55.30		60.83	2.30	2.53
24 L.H., Daily Totals		$1365.70		$2039.63	$56.90	$84.98

Crew Q-19	Hr.	Daily	Hr.	Daily	Bare Costs	Incl. O&P
1 Steamfitter	$58.50	$468.00	$88.35	$706.80	$52.88	$79.62
1 Steamfitter Apprentice	46.80	374.40	70.65	565.20		
1 Electrician	53.35	426.80	79.85	638.80		
24 L.H., Daily Totals		$1269.20		$1910.80	$52.88	$79.62

Crew Q-20	Hr.	Daily	Hr.	Daily	Bare Costs	Incl. O&P
1 Sheet Metal Worker	$54.70	$437.60	$83.55	$668.40	$50.05	$76.11
1 Sheet Metal Apprentice	43.75	350.00	66.80	534.40		
.5 Electrician	53.35	213.40	79.85	319.40		
20 L.H., Daily Totals		$1001.00		$1522.20	$50.05	$76.11

Crew Q-21	Hr.	Daily	Hr.	Daily	Bare Costs	Incl. O&P
2 Steamfitters	$58.50	$936.00	$88.35	$1413.60	$54.29	$81.80
1 Steamfitter Apprentice	46.80	374.40	70.65	565.20		
1 Electrician	53.35	426.80	79.85	638.80		
32 L.H., Daily Totals		$1737.20		$2617.60	$54.29	$81.80

Crew Q-22	Hr.	Daily	Hr.	Daily	Bare Costs	Incl. O&P
1 Plumber	$57.55	$460.40	$86.90	$695.20	$51.80	$78.22
1 Plumber Apprentice	46.05	368.40	69.55	556.40		
1 Hyd. Crane, 12 Ton		663.00		729.30	41.44	45.58
16 L.H., Daily Totals		$1491.80		$1980.90	$93.24	$123.81

Crew Q-22A	Hr.	Daily	Hr.	Daily	Bare Costs	Incl. O&P
1 Plumber	$57.55	$460.40	$86.90	$695.20	$47.63	$72.30
1 Plumber Apprentice	46.05	368.40	69.55	556.40		
1 Laborer	36.65	293.20	56.55	452.40		
1 Equip. Oper. (crane)	50.25	402.00	76.20	609.60		
1 Hyd. Crane, 12 Ton		663.00		729.30	20.72	22.79
32 L.H., Daily Totals		$2187.00		$3042.90	$68.34	$95.09

Crew Q-23	Hr.	Daily	Hr.	Daily	Bare Costs	Incl. O&P
1 Plumber Foreman (outside)	$59.55	$476.40	$89.90	$719.20	$55.33	$83.65
1 Plumber	57.55	460.40	86.90	695.20		
1 Equip. Oper. (medium)	48.90	391.20	74.15	593.20		
1 Lattice Boom Crane, 20 Ton		960.50		1056.55	40.02	44.02
24 L.H., Daily Totals		$2288.50		$3064.15	$95.35	$127.67

Crew R-1	Hr.	Daily	Hr.	Daily	Bare Costs	Incl. O&P
1 Electrician Foreman	$53.85	$430.80	$80.60	$644.80	$47.20	$71.26
3 Electricians	53.35	1280.40	79.85	1916.40		
2 Helpers	34.65	554.40	53.70	859.20		
48 L.H., Daily Totals		$2265.60		$3420.40	$47.20	$71.26

Crew R-1A	Hr.	Daily	Hr.	Daily	Bare Costs	Incl. O&P
1 Electrician	$53.35	$426.80	$79.85	$638.80	$44.00	$66.78
1 Helper	34.65	277.20	53.70	429.60		
16 L.H., Daily Totals		$704.00		$1068.40	$44.00	$66.78

Crew R-2	Hr.	Daily	Hr.	Daily	Bare Costs	Incl. O&P
1 Electrician Foreman	$53.85	$430.80	$80.60	$644.80	$47.64	$71.96
3 Electricians	53.35	1280.40	79.85	1916.40		
2 Helpers	34.65	554.40	53.70	859.20		
1 Equip. Oper. (crane)	50.25	402.00	76.20	609.60		
1 S.P. Crane, 4x4, 5 Ton		280.00		308.00	5.00	5.50
56 L.H., Daily Totals		$2947.60		$4338.00	$52.64	$77.46

Crew R-3	Hr.	Daily	Hr.	Daily	Bare Costs	Incl. O&P
1 Electrician Foreman	$53.85	$430.80	$80.60	$644.80	$52.93	$79.42
1 Electrician	53.35	426.80	79.85	638.80		
.5 Equip. Oper. (crane)	50.25	201.00	76.20	304.80		
.5 S.P. Crane, 4x4, 5 Ton		140.00		154.00	7.00	7.70
20 L.H., Daily Totals		$1198.60		$1742.40	$59.93	$87.12

Crew R-4	Hr.	Daily	Hr.	Daily	Bare Costs	Incl. O&P
1 Struc. Steel Foreman (outside)	$53.10	$424.80	$93.70	$749.60	$51.95	$88.83
3 Struc. Steel Workers	51.10	1226.40	90.20	2164.80		
1 Electrician	53.35	426.80	79.85	638.80		
1 Welder, Gas Engine, 300 amp		142.00		156.20	3.55	3.90
40 L.H., Daily Totals		$2220.00		$3709.40	$55.50	$92.73

Crew R-5	Bare Costs Hr.	Bare Costs Daily	Incl. Subs O&P Hr.	Incl. Subs O&P Daily	Cost Per Labor-Hour Bare Costs	Cost Per Labor-Hour Incl. O&P
1 Electrician Foreman	$53.85	$430.80	$80.60	$644.80	$46.60	$70.41
4 Electrician Linemen	53.35	1707.20	79.85	2555.20		
2 Electrician Operators	53.35	853.60	79.85	1277.60		
4 Electrician Groundmen	34.65	1108.80	53.70	1718.40		
1 Crew Truck		224.80		247.28		
1 Flatbed Truck, 20,000 GVW		253.40		278.74		
1 Pickup Truck, 3/4 Ton		157.00		172.70		
.2 Hyd. Crane, 55 Ton		227.40		250.14		
.2 Hyd. Crane, 12 Ton		132.60		145.86		
.2 Earth Auger, Truck-Mtd.		85.00		93.50		
1 Tractor w/Winch		434.60		478.06	17.21	18.93
88 L.H., Daily Totals		$5615.20		$7862.28	$63.81	$89.34

Crew R-6	Hr.	Daily	Hr.	Daily	Bare Costs	Incl. O&P
1 Electrician Foreman	$53.85	$430.80	$80.60	$644.80	$46.60	$70.41
4 Electrician Linemen	53.35	1707.20	79.85	2555.20		
2 Electrician Operators	53.35	853.60	79.85	1277.60		
4 Electrician Groundmen	34.65	1108.80	53.70	1718.40		
1 Crew Truck		224.80		247.28		
1 Flatbed Truck, 20,000 GVW		253.40		278.74		
1 Pickup Truck, 3/4 Ton		157.00		172.70		
.2 Hyd. Crane, 55 Ton		227.40		250.14		
.2 Hyd. Crane, 12 Ton		132.60		145.86		
.2 Earth Auger, Truck-Mtd.		85.00		93.50		
1 Tractor w/Winch		434.60		478.06		
3 Cable Trailers		586.80		645.48		
.5 Tensioning Rig		194.55		214.01		
.5 Cable Pulling Rig		1136.50		1250.15	39.01	42.91
88 L.H., Daily Totals		$7533.05		$9971.92	$85.60	$113.32

Crew R-7	Hr.	Daily	Hr.	Daily	Bare Costs	Incl. O&P
1 Electrician Foreman	$53.85	$430.80	$80.60	$644.80	$37.85	$58.18
5 Electrician Groundmen	34.65	1386.00	53.70	2148.00		
1 Crew Truck		224.80		247.28	4.68	5.15
48 L.H., Daily Totals		$2041.60		$3040.08	$42.53	$63.34

Crew R-8	Hr.	Daily	Hr.	Daily	Bare Costs	Incl. O&P
1 Electrician Foreman	$53.85	$430.80	$80.60	$644.80	$47.20	$71.26
3 Electrician Linemen	53.35	1280.40	79.85	1916.40		
2 Electrician Groundmen	34.65	554.40	53.70	859.20		
1 Pickup Truck, 3/4 Ton		157.00		172.70		
1 Crew Truck		224.80		247.28	7.95	8.75
48 L.H., Daily Totals		$2647.40		$3840.38	$55.15	$80.01

Crew R-9	Hr.	Daily	Hr.	Daily	Bare Costs	Incl. O&P
1 Electrician Foreman	$53.85	$430.80	$80.60	$644.80	$44.06	$66.87
1 Electrician Lineman	53.35	426.80	79.85	638.80		
2 Electrician Operators	53.35	853.60	79.85	1277.60		
4 Electrician Groundmen	34.65	1108.80	53.70	1718.40		
1 Pickup Truck, 3/4 Ton		157.00		172.70		
1 Crew Truck		224.80		247.28	5.97	6.56
64 L.H., Daily Totals		$3201.80		$4699.58	$50.03	$73.43

Crew R-10	Hr.	Daily	Hr.	Daily	Bare Costs	Incl. O&P
1 Electrician Foreman	$53.85	$430.80	$80.60	$644.80	$50.32	$75.62
4 Electrician Linemen	53.35	1707.20	79.85	2555.20		
1 Electrician Groundman	34.65	277.20	53.70	429.60		
1 Crew Truck		224.80		247.28		
3 Tram Cars		399.45		439.39	13.01	14.31
48 L.H., Daily Totals		$3039.45		$4316.27	$63.32	$89.92

Crew R-11	Hr.	Daily	Hr.	Daily	Bare Costs	Incl. O&P
1 Electrician Foreman	$53.85	$430.80	$80.60	$644.80	$50.59	$76.11
4 Electricians	53.35	1707.20	79.85	2555.20		
1 Equip. Oper. (crane)	50.25	402.00	76.20	609.60		
1 Common Laborer	36.65	293.20	56.55	452.40		
1 Crew Truck		224.80		247.28		
1 Hyd. Crane, 12 Ton		663.00		729.30	15.85	17.44
56 L.H., Daily Totals		$3721.00		$5238.58	$66.45	$93.55

Crew R-12	Hr.	Daily	Hr.	Daily	Bare Costs	Incl. O&P
1 Carpenter Foreman (inside)	$46.35	$370.80	$71.50	$572.00	$43.30	$67.73
4 Carpenters	45.85	1467.20	70.75	2264.00		
4 Common Laborers	36.65	1172.80	56.55	1809.60		
1 Equip. Oper. (medium)	48.90	391.20	74.15	593.20		
1 Steel Worker	51.10	408.80	90.20	721.60		
1 Dozer, 200 H.P.		1325.00		1457.50		
1 Pickup Truck, 3/4 Ton		157.00		172.70	16.84	18.52
88 L.H., Daily Totals		$5292.80		$7590.60	$60.15	$86.26

Crew R-13	Hr.	Daily	Hr.	Daily	Bare Costs	Incl. O&P
1 Electrician Foreman	$53.85	$430.80	$80.60	$644.80	$51.43	$77.18
3 Electricians	53.35	1280.40	79.85	1916.40		
.25 Equip. Oper. (crane)	50.25	100.50	76.20	152.40		
1 Equipment Oiler	43.55	348.40	66.00	528.00		
.25 Hydraulic Crane, 33 Ton		189.80		208.78	4.52	4.97
42 L.H., Daily Totals		$2349.90		$3450.38	$55.95	$82.15

Crew R-15	Hr.	Daily	Hr.	Daily	Bare Costs	Incl. O&P
1 Electrician Foreman	$53.85	$430.80	$80.60	$644.80	$52.38	$78.56
4 Electricians	53.35	1707.20	79.85	2555.20		
1 Equipment Oper. (light)	47.05	376.40	71.35	570.80		
1 Aerial Lift Truck, 40' Boom		325.00		357.50	6.77	7.45
48 L.H., Daily Totals		$2839.40		$4128.30	$59.15	$86.01

Crew R-18	Hr.	Daily	Hr.	Daily	Bare Costs	Incl. O&P
.25 Electrician Foreman	$53.85	$107.70	$80.60	$161.20	$41.88	$63.82
1 Electrician	53.35	426.80	79.85	638.80		
2 Helpers	34.65	554.40	53.70	859.20		
26 L.H., Daily Totals		$1088.90		$1659.20	$41.88	$63.82

Crew R-19	Hr.	Daily	Hr.	Daily	Bare Costs	Incl. O&P
.5 Electrician Foreman	$53.85	$215.40	$80.60	$322.40	$53.45	$80.00
2 Electricians	53.35	853.60	79.85	1277.60		
20 L.H., Daily Totals		$1069.00		$1600.00	$53.45	$80.00

Crew R-21	Hr.	Daily	Hr.	Daily	Bare Costs	Incl. O&P
1 Electrician Foreman	$53.85	$430.80	$80.60	$644.80	$53.36	$79.89
3 Electricians	53.35	1280.40	79.85	1916.40		
.1 Equip. Oper. (medium)	48.90	39.12	74.15	59.32		
.1 S.P. Crane, 4x4, 25 Ton		60.72		66.79	1.85	2.04
32.8 L.H., Daily Totals		$1811.04		$2687.31	$55.21	$81.93

Crew R-22	Hr.	Daily	Hr.	Daily	Bare Costs	Incl. O&P
.66 Electrician Foreman	$53.85	$284.33	$80.60	$425.57	$45.40	$68.73
2 Electricians	53.35	853.60	79.85	1277.60		
2 Helpers	34.65	554.40	53.70	859.20		
37.28 L.H., Daily Totals		$1692.33		$2562.37	$45.40	$68.73

Crew No.	Bare Costs		Incl. Subs O & P		Cost Per Labor-Hour	
Crew R-30	**Hr.**	**Daily**	**Hr.**	**Daily**	**Bare Costs**	**Incl. O&P**
.25 Electrician Foreman (outside)	$55.35	$110.70	$82.85	$165.70	$43.23	$65.74
1 Electrician	53.35	426.80	79.85	638.80		
2 Laborers, (Semi-Skilled)	36.65	586.40	56.55	904.80		
26 L.H., Daily Totals		$1123.90		$1709.30	$43.23	$65.74
Crew R-31	**Hr.**	**Daily**	**Hr.**	**Daily**	**Bare Costs**	**Incl. O&P**
1 Electrician	$53.35	$426.80	$79.85	$638.80	$53.35	$79.85
1 Core Drill, Electric, 2.5 H.P.		46.20		50.82	5.78	6.35
8 L.H., Daily Totals		$473.00		$689.62	$59.13	$86.20
Crew W-41E	**Hr.**	**Daily**	**Hr.**	**Daily**	**Bare Costs**	**Incl. O&P**
.5 Plumber Foreman (outside)	$59.55	$238.20	$89.90	$359.60	$49.59	$75.36
1 Plumber	57.55	460.40	86.90	695.20		
1 Laborer	36.65	293.20	56.55	452.40		
20 L.H., Daily Totals		$991.80		$1507.20	$49.59	$75.36

Historical Cost Indexes

The table below lists both the RSMeans® historical cost index based on Jan. 1, 1993 = 100 as well as the computed value of an index based on Jan. 1, 2014 costs. Since the Jan. 1, 2014 figure is estimated, space is left to write in the actual index figures as they become available through either the quarterly *RSMeans Construction Cost Indexes* or as printed in the *Engineering News-Record.* To compute the actual index based on Jan. 1, 2014 = 100, divide the historical cost index for a particular year by the actual Jan. 1, 2014 construction cost index. Space has been left to advance the index figures as the year progresses.

Year	Historical Cost Index Jan. 1, 1993 = 100		Current Index Based on Jan. 1, 2014 = 100		Year	Historical Cost Index Jan. 1, 1993 = 100	Current Index Based on Jan. 1, 2014 = 100		Year	Historical Cost Index Jan. 1, 1993 = 100	Current Index Based on Jan. 1, 2014 = 100	
	Est.	Actual	Est.	Actual		Actual	Est.	Actual		Actual	Est.	Actual
Oct 2014*					July 1999	117.6	58.0		July 1981	70.0	34.5	
July 2014*					1998	115.1	56.8		1980	62.9	31.0	
April 2014*					1997	112.8	55.6		1979	57.8	28.5	
Jan 2014*	202.7		100.0	100.0	1996	110.2	54.4		1978	53.5	26.4	
July 2013		201.2	99.3		1995	107.6	53.1		1977	49.5	24.4	
2012		194.6	96.0		1994	104.4	51.5		1976	46.9	23.1	
2011		191.2	94.3		1993	101.7	50.2		1975	44.8	22.1	
2010		183.5	90.5		1992	99.4	49.1		1974	41.4	20.4	
2009		180.1	88.9		1991	96.8	47.8		1973	37.7	18.6	
2008		180.4	89.0		1990	94.3	46.5		1972	34.8	17.2	
2007		169.4	83.6		1989	92.1	45.5		1971	32.1	15.8	
2006		162.0	79.9		1988	89.9	44.3		1970	28.7	14.2	
2005		151.6	74.8		1987	87.7	43.3		1969	26.9	13.3	
2004		143.7	70.9		1986	84.2	41.6		1968	24.9	12.3	
2003		132.0	65.1		1985	82.6	40.8		1967	23.5	11.6	
2002		128.7	63.5		1984	82.0	40.4		1966	22.7	11.2	
2001		125.1	61.7		1983	80.2	39.5		1965	21.7	10.7	
2000		120.9	59.6		1982	76.1	37.6		1964	21.2	10.5	

Adjustments to Costs

The "Historical Cost Index" can be used to convert national average building costs at a particular time to the approximate building costs for some other time.

Time Adjustment Using the Historical Cost Indexes:

$$\frac{\text{Index for Year A}}{\text{Index for Year B}} \times \text{Cost in Year B} = \text{Cost in Year A}$$

Example:

Estimate and compare construction costs for different years in the same city.

To estimate the national average construction cost of a building in 1970, knowing that it cost $900,000 in 2014:

INDEX in 1970 = 28.7

INDEX in 2014 = 202.7

$$\frac{\text{INDEX 1970}}{\text{INDEX 2014}} \times \text{Cost 2014} = \text{Cost 1970}$$

$$\frac{28.7}{202.7} \times \$900{,}000 = .142 \times \$900{,}000 = \$128{,}000$$

The construction cost of the building in 1970 is $128,000.

Note: The city cost indexes for Canada can be used to convert U.S. national averages to local costs in Canadian dollars.

*Historical Cost Index updates and other resources are provided on the following website.
http://www.reedconstructiondata.com/rsmeans/chgnotice/456321

How to Use the City Cost Indexes

What you should know before you begin

RSMeans City Cost Indexes (CCI) are an extremely useful tool to use when you want to compare costs from city to city and region to region.

This publication contains average construction cost indexes for 731 U.S. and Canadian cities covering over 930 three-digit zip code locations, as listed directly under each city.

Keep in mind that a City Cost Index number is a percentage ratio of a specific city's cost to the national average cost of the same item at a stated time period.

In other words, these index figures represent relative construction factors (or, if you prefer, multipliers) for Material and Installation costs, as well as the weighted average for Total In Place costs for each CSI MasterFormat division. Installation costs include both labor and equipment rental costs. When estimating equipment rental rates only, for a specific location, use 01 54 33 EQUIPMENT RENTAL COSTS in the Reference Section at the back of the book.

The 30 City Average Index is the average of 30 major U.S. cities and serves as a National Average.

Index figures for both material and installation are based on the 30 major city average of 100 and represent the cost relationship as of July 1, 2013. The index for each division is computed from representative material and labor quantities for that division. The weighted average for each city is a weighted total of the components listed above it, but does not include relative productivity between trades or cities.

As changes occur in local material prices, labor rates, and equipment rental rates (including fuel costs), the impact of these changes should be accurately measured by the change in the City Cost Index for each particular city (as compared to the 30 City Average).

Therefore, if you know (or have estimated) building costs in one city today, you can easily convert those costs to expected building costs in another city.

In addition, by using the Historical Cost Index, you can easily convert National Average building costs at a particular time to the approximate building costs for some other time. The City Cost Indexes can then be applied to calculate the costs for a particular city.

Quick Calculations

Location Adjustment Using the City Cost Indexes:

$$\frac{\text{Index for City A}}{\text{Index for City B}} \times \text{Cost in City B} = \text{Cost in City A}$$

Time Adjustment for the National Average Using the Historical Cost Index:

$$\frac{\text{Index for Year A}}{\text{Index for Year B}} \times \text{Cost in Year B} = \text{Cost in Year A}$$

Adjustment from the National Average:

$$\frac{\text{Index for City A}}{100} \times \text{National Average Cost} = \text{Cost in City A}$$

Since each of the other RSMeans publications contains many different items, any *one* item multiplied by the particular city index may give incorrect results. However, the larger the number of items compiled, the closer the results should be to actual costs for that particular city.

The City Cost Indexes for Canadian cities are calculated using Canadian material and equipment prices and labor rates, in Canadian dollars. Therefore, indexes for Canadian cities can be used to convert U.S. National Average prices to local costs in Canadian dollars.

How to use this section

1. Compare costs from city to city.

In using the RSMeans Indexes, remember that an index number is not a fixed number but a ratio: It's a percentage ratio of a building component's cost at any stated time to the National Average cost of that same component at the same time period. Put in the form of an equation:

$$\frac{\text{Specific City Cost}}{\text{National Average Cost}} \times 100 = \text{City Index Number}$$

Therefore, when making cost comparisons between cities, do not subtract one city's index number from the index number of another city and read the result as a percentage difference. Instead, divide one city's index number by that of the other city. The resulting number may then be used as a multiplier to calculate cost differences from city to city.

The formula used to find cost differences between cities for the purpose of comparison is as follows:

$$\frac{\text{City A Index}}{\text{City B Index}} \times \text{City B Cost (Known)} = \text{City A Cost (Unknown)}$$

In addition, you can use RSMeans CCI to calculate and compare costs division by division between cities using the same basic formula. (Just be sure that you're comparing similar divisions.)

2. Compare a specific city's construction costs with the National Average.

When you're studying construction location feasibility, it's advisable to compare a prospective project's cost index with an index of the National Average cost.

For example, divide the weighted average index of construction costs of a specific city by that of the 30 City Average, which = 100.

$$\frac{\text{City Index}}{100} = \text{\% of National Average}$$

As a result, you get a ratio that indicates the relative cost of construction in that city in comparison with the National Average.

3. Convert U.S. National Average to actual costs in Canadian City.

$$\frac{\text{Index for Canadian City}}{100} \times \text{National Average Cost} = \text{Cost in Canadian City in \$ CAN}$$

4. **Adjust construction cost data based on a National Average.**

When you use a source of construction cost data which is based on a National Average (such as RSMeans cost data publications), it is necessary to adjust those costs to a specific location.

$$\frac{\text{City Index}}{100} \times \frac{\text{"Book" Cost Based on}}{\text{National Average Costs}} = \frac{\text{City Cost}}{\text{(Unknown)}}$$

5. **When applying the City Cost Indexes to demolition projects, use the appropriate division installation index.** For example, for removal of existing doors and windows, use Division 8 (Openings) index.

What you might like to know about how we developed the Indexes

The information presented in the CCI is organized according to the Construction Specifications Institute (CSI) MasterFormat 2012 classification system.

To create a reliable index, RSMeans researched the building type most often constructed in the United States and Canada. Because it was concluded that no one type of building completely represented the building construction industry, nine different types of buildings were combined to create a composite model.

The exact material, labor, and equipment quantities are based on detailed analyses of these nine building types, and then each quantity is weighted in proportion to expected usage. These various material items, labor hours, and equipment rental rates are thus combined to form a composite building representing as closely as possible the actual usage of materials, labor, and equipment used in the North American building construction industry.

The following structures were chosen to make up that composite model:

1. Factory, 1 story
2. Office, 2–4 story
3. Store, Retail
4. Town Hall, 2–3 story
5. High School, 2–3 story
6. Hospital, 4–8 story
7. Garage, Parking
8. Apartment, 1–3 story
9. Hotel/Motel, 2–3 story

For the purposes of ensuring the timeliness of the data, the components of the index for the composite model have been streamlined. They currently consist of:

- specific quantities of 66 commonly used construction materials;
- specific labor-hours for 21 building construction trades; and
- specific days of equipment rental for 6 types of construction equipment (normally used to install the 66 material items by the 21 trades.) Fuel costs and routine maintenance costs are included in the equipment cost.

A sophisticated computer program handles the updating of all costs for each city on a quarterly basis. Material and equipment price quotations are gathered quarterly from 731 cities in the United States and Canada. These prices and the latest negotiated labor wage rates for 21 different building trades are used to compile the quarterly update of the City Cost Index.

The 30 major U.S. cities used to calculate the National Average are:

Atlanta, GA	Memphis, TN
Baltimore, MD	Milwaukee, WI
Boston, MA	Minneapolis, MN
Buffalo, NY	Nashville, TN
Chicago, IL	New Orleans, LA
Cincinnati, OH	New York, NY
Cleveland, OH	Philadelphia, PA
Columbus, OH	Phoenix, AZ
Dallas, TX	Pittsburgh, PA
Denver, CO	St. Louis, MO
Detroit, MI	San Antonio, TX
Houston, TX	San Diego, CA
Indianapolis, IN	San Francisco, CA
Kansas City, MO	Seattle, WA
Los Angeles, CA	Washington, DC

What the CCI does not indicate

The weighted average for each city is a total of the divisional components weighted to reflect typical usage, but it does not include the productivity variations between trades or cities.

In addition, the CCI does not take into consideration factors such as the following:

- managerial efficiency
- competitive conditions
- automation
- restrictive union practices
- unique local requirements
- regional variations due to specific building codes

DIVISION		UNITED STATES			ALABAMA															
		30 CITY			ANNISTON			BIRMINGHAM			BUTLER			DECATUR			DOTHAN			
		AVERAGE			362			350 - 352			369			356			363			
		MAT.	INST.	TOTAL	MAT.	INST.	TOTAL	MAT.	INST.	TOTAL	MAT.	INST.	TOTAL	MAT.	INST.	TOTAL	MAT.	INST.	TOTAL	
015433	CONTRACTOR EQUIPMENT		100.0	100.0		101.6	101.6		101.7	101.7		98.8	98.8		101.6	101.6		98.8	98.8	
0241, 31 - 34	SITE & INFRASTRUCTURE, DEMOLITION	100.0	100.0	100.0	87.7	94.3	92.3	88.8	94.9	93.1	101.0	88.3	92.1	81.3	92.8	89.3	98.7	88.3	91.4	
0310	Concrete Forming & Accessories	100.0	100.0	100.0	92.7	48.2	54.3	93.3	70.2	73.4	89.0	43.6	49.8	94.7	46.4	53.0	97.9	43.0	50.5	
0320	Concrete Reinforcing	100.0	100.0	100.0	87.0	87.4	87.2	92.9	87.8	90.4	92.0	47.4	69.7	87.0	79.5	83.3	92.0	46.8	69.4	
0330	Cast-in-Place Concrete	100.0	100.0	100.0	102.0	50.5	80.6	99.7	74.7	89.3	99.4	55.7	81.3	93.7	65.1	81.9	99.4	47.0	77.7	
03	CONCRETE	100.0	100.0	100.0	99.8	57.9	79.1	95.1	76.1	85.7	100.6	50.4	75.8	91.4	60.7	76.3	100.0	47.0	73.9	
04	MASONRY	100.0	100.0	100.0	90.5	67.9	76.5	89.9	76.2	81.4	95.3	49.4	66.9	88.4	52.1	66.0	96.6	40.4	61.8	
05	METALS	100.0	100.0	100.0	96.9	92.5	95.5	97.0	94.7	96.3	95.8	78.1	90.4	99.0	87.7	95.5	95.9	76.9	90.0	
06	WOOD, PLASTICS & COMPOSITES	100.0	100.0	100.0	92.7	43.7	65.1	96.3	68.3	80.5	87.3	43.4	62.5	98.0	43.9	67.5	99.6	43.4	67.9	
07	THERMAL & MOISTURE PROTECTION	100.0	100.0	100.0	100.1	53.8	81.3	95.9	81.4	90.0	100.1	57.6	82.9	93.1	59.3	79.4	100.1	49.9	79.7	
08	OPENINGS	100.0	100.0	100.0	93.9	50.9	83.8	95.8	72.3	90.3	94.0	45.2	82.4	98.8	51.6	87.6	94.0	45.2	82.5	
0920	Plaster & Gypsum Board	100.0	100.0	100.0	87.5	42.4	56.8	97.0	67.8	77.1	84.9	42.0	55.7	98.1	42.6	60.4	95.3	42.0	59.1	
0950, 0980	Ceilings & Acoustic Treatment	100.0	100.0	100.0	87.8	42.4	57.6	98.4	67.8	78.0	87.8	42.0	57.3	95.7	42.6	60.4	87.8	42.0	57.3	
0960	Flooring	100.0	100.0	100.0	104.1	41.0	85.2	109.0	76.2	99.2	109.2	57.0	93.6	106.0	51.5	89.7	114.9	28.0	88.9	
0970, 0990	Wall Finishes & Painting/Coating	100.0	100.0	100.0	91.4	32.1	55.8	93.5	56.8	71.4	91.4	54.5	69.2	89.3	68.2	76.7	91.4	54.5	69.2	
09	FINISHES	100.0	100.0	100.0	92.7	44.0	65.6	99.4	69.3	82.6	95.9	46.7	68.5	95.9	48.8	69.7	98.7	41.0	66.6	
COVERS	DIVS. 10 - 14, 25, 28, 41, 43, 44, 46	100.0	100.0	100.0	100.0	70.1	94.0	100.0	87.4	97.5	100.0	42.5	88.5	100.0	44.4	88.9	100.0	42.6	88.5	
21, 22, 23	FIRE SUPPRESSION, PLUMBING & HVAC	100.0	100.0	100.0	99.9	56.6	82.6	100.0	62.8	85.1	97.3	34.0	71.9	100.0	41.7	76.6	97.3	33.9	71.9	
26, 27, 3370	ELECTRICAL, COMMUNICATIONS & UTIL.	100.0	100.0	100.0	90.7	58.7	74.0	99.7	62.8	80.4	92.6	41.2	65.7	94.5	66.5	79.9	91.5	58.8	74.4	
MF2010	WEIGHTED AVERAGE	100.0	100.0	100.0	96.5	63.0	81.9	97.5	74.1	87.3	96.8	50.5	76.6	96.6	59.1	80.3	96.9	50.5	76.7	

DIVISION		ALABAMA																	
		EVERGREEN			GADSDEN			HUNTSVILLE			JASPER			MOBILE			MONTGOMERY		
		364			359			357 - 358			355			365 - 366			360 - 361		
		MAT.	INST.	TOTAL	MAT.	INST.	TOTAL	MAT.	INST.	TOTAL	MAT.	INST.	TOTAL	MAT.	INST.	TOTAL	MAT.	INST.	TOTAL
015433	CONTRACTOR EQUIPMENT		98.8	98.8		101.6	101.6		101.6	101.6		101.6	101.6		98.8	98.8		98.8	98.8
0241, 31 - 34	SITE & INFRASTRUCTURE, DEMOLITION	101.5	88.6	92.4	86.7	93.8	91.7	81.1	93.8	90.0	86.3	93.6	91.4	94.0	89.3	90.7	94.0	89.2	90.6
0310	Concrete Forming & Accessories	85.6	45.2	50.8	87.0	43.3	49.3	94.7	71.1	74.3	92.2	35.3	43.0	97.1	55.0	60.7	96.9	45.0	52.1
0320	Concrete Reinforcing	92.0	47.4	69.7	92.2	87.7	90.0	87.0	80.5	83.8	87.0	86.5	86.8	89.8	79.4	84.6	94.9	87.1	91.0
0330	Cast-in-Place Concrete	99.5	58.3	82.4	93.8	66.7	82.5	91.3	69.6	82.3	103.9	44.8	79.3	104.3	67.4	88.9	105.4	57.4	85.5
03	CONCRETE	100.9	52.0	76.8	95.7	61.4	78.8	90.3	73.3	81.9	98.9	50.1	74.8	97.0	65.3	81.4	98.3	58.9	78.8
04	MASONRY	95.4	54.1	69.8	86.7	62.3	71.6	89.7	67.7	76.1	84.2	42.0	58.1	93.8	55.1	69.9	90.4	50.5	65.7
05	METALS	95.9	78.1	90.4	96.9	94.0	96.0	99.0	90.5	96.4	96.9	91.1	95.1	97.9	90.6	95.6	97.0	92.8	95.7
06	WOOD, PLASTICS & COMPOSITES	83.6	43.4	60.9	87.8	39.6	60.6	98.0	74.3	84.6	94.8	32.3	59.6	98.2	54.1	73.3	97.7	43.4	67.0
07	THERMAL & MOISTURE PROTECTION	100.1	57.9	83.0	93.1	72.4	84.7	93.0	78.5	87.1	93.2	46.1	74.1	99.8	71.3	88.2	100.2	67.8	87.0
08	OPENINGS	94.0	45.2	82.4	95.4	49.8	84.6	98.8	68.2	91.5	95.3	48.7	84.3	97.2	58.6	88.1	98.3	54.0	87.8
0920	Plaster & Gypsum Board	84.2	42.0	55.5	89.8	38.2	54.7	98.1	73.9	81.7	94.2	30.7	51.0	92.1	53.1	65.5	91.5	42.0	57.8
0950, 0980	Ceilings & Acoustic Treatment	87.8	42.0	57.3	92.1	38.2	56.2	97.6	73.9	81.9	92.1	30.7	51.2	93.3	53.1	66.5	95.9	42.0	60.0
0960	Flooring	107.0	57.0	92.1	102.0	76.2	94.3	106.0	76.2	97.1	104.3	41.0	85.3	114.6	59.5	98.1	110.8	59.5	95.5
0970, 0990	Wall Finishes & Painting/Coating	91.4	54.5	69.2	89.3	62.0	72.9	89.3	59.7	71.5	89.3	39.4	59.3	95.0	55.4	71.2	91.2	54.5	69.1
09	FINISHES	95.3	47.9	68.9	93.3	49.4	68.8	96.4	71.1	82.3	94.5	35.2	61.5	99.0	54.7	74.3	99.0	47.1	70.1
COVERS	DIVS. 10 - 14, 25, 28, 41, 43, 44, 46	100.0	44.1	88.8	100.0	79.5	95.9	100.0	85.3	97.0	100.0	40.8	88.1	100.0	83.9	96.8	100.0	79.4	95.9
21, 22, 23	FIRE SUPPRESSION, PLUMBING & HVAC	97.3	36.3	72.9	102.0	36.9	75.9	100.0	52.9	81.1	102.0	53.8	82.6	100.0	61.5	84.5	100.1	34.8	73.9
26, 27, 3370	ELECTRICAL, COMMUNICATIONS & UTIL.	90.1	41.2	64.6	94.5	62.8	78.0	95.7	66.5	80.4	94.1	62.8	77.7	93.4	58.5	75.2	93.9	62.3	77.4
MF2010	WEIGHTED AVERAGE	96.5	51.9	77.1	96.7	60.8	81.1	96.7	70.8	85.4	97.0	56.7	79.4	97.8	65.9	83.9	97.9	58.1	80.5

DIVISION		ALABAMA									ALASKA								
		PHENIX CITY			SELMA			TUSCALOOSA			ANCHORAGE			FAIRBANKS			JUNEAU		
		368			367			354			995 - 996			997			998		
		MAT.	INST.	TOTAL	MAT.	INST.	TOTAL	MAT.	INST.	TOTAL	MAT.	INST.	TOTAL	MAT.	INST.	TOTAL	MAT.	INST.	TOTAL
015433	CONTRACTOR EQUIPMENT		98.8	98.8		98.8	98.8		101.6	101.6		113.5	113.5		113.5	113.5		113.5	113.5
0241, 31 - 34	SITE & INFRASTRUCTURE, DEMOLITION	105.2	89.3	94.0	98.6	89.1	92.0	81.6	94.0	90.3	125.5	128.8	127.8	116.4	128.8	125.1	132.0	128.8	129.8
0310	Concrete Forming & Accessories	92.6	39.9	47.1	90.3	43.5	49.9	94.6	48.8	55.1	127.8	119.0	120.2	133.0	118.7	120.7	129.2	118.9	120.3
0320	Concrete Reinforcing	91.9	63.3	77.7	92.0	86.2	89.1	87.0	87.7	87.3	148.0	111.5	129.8	142.1	111.5	126.8	122.7	111.5	117.1
0330	Cast-in-Place Concrete	99.5	57.9	82.2	99.4	47.2	77.8	95.1	68.5	84.1	131.0	117.4	125.3	129.2	117.7	124.4	136.1	117.4	128.3
03	CONCRETE	103.6	52.4	78.3	99.4	54.5	77.3	92.1	64.5	78.5	143.0	116.3	129.8	126.5	116.3	121.4	139.9	116.2	128.2
04	MASONRY	95.3	41.8	62.2	98.7	39.3	62.0	88.6	65.5	74.3	180.9	122.2	144.6	182.2	122.2	145.1	167.6	122.2	139.5
05	METALS	95.8	83.5	92.0	95.8	90.3	94.1	98.2	94.1	96.9	114.6	102.5	110.9	118.8	102.6	113.8	117.6	102.4	112.9
06	WOOD, PLASTICS & COMPOSITES	92.5	36.3	60.8	89.3	43.4	63.4	98.0	45.4	68.3	117.1	117.9	117.6	125.0	117.4	120.7	119.8	117.9	118.7
07	THERMAL & MOISTURE PROTECTION	100.4	64.9	86.0	100.0	53.0	80.9	93.1	74.4	85.5	169.9	115.5	147.8	169.2	116.6	147.9	163.4	115.5	143.9
08	OPENINGS	93.9	44.6	82.3	93.9	54.0	84.5	98.8	59.8	89.6	119.7	115.2	118.7	116.2	115.2	116.0	116.6	115.2	116.3
0920	Plaster & Gypsum Board	89.0	34.7	52.1	86.8	42.0	56.3	98.1	44.2	61.4	126.3	118.3	120.9	148.9	117.8	127.7	131.2	118.3	122.4
0950, 0980	Ceilings & Acoustic Treatment	87.8	34.7	52.5	87.8	42.0	57.3	97.6	44.2	62.1	126.5	118.3	121.0	121.7	117.8	119.1	123.5	118.3	120.0
0960	Flooring	111.3	59.5	95.8	109.8	29.0	85.6	106.0	76.2	97.1	150.1	128.3	143.6	140.7	128.3	137.0	142.1	128.3	138.0
0970, 0990	Wall Finishes & Painting/Coating	91.4	54.5	69.2	91.4	54.5	69.2	89.3	46.9	63.8	146.5	113.3	126.5	143.9	121.9	130.7	143.3	113.3	125.2
09	FINISHES	97.5	43.2	67.2	96.0	41.2	65.5	96.3	52.0	71.6	143.1	120.5	130.5	139.4	121.2	129.2	139.6	120.5	129.0
COVERS	DIVS. 10 - 14, 25, 28, 41, 43, 44, 46	100.0	79.0	95.8	100.0	42.5	88.5	100.0	81.3	96.2	100.0	112.5	102.5	100.0	112.5	102.5	100.0	112.5	102.5
21, 22, 23	FIRE SUPPRESSION, PLUMBING & HVAC	97.3	35.1	72.4	97.3	34.3	72.1	100.0	34.6	73.8	100.2	104.3	101.9	100.2	106.7	102.8	100.1	100.6	100.3
26, 27, 3370	ELECTRICAL, COMMUNICATIONS & UTIL.	92.1	71.2	81.2	91.2	41.2	65.1	95.1	62.8	78.2	120.5	115.6	117.9	134.7	115.6	124.7	125.5	115.6	120.3
MF2010	WEIGHTED AVERAGE	97.3	55.7	79.2	96.7	50.8	76.7	96.7	62.0	81.6	121.3	114.3	118.3	120.7	115.0	118.2	120.6	113.5	117.5

DIVISION		ALASKA			ARIZONA																
		KETCHIKAN			CHAMBERS			FLAGSTAFF			GLOBE			KINGMAN			MESA/TEMPE		
		999			865			860			855			864			852		
		MAT.	INST.	TOTAL	MAT.	INST.	TOTAL	MAT.	INST.	TOTAL	MAT.	INST.	TOTAL	MAT.	INST.	TOTAL	MAT.	INST.	TOTAL
015433	CONTRACTOR EQUIPMENT		113.5	113.5		92.6	92.6		92.6	92.6		94.2	94.2		92.6	92.6		94.2	94.2
0241, 31 - 34	SITE & INFRASTRUCTURE, DEMOLITION	170.9	128.8	141.3	67.2	96.6	87.9	84.8	96.8	93.3	100.7	98.4	99.1	67.2	96.8	88.0	91.4	98.6	96.4
0310	Concrete Forming & Accessories	124.5	118.9	119.7	98.2	59.2	64.5	103.7	66.4	71.5	96.3	59.3	64.3	96.5	66.4	70.5	99.3	69.6	73.6
0320	Concrete Reinforcing	111.6	111.5	111.5	100.0	83.5	91.7	99.8	83.5	91.7	110.1	83.5	96.8	100.1	83.5	91.8	110.8	83.5	97.2
0330	Cast-in-Place Concrete	261.0	117.4	201.4	90.3	73.5	83.3	90.4	73.8	83.5	93.8	73.6	85.4	90.0	73.8	83.3	94.6	73.9	86.0
03	CONCRETE	214.6	116.2	166.0	96.5	68.9	82.8	116.1	72.2	94.4	109.8	68.9	89.7	96.1	72.2	84.3	101.7	73.7	87.8
04	MASONRY	191.0	122.2	148.4	90.6	62.7	73.3	90.7	62.8	73.4	110.3	62.7	80.9	90.6	62.8	73.4	110.6	62.8	81.0
05	METALS	119.0	102.4	113.9	95.1	75.0	88.9	95.6	75.7	89.5	92.4	75.5	87.2	95.7	75.8	89.6	92.7	76.4	87.7
06	WOOD, PLASTICS & COMPOSITES	115.7	117.9	117.0	96.1	56.3	73.6	102.2	65.7	81.6	90.0	56.4	71.0	91.6	65.7	77.0	93.4	69.9	80.2
07	THERMAL & MOISTURE PROTECTION	172.3	115.5	149.2	95.1	65.1	82.9	96.7	68.2	85.1	99.3	64.6	85.2	95.0	65.7	83.1	98.6	66.6	85.6
08	OPENINGS	118.1	115.2	117.4	108.1	65.7	98.1	108.2	70.8	99.4	102.2	65.8	93.6	108.3	70.8	99.4	102.3	73.2	95.4
0920	Plaster & Gypsum Board	137.4	118.3	124.4	85.6	55.0	64.8	89.1	64.7	72.5	94.2	55.0	67.5	79.1	64.7	69.3	96.4	69.0	77.7
0950, 0980	Ceilings & Acoustic Treatment	114.7	118.3	117.1	107.8	55.0	72.7	108.7	64.7	79.4	96.4	55.0	68.8	108.7	64.7	79.4	96.4	69.0	78.1
0960	Flooring	140.7	128.3	137.0	92.1	40.3	76.6	94.1	40.4	78.1	100.0	40.3	82.1	91.1	54.7	80.2	101.1	50.4	86.0
0970, 0990	Wall Finishes & Painting/Coating	143.9	113.3	125.5	97.8	56.3	72.9	97.8	56.3	72.9	108.9	56.3	77.3	97.8	56.3	72.9	108.9	56.3	77.3
09	FINISHES	140.7	120.5	129.5	95.7	54.0	72.4	98.8	59.6	76.9	97.6	54.1	73.4	94.7	62.0	76.5	97.1	64.0	78.7
COVERS	DIVS. 10 - 14, 25, 28, 41, 43, 44, 46	100.0	112.5	102.5	100.0	82.3	96.5	100.0	83.3	96.7	100.0	82.5	96.5	100.0	83.3	96.7	100.0	84.0	96.8
21, 22, 23	FIRE SUPPRESSION, PLUMBING & HVAC	98.4	100.6	99.3	97.0	78.9	89.7	100.2	80.0	92.1	95.2	78.9	88.7	97.0	80.0	90.2	100.1	79.0	91.6
26, 27, 3370	ELECTRICAL, COMMUNICATIONS & UTIL.	134.6	115.6	124.7	102.2	72.6	86.7	101.1	65.8	82.7	98.7	65.8	81.6	102.2	65.8	83.2	95.3	65.8	79.9
MF2010	WEIGHTED AVERAGE	132.0	113.5	124.0	97.3	71.9	86.3	101.0	72.9	88.8	99.0	71.2	86.9	97.3	73.2	86.8	98.8	73.8	87.9

DIVISION		ARIZONA												ARKANSAS					
		PHOENIX			PRESCOTT			SHOW LOW			TUCSON			BATESVILLE			CAMDEN		
		850,853			863			859			856 - 857			725			717		
		MAT.	INST.	TOTAL	MAT.	INST.	TOTAL	MAT.	INST.	TOTAL	MAT.	INST.	TOTAL	MAT.	INST.	TOTAL	MAT.	INST.	TOTAL
015433	CONTRACTOR EQUIPMENT		94.7	94.7		92.6	92.6		94.2	94.2		94.2	94.2		89.0	89.0		89.0	89.0
0241, 31 - 34	SITE & INFRASTRUCTURE, DEMOLITION	91.9	98.8	96.7	73.5	96.6	89.7	102.8	98.4	99.7	87.1	98.6	95.1	73.4	84.9	81.5	73.3	84.5	81.2
0310	Concrete Forming & Accessories	100.2	69.7	73.9	99.8	53.5	59.8	103.2	69.3	74.0	99.8	69.5	73.6	85.2	45.9	51.3	83.9	31.4	38.6
0320	Concrete Reinforcing	109.0	83.7	96.3	99.8	83.5	91.6	110.8	83.5	97.2	91.8	83.5	87.6	86.7	65.8	76.3	88.2	65.7	76.9
0330	Cast-in-Place Concrete	94.7	74.0	86.1	90.3	73.5	83.3	93.9	73.6	85.5	97.4	73.9	87.6	81.1	46.1	66.6	81.2	39.0	63.7
03	CONCRETE	101.3	73.8	87.7	101.7	66.3	84.2	112.1	73.4	93.0	99.7	73.6	86.8	85.2	50.5	68.0	86.9	41.5	64.5
04	MASONRY	98.0	64.6	77.3	90.7	62.7	73.4	110.4	62.7	80.9	95.9	62.8	75.4	98.7	41.5	63.3	111.6	31.9	62.3
05	METALS	94.2	77.2	88.9	95.6	74.7	89.2	92.2	75.7	87.1	93.5	76.3	88.2	98.0	66.9	88.4	97.3	66.5	87.8
06	WOOD, PLASTICS & COMPOSITES	94.4	69.9	80.6	97.5	48.5	69.9	98.0	69.9	82.1	93.7	69.9	80.3	86.3	46.5	63.9	89.8	30.4	56.3
07	THERMAL & MOISTURE PROTECTION	98.2	68.7	86.2	95.5	64.2	82.8	99.5	66.5	86.1	99.6	65.8	85.9	100.2	44.8	77.7	96.1	36.4	71.9
08	OPENINGS	104.2	73.2	96.9	108.2	61.5	97.2	101.3	73.2	94.7	98.3	73.2	92.4	94.6	46.0	83.1	93.5	40.6	81.0
0920	Plaster & Gypsum Board	98.5	69.0	78.4	85.7	47.0	59.4	98.6	69.0	78.4	101.9	69.0	79.5	81.0	45.4	56.7	80.2	28.7	45.2
0950, 0980	Ceilings & Acoustic Treatment	103.5	69.0	80.5	106.9	47.0	67.1	96.4	69.0	78.1	97.3	69.0	78.4	90.6	45.4	60.5	86.2	28.7	47.9
0960	Flooring	101.4	52.7	86.8	92.9	40.3	77.1	102.5	45.9	85.5	91.9	45.0	77.9	104.6	60.1	91.3	111.2	40.3	90.0
0970, 0990	Wall Finishes & Painting/Coating	108.9	62.0	80.7	97.8	56.3	72.9	108.9	56.3	77.3	110.1	56.3	77.8	99.8	42.0	65.1	99.5	50.8	70.2
09	FINISHES	99.1	65.0	80.1	96.2	49.4	70.1	99.2	63.0	79.0	95.4	62.9	77.3	89.8	47.8	66.4	92.1	34.2	59.8
COVERS	DIVS. 10 - 14, 25, 28, 41, 43, 44, 46	100.0	84.0	96.8	100.0	81.4	96.3	100.0	84.0	96.8	100.0	84.0	96.8	100.0	40.6	88.1	100.0	36.5	87.3
21, 22, 23	FIRE SUPPRESSION, PLUMBING & HVAC	100.0	80.0	92.0	100.2	78.9	91.7	95.2	79.0	88.7	100.1	79.0	91.6	95.3	52.1	77.9	95.3	53.5	78.5
26, 27, 3370	ELECTRICAL, COMMUNICATIONS & UTIL.	102.2	68.3	84.5	100.7	65.8	82.5	95.7	65.8	80.1	97.6	65.8	81.0	95.0	64.0	78.8	91.5	62.0	76.1
MF2010	WEIGHTED AVERAGE	99.4	74.8	88.7	98.9	69.7	86.2	99.1	73.5	88.0	97.6	73.6	87.1	94.2	55.1	77.2	94.5	50.4	75.3

DIVISION		ARKANSAS																	
		FAYETTEVILLE			FORT SMITH			HARRISON			HOT SPRINGS			JONESBORO			LITTLE ROCK		
		727			729			726			719			724			720 - 722		
		MAT.	INST.	TOTAL	MAT.	INST.	TOTAL	MAT.	INST.	TOTAL	MAT.	INST.	TOTAL	MAT.	INST.	TOTAL	MAT.	INST.	TOTAL
015433	CONTRACTOR EQUIPMENT		89.0	89.0		89.0	89.0		89.0	89.0		89.0	89.0		107.5	107.5		89.0	89.0
0241, 31 - 34	SITE & INFRASTRUCTURE, DEMOLITION	72.8	86.4	82.4	77.7	86.3	83.8	77.9	84.9	82.8	76.4	85.9	83.0	96.9	100.3	99.3	87.9	86.3	86.8
0310	Concrete Forming & Accessories	80.6	39.4	45.1	100.2	57.7	63.5	89.7	45.8	51.8	81.2	36.5	42.6	88.7	49.2	54.6	95.3	60.4	65.2
0320	Concrete Reinforcing	86.7	71.6	79.2	87.7	72.3	80.0	86.3	65.7	76.0	86.5	65.8	76.2	83.9	64.8	74.4	92.7	66.5	79.6
0330	Cast-in-Place Concrete	81.1	50.2	68.3	92.8	70.4	83.5	90.0	44.2	71.0	83.0	39.5	64.9	88.3	55.9	74.9	91.5	70.5	82.8
03	CONCRETE	84.9	50.1	67.7	93.1	65.3	79.4	92.6	49.8	71.4	90.7	44.0	67.6	89.8	55.9	73.1	94.7	65.4	80.2
04	MASONRY	89.1	42.1	60.0	94.1	50.0	66.8	99.0	39.7	62.3	83.9	34.1	53.1	90.5	42.6	60.9	93.9	50.0	66.7
05	METALS	98.0	69.3	89.1	100.3	71.1	91.3	99.1	66.6	89.1	97.3	66.9	87.9	94.6	78.8	89.7	100.5	69.2	90.8
06	WOOD, PLASTICS & COMPOSITES	81.9	36.3	56.2	103.9	60.4	79.4	92.3	46.5	66.5	86.8	36.5	58.4	90.1	49.4	67.1	96.3	63.9	78.1
07	THERMAL & MOISTURE PROTECTION	101.0	46.2	78.8	101.4	54.5	82.4	100.5	43.5	77.4	96.3	38.6	72.9	105.2	50.0	82.8	97.7	54.9	80.3
08	OPENINGS	94.6	45.8	83.0	95.4	57.1	86.4	95.4	46.7	83.9	93.5	42.4	81.4	97.4	53.5	87.0	96.2	58.9	87.4
0920	Plaster & Gypsum Board	78.7	34.9	48.9	86.6	59.7	68.3	85.5	45.4	58.2	78.8	35.1	49.0	95.4	48.0	63.1	92.1	63.3	72.5
0950, 0980	Ceilings & Acoustic Treatment	90.6	34.9	53.5	92.4	59.7	70.6	92.4	45.4	61.1	86.2	35.1	52.2	93.8	48.0	63.3	95.8	63.3	74.2
0960	Flooring	102.0	60.1	89.5	111.0	61.9	96.3	106.9	60.1	92.9	110.0	60.1	95.1	74.5	54.4	68.5	109.1	61.9	95.0
0970, 0990	Wall Finishes & Painting/Coating	99.8	30.0	57.8	99.8	55.0	72.8	99.8	42.0	65.1	99.5	50.8	70.2	88.3	49.4	64.9	100.9	56.4	74.2
09	FINISHES	88.9	40.9	62.2	93.0	58.4	73.8	91.8	47.8	67.3	91.8	41.3	63.7	86.5	49.6	66.0	95.9	60.7	76.3
COVERS	DIVS. 10 - 14, 25, 28, 41, 43, 44, 46	100.0	50.0	90.0	100.0	76.8	95.4	100.0	50.5	90.1	100.0	37.5	87.5	100.0	46.3	89.3	100.0	77.2	95.4
21, 22, 23	FIRE SUPPRESSION, PLUMBING & HVAC	95.3	50.6	77.4	100.2	50.1	80.1	95.3	49.8	77.0	95.3	49.1	76.8	100.4	52.9	81.4	100.0	56.4	82.5
26, 27, 3370	ELECTRICAL, COMMUNICATIONS & UTIL.	88.8	52.4	69.8	92.3	66.1	78.6	93.5	38.0	64.5	93.5	69.4	81.0	99.0	64.0	80.7	99.3	69.5	83.7
MF2010	WEIGHTED AVERAGE	93.0	52.9	75.6	96.8	61.7	81.5	95.5	51.1	76.1	93.8	52.3	75.8	96.3	59.4	80.2	98.0	63.8	83.1

DIVISION		ARKANSAS												CALIFORNIA					
		PINE BLUFF			RUSSELLVILLE			TEXARKANA			WEST MEMPHIS			ALHAMBRA			ANAHEIM		
		716			728			718			723			917 - 918			928		
		MAT.	INST.	TOTAL	MAT.	INST.	TOTAL	MAT.	INST.	TOTAL	MAT.	INST.	TOTAL	MAT.	INST.	TOTAL	MAT.	INST.	TOTAL
015433	CONTRACTOR EQUIPMENT		89.0	89.0		89.0	89.0		89.7	89.7		107.5	107.5		98.9	98.9		100.0	100.0
0241, 31 - 34	SITE & INFRASTRUCTURE, DEMOLITION	78.5	86.3	84.0	74.5	84.9	81.8	90.2	86.9	87.9	103.2	100.3	101.2	96.4	109.7	105.7	96.5	107.7	104.4
0310	Concrete Forming & Accessories	80.9	60.3	63.2	85.9	56.0	60.1	87.6	41.0	47.4	94.4	49.5	55.6	115.1	116.9	116.6	105.5	123.1	120.7
0320	Concrete Reinforcing	88.1	66.5	77.3	87.3	65.6	76.5	87.7	65.8	76.8	83.9	64.9	74.4	107.4	114.9	111.1	93.2	114.8	103.9
0330	Cast-in-Place Concrete	83.0	70.4	77.8	85.0	46.0	68.8	90.5	44.0	71.2	92.6	56.0	77.4	94.7	119.1	104.8	91.3	121.7	104.0
03	CONCRETE	91.5	65.4	78.6	88.4	54.9	71.9	89.1	47.6	68.6	97.3	56.0	76.9	104.0	116.4	110.1	103.1	120.0	111.4
04	MASONRY	120.2	50.0	76.8	93.3	38.0	59.1	98.1	33.7	58.2	76.8	42.6	55.7	122.3	118.6	120.0	77.4	116.2	101.4
05	METALS	98.0	69.1	89.1	98.0	66.5	88.3	90.8	67.0	83.5	93.6	79.1	89.2	84.0	99.9	88.9	103.1	100.2	102.2
06	WOOD, PLASTICS & COMPOSITES	86.3	63.9	73.7	87.7	60.4	72.3	95.2	43.0	65.7	96.3	49.4	69.9	96.6	113.1	105.9	98.5	121.4	111.4
07	THERMAL & MOISTURE PROTECTION	96.4	54.9	79.6	101.2	44.7	78.3	97.1	45.3	76.1	105.6	50.0	83.1	95.2	117.2	104.2	97.9	121.3	107.4
08	OPENINGS	93.5	58.9	85.3	94.6	56.2	85.5	98.3	46.7	86.1	97.4	53.5	87.0	89.8	115.5	95.9	100.3	120.2	105.0
0920	Plaster & Gypsum Board	78.4	63.3	68.1	81.0	59.7	66.5	82.2	41.7	54.6	98.0	48.0	64.0	102.1	113.7	110.0	100.6	122.0	115.2
0950, 0980	Ceilings & Acoustic Treatment	86.2	63.3	71.0	90.6	59.7	70.0	89.7	41.7	57.8	91.8	48.0	62.7	102.5	113.7	109.9	106.7	122.0	116.9
0960	Flooring	109.7	61.9	95.4	104.3	60.1	91.1	111.9	51.9	94.0	76.7	54.4	70.0	93.0	109.2	97.8	105.1	114.1	107.8
0970, 0990	Wall Finishes & Painting/Coating	99.5	56.4	73.6	99.8	35.7	61.3	99.5	31.0	58.3	88.3	50.8	65.8	101.7	108.4	105.7	100.4	106.2	103.9
09	FINISHES	91.8	60.7	74.5	90.0	55.3	70.7	94.3	41.3	64.7	87.7	49.8	66.6	98.0	114.0	106.9	102.0	119.6	111.8
COVERS	DIVS. 10 - 14, 25, 28, 41, 43, 44, 46	100.0	77.2	95.4	100.0	42.1	88.4	100.0	33.5	86.7	100.0	46.4	89.3	100.0	110.1	102.0	100.0	111.3	102.3
21, 22, 23	FIRE SUPPRESSION, PLUMBING & HVAC	100.2	52.4	81.1	95.3	51.9	77.9	100.2	54.6	81.9	95.5	66.5	83.9	95.1	113.7	102.6	100.0	118.0	107.2
26, 27, 3370	ELECTRICAL, COMMUNICATIONS & UTIL.	91.6	69.5	80.1	92.2	45.7	68.0	93.4	36.8	63.9	100.6	65.8	82.4	120.2	119.9	120.0	91.4	105.9	99.0
MF2010	WEIGHTED AVERAGE	96.9	62.9	82.1	94.1	54.4	76.8	95.7	49.9	75.7	95.6	62.7	81.2	98.2	113.9	105.0	98.9	114.1	105.6

DIVISION		CALIFORNIA																	
		BAKERSFIELD			BERKELEY			EUREKA			FRESNO			INGLEWOOD			LONG BEACH		
		932 - 933			947			955			936 - 938			903 - 905			906 - 908		
		MAT.	INST.	TOTAL	MAT.	INST.	TOTAL	MAT.	INST.	TOTAL	MAT.	INST.	TOTAL	MAT.	INST.	TOTAL	MAT.	INST.	TOTAL
015433	CONTRACTOR EQUIPMENT		97.9	97.9		98.5	98.5		97.6	97.6		97.9	97.9		95.7	95.7		95.7	95.7
0241, 31 - 34	SITE & INFRASTRUCTURE, DEMOLITION	97.5	105.3	103.0	119.7	106.0	110.1	108.5	101.6	103.7	99.4	104.8	103.2	92.6	102.2	99.3	99.8	102.2	101.5
0310	Concrete Forming & Accessories	104.6	122.7	120.2	118.3	148.0	143.9	114.9	130.6	128.4	103.9	131.7	127.9	113.1	116.9	116.4	108.0	116.9	115.7
0320	Concrete Reinforcing	108.4	114.7	111.5	92.5	116.2	104.3	101.9	115.4	108.6	84.7	115.4	100.0	108.6	114.9	111.7	107.6	114.9	111.2
0330	Cast-in-Place Concrete	92.1	120.7	104.0	127.8	122.0	125.4	99.1	116.3	106.2	97.5	117.3	105.7	87.6	120.2	101.1	99.7	120.2	108.2
03	CONCRETE	101.1	119.4	110.1	110.4	131.2	120.6	114.4	121.4	117.8	101.5	122.4	111.8	95.7	116.8	106.1	105.0	116.8	110.8
04	MASONRY	93.2	115.7	107.1	119.2	131.9	127.0	99.4	123.4	114.3	96.3	121.3	111.7	75.8	118.7	102.4	84.5	118.7	105.7
05	METALS	105.8	99.4	103.8	106.5	100.3	104.6	102.9	97.8	101.3	105.9	100.4	104.2	93.8	100.6	95.9	93.7	100.6	95.8
06	WOOD, PLASTICS & COMPOSITES	93.8	121.5	109.4	110.4	154.0	135.0	112.3	135.9	125.6	105.3	135.9	122.5	106.0	113.2	110.1	99.5	113.2	107.2
07	THERMAL & MOISTURE PROTECTION	104.0	111.9	107.2	103.2	135.9	116.5	101.9	116.0	107.6	94.0	113.6	101.9	100.9	118.5	108.1	101.1	118.5	108.2
08	OPENINGS	96.5	116.2	101.2	96.8	140.5	107.1	99.8	112.7	102.8	96.7	124.0	103.1	91.0	115.5	96.8	91.0	115.5	96.8
0920	Plaster & Gypsum Board	98.9	122.0	114.7	102.8	155.2	138.4	106.0	136.9	127.0	97.5	136.9	124.3	103.9	113.7	110.5	99.8	113.7	109.2
0950, 0980	Ceilings & Acoustic Treatment	109.6	122.0	117.9	104.1	155.2	138.1	112.1	136.9	128.6	106.0	136.9	126.5	101.7	113.7	109.7	101.7	113.7	109.7
0960	Flooring	104.9	109.2	106.2	108.7	126.4	114.0	109.3	118.7	112.1	110.4	135.9	118.0	98.9	109.2	102.0	96.5	109.2	100.3
0970, 0990	Wall Finishes & Painting/Coating	100.8	104.2	102.8	102.7	142.1	126.4	102.0	47.2	69.1	115.0	108.6	111.1	102.1	108.4	105.9	102.1	108.4	105.9
09	FINISHES	102.9	118.7	111.7	103.9	145.5	127.1	107.9	122.2	115.9	104.3	131.9	119.7	101.4	114.1	108.4	100.6	114.1	108.1
COVERS	DIVS. 10 - 14, 25, 28, 41, 43, 44, 46	100.0	111.3	102.3	100.0	125.1	105.0	100.0	121.4	104.3	100.0	121.4	104.3	100.0	110.2	102.0	100.0	110.2	102.0
21, 22, 23	FIRE SUPPRESSION, PLUMBING & HVAC	100.1	108.3	103.4	95.2	144.7	115.0	95.1	99.1	96.7	100.2	109.6	104.0	94.7	113.7	102.4	94.7	113.7	102.4
26, 27, 3370	ELECTRICAL, COMMUNICATIONS & UTIL.	102.0	102.3	102.2	109.2	142.6	126.6	98.8	112.6	106.0	91.7	102.9	97.6	105.5	119.9	113.0	105.2	119.9	112.9
MF2010	WEIGHTED AVERAGE	100.8	110.6	105.1	103.4	133.1	116.4	101.6	111.6	106.0	100.0	114.5	106.3	95.7	113.4	103.4	97.1	113.4	104.2

DIVISION		CALIFORNIA																	
		LOS ANGELES			MARYSVILLE			MODESTO			MOJAVE			OAKLAND			OXNARD		
		900 - 902			959			953			935			946			930		
		MAT.	INST.	TOTAL	MAT.	INST.	TOTAL	MAT.	INST.	TOTAL	MAT.	INST.	TOTAL	MAT.	INST.	TOTAL	MAT.	INST.	TOTAL
015433	CONTRACTOR EQUIPMENT		99.2	99.2		97.6	97.6		97.6	97.6		97.9	97.9		98.5	98.5		96.8	96.8
0241, 31 - 34	SITE & INFRASTRUCTURE, DEMOLITION	100.3	105.7	104.1	104.8	104.5	104.6	99.6	104.5	103.0	93.1	105.3	101.7	126.2	106.0	112.0	99.7	103.4	102.3
0310	Concrete Forming & Accessories	109.9	123.2	121.4	105.0	131.8	128.2	101.3	131.8	127.7	116.3	113.5	113.9	107.2	148.0	142.4	107.7	123.1	121.0
0320	Concrete Reinforcing	109.4	115.0	112.2	101.9	115.3	108.6	105.5	115.5	110.5	102.0	114.6	108.3	94.6	116.2	105.4	100.1	114.8	107.4
0330	Cast-in-Place Concrete	94.6	120.9	105.5	110.7	117.6	113.6	99.1	117.6	106.8	84.4	120.7	99.4	121.2	122.0	121.5	98.2	121.0	107.6
03	CONCRETE	100.9	119.9	110.3	115.1	122.6	118.8	105.9	122.6	114.1	93.9	115.4	104.5	109.9	131.2	120.4	102.1	119.8	110.8
04	MASONRY	90.0	119.9	108.5	100.3	118.2	111.4	99.3	122.3	113.5	94.7	115.7	107.7	126.6	131.9	129.9	97.5	116.5	109.3
05	METALS	100.6	102.0	101.0	102.4	101.1	102.0	99.6	101.4	100.2	102.6	99.3	101.6	100.8	100.3	100.7	100.7	100.0	100.5
06	WOOD, PLASTICS & COMPOSITES	105.7	121.4	114.5	99.5	135.9	120.0	95.3	135.9	118.2	105.0	109.2	107.4	98.4	154.0	129.8	99.6	121.5	111.9
07	THERMAL & MOISTURE PROTECTION	100.3	121.1	108.7	101.4	119.3	108.6	101.0	118.5	108.1	100.1	110.2	104.2	101.2	135.9	115.3	103.3	119.4	109.8
08	OPENINGS	97.8	120.2	103.1	99.1	124.7	105.2	98.0	124.7	104.3	92.1	109.5	96.2	96.8	140.5	107.2	94.5	120.2	100.6
0920	Plaster & Gypsum Board	103.5	122.0	116.1	98.3	136.9	124.5	101.0	136.9	125.4	108.9	109.4	109.2	97.4	155.2	136.7	101.8	122.0	115.6
0950, 0980	Ceilings & Acoustic Treatment	111.6	122.0	118.5	111.2	136.9	128.3	106.7	136.9	126.8	107.4	109.4	108.7	107.0	155.2	139.1	109.9	122.0	118.0
0960	Flooring	96.8	114.1	101.9	105.2	111.3	107.0	105.6	113.5	107.9	110.4	109.2	110.0	104.4	126.4	111.0	102.9	114.1	106.2
0970, 0990	Wall Finishes & Painting/Coating	101.1	108.4	105.5	102.0	118.3	111.8	102.0	118.3	111.8	100.5	104.2	102.7	102.7	142.1	126.4	100.5	101.2	100.9
09	FINISHES	102.9	119.8	112.3	105.0	128.9	118.3	104.0	129.2	118.1	104.6	111.3	108.3	103.1	145.5	126.7	102.3	119.1	111.6
COVERS	DIVS. 10 - 14, 25, 28, 41, 43, 44, 46	100.0	111.3	102.3	100.0	121.4	104.3	100.0	121.4	104.3	100.0	110.0	102.0	100.0	125.1	105.0	100.0	111.5	102.3
21, 22, 23	FIRE SUPPRESSION, PLUMBING & HVAC	100.0	118.0	107.2	95.1	109.8	101.0	100.0	108.8	103.5	95.2	108.3	100.4	100.1	144.7	118.0	100.1	118.0	107.3
26, 27, 3370	ELECTRICAL, COMMUNICATIONS & UTIL.	103.6	120.6	112.5	95.4	109.9	103.0	97.9	106.5	102.4	90.5	102.3	96.6	108.3	142.6	126.2	96.2	109.9	103.3
MF2010	WEIGHTED AVERAGE	100.1	116.5	107.3	100.8	115.1	107.0	100.4	114.9	106.7	96.8	108.6	101.9	103.9	133.1	116.6	99.6	114.2	105.9

	DIVISION	CALIFORNIA																	
		PALM SPRINGS			PALO ALTO			PASADENA			REDDING			RICHMOND			RIVERSIDE		
		922			943			910 - 912			960			948			925		
		MAT.	INST.	TOTAL	MAT.	INST.	TOTAL	MAT.	INST.	TOTAL	MAT.	INST.	TOTAL	MAT.	INST.	TOTAL	MAT.	INST.	TOTAL
015433	CONTRACTOR EQUIPMENT		98.8	98.8		98.5	98.5		98.9	98.9		97.6	97.6		98.5	98.5		98.8	98.8
0241, 31 - 34	SITE & INFRASTRUCTURE, DEMOLITION	88.2	105.7	100.5	115.6	106.0	108.8	93.4	109.7	104.8	109.5	104.5	106.0	125.4	106.0	111.8	95.0	105.7	102.5
0310	Concrete Forming & Accessories	102.1	116.9	114.9	105.4	144.5	139.2	104.9	116.9	115.2	105.0	129.6	126.2	120.9	147.8	144.2	105.8	123.1	120.7
0320	Concrete Reinforcing	107.1	114.7	110.9	92.5	115.9	104.2	108.3	114.9	111.6	104.7	115.3	110.0	92.5	116.0	104.2	104.1	114.7	109.4
0330	Cast-in-Place Concrete	87.1	121.6	101.4	108.1	122.0	113.8	89.9	119.1	102.0	110.9	117.6	113.7	124.3	122.0	123.4	94.7	121.7	105.9
03	CONCRETE	97.7	117.2	107.3	99.3	129.6	114.2	99.4	116.4	107.8	115.2	121.5	118.3	112.0	131.0	121.4	103.8	120.0	111.8
04	MASONRY	74.9	115.9	100.3	101.7	127.8	117.8	106.8	118.6	114.1	110.3	118.2	115.2	119.0	127.8	124.4	75.9	115.9	100.7
05	METALS	103.8	99.9	102.6	98.3	99.8	98.8	84.1	99.9	88.9	100.5	101.1	100.7	98.4	99.8	98.8	103.4	100.1	102.4
06	WOOD, PLASTICS & COMPOSITES	93.2	113.3	104.5	95.8	149.6	126.1	83.3	113.1	100.1	97.8	132.8	117.5	114.2	154.0	136.7	98.5	121.4	111.4
07	THERMAL & MOISTURE PROTECTION	97.6	119.0	106.3	100.8	133.4	114.0	95.0	117.2	104.0	108.6	118.9	112.8	101.4	134.2	114.7	98.1	121.2	107.5
08	OPENINGS	96.6	115.5	101.1	96.8	137.5	106.5	89.8	115.5	95.9	102.0	123.0	107.0	96.9	139.9	107.1	99.1	120.2	104.1
0920	Plaster & Gypsum Board	95.5	113.7	107.9	95.6	150.6	133.0	95.0	113.7	107.7	104.5	133.7	124.4	104.9	155.2	139.1	99.9	122.0	115.0
0950, 0980	Ceilings & Acoustic Treatment	104.1	113.7	110.5	105.0	150.6	135.4	102.5	113.7	109.9	131.8	133.7	133.1	105.0	155.2	138.4	111.0	122.0	118.4
0960	Flooring	107.7	109.2	108.1	103.4	126.4	110.3	88.9	109.2	95.0	105.4	111.3	107.2	110.3	126.4	115.1	109.1	114.1	110.6
0970, 0990	Wall Finishes & Painting/Coating	98.7	113.7	107.7	102.7	142.1	126.4	101.7	108.4	105.7	104.1	118.3	112.6	102.7	142.1	126.4	98.7	106.2	103.2
09	FINISHES	100.7	114.7	108.5	101.4	142.9	124.5	95.7	114.0	105.9	110.9	127.1	119.9	105.4	145.5	127.7	103.5	119.6	112.5
COVERS	DIVS. 10 - 14, 25, 28, 41, 43, 44, 46	100.0	110.4	102.1	100.0	124.7	104.9	100.0	110.1	102.0	100.0	121.1	104.2	100.0	125.1	105.0	100.0	111.3	102.3
21, 22, 23	FIRE SUPPRESSION, PLUMBING & HVAC	95.1	113.6	102.5	95.2	143.3	114.5	95.1	113.7	102.6	100.1	109.8	104.0	95.2	144.7	115.0	100.0	118.0	107.2
26, 27, 3370	ELECTRICAL, COMMUNICATIONS & UTIL.	95.0	106.0	100.8	108.1	153.8	132.0	116.8	119.9	118.4	97.8	109.9	104.1	108.8	132.4	121.1	91.6	106.0	99.1
MF2010	WEIGHTED AVERAGE	96.7	111.6	103.2	99.5	133.1	114.2	96.3	113.9	104.0	103.6	114.6	108.4	102.5	131.2	115.0	98.9	114.0	105.5

	DIVISION	CALIFORNIA																	
		SACRAMENTO			SALINAS			SAN BERNARDINO			SAN DIEGO			SAN FRANCISCO			SAN JOSE		
		942,956 - 958			939			923 - 924			919 - 921			940 - 941			951		
		MAT.	INST.	TOTAL	MAT.	INST.	TOTAL	MAT.	INST.	TOTAL	MAT.	INST.	TOTAL	MAT.	INST.	TOTAL	MAT.	INST.	TOTAL
015433	CONTRACTOR EQUIPMENT		98.0	98.0		97.9	97.9		98.8	98.8		99.0	99.0		108.8	108.8		98.5	98.5
0241, 31 - 34	SITE & INFRASTRUCTURE, DEMOLITION	101.0	111.5	108.3	112.3	105.1	107.2	75.1	105.7	96.6	99.7	103.3	102.2	128.6	111.8	116.8	132.0	99.7	109.3
0310	Concrete Forming & Accessories	105.6	132.4	128.7	110.9	132.8	129.8	109.5	116.9	115.9	103.8	112.0	110.9	106.9	149.0	143.2	107.4	147.9	142.3
0320	Concrete Reinforcing	87.6	115.6	101.6	100.6	115.9	108.3	104.1	114.7	109.4	104.9	114.7	109.8	107.8	116.6	112.2	92.8	116.2	104.5
0330	Cast-in-Place Concrete	102.5	118.4	109.1	97.0	117.9	105.7	65.5	121.6	88.8	98.5	108.7	102.7	124.3	123.5	124.0	115.1	121.6	117.8
03	CONCRETE	99.8	122.7	111.1	112.4	123.2	117.7	77.7	117.2	97.2	104.6	110.7	107.6	113.5	132.7	123.0	111.8	131.4	121.5
04	MASONRY	103.4	122.3	115.1	94.4	127.4	114.8	82.4	115.9	103.1	97.1	114.2	107.7	127.2	136.2	132.8	126.8	131.9	130.0
05	METALS	96.4	95.7	96.2	105.3	102.5	104.4	103.3	99.9	102.3	99.6	101.3	100.1	107.1	110.1	108.0	97.8	107.1	100.6
06	WOOD, PLASTICS & COMPOSITES	92.3	136.1	117.0	104.4	136.1	122.3	102.3	113.3	108.5	97.0	108.8	103.6	98.4	154.2	129.9	107.5	153.8	133.6
07	THERMAL & MOISTURE PROTECTION	108.5	121.1	113.6	100.6	125.9	110.9	96.8	119.0	105.8	99.7	106.6	102.5	102.9	139.0	117.6	97.4	138.4	114.1
08	OPENINGS	110.7	124.8	114.1	95.6	130.7	103.9	96.6	115.5	101.1	96.8	111.4	100.3	101.4	140.6	110.7	89.5	140.3	101.5
0920	Plaster & Gypsum Board	92.5	136.9	122.7	102.9	137.1	126.2	101.7	113.7	109.9	102.3	108.9	106.8	99.8	155.2	137.5	96.2	155.2	136.3
0950, 0980	Ceilings & Acoustic Treatment	105.0	136.9	126.2	107.4	137.1	127.2	106.7	113.7	111.4	109.6	108.9	109.1	114.8	155.2	141.7	104.8	155.2	138.3
0960	Flooring	103.3	119.7	108.2	105.1	123.3	110.6	111.0	109.2	110.4	96.2	114.1	101.5	104.4	126.4	111.0	98.7	126.4	107.0
0970, 0990	Wall Finishes & Painting/Coating	100.3	118.3	111.1	101.6	142.1	126.0	98.7	106.2	103.2	98.3	106.4	103.2	102.7	152.6	132.7	102.3	142.1	126.2
09	FINISHES	99.8	130.4	116.9	104.2	133.7	120.6	101.8	113.9	108.6	101.0	111.3	106.7	105.2	146.8	128.4	102.7	145.3	126.4
COVERS	DIVS. 10 - 14, 25, 28, 41, 43, 44, 46	100.0	121.9	104.4	100.0	121.5	104.3	100.0	110.2	102.1	100.0	109.1	101.8	100.0	125.6	105.1	100.0	124.7	105.0
21, 22, 23	FIRE SUPPRESSION, PLUMBING & HVAC	100.0	121.9	108.7	95.2	115.3	103.2	95.1	113.8	102.6	100.0	116.3	106.5	100.0	171.6	128.7	100.0	143.3	117.3
26, 27, 3370	ELECTRICAL, COMMUNICATIONS & UTIL.	103.4	111.3	107.5	91.7	122.8	107.9	95.0	103.7	99.6	102.2	98.9	100.5	108.3	161.4	136.1	101.0	153.8	128.6
MF2010	WEIGHTED AVERAGE	101.2	118.6	108.8	100.2	120.3	108.9	94.6	111.2	101.9	100.2	109.1	104.1	106.1	143.8	122.5	102.2	134.6	116.3

	DIVISION	CALIFORNIA																	
		SAN LUIS OBISPO			SAN MATEO			SAN RAFAEL			SANTA ANA			SANTA BARBARA			SANTA CRUZ		
		934			944			949			926 - 927			931			950		
		MAT.	INST.	TOTAL	MAT.	INST.	TOTAL	MAT.	INST.	TOTAL	MAT.	INST.	TOTAL	MAT.	INST.	TOTAL	MAT.	INST.	TOTAL
015433	CONTRACTOR EQUIPMENT		97.9	97.9		98.5	98.5		98.5	98.5		98.8	98.8		97.9	97.9		98.5	98.5
0241, 31 - 34	SITE & INFRASTRUCTURE, DEMOLITION	104.7	105.3	105.1	122.8	106.0	111.0	113.7	111.6	112.2	86.7	105.7	100.0	99.6	105.3	103.6	131.6	99.5	109.0
0310	Concrete Forming & Accessories	118.0	116.9	117.0	111.2	147.9	142.9	115.2	147.8	143.4	109.7	116.9	115.9	108.6	123.1	121.1	107.5	133.0	129.5
0320	Concrete Reinforcing	102.0	114.8	108.4	92.5	116.1	104.3	93.1	116.1	104.6	107.6	114.7	111.2	100.1	114.8	107.4	115.0	115.9	115.5
0330	Cast-in-Place Concrete	104.2	120.8	111.1	120.4	122.0	121.0	140.1	121.2	132.2	83.6	121.6	99.4	97.8	120.9	107.4	114.3	119.8	116.6
03	CONCRETE	110.1	116.9	113.4	108.4	131.1	119.6	129.7	130.6	130.2	95.2	117.2	106.1	102.0	119.7	110.7	115.0	124.1	119.5
04	MASONRY	96.1	117.4	109.3	118.7	130.8	126.2	98.0	130.8	118.3	72.6	116.2	99.6	94.9	117.4	108.8	130.4	127.5	128.6
05	METALS	103.2	99.7	102.1	98.2	100.1	98.8	99.6	97.8	99.1	103.4	99.9	102.3	101.1	100.0	100.8	105.1	105.8	105.3
06	WOOD, PLASTICS & COMPOSITES	107.5	113.4	110.8	103.5	154.0	132.0	100.4	153.8	130.5	103.9	113.3	109.2	99.6	121.5	111.9	107.5	136.2	123.7
07	THERMAL & MOISTURE PROTECTION	100.8	117.5	107.6	101.2	136.2	115.4	105.2	132.9	116.5	97.9	119.1	106.5	100.2	119.0	107.9	97.4	128.5	110.0
08	OPENINGS	93.9	111.8	98.1	96.8	139.9	107.0	108.2	139.8	115.7	96.0	115.5	100.6	95.2	120.2	101.1	90.7	130.8	100.2
0920	Plaster & Gypsum Board	110.4	113.7	112.6	100.8	155.2	137.8	102.5	155.2	138.4	102.4	113.7	110.1	101.8	122.0	115.6	103.7	137.1	126.4
0950, 0980	Ceilings & Acoustic Treatment	107.4	113.7	111.6	105.0	155.2	138.4	113.0	155.2	141.1	106.7	113.7	111.4	109.9	122.0	118.0	108.5	137.1	127.5
0960	Flooring	111.3	109.2	110.7	106.0	126.4	112.1	115.1	121.3	117.0	111.5	109.2	110.8	104.4	109.2	105.8	102.8	123.3	108.9
0970, 0990	Wall Finishes & Painting/Coating	100.5	101.1	100.9	102.7	142.1	126.4	99.0	142.1	124.9	98.7	106.2	103.2	100.5	101.2	100.9	102.5	142.1	126.3
09	FINISHES	106.1	112.6	109.7	103.4	145.5	126.8	105.8	144.4	127.3	103.3	113.9	109.2	102.9	118.2	111.4	105.7	133.8	121.3
COVERS	DIVS. 10 - 14, 25, 28, 41, 43, 44, 46	100.0	120.4	104.1	100.0	125.2	105.0	100.0	124.5	104.9	100.0	110.4	102.1	100.0	111.5	102.3	100.0	121.8	104.4
21, 22, 23	FIRE SUPPRESSION, PLUMBING & HVAC	95.2	113.8	102.6	95.2	142.8	114.3	95.1	164.7	123.0	95.1	113.8	102.6	100.1	118.0	107.3	100.0	115.3	106.1
26, 27, 3370	ELECTRICAL, COMMUNICATIONS & UTIL.	90.5	106.0	98.6	108.1	143.6	126.7	105.1	116.4	111.0	95.0	105.9	100.7	89.4	111.8	101.1	99.9	122.8	111.9
MF2010	WEIGHTED AVERAGE	99.3	111.6	104.7	101.7	132.7	115.2	104.2	133.6	117.0	96.5	111.6	103.1	98.9	114.6	105.7	104.1	120.4	111.2

	DIVISION	CALIFORNIA															COLORADO		
		SANTA ROSA			STOCKTON			SUSANVILLE			VALLEJO			VAN NUYS			ALAMOSA		
		954			952			961			945			913 - 916			811		
		MAT.	INST.	TOTAL	MAT.	INST.	TOTAL	MAT.	INST.	TOTAL	MAT.	INST.	TOTAL	MAT.	INST.	TOTAL	MAT.	INST.	TOTAL
015433	CONTRACTOR EQUIPMENT		98.2	98.2		97.6	97.6		97.6	97.6		98.5	98.5		98.9	98.9		93.8	93.8
0241, 31 - 34	SITE & INFRASTRUCTURE, DEMOLITION	100.1	104.5	103.2	99.4	104.5	103.0	115.9	104.5	107.9	100.5	111.4	108.2	109.9	109.7	109.7	132.1	89.0	101.9
0310	Concrete Forming & Accessories	103.1	146.8	140.8	105.3	131.8	128.2	106.2	129.6	126.4	105.6	146.5	140.9	110.9	116.9	116.1	105.2	69.0	73.9
0320	Concrete Reinforcing	102.7	116.4	109.6	105.5	115.5	110.5	104.7	115.3	110.0	94.3	116.3	105.3	108.3	114.9	111.6	108.0	78.4	93.2
0330	Cast-in-Place Concrete	108.7	119.2	113.1	96.6	117.6	105.3	100.9	117.6	107.8	111.6	119.6	114.9	94.8	119.1	104.9	100.6	79.3	91.8
03	CONCRETE	114.9	130.0	122.4	104.9	122.6	113.6	117.6	121.5	119.5	105.2	129.5	117.2	113.8	116.4	115.1	115.5	74.6	95.3
04	MASONRY	98.6	133.7	120.3	99.2	122.3	113.5	108.3	118.2	114.4	74.6	132.4	110.4	122.3	118.6	120.0	123.2	72.7	92.0
05	METALS	103.7	104.3	103.9	99.8	101.4	100.3	99.6	101.1	100.1	99.6	97.6	99.0	83.2	99.9	88.3	97.2	81.1	92.2
06	WOOD, PLASTICS & COMPOSITES	94.6	153.6	127.9	100.8	135.9	120.6	99.5	132.8	118.3	89.8	153.8	125.9	91.6	113.1	103.8	93.8	69.0	79.8
07	THERMAL & MOISTURE PROTECTION	98.3	134.5	113.0	101.1	118.5	108.2	109.2	118.9	113.2	103.0	134.0	115.6	95.8	117.2	104.5	105.1	78.4	94.2
08	OPENINGS	97.4	140.2	107.6	98.0	124.7	104.3	101.9	123.0	106.9	110.0	140.3	117.2	89.7	115.5	95.8	98.4	75.3	92.9
0920	Plaster & Gypsum Board	97.2	155.2	136.7	101.0	136.9	125.4	105.4	133.7	124.7	97.3	155.2	136.7	99.9	113.7	109.3	75.8	67.9	70.4
0950, 0980	Ceilings & Acoustic Treatment	106.7	155.2	139.0	113.7	136.9	129.1	123.9	133.7	130.4	115.0	155.2	141.8	99.8	113.7	109.1	104.2	67.9	80.1
0960	Flooring	108.1	117.0	110.8	105.6	113.5	107.9	105.9	111.3	107.5	111.2	126.4	115.7	91.1	109.2	96.5	112.4	55.8	95.5
0970, 0990	Wall Finishes & Painting/Coating	98.7	142.1	124.8	102.0	118.3	111.8	104.1	118.3	112.6	99.9	142.1	125.3	101.7	108.4	105.7	113.0	24.1	59.6
09	FINISHES	102.6	143.0	125.1	105.5	129.2	118.7	110.4	127.1	119.7	102.8	144.7	126.1	97.7	114.0	106.8	104.9	61.2	80.5
COVERS	DIVS. 10 - 14, 25, 28, 41, 43, 44, 46	100.0	123.2	104.7	100.0	121.4	104.3	100.0	121.1	104.2	100.0	123.7	104.7	100.0	110.1	102.0	100.0	89.1	97.8
21, 22, 23	FIRE SUPPRESSION, PLUMBING & HVAC	95.1	163.4	122.5	100.0	108.8	103.5	95.2	109.8	101.0	100.0	128.8	111.6	95.1	113.7	102.6	95.2	72.6	86.1
26, 27, 3370	ELECTRICAL, COMMUNICATIONS & UTIL.	95.4	116.6	106.4	97.9	110.9	104.7	98.1	109.9	104.3	100.6	125.1	113.4	116.8	119.9	118.4	96.8	74.2	85.0
MF2010	WEIGHTED AVERAGE	100.3	133.4	114.7	100.5	115.5	107.0	102.6	114.6	107.8	100.6	127.2	112.2	99.1	113.9	105.5	102.0	74.6	90.1

	DIVISION	COLORADO																	
		BOULDER			COLORADO SPRINGS			DENVER			DURANGO			FORT COLLINS			FORT MORGAN		
		803			808 - 809			800 - 802			813			805			807		
		MAT.	INST.	TOTAL	MAT.	INST.	TOTAL	MAT.	INST.	TOTAL	MAT.	INST.	TOTAL	MAT.	INST.	TOTAL	MAT.	INST.	TOTAL
015433	CONTRACTOR EQUIPMENT		94.9	94.9		93.4	93.4		97.7	97.7		93.8	93.8		94.9	94.9		94.9	94.9
0241, 31 - 34	SITE & INFRASTRUCTURE, DEMOLITION	92.4	95.1	94.3	94.5	93.1	93.5	93.7	101.2	99.0	125.8	89.0	100.0	104.3	94.7	97.5	95.3	94.6	94.8
0310	Concrete Forming & Accessories	102.0	81.7	84.4	92.9	77.4	79.5	98.5	77.7	80.5	111.4	69.2	74.9	99.3	76.4	79.5	102.5	76.6	80.2
0320	Concrete Reinforcing	102.0	78.5	90.3	101.2	82.7	92.0	101.2	82.7	92.0	108.0	78.4	93.2	102.1	78.6	90.3	102.2	78.4	90.3
0330	Cast-in-Place Concrete	102.4	79.8	93.0	105.2	82.9	95.9	99.6	82.6	92.5	115.7	79.4	100.6	115.6	78.6	100.2	100.5	78.6	91.4
03	CONCRETE	104.7	80.6	92.8	108.1	80.6	94.6	102.4	80.6	91.7	117.8	74.8	96.5	115.3	77.8	96.8	103.2	77.8	90.7
04	MASONRY	93.1	73.2	80.8	94.1	73.0	81.0	96.1	73.3	82.0	110.9	72.4	87.1	110.4	76.9	89.7	107.5	72.9	86.1
05	METALS	97.7	84.4	93.6	100.8	86.5	96.4	103.4	86.7	98.2	97.2	81.2	92.2	98.9	81.3	93.5	97.5	81.1	92.4
06	WOOD, PLASTICS & COMPOSITES	93.3	84.9	88.6	83.8	79.1	81.2	91.0	79.0	84.3	102.6	69.0	83.7	90.8	79.0	84.1	93.3	79.0	85.2
07	THERMAL & MOISTURE PROTECTION	104.4	80.5	94.7	105.3	80.3	95.1	103.7	74.6	91.9	105.0	78.3	94.2	104.8	72.2	91.6	104.4	79.9	94.4
08	OPENINGS	101.5	83.9	97.3	105.8	82.0	100.2	106.5	81.9	100.7	105.3	75.3	98.2	101.4	80.7	96.5	101.4	80.7	96.5
0920	Plaster & Gypsum Board	104.0	84.6	90.8	86.1	78.6	81.0	99.6	78.6	85.3	88.1	67.9	74.3	98.0	78.6	84.8	104.0	78.6	86.7
0950, 0980	Ceilings & Acoustic Treatment	98.0	84.6	89.1	105.1	78.6	87.4	108.2	78.6	88.5	104.2	67.9	80.1	98.0	78.6	85.0	98.0	78.6	85.0
0960	Flooring	103.7	85.3	98.2	95.8	69.5	87.9	100.5	86.3	96.3	116.8	55.8	98.5	100.4	55.8	87.1	104.1	55.8	89.7
0970, 0990	Wall Finishes & Painting/Coating	100.8	68.0	81.1	100.5	41.4	65.0	100.8	77.3	86.7	113.0	24.1	59.6	100.8	41.1	64.9	100.8	54.7	73.1
09	FINISHES	102.6	81.2	90.7	99.4	72.0	84.2	103.1	79.0	89.7	106.9	61.2	81.5	101.4	68.9	83.3	102.6	70.4	84.7
COVERS	DIVS. 10 - 14, 25, 28, 41, 43, 44, 46	100.0	90.0	98.0	100.0	89.7	97.9	100.0	89.4	97.9	100.0	89.1	97.8	100.0	89.4	97.9	100.0	89.4	97.9
21, 22, 23	FIRE SUPPRESSION, PLUMBING & HVAC	95.2	77.6	88.1	100.2	72.7	89.2	100.0	79.9	91.9	95.2	84.9	91.0	100.1	77.5	91.0	95.2	77.5	88.1
26, 27, 3370	ELECTRICAL, COMMUNICATIONS & UTIL.	96.9	85.2	90.8	100.5	83.0	91.4	102.1	85.2	93.3	96.2	71.1	83.1	96.9	85.1	90.7	97.2	85.1	90.9
MF2010	WEIGHTED AVERAGE	98.6	82.0	91.3	101.3	79.3	91.7	101.6	82.6	93.3	102.4	76.8	91.2	102.1	79.6	92.3	99.1	79.6	90.6

	DIVISION	COLORADO																	
		GLENWOOD SPRINGS			GOLDEN			GRAND JUNCTION			GREELEY			MONTROSE			PUEBLO		
		816			804			815			806			814			810		
		MAT.	INST.	TOTAL	MAT.	INST.	TOTAL	MAT.	INST.	TOTAL	MAT.	INST.	TOTAL	MAT.	INST.	TOTAL	MAT.	INST.	TOTAL
015433	CONTRACTOR EQUIPMENT		96.5	96.5		94.9	94.9		96.5	96.5		94.9	94.9		95.1	95.1		93.8	93.8
0241, 31 - 34	SITE & INFRASTRUCTURE, DEMOLITION	140.8	95.9	109.3	104.4	94.9	97.7	125.5	95.5	104.5	91.7	94.1	93.4	134.6	92.1	104.8	117.7	91.5	99.3
0310	Concrete Forming & Accessories	102.3	76.7	80.2	95.2	76.5	79.1	110.1	76.2	80.8	97.5	80.2	82.5	101.8	76.4	79.9	107.3	77.8	81.8
0320	Concrete Reinforcing	106.9	78.4	92.7	102.2	78.4	90.3	107.3	78.3	92.8	102.0	77.0	89.5	106.8	78.3	92.6	103.3	82.7	93.1
0330	Cast-in-Place Concrete	100.6	78.8	91.5	100.5	78.6	91.4	111.4	78.1	97.5	96.6	58.5	80.8	100.6	78.7	91.5	99.9	83.7	93.2
03	CONCRETE	120.6	77.9	99.6	113.8	77.8	96.0	114.1	77.4	96.0	99.9	72.2	86.3	111.6	77.7	94.9	104.3	81.1	92.8
04	MASONRY	97.5	72.9	82.3	110.3	72.9	87.2	129.6	72.7	94.4	104.6	49.7	70.6	103.9	72.7	84.6	94.1	73.2	81.2
05	METALS	96.9	81.8	92.2	97.7	81.2	92.6	98.4	80.5	92.9	98.9	78.5	92.6	96.0	80.8	91.3	100.0	87.4	96.1
06	WOOD, PLASTICS & COMPOSITES	89.1	79.1	83.4	86.2	79.0	82.1	100.4	79.1	88.4	88.4	84.9	86.4	90.2	79.2	84.0	96.3	79.4	86.8
07	THERMAL & MOISTURE PROTECTION	105.0	79.5	94.6	105.3	74.2	92.7	104.1	68.8	89.8	104.3	65.0	88.3	105.1	79.5	94.7	103.6	80.0	94.0
08	OPENINGS	104.3	80.8	98.7	101.4	80.7	96.5	105.0	80.8	99.3	101.4	83.9	97.3	105.5	80.8	99.7	100.1	82.1	95.9
0920	Plaster & Gypsum Board	111.5	78.6	89.1	95.8	78.6	84.1	123.4	78.6	92.9	96.5	84.6	88.4	75.0	78.6	77.4	80.2	78.6	79.1
0950, 0980	Ceilings & Acoustic Treatment	103.4	78.6	86.9	98.0	78.6	85.0	103.4	78.6	86.9	98.0	84.6	89.1	104.2	78.6	87.2	113.1	78.6	90.1
0960	Flooring	111.7	51.4	93.7	98.3	55.8	85.6	116.2	55.8	98.1	99.5	55.8	86.4	114.6	46.2	94.2	113.4	86.3	105.3
0970, 0990	Wall Finishes & Painting/Coating	112.9	54.7	77.9	100.8	68.0	81.1	112.9	68.0	85.9	100.8	25.6	55.6	113.0	24.1	59.6	113.0	38.4	68.1
09	FINISHES	109.9	69.8	87.5	101.3	72.6	85.3	110.8	72.0	89.2	100.1	70.7	83.7	105.4	65.3	83.1	105.8	74.9	88.6
COVERS	DIVS. 10 - 14, 25, 28, 41, 43, 44, 46	100.0	89.5	97.9	100.0	89.4	97.9	100.0	89.5	97.9	100.0	90.0	98.0	100.0	89.8	98.0	100.0	90.3	98.1
21, 22, 23	FIRE SUPPRESSION, PLUMBING & HVAC	95.2	84.8	91.0	95.2	77.0	87.9	100.0	84.3	93.7	100.1	77.5	91.0	95.2	84.8	91.0	100.0	72.8	89.1
26, 27, 3370	ELECTRICAL, COMMUNICATIONS & UTIL.	93.5	71.1	81.8	97.2	85.1	90.9	95.9	55.1	74.6	96.9	85.1	90.7	95.9	57.7	76.0	96.9	74.3	85.1
MF2010	WEIGHTED AVERAGE	102.2	79.3	92.2	100.6	79.6	91.4	104.4	76.7	92.4	99.7	76.1	89.4	101.2	76.5	90.5	100.9	78.6	91.2

DIVISION		COLORADO			CONNECTICUT															
		SALIDA			BRIDGEPORT			BRISTOL			HARTFORD			MERIDEN			NEW BRITAIN		
		812			066			060			061			064			060		
		MAT.	INST.	TOTAL	MAT.	INST.	TOTAL	MAT.	INST.	TOTAL	MAT.	INST.	TOTAL	MAT.	INST.	TOTAL	MAT.	INST.	TOTAL
015433	CONTRACTOR EQUIPMENT		95.1	95.1		100.6	100.6		100.6	100.6		100.6	100.6		101.1	101.1		100.6	100.6
0241, 31 - 34	SITE & INFRASTRUCTURE, DEMOLITION	125.5	92.4	102.3	111.7	104.0	106.3	110.9	104.0	106.1	106.0	104.0	104.6	108.9	104.7	105.9	111.1	104.0	106.1
0310	Concrete Forming & Accessories	110.0	76.4	81.0	98.2	123.8	120.3	98.2	123.7	120.3	95.1	123.7	119.8	97.9	123.7	120.2	98.6	123.7	120.3
0320	Concrete Reinforcing	106.6	78.3	92.5	110.0	129.3	119.7	110.0	129.3	119.7	109.5	129.3	119.4	110.0	129.3	119.7	110.0	129.3	119.7
0330	Cast-in-Place Concrete	115.2	78.7	100.1	103.7	127.5	113.6	97.2	127.4	109.8	99.0	127.4	110.8	93.5	127.4	107.6	98.8	127.4	110.7
03	CONCRETE	112.8	77.7	95.5	105.3	125.8	115.4	102.2	125.7	113.8	102.8	125.8	114.1	100.5	125.7	112.9	103.0	125.7	114.2
04	MASONRY	131.2	72.7	95.0	98.1	134.8	120.8	90.8	134.8	118.0	94.7	134.8	119.5	90.4	134.8	117.9	92.1	134.8	118.6
05	METALS	95.7	80.8	91.1	102.0	125.1	109.1	102.0	125.1	109.1	107.8	125.1	113.1	99.5	125.1	107.4	98.6	125.1	106.8
06	WOOD, PLASTICS & COMPOSITES	97.4	79.2	87.2	99.1	122.9	112.5	99.1	122.9	112.5	92.6	122.9	109.7	99.1	122.9	112.5	99.1	122.9	112.5
07	THERMAL & MOISTURE PROTECTION	104.1	79.5	94.1	100.3	129.2	112.0	100.4	126.9	111.2	104.8	126.9	113.8	100.4	126.9	111.2	100.4	126.9	111.2
08	OPENINGS	98.5	80.8	94.3	107.0	132.2	113.0	107.0	132.2	113.0	106.1	132.2	112.2	109.4	132.2	114.8	107.0	132.2	113.0
0920	Plaster & Gypsum Board	75.4	78.6	77.5	99.9	123.0	115.6	99.9	123.0	115.6	95.4	123.0	114.2	101.5	123.0	116.1	99.9	123.0	115.6
0950, 0980	Ceilings & Acoustic Treatment	104.2	78.6	87.2	97.9	123.0	114.6	97.9	123.0	114.6	101.3	123.0	115.7	102.1	123.0	116.0	97.9	123.0	114.6
0960	Flooring	119.1	46.2	97.3	99.3	133.6	109.6	99.3	133.6	109.6	97.8	133.6	108.5	99.3	133.6	109.6	99.3	133.6	109.6
0970, 0990	Wall Finishes & Painting/Coating	113.0	24.1	59.6	96.4	122.3	112.0	96.4	122.3	112.0	97.3	122.3	112.3	96.4	122.3	112.0	96.4	122.3	112.0
09	FINISHES	105.6	65.3	83.2	100.8	125.1	114.3	100.8	125.1	114.3	99.7	125.1	113.8	101.9	125.1	114.8	100.8	125.1	114.3
COVERS	DIVS. 10 - 14, 25, 28, 41, 43, 44, 46	100.0	89.9	98.0	100.0	110.4	102.1	100.0	110.5	102.1	100.0	110.5	102.1	100.0	110.5	102.1	100.0	110.5	102.1
21, 22, 23	FIRE SUPPRESSION, PLUMBING & HVAC	95.2	72.6	86.1	100.0	116.0	106.4	100.0	116.0	106.4	100.0	116.0	106.4	95.1	116.0	103.5	100.0	116.0	106.4
26, 27, 3370	ELECTRICAL, COMMUNICATIONS & UTIL.	96.1	74.2	84.7	99.2	108.7	104.2	99.2	111.3	105.5	98.6	112.3	105.7	99.2	111.3	105.5	99.3	111.3	105.6
MF2010	WEIGHTED AVERAGE	101.7	76.2	90.6	101.8	120.2	109.8	101.1	120.4	109.5	102.0	120.6	110.1	99.6	120.5	108.7	100.8	120.4	109.3

DIVISION		CONNECTICUT																	
		NEW HAVEN			NEW LONDON			NORWALK			STAMFORD			WATERBURY			WILLIMANTIC		
		065			063			068			069			067			062		
		MAT.	INST.	TOTAL	MAT.	INST.	TOTAL	MAT.	INST.	TOTAL	MAT.	INST.	TOTAL	MAT.	INST.	TOTAL	MAT.	INST.	TOTAL
015433	CONTRACTOR EQUIPMENT		101.1	101.1		101.1	101.1		100.6	100.6		100.6	100.6		100.6	100.6		100.6	100.6
0241, 31 - 34	SITE & INFRASTRUCTURE, DEMOLITION	110.9	104.7	106.5	103.1	104.7	104.2	111.5	104.0	106.2	112.2	104.0	106.4	111.4	104.0	106.2	111.6	104.0	106.3
0310	Concrete Forming & Accessories	98.0	123.7	120.2	97.9	123.7	120.2	98.2	124.2	120.6	98.2	124.2	120.7	98.2	123.7	120.2	98.2	123.7	120.2
0320	Concrete Reinforcing	110.0	129.3	119.7	86.3	129.3	107.7	110.0	129.4	119.7	110.0	129.4	119.7	110.0	129.3	119.7	110.0	129.3	119.7
0330	Cast-in-Place Concrete	100.5	127.4	111.7	85.7	127.4	103.0	102.0	128.9	113.2	103.7	128.9	114.2	103.7	127.4	113.6	96.9	127.4	109.6
03	CONCRETE	117.2	125.7	121.4	90.2	125.7	107.7	104.5	126.5	115.3	105.3	126.5	115.8	105.3	125.7	115.4	102.1	125.7	113.7
04	MASONRY	91.0	134.8	118.1	89.4	134.8	117.5	90.6	134.8	118.0	91.3	134.8	118.2	91.3	134.8	118.2	90.6	134.8	118.0
05	METALS	98.8	125.1	106.9	98.5	125.1	106.7	102.0	125.5	109.3	102.0	125.6	109.3	102.0	125.0	109.1	101.8	125.0	108.9
06	WOOD, PLASTICS & COMPOSITES	99.1	122.9	112.5	99.1	122.9	112.5	99.1	122.9	112.5	99.1	122.9	112.5	99.1	122.9	112.5	99.1	122.9	112.5
07	THERMAL & MOISTURE PROTECTION	100.5	126.7	111.1	100.4	126.9	111.1	100.4	129.4	112.2	100.4	129.4	112.2	100.4	126.7	111.1	100.6	125.7	110.8
08	OPENINGS	107.0	132.2	113.0	109.9	132.2	115.2	107.0	132.2	113.0	107.0	132.2	113.0	107.0	132.2	113.0	110.0	132.2	115.2
0920	Plaster & Gypsum Board	99.9	123.0	115.6	99.9	123.0	115.6	99.9	123.0	115.6	99.9	123.0	115.6	99.9	123.0	115.6	99.9	123.0	115.6
0950, 0980	Ceilings & Acoustic Treatment	97.9	123.0	114.6	95.9	123.0	113.9	97.9	123.0	114.6	97.9	123.0	114.6	97.9	123.0	114.6	95.9	123.0	113.9
0960	Flooring	99.3	133.6	109.6	99.3	133.6	109.6	99.3	133.6	109.6	99.3	133.6	109.6	99.3	133.6	109.6	99.3	135.7	110.2
0970, 0990	Wall Finishes & Painting/Coating	96.4	122.3	112.0	96.4	122.3	112.0	96.4	122.3	112.0	96.4	122.3	112.0	96.4	122.3	112.0	96.4	122.3	112.0
09	FINISHES	100.8	125.1	114.4	99.8	125.1	113.9	100.8	125.1	114.4	100.9	125.1	114.4	100.7	125.1	114.3	100.5	125.4	114.4
COVERS	DIVS. 10 - 14, 25, 28, 41, 43, 44, 46	100.0	110.5	102.1	100.0	110.5	102.1	100.0	110.6	102.1	100.0	110.7	102.1	100.0	110.4	102.1	100.0	110.5	102.1
21, 22, 23	FIRE SUPPRESSION, PLUMBING & HVAC	100.0	116.0	106.4	95.1	116.0	103.5	100.0	116.1	106.4	100.0	116.1	106.4	100.0	116.0	106.4	100.0	116.0	106.4
26, 27, 3370	ELECTRICAL, COMMUNICATIONS & UTIL.	99.2	111.3	105.5	95.9	111.3	103.9	99.2	156.4	129.1	99.2	163.8	132.9	98.7	108.7	103.9	99.2	112.2	106.0
MF2010	WEIGHTED AVERAGE	102.2	120.5	110.2	97.7	120.5	107.6	101.4	126.9	112.5	101.5	127.9	113.0	101.4	120.1	109.6	101.4	120.6	109.7

DIVISION		D.C.			DELAWARE									FLORIDA					
		WASHINGTON			DOVER			NEWARK			WILMINGTON			DAYTONA BEACH			FORT LAUDERDALE		
		200 - 205			199			197			198			321			333		
		MAT.	INST.	TOTAL	MAT.	INST.	TOTAL	MAT.	INST.	TOTAL	MAT.	INST.	TOTAL	MAT.	INST.	TOTAL	MAT.	INST.	TOTAL
015433	CONTRACTOR EQUIPMENT		104.4	104.4		117.0	117.0		117.0	117.0		117.2	117.2		98.8	98.8		91.7	91.7
0241, 31 - 34	SITE & INFRASTRUCTURE, DEMOLITION	106.8	94.0	97.8	99.7	112.3	108.6	99.1	112.3	108.4	92.4	112.6	106.6	107.0	90.4	95.4	93.6	78.3	82.9
0310	Concrete Forming & Accessories	99.5	79.2	81.9	97.1	101.5	100.9	97.2	101.5	100.9	98.3	101.5	101.1	97.3	69.7	73.5	95.5	70.1	73.6
0320	Concrete Reinforcing	109.0	91.3	100.2	98.5	105.2	101.9	93.8	105.2	99.5	96.4	105.2	100.8	92.8	79.0	85.9	89.8	74.4	82.1
0330	Cast-in-Place Concrete	121.4	88.1	107.5	99.1	100.6	99.7	84.4	100.6	91.1	88.9	100.6	93.8	86.6	72.0	80.5	90.9	78.4	85.7
03	CONCRETE	109.6	85.8	97.9	99.9	102.8	101.3	92.1	102.8	97.4	94.8	102.8	98.7	88.2	73.4	80.9	90.7	75.0	82.9
04	MASONRY	95.9	80.0	86.1	103.3	97.2	99.5	99.9	97.2	98.2	103.6	97.2	99.6	92.3	66.8	76.5	89.9	69.4	77.3
05	METALS	100.1	107.3	102.3	103.1	117.0	107.4	103.7	117.0	107.8	102.9	117.0	107.3	96.4	92.8	95.3	95.5	91.5	94.2
06	WOOD, PLASTICS & COMPOSITES	103.5	77.2	88.7	95.0	101.1	98.4	97.6	101.1	99.6	98.2	101.1	99.8	99.7	70.8	83.4	96.0	68.7	80.6
07	THERMAL & MOISTURE PROTECTION	102.4	86.0	95.8	101.9	110.2	105.3	104.2	110.2	106.7	102.2	110.2	105.5	99.6	74.9	89.6	96.9	82.3	91.0
08	OPENINGS	100.3	87.6	97.3	94.3	110.5	98.1	93.4	110.5	97.4	96.6	110.5	99.9	98.9	69.1	91.8	96.9	67.1	89.9
0920	Plaster & Gypsum Board	105.8	76.5	85.9	105.9	101.0	102.6	103.4	101.0	101.8	104.8	101.0	102.2	95.1	70.3	78.3	98.5	68.2	77.9
0950, 0980	Ceilings & Acoustic Treatment	107.8	76.5	87.0	104.2	101.0	102.0	95.0	101.0	99.0	102.2	101.0	101.4	90.7	70.3	77.2	91.3	68.2	75.9
0960	Flooring	104.7	95.3	101.9	99.1	109.0	102.1	95.5	109.0	99.6	101.4	109.0	103.7	115.2	74.9	103.2	115.4	70.6	102.0
0970, 0990	Wall Finishes & Painting/Coating	104.6	84.0	92.2	97.3	105.3	102.1	99.5	105.3	103.0	97.3	105.3	102.1	99.4	75.7	85.2	96.5	71.7	81.6
09	FINISHES	101.6	81.7	90.5	101.4	102.8	102.2	97.4	102.8	100.4	102.0	102.8	102.4	100.8	70.9	84.1	99.4	69.6	82.8
COVERS	DIVS. 10 - 14, 25, 28, 41, 43, 44, 46	100.0	98.4	99.7	100.0	89.7	97.9	100.0	89.7	97.9	100.0	89.7	97.9	100.0	83.5	96.7	100.0	86.5	97.3
21, 22, 23	FIRE SUPPRESSION, PLUMBING & HVAC	100.1	92.9	97.2	100.0	115.8	106.3	100.2	115.8	106.5	100.2	115.8	106.4	100.0	77.8	91.1	100.0	68.0	87.2
26, 27, 3370	ELECTRICAL, COMMUNICATIONS & UTIL.	102.7	102.6	102.6	95.2	109.0	102.4	97.7	109.0	103.6	96.6	109.0	103.1	94.4	56.2	74.5	92.6	72.7	82.2
MF2010	WEIGHTED AVERAGE	101.6	91.6	97.2	99.7	108.1	103.3	98.8	108.1	102.8	99.4	108.1	103.2	97.4	74.4	87.3	96.4	73.9	86.6

	DIVISION	FLORIDA																	
		FORT MYERS			GAINESVILLE			JACKSONVILLE			LAKELAND			MELBOURNE			MIAMI		
		339,341			326,344			320,322			338			329			330 - 332,340		
		MAT.	INST.	TOTAL	MAT.	INST.	TOTAL	MAT.	INST.	TOTAL	MAT.	INST.	TOTAL	MAT.	INST.	TOTAL	MAT.	INST.	TOTAL
015433	CONTRACTOR EQUIPMENT		98.8	98.8		98.8	98.8		98.8	98.8		98.8	98.8		98.8	98.8		91.7	91.7
0241, 31 - 34	SITE & INFRASTRUCTURE, DEMOLITION	104.3	89.5	93.9	115.4	89.5	97.2	107.0	89.8	94.9	106.2	89.9	94.7	113.9	89.7	96.9	96.7	78.1	83.6
0310	Concrete Forming & Accessories	91.1	76.6	78.6	92.3	55.5	60.5	97.1	55.9	61.5	87.4	77.1	78.5	93.5	71.4	74.4	100.7	70.5	74.6
0320	Concrete Reinforcing	90.8	95.7	93.3	98.5	66.9	82.7	92.8	67.0	79.9	93.1	96.6	94.8	93.9	79.0	86.5	95.9	74.5	85.2
0330	Cast-in-Place Concrete	94.8	69.4	84.3	99.4	63.6	84.5	87.4	69.5	80.0	97.0	70.8	86.1	104.3	74.3	91.9	91.7	79.4	86.6
03	CONCRETE	91.3	78.6	85.1	98.7	62.0	80.6	88.6	64.2	76.6	93.0	79.5	86.3	99.0	75.0	87.1	92.4	75.5	84.0
04	MASONRY	84.1	62.9	71.0	106.7	62.3	79.2	92.3	62.3	73.7	100.3	77.1	85.9	91.2	71.0	78.7	90.3	73.2	79.7
05	METALS	97.6	98.6	97.9	95.3	86.6	92.7	95.0	87.0	92.5	97.5	99.5	98.1	104.4	93.0	100.9	95.9	90.3	94.1
06	WOOD, PLASTICS & COMPOSITES	92.5	79.2	85.0	93.2	53.0	70.5	99.7	53.0	73.4	87.5	79.2	82.8	94.9	70.8	81.3	103.4	68.7	83.8
07	THERMAL & MOISTURE PROTECTION	96.8	81.3	90.5	99.9	62.5	84.7	99.8	63.1	84.9	96.7	86.0	92.4	100.0	77.3	90.8	98.1	76.5	89.4
08	OPENINGS	97.9	76.2	92.8	97.1	54.0	86.9	98.9	57.2	89.0	97.9	76.9	92.9	98.1	72.4	92.0	99.1	67.0	91.5
0920	Plaster & Gypsum Board	93.9	79.0	83.8	90.8	51.9	64.4	95.1	51.9	65.8	90.5	79.0	82.7	90.8	70.3	76.9	96.2	68.2	77.1
0950, 0980	Ceilings & Acoustic Treatment	85.8	79.0	81.3	86.1	51.9	63.4	90.7	51.9	64.9	85.8	79.0	81.3	88.1	70.3	76.3	95.9	68.2	77.4
0960	Flooring	112.1	56.1	95.4	112.6	43.9	92.1	115.2	65.7	100.4	109.8	57.6	94.2	112.8	74.9	101.5	118.5	74.3	105.3
0970, 0990	Wall Finishes & Painting/Coating	101.6	62.1	77.8	99.4	62.1	77.0	99.4	66.3	79.5	101.6	62.1	77.8	99.4	93.9	96.1	93.5	71.7	80.4
09	FINISHES	98.9	71.5	83.6	99.4	52.9	73.5	100.9	57.7	76.8	97.9	71.8	83.3	99.6	73.9	85.3	101.9	70.8	84.5
COVERS	DIVS. 10 - 14, 25, 28, 41, 43, 44, 46	100.0	74.6	94.9	100.0	81.9	96.4	100.0	81.4	96.3	100.0	74.6	94.9	100.0	84.9	97.0	100.0	87.1	97.4
21, 22, 23	FIRE SUPPRESSION, PLUMBING & HVAC	97.4	65.7	84.7	98.8	65.5	85.5	100.0	65.5	86.2	97.4	82.5	91.5	100.0	79.9	92.0	100.0	67.9	87.1
26, 27, 3370	ELECTRICAL, COMMUNICATIONS & UTIL.	94.7	64.3	78.9	94.7	73.5	83.6	94.1	63.7	78.2	92.8	63.1	77.3	95.7	69.5	82.0	96.6	75.5	85.6
MF2010	WEIGHTED AVERAGE	96.4	74.1	86.7	98.7	68.0	85.3	97.2	67.7	84.3	97.1	79.3	89.4	99.9	77.9	90.3	97.7	74.6	87.6

	DIVISION	FLORIDA																	
		ORLANDO			PANAMA CITY			PENSACOLA			SARASOTA			ST. PETERSBURG			TALLAHASSEE		
		327 - 328,347			324			325			342			337			323		
		MAT.	INST.	TOTAL	MAT.	INST.	TOTAL	MAT.	INST.	TOTAL	MAT.	INST.	TOTAL	MAT.	INST.	TOTAL	MAT.	INST.	TOTAL
015433	CONTRACTOR EQUIPMENT		98.8	98.8		98.8	98.8		98.8	98.8		98.8	98.8		98.8	98.8		98.8	98.8
0241, 31 - 34	SITE & INFRASTRUCTURE, DEMOLITION	108.9	89.6	95.4	119.4	88.6	97.8	118.9	89.1	98.0	111.3	89.6	96.1	107.8	89.3	94.8	105.5	88.9	93.8
0310	Concrete Forming & Accessories	99.8	73.7	77.3	96.4	44.9	52.0	94.4	53.0	58.6	96.2	76.7	79.4	95.0	52.2	58.1	100.4	45.1	52.7
0320	Concrete Reinforcing	95.9	76.2	86.1	97.0	74.9	85.9	99.4	75.4	87.4	92.8	96.5	94.6	93.1	88.8	91.0	102.5	66.8	84.6
0330	Cast-in-Place Concrete	107.4	71.8	92.6	91.9	57.7	77.7	113.4	67.0	94.2	104.0	70.6	90.1	98.0	65.5	84.5	93.5	57.3	78.5
03	CONCRETE	99.0	74.6	86.9	96.8	56.7	77.0	106.3	63.6	85.2	97.2	79.2	88.3	94.6	65.2	80.1	94.3	55.2	75.0
04	MASONRY	97.1	66.8	78.4	97.0	48.1	66.7	114.9	55.4	78.1	92.6	77.1	83.0	138.8	49.9	83.8	93.8	54.8	69.6
05	METALS	95.2	91.3	94.0	96.2	88.2	93.7	97.3	89.6	94.9	99.8	99.1	99.6	98.4	94.6	97.2	93.2	86.2	91.1
06	WOOD, PLASTICS & COMPOSITES	99.3	76.9	86.7	98.5	42.7	67.0	96.1	52.7	71.6	99.4	79.2	88.0	97.4	51.3	71.4	98.3	42.3	66.7
07	THERMAL & MOISTURE PROTECTION	98.9	75.5	89.4	100.1	57.5	82.8	100.0	63.4	85.1	98.4	86.0	93.4	96.9	58.7	81.4	105.5	72.7	92.2
08	OPENINGS	100.0	70.9	93.1	96.8	47.6	85.1	96.7	58.3	87.6	99.6	76.0	94.0	97.9	62.4	89.5	102.0	48.5	89.3
0920	Plaster & Gypsum Board	99.8	76.7	84.0	93.7	41.4	58.1	94.8	51.7	65.5	95.7	79.0	84.3	96.5	50.2	65.0	105.5	40.9	61.6
0950, 0980	Ceilings & Acoustic Treatment	99.6	76.7	84.3	88.1	41.4	57.0	88.1	51.7	63.9	91.3	79.0	83.1	87.8	50.2	62.7	99.5	40.9	60.5
0960	Flooring	112.1	74.9	101.0	114.8	43.8	93.5	110.4	64.1	96.6	117.1	58.9	99.7	114.2	56.2	96.8	112.1	63.1	97.4
0970, 0990	Wall Finishes & Painting/Coating	96.2	71.7	81.5	99.4	66.3	79.5	99.4	66.3	79.5	102.2	62.1	78.1	101.6	62.1	77.8	96.4	66.3	78.3
09	FINISHES	103.1	74.1	86.9	101.3	45.8	70.4	99.8	55.7	75.3	102.6	72.0	85.6	100.4	52.9	74.0	103.5	49.3	73.3
COVERS	DIVS. 10 - 14, 25, 28, 41, 43, 44, 46	100.0	84.2	96.8	100.0	47.4	89.5	100.0	47.5	89.5	100.0	74.6	94.9	100.0	57.3	91.4	100.0	67.6	93.5
21, 22, 23	FIRE SUPPRESSION, PLUMBING & HVAC	100.0	57.8	83.1	100.0	53.4	81.3	100.0	53.6	81.4	99.9	64.1	85.5	100.0	59.7	83.9	100.0	39.8	75.9
26, 27, 3370	ELECTRICAL, COMMUNICATIONS & UTIL.	98.9	61.4	79.3	93.1	60.7	76.2	97.0	56.8	76.0	95.7	63.0	78.6	92.8	63.0	77.3	99.2	60.8	79.1
MF2010	WEIGHTED AVERAGE	99.3	71.4	87.1	98.5	59.1	81.3	100.8	62.4	84.1	99.3	75.3	88.8	100.2	64.7	84.7	98.7	58.0	81.0

	DIVISION	FLORIDA						GEORGIA											
		TAMPA			WEST PALM BEACH			ALBANY			ATHENS			ATLANTA			AUGUSTA		
		335 - 336,346			334,349			317,398			306			300 - 303,399			308 - 309		
		MAT.	INST.	TOTAL	MAT.	INST.	TOTAL	MAT.	INST.	TOTAL	MAT.	INST.	TOTAL	MAT.	INST.	TOTAL	MAT.	INST.	TOTAL
015433	CONTRACTOR EQUIPMENT		98.8	98.8		91.7	91.7		92.5	92.5		93.8	93.8		94.3	94.3		93.8	93.8
0241, 31 - 34	SITE & INFRASTRUCTURE, DEMOLITION	108.3	89.8	95.3	90.5	78.4	82.0	96.0	79.8	84.6	93.9	93.2	93.4	90.8	94.7	93.5	87.9	92.9	91.4
0310	Concrete Forming & Accessories	98.0	77.3	80.1	99.2	69.7	73.7	90.9	44.5	50.9	90.6	47.0	52.9	94.1	75.0	77.6	91.6	67.0	70.4
0320	Concrete Reinforcing	89.8	96.6	93.2	92.3	74.2	83.3	89.6	81.4	85.5	98.6	80.4	89.5	97.9	82.4	90.2	99.0	74.1	86.6
0330	Cast-in-Place Concrete	95.8	70.9	85.5	86.4	75.5	81.9	92.4	55.3	77.0	107.4	56.5	86.3	107.4	71.1	92.3	101.4	50.6	80.3
03	CONCRETE	93.2	79.6	86.5	87.8	73.7	80.9	91.0	56.7	74.1	103.8	57.3	80.8	101.3	75.1	88.4	96.8	63.0	80.1
04	MASONRY	90.9	77.1	82.4	89.5	67.7	76.0	93.8	49.9	66.6	77.7	54.7	63.5	90.9	66.9	76.1	91.2	44.0	62.0
05	METALS	97.4	99.7	98.1	94.6	90.9	93.5	96.6	86.4	93.5	90.8	74.6	85.8	91.6	77.9	87.4	90.5	72.7	85.0
06	WOOD, PLASTICS & COMPOSITES	101.7	79.2	89.0	101.3	68.7	82.9	93.0	38.9	62.5	90.3	41.9	63.0	94.4	77.5	84.9	91.7	72.6	80.9
07	THERMAL & MOISTURE PROTECTION	97.1	86.0	92.6	96.7	71.4	86.4	96.5	61.4	82.3	93.6	53.8	77.5	93.5	73.0	85.2	93.3	58.1	79.0
08	OPENINGS	99.1	79.2	94.4	96.2	67.1	89.3	98.8	45.0	86.0	90.4	46.9	80.1	95.8	73.4	90.5	90.4	63.7	84.1
0920	Plaster & Gypsum Board	99.2	79.0	85.5	103.2	68.2	79.4	100.1	37.4	57.5	96.9	40.4	58.4	99.1	77.2	84.2	98.0	72.1	80.4
0950, 0980	Ceilings & Acoustic Treatment	91.3	79.0	83.1	85.8	68.2	74.1	93.5	37.4	56.2	94.1	40.4	58.4	94.1	77.2	82.8	95.1	72.1	79.8
0960	Flooring	115.4	57.6	98.1	117.4	67.7	102.5	116.2	48.7	96.0	94.5	54.9	82.7	95.8	66.4	87.0	94.8	47.4	80.6
0970, 0990	Wall Finishes & Painting/Coating	101.6	62.1	77.8	96.5	71.7	81.6	97.6	53.9	71.3	96.1	46.9	66.6	96.1	85.9	89.9	96.1	46.9	66.6
09	FINISHES	102.0	71.8	85.2	99.1	69.0	82.4	101.6	44.2	69.6	93.5	46.8	67.5	93.8	74.8	83.2	93.3	61.9	75.8
COVERS	DIVS. 10 - 14, 25, 28, 41, 43, 44, 46	100.0	84.9	97.0	100.0	86.5	97.3	100.0	80.3	96.1	100.0	79.4	95.9	100.0	85.7	97.1	100.0	80.6	96.1
21, 22, 23	FIRE SUPPRESSION, PLUMBING & HVAC	100.0	82.6	93.0	97.4	64.0	84.0	100.0	69.5	87.7	95.1	71.1	85.5	100.0	71.9	88.7	100.0	62.6	85.0
26, 27, 3370	ELECTRICAL, COMMUNICATIONS & UTIL.	92.6	63.1	77.2	93.8	72.7	82.8	96.1	60.1	77.3	99.5	71.2	84.7	98.8	72.7	85.2	100.1	63.2	80.9
MF2010	WEIGHTED AVERAGE	97.9	79.8	90.0	95.3	72.3	85.3	97.5	62.5	82.3	94.7	65.1	81.8	96.8	75.3	87.5	95.6	64.8	82.2

	DIVISION	GEORGIA																	
		COLUMBUS			DALTON			GAINESVILLE			MACON			SAVANNAH			STATESBORO		
		318 - 319			307			305			310 - 312			313 - 314			304		
		MAT.	INST.	TOTAL	MAT.	INST.	TOTAL	MAT.	INST.	TOTAL	MAT.	INST.	TOTAL	MAT.	INST.	TOTAL	MAT.	INST.	TOTAL
015433	CONTRACTOR EQUIPMENT		92.5	92.5		105.5	105.5		93.8	93.8		102.8	102.8		93.5	93.5		93.2	93.2
0241, 31 - 34	SITE & INFRASTRUCTURE, DEMOLITION	95.9	79.9	84.6	94.2	97.5	96.5	93.7	93.2	93.3	96.6	94.8	95.3	97.6	81.2	86.1	95.0	77.6	82.7
0310	Concrete Forming & Accessories	90.8	55.6	60.4	83.8	47.8	52.7	93.8	43.9	50.7	90.3	53.1	58.2	92.5	51.0	56.7	78.7	52.6	56.2
0320	Concrete Reinforcing	89.8	81.8	85.8	98.1	76.3	87.2	98.4	80.3	89.4	91.0	81.5	86.2	96.8	74.2	85.5	97.7	43.1	70.4
0330	Cast-in-Place Concrete	92.1	54.8	76.6	104.3	51.0	82.2	112.8	53.8	88.3	90.9	66.7	80.8	100.2	55.8	81.8	107.1	60.7	87.9
03	CONCRETE	90.8	61.5	76.4	102.9	55.8	79.7	105.7	54.9	80.6	90.4	64.4	77.6	95.9	58.4	77.4	103.2	55.3	79.5
04	MASONRY	93.7	56.2	70.5	78.8	37.3	53.2	86.3	55.1	67.0	105.6	45.3	68.3	90.3	51.4	66.2	80.2	41.3	56.1
05	METALS	96.1	87.5	93.5	91.6	84.3	89.3	90.0	73.8	85.0	92.2	87.1	90.6	93.3	83.8	90.3	95.0	74.0	88.5
06	WOOD, PLASTICS & COMPOSITES	93.0	53.3	70.6	74.1	48.5	59.7	94.2	38.8	62.9	101.2	52.4	73.7	105.9	47.0	72.7	67.6	55.1	60.6
07	THERMAL & MOISTURE PROTECTION	96.5	64.6	83.5	95.6	52.5	78.1	93.6	55.8	78.2	94.9	63.2	82.1	95.6	57.4	80.1	94.2	52.1	77.1
08	OPENINGS	98.8	60.2	89.6	91.2	50.7	81.6	90.4	40.9	78.7	97.6	53.7	87.2	103.0	49.6	90.4	92.2	41.6	80.2
0920	Plaster & Gypsum Board	100.1	52.3	67.6	84.0	47.3	59.0	99.1	37.2	57.0	105.5	51.3	68.6	98.1	45.8	62.5	84.1	54.1	63.7
0950, 0980	Ceilings & Acoustic Treatment	93.5	52.3	66.1	106.9	47.3	67.2	94.1	37.2	56.2	88.4	51.3	63.7	101.5	45.8	64.4	102.2	54.1	70.2
0960	Flooring	116.2	51.3	96.8	94.7	47.8	80.7	95.7	47.4	81.3	91.7	48.7	78.8	113.7	47.2	93.8	112.7	45.7	92.7
0970, 0990	Wall Finishes & Painting/Coating	97.6	67.9	79.7	87.3	61.1	71.5	96.1	46.9	66.6	99.8	53.9	72.2	96.0	59.5	74.1	94.0	40.1	61.6
09	FINISHES	101.5	54.7	75.4	101.7	48.1	71.9	94.0	42.6	65.3	89.7	51.0	68.2	101.7	49.9	72.9	104.7	50.1	74.3
COVERS	DIVS. 10 - 14, 25, 28, 41, 43, 44, 46	100.0	81.9	96.4	100.0	23.4	84.6	100.0	40.5	88.1	100.0	80.3	96.1	100.0	80.0	96.0	100.0	44.7	88.9
21, 22, 23	FIRE SUPPRESSION, PLUMBING & HVAC	100.0	64.7	85.8	95.2	58.2	80.4	95.1	71.0	85.5	100.0	68.2	87.2	100.0	62.9	85.2	95.7	58.0	80.5
26, 27, 3370	ELECTRICAL, COMMUNICATIONS & UTIL.	96.2	71.3	83.2	109.8	69.3	88.7	99.5	71.2	84.7	95.2	62.7	78.2	101.0	58.7	78.9	99.6	58.7	78.3
MF2010	WEIGHTED AVERAGE	97.4	66.6	84.0	96.4	60.1	80.6	95.2	62.8	81.1	96.2	65.9	83.0	98.4	62.0	82.6	96.5	56.9	79.2

	DIVISION	GEORGIA						HAWAII									IDAHO		
		VALDOSTA			WAYCROSS			HILO			HONOLULU			STATES & POSS., GUAM			BOISE		
		316			315			967			968			969			836 - 837		
		MAT.	INST.	TOTAL	MAT.	INST.	TOTAL	MAT.	INST.	TOTAL	MAT.	INST.	TOTAL	MAT.	INST.	TOTAL	MAT.	INST.	TOTAL
015433	CONTRACTOR EQUIPMENT		92.5	92.5		92.5	92.5		98.6	98.6		98.6	98.6		162.1	162.1		97.9	97.9
0241, 31 - 34	SITE & INFRASTRUCTURE, DEMOLITION	104.6	79.9	87.3	101.5	78.6	85.4	126.9	106.4	112.5	136.4	106.4	115.3	165.7	103.6	122.1	82.4	97.1	92.7
0310	Concrete Forming & Accessories	81.4	45.0	50.0	83.1	66.3	68.6	108.1	135.3	131.5	119.2	135.3	133.1	110.3	64.7	70.9	100.7	79.2	82.2
0320	Concrete Reinforcing	91.7	78.8	85.3	91.7	75.0	83.3	109.0	121.6	115.3	116.9	121.6	119.2	188.7	31.1	110.0	101.9	81.2	91.6
0330	Cast-in-Place Concrete	90.5	56.5	76.4	102.2	49.0	80.1	186.3	126.3	161.4	153.8	126.3	142.4	161.5	106.6	138.7	91.1	89.4	90.4
03	CONCRETE	95.4	56.9	76.4	98.7	62.8	81.0	148.2	128.4	138.4	141.5	128.4	135.0	149.6	73.5	112.0	100.7	83.2	92.1
04	MASONRY	99.3	51.0	69.4	100.0	39.9	62.8	127.9	128.5	128.3	128.1	128.5	128.4	182.7	44.3	97.1	118.9	82.4	96.3
05	METALS	95.7	85.4	92.5	94.8	79.3	90.0	107.6	107.9	107.7	120.8	107.9	116.8	138.8	76.3	119.5	101.1	81.2	95.0
06	WOOD, PLASTICS & COMPOSITES	81.0	38.5	57.0	82.8	74.1	77.9	101.3	137.9	122.0	119.2	137.9	129.7	113.8	68.6	88.3	88.9	77.7	82.6
07	THERMAL & MOISTURE PROTECTION	96.7	63.3	83.1	96.5	51.2	78.1	108.0	124.8	114.8	123.3	124.8	123.9	124.5	67.7	101.5	95.0	81.2	89.4
08	OPENINGS	94.7	43.5	82.6	94.7	59.2	86.3	99.7	132.7	107.5	108.8	132.7	114.4	104.3	55.2	92.6	99.1	72.5	92.8
0920	Plaster & Gypsum Board	92.0	37.1	54.6	92.4	73.7	79.7	107.7	138.9	128.9	143.9	138.9	140.5	213.3	56.8	106.8	84.9	76.9	79.5
0950, 0980	Ceilings & Acoustic Treatment	90.8	37.1	55.1	88.9	73.7	78.8	121.3	138.9	133.0	132.4	138.9	136.7	240.6	56.8	118.3	107.2	76.9	87.0
0960	Flooring	110.1	48.7	91.7	111.2	30.4	87.1	125.5	138.2	129.3	144.2	138.2	142.4	147.9	47.2	117.8	95.5	85.2	92.4
0970, 0990	Wall Finishes & Painting/Coating	97.6	53.3	70.9	97.6	46.9	67.1	99.1	145.5	127.0	107.4	145.5	130.3	104.1	36.7	63.6	102.9	40.3	65.3
09	FINISHES	99.0	44.5	68.7	98.6	58.7	76.4	115.6	138.1	128.1	130.3	138.1	134.6	192.9	60.0	118.8	97.8	76.5	85.9
COVERS	DIVS. 10 - 14, 25, 28, 41, 43, 44, 46	100.0	79.1	95.8	100.0	52.9	90.6	100.0	116.4	103.3	100.0	116.4	103.3	100.0	76.3	95.3	100.0	89.8	98.0
21, 22, 23	FIRE SUPPRESSION, PLUMBING & HVAC	100.0	71.4	88.5	96.8	58.3	81.4	100.1	109.4	103.8	100.2	109.4	103.8	102.1	38.1	76.4	100.0	72.5	89.0
26, 27, 3370	ELECTRICAL, COMMUNICATIONS & UTIL.	94.2	57.6	75.1	98.7	58.7	77.8	105.2	124.0	115.0	106.5	124.0	115.7	151.3	42.2	94.3	96.6	74.6	85.1
MF2010	WEIGHTED AVERAGE	97.5	62.6	82.3	97.3	60.7	81.3	110.4	121.0	115.0	114.9	121.0	117.6	131.2	58.9	99.7	99.8	79.4	90.9

	DIVISION	IDAHO															ILLINOIS		
		COEUR D'ALENE			IDAHO FALLS			LEWISTON			POCATELLO			TWIN FALLS			BLOOMINGTON		
		838			834			835			832			833			617		
		MAT.	INST.	TOTAL	MAT.	INST.	TOTAL	MAT.	INST.	TOTAL	MAT.	INST.	TOTAL	MAT.	INST.	TOTAL	MAT.	INST.	TOTAL
015433	CONTRACTOR EQUIPMENT		92.5	92.5		97.9	97.9		92.5	92.5		97.9	97.9		97.9	97.9		101.4	101.4
0241, 31 - 34	SITE & INFRASTRUCTURE, DEMOLITION	79.9	92.3	88.6	80.3	96.9	91.9	86.3	93.1	91.1	83.1	97.1	92.9	89.4	98.0	95.5	96.3	97.8	97.3
0310	Concrete Forming & Accessories	113.8	82.2	86.5	94.9	79.0	81.1	118.9	83.1	88.0	100.9	78.9	81.9	101.8	56.0	62.3	85.5	119.2	114.6
0320	Concrete Reinforcing	109.5	96.9	103.2	103.9	81.0	92.4	109.5	97.1	103.3	102.3	81.2	91.8	104.2	81.0	92.6	95.8	111.2	103.5
0330	Cast-in-Place Concrete	98.5	86.4	93.5	86.8	75.0	81.9	102.3	86.8	95.9	93.7	89.2	91.8	96.1	64.4	82.9	98.7	113.8	104.9
03	CONCRETE	107.5	86.3	97.1	92.9	78.0	85.6	111.1	86.9	99.2	99.9	83.0	91.5	107.5	64.2	86.1	97.2	115.8	106.4
04	MASONRY	120.4	85.3	98.7	114.1	82.4	94.5	120.8	85.3	98.8	116.6	82.4	95.4	119.0	82.4	96.4	113.4	118.3	116.4
05	METALS	95.2	87.1	92.7	108.9	79.6	99.9	94.6	88.0	92.6	109.0	80.9	100.3	109.0	79.5	99.9	97.3	113.3	102.2
06	WOOD, PLASTICS & COMPOSITES	92.7	81.8	86.5	83.0	79.4	80.9	97.9	81.8	88.8	88.9	77.7	82.6	89.8	48.3	66.4	88.1	118.1	105.0
07	THERMAL & MOISTURE PROTECTION	147.5	81.5	120.7	94.6	71.7	85.3	147.7	81.5	120.8	95.0	73.9	86.4	95.7	75.2	87.4	100.3	113.3	105.6
08	OPENINGS	117.8	73.5	107.4	102.8	68.6	94.7	117.8	76.1	107.9	99.8	67.5	92.1	102.8	49.6	90.2	93.1	106.6	96.3
0920	Plaster & Gypsum Board	152.1	81.2	103.9	75.3	78.7	77.6	153.2	81.2	104.3	77.6	76.9	77.1	78.5	46.7	56.8	90.3	118.6	109.5
0950, 0980	Ceilings & Acoustic Treatment	138.9	81.2	100.5	105.1	78.7	87.5	138.9	81.2	100.5	113.1	76.9	89.0	107.8	46.7	67.1	94.3	118.6	110.5
0960	Flooring	137.5	46.7	110.3	95.7	43.9	80.2	140.2	92.5	126.0	98.7	85.2	94.6	99.7	43.9	83.0	98.0	120.9	104.8
0970, 0990	Wall Finishes & Painting/Coating	122.3	66.3	88.7	102.9	41.2	65.8	122.3	66.3	88.7	102.8	42.0	66.2	102.9	38.1	63.9	95.3	127.8	114.8
09	FINISHES	166.2	74.0	114.8	95.5	69.5	81.0	167.3	83.0	120.4	98.9	76.7	86.5	98.9	50.8	72.1	97.4	121.1	110.6
COVERS	DIVS. 10 - 14, 25, 28, 41, 43, 44, 46	100.0	89.7	97.9	100.0	48.9	89.8	100.0	89.7	97.9	100.0	89.8	98.0	100.0	45.4	89.1	100.0	103.9	100.8
21, 22, 23	FIRE SUPPRESSION, PLUMBING & HVAC	99.5	84.3	93.4	101.0	72.3	89.5	100.7	87.2	95.3	100.0	72.4	88.9	100.0	70.8	88.3	95.0	107.8	100.1
26, 27, 3370	ELECTRICAL, COMMUNICATIONS & UTIL.	88.7	79.4	83.9	88.7	70.9	79.4	86.7	80.8	83.6	94.1	70.9	82.0	90.2	61.6	75.2	93.5	96.8	95.3
MF2010	WEIGHTED AVERAGE	107.6	83.3	97.0	99.6	75.5	89.1	108.3	85.5	98.4	100.8	78.4	91.1	101.9	68.5	87.4	96.9	109.9	102.6

	DIVISION	ILLINOIS																	
		CARBONDALE 629			CENTRALIA 628			CHAMPAIGN 618 - 619			CHICAGO 606 - 608			DECATUR 625			EAST ST. LOUIS 620 - 622		
		MAT.	INST.	TOTAL	MAT.	INST.	TOTAL	MAT.	INST.	TOTAL	MAT.	INST.	TOTAL	MAT.	INST.	TOTAL	MAT.	INST.	TOTAL
015433	CONTRACTOR EQUIPMENT		107.7	107.7		107.7	107.7		102.1	102.1		93.9	93.9		102.1	102.1		107.7	107.7
0241, 31 - 34	SITE & INFRASTRUCTURE, DEMOLITION	100.2	98.6	99.0	100.5	99.2	99.6	105.2	98.4	100.4	105.3	95.4	98.4	92.2	98.6	96.7	102.6	99.2	100.2
0310	Concrete Forming & Accessories	95.5	107.9	106.2	97.4	114.8	112.4	91.6	117.1	113.7	96.6	159.8	151.1	97.7	116.9	114.2	92.8	115.0	112.0
0320	Concrete Reinforcing	89.8	100.4	95.1	89.8	113.3	101.5	95.8	105.1	100.5	95.5	163.2	129.3	87.9	105.9	96.9	89.7	113.4	101.5
0330	Cast-in-Place Concrete	91.8	101.1	95.7	92.3	118.5	103.2	114.4	109.2	112.2	103.4	149.6	122.6	100.2	111.9	105.1	93.8	118.5	104.1
03	CONCRETE	87.0	104.8	95.8	87.5	116.3	101.7	109.6	112.1	110.8	99.2	155.5	127.0	97.5	113.1	105.2	88.4	116.5	102.3
04	MASONRY	79.4	107.7	96.9	79.5	117.4	103.0	136.2	118.4	125.2	101.8	158.0	136.6	75.9	115.5	100.4	79.7	117.4	103.0
05	METALS	94.3	113.7	100.3	94.4	120.8	102.5	97.3	108.7	100.8	95.0	135.9	107.6	97.9	109.6	101.5	95.4	121.0	103.3
06	WOOD, PLASTICS & COMPOSITES	106.0	106.1	106.1	108.7	112.3	110.7	95.0	116.5	107.1	98.2	159.6	132.8	106.2	116.5	112.0	103.2	112.3	108.3
07	THERMAL & MOISTURE PROTECTION	98.0	101.6	99.5	98.0	112.3	103.8	100.9	114.5	106.5	99.2	147.5	118.8	104.0	109.6	106.3	98.0	111.9	103.6
08	OPENINGS	88.3	110.5	93.6	88.4	117.3	95.2	93.7	112.1	98.1	101.3	162.6	115.8	99.3	112.4	102.4	88.4	117.3	95.3
0920	Plaster & Gypsum Board	95.7	106.2	102.9	96.8	112.6	107.6	92.5	116.9	109.1	89.7	161.3	138.4	98.4	116.9	111.0	94.6	112.6	106.9
0950, 0980	Ceilings & Acoustic Treatment	96.2	106.2	102.9	96.2	112.6	107.1	94.3	116.9	109.4	101.5	161.3	141.3	102.4	116.9	112.1	96.2	112.6	107.1
0960	Flooring	121.3	122.4	121.6	122.3	117.9	120.9	100.7	122.4	107.1	93.1	150.7	110.3	108.6	118.8	111.7	120.2	117.9	119.5
0970, 0990	Wall Finishes & Painting/Coating	104.5	99.5	101.5	104.5	105.2	104.9	95.3	111.1	104.8	93.8	156.1	131.2	96.1	108.4	103.5	104.5	105.2	104.9
09	FINISHES	102.4	108.9	106.0	102.8	114.4	109.3	99.3	118.0	109.7	98.7	158.7	132.1	102.9	117.3	110.9	102.0	114.5	109.0
COVERS	DIVS. 10 - 14, 25, 28, 41, 43, 44, 46	100.0	104.1	100.8	100.0	106.5	101.3	100.0	103.1	100.6	100.0	125.0	105.0	100.0	103.0	100.6	100.0	106.5	101.3
21, 22, 23	FIRE SUPPRESSION, PLUMBING & HVAC	95.0	104.3	98.7	95.0	94.4	94.8	95.0	105.3	99.1	99.9	136.1	114.4	100.0	98.9	99.5	99.9	98.4	99.3
26, 27, 3370	ELECTRICAL, COMMUNICATIONS & UTIL.	94.0	106.7	100.6	95.5	104.3	100.1	96.8	94.5	95.6	96.1	137.1	117.5	98.4	91.3	94.7	95.0	104.3	99.9
MF2010	WEIGHTED AVERAGE	93.9	106.2	99.2	94.1	108.4	100.4	100.2	108.0	103.6	98.9	141.9	117.6	98.2	105.9	101.6	95.5	109.3	101.5

	DIVISION	ILLINOIS																	
		EFFINGHAM 624			GALESBURG 614			JOLIET 604			KANKAKEE 609			LA SALLE 613			NORTH SUBURBAN 600 - 603		
		MAT.	INST.	TOTAL	MAT.	INST.	TOTAL	MAT.	INST.	TOTAL	MAT.	INST.	TOTAL	MAT.	INST.	TOTAL	MAT.	INST.	TOTAL
015433	CONTRACTOR EQUIPMENT		102.1	102.1		101.4	101.4		92.0	92.0		92.0	92.0		101.4	101.4		92.0	92.0
0241, 31 - 34	SITE & INFRASTRUCTURE, DEMOLITION	97.5	98.2	98.0	98.7	97.7	98.0	105.2	94.7	97.8	98.9	93.9	95.4	98.1	98.4	98.3	104.4	93.7	96.8
0310	Concrete Forming & Accessories	102.2	115.5	113.7	91.5	119.0	115.3	98.4	163.3	154.4	92.4	141.3	134.6	105.1	125.1	122.4	97.8	153.4	145.8
0320	Concrete Reinforcing	90.7	101.0	95.9	95.4	111.1	103.2	95.5	155.7	125.6	96.3	146.5	121.4	95.5	143.7	119.6	95.5	157.3	126.4
0330	Cast-in-Place Concrete	99.9	107.3	103.0	101.6	107.9	104.2	103.3	149.4	122.4	96.4	132.5	111.4	101.5	123.6	110.7	103.4	143.5	120.0
03	CONCRETE	98.4	109.9	104.1	100.1	113.7	106.8	99.3	155.5	127.1	93.7	138.2	115.6	100.9	127.9	114.3	99.3	149.4	124.0
04	MASONRY	83.8	107.6	98.5	113.5	119.7	117.4	105.0	151.1	133.5	101.3	139.8	125.1	113.5	125.6	121.0	101.8	146.6	129.5
05	METALS	95.1	106.1	98.5	97.3	112.8	102.1	92.9	130.6	104.6	92.9	124.7	102.7	97.4	130.3	107.5	94.1	130.1	105.2
06	WOOD, PLASTICS & COMPOSITES	108.8	116.5	113.1	94.8	118.2	108.0	99.7	164.8	136.4	92.8	140.7	119.8	110.0	123.5	117.6	98.2	154.1	129.7
07	THERMAL & MOISTURE PROTECTION	103.6	107.6	105.2	100.5	109.9	104.3	99.1	144.5	117.5	98.4	135.6	113.5	100.6	119.4	108.2	99.5	140.8	116.3
08	OPENINGS	93.8	111.2	97.9	93.1	112.2	97.6	99.0	163.5	114.2	91.9	147.0	105.0	93.1	128.3	101.5	99.0	157.1	112.8
0920	Plaster & Gypsum Board	98.3	116.9	111.0	92.5	118.7	110.3	86.7	166.8	141.2	84.1	142.0	123.5	99.2	124.2	116.2	89.7	155.8	134.6
0950, 0980	Ceilings & Acoustic Treatment	96.2	116.9	110.0	94.3	118.7	110.5	101.5	166.8	144.9	101.5	142.0	128.4	94.3	124.2	114.2	101.5	155.8	137.6
0960	Flooring	109.7	122.4	113.5	100.6	120.9	106.7	92.7	142.9	107.7	90.1	134.3	103.3	106.5	124.8	112.0	93.1	142.9	108.0
0970, 0990	Wall Finishes & Painting/Coating	96.1	104.2	101.0	95.3	96.0	95.7	92.0	151.4	127.7	92.0	127.8	113.5	95.3	127.8	114.8	93.8	151.4	128.4
09	FINISHES	101.9	116.8	110.2	98.6	117.8	109.3	98.1	160.0	132.6	96.6	137.1	119.2	101.2	124.7	114.3	98.7	152.5	128.7
COVERS	DIVS. 10 - 14, 25, 28, 41, 43, 44, 46	100.0	71.7	94.3	100.0	103.8	100.8	100.0	125.4	105.1	100.0	117.6	103.5	100.0	104.4	100.9	100.0	120.4	104.1
21, 22, 23	FIRE SUPPRESSION, PLUMBING & HVAC	95.1	103.5	98.4	95.0	105.7	99.3	100.0	132.7	113.1	95.1	126.6	107.7	95.0	124.8	107.0	99.9	130.9	112.3
26, 27, 3370	ELECTRICAL, COMMUNICATIONS & UTIL.	96.0	106.6	101.6	94.5	87.3	90.7	95.4	141.2	119.3	90.1	140.4	116.4	91.6	140.4	117.1	95.2	131.5	114.2
MF2010	WEIGHTED AVERAGE	96.3	106.3	100.6	97.6	107.6	102.0	98.4	140.7	116.8	94.8	130.8	110.5	97.7	125.1	109.6	98.5	135.9	114.8

	DIVISION	ILLINOIS																	
		PEORIA 615 - 616			QUINCY 623			ROCK ISLAND 612			ROCKFORD 610 - 611			SOUTH SUBURBAN 605			SPRINGFIELD 626 - 627		
		MAT.	INST.	TOTAL	MAT.	INST.	TOTAL	MAT.	INST.	TOTAL	MAT.	INST.	TOTAL	MAT.	INST.	TOTAL	MAT.	INST.	TOTAL
015433	CONTRACTOR EQUIPMENT		101.4	101.4		102.1	102.1		101.4	101.4		101.4	101.4		92.0	92.0		102.1	102.1
0241, 31 - 34	SITE & INFRASTRUCTURE, DEMOLITION	99.2	97.8	98.2	96.3	98.1	97.6	96.8	96.8	96.8	98.5	98.8	98.7	104.4	93.7	96.8	99.2	98.6	98.8
0310	Concrete Forming & Accessories	94.5	119.9	116.4	100.0	113.8	111.9	93.2	101.3	100.2	98.9	130.1	125.8	97.8	153.4	145.8	98.1	117.3	114.6
0320	Concrete Reinforcing	92.9	111.2	102.0	90.4	105.6	98.0	95.4	104.2	99.8	87.8	138.2	113.0	95.5	157.3	126.4	92.8	106.1	99.4
0330	Cast-in-Place Concrete	98.5	116.0	105.8	100.0	104.2	101.8	99.4	98.4	99.0	100.9	126.4	111.5	103.4	143.5	120.0	95.2	108.5	100.7
03	CONCRETE	97.3	116.9	106.9	98.0	109.0	103.4	98.1	101.2	99.6	97.8	130.1	113.7	99.3	149.4	124.0	95.6	112.1	103.7
04	MASONRY	113.5	120.0	117.5	104.2	104.6	104.4	113.4	96.2	102.7	88.8	134.3	117.0	101.8	146.6	129.5	86.1	119.0	106.5
05	METALS	100.0	113.6	104.2	95.1	108.4	99.2	97.3	108.3	100.7	100.0	127.8	108.6	94.1	130.1	105.2	95.5	109.7	99.9
06	WOOD, PLASTICS & COMPOSITES	102.5	118.2	111.4	106.1	116.5	112.0	96.4	100.6	98.8	102.5	127.8	116.8	98.2	154.1	129.7	103.0	116.5	110.6
07	THERMAL & MOISTURE PROTECTION	101.2	113.8	106.3	103.6	104.1	103.8	100.4	98.5	99.7	103.5	128.4	113.6	99.5	140.8	116.3	105.2	112.8	108.3
08	OPENINGS	99.2	116.7	103.4	94.6	112.4	98.8	93.1	101.0	95.0	99.2	132.5	107.1	99.0	157.1	112.8	101.3	112.4	103.9
0920	Plaster & Gypsum Board	96.3	118.7	111.5	96.8	116.9	110.5	92.5	100.5	97.9	96.3	128.5	118.2	89.7	155.8	134.6	97.1	116.9	110.6
0950, 0980	Ceilings & Acoustic Treatment	100.5	118.7	112.6	96.2	116.9	110.0	94.3	100.5	98.4	100.5	128.5	119.2	101.5	155.8	137.6	106.6	116.9	113.5
0960	Flooring	103.8	120.9	108.9	108.6	110.1	109.1	101.7	107.5	103.4	103.8	124.8	110.1	93.1	142.9	108.0	113.1	109.8	112.1
0970, 0990	Wall Finishes & Painting/Coating	95.3	127.8	114.8	96.1	105.1	101.5	95.3	96.0	95.7	95.3	135.3	119.3	93.8	151.4	128.4	94.6	105.1	100.9
09	FINISHES	101.4	121.4	112.6	101.3	113.6	108.2	98.8	101.9	100.6	101.4	129.8	117.2	98.7	152.5	128.7	106.1	115.4	111.3
COVERS	DIVS. 10 - 14, 25, 28, 41, 43, 44, 46	100.0	104.3	100.9	100.0	72.7	94.5	100.0	97.2	99.4	100.0	111.6	102.3	100.0	120.4	104.1	100.0	103.3	100.7
21, 22, 23	FIRE SUPPRESSION, PLUMBING & HVAC	99.9	105.8	102.3	95.1	100.8	97.4	95.0	99.8	96.9	100.0	114.9	106.0	99.9	130.9	112.3	99.9	103.4	101.3
26, 27, 3370	ELECTRICAL, COMMUNICATIONS & UTIL.	95.5	97.4	96.5	93.2	81.3	87.0	86.5	95.7	91.3	95.8	127.0	112.1	95.2	131.5	114.2	101.0	92.8	96.8
MF2010	WEIGHTED AVERAGE	99.9	110.4	104.5	96.9	101.6	99.0	96.6	99.8	98.0	98.9	123.5	109.6	98.5	135.9	114.8	99.0	107.2	102.5

	DIVISION	INDIANA																	
		ANDERSON 460			BLOOMINGTON 474			COLUMBUS 472			EVANSVILLE 476 - 477			FORT WAYNE 467 - 468			GARY 463 - 464		
		MAT.	INST.	TOTAL	MAT.	INST.	TOTAL	MAT.	INST.	TOTAL	MAT.	INST.	TOTAL	MAT.	INST.	TOTAL	MAT.	INST.	TOTAL
015433	CONTRACTOR EQUIPMENT		96.7	96.7		87.2	87.2		87.2	87.2		116.1	116.1		96.7	96.7		96.7	96.7
0241, 31 - 34	SITE & INFRASTRUCTURE, DEMOLITION	90.5	96.0	94.4	82.7	94.8	91.2	79.4	94.6	90.1	87.8	123.5	112.9	91.5	96.0	94.6	91.0	99.2	96.8
0310	Concrete Forming & Accessories	98.8	80.8	83.3	101.9	81.1	83.9	95.9	78.9	81.2	95.4	81.9	83.7	97.1	75.4	78.3	98.9	115.2	113.0
0320	Concrete Reinforcing	98.0	83.7	90.9	87.1	83.4	85.2	87.5	83.4	85.4	95.4	76.7	86.1	98.0	76.8	87.4	98.0	110.1	104.0
0330	Cast-in-Place Concrete	104.8	78.1	93.7	100.8	78.0	91.3	100.4	72.5	88.8	96.3	87.7	92.8	111.3	83.7	99.9	109.5	112.4	110.7
03	CONCRETE	96.3	80.9	88.7	102.3	80.3	91.4	101.6	77.3	89.6	102.7	83.1	93.0	99.3	79.1	89.3	98.5	113.1	105.7
04	MASONRY	84.8	79.7	81.7	90.5	76.4	81.8	90.4	76.2	81.6	86.4	83.3	84.5	88.2	77.3	81.4	86.1	112.8	102.7
05	METALS	92.9	89.8	92.0	95.9	78.7	90.6	95.9	78.0	90.4	89.2	84.4	87.8	92.9	86.4	90.9	92.9	108.0	97.6
06	WOOD, PLASTICS & COMPOSITES	108.2	80.8	92.8	118.0	81.1	97.2	112.5	78.3	93.2	97.4	80.8	88.0	107.9	74.8	89.2	105.7	115.0	110.9
07	THERMAL & MOISTURE PROTECTION	105.4	77.1	93.9	95.3	79.7	88.9	94.9	79.4	88.6	99.3	84.7	93.4	105.2	79.0	94.6	104.0	109.0	106.0
08	OPENINGS	95.0	81.9	91.9	105.0	82.0	99.6	100.8	80.5	96.0	98.3	79.8	93.9	95.0	74.7	90.2	95.0	119.4	100.8
0920	Plaster & Gypsum Board	94.3	80.6	85.0	98.1	81.2	86.6	94.7	78.3	83.6	93.1	79.6	83.9	93.5	74.4	80.5	87.6	115.8	106.8
0950, 0980	Ceilings & Acoustic Treatment	89.6	80.6	83.7	83.9	81.2	82.1	83.9	78.3	80.2	88.8	79.6	82.7	89.6	74.4	79.5	89.6	115.8	107.1
0960	Flooring	99.9	86.6	96.0	105.8	76.4	97.0	101.4	76.4	93.9	100.8	82.2	95.2	99.9	80.2	94.0	99.9	123.6	107.0
0970, 0990	Wall Finishes & Painting/Coating	97.3	70.9	81.4	89.6	83.0	85.7	89.6	83.0	85.7	94.7	86.7	89.9	97.3	74.4	83.5	97.3	126.6	114.9
09	FINISHES	95.0	80.9	87.1	96.2	80.4	87.4	94.5	78.7	85.7	95.2	82.4	88.1	94.8	76.1	84.4	94.0	118.3	107.5
COVERS	DIVS. 10 - 14, 25, 28, 41, 43, 44, 46	100.0	91.5	98.3	100.0	89.1	97.8	100.0	88.7	97.7	100.0	94.1	98.8	100.0	89.5	97.9	100.0	105.3	101.1
21, 22, 23	FIRE SUPPRESSION, PLUMBING & HVAC	100.0	75.8	90.3	99.8	79.0	91.4	94.9	79.0	88.5	100.0	80.2	92.0	100.0	72.4	88.9	100.0	105.9	102.3
26, 27, 3370	ELECTRICAL, COMMUNICATIONS & UTIL.	85.9	88.4	87.2	98.2	83.5	90.5	97.3	85.2	91.0	94.5	86.3	90.2	86.6	78.4	82.3	97.7	110.2	104.2
MF2010	WEIGHTED AVERAGE	95.5	83.1	90.1	98.7	81.5	91.2	96.6	80.9	89.7	96.5	86.6	92.1	96.0	79.2	88.7	96.8	110.1	102.6

	DIVISION	INDIANA																	
		INDIANAPOLIS 461 - 462			KOKOMO 469			LAFAYETTE 479			LAWRENCEBURG 470			MUNCIE 473			NEW ALBANY 471		
		MAT.	INST.	TOTAL	MAT.	INST.	TOTAL	MAT.	INST.	TOTAL	MAT.	INST.	TOTAL	MAT.	INST.	TOTAL	MAT.	INST.	TOTAL
015433	CONTRACTOR EQUIPMENT		93.3	93.3		96.7	96.7		87.2	87.2		105.0	105.0		95.7	95.7		95.1	95.1
0241, 31 - 34	SITE & INFRASTRUCTURE, DEMOLITION	90.3	99.3	96.6	87.1	95.8	93.2	80.2	94.5	90.3	78.0	111.0	101.2	82.5	95.0	91.3	75.3	97.4	90.8
0310	Concrete Forming & Accessories	98.0	85.1	86.9	102.4	77.8	81.2	93.5	82.8	84.3	92.4	77.2	79.3	93.5	80.5	82.3	90.2	75.1	77.2
0320	Concrete Reinforcing	100.0	85.4	92.7	88.7	83.7	86.2	87.1	83.6	85.4	86.3	76.6	81.5	96.3	83.6	90.0	87.7	79.6	83.6
0330	Cast-in-Place Concrete	103.3	84.7	95.6	103.7	81.0	94.3	100.9	80.7	92.5	94.4	76.6	87.0	105.9	77.7	94.2	97.4	75.7	88.4
03	CONCRETE	99.5	84.8	92.3	93.1	80.5	86.9	101.8	81.9	92.0	95.0	77.4	86.3	101.1	80.6	91.0	100.2	76.5	88.5
04	MASONRY	90.3	81.3	84.7	84.4	80.0	81.7	95.5	79.8	85.8	75.6	76.0	75.9	92.4	79.8	84.6	82.3	69.5	74.4
05	METALS	93.6	81.7	89.9	89.5	89.3	89.4	94.3	78.6	89.5	91.1	85.2	89.3	97.7	89.9	95.3	93.0	82.7	89.8
06	WOOD, PLASTICS & COMPOSITES	105.3	85.3	94.0	111.6	76.6	91.9	109.5	83.2	94.7	95.8	77.0	85.2	111.4	80.6	94.0	97.9	76.0	85.5
07	THERMAL & MOISTURE PROTECTION	100.9	82.6	93.5	105.0	77.1	93.7	94.9	81.5	89.5	99.2	78.3	90.7	97.9	78.0	89.8	86.8	71.4	80.6
08	OPENINGS	103.1	84.8	98.8	90.0	79.6	87.5	99.2	83.1	95.4	100.4	76.8	94.8	98.5	81.7	94.5	97.9	78.2	93.2
0920	Plaster & Gypsum Board	88.5	85.1	86.2	99.8	76.3	83.8	91.8	83.3	86.0	72.6	76.8	75.5	93.1	80.6	84.6	91.2	75.6	80.6
0950, 0980	Ceilings & Acoustic Treatment	91.3	85.1	87.2	89.6	76.3	80.8	79.4	83.3	82.0	93.2	76.8	82.3	84.8	80.6	82.0	88.8	75.6	80.0
0960	Flooring	101.7	86.6	97.2	103.6	89.9	99.5	100.4	88.1	96.8	74.8	86.6	78.3	100.7	86.6	96.5	98.3	64.0	88.1
0970, 0990	Wall Finishes & Painting/Coating	99.5	83.0	89.6	97.3	73.3	82.9	89.6	95.1	92.9	90.6	73.2	80.1	89.6	70.9	78.4	94.7	84.5	88.6
09	FINISHES	95.4	85.4	89.8	96.6	79.3	87.0	92.9	85.2	88.6	85.8	78.9	82.0	94.0	80.7	86.6	94.5	74.3	83.2
COVERS	DIVS. 10 - 14, 25, 28, 41, 43, 44, 46	100.0	93.0	98.6	100.0	89.1	97.8	100.0	91.3	98.3	100.0	44.9	89.0	100.0	90.9	98.2	100.0	44.3	88.8
21, 22, 23	FIRE SUPPRESSION, PLUMBING & HVAC	99.9	80.1	92.0	95.1	79.2	88.7	94.9	80.2	89.0	95.6	75.7	87.6	99.8	75.8	90.2	95.1	76.9	87.8
26, 27, 3370	ELECTRICAL, COMMUNICATIONS & UTIL.	100.7	88.3	94.2	90.4	78.8	84.4	96.8	82.6	89.4	92.3	75.3	83.4	90.2	80.3	85.0	92.8	77.6	84.9
MF2010	WEIGHTED AVERAGE	98.3	85.1	92.6	93.3	82.0	88.4	96.3	82.9	90.5	93.3	79.3	87.2	97.3	81.8	90.6	94.4	77.0	86.8

	DIVISION	INDIANA									IOWA								
		SOUTH BEND 465 - 466			TERRE HAUTE 478			WASHINGTON 475			BURLINGTON 526			CARROLL 514			CEDAR RAPIDS 522 - 524		
		MAT.	INST.	TOTAL	MAT.	INST.	TOTAL	MAT.	INST.	TOTAL	MAT.	INST.	TOTAL	MAT.	INST.	TOTAL	MAT.	INST.	TOTAL
015433	CONTRACTOR EQUIPMENT		105.0	105.0		116.1	116.1		116.1	116.1		100.2	100.2		100.2	100.2		96.8	96.8
0241, 31 - 34	SITE & INFRASTRUCTURE, DEMOLITION	92.9	95.9	95.0	89.7	124.2	113.9	89.2	121.5	111.9	96.5	97.3	97.1	85.8	97.3	93.9	98.5	96.0	96.8
0310	Concrete Forming & Accessories	102.6	79.8	82.9	96.3	78.1	80.6	97.1	80.6	82.9	97.3	77.3	80.1	85.2	51.6	56.2	103.2	81.0	84.1
0320	Concrete Reinforcing	101.1	79.1	90.1	95.4	83.8	89.6	88.2	50.3	69.3	94.3	85.4	89.9	95.1	81.1	88.1	95.1	86.8	90.9
0330	Cast-in-Place Concrete	96.3	80.7	89.8	93.3	87.0	90.7	101.5	86.6	95.3	109.6	56.3	87.5	109.6	61.0	89.4	109.9	81.3	98.0
03	CONCRETE	90.8	81.3	86.1	105.6	82.6	94.2	111.1	77.1	94.3	101.7	72.5	87.3	100.5	61.7	81.3	101.8	82.8	92.4
04	MASONRY	90.7	79.1	83.5	93.6	80.5	85.5	86.4	82.3	83.9	97.2	66.0	77.9	99.0	73.5	83.2	102.8	77.5	87.1
05	METALS	93.0	99.8	95.1	89.9	88.1	89.4	84.7	69.6	80.0	87.0	94.9	89.4	87.0	90.4	88.1	89.3	93.8	90.7
06	WOOD, PLASTICS & COMPOSITES	105.0	79.2	90.4	99.5	76.4	86.5	100.0	80.7	89.1	95.3	78.4	85.7	81.8	45.9	61.6	101.8	80.5	89.8
07	THERMAL & MOISTURE PROTECTION	97.5	81.9	91.2	99.4	81.9	92.3	99.5	84.4	93.4	103.8	75.6	92.4	104.1	70.5	90.5	104.7	79.9	94.6
08	OPENINGS	91.0	78.9	88.1	98.9	79.4	94.3	95.6	69.2	89.3	92.4	72.1	87.6	96.6	54.7	86.7	97.6	81.7	93.8
0920	Plaster & Gypsum Board	87.0	78.9	81.5	93.1	75.0	80.8	93.0	79.5	83.8	100.7	77.7	85.1	95.9	44.3	60.8	105.3	80.1	88.2
0950, 0980	Ceilings & Acoustic Treatment	93.4	78.9	83.8	88.8	75.0	79.6	82.5	79.5	80.5	110.6	77.7	88.7	110.6	44.3	66.5	114.1	80.1	91.5
0960	Flooring	97.4	89.9	95.1	100.8	86.6	96.6	101.7	77.8	94.6	111.2	38.5	89.5	105.6	33.8	84.2	127.9	79.7	113.5
0970, 0990	Wall Finishes & Painting/Coating	87.6	87.4	87.5	94.7	83.7	88.1	94.7	85.2	89.0	99.0	80.2	87.7	99.0	79.9	87.5	100.4	73.7	84.3
09	FINISHES	95.3	82.4	88.1	95.2	79.9	86.6	94.3	81.4	87.1	107.2	70.4	86.7	103.3	49.4	73.3	113.2	80.0	94.7
COVERS	DIVS. 10 - 14, 25, 28, 41, 43, 44, 46	100.0	90.7	98.1	100.0	93.2	98.6	100.0	94.0	98.8	100.0	89.2	97.8	100.0	66.9	93.4	100.0	90.2	98.0
21, 22, 23	FIRE SUPPRESSION, PLUMBING & HVAC	99.9	77.4	90.9	100.0	80.4	92.1	95.1	79.3	88.7	95.3	76.9	87.9	95.3	74.2	86.8	100.2	80.7	92.4
26, 27, 3370	ELECTRICAL, COMMUNICATIONS & UTIL.	97.6	86.6	91.9	92.6	83.6	87.9	93.1	86.3	89.5	100.7	69.3	84.3	101.4	81.3	90.9	98.2	81.3	89.3
MF2010	WEIGHTED AVERAGE	95.7	84.2	90.7	97.2	85.8	92.2	95.0	83.3	89.9	96.6	76.9	88.0	96.4	72.2	85.9	99.4	83.4	92.4

		IOWA																	
	DIVISION	COUNCIL BLUFFS			CRESTON			DAVENPORT			DECORAH			DES MOINES			DUBUQUE		
		515			508			527 - 528			521			500 - 503,509			520		
		MAT.	INST.	TOTAL	MAT.	INST.	TOTAL	MAT.	INST.	TOTAL	MAT.	INST.	TOTAL	MAT.	INST.	TOTAL	MAT.	INST.	TOTAL
015433	CONTRACTOR EQUIPMENT		96.3	96.3		100.2	100.2		100.2	100.2		100.2	100.2		101.8	101.8		95.7	95.7
0241, 31 - 34	SITE & INFRASTRUCTURE, DEMOLITION	103.6	92.3	95.7	86.7	96.4	93.5	96.8	99.3	98.6	95.1	96.3	95.9	95.5	99.7	98.5	96.4	93.3	94.2
0310	Concrete Forming & Accessories	84.6	70.7	72.6	84.5	67.2	69.6	102.5	91.2	92.7	94.9	46.5	53.1	103.2	78.9	82.2	85.9	73.8	75.4
0320	Concrete Reinforcing	97.0	79.3	88.2	93.3	81.2	87.3	95.1	100.0	97.5	94.3	79.0	86.7	102.7	81.4	92.0	93.7	86.6	90.2
0330	Cast-in-Place Concrete	114.3	75.1	98.0	102.1	65.4	86.9	105.9	91.7	100.0	106.7	59.3	87.0	89.8	88.0	89.1	107.6	100.4	104.6
03	CONCRETE	104.1	74.7	89.6	96.9	70.2	83.7	99.9	93.5	96.7	99.6	58.4	79.3	95.0	83.2	89.2	98.6	86.2	92.4
04	MASONRY	104.5	75.9	86.8	96.9	81.4	87.3	100.0	85.5	91.1	117.7	72.7	89.8	90.7	80.4	84.3	103.7	73.7	85.1
05	METALS	94.3	90.2	93.1	89.4	90.6	89.8	89.3	103.9	93.8	87.1	87.7	87.3	95.2	92.9	94.5	88.0	93.5	89.7
06	WOOD, PLASTICS & COMPOSITES	80.6	69.7	74.4	79.1	63.1	70.1	101.9	90.3	95.3	92.3	37.9	61.6	98.8	77.4	86.8	82.2	71.7	76.3
07	THERMAL & MOISTURE PROTECTION	104.2	68.4	89.6	106.2	78.8	95.1	104.2	87.4	97.4	104.0	54.7	84.0	100.6	79.5	92.0	104.4	71.3	91.0
08	OPENINGS	96.6	75.5	91.6	106.6	64.4	96.6	97.6	91.4	96.1	95.2	49.8	84.5	101.9	82.8	97.4	96.6	79.3	92.5
0920	Plaster & Gypsum Board	95.9	69.0	77.6	94.1	62.0	72.2	105.3	90.0	94.9	99.6	36.0	56.4	91.9	76.8	81.6	95.9	71.1	79.0
0950, 0980	Ceilings & Acoustic Treatment	110.6	69.0	82.9	111.5	62.0	78.5	114.1	90.0	98.1	110.6	36.0	61.0	115.0	76.8	89.6	110.6	71.1	84.3
0960	Flooring	104.3	82.1	97.6	104.2	33.8	83.2	113.5	98.7	109.1	111.1	48.0	92.2	109.4	87.6	102.9	118.8	77.6	106.5
0970, 0990	Wall Finishes & Painting/Coating	95.3	66.9	78.2	95.8	79.9	86.3	99.0	97.6	98.1	99.0	33.6	59.7	91.2	79.9	84.4	99.4	65.1	78.8
09	FINISHES	104.1	72.0	86.2	102.4	61.5	79.6	109.1	93.2	100.2	106.9	43.6	71.6	105.5	80.4	91.5	108.4	72.9	88.6
COVERS	DIVS. 10 - 14, 25, 28, 41, 43, 44, 46	100.0	88.9	97.8	100.0	71.4	94.3	100.0	93.1	98.6	100.0	85.2	97.0	100.0	90.3	98.1	100.0	88.6	97.7
21, 22, 23	FIRE SUPPRESSION, PLUMBING & HVAC	100.2	75.8	90.4	95.3	80.3	89.2	100.2	93.2	97.4	95.3	74.5	86.9	100.0	79.8	91.9	100.2	75.4	90.2
26, 27, 3370	ELECTRICAL, COMMUNICATIONS & UTIL.	103.7	81.6	92.1	94.8	81.3	87.8	95.5	91.2	93.2	98.1	45.4	70.6	106.9	81.3	93.6	102.2	78.4	89.8
MF2010	WEIGHTED AVERAGE	100.2	78.8	90.9	96.7	77.9	88.5	98.4	93.4	96.2	97.4	65.6	83.5	99.4	83.9	92.7	98.5	80.4	90.6

		IOWA																	
	DIVISION	FORT DODGE			MASON CITY			OTTUMWA			SHENANDOAH			SIBLEY			SIOUX CITY		
		505			504			525			516			512			510 - 511		
		MAT.	INST.	TOTAL	MAT.	INST.	TOTAL	MAT.	INST.	TOTAL	MAT.	INST.	TOTAL	MAT.	INST.	TOTAL	MAT.	INST.	TOTAL
015433	CONTRACTOR EQUIPMENT		100.2	100.2		100.2	100.2		95.7	95.7		96.3	96.3		100.2	100.2		100.2	100.2
0241, 31 - 34	SITE & INFRASTRUCTURE, DEMOLITION	94.6	95.0	94.9	94.6	96.2	95.8	96.8	91.4	93.0	102.0	91.4	94.6	106.0	95.0	98.3	107.6	95.8	99.3
0310	Concrete Forming & Accessories	85.1	46.0	51.3	89.6	46.4	52.3	92.9	71.9	74.8	86.2	57.8	61.7	86.6	38.7	45.3	103.2	67.9	72.7
0320	Concrete Reinforcing	93.3	67.4	80.4	93.2	80.6	86.9	94.3	85.6	90.0	97.0	68.6	82.8	97.0	67.3	82.2	95.1	78.2	86.7
0330	Cast-in-Place Concrete	95.8	45.0	74.7	95.8	58.7	80.4	110.3	54.8	87.3	110.5	61.1	90.0	108.3	45.7	82.3	109.0	56.7	87.3
03	CONCRETE	92.6	51.1	72.1	92.9	58.5	75.9	101.2	69.4	85.5	101.5	62.1	82.1	100.5	48.1	74.6	101.2	66.9	84.3
04	MASONRY	95.9	38.5	60.4	108.7	69.4	84.4	100.3	57.1	73.6	104.0	75.9	86.6	122.1	38.8	70.6	97.2	55.6	71.4
05	METALS	89.5	81.9	87.2	89.5	88.8	89.3	86.9	92.0	88.5	93.3	84.3	90.5	87.2	81.4	85.4	89.3	88.2	89.0
06	WOOD, PLASTICS & COMPOSITES	79.6	46.2	60.8	83.9	37.9	58.0	89.4	78.2	83.1	82.2	54.0	66.3	83.1	36.7	56.9	101.8	68.2	82.9
07	THERMAL & MOISTURE PROTECTION	105.6	58.5	86.5	105.2	64.6	88.7	104.6	65.2	88.6	103.5	66.6	88.5	103.8	49.2	81.6	104.3	64.9	88.3
08	OPENINGS	100.8	50.1	88.8	93.3	50.2	83.1	96.6	76.2	91.8	88.1	55.5	80.4	93.4	44.9	81.9	97.6	66.9	90.3
0920	Plaster & Gypsum Board	94.1	44.6	60.4	94.1	36.0	54.6	97.0	77.7	83.9	95.9	52.8	66.6	95.9	34.7	54.3	105.3	67.2	79.4
0950, 0980	Ceilings & Acoustic Treatment	111.5	44.6	67.0	111.5	36.0	61.3	110.6	77.7	88.7	110.6	52.8	72.1	110.6	34.7	60.1	114.1	67.2	82.9
0960	Flooring	105.8	48.0	88.5	107.9	48.0	90.0	121.8	51.1	100.6	104.9	34.9	84.0	106.4	34.6	85.0	113.6	54.7	96.0
0970, 0990	Wall Finishes & Painting/Coating	95.8	63.1	76.1	95.8	30.7	56.7	99.4	79.9	87.7	95.3	66.9	78.2	99.0	63.1	77.4	99.0	63.1	77.4
09	FINISHES	104.4	47.1	72.5	105.0	42.9	70.4	109.5	69.2	87.0	104.2	53.4	75.9	106.7	38.9	68.9	110.7	64.8	85.1
COVERS	DIVS. 10 - 14, 25, 28, 41, 43, 44, 46	100.0	79.8	96.0	100.0	84.8	97.0	100.0	84.2	96.8	100.0	67.9	93.6	100.0	78.9	95.8	100.0	86.9	97.4
21, 22, 23	FIRE SUPPRESSION, PLUMBING & HVAC	95.3	65.8	83.4	95.3	71.9	85.9	95.3	71.3	85.7	95.3	75.1	87.2	95.3	69.5	84.9	100.2	76.3	90.6
26, 27, 3370	ELECTRICAL, COMMUNICATIONS & UTIL.	101.9	42.9	71.1	100.9	57.0	77.9	100.5	67.8	83.4	98.1	80.6	89.0	98.1	42.4	69.1	98.2	74.7	85.9
MF2010	WEIGHTED AVERAGE	96.6	58.8	80.2	96.5	66.6	83.5	97.3	73.0	86.7	97.1	72.1	86.2	97.7	57.5	80.2	99.1	73.6	88.0

		IOWA						KANSAS											
	DIVISION	SPENCER			WATERLOO			BELLEVILLE			COLBY			DODGE CITY			EMPORIA		
		513			506 - 507			669			677			678			668		
		MAT.	INST.	TOTAL	MAT.	INST.	TOTAL	MAT.	INST.	TOTAL	MAT.	INST.	TOTAL	MAT.	INST.	TOTAL	MAT.	INST.	TOTAL
015433	CONTRACTOR EQUIPMENT		100.2	100.2		100.2	100.2		103.1	103.1		103.1	103.1		103.1	103.1		101.4	101.4
0241, 31 - 34	SITE & INFRASTRUCTURE, DEMOLITION	106.1	95.0	98.3	99.5	96.1	97.1	111.0	94.7	99.5	109.4	94.8	99.1	111.9	94.5	99.7	102.8	92.2	95.3
0310	Concrete Forming & Accessories	92.8	38.7	46.1	101.6	48.1	55.4	95.0	55.3	60.7	102.2	59.6	65.4	95.3	59.5	64.4	86.5	66.7	69.4
0320	Concrete Reinforcing	97.0	67.3	82.2	94.0	86.5	90.2	99.6	56.1	77.9	101.2	56.1	78.7	98.8	56.1	77.5	98.2	56.7	77.5
0330	Cast-in-Place Concrete	108.3	45.7	82.3	102.6	55.0	82.9	118.9	56.9	93.2	114.3	57.0	90.6	116.4	56.7	91.6	114.9	54.8	90.0
03	CONCRETE	100.9	48.1	74.9	98.4	59.1	79.0	117.7	57.4	88.0	114.9	59.3	87.5	116.2	59.1	88.0	110.0	61.8	86.2
04	MASONRY	122.1	38.8	70.6	96.6	73.0	82.0	94.9	59.8	73.2	96.9	59.8	73.9	104.4	60.0	76.9	100.5	68.4	80.7
05	METALS	87.1	81.4	85.4	91.8	92.8	92.1	97.8	78.4	91.9	98.2	78.7	92.2	99.7	77.8	92.9	97.6	79.9	92.1
06	WOOD, PLASTICS & COMPOSITES	89.3	36.7	59.6	98.6	37.9	64.4	96.4	53.3	72.1	103.9	59.2	78.7	95.3	59.2	74.9	87.7	68.2	76.7
07	THERMAL & MOISTURE PROTECTION	104.6	49.2	82.1	105.4	71.4	91.6	98.4	62.3	83.7	101.9	62.9	86.1	101.9	62.3	85.8	97.0	77.4	89.1
08	OPENINGS	104.3	44.9	90.2	94.2	53.5	84.5	96.4	49.4	85.3	99.5	52.6	88.4	99.5	52.6	88.4	94.4	57.5	85.6
0920	Plaster & Gypsum Board	97.0	34.7	54.7	103.9	36.0	57.7	98.0	51.8	66.6	100.7	57.9	71.6	94.8	57.9	69.7	95.1	67.1	76.0
0950, 0980	Ceilings & Acoustic Treatment	110.6	34.7	60.1	115.0	36.0	62.5	91.6	51.8	65.1	94.3	57.9	70.1	94.3	57.9	70.1	91.6	67.1	75.3
0960	Flooring	109.1	34.6	86.8	113.3	76.9	102.4	106.4	39.9	86.5	105.7	39.9	86.0	102.2	39.9	83.6	101.8	37.1	82.4
0970, 0990	Wall Finishes & Painting/Coating	99.0	63.1	77.4	95.8	79.4	86.0	97.4	39.8	62.8	97.4	39.8	62.8	97.4	39.8	62.8	97.4	39.8	62.8
09	FINISHES	107.6	38.9	69.3	108.8	54.4	78.5	101.3	50.4	72.9	101.5	53.9	75.0	99.8	53.9	74.2	98.6	58.6	76.3
COVERS	DIVS. 10 - 14, 25, 28, 41, 43, 44, 46	100.0	78.9	95.8	100.0	85.3	97.1	100.0	41.8	88.3	100.0	42.5	88.5	100.0	42.5	88.5	100.0	41.5	88.3
21, 22, 23	FIRE SUPPRESSION, PLUMBING & HVAC	95.3	69.5	84.9	100.2	80.0	92.1	95.1	73.1	86.3	95.1	68.7	84.5	100.0	68.7	87.4	95.1	73.4	86.4
26, 27, 3370	ELECTRICAL, COMMUNICATIONS & UTIL.	99.9	42.4	69.9	97.0	57.0	76.1	93.3	66.0	79.1	97.9	66.0	81.3	94.7	73.4	83.6	90.8	73.1	81.5
MF2010	WEIGHTED AVERAGE	99.2	57.5	81.0	98.3	70.9	86.4	99.4	65.6	84.7	100.2	65.6	85.1	101.7	66.5	86.4	97.8	70.0	85.7

DIVISION		KANSAS																	
		FORT SCOTT			HAYS			HUTCHINSON			INDEPENDENCE			KANSAS CITY			LIBERAL		
		667			676			675			673			660 - 662			679		
		MAT.	INST.	TOTAL	MAT.	INST.	TOTAL	MAT.	INST.	TOTAL	MAT.	INST.	TOTAL	MAT.	INST.	TOTAL	MAT.	INST.	TOTAL
015433	CONTRACTOR EQUIPMENT		102.3	102.3		103.1	103.1		103.1	103.1		103.1	103.1		99.7	99.7		103.1	103.1
0241, 31 - 34	SITE & INFRASTRUCTURE, DEMOLITION	99.4	93.0	94.9	115.5	94.8	100.9	93.4	94.6	94.3	113.1	94.6	100.1	93.5	92.9	93.1	115.2	94.6	100.7
0310	Concrete Forming & Accessories	102.6	78.6	81.9	99.7	59.6	65.1	89.9	59.5	63.6	110.8	67.8	73.7	99.3	96.3	96.7	95.8	59.5	64.5
0320	Concrete Reinforcing	97.5	96.3	96.9	98.8	56.1	77.5	98.8	56.1	77.5	98.2	63.0	80.6	94.6	97.8	96.2	100.2	56.1	78.2
0330	Cast-in-Place Concrete	106.6	55.7	85.5	90.2	57.0	76.5	83.6	54.2	71.4	116.8	54.5	91.0	91.4	99.3	94.7	90.3	54.2	75.3
03	CONCRETE	105.2	74.6	90.1	106.7	59.3	83.3	89.1	58.3	73.9	117.4	63.3	90.7	97.3	97.9	97.6	108.9	58.3	83.9
04	MASONRY	101.0	61.4	76.5	104.3	59.8	76.8	96.4	59.6	73.6	94.0	65.7	76.5	103.6	99.5	101.1	102.4	59.6	75.9
05	METALS	97.5	94.4	96.5	97.8	78.7	91.9	97.6	77.7	91.5	97.5	81.3	92.5	104.9	100.6	103.5	98.1	77.7	91.8
06	WOOD, PLASTICS & COMPOSITES	106.5	85.4	94.6	100.6	59.2	77.2	89.9	59.2	72.6	114.7	69.6	89.3	102.4	96.7	99.1	95.9	59.2	75.2
07	THERMAL & MOISTURE PROTECTION	97.8	82.8	91.7	102.1	62.9	86.2	100.9	62.2	85.2	101.9	82.4	94.0	97.5	99.8	98.4	102.3	62.2	86.0
08	OPENINGS	94.4	80.1	91.0	99.4	52.6	88.3	99.4	52.6	88.3	97.2	59.8	88.4	95.6	90.5	94.4	99.5	52.6	88.4
0920	Plaster & Gypsum Board	100.6	84.9	89.9	98.1	57.9	70.7	93.6	57.9	69.3	107.8	68.6	81.1	93.5	96.5	95.5	95.5	57.9	69.9
0950, 0980	Ceilings & Acoustic Treatment	91.6	84.9	87.2	94.3	57.9	70.1	94.3	57.9	70.1	94.3	68.6	77.2	91.6	96.5	94.9	94.3	57.9	70.1
0960	Flooring	117.6	40.1	94.4	104.7	39.9	85.3	99.5	39.9	81.7	109.9	39.9	89.0	93.5	100.5	95.6	102.4	39.9	83.7
0970, 0990	Wall Finishes & Painting/Coating	99.2	39.8	63.5	97.4	39.8	62.8	97.4	39.8	62.8	97.4	39.8	62.8	105.6	69.2	83.7	97.4	39.8	62.8
09	FINISHES	104.0	68.4	84.2	101.4	53.9	74.9	97.3	53.9	73.1	103.8	60.0	79.4	98.5	93.1	95.4	100.7	53.9	74.6
COVERS	DIVS. 10 - 14, 25, 28, 41, 43, 44, 46	100.0	47.4	89.5	100.0	42.5	88.5	100.0	42.5	88.5	100.0	43.7	88.7	100.0	63.5	92.7	100.0	42.5	88.5
21, 22, 23	FIRE SUPPRESSION, PLUMBING & HVAC	95.1	69.3	84.7	95.1	68.7	84.5	95.1	68.7	84.5	95.1	71.1	85.5	99.9	96.8	98.6	95.1	68.7	84.5
26, 27, 3370	ELECTRICAL, COMMUNICATIONS & UTIL.	90.1	73.1	81.2	96.8	66.0	80.7	91.7	66.0	78.3	94.0	73.1	83.0	95.5	97.3	96.4	94.7	73.4	83.6
MF2010	WEIGHTED AVERAGE	97.7	74.3	87.5	99.6	65.6	84.8	95.8	65.4	82.5	99.9	70.3	87.0	99.4	95.7	97.7	99.5	66.4	85.1

DIVISION		KANSAS									KENTUCKY								
		SALINA			TOPEKA			WICHITA			ASHLAND			BOWLING GREEN			CAMPTON		
		674			664 - 666			670 - 672			411 - 412			421 - 422			413 - 414		
		MAT.	INST.	TOTAL	MAT.	INST.	TOTAL	MAT.	INST.	TOTAL	MAT.	INST.	TOTAL	MAT.	INST.	TOTAL	MAT.	INST.	TOTAL
015433	CONTRACTOR EQUIPMENT		103.1	103.1		101.4	101.4		103.1	103.1		98.0	98.0		95.1	95.1		101.5	101.5
0241, 31 - 34	SITE & INFRASTRUCTURE, DEMOLITION	102.2	94.8	97.0	97.9	91.2	93.2	98.7	93.3	94.9	108.9	87.0	93.5	75.6	97.7	91.1	83.8	98.8	94.3
0310	Concrete Forming & Accessories	91.8	55.7	60.6	97.2	41.8	49.4	97.8	51.7	58.0	89.0	105.7	103.4	86.5	84.0	84.4	90.4	84.6	85.4
0320	Concrete Reinforcing	98.2	62.6	80.4	97.7	97.2	97.5	97.5	75.8	86.7	89.1	106.1	97.6	86.4	89.5	88.0	87.3	105.0	96.1
0330	Cast-in-Place Concrete	101.0	54.7	81.8	96.1	46.5	75.6	94.7	52.7	77.3	88.7	102.9	94.6	88.0	97.5	91.9	98.1	75.2	88.6
03	CONCRETE	103.0	58.0	80.8	98.3	55.4	77.1	97.6	57.8	78.0	96.1	105.2	100.6	94.4	89.7	92.1	98.2	85.2	91.8
04	MASONRY	118.4	59.8	82.1	97.6	56.9	72.4	94.9	51.1	67.8	92.2	105.1	100.2	94.0	81.6	86.3	91.6	70.2	78.4
05	METALS	99.5	81.5	93.9	101.6	96.4	100.0	101.6	83.7	96.1	92.1	109.8	97.6	93.7	86.9	91.6	93.0	93.7	93.2
06	WOOD, PLASTICS & COMPOSITES	91.5	53.3	70.0	95.4	38.1	63.1	95.3	52.5	71.2	79.0	105.4	93.9	92.3	83.9	87.5	90.8	91.1	90.9
07	THERMAL & MOISTURE PROTECTION	101.4	62.6	85.7	100.6	69.1	87.8	100.0	57.6	82.8	91.3	99.3	94.5	86.7	82.1	84.8	99.6	69.4	87.3
08	OPENINGS	99.4	49.7	87.6	98.6	55.3	88.4	101.6	56.6	91.0	97.0	99.3	97.5	97.9	79.4	93.5	99.3	89.8	97.1
0920	Plaster & Gypsum Board	93.6	51.8	65.2	97.6	36.1	55.8	92.4	51.0	64.2	61.2	105.6	91.4	86.4	83.7	84.6	86.4	90.2	89.0
0950, 0980	Ceilings & Acoustic Treatment	94.3	51.8	66.0	97.6	36.1	56.7	99.4	51.0	67.2	83.8	105.6	98.3	88.8	83.7	85.4	88.8	90.2	89.7
0960	Flooring	100.7	63.1	89.5	106.9	75.0	97.3	108.3	64.1	95.1	80.1	104.2	87.3	96.5	87.6	93.9	98.6	42.0	81.7
0970, 0990	Wall Finishes & Painting/Coating	97.4	39.8	62.8	99.2	53.7	71.9	96.3	49.0	67.9	96.9	103.4	100.8	94.7	74.3	82.4	94.7	66.8	77.9
09	FINISHES	98.5	54.8	74.2	103.1	47.0	71.8	102.7	53.5	75.3	82.5	105.7	95.4	93.1	84.0	88.0	93.9	75.7	83.8
COVERS	DIVS. 10 - 14, 25, 28, 41, 43, 44, 46	100.0	54.2	90.8	100.0	48.4	89.7	100.0	51.1	90.2	100.0	96.3	99.3	100.0	59.8	91.9	100.0	57.3	91.4
21, 22, 23	FIRE SUPPRESSION, PLUMBING & HVAC	100.0	68.7	87.5	100.0	69.0	87.5	99.8	64.3	85.6	95.0	92.9	94.1	100.0	84.3	93.7	95.1	81.5	89.7
26, 27, 3370	ELECTRICAL, COMMUNICATIONS & UTIL.	94.5	73.3	83.4	94.2	73.1	83.2	99.0	73.3	85.6	90.7	99.6	95.4	93.3	82.1	87.4	90.6	67.7	78.6
MF2010	WEIGHTED AVERAGE	100.4	67.0	85.9	99.4	66.5	85.1	99.9	65.1	84.8	94.0	100.1	96.6	95.5	84.8	90.8	95.1	80.2	88.6

DIVISION		KENTUCKY																	
		CORBIN			COVINGTON			ELIZABETHTOWN			FRANKFORT			HAZARD			HENDERSON		
		407 - 409			410			427			406			417 - 418			424		
		MAT.	INST.	TOTAL	MAT.	INST.	TOTAL	MAT.	INST.	TOTAL	MAT.	INST.	TOTAL	MAT.	INST.	TOTAL	MAT.	INST.	TOTAL
015433	CONTRACTOR EQUIPMENT		101.5	101.5		105.0	105.0		95.1	95.1		101.5	101.5		101.5	101.5		116.1	116.1
0241, 31 - 34	SITE & INFRASTRUCTURE, DEMOLITION	83.0	98.5	93.9	79.4	112.3	102.5	70.4	97.8	89.6	87.1	99.7	95.9	81.8	100.1	94.6	78.4	124.1	110.5
0310	Concrete Forming & Accessories	88.9	72.3	74.6	86.0	92.9	92.0	81.6	82.4	82.3	99.5	73.4	77.0	87.1	85.4	85.6	93.4	83.5	84.8
0320	Concrete Reinforcing	87.0	72.1	79.5	85.9	94.0	90.0	86.9	93.8	90.3	99.8	93.7	96.8	87.7	105.0	96.3	86.6	87.6	87.1
0330	Cast-in-Place Concrete	92.2	59.9	78.8	93.9	99.6	96.3	79.6	74.6	77.5	87.2	70.6	80.3	94.4	84.2	90.2	77.8	92.5	83.9
03	CONCRETE	92.6	68.4	80.7	96.5	95.5	96.0	86.6	82.0	84.3	93.0	76.6	84.9	95.0	88.7	91.9	91.6	87.4	89.6
04	MASONRY	89.8	63.0	73.2	105.4	103.3	104.1	78.5	76.8	77.4	87.2	80.8	83.3	89.3	72.8	79.1	97.8	94.5	95.8
05	METALS	93.4	79.8	89.2	91.0	91.7	91.2	92.8	88.8	91.6	95.8	89.5	93.9	93.0	93.7	93.2	84.4	88.6	85.7
06	WOOD, PLASTICS & COMPOSITES	88.5	77.7	82.4	88.7	86.0	87.2	87.4	83.9	85.4	105.5	70.3	85.6	87.7	91.1	89.6	95.1	80.8	87.1
07	THERMAL & MOISTURE PROTECTION	98.8	65.6	85.3	99.4	95.4	97.8	86.4	78.8	83.3	98.8	77.1	90.0	99.5	71.3	88.1	99.0	93.6	96.8
08	OPENINGS	99.3	67.2	91.7	101.3	85.7	97.6	97.9	85.3	94.9	102.1	78.2	96.4	99.7	80.3	95.1	96.0	81.8	92.7
0920	Plaster & Gypsum Board	90.6	76.4	80.9	69.4	86.1	80.8	85.6	83.7	84.3	100.9	68.8	79.0	85.6	90.2	88.7	89.7	79.7	82.9
0950, 0980	Ceilings & Acoustic Treatment	86.1	76.4	79.6	92.3	86.1	88.2	88.8	83.7	85.4	96.8	68.8	78.1	88.8	90.2	89.7	82.5	79.7	80.6
0960	Flooring	100.6	42.0	83.1	72.7	86.2	76.7	94.1	87.6	92.2	108.0	55.4	92.3	97.1	42.7	80.9	100.1	87.6	96.4
0970, 0990	Wall Finishes & Painting/Coating	94.2	50.5	67.9	90.6	84.1	86.7	94.7	77.8	84.6	93.9	81.5	86.5	94.7	66.8	77.9	94.7	98.1	96.7
09	FINISHES	93.9	65.0	77.8	84.8	90.5	87.9	92.0	82.5	86.7	101.6	70.3	84.2	93.2	76.5	83.9	92.6	85.7	88.8
COVERS	DIVS. 10 - 14, 25, 28, 41, 43, 44, 46	100.0	49.7	89.9	100.0	102.8	100.6	100.0	86.6	97.3	100.0	67.2	93.4	100.0	58.1	91.6	100.0	67.1	93.4
21, 22, 23	FIRE SUPPRESSION, PLUMBING & HVAC	95.1	72.9	86.2	95.6	90.6	93.6	95.4	81.9	90.0	100.1	84.1	93.7	95.1	82.4	90.0	95.4	79.3	88.9
26, 27, 3370	ELECTRICAL, COMMUNICATIONS & UTIL.	90.5	82.1	86.1	94.6	79.6	86.8	90.4	82.1	86.1	99.0	89.2	93.9	90.6	59.7	74.5	92.7	82.2	87.2
MF2010	WEIGHTED AVERAGE	94.4	73.3	85.2	95.2	93.2	94.3	92.2	83.7	88.5	97.9	82.4	91.1	94.6	79.9	88.2	93.0	87.9	90.8

City Cost Indexes

DIVISION		KENTUCKY																	
		LEXINGTON			LOUISVILLE			OWENSBORO			PADUCAH			PIKEVILLE			SOMERSET		
		403 - 405			400 - 402			423			420			415 - 416			425 - 426		
		MAT.	INST.	TOTAL	MAT.	INST.	TOTAL	MAT.	INST.	TOTAL	MAT.	INST.	TOTAL	MAT.	INST.	TOTAL	MAT.	INST.	TOTAL
015433	**CONTRACTOR EQUIPMENT**		101.5	101.5		95.1	95.1		116.1	116.1		116.1	116.1		98.0	98.0		101.5	101.5
0241, 31 - 34	**SITE & INFRASTRUCTURE, DEMOLITION**	85.0	100.9	96.2	75.7	97.8	91.3	87.9	124.1	113.3	80.9	123.6	110.9	119.6	86.0	96.0	75.1	99.3	92.1
0310	Concrete Forming & Accessories	101.5	74.3	78.0	100.6	83.0	85.4	91.8	83.4	84.5	89.9	81.8	82.9	97.7	90.9	91.9	88.0	77.6	79.0
0320	Concrete Reinforcing	95.5	93.5	94.5	97.1	93.8	95.4	86.6	90.2	88.4	87.1	89.0	88.0	89.6	105.9	97.7	86.9	93.7	90.3
0330	Cast-in-Place Concrete	94.3	93.9	94.2	95.8	75.4	87.4	90.7	96.1	92.9	82.9	88.3	85.1	97.5	98.7	98.0	77.8	100.3	87.1
03	**CONCRETE**	95.8	84.9	90.5	96.7	82.5	89.7	103.4	89.1	96.4	96.0	85.5	90.8	109.6	97.1	103.5	81.8	88.6	85.2
04	**MASONRY**	88.4	73.4	79.1	88.9	78.2	82.3	90.5	92.6	91.8	93.5	90.0	91.3	90.0	98.5	95.2	84.7	78.1	80.6
05	**METALS**	95.8	89.9	93.9	97.6	88.9	94.9	85.9	90.2	87.2	83.0	88.4	84.6	92.0	108.5	97.1	92.9	89.3	91.8
06	**WOOD, PLASTICS & COMPOSITES**	105.1	70.3	85.5	103.1	83.9	92.3	93.0	80.8	86.1	90.6	80.9	85.2	88.6	90.4	89.6	88.3	77.7	82.3
07	**THERMAL & MOISTURE PROTECTION**	99.1	85.7	93.6	97.9	79.5	90.4	99.3	93.9	97.1	99.0	80.9	91.7	91.9	81.5	87.7	99.0	71.8	87.9
08	**OPENINGS**	99.5	75.3	93.8	99.0	85.3	95.8	96.0	83.9	93.1	95.2	79.4	91.5	97.7	88.4	95.5	98.7	79.6	94.2
0920	Plaster & Gypsum Board	100.6	68.8	79.0	93.2	83.7	86.7	88.2	79.7	82.4	87.1	79.8	82.1	64.5	90.2	82.0	85.6	76.4	79.4
0950, 0980	Ceilings & Acoustic Treatment	90.5	68.8	76.1	94.3	83.7	87.2	82.5	79.7	80.6	82.5	79.8	80.7	83.8	90.2	88.0	88.8	76.4	80.5
0960	Flooring	105.0	71.6	95.0	102.9	87.6	98.3	99.5	87.6	96.0	98.5	60.4	87.1	83.8	104.2	89.9	97.4	42.0	80.8
0970, 0990	Wall Finishes & Painting/Coating	94.2	82.8	87.3	93.9	77.8	84.2	94.7	98.1	96.7	94.7	81.4	86.7	96.9	83.0	88.5	94.7	77.8	84.6
09	**FINISHES**	97.5	73.6	84.2	97.9	83.5	89.9	92.9	85.2	88.6	92.1	77.8	84.1	84.9	93.0	89.4	92.6	70.9	80.5
COVERS	**DIVS. 10 - 14, 25, 28, 41, 43, 44, 46**	100.0	88.4	97.7	100.0	87.1	97.4	100.0	100.9	100.2	100.0	63.4	92.7	100.0	57.8	91.5	100.0	60.0	92.0
21, 22, 23	**FIRE SUPPRESSION, PLUMBING & HVAC**	100.0	80.6	92.2	100.0	82.6	93.0	100.0	78.4	91.3	95.4	82.9	90.4	95.0	89.0	92.6	95.4	79.1	88.8
26, 27, 3370	**ELECTRICAL, COMMUNICATIONS & UTIL.**	93.4	82.1	87.5	100.0	82.1	90.6	92.7	82.2	87.3	95.1	81.8	88.2	93.7	75.1	84.0	91.0	82.1	86.4
MF2010	**WEIGHTED AVERAGE**	97.0	82.4	90.7	97.8	84.2	91.9	95.6	88.9	92.7	93.2	86.2	90.2	96.2	90.0	93.5	92.7	81.5	87.8

DIVISION		LOUISIANA																	
		ALEXANDRIA			BATON ROUGE			HAMMOND			LAFAYETTE			LAKE CHARLES			MONROE		
		713 - 714			707 - 708			704			705			706			712		
		MAT.	INST.	TOTAL	MAT.	INST.	TOTAL	MAT.	INST.	TOTAL	MAT.	INST.	TOTAL	MAT.	INST.	TOTAL	MAT.	INST.	TOTAL
015433	**CONTRACTOR EQUIPMENT**		89.7	89.7		89.3	89.3		89.9	89.9		89.9	89.9		89.3	89.3		89.7	89.7
0241, 31 - 34	**SITE & INFRASTRUCTURE, DEMOLITION**	96.3	86.7	89.5	96.0	86.0	89.0	99.1	86.9	90.5	100.1	87.0	90.9	100.8	85.9	90.3	96.3	85.9	89.0
0310	Concrete Forming & Accessories	83.0	44.2	49.5	94.1	61.7	66.1	78.1	46.9	51.2	95.2	54.9	60.4	96.1	57.5	62.8	82.5	43.9	49.2
0320	Concrete Reinforcing	89.4	56.0	72.7	95.1	59.0	77.1	93.3	59.4	76.4	94.6	59.0	76.8	94.6	59.2	76.9	88.4	55.6	72.0
0330	Cast-in-Place Concrete	94.6	50.3	76.2	91.8	58.7	78.0	93.9	43.9	73.2	93.5	47.6	74.4	98.4	67.4	85.5	94.6	56.7	78.8
03	**CONCRETE**	94.7	49.7	72.5	95.9	60.7	78.5	93.3	49.1	71.5	94.4	53.8	74.4	96.7	61.8	79.5	94.5	51.5	73.3
04	**MASONRY**	116.0	54.2	77.8	97.6	54.7	71.0	99.0	51.8	69.8	99.0	53.0	70.5	98.3	59.5	74.3	110.9	48.4	72.2
05	**METALS**	89.5	71.1	83.9	106.0	70.1	94.9	97.6	69.5	88.9	96.7	70.0	88.5	96.7	70.5	88.7	89.5	68.5	83.0
06	**WOOD, PLASTICS & COMPOSITES**	89.8	42.7	63.2	98.5	65.3	79.8	84.8	48.1	64.1	106.3	57.0	78.5	104.4	59.1	78.8	89.1	43.2	63.2
07	**THERMAL & MOISTURE PROTECTION**	97.5	54.1	79.9	95.8	62.4	82.2	96.9	57.0	80.7	97.4	60.8	82.5	96.5	59.5	81.5	97.5	54.7	80.1
08	**OPENINGS**	99.8	45.8	87.0	98.6	60.0	89.5	92.7	52.5	83.2	96.3	53.9	86.3	96.3	55.7	86.7	99.7	48.8	87.7
0920	Plaster & Gypsum Board	79.0	41.4	53.4	99.4	64.6	75.7	100.1	46.8	63.8	109.6	56.0	73.2	109.6	58.1	74.6	78.6	41.9	53.6
0950, 0980	Ceilings & Acoustic Treatment	87.1	41.4	56.7	103.1	64.6	77.4	98.7	46.8	64.2	97.0	56.0	69.7	98.0	58.1	71.5	87.1	41.9	57.0
0960	Flooring	110.0	66.9	97.1	106.0	66.9	94.3	98.6	66.9	89.1	106.8	66.9	94.9	106.8	74.3	97.1	109.5	58.1	94.2
0970, 0990	Wall Finishes & Painting/Coating	99.5	65.8	79.2	96.1	47.3	66.8	103.4	49.3	70.9	103.4	59.2	76.8	103.4	51.0	71.9	99.5	49.1	69.2
09	**FINISHES**	93.2	50.0	69.1	100.6	61.6	78.9	97.8	50.9	71.7	101.1	57.5	76.8	101.3	59.9	78.2	93.0	46.5	67.1
COVERS	**DIVS. 10 - 14, 25, 28, 41, 43, 44, 46**	100.0	51.4	90.3	100.0	80.9	96.2	100.0	46.6	89.3	100.0	79.4	95.9	100.0	80.4	96.1	100.0	47.0	89.4
21, 22, 23	**FIRE SUPPRESSION, PLUMBING & HVAC**	100.2	56.5	82.7	100.0	59.5	83.8	95.2	43.7	74.5	100.1	60.7	84.3	100.1	61.7	84.7	100.2	54.1	81.7
26, 27, 3370	**ELECTRICAL, COMMUNICATIONS & UTIL.**	90.5	58.2	73.6	103.0	60.8	81.0	98.2	57.3	76.8	99.5	66.6	82.3	98.9	66.6	82.0	92.2	59.6	75.2
MF2010	**WEIGHTED AVERAGE**	96.8	57.8	79.8	100.4	63.6	84.4	96.3	54.9	78.3	98.5	62.7	82.9	98.6	65.0	84.0	96.7	56.4	79.2

DIVISION		LOUISIANA									MAINE								
		NEW ORLEANS			SHREVEPORT			THIBODAUX			AUGUSTA			BANGOR			BATH		
		700 - 701			710 - 711			703			043			044			045		
		MAT.	INST.	TOTAL	MAT.	INST.	TOTAL	MAT.	INST.	TOTAL	MAT.	INST.	TOTAL	MAT.	INST.	TOTAL	MAT.	INST.	TOTAL
015433	**CONTRACTOR EQUIPMENT**		90.2	90.2		89.7	89.7		89.9	89.9		100.6	100.6		100.6	100.6		100.6	100.6
0241, 31 - 34	**SITE & INFRASTRUCTURE, DEMOLITION**	105.2	88.4	93.4	100.8	86.0	90.4	101.4	87.0	91.3	89.2	100.8	97.3	91.0	101.0	98.0	88.9	100.8	97.2
0310	Concrete Forming & Accessories	97.7	66.5	70.8	98.8	44.7	52.2	90.1	64.3	67.8	99.2	97.0	97.3	93.6	98.8	98.1	89.7	97.2	96.2
0320	Concrete Reinforcing	96.7	60.1	78.4	92.6	56.1	74.4	93.3	59.2	76.3	96.4	110.1	103.2	90.1	111.5	100.8	89.2	111.2	100.2
0330	Cast-in-Place Concrete	99.7	70.1	87.4	99.4	53.4	80.3	100.9	51.3	80.3	97.9	60.3	82.3	80.2	113.0	93.8	80.2	60.3	72.0
03	**CONCRETE**	97.8	67.1	82.6	99.6	50.9	75.6	98.4	59.3	79.1	104.6	86.2	95.5	94.4	105.3	99.8	94.5	86.4	90.5
04	**MASONRY**	101.8	60.7	76.4	102.0	51.0	70.4	124.8	49.3	78.1	96.8	64.8	77.0	106.0	103.9	104.7	112.3	85.8	95.9
05	**METALS**	109.5	73.0	98.3	93.3	69.2	85.9	97.6	69.3	88.9	97.8	88.0	94.8	88.7	90.4	89.2	87.2	89.6	87.9
06	**WOOD, PLASTICS & COMPOSITES**	99.7	67.9	81.8	101.7	43.7	69.0	92.9	69.7	79.8	98.6	107.7	103.7	95.1	98.7	97.1	89.7	107.7	99.9
07	**THERMAL & MOISTURE PROTECTION**	96.1	69.0	85.1	97.1	52.2	78.9	96.7	54.6	79.6	104.3	66.8	89.1	100.2	86.0	94.5	100.2	73.2	89.2
08	**OPENINGS**	96.2	66.4	89.2	98.8	46.4	86.4	97.1	64.2	89.3	104.2	90.0	100.8	102.5	85.0	98.4	102.5	90.0	99.5
0920	Plaster & Gypsum Board	108.2	67.2	80.3	90.0	42.5	57.7	102.7	69.1	79.8	100.8	107.4	105.2	98.8	98.0	98.2	95.1	107.4	103.4
0950, 0980	Ceilings & Acoustic Treatment	98.7	67.2	77.8	91.3	42.5	58.8	98.7	69.1	79.0	101.5	107.4	105.4	91.6	98.0	95.9	89.6	107.4	101.4
0960	Flooring	105.3	66.9	93.8	113.4	62.4	98.1	104.3	44.9	86.5	103.1	56.3	89.1	98.9	113.0	103.1	97.3	56.3	85.0
0970, 0990	Wall Finishes & Painting/Coating	108.8	65.5	82.8	95.5	45.5	65.4	104.7	50.6	72.2	101.0	64.7	79.2	95.3	45.1	65.1	95.3	45.1	65.1
09	**FINISHES**	102.4	66.6	82.4	98.2	47.4	69.9	100.1	60.1	77.8	101.4	87.7	93.8	98.6	96.3	97.3	97.1	85.6	90.7
COVERS	**DIVS. 10 - 14, 25, 28, 41, 43, 44, 46**	100.0	83.2	96.6	100.0	77.3	95.5	100.0	81.3	96.3	100.0	103.3	100.7	100.0	110.0	102.0	100.0	103.3	100.7
21, 22, 23	**FIRE SUPPRESSION, PLUMBING & HVAC**	100.0	65.8	86.3	100.0	58.9	83.5	95.2	61.7	81.7	99.9	65.6	86.2	100.1	75.9	90.4	95.2	65.7	83.4
26, 27, 3370	**ELECTRICAL, COMMUNICATIONS & UTIL.**	103.2	72.9	87.4	97.7	67.2	81.8	96.7	72.8	84.2	99.3	80.0	89.2	98.5	77.2	87.4	96.4	80.0	87.9
MF2010	**WEIGHTED AVERAGE**	101.5	69.8	87.7	98.4	59.6	81.5	98.7	64.9	84.0	100.3	80.8	91.8	97.6	90.8	94.7	96.1	83.0	90.4

DIVISION		MAINE																	
		HOULTON			KITTERY			LEWISTON			MACHIAS			PORTLAND			ROCKLAND		
		047			039			042			046			040 - 041			048		
		MAT.	INST.	TOTAL	MAT.	INST.	TOTAL	MAT.	INST.	TOTAL	MAT.	INST.	TOTAL	MAT.	INST.	TOTAL	MAT.	INST.	TOTAL
015433	**CONTRACTOR EQUIPMENT**		100.6	100.6		100.6	100.6		100.6	100.6		100.6	100.6		100.6	100.6		100.6	100.6
0241, 31 - 34	**SITE & INFRASTRUCTURE, DEMOLITION**	90.7	101.8	98.5	85.6	100.8	96.3	88.2	101.0	97.2	90.1	101.8	98.3	86.9	101.0	96.8	86.3	101.8	97.2
0310	Concrete Forming & Accessories	97.1	103.6	102.7	90.4	97.7	96.7	98.9	98.9	98.9	94.4	103.5	102.3	95.8	98.9	98.4	95.4	103.5	102.4
0320	Concrete Reinforcing	90.1	109.9	100.0	84.5	111.3	97.8	110.8	111.5	111.1	90.1	109.9	100.0	107.5	111.5	109.5	90.1	109.9	100.0
0330	Cast-in-Place Concrete	80.3	72.6	77.1	81.4	61.2	73.0	81.9	113.0	94.8	80.2	72.0	76.8	95.5	113.0	102.7	81.9	72.0	77.8
03	**CONCRETE**	95.5	93.3	94.4	90.1	87.0	88.6	95.1	105.3	100.1	94.9	93.1	94.0	100.8	105.3	103.0	92.2	93.1	92.6
04	**MASONRY**	91.0	76.3	81.9	101.0	86.8	92.2	91.7	103.9	99.2	91.0	76.3	81.9	101.3	103.9	102.9	85.9	76.3	80.0
05	**METALS**	87.4	87.9	87.5	87.1	89.8	87.9	91.7	90.4	91.3	87.4	87.9	87.6	99.4	90.4	96.6	87.3	87.9	87.5
06	**WOOD, PLASTICS & COMPOSITES**	99.1	107.7	104.0	88.7	107.7	99.4	101.2	98.7	99.8	96.1	107.7	102.6	96.1	98.7	97.5	96.9	107.7	103.0
07	**THERMAL & MOISTURE PROTECTION**	100.3	74.0	89.6	103.2	72.5	90.7	100.1	86.0	94.4	100.3	73.0	89.2	101.4	86.0	95.2	100.0	73.0	89.1
08	**OPENINGS**	102.6	88.2	99.2	102.7	90.0	99.6	105.6	85.0	100.7	102.6	88.2	99.2	105.7	85.0	100.8	102.5	88.2	99.1
0920	Plaster & Gypsum Board	101.0	107.4	105.3	94.7	107.4	103.3	104.5	98.0	100.1	99.5	107.4	104.8	101.0	98.0	99.0	99.5	107.4	104.8
0950, 0980	Ceilings & Acoustic Treatment	89.6	107.4	101.4	90.6	107.4	101.8	101.4	98.0	99.1	89.6	107.4	101.4	104.4	98.0	100.1	89.6	107.4	101.4
0960	Flooring	100.1	52.7	85.9	97.5	58.7	85.9	101.6	113.0	105.0	99.3	52.7	85.4	101.9	113.0	105.2	99.6	52.7	85.6
0970, 0990	Wall Finishes & Painting/Coating	95.3	140.1	122.2	95.4	39.3	61.7	95.3	45.1	65.1	95.3	140.1	122.2	93.4	45.1	64.4	95.3	140.1	122.2
09	**FINISHES**	98.8	99.7	99.3	96.2	85.6	90.3	101.8	96.3	98.7	98.3	99.7	99.1	102.5	96.3	99.0	98.0	99.7	98.9
COVERS	**DIVS. 10 - 14, 25, 28, 41, 43, 44, 46**	100.0	108.9	101.8	100.0	103.6	100.7	100.0	110.0	102.0	100.0	108.9	101.8	100.0	110.0	102.0	100.0	108.9	101.8
21, 22, 23	**FIRE SUPPRESSION, PLUMBING & HVAC**	95.2	74.1	86.8	95.2	77.3	88.0	100.1	75.9	90.4	95.2	74.1	86.8	100.0	75.9	90.3	95.2	74.1	86.8
26, 27, 3370	**ELECTRICAL, COMMUNICATIONS & UTIL.**	100.6	80.0	89.9	98.0	80.0	88.6	100.6	80.1	89.9	100.6	80.0	89.9	103.3	80.1	91.2	100.5	80.0	89.8
MF2010	**WEIGHTED AVERAGE**	95.9	86.6	91.8	95.1	85.7	91.0	98.3	91.2	95.2	95.7	86.5	91.7	100.8	91.2	96.6	95.0	86.5	91.3

DIVISION		MAINE			MARYLAND														
		WATERVILLE			ANNAPOLIS			BALTIMORE			COLLEGE PARK			CUMBERLAND			EASTON		
		049			214			210 - 212			207 - 208			215			216		
		MAT.	INST.	TOTAL	MAT.	INST.	TOTAL	MAT.	INST.	TOTAL	MAT.	INST.	TOTAL	MAT.	INST.	TOTAL	MAT.	INST.	TOTAL
015433	**CONTRACTOR EQUIPMENT**		100.6	100.6		99.3	99.3		102.8	102.8		104.3	104.3		99.3	99.3		99.3	99.3
0241, 31 - 34	**SITE & INFRASTRUCTURE, DEMOLITION**	90.8	100.8	97.8	98.2	90.9	93.1	99.8	94.8	96.3	103.7	94.1	96.9	91.4	90.9	91.1	98.2	88.0	91.0
0310	Concrete Forming & Accessories	89.2	97.0	96.0	98.1	70.3	74.1	100.3	72.6	76.4	86.2	71.2	73.2	91.6	80.8	82.3	89.5	71.7	74.2
0320	Concrete Reinforcing	90.1	110.1	100.1	104.0	82.1	93.0	104.5	82.1	93.3	109.0	80.5	94.7	87.7	74.6	81.2	87.0	80.6	83.8
0330	Cast-in-Place Concrete	80.3	60.3	72.0	110.8	77.1	96.8	111.3	77.9	97.4	122.5	78.0	104.0	93.6	85.7	90.3	104.0	49.2	81.2
03	**CONCRETE**	96.0	86.2	91.1	103.7	75.9	90.0	104.2	77.2	90.9	109.4	76.6	93.2	89.4	82.2	85.9	97.0	66.6	82.0
04	**MASONRY**	100.7	64.8	78.5	96.1	72.9	81.7	98.4	72.9	82.6	108.4	69.4	84.3	94.6	84.4	88.3	108.1	43.7	68.2
05	**METALS**	87.3	88.0	87.5	104.7	93.8	101.3	101.5	94.3	99.3	87.0	97.5	90.2	99.2	91.2	96.7	99.4	88.1	95.9
06	**WOOD, PLASTICS & COMPOSITES**	89.2	107.7	99.6	97.6	70.6	82.4	97.3	73.7	84.0	87.6	70.8	78.1	86.3	79.8	82.6	84.1	79.1	81.3
07	**THERMAL & MOISTURE PROTECTION**	100.3	66.8	86.7	101.1	79.6	92.4	99.1	80.2	91.4	102.6	78.5	92.9	99.5	80.4	91.7	99.6	60.4	83.7
08	**OPENINGS**	102.6	90.0	99.6	99.3	78.1	94.3	97.7	79.8	93.5	94.7	75.2	90.1	96.0	78.7	91.9	94.4	72.9	89.3
0920	Plaster & Gypsum Board	95.1	107.4	103.4	100.5	70.0	79.7	103.4	73.0	82.8	97.1	70.0	78.6	100.7	79.5	86.3	100.3	78.8	85.7
0950, 0980	Ceilings & Acoustic Treatment	89.6	107.4	101.4	100.6	70.0	80.2	98.1	73.0	81.4	98.1	70.0	79.4	100.1	79.5	86.4	100.1	78.8	85.9
0960	Flooring	97.0	56.3	84.8	100.7	78.6	94.1	100.0	78.6	93.6	97.9	80.5	92.7	96.2	92.5	95.1	95.4	52.6	82.6
0970, 0990	Wall Finishes & Painting/Coating	95.3	64.7	76.9	91.7	80.9	85.2	94.6	80.9	86.4	104.6	77.5	88.3	94.4	75.0	82.7	94.4	77.5	84.3
09	**FINISHES**	97.2	87.7	91.9	97.2	72.1	83.2	99.9	73.9	85.4	96.3	72.8	83.2	97.6	82.3	89.1	97.8	70.5	82.6
COVERS	**DIVS. 10 - 14, 25, 28, 41, 43, 44, 46**	100.0	103.3	100.7	100.0	85.6	97.1	100.0	86.3	97.3	100.0	83.8	96.8	100.0	90.2	98.0	100.0	77.2	95.4
21, 22, 23	**FIRE SUPPRESSION, PLUMBING & HVAC**	95.2	65.6	83.3	100.0	81.7	92.6	100.0	81.7	92.7	95.3	85.0	91.2	95.1	73.0	86.2	95.1	68.6	84.5
26, 27, 3370	**ELECTRICAL, COMMUNICATIONS & UTIL.**	100.6	80.0	89.8	96.7	91.7	94.1	101.4	91.7	96.3	102.2	99.8	100.9	96.6	81.6	88.8	96.1	64.4	79.5
MF2010	**WEIGHTED AVERAGE**	96.2	80.8	89.5	100.3	81.8	92.3	100.5	82.8	92.8	97.6	83.9	91.7	96.0	81.9	89.9	97.4	69.3	85.2

DIVISION		MARYLAND															MASSACHUSETTS		
		ELKTON			HAGERSTOWN			SALISBURY			SILVER SPRING			WALDORF			BOSTON		
		219			217			218			209			206			020 - 022, 024		
		MAT.	INST.	TOTAL	MAT.	INST.	TOTAL	MAT.	INST.	TOTAL	MAT.	INST.	TOTAL	MAT.	INST.	TOTAL	MAT.	INST.	TOTAL
015433	**CONTRACTOR EQUIPMENT**		99.3	99.3		99.3	99.3		99.3	99.3		97.2	97.2		97.2	97.2		106.3	106.3
0241, 31 - 34	**SITE & INFRASTRUCTURE, DEMOLITION**	85.8	88.7	87.8	89.8	91.6	91.0	98.1	88.0	91.0	91.3	87.5	88.6	98.0	87.1	90.3	100.8	108.2	106.0
0310	Concrete Forming & Accessories	95.5	82.6	84.4	90.6	76.3	78.3	103.5	52.9	59.8	94.8	71.4	74.6	102.4	69.6	74.1	100.9	142.0	136.4
0320	Concrete Reinforcing	87.0	101.6	94.3	87.7	74.6	81.2	87.0	65.6	76.3	107.6	80.4	94.0	108.3	80.4	94.4	108.7	157.1	132.8
0330	Cast-in-Place Concrete	84.2	76.5	81.0	89.2	85.7	87.8	104.0	47.9	80.7	125.4	79.8	106.5	140.5	76.9	114.1	104.6	152.2	124.3
03	**CONCRETE**	82.6	84.7	83.7	86.0	80.2	83.1	97.9	55.0	76.7	108.0	77.1	92.7	118.7	75.3	97.3	108.3	147.2	127.5
04	**MASONRY**	94.1	59.9	73.0	100.3	84.4	90.4	107.7	48.1	70.9	107.3	72.7	85.9	92.7	67.6	77.2	108.6	162.1	141.7
05	**METALS**	99.5	99.2	99.4	99.3	91.3	96.8	99.5	82.4	94.2	91.2	93.9	92.1	91.2	93.8	92.0	100.8	132.1	110.5
06	**WOOD, PLASTICS & COMPOSITES**	91.3	88.8	89.9	85.4	73.5	78.7	101.5	56.3	76.0	94.5	70.1	80.7	102.3	70.1	84.1	100.3	141.7	123.6
07	**THERMAL & MOISTURE PROTECTION**	99.2	75.1	89.4	99.2	78.6	90.8	99.8	65.2	85.7	105.6	84.4	97.0	106.1	82.2	96.4	104.0	151.9	123.5
08	**OPENINGS**	94.4	81.4	91.3	94.4	74.5	89.6	94.6	62.8	87.1	86.8	74.7	84.0	87.5	74.7	84.5	99.2	144.4	109.9
0920	Plaster & Gypsum Board	103.6	88.8	93.6	100.3	73.0	81.8	110.0	55.3	72.8	102.9	70.0	80.5	106.2	70.0	81.6	104.2	142.4	130.2
0950, 0980	Ceilings & Acoustic Treatment	100.1	88.8	92.6	101.0	73.0	82.4	100.1	55.3	70.3	107.8	70.0	82.6	107.8	70.0	82.6	101.8	142.4	128.8
0960	Flooring	97.6	59.5	86.2	95.8	92.5	94.8	101.0	65.9	90.5	104.0	80.5	97.0	107.5	80.5	99.4	98.6	183.1	123.9
0970, 0990	Wall Finishes & Painting/Coating	94.4	77.5	84.3	94.4	77.5	84.3	94.4	77.5	84.3	111.6	77.5	91.1	111.6	77.5	91.1	98.6	150.8	130.0
09	**FINISHES**	98.0	78.4	87.1	97.5	78.9	87.1	100.7	57.3	76.5	96.8	72.3	83.2	98.5	71.7	83.6	102.1	151.0	129.3
COVERS	**DIVS. 10 - 14, 25, 28, 41, 43, 44, 46**	100.0	61.4	92.3	100.0	89.5	97.9	100.0	39.7	87.9	100.0	83.1	96.6	100.0	81.4	96.3	100.0	118.6	103.7
21, 22, 23	**FIRE SUPPRESSION, PLUMBING & HVAC**	95.1	79.3	88.8	100.0	81.4	92.5	95.1	61.2	81.5	95.3	86.3	91.7	95.3	83.7	90.6	100.1	127.0	110.9
26, 27, 3370	**ELECTRICAL, COMMUNICATIONS & UTIL.**	97.9	91.7	94.6	96.4	81.6	88.7	94.8	68.7	81.2	99.4	99.8	99.6	96.6	99.8	98.3	100.9	138.6	120.6
MF2010	**WEIGHTED AVERAGE**	95.1	81.9	89.4	96.9	82.8	90.7	97.8	63.3	82.8	96.9	83.8	91.2	97.5	82.2	90.9	101.8	138.1	117.6

	DIVISION	MASSACHUSETTS																	
		BROCKTON			BUZZARDS BAY			FALL RIVER			FITCHBURG			FRAMINGHAM			GREENFIELD		
		023			025			027			014			017			013		
		MAT.	INST.	TOTAL	MAT.	INST.	TOTAL	MAT.	INST.	TOTAL	MAT.	INST.	TOTAL	MAT.	INST.	TOTAL	MAT.	INST.	TOTAL
015433	CONTRACTOR EQUIPMENT		102.4	102.4		102.4	102.4		103.3	103.3		100.6	100.6		101.8	101.8		100.6	100.6
0241, 31 - 34	SITE & INFRASTRUCTURE, DEMOLITION	95.7	104.8	102.1	85.8	104.8	99.1	94.8	104.9	101.9	85.8	104.7	99.1	82.5	104.5	97.9	89.6	103.2	99.2
0310	Concrete Forming & Accessories	102.6	131.5	127.5	100.2	131.4	127.1	102.6	131.7	127.7	94.8	126.0	121.7	102.0	131.2	127.2	93.2	112.1	109.5
0320	Concrete Reinforcing	106.8	157.0	131.9	85.6	165.2	125.4	106.8	165.2	136.0	85.4	156.0	120.6	85.3	156.7	121.0	88.6	126.0	107.3
0330	Cast-in-Place Concrete	98.6	152.4	120.9	82.0	152.4	111.2	95.4	152.9	119.3	85.7	148.3	111.7	85.8	140.8	108.6	88.2	129.2	105.2
03	CONCRETE	104.8	142.5	123.4	88.7	144.0	115.9	103.3	144.3	123.5	86.8	138.3	112.2	89.7	138.4	113.7	90.6	119.9	105.1
04	MASONRY	102.4	159.6	137.8	95.0	159.6	134.9	103.3	159.5	138.1	93.0	155.2	131.5	98.6	155.3	133.7	96.9	133.2	119.4
05	METALS	99.4	129.7	108.8	94.4	132.9	106.3	99.4	133.3	109.9	95.6	126.2	105.0	95.6	129.0	105.9	97.8	109.2	101.3
06	WOOD, PLASTICS & COMPOSITES	102.0	129.4	117.5	98.9	129.4	116.1	102.0	129.6	117.6	95.7	122.7	111.0	102.2	129.1	117.4	93.5	109.8	102.7
07	THERMAL & MOISTURE PROTECTION	101.7	148.3	120.6	101.0	146.7	119.5	101.6	145.9	119.6	100.8	140.1	116.8	100.9	145.2	118.8	100.9	121.2	109.1
08	OPENINGS	98.1	137.7	107.4	94.0	135.4	103.8	98.1	135.5	106.9	104.5	133.9	111.5	94.7	137.6	104.8	104.5	114.7	106.9
0920	Plaster & Gypsum Board	95.4	129.7	118.7	91.1	129.7	117.3	95.4	129.7	118.7	97.3	122.8	114.6	99.9	129.7	120.1	98.0	109.4	105.8
0950, 0980	Ceilings & Acoustic Treatment	100.9	129.7	120.1	90.0	129.7	116.4	100.9	129.7	120.1	90.6	122.8	112.0	90.6	129.7	116.6	99.5	109.4	106.1
0960	Flooring	100.4	183.1	125.1	98.9	183.1	124.1	99.8	183.1	124.7	98.2	183.1	123.6	99.6	183.1	124.5	97.5	149.9	113.2
0970, 0990	Wall Finishes & Painting/Coating	96.0	135.1	119.5	96.0	135.1	119.5	96.0	135.1	119.5	95.1	135.1	119.2	96.1	135.1	119.6	95.1	108.5	103.1
09	FINISHES	100.0	141.4	123.0	96.0	141.4	121.3	99.9	141.6	123.1	96.1	137.4	119.1	96.7	141.2	121.5	98.3	118.7	109.7
COVERS	DIVS. 10 - 14, 25, 28, 41, 43, 44, 46	100.0	116.4	103.3	100.0	116.4	103.3	100.0	117.0	103.4	100.0	102.5	100.5	100.0	115.9	103.2	100.0	105.1	101.0
21, 22, 23	FIRE SUPPRESSION, PLUMBING & HVAC	100.1	111.2	104.5	95.2	111.2	101.6	100.1	111.3	104.6	95.8	113.6	103.0	95.8	120.4	105.7	95.8	100.8	97.8
26, 27, 3370	ELECTRICAL, COMMUNICATIONS & UTIL.	100.7	100.7	100.7	97.6	100.7	99.2	100.6	100.7	100.6	100.4	104.2	102.4	96.7	129.2	113.7	100.4	92.8	96.4
MF2010	WEIGHTED AVERAGE	100.4	126.3	111.7	94.9	126.7	108.8	100.2	126.8	111.8	96.3	124.6	108.6	95.5	131.0	111.0	97.5	110.2	103.0

	DIVISION	MASSACHUSETTS																	
		HYANNIS			LAWRENCE			LOWELL			NEW BEDFORD			PITTSFIELD			SPRINGFIELD		
		026			019			018			027			012			010 - 011		
		MAT.	INST.	TOTAL	MAT.	INST.	TOTAL	MAT.	INST.	TOTAL	MAT.	INST.	TOTAL	MAT.	INST.	TOTAL	MAT.	INST.	TOTAL
015433	CONTRACTOR EQUIPMENT		102.4	102.4		102.4	102.4		100.6	100.6		103.3	103.3		100.6	100.6		100.6	100.6
0241, 31 - 34	SITE & INFRASTRUCTURE, DEMOLITION	91.9	104.8	101.0	94.5	104.8	101.7	93.7	104.7	101.5	93.4	104.9	101.5	94.7	103.1	100.6	94.2	103.2	100.5
0310	Concrete Forming & Accessories	94.6	131.4	126.3	103.2	131.7	127.8	100.0	131.5	127.2	102.6	131.7	127.7	99.9	111.5	109.9	100.2	111.7	110.1
0320	Concrete Reinforcing	85.6	165.2	125.4	105.6	146.6	126.0	106.5	146.3	126.4	106.8	165.2	136.0	87.9	126.0	106.9	106.5	126.0	116.2
0330	Cast-in-Place Concrete	90.0	152.4	115.9	99.2	149.8	120.2	90.2	149.7	114.9	84.1	152.9	112.6	98.4	128.3	110.8	93.9	128.5	108.3
03	CONCRETE	95.0	144.0	119.2	104.9	139.8	122.1	96.3	139.5	117.6	97.9	144.3	120.8	97.2	119.3	108.1	98.0	119.4	108.6
04	MASONRY	101.4	159.6	137.4	103.3	159.6	138.1	92.4	155.2	131.3	101.3	159.5	137.3	93.0	131.5	116.8	92.7	131.9	117.0
05	METALS	95.9	132.9	107.3	98.2	125.8	106.7	98.1	123.0	105.8	99.4	133.3	109.9	98.0	109.2	101.4	100.6	109.2	103.2
06	WOOD, PLASTICS & COMPOSITES	91.9	129.4	113.1	102.7	129.4	117.7	101.9	129.4	117.4	102.0	129.6	117.6	101.9	109.8	106.4	101.9	109.8	106.4
07	THERMAL & MOISTURE PROTECTION	101.2	146.7	119.7	101.3	147.9	120.2	101.2	145.8	119.3	101.6	145.9	119.6	101.2	120.4	109.0	101.2	120.6	109.1
08	OPENINGS	94.6	135.4	104.3	98.4	134.9	107.1	105.7	134.9	112.6	98.1	135.5	106.9	105.7	114.7	107.8	105.7	114.7	107.8
0920	Plaster & Gypsum Board	86.7	129.7	115.9	102.8	129.7	121.1	102.8	129.7	121.1	95.4	129.7	118.7	102.8	109.4	107.3	102.8	109.4	107.3
0950, 0980	Ceilings & Acoustic Treatment	92.9	129.7	117.4	101.5	129.7	120.2	101.5	129.7	120.2	100.9	129.7	120.1	101.5	109.4	106.8	101.5	109.4	106.8
0960	Flooring	96.8	183.1	122.6	100.2	183.1	125.0	100.2	183.1	125.0	99.8	183.1	124.7	100.4	149.9	115.2	99.9	149.9	114.8
0970, 0990	Wall Finishes & Painting/Coating	96.0	135.1	119.5	95.2	135.1	119.2	95.1	135.1	119.2	96.0	135.1	119.5	95.1	108.5	103.1	96.5	108.5	103.7
09	FINISHES	95.8	141.4	121.2	100.4	141.4	123.2	100.3	141.4	123.2	99.7	141.6	123.0	100.4	118.3	110.4	100.3	118.4	110.4
COVERS	DIVS. 10 - 14, 25, 28, 41, 43, 44, 46	100.0	116.4	103.3	100.0	116.5	103.3	100.0	116.5	103.3	100.0	117.0	103.4	100.0	104.5	100.9	100.0	104.6	100.9
21, 22, 23	FIRE SUPPRESSION, PLUMBING & HVAC	100.1	111.2	104.5	100.0	120.4	108.2	100.0	118.1	107.3	100.1	111.3	104.6	100.0	100.0	100.0	100.0	100.2	100.1
26, 27, 3370	ELECTRICAL, COMMUNICATIONS & UTIL.	98.1	100.7	99.4	99.4	129.2	115.0	99.9	129.2	115.2	101.5	100.7	101.1	99.9	92.8	96.2	99.9	92.8	96.2
MF2010	WEIGHTED AVERAGE	97.6	126.7	110.3	100.1	131.3	113.7	99.4	130.1	112.8	99.6	126.8	111.4	99.6	109.6	103.9	100.0	109.8	104.3

	DIVISION	MASSACHUSETTS			MICHIGAN														
		WORCESTER			ANN ARBOR			BATTLE CREEK			BAY CITY			DEARBORN			DETROIT		
		015 - 016			481			490			487			481			482		
		MAT.	INST.	TOTAL	MAT.	INST.	TOTAL	MAT.	INST.	TOTAL	MAT.	INST.	TOTAL	MAT.	INST.	TOTAL	MAT.	INST.	TOTAL
015433	CONTRACTOR EQUIPMENT		100.6	100.6		110.7	110.7		102.6	102.6		110.7	110.7		110.7	110.7		98.7	98.7
0241, 31 - 34	SITE & INFRASTRUCTURE, DEMOLITION	94.1	104.7	101.5	78.7	97.2	91.6	89.4	87.9	88.4	70.2	96.4	88.6	78.4	97.4	91.7	92.0	99.0	96.9
0310	Concrete Forming & Accessories	100.5	126.1	122.6	97.8	111.4	109.6	97.5	87.1	88.5	97.9	89.0	90.2	97.7	116.2	113.6	99.5	116.2	113.9
0320	Concrete Reinforcing	106.5	155.2	130.8	95.7	123.7	109.7	92.2	96.0	94.1	95.7	122.8	109.3	95.7	123.8	109.8	97.2	123.8	110.5
0330	Cast-in-Place Concrete	93.3	148.3	116.2	86.2	109.5	95.9	95.1	100.0	97.1	82.5	92.0	86.5	84.3	112.5	96.0	91.4	112.5	100.1
03	CONCRETE	97.8	138.2	117.7	92.7	113.2	102.8	96.0	92.6	94.3	91.0	97.0	94.0	91.8	116.4	103.9	95.5	115.2	105.2
04	MASONRY	92.2	155.2	131.2	97.2	108.2	104.0	99.1	86.5	91.3	96.8	87.3	90.9	97.1	113.6	107.3	95.5	113.6	106.7
05	METALS	100.6	126.0	108.4	93.7	118.2	101.2	95.8	88.6	93.6	94.3	115.0	100.7	93.7	118.5	101.4	96.3	101.2	97.8
06	WOOD, PLASTICS & COMPOSITES	102.4	122.7	113.9	94.9	112.4	104.8	100.3	86.3	92.4	94.9	88.7	91.4	94.9	116.9	107.3	97.2	116.9	108.3
07	THERMAL & MOISTURE PROTECTION	101.2	140.1	117.0	100.0	108.5	103.4	96.7	85.5	92.1	97.8	94.2	96.3	98.5	113.5	104.6	97.3	113.5	103.9
08	OPENINGS	105.7	133.7	112.3	97.7	110.0	100.6	94.2	82.7	91.5	97.7	94.8	97.0	97.7	112.4	101.1	100.0	112.1	102.8
0920	Plaster & Gypsum Board	102.8	122.8	116.4	106.2	111.8	110.0	93.0	82.0	85.6	106.2	87.3	93.4	106.2	116.4	113.2	109.0	116.4	114.0
0950, 0980	Ceilings & Acoustic Treatment	101.5	122.8	115.7	93.1	111.8	105.5	96.5	82.0	86.9	94.1	87.3	89.6	93.1	116.4	108.6	96.8	116.4	109.8
0960	Flooring	100.2	183.1	125.0	96.1	117.6	102.5	103.7	93.6	100.7	96.1	81.9	91.8	95.7	116.0	101.8	97.1	116.0	102.7
0970, 0990	Wall Finishes & Painting/Coating	95.1	135.1	119.2	86.5	100.8	95.1	97.2	81.8	87.9	86.5	83.7	84.9	86.5	99.0	94.0	88.5	99.0	94.8
09	FINISHES	100.3	137.4	121.0	94.3	111.7	104.0	97.7	87.5	92.0	94.1	86.5	89.9	94.2	114.8	105.7	96.8	114.8	106.8
COVERS	DIVS. 10 - 14, 25, 28, 41, 43, 44, 46	100.0	108.6	101.7	100.0	106.1	101.2	100.0	98.2	99.6	100.0	92.6	98.5	100.0	107.7	101.5	100.0	107.7	101.5
21, 22, 23	FIRE SUPPRESSION, PLUMBING & HVAC	100.0	113.1	105.3	100.0	99.9	100.0	100.0	87.7	95.1	100.0	84.6	93.8	100.0	108.8	103.5	100.0	109.0	103.6
26, 27, 3370	ELECTRICAL, COMMUNICATIONS & UTIL.	99.9	104.2	102.1	96.0	110.0	103.4	90.3	80.4	85.1	95.0	91.1	92.9	96.0	105.2	100.8	97.2	105.1	101.3
MF2010	WEIGHTED AVERAGE	100.0	124.6	110.7	96.4	107.9	101.4	96.8	87.4	92.7	95.8	92.5	94.4	96.2	110.9	102.6	97.9	109.3	102.8

DIVISION		MICHIGAN																	
		FLINT			GAYLORD			GRAND RAPIDS			IRON MOUNTAIN			JACKSON			KALAMAZOO		
		484 - 485			497			493,495			498 - 499			492			491		
		MAT.	INST.	TOTAL	MAT.	INST.	TOTAL	MAT.	INST.	TOTAL	MAT.	INST.	TOTAL	MAT.	INST.	TOTAL	MAT.	INST.	TOTAL
015433	CONTRACTOR EQUIPMENT		110.7	110.7		105.1	105.1		102.6	102.6		94.2	94.2		105.1	105.1		102.6	102.6
0241, 31 - 34	SITE & INFRASTRUCTURE, DEMOLITION	68.2	96.5	88.1	83.4	85.7	85.0	89.5	88.0	88.4	92.1	94.1	93.5	105.8	88.0	93.3	89.7	87.9	88.4
0310	Concrete Forming & Accessories	100.6	91.4	92.7	95.4	77.7	80.1	97.0	84.7	86.4	87.5	86.1	86.3	92.4	88.3	88.9	97.5	86.8	88.2
0320	Concrete Reinforcing	95.7	123.2	109.5	85.9	109.3	97.6	98.6	95.9	97.2	85.7	95.4	90.6	83.5	123.1	103.3	92.2	93.6	92.9
0330	Cast-in-Place Concrete	86.8	94.7	90.1	94.9	87.8	92.0	100.1	97.6	99.1	112.3	87.2	101.9	94.8	93.8	94.4	97.0	99.8	98.2
03	CONCRETE	93.2	99.1	96.1	92.4	88.2	90.3	99.3	90.7	95.1	101.8	88.4	95.2	87.2	97.3	92.2	99.4	92.0	95.8
04	MASONRY	97.3	96.8	97.0	109.6	80.5	91.6	96.1	86.6	90.2	95.6	87.6	90.7	89.7	92.8	91.7	97.8	86.5	90.8
05	METALS	93.7	115.7	100.5	97.2	111.7	101.6	93.0	88.2	91.5	96.6	91.1	94.9	97.3	113.6	102.4	95.8	87.5	93.3
06	WOOD, PLASTICS & COMPOSITES	98.2	89.9	93.5	92.8	76.9	83.8	101.5	83.1	91.1	87.5	86.1	86.7	91.5	86.6	88.7	100.3	86.3	92.4
07	THERMAL & MOISTURE PROTECTION	97.8	97.9	97.9	95.1	76.2	87.4	99.5	77.4	90.5	98.6	83.1	92.3	94.5	94.5	94.5	96.7	85.5	92.1
08	OPENINGS	97.7	95.5	97.1	94.8	76.0	90.3	100.6	85.6	97.0	101.3	77.3	95.7	93.8	93.9	93.8	94.2	82.1	91.3
0920	Plaster & Gypsum Board	107.7	88.6	94.7	92.8	74.9	80.6	94.9	78.8	83.9	53.7	86.1	75.7	91.0	84.9	86.9	93.0	82.0	85.6
0950, 0980	Ceilings & Acoustic Treatment	93.1	88.6	90.1	95.7	74.9	81.9	105.1	78.8	87.6	94.6	86.1	88.9	95.7	84.9	88.5	96.5	82.0	86.9
0960	Flooring	96.1	94.9	95.7	95.7	93.9	95.1	103.6	84.3	97.8	119.6	95.6	112.4	94.4	84.3	91.4	103.7	77.1	95.8
0970, 0990	Wall Finishes & Painting/Coating	86.5	87.3	87.0	93.1	83.7	87.5	98.9	80.0	87.6	111.4	71.8	87.6	93.1	98.8	96.5	97.2	81.8	87.9
09	FINISHES	93.6	90.7	92.0	97.5	79.8	87.7	101.7	83.8	91.7	99.7	86.4	92.3	98.6	88.0	92.7	97.7	84.2	90.2
COVERS	DIVS. 10 - 14, 25, 28, 41, 43, 44, 46	100.0	93.9	98.8	100.0	90.7	98.1	100.0	98.2	99.6	100.0	89.5	97.9	100.0	93.7	98.7	100.0	98.2	99.6
21, 22, 23	FIRE SUPPRESSION, PLUMBING & HVAC	100.0	90.6	96.2	95.5	80.1	89.3	100.0	83.8	93.5	95.4	86.0	91.6	95.5	88.8	92.8	100.0	81.6	92.6
26, 27, 3370	ELECTRICAL, COMMUNICATIONS & UTIL.	96.0	98.3	97.2	88.4	78.9	83.4	95.2	72.3	83.3	94.5	86.3	90.2	92.2	110.0	101.5	90.2	83.7	86.8
MF2010	WEIGHTED AVERAGE	96.1	96.8	96.4	95.5	84.4	90.7	98.1	84.5	92.2	97.5	87.4	93.1	95.0	95.9	95.4	97.1	85.9	92.2

DIVISION		MICHIGAN															MINNESOTA		
		LANSING			MUSKEGON			ROYAL OAK			SAGINAW			TRAVERSE CITY			BEMIDJI		
		488 - 489			494			480,483			486			496			566		
		MAT.	INST.	TOTAL	MAT.	INST.	TOTAL	MAT.	INST.	TOTAL	MAT.	INST.	TOTAL	MAT.	INST.	TOTAL	MAT.	INST.	TOTAL
015433	CONTRACTOR EQUIPMENT		110.7	110.7		102.6	102.6		96.2	96.2		110.7	110.7		94.2	94.2		99.0	99.0
0241, 31 - 34	SITE & INFRASTRUCTURE, DEMOLITION	90.5	96.6	94.8	87.5	87.9	87.8	83.0	96.8	92.7	71.2	96.4	88.9	78.4	93.6	89.0	94.0	98.3	97.0
0310	Concrete Forming & Accessories	98.0	89.2	90.4	97.9	85.6	87.3	93.6	112.7	110.1	97.8	88.8	90.0	87.5	75.7	77.3	88.7	91.4	91.0
0320	Concrete Reinforcing	99.0	123.0	111.0	92.8	96.1	94.5	86.8	123.8	105.2	95.7	122.8	109.3	87.0	92.9	90.0	97.0	108.0	102.5
0330	Cast-in-Place Concrete	100.5	92.9	97.3	94.8	95.8	95.2	75.5	108.3	89.1	85.2	91.9	88.0	87.6	80.0	84.5	104.1	104.2	104.1
03	CONCRETE	100.0	97.4	98.7	94.3	90.6	92.5	79.9	112.2	95.8	92.2	96.9	94.5	83.9	80.8	82.4	96.9	100.0	98.4
04	MASONRY	95.3	93.3	94.1	96.3	89.4	92.0	92.0	112.1	104.5	98.7	87.3	91.6	93.7	82.3	86.7	97.2	104.8	101.9
05	METALS	91.4	115.1	98.7	93.6	88.5	92.0	96.8	99.4	97.6	93.8	114.8	100.3	96.5	89.7	94.4	90.2	121.0	99.7
06	WOOD, PLASTICS & COMPOSITES	96.4	88.0	91.6	96.5	85.2	90.1	90.4	113.1	103.2	91.2	88.7	89.8	87.5	75.2	80.6	71.5	87.5	80.6
07	THERMAL & MOISTURE PROTECTION	98.6	90.6	95.3	95.8	80.4	89.5	96.5	112.3	102.9	98.6	94.2	96.9	97.8	72.8	87.6	105.6	97.1	102.1
08	OPENINGS	98.7	94.4	97.7	93.4	86.5	91.8	97.6	110.4	100.6	95.6	94.8	95.4	101.3	71.4	94.3	97.5	109.2	100.2
0920	Plaster & Gypsum Board	92.2	86.6	88.4	74.3	80.9	78.8	103.2	112.5	109.5	106.2	87.3	93.4	53.7	74.9	68.1	100.6	87.4	91.6
0950, 0980	Ceilings & Acoustic Treatment	96.8	86.6	90.0	97.4	80.9	86.4	92.3	112.5	105.8	93.1	87.3	89.3	94.6	74.9	81.5	137.1	87.4	104.0
0960	Flooring	102.8	84.3	97.2	102.5	84.3	97.1	93.7	116.0	100.4	96.1	81.9	91.8	119.6	93.9	111.9	107.8	123.0	112.3
0970, 0990	Wall Finishes & Painting/Coating	96.3	81.5	87.4	95.6	80.0	86.3	87.9	98.8	94.5	86.5	83.7	84.9	111.4	45.6	71.9	95.3	97.9	96.9
09	FINISHES	98.2	87.0	91.9	94.3	84.0	88.6	93.3	112.2	103.8	93.9	86.5	89.8	98.6	74.3	85.1	111.3	97.1	103.4
COVERS	DIVS. 10 - 14, 25, 28, 41, 43, 44, 46	100.0	93.3	98.7	100.0	98.0	99.6	100.0	104.6	100.9	100.0	92.6	98.5	100.0	86.8	97.4	100.0	99.3	99.9
21, 22, 23	FIRE SUPPRESSION, PLUMBING & HVAC	99.9	88.9	95.5	99.9	83.1	93.1	95.4	108.0	100.5	100.0	84.2	93.7	95.4	79.4	89.0	95.5	84.8	91.2
26, 27, 3370	ELECTRICAL, COMMUNICATIONS & UTIL.	99.2	91.4	95.1	90.7	77.2	83.6	98.3	104.5	101.5	93.5	91.1	92.2	89.9	78.9	84.2	104.4	105.2	104.8
MF2010	WEIGHTED AVERAGE	97.7	94.1	96.1	95.6	85.5	91.2	94.2	107.5	100.0	95.7	92.4	94.2	94.6	81.0	88.6	97.7	99.5	98.5

DIVISION		MINNESOTA																	
		BRAINERD			DETROIT LAKES			DULUTH			MANKATO			MINNEAPOLIS			ROCHESTER		
		564			565			556 - 558			560			553 - 555			559		
		MAT.	INST.	TOTAL	MAT.	INST.	TOTAL	MAT.	INST.	TOTAL	MAT.	INST.	TOTAL	MAT.	INST.	TOTAL	MAT.	INST.	TOTAL
015433	CONTRACTOR EQUIPMENT		101.6	101.6		99.0	99.0		100.4	100.4		101.6	101.6		103.8	103.8		100.4	100.4
0241, 31 - 34	SITE & INFRASTRUCTURE, DEMOLITION	94.2	103.0	100.4	92.2	98.6	96.7	96.3	100.4	99.2	91.1	103.7	99.9	97.9	105.6	103.3	97.0	100.0	99.1
0310	Concrete Forming & Accessories	90.6	93.6	93.2	85.7	93.5	92.5	100.4	112.2	110.6	98.4	101.1	100.7	101.2	127.4	123.8	102.8	106.6	106.1
0320	Concrete Reinforcing	95.9	108.2	102.0	97.0	108.0	102.5	102.6	108.5	105.5	95.8	118.2	107.0	97.2	118.5	107.8	94.8	118.2	106.5
0330	Cast-in-Place Concrete	113.1	108.2	111.1	101.1	107.5	103.7	107.4	107.0	107.2	104.2	109.0	106.2	110.5	117.3	113.3	109.0	100.9	105.6
03	CONCRETE	100.8	102.4	101.6	94.5	102.1	98.2	101.3	110.3	105.7	96.5	107.9	102.1	103.4	122.6	112.9	100.7	107.7	104.1
04	MASONRY	119.6	113.5	115.8	119.9	110.8	114.2	104.6	117.0	112.3	108.6	110.5	109.8	114.7	125.3	121.3	107.0	112.5	110.4
05	METALS	91.3	121.0	100.5	90.2	120.9	99.7	93.4	123.3	102.6	91.2	126.7	102.1	98.6	130.0	108.3	95.2	128.4	105.4
06	WOOD, PLASTICS & COMPOSITES	89.5	87.3	88.3	68.5	87.5	79.3	103.6	111.9	108.3	99.0	98.7	98.8	109.0	126.0	118.6	109.0	105.2	106.9
07	THERMAL & MOISTURE PROTECTION	104.7	109.9	106.8	105.4	110.1	107.3	100.1	116.9	106.9	105.0	100.6	103.2	102.1	128.6	112.9	103.5	102.0	102.9
08	OPENINGS	84.8	109.0	90.5	97.4	109.2	100.2	98.3	118.2	103.0	89.1	117.7	95.9	94.9	133.8	104.1	93.5	122.5	100.4
0920	Plaster & Gypsum Board	84.6	87.4	86.5	99.5	87.4	91.3	91.1	112.6	105.8	89.5	99.1	96.0	99.3	127.0	118.2	99.5	105.8	103.8
0950, 0980	Ceilings & Acoustic Treatment	66.7	87.4	80.5	137.1	87.4	104.0	98.7	112.6	108.0	66.7	99.1	88.3	98.5	127.0	117.5	95.0	105.8	102.2
0960	Flooring	106.7	123.0	111.6	106.6	123.0	111.5	107.1	124.3	112.3	108.3	123.0	112.7	98.3	123.0	105.7	103.2	123.0	109.1
0970, 0990	Wall Finishes & Painting/Coating	89.6	97.9	94.6	95.3	97.9	96.9	91.0	111.3	103.2	101.0	105.3	103.6	96.5	128.1	115.5	91.3	105.3	99.7
09	FINISHES	93.1	98.5	96.1	110.7	98.6	104.0	101.4	114.7	108.8	94.8	105.3	100.7	100.5	127.2	115.4	99.4	109.5	105.0
COVERS	DIVS. 10 - 14, 25, 28, 41, 43, 44, 46	100.0	100.9	100.2	100.0	101.2	100.2	100.0	100.1	100.0	100.0	101.1	100.2	100.0	109.1	101.8	100.0	102.6	100.5
21, 22, 23	FIRE SUPPRESSION, PLUMBING & HVAC	94.8	87.8	92.0	95.5	87.6	92.3	99.9	97.2	98.8	94.8	88.5	92.3	99.9	116.9	106.7	100.0	96.0	98.4
26, 27, 3370	ELECTRICAL, COMMUNICATIONS & UTIL.	101.9	104.9	103.4	104.1	69.1	85.8	98.4	104.9	101.8	108.8	89.2	98.5	101.8	109.0	105.6	98.2	89.2	93.5
MF2010	WEIGHTED AVERAGE	96.3	102.3	98.9	98.3	96.7	97.6	99.0	108.6	103.2	96.6	102.4	99.1	100.6	119.9	109.0	98.8	104.9	101.5

	DIVISION	MINNESOTA															MISSISSIPPI		
		SAINT PAUL			ST. CLOUD			THIEF RIVER FALLS			WILLMAR			WINDOM			BILOXI		
		550 - 551			563			567			562			561			395		
		MAT.	INST.	TOTAL	MAT.	INST.	TOTAL	MAT.	INST.	TOTAL	MAT.	INST.	TOTAL	MAT.	INST.	TOTAL	MAT.	INST.	TOTAL
015433	CONTRACTOR EQUIPMENT		100.4	100.4		101.6	101.6		99.0	99.0		101.6	101.6		101.6	101.6		99.3	99.3
0241, 31 - 34	SITE & INFRASTRUCTURE, DEMOLITION	99.1	101.1	100.5	89.7	105.1	100.5	92.9	98.2	96.7	89.1	102.8	98.7	83.3	101.7	96.3	98.8	89.6	92.3
0310	Concrete Forming & Accessories	102.0	124.9	121.8	87.7	120.4	116.0	89.4	90.8	90.6	87.5	93.1	92.3	92.0	86.8	87.5	95.6	54.1	59.8
0320	Concrete Reinforcing	98.5	118.5	108.5	95.9	118.4	107.1	97.4	107.9	102.7	95.6	118.0	106.8	95.6	117.1	106.4	95.8	58.4	77.1
0330	Cast-in-Place Concrete	120.2	116.7	118.7	99.9	115.8	106.5	103.2	89.5	97.5	101.4	86.9	95.4	87.7	67.8	79.4	108.7	54.8	86.3
03	CONCRETE	110.6	121.3	115.9	92.3	118.9	105.4	95.6	94.6	95.1	92.2	96.7	94.4	83.2	87.2	85.2	100.0	56.8	78.7
04	MASONRY	115.6	125.3	121.6	105.3	122.1	115.7	97.2	104.8	101.9	109.7	113.5	112.0	119.4	91.2	102.0	90.3	45.1	62.4
05	METALS	94.1	129.8	105.1	92.0	128.0	103.1	90.3	120.3	99.6	91.1	125.8	101.8	91.0	123.5	101.0	91.9	82.4	89.0
06	WOOD, PLASTICS & COMPOSITES	102.8	122.7	114.0	86.8	117.6	104.2	72.5	87.5	81.0	86.6	87.5	87.1	90.8	84.8	87.4	102.3	57.4	77.0
07	THERMAL & MOISTURE PROTECTION	101.8	128.1	112.5	104.8	120.1	111.0	106.5	100.0	103.9	104.6	110.8	107.1	104.6	91.0	99.1	97.0	55.4	80.1
08	OPENINGS	97.1	132.0	105.3	89.1	129.2	98.6	97.5	109.2	100.2	86.7	93.5	88.3	90.2	92.0	90.6	97.2	58.6	88.0
0920	Plaster & Gypsum Board	87.3	123.8	112.1	84.6	118.6	107.7	100.2	87.4	91.5	84.6	87.6	86.7	84.6	84.8	84.7	104.4	56.5	71.8
0950, 0980	Ceilings & Acoustic Treatment	98.5	123.8	115.4	66.7	118.6	101.2	137.1	87.4	104.0	66.7	87.6	80.6	66.7	84.8	78.7	95.9	56.5	69.7
0960	Flooring	99.5	123.0	106.5	103.3	123.0	109.2	107.5	123.0	112.1	105.1	123.0	110.4	107.3	123.0	112.0	114.4	56.1	97.0
0970, 0990	Wall Finishes & Painting/Coating	99.9	127.1	116.3	101.0	128.1	117.3	95.3	97.9	96.9	95.3	97.9	96.9	95.3	105.3	101.3	94.7	44.4	64.5
09	FINISHES	99.5	125.1	113.8	92.6	121.9	108.9	111.1	97.1	103.3	92.7	96.2	94.6	92.8	93.2	93.0	100.8	53.6	74.5
COVERS	DIVS. 10 - 14, 25, 28, 41, 43, 44, 46	100.0	108.5	101.7	100.0	107.2	101.4	100.0	99.2	99.8	100.0	100.8	100.2	100.0	95.6	99.1	100.0	56.9	91.4
21, 22, 23	FIRE SUPPRESSION, PLUMBING & HVAC	99.9	114.9	105.9	99.7	116.3	106.3	95.5	84.5	91.1	94.8	108.4	100.2	94.8	85.3	91.0	100.0	50.1	80.0
26, 27, 3370	ELECTRICAL, COMMUNICATIONS & UTIL.	95.2	111.4	103.7	101.9	111.4	106.9	101.4	69.1	84.6	101.9	87.2	94.2	108.8	89.1	98.5	98.5	57.6	77.1
MF2010	WEIGHTED AVERAGE	100.2	118.8	108.3	96.3	117.8	105.7	97.3	93.8	95.7	94.9	103.0	98.4	95.3	93.3	94.4	97.7	59.1	80.9

	DIVISION	MISSISSIPPI																	
		CLARKSDALE			COLUMBUS			GREENVILLE			GREENWOOD			JACKSON			LAUREL		
		386			397			387			389			390 - 392			394		
		MAT.	INST.	TOTAL	MAT.	INST.	TOTAL	MAT.	INST.	TOTAL	MAT.	INST.	TOTAL	MAT.	INST.	TOTAL	MAT.	INST.	TOTAL
015433	CONTRACTOR EQUIPMENT		99.3	99.3		99.3	99.3		99.3	99.3		99.3	99.3		99.3	99.3		99.3	99.3
0241, 31 - 34	SITE & INFRASTRUCTURE, DEMOLITION	95.6	89.4	91.2	97.1	89.6	91.9	101.3	89.6	93.1	98.3	89.1	91.8	95.9	89.6	91.5	102.9	89.5	93.5
0310	Concrete Forming & Accessories	85.3	50.4	55.2	84.4	52.4	56.8	82.0	65.2	67.5	94.3	50.4	56.4	94.6	58.9	63.8	84.4	54.2	58.3
0320	Concrete Reinforcing	100.0	42.9	71.5	102.7	43.9	73.3	100.6	62.2	81.4	100.0	54.2	77.1	101.2	57.5	79.4	103.3	40.3	71.9
0330	Cast-in-Place Concrete	103.0	54.4	82.8	110.8	59.5	89.5	106.0	56.5	85.5	110.7	54.1	87.2	98.9	57.8	81.8	108.4	56.1	86.7
03	CONCRETE	95.9	52.1	74.3	101.6	54.9	78.6	101.1	63.0	82.3	102.1	53.9	78.3	96.1	59.7	78.2	104.0	53.9	79.3
04	MASONRY	93.5	45.8	64.0	114.7	52.5	76.2	138.0	50.3	83.8	94.1	45.6	64.1	95.7	50.3	67.6	110.9	49.0	72.6
05	METALS	90.5	74.9	85.7	88.9	75.8	84.9	91.5	84.1	89.3	90.5	76.4	86.2	98.5	82.3	93.5	89.0	74.8	84.6
06	WOOD, PLASTICS & COMPOSITES	85.6	52.4	66.9	87.2	53.4	68.1	82.2	69.7	75.1	98.7	52.4	72.6	99.6	61.0	77.8	88.3	55.9	70.0
07	THERMAL & MOISTURE PROTECTION	95.9	48.0	76.5	97.0	52.3	78.8	96.2	61.0	81.9	96.3	47.9	76.6	97.1	60.3	82.1	97.1	54.0	79.6
08	OPENINGS	96.8	50.6	85.8	96.5	50.7	85.7	96.8	62.9	88.8	96.8	51.9	86.2	101.0	56.8	90.5	93.5	51.5	83.6
0920	Plaster & Gypsum Board	87.7	51.4	63.0	92.7	52.4	65.3	87.3	69.2	75.0	98.9	51.4	66.6	97.4	60.2	72.1	92.7	54.9	67.0
0950, 0980	Ceilings & Acoustic Treatment	87.8	51.4	63.6	90.3	52.4	65.1	89.7	69.2	76.1	87.8	51.4	63.6	98.4	60.2	73.0	90.3	54.9	66.8
0960	Flooring	111.6	53.8	94.3	108.2	60.3	93.9	109.9	53.8	93.1	117.1	53.8	98.2	111.9	56.1	95.2	106.8	53.8	91.0
0970, 0990	Wall Finishes & Painting/Coating	97.4	47.1	67.2	94.7	54.1	70.3	97.4	67.3	79.3	97.4	47.1	67.2	91.7	67.3	77.0	94.7	62.2	75.2
09	FINISHES	96.5	51.0	71.1	96.3	54.4	73.0	97.0	64.3	78.8	99.9	51.0	72.7	100.6	59.6	77.8	96.4	55.5	73.6
COVERS	DIVS. 10 - 14, 25, 28, 41, 43, 44, 46	100.0	57.2	91.4	100.0	58.4	91.7	100.0	60.0	92.0	100.0	57.2	91.4	100.0	59.1	91.8	100.0	38.5	87.7
21, 22, 23	FIRE SUPPRESSION, PLUMBING & HVAC	97.9	42.2	75.6	97.3	38.3	73.7	100.0	48.6	79.4	97.9	40.2	74.7	100.0	58.3	83.3	97.4	46.8	77.1
26, 27, 3370	ELECTRICAL, COMMUNICATIONS & UTIL.	95.6	47.5	70.5	95.7	50.3	72.0	95.6	64.2	79.2	95.6	44.5	68.9	100.1	64.2	81.3	97.3	58.3	77.0
MF2010	WEIGHTED AVERAGE	95.8	53.8	77.5	97.1	55.1	78.8	99.4	63.1	83.6	97.0	53.3	78.0	99.1	63.5	83.6	97.2	57.1	79.7

	DIVISION	MISSISSIPPI									MISSOURI								
		MCCOMB			MERIDIAN			TUPELO			BOWLING GREEN			CAPE GIRARDEAU			CHILLICOTHE		
		396			393			388			633			637			646		
		MAT.	INST.	TOTAL	MAT.	INST.	TOTAL	MAT.	INST.	TOTAL	MAT.	INST.	TOTAL	MAT.	INST.	TOTAL	MAT.	INST.	TOTAL
015433	CONTRACTOR EQUIPMENT		99.3	99.3		99.3	99.3		99.3	99.3		106.6	106.6		106.6	106.6		102.1	102.1
0241, 31 - 34	SITE & INFRASTRUCTURE, DEMOLITION	90.3	89.2	89.6	94.3	89.9	91.2	93.1	89.3	90.4	90.1	95.1	93.6	91.7	95.0	94.1	105.0	94.6	97.7
0310	Concrete Forming & Accessories	84.4	52.2	56.6	81.7	64.8	67.1	82.5	52.5	56.6	94.2	88.8	89.5	87.3	75.1	76.8	89.9	95.2	94.5
0320	Concrete Reinforcing	103.9	42.3	73.1	102.7	57.5	80.1	97.9	54.2	76.1	103.2	100.8	102.0	104.5	88.8	96.7	97.4	110.0	103.7
0330	Cast-in-Place Concrete	96.2	53.9	78.7	102.9	60.1	85.2	103.0	55.8	83.4	92.6	85.4	89.6	91.6	87.3	89.8	97.9	87.8	93.7
03	CONCRETE	91.6	52.5	72.3	95.8	63.2	79.7	95.4	55.4	75.7	95.9	91.0	93.5	95.0	83.4	89.3	102.1	95.9	99.1
04	MASONRY	116.2	45.1	72.2	90.0	54.5	68.0	127.9	48.7	78.9	122.1	97.9	107.1	118.5	79.0	94.1	103.6	100.6	101.7
05	METALS	89.1	73.1	84.2	90.0	82.5	87.7	90.4	76.6	86.2	96.1	113.8	101.6	97.2	107.3	100.3	96.9	110.5	101.1
06	WOOD, PLASTICS & COMPOSITES	87.2	55.3	69.2	84.6	66.7	74.5	82.8	53.4	66.2	95.3	88.0	91.2	88.1	71.5	78.7	96.0	95.4	95.7
07	THERMAL & MOISTURE PROTECTION	96.6	47.2	76.6	96.8	62.6	82.9	95.9	53.8	78.8	103.2	98.8	101.4	103.1	83.5	95.1	97.8	97.0	97.4
08	OPENINGS	96.6	52.2	86.1	96.5	65.5	89.2	96.8	51.8	86.1	97.0	98.5	97.3	97.0	75.5	91.9	96.0	98.1	96.5
0920	Plaster & Gypsum Board	92.7	54.3	66.6	92.7	66.1	74.6	87.3	52.4	63.6	97.9	87.6	90.9	96.4	70.6	78.8	97.4	95.1	95.8
0950, 0980	Ceilings & Acoustic Treatment	90.3	54.3	66.4	92.3	66.1	74.8	87.8	52.4	64.2	97.1	87.6	90.8	97.1	70.6	79.5	92.2	95.1	94.1
0960	Flooring	108.2	53.8	91.9	106.7	56.1	91.6	110.1	53.8	93.3	98.4	98.5	98.4	95.2	85.6	92.3	96.0	104.4	98.5
0970, 0990	Wall Finishes & Painting/Coating	94.7	43.2	63.8	94.7	77.8	84.5	97.4	60.0	74.9	102.3	102.8	102.6	102.3	74.8	85.8	96.2	91.3	93.3
09	FINISHES	95.7	52.1	71.4	96.0	65.2	78.8	96.0	53.8	72.5	100.8	91.2	95.4	99.6	75.4	86.1	99.9	96.4	97.9
COVERS	DIVS. 10 - 14, 25, 28, 41, 43, 44, 46	100.0	60.7	92.1	100.0	61.1	92.2	100.0	58.4	91.7	100.0	90.2	98.0	100.0	92.3	98.5	100.0	90.9	98.2
21, 22, 23	FIRE SUPPRESSION, PLUMBING & HVAC	97.3	35.3	72.5	100.0	60.4	84.1	98.1	44.2	76.5	95.0	97.7	96.1	99.9	96.6	98.6	95.2	99.2	96.8
26, 27, 3370	ELECTRICAL, COMMUNICATIONS & UTIL.	94.1	50.6	71.4	97.3	58.2	76.9	95.4	50.4	71.9	98.0	80.8	89.1	98.0	103.9	101.1	93.7	80.5	86.8
MF2010	WEIGHTED AVERAGE	95.7	52.9	77.1	96.2	65.3	82.7	97.3	56.1	79.4	98.1	94.7	96.6	99.1	90.6	95.4	97.7	96.2	97.1

		MISSOURI																	
DIVISION		COLUMBIA			FLAT RIVER			HANNIBAL			HARRISONVILLE			JEFFERSON CITY			JOPLIN		
		652			636			634			647			650 - 651			648		
		MAT.	INST.	TOTAL	MAT.	INST.	TOTAL	MAT.	INST.	TOTAL	MAT.	INST.	TOTAL	MAT.	INST.	TOTAL	MAT.	INST.	TOTAL
015433	CONTRACTOR EQUIPMENT		107.7	107.7		106.6	106.6		106.6	106.6		102.1	102.1		107.7	107.7		105.8	105.8
0241, 31 - 34	SITE & INFRASTRUCTURE, DEMOLITION	100.0	96.6	97.6	92.8	95.2	94.5	88.0	95.1	93.0	96.1	95.6	95.8	99.2	96.6	97.4	105.6	99.4	101.3
0310	Concrete Forming & Accessories	85.5	80.0	80.7	100.1	86.2	88.1	92.4	79.3	81.1	87.3	100.9	99.1	95.6	80.0	82.1	101.6	73.7	77.5
0320	Concrete Reinforcing	98.1	116.3	107.2	104.5	108.0	106.3	102.7	100.8	101.7	97.0	110.1	103.5	100.4	113.1	106.7	100.6	101.7	101.1
0330	Cast-in-Place Concrete	89.8	87.1	88.7	95.6	94.2	95.0	87.6	85.2	86.6	100.3	106.1	102.7	95.3	87.1	91.9	106.0	79.2	94.9
03	CONCRETE	86.8	90.6	88.7	98.8	94.2	96.6	92.2	86.8	89.5	98.5	104.8	101.6	91.9	90.0	91.0	101.7	81.7	91.8
04	MASONRY	133.2	88.7	105.7	119.0	80.5	95.2	113.5	97.9	103.8	98.3	105.3	102.6	97.0	88.7	91.9	96.6	86.8	90.5
05	METALS	97.1	119.7	104.1	96.0	115.8	102.1	96.1	113.5	101.5	97.3	111.5	101.7	96.6	118.4	103.3	100.0	102.2	100.7
06	WOOD, PLASTICS & COMPOSITES	95.4	76.3	84.6	103.2	84.5	92.7	93.3	75.5	83.2	92.5	99.8	96.6	102.0	76.3	87.5	107.5	71.8	87.4
07	THERMAL & MOISTURE PROTECTION	95.5	88.1	92.5	103.4	93.8	99.5	103.0	96.9	100.5	97.1	105.8	100.6	100.8	88.1	95.7	97.1	79.6	90.0
08	OPENINGS	94.7	93.3	94.4	96.9	98.8	97.4	97.0	84.3	94.0	96.1	102.9	97.7	94.2	92.5	93.8	97.2	78.2	92.7
0920	Plaster & Gypsum Board	85.3	75.5	78.7	103.8	84.1	90.4	96.7	74.7	81.8	92.5	99.6	97.3	89.2	75.5	79.9	103.5	70.7	81.2
0950, 0980	Ceilings & Acoustic Treatment	97.9	75.5	83.0	97.1	84.1	88.4	97.1	74.7	82.2	92.2	99.6	97.1	102.1	75.5	84.4	93.0	70.7	78.2
0960	Flooring	109.7	97.5	106.1	101.3	85.6	96.6	97.7	98.5	98.0	91.7	104.4	95.5	116.0	97.5	110.5	121.0	76.3	107.6
0970, 0990	Wall Finishes & Painting/Coating	102.4	78.8	88.2	102.3	80.0	88.9	102.3	102.8	102.6	101.1	105.3	103.6	99.6	78.8	87.1	95.8	76.5	84.2
09	FINISHES	97.5	81.6	88.6	102.5	84.1	92.3	100.2	83.8	91.1	97.5	101.7	99.9	101.7	81.6	90.5	106.1	74.5	88.4
COVERS	DIVS. 10 - 14, 25, 28, 41, 43, 44, 46	100.0	95.6	99.1	100.0	94.3	98.8	100.0	88.8	97.8	100.0	93.1	98.6	100.0	95.6	99.1	100.0	89.6	97.9
21, 22, 23	FIRE SUPPRESSION, PLUMBING & HVAC	99.9	97.8	99.0	95.0	97.6	96.1	95.0	97.6	96.1	95.1	100.5	97.2	99.9	98.7	99.4	100.1	72.7	89.1
26, 27, 3370	ELECTRICAL, COMMUNICATIONS & UTIL.	98.4	86.6	92.2	102.4	104.0	103.2	96.8	82.9	89.6	100.4	99.7	100.1	104.3	86.6	95.1	91.5	69.7	80.1
MF2010	WEIGHTED AVERAGE	98.6	93.5	96.4	99.0	95.8	97.6	97.1	92.6	95.1	97.3	102.2	99.5	98.4	93.4	96.2	99.5	81.0	91.5

		MISSOURI																	
DIVISION		KANSAS CITY			KIRKSVILLE			POPLAR BLUFF			ROLLA			SEDALIA			SIKESTON		
		640 - 641			635			639			654 - 655			653			638		
		MAT.	INST.	TOTAL	MAT.	INST.	TOTAL	MAT.	INST.	TOTAL	MAT.	INST.	TOTAL	MAT.	INST.	TOTAL	MAT.	INST.	TOTAL
015433	CONTRACTOR EQUIPMENT		103.4	103.4		99.0	99.0		101.3	101.3		107.7	107.7		99.7	99.7		101.3	101.3
0241, 31 - 34	SITE & INFRASTRUCTURE, DEMOLITION	98.2	98.0	98.0	90.9	91.2	91.1	78.4	95.0	90.1	99.0	96.6	97.3	94.9	92.5	93.2	81.7	95.1	91.1
0310	Concrete Forming & Accessories	100.3	107.8	106.8	85.2	79.3	80.1	85.3	75.0	76.4	92.7	94.0	93.8	90.1	79.3	80.8	86.4	75.0	76.5
0320	Concrete Reinforcing	95.5	117.1	106.3	103.4	101.9	102.6	106.5	97.6	102.0	98.6	97.7	98.2	97.1	109.4	103.2	105.9	101.4	103.6
0330	Cast-in-Place Concrete	98.6	107.8	102.4	95.5	85.1	91.2	73.5	87.8	79.5	91.8	96.4	93.7	95.9	84.5	91.2	78.6	87.8	82.4
03	CONCRETE	97.8	109.7	103.7	110.2	86.4	98.5	86.0	84.6	85.3	88.6	96.5	92.5	101.1	87.6	94.4	89.6	85.3	87.5
04	MASONRY	100.4	109.1	105.8	125.6	88.7	102.8	117.1	79.1	93.6	109.4	89.4	97.0	115.2	86.6	97.5	116.8	79.1	93.5
05	METALS	107.7	114.9	109.9	95.7	105.2	98.6	96.3	102.9	98.3	96.5	111.9	101.3	95.1	109.3	99.5	96.6	104.4	99.0
06	WOOD, PLASTICS & COMPOSITES	106.7	107.5	107.1	81.1	76.3	78.4	80.3	71.5	75.3	103.0	94.9	98.5	95.4	76.3	84.6	81.9	71.5	76.0
07	THERMAL & MOISTURE PROTECTION	97.2	108.6	101.8	109.8	89.1	101.4	108.3	85.3	99.0	95.6	95.5	95.6	101.8	93.7	98.5	108.5	83.9	98.5
08	OPENINGS	103.5	109.8	105.0	102.0	85.6	98.1	103.0	77.9	97.1	94.7	90.4	93.7	99.0	87.3	96.2	103.0	78.9	97.3
0920	Plaster & Gypsum Board	100.2	107.5	105.1	91.8	75.5	80.7	92.3	70.6	77.5	87.2	94.7	92.3	80.4	75.5	77.1	94.1	70.6	78.1
0950, 0980	Ceilings & Acoustic Treatment	99.3	107.5	104.7	95.3	75.5	82.1	97.1	70.6	79.5	97.9	94.7	95.8	97.9	75.5	83.0	97.1	70.6	79.5
0960	Flooring	96.6	109.3	100.4	76.5	97.5	82.8	90.2	83.5	88.2	113.0	97.5	108.3	86.4	97.5	89.7	90.6	83.5	88.5
0970, 0990	Wall Finishes & Painting/Coating	101.1	112.6	108.0	97.6	80.4	87.2	97.1	74.8	83.7	102.4	102.8	102.7	102.4	102.4	102.4	97.1	74.8	83.7
09	FINISHES	101.4	108.5	105.4	100.0	81.7	89.8	99.6	75.1	85.9	98.8	95.5	96.9	95.7	83.7	89.0	100.3	75.1	86.2
COVERS	DIVS. 10 - 14, 25, 28, 41, 43, 44, 46	100.0	101.3	100.3	100.0	89.0	97.8	100.0	92.4	98.5	100.0	91.4	98.3	100.0	91.5	98.3	100.0	92.4	98.5
21, 22, 23	FIRE SUPPRESSION, PLUMBING & HVAC	100.0	106.7	102.7	95.1	97.1	95.9	95.1	96.0	95.4	95.0	98.0	96.2	94.9	93.8	94.5	95.1	96.1	95.5
26, 27, 3370	ELECTRICAL, COMMUNICATIONS & UTIL.	102.0	100.1	101.0	97.0	82.9	89.6	97.2	103.9	100.7	96.6	86.6	91.4	98.2	99.7	99.0	96.3	103.9	100.3
MF2010	WEIGHTED AVERAGE	101.6	106.7	103.8	100.3	90.0	95.8	97.0	90.3	94.1	96.3	95.6	96.0	98.1	92.6	95.7	97.5	90.6	94.5

		MISSOURI									MONTANA								
DIVISION		SPRINGFIELD			ST. JOSEPH			ST. LOUIS			BILLINGS			BUTTE			GREAT FALLS		
		656 - 658			644 - 645			630 - 631			590 - 591			597			594		
		MAT.	INST.	TOTAL	MAT.	INST.	TOTAL	MAT.	INST.	TOTAL	MAT.	INST.	TOTAL	MAT.	INST.	TOTAL	MAT.	INST.	TOTAL
015433	CONTRACTOR EQUIPMENT		102.3	102.3		102.1	102.1		107.6	107.6		99.3	99.3		99.0	99.0		99.0	99.0
0241, 31 - 34	SITE & INFRASTRUCTURE, DEMOLITION	97.2	94.4	95.2	99.8	93.7	95.5	91.9	98.4	96.5	95.9	97.1	96.8	103.3	96.9	98.8	107.1	96.9	100.0
0310	Concrete Forming & Accessories	98.2	74.7	77.9	100.2	91.1	92.3	100.2	103.6	103.1	101.3	71.6	75.7	88.4	73.3	75.4	101.1	71.6	75.6
0320	Concrete Reinforcing	94.3	116.0	105.1	94.4	109.8	102.1	95.5	114.0	104.7	93.9	82.8	88.3	101.9	82.9	92.4	94.0	82.8	88.4
0330	Cast-in-Place Concrete	97.5	78.8	89.7	98.5	103.6	100.6	91.6	105.0	97.2	120.2	68.9	98.9	132.3	70.3	106.6	139.8	65.0	108.7
03	CONCRETE	97.7	84.6	91.2	97.5	99.5	98.5	94.5	106.8	100.5	106.9	73.6	90.5	110.6	74.9	93.0	115.9	72.2	94.3
04	MASONRY	89.5	87.2	88.1	100.1	96.7	98.0	99.7	112.5	107.6	117.9	75.0	91.4	113.3	77.9	91.4	117.5	76.7	92.3
05	METALS	101.2	107.5	103.1	103.7	110.7	105.9	101.9	120.5	107.7	107.5	91.0	102.4	100.6	91.1	97.7	103.8	91.1	99.9
06	WOOD, PLASTICS & COMPOSITES	103.4	72.8	86.1	107.5	89.7	97.4	103.1	100.8	101.8	99.5	70.0	82.8	86.6	72.5	78.6	100.7	70.0	83.4
07	THERMAL & MOISTURE PROTECTION	99.9	81.1	92.2	97.7	97.4	97.5	103.2	107.6	105.0	108.1	71.2	93.1	108.0	72.5	93.6	108.5	69.8	92.8
08	OPENINGS	101.6	81.2	96.8	101.7	98.3	100.9	97.2	110.5	100.4	93.9	68.7	88.0	92.3	71.3	87.3	95.0	68.1	88.6
0920	Plaster & Gypsum Board	88.3	71.9	77.1	104.8	89.1	94.1	104.7	100.8	102.1	99.9	69.3	79.1	100.3	71.9	81.0	109.7	69.3	82.2
0950, 0980	Ceilings & Acoustic Treatment	97.9	71.9	80.6	98.4	89.1	92.2	102.4	100.8	101.3	99.8	69.3	79.5	107.0	71.9	83.6	108.8	69.3	82.5
0960	Flooring	110.5	76.3	100.3	99.8	104.4	101.2	101.1	99.4	100.6	106.0	71.5	95.7	103.9	75.5	95.4	111.0	75.5	100.4
0970, 0990	Wall Finishes & Painting/Coating	96.2	76.5	84.4	96.2	88.6	91.6	102.3	106.5	104.8	99.0	88.5	92.7	97.2	52.3	70.2	97.2	88.5	92.0
09	FINISHES	101.0	75.1	86.6	102.5	92.7	97.0	103.7	102.5	103.0	102.7	73.0	86.2	103.4	71.2	85.5	107.4	73.7	88.6
COVERS	DIVS. 10 - 14, 25, 28, 41, 43, 44, 46	100.0	93.0	98.6	100.0	96.9	99.4	100.0	102.3	100.5	100.0	96.1	99.2	100.0	96.4	99.3	100.0	96.1	99.2
21, 22, 23	FIRE SUPPRESSION, PLUMBING & HVAC	99.9	73.0	89.1	100.1	89.5	95.8	100.0	107.6	103.0	100.1	74.9	90.0	100.2	70.7	88.4	100.2	72.2	89.0
26, 27, 3370	ELECTRICAL, COMMUNICATIONS & UTIL.	102.9	70.3	85.9	100.5	80.5	90.1	100.5	107.1	103.9	96.8	73.9	84.8	104.4	72.2	87.6	96.9	72.2	84.0
MF2010	WEIGHTED AVERAGE	99.9	82.1	92.1	100.7	93.9	97.8	99.6	107.6	103.1	102.2	77.9	91.6	102.0	77.2	91.2	103.4	77.1	92.0

DIVISION		MONTANA																	
		HAVRE			HELENA			KALISPELL			MILES CITY			MISSOULA			WOLF POINT		
		595			596			599			593			598			592		
		MAT.	INST.	TOTAL	MAT.	INST.	TOTAL	MAT.	INST.	TOTAL	MAT.	INST.	TOTAL	MAT.	INST.	TOTAL	MAT.	INST.	TOTAL
015433	CONTRACTOR EQUIPMENT		99.0	99.0		99.0	99.0		99.0	99.0		99.0	99.0		99.0	99.0		99.0	99.0
0241, 31 - 34	SITE & INFRASTRUCTURE, DEMOLITION	111.0	96.7	100.9	100.1	96.6	97.7	93.4	96.8	95.8	99.9	96.3	97.4	85.8	96.6	93.4	118.0	96.4	102.9
0310	Concrete Forming & Accessories	81.6	69.1	70.8	104.4	66.7	71.9	91.6	71.5	74.3	99.6	67.6	72.0	91.6	71.5	74.2	92.2	67.8	71.2
0320	Concrete Reinforcing	102.7	82.8	92.8	106.8	82.8	94.8	104.6	83.4	94.0	102.3	82.8	92.6	103.6	83.4	93.5	103.7	82.8	93.3
0330	Cast-in-Place Concrete	142.5	61.8	109.0	106.9	64.3	89.2	114.9	64.0	93.8	125.8	61.6	99.1	97.5	66.2	84.5	140.8	62.0	108.1
03	CONCRETE	119.1	70.0	94.9	104.5	69.8	87.3	100.0	72.0	86.2	107.4	69.3	88.6	88.0	72.7	80.4	123.0	69.6	96.6
04	MASONRY	114.2	71.1	87.5	110.7	77.1	89.9	112.2	78.3	91.2	119.4	66.5	86.7	135.6	77.9	99.9	120.5	67.2	87.5
05	METALS	96.8	90.9	95.0	102.7	90.0	98.7	96.6	91.4	95.0	95.9	91.1	94.4	97.2	91.3	95.4	96.0	91.1	94.5
06	WOOD, PLASTICS & COMPOSITES	78.4	70.0	73.7	103.7	63.8	81.2	89.9	70.0	78.7	97.8	70.0	82.1	89.9	70.0	78.7	89.5	70.0	78.5
07	THERMAL & MOISTURE PROTECTION	108.3	69.4	92.5	105.4	72.2	91.9	107.7	74.3	94.1	108.0	66.8	91.3	107.4	80.6	96.5	108.7	67.1	91.8
08	OPENINGS	92.4	68.1	86.6	96.2	66.3	89.2	92.3	68.3	86.6	91.8	68.0	86.2	92.3	68.0	86.6	91.9	67.8	86.2
0920	Plaster & Gypsum Board	96.2	69.3	77.9	97.9	62.9	74.1	100.3	69.3	79.2	109.0	69.3	82.0	100.3	69.3	79.2	104.2	69.3	80.4
0950, 0980	Ceilings & Acoustic Treatment	107.0	69.3	81.9	109.7	62.9	78.5	107.0	69.3	81.9	106.1	69.3	81.6	107.0	69.3	81.9	106.1	69.3	81.6
0960	Flooring	101.3	75.5	93.6	106.1	60.2	92.4	105.7	75.5	96.7	110.7	71.5	99.0	105.7	75.5	96.7	107.1	71.5	96.5
0970, 0990	Wall Finishes & Painting/Coating	97.2	49.7	68.7	93.5	48.2	66.3	97.2	48.2	67.8	97.2	48.2	67.8	97.2	63.3	76.8	97.2	49.7	68.7
09	FINISHES	102.8	67.9	83.4	104.7	63.0	81.5	103.3	69.4	84.4	106.2	65.9	83.8	102.7	71.0	85.1	105.9	66.2	83.8
COVERS	DIVS. 10 - 14, 25, 28, 41, 43, 44, 46	100.0	72.6	94.5	100.0	95.6	99.1	100.0	96.2	99.2	100.0	92.6	98.5	100.0	96.1	99.2	100.0	92.9	98.6
21, 22, 23	FIRE SUPPRESSION, PLUMBING & HVAC	95.3	69.3	84.9	100.1	70.9	88.4	95.3	70.3	85.3	95.3	69.7	85.0	100.2	69.9	88.0	95.3	70.0	85.2
26, 27, 3370	ELECTRICAL, COMMUNICATIONS & UTIL.	96.9	72.7	84.3	103.8	72.7	87.6	101.2	71.0	85.4	96.9	77.4	86.7	102.2	69.3	85.0	96.9	77.4	86.7
MF2010	WEIGHTED AVERAGE	100.6	74.2	89.1	102.0	75.1	90.2	98.4	76.3	88.8	99.5	74.7	88.7	99.4	76.4	89.4	101.7	74.9	90.0

DIVISION		NEBRASKA																	
		ALLIANCE			COLUMBUS			GRAND ISLAND			HASTINGS			LINCOLN			MCCOOK		
		693			686			688			689			683 - 685			690		
		MAT.	INST.	TOTAL	MAT.	INST.	TOTAL	MAT.	INST.	TOTAL	MAT.	INST.	TOTAL	MAT.	INST.	TOTAL	MAT.	INST.	TOTAL
015433	CONTRACTOR EQUIPMENT		97.5	97.5		101.4	101.4		101.4	101.4		101.4	101.4		101.4	101.4		101.4	101.4
0241, 31 - 34	SITE & INFRASTRUCTURE, DEMOLITION	98.8	98.4	98.6	99.1	92.0	94.1	103.8	92.8	96.1	102.6	92.0	95.2	93.6	92.8	93.0	102.1	91.9	95.0
0310	Concrete Forming & Accessories	88.3	57.7	61.9	95.9	68.1	71.9	95.4	74.0	76.9	98.5	73.6	77.0	97.5	68.9	72.8	93.9	57.3	62.3
0320	Concrete Reinforcing	110.1	85.0	97.6	101.7	76.9	89.3	101.1	76.8	89.0	101.1	77.5	89.3	99.2	76.3	87.7	103.1	76.5	89.8
0330	Cast-in-Place Concrete	109.4	61.3	89.4	114.9	61.3	92.6	121.6	68.6	99.6	121.6	64.6	97.9	99.8	71.0	87.9	117.6	60.6	94.0
03	CONCRETE	120.2	64.5	92.7	106.9	68.2	87.8	111.7	73.4	92.8	111.9	71.9	92.1	98.0	71.9	85.1	109.5	63.1	86.6
04	MASONRY	109.7	74.9	88.2	112.6	79.0	91.8	105.7	77.4	88.2	114.2	89.5	98.9	95.9	68.4	78.9	105.5	74.7	86.5
05	METALS	101.2	79.1	94.4	91.9	85.2	89.8	93.6	87.0	91.5	94.4	86.5	91.9	95.6	86.2	92.7	96.0	84.2	92.4
06	WOOD, PLASTICS & COMPOSITES	88.3	53.4	68.6	99.7	67.2	81.4	99.0	73.9	84.8	102.7	73.9	86.4	98.3	67.2	80.7	97.1	53.4	72.5
07	THERMAL & MOISTURE PROTECTION	106.4	69.5	91.4	104.4	70.7	90.7	104.5	74.2	92.2	104.5	85.4	96.8	100.4	70.1	88.1	100.8	67.4	87.2
08	OPENINGS	92.8	59.4	84.9	94.1	66.7	87.6	94.2	70.9	88.7	94.2	70.3	88.5	103.2	63.5	93.8	93.2	57.5	84.7
0920	Plaster & Gypsum Board	81.0	51.9	61.2	88.7	66.1	73.3	88.0	73.0	77.8	89.9	73.0	78.4	90.1	66.1	73.8	92.5	51.9	64.9
0950, 0980	Ceilings & Acoustic Treatment	98.5	51.9	67.5	93.3	66.1	75.2	93.3	73.0	79.8	93.3	73.0	79.8	99.3	66.1	77.2	94.3	51.9	66.1
0960	Flooring	104.5	83.6	98.2	102.5	102.2	102.4	102.3	109.2	104.4	103.4	102.2	103.1	107.6	85.1	100.9	101.6	83.6	96.2
0970, 0990	Wall Finishes & Painting/Coating	174.1	56.0	103.1	91.9	65.1	75.8	91.9	69.1	78.2	91.9	65.1	75.8	100.0	79.6	87.7	95.3	48.8	67.3
09	FINISHES	102.8	60.8	79.4	98.5	72.9	84.3	98.7	78.8	87.6	99.2	76.9	86.8	102.3	72.7	85.8	99.6	59.9	77.5
COVERS	DIVS. 10 - 14, 25, 28, 41, 43, 44, 46	100.0	67.0	93.4	100.0	88.3	97.7	100.0	89.3	97.9	100.0	89.1	97.8	100.0	88.6	97.7	100.0	66.3	93.2
21, 22, 23	FIRE SUPPRESSION, PLUMBING & HVAC	95.3	72.0	85.9	95.1	77.1	87.9	100.0	78.5	91.4	95.1	77.2	87.9	99.9	78.5	91.3	95.0	77.1	87.8
26, 27, 3370	ELECTRICAL, COMMUNICATIONS & UTIL.	91.7	71.0	80.9	92.0	82.4	87.0	90.7	65.9	77.8	90.1	85.6	87.7	103.2	65.9	83.7	93.9	71.0	81.9
MF2010	WEIGHTED AVERAGE	100.4	71.6	87.9	97.5	77.9	88.9	99.1	77.8	89.9	98.5	81.2	90.9	99.5	75.3	89.0	98.1	72.1	86.8

DIVISION		NEBRASKA												NEVADA					
		NORFOLK			NORTH PLATTE			OMAHA			VALENTINE			CARSON CITY			ELKO		
		687			691			680 - 681			692			897			898		
		MAT.	INST.	TOTAL	MAT.	INST.	TOTAL	MAT.	INST.	TOTAL	MAT.	INST.	TOTAL	MAT.	INST.	TOTAL	MAT.	INST.	TOTAL
015433	CONTRACTOR EQUIPMENT		92.7	92.7		101.4	101.4		92.7	92.7		95.6	95.6		97.9	97.9		97.9	97.9
0241, 31 - 34	SITE & INFRASTRUCTURE, DEMOLITION	82.3	91.1	88.5	103.4	92.0	95.4	86.4	91.8	90.2	86.4	95.9	93.1	78.7	99.6	93.4	61.9	98.4	87.5
0310	Concrete Forming & Accessories	82.4	72.7	74.0	96.3	73.1	76.3	93.5	72.9	75.7	84.5	57.1	60.8	105.3	94.6	96.1	111.2	83.2	87.0
0320	Concrete Reinforcing	101.8	66.6	84.3	102.5	77.4	90.0	104.1	76.9	90.5	103.1	66.3	84.8	107.5	118.0	112.7	106.9	113.9	110.4
0330	Cast-in-Place Concrete	116.1	63.8	94.4	117.6	65.9	96.2	92.6	77.1	86.2	104.3	58.6	85.3	104.1	86.2	96.7	98.7	80.9	91.3
03	CONCRETE	105.7	68.7	87.5	109.6	72.1	91.1	94.6	75.4	85.1	107.0	60.0	83.8	104.1	96.0	100.1	98.6	88.3	93.5
04	MASONRY	118.7	78.7	93.9	93.2	89.5	90.9	98.9	79.9	87.1	105.3	74.8	86.4	116.6	89.3	99.7	114.9	73.1	89.0
05	METALS	95.2	74.1	88.7	95.1	86.1	92.3	95.4	79.5	90.5	107.3	74.1	97.1	90.1	104.0	94.4	92.8	99.5	94.8
06	WOOD, PLASTICS & COMPOSITES	82.8	73.4	77.5	99.1	73.9	84.9	91.2	72.1	80.5	82.8	52.9	65.9	91.6	96.6	94.4	99.4	83.7	90.6
07	THERMAL & MOISTURE PROTECTION	104.4	74.9	92.4	100.8	84.1	94.0	100.0	79.8	91.8	101.6	69.1	88.4	102.7	91.2	98.0	99.0	76.6	89.9
08	OPENINGS	95.9	67.7	89.2	92.5	70.3	87.3	101.4	71.9	94.4	95.0	56.5	85.9	99.6	109.4	101.9	100.6	87.3	97.4
0920	Plaster & Gypsum Board	88.7	73.0	78.0	92.5	73.0	79.3	99.5	71.7	80.6	94.7	51.9	65.6	91.3	96.4	94.8	93.3	83.1	86.4
0950, 0980	Ceilings & Acoustic Treatment	106.9	73.0	84.4	94.3	73.0	80.1	112.8	71.7	85.5	113.1	51.9	72.4	101.8	96.4	98.2	102.3	83.1	89.5
0960	Flooring	131.7	102.2	122.9	102.4	102.2	102.3	129.9	83.4	116.0	133.3	81.3	117.7	100.8	109.9	103.6	103.7	52.1	88.2
0970, 0990	Wall Finishes & Painting/Coating	164.9	65.1	104.9	95.3	65.1	77.1	148.8	68.2	100.4	171.4	67.5	108.9	102.7	82.5	90.6	104.2	97.9	100.4
09	FINISHES	120.0	76.9	96.0	99.9	76.9	87.1	119.6	74.0	94.2	123.7	61.1	88.8	100.4	96.2	98.1	99.4	78.9	88.0
COVERS	DIVS. 10 - 14, 25, 28, 41, 43, 44, 46	100.0	87.9	97.6	100.0	68.6	93.7	100.0	88.1	97.6	100.0	64.3	92.8	100.0	92.0	98.4	100.0	64.5	92.9
21, 22, 23	FIRE SUPPRESSION, PLUMBING & HVAC	94.9	76.9	87.7	99.9	77.1	90.8	99.8	78.0	91.1	94.7	76.9	87.6	100.0	81.4	92.6	97.7	81.2	91.1
26, 27, 3370	ELECTRICAL, COMMUNICATIONS & UTIL.	90.9	82.4	86.5	92.0	78.9	85.2	97.9	82.4	89.8	89.0	88.6	88.8	100.4	97.4	98.8	98.1	97.3	97.7
MF2010	WEIGHTED AVERAGE	99.2	77.5	89.8	98.5	79.6	90.2	99.6	79.3	90.7	100.6	73.6	88.9	99.1	93.8	96.8	97.6	86.1	92.6

DIVISION		NEVADA									NEW HAMPSHIRE								
		ELY			LAS VEGAS			RENO			CHARLESTON			CLAREMONT			CONCORD		
		893			889 - 891			894 - 895			036			037			032 - 033		
		MAT.	INST.	TOTAL	MAT.	INST.	TOTAL	MAT.	INST.	TOTAL	MAT.	INST.	TOTAL	MAT.	INST.	TOTAL	MAT.	INST.	TOTAL
015433	CONTRACTOR EQUIPMENT		97.9	97.9		97.9	97.9		97.9	97.9		100.6	100.6		100.6	100.6		100.6	100.6
0241, 31 - 34	SITE & INFRASTRUCTURE, DEMOLITION	67.0	99.9	90.1	69.7	101.6	92.1	67.0	99.6	89.9	87.4	99.2	95.7	81.2	99.2	93.9	94.5	101.4	99.3
0310	Concrete Forming & Accessories	104.2	105.9	105.6	105.4	111.0	110.3	100.5	94.5	95.3	88.1	82.9	83.6	93.8	82.9	84.4	95.9	94.3	94.5
0320	Concrete Reinforcing	105.7	115.2	110.4	97.5	121.8	109.6	99.9	121.4	110.6	84.5	95.2	89.8	84.5	95.2	89.8	92.9	95.8	94.4
0330	Cast-in-Place Concrete	105.9	106.7	106.2	102.6	107.5	104.6	112.0	86.2	101.3	97.4	72.1	86.9	89.5	72.1	82.3	111.5	92.2	103.5
03	CONCRETE	106.6	107.6	107.1	101.8	111.4	106.6	106.3	96.6	101.5	99.5	81.6	90.6	91.5	81.6	86.6	106.0	93.7	100.0
04	MASONRY	119.0	104.2	109.9	108.1	102.8	104.8	114.2	89.3	98.8	87.8	82.3	84.4	87.3	82.3	84.2	97.6	102.1	100.4
05	METALS	92.7	104.8	96.4	99.8	109.3	102.7	94.2	105.3	97.6	95.2	90.6	93.8	95.2	90.6	93.8	101.2	94.0	99.0
06	WOOD, PLASTICS & COMPOSITES	90.3	106.3	99.3	89.2	110.4	101.2	85.2	96.6	91.6	86.9	92.3	90.0	93.3	92.3	92.7	94.8	93.3	93.9
07	THERMAL & MOISTURE PROTECTION	99.6	95.9	98.1	112.5	102.0	108.2	99.1	91.2	95.9	102.8	72.0	90.3	102.7	72.0	90.2	104.9	91.4	99.4
08	OPENINGS	100.5	99.6	100.3	99.6	117.9	104.0	98.6	104.2	99.9	102.6	78.0	96.8	103.8	78.0	97.7	104.1	88.4	100.4
0920	Plaster & Gypsum Board	90.0	106.4	101.1	87.8	110.7	103.4	82.0	96.4	91.8	94.7	91.4	92.5	95.8	91.4	92.8	98.4	92.4	94.4
0950, 0980	Ceilings & Acoustic Treatment	102.3	106.4	105.0	111.1	110.7	110.8	107.6	96.4	100.1	90.6	91.4	91.2	90.6	91.4	91.2	95.9	92.4	93.6
0960	Flooring	101.5	58.2	88.6	93.3	109.9	98.3	98.8	109.9	102.1	96.3	32.7	77.3	98.2	32.7	78.6	97.1	112.2	101.6
0970, 0990	Wall Finishes & Painting/Coating	104.2	123.6	115.9	107.2	123.6	117.0	104.2	82.5	91.2	95.4	46.7	66.2	95.4	46.9	66.3	95.1	97.5	96.5
09	FINISHES	98.9	99.4	99.2	98.4	112.2	106.1	98.0	96.2	97.0	94.9	71.3	81.8	95.1	71.3	81.9	95.9	98.0	97.0
COVERS	DIVS. 10 - 14, 25, 28, 41, 43, 44, 46	100.0	65.0	93.0	100.0	106.3	101.3	100.0	92.0	98.4	100.0	90.0	98.0	100.0	90.0	98.0	100.0	104.8	101.0
21, 22, 23	FIRE SUPPRESSION, PLUMBING & HVAC	97.7	106.7	101.3	100.2	105.1	102.1	100.0	81.5	92.6	95.2	40.4	73.2	95.2	40.5	73.3	99.9	86.1	94.4
26, 27, 3370	ELECTRICAL, COMMUNICATIONS & UTIL.	98.4	109.7	104.3	102.7	120.6	112.0	98.8	97.4	98.0	99.6	54.1	75.8	99.6	54.1	75.8	99.6	80.4	89.6
MF2010	WEIGHTED AVERAGE	98.7	103.4	100.8	100.2	109.4	104.2	99.0	93.8	96.7	96.9	70.1	85.2	96.0	70.1	84.7	100.8	92.3	97.1

DIVISION		NEW HAMPSHIRE															NEW JERSEY		
		KEENE			LITTLETON			MANCHESTER			NASHUA			PORTSMOUTH			ATLANTIC CITY		
		034			035			031			030			038			082,084		
		MAT.	INST.	TOTAL	MAT.	INST.	TOTAL	MAT.	INST.	TOTAL	MAT.	INST.	TOTAL	MAT.	INST.	TOTAL	MAT.	INST.	TOTAL
015433	CONTRACTOR EQUIPMENT		100.6	100.6		100.6	100.6		100.6	100.6		100.6	100.6		100.6	100.6		98.8	98.8
0241, 31 - 34	SITE & INFRASTRUCTURE, DEMOLITION	95.2	99.5	98.2	81.3	99.5	94.1	93.6	101.4	99.1	96.5	101.4	100.0	90.1	101.5	98.1	98.0	104.4	102.5
0310	Concrete Forming & Accessories	92.6	84.9	85.9	103.4	84.9	87.4	98.4	94.8	95.3	100.3	94.8	95.5	89.4	95.1	94.3	110.1	128.4	125.9
0320	Concrete Reinforcing	84.5	95.3	89.9	85.2	95.3	90.3	106.7	95.8	101.3	105.5	95.8	100.7	84.5	95.9	90.2	79.1	121.6	100.3
0330	Cast-in-Place Concrete	97.9	74.4	88.1	87.9	74.4	82.3	111.9	115.3	113.3	92.6	115.3	102.0	87.9	115.4	99.3	84.2	132.2	104.1
03	CONCRETE	99.2	83.2	91.3	91.0	83.3	87.2	108.6	101.9	105.3	99.5	101.9	100.6	91.0	102.1	96.4	92.9	127.3	109.9
04	MASONRY	91.7	85.9	88.1	96.9	85.9	90.1	94.9	102.1	99.4	91.4	102.1	98.0	87.7	102.1	96.6	99.2	125.4	115.4
05	METALS	95.8	91.2	94.4	95.9	91.2	94.4	103.8	94.5	100.9	100.8	94.5	98.9	97.3	95.1	96.7	95.6	106.5	99.0
06	WOOD, PLASTICS & COMPOSITES	91.6	92.3	92.0	102.9	92.3	96.9	95.1	93.3	94.1	101.4	93.3	96.8	88.2	93.3	91.1	115.6	130.2	123.9
07	THERMAL & MOISTURE PROTECTION	103.2	75.0	91.8	102.8	73.5	90.9	106.9	94.9	102.0	103.6	94.9	100.1	103.3	115.6	108.3	104.6	123.2	112.2
08	OPENINGS	101.1	82.9	96.8	104.8	78.7	98.6	106.9	88.4	102.6	105.8	88.4	101.6	106.5	85.1	101.4	102.3	125.4	107.7
0920	Plaster & Gypsum Board	95.1	91.4	92.6	108.4	91.4	96.9	98.7	92.4	94.4	103.8	92.4	96.1	94.7	92.4	93.2	107.3	130.5	123.1
0950, 0980	Ceilings & Acoustic Treatment	90.6	91.4	91.2	90.6	91.4	91.2	98.6	92.4	94.5	101.4	92.4	95.4	91.6	92.4	92.2	88.9	130.5	116.6
0960	Flooring	97.9	53.6	84.7	106.2	32.7	84.2	98.2	112.2	102.4	100.9	112.2	104.2	96.5	112.2	101.2	102.2	150.6	116.6
0970, 0990	Wall Finishes & Painting/Coating	95.4	46.7	66.2	95.4	61.2	74.9	98.2	97.5	97.8	95.4	97.5	96.7	95.4	97.5	96.7	93.0	117.3	107.6
09	FINISHES	96.7	76.2	85.3	99.0	73.8	85.0	98.5	98.0	98.2	101.0	98.0	99.3	95.9	98.0	97.1	99.4	132.5	117.8
COVERS	DIVS. 10 - 14, 25, 28, 41, 43, 44, 46	100.0	91.2	98.2	100.0	99.2	99.8	100.0	104.8	101.0	100.0	104.8	101.0	100.0	104.8	101.0	100.0	110.5	102.1
21, 22, 23	FIRE SUPPRESSION, PLUMBING & HVAC	95.2	44.2	74.7	95.2	65.2	83.2	99.9	86.2	94.4	100.1	86.2	94.5	100.1	86.2	94.5	99.6	119.6	107.6
26, 27, 3370	ELECTRICAL, COMMUNICATIONS & UTIL.	99.6	64.7	81.4	100.8	57.3	78.1	101.6	80.4	90.6	102.2	80.4	90.8	100.1	80.4	89.8	92.7	141.4	118.1
MF2010	WEIGHTED AVERAGE	97.4	74.0	87.2	97.1	77.2	88.4	102.1	93.6	98.4	100.6	93.6	97.6	98.2	94.2	96.4	98.1	123.5	109.2

DIVISION		NEW JERSEY																	
		CAMDEN			DOVER			ELIZABETH			HACKENSACK			JERSEY CITY			LONG BRANCH		
		081			078			072			076			073			077		
		MAT.	INST.	TOTAL	MAT.	INST.	TOTAL	MAT.	INST.	TOTAL	MAT.	INST.	TOTAL	MAT.	INST.	TOTAL	MAT.	INST.	TOTAL
015433	CONTRACTOR EQUIPMENT		98.8	98.8		100.6	100.6		100.6	100.6		100.6	100.6		98.8	98.8		98.4	98.4
0241, 31 - 34	SITE & INFRASTRUCTURE, DEMOLITION	99.1	104.7	103.0	108.2	105.4	106.3	112.7	105.4	107.6	108.9	105.5	106.5	98.0	105.4	103.2	102.6	105.3	104.5
0310	Concrete Forming & Accessories	101.7	128.4	124.7	97.4	129.0	124.7	109.3	129.0	126.3	97.4	128.9	124.6	101.1	129.0	125.2	101.7	128.7	125.0
0320	Concrete Reinforcing	103.8	121.8	112.8	80.0	128.6	104.3	80.0	128.6	104.3	80.0	128.6	104.3	103.8	128.6	116.2	80.0	128.6	104.3
0330	Cast-in-Place Concrete	81.6	132.1	102.6	102.2	127.3	112.6	87.8	127.3	104.2	99.9	127.3	111.3	79.8	127.4	99.5	88.7	132.1	106.7
03	CONCRETE	94.1	127.3	110.4	99.7	127.4	113.3	95.3	127.4	111.1	97.8	127.3	112.4	93.2	127.2	110.0	97.1	128.7	112.7
04	MASONRY	89.7	125.4	111.8	87.3	125.9	111.2	101.2	125.9	116.5	90.8	125.9	112.5	80.8	125.9	108.7	94.4	125.4	113.6
05	METALS	100.7	106.6	102.5	95.6	113.6	101.2	97.1	113.6	102.2	95.7	113.4	101.2	100.8	111.2	104.0	95.7	111.1	100.5
06	WOOD, PLASTICS & COMPOSITES	104.2	130.2	118.9	101.5	130.2	117.7	117.1	130.2	124.5	101.5	130.2	117.7	102.3	130.2	118.0	103.4	130.1	118.5
07	THERMAL & MOISTURE PROTECTION	104.5	122.7	111.9	101.5	130.6	113.4	101.7	130.6	113.5	101.3	123.3	110.3	101.1	130.6	113.1	101.2	123.4	110.3
08	OPENINGS	104.6	125.4	109.5	108.8	127.9	113.4	106.9	127.9	111.9	106.2	127.9	111.4	104.5	127.9	110.0	100.7	127.9	107.1
0920	Plaster & Gypsum Board	102.8	130.5	121.6	99.9	130.5	120.7	107.7	130.5	123.2	99.9	130.5	120.7	103.2	130.5	121.8	101.7	130.5	121.3
0950, 0980	Ceilings & Acoustic Treatment	99.7	130.5	120.2	88.6	130.5	116.5	90.6	130.5	117.2	88.6	130.5	116.5	99.5	130.5	120.1	88.6	130.5	116.5
0960	Flooring	98.7	150.6	114.2	96.8	172.3	119.4	101.2	172.3	122.5	96.8	172.3	119.4	97.7	172.3	120.0	97.9	172.3	120.1
0970, 0990	Wall Finishes & Painting/Coating	93.0	117.3	107.6	94.6	119.3	109.4	94.6	119.3	109.4	94.6	119.3	109.4	94.7	119.3	109.5	94.7	117.3	108.3
09	FINISHES	100.0	132.5	118.1	97.1	135.5	118.5	100.4	135.5	119.9	96.9	135.5	118.4	99.8	135.5	119.7	97.9	136.0	119.2
COVERS	DIVS. 10 - 14, 25, 28, 41, 43, 44, 46	100.0	110.5	102.1	100.0	113.6	102.7	100.0	113.6	102.7	100.0	113.6	102.7	100.0	113.6	102.7	100.0	110.4	102.1
21, 22, 23	FIRE SUPPRESSION, PLUMBING & HVAC	100.0	119.7	107.9	99.6	123.4	109.1	100.0	122.8	109.1	99.6	123.3	109.1	100.0	123.4	109.3	99.6	122.6	108.8
26, 27, 3370	ELECTRICAL, COMMUNICATIONS & UTIL.	97.9	141.4	120.6	90.9	137.6	115.3	91.5	137.6	115.6	90.9	139.7	116.4	95.5	139.7	118.6	90.6	131.5	111.9
MF2010	WEIGHTED AVERAGE	99.4	123.6	109.9	98.7	125.4	110.3	99.5	125.3	110.7	98.4	125.5	110.2	98.5	125.5	110.2	97.8	124.1	109.2

DIVISION		NEW JERSEY																	
		NEW BRUNSWICK			NEWARK			PATERSON			POINT PLEASANT			SUMMIT			TRENTON		
		088 - 089			070 - 071			074 - 075			087			079			085 - 086		
		MAT.	INST.	TOTAL	MAT.	INST.	TOTAL	MAT.	INST.	TOTAL	MAT.	INST.	TOTAL	MAT.	INST.	TOTAL	MAT.	INST.	TOTAL
015433	CONTRACTOR EQUIPMENT		98.4	98.4		100.6	100.6		100.6	100.6		98.4	98.4		100.6	100.6		98.4	98.4
0241, 31 - 34	SITE & INFRASTRUCTURE, DEMOLITION	111.8	105.3	107.3	114.4	105.4	108.1	110.9	105.5	107.1	113.5	105.3	107.7	110.1	105.4	106.8	96.6	105.3	102.7
0310	Concrete Forming & Accessories	104.6	128.9	125.6	97.2	129.0	124.6	99.4	128.9	124.8	99.3	120.3	117.4	100.2	129.0	125.0	99.9	128.6	124.6
0320	Concrete Reinforcing	80.0	128.6	104.3	103.3	128.6	116.0	103.8	128.6	116.2	80.0	128.5	104.2	80.0	128.6	104.3	103.3	115.7	109.5
0330	Cast-in-Place Concrete	103.9	132.3	115.7	109.6	127.3	117.0	101.6	127.3	112.3	103.9	128.4	114.1	84.9	127.3	102.5	99.5	132.0	113.0
03	CONCRETE	109.8	128.9	119.3	106.9	127.4	117.0	103.4	127.3	115.2	109.5	123.7	116.5	92.3	127.4	109.6	102.3	126.3	114.1
04	MASONRY	96.6	125.9	114.7	93.4	125.9	113.5	87.8	125.9	111.4	86.4	124.6	110.0	89.3	125.9	112.0	92.1	125.4	112.7
05	METALS	95.7	111.2	100.4	102.8	113.6	106.1	95.7	113.5	101.2	95.7	110.7	100.3	95.6	113.6	101.1	100.4	105.7	102.0
06	WOOD, PLASTICS & COMPOSITES	108.9	130.1	120.9	99.0	130.2	116.6	104.1	130.2	118.8	101.6	119.1	111.5	105.3	130.2	119.3	101.4	130.1	117.6
07	THERMAL & MOISTURE PROTECTION	104.8	129.9	115.0	102.0	130.6	113.6	101.6	123.3	110.4	104.8	122.1	111.8	101.9	130.6	113.6	104.2	123.1	111.9
08	OPENINGS	97.2	127.9	104.5	107.5	127.9	112.3	111.6	127.9	115.4	99.1	122.2	104.6	113.6	127.9	117.0	106.7	124.4	110.9
0920	Plaster & Gypsum Board	103.9	130.5	122.0	99.9	130.5	120.7	103.2	130.5	121.8	99.4	119.1	112.8	101.7	130.5	121.3	99.9	130.5	120.7
0950, 0980	Ceilings & Acoustic Treatment	88.9	130.5	116.6	102.4	130.5	121.1	99.5	130.5	120.1	88.9	119.1	109.0	88.6	130.5	116.5	101.4	130.5	120.8
0960	Flooring	100.0	172.3	121.6	98.5	172.3	120.6	97.7	172.3	120.0	97.8	150.6	113.6	98.0	172.3	120.2	99.3	172.3	121.1
0970, 0990	Wall Finishes & Painting/Coating	93.0	119.3	108.8	96.0	119.3	110.0	94.6	119.3	109.4	93.0	117.3	107.6	94.6	119.3	109.4	98.3	117.3	109.7
09	FINISHES	99.7	135.4	119.6	98.6	135.5	119.1	100.0	135.5	119.8	98.5	125.9	113.7	98.0	135.5	118.9	100.1	136.0	120.1
COVERS	DIVS. 10 - 14, 25, 28, 41, 43, 44, 46	100.0	113.4	102.7	100.0	113.6	102.7	100.0	113.6	102.7	100.0	104.2	100.8	100.0	113.6	102.7	100.0	110.4	102.1
21, 22, 23	FIRE SUPPRESSION, PLUMBING & HVAC	99.6	122.9	108.9	100.0	122.8	109.1	100.0	123.4	109.4	99.6	122.5	108.7	99.6	122.8	108.9	100.0	122.5	109.0
26, 27, 3370	ELECTRICAL, COMMUNICATIONS & UTIL.	93.4	137.6	116.5	99.2	139.7	120.3	95.5	137.6	117.5	92.7	131.5	112.9	91.5	137.6	115.6	101.4	138.7	120.8
MF2010	WEIGHTED AVERAGE	99.7	125.2	110.8	101.9	125.6	112.2	100.2	125.2	111.1	99.2	121.4	108.8	98.7	125.3	110.3	100.8	124.0	110.9

DIVISION		NEW JERSEY			NEW MEXICO														
		VINELAND			ALBUQUERQUE			CARRIZOZO			CLOVIS			FARMINGTON			GALLUP		
		080,083			870 - 872			883			881			874			873		
		MAT.	INST.	TOTAL	MAT.	INST.	TOTAL	MAT.	INST.	TOTAL	MAT.	INST.	TOTAL	MAT.	INST.	TOTAL	MAT.	INST.	TOTAL
015433	CONTRACTOR EQUIPMENT		98.8	98.8		109.6	109.6		109.6	109.6		109.6	109.6		109.6	109.6		109.6	109.6
0241, 31 - 34	SITE & INFRASTRUCTURE, DEMOLITION	102.7	104.7	104.1	82.2	104.1	97.6	102.9	104.1	103.7	91.2	104.1	100.2	88.4	104.1	99.4	96.0	104.1	101.7
0310	Concrete Forming & Accessories	96.8	128.4	124.1	101.4	65.0	70.0	99.2	65.0	69.7	99.1	64.9	69.6	101.4	65.0	70.0	101.5	65.0	70.0
0320	Concrete Reinforcing	79.1	118.5	98.7	100.5	71.1	85.8	109.9	71.1	90.5	111.2	71.1	91.2	109.9	71.1	90.5	105.2	71.1	88.2
0330	Cast-in-Place Concrete	90.7	132.2	107.9	96.9	71.2	86.2	95.4	71.2	85.3	95.3	71.1	85.3	97.8	71.2	86.8	92.1	71.2	83.4
03	CONCRETE	97.6	126.7	111.9	101.9	69.4	85.9	119.1	69.4	94.6	107.6	69.3	88.7	105.7	69.4	87.8	112.5	69.4	91.2
04	MASONRY	87.9	125.4	111.1	99.5	60.6	75.4	101.6	60.6	76.2	101.7	60.6	76.2	106.3	60.6	78.0	95.4	60.6	73.8
05	METALS	95.6	106.0	98.8	103.8	87.9	98.9	98.4	87.9	95.2	98.1	87.8	94.9	101.4	87.9	97.2	100.6	87.9	96.6
06	WOOD, PLASTICS & COMPOSITES	98.5	130.2	116.4	92.8	65.5	77.4	88.9	65.5	75.7	88.9	65.5	75.7	92.9	65.5	77.4	92.9	65.5	77.4
07	THERMAL & MOISTURE PROTECTION	104.4	123.2	112.1	99.4	71.8	88.2	100.7	71.8	89.0	99.6	71.8	88.3	99.6	71.8	88.3	100.5	71.8	88.8
08	OPENINGS	98.6	125.0	104.9	101.4	68.5	93.6	98.6	68.5	91.5	98.8	68.5	91.6	104.0	68.5	95.6	104.1	68.5	95.7
0920	Plaster & Gypsum Board	98.0	130.5	120.1	88.3	64.1	71.8	75.4	64.1	67.7	75.4	64.1	67.7	81.6	64.1	69.7	81.6	64.1	69.7
0950, 0980	Ceilings & Acoustic Treatment	88.9	130.5	116.6	106.3	64.1	78.2	104.2	64.1	77.5	104.2	64.1	77.5	103.1	64.1	77.2	103.1	64.1	77.2
0960	Flooring	97.1	150.6	113.1	100.2	67.2	90.3	100.2	67.2	90.4	100.2	67.2	90.4	101.7	67.2	91.4	101.7	67.2	91.4
0970, 0990	Wall Finishes & Painting/Coating	93.0	117.3	107.6	109.1	68.1	84.4	102.9	68.1	82.0	102.9	68.1	82.0	102.9	68.1	82.0	102.9	68.1	82.0
09	FINISHES	97.2	132.5	116.8	98.3	65.5	80.0	99.0	65.5	80.3	97.6	65.5	79.7	96.9	65.5	79.4	98.1	65.5	79.9
COVERS	DIVS. 10 - 14, 25, 28, 41, 43, 44, 46	100.0	110.5	102.1	100.0	76.3	95.2	100.0	76.3	95.2	100.0	76.3	95.2	100.0	76.3	95.2	100.0	76.3	95.2
21, 22, 23	FIRE SUPPRESSION, PLUMBING & HVAC	99.6	119.6	107.6	100.2	70.7	88.4	97.2	70.7	86.6	97.2	70.5	86.5	100.1	70.7	88.3	97.1	70.7	86.6
26, 27, 3370	ELECTRICAL, COMMUNICATIONS & UTIL.	92.7	141.4	118.1	88.5	72.9	80.4	90.6	72.9	81.4	88.1	72.9	80.2	86.8	72.9	79.6	86.0	72.9	79.2
MF2010	WEIGHTED AVERAGE	97.5	123.4	108.8	99.2	73.6	88.0	100.1	73.6	88.5	98.1	73.5	87.4	99.7	73.6	88.3	99.3	73.6	88.1

DIVISION		NEW MEXICO																	
		LAS CRUCES			LAS VEGAS			ROSWELL			SANTA FE			SOCORRO			TRUTH/CONSEQUENCES		
		880			877			882			875			878			879		
		MAT.	INST.	TOTAL	MAT.	INST.	TOTAL	MAT.	INST.	TOTAL	MAT.	INST.	TOTAL	MAT.	INST.	TOTAL	MAT.	INST.	TOTAL
015433	CONTRACTOR EQUIPMENT		86.0	86.0		109.6	109.6		109.6	109.6		109.6	109.6		109.6	109.6		86.1	86.1
0241, 31 - 34	SITE & INFRASTRUCTURE, DEMOLITION	91.6	83.8	86.1	87.8	104.1	99.2	93.3	104.1	100.9	92.9	104.1	100.7	84.3	104.1	98.2	101.9	83.8	89.2
0310	Concrete Forming & Accessories	95.6	63.8	68.2	101.4	65.0	70.0	99.2	65.0	69.7	100.1	65.0	69.8	101.5	65.0	70.0	98.7	63.8	68.6
0320	Concrete Reinforcing	107.3	71.0	89.1	106.9	71.1	89.0	111.2	71.1	91.2	106.0	71.1	88.6	109.1	71.1	90.1	102.8	71.0	86.9
0330	Cast-in-Place Concrete	89.9	63.7	79.0	95.2	71.2	85.2	95.3	71.2	85.3	103.2	71.2	89.9	93.2	71.2	84.1	102.1	63.7	86.1
03	CONCRETE	86.1	65.9	76.1	103.1	69.4	86.4	108.3	69.4	89.1	105.6	69.4	87.7	102.0	69.4	85.9	95.1	65.9	80.7
04	MASONRY	97.5	60.2	74.4	95.6	60.6	73.9	112.2	60.6	80.3	99.5	60.6	75.4	95.5	60.6	73.9	93.1	60.2	72.8
05	METALS	97.1	81.3	92.2	100.3	87.9	96.4	99.3	87.9	95.8	97.7	87.9	94.7	100.6	87.9	96.6	100.1	81.4	94.3
06	WOOD, PLASTICS & COMPOSITES	78.8	64.4	70.7	92.9	65.5	77.4	88.9	65.5	75.7	94.6	65.5	78.2	92.9	65.5	77.4	84.6	64.4	73.2
07	THERMAL & MOISTURE PROTECTION	86.2	67.2	78.5	99.2	71.8	88.1	99.7	71.8	88.4	101.7	71.8	89.5	99.1	71.8	88.0	86.9	67.2	78.9
08	OPENINGS	91.5	67.9	85.9	100.3	68.5	92.8	98.6	68.5	91.5	102.4	68.5	94.4	100.1	68.5	92.7	93.3	67.9	87.3
0920	Plaster & Gypsum Board	74.2	64.1	67.3	81.6	64.1	69.7	75.4	64.1	67.7	91.6	64.1	72.9	81.6	64.1	69.7	83.4	64.1	70.3
0950, 0980	Ceilings & Acoustic Treatment	91.4	64.1	73.2	103.1	64.1	77.2	104.2	64.1	77.5	103.6	64.1	77.3	103.1	64.1	77.2	90.4	64.1	72.9
0960	Flooring	129.3	67.2	110.7	101.7	67.2	91.4	100.2	67.2	90.4	108.8	67.2	96.4	101.7	67.2	91.4	132.2	67.2	112.7
0970, 0990	Wall Finishes & Painting/Coating	90.1	68.1	76.8	102.9	68.1	82.0	102.9	68.1	82.0	107.7	68.1	83.9	102.9	68.1	82.0	93.7	68.1	78.3
09	FINISHES	107.1	64.6	83.4	96.8	65.5	79.3	97.7	65.5	79.8	101.9	65.5	81.6	96.7	65.5	79.3	109.0	64.6	84.3
COVERS	DIVS. 10 - 14, 25, 28, 41, 43, 44, 46	100.0	73.6	94.7	100.0	76.3	95.2	100.0	76.3	95.2	100.0	76.3	95.2	100.0	76.3	95.2	100.0	73.6	94.7
21, 22, 23	FIRE SUPPRESSION, PLUMBING & HVAC	100.4	70.5	88.4	97.1	70.7	86.6	100.0	70.7	88.3	100.1	70.7	88.3	97.1	70.7	86.6	97.0	70.5	86.4
26, 27, 3370	ELECTRICAL, COMMUNICATIONS & UTIL.	89.9	72.9	81.0	88.5	72.9	80.4	89.6	72.9	80.9	100.8	72.9	86.3	86.5	72.9	79.4	90.1	72.9	81.1
MF2010	WEIGHTED AVERAGE	95.8	70.4	84.7	97.7	73.6	87.2	99.8	73.6	88.4	100.5	73.6	88.8	97.3	73.6	87.0	97.0	70.4	85.4

DIVISION		NEW MEXICO			NEW YORK															
		TUCUMCARI			ALBANY			BINGHAMTON			BRONX			BROOKLYN			BUFFALO		
		884			120 - 122			137 - 139			104			112			140 - 142		
		MAT.	INST.	TOTAL	MAT.	INST.	TOTAL	MAT.	INST.	TOTAL	MAT.	INST.	TOTAL	MAT.	INST.	TOTAL	MAT.	INST.	TOTAL
015433	CONTRACTOR EQUIPMENT		109.6	109.6		113.5	113.5		114.4	114.4		112.8	112.8		114.1	114.1		97.0	97.0
0241, 31 - 34	SITE & INFRASTRUCTURE, DEMOLITION	90.8	104.1	100.1	75.0	107.1	97.5	92.8	94.3	93.9	110.9	123.1	119.5	117.0	128.4	125.0	98.8	98.3	98.4
0310	Concrete Forming & Accessories	99.1	64.9	69.6	100.7	96.1	96.7	102.2	83.3	85.9	103.3	171.7	162.3	108.6	171.2	162.6	97.4	115.5	113.1
0320	Concrete Reinforcing	109.0	71.1	90.1	105.6	106.3	106.0	93.5	96.4	94.9	96.7	184.6	140.6	94.9	184.6	139.7	102.1	103.0	102.5
0330	Cast-in-Place Concrete	95.3	71.1	85.3	88.1	100.9	93.4	100.9	127.9	112.1	94.6	173.8	127.5	103.0	172.0	131.7	105.5	119.2	111.2
03	CONCRETE	106.7	69.3	88.3	101.0	100.3	100.6	96.2	102.7	99.4	96.5	173.2	134.4	108.3	172.0	139.7	104.6	113.7	109.1
04	MASONRY	112.9	60.6	80.5	90.9	100.3	96.7	105.8	126.6	118.7	94.3	176.4	145.1	116.4	176.4	153.5	107.3	119.6	114.9
05	METALS	98.1	87.8	94.9	101.3	109.5	103.9	96.0	118.2	102.9	100.8	149.3	115.8	104.4	140.3	115.4	99.5	95.6	98.3
06	WOOD, PLASTICS & COMPOSITES	88.9	65.5	75.7	97.7	93.6	95.4	107.3	79.3	91.5	100.1	171.4	140.4	109.3	171.1	144.1	98.7	115.2	108.0
07	THERMAL & MOISTURE PROTECTION	99.5	71.8	88.3	99.4	96.7	98.3	106.9	96.3	102.6	108.5	163.2	130.7	108.9	162.7	130.8	101.5	110.4	105.1
08	OPENINGS	98.5	68.5	91.4	99.9	92.0	98.0	90.7	81.6	88.5	87.0	172.2	107.1	88.0	166.5	106.6	91.8	103.3	94.6
0920	Plaster & Gypsum Board	75.4	64.1	67.7	105.0	93.1	96.9	111.6	78.2	88.9	105.4	173.3	151.6	106.9	173.3	152.1	103.2	115.4	111.5
0950, 0980	Ceilings & Acoustic Treatment	104.2	64.1	77.5	95.9	93.1	94.0	94.6	78.2	83.7	82.8	173.3	143.0	90.2	173.3	145.5	101.6	115.4	110.8
0960	Flooring	100.2	67.2	90.4	97.0	105.2	99.4	106.2	93.1	102.3	96.6	185.6	123.2	113.2	185.6	134.8	101.0	120.1	106.7
0970, 0990	Wall Finishes & Painting/Coating	102.9	68.1	82.0	96.8	89.4	92.4	93.3	95.3	94.5	97.9	148.0	128.0	124.2	148.0	138.5	102.4	112.5	108.5
09	FINISHES	97.6	65.5	79.7	97.8	96.7	97.2	96.7	84.9	90.1	95.3	172.0	138.1	111.3	171.8	145.0	101.0	116.9	109.9
COVERS	DIVS. 10 - 14, 25, 28, 41, 43, 44, 46	100.0	76.3	95.2	100.0	96.0	99.2	100.0	93.5	98.7	100.0	132.7	106.6	100.0	132.0	106.4	100.0	106.1	101.2
21, 22, 23	FIRE SUPPRESSION, PLUMBING & HVAC	97.2	70.5	86.5	100.0	98.8	99.5	100.4	87.3	95.1	100.2	164.7	126.1	99.8	164.4	125.7	100.0	97.1	98.8
26, 27, 3370	ELECTRICAL, COMMUNICATIONS & UTIL.	90.6	72.9	81.4	96.0	99.3	97.7	98.9	107.9	103.6	96.1	175.0	137.3	98.7	175.0	138.5	97.5	101.9	99.8
MF2010	WEIGHTED AVERAGE	98.8	73.5	87.8	98.6	100.2	99.3	98.1	99.4	98.6	98.0	163.9	126.7	102.6	162.9	128.9	99.7	105.8	102.4

DIVISION		NEW YORK																	
		ELMIRA			FAR ROCKAWAY			FLUSHING			GLENS FALLS			HICKSVILLE			JAMAICA		
		148 - 149			116			113			128			115,117,118			114		
		MAT.	INST.	TOTAL	MAT.	INST.	TOTAL	MAT.	INST.	TOTAL	MAT.	INST.	TOTAL	MAT.	INST.	TOTAL	MAT.	INST.	TOTAL
015433	CONTRACTOR EQUIPMENT		116.0	116.0		114.1	114.1		114.1	114.1		113.5	113.5		114.1	114.1		114.1	114.1
0241, 31 - 34	SITE & INFRASTRUCTURE, DEMOLITION	93.3	94.3	94.0	120.2	128.4	125.9	120.2	128.4	126.0	65.2	106.7	94.3	110.1	127.3	122.2	114.5	128.4	124.3
0310	Concrete Forming & Accessories	83.9	86.3	86.0	94.9	171.2	160.8	98.7	171.2	161.3	87.1	86.9	86.9	91.5	144.5	137.3	98.7	171.2	161.3
0320	Concrete Reinforcing	100.5	90.6	95.6	94.9	184.6	139.7	96.5	184.6	140.5	97.0	94.1	95.5	94.9	184.4	139.6	94.9	184.6	139.7
0330	Cast-in-Place Concrete	94.7	96.4	95.4	111.4	172.0	136.6	111.4	172.0	136.6	81.7	96.9	88.0	94.7	165.6	124.1	103.0	172.0	131.7
03	CONCRETE	95.0	92.3	93.7	114.3	172.0	142.8	114.9	172.0	143.1	89.9	92.6	91.3	100.1	157.8	128.6	107.6	172.0	139.4
04	MASONRY	101.4	93.7	96.6	119.7	176.4	154.8	113.9	176.4	152.6	90.1	93.2	92.0	110.5	166.0	144.9	118.1	176.4	154.2
05	METALS	95.4	114.7	101.3	104.4	140.3	115.5	104.4	140.3	115.5	94.8	105.1	98.0	105.9	138.2	115.8	104.4	140.3	115.5
06	WOOD, PLASTICS & COMPOSITES	86.3	84.9	85.5	92.2	171.1	136.7	97.1	171.1	138.8	86.1	84.9	85.4	88.7	141.5	118.5	97.1	171.1	138.8
07	THERMAL & MOISTURE PROTECTION	102.2	84.1	94.8	108.8	162.7	130.7	108.9	162.7	130.7	93.0	92.8	92.9	108.5	154.3	127.1	108.7	162.7	130.6
08	OPENINGS	92.4	83.8	90.3	86.7	166.5	105.6	86.7	166.5	105.6	91.3	84.5	89.7	86.7	150.3	101.8	86.7	166.5	105.6
0920	Plaster & Gypsum Board	100.2	84.2	89.3	95.1	173.3	148.3	97.8	173.3	149.1	95.2	84.2	87.7	94.8	142.7	127.4	97.8	173.3	149.1
0950, 0980	Ceilings & Acoustic Treatment	93.4	84.2	87.2	78.7	173.3	141.6	78.7	173.3	141.6	85.9	84.2	84.8	77.7	142.7	121.0	78.7	173.3	141.6
0960	Flooring	92.9	93.1	93.0	108.5	185.6	131.6	110.0	185.6	132.6	86.1	105.2	91.8	107.6	145.3	118.9	110.0	185.6	132.6
0970, 0990	Wall Finishes & Painting/Coating	101.7	87.6	93.2	124.2	148.0	138.5	124.2	148.0	138.5	94.1	89.3	91.2	124.2	148.0	138.5	124.2	148.0	138.5
09	FINISHES	96.1	87.4	91.3	106.4	171.8	142.8	107.1	171.8	143.1	89.1	89.8	89.5	104.9	143.7	126.5	106.6	171.8	142.9
COVERS	DIVS. 10 - 14, 25, 28, 41, 43, 44, 46	100.0	93.4	98.7	100.0	132.0	106.4	100.0	132.0	106.4	100.0	92.7	98.5	100.0	125.1	105.0	100.0	132.0	106.4
21, 22, 23	FIRE SUPPRESSION, PLUMBING & HVAC	95.3	87.8	92.3	94.9	164.4	122.8	94.9	164.4	122.8	95.4	92.0	94.0	99.8	151.8	120.6	94.9	164.4	122.8
26, 27, 3370	ELECTRICAL, COMMUNICATIONS & UTIL.	95.5	96.9	96.2	106.1	175.0	142.1	106.1	175.0	142.1	91.1	99.3	95.4	98.0	135.8	117.7	96.9	175.0	137.7
MF2010	WEIGHTED AVERAGE	95.9	93.1	94.7	102.4	162.9	128.7	102.3	162.9	128.7	92.6	95.0	93.6	100.7	146.6	120.7	100.6	162.9	127.7

DIVISION		NEW YORK																	
		JAMESTOWN			KINGSTON			LONG ISLAND CITY			MONTICELLO			MOUNT VERNON			NEW ROCHELLE		
		147			124			111			127			105			108		
		MAT.	INST.	TOTAL	MAT.	INST.	TOTAL	MAT.	INST.	TOTAL	MAT.	INST.	TOTAL	MAT.	INST.	TOTAL	MAT.	INST.	TOTAL
015433	CONTRACTOR EQUIPMENT		94.0	94.0		114.1	114.1		114.1	114.1		114.1	114.1		112.8	112.8		112.8	112.8
0241, 31 - 34	SITE & INFRASTRUCTURE, DEMOLITION	94.5	93.8	94.0	122.1	124.3	123.7	118.1	128.4	125.3	117.5	124.2	122.2	117.9	118.8	118.6	117.1	118.8	118.3
0310	Concrete Forming & Accessories	83.9	81.2	81.6	88.7	106.0	103.6	103.1	171.2	161.9	95.7	105.0	103.7	93.5	128.5	123.7	108.6	128.5	125.8
0320	Concrete Reinforcing	100.7	98.6	99.6	97.4	140.6	119.0	94.9	184.6	139.7	96.6	140.4	118.5	95.2	183.2	139.2	95.3	183.2	139.2
0330	Cast-in-Place Concrete	98.3	91.3	95.4	110.4	131.3	119.1	106.3	172.0	133.6	103.4	126.8	113.1	105.7	128.9	115.3	105.7	128.9	115.3
03	CONCRETE	98.0	88.0	93.0	112.5	120.8	116.6	110.7	172.0	140.9	107.2	118.8	112.9	105.9	137.8	121.6	105.4	137.8	121.4
04	MASONRY	109.1	91.1	98.0	104.0	136.8	124.3	113.3	176.4	152.3	98.0	131.0	118.4	100.0	126.1	116.2	100.0	126.1	116.2
05	METALS	92.7	90.9	92.2	103.3	116.5	107.4	104.4	140.3	115.4	103.3	116.1	107.2	100.6	134.6	111.1	100.9	134.6	111.3
06	WOOD, PLASTICS & COMPOSITES	84.8	78.6	81.3	87.3	98.6	93.7	103.3	171.1	141.5	95.2	98.6	97.1	89.3	131.7	113.2	107.6	131.7	121.2
07	THERMAL & MOISTURE PROTECTION	101.6	84.0	94.5	114.3	136.3	123.3	108.8	162.7	130.7	114.0	134.0	122.1	109.4	135.8	120.1	109.4	135.8	120.2
08	OPENINGS	92.2	82.5	89.9	92.8	117.5	98.7	86.7	166.5	105.6	88.5	117.5	95.4	87.0	145.0	100.7	87.0	145.0	100.7
0920	Plaster & Gypsum Board	93.2	77.7	82.6	97.2	98.6	98.1	102.6	173.3	150.7	99.5	98.6	98.9	100.1	132.3	122.0	112.8	132.3	126.1
0950, 0980	Ceilings & Acoustic Treatment	90.7	77.7	82.0	74.3	98.6	90.5	78.7	173.3	141.6	74.3	98.6	90.5	81.1	132.3	115.2	81.1	132.3	115.2
0960	Flooring	95.9	89.7	94.1	103.1	73.8	94.4	111.4	185.6	133.6	105.4	73.8	95.9	89.0	180.7	116.4	95.2	180.7	120.8
0970, 0990	Wall Finishes & Painting/Coating	103.0	91.3	96.0	125.9	113.8	118.6	124.2	148.0	138.5	125.9	113.8	118.6	96.1	148.0	127.3	96.1	148.0	127.3
09	FINISHES	95.6	83.0	88.6	103.7	98.4	100.7	107.9	171.8	143.5	104.2	97.8	100.7	92.8	139.7	118.9	96.0	139.7	120.3
COVERS	DIVS. 10 - 14, 25, 28, 41, 43, 44, 46	100.0	94.8	99.0	100.0	109.7	101.9	100.0	132.0	106.4	100.0	108.9	101.8	100.0	117.8	103.6	100.0	117.8	103.6
21, 22, 23	FIRE SUPPRESSION, PLUMBING & HVAC	95.2	84.6	91.0	95.3	121.1	105.6	99.8	164.4	125.7	95.3	117.9	104.4	95.4	118.6	104.7	95.4	118.6	104.7
26, 27, 3370	ELECTRICAL, COMMUNICATIONS & UTIL.	94.3	90.4	92.3	92.7	113.7	103.6	97.5	175.0	137.9	92.7	113.7	103.6	94.4	158.9	128.1	94.4	158.9	128.1
MF2010	WEIGHTED AVERAGE	96.0	87.8	92.4	100.7	118.3	108.4	102.2	162.9	128.6	99.4	116.5	106.8	97.9	133.3	113.3	98.2	133.3	113.5

DIVISION		NEW YORK																	
		NEW YORK			NIAGARA FALLS			PLATTSBURGH			POUGHKEEPSIE			QUEENS			RIVERHEAD		
		100 - 102			143			129			125 - 126			110			119		
		MAT.	INST.	TOTAL	MAT.	INST.	TOTAL	MAT.	INST.	TOTAL	MAT.	INST.	TOTAL	MAT.	INST.	TOTAL	MAT.	INST.	TOTAL
015433	CONTRACTOR EQUIPMENT		113.2	113.2		94.0	94.0		99.2	99.2		114.1	114.1		114.1	114.1		114.1	114.1
0241, 31 - 34	SITE & INFRASTRUCTURE, DEMOLITION	119.6	123.8	122.5	96.6	95.4	95.8	89.9	102.7	98.9	118.3	124.0	122.3	113.4	128.4	123.9	111.3	127.3	122.5
0310	Concrete Forming & Accessories	107.7	179.3	169.5	83.9	113.2	109.2	91.9	83.8	84.9	88.7	148.2	140.0	91.7	171.2	160.3	96.0	144.5	137.9
0320	Concrete Reinforcing	102.4	184.7	143.5	99.3	99.3	99.3	101.3	93.6	97.5	97.4	140.9	119.1	96.5	184.6	140.5	96.7	184.4	140.5
0330	Cast-in-Place Concrete	106.4	179.2	136.7	101.6	118.7	108.7	100.1	92.7	97.0	106.9	127.3	115.4	97.9	172.0	128.7	96.3	165.6	125.0
03	CONCRETE	106.7	178.5	142.1	100.2	111.7	105.9	104.1	88.4	96.3	109.6	138.2	123.7	103.3	172.0	137.2	100.8	157.8	128.9
04	MASONRY	104.1	178.4	150.0	117.3	118.8	118.2	85.9	87.3	86.8	97.8	128.1	116.6	107.8	176.4	150.2	115.8	166.0	146.9
05	METALS	114.1	149.5	125.0	95.4	92.3	94.5	99.2	85.5	95.0	103.3	118.5	108.0	104.4	140.3	115.4	106.3	138.2	116.1
06	WOOD, PLASTICS & COMPOSITES	104.6	180.6	147.5	84.7	110.9	99.5	94.1	81.9	87.2	87.3	158.2	127.3	88.8	171.1	135.3	94.0	141.5	120.8
07	THERMAL & MOISTURE PROTECTION	108.5	165.3	131.5	101.7	108.0	104.3	108.4	89.8	100.9	114.3	138.8	124.2	108.5	162.7	130.5	109.2	154.3	127.5
08	OPENINGS	92.6	177.5	112.7	92.2	100.2	94.1	98.8	80.1	94.4	92.8	149.1	106.2	86.7	166.5	105.6	86.7	150.3	101.8
0920	Plaster & Gypsum Board	112.2	182.7	160.2	93.2	111.0	105.3	119.4	80.6	93.0	97.2	160.0	139.9	94.8	173.3	148.2	96.1	142.7	127.8
0950, 0980	Ceilings & Acoustic Treatment	101.5	182.7	155.5	90.7	111.0	104.2	100.8	80.6	87.4	74.3	160.0	131.3	78.7	173.3	141.6	78.5	142.7	121.3
0960	Flooring	97.6	185.6	123.9	95.9	113.8	101.3	109.9	105.2	108.5	103.1	148.0	116.6	107.6	185.6	130.9	108.5	145.3	119.5
0970, 0990	Wall Finishes & Painting/Coating	97.9	156.4	133.1	103.0	109.7	107.0	123.3	76.4	95.1	125.9	113.8	118.6	124.2	148.0	138.5	124.2	148.0	138.5
09	FINISHES	101.2	178.9	144.5	95.7	113.6	105.7	100.9	86.4	92.9	103.5	146.3	127.3	105.3	171.8	142.3	105.5	143.7	126.8
COVERS	DIVS. 10 - 14, 25, 28, 41, 43, 44, 46	100.0	137.1	107.4	100.0	104.7	100.9	100.0	48.9	89.8	100.0	113.3	102.7	100.0	132.0	106.4	100.0	125.1	105.0
21, 22, 23	FIRE SUPPRESSION, PLUMBING & HVAC	100.1	165.7	126.4	95.2	97.1	95.9	95.3	92.0	94.0	95.3	117.9	104.4	99.8	164.4	125.7	100.0	151.8	120.7
26, 27, 3370	ELECTRICAL, COMMUNICATIONS & UTIL.	104.2	175.0	141.1	92.9	99.3	96.3	89.2	90.4	89.8	92.7	113.7	103.6	98.0	175.0	138.2	99.7	135.8	118.5
MF2010	WEIGHTED AVERAGE	103.8	166.4	131.1	96.9	103.9	100.0	97.3	88.4	93.4	100.0	127.5	112.0	100.8	162.9	127.8	101.4	146.6	121.1

DIVISION		NEW YORK																	
		ROCHESTER			SCHENECTADY			STATEN ISLAND			SUFFERN			SYRACUSE			UTICA		
		144 - 146			123			103			109			130 - 132			133 - 135		
		MAT.	INST.	TOTAL	MAT.	INST.	TOTAL	MAT.	INST.	TOTAL	MAT.	INST.	TOTAL	MAT.	INST.	TOTAL	MAT.	INST.	TOTAL
015433	CONTRACTOR EQUIPMENT		116.4	116.4		113.5	113.5		112.8	112.8		112.8	112.8		113.5	113.5		113.5	113.5
0241, 31 - 34	SITE & INFRASTRUCTURE, DEMOLITION	77.8	109.6	100.2	74.9	107.1	97.5	122.7	123.1	123.0	113.8	119.1	117.5	91.8	106.6	102.2	70.8	104.5	94.4
0310	Concrete Forming & Accessories	104.8	92.4	94.1	103.1	96.1	97.1	93.0	171.7	160.9	102.1	131.2	127.2	101.1	91.2	92.5	102.4	82.2	85.0
0320	Concrete Reinforcing	101.9	90.7	96.3	96.0	106.3	101.2	96.7	184.6	140.6	95.3	140.9	118.1	94.4	94.8	94.6	94.4	91.9	93.2
0330	Cast-in-Place Concrete	97.6	98.9	98.1	98.6	100.9	99.5	105.7	173.8	134.0	102.5	134.8	115.9	93.8	105.4	98.6	85.8	98.3	91.0
03	CONCRETE	107.3	95.2	101.3	104.5	100.3	102.4	107.6	173.2	140.0	102.4	133.3	117.6	99.5	97.5	98.6	97.8	90.6	94.2
04	MASONRY	100.2	98.6	99.2	88.1	100.3	95.7	106.7	176.4	149.8	99.4	138.2	123.4	98.1	106.3	103.2	90.0	97.7	94.8
05	METALS	101.4	104.5	102.3	98.8	109.5	102.1	98.7	149.3	114.3	98.7	118.1	104.7	99.5	105.0	101.2	97.5	103.8	99.5
06	WOOD, PLASTICS & COMPOSITES	103.8	90.6	96.4	106.0	93.6	99.0	87.8	171.4	135.0	99.7	131.7	117.7	103.7	87.6	94.6	103.7	79.4	90.0
07	THERMAL & MOISTURE PROTECTION	100.5	98.0	99.5	94.2	96.7	95.3	108.8	163.2	130.9	109.3	140.9	122.1	102.5	97.9	100.6	90.9	94.4	92.3
08	OPENINGS	96.7	86.9	94.4	96.9	92.0	95.7	87.0	172.2	107.1	87.0	135.5	98.5	92.5	80.4	89.6	95.2	79.2	91.4
0920	Plaster & Gypsum Board	111.1	90.3	96.9	106.6	93.1	97.4	100.2	173.3	149.9	104.6	132.3	123.5	101.9	87.0	91.7	101.9	78.6	86.0
0950, 0980	Ceilings & Acoustic Treatment	99.4	90.3	93.3	92.0	93.1	92.8	82.8	173.3	143.0	81.1	132.3	115.2	94.6	87.0	89.5	94.6	78.6	83.9
0960	Flooring	92.3	104.9	96.1	92.6	105.2	96.3	92.9	185.6	120.6	92.0	145.3	107.9	93.6	94.9	94.0	90.9	95.0	92.1
0970, 0990	Wall Finishes & Painting/Coating	102.9	97.4	99.6	94.1	89.4	91.3	97.9	148.0	128.0	96.1	121.7	111.5	98.3	97.0	97.5	90.8	97.0	94.5
09	FINISHES	103.1	95.2	98.7	94.5	96.7	95.8	94.7	172.0	137.8	93.8	130.2	114.1	96.0	92.1	93.8	94.1	85.4	89.2
COVERS	DIVS. 10 - 14, 25, 28, 41, 43, 44, 46	100.0	97.0	99.4	100.0	96.0	99.2	100.0	132.7	106.6	100.0	121.8	104.4	100.0	96.3	99.3	100.0	91.9	98.4
21, 22, 23	FIRE SUPPRESSION, PLUMBING & HVAC	100.0	87.9	95.1	100.2	98.8	99.6	100.2	164.7	126.1	95.4	119.6	105.1	100.2	91.2	96.6	100.2	88.8	95.6
26, 27, 3370	ELECTRICAL, COMMUNICATIONS & UTIL.	99.2	94.3	96.7	95.5	99.3	97.5	96.1	175.0	137.3	102.3	121.5	112.3	99.1	97.7	98.4	97.3	97.7	97.5
MF2010	WEIGHTED AVERAGE	100.3	95.6	98.3	97.8	100.2	98.8	99.7	163.9	127.6	98.0	126.2	110.3	98.5	97.0	97.9	96.7	93.2	95.1

DIVISION		NEW YORK									NORTH CAROLINA								
		WATERTOWN			WHITE PLAINS			YONKERS			ASHEVILLE			CHARLOTTE			DURHAM		
		136			106			107			287 - 288			281 - 282			277		
		MAT.	INST.	TOTAL	MAT.	INST.	TOTAL	MAT.	INST.	TOTAL	MAT.	INST.	TOTAL	MAT.	INST.	TOTAL	MAT.	INST.	TOTAL
015433	CONTRACTOR EQUIPMENT		113.5	113.5		112.8	112.8		112.8	112.8		95.6	95.6		95.6	95.6		100.8	100.8
0241, 31 - 34	SITE & INFRASTRUCTURE, DEMOLITION	78.1	106.1	97.8	110.6	118.8	116.4	119.1	118.8	118.9	102.4	76.6	84.3	106.0	76.7	85.4	102.6	85.3	90.4
0310	Concrete Forming & Accessories	87.5	87.4	87.4	107.1	128.5	125.6	107.4	128.5	125.7	95.4	42.1	49.4	98.6	43.7	51.2	100.0	45.1	52.7
0320	Concrete Reinforcing	95.0	94.8	94.9	95.3	183.2	139.2	99.1	183.2	141.1	94.4	63.0	78.7	100.3	58.3	79.3	94.8	57.6	76.2
0330	Cast-in-Place Concrete	99.8	91.8	96.5	93.8	128.9	108.4	105.0	128.9	114.9	110.7	51.9	86.3	115.3	50.4	88.4	104.1	47.6	80.7
03	CONCRETE	110.2	91.2	100.8	96.1	137.8	116.7	105.6	137.9	121.5	102.3	51.2	77.1	104.9	50.5	78.1	98.8	50.1	74.8
04	MASONRY	91.0	100.2	96.7	98.9	126.1	115.7	103.6	126.1	117.5	81.7	44.5	58.7	88.9	52.1	66.1	85.5	38.6	56.4
05	METALS	97.6	104.8	99.8	100.3	134.6	110.9	109.9	134.6	117.5	94.2	82.7	90.7	95.0	81.3	90.8	108.6	80.2	99.9
06	WOOD, PLASTICS & COMPOSITES	85.0	86.3	85.8	105.6	131.7	120.3	105.4	131.7	120.3	96.4	41.2	65.2	101.6	42.8	68.4	99.5	45.7	69.1
07	THERMAL & MOISTURE PROTECTION	91.2	94.9	92.7	109.2	135.8	120.0	109.5	135.8	120.2	103.8	44.5	79.7	98.5	47.5	77.8	106.6	46.6	82.2
08	OPENINGS	95.2	83.2	92.4	87.0	145.0	100.7	90.0	145.3	103.1	94.7	44.6	82.9	99.9	45.5	87.1	97.8	47.7	85.9
0920	Plaster & Gypsum Board	92.2	85.7	87.7	107.6	132.3	124.4	111.8	132.3	125.8	97.5	39.2	57.8	97.8	40.9	59.1	108.8	43.8	64.6
0950, 0980	Ceilings & Acoustic Treatment	94.6	85.7	88.6	81.1	132.3	115.2	99.7	132.3	121.4	91.6	39.2	56.7	96.0	40.9	59.3	95.1	43.8	61.0
0960	Flooring	84.8	95.0	87.9	93.8	180.7	119.8	93.4	180.7	119.5	101.5	43.5	84.1	101.1	44.2	84.1	106.2	43.5	87.4
0970, 0990	Wall Finishes & Painting/Coating	90.8	91.4	91.2	96.1	148.0	127.3	96.1	148.0	127.3	107.1	41.1	67.4	107.3	50.4	73.1	103.4	38.2	64.2
09	FINISHES	91.7	88.8	90.1	94.3	139.7	119.5	99.4	139.7	121.8	96.8	41.5	66.0	97.0	43.8	67.4	100.2	43.8	68.8
COVERS	DIVS. 10 - 14, 25, 28, 41, 43, 44, 46	100.0	93.8	98.8	100.0	117.8	103.6	100.0	120.5	104.1	100.0	78.2	95.6	100.0	78.6	95.7	100.0	73.8	94.8
21, 22, 23	FIRE SUPPRESSION, PLUMBING & HVAC	100.2	81.5	92.7	100.4	118.6	107.7	100.4	118.6	107.7	100.4	53.5	81.6	100.0	55.3	82.1	100.3	53.1	81.4
26, 27, 3370	ELECTRICAL, COMMUNICATIONS & UTIL.	99.1	90.5	94.6	94.4	158.9	128.1	102.4	158.9	131.9	96.8	57.0	76.0	95.7	59.9	77.1	94.4	45.6	68.9
MF2010	WEIGHTED AVERAGE	98.1	91.9	95.4	98.0	133.3	113.4	102.5	133.4	116.0	97.6	55.9	79.4	98.7	57.6	80.8	100.2	54.4	80.2

DIVISION		NORTH CAROLINA																	
		ELIZABETH CITY			FAYETTEVILLE			GASTONIA			GREENSBORO			HICKORY			KINSTON		
		279			283			280			270,272 - 274			286			285		
		MAT.	INST.	TOTAL	MAT.	INST.	TOTAL	MAT.	INST.	TOTAL	MAT.	INST.	TOTAL	MAT.	INST.	TOTAL	MAT.	INST.	TOTAL
015433	CONTRACTOR EQUIPMENT		105.0	105.0		100.8	100.8		95.6	95.6		100.8	100.8		100.8	100.8		100.8	100.8
0241, 31 - 34	SITE & INFRASTRUCTURE, DEMOLITION	107.1	87.0	93.0	101.6	85.2	90.1	102.2	76.6	84.3	102.5	85.3	90.4	101.3	85.1	90.0	100.4	85.2	89.7
0310	Concrete Forming & Accessories	85.5	43.3	49.1	95.0	60.4	65.1	101.9	39.5	48.0	99.7	45.4	52.8	91.6	37.9	45.2	87.9	43.0	49.2
0320	Concrete Reinforcing	92.8	47.3	70.1	98.0	57.7	77.9	94.8	58.3	76.6	93.7	57.8	75.8	94.4	57.9	76.1	93.9	47.2	70.6
0330	Cast-in-Place Concrete	104.3	47.4	80.7	116.1	48.9	88.2	108.2	53.4	85.5	103.4	48.2	80.5	110.7	48.7	84.9	106.9	45.4	81.3
03	CONCRETE	98.8	47.3	73.4	104.4	57.3	81.2	100.9	49.7	75.6	98.3	50.4	74.7	102.0	47.3	75.0	98.9	46.4	73.0
04	MASONRY	97.5	48.8	67.4	84.7	39.6	56.8	85.8	51.7	64.7	82.9	42.1	57.6	72.2	44.4	55.0	77.5	49.0	59.8
05	METALS	96.1	76.9	90.2	113.2	80.4	103.1	94.9	81.2	90.7	101.9	80.4	95.3	94.3	79.9	89.9	93.2	76.0	87.9
06	WOOD, PLASTICS & COMPOSITES	82.0	44.5	60.9	95.9	66.6	79.4	105.5	37.4	67.1	99.2	45.9	69.1	90.9	36.0	59.9	87.7	43.9	63.0
07	THERMAL & MOISTURE PROTECTION	105.9	44.1	80.8	103.7	46.6	80.5	104.0	46.0	80.5	106.3	43.7	80.9	104.1	42.5	79.1	103.9	43.7	79.5
08	OPENINGS	95.0	39.3	81.8	94.8	59.1	86.4	98.4	42.0	85.1	97.8	47.8	86.0	94.8	37.6	81.2	94.9	44.2	82.9
0920	Plaster & Gypsum Board	100.1	42.0	60.6	102.8	65.4	77.3	105.1	35.3	57.6	110.5	44.0	65.3	97.5	33.9	54.2	97.6	42.0	59.8
0950, 0980	Ceilings & Acoustic Treatment	95.1	42.0	59.7	92.5	65.4	74.5	95.2	35.3	55.3	95.1	44.0	61.1	91.6	33.9	53.2	95.2	42.0	59.8
0960	Flooring	98.3	23.6	76.0	101.7	43.5	84.3	104.8	43.5	86.5	106.2	40.4	86.5	101.4	33.5	81.1	98.7	23.0	76.1
0970, 0990	Wall Finishes & Painting/Coating	103.4	42.9	67.0	107.1	33.9	63.1	107.1	41.1	67.4	103.4	32.3	60.7	107.1	41.1	67.4	107.1	40.1	66.8
09	FINISHES	97.1	39.5	65.0	97.7	55.5	74.2	99.5	39.4	66.0	100.5	42.8	68.3	96.9	36.2	63.1	96.8	38.8	64.5
COVERS	DIVS. 10 - 14, 25, 28, 41, 43, 44, 46	100.0	72.2	94.4	100.0	76.0	95.2	100.0	77.9	95.6	100.0	77.8	95.6	100.0	77.6	95.5	100.0	73.4	94.7
21, 22, 23	FIRE SUPPRESSION, PLUMBING & HVAC	95.3	52.2	78.0	100.3	52.8	81.3	100.4	53.5	81.6	100.2	53.3	81.4	95.5	53.3	78.6	95.5	52.3	78.2
26, 27, 3370	ELECTRICAL, COMMUNICATIONS & UTIL.	94.1	35.3	63.4	95.6	51.9	72.8	96.2	58.7	76.6	93.4	57.0	74.4	94.2	58.7	75.7	94.0	47.2	69.6
MF2010	WEIGHTED AVERAGE	96.9	52.2	77.4	100.9	58.5	82.5	98.4	56.1	80.0	98.8	56.3	80.3	95.6	54.9	77.9	95.3	53.7	77.2

DIVISION		NORTH CAROLINA															NORTH DAKOTA		
		MURPHY			RALEIGH			ROCKY MOUNT			WILMINGTON			WINSTON-SALEM			BISMARCK		
		289			275 - 276			278			284			271			585		
		MAT.	INST.	TOTAL	MAT.	INST.	TOTAL	MAT.	INST.	TOTAL	MAT.	INST.	TOTAL	MAT.	INST.	TOTAL	MAT.	INST.	TOTAL
015433	CONTRACTOR EQUIPMENT		95.6	95.6		100.8	100.8		100.8	100.8		95.6	95.6		100.8	100.8		99.0	99.0
0241, 31 - 34	SITE & INFRASTRUCTURE, DEMOLITION	103.6	76.5	84.5	103.7	85.3	90.8	105.0	85.3	91.2	103.6	76.8	84.8	102.8	85.3	90.5	100.7	97.6	98.5
0310	Concrete Forming & Accessories	102.6	40.2	48.7	98.7	47.6	54.6	92.1	45.2	51.6	96.8	49.9	56.3	101.6	49.0	56.2	103.6	41.3	49.8
0320	Concrete Reinforcing	93.9	46.9	70.4	100.0	57.8	78.9	92.8	57.6	75.2	95.1	57.6	76.4	93.7	57.6	75.7	105.6	83.2	94.4
0330	Cast-in-Place Concrete	114.5	45.0	85.7	109.3	52.0	85.5	102.0	48.3	79.7	110.3	50.2	85.3	105.9	50.6	83.0	109.5	48.4	84.1
03	CONCRETE	105.3	45.0	75.5	102.0	52.8	77.7	99.5	50.3	75.2	102.2	53.1	77.9	99.6	52.8	76.5	103.6	53.0	78.6
04	MASONRY	74.5	41.9	54.4	88.1	42.9	60.2	76.2	40.2	53.9	72.4	42.4	53.9	83.0	39.3	56.0	110.9	57.4	77.8
05	METALS	92.1	75.3	87.0	94.8	80.4	90.4	95.3	79.6	90.5	93.9	80.2	89.7	99.4	80.2	93.5	97.7	85.4	93.9
06	WOOD, PLASTICS & COMPOSITES	106.2	40.5	69.2	97.9	48.6	70.1	89.5	45.7	64.8	98.8	50.1	71.3	99.2	50.4	71.7	87.3	35.6	58.1
07	THERMAL & MOISTURE PROTECTION	104.0	41.6	78.6	100.6	47.9	79.2	106.3	41.6	80.1	103.8	47.9	81.1	106.3	44.3	81.2	109.3	50.8	85.6
08	OPENINGS	94.7	39.8	81.7	97.0	48.5	85.5	94.3	42.9	82.1	94.9	51.7	84.6	97.8	50.3	86.5	100.9	47.9	88.4
0920	Plaster & Gypsum Board	104.2	38.5	59.5	102.5	46.8	64.6	102.1	43.8	62.5	100.3	48.3	65.0	110.5	48.7	68.5	94.3	33.8	53.1
0950, 0980	Ceilings & Acoustic Treatment	91.6	38.5	56.3	95.9	46.8	63.2	92.4	43.8	60.1	92.5	48.3	63.1	95.1	48.7	64.2	136.9	33.8	68.3
0960	Flooring	105.2	24.5	81.1	103.4	43.5	85.4	101.8	22.2	78.0	102.3	45.3	85.2	106.2	43.5	87.4	109.8	73.4	98.9
0970, 0990	Wall Finishes & Painting/Coating	107.1	40.2	66.9	102.1	37.7	63.4	103.4	39.3	64.9	107.1	39.5	66.5	103.4	37.5	63.8	94.1	30.0	55.6
09	FINISHES	99.0	36.8	64.3	99.7	45.7	69.6	97.9	40.2	65.8	97.6	47.8	69.9	100.5	46.7	70.5	111.7	43.8	73.9
COVERS	DIVS. 10 - 14, 25, 28, 41, 43, 44, 46	100.0	77.5	95.5	100.0	74.5	94.9	100.0	74.4	94.9	100.0	75.7	95.1	100.0	79.3	95.9	100.0	84.8	97.0
21, 22, 23	FIRE SUPPRESSION, PLUMBING & HVAC	95.5	52.0	78.0	100.0	53.6	81.4	95.3	53.9	78.7	100.4	55.1	82.2	100.2	53.5	81.5	100.2	61.3	84.6
26, 27, 3370	ELECTRICAL, COMMUNICATIONS & UTIL.	97.8	30.0	62.4	96.5	41.3	67.7	96.1	40.6	67.1	97.0	51.9	73.4	93.4	57.0	74.4	102.2	73.3	87.1
MF2010	WEIGHTED AVERAGE	96.4	49.2	75.8	98.2	55.1	79.4	96.0	53.2	77.4	97.2	56.6	79.5	98.5	57.1	80.5	102.0	64.0	85.5

DIVISION		NORTH DAKOTA																	
		DEVILS LAKE			DICKINSON			FARGO			GRAND FORKS			JAMESTOWN			MINOT		
		583			586			580 - 581			582			584			587		
		MAT.	INST.	TOTAL	MAT.	INST.	TOTAL	MAT.	INST.	TOTAL	MAT.	INST.	TOTAL	MAT.	INST.	TOTAL	MAT.	INST.	TOTAL
015433	CONTRACTOR EQUIPMENT		99.0	99.0		99.0	99.0		99.0	99.0		99.0	99.0		99.0	99.0		99.0	99.0
0241, 31 - 34	SITE & INFRASTRUCTURE, DEMOLITION	103.3	95.2	97.6	111.5	93.7	99.0	98.4	97.7	97.9	107.3	93.7	97.8	102.3	93.7	96.3	104.8	97.6	99.8
0310	Concrete Forming & Accessories	102.6	36.0	45.1	91.7	34.9	42.7	102.5	42.2	50.5	95.6	34.8	43.1	93.2	34.6	42.6	91.4	66.0	69.5
0320	Concrete Reinforcing	101.9	83.5	92.7	102.8	41.0	71.9	94.5	83.3	88.9	100.3	83.3	91.8	102.5	48.3	75.4	103.8	83.6	93.7
0330	Cast-in-Place Concrete	122.9	47.0	91.4	111.2	45.2	83.8	100.9	50.3	79.9	111.2	45.1	83.8	121.4	45.0	89.7	111.2	47.5	84.7
03	CONCRETE	110.8	50.2	80.9	110.1	41.0	76.0	104.7	54.0	79.7	107.3	48.7	78.4	109.3	42.2	76.2	105.8	63.7	85.0
04	MASONRY	111.7	64.4	82.4	113.4	59.2	79.8	97.9	57.4	72.9	106.2	63.8	80.0	123.8	33.4	67.8	105.6	64.2	80.0
05	METALS	95.1	84.6	91.9	95.1	64.1	85.5	99.0	86.0	95.0	95.1	79.3	90.2	95.1	65.1	85.8	95.4	86.3	92.6
06	WOOD, PLASTICS & COMPOSITES	88.7	32.8	57.1	76.6	32.8	51.9	86.8	36.1	58.2	80.8	32.8	53.7	78.5	32.8	52.7	76.4	68.7	72.0
07	THERMAL & MOISTURE PROTECTION	106.7	49.4	83.4	107.1	47.3	82.8	108.7	51.8	85.6	106.9	49.4	83.5	106.6	41.5	80.1	106.6	56.0	86.1
08	OPENINGS	96.6	40.8	83.4	96.5	30.9	81.0	98.8	48.2	86.8	96.6	40.8	83.4	96.6	33.0	81.5	96.7	66.0	89.4
0920	Plaster & Gypsum Board	115.2	30.9	57.9	105.2	30.9	54.7	91.5	34.3	52.6	106.7	30.9	55.1	106.3	30.9	55.0	105.2	67.9	79.8
0950, 0980	Ceilings & Acoustic Treatment	142.1	30.9	68.1	142.1	30.9	68.1	138.7	34.3	69.2	142.1	30.9	68.1	142.1	30.9	68.1	142.1	67.9	92.7
0960	Flooring	117.0	36.5	92.9	110.5	36.5	88.4	109.8	73.4	98.9	112.4	36.5	89.7	111.2	36.5	88.9	110.3	86.4	103.1
0970, 0990	Wall Finishes & Painting/Coating	94.9	23.0	51.7	94.9	23.0	51.7	91.0	72.2	79.7	94.9	28.9	55.2	94.9	23.0	51.7	94.9	25.2	53.0
09	FINISHES	117.6	32.9	70.4	115.3	32.9	69.4	112.3	48.8	76.9	115.5	33.6	69.9	114.7	32.9	69.1	114.5	65.4	87.1
COVERS	DIVS. 10 - 14, 25, 28, 41, 43, 44, 46	100.0	33.3	86.6	100.0	33.5	86.7	100.0	84.9	97.0	100.0	33.4	86.7	100.0	82.8	96.6	100.0	88.5	97.7
21, 22, 23	FIRE SUPPRESSION, PLUMBING & HVAC	95.5	74.6	87.1	95.5	67.1	84.1	100.1	77.6	91.1	100.4	33.9	73.8	95.5	35.8	71.5	100.4	61.4	84.7
26, 27, 3370	ELECTRICAL, COMMUNICATIONS & UTIL.	99.3	36.7	66.6	109.1	74.2	90.9	104.8	67.7	85.5	103.3	54.8	78.0	99.3	36.6	66.6	106.9	77.5	91.5
MF2010	WEIGHTED AVERAGE	101.0	58.6	82.5	101.9	57.9	82.7	101.7	67.6	86.9	101.8	51.6	79.9	101.0	45.1	76.7	101.8	71.0	88.4

DIVISION		NORTH DAKOTA			OHIO														
		WILLISTON			AKRON			ATHENS			CANTON			CHILLICOTHE			CINCINNATI		
		588			442 - 443			457			446 - 447			456			451 - 452		
		MAT.	INST.	TOTAL	MAT.	INST.	TOTAL	MAT.	INST.	TOTAL	MAT.	INST.	TOTAL	MAT.	INST.	TOTAL	MAT.	INST.	TOTAL
015433	CONTRACTOR EQUIPMENT		99.0	99.0		95.3	95.3		90.7	90.7		95.3	95.3		100.3	100.3		100.0	100.0
0241, 31 - 34	SITE & INFRASTRUCTURE, DEMOLITION	105.3	93.7	97.2	96.8	102.4	100.7	111.6	92.6	98.3	96.9	102.0	100.4	96.5	104.0	101.8	93.6	103.5	100.5
0310	Concrete Forming & Accessories	97.3	34.9	43.5	100.2	92.2	93.3	94.1	85.8	86.9	100.2	82.3	84.7	96.2	92.1	92.6	98.1	80.6	83.0
0320	Concrete Reinforcing	104.8	41.0	72.9	93.9	93.3	93.6	90.1	86.6	88.4	93.9	76.5	85.2	87.0	81.0	84.0	92.2	81.7	87.0
0330	Cast-in-Place Concrete	111.2	45.2	83.8	92.4	97.4	94.5	111.9	92.2	103.7	93.3	95.2	94.1	101.6	96.1	99.3	93.6	93.6	93.6
03	CONCRETE	107.3	41.0	74.6	92.9	93.5	93.2	106.4	87.6	97.2	93.4	85.3	89.4	100.8	91.3	96.1	95.3	85.5	90.4
04	MASONRY	100.0	59.2	74.7	92.1	94.1	93.3	82.7	89.5	87.0	92.8	84.6	87.7	90.1	96.7	94.2	89.8	85.5	87.1
05	METALS	95.2	64.2	85.7	94.8	82.2	90.9	101.6	78.6	94.5	94.8	75.1	88.7	93.7	84.9	91.0	95.9	84.6	92.4
06	WOOD, PLASTICS & COMPOSITES	82.2	32.8	54.3	99.1	91.4	94.7	83.4	86.8	85.3	99.5	80.8	88.9	96.0	90.6	92.9	98.5	78.5	87.2
07	THERMAL & MOISTURE PROTECTION	106.8	47.3	82.7	111.2	95.8	104.9	99.7	91.9	96.6	112.3	91.6	103.9	101.2	94.3	98.4	99.5	89.9	95.6
08	OPENINGS	96.7	30.9	81.1	115.2	92.5	109.8	99.9	83.6	96.1	108.5	77.2	101.1	91.9	84.0	90.0	99.9	78.9	94.9
0920	Plaster & Gypsum Board	106.7	30.9	55.1	102.0	90.8	94.4	93.9	86.1	88.6	103.1	79.9	87.3	96.6	90.5	92.5	98.3	78.1	84.6
0950, 0980	Ceilings & Acoustic Treatment	142.1	30.9	68.1	98.7	90.8	93.5	102.0	86.1	91.5	98.7	79.9	86.2	96.3	90.5	92.5	97.2	78.1	84.5
0960	Flooring	113.2	36.5	90.3	98.6	92.5	96.7	124.5	100.8	117.4	98.7	82.0	93.7	101.5	100.8	101.3	102.5	90.0	98.7
0970, 0990	Wall Finishes & Painting/Coating	94.9	32.2	57.2	97.1	103.9	101.2	104.9	93.1	97.8	97.1	82.8	88.5	102.2	89.8	94.7	102.2	84.8	91.7
09	FINISHES	115.6	34.0	70.1	100.1	93.3	96.4	102.9	90.1	95.8	100.4	81.8	90.0	100.4	93.7	96.6	100.8	82.2	90.4
COVERS	DIVS. 10 - 14, 25, 28, 41, 43, 44, 46	100.0	33.5	86.7	100.0	95.3	99.1	100.0	53.3	90.6	100.0	93.1	98.6	100.0	93.1	98.6	100.0	91.4	98.3
21, 22, 23	FIRE SUPPRESSION, PLUMBING & HVAC	95.5	67.1	84.1	100.0	94.3	97.7	95.2	52.4	78.0	100.0	82.5	93.0	95.6	94.0	94.9	100.0	81.3	92.5
26, 27, 3370	ELECTRICAL, COMMUNICATIONS & UTIL.	103.7	74.2	88.3	99.3	92.8	95.9	97.8	98.9	98.3	98.5	87.8	92.9	97.3	85.4	91.1	96.1	79.7	87.6
MF2010	WEIGHTED AVERAGE	100.3	58.0	81.9	99.8	93.4	97.0	99.1	80.6	91.1	99.2	85.0	93.0	96.4	92.2	94.5	97.8	84.8	92.1

DIVISION		OHIO																	
		CLEVELAND			COLUMBUS			DAYTON			HAMILTON			LIMA			LORAIN		
		441			430 - 432			453 - 454			450			458			440		
		MAT.	INST.	TOTAL	MAT.	INST.	TOTAL	MAT.	INST.	TOTAL	MAT.	INST.	TOTAL	MAT.	INST.	TOTAL	MAT.	INST.	TOTAL
015433	CONTRACTOR EQUIPMENT		95.6	95.6		94.9	94.9		95.3	95.3		100.3	100.3		93.1	93.1		95.3	95.3
0241, 31 - 34	SITE & INFRASTRUCTURE, DEMOLITION	96.7	102.9	101.1	95.2	99.3	98.1	92.4	103.4	100.1	92.3	103.6	100.2	105.2	92.5	96.3	96.3	101.3	99.8
0310	Concrete Forming & Accessories	100.3	97.9	98.2	99.1	85.9	87.7	98.1	79.0	81.6	98.1	80.1	82.5	94.0	84.3	85.6	100.3	85.3	87.3
0320	Concrete Reinforcing	94.3	93.7	94.0	101.6	81.7	91.7	92.2	80.6	86.4	92.2	81.7	87.0	90.1	79.1	84.6	93.9	93.6	93.7
0330	Cast-in-Place Concrete	90.7	105.5	96.8	90.6	95.9	92.8	87.0	86.0	86.6	93.3	92.8	93.1	102.8	97.2	100.5	88.0	103.5	94.4
03	CONCRETE	92.2	99.0	95.5	95.8	88.3	92.1	92.2	81.5	86.9	95.1	85.0	90.1	99.3	87.6	93.5	90.8	92.6	91.7
04	MASONRY	96.6	105.1	101.9	97.9	94.8	96.0	89.3	81.7	84.6	89.6	84.0	86.1	112.9	81.7	93.6	88.8	98.5	94.8
05	METALS	96.3	85.2	92.9	95.9	80.5	91.2	95.2	77.8	89.8	95.2	85.1	92.1	101.6	80.4	95.1	95.4	83.3	91.6
06	WOOD, PLASTICS & COMPOSITES	98.2	94.8	96.2	97.5	83.7	89.7	99.7	77.3	87.1	98.5	78.5	87.2	83.3	84.0	83.7	99.1	82.0	89.4
07	THERMAL & MOISTURE PROTECTION	109.8	108.1	109.1	101.8	95.7	99.3	104.5	86.4	97.2	101.3	88.7	96.2	99.3	92.5	96.5	112.2	101.1	107.7
08	OPENINGS	104.5	94.3	102.1	103.0	80.4	97.7	99.3	78.0	94.2	96.9	78.9	92.6	99.9	78.3	94.8	108.5	87.4	103.5
0920	Plaster & Gypsum Board	101.2	94.3	96.5	97.6	83.2	87.8	98.3	76.8	83.7	98.3	78.1	84.6	93.9	83.2	86.6	102.0	81.2	87.8
0950, 0980	Ceilings & Acoustic Treatment	97.0	94.3	95.2	100.4	83.2	89.0	98.2	76.8	84.0	97.2	78.1	84.5	101.0	83.2	89.2	98.7	81.2	87.1
0960	Flooring	98.4	105.1	100.4	94.7	87.9	92.6	105.2	80.7	97.9	102.5	90.0	98.7	123.6	88.3	113.1	98.7	105.1	100.6
0970, 0990	Wall Finishes & Painting/Coating	97.1	105.2	102.0	100.9	93.3	96.3	102.2	85.2	92.0	102.2	84.8	91.7	104.9	83.8	92.2	97.1	105.2	102.0
09	FINISHES	99.7	99.6	99.6	97.5	86.6	91.4	101.7	79.5	89.3	100.7	81.9	90.2	101.9	84.7	92.3	100.2	90.4	94.7
COVERS	DIVS. 10 - 14, 25, 28, 41, 43, 44, 46	100.0	102.1	100.4	100.0	93.8	98.8	100.0	90.9	98.2	100.0	90.9	98.2	100.0	94.1	98.8	100.0	98.3	99.7
21, 22, 23	FIRE SUPPRESSION, PLUMBING & HVAC	100.0	99.8	99.9	100.0	92.3	96.9	100.8	81.7	93.1	100.5	80.5	92.5	95.2	84.7	91.0	100.0	85.2	94.0
26, 27, 3370	ELECTRICAL, COMMUNICATIONS & UTIL.	98.8	106.2	102.6	97.1	88.1	92.4	94.7	82.7	88.4	95.1	81.0	87.8	98.1	78.4	87.8	98.7	86.7	92.4
MF2010	WEIGHTED AVERAGE	98.9	100.0	99.4	98.6	89.7	94.7	97.5	83.2	91.3	97.4	84.5	91.8	99.6	84.4	93.0	98.8	90.4	95.2

DIVISION		OHIO																	
		MANSFIELD			MARION			SPRINGFIELD			STEUBENVILLE			TOLEDO			YOUNGSTOWN		
		448 - 449			433			455			439			434 - 436			444 - 445		
		MAT.	INST.	TOTAL	MAT.	INST.	TOTAL	MAT.	INST.	TOTAL	MAT.	INST.	TOTAL	MAT.	INST.	TOTAL	MAT.	INST.	TOTAL
015433	CONTRACTOR EQUIPMENT		95.3	95.3		94.5	94.5		95.3	95.3		99.3	99.3		97.3	97.3		95.3	95.3
0241, 31 - 34	SITE & INFRASTRUCTURE, DEMOLITION	92.7	102.2	99.3	91.1	97.8	95.8	92.7	102.1	99.3	134.7	107.7	115.8	94.4	99.8	98.2	96.7	103.0	101.1
0310	Concrete Forming & Accessories	90.3	83.5	84.4	95.7	81.3	83.3	98.1	83.3	85.4	97.6	87.9	89.2	99.0	92.5	93.4	100.2	85.9	87.9
0320	Concrete Reinforcing	85.4	75.6	80.5	93.7	81.5	87.6	92.2	78.9	85.6	91.1	85.9	88.5	101.6	87.2	94.4	93.9	86.1	90.0
0330	Cast-in-Place Concrete	85.6	94.5	89.3	82.8	92.9	87.0	89.4	85.9	87.9	89.9	94.7	91.9	90.6	100.2	94.6	91.5	97.1	93.9
03	CONCRETE	85.6	85.4	85.5	88.0	85.1	86.5	93.3	83.1	88.3	91.7	89.3	90.5	95.8	93.9	94.9	92.5	89.3	90.9
04	MASONRY	91.2	93.7	92.7	100.0	92.4	95.3	89.5	81.7	84.6	86.5	93.8	91.0	106.6	96.8	100.6	92.4	91.1	91.6
05	METALS	95.6	75.5	89.4	95.0	78.1	89.8	95.1	77.0	89.5	91.6	79.9	87.9	95.7	86.9	93.0	94.8	79.3	90.0
06	WOOD, PLASTICS & COMPOSITES	86.8	82.0	84.1	93.5	81.7	86.9	101.1	83.4	91.1	89.3	86.1	87.5	97.5	91.5	94.1	99.1	84.1	90.6
07	THERMAL & MOISTURE PROTECTION	111.6	93.6	104.3	101.4	82.0	93.5	104.4	87.0	97.4	113.5	96.6	106.6	103.4	100.9	102.4	112.4	92.8	104.4
08	OPENINGS	108.8	76.7	101.2	97.1	77.6	92.5	97.4	78.6	92.9	97.5	82.9	94.0	100.4	88.8	97.6	108.5	84.2	102.7
0920	Plaster & Gypsum Board	95.2	81.2	85.7	95.4	81.2	85.7	98.3	83.2	88.0	95.5	85.1	88.4	97.6	91.2	93.2	102.0	83.3	89.3
0950, 0980	Ceilings & Acoustic Treatment	99.6	81.2	87.4	100.4	81.2	87.6	98.2	83.2	88.2	97.3	85.1	89.2	100.4	91.2	94.3	98.7	83.3	88.5
0960	Flooring	94.6	104.7	97.6	93.5	71.0	86.8	105.2	80.7	97.9	122.7	99.8	115.9	93.9	101.5	96.2	98.7	92.1	96.7
0970, 0990	Wall Finishes & Painting/Coating	97.1	88.5	91.9	100.9	48.5	69.4	102.2	85.2	92.0	114.2	102.6	107.2	100.9	102.1	101.6	97.1	92.6	94.4
09	FINISHES	98.0	87.7	92.3	96.4	76.1	85.1	101.7	83.1	91.4	114.3	91.2	101.4	97.2	94.8	95.9	100.3	87.3	93.0
COVERS	DIVS. 10 - 14, 25, 28, 41, 43, 44, 46	100.0	93.2	98.6	100.0	54.9	91.0	100.0	91.6	98.3	100.0	95.2	99.0	100.0	96.3	99.3	100.0	94.0	98.8
21, 22, 23	FIRE SUPPRESSION, PLUMBING & HVAC	95.1	83.4	90.4	95.2	89.7	93.0	100.8	81.7	93.1	95.5	91.9	94.0	100.1	99.2	99.7	100.0	88.4	95.3
26, 27, 3370	ELECTRICAL, COMMUNICATIONS & UTIL.	96.2	77.7	86.5	91.2	77.7	84.1	94.7	85.6	90.0	86.4	112.7	100.1	97.2	101.6	99.5	98.7	82.9	90.5
MF2010	WEIGHTED AVERAGE	96.6	85.6	91.8	95.0	83.7	90.1	97.5	84.2	91.7	96.8	94.5	95.8	98.7	96.4	97.7	99.1	88.3	94.4

DIVISION		OHIO			OKLAHOMA														
		ZANESVILLE			ARDMORE			CLINTON			DURANT			ENID			GUYMON		
		437 - 438			734			736			747			737			739		
		MAT.	INST.	TOTAL	MAT.	INST.	TOTAL	MAT.	INST.	TOTAL	MAT.	INST.	TOTAL	MAT.	INST.	TOTAL	MAT.	INST.	TOTAL
015433	CONTRACTOR EQUIPMENT		94.5	94.5		82.5	82.5		81.6	81.6		81.6	81.6		81.6	81.6		81.6	81.6
0241, 31 - 34	SITE & INFRASTRUCTURE, DEMOLITION	93.9	99.5	97.9	94.7	91.4	92.4	96.1	90.0	91.8	92.5	89.7	90.6	97.6	90.0	92.2	99.9	89.7	92.7
0310	Concrete Forming & Accessories	92.8	82.6	84.0	94.4	41.3	48.6	92.9	47.7	53.9	85.7	45.1	50.6	96.6	34.8	43.3	100.3	45.9	53.3
0320	Concrete Reinforcing	93.1	85.3	89.2	87.6	80.2	83.9	88.1	80.2	84.2	92.3	80.6	86.4	87.5	80.2	83.9	88.1	80.2	84.1
0330	Cast-in-Place Concrete	87.2	93.3	89.8	99.3	44.2	76.4	95.9	45.9	75.2	92.1	43.5	71.9	95.9	47.2	75.7	96.0	43.9	74.3
03	CONCRETE	91.5	86.5	89.0	97.5	50.0	74.1	97.1	53.5	75.6	92.9	51.5	72.5	97.7	48.2	73.3	100.7	51.9	76.7
04	MASONRY	96.8	83.4	88.5	93.4	58.1	71.5	118.8	58.1	81.2	92.4	63.4	74.5	100.6	58.1	74.3	95.9	54.6	70.3
05	METALS	96.4	80.3	91.4	93.3	70.6	86.3	93.4	70.6	86.4	91.4	71.3	85.2	94.8	70.5	87.3	93.9	70.3	86.6
06	WOOD, PLASTICS & COMPOSITES	89.6	81.7	85.2	100.7	38.7	65.7	99.7	47.3	70.1	89.5	44.0	63.8	103.6	30.0	62.0	107.5	47.0	73.4
07	THERMAL & MOISTURE PROTECTION	101.5	91.1	97.3	106.1	61.0	87.8	106.1	62.0	88.2	99.9	62.3	84.6	106.2	60.3	87.6	106.5	58.4	87.0
08	OPENINGS	97.1	80.2	93.1	93.9	49.6	83.4	93.9	54.0	84.4	93.9	52.7	84.1	93.9	44.4	82.1	94.0	49.8	83.5
0920	Plaster & Gypsum Board	92.8	81.2	84.9	84.3	37.5	52.4	84.0	46.3	58.3	79.1	42.9	54.5	85.1	28.4	46.5	85.7	46.0	58.7
0950, 0980	Ceilings & Acoustic Treatment	100.4	81.2	87.6	86.2	37.5	53.8	86.2	46.3	59.7	85.4	42.9	57.1	86.2	28.4	47.8	88.9	46.0	60.4
0960	Flooring	92.0	86.9	90.5	110.5	43.9	90.6	109.4	41.9	89.2	106.9	63.1	93.8	111.2	41.9	90.5	112.9	24.8	86.5
0970, 0990	Wall Finishes & Painting/Coating	100.9	76.0	85.9	96.5	52.4	70.0	96.5	52.4	70.0	100.1	52.4	71.4	96.5	52.4	70.0	96.5	33.8	58.8
09	FINISHES	95.9	82.2	88.3	93.8	41.0	64.4	93.6	45.8	67.0	91.8	47.7	67.2	94.3	35.4	61.5	95.8	39.7	64.6
COVERS	DIVS. 10 - 14, 25, 28, 41, 43, 44, 46	100.0	90.0	98.0	100.0	78.0	95.6	100.0	79.0	95.8	100.0	78.3	95.7	100.0	77.1	95.4	100.0	77.8	95.5
21, 22, 23	FIRE SUPPRESSION, PLUMBING & HVAC	95.2	89.7	93.0	95.4	66.3	83.7	95.4	66.3	83.7	95.4	65.8	83.5	100.3	66.3	86.6	95.4	64.2	82.9
26, 27, 3370	ELECTRICAL, COMMUNICATIONS & UTIL.	91.5	78.2	84.6	89.7	70.4	79.6	90.7	70.4	80.1	94.7	70.4	82.0	90.7	70.4	80.1	92.3	63.0	77.0
MF2010	WEIGHTED AVERAGE	95.5	85.6	91.2	95.2	62.3	80.9	96.5	63.6	82.2	94.3	64.0	81.1	97.2	60.9	81.4	96.3	60.5	80.7

DIVISION		OKLAHOMA																	
		LAWTON			MCALESTER			MIAMI			MUSKOGEE			OKLAHOMA CITY			PONCA CITY		
		735			745			743			744			730 - 731			746		
		MAT.	INST.	TOTAL	MAT.	INST.	TOTAL	MAT.	INST.	TOTAL	MAT.	INST.	TOTAL	MAT.	INST.	TOTAL	MAT.	INST.	TOTAL
015433	CONTRACTOR EQUIPMENT		82.5	82.5		81.6	81.6		89.7	89.7		89.7	89.7		82.7	82.7		81.6	81.6
0241, 31 - 34	SITE & INFRASTRUCTURE, DEMOLITION	94.3	91.4	92.2	86.3	89.9	88.9	87.8	87.2	87.4	88.0	87.0	87.3	93.7	91.8	92.3	93.0	89.9	90.8
0310	Concrete Forming & Accessories	100.2	45.3	52.8	83.6	44.6	49.9	96.8	67.7	71.7	101.3	34.1	43.3	98.2	55.5	61.3	92.2	47.3	53.4
0320	Concrete Reinforcing	87.7	80.2	84.0	91.9	80.2	86.1	90.5	80.3	85.4	91.3	79.7	85.5	92.8	80.2	86.5	91.3	80.3	85.8
0330	Cast-in-Place Concrete	92.8	47.3	73.9	80.8	46.0	66.3	84.6	48.3	69.6	85.7	46.6	69.4	94.5	47.7	75.1	94.6	45.5	74.2
03	CONCRETE	93.8	52.9	73.6	83.4	52.1	68.0	88.2	64.0	76.2	90.0	48.3	69.4	97.3	57.5	77.7	95.0	53.1	74.3
04	MASONRY	95.9	58.1	72.5	110.7	57.1	77.5	95.2	57.2	71.7	113.0	46.0	71.5	98.7	57.4	73.1	87.8	57.1	68.8
05	METALS	98.5	70.6	89.9	91.3	70.5	84.9	91.3	82.8	88.6	92.7	81.0	89.1	92.0	70.7	85.4	91.3	70.8	84.9
06	WOOD, PLASTICS & COMPOSITES	106.4	44.1	71.3	86.9	44.0	62.7	102.4	74.6	86.7	107.0	31.4	64.3	100.3	58.2	76.5	97.6	47.0	69.1
07	THERMAL & MOISTURE PROTECTION	106.1	61.7	88.1	99.6	61.0	83.9	99.9	64.9	85.7	100.0	49.5	79.5	98.8	62.8	84.2	100.1	66.7	86.5
08	OPENINGS	95.4	52.1	85.1	93.9	52.5	84.1	93.9	69.3	88.1	93.9	42.8	81.8	94.8	59.7	86.5	93.9	53.8	84.4
0920	Plaster & Gypsum Board	87.1	43.0	57.1	77.9	42.9	54.1	85.0	74.3	77.7	87.2	29.7	48.1	92.9	57.5	68.8	83.5	46.0	58.0
0950, 0980	Ceilings & Acoustic Treatment	94.2	43.0	60.1	85.4	42.9	57.1	85.4	74.3	78.0	94.3	29.7	51.3	97.5	57.5	70.9	85.4	46.0	59.2
0960	Flooring	113.1	41.9	91.8	105.8	41.9	86.7	112.8	63.1	97.9	115.3	40.5	92.9	109.5	41.9	89.3	109.9	41.9	89.6
0970, 0990	Wall Finishes & Painting/Coating	96.5	52.4	70.0	100.1	38.6	63.1	100.1	79.1	87.5	100.1	34.9	60.9	97.9	52.4	70.6	100.1	52.4	71.4
09	FINISHES	96.4	43.8	67.1	90.8	42.0	63.6	93.7	68.8	79.8	96.7	33.7	61.6	98.6	52.0	72.6	93.4	45.3	66.6
COVERS	DIVS. 10 - 14, 25, 28, 41, 43, 44, 46	100.0	78.6	95.7	100.0	78.3	95.6	100.0	82.1	96.4	100.0	76.5	95.3	100.0	80.0	96.0	100.0	78.6	95.7
21, 22, 23	FIRE SUPPRESSION, PLUMBING & HVAC	100.3	66.3	86.7	95.4	61.5	81.8	95.4	61.8	81.9	100.3	60.5	84.3	100.0	66.0	86.4	95.4	61.7	81.9
26, 27, 3370	ELECTRICAL, COMMUNICATIONS & UTIL.	92.3	70.4	80.9	93.1	66.1	79.0	94.6	66.1	79.7	92.7	53.7	72.3	97.3	70.4	83.3	92.7	70.0	80.8
MF2010	WEIGHTED AVERAGE	97.5	63.3	82.6	93.7	61.1	79.5	94.0	68.3	82.8	96.6	56.3	79.0	97.3	65.4	83.4	94.3	62.5	80.5

DIVISION		OKLAHOMA												OREGON					
		POTEAU			SHAWNEE			TULSA			WOODWARD			BEND			EUGENE		
		749			748			740 - 741			738			977			974		
		MAT.	INST.	TOTAL	MAT.	INST.	TOTAL	MAT.	INST.	TOTAL	MAT.	INST.	TOTAL	MAT.	INST.	TOTAL	MAT.	INST.	TOTAL
015433	CONTRACTOR EQUIPMENT		89.0	89.0		81.6	81.6		89.7	89.7		81.6	81.6		97.9	97.9		97.9	97.9
0241, 31 - 34	SITE & INFRASTRUCTURE, DEMOLITION	74.3	85.9	82.4	96.0	89.9	91.7	94.2	87.4	89.4	96.3	90.0	91.9	106.8	102.8	103.9	96.9	102.7	101.0
0310	Concrete Forming & Accessories	90.0	41.9	48.5	85.5	44.7	50.3	101.4	41.1	49.4	93.1	47.8	54.0	111.2	100.6	102.0	107.6	100.4	101.4
0320	Concrete Reinforcing	92.4	80.3	86.3	91.3	80.2	85.8	91.6	80.2	85.9	87.5	80.3	83.9	95.7	101.5	98.6	99.7	101.5	100.6
0330	Cast-in-Place Concrete	84.6	45.7	68.5	97.6	43.7	75.2	93.3	47.0	74.1	95.9	46.1	75.2	94.2	104.1	98.3	91.2	104.0	96.5
03	CONCRETE	90.6	51.6	71.3	96.6	51.4	74.3	95.3	51.7	73.8	97.4	53.6	75.8	104.8	101.6	103.2	96.3	101.5	98.9
04	MASONRY	95.5	57.2	71.8	111.9	57.1	78.0	96.1	61.0	74.4	89.5	58.1	70.1	98.7	104.5	102.3	96.0	104.5	101.3
05	METALS	91.3	82.4	88.5	91.2	70.4	84.8	95.9	82.0	91.6	93.5	71.0	86.5	90.7	96.6	92.5	91.4	96.4	93.0
06	WOOD, PLASTICS & COMPOSITES	94.3	40.0	63.7	89.3	44.0	63.7	106.2	37.5	67.4	99.8	47.0	70.0	103.3	100.3	101.6	98.8	100.3	99.6
07	THERMAL & MOISTURE PROTECTION	100.0	61.0	84.2	100.0	60.3	83.9	100.0	62.6	84.8	106.2	67.1	90.3	104.6	96.3	101.2	103.9	93.4	99.6
08	OPENINGS	93.9	50.5	83.6	93.9	52.5	84.1	95.4	48.1	84.2	93.9	53.8	84.4	93.8	103.8	96.2	94.2	103.8	96.5
0920	Plaster & Gypsum Board	82.0	38.7	52.5	79.1	42.9	54.5	87.2	36.0	52.4	84.6	46.0	58.4	103.7	100.2	101.3	101.1	100.2	100.5
0950, 0980	Ceilings & Acoustic Treatment	85.4	38.7	54.3	85.4	42.9	57.1	94.3	36.0	55.5	88.9	46.0	60.4	92.3	100.2	97.5	93.3	100.2	97.9
0960	Flooring	109.3	63.1	95.4	106.9	33.3	84.9	114.0	43.6	92.9	109.4	43.9	89.8	107.6	105.3	106.9	106.1	105.3	105.9
0970, 0990	Wall Finishes & Painting/Coating	100.1	52.4	71.4	100.1	34.9	60.9	100.1	41.4	64.8	96.5	52.4	70.0	104.6	72.3	85.2	104.6	72.3	85.2
09	FINISHES	91.5	45.3	65.8	92.1	40.1	63.1	96.5	39.7	64.9	94.4	45.9	67.4	103.3	98.1	100.4	101.7	98.1	99.7
COVERS	DIVS. 10 - 14, 25, 28, 41, 43, 44, 46	100.0	78.1	95.6	100.0	78.3	95.7	100.0	79.3	95.8	100.0	79.0	95.8	100.0	101.9	100.4	100.0	101.9	100.4
21, 22, 23	FIRE SUPPRESSION, PLUMBING & HVAC	95.4	61.7	81.9	95.4	65.8	83.5	100.3	63.5	85.5	95.4	66.3	83.7	95.1	102.6	98.1	100.0	102.6	101.0
26, 27, 3370	ELECTRICAL, COMMUNICATIONS & UTIL.	92.8	66.1	78.9	94.8	70.4	82.1	94.7	66.1	79.8	92.2	70.4	80.8	95.8	96.3	96.1	94.5	96.3	95.4
MF2010	WEIGHTED AVERAGE	93.5	62.1	79.8	95.7	62.2	81.1	97.4	62.2	82.1	95.3	63.8	81.6	97.3	100.5	98.7	97.1	100.4	98.5

DIVISION		OREGON																	
		KLAMATH FALLS			MEDFORD			PENDLETON			PORTLAND			SALEM			VALE		
		976			975			978			970 - 972			973			979		
		MAT.	INST.	TOTAL	MAT.	INST.	TOTAL	MAT.	INST.	TOTAL	MAT.	INST.	TOTAL	MAT.	INST.	TOTAL	MAT.	INST.	TOTAL
015433	CONTRACTOR EQUIPMENT		97.9	97.9		97.9	97.9		95.4	95.4		97.9	97.9		97.9	97.9		95.4	95.4
0241, 31 - 34	SITE & INFRASTRUCTURE, DEMOLITION	110.8	102.7	105.1	104.5	102.7	103.3	103.2	96.4	98.4	99.4	102.8	101.7	92.9	102.8	99.8	91.4	96.3	94.9
0310	Concrete Forming & Accessories	104.1	100.3	100.9	103.2	100.3	100.7	104.4	100.7	101.2	108.8	100.6	101.7	107.1	100.6	101.5	110.9	99.7	101.2
0320	Concrete Reinforcing	95.7	101.5	98.6	97.3	101.5	99.4	95.1	101.5	98.3	100.5	101.5	101.0	106.1	101.5	103.8	92.7	101.4	97.0
0330	Cast-in-Place Concrete	94.2	104.0	98.3	94.2	104.0	98.3	95.0	105.4	99.3	93.7	104.1	98.0	88.4	104.1	94.9	74.9	105.0	87.4
03	CONCRETE	107.6	101.5	104.6	102.2	101.5	101.8	88.4	102.1	95.2	97.7	101.6	99.7	96.0	101.6	98.8	74.7	101.5	88.0
04	MASONRY	110.8	104.5	106.9	93.1	104.5	100.2	101.7	104.6	103.5	96.5	104.5	101.5	103.3	104.5	104.1	100.2	104.6	102.9
05	METALS	90.7	96.3	92.5	91.0	96.3	92.6	97.1	97.1	97.1	92.4	96.6	93.7	97.8	96.6	97.4	97.0	96.0	96.7
06	WOOD, PLASTICS & COMPOSITES	93.8	100.3	97.5	92.6	100.3	97.0	95.9	100.4	98.4	99.8	100.3	100.1	96.7	100.3	98.7	104.8	100.4	102.3
07	THERMAL & MOISTURE PROTECTION	104.7	94.8	100.7	104.4	94.8	100.5	97.4	95.4	96.6	103.8	98.1	101.5	101.7	96.3	99.5	96.9	94.8	96.1
08	OPENINGS	93.9	103.8	96.2	96.3	103.8	98.1	89.5	103.9	92.9	92.2	103.8	95.0	95.9	103.8	97.8	89.4	95.8	90.9
0920	Plaster & Gypsum Board	98.2	100.2	99.6	97.5	100.2	99.3	85.2	100.2	95.4	100.9	100.2	100.4	97.3	100.2	99.3	91.9	100.2	97.5
0950, 0980	Ceilings & Acoustic Treatment	100.3	100.2	100.2	107.3	100.2	102.6	64.6	100.2	88.3	95.3	100.2	98.5	100.7	100.2	100.4	64.6	100.2	88.3
0960	Flooring	105.0	105.3	105.1	104.4	105.3	104.7	72.5	105.3	82.3	103.6	105.3	104.1	104.3	105.3	104.6	74.2	105.3	83.5
0970, 0990	Wall Finishes & Painting/Coating	104.6	59.4	77.4	104.6	59.4	77.4	96.1	66.9	78.5	104.3	66.9	81.8	102.8	72.3	84.5	96.1	72.3	81.8
09	FINISHES	104.0	96.7	99.9	104.4	96.7	100.1	71.8	97.6	86.2	101.4	97.5	99.2	100.5	98.1	99.2	72.3	98.2	86.7
COVERS	DIVS. 10 - 14, 25, 28, 41, 43, 44, 46	100.0	101.8	100.4	100.0	101.8	100.4	100.0	92.2	98.4	100.0	101.9	100.4	100.0	101.9	100.4	100.0	102.1	100.4
21, 22, 23	FIRE SUPPRESSION, PLUMBING & HVAC	95.1	102.6	98.1	100.0	102.6	101.0	96.9	115.3	104.3	100.0	102.6	101.0	100.0	102.6	101.1	96.9	71.5	86.7
26, 27, 3370	ELECTRICAL, COMMUNICATIONS & UTIL.	94.6	81.3	87.6	97.9	81.3	89.2	87.3	96.5	92.1	94.7	103.2	99.1	102.1	96.3	99.1	87.3	73.9	80.3
MF2010	WEIGHTED AVERAGE	98.2	98.2	98.2	98.4	98.2	98.3	93.1	102.5	97.2	97.2	101.4	99.1	99.0	100.5	99.7	91.3	89.8	90.7

DIVISION		PENNSYLVANIA																	
		ALLENTOWN			ALTOONA			BEDFORD			BRADFORD			BUTLER			CHAMBERSBURG		
		181			166			155			167			160			172		
		MAT.	INST.	TOTAL	MAT.	INST.	TOTAL	MAT.	INST.	TOTAL	MAT.	INST.	TOTAL	MAT.	INST.	TOTAL	MAT.	INST.	TOTAL
015433	CONTRACTOR EQUIPMENT		113.5	113.5		113.5	113.5		110.5	110.5		113.5	113.5		113.5	113.5		112.7	112.7
0241, 31 - 34	SITE & INFRASTRUCTURE, DEMOLITION	90.7	105.4	101.0	94.2	105.5	102.1	103.7	102.2	102.7	89.5	104.8	100.3	85.7	106.9	100.6	85.1	102.7	97.4
0310	Concrete Forming & Accessories	100.5	114.2	112.3	85.7	80.4	81.1	85.2	82.7	83.0	87.9	82.5	83.2	87.2	95.7	94.5	89.5	81.8	82.9
0320	Concrete Reinforcing	94.4	111.5	103.0	91.5	106.0	98.7	89.7	83.5	86.6	93.5	106.4	99.9	92.2	112.5	102.3	88.7	106.3	97.5
0330	Cast-in-Place Concrete	85.0	101.5	91.8	94.7	87.5	91.7	109.6	71.2	93.6	90.4	92.9	91.5	83.7	98.1	89.6	94.9	71.0	85.0
03	CONCRETE	94.3	110.0	102.1	89.5	89.1	89.3	103.4	80.2	91.9	96.0	92.0	94.0	81.7	100.8	91.1	103.3	84.1	93.8
04	MASONRY	94.2	102.2	99.2	96.9	65.6	77.5	114.6	88.5	98.5	94.0	88.5	90.6	98.9	98.5	98.7	94.7	88.9	91.1
05	METALS	99.9	123.1	107.0	93.7	118.3	101.3	97.5	105.6	100.0	97.8	118.2	104.1	93.4	123.4	102.7	99.5	116.5	104.7
06	WOOD, PLASTICS & COMPOSITES	103.0	117.3	111.1	81.2	82.7	82.0	83.7	82.6	83.1	87.6	80.3	83.5	82.6	94.8	89.5	92.5	82.7	87.0
07	THERMAL & MOISTURE PROTECTION	102.5	120.4	109.7	101.8	92.0	97.8	104.0	88.6	97.8	102.4	92.3	98.3	101.5	100.5	101.1	102.8	74.0	91.1
08	OPENINGS	92.5	116.3	98.1	86.4	90.1	87.3	92.1	83.6	90.1	92.6	92.3	92.5	86.4	105.7	91.0	89.7	84.8	88.6
0920	Plaster & Gypsum Board	99.7	117.6	111.8	91.1	81.9	84.8	91.6	81.9	85.0	91.9	79.4	83.4	91.1	94.4	93.3	99.4	81.9	87.5
0950, 0980	Ceilings & Acoustic Treatment	85.7	117.6	106.9	90.1	81.9	84.6	95.7	81.9	86.5	88.4	79.4	82.4	91.1	94.4	93.3	90.3	81.9	84.7
0960	Flooring	93.6	97.6	94.8	87.4	99.3	91.0	96.9	89.8	94.7	88.5	105.8	93.6	88.2	82.5	86.5	93.7	47.0	79.7
0970, 0990	Wall Finishes & Painting/Coating	98.3	72.1	82.5	94.6	108.9	103.2	104.7	83.3	91.9	98.3	92.6	94.9	94.6	108.9	103.2	101.9	83.3	90.7
09	FINISHES	93.8	107.1	101.2	92.2	85.9	88.7	98.3	82.6	89.6	91.9	85.7	88.5	92.1	93.5	92.9	93.9	75.4	83.6
COVERS	DIVS. 10 - 14, 25, 28, 41, 43, 44, 46	100.0	104.5	100.9	100.0	95.3	99.1	100.0	98.6	99.7	100.0	99.3	99.9	100.0	102.3	100.5	100.0	95.2	99.0
21, 22, 23	FIRE SUPPRESSION, PLUMBING & HVAC	100.2	110.6	104.4	99.8	84.3	93.6	95.0	89.1	92.7	95.4	90.9	93.6	95.0	93.9	94.5	95.4	89.0	92.8
26, 27, 3370	ELECTRICAL, COMMUNICATIONS & UTIL.	98.2	98.9	98.6	88.3	111.1	100.2	93.7	111.1	102.8	91.7	111.1	101.8	88.9	110.1	100.0	89.4	88.3	88.9
MF2010	WEIGHTED AVERAGE	97.6	108.8	102.4	94.3	92.7	93.6	98.0	92.6	95.7	95.3	96.8	96.0	92.1	102.2	96.5	96.0	89.7	93.2

DIVISION		PENNSYLVANIA																	
		DOYLESTOWN			DUBOIS			ERIE			GREENSBURG			HARRISBURG			HAZLETON		
		189			158			164 - 165			156			170 - 171			182		
		MAT.	INST.	TOTAL	MAT.	INST.	TOTAL	MAT.	INST.	TOTAL	MAT.	INST.	TOTAL	MAT.	INST.	TOTAL	MAT.	INST.	TOTAL
015433	CONTRACTOR EQUIPMENT		94.6	94.6		110.5	110.5		113.5	113.5		110.5	110.5		112.7	112.7		113.5	113.5
0241, 31 - 34	SITE & INFRASTRUCTURE, DEMOLITION	103.5	91.4	95.1	108.7	102.5	104.4	91.1	105.9	101.5	99.8	104.7	103.2	85.2	104.5	98.7	84.1	105.7	99.3
0310	Concrete Forming & Accessories	84.7	129.1	123.0	84.7	83.9	84.0	99.9	88.3	89.9	91.5	95.6	95.1	99.8	89.4	90.8	82.3	90.2	89.1
0320	Concrete Reinforcing	91.3	135.0	113.1	89.1	106.7	97.9	93.5	106.5	100.0	89.1	112.5	100.8	98.1	106.7	102.4	91.7	110.5	101.0
0330	Cast-in-Place Concrete	80.3	87.6	83.4	105.7	93.5	100.6	93.1	82.6	88.7	101.7	97.5	100.0	97.1	96.8	97.0	80.3	92.7	85.5
03	CONCRETE	90.0	115.6	102.6	104.8	92.8	98.9	88.6	90.9	89.7	98.4	100.5	99.4	99.2	96.5	97.9	87.0	96.1	91.5
04	MASONRY	97.2	127.9	116.2	114.6	89.6	99.2	86.7	92.8	90.5	125.3	98.5	108.7	95.0	90.6	92.3	106.2	96.5	100.2
05	METALS	97.4	125.6	106.1	97.5	117.6	103.7	93.9	117.0	101.0	97.5	121.9	105.0	106.1	120.3	110.5	99.6	121.4	106.3
06	WOOD, PLASTICS & COMPOSITES	83.1	131.5	110.4	82.6	82.4	82.5	99.2	86.7	92.1	90.5	94.7	92.9	100.1	88.7	93.7	81.6	88.6	85.6
07	THERMAL & MOISTURE PROTECTION	100.6	132.8	113.7	104.3	95.5	100.7	102.3	93.4	98.7	103.9	100.1	102.4	106.2	112.2	108.6	102.0	107.6	104.2
08	OPENINGS	94.7	140.0	105.4	92.1	93.5	92.4	86.6	93.0	88.1	92.1	105.7	95.3	95.8	95.4	95.7	93.0	94.5	93.4
0920	Plaster & Gypsum Board	89.6	132.2	118.6	90.5	81.7	84.5	99.7	86.0	90.4	92.9	94.4	93.9	103.4	88.1	93.0	90.0	88.0	88.7
0950, 0980	Ceilings & Acoustic Treatment	84.9	132.2	116.4	95.7	81.7	86.4	85.7	86.0	85.9	94.8	94.4	94.5	99.4	88.1	91.9	86.7	88.0	87.6
0960	Flooring	78.6	136.8	96.0	96.7	105.8	99.4	92.3	93.5	92.7	99.8	69.3	90.7	98.7	93.4	97.1	86.2	96.3	89.2
0970, 0990	Wall Finishes & Painting/Coating	97.8	70.5	81.4	104.7	108.9	107.2	103.3	92.6	96.9	104.7	108.9	107.2	102.3	91.5	95.8	98.3	105.6	102.7
09	FINISHES	85.1	123.5	106.5	98.6	89.3	93.4	94.3	89.2	91.5	98.7	91.9	94.9	98.8	89.9	93.8	90.2	91.0	90.6
COVERS	DIVS. 10 - 14, 25, 28, 41, 43, 44, 46	100.0	71.9	94.4	100.0	98.6	99.7	100.0	100.9	100.2	100.0	102.1	100.4	100.0	96.4	99.3	100.0	99.1	99.8
21, 22, 23	FIRE SUPPRESSION, PLUMBING & HVAC	95.0	125.4	107.2	95.0	89.5	92.8	99.8	93.1	97.1	95.0	92.8	94.1	100.0	90.7	96.3	95.4	98.8	96.7
26, 27, 3370	ELECTRICAL, COMMUNICATIONS & UTIL.	91.1	126.9	109.8	94.3	111.1	103.1	90.1	87.4	88.7	94.3	111.1	103.1	96.8	88.4	92.4	92.6	91.5	92.0
MF2010	WEIGHTED AVERAGE	94.6	120.8	106.0	98.4	97.1	97.8	94.1	95.0	94.5	98.0	101.6	99.5	99.6	95.9	98.0	95.0	98.8	96.7

DIVISION		PENNSYLVANIA																	
		INDIANA			JOHNSTOWN			KITTANNING			LANCASTER			LEHIGH VALLEY			MONTROSE		
		157			159			162			175 - 176			180			188		
		MAT.	INST.	TOTAL	MAT.	INST.	TOTAL	MAT.	INST.	TOTAL	MAT.	INST.	TOTAL	MAT.	INST.	TOTAL	MAT.	INST.	TOTAL
015433	CONTRACTOR EQUIPMENT		110.5	110.5		110.5	110.5		113.5	113.5		112.7	112.7		113.5	113.5		113.5	113.5
0241, 31 - 34	SITE & INFRASTRUCTURE, DEMOLITION	97.9	103.3	101.7	104.2	103.9	104.0	88.2	106.8	101.3	77.7	104.4	96.5	87.9	105.1	99.9	86.6	103.4	98.4
0310	Concrete Forming & Accessories	85.8	86.3	86.2	84.7	84.2	84.3	87.2	95.8	94.6	91.6	88.9	89.3	94.2	113.6	110.9	83.3	88.5	87.8
0320	Concrete Reinforcing	88.4	112.6	100.5	89.7	112.2	100.9	92.2	112.6	102.4	88.4	106.6	97.5	91.7	111.4	101.5	96.0	108.4	102.2
0330	Cast-in-Place Concrete	99.7	97.3	98.7	110.6	92.9	103.2	86.9	98.0	91.5	80.8	97.7	87.8	86.9	100.9	92.7	85.2	90.9	87.6
03	CONCRETE	95.9	96.3	96.1	104.2	93.7	99.0	84.1	100.8	92.4	90.9	96.5	93.7	93.3	109.4	101.2	92.0	94.1	93.1
04	MASONRY	110.3	100.2	104.1	111.1	88.0	96.8	101.7	100.2	100.8	99.9	90.8	94.3	94.2	101.7	98.8	94.1	96.0	95.3
05	METALS	97.6	121.5	105.0	97.5	120.1	104.5	93.5	123.4	102.7	99.5	119.7	105.7	99.5	121.7	106.4	97.9	115.3	103.2
06	WOOD, PLASTICS & COMPOSITES	84.5	82.4	83.3	82.6	82.6	82.6	82.6	94.8	89.5	95.2	88.7	91.5	94.4	117.3	107.3	82.4	88.3	85.8
07	THERMAL & MOISTURE PROTECTION	103.8	97.9	101.4	104.0	93.7	99.8	101.6	100.9	101.3	102.4	98.5	100.8	102.3	103.6	102.8	102.0	92.4	98.1
08	OPENINGS	92.1	95.2	92.8	92.1	91.8	92.0	86.4	102.0	90.1	89.7	99.3	92.0	93.0	116.3	98.5	89.8	93.8	90.8
0920	Plaster & Gypsum Board	92.0	81.7	85.0	90.3	81.9	84.6	91.1	94.4	93.3	101.3	88.1	92.3	93.0	117.6	109.7	90.4	87.7	88.6
0950, 0980	Ceilings & Acoustic Treatment	95.7	81.7	86.4	94.8	81.9	86.2	91.1	94.4	93.3	90.3	88.1	88.8	86.7	117.6	107.2	88.4	87.7	88.0
0960	Flooring	97.4	105.8	99.9	96.7	100.0	97.7	88.2	105.8	93.5	94.5	93.4	94.2	90.9	97.6	92.9	86.7	58.9	78.4
0970, 0990	Wall Finishes & Painting/Coating	104.7	108.9	107.2	104.7	108.9	107.2	94.6	108.9	103.2	101.9	58.4	75.7	98.3	67.9	80.0	98.3	105.6	102.7
09	FINISHES	98.0	90.2	93.7	97.9	88.7	92.8	92.3	97.5	95.2	93.9	86.2	89.6	92.2	105.5	99.6	91.0	84.8	87.5
COVERS	DIVS. 10 - 14, 25, 28, 41, 43, 44, 46	100.0	100.7	100.1	100.0	98.8	99.8	100.0	102.3	100.5	100.0	96.3	99.3	100.0	104.1	100.8	100.0	99.1	99.8
21, 22, 23	FIRE SUPPRESSION, PLUMBING & HVAC	95.0	92.7	94.1	95.0	89.8	92.9	95.0	96.1	95.4	95.4	90.8	93.5	95.4	110.3	101.4	95.4	98.0	96.4
26, 27, 3370	ELECTRICAL, COMMUNICATIONS & UTIL.	94.3	111.1	103.1	94.3	111.1	103.1	88.3	111.1	100.2	90.9	45.1	67.0	92.6	133.5	113.9	91.7	97.2	94.5
MF2010	WEIGHTED AVERAGE	96.9	100.1	98.3	97.9	97.3	97.7	92.5	103.4	97.3	94.8	89.3	92.4	95.4	112.5	102.9	94.4	97.1	95.6

DIVISION		PENNSYLVANIA																	
		NEW CASTLE			NORRISTOWN			OIL CITY			PHILADELPHIA			PITTSBURGH			POTTSVILLE		
		161			194			163			190 - 191			150 - 152			179		
		MAT.	INST.	TOTAL	MAT.	INST.	TOTAL	MAT.	INST.	TOTAL	MAT.	INST.	TOTAL	MAT.	INST.	TOTAL	MAT.	INST.	TOTAL
015433	CONTRACTOR EQUIPMENT		113.5	113.5		98.0	98.0		113.5	113.5		97.5	97.5		111.8	111.8		112.7	112.7
0241, 31 - 34	SITE & INFRASTRUCTURE, DEMOLITION	86.1	106.9	100.7	92.4	99.8	97.6	84.7	105.0	99.0	99.9	99.2	99.4	103.4	106.7	105.7	80.4	104.6	97.4
0310	Concrete Forming & Accessories	87.2	97.4	96.0	83.5	130.8	124.3	87.2	83.5	84.0	98.7	139.9	134.3	99.0	98.7	98.7	83.2	90.0	89.1
0320	Concrete Reinforcing	91.0	93.3	92.2	92.5	142.2	117.3	92.2	93.1	92.6	100.5	142.2	121.3	90.1	112.9	101.5	87.7	103.8	95.7
0330	Cast-in-Place Concrete	84.4	96.9	89.6	79.7	127.5	99.6	82.0	96.5	88.1	94.8	131.7	110.1	105.7	97.7	102.4	86.0	97.9	91.0
03	CONCRETE	82.0	97.6	89.7	86.9	131.4	108.9	80.5	91.2	85.7	98.2	136.9	117.3	101.7	102.0	101.9	94.6	96.6	95.6
04	MASONRY	98.1	98.5	98.4	106.5	125.7	118.4	98.2	95.2	96.3	95.5	130.2	117.0	103.3	103.3	103.3	94.5	92.9	93.5
05	METALS	93.5	114.0	99.8	99.4	130.9	109.1	93.5	112.1	99.2	100.7	131.2	110.1	98.9	122.9	106.3	99.7	118.9	105.6
06	WOOD, PLASTICS & COMPOSITES	82.6	98.1	91.4	81.5	131.4	109.7	82.6	80.3	81.3	97.2	141.5	122.2	100.5	98.0	99.1	84.9	88.6	87.0
07	THERMAL & MOISTURE PROTECTION	101.5	97.0	99.7	103.9	132.6	115.6	101.4	94.9	98.8	102.7	135.7	116.1	104.1	102.0	103.3	102.5	107.1	104.3
08	OPENINGS	86.4	98.0	89.2	88.6	142.0	101.2	86.4	80.7	85.1	94.4	147.5	106.9	95.5	107.5	98.3	89.8	92.5	90.4
0920	Plaster & Gypsum Board	91.1	97.8	95.7	88.2	132.2	118.1	91.1	79.4	83.2	97.2	142.6	128.1	99.8	97.8	98.5	95.7	88.0	90.5
0950, 0980	Ceilings & Acoustic Treatment	91.1	97.8	95.6	90.5	132.2	118.3	91.1	79.4	83.3	91.6	142.6	125.6	95.7	97.8	97.1	90.3	88.0	88.8
0960	Flooring	88.2	58.5	79.4	91.6	139.5	105.9	88.2	105.8	93.5	98.5	139.5	110.7	102.8	108.0	104.3	91.1	93.4	91.8
0970, 0990	Wall Finishes & Painting/Coating	94.6	108.9	103.2	100.8	147.4	128.8	94.6	108.9	103.2	105.9	154.0	134.8	104.7	121.2	114.6	101.9	105.6	104.1
09	FINISHES	92.1	91.2	91.6	92.9	133.9	115.7	92.0	88.3	89.9	98.2	141.6	122.4	100.9	102.1	101.6	92.4	91.9	92.1
COVERS	DIVS. 10 - 14, 25, 28, 41, 43, 44, 46	100.0	102.6	100.5	100.0	111.7	102.3	100.0	100.4	100.1	100.0	120.5	104.1	100.0	102.5	100.5	100.0	96.7	99.3
21, 22, 23	FIRE SUPPRESSION, PLUMBING & HVAC	95.0	93.5	94.4	95.2	125.8	107.4	95.0	92.6	94.0	100.0	130.5	112.2	99.9	101.2	100.4	95.4	98.2	96.5
26, 27, 3370	ELECTRICAL, COMMUNICATIONS & UTIL.	88.9	88.7	88.8	92.0	148.7	121.6	90.8	110.1	100.9	98.6	148.7	124.7	96.7	111.1	104.2	89.0	91.1	90.1
MF2010	WEIGHTED AVERAGE	92.1	97.2	94.4	94.9	129.5	110.0	92.1	97.0	94.3	98.9	133.4	113.9	99.7	105.8	102.3	94.7	98.0	96.1

DIVISION		PENNSYLVANIA																	
		READING			SCRANTON			STATE COLLEGE			STROUDSBURG			SUNBURY			UNIONTOWN		
		195 - 196			184 - 185			168			183			178			154		
		MAT.	INST.	TOTAL	MAT.	INST.	TOTAL	MAT.	INST.	TOTAL	MAT.	INST.	TOTAL	MAT.	INST.	TOTAL	MAT.	INST.	TOTAL
015433	CONTRACTOR EQUIPMENT		117.0	117.0		113.5	113.5		112.7	112.7		113.5	113.5		113.5	113.5		110.5	110.5
0241, 31 - 34	SITE & INFRASTRUCTURE, DEMOLITION	96.9	111.5	107.1	91.1	105.9	101.5	81.6	104.2	97.5	85.6	103.5	98.2	90.9	105.6	101.2	98.5	104.7	102.8
0310	Concrete Forming & Accessories	99.0	90.5	91.7	100.7	90.1	91.5	85.4	80.1	80.9	88.7	89.5	89.4	95.8	88.9	89.9	79.0	95.7	93.4
0320	Concrete Reinforcing	93.8	103.9	98.8	94.4	108.7	101.6	92.7	106.7	99.7	94.7	115.2	104.9	90.1	106.7	98.4	89.1	112.5	100.8
0330	Cast-in-Place Concrete	71.5	98.2	82.6	88.7	92.9	90.5	85.6	66.6	77.7	83.6	73.4	79.4	94.0	96.5	95.0	99.7	97.5	98.8
03	CONCRETE	86.2	96.9	91.5	96.1	95.8	96.0	96.3	81.9	89.2	90.8	89.8	90.3	98.2	96.2	97.2	95.5	100.6	98.0
04	MASONRY	96.3	94.0	94.9	94.6	96.5	95.8	99.5	79.2	87.0	91.8	101.3	97.7	93.9	90.4	91.7	127.7	98.5	109.6
05	METALS	99.7	119.2	105.7	101.9	121.2	107.9	97.6	119.0	104.2	99.6	117.8	105.2	99.4	120.0	105.8	97.3	122.0	104.9
06	WOOD, PLASTICS & COMPOSITES	100.3	88.5	93.7	103.0	88.2	94.7	89.7	82.7	85.7	88.6	88.3	88.4	93.2	88.6	90.6	76.4	94.7	86.7
07	THERMAL & MOISTURE PROTECTION	104.3	113.6	108.1	102.4	96.3	99.9	101.6	92.1	97.7	102.1	85.8	95.5	103.5	105.8	104.4	103.7	100.2	102.3
08	OPENINGS	93.1	98.3	94.3	92.5	93.7	92.8	89.6	90.1	89.7	93.0	89.1	92.1	89.9	93.4	90.7	92.0	105.7	95.3
0920	Plaster & Gypsum Board	100.4	88.0	92.0	101.9	87.6	92.2	93.1	81.9	85.5	91.4	87.7	88.9	95.2	88.0	90.3	88.4	94.4	92.5
0950, 0980	Ceilings & Acoustic Treatment	81.5	88.0	85.8	94.6	87.6	89.9	85.8	81.9	83.2	84.9	87.7	86.8	86.8	88.0	87.6	94.8	94.4	94.5
0960	Flooring	95.5	93.4	94.9	93.6	108.9	98.2	91.4	99.3	93.8	89.0	52.3	78.0	91.8	89.3	91.0	94.4	105.8	97.8
0970, 0990	Wall Finishes & Painting/Coating	99.5	105.6	103.1	98.3	105.6	102.7	98.3	108.9	104.7	98.3	66.0	78.9	101.9	105.6	104.1	104.7	108.9	107.2
09	FINISHES	94.0	91.5	92.6	96.0	94.6	95.2	91.2	85.9	88.3	90.9	80.3	85.0	93.1	90.6	91.7	96.5	97.3	96.9
COVERS	DIVS. 10 - 14, 25, 28, 41, 43, 44, 46	100.0	99.0	99.8	100.0	99.1	99.8	100.0	91.8	98.4	100.0	63.5	92.7	100.0	95.9	99.2	100.0	102.1	100.4
21, 22, 23	FIRE SUPPRESSION, PLUMBING & HVAC	100.2	107.3	103.0	100.2	98.8	99.6	95.4	84.6	91.1	95.4	100.5	97.4	95.4	90.4	93.4	95.0	92.8	94.1
26, 27, 3370	ELECTRICAL, COMMUNICATIONS & UTIL.	98.7	91.1	94.7	98.2	97.2	97.7	90.8	111.1	101.4	92.6	142.5	118.6	89.3	93.0	91.2	91.3	111.1	101.6
MF2010	WEIGHTED AVERAGE	97.1	101.1	98.8	98.3	99.7	98.9	94.9	93.0	94.1	94.9	102.0	98.0	95.5	96.3	95.8	97.1	102.2	99.4

DIVISION		PENNSYLVANIA																	
		WASHINGTON			WELLSBORO			WESTCHESTER			WILKES-BARRE			WILLIAMSPORT			YORK		
		153			169			193			186 - 187			177			173 - 174		
		MAT.	INST.	TOTAL	MAT.	INST.	TOTAL	MAT.	INST.	TOTAL	MAT.	INST.	TOTAL	MAT.	INST.	TOTAL	MAT.	INST.	TOTAL
015433	CONTRACTOR EQUIPMENT		110.5	110.5		113.5	113.5		98.0	98.0		113.5	113.5		113.5	113.5		112.7	112.7
0241, 31 - 34	SITE & INFRASTRUCTURE, DEMOLITION	98.6	104.7	102.9	93.1	103.3	100.3	98.0	97.2	97.4	83.7	105.7	99.2	82.7	104.2	97.8	81.1	104.5	97.5
0310	Concrete Forming & Accessories	85.9	95.8	94.5	87.3	87.1	87.1	90.1	129.3	123.9	91.3	90.4	90.5	92.3	61.5	65.7	86.6	89.5	89.1
0320	Concrete Reinforcing	89.1	112.6	100.9	92.7	108.3	100.5	91.7	109.7	100.7	93.5	110.5	102.0	89.4	66.6	78.0	90.1	106.7	98.4
0330	Cast-in-Place Concrete	99.7	97.6	98.8	89.7	88.8	89.3	88.2	126.7	104.2	80.3	92.6	85.4	79.6	74.3	77.4	86.4	97.9	91.2
03	CONCRETE	96.0	100.6	98.3	98.5	92.8	95.7	94.3	124.4	109.1	87.9	96.2	92.0	86.0	68.8	77.5	95.8	96.9	96.4
04	MASONRY	109.8	100.2	103.9	100.2	90.4	94.2	101.2	125.7	116.3	106.6	96.1	100.1	86.4	94.2	91.2	94.9	90.8	92.4
05	METALS	97.2	122.3	105.0	97.7	115.4	103.1	99.4	114.9	104.2	97.8	121.8	105.2	99.5	100.2	99.7	101.1	120.3	107.0
06	WOOD, PLASTICS & COMPOSITES	84.6	94.7	90.3	87.1	88.6	87.9	88.8	131.4	112.8	91.0	88.6	89.7	89.6	52.2	68.5	88.7	88.7	88.7
07	THERMAL & MOISTURE PROTECTION	103.8	100.6	102.5	102.6	90.3	97.6	104.3	131.7	115.4	102.0	107.4	104.2	103.0	103.3	103.1	102.5	112.2	106.5
08	OPENINGS	92.0	105.7	95.3	92.5	94.0	92.9	88.6	125.7	97.4	89.8	100.4	92.3	89.9	54.8	81.6	89.8	95.4	91.1
0920	Plaster & Gypsum Board	91.8	94.4	93.5	91.3	88.0	89.1	90.0	132.2	118.7	92.3	88.0	89.4	95.7	50.5	64.9	96.8	88.1	90.9
0950, 0980	Ceilings & Acoustic Treatment	94.8	94.4	94.5	85.8	88.0	87.3	90.5	132.2	118.3	88.4	88.0	88.2	90.3	50.5	63.8	89.3	88.1	88.5
0960	Flooring	97.5	105.8	100.0	88.2	51.7	77.3	94.4	139.5	107.9	89.8	96.3	91.7	90.8	50.6	78.8	92.3	93.4	92.6
0970, 0990	Wall Finishes & Painting/Coating	104.7	106.8	106.0	98.3	105.6	102.7	100.8	147.4	128.8	98.3	105.6	102.7	101.9	105.6	104.1	101.9	91.5	95.6
09	FINISHES	97.8	97.3	97.6	91.6	82.6	86.6	94.4	131.9	115.3	91.9	92.3	92.1	92.9	63.1	76.3	92.7	89.9	91.1
COVERS	DIVS. 10 - 14, 25, 28, 41, 43, 44, 46	100.0	102.1	100.4	100.0	95.8	99.2	100.0	115.5	103.1	100.0	99.0	99.8	100.0	93.2	98.6	100.0	96.5	99.3
21, 22, 23	FIRE SUPPRESSION, PLUMBING & HVAC	95.0	96.0	95.4	95.4	89.6	93.0	95.2	125.5	107.3	95.4	98.7	96.7	95.4	91.4	93.8	100.2	90.8	96.4
26, 27, 3370	ELECTRICAL, COMMUNICATIONS & UTIL.	93.7	111.1	102.8	91.7	83.3	87.3	91.9	117.6	105.3	92.6	91.5	92.0	89.8	47.7	67.8	90.9	88.4	89.6
MF2010	WEIGHTED AVERAGE	96.7	103.2	99.5	95.9	92.3	94.3	95.8	121.7	107.1	94.7	99.2	96.7	93.6	79.3	87.4	96.5	96.0	96.3

DIVISION		PUERTO RICO			RHODE ISLAND						SOUTH CAROLINA								
		SAN JUAN			NEWPORT			PROVIDENCE			AIKEN			BEAUFORT			CHARLESTON		
		009			028			029			298			299			294		
		MAT.	INST.	TOTAL	MAT.	INST.	TOTAL	MAT.	INST.	TOTAL	MAT.	INST.	TOTAL	MAT.	INST.	TOTAL	MAT.	INST.	TOTAL
015433	CONTRACTOR EQUIPMENT		90.2	90.2		102.1	102.1		102.1	102.1		100.5	100.5		100.5	100.5		100.5	100.5
0241, 31 - 34	SITE & INFRASTRUCTURE, DEMOLITION	132.3	90.5	103.0	90.6	104.2	100.2	92.0	104.2	100.6	118.8	87.1	96.5	113.9	85.3	93.8	97.9	86.0	89.5
0310	Concrete Forming & Accessories	93.7	18.1	28.4	102.6	120.6	118.1	103.9	120.6	118.3	98.6	68.9	73.0	97.5	38.8	46.8	96.5	64.4	68.8
0320	Concrete Reinforcing	191.6	13.1	102.5	106.8	153.9	130.3	103.4	153.9	128.6	94.8	67.3	81.0	93.9	26.6	60.3	93.7	61.2	77.5
0330	Cast-in-Place Concrete	103.7	31.2	73.6	80.3	124.1	98.5	96.9	124.1	108.2	80.9	70.8	76.7	80.9	47.8	67.2	94.9	50.2	76.3
03	CONCRETE	110.5	22.7	67.2	96.1	127.4	111.6	103.6	127.4	115.3	101.3	70.4	86.0	98.6	41.7	70.5	94.1	60.2	77.3
04	MASONRY	83.6	17.2	42.5	96.4	128.9	116.5	100.6	128.9	118.1	73.9	62.5	66.8	87.4	33.8	54.3	88.6	41.6	59.6
05	METALS	117.2	34.9	91.8	99.4	126.8	107.8	106.5	126.8	112.8	92.9	85.1	90.5	92.9	69.3	85.6	94.8	81.6	90.7
06	WOOD, PLASTICS & COMPOSITES	97.0	17.9	52.4	101.9	119.8	112.0	104.2	119.8	113.0	99.2	70.1	82.8	97.6	38.7	64.3	96.4	70.0	81.5
07	THERMAL & MOISTURE PROTECTION	129.7	21.3	85.7	101.5	120.4	109.2	102.3	120.4	109.6	105.3	67.7	90.0	105.0	39.6	78.5	104.2	48.2	81.5
08	OPENINGS	150.1	15.1	118.2	98.1	128.0	105.1	100.9	128.0	107.3	92.5	66.3	86.3	92.6	37.5	79.5	96.0	64.8	88.6
0920	Plaster & Gypsum Board	154.5	15.3	59.8	94.7	119.8	111.8	97.0	119.8	112.5	104.6	69.1	80.4	108.1	36.6	59.4	109.8	68.9	82.0
0950, 0980	Ceilings & Acoustic Treatment	230.8	15.3	87.4	95.3	119.8	111.6	95.2	119.8	111.6	96.0	69.1	78.1	99.5	36.6	57.7	99.5	68.9	79.1
0960	Flooring	225.0	13.4	161.7	99.8	140.3	111.9	100.4	140.3	112.4	104.8	69.7	94.3	106.1	52.1	90.0	105.8	59.3	91.9
0970, 0990	Wall Finishes & Painting/Coating	203.3	17.2	91.4	96.0	124.2	112.9	94.4	124.2	112.3	105.7	71.9	85.4	105.7	35.4	63.5	105.7	68.3	83.2
09	FINISHES	214.4	17.4	104.6	98.6	124.7	113.2	98.0	124.7	112.9	102.4	69.5	84.0	103.5	40.7	68.5	101.6	64.9	81.2
COVERS	DIVS. 10 - 14, 25, 28, 41, 43, 44, 46	100.0	16.5	83.3	100.0	105.8	101.2	100.0	105.8	101.2	100.0	73.6	94.7	100.0	72.1	94.4	100.0	70.7	94.1
21, 22, 23	FIRE SUPPRESSION, PLUMBING & HVAC	102.8	14.2	67.3	100.1	108.4	103.4	99.9	108.4	103.3	95.5	64.5	83.0	95.5	37.3	72.1	100.4	54.9	82.1
26, 27, 3370	ELECTRICAL, COMMUNICATIONS & UTIL.	126.9	13.5	67.7	101.5	100.4	100.9	102.5	100.4	101.4	96.8	67.4	81.5	100.7	34.4	66.1	99.0	90.9	94.8
MF2010	WEIGHTED AVERAGE	122.1	24.6	79.7	99.0	116.6	106.6	101.5	116.6	108.1	96.4	70.4	85.1	97.1	45.7	74.7	97.7	66.4	84.1

DIVISION		SOUTH CAROLINA															SOUTH DAKOTA		
		COLUMBIA			FLORENCE			GREENVILLE			ROCK HILL			SPARTANBURG			ABERDEEN		
		290 - 292			295			296			297			293			574		
		MAT.	INST.	TOTAL	MAT.	INST.	TOTAL	MAT.	INST.	TOTAL	MAT.	INST.	TOTAL	MAT.	INST.	TOTAL	MAT.	INST.	TOTAL
015433	CONTRACTOR EQUIPMENT		100.5	100.5		100.5	100.5		100.5	100.5		100.5	100.5		100.5	100.5		99.0	99.0
0241, 31 - 34	SITE & INFRASTRUCTURE, DEMOLITION	98.0	85.9	89.5	107.9	86.0	92.5	102.9	85.6	90.7	100.4	84.6	89.3	102.7	85.6	90.7	94.3	94.5	94.4
0310	Concrete Forming & Accessories	95.4	46.1	52.9	84.6	46.3	51.5	96.0	46.1	52.9	94.5	39.6	47.1	99.0	46.2	53.5	98.9	37.5	45.9
0320	Concrete Reinforcing	96.8	61.0	78.9	93.3	61.2	77.3	93.2	45.9	69.6	94.1	45.6	69.8	93.2	58.9	76.1	100.5	39.3	69.9
0330	Cast-in-Place Concrete	97.9	50.7	78.3	80.9	50.0	68.1	80.9	49.9	68.0	80.9	44.5	65.8	80.9	50.0	68.1	110.5	44.5	83.1
03	CONCRETE	95.9	52.2	74.3	93.3	52.0	72.9	92.2	49.1	70.9	90.3	44.2	67.5	92.4	51.6	72.3	103.7	41.7	73.1
04	MASONRY	85.5	38.3	56.3	74.0	41.6	54.0	72.0	41.6	53.2	94.1	35.2	57.7	74.0	41.6	54.0	104.8	55.8	74.5
05	METALS	92.1	80.8	88.6	93.7	81.2	89.8	93.7	75.7	88.1	92.9	72.4	86.6	93.7	80.4	89.6	98.4	65.2	88.2
06	WOOD, PLASTICS & COMPOSITES	100.4	45.8	69.6	82.1	45.8	61.6	96.0	45.8	67.7	94.3	39.8	63.5	100.3	45.8	69.6	96.3	36.8	62.8
07	THERMAL & MOISTURE PROTECTION	100.0	44.2	77.4	104.5	45.7	80.6	104.4	45.7	80.6	104.3	40.5	78.4	104.5	45.7	80.6	104.3	48.8	81.8
08	OPENINGS	97.5	51.7	86.7	92.6	51.7	82.9	92.5	48.1	82.0	92.6	40.3	80.2	92.5	51.1	82.7	93.9	37.6	80.6
0920	Plaster & Gypsum Board	102.7	44.0	62.8	97.4	44.0	61.0	103.1	44.0	62.9	102.4	37.7	58.4	106.1	44.0	63.8	106.1	35.1	57.8
0950, 0980	Ceilings & Acoustic Treatment	101.2	44.0	63.1	96.9	44.0	61.7	96.0	44.0	61.4	96.0	37.7	57.2	96.0	44.0	61.4	108.2	35.1	59.6
0960	Flooring	99.7	44.4	83.1	98.4	44.4	82.3	103.7	58.2	90.1	102.9	45.3	85.7	105.0	58.2	91.0	111.1	51.2	93.2
0970, 0990	Wall Finishes & Painting/Coating	102.4	68.3	81.9	105.7	68.3	83.2	105.7	68.3	83.2	105.7	40.0	66.2	105.7	68.3	83.2	95.6	37.9	60.9
09	FINISHES	99.7	47.7	70.7	98.5	48.1	70.4	100.2	50.4	72.5	99.6	40.2	66.5	101.0	50.4	72.8	106.0	39.7	69.1
COVERS	DIVS. 10 - 14, 25, 28, 41, 43, 44, 46	100.0	67.9	93.6	100.0	67.9	93.6	100.0	67.9	93.6	100.0	66.2	93.2	100.0	67.9	93.6	100.0	41.7	88.3
21, 22, 23	FIRE SUPPRESSION, PLUMBING & HVAC	100.0	53.2	81.2	100.4	53.3	81.5	100.4	53.1	81.4	95.5	45.2	75.3	100.4	53.3	81.5	100.2	40.6	76.3
26, 27, 3370	ELECTRICAL, COMMUNICATIONS & UTIL.	99.0	61.1	79.3	96.8	61.1	78.2	99.0	58.0	77.6	99.0	57.9	77.6	99.0	58.0	77.6	99.2	52.4	74.7
MF2010	WEIGHTED AVERAGE	97.2	57.3	79.8	96.2	57.8	79.4	96.2	56.5	78.9	95.6	51.2	76.3	96.4	57.4	79.5	100.1	50.6	78.6

		SOUTH DAKOTA																	
	DIVISION	MITCHELL			MOBRIDGE			PIERRE			RAPID CITY			SIOUX FALLS			WATERTOWN		
		573			576			575			577			570 - 571			572		
		MAT.	INST.	TOTAL	MAT.	INST.	TOTAL	MAT.	INST.	TOTAL	MAT.	INST.	TOTAL	MAT.	INST.	TOTAL	MAT.	INST.	TOTAL
015433	CONTRACTOR EQUIPMENT		99.0	99.0		99.0	99.0		99.0	99.0		99.0	99.0		100.0	100.0		99.0	99.0
0241, 31 - 34	SITE & INFRASTRUCTURE, DEMOLITION	91.2	94.5	93.5	91.2	94.5	93.5	93.0	94.6	94.1	92.8	94.6	94.1	92.4	96.2	95.1	91.1	94.5	93.5
0310	Concrete Forming & Accessories	98.1	37.8	46.1	88.8	37.7	44.7	101.3	39.3	47.8	106.7	37.5	47.0	104.1	41.4	50.0	85.4	37.5	44.1
0320	Concrete Reinforcing	99.9	47.8	73.9	102.5	39.3	70.9	104.3	71.5	87.9	94.0	71.6	82.8	101.7	71.5	86.6	97.1	39.4	68.3
0330	Cast-in-Place Concrete	107.4	43.3	80.8	107.4	44.5	81.3	96.8	43.2	74.6	106.5	43.8	80.5	102.8	44.0	78.4	107.4	47.0	82.3
03	CONCRETE	101.3	42.9	72.5	101.1	41.8	71.8	97.3	48.1	73.0	100.6	47.6	74.4	99.2	49.3	74.6	100.0	42.6	71.7
04	MASONRY	93.9	54.2	69.4	102.5	55.8	73.6	103.2	54.4	73.0	102.6	57.4	74.6	97.5	54.6	70.9	127.0	58.5	84.6
05	METALS	97.4	65.4	87.6	97.5	65.2	87.5	99.7	80.1	93.6	100.2	80.7	94.2	101.4	80.3	94.9	97.4	65.8	87.7
06	WOOD, PLASTICS & COMPOSITES	95.2	37.4	62.6	83.9	37.1	57.5	98.2	37.6	64.0	100.1	34.1	62.9	96.2	40.2	64.6	80.3	36.8	55.8
07	THERMAL & MOISTURE PROTECTION	104.1	46.8	80.9	104.2	48.8	81.7	105.1	47.1	81.5	104.7	49.0	82.1	109.3	49.4	85.0	104.0	49.5	81.9
08	OPENINGS	92.7	38.2	79.8	95.1	37.2	81.4	101.6	47.3	88.7	97.5	45.4	85.2	98.9	48.7	87.0	92.7	37.3	79.6
0920	Plaster & Gypsum Board	104.5	35.7	57.7	97.2	35.4	55.1	97.2	35.9	55.5	105.6	32.3	55.8	97.3	38.6	57.4	95.2	35.1	54.3
0950, 0980	Ceilings & Acoustic Treatment	104.6	35.7	58.7	108.2	35.4	59.7	106.4	35.9	59.5	110.8	32.3	58.6	107.3	38.6	61.6	104.6	35.1	58.4
0960	Flooring	110.7	51.2	93.0	106.4	51.2	89.9	110.5	36.5	88.4	110.4	80.1	101.4	105.9	76.3	97.1	105.1	51.2	89.0
0970, 0990	Wall Finishes & Painting/Coating	95.6	41.5	63.1	95.6	42.6	63.7	99.1	45.5	66.9	95.6	45.5	65.5	93.9	45.5	64.8	95.6	37.9	60.9
09	FINISHES	104.8	40.4	68.9	103.3	40.3	68.2	105.5	38.2	68.0	106.3	44.9	72.1	103.8	47.5	72.4	101.9	39.7	67.2
COVERS	DIVS. 10 - 14, 25, 28, 41, 43, 44, 46	100.0	38.7	87.7	100.0	41.7	88.3	100.0	77.9	95.6	100.0	77.8	95.6	100.0	78.3	95.7	100.0	41.7	88.3
21, 22, 23	FIRE SUPPRESSION, PLUMBING & HVAC	95.3	40.3	73.2	95.3	40.8	73.4	100.0	64.6	85.8	100.2	65.1	86.1	100.1	39.4	75.8	95.3	40.8	73.4
26, 27, 3370	ELECTRICAL, COMMUNICATIONS & UTIL.	97.5	45.0	70.1	99.2	45.9	71.4	104.6	52.4	77.4	95.7	52.5	73.1	97.4	76.9	86.7	96.7	45.9	70.2
MF2010	WEIGHTED AVERAGE	97.5	49.5	76.6	98.1	49.8	77.1	100.8	59.2	82.7	100.0	60.3	82.7	100.0	58.8	82.1	98.5	50.2	77.5

		TENNESSEE																	
	DIVISION	CHATTANOOGA			COLUMBIA			COOKEVILLE			JACKSON			JOHNSON CITY			KNOXVILLE		
		373 - 374			384			385			383			376			377 - 379		
		MAT.	INST.	TOTAL	MAT.	INST.	TOTAL	MAT.	INST.	TOTAL	MAT.	INST.	TOTAL	MAT.	INST.	TOTAL	MAT.	INST.	TOTAL
015433	CONTRACTOR EQUIPMENT		104.8	104.8		98.8	98.8		98.8	98.8		104.9	104.9		98.7	98.7		98.7	98.7
0241, 31 - 34	SITE & INFRASTRUCTURE, DEMOLITION	103.4	99.6	100.7	86.7	88.1	87.7	92.0	87.6	88.9	95.1	97.9	97.1	110.7	88.4	95.1	89.5	88.9	89.1
0310	Concrete Forming & Accessories	97.9	56.6	62.3	82.3	63.5	66.1	82.5	36.3	42.6	89.0	46.0	51.9	84.6	39.4	45.6	96.4	42.4	49.8
0320	Concrete Reinforcing	89.9	63.2	76.6	87.7	63.2	75.4	87.7	63.2	75.4	87.6	64.0	75.8	90.5	62.1	76.3	89.9	62.2	76.1
0330	Cast-in-Place Concrete	100.0	60.4	83.5	91.8	52.4	75.5	104.0	44.3	79.2	101.6	50.1	80.2	80.4	61.0	72.4	93.9	48.8	75.2
03	CONCRETE	93.8	60.8	77.5	93.6	61.2	77.6	103.2	46.3	75.1	94.7	52.7	74.0	100.2	53.2	77.0	91.2	50.3	71.1
04	MASONRY	108.3	49.7	72.1	118.0	55.5	79.3	112.3	44.5	70.4	118.2	48.0	74.8	123.3	45.6	75.2	85.2	53.0	65.3
05	METALS	97.7	87.4	94.5	91.5	87.6	90.3	91.6	87.5	90.3	93.8	87.6	91.9	94.9	86.4	92.3	98.3	86.7	94.7
06	WOOD, PLASTICS & COMPOSITES	107.1	58.6	79.8	73.0	65.9	69.0	73.2	34.4	51.3	88.9	46.4	65.0	79.1	37.8	55.8	93.4	37.8	62.0
07	THERMAL & MOISTURE PROTECTION	100.1	57.6	82.8	93.7	60.2	80.1	94.1	46.5	74.8	96.0	55.0	79.3	95.2	53.5	78.3	93.3	58.2	79.0
08	OPENINGS	101.7	57.1	91.2	92.5	59.3	84.6	92.5	43.2	80.8	100.1	52.9	88.9	98.0	46.2	85.7	95.3	46.2	83.7
0920	Plaster & Gypsum Board	82.0	57.8	65.6	85.1	65.3	71.6	85.1	32.8	49.5	86.8	45.2	58.5	99.3	36.3	56.5	106.9	36.3	58.9
0950, 0980	Ceilings & Acoustic Treatment	97.1	57.8	71.0	81.5	65.3	70.7	81.5	32.8	49.1	90.3	45.2	60.3	93.3	36.3	55.4	94.2	36.3	55.7
0960	Flooring	98.2	57.7	86.1	94.5	20.9	72.5	94.5	59.7	84.1	91.6	41.5	76.6	97.2	41.4	80.5	101.3	56.8	88.0
0970, 0990	Wall Finishes & Painting/Coating	97.0	53.9	71.1	87.7	34.1	55.5	87.7	37.3	57.4	89.2	50.6	66.0	94.3	43.0	63.5	94.3	49.6	67.4
09	FINISHES	94.5	56.6	73.4	92.3	53.2	70.5	92.8	39.3	63.0	91.9	44.7	65.6	99.1	38.9	65.6	92.1	44.6	65.6
COVERS	DIVS. 10 - 14, 25, 28, 41, 43, 44, 46	100.0	42.4	88.5	100.0	49.2	89.8	100.0	42.1	88.4	100.0	44.2	88.8	100.0	44.3	88.8	100.0	46.8	89.3
21, 22, 23	FIRE SUPPRESSION, PLUMBING & HVAC	100.2	61.9	84.9	97.4	79.8	90.3	97.4	74.2	88.1	100.1	69.5	87.8	100.0	60.4	84.1	100.0	65.7	86.2
26, 27, 3370	ELECTRICAL, COMMUNICATIONS & UTIL.	101.9	69.8	85.1	92.6	58.8	75.0	94.4	63.1	78.0	99.5	59.9	78.8	91.9	47.4	68.7	97.8	57.3	76.6
MF2010	WEIGHTED AVERAGE	99.5	65.5	84.7	95.3	67.5	83.2	96.4	60.4	80.7	98.3	62.5	82.7	99.2	56.7	80.7	96.2	60.5	80.6

		TENNESSEE									TEXAS								
	DIVISION	MCKENZIE			MEMPHIS			NASHVILLE			ABILENE			AMARILLO			AUSTIN		
		382			375,380 - 381			370 - 372			795 - 796			790 - 791			786 - 787		
		MAT.	INST.	TOTAL	MAT.	INST.	TOTAL	MAT.	INST.	TOTAL	MAT.	INST.	TOTAL	MAT.	INST.	TOTAL	MAT.	INST.	TOTAL
015433	CONTRACTOR EQUIPMENT		98.8	98.8		103.2	103.2		105.2	105.2		89.7	89.7		89.7	89.7		89.0	89.0
0241, 31 - 34	SITE & INFRASTRUCTURE, DEMOLITION	91.7	87.7	88.9	93.0	94.6	94.1	97.7	100.3	99.6	94.0	86.7	88.8	92.7	87.7	89.2	95.8	86.8	89.5
0310	Concrete Forming & Accessories	90.1	39.0	46.0	95.4	64.7	68.9	99.3	65.7	70.3	100.3	38.9	47.3	101.2	52.5	59.2	96.9	54.3	60.2
0320	Concrete Reinforcing	87.8	63.9	75.9	95.5	64.5	80.0	97.2	63.6	80.4	89.4	51.7	70.6	93.2	51.7	72.5	91.1	49.4	70.2
0330	Cast-in-Place Concrete	101.8	58.3	83.7	92.4	61.9	79.8	89.4	66.0	79.7	100.1	47.4	78.2	92.5	53.9	76.5	93.3	61.7	80.2
03	CONCRETE	101.9	52.4	77.5	91.6	65.2	78.6	90.6	66.9	78.9	98.3	45.5	72.2	97.5	53.7	75.9	92.7	56.7	74.9
04	MASONRY	116.5	47.0	73.5	100.1	63.2	77.3	91.0	59.3	71.4	98.7	56.9	72.9	100.5	57.7	74.0	99.5	45.8	66.3
05	METALS	91.6	87.8	90.4	99.0	89.3	96.0	99.5	88.7	96.2	100.6	68.5	90.7	96.1	68.9	87.7	97.3	64.8	87.3
06	WOOD, PLASTICS & COMPOSITES	82.1	37.1	56.7	96.4	66.6	79.6	106.2	66.5	83.8	104.2	37.3	66.4	98.2	54.7	73.7	93.4	58.1	73.5
07	THERMAL & MOISTURE PROTECTION	94.1	49.5	76.0	95.5	64.0	82.7	97.1	61.7	82.7	101.2	49.6	80.2	101.8	53.5	82.2	101.1	49.6	80.2
08	OPENINGS	92.5	45.1	81.3	101.3	64.6	92.6	101.5	64.7	92.8	92.8	41.0	80.5	94.3	52.8	84.5	101.9	55.1	90.8
0920	Plaster & Gypsum Board	88.0	35.6	52.4	92.8	65.9	74.5	95.0	65.9	75.2	86.2	35.8	51.9	92.4	53.8	66.2	88.6	57.3	67.3
0950, 0980	Ceilings & Acoustic Treatment	81.5	35.6	51.0	95.4	65.9	75.8	92.1	65.9	74.6	93.3	35.8	55.1	102.3	53.8	70.0	99.4	57.3	71.4
0960	Flooring	97.2	40.1	80.1	98.9	36.7	80.3	97.4	65.4	87.9	113.4	77.9	102.8	108.3	77.9	99.2	105.5	40.2	86.0
0970, 0990	Wall Finishes & Painting/Coating	87.7	50.6	65.4	91.9	56.0	70.3	95.3	73.6	82.2	98.8	50.2	69.6	98.9	49.5	69.2	102.8	40.8	65.6
09	FINISHES	93.9	39.1	63.3	96.5	58.4	75.3	94.8	66.4	79.0	96.2	46.4	68.5	99.3	56.9	75.6	100.1	50.3	72.4
COVERS	DIVS. 10 - 14, 25, 28, 41, 43, 44, 46	100.0	28.2	85.6	100.0	82.1	96.4	100.0	82.6	96.5	100.0	76.1	95.2	100.0	67.7	93.5	100.0	78.3	95.7
21, 22, 23	FIRE SUPPRESSION, PLUMBING & HVAC	97.4	69.8	86.3	100.0	74.9	90.0	100.0	83.2	93.3	100.3	46.0	78.5	100.1	50.8	80.3	100.1	56.1	82.5
26, 27, 3370	ELECTRICAL, COMMUNICATIONS & UTIL.	94.1	64.4	78.6	100.9	64.9	82.1	97.6	63.1	79.6	94.9	48.7	70.7	98.7	64.6	80.9	97.0	62.1	78.8
MF2010	WEIGHTED AVERAGE	96.6	60.4	80.9	98.5	71.3	86.7	97.9	74.1	87.5	98.3	53.6	78.8	98.2	60.0	81.6	98.6	59.1	81.4

DIVISION		TEXAS																	
		BEAUMONT			BROWNWOOD			BRYAN			CHILDRESS			CORPUS CHRISTI			DALLAS		
		776 - 777			768			778			792			783 - 784			752 - 753		
		MAT.	INST.	TOTAL	MAT.	INST.	TOTAL	MAT.	INST.	TOTAL	MAT.	INST.	TOTAL	MAT.	INST.	TOTAL	MAT.	INST.	TOTAL
015433	**CONTRACTOR EQUIPMENT**		91.2	91.2		89.7	89.7		91.2	91.2		89.7	89.7		96.2	96.2		99.1	99.1
0241, 31 - 34	**SITE & INFRASTRUCTURE, DEMOLITION**	89.0	87.6	88.0	100.5	86.1	90.4	80.7	90.3	87.5	103.1	87.0	91.8	138.3	82.8	99.3	107.7	88.2	94.0
0310	Concrete Forming & Accessories	106.7	48.4	56.4	97.7	33.0	41.8	84.5	45.5	50.8	98.7	61.9	66.9	98.8	39.8	47.9	97.5	64.1	68.7
0320	Concrete Reinforcing	89.0	41.5	65.3	91.7	50.7	71.3	91.3	49.5	70.4	89.5	51.0	70.3	84.6	47.2	65.9	99.1	52.8	76.0
0330	Cast-in-Place Concrete	86.8	51.7	72.2	105.3	42.7	79.3	69.6	65.7	68.0	102.7	49.4	80.6	109.1	46.9	83.3	96.4	54.8	79.1
03	**CONCRETE**	96.2	49.3	73.1	105.2	40.8	73.4	80.7	54.5	67.8	107.8	56.2	82.3	99.0	45.5	72.6	99.6	60.1	80.1
04	**MASONRY**	99.8	57.3	73.5	130.0	48.3	79.5	137.2	56.3	87.2	102.9	49.8	70.0	88.6	51.4	65.6	101.9	56.5	73.8
05	**METALS**	95.2	64.6	85.8	97.3	62.8	86.6	94.8	70.5	87.3	98.1	65.7	88.1	92.7	75.0	87.2	100.6	79.6	94.1
06	**WOOD, PLASTICS & COMPOSITES**	114.0	48.4	77.0	101.5	29.2	60.7	79.0	39.6	56.8	103.5	67.5	83.2	116.8	38.2	72.4	102.3	67.6	82.7
07	**THERMAL & MOISTURE PROTECTION**	103.5	55.5	84.0	100.1	44.1	77.4	95.4	56.2	79.5	101.7	48.8	80.3	102.3	48.4	80.4	92.1	62.5	80.1
08	**OPENINGS**	94.4	45.2	82.8	91.6	37.0	78.7	97.4	45.9	85.2	90.0	56.6	82.1	106.7	39.5	90.8	103.9	58.4	93.2
0920	Plaster & Gypsum Board	97.4	47.3	63.4	87.8	27.5	46.8	84.7	38.3	53.1	85.8	66.9	73.0	93.9	36.7	54.9	96.7	66.9	76.4
0950, 0980	Ceilings & Acoustic Treatment	95.5	47.3	63.5	91.7	27.5	49.0	89.0	38.3	55.3	91.5	66.9	75.2	91.5	36.7	55.0	103.4	66.9	79.1
0960	Flooring	109.1	74.3	98.7	109.2	45.5	90.1	84.2	62.9	77.9	111.6	41.6	90.7	115.4	64.2	100.1	107.8	48.6	90.1
0970, 0990	Wall Finishes & Painting/Coating	96.1	48.6	67.6	96.3	29.8	56.3	92.6	62.5	74.5	98.8	38.6	62.7	115.0	57.7	80.6	105.5	51.5	73.1
09	**FINISHES**	90.2	52.7	69.4	95.3	34.0	61.2	79.5	48.8	62.4	96.4	56.8	74.4	103.7	45.6	71.3	100.4	60.4	78.1
COVERS	**DIVS. 10 - 14, 25, 28, 41, 43, 44, 46**	100.0	79.4	95.9	100.0	37.3	87.4	100.0	79.8	96.0	100.0	70.0	94.0	100.0	78.3	95.7	100.0	82.0	96.4
21, 22, 23	**FIRE SUPPRESSION, PLUMBING & HVAC**	100.2	59.7	83.9	95.3	49.1	76.8	95.3	65.9	83.5	95.4	52.3	78.1	100.2	40.0	76.0	100.0	62.1	84.8
26, 27, 3370	**ELECTRICAL, COMMUNICATIONS & UTIL.**	92.1	70.8	81.0	93.2	43.7	67.4	90.2	67.7	78.4	94.9	43.4	68.0	91.8	61.4	76.0	95.7	65.2	79.7
MF2010	**WEIGHTED AVERAGE**	96.7	61.2	81.2	98.5	48.4	76.7	94.1	63.0	80.5	97.9	56.9	80.0	99.7	53.7	79.6	100.1	65.8	85.2

DIVISION		TEXAS																	
		DEL RIO			DENTON			EASTLAND			EL PASO			FORT WORTH			GALVESTON		
		788			762			764			798 - 799,885			760 - 761			775		
		MAT.	INST.	TOTAL	MAT.	INST.	TOTAL	MAT.	INST.	TOTAL	MAT.	INST.	TOTAL	MAT.	INST.	TOTAL	MAT.	INST.	TOTAL
015433	**CONTRACTOR EQUIPMENT**		89.0	89.0		95.7	95.7		89.7	89.7		89.8	89.8		89.7	89.7		100.1	100.1
0241, 31 - 34	**SITE & INFRASTRUCTURE, DEMOLITION**	119.1	85.4	95.4	100.5	79.3	85.6	103.1	86.0	91.1	95.1	86.1	88.8	97.4	87.1	90.1	108.0	88.1	94.1
0310	Concrete Forming & Accessories	95.3	36.5	44.6	104.6	32.3	42.2	98.5	35.8	44.4	97.2	43.8	51.1	96.9	63.5	68.1	94.2	65.6	69.5
0320	Concrete Reinforcing	85.2	46.7	66.0	93.0	51.0	72.1	92.0	50.7	71.3	97.5	45.8	71.7	97.3	52.7	75.0	90.8	62.8	76.8
0330	Cast-in-Place Concrete	117.4	42.8	86.4	81.5	44.4	66.1	111.3	43.3	83.1	96.6	41.3	73.7	98.6	50.7	78.7	92.3	66.9	81.8
03	**CONCRETE**	119.8	41.7	81.3	82.8	42.0	62.7	110.0	42.4	76.6	99.9	44.3	72.5	100.7	57.7	79.5	97.2	67.1	82.3
04	**MASONRY**	103.6	50.0	70.4	138.8	47.1	82.1	99.9	48.2	67.9	96.5	46.9	65.8	98.4	56.4	72.4	95.9	56.4	71.5
05	**METALS**	92.5	60.8	82.7	96.9	75.5	90.3	97.1	63.2	86.6	95.3	62.6	85.2	98.8	68.5	89.5	96.3	88.8	94.0
06	**WOOD, PLASTICS & COMPOSITES**	97.0	34.6	61.8	112.9	31.9	67.2	108.3	32.8	65.7	99.9	45.8	69.4	99.5	67.5	81.4	96.4	66.4	79.4
07	**THERMAL & MOISTURE PROTECTION**	99.2	43.6	76.6	98.3	44.1	76.3	100.5	44.2	77.6	98.9	54.5	80.9	98.4	52.6	79.8	94.5	65.3	82.6
08	**OPENINGS**	99.3	33.9	83.9	107.8	37.5	91.1	72.9	39.0	64.9	91.2	43.9	80.0	98.7	58.3	89.2	101.8	64.0	92.9
0920	Plaster & Gypsum Board	89.7	33.0	51.2	92.3	30.2	50.0	87.8	31.2	49.3	93.7	44.6	60.3	92.0	66.9	74.9	92.0	65.7	74.1
0950, 0980	Ceilings & Acoustic Treatment	87.8	33.0	51.4	96.2	30.2	52.2	91.7	31.2	51.4	95.9	44.6	61.7	102.3	66.9	78.8	91.6	65.7	74.4
0960	Flooring	97.3	34.7	78.6	102.5	43.5	84.8	141.2	45.5	112.6	114.0	63.5	98.9	137.1	41.4	108.5	98.1	62.9	87.6
0970, 0990	Wall Finishes & Painting/Coating	101.3	38.4	63.5	106.2	29.6	60.2	97.8	30.5	57.4	93.2	34.9	58.2	95.2	51.1	68.7	104.3	47.0	69.9
09	**FINISHES**	95.8	36.0	62.5	92.0	33.5	59.4	104.8	36.3	66.6	99.2	46.5	69.9	107.4	58.8	80.3	88.8	63.0	74.4
COVERS	**DIVS. 10 - 14, 25, 28, 41, 43, 44, 46**	100.0	37.1	87.4	100.0	36.0	87.2	100.0	38.4	87.7	100.0	77.3	95.4	100.0	81.7	96.3	100.0	84.3	96.9
21, 22, 23	**FIRE SUPPRESSION, PLUMBING & HVAC**	95.2	46.9	75.8	95.3	37.9	72.3	95.3	49.4	76.9	100.0	34.2	73.6	100.0	57.7	83.1	95.2	66.0	83.5
26, 27, 3370	**ELECTRICAL, COMMUNICATIONS & UTIL.**	94.0	30.1	60.6	95.6	43.8	68.6	93.1	43.8	67.3	92.7	54.9	73.0	94.5	62.4	77.8	91.7	69.5	80.1
MF2010	**WEIGHTED AVERAGE**	99.4	46.2	76.3	98.2	46.6	75.7	96.5	49.1	75.9	97.2	50.6	76.9	99.6	62.5	83.5	96.3	69.7	84.7

DIVISION		TEXAS																	
		GIDDINGS			GREENVILLE			HOUSTON			HUNTSVILLE			LAREDO			LONGVIEW		
		789			754			770 - 772			773			780			756		
		MAT.	INST.	TOTAL	MAT.	INST.	TOTAL	MAT.	INST.	TOTAL	MAT.	INST.	TOTAL	MAT.	INST.	TOTAL	MAT.	INST.	TOTAL
015433	**CONTRACTOR EQUIPMENT**		89.0	89.0		96.7	96.7		100.0	100.0		91.2	91.2		89.0	89.0		91.5	91.5
0241, 31 - 34	**SITE & INFRASTRUCTURE, DEMOLITION**	105.0	85.4	91.3	101.0	83.5	88.7	106.7	87.9	93.5	95.8	87.9	90.3	99.1	86.8	90.5	97.3	90.3	92.4
0310	Concrete Forming & Accessories	93.0	55.5	60.7	89.1	27.3	35.7	96.3	65.8	69.9	91.8	55.4	60.4	95.3	40.0	47.5	85.5	28.1	35.9
0320	Concrete Reinforcing	85.8	46.8	66.3	99.6	24.9	62.3	90.6	63.3	77.0	91.5	50.5	71.0	85.2	47.6	66.4	98.4	22.7	60.6
0330	Cast-in-Place Concrete	99.5	42.9	76.0	87.8	37.7	67.0	89.5	66.9	80.1	95.6	48.2	75.9	84.1	59.3	73.8	102.1	35.0	74.2
03	**CONCRETE**	96.5	50.2	73.7	92.2	32.6	62.8	94.9	67.2	81.3	103.8	52.8	78.7	91.4	49.1	70.6	108.4	30.8	70.1
04	**MASONRY**	111.8	50.0	73.5	159.0	46.1	89.1	95.8	66.6	77.7	135.8	51.6	83.7	97.3	58.6	73.3	154.6	42.0	85.0
05	**METALS**	92.0	61.8	82.6	97.9	63.6	87.3	99.2	89.7	96.3	94.7	66.3	85.9	95.0	64.0	85.5	91.2	50.5	78.7
06	**WOOD, PLASTICS & COMPOSITES**	96.0	59.4	75.4	92.3	26.3	55.1	98.8	66.4	80.5	88.3	58.2	71.3	97.0	38.1	63.7	86.2	28.4	53.6
07	**THERMAL & MOISTURE PROTECTION**	99.9	47.7	78.7	92.2	39.5	70.8	94.1	68.1	83.6	96.3	48.2	76.8	98.2	51.9	79.4	93.5	34.9	69.7
08	**OPENINGS**	98.5	49.0	86.7	102.1	26.0	84.1	104.4	64.0	94.8	97.4	48.5	85.8	99.2	39.4	85.0	91.7	26.0	76.1
0920	Plaster & Gypsum Board	88.6	58.7	68.3	90.4	24.5	45.5	94.5	65.7	74.9	89.5	57.4	67.7	90.8	36.7	54.0	88.7	26.8	46.6
0950, 0980	Ceilings & Acoustic Treatment	87.8	58.7	68.4	99.0	24.5	49.4	97.0	65.7	76.2	89.0	57.4	67.9	92.2	36.7	55.2	95.4	26.8	49.7
0960	Flooring	97.8	34.7	79.0	103.7	47.8	87.0	99.1	62.9	88.3	87.4	34.7	71.7	97.1	64.2	87.3	108.1	35.1	86.2
0970, 0990	Wall Finishes & Painting/Coating	101.3	30.5	58.8	105.5	25.1	57.2	104.3	62.5	79.2	92.6	31.3	55.8	101.3	54.4	73.1	95.1	29.3	55.6
09	**FINISHES**	94.7	50.2	69.9	96.9	30.4	59.8	96.0	64.7	78.5	81.9	49.9	64.1	95.2	45.1	67.3	101.0	29.4	61.1
COVERS	**DIVS. 10 - 14, 25, 28, 41, 43, 44, 46**	100.0	42.9	88.6	100.0	34.4	86.9	100.0	84.3	96.9	100.0	43.3	88.6	100.0	76.3	95.2	100.0	21.9	84.4
21, 22, 23	**FIRE SUPPRESSION, PLUMBING & HVAC**	95.3	62.5	82.1	95.2	43.5	74.4	100.0	66.1	86.4	95.3	63.3	82.5	100.1	38.8	75.5	95.1	27.5	68.0
26, 27, 3370	**ELECTRICAL, COMMUNICATIONS & UTIL.**	90.8	43.4	66.1	92.3	27.1	58.3	93.7	67.8	80.2	90.2	43.4	65.8	94.2	59.9	76.3	92.7	53.2	72.0
MF2010	**WEIGHTED AVERAGE**	96.4	55.6	78.6	99.4	42.2	74.5	98.7	70.9	86.6	97.2	56.9	79.6	97.1	53.7	78.2	99.0	40.5	73.5

DIVISION		TEXAS																	
		LUBBOCK			LUFKIN			MCALLEN			MCKINNEY			MIDLAND			ODESSA		
		793 - 794			759			785			750			797			797		
		MAT.	INST.	TOTAL	MAT.	INST.	TOTAL	MAT.	INST.	TOTAL	MAT.	INST.	TOTAL	MAT.	INST.	TOTAL	MAT.	INST.	TOTAL
015433	CONTRACTOR EQUIPMENT		98.2	98.2		91.5	91.5		96.4	96.4		96.7	96.7		98.2	98.2		89.7	89.7
0241, 31 - 34	SITE & INFRASTRUCTURE, DEMOLITION	116.1	85.0	94.3	92.5	91.8	92.0	142.5	82.9	100.6	97.4	83.6	87.7	118.7	85.0	95.1	94.2	87.0	89.1
0310	Concrete Forming & Accessories	99.1	51.0	57.6	88.6	56.8	61.2	99.9	35.8	44.6	88.2	36.3	43.4	103.0	51.0	58.2	100.2	50.9	57.7
0320	Concrete Reinforcing	90.5	51.7	71.1	100.2	62.4	81.3	84.8	47.1	66.0	99.6	51.1	75.4	91.4	50.9	71.2	89.4	50.9	70.2
0330	Cast-in-Place Concrete	100.3	48.4	78.8	91.3	44.7	72.0	118.5	45.3	88.1	82.4	45.2	66.9	106.6	48.6	82.5	100.1	47.5	78.3
03	CONCRETE	96.9	52.0	74.7	100.8	54.3	77.8	106.6	43.1	75.3	87.6	44.0	66.0	101.6	51.9	77.1	98.3	50.7	74.8
04	MASONRY	98.1	55.7	71.8	118.7	52.3	77.6	103.2	51.4	71.1	171.0	48.4	95.1	114.8	42.0	69.7	98.7	42.0	63.6
05	METALS	104.1	80.1	96.7	98.0	69.1	89.1	92.4	74.5	86.9	97.8	74.4	90.6	102.4	79.0	95.2	99.9	67.1	89.8
06	WOOD, PLASTICS & COMPOSITES	104.1	54.9	76.3	94.3	59.4	74.6	115.4	33.6	69.2	91.0	36.4	60.2	109.1	54.9	78.5	104.2	54.7	76.3
07	THERMAL & MOISTURE PROTECTION	90.8	52.7	75.3	93.3	48.0	74.9	102.6	45.0	79.2	92.0	45.9	73.3	91.0	47.4	73.3	101.2	46.6	79.0
08	OPENINGS	103.6	49.4	90.8	69.6	53.4	65.8	103.4	36.8	87.6	102.0	39.8	87.3	102.5	49.1	89.9	92.8	49.0	82.4
0920	Plaster & Gypsum Board	86.7	53.8	64.3	87.7	58.7	68.0	94.9	31.9	52.0	89.6	34.8	52.4	88.3	53.8	64.8	86.2	53.8	64.2
0950, 0980	Ceilings & Acoustic Treatment	95.1	53.8	67.6	88.3	58.7	68.6	92.2	31.9	52.1	99.0	34.8	56.3	92.4	53.8	66.7	93.3	53.8	67.0
0960	Flooring	107.0	39.3	86.7	143.4	38.2	112.0	114.9	65.0	100.0	103.3	35.6	83.1	108.2	48.6	90.4	113.4	48.6	94.0
0970, 0990	Wall Finishes & Painting/Coating	110.2	49.5	73.7	95.1	37.8	60.6	115.0	28.9	63.2	105.5	41.2	66.9	110.2	49.5	73.7	98.8	49.5	69.2
09	FINISHES	98.8	48.8	70.9	109.1	52.0	77.3	104.3	39.8	68.4	96.4	36.7	63.1	99.2	50.7	72.1	96.2	50.6	70.8
COVERS	DIVS. 10 - 14, 25, 28, 41, 43, 44, 46	100.0	77.8	95.6	100.0	70.8	94.2	100.0	76.0	95.2	100.0	39.7	87.9	100.0	78.6	95.7	100.0	78.3	95.6
21, 22, 23	FIRE SUPPRESSION, PLUMBING & HVAC	99.8	45.2	77.9	95.1	63.5	82.4	95.3	30.8	69.4	95.2	41.4	73.6	94.9	45.3	75.0	100.3	45.3	78.2
26, 27, 3370	ELECTRICAL, COMMUNICATIONS & UTIL.	93.6	49.6	70.6	94.0	43.4	67.6	91.6	34.2	61.6	92.4	45.2	67.8	93.5	40.4	65.8	95.0	40.3	66.4
MF2010	WEIGHTED AVERAGE	100.0	56.3	81.0	95.9	59.1	79.9	99.7	46.6	76.6	99.3	48.9	77.3	99.9	53.6	79.7	98.2	52.5	78.3

DIVISION		TEXAS																	
		PALESTINE			SAN ANGELO			SAN ANTONIO			TEMPLE			TEXARKANA			TYLER		
		758			769			781 - 782			765			755			757		
		MAT.	INST.	TOTAL	MAT.	INST.	TOTAL	MAT.	INST.	TOTAL	MAT.	INST.	TOTAL	MAT.	INST.	TOTAL	MAT.	INST.	TOTAL
015433	CONTRACTOR EQUIPMENT		91.5	91.5		89.7	89.7		91.5	91.5		89.7	89.7		91.5	91.5		91.5	91.5
0241, 31 - 34	SITE & INFRASTRUCTURE, DEMOLITION	98.6	91.1	93.3	97.0	88.1	90.8	101.8	91.4	94.5	86.0	85.9	85.9	87.0	91.1	89.9	96.6	91.5	93.1
0310	Concrete Forming & Accessories	80.1	55.8	59.1	98.0	36.5	44.9	94.5	54.8	60.3	101.3	39.7	48.1	95.4	34.0	42.4	90.3	33.1	40.9
0320	Concrete Reinforcing	97.7	50.7	74.2	91.6	50.9	71.3	91.8	49.1	70.5	91.8	47.4	69.6	97.6	52.4	75.0	98.4	53.0	75.7
0330	Cast-in-Place Concrete	83.5	42.6	66.5	99.3	46.6	77.4	84.5	65.4	76.6	81.4	48.4	67.7	84.1	43.4	67.2	100.2	42.5	76.3
03	CONCRETE	102.9	50.9	77.2	100.4	43.9	72.5	92.6	58.3	75.7	86.0	45.2	65.9	93.6	42.0	68.1	108.0	41.4	75.1
04	MASONRY	113.5	48.5	73.3	126.6	41.6	74.0	98.9	66.5	78.8	137.9	48.4	82.5	175.4	44.7	94.5	164.8	54.4	96.5
05	METALS	97.7	62.7	86.9	97.4	66.5	87.9	96.1	68.7	87.6	97.1	62.8	86.5	91.1	66.9	83.6	97.6	68.0	88.5
06	WOOD, PLASTICS & COMPOSITES	84.7	59.4	70.4	101.8	36.1	64.7	96.7	52.8	71.9	111.0	39.4	70.6	98.1	32.3	61.0	96.0	30.5	59.0
07	THERMAL & MOISTURE PROTECTION	93.7	46.4	74.5	100.0	42.1	76.5	95.9	64.3	83.1	99.6	45.0	77.4	93.2	47.2	74.5	93.6	53.2	77.2
08	OPENINGS	69.5	50.7	65.1	91.7	38.8	79.2	99.9	51.7	88.5	69.6	41.8	63.0	91.7	37.9	78.9	69.5	36.9	61.8
0920	Plaster & Gypsum Board	84.4	58.7	66.9	87.8	34.6	51.6	88.8	51.8	63.6	87.8	38.0	53.9	93.2	30.7	50.7	87.7	28.9	47.7
0950, 0980	Ceilings & Acoustic Treatment	88.3	58.7	68.6	91.7	34.6	53.7	91.2	51.8	65.0	91.7	38.0	56.0	95.4	30.7	52.4	88.3	28.9	48.7
0960	Flooring	135.3	58.9	112.5	109.2	34.7	86.9	97.6	64.3	87.6	143.4	41.7	113.0	115.6	59.4	98.7	145.3	37.0	112.9
0970, 0990	Wall Finishes & Painting/Coating	95.1	30.5	56.3	96.3	49.5	68.2	98.5	54.4	72.0	97.8	38.5	62.2	95.1	39.4	61.6	95.1	39.4	61.6
09	FINISHES	106.9	54.4	77.7	95.0	36.4	62.4	94.8	55.4	72.8	104.1	39.6	68.2	103.1	38.1	66.9	109.9	32.6	66.9
COVERS	DIVS. 10 - 14, 25, 28, 41, 43, 44, 46	100.0	40.7	88.1	100.0	75.0	95.0	100.0	80.6	96.1	100.0	36.8	87.3	100.0	74.9	95.0	100.0	74.7	94.9
21, 22, 23	FIRE SUPPRESSION, PLUMBING & HVAC	95.1	58.5	80.4	95.3	44.6	75.0	100.0	64.5	85.8	95.3	48.2	76.4	95.1	32.9	70.2	95.1	57.7	80.1
26, 27, 3370	ELECTRICAL, COMMUNICATIONS & UTIL.	90.1	43.4	65.7	97.2	45.8	70.4	94.3	60.0	76.4	94.3	64.9	79.0	93.8	61.1	76.8	92.7	55.0	73.0
MF2010	WEIGHTED AVERAGE	95.4	55.8	78.2	98.2	49.5	77.0	97.5	64.5	83.1	95.0	52.7	76.6	98.4	49.7	77.2	98.9	54.6	79.6

DIVISION		TEXAS															UTAH		
		VICTORIA			WACO			WAXAHACKIE			WHARTON			WICHITA FALLS			LOGAN		
		779			766 - 767			751			774			763			843		
		MAT.	INST.	TOTAL	MAT.	INST.	TOTAL	MAT.	INST.	TOTAL	MAT.	INST.	TOTAL	MAT.	INST.	TOTAL	MAT.	INST.	TOTAL
015433	CONTRACTOR EQUIPMENT		99.0	99.0		89.7	89.7		96.7	96.7		100.1	100.1		89.7	89.7		97.2	97.2
0241, 31 - 34	SITE & INFRASTRUCTURE, DEMOLITION	112.7	82.3	91.4	94.2	87.0	89.2	99.0	84.4	88.8	118.4	85.5	95.3	94.9	86.7	89.1	93.4	96.2	95.4
0310	Concrete Forming & Accessories	94.0	35.1	43.2	99.9	39.6	47.9	88.2	59.5	63.5	88.7	36.7	43.9	99.9	37.2	45.8	104.8	58.1	64.5
0320	Concrete Reinforcing	87.2	30.3	58.8	91.5	49.4	70.5	99.6	51.1	75.4	90.7	50.2	70.5	91.5	51.7	71.6	102.3	80.2	91.3
0330	Cast-in-Place Concrete	103.6	41.0	77.7	88.1	65.8	78.8	86.9	41.7	68.1	106.5	51.2	83.5	94.1	44.7	73.6	86.9	72.8	81.0
03	CONCRETE	104.5	38.5	71.9	93.0	51.7	72.6	91.3	53.2	72.5	108.8	46.4	78.0	95.8	43.8	70.1	107.9	67.9	88.2
04	MASONRY	112.4	34.1	64.0	98.3	53.7	70.7	159.7	48.5	90.9	97.0	44.6	64.6	98.8	56.9	72.9	103.9	59.7	76.5
05	METALS	94.9	70.3	87.3	99.4	67.0	89.4	97.9	77.0	91.4	96.3	78.5	90.8	99.4	68.5	89.8	101.0	78.2	94.0
06	WOOD, PLASTICS & COMPOSITES	99.9	35.9	63.8	109.2	35.1	67.4	91.0	67.5	77.8	89.4	32.6	57.3	109.2	35.0	67.3	81.5	56.2	67.2
07	THERMAL & MOISTURE PROTECTION	98.4	39.5	74.5	100.5	46.9	78.8	92.1	46.3	73.5	94.8	49.3	76.3	100.5	50.5	80.2	97.9	61.8	83.3
08	OPENINGS	101.7	35.2	85.9	81.5	42.5	72.2	102.1	56.8	91.3	101.8	37.4	86.6	81.5	40.3	71.7	94.3	57.1	85.5
0920	Plaster & Gypsum Board	89.3	34.3	51.9	88.3	33.6	51.1	90.1	66.9	74.3	86.8	30.9	48.8	88.3	33.4	51.0	75.6	54.8	61.4
0950, 0980	Ceilings & Acoustic Treatment	92.5	34.3	53.8	93.5	33.6	53.6	100.7	66.9	78.2	91.6	30.9	51.2	93.5	33.4	53.5	105.1	54.8	71.6
0960	Flooring	97.8	38.2	80.0	142.3	45.1	113.3	103.3	38.2	83.8	96.1	45.5	80.9	143.5	77.9	123.9	100.2	48.6	84.8
0970, 0990	Wall Finishes & Painting/Coating	104.5	37.8	64.4	97.8	49.8	69.0	105.5	37.8	64.8	104.3	31.3	60.5	100.5	49.5	69.9	102.9	62.0	78.3
09	FINISHES	87.0	36.0	58.6	104.7	40.2	68.8	97.1	54.8	73.5	88.3	36.7	59.5	105.3	45.0	71.7	98.5	56.0	74.8
COVERS	DIVS. 10 - 14, 25, 28, 41, 43, 44, 46	100.0	36.6	87.3	100.0	78.2	95.6	100.0	40.6	88.1	100.0	38.7	87.7	100.0	65.2	93.0	100.0	54.1	90.8
21, 22, 23	FIRE SUPPRESSION, PLUMBING & HVAC	95.3	31.2	69.6	100.2	54.2	81.8	95.2	40.5	73.2	95.2	63.4	82.5	100.2	48.4	79.5	100.0	69.0	87.6
26, 27, 3370	ELECTRICAL, COMMUNICATIONS & UTIL.	96.5	39.7	66.9	97.4	65.1	80.5	92.4	65.2	78.2	95.3	43.8	68.4	99.1	61.0	79.2	95.4	74.3	84.4
MF2010	WEIGHTED AVERAGE	98.2	42.7	74.1	97.2	57.3	79.8	99.3	56.4	80.6	98.1	53.8	78.8	97.8	55.0	79.2	99.7	68.8	86.2

DIVISION		UTAH												VERMONT					
		OGDEN			PRICE			PROVO			SALT LAKE CITY			BELLOWS FALLS			BENNINGTON		
		842,844			845			846 - 847			840 - 841			051			052		
		MAT.	INST.	TOTAL	MAT.	INST.	TOTAL	MAT.	INST.	TOTAL	MAT.	INST.	TOTAL	MAT.	INST.	TOTAL	MAT.	INST.	TOTAL
015433	CONTRACTOR EQUIPMENT		97.2	97.2		96.3	96.3		96.3	96.3		97.2	97.2		100.6	100.6		100.6	100.6
0241, 31 - 34	SITE & INFRASTRUCTURE, DEMOLITION	82.2	96.2	92.1	90.8	94.6	93.5	89.8	94.6	93.2	81.8	96.2	91.9	82.6	100.1	94.9	82.0	100.1	94.7
0310	Concrete Forming & Accessories	104.8	58.1	64.5	107.1	49.4	57.3	106.3	58.0	64.6	107.3	58.0	64.7	100.2	87.5	89.3	97.8	109.1	107.6
0320	Concrete Reinforcing	102.0	80.2	91.1	109.8	80.1	95.0	110.7	80.2	95.5	104.3	80.2	92.3	83.0	87.3	85.2	83.0	87.3	85.2
0330	Cast-in-Place Concrete	88.2	72.8	81.8	87.0	60.0	75.8	87.0	72.7	81.1	96.4	72.7	86.6	89.5	113.1	99.3	89.5	113.1	99.3
03	CONCRETE	97.5	67.9	82.9	109.3	59.6	84.8	107.8	67.8	88.1	116.9	67.8	92.6	93.5	96.2	94.8	93.3	105.8	99.5
04	MASONRY	97.8	59.7	74.2	109.1	59.7	78.5	109.3	59.7	78.6	111.0	59.7	79.3	94.6	96.9	96.0	102.1	96.9	98.9
05	METALS	101.5	78.2	94.3	98.3	77.6	91.9	99.3	78.0	92.7	106.1	78.0	97.4	95.6	89.3	93.6	95.5	89.2	93.6
06	WOOD, PLASTICS & COMPOSITES	81.5	56.2	67.2	84.4	45.3	62.3	82.9	56.2	67.8	83.2	56.2	68.0	102.2	85.5	92.8	99.2	115.1	108.1
07	THERMAL & MOISTURE PROTECTION	96.9	61.8	82.6	99.5	59.4	83.3	99.5	61.8	84.2	103.8	61.8	86.8	98.0	84.8	92.7	98.0	83.1	92.0
08	OPENINGS	94.3	57.1	85.5	98.3	50.1	86.9	98.3	57.1	88.5	96.1	57.1	86.9	103.5	88.1	99.9	103.5	104.2	103.7
0920	Plaster & Gypsum Board	75.6	54.8	61.4	77.9	43.5	54.5	76.0	54.8	61.6	83.0	54.8	63.8	99.8	84.4	89.3	97.9	114.9	109.5
0950, 0980	Ceilings & Acoustic Treatment	105.1	54.8	71.6	105.1	43.5	64.2	105.1	54.8	71.6	99.2	54.8	69.6	92.3	84.4	87.1	92.3	114.9	107.3
0960	Flooring	98.0	48.6	83.3	101.2	36.1	81.8	100.9	48.6	85.3	102.2	48.6	86.2	101.6	106.5	103.1	100.8	106.5	102.5
0970, 0990	Wall Finishes & Painting/Coating	102.9	62.0	78.3	102.9	38.3	64.1	102.9	49.0	70.5	106.5	49.0	71.9	95.7	106.8	102.3	95.7	106.8	102.3
09	FINISHES	96.5	56.0	74.0	99.5	44.5	68.9	99.1	54.6	74.3	98.2	54.6	73.9	97.2	92.6	94.6	96.7	110.1	104.2
COVERS	DIVS. 10 - 14, 25, 28, 41, 43, 44, 46	100.0	54.1	90.8	100.0	47.0	89.4	100.0	54.0	90.8	100.0	54.0	90.8	100.0	100.3	100.1	100.0	103.5	100.7
21, 22, 23	FIRE SUPPRESSION, PLUMBING & HVAC	100.0	69.0	87.6	97.5	68.6	85.9	100.0	69.0	87.6	100.1	69.0	87.6	95.2	98.5	96.5	95.2	98.4	96.5
26, 27, 3370	ELECTRICAL, COMMUNICATIONS & UTIL.	95.7	74.3	84.5	101.2	74.3	87.1	96.2	61.2	77.9	98.6	61.2	79.1	98.6	91.5	94.9	98.6	62.6	79.8
MF2010	WEIGHTED AVERAGE	97.9	68.8	85.2	100.1	65.3	84.9	100.1	66.7	85.6	102.2	66.8	86.8	96.6	94.7	95.8	96.9	95.3	96.2

DIVISION		VERMONT																	
		BRATTLEBORO			BURLINGTON			GUILDHALL			MONTPELIER			RUTLAND			ST. JOHNSBURY		
		053			054			059			056			057			058		
		MAT.	INST.	TOTAL	MAT.	INST.	TOTAL	MAT.	INST.	TOTAL	MAT.	INST.	TOTAL	MAT.	INST.	TOTAL	MAT.	INST.	TOTAL
015433	CONTRACTOR EQUIPMENT		100.6	100.6		100.6	100.6		100.6	100.6		100.6	100.6		100.6	100.6		100.6	100.6
0241, 31 - 34	SITE & INFRASTRUCTURE, DEMOLITION	83.4	100.1	95.1	86.2	100.0	95.9	81.7	98.7	93.6	86.6	99.8	95.8	85.9	99.8	95.6	81.7	98.7	93.6
0310	Concrete Forming & Accessories	100.5	87.4	89.2	98.0	89.6	90.8	98.1	80.1	82.6	101.8	86.1	88.2	100.8	89.6	91.1	96.5	80.1	82.4
0320	Concrete Reinforcing	82.2	87.3	84.7	105.7	87.2	96.4	83.8	87.2	85.5	89.0	87.2	88.1	103.7	87.2	95.5	82.2	87.2	84.7
0330	Cast-in-Place Concrete	92.3	113.1	100.9	102.6	112.0	106.5	86.8	103.2	93.6	105.0	111.9	107.9	87.7	111.9	97.8	86.8	103.2	93.6
03	CONCRETE	95.5	96.2	95.9	104.0	96.7	100.4	91.0	89.5	90.3	102.8	95.2	99.0	96.9	96.7	96.8	90.7	89.5	90.1
04	MASONRY	101.1	96.9	98.5	104.8	95.4	99.0	101.3	79.6	87.8	100.1	95.4	97.2	85.8	95.4	91.7	124.9	79.6	96.9
05	METALS	95.5	89.2	93.6	104.0	88.4	99.2	95.6	88.3	93.3	100.2	88.3	96.5	101.0	88.4	97.1	95.6	88.3	93.3
06	WOOD, PLASTICS & COMPOSITES	102.6	85.5	92.9	98.3	90.3	93.8	98.5	85.5	91.2	96.4	85.5	90.3	102.7	90.3	95.7	93.6	85.5	89.0
07	THERMAL & MOISTURE PROTECTION	98.1	80.0	90.8	104.3	82.5	95.4	97.9	72.6	87.6	105.0	82.0	95.6	98.2	82.5	91.8	97.8	72.6	87.6
08	OPENINGS	103.5	88.1	99.9	107.5	86.4	102.5	103.5	83.8	98.8	103.8	83.8	99.1	106.6	86.4	101.8	103.5	83.8	98.8
0920	Plaster & Gypsum Board	99.8	84.4	89.3	100.5	89.3	92.9	106.1	84.4	91.3	105.4	84.4	91.1	100.5	89.3	92.9	108.0	84.4	91.9
0950, 0980	Ceilings & Acoustic Treatment	92.3	84.4	87.1	96.0	89.3	91.6	92.3	84.4	87.1	102.5	84.4	90.5	97.0	89.3	91.9	92.3	84.4	87.1
0960	Flooring	101.7	106.5	103.2	103.1	106.5	104.1	104.6	106.5	105.2	104.0	106.5	104.8	101.6	106.5	103.1	107.7	106.5	107.3
0970, 0990	Wall Finishes & Painting/Coating	95.7	106.8	102.3	100.4	85.3	91.4	95.7	85.3	89.4	100.5	85.3	91.4	95.7	85.3	89.4	95.7	85.3	89.4
09	FINISHES	97.3	92.6	94.7	99.1	92.7	95.5	98.8	85.9	91.6	101.0	89.8	94.8	98.4	92.7	95.2	99.9	85.9	92.1
COVERS	DIVS. 10 - 14, 25, 28, 41, 43, 44, 46	100.0	100.3	100.1	100.0	100.2	100.0	100.0	94.3	98.9	100.0	99.7	99.9	100.0	100.2	100.0	100.0	94.3	98.9
21, 22, 23	FIRE SUPPRESSION, PLUMBING & HVAC	95.2	98.4	96.5	99.9	73.3	89.3	95.2	65.3	83.2	95.0	73.3	86.3	100.1	73.3	89.3	95.2	65.3	83.2
26, 27, 3370	ELECTRICAL, COMMUNICATIONS & UTIL.	98.6	91.5	94.9	98.0	62.6	79.5	98.6	62.6	79.8	98.0	62.6	79.5	98.6	62.6	79.8	98.6	62.6	79.8
MF2010	WEIGHTED AVERAGE	97.2	94.5	96.0	101.5	85.1	94.4	96.8	79.2	89.1	99.2	84.3	92.7	99.2	85.1	93.0	97.9	79.2	89.8

DIVISION		VERMONT			VIRGINIA														
		WHITE RIVER JCT.			ALEXANDRIA			ARLINGTON			BRISTOL			CHARLOTTESVILLE			CULPEPER		
		050			223			222			242			229			227		
		MAT.	INST.	TOTAL	MAT.	INST.	TOTAL	MAT.	INST.	TOTAL	MAT.	INST.	TOTAL	MAT.	INST.	TOTAL	MAT.	INST.	TOTAL
015433	CONTRACTOR EQUIPMENT		100.6	100.6		102.1	102.1		100.8	100.8		100.8	100.8		105.1	105.1		100.8	100.8
0241, 31 - 34	SITE & INFRASTRUCTURE, DEMOLITION	85.7	98.8	94.9	113.5	89.3	96.5	123.7	87.3	98.2	107.8	85.7	92.3	112.6	87.7	95.1	111.2	87.2	94.3
0310	Concrete Forming & Accessories	95.0	80.7	82.7	93.3	73.6	76.3	92.5	73.0	75.7	88.4	43.4	49.6	86.8	49.6	54.7	83.9	72.3	73.9
0320	Concrete Reinforcing	83.0	87.2	85.1	85.0	87.0	86.0	95.9	87.0	91.5	95.9	67.3	81.6	95.3	73.0	84.2	95.9	86.9	91.4
0330	Cast-in-Place Concrete	92.3	104.1	97.2	107.5	82.2	97.0	104.6	81.4	95.0	104.1	49.1	81.3	108.3	56.1	86.6	107.1	68.8	91.2
03	CONCRETE	97.3	90.1	93.7	101.2	80.1	90.8	105.6	79.6	92.8	102.1	51.7	77.2	102.8	57.9	80.7	100.1	74.9	87.6
04	MASONRY	112.6	81.2	93.1	86.0	72.9	77.8	99.1	71.3	81.9	89.1	52.3	66.3	112.3	54.2	76.3	100.9	71.2	82.5
05	METALS	95.6	88.3	93.3	102.5	97.0	100.8	101.1	97.4	99.9	99.9	85.8	95.6	100.2	90.6	97.2	100.3	95.4	98.8
06	WOOD, PLASTICS & COMPOSITES	95.9	85.5	90.0	98.1	71.9	83.3	94.4	71.9	81.7	86.6	40.9	60.8	85.1	46.6	63.4	83.7	71.9	77.1
07	THERMAL & MOISTURE PROTECTION	98.2	73.3	88.1	103.2	82.4	94.7	105.1	81.2	95.4	104.7	61.3	87.1	104.3	69.0	90.0	104.5	78.7	94.0
08	OPENINGS	103.5	83.8	98.8	96.4	75.6	91.5	94.6	75.6	90.1	97.4	46.8	85.5	95.7	52.7	85.5	96.0	75.6	91.2
0920	Plaster & Gypsum Board	96.5	84.4	88.3	108.4	70.9	82.9	104.4	70.9	81.6	100.1	38.9	58.4	100.1	44.1	62.0	100.3	70.9	80.3
0950, 0980	Ceilings & Acoustic Treatment	92.3	84.4	87.1	97.7	70.9	79.9	95.1	70.9	79.0	94.2	38.9	57.4	94.2	44.1	60.8	95.1	70.9	79.0
0960	Flooring	99.7	106.5	101.8	106.7	84.5	100.0	105.2	82.7	98.5	101.9	67.7	91.7	100.6	67.7	90.7	100.6	82.7	95.2
0970, 0990	Wall Finishes & Painting/Coating	95.7	85.3	89.4	120.5	79.9	96.1	120.5	79.9	96.1	106.3	55.5	75.7	106.3	77.7	89.1	120.5	77.7	94.8
09	FINISHES	96.5	86.3	90.8	102.9	75.5	87.6	102.6	74.7	87.0	99.2	48.2	70.8	98.8	54.6	74.2	99.6	74.3	85.5
COVERS	DIVS. 10 - 14, 25, 28, 41, 43, 44, 46	100.0	94.9	99.0	100.0	88.5	97.7	100.0	86.0	97.2	100.0	69.5	93.9	100.0	78.0	95.6	100.0	86.0	97.2
21, 22, 23	FIRE SUPPRESSION, PLUMBING & HVAC	95.2	66.1	83.5	100.3	87.7	95.2	100.3	86.8	94.9	95.4	51.6	77.8	95.4	68.7	84.7	95.4	73.2	86.5
26, 27, 3370	ELECTRICAL, COMMUNICATIONS & UTIL.	98.6	62.6	79.8	95.8	99.2	97.6	93.4	99.2	96.5	95.5	36.2	64.6	95.4	70.8	82.6	98.0	99.2	98.6
MF2010	WEIGHTED AVERAGE	97.9	79.7	90.0	99.8	85.4	93.5	100.6	84.7	93.6	98.1	55.6	79.6	99.2	67.3	85.3	98.6	80.8	90.9

	DIVISION	VIRGINIA																	
		FAIRFAX			FARMVILLE			FREDERICKSBURG			GRUNDY			HARRISONBURG			LYNCHBURG		
		220 - 221			239			224 - 225			246			228			245		
		MAT.	INST.	TOTAL	MAT.	INST.	TOTAL	MAT.	INST.	TOTAL	MAT.	INST.	TOTAL	MAT.	INST.	TOTAL	MAT.	INST.	TOTAL
015433	CONTRACTOR EQUIPMENT		100.8	100.8		105.1	105.1		100.8	100.8		100.8	100.8		100.8	100.8		100.8	100.8
0241, 31 - 34	SITE & INFRASTRUCTURE, DEMOLITION	122.4	87.3	97.8	107.2	87.2	93.2	110.8	87.2	94.2	105.6	84.4	90.7	119.4	85.9	95.9	106.5	85.9	92.1
0310	Concrete Forming & Accessories	86.8	72.8	74.7	100.4	44.6	52.3	86.8	70.8	73.0	91.5	37.2	44.6	82.9	43.5	48.9	88.4	61.6	65.3
0320	Concrete Reinforcing	95.9	87.0	91.5	94.7	51.5	73.1	96.7	86.9	91.8	94.6	52.0	73.3	95.9	66.9	81.5	95.3	67.3	81.3
0330	Cast-in-Place Concrete	104.6	81.3	94.9	106.8	54.1	85.0	106.2	64.7	89.0	104.1	48.1	80.9	104.6	58.8	85.6	104.1	58.4	85.2
03	CONCRETE	105.3	79.4	92.5	101.4	50.9	76.5	99.8	72.8	86.5	100.9	45.3	73.5	103.2	54.9	79.3	100.8	63.0	82.1
04	MASONRY	99.0	71.3	81.8	95.6	45.6	64.7	99.9	71.2	82.1	90.9	48.1	64.4	97.4	51.2	68.8	104.9	54.1	73.5
05	METALS	100.4	97.0	99.3	100.0	78.2	93.3	100.3	96.2	99.0	99.9	71.0	91.0	100.2	85.1	95.6	100.1	86.3	95.9
06	WOOD, PLASTICS & COMPOSITES	86.6	71.9	78.3	101.9	45.0	69.8	86.6	69.6	77.0	89.6	35.8	59.3	82.7	39.8	58.5	86.6	64.0	73.9
07	THERMAL & MOISTURE PROTECTION	105.0	73.9	92.4	105.3	50.2	83.0	104.5	79.1	94.2	104.7	46.1	80.9	104.9	64.5	88.5	104.5	65.1	88.5
08	OPENINGS	94.6	75.6	90.1	94.9	44.0	82.9	95.7	74.0	90.6	97.4	35.2	82.7	96.0	50.6	85.3	96.0	59.4	87.3
0920	Plaster & Gypsum Board	100.3	70.9	80.3	109.3	42.5	63.8	100.3	68.5	78.7	100.1	33.7	54.9	100.1	37.8	57.7	100.1	62.7	74.7
0950, 0980	Ceilings & Acoustic Treatment	95.1	70.9	79.0	92.5	42.5	59.2	95.1	68.5	77.4	94.2	33.7	53.9	94.2	37.8	56.7	94.2	62.7	73.3
0960	Flooring	102.2	82.7	96.4	108.6	61.4	94.5	102.2	82.7	96.4	103.2	34.3	82.6	100.3	75.3	92.9	101.9	67.7	91.7
0970, 0990	Wall Finishes & Painting/Coating	120.5	79.9	96.1	107.7	39.6	66.7	120.5	77.7	94.8	106.3	35.9	64.0	120.5	55.5	81.4	106.3	55.5	75.7
09	FINISHES	101.2	74.7	86.4	101.8	47.0	71.3	100.1	72.9	85.0	99.3	35.9	64.0	100.0	48.9	71.5	99.0	61.5	78.1
COVERS	DIVS. 10 - 14, 25, 28, 41, 43, 44, 46	100.0	86.0	97.2	100.0	46.8	89.3	100.0	81.9	96.4	100.0	44.8	88.9	100.0	69.3	93.8	100.0	72.7	94.5
21, 22, 23	FIRE SUPPRESSION, PLUMBING & HVAC	95.4	86.8	91.9	95.3	44.3	74.9	95.4	86.5	91.8	95.4	63.5	82.6	95.4	67.1	84.1	95.4	68.4	84.6
26, 27, 3370	ELECTRICAL, COMMUNICATIONS & UTIL.	96.6	99.2	98.0	90.4	49.4	69.0	93.6	99.2	96.6	95.5	45.7	69.5	95.7	98.7	97.3	96.6	53.9	74.3
MF2010	WEIGHTED AVERAGE	99.3	84.4	92.8	97.8	53.3	78.5	98.2	83.0	91.6	98.0	53.4	78.6	98.8	68.1	85.5	98.6	66.1	84.5

	DIVISION	VIRGINIA																	
		NEWPORT NEWS			NORFOLK			PETERSBURG			PORTSMOUTH			PULASKI			RICHMOND		
		236			233 - 235			238			237			243			230 - 232		
		MAT.	INST.	TOTAL	MAT.	INST.	TOTAL	MAT.	INST.	TOTAL	MAT.	INST.	TOTAL	MAT.	INST.	TOTAL	MAT.	INST.	TOTAL
015433	CONTRACTOR EQUIPMENT		105.1	105.1		105.7	105.7		105.1	105.1		105.0	105.0		100.8	100.8		105.0	105.0
0241, 31 - 34	SITE & INFRASTRUCTURE, DEMOLITION	106.2	88.4	93.7	105.7	89.4	94.3	109.7	88.6	94.9	104.7	88.2	93.1	104.9	85.1	91.0	103.5	88.6	93.0
0310	Concrete Forming & Accessories	99.7	65.4	70.1	99.5	65.6	70.2	93.0	58.3	63.1	88.9	54.2	58.9	91.5	39.5	46.6	100.7	58.3	64.1
0320	Concrete Reinforcing	94.4	72.8	83.6	99.5	72.8	86.2	94.1	73.2	83.6	94.1	72.8	83.4	94.6	66.7	80.7	100.7	73.2	87.0
0330	Cast-in-Place Concrete	103.8	66.8	88.5	111.9	66.9	93.2	110.3	54.2	87.0	102.8	66.7	87.8	104.1	48.9	81.2	98.8	54.2	80.3
03	CONCRETE	98.6	68.6	83.8	103.3	68.7	86.2	103.6	61.2	82.7	97.3	63.5	80.7	100.9	49.6	75.6	97.3	61.2	79.5
04	MASONRY	90.6	54.9	68.5	99.1	54.9	71.7	103.3	53.6	72.5	96.0	54.9	70.6	85.9	49.1	63.1	92.1	53.5	68.2
05	METALS	102.2	90.4	98.6	102.0	90.5	98.4	100.1	91.1	97.3	101.2	89.7	97.7	100.0	83.8	95.0	106.2	91.1	101.5
06	WOOD, PLASTICS & COMPOSITES	100.6	67.7	82.0	97.0	67.7	80.5	91.4	59.8	73.6	87.6	52.6	67.8	89.6	37.6	60.3	101.6	59.8	78.0
07	THERMAL & MOISTURE PROTECTION	105.3	69.7	90.8	101.9	69.7	88.8	105.3	69.6	90.8	105.3	67.0	89.7	104.7	51.0	82.9	103.7	69.6	89.8
08	OPENINGS	95.3	63.3	87.7	94.2	63.3	86.9	94.6	59.9	86.4	95.4	55.0	85.8	97.5	43.2	84.6	99.0	59.9	89.8
0920	Plaster & Gypsum Board	110.2	65.8	80.0	106.3	65.8	78.8	101.7	57.7	71.8	102.0	50.2	66.8	100.1	35.5	56.1	104.7	57.7	72.8
0950, 0980	Ceilings & Acoustic Treatment	96.1	65.8	75.9	97.7	65.8	76.5	93.4	57.7	69.7	96.1	50.2	65.6	94.2	35.5	55.1	100.4	57.7	72.0
0960	Flooring	108.6	67.7	96.4	103.5	67.7	92.8	104.7	72.5	95.1	101.7	67.7	91.5	103.2	67.7	92.6	109.4	72.5	98.4
0970, 0990	Wall Finishes & Painting/Coating	107.7	42.9	68.8	105.8	77.7	88.9	107.7	77.7	89.7	107.7	77.7	89.7	106.3	55.5	75.7	107.0	77.7	89.4
09	FINISHES	102.5	63.5	80.8	100.7	67.4	82.2	100.1	62.5	79.1	99.5	57.9	76.3	99.3	44.7	68.9	104.4	62.5	81.1
COVERS	DIVS. 10 - 14, 25, 28, 41, 43, 44, 46	100.0	75.4	95.1	100.0	75.4	95.1	100.0	78.2	95.6	100.0	73.7	94.7	100.0	45.0	89.0	100.0	78.2	95.6
21, 22, 23	FIRE SUPPRESSION, PLUMBING & HVAC	100.2	64.1	85.7	100.0	64.7	85.8	95.3	66.9	83.9	100.2	64.7	86.0	95.4	63.4	82.5	100.0	66.9	86.7
26, 27, 3370	ELECTRICAL, COMMUNICATIONS & UTIL.	93.1	60.3	76.0	96.9	59.7	77.4	93.3	70.8	81.5	91.4	59.7	74.9	95.5	55.7	74.7	96.2	70.8	83.0
MF2010	WEIGHTED AVERAGE	99.2	68.2	85.7	100.0	68.8	86.4	98.6	68.9	85.6	98.6	66.1	84.5	97.7	58.3	80.6	100.4	68.9	86.7

	DIVISION	VIRGINIA									WASHINGTON								
		ROANOKE			STAUNTON			WINCHESTER			CLARKSTON			EVERETT			OLYMPIA		
		240 - 241			244			226			994			982			985		
		MAT.	INST.	TOTAL	MAT.	INST.	TOTAL	MAT.	INST.	TOTAL	MAT.	INST.	TOTAL	MAT.	INST.	TOTAL	MAT.	INST.	TOTAL
015433	CONTRACTOR EQUIPMENT		100.8	100.8		105.1	105.1		100.8	100.8		90.4	90.4		102.1	102.1		102.1	102.1
0241, 31 - 34	SITE & INFRASTRUCTURE, DEMOLITION	105.2	85.9	91.7	109.1	87.3	93.8	118.1	87.2	96.4	97.1	90.0	92.1	89.4	110.3	104.1	90.2	110.4	104.4
0310	Concrete Forming & Accessories	97.9	61.8	66.8	91.1	51.1	56.6	85.2	70.8	72.8	114.8	68.8	75.1	114.9	99.6	101.7	103.1	99.8	100.2
0320	Concrete Reinforcing	95.6	67.4	81.5	95.3	50.5	72.9	95.3	86.3	90.8	105.7	87.4	96.6	110.2	100.2	105.2	116.1	100.2	108.2
0330	Cast-in-Place Concrete	118.3	58.5	93.5	108.3	52.0	84.9	104.6	64.7	88.1	97.0	83.2	91.3	99.8	103.8	101.5	89.2	105.4	96.0
03	CONCRETE	105.5	63.1	84.6	102.3	53.0	78.0	102.6	72.7	87.9	110.7	77.4	94.3	98.7	100.7	99.7	96.9	101.3	99.1
04	MASONRY	92.1	54.1	68.6	100.6	52.3	70.7	95.1	60.0	73.3	97.2	80.7	87.0	111.9	99.0	103.9	103.1	99.9	101.1
05	METALS	102.2	86.5	97.4	100.2	81.7	94.5	100.3	94.8	98.6	88.1	81.2	86.0	104.3	91.7	100.4	104.2	91.6	100.3
06	WOOD, PLASTICS & COMPOSITES	99.0	64.0	79.3	89.6	51.0	67.8	85.1	69.6	76.4	103.3	65.2	81.8	112.1	99.3	104.9	94.6	99.3	97.2
07	THERMAL & MOISTURE PROTECTION	104.3	67.2	89.3	104.3	53.0	83.5	105.0	76.2	93.3	142.7	76.9	116.0	100.5	100.5	100.5	101.6	97.8	100.1
08	OPENINGS	96.4	59.4	87.6	96.0	49.2	84.9	97.5	74.7	92.1	106.7	69.7	98.0	101.9	99.8	101.4	103.5	99.8	102.6
0920	Plaster & Gypsum Board	108.4	62.7	77.3	100.1	48.6	65.1	100.3	68.5	78.7	130.6	64.0	85.3	110.7	99.2	102.9	103.1	99.2	100.5
0950, 0980	Ceilings & Acoustic Treatment	97.7	62.7	74.5	94.2	48.6	63.9	95.1	68.5	77.4	97.9	64.0	75.3	99.9	99.2	99.5	106.3	99.2	101.6
0960	Flooring	106.7	67.7	95.0	102.7	40.2	84.0	101.6	82.7	96.0	98.3	48.0	83.2	117.9	93.3	110.6	109.2	93.4	104.5
0970, 0990	Wall Finishes & Painting/Coating	106.3	55.5	75.7	106.3	35.3	63.6	120.5	89.1	101.6	106.4	59.8	78.4	104.8	90.1	96.0	110.3	90.1	98.1
09	FINISHES	101.8	62.4	79.8	99.2	47.6	70.4	100.5	74.2	85.9	114.8	62.6	85.7	109.0	97.1	102.4	104.6	97.4	100.6
COVERS	DIVS. 10 - 14, 25, 28, 41, 43, 44, 46	100.0	72.7	94.5	100.0	72.6	94.5	100.0	85.7	97.1	100.0	77.3	95.5	100.0	99.6	99.9	100.0	100.1	100.0
21, 22, 23	FIRE SUPPRESSION, PLUMBING & HVAC	100.3	65.8	86.5	95.4	60.1	81.2	95.4	86.8	91.9	95.4	84.7	91.1	100.1	100.3	100.2	100.0	100.8	100.3
26, 27, 3370	ELECTRICAL, COMMUNICATIONS & UTIL.	95.5	55.7	74.7	94.3	74.6	84.0	94.1	99.2	96.8	94.9	94.4	94.6	103.0	98.5	100.7	101.0	99.7	100.3
MF2010	WEIGHTED AVERAGE	100.2	66.0	85.3	98.4	62.7	82.9	98.7	82.1	91.5	100.6	80.7	92.0	102.1	99.6	101.0	101.0	100.0	100.6

	DIVISION	WASHINGTON																	
		RICHLAND			SEATTLE			SPOKANE			TACOMA			VANCOUVER			WENATCHEE		
		993			980 - 981,987			990 - 992			983 - 984			986			988		
		MAT.	INST.	TOTAL	MAT.	INST.	TOTAL	MAT.	INST.	TOTAL	MAT.	INST.	TOTAL	MAT.	INST.	TOTAL	MAT.	INST.	TOTAL
015433	CONTRACTOR EQUIPMENT		90.4	90.4		102.0	102.0		90.4	90.4		102.1	102.1		97.1	97.1		102.1	102.1
0241, 31 - 34	SITE & INFRASTRUCTURE, DEMOLITION	99.6	90.8	93.4	94.5	108.6	104.4	98.7	90.8	93.2	92.4	110.4	105.0	102.3	98.1	99.3	101.4	109.2	106.9
0310	Concrete Forming & Accessories	114.9	78.4	83.4	106.9	100.3	101.2	119.8	78.0	83.7	105.5	99.8	100.6	106.2	92.1	94.0	107.2	77.5	81.6
0320	Concrete Reinforcing	101.4	87.6	94.5	112.2	100.3	106.3	102.1	87.6	94.9	109.0	100.2	104.6	109.9	100.0	104.9	109.8	89.3	99.6
0330	Cast-in-Place Concrete	97.2	85.5	92.4	103.6	105.6	104.5	101.0	85.3	94.5	102.6	105.4	103.8	114.6	99.5	108.3	104.8	77.4	93.4
03	CONCRETE	110.3	82.6	96.6	102.0	101.7	101.8	112.6	82.3	97.6	100.5	101.3	100.9	110.5	95.9	103.3	108.4	79.8	94.3
04	MASONRY	99.3	83.2	89.3	109.2	99.9	103.4	99.9	83.8	89.9	106.7	99.9	102.5	107.3	96.7	100.7	109.1	91.5	98.2
05	METALS	88.5	84.1	87.1	104.5	93.4	101.1	90.8	83.7	88.6	106.0	91.6	101.6	103.4	92.1	99.9	103.5	86.1	98.1
06	WOOD, PLASTICS & COMPOSITES	103.4	75.7	87.8	102.2	99.3	100.6	112.1	75.7	91.6	101.3	99.3	100.2	93.1	91.9	92.5	102.9	75.9	87.6
07	THERMAL & MOISTURE PROTECTION	144.1	79.7	117.9	97.6	100.9	99.0	140.6	79.1	115.6	100.2	98.7	99.6	100.6	92.5	97.3	100.8	82.2	93.3
08	OPENINGS	109.2	71.2	100.2	105.2	99.8	103.9	109.7	71.7	100.7	102.7	99.8	102.0	98.5	94.4	97.5	102.2	71.5	94.9
0920	Plaster & Gypsum Board	130.6	74.8	92.6	106.9	99.2	101.7	121.5	74.8	89.7	111.5	99.2	103.1	109.4	91.9	97.5	114.0	75.1	87.5
0950, 0980	Ceilings & Acoustic Treatment	104.8	74.8	84.8	101.8	99.2	100.1	99.8	74.8	83.2	103.5	99.2	100.6	98.4	91.9	94.1	95.3	75.1	81.8
0960	Flooring	98.7	81.6	93.5	110.9	101.0	108.0	98.2	86.9	94.8	111.4	93.4	106.0	118.1	104.0	113.9	114.1	63.8	99.1
0970, 0990	Wall Finishes & Painting/Coating	106.4	66.2	82.2	104.2	90.1	95.7	106.6	72.4	86.0	104.8	90.1	96.0	107.2	68.4	83.9	104.8	66.2	81.6
09	FINISHES	116.6	76.7	94.4	108.1	98.9	102.9	114.3	78.5	94.3	108.2	97.4	102.2	105.6	91.9	98.0	108.5	72.9	88.7
COVERS	DIVS. 10 - 14, 25, 28, 41, 43, 44, 46	100.0	87.2	97.4	100.0	100.1	100.0	100.0	87.1	97.4	100.0	100.1	100.0	100.0	62.0	92.4	100.0	78.5	95.7
21, 22, 23	FIRE SUPPRESSION, PLUMBING & HVAC	100.3	109.7	104.1	100.0	117.1	106.8	100.2	85.9	94.5	100.1	100.8	100.4	100.2	88.6	95.5	95.2	87.9	92.3
26, 27, 3370	ELECTRICAL, COMMUNICATIONS & UTIL.	92.1	94.4	93.3	102.4	107.0	104.8	90.5	79.1	84.5	102.8	99.7	101.2	107.5	101.0	104.1	103.7	98.5	101.0
MF2010	WEIGHTED AVERAGE	102.2	89.6	96.7	102.5	104.8	103.5	102.5	82.7	93.8	102.3	100.0	101.3	103.1	93.2	98.8	102.0	87.1	95.5

	DIVISION	WASHINGTON			WEST VIRGINIA														
		YAKIMA			BECKLEY			BLUEFIELD			BUCKHANNON			CHARLESTON			CLARKSBURG		
		989			258 - 259			247 - 248			262			250 - 253			263 - 264		
		MAT.	INST.	TOTAL	MAT.	INST.	TOTAL	MAT.	INST.	TOTAL	MAT.	INST.	TOTAL	MAT.	INST.	TOTAL	MAT.	INST.	TOTAL
015433	CONTRACTOR EQUIPMENT		102.1	102.1		100.8	100.8		100.8	100.8		100.8	100.8		100.8	100.8		100.8	100.8
0241, 31 - 34	SITE & INFRASTRUCTURE, DEMOLITION	94.7	109.8	105.3	97.9	89.2	91.8	99.6	89.2	92.3	105.8	89.2	94.1	99.0	90.0	92.7	106.3	89.2	94.3
0310	Concrete Forming & Accessories	106.0	95.3	96.8	86.3	92.0	91.2	88.3	91.9	91.4	87.7	92.2	91.6	99.6	93.4	94.2	85.3	92.1	91.1
0320	Concrete Reinforcing	109.5	88.8	99.2	92.3	89.4	90.9	94.2	84.9	89.6	94.8	85.0	89.9	100.6	89.5	95.1	94.8	84.4	89.6
0330	Cast-in-Place Concrete	109.6	84.4	99.2	102.0	102.1	102.0	101.9	102.1	102.0	101.6	100.3	101.0	100.6	102.7	101.4	111.3	98.4	105.9
03	CONCRETE	105.1	90.0	97.7	96.5	95.5	96.0	96.9	94.7	95.8	99.7	94.2	97.0	98.1	96.4	97.2	103.7	93.4	98.6
04	MASONRY	100.4	81.6	88.8	90.8	93.4	92.4	87.5	93.4	91.1	97.4	92.3	94.2	89.2	95.3	93.0	100.7	92.3	95.5
05	METALS	104.2	85.7	98.5	101.2	98.2	100.3	100.2	96.5	99.1	100.4	97.3	99.4	99.8	98.9	99.5	100.4	96.9	99.3
06	WOOD, PLASTICS & COMPOSITES	101.7	99.3	100.4	85.1	92.8	89.4	88.6	92.8	91.0	87.9	92.8	90.6	101.1	92.8	96.4	84.4	92.8	89.1
07	THERMAL & MOISTURE PROTECTION	100.3	81.6	92.7	107.1	89.4	99.9	104.4	89.4	98.4	104.7	90.0	98.7	103.6	90.2	98.2	104.6	88.9	98.3
08	OPENINGS	102.2	81.0	97.2	96.1	85.0	93.5	97.9	83.9	94.6	97.9	83.9	94.6	98.1	85.0	95.0	97.9	85.1	94.9
0920	Plaster & Gypsum Board	111.1	99.2	103.0	93.3	92.4	92.7	99.3	92.4	94.6	99.7	92.4	94.7	97.5	92.4	94.0	97.5	92.4	94.0
0950, 0980	Ceilings & Acoustic Treatment	97.3	99.2	98.6	93.4	92.4	92.7	92.4	92.4	92.4	94.2	92.4	93.0	102.2	92.4	95.7	94.2	92.4	93.0
0960	Flooring	112.4	59.1	96.4	100.5	114.1	104.6	99.8	114.1	104.1	99.5	114.1	103.9	104.4	114.1	107.3	98.6	114.1	103.2
0970, 0990	Wall Finishes & Painting/Coating	104.8	72.4	85.3	107.1	93.5	98.9	106.3	93.5	98.6	106.3	93.8	98.8	106.6	93.5	98.7	106.3	93.8	98.8
09	FINISHES	107.2	86.5	95.7	97.8	96.9	97.3	97.5	96.9	97.2	98.4	96.7	97.4	102.4	97.4	99.6	97.7	96.7	97.1
COVERS	DIVS. 10 - 14, 25, 28, 41, 43, 44, 46	100.0	97.3	99.5	100.0	56.6	91.3	100.0	56.6	91.3	100.0	96.7	99.3	100.0	97.7	99.5	100.0	96.7	99.3
21, 22, 23	FIRE SUPPRESSION, PLUMBING & HVAC	100.1	109.1	103.7	95.5	91.5	93.9	95.4	91.5	93.8	95.4	92.3	94.2	100.1	92.9	97.2	95.4	92.2	94.1
26, 27, 3370	ELECTRICAL, COMMUNICATIONS & UTIL.	105.9	94.4	99.9	91.2	91.9	91.6	94.4	91.9	93.1	95.8	96.8	96.3	96.3	91.9	94.0	95.8	96.8	96.3
MF2010	WEIGHTED AVERAGE	102.5	94.4	99.0	96.9	92.0	94.8	97.1	91.7	94.7	98.2	93.6	96.2	99.0	94.1	96.9	98.8	93.5	96.5

	DIVISION	WEST VIRGINIA																	
		GASSAWAY			HUNTINGTON			LEWISBURG			MARTINSBURG			MORGANTOWN			PARKERSBURG		
		266			255 - 257			249			254			265			261		
		MAT.	INST.	TOTAL	MAT.	INST.	TOTAL	MAT.	INST.	TOTAL	MAT.	INST.	TOTAL	MAT.	INST.	TOTAL	MAT.	INST.	TOTAL
015433	CONTRACTOR EQUIPMENT		100.8	100.8		100.8	100.8		100.8	100.8		100.8	100.8		100.8	100.8		100.8	100.8
0241, 31 - 34	SITE & INFRASTRUCTURE, DEMOLITION	103.2	89.2	93.3	102.4	90.4	94.0	115.3	89.2	97.0	101.4	89.7	93.2	100.6	89.8	93.0	108.8	89.9	95.5
0310	Concrete Forming & Accessories	87.2	92.2	91.5	98.2	96.0	96.3	85.6	91.8	91.0	86.3	82.4	83.0	85.6	91.8	91.0	89.8	90.0	90.0
0320	Concrete Reinforcing	94.8	85.0	89.9	93.7	90.6	92.2	94.8	84.9	89.9	92.3	83.0	87.6	94.8	84.3	89.6	94.2	84.8	89.5
0330	Cast-in-Place Concrete	106.3	101.0	104.1	111.1	103.5	108.0	101.9	102.1	102.0	106.8	96.0	102.3	101.5	100.1	101.0	103.8	97.1	101.0
03	CONCRETE	100.2	94.4	97.3	101.8	98.0	99.9	105.8	94.6	100.3	99.9	88.0	94.1	96.6	93.8	95.3	101.6	92.1	96.9
04	MASONRY	101.5	92.3	95.8	89.8	99.9	96.0	90.3	93.4	92.2	91.7	85.6	87.9	117.9	92.3	102.0	78.3	90.7	86.0
05	METALS	100.3	97.2	99.4	103.6	99.5	102.3	100.3	96.5	99.1	101.6	94.9	99.5	100.4	96.7	99.3	101.0	96.8	99.7
06	WOOD, PLASTICS & COMPOSITES	87.0	92.8	90.2	97.2	95.4	96.2	84.7	92.8	89.3	85.1	81.3	82.9	84.7	92.8	89.3	88.0	88.8	88.4
07	THERMAL & MOISTURE PROTECTION	104.4	90.0	98.6	107.3	91.4	100.8	105.2	89.4	98.8	107.4	78.2	95.5	104.5	88.9	98.2	104.4	89.6	98.4
08	OPENINGS	96.1	83.9	93.2	95.4	86.6	93.4	97.9	83.9	94.6	98.1	73.9	92.4	99.1	85.1	95.8	96.9	81.8	93.3
0920	Plaster & Gypsum Board	99.0	92.4	94.5	102.1	95.1	97.4	97.5	92.4	94.0	93.9	80.5	84.8	97.5	92.4	94.0	100.1	88.3	92.0
0950, 0980	Ceilings & Acoustic Treatment	94.2	92.4	93.0	96.1	95.1	95.4	94.2	92.4	93.0	96.1	80.5	85.7	94.2	92.4	93.0	94.2	88.3	90.3
0960	Flooring	99.3	114.1	103.8	107.9	120.3	111.6	98.7	114.1	103.3	100.5	114.1	104.6	98.7	114.1	103.3	102.4	114.1	105.9
0970, 0990	Wall Finishes & Painting/Coating	106.3	93.5	98.6	107.1	93.5	98.9	106.3	77.0	88.7	107.1	42.2	68.1	106.3	93.8	98.8	106.3	93.5	98.6
09	FINISHES	97.9	96.6	97.2	101.7	100.5	101.0	98.8	95.1	96.7	98.7	82.3	89.6	97.3	96.7	96.9	99.3	94.8	96.8
COVERS	DIVS. 10 - 14, 25, 28, 41, 43, 44, 46	100.0	96.7	99.3	100.0	98.6	99.7	100.0	56.6	91.3	100.0	95.4	99.1	100.0	65.4	93.1	100.0	97.1	99.4
21, 22, 23	FIRE SUPPRESSION, PLUMBING & HVAC	95.4	91.4	93.8	100.4	89.3	95.9	95.4	91.5	93.8	95.5	86.7	91.9	95.4	92.2	94.1	100.2	88.1	95.4
26, 27, 3370	ELECTRICAL, COMMUNICATIONS & UTIL.	95.8	91.9	93.8	94.9	96.1	95.5	91.8	91.9	91.8	96.9	82.8	89.6	96.0	96.8	96.4	96.0	92.8	94.3
MF2010	WEIGHTED AVERAGE	98.2	92.8	95.8	99.8	95.2	97.8	98.4	91.4	95.4	98.3	86.1	93.0	98.8	92.6	96.1	98.9	91.4	95.6

DIVISION		WEST VIRGINIA									WISCONSIN								
		PETERSBURG			ROMNEY			WHEELING			BELOIT			EAU CLAIRE			GREEN BAY		
		268			267			260			535			547			541 - 543		
		MAT.	INST.	TOTAL	MAT.	INST.	TOTAL	MAT.	INST.	TOTAL	MAT.	INST.	TOTAL	MAT.	INST.	TOTAL	MAT.	INST.	TOTAL
015433	CONTRACTOR EQUIPMENT		100.8	100.8		100.8	100.8		100.8	100.8		102.6	102.6		101.2	101.2		99.0	99.0
0241, 31 - 34	SITE & INFRASTRUCTURE, DEMOLITION	99.9	89.8	92.8	102.6	89.8	93.6	109.5	89.9	95.7	95.8	108.6	104.8	94.6	103.6	100.9	98.2	99.6	99.2
0310	Concrete Forming & Accessories	88.7	91.6	91.2	84.8	91.6	90.7	91.5	92.6	92.4	100.1	102.1	101.9	99.7	101.3	101.1	107.8	100.5	101.5
0320	Concrete Reinforcing	94.2	84.9	89.6	94.8	76.3	85.6	93.6	84.3	89.0	96.6	117.1	106.9	93.7	100.9	97.3	91.9	93.2	92.6
0330	Cast-in-Place Concrete	101.5	96.3	99.4	106.3	96.3	102.1	103.8	100.9	102.6	103.7	101.9	102.9	102.1	99.5	101.0	105.6	100.9	103.7
03	CONCRETE	96.8	92.5	94.7	100.0	91.0	95.5	101.6	94.4	98.1	99.9	104.9	102.3	98.8	100.7	99.7	101.9	99.5	100.7
04	MASONRY	92.9	92.3	92.5	90.7	92.3	91.7	100.2	91.7	94.9	98.5	104.8	102.4	89.8	102.7	97.8	120.3	101.9	108.9
05	METALS	100.5	96.6	99.3	100.5	93.7	98.4	101.2	96.9	99.9	94.2	108.4	98.6	93.1	102.7	96.0	95.4	99.1	96.5
06	WOOD, PLASTICS & COMPOSITES	88.9	92.8	91.1	83.8	92.8	88.9	89.6	92.8	91.4	98.2	101.6	100.1	105.2	101.6	103.2	110.0	101.6	105.3
07	THERMAL & MOISTURE PROTECTION	104.5	84.5	96.4	104.6	81.2	95.1	104.7	89.5	98.5	103.4	97.9	101.2	103.2	88.1	97.1	105.0	87.1	97.7
08	OPENINGS	99.1	83.9	95.5	99.0	82.0	95.0	97.6	85.1	94.7	103.6	110.2	105.1	101.0	98.1	100.3	97.4	97.8	97.5
0920	Plaster & Gypsum Board	99.7	92.4	94.7	97.1	92.4	93.9	100.1	92.4	94.8	92.8	101.9	99.0	101.4	101.9	101.8	96.6	101.9	100.2
0950, 0980	Ceilings & Acoustic Treatment	94.2	92.4	93.0	94.2	92.4	93.0	94.2	92.4	93.0	92.5	101.9	98.8	101.7	101.9	101.9	92.7	101.9	98.9
0960	Flooring	100.3	114.1	104.5	98.5	114.1	103.2	103.2	114.1	106.4	101.9	120.3	107.4	94.2	116.2	100.8	113.3	116.2	114.2
0970, 0990	Wall Finishes & Painting/Coating	106.3	93.8	98.8	106.3	93.8	98.8	106.3	93.8	98.8	91.8	99.3	96.3	88.0	88.3	88.2	98.1	85.0	90.2
09	FINISHES	98.1	96.7	97.3	97.4	96.7	97.0	99.5	96.9	98.1	99.7	105.1	102.7	99.2	103.2	101.4	103.3	102.6	102.9
COVERS	DIVS. 10 - 14, 25, 28, 41, 43, 44, 46	100.0	65.3	93.1	100.0	96.7	99.3	100.0	97.1	99.4	100.0	98.2	99.6	100.0	98.5	99.7	100.0	98.1	99.6
21, 22, 23	FIRE SUPPRESSION, PLUMBING & HVAC	95.4	91.2	93.7	95.4	90.8	93.6	100.3	92.8	97.3	100.0	100.0	100.0	100.1	87.5	95.1	100.4	84.3	93.9
26, 27, 3370	ELECTRICAL, COMMUNICATIONS & UTIL.	99.2	82.9	90.7	98.5	82.9	90.3	93.1	96.8	95.0	102.8	87.3	94.7	103.9	86.9	95.0	98.2	82.9	90.2
MF2010	WEIGHTED AVERAGE	98.0	90.1	94.6	98.2	90.4	94.8	99.8	93.8	97.2	99.6	101.9	100.6	98.7	96.4	97.7	100.5	94.1	97.7

DIVISION		WISCONSIN																	
		KENOSHA			LA CROSSE			LANCASTER			MADISON			MILWAUKEE			NEW RICHMOND		
		531			546			538			537			530,532			540		
		MAT.	INST.	TOTAL	MAT.	INST.	TOTAL	MAT.	INST.	TOTAL	MAT.	INST.	TOTAL	MAT.	INST.	TOTAL	MAT.	INST.	TOTAL
015433	CONTRACTOR EQUIPMENT		100.5	100.5		101.2	101.2		102.6	102.6		102.6	102.6		90.8	90.8		101.6	101.6
0241, 31 - 34	SITE & INFRASTRUCTURE, DEMOLITION	101.8	105.5	104.4	88.5	103.6	99.1	95.0	108.5	104.4	91.3	108.6	103.4	98.0	98.7	98.5	91.7	104.0	100.3
0310	Concrete Forming & Accessories	106.8	102.5	103.1	86.7	100.8	98.9	99.5	99.7	99.7	103.7	101.7	102.0	102.2	120.1	117.6	95.2	102.8	101.8
0320	Concrete Reinforcing	96.4	98.6	97.5	93.4	92.9	93.1	97.9	92.8	95.4	102.3	93.1	97.7	100.5	98.9	99.7	91.0	100.7	95.9
0330	Cast-in-Place Concrete	112.8	106.0	109.9	91.8	97.0	94.0	103.0	101.0	102.2	94.3	101.8	97.4	102.3	110.6	105.7	106.3	83.6	96.9
03	CONCRETE	104.6	103.0	103.8	90.1	98.1	94.1	99.6	98.9	99.2	97.1	100.1	98.6	100.3	111.9	106.0	96.6	95.9	96.3
04	MASONRY	95.9	115.5	108.1	89.0	102.7	97.5	98.6	104.9	102.5	99.4	104.8	102.8	102.2	120.5	113.5	115.4	102.8	107.6
05	METALS	95.0	101.6	97.0	93.0	98.9	94.8	91.8	96.1	93.1	93.5	98.1	94.9	98.1	94.9	97.1	93.2	101.5	95.8
06	WOOD, PLASTICS & COMPOSITES	101.8	98.7	100.1	90.2	101.6	96.7	97.5	101.6	99.8	97.0	101.6	99.6	101.1	121.3	112.5	94.0	104.9	100.2
07	THERMAL & MOISTURE PROTECTION	103.4	108.4	105.4	102.7	87.2	96.4	103.1	83.5	95.2	106.5	94.0	101.4	99.9	113.0	105.2	104.9	93.8	100.4
08	OPENINGS	97.6	102.8	98.9	101.0	86.7	97.6	99.2	87.5	96.4	105.1	101.6	104.3	106.2	115.1	108.3	86.9	99.9	90.0
0920	Plaster & Gypsum Board	83.2	99.0	93.9	95.9	101.9	100.0	91.5	101.9	98.6	94.2	101.9	99.5	99.9	122.1	115.0	85.5	105.5	99.1
0950, 0980	Ceilings & Acoustic Treatment	92.5	99.0	96.8	100.8	101.9	101.6	87.2	101.9	97.0	102.4	101.9	102.1	96.1	122.1	113.4	65.8	105.5	92.2
0960	Flooring	120.9	112.4	118.4	88.3	116.2	96.7	101.6	112.6	104.9	103.5	112.6	106.2	101.6	116.1	105.9	106.5	116.2	109.4
0970, 0990	Wall Finishes & Painting/Coating	101.5	115.2	109.7	88.0	75.5	80.5	91.8	67.8	77.4	89.8	99.3	95.5	93.2	125.6	112.7	101.0	88.3	93.4
09	FINISHES	104.6	105.5	105.1	96.1	101.8	99.3	98.3	98.4	98.3	101.0	103.9	102.6	102.5	121.0	112.8	93.7	103.3	99.0
COVERS	DIVS. 10 - 14, 25, 28, 41, 43, 44, 46	100.0	101.6	100.3	100.0	98.5	99.7	100.0	50.6	90.1	100.0	98.2	99.6	100.0	104.8	101.0	100.0	98.4	99.7
21, 22, 23	FIRE SUPPRESSION, PLUMBING & HVAC	100.2	96.9	98.9	100.1	87.4	95.0	95.1	87.2	92.0	100.0	96.3	98.5	100.0	102.5	101.0	94.8	86.8	91.6
26, 27, 3370	ELECTRICAL, COMMUNICATIONS & UTIL.	103.4	99.6	101.4	104.2	87.1	95.3	102.6	87.1	94.5	103.6	94.3	98.8	102.9	101.9	102.4	102.0	87.1	94.3
MF2010	WEIGHTED AVERAGE	100.2	102.9	101.4	97.2	95.0	96.3	97.4	93.5	95.7	99.5	99.8	99.6	100.9	107.9	104.0	96.2	95.9	96.1

DIVISION		WISCONSIN																	
		OSHKOSH			PORTAGE			RACINE			RHINELANDER			SUPERIOR			WAUSAU		
		549			539			534			545			548			544		
		MAT.	INST.	TOTAL	MAT.	INST.	TOTAL	MAT.	INST.	TOTAL	MAT.	INST.	TOTAL	MAT.	INST.	TOTAL	MAT.	INST.	TOTAL
015433	CONTRACTOR EQUIPMENT		99.0	99.0		102.6	102.6		102.6	102.6		99.0	99.0		101.6	101.6		99.0	99.0
0241, 31 - 34	SITE & INFRASTRUCTURE, DEMOLITION	90.0	99.6	96.7	85.8	108.4	101.6	95.6	109.3	105.2	102.1	99.6	100.3	88.6	104.0	99.4	85.9	99.6	95.5
0310	Concrete Forming & Accessories	91.0	99.9	98.7	92.1	100.7	99.5	100.4	102.5	102.2	88.4	99.9	98.3	93.4	94.0	94.0	90.3	100.5	99.1
0320	Concrete Reinforcing	92.0	93.1	92.6	98.0	93.0	95.5	96.6	98.6	97.6	92.2	92.2	92.2	91.0	92.7	91.8	92.2	92.9	92.5
0330	Cast-in-Place Concrete	97.9	100.8	99.1	88.5	103.0	94.5	101.7	105.7	103.3	111.0	101.0	106.9	100.1	100.0	100.1	91.3	100.9	95.3
03	CONCRETE	91.7	99.1	95.3	87.5	100.1	93.7	98.9	102.9	100.9	103.6	99.0	101.3	91.2	96.1	93.6	86.8	99.4	93.0
04	MASONRY	102.5	102.2	102.3	97.5	104.8	102.1	98.6	115.5	109.0	118.7	102.7	108.8	114.7	106.3	109.5	102.1	101.9	102.0
05	METALS	93.4	98.4	94.9	92.5	97.1	93.9	95.9	101.5	97.6	93.3	97.8	94.7	94.1	99.4	95.7	93.2	98.9	94.9
06	WOOD, PLASTICS & COMPOSITES	89.8	101.6	96.5	88.4	101.6	95.8	98.4	98.7	98.6	87.4	101.6	95.4	92.4	92.2	92.3	89.3	101.6	96.3
07	THERMAL & MOISTURE PROTECTION	104.2	87.2	97.3	102.6	93.8	99.0	103.4	107.9	105.2	105.0	85.2	96.9	104.5	94.3	100.4	104.1	85.8	96.6
08	OPENINGS	93.4	97.8	94.5	99.3	94.2	98.1	103.6	102.8	103.4	93.5	88.8	92.4	86.4	92.5	87.8	93.6	97.7	94.6
0920	Plaster & Gypsum Board	83.9	101.9	96.2	85.4	101.9	96.7	92.8	99.0	97.0	83.9	101.9	96.2	85.4	92.4	90.2	83.9	101.9	96.2
0950, 0980	Ceilings & Acoustic Treatment	92.7	101.9	98.9	89.9	101.9	97.9	92.5	99.0	96.8	92.7	101.9	98.9	66.7	92.4	83.8	92.7	101.9	98.9
0960	Flooring	105.1	116.2	108.4	98.2	120.3	104.8	101.9	112.4	105.0	104.3	116.2	107.9	107.8	124.7	112.9	104.9	116.2	108.3
0970, 0990	Wall Finishes & Painting/Coating	95.3	85.0	89.1	91.8	67.8	77.4	91.8	115.3	105.9	95.3	61.7	75.1	89.6	103.7	98.1	95.3	85.0	89.1
09	FINISHES	98.4	100.9	99.8	96.4	100.5	98.7	99.7	105.5	103.0	99.3	99.1	99.2	93.1	100.7	97.3	98.1	102.6	100.6
COVERS	DIVS. 10 - 14, 25, 28, 41, 43, 44, 46	100.0	66.1	93.2	100.0	64.8	92.9	100.0	101.6	100.3	100.0	56.4	91.3	100.0	96.7	99.3	100.0	98.1	99.6
21, 22, 23	FIRE SUPPRESSION, PLUMBING & HVAC	95.5	83.4	90.6	95.1	96.1	95.5	100.0	96.9	98.8	95.5	87.0	92.1	94.8	90.4	93.0	95.5	86.3	91.8
26, 27, 3370	ELECTRICAL, COMMUNICATIONS & UTIL.	102.6	81.0	91.3	106.8	94.3	100.3	102.6	95.2	98.7	101.9	82.6	91.8	107.2	99.3	103.1	103.8	82.6	92.7
MF2010	WEIGHTED AVERAGE	96.3	92.4	94.6	96.1	97.9	96.9	99.7	102.6	101.0	98.7	92.4	95.9	96.1	97.6	96.7	95.7	94.4	95.2

	DIVISION	WYOMING																	
		CASPER			CHEYENNE			NEWCASTLE			RAWLINS			RIVERTON			ROCK SPRINGS		
		826			820			827			823			825			829 - 831		
		MAT.	INST.	TOTAL	MAT.	INST.	TOTAL	MAT.	INST.	TOTAL	MAT.	INST.	TOTAL	MAT.	INST.	TOTAL	MAT.	INST.	TOTAL
015433	CONTRACTOR EQUIPMENT		97.9	97.9		97.9	97.9		97.9	97.9		97.9	97.9		97.9	97.9		97.9	97.9
0241, 31 - 34	SITE & INFRASTRUCTURE, DEMOLITION	97.2	95.2	95.8	90.1	95.2	93.7	82.0	94.8	91.0	95.8	94.8	95.1	89.3	94.8	93.2	85.8	94.7	92.1
0310	Concrete Forming & Accessories	104.4	43.5	51.9	106.6	54.7	61.8	97.1	63.5	68.1	101.3	63.5	68.6	96.0	63.0	67.5	103.5	62.9	68.4
0320	Concrete Reinforcing	108.6	70.5	89.6	100.8	71.3	86.1	108.4	71.7	90.1	108.1	71.7	89.9	109.1	71.4	90.2	109.1	71.4	90.3
0330	Cast-in-Place Concrete	98.6	72.6	87.8	91.1	72.7	83.5	92.0	62.1	79.6	92.1	62.1	79.6	92.0	52.6	75.7	92.0	52.3	75.5
03	CONCRETE	103.0	59.4	81.5	100.1	64.6	82.6	100.6	64.8	82.9	114.3	64.8	89.9	108.9	61.3	85.4	101.1	61.2	81.4
04	MASONRY	100.0	34.5	59.5	95.8	47.6	65.9	92.6	55.7	69.7	92.6	55.7	69.7	92.6	49.9	66.1	147.5	48.7	86.4
05	METALS	98.1	71.0	89.7	101.5	72.4	92.5	96.6	72.4	89.2	96.7	72.4	89.2	96.8	71.6	89.0	97.5	71.9	89.6
06	WOOD, PLASTICS & COMPOSITES	96.3	41.0	65.1	95.6	55.6	73.0	85.6	68.4	75.9	89.6	68.4	77.6	84.5	68.4	75.5	93.6	68.4	79.4
07	THERMAL & MOISTURE PROTECTION	105.2	49.4	82.5	99.7	54.4	81.3	101.2	55.9	82.8	102.7	55.9	83.7	102.1	59.7	84.9	101.3	59.3	84.2
08	OPENINGS	102.1	48.9	89.5	102.3	56.9	91.6	106.2	60.8	95.5	105.9	60.8	95.2	106.1	60.8	95.3	106.5	60.8	95.7
0920	Plaster & Gypsum Board	93.1	39.1	56.3	84.6	54.2	63.9	80.3	67.4	71.5	80.7	67.4	71.6	80.3	67.4	71.5	90.0	67.4	74.6
0950, 0980	Ceilings & Acoustic Treatment	108.9	39.1	62.4	100.3	54.2	69.6	102.3	67.4	79.0	102.3	67.4	79.0	102.3	67.4	79.0	102.3	67.4	79.0
0960	Flooring	109.6	33.6	86.8	107.4	54.0	91.5	101.0	52.6	86.5	103.8	52.6	88.5	100.6	52.6	86.2	105.9	52.6	89.9
0970, 0990	Wall Finishes & Painting/Coating	100.0	44.0	66.3	106.9	44.0	69.1	102.6	56.6	74.9	102.6	56.6	74.9	102.6	56.6	74.9	102.6	56.6	74.9
09	FINISHES	106.4	40.1	69.5	103.2	52.8	75.1	97.9	60.6	77.1	100.2	60.6	78.2	98.7	60.6	77.5	100.6	60.3	78.2
COVERS	DIVS. 10 - 14, 25, 28, 41, 43, 44, 46	100.0	88.9	97.8	100.0	90.6	98.1	100.0	91.5	98.3	100.0	91.5	98.3	100.0	58.0	91.6	100.0	82.7	96.5
21, 22, 23	FIRE SUPPRESSION, PLUMBING & HVAC	100.0	65.4	86.1	100.0	65.5	86.2	97.5	64.8	84.3	97.5	64.8	84.3	97.5	64.7	84.3	100.0	64.2	85.6
26, 27, 3370	ELECTRICAL, COMMUNICATIONS & UTIL.	96.0	65.6	80.1	99.1	70.2	84.0	96.8	70.6	83.2	96.8	70.6	83.2	96.8	64.4	79.9	95.0	64.4	79.0
MF2010	WEIGHTED AVERAGE	100.4	60.7	83.1	100.1	65.7	85.2	98.2	67.8	84.9	100.2	67.8	86.1	99.3	64.9	84.3	101.8	65.4	85.9

	DIVISION	WYOMING												CANADA					
		SHERIDAN			WHEATLAND			WORLAND			YELLOWSTONE NAT'L PA			BARRIE, ONTARIO			BATHURST, NEW BRUNSWICK		
		828			822			824			821								
		MAT.	INST.	TOTAL	MAT.	INST.	TOTAL	MAT.	INST.	TOTAL	MAT.	INST.	TOTAL	MAT.	INST.	TOTAL	MAT.	INST.	TOTAL
015433	CONTRACTOR EQUIPMENT		97.9	97.9		97.9	97.9		97.9	97.9		97.9	97.9		101.6	101.6		101.9	101.9
0241, 31 - 34	SITE & INFRASTRUCTURE, DEMOLITION	89.8	95.2	93.6	86.4	94.8	92.3	83.9	94.7	91.5	84.0	94.9	91.7	115.3	101.0	105.3	98.4	97.3	97.6
0310	Concrete Forming & Accessories	104.3	51.6	58.8	99.1	51.1	57.7	99.2	62.7	67.7	99.2	64.0	68.8	125.1	91.5	96.1	103.7	65.1	70.4
0320	Concrete Reinforcing	109.1	71.8	90.4	108.4	71.7	90.1	109.1	71.1	90.1	111.0	71.1	91.1	174.2	86.5	130.4	139.2	58.3	98.8
0330	Cast-in-Place Concrete	95.1	72.7	85.8	96.0	61.8	81.8	92.0	52.2	75.5	92.0	54.2	76.3	166.0	90.4	134.6	136.1	62.8	105.7
03	CONCRETE	108.8	63.3	86.3	105.4	59.2	82.6	100.8	61.0	81.2	101.1	62.3	81.9	151.0	90.3	121.1	130.9	63.7	97.7
04	MASONRY	92.8	57.2	70.8	92.9	55.2	69.6	92.6	44.9	63.1	92.6	48.5	65.3	165.2	99.8	124.7	165.6	66.1	104.0
05	METALS	100.3	72.9	91.9	96.6	72.2	89.1	96.8	71.2	88.9	97.4	71.2	89.3	108.2	91.4	103.0	104.4	72.6	94.6
06	WOOD, PLASTICS & COMPOSITES	97.0	51.6	71.4	87.5	52.1	67.5	87.5	68.4	76.8	87.5	68.4	76.8	120.6	90.1	103.4	97.8	65.4	79.5
07	THERMAL & MOISTURE PROTECTION	102.4	56.5	83.7	101.5	55.0	82.6	101.3	56.0	82.9	100.8	57.5	83.2	112.4	91.0	103.7	104.6	63.6	87.9
08	OPENINGS	106.8	52.5	93.9	104.8	54.7	93.0	106.4	60.8	95.6	99.8	60.8	90.6	92.5	88.6	91.5	89.2	57.2	81.6
0920	Plaster & Gypsum Board	105.6	50.0	67.8	80.3	50.5	60.0	80.3	67.4	71.5	80.5	67.4	71.6	151.2	89.8	109.4	151.8	64.2	92.2
0950, 0980	Ceilings & Acoustic Treatment	105.2	50.0	68.5	102.3	50.5	67.8	102.3	67.4	79.0	103.1	67.4	79.3	95.2	89.8	91.6	100.9	64.2	76.5
0960	Flooring	105.0	52.6	89.3	102.5	51.7	87.3	102.5	52.6	87.6	102.5	52.6	87.6	130.7	100.7	121.7	117.8	48.0	96.9
0970, 0990	Wall Finishes & Painting/Coating	105.1	44.0	68.4	102.6	35.2	62.1	102.6	56.6	74.9	102.6	56.6	74.9	111.9	91.8	99.8	108.4	52.7	74.9
09	FINISHES	106.2	49.7	74.7	98.6	48.3	70.6	98.3	60.5	77.2	98.5	61.4	77.8	117.3	93.5	104.0	113.4	60.9	84.2
COVERS	DIVS. 10 - 14, 25, 28, 41, 43, 44, 46	100.0	90.1	98.0	100.0	82.0	96.4	100.0	57.8	91.5	100.0	59.0	91.8	139.2	76.2	126.6	131.1	66.9	118.3
21, 22, 23	FIRE SUPPRESSION, PLUMBING & HVAC	97.5	65.5	84.6	97.5	64.5	84.2	97.5	64.4	84.2	97.5	66.2	84.9	102.2	101.5	101.9	102.5	71.3	90.0
26, 27, 3370	ELECTRICAL, COMMUNICATIONS & UTIL.	99.5	64.5	81.2	96.8	70.2	82.9	96.8	64.4	79.9	95.7	64.4	79.4	116.5	92.6	104.0	114.2	63.3	87.6
MF2010	WEIGHTED AVERAGE	100.9	65.2	85.4	98.7	64.6	83.9	98.3	64.1	83.4	97.7	65.2	83.5	117.3	94.9	107.5	112.1	68.6	93.1

	DIVISION	CANADA																	
		BRANDON, MANITOBA			BRANTFORD, ONTARIO			BRIDGEWATER, NOVA SCOTIA			CALGARY, ALBERTA			CAP-DE-LA-MADELEINE, QUEBEC			CHARLESBOURG, QUEBEC		
		MAT.	INST.	TOTAL	MAT.	INST.	TOTAL	MAT.	INST.	TOTAL	MAT.	INST.	TOTAL	MAT.	INST.	TOTAL	MAT.	INST.	TOTAL
015433	CONTRACTOR EQUIPMENT		103.9	103.9		101.6	101.6		101.5	101.5		106.4	106.4		102.5	102.5		102.5	102.5
0241, 31 - 34	SITE & INFRASTRUCTURE, DEMOLITION	117.9	99.9	105.3	114.9	101.3	105.4	99.9	98.9	99.2	124.6	105.7	111.3	95.9	100.6	99.2	95.9	100.6	99.2
0310	Concrete Forming & Accessories	126.8	74.1	81.3	124.9	99.1	102.6	97.1	73.2	76.5	127.7	96.8	101.0	130.3	90.0	95.5	130.3	90.0	95.5
0320	Concrete Reinforcing	189.5	54.9	122.3	167.9	85.2	126.6	142.9	48.2	95.6	130.9	74.1	102.5	142.9	81.4	112.2	142.9	81.4	112.2
0330	Cast-in-Place Concrete	146.1	77.7	117.7	162.0	112.0	141.3	166.5	73.2	127.8	201.5	106.2	162.0	130.8	99.3	117.7	130.8	99.3	117.7
03	CONCRETE	138.9	72.4	106.1	147.1	100.9	124.3	144.7	69.3	107.5	167.1	95.9	132.0	128.7	91.7	110.5	128.7	91.7	110.5
04	MASONRY	186.5	68.8	113.7	167.9	104.4	128.6	163.2	73.2	107.5	203.4	90.7	133.7	163.7	88.6	117.2	163.7	88.6	117.2
05	METALS	117.8	77.8	105.5	105.5	92.2	101.4	104.6	75.3	95.6	130.4	89.2	117.7	103.8	88.8	99.2	103.8	88.8	99.2
06	WOOD, PLASTICS & COMPOSITES	124.4	75.1	96.6	119.8	97.9	107.4	89.2	72.6	79.9	107.8	96.5	101.4	131.6	89.9	108.1	131.6	89.9	108.1
07	THERMAL & MOISTURE PROTECTION	109.1	73.3	94.6	112.6	96.6	106.1	107.5	71.4	92.9	121.9	94.1	110.6	106.2	91.8	100.3	106.2	91.8	100.3
08	OPENINGS	97.0	66.3	89.7	93.6	94.9	93.9	86.6	66.4	81.8	94.2	86.1	92.3	94.7	83.0	91.9	94.7	83.0	91.9
0920	Plaster & Gypsum Board	129.5	73.9	91.7	142.7	97.8	112.1	141.8	71.7	94.1	140.9	95.9	110.3	172.3	89.4	115.9	172.3	89.4	115.9
0950, 0980	Ceilings & Acoustic Treatment	102.5	73.9	83.5	97.4	97.8	97.6	97.4	71.7	80.3	145.9	95.9	112.6	97.4	89.4	92.1	97.4	89.4	92.1
0960	Flooring	136.3	71.5	116.9	130.5	100.7	121.6	113.1	68.5	99.8	133.3	94.3	121.7	130.5	101.4	121.8	130.5	101.4	121.8
0970, 0990	Wall Finishes & Painting/Coating	113.1	60.0	81.2	109.0	100.8	104.1	109.0	65.5	82.9	114.2	112.4	113.1	109.0	93.7	99.8	109.0	93.7	99.8
09	FINISHES	118.8	72.5	93.0	115.6	100.1	107.0	109.9	71.8	88.7	129.1	98.5	112.0	118.9	92.8	104.4	118.9	92.8	104.4
COVERS	DIVS. 10 - 14, 25, 28, 41, 43, 44, 46	131.1	69.1	118.7	131.1	78.4	120.6	131.1	68.2	118.5	131.1	98.4	124.6	131.1	87.3	122.4	131.1	87.3	122.4
21, 22, 23	FIRE SUPPRESSION, PLUMBING & HVAC	102.5	85.9	95.9	102.5	104.5	103.3	102.5	86.3	96.0	101.3	92.2	97.6	102.9	93.0	98.9	102.9	93.0	98.9
26, 27, 3370	ELECTRICAL, COMMUNICATIONS & UTIL.	120.0	71.3	94.6	112.9	92.1	102.0	118.7	65.9	91.1	114.2	100.8	107.2	112.6	74.3	92.6	112.6	74.3	92.6
MF2010	WEIGHTED AVERAGE	118.7	77.3	100.7	115.5	98.8	108.2	113.4	76.1	97.2	124.7	95.4	111.9	112.8	89.4	102.6	112.8	89.4	102.6

DIVISION		CANADA																	
		CHARLOTTETOWN, PRINCE EDWARD ISLAND			CHICOUTIMI, QUEBEC			CORNER BROOK, NEWFOUNDLAND			CORNWALL, ONTARIO			DALHOUSIE, NEW BRUNSWICK			DARTMOUTH, NOVA SCOTIA		
		MAT.	INST.	TOTAL	MAT.	INST.	TOTAL	MAT.	INST.	TOTAL	MAT.	INST.	TOTAL	MAT.	INST.	TOTAL	MAT.	INST.	TOTAL
015433	CONTRACTOR EQUIPMENT		101.5	101.5		102.5	102.5		103.1	103.1		101.6	101.6		101.9	101.9		101.5	101.5
0241, 31 - 34	SITE & INFRASTRUCTURE, DEMOLITION	115.8	96.4	102.2	96.1	99.7	98.6	120.7	97.9	104.7	113.0	100.8	104.5	97.6	97.3	97.4	111.0	98.9	102.5
0310	Concrete Forming & Accessories	105.6	59.5	65.8	131.4	92.8	98.1	106.9	63.0	69.0	122.8	91.6	95.9	103.2	65.3	70.5	98.9	73.2	76.8
0320	Concrete Reinforcing	146.3	48.3	97.3	105.0	96.4	100.7	173.8	50.1	112.0	167.9	84.9	126.5	144.1	58.4	101.3	181.7	48.2	115.0
0330	Cast-in-Place Concrete	164.6	62.0	122.0	128.2	99.2	116.1	170.1	71.7	129.2	145.7	102.2	127.7	137.6	62.8	106.6	164.2	73.2	126.4
03	CONCRETE	153.1	59.1	106.7	120.1	95.6	108.0	167.6	64.5	116.8	139.3	94.1	117.0	135.2	63.8	100.0	152.9	69.3	111.6
04	MASONRY	182.6	63.7	109.0	167.4	94.1	122.0	183.2	65.3	110.2	166.7	95.7	122.7	158.4	66.1	101.3	196.5	73.2	120.2
05	METALS	127.0	68.6	108.9	104.2	92.3	100.6	118.3	74.0	104.6	105.5	90.8	101.0	101.7	72.8	92.8	118.3	75.3	105.0
06	WOOD, PLASTICS & COMPOSITES	88.5	59.1	71.9	132.6	93.2	110.4	104.3	62.1	80.5	118.6	91.0	103.0	97.7	65.4	79.5	94.6	72.6	82.2
07	THERMAL & MOISTURE PROTECTION	119.5	61.7	96.1	105.0	95.9	101.3	112.7	63.2	92.6	112.4	91.1	103.8	110.8	63.6	91.6	110.4	71.4	94.6
08	OPENINGS	93.4	51.1	83.4	93.8	80.5	90.6	102.9	58.5	92.4	94.7	88.2	93.1	90.3	57.2	82.4	88.3	66.4	83.1
0920	Plaster & Gypsum Board	144.0	57.7	85.3	176.1	92.8	119.4	160.5	60.8	92.7	208.0	90.7	128.2	148.2	64.2	91.1	156.0	71.7	98.6
0950, 0980	Ceilings & Acoustic Treatment	121.3	57.7	79.0	100.1	92.8	95.2	103.4	60.8	75.0	100.9	90.7	94.1	100.2	64.2	76.3	110.5	71.7	84.7
0960	Flooring	116.1	64.0	100.5	132.3	101.4	123.0	122.1	58.3	103.0	130.5	99.2	121.1	116.3	73.4	103.5	118.0	68.5	103.2
0970, 0990	Wall Finishes & Painting/Coating	112.9	43.9	71.4	108.4	105.7	106.7	113.0	63.1	83.0	109.0	93.9	99.9	109.9	52.7	75.5	113.0	65.5	84.4
09	FINISHES	118.9	59.1	85.6	120.3	96.5	107.0	119.5	62.2	87.6	125.3	93.6	107.6	112.7	65.9	86.6	118.0	71.8	92.3
COVERS	DIVS. 10 - 14, 25, 28, 41, 43, 44, 46	131.1	66.4	118.2	131.1	88.0	122.5	131.1	67.2	118.3	131.1	75.8	120.1	131.1	66.9	118.3	131.1	68.2	118.5
21, 22, 23	FIRE SUPPRESSION, PLUMBING & HVAC	102.7	65.2	87.7	102.5	93.7	99.0	102.5	73.4	90.8	102.9	102.3	102.6	102.5	71.3	90.0	102.5	86.3	96.0
26, 27, 3370	ELECTRICAL, COMMUNICATIONS & UTIL.	111.9	53.6	81.5	111.0	86.5	98.2	117.4	59.7	87.2	113.9	93.1	103.0	115.3	59.8	86.3	122.2	65.9	92.8
MF2010	WEIGHTED AVERAGE	120.5	64.0	95.9	111.9	93.1	103.7	122.2	69.0	99.0	115.5	95.1	106.7	112.1	68.8	93.2	119.7	76.1	100.7

DIVISION		CANADA																	
		EDMONTON, ALBERTA			FORT MCMURRAY, ALBERTA			FREDERICTON, NEW BRUNSWICK			GATINEAU, QUEBEC			GRANBY, QUEBEC			HALIFAX, NOVA SCOTIA		
		MAT.	INST.	TOTAL	MAT.	INST.	TOTAL	MAT.	INST.	TOTAL	MAT.	INST.	TOTAL	MAT.	INST.	TOTAL	MAT.	INST.	TOTAL
015433	CONTRACTOR EQUIPMENT		106.4	106.4		103.7	103.7		101.9	101.9		102.5	102.5		102.5	102.5		101.5	101.5
0241, 31 - 34	SITE & INFRASTRUCTURE, DEMOLITION	146.0	105.7	117.7	120.4	101.9	107.4	100.8	97.3	98.3	95.7	100.5	99.1	96.2	100.5	99.2	105.2	98.8	100.7
0310	Concrete Forming & Accessories	127.2	96.8	101.0	123.0	94.3	98.2	122.7	65.6	73.5	130.3	89.8	95.4	130.3	89.8	95.3	104.9	78.3	81.9
0320	Concrete Reinforcing	140.3	74.1	107.2	155.3	74.1	114.7	136.1	58.5	97.4	151.1	81.3	116.3	151.1	81.3	116.3	147.6	62.8	105.3
0330	Cast-in-Place Concrete	201.0	106.2	161.7	215.2	104.8	169.4	134.8	62.9	104.9	129.0	99.3	116.7	133.3	99.2	119.2	165.6	81.0	130.5
03	CONCRETE	168.4	95.9	132.6	170.0	94.3	132.7	133.1	64.0	99.0	129.2	91.7	110.7	131.2	91.6	111.7	146.0	76.9	111.9
04	MASONRY	193.2	90.7	129.8	207.6	89.6	134.6	192.4	67.6	115.2	163.6	88.6	117.2	163.9	88.6	117.3	193.7	83.4	125.5
05	METALS	130.4	89.2	117.7	130.8	89.0	117.9	123.0	73.3	107.7	103.8	88.7	99.2	103.8	88.6	99.1	130.3	81.6	115.3
06	WOOD, PLASTICS & COMPOSITES	108.6	96.5	101.8	113.7	93.8	102.5	112.9	65.4	86.1	131.6	89.9	108.1	131.6	89.9	108.1	88.7	77.4	82.3
07	THERMAL & MOISTURE PROTECTION	127.2	94.1	113.8	120.1	92.6	108.9	117.3	64.6	95.9	106.2	91.8	100.3	106.2	90.2	99.7	118.6	77.9	102.1
08	OPENINGS	94.4	86.1	92.4	94.7	84.6	92.3	89.3	56.1	81.5	94.7	78.2	90.8	94.7	78.2	90.8	86.5	72.0	83.1
0920	Plaster & Gypsum Board	145.5	95.9	111.8	140.9	93.3	108.5	153.7	64.2	92.8	140.5	89.4	105.8	143.0	89.4	106.6	140.8	76.6	97.1
0950, 0980	Ceilings & Acoustic Treatment	150.5	95.9	114.2	107.2	93.3	97.9	113.3	64.2	80.6	97.4	89.4	92.1	97.4	89.4	92.1	116.1	76.6	89.8
0960	Flooring	135.0	94.3	122.8	130.5	94.3	119.7	128.5	76.7	113.0	130.5	101.4	121.8	130.5	101.4	121.8	118.2	90.0	109.8
0970, 0990	Wall Finishes & Painting/Coating	114.6	112.4	113.3	109.1	97.5	102.2	112.6	67.2	85.3	109.0	93.7	99.8	109.0	93.7	99.8	113.9	82.0	94.7
09	FINISHES	133.3	98.5	113.9	118.9	95.0	105.6	121.4	68.1	91.7	114.5	92.8	102.4	114.9	92.8	102.6	117.2	81.2	97.2
COVERS	DIVS. 10 - 14, 25, 28, 41, 43, 44, 46	131.1	98.4	124.6	131.1	97.5	124.4	131.1	66.9	118.3	131.1	87.3	122.4	131.1	87.3	122.4	131.1	69.6	118.8
21, 22, 23	FIRE SUPPRESSION, PLUMBING & HVAC	101.2	92.2	97.6	102.9	101.0	102.2	102.6	81.1	94.0	102.9	93.0	98.9	102.5	93.0	98.7	100.7	77.7	91.5
26, 27, 3370	ELECTRICAL, COMMUNICATIONS & UTIL.	110.3	100.8	105.4	107.1	87.5	96.8	117.2	78.9	97.2	112.6	74.3	92.6	113.2	74.3	92.9	120.1	80.5	99.4
MF2010	WEIGHTED AVERAGE	125.0	95.4	112.1	124.1	94.2	111.1	118.1	74.0	98.9	112.5	89.2	102.3	112.8	89.1	102.4	119.9	80.6	102.8

DIVISION		CANADA																	
		HAMILTON, ONTARIO			HULL, QUEBEC			JOLIETTE, QUEBEC			KAMLOOPS, BRITISH COLUMBIA			KINGSTON, ONTARIO			KITCHENER, ONTARIO		
		MAT.	INST.	TOTAL	MAT.	INST.	TOTAL	MAT.	INST.	TOTAL	MAT.	INST.	TOTAL	MAT.	INST.	TOTAL	MAT.	INST.	TOTAL
015433	CONTRACTOR EQUIPMENT		108.3	108.3		102.5	102.5		102.5	102.5		105.6	105.6		103.9	103.9		103.8	103.8
0241, 31 - 34	SITE & INFRASTRUCTURE, DEMOLITION	117.0	112.7	114.0	95.7	100.5	99.1	96.3	100.6	99.3	118.4	104.0	108.3	113.0	104.5	107.1	101.2	105.1	104.0
0310	Concrete Forming & Accessories	127.9	94.2	98.8	130.3	89.8	95.4	130.3	90.0	95.5	123.5	92.8	97.0	122.9	91.7	96.0	118.0	86.6	90.9
0320	Concrete Reinforcing	153.0	93.1	123.1	151.1	81.3	116.3	142.9	81.4	112.2	112.0	79.2	95.6	167.9	84.9	126.5	103.3	93.0	98.2
0330	Cast-in-Place Concrete	151.7	100.2	130.3	129.0	99.3	116.7	134.4	99.3	119.8	116.3	103.1	110.8	145.7	102.2	127.6	140.7	90.1	119.7
03	CONCRETE	140.5	96.0	118.6	129.2	91.7	110.7	130.4	91.7	111.3	136.9	93.9	115.7	141.2	94.1	118.0	118.8	89.2	104.2
04	MASONRY	193.0	99.7	135.3	163.6	88.6	117.2	164.0	88.6	117.3	170.7	96.6	124.9	173.6	95.7	125.4	159.0	95.3	119.6
05	METALS	121.5	92.7	112.6	103.8	88.7	99.2	103.8	88.8	99.2	106.2	88.7	100.8	107.2	90.7	102.1	113.9	92.5	107.3
06	WOOD, PLASTICS & COMPOSITES	105.6	93.7	98.9	131.6	89.9	108.1	131.6	89.9	108.1	101.3	91.3	95.7	118.6	91.1	103.1	109.7	85.3	96.0
07	THERMAL & MOISTURE PROTECTION	124.3	95.0	112.4	106.2	91.8	100.3	106.2	91.8	100.3	121.2	88.6	108.0	112.4	92.2	104.2	112.1	91.6	103.8
08	OPENINGS	91.9	92.1	91.9	94.7	78.2	90.8	94.7	83.0	91.9	91.0	87.7	90.2	94.7	87.9	93.1	85.7	85.8	85.7
0920	Plaster & Gypsum Board	172.8	93.5	118.9	140.5	89.4	105.8	172.3	89.4	115.9	124.3	90.6	101.4	211.3	90.8	129.4	141.4	84.9	102.9
0950, 0980	Ceilings & Acoustic Treatment	130.0	93.5	105.7	97.4	89.4	92.1	97.4	89.4	92.1	97.4	90.6	92.9	114.3	90.8	98.7	105.3	84.9	91.7
0960	Flooring	132.6	104.1	124.1	130.5	101.4	121.8	130.5	101.4	121.8	129.8	56.6	107.9	130.5	99.2	121.1	124.4	104.1	118.3
0970, 0990	Wall Finishes & Painting/Coating	114.2	102.5	107.2	109.0	93.7	99.8	109.0	93.7	99.8	109.0	85.8	95.0	109.0	86.8	95.6	107.9	92.5	98.6
09	FINISHES	129.3	97.1	111.3	114.5	92.8	102.4	118.9	92.8	104.4	115.0	86.0	98.9	128.7	92.9	108.7	114.3	90.1	100.8
COVERS	DIVS. 10 - 14, 25, 28, 41, 43, 44, 46	131.1	100.4	125.0	131.1	87.3	122.4	131.1	87.3	122.4	131.1	97.9	124.5	131.1	75.8	120.1	131.1	98.3	124.6
21, 22, 23	FIRE SUPPRESSION, PLUMBING & HVAC	101.4	85.5	95.1	102.5	93.0	98.7	102.5	93.0	98.7	102.5	96.5	100.1	102.9	102.5	102.7	100.5	87.6	95.3
26, 27, 3370	ELECTRICAL, COMMUNICATIONS & UTIL.	111.6	96.0	103.5	114.5	74.3	93.5	113.2	74.3	92.9	116.7	83.9	99.6	113.9	91.7	102.3	114.4	93.6	103.6
MF2010	WEIGHTED AVERAGE	119.3	95.3	108.8	112.6	89.2	102.4	113.0	89.4	102.7	114.9	92.4	105.1	116.6	95.2	107.3	111.6	92.0	103.1

	DIVISION	CANADA																	
		LAVAL, QUEBEC			LETHBRIDGE, ALBERTA			LLOYDMINSTER, ALBERTA			LONDON, ONTARIO			MEDICINE HAT, ALBERTA			MONCTON, NEW BRUNSWICK		
		MAT.	INST.	TOTAL	MAT.	INST.	TOTAL	MAT.	INST.	TOTAL	MAT.	INST.	TOTAL	MAT.	INST.	TOTAL	MAT.	INST.	TOTAL
015433	CONTRACTOR EQUIPMENT		102.5	102.5		103.7	103.7		103.7	103.7		103.9	103.9		103.7	103.7		101.9	101.9
0241, 31 - 34	SITE & INFRASTRUCTURE, DEMOLITION	96.1	100.5	99.2	113.0	102.5	105.6	112.9	101.9	105.2	116.2	105.3	108.5	111.6	101.9	104.8	97.8	97.5	97.6
0310	Concrete Forming & Accessories	130.5	89.8	95.4	124.2	94.4	98.5	122.4	84.4	89.6	126.1	88.0	93.2	124.1	84.3	89.8	103.7	65.9	71.0
0320	Concrete Reinforcing	151.1	81.3	116.3	155.3	74.1	114.8	155.3	74.0	114.7	130.9	93.0	112.0	155.3	74.0	114.7	139.2	61.5	100.4
0330	Cast-in-Place Concrete	133.3	99.3	119.2	161.4	104.8	137.9	149.7	101.0	129.5	161.6	98.1	135.2	149.7	101.0	129.5	131.1	79.8	109.8
03	CONCRETE	131.2	91.7	111.7	144.7	94.4	119.9	139.1	88.6	114.2	141.5	92.5	117.3	139.2	88.6	114.2	128.5	70.8	100.1
04	MASONRY	163.8	88.6	117.3	181.6	89.6	124.7	163.2	82.9	113.5	196.0	97.0	134.7	163.2	82.9	113.5	165.2	66.2	103.9
05	METALS	103.9	88.7	99.2	124.2	89.0	113.4	106.4	88.9	101.0	122.9	92.0	113.4	106.4	88.8	100.9	104.4	81.5	97.4
06	WOOD, PLASTICS & COMPOSITES	131.8	89.9	108.2	117.3	93.8	104.1	113.7	83.8	96.9	108.5	86.4	96.0	117.3	83.8	98.4	97.8	65.4	79.5
07	THERMAL & MOISTURE PROTECTION	106.8	91.8	100.7	117.4	92.6	107.3	114.3	88.0	103.7	125.2	92.8	112.0	120.9	88.0	107.6	108.9	65.8	91.4
08	OPENINGS	94.7	78.2	90.8	94.7	84.6	92.3	94.7	79.2	91.0	91.1	87.0	90.1	94.7	79.2	91.0	89.2	62.4	82.8
0920	Plaster & Gypsum Board	143.3	89.4	106.7	131.4	93.3	105.5	126.2	83.0	96.8	173.0	86.0	113.8	129.2	83.0	97.7	151.8	64.2	92.2
0950, 0980	Ceilings & Acoustic Treatment	97.4	89.4	92.1	106.3	93.3	97.6	97.4	83.0	87.8	130.1	86.0	100.7	97.4	83.0	87.8	100.9	64.2	76.5
0960	Flooring	130.5	101.4	121.8	130.5	94.3	119.7	130.5	94.3	119.7	132.6	104.1	124.1	130.5	94.3	119.7	117.8	73.4	104.5
0970, 0990	Wall Finishes & Painting/Coating	109.0	93.7	99.8	109.0	106.5	107.5	109.1	83.0	93.4	112.6	99.3	104.6	109.0	83.0	93.4	108.4	52.7	74.9
09	FINISHES	114.9	92.8	102.6	116.5	95.9	105.0	113.9	85.7	98.2	129.0	91.9	108.3	114.2	85.7	98.3	113.4	65.9	86.9
COVERS	DIVS. 10 - 14, 25, 28, 41, 43, 44, 46	131.1	87.3	122.4	131.1	97.5	124.4	131.1	94.1	123.7	131.1	99.0	124.7	131.1	94.1	123.7	131.1	66.9	118.3
21, 22, 23	FIRE SUPPRESSION, PLUMBING & HVAC	100.5	93.0	97.5	102.7	97.6	100.7	102.9	97.6	100.8	101.5	82.8	94.0	102.5	94.2	99.2	102.5	71.6	90.1
26, 27, 3370	ELECTRICAL, COMMUNICATIONS & UTIL.	114.4	74.3	93.5	108.4	87.5	97.5	106.1	87.5	96.4	109.7	93.6	101.3	106.1	87.5	96.4	118.8	63.3	89.8
MF2010	WEIGHTED AVERAGE	112.4	89.2	102.3	118.7	93.7	107.8	113.9	90.3	103.6	119.5	91.9	107.5	114.0	89.6	103.4	112.3	71.4	94.5

	DIVISION	CANADA																	
		MONTREAL, QUEBEC			MOOSE JAW, SASKATCHEWAN			NEW GLASGOW, NOVA SCOTIA			NEWCASTLE, NEW BRUNSWICK			NORTH BAY, ONTARIO			OSHAWA, ONTARIO		
		MAT.	INST.	TOTAL	MAT.	INST.	TOTAL	MAT.	INST.	TOTAL	MAT.	INST.	TOTAL	MAT.	INST.	TOTAL	MAT.	INST.	TOTAL
015433	CONTRACTOR EQUIPMENT		104.0	104.0		100.2	100.2		101.5	101.5		101.9	101.9		101.6	101.6		103.8	103.8
0241, 31 - 34	SITE & INFRASTRUCTURE, DEMOLITION	107.7	100.0	102.3	112.7	96.5	101.3	104.4	98.9	100.5	98.4	97.3	97.6	119.5	100.4	106.1	112.9	104.5	107.0
0310	Concrete Forming & Accessories	131.4	93.1	98.4	107.3	61.9	68.1	98.8	73.2	76.8	103.7	65.3	70.6	127.4	89.0	94.3	123.8	91.1	95.6
0320	Concrete Reinforcing	135.8	96.4	116.1	109.6	63.8	86.7	173.8	48.2	111.1	139.2	58.4	98.8	205.4	84.4	145.0	163.7	87.1	125.4
0330	Cast-in-Place Concrete	174.5	100.5	143.8	145.3	72.1	114.9	164.2	73.2	126.4	136.1	62.9	105.7	158.1	88.4	129.1	162.7	89.2	132.2
03	CONCRETE	148.7	96.2	122.8	122.3	66.5	94.8	151.6	69.3	111.0	130.9	63.8	97.8	154.2	88.1	121.6	145.4	89.9	118.0
04	MASONRY	172.0	94.1	123.8	161.7	65.5	102.2	182.8	73.2	115.0	165.6	66.1	104.0	188.0	91.5	128.3	162.1	96.0	121.2
05	METALS	124.5	92.8	114.7	102.7	74.9	94.1	116.0	75.3	103.4	104.4	72.8	94.7	117.0	90.4	108.8	105.5	92.6	101.5
06	WOOD, PLASTICS & COMPOSITES	120.2	93.5	105.1	99.0	60.4	77.2	94.6	72.6	82.2	97.8	65.4	79.5	127.3	89.6	106.0	117.2	90.1	101.9
07	THERMAL & MOISTURE PROTECTION	118.4	96.5	109.5	105.6	65.5	89.3	110.4	71.4	94.6	108.9	63.6	90.5	115.7	87.4	104.2	113.1	89.2	103.4
08	OPENINGS	92.8	82.4	90.3	90.1	58.1	82.5	88.3	66.4	83.1	89.2	57.2	81.6	95.9	85.7	93.5	92.2	89.9	91.7
0920	Plaster & Gypsum Board	144.9	92.8	109.5	121.3	59.1	79.0	154.0	71.7	98.0	151.8	64.2	92.2	151.5	89.2	109.1	146.6	89.8	108.0
0950, 0980	Ceilings & Acoustic Treatment	115.8	92.8	100.5	97.4	59.1	71.9	102.5	71.7	82.0	100.9	64.2	76.5	102.5	89.2	93.6	100.8	89.8	93.5
0960	Flooring	138.7	101.4	127.5	120.8	64.4	104.0	118.0	68.5	103.2	117.8	73.4	104.5	136.3	99.2	125.2	127.7	106.8	121.5
0970, 0990	Wall Finishes & Painting/Coating	112.9	105.7	108.6	109.0	68.2	84.5	113.0	65.5	84.4	108.4	52.7	74.9	113.0	93.1	101.1	107.9	106.6	107.1
09	FINISHES	123.1	96.7	108.4	110.3	62.7	83.8	116.0	71.8	91.4	113.4	65.9	86.9	121.6	91.8	105.0	115.7	95.7	104.5
COVERS	DIVS. 10 - 14, 25, 28, 41, 43, 44, 46	131.1	88.6	122.6	131.1	66.1	118.1	131.1	68.2	118.5	131.1	66.9	118.3	131.1	74.5	119.8	131.1	99.7	124.8
21, 22, 23	FIRE SUPPRESSION, PLUMBING & HVAC	101.3	93.8	98.3	102.9	77.8	92.8	102.5	86.3	96.0	102.5	71.3	90.0	102.5	100.3	101.6	100.5	102.9	101.5
26, 27, 3370	ELECTRICAL, COMMUNICATIONS & UTIL.	113.8	86.5	99.6	115.4	64.3	88.7	117.7	65.9	90.7	113.4	63.3	87.3	117.7	93.0	104.8	115.6	92.6	103.6
MF2010	WEIGHTED AVERAGE	119.2	93.4	107.9	111.2	70.8	93.6	117.7	76.1	99.6	112.1	69.3	93.5	120.5	92.9	108.5	114.6	96.1	106.5

	DIVISION	CANADA																	
		OTTAWA, ONTARIO			OWEN SOUND, ONTARIO			PETERBOROUGH, ONTARIO			PORTAGE LA PRAIRIE, MANITOBA			PRINCE ALBERT, SASKATCHEWAN			PRINCE GEORGE, BRITISH COLUMBIA		
		MAT.	INST.	TOTAL	MAT.	INST.	TOTAL	MAT.	INST.	TOTAL	MAT.	INST.	TOTAL	MAT.	INST.	TOTAL	MAT.	INST.	TOTAL
015433	CONTRACTOR EQUIPMENT		103.8	103.8		101.6	101.6		101.6	101.6		103.9	103.9		100.2	100.2		105.6	105.6
0241, 31 - 34	SITE & INFRASTRUCTURE, DEMOLITION	110.8	105.0	106.7	115.3	100.9	105.2	114.9	100.8	105.0	113.7	99.9	104.0	107.8	96.7	100.0	121.8	104.0	109.3
0310	Concrete Forming & Accessories	126.4	90.9	95.8	125.1	87.5	92.6	124.9	90.1	94.9	124.3	73.6	80.6	107.3	61.7	67.9	113.0	87.5	91.0
0320	Concrete Reinforcing	144.0	93.0	118.5	174.2	86.5	130.4	167.9	84.9	126.5	155.3	54.9	105.2	114.6	63.8	89.2	112.0	79.2	95.6
0330	Cast-in-Place Concrete	161.6	99.0	135.6	166.0	84.3	132.1	162.0	90.1	132.2	149.7	77.1	119.6	131.6	72.0	106.9	145.8	103.1	128.0
03	CONCRETE	143.6	94.1	119.2	151.0	86.4	119.2	147.1	89.3	118.6	132.7	72.0	102.7	116.7	66.3	91.9	150.0	91.5	121.2
04	MASONRY	175.1	97.0	126.8	165.2	97.3	123.2	167.9	98.4	124.9	166.3	67.7	105.3	160.8	65.5	101.8	172.9	96.6	125.7
05	METALS	123.6	91.9	113.8	108.2	91.2	103.0	105.5	90.9	101.0	106.4	77.7	97.5	102.8	74.7	94.1	106.2	88.8	100.8
06	WOOD, PLASTICS & COMPOSITES	108.8	90.5	98.5	120.6	86.2	101.2	119.8	88.2	102.0	117.1	75.1	93.4	99.0	60.4	77.2	101.3	83.9	91.5
07	THERMAL & MOISTURE PROTECTION	127.5	92.7	113.4	112.4	88.2	102.6	112.6	93.1	104.7	106.1	72.8	92.6	105.5	64.4	88.8	115.3	87.9	104.1
08	OPENINGS	93.8	89.2	92.7	92.5	85.0	90.7	93.6	87.5	92.2	94.7	66.3	88.0	88.9	58.1	81.6	91.0	83.7	89.3
0920	Plaster & Gypsum Board	192.8	90.2	123.0	151.2	85.7	106.7	142.7	87.8	105.4	123.6	73.9	89.8	121.3	59.1	79.0	124.3	83.0	96.2
0950, 0980	Ceilings & Acoustic Treatment	134.7	90.2	105.1	95.2	85.7	88.9	97.4	87.8	91.0	97.4	73.9	81.8	97.4	59.1	71.9	97.4	83.0	87.8
0960	Flooring	126.9	99.2	118.6	130.7	100.7	121.7	130.5	99.2	121.1	130.5	71.5	112.8	120.8	64.4	104.0	126.1	77.6	111.6
0970, 0990	Wall Finishes & Painting/Coating	112.4	94.1	101.4	111.9	91.8	99.8	109.0	95.5	100.9	109.1	60.0	79.6	109.0	58.2	78.5	109.0	85.8	95.0
09	FINISHES	130.7	92.9	109.7	117.3	90.6	102.4	115.6	92.5	102.7	113.4	72.3	90.5	110.3	61.6	83.2	113.9	85.1	97.9
COVERS	DIVS. 10 - 14, 25, 28, 41, 43, 44, 46	131.1	97.1	124.3	139.2	75.0	126.4	131.1	76.0	120.1	131.1	68.7	118.6	131.1	66.1	118.1	131.1	97.1	124.3
21, 22, 23	FIRE SUPPRESSION, PLUMBING & HVAC	101.5	83.2	94.2	102.2	100.2	101.4	102.5	103.9	103.1	102.5	85.4	95.6	102.9	70.1	89.7	102.5	96.5	100.1
26, 27, 3370	ELECTRICAL, COMMUNICATIONS & UTIL.	109.0	94.4	101.4	117.9	91.7	104.2	112.9	92.6	102.3	115.1	61.5	87.1	115.4	64.3	88.7	113.1	83.9	97.8
MF2010	WEIGHTED AVERAGE	119.2	92.5	107.5	117.4	93.0	106.8	115.5	94.9	106.5	113.9	75.6	97.2	110.3	69.0	92.3	115.9	91.7	105.3

	DIVISION	CANADA																	
		QUEBEC, QUEBEC			RED DEER, ALBERTA			REGINA, SASKATCHEWAN			RIMOUSKI, QUEBEC			ROUYN-NORANDA, QUEBEC			SAINT HYACINTHE, QUEBEC		
		MAT.	INST.	TOTAL	MAT.	INST.	TOTAL	MAT.	INST.	TOTAL	MAT.	INST.	TOTAL	MAT.	INST.	TOTAL	MAT.	INST.	TOTAL
015433	CONTRACTOR EQUIPMENT		104.4	104.4		103.7	103.7		100.2	100.2		102.5	102.5		102.5	102.5		102.5	102.5
0241, 31 - 34	SITE & INFRASTRUCTURE, DEMOLITION	103.7	100.0	101.1	111.6	101.9	104.8	123.5	98.4	105.9	96.1	99.7	98.6	95.7	100.5	99.1	96.1	100.5	99.2
0310	Concrete Forming & Accessories	131.8	93.4	98.6	139.3	84.3	91.9	114.1	90.4	93.7	130.3	92.8	98.0	130.3	89.8	95.4	130.3	89.8	95.4
0320	Concrete Reinforcing	136.6	96.4	116.5	155.3	74.0	114.7	132.1	83.4	107.8	107.7	96.4	102.1	151.1	81.3	116.3	151.1	81.3	116.3
0330	Cast-in-Place Concrete	142.8	100.9	125.4	149.7	101.0	129.5	173.6	94.9	140.9	135.7	99.2	120.5	129.0	99.3	116.7	133.3	99.3	119.2
03	CONCRETE	133.9	96.4	115.4	140.3	88.6	114.8	150.3	90.6	120.9	125.3	95.6	110.7	129.2	91.7	110.7	131.2	91.7	111.7
04	MASONRY	172.0	94.1	123.8	163.2	82.9	113.5	184.7	94.4	128.8	163.4	94.1	120.5	163.6	88.6	117.2	163.8	88.6	117.3
05	METALS	124.4	93.0	114.7	106.4	88.8	100.9	125.6	85.7	113.3	103.4	92.3	100.0	103.8	88.7	99.2	103.8	88.7	99.2
06	WOOD, PLASTICS & COMPOSITES	119.9	93.5	105.0	117.3	83.8	98.4	98.3	91.3	94.4	131.6	93.2	110.0	131.6	89.9	108.1	131.6	89.9	108.1
07	THERMAL & MOISTURE PROTECTION	118.9	96.6	109.9	130.7	88.0	113.4	117.9	85.6	104.8	106.2	95.9	102.0	106.2	91.8	100.3	106.6	91.8	100.6
08	OPENINGS	94.0	90.3	93.1	94.7	79.2	91.0	90.8	80.2	88.3	94.2	80.5	91.0	94.7	78.2	90.8	94.7	78.2	90.8
0920	Plaster & Gypsum Board	164.8	92.8	115.8	129.2	83.0	97.7	147.8	91.0	109.2	172.1	92.8	118.1	140.3	89.4	105.7	142.7	89.4	106.5
0950, 0980	Ceilings & Acoustic Treatment	122.8	92.8	102.8	97.4	83.0	87.8	123.9	91.0	102.0	96.5	92.8	94.0	96.5	89.4	91.8	96.5	89.4	91.8
0960	Flooring	134.3	101.4	124.5	133.2	94.3	121.6	132.5	66.7	112.8	130.5	101.4	121.8	130.5	101.4	121.8	130.5	101.4	121.8
0970, 0990	Wall Finishes & Painting/Coating	115.4	105.7	109.5	109.0	83.0	93.4	112.9	90.9	99.7	109.0	105.7	107.0	109.0	93.7	99.8	109.0	93.7	99.8
09	FINISHES	126.3	96.7	109.8	115.0	85.7	98.7	125.5	86.8	103.9	118.7	96.5	106.3	114.3	92.8	102.3	114.6	92.8	102.5
COVERS	DIVS. 10 - 14, 25, 28, 41, 43, 44, 46	131.1	88.7	122.6	131.1	94.1	123.7	131.1	73.5	119.6	131.1	88.0	122.5	131.1	87.3	122.4	131.1	87.3	122.4
21, 22, 23	FIRE SUPPRESSION, PLUMBING & HVAC	101.2	93.8	98.3	102.5	94.2	99.2	100.7	84.2	94.1	102.5	93.7	99.0	102.5	93.0	98.7	98.9	93.0	96.5
26, 27, 3370	ELECTRICAL, COMMUNICATIONS & UTIL.	110.9	86.5	98.2	106.1	87.5	96.4	115.4	91.4	102.9	113.2	86.5	99.3	113.2	74.3	92.9	114.0	74.3	93.3
MF2010	WEIGHTED AVERAGE	117.5	93.8	107.2	114.5	89.6	103.6	120.3	88.4	106.4	112.3	93.1	103.9	112.5	89.2	102.3	111.9	89.2	102.0

	DIVISION	CANADA																	
		SAINT JOHN, NEW BRUNSWICK			SARNIA, ONTARIO			SASKATOON, SASKATCHEWAN			SAULT STE MARIE, ONTARIO			SHERBROOKE, QUEBEC			SOREL, QUEBEC		
		MAT.	INST.	TOTAL	MAT.	INST.	TOTAL	MAT.	INST.	TOTAL	MAT.	INST.	TOTAL	MAT.	INST.	TOTAL	MAT.	INST.	TOTAL
015433	CONTRACTOR EQUIPMENT		101.9	101.9		101.6	101.6		100.2	100.2		101.6	101.6		102.5	102.5		102.5	102.5
0241, 31 - 34	SITE & INFRASTRUCTURE, DEMOLITION	98.5	98.5	98.5	113.4	100.9	104.6	108.7	98.4	101.5	103.7	100.4	101.4	96.2	100.5	99.2	96.3	100.6	99.3
0310	Concrete Forming & Accessories	123.2	66.3	74.1	123.7	97.4	101.0	107.4	90.4	92.7	113.0	87.0	90.6	130.3	89.8	95.4	130.3	90.0	95.5
0320	Concrete Reinforcing	139.2	61.9	100.6	119.4	86.3	102.9	115.5	83.4	99.5	107.8	84.6	96.2	151.1	81.3	116.3	142.9	81.4	112.2
0330	Cast-in-Place Concrete	134.1	80.7	111.9	149.5	103.7	130.5	141.4	94.9	122.1	134.3	88.5	115.3	133.3	99.3	119.2	134.4	99.3	119.8
03	CONCRETE	131.4	71.4	101.8	133.3	97.4	115.6	122.5	90.6	106.7	117.3	87.3	102.5	131.2	91.7	111.7	130.4	91.7	111.3
04	MASONRY	185.8	72.2	115.5	179.3	101.2	131.0	167.2	94.4	122.2	164.3	94.5	121.1	163.9	88.6	117.3	164.0	88.6	117.3
05	METALS	104.3	82.6	97.6	105.5	91.3	101.1	101.9	85.7	96.9	104.7	90.8	100.4	103.8	88.7	99.2	103.8	88.8	99.2
06	WOOD, PLASTICS & COMPOSITES	121.4	64.5	89.3	118.9	96.8	106.4	98.0	91.3	94.3	106.0	86.5	95.0	131.6	89.9	108.1	131.6	89.9	108.1
07	THERMAL & MOISTURE PROTECTION	109.2	69.6	93.1	112.6	96.7	106.2	106.8	85.6	98.2	111.5	89.4	102.5	106.2	91.8	100.3	106.2	91.8	100.3
08	OPENINGS	89.1	60.8	82.4	95.7	91.5	94.7	90.1	80.2	87.7	87.2	85.3	86.8	94.7	78.2	90.8	94.7	83.0	91.9
0920	Plaster & Gypsum Board	169.4	63.3	97.2	164.7	96.7	118.4	129.7	91.0	103.4	131.7	86.0	100.6	142.7	89.4	106.5	172.1	89.4	115.9
0950, 0980	Ceilings & Acoustic Treatment	107.2	63.3	78.0	102.7	96.7	98.7	119.7	91.0	100.6	97.4	86.0	89.8	96.5	89.4	91.8	96.5	89.4	91.8
0960	Flooring	128.4	73.4	112.0	130.5	108.5	123.9	120.7	66.7	104.6	123.6	102.7	117.3	130.5	101.4	121.8	130.5	101.4	121.8
0970, 0990	Wall Finishes & Painting/Coating	108.4	78.8	90.6	109.0	108.0	108.4	109.9	90.9	98.5	109.0	100.0	103.6	109.0	93.7	99.8	109.0	93.7	99.8
09	FINISHES	120.2	68.5	91.4	119.8	101.1	109.4	116.5	86.8	99.9	111.4	91.3	100.2	114.6	92.8	102.5	118.7	92.8	104.3
COVERS	DIVS. 10 - 14, 25, 28, 41, 43, 44, 46	131.1	67.3	118.3	131.1	77.5	120.4	131.1	73.5	119.6	131.1	98.4	124.6	131.1	87.3	122.4	131.1	87.3	122.4
21, 22, 23	FIRE SUPPRESSION, PLUMBING & HVAC	102.5	81.0	93.9	102.5	110.2	105.6	100.6	84.2	94.0	102.5	96.8	100.2	102.9	93.0	98.9	102.5	93.0	98.7
26, 27, 3370	ELECTRICAL, COMMUNICATIONS & UTIL.	121.9	90.3	105.4	115.7	94.9	104.8	117.1	91.4	103.7	114.4	93.0	103.2	113.2	74.3	92.9	113.2	74.3	92.9
MF2010	WEIGHTED AVERAGE	114.6	78.3	98.8	115.3	99.5	108.4	111.3	88.4	101.3	110.7	93.1	103.0	112.8	89.2	102.5	113.0	89.4	102.7

	DIVISION	CANADA																	
		ST CATHARINES, ONTARIO			ST JEROME, QUEBEC			ST JOHNS, NEWFOUNDLAND			SUDBURY, ONTARIO			SUMMERSIDE, PRINCE EDWARD ISLAND			SYDNEY, NOVA SCOTIA		
		MAT.	INST.	TOTAL	MAT.	INST.	TOTAL	MAT.	INST.	TOTAL	MAT.	INST.	TOTAL	MAT.	INST.	TOTAL	MAT.	INST.	TOTAL
015433	CONTRACTOR EQUIPMENT		101.6	101.6		102.5	102.5		103.1	103.1		101.6	101.6		101.5	101.5		101.5	101.5
0241, 31 - 34	SITE & INFRASTRUCTURE, DEMOLITION	101.8	101.8	101.8	95.7	100.5	99.1	116.4	99.9	104.8	101.9	101.3	101.5	112.2	96.4	101.1	100.1	98.9	99.2
0310	Concrete Forming & Accessories	115.9	93.9	96.9	130.3	89.8	95.4	113.8	82.4	86.7	111.9	87.4	90.7	99.1	59.6	65.0	98.8	73.2	76.8
0320	Concrete Reinforcing	104.3	93.0	98.7	151.1	81.3	116.3	161.9	74.4	118.2	105.1	93.0	99.0	171.5	48.3	110.0	173.8	48.2	111.1
0330	Cast-in-Place Concrete	134.4	100.1	120.1	129.0	99.3	116.7	166.6	87.6	133.8	135.5	96.2	119.2	155.8	62.0	116.9	126.6	73.2	104.5
03	CONCRETE	115.8	95.8	105.9	129.2	91.7	110.7	159.3	83.0	121.6	116.2	91.6	104.0	156.5	59.1	108.5	133.8	69.3	102.0
04	MASONRY	158.6	99.7	122.1	163.6	88.6	117.2	187.1	89.8	126.9	158.6	94.9	119.2	181.9	63.7	108.8	180.0	73.2	113.9
05	METALS	104.6	92.4	100.8	103.8	88.7	99.2	126.2	83.4	113.0	104.6	91.9	100.7	116.0	68.6	101.4	116.0	75.3	103.4
06	WOOD, PLASTICS & COMPOSITES	107.4	93.6	99.6	131.6	89.9	108.1	101.4	80.9	89.8	103.5	86.5	93.9	94.9	59.1	74.7	94.6	72.6	82.2
07	THERMAL & MOISTURE PROTECTION	112.1	95.9	105.5	106.2	91.8	100.3	118.6	84.9	105.0	111.5	90.3	102.9	109.9	62.5	90.7	110.4	71.4	94.6
08	OPENINGS	85.2	90.7	86.5	94.7	78.2	90.8	99.3	73.4	93.2	86.0	85.3	85.8	99.1	51.1	87.8	88.3	66.4	83.1
0920	Plaster & Gypsum Board	129.0	93.4	104.8	140.3	89.4	105.7	154.5	80.1	103.9	133.6	86.0	101.2	154.4	57.7	88.6	154.0	71.7	98.0
0950, 0980	Ceilings & Acoustic Treatment	100.8	93.4	95.9	96.5	89.4	91.8	116.8	80.1	92.4	95.5	86.0	89.2	102.5	57.7	72.7	102.5	71.7	82.0
0960	Flooring	123.0	100.7	116.3	130.5	101.4	121.8	122.3	60.3	103.8	120.9	102.7	115.4	118.1	64.0	101.9	118.0	68.5	103.2
0970, 0990	Wall Finishes & Painting/Coating	107.9	102.5	104.7	109.0	93.7	99.8	112.8	75.8	90.6	107.9	93.5	99.3	113.0	43.9	71.4	113.0	65.5	84.4
09	FINISHES	111.2	96.4	103.0	114.3	92.8	102.3	121.1	78.0	97.1	110.1	90.7	99.3	117.0	59.1	84.7	116.0	71.8	91.4
COVERS	DIVS. 10 - 14, 25, 28, 41, 43, 44, 46	131.1	76.3	120.1	131.1	87.3	122.4	131.1	73.2	119.5	131.1	98.5	124.6	131.1	66.4	118.2	131.1	68.2	118.5
21, 22, 23	FIRE SUPPRESSION, PLUMBING & HVAC	100.5	84.4	94.1	102.5	93.0	98.7	100.8	73.9	90.0	102.5	85.8	95.8	102.5	65.2	87.6	102.5	86.3	96.0
26, 27, 3370	ELECTRICAL, COMMUNICATIONS & UTIL.	116.0	94.5	104.7	113.8	74.3	93.2	115.9	72.4	93.2	113.1	94.8	103.6	116.7	53.6	83.7	117.7	65.9	90.7
MF2010	WEIGHTED AVERAGE	109.7	93.0	102.4	112.5	89.2	102.3	122.0	80.4	103.9	109.9	91.7	101.9	119.5	64.1	95.4	115.6	76.1	98.4

DIVISION		CANADA																	
		THUNDER BAY, ONTARIO			TIMMINS, ONTARIO			TORONTO, ONTARIO			TROIS RIVIERES, QUEBEC			TRURO, NOVA SCOTIA			VANCOUVER, BRITISH COLUMBIA		
		MAT.	INST.	TOTAL	MAT.	INST.	TOTAL	MAT.	INST.	TOTAL	MAT.	INST.	TOTAL	MAT.	INST.	TOTAL	MAT.	INST.	TOTAL
015433	**CONTRACTOR EQUIPMENT**		101.6	101.6		101.6	101.6		103.9	103.9		102.5	102.5		101.5	101.5		112.1	112.1
0241, 31 - 34	**SITE & INFRASTRUCTURE, DEMOLITION**	106.8	101.7	103.2	114.9	100.4	104.7	137.0	105.8	115.1	100.9	100.6	100.7	100.1	98.9	99.2	120.3	108.2	111.8
0310	Concrete Forming & Accessories	123.7	90.8	95.3	124.9	89.0	93.9	128.5	99.5	103.4	133.2	90.0	95.9	97.1	73.3	76.5	122.0	91.6	95.8
0320	Concrete Reinforcing	93.2	91.9	92.5	167.9	84.4	126.2	151.3	93.2	122.3	173.8	81.4	127.6	142.9	48.2	95.6	142.2	81.7	112.0
0330	Cast-in-Place Concrete	147.9	98.5	127.4	162.0	88.4	131.5	150.7	108.2	133.1	131.1	99.3	117.9	168.3	73.3	128.8	157.7	99.2	133.4
03	**CONCRETE**	124.2	93.7	109.2	147.1	88.1	118.0	139.8	101.1	120.7	136.8	91.7	114.6	145.5	69.3	107.9	148.1	92.7	120.8
04	**MASONRY**	159.3	99.0	122.0	167.9	91.5	120.6	194.2	104.1	138.4	183.8	88.6	124.9	163.4	73.2	107.6	182.3	93.0	127.0
05	**METALS**	104.4	91.3	100.4	105.5	90.4	100.9	125.1	93.1	115.2	115.1	88.8	107.0	104.6	75.3	95.6	141.4	91.4	126.0
06	**WOOD, PLASTICS & COMPOSITES**	117.2	89.5	101.6	119.8	89.6	102.7	111.5	98.7	104.3	139.8	89.9	111.7	89.2	72.6	79.9	104.1	91.7	97.1
07	**THERMAL & MOISTURE PROTECTION**	112.4	93.0	104.5	112.6	87.4	102.4	124.3	99.6	114.2	109.3	91.8	102.2	107.5	71.4	92.9	130.0	88.5	113.2
08	**OPENINGS**	84.3	87.8	85.1	93.6	85.7	91.7	91.8	96.8	93.0	97.0	83.0	93.7	86.6	66.4	81.8	94.5	87.9	93.0
0920	Plaster & Gypsum Board	162.4	89.2	112.6	142.7	89.2	106.3	155.3	98.6	116.7	186.1	89.4	120.3	141.8	71.7	94.1	140.0	90.9	106.6
0950, 0980	Ceilings & Acoustic Treatment	95.5	89.2	91.3	97.4	89.2	91.9	125.4	98.6	107.6	101.6	89.4	93.5	97.4	71.7	80.3	150.0	90.9	110.7
0960	Flooring	127.7	107.8	121.7	130.5	99.2	121.1	132.8	110.4	126.1	136.3	101.4	125.8	113.1	68.5	99.8	129.2	97.7	119.8
0970, 0990	Wall Finishes & Painting/Coating	107.9	92.1	98.4	109.0	93.1	99.5	114.8	106.6	109.9	113.0	93.7	101.4	109.0	65.5	82.9	114.0	97.7	104.2
09	**FINISHES**	116.4	94.2	104.0	115.6	91.8	102.4	127.3	102.5	113.5	125.3	92.8	107.2	109.9	71.8	88.7	129.1	93.6	109.3
COVERS	**DIVS. 10 - 14, 25, 28, 41, 43, 44, 46**	131.1	76.0	120.1	131.1	74.5	119.8	131.1	102.1	125.3	131.1	87.3	122.4	131.1	68.2	118.5	131.1	96.9	124.3
21, 22, 23	**FIRE SUPPRESSION, PLUMBING & HVAC**	100.5	84.3	94.0	102.5	100.3	101.6	101.2	94.6	98.6	102.5	93.0	98.7	102.5	86.3	96.0	101.3	82.3	93.7
26, 27, 3370	**ELECTRICAL, COMMUNICATIONS & UTIL.**	114.4	93.0	103.2	114.4	93.0	103.2	111.5	96.6	103.7	118.0	74.3	95.2	112.9	65.9	88.4	111.3	81.3	95.6
MF2010	**WEIGHTED AVERAGE**	110.9	91.7	102.6	115.6	92.9	105.7	120.2	99.0	111.0	117.9	89.4	105.5	113.0	76.1	96.9	123.3	90.0	108.8

DIVISION		CANADA																	
		VICTORIA, BRITISH COLUMBIA			WHITEHORSE, YUKON			WINDSOR, ONTARIO			WINNIPEG, MANITOBA			YARMOUTH, NOVA SCOTIA			YELLOWKNIFE, NWT		
		MAT.	INST.	TOTAL	MAT.	INST.	TOTAL	MAT.	INST.	TOTAL	MAT.	INST.	TOTAL	MAT.	INST.	TOTAL	MAT.	INST.	TOTAL
015433	**CONTRACTOR EQUIPMENT**		108.7	108.7		101.7	101.7		101.6	101.6		105.9	105.9		101.5	101.5		101.6	101.6
0241, 31 - 34	**SITE & INFRASTRUCTURE, DEMOLITION**	121.5	107.7	111.8	123.0	97.4	105.0	97.3	101.6	100.3	119.5	101.3	106.7	104.2	98.9	100.5	138.7	101.7	112.8
0310	Concrete Forming & Accessories	112.9	91.0	94.0	117.7	60.9	68.7	123.7	90.0	94.6	131.5	68.1	76.8	98.8	73.2	76.8	116.1	82.0	86.7
0320	Concrete Reinforcing	113.0	81.6	97.3	153.9	61.9	107.9	102.2	93.0	97.6	141.0	57.4	99.3	173.8	48.2	111.1	154.5	64.7	109.6
0330	Cast-in-Place Concrete	146.4	98.6	126.6	191.7	71.3	141.7	137.7	100.3	122.1	181.5	73.8	136.8	162.4	73.2	125.4	231.5	89.4	172.5
03	**CONCRETE**	152.5	91.9	122.6	158.2	65.4	112.4	117.6	94.1	106.0	153.4	68.7	111.6	150.8	69.3	110.6	176.9	81.7	129.9
04	**MASONRY**	170.1	93.0	122.4	205.6	63.9	117.9	158.7	97.8	121.0	179.6	69.3	111.4	182.6	73.2	114.9	209.1	76.2	126.8
05	**METALS**	104.3	87.4	99.1	129.3	74.6	112.5	104.5	91.8	100.6	143.3	74.6	122.1	116.0	75.3	103.4	129.3	79.8	114.0
06	**WOOD, PLASTICS & COMPOSITES**	101.6	91.6	96.0	100.8	59.6	77.5	117.2	89.1	101.4	111.1	69.0	87.3	94.6	72.6	82.2	100.3	83.7	90.9
07	**THERMAL & MOISTURE PROTECTION**	115.8	88.3	104.7	125.8	63.4	100.5	112.2	92.2	104.0	118.7	70.7	99.2	110.4	71.4	94.6	122.1	80.0	105.0
08	**OPENINGS**	91.8	84.1	90.0	95.4	56.8	86.3	84.1	88.1	85.0	93.1	61.6	85.6	88.3	66.4	83.1	94.3	70.5	88.7
0920	Plaster & Gypsum Board	121.7	90.9	100.7	146.1	58.2	86.3	147.7	88.8	107.6	139.7	67.5	90.6	154.0	71.7	98.0	176.0	83.1	112.8
0950, 0980	Ceilings & Acoustic Treatment	99.3	90.9	93.7	127.7	58.2	81.4	95.5	88.8	91.0	143.6	67.5	93.0	102.5	71.7	82.0	133.5	83.1	99.9
0960	Flooring	125.9	77.6	111.5	140.3	62.5	117.0	127.7	105.0	120.9	131.3	74.0	114.1	118.0	68.5	103.2	134.7	94.2	122.6
0970, 0990	Wall Finishes & Painting/Coating	109.9	97.7	102.6	124.2	57.2	83.9	107.9	95.4	100.4	112.9	55.4	78.4	113.0	65.5	84.4	121.6	82.0	97.8
09	**FINISHES**	114.1	90.1	100.7	130.9	60.6	91.7	114.0	93.5	102.6	127.8	68.3	94.6	116.0	71.8	91.4	134.9	83.7	106.4
COVERS	**DIVS. 10 - 14, 25, 28, 41, 43, 44, 46**	131.1	72.9	119.5	131.1	65.4	118.0	131.1	75.4	120.0	131.1	67.9	118.5	131.1	68.2	118.5	131.1	68.7	118.6
21, 22, 23	**FIRE SUPPRESSION, PLUMBING & HVAC**	102.5	82.2	94.4	102.9	76.4	92.3	100.5	83.9	93.9	101.2	65.7	87.0	102.5	86.3	96.0	103.0	96.0	100.2
26, 27, 3370	**ELECTRICAL, COMMUNICATIONS & UTIL.**	113.8	81.3	96.8	134.3	63.7	97.5	119.3	94.5	106.4	112.9	66.9	88.9	117.7	65.9	90.7	132.1	87.2	108.7
MF2010	**WEIGHTED AVERAGE**	115.9	88.1	103.8	126.3	69.8	101.7	110.2	91.7	102.2	123.6	70.8	100.6	117.6	76.1	99.6	128.8	85.8	110.1

Location Factors

Costs shown in RSMeans cost data publications are based on national averages for materials and installation. To adjust these costs to a specific location, simply multiply the base cost by the factor and divide by 100 for that city. The data is arranged alphabetically by state and postal zip code numbers. For a city not listed, use the factor for a nearby city with similar economic characteristics.

STATE/ZIP	CITY	MAT.	INST.	TOTAL
ALABAMA				
350-352	Birmingham	97.5	74.1	87.3
354	Tuscaloosa	96.7	62.0	81.6
355	Jasper	97.0	56.7	79.4
356	Decatur	96.6	59.1	80.3
357-358	Huntsville	96.7	70.8	85.4
359	Gadsden	96.7	60.8	81.1
360-361	Montgomery	97.9	58.1	80.5
362	Anniston	96.5	63.0	81.9
363	Dothan	96.9	50.5	76.7
364	Evergreen	96.5	51.9	77.1
365-366	Mobile	97.8	65.9	83.9
367	Selma	96.7	50.8	76.7
368	Phenix City	97.3	55.7	79.2
369	Butler	96.8	50.5	76.6
ALASKA				
995-996	Anchorage	121.3	114.3	118.3
997	Fairbanks	120.7	115.0	118.2
998	Juneau	120.6	113.5	117.5
999	Ketchikan	132.0	113.5	124.0
ARIZONA				
850,853	Phoenix	99.4	74.8	88.7
851,852	Mesa/Tempe	98.8	73.8	87.9
855	Globe	99.0	71.2	86.9
856-857	Tucson	97.6	73.6	87.1
859	Show Low	99.1	73.5	88.0
860	Flagstaff	101.0	72.9	88.8
863	Prescott	98.9	69.7	86.2
864	Kingman	97.3	73.2	86.8
865	Chambers	97.3	71.9	86.3
ARKANSAS				
716	Pine Bluff	96.9	62.9	82.1
717	Camden	94.5	50.4	75.3
718	Texarkana	95.7	49.9	75.7
719	Hot Springs	93.8	52.3	75.8
720-722	Little Rock	98.0	63.8	83.1
723	West Memphis	95.6	62.7	81.2
724	Jonesboro	96.3	59.4	80.2
725	Batesville	94.2	55.1	77.2
726	Harrison	95.5	51.1	76.1
727	Fayetteville	93.0	52.9	75.6
728	Russellville	94.1	54.4	76.8
729	Fort Smith	96.8	61.7	81.5
CALIFORNIA				
900-902	Los Angeles	100.1	116.5	107.3
903-905	Inglewood	95.7	113.4	103.4
906-908	Long Beach	97.1	113.4	104.2
910-912	Pasadena	96.3	113.9	104.0
913-916	Van Nuys	99.1	113.9	105.5
917-918	Alhambra	98.2	113.9	105.0
919-921	San Diego	100.2	109.1	104.1
922	Palm Springs	96.7	111.6	103.2
923-924	San Bernardino	94.6	111.2	101.9
925	Riverside	98.9	114.0	105.5
926-927	Santa Ana	96.5	111.6	103.1
928	Anaheim	98.9	114.1	105.6
930	Oxnard	99.6	114.2	105.9
931	Santa Barbara	98.9	114.6	105.7
932-933	Bakersfield	100.8	110.6	105.1
934	San Luis Obispo	99.3	111.6	104.7
935	Mojave	96.8	108.6	101.9
936-938	Fresno	100.0	114.5	106.3
939	Salinas	100.2	120.3	108.9
940-941	San Francisco	106.1	143.8	122.5
942,956-958	Sacramento	101.2	118.6	108.8
943	Palo Alto	99.5	133.1	114.2
944	San Mateo	101.7	132.7	115.2
945	Vallejo	100.6	127.2	112.2
946	Oakland	103.9	133.1	116.6
947	Berkeley	103.4	133.1	116.4
948	Richmond	102.5	131.2	115.0
949	San Rafael	104.2	133.6	117.0
950	Santa Cruz	104.1	120.4	111.2
CALIFORNIA (CONT'D)				
951	San Jose	102.2	134.6	116.3
952	Stockton	100.5	115.5	107.0
953	Modesto	100.4	114.9	106.7
954	Santa Rosa	100.3	133.4	114.7
955	Eureka	101.6	111.6	106.0
959	Marysville	100.8	115.1	107.0
960	Redding	103.6	114.6	108.4
961	Susanville	102.6	114.6	107.8
COLORADO				
800-802	Denver	101.6	82.6	93.3
803	Boulder	98.6	82.0	91.3
804	Golden	100.6	79.6	91.4
805	Fort Collins	102.1	79.6	92.3
806	Greeley	99.7	76.1	89.4
807	Fort Morgan	99.1	79.6	90.6
808-809	Colorado Springs	101.3	79.3	91.7
810	Pueblo	100.9	78.6	91.2
811	Alamosa	102.0	74.6	90.1
812	Salida	101.7	76.2	90.6
813	Durango	102.4	76.8	91.2
814	Montrose	101.2	76.5	90.5
815	Grand Junction	104.4	76.7	92.4
816	Glenwood Springs	102.2	79.3	92.2
CONNECTICUT				
060	New Britain	100.8	120.4	109.3
061	Hartford	102.0	120.6	110.1
062	Willimantic	101.4	120.6	109.7
063	New London	97.7	120.5	107.6
064	Meriden	99.6	120.5	108.7
065	New Haven	102.2	120.5	110.2
066	Bridgeport	101.8	120.2	109.8
067	Waterbury	101.4	120.1	109.6
068	Norwalk	101.4	126.9	112.5
069	Stamford	101.5	127.9	113.0
D.C.				
200-205	Washington	101.6	91.6	97.2
DELAWARE				
197	Newark	98.8	108.1	102.8
198	Wilmington	99.4	108.1	103.2
199	Dover	99.7	108.1	103.3
FLORIDA				
320,322	Jacksonville	97.2	67.7	84.3
321	Daytona Beach	97.4	74.4	87.3
323	Tallahassee	98.7	58.0	81.0
324	Panama City	98.5	59.1	81.3
325	Pensacola	100.8	62.4	84.1
326,344	Gainesville	98.7	68.0	85.3
327-328,347	Orlando	99.3	71.4	87.1
329	Melbourne	99.9	77.9	90.3
330-332,340	Miami	97.7	74.6	87.6
333	Fort Lauderdale	96.4	73.9	86.6
334,349	West Palm Beach	95.3	72.3	85.3
335-336,346	Tampa	97.9	79.8	90.0
337	St. Petersburg	100.2	64.7	84.7
338	Lakeland	97.1	79.3	89.4
339,341	Fort Myers	96.4	74.1	86.7
342	Sarasota	99.3	75.3	88.8
GEORGIA				
300-303,399	Atlanta	96.8	75.3	87.5
304	Statesboro	96.5	56.9	79.2
305	Gainesville	95.2	62.8	81.1
306	Athens	94.7	65.1	81.8
307	Dalton	96.4	60.1	80.6
308-309	Augusta	95.6	64.8	82.2
310-312	Macon	96.2	65.9	83.0
313-314	Savannah	98.4	62.0	82.6
315	Waycross	97.3	60.7	81.3
316	Valdosta	97.5	62.6	82.3
317,398	Albany	97.5	62.5	82.3
318-319	Columbus	97.4	66.6	84.0

STATE/ZIP	CITY	MAT.	INST.	TOTAL
HAWAII				
967	Hilo	110.4	121.0	115.0
968	Honolulu	114.9	121.0	117.6
STATES & POSS.				
969	Guam	131.2	58.9	99.7
IDAHO				
832	Pocatello	100.8	78.4	91.1
833	Twin Falls	101.9	68.5	87.4
834	Idaho Falls	99.6	75.5	89.1
835	Lewiston	108.3	85.5	98.4
836-837	Boise	99.8	79.4	90.9
838	Coeur d'Alene	107.6	83.3	97.0
ILLINOIS				
600-603	North Suburban	98.5	135.9	114.8
604	Joliet	98.4	140.7	116.8
605	South Suburban	98.5	135.9	114.8
606-608	Chicago	98.9	141.9	117.6
609	Kankakee	94.8	130.8	110.5
610-611	Rockford	98.9	123.5	109.6
612	Rock Island	96.6	99.8	98.0
613	La Salle	97.7	125.1	109.6
614	Galesburg	97.6	107.6	102.0
615-616	Peoria	99.9	110.4	104.5
617	Bloomington	96.9	109.9	102.6
618-619	Champaign	100.2	108.0	103.6
620-622	East St. Louis	95.5	109.3	101.5
623	Quincy	96.9	101.6	99.0
624	Effingham	96.3	106.3	100.6
625	Decatur	98.2	105.9	101.6
626-627	Springfield	99.0	107.2	102.5
628	Centralia	94.1	108.4	100.4
629	Carbondale	93.9	106.2	99.2
INDIANA				
460	Anderson	95.5	83.1	90.1
461-462	Indianapolis	98.3	85.1	92.6
463-464	Gary	96.8	110.1	102.6
465-466	South Bend	95.7	84.2	90.7
467-468	Fort Wayne	96.0	79.2	88.7
469	Kokomo	93.3	82.0	88.4
470	Lawrenceburg	93.3	79.3	87.2
471	New Albany	94.4	77.0	86.8
472	Columbus	96.6	80.9	89.7
473	Muncie	97.3	81.8	90.6
474	Bloomington	98.7	81.5	91.2
475	Washington	95.0	83.3	89.9
476-477	Evansville	96.5	86.6	92.1
478	Terre Haute	97.2	85.8	92.2
479	Lafayette	96.3	82.9	90.5
IOWA				
500-503,509	Des Moines	99.4	83.9	92.7
504	Mason City	96.5	66.6	83.5
505	Fort Dodge	96.6	58.8	80.2
506-507	Waterloo	98.3	70.9	86.4
508	Creston	96.7	77.9	88.5
510-511	Sioux City	99.1	73.6	88.0
512	Sibley	97.7	57.5	80.2
513	Spencer	99.2	57.5	81.0
514	Carroll	96.4	72.2	85.9
515	Council Bluffs	100.2	78.8	90.9
516	Shenandoah	97.1	72.1	86.2
520	Dubuque	98.5	80.4	90.6
521	Decorah	97.4	65.6	83.5
522-524	Cedar Rapids	99.4	83.4	92.4
525	Ottumwa	97.3	73.0	86.7
526	Burlington	96.6	76.9	88.0
527-528	Davenport	98.4	93.4	96.2
KANSAS				
660-662	Kansas City	99.4	95.7	97.7
664-666	Topeka	99.4	66.5	85.1
667	Fort Scott	97.7	74.3	87.5
668	Emporia	97.8	70.0	85.7
669	Belleville	99.4	65.6	84.7
670-672	Wichita	99.9	65.1	84.8
673	Independence	99.9	70.3	87.0
674	Salina	100.4	67.0	85.9
675	Hutchinson	95.8	65.4	82.5
676	Hays	99.6	65.6	84.8
677	Colby	100.2	65.6	85.1
KANSAS (CONT'D)				
678	Dodge City	101.7	66.5	86.4
679	Liberal	99.5	66.4	85.1
KENTUCKY				
400-402	Louisville	97.8	84.2	91.9
403-405	Lexington	97.0	82.4	90.7
406	Frankfort	97.9	82.4	91.1
407-409	Corbin	94.4	73.3	85.2
410	Covington	95.2	93.2	94.3
411-412	Ashland	94.0	100.1	96.6
413-414	Campton	95.1	80.2	88.6
415-416	Pikeville	96.2	90.0	93.5
417-418	Hazard	94.6	79.9	88.2
420	Paducah	93.2	86.2	90.2
421-422	Bowling Green	95.5	84.8	90.8
423	Owensboro	95.6	88.9	92.7
424	Henderson	93.0	87.9	90.8
425-426	Somerset	92.7	81.5	87.8
427	Elizabethtown	92.2	83.7	88.5
LOUISIANA				
700-701	New Orleans	101.5	69.8	87.7
703	Thibodaux	98.7	64.9	84.0
704	Hammond	96.3	54.9	78.3
705	Lafayette	98.5	62.7	82.9
706	Lake Charles	98.6	65.0	84.0
707-708	Baton Rouge	100.4	63.6	84.4
710-711	Shreveport	98.4	59.6	81.5
712	Monroe	96.7	56.4	79.2
713-714	Alexandria	96.8	57.8	79.8
MAINE				
039	Kittery	95.1	85.7	91.0
040-041	Portland	100.8	91.2	96.6
042	Lewiston	98.3	91.2	95.2
043	Augusta	100.3	80.8	91.8
044	Bangor	97.6	90.8	94.7
045	Bath	96.1	83.0	90.4
046	Machias	95.7	86.5	91.7
047	Houlton	95.9	86.6	91.8
048	Rockland	95.0	86.5	91.3
049	Waterville	96.2	80.8	89.5
MARYLAND				
206	Waldorf	97.5	82.2	90.9
207-208	College Park	97.6	83.9	91.7
209	Silver Spring	96.9	83.8	91.2
210-212	Baltimore	100.5	82.8	92.8
214	Annapolis	100.3	81.8	92.3
215	Cumberland	96.0	81.9	89.9
216	Easton	97.4	69.3	85.2
217	Hagerstown	96.9	82.8	90.7
218	Salisbury	97.8	63.3	82.8
219	Elkton	95.1	81.9	89.4
MASSACHUSETTS				
010-011	Springfield	100.0	109.8	104.3
012	Pittsfield	99.6	109.6	103.9
013	Greenfield	97.5	110.2	103.0
014	Fitchburg	96.3	124.6	108.6
015-016	Worcester	100.0	124.6	110.7
017	Framingham	95.5	131.0	111.0
018	Lowell	99.4	130.1	112.8
019	Lawrence	100.1	131.3	113.7
020-022, 024	Boston	101.8	138.1	117.6
023	Brockton	100.4	126.3	111.7
025	Buzzards Bay	94.9	126.7	108.8
026	Hyannis	97.6	126.7	110.3
027	New Bedford	99.6	126.8	111.4
MICHIGAN				
480,483	Royal Oak	94.2	107.5	100.0
481	Ann Arbor	96.4	107.9	101.4
482	Detroit	97.9	109.3	102.8
484-485	Flint	96.1	96.8	96.4
486	Saginaw	95.7	92.4	94.2
487	Bay City	95.8	92.5	94.4
488-489	Lansing	97.7	94.1	96.1
490	Battle Creek	96.8	87.4	92.7
491	Kalamazoo	97.1	85.9	92.2
492	Jackson	95.0	95.9	95.4
493,495	Grand Rapids	98.1	84.5	92.2
494	Muskegon	95.6	85.5	91.2

STATE/ZIP	CITY	MAT.	INST.	TOTAL
MICHIGAN (CONT'D)				
496	Traverse City	94.6	81.0	88.6
497	Gaylord	95.5	84.4	90.7
498-499	Iron mountain	97.5	87.4	93.1
MINNESOTA				
550-551	Saint Paul	100.2	118.8	108.3
553-555	Minneapolis	100.6	119.9	109.0
556-558	Duluth	99.0	108.6	103.2
559	Rochester	98.8	104.9	101.5
560	Mankato	96.6	102.4	99.1
561	Windom	95.3	93.3	94.4
562	Willmar	94.9	103.0	98.4
563	St. Cloud	96.3	117.8	105.7
564	Brainerd	96.3	102.3	98.9
565	Detroit Lakes	98.3	96.7	97.6
566	Bemidji	97.7	99.5	98.5
567	Thief River Falls	97.3	93.8	95.7
MISSISSIPPI				
386	Clarksdale	95.8	53.8	77.5
387	Greenville	99.4	63.1	83.6
388	Tupelo	97.3	56.1	79.4
389	Greenwood	97.0	53.3	78.0
390-392	Jackson	99.1	63.5	83.6
393	Meridian	96.2	65.3	82.7
394	Laurel	97.2	57.1	79.7
395	Biloxi	97.7	59.1	80.9
396	Mccomb	95.7	52.9	77.1
397	Columbus	97.1	55.1	78.8
MISSOURI				
630-631	St. Louis	99.6	107.6	103.1
633	Bowling Green	98.1	94.7	96.6
634	Hannibal	97.1	92.6	95.1
635	Kirksville	100.3	90.0	95.8
636	Flat River	99.0	95.8	97.6
637	Cape Girardeau	99.1	90.6	95.4
638	Sikeston	97.5	90.6	94.5
639	Poplar Bluff	97.0	90.3	94.1
640-641	Kansas City	101.6	106.7	103.8
644-645	St. Joseph	100.7	93.9	97.8
646	Chillicothe	97.7	96.2	97.1
647	Harrisonville	97.3	102.2	99.5
648	Joplin	99.5	81.0	91.5
650-651	Jefferson City	98.4	93.4	96.2
652	Columbia	98.6	93.5	96.4
653	Sedalia	98.1	92.6	95.7
654-655	Rolla	96.3	95.6	96.0
656-658	Springfield	99.9	82.1	92.1
MONTANA				
590-591	Billings	102.2	77.9	91.6
592	Wolf Point	101.7	74.9	90.0
593	Miles City	99.5	74.7	88.7
594	Great Falls	103.4	77.1	92.0
595	Havre	100.6	74.2	89.1
596	Helena	102.0	75.1	90.2
597	Butte	102.0	77.2	91.2
598	Missoula	99.4	76.4	89.4
599	Kalispell	98.4	76.3	88.8
NEBRASKA				
680-681	Omaha	99.6	79.3	90.7
683-685	Lincoln	99.5	75.3	89.0
686	Columbus	97.5	77.9	88.9
687	Norfolk	99.2	77.5	89.8
688	Grand Island	99.1	77.8	89.9
689	Hastings	98.5	81.2	90.9
690	Mccook	98.1	72.1	86.8
691	North Platte	98.5	79.6	90.2
692	Valentine	100.6	73.6	88.9
693	Alliance	100.4	71.6	87.9
NEVADA				
889-891	Las Vegas	100.2	109.4	104.2
893	Ely	98.7	103.4	100.8
894-895	Reno	99.0	93.8	96.7
897	Carson City	99.1	93.8	96.8
898	Elko	97.6	86.1	92.6
NEW HAMPSHIRE				
030	Nashua	100.6	93.6	97.6
031	Manchester	102.1	93.6	98.4

STATE/ZIP	CITY	MAT.	INST.	TOTAL
NEW HAMPSHIRE (CONT'D)				
032-033	Concord	100.8	92.3	97.1
034	Keene	97.4	74.0	87.2
035	Littleton	97.1	77.2	88.4
036	Charleston	96.9	70.1	85.2
037	Claremont	96.0	70.1	84.7
038	Portsmouth	98.2	94.2	96.4
NEW JERSEY				
070-071	Newark	101.9	125.6	112.2
072	Elizabeth	99.5	125.3	110.7
073	Jersey City	98.5	125.5	110.2
074-075	Paterson	100.2	125.2	111.1
076	Hackensack	98.4	125.5	110.2
077	Long Branch	97.8	124.1	109.2
078	Dover	98.7	125.4	110.3
079	Summit	98.7	125.3	110.3
080,083	Vineland	97.5	123.4	108.8
081	Camden	99.4	123.6	109.9
082,084	Atlantic City	98.1	123.5	109.2
085-086	Trenton	100.8	124.0	110.9
087	Point Pleasant	99.2	121.4	108.8
088-089	New Brunswick	99.7	125.2	110.8
NEW MEXICO				
870-872	Albuquerque	99.2	73.6	88.0
873	Gallup	99.3	73.6	88.1
874	Farmington	99.7	73.6	88.3
875	Santa Fe	100.5	73.6	88.8
877	Las Vegas	97.7	73.6	87.2
878	Socorro	97.3	73.6	87.0
879	Truth/Consequences	97.0	70.4	85.4
880	Las Cruces	95.8	70.4	84.7
881	Clovis	98.1	73.5	87.4
882	Roswell	99.8	73.6	88.4
883	Carrizozo	100.1	73.6	88.5
884	Tucumcari	98.8	73.5	87.8
NEW YORK				
100-102	New York	103.8	166.4	131.1
103	Staten Island	99.7	163.9	127.6
104	Bronx	98.0	163.9	126.7
105	Mount Vernon	97.9	133.3	113.3
106	White Plains	98.0	133.3	113.4
107	Yonkers	102.5	133.4	116.0
108	New Rochelle	98.2	133.3	113.5
109	Suffern	98.0	126.2	110.3
110	Queens	100.8	162.9	127.8
111	Long Island City	102.2	162.9	128.6
112	Brooklyn	102.6	162.9	128.9
113	Flushing	102.3	162.9	128.7
114	Jamaica	100.6	162.9	127.7
115,117,118	Hicksville	100.7	146.6	120.7
116	Far Rockaway	102.4	162.9	128.7
119	Riverhead	101.4	146.6	121.1
120-122	Albany	98.6	100.2	99.3
123	Schenectady	97.8	100.2	98.8
124	Kingston	100.7	118.3	108.4
125-126	Poughkeepsie	100.0	127.5	112.0
127	Monticello	99.4	116.5	106.8
128	Glens Falls	92.6	95.0	93.6
129	Plattsburgh	97.3	88.4	93.4
130-132	Syracuse	98.5	97.0	97.9
133-135	Utica	96.7	93.2	95.1
136	Watertown	98.1	91.9	95.4
137-139	Binghamton	98.1	99.4	98.6
140-142	Buffalo	99.7	105.8	102.4
143	Niagara Falls	96.9	103.9	100.0
144-146	Rochester	100.3	95.6	98.3
147	Jamestown	96.0	87.8	92.4
148-149	Elmira	95.9	93.1	94.7
NORTH CAROLINA				
270,272-274	Greensboro	98.8	56.3	80.3
271	Winston-Salem	98.5	57.1	80.5
275-276	Raleigh	98.2	55.1	79.4
277	Durham	100.2	54.4	80.2
278	Rocky Mount	96.0	53.2	77.4
279	Elizabeth City	96.9	52.2	77.4
280	Gastonia	98.4	56.1	80.0
281-282	Charlotte	98.7	57.6	80.8
283	Fayetteville	100.9	58.5	82.5
284	Wilmington	97.2	56.6	79.5
285	Kinston	95.3	53.7	77.2

Location Factors

STATE/ZIP	CITY	MAT.	INST.	TOTAL
NORTH CAROLINA (CONT'D)				
286	Hickory	95.6	54.9	77.9
287-288	Asheville	97.6	55.9	79.4
289	Murphy	96.4	49.2	75.8
NORTH DAKOTA				
580-581	Fargo	101.7	67.6	86.9
582	Grand Forks	101.8	51.6	79.9
583	Devils Lake	101.0	58.6	82.5
584	Jamestown	101.0	45.1	76.7
585	Bismarck	102.0	64.0	85.5
586	Dickinson	101.9	57.9	82.7
587	Minot	101.8	71.0	88.4
588	Williston	100.3	58.0	81.9
OHIO				
430-432	Columbus	98.6	89.7	94.7
433	Marion	95.0	83.7	90.1
434-436	Toledo	98.7	96.4	97.7
437-438	Zanesville	95.5	85.6	91.2
439	Steubenville	96.8	94.5	95.8
440	Lorain	98.8	90.4	95.2
441	Cleveland	98.9	100.0	99.4
442-443	Akron	99.8	93.4	97.0
444-445	Youngstown	99.1	88.3	94.4
446-447	Canton	99.2	85.0	93.0
448-449	Mansfield	96.6	85.6	91.8
450	Hamilton	97.4	84.5	91.8
451-452	Cincinnati	97.8	84.8	92.1
453-454	Dayton	97.5	83.2	91.3
455	Springfield	97.5	84.2	91.7
456	Chillicothe	96.4	92.2	94.5
457	Athens	99.1	80.6	91.1
458	Lima	99.6	84.4	93.0
OKLAHOMA				
730-731	Oklahoma City	97.3	65.4	83.4
734	Ardmore	95.2	62.3	80.9
735	Lawton	97.5	63.3	82.6
736	Clinton	96.5	63.6	82.2
737	Enid	97.2	60.9	81.4
738	Woodward	95.3	63.8	81.6
739	Guymon	96.3	60.5	80.7
740-741	Tulsa	97.4	62.2	82.1
743	Miami	94.0	68.3	82.8
744	Muskogee	96.6	56.3	79.0
745	Mcalester	93.7	61.1	79.5
746	Ponca City	94.3	62.5	80.5
747	Durant	94.3	64.0	81.1
748	Shawnee	95.7	62.2	81.1
749	Poteau	93.5	62.1	79.8
OREGON				
970-972	Portland	97.2	101.4	99.1
973	Salem	99.0	100.5	99.7
974	Eugene	97.1	100.4	98.5
975	Medford	98.4	98.2	98.3
976	Klamath Falls	98.2	98.2	98.2
977	Bend	97.3	100.5	98.7
978	Pendleton	93.1	102.5	97.2
979	Vale	91.3	89.8	90.7
PENNSYLVANIA				
150-152	Pittsburgh	99.7	105.8	102.3
153	Washington	96.7	103.2	99.5
154	Uniontown	97.1	102.2	99.4
155	Bedford	98.0	92.6	95.7
156	Greensburg	98.0	101.6	99.5
157	Indiana	96.9	100.1	98.3
158	Dubois	98.4	97.1	97.8
159	Johnstown	97.9	97.3	97.7
160	Butler	92.1	102.2	96.5
161	New Castle	92.1	97.2	94.4
162	Kittanning	92.5	103.4	97.3
163	Oil City	92.1	97.0	94.3
164-165	Erie	94.1	95.0	94.5
166	Altoona	94.3	92.7	93.6
167	Bradford	95.3	96.8	96.0
168	State College	94.9	93.0	94.1
169	Wellsboro	95.9	92.3	94.3
170-171	Harrisburg	99.6	95.9	98.0
172	Chambersburg	96.0	89.7	93.2
173-174	York	96.5	96.0	96.3
175-176	Lancaster	94.8	89.3	92.4

STATE/ZIP	CITY	MAT.	INST.	TOTAL
PENNSYLVANIA (CONT'D)				
177	Williamsport	93.6	79.3	87.4
178	Sunbury	95.5	96.3	95.8
179	Pottsville	94.7	98.0	96.1
180	Lehigh Valley	95.4	112.5	102.9
181	Allentown	97.6	108.8	102.4
182	Hazleton	95.0	98.8	96.7
183	Stroudsburg	94.9	102.0	98.0
184-185	Scranton	98.3	99.7	98.9
186-187	Wilkes-Barre	94.7	99.2	96.7
188	Montrose	94.4	97.1	95.6
189	Doylestown	94.6	120.8	106.0
190-191	Philadelphia	98.9	133.4	113.9
193	Westchester	95.8	121.7	107.1
194	Norristown	94.9	129.5	110.0
195-196	Reading	97.1	101.1	98.8
PUERTO RICO				
009	San Juan	122.1	24.6	79.7
RHODE ISLAND				
028	Newport	99.0	116.6	106.6
029	Providence	101.5	116.6	108.1
SOUTH CAROLINA				
290-292	Columbia	97.2	57.3	79.8
293	Spartanburg	96.4	57.4	79.5
294	Charleston	97.7	66.4	84.1
295	Florence	96.2	57.8	79.4
296	Greenville	96.2	56.5	78.9
297	Rock Hill	95.6	51.2	76.3
298	Aiken	96.4	70.4	85.1
299	Beaufort	97.1	45.7	74.7
SOUTH DAKOTA				
570-571	Sioux Falls	100.0	58.8	82.1
572	Watertown	98.5	50.2	77.5
573	Mitchell	97.5	49.5	76.6
574	Aberdeen	100.1	50.6	78.6
575	Pierre	100.8	59.2	82.7
576	Mobridge	98.1	49.8	77.1
577	Rapid City	100.0	60.3	82.7
TENNESSEE				
370-372	Nashville	97.9	74.1	87.5
373-374	Chattanooga	99.5	65.5	84.7
375,380-381	Memphis	98.5	71.3	86.7
376	Johnson City	99.2	56.7	80.7
377-379	Knoxville	96.2	60.5	80.6
382	Mckenzie	96.6	60.4	80.9
383	Jackson	98.3	62.5	82.7
384	Columbia	95.3	67.5	83.2
385	Cookeville	96.4	60.4	80.7
TEXAS				
750	Mckinney	99.3	48.9	77.3
751	Waxahackie	99.3	56.4	80.6
752-753	Dallas	100.1	65.8	85.2
754	Greenville	99.4	42.2	74.5
755	Texarkana	98.4	49.7	77.2
756	Longview	99.0	40.5	73.5
757	Tyler	98.9	54.6	79.6
758	Palestine	95.4	55.8	78.2
759	Lufkin	95.9	59.1	79.9
760-761	Fort Worth	99.6	62.5	83.5
762	Denton	98.2	46.6	75.7
763	Wichita Falls	97.8	55.0	79.2
764	Eastland	96.5	49.1	75.9
765	Temple	95.0	52.7	76.6
766-767	Waco	97.2	57.3	79.8
768	Brownwood	98.5	48.4	76.7
769	San Angelo	98.2	49.5	77.0
770-772	Houston	98.7	70.9	86.6
773	Huntsville	97.2	56.9	79.6
774	Wharton	98.1	53.8	78.8
775	Galveston	96.3	69.7	84.7
776-777	Beaumont	96.7	61.2	81.2
778	Bryan	94.1	63.0	80.5
779	Victoria	98.2	42.7	74.1
780	Laredo	97.1	53.7	78.2
781-782	San Antonio	97.5	64.5	83.1
783-784	Corpus Christi	99.7	53.7	79.6
785	Mc Allen	99.7	46.6	76.6
786-787	Austin	98.6	59.1	81.4

STATE/ZIP	CITY	MAT.	INST.	TOTAL
TEXAS (CONT'D)				
788	Del Rio	99.4	46.2	76.3
789	Giddings	96.4	55.6	78.6
790-791	Amarillo	98.2	60.0	81.6
792	Childress	97.9	56.9	80.0
793-794	Lubbock	100.0	56.3	81.0
795-796	Abilene	98.3	53.6	78.8
797	Midland	99.9	53.6	79.7
798-799,885	El Paso	97.2	50.6	76.9
UTAH				
840-841	Salt Lake City	102.2	66.8	86.8
842,844	Ogden	97.9	68.8	85.2
843	Logan	99.7	68.8	86.2
845	Price	100.1	65.3	84.9
846-847	Provo	100.1	66.7	85.6
VERMONT				
050	White River Jct.	97.9	79.7	90.0
051	Bellows Falls	96.6	94.7	95.8
052	Bennington	96.9	95.3	96.2
053	Brattleboro	97.2	94.5	96.0
054	Burlington	101.5	85.1	94.4
056	Montpelier	99.2	84.3	92.7
057	Rutland	99.2	85.1	93.0
058	St. Johnsbury	97.9	79.2	89.8
059	Guildhall	96.8	79.2	89.1
VIRGINIA				
220-221	Fairfax	99.3	84.4	92.8
222	Arlington	100.6	84.7	93.6
223	Alexandria	99.8	85.4	93.5
224-225	Fredericksburg	98.2	83.0	91.6
226	Winchester	98.7	82.1	91.5
227	Culpeper	98.6	80.8	90.9
228	Harrisonburg	98.8	68.1	85.5
229	Charlottesville	99.2	67.3	85.3
230-232	Richmond	100.4	68.9	86.7
233-235	Norfolk	100.0	68.8	86.4
236	Newport News	99.2	68.2	85.7
237	Portsmouth	98.6	66.1	84.5
238	Petersburg	98.6	68.9	85.6
239	Farmville	97.8	53.3	78.5
240-241	Roanoke	100.2	66.0	85.3
242	Bristol	98.1	55.6	79.6
243	Pulaski	97.7	58.3	80.6
244	Staunton	98.4	62.7	82.9
245	Lynchburg	98.6	66.1	84.5
246	Grundy	98.0	53.4	78.6
WASHINGTON				
980-981,987	Seattle	102.5	104.8	103.5
982	Everett	102.1	99.6	101.0
983-984	Tacoma	102.3	100.0	101.3
985	Olympia	101.0	100.0	100.6
986	Vancouver	103.1	93.2	98.8
988	Wenatchee	102.0	87.1	95.5
989	Yakima	102.5	94.4	99.0
990-992	Spokane	102.5	82.7	93.8
993	Richland	102.2	89.6	96.7
994	Clarkston	100.6	80.7	92.0
WEST VIRGINIA				
247-248	Bluefield	97.1	91.7	94.7
249	Lewisburg	98.4	91.4	95.4
250-253	Charleston	99.0	94.1	96.9
254	Martinsburg	98.3	86.1	93.0
255-257	Huntington	99.8	95.2	97.8
258-259	Beckley	96.9	92.0	94.8
260	Wheeling	99.8	93.8	97.2
261	Parkersburg	98.9	91.4	95.6
262	Buckhannon	98.2	93.6	96.2
263-264	Clarksburg	98.8	93.5	96.5
265	Morgantown	98.8	92.6	96.1
266	Gassaway	98.2	92.8	95.8
267	Romney	98.2	90.4	94.8
268	Petersburg	98.0	90.1	94.6
WISCONSIN				
530,532	Milwaukee	100.9	107.9	104.0
531	Kenosha	100.2	102.9	101.4
534	Racine	99.7	102.6	101.0
535	Beloit	99.6	101.9	100.6
537	Madison	99.5	99.8	99.6

STATE/ZIP	CITY	MAT.	INST.	TOTAL
WISCONSIN (CONT'D)				
538	Lancaster	97.4	93.5	95.7
539	Portage	96.1	97.9	96.9
540	New Richmond	96.2	95.9	96.1
541-543	Green Bay	100.5	94.1	97.7
544	Wausau	95.7	94.4	95.2
545	Rhinelander	98.7	92.4	95.9
546	La Crosse	97.2	95.0	96.3
547	Eau Claire	98.7	96.4	97.7
548	Superior	96.1	97.6	96.7
549	Oshkosh	96.3	92.4	94.6
WYOMING				
820	Cheyenne	100.1	65.7	85.2
821	Yellowstone Nat'l Park	97.7	65.2	83.5
822	Wheatland	98.7	64.6	83.9
823	Rawlins	100.2	67.8	86.1
824	Worland	98.3	64.1	83.4
825	Riverton	99.3	64.9	84.3
826	Casper	100.4	60.7	83.1
827	Newcastle	98.2	67.8	84.9
828	Sheridan	100.9	65.2	85.4
829-831	Rock Springs	101.8	65.4	85.9
CANADIAN FACTORS (reflect Canadian currency)				
ALBERTA				
	Calgary	124.7	95.4	111.9
	Edmonton	125.0	95.4	112.1
	Fort McMurray	124.1	94.2	111.1
	Lethbridge	118.7	93.7	107.8
	Lloydminster	113.9	90.3	103.6
	Medicine Hat	114.0	89.6	103.4
	Red Deer	114.5	89.6	103.6
BRITISH COLUMBIA				
	Kamloops	114.9	92.4	105.1
	Prince George	115.9	91.7	105.3
	Vancouver	123.3	90.0	108.8
	Victoria	115.9	88.1	103.8
MANITOBA				
	Brandon	118.7	77.3	100.7
	Portage la Prairie	113.9	75.6	97.2
	Winnipeg	123.6	70.8	100.6
NEW BRUNSWICK				
	Bathurst	112.1	68.6	93.1
	Dalhousie	112.1	68.8	93.2
	Fredericton	118.1	74.0	98.9
	Moncton	112.3	71.4	94.5
	Newcastle	112.1	69.3	93.5
	St. John	114.6	78.3	98.8
NEWFOUNDLAND				
	Corner Brook	122.2	69.0	99.0
	St Johns	122.0	80.4	103.9
NORTHWEST TERRITORIES				
	Yellowknife	128.8	85.8	110.1
NOVA SCOTIA				
	Bridgewater	113.4	76.1	97.2
	Dartmouth	119.7	76.1	100.7
	Halifax	119.9	80.6	102.8
	New Glasgow	117.7	76.1	99.6
	Sydney	115.6	76.1	98.4
	Truro	113.0	76.1	96.9
	Yarmouth	117.6	76.1	99.6
ONTARIO				
	Barrie	117.3	94.9	107.5
	Brantford	115.5	98.8	108.2
	Cornwall	115.5	95.1	106.7
	Hamilton	119.3	95.3	108.8
	Kingston	116.6	95.2	107.3
	Kitchener	111.6	92.0	103.1
	London	119.5	91.9	107.5
	North Bay	120.5	92.9	108.5
	Oshawa	114.6	96.1	106.5
	Ottawa	119.2	92.5	107.5
	Owen Sound	117.4	93.0	106.8
	Peterborough	115.5	94.9	106.5
	Sarnia	115.3	99.5	108.4

STATE/ZIP	CITY	MAT.	INST.	TOTAL
ONTARIO (CONT'D)				
	Sault Ste Marie	110.7	93.1	103.0
	St. Catharines	109.7	93.0	102.4
	Sudbury	109.9	91.7	101.9
	Thunder Bay	110.9	91.7	102.6
	Timmins	115.6	92.9	105.7
	Toronto	120.2	99.0	111.0
	Windsor	110.2	91.7	102.2
PRINCE EDWARD ISLAND				
	Charlottetown	120.5	64.0	95.9
	Summerside	119.5	64.1	95.4
QUEBEC				
	Cap-de-la-Madeleine	112.8	89.4	102.6
	Charlesbourg	112.8	89.4	102.6
	Chicoutimi	111.9	93.1	103.7
	Gatineau	112.5	89.2	102.3
	Granby	112.8	89.1	102.4
	Hull	112.6	89.2	102.4
	Joliette	113.0	89.4	102.7
	Laval	112.4	89.2	102.3
	Montreal	119.2	93.4	107.9
	Quebec	117.5	93.8	107.2
	Rimouski	112.3	93.1	103.9
	Rouyn-Noranda	112.5	89.2	102.3
	Saint Hyacinthe	111.9	89.2	102.0
	Sherbrooke	112.8	89.2	102.5
	Sorel	113.0	89.4	102.7
	St Jerome	112.5	89.2	102.3
	Trois Rivieres	117.9	89.4	105.5
SASKATCHEWAN				
	Moose Jaw	111.2	70.8	93.6
	Prince Albert	110.3	69.0	92.3
	Regina	120.3	88.4	106.4
	Saskatoon	111.3	88.4	101.3
YUKON				
	Whitehorse	126.3	69.8	101.7

R011105-05 Tips for Accurate Estimating

1. Use pre-printed or columnar forms for orderly sequence of dimensions and locations and for recording telephone quotations.
2. Use only the front side of each paper or form except for certain pre-printed summary forms.
3. Be consistent in listing dimensions: For example, length x width x height. This helps in rechecking to ensure that, the total length of partitions is appropriate for the building area.
4. Use printed (rather than measured) dimensions where given.
5. Add up multiple printed dimensions for a single entry where possible.
6. Measure all other dimensions carefully.
7. Use each set of dimensions to calculate multiple related quantities.
8. Convert foot and inch measurements to decimal feet when listing. Memorize decimal equivalents to .01 parts of a foot (1/8″ equals approximately .01′).
9. Do not "round off" quantities until the final summary.
10. Mark drawings with different colors as items are taken off.
11. Keep similar items together, different items separate.
12. Identify location and drawing numbers to aid in future checking for completeness.
13. Measure or list everything on the drawings or mentioned in the specifications.
14. It may be necessary to list items not called for to make the job complete.
15. Be alert for: Notes on plans such as N.T.S. (not to scale); changes in scale throughout the drawings; reduced size drawings; discrepancies between the specifications and the drawings.
16. Develop a consistent pattern of performing an estimate. For example:
    a. Start the quantity takeoff at the lower floor and move to the next higher floor.
    b. Proceed from the main section of the building to the wings.
    c. Proceed from south to north or vice versa, clockwise or counterclockwise.
    d. Take off floor plan quantities first, elevations next, then detail drawings.
17. List all gross dimensions that can be either used again for different quantities, or used as a rough check of other quantities for verification (exterior perimeter, gross floor area, individual floor areas, etc.).
18. Utilize design symmetry or repetition (repetitive floors, repetitive wings, symmetrical design around a center line, similar room layouts, etc.). Note: Extreme caution is needed here so as not to omit or duplicate an area.
19. Do not convert units until the final total is obtained. For instance, when estimating concrete work, keep all units to the nearest cubic foot, then summarize and convert to cubic yards.
20. When figuring alternatives, it is best to total all items involved in the basic system, then total all items involved in the alternates. Therefore you work with positive numbers in all cases. When adds and deducts are used, it is often confusing whether to add or subtract a portion of an item; especially on a complicated or involved alternate.

R011105-40 Weather Data and Design Conditions

City	Latitude (1)		Winter Temperatures (1)			Winter Degree Days (2)	Summer (Design Dry Bulb) Temperatures and Relative Humidity		
	0	1′	Med. of Annual Extremes	99%	97½%		1%	2½%	5%
UNITED STATES									
Albuquerque, NM	35	0	5.1	12	16	4,400	96/61	94/61	92/61
Atlanta, GA	33	4	11.9	17	22	3,000	94/74	92/74	90/73
Baltimore, MD	39	2	7	14	17	4,600	94/75	91/75	89/74
Birmingham, AL	33	3	13	17	21	2,600	96/74	94/75	92/74
Bismarck, ND	46	5	-32	-23	-19	8,800	95/68	91/68	88/67
Boise, ID	43	3	1	3	10	5,800	96/65	94/64	91/64
Boston, MA	42	2	-1	6	9	5,600	91/73	88/71	85/70
Burlington, VT	44	3	-17	-12	-7	8,200	88/72	85/70	82/69
Charleston, WV	38	2	3	7	11	4,400	92/74	90/73	87/72
Charlotte, NC	35	1	13	18	22	3,200	95/74	93/74	91/74
Casper, WY	42	5	-21	-11	-5	7,400	92/58	90/57	87/57
Chicago, IL	41	5	-8	-3	2	6,600	94/75	91/74	88/73
Cincinnati, OH	39	1	0	1	6	4,400	92/73	90/72	88/72
Cleveland, OH	41	2	-3	1	5	6,400	91/73	88/72	86/71
Columbia, SC	34	0	16	20	24	2,400	97/76	95/75	93/75
Dallas, TX	32	5	14	18	22	2,400	102/75	100/75	97/75
Denver, CO	39	5	-10	-5	1	6,200	93/59	91/59	89/59
Des Moines, IA	41	3	-14	-10	-5	6,600	94/75	91/74	88/73
Detroit, MI	42	2	-3	3	6	6,200	91/73	88/72	86/71
Great Falls, MT	47	3	-25	-21	-15	7,800	91/60	88/60	85/59
Hartford, CT	41	5	-4	3	7	6,200	91/74	88/73	85/72
Houston, TX	29	5	24	28	33	1,400	97/77	95/77	93/77
Indianapolis, IN	39	4	-7	-2	2	5,600	92/74	90/74	87/73
Jackson, MS	32	2	16	21	25	2,200	97/76	95/76	93/76
Kansas City, MO	39	1	-4	2	6	4,800	99/75	96/74	93/74
Las Vegas, NV	36	1	18	25	28	2,800	108/66	106/65	104/65
Lexington, KY	38	0	-1	3	8	4,600	93/73	91/73	88/72
Little Rock, AR	34	4	11	15	20	3,200	99/76	96/77	94/77
Los Angeles, CA	34	0	36	41	43	2,000	93/70	89/70	86/69
Memphis, TN	35	0	10	13	18	3,200	98/77	95/76	93/76
Miami, FL	25	5	39	44	47	200	91/77	90/77	89/77
Milwaukee, WI	43	0	-11	-8	-4	7,600	90/74	87/73	84/71
Minneapolis, MN	44	5	-22	-16	-12	8,400	92/75	89/73	86/71
New Orleans, LA	30	0	28	29	33	1,400	93/78	92/77	90/77
New York, NY	40	5	6	11	15	5,000	92/74	89/73	87/72
Norfolk, VA	36	5	15	20	22	3,400	93/77	91/76	89/76
Oklahoma City, OK	35	2	4	9	13	3,200	100/74	97/74	95/73
Omaha, NE	41	2	-13	-8	-3	6,600	94/76	91/75	88/74
Philadelphia, PA	39	5	6	10	14	4,400	93/75	90/74	87/72
Phoenix, AZ	33	3	27	31	34	1,800	109/71	107/71	105/71
Pittsburgh, PA	40	3	-1	3	7	6,000	91/72	88/71	86/70
Portland, ME	43	4	-10	-6	-1	7,600	87/72	84/71	81/69
Portland, OR	45	4	18	17	23	4,600	89/68	85/67	81/65
Portsmouth, NH	43	1	-8	-2	2	7,200	89/73	85/71	83/70
Providence, RI	41	4	-1	5	9	6,000	89/73	86/72	83/70
Rochester, NY	43	1	-5	1	5	6,800	91/73	88/71	85/70
Salt Lake City, UT	40	5	0	3	8	6,000	97/62	95/62	92/61
San Francisco, CA	37	5	36	38	40	3,000	74/63	71/62	69/61
Seattle, WA	47	4	22	22	27	5,200	85/68	82/66	78/65
Sioux Falls, SD	43	4	-21	-15	-11	7,800	94/73	91/72	88/71
St. Louis, MO	38	4	-3	3	8	5,000	98/75	94/75	91/75
Tampa, FL	28	0	32	36	40	680	92/77	91/77	90/76
Trenton, NJ	40	1	4	11	14	5,000	91/75	88/74	85/73
Washington, DC	38	5	7	14	17	4,200	93/75	91/74	89/74
Wichita, KS	37	4	-3	3	7	4,600	101/72	98/73	96/73
Wilmington, DE	39	4	5	10	14	5,000	92/74	89/74	87/73
ALASKA									
Anchorage	61	1	-29	-23	-18	10,800	71/59	68/58	66/56
Fairbanks	64	5	-59	-51	-47	14,280	82/62	78/60	75/59
CANADA									
Edmonton, Alta.	53	3	-30	-29	-25	11,000	85/66	82/65	79/63
Halifax, N.S.	44	4	-4	1	5	8,000	79/66	76/65	74/64
Montreal, Que.	45	3	-20	-16	-10	9,000	88/73	85/72	83/71
Saskatoon, Sask.	52	1	-35	-35	-31	11,000	89/68	86/66	83/65
St. John's, N.F.	47	4	1	3	7	8,600	77/66	75/65	73/64
Saint John, N.B.	45	2	-15	-12	-8	8,200	80/67	77/65	75/64
Toronto, Ont.	43	4	-10	-5	-1	7,000	90/73	87/72	85/71
Vancouver, B.C.	49	1	13	15	19	6,000	79/67	77/66	74/65
Winnipeg, Man.	49	5	-31	-30	-27	10,800	89/73	86/71	84/70

(1) Handbook of Fundamentals, ASHRAE, Inc., NY 1989
(2) Local Climatological Annual Survey, USDC Env. Science Services Administration, Asheville, NC

R011105-50 Metric Conversion Factors

Description: This table is primarily for converting customary U.S. units in the left hand column to SI metric units in the right hand column. In addition, conversion factors for some commonly encountered Canadian and non-SI metric units are included.

	If You Know		Multiply By		To Find
Length	Inches	x	25.4[a]	=	Millimeters
	Feet	x	0.3048[a]	=	Meters
	Yards	x	0.9144[a]	=	Meters
	Miles (statute)	x	1.609	=	Kilometers
Area	Square inches	x	645.2	=	Square millimeters
	Square feet	x	0.0929	=	Square meters
	Square yards	x	0.8361	=	Square meters
Volume (Capacity)	Cubic inches	x	16,387	=	Cubic millimeters
	Cubic feet	x	0.02832	=	Cubic meters
	Cubic yards	x	0.7646	=	Cubic meters
	Gallons (U.S. liquids)[b]	x	0.003785	=	Cubic meters[c]
	Gallons (Canadian liquid)[b]	x	0.004546	=	Cubic meters[c]
	Ounces (U.S. liquid)[b]	x	29.57	=	Milliliters[c, d]
	Quarts (U.S. liquid)[b]	x	0.9464	=	Liters[c, d]
	Gallons (U.S. liquid)[b]	x	3.785	=	Liters[c, d]
Force	Kilograms force[d]	x	9.807	=	Newtons
	Pounds force	x	4.448	=	Newtons
	Pounds force	x	0.4536	=	Kilograms force[d]
	Kips	x	4448	=	Newtons
	Kips	x	453.6	=	Kilograms force[d]
Pressure, Stress, Strength (Force per unit area)	Kilograms force per square centimeter[d]	x	0.09807	=	Megapascals
	Pounds force per square inch (psi)	x	0.006895	=	Megapascals
	Kips per square inch	x	6.895	=	Megapascals
	Pounds force per square inch (psi)	x	0.07031	=	Kilograms force per square centimeter[d]
	Pounds force per square foot	x	47.88	=	Pascals
	Pounds force per square foot	x	4.882	=	Kilograms force per square meter[d]
Bending Moment Or Torque	Inch-pounds force	x	0.01152	=	Meter-kilograms force[d]
	Inch-pounds force	x	0.1130	=	Newton-meters
	Foot-pounds force	x	0.1383	=	Meter-kilograms force[d]
	Foot-pounds force	x	1.356	=	Newton-meters
	Meter-kilograms force[d]	x	9.807	=	Newton-meters
Mass	Ounces (avoirdupois)	x	28.35	=	Grams
	Pounds (avoirdupois)	x	0.4536	=	Kilograms
	Tons (metric)	x	1000	=	Kilograms
	Tons, short (2000 pounds)	x	907.2	=	Kilograms
	Tons, short (2000 pounds)	x	0.9072	=	Megagrams[e]
Mass per Unit Volume	Pounds mass per cubic foot	x	16.02	=	Kilograms per cubic meter
	Pounds mass per cubic yard	x	0.5933	=	Kilograms per cubic meter
	Pounds mass per gallon (U.S. liquid)[b]	x	119.8	=	Kilograms per cubic meter
	Pounds mass per gallon (Canadian liquid)[b]	x	99.78	=	Kilograms per cubic meter
Temperature	Degrees Fahrenheit	(F-32)/1.8		=	Degrees Celsius
	Degrees Fahrenheit	(F+459.67)/1.8		=	Degrees Kelvin
	Degrees Celsius	C+273.15		=	Degrees Kelvin

[a]The factor given is exact
[b]One U.S. gallon = 0.8327 Canadian gallon
[c]1 liter = 1000 milliliters = 1000 cubic centimeters
1 cubic decimeter = 0.001 cubic meter
[d]Metric but not SI unit
[e]Called "tonne" in England and "metric ton" in other metric countries

R011105-60 Weights and Measures

Measures of Length
1 Mile = 1760 Yards = 5280 Feet
1 Yard = 3 Feet = 36 inches
1 Foot = 12 Inches
1 Mil = 0.001 Inch
1 Fathom = 2 Yards = 6 Feet
1 Rod = 5.5 Yards = 16.5 Feet
1 Hand = 4 Inches
1 Span = 9 Inches
1 Micro-inch = One Millionth Inch or 0.000001 Inch
1 Micron = One Millionth Meter + 0.00003937 Inch

Surveyor's Measure
1 Mile = 8 Furlongs = 80 Chains
1 Furlong = 10 Chains = 220 Yards
1 Chain = 4 Rods = 22 Yards = 66 Feet = 100 Links
1 Link = 7.92 Inches

Square Measure
1 Square Mile = 640 Acres = 6400 Square Chains
1 Acre = 10 Square Chains = 4840 Square Yards = 43,560 Sq. Ft.
1 Square Chain = 16 Square Rods = 484 Square Yards = 4356 Sq. Ft.
1 Square Rod = 30.25 Square Yards = 272.25 Square Feet = 625 Square Lines
1 Square Yard = 9 Square Feet
1 Square Foot = 144 Square Inches
An Acre equals a Square 208.7 Feet per Side

Cubic Measure
1 Cubic Yard = 27 Cubic Feet
1 Cubic Foot = 1728 Cubic Inches
1 Cord of Wood = 4 x 4 x 8 Feet = 128 Cubic Feet
1 Perch of Masonry = 16½ x 1½ x 1 Foot = 24.75 Cubic Feet

Avoirdupois or Commercial Weight
1 Gross or Long Ton = 2240 Pounds
1 Net or Short ton = 2000 Pounds
1 Pound = 16 Ounces = 7000 Grains
1 Ounce = 16 Drachms = 437.5 Grains
1 Stone = 14 Pounds

Shipping Measure
For Measuring Internal Capacity of a Vessel:
1 Register Ton = 100 Cubic Feet

For Measurement of Cargo:
Approximately 40 Cubic Feet of Merchandise is considered a Shipping Ton, unless that bulk would weigh more than 2000 Pounds, in which case Freight Charge may be based upon weight.

40 Cubic Feet = 32.143 U.S. Bushels = 31.16 Imp. Bushels

Liquid Measure
1 Imperial Gallon = 1.2009 U.S. Gallon = 277.42 Cu. In.
1 Cubic Foot = 7.48 U.S. Gallons

R012909-80 Sales Tax by State

State sales tax on materials is tabulated below (5 states have no sales tax). Many states allow local jurisdictions, such as a county or city, to levy additional sales tax.

Some projects may be sales tax exempt, particularly those constructed with public funds.

State	Tax (%)	State	Tax (%)	State	Tax (%)	State	Tax (%)
Alabama	4	Illinois	6.25	Montana	0	Rhode Island	7
Alaska	0	Indiana	7	Nebraska	5.5	South Carolina	6
Arizona	5.6	Iowa	6	Nevada	6.85	South Dakota	4
Arkansas	6	Kansas	6.3	New Hampshire	0	Tennessee	7
California	7.25	Kentucky	6	New Jersey	7	Texas	6.25
Colorado	2.9	Louisiana	4	New Mexico	5.13	Utah	5.95
Connecticut	6.35	Maine	5	New York	4	Vermont	6
Delaware	0	Maryland	6	North Carolina	4.75	Virginia	5
District of Columbia	6	Massachusetts	6.25	North Dakota	5	Washington	6.5
Florida	6	Michigan	6	Ohio	5.5	West Virginia	6
Georgia	4	Minnesota	6.88	Oklahoma	4.5	Wisconsin	5
Hawaii	4	Mississippi	7	Oregon	0	Wyoming	4
Idaho	6	Missouri	4.23	Pennsylvania	6	Average	5.06 %

Sales Tax by Province (Canada)

GST - a value-added tax, which the government imposes on most goods and services provided in or imported into Canada. PST - a retail sales tax, which three of the provinces impose on the price of most goods and some services. QST - a value-added tax, similar to the federal GST, which Quebec imposes. HST - Five provinces have combined their retail sales tax with the federal GST into one harmonized tax.

Province	PST (%)	QST (%)	GST(%)	HST(%)
Alberta	0	0	5	0
British Columbia	7	0	5	0
Manitoba	7	0	5	0
New Brunswick	0	0	0	13
Newfoundland	0	0	0	13
Northwest Territories	0	0	5	0
Nova Scotia	0	0	0	15
Ontario	0	0	0	13
Prince Edward Island	0	0	5	14
Quebec	0	9.975	5	0
Saskatchewan	5	0	5	0
Yukon	0	0	5	0

General Requirements — R0129 Payment Procedures

R012909-85 Unemployment Taxes and Social Security Taxes

State unemployment tax rates vary not only from state to state, but also with the experience rating of the contractor. The federal unemployment tax rate is 6.0% of the first $7,000 of wages. This is reduced by a credit of up to 5.4% for timely payment to the state. The minimum federal unemployment tax is 0.6% after all credits.

Social security (FICA) for 2014 is estimated at time of publication to be 7.65% of wages up to $115,800.

R012909-90 Overtime

One way to improve the completion date of a project or eliminate negative float from a schedule is to compress activity duration times. This can be achieved by increasing the crew size or working overtime with the proposed crew.

To determine the costs of working overtime to compress activity duration times, consider the following examples. Below is an overtime efficiency and cost chart based on a five, six, or seven day week with an eight through twelve hour day. Payroll percentage increases for time and one half and double time are shown for the various working days.

Days per Week	Hours per Day	Production Efficiency					Payroll Cost Factors	
		1st Week	2nd Week	3rd Week	4th Week	Average 4 Weeks	@ 1-1/2 Times	@ 2 Times
5	8	100%	100%	100%	100%	100 %	100 %	100 %
	9	100	100	95	90	96.25	105.6	111.1
	10	100	95	90	85	91.25	110.0	120.0
	11	95	90	75	65	81.25	113.6	127.3
	12	90	85	70	60	76.25	116.7	133.3
6	8	100	100	95	90	96.25	108.3	116.7
	9	100	95	90	85	92.50	113.0	125.9
	10	95	90	85	80	87.50	116.7	133.3
	11	95	85	70	65	78.75	119.7	139.4
	12	90	80	65	60	73.75	122.2	144.4
7	8	100	95	85	75	88.75	114.3	128.6
	9	95	90	80	70	83.75	118.3	136.5
	10	90	85	75	65	78.75	121.4	142.9
	11	85	80	65	60	72.50	124.0	148.1
	12	85	75	60	55	68.75	126.2	152.4

General Requirements — R0131 Project Management & Coordination

R013113-40 Builder's Risk Insurance

Builder's Risk Insurance is insurance on a building during construction. Premiums are paid by the owner or the contractor. Blasting, collapse and underground insurance would raise total insurance costs above those listed. Floater policy for materials delivered to the job runs $.75 to $1.25 per $100 value. Contractor equipment insurance runs $.50 to $1.50 per $100 value. Insurance for miscellaneous tools to $1,500 value runs from $3.00 to $7.50 per $100 value.

Tabulated below are New England Builder's Risk insurance rates in dollars per $100 value for $1,000 deductible. For $25,000 deductible, rates can be reduced 13% to 34%. On contracts over $1,000,000, rates may be lower than those tabulated. Policies are written annually for the total completed value in place. For "all risk" insurance (excluding flood, earthquake and certain other perils) add $.025 to total rates below.

Coverage	Frame Construction (Class 1)		Brick Construction (Class 4)		Fire Resistive (Class 6)	
	Range	Average	Range	Average	Range	Average
Fire Insurance	$.350 to $ .850	$.600	$.158 to $.189	$.174	$.052 to $.080	$.070
Extended Coverage	.115 to .200	.158	.080 to .105	.101	.081 to .105	.100
Vandalism	.012 to .016	.014	.008 to .011	.011	.008 to .011	.010
Total Annual Rate	$.477 to $1.066	$.772	$.246 to $.305	$.286	$.141 to $.196	$.180

R013113-50 General Contractor's Overhead

There are two distinct types of overhead on a construction project: Project Overhead and Main Office Overhead. Project Overhead includes those costs at a construction site not directly associated with the installation of construction materials. Examples of Project Overhead costs include the following:

1. Superintendent
2. Construction office and storage trailers
3. Temporary sanitary facilities
4. Temporary utilities
5. Security fencing
6. Photographs
7. Clean up
8. Performance and payment bonds

The above Project Overhead items are also referred to as General Requirements and therefore are estimated in Division 1. Division 1 is the first division listed in the CSI MasterFormat but it is usually the last division estimated. The sum of the costs in Divisions 1 through 49 is referred to as the sum of the direct costs.

All construction projects also include indirect costs. The primary components of indirect costs are the contractor's Main Office Overhead and profit. The amount of the Main Office Overhead expense varies depending on the the following:

1. Owner's compensation
2. Project managers and estimator's wages
3. Clerical support wages
4. Office rent and utilities
5. Corporate legal and accounting costs
6. Advertising
7. Automobile expenses
8. Association dues
9. Travel and entertainment expenses

These costs are usually calculated as a percentage of annual sales volume. This percentage can range from 35% for a small contractor doing less than $500,000 to 5% for a large contractor with sales in excess of $100 million.

R013113-60 Workers' Compensation Insurance Rates by Trade

The table below tabulates the national averages for workers' compensation insurance rates by trade and type of building. The average "Insurance Rate" is multiplied by the "% of Building Cost" for each trade. This produces the "Workers' Compensation" cost by % of total labor cost, to be added for each trade by building type to determine the weighted average workers' compensation rate for the building types analyzed.

Trade	Insurance Rate (% Labor Cost)		% of Building Cost			Workers' Compensation		
	Range	Average	Office Bldgs.	Schools & Apts.	Mfg.	Office Bldgs.	Schools & Apts.	Mfg.
Excavation, Grading, etc.	3.9 % to 18.7%	9.7%	4.8%	4.9%	4.5%	0.47%	0.48%	0.44%
Piles & Foundations	5.8 to 37	14.7	7.1	5.2	8.7	1.04	0.76	1.28
Concrete	4.6 to 38.6	13.1	5.0	14.8	3.7	0.66	1.94	0.48
Masonry	5.1 to 35	13.5	6.9	7.5	1.9	0.93	1.01	0.26
Structural Steel	5.8 to 108.2	34.6	10.7	3.9	17.6	3.70	1.35	6.09
Miscellaneous & Ornamental Metals	5.3 to 31.2	12.0	2.8	4.0	3.6	0.34	0.48	0.43
Carpentry & Millwork	5.5 to 34.9	15.4	3.7	4.0	0.5	0.57	0.62	0.08
Metal or Composition Siding	5.8 to 68.3	18.2	2.3	0.3	4.3	0.42	0.05	0.78
Roofing	5.8 to 93.3	31.3	2.3	2.6	3.1	0.72	0.81	0.97
Doors & Hardware	4.2 to 34.8	11.1	0.9	1.4	0.4	0.10	0.16	0.04
Sash & Glazing	5.3 to 36.8	13.1	3.5	4.0	1.0	0.46	0.52	0.13
Lath & Plaster	3.7 to 35	11.6	3.3	6.9	0.8	0.38	0.80	0.09
Tile, Marble & Floors	3.3 to 27.1	8.5	2.6	3.0	0.5	0.22	0.26	0.04
Acoustical Ceilings	3.4 to 36.7	8.1	2.4	0.2	0.3	0.19	0.02	0.02
Painting	5 to 33	11.5	1.5	1.6	1.6	0.17	0.18	0.18
Interior Partitions	5.5 to 34.9	15.4	3.9	4.3	4.4	0.60	0.66	0.68
Miscellaneous Items	2.5 to 150	14.4	5.2	3.7	9.7	0.75	0.53	1.40
Elevators	1.6 to 14.4	5.7	2.1	1.1	2.2	0.12	0.06	0.13
Sprinklers	3.2 to 18.3	7.2	0.5	—	2.0	0.04	—	0.14
Plumbing	2.6 to 17.5	7.1	4.9	7.2	5.2	0.35	0.51	0.37
Heat., Vent., Air Conditioning	4 to 18.8	8.8	13.5	11.0	12.9	1.19	0.97	1.14
Electrical	2.8 to 13.2	5.8	10.1	8.4	11.1	0.59	0.49	0.64
Total	1.6 % to 150%	—	100.0%	100.0%	100.0%	14.01%	12.66%	15.81%
		Overall Weighted Average	14.16%					

Workers' Compensation Insurance Rates by States

The table below lists the weighted average workers' compensation base rate for each state with a factor comparing this with the national average of 14%.

State	Weighted Average	Factor	State	Weighted Average	Factor	State	Weighted Average	Factor
Alabama	20.3%	145	Kentucky	14.9%	106	North Dakota	9.6%	69
Alaska	14.3	102	Louisiana	19.3	138	Ohio	8.6	61
Arizona	13.1	94	Maine	12.7	91	Oklahoma	12.9	92
Arkansas	8.6	61	Maryland	14.6	104	Oregon	12.3	88
California	26.8	191	Massachusetts	12.0	86	Pennsylvania	18.1	129
Colorado	7.5	54	Michigan	16.9	121	Rhode Island	12.6	90
Connecticut	24.6	176	Minnesota	23.5	168	South Carolina	19.6	140
Delaware	12.4	89	Mississippi	13.8	99	South Dakota	15.5	111
District of Columbia	11.5	82	Missouri	15.4	110	Tennessee	12.5	89
Florida	11.1	79	Montana	8.9	64	Texas	9.5	68
Georgia	29.2	209	Nebraska	18.3	131	Utah	9.1	65
Hawaii	8.4	60	Nevada	9.0	64	Vermont	13.6	97
Idaho	10.7	76	New Hampshire	20.8	149	Virginia	9.9	71
Illinois	28.1	201	New Jersey	15.0	107	Washington	10.9	78
Indiana	5.7	41	New Mexico	15.0	107	West Virginia	8.3	59
Iowa	15.1	108	New York	17.9	128	Wisconsin	13.6	97
Kansas	8.7	62	North Carolina	17.4	124	Wyoming	6.3	45
			Weighted Average for U.S. is	14.2% of payroll = 100%				

Rates in the following table are the base or manual costs per $100 of payroll for workers' compensation in each state. Rates are usually applied to straight time wages only and not to premium time wages and bonuses.

The weighted average skilled worker rate for 35 trades is 14%. For bidding purposes, apply the full value of workers' compensation directly to total labor costs, or if labor is 38%, materials 42% and overhead and profit 20% of total cost, carry 38/80 × 14% =6.6% of cost (before overhead and profit) into overhead. Rates vary not only from state to state but also with the experience rating of the contractor.

Rates are the most current available at the time of publication.

R013113-60 Workers' Compensation Insurance Rates by Trade and State (cont.)

State	Carpentry — 3 stories or less 5651	Carpentry — interior cab. work 5437	Carpentry — general 5403	Concrete Work — NOC 5213	Concrete Work — flat (flr., sdwk.) 5221	Electrical Wiring — inside 5190	Excavation — earth NOC 6217	Excavation — rock 6217	Glaziers 5462	Insulation Work 5479	Lathing 5443	Masonry 5022	Painting & Decorating 5474	Pile Driving 6003	Plastering 5480	Plumbing 5183
AL	34.03	17.01	24.59	13.79	10.37	8.87	16.42	16.42	19.75	19.84	9.24	23.79	15.91	18.57	15.88	8.73
AK	17.14	9.44	11.58	10.41	9.44	5.63	8.23	8.23	32.03	14.36	11.73	10.86	10.79	21.68	11.31	6.20
AZ	21.40	9.27	20.28	10.31	6.93	6.08	7.07	7.07	10.80	14.99	5.47	13.01	12.78	10.80	8.64	6.84
AR	12.80	4.77	9.18	6.63	5.16	4.03	6.82	6.82	8.14	7.67	3.99	8.30	7.08	8.10	13.00	4.55
CA	34.87	34.83	34.86	16.55	16.55	13.24	18.65	18.65	24.16	16.48	17.83	25.13	23.59	23.18	35.04	17.48
CO	8.95	4.93	6.07	6.68	4.75	2.92	5.49	5.49	6.00	7.27	3.95	8.29	6.56	6.51	5.69	4.16
CT	30.95	18.48	31.07	26.49	14.43	7.89	16.02	16.02	20.24	20.60	9.92	31.53	18.60	20.17	18.75	11.28
DE	12.46	12.46	9.93	10.07	8.80	4.53	7.70	7.70	10.90	9.93	6.60	11.53	13.39	15.03	10.90	6.60
DC	8.92	7.25	6.94	12.65	7.38	4.69	8.02	8.02	9.61	6.63	9.83	9.32	5.82	17.36	8.00	7.80
FL	11.32	8.62	11.92	12.51	6.20	6.17	7.30	7.30	10.56	10.45	5.53	10.92	10.23	17.84	9.68	5.63
GA	68.34	20.32	28.40	21.36	15.09	12.05	17.32	17.32	19.39	18.80	11.94	35.11	33.04	26.44	19.65	12.76
HI	8.51	6.92	12.69	7.05	4.02	4.47	5.40	5.40	8.22	5.20	6.26	7.01	6.19	7.32	8.33	5.31
ID	14.63	6.84	12.43	11.48	5.28	4.43	7.25	7.25	8.81	9.06	4.46	10.55	9.38	7.65	8.43	5.10
IL	33.34	21.75	28.57	38.59	16.83	10.03	14.53	14.53	25.55	21.41	14.32	28.94	18.57	24.53	32.84	12.51
IN	8.36	4.21	6.98	4.56	4.15	2.83	3.92	3.92	5.29	6.15	3.85	5.14	4.97	7.23	3.71	2.56
IA	15.46	11.35	19.21	14.85	11.16	5.20	11.01	11.01	15.76	9.89	6.48	14.30	10.13	11.38	21.73	7.80
KS	14.92	7.39	9.74	8.07	5.70	4.46	5.55	5.55	7.54	8.17	4.09	8.38	7.75	7.56	6.20	5.23
KY	18.75	13.00	20.00	8.50	6.50	6.00	14.00	14.00	22.12	11.15	7.60	7.50	9.00	16.03	12.93	5.00
LA	31.05	16.86	25.02	13.35	12.37	7.85	14.11	14.11	19.19	15.06	7.99	17.31	17.64	18.85	14.84	7.10
ME	12.24	8.87	16.18	16.43	8.70	4.74	12.34	12.34	14.82	12.79	5.09	10.88	12.44	9.00	9.99	5.90
MD	11.40	15.89	15.00	19.22	6.00	5.23	8.11	8.11	13.77	12.42	6.15	10.11	6.40	15.19	9.17	6.65
MA	8.68	5.23	9.61	18.85	6.24	2.84	4.35	4.35	9.58	7.78	5.27	10.55	5.09	12.92	4.68	3.50
MI	18.41	10.45	18.41	18.27	12.58	5.82	13.28	13.28	11.67	10.92	10.45	18.82	13.49	37.00	10.88	7.18
MN	20.27	20.82	34.13	11.58	13.15	5.83	11.00	11.00	36.85	18.37	11.55	15.59	17.55	27.58	11.58	8.09
MS	22.76	9.99	13.34	9.60	6.67	6.18	9.49	9.49	14.94	10.15	5.94	13.12	11.01	13.14	10.43	8.17
MO	19.66	11.10	12.81	12.38	10.34	6.15	9.96	9.96	9.47	11.55	7.44	15.13	11.42	13.38	12.15	9.00
MT	9.67	6.59	10.80	6.22	5.67	3.71	7.41	7.41	7.50	8.67	4.15	7.62	8.22	7.95	6.80	4.79
NE	21.80	13.15	21.48	18.23	11.20	8.03	17.38	17.38	12.05	21.53	8.35	18.73	13.00	13.93	13.68	8.73
NV	10.41	6.82	12.39	10.83	5.26	4.33	8.75	8.75	6.55	7.55	3.41	6.63	7.10	8.43	6.21	6.76
NH	24.24	11.06	19.63	28.40	16.47	5.63	12.61	12.61	14.12	12.63	7.30	16.19	20.93	18.53	11.20	9.49
NJ	18.69	13.18	18.69	15.51	11.69	5.13	10.83	10.83	11.21	12.63	15.13	18.71	11.28	15.79	15.13	6.96
NM	25.01	9.56	18.92	11.73	9.43	6.05	7.22	7.22	10.95	10.56	6.11	15.74	12.91	12.66	14.20	6.70
NY	12.54	10.74	19.18	23.71	17.15	7.85	12.16	12.16	18.72	10.88	12.86	21.24	14.99	20.13	10.02	10.83
NC	30.12	9.54	15.76	16.11	9.31	10.56	15.41	15.41	14.28	13.03	8.11	13.05	14.08	16.61	14.12	10.73
ND	9.54	4.31	9.54	5.43	5.43	3.75	4.13	4.13	10.50	10.50	9.93	6.52	6.00	10.59	9.93	6.11
OH	7.73	4.31	5.52	4.67	4.92	3.43	4.79	4.79	8.83	11.38	36.65	6.93	8.15	6.99	7.80	3.52
OK	18.82	10.32	12.49	13.50	7.66	6.82	9.48	9.48	11.79	11.31	5.13	14.63	10.66	11.24	14.01	6.82
OR	22.30	8.52	11.92	10.51	10.42	5.35	9.08	9.08	12.64	10.88	6.09	14.37	11.81	11.44	10.81	6.35
PA	19.22	19.22	15.60	19.54	13.58	7.39	11.11	11.11	14.34	15.60	14.34	16.04	17.13	19.44	14.34	9.19
RI	9.84	11.55	13.75	12.56	9.68	4.58	8.16	8.16	10.61	12.48	5.17	10.07	8.82	19.30	8.69	6.75
SC	29.32	13.80	17.10	12.84	8.21	9.82	12.01	12.01	12.92	14.13	9.57	15.39	16.10	15.67	14.47	9.40
SD	20.86	17.96	20.35	16.10	10.49	5.08	9.75	9.75	10.84	14.71	6.95	14.65	11.22	12.63	10.35	10.02
TN	22.71	10.14	11.54	10.75	7.08	5.49	10.05	10.05	11.07	10.34	6.60	11.69	8.27	12.08	11.04	5.17
TX	10.12	7.33	10.12	8.08	5.99	5.98	7.60	7.60	8.25	9.27	4.15	8.69	7.28	10.26	7.28	5.27
UT	14.72	6.59	7.50	7.34	6.57	3.56	6.79	6.79	7.45	7.15	6.41	8.96	7.73	6.72	5.56	4.17
VT	15.69	10.21	13.77	13.60	11.57	4.37	11.83	11.83	12.40	15.34	5.79	11.92	8.78	10.43	9.35	7.38
VA	12.38	8.14	8.89	10.51	5.76	4.50	7.40	7.40	8.02	7.27	5.45	7.98	8.93	8.31	6.69	5.07
WA	9.18	9.18	9.18	8.55	8.55	3.18	6.68	6.68	13.99	6.79	9.18	13.26	12.26	21.68	10.43	4.33
WV	14.85	5.54	8.34	7.68	4.51	3.53	6.46	6.46	6.74	6.29	4.33	8.39	6.03	8.09	6.54	3.52
WI	12.52	13.14	17.06	11.98	7.59	5.08	7.84	7.84	13.88	11.22	4.00	13.69	12.43	28.24	10.16	5.95
WY	5.82	5.82	5.82	5.82	5.82	5.82	5.82	5.82	5.82	5.82	5.82	5.82	5.82	5.82	5.82	5.82
AVG.	18.19	11.07	15.38	13.15	8.92	5.83	9.69	9.69	13.15	11.67	8.12	13.49	11.54	14.65	11.55	7.08

R013113-60 Workers' Compensation Insurance Rates by Trade and State (cont.)

State	Roofing 5551	Sheet Metal Work (HVAC) 5538	Steel Erection — door & sash 5102	Steel Erection — inter., ornam. 5102	Steel Erection — structure 5040	Steel Erection — NOC 5057	Tile Work — (interior ceramic) 5348	Waterproofing 9014	Wrecking 5701	Boiler- making 3726	Mill- wrights 3724	Structural Painters 5037	Elevator Constructor 5160	Truck Drivers 7219	Sprinkler Installer 5188
AL	54.71	10.34	12.52	12.52	40.75	21.56	12.35	9.27	40.75	11.93	12.83	55.64	10.09	23.31	8.40
AK	25.49	8.05	7.82	7.82	29.28	26.40	5.33	5.77	29.28	8.56	8.24	31.37	4.06	11.75	5.13
AZ	29.22	10.31	13.61	13.61	26.91	12.89	4.50	6.21	26.91	9.07	7.41	59.76	5.09	8.01	5.48
AR	16.03	4.74	8.34	8.34	15.40	11.28	6.69	2.58	15.40	4.94	5.76	25.69	2.73	10.96	5.72
CA	71.13	18.78	18.41	18.41	35.28	19.69	14.59	23.59	19.69	8.98	13.58	35.28	5.07	27.13	18.32
CO	15.60	4.16	5.36	5.36	23.64	10.59	4.10	3.23	18.40	4.37	3.87	22.88	2.78	7.22	3.85
CT	58.74	12.93	21.24	21.24	65.13	29.83	15.63	6.43	65.13	19.44	14.12	66.07	11.07	22.67	10.49
DE	26.71	5.55	13.16	13.16	23.27	13.16	8.00	11.53	23.27	6.17	6.19	23.27	6.19	11.13	6.60
DC	17.05	6.45	23.33	23.33	26.02	7.91	6.08	4.38	26.02	15.31	7.62	35.16	9.65	14.34	4.00
FL	18.17	7.69	9.91	9.91	23.04	12.59	5.63	5.26	23.04	9.07	6.05	39.44	3.54	9.24	6.41
GA	93.33	16.61	22.17	22.17	60.89	25.88	13.50	11.64	60.89	19.54	14.79	72.60	7.53	19.39	15.17
HI	17.24	4.93	7.60	7.60	21.77	6.88	6.52	5.40	21.77	4.82	7.71	24.83	2.28	10.54	3.49
ID	24.73	7.46	12.31	12.31	25.42	10.44	6.15	4.84	25.42	5.30	5.61	29.98	3.26	9.46	4.65
IL	47.55	15.21	31.16	31.16	84.68	19.36	27.09	7.88	84.68	23.23	15.10	101.23	14.42	19.73	16.65
IN	11.57	4.32	5.32	5.32	11.59	4.80	3.26	2.77	11.59	3.85	3.70	13.20	1.58	5.75	3.20
IA	28.01	8.53	9.15	9.15	52.79	14.30	9.38	6.15	25.61	14.19	8.59	87.69	6.14	13.60	7.75
KS	16.26	5.64	6.71	6.71	18.97	11.93	5.23	4.63	11.96	6.01	7.60	26.96	3.22	7.88	5.30
KY	34.00	12.00	11.72	11.72	36.00	11.02	16.73	3.50	36.00	12.15	8.35	150.00	9.56	10.00	10.32
LA	43.53	14.84	20.14	20.14	44.26	12.79	9.99	7.91	44.20	15.48	10.94	55.57	6.11	23.25	11.22
ME	22.72	6.77	8.92	8.92	37.18	12.21	5.76	4.64	37.18	8.08	6.58	54.29	4.39	8.15	6.70
MD	29.05	11.90	11.92	11.92	59.02	17.92	6.49	5.84	32.91	14.58	9.95	43.60	4.46	16.02	11.88
MA	30.99	5.72	6.89	6.89	54.08	33.00	5.81	2.48	23.75	15.42	5.98	23.51	4.21	8.28	4.13
MI	41.18	9.83	12.28	12.28	37.00	11.27	11.73	6.46	37.00	9.41	8.67	11.27	5.13	9.09	8.21
MN	67.71	17.30	9.82	9.82	108.16	7.70	14.69	7.95	7.70	10.92	12.24	36.60	6.41	11.39	7.62
MS	31.61	7.46	15.08	15.08	38.45	7.53	8.45	5.85	38.45	7.77	9.97	38.74	4.78	52.03	7.53
MO	36.73	8.64	12.33	12.33	47.55	24.29	10.41	5.92	47.55	14.45	9.66	60.85	6.20	12.44	8.43
MT	29.42	4.80	6.98	6.98	18.72	5.89	5.47	4.64	5.89	5.03	5.91	28.63	3.01	6.06	3.67
NE	35.08	12.88	14.80	14.80	59.03	14.13	8.68	6.73	67.83	16.33	11.35	66.18	7.13	15.98	10.30
NV	16.59	7.34	7.06	7.06	19.32	12.20	4.42	4.23	12.20	4.79	6.15	28.67	4.30	10.30	5.47
NH	36.82	8.93	14.28	14.28	95.03	28.57	12.35	7.17	95.03	8.91	9.04	54.18	6.29	36.82	10.68
NJ	40.10	7.22	12.87	12.87	17.25	13.36	9.55	6.63	16.86	4.18	9.20	23.37	7.00	15.75	5.89
NM	35.07	9.63	11.79	11.79	45.52	16.76	7.73	7.53	45.52	8.00	8.00	42.38	8.41	14.60	7.75
NY	35.96	13.56	19.25	19.25	35.26	19.08	10.35	7.49	18.10	21.41	10.22	42.40	12.20	14.37	6.45
NC	35.39	11.87	14.43	14.43	45.05	22.54	7.62	6.10	45.05	18.41	9.71	78.32	8.90	17.84	10.60
ND	17.63	6.11	10.59	10.59	10.90	10.59	10.50	17.63	8.18	3.33	10.50	10.59	3.33	8.79	6.11
OH	14.89	4.00	6.58	6.58	11.33	5.84	4.85	4.58	5.84	2.76	4.52	24.79	2.13	10.02	3.68
OK	22.44	9.62	10.98	10.98	29.21	15.15	6.69	6.41	32.94	9.25	6.21	52.38	3.88	11.26	6.40
OR	27.22	6.85	7.08	7.08	22.04	13.44	9.10	6.73	22.04	9.10	9.28	11.81	3.70	14.41	6.02
PA	38.35	8.92	19.10	19.10	30.01	19.10	11.69	16.04	30.01	7.95	8.85	30.01	8.85	16.69	9.19
RI	23.59	8.48	7.37	7.37	35.11	24.46	5.33	4.98	22.15	11.33	6.32	39.60	4.85	10.13	5.85
SC	46.45	10.88	15.24	15.24	31.33	24.73	7.98	5.42	31.33	11.24	10.59	66.07	8.64	15.07	10.12
SD	34.39	7.29	14.74	14.74	33.52	23.05	6.19	5.30	33.52	22.67	11.38	66.74	6.69	10.62	8.14
TN	29.78	8.13	13.42	13.42	16.07	18.88	7.46	4.09	16.07	8.10	7.46	40.34	4.92	10.23	6.10
TX	16.22	11.69	7.60	7.60	22.99	9.48	4.68	5.57	7.13	4.30	5.95	7.34	3.34	12.75	5.27
UT	23.93	5.78	5.71	5.71	24.61	11.81	5.75	4.31	20.14	5.91	4.77	21.36	3.71	7.05	4.16
VT	27.01	7.30	12.01	12.01	34.39	18.99	7.01	7.90	34.39	8.67	7.28	48.29	6.07	13.40	6.27
VA	25.97	5.47	7.46	7.46	29.39	10.80	5.76	3.30	29.39	4.92	7.38	27.90	5.39	9.03	5.30
WA	21.14	4.47	8.09	8.09	8.09	8.09	7.86	21.14	8.09	2.77	6.15	12.26	2.46	8.52	4.70
WV	19.62	5.62	7.13	7.13	19.39	9.55	4.86	3.11	20.70	5.62	6.64	35.86	4.23	8.93	3.76
WI	28.04	9.43	12.08	12.08	20.72	16.12	13.78	5.07	20.72	9.68	8.34	24.08	4.11	10.15	4.64
WY	5.82	5.82	5.82	5.82	5.82	5.82	5.82	5.82	5.82	5.82	5.82	5.82	5.82	5.82	5.82
AVG.	31.29	8.79	11.95	11.95	34.64	15.21	8.54	6.86	29.21	9.87	8.39	41.47	5.67	13.67	7.24

R013113-60 Workers' Compensation (cont.) (Canada in Canadian dollars)

Province		Alberta	British Columbia	Manitoba	Ontario	New Brunswick	Newfndld. & Labrador	Northwest Territories	Nova Scotia	Prince Edward Island	Quebec	Saskat-chewan	Yukon
Carpentry—3 stories or less	Rate	5.71	3.27	3.98	4.55	3.17	7.81	4.99	6.85	6.42	12.03	3.43	6.63
	Code	42143	721028	40102	723	238130	4226	4-41	4226	403	80110	B1226	202
Carpentry—interior cab. work	Rate	1.89	2.80	3.98	4.55	3.17	4.07	4.99	5.87	3.94	12.03	3.43	6.63
	Code	42133	721021	40102	723	238350	4279	4-41	4274	402	80110	B1127	202
CARPENTRY—general	Rate	5.71	3.27	3.98	4.55	3.17	4.07	4.99	6.85	6.42	12.03	3.43	6.63
	Code	42143	721028	40102	723	238130	4299	4-41	4226	403	80110	B1202	202
CONCRETE WORK—NOC	Rate	3.88	3.57	6.50	18.31	3.17	7.81	4.99	5.75	3.94	1.92	2.99	6.63
	Code	42104	721010	40110	748	238110	4224	4-41	4224	402	80100	B1314	203
CONCRETE WORK—flat	Rate	3.88	3.57	6.50	18.31	3.17	7.81	4.99	5.75	3.94	11.92	2.99	6.63
(flr. sidewalk)	Code	42104	721010	40110	748	238110	4224	4-41	4224	402	80100	B1314	203
ELECTRICAL Wiring—inside	Rate	1.56	1.18	1.65	3.69	1.28	2.29	2.44	1.97	3.94	5.04	1.83	3.89
	Code	42124	721019	40203	704	238210	4261	4-46	4261	402	80170	B1105	206
EXCAVATION—earth NOC	Rate	1.86	3.79	4.00	5.29	3.17	2.93	3.88	3.89	3.94	6.40	2.19	3.89
	Code	40604	721031	40706	711	238190	4214	4-43	4214	402	80030	R1106	207
EXCAVATION—rock	Rate	1.86	3.79	4.00	5.29	3.17	2.93	3.88	3.89	3.94	6.40	2.19	3.89
	Code	40604	721031	40706	711	238190	4214	4-43	4214	402	80030	R1106	207
GLAZIERS	Rate	2.23	3.79	3.98	10.25	3.17	5.10	4.99	5.22	3.94	15.02	2.99	3.89
	Code	42121	715020	40109	751	238150	4233	4-41	4233	402	80150	B1304	212
INSULATION WORK	Rate	1.62	9.87	3.98	10.25	3.17	5.10	4.99	5.22	3.94	12.03	3.43	6.63
	Code	42184	721029	40102	751	238310	4234	4-41	4234	402	80110	B1207	202
LATHING	Rate	4.20	7.84	3.98	4.55	3.17	4.07	4.99	5.87	3.94	12.03	2.99	6.63
	Code	42135	721042	40102	723	238390	4279	4-41	4271	402	80110	B1316	202
MASONRY	Rate	3.88	3.30	3.98	12.70	3.17	5.10	4.99	5.22	3.94	11.92	2.99	6.63
	Code	42102	721037	40102	741	238140	4231	4-41	4231	402	80100	B1318	202
PAINTING & DECORATING	Rate	3.16	3.62	3.11	7.51	3.17	4.07	4.99	5.87	3.94	12.03	3.43	6.63
	Code	42111	721041	40105	719	238320	4275	4-41	4275	402	80110	B1201	202
PILE DRIVING	Rate	3.88	3.42	4.00	7.03	3.17	2.93	3.88	5.75	6.42	6.40	2.99	6.63
	Code	42159	722004	40706	732	238190	4129	4-43	4221	403	80030	B1310	202
PLASTERING	Rate	4.20	7.84	4.45	7.51	3.17	4.07	4.99	5.87	3.94	12.03	3.43	6.63
	Code	42135	721042	40108	719	238390	4271	4-41	4271	402	80110	B1221	202
PLUMBING	Rate	1.56	2.55	2.78	4.16	1.28	2.47	2.44	2.81	3.94	5.89	1.83	3.89
	Code	42122	721043	40204	707	238220	4241	4-46	4241	401	80160	B1101	214
ROOFING	Rate	4.71	5.42	7.22	14.80	3.17	7.81	4.99	9.43	3.94	18.52	2.99	6.63
	Code	42118	721036	40403	728	238160	4236	4-41	4236	402	80130	B1320	202
SHEET METAL WORK (HVAC)	Rate	1.56	2.55	7.22	4.16	1.28	2.47	2.44	2.81	3.94	5.89	1.83	3.89
	Code	42117	721043	40402	707	238220	4244	4-46	4244	401	80160	B1107	208
STEEL ERECTION—door & sash	Rate	1.62	10.62	6.82	18.31	3.17	7.81	4.99	5.22	3.94	16.70	2.99	6.63
	Code	42106	722005	40502	748	238120	4227	-37	4227	402	80080	B1322	202
STEEL ERECTION—inter., ornam.	Rate	1.62	10.62	6.82	18.31	3.17	7.81	4.99	5.22	3.94	16.70	2.99	6.63
	Code	42106	722005	40502	748	238120	4227	4-41	4227	402	80080	B1322	202
STEEL ERECTION—structure	Rate	1.62	10.62	6.82	18.31	3.17	7.81	4.99	5.22	3.94	16.70	2.99	6.63
	Code	42106	722005	40502	748	238120	4227	4-41	4227	402	80080	B1322	202
STEEL ERECTION—NOC	Rate	1.62	10.62	6.82	18.31	3.17	7.81	4.99	5.22	3.94	16.70	2.99	6.63
	Code	42106	722005	40502	748	238120	4227	4-41	4227	402	80080	B1322	202
TILE WORK—inter. (ceramic)	Rate	3.02	3.65	1.86	7.51	3.17	4.07	4.99	5.87	3.94	12.03	2.99	6.63
	Code	42113	721054	40103	719	238340	4276	4-41	4276	402	80110	B1301	202
WATERPROOFING	Rate	3.16	3.62	3.98	4.55	3.17	4.07	4.99	5.22	3.94	18.52	3.13	6.63
	Code	42139	721016	40102	723	238190	4299	4-41	4239	402	80130	B1217	202
WRECKING	Rate	1.86	5.64	6.90	18.31	3.17	2.93	3.88	3.89	3.94	12.03	2.19	6.63
	Code	40604	721005	40106	748	238190	4211	4-43	4211	402	80110	R1125	202

General Requirements — R0131 Project Management & Coordination

R013113-80 Performance Bond

This table shows the cost of a Performance Bond for a construction job scheduled to be completed in 12 months. Add 1% of the premium cost per month for jobs requiring more than 12 months to complete. The rates are "standard" rates offered to contractors that the bonding company considers financially sound and capable of doing the work. Preferred rates are offered by some bonding companies based upon financial strength of the contractor. Actual rates vary from contractor to contractor and from bonding company to bonding company. Contractors should prequalify through a bonding agency before submitting a bid on a contract that requires a bond.

Contract Amount	Building Construction Class B Projects	Highways & Bridges: Class A New Construction	Highways & Bridges: Class A-1 Highway Resurfacing
First $ 100,000 bid	$25.00 per M	$15.00 per M	$9.40 per M
Next 400,000 bid	$ 2,500 plus $15.00 per M	$ 1,500 plus $10.00 per M	$ 940 plus $7.20 per M
Next 2,000,000 bid	8,500 plus 10.00 per M	5,500 plus 7.00 per M	3,820 plus 5.00 per M
Next 2,500,000 bid	28,500 plus 7.50 per M	19,500 plus 5.50 per M	15,820 plus 4.50 per M
Next 2,500,000 bid	47,250 plus 7.00 per M	33,250 plus 5.00 per M	28,320 plus 4.50 per M
Over 7,500,000 bid	64,750 plus 6.00 per M	45,750 plus 4.50 per M	39,570 plus 4.00 per M

General Requirements — R0154 Construction Aids

R015423-10 Steel Tubular Scaffolding

On new construction, tubular scaffolding is efficient up to 60′ high or five stories. Above this it is usually better to use a hung scaffolding if construction permits. Swing scaffolding operations may interfere with tenants. In this case, the tubular is more practical at all heights.

In repairing or cleaning the front of an existing building the cost of tubular scaffolding per S.F. of building front increases as the height increases above the first tier. The first tier cost is relatively high due to leveling and alignment.

The minimum efficient crew for erecting and dismantling is three workers. They can set up and remove 18 frame sections per day up to 5 stories high. For 6 to 12 stories high, a crew of four is most efficient. Use two or more on top and two on the bottom for handing up or hoisting. They can also set up and remove 18 frame sections per day. At 7′ horizontal spacing, this will run about 800 S.F. per day of erecting and dismantling. Time for placing and removing planks must be added to the above. A crew of three can place and remove 72 planks per day up to 5 stories. For over 5 stories, a crew of four can place and remove 80 planks per day.

The table below shows the number of pieces required to erect tubular steel scaffolding for 1000 S.F. of building frontage. This area is made up of a scaffolding system that is 12 frames (11 bays) long by 2 frames high.

For jobs under twenty-five frames, add 50% to rental cost. Rental rates will be lower for jobs over three months duration. Large quantities for long periods can reduce rental rates by 20%.

Description of Component	Number of Pieces for 1000 S.F. of Building Front	Unit
5′ Wide Standard Frame, 6′-4″ High	24	Ea.
Leveling Jack & Plate	24	↓
Cross Brace	44	↓
Side Arm Bracket, 21″	12	↓
Guardrail Post	12	↓
Guardrail, 7′ section	22	↓
Stairway Section	2	↓
Stairway Starter Bar	1	↓
Stairway Inside Handrail	2	↓
Stairway Outside Handrail	2	↓
Walk-Thru Frame Guardrail	2	↓

Scaffolding is often used as falsework over 15′ high during construction of cast-in-place concrete beams and slabs. Two foot wide scaffolding is generally used for heavy beam construction. The span between frames depends upon the load to be carried with a maximum span of 5′.

Heavy duty shoring frames with a capacity of 10,000#/leg can be spaced up to 10′ O.C. depending upon form support design and loading.

Scaffolding used as horizontal shoring requires less than half the material required with conventional shoring.

On new construction, erection is done by carpenters.

Rolling towers supporting horizontal shores can reduce labor and speed the job. For maintenance work, catwalks with spans up to 70′ can be supported by the rolling towers.

R015433-10 Contractor Equipment

Rental Rates shown elsewhere in the book pertain to late model high quality machines in excellent working condition, rented from equipment dealers. Rental rates from contractors may be substantially lower than the rental rates from equipment dealers depending upon economic conditions; for older, less productive machines, reduce rates by a maximum of 15%. Any overtime must be added to the base rates. For shift work, rates are lower. Usual rule of thumb is 150% of one shift rate for two shifts; 200% for three shifts.

For periods of less than one week, operated equipment is usually more economical to rent than renting bare equipment and hiring an operator.

Costs to move equipment to a job site (mobilization) or from a job site (demobilization) are not included in rental rates, nor in any Equipment costs on any Unit Price line items or crew listings. These costs can be found elsewhere. If a piece of equipment is already at a job site, it is not appropriate to utilize mob/demob costs in an estimate again.

Rental rates vary throughout the country with larger cities generally having lower rates. Lease plans for new equipment are available for periods in excess of six months with a percentage of payments applying toward purchase.

Rental rates can also be treated as reimbursement costs for contractor-owned equipment. Owned equipment costs include depreciation, loan payments, interest, taxes, insurance, storage, and major repairs.

Monthly rental rates vary from 2% to 5% of the cost of the equipment depending on the anticipated life of the equipment and its wearing parts. Weekly rates are about 1/3 the monthly rates and daily rental rates about 1/3 the weekly rate.

The hourly operating costs for each piece of equipment include costs to the user such as fuel, oil, lubrication, normal expendables for the equipment, and a percentage of mechanic's wages chargeable to maintenance. The hourly operating costs listed do not include the operator's wages.

The daily cost for equipment used in the standard crews is figured by dividing the weekly rate by five, then adding eight times the hourly operating cost to give the total daily equipment cost, not including the operator. This figure is in the right hand column of the Equipment listings under Equipment Cost/Day.

Pile Driving rates shown for pile hammer and extractor do not include leads, crane, boiler or compressor. Vibratory pile driving requires an added field specialist during set-up and pile driving operation for the electric model. The hydraulic model requires a field specialist for set-up only. Up to 125 reuses of sheet piling are possible using vibratory drivers. For normal conditions, crane capacity for hammer type and size are as follows.

Crane Capacity	Hammer Type and Size		
	Air or Steam	Diesel	Vibratory
25 ton	to 8,750 ft.-lb.		70 H.P.
40 ton	15,000 ft.-lb.	to 32,000 ft.-lb.	170 H.P.
60 ton	25,000 ft.-lb.		300 H.P.
100 ton		112,000 ft.-lb.	

Cranes should be specified for the job by size, building and site characteristics, availability, performance characteristics, and duration of time required.

Backhoes & Shovels rent for about the same as equivalent size cranes but maintenance and operating expense is higher. Crane operators rate must be adjusted for high boom heights. Average adjustments: for 150′ boom add 2% per hour; over 185′, add 4% per hour; over 210′, add 6% per hour; over 250′, add 8% per hour and over 295′, add 12% per hour.

Tower Cranes of the climbing or static type have jibs from 50′ to 200′ and capacities at maximum reach range from 4,000 to 14,000 pounds. Lifting capacities increase up to maximum load as the hook radius decreases.

Typical rental rates, based on purchase price are about 2% to 3% per month.

Erection and dismantling runs between 500 and 2000 labor hours. Climbing operation takes 10 labor hours per 20′ climb. Crane dead time is about 5 hours per 40′ climb. If crane is bolted to side of the building add cost of ties and extra mast sections. Climbing cranes have from 80′ to 180′ of mast while static cranes have 80′ to 800′ of mast.

Truck Cranes can be converted to tower cranes by using tower attachments. Mast heights over 400′ have been used.

A single 100′ high material **Hoist and Tower** can be erected and dismantled in about 400 labor hours; a double 100′ high hoist and tower in about 600 labor hours. Erection times for additional heights are 3 and 4 labor hours per vertical foot respectively up to 150′, and 4 to 5 labor hours per vertical foot over 150′ high. A 40′ high portable Buck hoist takes about 160 labor hours to erect and dismantle. Additional heights take 2 labor hours per vertical foot to 80′ and 3 labor hours per vertical foot for the next 100′. Most material hoists do not meet local code requirements for carrying personnel.

A 150′ high **Personnel Hoist** requires about 500 to 800 labor hours to erect and dismantle. Budget erection time at 5 labor hours per vertical foot for all trades. Local code requirements or labor scarcity requiring overtime can add up to 50% to any of the above erection costs.

Earthmoving Equipment: The selection of earthmoving equipment depends upon the type and quantity of material, moisture content, haul distance, haul road, time available, and equipment available. Short haul cut and fill operations may require dozers only, while another operation may require excavators, a fleet of trucks, and spreading and compaction equipment. Stockpiled material and granular material are easily excavated with front end loaders. Scrapers are most economically used with hauls between 300′ and 1-1/2 miles if adequate haul roads can be maintained. Shovels are often used for blasted rock and any material where a vertical face of 8′ or more can be excavated. Special conditions may dictate the use of draglines, clamshells, or backhoes. Spreading and compaction equipment must be matched to the soil characteristics, the compaction required and the rate the fill is being supplied.

R015433-15 Heavy Lifting

Hydraulic Climbing Jacks

The use of hydraulic heavy lift systems is an alternative to conventional type crane equipment. The lifting, lowering, pushing, or pulling mechanism is a hydraulic climbing jack moving on a square steel jackrod from 1-5/8″ to 4″ square, or a steel cable. The jackrod or cable can be vertical or horizontal, stationary or movable, depending on the individual application. When the jackrod is stationary, the climbing jack will climb the rod and push or pull the load along with itself. When the climbing jack is stationary, the jackrod is movable with the load attached to the end and the climbing jack will lift or lower the jackrod with the attached load. The heavy lift system is normally operated by a single control lever located at the hydraulic pump.

The system is flexible in that one or more climbing jacks can be applied wherever a load support point is required, and the rate of lift synchronized.

Economic benefits have been demonstrated on projects such as: erection of ground assembled roofs and floors, complete bridge spans, girders and trusses, towers, chimney liners and steel vessels, storage tanks, and heavy machinery. Other uses are raising and lowering offshore work platforms, caissons, tunnel sections and pipelines.

R024119-10 Demolition Defined

Whole Building Demolition - Demolition of the whole building with no concern for any particular building element, component, or material type being demolished. This type of demolition is accomplished with large pieces of construction equipment that break up the structure, load it into trucks and haul it to a disposal site, but disposal or dump fees are not included. Demolition of below-grade foundation elements, such as footings, foundation walls, grade beams, slabs on grade, etc., is not included. Certain mechanical equipment containing flammable liquids or ozone-depleting refrigerants, electric lighting elements, communication equipment components, and other building elements may contain hazardous waste, and must be removed, either selectively or carefully, as hazardous waste before the building can be demolished.

Foundation Demolition - Demolition of below-grade foundation footings, foundation walls, grade beams, and slabs on grade. This type of demolition is accomplished by hand or pneumatic hand tools, and does not include saw cutting, or handling, loading, hauling, or disposal of the debris.

Gutting - Removal of building interior finishes and electrical/mechanical systems down to the load-bearing and sub-floor elements of the rough building frame, with no concern for any particular building element, component, or material type being demolished. This type of demolition is accomplished by hand or pneumatic hand tools, and includes loading into trucks, but not hauling, disposal or dump fees, scaffolding, or shoring. Certain mechanical equipment containing flammable liquids or ozone-depleting refrigerants, electric lighting elements, communication equipment components, and other building elements may contain hazardous waste, and must be removed, either selectively or carefully, as hazardous waste, before the building is gutted.

Selective Demolition - Demolition of a selected building element, component, or finish, with some concern for surrounding or adjacent elements, components, or finishes (see the first Subdivision (s) at the beginning of appropriate Divisions). This type of demolition is accomplished by hand or pneumatic hand tools, and does not include handling, loading, storing, hauling, or disposal of the debris, scaffolding, or shoring. "Gutting" methods may be used in order to save time, but damage that is caused to surrounding or adjacent elements, components, or finishes may have to be repaired at a later time.

Careful Removal - Removal of a piece of service equipment, building element or component, or material type, with great concern for both the removed item and surrounding or adjacent elements, components or finishes. The purpose of careful removal may be to protect the removed item for later re-use, preserve a higher salvage value of the removed item, or replace an item while taking care to protect surrounding or adjacent elements, components, connections, or finishes from cosmetic and/or structural damage. An approximation of the time required to perform this type of removal is 1/3 to 1/2 the time it would take to install a new item of like kind. This type of removal is accomplished by hand or pneumatic hand tools, and does not include loading, hauling, or storing the removed item, scaffolding, shoring, or lifting equipment.

Cutout Demolition - Demolition of a small quantity of floor, wall, roof, or other assembly, with concern for the appearance and structural integrity of the surrounding materials. This type of demolition is accomplished by hand or pneumatic hand tools, and does not include saw cutting, handling, loading, hauling, or disposal of debris, scaffolding, or shoring.

Rubbish Handling - Work activities that involve handling, loading or hauling of debris. Generally, the cost of rubbish handling must be added to the cost of all types of demolition, with the exception of whole building demolition.

Minor Site Demolition - Demolition of site elements outside the footprint of a building. This type of demolition is accomplished by hand or pneumatic hand tools, or with larger pieces of construction equipment, and may include loading a removed item onto a truck (check the Crew for equipment used). It does not include saw cutting, hauling or disposal of debris, and, sometimes, handling or loading.

R024119-20 Dumpsters

Dumpster rental costs on construction sites are presented in two ways.

The cost per week rental includes the delivery of the dumpster; its pulling or emptying once per week, and its final removal. The assumption is made that the dumpster contractor could choose to empty a dumpster by simply bringing in an empty unit and removing the full one. These costs also include the disposal of the materials in the dumpster.

The Alternate Pricing can be used when actual planned conditions are not approximated by the weekly numbers. For example, these lines can be used when a dumpster is needed for 4 weeks and will need to be emptied 2 or 3 times per week. Conversely the Alternate Pricing lines can be used when a dumpster will be rented for several weeks or months but needs to be emptied only a few times over this period.

R024119-30 Rubbish Handling Chutes

To correctly estimate the cost of rubbish handling chute systems, the individual components must be priced separately. First choose the size of the system; a 30-inch diameter chute is quite common, but the sizes range from 18 to 36 inches in diameter. The 30-inch chute comes in a standard weight and two thinner weights. The thinner weight chutes are sometimes chosen for cost savings, but they are more easily damaged.

There are several types of major chute pieces that make up the chute system. The first component to consider is the top chute section (top intake hopper) where the material is dropped into the chute at the highest point. After determining the top chute, the intermediate chute pieces called the regular chute sections are priced. Next, the number of chute control door sections (intermediate intake hoppers) must be determined. In the more complex systems, a chute control door section is provided at each floor level. The last major component to consider is bolt down frames; these are usually provided at every other floor level.

There are a number of accessories to consider for safe operation and control. There are covers for the top chute and the chute door sections. The top chute can have a trough that allows for better loading of the chute. For the safest operation, a chute warning light system can be added that will warn the other chute intake locations not to load while another is being used. There are dust control devices that spray a water mist to keep down the dust as the debris is loaded into a Dumpster. There are special breakaway cords that are used to prevent damage to the chute if the Dumpster is removed without disconnecting from the chute. There are chute liners that can be installed to protect the chute structure from physical damage from rough abrasive materials. Warning signs can be posted at each floor level that is provided with a chute control door section.

In summary, a complete rubbish handling chute system will include one top section, several intermediate regular sections, several intermediate control door (intake hopper) sections and bolt down frames at every other floor level starting with the top floor. If so desired, the system can also include covers and a light warning system for a safer operation. The bottom of the chute should always be above the Dumpster and should be tied off with a breakaway cord to the Dumpster.

R026510-20 Underground Storage Tank Removal

Underground Storage Tank Removal can be divided into two categories: Non-Leaking and Leaking. Prior to removing an underground storage tank, tests should be made, with the proper authorities present, to determine whether a tank has been leaking or the surrounding soil has been contaminated.

To safely remove Liquid Underground Storage Tanks:

1. Excavate to the top of the tank.
2. Disconnect all piping.
3. Open all tank vents and access ports.
4. Remove all liquids and/or sludge.
5. Purge the tank with an inert gas.
6. Provide access to the inside of the tank and clean out the interior using proper personal protective equipment (PPE).
7. Excavate soil surrounding the tank using proper PPE for on-site personnel.
8. Pull and properly dispose of the tank.
9. Clean up the site of all contaminated material.
10. Install new tanks or close the excavation.

R031113-10 Wall Form Materials

Aluminum Forms

Approximate weight is 3 lbs. per S.F.C.A. Standard widths are available from 4″ to 36″ with 36″ most common. Standard lengths of 2′, 4′, 6′ to 8′ are available. Forms are lightweight and fewer ties are needed with the wider widths. The form face is either smooth or textured.

Metal Framed Plywood Forms

Manufacturers claim over 75 reuses of plywood and over 300 reuses of steel frames. Many specials such as corners, fillers, pilasters, etc. are available. Monthly rental is generally about 15% of purchase price for first month and 9% per month thereafter with 90% of rental applied to purchase for the first month and decreasing percentages thereafter. Aluminum framed forms cost 25% to 30% more than steel framed.

After the first month, extra days may be prorated from the monthly charge. Rental rates do not include ties, accessories, cleaning, loss of hardware or freight in and out. Approximate weight is 5 lbs. per S.F. for steel; 3 lbs. per S.F. for aluminum.

Forms can be rented with option to buy.

Plywood Forms, Job Fabricated

There are two types of plywood used for concrete forms.

1. Exterior plyform which is completely waterproof. This is face oiled to facilitate stripping. Ten reuses can be expected with this type with 25 reuses possible.
2. An overlaid type consists of a resin fiber fused to exterior plyform. No oiling is required except to facilitate cleaning. This is available in both high density (HDO) and medium density overlaid (MDO). Using HDO, 50 reuses can be expected with 200 possible.

Plyform is available in 5/8″ and 3/4″ thickness. High density overlaid is available in 3/8″, 1/2″, 5/8″ and 3/4″ thickness.

5/8″ thick is sufficient for most building forms, while 3/4″ is best on heavy construction.

Plywood Forms, Modular, Prefabricated

There are many plywood forming systems without frames. Most of these are manufactured from 1-1/8″ (HDO) plywood and have some hardware attached. These are used principally for foundation walls 8′ or less high. With care and maintenance, 100 reuses can be attained with decreasing quality of surface finish.

Steel Forms

Approximate weight is 6-1/2 lbs. per S.F.C.A. including accessories. Standard widths are available from 2″ to 24″, with 24″ most common. Standard lengths are from 2′ to 8′, with 4′ the most common. Forms are easily ganged into modular units.

Forms are usually leased for 15% of the purchase price per month prorated daily over 30 days.

Rental may be applied to sale price, and usually rental forms are bought. With careful handling and cleaning 200 to 400 reuses are possible.

Straight wall gang forms up to 12′ x 20′ or 8′ x 30′ can be fabricated. These crane handled forms usually lease for approx. 9% per month.

Individual job analysis is available from the manufacturer at no charge.

R031113-30 Slipforms

The slipform method of forming may be used for forming circular silo and multi-celled storage bin type structures over 30′ high, and building core shear walls over eight stories high. The shear walls, usually enclose elevator shafts, stairwells, mechanical spaces, and toilet rooms. Reuse of the form on duplicate structures will reduce the height necessary and spread the cost of building the form. Slipform systems can be used to cast chimneys, towers, piers, dams, underground shafts or other structures capable of being extruded.

Slipforms are usually 4′ high and are raised semi-continuously by jacks climbing on rods which are embedded in the concrete. The jacks are powered by a hydraulic, pneumatic, or electric source and are available in 3, 6, and 22 ton capacities. Interior work decks and exterior scaffolds must be provided for placing inserts, embedded items, reinforcing steel, and concrete. Scaffolds below the form for finishers may be required. The interior work decks are often used as roof slab forms on silos and bin work. Form raising rates will range from 6″ to 20″ per hour for silos; 6″ to 30″ per hour for buildings; and 6″ to 48″ per hour for shaft work.

Reinforcing bars and stressing strands are usually hoisted by crane or gin pole, and the concrete material can be hoisted by crane, winch-powered skip, or pumps. The slipform system is operated on a continuous 24-hour day when a monolithic structure is desired. For least cost, the system is operated only during normal working hours.

Placing concrete will range from 0.5 to 1.5 labor-hours per C.Y. Bucks, blockouts, keyways, weldplates, etc. are extra.

R031113-40 Forms for Reinforced Concrete

Design Economy

Avoid many sizes in proportioning beams and columns.

From story to story avoid changing column dimensions. Gain strength by adding steel or using a richer mix. If a change in size of column is necessary, vary one dimension only to minimize form alterations. Keep beams and columns the same width.

From floor to floor in a multi-story building vary beam depth, not width, as that will leave slab panel form unchanged. It is cheaper to vary the strength of a beam from floor to floor by means of steel area than by 2″ changes in either width or depth.

Cost Factors

Material includes the cost of lumber, cost of rent for metal pans or forms if used, nails, form ties, form oil, bolts and accessories.

Labor includes the cost of carpenters to make up, erect, remove and repair, plus common labor to clean and move. Having carpenters remove forms minimizes repairs.

Improper alignment and condition of forms will increase finishing cost. When forms are heavily oiled, concrete surfaces must be neutralized before finishing. Special curing compounds will cause spillages to spall off in first frost. Gang forming methods will reduce costs on large projects.

Materials Used

Boards are seldom used unless their architectural finish is required. Generally, steel, fiberglass and plywood are used for contact surfaces. Labor on plywood is 10% less than with boards. The plywood is backed up with 2 x 4′s at 12″ to 32″ O.C. Walers are generally 2 - 2 x 4′s. Column forms are held together with steel yokes or bands. Shoring is with adjustable shoring or scaffolding for high ceilings.

Reuse

Floor and column forms can be reused four or possibly five times without excessive repair. Remember to allow for 10% waste on each reuse.

When modular sized wall forms are made, up to twenty uses can be expected with exterior plyform.

When forms are reused, the cost to erect, strip, clean and move will not be affected. 10% replacement of lumber should be included and about one hour of carpenter time for repairs on each reuse per 100 S.F.

The reuse cost for certain accessory items normally rented on a monthly basis will be lower than the cost for the first use.

After fifth use, new material required plus time needed for repair prevent form cost from dropping further and it may go up. Much depends on care in stripping, the number of special bays, changes in beam or column sizes and other factors.

Costs for multiple use of formwork may be developed as follows:

2 Uses

$$\frac{\text{(1st Use + Reuse)}}{2} = \text{avg. cost/2 uses}$$

3 Uses

$$\frac{\text{(1st Use + 2 Reuse)}}{3} = \text{avg. cost/3 uses}$$

4 Uses

$$\frac{\text{(1st use + 3 Reuse)}}{4} = \text{avg. cost/4 uses}$$

R031113-60 Formwork Labor-Hours

Item	Hours Required				Total Hours	Multiple Use		
	Unit	Fabricate	Erect & Strip	Clean & Move	1 Use	2 Use	3 Use	4 Use
Beam and Girder, interior beams, 12″ wide	100 S.F.	6.4	8.3	1.3	16.0	13.3	12.4	12.0
Hung from steel beams		5.8	7.7	1.3	14.8	12.4	11.6	11.2
Beam sides only, 36″ high		5.8	7.2	1.3	14.3	11.9	11.1	10.7
Beam bottoms only, 24″ wide		6.6	13.0	1.3	20.9	18.1	17.2	16.7
Box out for openings		9.9	10.0	1.1	21.0	16.6	15.1	14.3
Buttress forms, to 8′ high		6.0	6.5	1.2	13.7	11.2	10.4	10.0
Centering, steel, 3/4″ rib lath			1.0		1.0			
3/8″ rib lath or slab form			0.9		0.9			
Chamfer strip or keyway	100 L.F.		1.5		1.5	1.5	1.5	1.5
Columns, fiber tube 8″ diameter			20.6		20.6			
12″			21.3		21.3			
16″			22.9		22.9			
20″			23.7		23.7			
24″			24.6		24.6			
30″			25.6		25.6			
Columns, round steel, 12″ diameter			22.0		22.0	22.0	22.0	22.0
16″			25.6		25.6	25.6	25.6	25.6
20″			30.5		30.5	30.5	30.5	30.5
24″			37.7		37.7	37.7	37.7	37.7
Columns, plywood 8″ x 8″	100 S.F.	7.0	11.0	1.2	19.2	16.2	15.2	14.7
12″ x 12″		6.0	10.5	1.2	17.7	15.2	14.4	14.0
16″ x 16″		5.9	10.0	1.2	17.1	14.7	13.8	13.4
24″ x 24″		5.8	9.8	1.2	16.8	14.4	13.6	13.2
Columns, steel framed plywood 8″ x 8″			10.0	1.0	11.0	11.0	11.0	11.0
12″ x 12″			9.3	1.0	10.3	10.3	10.3	10.3
16″ x 16″			8.5	1.0	9.5	9.5	9.5	9.5
24″ x 24″			7.8	1.0	8.8	8.8	8.8	8.8
Drop head forms, plywood		9.0	12.5	1.5	23.0	19.0	17.7	17.0
Coping forms		8.5	15.0	1.5	25.0	21.3	20.0	19.4
Culvert, box			14.5	4.3	18.8	18.8	18.8	18.8
Curb forms, 6″ to 12″ high, on grade		5.0	8.5	1.2	14.7	12.7	12.1	11.7
On elevated slabs		6.0	10.8	1.2	18.0	15.5	14.7	14.3
Edge forms to 6″ high, on grade	100 L.F.	2.0	3.5	0.6	6.1	5.6	5.4	5.3
7″ to 12″ high	100 S.F.	2.5	5.0	1.0	8.5	7.8	7.5	7.4
Equipment foundations		10.0	18.0	2.0	30.0	25.5	24.0	23.3
Flat slabs, including drops		3.5	6.0	1.2	10.7	9.5	9.0	8.8
Hung from steel		3.0	5.5	1.2	9.7	8.7	8.4	8.2
Closed deck for domes		3.0	5.8	1.2	10.0	9.0	8.7	8.5
Open deck for pans		2.2	5.3	1.0	8.5	7.9	7.7	7.6
Footings, continuous, 12″ high		3.5	3.5	1.5	8.5	7.3	6.8	6.6
Spread, 12″ high		4.7	4.2	1.6	10.5	8.7	8.0	7.7
Pile caps, square or rectangular		4.5	5.0	1.5	11.0	9.3	8.7	8.4
Grade beams, 24″ deep		2.5	5.3	1.2	9.0	8.3	8.0	7.9
Lintel or Sill forms		8.0	17.0	2.0	27.0	23.5	22.3	21.8
Spandrel beams, 12″ wide		9.0	11.2	1.3	21.5	17.5	16.2	15.5
Stairs			25.0	4.0	29.0	29.0	29.0	29.0
Trench forms in floor		4.5	14.0	1.5	20.0	18.3	17.7	17.4
Walls, Plywood, at grade, to 8′ high		5.0	6.5	1.5	13.0	11.0	9.7	9.5
8′ to 16′		7.5	8.0	1.5	17.0	13.8	12.7	12.1
16′ to 20′		9.0	10.0	1.5	20.5	16.5	15.2	14.5
Foundation walls, to 8′ high		4.5	6.5	1.0	12.0	10.3	9.7	9.4
8′ to 16′ high		5.5	7.5	1.0	14.0	11.8	11.0	10.6
Retaining wall to 12′ high, battered		6.0	8.5	1.5	16.0	13.5	12.7	12.3
Radial walls to 12′ high, smooth		8.0	9.5	2.0	19.5	16.0	14.8	14.3
2′ chords		7.0	8.0	1.5	16.5	13.5	12.5	12.0
Prefabricated modular, to 8′ high		—	4.3	1.0	5.3	5.3	5.3	5.3
Steel, to 8′ high		—	6.8	1.2	8.0	8.0	8.0	8.0
8′ to 16′ high		—	9.1	1.5	10.6	10.3	10.2	10.2
Steel framed plywood to 8′ high		—	6.8	1.2	8.0	7.5	7.3	7.2
8′ to 16′ high		—	9.3	1.2	10.5	9.5	9.2	9.0

R032110-10 Reinforcing Steel Weights and Measures

Bar Designation No.**	Nominal Weight Lb./Ft.	U.S. Customary Units Nominal Dimensions*			SI Units Nominal Dimensions*			
		Diameter in.	Cross Sectional Area, in.^{2}	Perimeter in.	Nominal Weight kg/m	Diameter mm	Cross Sectional Area, cm^2	Perimeter mm
3	.376	.375	.11	1.178	.560	9.52	.71	29.9
4	.668	.500	.20	1.571	.994	12.70	1.29	39.9
5	1.043	.625	.31	1.963	1.552	15.88	2.00	49.9
6	1.502	.750	.44	2.356	2.235	19.05	2.84	59.8
7	2.044	.875	.60	2.749	3.042	22.22	3.87	69.8
8	2.670	1.000	.79	3.142	3.973	25.40	5.10	79.8
9	3.400	1.128	1.00	3.544	5.059	28.65	6.45	90.0
10	4.303	1.270	1.27	3.990	6.403	32.26	8.19	101.4
11	5.313	1.410	1.56	4.430	7.906	35.81	10.06	112.5
14	7.650	1.693	2.25	5.320	11.384	43.00	14.52	135.1
18	13.600	2.257	4.00	7.090	20.238	57.33	25.81	180.1

* The nominal dimensions of a deformed bar are equivalent to those of a plain round bar having the same weight per foot as the deformed bar.
** Bar numbers are based on the number of eighths of an inch included in the nominal diameter of the bars.

R032110-20 Metric Rebar Specification - ASTM A615-81

Grade 300 (300 MPa* = 43,560 psi; +8.7% vs. Grade 40)				
Grade 400 (400 MPa* = 58,000 psi; –3.4% vs. Grade 60)				
Bar No.	Diameter mm	Area mm^2	Equivalent in.^{2}	Comparison with U.S. Customary Bars
10M	11.3	100	.16	Between #3 & #4
15M	16.0	200	.31	#5 (.31 in.^{2})
20M	19.5	300	.47	#6 (.44 in.^{2})
25M	25.2	500	.78	#8 (.79 in.^{2})
30M	29.9	700	1.09	#9 (1.00 in.^{2})
35M	35.7	1000	1.55	#11 (1.56 in.^{2})
45M	43.7	1500	2.33	#14 (2.25 in.^{2})
55M	56.4	2500	3.88	#18 (4.00 in.^{2})

* MPa = megapascals

R032110-25 Comparison of U.S. Customary Units and SI Units for Reinforcing Bars

U.S. Customary Units							
		Nominal Dimensions[a]			Deformation Requirements, in.		
Bar Designation No.[b]	Nominal Weight, lb/ft	Diameter in.	Cross Sectional Area, in.^{2}	Perimeter in.	Maximum Average Spacing	Minimum Average Height	Maximum Gap (Chord of 12-1/2% of Nominal Perimeter)
3	0.376	0.375	0.11	1.178	0.262	0.015	0.143
4	0.668	0.500	0.20	1.571	0.350	0.020	0.191
5	1.043	0.625	0.31	1.963	0.437	0.028	0.239
6	1.502	0.750	0.44	2.356	0.525	0.038	0.286
7	2.044	0.875	0.60	2.749	0.612	0.044	0.334
8	2.670	1.000	0.79	3.142	0.700	0.050	0.383
9	3.400	1.128	1.00	3.544	0.790	0.056	0.431
10	4.303	1.270	1.27	3.990	0.889	0.064	0.487
11	5.313	1.410	1.56	4.430	0.987	0.071	0.540
14	7.65	1.693	2.25	5.32	1.185	0.085	0.648
18	13.60	2.257	4.00	7.09	1.58	0.102	0.864
SI UNITS							
		Nominal Dimensions[a]			Deformation Requirements, mm		
Bar Designation No.[b]	Nominal Weight kg/m	Diameter, mm	Cross Sectional Area, cm^2	Perimeter, mm	Maximum Average Spacing	Minimum Average Height	Maximum Gap (Chord of 12-1/2% of Nominal Perimeter)
3	0.560	9.52	0.71	29.9	6.7	0.38	3.5
4	0.994	12.70	1.29	39.9	8.9	0.51	4.9
5	1.552	15.88	2.00	49.9	11.1	0.71	6.1
6	2.235	19.05	2.84	59.8	13.3	0.96	7.3
7	3.042	22.22	3.87	69.8	15.5	1.11	8.5
8	3.973	25.40	5.10	79.8	17.8	1.27	9.7
9	5.059	28.65	6.45	90.0	20.1	1.42	10.9
10	6.403	32.26	8.19	101.4	22.6	1.62	11.4
11	7.906	35.81	10.06	112.5	25.1	1.80	13.6
14	11.384	43.00	14.52	135.1	30.1	2.16	16.5
18	20.238	57.33	25.81	180.1	40.1	2.59	21.9

[a]Nominal dimensions of a deformed bar are equivalent to those of a plain round bar having the same weight per foot as the deformed bar.

[b]Bar numbers are based on the number of eighths of an inch included in the nominal diameter of the bars.

R032110-40 Weight of Steel Reinforcing Per Square Foot of Wall (PSF)

Reinforced Weights: The table below suggests the weights per square foot for reinforcing steel in walls. Weights are approximate and will be the same for all grades of steel bars. For bars in two directions, add weights for each size and spacing.

C/C Spacing in Inches	Bar Size								
	#3 Wt. (PSF)	#4 Wt. (PSF)	#5 Wt. (PSF)	#6 Wt. (PSF)	#7 Wt. (PSF)	#8 Wt. (PSF)	#9 Wt. (PSF)	#10 Wt. (PSF)	#11 Wt. (PSF)
2″	2.26	4.01	6.26	9.01	12.27				
3″	1.50	2.67	4.17	6.01	8.18	10.68	13.60	17.21	21.25
4″	1.13	2.01	3.13	4.51	6.13	8.10	10.20	12.91	15.94
5″	.90	1.60	2.50	3.60	4.91	6.41	8.16	10.33	12.75
6″	.752	1.34	2.09	3.00	4.09	5.34	6.80	8.61	10.63
8″	.564	1.00	1.57	2.25	3.07	4.01	5.10	6.46	7.97
10″	.451	.802	1.25	1.80	2.45	3.20	4.08	5.16	6.38
12″	.376	.668	1.04	1.50	2.04	2.67	3.40	4.30	5.31
18″	.251	.445	.695	1.00	1.32	1.78	2.27	2.86	3.54
24″	.188	.334	.522	.751	1.02	1.34	1.70	2.15	2.66
30″	.150	.267	.417	.600	.817	1.07	1.36	1.72	2.13
36″	.125	.223	.348	.501	.681	.890	1.13	1.43	1.77
42″	.107	.191	.298	.429	.584	.753	.97	1.17	1.52
48″	.094	.167	.261	.376	.511	.668	.85	1.08	1.33

R032110-50 Minimum Wall Reinforcement Weight (PSF)

This table lists the approximate minimum wall reinforcement weights per S.F. according to the specification of .12% of gross area for vertical bars and .20% of gross area for horizontal bars.

Location	Wall Thickness	Bar Size	Horizontal Steel Spacing C/C	Sq. In. Req'd per S.F.	Total Wt. per S.F.	Bar Size	Vertical Steel Spacing C/C	Sq. In. Req'd per S.F.	Total Wt. per S.F.	Horizontal & Vertical Steel Total Weight per S.F.
Both Faces	10″	#4	18″	.24	.89#	#3	18″	.14	.50#	1.39#
	12″	#4	16″	.29	1.00	#3	16″	.17	.60	1.60
	14″	#4	14″	.34	1.14	#3	13″	.20	.69	1.84
	16″	#4	12″	.38	1.34	#3	11″	.23	.82	2.16
	18″	#5	17″	.43	1.47	#4	18″	.26	.89	2.36
One Face	6″	#3	9″	.15	.50	#3	18″	.09	.25	.75
	8″	#4	12″	.19	.67	#3	11″	.12	.41	1.08
	10″	#5	15″	.24	.83	#4	16″	.14	.50	1.34

R032110-70 Bend, Place and Tie Reinforcing

Placing and tying by rodmen for footings and slabs runs from nine hrs. per ton for heavy bars to fifteen hrs. per ton for light bars. For beams, columns, and walls, production runs from eight hrs. per ton for heavy bars to twenty hrs. per ton for light bars. Overall average for typical reinforced concrete buildings is about fourteen hrs. per ton. These production figures include the time for placing of accessories and usual inserts, but not their material cost (allow 15% of the cost of delivered bent rods). Equipment handling is necessary for the larger-sized bars so that installation costs for the very heavy bars will not decrease proportionately.

Installation costs for splicing reinforcing bars include allowance for equipment to hold the bars in place while splicing as well as necessary scaffolding for iron workers.

R032110-80 Shop-Fabricated Reinforcing Steel

The material prices for reinforcing, shown in the unit cost sections of the book, are for 50 tons or more of shop-fabricated reinforcing steel and include:

1. Mill base price of reinforcing steel
2. Mill grade/size/length extras
3. Mill delivery to the fabrication shop
4. Shop storage and handling
5. Shop drafting/detailing
6. Shop shearing and bending
7. Shop listing
8. Shop delivery to the job site

Both material and installation costs can be considerably higher for small jobs consisting primarily of smaller bars, while material costs may be slightly lower for larger jobs.

R032205-30 Common Stock Styles of Welded Wire Fabric

This table provides some of the basic specifications, sizes, and weights of welded wire fabric used for reinforcing concrete.

	New Designation	Old Designation		Steel Area per Foot				Approximate Weight per 100 S.F.	
	Spacing — Cross Sectional Area (in.) — (Sq. in. 100)	Spacing — Wire Gauge (in.) — (AS & W)		Longitudinal		Transverse			
				in.	cm	in.	cm	lbs	kg
Rolls	6 x 6 — W1.4 x W1.4	6 x 6 — 10 x 10		.028	.071	.028	.071	21	9.53
	6 x 6 — W2.0 x W2.0	6 x 6 — 8 x 8	1	.040	.102	.040	.102	29	13.15
	6 x 6 — W2.9 x W2.9	6 x 6 — 6 x 6		.058	.147	.058	.147	42	19.05
	6 x 6 — W4.0 x W4.0	6 x 6 — 4 x 4		.080	.203	.080	.203	58	26.91
	4 x 4 — W1.4 x W1.4	4 x 4 — 10 x 10		.042	.107	.042	.107	31	14.06
	4 x 4 — W2.0 x W2.0	4 x 4 — 8 x 8	1	.060	.152	.060	.152	43	19.50
	4 x 4 — W2.9 x W2.9	4 x 4 — 6 x 6		.087	.227	.087	.227	62	28.12
	4 x 4 — W4.0 x W4.0	4 x 4 — 4 x 4		.120	.305	.120	.305	85	38.56
Sheets	6 x 6 — W2.9 x W2.9	6 x 6 — 6 x 6		.058	.147	.058	.147	42	19.05
	6 x 6 — W4.0 x W4.0	6 x 6 — 4 x 4		.080	.203	.080	.203	58	26.31
	6 x 6 — W5.5 x W5.5	6 x 6 — 2 x 2	2	.110	.279	.110	.279	80	36.29
	4 x 4 — W1.4 x W1.4	4 x 4 — 4 x 4		.120	.305	.120	.305	85	38.56

NOTES: 1. Exact W—number size for 8 gauge is W2.1
2. Exact W—number size for 2 gauge is W5.4

R033053-10 Spread Footings

General: A spread footing is used to convert a concentrated load (from one superstructure column, or substructure grade beams) into an allowable area load on supporting soil.

Because of punching action from the column load, a spread footing is usually thicker than strip footings which support wall loads. One or two story commercial or residential buildings should have no less than 1′ thick spread footings. Heavier loads require no less than 2′ thick. Spread footings may be square, rectangular or octagonal in plan.

Spread footings tend to minimize excavation and foundation materials, as well as labor and equipment. Another advantage is that footings and soil conditions can be readily examined. They are the most widely used type of footing, especially in mild climates and for buildings of four stories or under. This is because they are usually more economical than other types, if suitable soil and site conditions exist.

They are used when suitable supporting soil is located within several feet of the surface or line of subsurface excavation. Suitable soil types include sands and gravels, gravels with a small amount of clay or silt, hardpan, chalk, and rock. Pedestals may be used to bring the column base load down to the top of footing. Alternately, undesirable soil between underside of footing and top of bearing level can be removed and replaced with lean concrete mix or compacted granular material.

Depth of footing should be below topsoil, uncompacted fill, muck, etc. It must be lower than frost penetration but should be above the water table. It must not be at the ground surface because of potential surface erosion. If the ground slopes, approximately three horizontal feet of edge protection must remain. Differential footing elevations may overlap soil stresses or cause excavation problems if clear spacing between footings is less than the difference in depth.

Other footing types are usually used for the following reasons:

A. Bearing capacity of soil is low.

B. Very large footings are required, at a cost disadvantage.

C. Soil under footing (shallow or deep) is very compressible, with probability of causing excessive or differential settlement.

D. Good bearing soil is deep.

E. Potential for scour action exists.

F. Varying subsoil conditions within building perimeter.

Cost of spread footings for a building is determined by:

1. The soil bearing capacity.
2. Typical bay size.
3. Total load (live plus dead) per S.F. for roof and elevated floor levels.
4. The size and shape of the building.
5. Footing configuration. Does the building utilize outer spread footings or are there continuous perimeter footings only or a combination of spread footings plus continuous footings?

Soil Bearing Capacity in Kips per S.F.

Bearing Material	Typical Allowable Bearing Capacity
Hard sound rock	120 KSF
Medium hard rock	80
Hardpan overlaying rock	24
Compact gravel and boulder-gravel; very compact sandy gravel	20
Soft rock	16
Loose gravel; sandy gravel; compact sand; very compact sand-inorganic silt	12
Hard dry consolidated clay	10
Loose coarse to medium sand; medium compact fine sand	8
Compact sand-clay	6
Loose fine sand; medium compact sand-inorganic silts	4
Firm or stiff clay	3
Loose saturated sand-clay; medium soft clay	2

R033053-50 Industrial Chimneys

Foundation requirements in C.Y. of concrete for various sized chimneys.

Size Chimney	2 Ton Soil	3 Ton Soil	Size Chimney	2 Ton Soil	3 Ton Soil	Size Chimney	2 Ton Soil	3 Ton Soil
75′ x 3′-0″	13 C.Y.	11 C.Y.	160′ x 6′-6″	86 C.Y.	76 C.Y.	300′ x 10′-0″	325 C.Y.	245 C.Y.
85′ x 5′-6″	19	16	175′ x 7′-0″	108	95	350′ x 12′-0″	422	320
100′ x 5′-0″	24	20	200′ x 6′-0″	125	105	400′ x 14′-0″	520	400
125′ x 5′-6″	43	36	250′ x 8′-0″	230	175	500′ x 18′-0″	725	575

R033053-60 Maximum Depth of Frost Penetration in Inches

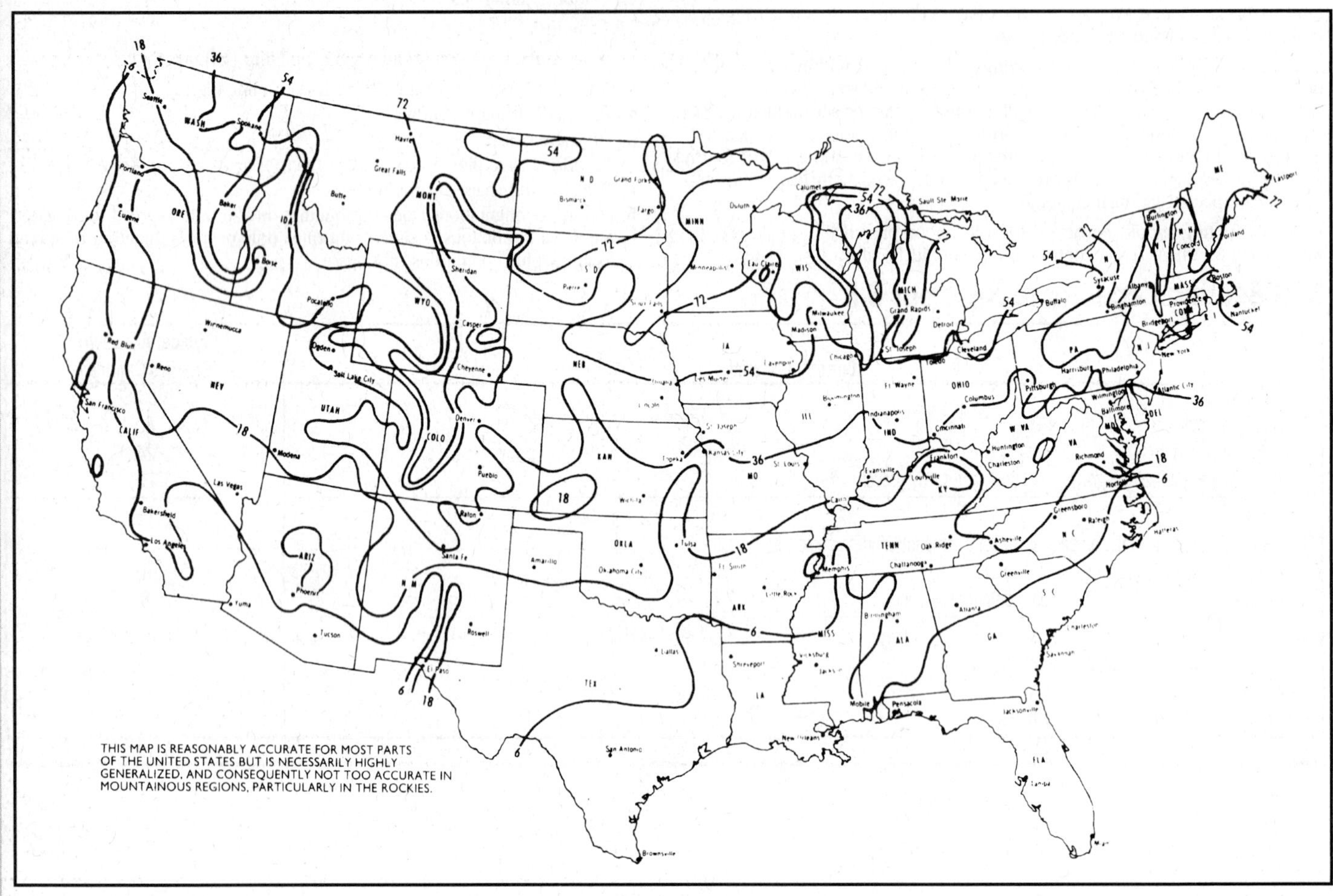

R033105-10 Proportionate Quantities

The tables below show both quantities per S.F. of floor areas as well as form and reinforcing quantities per C.Y. Unusual structural requirements would increase the ratios below. High strength reinforcing would reduce the steel weights. Figures are for 3000 psi concrete and 60,000 psi reinforcing unless specified otherwise.

Type of Construction	Live Load	Span	Per S.F. of Floor Area				Per C.Y. of Concrete		
			Concrete	Forms	Reinf.	Pans	Forms	Reinf.	Pans
Flat Plate	50 psf	15 Ft.	.46 C.F.	1.06 S.F.	1.71lb.		62 S.F.	101 lb.	
		20	.63	1.02	2.40		44	104	
		25	.79	1.02	3.03		35	104	
	100	15	.46	1.04	2.14		61	126	
		20	.71	1.02	2.72		39	104	
		25	.83	1.01	3.47		33	113	
Flat Plate (waffle construction) 20″ domes	50	20	.43	1.00	2.10	.84 S.F.	63	135	53 S.F.
		25	.52	1.00	2.90	.89	52	150	46
		30	.64	1.00	3.70	.87	42	155	37
	100	20	.51	1.00	2.30	.84	53	125	45
		25	.64	1.00	3.20	.83	42	135	35
		30	.76	1.00	4.40	.81	36	160	29
Waffle Construction 30″ domes	50	25	.69	1.06	1.83	.68	42	72	40
		30	.74	1.06	2.39	.69	39	87	39
		35	.86	1.05	2.71	.69	33	85	39
		40	.78	1.00	4.80	.68	35	165	40
Flat Slab (two way with drop panels)	50	20	.62	1.03	2.34		45	102	
		25	.77	1.03	2.99		36	105	
		30	.95	1.03	4.09		29	116	
	100	20	.64	1.03	2.83		43	119	
		25	.79	1.03	3.88		35	133	
		30	.96	1.03	4.66		29	131	
	200	20	.73	1.03	3.03		38	112	
		25	.86	1.03	4.23		32	133	
		30	1.06	1.03	5.30		26	135	
One Way Joists 20″ Pans	50	15	.36	1.04	1.40	.93	78	105	70
		20	.42	1.05	1.80	.94	67	120	60
		25	.47	1.05	2.60	.94	60	150	54
	100	15	.38	1.07	1.90	.93	77	140	66
		20	.44	1.08	2.40	.94	67	150	58
		25	.52	1.07	3.50	.94	55	185	49
One Way Joists 8″ x 16″ filler blocks	50	15	.34	1.06	1.80	.81 Ea.	84	145	64 Ea.
		20	.40	1.08	2.20	.82	73	145	55
		25	.46	1.07	3.20	.83	63	190	49
	100	15	.39	1.07	1.90	.81	74	130	56
		20	.46	1.09	2.80	.82	64	160	48
		25	.53	1.10	3.60	.83	56	190	42
One Way Beam & Slab	50	15	.42	1.30	1.73		84	111	
		20	.51	1.28	2.61		68	138	
		25	.64	1.25	2.78		53	117	
	100	15	.42	1.30	1.90		84	122	
		20	.54	1.35	2.69		68	154	
		25	.69	1.37	3.93		54	145	
	200	15	.44	1.31	2.24		80	137	
		20	.58	1.40	3.30		65	163	
		25	.69	1.42	4.89		53	183	
Two Way Beam & Slab	100	15	.47	1.20	2.26		69	130	
		20	.63	1.29	3.06		55	131	
		25	.83	1.33	3.79		43	123	
	200	15	.49	1.25	2.70		41	149	
		20	.66	1.32	4.04		54	165	
		25	.88	1.32	6.08		41	187	

R033105-10 Proportionate Quantities (cont.)

4000 psi Concrete and 60,000 psi Reinforcing—Form and Reinforcing Quantities per C.Y.

Item	Size	Forms	Reinforcing	Minimum	Maximum
Columns (square tied)	10″ x 10″	130 S.F.C.A.	#5 to #11	220 lbs.	875 lbs.
	12″ x 12″	108	#6 to #14	200	955
	14″ x 14″	92	#7 to #14	190	900
	16″ x 16″	81	#6 to #14	187	1082
	18″ x 18″	72	#6 to #14	170	906
	20″ x 20″	65	#7 to #18	150	1080
	22″ x 22″	59	#8 to #18	153	902
	24″ x 24″	54	#8 to #18	164	884
	26″ x 26″	50	#9 to #18	169	994
	28″ x 28″	46	#9 to #18	147	864
	30″ x 30″	43	#10 to #18	146	983
	32″ x 32″	40	#10 to #18	175	866
	34″ x 34″	38	#10 to #18	157	772
	36″ x 36″	36	#10 to #18	175	852
	38″ x 38″	34	#10 to #18	158	765
	40″ x 40″	32	#10 to #18	143	692

Item	Size	Form	Spiral	Reinforcing	Minimum	Maximum
Columns (spirally reinforced)	12″ diameter	34.5 L.F.	190 lbs.	#4 to #11	165 lbs.	1505 lb.
		34.5	190	#14 & #18	—	1100
	14″	25	170	#4 to #11	150	970
		25	170	#14 & #18	800	1000
	16″	19	160	#4 to #11	160	950
		19	160	#14 & #18	605	1080
	18″	15	150	#4 to #11	160	915
		15	150	#14 & #18	480	1075
	20″	12	130	#4 to #11	155	865
		12	130	#14 & #18	385	1020
	22″	10	125	#4 to #11	165	775
		10	125	#14 & #18	320	995
	24″	9	120	#4 to #11	195	800
		9	120	#14 & #18	290	1150
	26″	7.3	100	#4 to #11	200	729
		7.3	100	#14 & #18	235	1035
	28″	6.3	95	#4 to #11	175	700
		6.3	95	#14 & #18	200	1075
	30″	5.5	90	#4 to #11	180	670
		5.5	90	#14 & #18	175	1015
	32″	4.8	85	#4 to #11	185	615
		4.8	85	#14 & #18	155	955
	34″	4.3	80	#4 to #11	180	600
		4.3	80	#14 & #18	170	855
	36″	3.8	75	#4 to #11	165	570
		3.8	75	#14 & #18	155	865
	40″	3.0	70	#4 to #11	165	500
		3.0	70	#14 & #18	145	765

R033105-10 Proportionate Quantities (cont.)

3000 psi Concrete and 60,000 psi Reinforcing—Form and Reinforcing Quantities per C.Y.						
Item	Type	Loading	Height	C.Y./L.F.	Forms/C.Y.	Reinf./C.Y.
Retaining Walls	Cantilever	Level Backfill	4 Ft.	0.2 C.Y.	49 S.F.	35 lbs.
			8	0.5	42	45
			12	0.8	35	70
			16	1.1	32	85
			20	1.6	28	105
		Highway Surcharge	4	0.3	41	35
			8	0.5	36	55
			12	0.8	33	90
			16	1.2	30	120
			20	1.7	27	155
		Railroad Surcharge	4	0.4	28	45
			8	0.8	25	65
			12	1.3	22	90
			16	1.9	20	100
			20	2.6	18	120
	Gravity, with Vertical Face	Level Backfill	4	0.4	37	None
			7	0.6	27	↓
			10	1.2	20	↓
		Sloping Backfill	4	0.3	31	↓
			7	0.8	21	↓
			10	1.6	15	↓

Item	Span	Live Load in Kips per Linear Foot							
		Under 1 Kip		2 to 3 Kips		4 to 5 Kips		6 to 7 Kips	
		Forms	Reinf.	Forms	Reinf.	Forms	Reinf.	Forms	Reinf.
Beams	10 Ft.	—	—	90 S.F.	170 #	85 S.F.	175 #	75 S.F.	185 #
	16	130 S.F.	165 #	85	180	75	180	65	225
	20	110	170	75	185	62	200	51	200
	26	90	170	65	215	62	215	—	—
	30	85	175	60	200	—	—	—	—

Item	Size	Type	Forms per C.Y.	Reinforcing per C.Y.
Spread Footings	Under 1 C.Y.	1,000 psf soil	24 S.F.	44 lbs.
		5,000	24	42
		10,000	24	52
	1 C.Y. to 5 C.Y.	1,000	14	49
		5,000	14	50
		10,000	14	50
	Over 5 C.Y.	1,000	9	54
		5,000	9	52
		10,000	9	56
Pile Caps (30 Ton Concrete Piles)	Under 5 C.Y.	shallow caps	20	65
		medium	20	50
		deep	20	40
	5 C.Y. to 10 C.Y.	shallow	14	55
		medium	15	45
		deep	15	40
	10 C.Y. to 20 C.Y.	shallow	11	60
		medium	11	45
		deep	12	35
	Over 20 C.Y.	shallow	9	60
		medium	9	45
		deep	10	40

R033105-10 Proportionate Quantities (cont.)

3000 psi Concrete and 60,000 psi Reinforcing — Form and Reinforcing Quantities per C.Y.

Item	Size	Pile Spacing	50 T Pile	100 T Pile	50 T Pile	100 T Pile
Pile Caps (Steel H Piles)	Under 5 C.Y.	24″ O.C.	24 S.F.	24 S.F.	75 lbs.	90 lbs.
		30″	25	25	80	100
		36″	24	24	80	110
	5 C.Y. to 10 C.Y.	24″	15	15	80	110
		30″	15	15	85	110
		36″	15	15	75	90
	Over 10 C.Y.	24″	13	13	85	90
		30″	11	11	85	95
		36″	10	10	85	90

Item	Height	8″ Thick Forms	8″ Thick Reinf.	10″ Thick Forms	10″ Thick Reinf.	12″ Thick Forms	12″ Thick Reinf.	15″ Thick Forms	15″ Thick Reinf.
Basement Walls	7 Ft.	81 S.F.	44 lbs.	65 S.F.	45 lbs.	54 S.F.	44 lbs.	41 S.F.	43 lbs.
	8	↓	44	↓	45	↓	44	↓	43
	9	↓	46	↓	45	↓	44	↓	43
	10	↓	57	↓	45	↓	44	↓	43
	12	↓	83	↓	50	↓	52	↓	43
	14	↓	116	↓	65	↓	64	↓	51
	16	↓		↓	86	↓	90	↓	65
	18	↓		↓		↓	106	↓	70

R033105-20 Materials for One C.Y. of Concrete

This is an approximate method of figuring quantities of cement, sand and coarse aggregate for a field mix with waste allowance included.

With crushed gravel as coarse aggregate, to determine barrels of cement required, divide 10 by total mix; that is, for 1:2:4 mix, 10 divided by 7 = 1-3/7 barrels. If the coarse aggregate is crushed stone, use 10-1/2 instead of 10 as given for gravel.

To determine tons of sand required, multiply barrels of cement by parts of sand and then by 0.2; that is, for the 1:2:4 mix, as above, 1-3/7 x 2 x .2 = .57 tons.

Tons of crushed gravel are in the same ratio to tons of sand as parts in the mix, or 4/2 x .57 = 1.14 tons.

1 bag cement = 94#
4 bags = 1 barrel

1 C.Y. sand or crushed gravel = 2700#
1 ton sand or crushed gravel = 20 C.F.

1 C.Y. crushed stone = 2575#
1 ton crushed stone = 21 C.F.

Average carload of cement is 692 bags; of sand or gravel is 56 tons.

Do not stack stored cement over 10 bags high.

R033105-30 Metric Equivalents of Cement Content for Concrete Mixes

94 Pound Bags per Cubic Yard	Kilograms per Cubic Meter	94 Pound Bags per Cubic Yard	Kilograms per Cubic Meter
1.0	55.77	7.0	390.4
1.5	83.65	7.5	418.3
2.0	111.5	8.0	446.2
2.5	139.4	8.5	474.0
3.0	167.3	9.0	501.9
3.5	195.2	9.5	529.8
4.0	223.1	10.0	557.7
4.5	251.0	10.5	585.6
5.0	278.8	11.0	613.5
5.5	306.7	11.5	641.3
6.0	334.6	12.0	669.2
6.5	362.5	12.5	697.1

a. If you know the cement content in pounds per cubic yard, multiply by .5933 to obtain kilograms per cubic meter.

b. If you know the cement content in 94 pound bags per cubic yard, multiply by 55.77 to obtain kilograms per cubic meter.

R033105-40 Metric Equivalents of Common Concrete Strengths (to convert other psi values to megapascals, multiply by 0.006895)

U.S. Values psi	SI Value Megapascals	Non-SI Metric Value kgf/cm²*
2000	14	140
2500	17	175
3000	21	210
3500	24	245
4000	28	280
4500	31	315
5000	34	350
6000	41	420
7000	48	490
8000	55	560
9000	62	630
10,000	69	705

* kilograms force per square centimeter

R033105-50 Quantities of Cement, Sand and Stone for One C.Y. of Concrete per Various Mixes

This table can be used to determine the quantities of the ingredients for smaller quantities of site mixed concrete.

	Mix = 1:1:1-3/4			Mix = 1:2:2.25			Mix = 1:2.25:3			Mix = 1:3:4		
Concrete (C.Y.)	Cement (sacks)	Sand (C.Y.)	Stone (C.Y.)	Cement (sacks)	Sand (C.Y.)	Stone (C.Y.)	Cement (sacks)	Sand (C.Y.)	Stone (C.Y.)	Cement (sacks)	Sand (C.Y.)	Stone (C.Y.)
1	10	.37	.63	7.75	.56	.65	6.25	.52	.70	5	.56	.74
2	20	.74	1.26	15.50	1.12	1.30	12.50	1.04	1.40	10	1.12	1.48
3	30	1.11	1.89	23.25	1.68	1.95	18.75	1.56	2.10	15	1.68	2.22
4	40	1.48	2.52	31.00	2.24	2.60	25.00	2.08	2.80	20	2.24	2.96
5	50	1.85	3.15	38.75	2.80	3.25	31.25	2.60	3.50	25	2.80	3.70
6	60	2.22	3.78	46.50	3.36	3.90	37.50	3.12	4.20	30	3.36	4.44
7	70	2.59	4.41	54.25	3.92	4.55	43.75	3.64	4.90	35	3.92	5.18
8	80	2.96	5.04	62.00	4.48	5.20	50.00	4.16	5.60	40	4.48	5.92
9	90	3.33	5.67	69.75	5.04	5.85	56.25	4.68	6.30	45	5.04	6.66
10	100	3.70	6.30	77.50	5.60	6.50	62.50	5.20	7.00	50	5.60	7.40
11	110	4.07	6.93	85.25	6.16	7.15	68.75	5.72	7.70	55	6.16	8.14
12	120	4.44	7.56	93.00	6.72	7.80	75.00	6.24	8.40	60	6.72	8.88
13	130	4.82	8.20	100.76	7.28	8.46	81.26	6.76	9.10	65	7.28	9.62
14	140	5.18	8.82	108.50	7.84	9.10	87.50	7.28	9.80	70	7.84	10.36
15	150	5.56	9.46	116.26	8.40	9.76	93.76	7.80	10.50	75	8.40	11.10
16	160	5.92	10.08	124.00	8.96	10.40	100.00	8.32	11.20	80	8.96	11.84
17	170	6.30	10.72	131.76	9.52	11.06	106.26	8.84	11.90	85	9.52	12.58
18	180	6.66	11.34	139.50	10.08	11.70	112.50	9.36	12.60	90	10.08	13.32
19	190	7.04	11.98	147.26	10.64	12.36	118.76	9.84	13.30	95	10.64	14.06
20	200	7.40	12.60	155.00	11.20	13.00	125.00	10.40	14.00	100	11.20	14.80
21	210	7.77	13.23	162.75	11.76	13.65	131.25	10.92	14.70	105	11.76	15.54
22	220	8.14	13.86	170.05	12.32	14.30	137.50	11.44	15.40	110	12.32	16.28
23	230	8.51	14.49	178.25	12.88	14.95	143.75	11.96	16.10	115	12.88	17.02
24	240	8.88	15.12	186.00	13.44	15.60	150.00	12.48	16.80	120	13.44	17.76
25	250	9.25	15.75	193.75	14.00	16.25	156.25	13.00	17.50	125	14.00	18.50
26	260	9.64	16.40	201.52	14.56	16.92	162.52	13.52	18.20	130	14.56	19.24
27	270	10.00	17.00	209.26	15.12	17.56	168.76	14.04	18.90	135	15.02	20.00
28	280	10.36	17.64	217.00	15.68	18.20	175.00	14.56	19.60	140	15.68	20.72
29	290	10.74	18.28	224.76	16.24	18.86	181.26	15.08	20.30	145	16.24	21.46

R033105-65 Field-Mix Concrete

Presently most building jobs are built with ready-mixed concrete except at isolated locations and some larger jobs requiring over 10,000 C.Y. where land is readily available for setting up a temporary batch plant.

The most economical mix is a controlled mix using local aggregate proportioned by trial to give the required strength with the least cost of material.

R033105-80 Slab on Grade

General: Ground slabs are classified on the basis of use. Thickness is generally controlled by the heaviest concentrated load supported. If load area is greater than 80 sq. in., soil bearing may be important. The base granular fill must be a uniformly compacted material of limited capillarity, such as gravel or crushed rock. Concrete is placed on this surface of the vapor barrier on top of base.

Ground slabs are either single or two course floors. Single course are widely used. Two course floors have a subsequent wear resistant topping.

Reinforcement is provided to maintain tightly closed cracks.

Control joints limit crack locations and provide for differential horizontal movement only. Isolation joints allow both horizontal and vertical differential movement.

Use of Table: Determine appropriate type of slab (A, B, C, or D) by considering type of use or amount of abrasive wear of traffic type.

Determine thickness by maximum allowable wheel load or uniform load, opposite 1st column, thickness. Increase the controlling thickness if details require, and select either plain or reinforced slab thickness and type.

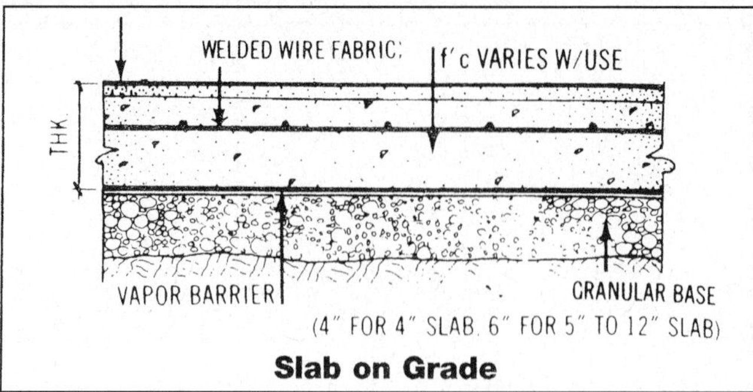

Slab on Grade

Thickness and Loading Assumptions by Type of Use

SLAB THICKNESS (IN.)	TYPE	A	B	C	D	◄ Slab I.D.
		Non	Light	Normal	Heavy	◄ Industrial
		Little	Light	Moderate	Severe	◄ Abrasion
		Foot Only	Pneumatic Wheels	Solid Rubber Wheels	Steel Tires	◄ Type of Traffic
		Load* (K)	Load* (K)	Load* (K)	Load* (K)	Max. Uniform Load to Slab ▼ (PSF)
4"	Reinf. Plain	4K				100
5"	Reinf. Plain	6K	4K			200
6"	Reinf. Plain		8K	6K	6K	500 to 800
7"	Reinf. Plain			9K	8K	1,500
8"	Reinf. Plain				11K	* Max. Wheel Load in Kips (incl. impact)
10"	Reinf. Plain				14K	
12"	Reinf. Plain					
DESIGN ASSUMPTIONS	Concrete, Chuted	f'c = 3.5 KSI	4 KSI	4.5 KSI	Slab @ 3.5 KSI	ASSUMPTIONS BY SLAB TYPE
	Toppings			1" Integral	1" Bonded	
	Finish	Steel Trowel	Steel Trowel	Steel Trowel	Screed & Steel Trowel	
	Compacted Granular Base	4" deep for 4" slab thickness 6" deep for 5" slab thickness & greater				ASSUMPTIONS FOR ALL SLAB TYPES
	Vapor Barrier	6 mil polyethylene				
	Forms & Joints	Allowances included				
	Rein-forcement	WWF as required ≥ 60,000 psi				

Concrete R0331 Structural Concrete

R033105-85 Lift Slabs

The cost advantage of the lift slab method is due to placing all concrete, reinforcing steel, inserts and electrical conduit at ground level and in reduction of formwork. Minimum economical project size is about 30,000 S.F. Slabs may be tilted for parking garage ramps.

It is now used in all types of buildings and has gone up to 22 stories high in apartment buildings. Current trend is to use post-tensioned flat plate slabs with spans from 22′ to 35′. Cylindrical void forms are used when deep slabs are required. One pound of prestressing steel is about equal to seven pounds of conventional reinforcing.

To be considered cured for stressing and lifting, a slab must have attained 75% of design strength. Seven days are usually sufficient with four to five days possible if high early strength cement is used. Slabs can be stacked using two coats of a non-bonding agent to insure that slabs do not stick to each other. Lifting is done by companies specializing in this work. Lift rate is 5′ to 15′ per hour with an average of 10′ per hour. Total areas up to 33,000 S.F. have been lifted at one time. 24 to 36 jacking columns are common. Most economical bay sizes are 24′ to 28′ with four to fourteen stories most efficient. Continuous design reduces reinforcing steel cost. Use of post-tensioned slabs allows larger bay sizes.

Concrete R0341 Precast Structural Concrete

R034105-30 Prestressed Precast Concrete Structural Units

Type	Location	Depth	Span in Ft.		Live Load Lb. per S.F.
Double Tee (8′ to 10′)	Floor	28″ to 34″	60 to 80		50 to 80
	Roof	12″ to 24″	30 to 50		40
	Wall	Width 8′	Up to 55′ high		Wind
Multiple Tee (8′)	Roof	8″ to 12″	15 to 40		40
	Floor	8″ to 12″	15 to 30		100
Plank	Roof		Roof	Floor	
		4″	13	12	40 for Roof
		6″	22	18	
or	or	8″	26	25	
		10″	33	29	100 for Floor
	Floor	12″	42	32	
Single Tee (8′ to 10′)	Roof	28″	40		40
		32″	80		
		36″	100		
		48″	120		
AASHO Girder	Bridges	Type 4	100		Highway
		5	110		
		6	125		
Box Beam (4′)	Bridges	15″ 27″ 33″	40 to 100		Highway

The majority of precast projects today utilize double tees rather than single tees because of speed and ease of installation. As a result casting beds at manufacturing plants are normally formed for double tees. Single tee projects will therefore require an initial set up charge to be spread over the individual single tee costs.

For floors, a 2″ to 3″ topping is field cast over the shapes. For roofs, insulating concrete or rigid insulation is placed over the shapes.

Member lengths up to 40′ are standard haul, 40′ to 60′ require special permits and lengths over 60′ must be escorted. Over width and/or over length can add up to 100% on hauling costs.

Large heavy members may require two cranes for lifting which would increase erection costs by about 45%. An eight man crew can install 12 to 20 double tees, or 45 to 70 quad tees or planks per day.

Grouting of connections must also be included.

Several system buildings utilizing precast members are available. Heights can go up to 22 stories for apartment buildings. Optimum design ratio is 3 S.F. of surface to 1 S.F. of floor area.

R034136-90 Prestressed Concrete, Post-Tensioned

In post-tensioned concrete the steel tendons are tensioned after the concrete has reached about 3/4 of its ultimate strength. The cableways are grouted after tensioning to provide bond between the steel and concrete. If bond is to be prevented, the tendons are coated with a corrosion-preventative grease and wrapped with waterproofed paper or plastic. Bonded tendons are usually used when ultimate strength (beams & girders) are controlling factors.

High strength concrete is used to fully utilize the steel, thereby reducing the size and weight of the member. A plasticizing agent may be added to reduce water content. Maximum size aggregate ranges from 1/2″ to 1-1/2″ depending on the spacing of the tendons.

The types of steel commonly used are bars and strands. Job conditions determine which is best suited. Bars are best for vertical prestresses since they are easy to support. The trend is for steel manufacturers to supply a finished package, cut to length, which reduces field preparation to a minimum.

Bars vary from 3/4″ to 1-3/8″ diameter. Table below gives time in labor-hours per tendon for placing, tensioning and grouting (if required) a 75′ beam. Tendons used in buildings are not usually grouted; tendons for bridges usually are grouted. For strands the table indicates the labor-hours per pound for typical prestressed units 100′ long. Simple span beams usually require one-end stressing regardless of lengths. Continuous beams are usually stressed from two ends. Long slabs are poured from the center outward and stressed in 75′ increments after the initial 150′ center pour.

Labor Hours per Tendon and per Pound of Prestressed Steel						
Length	**100′ Beam**		**75′ Beam**		**100′ Slab**	
Type Steel	**Strand**		**Bars**		**Strand**	
Diameter	**0.5″**		**3/4″**	**1-3/8″**	**0.5″**	**0.6″**
Number	**4**	**12**	**1**	**1**	**1**	**1**
Force in Kips	**100**	**300**	**42**	**143**	**25**	**35**
Preparation & Placing Cables	3.6	7.4	0.9	2.9	0.9	1.1
Stressing Cables	2.0	2.4	0.8	1.6	0.5	0.5
Grouting, if required	2.5	3.0	0.6	1.3		
Total Labor Hours	8.1	12.8	2.3	5.8	1.4	1.6
Prestressing Steel Weights (Lbs.)	215	640	115	380	53	74
Labor-hours per Lb. Bonded	0.038	0.020	0.020	0.015		
Non-bonded					0.026	0.022

Flat Slab construction — 4000 psi concrete with span-to-depth ratio between 36 and 44. Two way post-tensioned steel averages 1.0 lb. per S.F. for 24′ to 28′ bays (usually strand) and additional reinforcing steel averages .5 lb. per S.F.

Pan and Joist construction — 4000 psi concrete with span-to-depth ratio 28 to 30. Post-tensioned steel averages .8 lb. per S.F. and reinforcing steel about 1.0 lb. per S.F. Placing and stressing averages 40 hours per ton of total material.

Beam construction — 4000 to 5000 psi concrete. Steel weights vary greatly.

Labor cost per pound goes down as the size and length of the tendon increase. The primary economic consideration is the cost per kip for the member.

Post-tensioning becomes feasible for beams and girders over 30′ long; for continuous two-way slabs over 20′ clear; also in transferring upper building loads over longer spans at lower levels. Post-tension suppliers will provide engineering services at no cost to the user. Substantial economies are possible by using post-tensioned Lift Slabs.

R034713-20 Tilt Up Concrete Panels

The advantage of tilt up construction is in the low cost of forms and the placing of concrete and reinforcing. Panels up to 75′ high and 5-1/2″ thick have been tilted using strongbacks. Tilt up has been used for one to five story buildings and is well-suited for warehouses, stores, offices, schools and residences.

The panels are cast in forms on the floor slab. Most jobs use 5-1/2″ thick solid reinforced concrete panels. Sandwich panels with a layer of insulating materials are also used. Where dampness is a factor, lightweight aggregate is used. Optimum panel size is 300 to 500 S.F.

Slabs are usually poured with 3000 psi concrete which permits tilting seven days after pouring. Slabs may be stacked on top of each other and are separated from each other by either two coats of bond breaker or a film of polyethylene. Use of high early-strength cement allows tilting two days after a pour. Tilting up is done with a roller outrigger crane with a capacity of at least 1-1/2 times the weight of the panel at the required reach. Exterior precast columns can be set at the same time as the panels; interior precast columns can be set first and the panels clipped directly to them. The use of cast-in-place concrete columns is diminishing due to shrinkage problems. Structural steel columns are sometimes used if crane rails are planned. Panels can be clipped to the columns or lowered between the flanges. Steel channels with anchors may be used as edge forms for the slab. When the panels are lifted the channels form an integral steel column to take structural loads. Roof loads can be carried directly by the panels for wall heights to 14′.

Requirements of local building codes may be a limiting factor and should be checked. Building floor slabs should be poured first and should be a minimum of 5″ thick with 100% compaction of soil or 6″ thick with less than 100% compaction.

Setting times as fast as nine minutes per panel have been observed, but a safer expectation would be four panels per hour with a crane and a four-man setting crew. If crane erects from inside building, some provision must be made to get crane out after walls are erected. Good yarding procedure is important to minimize delays. Equalizing three-point lifting beams and self-releasing pick-up hooks speed erection. If panels must be carried to their final location, setting time per panel will be increased and erection costs may approach the erection cost range of architectural precast wall panels. Placing panels into slots formed in continuous footers will speed erection.

Reinforcing should be with #5 bars with vertical bars on the bottom. If surface is to be sandblasted, stainless steel chairs should be used to prevent rust staining.

Use of a broom finish is popular since the unavoidable surface blemishes are concealed.

Precast columns run from three to five times the C.Y. price of the panels only.

R035216-10 Lightweight Concrete

Lightweight aggregate concrete is usually purchased ready mixed, but it can also be field mixed.

Vermiculite or Perlite comes in bags of 4 C.F. under various trade names. Weight is about 8 lbs. per C.F. For insulating roof fill use 1:6 mix. For structural deck use 1:4 mix over gypsum boards, steeltex, steel centering, etc., supported by closely spaced joists or bulb trees. For structural slabs use 1:3:2 vermiculite sand concrete over steeltex, metal lath, steel centering, etc., on joists spaced 2′-0″ O.C. for maximum L.L. of 80 P.S.F. Use same mix for slab base fill over steel flooring or regular reinforced concrete slab when tile, terrazzo or other finish is to be laid over.

For slabs on grade use 1:3:2 mix when tile, etc., finish is to be laid over. If radiant heating units are installed use a 1:6 mix for a base. After coils are in place, cover with a regular granolithic finish (mix 1:3:2) to a minimum depth of 1-1/2″ over top of units.

Reinforce all slabs with 6 x 6 or 10 x 10 welded wire mesh.

R050516-30 Coating Structural Steel

On field-welded jobs, the shop-applied primer coat is necessarily omitted. All painting must be done in the field and usually consists of red oxide rust inhibitive paint or an aluminum paint. The table below shows paint coverage and daily production for field painting.

See Division 05 01 10.51 for steel surface preparation treatments such as wire brushing, pressure washing and sand blasting.

Type Construction	Surface Area per Ton	Coat	One Gallon Covers		In 8 Hrs. Person Covers		Average per Ton Spray	
			Brush	Spray	Brush	Spray	Gallons	Labor-hours
Light Structural	300 S.F. to 500 S.F.	1st	500 S.F.	455 S.F.	640 S.F.	2000 S.F.	0.9 gals.	1.6 L.H.
		2nd	450	410	800	2400	1.0	1.3
		3rd	450	410	960	3200	1.0	1.0
Medium	150 S.F. to 300 S.F.	All	400	365	1600	3200	0.6	0.6
Heavy Structural	50 S.F. to 150 S.F.	1st	400	365	1920	4000	0.2	0.2
		2nd	400	365	2000	4000	0.2	0.2
		3rd	400	365	2000	4000	0.2	0.2
Weighted Average	225 S.F.	All	400	365	1350	3000	0.6	0.6

R050521-20 Welded Structural Steel

Usual weight reductions with welded design run 10% to 20% compared with bolted or riveted connections. This amounts to about the same total cost compared with bolted structures since field welding is more expensive than bolts. For normal spans of 18′ to 24′ figure 6 to 7 connections per ton.

Trusses — For welded trusses add 4% to weight of main members for connections. Up to 15% less steel can be expected in a welded truss compared to one that is shop bolted. Cost of erection is the same whether shop bolted or welded.

General — Typical electrodes for structural steel welding are E6010, E6011, E60T and E70T. Typical buildings vary between 2# to 8# of weld rod per ton of steel. Buildings utilizing continuous design require about three times as much welding as conventional welded structures. In estimating field erection by welding, it is best to use the average linear feet of weld per ton to arrive at the welding cost per ton. The type, size and position of the weld will have a direct bearing on the cost per linear foot. A typical field welder will deposit 1.8# to 2# of weld rod per hour manually. Using semiautomatic methods can increase production by as much as 50% to 75%.

R051223-10 Structural Steel

The bare material prices for structural steel, shown in the unit cost sections of the book, are for 100 tons of shop-fabricated structural steel and include:

1. Mill base price of structural steel
2. Mill scrap/grade/size/length extras
3. Mill delivery to a metals service center (warehouse)
4. Service center storage and handling
5. Service center delivery to a fabrication shop
6. Shop storage and handling
7. Shop drafting/detailing
8. Shop fabrication
9. Shop coat of primer paint
10. Shop listing
11. Shop delivery to the job site

In unit cost sections of the book that contain items for field fabrication of steel components, the bare material cost of steel includes:

1. Mill base price of structural steel
2. Mill scrap/grade/size/length extras
3. Mill delivery to a metals service center (warehouse)
4. Service center storage and handling
5. Service center delivery to the job site

R051223-20 Steel Estimating Quantities

One estimate on erection is that a crane can handle 35 to 60 pieces per day. Say the average is 45. With usual sizes of beams, girders, and columns, this would amount to about 20 tons per day. The type of connection greatly affects the speed of erection. Moment connections for continuous design slow down production and increase erection costs.

Short open web bar joists can be set at the rate of 75 to 80 per day, with 50 per day being the average for setting long span joists.

After main members are calculated, add the following for usual allowances: base plates 2% to 3%; column splices 4% to 5%; and miscellaneous details 4% to 5%, for a total of 10% to 13% in addition to main members.

The ratio of column to beam tonnage varies depending on type of steels used, typical spans, story heights and live loads.

It is more economical to keep the column size constant and to vary the strength of the column by using high strength steels. This also saves floor space. Buildings have recently gone as high as ten stories with 8″ high strength columns. For light columns under W8X31 lb. sections, concrete filled steel columns are economical.

High strength steels may be used in columns and beams to save floor space and to meet head room requirements. High strength steels in some sizes sometimes require long lead times.

Round, square and rectangular columns, both plain and concrete filled, are readily available and save floor area, but are higher in cost per pound than rolled columns. For high unbraced columns, tube columns may be less expensive.

Below are average minimum figures for the weights of the structural steel frame for different types of buildings using A36 steel, rolled shapes and simple joints. For economy in domes, rise to span ratio = .13. Open web joist framing systems will reduce weights by 10% to 40%. Composite design can reduce steel weight by up to 25% but additional concrete floor slab thickness may be required. Continuous design can reduce the weights up to 20%. There are many building codes with different live load requirements and different structural requirements, such as hurricane and earthquake loadings which can alter the figures.

Structural Steel Weights per S.F. of Floor Area

Type of Building	No. of Stories	Avg. Spans	L.L. #/S.F.	Lbs. Per S.F.	Type of Building	No. of Stories	Avg. Spans	L.L. #/S.F.	Lbs. Per S.F.
Steel Frame Mfg.	1	20′x20′	40	8	Apartments	2-8	20′x20′	40	8
		30′x30′		13		9-25			14
		40′x40′		18	Office	to 10	Various	80	10
Parking garage	4	Various	80	8.5		20			18
Domes (Schwedler)*	1	200′	30	10		30			26
		300′		15		over 50			35

R051223-25 Common Structural Steel Specifications

ASTM A992 (formerly A36, then A572 Grade 50) is the all-purpose carbon grade steel widely used in building and bridge construction.

The other high-strength steels listed below may each have certain advantages over ASTM A992 stuctural carbon steel, depending on the application. They have proven to be economical choices where, due to lighter members, the reduction of dead load and the associated savings in shipping cost can be significant.

ASTM A588 atmospheric weathering, high-strength low-alloy steels can be used in the bare (uncoated) condition, where exposure to normal atmosphere causes a tightly adherant oxide to form on the surface protecting the steel from further oxidation. ASTM A242 corrosion-resistant, high-strength low-alloy steels have enhanced atmospheric corrosion resistance of at least two times that of carbon structural steels with copper, or four times that of carbon structural steels without copper. The reduction or elimination of maintenance resulting from the use of these steels often offsets their higher initial cost.

Steel Type	ASTM Designation	Minimum Yield Stress in KSI	Shapes Available
Carbon	A36	36	All structural shape groups, and plates & bars up thru 8″ thick
	A529	50	Structural shape group 1, and plates & bars up thru 2″ thick
		50	All structural shape groups
High-Strength Low-Alloy	A913	60	
Quenched & Self-Tempered		65	
		70	
		42	All structural shape groups, and plates & bars up thru 6″ thick
High-Strength Low-Alloy	A572	50	All structural shape groups, and plates & bars up thru 4″ thick
Columbium-Vanadium		55	Structural shape groups 1 & 2, and plates & bars up thru 2″ thick
		60	Structural shape groups 1 & 2, and plates & bars up thru 1-1/4″ thick
		65	Structural shape group 1, and plates & bars up thru 1-1/4″ thick
High-Strength Low-Alloy	A992	50	All structural shape groups
Columbium-Vanadium			
Weathering		42	Structural shape groups 4 & 5, and plates & bars over 1-1/2″ up thru 4″ thick
High-Strength Low-Alloy	A242	46	Structural shape group 3, and plates & bars over 3/4″ up thru 1-1/2″ thick
		50	Structural shape groups 1 & 2, and plates & bars up thru 3/4″ thick
Weathering		42	Plates & bars over 5″ up thru 8″ thick
High-Strength Low-Alloy	A588	46	Plates & bars over 4″ up thru 5″ thick
		50	All structural shape groups, and plates & bars up thru 4″ thick
Quenched and Tempered	A852	70	Plates & bars up thru 4″ thick
Low-Alloy			
Quenched and Tempered	A514	90	Plates & bars over 2-1/2″ up thru 6″ thick
Alloy		100	Plates & bars up thru 2-1/2″ thick

R051223-30 High Strength Steels

The mill price of high strength steels may be higher than A992 carbon steel but their proper use can achieve overall savings thru total reduced weights. For columns with L/r over 100, A992 steel is best; under 100, high strength steels are economical. For heavy columns, high strength steels are economical when cover plates are eliminated. There is no economy using high strength steels for clip angles or supports or for beams where deflection governs. Thinner members are more economical than thick.

The per ton erection and fabricating costs of the high strength steels will be higher than for A992 since the same number of pieces, but less weight, will be installed.

R051223-35 Common Steel Sections

The upper portion of this table shows the name, shape, common designation and basic characteristics of commonly used steel sections. The lower portion explains how to read the designations used for the above illustrated common sections.

Shape & Designation	Name & Characteristics	Shape & Designation	Name & Characteristics
W	W Shape Parallel flange surfaces	MC	Miscellaneous Channel Infrequently rolled by some producers
S	American Standard Beam (I Beam) Sloped inner flange	L	Angle Equal or unequal legs, constant thickness
M	Miscellaneous Beams Cannot be classified as W, HP or S; infrequently rolled by some producers	T	Structural Tee Cut from W, M or S on center of web
C	American Standard Channel Sloped inner flange	HP	Bearing Pile Parallel flanges and equal flange and web thickness

Common drawing designations follow:

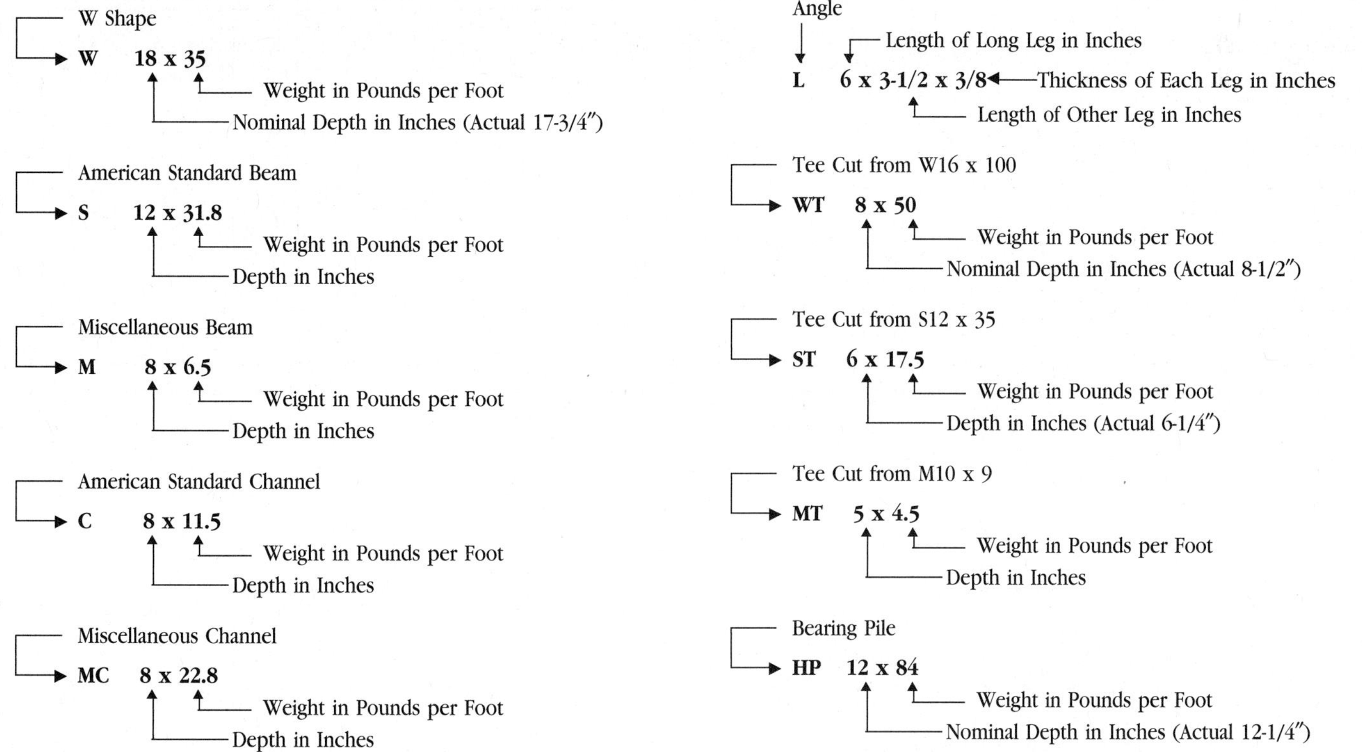

R051223-45 Installation Time for Structural Steel Building Components

The following tables show the expected average installation times for various structural steel shapes. Table A presents installation times for columns, Table B for beams, Table C for light framing and bolts, and Table D for structural steel for various project types.

Table A

Description	Labor-Hours	Unit
Columns		
Steel, Concrete Filled		
3-1/2" Diameter	.933	Ea.
6-5/8" Diameter	1.120	Ea.
Steel Pipe		
3" Diameter	.933	Ea.
8" Diameter	1.120	Ea.
12" Diameter	1.244	Ea.
Structural Tubing		
4" x 4"	.966	Ea.
8" x 8"	1.120	Ea.
12" x 8"	1.167	Ea.
W Shape 2 Tier		
W8 x 31	.052	L.F.
W8 x 67	.057	L.F.
W10 x 45	.054	L.F.
W10 x 112	.058	L.F.
W12 x 50	.054	L.F.
W12 x 190	.061	L.F.
W14 x 74	.057	L.F.
W14 x 176	.061	L.F.

Table B

Description	Labor-Hours	Unit	Labor-Hours	Unit
Beams, W Shape				
W6 x 9	.949	Ea.	.093	L.F.
W10 x 22	1.037	Ea.	.085	L.F.
W12 x 26	1.037	Ea.	.064	L.F.
W14 x 34	1.333	Ea.	.069	L.F.
W16 x 31	1.333	Ea.	.062	L.F.
W18 x 50	2.162	Ea.	.088	L.F.
W21 x 62	2.222	Ea.	.077	L.F.
W24 x 76	2.353	Ea.	.072	L.F.
W27 x 94	2.581	Ea.	.067	L.F.
W30 x 108	2.857	Ea.	.067	L.F.
W33 x 130	3.200	Ea.	.071	L.F.
W36 x 300	3.810	Ea.	.077	L.F.

Table C

Description	Labor-Hours	Unit
Light Framing		
Angles 4" and Larger	.055	lbs.
Less than 4"	.091	lbs.
Channels 8" and Larger	.048	lbs.
Less than 8"	.072	lbs.
Cross Bracing Angles	.055	lbs.
Rods	.034	lbs.
Hanging Lintels	.069	lbs.
High Strength Bolts in Place		
3/4" Bolts	.070	Ea.
7/8" Bolts	.076	Ea.

Table D

Description	Labor-Hours	Unit	Labor-Hours	Unit
Apartments, Nursing Homes, etc.				
1-2 Stories	4.211	Piece	7.767	Ton
3-6 Stories	4.444	Piece	7.921	Ton
7-15 Stories	4.923	Piece	9.014	Ton
Over 15 Stories	5.333	Piece	9.209	Ton
Offices, Hospitals, etc.				
1-2 Stories	4.211	Piece	7.767	Ton
3-6 Stories	4.741	Piece	8.889	Ton
7-15 Stories	4.923	Piece	9.014	Ton
Over 15 Stories	5.120	Piece	9.209	Ton
Industrial Buildings				
1 Story	3.478	Piece	6.202	Ton

R051223-80 Dimensions and Weights of Sheet Steel

Gauge No.	Approximate Thickness				Weight		
	Inches (in fractions)	Inches (in decimal parts)		Millimeters			
	Wrought Iron	Wrought Iron	Steel	Steel	per S.F. in Ounces	per S.F. in Lbs.	per Square Meter in Kg.
0000000	1/2″	.5	.4782	12.146	320	20.000	97.650
000000	15/32″	.46875	.4484	11.389	300	18.750	91.550
00000	7/16″	.4375	.4185	10.630	280	17.500	85.440
0000	13/32″	.40625	.3886	9.870	260	16.250	79.330
000	3/8″	.375	.3587	9.111	240	15.000	73.240
00	11/32″	.34375	.3288	8.352	220	13.750	67.130
0	5/16″	.3125	.2989	7.592	200	12.500	61.030
1	9/32″	.28125	.2690	6.833	180	11.250	54.930
2	17/64″	.265625	.2541	6.454	170	10.625	51.880
3	1/4″	.25	.2391	6.073	160	10.000	48.820
4	15/64″	.234375	.2242	5.695	150	9.375	45.770
5	7/32″	.21875	.2092	5.314	140	8.750	42.720
6	13/64″	.203125	.1943	4.935	130	8.125	39.670
7	3/16″	.1875	.1793	4.554	120	7.500	36.320
8	11/64″	.171875	.1644	4.176	110	6.875	33.570
9	5/32″	.15625	.1495	3.797	100	6.250	30.520
10	9/64″	.140625	.1345	3.416	90	5.625	27.460
11	1/8″	.125	.1196	3.038	80	5.000	24.410
12	7/64″	.109375	.1046	2.657	70	4.375	21.360
13	3/32″	.09375	.0897	2.278	60	3.750	18.310
14	5/64″	.078125	.0747	1.897	50	3.125	15.260
15	9/128″	.0713125	.0673	1.709	45	2.813	13.730
16	1/16″	.0625	.0598	1.519	40	2.500	12.210
17	9/160″	.05625	.0538	1.367	36	2.250	10.990
18	1/20″	.05	.0478	1.214	32	2.000	9.765
19	7/160″	.04375	.0418	1.062	28	1.750	8.544
20	3/80″	.0375	.0359	.912	24	1.500	7.324
21	11/320″	.034375	.0329	.836	22	1.375	6.713
22	1/32″	.03125	.0299	.759	20	1.250	6.103
23	9/320″	.028125	.0269	.683	18	1.125	5.490
24	1/40″	.025	.0239	.607	16	1.000	4.882
25	7/320″	.021875	.0209	.531	14	.875	4.272
26	3/160″	.01875	.0179	.455	12	.750	3.662
27	11/640″	.0171875	.0164	.417	11	.688	3.357
28	1/64″	.015625	.0149	.378	10	.625	3.052

R053100-10 Decking Descriptions

General - All Deck Products

Steel deck is made by cold forming structural grade sheet steel into a repeating pattern of parallel ribs. The strength and stiffness of the panels are the result of the ribs and the material properties of the steel. Deck lengths can be varied to suit job conditions, but because of shipping considerations, are usually less than 40 feet. Standard deck width varies with the product used but full sheets are usually 12″, 18″, 24″, 30″, or 36″. Deck is typically furnished in a standard width with the ends cut square. Any cutting for width, such as at openings or for angular fit, is done at the job site.

Deck is typically attached to the building frame with arc puddle welds, self-drilling screws, or powder or pneumatically driven pins. Sheet to sheet fastening is done with screws, button punching (crimping), or welds.

Composite Floor Deck

After installation and adequate fastening, floor deck serves several purposes. It (a) acts as a working platform, (b) stabilizes the frame, (c) serves as a concrete form for the slab, and (d) reinforces the slab to carry the design loads applied during the life of the building. Composite decks are distinguished by the presence of shear connector devices as part of the deck. These devices are designed to mechanically lock the concrete and deck together so that the concrete and the deck work together to carry subsequent floor loads. These shear connector devices can be rolled-in embossments, lugs, holes, or wires welded to the panels. The deck profile can also be used to interlock concrete and steel.

Composite deck finishes are either galvanized (zinc coated) or phosphatized/painted. Galvanized deck has a zinc coating on both the top and bottom surfaces. The phosphatized/painted deck has a bare (phosphatized) top surface that will come into contact with the concrete. This bare top surface can be expected to develop rust before the concrete is placed. The bottom side of the deck has a primer coat of paint.

Composite floor deck is normally installed so the panel ends do not overlap on the supporting beams. Shear lugs or panel profile shape often prevent a tight metal to metal fit if the panel ends overlap; the air gap caused by overlapping will prevent proper fusion with the structural steel supports when the panel end laps are shear stud welded.

Adequate end bearing of the deck must be obtained as shown on the drawings. If bearing is actually less in the field than shown on the drawings, further investigation is required.

Roof Deck

Roof deck is not designed to act compositely with other materials. Roof deck acts alone in transferring horizontal and vertical loads into the building frame. Roof deck rib openings are usually narrower than floor deck rib openings. This provides adequate support of rigid thermal insulation board.

Roof deck is typically installed to endlap approximately 2″ over supports. However, it can be butted (or lapped more than 2″) to solve field fit problems. Since designers frequently use the installed deck system as part of the horizontal bracing system (the deck as a diaphragm), any fastening substitution or change should be approved by the designer. Continuous perimeter support of the deck is necessary to limit edge deflection in the finished roof and may be required for diaphragm shear transfer.

Standard roof deck finishes are galvanized or primer painted. The standard factory applied paint for roof deck is a primer paint and is not intended to weather for extended periods of time. Field painting or touching up of abrasions and deterioration of the primer coat or other protective finishes is the responsibility of the contractor.

Cellular Deck

Cellular deck is made by attaching a bottom steel sheet to a roof deck or composite floor deck panel. Cellular deck can be used in the same manner as floor deck. Electrical, telephone, and data wires are easily run through the chase created between the deck panel and the bottom sheet.

When used as part of the electrical distribution system, the cellular deck must be installed so that the ribs line up and create a smooth cell transition at abutting ends. The joint that occurs at butting cell ends must be taped or otherwise sealed to prevent wet concrete from seeping into the cell. Cell interiors must be free of welding burrs, or other sharp intrusions, to prevent damage to wires.

When used as a roof deck, the bottom flat plate is usually left exposed to view. Care must be maintained during erection to keep good alignment and prevent damage.

Cellular deck is sometimes used with the flat plate on the top side to provide a flat working surface. Installation of the deck for this purpose requires special methods for attachment to the frame because the flat plate, now on the top, can prevent direct access to the deck material that is bearing on the structural steel. It may be advisable to treat the flat top surface to prevent slipping.

Cellular deck is always furnished galvanized or painted over galvanized.

Form Deck

Form deck can be any floor or roof deck product used as a concrete form. Connections to the frame are by the same methods used to anchor floor and roof deck. Welding washers are recommended when welding deck that is less than 20 gauge thickness.

Form deck is furnished galvanized, prime painted, or uncoated. Galvanized deck must be used for those roof deck systems where form deck is used to carry a lightweight insulating concrete fill.

Wood, Plastics & Comp. R0611 Wood Framing

R061110-30 Lumber Product Material Prices

The price of forest products fluctuates widely from location to location and from season to season depending upon economic conditions. The bare material prices in the unit cost sections of the book show the National Average material prices in effect Jan. 1 of this book year. It must be noted that lumber prices in general may change significantly during the year.

Availability of certain items depends upon geographic location and must be checked prior to firm-price bidding.

R133419-10 Pre-Engineered Steel Buildings

These buildings are manufactured by many companies and normally erected by franchised dealers throughout the U.S. The four basic types are: Rigid Frames, Truss type, Post and Beam and the Sloped Beam type. Most popular roof slope is low pitch of 1″ in 12″. The minimum economical area of these buildings is about 3000 S.F. of floor area. Bay sizes are usually 20′ to 24′ but can go as high as 30′ with heavier girts and purlins. Eave heights are usually 12′ to 24′ with 18′ to 20′ most typical.

Material prices shown in the Unit Price section are bare costs for the building shell only and do not include floors, foundations, anchor bolts, interior finishes or utilities. Costs assume at least three bays of 24′ each, a 1″ in 12″ roof slope, and they are based on 30 psf roof load and 20 psf wind load (wind load is a function of wind speed, building height, and terrain characteristics; this should be determined by a Registered Structural Engineer) and no unusual requirements. Costs include the structural frame, 26 ga. non-insulated colored corrugated or ribbed roofing and siding panels, fasteners, closures, trim and flashing but no allowance for insulation, doors, windows, skylights, gutters or downspouts. Very large projects would generally cost less for materials than the prices shown. For roof panel substitutions and wall panel substitutions, see appropriate Unit Price sections.

Conditions at the site, weather, shape and size of the building, and labor availability will affect the erection cost of the building.

R220102-20 Labor Adjustment Factors

Labor Adjustment Factors are provided for Divisions 21, 22, and 23 to assist the mechanical estimator account for the various complexities and special conditions of any particular project. While a single percentage has been entered on each line of Division 22 01 02.20, it should be understood that these are just suggested midpoints of ranges of values commonly used by mechanical estimators. They may be increased or decreased depending on the severity of the special conditions.

The group for ″existing occupied buildings″, has been the subject of requests for explanation. Actually there are two stages to this group: buildings that are existing and ″finished″ but unoccupied, and those that also are occupied. Buildings that are ″finished″ may result in higher labor costs due to the workers having to be more careful not to damage finished walls, ceilings, floors, etc. and may necessitate special protective coverings and barriers. Also corridor bends and doorways may not accommodate long pieces of pipe or larger pieces of equipment. Work above an already hung ceiling can be very time consuming. The addition of occupants may force the work to be done on premium time (nights and/or weekends), eliminate the possible use of some preferred tools such as pneumatic drivers, powder charged drivers etc. The estimator should evaluate the access to the work area and just how the work is going to be accomplished to arrive at an increase in labor costs over ″normal″ new construction productivity.

R220523-80 Valve Materials

VALVE MATERIALS

Bronze:
Bronze is one of the oldest materials used to make valves. It is most commonly used in hot and cold water systems and other non-corrosive services. It is often used as a seating surface in larger iron body valves to ensure tight closure.

Carbon Steel:
Carbon steel is a high strength material. Therefore, valves made from this metal are used in higher pressure services, such as steam lines up to 600 psi at 850° F. Many steel valves are available with butt-weld ends for economy and are generally used in high pressure steam service as well as other higher pressure non-corrosive services.

Forged Steel:
Valves from tough carbon steel are used in service up to 2000 psi and temperatures up to 1000° F in Gate, Globe and Check valves.

Iron:
Valves are normally used in medium to large pipe lines to control non-corrosive fluid and gases, where pressures do not exceed 250 psi at 450° F or 500 psi cold water, oil or gas.

Stainless Steel:
Developed steel alloys can be used in over 90% corrosive services.

Plastic PVC:
This is used in a great variety of valves generally in high corrosive service with lower temperatures and pressures.

VALVE SERVICE PRESSURES

Pressure ratings on valves provide an indication of the safe operating pressure for a valve at some elevated temperature. This temperature is dependent upon the materials used and the fabrication of the valve. When specific data is not available, a good "rule-of-thumb" to follow is the temperature of saturated steam on the primary rating indicated on the valve body. Example: The valve has the number 150S printed on the side indicating 150 psi and hence, a maximum operating temperature of 367° F (temperature of saturated steam and 150 psi).

DEFINITIONS

1. "WOG" – Water, oil, gas (cold working pressures).
2. "SWP" – Steam working pressure.
3. 100% area (full port) – means the area through the valve is equal to or greater than the area of standard pipe.
4. "Standard Opening" – means that the area through the valve is less than the area of standard pipe and therefore these valves should be used only where restriction of flow is unimportant.
5. "Round Port" – means the valve has a full round opening through the plug and body, of the same size and area as standard pipe.
6. "Rectangular Port" – valves have rectangular shaped ports through the plug body. The area of the port is either equal to 100% of the area of standard pipe, or restricted (standard opening). In either case it is clearly marked.
7. "ANSI" – American National Standards Institute.

R220523-90 Valve Selection Considerations

INTRODUCTION: In any piping application, valve performance is critical. Valves should be selected to give the best performance at the lowest cost.

The following is a list of performance characteristics generally expected of valves.

1. Stopping flow or starting it.
2. Throttling flow (Modulation).
3. Flow direction changing.
4. Checking backflow (Permitting flow in only one direction).
5. Relieving or regulating pressure.

In order to properly select the right valve, some facts must be determined.

A. What liquid or gas will flow through the valve?

B. Does the fluid contain suspended particles?

C. Does the fluid remain in liquid form at all times?

D. Which metals does fluid corrode?

E. What are the pressure and temperature limits? (As temperature and pressure rise, so will the price of the valve.)

F. Is there constant line pressure?

G. Is the valve merely an on-off valve?

H. Will checking of backflow be required?

I. Will the valve operate frequently or infrequently?

Valves are classified by design type into such classifications as Gate, Globe, Angle, Check, Ball, Butterfly and Plug. They are also classified by end connection, stem, pressure restrictions and material such as bronze, cast iron, etc. Each valve has a specific use. A quality valve used correctly will provide a lifetime of trouble- free service, but a high quality valve installed in the wrong service may require frequent attention.

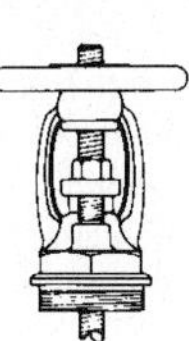

STEM TYPES
(OS & Y)—Rising Stem-Outside Screw and Yoke

Offers a visual indication of whether the valve is open or closed. Recommended where high temperatures, corrosives, and solids in the line might cause damage to inside-valve stem threads. The stem threads are engaged by the yoke bushing so the stem rises through the hand wheel as it is turned.

(R.S.)—Rising Stem-Inside Screw

Adequate clearance for operation must be provided because both the hand wheel and the stem rise.
The valve wedge position is indicated by the position of the stem and hand wheel.

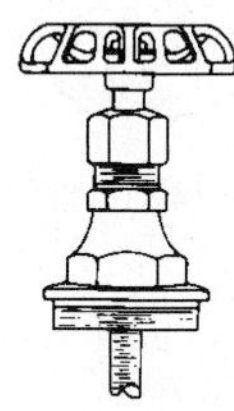

(N.R.S.)—Non-Rising Stem-Inside Screw

A minimum clearance is required for operating this type of valve. Excessive wear or damage to stem threads inside the valve may be caused by heat, corrosion, and solids. Because the hand wheel and stem do not rise, wedge position cannot be visually determined.

VALVE TYPES
Gate Valves

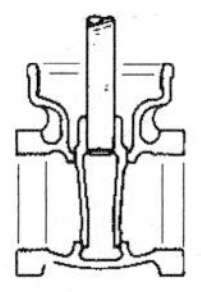

Provide full flow, minute pressure drop, minimum turbulence and minimum fluid trapped in the line.
They are normally used where operation is infrequent.

Globe Valves

Globe valves are designed for throttling and/or frequent operation with positive shut-off. Particular attention must be paid to the several types of seating materials available to avoid unnecessary wear. The seats must be compatible with the fluid in service and may be composition or metal. The configuration of the Globe valve opening causes turbulence which results in increased resistance. Most bronze Globe valves are rising stem-inside screw, but they are also available on O.S. & Y.

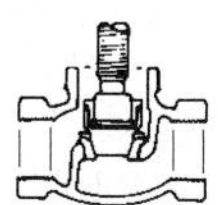

Angle Valves

The fundamental difference between the Angle valve and the Globe valve is the fluid flow through the Angle valve. It makes a 90° turn and offers less resistance to flow than the Globe valve while replacing an elbow. An Angle valve thus reduces the number of joints and installation time.

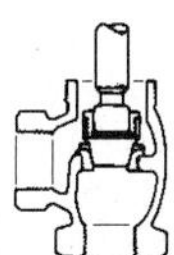

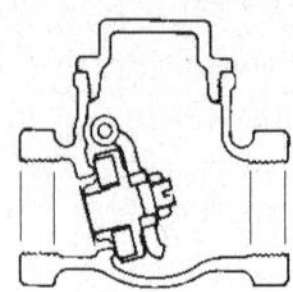

Check Valves

Check valves are designed to prevent backflow by automatically seating when the direction of fluid is reversed.
Swing Check valves are generally installed with Gate-valves, as they provide comparable full flow. Usually recommended for lines where flow velocities are low and should not be used on lines with pulsating flow. Recommended for horizontal installation, or in vertical lines only where flow is upward.

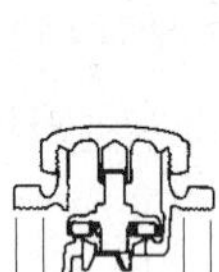

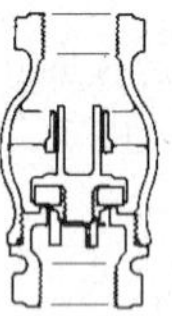

Lift Check Valves

These are commonly used with Globe and Angle valves since they have similar diaphragm seating arrangements and are recommended for preventing backflow of steam, air, gas and water, and on vapor lines with high flow velocities. For horizontal lines, horizontal lift checks should be used and vertical lift checks for vertical lines.

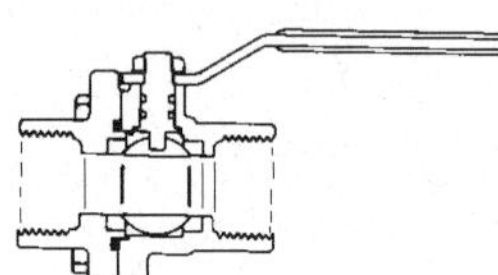

Ball Valves

Ball valves are light and easily installed, yet because of modern elastomeric seats, provide tight closure. Flow is controlled by rotating up to 90° a drilled ball which fits tightly against resilient seals. This ball seats with flow in either direction, and valve handle indicates the degree of opening. Recommended for frequent operation readily adaptable to automation, ideal for installation where space is limited.

Butterfly Valves

Butterfly valves provide bubble-tight closure with excellent throttling characteristics. They can be used for full-open, closed and for throttling applications.

The Butterfly valve consists of a disc within the valve body which is controlled by a shaft. In its closed position, the valve disc seals against a resilient seat. The disc position throughout the full 90° rotation is visually indicated by the position of the operator.

A Butterfly valve is only a fraction of the weight of a Gate valve and requires no gaskets between flanges in most cases. Recommended for frequent operation and adaptable to automation where space is limited.

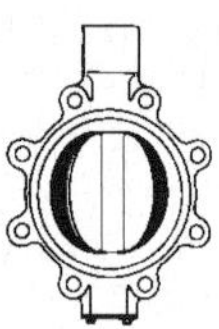

Wafer and Lug type bodies when installed between two pipe flanges, can be easily removed from the line. The pressure of the bolted flanges holds the valve in place.
Locating lugs makes installation easier.

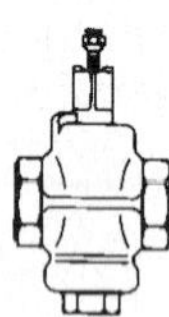

Plug Valves

Lubricated plug valves, because of the wide range of service to which they are adapted, may be classified as all purpose valves. They can be safely used at all pressure and vacuums, and at all temperatures up to the limits of available lubricants. They are the most satisfactory valves for the handling of gritty suspensions and many other destructive, erosive, corrosive and chemical solutions.

R221113-50 Pipe Material Considerations

1. Malleable fittings should be used for gas service.
2. Malleable fittings are used where there are stresses/strains due to expansion and vibration.
3. Cast fittings may be broken as an aid to disassembling of heating lines frozen by long use, temperature and minerals.
4. Cast iron pipe is extensively used for underground and submerged service.
5. Type M (light wall) copper tubing is available in hard temper only and is used for nonpressure and less severe applications than K and L.
6. Type L (medium wall) copper tubing, available hard or soft for interior service.
7. Type K (heavy wall) copper tubing, available in hard or soft temper for use where conditions are severe. For underground and interior service.
8. Hard drawn tubing requires fewer hangers or supports but should not be bent. Silver brazed fittings are recommended, however soft solder is normally used.
9. Type DMV (very light wall) copper tubing designed for drainage, waste and vent plus other non-critical pressure services.

Domestic/Imported Pipe and Fittings Cost

The prices shown in this publication for steel/cast iron pipe and steel, cast iron, malleable iron fittings are based on domestic production sold at the normal trade discounts. The above listed items of foreign manufacture may be available at prices of 1/3 to 1/2 those shown. Some imported items after minor machining or finishing operations are being sold as domestic to further complicate the system.

Caution: Most pipe prices in this book also include a coupling and pipe hangers which for the larger sizes can add significantly to the per foot cost and should be taken into account when comparing "book cost" with quoted supplier's cost.

R260519-94 Size Required and Weight (Lbs./1000 L.F.) of Aluminum and Copper THW Wire by Ampere Load

Amperes	Copper Size	Aluminum Size	Copper Weight	Aluminum Weight
15	14	12	24	11
20	12	10	33	17
30	10	8	48	39
45	8	6	77	52
65	6	4	112	72
85	4	2	167	101
100	3	1	205	136
115	2	1/0	252	162
130	1	2/0	324	194
150	1/0	3/0	397	233
175	2/0	4/0	491	282
200	3/0	250	608	347
230	4/0	300	753	403
255	250	400	899	512
285	300	500	1068	620
310	350	500	1233	620
335	400	600	1396	772
380	500	750	1732	951

R260533-70 Pull Boxes and Cabinets

List cabinets and pull boxes by NEMA type and size.

Example:	TYPE	SIZE
	NEMA 1	6″W x 6″H x 4″D
	NEMA 3R	6″W x 6″H x 4″D

Labor-hours for wall mount (indoor or outdoor) installations include:
1. Unloading and uncrating
2. Handling of enclosures up to 200′ from loading dock using a dolly or pipe rollers
3. Measuring and marking
4. Drilling (4) anchor type lead fasteners using a hammer drill
5. Mounting and leveling boxes

Note: A plywood backboard is not included.

Labor-hours for ceiling mounting include:
1. Unloading and uncrating
2. Handling boxes up to 100′ from loading dock
3. Measuring and marking
4. Drilling (4) anchor type lead fasteners using a hammer drill
5. Installing and leveling boxes to a height of 15′ using rolling staging

Labor-hours for free standing cabinets include:
1. Unloading and uncrating
2. Handling of cabinets up to 200′ from loading dock using a dolly or pipe rollers
3. Marking of floor
4. Drilling (4) anchor type lead fasteners using a hammer drill
5. Leveling and shimming

Labor-hours for telephone cabinets include:
1. Unloading and uncrating
2. Handling cabinets up to 200′ using a dolly or pipe rollers
3. Measuring and marking
4. Mounting and leveling, using (4) lead anchor type fasteners

R260533-75 Weight Comparisons of Common Size Cast Boxes in Lbs.

Size NEMA 4 or 9	Cast Iron	Cast Aluminum	Size NEMA 7	Cast Iron	Cast Aluminum
6″ x 6″ x 6″	17	7	6″ x 6″ x 6″	40	15
8″ x 6″ x 6″	21	8	8″ x 6″ x 6″	50	19
10″ x 6″ x 6″	23	9	10″ x 6″ x 6″	55	21
12″ x 12″ x 6″	52	20	12″ x 6″ x 6″	100	37
16″ x 16″ x 6″	97	36	16″ x 16″ x 6″	140	52
20″ x 20″ x 6″	133	50	20″ x 20″ x 6″	180	67
24″ x 18″ x 8″	149	56	24″ x 18″ x 8″	250	93
24″ x 24″ x 10″	238	88	24″ x 24″ x 10″	358	133
30″ x 24″ x 12″	324	120	30″ x 24″ x 10″	475	176
36″ x 36″ x 12″	500	185	30″ x 24″ x 12″	510	189

R260590-05 Typical Overhead Service Entrance

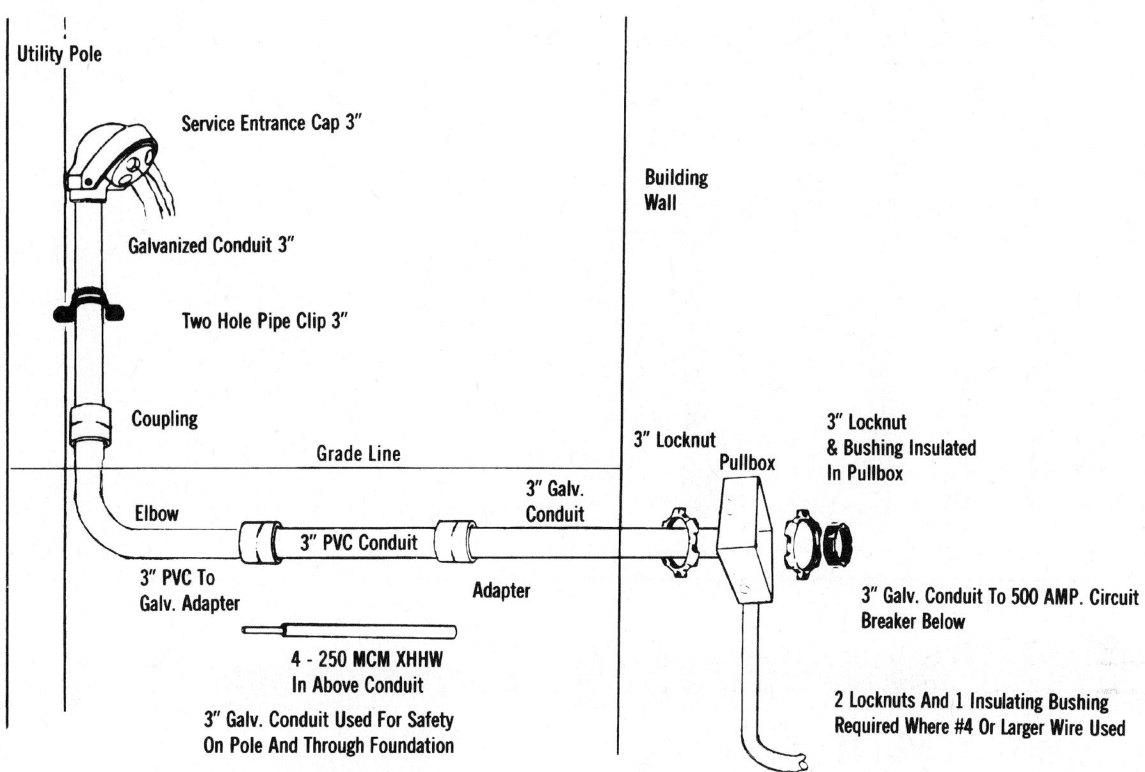

R262213-60 Oil Filled Transformers

Transformers in this section include:
1. Rigging (as required)
2. Rental of crane and operator
3. Setting of oil filled transformer
4. (4) Anchor bolts, nuts and washers in concrete pad

Price does not include:
1. Primary and secondary terminations
2. Transformer pad
3. Equipment grounding
4. Cable
5. Conduit locknuts or bushings

Transformers in Unit Price sections for dry type, back-boost and isolating transformers include:
1. Unloading and uncrating
2. Hauling transformer to within 200′ of loading dock
3. Setting in place
4. Wall mounting hardware
5. Testing

Price does not include:
1. Structural supports
2. Suspension systems
3. Welding or fabrication
4. Primary & secondary terminations

Add the following percentages to the labor for ceiling mounted transformers:

10' to 15′	= + 15%
15′ to 25′	= + 30%
Over 25′	= + 35%

Job Conditions: Productivities are based on new construction. Installation is assumed to be on the first floor, in an obstructed area to a height of 10′. Material staging area is within 100′ of final transformer location.

Electrical — R2622 Low Voltage Transformers

R262213-65 Transformer Weight (Lbs.) by kVA

Oil Filled 3 Phase 5/15 KV To 480/277			
kVA	Lbs.	kVA	Lbs.
150	1800	1000	6200
300	2900	1500	8400
500	4700	2000	9700
750	5300	3000	15000
Dry 240/480 To 120/240 Volt			
1 Phase		3 Phase	
kVA	Lbs.	kVA	Lbs.
1	23	3	90
2	36	6	135
3	59	9	170
5	73	15	220
7.5	131	30	310
10	149	45	400
15	205	75	600
25	255	112.5	950
37.5	295	150	1140
50	340	225	1575
75	550	300	1870
100	670	500	2850
167	900	750	4300

Electrical — R2625 Enclosed Bus Assemblies

R262513-15 Weight (Lbs./L.F.) of 4 Pole Aluminum and Copper Bus Duct by Ampere Load

Amperes	Aluminum Feeder	Copper Feeder	Aluminum Plug-In	Copper Plug-In
225			7	7
400			8	13
600	10	10	11	14
800	10	19	13	18
1000	11	19	16	22
1350	14	24	20	30
1600	17	26	25	39
2000	19	30	29	46
2500	27	43	36	56
3000	30	48	42	73
4000	39	67		
5000		78		

R263213-45 Generator Weight (Lbs.) by kW

3 Phase 4 Wire /480 Volt			
Gas		Diesel	
kW	Lbs.	kW	Lbs.
7.5	600	30	1800
10	630	50	2230
15	960	75	2250
30	1500	100	3840
65	2350	125	4030
85	2570	150	5500
115	4310	175	5650
170	6530	200	5930
		250	6320
		300	7840
		350	8220
		400	10750
		500	11900

R263413-31 Ampere Values Determined by Horsepower, Voltage and Phase Values

H.P.	Amperes							
	Single Phase			Three Phase				
	115V	208V	230V	200V	208V	230V	460V	575V
1/6	4.4A	2.4A	2.2A					
1/4	5.8	3.2	2.9					
1/3	7.2	4.0	3.6					
1/2	9.8	5.4	4.9	2.5A	2.4A	2.2A	1.1A	0.9A
3/4	13.8	7.6	6.9	3.7	3.5	3.2	1.6	1.3
1	16	8.8	8	4.8	4.6	4.2	2.1	1.7
1-1/2	20	11	10	6.9	6.6	6.0	3.0	2.4
2	24	13.2	12	7.8	7.5	6.8	3.4	2.7
3	34	18.7	17	11.0	10.6	9.6	4.8	3.9
5	56	30.8	28	17.5	16.7	15.2	7.6	6.1
7-1/2	80	44	40	25.3	24.2	22	11	9
10	100	55	50	32.2	30.8	28	14	11
15				48.3	46.2	42	21	17
20				62.1	59.4	54	27	22
25				78.2	74.8	68	34	27
30				92	88	80	40	32
40				120	114	104	52	41
50				150	143	130	65	52
60				177	169	154	77	62
75				221	211	192	96	77
100				285	273	248	124	99
125				359	343	312	156	125
150				414	396	360	180	144
200				552	528	480	240	192
250							302	242
300							361	289
350							414	336
400							477	382

R312316-40 Excavating

The selection of equipment used for structural excavation and bulk excavation or for grading is determined by the following factors.

1. Quantity of material
2. Type of material
3. Depth or height of cut
4. Length of haul
5. Condition of haul road
6. Accessibility of site
7. Moisture content and dewatering requirements
8. Availability of excavating and hauling equipment

Some additional costs must be allowed for hand trimming the sides and bottom of concrete pours and other excavation below the general excavation.

When planning excavation and fill, the following should also be considered.

1. Swell factor
2. Compaction factor
3. Moisture content
4. Density requirements

A typical example for scheduling and estimating the cost of excavation of a 15′ deep basement on a dry site when the material must be hauled off the site is outlined below.

Assumptions:

1. Swell factor, 18%
2. No mobilization or demobilization
3. Allowance included for idle time and moving on job
4. No dewatering, sheeting, or bracing
5. No truck spotter or hand trimming

Number of B.C.Y. per truck = 1.5 C.Y. bucket x 8 passes = 12 loose C.Y.

$$= 12 \times \frac{100}{118} = 10.2 \text{ B.C.Y. per truck}$$

Truck Haul Cycle:

Load truck, 8 passes	=	4 minutes
Haul distance, 1 mile	=	9 minutes
Dump time	=	2 minutes
Return, 1 mile	=	7 minutes
Spot under machine	=	1 minute
		23 minute cycle

Fleet Haul Production per day in B.C.Y.

$$4 \text{ trucks} \times \frac{50 \text{ min. hour}}{23 \text{ min. haul cycle}} \times 8 \text{ hrs.} \times 10.2 \text{ B.C.Y.}$$

$$= 4 \times 2.2 \times 8 \times 10.2 = 718 \text{ B.C.Y./day}$$

Add the mobilization and demobilization costs to the total excavation costs. When equipment is rented for more than three days, there is often no mobilization charge by the equipment dealer. On larger jobs outside of urban areas, scrapers can move earth economically provided a dump site or fill area and adequate haul roads are available. Excavation within sheeting bracing or cofferdam bracing is usually done with a clamshell and production is low, since the clamshell may have to be guided by hand between the bracing. When excavating or filling an area enclosed with a wellpoint system, add 10% to 15% to the cost to allow for restricted access. When estimating earth excavation quantities for structures, allow work space outside the building footprint for construction of the foundation and a slope of 1:1 unless sheeting is used.

R312316-45 Excavating Equipment

The table below lists THEORETICAL hourly production in C.Y./hr. bank measure for some typical excavation equipment. Figures assume 50 minute hours, 83% job efficiency, 100% operator efficiency, 90° swing and properly sized hauling units, which must be modified for adverse digging and loading conditions. Actual production costs in the front of the book average about 50% of the theoretical values listed here.

Equipment	Soil Type	B.C.Y. Weight	% Swell	1 C.Y.	1-1/2 C.Y.	2 C.Y.	2-1/2 C.Y.	3 C.Y.	3-1/2 C.Y.	4 C.Y.
Hydraulic Excavator "Backhoe" 15′ Deep Cut	Moist loam, sandy clay	3400 lb.	40%	85	125	175	220	275	330	380
	Sand and gravel	3100	18	80	120	160	205	260	310	365
	Common earth	2800	30	70	105	150	190	240	280	330
	Clay, hard, dense	3000	33	65	100	130	170	210	255	300
Power Shovel Optimum Cut (Ft.)	Moist loam, sandy clay	3400	40	170	245	295	335	385	435	475
				(6.0)	(7.0)	(7.8)	(8.4)	(8.8)	(9.1)	(9.4)
	Sand and gravel	3100	18	165	225	275	325	375	420	460
				(6.0)	(7.0)	(7.8)	(8.4)	(8.8)	(9.1)	(9.4)
	Common earth	2800	30	145	200	250	295	335	375	425
				(7.8)	(9.2)	(10.2)	(11.2)	(12.1)	(13.0)	(13.8)
	Clay, hard, dense	3000	33	120	175	220	255	300	335	375
				(9.0)	(10.7)	(12.2)	(13.3)	(14.2)	(15.1)	(16.0)
Drag Line Optimum Cut (Ft.)	Moist loam, sandy clay	3400	40	130	180	220	250	290	325	385
				(6.6)	(7.4)	(8.0)	(8.5)	(9.0)	(9.5)	(10.0)
	Sand and gravel	3100	18	130	175	210	245	280	315	375
				(6.6)	(7.4)	(8.0)	(8.5)	(9.0)	(9.5)	(10.0)
	Common earth	2800	30	110	160	190	220	250	280	310
				(8.0)	(9.0)	(9.9)	(10.5)	(11.0)	(11.5)	(12.0)
	Clay, hard, dense	3000	33	90	130	160	190	225	250	280
				(9.3)	(10.7)	(11.8)	(12.3)	(12.8)	(13.3)	(12.0)
				Wheel Loaders				Track Loaders		
				3 C.Y.	4 C.Y.	6 C.Y.	8 C.Y.	2-1/4 C.Y.	3 C.Y.	4 C.Y.
Loading Tractors	Moist loam, sandy clay	3400	40	260	340	510	690	135	180	250
	Sand and gravel	3100	18	245	320	480	650	130	170	235
	Common earth	2800	30	230	300	460	620	120	155	220
	Clay, hard, dense	3000	33	200	270	415	560	110	145	200
	Rock, well-blasted	4000	50	180	245	380	520	100	130	180

R312319-90 Wellpoints

A single stage wellpoint system is usually limited to dewatering an average 15′ depth below normal ground water level. Multi-stage systems are employed for greater depth with the pumping equipment installed only at the lowest header level. Ejectors with unlimited lift capacity can be economical when two or more stages of wellpoints can be replaced or when horizontal clearance is restricted, such as in deep trenches or tunneling projects, and where low water flows are expected. Wellpoints are usually spaced on 2-1/2′ to 10′ centers along a header pipe. Wellpoint spacing, header size, and pump size are all determined by the expected flow as dictated by soil conditions.

In almost all soils encountered in wellpoint dewatering, the wellpoints may be jetted into place. Cemented soils and stiff clays may require sand wicks about 12″ in diameter around each wellpoint to increase efficiency and eliminate weeping into the excavation. These sand wicks require 1/2 to 3 C.Y. of washed filter sand and are installed by using a 12″ diameter steel casing and hole puncher jetted into the ground 2′ deeper than the wellpoint. Rock may require predrilled holes.

Labor required for the complete installation and removal of a single stage wellpoint system is in the range of 3/4 to 2 labor-hours per linear foot of header, depending upon jetting conditions, wellpoint spacing, etc.

Continuous pumping is necessary except in some free draining soil where temporary flooding is permissible (as in trenches which are backfilled after each day's work). Good practice requires provision of a stand-by pump during the continuous pumping operation.

Systems for continuous trenching below the water table should be installed three to four times the length of expected daily progress to ensure uninterrupted digging, and header pipe size should not be changed during the job.

For pervious free draining soils, deep wells in place of wellpoints may be economical because of lower installation and maintenance costs. Daily production ranges between two to three wells per day, for 25′ to 40′ depths, to one well per day for depths over 50′.

Detailed analysis and estimating for any dewatering problem is available at no cost from wellpoint manufacturers. Major firms will quote "sufficient equipment" quotes or their affiliates offer lump sum proposals to cover complete dewatering responsibility.

Description for 200′ System with 8″ Header		Quantities
Equipment & Material	Wellpoints 25′ long, 2″ diameter @ 5′ O.C.	40 Each
	Header pipe, 8″ diameter	200 L.F.
	Discharge pipe, 8″ diameter	100 L.F.
	8″ valves	3 Each
	Combination jetting & wellpoint pump (standby)	1 Each
	Wellpoint pump, 8″ diameter	1 Each
	Transportation to and from site	1 Day
	Fuel for 30 days x 60 gal./day	1800 Gallons
	Lubricants for 30 days x 16 lbs./day	480 Lbs.
	Sand for points	40 C.Y.
Labor	Technician to supervise installation	1 Week
	Labor for installation and removal of system	300 Labor-hours
	4 Operators straight time 40 hrs./wk. for 4.33 wks.	693 Hrs.
	4 Operators overtime 2 hrs./wk. for 4.33 wks.	35 Hrs.

R312323-30 Compacting Backfill

Compaction of fill in embankments, around structures, in trenches, and under slabs is important to control settlement. Factors affecting compaction are:

1. Soil gradation
2. Moisture content
3. Equipment used
4. Depth of fill per lift
5. Density required

Production Rate:

$$\frac{1.75' \text{ plate width x 50 F.P.M. x 50 min./hr. x .67' lift}}{27 \text{ C.F. per C.Y.}} = 108.5 \text{ C.Y./hr.}$$

Production Rate for 4 Passes:

$$\frac{108.5 \text{ C.Y.}}{4 \text{ passes}} = 27.125 \text{ C.Y./hr. x 8 hrs.} = 217 \text{ C.Y./day}$$

Example:

Compact granular fill around a building foundation using a 21″ wide x 24″ vibratory plate in 8″ lifts. Operator moves at 50 F.P.M. working a 50 minute hour to develop 95% Modified Proctor Density with 4 passes.

R314116-40 Wood Sheet Piling

Wood sheet piling may be used for depths to 20′ where there is no ground water. If moderate ground water is encountered Tongue & Groove sheeting will help to keep it out. When considerable ground water is present, steel sheeting must be used.

For estimating purposes on trench excavation, sizes are as follows:

Depth	Sheeting	Wales	Braces	B.F. per S.F.
To 8′	3 x 12′s	6 x 8′s, 2 line	6 x 8′s, @ 10′	4.0 @ 8′
8′ x 12′	3 x 12′s	10 x 10′s, 2 line	10 x 10′s, @ 9′	5.0 average
12′ to 20′	3 x 12′s	12 x 12′s, 3 line	12 x 12′s, @ 8′	7.0 average

Sheeting to be toed in at least 2′ depending upon soil conditions. A five person crew with an air compressor and sheeting driver can drive and brace 440 SF/day at 8′ deep, 360 SF/day at 12′ deep, and 320 SF/day at 16′ deep. For normal soils, piling can be pulled in 1/3 the time to install. Pulling difficulty increases with the time in the ground. Production can be increased by high pressure jetting.

R314116-45 Steel Sheet Piling

Limiting weights are 22 to 38#/S.F. of wall surface with 27#/S.F. average for usual types and sizes. (Weights of piles themselves are from 30.7#/L.F. to 57#/L.F. but they are 15″ to 21″ wide.) Lightweight sections 12″ to 28″ wide from 3 ga. to 12 ga. thick are also available for shallow excavations. Piles may be driven two at a time with an impact or vibratory hammer (use vibratory to pull) hung from a crane without leads. A reasonable estimate of the life of steel sheet piling is 10 uses with up to 125 uses possible if a vibratory hammer is used. Used piling costs from 50% to 80% of new piling depending on location and market conditions. Sheet piling and H piles can be rented for about 30% of the delivered mill price for the first month and 5% per month thereafter. Allow 1 labor-hour per pile for cleaning and trimming after driving. These costs increase with depth and hydrostatic head. Vibratory drivers are faster in wet granular soils and are excellent for pile extraction. Pulling difficulty increases with the time in the ground and may cost more than driving. It is often economical to abandon the sheet piling, especially if it can be used as the outer wall form. Allow about 1/3 additional length or more for toeing into ground. Add bracing, waler and strut costs. Waler costs can equal the cost per ton of sheeting.

Earthwork R3145 Vibroflotation & Densification

R314513-90 Vibroflotation and Vibro Replacement Soil Compaction

Vibroflotation is a proprietary system of compacting sandy soils in place to increase relative density to about 70%. Typical bearing capacities attained will be 6000 psf for saturated sand and 12,000 psf for dry sand. Usual range is 4000 to 8000 psf capacity. Costs in the front of the book are for a vertical foot of compacted cylinder 6′ to 10′ in diameter.

Vibro replacement is a proprietary system of improving cohesive soils in place to increase bearing capacity. Most silts and clays above or below the water table can be strengthened by installation of stone columns.

The process consists of radial displacement of the soil by vibration. The created hole is then backfilled in stages with coarse granular fill which is thoroughly compacted and displaced into the surrounding soil in the form of a column.

The total project cost would depend on the number and depth of the compacted cylinders. The installing company guarantees relative soil density of the sand cylinders after compaction and the bearing capacity of the soil after the replacement process. Detailed estimating information is available from the installer at no cost.

R316000-20 Pile Caps, Piles and Caissons

General: The function of a reinforced concrete pile cap is to transfer superstructure load from isolated column or pier to each pile in its supporting cluster. To do this, the cap must be thick and rigid, with all piles securely embedded into and bonded to it.

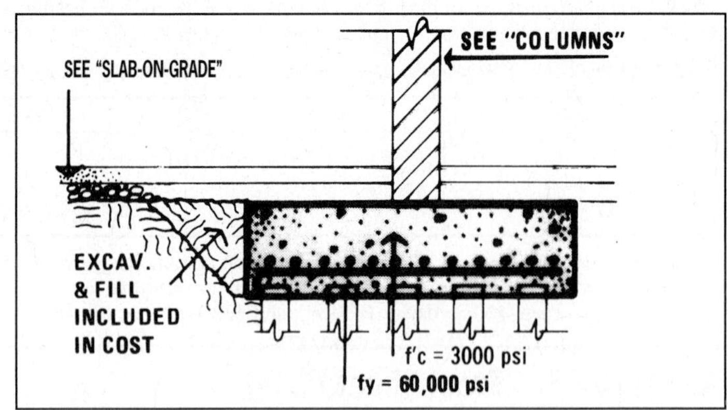

Figure 1.1-331 Section Through Pile Cap

Table 1.1-332 Concrete Quantities for Pile Caps

Load	Number of Piles @ 3'-0" O.C. Per Footing Cluster									
Working (K)	2 (CY)	4 (CY)	6 (CY)	8 (CY)	10 (CY)	12 (CY)	14 (CY)	16 (CY)	18 (CY)	20 (CY)
50	(.9)	(1.9)	(3.3)	(4.9)	(5.7)	(7.8)	(9.9)	(11.1)	(14.4)	(16.5)
100	(1.0)	(2.2)	(3.3)	(4.9)	(5.7)	(7.8)	(9.9)	(11.1)	(14.4)	(16.5)
200	(1.0)	(2.2)	(4.0)	(4.9)	(5.7)	(7.8)	(9.9)	(11.1)	(14.4)	(16.5)
400	(1.1)	(2.6)	(5.2)	(6.3)	(7.4)	(8.2)	(13.7)	(11.1)	(14.4)	(16.5)
800		(2.9)	(5.8)	(7.5)	(9.2)	(13.6)	(17.6)	(15.9)	(19.7)	(22.1)
1200			(5.8)	(8.3)	(9.7)	(14.2)	(18.3)	(20.4)	(21.2)	(22.7)
1600				(9.8)	(11.4)	(14.5)	(19.5)	(20.4)	(24.6)	(27.2)
2000				(9.8)	(11.4)	(16.6)	(24.1)	(21.7)	(26.0)	(28.8)
3000						(17.5)		(26.5)	(30.3)	(32.9)
4000								(30.2)	(30.7)	(36.5)

Table 1.1-333 Concrete Quantities for Pile Caps

Load	Number of Piles @ 4'-6" O.C. Per Footing					
Working (K)	2 (CY)	3 (CY)	4 (CY)	5 (CY)	6 (CY)	7 (CY)
50	(2.3)	(3.6)	(5.6)	(11.0)	(13.7)	(12.9)
100	(2.3)	(3.6)	(5.6)	(11.0)	(13.7)	(12.9)
200	(2.3)	(3.6)	(5.6)	(11.0)	(13.7)	(12.9)
400	(3.0)	(3.6)	(5.6)	(11.0)	(13.7)	(12.9)
800			(6.2)	(11.5)	(14.0)	(12.9)
1200				(13.0)	(13.7)	(13.4)
1600						(14.0)

R316000-20 Pile Caps, Piles and Caissons (cont.)

General: Piles are column-like shafts which receive superstructure loads, overturning forces, or uplift forces. They receive these loads from isolated column or pier foundations (pile caps), foundation walls, grade beams, or foundation mats. The piles then transfer these loads through shallower poor soil strata to deeper soil of adequate support strength and acceptable settlement with load.

Be sure that other foundation types aren't better suited to the job. Consider ground and settlement, as well as loading, when reviewing. Piles usually are associated with difficult foundation problems and substructure condition. Ground conditions determine type of pile (different pile types have been developed to suit ground conditions.) Decide each case by technical study, experience and sound engineering judgment—not rules of thumb. A full investigation of ground conditions, early, is essential to provide maximum information for professional foundation engineering and an acceptable structure.

Piles support loads by end bearing and friction. Both are generally present; however, piles are designated by their principal method of load transfer to soil.

Boring should be taken at expected pile locations. Ground strata (to bedrock or depth of 1-1/2 building width) must be located and identified with appropriate strengths and compressibilities. The sequence of strata determines if end bearing or friction piles are best suited.

End bearing piles have shafts which pass through soft strata or thin hard strata and tip bear on bedrock or penetrate some distance into a dense, adequate soil (sand or gravel.)

Friction piles have shafts which may be entirely embedded in cohesive soil (moist clay), and develop required support mainly by adhesion or "skin-friction" between soil and shaft area.

Piles pass through soil by either one of two ways:

1. Displacement piles force soil out of the way. This may cause compaction, ground heaving, remolding of sensitive soils, damage to adjacent structures, or hard driving.
2. Non-displacement piles have either a hole bored and the pile cast or placed in hole, or open ended pipe (casing) driven and the soil core removed. They tend to eliminate heaving or lateral pressure damage to adjacent structures of piles. Steel "HP" piles are considered of small displacement.

Placement of piles (attitude) is most often vertical; however, they are sometimes battered (placed at a small angle from vertical) to advantageously resist lateral loads. Seldom are piles installed singly but rather in clusters (or groups). Codes require a minimum of three piles per major column load or two per foundation wall or grade beam. Single pile capacity is limited by pile structural strength or support strength of soil. Support capacity of a pile cluster is almost always less than the sum of its individual pile capacities due to overlapping of bearing the friction stresses.

Large rigs for heavy, long piles create large soil surface loads and additional expense on weak ground.

Fewer piles create higher costs per pile.

Pile load tests are frequently required by code, ground situation, or pile type.

R316000-20 Pile Caps, Piles and Caissons (cont.)

General: Caissons, as covered in this section, are drilled cylindrical foundation shafts which function primarily as short column-like compression members. They transfer superstructure loads through inadequate soils to bedrock or hard stratum. They may be either reinforced or unreinforced and either straight or belled out at the bearing level.

Shaft diameters range in size from 20″ to 84″ with the most usual sizes beginning at 34″. If inspection of bottom is required, the minimum diameter practical is 30″. If handwork is required (in addition to mechanical belling, etc.) the minimum diameter is 32″. The most frequently used shaft diameter is probably 36″ with a 5′ or 6′ bell diameter. The maximum bell diameter practical is three times the shaft diameter.

Plain concrete is commonly used, poured directly against the excavated face of soil. Permanent casings add to cost and economically should be avoided. Wet or loose strata are undesirable. The associated installation sometimes involves a mudding operation with bentonite clay slurry to keep walls of excavation stable (costs not included here).

Reinforcement is sometimes used, especially for heavy loads. It is required if uplift, bending moment, or lateral loads exist. A small amount of reinforcement is desirable at the top portion of each caisson, even if the above conditions theoretically are not present. This will provide for construction eccentricities and other possibilities. Reinforcement, if present, should extend below the soft strata. Horizontal reinforcement is not required for belled bottoms.

There are three basic types of caisson bearing details:

1. Belled, which are generally recommended to provide reduced bearing pressure on soil. These are not for shallow depths or poor soils. Good soils for belling include most clays, hardpan, soft shale, and decomposed rock.

   Soils requiring handwork include hard shale, limestone, and sandstone.

   Soils not recommended include sand, gravel, silt, and igneous rock. Compact sand and gravel above water table may stand. Water in the bearing strata is undesirable.
2. Straight shafted, which have no bell but the entire length is enlarged to permit safe bearing pressures. They are most economical for light loads on high bearing capacity soil.
3. Socketed (or keyed), which are used for extremely heavy loads. They involve sinking the shaft into rock for combined friction and bearing support action. Reinforcement of shaft is usually necessary. Wide flange cores are frequently used here.

Advantages include:

A. Shafts can pass through soils that piles cannot
B. No soil heaving or displacement during installation
C. No vibration during installation
D. Less noise than pile driving
E. Bearing strata can be visually inspected & tested

Uses include:

A. Situations where unsuitable soil exists to moderate depth
B. Tall structures
C. Heavy structures
D. Underpinning (extensive use)

See R033053-10 for Soil Bearing Capacities.

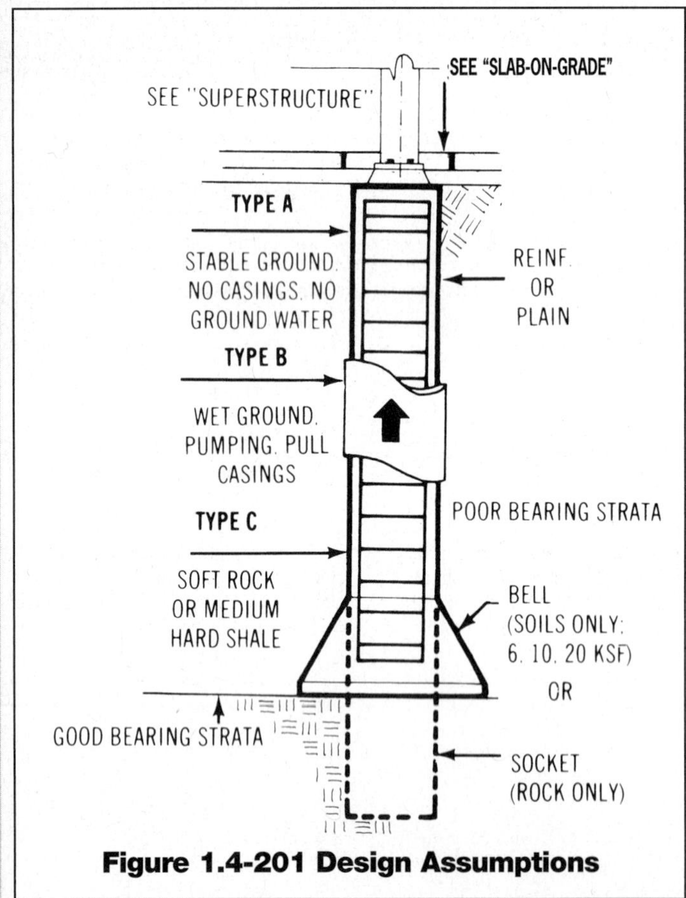

Figure 1.4-201 Design Assumptions

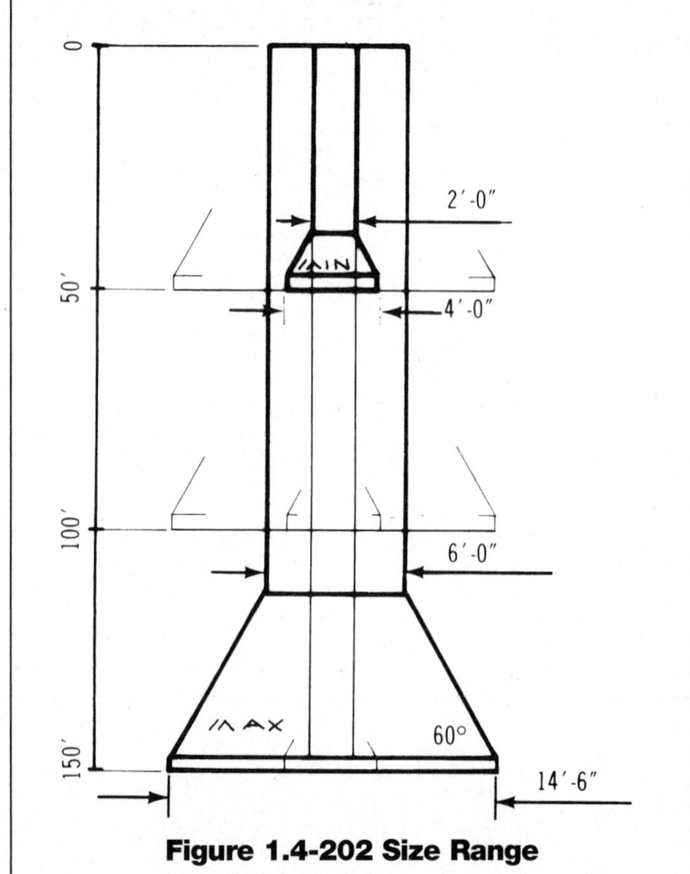

Figure 1.4-202 Size Range

Earthwork — R3171 Tunnel Excavation

R317100-10 Tunnel Excavation

Bored tunnel excavation is common in rock for diameters from 4 feet for sewer and utilities, to 60 feet for vehicles. Production varies from a few linear feet per day to over 200 linear feet per day. In the smaller diameters, the productivity is limited by the restricted area for mucking or the removal of excavated material.

Most of the tunnels in rock today are excavated by boring machines called moles. Preparation for starting the excavation or setting up the mole is very costly. Shafts must be excavated to the invert of the proposed tunnel and the mole must be lowered into the shaft. If excavating a portal tunnel, that is starting at an open face, the cost is reduced considerably both for mobilization and mucking.

In soft ground and mixed material, special bucket excavators and rotary excavators are used inside a shield. Tunnel liners must follow directly behind the shield to support the earth and prevent cave-ins.

Traditional muck haulage operations are performed by rail with locomotives and muck cars. Sometimes conveyors are more economical and require less ventilation of the tunnel.

Ventilation and air compression are other important cost factors to consider in tunnel excavation. Continuous ventilation ducts are sometimes fabricated at the tunnel site.

Tunnel linings are steel, cast in place reinforced concrete, shotcrete, or a combination of these. When required, contact grouting is performed by pumping grout between the lining and the excavation. Intermittent holes are drilled into the lining and separate costs are determined for drilling per hole, grout pump connecting per hole, and grout per cubic foot.

Consolidation grouting and roof bolts may also be required where the excavation is unstable or faulting occurs.

Tunnel boring is usually done 24 hours per day. A typical crew for rock boring is:

Tunneling Crew based on three 8 hour shifts

1 Shifter
1 Walker
1 Machine Operator for mole
1 Oiler
1 Mechanic
3 Locomotives with operators
5 Miners for rails, vent ducts, and roof bolts
1 Electrician
2 Pumps
2 Laborers for hoisting
1 Hoist operator for muck removal
1 Oiler

Surface Crew Based on normal 8 hour shift

2 Shop Mechanics
1 Electrician
1 Shifter
2 Laborers
1 Operator with 18 ton cherry picker
1 Operator with front end loader

Exterior Improvements — R3201 Flexible Paving Surface Treatment

R320113-70 Pavement Maintenance

Routine pavement maintenance should be performed to keep a paved surface from deteriorating under the normal forces of nature and traffic.

The msot important maintenance function is the early detection and repair of minor pavement defects. Cracks and other surface breaks can develop into serious defects if not repaired in their earliest stages. For these reasons a pavement preventive maintenance program should include frequent close inspections of pavement surfaces. When suspicious areas are detected, a detailed investigation should be undertaken to determine the appropriate repair. Where subsurface or pavement deterioration is detected, the Benkelman Beam can be used to make deflection measurements under normal traffic stresses. This is done to determine the extent of the affected area.

Patching or resurfacing work should be done during warm (10°C and above) and dry weather. Adequate compaction is dificult to achieve when hot or warm mixtures are placed on cold pavements. Asphalt and asphalt mixtures usually do not bond well to damp surfaces.

Exterior Improvements — R3292 Turf & Grasses

R329219-50 Seeding

The type of grass is determined by light, shade and moisture content of soil plus intended use. Fertilizer should be disked 4″ before seeding. For steep slopes disk five tons of mulch and lay two tons of hay or straw on surface per acre after seeding. Surface mulch can be staked, lightly disked or tar emulsion sprayed. Material for mulch can be wood chips, peat moss, partially rotted hay or straw, wood fibers and sprayed emulsions. Hemp seed blankets with fertilizer are also available. For spring seeding, watering is necessary. Late fall seeding may have to be reseeded in the spring. Hydraulic seeding, power mulching, and aerial seeding can be used on large areas.

R329343-20 Trees and Plants by Environment and Purposes

Dry, Windy, Exposed Areas
Barberry
Junipers, all varieties
Locust
Maple
Oak
Pines, all varieties
Poplar, Hybrid
Privet
Spruce, all varieties
Sumac, Staghorn

Lightly Wooded Areas
Dogwood
Hemlock
Larch
Pine, White
Rhododendron
Spruce, Norway
Redbud

Total Shade Areas
Hemlock
Ivy, English
Myrtle
Pachysandra
Privet
Spice Bush
Yews, Japanese

Cold Temperatures of Northern U.S. and Canada
Arborvitae, American
Birch, White
Dogwood, Silky
Fir, Balsam
Fir, Douglas
Hemlock
Juniper, Andorra
Juniper, Blue Rug
Linden, Little Leaf
Maple, Sugar
Mountain Ash
Myrtle
Olive, Russian
Pine, Mugho
Pine, Ponderosa
Pine, Red
Pine, Scotch
Poplar, Hybrid
Privet
Rosa Rugosa
Spruce, Dwarf Alberta
Spruce, Black Hills
Spruce, Blue
Spruce, Norway
Spruce, White, Engelman
Yellow Wood

Wet, Swampy Areas
American Arborvitae
Birch, White
Black Gum
Hemlock
Maple, Red
Pine, White
Willow

Poor, Dry, Rocky Soil
Barberry
Crownvetch
Eastern Red Cedar
Juniper, Virginiana
Locust, Black
Locust, Bristly
Locust, Honey
Olive, Russian
Pines, all varieties
Privet
Rosa Rugosa
Sumac, Staghorn

Seashore Planting
Arborvitae, American
Juniper, Tamarix
Locust, Black
Oak, White
Olive, Russian
Pine, Austrian
Pine, Japanese Black
Pine, Mugho
Pine, Scotch
Privet, Amur River
Rosa Rugosa
Yew, Japanese

City Planting
Barberry
Fir, Concolor
Forsythia
Hemlock
Holly, Japanese
Ivy, English
Juniper, Andorra
Linden, Little Leaf
Locust, Honey
Maple, Norway, Silver
Oak, Pin, Red
Olive, Russian
Pachysandra
Pine, Austrian
Pine, White
Privet
Rosa Rugosa
Sumac, Staghorn
Yew, Japanese

Bonsai Planting
Azaleas
Birch, White
Ginkgo
Junipers
Pine, Bristlecone
Pine, Mugho
Spruce,k Engleman
Spruce, Dwarf Alberta

Street Planting
Linden, Little Leaf
Oak, Pin
Ginkgo

Fast Growth
Birch, White
Crownvetch
Dogwood, Silky
Fir, Douglas
Juniper, Blue Pfitzer
Juniper, Blue Rug
Maple, Silver
Olive, Autumn
Pines, Austrian, Ponderosa, Red Scotch and White
Poplar, Hybrid
Privet
Spruce, Norway
Spruce, Serbian
Texus, Cuspidata, Hicksi
Willow

Dense, Impenetrable Hedges
Field Plantings:
- Locust, Bristly,
- Olive, Autumn
- Sumac

Residential Area:
- Barberry, Red or Green
- Juniper, Blue Pfitzer
- Rosa Rugosa

Food for Birds
Ash, Mountain
Barberry
Bittersweet
Cherry, Manchu
Dogwood, Silky
Honesuckle, Rem Red
Hawthorn
Oaks
Olive, Autumn, Russian
Privet
Rosa Rugosa
Sumac

Erosion Control
Crownvetch
Locust, Bristly
Willow

R329343-30 Zones of Plant Hardiness

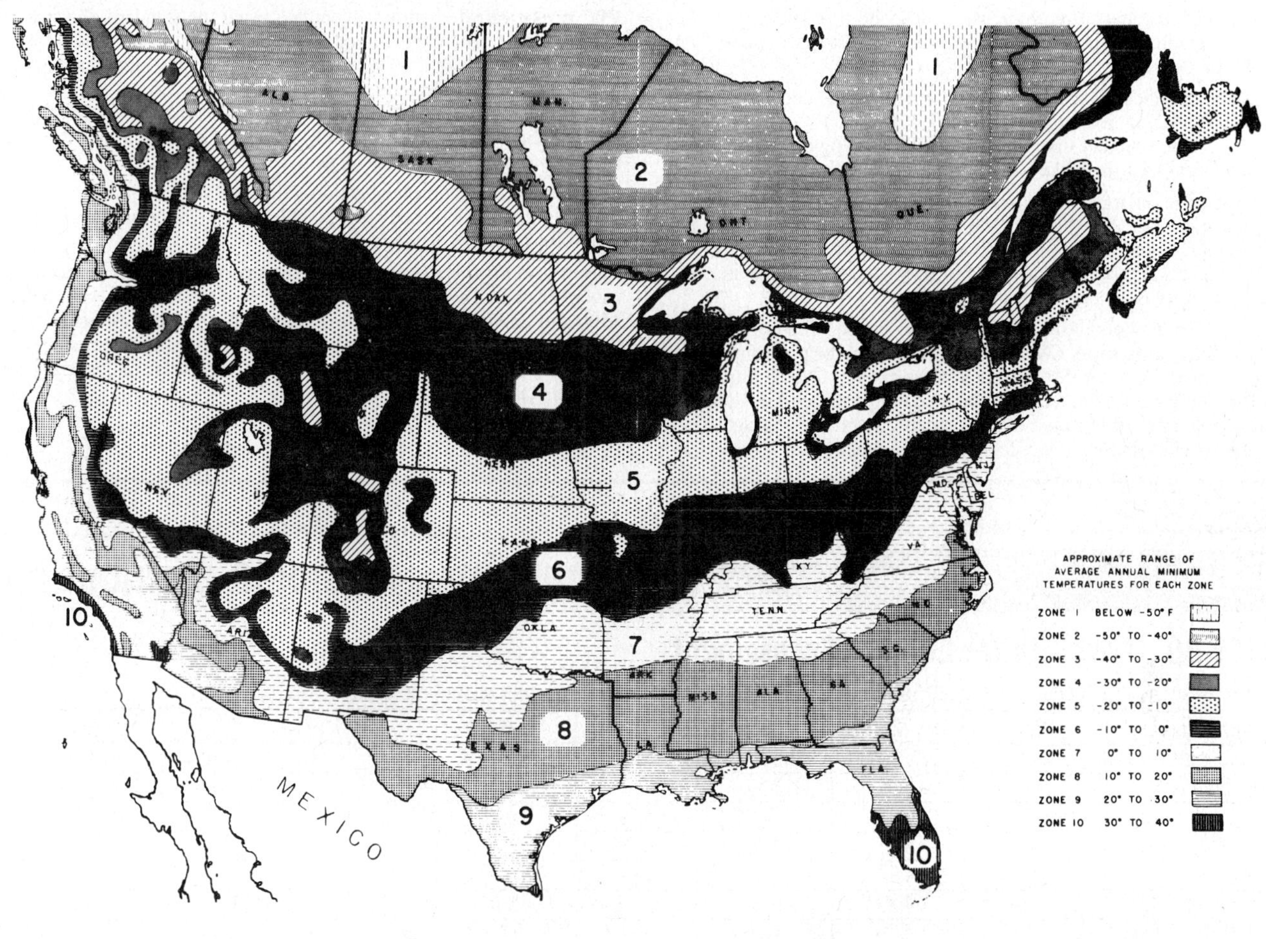

Utilities — R3311 Water Utility Distribution Piping

R331113-80 Piping Designations

There are several systems currently in use to describe pipe and fittings. The following paragraphs will help to identify and clarify classifications of piping systems used for water distribution.

Piping may be classified by schedule. Piping schedules include 5S, 10S, 10, 20, 30, Standard, 40, 60, Extra Strong, 80, 100, 120, 140, 160 and Double Extra Strong. These schedules are dependent upon the pipe wall thickness. The wall thickness of a particular schedule may vary with pipe size.

Ductile iron pipe for water distribution is classified by Pressure Classes such as Class 150, 200, 250, 300 and 350. These classes are actually the rated water working pressure of the pipe in pounds per square inch (psi). The pipe in these pressure classes is designed to withstand the rated water working pressure plus a surge allowance of 100 psi.

The American Water Works Association (AWWA) provides standards for various types of **plastic pipe.** C-900 is the specification for polyvinyl chloride (PVC) piping used for water distribution in sizes ranging from 4″ through 12″. C-901 is the specification for polyethylene (PE) pressure pipe, tubing and fittings used for water distribution in sizes ranging from 1/2″ through 3″. C-905 is the specification for PVC piping sizes 14″ and greater.

PVC pressure-rated pipe is identified using the standard dimensional ratio (SDR) method. This method is defined by the American Society for Testing and Materials (ASTM) Standard D 2241. This pipe is available in SDR numbers 64, 41, 32.5, 26, 21, 17, and 13.5. Pipe with an SDR of 64 will have the thinnest wall while pipe with an SDR of 13.5 will have the thickest wall. When the pressure rating (PR) of a pipe is given in psi, it is based on a line supplying water at 73 degrees F.

The National Sanitation Foundation (NSF) seal of approval is applied to products that can be used with potable water. These products have been tested to ANSI/NSF Standard 14.

Valves and strainers are classified by American National Standards Institute (ANSI) Classes. These Classes are 125, 150, 200, 250, 300, 400, 600, 900, 1500 and 2500. Within each class there is an operating pressure range dependent upon temperature. Design parameters should be compared to the appropriate material dependent, pressure-temperature rating chart for accurate valve selection.

Utilities — R3371 Elec. Utility Transmission & Distribution

R337119-30 Concrete for Conduit Encasement

Table below lists C.Y. of concrete for 100 L.F. of trench. Conduits separation center to center should meet 7.5″ (N.E.C.).

Number of Conduits	1	2	3	4	6	8	9	Number of Conduits
Trench Dimension	11.5″ x 11.5″	11.5″ x 19″	11.5″ x 27″	19″ x 19″	19″ x 27″	19″ x 38″	27″ x 27″	Trench Dimension
Conduit Diameter 2.0″	3.29	5.39	7.64	8.83	12.51	17.66	17.72	Conduit Diameter 2.0″
2.5″	3.23	5.29	7.49	8.62	12.19	17.23	17.25	2.5″
3.0″	3.15	5.13	7.24	8.29	11.71	16.59	16.52	3.0″
3.5″	3.08	4.97	7.02	7.99	11.26	15.98	15.84	3.5″
4.0″	2.99	4.80	6.76	7.65	10.74	15.30	15.07	4.0″
5.0″	2.78	4.37	6.11	6.78	9.44	13.57	13.12	5.0″
6.0″	2.52	3.84	5.33	5.74	7.87	11.48	10.77	6.0″

R347216-10 Single Track R.R. Siding

The costs for a single track RR siding in the Unit Price section include the components shown in the table below.

Description of Component	Qty. per L.F. of Track	Unit
Ballast, 1-1/2″ crushed stone	.667	C.Y.
6″ x 8″ x 8′-6″ Treated timber ties, 22″ O.C.	.545	Ea.
Tie plates, 2 per tie	1.091	Ea.
Track rail	2.000	L.F.
Spikes, 6″, 4 per tie	2.182	Ea.
Splice bars w/ bolts, lock washers & nuts, @ 33′ O.C.	.061	Pair
Crew B-14 @ 57 L.F./Day	.018	Day

R347216-20 Single Track, Steel Ties, Concrete Bed

The costs for a R.R. siding with steel ties and a concrete bed in the Unit Price section include the components shown in the table below.

Description of Component	Qty. per L.F. of Track	Unit
Concrete bed, 9′ wide, 10″ thick	.278	C.Y.
Ties, W6x16 x 6′-6″ long, @ 30″ O.C.	.400	Ea.
Tie plates, 4 per tie	1.600	Ea.
Track rail	2.000	L.F.
Tie plate bolts, 1″, 8 per tie	3.200	Ea.
Splice bars w/bolts, lock washers & nuts, @ 33′ O.C.	.061	Pair
Crew B-14 @ 22 L.F./Day	.045	Day

Change Orders

Change Order Considerations

A change order is a written document, usually prepared by the design professional, and signed by the owner, the architect/engineer, and the contractor. A change order states the agreement of the parties to: an addition, deletion, or revision in the work; an adjustment in the contract sum, if any; or an adjustment in the contract time, if any. Change orders, or "extras" in the construction process occur after execution of the construction contract and impact architects/engineers, contractors, and owners.

Change orders that are properly recognized and managed can ensure orderly, professional, and profitable progress for all who are involved in the project. There are many causes for change orders and change order requests. In all cases, change orders or change order requests should be addressed promptly and in a precise and prescribed manner. The following paragraphs include information regarding change order pricing and procedures.

The Causes of Change Orders

Reasons for issuing change orders include:

- Unforeseen field conditions that require a change in the work
- Correction of design discrepancies, errors, or omissions in the contract documents
- Owner-requested changes, either by design criteria, scope of work, or project objectives
- Completion date changes for reasons unrelated to the construction process
- Changes in building code interpretations, or other public authority requirements that require a change in the work
- Changes in availability of existing or new materials and products

Procedures

Properly written contract documents must include the correct change order procedures for all parties—owners, design professionals and contractors—to follow in order to avoid costly delays and litigation.

Being "in the right" is not always a sufficient or acceptable defense. The contract provisions requiring notification and documentation must be adhered to within a defined or reasonable time frame.

The appropriate method of handling change orders is by a written proposal and acceptance by all parties involved. Prior to starting work on a project, all parties should identify their authorized agents who may sign and accept change orders, as well as any limits placed on their authority.

Time may be a critical factor when the need for a change arises. For such cases, the contractor might be directed to proceed on a "time and materials" basis, rather than wait for all paperwork to be processed—a delay that could impede progress. In this situation, the contractor must still follow the prescribed change order procedures including, but not limited to, notification and documentation.

All forms used for change orders should be dated and signed by the proper authority. Lack of documentation can be very costly, especially if legal judgments are to be made, and if certain field personnel are no longer available. For time and material change orders, the contractor should keep accurate daily records of all labor and material allocated to the change.

Owners or awarding authorities who do considerable and continual building construction (such as the federal government) realize the inevitability of change orders for numerous reasons, both predictable and unpredictable. As a result, the federal government, the American Institute of Architects (AIA), the Engineers Joint Contract Documents Committee (EJCDC) and other contractor, legal, and technical organizations have developed standards and procedures to be followed by all parties to achieve contract continuance and timely completion, while being financially fair to all concerned.

In addition to the change order standards put forth by industry associations, there are also many books available on the subject.

Pricing Change Orders

When pricing change orders, regardless of their cause, the most significant factor is when the change occurs. The need for a change may be perceived in the field or requested by the architect/engineer *before* any of the actual installation has begun, or may evolve or appear *during* construction when the item of work in question is partially installed. In the latter cases, the original sequence of construction is disrupted, along with all contiguous and supporting systems. Change orders cause the greatest impact when they occur *after* the installation has been completed and must be uncovered, or even replaced. Post-completion changes may be caused by necessary design changes, product failure, or changes in the owner's requirements that are not discovered until the building or the systems begin to function.

Specified procedures of notification and record keeping must be adhered to and enforced regardless of the stage of construction: *before, during,* or *after* installation. Some bidding documents anticipate change orders by requiring that unit prices including overhead and profit percentages—for additional as well as deductible changes—be listed. Generally these unit prices do not fully take into account the ripple effect, or impact on other trades, and should be used for general guidance only.

When pricing change orders, it is important to classify the time frame in which the change occurs. There are two basic time frames for change orders: *pre-installation change orders*, which occur before the start of construction, and *post-installation change orders*, which involve reworking after the original installation. Change orders that occur between these stages may be priced according to the extent of work completed using a combination of techniques developed for pricing *pre-* and *post-installation* changes.

The following factors are the basis for a check list to use when preparing a change order estimate.

Factors To Consider When Pricing Change Orders

As an estimator begins to prepare a change order, the following questions should be reviewed to determine their impact on the final price.

General

- *Is the change order work* pre-installation *or* post-installation?

  Change order work costs vary according to how much of the installation has been completed. Once workers have the project scoped in their minds, even though they have not started, it can be difficult to refocus. Consequently they may spend more than the normal amount of time understanding the change. Also, modifications to work in place, such as trimming or refitting, usually take more time than was initially estimated. The greater the amount of work in place, the more reluctant workers are to change it. Psychologically they may resent the change and as a result the rework takes longer than normal. Post-installation change order estimates must include demolition of existing work as required to accomplish the change. If the work is performed at a later time, additional obstacles, such as building finishes, may be present which must be protected. Regardless of whether the change occurs pre-installation or post-installation, attempt to isolate the identifiable factors and price them separately. For example, add shipping costs that may be required pre-installation or any demolition required post-installation. Then analyze the potential impact on productivity of psychological and/or learning curve factors and adjust the output rates accordingly. One approach is to break down the typical workday into segments and quantify the impact on each segment. The following chart may be useful as a guide:

Activities (Productivity) Expressed as Percentages of a Workday			
Task	**Means Mechanical Cost Data (for New Construction)**	**Pre-Installation Change Orders**	**Post-Installation Change Orders**
1. Study plans	3%	6%	6%
2. Material procurement	3%	3%	3%
3. Receiving and storing	3%	3%	3%
4. Mobilization	5%	5%	5%
5. Site movement	5%	5%	8%
6. Layout and marking	8%	10%	12%
7. Actual installation	64%	59%	54%
8. Clean-up	3%	3%	3%
9. Breaks—non-productive	6%	6%	6%
Total	100%	100%	100%

Change Order Installation Efficiency

The labor-hours expressed (for new construction) are based on average installation time, using an efficiency level of approximately 60%-65%. For change order situations, adjustments to this efficiency level should reflect the daily labor-hour allocation for that particular occurrence.

If any of the specific percentages expressed in the above chart do not apply to a particular project situation, then those percentage points should be reallocated to the appropriate task(s). Example: Using data for new construction, assume there is no new material being utilized. The percentages for Tasks 2 and 3 would therefore be reallocated to other tasks. If the time required for Tasks 2 and 3 can now be applied to installation, we can add the time allocated for *Material Procurement* and *Receiving and Storing* to the *Actual Installation* time for new construction, thereby increasing the Actual Installation percentage.

This chart shows that, due to reduced productivity, labor costs will be higher than those for new construction by 5%-15% for pre-installation change orders and by 15%-25% for post-installation change orders. Each job and change order is unique and must be examined individually. Many factors, covered elsewhere in this section, can each have a significant impact on productivity and change order costs. All such factors should be considered in every case.

- *Will the change substantially delay the original completion date?*

  A significant change in the project may cause the original completion date to be extended. The extended schedule may subject the contractor to new wage rates dictated by relevant labor contracts. Project supervision and other project overhead must also be extended beyond the original completion date. The schedule extension may also put installation into a new weather season. For example, underground piping scheduled for October installation was delayed until January. As a result, frost penetrated the trench area, thereby changing the degree of difficulty of the task. Changes and delays may have a ripple effect throughout the project. This effect must be analyzed and negotiated with the owner.

- *What is the net effect of a deduct change order?*

  In most cases, change orders resulting in a deduction or credit reflect only bare costs. The contractor may retain the overhead and profit based on the original bid.

Materials

- *Will you have to pay more or less for the new material, required by the change order, than you paid for the original purchase?*

  The same material prices or discounts will usually apply to materials purchased for change orders as new construction. In some instances, however, the contractor may forfeit the advantages of competitive pricing for change orders. Consider the following example:

  A contractor purchased over $20,000 worth of fan coil units for an installation, and obtained the maximum discount. Some time later it was determined the project required an additional matching unit. The contractor has to purchase this unit from the original supplier to ensure a match. The supplier at this time may not discount the unit because of the small quantity, and the fact that he is no longer in a competitive situation. The impact of quantity on purchase can add between 0% and 25% to material prices and/or subcontractor quotes.

- *If materials have been ordered or delivered to the job site, will they be subject to a cancellation charge or restocking fee?*

  Check with the supplier to determine if ordered materials are subject to a cancellation charge. Delivered materials not used as result of a change order may be subject to a restocking fee if returned to the supplier. Common restocking charges run between 20% and 40%. Also, delivery charges to return the goods to the supplier must be added.

Labor

- *How efficient is the existing crew at the actual installation?*

  Is the same crew that performed the initial work going to do the change order? Possibly the change consists of the installation of a unit identical to one already installed; therefore, the change should take less time. Be sure to consider this potential productivity increase and modify the productivity rates accordingly.

- *If the crew size is increased, what impact will that have on supervision requirements?*

  Under most bargaining agreements or management practices, there is a point at which a working foreman is replaced by a nonworking foreman. This replacement increases project overhead by adding a nonproductive worker. If additional workers are added to accelerate the project or to perform changes while maintaining the schedule, be sure to add additional supervision time if warranted. Calculate the hours involved and the additional cost directly if possible.

- *What are the other impacts of increased crew size?*

  The larger the crew, the greater the potential for productivity to decrease. Some of the factors that cause this productivity loss are: overcrowding (producing restrictive conditions in the working space), and possibly a shortage of any special tools and equipment required. Such factors affect not only the crew working on the elements directly involved in the change order, but other crews whose movement may also be hampered.

  As the crew increases, check its basic composition for changes by the addition or deletion of apprentices or nonworking foreman, and quantify the potential effects of equipment shortages or other logistical factors.

- *As new crews, unfamiliar with the project, are brought onto the site, how long will it take them to become oriented to the project requirements?*

  The orientation time for a new crew to become 100% effective varies with the site and type of project. Orientation is easiest at a new construction site, and most difficult at existing, very restrictive renovation sites. The type of work also affects orientation time. When all elements of the work are exposed, such as concrete or masonry work, orientation is decreased. When the work is concealed or less visible, such as existing electrical systems, orientation takes longer. Usually orientation can be accomplished in one day or less. Costs for added orientation should be itemized and added to the total estimated cost.

- *How much actual production can be gained by working overtime?*

  Short term overtime can be used effectively to accomplish more work in a day. However, as overtime is scheduled to run beyond several weeks, studies have shown marked decreases in output. The following chart shows the effect of long term overtime on worker efficiency. If the anticipated change requires extended overtime to keep the job on schedule, these factors can be used as a guide to predict the impact on time and cost. Add project overhead, particularly supervision, that may also be incurred.

Days per Week	Hours per Day	Production Efficiency: 1 Week	2 Weeks	3 Weeks	4 Weeks	Average 4 Weeks	Payroll Cost Factors: @ 1-1/2 Times	@ 2 Times
5	8	100%	100%	100%	100%	100%	100%	100%
	9	100	100	95	90	96.25	105.6	111.1
	10	100	95	90	85	91.25	110.0	120.0
	11	95	90	75	65	81.25	113.6	127.3
	12	90	85	70	60	76.25	116.7	133.3
6	8	100	100	95	90	96.25	108.3	116.7
	9	100	95	90	85	92.50	113.0	125.9
	10	95	90	85	80	87.50	116.7	133.3
	11	95	85	70	65	78.75	119.7	139.4
	12	90	80	65	60	73.75	122.2	144.4
7	8	100	95	85	75	88.75	114.3	128.6
	9	95	90	80	70	83.75	118.3	136.5
	10	90	85	75	65	78.75	121.4	142.9
	11	85	80	65	60	72.50	124.0	148.1
	12	85	75	60	55	68.75	126.2	152.4

Effects of Overtime

Caution: Under many labor agreements, Sundays and holidays are paid at a higher premium than the normal overtime rate.

The use of long-term overtime is counterproductive on almost any construction job; that is, the longer the period of overtime, the lower the actual production rate. Numerous studies have been conducted, and while they have resulted in slightly different numbers, all reach the same conclusion. The figure above tabulates the effects of overtime work on efficiency.

As illustrated, there can be a difference between the *actual* payroll cost per hour and the *effective* cost per hour for overtime work. This is due to the reduced production efficiency with the increase in weekly hours beyond 40. This difference between actual and effective cost results from overtime work over a prolonged period. Short-term overtime work does not result in as great a reduction in efficiency and, in such cases, effective cost may not vary significantly from the actual payroll cost. As the total hours per week are increased on a regular basis, more time is lost due to fatigue, lowered morale, and an increased accident rate.

As an example, assume a project where workers are working 6 days a week, 10 hours per day. From the figure above (based on productivity studies), the average effective productive hours over a 4-week period are:

$$0.875 \times 60 = 52.5$$

Depending upon the locale and day of week, overtime hours may be paid at time and a half or double time. For time and a half, the overall (average) *actual* payroll cost (including regular and overtime hours) is determined as follows:

$$\frac{40 \text{ reg. hrs.} + (20 \text{ overtime hrs.} \times 1.5)}{60 \text{ hrs.}} = 1.167$$

Based on 60 hours, the payroll cost per hour will be 116.7% of the normal rate at 40 hours per week. However, because the effective production (efficiency) for 60 hours is reduced to the equivalent of 52.5 hours, the effective cost of overtime is calculated as follows:

For time and a half:

$$\frac{40 \text{ reg. hrs.} + (20 \text{ overtime hrs.} \times 1.5)}{52.5 \text{ hrs.}} = 1.33$$

Installed cost will be 133% of the normal rate (for labor).

Thus, when figuring overtime, the actual cost per unit of work will be higher than the apparent overtime payroll dollar increase, due to the reduced productivity of the longer workweek. These efficiency calculations are true only for those cost factors determined by hours worked. Costs that are applied weekly or monthly, such as equipment rentals, will not be similarly affected.

Equipment

- *What equipment is required to complete the change order?*

  Change orders may require extending the rental period of equipment already on the job site, or the addition of special equipment brought in to accomplish the change work. In either case, the additional rental charges and operator labor charges must be added.

Summary

The preceding considerations and others you deem appropriate should be analyzed and applied to a change order estimate. The impact of each should be quantified and listed on the estimate to form an audit trail.

Change orders that are properly identified, documented, and managed help to ensure the orderly, professional and profitable progress of the work. They also minimize potential claims or disputes at the end of the project.

Estimating Tips

- The cost figures in this Square Foot Cost section were derived from approximately 11,000 projects contained in the RSMeans database of completed construction projects. They include the contractor's overhead and profit, but do not generally include architectural fees or land costs. The figures have been adjusted to January of the current year. New projects are added to our files each year, and outdated projects are discarded. For this reason, certain costs may not show a uniform annual progression. In no case are all subdivisions of a project listed.
- These projects were located throughout the U.S. and reflect a tremendous variation in square foot (S.F.) and cubic foot (C.F.) costs. This is due to differences, not only in labor and material costs, but also in individual owners' requirements. For instance, a bank in a large city would have different features than one in a rural area. This is true of all the different types of buildings analyzed. Therefore, caution should be exercised when using these square foot costs. For example, for courthouses, costs in the database are local courthouse costs and will not apply to the larger, more elaborate federal courthouses. As a general rule, the projects in the 1/4 column do not include any site work or equipment, while the projects in the 3/4 column may include both equipment and site work. The median figures do not generally include site work.
- None of the figures "go with" any others. All individual cost items were computed and tabulated separately. Thus, the sum of the median figures for plumbing, HVAC and electrical will not normally total up to the total mechanical and electrical costs arrived at by separate analysis and tabulation of the projects.
- Each building was analyzed as to total and component costs and percentages. The figures were arranged in ascending order with the results tabulated as shown. The 1/4 column shows that 25% of the projects had lower costs and 75% had higher. The 3/4 column shows that 75% of the projects had lower costs and 25% had higher. The median column shows that 50% of the projects had lower costs and 50% had higher.
- There are two times when square foot costs are useful. The first is in the conceptual stage when no details are available. Then square foot costs make a useful starting point. The second is after the bids are in and the costs can be worked back into their appropriate categories for information purposes. As soon as details become available in the project design, the square foot approach should be discontinued and the project priced as to its particular components. When more precision is required, or for estimating the replacement cost of specific buildings, the current edition of *RSMeans Square Foot Costs* should be used.
- In using the figures in this section, it is recommended that the median column be used for preliminary figures if no additional information is available. The median figures, when multiplied by the total city construction cost index figures (see city cost indexes) and then multiplied by the project size modifier at the end of this section, should present a fairly accurate base figure, which would then have to be adjusted in view of the estimator's experience, local economic conditions, code requirements, and the owner's particular requirements. There is no need to factor the percentage figures, as these should remain constant from city to city. All tabulations mentioning air conditioning had at least partial air conditioning.
- The editors of this book would greatly appreciate receiving cost figures on one or more of your recent projects, which would then be included in the averages for next year. All cost figures received will be kept confidential, except that they will be averaged with other similar projects to arrive at square foot cost figures for next year's book. See the last page of the book for details and the discount available for submitting one or more of your projects.

50 17 00 | S.F. Costs

			UNIT	UNIT COSTS 1/4	UNIT COSTS MEDIAN	UNIT COSTS 3/4	% OF TOTAL 1/4	% OF TOTAL MEDIAN	% OF TOTAL 3/4	
01	0010	**APARTMENTS Low Rise (1 to 3 story)**	S.F.	73	92.50	123				01
	0020	Total project cost	C.F.	6.55	8.70	10.75				
	0100	Site work	S.F.	5.35	8.55	15	6.05%	10.55%	13.95%	
	0500	Masonry		1.44	3.55	5.80	1.54%	3.92%	6.50%	
	1500	Finishes		7.75	10.65	13.15	9.05%	10.75%	12.85%	
	1800	Equipment		2.40	3.63	5.40	2.71%	3.99%	5.95%	
	2720	Plumbing		5.70	7.30	9.30	6.65%	8.95%	10.05%	
	2770	Heating, ventilating, air conditioning		3.63	4.47	6.55	4.20%	5.60%	7.60%	
	2900	Electrical		4.25	5.65	7.65	5.20%	6.65%	8.35%	
	3100	Total: Mechanical & Electrical		15.10	19.60	24	16.05%	18.20%	23%	
	9000	Per apartment unit, total cost	Apt.	68,000	104,000	153,500				
	9500	Total: Mechanical & Electrical	″	12,900	20,300	26,500				
02	0010	**APARTMENTS Mid Rise (4 to 7 story)**	S.F.	97	117	144				02
	0020	Total project costs	C.F.	7.55	10.45	14.30				
	0100	Site work	S.F.	3.88	7.85	15.15	5.25%	6.70%	9.20%	
	0500	Masonry		4.05	8.90	12.15	5.10%	7.25%	10.50%	
	1500	Finishes		12.70	17	20	10.70%	13.50%	17.70%	
	1800	Equipment		2.71	4.22	5.55	2.54%	3.47%	4.31%	
	2500	Conveying equipment		2.20	2.70	3.26	2.05%	2.27%	2.69%	
	2720	Plumbing		5.70	9.10	9.65	5.70%	7.20%	8.95%	
	2900	Electrical		6.40	8.70	10.55	6.35%	7.20%	8.95%	
	3100	Total: Mechanical & Electrical		20.50	25.50	31	18.25%	21%	23%	
	9000	Per apartment unit, total cost	Apt.	109,500	129,500	214,000				
	9500	Total: Mechanical & Electrical	″	20,700	24,000	25,100				
03	0010	**APARTMENTS High Rise (8 to 24 story)**	S.F.	110	127	152				03
	0020	Total project costs	C.F.	10.70	12.45	14.65				
	0100	Site work	S.F.	3.99	6.45	9	2.58%	4.84%	6.15%	
	0500	Masonry		6.35	11.55	14.40	4.74%	9.65%	11.05%	
	1500	Finishes		12.20	15.25	18	9.75%	11.80%	13.70%	
	1800	Equipment		3.54	4.35	5.75	2.78%	3.49%	4.35%	
	2500	Conveying equipment		2.50	3.79	5.15	2.23%	2.78%	3.37%	
	2720	Plumbing		7.05	9.55	11.70	6.80%	7.20%	10.45%	
	2900	Electrical		7.55	9.55	12.90	6.45%	7.65%	8.80%	
	3100	Total: Mechanical & Electrical		22.50	29	35	17.95%	22.50%	24.50%	
	9000	Per apartment unit, total cost	Apt.	114,000	126,000	174,500				
	9500	Total: Mechanical & Electrical	″	22,700	28,200	29,900				
04	0010	**AUDITORIUMS**	S.F.	115	158	230				04
	0020	Total project costs	C.F.	7.15	9.95	14.30				
	2720	Plumbing	S.F.	6.80	10	11.95	5.85%	7.20%	8.70%	
	2900	Electrical		9.20	13.30	22	6.85%	9.50%	11.40%	
	3100	Total: Mechanical & Electrical		61	81.50	99	24.50%	27.50%	31%	
05	0010	**AUTOMOTIVE SALES**	S.F.	84.50	115	143				05
	0020	Total project costs	C.F.	5.55	6.70	8.65				
	2720	Plumbing	S.F.	3.86	6.70	7.30	2.89%	6.05%	6.50%	
	2770	Heating, ventilating, air conditioning		5.95	9.10	9.80	4.61%	10%	10.35%	
	2900	Electrical		6.80	10.70	15.70	7.25%	8.80%	12.15%	
	3100	Total: Mechanical & Electrical		19	30.50	36.50	17.30%	20.50%	22%	
06	0010	**BANKS**	S.F.	166	207	263				06
	0020	Total project costs	C.F.	11.85	16.10	21				
	0100	Site work	S.F.	19	30	42.50	7.90%	12.95%	17%	
	0500	Masonry		7.55	16.15	29	3.36%	6.90%	10.05%	
	1500	Finishes		14.85	22.50	28	5.85%	8.65%	11.70%	
	1800	Equipment		6.25	13.75	28	1.34%	5.55%	10.50%	
	2720	Plumbing		5.20	7.45	10.85	2.82%	3.90%	4.93%	
	2770	Heating, ventilating, air conditioning		9.85	13.20	17.60	4.86%	7.15%	8.50%	
	2900	Electrical		15.75	21	27.50	8.25%	10.20%	12.20%	
	3100	Total: Mechanical & Electrical		38	50.50	61	16.55%	19.45%	24%	
	3500	See also division 11020 & 11030 (MF2004 11 16 00 & 11 17 00)								

50 17 00 | S.F. Costs

			UNIT	UNIT COSTS			% OF TOTAL			
				1/4	MEDIAN	3/4	1/4	MEDIAN	3/4	
13	0010	**CHURCHES**	S.F.	112	143	185				13
	0020	Total project costs	C.F.	6.90	8.75	11.50				
	1800	Equipment	S.F.	1.21	3.17	6.80	.93%	2.11%	4.31%	
	2720	Plumbing		4.35	6.10	8.95	3.51%	4.96%	6.25%	
	2770	Heating, ventilating, air conditioning		10.15	13.25	18.75	7.50%	10%	12%	
	2900	Electrical		9.55	12.95	17.65	7.30%	8.80%	10.90%	
	3100	Total: Mechanical & Electrical	↓	29.50	39	51.50	18.30%	22%	24.50%	
	3500	See also division 11040 (MF2004 11 91 00)								
15	0010	**CLUBS, COUNTRY**	S.F.	120	144	182				15
	0020	Total project costs	C.F.	9.65	11.75	16.25				
	2720	Plumbing	S.F.	7.25	10.75	24.50	5.60%	7.90%	10%	
	2900	Electrical		9.40	12.90	16.80	7%	8.95%	11%	
	3100	Total: Mechanical & Electrical	↓	50	62.50	66	19%	26.50%	29.50%	
17	0010	**CLUBS, SOCIAL Fraternal**	S.F.	101	138	184				17
	0020	Total project costs	C.F.	5.95	9.05	10.80				
	2720	Plumbing	S.F.	6	7.50	11.30	5.60%	6.90%	8.55%	
	2770	Heating, ventilating, air conditioning		7.05	10.50	13.45	8.20%	9.25%	14.40%	
	2900	Electrical		6.85	11.80	13.55	5.95%	9.30%	10.55%	
	3100	Total: Mechanical & Electrical	↓	37	40	51	21%	23%	23.50%	
18	0010	**CLUBS, Y.M.C.A.**	S.F.	132	164	247				18
	0020	Total project costs	C.F.	5.55	9.25	13.80				
	2720	Plumbing	S.F.	7.60	15.10	16.95	5.65%	7.60%	10.85%	
	2900	Electrical		9.85	12.75	22	6.45%	8%	9.20%	
	3100	Total: Mechanical & Electrical	↓	40	45.50	74.50	20.50%	21.50%	28.50%	
19	0010	**COLLEGES Classrooms & Administration**	S.F.	116	164	212				19
	0020	Total project costs	C.F.	6.95	12.35	19.85				
	0500	Masonry	S.F.	9.15	16.85	20.50	5.10%	8.05%	10.50%	
	2720	Plumbing		5.20	12.65	23.50	5.10%	6.60%	8.95%	
	2900	Electrical		9.85	15.55	21.50	7.70%	9.85%	12%	
	3100	Total: Mechanical & Electrical	↓	41.50	57	76	23%	28%	31.50%	
21	0010	**COLLEGES Science, Engineering, Laboratories**	S.F.	225	264	305				21
	0020	Total project costs	C.F.	12.95	18.85	21.50				
	1800	Equipment	S.F.	6.10	28.50	31	2%	6.45%	12.65%	
	2900	Electrical		18.60	26.50	40.50	7.10%	9.40%	12.10%	
	3100	Total: Mechanical & Electrical	↓	69	82	127	28.50%	31.50%	41%	
	3500	See also division 11600 (MF2004 11 53 00)								
23	0010	**COLLEGES Student Unions**	S.F.	144	195	236				23
	0020	Total project costs	C.F.	8	10.50	13				
	3100	Total: Mechanical & Electrical	S.F.	54	58.50	69	23.50%	26%	29%	
25	0010	**COMMUNITY CENTERS**	S.F.	120	147	199				25
	0020	Total project costs	C.F.	7.75	11.05	14.35				
	1800	Equipment	S.F.	2.40	4.70	7.75	1.47%	3.01%	5.30%	
	2720	Plumbing		5.60	9.85	13.45	4.85%	7%	8.95%	
	2770	Heating, ventilating, air conditioning		8.45	13.10	18.80	6.80%	10.35%	12.90%	
	2900	Electrical		10.05	13.25	19.50	7.15%	8.90%	10.40%	
	3100	Total: Mechanical & Electrical	↓	33	42	60	18.80%	23%	30%	
28	0010	**COURT HOUSES**	S.F.	171	202	278				28
	0020	Total project costs	C.F.	13.15	15.70	19.80				
	2720	Plumbing	S.F.	8.15	11.35	12.90	5.95%	7.45%	8.20%	
	2900	Electrical		18.20	20.50	30	9.05%	10.65%	12.15%	
	3100	Total: Mechanical & Electrical	↓	51	66.50	73	22.50%	26.50%	30%	
30	0010	**DEPARTMENT STORES**	S.F.	63.50	85.50	108				30
	0020	Total project costs	C.F.	3.39	4.40	5.95				
	2720	Plumbing	S.F.	1.97	2.49	3.78	1.82%	4.21%	5.90%	
	2770	Heating, ventilating, air conditioning	↓	4.79	8.90	13.35	8.20%	9.10%	14.80%	

	50 17 00 | S.F. Costs		UNIT	UNIT COSTS 1/4	UNIT COSTS MEDIAN	UNIT COSTS 3/4	% OF TOTAL 1/4	% OF TOTAL MEDIAN	% OF TOTAL 3/4	
30	2900	Electrical	S.F.	7.25	9.95	11.75	9.05%	12.15%	14.95%	30
	3100	Total: Mechanical & Electrical		12.80	16.35	28.50	13.20%	21.50%	50%	
31	0010	**DORMITORIES Low Rise (1 to 3 story)**	S.F.	120	166	208				31
	0020	Total project costs	C.F.	6.80	10.95	16.45				
	2720	Plumbing	S.F.	7.20	9.65	12.20	8.05%	9%	9.65%	
	2770	Heating, ventilating, air conditioning		7.65	9.15	12.20	4.61%	8.05%	10%	
	2900	Electrical		7.95	12.10	16.55	6.40%	8.65%	9.50%	
	3100	Total: Mechanical & Electrical		41.50	44.50	69	22%	25%	27%	
	9000	Per bed, total cost	Bed	51,000	56,500	121,000				
32	0010	**DORMITORIES Mid Rise (4 to 8 story)**	S.F.	147	192	237				32
	0020	Total project costs	C.F.	16.25	17.85	21.50				
	2900	Electrical	S.F.	15.65	17.80	24	8.20%	10.20%	11.95%	
	3100	Total: Mechanical & Electrical	"	44	87.50	88.50	25%	30.50%	35.50%	
	9000	Per bed, total cost	Bed	21,000	47,900	277,000				
34	0010	**FACTORIES**	S.F.	55.50	83	128				34
	0020	Total project costs	C.F.	3.58	5.35	8.85				
	0100	Site work	S.F.	6.35	11.60	18.40	6.95%	11.45%	17.95%	
	2720	Plumbing		3.01	5.60	9.25	3.73%	6.05%	8.10%	
	2770	Heating, ventilating, air conditioning		5.85	8.40	11.35	5.25%	8.45%	11.35%	
	2900	Electrical		6.95	10.95	16.75	8.10%	10.50%	14.20%	
	3100	Total: Mechanical & Electrical		19.80	30.50	40	21%	28.50%	35.50%	
36	0010	**FIRE STATIONS**	S.F.	111	153	209				36
	0020	Total project costs	C.F.	6.45	8.90	11.85				
	0500	Masonry	S.F.	15.10	29	37.50	7.50%	10.95%	15.55%	
	1140	Roofing		3.61	9.75	11.10	1.90%	4.94%	5.05%	
	1580	Painting		2.80	4.18	4.28	1.37%	1.57%	2.07%	
	1800	Equipment		1.38	2.64	4.87	.62%	1.63%	3.42%	
	2720	Plumbing		6.20	9.85	13.95	5.85%	7.35%	9.45%	
	2770	Heating, ventilating, air conditioning		6.15	9.95	15.35	5.15%	7.40%	9.40%	
	2900	Electrical		7.95	14.10	19	6.90%	8.60%	10.60%	
	3100	Total: Mechanical & Electrical		40.50	52	59	18.40%	23%	27%	
37	0010	**FRATERNITY HOUSES & Sorority Houses**	S.F.	111	142	195				37
	0020	Total project costs	C.F.	11	11.50	13.85				
	2720	Plumbing	S.F.	8.35	9.60	17.55	6.80%	8%	10.85%	
	2900	Electrical		7.30	15.80	19.35	6.60%	9.90%	10.65%	
38	0010	**FUNERAL HOMES**	S.F.	117	159	289				38
	0020	Total project costs	C.F.	11.90	13.25	25.50				
	2900	Electrical	S.F.	5.15	9.45	10.35	3.58%	4.44%	5.95%	
39	0010	**GARAGES, COMMERCIAL (Service)**	S.F.	66	102	141				39
	0020	Total project costs	C.F.	4.35	6.40	9.35				
	1800	Equipment	S.F.	3.73	8.40	13.05	2.21%	4.62%	6.80%	
	2720	Plumbing		4.58	7.05	12.85	5.45%	7.85%	10.65%	
	2730	Heating & ventilating		4.51	7.80	9.95	5.25%	6.85%	8.20%	
	2900	Electrical		6.30	9.60	13.80	7.15%	9.25%	10.85%	
	3100	Total: Mechanical & Electrical		11.85	26.50	39	12.35%	17.40%	26%	
40	0010	**GARAGES, MUNICIPAL (Repair)**	S.F.	97	129	182				40
	0020	Total project costs	C.F.	6.05	7.70	13.25				
	0500	Masonry	S.F.	5.10	17.85	27.50	4.03%	9.15%	12.50%	
	2720	Plumbing		4.36	8.35	15.75	3.59%	6.70%	7.95%	
	2730	Heating & ventilating		7.45	10.80	21	6.15%	7.45%	13.50%	
	2900	Electrical		7.15	11.30	16.40	6.65%	8.15%	11.15%	
	3100	Total: Mechanical & Electrical		33	46	66.50	21.50%	23%	28.50%	
41	0010	**GARAGES, PARKING**	S.F.	38.50	55	95				41
	0020	Total project costs	C.F.	3.55	4.82	7				

	50 17 00 | S.F. Costs		UNIT	UNIT COSTS 1/4	UNIT COSTS MEDIAN	UNIT COSTS 3/4	% OF TOTAL 1/4	% OF TOTAL MEDIAN	% OF TOTAL 3/4	
41	2720	Plumbing	S.F.	.65	1.66	2.56	1.72%	2.70%	3.85%	41
	2900	Electrical		2.08	2.74	4	4.52%	5.40%	6.35%	
	3100	Total: Mechanical & Electrical		4.03	5.90	7.35	7%	8.90%	11.05%	
	3200									
	9000	Per car, total cost	Car	16,000	20,000	25,500				
43	0010	**GYMNASIUMS**	S.F.	105	140	193				43
	0020	Total project costs	C.F.	5.25	7.10	8.75				
	1800	Equipment	S.F.	2.50	4.68	8.45	1.76%	3.26%	6.70%	
	2720	Plumbing		5.75	7.75	10.20	4.65%	6.40%	7.75%	
	2770	Heating, ventilating, air conditioning		6.25	10.90	22	5.15%	9.05%	11.10%	
	2900	Electrical		8.35	11	14.65	6.75%	8.50%	10.30%	
	3100	Total: Mechanical & Electrical		28.50	40	48.50	19.75%	23.50%	29%	
	3500	See also division 11480 (MF2004 11 67 00)								
46	0010	**HOSPITALS**	S.F.	202	253	350				46
	0020	Total project costs	C.F.	15.30	19.05	27				
	1800	Equipment	S.F.	5.10	9.85	17	.80%	2.53%	4.80%	
	2720	Plumbing		17.40	24.50	31.50	7.60%	9.10%	10.85%	
	2770	Heating, ventilating, air conditioning		25.50	32.50	45.50	7.80%	12.95%	16.65%	
	2900	Electrical		22.50	30	44.50	10%	11.75%	14.10%	
	3100	Total: Mechanical & Electrical		65	90.50	134	28%	33.50%	37%	
	9000	Per bed or person, total cost	Bed	233,500	322,000	371,000				
	9900	See also division 11700 (MF2004 11 71 00)								
48	0010	**HOUSING For the Elderly**	S.F.	99	126	154				48
	0020	Total project costs	C.F.	7.10	9.85	12.55				
	0100	Site work	S.F.	6.90	10.75	15.75	5.05%	7.90%	12.10%	
	0500	Masonry		1.87	11.30	16.50	1.30%	6.05%	11%	
	1800	Equipment		2.40	3.30	5.25	1.88%	3.23%	4.43%	
	2510	Conveying systems		2.24	3.24	4.40	1.78%	2.20%	2.81%	
	2720	Plumbing		7.40	9.40	11.85	8.15%	9.55%	10.50%	
	2730	Heating, ventilating, air conditioning		3.78	5.35	8	3.30%	5.60%	7.25%	
	2900	Electrical		7.40	10.05	12.85	7.30%	8.50%	10.25%	
	3100	Total: Mechanical & Electrical		25.50	31.50	40.50	18.10%	22.50%	29%	
	9000	Per rental unit, total cost	Unit	92,500	108,000	120,000				
	9500	Total: Mechanical & Electrical	"	20,600	23,600	27,600				
50	0010	**HOUSING Public (Low Rise)**	S.F.	83.50	116	151				50
	0020	Total project costs	C.F.	6.60	9.25	11.50				
	0100	Site work	S.F.	10.60	15.30	25	8.35%	11.75%	16.50%	
	1800	Equipment		2.27	3.70	5.65	2.26%	3.03%	4.24%	
	2720	Plumbing		6	7.95	10.05	7.15%	9.05%	11.60%	
	2730	Heating, ventilating, air conditioning		3.02	5.85	6.40	4.26%	6.05%	6.45%	
	2900	Electrical		5.05	7.50	10.45	5.10%	6.55%	8.25%	
	3100	Total: Mechanical & Electrical		24	31	34.50	14.50%	17.55%	26.50%	
	9000	Per apartment, total cost	Apt.	91,500	104,000	131,000				
	9500	Total: Mechanical & Electrical	"	19,500	24,100	26,700				
51	0010	**ICE SKATING RINKS**	S.F.	71.50	167	183				51
	0020	Total project costs	C.F.	5.25	5.35	6.20				
	2720	Plumbing	S.F.	2.66	5	5.10	3.12%	3.23%	5.65%	
	2900	Electrical		7.65	11.70	12.40	6.30%	10.15%	15.05%	
52	0010	**JAILS**	S.F.	185	280	360				52
	0020	Total project costs	C.F.	19.60	27.50	32				
	1800	Equipment	S.F.	8.45	25	42.50	2.80%	6.95%	9.80%	
	2720	Plumbing		21	28	37	7%	8.90%	13.35%	
	2770	Heating, ventilating, air conditioning		19.60	26	50.50	7.50%	9.45%	17.75%	
	2900	Electrical		23	31	39	9.40%	11.70%	15.25%	
	3100	Total: Mechanical & Electrical		61	108	128	28%	30%	34%	
53	0010	**LIBRARIES**	S.F.	140	186	243				53
	0020	Total project costs	C.F.	9.40	11.75	15				

50 17 00 | S.F. Costs

			UNIT	UNIT COSTS 1/4	UNIT COSTS MEDIAN	UNIT COSTS 3/4	% OF TOTAL 1/4	% OF TOTAL MEDIAN	% OF TOTAL 3/4	
53	0500	Masonry	S.F.	9.30	19	31.50	5.60%	7.15%	10.95%	53
	1800	Equipment		1.88	5.05	7.60	.28%	1.39%	4.07%	
	2720	Plumbing		5.05	7.20	10	3.38%	4.60%	5.70%	
	2770	Heating, ventilating, air conditioning		11.20	19	23	7.80%	10.95%	12.80%	
	2900	Electrical		14.10	18.65	25	8.40%	10.55%	12%	
	3100	Total: Mechanical & Electrical		46	56	65.50	21%	24%	28%	
54	0010	**LIVING, ASSISTED**	S.F.	127	151	176				54
	0020	Total project costs	C.F.	10.70	12.50	14.20				
	0500	Masonry	S.F.	3.68	4.44	5.45	2.36%	3.16%	3.86%	
	1800	Equipment		2.81	3.35	4.30	2.12%	2.45%	2.87%	
	2720	Plumbing		10.65	14.25	14.75	6.05%	8.15%	10.60%	
	2770	Heating, ventilating, air conditioning		12.65	13.20	14.45	7.95%	9.35%	9.70%	
	2900	Electrical		12.45	14	16.15	9%	9.95%	10.65%	
	3100	Total: Mechanical & Electrical		34	40.50	47	24%	28.50%	31.50%	
55	0010	**MEDICAL CLINICS**	S.F.	129	159	203				55
	0020	Total project costs	C.F.	9.45	12.25	16.30				
	1800	Equipment	S.F.	1.65	7.35	11.40	1.05%	2.94%	6.35%	
	2720	Plumbing		8.60	12.10	16.15	6.15%	8.40%	10.10%	
	2770	Heating, ventilating, air conditioning		10.25	13.40	19.70	6.65%	8.85%	11.35%	
	2900	Electrical		11.10	15.75	20.50	8.10%	10%	12.20%	
	3100	Total: Mechanical & Electrical		35.50	48	66	22%	27%	33.50%	
	3500	See also division 11700 (MF2004 11 71 00)								
57	0010	**MEDICAL OFFICES**	S.F.	121	151	185				57
	0020	Total project costs	C.F.	9.05	12.30	16.60				
	1800	Equipment	S.F.	4.01	7.95	11.15	.66%	4.87%	6.50%	
	2720	Plumbing		6.70	10.35	13.95	5.60%	6.80%	8.50%	
	2770	Heating, ventilating, air conditioning		8.10	11.70	15.45	6.10%	8%	9.70%	
	2900	Electrical		9.70	14.20	19.75	7.50%	9.80%	11.70%	
	3100	Total: Mechanical & Electrical		26.50	38.50	55	19.30%	23%	28.50%	
59	0010	**MOTELS**	S.F.	76.50	111	145				59
	0020	Total project costs	C.F.	6.80	9.15	14.95				
	2720	Plumbing	S.F.	7.75	9.90	11.80	9.45%	10.60%	12.55%	
	2770	Heating, ventilating, air conditioning		4.74	7.05	12.65	5.60%	5.60%	10%	
	2900	Electrical		7.25	9.15	11.40	7.45%	9.05%	10.45%	
	3100	Total: Mechanical & Electrical		21.50	31	53	18.50%	24%	25.50%	
	5000									
	9000	Per rental unit, total cost	Unit	39,000	74,000	80,000				
	9500	Total: Mechanical & Electrical	"	7,600	11,500	13,400				
60	0010	**NURSING HOMES**	S.F.	120	155	193				60
	0020	Total project costs	C.F.	9.45	11.80	16.15				
	1800	Equipment	S.F.	3.28	5	8.40	2%	3.62%	4.99%	
	2720	Plumbing		10.30	15.60	18.80	8.75%	10.10%	12.70%	
	2770	Heating, ventilating, air conditioning		10.85	16.45	22	9.70%	11.45%	11.80%	
	2900	Electrical		11.90	14.95	20.50	9.40%	10.60%	12.50%	
	3100	Total: Mechanical & Electrical		28.50	39.50	66.50	26%	28%	30%	
	9000	Per bed or person, total cost	Bed	53,500	67,000	86,000				
61	0010	**OFFICES Low Rise (1 to 4 story)**	S.F.	101	132	172				61
	0020	Total project costs	C.F.	7.25	9.95	13.10				
	0100	Site work	S.F.	8.15	14.20	21	5.95%	9.70%	13.55%	
	0500	Masonry		3.92	7.90	14.40	2.61%	5.40%	8.45%	
	1800	Equipment		.91	2.11	5.75	.57%	1.50%	3.42%	
	2720	Plumbing		3.62	5.60	8.20	3.66%	4.50%	6.10%	
	2770	Heating, ventilating, air conditioning		8	11.15	16.35	7.20%	10.30%	11.70%	
	2900	Electrical		8.30	11.85	16.85	7.45%	9.65%	11.40%	
	3100	Total: Mechanical & Electrical		23	31.50	48	18.20%	22%	27%	
62	0010	**OFFICES Mid Rise (5 to 10 story)**	S.F.	107	130	176				62
	0020	Total project costs	C.F.	7.60	9.70	13.75				

	50 17 00 | S.F. Costs		UNIT	UNIT COSTS 1/4	MEDIAN	3/4	% OF TOTAL 1/4	MEDIAN	3/4	
62	2720	Plumbing	S.F.	3.24	5	7.20	2.83%	3.74%	4.50%	62
	2770	Heating, ventilating, air conditioning		8.15	11.65	18.60	7.65%	9.40%	11%	
	2900	Electrical		7.95	10.20	15.40	6.35%	7.80%	10%	
	3100	Total: Mechanical & Electrical	↓	20.50	26.50	50.50	18.95%	21%	27.50%	
63	0010	**OFFICES High Rise (11 to 20 story)**	S.F.	132	166	204				63
	0020	Total project costs	C.F.	9.20	11.50	16.50				
	2900	Electrical	S.F.	8	9.75	14.50	5.80%	7.85%	10.50%	
	3100	Total: Mechanical & Electrical	↓	26	34.50	58.50	16.90%	23.50%	34%	
64	0010	**POLICE STATIONS**	S.F.	158	208	265				64
	0020	Total project costs	C.F.	12.60	15.40	21				
	0500	Masonry	S.F.	15.80	26	33	7.80%	9.10%	11.35%	
	1800	Equipment		2.27	11.05	17.50	.98%	3.35%	6.70%	
	2720	Plumbing		8.85	17.60	22	5.65%	6.90%	10.75%	
	2770	Heating, ventilating, air conditioning		12.95	18.35	26	5.85%	10.55%	11.70%	
	2900	Electrical		17.25	25	32.50	9.80%	11.70%	14.50%	
	3100	Total: Mechanical & Electrical	↓	65.50	69	92	28.50%	31.50%	32.50%	
65	0010	**POST OFFICES**	S.F.	125	154	196				65
	0020	Total project costs	C.F.	7.50	9.50	10.80				
	2720	Plumbing	S.F.	5.05	6.95	8.80	4.24%	5.30%	5.60%	
	2770	Heating, ventilating, air conditioning		7.85	10.85	12.05	6.65%	7.15%	9.35%	
	2900	Electrical		10.30	14.50	17.20	7.25%	9%	11%	
	3100	Total: Mechanical & Electrical	↓	29.50	39	44	16.25%	18.80%	22%	
66	0010	**POWER PLANTS**	S.F.	865	1,150	2,100				66
	0020	Total project costs	C.F.	24	52	111				
	2900	Electrical	S.F.	61	130	193	9.30%	12.75%	21.50%	
	8100	Total: Mechanical & Electrical	↓	152	495	1,100	32.50%	32.50%	52.50%	
67	0010	**RELIGIOUS EDUCATION**	S.F.	101	131	161				67
	0020	Total project costs	C.F.	5.60	8	10				
	2720	Plumbing	S.F.	4.20	5.95	8.40	4.40%	5.30%	7.10%	
	2770	Heating, ventilating, air conditioning		10.60	12	16.95	10.05%	11.45%	12.35%	
	2900	Electrical		7.40	11.25	14.85	7.60%	9.05%	10.35%	
	3100	Total: Mechanical & Electrical	↓	32.50	42.50	52	22%	23%	27%	
69	0010	**RESEARCH Laboratories & Facilities**	S.F.	154	217	315				69
	0020	Total project costs	C.F.	11.20	22	26.50				
	1800	Equipment	S.F.	6.70	13.15	32	.94%	4.58%	9.10%	
	2720	Plumbing		10.95	19.30	31	6.15%	8.30%	10.80%	
	2770	Heating, ventilating, air conditioning		13.55	45.50	54	7.25%	16.50%	17.50%	
	2900	Electrical		18.10	29	47.50	9.55%	11.15%	14.60%	
	3100	Total: Mechanical & Electrical	↓	57.50	102	144	29.50%	37%	41%	
70	0010	**RESTAURANTS**	S.F.	146	188	245				70
	0020	Total project costs	C.F.	12.25	16.10	21				
	1800	Equipment	S.F.	8.65	23	35	6.10%	13%	15.65%	
	2720	Plumbing		11.55	14	18.40	6.10%	8.15%	9%	
	2770	Heating, ventilating, air conditioning		14.70	20.50	24.50	9.20%	12%	12.40%	
	2900	Electrical		15.40	19	24.50	8.35%	10.55%	11.55%	
	3100	Total: Mechanical & Electrical	↓	47.50	51	66	21%	25%	29.50%	
	9000	Per seat unit, total cost	Seat	5,350	7,125	8,450				
	9500	Total: Mechanical & Electrical	"	1,350	1,775	2,100				
72	0010	**RETAIL STORES**	S.F.	68	91.50	121				72
	0020	Total project costs	C.F.	4.62	6.60	9.15				
	2720	Plumbing	S.F.	2.47	4.12	7.05	3.26%	4.60%	6.80%	
	2770	Heating, ventilating, air conditioning		5.35	7.30	10.95	6.75%	8.75%	10.15%	
	2900	Electrical		6.15	8.40	12.10	7.25%	9.90%	11.60%	
	3100	Total: Mechanical & Electrical	↓	16.30	21	28	17.05%	21%	23.50%	
74	0010	**SCHOOLS Elementary**	S.F.	111	138	169				74
	0020	Total project costs	C.F.	7.25	9.30	12.05				
	0500	Masonry	S.F.	7.85	16.95	25	4.89%	10.50%	14%	
	1800	Equipment	↓	2.58	4.93	9.30	1.83%	3.13%	4.61%	

50 17 00 | S.F. Costs

		50 17 00 | S.F. Costs	UNIT	UNIT COSTS 1/4	UNIT COSTS MEDIAN	UNIT COSTS 3/4	% OF TOTAL 1/4	% OF TOTAL MEDIAN	% OF TOTAL 3/4	
74	2720	Plumbing	S.F.	6.40	9.05	12.05	5.70%	7.15%	9.35%	74
	2730	Heating, ventilating, air conditioning		9.60	15.25	21.50	8.15%	10.80%	14.90%	
	2900	Electrical		10.50	14	17.85	8.45%	10.05%	11.85%	
	3100	Total: Mechanical & Electrical		38	46.50	59	25%	27.50%	30%	
	9000	Per pupil, total cost	Ea.	12,800	19,000	42,500				
	9500	Total: Mechanical & Electrical	"	3,600	4,550	11,500				
76	0010	**SCHOOLS Junior High & Middle**	S.F.	115	142	173				76
	0020	Total project costs	C.F.	7.30	9.40	10.55				
	0500	Masonry	S.F.	12.30	18.40	22.50	7.55%	11.10%	14.30%	
	1800	Equipment		3.14	5.90	8.65	1.80%	3.03%	4.80%	
	2720	Plumbing		6.70	8.25	10.20	5.30%	6.80%	7.25%	
	2770	Heating, ventilating, air conditioning		13.35	16.20	28.50	8.90%	11.55%	14.20%	
	2900	Electrical		11.70	14.45	18.25	8.05%	9.55%	10.60%	
	3100	Total: Mechanical & Electrical		36.50	47	58.50	23%	25.50%	29.50%	
	9000	Per pupil, total cost	Ea.	14,500	19,100	25,600				
78	0010	**SCHOOLS Senior High**	S.F.	120	147	184				78
	0020	Total project costs	C.F.	7.20	10.10	17.05				
	1800	Equipment	S.F.	3.17	7.40	10.25	1.88%	2.67%	4.30%	
	2720	Plumbing		6.70	10.05	18.35	5.60%	6.90%	8.30%	
	2770	Heating, ventilating, air conditioning		11.50	15.65	24	8.95%	11.60%	15%	
	2900	Electrical		12.05	16.05	23.50	8.70%	10.35%	12.50%	
	3100	Total: Mechanical & Electrical		39.50	47	76	24%	26.50%	28.50%	
	9000	Per pupil, total cost	Ea.	11,200	22,900	28,600				
80	0010	**SCHOOLS Vocational**	S.F.	96.50	140	173				80
	0020	Total project costs	C.F.	6	8.60	11.90				
	0500	Masonry	S.F.	4.24	14	21.50	3.20%	4.61%	10.95%	
	1800	Equipment		3.02	7.50	10.35	1.24%	3.10%	4.26%	
	2720	Plumbing		6.15	9.20	13.55	5.40%	6.90%	8.55%	
	2770	Heating, ventilating, air conditioning		8.65	16.05	27	8.60%	11.90%	14.65%	
	2900	Electrical		10.05	13.75	18.80	8.45%	11%	13.20%	
	3100	Total: Mechanical & Electrical		38	66.50	85	27.50%	29.50%	31%	
	9000	Per pupil, total cost	Ea.	13,400	36,000	53,500				
83	0010	**SPORTS ARENAS**	S.F.	84.50	113	174				83
	0020	Total project costs	C.F.	4.58	8.20	10.60				
	2720	Plumbing	S.F.	4.22	7.40	15.65	4.35%	6.35%	9.40%	
	2770	Heating, ventilating, air conditioning		10.55	12.45	17.30	8.80%	10.20%	13.55%	
	2900	Electrical		7.80	11.90	15.40	7.65%	9.75%	11.90%	
85	0010	**SUPERMARKETS**	S.F.	78	90	106				85
	0020	Total project costs	C.F.	4.34	5.25	7.95				
	2720	Plumbing	S.F.	4.35	5.50	6.40	5.40%	6%	7.45%	
	2770	Heating, ventilating, air conditioning		6.40	8.50	10.35	8.60%	8.65%	9.60%	
	2900	Electrical		9.15	11.15	13.25	10.40%	12.45%	13.60%	
	3100	Total: Mechanical & Electrical		25	26.50	35.50	23.50%	27.50%	28.50%	
86	0010	**SWIMMING POOLS**	S.F.	126	295	450				86
	0020	Total project costs	C.F.	10.10	12.60	13.75				
	2720	Plumbing	S.F.	11.65	13.35	17.90	4.80%	9.70%	20.50%	
	2900	Electrical		9.60	15.35	34	6.05%	6.95%	7.60%	
	3100	Total: Mechanical & Electrical		60	79	104	11.15%	14.10%	23.50%	
87	0010	**TELEPHONE EXCHANGES**	S.F.	157	238	310				87
	0020	Total project costs	C.F.	10.40	16.75	23				
	2720	Plumbing	S.F.	7.05	10.90	16	4.52%	5.80%	6.90%	
	2770	Heating, ventilating, air conditioning		16.40	33	41	11.80%	16.05%	18.40%	
	2900	Electrical		17.05	27	48	10.90%	14%	17.85%	
	3100	Total: Mechanical & Electrical		50.50	95.50	135	29.50%	33.50%	44.50%	
91	0010	**THEATERS**	S.F.	99.50	135	199				91
	0020	Total project costs	C.F.	4.87	7.20	10.55				

	50 17 00 | S.F. Costs		UNIT	UNIT COSTS 1/4	UNIT COSTS MEDIAN	UNIT COSTS 3/4	% OF TOTAL 1/4	% OF TOTAL MEDIAN	% OF TOTAL 3/4	
91	2720	Plumbing	S.F.	3.28	3.80	15.55	2.92%	4.70%	6.80%	91
	2770	Heating, ventilating, air conditioning		10.25	12.40	15.35	8%	12.25%	13.40%	
	2900	Electrical		9.20	12.45	25.50	8.05%	9.35%	12.25%	
	3100	Total: Mechanical & Electrical	↓	23.50	32	37.50	21.50%	25.50%	27.50%	
94	0010	**TOWN HALLS City Halls & Municipal Buildings**	S.F.	110	149	195				94
	0020	Total project costs	C.F.	7.95	12.40	18.10				
	2720	Plumbing	S.F.	4.91	9.20	16.10	4.31%	5.95%	7.95%	
	2770	Heating, ventilating, air conditioning		8.85	17.60	26	7.05%	9.05%	13.45%	
	2900	Electrical		11.15	15.40	21	8.05%	9.45%	11.65%	
	3100	Total: Mechanical & Electrical	↓	38.50	49	75	22%	26.50%	31%	
97	0010	**WAREHOUSES & Storage Buildings**	S.F.	44	63	93				97
	0020	Total project costs	C.F.	2.30	3.60	5.95				
	0100	Site work	S.F.	4.53	9	13.55	6.05%	12.95%	19.55%	
	0500	Masonry		2.49	6.25	13.45	3.60%	7.35%	12%	
	1800	Equipment		.67	1.52	8.55	.72%	1.69%	5.55%	
	2720	Plumbing		1.46	2.62	4.91	2.90%	4.80%	6.55%	
	2730	Heating, ventilating, air conditioning		1.67	4.71	6.30	2.41%	5%	8.90%	
	2900	Electrical		2.60	4.89	8	5.15%	7.20%	10.05%	
	3100	Total: Mechanical & Electrical	↓	7.25	11.10	22	13.30%	18.90%	26%	
99	0010	**WAREHOUSE & OFFICES Combination**	S.F.	54	72	99				99
	0020	Total project costs	C.F.	2.76	4.01	5.90				
	1800	Equipment	S.F.	.93	1.81	2.69	.42%	1.19%	2.40%	
	2720	Plumbing		2.08	3.63	5.40	3.74%	4.76%	6.30%	
	2770	Heating, ventilating, air conditioning		3.29	5.15	7.20	5%	5.65%	10.05%	
	2900	Electrical		3.68	5.45	8.50	5.85%	8%	10%	
	3100	Total: Mechanical & Electrical	↓	10.25	15.55	24.50	14.55%	19.95%	24.50%	

Square Foot Project Size Modifier

One factor that affects the S.F. cost of a particular building is the size. In general, for buildings built to the same specifications in the same locality, the larger building will have the lower S.F. cost. This is due mainly to the decreasing contribution of the exterior walls, plus the economy of scale usually achievable in larger buildings. The area conversion scale shown below will give a factor to convert costs for the typical size building to an adjusted cost for the particular project.

The square foot base size table lists the median costs, most typical project size in our accumulated data, and the range in size of the projects.

The size factor for your project is determined by dividing your project area in S.F. by the typical project size for the particular building type. With this factor, enter the area conversion scale at the appropriate size factor and determine the appropriate cost multiplier for your building size.

Example: Determine the cost per S.F. for a 100,000 S.F. Mid-rise apartment building.

$$\frac{\text{Proposed building area} = 100{,}000 \text{ S.F.}}{\text{Typical size from below} = 50{,}000 \text{ S.F.}} = 2.00$$

Enter area conversion scale at 2.0, intersect curve, read horizontally the appropriate cost multiplier of .94. Size adjusted cost becomes .94 × $117.00 = $110.00 based on national average costs.

Note: For size factors less than .50, the cost multiplier is 1.1
For size factors greater than 3.5, the cost multiplier is .90

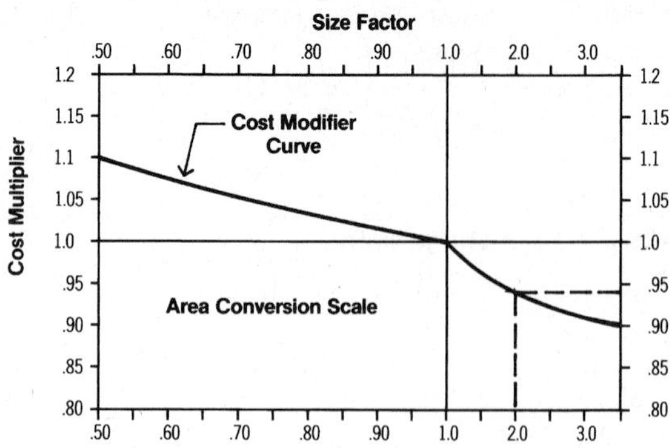

Square Foot Base Size

Building Type	Median Cost per S.F.	Typical Size Gross S.F.	Typical Range Gross S.F.	Building Type	Median Cost per S.F.	Typical Size Gross S.F.	Typical Range Gross S.F.
Apartments, Low Rise	$ 92.50	21,000	9,700 - 37,200	Jails	$ 280.00	40,000	5,500 - 145,000
Apartments, Mid Rise	117.00	50,000	32,000 - 100,000	Libraries	186.00	12,000	7,000 - 31,000
Apartments, High Rise	127.00	145,000	95,000 - 600,000	Living, Assisted	151.00	32,300	23,500 - 50,300
Auditoriums	158.00	25,000	7,600 - 39,000	Medical Clinics	159.00	7,200	4,200 - 15,700
Auto Sales	115.00	20,000	10,800 - 28,600	Medical Offices	151.00	6,000	4,000 - 15,000
Banks	207.00	4,200	2,500 - 7,500	Motels	111.00	40,000	15,800 - 120,000
Churches	143.00	17,000	2,000 - 42,000	Nursing Homes	155.00	23,000	15,000 - 37,000
Clubs, Country	144.00	6,500	4,500 - 15,000	Offices, Low Rise	132.00	20,000	5,000 - 80,000
Clubs, Social	138.00	10,000	6,000 - 13,500	Offices, Mid Rise	130.00	120,000	20,000 - 300,000
Clubs, YMCA	164.00	28,300	12,800 - 39,400	Offices, High Rise	166.00	260,000	120,000 - 800,000
Colleges (Class)	164.00	50,000	15,000 - 150,000	Police Stations	208.00	10,500	4,000 - 19,000
Colleges (Science Lab)	264.00	45,600	16,600 - 80,000	Post Offices	154.00	12,400	6,800 - 30,000
College (Student Union)	195.00	33,400	16,000 - 85,000	Power Plants	1150.00	7,500	1,000 - 20,000
Community Center	147.00	9,400	5,300 - 16,700	Religious Education	131.00	9,000	6,000 - 12,000
Court Houses	202.00	32,400	17,800 - 106,000	Research	217.00	19,000	6,300 - 45,000
Dept. Stores	85.50	90,000	44,000 - 122,000	Restaurants	188.00	4,400	2,800 - 6,000
Dormitories, Low Rise	166.00	25,000	10,000 - 95,000	Retail Stores	91.50	7,200	4,000 - 17,600
Dormitories, Mid Rise	192.00	85,000	20,000 - 200,000	Schools, Elementary	138.00	41,000	24,500 - 55,000
Factories	83.00	26,400	12,900 - 50,000	Schools, Jr. High	142.00	92,000	52,000 - 119,000
Fire Stations	153.00	5,800	4,000 - 8,700	Schools, Sr. High	147.00	101,000	50,500 - 175,000
Fraternity Houses	142.00	12,500	8,200 - 14,800	Schools, Vocational	140.00	37,000	20,500 - 82,000
Funeral Homes	159.00	10,000	4,000 - 20,000	Sports Arenas	113.00	15,000	5,000 - 40,000
Garages, Commercial	102.00	9,300	5,000 - 13,600	Supermarkets	90.00	44,000	12,000 - 60,000
Garages, Municipal	129.00	8,300	4,500 - 12,600	Swimming Pools	295.00	20,000	10,000 - 32,000
Garages, Parking	55.00	163,000	76,400 - 225,300	Telephone Exchange	238.00	4,500	1,200 - 10,600
Gymnasiums	140.00	19,200	11,600 - 41,000	Theaters	135.00	10,500	8,800 - 17,500
Hospitals	253.00	55,000	27,200 - 125,000	Town Halls	149.00	10,800	4,800 - 23,400
House (Elderly)	126.00	37,000	21,000 - 66,000	Warehouses	63.00	25,000	8,000 - 72,000
Housing (Public)	116.00	36,000	14,400 - 74,400	Warehouse & Office	72.00	25,000	8,000 - 72,000
Ice Rinks	167.00	29,000	27,200 - 33,600				

Abbreviations

A	Area Square Feet; Ampere
AAFES	Army and Air Force Exchange Service
ABS	Acrylonitrile Butadiene Stryrene; Asbestos Bonded Steel
A.C., AC	Alternating Current; Air-Conditioning; Asbestos Cement; Plywood Grade A & C
ACI	American Concrete Institute
ACR	Air Conditioning Refrigeration
ADA	Americans with Disabilities Act
AD	Plywood, Grade A & D
Addit.	Additional
Adj.	Adjustable
af	Audio-frequency
AFUE	Annual Fuel Utilization Efficiency
AGA	American Gas Association
Agg.	Aggregate
A.H., Ah	Ampere Hours
A hr.	Ampere-hour
A.H.U., AHU	Air Handling Unit
A.I.A.	American Institute of Architects
AIC	Ampere Interrupting Capacity
Allow.	Allowance
alt., alt	Alternate
Alum.	Aluminum
a.m.	Ante Meridiem
Amp.	Ampere
Anod.	Anodized
ANSI	American National Standards Institute
APA	American Plywood Association
Approx.	Approximate
Apt.	Apartment
Asb.	Asbestos
A.S.B.C.	American Standard Building Code
Asbe.	Asbestos Worker
ASCE.	American Society of Civil Engineers
A.S.H.R.A.E.	American Society of Heating, Refrig. & AC Engineers
ASME	American Society of Mechanical Engineers
ASTM	American Society for Testing and Materials
Attchmt.	Attachment
Avg., Ave.	Average
AWG	American Wire Gauge
AWWA	American Water Works Assoc.
Bbl.	Barrel
B&B, BB	Grade B and Better; Balled & Burlapped
B&S	Bell and Spigot
B.&W.	Black and White
b.c.c.	Body-centered Cubic
B.C.Y.	Bank Cubic Yards
BE	Bevel End
B.F.	Board Feet
Bg. cem.	Bag of Cement
BHP	Boiler Horsepower; Brake Horsepower
B.I.	Black Iron
bidir.	bidirectional
Bit., Bitum.	Bituminous
Bit., Conc.	Bituminous Concrete
Bk.	Backed
Bkrs.	Breakers
Bldg., bldg	Building
Blk.	Block
Bm.	Beam
Boil.	Boilermaker
bpm	Blows per Minute
BR	Bedroom
Brg.	Bearing
Brhe.	Bricklayer Helper
Bric.	Bricklayer
Brk., brk	Brick
brkt	Bracket
Brng.	Bearing
Brs.	Brass
Brz.	Bronze
Bsn.	Basin
Btr.	Better
Btu	British Thermal Unit
BTUH	BTU per Hour
Bu.	bushels
BUR	Built-up Roofing
BX	Interlocked Armored Cable
°C	degree centegrade
c	Conductivity, Copper Sweat
C	Hundred; Centigrade
C/C	Center to Center, Cedar on Cedar
C-C	Center to Center
Cab	Cabinet
Cair.	Air Tool Laborer
Cal.	caliper
Calc	Calculated
Cap.	Capacity
Carp.	Carpenter
C.B.	Circuit Breaker
C.C.A.	Chromate Copper Arsenate
C.C.F.	Hundred Cubic Feet
cd	Candela
cd/sf	Candela per Square Foot
CD	Grade of Plywood Face & Back
CDX	Plywood, Grade C & D, exterior glue
Cefi.	Cement Finisher
Cem.	Cement
CF	Hundred Feet
C.F.	Cubic Feet
CFM	Cubic Feet per Minute
CFRP	Carbon Fiber Reinforced Plastic
c.g.	Center of Gravity
CHW	Chilled Water; Commercial Hot Water
C.I., CI	Cast Iron
C.I.P., CIP	Cast in Place
Circ.	Circuit
C.L.	Carload Lot
CL	Chain Link
Clab.	Common Laborer
Clam	Common maintenance laborer
C.L.F.	Hundred Linear Feet
CLF	Current Limiting Fuse
CLP	Cross Linked Polyethylene
cm	Centimeter
CMP	Corr. Metal Pipe
CMU	Concrete Masonry Unit
CN	Change Notice
Col.	Column
CO_2	Carbon Dioxide
Comb.	Combination
comm.	Commercial, Communication
Compr.	Compressor
Conc.	Concrete
Cont., cont	Continuous; Continued, Container
Corr.	Corrugated
Cos	Cosine
Cot	Cotangent
Cov.	Cover
C/P	Cedar on Paneling
CPA	Control Point Adjustment
Cplg.	Coupling
CPM	Critical Path Method
CPVC	Chlorinated Polyvinyl Chloride
C.Pr.	Hundred Pair
CRC	Cold Rolled Channel
Creos.	Creosote
Crpt.	Carpet & Linoleum Layer
CRT	Cathode-ray Tube
CS	Carbon Steel, Constant Shear Bar Joist
Csc	Cosecant
C.S.F.	Hundred Square Feet
CSI	Construction Specifications Institute
CT	Current Transformer
CTS	Copper Tube Size
Cu	Copper, Cubic
Cu. Ft.	Cubic Foot
cw	Continuous Wave
C.W.	Cool White; Cold Water
Cwt.	100 Pounds
C.W.X.	Cool White Deluxe
C.Y.	Cubic Yard (27 cubic feet)
C.Y./Hr.	Cubic Yard per Hour
Cyl.	Cylinder
d	Penny (nail size)
D	Deep; Depth; Discharge
Dis., Disch.	Discharge
Db	Decibel
Dbl.	Double
DC	Direct Current
DDC	Direct Digital Control
Demob.	Demobilization
d.f.t.	Dry Film Thickness
d.f.u.	Drainage Fixture Units
D.H.	Double Hung
DHW	Domestic Hot Water
DI	Ductile Iron
Diag.	Diagonal
Diam., Dia	Diameter
Distrib.	Distribution
Div.	Division
Dk.	Deck
D.L.	Dead Load; Diesel
DLH	Deep Long Span Bar Joist
dlx	Deluxe
Do.	Ditto
DOP	Dioctyl Phthalate Penetration Test (Air Filters)
Dp., dp	Depth
D.P.S.T.	Double Pole, Single Throw
Dr.	Drive
DR	Dimension Ratio
Drink.	Drinking
D.S.	Double Strength
D.S.A.	Double Strength A Grade
D.S.B.	Double Strength B Grade
Dty.	Duty
DWV	Drain Waste Vent
DX	Deluxe White, Direct Expansion
dyn	Dyne
e	Eccentricity
E	Equipment Only; East; emissivity
Ea.	Each
EB	Encased Burial
Econ.	Economy
E.C.Y	Embankment Cubic Yards
EDP	Electronic Data Processing
EIFS	Exterior Insulation Finish System
E.D.R.	Equiv. Direct Radiation
Eq.	Equation
EL	elevation
Elec.	Electrician; Electrical
Elev.	Elevator; Elevating
EMT	Electrical Metallic Conduit; Thin Wall Conduit
Eng.	Engine, Engineered
EPDM	Ethylene Propylene Diene Monomer
EPS	Expanded Polystyrene
Eqhv.	Equip. Oper., Heavy
Eqlt.	Equip. Oper., Light
Eqmd.	Equip. Oper., Medium
Eqmm.	Equip. Oper., Master Mechanic
Eqol.	Equip. Oper., Oilers
Equip.	Equipment
ERW	Electric Resistance Welded

Abbreviations

E.S.	Energy Saver
Est.	Estimated
esu	Electrostatic Units
E.W.	Each Way
EWT	Entering Water Temperature
Excav.	Excavation
excl	Excluding
Exp., exp	Expansion, Exposure
Ext., ext	Exterior; Extension
Extru.	Extrusion
f.	Fiber stress
F	Fahrenheit; Female; Fill
Fab., fab	Fabricated; fabric
FBGS	Fiberglass
F.C.	Footcandles
f.c.c.	Face-centered Cubic
f'c.	Compressive Stress in Concrete; Extreme Compressive Stress
F.E.	Front End
FEP	Fluorinated Ethylene Propylene (Teflon)
F.G.	Flat Grain
F.H.A.	Federal Housing Administration
Fig.	Figure
Fin.	Finished
FIPS	Female Iron Pipe Size
Fixt.	Fixture
FJP	Finger jointed and primed
Fl. Oz.	Fluid Ounces
Flr.	Floor
FM	Frequency Modulation; Factory Mutual
Fmg.	Framing
FM/UL	Factory Mutual/Underwriters Labs
Fdn.	Foundation
FNPT	Female National Pipe Thread
Fori.	Foreman, Inside
Foro.	Foreman, Outside
Fount.	Fountain
fpm	Feet per Minute
FPT	Female Pipe Thread
Fr	Frame
F.R.	Fire Rating
FRK	Foil Reinforced Kraft
FSK	Foil/scrim/kraft
FRP	Fiberglass Reinforced Plastic
FS	Forged Steel
FSC	Cast Body; Cast Switch Box
Ft., ft	Foot; Feet
Ftng.	Fitting
Ftg.	Footing
Ft lb.	Foot Pound
Furn.	Furniture
FVNR	Full Voltage Non-Reversing
FVR	Full Voltage Reversing
FXM	Female by Male
Fy.	Minimum Yield Stress of Steel
g	Gram
G	Gauss
Ga.	Gauge
Gal., gal.	Gallon
gpm, GPM	Gallon per Minute
Galv., galv	Galvanized
GC/MS	Gas Chromatograph/Mass Spectrometer
Gen.	General
GFI	Ground Fault Interrupter
GFRC	Glass Fiber Reinforced Concrete
Glaz.	Glazier
GPD	Gallons per Day
gpf	Gallon per flush
GPH	Gallons per Hour
GPM	Gallons per Minute
GR	Grade
Gran.	Granular
Grnd.	Ground
GVW	Gross Vehicle Weight
GWB	Gypsum wall board
H	High Henry
HC	High Capacity
H.D., HD	Heavy Duty; High Density
H.D.O.	High Density Overlaid
HDPE	High density polyethelene plastic
Hdr.	Header
Hdwe.	Hardware
H.I.D., HID	High Intensity Discharge
Help.	Helper Average
HEPA	High Efficiency Particulate Air Filter
Hg	Mercury
HIC	High Interrupting Capacity
HM	Hollow Metal
HMWPE	high molecular weight polyethylene
HO	High Output
Horiz.	Horizontal
H.P., HP	Horsepower; High Pressure
H.P.F.	High Power Factor
Hr.	Hour
Hrs./Day	Hours per Day
HSC	High Short Circuit
Ht.	Height
Htg.	Heating
Htrs.	Heaters
HVAC	Heating, Ventilation & Air-Conditioning
Hvy.	Heavy
HW	Hot Water
Hyd.; Hydr.	Hydraulic
Hz	Hertz (cycles)
I.	Moment of Inertia
IBC	International Building Code
I.C.	Interrupting Capacity
ID	Inside Diameter
I.D.	Inside Dimension; Identification
I.F.	Inside Frosted
I.M.C.	Intermediate Metal Conduit
In.	Inch
Incan.	Incandescent
Incl.	Included; Including
Int.	Interior
Inst.	Installation
Insul., insul	Insulation/Insulated
I.P.	Iron Pipe
I.P.S., IPS	Iron Pipe Size
IPT	Iron Pipe Threaded
I.W.	Indirect Waste
J	Joule
J.I.C.	Joint Industrial Council
K	Thousand; Thousand Pounds; Heavy Wall Copper Tubing, Kelvin
K.A.H.	Thousand Amp. Hours
kcmil	Thousand Circular Mils
KD	Knock Down
K.D.A.T.	Kiln Dried After Treatment
kg	Kilogram
kG	Kilogauss
kgf	Kilogram Force
kHz	Kilohertz
Kip	1000 Pounds
KJ	Kiljoule
K.L.	Effective Length Factor
K.L.F.	Kips per Linear Foot
Km	Kilometer
KO	Knock Out
K.S.F.	Kips per Square Foot
K.S.I.	Kips per Square Inch
kV	Kilovolt
kVA	Kilovolt Ampere
kVAR	Kilovar (Reactance)
KW	Kilowatt
KWh	Kilowatt-hour
L	Labor Only; Length; Long; Medium Wall Copper Tubing
Lab.	Labor
lat	Latitude
Lath.	Lather
Lav.	Lavatory
lb.; #	Pound
L.B., LB	Load Bearing; L Conduit Body
L. & E.	Labor & Equipment
lb./hr.	Pounds per Hour
lb./L.F.	Pounds per Linear Foot
lbf/sq.in.	Pound-force per Square Inch
L.C.L.	Less than Carload Lot
L.C.Y.	Loose Cubic Yard
Ld.	Load
LE	Lead Equivalent
LED	Light Emitting Diode
L.F.	Linear Foot
L.F. Nose	Linear Foot of Stair Nosing
L.F. Rsr	Linear Foot of Stair Riser
Lg.	Long; Length; Large
L & H	Light and Heat
LH	Long Span Bar Joist
L.H.	Labor Hours
L.L., LL	Live Load
L.L.D.	Lamp Lumen Depreciation
lm	Lumen
lm/sf	Lumen per Square Foot
lm/W	Lumen per Watt
LOA	Length Over All
log	Logarithm
L-O-L	Lateralolet
long.	longitude
L.P., LP	Liquefied Petroleum; Low Pressure
L.P.F.	Low Power Factor
LR	Long Radius
L.S.	Lump Sum
Lt.	Light
Lt. Ga.	Light Gauge
L.T.L.	Less than Truckload Lot
Lt. Wt.	Lightweight
L.V.	Low Voltage
M	Thousand; Material; Male; Light Wall Copper Tubing
M^{2}CA	Meters Squared Contact Area
m/hr.; M.H.	Man-hour
mA	Milliampere
Mach.	Machine
Mag. Str.	Magnetic Starter
Maint.	Maintenance
Marb.	Marble Setter
Mat; Mat'l.	Material
Max.	Maximum
MBF	Thousand Board Feet
MBH	Thousand BTU's per hr.
MC	Metal Clad Cable
MCC	Motor Control Center
M.C.F.	Thousand Cubic Feet
MCFM	Thousand Cubic Feet per Minute
M.C.M.	Thousand Circular Mils
MCP	Motor Circuit Protector
MD	Medium Duty
MDF	Medium-density fibreboard
M.D.O.	Medium Density Overlaid
Med.	Medium
MF	Thousand Feet
M.F.B.M.	Thousand Feet Board Measure
Mfg.	Manufacturing
Mfrs.	Manufacturers
mg	Milligram
MGD	Million Gallons per Day
MGPH	Thousand Gallons per Hour
MH, M.H.	Manhole; Metal Halide; Man-Hour
MHz	Megahertz
Mi.	Mile
MI	Malleable Iron; Mineral Insulated
MIPS	Male Iron Pipe Size
mj	Mechanical Joint
m	Meter
mm	Millimeter
Mill.	Millwright
Min., min.	Minimum, minute

Misc.	Miscellaneous
ml	Milliliter, Mainline
M.L.F.	Thousand Linear Feet
Mo.	Month
Mobil.	Mobilization
Mog.	Mogul Base
MPH	Miles per Hour
MPT	Male Pipe Thread
MRGWB	Moisture Resistant Gypsum Wallboard
MRT	Mile Round Trip
ms	Millisecond
M.S.F.	Thousand Square Feet
Mstz.	Mosaic & Terrazzo Worker
M.S.Y.	Thousand Square Yards
Mtd., mtd., mtd	Mounted
Mthe.	Mosaic & Terrazzo Helper
Mtng.	Mounting
Mult.	Multi; Multiply
M.V.A.	Million Volt Amperes
M.V.A.R.	Million Volt Amperes Reactance
MV	Megavolt
MW	Megawatt
MXM	Male by Male
MYD	Thousand Yards
N	Natural; North
nA	Nanoampere
NA	Not Available; Not Applicable
N.B.C.	National Building Code
NC	Normally Closed
NEMA	National Electrical Manufacturers Assoc.
NEHB	Bolted Circuit Breaker to 600V.
NFPA	National Fire Protection Association
NLB	Non-Load-Bearing
NM	Non-Metallic Cable
nm	Nanometer
No.	Number
NO	Normally Open
N.O.C.	Not Otherwise Classified
Nose.	Nosing
NPT	National Pipe Thread
NQOD	Combination Plug-on/Bolt on Circuit Breaker to 240V.
N.R.C., NRC	Noise Reduction Coefficient/ Nuclear Regulator Commission
N.R.S.	Non Rising Stem
ns	Nanosecond
nW	Nanowatt
OB	Opposing Blade
OC	On Center
OD	Outside Diameter
O.D.	Outside Dimension
ODS	Overhead Distribution System
O.G.	Ogee
O.H.	Overhead
O&P	Overhead and Profit
Oper.	Operator
Opng.	Opening
Orna.	Ornamental
OSB	Oriented Strand Board
OS&Y	Outside Screw and Yoke
OSHA	Occupational Safety and Health Act
Ovhd.	Overhead
OWG	Oil, Water or Gas
Oz.	Ounce
P.	Pole; Applied Load; Projection
p.	Page
Pape.	Paperhanger
P.A.P.R.	Powered Air Purifying Respirator
PAR	Parabolic Reflector
P.B., PB	Push Button
Pc., Pcs.	Piece, Pieces
P.C.	Portland Cement; Power Connector
P.C.F.	Pounds per Cubic Foot
PCM	Phase Contrast Microscopy
PDCA	Painting and Decorating Contractors of America
P.E., PE	Professional Engineer; Porcelain Enamel; Polyethylene; Plain End
P.E.C.I.	Porcelain Enamel on Cast Iron
Perf.	Perforated
PEX	Cross linked polyethylene
Ph.	Phase
P.I.	Pressure Injected
Pile.	Pile Driver
Pkg.	Package
Pl.	Plate
Plah.	Plasterer Helper
Plas.	Plasterer
plf	Pounds Per Linear Foot
Pluh.	Plumbers Helper
Plum.	Plumber
Ply.	Plywood
p.m.	Post Meridiem
Pntd.	Painted
Pord.	Painter, Ordinary
pp	Pages
PP, PPL	Polypropylene
P.P.M.	Parts per Million
Pr.	Pair
P.E.S.B.	Pre-engineered Steel Building
Prefab.	Prefabricated
Prefin.	Prefinished
Prop.	Propelled
PSF, psf	Pounds per Square Foot
PSI, psi	Pounds per Square Inch
PSIG	Pounds per Square Inch Gauge
PSP	Plastic Sewer Pipe
Pspr.	Painter, Spray
Psst.	Painter, Structural Steel
P.T.	Potential Transformer
P. & T.	Pressure & Temperature
Ptd.	Painted
Ptns.	Partitions
Pu	Ultimate Load
PVC	Polyvinyl Chloride
Pvmt.	Pavement
PRV	Pressure Relief Valve
Pwr.	Power
Q	Quantity Heat Flow
Qt.	Quart
Quan., Qty.	Quantity
Q.C.	Quick Coupling
r	Radius of Gyration
R	Resistance
R.C.P.	Reinforced Concrete Pipe
Rect.	Rectangle
recpt.	receptacle
Reg.	Regular
Reinf.	Reinforced
Req'd.	Required
Res.	Resistant
Resi.	Residential
RF	Radio Frequency
RFID	Radio-frequency identification
Rgh.	Rough
RGS	Rigid Galvanized Steel
RHW	Rubber, Heat & Water Resistant; Residential Hot Water
rms	Root Mean Square
Rnd.	Round
Rodm.	Rodman
Rofc.	Roofer, Composition
Rofp.	Roofer, Precast
Rohe.	Roofer Helpers (Composition)
Rots.	Roofer, Tile & Slate
R.O.W.	Right of Way
RPM	Revolutions per Minute
R.S.	Rapid Start
Rsr	Riser
RT	Round Trip
S.	Suction; Single Entrance; South
SBS	Styrene Butadiere Styrene
SC	Screw Cover
SCFM	Standard Cubic Feet per Minute
Scaf.	Scaffold
Sch., Sched.	Schedule
S.C.R.	Modular Brick
S.D.	Sound Deadening
SDR	Standard Dimension Ratio
S.E.	Surfaced Edge
Sel.	Select
SER, SEU	Service Entrance Cable
S.F.	Square Foot
S.F.C.A.	Square Foot Contact Area
S.F. Flr.	Square Foot of Floor
S.F.G.	Square Foot of Ground
S.F. Hor.	Square Foot Horizontal
SFR	Square Feet of Radiation
S.F. Shlf.	Square Foot of Shelf
S4S	Surface 4 Sides
Shee.	Sheet Metal Worker
Sin.	Sine
Skwk.	Skilled Worker
SL	Saran Lined
S.L.	Slimline
Sldr.	Solder
SLH	Super Long Span Bar Joist
S.N.	Solid Neutral
SO	Stranded with oil resistant inside insulation
S-O-L	Socketolet
sp	Standpipe
S.P.	Static Pressure; Single Pole; Self-Propelled
Spri.	Sprinkler Installer
spwg	Static Pressure Water Gauge
S.P.D.T.	Single Pole, Double Throw
SPF	Spruce Pine Fir; Sprayed Polyurethane Foam
S.P.S.T.	Single Pole, Single Throw
SPT	Standard Pipe Thread
Sq.	Square; 100 Square Feet
Sq. Hd.	Square Head
Sq. In.	Square Inch
S.S.	Single Strength; Stainless Steel
S.S.B.	Single Strength B Grade
sst, ss	Stainless Steel
Sswk.	Structural Steel Worker
Sswl.	Structural Steel Welder
St.; Stl.	Steel
STC	Sound Transmission Coefficient
Std.	Standard
Stg.	Staging
STK	Select Tight Knot
STP	Standard Temperature & Pressure
Stpi.	Steamfitter, Pipefitter
Str.	Strength; Starter; Straight
Strd.	Stranded
Struct.	Structural
Sty.	Story
Subj.	Subject
Subs.	Subcontractors
Surf.	Surface
Sw.	Switch
Swbd.	Switchboard
S.Y.	Square Yard
Syn.	Synthetic
S.Y.P.	Southern Yellow Pine
Sys.	System
t.	Thickness
T	Temperature; Ton
Tan	Tangent
T.C.	Terra Cotta
T & C	Threaded and Coupled
T.D.	Temperature Difference
Tdd	Telecommunications Device for the Deaf
T.E.M.	Transmission Electron Microscopy
temp	Temperature, Tempered, Temporary
TFFN	Nylon Jacketed Wire

Abbreviations

TFE	Tetrafluoroethylene (Teflon)
T. & G.	Tongue & Groove; Tar & Gravel
Th.,Thk.	Thick
Thn.	Thin
Thrded	Threaded
Tilf.	Tile Layer, Floor
Tilh.	Tile Layer, Helper
THHN	Nylon Jacketed Wire
THW.	Insulated Strand Wire
THWN	Nylon Jacketed Wire
T.L.,TL	Truckload
T.M.	Track Mounted
Tot.	Total
T-O-L	Threadolet
tmpd	Tempered
TPO	Thermoplastic Polyolefin
T.S.	Trigger Start
Tr.	Trade
Transf.	Transformer
Trhv.	Truck Driver, Heavy
Trlr	Trailer
Trlt.	Truck Driver, Light
TTY	Teletypewriter
TV	Television
T.W.	Thermoplastic Water Resistant Wire
UCI	Uniform Construction Index
UF	Underground Feeder
UGND	Underground Feeder
UHF	Ultra High Frequency
U.I.	United Inch
U.L., UL	Underwriters Laboratory
Uld.	unloading
Unfin.	Unfinished
UPS	Uninterruptible Power Supply
URD	Underground Residential Distribution
US	United States
USGBC	U.S. Green Building Council
USP	United States Primed
UTMCD	Uniform Traffic Manual For Control Devices
UTP	Unshielded Twisted Pair
V	Volt
VA	Volt Amperes
V.C.T.	Vinyl Composition Tile
VAV	Variable Air Volume
VC	Veneer Core
VDC	Volts Direct Current
Vent.	Ventilation
Vert.	Vertical
V.F.	Vinyl Faced
V.G.	Vertical Grain
VHF	Very High Frequency
VHO	Very High Output
Vib.	Vibrating
VLF	Vertical Linear Foot
VOC	Volitile Organic Compound
Vol.	Volume
VRP	Vinyl Reinforced Polyester
W	Wire; Watt; Wide; West
w/	With
W.C., WC	Water Column; Water Closet
W.F.	Wide Flange
W.G.	Water Gauge
Wldg.	Welding
W. Mile	Wire Mile
W-O-L	Weldolet
W.R.	Water Resistant
Wrck.	Wrecker
W.S.P.	Water, Steam, Petroleum
WT., Wt.	Weight
WWF	Welded Wire Fabric
XFER	Transfer
XFMR	Transformer
XHD	Extra Heavy Duty
XHHW, XLPE	Cross-Linked Polyethylene Wire Insulation
XLP	Cross-linked Polyethylene
Xport	Transport
Y	Wye
yd	Yard
yr	Year
Δ	Delta
%	Percent
~	Approximately
Ø	Phase; diameter
@	At
#	Pound; Number
<	Less Than
>	Greater Than
Z	zone

D

E

Q

R

S

T

X

Notes

Notes

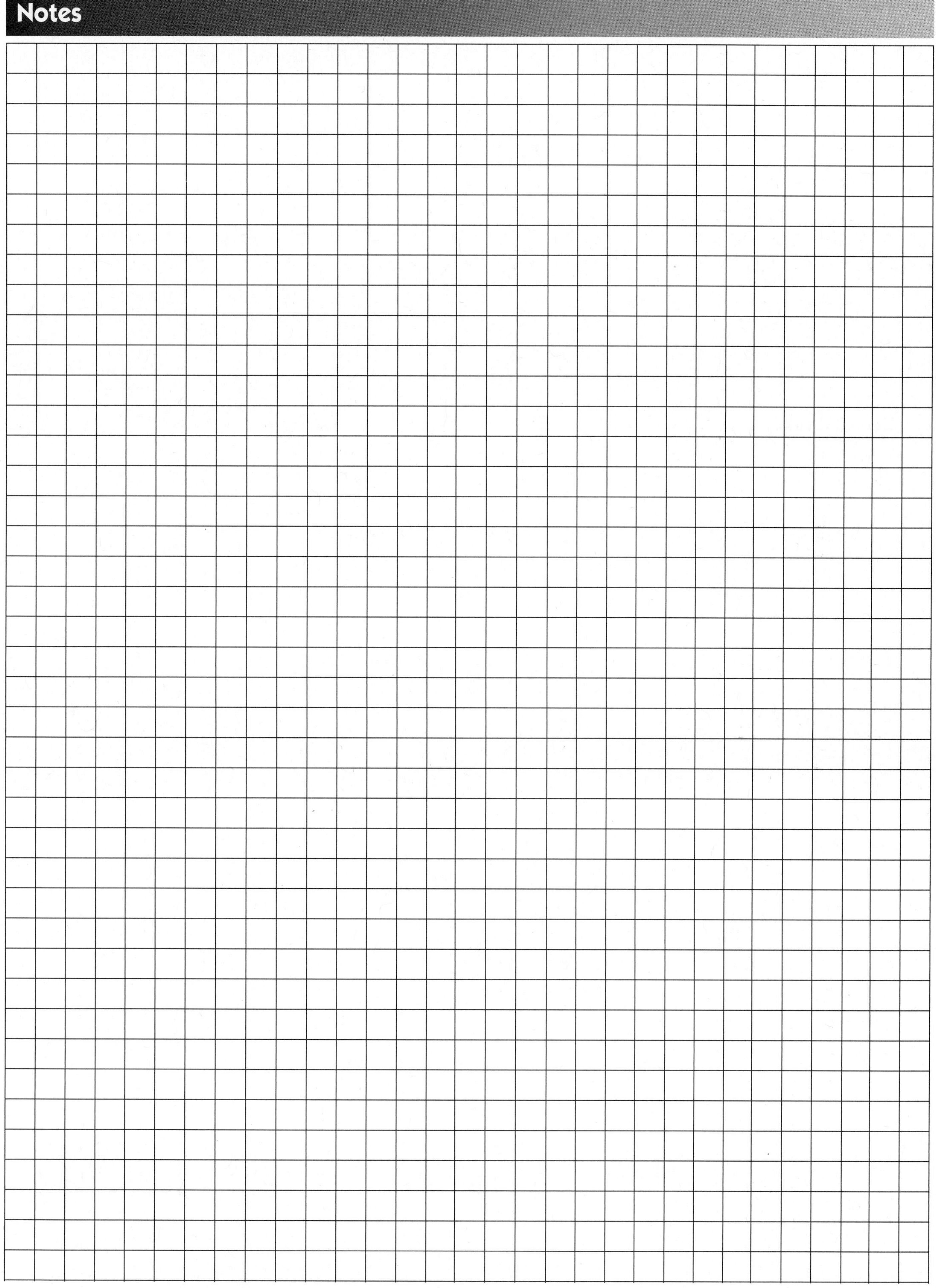

Notes

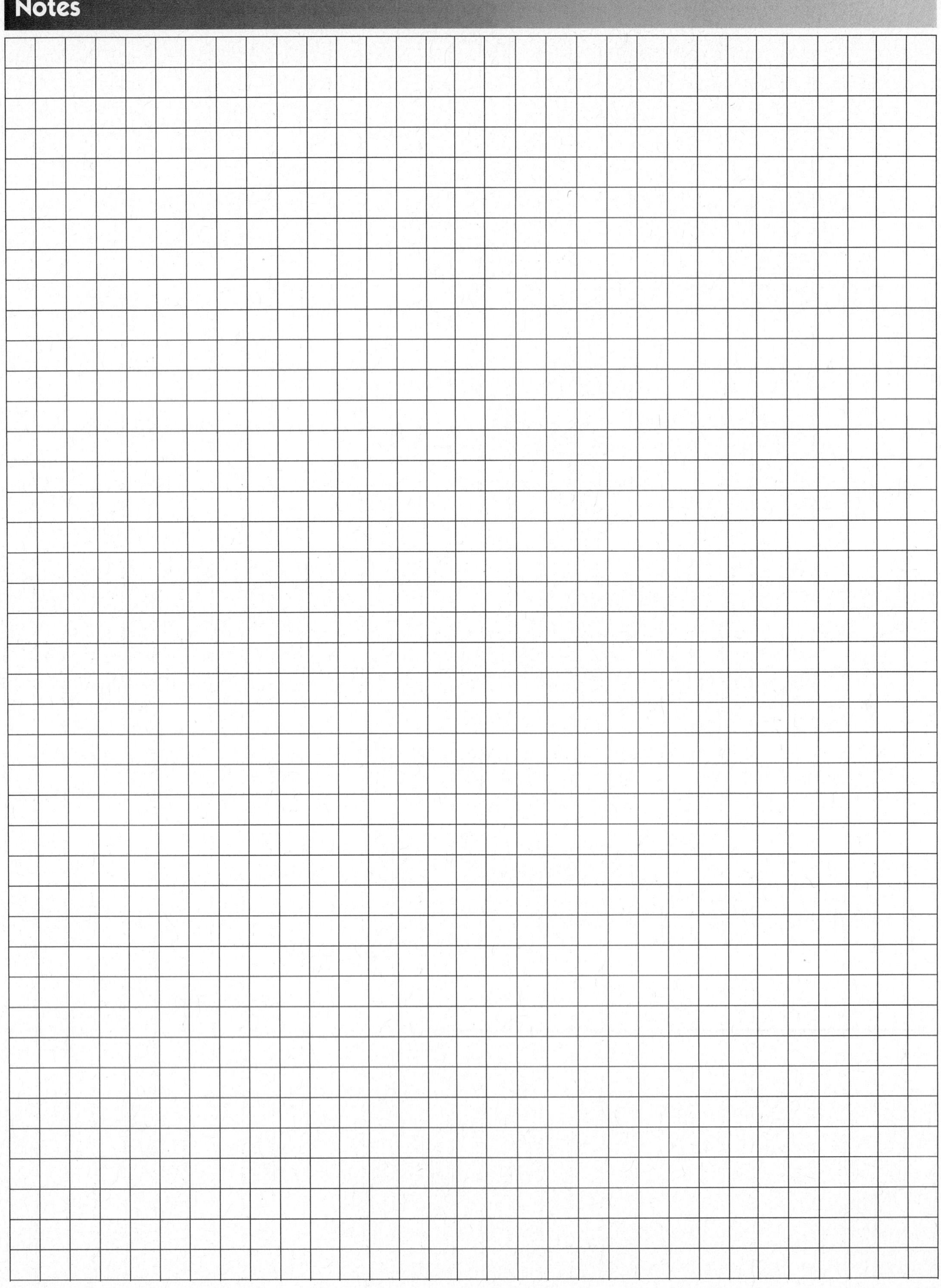

Notes

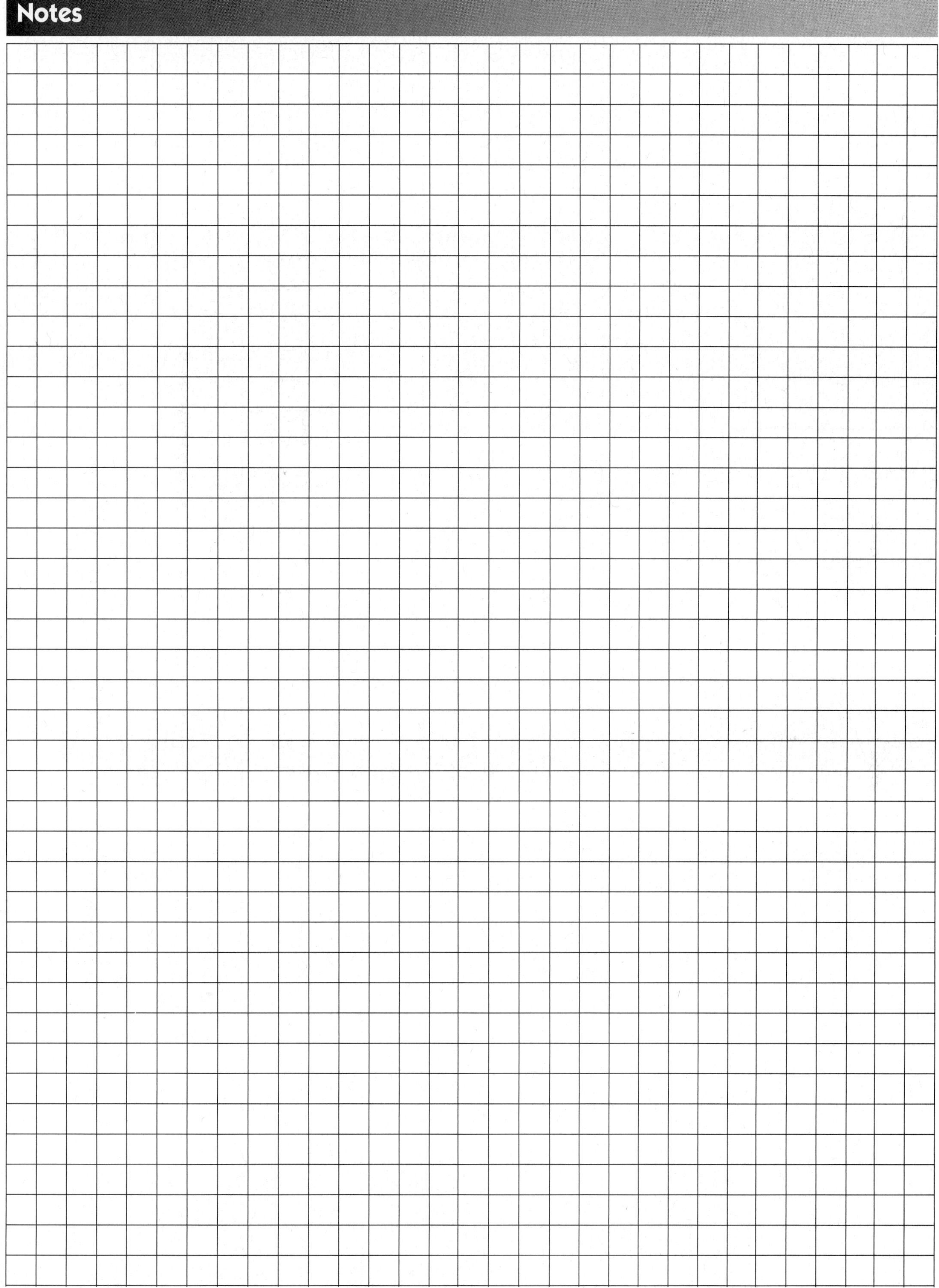

Notes

Notes

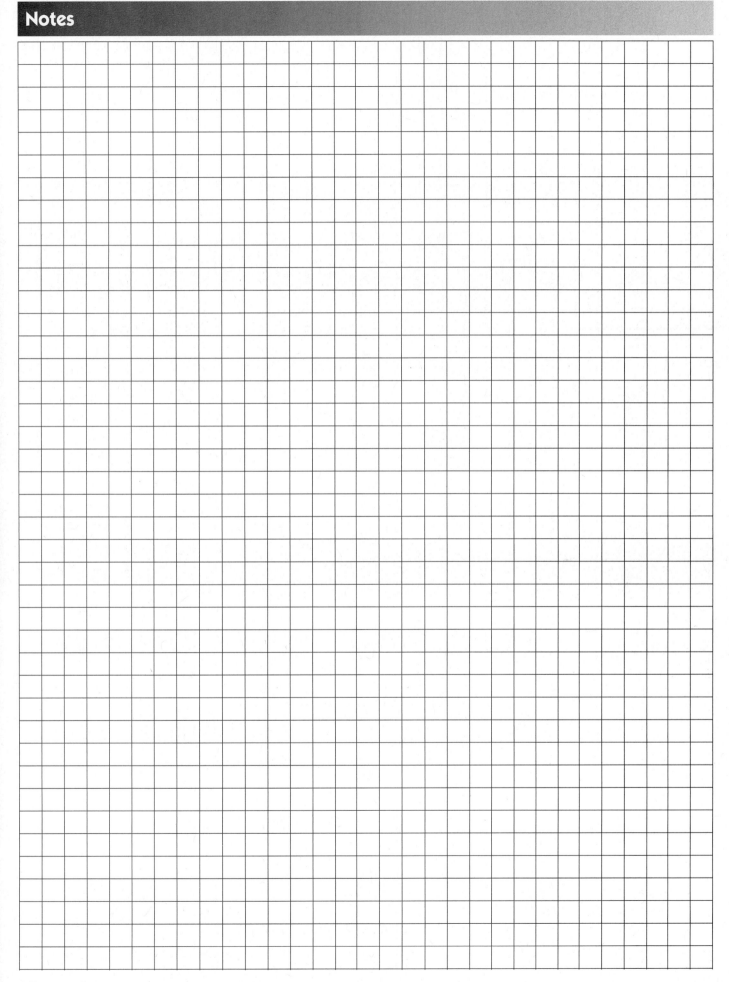

Notes

Reed Construction Data/RSMeans—a tradition of excellence in construction cost information and services since 1942

more information visit the RSMeans website at **www.rsmeans.com**

Unit prices according to the latest MasterFormat!

ook Selection Guide The following table provides definitive information on the content of each cost data publication. The number of lines of data provided each unit price or assemblies division, as well as the number of reference tables and crews, is listed for each book. The presence of other elements such as equipnt rental costs, historical cost indexes, city cost indexes, square foot models, or cross-referenced indexes is also indicated. You can use the table to help select the Means book that has the quantity and type of information you most need in your work.

it Cost visions	Building Construction	Mechanical	Electrical	Commercial Renovation	Square Foot	Site Work Landsc.	Green Building	Interior	Concrete Masonry	Open Shop	Heavy Construction	Light Commercial	Facilities Construction	Plumbing	Residential
1	582	395	411	508		521	196	319	463	581	511	244	1054	411	180
2	789	279	85	741		1005	180	399	222	788	746	481	1233	289	257
3	1689	347	231	1087		1477	995	329	2040	1688	1685	479	1787	317	385
4	963	21	0	932		729	179	620	1159	931	621	521	1176	0	435
5	1873	158	155	1068		811	1779	1073	716	1873	1034	954	1874	204	723
6	2404	18	18	2062		110	560	1523	278	2400	126	2085	2076	22	2612
7	1531	215	128	1569		584	747	516	509	1528	26	1282	1633	227	1018
8	2107	81	45	2530		268	1140	1768	112	2093	0	2109	2842	0	1511
9	1929	72	26	1726		313	449	2014	387	1870	15	1644	2179	54	1430
10	1006	17	10	623		215	27	818	158	1006	29	511	1098	235	224
11	1026	208	165	489		126	54	878	28	1010	0	230	1046	169	110
12	557	0	2	327		248	151	1669	12	545	19	364	1718	23	317
13	729	117	114	246		347	128	253	66	728	253	82	756	71	80
14	277	36	0	224		0	0	261	0	277	0	12	297	16	6
21	90	0	16	37		0	0	250	0	87	0	68	427	432	220
22	1159	7329	156	1170		1578	1099	849	20	1123	1687	861	7171	9082	712
23	1182	6901	586	886		160	898	741	38	1106	111	759	5013	1830	434
26	1379	499	10134	1039		816	661	1142	55	1351	562	1305	10083	444	621
27	72	0	298	34		13	0	60	0	72	39	52	280	0	4
28	97	58	137	72		0	21	79	0	100	0	41	152	44	26
31	1498	735	610	803		3206	289	7	1208	1443	3285	600	1560	660	611
32	782	54	8	859		4396	352	362	286	753	1824	407	1660	166	455
33	509	1020	542	209		2036	41	0	231	508	2024	120	1585	1211	118
34	109	0	47	4		192	0	0	31	64	195	0	130	0	0
35	18	0	0	0		327	0	0	0	18	442	0	83	0	0
41	60	0	0	32		0	0	22	0	60	30	0	67	14	0
44	75	83	0	0		0	0	0	0	0	0	0	75	75	0
46	23	16	0	0		274	151	0	0	23	264	0	33	33	0
48	12	0	27	0		0	27	0	0	12	19	12	12	0	12
Totals	24527	18659	13951	19277		19752	10124	15952	8019	24038	15547	15223	49100	16029	12501

ssem Div	Building Construction	Mechanical	Elec-trical	Commercial Renovation	Square Foot	Site Work Landscape	Assemblies	Green Building	Interior	Concrete Masonry	Heavy Construction	Light Commercial	Facilities Construction	Plumbing	Asm Div	Residential
A		15	0	188	150	577	598	0	0	536	571	154	24	0	1	368
B		0	0	848	2498	0	5658	56	329	1975	368	2089	174	0	2	211
C		0	0	647	872	0	1251	0	1576	146	0	762	249	0	3	588
D		1067	943	712	1856	72	2532	330	822	0	0	1342	1106	1081	4	851
E		0	0	86	260	0	300	0	5	0	0	257	5	0	5	392
F		0	0	0	114	0	114	0	0	0	0	114	3	0	6	357
G		522	442	318	262	3740	663	0	0	534	1558	199	288	758	7	307
															8	760
															9	80
															10	0
															11	0
															12	0
tals		1604	1385	2799	6012	4389	11116	386	2732	3191	2497	4917	1849	1839		3914

eference Section	Building Construction Costs	Mechanical	Electrical	Commercial Renovation	Square Foot	Site Work Landscape	Assem.	Green Building	Interior	Concrete Masonry	Open Shop	Heavy Construction	Light Commercial	Facilities Construction	Plumbing	Resi.
eference bles	yes	yes	yes	yes	no	yes	yes	yes	yes	yes	yes	yes	yes	yes	yes	yes
odels					111			25					50			28
ews	561	561	561	540		561		558	561	561	538	561	538	540	561	538
uipment ental osts	yes	yes	yes	yes		yes		yes	yes	yes	yes	yes	yes	yes	yes	yes
storical ost dexes	yes	yes	yes	yes	yes	yes	yes	yes	yes	yes	yes	yes	yes	yes	yes	no
ity Cost dexes	yes	yes	yes	yes	yes	yes	yes	yes	yes	yes	yes	yes	yes	yes	yes	yes

Annual Cost Guides

Cost guides also available in CD an at rsmeansonline.cor

For more information visit the RSMeans website at **www.rsmeans.com**

Unit prices according to the latest MasterForma

RSMeans Building Construction Cost Data 2014

Offers you unchallenged unit price reliability in an easy-to-use format. Whether used for verifying complete, finished estimates or for periodic checks, it supplies more cost facts better and faster than any comparable source. More than 24,000 unit prices have been updated for 2014. The City Cost Indexes and Location Factors cover more than 930 areas, for indexing to any project location in North America. Order and get *RSMeans Quarterly Update Service* FREE.

$194.95 | Available Sept. 2013 | Catalog no. 60014

RSMeans Green Building Cost Data 2014

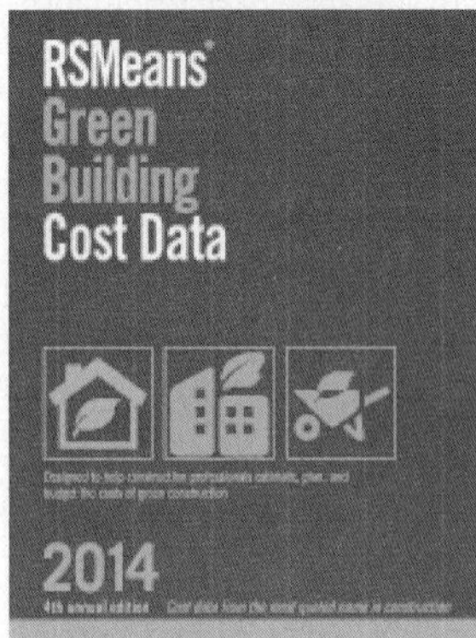

Estimate, plan, and budget the costs of green building for both new commercial construction and renovation work with this fourth edition of *RSMeans Green Building Cost Data.* More than 10,000 unit costs for a wide array of green building products plus assemblies costs. Easily identified cross references to LEED and Green Globes building rating systems criteria.

$160.95 | Available Nov. 2013 | Catalog no. 60554

RSMeans Mechanical Cost Data 2014

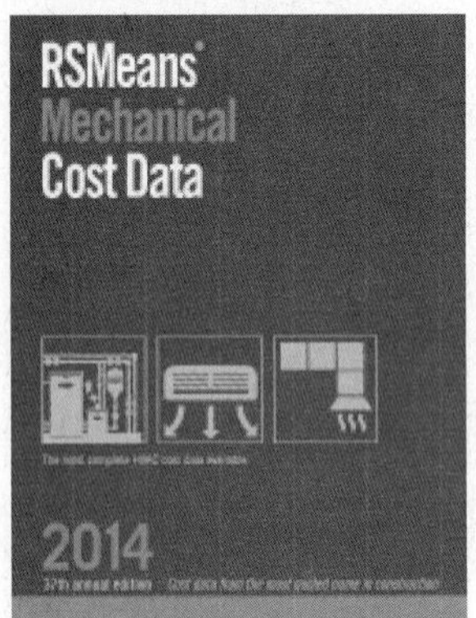

Total unit and systems price guidance for mechanical construction . . . materials, parts, fittings, and complete labor cost information. Includes prices for piping, heating, air conditioning, ventilation, and all related construction.

Plus new 2014 unit costs for:

- Thousands of installed HVAC/controls assemblies components
- "On-site" Location Factors for more than 930 cities and towns in the U.S. and Canada
- Crews, labor, and equipment

$191.95 | Available Oct. 2013 | Catalog no. 60024

RSMeans Facilities Construction Cost Data 2014

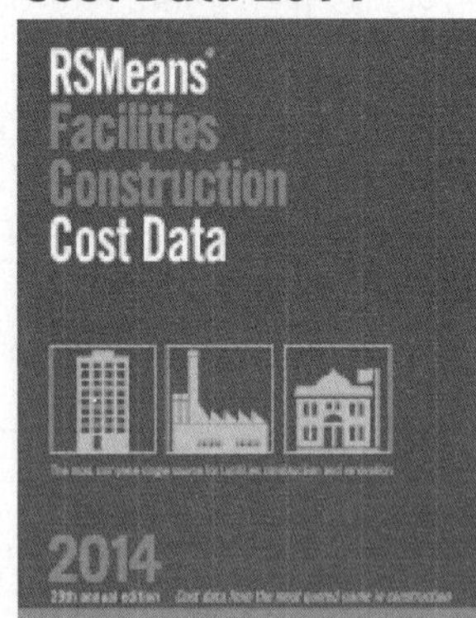

For the maintenance and construction o commercial, industrial, municipal, and institutional properties. Costs are show for new and remodeling construction an are broken down into materials, labor, equipment, and overhead and profit. Special emphasis is given to sections on mechanical, electrical, furnishings, site work, building maintenance, finish work, and demolition.

More than 49,000 unit costs, plus assemblies costs and a comprehensive Reference Section are included.

$496.95 | Available Nov. 2013 | Catalog no. 60204

RSMeans Square Foot Costs 2014

Accurate and Easy-to-Use

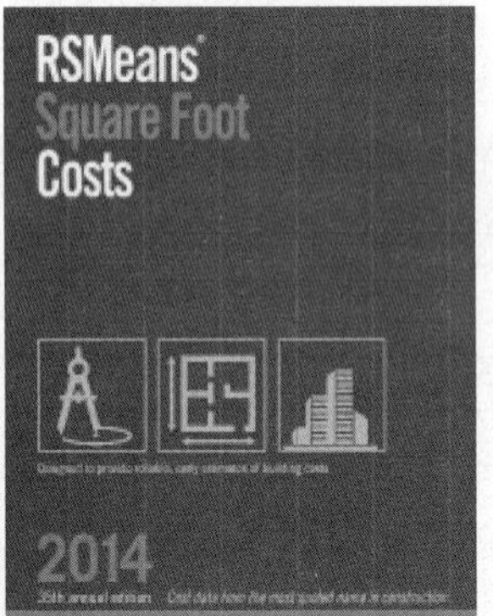

- **Updated price information** based on nationwide figures from suppliers, estimators, labor experts, and contractors
- New green building models
- Realistic graphics, offering true-to-life illustrations of building projects
- Extensive information on using square foot cost data, including sample estimates and alternate pricing methods

$205.95 | Available Oct. 2013 | Catalog no. 60054

RSMeans Commercial Renovation Cost Data 2014

Commercial/Multifamily Residential

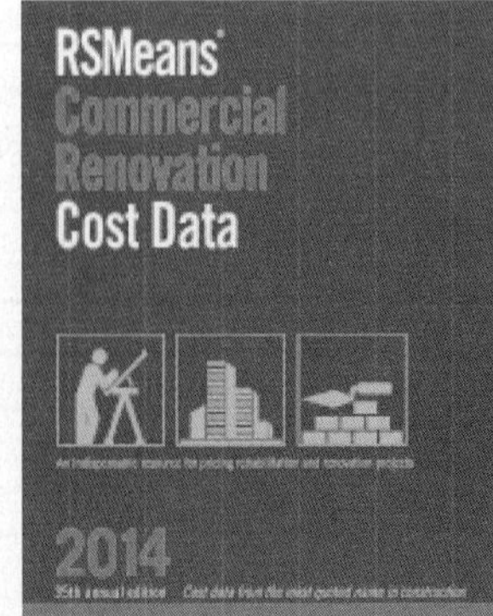

Use this valuable tool to estimate commercial and multifamily residential renovation and remodeling.

Includes: New costs for hundreds of unique methods, materials, and conditions that only come up in repair and remodeling, PLUS:

- Unit costs for more than 19,000 construction components
- Installed costs for more than 2,000 assemblies
- More than 930 "on-site" localization factors for the U.S. and Canada

$156.95 | Available Oct. 2013 | Catalog no. 60044

Annual Cost Guides

Cost guides also available in CD and at rsmeansonline.com

For more information visit the RSMeans website at **www.rsmeans.com**

Unit prices according to the latest MasterFormat!

RSMeans Electrical Cost Data 2014

Pricing information for every part of electrical cost planning. More than 13,000 unit and systems costs with design tables; clear specifications and drawings; engineering guides; illustrated estimating procedures; complete labor-hour and materials costs for better scheduling and procurement; and the latest electrical products and construction methods.

- A variety of special electrical systems, including cathodic protection
- Costs for maintenance, demolition, HVAC/mechanical, specialties, equipment, and more

196.95 | Available Oct. 2013 | Catalog no. 60034

RSMeans Electrical Change Order Cost Data 2014

RSMeans Electrical Change Order Cost Data provides you with electrical unit prices exclusively for pricing change orders—based on the recent, direct experience of contractors and suppliers. Analyze and check your own change order estimates against the experience others have had doing the same work. It also covers productivity analysis and change order cost justifications. With useful information for calculating the effects of change orders and dealing with their administration.

$187.95 | Available Dec. 2013 | Catalog no. 60234

RSMeans Assemblies Cost Data 2014

RSMeans Assemblies Cost Data takes the guesswork out of preliminary or conceptual estimates. Now you don't have to try to calculate the assembled cost by working up individual component costs. We've done all the work for you.

Presents detailed illustrations, descriptions, specifications, and costs for every conceivable building assembly—over 350 types in all—arranged in the easy-to-use UNIFORMAT II system. Each illustrated "assembled" cost includes a complete grouping of materials and associated installation costs, including the installing contractor's overhead and profit.

315.95 | Available Sept. 2013 | Catalog no. 60064

RSMeans Open Shop Building Construction Cost Data 2014

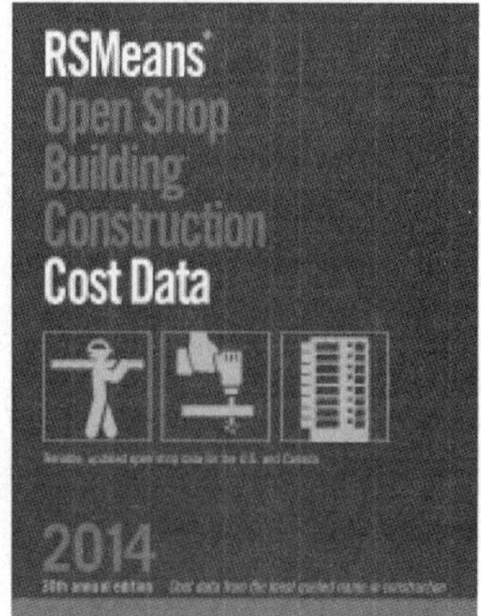

The latest costs for accurate budgeting and estimating of new commercial and residential construction . . . renovation work . . . change orders . . . cost engineering.

RSMeans Open Shop "BCCD" will assist you to:

- Develop benchmark prices for change orders
- Plug gaps in preliminary estimates and budgets
- Estimate complex projects
- Substantiate invoices on contracts
- Price ADA-related renovations

$165.95 | Available Dec. 2013 | Catalog no. 60154

RSMeans Residential Cost Data 2014

Contains square foot costs for 28 basic home models with the look of today, plus hundreds of custom additions and modifications you can quote right off the page. Includes costs for more than 700 residential systems. Complete with blank estimating forms, sample estimates, and step-by-step instructions.

Contains line items for cultured stone and brick, PVC trim lumber, and TPO roofing.

139.95 | Available Oct. 2013 | Catalog no. 60174

RSMeans Site Work & Landscape Cost Data 2014

Includes unit and assemblies costs for earthwork, sewerage, piped utilities, site improvements, drainage, paving, trees and shrubs, street openings/repairs, underground tanks, and more. Contains 78 types of assemblies costs for accurate conceptual estimates.

Includes:

- Estimating for infrastructure improvements
- Environmentally-oriented construction
- ADA-mandated handicapped access
- Hazardous waste line items

$187.95 | Available Dec. 2013 | Catalog no. 60284

RSMeans Facilities Maintenance & Repair Cost Data 2014

RSMeans Facilities Maintenance & Repair Cost Data gives you a complete system to manage and plan your facility repair and maintenance costs and budget efficiently. Guidelines for auditing a facility and developing an annual maintenance plan. Budgeting is included, along with reference tables on cost and management, and information on frequency and productivity of maintenance operations.

The only nationally recognized source of maintenance and repair costs. Developed in cooperation with the Civil Engineering Research Laboratory (CERL) of the Army Corps of Engineers.

$434.95 | Available Nov. 2013 | Catalog no. 60304

RSMeans Light Commercial Cost Data 2014

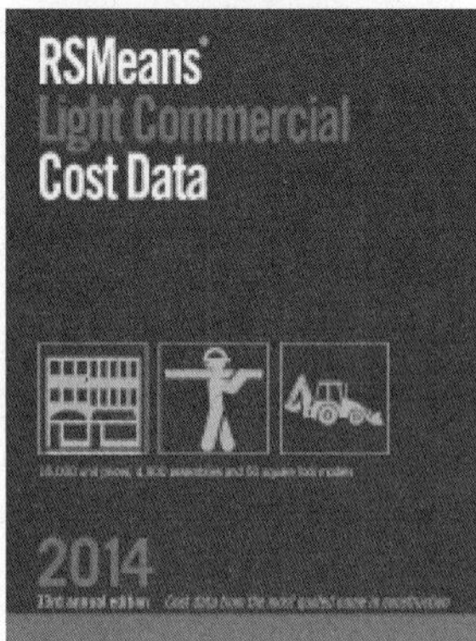

Specifically addresses the light commercial market, which is a specialized niche in the construction industry. Aids you, the owner/designer/contractor, in preparing all types of estimates—from budgets to detailed bids. Includes new advances in methods and materials.

Assemblies Section allows you to evaluate alternatives in the early stages of design/planning.

More than 15,000 unit costs ensure that you have the prices you need . . . when you need them.

$144.95 | Available Nov. 2013 | Catalog no. 60184

RSMeans Concrete & Masonry Cost Data 2014

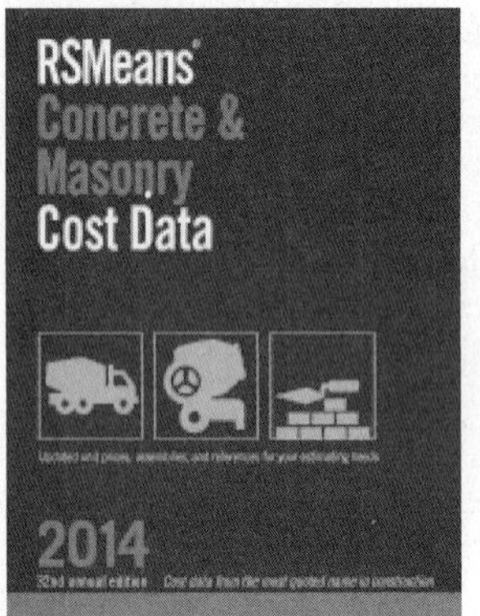

Provides you with cost facts for virtually all concrete/masonry estimating needs, from complicated formwork to various sizes and face finishes of brick and block—all in great detail. The comprehensive Unit Price Section contains more than 8,000 selected entries. Also contains an Assemblies [Cost] Section, and a detailed Reference Section that supplements the cost data.

$177.95 | Available Dec. 2013 | Catalog no. 60114

RSMeans Labor Rates for the Construction Industry 2014

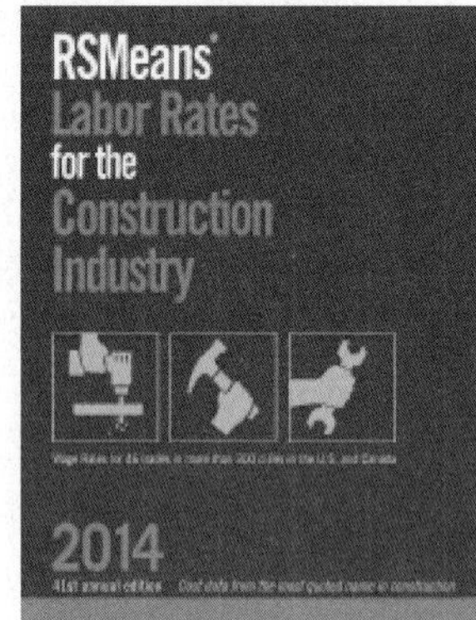

Complete information for estimating labor costs, making comparisons, and negotiating wage rates by trade for more than 300 U.S. and Canadian cities. With 46 construction trades listed by local union number in each city, and historical wage rates included for comparison. Each city chart lists the county and is alphabetically arranged with handy visual flip tabs for quick reference.

$424.95 | Available Dec. 2013 | Catalog no. 60124

RSMeans Construction Cost Indexes 2014

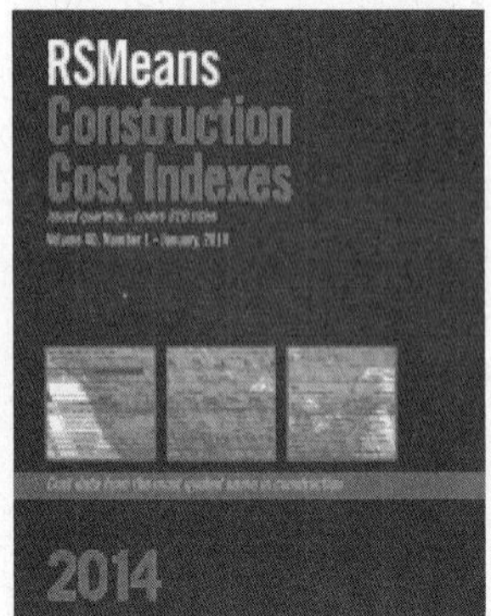

What materials and labor costs will change unexpectedly this year? By how much?

- Breakdowns for 318 major cities
- National averages for 30 key cities
- Expanded five major city indexes
- Historical construction cost indexes

$362.00 per year (subscription) | Catalog no. 50144
$93.50 individual quarters | Catalog no. 60144 A,B,C,D

RSMeans Interior Cost Data 2014

Provides you with prices and guidance needed to make accurate interior work estimates. Contains costs on materials, equipment, hardware, custom installations, furnishings, and labor costs . . . for new and remodel commercial and industrial interior construction, including updated information on office furnishings, and reference information.

$196.95 | Available Nov. 2013 | Catalog no. 60094

Annual Cost Guides

Cost guides also available in CD and at rsmeansonline.com

For more information visit the RSMeans website at **www.rsmeans.com**

Unit prices according to the latest MasterFormat!

RSMeans Heavy Construction Cost Data 2014

A comprehensive guide to heavy construction costs. Includes costs for highly specialized projects such as tunnels, dams, highways, airports, and waterways. Information on labor rates, equipment, and materials costs is included. Features unit price costs, systems costs, and numerous reference tables for costs and design.

$192.95 | Available Dec. 2013 | Catalog no. 60164

RSMeans Plumbing Cost Data 2014

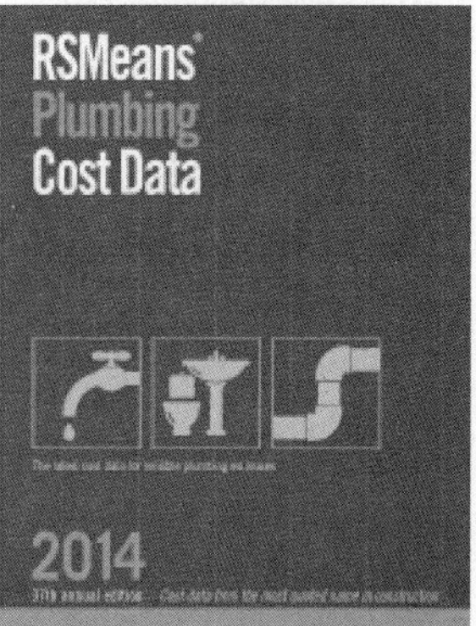

Comprehensive unit prices and assemblies for plumbing, irrigation systems, commercial and residential fire protection, point-of-use water heaters, and the latest approved materials. This publication and its companion, *RSMeans Mechanical Cost Data*, provide full-range cost estimating coverage for all the mechanical trades.

Now contains updated costs for potable water and radiant heat systems and high efficiency fixtures.

$194.95 | Available Oct. 2013 | Catalog no. 60214

2014 RSMeans Seminar Schedule

Note: call for exact dates and details.

Location	Dates
Seattle, WA	January and August
Dallas/Ft. Worth, TX	January
Austin, TX	February
Anchorage, AK	March and September
Las Vegas, NV	March and October
Washington, DC	April and September
Phoenix, AZ	April
Kansas City, MO	April
Toronto	May
Denver, CO	May
San Francisco, CA	June
Bethesda, MD	June
Columbus, GA	June
El Segundo, CA	August

Location	Dates
Jacksonville FL	September
Dallas, TX	September
Charleston, SC	October
Houston, TX	October
Atlanta, GA	November
Baltimore, MD	November
Orlando, FL	November
San Diego, CA	December
San Antonio, TX	December
Raleigh, NC	December

1-800-334-3509, Press 1

Professional Development

For more information visit the RSMeans website at **www.rsmeans.com**

eLearning Training Sessions

Learn how to use *RSMeans Online™* or *RSMeans CostWorks®* CD from the convenience of your home or office. Our eLearning training sessions let you join a training conference call and share the instructors' desktops, so you can view the presentation and step-by-step instructions on your own computer screen. The live webinars are held from 9 a.m. to 4 p.m. eastern standard time, with a one-hour break for lunch.

For these sessions, you must have a computer with high speed Internet access and a compatible Web browser. Learn more at www.rsmeansonline.com or call for a schedule: 1-800-334-3509 and press 1. Webinars are generally held on selected Wednesdays each month.

RSMeans Online™
$299 per person

RSMeans CostWorks® CD
$299 per person

Professional Development

For more information visit the RSMeans website at **www.rsmeans.com**

RSMeans Online™ Training

Construction estimating is vital to the decision-making process at each state of every project. RSMeansOnline works the way you do. It's systematic, flexible and intuitive. In this one day class you will see how you can estimate any phase of any project faster and better.

Some of what you'll learn:
- Customizing RSMeansOnline
- Making the most of RSMeans "Circle Reference" numbers
- How to integrate your cost data
- Generate reports, exporting estimates to MS Excel, sharing, collaborating and more

Also available: RSMeans Online™ training webinar

Maintenance & Repair Estimating for Facilities

This *two-day* course teaches attendees how to plan, budget, and estimate the cost of ongoing and preventive maintenance and repair for existing buildings and grounds.

Some of what you'll learn:
- The most financially favorable maintenance, repair, and replacement scheduling and estimating
- Auditing and value engineering facilities
- Preventive planning and facilities upgrading
- Determining both in-house and contract-out service costs
- Annual, asset-protecting M&R plan

Who should attend: facility managers, maintenance supervisors, buildings and grounds superintendents, plant managers, planners, estimators, and others involved in facilities planning and budgeting.

Facilities Construction Estimating

In this *two-day* course, professionals working in facilities management can get help with their daily challenges to establish budgets for all phases of a project.

Some of what you'll learn:
- Determining the full scope of a project
- Identifying the scope of risks and opportunities
- Creative solutions to estimating issues
- Organizing estimates for presentation and discussion
- Special techniques for repair/remodel and maintenance projects
- Negotiating project change orders

Who should attend: facility managers, engineers, contractors, facility tradespeople, planners, and project managers.

Practical Project Management for Construction Professionals

In this *two-day* course, acquire the essential knowledge and develop the skills to effectively and efficiently execute the day-to-day responsibilities of the construction project manager.

Covers:
- General conditions of the construction contract
- Contract modifications: change orders and construction change directives
- Negotiations with subcontractors and vendors
- Effective writing: notification and communications
- Dispute resolution: claims and liens

Who should attend: architects, engineers, owner's representatives, project managers.

Scheduling with Microsoft Project

Two days of hands-on training gives you a look at the basics to putting it all together with MS Project to create a complete project schedule. Learn to work better and faster, manage changes and updates, enhance tracking, generate actionable reports, and boost control of your time and budget.

Some of what you'll learn:
- Defining task/activity relationships: link tasks, establish predecessor/successor relationships and create logs
- Baseline scheduling: the essential what/when
- Using MS Project to determine the critical path
- Integrate RSMeans data with MS Project
- The dollar-loaded schedule: enhance the reliability of this "must-know" technique for scheduling public projects.

Who should attend: contractors' project-management teams, architects and engineers, project owners and their representatives, and those interested in improving their project planning and scheduling and management skills.

Mechanical & Electrical Estimating

This *two-day* course teaches attendees how to prepare more accurate and complete mechanical/electrical estimates, avoiding the pitfalls of omission and double-counting, while understanding the composition and rationale within the RSMeans mechanical/electrical database.

Some of what you'll learn:
- The unique way mechanical and electrical systems are interrelated
- M&E estimates—conceptual, planning, budgeting, and bidding stages
- Order of magnitude, square foot, assemblies, and unit price estimating
- Comparative cost analysis of equipment and design alternatives

Who should attend: architects, engineers, facilities managers, mechanical and electrical contractors, and others who need a highly reliable method for developing, understanding, and evaluating mechanical and electrical contracts.

Professional Development

For more information visit the RSMeans website at **www.rsmeans.com**

Unit Price Estimating

This interactive *two-day* seminar teaches attendees how to interpret project information and process it into final, detailed estimates with the greatest accuracy level.

The most important credential an estimator can take to the job is the ability to visualize construction and estimate accurately.

Some of what you'll learn:
- Interpreting the design in terms of cost
- The most detailed, time-tested methodology for accurate pricing
- Key cost drivers—material, labor, equipment, staging, and subcontracts
- Understanding direct and indirect costs for accurate job cost accounting and change order management

Who should attend: corporate and government estimators and purchasers, architects, engineers, and others who need to produce accurate project estimates.

Conceptual Estimating Using RSMeans CostWorks® CD

This *two day* class uses the leading industry data and a powerful software package to develop highly accurate conceptual estimates for your construction projects. All attendees must bring a laptop computer loaded with the current year *Square Foot Model Costs* and the *Assemblies Cost Data* CostWorks titles.

Some of what you'll learn:
- Introduction to conceptual estimating
- Types of conceptual estimates
- Helpful hints
- Order of magnitude estimating
- Square foot estimating
- Assemblies estimating

Who should attend: architects, engineers, contractors, construction estimators, owner's representatives, and anyone looking for an electronic method for performing square foot estimating.

RSMeans CostWorks® CD Training

This one-day course helps users become more familiar with the functionality of *RSMeans CostWorks* program. Each menu, icon, screen, and function found in the program is explained in depth. Time is devoted to hands-on estimating exercises.

Some of what you'll learn:
- Searching the database using all navigation methods
- Exporting RSMeans data to your preferred spreadsheet format
- Viewing crews, assembly components, and much more
- Automatically regionalizing the database

This training session requires you to bring a laptop computer to class.

When you register for this course you will receive an outline for your laptop requirements.

Also offering web training for CostWorks CD!

Assessing Scope of Work for Facility Construction Estimating

This *two-day* practical training program addresses the vital importance of understanding the SCOPE of projects in order to produce accurate cost estimates for facility repair and remodeling.

Some of what you'll learn:
- Discussions of site visits, plans/specs, record drawings of facilities, and site-specific lists
- Review of CSI divisions, including means, methods, materials, and the challenges of scoping each topic
- Exercises in SCOPE identification and SCOPE writing for accurate estimating of projects
- Hands-on exercises that require SCOPE, take-off, and pricing

Who should attend: corporate and government estimators, planners, facility managers, and others who need to produce accurate project estimates.

Facilities Est. Using RSMeans CostWorks® CD

Combines hands-on skill building with best estimating practices and real-life problems. Brings you up-to-date with key concepts, and provides tips, pointers, and guidelines to save time and avoid cost oversights and errors.

Some of what you'll learn:
- Estimating process concepts
- Customizing and adapting RSMeans cost data
- Establishing scope of work to account for all known variables
- Budget estimating: when, why, and how
- Site visits: what to look for—what you can't afford to overlook
- How to estimate repair and remodeling variables

This training session requires you to bring a laptop computer to class.

Who should attend: facility managers, architects, engineers, contractors, facility tradespeople, planners, project managers and anyone involved with JOC, SABRE, or IDIQ.

Unit Price Est. Using RSMeans CostWorks® CD

Step-by-step instruction and practice problems to identify and track key cost drivers—material, labor, equipment, staging, and subcontractors—for each specific task. Learn the most detailed, time-tested methodology for accurately "pricing" these variables, their impact on each other and on total cost.

Some of what you'll learn:
- Unit price cost estimating
- Order of magnitude, square foot, and assemblies estimating
- Quantity takeoff
- Direct and indirect construction costs
- Development of contractor's bill rates
- How to use *RSMeans Building Construction Cost Data*

This training session requires you to bring a laptop computer to class.

Who should attend: architects, engineers, corporate and government estimators, facility managers, and government procurement staff.

Registration Information

Register early . . . Save up to $100! Register 30 days before the start date of a seminar and save $100 off your total fee. *Note: This discount can be applied only once per order. It cannot be applied to team discount registrations or any other special offer.*

How to register Register by phone today! The RSMeans toll-free number for making reservations is **1-800-334-3509, Press 1.**

Two day seminar registration fee - $935. One-day *RSMeans CostWorks*® training registration fee - $375. To register by mail, complete the registration form and return, with your full fee, to: RSMeans Seminars, 700 Longwater Drive, Norwell, MA 02061.

Government pricing All federal government employees save off the regular seminar price. Other promotional discounts cannot be combined with the government discount.

Team discount program for two to four seminar registrations. Call for pricing: 1-800-334-3509, Press 1

Multiple course discounts When signing up for two or more courses, call for pricing.

Refund policy Cancellations will be accepted up to ten business days prior to the seminar start. There are no refunds for cancellations received later than ten working days prior to the first day of the seminar. A $150 processing fee will be applied for all cancellations. Written notice of cancellation is required. Substitutions can be made at any time before the session starts. **No-shows are subject to the full seminar fee.**

AACE approved courses Many seminars described and offered here have been approved for 14 hours (1.4 recertification credits) of credit by the AACE International Certification Board toward meeting the continuing education requirements for recertification as a Certified Cost Engineer/Certified Cost Consultant.

AIA Continuing Education We are registered with the AIA Continuing Education System (AIA/CES) and are committed to developing quality learning activities in accordance with the CES criteria. Many seminars meet the AIA/CES criteria for Quality Level 2. AIA members may receive (14) learning units (LUs) for each two-day RSMeans course.

Daily course schedule The first day of each seminar session begins at 8:30 a.m. and ends at 4:30 p.m. The second day begins at 8:00 a.m. and ends at 4:00 p.m. Participants are urged to bring a hand-held calculator, since many actual problems will be worked out in each session.

Continental breakfast Your registration includes the cost of a continental breakfast, and a morning and afternoon refreshment break. These informal segments allow you to discuss topics of mutual interest with other seminar attendees. (You are free to make your own lunch and dinner arrangements.)

Hotel/transportation arrangements RSMeans arranges to hold a block of rooms at most host hotels. To take advantage of special group rates when making your reservation, be sure to mention that you are attending the RSMeans seminar. You are, of course, free to stay at the lodging place of your choice. **(Hotel reservations and transportation arrangements should be made directly by seminar attendees.)**

Important Class sizes are limited, so please register as soon as possible.

Note: *Pricing subject to change.*

Registration Form

ADDS-1000

Call 1-800-334-3509, Press 1 to register or FAX this form 1-800-632-6732. Visit our website: www.rsmeans.com

Please register the following people for the RSMeans construction seminars as shown here. We understand that we must make our own hotel reservations if overnight stays are necessary.

☐ Full payment of $____________________ enclosed.

☐ Bill me.

Please print name of registrant(s).

(To appear on certificate of completion)

____

____

____

____

P.O. #: ________________________

GOVERNMENT AGENCIES MUST SUPPLY PURCHASE ORDER NUMBER OR TRAINING FORM.

Firm name______________________________

Address______________________________

City/State/Zip______________________________

Telephone no.____________________ Fax no.________________

E-mail address______________________________

Charge registration(s) to: ☐ MasterCard ☐ VISA ☐ American Express

Account no.____________________ Exp. date________________

Cardholder's signature______________________________

Seminar name______________________________

Seminar City ______________________________

Please mail check to: RSMeans Seminars, 700 Longwater Drive, Norwell, MA 02061 USA

New Titles - Reference Books

For more information visit the RSMeans website at **www.rsmeans.com**

Solar Energy: Technologies & Project Delivery for Buildings

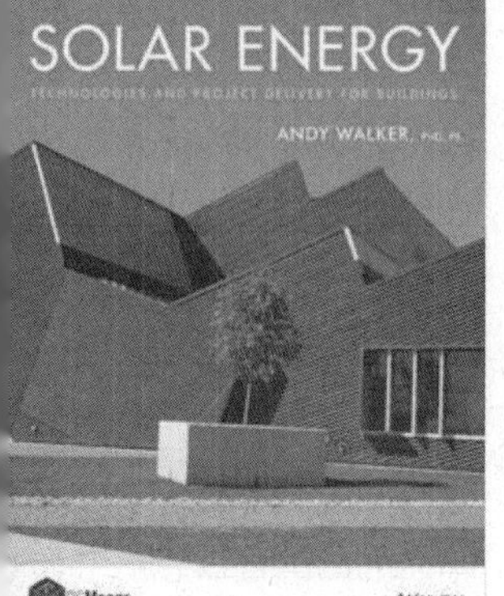

by Andy Walker

An authoritative reference on the design of solar energy systems in building projects, with applications, operating principles, and simple tools for the construction, engineering, and design professionals, the book simplifies the solar design and engineering process and provides sample documentation for the complete design of a solar energy system for buildings.

$85.00 | 320 pages, hardover | Catalog no. 67365

Project Control: Integrating Cost & Schedule in Construction

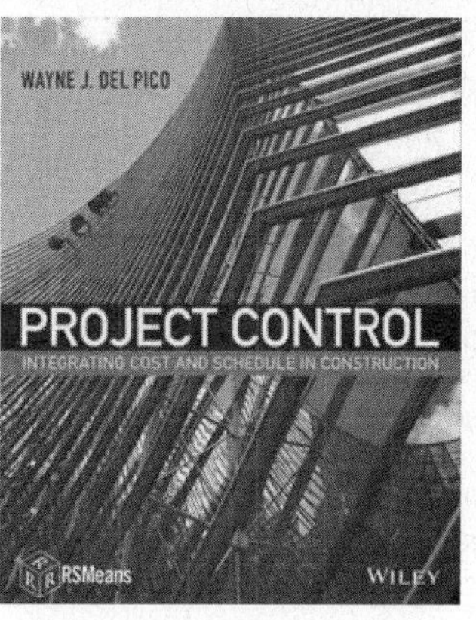

by Wayne J. Del Pico

Written by a seasoned professional in the field, *Project Control: Integrating Cost and Schedule in Construction* fills a void in the area of project control as applied in the construction industry today. It demonstrates how productivity models for an individual project are created, monitored, and controlled, and how corrective actions are implemented as deviations from the baseline occur.

$95.00 | 240 pages, softcover | Catalog no. 67366

Project Scheduling & Management for Construction, 4th Edition

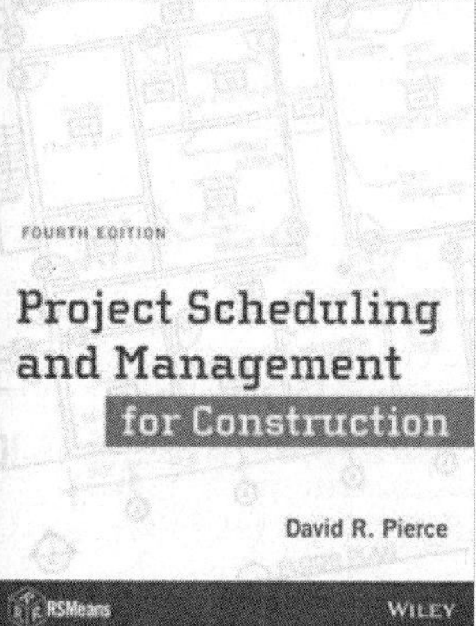

by David R. Pierce

First published in 1988 by RSMeans, the new edition of *Project Scheduling and Management for Construction* has been substantially revised for both professionals and students enrolled in construction management and civil engineering programs. While retaining its emphasis on developing practical, professional-level scheduling skills, the new edition is a relatable, real world case study.

$95.00 | 272 pages, softcover | Catalog no. 67367

How to Estimate with Means Data & CostWorks, 4th Edition

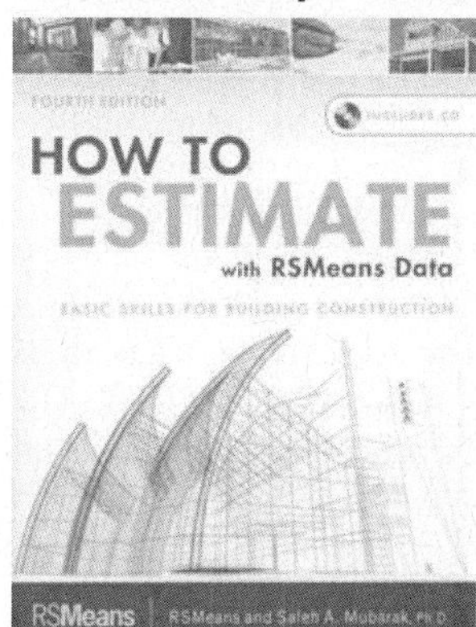

by RSMeans and Saleh A. Mubarak, Ph.D.

This step-by-step guide takes you through all the major construction items with extensive coverage of site work, concrete and masonry, wood and metal framing, doors and windows, and other divisions. The only construction cost estimating handbook that uses the most popular source of construction data, RSMeans, this indispensible guide features access to the instructional version of CostWorks in electronic form, enabling you to practice techniques to solve real-world estimating problems.

$75.00 | 292 pages, softcover | Includes CostWorks CD
Catalog no. 67324C

Estimating Building Costs, 2nd Edition

For the Residential & Light Commercial Construction Professional

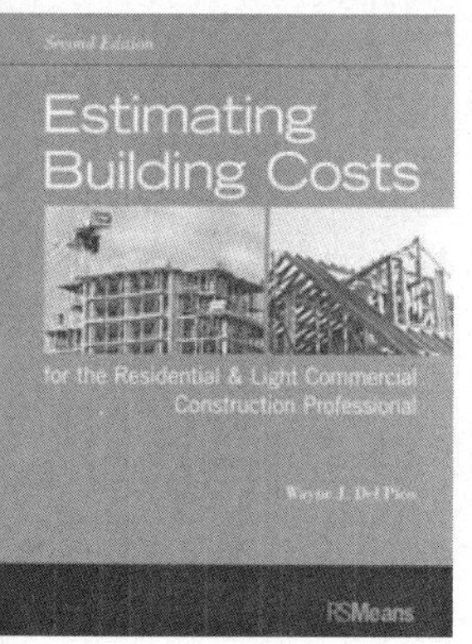

by Wayne J. Del Pico

This book guides readers through the entire estimating process, explaining in detail how to put together a reliable estimate that can be used not only for budgeting, but also for developing a schedule, managing a project, dealing with contingencies, and ultimately making a profit.

Completely revised and updated to reflect the CSI MasterFormat 2010 system, this practical guide describes estimating techniques for each building system and how to apply them according to the latest industry standards.

$75.00 | 398 pages, softcover | Catalog no. 67343A

RSMeans Cost Data, Student Edition

This book provides a thorough introduction to cost estimating in a self-contained print and online package. With clear explanations and a hands-on, example-driven approach, it is the ideal reference for students and new professionals.

Features include:

- Commercial and residential construction cost data in print and online formats
- Complete how-to guidance on the essentials of cost estimating
- A supplemental website with plans, problem sets, and a full sample estimate

$99.00 | 512 pages, softcover | Catalog no. 67363

Reference Books

For more information visit the RSMeans website at **www.rsmeans.com**

Value Engineering: Practical Applications

For Design, Construction, Maintenance & Operations

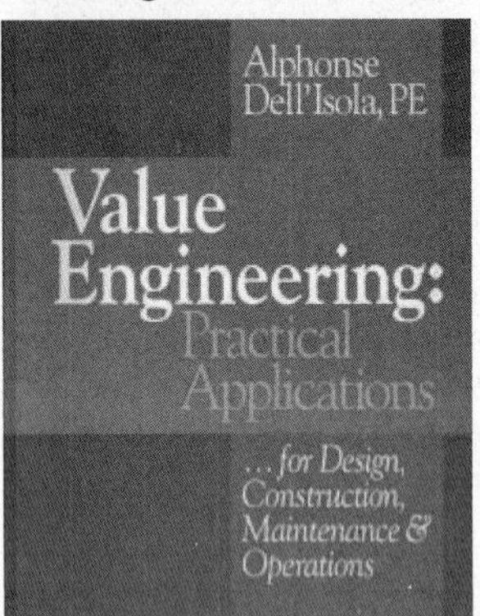

by Alphonse Dell'Isola, PE

A tool for immediate application—for engineers, architects, facility managers, owners, and contractors. Includes making the case for VE—the management briefing; integrating VE into planning, budgeting, and design; conducting life cycle costing; using VE methodology in design review and consultant selection; case studies; VE workbook; and a life cycle costing program on disk.

$85.00 | Over 450 pages, illustrated, softcover | Catalog no. 67319A

Complete Book of Framing, 2nd Edition

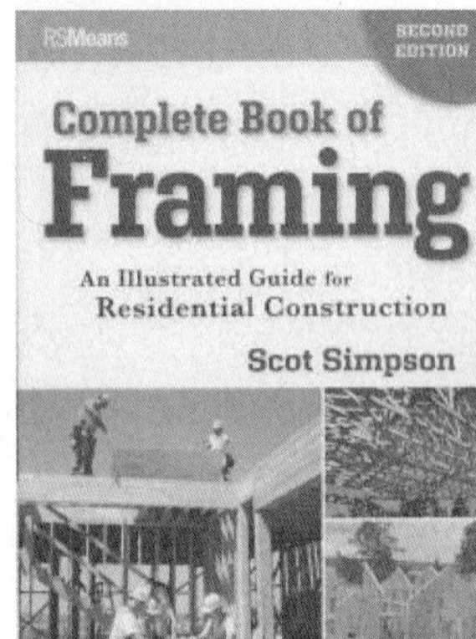

by Scot Simpson

This updated, easy-to-learn guide to rough carpentry and framing is written by an expert with more than thirty years of framing experience. Starting with the basics, this book begins with types of lumber, nails, and what tools are needed, followed by detailed, fully illustrated steps for framing each building element. Framer-Friendly Tips throughout the book show how to get a task done right—and more easily.

$29.95 | 352 pages, softcover | Catalog no. 67353A

Risk Management for Design & Construction

Introduces risk as a central pillar of project management and shows how a project manager can prepare to deal with uncertainty. Experts in the field use clear, straightforward terminology to demystify the concepts of project uncertainty and risk. Includes:

- Integrated cost and schedule risk analysis
- An introduction to a ready-to-use system of analyzing a project's risks and tools to proactively manage risks
- A methodology that was developed and used by the Washington State Department of Transportation
- Case studies and examples on the proper application of principles
- Information about combining value analysis with risk analysis

$125.00 | Over 250 pages, softcover | Catalog no. 67359

Green Home Improvement

by Daniel D. Chiras, PhD

With energy costs rising and environmental awareness increasing, people are looking to make their homes greener. This book, with 65 projects and actual costs and projected savings help homeowners prioritize their green improvements.

Projects range from simple water savers that cost only a few dollars, to bigger-ticket items such as HVAC systems. With color photos and cost estimates, each project compares options and describes the work involved, the benefits, and the savings.

$34.95 | 320 pages, illustrated, softcover | Catalog no. 67355

How Your House Works, 2nd Edition

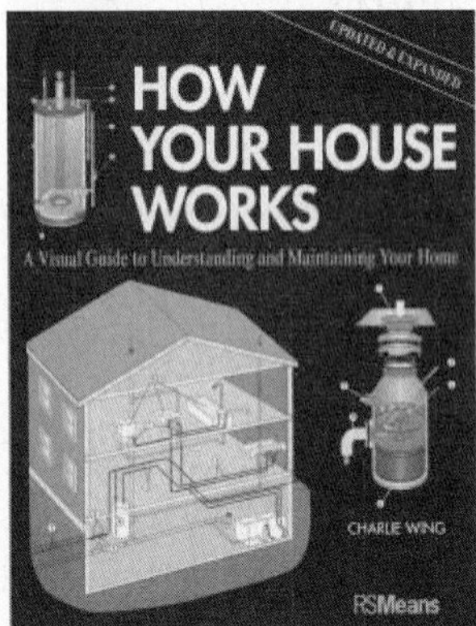

by Charlie Wing

Knowledge of your home's systems helps you control repair and construction costs and makes sure the correct elements are being installed or replaced. This book uncovers the mysteries behind just about every major appliance and building element in your house. See-through, cross-section drawings in full color show you exactly how these things should be put together and how they function, including what to check if they don't work. It just might save you having to call in a professional.

$22.95 | 192 pages, softcover | Catalog no. 67351A

Life Cycle Costing for Facilities

by Alphonse Dell'Isola and Dr. Steven Kirk

Guidance for achieving higher quality design and construction projects at lower costs! Cost-cutting efforts often sacrifice quality to yield the cheapest product. Life cycle costing enables building designers and owners to achieve both. The authors of this book show how LCC can work for a variety of projects — from roads to HVAC upgrades to different types of buildings.

$99.95 | 396 pages, hardcover | Catalog no. 67341

Builder's Essentials: Plan Reading & Material Takeoff

For Residential and Light Commercial Construction

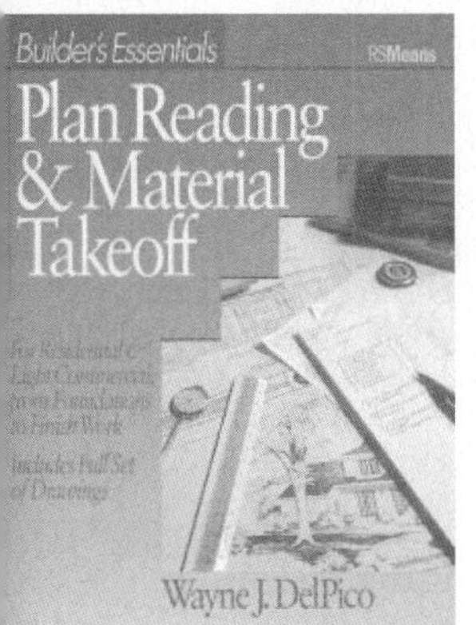

by Wayne J. DelPico

A valuable tool for understanding plans and specs, and accurately calculating material quantities. Step-by-step instructions and takeoff procedures based on a full set of working drawings.

$45.00 | Over 420 pages, softcover | Catalog no. 67307

Means Unit Price Estimating Methods, 4th Edition

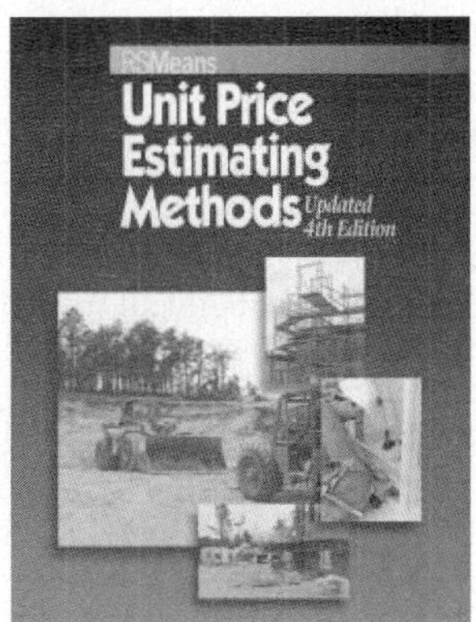

Includes cost data and estimating examples, updated to reflect changes to the CSI numbering system and new features of RSMeans cost data. It describes the most productive, universally accepted ways to estimate, and uses checklists and forms to illustrate shortcuts and timesavers. A model estimate demonstrates procedures. A new chapter explores computer estimating alternatives.

$75.00 | Over 350 pages, illustrated, softcover | Catalog no. 67303B

Concrete Repair & Maintenance Illustrated

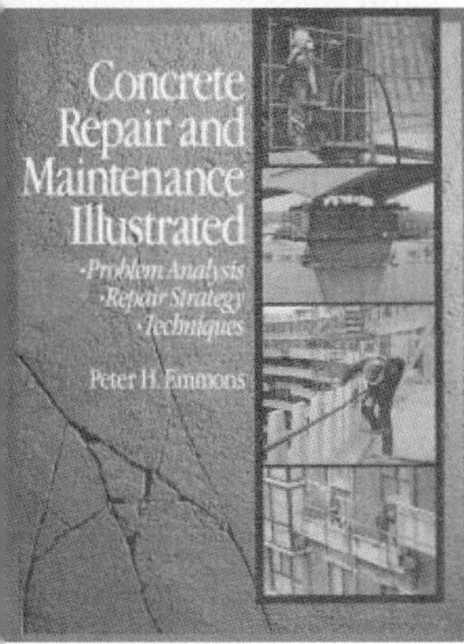

by Peter Emmons

Hundreds of illustrations show users how to analyze, repair, clean, and maintain concrete structures for optimal performance and cost effectiveness. From parking garages to roads and bridges to structural concrete, this comprehensive book describes the causes, effects, and remedies for concrete wear and failure. Invaluable for planning jobs, selecting materials, and training employees, this book is a must-have for concrete specialists, general contractors, facility managers, civil and structural engineers, and architects.

$75.00 | 300 pages, illustrated, softcover | Catalog no. 67146

Square Foot & UNIFORMAT Assemblies Estimating Methods, 3rd Edition

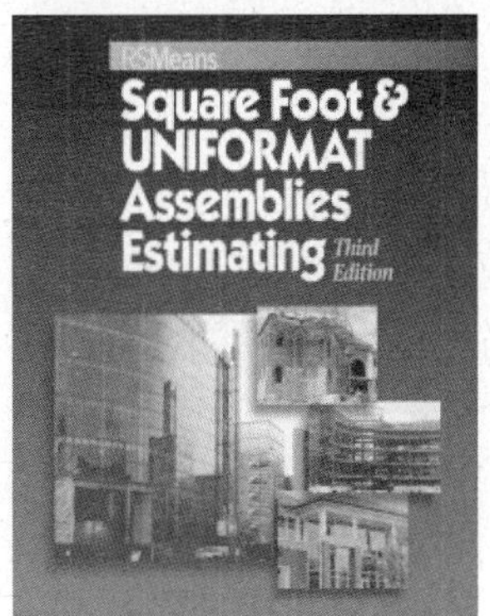

Develop realistic square foot and assemblies costs for budgeting and construction funding. The new edition features updated guidance on square foot and assemblies estimating using UNIFORMAT II. An essential reference for anyone who performs conceptual estimates.

$75.00 | Over 300 pages, illustrated, softcover | Catalog no. 67145B

Electrical Estimating Methods, 3rd Edition

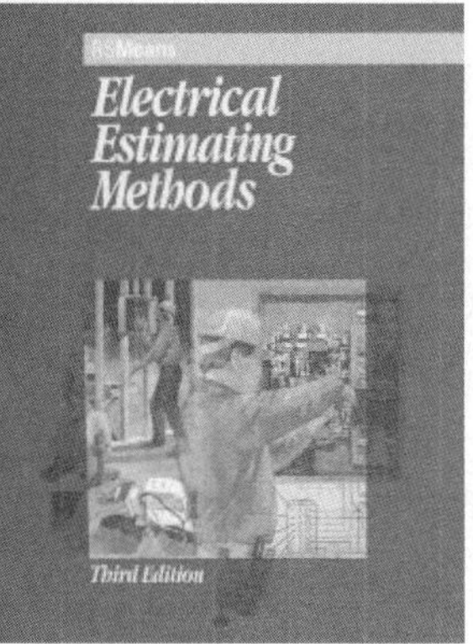

Expanded edition includes sample estimates and cost information in keeping with the latest version of the CSI MasterFormat and UNIFORMAT II. Complete coverage of fiber optic and uninterruptible power supply electrical systems, broken down by components, and explained in detail. Includes a new chapter on computerized estimating methods. A practical companion to *RSMeans Electrical Cost Data.*

$75.00 | Over 325 pages, hardcover | Catalog no. 67230B

Residential & Light Commercial Construction Standards, 3rd Edition

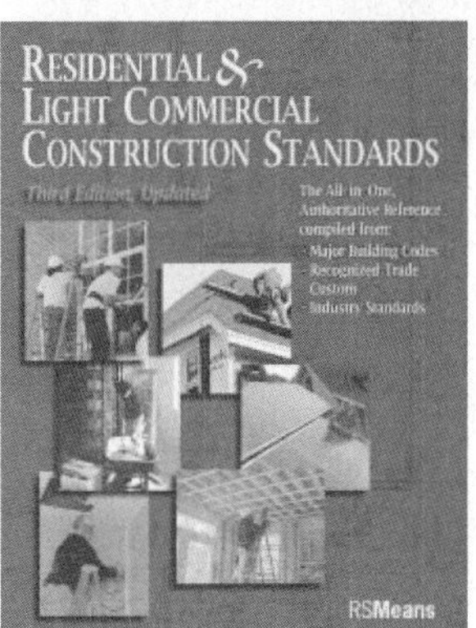

Updated by RSMeans and contributing authors

This book provides authoritative requirements and recommendations compiled from leading professional associations, industry publications, and building code organizations. This all-in-one reference helps establish a standard for workmanship, quickly resolve disputes, and avoid defect claims. Updated third edition includes new coverage of green building, seismic, hurricane, and mold-resistant construction.

$65.00 | Over 550 pages, illustrated, softcover | Catalog no. 67322B

Reference Books

For more information visit the RSMeans website at **www.rsmeans.com**

Construction Business Management

by Nick Ganaway

Only 43% of construction firms stay in business after four years. Make sure your company thrives with valuable guidance from a pro with 25 years of success as a commercial contractor. Find out what it takes to build all aspects of a business that is profitable, enjoyable, and enduring. With a bonus chapter on retail construction.

$55.00 | 200 pages, softcover | Catalog no. 67352

Project Scheduling & Management for Construction, 3rd Edition

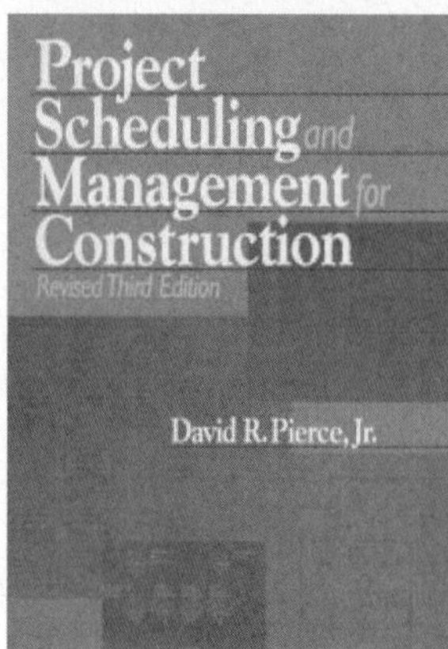

by David R. Pierce, Jr.

A comprehensive yet easy-to-follow guide to construction project scheduling and control—from vital project management principles through the latest scheduling, tracking, and controlling techniques. The author is a leading authority on scheduling, with years of field and teaching experience at leading academic institutions. Spend a few hours with this book and come away with a solid understanding of this essential management topic.

$64.95 | Over 300 pages, illustrated, hardcover | Catalog no. 67247B

The Practice of Cost Segregation Analysis

by Bruce A. Desrosiers and Wayne J. DelPico

This expert guide walks you through the practice of cost segregation analysis, which enables property owners to defer taxes and benefit from "accelerated cost recovery" through depreciation deductions on assets that are properly identified and classified.

With a glossary of terms, sample cost segregation estimates for various building types, key information resources, and updates via a dedicated website, this book is a critical resource for anyone involved in cost segregation analysis.

$105.00 | Over 225 pages | Catalog no. 67345

Preventive Maintenance for Multi-Family Housing

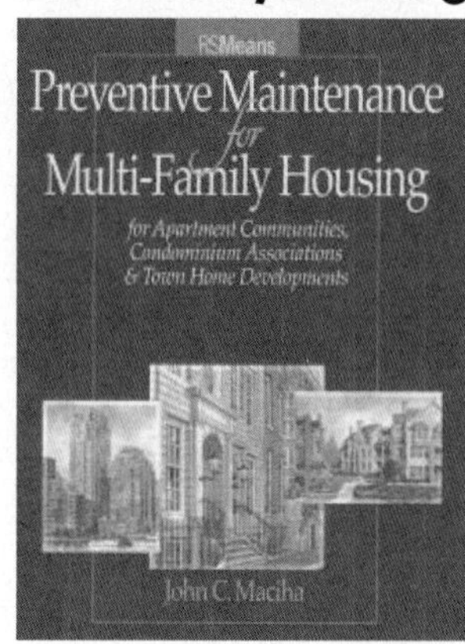

by John C. Maciha

Prepared by one of the nation's leading experts on multi-family housing.

This complete PM system for apartment and condominium communities features expert guidance, checklists for buildings and grounds maintenance tasks and their frequencies, a reusable wall chart to track maintenance, and a dedicated website featuring customizable electronic forms. A must-have for anyone involved with multi-family housing maintenance and upkeep.

$95.00 | 225 pages | Catalog no. 67346

Green Building: Project Planning & Cost Estimating, 3rd Edition

Since the widely read first edition of this book, green building has gone from a growing trend to a major force in design and construction.

This new edition has been updated with the latest in green building technologies, design concepts, standards, and costs. Full-color with all new case studies—plus a new chapter on commercial real estate.

$110.00 | Over 450 pages, softcover | Catalog no. 67338B

Job Order Contracting: Expediting Construction Project Delivery

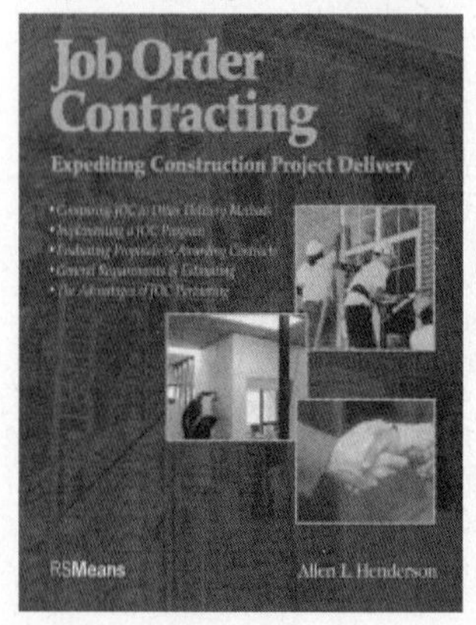

by Allen L. Henderson

Expert guidance to help you implement JOC—fast becoming the preferred project delivery method for repair and renovation, minor new construction, and maintenance projects in the public sector and in many states and municipalities. The author, a leading JOC expert and practitioner, shows how to:

- Establish a JOC program
- Evaluate proposals & award contracts
- Handle general requirements and estimating
- Partner for maximum benefits

$89.95 | 192 pages, illustrated, hardcover | Catalog no. 67348

Construction Supervision

This brand new title inspires supervisory excellence with proven tactics and techniques applied by thousands of construction supervisors over the past decade. Recognizing the unique and critical role the supervisor plays in project success, the book's leadership guidelines carve out a practical blueprint for motivating work performance and increasing productivity through effective communication.

Features:

- A unique focus on field supervision and crew management
- Coverage of supervision from the foreman to the superintendent level
- An overview of technical skills whose mastery will build confidence and success for the supervisor
- A detailed view of "soft" management and communication skills

$95.00 | Over 450 pages, softcover | Catalog no. 67358

Building & Renovating Schools

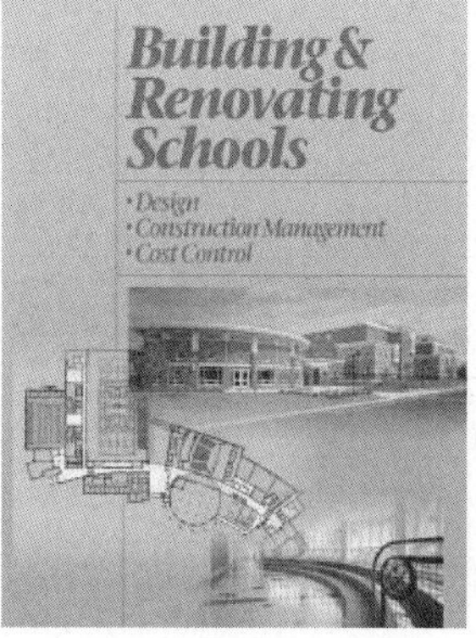

This all-inclusive guide covers every step of the school construction process—from initial planning, needs assessment, and design, right through moving into the new facility. A must-have resource for anyone concerned with new school construction or renovation. With square foot cost models for elementary, middle, and high school facilities, and real-life case studies of recently completed school projects.

The contributors to this book—architects, construction project managers, contractors, and estimators who specialize in school construction—provide start-to-finish, expert guidance on the process.

$99.95 | Over 425 pages, hardcover | Catalog no. 67342

The Homeowner's Guide to Mold

By Michael Pugliese

Expert guidance to protect your health and your home.

Mold, whether caused by leaks, humidity or flooding, is a real health and financial issue—for homeowners and contractors.

This full-color book explains:

- Construction and maintenance practices to prevent mold
- How to inspect for and remove mold
- Mold remediation procedures and costs
- What to do after a flood
- How to deal with insurance companies

$21.95 | 144 pages, softcover | Catalog no. 67344

Universal Design Ideas for Style, Comfort & Safety

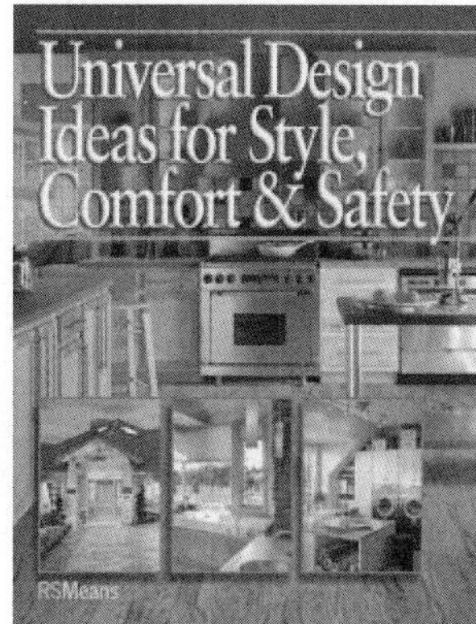

by RSMeans and Lexicon Consulting, Inc.

Incorporating universal design when building or remodeling helps people of any age and physical ability more fully and safely enjoy their living spaces. This book shows how universal design can be artfully blended into the most attractive homes. It discusses specialized products like adjustable countertops and chair lifts, as well as simple ways to enhance a home's safety and comfort. With color photos and expert guidance, every area of the home is covered. Includes budget estimates that give an idea how much projects will cost.

$21.95 | 160 pages, illustrated, softcover | Catalog no. 67354

Landscape Estimating Methods, 5th Edition

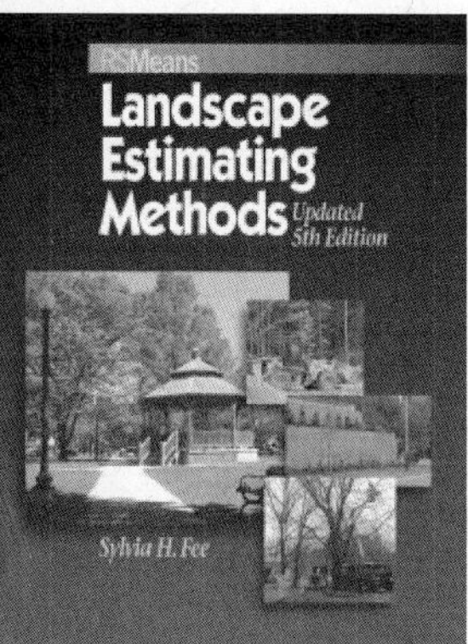

Answers questions about preparing competitive landscape construction estimates, with up-to-date cost estimates and the new MasterFormat classification system. Expanded and revised to address the latest materials and methods, including new coverage on approaches to green building. Includes:

- Step-by-step explanation of the estimating process
- Sample forms and worksheets that save time and prevent errors

$75.00 | Over 350 pages, softcover | Catalog no. 67295C

Plumbing Estimating Methods, 3rd Edition

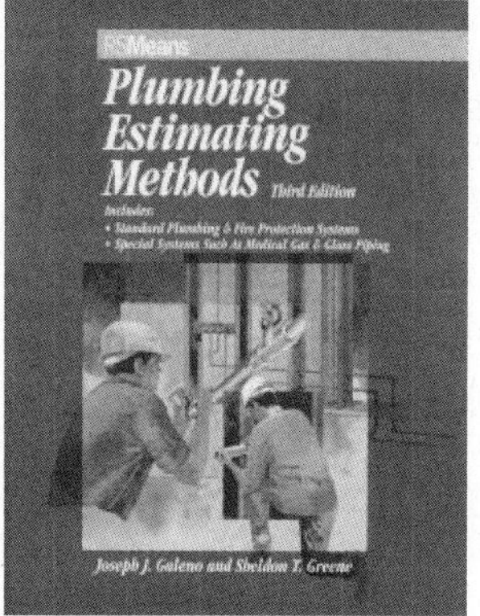

by Joseph J. Galeno and Sheldon T. Greene

Updated and revised! This practical guide walks you through a plumbing estimate, from basic materials and installation methods through change order analysis. *Plumbing Estimating Methods* covers residential, commercial, industrial, and medical systems, and features sample takeoff and estimate forms and detailed illustrations of systems and components.

$65.00 | 330+ pages, softcover | Catalog no. 67283B

Reference Books

For more information visit the RSMeans website at **www.rsmeans.com**

Understanding & Negotiating Construction Contracts

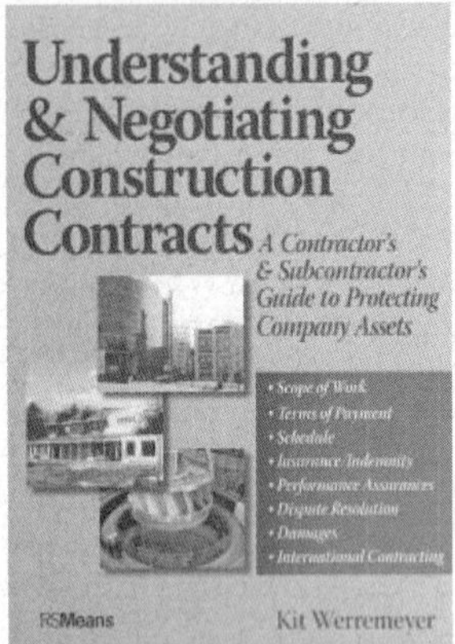

by Kit Werremeyer

Take advantage of the author's 30 years' experience in small-to-large (including international) construction projects. Learn how to identify, understand, and evaluate high risk terms and conditions typically found in all construction contracts—then negotiate to lower or eliminate the risk, improve terms of payment, and reduce exposure to claims and disputes. The author avoids "legalese" and gives real-life examples from actual projects.

$80.00 | 300 pages, softcover | Catalog no. 67350

Means Illustrated Construction Dictionary

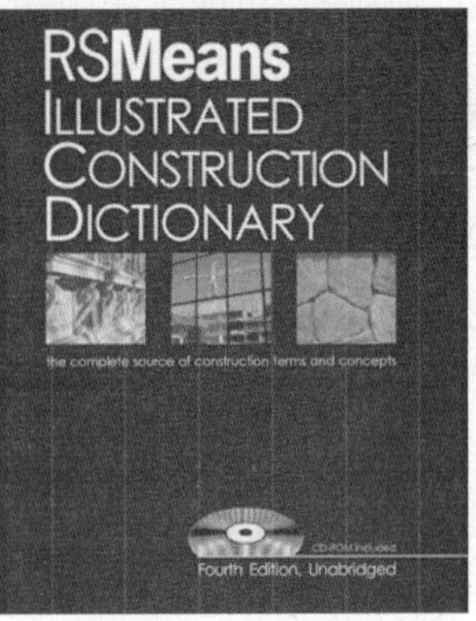

Unabridged 4th Edition, with CD-ROM

Long regarded as the industry's finest, *Means Illustrated Construction Dictionary* is now even better. With nearly 20,000 terms and more than 1,400 illustrations and photos, it is the clear choice for the most comprehensive and current information. The companion CD-ROM that comes with this new edition adds extra features such as: larger graphics and expanded definitions.

$99.95 | Over 790 pages, illust., hardcover | Catalog no. 67292B

The Gypsum Construction Handbook

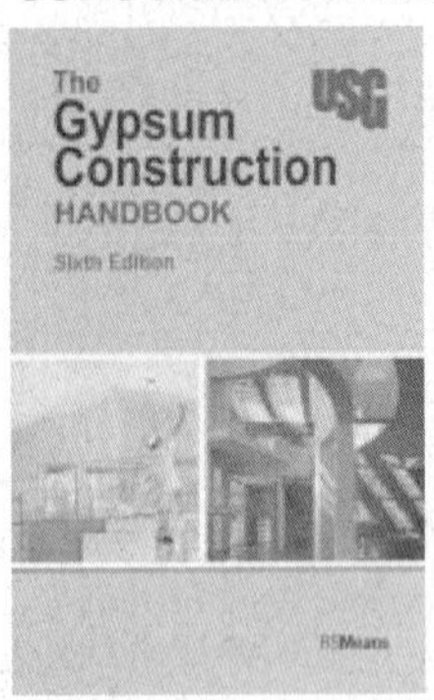

by USG

An invaluable reference of construction procedures for gypsum drywall, cement board, veneer plaster, and conventional plaster. This new edition includes the newest product developments, installation methods, fire- and sound-rated construction information, illustrated framing-to-finish application instructions, estimating and planning information, and more. Great for architects and engineers; contractors, builders, and dealers; apprentices and training programs; building inspectors and code officials; and anyone interested in gypsum construction. Features information on tools and safety practices, a glossary of construction terms, and a list of important agencies and associations. Also available in Spanish.

$40.00 | Over 575 pages, illustrated, softcover | Catalog no. 67357A

RSMeans Estimating Handbook, 3rd Edition

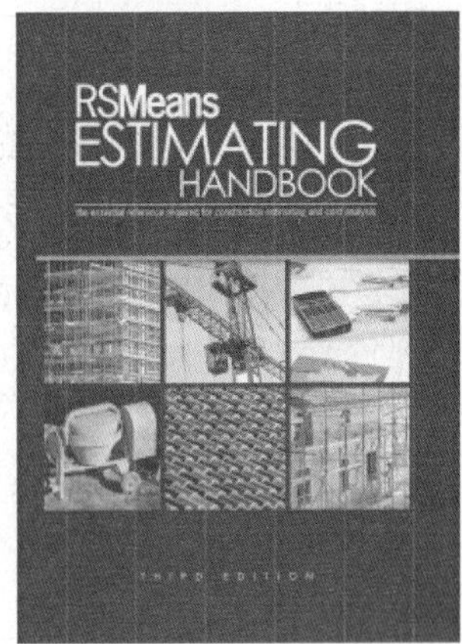

Widely used in the industry for tasks ranging from routine estimates to special cost analysis projects, this handbook has been completely updated and reorganized with new and expanded technical information.

RSMeans Estimating Handbook will help construction professionals:

- Evaluate architectural plans and specifications
- Prepare accurate quantity takeoffs
- Compare design alternatives and costs
- Perform value engineering
- Double-check estimates and quotes
- Estimate change orders

$110.00 | Over 900 pages, hardcover | Catalog No. 67276B

Mechanical Estimating Methods, 4th Edition

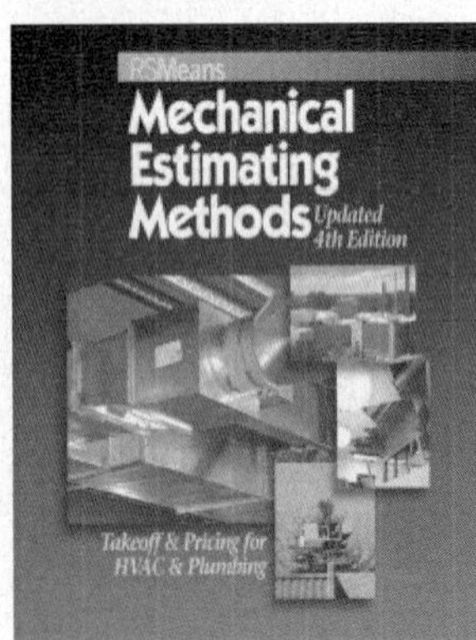

Completely updated, this guide assists you in making a review of plans, specs, and bid packages, with suggestions for takeoff procedures, listings, substitutions, and pre-bid scheduling for all components of HVAC. Includes suggestions for budgeting labor and equipment usage. Compares materials and construction methods to allow you to select the best option.

$70.00 per copy | Over 350 pages, illustrated, softcover | Catalog no. 67294B